Preparation Strategy & Analysis Videos

AF819595

# 24 Years

2025-2013 JEE Main & 2012-2002 AIEEE

# JEE Main

## Chapterwise Topicwise Solved Papers

# MATHEMATICS

**Authors**
Brijesh Dwivedi
Anjali Thakur

arihant
SUCCESS AND NOTHING LESS

arihant
SUCCESS AND NOTHING LESS

**ARIHANT PRAKASHAN (Series), MEERUT**



卐 **Administrative & Production Offices**

**Regd. Office**

'Ramchhaya' 4577/15, Agarwal Road, Darya Ganj, New Delhi -110002
Tele: 011- 47630600, 43518550

卐 **Head Office**

Kalindi, TP Nagar, Meerut (UP) - 250002
Tel: 0121-7156203, 7156204

卐 **Sales & Support Offices**

Agra, Ahmedabad, Bengaluru, Bareilly, Chennai, Delhi, Guwahati, Hyderabad, Jaipur, Jhansi, Kolkata, Lucknow, Nagpur & Pune.

卐 **PRICE** ₹810.00

卐 **PO No :** TXT-XX-XXXXXXX-X-XX

Published by Arihant Publications (India) Ltd.

For further information about the books published by Arihant, log on to www.arihantbooks.com or e-mail at info@arihantbooks.com

Follow us on

# Preface

JEE Main is a gateway examination for candidates expecting to seek admission in Bachelor in Engineering (BE), Bachelor of Technology (B.Tech) and Bachelor of Architecture (B.Arch) at Indian Institutes of Information Technology (IIITs), National Institutes of Technology (NITs), Delhi Technological University and other Centrally Funded Technical Institutes (CFTIs).

JEE Main is also an examination which is like screening examination for JEE Advanced (The gateway examination to India's most reputed Technical Institutes, Indian Institutes of Technology— IITs).

To make the students well-versed with the pattern as well as the level of the questions asked in the exam, this book contains Chapterwise Topicwise Solutions of Questions asked in Last 24 Years' Examinations of JEE Main (formerly known as AIEEE).

**The Unique and exclusive features of this book are:**

- Questions are further sorted according to their difficulty level in each topic.
- Questions on frequently asked concepts are marked with concept name to get edge over others.

Solutions to all the questions have been kept very detailed and accurate with ingredient of special features like Key idea, Concept Enhancer & Alternate/Time saver techniques.

We hope this book would be highly beneficial for the students. We would be grateful if any discrepancy or mistake in the questions or answers is brought to our notice so that these could be rectified in subsequent editions.

Publisher

# Contents

## Class 11

## Class 12

# SYLLABUS

## MATHEMATICS

### UNIT 1 Sets, Relations and Functions

Sets and their representation: Union, intersection, and complement of sets and their algebraic properties; Power set; Relation, Type of relations, equivalence relations, functions; one-one, into and onto functions, the composition of functions

### UNIT 2 Complex Numbers and Quadratic Equations

Complex numbers as ordered pairs of reals, Representation of complex numbers in the form a + ib and their representation in a plane, Argand diagram, algebra of complex number, modulus, and argument (or amplitude) of a complex number, Quadratic equations in real and complex number system and their solutions, Relations between roots and co-efficient, nature of roots, the formation of quadratic equations with given roots

### UNIT 3 Matrices and Determinants

Matrices, algebra of matrices, type of matrices, determinants, and matrices of order two and three, evaluation of determinants, area of triangles using determinants, Adjoint and evaluation of inverse of a square matrix using determinates and Test of consistency and solution of simultaneous linear equations in two or three variables using matrices

### UNIT 4 Permutations and Combinations

The fundamental principle of counting, permutation as an arrangement and combination as section, Meaning of P (n,r) and C (n,r), simple applications

### UNIT 5 Binomial Theorem and Its Simple Applications

Binomial theorem for a positive integral index, general term and middle term, and simple applications

### UNIT 6 Sequence and Series

Arithmetic and Geometric progressions, insertion of arithmetic, geometric means between two given numbers, Relation between A.M and G.M

### UNIT 7 Limit, Continuity and Differentiability

Real–valued functions, algebra of functions, polynomials, rational, trigonometric, logarithmic, and exponential functions, inverse function. Graphs of simple functions. Limits, continuity and differentiability. Differentiation of the sum, difference, product, and quotient of two functions. Differentiation of trigonometric, inverse trigonometric, logarithmic, exponential, composite, and implicit functions; derivatives of order up to two, Applications of derivatives:

Rate of change of quantities, monotonic-Increasing and decreasing functions, Maxima and minima of functions of one variable,

### UNIT 8

Integral Calculus - Integral as an anti-derivative, Fundamental integral involving algebraic, trigonometric, exponential, and logarithmic functions. Integrations by substitution, by parts and by partial functions. Integration using trigonometricidentities. Evaluation of simple integrals of the type

$$\int \frac{dx}{x^2 \pm a^2},\ \int \frac{dx}{\sqrt{x^2 \pm a^2}},\ \int \frac{dx}{a^2 - x^2},\ \int \frac{dx}{\sqrt{a^2 - x^2}},\ \int \frac{dx}{ax^2+bx+c},\ \int \frac{dx}{\sqrt{ax^2+bx+c}},\ \int \frac{(px+q)dx}{ax^2+bx+c},\ \int \frac{(px+q)dx}{\sqrt{ax^2+bx+c}},\ \int \sqrt{a^2 \pm x^2}\,dx,\ \int \sqrt{x^2 - a^2}\,dx$$

The fundamental theorem of calculus, properties of definite integrals. Evaluation of definite integrals, determining areas of the regions bounded by simple curves in standard form

**UNIT 9 Diffrential Equations**

Ordinary differential equations, their order, and degree, the solution of differential equation by the method of separation of variables, solution of a homogeneous and linear differential equation of the type $\frac{dy}{dx} + p(x)y = q(x)$

**UNIT 10 Co-ordinate Geometry**

Cartesian system of rectangular coordinates in a plane, distance formula, sections formula, locus and its equation, the slope of a line, parallel and perpendicular lines, intercepts of a line on the co-ordinate axis.

Straight line - Various forms of equations of a line, intersection of lines, angles between two lines, conditions for concurrence of three lines, the distance of a point form a line, co-ordinate of the centroid, orthocentre, and circumcentre of a triangle,

Circle, conic sections - A standard form of equations of a circle, the general form of the equation of a circle, its radius and central, equation of a circle when the endpoints of a diameter are given, points of intersection of a line and a circle with the centre at the origin and sections of conics, equations of conic sections (parabola, ellipse, and hyperbola) in standard forms,

**UNIT 11 Three Dimensional Geometry**

Coordinates of a point in space, the distance between two points, section formula, directions ratios, and direction cosines, and the angle between two intersecting lines. Skew lines, the shortest distance between them, and its equation. Equations of a line

**UNIT 12 Vector Algebra**

Vectors and scalars, the addition of vectors, components of a vector in two dimensions and three-dimensional space, scalar and vector products

**UNIT 13 Statistics and Probability**

Measures of discretion; calculation of mean, median, mode of grouped and ungrouped data calculation of standard deviation, variance, and mean deviation for grouped and ungrouped data.

Probability: Probability of an event, addition and multiplication theorems of probability, Baye's theorem, probability distribution of a random variate.

**UNIT 14 Trigonometry**

Trigonometrical identities and trigonometrical functions, inverse trigonometrical functions, and their properties.

# C-01

JEE Main Chapterwise Topicwise Mathematics

In JEE Main there is usually 0-1 question per paper are visible from this chapter. In past 3 years, 8 conceptual and 6 numerical type questions of moderate level were asked.

**For detailed analysis scan QR code**

# SETS

## Topic 1 Sets, Types of Sets, Operations on Sets, De-Morgan's Laws and Venn Diagrams

### Level 1

**01** Let $A = \{1, 6, 11, 16, ...\}$ and $B = \{9, 16, 23, 30, ....\}$ be the sets consisting of the first 2025 terms of two arithmetic progressions. Then, $n(A \cup B)$ is *[2025, 4 April Shift-I]*

(a) 3761 (b) 4003 (c) 3814 (d) 4027

**02** Let $A = \{1, 2, 3, ....., 10\}$ and $B = \left\{\frac{m}{n} : m, n \in A, m < n \text{ and } \gcd(m, n) = 1\right\}$.

Then, $n(B)$ is equal to *[2025, 22 Jan Shift-I]*

(a) 29 (b) 36 (c) 31 (d) 37

**Imp.** De-Morgan's Laws & Venn Diagram

**03** Let $A = \{n \in [100, 700] \cap N : n$ is neither a multiple of 3 nor a multiple of 4$\}$. Then, the number of elements in $A$ is *[2024, 6 April Shift-I]*

(a) 300 (b) 280 (c) 310 (d) 290

**04** Let $S = \{x \in R : (\sqrt{3} + \sqrt{2})^x + (\sqrt{3} - \sqrt{2})^x = 10\}$. Then, the number of elements in $S$, is *[2024, 1 Feb Shift-I]*

(a) 0 (b) 1 (c) 4 (d) 2

**05** An organisation awarded 48 medals in event '$A$', 25 in event '$B$' and 18 in event '$C$'. If these medals went to total 60 men and only five men got medals in all the three events, then how many received medals in exactly two of three events? *[2023, 11 April Shift-I]*

(a) 10 (b) 9 (c) 21 (d) 15

**06** Let $A = \{x \in R : |x + 1| < 2\}$ and $B = \{x \in R : |x - 1| \geq 2\}$. Then, which one of the following statements is not true? *[2022, 25 June Shift-II]*

(a) $A - B = (-1, 1)$ (b) $B - A = R - (-3, 1)$

(c) $A \cap B = (-3, -1]$ (d) $A \cup B = R - [1, 3)$

**07** Out of all the patients in a hospital 89% are found to be suffering from heart ailment and 98% are suffering from lungs infection. If $K$% of them are suffering from both ailments, then $K$ can not belong to the set *[2021, 26 Aug Shift-I]*

(a) {80, 83, 86, 89} (b) {84, 86, 88, 90}

(c) {79, 81, 83, 85} (d) {84, 87, 90, 93}

**08** Let $S_1, S_2$ and $S_3$ be three sets defined as

$S_1 = \{z \in C : |z - 1| \leq \sqrt{2}\}$, $S_2 = \{z \in C : \text{Re}((1 - i)z) \geq 1\}$

and $S_3 = \{z \in C : \text{Im}(z) \leq 1\}$.

Then, the set $S_1 \cap S_2 \cap S_3$ *[2021, 17 March Shift-II]*

(a) has exactly two elements

(b) has infinitely many elements

(c) has exactly three elements

(d) is a singleton

**09** In a school, there are three types of games to be played. Some of the students play two types of games, but none plays all the three games. Which Venn diagram can justify the above statement? *[2021, 17 March Shift-I]*

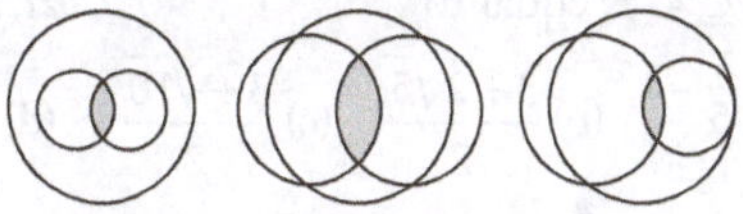

(P) (Q) (R)

(a) $P$ and $Q$ (b) $P$ and $R$

(c) None of these (d) $Q$ and $R$

**10** Let $Z$ be the set of integers. If $A = \{x \in Z : 2^{(x+2)(x^2 - 5x + 6)} = 1\}$ and $B = \{x \in Z : -3 < 2x - 1 < 9\}$, then the number of subsets of the set $A \times B$, is *[2019, 12 Jan Shift-II]*

(a) $2^{12}$ (b) $2^{18}$ (c) $2^{15}$ (d) $2^{10}$

**11** Two sets $A$ and $B$ as under :

$A = \{(a, b) \in R \times R : |a - 5| < 1 \text{ and } |b - 5| < 1\}$;

$B = \{(a, b) \in R \times R : 4(a - 6)^2 + 9(b - 5)^2 \leq 36\}$.

Then, *[JEE Main 2018]*

(a) neither $A \subset B$ nor $B \subset A$ (b) $B \subset A$

(c) $A \subset B$ (d) $A \cap B = \phi$ (an empty set)

**12** Let $P = \{\theta : \sin\theta - \cos\theta = \sqrt{2}\cos\theta\}$ and $Q = \{\theta : \sin\theta + \cos\theta = \sqrt{2}\sin\theta\}$ be two sets. Then **[JEE Main 2016]**

(a) $P \subset Q$ and $Q - P \neq \phi$ (b) $Q \not\subset P$
(c) $P = Q$ (d) $P \not\subset Q$

**13** If $X = \{4^n - 3n - 1 : n \in N\}$ and $Y = \{9(n-1) : n \in N\}$, where $N$ is the set of natural numbers, then $X \cup Y$ is equal to **[JEE Main 2014]**

(a) $N$ (b) $Y - X$ (c) $X$ (d) $Y$

**14** If $A, B$ and $C$ are three sets such that $A \cap B = A \cap C$ and $A \cup B = A \cup C$, then **[AIEEE 2009]**

(a) $A = C$ (b) $B = C$ (c) $A \cap B = \phi$ (d) $A = B$

## Level 2

**15** Let

$$A = \left\{ x \in (0, \pi) - \left\{\frac{\pi}{2}\right\} : \log_{(2/\pi)} |\sin x| + \log_{(2/\pi)} |\cos x| = 2 \right\}$$ and

$B = \{x \geq 0 : \sqrt{x}(\sqrt{x} - 4) - 3|\sqrt{x} - 2| + 6 = 0\}$. Then $n(A \cup B)$ is equal to **[2025, 24 Jan Shift -II]**

(a) 2 (b) 4 (c) 6 (d) 8

**16** Let $A = \{(x, y) \in R \times R : |x + y| \geq 3\}$ and $B = \{(x, y) \in R \times R : |x| + |y| \leq 3\}$.
If $C = \{(x, y) \in A \cap B : x = 0 \text{ or } y = 0\}$, then $\sum_{(x, y) \in C} |x + y|$ is **[2025, 23 Jan Shift -II]**

(a) 12 (b) 24 (c) 18 (d) 15

**17** Let $A = \{(x, y) \in R \times R \mid 2x^2 + 2y^2 - 2x - 2y = 1\}$, $B = \{(x, y) \in R \times R \mid 4x^2 + 4y^2 - 16y + 7 = 0\}$ and $C = \{(x, y) \in R \times R \mid x^2 + y^2 - 4x - 2y + 5 \leq r^2\}$. Then, the minimum value of $|r|$ such that $A \cup B \subseteq C$ is equal to **[2021, 27 July Shift-I]**

(a) $1 + \sqrt{5}$ (b) $\frac{3 + 2\sqrt{5}}{2}$ (c) $\frac{3 + \sqrt{10}}{2}$ (d) $\frac{2 + \sqrt{10}}{2}$

**18** Let $\bigcup_{i=1}^{50} X_i = \bigcup_{i=1}^{n} Y_i = T$, where each $X_i$ contains 10 elements and each $Y_i$ contains 5 elements. If each element of the set $T$ is an element of exactly 20 of sets $X_i$'s and exactly 6 of sets $Y_i$'s, then $n$ is equal to **[2020, 4 Sep Shift-II]**

(a) 50 (b) 15
(c) 45 (d) 30

**19** Consider the two sets $A = \{m \in R :$ both the roots of $x^2 - (m+1)x + m + 4 = 0$ are real$\}$ and $B = [-3, 5)$. Which of the following is not true? **[2020, 3 Sep Shift-I]**

(a) $A - B = (-\infty, -3) \cup [5, \infty)$
(b) $A \cap B = \{-3\}$
(c) $B - A = (-3, 5]$
(d) $A \cup B = R$

**20** Let $S = \{x \in R : x \geq 0$ and $2|\sqrt{x} - 3| + \sqrt{x}(\sqrt{x} - 6) + 6 = 0\}$. Then, $S$ **[JEE Main 2018]**

(a) is an empty set.
(b) contains exactly one element.
(c) contains exactly two elements.
(d) contains exactly four elements.

## Level 3

**21** Let $S = \left\{ x \in [-6, 3] - \{-2, 2\} : \frac{|x + 3| - 1}{|x| - 2} \geq 0 \right\}$ and $T = \{x \in Z : x^2 - 7|x| + 9 \leq 0\}$. Then, the number of elements in $S \cap T$ is **[2022, 28 July Shift-II]**

(a) 7 (b) 5
(c) 4 (d) 3

**22** Let $A, B$ and $C$ be sets such that $\phi \neq A \cap B \subseteq C$. Then, which of the following statements is not true? **[2019, 12 April Shift-II]**

(a) $B \cap C \neq \phi$
(b) If $(A - B) \subseteq C$, then $A \subseteq C$
(c) $(C \cup A) \cap (C \cup B) = C$
(d) If $(A - C) \subseteq B$, then $A \subseteq B$

## Numerical Value Type Questions

**23** For $n \geq 2$, let $S_n$ denote the set of all subsets of $\{1, 2, ..., n\}$ with no two consecutive numbers. For example $\{1, 3, 5\} \in S_6$ but $\{1, 2, 4\} \notin S_6$. Then, $n(S_5)$ is equal to ........... **[2025, 7 April Shift -I]**

**Imp.** Cartesian Product of Sets

**24** Let $S = [p_1, p_2, ...., p_{10}]$ be the set of first ten prime number. Let $A = S \cup P$, where $P$ is the set of all possible products of distinct elements of $S$. Then, the number of all ordered pair $(x, y), x \in S, y \in A$ such that $x$ divides $y$ is ........... **[2025, 24 Jan Shift -I]**

**25** The number of elements in the set $\{n \in N : 10 \leq n \leq 100$ and $3^n - 3$ is a multiple of 7$\}$ is ............ . **[2023, 15 April Shift -I]**

**26** Let $\lambda \in R$ and let the equation $E$ be $|x|^2 - 2|x| + |\lambda - 3| = 0$. Then, the largest element in the set $S = \{x + \lambda : x$ is an integer solution of $E\}$ is ....... . **[2023, 24 Jan Shift - I]**

**27** Let $S = \{4, 6, 9\}$ and $T = \{9, 10, 11, ...., 1000\}$. If $A = \{a_1 + a_2 + .... + a_k : k \in N, a_1, a_2, a_3, ...., a_k \in S\}$, then the sum of all the elements in the set $T - A$ is equal to ...... . **[2022, 29 July Shift-I]**

**28** Let $A = \{1, 2, 3, 4, 5, 6, 7\}$ and $B = \{3, 6, 7, 9\}$, Then, the number of elements in the set $\{C \subseteq A : C \cap B \neq \phi\}$ is ............ . **[2022, 26 July Shift-II]**

**29** Let A : { 1, 2, 3, 4, 5, 6, 7}. Define $B = \{T \subseteq A$ : either $1 \notin T$ or $2 \in T\}$ and $C = \{T \subseteq A : T$ the sum of all the elements of $T$ is a prime number }. Then, the number of elements in the set $B \cup C$ is ......... .
*[2022, 25 July Shift-II]*

**Imp.** Sum of $n$ terms of Natural Numbers

**30** Let $A = \{n \in N : \text{HCF}(n, 45) = 1\}$ and $B = \{2k : k \in \{1, 2, ..., 100\}\}$. Then, the sum of all the elements of $A \cap B$ is ...... . *[2022, 26 June Shift-I]*

**31** The sum of all the elements of the set $\{\alpha \in \{1, 2, ..., 100\} : \text{HCF}(\alpha, 24) = 1\}$ is ...... .
*[2022, 24 June Shift-II]*

**32** If $A = \{x \in R : |x-2| > 1\}$, $B = \{x \in R : \sqrt{x^2 - 3} > 1\}$ and $C = \{x \in R : |x-4| \geq 2\}$ and $Z$ is the set of all integers, then the number of subsets of the set $(A \cap B \cap C)^C \cap Z$ is ...... . *[2021, 27 Aug Shift-I]*

**33** Let $A = \{n \in N \mid n^2 \leq n + 10000\}$, $B = \{3k + 1 \mid k \in N\}$ and $C = \{2k \mid k \in N\}$, then the sum of all the elements of the set $A \cap (B - C)$ is equal to ........... .
*[2021, 27 July Shift-II]*

**34** Let $A = \{n \in N : n$ is a 3-digit number$\}$, $B = \{9k + 2 : k \in N\}$ and $C = \{9k + l : k \in N\}$ for some $l (0 < l < 9)$.
If the sum of all the elements of the set $A \cap (B \cup C)$ is $274 \times 400$, then $l$ is equal to .......... . *[2021, 24 Feb Shift-I]*

## Topic 2 Practical Problems

**01** Let $A$ and $B$ be two finite sets with $m$ and $n$ elements, respectively. The total number of subsets of the set $A$ is 56 more than the total number of subsets of $B$. Then, the distance of the point $P(m, n)$ from the point $Q(-2, -3)$ is *[2024, 27 Jan Shift-II]*
(a) 6 (b) 10
(c) 4 (d) 8

**02** A survey shows that 63% of the people in a city read newspaper $A$ whereas 76% read newspaper $(B)$. If $x$% of the people read both the newspapers, then a possible value of $x$ can be *[2020, 4 Sep Shift-I]*
(a) 55 (b) 29 (c) 65 (d) 37

**03** In a class of 140 students numbered 1 to 140, all even numbered students opted Mathematics course, those whose number is divisible by 3 opted Physics course and those whose number is divisible by 5 opted Chemistry course. Then, the number of students who did not opt for any of the three courses is
*[2019, 10 Jan. Shift-I]*
(a) 42 (b) 102 (c) 38 (d) 1

### Numerical Value Type Questions

**04** In a survey of 220 students of a higher secondary school, it was found that atleast 125 and atmost 130 students studied Mathematics; atleast 85 and atmost 95 studied Physics; atleast 75 and atmost 90 studied Chemistry; 30 studied both Physics and Chemistry; 50 studied both Chemistry and Mathematics; 40 studied both Mathematics and Physics and 10 studied none of these subjects. Let $m$ and $n$ respectively be the least and the most number of students who studied all the three subjects. Then, $m + n$ is equal to ...... . *[2024, 4 April Shift-I]*

**05** A group of 40 students appeared in an examination of 3 subjects — Mathematics, Physics and Chemistry. It was found that all students passed in atleast one of the subjects, 20 students passed in Mathematics, 25 students passed in Physics, 16 students passed in Chemistry, atmost 11 students passed in both Mathematics and Physics, atmost 15 students passed in both Physics and Chemistry, atmost 15 students passed in both Mathematics and Chemistry. The maximum number of students passed in all the three subjects is ......... . *[2024, 30 Jan Shift-I]*

## • Key Index

### Topic 1

| | | | | | | | | | |
|---|---|---|---|---|---|---|---|---|---|
| **1** (a) | **2** (c) | **3** (a) | **4** (d) | **5** (c) | **6** (b) | **7** (c) | **8** (b) | **9** (c) | **10** (c) |
| **11** (c) | **12** (c) | **13** (d) | **14** (b) | **15** (d) | **16** (a) | **17** (b) | **18** (d) | **19** (c) | **20** (c) |
| **21** (d) | **22** (d) | **23** (13) | **24** (5120) | **25** (15) | **26** (5) | **27** (11) | **28** (112) | **29** (107) | **30** (5264) |
| **31** (1633) | **32** (256) | **33** (832) | **34** (5) | | | | | | |

### Topic 2

| | | | | |
|---|---|---|---|---|
| **1** (b) | **2** (a) | **3** (c) | **4** (45) | **5** (10) |

# Detailed **Explanations**

## Topic 1

**01.** (a) $A = \{1, 6, 11, 16, ...\} \rightarrow 5n-4$

$B = \{9, 16, 23, 30, ...\} \rightarrow 7n+2$

$(a_{2025})_A = 5(2025) - 4 = 10121$

$(a_{2025})_B = 7(2025) + 2 = 14177$

Common terms : $(A \cap B)$

$= 16, 51, 86, ...$

Its general term $\le 10121$

$$16 + (n-1)35 \le 10121$$

$$n \le \frac{10121 - 16}{35} + 1$$

$$n \le 289.71$$

$\therefore \quad n = 289$

$\therefore$Required value $= 2025 + 2025 - 289$

$= 3761$

**02.** (c) $A = \{1, 2, 3, ..., 10\}$

$B = \left\{\frac{m}{n} : m, n \in A, m < n \text{ and } \gcd(m, n) = 1\right\}$

$m = 1 \Rightarrow n = 2, 3, 4, ..., 10 \quad \rightarrow 9$ elements

$m = 2 \Rightarrow n = 3, 5, 7, 9 \quad \rightarrow 4$ elements

$m = 3 \Rightarrow n = 4, 5, 7, 8, 10 \quad \rightarrow 5$ elements

$m = 4 \Rightarrow n = 5, 7, 9 \quad \rightarrow 3$ elements

$m = 5 \Rightarrow n = 6, 7, 8, 9 \quad \rightarrow 4$ elements

$m = 6 \Rightarrow n = 7 \quad \rightarrow 1$ element

$m = 7 \Rightarrow n = 8, 9, 10 \quad \rightarrow 3$ elements

$m = 8 \Rightarrow n = 9 \quad \rightarrow 1$ element

$m = 9 \Rightarrow n = 10 \quad \rightarrow 1$ element

$\therefore n(B) = 31$

**03.** (a) Total element $= 601$

(Natural number between [100, 700])

Let set $P$ denotes number of multiple of 3 and set $Q$ denotes number of multiple of 4.

Now, according to the question, we need

$$n(\overline{P} \cap \overline{Q})$$

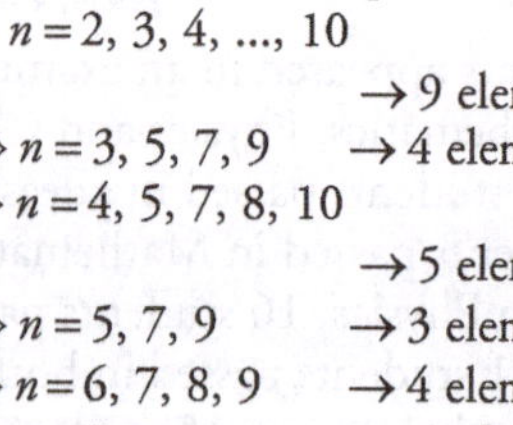


$\Rightarrow$ By De-morgan's law

$$n(\overline{P} \cap \overline{Q}) = n(\overline{P \cup Q})$$

Now, $\quad n(P) = 200$

$[\because 700 \ge (102 + (n-1)\,3]$

$n(Q) = 151$

$[\because 700 \ge (100 + (m-1)\,4)]$

$n(P \cap Q) = 50$

$[\because 700 \ge (104 + (r-1)12)]$

$\Rightarrow n(P \cup Q) = n(P) + n(Q) - n(P \cap Q)$

$= 200 + 151 - 50 = 301$

$\Rightarrow n(\overline{P \cup Q}) = 601 - 301$

$= 300 = n(\overline{P} \cap \overline{Q})$

$\therefore$Number of elements in set $A$ is 300.

**04.** (d) Given,

$(\sqrt{3} + \sqrt{2})^x + (\sqrt{3} - \sqrt{2})^x = 10$

Let $(\sqrt{3} + \sqrt{2})^x = t$

$\Rightarrow (\sqrt{3} - \sqrt{2})^x = \frac{1}{t} \Rightarrow t + \frac{1}{t} = 10$

$\Rightarrow t^2 - 10t + 1 = 0$

$\Rightarrow \quad t = \frac{10 \pm \sqrt{96}}{2} = 5 \pm 2\sqrt{6}$

$\Rightarrow (\sqrt{3} + \sqrt{2})^x = (\sqrt{3} \pm \sqrt{2})^2$

$x = \pm 2$

$\therefore$ Number of elements in $S = 2$

**05.** (c) We have, $n(A) = 48$,

$n(B) = 25, n(C) = 18$

$n(A \cup B \cup C) = 60$

and $\quad n(A \cap B \cap C) = 5$

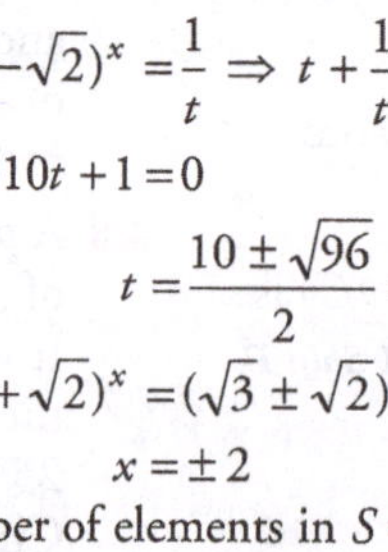


$e = 5,$

$a + c + e + d = 48$

$\Rightarrow \quad a + c + d = 43 \quad$ ...(i)

$b + c + e + f = 25$

$\Rightarrow \quad b + c + f = 20 \quad$ ...(ii)

$d + e + f + g = 18$

$\Rightarrow \quad d + f + g = 13 \quad$ ...(iii)

$a + b + c + d + e + f + g = 60$

$\Rightarrow \quad a + b + c + d + f + g = 55 \quad$ ...(iv)

Adding Eqs. (i), (ii) and (iii), we get

$a + b + 2c + 2d + 2f + g = 76 \quad$ ...(v)

Subtract Eq. (iv) from Eq. (v), we get

$c + d + f = 21$

$\therefore$Number of men who received exactly 2 medals $= c + d + f = 21$

> **Alternate Solution**
>
> We have, $n(A) = 48, n(B) = 25, n(C) = 18$
>
> $n(A \cup B \cup C) = 60$ and $n(A \cap B \cap C) = 5$
>
> Number of medals in exactly two of three events
>
> $= n(A) + n(B) + n(C) - n(A \cup B \cup C) - 2n(A \cap B \cap C)$
>
> $= 48 + 25 + 18 - 60 - 2 \times 5 = 21$

**06.** (b) For set $A$, $-2 < x + 1 < 2$

$\Rightarrow \quad -3 < x < 1$

$\Rightarrow \quad x \in (-3, 1)$

For set $B$, $x - 1 \le -2$ or $x - 1 \ge 2$

$\Rightarrow \quad x \le -1$ or $x \ge 3$

$\Rightarrow \quad x \in (-\infty, -1] \cup [3, \infty)$

$A - B = (-1, 1)$

$B - A = R - (-3, 3)$

$A \cap B = (-3, -1]$

$A \cup B = R - [1, 3)$

**07.** (c) Let $A$ = Set of patients suffering from heart ailment and $B$ = Set of patients suffering from lungs infection

Given, $n(A) = 89\%$ and $n(B) = 98\%$

$n(A \cup B) \ge n(A) + n(B) - n(A \cap B)$

$\Rightarrow \quad 100 \ge 89 + 98 - n(A \cap B)$

$\Rightarrow \quad 87 \le n(A \cap B)$

Also, $n(A \cap B) = \min\{n(A), n(B)\}$

$\Rightarrow \quad n(A \cap B) \le 89$

$\therefore \quad 87 \le n(A \cap B) \le 89$

So, $n(A \cap B) \notin \{79, 81, 83, 85\}$.

> **Alternate Solution**
>
> $n(A \cup B) \ge n(A) + n(B) - n(A \cap B)$
>
> $\Rightarrow 100 \ge 89 + 98 - n(A \cap B)$
>
> $\Rightarrow 87 \le n(A \cap B)$
>
> $n(A \cap B) \le 89$
>
> So, $n(A \cap B) \notin \{79, 81, 83, 85\}$.

**08.** *(b)*

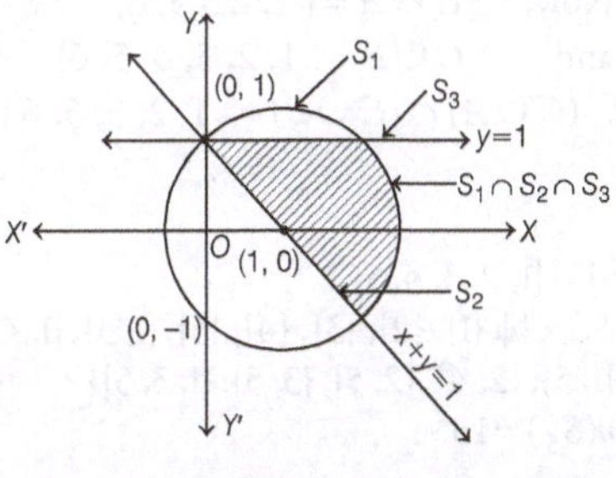


Here, $|z-1| \le \sqrt{2}$, represents a circle $S_1$ whose centre is (1, 0) and radius $\sqrt{2}$ units and $z$ lies on and inside the circle.

Re $((1-i)z) \ge 1$, represents

$$S_2 : x + y \ge 1$$

and Im $(z) \le 1$, represents $S_3 : y \le 1$.

Thus, from figure, $S_1 \cap S_2 \cap S_3$ has infinitely many elements in the shaded region.

**09.** *(c)*

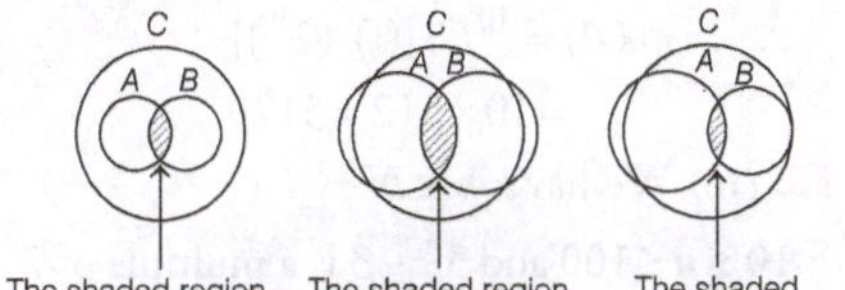


The shaded region of this Venn diagram represents the students who play all three types of games.

The shaded region of this Venn diagram represent the students who play all three type of games.

The shaded region of this Venn diagram represent the students who play all three type of games.

As none plays all the three games, the intersection of all three circles must be zero.

**10.** *(c)* Given, set

$$A = \{x \in Z : 2^{(x+2)(x^2-5x+6)} = 1\}$$

Consider,

$$2^{(x+2)(x^2-5x+6)} = 1 = 2^0$$

$\Rightarrow \quad (x+2)(x-3)(x-2) = 0$

$\Rightarrow \quad x = -2, 2, 3$

$\Rightarrow \quad A = \{-2, 2, 3\}$

Also, we have set

$$B = \{x \in Z : -3 < 2x - 1 < 9\}$$

Consider,

$$-3 < 2x - 1 < 9, x \in Z$$

$\Rightarrow \quad -2 < 2x < 10, x \in Z$

$\Rightarrow \quad -1 < x < 5, x \in Z$

$\Rightarrow \quad B = \{0, 1, 2, 3, 4\}$

So, $A \times B$ has 15 elements.

$\therefore$ Number of subsets of $A \times B = 2^{15}$

[$\because$ if $n(A) = m$, the number of possible subsets $= 2^m$]

**11.** *(c)*

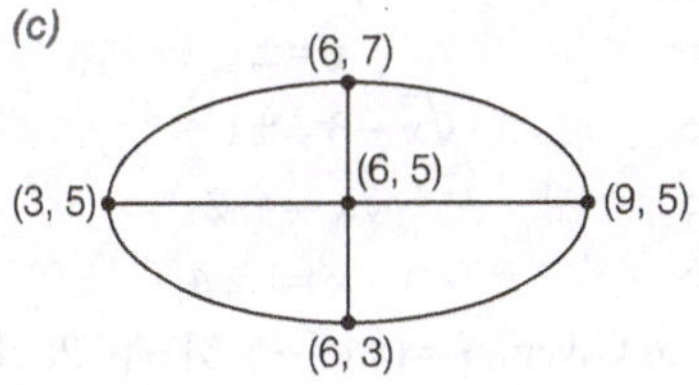


From set A

$$|a-5| < 1$$

$$-1 < a - 5 < 1$$

$$4 < a < 6 \text{ and } |b-5| < 1$$

$$4 < b < 6$$

$$A = \{4 < a, b < 6\}$$

From set B

$$\frac{(a-6)^2}{9} + \frac{(b-5)^2}{4} \le 1$$

$$B = \{3 < a, b < 7\}$$

$$A \subset B.$$

**12.** *(c)* $\because \sin\theta - \cos\theta = \sqrt{2}\cos\theta$

$$\sin\theta = \cos\theta + \sqrt{2}\cos\theta$$

$$= (\sqrt{2}+1)\cos\theta$$

$\Rightarrow \quad \tan\theta = \sqrt{2} + 1$

Similarly, $(\sqrt{2}-1)\sin\theta = \cos\theta$

$$\Rightarrow \tan\theta = \frac{1}{\sqrt{2}-1} \times \frac{\sqrt{2}+1}{\sqrt{2}+1}$$

$$\tan\theta = \sqrt{2} + 1$$

$\therefore \quad P = Q$

**13.** *(d)* $\because X = \{4^n - 3n - 1 : n \in N\}$

$$X = \{0, 9, 54, 243, \ldots\}$$

[put $n = 1, 2, 3, \ldots$]

$$Y = \{9(n-1) : n \in N\}$$

$$Y = \{0, 9, 18, 27, \ldots\}$$

It is clear that $X \subset Y$.

$\therefore \quad X \cup Y = Y$

**14.** *(b)* Given, $A \cap B = A \cap C$

and $\quad A \cup B = A \cup C$

$\therefore \quad B = C$

**15.** *(d)* $\log_{2/\pi} |\sin x| + \log_{2/\pi} |\cos x| = 2$

$\Rightarrow \quad \log_{2/\pi} |\sin x \cos x| = 2$

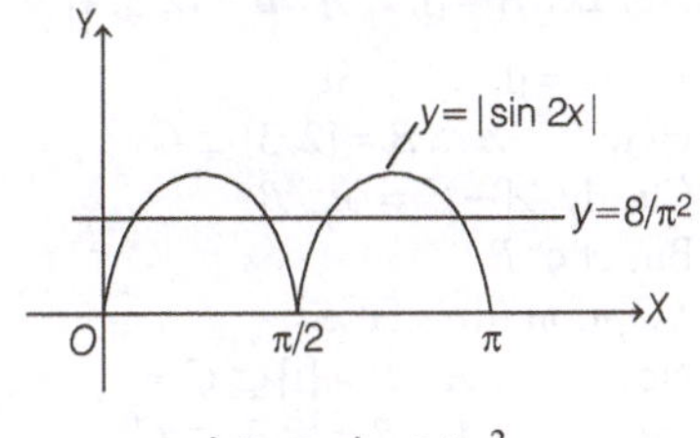


$\Rightarrow \quad |\sin 2x| = 8/\pi^2$

$$\left[\because \sin x \cos x = \frac{\sin 2x}{2}\right]$$

$\therefore$ Number of solution = 4,

$\therefore n(A) = 4$

Now,

$$\sqrt{x}(\sqrt{x} - 4) - 3|\sqrt{x} - 2| + 6 = 0 \quad \ldots\text{(i)}$$

Case I Let $\sqrt{x} = y < 2$

Then, Eq. (i) becomes

$$y(y-4) + 3|y-2| + 6 = 0$$

$\Rightarrow \quad y^2 - 4y + 3y - 6 + 6 = 0$

$\Rightarrow \quad y^2 - y = 0 \Rightarrow y = 0, 1$

$\Rightarrow \quad \sqrt{x} = 0, 1 \Rightarrow x = 0, 1$

Case II Again, let $\sqrt{x} = y > 2$

Then, $y^2 - 4y - 3y + 6 + 6 = 0$

$\Rightarrow \quad y^2 - 7y + 12 = 0$

$\Rightarrow \quad y = 3, 4$

$\Rightarrow \quad \sqrt{x} = 3, 4$

$\Rightarrow \quad x = 9, 16$

So, total number of solutions = 4

$\therefore \quad n(B) = 4$

Hence, $n(A \cup B) = 4 + 4 = 8$

**16.** *(a)* $A = \{(x, y) \in R \times R : |x + y| \ge 3\}$

$\Rightarrow x + y \ge 3 \quad \text{or} \quad x + y \le -3 \quad \ldots\text{(i)}$

$B = \{(x, y) \in R \times R : |x| + |y| \le 3\}$

I quadrant : $x + y \le 3$

II quadrant : $-x + y \le 3 \quad \ldots\text{(ii)}$

III quadrant : $-x - y \le 3$

IV quadrant : $x - y \le 3$

$C : \{(x, y) \; A \cap B : x = 0 \text{ or } y = 0\}$

Put $x = 0$ in conditions (i) and (ii), we get common solution $y = -3, 3$.

Similarly, $y = 0$ we get $x = -3, 3$

$\therefore C = \{(0, 3), (0, -3), (3, 0), (-3, 0)\}$

$$\therefore \sum_{(x, y) \in C} |x + y| = 3 + 3 + 3 + 3 = 12$$

**17.** *(b)* Let $O$ be the centre of the circle $C$. $C_1$ and $C_2$ are the centres of circles $A$ and $B$, respectively.

So, $\quad C_1 = \left(\frac{1}{2}, \frac{1}{2}\right), R_1 = 1$

and $C_2 = (0, 2), R_2 = \frac{3}{2}$

$O = (2, 1)$, radius $r_3 = |r|$

Now, for $A \cup B \subseteq C$, we have

$$r \ge OC_2 + R_2$$

$$= \sqrt{2^2 + 1^2} + \frac{3}{2} = \frac{2\sqrt{5} + 3}{2}$$

$$\Rightarrow r \ge \frac{2\sqrt{5} + 3}{2}$$

**18.** *(d)* According to the given information number of distinct elements in

$$\bigcup_{i=1}^{50} X_i = \frac{50\times 10}{20} = 25$$

and number of distinct elements in

$$\bigcup_{i=1}^{n} Y_i = \frac{n\times 5}{6}$$

$\because \quad \bigcup_{i=1}^{50} X_i = \bigcup_{i=1}^{n} Y_i = T$ [given]

$\Rightarrow \quad 25 = \frac{n\times 5}{6}$

$\Rightarrow \quad n = 30$

**19.** *(c)* Given sets $A = \{m \in R$ : both the roots of $x^2 - (m+1)x + m + 4 = 0$ are real$\}$ and $B = [-3, 5)$

$\because$ Roots of $x^2 - (m+1)x + m + 4 = 0$ are real, $m \in R$.

$\therefore \; D \ge 0 \Rightarrow (m+1)^2 - 4(m+4) \ge 0$

$\Rightarrow \; m^2 - 2m - 15 \ge 0$

$\Rightarrow \; m^2 - 5m + 3m - 15 \ge 0$

$\Rightarrow \; m(m-5) + 3(m-5) \ge 0$

$\Rightarrow \; (m+3)(m-5) \ge 0$

$\Rightarrow \; m \in (-\infty, -3] \cup [5, \infty)$

$\therefore \; A = (-\infty, -3] \cup [5, \infty)$

$\therefore \; A - B = (-\infty, -3) \cup [5, \infty)$

$A \cap B = \{-3\}$,

$B - A = (-3, 5)$

and $A \cup B = R$

Hence, option (c) is correct.

**20.** *(c)*

**key idea** (i) Let $\sqrt{x} - 3 = y$ (where, $y \in R$)

(ii) Solve it further to obtain value of $x$.

(iii) Number of element in $S$ = Number of values of $x$.

We have,

$2|\sqrt{x} - 3| + \sqrt{x}(\sqrt{x} - 6) + 6 = 0$

Let $\quad \sqrt{x} - 3 = y$

$\Rightarrow \quad \sqrt{x} = y + 3$

$\therefore \; 2|y| + (y+3)(y-3) + 6 = 0$

$\Rightarrow \; 2|y| + y^2 - 3 = 0$

$\Rightarrow \; |y|^2 + 2|y| - 3 = 0$

$\Rightarrow \; (|y|+3)(|y|-1) = 0$

$\Rightarrow \; |y| \ne -3$

$\Rightarrow \; |y| = 1$

$\Rightarrow \; y = \pm 1$

$\Rightarrow \; \sqrt{x} - 3 = \pm 1$

$\Rightarrow \; \sqrt{x} = 4, 2$

$\Rightarrow \; x = 16, 4$

**21.** *(d)* Given, $S = \{x \in [-6, 3] - \{-2, 2\} : \frac{|x+3| - 1}{|x| - 2}\} \ge 0$

and $\quad T = \{x \in Z : x^2 - 7|x| + 9 \le 0\}$

Here, $x^2 - 7|x| + 9 \le 0$

$\Rightarrow |x|^2 - 7|x| + 9 \le 0 \quad [\because |x|^2 = x^2]$

$$\Rightarrow \left[|x| - \left(\frac{7+\sqrt{49-36}}{2}\right)\right] \left[|x| - \left(\frac{7-\sqrt{49-36}}{2}\right)\right] \le 0$$

$$\Rightarrow \quad |x| \in \left[\frac{7-\sqrt{13}}{2}, \frac{7+\sqrt{13}}{2}\right]$$

But $x$ is an integer,

$|x| = 2, 3, 4, 5$ and $x = \pm 2, \pm 3, \pm 4, \pm 5$

$T = \{\pm 2, \pm 3, \pm 4, \pm 5\}$

$S \cap T \ne \{\pm 2, 4, 5\}$ as $S \notin \{\pm 2, 4, 5\}$

When $x = -3$, then $\frac{|-3+3| - 1}{|-3| - 2} = \frac{-1}{1} < 0$

When $x = -4$, then $\frac{|-4+3| - 1}{|-4| - 2} = 0$

When $x = -5$, then $\frac{|-5+3| - 1}{|-5| - 2} = \frac{1}{3} \ge 0$

When $x = 3$, then $\frac{|3+3| - 1}{|3| - 2} = 5 \ge 0$

Hence, $S \cap T = \{-5, -4, 3\}$

$\Rightarrow n(S \cap T) = 3$

**22.** *(d)* Let $A = \{1, 2, 3\}$, $B = \{2, 3, 4\}$ and $C = \{1, 2, 3, 5, 6\}$

Here $\quad A \cap B = \{2, 3\} \subseteq C$

Clearly, $A - C = \phi \subseteq B$

But $A \not\subset B$

$\therefore$ Option (d) is false.

Now, $\quad A - B = \{1\} \subseteq C$

and $\quad A \cap B = \{2, 3\} \subseteq C$

$\therefore \quad A \subseteq C$

Thus, option (b) is correct.

Now, $\quad B \cap C = \{2, 3\} \ne \phi$

Thus, option (a) is correct.

Now, $\quad C \cup A = \{1, 2, 3, 5, 6\}$

and $\quad C \cup B = \{1, 2, 3, 4, 5, 6\}$

$\therefore (C \cup A) \cap (C \cup B) = \{1, 2, 3, 5, 6\} = C$

**23.** *(13)*

$A = \{1, 2, 3, 4, 5\}$

$S_5 = \{\phi, \{1\}, \{2\}, \{3\}, \{4\}, \{5\}, \{1, 3\}, \{1, 4\}, \{1, 5\}, \{2, 4\}, \{2, 5\}, \{3, 5\}, \{1, 3, 5\}\}$

$n(S_5) = 13$

**24.** *(5120)* Given,

$S = \{p_1, p_2, p_3, \ldots, p_{10}\}$

and $A$ = {product of distinct element of $S$}

$P = \{(x, y) : x \in S, y \in A$ and $y$ is divided by $x\}$

Since, $x \in S$

Let $x = P_i \in S$; selected from $S$ is ${}^{10}C_1$ ways

Now, $x \mid y$ iff $y$ contains $P_i$ in product. i.e., $P_i$ is there and at rest one other element as a product

$\therefore \quad n(P) = {}^{10}C_1[(1)\cdot(2^9)]$

$= 10 \times 512 = 5120$

**25.** *(15)* We have, $n \in N$

$10 \le n \le 100$ and $3^n - 3$ is a multiple of 7

$\therefore 3^n - 3 = 7k, \forall k \in I$

$3^n = 7k + 3$

Now, $\; 3 = 3 \pmod 7$

$3^2 = 2 \pmod 7$

$3^3 = -1 \pmod 7$ or $6 \pmod 7$

$3^6 = 1 \pmod 7$

$3^7 = 3 \pmod 7$

Since, $3^{13} = 3 \pmod 7$

$\vdots$

$\therefore n$ can be 13, 19, ... 97

$\therefore$ Total $n = 15$

**26.** *(5)* Given,

$|x|^2 - 2|x| + |\lambda - 3| = 0$

$\Rightarrow |x|^2 - 2|x| + 1 + |\lambda - 3| - 1 = 0$

$\Rightarrow (|x| - 1)^2 + |\lambda - 3| = 1$

Since, $\lambda \in R$

At $\lambda = 3$, $x = 0, 2$

At $\lambda = 4$ or 2, $x = 1$ or $-1$

So, maximum value of $x + \lambda = 5$

Hence, the largest element in the set $S$ is 5.

**27.** *(11)* $S = \{4, 6, 9\}$,

$T = \{9, 10, 11, \ldots\ldots, 1000\}$

$A = \{a_1 + a_2 + ..... + a_k : k \in N,$
$a_1, a_2, a_3 ...., a_k \in S\}$

$A = \{a; a = 4x + 6y + 9z, \ x, y, z$
$\{0, 1, 2, ..., \}\}$

$\therefore A = \{9, 10, 12, 13, 14, 15, 16, 17, 18, ...\}$

$T - A = \{11\}$

**28.** *(112)* Given, $A = \{1, 2, 3, 4, 5, 6, 7\}$ and $B = \{3, 6, 7, 9\}$

Total number of subset of $A = 2^7 = 128$

As, $C \subseteq A$ and $C \cap B = \phi$, so the set $C$ contains only 4 elements

i.e. $\{1, 2, 4, 5\}$

$\therefore$ The number of elements in the set $\{C \subseteq A : C \cap B \neq \phi\}$

= Total number of sets − Number of set when $C \subseteq A$ and $C \cap B = \phi$

$= 128 - 2^4$
$= 128 - 16$
$= 112$

**29.** *(107)* Apply De-morgan's law

$(B \cup C)' = B' \cap C'$

$B'$ containing subsets of $A$ which do not contains 2 but element 1.

And $C'$ is a set containing subsets of $A$ whose sum of elements is not prime.

So, we need to calculate number of subsets of {3,4,5,6,7} whose sum of elements plus 1 is composite.

Number of 5 elements subset = 1

Number of 4 elements subset = 3 (except selecting 3 or 7)

Number of 3 elements subset = 6 (except selecting {3,4,5} {3,6,7} {4,5,7} or {5,6,7}

Number of such 2 elements subset = 7 (except selecting {3,7},{4,6} {5,7})

Number of such 1 element subset = 3 (except selecting {4} or {6})

Number of such 0 elements subset = 1

Here,

$n(B' \cap C') = 1 + 3 + 6 + 7 + 3 + 1 = 21$

Therefore $n(B \cup C) = 2^7 - 21 = 107$

> **Concept Enhancer**
>
> According to De-Morgan's first law, the complement of the union of two sets A and B is equal to the intersection of the complement of the sets A and B.
>
> $(A \cup B)' = A' \cap B'$
>
> where, $A'$ and $B'$ denotes the complement.

**30.** *(5264)* $B = \{2, 4, 6, 8, ..., 200\}$ and

$45 = 3^2 \cdot 5$

$n$ (multiples of 6) = 33

$n$ (multiples of 10) = 20

$n$ (multiples of 30) = 6

$$\text{Sum} = S_6 = 6\,(1 + 2 + ... + 33) = \frac{6\,(33 \times 34)}{2}$$

$$S_{10} = 10\,(1 + 2 + ... + 20) = \frac{10\,(20 \times 21)}{2}$$

$$S_{30} = 30\,(1 + 2 + ... + 6) = \frac{30\,(6 \times 7)}{2}$$

$$\text{Total sum, } S_{200} = 2\,(1 + 2 + ... + 100) = \frac{2\,(100 \times 101)}{2}$$

Sum of elements of $A \cap B$

$$= 2\left[\frac{100 \times 101}{2} - 3\,\frac{(33 \times 34)}{2} - 5\left(\frac{20 \times 21}{2}\right) + 15\left(6 \times \frac{7}{2}\right)\right]$$

$= 10100 - 3366 - 2100 + 630 = 5264$

> **Alternate Solution**
>
> $$S_6 = \frac{6(33 \times 34)}{2} = 3366$$
> $$S_{10} = \frac{10(20 \times 21)}{2} = 2100$$
> $$S_{30} = \frac{30(6 \times 7)}{2} = 630$$
> $$\therefore S_{200} = \frac{2(100 \times 101)}{2} = 10100$$
>
> Sum of natural numbers up to 100 which are neither divisible by 3 nor by 5.
>
> $= 10100 - 3366 - 2100 + 630 = 5264$

**31.** *(1633)*

> **key idea** Let $S(\alpha)$ is the sum of all element of the set.
>
> $S(\alpha) = \Sigma(\alpha) - \Sigma$ (multiple of 2) $- \Sigma$ (multiple of 3) $+ \Sigma$ (multiple of 6)

Let $S(\alpha)$ is the sum of all elements of the set

$\{\alpha \in \{1, 2, 3, ..., 100\};$ HCF $(24, \alpha) = 1\}$

Since, HCF $(24, \alpha) = 1$

$\Rightarrow S(\alpha) = (1 + 2 + 3 ... + 100) - \Sigma(\text{multiple of } 3) - \Sigma(\text{multiple of } 2) + \Sigma(\text{multiple of } 6)$

$\Rightarrow S(\alpha) = (1 + 2 + 3 ... + 100) - (2 + 4 + 6 + ... + 100) - (3 + 6 + 9 + ... + 99) + (6 + 12 + 18 + ... 96)$

$$\Rightarrow S(\alpha) = \frac{100 \cdot (100 + 1)}{2} - 2 \cdot (1 + 2 + 3 + ... + 50) - 3(1 + 2 + 3 + ... + 33) + 6(1 + 2 + 3 + ... 16)$$

$$= \frac{100 \cdot 101}{2} - 2 \cdot \frac{50 \cdot (50 + 1)}{2} - 3 \cdot \frac{33 \cdot (33 + 1)}{2} + 6 \cdot \frac{16 \cdot (16 + 1)}{2}$$

$= 5050 - 2550 - 1683 + 816$

$= 1633$

**32.** *(256)* $A = \{x \in R : |x - 2| > 1\}$

$\Rightarrow \quad A = (-\infty, 1) \cup (3, \infty)$

$B = \{x \in R : \sqrt{x^2 - 3} > 1\}$

$\Rightarrow \quad B = (-\infty, -2) \cup (2, \infty)$

$C = \{x \in R : |x - 4| \geq 2\}$

$\Rightarrow \quad C = (-\infty, 2] \cup [6, \infty)$

$\Rightarrow \quad A \cap B \cap C = (-\infty, -2) \cup [6, \infty)$

$\Rightarrow (A \cap B \cap C)^C = [-2, 6)$

$\therefore (A \cap B \cap C)^C \cap Z$

$= \{-2, -1, 0, 1, 2, 3, 4, 5\}$

Number of subsets of

$(A \cap B \cap C)^C \cap Z = 2^8 = 256$

**33.** *(832)* Let $A = \{n \in N \mid n^2 \leq n + 10000\}$

$n^2 \leq n + 10000$

$n^2 - n \leq 10000$

$\Rightarrow \quad n(n - 1) \leq 100 \times 100$

$\Rightarrow \quad A = \{1, 2, 3, ......, 100\} \quad [\because n \in N]$

Now, $B = \{3k + 1 \mid k \in N\}$

$B = \{4, 7, 10, 13, ...\}$

and $\quad C = \{2k \mid k \in N\}$

$C = \{2, 4, 6, 8, ...\}$

So, $B - C = \{7, 13, 19, ......, 97, ...\}$

So, $A \cap (B - C) = \{7, 13, 19, ......, 97\}$

This form an AP with common difference $(d = 6)$

$\Rightarrow \quad 97 = 7 + (n - 1)6$

$$n = \frac{97 - 7}{6} + 1 = 16$$

$[\because a_n = a + (n - 1)d]$

$$\text{Hence, sum} = \frac{16}{2}[7 + 97]$$

$$= 832 \quad \left[\because S_n = \frac{n}{2}(a + l)\right]$$

**34.** *(5)* Given, $A = \{n \in N : n \text{ is a 3-digit number}\}$

$B = \{9k + 2 : k \in N\}$

and $\quad C = \{9k + l : k \in N\}$

$\because$ 3 digit number of the form $9k+2$ are $\{101, 110, \ldots 992\}$

$\Rightarrow \quad \text{Sum} = \frac{100}{2}[101+992]$

$= \frac{100 \times 1093}{2}$

Similarly, 3-digit number of the form $9k+5$ are $\{104, 113, 122, \ldots, 995\}$

$\Rightarrow \text{Sum} = \frac{100}{2}[104+995] = \frac{100 \times 1099}{2}$

Their sum

$= \frac{100 \times 1093}{2} + \frac{100 \times 1099}{2}$

$= 100 \times 1096 = 400 \times 274$

Hence, we can say the value of $l = 5$ as the second series of numbers obtained by set $C$ is of the form $9k+5$.

$\therefore$ Required value of $l = 5$

## Topic 2

**01.** *(b)* Given, $n(A) = m$ and $n(B) = n$

$\therefore$ Number of subsets of $A = 2^m$

and number of subsets of $B = 2^n$

According to the question,

$2^m - 2^n = 56$

$\Rightarrow \quad 2^m - 2^n = 2^6 - 2^3$

$\Rightarrow \quad m = 6$ and $n = 3$

So, distance between the points $P(m, n)$, i.e. $P(6, 3)$ and $Q(-2, -3)$ is given by

$\sqrt{(6+2)^2 + (3+3)^2} = \sqrt{64+36}$

$= \sqrt{100} = 10$

**02.** *(a)* Let $n(A)$ = Number of people read newspaper $A = 63\%$

$n(B)$ = Number of people read newspaper $B = 76\%$

and $n(A \cap B)$ = Number of people read both $= x\%$

$\because n(A \cup B) = n(A) + n(B) - n(A \cap B)$

and $76 \le n(A \cup B) \le 100$

$\therefore \quad 76 \le 63 + 76 - x \le 100$

$\Rightarrow \quad 0 \le 63 - x \le 24$

$\Rightarrow \quad 39 \le x \le 63$

Hence, option (a) is correct.

**03.** *(c)*

> **key idea** (i) First find the number of students who have opted any of the 3 courses (i.e., $n(A \cup B \cup C)$), where, $n(A)$, $n(B)$ and $n(C)$ is number of students opted for Mathematics, Physics and Chemistry, respectively.
>
> (ii) $n(A \cup B \cup C) = \Sigma n(A) - \Sigma n(A \cap B) + n(A \cap B \cap C)$
>
> (iii) Number of students who did not opt any subjects = Total students $- n(A \cup B \cup C)$

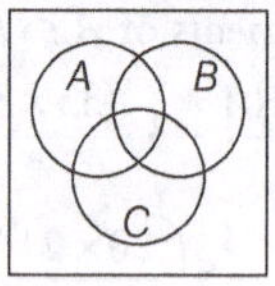


Let $A$ be the set of even numbered students then

$$n(A) = \left[\frac{140}{2}\right] = 70$$

([.] denotes greatest integer function)

Let $B$ be the set of those students whose number is divisible by 3, then

$$n(B) = \left[\frac{140}{3}\right] = 46$$

Let $C$ be the set of those students whose number is divisible by 5,

then $n(C) = \left[\frac{140}{5}\right] = 28$

Now, $n(A \cap B) = \left[\frac{140}{6}\right] = 23$

(numbers divisible by both 2 and 3)

$$n(B \cap C) = \left[\frac{140}{15}\right] = 9$$

(numbers divisible by both 3 and 5)

$$n(C \cap A) = \left[\frac{140}{10}\right] = 14$$

(numbers divisible by both 2 and 5)

$$n(A \cap B \cap C) = \left[\frac{140}{30}\right] = 4$$

(numbers divisible by 2, 3 and 5)

and $n(A \cup B \cup C)$

$= \Sigma n(A) - \Sigma n(A \cap B) + n(A \cap B \cap C)$

$= (70 + 46 + 28) - (23 + 9 + 14) + 4 = 102$

$\therefore$ Number of students who did not opt any of the three courses

$= \text{Total students} - n(A \cup B \cup C)$

$= 140 - 102 = 38$

**04.** *(45)* According to the question,

$125 \le m + 90 - x \le 130$

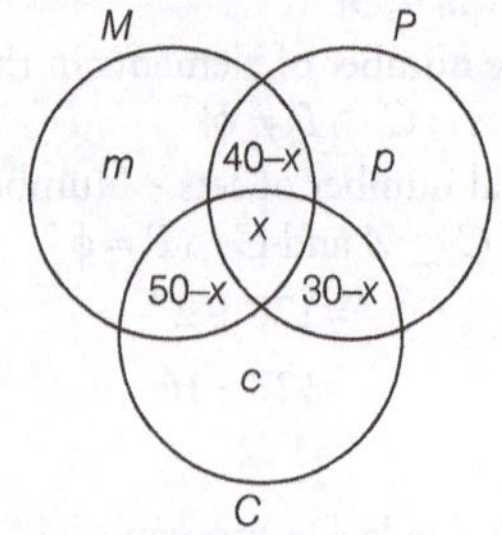


$85 \le p + 70 - x \le 95$

$75 \le c + 80 - x \le 90$

$m + p + c + 120 - 2x = 210$

$\Rightarrow \quad 15 \le x \le 45$

and $\quad 30 - x \ge 0$

$\Rightarrow \quad 15 \le x \le 30$

Hence, $30 + 15 = 45$

**05.** *(10)* Given, $n(M \cup P \cup C) = 40$,

We have,

$x = n(P \cap M \cap C)$

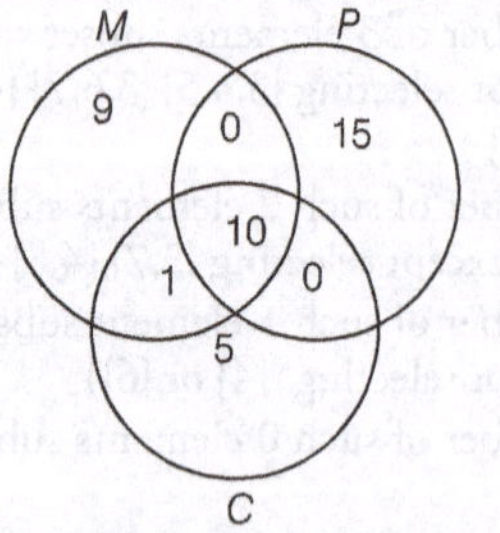


Now, $11 - n(P \cap M \cap C) \ge 0$

$\Rightarrow n(P \cap M \cap C) = 11$

Which does not satisfy.

So, $n(P \cap M \cap C) = 10$

Hence, maximum number of students passed in all the three subjects is 10.

# C-02

JEE Main Chapterwise Topicwise Mathematics

In JEE Main, there is usually 0-1 question per paper is visible from this chapter. In past 3 years, 25 conceptual and 4 numerical value type questions of moderate to hard difficulty level (most of these are from moderate difficulty level) were asked.

**For detailed analysis scan QR code**

# RELATIONS AND FUNCTIONS

## Topic 1 Domain, Co-domain and Range of Relations and Functions

### Level 1

**01** If the domain of the function

$$f(x)=\frac{1}{\sqrt{10+3x-x^2}}+\frac{1}{\sqrt{x+|x|}}$$ is $(a, b)$, then

$(1+a)^2+b^2$ is equal to **[2025, 2 April Shift-II]**

(a) 26 (b) 30 (c) 25 (d) 29

**Imp.** Domain of the Function

**02** If the domain of the function

$$f(x)=\log_e\left(\frac{2x-3}{5+4x}\right)+\sin^{-1}\left(\frac{4+3x}{2-x}\right)$$ is $[\alpha, \beta)$, then

$\alpha^2+4\beta$ is equal to **[2025, 3 April Shift-I]**

(a) 7 (b) 4 (c) 5 (d) 3

**03** If the range of the function $f(x)=\frac{5-x}{x^2-3x+2}$, $x\neq 1, 2$ is $(-\infty, \alpha]\cup[\beta, \infty)$, then $\alpha^2+\beta^2$ is equal to **[2025, 7 April Shift-II]**

(a) 192 (b) 194 (c) 190 (d) 188

**04** Let the sets $A$ and $B$ denote the domain and range respectively of the function $f(x)=\frac{1}{\sqrt{\lceil x\rceil-x}}$, where $\lceil x\rceil$ denotes the smallest integer greater than or equal to $x$. Then, among the statements **[2023, 6 April Shift-II]**

(S1): $A\cap B=(1, \infty)-N$ and (S2) : $A\cup B=(1, \infty)$

(a) Both (S1) and (S2) are true

(b) Neither (S1) nor (S2) is true

(c) only (S1) is true

(d) only (S2) is true

**05** If the domain of the function $f(x)=\frac{[x]}{1+x^2}$, where $[x]$ is greatest integer $\leq x$, is $[2, 6)$, then its range is **[2023, 31 Jan Shift-I]**

(a) $\left(\frac{5}{37}, \frac{2}{5}\right]-\left\{\frac{9}{29}, \frac{27}{109}, \frac{18}{89}, \frac{9}{53}\right\}$

(b) $\left(\frac{5}{26}, \frac{2}{5}\right]-\left\{\frac{9}{29}, \frac{27}{109}, \frac{18}{89}, \frac{9}{53}\right\}$

(c) $\left(\frac{5}{26}, \frac{2}{5}\right]$

(d) $\left(\frac{5}{37}, \frac{2}{5}\right]$

**06** If $R=\{(x, y): x, y\in Z, x^2+3y^2\leq 8\}$ is a relation on the set of integers $Z$, then the domain of $R^{-1}$ is **[2020, 2 Sep Shift-I]**

(a) $\{-1, 0, 1\}$ (b) $\{-2, -1, 1, 2\}$

(c) $\{-2, -1, 0, 1, 2\}$ (d) $\{0, 1\}$

**07** The domain of the function $f(x)=\frac{1}{\sqrt{|x|-x}}$ is **[AIEEE 2011]**

(a) $(0, \infty)$ (b) $(-\infty, 0)$

(c) $(-\infty, \infty)-\{0\}$ (d) $(-\infty, \infty)$

**08** The domain of definition of the function

$$f(x)=\sqrt{\log_{10}\left(\frac{5x-x^2}{4}\right)}$$ is **[AIEEE 2002]**

(a) $[1, 4]$ (b) $[1, 0]$ (c) $[0, 5]$ (d) $[5, 0]$

### Level 2

**09** Let $A=\{1, 2, 3, \ldots, 100\}$ and $R$ be a relation on $A$ such that $R=\{(a, b): a=2b+1\}$. Let $(a_1, a_2), (a_2, a_3), (a_3, a_4), \ldots, (a_k, a_{k+1})$ be a sequence of $k$ elements of $R$ such that the second entry of an ordered pair is equal to the first entry of the next ordered pair. Then, the largest integer $k$, for which such a sequence exists, is equal to **[2025, 2 April Shift-II]**

(a) 8 (b) 5

(c) 6 (d) 7

**10** If the domain of the function $f(x) = \log_7(1 - \log_4(x^2 - 9x + 18))$ is $(\alpha, \beta) \cup (\gamma, \delta)$, then $\alpha + \beta + \gamma + \delta$ is equal to *[2025, 3 April Shift-II]*

(a) 18 (b) 17 (c) 16 (d) 15

**11** Let the domains of the functions $f(x) = \log_4(\log_3(\log_7(8 - \log_2(x^2 + 4x + 5))))$ and $g(x) = \sin^{-1}\left(\dfrac{7x + 10}{x - 2}\right)$ be $(\alpha, \beta)$ and $[\gamma, \delta]$, respectively. Then, $\alpha^2 + \beta^2 + \gamma^2 + \delta^2$ is equal to *[2025, 4 April Shift-II]*

(a) 13 (b) 15 (c) 14 (d) 16

**12** Let $A = \{(\alpha, \beta) \in R \times R : |\alpha - 1| \le 4 \text{ and } |\beta - 5| \le 6\}$ and $B = \{(\alpha, \beta) \in R \times R : 16(\alpha - 2)^2 + 9(\beta - 6)^2 \le 144\}$.

Then, *[2025, 7 April Shift-II]*

(a) $A \cup B = \{(x, y) : -4 \le x \le 4, -1 \le y \le 11\}$

(b) $A \subset B$

(c) $B \subset A$

(d) neither $A \subset B$ nor $B \subset A$

**13** If the domain of the function $\log_5(18x - x^2 - 77)$ is $(\alpha, \beta)$ and the domain of the function $\log_{(x-1)}\left(\dfrac{2x^2 + 3x - 2}{x^2 - 3x - 4}\right)$ is $(\gamma, \delta)$, then $\alpha^2 + \beta^2 + \gamma^2$ is equal to *[2025, 29 Jan Shift-II]*

(a) 195 (b) 179

(c) 186 (d) 174

**14** Let $S = N \cup \{0\}$. Define a relation $R$ from $S$ to $R$ by

$$R = \left\{(x, y) : \log_e y = x \log_e\left(\frac{2}{5}\right), x \in S, y \in R\right\}$$

Then, the sum of all the elements in the range of $R$ is equal to *[2025, 29 Jan Shift-II]*

(a) $\dfrac{10}{9}$ (b) $\dfrac{3}{2}$ (c) $\dfrac{5}{2}$ (d) $\dfrac{5}{3}$

**15** Let $f(x) = \log_e x$ and

$$g(x) = \frac{x^4 - 2x^3 + 3x^2 - 2x + 2}{2x^2 - 2x + 1}$$

Then, the domain of *fog* is *[2025, 23 Jan Shift-I]*

(a) $R$ (b) $[0, \infty)$ (c) $(0, \infty)$ (d) $[1, \infty)$

**16** If the domain of the function

$$f(x) = \frac{\sqrt{x^2 - 25}}{(4 - x^2)} + \log_{10}(x^2 + 2x - 15)$$ is

$(-\infty, \alpha) \cup [\beta, \infty)$, then $\alpha^2 + \beta^3$ is equal to *[2024, 1 Feb Shift-II]*

(a) 175 (b) 150

(c) 125 (d) 140

**17** The domain of the function $f(x) = \dfrac{1}{\sqrt{[x]^2 - 3[x] - 10}}$ is (Where, $[x]$ denotes the greatest integer less than or equal to $x$) *[2023, 11 April Shift-II]*

(a) $(-\infty, -2) \cup (5, \infty)$ (b) $(-\infty, -3] \cup [6, \infty)$

(c) $(-\infty, -3] \cup (5, \infty)$ (d) $(-\infty, -2) \cup [6, \infty)$

**18** Let $f : R - \{2, 6\} \to R$ be real valued function defined as $f(x) = \dfrac{x^2 + 2x + 1}{x^2 - 8x + 12}$. Then, range of $f$ is *[2023, 31 Jan Shift-II]*

(a) $\left(-\infty, -\dfrac{21}{4}\right] \cup [1, \infty)$ (b) $\left(-\infty, -\dfrac{21}{4}\right] \cup [0, \infty)$

(c) $\left(-\infty, -\dfrac{21}{4}\right] \cup \left[\dfrac{21}{4}, \infty\right)$ (d) $\left(-\infty, -\dfrac{21}{4}\right) \cup (0, \infty)$

**19** The domain of $f(x) = \dfrac{\log_{(x+1)}(x - 2)}{e^{2\log_e x} - (2x + 3)}$, $x \in R$ is *[2023, 29 Jan Shift-I]*

(a) $R - \{3\}$ (b) $(2, \infty) - \{3\}$

(c) $(-1, \infty) - \{3\}$ (d) $R - \{-1, 3)$

**20** Let $[x]$ denote the greatest integer $\le x$, where $x \in R$. If the domain of the real valued function $f(x) = \sqrt{\dfrac{|[x]| - 2}{|[x]| - 3}}$ is $(-\infty, a) \cup [b, c) \cup [4, \infty)$, $a < b < c$, then the value of $a + b + c$ is *[2021, 20 July Shift-I]*

(a) 8 (b) 1 (c) −2 (d) −3

**21** If the functions are defined as $f(x) = \sqrt{x}$ and $g(x) = \sqrt{1 - x}$, then what is the common domain of the following functions? $f + g, f - g, f / g, g / f, g - f$, where, $(f / g)(x) = \dfrac{f(x)}{g(x)}$ *[2021, 18 March, Shift-I]*

(a) $0 \le x \le 1$ (b) $0 \le x < 1$ (c) $0 < x < 1$ (d) $0 < x \le 1$

## Topic 2 Types of Functions and Algebraic Operations on Real Functions

### Level 1

**01** Let $f$ be a function such that $f(x) + 3f\left(\dfrac{24}{x}\right) = 4x$, $x \ne 0$. Then, $f(3) + f(8)$ is equal to *[2025, 3 April Shift-II]*

(a) 11 (b) 10 (c) 13 (d) 12

**02** Consider the sets $A = \{(x, y) \in R \times R : x^2 + y^2 = 25\}$, $B = \{(x, y) \in R \times R : x^2 + 9y^2 = 144\}$, $C = \{(x, y) \in Z \times Z : x^2 + y^2 \le 4\}$ and $D = A \cap B$. The total number of one-one function from the set $D$ to the set $C$ is *[2025, 4 April Shift-I]*

(a) 15120 (b) 17160 (c) 19320 (d) 18290

**03** If $[x]$ be the greatest integer less than or equal to $x$, then $\sum_{n=8}^{100}\left[\frac{(-1)^n n}{2}\right]$ is equal to *[2021, 25 July Shift-II]*

(a) 0 (b) 4
(c) −2 (d) 2

**04** The graph of the function $y = f(x)$ is symmetrical about the line $x = 2$, then *[AIEEE 2004]*

(a) $f(x+2) = f(x-2)$ (b) $f(2+x) = f(2-x)$
(c) $f(x) = f(-x)$ (d) $f(x) = -f(-x)$

## Level 2

**Imp.** Algebraic Operations on Real Function

**05** Let $f(x) = \frac{2^{x+2}+16}{2^{2x+1}+2^{x+4}+32}$. Then, the value of $8\left(f\left(\frac{1}{15}\right)+f\left(\frac{2}{15}\right)+\ldots+f\left(\frac{59}{15}\right)\right)$ is equal to *[2025, 24 Jan Shift-I]*

(a) 118 (b) 102
(c) 108 (d) 92

**06** Let $f: R-\{0,1\} \to R$ be a function such that $f(x)+f\left(\frac{1}{1-x}\right) = 1+x$. Then, $f(2)$ is equal to *[2023, 1 Feb Shift-II]*

(a) $\frac{9}{2}$ (b) $\frac{9}{4}$ (c) $\frac{7}{3}$ (d) $\frac{7}{4}$

**07** Consider a function $f: N \to R$, satisfying $f(1)+2f(2)+3f(3)+\ldots+xf(x) = x(x+1)f(x)$; $x \ge 2$ with $f(1) = 1$. Then, $\frac{1}{f(2022)}+\frac{1}{f(2028)}$ is equal to *[2023, 29 Jan Shift-II]*

(a) 8200 (b) 8000 (c) 8400 (d) 8100

**08** Let $f(x) = 2x^n + \lambda$, $\lambda \in R$, $n \in N$ and $f(4) = 133$, $f(5) = 255$. Then, the sum of all the positive integer divisors of $(f(3)-f(2))$ is *[2023, 25 Jan Shift-II]*

(a) 60 (b) 59 (c) 58 (d) 61

**09** If $f(x) = \frac{2^{2x}}{2^{2x}+2}$, $x \in R$, then $f\left(\frac{1}{2023}\right)+f\left(\frac{2}{2023}\right)+\ldots+f\left(\frac{2022}{2023}\right)$ is equal to *[2023, 24 Jan Shift-II]*

(a) 2010 (b) 2011 (c) 1011 (d) 1010

**10** Let $f: N \to N$ be a function such that $f(m+n) = f(m)+f(n)$ for every $m, n \in N$. If $f(6) = 18$, then $f(2)\cdot f(3)$ is equal to *[2021, 31 Aug Shift-II]*

(a) 6 (b) 54
(c) 18 (d) 36

**11** A function $f(x)$ is given by $f(x) = \frac{5^x}{5^x+5}$, then the sum of the series $f\left(\frac{1}{20}\right)+f\left(\frac{2}{20}\right)+f\left(\frac{3}{20}\right)+\ldots+f\left(\frac{39}{20}\right)$ is equal to *[2021, 25 Feb Shift-II]*

(a) 29/2 (b) 49/2 (c) 39/2 (d) 19/2

## Level 3

**12** Let $A = \{x \in R : [x+3]+[x+4] \le 3\}$,
$$B = \left\{x \in R : 3^x\left(\sum_{r=1}^{\infty}\frac{3}{10^r}\right)^{x-3} < 3^{-3x}\right\},$$ where $[t]$ denotes greatest integer function. Then, *[2023, 6 April Shift-I]*

(a) $A = B$ (b) $A \cap B = \phi$
(c) $B \subset C, A \ne B$ (d) $A \subset B, A \ne B$

**13** The number of functions $f:\{1,2,3,4\} \to \{a \in Z; |a| \le 8\}$ satisfying $f(n)+\frac{1}{n}f(n+1) = 1, \forall n \in \{1,2,3\}$ is *[2023, 25 Jan Shift-II]*

(a) 3 (b) 2
(c) 1 (d) 4

### Numerical Value Type Questions

**14** Let the domain of the function $f(x) = \cos^{-1}\left(\frac{4x+5}{3x-7}\right)$ be $[\alpha, \beta]$ and the domain of $g(x) = \log_2(2-6\log_{27}(2x+5))$ be $(\gamma, \delta)$. Then, $|7(\alpha+\beta)+4(\gamma+\delta)|$ is equal to ........ *[2025, 8 April Shift-II]*

**15** If $S = \{a \in R : |2a-1| = 3[a]+2\{a\}\}$, where $[t]$ denotes the greatest integer less than or equal to $t$ and $\{t\}$ represents the fractional part of $t$, then $72\sum_{a \in S} a$ is equal to ......... *[2024, 05 April Shift-I]*

**16** Let $A = \{1, 2, 3, \ldots, 7\}$ and let $P(A)$ denote the power set of $A$. If the number of functions $f: A \to P(A)$ such that $a \in f(a)$, $\forall a \in A$ is $m^n$, $m$ and $n \in N$ and $m$ is least, then $m+n$ is equal to ...... *[2024, 30 Jan Shift-I]*

**17** Let $A = \{1, 2, 3, 5, 8, 9\}$. Then, the number of possible functions $f: A \to A$ such that $f(m\cdot n) = f(m)\cdot f(n), \forall m, n \in A$ with $m \cdot n \in A$ is equal to ........ *[2023, 30 Jan Shift-II]*

## Key Index

### Topic 1

| | | | | | | | | | |
|---|---|---|---|---|---|---|---|---|---|
| **1** (a) | **2** (b) | **3** (b) | **4** (c) | **5** (d) | **6** (a) | **7** (b) | **8** (a) | **9** (b) | **10** (a) |
| **11** (b) | **12** (c) | **13** (c) | **14** (d) | **15** (a) | **16** (b) | **17** (d) | **18** (b) | **19** (b) | **20** (c) |
| **21** (c) | | | | | | | | | |

### Topic 2

| | | | | | | | | | |
|---|---|---|---|---|---|---|---|---|---|
| **1** (a) | **2** (b) | **3** (b) | **4** (b) | **5** (a) | **6** (b) | **7** (d) | **8** (a) | **9** (c) | **10** (b) |
| **11** (c) | **12** (a) | **13** (b) | **14** (96) | **15** (18) | **16** (51) | **17** (432) | | | |

# Detailed **Explanations**

### Topic 1

**01.** *(a)* $x+|x|=\begin{cases}2x & x\geq 0\\ 0 & x<0\end{cases}$

$\Rightarrow \dfrac{1}{\sqrt{x+|x|}}$, domain is $x>0 \Rightarrow (0,\infty)$

Similarly, $\dfrac{1}{\sqrt{3x+10-x^2}}$ is defined

when $3x+10-x^2>0$

$\Rightarrow x^2-3x-10<0$

$\Rightarrow (x-5)(x+2)<0 \Rightarrow x\in(-2,5)$

$\Rightarrow$ Domain will be $(0,\infty)\cap(-2,5)$

$=(0,5)$

$\therefore (1+a)^2+b^2=1+25=26$

**02.** *(b)* $\dfrac{2x-3}{5+4x}>0$ ...(i)

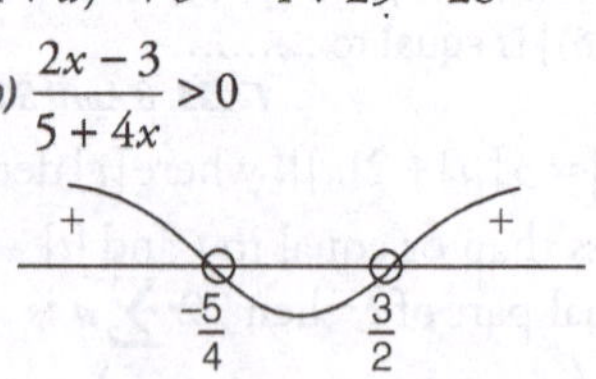


$\Rightarrow x<-\dfrac{5}{4}$ or $x>\dfrac{3}{2}$

$-1\leq\dfrac{4+3x}{2-x}\Rightarrow -3\leq x<2$ ...(ii)

and for $\dfrac{4+3x}{2-x}\leq 1 \Rightarrow x\in\left(-\infty,-\dfrac{1}{2}\right]$

$\cup(2,\infty)$ ...(iii)

From Eqs. (i), (ii), (iii), we get

$-3\leq x\leq -\dfrac{1}{2}$ or $\left(x<-\dfrac{5}{4}\text{ or } x>\dfrac{3}{2}\right)$

$\Rightarrow x\in\left[-3,\dfrac{-5}{4}\right)$

$\alpha=-3, \beta=-\dfrac{5}{4}$

$\therefore \quad \alpha^2+4\beta=9-5=4$

**03.** *(b)* To find range put $y=f(x)$ and solve for $x$ in terms of $y$.

$y=\dfrac{5-x}{x^2-3x+2}$

$x^2y-3xy+2y=5-x$

$x^2y-3xy+x+2y-5=0$

$x^2y-(3y+1)x+2y-5=0\Rightarrow D\geq 0$

$(3y-1)^2-4y(2y-5)\geq 0$

$9y^2-6y+1-8y^2+20y\geq 0$

$y^2+14y+1\geq 0$

$y=\dfrac{-14\pm\sqrt{196-4}}{2}=\dfrac{-14\pm\sqrt{192}}{2}$

$y=\dfrac{-14\pm 8\sqrt{3}}{2}=-7\pm 4\sqrt{3}$

Thus, $\alpha=-7-4\sqrt{3}$ and

$\beta=-7+4\sqrt{3}$

Then, range is $(-\infty,\alpha]\cup[\beta,\infty)$

$=(-\infty,-7-4\sqrt{3}]\cup[-7+4\sqrt{3},\infty)$,

$\alpha^2+\beta^2=(-7-4\sqrt{3})^2+(-7+4\sqrt{3})^2$

$=49+48+56\sqrt{3}$

$+49+48-56\sqrt{3}=194$

**04.** *(c)* We have, $f(x)=\dfrac{1}{\sqrt{[x]-x}}$

$\therefore$ Domain of $f(x)$, $A=R-I$ and

Range of $f(x)$, $B=(1,\infty)$

$\therefore A\cap B=(1,\infty)-N$ and $A\cup B\neq(1,\infty)$

$\therefore$Only $S_1$ is true.

**05.** *(d)* We have,

$f(x)=\dfrac{[x]}{1+x^2}, 2\leq x<6$

$=\begin{cases}\dfrac{2}{1+x^2}, & 2\leq x<3\\ \dfrac{3}{1+x^2}, & 3\leq x<4\\ \dfrac{4}{1+x^2}, & 4\leq x<5\\ \dfrac{5}{1+x^2}, & 5\leq x<6\end{cases}$

Since, $f(x)$ is a decreasing function.

So, range of $f(x)=(f(6), f(2)]$

Now, $f(2)=\dfrac{2}{5}$

and $f(6)=\dfrac{5}{37}$

$\therefore$ Range of $f(x)$ is $\left(\dfrac{5}{37},\dfrac{2}{5}\right]$.

**06.** *(a)* Given,

$R=\{(x,y)|x,y\in Z, x^2+3y^2\leq 8\}$

For $x=0$,

$3y^2\leq 8\Rightarrow y^2\leq\dfrac{8}{3}$

$\Rightarrow \quad y=0,\pm1$

For $x=\pm1$, $1+3y^2\leq 8$

$\Rightarrow \quad y^2\leq\dfrac{7}{3}$

$y=0,\pm1$

For $x=\pm2$

$4+3y^2\leq 8\Rightarrow y^2\leq\dfrac{4}{3}\Rightarrow y=0$

$\therefore$ The relation

$R = \{(0, 0), (0, -1), (0, 1), (1, 0), (1, -1), (1, 1), (-1, 0), (-1, 1), (2, 0), (-2, 0)\}$

Thus, $R$ can be written as

$\{-2, -1, 0, 1, 2\} \rightarrow \{-1, 0, 1\}$

where, range of $R$ is $\{-1, 0, 1\}$.

Since, the domain of $R^{-1}$ is the range of $R$, then $R^{-1}$ can be written as

$\{-1, 0, 1\} \rightarrow \{-2, -1, 0, 1, 2\}$

$\therefore$ Domain of $R^{-1}$ is $\{-1, 0, 1\}$.

**07.** (b) $y = \dfrac{1}{\sqrt{|x| - x}}$

For domain, $|x| - x > 0 \Rightarrow |x| > x$

i.e., only possible, if $x < 0$.

$\therefore\ x \in (-\infty, 0)$

**08.** (a) Given that

$$f(x) = \sqrt{\log_{10}\left(\frac{5x - x^2}{4}\right)}$$

For domain of $f(x)$,

$$\log_{10}\left(\frac{5x - x^2}{4}\right) \geq 0$$

$\Rightarrow \quad \dfrac{5x - x^2}{4} \geq 1$

$\Rightarrow \quad x^2 - 5x + 4 \leq 0$

$\Rightarrow \quad (x - 1)(x - 4) \leq 0$

$\Rightarrow \quad x \in [1, 4]$

**09.** (b) The sequence of $k$ terms be $(a_1, a_2), (a_2, a_3) \ldots, (a_k, a_{k+1})$ where $a_i = 2a_{i+1} + 1$, using $A$ on $A$ relation

$a_1 = 2a_2 + 1$, $a_1$ will be odd

$a_2 = 2a_3 + 1 \Rightarrow$

$a_1 = 2(2a_3 + 1) + 1 = 4a_3 + 3$

$a_3 = 2a_4 + 1 \Rightarrow$

$a_1 = 4(2a_4 + 1) + 3 = 8a_4 + 7$

$\vdots$

$a_k = 2(a_{k+1}) + 1 \Rightarrow$

$a_1 = 2^k \cdot a_{k+1} + 2^k - 1 \in A$

$\Rightarrow \dfrac{a_1 + 1 - 2^k}{2^k} = a_{k+1} \Rightarrow 2^k \mid (a_1 + 1)$

for highest $k$, $a_1 + 1 \in \{2, \ldots, 101\}$

$k$ maximum when $k = 6$, as $k = 7$, $2^k = 128$

$|128 \mid a_i\ \forall\ a_1 \in A \Rightarrow a_1 = 95$ and $k = 6$

(95, 47), (47, 23), (23, 11), (11, 5), (5, 2) will be sequence.

**10.** (a) $\log_4(x^2 - 9x + 18) > 0$

Case I $\log_4(x^2 - 9x + 18) > 0$

$x^2 - 9x + 18 < 4$

$x^2 - 9x + 14 < 0$

$(x - 2)(x - 7) < 0$

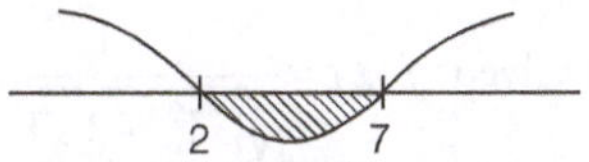


Case II $x^2 - 9x + 18 > 0$

$(x - 3)(x - 6) > 0$

$x < 0$ or $x > 6$

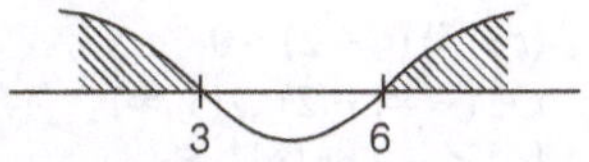


Common domain

$x \in (2, 3) \cup (6, 7)$

$\alpha = 2, \beta = 3$, $\gamma = 6$ and $\delta = 7$

$\therefore$ Sum = 18

**11.** (b) $f(x) = \log_4(\log_3(\log_7(8 - \log_2(x^2 + 4x + 5))))$

$f(x)$ is define when

$\Rightarrow \log_3(\log_7(8 - \log_2(x^2 + 4x + 5))) > 0$

$\Rightarrow \log_7(8 - \log_2(x^2 + 4x + 5)) > 3^0$ (1)

$\Rightarrow 8 - \log_2(x^2 + 4x + 5) > 7^1$

$\Rightarrow 8 - 7 > \log_2(x^2 + 4x + 5)$

$\Rightarrow \log_2(x^2 + 4x + 5) < 1$

$\Rightarrow x^2 + 4x + 5 < 2^1$ also

$x^2 + 4x + 5 > 0$

$\Rightarrow x^2 + 4x + 3 < 0,\ x^2 + 4x + 5 > 0$

$\Rightarrow (x + 3)(x + 1) < 0,\ (x + 2)^2 + 1 > 0$

$x \in R$

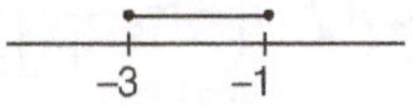


$\therefore x \in (-3, -1) \Rightarrow \alpha = -3, \beta = -1$

and $g(x) = \sin^{-1}\left(\dfrac{7x + 10}{x - 2}\right)$

$g(x)$ defines when $-1 \leq \dfrac{7x + 10}{x - 2} \leq 1$

$\dfrac{7x + 10}{x - 2} \geq -1$ and $\dfrac{7x + 10}{x - 2} \leq 1$

$\dfrac{7x + 10 + x - 2}{x - 2} \geq 0$ and

$\dfrac{7x + 10 - x + 2}{x - 2} \leq 0$

$\dfrac{8x + 8}{x - 2} \geq 0$ and $\dfrac{6x + 12}{x - 2} \leq 0$

$\dfrac{x + 1}{x - 2} \geq 0$ and $\dfrac{x + 2}{x - 2} \leq 0$

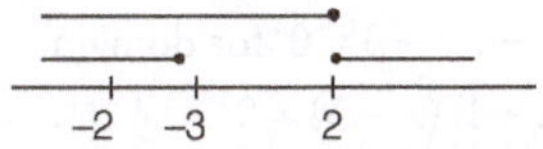


$\therefore x \in [-2, -1] \Rightarrow \gamma = -2, \delta = -1$

$\therefore \alpha^2 + \beta^2 + \gamma^2 + \delta^2 = (-3)^2 + (-1)^2 + (-2)^2 + (-1)^2$

$= 9 + 1 + 4 + 1 = 15$

**12.** (c) Apply $|x| \leq a \Rightarrow -a \leq x \leq a$

$|\alpha - 1| \leq 4 \Rightarrow -4 \leq \alpha - 1 \leq 4$

$\Rightarrow -3 \leq \alpha \leq 5$

And $|\beta - 5| \leq 6$

$\Rightarrow \ -1 \leq \beta \leq 11$

Also, $16(\alpha - 2)^2 + 9(\beta - 6)^2 \leq 144$

$\Rightarrow \dfrac{16(\alpha - 2)^2}{144} + \dfrac{9(\beta - 6)^2}{144} \leq \dfrac{144}{144}$

$\dfrac{(\alpha - 2)^2}{9} + \dfrac{(\beta - 6)^2}{16} \leq 1$

Ellipse centred at (2, 6) with semi-major axis is $a = 3$, semi-minor axis $b = 4$

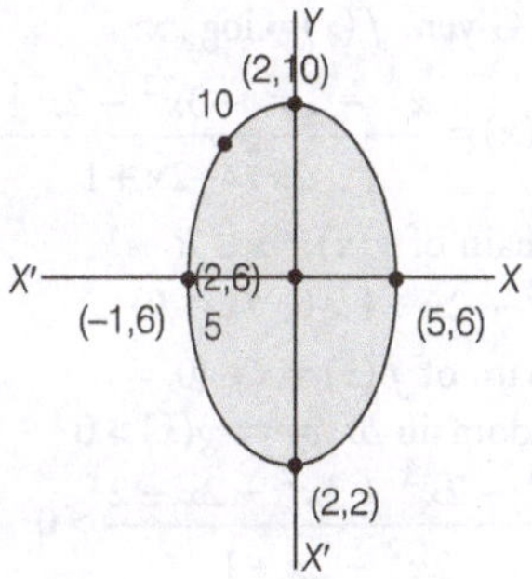


Set $A$ is a rectangle defined by $-3 \leq \alpha \leq 5$ and $-1 \leq \beta \leq 11$.

Set $B$ is an ellipse contained within the rectangle defined by $-1 \leq \alpha \leq 6$ and $2 \leq \beta \leq 10$.

Thus, $B \subset A$

**13.** (c) Let $f_1(x) = \log_5(18x - x^2 - 77)$

$\therefore 18x - x^2 - 77 > 0$

$x^2 - 18x + 77 < 0$

$x \in (7, 11)$

$\Rightarrow \alpha = 7, \beta = 11$

and $f_2(x) = \log_{(x-1)}\left(\dfrac{2x^2 + 3x - 2}{x^2 - 3x - 4}\right)$

$\therefore x - 1 > 0,\ x - 1 \neq 1,\ \dfrac{2x^2 + 3x - 2}{x^2 - 3x - 4} > 0$

$x > 1,\ x \neq 2,\ \dfrac{(2x - 1)(x + 2)}{(x - 4)(x + 1)} > 0$

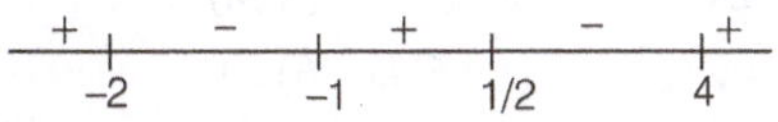


$\Rightarrow \quad x \in [4, \infty) \Rightarrow \gamma = 4$

$\therefore \alpha^2 + \beta^2 + \gamma^2 = 49 + 121 + 16$

$= 186$

**14.** (d) $S = \{0, 1, 2, 3, \ldots\}$

$\log_e y = x \log_e\left(\frac{2}{5}\right) \Rightarrow y = \left(\frac{2}{5}\right)^x$

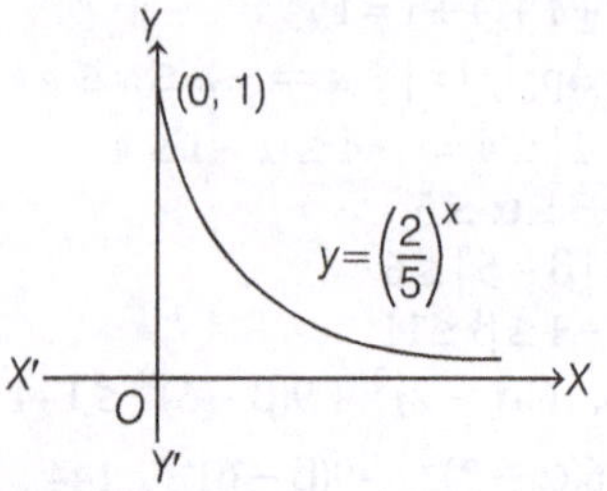


$\therefore$ Required sum

$= 1 + \left(\frac{2}{5}\right)^1 + \left(\frac{2}{5}\right)^2 + \left(\frac{2}{5}\right)^3 + \ldots$

$= \frac{1}{1 - \frac{2}{5}} = \frac{5}{3}$

**15.** (a) Given, $f(x) = \log_e x$

$$g(x) = \frac{x^4 - 2x^3 + 3x^2 - 2x + 2}{2x^2 - 2x + 1}$$

Domain of $g(x) = x \in R$

$[2x^2 - 2x + 1 > 0, \forall x \in R]$

domain of $f(x) = x \in (0, \infty)$

For domain of $fog \Rightarrow g(x) > 0$

$\therefore \frac{x^4 - 2x^3 + 3x^2 - 2x + 2}{2x^2 - 2x + 1} > 0$

$[2x^2 + 2x + 1 > 0, \forall x \in R, D = 4 - 8 < 0]$

$\Rightarrow x^4 - 2x^3 + 3x^2 - 2x + 2 > 0$

$\Rightarrow (x^2 + 1)(x^2 - 2x + 2) > 0, \forall x \in R$

$\therefore$ Domain of $fog = x \in R$

**16.** (b) We have, $f(x) = \frac{\sqrt{x^2 - 25}}{4 - x^2} + \log_{10}(x^2 + 2x - 15)$

Now, for $\frac{\sqrt{x^2 - 25}}{4 - x^2}$

$x^2 - 25 \geq 0$

$(x - 5)(x + 5) \geq 0$

$\Rightarrow x \in (-\infty, -5] \cup [5, \infty)$ ...(i)

and $4 - x^2 \neq 0$ ...(ii)

Now, for $\log_{10}(x^2 + 2x - 15)$

$\Rightarrow x^2 + 2x - 15 > 0$

$\Rightarrow (x + 5)(x - 3) > 0$

$\Rightarrow x \in (-\infty, -5) \cup (3, \infty)$ ...(iii)

From Eqs. (i), (ii) and (iii), we get

$x \in (-\infty, -5) \cup [5, \infty)$

So, $\alpha = -5, \beta = 5$, then

$\alpha^2 + \beta^3 = 25 + 125 = 150$

**17.** (d) Given, $f(x) = \frac{1}{\sqrt{[x]^2 - 3[x] - 10}}$

For domain of $f(x)$,

$[x]^2 - 3[x] - 10 > 0$

Let $[x] = t$

Then, $t^2 - 3t - 10 > 0$

$\Rightarrow (t - 5)(t + 2) > 0$

$\Rightarrow t \in (-\infty, -2) \cup (5, \infty)$

$\Rightarrow [x] < -2$ or $[x] > 5$

But $[x]$ is an integer value.

$\Rightarrow x < -2$ or $x \geq 6$

Hence, domain is

$x \in (-\infty, -2) \cup [6, \infty)$.

**18.** (b) Let $y = \frac{x^2 + 2x + 1}{x^2 - 8x + 12}$

$\Rightarrow (y - 1)x^2 - (8y + 2)x + 12y - 1 = 0$

Since, $x \in R$.

$\therefore D \geq 0$

$\Rightarrow (8y + 2)^2 - 4(y - 1)(12y - 1) \geq 0$

$\Rightarrow 16y^2 + 8y + 1 - 12y^2 + 12y + y - 1 \geq 0$

$\Rightarrow y(4y + 21) \geq 0$

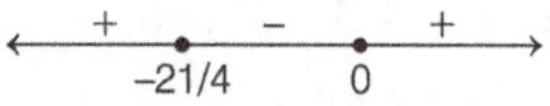


$\therefore y \in \left(-\infty, \frac{-21}{4}\right] \cup [0, \infty)$

So, range of $f$ is $\left(-\infty, \frac{-21}{4}\right] \cup [0, \infty)$.

**19.** (b) We have, $f(x) = \frac{\log_{(x+1)}(x - 2)}{e^{2\log_e x} - 2x - 3}$

For domain, $x - 2 > 0$ ...(i)

$x + 1 > 0$ ...(ii)

$x + 1 \neq 1$ ...(iii)

$x > 0$ ...(iv)

$\Rightarrow x > 2$ and $x > -1$ and $x > 0$

and $x \neq 0$

$\Rightarrow x \in (2, \infty)$ ... (v)

Also, $f(x) = \frac{\log_{(x+1)}(x - 2)}{e^{\log_e x^2} - 2x - 3}$

$(\because a \log x = \log x^a)$

$\Rightarrow f(x) = \frac{\log_{(x+1)}(x - 2)}{x^2 - 2x - 3}$

$(\because a^{\log_a x} = x)$

$\therefore x^2 - 2x - 3 \neq 0$ for domain

$\Rightarrow (x + 1)(x - 3) \neq 0$

$\Rightarrow x \neq -1, 3$ ...(vi)

From Eqs. (v) and (vi), we get

$\therefore$ Required domain is $(2, \infty) - \{3\}$.

**20.** (c) $f(x) = \sqrt{\frac{|[x]| - 2}{|[x]| - 3}}, \frac{|[x]| - 2}{|[x]| - 3} \geq 0$

Let $|[x]| = t$

$\Rightarrow \frac{t - 2}{t - 3} \geq 0$

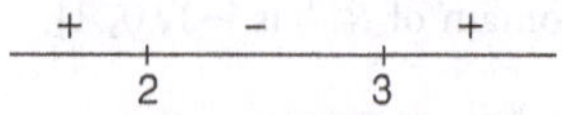


$t \in (-\infty, 2] \cup (3, \infty)$

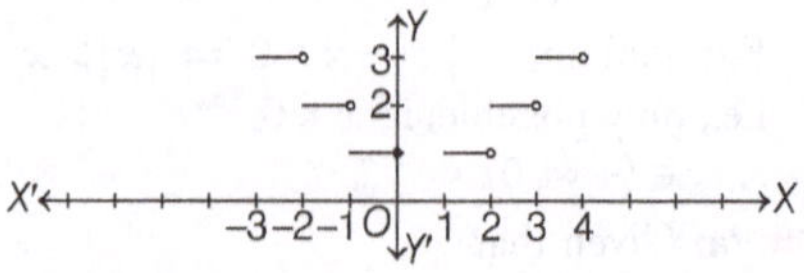


$|[x]| = 3 \Rightarrow x \in [-3, -2) \cup [3, 4)$

Domain of $A$

$x = (-\infty, -3) \cup [-2, 3) \cup [4, \infty)$

$a = -3;$

$b = -2;$

$c = 3$

$\therefore a + b + c = -3 + (-2) + 3 = -2$

> **Alternate Solution**
>
> $\frac{|[x]| - 2}{|[x]| - 3} \geq 0$
>
> $|[x]| \leq 2$ or $|[x]| > 3$
>
> $\Rightarrow -2 \leq [x] \leq 2$ or $[x] < -3$ or $[x] > 3$
>
> $\Rightarrow -2 \leq x < 3$ or $x < -3$ or $x \geq 4$
>
> $\Rightarrow x \in (-\infty, -3) \cup [-2, 3) \cup [4, \infty)$
>
> $\therefore a = -3, b = -2, c = 3$
>
> Therefore, $a + b + c = -2$

**21.** (c) Given, $f(x) = \sqrt{x}$

and $g(x) = \sqrt{1 - x}$

$\therefore$ Domain of $f(x) = D_1$ is $x \geq 0$

i.e. $D_1 : x \in [0, \infty)$

and domain of $g(x) = D_2$ is

$1 - x \geq 0$

$\Rightarrow x \leq 1$

i.e. $D_2 : x \in (-\infty, 1]$

As, we know that, the domain of $f + g, f - g, g - f$ will be $D_1 \cap D_2$ as well as the domain for $f/g$ is $D_1 \cap D_2$ except all those value(s) of $x$, such that $g(x) = 0$. Similarly, for $g/f$ is $D_1 \cap D_2$ but $f(x) \neq 0$.

Hence, common domain for

$(f + g), (f - g), \left(\frac{f}{g}\right), \left(\frac{g}{f}\right)$

and $(g - f)$ will be $0 < x < 1$.

## Topic 2

**01.** (a) $f(x)+3f\left(\frac{24}{x}\right)=4x$

put $x=3$

$$f(3)+3f\left(\frac{24}{3}\right)=4(3)$$

$$f(3)+3f(8)=12 \quad \ldots(i)$$

put $x=8$

$$f(8)+3f\left(\frac{24}{8}\right)=4(8)$$

$$f(8)+3f(3)=32$$

$$3f(8)+9f(3)=96 \quad \ldots(ii)$$

Subtracting Eqs (i) and (ii), we get

$$8f(3)=84$$

$$f(3)=\frac{84}{8}=\frac{21}{2}$$

$$f(3)+3f(8)=12 \quad \text{[using Eq. (i)]}$$

$$\frac{21}{2}+3f(8)=12$$

$$3f(8)=\frac{3}{2} \Rightarrow f(8)=\frac{1}{2}$$

$$\therefore f(3)+f(8)=\frac{21}{2}+\frac{1}{2}=\frac{22}{2}=11$$

**02.** (b) $A \to x^2+y^2=25$

$B \to x^2+9y^2=144 \Rightarrow \frac{x^2}{144}+\frac{y^2}{16}=1$

$C \to x^2+y^2 \le 4$

$D=A\cap B$

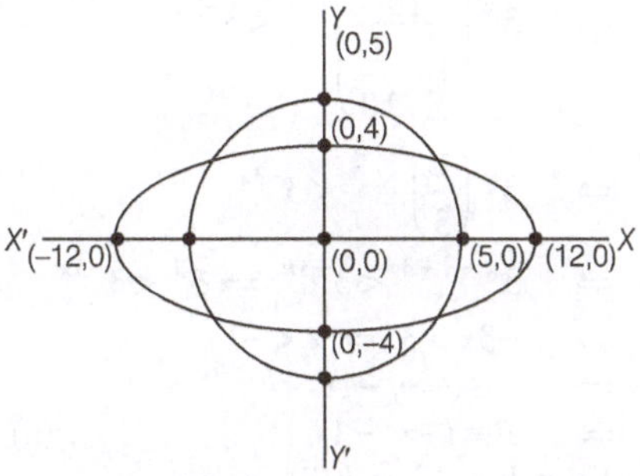


$\therefore n(A\cap B)=4$

For $n(C)$

$\therefore n(C)=12+1=13$

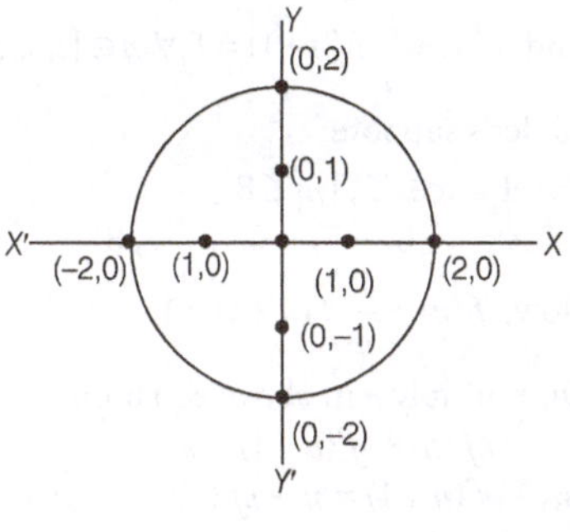


$\therefore$ Number of one-one function from $D$ to $C$

$$={}^{13}P_4$$

$$=13\times12\times11\times10=17160$$

**03.** (b) We have, $\sum_{n=8}^{100}\left[\frac{(-1)^n n}{2}\right]$

($\because [x]$ is the greatest integer function)

On substituting the values of $n$, we get

$$[4]+[-4.5]+[5]+[-5.5]+\ldots+[-49.5]+[50]$$

$$=4-5+5-6+\ldots-50+50$$

$$=4$$

**04.** (b) Given, graph is symmetrical about the line $x=2$.

$\therefore f(2+x)=f(2-x)$

**05.** (a) Given, $f(x)=\frac{2^{x+2}+16}{2^{2x+1}+2^{x+4}+32}$

$$\Rightarrow f(x)=\frac{2^x\cdot2^2+2^4}{(2^x)^2\cdot2+2^x\cdot2^4+2^5}$$

$$=\frac{2(2\cdot2^x+2^3)}{2((2^x)^2+8\cdot2^x+16)}$$

$$\Rightarrow f(x)=\frac{2(2^x+4)}{(2^x+4)^2}=\frac{2}{2^x+4} \quad \ldots(i)$$

Also, $f(4-x)=\frac{2}{2^{4-x}+4}=\frac{2}{\frac{2^4+4\cdot2^x}{2^x}}$

$$\Rightarrow f(4-x)=\frac{2\cdot2^x}{4(2^x+4)}=\frac{2^x}{2(2^x+4)} \quad \ldots(ii)$$

On adding Eqs. (i) and (ii), we get

$$f(x)+f(4-x)=\frac{2}{2^x+4}+\frac{2^x}{2(2^x+4)}$$

$$=\frac{(4+2^x)}{2(2^x+4)}$$

$$\Rightarrow f(x)+f(4-x)=\frac{1}{2} \quad \ldots(iii)$$

Now, $f(1/15)+f\left(4-\frac{1}{15}\right)$

$$=f(1/15)+f\left(\frac{59}{15}\right)=\frac{1}{2}$$

Similarly, $f(2/15)+f(58/15)=\frac{1}{2}$

$$f(3/15)+f\left(\frac{57}{15}\right)=\frac{1}{2}$$

and soon.

$f(29/15)+f(31/15)=1/2$ and

$$f(30/15)=f(2)=\frac{2}{2^2+4}=\frac{2}{8}=\frac{1}{4}$$

Hence,

$8[f(1/15)+f(2/15)+\ldots+f(59/15)]$

$$=8\left[\underbrace{1/2+1/2+1/2+\ldots+1/2}_{29\text{ terms}}+f(30/15)\right]$$

$$=8[29\times1/2+1/4]=118$$

**06.** (b) We have, $f(x)+f\left(\frac{1}{1-x}\right)=1+x$

Put $x=2$, then $f(2)+f(-1)=3 \quad \ldots(i)$

Put $x=1/2$, then

$$f\left(\frac{1}{2}\right)+f(2)=\frac{3}{2} \quad \ldots(ii)$$

Put $x=-1$, then

$$f(-1)+f\left(\frac{1}{2}\right)=0 \quad \ldots(iii)$$

On subtracting Eq. (iii) from Eq. (i), we get

$$f(2)-f\left(\frac{1}{2}\right)=3 \quad \ldots(iv)$$

On adding Eqs. (ii) and (iv), we get

$$2f(2)=\frac{3}{2}+3$$

$$\Rightarrow \quad 2f(2)=\frac{9}{2} \Rightarrow f(2)=\frac{9}{4}$$

**Alternate Solution**

We have, $f(x)+f\left(\frac{1}{1-x}\right)=1+x \quad \ldots(i)$

Replacing $x$ by $\frac{1}{1-x}$ in Eq. (i), we get

$$f\left(\frac{1}{1-x}\right)+f\left(\frac{1}{1-\frac{1}{1-x}}\right)=1+\frac{1}{1-x}$$

$$\Rightarrow f\left(\frac{1}{1-x}\right)+f\left(\frac{1-x}{1-x-1}\right)=\frac{1-x+1}{1-x}$$

$$\Rightarrow f\left(\frac{1}{1-x}\right)+f\left(\frac{1-x}{-x}\right)=\frac{2-x}{1-x} \quad \ldots(ii)$$

Now, replacing $x$ by $\frac{x-1}{x}$ in Eq. (i),

we get

$$f\left(\frac{x-1}{x}\right)+f(x)=\frac{2x-1}{x} \quad \ldots(iii)$$

Putting $x=2$ in Eqs. (i), (ii) and (iii), we get

$$f(2)+f(-1)=3 \quad \ldots(iv)$$

$$f(-1)+f\left(\frac{1}{2}\right)=0 \quad \ldots(v)$$

and $$f\left(\frac{1}{2}\right)+f(2)=3/2 \quad \ldots(vi)$$

On solving Eqs. (iv), (v) and (vi), we get

$$f(2)=\frac{9}{4}$$

**07.** *(d)* Let us consider a function $f: N \to R$ satisfying

$$f(1)+2f(2)+3f(3)\ldots+xf(x) = x(x+1)f(x) \quad \ldots(i)$$

We have for $x \ge 2$

$$f(1)+2f(2)+3f(3)+\ldots+xf(x) = x(x+1)f(x) \quad \ldots(ii)$$

Now we will replace $x$ by $(x+1)$ and subtracting Eq. (i) from Eq. (ii),

$$x(x+1)f(x)+(x+1)f(x+1) = (x+1)(x+2)f(x+1)$$

$$\Rightarrow \quad \frac{x}{f(x+1)}+\frac{1}{f(x)}=\frac{(x+2)}{f(x)}$$

$$\Rightarrow \quad xf(x)=(x+1)f(x+1) \text{ for } x \ge 2$$

$$f(2)=\frac{1}{4};\ f(4)=\frac{1}{8}$$

Now, $f(2022)=\dfrac{1}{4044}$

Similarly, $f(2028)=\dfrac{1}{4056}$

So $\dfrac{1}{f(2022)}+\dfrac{1}{f(2028)}$

$$=4044+4056=8100$$

**08.** *(a)* Given, $f(x)=2x^n+\lambda$, $\lambda \in R, n \in N$

and put $x=4$

$$f(4)=133,\ f(5)=255$$

$$f(4)=2x^n+\lambda$$

So, $133=2(4)^n+\lambda \quad \ldots(i)$

and put $x=5$

$$255=2(5)^n+\lambda \quad \ldots(ii)$$

On subtracting Eq. (i) from Eq. (ii), we get

$$122=2[5^n-4^n] \text{ or } 5^n-4^n=61$$

By hit and trial method for $n=3$,

$$5^3-4^3=61$$

$n=3$, satisfy the equation

Now, putting the value of $n$ in Eq. (i), we get

$$133=2(4)^3+\lambda$$

$$\lambda=133-128$$

$$\Rightarrow \quad \lambda=5$$

Now, $f(x)=2x^3+5$

So, $f(3)=2(3)^3+5=59$

$$f(2)=2(2)^3+5=21$$

So, $f(3)-f(2)=59-21=38$

Positive integer divisor of 38

$$=1, 2, 19, 38.$$

So, sum of positive integer divisors

$$=1+2+19+38$$
$$=60$$

**09.** *(c)* Given, $f(x)=\dfrac{2^{2x}}{2^{2x}+2}$

$$\therefore f(1-x)=\frac{2^{2(1-x)}}{2^{2(1-x)}+2}=\frac{\dfrac{2^2}{2^{2x}}}{\dfrac{2^2}{2^{2x}}+2}$$

$$=\frac{4}{4+2\cdot 2^{2x}}=\frac{2}{2+2^{2x}}$$

$\therefore f(x)+f(1-x)$

$$=\frac{2^{2x}}{2^{2x}+2}+\frac{2}{2+2^{2x}}$$

$$=\frac{2^{2x}+2}{2^{2x}+2}=1$$

$$\therefore f(x)+f(1-x)=1$$

$$\therefore \quad f\left(\frac{1}{2023}\right)+f\left(1-\frac{1}{2023}\right)=1$$

$$\Rightarrow f\left(\frac{1}{2023}\right)+f\left(\frac{2022}{2023}\right)=1$$

So, $\left\{f\left(\dfrac{1}{2023}\right)+f\left(\dfrac{2022}{2023}\right)\right\}$

$$+\left\{f\left(\frac{2}{2023}\right)+f\left(\frac{2021}{2023}\right)\right\}$$

$$+\ldots 1011 \text{ times}$$

$$=1+1+\ldots\ldots 1011 \text{ times}=1011$$

**10.** *(b)* $f(m+n)=f(m)+f(n),\ m, n \in N$

$\therefore \quad f(3+3)=f(3)+f(3)$

$\Rightarrow \quad f(6)=2f(3)=18 \quad [\because f(6)=18]$

Also $f(3)=f(2+1)=f(2)+f(1)$

$$=f(1+1)+f(1)$$

$$f(3)=f(1)+f(1)+f(1)$$

$\Rightarrow \quad 9=3f(1)$

$\Rightarrow \quad f(1)=3$

$\therefore \quad f(2)=f(1+1)=f(1)+f(1)$

$$=3+3=6$$

Hence, $f(2)\cdot f(3)=6\cdot 9=54$

**11.** *(c)* Given, $f(x)=\dfrac{5^x}{5^x+5}$, then

$$f(2-x)=\frac{5^{2-x}}{5^{2-x}+5}=\frac{5}{5^x+5}$$

This gives,

$$f(x)+f(2-x)=\frac{5^x+5}{5^x+5}=1$$

$$\Rightarrow f\left(\frac{1}{20}\right)+f\left(2-\frac{1}{20}\right)$$

$$=f\left(\frac{1}{20}\right)+f\left(\frac{39}{20}\right)=1$$

Similarly,

$$f\left(\frac{2}{20}\right)+f\left(\frac{38}{20}\right)=1 \text{ and so on.}$$

$$\therefore f\left(\frac{1}{20}\right)+f\left(\frac{2}{20}\right)+\ldots$$

$$+f\left(\frac{38}{20}\right)+f\left(\frac{39}{20}\right)$$

$$=1+1+\ldots+1+f\left(\frac{20}{20}\right)$$

$$=19+f(1)=19+\frac{1}{2}=\frac{39}{2}$$

**12.** *(a)* We have,

$$A=\{x \in R: [x+3]+[x+4] \le 3\}$$

Here, $[x+3]+[x+4] \le 3$

$\Rightarrow \quad [x]+3+[x]+4 \le 3$

$$(\because [x+n]=[x]+n, n \in I)$$

$\Rightarrow \quad 2[x]+4 \le 0$

$\Rightarrow \quad [x] \le -2$

$\Rightarrow \quad x \in (-\infty, -1)$

$$A \equiv (-\infty, -1) \quad \ldots(i)$$

Also,

$$B=\left\{x \in R: 3^x\left(\sum_{r=1}^{\infty}\frac{3}{10^r}\right)^{x-3}<3^{-3x}\right\}$$

Here,

$$3^x\left(\frac{3}{10}+\frac{3}{10^2}+\frac{3}{10^3}+\ldots\right)^{x-3}<3^{-3x}$$

$$\Rightarrow \quad 3^x\left(\frac{\dfrac{3}{10}}{1-\dfrac{1}{10}}\right)^{x-3}<3^{-3x}$$

$$\Rightarrow \quad 3^x\left(\frac{1}{3}\right)^{x-3}<3^{-3x}$$

$\Rightarrow \quad 3^{x-x+3}<3^{-3x} \Rightarrow 3^3<3^{-3x}$

$\Rightarrow \quad -3x>3 \Rightarrow x<-1$

$\Rightarrow \quad x \in (-\infty, -1)$

$\Rightarrow \quad B \equiv (-\infty, -1) \quad \ldots(ii)$

From Eqs. (i) and (ii), we get

$$A=B$$

**13.** *(b)* Given,

$$f:\{1, 2, 3, 4\} \to \{a \in Z; |a| \le 8\}$$

And $f(n)+\dfrac{1}{n}f(n+1)=1, \forall n \in \{1, 2, 3\}$

So, let's suppose

$$A=a \in Z; |a| \le 8$$

So, $A=\{-8, -7, -6, \ldots, 7, 8\}$

Now, $f(n)+\dfrac{1}{n}f(n+1)=1$

On multiply $n$ in above equation

$$nf(n)+f(n+1)=n$$

as, $\quad f(n+1)=n-nf(n)$

Now, for $n = 1$,

$$f(2) = 1 - f(1)$$

For $n = 2$,

$$f(3) = 2 - 2f(2) = 2 - 2(1 - f(1))$$

or, $f(3) = 2f(1)$

For $n = 3$

$$f(4) = 3 - 3f(3) = 3 - 3(2f(1))$$

or $f(4) = 3 - 6f(1)$

We know that

$$f(1) \in \{-8, -7, \ldots, 7, 8\} \quad \ldots\text{(i)}$$

Now, $-8 \le f(2) \le 8$

$$-8 \le 1 - f(1) \le 8$$

Subtract 1 from the inequality

$$-9 \le -f(1) \le 7$$

Now, multiply $(-1)$ in the inequality

$$9 \ge f(1) \ge -7$$

or $\quad f(1) \in \{-7, -6 \ldots, 8, 9\} \quad \ldots\text{(ii)}$

Now, $-8 \le f(3) \le 8$; $-8 \le 2f(1) \le 8$

Divide the inequality by 2

$$-4 \le f(1) \le 4$$

or $\quad f(1) \in \{-4, -3, -2, -1, 0, 1, 2, 3, 4\} \quad \ldots\text{(iii)}$

Now, $-8 \le f(4) \le 8$

$$-8 \le 3 - 6f(1) \le 8$$

Subtract 3 from inequality

$$-11 \le -6f(1) \le 5$$

Now, divide the inequality by $(-6)$

$$\frac{11}{6} \ge f(1) \ge \frac{-5}{6}$$

But, $f(1) \in I$

So, $\quad f(1) \in \{0, 1\} \quad \ldots\text{(iv)}$

From Eqs. (i), (ii), (iii) and (iv),

$$f(1) = 0, 1$$

If $f(1) = 0$, then $f(2) = 1$, $f(3) = 0$, $f(4) = 3$

If $f(1) = 1$, then $f(2) = 0$, $f(3) = 2$, $f(4) = -3$

So, number of such functions = 2

**14.** *(96)* $-1 \le \dfrac{4x+5}{3x-7} \le 1$

$$\frac{4x+5}{3x-7} + 1 \ge 0 \text{ and } \frac{4x+5}{3x-7} - 1 \le 0$$

$$\frac{7x-2}{3x-7} \ge 0 \text{ and } \frac{x+12}{3x-7} \le 0'$$

$$x \ge \frac{7}{3} \text{ or } x \le \frac{2}{7} \text{ and } -12 \le x \le \frac{7}{3}$$

Domain =

$$\left(-\infty, +\frac{2}{7}\right] \cup \left(\frac{7}{3}, \infty\right) \cap \left[-12, \frac{7}{3}\right]$$

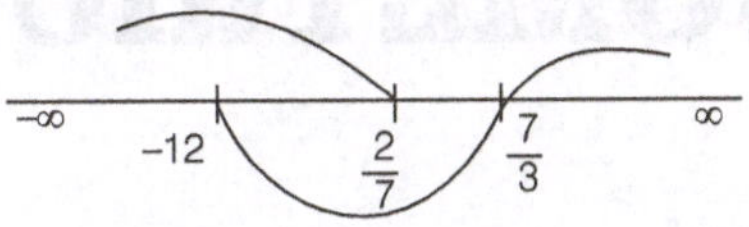


$$\text{Domain} = \left[-12, \frac{2}{7}\right]$$

$$\Rightarrow \quad \alpha = -12, \quad \beta = \frac{2}{7}$$

$$g(x) = \log_2 (2 - 6\log_{27}(2x+5))$$
$$= \log_2 (2 - 2\log_3 (2x+5))$$

For $g(x)$ to be well defined

$2x + 5 > 0$

$$\Rightarrow \quad x > \frac{-5}{2}$$

$$2 - 2\log_3(2x+5) > 0$$
$$2 > 2\log_3(2x+5)$$
$$\log_3(2x+5) < 1$$
$$2x + 5 < 3$$
$$x < -1$$

$$\text{Domain} = \left(\frac{-5}{2}, -1\right)$$

$$\gamma = \frac{-5}{2} \Rightarrow \delta = -1$$

So, $|7(\alpha+\beta) + 4(\gamma+\delta)|$

$$\left|7\left(-12 + \frac{2}{7}\right) + 4\left(\frac{-5}{2} - 1\right)\right|$$

$$|-84 + 2 - 10 - 4|$$
$$= |-96| = 96$$

**15.** *(18)* We have,

$$|2a - 1| = 3[a] + 2\{a\}$$

$$\Rightarrow \quad |2a - 1| = 3[a] + 2(a - [a])$$

$$\Rightarrow \quad |2a - 1| = [a] + 2a$$

When $a > \dfrac{1}{2}$, then

$$2a - 1 = [a] + 2a$$

$$\Rightarrow \quad [a] = -1$$

$\Rightarrow a \in [-1, 0)$ which is not satisfied.

When $a < \dfrac{1}{2}$, then

$$-2a + 1 = [a] + 2a$$

On putting $\quad a = I + f$

$$-2(I + f) + 1 = I + 2(I + f)$$

$$\Rightarrow \quad -2I - 2f + 1 = I + 2I + 2f$$

$$\Rightarrow \quad 5I + 4f = 1$$

$$\Rightarrow \quad I = 0, f = \frac{1}{4}$$

$$\therefore \quad a = 1/4$$

$$\therefore \quad 72\,\Sigma a = 72 \times \frac{1}{4} = 18$$

**16.** *(51)* Given, $A = \{1, 2, 3, \ldots, 7\}$

Then, $n(P(A)) = 2^7 = 128$

$$f : A \to P(A)$$

So, number of function

$$= 128 \times 128 \times \ldots \times 128$$
$$= 128^7 = (2^7)^7 = 2^{49}$$

$$\Rightarrow \quad m^n = 2^{49}$$

$$\therefore \quad m = 2 \text{ and } n = 49$$

$$\therefore \quad m + n = 51$$

**17.** *(432)* $f(1 \cdot n) = f(1) \cdot f(n) \Rightarrow f(1) = 1$

$$f(3.3) = (f(3))^2$$

$$\Rightarrow \quad f(9) = f(3) \times f(3)$$

So, $\quad f(3) = 1 \text{ or } 3$

$f(2)$, $f(5)$ and $f(8)$ can be chosen in 6 ways each.

Therefore, total number of functions

$$= 1 \times 6 \times 2 \times 6 \times 6 \times 1 = 432$$

# C-03



In JEE Main, there is usually 0-1 question per paper is visible from this chapter.
In past 3 years, 29 conceptual and 6 numerical value type questions of easy to moderate level were asked from this chapter.
**For detailed analysis scan QR code**

# TRIGONOMETRIC FUNCTIONS

## Topic 1 Trigonometric Ratios and Identities, Domain & Range of Trigonometric Functions

### Level 1

**01** If $\sin x = \frac{-3}{5}$, where $\pi < x < \frac{3\pi}{2}$, then $80(\tan^2 x - \cos x)$ is equal to *[2024, 8 April Shift-I]*

(a) 108 (b) 109
(c) 18 (d) 19

**Imp.** Domain & Range of Trigonometric Function

**02** Let $f(x) = \frac{1}{7 - \sin 5x}$ be a function defined on $R$. Then, the range of the function $f(x)$ is equal to *[2024, 6 April Shift-I]*

(a) $\left[\frac{1}{7}, \frac{1}{6}\right]$ (b) $\left[\frac{1}{7}, \frac{1}{5}\right]$ (c) $\left[\frac{1}{8}, \frac{1}{6}\right]$ (d) $\left[\frac{1}{8}, \frac{1}{5}\right]$

**03** For $\alpha, \beta \in \left(0, \frac{\pi}{2}\right)$, let $3\sin(\alpha + \beta) = 2\sin(\alpha - \beta)$ and a real number $k$ be such that $\tan\alpha = k\tan\beta$. Then, the value of $k$ is equal to *[2024, 30 Jan Shift-II]*

(a) $\frac{-2}{3}$ (b) $-5$ (c) $\frac{2}{3}$ (d) 5

**04** If $\cot\alpha = 1$ and $\sec\beta = -\frac{5}{3}$, where $\pi < \alpha < \frac{3\pi}{2}$ and $\frac{\pi}{2} < \beta < \pi$, then the value of $\tan(\alpha + \beta)$ and the quadrant in which $\alpha + \beta$ lies, respectively are *[2022, 28 June Shift-II]*

(a) $-\frac{1}{7}$ and IVth quadrant (b) 7 and Ist quadrant
(c) $-7$ and IVth quadrant (d) $\frac{1}{7}$ and Ist quadrant

**05** $16\sin(20°)\sin(40°)\sin(80°)$ is equal to *[2022, 26 June Shift-II]*

(a) $\sqrt{3}$ (b) $2\sqrt{3}$ (c) 3 (d) $4\sqrt{3}$

**06** cosec 18° is a root of the equation *[2021, 31 Aug Shift-I]*

(a) $x^2 + 2x - 4 = 0$ (b) $4x^2 + 2x - 1 = 0$
(c) $x^2 - 2x + 4 = 0$ (d) $x^2 - 2x - 4 = 0$

**07** The value of $\sin 10° \sin 30° \sin 50° \sin 70°$ is *[2019, 9 April Shift-II]*

(a) $\frac{1}{36}$ (b) $\frac{1}{32}$ (c) $\frac{1}{16}$ (d) $\frac{1}{18}$

**08** The expression $\frac{\tan A}{1 - \cot A} + \frac{\cot A}{1 - \tan A}$ can be written as *[JEE Main 2013]*

(a) $\sin A \cos A + 1$ (b) $\sec A \operatorname{cosec} A + 1$
(c) $\tan A + \cot A$ (d) $\sec A + \operatorname{cosec} A$

**09** Let $\cos(\alpha + \beta) = \frac{4}{5}$ and $\sin(\alpha - \beta) = \frac{5}{13}$, where $0 \le \alpha, \beta \le \frac{\pi}{4}$. Then, $\tan 2\alpha$ is equal to *[AIEEE 2010]*

(a) $\frac{25}{16}$ (b) $\frac{56}{33}$ (c) $\frac{19}{12}$ (d) $\frac{20}{7}$

**10** The value of $\frac{1 - \tan^2 15°}{1 + \tan^2 15°}$ is *[AIEEE 2002]*

(a) 1 (b) $\sqrt{3}$ (c) $\sqrt{3}/2$ (d) 2

**11** If $\tan\theta = -\frac{4}{3}$, then $\sin\theta$ is *[AIEEE 2002]*

(a) $-\frac{4}{5}$ but not $\frac{4}{5}$ (b) $-\frac{4}{5}$ or $\frac{4}{5}$
(c) $\frac{4}{5}$ but not $-\frac{4}{5}$ (d) None of these

**12** If $y = \sin^2\theta + \operatorname{cosec}^2\theta, \theta \ne 0$, then *[AIEEE 2002]*

(a) $y = 0$ (b) $y \le 2$
(c) $y \ge -2$ (d) $y \ge 2$

## Level 2

**13** If $10\sin^4\theta + 15\cos^4\theta = 6$, then the value of $\frac{27\,\text{cosec}^6\theta + 8\sec^6\theta}{16\sec^8\theta}$ is **[2025, 4 April Shift-I]**

(a) $\frac{1}{5}$ (b) $\frac{3}{4}$ (c) $\frac{3}{5}$ (d) $\frac{2}{5}$

**14** The value of $(\sin 70°)(\cot 10° \cot 70° - 1)$ is **[2025, 23 Jan Shift-I]**

(a) $\frac{2}{3}$ (b) 0 (c) $\frac{3}{2}$ (d) 1

**15** Let $|\cos\theta\cos(60-\theta)\cos(60+\theta)| \le \frac{1}{8}$, $\theta \in [0, 2\pi]$. Then, the sum of all $\theta \in [0, 2\pi]$, where $\cos 3\theta$ attains its maximum value, is **[2024, 9 April Shift-I]**

(a) $9\pi$ (b) $6\pi$ (c) $15\pi$ (d) $18\pi$

**16** If the value of $\frac{3\cos 36° + 5\sin 18°}{5\cos 36° - 3\sin 18°}$ is $\frac{a\sqrt{5}-b}{c}$, where $a, b, c$ are natural numbers and $\gcd(a, c) = 1$, then $a + b + c$ is equal to **[2024, 8 April Shift-II]**

(a) 40 (b) 50 (c) 54 (d) 52

**17** If $\tan A = \frac{1}{\sqrt{x(x^2+x+1)}}$, $\tan B = \frac{\sqrt{x}}{\sqrt{x^2+x+1}}$ and $\tan C = (x^{-3} + x^{-2} + x^{-1})^{\frac{1}{2}}$, $0 < A, B, C < \frac{\pi}{2}$, then $A + B$ is equal to **[2024, 1 Feb Shift-I]**

(a) $\frac{\pi}{2} - C$ (b) $C$

(c) $\pi - C$ (d) $2\pi - C$

**18** The value of $36(4\cos^2 9° - 1)(4\cos^2 27° - 1)(4\cos^2 81° - 1)(4\cos^2 243° - 1)$ is **[2023, 8 April Shift-I]**

(a) 27 (b) 36 (c) 18 (d) 54

**19** $\alpha = \sin 36°$ is a root of which of the following equation? **[2022, 27 June Shift-II]**

(a) $16x^4 - 10x^2 - 5 = 0$ (b) $16x^4 + 20x^2 - 5 = 0$

(c) $16x^4 - 20x^2 + 5 = 0$ (d) $16x^4 - 10x^2 + 5 = 0$

**20** The value of $2\sin(12°) - \sin(72°)$ is **[2022, 25 June Shift-II]**

(a) $\frac{\sqrt{5}(1-\sqrt{3})}{4}$ (b) $\frac{1-\sqrt{5}}{8}$

(c) $\frac{\sqrt{3}(1-\sqrt{5})}{2}$ (d) $\frac{\sqrt{3}(1-\sqrt{5})}{4}$

**21** If $\sin\theta + \cos\theta = \frac{1}{2}$, then $16[\sin(2\theta) + \cos(4\theta) + \sin(6\theta)]$ is equal to **[2021, 27 July Shift-I]**

(a) 23 (b) $-27$ (c) $-23$ (d) 27

**22** If $15\sin^4\alpha + 10\cos^4\alpha = 6$, for some $\alpha \in R$, then the value of $27\sec^6\alpha + 8\,\text{cosec}^6\alpha$ is equal to **[2021, 18 March Shift-II]**

(a) 350 (b) 500 (c) 400 (d) 250

**23** If for $x \in \left(0, \frac{\pi}{2}\right)$, $\log_{10}\sin x + \log_{10}\cos x = -1$ and $\log_{10}(\sin x + \cos x) = \frac{1}{2}(\log_{10} n - 1)$, $n > 0$, then the value of $n$ is equal to **[2021, 16 March Shift-I]**

(a) 20 (b) 12 (c) 9 (d) 16

**24** All possible values of $\theta \in [0, 2\pi]$ for which $\sin 2\theta + \tan 2\theta > 0$ lie in **[2021, 25 Feb Shift-I]**

(a) $\left(0, \frac{\pi}{2}\right) \cup \left(\pi, \frac{3\pi}{2}\right)$

(b) $\left(0, \frac{\pi}{2}\right) \cup \left(\frac{\pi}{2}, \frac{3\pi}{4}\right) \cup \left(\pi, \frac{7\pi}{6}\right)$

(c) $\left(0, \frac{\pi}{4}\right) \cup \left(\frac{\pi}{2}, \frac{3\pi}{4}\right) \cup \left(\frac{3\pi}{2}, \frac{11\pi}{6}\right)$

(d) $\left(0, \frac{\pi}{4}\right) \cup \left(\frac{\pi}{2}, \frac{3\pi}{4}\right) \cup \left(\pi, \frac{5\pi}{4}\right) \cup \left(\frac{3\pi}{2}, \frac{7\pi}{4}\right)$

**25** If $L = \sin^2\left(\frac{\pi}{16}\right) - \sin^2\left(\frac{\pi}{8}\right)$ and $M = \cos^2\left(\frac{\pi}{16}\right) - \sin^2\left(\frac{\pi}{8}\right)$, then **[2020, 5 Sep. Shift-II]**

(a) $L = -\frac{1}{2\sqrt{2}} + \frac{1}{2}\cos\frac{\pi}{8}$ (b) $L = \frac{1}{4\sqrt{2}} - \frac{1}{4}\cos\frac{\pi}{8}$

(c) $M = \frac{1}{4\sqrt{2}} + \frac{1}{4}\cos\frac{\pi}{8}$ (d) $M = \frac{1}{2\sqrt{2}} + \frac{1}{2}\cos\frac{\pi}{8}$

**Imp.** AM-GM inequality

**26** The minimum value of $2^{\sin x} + 2^{\cos x}$ is **[2020, 4 Sep Shift-II]**

(a) $2^{-1+\sqrt{2}}$ (b) $2^{-1+\frac{1}{\sqrt{2}}}$

(c) $2^{1-\frac{1}{\sqrt{2}}}$ (d) $2^{1-\sqrt{2}}$

**27** The value of $\cos^3\left(\frac{\pi}{8}\right)\cdot\cos\left(\frac{3\pi}{8}\right) + \sin^3\left(\frac{\pi}{8}\right)\cdot\sin\left(\frac{3\pi}{8}\right)$ is **[2020, 9 Jan Shift-I]**

(a) $\frac{1}{4}$ (b) $\frac{1}{2\sqrt{2}}$ (c) $\frac{1}{2}$ (d) $\frac{1}{\sqrt{2}}$

**28** The equation $y = \sin x\sin(x+2) - \sin^2(x+1)$ represents a straight line lying in **[2019, 12 April Shift-I]**

(a) second and third quadrants only

(b) first, second and fourth quadrants

(c) first, third and fourth quadrants

(d) third and fourth quadrants only

**29** All the pairs $(x, y)$ that satisfy the inequality $2^{\sqrt{\sin^2 x - 2\sin x + 5}} \cdot \dfrac{1}{4^{\sin^2 y}} \le 1$ also satisfy the equation **[2019, 10 April Shift-I]**

(a) $2|\sin x| = 3\sin y$ (b) $\sin x = |\sin y|$
(c) $\sin x = 2\sin y$ (d) $2\sin x = \sin y$

**30** If an angle $A$ of a $\Delta ABC$ satisfies $5\cos A + 3 = 0$, then the roots of the quadratic equation, $9x^2 + 27x + 20 = 0$ are **[JEE Main 2018]**

(a) $\sin A, \sec A$ (b) $\sec A, \cot A$
(c) $\sec A, \tan A$ (d) $\tan A, \cos A$

**31** If $5(\tan^2 x - \cos^2 x) = 2\cos 2x + 9$, then the value of $\cos 4x$ is **[JEE Main 2017]**

(a) $\dfrac{1}{3}$ (b) $\dfrac{2}{9}$ (c) $-\dfrac{7}{9}$ (d) $-\dfrac{3}{5}$

**32** If $A > 0, B > 0$ and $A + B = \pi/6$, then the minimum value of $\tan A + \tan B$ is **[JEE Main 2016]**

(a) $\sqrt{3} - \sqrt{2}$ (b) $4 - 2\sqrt{3}$ (c) $2/\sqrt{3}$ (d) $2 - \sqrt{3}$

**33** If $f_k(x) = \dfrac{1}{k}(\sin^k x + \cos^k x)$, where $x \in R, k \ge 1$, then $f_4(x) - f_6(x)$ is equal to **[JEE Main 2014]**

(a) 1/6 (b) 1/3 (c) 1/4 (d) 1/12

**34** If $A = \sin^2 x + \cos^4 x$, then for all real $x$ **[AIEEE 2011]**

(a) $1 \le A \le 2$ (b) $\dfrac{3}{4} \le A \le \dfrac{13}{16}$
(c) $\dfrac{3}{4} \le A \le 1$ (d) $\dfrac{13}{16} \le A \le 1$

**35** Let $\alpha$ and $\beta$ be such that $\pi < \alpha - \beta < 3\pi$. If $\sin\alpha + \sin\beta = -\dfrac{21}{65}$ and $\cos\alpha + \cos\beta = -\dfrac{27}{65}$, then the value of $\cos\left(\dfrac{\alpha - \beta}{2}\right)$ is **[AIEEE 2004]**

(a) $-\dfrac{3}{\sqrt{130}}$ (b) $\dfrac{3}{\sqrt{130}}$ (c) $\dfrac{6}{65}$ (d) $-\dfrac{6}{65}$

**36** If $\sin(\alpha + \beta) = 1$ and $\sin(\alpha - \beta) = \dfrac{1}{2}$, then $\tan(\alpha + 2\beta)\tan(2\alpha + \beta)$ is equal to **[AIEEE 2002]**

(a) 1 (b) $-1$
(c) zero (d) None of these

## Level 3

**37** If $\sum_{r=1}^{13}\left\{\dfrac{1}{\sin\left(\dfrac{\pi}{4} + (r-1)\dfrac{\pi}{6}\right)\sin\left(\dfrac{\pi}{4} + \dfrac{r\pi}{6}\right)}\right\} = a\sqrt{3} + b, a, b \in Z$, then $a^2 + b^2$ is equal to **[2025, 28 Jan Shift-II]**

(a) 2 (b) 8 (c) 10 (d) 4

**38** Let $S = \left\{x \in \left(-\dfrac{\pi}{2}, \dfrac{\pi}{2}\right) : 9^{1 - \tan^2 x} + 9^{\tan^2 x} = 10\right\}$ and $\beta = \sum_{x \in S} \tan^2\left(\dfrac{x}{3}\right)$, then $\dfrac{1}{6}(\beta - 14)^2$ is equal to **[2023, 10 April Shift-II]**

(a) 8 (b) 16 (c) 64 (d) 32

**39** $96\cos\dfrac{\pi}{33}\cos\dfrac{2\pi}{33}\cos\dfrac{4\pi}{33}\cos\dfrac{8\pi}{33}\cos\dfrac{16\pi}{33}$ is equal to **[2023, 10 April Shift-I]**

(a) 4 (b) 3 (c) 2 (d) 1

**40** $2\sin\left(\dfrac{\pi}{22}\right)\sin\left(\dfrac{3\pi}{22}\right)\sin\left(\dfrac{5\pi}{22}\right)\sin\left(\dfrac{7\pi}{22}\right)\sin\left(\dfrac{9\pi}{22}\right)$ is equal to **[2022, 25 July Shift-II]**

(a) 3/16 (b) 1/16 (c) 1/32 (d) 9/32

**41** The range of the function $f(x) = \log_{\sqrt{5}}\left(3 + \cos\left(\dfrac{3\pi}{4} + x\right) + \cos\left(\dfrac{\pi}{4} + x\right) + \cos\left(\dfrac{\pi}{4} - x\right) - \cos\left(\dfrac{3\pi}{4} - x\right)\right)$ is **[2021, 1 Sep Shift-II]**

(a) $\left[\dfrac{1}{\sqrt{5}}, \sqrt{5}\right]$ (b) $[0, 2]$ (c) $(0, \sqrt{5})$ (d) $[-2, 2]$

**42** The value of $2\sin\left(\dfrac{\pi}{8}\right)\sin\left(\dfrac{2\pi}{8}\right)\sin\left(\dfrac{3\pi}{8}\right)\sin\left(\dfrac{5\pi}{8}\right)\sin\left(\dfrac{6\pi}{8}\right)\sin\left(\dfrac{7\pi}{8}\right)$ is **[2021, 26 Aug Shift-II]**

(a) $\dfrac{1}{4\sqrt{2}}$ (b) $\dfrac{1}{4}$ (c) $\dfrac{1}{8}$ (d) $\dfrac{1}{8\sqrt{2}}$

**43** Let $f : R \to R$ be defined as $f(x + y) + f(x - y) = 2f(x)f(y), f\left(\dfrac{1}{2}\right) = -1$. Then, the value of $\sum_{k=1}^{20}\dfrac{1}{\sin(k)\sin[k + f(k)]}$ is equal to **[2021, 27 July Shift-I]**

(a) $\text{cosec}^2(21)\cos(20)\cos(2)$
(b) $\sec^2(1)\sec(21)\cos(20)$
(c) $\text{cosec}^2(1)\text{cosec}(21)\sin(20)$
(d) $\sec^2(21)\sin(20)\sin(2)$

**44** If $0 < x, y < \pi$ and $\cos x + \cos y - \cos(x + y) = 3/2$, then $\sin x + \cos y$ is equal to **[2021, 25 Feb Shift-II]**

(a) $\dfrac{1}{2}$ (b) $\dfrac{\sqrt{3}}{2}$ (c) $\dfrac{1 - \sqrt{3}}{2}$ (d) $\dfrac{1 + \sqrt{3}}{2}$

**45** If $\cos(\alpha + \beta) = \dfrac{3}{5}$, $\sin(\alpha - \beta) = \dfrac{5}{13}$ and $0 < \alpha, \beta < \dfrac{\pi}{4}$, then $\tan(2\alpha)$ is equal to **[2019, 8 April Shift-I]**

(a) $\dfrac{63}{52}$ (b) $\dfrac{63}{16}$ (c) $\dfrac{21}{16}$ (d) $\dfrac{33}{52}$

**46** For any $\theta \in \left(\frac{\pi}{4}, \frac{\pi}{2}\right)$, the expression

$3(\sin\theta - \cos\theta)^4 + 6(\sin\theta + \cos\theta)^2 + 4\sin^6\theta$ equals *[2019, 9 Jan Shift-I]*

(a) $13 - 4\cos^4\theta + 2\sin^2\theta\cos^2\theta$
(b) $13 - 4\cos^2\theta + 6\cos^4\theta$
(c) $13 - 4\cos^2\theta + 6\sin^2\theta\cos^2\theta$
(d) $13 - 4\cos^6\theta$

**47** If $\tan A$ and $\tan B$ are the roots of the quadratic equation, $3x^2 - 10x - 25 = 0$, then the value of $3\sin^2(A+B) - 10\sin(A+B)\cdot\cos(A+B) - 25\cos^2(A+B)$ is *[JEE Main 2018]*

(a) 10 (b) −10 (c) 25 (d) −25

**48** If $\cos\alpha + \cos\beta = \frac{3}{2}$ and $\sin\alpha + \sin\beta = \frac{1}{2}$ and $\theta$ is the arithmetic mean of $\alpha$ and $\beta$, then $\sin 2\theta + \cos 2\theta$ is equal to *[JEE Main 2015]*

(a) $\frac{3}{5}$ (b) $\frac{4}{5}$ (c) $\frac{7}{5}$ (d) $\frac{8}{5}$

## Numerical Value Type Questions

**49** The value of $\tan 9° - \tan 27° - \tan 63° + \tan 81°$ is ...... . *[2023, 6 April Shift-II]*

**50** If $\sin^2(10°)\sin(20°)\sin(40°)\sin(50°)\sin(70°) = \alpha - \frac{1}{16}\sin(10°)$, then $16 + \alpha^{-1}$ is equal to ........ . *[2021, 26 June Shift-I]*

**51** If $\frac{\sqrt{2}\sin\alpha}{\sqrt{1+\cos 2\alpha}} = \frac{1}{7}$ and $\sqrt{\frac{1-\cos 2\beta}{2}} = \frac{1}{\sqrt{10}}$, $\alpha, \beta \in \left(0, \frac{\pi}{2}\right)$, then $\tan(\alpha + 2\beta)$ is equal to...... . *[2020, 8 Jan Shift-II]*

## *Topic 2 Trigonometric Equations

**(This topic is not included in latest JEE Main syllabus.)*

### Level 1

**01** If $\theta \in [-2\pi, 2\pi]$, then the number of solutions of $2\sqrt{2}\cos^2\theta + (2-\sqrt{6})\cos\theta - \sqrt{3} = 0$, is equal to *[2025, 2 April Shift-I]*

(a) 12 (b) 8 (c) 6 (d) 10

**02** If $\theta \in \left[-\frac{7\pi}{6}, \frac{4\pi}{3}\right]$, then the number of solutions of $\sqrt{3}\operatorname{cosec}^2\theta - 2(\sqrt{3}-1)\operatorname{cosec}\theta - 4 = 0$ is equal to *[2025, 2 April Shift-II]*

(a) 7 (b) 6 (c) 10 (d) 8

**03** The number of solutions of the equation $2x + 3\tan x = \pi$, $x \in [-2\pi, 2\pi] - \left\{\pm\frac{\pi}{2}, \pm\frac{3\pi}{2}\right\}$ is *[2025, 3 April Shift-I]*

(a) 3 (b) 5 (c) 4 (d) 6

**04** If $\sin x + \sin^2 x = 1, x \in \left(0, \frac{\pi}{2}\right)$, then $(\cos^{12}x + \tan^{12}x) + 3(\cos^{10}x + \tan^{10}x + \cos^8 x + \tan^8 x) + (\cos^6 x + \tan^6 x)$ is equal to *[2025, 29 Jan Shift-II]*

(a) 4 (b) 1 (c) 3 (d) 2

**05** The sum of all values of $\theta \in [0, 2\pi]$ satisfying $2\sin^2\theta = \cos 2\theta$ and $2\cos^2\theta = 3\sin\theta$ is *[2025, 22 Jan Shift-II]*

(a) $4\pi$ (b) $\frac{5\pi}{6}$ (c) $\frac{\pi}{2}$ (d) $\pi$

**06** The number of solutions of the equation $4\sin^2 x - 4\cos^3 x + 9 - 4\cos x = 0$ $x \in |-2\pi, 2\pi|$ is *[2024, 1 Feb Shift-II]*

(a) 1 (b) 3 (c) 2 (d) 0

**07** The number of solutions of the equation $32^{\tan^2 x} + 32^{\sec^2 x} = 81, 0 \le x \le \frac{\pi}{4}$ is *[2021, 31 Aug Shift-II]*

(a) 3 (b) 1 (c) 0 (d) 2

**08** The possible values of $\theta \in (0, \pi)$ such that $\sin(\theta) + \sin(4\theta) + \sin(7\theta) = 0$ are *[AIEEE 2011]*

(a) $\frac{2\pi}{9}, \frac{\pi}{4}, \frac{4\pi}{9}, \frac{\pi}{2}, \frac{3\pi}{4}, \frac{8\pi}{9}$
(b) $\frac{\pi}{4}, \frac{5\pi}{12}, \frac{\pi}{2}, \frac{2\pi}{3}, \frac{3\pi}{4}, \frac{8\pi}{9}$
(c) $\frac{2\pi}{9}, \frac{\pi}{4}, \frac{\pi}{2}, \frac{2\pi}{3}, \frac{3\pi}{4}, \frac{35\pi}{36}$
(d) $\frac{2\pi}{9}, \frac{\pi}{4}, \frac{\pi}{2}, \frac{2\pi}{3}, \frac{3\pi}{4}, \frac{8\pi}{9}$

**09** The number of values of $x$ in the interval $[0, 3\pi]$ satisfying the equation $2\sin^2 x + 5\sin x - 3 = 0$ is *[AIEEE 2006]*

(a) 6 (b) 1 (c) 2 (d) 4

**10** If $f: R \to S$, defined by $f(x) = \sin x - \sqrt{3}\cos x + 1$, is onto, then the interval of $S$ is *[AIEEE 2004]*

(a) [0, 3] (b) [−1, 1] (c) [0, 1] (d) [−1, 3]

### Level 2

**11** The number of solutions of the equation $\cos 2\theta \cos\frac{\theta}{2} + \cos\frac{5\theta}{2} = 2\cos^3\frac{5\theta}{2}$ in $\left[-\frac{\pi}{2}, \frac{\pi}{2}\right]$ is *[2025, 7 April Shift-II]*

(a) 9 (b) 5 (c) 6 (d) 7

**12** Suppose $\theta \in \left[0, \frac{\pi}{4}\right]$ is a solution of $4\cos\theta - 3\sin\theta = 1$. Then, $\cos\theta$ is equal to *[2024, 5 April Shift-I]*
(a) $\frac{6-\sqrt{6}}{(3\sqrt{6}-2)}$ (b) $\frac{4}{(3\sqrt{6}+2)}$ (c) $\frac{4}{(3\sqrt{6}-2)}$ (d) $\frac{6+\sqrt{6}}{(3\sqrt{6}+2)}$

**13** The number of solutions of the equation $e^{\sin x} - 2e^{-\sin x} = 2$, is *[2024, 31 Jan Shift-II]*
(a) 1 (b) more than 2
(c) 2 (d) 0

**14** If $\alpha, -\frac{\pi}{2} < \alpha < \frac{\pi}{2}$ is the solution of $4\cos\theta + 5\sin\theta = 1$, then the value of $\tan\alpha$ is *[2024, 29 Jan Shift-I]*
(a) $\frac{10-\sqrt{10}}{12}$ (b) $\frac{\sqrt{10}-10}{6}$ (c) $\frac{\sqrt{10}-10}{12}$ (d) $\frac{10-\sqrt{10}}{6}$

**15** The number of elements in the set $S = \left\{x \in R : 2\cos\left(\frac{x^2+x}{6}\right) = 4^x + 4^{-x}\right\}$ is *[2022, 29 July Shift-II]*
(a) 1 (b) 3 (c) 0 (d) infinite

**16** The number of solutions of $|\cos x| = \sin x$, such that $-4\pi \le x \le 4\pi$ is *[2022, 25 July Shift-I]*
(a) 4 (b) 6 (c) 8 (d) 12

**17** The number of solutions of the equation $\cos\left(x+\frac{\pi}{3}\right)\cos\left(\frac{\pi}{3}-x\right) = \frac{1}{4}\cos^2 2x,\ x \in [-3\pi, 3\pi]$ is *[2022, 24 June Shift-II]*
(a) 8 (b) 5 (c) 6 (d) 7

**18** If $n$ is the number of solutions of the equation $2\cos x\left(4\sin\left(\frac{\pi}{4}+x\right)\sin\left(\frac{\pi}{4}-x\right)-1\right) = 1, x \in [0, \pi]$ and $S$ is the sum of all these solutions, then the ordered pair $(n, S)$ is *[2021, 1 Sep Shift-II]*
(a) $(3, 13\pi/9)$ (b) $(2, 2\pi/3)$ (c) $(2, 8\pi/9)$ (d) $(3, 5\pi/3)$

**Imp.** Solution of Trigonometric Equation

**19** The number of solutions of $\sin^7 x + \cos^7 x = 1$, $x \in [0, 4\pi]$ is equal to *[2021, 22 July Shift-II]*
(a) 11 (b) 7 (c) 5 (d) 9

**20** If the equation $\cos^4\theta + \sin^4\theta + \lambda = 0$ has real solutions for $\theta$, then $\lambda$ lies in the interval *[2020, 2 Sep Shift-II]*
(a) $\left(-\frac{5}{4}, -1\right)$ (b) $\left[-1, -\frac{1}{2}\right]$
(c) $\left(-\frac{1}{2}, -\frac{1}{4}\right]$ (d) $\left[-\frac{3}{2}, -\frac{5}{4}\right]$

**21** The number of solutions of the equation $1 + \sin^4 x = \cos^2 3x, x \in \left[-\frac{5\pi}{2}, \frac{5\pi}{2}\right]$ is *[2019, 12 April Shift-I]*
(a) 3 (b) 5 (c) 7 (d) 4

**22** The maximum value of $3\cos\theta + 5\sin\left(\theta - \frac{\pi}{6}\right)$ for any real value of $\theta$ is *[2019, 12 Jan Shift-I]*
(a) $\frac{\sqrt{79}}{2}$ (b) $\sqrt{34}$ (c) $\sqrt{31}$ (d) $\sqrt{19}$

**23** If sum of all the solutions of the equation $8\cos x \cdot \left(\cos\left(\frac{\pi}{6}+x\right)\cdot\cos\left(\frac{\pi}{6}-x\right)-\frac{1}{2}\right) = 1$ in $[0, \pi]$ is $k\pi$, then $k$ is equal to *[JEE Main 2018]*
(a) 2/3 (b) 13/9 (c) 8/9 (d) 20/9

**24** The number of solutions of $\sin 3x = \cos 2x$, in the interval $\left(\frac{\pi}{2}, \pi\right)$ is *[JEE Main 2018]*
(a) 2 (b) 4 (c) 3 (d) 1

**25** If $\alpha$ is a root of $25\cos^2\theta + 5\cos\theta - 12 = 0, \frac{\pi}{2} < \alpha < \pi$, then $\sin 2\alpha$ is equal to *[AIEEE 2002]*
(a) $\frac{24}{25}$ (b) $-\frac{24}{25}$
(c) $\frac{13}{18}$ (d) $-\frac{13}{18}$

**26** The equation $a\sin x + b\cos x = c$, where $|c| > \sqrt{a^2+b^2}$ has *[AIEEE 2002]*
(a) a unique solution (b) infinite number of solutions
(c) no solution (d) None of these

## Level 3

**27** If $2\sin^3 x + \sin 2x \cos x + 4\sin x - 4 = 0$ has exactly 3 solutions in the interval $\left[0, \frac{n\pi}{2}\right], n \in N$, then the roots of the equation $x^2 + nx + (n-3) = 0$ belong to *[2024, 30 Jan Shift-I]*
(a) $\left(-\frac{\sqrt{17}}{2}, \frac{\sqrt{17}}{2}\right)$ (b) $Z$
(c) $(0, \infty)$ (d) $(-\infty, 0)$

**28** The sum of the solutions $x \in R$ of the equation $\frac{3\cos 2x + \cos^3 2x}{\cos^6 x - \sin^6 x} = x^3 - x^2 + 6$ is *[2024, 29 Jan Shift-II]*
(a) 0 (b) −1
(c) 3 (d) 1

**29** The number of elements in the set $S=\{\theta\in[0,2\pi]:3\cos^4\theta-5\cos^2\theta-2\sin^6\theta+2=0\}$ is *[2023, 11 April Shift-I]*

(a) 12 (b) 8 (c) 9 (d) 10

**30** If $\tan 15^\circ+\frac{1}{\tan 75^\circ}+\frac{1}{\tan 105^\circ}+\tan 195^\circ=2a$, then the value of $\left(a+\frac{1}{a}\right)$ is *[2023, 30 Jan Shift-I]*

(a) $5-\frac{3}{2}\sqrt{3}$ (b) $4-2\sqrt{3}$ (c) 4 (d) 2

**31** If the solution of the equation $\log_{\cos x}\cot x+4\log_{\sin x}\tan x=1,\ x\in\left(0,\frac{\pi}{2}\right)$, is $\sin^{-1}\left(\frac{\alpha+\sqrt{\beta}}{2}\right)$, where $\alpha,\beta$ are integers, then $\alpha+\beta$ is equal to *[2023, 30 Jan Shift-I]*

(a) 3 (b) 5 (c) 6 (d) 4

**32** The set of all values of $\lambda$ for which the equation $\cos^2 2x-2\sin^4 x-2\cos^2 x=\lambda$ has a real solution $x$, is *[2023, 29 Jan Shift-II]*

(a) $\left[-\frac{3}{2},-1\right]$ (b) $[-2,-1]$ (c) $\left[-2,-\frac{3}{2}\right]$ (d) $\left[-1,-\frac{1}{2}\right]$

**33** Let $f(\theta)=3\left(\sin^4\left(\frac{3\pi}{2}-\theta\right)+\sin^4(3\pi+\theta)\right)-2(1-\sin^2 2\theta)$ and $S=\left\{\theta\in[0,\pi]:f'(\theta)=-\frac{\sqrt{3}}{2}\right\}$. If $4\beta=\sum_{\theta\in S}\theta$, then $f(\beta)$ is equal to *[2023, 29 Jan Shift-I]*

(a) $\frac{5}{4}$ (b) $\frac{3}{2}$ (c) $\frac{11}{8}$ (d) $\frac{9}{8}$

**34** Let $S=\left\{\theta\in\left(0,\frac{\pi}{2}\right):\sum_{m=1}^{9}\sec\left(\theta+(m-1)\frac{\pi}{6}\right)\sec\left(\theta+\frac{m\pi}{6}\right)=-\frac{8}{\sqrt{3}}\right\}$. Then, *[2022, 27 July Shift-II]*

(a) $S=\left\{\frac{\pi}{12}\right\}$ (b) $S=\left\{\frac{2\pi}{3}\right\}$

(c) $\sum_{\theta\in S}\theta=\frac{\pi}{2}$ (d) $\sum_{\theta\in S}\theta=\frac{3\pi}{4}$

**35** Let $S=\{\theta\in[0,2\pi]:8^{2\sin^2\theta}+8^{2\cos^2\theta}=16\}$. Then, $n(S)+\sum_{\theta\in S}\left(\sec\left(\frac{\pi}{4}+2\theta\right)\text{cosec}\left(\frac{\pi}{4}+2\theta\right)\right)$ is equal to *[2022, 26 July Shift-I]*

(a) 0 (b) −2 (c) −4 (d) 12

**36** Let $S=\left\{\theta\in[-\pi,\pi]-\left\{\pm\frac{\pi}{2}\right\}:\sin\theta\tan\theta+\tan\theta=\sin 2\theta\right\}$. If $T=\sum_{\theta\in S}\cos 2\theta$, then $T+n(S)$ is equal to *[2022, 24 June Shift-I]*

(a) $7+\sqrt{3}$ (b) 9

(c) $8+\sqrt{3}$ (d) 10

**37** The sum of solutions of the equation $\frac{\cos x}{1+\sin x}=|\tan 2x|,\ x\in\left(-\frac{\pi}{2},\frac{\pi}{2}\right)-\left\{\frac{\pi}{4},-\frac{\pi}{4}\right\}$ is *[2021, 26 Aug Shift-I]*

(a) $-\frac{11\pi}{30}$ (b) $\frac{\pi}{10}$

(c) $-\frac{7\pi}{30}$ (d) $-\frac{\pi}{15}$

**38** Let $\alpha=\max_{x\in R}\{8^{2\sin 3x}\cdot 4^{4\cos 3x}\}$ and $\beta=\min_{x\in R}\{8^{2\sin 3x}\cdot 4^{4\cos 3x}\}$. If $8x^2+bx+c=0$ is a quadratic equation whose roots are $\alpha^{1/5}$ and $\beta^{1/5}$, then the value of $c-b$ is equal to *[2021, 27 July Shift-II]*

(a) 42 (b) 47

(c) 43 (d) 50

**39** The number of solutions of the equation $x+2\tan x=\frac{\pi}{2}$ in the interval $[0,2\pi]$ is *[2021, 17 March Shift-II]*

(a) 3 (b) 4

(c) 2 (d) 5

**40** The number of roots of the equation, $(81)^{\sin^2 x}+(81)^{\cos^2 x}=30$ in the interval $[0,\pi]$ is equal to *[2021, 16 March Shift-I]*

(a) 3 (b) 4

(c) 8 (d) 2

**41** If $[x]$ denotes the greatest integer $\le x$, then the system of linear equations $[\sin\theta]x+[-\cos\theta]y=0$, $[\cot\theta]x+y=0$ *[2019, 12 April Shift-II]*

(a) have infinitely many solutions if $\theta\in\left(\frac{\pi}{2},\frac{2\pi}{3}\right)$ and has a unique solution if $\theta\in\left(\pi,\frac{7\pi}{6}\right)$.

(b) has a unique solution if $\theta\in\left(\frac{\pi}{2},\frac{2\pi}{3}\right)\cup\left(\pi,\frac{7\pi}{6}\right)$

(c) has a unique solution if $\theta\in\left(\frac{\pi}{2},\frac{2\pi}{3}\right)$ and have infinitely many solutions if $\theta\in\left(\pi,\frac{7\pi}{6}\right)$

(d) have infinitely many solutions if $\theta\in\left(\frac{\pi}{2},\frac{2\pi}{3}\right)\cup\left(\pi,\frac{7\pi}{6}\right)$

**42** If $5(\tan^2 x - \cos^2 x) = 2\cos 2x + 9$, then the value of $\cos 4x$ is **[JEE Main 2017]**

(a) $-\frac{3}{5}$ (b) $\frac{1}{3}$

(c) $\frac{2}{9}$ (d) $-\frac{7}{9}$

**43** If $0 \le x < 2\pi$, then the number of real values of $x$, which satisfy the equation $\cos x + \cos 2x + \cos 3x + \cos 4x = 0$, is **[JEE Main 2016]**

(a) 3 (b) 5

(c) 7 (d) 9

**44** The number of $x \in [0, 2\pi]$ for which

$\left|\sqrt{2\sin^4 x + 18\cos^2 x} - \sqrt{2\cos^4 x + 18\sin^2 x}\right| = 1$, is **[JEE Main 2016]**

(a) 2 (b) 6 (c) 4 (d) 8

**45** If $0 < x < \pi$ and $\cos x + \sin x = 1/2$, then $\tan x$ is **[AIEEE 2006]**

(a) $\frac{(4-\sqrt{7})}{3}$ (b) $-\frac{(4+\sqrt{7})}{3}$ (c) $\frac{(1+\sqrt{7})}{4}$ (d) $\frac{(1-\sqrt{7})}{4}$

## Numerical Value Type Questions

**46** The number of solutions of $\sin^2 x + (2 + 2x - x^2)\sin x - 3(x-1)^2 = 0$, where $-\pi \le x \le \pi$, is ........... **[2024, 5 April Shift-II]**

**47** Let $S = \{\sin^2 2\theta : (\sin^4\theta + \cos^4\theta)x^2 + (\sin 2\theta)x + (\sin^6\theta + \cos^6\theta) = 0$ has real roots$\}$.

If $\alpha$ and $\beta$ are the smallest and largest elements of the set $S$, respectively, then $3((\alpha-2)^2 + (\beta-1)^2)$ equals...... **[2024, 4 April Shift-II]**

**48** Let the set of all $a \in R$ such that the equation $\cos 2x + a\sin x = 2a - 7$ has a solution be $[p, q]$ and $r = \tan 9° - \tan 27° - \frac{1}{\cot 63°} + \tan 81°$, then $pqr$ is equal to ...... . **[2024, 27 Jan Shift-I]**

**Imp.** Trigonometric Function of Compound Angle

**49** If $m$ and $n$ respectively are the numbers of positive and negative values of $\theta$ in the interval $[-\pi, \pi]$ that satisfy the equation $\cos 2\theta \cos\frac{\theta}{2} = \cos 3\theta \cos\frac{9\theta}{2}$, then $mn$ is equal to ........ **[2023, 25 Jan Shift-II]**

**50** Let $S = \{\theta \in [0, 2\pi) : \tan(\pi\cos\theta) + \tan(\pi\sin\theta) = 0\}$.

Then, $\sum_{\theta \in S} \sin^2\left(\theta + \frac{\pi}{4}\right)$ is equal to ........ **[2023, 24 Jan Shift-II]**

**51** If the sum of solutions of the system of equations $2\sin^2\theta - \cos 2\theta = 0$ and $2\cos^2\theta + 3\sin\theta = 0$ in the interval $[0, 2\pi]$ is $k\pi$, then $k$ is equal to .............. **[2022, 26 July Shift-II]**

**52** The number of solutions of the equation $\sin x = \cos^2 x$ in the interval $(0,10)$ is ........... . **[2022, 29 June Shift-II]**

**53** The number of elements in the set $S = \{\theta \in [-4\pi, 4\pi] : 3\cos^2 2\theta + 6\cos 2\theta - 10\cos^2\theta + 5 = 0\}$ is ...... . **[2022, 29 June Shift-I]**

**54** The number of solutions of the equation $2\theta - \cos^2\theta + \sqrt{2} = 0$ in $R$ is equal to ...... . **[2022, 29 June Shift-I]**

**55** The number of value of $x$ in the interval $\left(\frac{\pi}{4}, \frac{7\pi}{4}\right)$ for which $14\operatorname{cosec}^2 x - 2\sin^2 x = 21 - 4\cos^2 x$ holds, is ............ **[2022, 25 June Shift-I]**

**56** Let $S$ be the sum of all solutions (in radians) of the equation $\sin^4\theta + \cos^4\theta - \sin\theta\cos\theta = 0$ in $[0, 4\pi]$. Then, $8S/\pi$ is equal to ......... . **[2021, 27 Aug Shift-II]**

**57** The number of solutions of the equation $|\cot x| = \cot x + \frac{1}{\sin x}$ in the interval $[0, 2\pi]$ is ........... . **[2021, 18 March Shift-I]**

## Key Index

### Topic 1

| | | | | | | | | | |
|---|---|---|---|---|---|---|---|---|---|
| 1 (b) | 2 (c) | 3 (b) | 4 (a) | 5 (b) | 6 (d) | 7 (c) | 8 (b) | 9 (b) | 10 (c) |
| 11 (b) | 12 (d) | 13 (d) | 14 (d) | 15 (b) | 16 (d) | 17 (b) | 18 (b) | 19 (c) | 20 (d) |
| 21 (c) | 22 (d) | 23 (b) | 24 (d) | 25 (d) | 26 (c) | 27 (b) | 28 (d) | 29 (b) | 30 (c) |
| 31 (c) | 32 (b) | 33 (d) | 34 (c) | 35 (a) | 36 (a) | 37 (b) | 38 (d) | 39 (b) | 40 (b) |
| 41 (b) | 42 (c) | 43 (c) | 44 (d) | 45 (b) | 46 (d) | 47 (d) | 48 (c) | 49 (4) | 50 (80) |
| 51 (1) | | | | | | | | | |

### Topic 2

| | | | | | | | | | |
|---|---|---|---|---|---|---|---|---|---|
| 1 (b) | 2 (b) | 3 (b) | 4 (d) | 5 (d) | 6 (d) | 7 (b) | 8 (a) | 9 (d) | 10 (d) |
| 11 (d) | 12 (c) | 13 (d) | 14 (c) | 15 (a) | 16 (c) | 17 (d) | 18 (a) | 19 (c) | 20 (b) |
| 21 (b) | 22 (d) | 23 (b) | 24 (d) | 25 (b) | 26 (c) | 27 (d) | 28 (d) | 29 (c) | 30 (c) |
| 31 (d) | 32 (a) | 33 (a) | 34 (c) | 35 (c) | 36 (b) | 37 (a) | 38 (a) | 39 (a) | 40 (b) |
| 41 (a) | 42 (d) | 43 (c) | 44 (d) | 45 (b) | 46 (2) | 47 (4) | 48 (48) | 49 (25) | 50 (2) |
| 51 (3) | 52 (4) | 53 (32) | 54 (1) | 55 (4) | 56 (56) | 57 (1) | | | |

# Detailed Explanations

## Topic 1

**01.** *(b)* $\sin x = -\frac{3}{5}$ and $\pi < x < \frac{3\pi}{2}$

So, $\cos x = -\frac{4}{5}$ and $\tan x = \frac{3}{4}$

Then,

$$80(\tan^2 x - \cos x) = 80\left[\left(\frac{3}{4}\right)^2 - \left(-\frac{4}{5}\right)\right]$$

$$= 80\left[\frac{9}{16} + \frac{4}{5}\right] = 80\left[\frac{45+64}{80}\right]$$

$$= 80 \times \frac{109}{80} = 109$$

**02.** *(c)* We know, $-1 \le \sin nx \le 1$

So, $-1 \le \sin 5x \le 1$

Multiplying $-1$,

$$-1 \le -\sin 5x \le 1$$

Adding 7,

$$6 \le 7 - \sin 5x \le 8$$

Now, taking reciprocal

$$\frac{1}{8} \le \frac{1}{7 - \sin 5x} \le \frac{1}{6}$$

$\therefore$ Range $= \left[\frac{1}{8}, \frac{1}{6}\right]$

**03.** *(b)* Given, $3\sin(\alpha+\beta) = 2\sin(\alpha-\beta)$

$\Rightarrow 3\sin\alpha\cos\beta + 3\cos\alpha\sin\beta$
$= 2\sin\alpha\cos\beta - 2\cos\alpha\sin\beta$

$\Rightarrow \sin\alpha\cos\beta = -5\cos\alpha\sin\beta$

$\Rightarrow \tan\alpha = -5\tan\beta$

On comparing,

$\tan\alpha = k\tan\beta \Rightarrow k = -5$

**04.** *(a)* Given, $\cot\alpha = 1$, $\alpha \in \left(\pi, \frac{3\pi}{2}\right)$

Then, $\tan\alpha = 1$

and $\sec\beta = \frac{-5}{3}$, $\beta \in \left(\frac{\pi}{2}, \pi\right)$

Then, $\tan\beta = -\frac{4}{3}$

$$\therefore \tan(\alpha+\beta) = \frac{\tan\alpha + \tan\beta}{1 - \tan\alpha \cdot \tan\beta}$$

$$= \frac{1 - \frac{4}{3}}{1 + \frac{4}{3}} = -\frac{1}{7}$$

$\alpha + \beta \in (3\pi/2, 2\pi)$ i.e. fourth quadrant.

**05.** *(b)* $16\sin 20°\sin 40°\sin 80°$

We know that

$$\sin\theta\sin(60° - \theta)\sin(60° + \theta) = \frac{1}{4}\sin 3\theta$$

In the given question, let $\theta = 20°$

$\therefore 16\sin 20°\sin(60° - 20°)\sin(60° + 20°)$

$$= 16 \times \frac{1}{4}\sin 60° = \frac{4\sqrt{3}}{2} = 2\sqrt{3}$$

**06.** *(d)* $\operatorname{cosec} 18° = \frac{1}{\sin 18°} = \frac{4}{\sqrt{5}-1}$

$$= \sqrt{5} + 1$$

If $x = \sqrt{5} + 1$, then $(x-1)^2 = 5$

$\Rightarrow x^2 - 2x - 4 = 0$

**07.** *(c)* We have,

$\sin 10°\sin 30°\sin 50°\sin 70°$

$= \sin(30°)[\sin(10°)\sin(50°)\sin(70°)]$

$= \frac{1}{2}[\sin(10°)\sin(60° - 10°)\sin(60° + 10°)]$

$= \frac{1}{2}\left[\frac{1}{4}\sin(3(10°))\right]$

$[\because \sin\theta\sin(60° - \theta)\sin(60° + \theta) = 1/4\sin 3\theta]$

$= \frac{1}{8}\sin 30°$

$= \frac{1}{8} \times \frac{1}{2} = \frac{1}{16}$

**08.** *(b)* Given expression is

$$\frac{\tan A}{1-\cot A}+\frac{\cot A}{1-\tan A}$$

$$=\frac{\sin A}{\cos A}\times\frac{\sin A}{\sin A-\cos A}+\frac{\cos A}{\sin A}\times\frac{\cos A}{\cos A-\sin A}$$

$$=\frac{1}{\sin A-\cos A}\left\{\frac{\sin^3 A-\cos^3 A}{\cos A\sin A}\right\}$$

$$=\frac{\sin^2 A+\sin A\cos A+\cos^2 A}{\sin A\cos A}$$

$$=\frac{1+\sin A\cos A}{\sin A\cos A}=1+\sec A\operatorname{cosec} A$$

**09.** *(b)* $\cos(\alpha+\beta)=\frac{4}{5}$

$\Rightarrow\ \alpha+\beta\in$ Ist quadrant

and $\sin(\alpha-\beta)=\frac{5}{13}$

$\Rightarrow\ \alpha-\beta\in$ Ist quadrant

Now, $2\alpha=(\alpha+\beta)+(\alpha-\beta)$

$$\therefore \tan 2\alpha=\frac{\tan(\alpha+\beta)+\tan(\alpha-\beta)}{1-\tan(\alpha+\beta)\tan(\alpha-\beta)}$$

$$=\frac{\frac{3}{4}+\frac{5}{12}}{1-\frac{3}{4}\cdot\frac{5}{12}}=\frac{56}{33}$$

**10.** *(c)* We know that

$$\frac{1-\tan^2 15^\circ}{1+\tan^2 15^\circ}=\cos 30^\circ=\frac{\sqrt3}{2}$$

**11.** *(b)* $\because\ \tan\theta=-\frac{4}{3}$

$$\therefore\quad \sin\theta=\frac{BC}{AC}=\frac{4}{5}$$

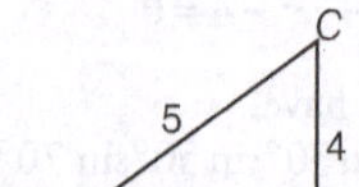


But, $\tan\theta$ is negative which is possible only, if $\theta$ lies in IInd and IVth quadrants.

So, $\sin\theta$ may be $\frac{4}{5}$ or $-\frac{4}{5}$.

**12.** *(d)* Given that

$y=\sin^2\theta+\operatorname{cosec}^2\theta$

$\therefore\ \ y=(\sin\theta-\operatorname{cosec}\theta)^2+2$

$\Rightarrow\qquad y\geq 2,\quad \theta\neq 0$

**13.** *(d)* $\because 10\sin^4\theta+15\cos^4\theta=6$

Let $\sin^2\theta=x$

$\Rightarrow 10x^2+15(1-x)^2=6$

$\Rightarrow 10x^2+15x^2-30x+15-6=0$

$\Rightarrow 25x^2-30x+9=0$

$\Rightarrow (5x-3)^2=0 \Rightarrow x=3/5$

$\therefore \sin^2\theta=\frac{3}{5},\ \cos^2\theta=\frac{2}{5}$

$$\therefore \frac{27\operatorname{cosec}^6\theta+8\sec^6\theta}{16\sec^8\theta}$$

$$=\frac{27\times\frac{125}{27}+8\times\frac{125}{8}}{16\times\frac{625}{16}}=\frac{250}{625}=\frac{2}{5}$$

**14.** *(d)* $\sin 70^\circ(\cot 10^\circ\cot 70^\circ-1)$

$$=\sin 70^\circ\left[\frac{\cos 10^\circ\cos 70^\circ-\sin 10^\circ\sin 70^\circ}{\sin 10^\circ\sin 70^\circ}\right]$$

$$=\frac{\cos 80^\circ}{\sin 10^\circ}=1$$

$\because \cos a\cos b-\sin a\sin b=\cos(a+b)$

and $\cos(90^\circ-\theta)=\sin\theta$

**15.** *(b)* As we know that

$\cos\theta\cdot\cos(60^\circ-\theta)\cdot\cos(60^\circ+\theta)$

$$=\frac{1}{4}\cos 3\theta$$

So, given equation reduces to

$$\left|\frac{1}{4}\cos 3\theta\right|\leq\frac{1}{8}\Rightarrow|\cos 3\theta|\leq\frac{1}{2}$$

$$\Rightarrow\qquad -\frac{1}{2}\leq\cos 3\theta\leq\frac{1}{2}$$

So, maximum value of $\cos 3\theta=1/2$

$$\Rightarrow\qquad 3\theta=2n\pi\pm\frac{\pi}{3},\ n\in I$$

$$\Rightarrow\qquad \theta=\frac{2n\pi}{3}\pm\frac{\pi}{9},\ n\in I$$

As, $\theta\in[0,2\pi]$, then possible value are

$$\theta=\left\{\frac{\pi}{9},\frac{5\pi}{9},\frac{7\pi}{9},\frac{11\pi}{9},\frac{13\pi}{9},\frac{17\pi}{9}\right\}$$

and its sum $=\frac{54\pi}{9}=6\pi$

**16.** *(d)* $$\frac{3\cos 36^\circ+5\sin 18^\circ}{5\cos 36^\circ-3\sin 18^\circ}$$

$$=\frac{3\left(\frac{\sqrt5+1}{4}\right)+5\left(\frac{\sqrt5-1}{4}\right)}{5\left(\frac{\sqrt5+1}{4}\right)-3\left(\frac{\sqrt5-1}{4}\right)}$$

$$=\frac{3\sqrt5+3+5\sqrt5-5}{5\sqrt5+5-3\sqrt5+3}=\frac{8\sqrt5-2}{2\sqrt5+8}$$

$$=\frac{4\sqrt5-1}{\sqrt5+4}\times\frac{\sqrt5-4}{\sqrt5-4}$$

$$=\frac{20-16\sqrt5-\sqrt5+4}{-11}$$

$$=\frac{24-17\sqrt5}{-11}=\frac{17\sqrt5-24}{11}$$

Now, $\frac{17\sqrt5-24}{11}=\frac{a\sqrt5-b}{c}$

So, $a=17, b=24$ and $c=11$

Therefore, $a+b+c=52$

**17.** *(b)* Given, $\tan A=\frac{1}{\sqrt{x(x^2+x+1)}}$,

$$\tan B=\frac{\sqrt x}{\sqrt{x^2+x+1}}$$

$$\tan C=\sqrt{\left(\frac{1}{x^3}+\frac{1}{x^2}+\frac{1}{x}\right)}$$

$$=\sqrt{\frac{1+x+x^2}{x^3}}$$

and $$\tan C=\frac{\sqrt{x^2+x+1}}{x\sqrt x}$$

$$\tan(A+B)=\frac{\tan A+\tan B}{1-\tan A\tan B}$$

$$=\left(\frac{\frac{1}{\sqrt x\sqrt{x^2+x+1}}+\frac{\sqrt x}{\sqrt{x^2+x+1}}}{1-\frac{1}{\sqrt x\sqrt{x^2+x+1}}\cdot\frac{\sqrt x}{\sqrt{x^2+x+1}}}\right)$$

$$=\left(\frac{1+x}{x^2+x+1-1}\right)\frac{\sqrt{x^2+x+1}}{\sqrt x}$$

$$=\frac{(1+x)\sqrt{x^2+x+1}}{x(x+1)\sqrt x}=\frac{\sqrt{x^2+x+1}}{x\sqrt x}$$

$\tan(A+B)=\tan C$

$\therefore\qquad A+B=C$

**18.** *(b)* $4\cos^2\theta-1=4-4\sin^2\theta-1$

$$=3-4\sin^2\theta$$

$$=\frac{3\sin\theta-4\sin^3\theta}{\sin\theta}=\frac{\sin 3\theta}{\sin\theta}$$

$\therefore$ Given expression

$$=36\cdot\frac{\sin(3\times 9^\circ)}{\sin 9^\circ}\times\frac{\sin(3\times 27^\circ)}{\sin 27^\circ}$$

$$\times\frac{\sin(3\times 81^\circ)}{\sin 81^\circ}\times\frac{\sin(3\times 243^\circ)}{\sin 243^\circ}$$

$$= 36 \cdot \frac{\sin 729^\circ}{\sin 9^\circ} = 36 \times \frac{\sin(2 \times 360^\circ + 9^\circ)}{\sin 9^\circ}$$

$$= 36 \times \frac{\sin 9^\circ}{\sin 9^\circ} = 36$$

**19.** *(c)* Given, $\alpha = \sin 36^\circ = \frac{\sqrt{10 - 2\sqrt{5}}}{4}$

$\Rightarrow \quad (4\alpha)^2 = 10 - 2\sqrt{5}$

$\Rightarrow \quad (16\alpha^2 - 10)^2 = (-2\sqrt{5})^2 = 20$

$\Rightarrow \quad 4(8\alpha^2 - 5)^2 = 20$

$\Rightarrow \quad 64\alpha^4 - 80\alpha^2 + 25 = 5$

$\Rightarrow \quad 64\alpha^4 - 80\alpha^2 + 20 = 0$

$\Rightarrow \quad 16\alpha^4 - 20\alpha^2 + 5 = 0$

$\therefore$ $\alpha$ is a root of the equation $16x^4 - 20x^2 + 5 = 0$

**20.** *(d)* $2 \sin (12^\circ) - \sin (72^\circ)$

$= \sin 12^\circ + \sin 12^\circ - \sin 72^\circ$

$= \sin 12^\circ + 2 \cos 42^\circ \sin (-30^\circ)$

$= \sin 12^\circ - 2 \times \frac{1}{2} \cos 42^\circ$

$= \sin 12^\circ - \sin 48^\circ$

$[\because \sin (\theta) = \cos (90^\circ - \theta)]$

$= 2 \cos 30^\circ \sin(-18^\circ)$

$$= -\frac{2\sqrt{3}}{2} \times \left[\frac{\sqrt{5} - 1}{4}\right] = \frac{\sqrt{3}(1 - \sqrt{5})}{4}$$

**21.** *(c)* Given, $\sin \theta + \cos \theta = \frac{1}{2}$

$\Rightarrow \sin^2 \theta + \cos^2 \theta + 2 \sin \theta \cos \theta = \frac{1}{4}$

$\Rightarrow \sin 2\theta = \frac{1}{4} - 1 = -\frac{3}{4}$

$[\because \sin 2A = 2 \sin A \cos A]$

$\therefore \cos 2\theta = \sqrt{1 - \sin^2 2\theta}$

$$= \sqrt{1 - \frac{9}{16}} = \pm \frac{\sqrt{7}}{4}$$

$1 + \cos 4\theta = 2 \cos^2 2\theta$

$[\because 1 + \cos 2A = 2 \cos^2 A]$

$$\cos 4\theta = 2\left(\frac{7}{16}\right) - 1 = \frac{-2}{16} = \frac{-1}{8}$$

$\therefore 16(\sin 2\theta + \cos 4\theta + \sin 6\theta)$

$= 16 [(\sin 2\theta + \sin 6\theta) + \cos 4\theta]$

$= 16(2 \sin 4\theta \cos 2\theta + \cos 4\theta)$

$= 16(4 \sin 2\theta \cos^2 2\theta + \cos 4\theta)$

$$= 16\left[4\left(\frac{-3}{4}\right)\left(\frac{7}{16}\right) - \frac{1}{8}\right]$$

$= -21 - 2 = -23$

**22.** *(d)* Given, $15 \sin^4 \alpha + 10 \cos^4 \alpha = 6$

$\Rightarrow 15 \sin^4 \alpha + 10 \cos^4 \alpha = 6(\sin^2 \alpha + \cos^2 \alpha)^2$

$\Rightarrow 15 \sin^4 \alpha + 10 \cos^4 \alpha = 6 (\sin^4 \alpha + \cos^4 \alpha + 2 \sin^2 \alpha \cos^2 \alpha)$

$\Rightarrow 9 \sin^4 \alpha + 4 \cos^4 \alpha - 12 \sin^2 \alpha \cos^2 \alpha = 0$

$\Rightarrow (3 \sin^2 \alpha - 2 \cos^2 \alpha)^2 = 0$

$\Rightarrow 3 \sin^2 \alpha - 2 \cos^2 \alpha = 0$

$\Rightarrow 3 \sin^2 \alpha = 2 \cos^2 \alpha \Rightarrow \tan^2 \alpha = 2/3$

$\therefore \quad \cot^2 \alpha = 3/2$

Now, $27 \sec^6 \alpha + 8 \operatorname{cosec}^6 \alpha$

$= 27(\sec^2 \alpha)^3 + 8(\operatorname{cosec}^2 \alpha)^3$

$= 27 (1 + \tan^2 \alpha)^3 + 8(1 + \cot^2 \alpha)^3$

$$= 27\left(1 + \frac{2}{3}\right)^3 + 8\left(1 + \frac{3}{2}\right)^3 = 250$$

**23.** *(b)* $\log_{10} \sin x + \log_{10} \cos x = -1$, $x \in (0, \pi/2)$

$\log_{10} (\sin x \cos x) = -1$

$\Rightarrow \sin x \cos x = 10^{-1} = 1/10$

$\log_{10} (\sin x + \cos x) = \frac{1}{2} (\log_{10} n - 1), n > 0$

$2 \log_{10} (\sin x + \cos x) = (\log_{10} n - \log_{10} 10)$

$\Rightarrow \log_{10} (\sin x + \cos x)^2 = \log_{10} (n/10)$

$\Rightarrow \quad (\sin x + \cos x)^2 = n/10$

$\Rightarrow \sin^2 x + \cos^2 x + 2 \sin x \cos x = \frac{n}{10}$

$\Rightarrow 1 + 2(1/10) = n/10 \Rightarrow 12/10 = n/10$

$\therefore \quad n = 12$

**24.** *(d)* $\sin 2\theta + \tan 2\theta > 0$

$\Rightarrow \quad \sin 2\theta + \frac{\sin 2\theta}{\cos 2\theta} > 0$

$\Rightarrow \quad \sin 2\theta\left(1 + \frac{1}{\cos 2\theta}\right) > 0$

$\Rightarrow \quad \sin 2\theta\left(\frac{\cos 2\theta + 1}{\cos 2\theta}\right) > 0$

$\Rightarrow \quad \tan 2\theta(2 \cos^2 \theta) > 0$

$\because \quad \cos 2\theta \neq 0 \Rightarrow 1 - 2 \sin^2 \theta \neq 0$

$\therefore \quad \sin \theta \neq \pm \frac{1}{\sqrt{2}}$

Now, $\tan 2\theta(1 + \cos 2\theta) > 0$

$\Rightarrow \tan 2\theta > 0$ as $1 + \cos 2\theta > 0$

$$\Rightarrow 2\theta \in \left(0, \frac{\pi}{2}\right) \cup \left(\pi, \frac{3\pi}{2}\right) \cup \left(2\pi, \frac{5\pi}{2}\right) \cup \left(3\pi, \frac{7\pi}{2}\right)$$

$$\therefore \theta \in \left(0, \frac{\pi}{4}\right) \cup \left(\frac{\pi}{2}, \frac{3\pi}{4}\right) \cup \left(\pi, \frac{5\pi}{4}\right) \cup \left(\frac{3\pi}{2}, \frac{7\pi}{4}\right)$$

**25.** *(d)* Given, $L = \sin^2 \frac{\pi}{16} - \sin^2 \frac{\pi}{8}$

$$= \sin\left(\frac{\pi}{16} + \frac{\pi}{8}\right) \sin\left(\frac{\pi}{16} - \frac{\pi}{8}\right)$$

$$= \sin \frac{3\pi}{16} \sin\left(-\frac{\pi}{16}\right)$$

$$= \frac{1}{2}\left[2 \sin \frac{3\pi}{16} \sin\left(-\frac{\pi}{16}\right)\right]$$

$$= \frac{1}{2}\left[\cos \frac{\pi}{4} - \cos \frac{2\pi}{16}\right] = \frac{1}{2\sqrt{2}} - \frac{1}{2} \cos \frac{\pi}{8}$$

and $M = \cos^2 \frac{\pi}{16} - \sin^2 \frac{\pi}{8}$

$$= \cos\left(\frac{\pi}{16} + \frac{\pi}{8}\right) \cos\left(\frac{\pi}{16} - \frac{\pi}{8}\right)$$

$$= \cos \frac{3\pi}{16} \cos\left(-\frac{\pi}{16}\right) = \frac{1}{2}\left[2 \cos \frac{3\pi}{16} \cos \frac{\pi}{16}\right]$$

$$= \frac{1}{2}\left[\cos \frac{\pi}{4} + \cos \frac{2\pi}{16}\right] = \frac{1}{2\sqrt{2}} + \frac{1}{2} \cos \frac{\pi}{8}$$

**26.** *(c)* $2^{\sin x} + 2^{\cos x}$ (given)

Using AM-GM inequality, we get

$$\frac{2^{\sin x} + 2^{\cos x}}{2} \geq \sqrt{2^{\sin x} \times 2^{\cos x}}$$

$$\Rightarrow \frac{2^{\sin x} + 2^{\cos x}}{2} \geq 2^{\frac{\sin x + \cos x}{2}} \quad \text{...(i)}$$

We know that

$-\sqrt{2} \leq \sin x + \cos x \leq \sqrt{2}$

$$\Rightarrow \frac{-1}{\sqrt{2}} \leq \frac{\sin x + \cos x}{2} \leq \frac{1}{\sqrt{2}}$$

(dividing by 2)

$$\Rightarrow 2^{(-1/\sqrt{2})} \leq 2^{\frac{\sin x + \cos x}{2}} \leq 2^{(1/\sqrt{2})}$$

The minimum value of

$$2^{\frac{\sin x + \cos x}{2}} = 2^{(-1/\sqrt{2})}$$

$\Rightarrow$ Minimum value of

$$\frac{2^{\sin x} + 2^{\cos x}}{2} = 2^{(-1/\sqrt{2})} \text{ [from Eq. (i)]}$$

i.e., minimum value of

$$2^{\sin x} + 2^{\cos x} = 2 \cdot 2^{(-1/\sqrt{2})} = 2^{1 - \frac{1}{\sqrt{2}}}$$

**27.** *(b)* Given, trigonometric expression

$$\cos^3\left(\frac{\pi}{8}\right) \cos\left(\frac{3\pi}{8}\right) + \sin^3\left(\frac{\pi}{8}\right) \sin \frac{3\pi}{8}$$

$$= \cos^3\left(\frac{\pi}{8}\right)\cos\left(\frac{\pi}{2}-\frac{\pi}{8}\right)+\sin^3\left(\frac{\pi}{8}\right)\sin\left(\frac{\pi}{2}-\frac{\pi}{8}\right)$$

$$= \cos^3\left(\frac{\pi}{8}\right)\sin\left(\frac{\pi}{8}\right)+\sin^3\left(\frac{\pi}{8}\right)\cos\left(\frac{\pi}{8}\right)$$

$$= \sin\left(\frac{\pi}{8}\right)\cos\left(\frac{\pi}{8}\right)\left[\cos^2\left(\frac{\pi}{8}\right)+\sin^2\left(\frac{\pi}{8}\right)\right]$$

$$= \frac{1}{2}\times 2\sin\left(\frac{\pi}{8}\right)\cos\left(\frac{\pi}{8}\right)$$

$$= \frac{1}{2}\sin\frac{\pi}{4} = \frac{1}{2\sqrt{2}}$$

**28.** *(d)*

> **key idea** Use formulae : $2\sin A\sin B = \cos(A-B)-\cos(A+B)$ and $\cos 2\theta = 1-2\sin^2\theta$

Given equation is

$y = \sin x\sin(x+2)-\sin^2(x+1)$

$$= \frac{1}{2}[\cos 2-\cos(2x+2)]-\frac{1}{2}[1-\cos(2x+2)]$$

$[\because 2\sin A\sin B = \cos(A-B) - \cos(A+B)$ and $\cos 2\theta = 1-2\sin^2\theta \Rightarrow 2\sin^2\theta = 1-\cos 2\theta]$

$$= \frac{1}{2}\cos 2-\frac{1}{2}\cos(2x+2)-\frac{1}{2}+\frac{1}{2}\cos(2x+2)$$

$$= \frac{1}{2}(\cos(2)-1) = -\frac{1}{2}(2\sin^2(1))$$

$= -\sin^2(1) < 0 \Rightarrow y < 0$

and as we know that $y<0$, is in third and fourth quadrants only.

**29.** *(b)* Given, inequality is

$$2^{\sqrt{\sin^2 x-2\sin x+5}}\cdot\frac{1}{4^{\sin^2 y}} \le 1$$

$$\Rightarrow 2^{\sqrt{(\sin x-1)^2+4}}\cdot 2^{-2\sin^2 y} \le 1$$

$$\Rightarrow 2^{\sqrt{(\sin x-1)^2+4}} \le 2^{2\sin^2 y}$$

$$\Rightarrow \sqrt{(\sin x-1)^2+4} \le 2\sin^2 y$$

[If $a>1$ and $a^m \le a^n \Rightarrow m \le n$]

$\because$ Range of $\sqrt{(\sin x-1)^2+4}$ is $[2, 2\sqrt{2}]$ and range of $2\sin^2 y$ is $[0, 2]$.

$\therefore$ The above inequality holds, if

$$\sqrt{(\sin x-1)^2+4} = 2 = 2\sin^2 y$$

$\Rightarrow \sin x = 1$ and $\sin^2 y = 1$

$\Rightarrow \sin x = |\sin y|$ [from the options]

**30.** *(c)* $5\cos A+3 = 0$

$$\Rightarrow \cos A = \frac{-3}{5} \Rightarrow \sec A = \frac{-5}{3}$$

Now, $9x^2+27x+20 = 0$

$\Rightarrow (3x+4)(3x+5) = 0$

$$\Rightarrow x = \frac{-4}{3} \text{ or } x = \frac{-5}{3}$$

Now, $\tan^2 A+1 = \sec^2 A$

$$\Rightarrow \tan^2 A = \frac{25}{9}-1 \Rightarrow \tan^2 A = \frac{16}{9}$$

$$\Rightarrow \tan A = \pm\frac{4}{3} \Rightarrow \tan A = -\frac{4}{3}$$

So, the roots are $\tan A$ and $\sec A$.

**31.** *(c)* $5(\tan^2 x-\cos^2 x) = 2\cos 2x+9$

Let $u = \tan^2 x$,

From given equation,

$$5\left(u-\frac{1}{1+u}\right) = 2\left(\frac{1-u}{1+u}\right)+9$$

$\Rightarrow 5(u^2+u-1) = 2-2u+9+9u$

$\therefore 5u^2-2u-16 = 0$

$\Rightarrow (5u+8)(u-2) = 0$

But, $u$ is positive

$\therefore u = 2$

Now, $\tan^2 x = 2$

$$\Rightarrow \cos 2x = \frac{1-\tan^2 x}{1+\tan^2 x} = \frac{1-2}{1+2} = \frac{-1}{3}$$

$$\Rightarrow \cos 4x = 2\cos^2 2x-1 = 2\left(\frac{1}{9}\right)-1 = \frac{-7}{9}$$

**32.** *(b)* We know that

$$\tan(A+B) = \frac{\tan A+\tan B}{1-\tan A\tan B}$$

$$\Rightarrow \frac{1}{\sqrt{3}} = \frac{y}{1-\tan A\tan B},$$

where $y = \tan A+\tan B$

$\Rightarrow \tan A\tan B = 1-\sqrt{3}y$

$\because$ AM $\ge$ GM

$$\Rightarrow \frac{\tan A+\tan B}{2} \ge \sqrt{\tan A\tan B}$$

$\Rightarrow y \ge 2\sqrt{1-\sqrt{3}y} \Rightarrow y^2 \ge 4-4\sqrt{3}y$

$\Rightarrow y^2+4\sqrt{3}y-4 \ge 0$

$\Rightarrow y \le -2\sqrt{3}-4$ or $y \ge -2\sqrt{3}+4$

($y \le -2\sqrt{3}-4$ is not possible as $\tan A\cdot\tan B < 0$)

**33.** *(d)* $f_k(x) = \frac{1}{k}(\sin^k x+\cos^k x)$,

Where, $x \in R$ and $k \ge 1$

Now, $f_4(x)-f_6(x)$

$$= \frac{1}{4}(\sin^4 x+\cos^4 x)-\frac{1}{6}(\sin^6 x+\cos^6 x)$$

$$= \frac{1}{4}(1-2\sin^2 x\cdot\cos^2 x)-\frac{1}{6}(1-3\sin^2 x\cdot\cos^2 x)$$

$$= \frac{1}{4}-\frac{1}{6} = \frac{1}{12}$$

**34.** *(c)* $A = \sin^2 x+\cos^4 x$

We have, $\cos^4 x \le \cos^2 x$

and $\sin^2 x = \sin^2 x$

On adding, we get

$\sin^2 x+\cos^4 x \le \sin^2 x+\cos^2 x$

$\therefore A \le 1$

Again, $A: t+(1-t)^2 = t^2-t+1, t \ge 0$, where minimum is $\frac{3}{4}$.

[let $t = \sin^2 x$]

Thus, $\frac{3}{4} \le A \le 1$.

**35.** *(a)* Given that

$$\sin\alpha+\sin\beta = -\frac{21}{65} \quad \text{...(i)}$$

$$\text{and } \cos\alpha+\cos\beta = -\frac{27}{65} \quad \text{...(ii)}$$

On squaring and adding Eqs. (i) and (ii), we get

$$\sin^2\alpha+\sin^2\beta+2\sin\alpha\sin\beta+\cos^2\alpha+\cos^2\beta+2\cos\alpha\cos\beta = \left(-\frac{21}{65}\right)^2+\left(-\frac{27}{65}\right)^2$$

$$\Rightarrow 2+2(\cos\alpha\cos\beta+\sin\alpha\sin\beta) = \frac{441}{4225}+\frac{729}{4225}$$

$$\Rightarrow 2[1+\cos(\alpha-\beta)] = \frac{1170}{4225}$$

$$\Rightarrow \cos^2\left(\frac{\alpha-\beta}{2}\right) = \frac{9}{130}$$

$$\therefore \cos\left(\frac{\alpha-\beta}{2}\right) = -\frac{3}{\sqrt{130}}$$

$$\left[\because \pi < \alpha-\beta < 3\pi \Rightarrow \frac{\pi}{2} < \frac{\alpha-\beta}{2} < \frac{3\pi}{2}\right]$$

**36.** *(a)* $\because \sin(\alpha+\beta) = 1$

$$\Rightarrow \alpha+\beta = \frac{\pi}{2} \quad \text{...(i)}$$

$$\text{and } \sin(\alpha-\beta) = \frac{1}{2}$$

$\Rightarrow \qquad \alpha - \beta = \pi/6 \qquad \ldots$(ii)

On solving Eqs. (i) and (ii), we get

$$\alpha = \frac{\pi}{3} \text{ and } \beta = \frac{\pi}{6}$$

$\therefore \tan(\alpha + 2\beta)\tan(2\alpha + \beta)$

$$= \tan\left(\frac{2\pi}{3}\right)\tan\left(\frac{5\pi}{6}\right)$$

$$= (-\sqrt{3}) \times \left(-\frac{1}{\sqrt{3}}\right) = 1$$

**37.** *(b)* $\displaystyle\sum_{r=1}^{13} \frac{1}{\sin\left(\frac{\pi}{4} + (r-1)\frac{\pi}{6}\right)\sin\left(\frac{\pi}{4} + \frac{r\pi}{6}\right)}$

$$= 2\sum_{r=1}^{13} \frac{\sin\left(\left(\frac{\pi}{4} + \frac{r\pi}{6}\right) - \left(\frac{\pi}{4} + (r-1)\frac{\pi}{6}\right)\right)}{\sin\left(\frac{\pi}{4} + (r-1)\frac{\pi}{6}\right)\sin\left(\frac{\pi}{4} + \frac{r\pi}{6}\right)}$$

$$= 2\sum_{r=1}^{13}\left[\cot\left(\frac{\pi}{4} + (r-1)\frac{\pi}{6}\right) - \cot\left(\frac{\pi}{4} + \frac{r\pi}{6}\right)\right]$$

$$= 2\left[\left\{\cot\frac{\pi}{4} - \cot\left(\frac{\pi}{4} + \frac{\pi}{6}\right)\right\} + \left\{\cot\left(\frac{\pi}{4} + \frac{\pi}{6}\right) - \cot\left(\frac{\pi}{4} + \frac{2\pi}{6}\right)\right\} + \ldots \ldots + \left\{\cot\left(\frac{\pi}{4} + \frac{12\pi}{6}\right) - \cot\left(\frac{\pi}{4} + \frac{13\pi}{6}\right)\right\}\right]$$

$$= 2\left[1 - \cot\left(\frac{\pi}{4} + \left(2\pi + \frac{\pi}{6}\right)\right)\right]$$

$$= 2\left[1 - \cot\left(\frac{\pi}{4} + \frac{\pi}{6}\right)\right]$$

$$= 2[1 - 2 + \sqrt{3}] = 2(\sqrt{3} - 1) = a\sqrt{3} + b \quad \text{[given]}$$

$\Rightarrow \quad a = 2$ and $b = -2$

$\Rightarrow a^2 + b^2 = 4 + 4 = 8$

**38.** *(d)* We have,

$$S = \left\{x \in \left(-\frac{\pi}{2}, \frac{\pi}{2}\right) : 9^{1-\tan^2 x} + 9^{\tan^2 x} = 10\right\}$$

and $\beta = \displaystyle\sum_{x \in S} \tan^2\left(\frac{x}{3}\right)$

Now, $9^{1-\tan^2 x} + 9^{\tan^2 x} = 10$

$$\frac{9}{9^{\tan^2 x}} + 9^{\tan^2 x} = 10$$

Let $\qquad 9^{\tan^2 x} = t$

$$\frac{9}{t} + t = 10$$

$\Rightarrow t^2 - 10t + 9 = 0 \Rightarrow \quad t = 9$ or $1$

If $\ t = 9, 9^{\tan^2 x} = 9 \Rightarrow x = \pm\dfrac{\pi}{4}$

If $t = 1, 9^{\tan^2 x} = 1 \quad \Rightarrow \quad x = 0$

Also, $\beta = \displaystyle\sum_{x \in S} \tan^2\left(\frac{x}{3}\right)$

$$= \tan^2\left(\frac{0}{3}\right) + \tan^2\left(\frac{\pi}{12}\right) + \tan^2\left(-\frac{\pi}{12}\right)$$

$$= 0 + 2\tan^2\frac{\pi}{12} = 2[2 - \sqrt{3}]^2$$

$$= 2[4 + 3 - 4\sqrt{3}] = 14 - 8\sqrt{3}$$

So, $\dfrac{1}{6}(\beta - 14)^2 = \dfrac{1}{6}[14 - 8\sqrt{3} - 14]^2$

$$= \frac{1}{6} \times 64 \times 3 = 32$$

**39.** *(b)* Let $A = 96\cos\dfrac{\pi}{33}\cos\dfrac{2\pi}{33}\cos\dfrac{4\pi}{33}\cos\dfrac{8\pi}{33}\cos\dfrac{16\pi}{33}$

$$\Rightarrow 2A = 96 \times 2\left(\cos\frac{\pi}{33}\cos\frac{2\pi}{33}\cos\frac{4\pi}{33}\cos\frac{8\pi}{33}\cos\frac{16\pi}{33}\right)$$

$$\Rightarrow 2A \times \sin\frac{\pi}{33} = 96 \times \left(2\sin\frac{\pi}{33}\cos\frac{\pi}{33}\cos\frac{2\pi}{33}\cos\frac{4\pi}{33}\cdot\cos\frac{8\pi}{33}\cos\frac{16\pi}{33}\right)$$

$$\Rightarrow 2A \times \sin\frac{\pi}{33} = 6 \times \sin\frac{32\pi}{33} = 6 \times \sin\frac{\pi}{33}$$

$\Rightarrow \qquad 2A = 6 \Rightarrow A = 3$

Thus, the required answer is 3.

**40.** *(b)* $2\sin\dfrac{\pi}{22}\cdot\sin\dfrac{3\pi}{22}\cdot\sin\dfrac{5\pi}{22}\cdot\sin\dfrac{7\pi}{22}\cdot\sin\dfrac{9\pi}{22}$

$$= 2\sin\left(\frac{\pi}{2} - \frac{10\pi}{22}\right)\sin\left(\frac{\pi}{2} - \frac{8\pi}{22}\right)\sin\left(\frac{\pi}{2} - \frac{6\pi}{22}\right)\sin\left(\frac{\pi}{2} - \frac{4\pi}{22}\right)\sin\left(\frac{\pi}{2} - \frac{2\pi}{22}\right)$$

$$= 2\cdot\cos\frac{10\pi}{22}\cdot\cos\frac{8\pi}{22}\cos\frac{6\pi}{22}\cdot\cos\frac{4\pi}{22}\cdot\cos\frac{2\pi}{22}$$

$$= 2\cdot\cos\frac{\pi}{11}\cdot\cos\frac{2\pi}{11}\cos\frac{3\pi}{11}\cdot\cos\frac{4\pi}{11}\cdot\cos\frac{5\pi}{11}$$

On multiplying and dividing by $\sin\left(\dfrac{\pi}{11}\right)$, we get

$$\frac{2\sin\frac{32\pi}{11}}{2^5\sin\frac{\pi}{11}}$$

$$\left[\because \cos\left(\frac{3\pi}{11}\right) = -\cos\left(\frac{8\pi}{11}\right), \cos\left(\frac{5\pi}{11}\right) = -\cos\left(\frac{16\pi}{11}\right) \text{ and } \sin 2\theta = 2\sin\theta\cos\theta\right]$$

$$= \frac{2}{32}\frac{\sin\left(3\pi - \frac{\pi}{11}\right)}{\sin\frac{\pi}{11}} = \frac{1}{16}$$

**41.** *(b)* Consider

$$\cos\left(\frac{3\pi}{4} + x\right) - \cos\left(\frac{3\pi}{4} - x\right)$$

$$= -2 \times \sin\frac{3\pi}{4}\sin(x) = -\sqrt{2}\sin x$$

Now, $\cos\left(\dfrac{\pi}{4} + x\right) + \cos\left(\dfrac{\pi}{4} - x\right)$

$$= 2 \times \cos\left(\frac{\pi}{4}\right)\cos x = \sqrt{2}\cos x$$

$$\therefore \cos\left(\frac{3\pi}{4} + x\right) + \cos\left(\frac{\pi}{4} + x\right) + \cos\left(\frac{\pi}{4} - x\right) - \cos\left(\frac{3\pi}{4} - x\right)$$

$= \sqrt{2}(\cos x - \sin x)$

Since, $-2 \le \sqrt{2}(\cos x - \sin x) \le 2$

$\Rightarrow 1 \le 3 + \sqrt{2}(\cos x - \sin x) \le 5$

$$\Rightarrow 1 \le 3 + \cos\left(\frac{3\pi}{4} + x\right) + \cos\left(\frac{\pi}{4} + x\right) + \cos\left(\frac{\pi}{4} - x\right) - \cos\left(\frac{3\pi}{4} - x\right) \le 5$$

$$\Rightarrow 0 \le \log_{\sqrt{5}}\left[3 + \cos\left(\frac{3\pi}{4} + x\right) + \cos\left(\frac{\pi}{4} + x\right) + \cos\left(\frac{\pi}{4} - x\right) - \cos\left(\frac{3\pi}{4} - x\right)\right] \le 2$$

$\therefore$ Range of the function is [0, 2].

**42.** *(c)* Given, $2\sin\left(\dfrac{\pi}{8}\right)\sin\left(\dfrac{2\pi}{8}\right)\sin\left(\dfrac{3\pi}{8}\right)\sin\left(\dfrac{5\pi}{8}\right)\sin\left(\dfrac{6\pi}{8}\right)\sin\left(\dfrac{7\pi}{8}\right)$

$$= 2\sin\frac{\pi}{8}\sin\frac{2\pi}{8}\sin\frac{3\pi}{8}\sin\left(\pi-\frac{3\pi}{8}\right)\sin\left(\pi-\frac{2\pi}{8}\right)\sin\left(\pi-\frac{\pi}{8}\right)$$

$$= 2\sin\frac{\pi}{8}\sin\frac{2\pi}{8}\sin\frac{3\pi}{8}\sin\frac{3\pi}{8}\sin\frac{2\pi}{8}\sin\frac{\pi}{8}$$

$$= 2\sin^2\left(\frac{\pi}{8}\right)\sin^2\left(\frac{2\pi}{8}\right)\sin^2\left(\frac{3\pi}{8}\right)$$

$$= 2\left(\sin^2\frac{\pi}{8}\right)\left(\frac{1}{\sqrt{2}}\right)^2\sin^2\left(\frac{\pi}{2}-\frac{\pi}{8}\right)$$

$$= 2\left(\sin^2\frac{\pi}{8}\right)\times\frac{1}{2}\times\cos^2\frac{\pi}{8}$$

$$= \sin^2\left(\frac{\pi}{8}\right)\cos^2\left(\frac{\pi}{8}\right)$$

$$= \frac{1}{4}\left(2\sin\frac{\pi}{8}\cos\frac{\pi}{8}\right)^2 = \frac{1}{4}\sin^2\left(\frac{\pi}{4}\right)$$

$$= \frac{1}{4}\left(\frac{1}{\sqrt{2}}\right)^2 = \frac{1}{8}$$

**43.** *(c)* $f: R \to R$ is defined as

$$f(x+y)+f(x-y)=2f(x)f(y)$$

Let $f(x)=\cos\lambda x$

$\because f\left(\frac{1}{2}\right)=-1 \Rightarrow \cos\frac{\lambda}{2}=-1$

$\Rightarrow \lambda = 2\pi$

Thus, $f(x)=\cos 2\pi x$

Now, $k$ is a natural number.

Thus, $f(k)=1$ $\quad(\because \cos 2n\pi = 1, n\in Z)$

Now, $\displaystyle\sum_{k=1}^{20}\frac{1}{\sin k\cdot\sin[k+f(k)]}$

$$= \sum_{k=1}^{20}\frac{1}{\sin k\cdot\sin(k+1)}$$

Multiply and divide by sin 1,

$$\frac{1}{\sin 1}\sum_{k=1}^{20}\frac{\sin 1}{\sin k\cdot\sin(k+1)}$$

$$= \frac{1}{\sin 1}\sum_{k=1}^{20}\frac{\sin[(k+1)-k]}{\sin k\cdot\sin(k+1)}$$

$$= \frac{1}{\sin 1}\sum_{k=1}^{20}[\cot k - \cot(k+1)]$$

$$= \frac{1}{\sin 1}(\cot 1 - \cot 21)$$

$$= \frac{1}{\sin 1}\left(\frac{\cos 1}{\sin 1}-\frac{\cos 21}{\sin 21}\right)$$

$$= \frac{1}{\sin 1}\left(\frac{\cos 1\sin 21 - \cos 21\sin 1}{\sin 1\cdot\sin 21}\right)$$

$$= \frac{1}{\sin 1}\left(\frac{\sin 20}{\sin 1\cdot\sin 21}\right)$$

$$= \operatorname{cosec}^2 1\cdot\operatorname{cosec} 21\cdot\sin 20$$

**44.** *(d)* Given, $\cos x+\cos y-\cos(x+y)=\frac{3}{2}$

$$\Rightarrow 2\cos\left(\frac{x+y}{2}\right)\cos\left(\frac{x-y}{2}\right)-\left[2\cos^2\left(\frac{x+y}{2}\right)-1\right]=\frac{3}{2}$$

$\left[\text{Use formula } \cos a+\cos b = 2\cos\left(\frac{a+b}{2}\right)\cos\left(\frac{a-b}{2}\right), \cos 2x = 2\cos^2 x - 1\right]$

$$\Rightarrow 2\cos\left(\frac{x+y}{2}\right)\cos\left(\frac{x-y}{2}\right)-2\cos^2\left(\frac{x+y}{2}\right)=\frac{3}{2}-1=\frac{1}{2}$$

$$\Rightarrow 4\cos\left(\frac{x+y}{2}\right)\cos\left(\frac{x-y}{2}\right)-4\cos^2\left(\frac{x+y}{2}\right)=\frac{1}{2}\times 2=1$$

$$=\cos^2\frac{(x-y)}{2}+\sin^2\frac{(x-y)}{2}$$

$$\Rightarrow \left[\cos\left(\frac{x-y}{2}\right)-2\cos\left(\frac{x+y}{2}\right)\right]^2+\sin^2\left(\frac{x-y}{2}\right)=0$$

$$\Rightarrow \sin\left(\frac{x-y}{2}\right)=0$$

and $\cos\left(\frac{x-y}{2}\right)-2\cos\left(\frac{x+y}{2}\right)=0$

$\Rightarrow x=y$ and $\cos 0 - 2\cos x = 0$ [using $x=y$]

Given, $\cos x = \frac{1}{2} = \cos y$

$$\therefore \sin x = \sqrt{1-\cos^2 x}=\sqrt{1-\frac{1}{4}}=\frac{\sqrt{3}}{2}$$

$$\therefore \sin x + \cos y = \frac{\sqrt{3}}{2}+\frac{1}{2}=\frac{1+\sqrt{3}}{2}$$

**45.** *(b)* Given, $\sin(\alpha-\beta)=\frac{5}{13}$

and $\cos(\alpha+\beta)=\frac{3}{5}$, where $\alpha,\beta\in\left(0,\frac{\pi}{4}\right)$

Since, $0<\alpha<\frac{\pi}{4}$ and $0<\beta<\frac{\pi}{4}$

$$\therefore 0<\alpha+\beta<\frac{\pi}{4}+\frac{\pi}{4}=\frac{\pi}{2}$$

$$\Rightarrow 0<\alpha+\beta<\frac{\pi}{2}$$

Also, $-\frac{\pi}{4}<-\beta<0$

$$\therefore 0-\frac{\pi}{4}<\alpha-\beta<\frac{\pi}{4}+0$$

$$\Rightarrow -\frac{\pi}{4}<\alpha-\beta<\frac{\pi}{4}$$

$\therefore \alpha+\beta\in\left(0,\frac{\pi}{2}\right)$ and $\alpha-\beta\in\left(-\frac{\pi}{4},\frac{\pi}{4}\right)$

But, $\sin(\alpha-\beta)>0$, therefore

$\alpha-\beta\in\left(0,\frac{\pi}{4}\right)$.

Now, $\sin(\alpha-\beta)=5/13$

$\Rightarrow \tan(\alpha-\beta)=5/12$ ...(i)

and $\cos(\alpha+\beta)=3/5$

$\Rightarrow \tan(\alpha+\beta)=4/3$ ...(ii)

Now, $\tan(2\alpha)=\tan[(\alpha+\beta)+(\alpha-\beta)]$

$$=\frac{\tan(\alpha+\beta)+\tan(\alpha-\beta)}{1-\tan(\alpha+\beta)\tan(\alpha-\beta)}$$

$$=\frac{\frac{4}{3}+\frac{5}{12}}{1-\frac{4}{3}\times\frac{5}{12}} \quad \text{[from Eqs. (i) and (ii)]}$$

$$=\frac{48+15}{36-20}=\frac{63}{16}$$

**46.** *(d)* Given expression

$$=3(\sin\theta-\cos\theta)^4+6(\sin\theta+\cos\theta)^2+4\sin^6\theta$$

$$=3((\sin\theta-\cos\theta)^2)^2+6(\sin\theta+\cos\theta)^2+4(\sin^2\theta)^3$$

$$=3(1-\sin 2\theta)^2+6(1+\sin 2\theta)+4(1-\cos^2\theta)^3$$

$[\because 1+\sin 2\theta=(\cos\theta+\sin\theta)^2$ and $1-\sin 2\theta=(\cos\theta-\sin\theta)^2]$

$$=3(1^2+\sin^2 2\theta-2\sin 2\theta)+6(1+\sin 2\theta)+4(1-\cos^6\theta-3\cos^2\theta+3\cos^4\theta)$$

$[\because (a-b)^2=a^2+b^2-2ab$ and $(a-b)^3=a^3-b^3-3a^2b+3ab^2]$

$$=3+3\sin^2 2\theta-6\sin 2\theta+6+6\sin 2\theta+4-4\cos^6\theta-12\cos^2\theta+12\cos^4\theta$$

$$=13+3\sin^2 2\theta-4\cos^6\theta-12\cos^2\theta+12\cos^4\theta$$
$$=13+3(2\sin\theta\cos\theta)^2-4\cos^6\theta-12\cos^2\theta(1-\cos^2\theta)$$
$$=13+12\sin^2\theta\cos^2\theta-4\cos^6\theta-12\cos^2\theta\sin^2\theta$$
$$=13-4\cos^6\theta$$

**47.** *(d)* We have,
$$3x^2-10x-25=0 \quad \text{...(i)}$$
∵ $\tan A$ and $\tan B$ are roots of Eq. (i).
∴ $\tan A+\tan B=10/3$
and $\tan A\tan B=\dfrac{-25}{3}$
Now, $\tan(A+B)=\dfrac{\tan A+\tan B}{1-\tan A\tan B}=\dfrac{10/3}{1+\dfrac{25}{3}}=\dfrac{5}{14}$
∴ $\sin(A+B)=5/\sqrt{221}$
and $\cos(A+B)=\dfrac{14}{\sqrt{221}}$
Now, $3\sin^2(A+B)-10\sin(A+B)\cos(A+B)-25\cos^2(A+B)$
$$=\frac{3\times 25}{221}-\left(\frac{10\times5\times14}{221}\right)-\left(\frac{25\times14\times14}{221}\right)$$
$$=\frac{25}{221}(3-28-196)=-25$$

**48.** *(c)* ∵ $\cos\alpha+\cos\beta=\dfrac{3}{2}$
$$\Rightarrow 2\cos\left(\frac{\alpha+\beta}{2}\right)\cos\left(\frac{\alpha-\beta}{2}\right)=\frac{3}{2} \quad \text{...(i)}$$
Also, $\sin\alpha+\sin\beta=\dfrac{1}{2}$
$$\Rightarrow 2\sin\left(\frac{\alpha+\beta}{2}\right)\cos\left(\frac{\alpha-\beta}{2}\right)=\frac{1}{2} \quad \text{...(ii)}$$
On dividing Eq. (ii) by Eq. (i), we get
$$\tan\left(\frac{\alpha+\beta}{2}\right)=\frac{1}{3}$$
Since, $\theta$ is the arithmetic mean of $\alpha$ and $\beta$
$$\theta=\frac{\alpha+\beta}{2}\Rightarrow \tan\theta=\frac{1}{3}$$

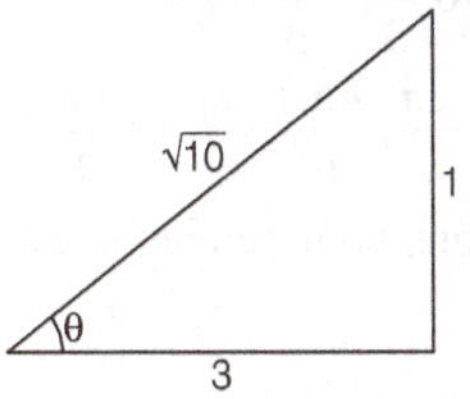


$$\Rightarrow \sin\theta=\frac{1}{\sqrt{10}} \text{ and } \cos\theta=\frac{3}{\sqrt{10}}$$
Now, $\sin 2\theta+\cos 2\theta=2\sin\theta\cos\theta+2\cos^2\theta-1$
$$=2\times\frac{1}{\sqrt{10}}\times\frac{3}{\sqrt{10}}+2\left(\frac{9}{10}\right)-1$$
$$=\frac{6}{10}+\frac{18}{10}-1=\frac{7}{5}$$

**49.** *(4)* Given,
$$\tan 9^\circ-\tan 27^\circ-\tan 63^\circ+\tan 81^\circ$$
$$=\tan 9^\circ+\tan(90^\circ-9^\circ)-\tan 27^\circ-\tan(90^\circ-27^\circ)$$
$$=\tan 9^\circ+\cot 9^\circ-\tan 27^\circ-\cot 27^\circ$$
$$=\frac{\sin 9^\circ}{\cos 9^\circ}+\frac{\cos 9^\circ}{\sin 9^\circ}-\left(\frac{\sin 27^\circ}{\cos 27^\circ}+\frac{\cos 27^\circ}{\sin 27^\circ}\right)$$
$$=\frac{\sin^2 9^\circ+\cos^2 9^\circ}{\sin 9^\circ\cos 9^\circ}-\left(\frac{\sin^2 27^\circ+\cos^2 27^\circ}{\cos 27^\circ\sin 27^\circ}\right)$$
$$=\frac{2}{\sin 18^\circ}-\frac{2}{\sin 54^\circ}=\frac{2\times4}{\sqrt5-1}-\frac{2\times4}{\sqrt5+1}$$
$$=8\left(\frac{\sqrt5+1-\sqrt5+1}{5-1}\right)=4$$

**50.** *(80)* Given,
$$\sin^2(10^\circ)\sin(20^\circ)\sin(40^\circ)\sin(50^\circ)\sin(70^\circ)$$
$$=\frac{1}{2}\sin 10^\circ\left(\cos 20^\circ+\frac{1}{2}\right)\left[\frac{1}{2}\sin 10^\circ\left(\cos 20^\circ-\frac{1}{2}\right)\right]$$
$$=\frac{1}{2}\left[\sin 10^\circ\cdot\cos 20^\circ+\frac{1}{2}\sin 10^\circ\right]\left[\frac{1}{2}\left\{\sin 10^\circ\cdot\cos 20^\circ-\frac{1}{2}\sin 10^\circ\right\}\right]$$
$$=\frac{1}{2}\left[\frac{1}{2}(\sin 30^\circ-\sin 10^\circ)+\frac{1}{2}\sin 10^\circ\right]\left[\frac{1}{2}\left\{\frac{1}{2}(\sin 30^\circ-\sin 10^\circ)-\frac{1}{2}\sin 10^\circ\right\}\right]$$
$$=\frac{1}{4}\times\frac{1}{2}\times\frac{1}{2}\times\frac{1}{2}[\sin 30^\circ-2\sin 10^\circ]$$
$$=\frac{1}{32}\left[\frac{1}{2}-2\sin 10^\circ\right]=\frac{1}{64}-\frac{1}{16}\sin 10^\circ$$
$$\therefore\ 16+\alpha^{-1}=16+64=80$$

**51.** *(1)* It is given that for $\alpha,\beta\in\left(0,\dfrac{\pi}{2}\right)$,
$$\frac{\sqrt2\sin\alpha}{\sqrt{1+\cos 2\alpha}}=\frac{1}{7}\Rightarrow\frac{\sqrt2\sin\alpha}{\sqrt2|\cos\alpha|}=\frac{1}{7}$$
$$\Rightarrow \tan\alpha=1/7$$
$$\because\ \alpha\in\left(0,\frac{\pi}{2}\right) \quad \text{...(i)}$$
and $\sqrt{\dfrac{1-\cos 2\beta}{2}}=\dfrac{1}{\sqrt{10}}$
$$\Rightarrow |\sin\beta|=\frac{1}{\sqrt{10}}\Rightarrow\sin\beta=\frac{1}{\sqrt{10}}$$
$$\because\ \beta\in\left(0,\frac{\pi}{2}\right)$$
$$\therefore\ \tan\beta=1/3 \quad \text{...(ii)}$$
$$\therefore \tan 2\beta=\frac{2\tan\beta}{1-\tan^2\beta}=\frac{\frac{2}{3}}{1-\frac{1}{9}}=\frac{6}{8}=\frac{3}{4}$$
Now, $\tan(\alpha+2\beta)=\dfrac{\tan\alpha+\tan 2\beta}{1-\tan\alpha\tan 2\beta}$
$$=\frac{\frac{1}{7}+\frac{3}{4}}{1-\left(\frac{1}{7}\times\frac{3}{4}\right)}=\frac{4+21}{28-3}=\frac{25}{25}=1$$

## Topic 2

**01.** *(b)* $\cos\theta=\dfrac{\sqrt6-2\pm\sqrt{(\sqrt6-2)^2+8\sqrt6}}{4\sqrt2}$
$$=\frac{\sqrt3}{2},-\frac{1}{\sqrt2}$$

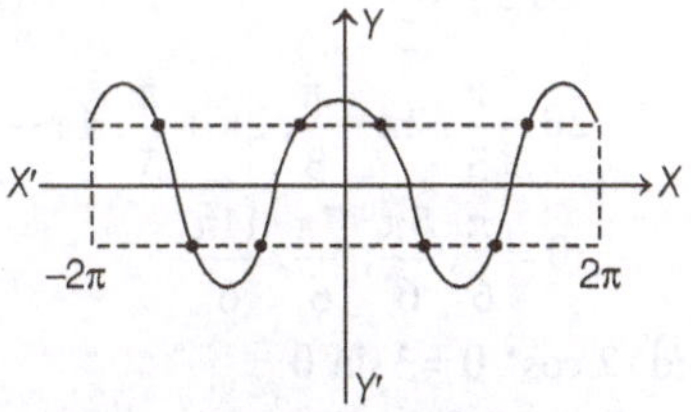


∴ Number of solutions $=8$
i.e. $\dfrac{-11\pi}{6},\dfrac{-5\pi}{4},\dfrac{-3\pi}{4},\dfrac{-\pi}{6},\dfrac{\pi}{6},\dfrac{3\pi}{4},\dfrac{5\pi}{4},\dfrac{11\pi}{6}$

**02.** *(b)*
$$\operatorname{cosec}\theta=\frac{2\sqrt3-2\pm\sqrt{(2(\sqrt3-1))^2+4\sqrt3(4)}}{2\sqrt3}$$
$$\Rightarrow \operatorname{cosec}\theta=-\frac{2}{\sqrt3},2$$
$$\theta=-\frac{7\pi}{6},\frac{-2\pi}{3},\frac{-\pi}{3},\frac{\pi}{6},\frac{5\pi}{6},\frac{4\pi}{3}$$
Total number of solution $=6$

**03.** *(b)* $2x+3\tan x=\pi$
$$3\tan x=\pi-2x$$

$\tan x = \dfrac{\pi - 2x}{3}$

5 solutions in $(-2\pi, 2\pi)$ $\tan x$ is increasing so we get 1 solution.

$$\left.\begin{matrix}\left(\dfrac{-3\pi}{2}, \dfrac{-\pi}{2}\right)\\ \left(\dfrac{-\pi}{2}, \dfrac{\pi}{2}\right)\\ \left(\dfrac{\pi}{2}, \dfrac{3\pi}{2}\right)\\ \left(\dfrac{3\pi}{2}, 2\pi\right)\\ \left(-2\pi, \dfrac{-3\pi}{2}\right)\end{matrix}\right\} \text{5 solutions}$$

**04.** *(d)* We have, $\sin x + \sin^2 x = 1$

$\Rightarrow \sin x = \cos^2 x \Rightarrow \tan x = \cos x$

$\therefore$ Given expression

$= 2\cos^{12} x + 6[\cos^{10} x + \cos^8 x] + 2\cos^6 x$

$= 2[\sin^6 x + 3\sin^5 x + 3\sin^4 x + \sin^3 x]$

$= 2\sin^3 x[(\sin x + 1)^3]$

$= 2[\sin^2 x + \sin x]^3 = 2$

**05.** *(d)* $2\sin^2\theta = \cos 2\theta$

$\Rightarrow 1 - \cos 2\theta = \cos 2\theta$

$\Rightarrow 2\cos 2\theta = 1$

$\cos 2\theta = \dfrac{1}{2}$

$2\theta = \dfrac{\pi}{3}, 2\pi - \dfrac{\pi}{3}, 2\pi + \dfrac{\pi}{3}, 4\pi - \dfrac{\pi}{3}$

$\theta = \dfrac{\pi}{6}, \dfrac{5\pi}{6}, \dfrac{7\pi}{6}, \dfrac{11\pi}{6}$

and $2\cos^2\theta = 3\sin\theta$

$\Rightarrow 2(1 - \sin^2\theta) = 3\sin\theta$

$\Rightarrow 2\sin^2\theta + 3\sin\theta - 2 = 0$

$\Rightarrow 2\sin^2\theta + 4\sin\theta - \sin\theta - 2 = 0$

$\Rightarrow (2\sin\theta - 1)(\sin\theta + 2) = 0$

$\sin\theta = \dfrac{1}{2}$ $\quad (\because \sin\theta \neq -2)$

$\theta = \dfrac{\pi}{6}, \dfrac{5\pi}{6}$

$\therefore$ Sum of common solutions is

$= \dfrac{\pi}{6} + \dfrac{5\pi}{6} = \pi$

**06.** *(d)* We have,

$4\sin^2 x - 4\cos^3 x + 9 - 4\cos x = 0$

$\Rightarrow 4(1 - \cos^2 x) - 4\cos^3 x + 9 - 4\cos x = 0$

$\Rightarrow 4\cos^3 x + 4\cos^2 x + 4\cos x - 13 = 0$

We know maximum value of $\cos x = 1$

So, $4\cos^3 x + 4\cos^2 x + 4\cos x$ can never be greater than 12, therefore no solution.

**07.** *(b)* $32^{\tan^2 x} + 32^{\sec^2 x} = 81$

$\Rightarrow 32^{\tan^2 x} + 32^{1 + \tan^2 x} = 81$

$\Rightarrow 33 \times 32^{\tan^2 x} = 81 \Rightarrow 32^{\tan^2 x} = \dfrac{27}{11}$

$\Rightarrow \tan^2 x = \ln_{32}\left(\dfrac{27}{11}\right)$

$\Rightarrow \tan x = \sqrt{\ln_{32}\left(\dfrac{27}{11}\right)} \in (0, 1)$

$\Rightarrow$ One solution in $\left[0, \dfrac{\pi}{4}\right]$.

**08.** *(a)* $\sin\theta + \sin 4\theta + \sin 7\theta = 0$

$\Rightarrow \sin 4\theta + (\sin\theta + \sin 7\theta) = 0$

$\Rightarrow \sin 4\theta + 2\sin 4\theta \cdot \cos 3\theta = 0$

$\Rightarrow \sin 4\theta\{1 + 2\cos 3\theta\} = 0$

$\Rightarrow \sin 4\theta = 0$ or $\cos 3\theta = -\dfrac{1}{2}$

Now, $\sin 4\theta = 0$

$\Rightarrow \theta = \dfrac{\pi}{4}, \dfrac{\pi}{2}, \dfrac{3\pi}{4}$ $\quad [\because \theta \in (0, \pi)]$

and $\cos 3\theta = -\dfrac{1}{2}$

$\Rightarrow \theta = \dfrac{2\pi}{9}, \dfrac{4\pi}{9}, \dfrac{8\pi}{9}$ $\quad [\because \theta \in (0, \pi)]$

So, $\theta = \dfrac{\pi}{4}, \dfrac{\pi}{2}, \dfrac{3\pi}{4}, \dfrac{2\pi}{9}, \dfrac{4\pi}{9}, \dfrac{8\pi}{9}$

**09.** *(d)* Given equation is

$2\sin^2 x + 5\sin x - 3 = 0.$

$\Rightarrow (2\sin x - 1)(\sin x + 3) = 0$

$\Rightarrow \sin x = \dfrac{1}{2}$ $\quad [\because \sin x \neq -3]$

Since, $-1 \le \sin x \le 1$

Thus, $x = \dfrac{\pi}{6}, \dfrac{5\pi}{6}, \dfrac{13\pi}{6}, \dfrac{17\pi}{6}$

Hence, the number of values of $x$ in the interval $[0, 3\pi]$ are 4.

**10.** *(d)* Given that

$f(x) = \sin x - \sqrt{3}\cos x + 1$

$\because -2 \le \sin x - \sqrt{3}\cos x \le 2$

$\left[\because -\sqrt{a^2 + b^2} \le a\sin x + b\cos x \le \sqrt{a^2 + b^2}\right]$

$\Rightarrow -1 \le \sin x - \sqrt{3}\cos x + 1 \le 3$

$\therefore$ Range of $f(x) = [-1, 3]$

**11.** *(d)* We have,

$\cos 2\theta\cos\dfrac{\theta}{2} + \cos\dfrac{5\theta}{2} = 2\cos^3\dfrac{5\theta}{2}$

$\cos 2\theta\cos\dfrac{\theta}{2} = 2\cos^3\dfrac{5\theta}{2} - \cos\dfrac{5\theta}{2}$

$\cos 2\theta\cos\dfrac{\theta}{2} = \cos\dfrac{5\theta}{2}\left(2\cos^2\dfrac{5\theta}{2} - 1\right)$

Use formula $2\cos^2 x - 1 = \cos 2x$

$\cos 2\theta\cos\dfrac{\theta}{2} = \cos\dfrac{5\theta}{2}(\cos 5\theta)$

Multiply both sides of the equation by 2

$2\cos 2\theta\cos\dfrac{\theta}{2} = 2\cos\dfrac{5\theta}{2}(\cos 5\theta)$

Use formula $2\cos x \times \cos y = \cos(x + y) + \cos(x - y)$

$\cos\left(2\theta + \dfrac{\theta}{2}\right) + \cos\left(2\theta - \dfrac{\theta}{2}\right)$

$= \cos\left(\dfrac{5\theta}{2} + 5\theta\right) + \cos\left(\dfrac{5\theta}{2} - 5\theta\right)$

$\cos\dfrac{5\theta}{2} + \cos\dfrac{3\theta}{2} = \cos\dfrac{15\theta}{2} + \cos\dfrac{5\theta}{2}$

$\cos\dfrac{15\theta}{2} - \cos\dfrac{3\theta}{2} = 0$

Use Formula

$\cos x - \cos y = -2\sin\dfrac{x + y}{2}\sin\dfrac{x - y}{2}$

$-2\sin\left(\dfrac{\dfrac{15\theta}{2} + \dfrac{3\theta}{2}}{2}\right)\sin\left(\dfrac{\dfrac{15\theta}{2} - \dfrac{3\theta}{2}}{2}\right) = 0$

$2\sin\left(\dfrac{9\theta}{2}\right)\sin 3\theta = 0$

This gives $\sin\dfrac{9\theta}{2} = 0$ or $\sin 3\theta = 0$

The general solution of $\sin\theta = 0$ is $\theta = n\pi$

Here $\theta \in \left[-\dfrac{\pi}{2}, \dfrac{\pi}{2}\right]$

$\dfrac{9\theta}{2} = k\pi$ or $3\theta = n\pi$

$\theta = \dfrac{2k\pi}{9}$ or $\theta = \dfrac{n\pi}{3}$

$\theta = 0, \dfrac{\pi}{3}, -\dfrac{\pi}{3}$

$\theta = \dfrac{2\pi}{9}, \dfrac{4\pi}{9}, -\dfrac{2\pi}{9}, -\dfrac{4\pi}{9}$

Therefore, total number of solutions are 7.

**12.** *(c)*

> **key idea** First convert the given equation into quadratic equation in $\cos\theta$ as $a\cos^2\theta + b\cos\theta + c = 0$, then $\cos\theta = \frac{-b \pm \sqrt{b^2 - 4ac}}{2a}$ and keep in mind that $\theta \in [0, \pi/4]$

Given, $4\cos\theta - 3\sin\theta = 1$

$\Rightarrow \quad 4\cos\theta - 1 = 3\sin\theta$

On squaring both sides, we get

$16\cos^2\theta + 1 - 8\cos\theta = 9\sin^2\theta$

$\Rightarrow 16\cos^2\theta + 1 - 8\cos\theta - 9 + 9\cos^2\theta = 0$

$\Rightarrow \quad 25\cos^2\theta - 8\cos\theta - 8 = 0$

$\Rightarrow \quad \cos\theta = \frac{8 \pm \sqrt{64 + 4 \times 25 \times 8}}{2 \times 25}$

$\Rightarrow \quad \cos\theta = \frac{8 \pm 4\sqrt{4 + 50}}{2 \times 25} = \frac{4 \pm 2\sqrt{54}}{25}$

Since, $\theta \in \left[0, \frac{\pi}{4}\right]$

$\therefore \cos\theta = \frac{4 + 2\sqrt{54}}{25} = \frac{4 + 6\sqrt{6}}{25}$

$= \frac{4 + 6\sqrt{6}}{25} \times \frac{6\sqrt{6} - 4}{6\sqrt{6} - 4} = \frac{4}{3\sqrt{6} - 2}$

**13.** *(d)* We have,

$$e^{\sin x} - \frac{2}{e^{\sin x}} = 2$$

Let $e^{\sin x} = t \ (t > 0)$

$t - \frac{2}{t} = 2 \Rightarrow t^2 - 2t - 2 = 0$

$t = \frac{2 \pm \sqrt{4 + 8}}{2} = 1 \pm \sqrt{3}$

$t = 1 - \sqrt{3}$ (rejected as $t > 0$)

$\therefore \quad t = 1 + \sqrt{3} \Rightarrow e^{\sin x} = 1 + \sqrt{3}$

$\sin x = \log_e(1 + \sqrt{3})$

$= \log_e 2.73 > 1$

So, no solution.

**14.** *(c)* We have, $4\cos\theta + 5\sin\theta = 1$

$\Rightarrow 16\cos^2\theta + 25\sin^2\theta + 40\sin\theta\cos\theta = 1$

$\Rightarrow 16 + 9\sin^2\theta + 20\sin 2\theta = 1$

$\Rightarrow 16 + 9\left(\frac{1 - \cos 2\theta}{2}\right) + 20\sin 2\theta = 1$

$\Rightarrow \frac{-9}{2}\cos 2\theta + 20\sin 2\theta = \frac{-39}{2}$

$\Rightarrow -9\cos 2\theta + 40\sin 2\theta = -39$

$\Rightarrow -9\left(\frac{1 - \tan^2\theta}{1 + \tan^2\theta}\right) + 40\left(\frac{2\tan\theta}{1 + \tan^2\theta}\right) = -39$

$\Rightarrow 48\tan^2\theta + 80\tan\theta + 30 = 0$

$\Rightarrow 24\tan^2\theta + 40\tan\theta + 15 = 0$

$\Rightarrow \tan\theta = \frac{-40 \pm \sqrt{160}}{2 \times 24}$

$\Rightarrow \tan\theta = \frac{\sqrt{10} - 10}{12}, \frac{-\sqrt{10} - 10}{12}$

Since, $\alpha \in \left(-\frac{\pi}{2}, \frac{\pi}{2}\right)$

$\therefore \ \tan\alpha = \frac{\sqrt{10} - 10}{12}$

**15.** *(a)* $S = \left\{x \in R : 2\cos\left(\frac{x^2 + x}{6}\right) = 4^x + 4^{-x}\right\}$

$\Rightarrow 2\cos\left(\frac{x^2 + x}{6}\right) = 4^x + 4^{-x}$

$\Rightarrow 2\cos\left(\frac{x^2 + x}{6}\right) \le 2$

and $4^x + 4^{-x} \ge 2$

So, $2\cos\left(\frac{x^2 + x}{6}\right) = 4^x + 4^{-x} = 2$

$\Rightarrow \ 2\cos\left(\frac{x^2 + x}{6}\right) = 2, \ 4^x + 4^{-x} = 2$

Both satisfy $x = 0$ only.

Hence, there is only one solution.

**16.** *(c)*

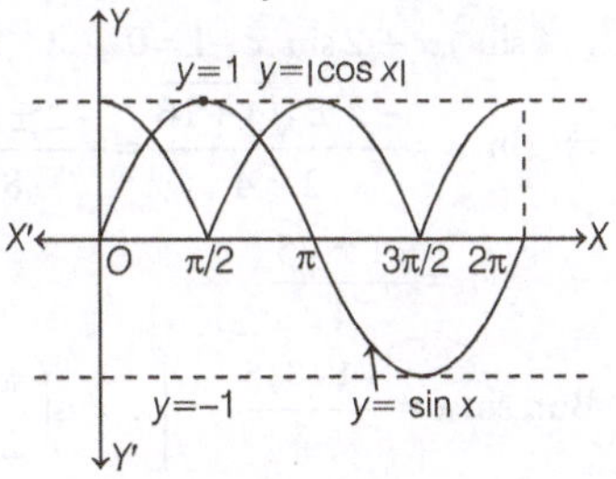


The number of solution of the equation $|\cos x| = \sin x, \ \forall \ x \in [-4\pi, 4\pi]$ will be equal to 4 times the number of solutions of the same equation $|\cos x| = \sin x$ for $x \in [0, 2\pi]$. Clearly, from the above diagram, two solution of the equation $|\cos x| = \sin x$ in $[0, 2\pi]$. Thus, total 8 solutions in $[-4\pi, 4\pi]$.

**17.** *(d)* Given, equation is

$\cos\left(x + \frac{\pi}{3}\right)\cos\left(\frac{\pi}{3} - x\right) = \frac{1}{4}\cos^2 2x, \ x \in [-3\pi, 3\pi]$

Since, we know that

$\cos(\alpha - \beta)\cos(\alpha + \beta) = \cos^2\alpha - \sin^2\beta$

$\Rightarrow \quad \cos^2\frac{\pi}{3} - \sin^2 x = \frac{1}{4}\cos^2 2x$

$\Rightarrow \quad (1 - 4\sin^2 x) = \cos^2 2x$

$\Rightarrow \quad (1 - 4\sin^2 x) = (1 - 2\sin^2 x)^2$

$\Rightarrow \quad 1 - 4\sin^2 x = 1 - 4\sin^2 x + 4\sin^4 x$

$\Rightarrow \quad 4\sin^4 x = 0 \Rightarrow \sin x = 0$

$\Rightarrow \ x = -3\pi, -2\pi, -\pi, 0, \pi, 2\pi, 3\pi$

$[\because x \in [-3\pi, 3\pi]]$

Thus, number of solution is 7.

**18.** *(a)* Given,

$2\cos x\left(4\sin\left(\frac{\pi}{4} + x\right)\sin\left(\frac{\pi}{4} - x\right) - 1\right) = 1$

$\Rightarrow 2\cos x\left(2\cos(2x) - 2\cos\left(\frac{\pi}{2}\right) - 1\right) = 1$

$\Rightarrow 2\cos x(4\cos^2 x - 3) = 1$

$\Rightarrow \cos 3x = \frac{1}{2} \Rightarrow 3x = \frac{\pi}{3}, \frac{5\pi}{3}, \frac{7\pi}{3}$

$\Rightarrow \quad x = \frac{\pi}{9}, \frac{5\pi}{9}, \frac{7\pi}{9}$

Number of solutions $= n = 3$

Sum of solutions $= S = \frac{13\pi}{9}$

**19.** *(c)* Given, $\sin^7 x + \cos^7 x = 1$ ...(i)

As, $\sin^2 x + \cos^2 x = 1$ ...(ii)

Now, $\sin^7 x \le \sin^2 x$ and $\cos^7 x \le \cos^2 x$.

But, according to question, Eqs. (i) and (ii), it is only possible, when $\sin^7 x = \sin^2 x$ and $\cos^7 x = \cos^2 x$

So, $\sin^2 x + \cos^2 x = 1$

When, $\sin x = 0$ and $\cos x = 1$

or $\sin x = 1$ and $\cos x = 0$

$\Rightarrow \quad x = 0, 2\pi, 4\pi, \frac{\pi}{2}, \frac{5\pi}{2}$

$\therefore$ 5 solutions.

**20.** *(b)* The expression, $\cos^4\theta + \sin^4\theta$

$= (\cos^2\theta + \sin^2\theta)^2 - 2\sin^2\theta\cos^2\theta$

$= 1 - \frac{1}{2}\sin^2(2\theta)$

$\because \quad \sin^2(2\theta) \in [0, 1]$

$\Rightarrow \quad -\frac{1}{2}\sin^2(2\theta) \in \left[-\frac{1}{2}, 0\right]$

$\Rightarrow \quad 1-\frac{1}{2}\sin^2(2\theta) \in \left[\frac{1}{2}, 1\right]$

Now, as $\cos^4\theta+\sin^4\theta+\lambda=0$

$\Rightarrow \lambda = -(\cos^4\theta+\sin^4\theta)$

$\Rightarrow \lambda \in [-1, -1/2]$ for real solution of the given equation $\cos^4\theta+\sin^4\theta+\lambda=0$ for $\theta$.

Hence, option (b) is correct.

**21.** *(b)* Given equation is

$1+\sin^4 x = \cos^2(3x)$

Since, range of $(1+\sin^4 x) = [1, 2]$

and range of $\cos^2(3x) = [0, 1]$

So, the given equation holds if

$$1+\sin^4 x = 1 = \cos^2(3x)$$

$\Rightarrow \quad \sin^4 x = 0$ and $\cos^2 3x = 1$

Since, $\quad x \in \left[-\frac{5\pi}{2}, \frac{5\pi}{2}\right]$

$\therefore \quad x = -2\pi, -\pi, 0, \pi, 2\pi.$

Thus, there are five different values of $x$ is possible.

**22.** *(d)* Given expression

$3\cos\theta+5\sin\left(\theta-\frac{\pi}{6}\right)$

$=3\cos\theta+5\left(\sin\theta\cos\frac{\pi}{6}-\sin\frac{\pi}{6}\cos\theta\right)$

$=3\cos\theta+5\left(\frac{\sqrt{3}}{2}\sin\theta-\frac{1}{2}\cos\theta\right)$

$=3\cos\theta-\frac{5}{2}\cos\theta+\frac{5\sqrt{3}}{2}\sin\theta$

$=\frac{1}{2}\cos\theta+\frac{5\sqrt{3}}{2}\sin\theta$

$\because$ The maximum value of $a\cos\theta+b\sin\theta$ is $\sqrt{a^2+b^2}$.

So, maximum value of

$$\frac{1}{2}\cos\theta+\frac{5\sqrt{3}}{2}\sin\theta \text{ is}$$

$$=\sqrt{\left(\frac{1}{2}\right)^2+\left(\frac{5\sqrt{3}}{2}\right)^2}=\sqrt{\frac{1}{4}+\frac{75}{4}}$$

$$=\sqrt{\frac{76}{4}}=\sqrt{19}$$

**23.** *(b)*

> **key idea** Apply the identities
> $\cos(x+y)\cos(x-y) = \cos^2 x - \sin^2 y$
> and $\cos 3x = 4\cos^3 x - 3\cos x$

We have,

$$8\cos x\left(\cos\left(\frac{\pi}{6}+x\right)\cos\left(\frac{\pi}{6}-x\right)-\frac{1}{2}\right)=1$$

$$\Rightarrow 8\cos x\left(\cos^2\frac{\pi}{6}-\sin^2 x-\frac{1}{2}\right)=1$$

$$\Rightarrow \quad 8\cos x\left(\frac{3}{4}-\sin^2 x-\frac{1}{2}\right)=1$$

$$\Rightarrow \quad 8\cos x\left(\frac{3}{4}-\frac{1}{2}-1+\cos^2 x\right)=1$$

$$\Rightarrow \quad 8\cos x\left(\frac{-3+4\cos^2 x}{4}\right)=1$$

$$\Rightarrow \quad 2(4\cos^3 x-3\cos x)=1$$

$$\Rightarrow \quad 2\cos 3x = 1 \Rightarrow \cos 3x = \frac{1}{2}$$

$$\Rightarrow \quad 3x = \frac{\pi}{3}, \frac{5\pi}{3}, \frac{7\pi}{3} \qquad [0 \le 3x \le 3\pi]$$

$$\Rightarrow \quad x = \frac{\pi}{9}, \frac{5\pi}{9}, \frac{7\pi}{9}$$

$$\text{Sum} = \frac{\pi}{9}+\frac{5\pi}{9}+\frac{7\pi}{9}=\frac{13\pi}{9}$$

$$\Rightarrow \quad k\pi = \frac{13\pi}{9}$$

Hence, $k = \frac{13}{9}$

**24.** *(d)* Given, $\sin 3x = \cos 2x$

$\Rightarrow \quad 3\sin x - 4\sin^3 x = 1-2\sin^2 x$

$\Rightarrow 4\sin^3 x - 2\sin^2 x - 3\sin x + 1 = 0$

$\Rightarrow (\sin x - 1)(4\sin^2 x + 2\sin x - 1) = 0$

Now, $\sin x \ne 1 \qquad \left[\because x \in \left(\frac{\pi}{2}, \pi\right)\right]$

$\therefore 4\sin^2 x + 2\sin x - 1 = 0$

$$\Rightarrow \sin x = \frac{-2\pm\sqrt{4+16}}{2\times 4} = \frac{-2\pm 2\sqrt{5}}{8}$$

$$= \frac{-1\pm\sqrt{5}}{4}$$

But, $\sin x \ne \frac{-1-\sqrt{5}}{4} \qquad \left[\because x \in \left(\frac{\pi}{2}, \pi\right)\right]$

$\therefore \sin x = \frac{-1+\sqrt{5}}{4},$

So, it has only one solution.

**25.** *(b)* Since, $\alpha$ is a root of

$$25\cos^2\theta + 5\cos\theta - 12 = 0$$

$\therefore \quad 25\cos^2\alpha + 5\cos\alpha - 12 = 0$

$\Rightarrow \quad (5\cos\alpha - 3)(5\cos\alpha + 4) = 0$

$\Rightarrow \quad \cos\alpha = -\frac{4}{5}$ or $\frac{3}{5}$

But, $\frac{\pi}{2} < \alpha < \pi$ i.e., in second quadrant.

$\therefore \quad \cos\alpha = -\frac{4}{5} \Rightarrow \sin\alpha = \frac{3}{5}$

Now, $\sin 2\alpha = 2\sin\alpha\cos\alpha$

$$= 2\times\frac{3}{5}\times\left(-\frac{4}{5}\right) = -\frac{24}{25}$$

**26.** *(c)* Since, $|c| > \sqrt{a^2+b^2}$

$\Rightarrow \quad c < -\sqrt{a^2+b^2}$

and $\quad c > \sqrt{a^2+b^2}$

But, $-\sqrt{a^2+b^2} \le a\sin x + b\cos x$

$\le \sqrt{a^2+b^2} \quad$ ...(i)

and $\quad a\sin x + b\cos x = c \quad$ ...(ii)

From Eqs. (i) and (ii), we can see that no solution exists.

**27.** *(d)*

$2\sin^3 x + \sin 2x\cos x + 4\sin x - 4 = 0$

$\Rightarrow 2\left(\frac{3\sin x - \sin 3x}{4}\right)$

$+ \sin 2x\cos x + 4\sin x - 4 = 0$

$\Rightarrow \frac{1}{2}(3\sin x - \sin 3x) + \frac{1}{2}$

$\{\sin(2x+x) + \sin(2x-x)\}$

$+ 4\sin x - 4 = 0$

$\Rightarrow \frac{3}{2}\sin x - \frac{1}{2}\sin 3x + \frac{1}{2}\sin 3x$

$+ \frac{1}{2}\sin x + 4\sin x - 4 = 0$

$\Rightarrow \quad \sin x\left(\frac{3}{2}+\frac{1}{2}+4\right) - 4 = 0$

$\Rightarrow \quad 6\sin x - 4 = 0$

$\Rightarrow \quad \sin x = \frac{2}{3}$

$\Rightarrow \quad \sin x = 0.66, \ x \in \left[0, \frac{n\pi}{2}\right]$

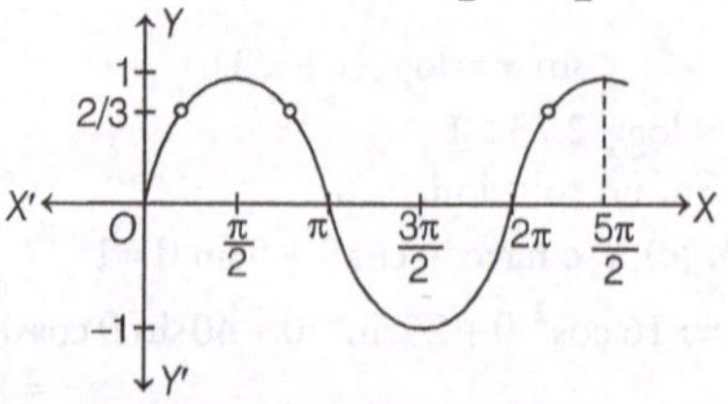


For exacts 3 solutions, $n = 5$

$\therefore x^2 + 5x + 2 = 0$

$$\Rightarrow x = \frac{-5\pm\sqrt{25-8}}{2} = \frac{-5\pm\sqrt{17}}{2}$$

$$= -0.438, -4.56 \in (-\infty, 0)$$

**28.** *(d)* $\dfrac{3\cos 2x+\cos^3 2x}{\cos^6 x-\sin^6 x}=x^3-x^2+6$

$$\text{LHS}=\frac{\cos 2x\,(3+\cos^2 2x)}{(\cos^2 x-\sin^2 x)(\cos^4 x+\sin^4 x+\cos^2 x\sin^2 x)}$$

$$=\frac{\cos 2x\,(3+\cos^2 2x)}{\cos 2x\,((\cos^2 x+\sin^2 x)^2-\cos^2 x\sin^2 x)}$$

$[\because \cos 2x=\cos^2 x-\sin^2 x]$

$$=\frac{3+\cos^2 2x}{1-\frac{1}{4}\sin^2 2x}$$

$[\because \sin 2x=2\sin x\cos x]$

$$=\frac{4\,(3+1-\sin^2 2x)}{4-\sin^2 2x}$$

$$=\frac{4(4-\sin^2 2x)}{4-\sin^2 2x}=4$$

Now, $x^3-x^2+6=4$

$x^3-x^2+2=0$

$\therefore$ Sum of solution $=\dfrac{-(b)}{a}=1$

> **Alternate Solution**
>
> $$\text{LHS}=\frac{\cos 2x(3+\cos^2 2x)}{(\cos^2 x-\sin^2 x)(\cos^4 x+\sin^4 x+\cos^2 x\sin^2 x)}$$
>
> $$=\frac{\cos 2x(3+\cos^2 2x)}{\cos 2x(1-\sin^2 x\cos^2 x)}$$
>
> $[\because \sin^4 x+\cos^4 x=1-2\sin^2 x\cos^2 x]$
>
> $$=\frac{3+\cos^2 2X}{1-\frac{1}{4}\sin^2 2x}=\frac{4(3+1-\sin^2 2x)}{4-\sin^2 2x}=4$$
>
> So, $x^3-x^2+6=4$
>
> $x^3-x^2+2=0$
>
> $\therefore$ Sum of solution $=-\dfrac{b}{a}=1$

**29.** *(c)* We have,

$S=\{\theta\in[0,2\pi]:3\cos^4\theta-5\cos^2\theta-2\sin^6\theta+2=0\}$

$\because 3\cos^4\theta-5\cos^2\theta-2\sin^6\theta+2=0$

$\Rightarrow 3\cos^4\theta-3\cos^2\theta-2\cos^2\theta-2\sin^6\theta+2=0$

$\Rightarrow 3\cos^2\theta(\cos^2\theta-1)-2(\cos^2\theta-1)-2\sin^6\theta=0$

$\Rightarrow -3\cos^2\theta\sin^2\theta+2\sin^2\theta-2\sin^6\theta=0$

$\Rightarrow -3\cos^2\theta\sin^2\theta+2\sin^2\theta(1-\sin^4\theta)=0$

$\Rightarrow -3\cos^2\theta\sin^2\theta+2\sin^2\theta(1-\sin^2\theta)(1+\sin^2\theta)=0$

$\Rightarrow -3\cos^2\theta\sin^2\theta+2\sin^2\theta\cos^2\theta(1+\sin^2\theta)=0$

$\Rightarrow \cos^2\theta\sin^2\theta[-3+2+2\sin^2\theta]=0$

$\Rightarrow \cos^2\theta\sin^2\theta(2\sin^2\theta-1)=0$

Either $\cos^2\theta=0$ or $\sin^2\theta=0$ or $2\sin^2\theta-1=0$

If $\sin^2\theta=0$, then $\theta=\{0,\pi,2\pi\}$

If $\cos^2\theta=0$, then $\theta=\left\{\dfrac{\pi}{2},\dfrac{3\pi}{2}\right\}$

If $2\sin^2\theta-1=0$, then

$$\theta=\left\{\frac{\pi}{4},\frac{3\pi}{4},\frac{5\pi}{4},\frac{7\pi}{4}\right\}$$

Hence, there are total 9 solutions of the given set.

**30.** *(c)* Given, $\tan 15^\circ+\dfrac{1}{\tan 75^\circ}+\dfrac{1}{\tan 105^\circ}+\tan 195^\circ=2a$ ... (i)

Since, $\tan 15^\circ=2-\sqrt{3}$

$\dfrac{1}{\tan 75^\circ}=\cot 75^\circ$

$=\tan 15^\circ=2-\sqrt{3}$

$[\because \cot\theta=\tan(90^\circ-\theta)]$

$\dfrac{1}{\tan 105^\circ}=\cot(105^\circ)=-\cot 75^\circ$

$=\sqrt{3}-2$

and $\tan(195^\circ)=\tan(180^\circ+15^\circ)$

$=\tan 15^\circ=2-\sqrt{3}$

From Eq. (i),

$2a=(2-\sqrt{3})+(2-\sqrt{3})+(\sqrt{3}-2)+(2-\sqrt{3})$

$\Rightarrow 2a=2(2-\sqrt{3})\Rightarrow a=2-\sqrt{3}$

Hence, $a+\dfrac{1}{a}=(2-\sqrt{3})+\dfrac{1}{(2-\sqrt{3})}$

$$=(2-\sqrt{3})+\frac{2+\sqrt{3}}{(2-\sqrt{3})(2+\sqrt{3})}$$

$$=(2-\sqrt{3})+\frac{(2+\sqrt{3})}{4-3}$$

$=2-\sqrt{3}+2+\sqrt{3}=4$

> **Alternate Solution**
>
> $\tan 15^\circ+\dfrac{1}{\tan 75^\circ}+\dfrac{1}{\tan 105^\circ}+\tan 195^\circ=2a$
>
> $\Rightarrow \tan 15^\circ+\tan 15^\circ-\tan 15^\circ+\tan 15^\circ=2a$
>
> $\Rightarrow a=\tan 15^\circ$
>
> Now, $a+\dfrac{1}{a}=\tan 15^\circ+\dfrac{1}{\tan 15^\circ}$
>
> $$=\frac{1+\tan^2 15^\circ}{\tan 15^\circ}=\frac{2}{\sin 30^\circ}=4$$

**31.** *(d)* Given equation is

$\log_{\cos x}\cot x+4\log_{\sin x}\tan x=1$, where $x\in\left(0,\dfrac{\pi}{2}\right)$

$$\Rightarrow \frac{\log_e\cot x}{\log_e\cos x}+4\frac{\log_e\tan x}{\log_e\sin x}=1$$

[Using base change formula]

$$\Rightarrow \frac{\log\cos x-\log\sin x}{\log\cos x}+4\left(\frac{\log\sin x-\log\cos x}{\log\sin x}\right)=1$$

$\Rightarrow (\log\cos x)(\log\sin x)-(\log(\sin x))^2+4(\log\sin x)(\log\cos x)-4(\log(\cos x))^2=\log\sin x\cdot\log\cos x$

$\Rightarrow (\log\sin x)^2-4(\log\sin x)(\log\cos x)+4(\log\cos x)^2=0$

$\Rightarrow (\log\sin x-2\log\cos x)^2=0$

$\Rightarrow \log\sin x=2\log\cos x$

$\Rightarrow \log\sin x=\log(\cos x)^2$

$\Rightarrow \sin x=\cos^2 x\Rightarrow 1-\sin^2 x=\sin x$

$\Rightarrow \sin^2 x+\sin x-1=0$

$\Rightarrow \sin x=\dfrac{-1\pm\sqrt{5}}{2}$

But, $x\in(0,\pi/2)$

So, $\sin x>0$

$\therefore \sin x=\dfrac{-1-\sqrt{5}}{2}<0$ (rejected)

Hence,

$\sin x=\dfrac{-1+\sqrt{5}}{2}=\dfrac{\alpha+\sqrt{\beta}}{2}$ [given]

Therefore, $\alpha+\beta=-1+5=4$

**32.** *(a)* $\cos^2 2x-2\sin^4 x-2\cos^2 x=\lambda$

$$\Rightarrow \cos^2 2x-\frac{1-2\cos 2x+\cos^2 2x}{2}-2\cos^2 x=\lambda$$

$$\Rightarrow \cos^2 2x+\frac{-\left(\frac{2}{2}-\frac{4\cos 2x}{2}+\frac{\cos 4x+1}{2}\right)}{2}-2\cos^2 x=\lambda$$

$$\Rightarrow \cos^2 2x+\frac{-\frac{(\cos 4x-4\cos 2x+(2+1))}{2}}{2}-2\cos^2 x=\lambda$$

$\Rightarrow \cos^2 2x + \frac{-3+4\cos 2x - \cos 4x}{4} - 2\left(\frac{\cos 2x + 1}{2}\right) = \lambda$

$\Rightarrow \frac{\cos 4x + (4\cos 2x - 4\cos 2x) + (2-7)}{4} = \lambda$

$\Rightarrow \frac{\cos 4x - 5}{4} = \lambda$

As we know, $-1 \le \cos 4x \le 1$

$\Rightarrow -6 \le \cos 4x - 5 \le -4$

$\Rightarrow \frac{-6}{4} \le \frac{\cos 4x - 5}{4} \le \frac{-4}{4}$

$\Rightarrow -\frac{3}{2} \le \lambda \le -1$

So, $\lambda \in \left[-\frac{3}{2}, -1\right]$

**33.** *(a)* We have, $f(\theta) = 3\left[\sin^4\left(\frac{3\pi}{2} - \theta\right) + \sin^4(3\pi + \theta)\right] - 2(1 - \sin^2(2\theta))$

(Using $\sin\left(\frac{3\pi}{2} - \theta\right) = -\cos\theta$ and $\sin(3\pi + \theta) = \sin\theta$), we get

$f(\theta) = 3(\cos^4\theta + \sin^4\theta) - 2(1 - \sin^2(2\theta))$

$= 3[(\cos^2\theta + \sin^2\theta)^2 - 2\sin^2\theta\cos^2\theta] - 2[(1 - \sin^2(2\theta)]$

$= 3\left[1 - \frac{(2\sin\theta\cos\theta)^2}{2}\right] - 2 + 2\sin^2(2\theta)$

$= 3 - \frac{3\sin^2(2\theta)}{2} - 2 + 2\sin^2(2\theta)$

$= 1 + \frac{\sin^2(2\theta)}{2}$ ... (i)

Now, $S = \left\{\theta \in [0, \pi] : f'(\theta) = -\frac{\sqrt{3}}{2}\right\}$

$f'(\theta) = \left[1 + \frac{\sin^2(2\theta)}{2}\right]' = \frac{1}{2} \times [\sin^2(2\theta)]'$

$= \sin(2\theta)\cdot\cos(2\theta)\cdot 2 = \sin(4\theta)$

$\theta \in [0, \pi] \Rightarrow 4\theta \in [0, 4\pi]$

Now, $\sin(4\theta) = -\frac{\sqrt{3}}{2}$ in $[0, 4\pi]$

$\Rightarrow 4\theta = \frac{4\pi}{3}, \frac{5\pi}{3}, \frac{10\pi}{3}, \frac{11\pi}{3}$

$\Rightarrow \theta = \frac{\pi}{3}, \frac{5\pi}{12}, \frac{5\pi}{6}, \frac{11\pi}{12}$

Now, $4\beta = \sum_{\theta \in S} \theta = \frac{\pi}{3} + \frac{5\pi}{12} + \frac{5\pi}{6} + \frac{11\pi}{12}$

$= \frac{4\pi + 5\pi + 10\pi + 11\pi}{12} = \frac{5\pi}{2}$

$\Rightarrow \beta = \frac{5\pi}{8}$

$\therefore f(\beta) = 1 + \frac{\sin^2(2\beta)}{2} = 1 + \frac{\sin^2\left(\frac{5\pi}{4}\right)}{2}$

$= 1 + \frac{\sin^2\left(\pi + \frac{\pi}{4}\right)}{2}$

[Using $\sin(\pi + x) = -\sin x$]

$= 1 + \frac{\sin^2\left(\frac{\pi}{4}\right)}{2} = 1 + \frac{\left(\frac{1}{\sqrt{2}}\right)^2}{2} = \frac{5}{4}$

**34.** *(c)* Given,

$\sum_{m=1}^{9} \frac{1}{\cos\left(\theta + (m-1)\frac{\pi}{6}\right) \times \cos\left(\theta + \frac{m\pi}{6}\right)} = \frac{-8}{\sqrt{3}}$

$\Rightarrow 2\sum_{m}^{9} \frac{\sin\left(\left(\theta + \frac{m\pi}{6}\right) - \left(\theta + (m-1)\frac{\pi}{6}\right)\right)}{\cos\left(\theta + (m-1)\frac{\pi}{6}\right)\cdot\cos\left(\theta + \frac{m\pi}{6}\right)} = \frac{-8}{\sqrt{3}}$

$\Rightarrow 2\sum_{m=1}^{9} \tan\left(\theta + \frac{m\pi}{6}\right) - \tan\left(\theta + (m-1)\frac{\pi}{6}\right) = \frac{-8}{\sqrt{3}}$

$\Rightarrow 2\left[\tan\left(\theta + \frac{3\pi}{2}\right) - \tan\theta\right] = \frac{-8}{\sqrt{3}}$

$\Rightarrow \cot\theta + \tan\theta = \frac{4}{\sqrt{3}}$

$\Rightarrow \tan\theta = \sqrt{3}$ or $\frac{1}{\sqrt{3}}$

$\Rightarrow \theta = \frac{\pi}{6}$ or $\frac{\pi}{3}$ as $\theta \in \left(0, \frac{\pi}{2}\right)$

$\therefore S = \left\{\frac{\pi}{6}, \frac{\pi}{3}\right\}, \sum_{\theta \in S} \theta = \frac{\pi}{6} + \frac{\pi}{3} = \frac{\pi}{2}$

**35.** *(c)* Given, $S = \{\theta \in [0, 2\pi] : 8^{2\sin^2\theta} + 8^{2\cos^2\theta} = 16\}$

$8^{2\sin^2\theta} + 8^{2(1-\sin^2\theta)} = 16$ [$\because \cos^2\theta = 1 - \sin^2\theta$]

$\Rightarrow 8^{2\sin^2\theta} + \frac{8^2}{8^{2\sin^2\theta}} = 16$

Let $8^{2\sin^2\theta} = y$

$\Rightarrow y + \frac{64}{y} = 16$

$\Rightarrow (y-8)^2 = 0$

$\Rightarrow y = 8 \Rightarrow 8^{2\sin^2\theta} = 8$

$\Rightarrow \sin^2\theta = 1/2$

As $\theta \in [0, 2\pi]$, so there are 4 values of $\theta$.

$\therefore \theta = \frac{\pi}{4}, \frac{3\pi}{4}, \frac{5\pi}{4}, \frac{7\pi}{4}$ $n(S) = 4$

Now, $n(S) + \sum_{\theta \in S}\left[\sec\left(\frac{\pi}{4} + 2\theta\right)\operatorname{cosec}\left(\frac{\pi}{4} + 2\theta\right)\right]$

$= 4 + \sum_{\theta \in S}\left[\frac{2}{2\cos\left(\frac{\pi}{4} + 2\theta\right)\sin\left(\frac{\pi}{4} + 2\theta\right)}\right]$

$= 4 + \sum_{\theta \in S}\left[\frac{2}{\sin\left(\frac{\pi}{2} + 4\theta\right)}\right]$

$= 4 + \sum_{\theta \in S}\frac{2}{\cos(4\theta)}$

$= 4 + \sum_{\theta \in S} 2\sec 4\theta$

$= 4 + 2[\sec\pi + \sec 3\pi + \sec 5\pi + \sec 7\pi]$

$= 4 + 2[-1 - 1 - 1 - 1]$

$= 4 - 8 = -4$

**36.** *(b)* $\because \sin\theta\tan\theta + \tan\theta = \sin 2\theta$

$\Rightarrow \tan\theta(\sin\theta + 1) - \sin 2\theta = 0$

$\Rightarrow \tan\theta(\sin\theta + 1 - 2\cos^2\theta) = 0$

$\Rightarrow \tan\theta[\sin\theta + 1 - 2(1 - \sin^2\theta)] = 0$

$\Rightarrow \tan\theta(2\sin^2\theta + \sin\theta - 1) = 0$

either $\tan\theta = 0 \Rightarrow \theta = 0, \pi, -\pi$

or $2\sin^2\theta + \sin\theta - 1 = 0$

$\Rightarrow (2\sin\theta - 1)(\sin\theta + 1) = 0$

$\Rightarrow \sin\theta = \frac{1}{2}$ and $\sin\theta = -1$

$\sin\theta = -1 \Rightarrow \theta = \frac{-\pi}{2}$, which is not possible thus the possible values of $\theta$ are $0, \pi, -\pi, \frac{5\pi}{6}$ and $\frac{\pi}{6}$.

Thus, $n(S) = 5$

$T = \sum_{\theta \in S}\cos 2\theta = \cos 0 + \cos 2\pi + \cos(-2\pi) + \cos\left(\frac{5\pi}{3}\right) + \cos\left(\frac{\pi}{3}\right)$

$$=1+1+1+\frac{1}{2}+\frac{1}{2}=4$$

$\therefore \quad T=4$

Thus, $T+n(S)=4+5=9$

**37.** *(a)* We have, $\frac{\cos x}{1+\sin x}=|\tan 2x|$

$$\Rightarrow \frac{\cos^2\left(\frac{x}{2}\right)-\sin^2\left(\frac{x}{2}\right)}{\cos^2\frac{x}{2}+\sin^2\frac{x}{2}+2\cos\frac{x}{2}\sin\frac{x}{2}}=|\tan 2x|$$

$$\Rightarrow \frac{\left[\cos\left(\frac{x}{2}\right)-\sin\left(\frac{x}{2}\right)\right]\left[\cos\frac{x}{2}+\sin\frac{x}{2}\right]}{\left(\cos\frac{x}{2}+\sin\frac{x}{2}\right)^2}=|\tan 2x|$$

$$\Rightarrow \frac{1-\tan\frac{x}{2}}{1+\tan\frac{x}{2}}=|\tan 2x|$$

$$\Rightarrow \tan\left(\frac{\pi}{4}-\frac{x}{2}\right)=|\tan 2x|$$

$$\Rightarrow \tan^2\left(\frac{\pi}{4}-\frac{x}{2}\right)=\tan^2 2x$$

$$\Rightarrow 2x=n\pi\pm\left(\frac{\pi}{4}-\frac{x}{2}\right)$$

$$\Rightarrow \frac{5x}{2}=\left(n+\frac{1}{4}\right)\pi \text{ or } \frac{3x}{2}=\left(n-\frac{1}{4}\right)\pi$$

$$\Rightarrow \frac{-\pi}{2}<\frac{2}{5}\left(n+\frac{1}{4}\right)\pi<\frac{\pi}{2}$$

$$\text{or } \frac{-\pi}{2}<\frac{2}{3}\left(n-\frac{1}{4}\right)\pi<\frac{\pi}{2}$$

$$\Rightarrow -\frac{5}{4}<n+\frac{1}{4}<\frac{5}{4} \text{ or } \frac{-3}{4}<n-\frac{1}{4}<\frac{3}{4}$$

$$\Rightarrow \frac{-6}{4}<n<1 \text{ or } \frac{-1}{2}<n<1$$

$\Rightarrow n=-1, 0 \quad \text{or} \quad n=0$

When, $n=-1$, $x=\left(\frac{-3}{10}\right)\pi$

or when $n=0$, $x=-\frac{\pi}{6}$

$$n=0, x=\left(\frac{1}{10}\right)\pi$$

$\therefore$ Required sum

$$=\left(\frac{-3}{10}\right)\pi+\left(\frac{1}{10}\right)\pi+\left(\frac{-1}{6}\right)\pi$$

$$=\left(\frac{-11}{30}\right)\pi$$

**38.** *(a)* Given, $\alpha=\max(8^{2\sin 3x}\cdot 4^{4\cos 3x})$

$$=\max(2^{6\sin 3x}\cdot 2^{8\cos 3x})$$

$$=\max(2^{6\sin 3x+8\cos 3x})$$

and $\beta=\min(8^{2\sin 3x}\cdot 4^{4\cos 3x})$

$$=\min(2^{6\sin 3x}\cdot 2^{8\cos 3x})$$

$$=\min(2^{6\sin 3x+8\cos 3x})$$

Now, determine the range of $6\sin 3x+8\cos 3x$

$\because$ Range of $a\sin x+b\cos x$

$$=[-\sqrt{a^2+b^2}, \sqrt{a^2+b^2}]$$

So, range of $6\sin 3x+8\cos 3x$

$$=[-\sqrt{6^2+8^2}, \sqrt{6^2+8^2}]$$

$$=[-10, 10]$$

So, $\alpha=2^{10}$ and $\beta=2^{-10}$

Now, $\alpha^{1/5}=(2^{10})^{1/5}=4$

$$\beta^{1/5}=(2^{-10})^{1/5}=\frac{1}{4}$$

Quadratic equation with roots 4 and $\frac{1}{4}$

is $x^2-\left(4+\frac{1}{4}\right)x+4\times\frac{1}{4}=0$

$$\Rightarrow x^2-\frac{17}{4}x+1=0$$

On multiplying both sides by 8, we get

$$8x^2-34x+8=0$$

On comparing $8x^2+bx+c$, we get

$$b=-34 \text{ and } c=8$$

So, $c-b=8-(-34)=42$

**39.** *(a)* Given, $x+2\tan x=\frac{\pi}{2}$

$$\Rightarrow 2\tan x=\frac{\pi}{2}-x \Rightarrow \tan x=\frac{\pi}{4}-\frac{x}{2}$$

$$\Rightarrow \tan x=\left(-\frac{1}{2}\right)x+\frac{\pi}{4} \quad \text{...(i)}$$

**Approach** In this type of problem solving, graphical approach is best because we have to find only number of solutions, not the solution (i.e. not the value(s) of $x$).

**Concept** To find the number of solution(s) for Eq. (i), first of all, let

$$y=\tan x \quad \text{... (ii)}$$

and $$y=\left(\frac{-1}{2}\right)x+\frac{\pi}{4} \quad \text{...(iii)}$$

and then draw the graph of Eqs. (ii) and (iii).

Now, total number of solution(s) = Total number of point(s) of intersection of the graph of Eqs. (ii) and (iii).

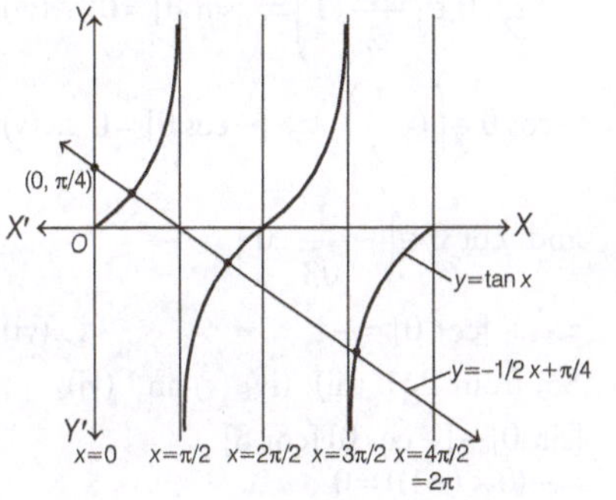


From above, we see that the red line i.e., $y=-\frac{1}{2}x+\frac{\pi}{4}$ intersects the black curve i.e., $y=\tan x$ at three distinct points in $[0, 2\pi]$.

$\therefore$Total number of solutions $=3$

**40.** *(b)* Given, $81^{\sin^2 x}+81^{\cos^2 x}=30$

$$\Rightarrow 81^{\sin^2 x}+81^{(1-\sin^2 x)}=30$$

$$\Rightarrow 81^{\sin^2 x}+\frac{81}{81^{\sin^2 x}}=30$$

Let $81^{\sin^2 x}=y$.

$$\therefore y+\frac{81}{y}=30 \Rightarrow y^2-30y+81=0$$

$\Rightarrow (y-27)(y-3)=0$

$\Rightarrow y=3$ or $y=27$

$81^{\sin^2 x}=3$ or $81^{\sin^2 x}=27$

$3^{4\sin^2 x}=3$ or $3^{4\sin^2 x}=3^3$

$\Rightarrow 4\sin^2 x=1$ or $4\sin^2 x=3$

$\Rightarrow \sin^2 x=1/4$ or $\sin^2 x=3/4$

$\Rightarrow \sin^2 x=\sin^2(\pi/6)$

or $\sin^2 x=\sin^2(\pi/3)$

$\Rightarrow x=n\pi\pm\pi/6$

or $x=n\pi\pm\pi/3$ From $[0, \pi]$,

$x=\pi/6, 5\pi/6$

or $x=\pi/3, 2\pi/3$

Hence, the total number of solutions = 4

**41.** *(a)* Given system of linear equations is

$[\sin\theta]x+[-\cos\theta]y=0$ ...(i)

and $[\cot\theta]x+y=0$ ...(ii)

Where, $[x]$ denotes the greatest integer $\le x$.

Here, $\Delta=\frac{[\sin\theta]}{[\cot\theta]}=\frac{[-\cos\theta]}{1}$

$$=[\sin\theta]-[-\cos\theta][\cot\theta] \text{ ...(iii)}$$

When, $\theta \in \left(\frac{\pi}{2}, \frac{2\pi}{3}\right)$

$\sin\theta \in \left(\frac{\sqrt{3}}{2}, 1\right) \Rightarrow [\sin\theta] = 0$ ...(iv)

$-\cos\theta \in \left(0, \frac{1}{2}\right) \Rightarrow [-\cos\theta] = 0$ ...(v)

and $\cot\theta \in \left(-\frac{1}{\sqrt{3}}, 0\right)$

$\Rightarrow [\cot\theta] = -1$ ...(vi)

So, from Eqs. (iii), (iv) (v) and (vi),

$[\sin\theta] - [-\cos\theta][\cot\theta]$
$= -(0 \times (-1)) = 0$

Thus, for $\theta \in \left(\frac{\pi}{2}, \frac{2\pi}{3}\right)$, the given system have infinitely many solutions.

When, $\theta \in \left(\pi, \frac{7\pi}{6}\right)$,

$\sin\theta \in \left(-\frac{1}{2}, 0\right)$

$\Rightarrow [\sin\theta] = -1$

$-\cos\theta \in \left(\frac{\sqrt{3}}{2}, 1\right)$

$\Rightarrow [\cos\theta] = 0$ and $\cot\theta \in (\sqrt{3}, \infty)$

$\Rightarrow [\cot\theta] = n, n \in N.$

So, $\Delta = -1 - (0 \times n) = -1$

Thus, for $\theta \in \left(\pi, \frac{7\pi}{6}\right)$, the given system has a unique solution.

**42.** *(d)* Given,

$5(\tan^2 x - \cos^2 x) = 2\cos 2x + 9$

$\Rightarrow 5\left(\frac{1-\cos 2x}{1+\cos 2x} - \frac{1+\cos 2x}{2}\right) = 2\cos 2x + 9$

On putting $\cos 2x = y$, we get

$5\left(\frac{1-y}{1+y} - \frac{1+y}{2}\right) = 2y + 9$

$\Rightarrow 5(2 - 2y - 1 - y^2 - 2y) = 2(1+y)(2y+9)$

$\Rightarrow 5(1 - 4y - y^2) = 2(2y + 9 + 2y^2 + 9y)$

$\Rightarrow 9y^2 + 42y + 13 = 0$

$\Rightarrow (3y+1)(3y+13) = 0$

$\Rightarrow y = -\frac{1}{3}, -\frac{13}{3}$

$\therefore \cos 2x = -\frac{1}{3}, -\frac{13}{3}$

$\Rightarrow \cos 2x = -\frac{1}{3}$ $\left[\because \cos 2x \neq -\frac{13}{3}\right]$

Now, $\cos 4x = 2\cos^2 2x - 1$

$= 2\left(-\frac{1}{3}\right)^2 - 1 = \frac{2}{9} - 1 = -\frac{7}{9}$

**43.** *(c)* Given equation is

$\cos x + \cos 2x + \cos 3x + \cos 4x = 0$

$\Rightarrow (\cos x + \cos 3x) + (\cos 2x + \cos 4x) = 0$

$\Rightarrow 2\cos 2x \cos x + 2\cos 3x \cos x = 0$

$\Rightarrow 2\cos x(\cos 2x + \cos 3x) = 0$

$\Rightarrow 2\cos x\left(2\cos\frac{5x}{2}\cos\frac{x}{2}\right) = 0$

$\Rightarrow \cos x \cdot \cos\frac{5x}{2} \cdot \cos\frac{x}{2} = 0$

$\Rightarrow \cos x = 0$ or $\cos\frac{5x}{2} = 0$ or $\cos\frac{x}{2} = 0$

Now, $\cos x = 0 \Rightarrow x = \frac{\pi}{2}, \frac{3\pi}{2}$

$[\because 0 \le x < 2\pi]$

$\cos\frac{5x}{2} = 0$

$\Rightarrow \frac{5x}{2} = \frac{\pi}{2}, \frac{3\pi}{2}, \frac{5\pi}{2}, \frac{7\pi}{2}, \frac{9\pi}{2}, \frac{11\pi}{2}, \ldots,$

$\Rightarrow x = \frac{\pi}{5}, \frac{3\pi}{5}, \pi, \frac{7\pi}{5}, \frac{9\pi}{5}$

$[\because 0 \le x < 2\pi]$

and $\cos\frac{x}{2} = 0$

$\Rightarrow \frac{x}{2} = \frac{\pi}{2}, \frac{3\pi}{2}, \frac{5\pi}{2}, \ldots$

$\Rightarrow x = \pi$ $[\because 0 \le x < 2\pi]$

Hence, $x = \frac{\pi}{2}, \frac{3\pi}{2}, \pi, \frac{\pi}{5}, \frac{3\pi}{5}, \frac{7\pi}{5}, \frac{9\pi}{5}$

**44.** *(d)* Let $a = 2\sin^4 x + 18\cos^2 x$

and $b = 2\cos^4 x + 18\sin^2 x$

Now, $a - b = 2(\sin^4 x - \cos^4 x) + 18(\cos^2 x - \sin^2 x)$

$= 2(\sin^2 x - \cos^2 x) + 18(\cos^2 x - \sin^2 x) = 16\cos 2x$

and $a + b = 2(\sin^4 x + \cos^4 x) + 18(\cos^2 x + \sin^2 x)$

$= 2\{(1 - 2\sin^2 x\cos^2 x)\} + 18$

$= 20 - \sin^2 2x = 19 + \cos^2 2x$

From given, equation is, $|\sqrt{a} - \sqrt{b}| = 1$

On squaring both sides, we get

$a + b - 2\sqrt{ab} = 1$

$\Rightarrow (a + b - 1)^2 = 4ab$

$\Rightarrow (a+b)^2 - 2(a+b) + 1 = 4ab$

$\Rightarrow (a-b)^2 - 2(a+b) + 1 = 0$

$\Rightarrow 256\cos^2 2x - 2(19 + \cos^2 2x) + 1 = 0$

$\Rightarrow 254\cos^2 2x - 37 = 0$

$\cos^2 2x = \frac{37}{254} = \lambda$ (say)

where $|\lambda| \le 1$

So, $\cos 2x = \pm\sqrt{\lambda}$

We have 4 values in the first cycle and the four again in the next cycle. Thus, total values of $x$ is 8.

**45.** *(b)* Given, $\cos x + \sin x = 1/2$

$\therefore \frac{1 - \tan^2\frac{x}{2}}{1 + \tan^2\frac{x}{2}} + \frac{2\tan\frac{x}{2}}{1 + \tan^2\frac{x}{2}} = \frac{1}{2}$

Let $\tan\frac{x}{2} = t$

$\Rightarrow \frac{1-t^2}{1+t^2} + \frac{2t}{1+t^2} = \frac{1}{2}$

$\Rightarrow 2(1 - t^2 + 2t) = 1 + t^2$

$\Rightarrow 3t^2 - 4t - 1 = 0 \Rightarrow t = \frac{2 \pm \sqrt{7}}{3}$

As $0 < x < \pi \Rightarrow 0 < \frac{x}{2} < \frac{\pi}{2}$

So, $\tan\frac{x}{2}$ is positive.

$\therefore t = \tan\frac{x}{2} = \frac{2+\sqrt{7}}{3}$

Now, $\tan x = \frac{2\tan\frac{x}{2}}{1 - \tan^2\frac{x}{2}} = \frac{2t}{1-t^2}$

$\Rightarrow \tan x = \frac{2\left(\frac{2+\sqrt{7}}{3}\right)}{1 - \left(\frac{2+\sqrt{7}}{3}\right)^2}$

$\Rightarrow \tan x = -\left(\frac{4+\sqrt{7}}{3}\right)$

**46.** *(2)* $\sin^2 x - (x^2 - 2x - 2)\sin x - 3(x-1)^2 = 0$

$\sin^2 x - (x-1)^2\sin x + 3\sin x - 3(x-1)^2 = 0$

$\sin x[\sin x - (x-1)^2] + 3[\sin x - (x-1)^2] = 0$

$[\sin x-(x-1)^2](\sin x+3)=0$

$\sin x=(x-1)^2$ and $\sin x=-3$ (reject)

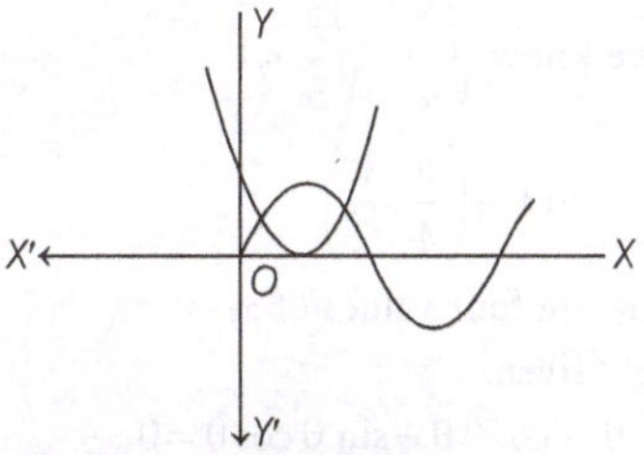


Point of intersection = 2

**Concept Enhancer**

(i) Range of $\sin x$ is $[-1, 1]$.

(ii) Number of solution of trigonometric function and algebraic can be found by plotting there curves and finding the number of intersection points of curves.

**47.** *(4)* $D=\sin^2 2\theta-4\left(1-\dfrac{\sin^2 2\theta}{2}\right)\left(1-\dfrac{3}{4}\sin^2 2\theta\right)$

$=(\sin 2\theta)^2-4\left(1-\dfrac{5}{4}\sin^2 2\theta+\dfrac{3}{8}\sin^4 2\theta\right)$

$\Rightarrow D=-\dfrac{3}{2}\sin^4 2\theta+6\sin^2 2\theta-4>0$

$\Rightarrow 3\sin^4 2\theta-12\sin^2 2\theta+8<0$

Critical point

$\sin^2 2\theta=\dfrac{12\pm\sqrt{(12)^2-4(3)(8)}}{2\times 3}$

$=\dfrac{12\pm 4\sqrt{3}}{6}=\dfrac{6\pm 2\sqrt{3}}{3}$

$\sin^2 2\theta=\dfrac{6\pm 2\sqrt{3}}{3}$ $\quad[\because \sin^2 2\theta\in[0,1]]$

$\therefore \sin^2 2\theta\in\left[2-\dfrac{2}{\sqrt{3}},1\right]$

$\alpha=2-\dfrac{2}{\sqrt{3}}, \beta=1$

$(\alpha-2)^2=\dfrac{4}{3}, (\beta-1)^2=0$

$\therefore 3(\alpha-2)^2+(\beta-1)^2=4$

**48.** *(48)* Given, equation is

$\cos 2x+a\sin x=2a-7$

$\Rightarrow a\sin x-2a=-7-\cos 2x$

$=-7-(1-2\sin^2 x)$

$=-8+2\sin^2 x=2(\sin^2 x-4)$

$\Rightarrow a(\sin x-2)=2(\sin x-2)(\sin x+2)$

$\Rightarrow (\sin x-2)[a-2(\sin x+2)]=0$

$\Rightarrow \sin x=2$ or $a=2(\sin x+2)$

But $\sin x=2$ [not possible]

Also, $-1\le\sin x\le 1$

$\Rightarrow 1\le\sin x+2\le 3\Rightarrow 2\le 2(\sin x+2)\le 6$

$\Rightarrow 2\le a\le 6\Rightarrow a\in[2,6]$

On comparing this with $a\in[p,q]$

we get $\quad p=2$ and $q=6$

Now, given

$r=\tan 9^\circ-\tan 27^\circ-\dfrac{1}{\cot 63^\circ}+\tan 81^\circ$

$=(\tan 9^\circ+\cot 9^\circ)-\tan 27^\circ-\cot 27^\circ$

$[\because \tan 81^\circ=\cot 9^\circ$ and $\cot 63^\circ=\tan 27^\circ]$

$=\dfrac{1}{\sin 9^\circ\cdot\cos 9^\circ}-\dfrac{1}{\sin 27^\circ\cdot\cos 27^\circ}$

$=2\left[\dfrac{4}{\sqrt{5}-1}-\dfrac{4}{\sqrt{5}+1}\right]$

$\Rightarrow r=4$

Hence, $pqr=2\times 6\times 4=48$

**49.** *(25)* Given, $m$ = number of positive values of $\theta$ in $[-\pi,\pi]$

$n$ = number of negative values of $\theta$ in $[-\pi,\pi]$

$\because \cos 2\theta\cdot\cos\dfrac{\theta}{2}=\cos 3\theta\cdot\cos\left(\dfrac{9\theta}{2}\right)$

On multiplying by 2 both the side, we get

$2\cos 2\theta\cdot\cos\dfrac{\theta}{2}=2\cos 3\theta\cdot\cos\left(\dfrac{9\theta}{2}\right)$

As we know that

$2\cos A\cdot\cos B$

$=\cos(A+B)+\cos(A-B)$

So, $\cos\left(2\theta+\dfrac{\theta}{2}\right)+\cos\left(2\theta-\dfrac{\theta}{2}\right)$

$=\cos\left(3\theta+\dfrac{9\theta}{2}\right)+\cos\left(3\theta-\dfrac{9\theta}{2}\right)$

$\Rightarrow \cos\dfrac{5\theta}{2}+\cos\dfrac{3\theta}{2}=\cos\dfrac{15\theta}{2}+\cos\left(-\dfrac{3\theta}{2}\right)$

So, $\cos\dfrac{5\theta}{2}+\cos\dfrac{3\theta}{2}=\cos\dfrac{15\theta}{2}+\cos\dfrac{3\theta}{2}$

$[\because \cos(-x)=\cos x]$

or $\cos\dfrac{5\theta}{2}=\cos\dfrac{15\theta}{2}$

$\because$ We know that, if $\cos\theta=\cos\alpha$

Then, $\theta=2n\pi\pm\alpha, n\in I$

So, $\dfrac{5\theta}{2}=2n\pi\pm\dfrac{15\theta}{2}$

On multiplying by 2 both the sides, we get

$5\theta=4n\pi\pm 15\theta$

Case I Taking (+)

$5\theta=4n\pi+15\theta \Rightarrow 10\theta=-4n\pi$

$\Rightarrow \theta=\dfrac{-2n\pi}{5}$ ...(i)

Case II Taking (–)

$5\theta=4n\pi-15\theta\Rightarrow 20\theta=4n\pi$

$\Rightarrow \theta=\dfrac{n\pi}{5}$ ...(ii)

From Eqs. (i) and (ii), solution set of Eq. (i) is subset of solution set of Eq. (ii)

So, $\theta=\dfrac{n\pi}{5}$

For $n=1, \theta=\dfrac{\pi}{5}$; $\quad n=-1, \theta=\dfrac{-\pi}{5}$

$n=2, \theta=\dfrac{2\pi}{5}$; $\quad n=-2, \theta=\dfrac{-2\pi}{5}$

$n=3, \theta=\dfrac{3\pi}{5}$; $\quad n=-3, \theta=\dfrac{-3\pi}{5}$

$n=4, \theta=\dfrac{4\pi}{5}$; $\quad n=-4, \theta=\dfrac{-4\pi}{5}$,

$n=5, \theta=\dfrac{5\pi}{5}$; $\quad n=-5, \theta=\dfrac{-5\pi}{5}$

So, $m=5$ and $n=5$

Now, $mn=5\times 5=25$

**50.** *(2)* Given,

$\tan(\pi\cos\theta)+\tan(\pi\sin\theta)=0$

$\Rightarrow \tan(\pi\cos\theta)=-\tan(\pi\sin\theta)$

$\Rightarrow \tan(\pi\cos\theta)=\tan(-\pi\sin\theta)$

$\Rightarrow \pi\cos\theta=n\pi-\pi\sin\theta$

$\Rightarrow \pi(\sin\theta+\cos\theta)=n\pi$

$\Rightarrow \sin\theta+\cos\theta=n, n\in I$

Since, $-\sqrt{2}\le\sin\theta+\cos\theta\le\sqrt{2}$

$\therefore$ Possible value of $n$ are $-1, 0, 1$.

$\therefore \quad \theta\in\left\{0,\dfrac{\pi}{2},\dfrac{3\pi}{4},\dfrac{7\pi}{4},\pi,\dfrac{3\pi}{2}\right\}$

$\therefore \sum_{\theta\in S}\sin^2\left(\theta+\dfrac{\pi}{4}\right)=\sin^2\left(\dfrac{\pi}{4}\right)+\sin^2\left(\dfrac{\pi}{2}+\dfrac{\pi}{4}\right)$

$\sin^2\left(\dfrac{3\pi}{4}+\dfrac{\pi}{4}\right)+\sin^2\left(\dfrac{7\pi}{4}+\dfrac{\pi}{4}\right)$

$+\sin^2\left(\pi+\dfrac{\pi}{4}\right)+\sin^2\left(\dfrac{3\pi}{2}+\dfrac{\pi}{4}\right)$

$=\left(\dfrac{1}{\sqrt{2}}\right)^2+\left(\dfrac{1}{\sqrt{2}}\right)^2+(0)^2+(0)^2+\left(-\dfrac{1}{\sqrt{2}}\right)^2+\left(\dfrac{-1}{\sqrt{2}}\right)^2$

$=\dfrac{1}{2}+\dfrac{1}{2}+0+0+\dfrac{1}{2}+\dfrac{1}{2}=2$

**51.** *(3)* Given, $2\sin^2\theta - \cos 2\theta = 0$

$$2\sin^2\theta - (1 - 2\sin^2\theta) = 0$$
$$4\sin^2\theta - 1 = 0$$
$$\Rightarrow \quad \sin\theta = \pm\frac{1}{2}$$

Also, $\quad 2\cos^2\theta + 3\sin\theta = 0$

$$2(1 - \sin^2\theta) + 3\sin\theta = 0$$
$$\Rightarrow \quad 2\sin^2\theta - 3\sin\theta - 2 = 0$$
$$\Rightarrow \quad (\sin\theta - 2)(2\sin\theta + 1) = 0$$
$$\Rightarrow \quad \sin\theta = -\frac{1}{2} \quad [\because \sin\theta \neq 2]$$
$$\therefore \sin\theta = -\frac{1}{2} \Rightarrow \theta = \frac{7\pi}{6}, \frac{11\pi}{6}$$

Now, sum of the solutions of the system of equation $= \frac{7\pi}{6} + \frac{11\pi}{6}$

$$\Rightarrow \frac{18\pi}{6} = k\pi \Rightarrow 3\pi = k\pi$$

Hence, $k = 3$

**52.** *(4)*

**key idea** Simplify the given equation using $\cos^2 x = 1 - \sin^2 x$ and solve further.

$$\sin x = \cos^2 x, x \in (0, 10)$$
$$\sin x = 1 - \sin^2 x$$
$$\Rightarrow \sin^2 x + \sin x - 1 = 0$$
$$\Rightarrow \quad \sin x = \frac{-1 \pm \sqrt{5}}{2}$$
$$\therefore \sin x = \frac{\sqrt{5} - 1}{2} \text{ or } \frac{-1 - \sqrt{5}}{2}$$

(Not possible)

$$\Rightarrow \quad y = \sin x$$

and $y = \frac{\sqrt{5} - 1}{2}$

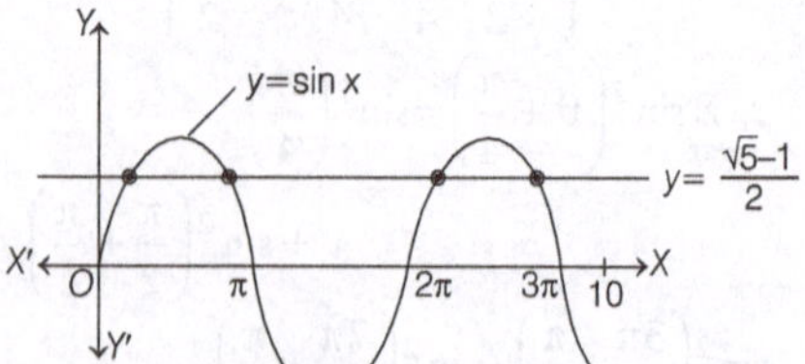


Circled dot represents the solutions. Clearly, number of solutions is 4.

**53.** *(32)* $3\cos^2 2\theta + 6\cos 2\theta - 10\cos^2\theta + 5 = 0$

$$\Rightarrow 3\cos^2 2\theta + 6\cos 2\theta - 5(2\cos^2\theta - 1) = 0$$
$$\Rightarrow 3\cos^2 2\theta + 6\cos 2\theta - 5\cos 2\theta = 0$$
$$\Rightarrow \quad 3\cos^2 2\theta + \cos 2\theta = 0$$
$$\Rightarrow \quad \cos 2\theta[3\cos 2\theta + 1] = 0$$
$$\cos 2\theta = 0 \quad \text{or} \quad \cos 2\theta = \frac{-1}{3}$$
$$2\theta = (2n+1)\cdot\frac{\pi}{2}$$
$$\therefore \quad \theta = \pm\frac{\pi}{4}, \pm\frac{3\pi}{4}, \ldots \pm\frac{15\pi}{4}$$
$$[\because \theta \in [-4\pi, 4\pi]]$$

Similarly, $\cos 2\theta = \frac{-1}{3}$ gives 16 solution.

∴ The number of elements in the set $S$ are $16 + 16 = 32$

**54.** *(1)* $2\theta - \cos^2\theta + \sqrt{2} = 0$

$$\Rightarrow 2\theta + \sqrt{2} = \cos^2\theta$$
$$\Rightarrow \; y = 2\theta + \sqrt{2} \text{ and } y = \cos^2\theta$$

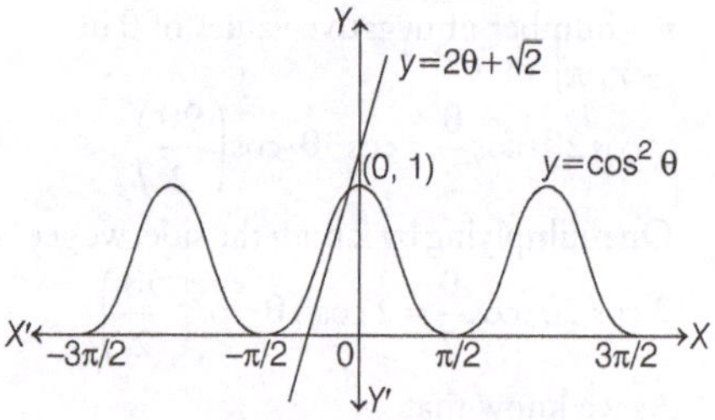


We can see that $y = \cos^2\theta$ and $y = 2\theta + \sqrt{2}$ intersect at one point.

∴Number of solutions is 1.

**55.** *(4)* Given, $14\,\text{cosec}^2 x - 2\sin^2 x = 21 - 4\cos^2 x$

$$\Rightarrow \frac{14}{\sin^2 x} - 2\sin^2 x = 21 - 4(1 - \sin^2 x)$$
$$\Rightarrow \frac{14 - 2\sin^4 x}{\sin^2 x} = 17 + 4\sin^2 x$$
$$\Rightarrow 14 - 2\sin^4 x = 17\sin^2 x + 4\sin^4 x$$
$$\Rightarrow 6\sin^4 x + 17\sin^2 x - 14 = 0$$
$$\Rightarrow (3\sin^2 x - 2)(2\sin^2 x + 7) = 0$$
$$\Rightarrow \quad \sin^2 x = \frac{2}{3} \quad \text{or} \quad \sin^2 x = -\frac{7}{2}$$

[not possible]

$$\Rightarrow \quad \sin x = \pm\sqrt{\frac{2}{3}} \text{ or } \pm\frac{\sqrt{2}}{\sqrt{3}}$$

As we know $\sqrt{\frac{2}{4}} < \sqrt{\frac{2}{3}} < 1$

$$\Rightarrow \quad \sin x \in \left(\frac{\pi}{4}, \frac{\pi}{2}\right)$$

There are four values of $x$.

**56.** *(56)* Given,

$$\sin^4\theta + \cos^4\theta - \sin\theta\cos\theta = 0$$
$$\theta \in [0, 4\pi]$$
$$\Rightarrow \; 1 - 2\sin^2\theta\cos^2\theta - \sin\theta\cos\theta = 0$$
$$\Rightarrow \quad 2 - \sin^2 2\theta - \sin 2\theta = 0$$
$$\Rightarrow \quad \sin^2 2\theta + \sin 2\theta - 2 = 0$$
$$\Rightarrow \quad (\sin 2\theta + 2)(\sin 2\theta - 1) = 0$$
$$\Rightarrow \quad \sin 2\theta = 1, 2\theta \in [0, 8\pi]$$
$$\theta = \frac{\pi}{4}, \frac{5\pi}{4}, \frac{9\pi}{4}, \frac{13\pi}{4}$$

Sum of solutions $S = \frac{28\pi}{4}$

Then, $\frac{8S}{\pi} = \frac{8}{\pi} \times \frac{28\pi}{4} = 56$

**57.** *(1)* Given, $|\cot x| = \cot x + \frac{1}{\sin x}$ ...(i)

If $\cot x > 0$, then $|\cot x| = \cot x$

From Eq. (i), $\cot x = \cot x + \frac{1}{\sin x}$

$$\Rightarrow \quad \frac{1}{\sin x} = 0 \quad \text{[not possible]}$$

If $\cot x < 0$, then $|\cot x| = -\cot x$

From Eq. (ii), we get

$$-\cot x = \cot x + \frac{1}{\sin x}$$
$$\Rightarrow 2\cot x + \frac{1}{\sin x} = 0$$
$$\Rightarrow 2\cos x = -1$$
$$\Rightarrow \quad x = \frac{2\pi}{3} \text{ or } \frac{4\pi}{3}$$

Here, $x = \frac{4\pi}{3}$ rejected because $\frac{4\pi}{3} \in$ third quadrant and in third quadrant $\cot x$ is positive. Since, we considered $\cot x < 0$.

$\therefore x = \frac{2\pi}{3}$ is the only one solution.

# C-04

JEE Main Chapterwise Topicwise Mathematics

In JEE Main, there are usually 1-2 questions per paper are visible from this chapter. In past 3 years, 84 conceptual and 32 numerical questions were asked from this chapter.

**For detailed analysis scan QR code**

# COMPLEX NUMBERS AND QUADRATIC EQUATIONS

## Topic 1 Algebraic Operations on Complex Number, Conjugate, Modules and Argument of Complex Numbers

### Level 1

**01** Let $z$ be a complex number such that $|z|=1$. If $\dfrac{2+k^2 z}{k+\bar{z}}=kz, k \in R$, then the maximum distance of $k+ik^2$ from the circle $|z-(1+2i)|=1$ is **[2025, 2 April Shift-I]**

(a) $\sqrt{5}+1$ (b) 3 (c) 2 (d) $\sqrt{3}+1$

**02** Let $z$ be a complex number such that the real part of $\dfrac{z-2i}{z+2i}$ is zero. Then, the maximum value of $|z-(6+8i)|$ is equal to **[2024, 9 April Shift-II]**

(a) 10 (b) 8 (c) 12 (d) $\infty$

**03** The sum of all possible value of $\theta \in [-\pi, 2\pi]$, for which $\dfrac{1+i\cos\theta}{1-2i\cos\theta}$ is purely imaginary, is equal to **[2024, 8 April Shift-II]**

(a) $4\pi$ (b) $5\pi$ (c) $3\pi$ (d) $2\pi$

**04** Let $z$ be a complex number such that $|z+2|=1$ and $\text{lm}\left(\dfrac{z+1}{z+2}\right)=\dfrac{1}{5}$. Then, the value of $|\text{Re}\,(\overline{z+2})|$ is **[2024, 8 April Shift-I]**

(a) $\dfrac{24}{5}$ (b) $\dfrac{2\sqrt{6}}{5}$ (c) $\dfrac{1+\sqrt{6}}{5}$ (d) $\dfrac{\sqrt{6}}{5}$

**05** If $z=x+iy, xy \neq 0$, satisfies the equation $z^2+i\bar{z}=0$, then $|z^2|$ is equal to **[2024, 30 Jan Shift-I]**

(a) 4 (b) 1/4 (c) 9 (d) 1

**06** If $z=\dfrac{1}{2}-2i$ is such that $|z+1|=\alpha z+\beta(1+i)$, $i=\sqrt{-1}$ and $\alpha, \beta \in R$, then $\alpha+\beta$ is equal to **[2024, 29 Jan Shift-II]**

(a) $-4$ (b) $-1$ (c) 2 (d) 3

**07** Let $A=\left\{\theta \in (0, 2\pi): \dfrac{1+2i\sin\theta}{1-i\sin\theta} \text{ is purely imaginary}\right\}$. Then, the sum of the elements in $A$ is **[2023, 8 April Shift-II]**

(a) $2\pi$ (b) $\pi$ (b) $3\pi$ (d) $4\pi$

**08** If $z=2+3i$, then $z^5+(\bar{z})^5$ is equal to **[2022, 29 July Shift-I]**

(a) 244 (b) 224 (c) 245 (d) 265

**09** The real part of the complex number $\dfrac{(1+2i)^8 \cdot (1-2i)^2}{(3+2i)\cdot\overline{(4-6i)}}$ is equal to **[2022, 30 June Shift-I]**

(a) 500/13 (b) 110/13 (c) 55/6 (d) 550/13

**10** Let a complex number $z, |z| \neq 1$, satisfy $\log_{\frac{1}{\sqrt{2}}}\left(\dfrac{|z|+11}{(|z|-1)^2}\right) \leq 2$. Then, the largest value of $|z|$ is equal to **[2021, 16 March Shift-I]**

(a) 5 (b) 7 (c) 8 (d) 6

**11** Let $\left(-2-\dfrac{1}{3}i\right)^3=\dfrac{x+iy}{27}\ (i=\sqrt{-1})$, where $x$ and $y$ are real numbers, then $y-x$ equals **[2019, 11 Jan Shift-I]**

(a) 91 (b) 85 (c) $-85$ (d) $-91$

**12** For all complex numbers $z$ of the form $1+i\alpha, \alpha \in R$, if $z^2=x+iy$, then **[JEE Main 2014]**

(a) $y^2-4x+2=0$ (b) $y^2+4x-4=0$

(c) $y^2-4x-4=0$ (d) $y^2+4x+2=0$

**13** If $z$ is a complex number of unit modulus at argument $\theta$, then $\arg\left(\frac{1+z}{1+\bar{z}}\right)$ equals *[JEE Main 2013]*

(a) $\frac{\pi}{2}-\theta$ (b) $\theta$

(c) $\pi-\theta$ (d) $-\theta$

**14** If $\left|z-\frac{4}{z}\right|=2$, then the maximum value of $|z|$ is equal to *[AIEEE 2009]*

(a) $\sqrt{3}+1$ (b) $\sqrt{5}+1$

(c) 2 (d) $2+\sqrt{2}$

**15** If $\left(\frac{1+i}{1-i}\right)^x=1$, then *[AIEEE 2003]*

(a) $x=4n$, where $n$ is any positive integer
(b) $x=2n$, where $n$ is any positive integer
(c) $x=4n+1$, where $n$ is any positive integer
(d) $x=2n+1$, where $n$ is any positive integer

## Level 2

**16** Let $A=\left\{\theta\in[0,2\pi]:1+10\,\mathrm{Re}\left(\frac{2\cos\theta+i\sin\theta}{\cos\theta-3i\sin\theta}\right)=0\right\}$. Then, $\sum_{\theta\in A}\theta^2$ is equal to *[2025, 8 April Shift-II]*

(a) $\frac{27}{4}\pi^2$ (b) $8\pi^2$ (c) $6\pi^2$ (d) $\frac{21}{4}\pi^2$

**17** Let $|z_1-8-2i|\le 1$ and $|z_2-2+6i|\le 2$, $z_1, z_2\in C$. Then, the minimum value of $|z_1-z_2|$ is *[2025, 29 Jan Shift-I]*

(a) 10 (b) 3 (c) 7 (d) 13

**18** The number of complex numbers $z$ satisfying $|z|=1$ and $\left|\frac{z}{\bar{z}}+\frac{\bar{z}}{z}\right|=1$, is *[2025, 23 Jan Shift-II]*

(a) 10 (b) 6 (c) 8 (d) 4

**19** Let $z_1$, $z_2$ and $z_3$ be three complex numbers on the circle $|z|=1$ with $\arg(z_1)=\frac{-\pi}{4}$, $\arg(z_2)=0$ and $\arg(z_3)=\frac{\pi}{4}$. If $|z_1\bar{z}_2+z_2\bar{z}_3+z_3\bar{z}_1|^2=\alpha+\beta\sqrt{2}$; $\alpha$, $\beta\in Z$, then the value of $\alpha^2+\beta^2$ is *[2025, 22 Jan Shift-I]*

(a) 41 (b) 29

(c) 31 (d) 24

**20** If the set $R=\{(a,b):a+5b=42,\ a,b\in N\}$ has $m$ elements and $\sum_{n=1}^{m}(1-i^{n!})=x+iy$, where $i=\sqrt{-1}$, then the value of $m+x+y$ is *[2024, 8 April Shift-I]*

(a) 12 (b) 5

(c) 4 (d) 8

**21** Let $\alpha$ and $\beta$ be the sum and the product of all the non-zero solutions of the equation $(\bar{z})^2+|z|=0$, $z\in C$. Then, $4(\alpha^2+\beta^2)$ is equal to *[2024, 4 April Shift-I]*

(a) 6 (b) 2 (c) 4 (d) 8

**22** Let $S=\{z\in C:|z-1|=1$ and $(\sqrt{2}-1)(z+\bar{z})-i(z-\bar{z})=2\sqrt{2}\}$. Let $z_1, z_2\in S$ be such that $|z_1|=\max_{z\in S}|z|$ and $|z_2|=\min_{z\in S}|z|$. Then, $|\sqrt{2}z_1-z_2|^2$ equals *[2024, 1 Feb Shift-I]*

(a) 4 (b) 3 (c) 2 (d) 1

**23** Let $z_1$ and $z_2$ be two complex numbers such that $z_1+z_2=5$ and $z_1^3+z_2^3=20+15i$. Then, $|z_1^4+z_2^4|$ equals *[2024, 31 Jan Shift-II]*

(a) 75 (b) $25\sqrt{3}$ (c) $30\sqrt{3}$ (d) $15\sqrt{15}$

**24** Let $w_1$ be the point obtained by the rotation of $z_1=5+4i$ about the origin through a right angle in the anti-clockwise direction, and $w_2$ be the point obtained by the rotation of $z_2=3+5i$ about the origin through a right angle in the clockwise direction. Then, the principal argument of $w_1-w_2$ is equal to *[2023, 11 April Shift-I]*

(a) $\pi+\tan^{-1}\frac{8}{9}$ (b) $\pi-\tan^{-1}\frac{8}{9}$

(c) $\pi-\tan^{-1}\frac{33}{5}$ (d) $-\pi+\tan^{-1}\frac{33}{5}$

**25** For $z=\alpha+i\beta$, $|z+2|=z+4(1+i)$, then $\alpha+\beta$ and $\alpha\beta$ are the roots of the equation *[2023, 8 April Shift-I]*

(a) $x^2+2x-3=0$ (b) $x^2+7x+12=0$

(c) $x^2+3x-4=0$ (d) $x^2+x-12=0$

**Imp.** Properties of Modulus of Complex Number

**26** If $z\ne 0$ be a complex number such that $\left|z-\frac{1}{z}\right|=2$, then the maximum value of $|z|$ is *[2022, 29 July Shift-II]*

(a) $\sqrt{2}$ (b) 1 (c) $\sqrt{2}-1$ (d) $\sqrt{2}+1$

**27** Let $z_1$ and $z_2$ be two complex numbers such that $\overline{z_1}=i\,\overline{z_2}$ and $\arg\left(\frac{z_1}{\overline{z_2}}\right)=\pi$. Then, *[2022, 25 June Shift-II]*

(a) $\arg z_2=\frac{\pi}{4}$ (b) $\arg z_2=-\frac{3\pi}{4}$

(c) $\arg z_1=\frac{\pi}{4}$ (d) $\arg z_1=-\frac{3\pi}{4}$

**28** Let $A=\{z\in C:1\le|z-(1+i)|\le 2\}$ and $B=\{z\in A:|z-(1-i)|=1\}$. Then, $B$ *[2022, 24 June Shift-I]*

(a) is an empty set
(b) contains exactly two elements
(c) contains exactly three elements
(d) is an infinite set

**29** If $z$ and $w$ are two complex numbers such that $|zw|=1$ and $\arg(z)-\arg(w)=3\pi/2$, then $\arg\left(\dfrac{1-2\bar{z}w}{1+3\bar{z}w}\right)$ is

(Here, arg ($z$) denotes the principal argument of complex number $z$) *[2021, 20 July Shift-I]*

(a) $\frac{\pi}{4}$ (b) $-\frac{3\pi}{4}$ (c) $-\frac{\pi}{4}$ (d) $\frac{3\pi}{4}$

**30** If $z_1, z_2$ are complex numbers such that $\text{Re}(z_1)=|z_1-1|$, $\text{Re}(z_2)=|z_2-1|$ and $\arg(z_1-z_2)=\frac{\pi}{6}$, then $\text{Im}(z_1+z_2)$ is equal to *[2020, 3 Sep Shift-II]*

(a) $\frac{\sqrt{3}}{2}$ (b) $\frac{1}{\sqrt{3}}$ (c) $\frac{2}{\sqrt{3}}$ (d) $2\sqrt{3}$

**31** Let $z \in C$ with $\text{Im}(z)=10$ and it satisfies $\dfrac{2z-n}{2z+n}=2i-1$ for some natural number $n$, then *[2019, 12 April Shift-II]*

(a) $n=20$ and $\text{Re}(z)=-10$ (b) $n=40$ and $\text{Re}(z)=10$
(c) $n=40$ and $\text{Re}(z)=-10$ (d) $n=20$ and $\text{Re}(z)=10$

**32** Let $z_1$ and $z_2$ be any two non-zero complex numbers such that $3|z_1|=4|z_2|$. If $z=\dfrac{3z_1}{2z_2}+\dfrac{2z_2}{3z_1}$, then *[2019, 10 Jan Shift-I]*

(a) $|z|=\frac{1}{2}\sqrt{\frac{17}{2}}$ (b) $\text{Im}(z)=0$

(c) $\text{Re}(z)=0$ (d) $|z|=\sqrt{\frac{5}{2}}$

**33** Let $A=\left\{\theta\in\left(-\frac{\pi}{2},\pi\right):\dfrac{3+2i\sin\theta}{1-2i\sin\theta}\text{ is purely imaginary}\right\}$. Then, the sum of the elements in $A$ is *[2019, 9 Jan Shift-I]*

(a) $\frac{3\pi}{4}$ (b) $\frac{5\pi}{6}$ (c) $\pi$ (d) $\frac{2\pi}{3}$

**34** The set of all $\alpha \in R$, for which $\omega=\dfrac{1+(1-8\alpha)z}{1-z}$ is a purely imaginary number, for all $z \in C$ satisfying $|z|=1$ and $\text{Re}\, z \neq 1$, is *[JEE Main 2018]*

(a) an empty set (b) equal to $R$

(c) $\{0\}$ (d) $\left\{0, \frac{1}{4}, -\frac{1}{4}\right\}$

**35** A value of $\theta$ for which $\dfrac{2+3i\sin\theta}{1-2i\sin\theta}$ is purely imaginary, is *[JEE Main 2016]*

(a) $\frac{\pi}{3}$ (b) $\frac{\pi}{6}$ (c) $\sin^{-1}\left(\frac{\sqrt{3}}{4}\right)$ (d) $\sin^{-1}\left(\frac{1}{\sqrt{3}}\right)$

**36** Let $w(\text{Im}\, w \neq 0)$ be a complex number. Then, the set of all complex number $z$ satisfying the equation $w-\overline{w}z=k(1-z)$, for some real number $k$, is *[JEE Main 2014]*

(a) $\{z:|z|=1\}$ (b) $\{z:z=\bar{z}\}$
(c) $\{z:z\neq 1\}$ (d) $\{z:|z|=1, z\neq 1\}$

**37** Let $a=\text{Im}\left(\dfrac{1+z^2}{2iz}\right)$, where $z$ is any non-zero complex number. The set $A=\{a:|z|=1 \text{ and } z\neq \pm 1\}$ is equal to *[JEE Main 2013]*

(a) $(-1, 1)$ (b) $[-1, 1]$ (c) $[0, 1)$ (d) $(-1, 0]$

**38** If a complex number $z$ satisfies the equation $z+\sqrt{2}|z+1|+i=0$, then $|z|$ is equal to *[JEE Main 2013]*

(a) 2 (b) $\sqrt{3}$ (c) $\sqrt{5}$ (d) 1

**39** $|z_1+z_2|^2+|z_1-z_2|^2$ is equal to *[AIEEE 2012]*

(a) $2(|z_1|+|z_2|)$ (b) $2(|z_1|^2+|z_2|^2)$
(c) $|z_1||z_2|$ (d) $|z_1|^2+|z_2|^2$

**40** If $z_1$ and $z_2$ are two non-zero complex numbers such that $|z_1+z_2|=|z_1|+|z_2|$, then $\arg(z_1)-\arg(z_2)$ is equal to *[AIEEE 2005]*

(a) $-\pi/2$ (b) 0 (c) $-\pi$ (d) $\pi/2$

**41** If $z=x-iy$ and $z^{1/3}=p+iq$, then $\left(\dfrac{x}{p}+\dfrac{y}{q}\right)\Big/(p^2+q^2)$ is equal to *[AIEEE 2004]*

(a) 1 (b) $-1$ (c) 2 (d) $-2$

**42** If $z$ and $w$ are two non-zero complex numbers such that $|zw|=1$ and $\arg(z)-\arg(w)=\pi/2$, then $\bar{z}(w)$ is equal to *[AIEEE 2003]*

(a) 1 (b) $-1$ (c) $i$ (d) $-i$

## Level 3

**43** Consider the following two statements

**Statement I** For any two non-zero complex numbers $z_1, z_2$

$$(|z_1|+|z_2|)\left|\frac{z_1}{|z_1|}+\frac{z_2}{|z_2|}\right|\leq 2(|z_1|+|z_2|) \text{ and}$$

**Statement II** If $x, y, z$ are three distinct complex numbers and $a, b, c$ are three positive real numbers such that $\dfrac{a}{|y-z|}=\dfrac{b}{|z-x|}=\dfrac{c}{|x-y|}$, then

$$\frac{a^2}{y-z}+\frac{b^2}{z-x}+\frac{c^2}{x-y}=1.$$

*[2023, 5 April Shift-I]*

Between the above two statements.

(a) Both Statement I and Statement II are incorrect.
(b) Statement I is incorrect but Statement II is correct.
(c) Both Statement I and Statement II are correct.
(d) Statement I is correct but Statement II is incorrect.

**44** If the set $\left\{\text{Re}\left(\frac{z-\bar{z}+z\bar{z}}{2-3z+5\bar{z}}\right): z \in C, \text{Re}(z)=3\right\}$ is equal to the interval $(\alpha, \beta]$, then $24(\beta-\alpha)$ is equal to **[2023, 15 April Shift-I]**

(a) 36 (b) 30
(c) 27 (d) 42

**45** Let $S=\{z \in C: \bar{z}=i(z^2+\text{Re}(\bar{z}))\}$. Then, $\sum_{z \in S}|z|^2$ is equal to **[2023, 13 April Shift-II]**

(a) 3 (b) 7/2
(c) 5/2 (d) 4

**46** For $a \in C$, let $A=\{z \in C: \text{Re}(a+\bar{z})>\text{Im}(\bar{a}+z)\}$ and $B=\{z \in C: \text{Re}(a+\bar{z})<\text{Im}(\bar{a}+z)\}$. Then, among the two statements :

(S1) If $\text{Re}(a), \text{Im}(a)>0$, then the set $A$ contains all the real numbers.

(S2) If $\text{Re}(a), \text{Im}(a)<0$, then the set $B$ contains all the real numbers. **[2023, 11 April Shift-II]**

(a) both are true (b) both are false
(c) only (S2) is true (d) only (S1) is true

**47** Let $S=\left\{z=x+iy: \frac{2z-3i}{4z+2i} \text{ is a real number}\right\}$.

Then, which of the following is not correct? **[2023, 10 April Shift-II]**

(a) $(x, y)=\left(0,-\frac{1}{2}\right)$ (b) $y+x^2+y^2 \neq -\frac{1}{4}$
(c) $x=0$ (d) $y \in\left(-\infty,-\frac{1}{2}\right) \cup\left(-\frac{1}{2}, \infty\right)$

**48** Let the complex number $z=x+iy$ be such that $\frac{2z-3i}{2z+i}$ is purely imaginary. If $x+y^2=0$, then $y^4+y^2-y$ is equal to **[2023, 10 April Shift-I]**

(a) 2/3 (b) 3/2
(c) 3/4 (d) 4/5

**49** Let $a \neq b$ two non-zero real numbers. Then, the number of elements in the set $X=\{z \in C: \text{Re}(az^2+bz)=a \text{ and } \text{Re}(bz^2+az)=b\}$ is equal to **[2023, 6 April Shift-II]**

(a) 2 (b) 1
(c) 3 (d) 0

**50** Let $a, b$ be two real numbers such that $ab<0$. If the complex number $\frac{1+ai}{b+i}$ is of unit modulus and $a+ib$ lies on the circle $|z-1|=|2z|$, then a possible value of $\frac{1+[a]}{4b}$, where $[t]$ is greatest integer function is **[2023, 1 Feb Shift-II]**

(a) $-\frac{1}{2}$ (b) $\frac{1}{2}$ (c) $-1$ (d) 1

**51** If the centre and radius of the circle $\left|\frac{z-2}{z-3}\right|=2$ are respectively, $(\alpha, \beta)$ and $\gamma$, then $3(\alpha+\beta+\gamma)$ is equal to **[2023, 1 Feb Shift-I]**

(a) 9 (b) 10
(c) 11 (d) 12

**52** For two non-zero complex numbers $z_1$ and $z_2$, if $\text{Re}(z_1 z_2)=0$ and $\text{Re}(z_1+z_2)=0$, then which of the following are possible? **[2023, 29 Jan Shift I]**

A. $\text{Im}(z_1)>0$ and $\text{Im}(z_2)>0$
B. $\text{Im}(z_1)<0$ and $\text{Im}(z_2)>0$
C. $\text{Im}(z_1)>0$ and $\text{Im}(z_2)<0$
D. $\text{Im}(z_1)<0$ and $\text{Im}(z_2)<0$

Choose the correct answer from the options given below.

(a) B and D (b) A and B
(c) A and C (d) B and C

**53** Let $S$ be the set of all $(\alpha, \beta)$, $\pi<\alpha, \beta<2\pi$, for which the complex number $\frac{1-i \sin \alpha}{1+2i \sin \alpha}$ is purely imaginary and $\frac{1+i \cos \beta}{1-2i \cos \beta}$ is purely real. Let $Z_{\alpha\beta}=\sin 2\alpha+i \cos 2\beta$, $(\alpha, \beta) \in S$. Then, $\sum_{(\alpha, \beta) \in S}\left(iz_{\alpha\beta}+\frac{1}{i\bar{z}_{\alpha\beta}}\right)$ is equal to **[2022, 27 July Shift-II]**

(a) 3 (b) $3i$
(c) 1 (d) $2-i$

**54** For $n \in N$, let $S_n=\left\{z \in C:|z-3+2i|=\frac{n}{4}\right\}$ and $T_n=\left\{z \in C:|z-2+3i|=\frac{1}{n}\right\}$. Then, the number of elements in the set $\{n \in N: S_n \cap T_n=\phi\}$ is **[2022, 25 July Shift-I]**

(a) 0 (b) 2
(c) 3 (d) 4

**55** Let $n$ denote the number of solutions of the equation $z^2+3\bar{z}=0$, where $z$ is a complex number. Then, the value of $\sum_{k=0}^{\infty} \frac{1}{n^k}$ is equal to **[2021, 22 July Shift-II]**

(a) 1 (b) 4/3
(c) 3/2 (d) 2

**56** If the real part of the complex number $(1-\cos\theta+2i \sin\theta)^{-1}$ is 1/5 for $\theta \in (0, \pi)$, then the value of the integral $\int_0^\theta \sin x \, dx$ is equal to **[2021, 20 July Shift-II]**

(a) 1 (b) 2
(c) $-1$ (d) 0

**57** Let $u=\dfrac{2z+i}{z-ki}, z=x+iy$ and $k>0$. If the curve represented by $\text{Re}(u)+\text{Im}(u)=1$ intersects the $Y$-axis at the points $P$ and $Q$ where $PQ=5$, then value of $k$ is *[2020, 4 Sep Shift-I]*

(a) 1/2 (b) 4
(c) 2 (d) 3/2

**58** If $\dfrac{3+i\sin\theta}{4-i\cos\theta}, \theta\in[0,2\pi]$, is a real number, then an argument of $\sin\theta+i\cos\theta$ is *[2020, 7 Jan Shift-II]*

(a) $\pi-\tan^{-1}\left(\dfrac{4}{3}\right)$ (b) $-\tan^{-1}\left(\dfrac{3}{4}\right)$
(c) $-\tan^{-1}\left(\dfrac{4}{3}\right)$ (d) $\pi-\tan^{-1}\left(\dfrac{3}{4}\right)$

**59** If $z$ and $w$ are two complex numbers such that $|zw|=1$ and $\arg(z)-\arg(w)=\pi/2$, then *[2019, 10 April Shift-II]*

(a) $\bar{z}w=-i$ (b) $z\bar{w}=\dfrac{1-i}{\sqrt{2}}$
(c) $\bar{z}w=i$ (d) $z\bar{w}=\dfrac{-1+i}{\sqrt{2}}$

**Imp.** Euler's form of Complex number

**60** Let $z=\left(\dfrac{\sqrt{3}}{2}+\dfrac{i}{2}\right)^5+\left(\dfrac{\sqrt{3}}{2}-\dfrac{i}{2}\right)^5$. If $R(z)$ and $I(z)$ respectively denote the real and imaginary parts of $z$, then *[2019, 10 Jan Shift-II]*

(a) $\text{Re}(z)>0$ and $\text{Im}(z)>0$ (b) $\text{Im}(z)=0$
(c) $\text{Re}(z)<0$ and $\text{Im}\ z)>0$ (d) $\text{Re}(z)=-3$

## Numerical Value Type Questions

**61** If $\alpha$ denotes the number of solutions of $|1-i|^x=2^x$ and $\beta=\left(\dfrac{|z|}{\arg(z)}\right)$, where $z=\dfrac{\pi}{4}(1+i)^4\left[\dfrac{1-\sqrt{\pi}i}{\sqrt{\pi}+i}+\dfrac{\sqrt{\pi}-i}{1+\sqrt{\pi}i}\right]$, $i=\sqrt{-1}$, then the distance of the point $(\alpha,\beta)$ from the line $4x-3y=7$ is ........ . *[2024, 31 Jan Shift-I]*

**62** Let $S=\left\{z\in C-\{i,2i\}:\dfrac{z^2+8iz-15}{z^2-3iz-2}\in R\right\}$. If $\alpha-\dfrac{13}{11}i\in S, \alpha\in R-\{0\}$, then $242\alpha^2$ is equal to ............ . *[2023, 11 April Shift-II]*

**63** Let $z=1+i$ and $z_1=\dfrac{1+i\bar{z}}{\bar{z}(1-z)+\dfrac{1}{z}}$. Then, $\dfrac{12}{\pi}\arg(z_1)$ is equal to .......... *[2023, 30 Jan Shift-I]*

**64** Let $\alpha=8-14i$, $A=\left\{z\in c:\dfrac{\alpha z-\bar{\alpha}\bar{z}}{z^2-(\bar{z})^2-112i}=1\right\}$ and $B=\{z\in c:|z+3i|=4\}$. Then, $\sum\limits_{z\in A\cap B}(\text{Re}\,z-\text{Im}\,z)$ is equal to ........ *[2023, 29 Jan Shift-II]*

**65** Let $S=\{z\in C:z^2+\bar{z}=0\}$. Then, $\sum\limits_{z\in S}(\text{Re}(z)+\text{Im}(z))$ is equal to ... . *[2022, 27 July Shift-I]*

**66** Sum of squares of modulus of all the complex numbers $z$ satisfying $\bar{z}=iz^2+z^2-z$ is equal to .......... . *[2022, 28 June Shift-II]*

**67** The number of elements in the set $\{z=a+ib\in C: a,b\in z$ and $1<|z-3+2i|<4\}$ is ......... *[2022, 28 June Shift-I]*

**68** If for the complex numbers $z$ satisfying $|z-2-2i|\le 1$ the maximum value of $|3iz+6|$ is attained at $a+ib$, then $a+b$ is equal to .......... . *[2021, 1 Sep Shift-II]*

**69** Let $Z_1$ and $Z_2$ be two complex numbers such that $\arg(Z_1-Z_2)=\dfrac{\pi}{4}$ and $Z_1, Z_2$ satisfy the equation $|Z-3|=\text{Re}(Z)$. Then, the imaginary part of $Z_1+Z_2$ is equal to ......... . *[2021, 27 Aug Shift-II]*

**70** The least positive integer $n$ such that $\dfrac{(2i)^n}{(1-i)^{n-2}}, i=\sqrt{-1}$, is a positive integer, is ......... . *[2021, 26 Aug Shift-II]*

**71** If the real part of the complex number $z=\dfrac{3+2i\cos\theta}{1-3i\cos\theta}, \theta\in\left(0,\dfrac{\pi}{2}\right)$ is zero, then the value of $\sin^2 3\theta+\cos^2\theta$ is equal to ....... *[2021, 27 July Shift-II]*

**72** Let $z$ and $w$ be two complex numbers, such that $w=z\bar{z}-2z+2, \left|\dfrac{z+i}{z-3i}\right|=1$ and $\text{Re}(w)$ has minimum value. Then, the minimum value of $n\in N$ for which $w^n$ is real, is equal to ...... . *[2021, 16 March Shift-I]*

**73** If the least and the largest real values of $\alpha$, for which the equation $z+\alpha|z-1|+2i=0$ ($z\in C$ and $i=\sqrt{-1}$ has a solution), are $p$ and $q$ respectively, then $4(p^2+q^2)$ is equal to ........... . *[2021, 24 Feb Shift-I]*

## Topic 2 Cube Roots of Unity, De-Moivre's Theorem and *n*th Roots of Unity

### Level 1

**01** Let $z = \dfrac{1 - i\sqrt{3}}{2}$, $i = \sqrt{-1}$. Then, the value of

$$21 + \left(z + \frac{1}{z}\right)^3 + \left(z^2 + \frac{1}{z^2}\right)^3 + \left(z^3 + \frac{1}{z^3}\right)^3 + \ldots + \left(z^{21} + \frac{1}{z^{21}}\right)^3$$

**[2021, 26 Jan Shift-I]**

(a) 12 (b) 13
(c) 18 (d) 20

**02** If $\alpha, \beta \in C$ are the distinct roots of the equation $x^2 - x + 1 = 0$, then $\alpha^{101} + \beta^{107}$ is equal to **[JEE Main 2018]**

(a) $-1$ (b) 0
(c) 1 (d) 2

**Imp.** Cube Root of Unity

**03** If $\omega (\neq 1)$ is a cube root of unity $(1 + \omega)^7 = A + B\omega$. Then, $(A, B)$ is equal to **[AIEEE 2011]**

(a) (1, 1) (b) (1, 0)
(c) $(-1, 1)$ (d) (0, 1)

**04** The value of $\displaystyle\sum_{k=1}^{10}\left(\sin\frac{2k\pi}{11} + i\cos\frac{2k\pi}{11}\right)$ is **[AIEEE 2006]**

(a) 1 (b) $-1$ (c) $-i$ (d) $i$

**05** If $\omega$ is an imaginary cube root of unity, then $(1 + \omega - \omega^2)^7$ equal to **[AIEEE 2002]**

(a) $128\,\omega$ (b) $-128\,\omega$
(c) $128\,\omega^2$ (d) $-128\,\omega^2$

### Level 2

**06** If $z$ is a complex number, then the number of common roots of the equations $z^{1985} + z^{100} + 1 = 0$ and $z^3 + 2z^2 + 2z + 1 = 0$ is equal to **[2024, 30 Jan Shift-II]**

(a) 2 (b) 1 (c) 3 (d) 0

**07** The complex number $z = \dfrac{i - 1}{\cos\frac{\pi}{3} + i\sin\frac{\pi}{3}}$ is equal to **[2023, 31 Jan Shift-II]**

(a) $\sqrt{2}\left(\cos\frac{5\pi}{12} + i\sin\frac{5\pi}{12}\right)$

(b) $\cos\frac{\pi}{12} - i\sin\frac{\pi}{12}$

(c) $\sqrt{2}\left(\cos\frac{\pi}{12} + i\sin\frac{\pi}{12}\right)$

(d) $\sqrt{2}i\left(\cos\frac{5\pi}{12} - i\sin\frac{5\pi}{12}\right)$

**08** Let $\alpha$ be a root of the equation $1 + x^2 + x^4 = 0$. Then, the value of $\alpha^{1011} + \alpha^{2022} - \alpha^{3033}$ is equal to **[2022, 29 June Shift-II]**

(a) 1 (b) $\alpha$
(c) $1 + \alpha$ (d) $1 + 2\alpha$

**09** If $a$ and $b$ are real numbers such that $(2 + \alpha)^4 = a + b\alpha$, where $\alpha = \dfrac{-1 + i\sqrt{3}}{2}$, then $a + b$ is equal to **[2020, 4 Sep Shift-II]**

(a) 24 (b) 33
(c) 9 (d) 57

**10** If $z$ is a non-real complex number, then the minimum value of $\dfrac{\text{Im } z^5}{(\text{Im } z)^5}$ is **[JEE Main 2015]**

(a) $-1$ (b) $-2$
(c) $-4$ (d) $-5$

**11** If the cube roots of unity are $1, \omega$ and $\omega^2$, then the roots of the equation $(x - 1)^3 + 8 = 0$, are **[AIEEE 2005]**

(a) $-1, 1 + 2\omega, 1 + 2\omega^2$
(b) $-1, 1 - 2\omega, 1 - 2\omega^2$
(c) $-1, -1, -1$
(d) $-1, -1 + 2\omega, -1 - 2\omega^2$

### Level 3

**Imp.** Application of Complex Number

**12** Let $O$ be the origin, the point $A$ be $z_1 = \sqrt{3} + 2\sqrt{2}i$, the point $B(z_2)$ be such that $\sqrt{3}|z_2| = |z_1|$ and $\arg(z_2) = \arg(z_1) + \frac{\pi}{6}$. Then, **[2025, 28 Jan Shift-I]**

(a) area of $\Delta ABO$ is $\frac{11}{4}$
(b) $ABO$ is a scalene triangle
(c) area of $\Delta ABO$ is $\frac{11}{\sqrt{3}}$
(d) $ABO$ is an obtuse angled isosceles triangle

**13** Let $p, q \in R$ and $(1 - \sqrt{3}i)^{200} = 2^{199}(p + iq)$, $i = \sqrt{-1}$. Then, $p + q + q^2$ and $p - q + q^2$ are roots of the equation **[2023, 24 Jan Shift-I]**

(a) $x^2 - 4x + 1 = 0$
(b) $x^2 - 4x - 1 = 0$
(c) $x^2 + 4x - 1 = 0$
(d) $x^2 + 4x + 1 = 0$

**14** The value of $\left(\dfrac{1+\sin\dfrac{2\pi}{9}+i\cos\dfrac{2\pi}{9}}{1+\sin\dfrac{2\pi}{9}-i\cos\dfrac{2\pi}{9}}\right)^3$ is *[2020, 2 Sep Shift-I]*

(a) $-\frac{1}{2}(\sqrt{3}-i)$ (b) $-\frac{1}{2}(1-i\sqrt{3})$
(c) $\frac{1}{2}(\sqrt{3}-i)$ (d) $\frac{1}{2}(1-i\sqrt{3})$

**15** Let $z \in C$ be such that $|z|<1$. If $\omega=\dfrac{5+3z}{5(1-z)}$, then *[2019, 9 April Shift-II]*

(a) $4\,\text{Im}(\omega)>5$ (b) $5\,\text{Re}(\omega)>1$
(c) $5\,\text{Im}(\omega)<1$ (d) $5\,\text{Re}(\omega)>4$

**16** If $z=\dfrac{\sqrt{3}}{2}+\dfrac{i}{2}(i=\sqrt{-1})$, then $(1+iz+z^5+iz^8)^9$ is equal to *[2019, 8 April Shift-II]*

(a) 1 (b) $(-1+2i)^9$ (c) $-1$ (d) 0

**17** Let $\alpha$ and $\beta$ be two roots of the equation $x^2+2x+2=0$, then $\alpha^{15}+\beta^{15}$ is equal to *[2019, 9 Jan Shift-I]*

(a) 256 (b) 512 (c) $-256$ (d) $-512$

**18** Let $z=1+ai$ be a complex number, $a>0$, such that $z^3$ is a real number. Then, the sum $1+z+z^2 \,.....+z^{11}$ is equal to *[JEE Main 2016]*

(a) $1365\sqrt{3}i$ (b) $-1365\sqrt{3}i$
(c) $-1250\sqrt{3}i$ (d) $1250\sqrt{3}i$

## Numerical Value Type Questions

**19** Let integers $a, b \in [-3, 3]$ be such that $a+b \neq 0$. Then, the number of all possible ordered pairs $(a, b)$ for which $\left|\dfrac{z-a}{z+b}\right|=1$ and

$$\begin{vmatrix} z+1 & \omega & \omega^2 \\ \omega & z+\omega^2 & 1 \\ \omega^2 & 1 & z+\omega \end{vmatrix}=1, z \in C,$$ where $\omega$ and $\omega^2$ are the roots of $x^2+x+1=0$ is equal to....... *[2025, 29 Jan Shift-II]*

**20** Let $\alpha, \beta$ be the roots of the equation $x^2-\sqrt{6}x+3=0$ such that $\text{Im}(\alpha)>\text{Im}(\beta)$. Let $a$, $b$ be integers not divisible by 3 and $n$ be a natural number such that $\dfrac{\alpha^{99}}{\beta}+\alpha^{98}=3^n(a+ib), i=\sqrt{-1}$.
Then, $n+a+b$ is equal to ....... *[2024, 29 Jan Shift-II]*

**21** Let $z=a+ib, b \neq 0$ be complex numbers satisfying $z^2=\bar{z}\cdot 2^{1-|Z|}$. Then, the least value of $n \in N$, such that $z^n=(z+1)^n$ is equal to ............... . *[2022, 28 July Shift-II]*

**22** The sum of 162th power of the root of the equation $x^3-2x^2+2x-1=0$ is *[2021, 26 Feb Shift-I]*

## Topic 3 Application of Complex Number in Geometry

### Level 1

**01** Among the statements

(S1) The set $\{z \in C-\{-i\}:|z|=1$ and $\dfrac{z-i}{z+i}$ is purely real$\}$ contains exactly two elements and

(S2) The set $\{z \in C-\{-1\}:|z|=1$ and $\dfrac{z-1}{z+1}$ is purely imaginary$\}$ contains infinitely many elements. *[2025, 7 April Shift-I]*

(a) only (S2) is correct
(b) both are correct
(c) both are incorrect
(d) only (S1) is correct

**02** If $z_1$ and $z_2$ are two distinct complex number such that $\left|\dfrac{z_1-2z_2}{\frac{1}{2}-z_1\bar{z}_2}\right|=2$, then *[2024, 6 April Shift-II]*

(a) both $z_1$ and $z_2$ lie on the same circle
(b) $z_1$ lies on a circle of radius $\frac{1}{2}$ and $z_2$ lies on a circle of radius 1.
(c) either $z_1$ lies on a circle of radius $\frac{1}{2}$ or $z_2$ lies on a circle of radius 1.
(d) either $z_1$ lies on a circle of radius 1 or $z_2$ lies on a circle of radius $\frac{1}{2}$

**03** If $z$ is a complex number such that $|z| \geq 1$, then the minimum value of $\left|z+\frac{1}{2}(3+4i)\right|$ is *[2024, 1 Feb Shift-II]*

(a) 5/2 (b) 3
(c) 2 (d) 3/2

**04** If $S=\{z \in C:|z-i|=|z+i|=|z-1|\}$, then $n(S)$ is *[2024, 27 Jan Shift-I]*

(a) 0 (b) 2
(c) 1 (d) 3

**05** For all $z \in C$ on the curve $C_1:|z|=4$, let the locus of the point $z+\dfrac{1}{z}$ be the curve $C_2$. Then, *[2023, 31 Jan Shift-I]*

(a) the curves $C_1$ and $C_2$ intersect at 4 points
(b) the curves $C_1$ and $C_2$ intersect at 2 points
(c) the curve $C_2$ lies inside $C_1$
(d) the curve $C_1$ lies inside $C_2$

**06** Let $z$ be a complex number such that $\left|\frac{z-2i}{z+i}\right|=2$, $z \neq -i$. Then, $z$ lies on the circle of radius 2 and centre **[2023, 25 Jan Shift-II]**

(a) (2, 0) (b) (0, 0) (c) (0, − 2) (d) (0, 2)

**07** Let a complex number be $w = 1-\sqrt{3}\,i$. Let another complex number $z$ be such that $|zw|=1$ and $\arg(z)-\arg(w)=\pi/2$. Then the area of the triangle with vertices origin, $z$ and $w$, is equal to **[2021, 18 March Shift-II]**

(a) 4 (b) 1/2 (c) 1/4 (d) 2

**08** If the four complex numbers $z, \bar{z}, \bar{z}-2\,\text{Re}(\bar{z})$ and $z-2\,\text{Re}(z)$ represent the vertices of a square of side 4 units in the argand plane, then $|z|$ is equal to **[2020, 5 Sep Shift-I]**

(a) $4\sqrt{2}$ (b) 2 (c) $2\sqrt{2}$ (d) 4

**09** If $|z+4| \leq 3$, then the maximum value of $|z+1|$ is **[AIEEE 2007]**

(a) 4 (b) 10 (c) 6 (d) 0

**10** If $w = \frac{z}{z-\frac{i}{3}}$ and $|w|=1$, then $z$ lies on **[AIEEE 2005]**

(a) a parabola (b) a straight line
(c) a circle (d) an ellipse

**11** If $|z^2-1| = |z|^2+1$, then $z$ lies on **[AIEEE 2004]**

(a) the real axis (b) the imaginary axis
(c) a circle (d) an ellipse

## Level 2

**12** If $z_1, z_2, z_3 \in C$ are the vertices of an equilateral triangle, whose centroid is $z_0$, then $\sum_{k=1}^{3}(z_k-z_0)^2$ is equal to **[2025, 3 April Shift-II]**

(a) $i$ (b) 0 (c) 1 (d) $-i$

**13** Let the product of $\omega_1 = (8+i)\sin\theta + (7+4i)\cos\theta$ and $\omega_2 = (1+8i)\sin\theta + (4+7i)\cos\theta$ be $\alpha + i\beta, i=\sqrt{-1}$. Let $p$ and $q$ be the maximum and the minimum values of $\alpha+\beta$ respectively. Then, $p+q$ is equal to **[2025, 4 April Shift-II]**

(a) 130 (b) 150 (c) 160 (d) 140

**14** If the locus of $z \in C$, such that $\text{Re}\left(\frac{z-1}{2z+i}\right)+\text{Re}\left(\frac{\bar{z}-1}{2\bar{z}-i}\right)=2$, is a circle of radius $r$ and centre $(a, b)$ then $\frac{15ab}{r^2}$ is equal to **[2025, 7 April Shift-II]**

(a) 12 (b) 18 (c) 16 (d) 24

**15** Let $\left|\frac{\bar{z}-i}{2\bar{z}+i}\right|=\frac{1}{3}$, $z \in C$ be the equation of a circle with centre at $C$. If the area of the triangle, whose vertices are at the points (0, 0), $C$ and $(\alpha, 0)$ is 11 sq. units, then $\alpha^2$ equals **[2025, 23 Jan Shift-I]**

(a) 81/25 (b) 121/25
(c) 100 (d) 50

**16** Let the curve $z(1+i)+\bar{z}(1-i)=4, z \in C$ divide the region $|z-3| \leq 1$ into two parts of areas $\alpha$ and $\beta$. Then, $|\alpha-\beta|$ equals **[2025, 23 Jan Shift-II]**

(a) $1+\frac{\pi}{4}$ (b) $1+\frac{\pi}{2}$ (c) $1+\frac{\pi}{3}$ (d) $1+\frac{\pi}{6}$

**17** Let $S_1 = \{z \in C : |z| \leq 5\}$, $S_2 = \left\{z \in C : \text{Im}\left(\frac{z+1-\sqrt{3}i}{1-\sqrt{3}i}\right) \geq 0\right\}$ and $S_3 = \{z \in C : \text{Re}(z) \geq 0\}$. Then, the area of the region $S_1 \cap S_2 \cap S_3$ is **[2024, 5 April Shift-II]**

(a) $\frac{125\pi}{24}$ (b) $\frac{125\pi}{12}$ (c) $\frac{125\pi}{4}$ (d) $\frac{125\pi}{6}$

**18** The area (in sq units) of the region $S = \{z \in C: |z-1| \leq 2; (z+\bar{z})+i(z-\bar{z}) \leq 2, \text{Im}(z) \geq 0\}$ is **[2024, 4 April Shift-II]**

(a) $\frac{7\pi}{4}$ (b) $\frac{3\pi}{2}$ (c) $\frac{17\pi}{8}$ (d) $\frac{7\pi}{3}$

**19** Let the minimum value $v_0$ of $v = |z|^2+|z-3|^2+|z-6i|^2, z \in C$ is attained at $z=z_0$. Then, $|2z_0^2-\bar{z}_0^3+3|^2+v_0^2$ is equal to **[2022, 27 July Shift-I]**

(a) 1000 (b) 1024 (c) 1105 (d) 1196

**20** For $z \in C$ if the minimum value of $(|z-3\sqrt{2}|+|z-p\sqrt{2}i|)$ is $5\sqrt{2}$, then a value of $p$ is **[2022, 25 July shift-II]**

(a) 3 (b) 7/2 (c) 4 (d) 9/2

**21** The number of points of intersection of $|z-(4+3i)|=2$ and $|z|+|z-4|=6, z \in C$ is **[2022, 27 June Shift-II]**

(a) 0 (b) 1 (c) 2 (d) 3

**22** Let a circle $C$ in complex plane pass through the points $z_1 = 3+4i, z_2 = 4+3i$ and $z_3 = 5i$. If $z (\neq z_1)$ is a point on $C$ such that the line through $z$ and $z_1$ is perpendicular to the line through $z_2$ and $z_3$, then arg $(z)$ is equal to **[2022, 25 June shift-I]**

(a) $\tan^{-1}\left(\frac{2}{\sqrt{5}}\right)-\pi$ (b) $\tan^{-1}\left(\frac{24}{7}\right)-\pi$

(c) $\tan^{-1}(3)-\pi$ (d) $\tan^{-1}\left(\frac{3}{4}\right)-\pi$

**23** If $z$ is a complex number such that $\frac{z-i}{z-1}$ is purely imaginary, then the minimum value of $|z-(3+3i)|$ is *[2021, 2019, 31 Aug Shift-II]*
(a) $2\sqrt{2}-1$ (b) $3\sqrt{2}$ (c) $6\sqrt{2}$ (d) $2\sqrt{2}$

**Imp.** Properties of Argument of Complex Number

**24** A point $z$ moves in the complex plane such that $\arg\left(\frac{z-2}{z+2}\right)=\frac{\pi}{4}$, then the minimum value of $|z-9\sqrt{2}-2i|^2$ equal to *[2021, 31 Aug Shift-I]*
(a) 100 (b) 98
(c) 95 (d) 102

**25** The equation $\arg\left(\frac{z-1}{z+1}\right)=\frac{\pi}{4}$ represents a circle with *[2021, 26 Aug Shift-I]*
(a) centre at (0, −1) and radius $\sqrt{2}$
(b) centre at (0, 1) and radius $\sqrt{2}$
(c) centre at (0, 0) and radius $\sqrt{2}$
(d) centre at (0, 1) and radius 2

**26** The point $P(a, b)$ undergoes the following three transformations successively
(A) Reflection about the line $y=x$.
(B) Translation through 2 units along the positive direction of $X$-axis.
(C) Rotation through angle $\pi/4$ about the origin in the anti-clockwise direction.

If the coordinates of the final position of the point $P$ are $\left(-\frac{1}{\sqrt{2}}, \frac{7}{\sqrt{2}}\right)$, then the value of $2a+b$ is equal to *[2021, 27 July Shift-II]*
(a) 13 (b) 9
(c) 5 (d) 7

**27** Let $z$ be a complex number such that $\left|\frac{z-i}{z+2i}\right|=1$ and $|z|=\frac{5}{2}$. Then, the value of $|z+3i|$ is *[2020, 9 Jan Shift-I]*
(a) $\sqrt{10}$ (b) $\frac{7}{2}$
(c) $\frac{15}{4}$ (d) $2\sqrt{3}$

**28** If $\text{Re}\left(\frac{z-1}{2z+i}\right)=1$, where $z=x+iy$, then the point $(x, y)$ lies on a *[2020, 7 Jan Shift-I]*
(a) straight line whose slope is $-\frac{2}{3}$.
(b) circle whose centre is at $\left(-\frac{1}{2}, -\frac{3}{2}\right)$.
(c) straight line whose slope is $\frac{3}{2}$.
(d) circle whose diameter is $\frac{\sqrt{5}}{2}$.

**29** All the points in the set $S=\left\{\frac{\alpha+i}{\alpha-i} : \alpha \in R\right\}$ $(i=\sqrt{-1})$ lie on a *[2019, 9 April Shift-I]*
(a) circle whose radius is $\sqrt{2}$.
(b) straight line whose slope is −1.
(c) circle whose radius is 1.
(d) straight line whose slope is 1.

**30** Let $z \in C$, the set of complex numbers. Then, the equation, $2|z+3i|-|z-i|=0$ represents *[JEE Main 2017]*
(a) a circle with radius 8/3.
(b) a circle with diameter 10/3.
(c) an ellipse with length of major axis 16/3.
(d) an ellipse with length of minor axis 16/9.

**31** The point represented by $2+i$ in the Argand plane moves 1 unit eastwards, then 2 units northwards and finally from there $2\sqrt{2}$ units in the south-westwards direction. Then its new position in the Argand plane is at the point represented by *[JEE Main 2016]*
(a) $1+i$ (b) $2+2i$ (c) $-2-2i$ (d) $-1-i$

**32** A complex number $z$ is said to be unimodular, if $|z|=1$. Suppose $z_1$ and $z_2$ are complex numbers such that $\frac{z_1-2z_2}{2-z_1\bar{z}_2}$ is unimodular and $z_2$ is not unimodular. Then, the point $z_1$ lies on a *[JEE Main 2015]*
(a) straight line parallel to $X$-axis
(b) straight line parallel to $Y$-axis
(c) circle of radius 2
(d) circle of radius $\sqrt{2}$

**33** If $z$ is a complex number such that $|z|\geq 2$, then the minimum value of $\left|z+\frac{1}{2}\right|$ *[JEE Main 2014]*
(a) is equal to 5/2
(b) lies in the interval (1, 2)
(c) is strictly greater than 5/2
(d) is strictly greater than 3/2 but less than 5/2

**34** The number of complex numbers $z$ such that $|z-1|=|z+1|=|z-i|$ is equal to *[AIEEE 2010]*
(a) 0 (b) 1 (c) 2 (d) ∞

**35** Let $z_1$ and $z_2$ be two roots of the equation $z^2+az+b=0$, $z$ being complex. Further, assume that the origin, $z_1$ and $z_2$ form an equilateral triangle. Then, *[AIEEE 2003]*
(a) $a^2=b$ (b) $a^2=2b$
(c) $a^2=3b$ (d) $a^2=4b$

## Level 3

**36** Let $C$ be the circle in the complex plane with centre $z_0 = \frac{1}{2}(1+3i)$ and radius $r = 1$. Let $z_1 = 1+i$ and the complex number $z_2$ be outside the circle $C$ such that $|z_1 - z_0||z_2 - z_0| = 1$. If $z_0$, $z_1$ and $z_2$ are collinear, then the smaller value of $|z_2|^2$ is equal to **[2023, 12 April Shift-I]**

(a) $\frac{13}{2}$ (b) $\frac{7}{2}$

(c) $\frac{5}{2}$ (d) $\frac{3}{2}$

**37** Let $z_1 = 2+3i$ and $z_2 = 3+4i$. The set $S = \{z \in C : |z - z_1|^2 - |z - z_2|^2 = |z_1 - z_2|^2\}$ represents a **[2023, 25 Jan Shift-I]**

(a) hyperbola with the length of the transverse axis 7
(b) straight line with the sum of its intercepts on the co-ordinate axes equals 18
(c) straight line with the sum of its intercepts on the co-ordinate axes equals 14
(d) hyperbola with eccentricity 2

**38** Let $S = \{z = x+iy : |z - 1 + i| \geq |z|, |z| < 2, |z+i| = |z-1|\}$. Then, the set of all values of $x$, for which $w = 2x + iy \in S$ for some $y \in R$, is **[2022, 29 July Shift-II]**

(a) $\left(-\sqrt{2}, \frac{1}{2\sqrt{2}}\right]$ (b) $\left(-\frac{1}{\sqrt{2}}, \frac{1}{4}\right]$

(c) $\left(-\sqrt{2}, \frac{1}{2}\right]$ (d) $\left(-\frac{1}{\sqrt{2}}, \frac{1}{2\sqrt{2}}\right]$

**39** Let arg $(z)$ represent the principal argument of the complex number $z$. Then, $|z| = 3$ and arg $(z-1)$ − arg $(z+1) = \pi/4$ intersect **[2022, 29 June Shift-II]**

(a) exactly at one point (b) exactly at two points
(c) nowhere (d) at infinitely many points

**40** The area of the polygon, whose vertices are the non-real roots of the equation $\bar{z} = iz^2$ is **[2022, 27 June shift-I]**

(a) $\frac{3\sqrt{3}}{4}$ (b) $\frac{3\sqrt{3}}{2}$

(c) $\frac{3}{2}$ (d) $\frac{3}{4}$

**41** Let $A = \left\{z \in C : \left|\frac{z+1}{z-1}\right| < 1\right\}$ and

$B = \left\{z \in C : \arg\left(\frac{z-1}{z+1}\right) = \frac{2\pi}{3}\right\}$

Then, $A \cap B$ is **[2022, 26 June Shift-I]**

(a) a portion of a circle centred at $\left(0, -\frac{1}{\sqrt{3}}\right)$ that lies in the second and third quadrants only
(b) a portion of a circle centred at $\left(0, -\frac{1}{\sqrt{3}}\right)$ that lies in the second quadrant only
(c) an empty set
(d) a portion of a circle of radius $\frac{2}{\sqrt{3}}$ that lies in the third quadrant only

**42** Let $C$ be the set of all complex numbers.
Let $S_1 = \{z \in C : |z - 2| \leq 1\}$ and
$S_2 = \{z \in C : z(1+i) + \bar{z}(1-i) \geq 4\}$.
Then, the maximum value of $\left|z - \frac{5}{2}\right|^2$ for $z \in S_1 \cap S_2$ is equal to **[2021, 27 July Shift-II]**

(a) $\frac{3+2\sqrt{2}}{4}$ (b) $\frac{5+2\sqrt{2}}{2}$ (c) $\frac{3+2\sqrt{2}}{2}$ (d) $\frac{5+2\sqrt{2}}{4}$

**43** Let $S_1$, $S_2$ and $S_3$ be three sets defined as
$S_1 = \{z \in C : |z - 1| \leq \sqrt{2}\}$
$S_2 = \{z \in C : \text{Re}[(1-i)z] \geq 1\}$
$S_3 = \{z \in C : \text{Im}(z) \leq 1\}$
Then, the set $S_1 \cap S_2 \cap S_3$ **[2021, 17 March Shift-II]**

(a) is a singleton
(b) has exactly two elements
(c) has infinitely many elements
(d) has exactly three elements

**Imp.** Application of Complex Number

**44** If $z \neq 1$ and $\frac{z^2}{z-1}$ is real, then the point represented by the complex number $z$ lies **[AIEEE 2012]**

(a) either on the real axis or on a circle passing through the origin
(b) on a circle with centre at the origin
(c) either on the real axis or on a circle not passing through the origin
(d) on the imaginary axis

## Numerical Value Type Questions

**45** Let $A = \{z \in C : |z - 2 - i| = 3\}$, $B = \{z \in C : \text{Re}(z - iz) = 2\}$ and $S = A \cap B$. Then $\sum_{z \in S} |z|^2$ is equal to ........... **[2025, 4 April Shift-I]**

**46** The sum of the square of the modulus of the elements in the set $\{z = a + ib : a, b \in z, z \in C, |z - 1| \leq 1, |z - 5| \leq |z - 5i|\}$ is .......... . **[2024, 9 April Shift-I]**

**47** Let $P = \{z \in C : |z + 2 - 3i| \le 1\}$ and $Q = \{z \in C : z(1+i) + \bar{z}(1-i) \le -8\}$. Let in $P \cap Q$, $|z - 3 + 2i|$ be maximum and minimum at $z_1$ and $z_2$, respectively. If $|z_1|^2 + 2|z_2|^2 = \alpha + \beta\sqrt{2}$, where $\alpha$, $\beta$ are integers, then $\alpha + \beta$ equals .......... . *[2024, 1 Feb Shift-I]*

**48** Let the complex numbers $\alpha$ and $1/\bar{\alpha}$ lie on the circles $|z - z_0|^2 = 4$ and $|z - z_0|^2 = 16$ respectively, where $z_0 = 1 + i$. Then, the value of $100|\alpha|^2$ is ......... . *[2024, 27 Jan Shift-II]*

**49** For $\alpha, \beta, z \in C$ and $\lambda > 1$, if $\sqrt{\lambda - 1}$ is the radius of the circle $|z - \alpha|^2 + |z - \beta|^2 = 2\lambda$, then $|\alpha - \beta|$ is equal to ........ *[2023, 6 April Shift-II]*

**50** Let $S = \{z \in C : |z - 2| \le 1, z(1+i) + \bar{z}(1-i) \le 2\}$. Let $|z - 4i|$ attains minimum and maximum values, respectively, at $z_1 \in S$ and $z_2 \in S$. If $5(|z_1|^2 + |z_2|^2) = \alpha + \beta\sqrt{5}$, where $\alpha$ and $\beta$ are integers, then the value of $\alpha + \beta$ is equal to ...... . *[2022, 29 June Shift-I]*

**51** Let $S = \{z \in C : |z - 3| \le 1$ and $z(4 + 3i) + \bar{z}(4 - 3i) \le 24\}$. If $\alpha + i\beta$ is the point in $S$ which is closest to $4i$, then $25(\alpha + \beta)$ is equal to .... . *[2022, 24 June shift-I]*

**52** Let $z_1, z_2$ be the roots of the equation $z^2 + az + 12 = 0$ and $z_1, z_2$ form an equilateral triangle with origin. Then, the value of $|a|$ is ...... . *[2021, 18 March Shift-I]*

**53** Let $z$ be those complex numbers which satisfy $|z + 5| \le 4$ and $z(1+i) + \bar{z}(1-i) \ge -10, i = \sqrt{-1}$. If the maximum value of $|z + 1|^2$ is $\alpha + \beta\sqrt{2}$, then the value of $(\alpha + \beta)$ is ......... . *[2021, 26 Feb Shift-II]*

## Topic 4 Quadratic Equations in Real & Complex Number System & Their Solutions, Relation between Roots & Coefficients, Nature of Roots, Formation of Quadratic Equation with given Roots

### Level 1

**01** Let $P_n = \alpha^n + \beta^n, n \in N$. If $P_{10} = 123$, $P_9 = 76$, $P_8 = 47$ and $P_1 = 1$, then the quadratic equation having roots $\frac{1}{\alpha}$ and $\frac{1}{\beta}$ is *[2025, 2 April Shift-I]*

(a) $x^2 - x - 1 = 0$ (b) $x^2 - x + 1 = 0$
(c) $x^2 + x + 1 = 0$ (d) $x^2 + x - 1 = 0$

**02** Consider the equation $x^2 + 4x - n = 0$, where $n \in [20, 100]$ is a natural number. Then, the number of all distinct values of $n$, for which the given equation has integral roots, is equal to *[2025, 4 April Shift-I]*

(a) 7 (b) 6 (c) 8 (d) 5

**03** Let the set of all values of $p \in R$, for which both the roots of the equation $x^2 - (p+2)x + (2p+9) = 0$ are negative real numbers be the interval $(\alpha, \beta]$. Then, $\beta - 2\alpha$ is equal to *[2025, 7 April Shift-I]*

(a) 9 (b) 0 (c) 5 (d) 20

**04** The number of solutions of the equation $\left(\frac{9}{x} - \frac{9}{\sqrt{x}} + 2\right)\left(\frac{2}{x} - \frac{7}{\sqrt{x}} + 3\right) = 0$ is *[2025, 29 Jan Shift-I]*

(a) 4 (b) 1
(c) 3 (d) 2

**05** The number of real solution(s) of the equation $x^2 + 3x + 2 = \min\{|x - 3|, |x + 2|\}$ is *[2025, 24 Jan Shift-II]*

(a) 2 (b) 3
(c) 0 (d) 1

**06** The product of all solutions of the equation $e^{5(\log_e x)^2 + 3} = x^8, x > 0$ is *[2025, 22 Jan Shift-I]*

(a) $e^{\frac{6}{5}}$ (b) $e$ (c) $e^{\frac{8}{5}}$ (d) $e^2$

**07** If 2 and 6 are the roots of the equation $ax^2 + bx + 1 = 0$, then the quadratic equation whose roots are $\frac{1}{2a+b}$ and $\frac{1}{6a+b}$, is *[2024, 4 April Shift-I]*

(a) $x^2 + 8x + 12 = 0$ (b) $2x^2 + 11x + 12 = 0$
(c) $x^2 + 10x + 16 = 0$ (d) $4x^2 + 14x + 12 = 0$

**08** Let $S = \{x \in R : (\sqrt{3} + \sqrt{2})^x + (\sqrt{3} - \sqrt{2})^x = 10\}$. Then, the number of elements in $S$, is *[2024, 1 Feb Shift-I]*

(a) 0 (b) 1 (c) 4 (d) 2

**09** Let $S$ be the set of positive integral values of $a$ for which $\frac{ax^2 + 2(a+1)x + 9a + 4}{x^2 - 8x + 32} < 0, \forall x \in R$. Then, the number of elements in $S$ is *[2024, 31 Jan Shift-I]*

(a) 1 (b) 3 (c) $\infty$ (d) 0

**10** The set of all $a \in R$ for which the equation $x|x - 1| + |x + 2| + a = 0$ has exactly one real root, is *[2023, 13 April Shift-I]*

(a) $(-\infty, \infty)$ (b) $(-6, -3)$
(c) $(-6, \infty)$ (d) $(-\infty, -3)$

**11** The number of integral values of $k$, for which one root of the equation $2x^2 - 8x + k = 0$ lies in the interval (1, 2) and its other root lies in the interval (2, 3), is *[2023, 1 Feb Shift-II]*

(a) 3 (b) 2
(c) 0 (d) 1

**12** The equation $x^2 - 4x + [x] + 3 = x[x]$, where $[x]$ denotes the greatest integer function, has **[2023, 24 Jan Shift-I]**

(a) a unique solution in $(-\infty, \infty)$
(b) no solution
(c) a unique solution in $(-\infty, 1)$
(d) exactly two solutions in $(-\infty, \infty)$

**13** Let $f(x) = ax^2 + bx + c$ be such that $f(1) = 3, f(-2) = \lambda$ and $f(3) = 4$. If $f(0) + f(1) + f(-2) + f(3) = 14$, then $\lambda$ is equal to **[2022, 28 July Shift-II]**

(a) $-4$ (b) $\frac{13}{2}$ (c) $\frac{23}{2}$ (d) 4

**14** The number of real solutions of $x^7 + 5x^3 + 3x + 1 = 0$ is equal to **[2022, 28 June Shift-I]**

(a) 0 (b) 1 (c) 3 (d) 5

**15** Let $a, b \in R$ be such that the equation $ax^2 - 2bx + 15 = 0$ has a repeated root $\alpha$. If $\alpha$ and $\beta$ are the roots of the equation $x^2 - 2bx + 21 = 0$, then $\alpha^2 + \beta^2$ is equal to **[2022, 25 June Shift-II]**

(a) 37 (b) 58 (c) 68 (d) 92

**16** The number of distinct real roots of the equation $x^7 - 7x - 2 = 0$ is **[2022, 24 June Shift-II]**

(a) 5 (b) 7 (c) 1 (d) 3

**17** The number of real solutions of the equation $x^2 - |x| - 12 = 0$ is **[2021, 25 July Shift-II]**

(a) 2 (b) 3 (c) 1 (d) 4

**18** Let $[x]$ denote the greatest integer less than or equal to $x$. Then, the values of $x \in R$ satisfying the equation $[e^x]^2 + [e^x + 1] - 3 = 0$ lie in the interval **[2021, 22 July Shift-II]**

(a) $\left[0, \frac{1}{e}\right)$ (b) $[\log_e 2, \log_e 3)$
(c) $[1, e)$ (d) $[0, \log_e 2)$

**19** The value of $3 + \cfrac{1}{4 + \cfrac{1}{3 + \cfrac{1}{4 + \cfrac{1}{3 + \dots\infty}}}}$ is equal to **[2021, 18 March shift-II]**

(a) $1.5 + \sqrt{3}$ (b) $2 + \sqrt{3}$ (c) $3 + 2\sqrt{3}$ (d) $4 + \sqrt{3}$

**20** If $\alpha, \beta \in R$ are such that $1 - 2i$ (here $i^2 = -1$) is a root of $z^2 + \alpha z + \beta = 0$, then $(\alpha - \beta)$ is equal to **[2021, 25 Feb Shift-II]**

(a) 3 (b) $-3$ (c) 7 (d) $-7$

**21** The integer '$k$', for which the inequality $x^2 - 2(3k - 1)x + 8k^2 - 7 > 0$ is valid for every $x$ in $R$, is **[2021, 25 Feb Shift-I]**

(a) 3 (b) 2 (c) 0 (d) 4

**22** If $\alpha$ and $\beta$ are the roots of the equation, $7x^2 - 3x - 2 = 0$, then the value of $\frac{\alpha}{1-\alpha^2} + \frac{\beta}{1-\beta^2}$ is equal to **[2020, 5 Sep Shift-II]**

(a) 27/32 (b) 1/24 (c) 3/8 (d) 27/16

**23** Let $[t]$ denote the greatest integer $\leq t$. Then, the equation in $x, [x]^2 + 2[x + 2] - 7 = 0$ has **[2020, 4 Sep Shift-I]**

(a) infinitely many solutions
(b) exactly four integral solutions
(c) no integral solution
(d) exactly two solutions

**24** If the equation $x^2 + bx + 45 = 0$ $(b \in R)$ has conjugate complex roots and they satisfy $|z + 1| = 2\sqrt{10}$, then **[2020, 8 Jan Shift-I]**

(a) $b^2 + b = 72$ (b) $b^2 + b = 12$
(c) $b^2 - b = 30$ (d) $b^2 - b = 42$

**25** Let $\alpha$ and $\beta$ be two real roots of the equation $(k + 1)\tan^2 x - \sqrt{2} \cdot \lambda \tan x = (1 - k)$, where $k(\neq -1)$ and $\lambda$ are real numbers. If $\tan^2(\alpha + \beta) = 50$, then a value of $\lambda$ is **[2020, 7 Jan Shift-I]**

(a) $5\sqrt{2}$ (b) 10 (c) $10\sqrt{2}$ (d) 5

**26** The sum of the solutions of the equation $|\sqrt{x} - 2| + \sqrt{x}(\sqrt{x} - 4) + 2 = 0$ $(x > 0)$ is equal to **[2019, 8 April Shift-I]**

(a) 9 (b) 12
(c) 4 (d) 10

**27** If an $\angle A$ of a $\Delta ABC$ satisfies $5\cos A + 3 = 0$, then the roots of the quadratic equation $9x^2 + 27x + 20 = 0$ are **[JEE Main 2018]**

(a) $\sin A, \sec A$ (b) $\sec A, \tan A$
(c) $\tan A, \cos A$ (d) $\sec A, \cot A$

**28** Let $p(x)$ be a quadratic polynomial such that $p(0) = 1$. If $p(x)$ leaves remainder 4 when divided by $x - 1$ and it leaves remainder 6 when divided by $x + 1$; then **[JEE Main 2017]**

(a) $p(2) = 11$ (b) $p(2) = 19$ (c) $p(-2) = 19$ (d) $p(-2) = 11$

**29** If $2 + 3i$ is one of the roots of the equation $2x^3 - 9x^2 + kx - 13 = 0, k \in R$, then the real root of this equation **[JEE Main 2015]**

(a) exists and is equal to $-\frac{1}{2}$ (b) exists and is equal to $\frac{1}{2}$
(c) exists and is equal to $\frac{1}{2}$ (d) does not exist

**30** The equation $\sqrt{3x^2 + x + 5} = x - 3$, where $x$ is real, has **[JEE Main 2014]**

(a) no solution (b) exactly one solution
(c) exactly two solution (d) exactly four solution

**31** If $\alpha$ and $\beta$ are roots of the equation $x^2 + px + \frac{3p}{4} = 0$, such that $|\alpha - \beta| = \sqrt{10}$, then $p$ belongs to the set **[JEE Main 2013]**

(a) $\{2, -5\}$ (b) $\{-3, 2\}$ (c) $\{-2, 5\}$ (d) $\{3, -5\}$

**32** The least integral value $\alpha$ of $x$ such that $\frac{x-5}{x^2+5x-14} > 0$, satisfies **[JEE Main 2013]**

(a) $\alpha^2 + 3\alpha - 4 = 0$ (b) $\alpha^2 - 5\alpha + 4 = 0$
(c) $\alpha^2 - 7\alpha + 6 = 0$ (d) $\alpha^2 + 5\alpha - 6 = 0$

**33** The values of '$a$' for which one root of the equation $x^2 - (a+1)x + a^2 + a - 8 = 0$ exceeds 2 and the other is lesser than 2, are given by **[JEE Main 2013, AIEEE 2006(S)]**

(a) $3 < a < 10$ (b) $a \geq 10$
(c) $-2 < a < 3$ (d) $a \leq -2$

**34** Let $p, q, r \in R$ and $r > p > 0$. If the quadratic equation $px^2 + qx + r = 0$ has two complex roots $\alpha$ and $\beta$, then $|\alpha| + |\beta|$ is **[AIEEE 2012]**

(a) equal to 1 (b) less than 2 but not equal to 1
(c) greater than 2 (d) equal to 2

**35** Sachin and Rahul attempted to solve a quadratic equation. Sachin made a mistake in writing down the constant term and ended up in roots (4, 3). Rahul made a mistake in writing down coefficient of $x$ to get roots (3, 2). The correct roots of equation are **[AIEEE 2011]**

(a) $-4, -3$ (b) 6, 1 (c) 4, 3 (d) $-6, -1$

**36** If $\alpha$ and $\beta$ are roots of equation $x^2 - x + 1 = 0$, then $\alpha^{2009} + \beta^{2009}$ is equal to **[AIEEE 2010]**

(a) $-2$ (b) $-1$ (c) 1 (d) 2

**37** The quadratic equations $x^2 - 6x + a = 0$ and $x^2 - cx + 6 = 0$ have one root in common. The other roots of the first and second equations are integers in the ratio 4 : 3. Then, the common root is **[AIEEE 2008]**

(a) 2 (b) 1 (c) 4 (d) 3

**38** If the roots of the quadratic equation $x^2 + px + q = 0$ are $\tan 30°$ and $\tan 15°$, respectively, then the value of $2 + q - p$ is **[AIEEE 2006, 2005(S)]**

(a) 2 (b) 3 (c) 0 (d) 1

**39** The value of $a$ for which the sum of the squares of the roots of the equation $x^2 - (a-2)x - a - 1 = 0$ assume the least value is **[AIEEE 2005]**

(a) 2 (b) 3 (c) 0 (d) 1

**40** Let two numbers have arithmetic mean 9 and geometric mean 4. Then, these numbers are the roots of the quadratic equation **[AIEEE 2004]**

(a) $x^2 + 18x + 16 = 0$ (b) $x^2 - 18x + 16 = 0$
(c) $x^2 + 18x - 16 = 0$ (d) $x^2 - 18x - 16 = 0$

**41** The number of real roots of $3^{2x^2 - 7x + 7} = 9$ is **[AIEEE 2002]**

(a) 0 (b) 2 (c) 1 (d) 4

## Level 2

**42** Let $\alpha$ and $\beta$ be the roots of $x^2 + \sqrt{3}x - 16 = 0$ and $\gamma$ and $\delta$ be the roots of $x^2 + 3x - 1 = 0$. If $P_n = \alpha^n + \beta^n$ and $Q_n = \gamma^n + \delta^n$, then $\frac{P_{25} + \sqrt{3}P_{24}}{2P_{23}} + \frac{Q_{25} - Q_{23}}{Q_{24}}$ is equal to **[2025, 3 April shift-I]**

(a) 7 (b) 5 (c) 4 (d) 3

**43** The number of real roots of the equation $x|x-2| + 3|x-3| + 1 = 0$ is **[2025, 7 April shift-II]**

(a) 4 (b) 3 (c) 1 (d) 2

**44** The sum of the squares of the roots of $|x-2|^2 + |x-2| - 2 = 0$ and the squares of the roots of $x^2 - 2|x-3| - 5 = 0$, is **[2025, 8 April shift-II]**

(a) 26 (b) 36 (c) 24 (d) 30

**45** If the set of all $a \in R$ for which the equation $2x^2 + (a-5)x + 15 = 3a$ has no real roots is the interval $(\alpha, \beta)$ and $X = (x \in Z : \alpha < x < \beta)$, then $\sum_{x \in X} x^2$ is equal to **[2025, 29 Jan Shift-II]**

(a) 2109 (b) 2129
(c) 2119 (d) 2139

**46** If $\alpha + i\beta$ and $\gamma + i\delta$ are the roots of $x^2 - (3-2i)x - (2i-2) = 0$, $i = \sqrt{-1}$, then $\alpha\gamma + \beta\delta$ is equal to **[2025, 28 Jan Shift-II]**

(a) 6 (b) $-6$
(c) $-2$ (d) 2

**47** The sum of the squares of all the roots of the equation $x^2 + |2x-3| - 4 = 0$, is **[2025, 28 Jan Shift-I]**

(a) $6(3-\sqrt{2})$ (b) $3(2-\sqrt{2})$
(c) $3(3-\sqrt{2})$ (d) $6(2-\sqrt{2})$

**48** If $\alpha$ and $\beta$ are the roots of the equation $2z^2 - 3z - 2i = 0$, where $i = \sqrt{-1}$, then $16 \cdot \text{Re}\left(\frac{\alpha^{19} + \beta^{19} + \alpha^{11} + \beta^{11}}{\alpha^{15} + \beta^{15}}\right) \cdot \text{Im}\left(\frac{\alpha^{19} + \beta^{19} + \alpha^{11} + \beta^{11}}{\alpha^{15} + \beta^{15}}\right)$ is equal to **[2025, 24 Jan Shift-I]**

(a) 409 (b) 441
(c) 312 (d) 398

**49** The product of all the rational roots of the equation $(x^2 - 9x + 11)^2 - (x-4)(x-5) = 3$ is equal to **[2025, 24 Jan Shift-I]**

(a) 28 (b) 14 (c) 7 (d) 21

**50** Let $\alpha_\theta$ and $\beta_\theta$ be the distinct roots of $2x^2+(\cos\theta)x-1=0, \theta \in (0, 2\pi)$. If $m$ and $M$ are the minimum and the maximum values of $\alpha_\theta^4+\beta_\theta^4$, then $16(M+m)$ equals **[2025, 22 Jan Shift-II]**

(a) 24 (b) 25 (c) 17 (d) 27

**51** Let $\alpha, \beta : \alpha > \beta$ be the roots of the equation $x^2-\sqrt{2}x-\sqrt{3}=0$. Let $P_n=\alpha^n-\beta^n, n \in N$. Then, $(11\sqrt{3}-10\sqrt{2})P_{10}+(11\sqrt{2}+10)P_{11}-11P_{12}$ is equal to **[2024, 9 April Shift-II]**

(a) $11\sqrt{3}\,P_9$ (b) $10\sqrt{3}P_9$

(c) $11\sqrt{2}\,P_9$ (d) $10\sqrt{2}\,P_9$

**52** Let $\alpha, \beta$ be the roots of the equation $x^2+2\sqrt{2}x-1=0$. The quadratic equation, whose roots are $\alpha^4+\beta^4$ and $\frac{1}{10}(\alpha^6+\beta^6)$, is **[2024, 9 April Shift-I]**

(a) $x^2-195x+9506=0$ (b) $x^2-190x+9466=0$

(c) $x^2-180x+9506=0$ (d) $x^2-195x+9466=0$

**53** Let $\alpha, \beta, \gamma$ be the three roots of the equation $x^3+bx+c=0$. If $\beta\gamma=1=-\alpha$, then $b^3+2c^3-3\alpha^3-6\beta^3-8\gamma^3$ is equal to **[2023, 8 April Shift-I]**

(a) 21 (b) 19

(c) $\frac{169}{8}$ (d) $\frac{155}{8}$

**54** The number of real solutions of the equation $3\left(x^2+\frac{1}{x^2}\right)-2\left(x+\frac{1}{x}\right)+5=0$, is **[2023, 24 Jan Shift-I]**

(a) 2 (b) 0 (c) 4 (d) 3

**55** The sum of all the real roots of the equation $(e^{2x}-4)(6e^{2x}-5e^x+1)=0$ is **[2022, 24 June Shift-II]**

(a) $\log_e 3$ (b) $-\log_e 3$ (c) $\log_e 6$ (d) $-\log_e 6$

**56** Let $\alpha, \beta$ be two roots of the equation $x^2+(20)^{1/4}x+(5)^{1/2}=0$. Then, $\alpha^8+\beta^8$ is equal to **[2021, 27 July Shift-I]**

(a) 10 (b) 100 (c) 50 (d) 160

**57** The number of real roots of the equation $e^{6x}-e^{4x}-2e^{3x}-12e^{2x}+e^x+1=0$ is **[2021, 25 July Shift-I]**

(a) 2 (b) 4 (c) 6 (d) 1

**58** If $\alpha$ and $\beta$ are the distinct roots of the equation $x^2+(3)^{1/4}x+3^{1/2}=0$, then the value of $\alpha^{96}(\alpha^{12}-1)+\beta^{96}(\beta^{12}-1)$ is equal to **[2021, 20 July Shift-I]**

(a) $56\times3^{25}$ (b) $56\times3^{24}$ (c) $52\times3^{24}$ (d) $28\times3^{25}$

**59** Let $\alpha$ and $\beta$ be the roots of $x^2-6x-2=0$. If $a_n=\alpha^n-\beta^n$ for $n \geq 1$, then the value of $\frac{a_{10}-2a_8}{3a_9}$ is **[2021, 25 Feb Shift-II]**

(a) 4 (b) 3 (c) 2 (d) 1

**60** Let $p$ and $q$ be two positive numbers, such that $p+q=2$ and $p^4+q^4=272$. Then, $p$ and $q$ are roots of the equation **[2021, 24 Feb Shift-I]**

(a) $x^2-2x+136=0$ (b) $x^2-2x+16=0$

(c) $x^2-2x+8=0$ (d) $x^2-2x+2=0$

**61** Suppose, $f(x)$ is a polynomial of degree four, having critical points at $-1, 0, 1$. If $T=\{x \in R\}\ f(x)=f(0)$, then the sum of squares of all the elements of $T$ is **[2020, 3 Sep Shift-II]**

(a) 2 (b) 4 (c) 8 (d) 6

**62** The set of all real values of $\lambda$ for which the quadratic equations, $(\lambda^2+1)x^2-4\lambda x+2=0$ always have exactly one root in the interval (0, 1) is **[2020, 3 Sep Shift-II]**

(a) (0, 2) (b) (−3, −1)

(c) (2, 4] (d) (1, 3]

**63** Let $\alpha$ and $\beta$ be the roots of the equation $x^2-x-1=0$. If $p_k=(\alpha)^k+(\beta)^k, k \geq 1$, then which one of the following statements is not true? **[2020, 7 Jan Shift-II]**

(a) $p_3=p_5-p_4$

(b) $(p_1+p_2+p_3+p_4+p_5)=26$

(c) $p_5=p_2 \cdot p_3$

(d) $p_5=11$

**64** The number of real roots of the equation $5+|2^x-1|=2^x(2^x-2)$ is **[2019, 10 April Shift-II]**

(a) 1 (b) 3 (c) 4 (d) 2

**65** If $\alpha$ and $\beta$ are the roots of the quadratic equation, $x^2+x\sin\theta-2\sin\theta=0, \theta \in \left(0, \frac{\pi}{2}\right)$, then $\frac{\alpha^{12}+\beta^{12}}{(\alpha^{-12}+\beta^{-12})(\alpha-\beta)^{24}}$ is equal to **[2019, 10 April Shift-I]**

(a) $\frac{2^{12}}{(\sin\theta+8)^{12}}$ (b) $\frac{2^6}{(\sin\theta+8)^{12}}$

(c) $\frac{2^{12}}{(\sin\theta-4)^{12}}$ (d) $\frac{2^{12}}{(\sin\theta-8)^6}$

**66** The number of integral values of $m$ for which equation $(1+m^2)x^2-2(1+3m)x+(1+8m)=0$, has no real root is **[2019, 8 April Shift-I]**

(a) 3 (b) infinitely many

(c) 1 (d) 2

**67** If $\alpha$ and $\beta$ are the roots of the equation $x^2 - 2x + 2 = 0$, then the least value of $n$ for which $(\alpha/\beta)^n = 1$ is **[2019, 8 April Shift-I]**

(a) 2 (b) 5 (c) 4 (d) 3

**68** Let $\alpha$ and $\beta$ be the roots of the quadratic equation $x^2 \sin\theta - x(\sin\theta\cos\theta + 1) + \cos\theta = 0\ (0 < \theta < 45°)$ and $\alpha < \beta$. Then, $\sum_{n=0}^{\infty}\left(\alpha^n + \frac{(-1)^n}{\beta^n}\right)$ is equal to **[2019, 11 Jan Shift-II]**

(a) $\frac{1}{1-\cos\theta} - \frac{1}{1+\sin\theta}$ (b) $\frac{1}{1-\cos\theta} + \frac{1}{1+\sin\theta}$

(c) $\frac{1}{1+\cos\theta} - \frac{1}{1-\sin\theta}$ (d) $\frac{1}{1+\cos\theta} + \frac{1}{1-\sin\theta}$

**69** Let $p, q$ and $r$ be real numbers ($p \neq q, r \neq 0$), such that the roots of the equation $\frac{1}{x+p} + \frac{1}{x+q} = \frac{1}{r}$ are equal in magnitude but opposite in sign, then the sum of squares of these roots is equal to **[JEE Main 2018]**

(a) $p^2 + q^2 + r^2$ (b) $p^2 + q^2$ (c) $2(p^2 + q^2)$ (d) $\frac{p^2 + q^2}{2}$

**70** For a positive integer $n$, if the quadratic equation, $x(x+1) + (x+1)(x+2) + \ldots + (x+\overline{n-1})(x+n) = 10n$ has two consecutive integral solutions, then $n$ is equal to **[JEE Main 2017]**

(a) 12 (b) 9 (c) 10 (d) 11

**71** If the two roots of the equation, $(a-1)(x^4 + x^2 + 1) + (a+1)(x^2 + x + 1)^2 = 0$ are real and distinct, then the set of all values of '$a$' is **[JEE Main 2015]**

(a) $\left(0, \frac{1}{2}\right)$ (b) $\left(-\frac{1}{2}, 0\right) \cup \left(0, \frac{1}{2}\right)$

(c) $\left(-\frac{1}{2}, 0\right)$ (d) $(-\infty, -2) \cup (2, \infty)$

**72** If $\frac{1}{\sqrt{\alpha}}$ and $\frac{1}{\sqrt{\beta}}$ are the roots of the equation, $ax^2 + bx + 1 = 0\ (a \neq 0, a, b \in R)$, then the equation, $x(x + b^3) + (a^3 - 3abx) = 0$ as roots **[JEE Main 2014]**

(a) $\alpha^{\frac{3}{2}}$ and $\beta^{\frac{3}{2}}$ (b) $\alpha\beta^{\frac{1}{2}}$ and $\alpha^{\frac{1}{2}}\beta$

(c) $\sqrt{\alpha\beta}$ and $\alpha\beta$ (d) $\alpha^{-\frac{3}{2}}$ and $\beta^{-\frac{3}{2}}$

**73** If $p$ and $q$ are non-zero real numbers and $\alpha^3 + \beta^3 = -p$, $\alpha\beta = q$, then a quadratic equation whose roots are $\frac{\alpha^2}{\beta}, \frac{\beta^2}{\alpha}$ is **[JEE Main 2013]**

(a) $px^2 - qx + p^2 = 0$ (b) $qx^2 + px + q^2 = 0$

(c) $px^2 + qx + p^2 = 0$ (d) $qx^2 - px + q^2 = 0$

**74** If the sum of the square of the roots of the equation $x^2 - (\sin\alpha - 2)x - (1 + \sin\alpha) = 0$ is least, then $\alpha$ is equal to **[AIEEE 2012]**

(a) $\pi/6$ (b) $\pi/4$ (c) $\pi/3$ (d) $\pi/2$

**75** If the roots of the equation $bx^2 + cx + a = 0$ is imaginary, then for all real values of $x$, the expression $3b^2x^2 + 6bcx + 2c^2$ is **[AIEEE 2009]**

(a) greater than $4ab$ (b) less than $4ab$

(c) greater than $-4ab$ (d) less than $-4ab$

**76** If both the roots of the quadratic equation $x^2 - 2kx + k^2 + k - 5 = 0$ are less than 5, then $k$ lies in the interval **[AIEEE 2005]**

(a) $[4, 5]$ (b) $(-\infty, 4)$ (c) $(6, \infty)$ (d) $(5, 6]$

**77** If the equation $a_n x^n + a_{n-1}x^{n-1} + \ldots + a_1 x = 0$, $a_1 \neq 0, n \geq 2$, has a positive root $x = \alpha$, then the equation $na_n x^{n-1} + (n-1)a_{n-1}x^{n-2} + \ldots + a_1 = 0$ has a positive root, which is **[AIEEE 2005]**

(a) equal to $\alpha$ (b) greater than or equal to $\alpha$

(c) smaller than $\alpha$ (d) greater than $\alpha$

**78** If one root of the equation $x^2 + px + 12 = 0$ is 4, while the equation $x^2 + px + q = 0$ has equal roots, then the value of $q$ is **[AIEEE 2004]**

(a) 49/4 (b) 12 (c) 3 (d) 4

**79** If the sum of the roots of the quadratic equation $ax^2 + bx + c = 0$ is equal to the sum of the squares of their reciprocals, then $\frac{a}{c}, \frac{b}{a}$ and $\frac{c}{b}$ are in **[AIEEE 2003]**

(a) arithmetic progression
(b) geometric progression
(c) harmonic progression
(d) arithmetic-geometric progression

**80** The value of $a$ for which one root of the quadratic equation $(a^2 - 5a + 3)x^2 + (3a - 1)x + 2 = 0$ is twice as large as the other, is **[AIEEE 2003]**

(a) 2/3 (b) −2/3 (c) 1/3 (d) −1/3

**81** If $\alpha \neq \beta$ and $\alpha^2 = 5\alpha - 3, \beta^2 = 5\beta - 3$, then the equation having $\alpha/\beta$ and $\beta/\alpha$ as its roots, is **[AIEEE 2002]**

(a) $3x^2 + 19x + 3 = 0$ (b) $3x^2 - 19x + 3 = 0$

(c) $3x^2 - 19x - 3 = 0$ (d) $x^2 - 16x + 1 = 0$

## Level 3

**82** Let $\alpha, \beta$ be the distinct roots of the equation $x^2 - (t^2 - 5t + 6)x + 1 = 0, t \in R$ and $a_n = \alpha^n + \beta^n$. Then, the minimum value of $\frac{a_{2023} + a_{2025}}{a_{2024}}$ is **[2024, 6 April Shift-I]**

(a) $\frac{-1}{4}$ (b) $\frac{1}{2}$ (c) $\frac{-1}{2}$ (d) $\frac{1}{4}$

**83** Let α and β be the roots of the equation $px^2 + qx - r = 0$, where $p \neq 0$. If $p$, $q$ and $r$ are the consecutive terms of a non-constant GP and $\frac{1}{\alpha} + \frac{1}{\beta} = \frac{3}{4}$, then the value of $(\alpha - \beta)^2$ is *[2024, 1 Feb Shift-II]*

(a) 9 (b) $\frac{20}{3}$ (c) 8 (d) $\frac{80}{9}$

**84** For $0 < c < b < a$, let $(a + b - 2c)x^2 + (b + c - 2a)x + (c + a - 2b) = 0$ and $\alpha \neq 1$ be one of its root. Then, among the two statements. *[2024, 31 Jan Shift-I]*

I. If $\alpha \in (-1, 0)$, then $b$ cannot be the geometric mean of $a$ and $c$.

II. If $\alpha \in (0, 1)$, then $b$ may be the geometric mean of $a$ and $c$.

(a) Neither I nor II is true (b) Only II is true
(c) Only I is true (d) Both I and II are true

**85** If α and β are the roots of the equation $x^2 - x - 1 = 0$ and $S_n = 2023\alpha^n + 2024\beta^n$, then *[2024, 27 Jan Shift-II]*

(a) $2S_{11} = S_{12} + S_{10}$ (b) $S_{12} = S_{11} + S_{10}$
(c) $S_{11} = S_{10} + S_{12}$ (d) $2S_{12} = S_{11} + S_{10}$

**86** The number of real roots of the equation $x|x| - 5|x + 2| + 6 = 0$, is *[2023, 15 April Shift-I]*

(a) 4 (b) 3 (c) 6 (d) 5

**87** Let α, β be the roots of the equation $x^2 - \sqrt{2}x + 2 = 0$. Then, $\alpha^{14} + \beta^{14}$ is equal to *[2023, 13 April Shift-II]*

(a) $-128$ (b) $-128\sqrt{2}$ (c) $-64\sqrt{2}$ (d) $-64$

**88** The sum of all the roots of the equation $|x^2 - 8x + 15| - 2x + 7 = 0$ is *[2023, 6 April Shift-I]*

(a) $9 + \sqrt{3}$ (b) $11 - \sqrt{3}$ (c) $11 + \sqrt{3}$ (d) $9 - \sqrt{3}$

**89** Let α, β be the roots of the quadratic equation $x^2 + \sqrt{6}x + 3 = 0$. Then, $\frac{\alpha^{23} + \beta^{23} + \alpha^{14} + \beta^{14}}{\alpha^{15} + \beta^{15} + \alpha^{10} + \beta^{10}}$ is equal to *[2023, 4 April Shift-I]*

(a) 81 (b) 72 (c) 9 (d) 729

**90** Let $S = \{x : x \in R \text{ and } (\sqrt{3} + \sqrt{2})^{x^2 - 4} + (\sqrt{3} - \sqrt{2})^{x^2 - 4} = 10\}$. Then, $n(S)$ is equal to *[2023 1 Feb .Shift-I]*

(a) 4 (b) 2 (c) 0 (d) 6

**91** The equation $e^{4x} + 8e^{3x} + 13e^{2x} - 8e^x + 1 = 0$, $x \in R$ has *[2023, 31 Jan Shift-II]*

(a) two solutions and only one of them is negative
(b) two solutions and both are negative
(c) no solution
(d) four solutions and two of which are negative

**92** The number of real roots of the equation $\sqrt{x^2 - 4x + 3} + \sqrt{x^2 - 9} = \sqrt{4x^2 - 14x + 6}$, is *[2023, 31 Jan Shift-I]*

(a) 3 (b) 2 (c) 0 (d) 1

**93** If the solution of the equation $\log_{\cos x} \cot x + 4\log_{\sin x} \tan x = 1$, $x \in \left(0, \frac{\pi}{2}\right)$, is $\sin^{-1}\left(\frac{\alpha + \sqrt{\beta}}{2}\right)$, where α, β are integers, then α + β is equal to *[2023, 30 Jan Shift-I]*

(a) 3 (b) 5 (c) 6 (d) 4

**94** Let $\lambda \neq 0$ be a real number. Let α, β be the roots of the equation $14x^2 - 31x + 3\lambda = 0$ and α, γ be the roots of the equation $35x^2 - 53x + 4\lambda = 0$. Then, $\frac{3\alpha}{\beta}$ and $\frac{4\alpha}{\gamma}$ are the roots of the equation *[2023, 29 Jan Shift-I]*

(a) $7x^2 - 245x + 250 = 0$
(b) $7x^2 + 245x - 250 = 0$
(c) $49x^2 - 245x + 250 = 0$
(d) $49x^2 + 245x + 250 = 0$

**95** Let α and β be the roots of the equation $x^2 - \sqrt{2}x + \sqrt{6} = 0$ and $\frac{1}{\alpha^2} + 1, \frac{1}{\beta^2} + 1$ be the roots of the equation $x^2 + ax + b = 0$. Then, the roots of the equation $x^2 - (a + b - 2)x + (a + b + 2) = 0$ are *[2022, 28 July Shift-II]*

(a) non-real complex number
(b) real and both negative
(c) real and both positive
(d) real and exactly one of them is positive

**96** If α, β, γ, δ are the roots of the equation $x^4 + x^3 + x^2 + x + 1 = 0$, then $\alpha^{2021} + \beta^{2021} + \gamma^{2021} + \delta^{2021}$ is equal to *[2022, 25 July Shift-I]*

(a) $-4$ (b) $-1$ (c) 1 (d) 4

**97** If the sum of the squares of the reciprocals of the roots α and β of the equation $3x^2 + \lambda x - 1 = 0$ is 15, then $6(\alpha^3 + \beta^3)^2$ is equal to *[2022, 24 June shift-I]*

(a) 18 (b) 24 (c) 36 (d) 96

**98** The numbers of pairs $(a, b)$ of real numbers, such that whenever α is a root of the equation $x^2 + ax + b = 0$, $\alpha^2 - 2$ is also a root of this equation, is *[2021, 01 Sep Shift-II]*

(a) 6 (b) 2 (c) 4 (d) 8

**99** The sum of the roots of the equation $x+1-2\log_2(3+2^x)+2\log_4(10-2^{-x})=0$ is **[2021, 31 Aug Shift-II]**

(a) $\log_2 14$ (b) $\log_2 11$
(c) $\log_2 12$ (d) $\log_2 13$

**100** The set of all values of $k>-1$, for which the equation $(3x^2+4x+3)^2-(k+1)(3x^2+4x+3)(3x^2+4x+2)+k(3x^2+4x+2)^2=0$ has real roots, is **[2021, 27 Aug Shift-II]**

(a) $\left(1, \frac{5}{2}\right]$ (b) $[2, 3)$ (c) $\left[-\frac{1}{2}, 1\right)$ (d) $\left(\frac{1}{2}, \frac{3}{2}\right]-\{1\}$

**101** Let $P(x)=x^2+bx+c$ be a quadratic polynomial with real coefficients, such that $\int_0^1 P(x)dx=1$ and $P(x)$ leaves remainder 5 when it is divided by $(x-2)$. Then, the value of $9(b+c)$ is equal to **[2021, 16 March Shift-II]**

(a) 9 (b) 15 (c) 7 (d) 11

**102** The number of elements in the set $\{x \in R: (|x|-3)|x+4|=6\}$ is equal to **[2021, 16 March Shift-I]**

(a) 3 (b) 2 (c) 4 (d) 1

**103** Let $a, b, c$ be in an arithmetic progression. Let the centroid of the triangle with vertices $(a, c)$, $(2, b)$ and $(a, b)$ be $\left(\frac{10}{3}, \frac{7}{3}\right)$. If $\alpha, \beta$ are the roots of the equation $ax^2+bx+1=0$, then the value of $\alpha^2+\beta^2-\alpha\beta$ is **[2021, 24 Feb Shift-II]**

(a) $\frac{71}{256}$ (b) $\frac{69}{256}$
(c) $-\frac{69}{256}$ (d) $-\frac{71}{256}$

**104** If $e^{(\cos^2 x+\cos^4 x+\cos^6 x+\ldots\infty)\log_e 2}$ satisfies the equation $t^2-9t+8=0$, then the value of $\frac{2\sin x}{\sin x+\sqrt{3}\cos x}, \left(0<x<\frac{\pi}{2}\right)$ is **[2021, 24 Feb Shift-I]**

(a) $\frac{1}{2}$ (b) $\sqrt{3}$ (c) $\frac{3}{2}$ (d) $2\sqrt{3}$

**105** Let $u=\frac{2z+i}{z-ki}$, $z=x+iy$ and $k>0$. If the curve represented by $\text{Re}(u)+\text{Im}(u)=1$ intersects the $Y$-axis at the points $P$ and $Q$ where $PQ=5$, then the value of $k$ is **[2020, 04 Sep Shift-II]**

(a) 3/2 (b) 1/2 (c) 4 (d) 2

**106** Let $a, b \in R$, $a \neq 0$ be such that the equation, $ax^2-2bx+5=0$ has a repeated root $\alpha$, which is also a root of the equation, $x^2-2bx-10=0$. If $\beta$ is the other root of this equation, then $\alpha^2+\beta^2$ is equal to **[2020, 9 Jan Shift-II]**

(a) 26 (b) 24 (c) 28 (d) 25

**107** The number of real roots of the equation, $e^{4x}+e^{3x}-4e^{2x}+e^x+1=0$ is **[2020, 9 Jan Shift-I]**

(a) 3 (b) 4 (c) 1 (d) 2

**108** Let $S$ be the set of all real roots of the equation, $3^x(3^x-1)+2=|3^x-1|+|3^x-2|$. Then, $S$ **[2020, 8 Jan Shift-II]**

(a) is a singleton.
(b) is an empty set.
(c) contains at least four elements.
(d) contains exactly two elements.

**109** If $\alpha$ and $\beta$ are the roots of the equation $375x^2-25x-2=0$, then $\lim_{n\to\infty}\sum_{r=1}^{n}\alpha^r+\lim_{n\to\infty}\sum_{r=1}^{n}\beta^r$ is equal to **[2019, 12 April Shift-I]**

(a) $\frac{21}{346}$ (b) $\frac{29}{358}$ (c) $\frac{1}{12}$ (d) $\frac{7}{116}$

**110** If both the roots of the quadratic equation $x^2-mx+4=0$ are real and distinct and they lie in the interval [1, 5] then $m$ lies in the interval **[2019, 9 Jan Shift-II]**

(a) (4, 5) (b) (−5, −4) (c) (5, 6) (d) (3, 4)

**111** The number of all possible positive integral values of $\alpha$ for which the roots of the quadratic equation, $6x^2-11x+\alpha=0$ are rational numbers is **[2019, 9 Jan Shift-II]**

(a) 5 (b) 2 (c) 4 (d) 3

**112** The sum of all real values of $x$ satisfying the equation $(x^2-5x+5)^{x^2+4x-60}=1$ is **[JEE Main 2016]**

(a) 3 (b) −4 (c) 6 (d) 5

**113** Let $\alpha$ and $\beta$ be the roots of equation $x^2-6x-2=0$. If $a_n=\alpha^n-\beta^n$, for $n \geq 1$, then the value of $a_{10}-2a_8/2a_9$ is equal to **[JEE Main 2015]**

(a) 6 (b) −6 (c) 3 (d) −3

**114** Let $\alpha$ and $\beta$ be the roots of equation $px^2+qx+r=0$, $p \neq 0$. If $p$, $q$ and $r$ are in AP and $\frac{1}{\alpha}+\frac{1}{\beta}=4$, then the value of $|\alpha-\beta|$ is **[JEE Main 2014]**

(a) $\frac{\sqrt{61}}{9}$ (b) $\frac{2\sqrt{17}}{9}$ (c) $\frac{\sqrt{34}}{9}$ (d) $\frac{2\sqrt{13}}{9}$

**115** If $\alpha$ and $\beta$ are roots of the equation, $x^2-4\sqrt{2}kx+2e^{4\ln k}-1=0$ for some $k$, and $\alpha^2+\beta^2=66$, then $\alpha^3+\beta^3$ is equal to **[JEE Main 2014]**

(a) $248\sqrt{2}$ (b) $280\sqrt{2}$
(c) $-32\sqrt{2}$ (d) $-280\sqrt{2}$

**116** The equation $e^{\sin x} - e^{-\sin x} - 4 = 0$ has *[AIEEE 2012]*
(a) infinite number of real roots
(b) no real root
(c) exactly one real root
(d) exactly four real roots

**117** Let for $a \neq a_1 \neq 0$, $f(x) = ax^2 + bx + c$, $g(x) = a_1x^2 + b_1 x + c_1$ and $p(x) = f(x) - g(x)$. If $p(x) = 0$ only for $x = -1$ and $p(-2) = 2$, then the value of $p(2)$ is *[AIEEE 2011]*
(a) 18 (b) 3
(c) 9 (d) 6

**118** All the values of $m$ for which both roots of the equation $x^2 - 2mx + m^2 - 1 = 0$ are greater than $-2$ but less than 4 lie in the interval *[AIEEE 2006]*
(a) $m > 3$ (b) $-1 < m < 3$
(c) $1 < m < 4$ (d) $-2 < m < 0$

## Numerical Value Type Questions

**119** If the set of all $a \in R - [1]$, for which the roots of the equation $(1-a)x^2 + 2(a-3)x + 9 = 0$ are positive is $(-\infty, -\alpha] \cup [\beta, \gamma)$, then $2\alpha + \beta + \gamma$ is equal to ......... . *[2025, 2 April Shift-II]*

**120** If $\alpha$ is a root of the equation $x^2 + x + 1 = 0$ and $\sum_{k=1}^{n}\left(\alpha^k + \frac{1}{\alpha^k}\right)^2 = 20$, then $n$ is equal to ........... . *[2025, 4 April Shift-II]*

**121** Let $\alpha, \beta$ be the roots of the equation $x^2 - ax - b = 0$ with Im $(\alpha) <$ Im $(\beta)$. Let $P_n = \alpha^n - \beta^n$. If $P_3 = -5\sqrt{7}\,i$, $P_4 = -3\sqrt{7}\,i$, $P_5 = 11\sqrt{7}i$ and $P_6 = 45\sqrt{7}i$, then $|\alpha^4 + \beta^4|$ is equal to....... *[2025, 23 Jan Shift-II]*

**122** If the equation $a(b-c)x^2 + b(c-a)x + c(a-b) = 0$ has equal roots, where $a + c = 15$ and $b = \frac{36}{5}$, then $a^2 + c^2$ is equal to ........... . *[2025, 23 Jan Shift-I]*

**123** If the set of all values of $a$, for which the equation $5x^3 - 15x - a = 0$ has three distinct real roots, the interval $(\alpha, \beta)$, then $\beta - 2\alpha$ is equal to ........... . *[2025, 23 Jan Shift-I]*

**124** The number of distinct real roots of the equation $|x+1||x+3| - 4|x+2| + 5 = 0$, is .............. . *[2024, 8 April Shift-II]*

**125** Let $\alpha$ and $\beta$ be roots of $x^2 + \sqrt{2}x - 8 = 0$. If $U_n = \alpha^n + \beta^n$, then $\frac{U_{10} + \sqrt{2}U_9}{2U_8}$ is equal to ...... . *[2024, 6 April Shift-II]*

**126** Let $x_1, x_2, x_3, x_4$ be the solution of the equation $4x^4 + 8x^3 - 17x^2 - 12x + 9 = 0$ and $(4 + x_1^2)(4 + x_2^2)(4 + x_3^2)(4 + x_4^2) = \frac{125}{16}m$. Then, the value of $m$ is .......... *[2024, 6 April Shift-I]*

**127** The number of real solutions of the equation $x|x+5| + 2|x+7| - 2 = 0$ is ...... *[2024, 5 April Shift-II]*

**128** The number of distinct real roots of the equation $|x||x+2| - 5|x+1| - 1 = 0$ is ............ . *[2024, 5 April Shift-I]*

**129** The number of real solutions of the equation $x(x^2 + 3|x| + 5|x-1| + 6|x-2|) = 0$ is ......... . *[2024, 30 Jan Shift-II]*

**130** Let $\alpha, \beta \in N$ be roots of the equation $x^2 - 70x + \lambda = 0$, where $\frac{\lambda}{2}, \frac{\lambda}{3} \notin N$. If $\lambda$ assumes the minimum possible value, then $\frac{(\sqrt{\alpha - 1} + \sqrt{\beta - 1})(\lambda + 35)}{|\alpha - \beta|}$ is equal to ............ . *[2024, 30 Jan Shift-I]*

**131** Let $\alpha, \beta$ be the roots of the equation $x^2 - x + 2 = 0$ with Im $(\alpha) >$ Im $(\beta)$. Then, $\alpha^6 + \alpha^4 + \beta^4 - 5\alpha^2$ is equal to ............... . *[2024, 29 Jan Shift-I]*

**132** If $\alpha$ satisfies the equation $x^2 + x + 1 = 0$ and $(1+\alpha)^7 = A + B\alpha + C\alpha^2$, $A, B, C \geq 0$, then $5(3A - 2B - C)$ is equal to ......... . *[2024, 27 Jan Shift-I]*

**133** The number of points, where the curve $f(x) = e^{8x} - e^{6x} - 3e^{4x} - e^{2x} + 1$, $x \in R$ cuts $X$-axis, is equal to ........... . *[2023, 11 April Shift-II]*

**134** If $a$ and $b$ are the roots of the equation $x^2 - 7x - 1 = 0$, then the value of $\frac{a^{21} + b^{21} + a^{17} + b^{17}}{a^{19} + b^{19}}$ is equal to ........... . *[2023, 11 April Shift-I]*

**135** Let $a, b, c$ be three distinct positive real numbers such that $(2a)^{\log_e a} = (bc)^{\log_e b}$ and $b^{\log_e 2} = a^{\log_e c}$. Then, $6a + 5bc$ is equal to ......... *[2023, 10 April Shift-I]*

**136** Let $m$ and $n$ be the numbers of real roots of the quadratic equations $x^2 - 12x + [x] + 31 = 0$ and $x^2 - 5|x+2| - 4 = 0$ respectively, where $[x]$ denotes the greatest integer $\leq x$. Then, $m^2 + mn + n^2$ is equal to ......... *[2023, 8 April Shift-II]*

**137** If the value of real number $a > 0$, for which $x^2 - 5ax + 1 = 0$ and $x^2 - ax - 5 = 0$ have a common real root is $\frac{3}{\sqrt{2\beta}}$, then $\beta$ is equal to ..........

*[2023, 30 Jan Shift-II]*

**138** Let $\alpha_1, \alpha_2, ...., \alpha_7$ be the roots of the equation $x^7 + 3x^5 - 13x^3 - 15x = 0$ and $|\alpha_1| \geq |\alpha_2| \geq ......$ $\geq |a_7|$. Then, $\alpha_1\alpha_2 - \alpha_3\alpha_4 + \alpha_5\alpha_6$ is equal to ....... .

*[2023, 29 Jan Shift-II]*

**139** Let $a \in R$ and let $\alpha, \beta$ be the roots of the equation $x^2 + 60^{\frac{1}{4}} x + a = 0$. If $\alpha^4 + \beta^4 = -30$, then the product of all possible values of $a$ is........... .

*[2023, 25 Jan Shfit-II]*

**140** Let $\alpha$ and $\beta$ $(\alpha > \beta)$ be the roots of the quadratic equation $x^2 - x - 4 = 0$.

If $P_n = \alpha^n - \beta^n, n \in N$, then $\frac{P_{15}P_{16} - P_{14}P_{16} - P_{15}^2 + P_{14}P_{15}}{P_{13}P_{14}}$ is equal to ........... .

*[2022, 29 July Shift-II]*

**141** If for some $p, q, r \in R$, not all have same sign, one of the roots of the equation $(p^2 + q^2)x^2 - 2q(p + r)x + q^2 + r^2 = 0$ is also a root of the equation $x^2 + 2x - 8 = 0$, then $\frac{q^2 + r^2}{p^2}$ is equal to ........... .

*[2022, 26 July Shift-I]*

**142** The number of real solutions of the equation $e^{4x} + 4e^{3x} - 58e^{2x} + 4e^x + 1 = 0$ is ...... .

*[2022, 28 June Shift-I]*

**143** The sum of the cubes of all the roots of the equation $x^4 - 3x^3 - 2x^2 + 3x + 1 = 0$ is ......... .

*[2022, 26 June Shift-I]*

**144** The number of distinct real roots of the equation $3x^4 + 4x^3 - 12x^2 + 4 = 0$ is ........... .

*[2021, 27 Aug Shift-I]*

**145** The sum of all integral values of $k$ $(k \neq 0)$ for which the equation $\frac{2}{x-1} - \frac{1}{x-2} = \frac{2}{k}$ in $x$ has no real roots, is

*[2021, 26 Aug Shift-I]*

**146** The number of real roots of the equation $e^{4x} - e^{3x} - 4e^{2x} - e^x + 1 = 0$ is equal to ...... .

*[2021, 27 July Shift-II]*

**147** The least positive value of '$a$' for which the equation, $2x^2 + (a - 10)x + 33/2 = 2a$ has real roots is ...... .

*[2020, 8 Jan Shift-I]*

# • Key Index

## Topic 1

| | | | | | | | | | |
|---|---|---|---|---|---|---|---|---|---|
| 1 (a) | 2 (c) | 3 (c) | 4 (b) | 5 (d) | 6 (d) | 7 (d) | 8 (a) | 9 (d) | 10 (b) |
| 11 (a) | 12 (b) | 13 (b) | 14 (b) | 15 (a) | 16 (d) | 17 (c) | 18 (c) | 19 (b) | 20 (a) |
| 21 (c) | 22 (c) | 23 (a) | 24 (b) | 25 (b) | 26 (d) | 27 (c) | 28 (d) | 29 (b) | 30 (d) |
| 31 (c) | 32 (*) | 33 (d) | 34 (c) | 35 (d) | 36 (d) | 37 (a) | 38 (c) | 39 (b) | 40 (b) |
| 41 (d) | 42 (d) | 43 (d) | 44 (b) | 45 (d) | 46 (b) | 47 (a) | 48 (c) | 49 (d) | 50 (*) |
| 51 (d) | 52 (d) | 53 (c) | 54 (d) | 55 (b) | 56 (a) | 57 (c) | 58 (c) | 59 (a) | 60 (b) |
| 61 (3) | 62 (1680) | 63 (9) | 64 (14) | 65 (0) | 66 (2) | 67 (40) | 68 (5) | 69 (6) | 70 (6) |
| 71 (1) | 72 (4) | 73 (10) | | | | | | | |

## Topic 2

| | | | | | | | | | |
|---|---|---|---|---|---|---|---|---|---|
| 1 (b) | 2 (c) | 3 (a) | 4 (c) | 5 (d) | 6 (a) | 7 (a) | 8 (a) | 9 (c) | 10 (c) |
| 11 (b) | 12 (d) | 13 (a) | 14 (a) | 15 (b) | 16 (c) | 17 (c) | 18 (b) | 19 (10) | 20 (49) |
| 21 (6) | 22 (3) | | | | | | | | |

## Topic 3

| | | | | | | | | | |
|---|---|---|---|---|---|---|---|---|---|
| 1 (a) | 2 (d) | 3 (d) | 4 (c) | 5 (a) | 6 (c) | 7 (b) | 8 (c) | 9 (c) | 10 (b) |
| 11 (b) | 12 (b) | 13 (d) | 14 (b) | 15 (c) | 16 (b) | 17 (b) | 18 (b) | 19 (a) | 20 (c) |
| 21 (c) | 22 (b) | 23 (d) | 24 (b) | 25 (b) | 26 (b) | 27 (b) | 28 (d) | 29 (c) | 30 (a) |
| 31 (a) | 32 (c) | 33 (b) | 34 (b) | 35 (c) | 36 (c) | 37 (c) | 38 (b) | 39 (c) | 40 (a) |
| 41 (b) | 42 (d) | 43 (c) | 44 (a) | 45 (22) | 46 (9) | 47 (36) | 48 (20) | 49 (2) | 50 (26) |
| 51 (80) | 52 (6) | 53 (48) | | | | | | | |

## Topic 4

| | | | | | | | | | |
|---|---|---|---|---|---|---|---|---|---|
| 1 (d) | 2 (b) | 3 (c) | 4 (a) | 5 (a) | 6 (c) | 7 (a) | 8 (d) | 9 (d) | 10 (a) |
| 11 (d) | 12 (a) | 13 (d) | 14 (b) | 15 (b) | 16 (d) | 17 (a) | 18 (d) | 19 (a) | 20 (d) |
| 21 (a) | 22 (d) | 23 (a) | 24 (c) | 25 (b) | 26 (d) | 27 (b) | 28 (c) | 29 (b) | 30 (a) |
| 31 (c) | 32 (d) | 33 (c) | 34 (c) | 35 (b) | 36 (c) | 37 (a) | 38 (b) | 39 (d) | 40 (b) |
| 41 (b) | 42 (b) | 43 (c) | 44 (b) | 45 (d) | 46 (d) | 47 (d) | 48 (b) | 49 (b) | 50 (b) |
| 51 (b) | 52 (a) | 53 (b) | 54 (b) | 55 (b) | 56 (c) | 57 (a) | 58 (c) | 59 (c) | 60 (b) |
| 61 (b) | 62 (d) | 63 (c) | 64 (a) | 65 (a) | 66 (b) | 67 (c) | 68 (b) | 69 (b) | 70 (d) |
| 71 (b) | 72 (a) | 73 (b) | 74 (d) | 75 (c) | 76 (b) | 77 (c) | 78 (a) | 79 (c) | 80 (a) |
| 81 (b) | 82 (a) | 83 (d) | 84 (d) | 85 (b) | 86 (b) | 87 (a) | 88 (a) | 89 (a) | 90 (a) |
| 91 (b) | 92 (d) | 93 (d) | 94 (c) | 95 (b) | 96 (b) | 97 (b) | 98 (a) | 99 (b) | 100 (a) |
| 101 (c) | 102 (b) | 103 (d) | 104 (a) | 105 (d) | 106 (d) | 107 (c) | 108 (a) | 109 (c) | 110 (a) |
| 111 (d) | 112 (a) | 113 (c) | 114 (d) | 115 (d) | 116 (b) | 117 (a) | 118 (b) | 119 (7) | 120 (11) |
| 121 (31) | 122 (117) | 123 (30) | 124 (2) | 125 (4) | 126 (221) | 127 (3) | 128 (3) | 129 (1) | 130 (60) |
| 131 (13) | 132 (5) | 133 (2) | 134 (51) | 135 (8) | 136 (9) | 137 (13) | 138 (9) | 139 (45) | 140 (16) |
| 141 (272) | 142 (2) | 143 (36) | 144 (4) | 145 (66) | 146 (2) | 147 (8) | | | |

# Detailed **Explanations**

## Topic 1

**01.** *(a)*

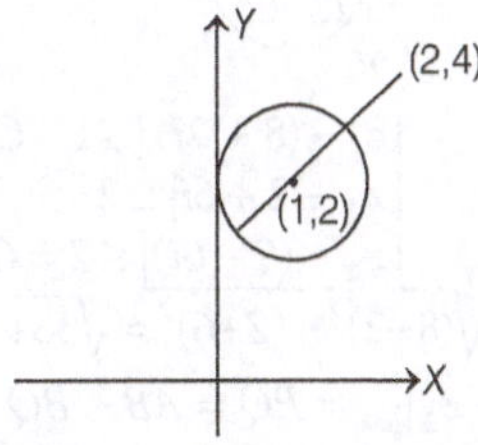


$$\frac{2+k^2 z}{k+\bar{z}} = kz$$

$\Rightarrow 2+k^2 z = k^2 z + kz\cdot\bar{z} \quad (\because z\bar{z} = |z|^2 = 1)$

$\Rightarrow \quad k = 2$

$\therefore \quad k + k^2 i = 2+4i$

$\therefore$ The maximum distance is

$\sqrt{(4-2)^2+(2-1)^2}$ + radius

$=\sqrt{5}+1$

**02.** *(c)* $\text{Re}\left(\frac{z-2i}{z+2i}\right)=0$

$$\Rightarrow \frac{z-2i}{z+2i}+\overline{\left(\frac{z-2i}{z+2i}\right)}=0$$

$$\Rightarrow \frac{(z-2i)(\bar{z}-2i)+(z+2i)(\bar{z}+2i)}{(z+2i)(\bar{z}-2i)}=0$$

$$\Rightarrow z\bar{z}-2i\bar{z}-2iz-4+z\bar{z}+2iz$$
$$+2\bar{z}i-4=0$$

$\Rightarrow \quad 2|z|^2 = 8 \Rightarrow |z|^2 = 4$

$|z| = 2$

$\therefore |z-(6+8i)|_{max} = 10+2 = 12$

**03.** *(c)* $z = \frac{1+i\cos\theta}{1-2i\cos\theta}$

$$=\frac{1+i\cos\theta}{1-2i\cos\theta}\times\frac{1+2i\cos\theta}{1+2i\cos\theta}$$

$$=\frac{1+2i\cos\theta+i\cos\theta-2\cos^2\theta}{1+4\cos^2\theta}$$

$$=\frac{1-2\cos^2\theta}{1+4\cos^2\theta}+i\frac{3\cos\theta}{1+4\cos^2\theta}$$

$\because z$ is purely imaginary

So, $\frac{1-2\cos^2\theta}{1+4\cos^2\theta}=0$

$1-2\cos^2\theta = 0 \qquad [\because 1+4\cos^2\theta \neq 0]$

$\Rightarrow \cos\theta = \pm\frac{1}{\sqrt{2}}$

So, $\theta = -\frac{3\pi}{4}, -\frac{\pi}{4}, \frac{\pi}{4}, \frac{3\pi}{4}, \frac{5\pi}{4}, \frac{7\pi}{4}$

Sum of all values $=\frac{12\pi}{4}=3\pi$

**04.** *(b)* Let $z = x+iy$

Given, $\text{Im}\left(\frac{z+1}{z+2}\right)=\frac{1}{5}$

$$\text{Im}\left(\frac{(x+1)+iy}{(x+2)+iy}\right)=\frac{1}{5}$$

$$\Rightarrow \text{Im}\left(\frac{((x+1)+iy)\times((x+2)-iy)}{((x+2)+iy)\,((x+2)-iy)}\right)=\frac{1}{5}$$

$$\Rightarrow \frac{y(x+2)-y(x+1)}{(x+2)^2+y^2}=\frac{1}{5}$$

and $\quad |z+2|=1$

$\Rightarrow \quad (x+2)^2+y^2=1$

So, $y(x+2)-y(x+1)=\frac{1}{5} \Rightarrow y=\frac{1}{5}$

$\because \quad (x+2)^2+y^2=1$

$\Rightarrow \quad (x+2)^2 = 1-\frac{1}{25}=\frac{24}{25}$

$\Rightarrow \quad x+2=\frac{2\sqrt{6}}{5} \qquad \ldots\text{(i)}$

$|\text{Re}(\overline{z+2})| = |\text{Re}(\bar{z}+2)| = (x+2)$

By using Eq. (i),

$|\text{Re}(\overline{z+2})| = \frac{2\sqrt{6}}{5}$

**05.** *(d)* We have,

$z^2 + i\bar{z} = 0$

$\Rightarrow \quad z^2 = -i\bar{z} \Rightarrow |z^2| = |-iz|$

$\Rightarrow \quad |z^2| = |z|$

$\therefore \quad |z|^2 - |z| = 0$

$\Rightarrow \quad |z|(|z|-1)=0$

$|z| = 0 \quad$ (not comparable)

$\therefore \quad |z| = 1$

So, $\quad |z^2| = 1$

**06.** *(d)* We have, $z=\frac{1}{2}-2i$

and $|z+1| = \alpha z + \beta(1+i)$

$$\Rightarrow \left|\frac{1}{2}-2i+1\right| = \alpha\left(\frac{1}{2}-2i\right)+\beta(1+i)$$

$$\Rightarrow \left|\frac{3}{2}-2i\right| = \frac{\alpha}{2}-2\alpha i+\beta+\beta i$$

$$\Rightarrow \left|\frac{3}{2}-2i\right| = \left(\frac{\alpha}{2}+\beta\right)+(\beta-2\alpha)i$$

$$\Rightarrow \sqrt{\frac{9}{4}+4} = \left(\frac{\alpha}{2}+\beta\right)+(\beta-2\alpha)i$$

$\Rightarrow \beta - 2\alpha = 0$ and $\frac{\alpha}{2}+\beta=\frac{5}{2}$

$\Rightarrow \beta = 2\alpha$ and $\frac{\alpha}{2}+2\alpha=\frac{5}{2} \Rightarrow \alpha = 1$

Also, $\beta = 2$

$\therefore \quad \alpha+\beta = 1+2 = 3$

**07.** *(d)* $z = \frac{1+2i\sin\theta}{1-i\sin\theta}\times\frac{1+i\sin\theta}{1+i\sin\theta}$

$$=\frac{(1-2\sin^2\theta)+i(3\sin\theta)}{1+\sin^2\theta}$$

According to the question,

$\text{Re}(z) = 0$

$\Rightarrow \quad 1-2\sin^2\theta = 0 \Rightarrow \sin\theta = \pm\frac{1}{\sqrt{2}}$

$\because \quad \theta \in (0, 2\pi)$

$\therefore \quad \theta = \frac{\pi}{4}, \frac{3\pi}{4}, \frac{5\pi}{4}, \frac{7\pi}{4}$

$\Rightarrow \quad A = \left\{\frac{\pi}{4}, \frac{3\pi}{4}, \frac{5\pi}{4}, \frac{7\pi}{4}\right\}$

$\therefore \quad$ Sum $= 4\pi$

**08.** *(a)* $z = 2+3i$

$z^5 + (\bar{z})^5 = (2+3i)^5 + (2-3i)^5$

$= (T_1+T_2+T_3+T_4+T_5+T_6)$
$+ (T_1-T_2+T_3-T_4+T_5-T_6)$

$= 2(T_1+T_3+T_5)$

$= 2({}^5C_0 2^5(3i)^0 + {}^5C_2 2^3(3i)^2$
$+ {}^5C_4 2(3i)^4)$

$= 2(32-10\times8\times9+5\times2\times81)$

$= 2(842-720) = 2\times122 = 244$

**09.** *(d)* We have, $\frac{(1+2i)^8\cdot(1-2i)^2}{(3+2i)\cdot(4-6i)}$

$$=\frac{(1+2i)^2(1-2i)^2(1+2i)^6}{26i} \qquad [\because i^2=-1]$$

$$=\frac{25(1+2i)^6}{26i}$$

$$=(-i)\frac{25}{26}(7+24i)(3-4i)$$

$$=\frac{(72-28)25}{26}=\frac{44\times25}{26}=\frac{550}{13}$$

**10.** *(b)* We have, $\log_{\frac{1}{\sqrt{2}}}\left(\frac{|z|+11}{(|z|-1)^2}\right) \le 2$

$\Rightarrow \quad \frac{|z|+11}{(|z|-1)^2} \ge \left(\frac{1}{\sqrt{2}}\right)^2$

$\Rightarrow \quad 2(|z|+11) \ge (|z|-1)^2$

$\Rightarrow \quad (|z|-7)(|z|+3) \le 0 \Rightarrow |z| \le 7$

($\because |z|$ can't be negative)

$\therefore$ Maximum value of $|z| = 7$

**11.** *(a)*

> **key idea** If $z = x + iy$, then $\bar{z} = x - iy$ and $i^2 = -1,\ i^3 = -i$

We have, $\frac{x+iy}{27} = \left(-2 - \frac{1}{3}i\right)^3$

$= \left[\frac{-1}{3}(6+i)\right]^3$

$\Rightarrow \quad \frac{x+iy}{27} = -\frac{1}{27}$

$(216 + 108i + 18i^2 + i^3)$

$= -\frac{1}{27}(198 + 107i)$

On comparing real and imaginary part, we get

$x = -198$ and $y = -107$

$\Rightarrow \quad y - x = -107 + 198 = 91$

**12.** *(b)* Let $z = 1 + i\alpha,\ \alpha \in R$

$z^2 = (1+i\alpha)(1+i\alpha)$

$\Rightarrow \quad x + iy = (1 + 2i\alpha - \alpha^2)$

$[\because z^2 = x + iy]$

On comparing real and imaginary parts, we get

$x = 1 - \alpha^2,\ y = 2\alpha$

Now, consider option (b), which is

$y^2 + 4x - 4 = 0$

LHS

$y^2 + 4x - 4 = (2\alpha)^2 + 4(1-\alpha^2) - 4$

$= 4\alpha^2 + 4 - 4\alpha^2 - 4 = 0$

$=$ RHS

Hence, $y^2 + 4x - 4 = 0$

**13.** *(b)* We have, $\frac{1+z}{1+\bar{z}} = \frac{1+z}{1+\frac{1}{z}} = z$

$|z^2| = 1 = z\bar{z}$

Then, the arg of the complex number $\frac{1+z}{1+\bar{z}}$ is just the argument of $z$ and that is, $\theta$.

**14.** *(b)* $|z| = \left|\left(z - \frac{4}{z}\right) + \frac{4}{z}\right|$

$\Rightarrow \quad |z| \le \left|z - \frac{4}{z}\right| + \frac{4}{|z|}$

$\Rightarrow \quad |z| \le 2 + \frac{4}{|z|}$

$\Rightarrow \quad |z|^2 - 2|z| - 4 \le 0$

$\Rightarrow (|z| - (\sqrt{5}+1))(|z| - (1-\sqrt{5})) \le 0$

$\Rightarrow \quad 1 - \sqrt{5} \le |z| \le \sqrt{5} + 1$

**15.** *(a)* Now, $\left(\frac{1+i}{1-i}\right)^x = \left[\frac{(1+i)(1+i)}{(1-i)(1+i)}\right]^x$

$= \left[\frac{(1+i)^2}{1-i^2}\right]^x = \left[\frac{1-1+2i}{2}\right]^x$

$\Rightarrow \quad \left(\frac{1+i}{1-i}\right)^x = (i)^x = 1$ [given]

$\Rightarrow \quad (i)^x = (i)^{4n}$

where, $n$ is any positive integer.

$\therefore \quad x = 4n$

**16.** *(d)* Given,

$1 + 10\,\text{Re}\left(\frac{2\cos\theta + i\sin\theta}{\cos\theta - 3i\sin\theta}\right) = 0$

$\Rightarrow \quad \text{Re}\left(\frac{2\cos\theta + i\sin\theta}{\cos\theta - 3i\sin\theta}\right) = -\frac{1}{10}$

$\Rightarrow \text{Re}\left[\frac{(2\cos\theta + i\sin\theta)(\cos\theta + 3i\sin\theta)}{\cos^2\theta + 9\sin^2\theta}\right]$

$= -\frac{1}{10}$

$\Rightarrow \quad \frac{2\cos^2\theta - 3\sin^2\theta}{\cos^2\theta + 9\sin^2\theta} = -\frac{1}{10}$

$\Rightarrow$

$20\cos^2\theta - 30\sin^2\theta = -\cos^2\theta - 9\sin^2\theta$

$\Rightarrow \quad 21\cos^2\theta = 21\sin^2\theta$

$\Rightarrow \quad \sin^2\theta = 1 - \sin^2\theta$

$\Rightarrow \quad \sin^2\theta = \sin^2\frac{\pi}{4}$

$\Rightarrow \quad \theta = n\pi \pm \frac{\pi}{4}$

So, possible values of $\theta$ are

$\theta = \frac{\pi}{4}, \frac{3\pi}{4}, \frac{5\pi}{4}, \frac{7\pi}{4}$

$\sum_{\theta \in A} \theta^2 = \frac{\pi^2}{16}(1^2 + 3^2 + 7^2 + 5^2)$

$= \frac{84\pi^2}{16} = \frac{21\pi^2}{4}$

**17.** *(c)* Given, $|z_1 - 8 - 2i| \le 1$

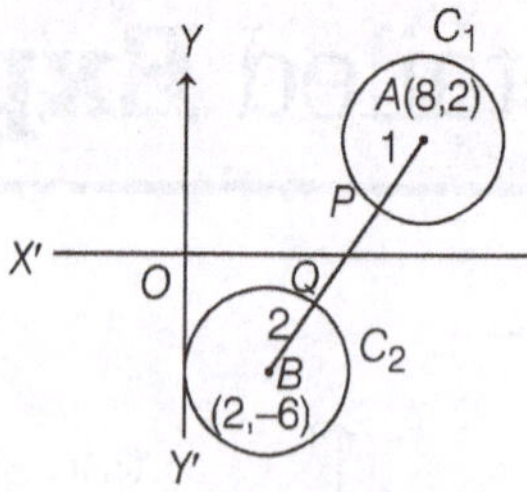


$\Rightarrow \quad |z_1 - (8+2i)| \le 1 = C_1$

and $\quad |z_2 - 2 + 6i| \le 2$

$\Rightarrow \quad |z_2 - (2-6i)| \le 2 = C_2$

$\therefore AB = \sqrt{(8-2)^2 + (2+6)^2} = \sqrt{36+64} = 10$

$\therefore |z_1 - z_2|_{\min} = PQ = AB - BQ - AP$

$= 10 - 2 - 1 = 7$

**18.** *(c)* $|z| = 1$

Let $z = e^{i\theta}$

$\bar{z} = e^{-i\theta} \Rightarrow \frac{z}{\bar{z}} = \frac{e^{i\theta}}{e^{-i\theta}} = e^{2i\theta}$

$= \cos 2\theta + i\sin 2\theta$

$\frac{\bar{z}}{z} = \frac{e^{-i\theta}}{e^{i\theta}} = \frac{1}{e^{2i\theta}} = e^{-2i\theta}$

$= \cos 2\theta - i\sin 2\theta$

$\frac{z}{\bar{z}} + \frac{\bar{z}}{z} = (\cos 2\theta + i\sin 2\theta)$

$+ (\cos 2\theta - i\sin 2\theta)$

$= 2\cos 2\theta$

$\because \quad \left|\frac{z}{\bar{z}} + \frac{\bar{z}}{z}\right| = 1 \Rightarrow |\cos 2\theta| = \frac{1}{2}$

$\therefore \quad \cos 2\theta = \pm\frac{1}{2}$

$\therefore$ Number of solutions $= 4 + 4 = 8$

> **Concept Enhancer**
> A complex number $z$ is written in the following form
> $z = a + ib = re^{i\theta} = r(\cos\theta + i\sin\theta)$

**19.** *(b)* Given, $z_1, z_2, z_3$ are three complex numbers lie on circle $|z| = 1$

$\arg(z_1) = -\frac{\pi}{4},\ \arg(z_2) = 0$

and $\arg(z_3) = \pi/4$

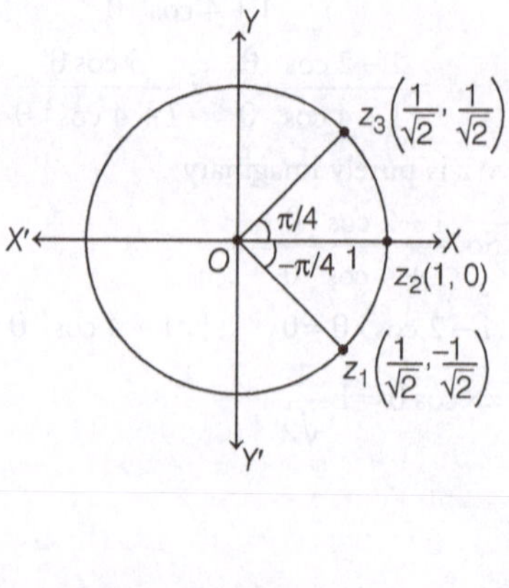

$\therefore \quad z_1 = \frac{1}{\sqrt{2}}(1-i), \bar{z}_1 = \frac{1}{\sqrt{2}}(1+i)$

$z_2 = 1 + 0i, \bar{z}_2 = 1$

$z_3 = \frac{1}{\sqrt{2}}(1+i), \bar{z}_3 = \frac{1}{\sqrt{2}}(1-i)$

$\therefore |z_1\bar{z}_2 + z_2\bar{z}_3 + z_3\bar{z}_1|^2$

$= \left|\frac{1}{\sqrt{2}}(1-i) + \frac{1}{\sqrt{2}}(1-i) + \frac{1}{2}(1+i)^2\right|^2$

$= \left|\sqrt{2} - \sqrt{2}i + \frac{1}{2}(2i)\right|^2$

$= |\sqrt{2} + (1-\sqrt{2})i|^2 = 2 + (1-\sqrt{2})^2$

$= 2 + 1 + 2 - 2\sqrt{2} = 5 - 2\sqrt{2}$

$\therefore \quad \alpha = 5, \beta = -2$

$\Rightarrow \alpha^2 + \beta^2 = 25 + 4 = 29$

**20.** (a) $a + 5b = 42 \Rightarrow b = \frac{42-a}{5}$

So, $(a, b) = \{(2, 8), (7, 7), (12, 6), (17, 5), (22, 4), (27, 3), (32, 2) (37, 1)\}$

$\Rightarrow$ 8 pairs

So, $m = 8$

So, $\sum_{n=1}^{8}(1 - i^{n!}) = \sum_{n=1}^{8} 1 - \sum_{n=1}^{8} i^{n!}$

$= 8 - (i + (-1) + (-1) + 1 + 1 + 1 + 1 + 1)$

$= 8 - (i + 3) = 5 - i = x + iy$

So, $x = 5, y = -1$

Therefore, $x + y + m = 5 - 1 + 8 = 12$

**21.** (c) Given $(\bar{z})^2 + |z| = 0$ ... (i)

Let $\quad z = x + iy$

So, $\quad \bar{z} = x - iy$

From Eq. (i), $(x - iy)^2 + |x + iy| = 0$

$\Rightarrow (x^2 - y^2) + \sqrt{x^2 + y^2} - 2xyi = 0$

On equating real and imaginary part from both sides, we get

$2xy = 0$ ... (ii)

and $x^2 - y^2 + \sqrt{x^2 + y^2} = 0$ ... (iii)

From Eq. (i), there are two cases arises

**Case I** When $x = 0$ and $y \neq 0$

From Eq. (iii),

$-y^2 + |y| = 0 \Rightarrow |y| = y^2 = |y|^2$

$\Rightarrow y = 0, \pm 1$ but $y \neq 0$

So, $y = \pm 1$

$\therefore \quad z = 0 + i$ or $z = 0 - i$ i.e. $z = i$

or $\quad z = -i$

**Case II** When $x \neq 0$ and $y = 0$

From Eq. (iii), $x^2 + |x| = 0$

$\Rightarrow \quad x = 0$ only which is not possible because $x \neq 0$.

Hence, $z = \pm i$ (finally)

Now, $\alpha = i - i = 0$ and

$\beta = (i)(-i) = -i^2 = -(-1) = 1$

Therefore, $4(\alpha^2 + \beta^2) = 4(0 + 1) = 4$

**22.** (c) Given $|z - 1| = 1$

$(x-1)^2 + y^2 = 1$ ...(i)

and $(\sqrt{2} - 1)(z + \bar{z}) - i(z - \bar{z}) = 2\sqrt{2}$

$(\sqrt{2} - 1)(2x) - i(2iy) = 2\sqrt{2}$

$[\because z + \bar{z} = 2x, z - \bar{z} = 2yi]$

$2(\sqrt{2} - 1)x + 2y = 2\sqrt{2}$

$(\sqrt{2} - 1)x + y = \sqrt{2}$ ...(ii)

On solving Eqs. (i) and (ii), we get

$x = 1, \frac{1}{2 - \sqrt{2}}$

When $x = 1$, then $y = 1$

$\therefore \quad z_2 = 1 + i$

and when $x = \frac{1}{2 - \sqrt{2}}$, then

$y = \sqrt{2} - \frac{1}{\sqrt{2}}$

$\therefore \quad z_1 = 1 + \frac{1}{\sqrt{2}} + \frac{i}{\sqrt{2}}$

$\therefore |\sqrt{2}z_1 - z_2|^2$

$= \left|\sqrt{2}\left(1 + \frac{1}{\sqrt{2}}\right) + i - 1 - i\right|^2$

$= (\sqrt{2})^2 = 2$

**23.** (a) Given, $z_1 + z_2 = 5$

$z_1^3 + z_2^3 = 20 + 15i$

$\Rightarrow z_1^3 + z_2^3 = (z_1 + z_2)((z_1 + z_2)^2 - 3z_1z_2)$

$\Rightarrow \quad 20 + 15i = 5(5^2 - 3z_1z_2)$

$\Rightarrow \quad 4 + 3i = 25 - 3z_1z_2$

$\Rightarrow \quad z_1z_2 = 7 - i$

$z_1 + z_2 = 5$

$\Rightarrow (z_1 + z_2)^2 = 5^2$

$z_1^2 + z_2^2 + 2z_1z_2 = 25$

$z_1^2 + z_2^2 + 14 - 2i = 25$

$z_1^2 + z_2^2 = 11 + 2i$

Squaring both sides,

$(z_1^2 + z_2^2)^2 = (11 + 2i)^2$

$z_1^4 + z_2^4 + 2(z_1z_2)^2$

$= 121 - 4 + 44i$

$z_1^4 + z_2^4 + 2(7 - i)^2 = 117 + 44i$

$z_1^4 + z_2^4 + 2(49 - 1 - 14i) = 117 + 44i$

$z_1^4 + z_2^4 = 21 + 72i$

$\therefore |z_1^4 + z_2^4| = \sqrt{21^2 + 72^2} = 75$

**24.** (b) We have, point $w_1$ obtained by rotation of $z_1$ about origin through 90° in anti-clockwise direction and point $w_2$ obtained by rotation of $z_2$ about origin through 90° in clockwise direction.

$\Rightarrow \quad w_1 = z_1 i$

and $\quad w_2 = z_2(-i)$

$w_1 = (5 + 4i)i$

and $\quad w_2 = (3 + 5i)(-i)$

$\Rightarrow \quad w_1 = -4 + 5i$ and $w_2 = 5 - 3i$

$w_1 - w_2 = -4 + 5i - 5 + 3i = -9 + 8i$

$\therefore$ Principal argument $= \pi - \tan^{-1}(8/9)$.

**25.** (b) Given that, $z = \alpha + i\beta$

and $|z + 2| = z + 4(1 + i)$

$\Rightarrow |\alpha + i\beta + 2| = \alpha + i\beta + 4 + 4i$

$\Rightarrow \sqrt{(\alpha + 2)^2 + \beta^2} = (\alpha + 4) + i(\beta + 4)$

On comparing real and imaginary parts, we get

$\beta + 4 = 0$

$\Rightarrow \quad \beta = -4$

and $\sqrt{(\alpha + 2)^2 + \beta^2} = \alpha + 4$

$\Rightarrow \quad \sqrt{(\alpha + 2)^2 + 16} = \alpha + 4$

$\Rightarrow \quad \sqrt{\alpha^2 + 4 + 4\alpha + 16} = \alpha + 4$

$\Rightarrow \quad \alpha^2 + 4\alpha + 20 = \alpha^2 + 16 + 8\alpha$

$\Rightarrow \quad \alpha = 1$

$\therefore \quad \alpha = 1, \beta = -4$

$\alpha + \beta = 1 - 4 = -3$

$\alpha\beta = 1(-4) = -4$

Hence, equation is

$x^2 - (-3 - 4)x + (-3)(-4) = 0$

$\Rightarrow x^2 + 7x + 12 = 0$

**26.** (d)

**key idea** Use properties of modulus of complex numbers $||z_1| - |z_2|| \leq |z_1 - z_2|$.

Given, $\left|z - \frac{1}{z}\right| = 2$

We know that, $||z_1| - |z_2|| \leq |z_1 - z_2|$

On putting $z_1 = z$ and $z_2 = \frac{1}{z}$, we get

$\left||z| - \frac{1}{|z|}\right| \leq \left|z - \frac{1}{z}\right| \Rightarrow \left||z| - \frac{1}{|z|}\right| \leq 2$

Let $|z| = t$

$\left|t - \frac{1}{t}\right| \leq 2 \Rightarrow |t^2 - 1| \leq 2|t|$

On squaring both sides, we get

$t^4 - 2t^2 + 1 \leq 4t^2$

$\Rightarrow \quad t^4 - 6t^2 + 9 - 8 \le 0$

$\Rightarrow \quad (t^2 - 3)^2 - (2\sqrt{2})^2 \le 0$

$\Rightarrow \quad t^2 \in [3 - 2\sqrt{2}, 3 + 2\sqrt{2}]$

$\Rightarrow \quad t^2 \in [(\sqrt{2} - 1)^2, (\sqrt{2} + 1)^2]$

$\Rightarrow \quad t \in [\sqrt{2} - 1, \sqrt{2} + 1]$

$\Rightarrow \quad |z| \in [\sqrt{2} - 1, \sqrt{2} + 1]$

So, maximum value of $|z| = \sqrt{2} + 1$

**27.** *(c)* Given, $\bar{z}_1 = i\bar{z}_2$

$\Rightarrow \quad \bar{z}_1 / \bar{z}_2 = i$

$\Rightarrow \quad z_1 / z_2 = -i$

$\Rightarrow \quad z_1 = -iz_2$

$\Rightarrow \quad \arg(z_1) = -\frac{\pi}{2} + \arg(z_2) \quad \ldots \text{(i)}$

Also, $\arg\left(\frac{z_1}{\bar{z}_2}\right) = \pi$

$\arg(z_1) - \arg(\bar{z}_2) = \pi$

$\arg(z_1) + \arg(z_2) = \pi \quad \ldots \text{(ii)}$

On substituting value of $\arg(z_1)$ in Eq. (ii), we get

$$-\frac{\pi}{2} + \arg(z_2) + \arg(z_2) = \pi$$

$$\Rightarrow \arg(z_2) = \frac{3\pi}{4}; \ \arg(z_1) = -\frac{\pi}{2} + \frac{3\pi}{4}$$

$$= \frac{\pi}{4}$$

**28.** *(d)* Given that

$$A = \{z \in C : 1 \le |z - (1 + i)| \le 2\}$$

and $B = \{z \in A : |z - (1 - i)| = 1\}$.

Here, $A$ represents all those points lying inside the region of circle having radius greater than or equal to 1 or less than or equal to 2 and $B$ represents all these points lying inside the circle having radius 1.

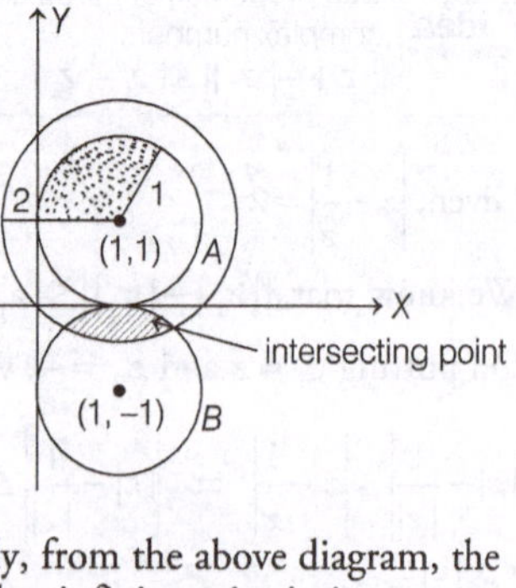


Clearly, from the above diagram, the set $B$ has infinite point in it.

So, $B$ is an infinite set.

**29.** *(b)* $|zw| = 1, \arg(z) - \arg(w) = \frac{3\pi}{2}$

Let $z = re^{i\theta} \Rightarrow \bar{z} = re^{-i\theta}$

$$w = \frac{1}{r} e^{i\left(\theta - \frac{3\pi}{2}\right)}$$

$$\Rightarrow w\bar{z} = \frac{1}{r} e^{i\left(\theta - \frac{3\pi}{2}\right)} \cdot re^{-i\theta}$$

$$\Rightarrow w\bar{z} = e^{i\left(\theta - \frac{3\pi}{2} - \theta\right)} = e^{-i\frac{3\pi}{2}}$$

$$\Rightarrow w\bar{z} = \cos\left(\frac{-3\pi}{2}\right) + i\sin\left(\frac{-3\pi}{2}\right)$$

$$\Rightarrow w\bar{z} = 0 + i$$

$$\Rightarrow w\bar{z} = i$$

$$\Rightarrow \left(\frac{1 - 2w\bar{z}}{1 + 3w\bar{z}}\right) = \left(\frac{1 - 2i}{1 + 3i} \times \frac{1 - 3i}{1 - 3i}\right)$$

$$= \frac{1 - 2i - 3i + 6i^2}{10} = \frac{-5 - 5i}{10}$$

$$\therefore \arg\left(\frac{1 - 2w\bar{z}}{1 + 3w\bar{z}}\right) = -\left(\pi - \frac{\pi}{4}\right) = -\frac{3\pi}{4}$$

**30.** *(d)* Let the complex numbers

$z_1 = x_1 + iy_1$ and $z_2 = x_2 + iy_2$

Now, it is given that

$$\text{Re}(z_1) = |z_1 - 1|$$

$$\Rightarrow x_1^2 = (x_1 - 1)^2 + y_1^2$$

$$\Rightarrow y_1^2 + 1 = 2x_1 \quad \ldots \text{(i)}$$

and $\text{Re}(z_2) = |z_2 - 1|$

$$\Rightarrow x_2^2 = (x_2 - 1)^2 + y_2^2$$

$$\Rightarrow y_2^2 + 1 = 2x_2 \quad \ldots \text{(ii)}$$

and $\arg(z_1 - z_2) = \frac{\pi}{6}$

$$\Rightarrow \frac{y_1 - y_2}{x_1 - x_2} = \frac{1}{\sqrt{3}} \quad \ldots \text{(iii)}$$

From Eqs. (i) and (ii), we get

$$y_1^2 - y_2^2 = 2(x_1 - x_2)$$

$$\Rightarrow \frac{y_1 - y_2}{x_1 - x_2}(y_1 + y_2) = 2$$

$$\Rightarrow \frac{1}{\sqrt{3}}(y_1 + y_2) = 2$$

[$\because$ Using Eq. (iii)]

$$\Rightarrow y_1 + y_2 = 2\sqrt{3}$$

$$\Rightarrow \text{Im}(z_1 + z_2) = 2\sqrt{3}$$

**31.** *(c)* Let $z = x + 10i$, as $\text{Im}(z) = 10$ [given]

Since, $z$ satisfies,

$$\frac{2z - n}{2z + n} = 2i - 1, n \in N,$$

$\therefore (2x + 20i - n) = (2i - 1)(2x + 20i + n)$

$\Rightarrow (2x - n) + 20i = (-2x - n - 40) + (4x + 2n - 20)i$

On comparing real and imaginary parts, we get

$$2x - n = -2x - n - 40$$

and $20 = 4x + 2n - 20$

$\Rightarrow 4x = -40$

and $4x + 2n = 40$

$\Rightarrow x = -10$

and $-40 + 2n = 40$

$\Rightarrow n = 40$

and $x = \text{Re}(z) = -10$

**32.** *(*)* Given, $3|z_1| = 4|z_2|$

$$\Rightarrow \frac{|z_1|}{|z_2|} = \frac{4}{3} \quad [\because z_2 \ne 0 \Rightarrow |z_2| \ne 0]$$

$$\therefore \frac{z_1}{z_2} = \left|\frac{z_1}{z_2}\right| e^{i\theta}$$

and $\frac{z_2}{z_1} = \left|\frac{z_2}{z_1}\right| e^{-i\theta}$

$[\because z = |z|(\cos\theta + i\sin\theta) = |z|e^{i\theta}]$

$$\Rightarrow \frac{z_1}{z_2} = \frac{4}{3}e^{i\theta}$$

and $\frac{z_2}{z_1} = \frac{3}{4}e^{-i\theta}$

$$\Rightarrow \frac{3}{2}\frac{z_1}{z_2} = 2e^{i\theta} \quad \ldots \text{(i)}$$

and $\frac{2}{3}\frac{z_2}{z_1} = \frac{1}{2}e^{-i\theta} \quad \ldots \text{(ii)}$

On adding Eqs. (i) and (ii), we get

$$z = \frac{3}{2}\frac{z_1}{z_2} + \frac{2}{3}\frac{z_2}{z_1} = 2e^{i\theta} + \frac{1}{2}e^{-i\theta}$$

$$= 2\cos\theta + 2i\sin\theta + \frac{1}{2}\cos\theta - \frac{1}{2}i\sin\theta$$

$[\because e^{\pm i\theta} = (\cos\theta \pm i\sin\theta)]$

$$= \frac{5}{2}\cos\theta + \frac{3}{2}i\sin\theta$$

$$\Rightarrow |z| = \sqrt{\left(\frac{5}{2}\right)^2 + \left(\frac{3}{2}\right)^2}$$

$$= \sqrt{\frac{34}{4}} = \frac{\sqrt{17}}{2}$$

**33.** *(d)* Let

$$z = \left(\frac{3 + 2i\sin\theta}{1 - 2i\sin\theta}\right) \times \left(\frac{1 + 2i\sin\theta}{1 + 2i\sin\theta}\right)$$

(rationalising the denominator)

$$= \frac{3 - 4\sin^2\theta + 8i\sin\theta}{1 + 4\sin^2\theta}$$

$[\because a^2 - b^2 = (a + b)(a - b) \text{ and } i^2 = -1]$

$$=\left(\frac{3-4\sin^2\theta}{1+4\sin^2\theta}\right)+\left(\frac{8\sin\theta}{1+4\sin^2\theta}\right)i$$

Since, real part of $z=0$

$\therefore \quad \dfrac{3-4\sin^2\theta}{1+4\sin^2\theta}=0$

$\Rightarrow \quad \sin^2\theta=3/4$

$\Rightarrow \quad \sin\theta=\pm\sqrt{3}/2$

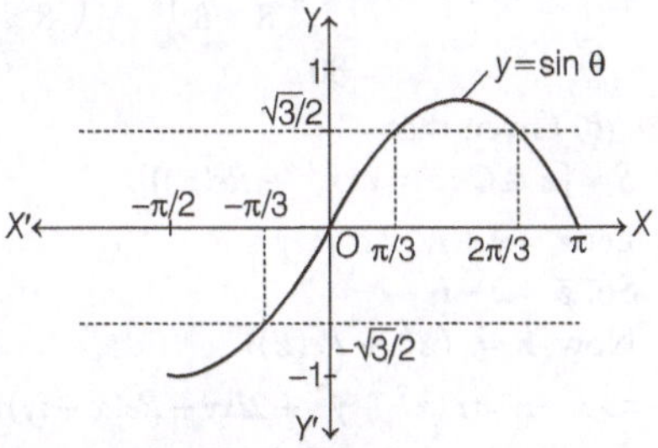


$\Rightarrow \quad \theta\in\left\{-\dfrac{\pi}{3},\dfrac{\pi}{3},\dfrac{2\pi}{3}\right\}$

Sum of values of $\theta=2\pi/3$.

**34.** *(c)* Given, $\omega=\dfrac{1+(1-8\alpha)z}{1-z}$

[$\because$ $\omega$ to be purely imaginary]

We know that $\omega+\overline{\omega}=0$

i.e., $\dfrac{1+(1-8\alpha)z}{1-z}+\dfrac{1+(1-8\alpha)\bar{z}}{1-\bar{z}}=0$

$\Rightarrow [1+(1-8\alpha)z][1-\bar{z}]+[1+(1-8\alpha)\bar{z}][1-z]=0$

$\Rightarrow [1-\bar{z}+(1-8\alpha)z-(1-8\alpha)z\bar{z}]+[1-z+(1-8\alpha)\bar{z}-(1-8\alpha)z\bar{z}]=0$

$\Rightarrow 2-(z+\bar{z})+(1-8\alpha)(z+\bar{z})-2(1-8\alpha)=0 \quad (\because z\bar{z}=|z|^2=1)$

$\Rightarrow 2-(z+\bar{z})+(z+\bar{z})-8\alpha(z+\bar{z})-2+16\alpha=0$

$\Rightarrow 16\alpha=8\alpha(z+\bar{z})$

Either $z+\bar{z}=2$ or $\alpha=\{0\}$. But $z+\bar{z}=2$ is not possible.

$\therefore \quad \alpha=\{0\}$

**35.** *(d)* Let $z=\dfrac{2+3i\sin\theta}{1-2i\sin\theta}$ is purely imaginary.

Then, we have $\text{Re}(z)=0$

Now, consider $z=\dfrac{2+3i\sin\theta}{1-2i\sin\theta}$

$$=\frac{(2+3i\sin\theta)(1+2i\sin\theta)}{(1-2i\sin\theta)(1+2i\sin\theta)}$$

$$=\frac{2+7i\sin\theta-6\sin^2\theta}{1+4\sin^2\theta}$$

$$=\frac{2-6\sin^2\theta}{1+4\sin^2\theta}+i\frac{7\sin\theta}{1+4\sin^2\theta}$$

$\because \text{Re}(z)=0$

$\therefore \quad \dfrac{2-6\sin^2\theta}{1+4\sin^2\theta}=0$

$\Rightarrow \quad 2=6\sin^2\theta$

$\Rightarrow \quad \sin^2\theta=\dfrac{1}{3}$

$\Rightarrow \quad \theta=\sin^{-1}\left(\pm\dfrac{1}{\sqrt{3}}\right)=\pm\sin^{-1}\left(\dfrac{1}{\sqrt{3}}\right)$

**36.** *(d)* Consider the equation

$w-\overline{w}z=k(1-z),\ k\in R$

Clearly $z\neq 1$ and $\dfrac{w-\overline{w}z}{1-z}$ is purely real.

$\therefore \quad \overline{\left(\dfrac{w-\overline{w}z}{1-z}\right)}=\dfrac{w-\overline{w}z}{1-z}$

$\Rightarrow \quad \dfrac{\overline{w}-w\bar{z}}{1-\bar{z}}=\dfrac{w-\overline{w}z}{1-z}$

$\Rightarrow \quad \overline{w}-\overline{w}z-w\bar{z}+wz\bar{z}=w-w\bar{z}-\overline{w}z+\overline{w}z\bar{z}$

$\Rightarrow \quad \overline{w}+w|z|^2=w+\overline{w}|z|^2$

$\Rightarrow \quad (w-\overline{w})(|z|^2)=w-\overline{w}$

$\Rightarrow \quad |z|^2=1 \quad [\because \text{Im } w\neq 0]$

$\Rightarrow \quad |z|=1$ and $z\neq 1$

$\therefore$ The required set is $\{z:|z|=1,\ z\neq 1\}$.

**37.** *(a)* Let $z=x+iy$

$\Rightarrow \quad z^2=x^2-y^2+2ixy$

Now, $\dfrac{1+z^2}{2iz}=\dfrac{1+x^2-y^2+2ixy}{2i(x+iy)}$

$$=\frac{(x^2-y^2+1)+2ixy}{2ix-2y}$$

$$=\frac{(x^2-y^2+1)+2ixy}{-2y+2ix}\times\frac{-2y-2ix}{-2y-2ix}$$

$$=\frac{y(x^2+y^2-1)+x(x^2+y^2+1)i}{2(x^2+y^2)}$$

On taking imaginary part,

$$a=\frac{x(x^2+y^2+1)}{2(x^2+y^2)}\qquad\left[\because a=\text{Im}\left(\frac{1+z^2}{2iz}\right)\right]$$

$\because \quad |z|=1\Rightarrow\sqrt{x^2+y^2}=1$

$\Rightarrow \quad x^2+y^2=1 \ \therefore a=\dfrac{x(1+1)}{2\times 1}=x$

Also, $z\neq 1\Rightarrow x+iy\neq 1$

$\therefore A=(-1,1)$

**38.** *(c)* Given equation is

$z+\sqrt{2}|z+1|+i=0$

Put $z=x+iy$ in the given equation.

$(x+iy)+\sqrt{2}|x+iy+1|+i=0$

$\Rightarrow x+iy+\sqrt{2}\,[\sqrt{(x+1)^2+y^2}\,]+i=0$

Now, equating real and imaginary part, we get

$x+\sqrt{2}\sqrt{(x+1)^2+y^2}=0$

and $y+1=0\Rightarrow y=-1$

$\Rightarrow x+\sqrt{2}\sqrt{(x+1)^2+(-1)^2}=0 \quad (\because y=-1)$

$\Rightarrow \sqrt{2}\sqrt{(x+1)^2+1}=-x$

$\Rightarrow 2[(x+1)^2+1]=x^2$

$\Rightarrow x^2+4x+4=0\Rightarrow x=-2$

Thus, $z=-2+i(-1)\Rightarrow |z|=\sqrt{5}$

**39.** *(b)* We have, $|z_1+z_2|^2+|z_1-z_2|^2$

$=|z_1|^2+|z_2|^2+z_1\bar{z}_2+\bar{z}_1z_2+|z_1|^2+|z_2|^2-z_1\bar{z}_2-\bar{z}_1z_2$

$[\because |z_1+z_2|^2=|z_1|^2+|z_2|^2+z_1\bar{z}_2+\bar{z}_1z_2$ and $|z_1-z_2|^2=|z_1|^2+|z_2|^2-z_1\bar{z}_2-\bar{z}_1z_2]$

$=2|z_1|^2+2|z_2|^2$

$=2(|z_1|^2+|z_2|^2)$

**40.** *(b)* Let $z_1=x_1+iy_1$

and $\quad z_2=x_2+iy_2$

Given that, $|z_1+z_2|=|z_1|+|z_2|$

$\therefore \quad \sqrt{(x_1+x_2)^2+(y_1+y_2)^2}=\sqrt{x_1^2+y_1^2}+\sqrt{x_2^2+y_2^2}$

On squaring both sides, we get

$x_1^2+x_2^2+2x_1x_2+y_1^2+y_2^2+2y_1y_2=x_1^2+y_1^2+x_2^2+y_2^2+2\sqrt{(x_1^2+y_1^2)(x_2^2+y_2^2)}$

$\Rightarrow x_1x_2+y_1y_2=\sqrt{(x_1^2+y_1^2)(x_2^2+y_2^2)}$

Again squaring, we get

$x_1^2x_2^2+y_1^2y_2^2+2x_1x_2y_1y_2=x_1^2x_2^2+y_1^2y_2^2+x_1^2y_2^2+y_1^2x_2^2$

$\Rightarrow (x_1y_2-y_1x_2)^2=0$

$\Rightarrow \quad \dfrac{y_1}{x_1}=\dfrac{y_2}{x_2}$

$\Rightarrow \quad \tan^{-1}\left(\dfrac{y_1}{x_1}\right)=\tan^{-1}\left(\dfrac{y_2}{x_2}\right)$

$\Rightarrow \quad \arg(z_1)=\arg(z_2)$

$\Rightarrow \arg(z_1)-\arg(z_2)=0$

**Alternate Solution**

Given that, $|z_1+z_2|=|z_1|+|z_2|$

On squaring both sides, we get

$|z_1|^2+|z_2|^2+2\text{Re}(z_1\bar{z}_2)=|z_1|^2+|z_2|^2+2|z_1||z_2|$

$\Rightarrow \text{Re}(z_1\bar{z}_2)=|z_1||z_2|$

$\Rightarrow |z_1||z_2|\cos(\theta_1-\theta_2)=|z_1||z_2|$

$\Rightarrow \theta_1-\theta_2=0$

$\Rightarrow \arg(z_1)-\arg(z_2)=0$

**41.** (d) $\because z^{1/3}=p+iq$

$\therefore z=(p+iq)^3$

$=p^3-iq^3+3ip^2q-3pq^2$

Given that, $z=x-iy$

$\therefore x-iy=p^3-3pq^2+i(3p^2q-q^3)$

$\Rightarrow x=p^3-3pq^2$ and $-y=3p^2q-q^3$

$\Rightarrow \dfrac{x}{p}=p^2-3q^2$

and $y/q=-3p^2+q^2$

$\therefore \dfrac{x}{p}+\dfrac{y}{q}=-2p^2-2q^2$

$\Rightarrow \dfrac{1}{(p^2+q^2)}\left(\dfrac{x}{p}+\dfrac{y}{q}\right)=-2$

**42.** (d) Let $z=r_1e^{i\theta}$ and $w=r_2e^{i\phi}$

$\Rightarrow \bar{z}=r_1e^{-i\theta}$

Given, $|zw|=1$

$\Rightarrow |r_1e^{i\theta}\cdot r_2e^{i\phi}|=1 \Rightarrow r_1r_2=1$ ...(i)

and $\arg(z)-\arg(w)=\pi/2$

$\Rightarrow \theta-\phi=\dfrac{\pi}{2}$ ...(ii)

Now, $\bar{z}w=r_1e^{-i\theta}\cdot r_2e^{i\phi}=r_1r_2e^{-i(\theta-\phi)}$

$=1\cdot e^{-i\pi/2}$ [from Eqs. (i) and (ii)]

$=\cos\dfrac{\pi}{2}-i\sin\dfrac{\pi}{2}\Rightarrow \bar{z}w=-i$

**43.** (d) **Statement I**

We know that

$$\left|\frac{z_1}{|z_1|}+\frac{z_2}{|z_2|}\right|\le\left|\frac{z_1}{|z_1|}\right|+\left|\frac{z_2}{|z_2|}\right|$$

$$\Rightarrow \left|\frac{z_1}{|z_1|}+\frac{z_2}{|z_2|}\right|\le\frac{|z_1|}{|z_1|}+\frac{|z_2|}{|z_2|}$$

$$\Rightarrow \left|\frac{z_1}{|z_1|}+\frac{z_2}{|z_2|}\right|\le 2$$

$$\Rightarrow (|z_1|+|z_2|)\left|\frac{z_1}{|z_1|}+\frac{z_2}{|z_2|}\right|\le 2(|z_1|+|z_2|)$$

$\therefore$ Statement I is correct.

**Statement II**

We have, $\dfrac{a}{|y-z|}=\dfrac{b}{|z-x|}=\dfrac{c}{|x-y|}$

$$\Rightarrow \frac{a^2}{|y-z|^2}=\frac{b^2}{|z-x|^2}=\frac{c^2}{|x-y|^2}=k$$

$\Rightarrow a^2=k(|y-z|^2)=k(y-z)\overline{(y-z)}$

$=k(y-z)(\bar{y}-\bar{z})$

$b^2=k(|z-x|^2)=k(z-x)(\bar{z}-\bar{x})$

and $c^2=k(|x-y|^2)=k(x-y)(\bar{x}-\bar{y})$

$$\therefore \frac{a^2}{y-z}+\frac{b^2}{z-x}+\frac{c^2}{x-y}$$

$=k(\bar{y}-\bar{z})+k(\bar{z}-\bar{x})+k(\bar{x}-\bar{y})$

$=k(\bar{y}-\bar{z}+\bar{z}-\bar{x}+\bar{x}-\bar{y})=k(0)=0$

$\therefore$ Statement II is incorrect.

**44.** (b) Let $w=\left(\dfrac{z-\bar{z}+z\bar{z}}{2-3z+5\bar{z}}\right), z\in C$

where $z=x+iy$

Since, $\text{Re}(z)=3$

$\Rightarrow z=3+iy \Rightarrow \bar{z}=3-iy$

$$w=\left(\frac{z-\bar{z}+z\bar{z}}{2-3z+5\bar{z}}\right)$$

$$=\left(\frac{(3+iy)-(3-iy)+(3+iy)(3-iy)}{2-3(3+iy)+5(3-iy)}\right)$$

$$=\left(\frac{2iy+9+y^2}{2-9-3iy+15-5iy}\right)$$

$$=\left(\frac{(9+y^2)+i(2y)}{8-8iy}\right)$$

$$=\frac{[(9+y^2)+i(2y)](1+iy)}{8(1-iy)(1+iy)}$$

$$=\frac{(9+y^2)+i(2y)+iy(9+y^2)-2y^2}{8(1+y^2)}$$

$$\therefore w=\frac{(9+y^2-2y^2)+i\{2y+9y+y^3\}}{8(1+y^2)}$$

$$\text{Re}(w)=\frac{9-y^2}{8(1+y^2)}$$

$$=\frac{1}{8}\left[\frac{10-(1+y^2)}{1+y^2}\right]=\frac{1}{8}\left[\frac{10}{1+y^2}-1\right]$$

Since, $1+y^2\in[1,\infty)$

$\Rightarrow \dfrac{1}{1+y^2}\in(0,1]\Rightarrow \dfrac{10}{1+y^2}\in(0,10]$

$\Rightarrow \dfrac{10}{1+y^2}-1\in(-1,9]$

$\Rightarrow \dfrac{1}{8}\left(\dfrac{10}{1+y^2}-1\right)\in\left(-\dfrac{1}{8},\dfrac{9}{8}\right]$

$\because \dfrac{1}{8}\left(\dfrac{10}{1-y^2}-1\right)\in(\alpha,\beta]$ [given]

$\therefore \alpha=-\dfrac{1}{8},\beta=\dfrac{9}{8}$

Now, $24(\beta-\alpha)=24\left(\dfrac{9}{8}+\dfrac{1}{8}\right)=24\left(\dfrac{10}{8}\right)$

$=30$

**45.** (d) Given that

$S=\{z\in C:\bar{z}=i(z^2+Re(\bar{z}))\}$

Let $z=x+iy, x\in R, y\in R$

So, $\bar{z}=x-iy$

Now, $\bar{z}=i(z^2+Re(\bar{z}))$

$\Rightarrow x-iy=i(x^2-y^2+2ixy+Re(x-iy))$

$\Rightarrow x-iy=i(x^2-y^2+x+i(2xy))$

Now, comparing both sides, we get

$x=-2xy$ ...(i)

and $-y=x^2-y^2+x$ ...(ii)

From Eq. (i), we get

$x(1+2y)=0\Rightarrow x=0,\ y=-\dfrac{1}{2}$

On putting $x=0$ in Eq. (ii), we get

$y^2=y\Rightarrow y=0,1$

If $y=-\dfrac{1}{2}\Rightarrow -\left(-\dfrac{1}{2}\right)=x^2-\dfrac{1}{4}+x$

$\Rightarrow \dfrac{1}{4}+\dfrac{1}{2}=x^2+x\Rightarrow \dfrac{3}{4}=x^2+x$

$\Rightarrow 4x^2+4x-3=0$

$\Rightarrow (2x+3)(2x-1)=0$

$\Rightarrow x=-\dfrac{3}{2}$ and $x=\dfrac{1}{2}$

Thus, $(0,0),(0,1),\left(-\dfrac{3}{2},-\dfrac{1}{2}\right)\left(\dfrac{1}{2},-\dfrac{1}{2}\right)$

All the points belongs to $S$.

So, $\sum_{z\in S}|z|^2$

$=0+1+\left(\dfrac{9}{4}+\dfrac{1}{4}\right)+\left(\dfrac{1}{4}+\dfrac{1}{4}\right)$

$=1+\dfrac{5}{2}+\dfrac{1}{2}=\dfrac{8}{2}=4$

$\therefore \sum_{z\in S}|z|^2=4$

**46.** (b) We have, $A=\{z\in C:\text{Re}(a+\bar{z})>\text{Im}(\bar{a}+z)\}, a\in C$

and $B=\{z\in C:\text{Re}(a+\bar{z})<\text{Im}(\bar{a}+z)\}$

Let $a=x_1+iy_1$ and $z=x+iy$

For set $A$, $\text{Re}(a+\bar{z}) > \text{Im}(\bar{a}+z)$

$\Rightarrow$ $\text{Re}(x_1+iy_1+x-iy) > \text{Im}(x_1-iy_1+x+iy)$

$\Rightarrow$ $\text{Re}((x_1+x)+i(y_1-y)) > \text{Im}((x_1+x)+i(y-y_1))$

$\Rightarrow$ $(x_1+x) > (y-y_1)$

Take, $x_1=2$, $y_1=10$, $x=-12$, $y=0$, for these values given inequality is not valid.

So, $S1$ is false.

Again for set $B$,

$\text{Re}(a+\bar{z}) < \text{Im}(\bar{a}+z)$

$\Rightarrow$ $\text{Re}(x_1+iy_1+x-iy) < \text{Im}(x_1-iy_1+x+iy)$

$\Rightarrow$ $\text{Re}((x_1+x)+i(y_1-y)) < \text{Im}((x_1+x)+(y-y_1))$

$\Rightarrow$ $x_1+x < y-y_1$

For $x_1=-2$, $y_1=-10$, $x=12$, $y=0$, the given inequality is not valid.

So, $S2$ is also false.

Thus, $S1$ and $S2$ both false.

**47.** (a) We have,

$$S=\left\{z=x+iy : \frac{2z-3i}{4z+2i} \text{ is a real number}\right\}$$

$$\frac{2z-3i}{4z+2i}\in R \Rightarrow \frac{2(x+iy)-3i}{4(x+iy)+2i}\in R$$

$$\frac{2x+i(2y-3)}{4x+i(4y+2)}\in R$$

$$\frac{2x+i(2y-3)}{4x+i(4y+2)}\times\frac{4x-i(4y+2)}{4x-i(4y+2)}\in R$$

$$\frac{8x^2+8y^2-8y-6+i[-2x(4y+2)+4x(2y-3)]}{(4x)^2+(4y+2)^2}\in R$$

The expression will be real if imaginary part is 0.

$-2x(4y+2)+4x(2y-3)=0$

$\Rightarrow x=0$

Also, $(4x)^2+(4y+2)^2\neq 0$

$\Rightarrow$ $y+x^2+y^2\neq\dfrac{-1}{4}$

If $x=0$, then $y\neq-\dfrac{1}{2}$

$$y\in\left(-\infty,-\frac{1}{2}\right)\cup\left(-\frac{1}{2},\infty\right)$$

**48.** (c) We have, $z=x+iy$ such that

$\dfrac{2z-3i}{2z+i}$ is purely imaginary

Now, $\left(\dfrac{2z-3i}{2z+i}\right)+\left(\overline{\dfrac{2z-3i}{2z+i}}\right)=0$

$\Rightarrow$ $\dfrac{2z-3i}{2z+i}+\dfrac{2\bar{z}+3i}{2\bar{z}-i}=0$

$\Rightarrow$ $(2z-3i)(2\bar{z}-i)+(2\bar{z}+3i)(2z+i)=0$

$\Rightarrow$ $\{4|z|^2-6i\bar{z}-2iz-3\}+\{4|z|^2+6iz+2i\bar{z}-3\}=0$

$\Rightarrow$ $8|z|^2-4i\bar{z}+4iz-6=0$

$\Rightarrow$ $8(x^2+y^2)-4i(x-iy)+4i(x+iy)-6=0$

$\Rightarrow$ $8(x^2+y^2)-4ix-4y+4ix-4y-6=0$

$\Rightarrow$ $8(x^2+y^2)-8y-6=0$

$\Rightarrow$ $4(x^2+y^2)-4y-3=0$

$\Rightarrow$ $4(y^4+y^2)-4y=3$

[given that $x+y^2=0 \Rightarrow x=-y^2$]

$\Rightarrow$ $y^4+y^2-y=3/4$

Hence, required answer is 3/4.

**49.** (d) Let $z=x+iy$

$\Rightarrow$ $z^2=x^2-y^2+2ixy$

$\therefore az^2+bz$

$=a(x^2-y^2+2ixy)+b(x+iy)$

$=a(x^2-y^2)+bx+2aixy+biy$

$\because$ $\text{Re}(az^2+bz)=a$

$\Rightarrow$ $a(x^2-y^2)+bx=a$

$\Rightarrow$ $x^2-y^2+\dfrac{b}{a}x=1$ ...(i)

and $bz^2+az=$

$b(x^2-y^2+2ixy)+a(x+iy)$

$=b(x^2-y^2)+ax+2bixy+aiy$

$\because$ $\text{Re}(bz^2+az)=b$

$\Rightarrow$ $b(x^2-y^2)+ax=b$

$\Rightarrow$ $x^2-y^2+\dfrac{a}{b}x=1$ ...(ii)

On subtracting Eq. (ii) from Eq. (i) we get

$$\frac{b}{a}x-\frac{a}{b}x=0$$

$\Rightarrow$ $\left(\dfrac{b}{a}-\dfrac{a}{b}\right)x=0$

$\Rightarrow$ $x=0$ or $\dfrac{b}{a}-\dfrac{a}{b}=0$

$\Rightarrow$ $b^2-a^2=0\Rightarrow a=\pm b$

$\Rightarrow$ $a=-b$ (since $a\neq b$)

From Eq. (i), when $x=0$,

then $0-y^2=1\Rightarrow y^2=-1$

$\Rightarrow$ $y\in\phi\Rightarrow z\in\phi$

$\Rightarrow$ has no solution

when, $a=-b$, then $x^2-y^2-x=1$

$\Rightarrow$ has infinite many solutions.

$\therefore$ $X=0$

**50.** (*) We have, $\left|\dfrac{1+ai}{b+i}\right|=1$

$\Rightarrow$ $|1+ai|=|b+i|$

$\Rightarrow$ $\sqrt{1+a^2}=\sqrt{b^2+1}$

$\Rightarrow$ $1+a^2=b^2+1$

$\Rightarrow$ $a^2=b^2\Rightarrow a=\pm b$

$\Rightarrow$ $a=-b$ [$\because ab<0$] ...(i)

Again, $a+ib$ lies on $|z-1|=|2z|$

$\therefore |a+ib-1|=|2(a+ib)|$

$\Rightarrow |a-ia-1|=|2(a-ia)|$

$\Rightarrow |(a-1)-ia|=2|a-ia|$

$\Rightarrow \sqrt{(a-1)^2+(-a)^2}=2\sqrt{(a)^2+(-a)^2}$

$\Rightarrow a^2-2a+1+a^2=8a^2$

$\Rightarrow 6a^2+2a-1=0$

$\Rightarrow a=\dfrac{-2\pm\sqrt{4+24}}{12}$

$=\dfrac{-2\pm2\sqrt{7}}{12}=\dfrac{-1\pm\sqrt{7}}{6}$

Now, when $a=\dfrac{-1+\sqrt{7}}{6}$, then $[a]=0$

and when $a=\dfrac{-1-\sqrt{7}}{6}$, then $[a]=-1$.

$\therefore$ $\dfrac{1+[a]}{4b}=\dfrac{1+0}{4b}$ or $\dfrac{1-1}{4b}=\dfrac{1}{4b}$ or 0

**51.** (d) Let $z=x+iy$

We have, $\left|\dfrac{z-2}{z-3}\right|=2\Rightarrow\left|\dfrac{x+iy-2}{x+iy-3}\right|=2$

$\Rightarrow$ $\dfrac{\sqrt{(x-2)^2+y^2}}{\sqrt{(x-3)^2+y^2}}=2$

$\Rightarrow (x-2)^2+y^2=4[(x-3)^2+y^2]$

$\Rightarrow$ $x^2-4x+4+y^2=4[x^2-6x+9+y^2]$

$\Rightarrow$ $x^2+y^2-\dfrac{20}{3}x+\dfrac{32}{3}=0$

$\therefore$ Centre of circle is

$$\left(-\frac{1}{2}\times\left(\frac{-20}{3}\right),-\frac{1}{2}\times(0)\right)$$

i.e. $\left(\dfrac{10}{3},0\right)$

And radius of circle

$$= \sqrt{\left(\frac{10}{3}\right)^2 + (0)^2 - \frac{32}{3}}$$

$$= \sqrt{\frac{100}{9} - \frac{32}{3}}$$

$$= \sqrt{\frac{100-96}{9}} = \frac{2}{3}$$

$\therefore \quad \alpha = \frac{10}{3}, \beta = 0, \gamma = \frac{2}{3}$

So, $3(\alpha + \beta + \gamma) = 3\left(\frac{10}{3} + 0 + \frac{2}{3}\right)$

$= 10 + 2 = 12$

**52.** *(d)* We have non-zero complex numbers $z_1$ and $z_2$.

Let $z_1 = x_1 + iy_1$; $z_2 = x_2 + iy_2$, $x_1, x_2, y_1, y_2 \in R$

$\therefore z_1 z_2 = (x_1 + iy_1)(x_2 + iy_2)$

$= x_1x_2 + ix_1y_2 + iy_1x_2 + i^2 y_1y_2$

$= (x_1x_2 - y_1y_2) + i(x_1y_2 + y_1x_2)$ $\quad [\because i^2 = -1]$

Now, $\text{Re}(z_1 \cdot z_2) = 0$

$\Rightarrow \quad x_1 x_2 - y_1 y_2 = 0$

$\Rightarrow \quad x_1 x_2 = y_1 y_2 \quad \ldots$(i)

And $z_1 + z_2 = (x_1 + iy_1) + (x_2 + iy_2)$

$= (x_1 + x_2) + i(y_1 + y_2)$

Now, $\text{Re}(z_1 + z_2) = 0$

$\Rightarrow \quad x_1 + x_2 = 0$

$\Rightarrow \quad x_1 = -x_2 \quad \ldots$ (ii)

From Eqs. (i) and (ii), we get

$y_1 y_2 = -x_2^2$; $\because x_2 \in R$; $\therefore -x_2^2 < 0$

$\Rightarrow \quad y_1, y_2 < 0$

i.e. $y_1 > 0$ and $y_2 < 0$ or $y_1 < 0$ and $y_2 > 0$

$\Rightarrow \text{Im}(z_1) > 0$ and $\text{Im}(z_2) < 0$

or $\text{Im}(z_1) < 0$ and $\text{Im}(z_2) > 0$

**53.** *(c)* $\alpha \in (\pi, 2\pi)$ and $\beta \in (\pi, 2\pi)$

$$\frac{1 - i\sin\alpha}{1 + 2i\sin\alpha} + \frac{1 + i\sin\alpha}{1 - 2i\sin\alpha} = 0$$

$\left[\because \frac{1 - i\sin\alpha}{1 + 2i\sin\alpha} \text{ is purely imaginary}\right]$

$\Rightarrow 1 - 2\sin^2\alpha = 0$

$\Rightarrow \alpha = \frac{5\pi}{4}$ or $\frac{7\pi}{4}$

$$\therefore \frac{1 + i\cos\beta}{1 - 2i\cos\beta} - \frac{1 - i\cos\beta}{1 + 2i\cos\beta} = 0$$

$\left[\because \frac{1 + i\cos\beta}{1 - 2i\cos\beta} \text{ is purely real}\right]$

$\Rightarrow \quad 2\cos\beta + \cos\beta = 0$

$\therefore \quad \beta = \frac{3\pi}{2}$

$\therefore \quad (\alpha, \beta) = \left(\frac{5\pi}{4}, \frac{3\pi}{2}\right)$ or $\left(\frac{7\pi}{4}, \frac{3\pi}{2}\right)$

$\therefore \quad z_{\alpha\beta} = (1 - i)$ or $(-1 - i)$

$$\therefore \sum_{(\alpha,\beta)\in S} \left(iz_{\alpha\beta} + \frac{1}{i\bar{z}_{\alpha\beta}}\right)$$

$$= i + 1 + \frac{1}{i - 1} + 1 - i - \frac{1}{i + 1}$$

$$= 2 + \frac{-i + 1 + 1 + i}{-2} = 1$$

**54.** *(d)*

> **key idea** The equation of circle whose centre is $(x_0, y_0)$ i.e. , $z_0$ and radius $r$ is $(x - x_0)^2 + (y - y_0)^2 = r^2$ or $|z - z_0| = r$
>
> $\therefore$ If $z = a + ib$, then $\arg z = \theta = \tan^{-1}\left(\frac{b}{a}\right)$

Given, $S_n = \left\{ z \in C : |z - 3 + 2i| = \frac{n}{4} \right\}$

represents a circle with centre $C_1(3, -2)$ and radius $r_1 = n/4$.

Also, $T_n = \left\{ z \in C : |z - 2 + 3i| = \frac{1}{n} \right\}$

also represent a circle with centre $C_2$ $(2, -3)$ and radius $r_2 = 1/n$

**Case I** Since, $S_n \cap T_n = \phi$ $\quad$ [given]

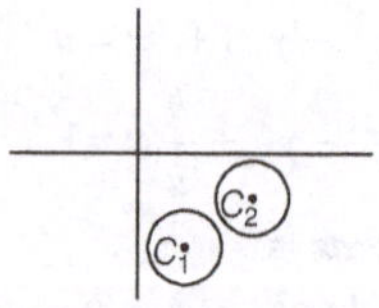


Thus, $C_1 C_2 > r_1 + r_2$

$\Rightarrow \quad \sqrt{(3 - 2)^2 + (-2 + 3)^2} > \frac{n}{4} + \frac{1}{n}$

$\Rightarrow \sqrt{2} > \frac{n^2 + 4}{4n}$

$\Rightarrow n^2 - 4\sqrt{2}n + 4 < 0 \Rightarrow n = 2\sqrt{2} \pm 2$

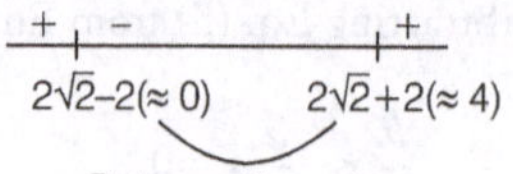


Thus, we get 4 values of $n$.

**Case II**

Since, $S_n \cap T_n = \phi$

$\Rightarrow \quad C_1 C_2 < |r_1 - r_2| \Rightarrow \sqrt{2} < \left|\frac{n}{4} - \frac{1}{n}\right|$

$\Rightarrow \quad 4\sqrt{2} < \frac{|n^2 - 4|}{n} \quad [\because 4\sqrt{2} \approx 5.6]$

Clearly, the above relation hold for $n \geq N$.

Thus, here $n$ may take infinite values $\geq N$.

**55.** *(b)* $z^2 + 3\bar{z} = 0$

On putting $z = x + iy$, we get

$$(x^2 - y^2) + i(2xy) + 3(x - iy) = 0$$

$\Rightarrow \quad (x^2 - y^2 + 3x) + i(2xy - 3y) = 0$

$$\begin{cases} x^2 - y^2 + 3x = 0 \\ y(2x - 3) = 0 \end{cases}$$

$y = 0$ or $x = 3/2$

If $y = 0$, then $x^2 + 3x = 0$

$\Rightarrow x = 0, -3$

So, (0, 0) and (−3, 0) are solutions, when $y = 0$.

If $x = \frac{3}{2}$, then

$$\frac{9}{4} - y^2 + \frac{9}{2} = 0$$

$\Rightarrow \quad y = \pm\frac{3\sqrt{3}}{2}$

$\therefore \quad (x, y) = \left(\frac{3}{2}, \frac{3\sqrt{3}}{2}\right)$ and $\left(\frac{3}{2}, -\frac{3\sqrt{3}}{2}\right)$

There are 4 solutions.

$$\sum_{k=0}^{\infty}\left(\frac{1}{n^k}\right) = 1 + \frac{1}{4} + \frac{1}{4^2} + \ldots \infty$$

$$= \frac{1}{1 - 1/4} = \frac{4}{3}$$

**56.** *(a)* Let $z = (1 - \cos\theta + 2i\sin\theta)^{-1}$

$$\Rightarrow z = \frac{1}{1 - \cos\theta + 2i\sin\theta}$$

$$= \frac{1}{1 - \cos\theta + 2i\sin\theta} \times \frac{1 - \cos\theta - 2i\sin\theta}{1 - \cos\theta - 2i\sin\theta}$$

$$= \frac{(1 - \cos\theta) - 2i\sin\theta}{(1 - \cos\theta)^2 - (2i\sin\theta)^2}$$

$$= \frac{2\sin^2\frac{\theta}{2} - 4i\sin\frac{\theta}{2}\cos\frac{\theta}{2}}{4\sin^4\frac{\theta}{2} + 16\sin^2\frac{\theta}{2}\cos^2\frac{\theta}{2}}$$

$$= \frac{\sin\frac{\theta}{2} - 2i\cos\frac{\theta}{2}}{2\sin\frac{\theta}{2}\left(\sin^2\frac{\theta}{2} + 4\cos^2\frac{\theta}{2}\right)}$$

Now,

$$\text{Re}(z)=\frac{\sin\frac{\theta}{2}}{2\sin\frac{\theta}{2}\left(\sin^2\frac{\theta}{2}+4\cos^2\frac{\theta}{2}\right)}$$

$$=\frac{1}{2\left(1+3\cos^2\frac{\theta}{2}\right)}$$

Given, $\text{Re}(z)=\frac{1}{5}$

$$\Rightarrow \frac{1}{2\left(1+3\cos^2\frac{\theta}{2}\right)}=\frac{1}{5}$$

$$\Rightarrow \cos\frac{\theta}{2}=\pm\frac{1}{\sqrt{2}}$$

$$\therefore \frac{\theta}{2}=n\pi\pm\frac{\pi}{4}$$

$$\therefore \theta=2n\pi\pm\frac{\pi}{2}$$

Given, range is $\theta\in(0,\pi)$.

$$\therefore \theta=\frac{\pi}{2}$$

Now, $\int_0^{\theta}\sin x\,dx=\int_0^{\frac{\pi}{2}}\sin x\,dx$

$$\int_0^{\frac{\pi}{2}}\sin x\,dx=[-\cos x]_0^{\pi/2}$$

$$=-(0-1)=1$$

**57.** *(c)* It is given $u=\frac{2z+i}{z-ki}$, $z=x+iy$, $k>0$

$$\Rightarrow u=\frac{2x+i(2y+1)}{x+i(y-k)}\times\frac{x-i(y-k)}{x-i(y-k)}$$

$$=\frac{[2x^2+(2y+1)(y-k)]+i[x(2y+1)-2x(y-k)]}{x^2+(y-k)^2}$$

$\therefore \text{Re}(u)+\text{Im}(u)$

$$\Rightarrow \frac{2x^2+(2y+1)(y-k)+x(2y+1)-2x(y-k)}{x^2+(y-k)^2}=1 \quad \text{[given]}$$

$$2x^2+(2y+1)(y-k)+x+2kx=x^2+(y-k)^2$$

$\because$ The above curve intersect the $Y$-axis.

So, on putting $x=0$, we get

$$2y^2+(1-2k)y-k=y^2-2ky+k^2$$

$$\Rightarrow y^2+y-(k^2+k)=0 \quad \ldots\text{(i)}$$

Now, the $y$-coordinates of points $P$ and $Q$ are roots of the Eq. (i).

So, $PQ$ = Difference of roots

$$=\sqrt{D}/|a|=\sqrt{1+4(k^2+k)}=5 \quad \text{[given]}$$

$$\Rightarrow 4k^2+4k+1=25$$

$$\Rightarrow k^2+k-6=0$$

$$\Rightarrow k=-3, 2 \Rightarrow k=2 \quad [\because k>0]$$

**58.** *(c)* Given complex expression

$$\frac{3+i\sin\theta}{4-i\cos\theta},\ \theta\in[0,2\pi]$$

$$=\frac{3+i\sin\theta}{4-i\cos\theta}\times\frac{4+i\cos\theta}{4+i\cos\theta}$$

[on rationalization]

$$=\frac{(12-\sin\theta\cos\theta)+i(4\sin\theta+3\cos\theta)}{16+\cos^2\theta}$$

$\because \frac{3+i\sin\theta}{4-i\cos\theta}$ is a real number.

So, $4\sin\theta+3\cos\theta=0$ ...(i)

Now, argument of $\sin\theta+i\cos\theta$ lies in either quadrant second or fourth because according to Eq. (i) $\sin\theta$ and $\cos\theta$ are opposite in sign, so either,

$$\arg(\sin\theta+i\cos\theta)=\pi-\tan^{-1}\left|\frac{\cos\theta}{\sin\theta}\right|$$

(for second quadrant)

$$=\pi-\tan^{-1}\left(\frac{4}{3}\right)$$

$$\left(\text{from Eq. (i), } \frac{\cos\theta}{\sin\theta}=-\frac{4}{3}\right)$$

or, $\arg(\sin\theta+i\cos\theta)=-\tan^{-1}\left|\frac{\cos\theta}{\sin\theta}\right|$

$$=-\tan^{-1}\left(\frac{4}{3}\right)$$

**59.** *(a)* It is given that, there are two complex numbers $z$ and $w$, such that $|zw|=1$

and $\arg(z)-\arg(w)=\pi/2$

$\therefore |z||w|=1 \quad [\because |z_1 z_2|=|z_1||z_2|]$

and $\arg(z)=\frac{\pi}{2}+\arg(w)$

Let $|z|=r$, then $|w|=\frac{1}{r}$ ...(i)

and let $\arg(w)=\theta$,

then $\arg(z)=\frac{\pi}{2}+\theta$ ...(ii)

So, we can assume

$$z=re^{i(\pi/2+\theta)} \quad \ldots\text{(iii)}$$

[$\because$ if $z=x+iy$ is a complex number, then it can be written as $z=re^{i\theta}$ where, $r=|z|$ and $\theta=\arg(z)$]

and $w=\frac{1}{r}e^{i\theta}$ ...(iv)

Now, $\bar{z}\cdot w=re^{-i(\pi/2+\theta)}\cdot\frac{1}{r}e^{i\theta}$

$$=e^{i(-\pi/2-\theta+\theta)}=e^{-i(\pi/2)}=-i$$

$[\because e^{-i\theta}=\cos\theta-i\sin\theta]$

and $z\bar{w}=re^{i(\pi/2+\theta)}\cdot\frac{1}{r}e^{-i\theta}$

$$=e^{i(\pi/2+\theta-\theta)}=e^{i(\pi/2)}=i$$

**• Concept Enhancer**

Euler's law states that for any real number $x$, $e^{ix}=\cos x+i\sin x$
where $e$ = base of natural logarithm
$i$ = imaginary unit
$x$ = angle in radians.

**60.** *(b)* Given,

$$z=\left(\frac{\sqrt{3}}{2}+\frac{i}{2}\right)^5+\left(\frac{\sqrt{3}}{2}-\frac{i}{2}\right)^5$$

$\because$ Euler's form of

$$\frac{\sqrt{3}}{2}+\frac{i}{2}=\left(\cos\frac{\pi}{6}+i\sin\frac{\pi}{6}\right)=e^{i(\pi/6)}$$

and $\frac{\sqrt{3}}{2}-\frac{i}{2}=\cos\left(\frac{-\pi}{6}\right)+i\sin\left(-\frac{\pi}{6}\right)$

$$=e^{-i\pi/6}$$

So, $z=(e^{i\pi/6})^5+(e^{-i\pi/6})^5$

$$=e^{i\frac{5\pi}{6}}+e^{-i\frac{5\pi}{6}}$$

$$=\left(\cos\frac{5\pi}{6}+i\sin\frac{5\pi}{6}\right)+\left(\cos\frac{5\pi}{6}-i\sin\frac{5\pi}{6}\right)$$

$[\because e^{\pm i\theta}=\cos\theta\pm i\sin\theta]$

$$=2\cos\frac{5\pi}{6};\ \therefore\ I(z)=0$$

and $R(z)=-2\cos\frac{\pi}{6}=-\sqrt{3}<0$

$$\left[\because \cos\frac{5\pi}{6}=\cos\left(\pi-\frac{\pi}{6}\right)=-\cos\frac{\pi}{6}\right]$$

**61.** *(3)* $|1-i|^x=2^x$

$$(\sqrt{2})^x=2^x$$

$$2^{\frac{x}{2}}=2^x\Rightarrow x=0$$

Number of solution $(\alpha)=1$.

Now, $z=\frac{\pi}{4}(1+i)^4\left(\frac{1-\sqrt{\pi}i}{\sqrt{\pi}+i}+\frac{\sqrt{\pi}-i}{1+\sqrt{\pi}i}\right)$

$$=\frac{\pi}{4}(1+i)^4\left(\frac{1+\pi+\pi+1}{(\sqrt{\pi}+i)(1+\sqrt{\pi}i)}\right)$$

$$=\frac{\pi}{4}(1+4i+6i^2+4i^3+1)\left(\frac{2+2\pi}{(\pi+1)}\right)$$

$$\frac{\pi}{2}\frac{(1+4i-6-4i+1)}{i}$$

$$=\frac{\pi}{2}\frac{(-4)}{i}=\frac{\pi}{2}\left(\frac{4i^2}{i}\right)=2\pi i \quad [\because i^2=-1]$$

$$z=2\pi i$$

$$\beta=\frac{|z|}{\arg(z)}=\frac{2\pi}{\frac{\pi}{2}}=4$$

Point $(\alpha, \beta)=(1, 4)$

Distance between $(\alpha, \beta)$ and $4x-3y-7=0$

$$D=\left|\frac{4-12-7}{5}\right|=3$$

**62.** *(1680)* Given that

$$S=\left\{z \in C-\{i, 2i\}: \frac{z^2+8iz-15}{z^2-3iz-2} \in R\right\}$$

Now, $\frac{z^2+8iz-15}{z^2-3iz-2} \in R$

$$\Rightarrow 1+\frac{11iz-13}{z^2-3iz-2} \in R$$

Let, $z=\alpha-\frac{13}{11}i \in S$ (given)

So, this implies,

$(z^2-3iz-2)$ is imaginary

Now, put, $z=x+iy$

$\Rightarrow \{x^2-y^2+3y-2+(2xy-3x)i\}=0$

[If $z$ is imaginary $\Rightarrow$ Re $z=0$]

$\therefore x^2-y^2+3y-2=0$

$\Rightarrow x^2=y^2-3y+2=(y-2)(y-1)$

Since, $z=\alpha-\frac{13}{11}i$

So, $x=\alpha$, $y=-\frac{13}{11}$

$$\therefore \quad \alpha^2=\left(-\frac{13}{11}-2\right)\left(\frac{-13}{11}-1\right)$$

$$=\left(\frac{-35}{11}\right)\times\left(\frac{-24}{11}\right)$$

$$\alpha^2=\frac{24\times 35}{121}$$

$$242\alpha^2=242\times\frac{24\times 35}{121}$$

$$=48\times 35=1680$$

**63.** *(9)* Given, $z=1+i \Rightarrow \bar{z}=1-i$

and $z_1=\frac{1+i\bar{z}}{\bar{z}(1-z)+\frac{1}{z}}$

$$=\frac{1+i(1-i)}{(1-i)(1-1-i)+\frac{1}{1+i}}$$

$$=\frac{2+i}{\frac{-3i-1}{2}}=\frac{4+2i}{-3i-1}$$

$$z_1=\frac{-(4+2i)(3i-1)}{(3i)^2-(1)^2}$$

$$\Rightarrow z_1=\frac{-12i+4-6i^2+2i}{-9-1}$$

$$=\frac{-10i+10}{-10}=\frac{-10(i-1)}{-10}$$

$\Rightarrow z_1=i-1=-1+i$

$\therefore \arg(z_1)=\frac{3\pi}{4}\left[\because \arg(z_1)=\tan^{-1}\left(\frac{b}{a}\right)\right]$

Hence, $\frac{12}{\pi}\arg(z_1)=\frac{12}{\pi}\times\frac{3\pi}{4}=9$

**64.** *(14)* Let $\alpha=8-14i$,

$$A=\left\{z \in c: \frac{\alpha z-\overline{\alpha z}}{z^2-(\bar{z})^2-112i}=1\right\}$$

and $B=\{z \in c: |z+3i|=4\}$

Let $z=a+ib$

Now, on putting $z=a+ib$ and $8-14i=\alpha$ in $A$, we get

$(8-14i)(a+ib)-(8+14i)(a-ib)$

$=(a+ib)^2-(a-ib)^2-112i$

$\Rightarrow i4(4b-7a)=i(4ab-112)$

$\Rightarrow 4b-7a=ab-28$ ...(i)

Rewriting Eq. (i) as,

$4b-ab=7a-28 \Rightarrow b=-7$

Again, if we rewrite Eq. (i) as,

$4b-ab=7a-28$

$\Rightarrow \quad 4b+28=7a+ab \Rightarrow a=4$

So, elements of $A$ can be seen as $4+ib$ and $a-7i$ types.

Elements of $B=\{z \in c: |z+3i|=4\}$

So, elements from the set $A \cap B$ can be seen as, $4-3i$ and $-7i$

Because, $B=\{z \notin c: |z+3i|=4\}$

$\Rightarrow |a+ib+3i|=4$

$\Rightarrow a^2+(b+3)^2=16$

For, $a=4$, $b=-3$ $b=-7$, $a=0$

So, $\sum_{z \in A \cap B}(\text{Re}\, z-\text{Im}\, z)=[4-(-3)]$

$+[0-(-7)]=7+7=14$

**65.** *(0)* $S=\{z \in C: z^2+\bar{z}=0\}$

Now, let $z=x+iy$

Then, $(x+iy)^2+(x-iy)=0$

$\Rightarrow \quad (x^2-y^2+x)+(2xy-y)i=0$

$\therefore x^2-y^2+x=0$ and $2xy-y=0$

$\Rightarrow \quad y(2x-1)=0$

$\therefore \quad y=0$ or $x=1/2$

(i) When $y=0 \quad \Rightarrow x^2+x=0$

$\Rightarrow \quad x=0$ or $-1$

$\therefore \quad z=0$ or $z=-1$

(ii) When $x=\frac{1}{2}$

$$\Rightarrow \frac{1}{4}-y^2+\frac{1}{2}=0 \Rightarrow y=\pm\frac{\sqrt{3}}{2}$$

$$\therefore z=\frac{1}{2}+\frac{\sqrt{3}}{2}i \text{ or } z=\frac{1}{2}-\frac{\sqrt{3}}{2}i$$

Now, $\sum_{z \in S}(\text{Re}(z)+\text{Im}(z))$

$$=0-1+\frac{1}{2}+\frac{1}{2}+0+0+\frac{\sqrt{3}}{2}-\frac{\sqrt{3}}{2}=0$$

**66.** *(2)* Given, $\bar{z}=iz^2+z^2-z$

Let $z=x+iy$

So, $2x=(1+i)(x^2-y^2+2xyi)$

$\Rightarrow \quad 2x=x^2-y^2-2xy$ ... (i)

and $x^2-y^2+2xy=0$ ...(ii)

From Eqs. (i) and (ii), we get

$x=0$ and $y=-\frac{1}{2}$

When $x=0$, then $y=0$

When $y=-\frac{1}{2}$, then $x^2-x-\frac{1}{4}=0$

$$\Rightarrow \quad x=\frac{-1\pm\sqrt{2}}{2}$$

So, there will be total 3 possible values of $z$, which are 0, $\left(\frac{-1+\sqrt{2}}{2}\right)-\frac{1}{2}i$

and $\left(\frac{-1-\sqrt{2}}{2}\right)-\frac{1}{2}i$

Sum of squares of modulus

$$=0+\left(\frac{\sqrt{2}-1}{2}\right)^2+\frac{1}{4}$$

$$+\left(\frac{\sqrt{2}+1}{2}\right)^2+\frac{1}{4}=2$$

**67.** *(40)* Given a relation

$1<|z-3+2i|<4$

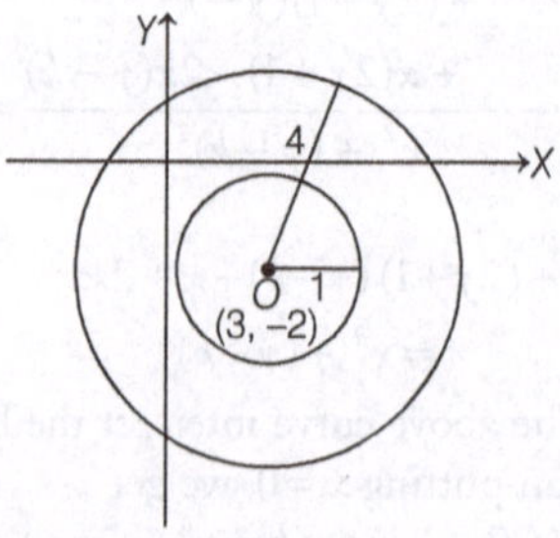


Let $z=a+ib$

$1<(a-3)^2+(b+2)^2<16$

$(0, \pm 2)$, $(\pm 2, 0)$, $(\pm 1, \pm 2)$, $(\pm 2, \pm 1)$

$(\pm 2, \pm 3)$, $(\pm 3, \pm 2)$, $(\pm 1, \pm 1)$, $(\pm 2, \pm 2)$

$(\pm 3, 0)$, $(0, \pm 3)$, $(\pm 3, \pm 1)$, $(\pm 1, \pm 3)$

Total number of points

$= 2 \times 4 + 4 \times 8 = 40$

**68.** *(5)* $|z - 2 - 2i| \le 1$

$\Rightarrow z$ lies inside the circle with centre at $2 + 2i$ and radius $= 1$, as shown in figure.

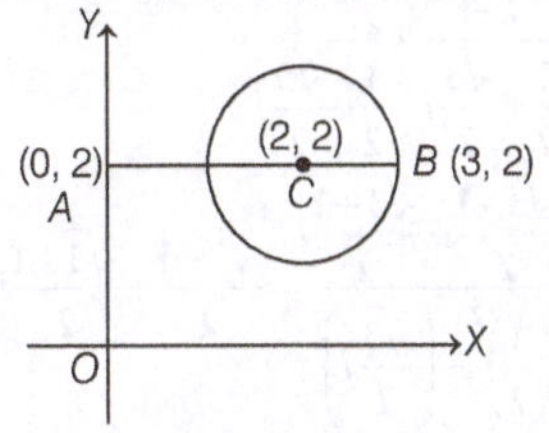


$$|3iz + 6| = |3i|\left|z + \frac{6}{3i}\right| = 3|z - 2i|$$

This is distance of $z$ from $2i$.

$\therefore z = 3 + 2i = a + ib \Rightarrow a + b = 5$

**69.** *(6)* Let $Z_1 = a_1 + ib_1, Z_2 = a_2 + ib_2$

$Z_1 - Z_2 = (a_1 - a_2) + i(b_1 - b_2)$

$\arg(Z_1 - Z_2) = \frac{\pi}{4}$

$\Rightarrow \tan^{-1}\left(\frac{b_1 - b_2}{a_1 - a_2}\right) = \frac{\pi}{4}$

$\Rightarrow b_1 - b_2 = a_1 - a_2$ ... (i)

Also, $|Z_1 - 3| = \text{Re}(Z_1)$

$\Rightarrow (a_1 - 3)^2 + b_1^2 = a_1^2$

and $|Z_2 - 3| = \text{Re}(Z_2)$

$\Rightarrow (a_2 - 3)^2 + b_2^2 = a_2^2$

$\Rightarrow (a_1 - 3)^2 - (a_2 - 3)^2 + b_1^2 - b_2^2 = a_1^2 - a_2^2$

$\Rightarrow (a_1 - a_2)(a_1 + a_2 - 6) + (b_1 - b_2)(b_1 + b_2) = (a_1 - a_2)(a_1 + a_2)$

$\Rightarrow a_1 + a_2 - 6 + b_1 + b_2 = a_1 + a_2$ [Using Eq. (i)]

$\Rightarrow b_1 + b_2 = 6$

$\Rightarrow \text{Im}(Z_1 + Z_2) = 6$

**70.** *(6)* We have,

$$\frac{(2i)^n}{(1-i)^{n-2}} = \frac{(2i)^n}{(1-i)^n(1-i)^{-2}}$$

$$= \left(\frac{2i}{1-i}\right)^n (1-i)^2$$

$$= \left[\frac{2i(1+i)}{(1-i)(1+i)}\right]^n (1 + i^2 - 2i)$$

$$= \left(\frac{2i-2}{2}\right)^n (1 - 1 - 2i) = (i-1)^n (-2i)$$

If $n = 1$, $(i-1)(-2i) = -2i^2 + 2i = 2 + 2i$

If $n = 2$, $-2i(i-1)^2 = -2i(-2i) = -4$

If $n = 4$,

$-2i(i-1)^4 = -2i(-2i)(-2i) = 8i$

If $n = 6$, $-2i(i-1)^6 = -2i(-2i)(-2i)(-2i) = 16$

So, least value of $n$ for which given complex is positive is 6.

**71.** *(1)* We have,

$$z = \frac{3 + 2i\cos\theta}{1 - 3i\cos\theta}$$

$$= \frac{3 + 2i\cos\theta}{1 - 3i\cos\theta} \times \frac{1 + 3i\cos\theta}{1 + 3i\cos\theta}$$

$$= \frac{(3 - 6\cos^2\theta) + i(9\cos\theta + 2\cos\theta)}{1 + 9\cos^2\theta}$$

$$z = \frac{(3 - 6\cos^2\theta) + (11\cos\theta)i}{1 + 9\cos^2\theta}$$

Given, $\text{Re}(z) = 0$

$$\Rightarrow \frac{3 - 6\cos^2\theta}{1 + 9\cos^2\theta} = 0$$

$\Rightarrow 3 - 6\cos^2\theta = 0$

$$\Rightarrow \cos\theta = \frac{1}{\sqrt{2}} \Rightarrow \theta = \frac{\pi}{4}, \left\{\theta \in \left(0, \frac{\pi}{2}\right)\right\}$$

Hence, $\sin^2 3\theta + \cos^2\theta$

$$= \sin^2\frac{3\pi}{4} + \cos^2\frac{\pi}{4}$$

$$= \frac{1}{2} + \frac{1}{2} = 1$$

**72.** *(4)* Given, $w = z\bar{z} - 2z + 2$

$$\left|\frac{z + i}{z - 3i}\right| = 1$$

$\Rightarrow |z + i| = |z - 3i|$

Let $z = x + iy$

$\Rightarrow |x + i(y+1)| = |x + i(y-3)|$

$\Rightarrow x^2 + (y+1)^2 = x^2 + (y-3)^2$

$\Rightarrow 2y + 1 = -6y + 9$

$\therefore y = 1$

Now, $w = z\bar{z} - 2z + 2$

$w = |z|^2 - 2z + 2$ $[\because z\bar{z} = |z|^2]$

$\Rightarrow w = x^2 + y^2 - 2(x + iy) + 2$

$\Rightarrow w = (x^2 + y^2 - 2x + 2) + i(-2y)$

$\Rightarrow w = (x^2 + 1 - 2x + 2) + i(-2)$

$w = (x-1)^2 + 2 - 2i$

$\text{Re}(w)$ has minimum value.

So, $(x-1)^2 + 2$ is minimum when $x = 1$

$\therefore w = 2 - 2i = 2(1 - i)$

$$= 2\sqrt{2}\left(\frac{1}{\sqrt{2}} - \frac{i}{\sqrt{2}}\right)$$

$w = 2\sqrt{2}\, e^{-i\pi/4}$

Now, $w^n = (2\sqrt{2})^n e^{\frac{-in\pi}{4}}$

$$= (2\sqrt{2})^n \left[\cos\left(\frac{n\pi}{4}\right) - i\underbrace{\sin\left(\frac{n\pi}{4}\right)}\right]$$

This has to be zero for $w^n$ to be real.

So, $\sin\left(\frac{n\pi}{4}\right) = 0$

$\Rightarrow \frac{n\pi}{4} = 0, \pi, 2\pi, 3\pi \ldots$

$\Rightarrow n = 0, 4, 8, 12 \ldots$

The minimum value of $n$ is 4 $(n \in N)$.

**73.** *(10)* The given equation is

$z + \alpha|z - 1| + 2i = 0$

Putting $z = x + iy$, we get

$x + iy + \alpha|x + iy - 1| + 2i = 0$

$x + \alpha\sqrt{(x-1)^2 + y^2} + i(y + 2) = 0 + 0i$

On comparing both sides,

$y + 2 = 0$

and $x + \alpha\sqrt{(x-1)^2 + y^2} = 0$

$$\Rightarrow y = -2 \text{ and } \alpha^2 = \frac{x^2}{x^2 - 2x + 5}$$

$$\text{Now, } \frac{x^2}{x^2 - 2x + 5} \in \left[0, \frac{5}{4}\right]$$

$$\therefore \alpha^2 \in \left[0, \frac{5}{4}\right] \Rightarrow \alpha \in \left[-\frac{\sqrt{5}}{2}, \frac{\sqrt{5}}{2}\right]$$

$$\therefore p = -\frac{\sqrt{5}}{2}; q = \frac{\sqrt{5}}{2}$$

$$\Rightarrow 4(p^2 + q^2) = 4\left(\frac{5}{4} + \frac{5}{4}\right) = 10$$

## Topic 2

**01.** *(b)* $21 + \left(z + \frac{1}{z}\right)^3 + \left(z^2 + \frac{1}{z^2}\right)^3 + \ldots + \left(z^{21} + \frac{1}{z^{21}}\right)^3$

Let $z = -\omega = \frac{1 - i\sqrt{3}}{2}$

$$= 21 + \left(-\omega - \frac{1}{\omega}\right)^3 + \left(\omega^2 + \frac{1}{\omega^2}\right)^3 + \ldots + \left(-\omega^{21} - \frac{1}{\omega^{21}}\right)^3$$

$[\because \omega^2 + \omega + 1 = 0 \text{ and } \omega^3 = 1]$

$= 21 - 8 = 13$

**02.** *(c)* We have, $\alpha, \beta$ are the roots of $x^2 - x + 1 = 0$

$\because$ Roots of $x^2 - x + 1 = 0$ are $-\omega, -\omega^2$

$\therefore$ Let $\alpha = -\omega$ and $\beta = -\omega^2$

$$\Rightarrow \alpha^{101} + \beta^{107} = (-\omega)^{101} + (-\omega^2)^{107}$$

$$= -(\omega^{101} + \omega^{214}) = -(\omega^2 + \omega) \quad [\because \omega^{3n} = 1]$$

$$= -(-1) = 1 \quad [\because 1 + \omega + \omega^2 = 0]$$

**03.** *(a)* $(1+\omega)^7 = A + B\omega,$

we know

$$1 + \omega + \omega^2 = 0$$

$$\therefore \quad 1 + \omega = -\omega^2$$

$$\Rightarrow \quad (-\omega^2)^7 = A + B\omega$$

$$\Rightarrow \quad -\omega^{14} = A + B\omega$$

$$\Rightarrow \quad -\omega^2 = A + B\omega \quad [\because \omega^{14} = \omega^{12} \cdot \omega^2 = \omega^2]$$

$$\Rightarrow \quad 1 + \omega = A + B\omega \quad [\because \omega^2 + \omega + 1 = 0]$$

On comparing, we get

$$A = 1, B = 1$$

**04.** *(c)* $\sum_{k=1}^{10}\left(\sin\frac{2k\pi}{11} + i\cos\frac{2k\pi}{11}\right)$

$$= i\sum_{k=1}^{10}\left(\cos\frac{2k\pi}{11} - i\sin\frac{2k\pi}{11}\right)$$

$$= i\sum_{k=1}^{10}\left(e^{-\frac{2k\pi}{11}i}\right)$$

$$= i\left\{\sum_{k=0}^{10}\left(e^{-\frac{2k\pi}{11}i}\right) - 1\right\}$$

$$= -i$$

$\left[\because \text{ sum of roots of unity is zero } \Rightarrow \sum_{k=0}^{10} e^{-\frac{2k\pi}{11}i} = 0\right]$

**● Concept Enhancer**

If $z^n = 1$, therefore $\omega$ is an $n$th roots of unity, the complex number $1, \omega, \omega^2, \ldots, \omega^{n-1}$ are $n$th roots of unity.

$\because$ The $n$th roots of unity $1, \omega, \omega^2, \ldots \omega^{n-1}$ are in GP with common ratio $\omega$.

So, the sum of all the $n$th roots of unity is $1 + \omega + \ldots + \omega^{n-1} = 0$

**05.** *(d)* Now, $(1 + \omega - \omega^2)^7 = (-\omega^2 - \omega^2)^7$

$$[\because 1 + \omega + \omega^2 = 0]$$

$$= (-2\omega^2)^7 = -2^7 \cdot \omega^{14}$$

$$= -128(\omega^3)^4 \omega^2$$

$$= -128\omega^2 \quad [\because \omega^3 = 1]$$

**● Concept Enhancer**

If $1, \omega, \omega^2$ are three cube roots of unity, then $1 + \omega + \omega^2 = 0$.

**06.** *(a)* Given, $z^3 + 2z^2 + 2z + 1 = 0 \quad \ldots(i)$

$$(z+1)(z^2 + z + 1) = 0$$

$$\Rightarrow \quad z = -1,\ z^2 + z + 1 = 0$$

$$z = \omega, \omega^2$$

$\because$ Eq. (i) have common roots with

$$z^{1985} + z^{100} + 1 = 0$$

Put $z = -1$

$\Rightarrow -1 + 1 + 1 = 0$

$\Rightarrow \ 1 = 0$ [false]

$\therefore$ $-1$ is not common root.

Put $z = \omega$

$\Rightarrow \quad \omega^{1985} + \omega^{100} + 1 = 0$

$\Rightarrow \quad \omega^2 + \omega + 1 = 0$ [true]

$\therefore$ $\omega$ is common root.

Put $z = \omega^2$

$\Rightarrow \quad \omega^{3970} + \omega^{200} + 1 = 0$

$\Rightarrow \quad \omega + \omega^2 + 1 = 0$ [true]

$\therefore$ $\omega^2$ is also common root.

Now, number of common roots = 2

**07.** *(a)* We have,

$$z = \frac{i - 1}{\cos\frac{\pi}{3} + i\sin\frac{\pi}{3}}$$

$$= \frac{-1 + i}{\cos\frac{\pi}{3} + i\sin\frac{\pi}{3}}$$

$$= \frac{\sqrt{2}\left(-\frac{1}{\sqrt{2}} + \frac{1}{\sqrt{2}}i\right)}{\cos\frac{\pi}{3} + i\sin\frac{\pi}{3}}$$

$$= \frac{\sqrt{2}\left(\cos\frac{3\pi}{4} + i\sin\frac{3\pi}{4}\right)}{\cos\frac{\pi}{3} + i\sin\frac{\pi}{3}}$$

$$= \frac{\sqrt{2}\,e^{i3\pi/4}}{e^{i\pi/3}} = \sqrt{2}\,e^{i\left(\frac{3\pi}{4} - \frac{\pi}{3}\right)}$$

$$= \sqrt{2}\,e^{\frac{i5\pi}{12}}$$

$$= \sqrt{2}\left(\cos\frac{5\pi}{12} + i\sin\frac{5\pi}{12}\right)$$

**Alternate Solution**

$$z = \frac{i - 1}{\cos\frac{\pi}{3} + i\sin\frac{\pi}{3}}$$

$$= \frac{i - 1}{\frac{1}{2} + \frac{\sqrt{3}}{2}i} \times \frac{\frac{1}{2} - \frac{\sqrt{3}}{2}i}{\frac{1}{2} - \frac{\sqrt{3}}{2}i}$$

$$= \frac{\frac{\sqrt{3}-1}{2} + \frac{\sqrt{3}+1}{2}i}{\left(\frac{1}{2}\right)^2 - \left(\frac{\sqrt{3}}{2}i\right)^2} = \frac{\sqrt{3}-1}{2} + \frac{\sqrt{3}+1}{2}i$$

Here, $r\cos\theta = \frac{\sqrt{3}-1}{2}, r\sin\theta = \frac{\sqrt{3}+1}{2}$

$$\therefore \tan\theta = \frac{\sqrt{3}+1}{\sqrt{3}-1} \Rightarrow \theta = \frac{5\pi}{12}$$

and $r^2 = 2 \Rightarrow r = \sqrt{2}$

$$\therefore \quad z = \sqrt{2}\left(\cos\frac{5\pi}{12} + i\sin\frac{5\pi}{12}\right)$$

**08.** *(a)*

**key idea** For cube roots of unity

$\because x^3 = 1 \Rightarrow x^3 - 1 = 0$

$(x - 1)(x^2 + x + 1) = 0 \Rightarrow x = 1$ or $x^2 + x + 1 = 0$

$x = \frac{-1 \pm \sqrt{1-4}}{2}$; $x = \frac{-1 + i\sqrt{3}}{2}$, $\frac{-1 - i\sqrt{3}}{2}$ and 1

$\omega = \frac{-1 + i\sqrt{3}}{2}$, $\omega^2 = \frac{-1 - i\sqrt{3}}{2}$

and $\omega^2 + \omega = -1$

Given, $x^4 + x^2 + 1 = 0$

$$\Rightarrow \quad (x^2 + x + 1)(x^2 - x + 1) = 0$$

$$\Rightarrow \quad x^2 + x + 1 = 0 \text{ or } x^2 - x + 1 = 0$$

$$\Rightarrow \quad x = \frac{-1 \pm i\sqrt{3}}{2} \text{ or } x = \frac{1 \pm i\sqrt{3}}{2}$$

$$\Rightarrow \quad x = \omega, \omega^2 \text{ or } x = -\omega, -\omega^2$$

$$\therefore \quad x = \pm\omega, \pm\omega^2$$

Therefore, $\alpha^{1011} + \alpha^{2022} - \alpha^{3033}$

$$= 1 + 1 - 1 = 1$$

$$[\text{when } x = \alpha = \omega \text{ or } \omega^2, \because \omega^3 = 1]$$

$$\alpha^{1011} + \alpha^{2022} - \alpha^{3033}$$

$$= -1 + 1 - (-1)$$

$$[\text{when } x = \alpha = -\omega \text{ or } -\omega^2] = 1$$

**09.** *(c)* It is given that, $(2 + \alpha)^4 = a + b\alpha$, where $a, b \in R$ and $\alpha = \frac{-1 + i\sqrt{3}}{2}$.

$\therefore \alpha = \dfrac{-1+i\sqrt{3}}{2}$ is non-real complex root of unity and we know that $\alpha^3 = 1$ and $1+\alpha+\alpha^2 = 0$, so

$(2+\alpha)^4 = 16+32\alpha+24\alpha^2+8\alpha^3+\alpha^4$

$= 16+8\alpha+24(\alpha+\alpha^2)+8+\alpha$ $[\because \alpha^3 = 1]$

$= 24+9\alpha-24$ $[\because 1+\alpha+\alpha^2 = 0]$

$9\alpha = a+b\alpha$ [given]

$\therefore a = 0$ and $b = 9$

$\therefore a+b = 9$

**10.** *(c)* Let $z = re^{i\theta}$

$\Rightarrow \dfrac{\text{Im } z^5}{(\text{Im } z)^5} = \dfrac{r^5(\sin 5\theta)}{r^5(\sin\theta)^5} = \dfrac{\sin 5\theta}{\sin^5\theta}$

On differentiating w.r.t. 'θ', we get

$\dfrac{dz}{d\theta} = \dfrac{\sin^5\theta \cdot 5\cos 5\theta - 5\sin 5\theta \sin^4\theta\cos\theta}{(\sin^5\theta)^2}$

Putting $\dfrac{dz}{d\theta} = 0$

$\Rightarrow 5\sin^4\theta(\sin\theta\cos 5\theta - \cos\theta\sin 5\theta) = 0$

$\Rightarrow \sin\theta = 0$ or $\sin(-4\theta) = 0$

$\Rightarrow \theta = n\pi$ or $\theta = n\pi/4$, where $n \in z$

As $z$ is non-real complex number.

$\therefore$Only $\theta = n\pi/4$ is possible.

$$f\left(\frac{\pi}{4}\right) = \frac{\sin\frac{5\pi}{4}}{\left(\sin\frac{\pi}{4}\right)^5} = -4$$

**11.** *(b)* Given that, $(x-1)^3 + 8 = 0$

$\Rightarrow (x-1)^3 = (-2)^3 \Rightarrow \left(\dfrac{x-1}{-2}\right)^3 = 1$

$\Rightarrow \left(\dfrac{x-1}{-2}\right) = (1)^{1/3}$

$\therefore$ Cube roots of $\left(\dfrac{x-1}{-2}\right)$ are 1, $\omega$ and $\omega^2$.

Cube roots of $(x-1)$ are $-2$, $-2\omega$ and $-2\omega^2$.

Cube roots of $x$ are $-1, 1-2\omega$ and $1-2\omega^2$.

**12.** *(d)* $z_1 = \sqrt{3}+2\sqrt{2}i$

$\sqrt{3}|z_2| = |z_1| \Rightarrow \left|\dfrac{z_2}{z_1}\right| = \dfrac{1}{\sqrt{3}}$

Also, $\arg(z_2) = \arg(z_1) + \dfrac{\pi}{6}$

$\arg\left(\dfrac{z_2}{z_1}\right) = \dfrac{\pi}{6}$

$\therefore \quad z_2 = \left|\dfrac{z_2}{z_1}\right| \cdot z_1 e^{i(\pi/6)}$

$z_2 = \dfrac{(\sqrt{3}+2\sqrt{2}i)(\sqrt{3}+i)}{2\sqrt{3}}$

$z_2 = \dfrac{(3-2\sqrt{2})+i(2\sqrt{6}+\sqrt{3})}{2\sqrt{3}}$

$z_1 - z_2 = \dfrac{3+2\sqrt{2}+i(2\sqrt{6}-\sqrt{3})}{2\sqrt{3}}$

$|z_1 - z_2| = |z_2|$

$\therefore \Delta ABO$ = isosceles triangle and angles are $\dfrac{\pi}{6}, \dfrac{\pi}{6}, \dfrac{2\pi}{3}$

$\therefore$ Obtuse angle triangle.

**13.** *(a)* Given, $(1-\sqrt{3}i)^{200} = 2^{199}(p+iq)$

$\Rightarrow \left[2\left(\dfrac{1}{2}-\dfrac{\sqrt{3}}{2}i\right)\right]^{200} = 2^{199}(p+iq)$

$\Rightarrow \dfrac{2^{200}}{2^{199}}\left[\cos\dfrac{\pi}{3} - i\sin\dfrac{\pi}{3}\right]^{200} = p+iq$

$\Rightarrow 2\left[\cos\dfrac{200\pi}{3} + i\sin\left(-\dfrac{200\pi}{3}\right)\right] = p+iq$

(using De-moivre's theorem)

$[\because \sin(-\theta) = -\sin\theta]$

$\Rightarrow 2\left[-\dfrac{1}{2} - i\dfrac{\sqrt{3}}{2}\right] = p+iq$

$\Rightarrow -1-\sqrt{3}i = p+iq$

So, $p = -1$, $q = -\sqrt{3}$

Let $x_1 = p+q+q^2 = (-1)+(-\sqrt{3}) + (-\sqrt{3})^2 = 2-\sqrt{3}$

and $x_2 = p-q+q^2 = (-1)-(-\sqrt{3}) + (-\sqrt{3})^2$

$= 2+\sqrt{3}$

Now, sum of roots $= x_1 + x_2 = 4$

and product of roots $= x_1 x_2 = 1$

Hence, the required equation whose roots are $x_1$ and $x_2$ is given by

$x^2 - (x_1+x_2)x + x_1x_2 = 0$

$\Rightarrow x^2 - 4x + 1 = 0$

> **Concept Enhancer**
>
> By De-moivre's formula, mathematical statement for any real number $x$, we have $(\cos x + \sin x)^n = \cos(nx) + i\sin(nx)$

**14.** *(a)* Given expression,

$$\left(\frac{1+\sin\frac{2\pi}{9}+i\cos\frac{2\pi}{9}}{1+\sin\frac{2\pi}{9}-i\cos\frac{2\pi}{9}}\right)^3$$

$$= \left[\frac{\left(\sin^2\frac{2\pi}{9} - i^2\cos^2\frac{2\pi}{9}\right) + \left(\sin\frac{2\pi}{9}+i\cos\frac{2\pi}{9}\right)}{1+\sin\frac{2\pi}{9}-i\cos\frac{2\pi}{9}}\right]^3$$

$$= \left[\frac{\left(\sin\frac{2\pi}{9}+i\cos\frac{2\pi}{9}\right)\cdot\left\{\sin\frac{2\pi}{9}-i\cos\frac{2\pi}{9}+1\right\}}{1+\sin\frac{2\pi}{9}-i\cos\frac{2\pi}{9}}\right]^3$$

$= \left(\sin\dfrac{2\pi}{9}+i\cos\dfrac{2\pi}{9}\right)^3$

$= \left(-i^2\sin\dfrac{2\pi}{9}+i\cos\dfrac{2\pi}{9}\right)^3$

$= i^3\left[\cos\dfrac{2\pi}{9} - i\sin\dfrac{2\pi}{9}\right]^3$

$= -i\left[\cos\left(-\dfrac{2\pi}{9}\right) + i\sin\left(-\dfrac{2\pi}{9}\right)\right]^3$

$= -i\left[\cos\dfrac{2\pi}{3} - i\sin\dfrac{2\pi}{3}\right]$

[According the De-Moivre's theorem]

$= -i\left[\left(-\dfrac{1}{2}\right) - i\dfrac{\sqrt{3}}{2}\right]$

$= -\dfrac{\sqrt{3}}{2} + \dfrac{i}{2}$

$= -\dfrac{1}{2}(\sqrt{3}-i)$

> **Concept Enhancer**
>
> De Moivre's theorem : It states that for any integer $n$, $(\cos\theta + i\sin\theta)^n = \cos(n\theta) + i\sin(n\theta)$ where $n$ is positive integer and '$i$' is the imaginary unit and $n$ is a rational number.

**15.** *(b)* Given complex number

$$\omega=\frac{5+3z}{5(1-z)}$$

$\Rightarrow\ 5\omega-5\omega z=5+3z$

$\Rightarrow\ (3+5\omega)z=5\omega-5$

$\Rightarrow\ |3+5\omega||z|=|5\omega-5|$ ...(i)

[applying modulus both sides and $|z_1z_2|=|z_1||z_2|$]

$\because\ |z|<1$

$\therefore\ |3+5\omega|>|5\omega-5|$ [from Eq. (i)]

$\Rightarrow\ \left|\omega+\frac{3}{5}\right|>|\omega-1|$

Let $\omega=x+iy$, then

$$\left(x+\frac{3}{5}\right)^2+y^2>(x-1)^2+y^2$$

$\Rightarrow\ x^2+\frac{9}{25}+\frac{6}{5}x>x^2+1-2x$

$\Rightarrow\ \frac{16x}{5}>\frac{16}{25}$

$\Rightarrow\ x>\frac{1}{5}$

$\Rightarrow\ 5x>1\Rightarrow 5\,\text{Re}(\omega)>1$

**16.** *(c)*

> **key idea** Use Euler formula, $e^{i\theta}=\cos\theta+i\sin\theta$

Given, $z=\frac{\sqrt{3}}{2}+\left(\frac{1}{2}\right)i$

$$=\cos\frac{\pi}{6}+i\sin\frac{\pi}{6}=e^{i\frac{\pi}{6}}$$

So, $(1+iz+z^5+iz^8)^9$

$$=\left(1+ie^{i\frac{\pi}{6}}+e^{i\frac{5\pi}{6}}+ie^{i\frac{8\pi}{6}}\right)^9$$

$$=\left(1+e^{i\frac{\pi}{2}}\cdot e^{i\frac{\pi}{6}}+e^{i\frac{5\pi}{6}}+e^{i\frac{\pi}{2}}\cdot e^{i\frac{4\pi}{3}}\right)^9$$

$$\left[\because i=e^{i\frac{\pi}{2}}\right]$$

$$=\left(1+e^{i\frac{2\pi}{3}}+e^{i\frac{5\pi}{6}}+e^{i\frac{11\pi}{6}}\right)^9$$

$$=\left[1+\left(\cos\frac{2\pi}{3}+i\sin\frac{2\pi}{3}\right)+\left(\cos\frac{5\pi}{6}+i\sin\frac{5\pi}{6}\right)+\left(\cos\frac{11\pi}{6}+i\sin\frac{11\pi}{6}\right)\right]^9$$

$$=\left(1-\frac{1}{2}+\frac{i\sqrt{3}}{2}-\frac{\sqrt{3}}{2}+\frac{1}{2}i+\frac{\sqrt{3}}{2}-\frac{i}{2}\right)^9$$

$$=\left(\frac{1}{2}+\frac{\sqrt{3}i}{2}\right)^9=\left(\cos\frac{\pi}{3}+i\sin\frac{\pi}{3}\right)^9$$

$=\cos 3\pi+i\sin 3\pi$

[$\because$ for any natural number '$n$' $(\cos\theta+i\sin\theta)^n=\cos(n\theta)+i\sin(n\theta)$] $=-1$

**17.** *(c)* We have, $x^2+2x+2=0$

$\Rightarrow\ x=\frac{-2\pm\sqrt{4-8}}{2}$

[$\because$ roots of $ax^2+bx+c=0$ are given by $x=\frac{-b\pm\sqrt{b^2-4ac}}{2a}$]

$\Rightarrow\ x=-1\pm i$

Let $\alpha=-1+i$ and $\beta=-1-i$.

Then, $\alpha^{15}+\beta^{15}=(-1+i)^{15}+(-1-i)^{15}$

$=-[(1-i)^{15}+(1+i)^{15}]$

$$=-\left[\left\{\sqrt{2}\left(\frac{1}{\sqrt{2}}-\frac{i}{\sqrt{2}}\right)\right\}^{15}+\left\{\sqrt{2}\left(\frac{1}{\sqrt{2}}+\frac{i}{\sqrt{2}}\right)\right\}^{15}\right]$$

$$=-\left[\left\{\sqrt{2}\left(\cos\frac{\pi}{4}-i\sin\frac{\pi}{4}\right)\right\}^{15}+\left\{\sqrt{2}\left(\cos\frac{\pi}{4}+i\sin\frac{\pi}{4}\right)\right\}^{15}\right]$$

$$=-(\sqrt{2})^{15}\left[\left(\cos\frac{15\pi}{4}-i\sin\frac{15\pi}{4}\right)+\left(\cos\frac{15\pi}{4}+i\sin\frac{15\pi}{4}\right)\right]$$

[using De Moivre's theorem $(\cos\theta\pm i\sin\theta)^n=\cos n\theta\pm i\sin n\theta$, $n\in Z$]

$$=-(\sqrt{2})^{15}\left[2\cos\frac{15\pi}{4}\right]$$

$$=-(\sqrt{2})^{15}\left[2\times\frac{1}{\sqrt{2}}\right]$$

$$\left[\because\cos\frac{15\pi}{4}=\cos\left(4\pi-\frac{\pi}{4}\right)=\cos\frac{\pi}{4}=\frac{1}{\sqrt{2}}\right]$$

$=-(\sqrt{2})^{16}=-2^8$

$=-256$

> **Alternate Solution**
>
> $\alpha^{15}+\beta^{15}=(-1+i)^{15}+(-1-i)^{15}$
>
> $=-[(1-i)^{15}+(1+i)^{15}]$
>
> $$=-\left[\frac{(1-i)^{16}}{1-i}+\frac{(1+i)^{16}}{1+i}\right]$$
>
> $$=-\left[\frac{[(1-i)^2]^8}{1-i}+\frac{[(1+i)^2]^8}{1+i}\right]$$
>
> $$=-\left[\frac{[1+i^2-2i]^8}{1-i}+\frac{[1+i^2+2i]^8}{1+i}\right]$$
>
> $$=-\left[\frac{(-2i)^8}{1-i}+\frac{(2i)^8}{1+i}\right]=-2^8\left[\frac{1}{1-i}+\frac{1}{1+i}\right]$$
>
> [$\because i^{4n}=1, n\in Z$]
>
> $$=-256\left[\frac{2}{1-(i)^2}\right]=-256\left[\frac{2}{2}\right]=-256$$

**18.** *(b)* Here, $z=1+ai$

$\Rightarrow z^2=1-a^2+2ai$

So, $z^2\cdot z=\{(1-a^2)+2ai\}\{1+ai\}$

$=(1-a^2)+2ai+(1-a^2)ai-2a^2$

$\because z^3$ is real

$\Rightarrow 2a+(1-a^2)a=0\Rightarrow a(3-a^2)=0$

$\Rightarrow a=\sqrt{3}$ [$\because a>0$]

Now, $1+z+z^2+\ldots+z^{11}$

$$=\frac{(1+\sqrt{3}i)^{12}-1}{1+\sqrt{3}i-1}=\frac{(1+\sqrt{3}i)^{12}-1}{\sqrt{3}i}$$

[$\because$ put the value $z=1+\sqrt{3}i$]

Now, $(1+\sqrt{3}i)^{12}=2^{12}\left(\frac{1}{2}+\frac{\sqrt{3}}{2}i\right)^{12}$

$$=2^{12}\left(\cos\frac{\pi}{3}+i\sin\frac{\pi}{3}\right)^{12}$$

$=2^{12}(\cos 4\pi+i\sin 4\pi)=2^{12}$

$\therefore 1+z+z^2+\ldots+z^{11}$

$$=\frac{2^{12}-1}{\sqrt{3}i}=\frac{4095}{\sqrt{3}i}$$

$$=-\frac{4095}{3}\sqrt{3}i=-1365\sqrt{3}i$$

**19.** *(10)* We have,

$a,b\in I,\ -3\le a,b\le 3,\ a+b\ne 0$

Given, $\left|\frac{2-a}{2+b}\right|=1\Rightarrow |z-a|=|z+b|$

and $$\begin{vmatrix} z+1 & \omega & \omega^2\\ \omega & z+\omega^2 & 1\\ \omega^2 & 1 & z+\omega\end{vmatrix}=1$$

$R_1\to R_1+R_2+R_3$

$$\Rightarrow \begin{vmatrix} z & z & z \\ \omega & z+\omega^2 & 1 \\ \omega^2 & 1 & z+\omega \end{vmatrix} = 1$$

$$\Rightarrow z\begin{vmatrix} 1 & 1 & 1 \\ \omega & z+\omega^2 & 1 \\ \omega^2 & 1 & z+\omega \end{vmatrix} = 1$$

$$\Rightarrow z\begin{vmatrix} 1 & 0 & 0 \\ \omega & z+\omega^2-\omega & 1-\omega \\ \omega^2 & 1-\omega^2 & z+\omega-\omega^2 \end{vmatrix} = 1$$

$\Rightarrow \quad z^3 = 1 \Rightarrow z = w, w^2, 1$

Now, $|1-a| = |1+b|$

$\Rightarrow$ 10 pairs

**20.** *(49)* $\alpha$ and $\beta$ are roots of equation

$$x^2 - \sqrt{6}x + 3 = 0$$

$$x = \frac{\sqrt{6} \pm \sqrt{6-12}}{2} = \frac{\sqrt{6} \pm \sqrt{6}i}{2}$$

$$= \frac{\sqrt{6}}{2}(1 \pm i)$$

Im $(\alpha) >$ Im $(\beta)$

$$\alpha = \frac{\sqrt{6}}{2}(1+i), \beta = \frac{\sqrt{6}}{2}(1-i)$$

$$\alpha = \frac{\sqrt{6}}{2}\sqrt{2}\cdot e^{i\frac{\pi}{4}} = \sqrt{3}\,e^{i\frac{\pi}{4}}$$

$$\alpha^{98} = (\sqrt{3})^{98}\,(e^{i\,98\times\frac{\pi}{4}})$$

$$= 3^{49}\,[e^{i\left(49\frac{\pi}{2}\right)}]$$

$$= 3^{49}\left(0 + i\sin 49\frac{\pi}{2}\right) = i\,(3^{49})$$

$$\therefore \frac{\alpha^{99}}{\beta} + \alpha^{98} = \alpha^{98}\left(\frac{\alpha}{\beta} + 1\right)$$

$$\Rightarrow i\left[3^{49}\left(\frac{1+i}{1-i} + 1\right)\right]$$

$$= i\,(3)^{49}\left(\frac{1+i+1-i}{1-i}\right)$$

$$= \frac{i}{1-i}(3)^{49}(2)$$

$$= \frac{i}{1-i} \times \left(\frac{1+i}{1+i}\right)(3)^{49}\,(2)$$

$$= \frac{i-1}{1+1}\cdot(3^{49})\cdot 2 = 3^{49}\,(-1+i)$$

$\therefore \quad a = -1, b = 1, n = 49$

$n + a + b = 49$

**21.** *(6)* Given, $z^2 = \bar{z}\cdot 2^{1-|z|}$

$$|z^2| = |\bar{z}|\cdot 2^{1-|z|}$$

$$|z|^2 = |\bar{z}| \times 2^{1-|z|}$$

$$|z| = 2^{1-|z|}$$

This is possible, when $|z| = 1$, as $|z| \neq 0$

$$\sqrt{a^2 + b^2} = 1 \Rightarrow a^2 + b^2 = 1$$

Again, $z^2 = \bar{z}\cdot 2^{1-|z|}$

$$(a+ib)^2 = (a-ib) \times 2^{1-1}$$

$[\because |z| = 1]$

$\Rightarrow \quad a^2 + i^2b^2 + 2abi = a - ib$

$\Rightarrow \quad a^2 - b^2 + 2abi = a - ib$

Comparing the real parts and imaginary part, we get

$$a^2 - b^2 = a \text{ and } 2ab = -b$$

$\Rightarrow \quad 2a = -1$ or $a = -\frac{1}{2} \quad [\because b \neq 0]$

Now, $\frac{1}{4} - b^2 = -\frac{1}{2}$

$\Rightarrow \quad b^2 = 3/4 \quad b = \pm\frac{\sqrt{3}}{2}$

$\therefore z = -\frac{1}{2} \pm \frac{\sqrt{3}}{2}i$, i.e. $z = \omega$ and $\omega^2$

$$z^n = (z+1)^n \Rightarrow \left(\frac{z+1}{z}\right)^n = 1$$

$$\Rightarrow \left(1 + \frac{1}{z}\right)^n = 1 \Rightarrow \left(1 + \frac{1}{\omega}\right)^n = 1$$

$$\Rightarrow \quad (1+\omega^2)^n = 1 \Rightarrow (-\omega)^n = 1$$

$[\because 1 + \omega + \omega^2 = 0]$

$\therefore$ The minimum value of $n$ is 6.

**22.** *(3)* Given, $x^3 - 2x^2 + 2x - 1 = 0$

i.e. $(x^3 - 1) - (2x^2 - 2x) = 0$

$\Rightarrow (x-1)(x^2 + x + 1) - 2x(x-1) = 0$

$\Rightarrow \quad (x-1)(x^2 + x + 1 - 2x) = 0$

$\Rightarrow \quad (x-1)(x^2 - x + 1) = 0$

$$\therefore x = 1 \text{ and } x = \frac{-(-1) \pm \sqrt{1-4}}{2}$$

$$= \frac{1 \pm \sqrt{3}i}{2}$$

$\therefore$ Roots are $1, -\omega, -\omega^2$.

Then, sum of 162· power of the roots

$$= (1)^{162} + (-\omega)^{162} + (-\omega^2)^{162}$$

$$= 1 + \omega^{162} + \omega^{324}$$

$$= 1 + (\omega^3)^{54} + (\omega^3)^{108}$$

$$= 1 + (1)^{54} + (1)^{108} \quad [\because \omega^3 = 1]$$

$$= 1 + 1 + 1 = 3$$

## Topic 3

**01.** *(a)* $\frac{z-i}{z+i} = \frac{\bar{z}+i}{\bar{z}-i}$

$\Rightarrow z\bar{z} - i\bar{z} - iz - 1 = z\bar{z} + zi + i\bar{z} - 1$

$\Rightarrow \quad \bar{z} + z = 0$

$\Rightarrow \quad x = 0 \quad$ ($Y$-axis)

$|z| = 1$

$\therefore \quad z = i \quad (z \neq -i$ is given)

Statement I in incorrect

$$\frac{z-1}{\bar{z}+1} + \frac{\bar{z}-1}{z+1} = 0$$

$\Rightarrow z\bar{z} + z - \bar{z} - 1 + z\bar{z} - z + \bar{z} - 1 = 0$

$\Rightarrow \quad z\bar{z} = 1 \Rightarrow |\bar{z}| = 1$

Statement II is correct.

**02.** *(d)* According to the question,

$$\left|\frac{z_1 - 2z_2}{\frac{1}{2} - z_1\bar{z}_2}\right| = 2 \Rightarrow \frac{|z_1 - 2z_2|}{\left|\frac{1}{2} - z_1\bar{z}_2\right|} = 2$$

$(\because$ modulus is distributive)

$$\Rightarrow |z_1 - 2z_2| = 2\left|\frac{1}{2} - z_1\bar{z}_2\right|$$

Squaring both sides,

$$|z_1 - 2z_2|^2 = |1 - 2z_1\bar{z}_2|^2$$

$$\Rightarrow (z_1 - 2z_2)(\bar{z}_1 - 2\bar{z}_2)$$

$$= (1 - 2z_1\bar{z}_2)(1 - 2\bar{z}_1z_2)$$

$(\because |z|^2 = z\bar{z})$

$$\Rightarrow |z_1|^2 + 4|z_2|^2 - 2\bar{z}_1z_2 - 2z_1\bar{z}_2$$

$$= 1 - 2\bar{z}_1z_2 - 2z_1\bar{z}_2 + 4|z_1|^2|z_2|^2$$

$$\Rightarrow |z_1|^2 + 4|z_2|^2 - 4|z_1|^2|z_2|^2 - 1 = 0$$

$$\Rightarrow (|z_1|^2 - 1)(1 - 4|z_2|^2) = 0$$

$$\Rightarrow \quad |z_1| = 1 \text{ and } |z_2| = \frac{1}{2}$$

**03.** *(d)* If $|z| \geq 1 \Rightarrow z$ lie out side or on circle $|z| = 1$

$\left|z + \frac{3}{2} + 2i\right| \Rightarrow$ distance of $z$ from $\left(-\frac{3}{2}, -2\right)$

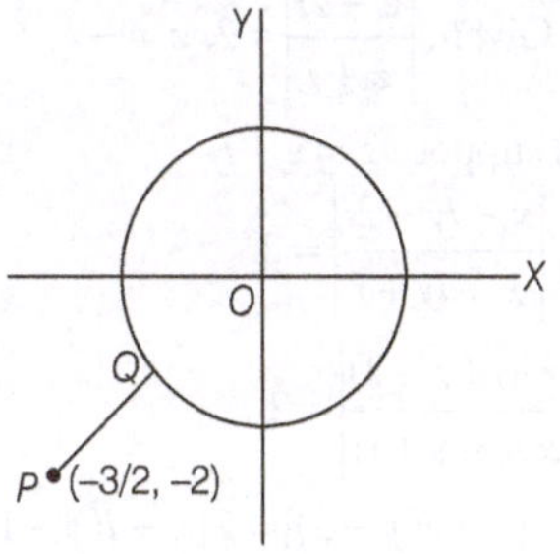


So, minimum value of $PQ$

$$= PO - QO$$

$$= \sqrt{\left(\frac{-3}{2} - 0\right)^2 + (-2-0)^2} - 1$$

$$= \sqrt{\frac{9}{4} + 4} - 1 \quad [\because \cos\theta = 1]$$

$$=\sqrt{\frac{25}{4}}-1=\frac{5}{2}-1=\frac{3}{2}$$

So, minimum value of $\left|z+\frac{1}{2}(3+4i)\right|$ is $\frac{3}{2}$.

**04.** *(c)* Given, $|z-i|=|z+i|=|z-1|$ ...(i)

From Eq. (i), we observe that, geometrically, $z$ will be circumcentre of triangle with vertices $A(1)$, $B(i)$, $C(-i)$ as shown in figure below

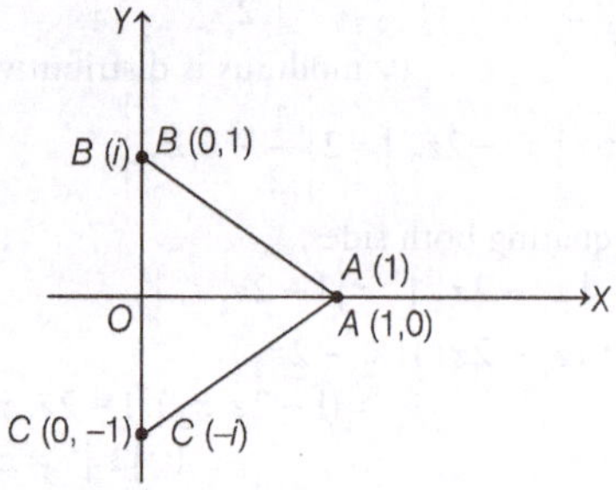


So, only one element exists in the set $S$. Hence, $n(S)=1$

**05.** *(a)* Let $w=z+\frac{1}{z}=4e^{i\theta}+\frac{1}{4}e^{-i\theta}$

$$\Rightarrow w=\frac{17}{4}\cos\theta+i\frac{15}{4}\sin\theta$$

So, locus of $w$ is ellipse

$$\frac{x^2}{\left(\frac{17}{4}\right)^2}+\frac{y^2}{\left(\frac{15}{4}\right)^2}=1$$

Locus of $z$ is circle $x^2+y^2=16$

So, intersect at 4 points.

**06.** *(c)* Given, $\left|\frac{z-2i}{z+i}\right|=2,\ z\neq -i$

Let's suppose $z=x+iy$

So, $\left|\frac{x+iy-2i}{x+iy+i}\right|=2$

$$\Rightarrow \left|\frac{x+i(y-2)}{x+i(y+1)}\right|=2$$

$$\Rightarrow |x+i(y-2)|=2|x+i(y+1)|$$

So, $\sqrt{x^2+(y-2)^2}=2\sqrt{x^2+(y+1)^2}$

Now, squaring both the sides, we get

$$x^2+y^2-4y+4=4(x^2+y^2+2y+1)$$

$$\Rightarrow 3x^2+3y^2+12y=0$$

or $x^2+y^2+4y=0$

$$\Rightarrow x^2+y^2+4y+4=4$$

$$\Rightarrow x^2+(y+2)^2=4$$

$\therefore$ Centre $=(0,-2)$

**07.** *(b)* Given, $w=1-\sqrt{3}i$

$$\Rightarrow |w|=\sqrt{(1)^2+(-\sqrt{3})^2}=\sqrt{1+3}=2$$

and $|zw|=1\Rightarrow |z||w|=1$

$$\Rightarrow |z|=\frac{1}{|w|}=\frac{1}{2}$$

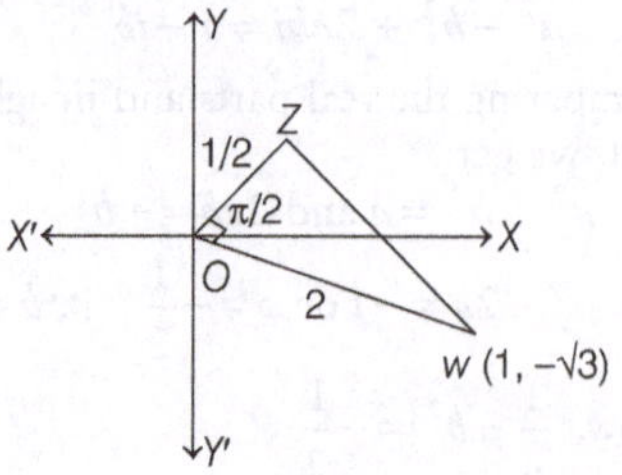


$$\therefore \text{Area of triangle}=\frac{1}{2}\times\frac{1}{2}\times 2=\frac{1}{2}$$

**08.** *(c)* It is given that the $z$, $\bar{z}$, $\bar{z}-2\text{Re}(\bar{z})$ and $z-2\text{Re}(z)$ are vertices of a square of side 4 units, so

$$|z-\bar{z}|=4 \quad \text{...(i)}$$

Now, let $z=x+iy$, so $\bar{z}=x-iy$ and $\text{Re}(z)=x$

Now, from Eq. (i), we have

$$2|y|=4\Rightarrow |y|=2$$

and $|z-(z-2\text{Re}(z))|=4$

$$\Rightarrow 2|x|=4\Rightarrow |x|=2$$

$$\therefore |z|=\sqrt{x^2+y^2}=\sqrt{4+4}=2\sqrt{2}$$

**09.** *(c)* $|z+4|\leq 3$ represents the interior and boundary of the circle with centre at $(-4, 0)$ and radius $=3$.

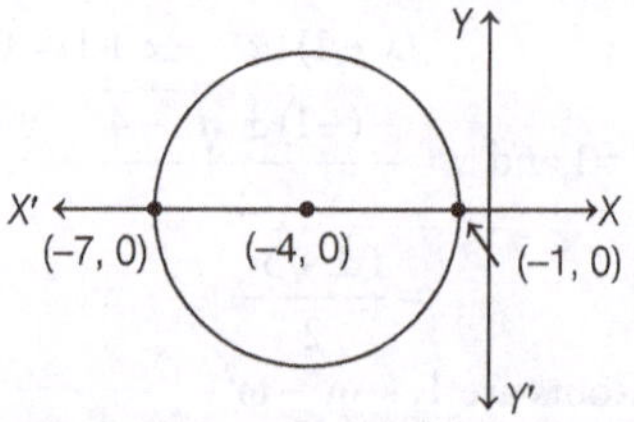


As $-1$ is an end point of a diameter of the circle, hence maximum possible value of $|z+1|$ is 6.

**10.** *(b)* Given that $w=\frac{z}{z-i/3}$

and $|w|=1$

$$\Rightarrow \left|\frac{z}{z-\frac{i}{3}}\right|=1\Rightarrow |z|=\left|z-\frac{i}{3}\right|$$

So, $z$ lies on perpendicular bisector of $(0, 0)$ and $(0, 1/3)$.

Hence, $z$ lies on a straight line.

**11.** *(b)* Using the relation, if

$$|z_1+z_2|=|z_1|+|z_2|$$

Then, $\arg(z_1)=\arg(z_2)$

Since, $|z^2+(-1)|=|z^2|+|-1|$

Then, $\arg(z^2)=\arg(-1)$

$\Rightarrow 2\arg(z)=\pi \quad [\because \arg(-1)=\pi]$

$\Rightarrow \arg(z)=\pi/2$

So, $z$ lies on $Y$-axis (imaginary axis).

**12.** *(b)* Centroid of triangle,

$$z_0=\frac{z_1+z_2+z_3}{3}$$

$z_1, z_2, z_3$ form an equilateral triangle

$$(z_1-z_2)^2+(z_2-z_3)^2+(z_3-z_1)^2=0 \quad \text{...(i)}$$

$$\sum_{k=1}^{3}(z_k-z_0)^2=\left(z_1-\frac{z_1+z_2+z_3}{3}\right)^2+\left(z_2-\frac{z_1+z_2+z_3}{3}\right)^2+\left(z_3-\frac{z_1+z_2+z_3}{3}\right)^2$$

$$\sum_{k=1}^{3}(z_k-z_0)^2=\frac{1}{9}(4z_1^2+z_2^2+z_3^2-4z_1z_2+2z_2z_3-4z_1z_3+4z_2^2+z_1^2+z_3^2-4z_1z_2-4z_2z_3+2z_1z_3+4z_3^2+z_1^2+z_2^2-4z_1z_3-4z_2z_3+2z_1z_2)$$

$$\sum_{k=1}^{3}(z_k-z_0)^2=\frac{1}{9}[6(z_1^2+z_2^2+z_3^2)-6(z_1z_2+z_1z_3+z_2z_3)]$$

$$\sum_{k=1}^{3}(z_k-z_0)^2=\frac{2}{3}[z_1^2+z_2^2+z_3^2-(z_1z_2+z_1z_3+z_2z_3)] \quad \text{...(ii)}$$

From Eq. (i), we get

$$(z_1-z_2)^2+(z_2-z_3)^2+(z_3-z_1)^2=0$$

$$2(z_1^2+z_2^2+z_3^2)-2(z_1z_2+z_2z_3+z_3z_1)=0$$

$$z_1^2+z_2^2+z_3^2=z_1z_2+z_2z_3+z_3z_1$$

Put in Eq. (ii), we get

$$\sum_{k=1}^{3}(z_k-z_0)^2=\frac{2}{3}[z_1^2+z_2^2+z_3^2-(z_1^2+z_2^2+z_3^2)]=0$$

**13.** *(d)* $\omega_1=(8+i)\sin\theta+(7+4i)\cos\theta$

$$=(8\sin\theta+7\cos\theta)+i(\sin\theta+4\cos\theta)$$

$$\omega_2=(1+8i)\sin\theta+(4+7i)\cos\theta$$

$$=(\sin\theta+4\cos\theta)+i(8\sin\theta+7\cos\theta)$$

$$\omega_1\omega_2=\alpha+i\beta$$

$$\therefore \alpha=(8\sin\theta+7\cos\theta)(\sin\theta+4\cos\theta)-(\sin\theta+4\cos\theta)(8\sin\theta+7\cos\theta)$$

$= 8\sin^2\theta + 39\sin\theta\cos\theta + 28\cos^2\theta$
$-8\sin^2\theta - 39\sin\theta\cos\theta - 28\cos^2\theta$
$= 0$

$\beta = (8\sin\theta + 7\cos\theta)^2 + (\sin\theta + 4\cos\theta)^2$
$= 64\sin^2\theta + 49\cos^2\theta + 112\sin\theta\cos\theta$
$\quad + \sin^2\theta + 16\cos^2\theta + 8\sin\theta\cos\theta$
$= 65\sin^2\theta + 65\cos^2\theta + 120\sin\theta\cos\theta$
$= 65 + 60\sin 2\theta$

$\therefore$ Maximum value of
$\alpha + \beta = 0 + 65 + 60 = 135 = p$
Minimum value of
$\alpha + \beta = 0 + 65 - 60 = 5 = q$
$\therefore p + q = 135 + 5 = 140$

**14.** *(b)* We know that $z = x + iy$ then $\bar{z} = x - iy$ this implies that $\text{Re}(z) = \text{Re}(\bar{z})$

$\text{Re}\left(\frac{z-1}{2z+i}\right) + \text{Re}\left(\frac{\bar{z}-1}{2\bar{z}-i}\right) = 2$ can be written as $2\,\text{Re}\left(\frac{z-1}{2z+i}\right) = 2$

$$2\,\text{Re}\left(\frac{x+iy-1}{2(x+iy)+i}\right) = 2$$

$$\text{Re}\left(\frac{z-1}{2z+i}\right) = 1$$

$$\text{Re}\left(\frac{x+iy-1}{2(x+iy)+i}\right) = 1$$

Now, solve

$$\left(\frac{x+iy-1}{2(x+iy)+i}\right) = \left[\frac{(x-1)+iy}{2x+i(2y+1)}\right]$$

$$= \frac{(x-1)+iy}{2x+i(2y+1)} \times \frac{2x-i(2y+1)}{2x-i(2y+1)}$$

$$= \frac{[(x+1)+iy][2x-i(2y+1)]}{4x^2 - i^2(2y+1)^2}$$

$$= \frac{(x-1+iy)(2x-2iy-i)}{4x^2+(2y+1)^2}$$

$$= \frac{2x^2 - 2ixy - xi - 2x + 2iy + i + 2ixy - 2i^2y^2 - i^2y}{4x^2+(2y+1)^2}$$

$$= \frac{2x^2 + 2y^2 - 2x + y - i(-x+2y+1)}{4x^2+(2y+1)^2}$$

So, $\text{Re}\left(\frac{x+iy-1}{2(x+iy)+i}\right) = 1$ becomes

$$\text{Re}\left(\frac{2x^2+2y^2-2x+y-i(-x+2y+1)}{4x^2+(2y+1)^2}\right) = 1$$

Taking real part,

$$\frac{2x^2+2y^2-2x+y}{4x^2+(2y+1)^2} = 1$$

$$2x^2+2y^2-2x+y = 4x^2+4y^2+4y+1$$

$$2x^2+2y^2+2x+3y+1 = 0$$

Dividing both sides of the equation by 2

$$x^2+y^2+x+\frac{3}{2}y+\frac{1}{2} = 0$$

Compare with

$$x^2+y^2+2gx+2fy+c = 0$$

The centre is $(-g, -f) = \left(-\frac{1}{2}, -\frac{3}{4}\right)$
$= (a, b)$ given and radius

$$r = \sqrt{g^2+f^2-c}$$

$$r = \sqrt{\frac{1}{4}+\frac{9}{16}-\frac{1}{2}}$$

$$\Rightarrow \sqrt{\frac{4}{16}+\frac{9}{16}-\frac{8}{16}} = \frac{\sqrt{5}}{4}$$

Thus, $\frac{15ab}{r^2} = \frac{15 \times -\frac{1}{2} \times -\frac{3}{4}}{\frac{5}{16}} = \frac{45/8}{1/16}$

$$= \frac{45}{8} \times \frac{16}{5} = 18$$

**15.** *(c)* Given that $\left|\frac{\bar{z}-i}{2\bar{z}+i}\right| = \frac{1}{3}$

$$\left|\frac{\bar{z}-i}{\bar{z}+1/2}\right| = \frac{2}{3}$$

Let $z = x + iy$, $x, y \in R$
$\Rightarrow \bar{z} = x - iy$

$$3|x-iy-i| = 2\left|x-iy+\frac{1}{2}\right|$$

$$\Rightarrow 9[x^2+(y+1)^2] = 4\left[x^2+\left(y-\frac{1}{3}\right)^2\right]$$

$$\Rightarrow 5x^2+5y^2+22y+8 = 0$$

$$\text{Centre} \equiv \left(0, \frac{-11}{5}\right)$$

$$\text{Area of } \Delta = \frac{1}{2}\begin{vmatrix} 0 & 0 & 1 \\ 0 & -\frac{11}{5} & 1 \\ \alpha & 0 & 1 \end{vmatrix} = 11$$

$$\left(-\frac{11}{5}\alpha\right)^2 = (11 \times 2)^2 \Rightarrow \alpha^2 = 100$$

**16.** *(b)* Let $z = x + iy$

$z(1+i) + \bar{z}(1-i) = 4, \quad z \in C$
$\Rightarrow (x+iy)(1+i) + (x-iy)(1-i) = 4$
$\Rightarrow x + ix + iy - y + x - ix - iy - y = 4$
$\Rightarrow \quad x - y = 2$
and $\quad |z-3| \le 1$
$\Rightarrow \quad (x-3)^2 + y^2 \le 1$

Solving $x - y = 2$ and $(x-3)^2 + y^2 = 1$ we get $x = 2$ or $3$

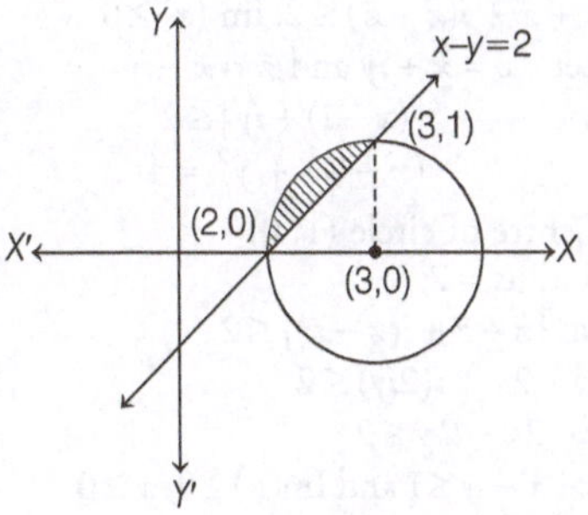


$$\text{Shaded area} = \pi(1)^2\left(\frac{90°}{360°}\right) - \frac{1}{2} \times 1 \times 1$$

$$\alpha = \frac{\pi}{4} - \frac{1}{2}$$

Unshaded area

$$= \beta = \pi - \left(\frac{\pi}{4} - \frac{1}{2}\right) = \frac{3\pi}{4} + \frac{1}{2}$$

$$|\alpha - \beta| = \frac{\pi}{2} + 1$$

**17.** *(b)* $S_1 : \{z \in C : |z| \le 5\}$
$\Rightarrow x^2 + y^2 \le 25$

$$S_2 = \left\{z \in C : \text{Im}\,\frac{(z+1-\sqrt{3}i)}{1-\sqrt{3}i} \ge 0\right\}$$

$$\text{Im}\left(\frac{z}{1-\sqrt{3}i} + \frac{(1-\sqrt{3}i)}{1-\sqrt{3}i}\right) \ge 0$$

$$\text{Im}\left(\frac{z}{1-\sqrt{3}i} + 1\right) \ge 0$$

$$\text{Im}\left(\frac{z}{1-\sqrt{3}i} \times \frac{1+\sqrt{3}i}{1+\sqrt{3}i}\right) \ge 0$$

$$\text{Im}\,(x+iy)(1+\sqrt{3}i) \ge 0$$

$$\Rightarrow \sqrt{3}x + y \ge 0, \theta \Rightarrow 150°$$

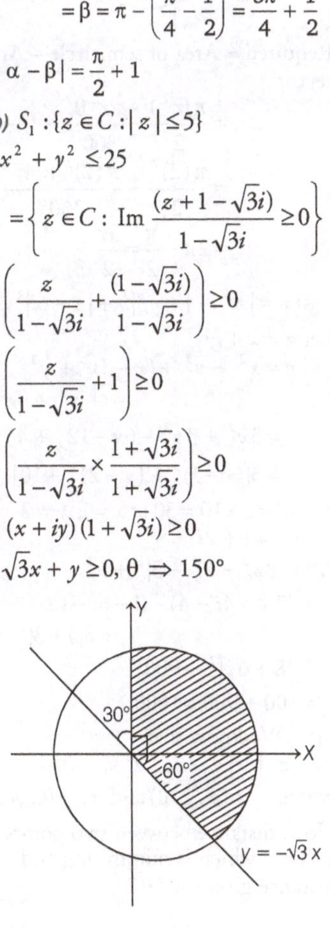

$S_3 : \{z \in C : \text{Re}(z) \ge 0\}$

$\Rightarrow \quad x \ge 0$

$\therefore$ Area $= \dfrac{\pi r^2 (150°)}{360} = \pi \times \dfrac{5\times 5\times 150}{360}$

$= \pi \dfrac{5\times 5\times 5}{12} = \dfrac{125\pi}{12}$

**18.** (b) $|z-1| \le 2,$

$z + \bar{z} + i(z - \bar{z}) \le 2$, Im $(z) \ge 0$

Let $z = x + iy$ and $\bar{z} = x - iy$

$|(x-1) + iy| \le 2$

$(x-1)^2 + y^2 = 4$

Centre of circle (1, 0)

Radius = 2;

Im $\{z + \bar{z} + i(z - \bar{z})\} \le 2$

$\Rightarrow \quad 2x + i(2iy) \le 2$

$\Rightarrow \quad 2x - 2y \le 2$

$\Rightarrow \quad x - y \le 1$ and Im$(z) \ge 0$, $y \ge 0$

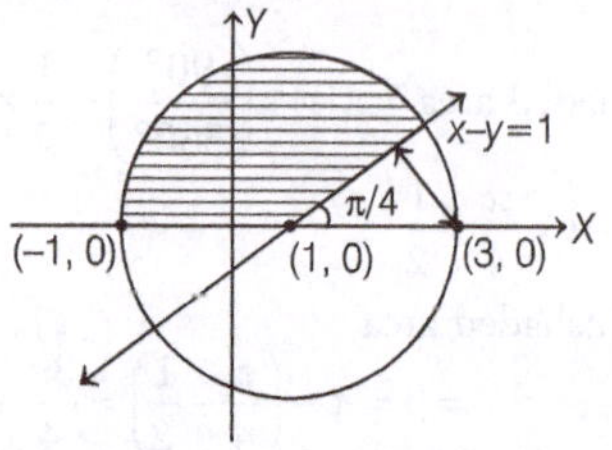


Required = Area of semicircle – Area of sector

$= \dfrac{\pi (r^2)}{2} - \dfrac{\pi r^2 \theta}{360}$

$= \dfrac{\pi (2)^2}{2} - \dfrac{\pi (2)^2 \times 45°}{360°}$

$= 2\pi - \dfrac{\pi}{2} = \dfrac{3\pi}{2}$

**19.** (a) $v = |z|^2 + |z-3|^2 + |z-6i|^2$

Let $z = x + iy$

$v = x^2 + y^2 + (x-3)^2 + y^2 + x^2 + (y-6)^2$

$= 3x^2 + 3y^2 - 6x - 12y + 45$

$= 3[(x-1)^2 + (y-2)^2 + 10]$

$\therefore \quad v_0 = 3 \times 10 = 30$ at $x = 1, y = 2$

$\therefore \quad z_0 = 1 + 2i$

$\therefore \quad |2z_0^2 - \bar{z}_0^3 + 3|^2 + v_0^2$

$= |2(1 + 4i - 4) - (1 - 6i - 12 + 8i) + 3|^2 + 900$

$= |8 + 6i|^2 + 900$

$= 100 + 900 = 1000$

**20.** (c) We know that,

$|z - z_1| + |z - z_2| \ge |z_1 - z_2|$

where $z_1 = (3\sqrt{2}, 0)$ and $z_2 = (0, p\sqrt{2})$

Now, distance between two points $z_1$ and $z_2$, which is minimising and distance given by $5\sqrt{2}$.

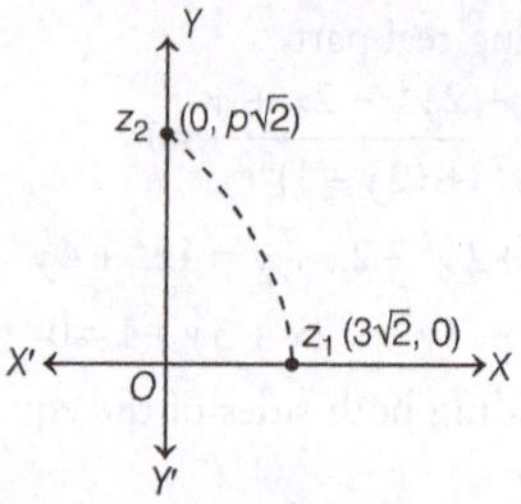


According to the question,

$|z_1 - z_2| = 5\sqrt{2}$

$\Rightarrow \sqrt{(3\sqrt{2})^2 + (p\sqrt{2})^2} = 5\sqrt{2}$

$\Rightarrow \sqrt{18 + 2p^2} = 5\sqrt{2} \Rightarrow 18 + 2p^2 = 50$

[On squaring both sides]

$\Rightarrow \quad p^2 = 16 \Rightarrow p = \pm 4$

Thus, the value of $p$ is 4.

**21.** (c)

| | |
|---|---|
| $\lvert z - (4+3i)\rvert = 2$ respresents a circle with centre (4, 3) and radius 2 units. | $\lvert z - (0 + i0)\rvert + \lvert z - (4 + i0)\rvert = 6$ represents an ellipse with foci (0, 0) and (4, 0) and the length of major axis equals 6 and length of semi-major axis $2\sqrt{5}$ and (4, 2) lies inside the both curve. |

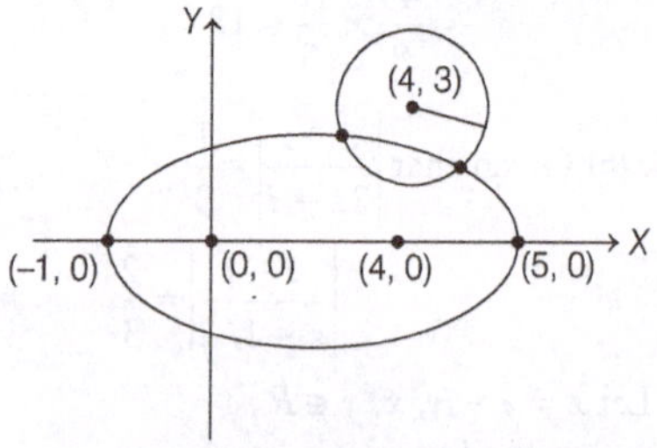


$\therefore$ Number of point of intersection = 2

**22.** (b) The equation of circle that passes through the points $z_1, z_2$ and $z_3$ is $x^2 + y^2 = 25$

Coordinates of $z_1$, $z_2$ and $z_3$ are (3, 4), (4, 3) and (0, 5) respectively.

Slope of $z_2\ z_3 = \dfrac{5-3}{0-4} = -\dfrac{1}{2}$

Slope of $z\ z_1 \times$ slope of $z_2\ z_3 = -1$

[$\because zz_1$ is perpendicular to $z_2\ z_3$]

$\therefore \quad$ Slope of $zz_1 = 2$

Equation of line $zz_1 : y - 4 = 2(x-3)$

$\Rightarrow \quad y = 2x - 2$

Substitute the value of $y$ in equation of circle $x^2 + (2x-2)^2 = 25$

$\Rightarrow \; x^2 + 4x^2 + 4 - 8x = 25$

$\Rightarrow \; (x-3)(5x+7) = 0 \Rightarrow x = 3$ or $-\dfrac{7}{5}$

If $x = 3$, then $y = 4$

(Not possible as $z \ne z_1$)

If $x = -\dfrac{7}{5}$, then $y = -\dfrac{14}{5} - 2 \Rightarrow \dfrac{-24}{5}$

Coordinate of $z$ is $\left(-\dfrac{7}{5}, -\dfrac{24}{5}\right)$.

Argument $= -\left[\pi - \tan^{-1}\left|\dfrac{y}{x}\right|\right]$

$= -\pi + \tan^{-1}\left|\dfrac{24}{7}\right|$

**23.** (d) Let $z = x + iy$

$\dfrac{z-i}{z-1} = \dfrac{x + i(y-1)}{(x-1) + iy} \times \dfrac{(x-1) - iy}{(x-1) - iy}$

$= \dfrac{x(x-1) + y(y-1)}{(x-1)^2 + y^2} + i\left[\dfrac{(x-1)(y-1) - xy}{(x-1)^2 + y^2}\right]$

As $\dfrac{z-i}{z-1}$ is purely imaginary,

$x^2 + y^2 - x - y = 0$

$\Rightarrow \left(x - \dfrac{1}{2}\right)^2 + \left(y - \dfrac{1}{2}\right)^2 = \dfrac{1}{2}$

This is a circle with centre (1/2, 1/2), radius $= 1/\sqrt{2}$ which passes through origin as shown in the figure.

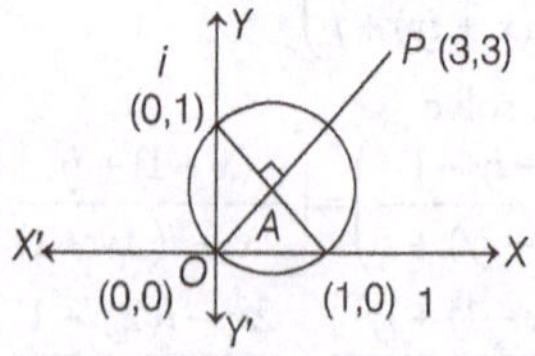


Minimum $|z - (3+3i)| = OP - \dfrac{1}{\sqrt{2}}$

$= \sqrt{\left(3 - \dfrac{1}{2}\right)^2 + \left(3 - \dfrac{1}{2}\right)^2} - \dfrac{1}{\sqrt{2}}$

$= \dfrac{5}{\sqrt{2}} - \dfrac{1}{\sqrt{2}} = \dfrac{4}{\sqrt{2}} = 2\sqrt{2}$

**24.** (b)

> **key idea** Use property of Argument
>
> $\therefore \arg\left(\dfrac{z_1}{z_2}\right) = \arg(z_1) - \arg(z_2)$

$\arg\left(\dfrac{z-2}{z+2}\right) = \dfrac{\pi}{4}$

If $z = x + iy$, $\arg[(x-2) + iy] - \arg[(x+2) + iy] = \dfrac{\pi}{4}$

$$\Rightarrow \tan^{-1}\left(\frac{y}{x-2}\right) - \tan^{-1}\left(\frac{y}{x+2}\right) = \frac{\pi}{4}$$

$$\Rightarrow \frac{\frac{y}{x-2} - \frac{y}{x+2}}{1 + \frac{y}{x-2}\cdot\frac{y}{x+2}} = \tan\left(\frac{\pi}{4}\right)$$

$$\Rightarrow \frac{xy + 2y - xy + 2y}{x^2 + y^2 - 4} = 1$$

$$\Rightarrow 4y = x^2 + y^2 - 4$$

$$\Rightarrow x^2 + y^2 - 4y - 4 = 0$$

$z$ is a circle with

Centre $= (0, 2)$ and Radius $= (2\sqrt{2})$

$|z - 9\sqrt{2} - 2i|^2$ is the distance of $(9\sqrt{2}, 2)$ from any point on circle.

Distance will be minimum when $(9\sqrt{2}, 2)$ will lie on the line joining the centre.

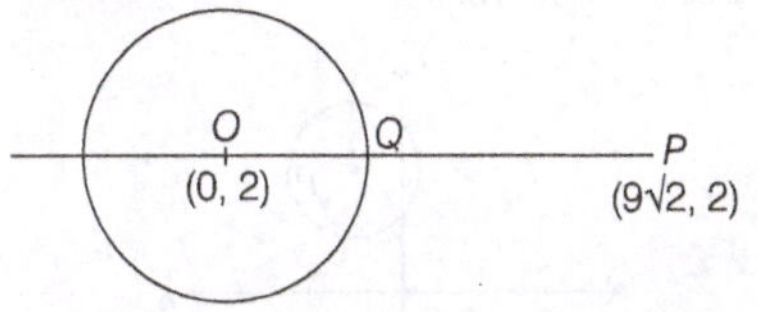


$$PQ = OP - OQ = 9\sqrt{2} - 2\sqrt{2} = 7\sqrt{2}$$

$$PQ^2 = (7\sqrt{2})^2 = 98$$

**25.** *(b)* We have,

$$\arg\left(\frac{z-1}{z+1}\right) = \frac{\pi}{4}$$

$$\Rightarrow \arg(z-1) - \arg(z+1) = \frac{\pi}{4}$$

Let $z = x + iy$

$$\arg[(x-1) + iy] - \arg[(x+1) + iy] = \frac{\pi}{4}$$

$$\Rightarrow \tan^{-1}\left(\frac{y}{x-1}\right) - \tan^{-1}\left(\frac{y}{x+1}\right) = \frac{\pi}{4}$$

$$\Rightarrow \left(\frac{\frac{y}{x-1} - \frac{y}{x+1}}{1 + \frac{y}{x-1}\cdot\frac{y}{x+1}}\right) = \tan\frac{\pi}{4}$$

$$\Rightarrow \frac{y(x+1) - y(x-1)}{(x^2 - 1) + y^2} = 1$$

$$\Rightarrow 2y = x^2 + y^2 - 1$$

$$\Rightarrow x^2 + y^2 - 2y - 1 = 0$$

$$\Rightarrow x^2 + (y-1)^2 = 2$$

$$\Rightarrow x^2 + (y-1)^2 = (\sqrt{2})^2$$

which is a circle with Centre (0, 1) and Radius $= \sqrt{2}$ units

**26.** *(b)* The image of $P(a, b)$ along $y = x$ is $Q(b, a)$.

Translating it 2 units along the positive direction of $X$-axis, it becomes $R(b + 2, a)$.

Then, rotation through $\frac{\pi}{4}$ about the origin in the anticlockwise direction, the final position of the point $P$ is $\left(-\frac{1}{\sqrt{2}}, \frac{7}{\sqrt{2}}\right)$

Now, applying rotational theorem,

$$-\frac{1}{\sqrt{2}} + \frac{7}{\sqrt{2}}i = [(b+2) + ai] \cdot \left(\cos\frac{\pi}{4} + i\sin\frac{\pi}{4}\right)$$

$$\Rightarrow -\frac{1}{\sqrt{2}} + \frac{7}{\sqrt{2}}i = \left(\frac{b+2}{\sqrt{2}} - \frac{a}{\sqrt{2}}\right) + i\left(\frac{b+2}{\sqrt{2}} + \frac{a}{\sqrt{2}}\right)$$

$$\Rightarrow -\frac{1}{\sqrt{2}} + \frac{7}{\sqrt{2}}i = \left(\frac{b-a+2}{\sqrt{2}}\right) + i\left(\frac{a+b+2}{\sqrt{2}}\right)$$

So, $\frac{b-a+2}{\sqrt{2}} = -\frac{1}{\sqrt{2}}$

$$\Rightarrow b - a = -3 \quad \ldots\text{(i)}$$

and $\frac{a+b+2}{\sqrt{2}} = \frac{7}{\sqrt{2}}$

$$\Rightarrow a + b = 5 \quad \ldots\text{(ii)}$$

Adding Eqs. (i) and (ii), we get

$$2b = 2$$

$$\Rightarrow b = 1$$

Substitute the value of $b$ in Eq. (ii), $a = 4$

Now, $2a + b = 2 \times 4 + 1 = 9$

> **Concept Enhancer**
>
> A complex rotation can be denoted as $|z|e^{i\theta}$ where $\theta$ is a real number in radius. $\theta$ denotes the rotation of the complex number in an anti-clockwise or counter clock wise direction and $-\theta$ for the clockwise direction.

**27.** *(b)* It is given for complex number $z$,

$$\left|\frac{z-i}{z+2i}\right| = 1 \quad \ldots\text{(i)}$$

and $$|z| = \frac{5}{2} \quad \ldots\text{(ii)}$$

From Eq. (i),

$$|z - i| = |z + 2i|$$

$\Rightarrow$ Locus of $z$ is a straight line and it is perpendicular bisector of line joining points $(0, -2)$ and $(0, 1)$, so locus of $z$ is $y = -\frac{1}{2}$, where $z = x + iy$.

From Eq. (ii), $\sqrt{x^2 + y^2} = 5/2$

$$\Rightarrow x^2 + \left(\frac{-1}{2}\right)^2 = \left(\frac{5}{2}\right)^2$$

$$\Rightarrow x^2 + \frac{1}{4} = \frac{25}{4}$$

$$\Rightarrow x^2 = 6$$

Now, $|z + 3i| = |x + (y+3)i|$

$$= \sqrt{x^2 + (y+3)^2} = \sqrt{6 + \left(-\frac{1}{2} + 3\right)^2}$$

$$(\because x^2 = 6 \text{ and } y = -1/2)$$

$$= \sqrt{6 + \frac{25}{4}} = \sqrt{\frac{24+25}{4}} = \sqrt{\frac{49}{4}} = \frac{7}{2}$$

**28.** *(d)* For $z = x + iy$,

$$\frac{z-1}{2z+i} = \frac{(x-1) + iy}{2x + (2y+1)i}$$

$$= \frac{[(x-1) + iy][2x - i(2y+1)]}{(2x)^2 + (2y+1)^2}$$

(on rationalization)

$$\therefore \text{Re}\left(\frac{z-1}{2z+i}\right) = \frac{2x(x-1) + y(2y+1)}{4x^2 + (2y+1)^2}$$

Now, it is given that $\text{Re}\left(\frac{z-1}{2z+i}\right) = 1$

$$\Rightarrow \frac{2x(x-1) + y(2y+1)}{4x^2 + (2y+1)^2} = 1$$

$\Rightarrow 2x^2 + 2y^2 + 2x + 3y + 1 = 0$, is a circle whose centre is $\left(-\frac{1}{2}, -\frac{3}{4}\right)$ and radius is $\sqrt{\frac{1}{4} + \frac{9}{16} - \frac{1}{2}} = \frac{\sqrt{5}}{4}$, so diameter is $\frac{\sqrt{5}}{2}$.

**29.** (c) Let $x+iy=\dfrac{\alpha+i}{\alpha-i}$

$\Rightarrow \quad x+iy=\dfrac{(\alpha+i)^2}{\alpha^2+1}=\dfrac{(\alpha^2-1)+(2\alpha)i}{\alpha^2+1}$

$=\dfrac{\alpha^2-1}{\alpha^2+1}+\left(\dfrac{2\alpha}{\alpha^2+1}\right)i$

On comparing real and imaginary parts, we get $x=\dfrac{\alpha^2-1}{\alpha^2+1}$ and $y=\dfrac{2\alpha}{\alpha^2+1}$

Now, $x^2+y^2=\left(\dfrac{\alpha^2-1}{\alpha^2+1}\right)^2+\left(\dfrac{2\alpha}{\alpha^2+1}\right)^2$

$=\dfrac{\alpha^4+1-2\alpha^2+4\alpha^2}{(\alpha^2+1)^2}=\dfrac{(\alpha^2+1)^2}{(\alpha^2+1)^2}=1$

$\Rightarrow x^2+y^2=1$

So, $S=\left\{\dfrac{\alpha+i}{\alpha-i};\alpha\in R\right\}$ lies on a circle with radius 1.

**30.** (a) We have, $2|z+3i|-|z-i|=0$

$\Rightarrow 2|x+i(y+3)|=|x+i(y-1)|$ $\quad(\because z=x+iy)$

$\Rightarrow 2\sqrt{x^2+(y+3)^2}=\sqrt{x^2+(y-1)^2}$

$\Rightarrow 4(x^2+(y+3)^2)=x^2+(y-1)^2$

$\Rightarrow 3x^2=y^2-2y+1-4y^2-24y-36$

$\Rightarrow 3x^2+3y^2+26y+35=0$

$\Rightarrow x^2+y^2+\dfrac{26}{3}y+\dfrac{35}{3}=0$

This is the equation of circle with radius,

$r=\sqrt{0^2+\left(\dfrac{13}{3}\right)^2-\dfrac{35}{3}}=\sqrt{\dfrac{64}{9}}=\dfrac{8}{3}$

$[\because r=\sqrt{g^2+f^2-c}]$

**31.** (a)

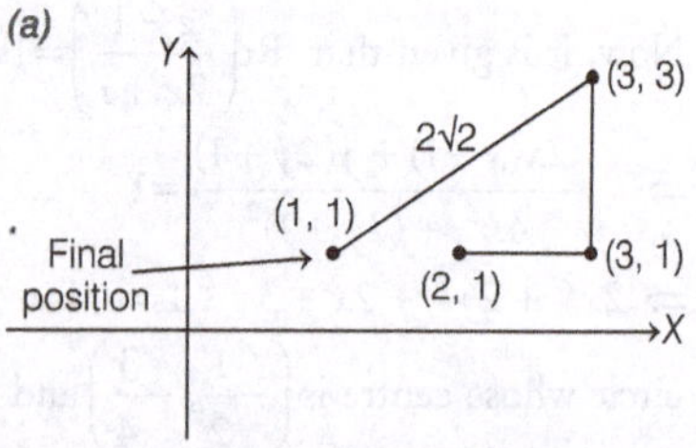


Hence, the final position of the point is represented by $1+i$.

**32.** (c)

**key idea** If z is unimodular, then $|z|=1$.

Also, use property of modulus i.e. $z\bar{z}=|z|^2$.

Given, $z_2$ is not unimodular i.e. $|z_2|\neq 1$ and $\dfrac{z_1-2z_2}{2-z_1\bar{z}_2}$ is unimodular

$\Rightarrow \left|\dfrac{z_1-2z_2}{2-z_1\bar{z}_2}\right|=1$

$\Rightarrow |z_1-2z_2|^2=|2-z_1\bar{z}_2|^2$

$\Rightarrow (z_1-2z_2)(\bar{z}_1-2\bar{z}_2)$

$=(2-z_1\bar{z}_2)(2-\bar{z}_1z_2)$ $\quad(\because z\bar{z}=|z|^2)$

$\Rightarrow |z_1|^2+4|z_2|^2-2\bar{z}_1z_2-2z_1\bar{z}_2$

$=4+|z_1|^2|z_2|^2-2\bar{z}_1z_2-2z_1\bar{z}_2$

$\Rightarrow (|z_2|^2-1)(|z_1|^2-4)=0$ $(\because |z_2|\neq 1)$

$\therefore |z_1|=2$

Let $z_1=x+iy\Rightarrow x^2+y^2=(2)^2$

$\therefore$ Point $z_1$ lies on a circle of radius 2.

**33.** (b) $|z|\geq 2$ is the region on or outside circle whose centre is (0, 0) and radius is 2.

Minimum $\left|z+\dfrac{1}{2}\right|$ is distance of $z$, which lie on the circle $|z|=2$ from $\left(\dfrac{-1}{2},0\right)$.

$\therefore$Minimum $\left|z+\dfrac{1}{2}\right|=$ Distance of $\left(-\dfrac{1}{2},0\right)$ from $(-2,0)$

$=\sqrt{\left(-2+\dfrac{1}{2}\right)^2+0}=\dfrac{3}{2}$

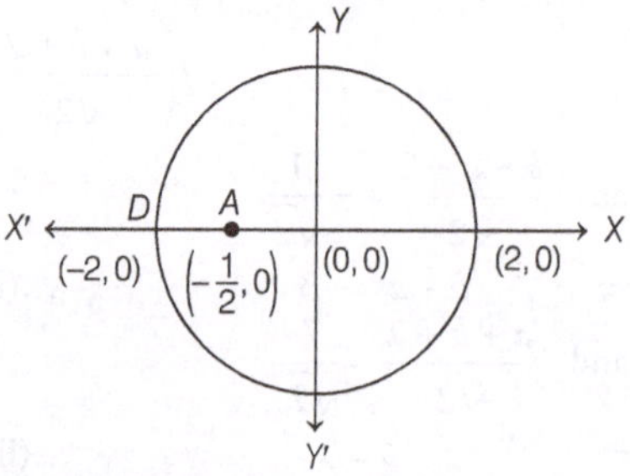


**Geometrically**

Min $\left|z+\dfrac{1}{2}\right|=AD=\sqrt{\left(\dfrac{-1}{2}+2\right)^2+0}=\dfrac{3}{2}$

**34.** (b) We have,

$|z-1|=|z+1|=|z-i|$

Clearly, $z$ is the circumcentre of the triangle formed by the vertices (1, 0) and (0, 1) and (–1, 0) which is unique.

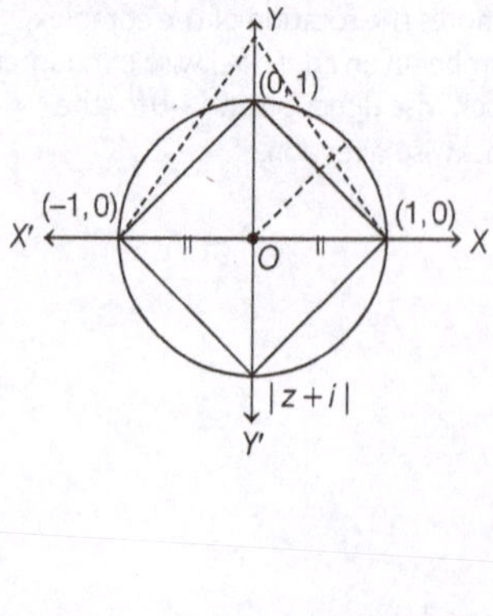


Hence, the number of complex number $z$ is one.

**35.** (c) Since, origin $z_1$ and $z_2$ are the vertices of an equilateral triangle, then

$z_1^2+z_2^2=z_1z_2$

$\Rightarrow \quad (z_1+z_2)^2=3z_1z_2 \quad$ ...(i)

Again, $z_1, z_2$ are the roots of the equation

Then, $z_1+z_2=-a$

and $\quad z_1z_2=b$

On putting these values in Eq. (i), we get

$(-a)^2=3b$

$\Rightarrow \quad a^2=3b$

**36.** (c) In cartesian plane, circle has centre $A\left(\dfrac{1}{2},\dfrac{3}{2}\right)$ and radius 1 units and $P(z_1)=(1,1)$

Let $Q(z_2)=(h,k)$

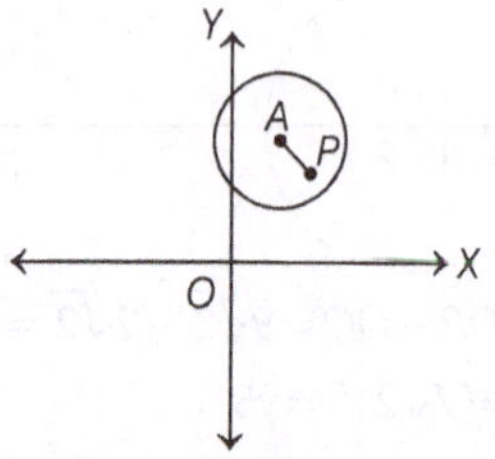


Now, $|z_1-z_0|=AP$

$=\sqrt{\left(1-\dfrac{1}{2}\right)^2+\left(1-\dfrac{3}{2}\right)^2}=\dfrac{1}{\sqrt{2}}$

$\therefore \quad |z_2-z_0|=\dfrac{1}{|z_1-z_0|}=\sqrt{2}$

Now, slope of AP, $\tan\theta=\dfrac{\dfrac{3}{2}-1}{\dfrac{1}{2}-1}=-1$

$\therefore \sin\theta=\dfrac{1}{\sqrt{2}},\ \cos\theta=\dfrac{-1}{\sqrt{2}}$

Since $b, z_0, z_1, z_2$ are collinear.

$\therefore A, P, Q$ are collinear.

Using parametric form of line for $Q$,

$\dfrac{h-\dfrac{1}{2}}{\cos\theta}=\pm\sqrt{2},\ \dfrac{k-\dfrac{3}{2}}{\sin\theta}=\pm\sqrt{2}$

$\Rightarrow \quad \dfrac{h-\dfrac{1}{2}}{\left(-\dfrac{1}{\sqrt{2}}\right)}=\pm\sqrt{2},\ \dfrac{k-\dfrac{3}{2}}{\left(\dfrac{1}{\sqrt{2}}\right)}=\pm\sqrt{2}$

$\Rightarrow \quad h=-\dfrac{1}{2}$ or $\dfrac{3}{2}$ and $k=\dfrac{5}{2}$ or $\dfrac{1}{2}$

$\therefore \quad Q = \left(-\frac{1}{2}, \frac{5}{2}\right) \text{ or } \left(\frac{3}{2}, \frac{1}{2}\right)$

Now, $(OQ)^2 = \left(\frac{-1}{2}\right)^2 + \left(\frac{5}{2}\right)^2$

or $\left(\frac{3}{2}\right)^2 + \left(\frac{1}{2}\right)^2$

i.e. $(OQ)^2 = \frac{13}{2} \text{ or } \frac{5}{2}$

$\Rightarrow |z_2|^2 = \frac{13}{2} \text{ or } \frac{5}{2}$

$\therefore$ Similar value of $|z_2|^2 = \frac{5}{2}$

**37.** *(c)* Given, $z_1 = 2 + 3i, z_2 = 3 + 4i$

$S = \{z \in C : |z - z_1|^2 - |z - z_2|^2 = |z_1 - z_2|^2\}$

Let $z = x + iy$

$\therefore |z - z_1| = |(x + iy) - (2 + 3i)| = \sqrt{(x-2)^2 + (y-3)^2}$

$\therefore |z - z_2| = |(x + iy) - (3 + 4i)| = \sqrt{(x-3)^2 + (y-4)^2}$

and $|z_1 - z_2| = |(2 + 3i) - (3 + 4i)| = |-1 - i| = \sqrt{1+1} = \sqrt{2}$

$\therefore |z - z_1|^2 - |z - z_2|^2 = |z_1 - z_2|^2$

$\Rightarrow [(x-2)^2 + (y-3)^2] - [(x-3)^2 + (y-4)^2] = 2$

$\Rightarrow (x^2 + y^2 - 4x - 6y + 13) - (x^2 + y^2 - 6x - 8y + 25) = 2$

$\Rightarrow \quad 2x + 2y = 14 \Rightarrow x + y = 7$

$\Rightarrow \quad \frac{x}{7} + \frac{y}{7} = 1$

$x$-intercept $= y$-intercept $= 7$

Sum of intercept on axes $= 14$

**38.** *(b)*

$S = \{z = x + iy : |z - 1 + i| \geq |z|, |z| < 2, |z + i| = |z - 1|\}$

**Case I** $|z - 1 + i| \geq |z|$

$\Rightarrow \quad |x + iy - 1 + i| \geq |x + iy|$

$\Rightarrow \quad |x - 1 + i(y+1)| \geq |x + iy|$

$\Rightarrow \quad \sqrt{(x-1)^2 + (y+1)^2} \geq \sqrt{x^2 + y^2}$

$\Rightarrow -2x + 2y + 2 \geq 0 \Rightarrow -x + y + 1 \geq 0$

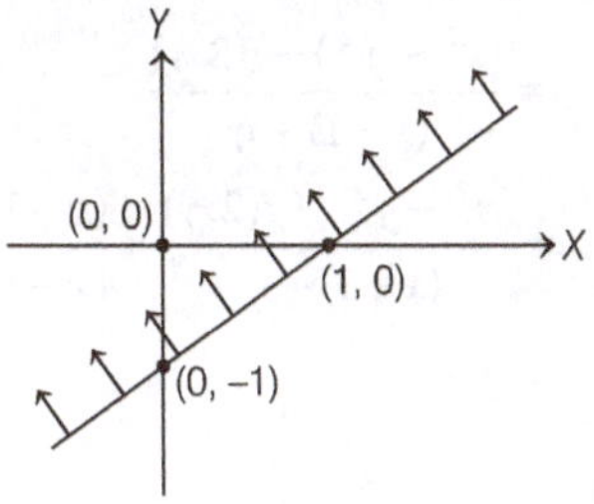


**Case II** $|z| < 2 \Rightarrow |x + iy| < 2$

$\Rightarrow \sqrt{x^2 + y^2} < 2 \Rightarrow x^2 + y^2 < 4$

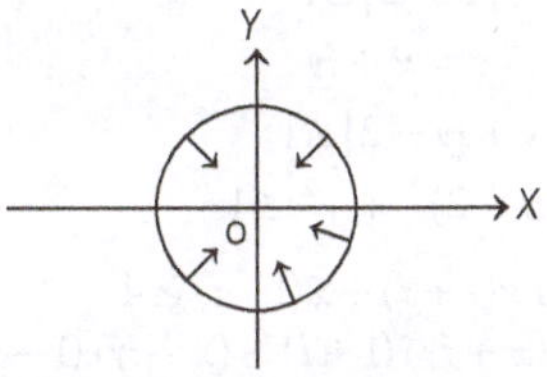


**Case III** $|z + i| = |z - 1|$

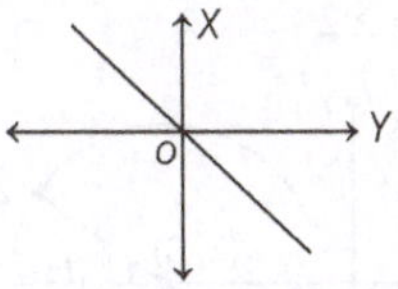


$\Rightarrow \quad |x + iy + i| = |x + iy - 1|$

$\Rightarrow \quad \sqrt{x^2 + (y+1)^2} = \sqrt{(x-1)^2 + y^2}$

$\Rightarrow \quad 2y = -2x \Rightarrow x + y = 0$

From Cases I, II and III,

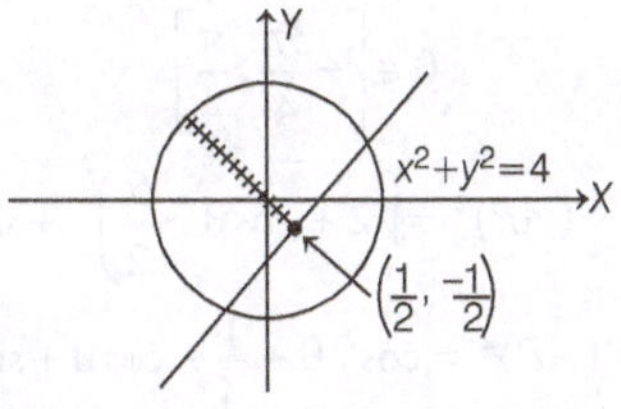


So, $w = 2x + iy \in S$

$\therefore \quad 2x \leq \frac{1}{2} \quad \Rightarrow \quad x \leq \frac{1}{4} \quad \ldots\text{(i)}$

Now, $(2x)^2 + (2x)^2 < 4$

$x^2 < \frac{1}{2} \Rightarrow x^2 - \left(\frac{1}{\sqrt{2}}\right)^2 < 0$

$\Rightarrow \quad x \in \left(\frac{-1}{\sqrt{2}}, \frac{1}{\sqrt{2}}\right) \quad \ldots\text{(ii)}$

From Eqs. (i) and (ii), we get

$x \in \left(-\frac{1}{\sqrt{2}}, \frac{1}{4}\right]$

**39.** *(c)* Let $\arg(z-1) - \arg(z+1) = \pi/4$

$\therefore |z| = 3$ represents a circle having radius 3 and centre at $z = 0$, i.e. at (0, 0). Now, $\arg(z-1) - \arg(z+1) = \pi/4$

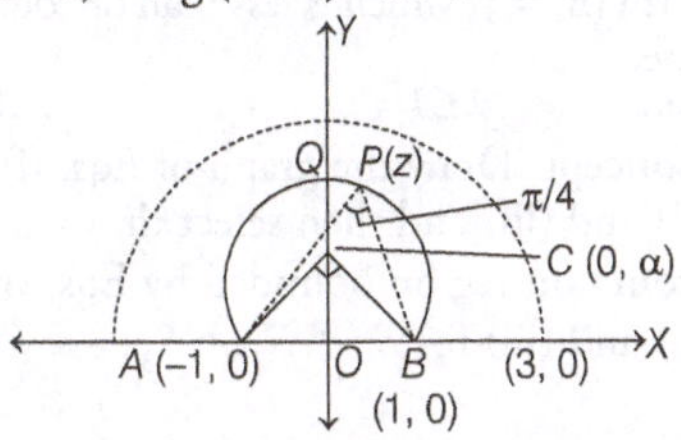


$z$ is on major arc of a circle having $AB$ as chord as shown in above figure.

$\angle ACB = 2\angle APB$

and $\angle ACO = \angle OCB$

$\therefore \quad \angle OCB = \angle APB = \pi/4$

$\Rightarrow \quad OC = OB = 1 = \alpha$

In right $\Delta OCB$, radius $= BC = \sqrt{2}$

$\therefore \ OQ = \alpha + \text{radius} = 1 + \sqrt{2}$

Another circle of radius 3 and centre (0, 0) does not intersect the circle of radius $\sqrt{2}$ and centre $(0, \sqrt{2})$.

Hence, there is no common point.

**40.** *(a)* Let $\quad z = x + iy, x, y \in R$

Given, $\quad \bar{z} = iz^2$

$\Rightarrow \quad x - iy = i(x + iy)^2$

$\Rightarrow \quad x - iy = i(x^2 - y^2 + 2ixy)$

$\Rightarrow \quad x - iy = -2xy + i(x^2 - y^2)$

On equating real and imaginary parts from both sides, we get

$x = -2xy, -y = x^2 - y^2$

$\Rightarrow x(1 + 2y) = 0 \Rightarrow x = 0 \text{ or } y = -1/2$

At $x = 0, \quad -y = 0 - y^2$

$\Rightarrow y - y^2 = 0 \quad \Rightarrow y = 0, 1$

Coordinates (0, 0) (0, 1).

At $y = \frac{-1}{2}, \frac{1}{2} = x^2 - \frac{1}{4}$

$\Rightarrow x^2 = \frac{3}{4} \Rightarrow x = \pm\frac{\sqrt{3}}{2}$

Coordinates $\left(\frac{\sqrt{3}}{2}, \frac{-1}{2}\right)\left(\frac{-\sqrt{3}}{2}, \frac{-1}{2}\right)$

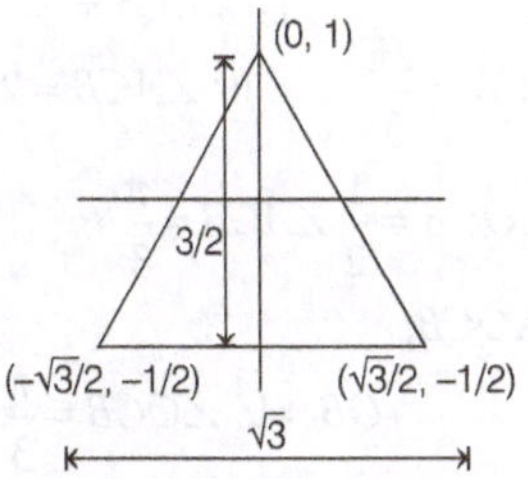


$\therefore$ Area $= \frac{1}{2} \times b \times h = \frac{1}{2} \times \sqrt{3} \times \frac{3}{2}$

$= \frac{3\sqrt{3}}{4}$ sq. units

**41.** *(b)*

> **key idea** If $z = x + iy$, then $|z|^2 = x^2 + y^2$

Given, $A = \left\{z \in C : \left|\frac{z+1}{z-1}\right| < 1\right\}$

and $B = \left\{z \in C : \arg\left(\frac{z-1}{z+1}\right) = \frac{2\pi}{3}\right\}$

Then, $\left|\dfrac{z+1}{z-1}\right| < 1$

$\Rightarrow \quad |z+1| < |z-1|$

$\Rightarrow \quad (x+1)^2 + y^2 < (x-1)^2 + y^2$ $\quad [\because z = x + iy]$

$\Rightarrow \quad (x+1)^2 - (x-1)^2 < 0$

$\Rightarrow \quad (x+1+x-1)(x+1-x+1) < 0$

$\Rightarrow \quad (2x)(2) < 0$

$\Rightarrow \quad x < 0$ $\quad \ldots$(i)

Now, $\arg\left(\dfrac{z-1}{z+1}\right) = \dfrac{2\pi}{3}$

It represents the part of a circle, where end points of the chord are (1, 0) and (–1, 0) i.e., lying on the $X$-axis.

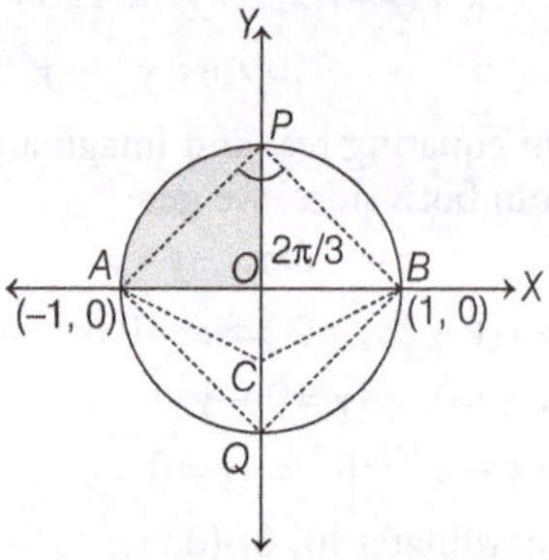


$\angle APB = \dfrac{2\pi}{3}$ because in this case the obtuse angle lies in the minor segment of the circle.

$O$ is the mid-point of $AB$.

$\angle AQB = \pi - \dfrac{2\pi}{3} = \dfrac{\pi}{3}$

$[\because \angle APB + \angle AQB = \pi]$

$\angle ACB = \dfrac{2\pi}{3}$ $\quad [\because \angle ACB = 2\angle AQB]$

$\therefore \angle OCB = \dfrac{1}{2}\angle ACB = \dfrac{\pi}{3}$

In $\Delta OCB$,

$OB = 1, \ \angle OCB = \dfrac{\pi}{3}$

$\tan\dfrac{\pi}{3} = \dfrac{OB}{OC}$

$\Rightarrow \quad OC = \dfrac{1}{\sqrt{3}}$

$\therefore \quad C \equiv \left(0, -\dfrac{1}{\sqrt{3}}\right)$

Thus, from Eq. (i), we can say that $A \cap B$ is a portion of a circle centred at $\left(0, -\dfrac{1}{\sqrt{3}}\right)$ that lies in second quadrant only.

**42.** (*d*) Let $S_1 = \{z \in C : |z-2| \le 1\}$ and $S_2 = \{z \in C : z(1+i) + \bar{z}(1-i) \ge 4\}$

Now, $|z-2| \le 1$

Let $\quad z = x + iy$

$\Rightarrow |x + iy - 2| \le 1$

$\Rightarrow (x-2)^2 + y^2 \le 1$

Also $z(1+i) + \bar{z}(1-i) \ge 4$

$\Rightarrow (x+iy)(1+i) + (x-iy)(1-i) \ge 4$

$\Rightarrow 2x - 2y \ge 4$

$\Rightarrow x - y \ge 2$

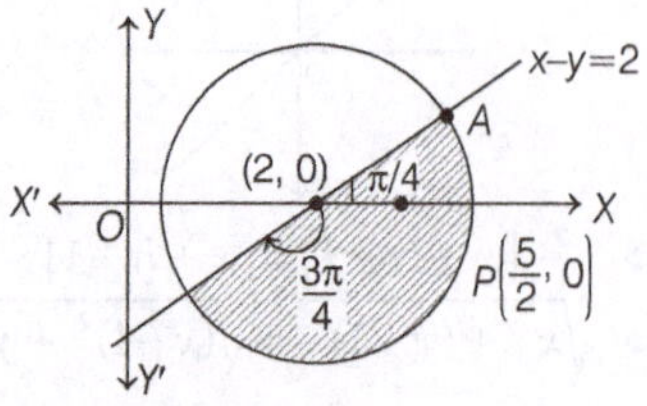


Let point on circle be $A(2 + \cos\theta, \sin\theta)$,

$$\theta \in \left[-\frac{3\pi}{4}, \frac{\pi}{4}\right]$$

$$(AP)^2 = \left(2 + \cos\theta - \frac{5}{2}\right)^2 + \sin^2\theta$$

$$\Rightarrow (AP)^2 = \cos^2\theta + \frac{1}{4} - \cos\theta + \sin^2\theta$$

$$\Rightarrow (AP)^2 = \frac{5}{4} - \cos\theta$$

For $(AP)^2$ to be maximum, $\theta = -\dfrac{3\pi}{4}$

$$(AP)^2 = \frac{5}{4} + \frac{1}{\sqrt{2}} = \frac{5 + 2\sqrt{2}}{4}$$

**43.** (*c*) For $|z-1| \le \sqrt{2}$ $\quad \ldots$(i)

$z$ lies on and inside the circle of radius $\sqrt{2}$ units and centre (1, 0).

For $S_2$, let $z = x + iy$

Now, $(1-i)(z) = (1-i)(x+iy)$

$= x + iy - ix + y = (x+y) + i(y-x)$

$\therefore \text{Re}[(1-i)z] = (x+y)$, which is greater than or equal to one.

i.e., $\quad x + y \ge 1$ $\quad \ldots$(ii)

Also, for $S_3$, Let $z = x + iy$

$\therefore \text{Im}(z) = y$, which is less than or equal to one.

i.e., $\quad y \le 1$ $\quad \ldots$(iii)

**Concept** Draw the graph of Eqs. (i), (ii) and (iii) and then select the common region bounded by Eqs. (i), (ii) and (iii) for $S_1 \cap S_2 \cap S_3$.

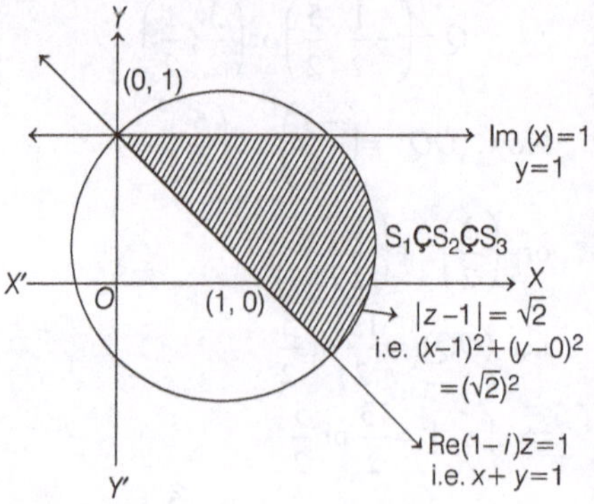


$\therefore S_1 \cap S_2 \cap S_3$ has infinitely many elements.

**44.** (*a*) **Given** A complex number $\dfrac{z^2}{z-1}$, $(z \ne 1)$ is purely real.

**To find** The locus of the complex number $z$.

**Method 1** Since, $\dfrac{z^2}{z-1}$, $(z \ne 1)$ is purely real.

Then, $\quad \dfrac{z^2}{z-1} = \dfrac{\bar{z}^2}{\bar{z}-1}$

$\Rightarrow \quad z^2(\bar{z}-1) = \bar{z}^2(z-1)$

$\Rightarrow \quad z^2\bar{z} - z^2 = \bar{z}^2 z - \bar{z}^2$

$\Rightarrow \quad zz\bar{z} - z^2 = \bar{z}\bar{z} \cdot z - \bar{z}^2$

$\Rightarrow \quad z|z|^2 - z^2 = \bar{z}|z|^2 - \bar{z}^2$

Rearranging the terms, we get

$z|z|^2 - \bar{z}|z|^2 = z^2 - \bar{z}^2$

$\Rightarrow \quad |z|^2(z - \bar{z}) = (z - \bar{z})(\bar{z} + z)$

$\Rightarrow |z|^2(z-\bar{z}) - (z-\bar{z})(\bar{z}+z) = 0$

$\Rightarrow (z - \bar{z})[|z|^2 - (\bar{z} + z)] = 0$

Either $(z - \bar{z}) = 0$ or $[|z|^2 - (\bar{z}+z)] = 0$

Now, $z = \bar{z}$

$\Rightarrow$ Locus of '$z$' is real axis and $\{|z|^2 - (\bar{z} + z)\} = 0$

$\Rightarrow \quad z\bar{z} - (\bar{z} + z) = 0$

Hence, locus of '$z$' is a circle passing through origin.

**Method 2** Put $z = x + iy$, then

$$\frac{z^2}{z-1} = \frac{(x+iy)^2}{(x+iy)-1}$$

$$= \frac{(x^2 - y^2) + i(2xy)}{(x-1) + iy}$$

$$= \frac{(x^2 - y^2) + i(2xy)}{(x-1) + iy} \times \frac{(x-1) - iy}{(x-1) - iy}$$

Since, $\frac{z^2}{z-1}$, $(z\neq 1)$ is purely real, hence its imaginary part should be equal to zero.

$\Rightarrow (x^2-y^2)(-y)+(2xy)(x-1)=0$

$\Rightarrow y(x^2-y^2+2x-2x^2)=0$

$\Rightarrow y(x^2+y^2-2x)=0$

Either $y=0$ or $x^2+y^2-2x=0$

Now, $y=0$

$\Rightarrow$ Locus of '$z$' is real axis and $x^2+y^2-2x=0.$

$\Rightarrow$ Locus of '$z$' is a circle passing through origin.

Hence, locus of '$z$' is either real axis or a circle passing through origin.

**45.** *(22)* $A=\{z\in C\,|\,|z-2-i|=3y\}$

Let $z=x+iy \Rightarrow (x-2)^2+(y-1)^2=9$ ...(i)

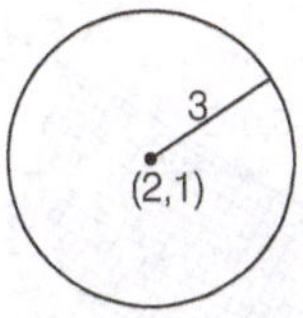


$B=\{z\in C: \text{Re}(z-iz)=2y\}$

Let $z=x+iy$

$iz=ix-y$

$z-iz=(x+y)+i(y-x)$

$\text{Re}(z-iz)=x+y$

$x+y=2$ ...(ii)

Solving Eqs. (i) and (ii), we get

$(x-2)^2+(2-x-1)^2=9$

$x^2-4x+4+x^2-2x+1=9$

$2x^2-6x-4=0$

$x^2-3x-2=0$

$x=\frac{3\pm\sqrt{17}}{2},\ y=\frac{1\mp\sqrt{17}}{2}$

$\Sigma|z|^2=\frac{1}{4}(2\times 26+2\times 18)$

$=22$

**46.** *(9)* Given,

$|z-1|\le 1$

Let $z=x+iy$

So, $|(x-1)+iy|\le 1$

$\Rightarrow \sqrt{(x-1)^2+y^2}\le 1$

$\Rightarrow (x-1)^2+y^2\le 1$ ...(i)

Also, $|z-5|\le|z-5i|$

$\Rightarrow (x-5)^2+(y)^2\le x^2+(y-5)^2$

$\Rightarrow -10x\le -10y \Rightarrow x\ge y$ ...(ii)

On solving Eqs. (i) and (ii), we get

$(x-1)^2+x^2=1$

$\Rightarrow 2x^2-2x=0$

$\Rightarrow x(x-1)=0$

$\Rightarrow x=0$ or $1$

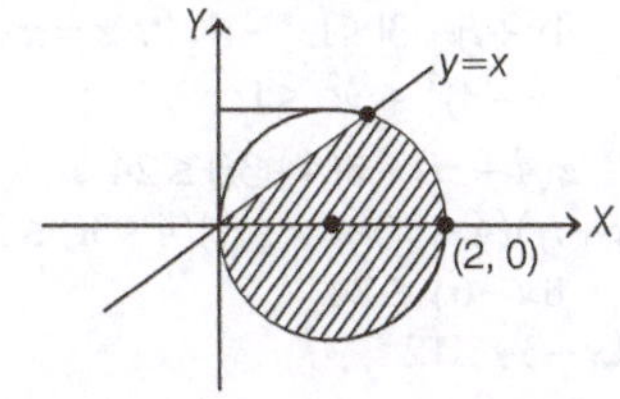


$\therefore y=0$ or $1$

Points (0, 0), (1, 0), (2, 0), (1, 1), (1, −1)

So, $0+1+4+1+1+1+1=9$

(as, $z_1=0+0i,\ z_2=1+0i,\ z_3=2+0i,\ z_4=1+i,\ z_5=1-i$)

**47.** *(36)* $(x+2)^2+(y-3)^2=1$

$L_1: x+y-1=0$

$L_2: x-y+4=0$

On solving circle and $L_1$, we get

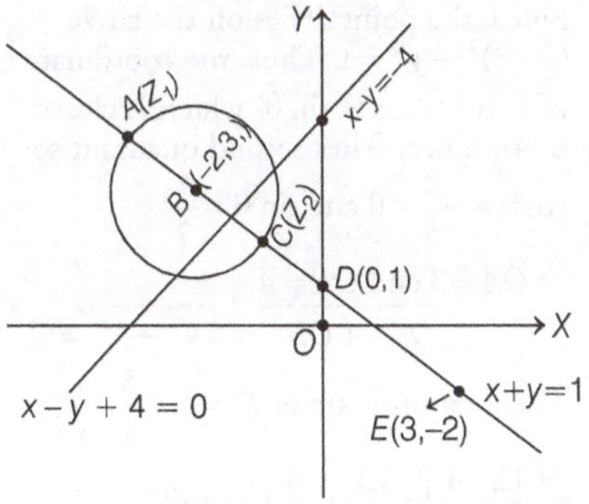


$A\left(0+(2\sqrt{2}+1)\left(\frac{-1}{\sqrt{2}}\right), 1+(2\sqrt{2}+1)\frac{1}{\sqrt{2}}\right)$

$A\left(\frac{-(2\sqrt{2}+1)}{\sqrt{2}}, \frac{3\sqrt{2}+1}{\sqrt{2}}\right)$

On solving circle and $C_2$, we get

$C\left(\frac{-3}{2},\frac{5}{2}\right)$

$|z_1|^2+2|z_2|^2=\frac{(2\sqrt{2}+1)^2}{2}+\frac{(3\sqrt{2}+1)^2}{2}+2\left(\frac{9}{4}+\frac{25}{4}\right)$

$=\frac{28+10\sqrt{2}+34}{2}$

$=31+5\sqrt{2}$

$\alpha=31, \beta=5$

Hence, $\alpha+\beta=36$

**48.** *(20)* Given, $\alpha$ and $\frac{1}{\alpha}$ are two complex numbers.

Also, $\alpha$ lie on $|z-z_0|^2=4$

and $\frac{1}{\alpha}$ lie on $|z-z_0|^2=16$

Now, $\alpha$ satisfies $|z-z_0|^2=4$

$\Rightarrow |\alpha-z_0|^2=4$

$\Rightarrow (\alpha-z_0)\overline{(\alpha-z_0)}=4$ $\quad [\because z\bar{z}=|z|^2]$

$\Rightarrow (\alpha-z_0)(\bar{\alpha}-\bar{z}_0)=4$

$[\because \overline{z_1\pm z_2}=\bar{z}_1\pm\bar{z}_2]$

$\Rightarrow \alpha\bar{\alpha}-\alpha\bar{z}_0-\bar{\alpha}z_0-z_0\bar{z}_0=4$

$\Rightarrow |\alpha|^2-\alpha\bar{z}_0-\bar{\alpha}z_0+|\bar{z}_0|^2=4$

$\Rightarrow |\alpha|^2-\alpha\bar{z}_0-\bar{\alpha}z_0+2=4$

[Given, $z_0=1+i$;

$\therefore |z_0|^2=1^2+1^2=2$]

$\Rightarrow \alpha\bar{z}_0+\alpha z_0=|\alpha|^2-2$ ...(i)

Also, $\frac{1}{\alpha}$ satisfies $|z-z_0|^2=16$

$\Rightarrow \left|\frac{1}{\bar{\alpha}}-z_0\right|^2=16$

$\Rightarrow \left|\frac{1-\bar{\alpha}z_0}{\bar{\alpha}}\right|^2=16$

$\Rightarrow (1-\bar{\alpha}z_0)\overline{(1-\bar{\alpha}z_0)}=16|\bar{\alpha}|^2$

$\Rightarrow (1-\bar{\alpha}z_0)(1-\bar{\bar{\alpha}}\bar{z}_0)=16|\alpha|^2$

$[\because |z|^2=|\bar{z}|^2]$

$\Rightarrow (1-\bar{\alpha}z_0)(1-\alpha\bar{z}_0)=16|\alpha|^2$

$[\because \bar{\bar{z}}=z]$

$\Rightarrow (1-\alpha\bar{z}_0-\bar{\alpha}z_0+\alpha\bar{\alpha}\,z_0\bar{z})=16|\alpha|^2$

$\Rightarrow 1-16|\alpha|^2+|\alpha|^2\cdot|z_0|^2=\alpha\bar{z}_0+\bar{\alpha}z_0$

$\Rightarrow 1-16|\alpha|^2+|\alpha|^2\cdot 2=|\alpha|^2-2$

[from Eq. (i)]

$\Rightarrow 1-14|\alpha|^2=|\alpha|^2-2$

$\Rightarrow 15|\alpha|^2=3 \Rightarrow |\alpha|^2=\frac{1}{5}$

Hence, $100|\alpha|^2=100\times\frac{1}{5}=20$

**49.** *(2)* Given, equation of circle is

$|z-\alpha|^2+|z-\beta|^2=2\lambda$

$\therefore\ 2\lambda=|\alpha-\beta|^2$ ...(i)

For circle,

$|z-z_1|^2+|z-z_2|^2=|z_1-z_2|^2$

$\therefore$ Radius, $r=\frac{|z_1-z_2|}{2}=\frac{|\alpha-\beta|}{2}$

$=\sqrt{\lambda-1}$

$\Rightarrow |\alpha-\beta| = 2\sqrt{\lambda-1}$

$\Rightarrow |\alpha-\beta|^2 = 4(\lambda-1)$

$\Rightarrow 2\lambda = 4(\lambda-1)$ [Using Eq. (i)]

$\Rightarrow 2\lambda = 4 \Rightarrow \lambda = 2$

$\Rightarrow |\alpha-\beta|^2 = 4 \Rightarrow |\alpha-\beta| = 2$

**50.** *(26)* $S = \{z \in C : |z-2| \le 1,\ z(1+i) + \bar{z}(1-i) \le 2\}$

and $5(|z_1|^2 + |z_2|^2) = \alpha + \beta\sqrt{5}$

Converting to cartesian form

$|z-2| \le 1 \equiv (x-2)^2 + (y-0)^2 \le 1$

$\Rightarrow$ Circle centred at (2, 0) and radius = 1

$z(1+i) + \bar{z}(1-i) \le 2$

$\Rightarrow (x+iy)(1+i) + (x-iy)(1-i) \le 2$

$\Rightarrow x + ix + iy - y + x - ix - iy - y \le 2$

$\Rightarrow 2x - 2y \le 2 \Rightarrow x - y \le 1$

$\Rightarrow y \ge x - 1$

$\Rightarrow$ Region above line $y = x - 1$.

The required common region.

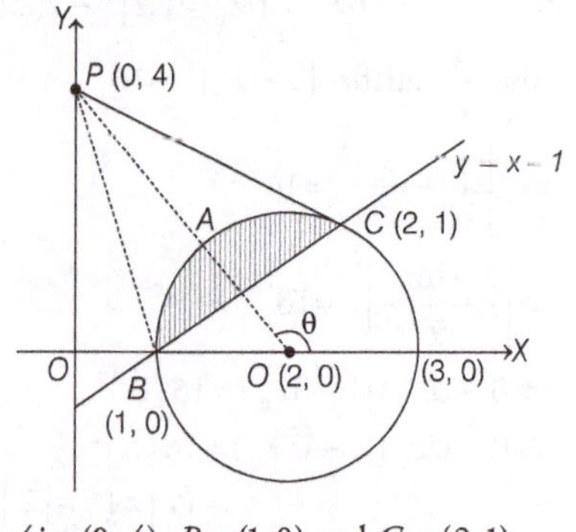


$4i \equiv (0, 4)$, $B \equiv (1, 0)$ and $C \equiv (2, 1)$

$PB = \sqrt{17}$, $PC = \sqrt{13}$

$\therefore$ Maximum is $PB$ and minimum is $PA$.

Line $PA : y - 0 = \dfrac{0-4}{2-0}(x-2)$

$\Rightarrow \quad y = -2(x-2)$

On solving $y = -2(x-2)$ and $(x-2)^2 + y^2 = 1$, we get

$(x-2)^2 + 4(x-2)^2 = 1$

$\Rightarrow \quad (x-2)^2 = \dfrac{1}{5} \Rightarrow x = 2 \pm \dfrac{1}{\sqrt{5}}$

$\therefore \quad A \equiv \left(2 - \dfrac{1}{\sqrt{5}}, \dfrac{2}{\sqrt{5}}\right)$

Minimum : $z_1 = \left(2 - \dfrac{1}{\sqrt{5}}\right) + \dfrac{2i}{\sqrt{5}}$

Maximum : $z_2 = 1 + 0i$

$|z_1|^2 = \dfrac{25 - 4\sqrt{5}}{5} = 5 - \dfrac{4}{\sqrt{5}}$

and $|z_2|^2 = 1$

$\therefore \quad 5(|z_1|^2 + |z_2|^2) = 5\left[5 - \dfrac{4}{\sqrt{5}} + 1\right]$

$= 30 - 4\sqrt{5} = \alpha + \beta\sqrt{5}$

$\therefore \quad \alpha + \beta = 30 - 4 = 26$

**51.** *(80)* Given, $|z-3| \le 1$

$\Rightarrow \quad |x + iy - 3| \le 1 \quad [\because z = x + iy]$

$\Rightarrow \quad (x-3)^2 + y^2 \le 1$

and $\quad z(4+3i) + \bar{z}(4-3i) \le 24$

$\Rightarrow (x+iy)(4+3i) + (x-iy)(4-3i) \le 24$

$\Rightarrow \quad 8x - 6y \le 24$

$\Rightarrow \quad 4x - 3y \le 12$

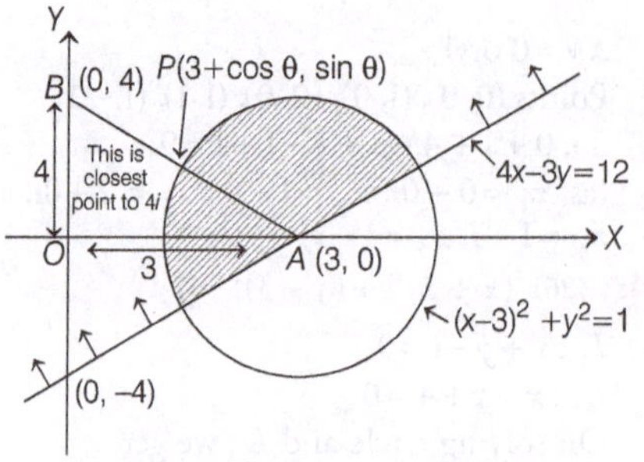


Since, the point $P$ lies on the circle $(x-3)^2 + y^2 = 1$. Thus, the coordinate of $P$ is $(3+\cos\theta, \sin\theta)$ which is closest to $4i$. Since, $\theta$ lies second quadrant so

$\cos\theta = -\dfrac{3}{5} < 0$ and $\sin\theta = \dfrac{4}{5}$

$[\because OA = 3$ and $OB = 4$, $\sqrt{OA^2 + OB^2} = \sqrt{4^2 + 3^2} = 5]$

Then, co-ordinate of $P = \left(3 - \dfrac{3}{5}, \dfrac{4}{5}\right)$

$= \left(\dfrac{12}{5}, \dfrac{4}{5}\right) = \dfrac{12}{5} + i\dfrac{4}{5} = \alpha + i\beta$

$\therefore \quad \alpha = \dfrac{12}{5}$ and $\beta = \dfrac{4}{5}$; $\therefore$ $25(\alpha + \beta)$

$= 25\left(\dfrac{12}{5} + \dfrac{4}{5}\right) = 25\left(\dfrac{16}{5}\right) = 5 \times 16 = 80$

**52.** *(6)* Given, $z_1, z_2$ are the roots of $z^2 + az + 12 = 0$

$\therefore z_1 + z_2 = \dfrac{-a}{1} = -a$

and $z_1 z_2 = \dfrac{12}{1} = 12$

Now, $z_1, z_2$ and origin forms an equilateral triangle.

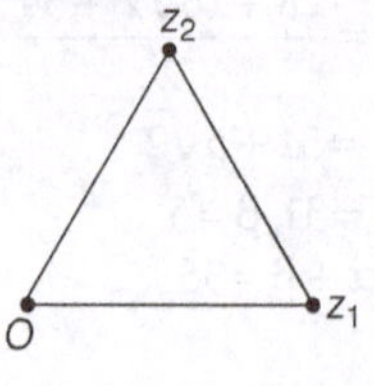


$\therefore z_1^2 + z_2^2 + 0^2 = z_1 z_2 + 0 + 0$

$\Rightarrow z_1^2 + z_2^2 = z_1 z_2$

$\Rightarrow z_1^2 + z_2^2 + 2z_1 z_2 = z_1 z_2 + 2z_1 z_2$

$\Rightarrow (z_1 + z_2)^2 = 3z_1 z_2$

$\Rightarrow \quad (-a)^2 = 3 \times (12)$

$\Rightarrow \quad a^2 = 36$

$\Rightarrow \quad |a|^2 = 36$

$\Rightarrow \quad |a| = \pm 6$ But $|a| \ge 0$

$\therefore \quad |a| = 6$

**53.** *(48)* Given, $|z+5| \le 4$, which is equation of circle.

$|z+5| \le 4 \Rightarrow (x+5)^2 + y^2 \le 16 \quad \ldots$(i)

and $z(1+i) + \bar{z}(1-i) \ge -10$

$\Rightarrow \quad (z + \bar{z}) + i(z - \bar{z}) \ge -10$

$\Rightarrow \quad x - y + 5 \ge 0 \ \ldots$(ii)

From Eqs. (i) and (ii), region bounded by inequalities are

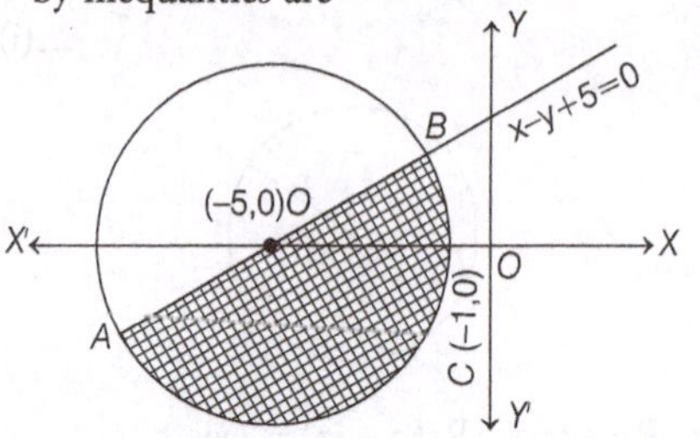


Now, $|z+1|^2 = |z-(-1)|^2$

Maximum value of $|z+1|^2$ will be equal to $(AC)^2$.

Now, $(x+5)^2 + y^2 = 16$

and $x - y + 5 = 0$

Given, $y = \pm 2\sqrt{2}$ and $x = \pm 2\sqrt{2} - 5$

$\therefore$ Coordinates are

$A(-2\sqrt{2} - 5, -2\sqrt{2})$,

$B(2\sqrt{2} - 5, 2\sqrt{2})$

and $\quad C(-1, 0)$

Then, $AC^2 = (2\sqrt{2} + 4)^2 + (2\sqrt{2})^2$

$= 32 + 16\sqrt{2}$

Given, that maximum value of $|z+1|^2$ is $\alpha + \beta\sqrt{2}$

$\Rightarrow \quad \alpha + \beta\sqrt{2} = 32 + 16\sqrt{2}$

$\Rightarrow \quad \alpha = 32, \beta = 16$

$\therefore \quad \alpha + \beta = 32 + 16 = 48$

## Topic 4

**01.** *(d)* $P_{10} = P_9 + P_8 \Rightarrow P_{10} - P_9 - P_8 = 0$

By Newton's method, therefore, the equation is

$x^2 - x - 1 = 0$ as $P_1 = 1$

$\alpha+\beta=1$ and $\alpha\beta=-1$

$\therefore$ Quadratic equation whose roots are $\frac{1}{\alpha}, \frac{1}{\beta}$ is

$$x^2-\left(\frac{1}{\alpha}+\frac{1}{\beta}\right)x+\frac{1}{\alpha\beta}=0$$

$$\Rightarrow \quad x^2+x-1=0$$

Option (d) is correct.

**02.** *(b)* $x^2+4x-n=0$

$n\in[20,100]$

$\Rightarrow x^2+4x=n$

$\Rightarrow x^2+4x+4=n+4$

$\Rightarrow (x+2)^2=n+4$

$\because n\in[20,100]$

$n+4\in[24,104]$

$\because n+4$ will be a perfect square.

$\therefore n+4=25, 36, 49, 64, 81, 100$

$\therefore$ 6 integral values

**03.** *(c)* $x^2-(p+2)x+(2p+9)=0$

$D\geq 0$

$(p+2)^2-4(2p+9)\geq 0$

$\Rightarrow \quad p^2-4p-32\geq 0$

$\Rightarrow \quad (p-8)(p+4)\geq 0$

$P\in(-\infty,-4]\cup[8,\infty)$ ...(i)

Sum $p+2<0$

$p<-2$ ...(ii)

Product $>0$

$2p+9>0$

$p>-\frac{9}{2}$ ...(iii)

From Eqs. (i), (ii), and (iii), we get

$$p\in\left(-\frac{9}{2},-4\right]$$

$\beta-2\alpha=-4+9$

$=5$

**04.** *(a)* Given equation is

$$\left(\frac{9}{x}-\frac{9}{\sqrt{x}}+2\right)\left(\frac{2}{x}-\frac{7}{\sqrt{x}}+3\right)=0 \quad \text{...(i)}$$

Let $\frac{1}{\sqrt{x}}=t$, so that $t>0$

From Eq. (i), we get

$(9t^2-9t+2)(2t^2-7t+3)=0$

$\Rightarrow (3t-2)(3t-1)(t-3)(2t-17)=0$

$\Rightarrow \quad t=\frac{1}{3},\frac{1}{2},\frac{2}{3},3$

$\therefore \quad x=9, 4, \frac{9}{4}, \frac{1}{9}$

Hence, number of solutions $=4$

**05.** *(a)* Graphical method is best solution for this question

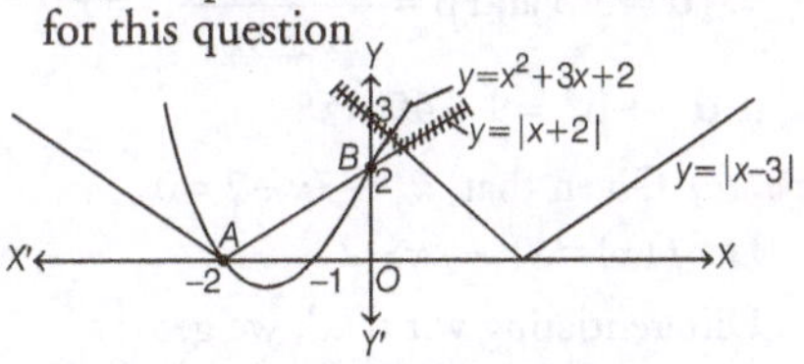


From graph, it is clear that total number of solutions $=2$ (Scratched graph rejected)

**06.** *(c)* $e^{5(\ln x)^2+3}=x^8$

On taking ln both sides, we get

$5(\ln x)^2+3=8\ln x$

$\Rightarrow 5(\ln x)^2-8\ln x+3=0$

$\Rightarrow 5(\ln x)^2-5\ln x-3\ln x+3=0$

$\Rightarrow \quad (5\ln x-3)(\ln x-1)=0$

$\Rightarrow \quad \ln x=\frac{3}{5}, 1 \Rightarrow x=e^{\frac{3}{5}}, e^1$

$\therefore \quad$ Product $=e^{3/5}\cdot e^1=e^{8/5}$

**07.** *(a)* Given 2 and 6 are the roots of the equation $ax^2+bx+1=0$

$\therefore$ Sum of roots $=-\frac{b}{a}=8$

Product of roots $=\frac{1}{a}=12$

$\Rightarrow \quad a=\frac{1}{12}$ so, $b=-\frac{2}{3}$

Now, $2a+b=\frac{2}{12}-\frac{2}{3}=-\frac{1}{2}$

and $6a+b=\frac{6}{12}-\frac{2}{3}=-\frac{1}{6}$

Now, $\frac{1}{2a+b}+\frac{1}{6a+b}=-2-6=-8$

and $\left(\frac{1}{2a+b}\right)\left(\frac{1}{6a+b}\right)=12$

$\therefore$ Required quadratic equation is

$x^2+8x+12=0$

**08.** *(d)* Given,

$(\sqrt{3}+\sqrt{2})^x+(\sqrt{3}-\sqrt{2})^x=10$

Let $(\sqrt{3}+\sqrt{2})^x=t\Rightarrow(\sqrt{3}-\sqrt{2})^x=\frac{1}{t}$

$t+\frac{1}{t}=10$

$t^2-10t+1=0$

$t=\frac{10\pm\sqrt{96}}{2}=5\pm 2\sqrt{6}$

$(\sqrt{3}+\sqrt{2})^x=(\sqrt{3}\pm\sqrt{2})^2$

$x=\pm 2$

$\therefore$ Number of solution $=2$

**09.** *(d)* $\frac{ax^2+2(a+1)x+9a+4}{x^2-8x+32}<0, \quad \forall x\in R$

$\Rightarrow \frac{ax^2+2(a+1)x+9a+4}{(x-4)^2+16}<0, \forall x\in R$

$\Rightarrow ax^2+2(a+1)x+9a+4<0, \forall x\in R$

coefficient of $x^2<0$ and $D<0$

$a<0$ and $D<0$

$\therefore$ No positive integral values of $a$

Hence, $\quad |S|=0$

**10.** *(a)* Let $f(x)=x|x-1|+|x+2|+\alpha$

$\therefore f(x)=0$

$\Rightarrow \quad x|x-1|+|x+2|+\alpha=0$

$\Rightarrow \quad x|x-1|+|x+2|=-\alpha$

Now, $x|x-1|+|x+2|$

$$=\begin{cases} x^2+2 & x\geq 1 \\ -x^2+2x+2 & -2\leq x<1 \\ -x^2-2 & x<-2 \end{cases}$$

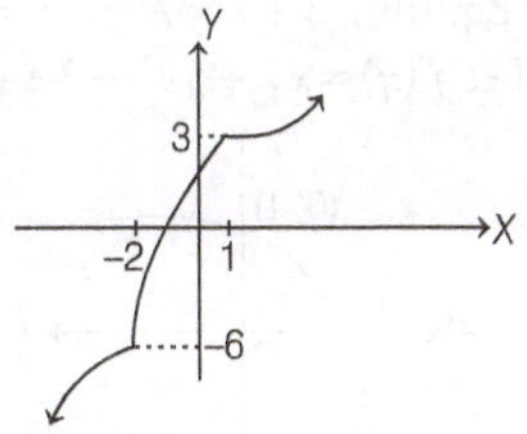


Thus, $-\alpha\in IR$

Hence, the set of all $\alpha\in R$ for which the equtіon $x|x-1|+|x+2|+a=0$ has exactly one real root is $(-\infty,\infty)$.

**11.** *(d)* Let $f(x)=2x^2-8x+k$.

Since, one root lies in the interval (1, 2) and other in (2, 3), then $f(1)>0$, $f(2)<0$, $f(3)>0$

Now, $f(1)>0$

$\Rightarrow k>6$, $f(2)<0$

$\Rightarrow k<8$ and $f(3)>0 \Rightarrow k>6$

$\therefore k\in(6,8)$

So, integral value of $k=7$.

Hence, there is only one integral value of $k$.

**12.** *(a)* Given equation involving GIF is

$x^2-4x+[x]+3=x[x]$

$\Rightarrow x^2-4x+3=x[x]-[x]$

$\Rightarrow (x-1)(x-3)=[x](x-1)$

$\Rightarrow (x-1)\{(x-3)-[x]\}=0$
$\Rightarrow (x-1)=0$ or $(x-3)=[x]$
$\Rightarrow x=1$ or $x-3=[x] \Rightarrow x-[x]=3$
$\Rightarrow \{x\}=3$, where $\{\cdot\}$ represents fractional part function.
As we know that $0 \le \{x\} < 1$
So, $\{x\}=3$ is not possible.
So, only one solution $x=1$ in $(-\infty, \infty)$.

**13.** *(d)* $f(x)=ax^2+bx+c$

$f(1)=a+b+c=3$ ...(i)

$f(3)=9a+3b+c=4$ ...(ii)

$f(0)+f(1)+f(-2)+f(3)=14$

$\Rightarrow c+3+\lambda+4=14$

$c+\lambda=7$ ...(iii)

Also, $c+a+b+c+4a-2b+c+9a+3b+c=14$

$\Rightarrow 14a+2b+4c=14$

$\Rightarrow 7a+b+2c=7$ ...(iv)

From Eqs. (i) and (iv),

$6a+c=4$ ...(v)

From Eqs. (i) and (ii),

$6a-2c=-5$ ...(vi)

From Eqs. (v) and (vi),

$3c=9 \Rightarrow c=3$

From Eq. (iii), $3+\lambda=7 \Rightarrow \lambda=4$

**14.** *(b)* Let $f(x)=x^7+5x^3+3x+1$

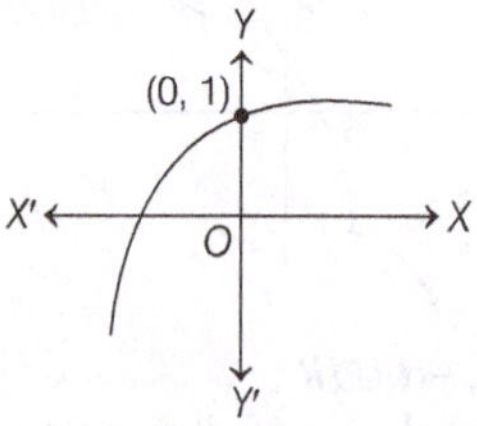


$\Rightarrow f'(x)=7x^6+15x^2+3>0, \forall x \in R$
$\Rightarrow f(x)$ is always increasing.
So, clearly it intersects $X$-axis at only one point.
$\therefore$ Number of real solution $=1$

**15.** *(b)* $ax^2-2bx+15=0$ has repeated roots $\alpha$.

So, $2\alpha=\frac{2b}{a}, \alpha^2=\frac{15}{a} \Rightarrow \alpha=\frac{15}{b}$

As $\alpha$ and $\beta$ are the roots of equation $x^2-2bx+21=0$

$\alpha+\beta=2b \Rightarrow \frac{15}{b}+\beta=2b$

$\Rightarrow \beta=2b-\frac{15}{b}=\frac{2b^2-15}{b}$

Also, $\alpha\beta=21 \Rightarrow \frac{15}{b} \times \frac{(2b^2-15)}{b}=21$

$\Rightarrow 10b^2-75=7b^2$

$\Rightarrow b=\pm 5$

$\Rightarrow \alpha=\pm 3$ and $\beta=\frac{2\times 25-15}{\pm 5}=\pm 7$

$\therefore \alpha^2+\beta^2=9+49=58$

**16.** *(d)* Given that $x^7-7x-2=0$
Let $f(x)=x^7-7x-2$
Differentiating w.r.t. '$x$', we get

$f'(x)=7x^6-7=7(x^6-1)$

Now, $f'(x)=0 \Rightarrow 7(x^6-1)=0$
$\Rightarrow x=\pm 1$
Now, $f(-1)=-1+7-2=4>0$ and
$f(1)=1-7-2=-8<0$
So, roughly sketch of $f(x)$ will be

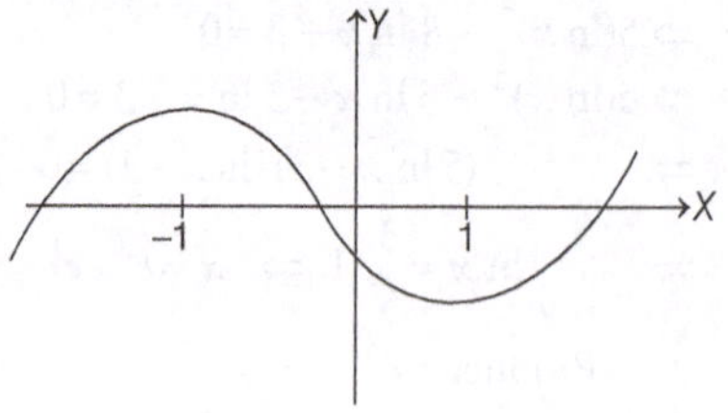


So, the number of real roots of $f(x)=0$ is 3.

**17.** *(a)* Given equation, $x^2-|x|-12=0$

$\Rightarrow |x^2|-|x|-12=0$

$\Rightarrow |x|^2-4|x|+3|x|-12=0$

$\Rightarrow (|x|-4)(|x|+3)=0$

So $|x|-4=0$ or $|x|+3=0$

$|x|=4$ or $|x|=-3$ (not possible)

$x=\pm 4$

Hence, the number of real solutions $=2$

**18.** *(d)* $[e^x]^2+[e^x+1]-3=0$

$\Rightarrow [e^x]^2+[e^x]+1-3=0$

$\Rightarrow [e^x]^2+[e^x]-2=0$

$\Rightarrow ([e^x]-1)([e^x]+2)=0$

$[e^x]=1$ or $[e^x]=-2$

$[e^x]=-2$ is not possible as $e^x>0$.

$\Rightarrow [e^x]=1$

$\Rightarrow 1 \le e^x < 2$

$\Rightarrow 0 \le x < \log_e 2$

**19.** *(a)* Let $x=3+\cfrac{1}{4+\cfrac{1}{3+\cfrac{1}{4+\cfrac{1}{3+\ldots\infty}}}}$

So, $x=3+\cfrac{1}{4+\cfrac{1}{x}}=3+\cfrac{1}{\frac{4x+1}{x}}$

$=3+\frac{x}{4x+1}$

$\Rightarrow (x-3)=\frac{x}{4x+1}$

$\Rightarrow 4x^2-12x-3=0$

$\Rightarrow x=\frac{3\pm 2\sqrt{3}}{2}$

$\Rightarrow x=\frac{3}{2}\pm\sqrt{3}=1.5\pm\sqrt{3}$

But from above, $x>0$.
$\therefore$ Only positive value of $x$ is accepted.
$\therefore x=1.5+\sqrt{3}$

**20.** *(d)* Given, root of $z^2+\alpha z+\beta=0$ is $1-2i$.
Since, it is quadratic equation and one root is complex in nature, its another root is complex conjugate.
$\therefore$ Two roots are $1-2i$ and $1+2i$.

Now, sum of roots $=-\frac{\alpha}{1}=-\alpha$

$=(1-2i)+(1+2i)=2$

Gives, $\alpha=-2$

Product of roots $=\frac{\beta}{1}=\beta$

$=(1-2i)(1+2i)$

$=1+4=5$

Gives, $\beta=5$

$\therefore \alpha-\beta=-2-5=-7$

**21.** *(a)* Given,
$x^2-2(3k-1)x+8k^2-7>0$
Here, $a>0$

$\therefore D<0$

$\Rightarrow [2(3k-1)]^2-4(8k^2-7)<0$

$\Rightarrow 4(9k^2+1-6k)-4(8k^2-7)<0$

$\Rightarrow k^2-6k+8<0$

$\Rightarrow (k-4)(k-2)<0$

$\therefore$

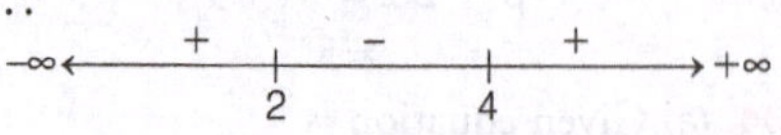


$k \in (2, 4)$

$\therefore$ Required integer, $k=3$

**22.** *(d)* Given quadratic equation $7x^2-3x-2=0$ has roots $\alpha$ and $\beta$.

So, $\alpha+\beta=\frac{3}{7}$ and $\alpha\beta=-\frac{2}{7}$.

Now, $\frac{\alpha}{1-\alpha^2}+\frac{\beta}{1-\beta^2}$

$=\frac{\alpha-\alpha\beta^2+\beta-\beta\alpha^2}{1-(\alpha^2+\beta^2)+\alpha^2\beta^2}$

$=\frac{(\alpha+\beta)-\alpha\beta(\alpha+\beta)}{1-[(\alpha+\beta)^2-2\alpha\beta]+(\alpha\beta)^2}$

$$= \frac{\frac{3}{7}+\frac{2}{7}\left(\frac{3}{7}\right)}{1-\left[\left(\frac{3}{7}\right)^2+\frac{4}{7}\right]+\frac{4}{49}}$$

$$= \frac{21+6}{49-37+4} = \frac{27}{16}$$

**23.** *(a)* Given equation in $x$ is

$[x]^2 + 2[x+2] - 7 = 0$

$\because$ For any integer $n$, $[x+n] = [x] + n$, $\forall\, x \in R$.

So, $[x]^2 + 2[x] + 4 - 7 = 0$

$\Rightarrow [x]^2 + 2[x] - 3 = 0$

$\Rightarrow [x]^2 + 3[x] - [x] - 3 = 0$

$\Rightarrow [x] = -3$ or $1$

$\therefore\ x \in [-3, -2) \cup [1, 2)$

$\therefore$ There are infinitely many solutions.

**24.** *(c)* It is given that roots of quadratic equation $x^2 + bx + 45 = 0,\ b \in R$ has conjugate complex roots, now let roots are $\alpha + i\beta$ and $\alpha - i\beta$.

$\therefore$ Sum of roots $= 2\alpha = -b$ ...(i)

and product of roots $= \alpha^2 + \beta^2 = 45$ ...(ii)

$\because$ Roots $\alpha \pm i\beta$ satisfy $|z+1| = 2\sqrt{10}$,

so $(\alpha+1)^2 + \beta^2 = 40$

$\Rightarrow \alpha^2 + 2\alpha + 1 + \beta^2 = 40$

From Eq. (ii),

On putting the value of $\alpha^2 + \beta^2$, we get

$45 + 2\alpha + 1 = 40$

$\Rightarrow 2\alpha = -6 \Rightarrow \alpha = -3$

From Eq. (i), we get

$-b = 2(-3)$ $[\because \alpha = -3]$

$\Rightarrow\ b = 6$

$\therefore\ b^2 + b = 36 + 6 = 42$

and $b^2 - b = 36 - 6 = 30$

**25.** *(b)* Given equation

$(k+1)\tan^2 x - \sqrt{2}\lambda \tan x = 1 - k$, where $k \neq -1$ and $\lambda \in R$ having roots $\alpha$ and $\beta$, so

$$\tan\alpha + \tan\beta = \frac{\sqrt{2}\lambda}{k+1}$$

and $$\tan\alpha \cdot \tan\beta = \frac{k-1}{k+1}$$

$\because$ $$\tan(\alpha+\beta) = \frac{\tan\alpha + \tan\beta}{1 - \tan\alpha\tan\beta}$$

$$= \frac{\frac{\sqrt{2}\lambda}{k+1}}{1 - \frac{k-1}{k+1}}$$

$\Rightarrow\ \tan(\alpha+\beta) = \frac{\sqrt{2}\lambda}{2} = \frac{\lambda}{\sqrt{2}}$

$\because\ \tan^2(\alpha+\beta) = 50 \Rightarrow \lambda^2 = 100$

$\Rightarrow\ \lambda = \pm 10$

**26.** *(d)*

> **key idea** Reduce the given equation into quadratic equation.

Given equation is

$|\sqrt{x}-2| + \sqrt{x}(\sqrt{x}-4) + 2 = 0$

$\Rightarrow |\sqrt{x}-2| + x - 4\sqrt{x} + 4 = 2$

$\Rightarrow |\sqrt{x}-2| + (\sqrt{x}-2)^2 = 2$

$\Rightarrow (|\sqrt{x}-2|)^2 + |\sqrt{x}-2| - 2 = 0$

Let $|\sqrt{x}-2| = y$, then above equation reduced to

$y^2 + y - 2 = 0$

$\Rightarrow\ (y+2)(y-1) = 0$

$\Rightarrow\ y = 1, -2$

$\therefore\ y = 1$ $[\because y = |\sqrt{x}-2| \geq 0]$

$\Rightarrow\ |\sqrt{x}-2| = 1 \Rightarrow \sqrt{x} - 2 = \pm 1$

$\Rightarrow\ \sqrt{x} = 3$ or $1 \Rightarrow\ x = 9$ or $1$

$\therefore$ Sum of roots $= 9 + 1 = 10$

**27.** *(b)* Here, $9x^2 + 27x + 20 = 0$

$$\therefore x = \frac{-b \pm \sqrt{b^2 - 4ac}}{2a}$$

$$\Rightarrow x = \frac{-27 \pm \sqrt{27^2 - 4 \times 9 \times 20}}{2 \times 9}$$

$\Rightarrow x = -\frac{4}{3}, -\frac{5}{3}$

Given, $\cos A = -\frac{3}{5}$;

$\therefore \sec A = \frac{1}{\cos A} = -\frac{5}{3}$

Here, $A$ is an obtuse angle.

$\therefore \tan A = -\sqrt{\sec^2 A - 1} = -\frac{4}{3}$.

Hence, roots of the equation are $\sec A$ and $\tan A$.

**28.** *(c)* Let $p(x) = ax^2 + bx + c$

$\because p(0) = 1$

$\Rightarrow\ c = 1$

Also, $p(1) = 4$ and $p(-1) = 6$

$\Rightarrow a + b + 1 = 4$ and $a - b + 1 = 6$

$\Rightarrow a + b = 3$ and $a - b = 5$

$\Rightarrow a = 4$

and $b = -1$

$p(x) = 4x^2 - x + 1$;

$p(2) = 16 - 2 + 1 = 15$

$p(-2) = 16 + 2 + 1 = 19$

**29.** *(b)* $\alpha = 2 + 3i;\ \beta = 2 - 3i,\ \gamma = ?$

$\alpha\beta\gamma = \frac{13}{2}$

$\left[\because \text{product of three roots} = -\frac{d}{a}\right]$

$\Rightarrow\ (4+9)\gamma = \frac{13}{2}$

$\Rightarrow\ \gamma = \frac{1}{2}$

**30.** *(a)* Here, $\sqrt{3x^2 + x + 5} = x - 3$

Squaring both the sides, we get

$3x^2 + x + 5 = (x-3)^2$

$\Rightarrow 3x^2 + x + 5 = x^2 + 9 - 6x$

$\Rightarrow 2x^2 + 7x - 4 = 0$

$\Rightarrow (2x-1)(x+4) = 0$

$\Rightarrow x = \frac{1}{2}$ or $x = -4$;

For $x = \frac{1}{2}$ and $x = -4$

LHS $\neq$ RHS of equation,

$\sqrt{3x^2 + x + 5} = x - 3$

Also, for every $x \in R$, LHS $\neq$ RHS of the given equation.

Hence, given equation has no solution.

**31.** *(c)* Given quadratic equation is

$x^2 + px + \frac{3p}{4} = 0$

So, $\alpha + \beta = -p,\ \alpha\beta = \frac{3p}{4}$

Now, given $|\alpha - \beta| = \sqrt{10}$

$\Rightarrow \alpha - \beta = \pm\sqrt{10}$

$\Rightarrow (\alpha - \beta)^2 = 10$

$\Rightarrow \alpha^2 + \beta^2 - 2\alpha\beta = 10$

$\Rightarrow (\alpha+\beta)^2 - 4\alpha\beta = 10$

$\Rightarrow p^2 - 4 \times \frac{3p}{4} = 10$

$\Rightarrow p = -2, 5 \Rightarrow p \in \{-2, 5\}$

**32.** *(d)* $\frac{x-5}{x^2 + 5x - 14} > 0$

$\Rightarrow\ \frac{x-5}{(x-2)(x+7)} > 0$

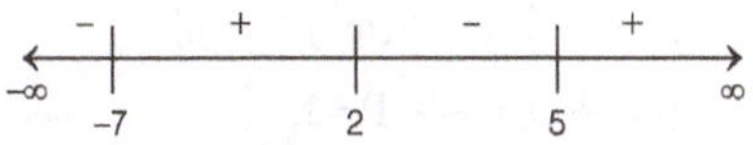


$\alpha = -6$ satisfy the option (d)

$\alpha^2 + 5\alpha - 6 = 0$

**33.** *(c)* $x^2 - (a+1)x + a^2 + a - 8 = 0$

Since, 2 lies between both roots

$\therefore f(2) < 0$

$\Rightarrow 4-2(a+1)+a^2+a-8<0$

$\Rightarrow a^2-a-6<0$

$\Rightarrow (a+2)(a-3)<0$

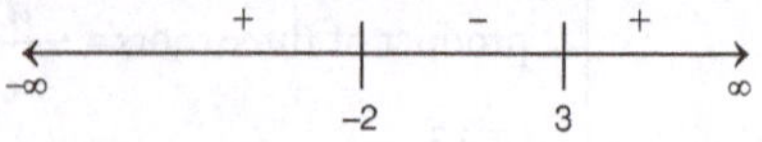


$\Rightarrow -2<a<3$

**34.** *(c)* Given quadratic equation is

$$px^2+qx+r=0 \quad \text{...(i)}$$

Since, α and β are two complex root.

$\therefore \beta=\bar{\alpha} \Rightarrow |\beta|=|\bar{\alpha}| \Rightarrow |\beta|=|\alpha|$ $(\because |\bar{\alpha}|=|\alpha|)$

Consider

$|\alpha|+|\beta|=|\alpha|+|\alpha|$ $(\because |\beta|=|\alpha|)$

$=2|\alpha|>2\cdot 1=2$ $(\because |\alpha|>1)$

Hence, $|\alpha|+|\beta|$ is greater than 2.

**35.** *(b)* Let the quadratic equation be

$$ax^2+bx+c=0$$

Sachin made a mistake in writing down constant terms.

So, sum of roots is correct.

i.e., $\alpha+\beta=7$

Rahul made mistake in writing down coefficient of $x$.

So, product of roots is correct.

i.e., $\alpha\beta=6$

Correct quadratic equation is

$$x^2-(\alpha+\beta)x+\alpha\beta=0$$

$x^2-7x+6=0$ having roots 1 and 6.

**36.** *(c)* Since, α and β are roots of the equation

$$x^2-x+1=0.$$

$\Rightarrow \quad \alpha+\beta=1, \ \alpha\beta=1$

$\Rightarrow \quad x=\dfrac{1\pm\sqrt{3}i}{2}$

$\Rightarrow \quad x=\dfrac{1+\sqrt{3}i}{2}$ or $\dfrac{1-\sqrt{3}i}{2}$

$\Rightarrow \quad x=-\omega^2$ or $-\omega$

Thus, $\alpha=-\omega^2$

Then, $\beta=-\omega$ or $\alpha=-\omega$, then $\beta=-\omega^2$

[where, $\omega^3=1$]

Hence,

$\alpha^{2009}+\beta^{2009}=(-\omega)^{2009}+(-\omega^2)^{2009}$

$=-[(\omega^3)^{669}\cdot\omega^2+(\omega^3)^{1339}\cdot\omega]$

$=-[\omega^2+\omega]=-(-1)=1$

**37.** *(a)* Let the roots of $x^2-6x+a=0$ be α, 4β and that of $x^2-cx+6=0$ be α and 3β.

$\therefore \quad \alpha+4\beta=6, \ 4\alpha\beta=a$

and $\alpha+3\beta=c, \ 3\alpha\beta=6$

$\Rightarrow \quad \dfrac{a}{6}=\dfrac{4}{3} \Rightarrow a=8$

$\therefore x^2-6x+8=0$

$\Rightarrow (x-4)(x-2)=0 \Rightarrow x=2, 4$

and $\quad x^2-cx+6=0$

$\Rightarrow \quad 2^2-2c+6=0 \Rightarrow c=5$

$\therefore \quad x^2-5x+6=0 \Rightarrow x=2, 3$

Hence, common root is 2.

**38.** *(b)* Given that, $x^2+px+q=0$

Sum of roots $=\tan 30°+\tan 15°=-p$

Product of roots $=\tan 30°\cdot\tan 15°=q$

$$\tan 45°=\frac{\tan 30°+\tan 15°}{1-\tan 30°\cdot\tan 15°}$$

$\Rightarrow \quad \dfrac{-p}{1-q}=1 \Rightarrow -p=1-q \Rightarrow q-p=1$

$\therefore 2+q-p=3$

**39.** *(d)* Let α and β be the roots of equation

$$x^2-(a-2)x-a-1=0$$

Then, $\alpha+\beta=a-2$ and $\alpha\beta=-a-1$

Now, $\quad \alpha^2+\beta^2=(\alpha+\beta)^2-2\alpha\beta$

$\Rightarrow \quad \alpha^2+\beta^2=(a-2)^2+2(a+1)$

$\Rightarrow \quad \alpha^2+\beta^2=a^2-2a+6$

$\Rightarrow \quad \alpha^2+\beta^2=(a-1)^2+5$

The value of $\alpha^2+\beta^2$ will be least, if

$a-1=0.$

$\Rightarrow a=1$

> **Alternate Solution**
>
> Since, $\alpha+\beta=(a-2)$ and $\alpha\beta=-a-1$
>
> Let $\quad f(a)=\alpha^2+\beta^2=(\alpha+\beta)^2-2\alpha\beta$
>
> $=(a-2)^2+2(a+1)=a^2-2a+6$
>
> $\Rightarrow \quad f'(a)=2a-2$
>
> For maxima or minima, put $f'(a)=0$.
>
> $\therefore \quad 2a-2=0 \Rightarrow a=1$
>
> Now, $f''(a)=2 \Rightarrow f''(1)=2>0$
>
> So, $f(a)$ is minimum at $a=1$

**40.** *(b)* Let α and β be two numbers whose arithmetic mean is 9 and geometric mean is 4.

$\therefore \quad \alpha+\beta=18$ and $\alpha\beta=16$

$\therefore$ Required equation

$$x^2-(\alpha+\beta)x+(\alpha\beta)=0$$

$\Rightarrow \quad x^2-18x+16=0$

**41.** *(b)* Given that, $3^{2x^2-7x+7}=3^2$

$\Rightarrow \quad 2x^2-7x+7=2$

$\Rightarrow \quad 2x^2-7x+5=0$

Now, $D=b^2-4ac$

$=49-40=9>0$

Hence, it has two real roots.

**42.** *(b)* $x^2+3x-1=0$ (roots γ, δ)

$\Rightarrow x^2-1=-3x$

$\Rightarrow \quad Q_n=\gamma^n+\delta^n$

$Q_{25}-Q_{23}=(\gamma^{25}-\gamma^{23})-(\delta^{25}-\delta^{23})$

$=\gamma^{23}(\gamma^2-1)+\delta^{23}(\delta^2-1)$

$=\gamma^{23}(-3\gamma)+\delta^{23}(-3\delta)$

$=-3[\gamma^{24}+\delta^{24}]$

$\Rightarrow \dfrac{Q_{25}-Q_{23}}{Q_{24}}=-3$

$x^2+\sqrt{3}x-16=0$ (roots α, β)

$\Rightarrow P_n=\alpha^n+\beta^n$

$P_{25}+\sqrt{3}P_{24}$

$=(\alpha^{25}+\sqrt{3}\alpha^{24})+(\beta^{25}+\sqrt{3}\beta^{24})$

$=\alpha^{23}(\alpha^2+\sqrt{3}\alpha)+\beta^{23}(\beta^2+\sqrt{3}\beta)$

$=\alpha^{23}(16)+16\beta^{23}$

$\Rightarrow \dfrac{P_{25}+\sqrt{3}P_{24}}{2\cdot P_{23}}=\dfrac{16(\alpha^{23}+\beta^{23})}{2(\alpha^{23}+\beta^{23})}=8$

$\Rightarrow \dfrac{P_{25}+\sqrt{3}P_{24}}{2P_{23}}+\dfrac{Q_{25}-Q_{23}}{Q_{24}}$

$=8+(-3)=5$

**43.** *(c)* We know that, $|x|=\begin{cases}-x, & x<0\\ x, & x\geq 0\end{cases}$

Similarly, $|x-2|=\begin{cases}-(x-2), & x<2\\ (x-2), & x\geq 2\end{cases}$

and $|x-3|=\begin{cases}-(x-3), & x<3\\ (x-3), & x\geq 3\end{cases}$

The zeroes of $|x-2|$ is 2 and that of $|x-3|$ is 3.

2 and 3 divide the number line in the intervals $(-\infty, 2)$, $[2, 3)$, $[3, \infty)$

In $(-\infty, 2]$

$-x(x-2)-3(x-3)+1=0$

$\Rightarrow \quad -x^2+2x-3x+9+1=0$

$\Rightarrow \quad -x^2-x+10=0$

$\Rightarrow \quad x^2+x-10=0$

$\Rightarrow \quad x=\dfrac{-1\pm\sqrt{41}}{2}\approx 2.7$ or $-3.7$

But we have case $(-\infty, 2]$, so

$x=\dfrac{-1-\sqrt{41}}{2}$

In $[2, 3)$, $x|x-2|+3|x-3|+1=0$

$x(x-2)-3(x-3)+1=0$

$x^2-2x-3x+9+1=0$

$x^2-5x+10=0$

$x = \dfrac{5 \pm \sqrt{-15}}{2}$ No real root exists.

In $[3, \infty)$

$$x|x-2| + 3|x-3| + 1 = 0$$
$$x(x-2) + 3(x-3) + 1 = 0$$
$$x^2 - 2x + 3x - 9 + 1 = 0$$
$$x^2 - x - 8 = 0$$
$$= \frac{1 \pm \sqrt{33}}{2} \approx 2.4 \text{ or } -3.4$$

But in this case $[3, \infty)$, neither solution is valid.

Now, count the number of real solutions, we found only one real root $\dfrac{-1-\sqrt{41}}{2}$.

The number of real roots is 1.

**44.** *(b)* $|x-2|^2 + |x-2| - 2 = 0$

Let $|x-2| = t$

$t^2 + t - 2 = 0$

$$(t+2)(t-1) = 0$$
$$\Rightarrow \quad |x-2| = 1 \quad (\because |x-2| \geq 0)$$
$$\Rightarrow \quad x - 2 = \pm 1$$
$$\Rightarrow \quad x = 3 \text{ or } 1$$

Also, $x^2 - 2|x-3| - 5 = 0$

$$x^2 - 2(x-3) - 5 = 0 \quad (x \geq 3)$$
$$x^2 - 2x + 1 = 0$$

$x = 1$ Not possible

$$x^2 + 2(x-3) - 5 = 0\ (x < 3)$$
$$x^2 + 2x - 11 = 0$$
$$\alpha \text{ or } \beta = -\frac{2 \pm \sqrt{48}}{2}$$
$$= \frac{-2 \pm 4\sqrt{3}}{2}$$
$$= -1 \pm 2\sqrt{3}$$

$\alpha = -1 - 2\sqrt{3}$ and $\beta = -1 + 2\sqrt{3}$

So, sum of squares of both the roots of equation is

$= 3^2 + 1^2 + (2\sqrt{3}-1)^2 + (2\sqrt{3}+1)^2$

$= 10 + 2(12+1) = 36$

**45.** *(d)* We have,

$$2x^2 + (a-5)x + 15 = 3a$$

has no real root

So, $(a-5)^2 - 8(15-3a) < 0$

$$= a^2 + 14a + 25 - 120 < 0$$
$$= a^2 + 14a - 95 < 0$$
$$= (a+19)(a-5) < 0$$
$$= a \in (-19, 5)$$

$$\therefore \quad \sum_{x \in X} x^2 = (1^2 + 2^2 + \ldots.. + 4^2)$$
$$+ (1^2 + 2^2 + \ldots.. + 18^2)$$
$$= \frac{4 \times 5 \times 9}{6} + \frac{18 \times 19 \times 37}{6}$$
$$= 30 + 2109 = 2139$$

> • Concept **Enhancer**
>
> $$\sum_{r=1}^{n} r^2 = 1^2 + 2^2 + 3^2 + \ldots + n^2$$
> $$= \frac{n(n+1)(2n+1)}{6}$$

**46.** *(d)* $x^2 - (3-2i)x - (2i-2) = 0$

$$x = \frac{(3-2i) \pm \sqrt{(3-2i)^2 + 4 \times 1(2i-2)}}{2 \times 1}$$
$$= \frac{(3-2i) \pm \sqrt{9 - 12i + 4i^2 + 8i - 8}}{2}$$
$$= \frac{(3-2i) \pm \sqrt{-3-4i}}{2}$$
$$= \frac{(3-2i) \pm \sqrt{(1-2i)^2}}{2}$$
$$= \frac{3-2i+1-2i}{2}, \frac{3-2i-1+2i}{2}$$
$$= \frac{4-4i}{2}, \frac{2}{2} = 2-2i, 1$$

Let $\alpha + i\beta = 2 - 2i$ and $\gamma + i\delta = 1$

$$\Rightarrow \quad \alpha = 2, \beta = -2, \gamma = 1, \delta = 0$$
$$\therefore \quad \alpha\gamma + \beta\delta = 2 \times 1 + 0 = 2$$

**47.** *(d)* If $x \geq 3/2$

$$\Rightarrow \quad x^2 + 2x - 3 - 4 = 0$$
$$\Rightarrow \quad x^2 + 2x - 7 = 0$$
$$\Rightarrow \alpha = \frac{-2 \pm \sqrt{4+28}}{2} = -1 \pm 2\sqrt{2}$$
$$\therefore \quad x = 2\sqrt{2} - 1 \quad [\text{as } x \geq 3/2]$$

and $x \leq 3/2 \Rightarrow x^2 - 2x + 3 - 4 = 0$

$$\Rightarrow \quad x^2 - 2x - 1 = 0$$
$$\Rightarrow \quad x = \frac{2 \pm \sqrt{4 + 4 \times 1}}{2} = \frac{2 \pm 2\sqrt{2}}{2}$$
$$= 1 \pm \sqrt{2}$$
$$\Rightarrow \quad x = 1 - \sqrt{2} \quad [\text{as } x \leq 3/2]$$

$\therefore$ Sum of the squares of all roots

$$= (2\sqrt{2}-1)^2 + (1-\sqrt{2})^2$$
$$= 8 + 1 - 2 \cdot 2\sqrt{2} + 1 + 2 - 2\sqrt{2}$$
$$= 12 - 6\sqrt{2} = 6(2 - \sqrt{2})$$

**48.** *(b)* Given quadratic equation is

$$2z^2 - 3z - 2i = 0 \quad \ldots\text{(i)}$$
$$\Rightarrow 2z^2 - 2i = 3z$$
$$\Rightarrow 2z\left(z - \frac{i}{z}\right) = 3z$$
$$\Rightarrow 2\left(z - \frac{i}{z}\right) = 3$$
$$\Rightarrow \quad z - \frac{i}{z} = \frac{3}{2}$$

$\because \alpha, \beta$ are the roots of Eq. (i)

So, $\left(\alpha - \dfrac{i}{\alpha}\right) = \dfrac{3}{2}$

On squaring both sides, we get

$$\alpha^2 - \frac{1}{\alpha^2} - 2i = 9/4$$
$$\Rightarrow \quad \alpha^2 - \frac{1}{\alpha^2} = \frac{9}{4} + 2i$$

Again, squaring both sides, we get

$$\alpha^4 + \frac{1}{\alpha^4} - 2 = \frac{81}{16} - 4 + 9i$$
$$\Rightarrow \quad \alpha^4 + \frac{1}{\alpha^4} = \frac{49}{16} + 9i \quad \ldots\text{(ii)}$$

Similarly, we can find

$$\beta^4 + \frac{1}{\beta^4} = \frac{49}{16} + 9i \quad \ldots\text{(iii)}$$

($\because \beta$ is also a root of Eq. (i))

Now, $\dfrac{\alpha^{19} + \beta^{19} + \alpha^{11} + \beta^{11}}{\alpha^{15} + \beta^{15}}$

$$= \frac{\alpha^{15}\left(\alpha^4 + \frac{1}{\alpha^4}\right) + \beta^{15}\left(\beta^4 + \frac{1}{\beta^4}\right)}{\alpha^{15} + \beta^{15}}$$
$$= \frac{(\alpha^{15} + \beta^{15})\left(\frac{49}{16} + 9i\right)}{(\alpha^{15} + \beta^{15})}$$

(using Eqs. (ii) and (iii))

$$= \frac{49}{16} + 9i$$

$\therefore$ Real part $= \dfrac{49}{16}$ and imaginary part $= 9$

Hence, 16[(real part) × (imaginary part]

$$= \left[\frac{49}{16} \times 9\right] \times 16$$
$$= 441$$

**49.** *(b)* Given equation is

$(x^2 - 9x + 11)^2 - (x-4)(x-5) = 3 \ldots$(i)

Let $x^2 - 9x = y$, then

Eq. (i) becomes, $(y+11)^2 - (y+20) = 3$

$$(\because (x-4)(x-5) = x^2 - 9x + 20)$$
$$\Rightarrow y^2 + 121 + 22y - y - 20 - 3 = 0$$
$$\Rightarrow y^2 + 21y + 98 = 0$$
$$\Rightarrow (y+14)(y+7) = 0$$
$$\Rightarrow y = -14, -7$$
$$\Rightarrow x^2 - 9x = -14 \text{ or } x^2 - 9x = -7$$
$$\Rightarrow x^2 - 9x + 14 = 0 \text{ or } x^2 - 9x + 7 = 0$$

$$\Rightarrow x=\frac{9\pm\sqrt{81-4\times 14}}{2}$$

$$\text{or } x=\frac{9\pm\sqrt{81-4\times 7}}{2}$$

$$\Rightarrow x=\frac{9\pm 5}{2} \text{ or } x=\frac{9\pm\sqrt{53}}{2}$$

$$\Rightarrow x=7,2 \text{ or } x=\frac{9+\sqrt{53}}{2},\ \frac{9-\sqrt{53}}{2}$$

(Rational roots) (Irrational roots)

∴Product of all rational roots

$=7\times 2=14$

**50.** *(b)* $2x^2+(\cos\theta)x-1=0$ $\begin{matrix}\alpha_\theta\\ \beta_\theta\end{matrix}$ ...(i)

$\theta\in(0,2\pi)$

$\alpha_\theta$ and $\beta_\theta$ be roots of Eq. (i), we get

$$\alpha_\theta+\beta_\theta=-\frac{\cos\theta}{2}$$

$$\alpha_\theta\beta_\theta=-\frac{1}{2}$$

$$\therefore\ \alpha_\theta^4+\beta_\theta^4=(\alpha_\theta^2+\beta_\theta^2)^2-2\alpha_\theta^2\beta_\theta^2$$

$$=((\alpha_\theta+\beta_\theta)^2-2\alpha_\theta\beta_\theta)^2-2\left(-\frac{1}{2}\right)^2$$

$$=\left(\frac{\cos^2\theta}{4}+1\right)^2-\frac{1}{2}$$

$$\therefore\quad m=1-\frac{1}{2}=\frac{1}{2}$$

$$M=\frac{25}{16}-\frac{1}{2}$$

$$\therefore\quad 16(M+m)=25$$

**51.** *(b)* Given $\alpha$ and $\beta$ are the roots of

$$x^2-\sqrt{2}x-\sqrt{3}=0$$

$$\therefore\ \alpha^2-\sqrt{2}\alpha-\sqrt{3}=0$$

On multiplying both sides by $\alpha^n$, we get

$$\alpha^{n+2}-\sqrt{2}\alpha^{n+1}-\sqrt{3}\,\alpha^n=0 \quad\text{...(i)}$$

Similarly,

$$\beta^{n+2}-\sqrt{2}\beta^{n+1}-\sqrt{3}\beta^n=0 \quad\text{...(ii)}$$

Subtracting

$$(\alpha^{n+2}-\beta^{n+2})-\sqrt{2}(\alpha^{n+1}-\beta^{n+1})-\sqrt{3}(\alpha^n-\beta^n)=0$$

$$\therefore\ P_{n+2}-\sqrt{2}P_{n+1}-\sqrt{3}P_n=0$$

Put $n=10$, $P_{12}-\sqrt{2}P_{11}-\sqrt{3}P_{10}=0$ ...(iii)

Put $n=9$, $P_{11}-\sqrt{2}P_{10}-\sqrt{3}P_9=0$ ...(iv)

On multiplying (−11) Eq. (iii) and (−10) Eq. (iv) and subtracting, we get

$$11(\sqrt{3}P_{10}+\sqrt{2}P_{11}-P_{12})-10(\sqrt{2}P_{10}-P_{11}+\sqrt{3}P_9)=0$$

$$\Rightarrow(11\sqrt{3}-10\sqrt{2})P_{10}+(11\sqrt{2}+10)P_{11}-11P_{12}-10\sqrt{3}P_9=0$$

$$\therefore\text{Required value}=10\sqrt{3}\,P_9$$

**52.** *(a)*

> **key idea** First use
> $\alpha^4+\beta^4=(\alpha^2+\beta^2)^2-2\alpha^2\beta^2$
> $=((\alpha+\beta)^2-2\alpha\beta)^2-2\alpha^2\beta^2$
> and then
> $\alpha^6+\beta^6=(\alpha^3+\beta^3)^2-2\alpha^3\beta^3$
> $(\alpha+\beta)[\{(\alpha+\beta)^2-3\alpha\beta\}]^2-2(\alpha\beta)^3$

Given $\alpha$ and $\beta$ are the roots of the given quadratic equation.

$$x^2+2\sqrt{2}x-1=0$$

$$\alpha+\beta=-2\sqrt{2},\ \alpha\beta=-1$$

$$\text{Now, }\alpha^4+\beta^4=(\alpha^2+\beta^2)^2-2\alpha^2\beta^2$$

$$=\{(\alpha+\beta)^2-2\alpha\beta\}^2-2(\alpha\beta)^2$$

$$=(8+2)^2-2(-1)^2=100-2=98$$

$$\text{and }\alpha^6+\beta^6=(\alpha^3+\beta^3)^2-2\alpha^3\beta^3$$

$$=[(\alpha+\beta)\{(\alpha+\beta)^2-3\alpha\beta\}]^2-2(\alpha\beta)^3$$

$$=[(-2\sqrt{2})(8+3)]^2-2(-1)^3$$

$$=(8)(121)+2=970$$

$$\therefore\quad \frac{\alpha^6+\beta^6}{10}=97$$

Hence, required quadratic equation is

$$x^2-(98+97)x+(98)(97)=0$$

$$\Rightarrow\quad x^2-195x+9506=0$$

**53.** *(b)* $x^3+bx+c=0$ [given] ...(i)

and $\beta\gamma=1=-\alpha$ ...(ii)

Let $\alpha,\beta,\gamma$ be the roots, then

$$\alpha+\beta+\gamma=0 \quad\text{...(iii)}$$

$$\alpha\beta+\beta\gamma+\gamma\alpha=-b \quad\text{...(iv)}$$

$$\alpha\beta\gamma=-c \quad\text{...(v)}$$

$\because\quad \beta\gamma=1$ [from (ii)]

$\Rightarrow\quad \alpha=-c$ [from (v)]

also $\alpha=-1$

$\therefore\quad c=1$

Now, from Eq. (iii), using Eq. (ii), we have

$$-1+\beta+\gamma=0$$

$$\Rightarrow\quad \beta+\gamma=1$$

and from Eq. (iv)

$$-\beta-\gamma+\beta\gamma=-b$$

$$\Rightarrow\quad -1+1=-b\Rightarrow b=0$$

∴ Eq. (i) becomes

$$x^3+1=0$$

roots are $x=-1,-\omega-\omega^2$

$$\alpha=-1,\beta=-\omega,\gamma=-\omega^2$$

$$b=0,\ c=1$$

Now, $b^3+2c^3-3\alpha^3-6\beta^3-8\gamma^3$

$=0+2+3+6+8=19$

**54.** *(b)* Given equation is

$$3\left(x^2+\frac{1}{x^2}\right)-2\left(x+\frac{1}{x}\right)+5=0$$

$$3\left\{\left(x+\frac{1}{x}\right)^2-2\right\}-2\left(x+\frac{1}{x}\right)+5=0$$

Let $x+\frac{1}{x}=t$; $\therefore 3(t^2-2)-2t+5=0$

$$\Rightarrow 3t^2-6-2t+5=0$$

$$\Rightarrow 3t^2-2t-1=0\Rightarrow(3t+1)(t-1)=0$$

$$\therefore\ t=1,-\frac{1}{3}$$

$$\therefore\ x+\frac{1}{x}=1,\ x+\frac{1}{x}=\frac{-1}{3}$$

$$x^2-x+1=0;\ x^2+1=-\frac{x}{3}$$

$$\Rightarrow 3x^2+3+x=0\ \because D=b^2-4ac,$$

$$\therefore D_1=(-1)^2-4\times1\times1,=1-4=-3<0$$

$$\text{and } D_2=1^2-4\times3\times3$$

$$=1-36=-35<0$$

No real solution.

Hence, there will not be any real solution.

**55.** *(b)* Given that

$$(e^{2x}-4)(6e^{2x}-5e^x+1)=0$$

Let $e^x=p$

Thus, $(p^2-4)(6p^2-5p+1)=0$

$$\Rightarrow(p+2)(p-2)(2p-1)(3p-1)=0$$

$$\Rightarrow\quad p=-2,2,1/2,1/3$$

| $\therefore e^x=-2$ | $e^x=2$ | $e^x=\frac{1}{2}$ | $e^x=\frac{1}{3}$ |
|---|---|---|---|
| This is not possible to take. | $x=\log_e 2$ | $x=-\log_e 2$ | $x=-\log_e 3$ |

Thus, all the real roots of the above equation are $\log_e 2,-\log_e 2,-\log_e 3$

Sum of these real roots

$$=\log_e 2-\log_e 2-\log_e 3$$

$$=-\log_e 3=-\log_e 3$$

**56.** *(c)* $x^2+(20)^{1/4}x+5^{1/2}=0$

$$\alpha+\beta=-20^{1/4}\quad \alpha\beta=5^{1/2}$$

$$(\alpha+\beta)^2=\alpha^2+\beta^2+2\alpha\beta$$

$$\Rightarrow 20^{1/2}=\alpha^2+\beta^2+2\cdot5^{1/2}$$

$$\Rightarrow 2\sqrt{5}=\alpha^2+\beta^2+2\sqrt{5}$$

$$\Rightarrow\ \alpha^2+\beta^2=0$$

Now, $(\alpha^2+\beta^2)^2=\alpha^4+\beta^4+2\alpha^2\beta^2$

$\Rightarrow 0=\alpha^4+\beta^4+2\cdot 5$

$\Rightarrow \alpha^4+\beta^4=-10$

Now, $(\alpha^4+\beta^4)^2=\alpha^8+\beta^8+2\alpha^4\beta^4$

$\Rightarrow 100=\alpha^8+\beta^8+2(5^{1/2})^4$

$\therefore \alpha^8+\beta^8=50$

**57.** (a) $e^{6x}-e^{4x}-2e^{3x}-12e^{2x}+e^x+1=0$

$\Rightarrow (e^{3x}-1)^2-e^x(e^{3x}-1)=12e^{2x}$

$\Rightarrow (e^{3x}-1)(e^{3x}-e^x-1)=12e^{2x}$

$\Rightarrow (e^{3x}-1)(e^x-e^{-x}-e^{-2x})=12$

$\Rightarrow e^x-e^{-x}-e^{-2x}=\dfrac{12}{e^{3x}-1}$

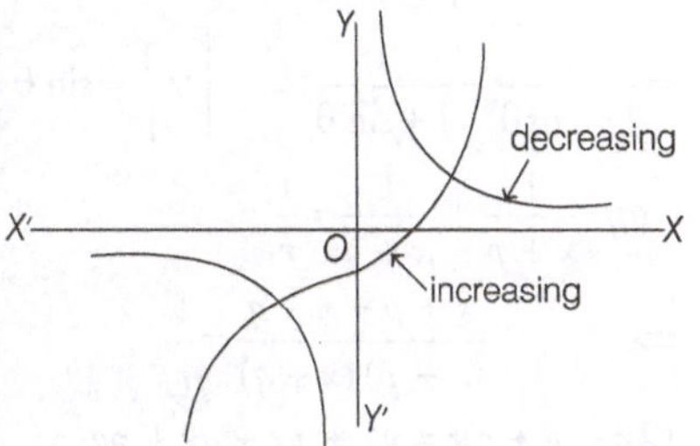


Hence, the number of real roots is 2.

**58.** (c) $x^2+3^{\frac{1}{4}}x+3^{\frac{1}{2}}=0$

$\therefore x=\dfrac{-3^{1/4}\pm\sqrt{3^{1/2}-4\cdot 3^{1/2}}}{2}$

$=\dfrac{3^{1/4}(-1\pm\sqrt{3}i)}{2}=3^{1/4}\left(\dfrac{-1+\sqrt{3}i}{2}\right)$

or $3^{1/4}\left(\dfrac{-1-\sqrt{3}i}{2}\right)=3^{1/4}\omega$ or $3^{1/4}\omega^2$

Now, $\alpha^{96}(\alpha^{12}-1)+\beta^{96}(\beta^{12}-1)$

$=\alpha^{108}-\alpha^{96}+\beta^{108}-\beta^{96}$

$=(\alpha^{108}+\beta^{108})-(\alpha^{96}+\beta^{96})$

$=\{(3^{1/4}\omega)^{108}+(3^{1/4}\omega^2)^{108}\}$

$-\{(3^{1/4}\omega)^{96}+(3^{1/4}\omega^2)^{96}\}$

$=3^{27}(\omega^{108}+\omega^{216})-3^{24}(\omega^{96}+\omega^{192})$

$=3^{27}(2)-3^{24}(2)=3^{24}(54)-3^{24}(2)$

$=3^{24}(52)=52\times 3^{24}$

**59.** (c) We have,

$$x^2-6x-2=0$$

Given, $\alpha$ and $\beta$ are roots of above quadratic equation, then

$\alpha^2-6\alpha-2=0$ ...(i)

$\beta^2-6\beta-2=0$ ...(ii)

Also, given $a_n=\alpha^n-\beta^n$, then

$\dfrac{a_{10}-2a_8}{3a_9}$

$=\dfrac{(\alpha^{10}-\beta^{10})-2(\alpha^8-\beta^8)}{3(\alpha^9-\beta^9)}$

$=\dfrac{\alpha^{10}-2\alpha^8-\beta^{10}+2\beta^8}{3(\alpha^9-\beta^9)}$

$=\dfrac{\alpha^8(\alpha^2-2)-\beta^8(\beta^2-2)}{3(\alpha^9-\beta^9)}$

[from Eqs. (i) and (ii) $\alpha^2-2=6\alpha$, $\beta^2-2=6\beta$]

$=\dfrac{\alpha^8(6\alpha)-\beta^8(6\beta)}{3(\alpha^9-\beta^9)}=\dfrac{6\alpha^9-6\beta^9}{3(\alpha^9-\beta^9)}=2$

**60.** (b) Given, $p, q$ are positive numbers.

$p+q=2$ ...(i)

$p^4+q^4=272$

$\Rightarrow (p^2+q^2)^2-2p^2q^2=272$

$\Rightarrow [(p+q)^2-2pq]^2-2p^2q^2=272$

$\Rightarrow [(2)^2-2pq]^2-2p^2q^2=272$

[from Eq. (i)]

$\Rightarrow (4-2pq)^2-2p^2q^2=272$

$\Rightarrow 16+4p^2q^2-16pq-2p^2q^2=272$

$\Rightarrow 2p^2q^2-16pq-256=0$

$\Rightarrow p^2q^2-8pq-128=0$

$\therefore pq=\dfrac{8\pm\sqrt{64+4\times128}}{2\times1}$

$=\dfrac{8\pm\sqrt{576}}{2}=\dfrac{8\pm24}{2}$

$\therefore pq=16,-8$

Here, $pq=-8$ is not possible as $p$ and $q$ are positive.

$\therefore pq=16$

Now, the equation whose roots are $p$ and $q$ is

$x^2-(p+q)x+pq=0$

$\Rightarrow x^2-2x+16=0$

**61.** (b) Let a polynomial $f(x)$ of degree four is

$f(x)=ax^4+bx^3+cx^2+dx+e$

$\therefore f'(x)=4ax^3+3bx^2+2cx+d=0$

$\Rightarrow x=-1, 0, 1$

$\therefore$ Sum of roots $=0\Rightarrow -\dfrac{3b}{4a}=0\Rightarrow b=0$

and product of roots $=0\Rightarrow -\dfrac{d}{4a}=0$

$\Rightarrow d=0$ and product of roots taking two at a time $=-1$

$\Rightarrow 2c/4a=-1\Rightarrow c=-2a$

$\therefore f(x)=ax^4-2ax^2+e$

Now, $f(x)=f(0)$ [given]

$\Rightarrow ax^4-2ax^2+e=e$

$\Rightarrow ax^2(x^2-2)=0$

$\Rightarrow x=0, 0, -\sqrt{2}, \sqrt{2}\in T$

$=\{x\in R\,|\,f(x)=f(0)\}$

$\therefore$ Sum of squares of all the element of $T=2+2=4$

**62.** (d) Given quadratic equations $f(x)=(\lambda^2+1)x^2-4\lambda x+2=0$ have exactly one root in the interval (0, 1).

So, $D>0\Rightarrow 16\lambda^2-4(\lambda^2+1)2>0$

$\Rightarrow 8\lambda^2-8>0$

$\Rightarrow \lambda^2>1$

$\Rightarrow \lambda\in(-\infty,-1)\cup(1,\infty)$ ...(i)

and $f(0)f(1)<0$

$\Rightarrow 2(\lambda^2+1-4\lambda+2)<0$

$\Rightarrow \lambda^2-4\lambda+3<0$

$\Rightarrow (\lambda-3)(\lambda-1)<0$

$\Rightarrow \lambda\in(1,3)$ ...(ii)

From Eqs, (i) and (ii), we get

$\lambda\in(1,3)$

And if $\lambda=3$, then the quadratic equation is $10x^2-12x+2=0$

$\Rightarrow 5x^2-6x+1=0$

$\Rightarrow x=1, 1/5$

$\because$ The root $x=1/5\in(0,1)$ for $\lambda=3$

$\therefore \lambda\in(1,3]$.

**63.** (c) It is given that $\alpha$ and $\beta$ are roots of quadratic equation $x^2-x-1=0$, so sum of roots $=\alpha+\beta=1$, and product of roots $=\alpha\beta=-1$

and $p_k=\alpha^k+\beta^k, k\geq1$

So, $p_1=\alpha+\beta=1$

$p_2=\alpha^2+\beta^2$

$p_2=(\alpha+\beta)^2-2\alpha\beta=1+2=3$

$p_3=(\alpha+\beta)^3-3\alpha\beta(\alpha+\beta)=1+3=4$

$p_4=\alpha^4+\beta^4$

$=(\alpha+\beta)^4-4\alpha\beta(\alpha^2+\beta^2)-6\alpha^2\beta^2$

$=1+12-6=7$ and $p_5=\alpha^5+\beta^5$

$p_5=(\alpha+\beta)^5-5\alpha\beta(\alpha^3+\beta^3)$

$-10\alpha^2\beta^2(\alpha+\beta)$

$=1+20-10=11$

$\because p_3=p_5-p_4=4$

$p_1+p_2+p_3+p_4+p_5$

$=1+3+4+7+11=26$

but $p_5\neq p_2\cdot p_3$

**64.** (a) Given equation

$5+|2^x-1|=2^x(2^x-2)$

**Case I**

If $2^x-1\geq0\Rightarrow x\geq0$,

then $5+2^x-1=2^x(2^x-2)$

Put $2^x=t$, then $5+t-1=t^2-2t$

$\Rightarrow \quad t^2-3t-4=0$

$\Rightarrow t=4$ or $-1 \Rightarrow t=4$ $\quad (\because t=2^x>0)$

$\Rightarrow 2^x=4 \Rightarrow x=2>0$

$\Rightarrow \quad x=2$ is the solution.

**Case II**

If $2^x-1<0 \Rightarrow x<0$, then

$5+1-2^x=2^x(2^x-2)$

Put $2^x=y$, then $6-y=y^2-2y$

$\Rightarrow \quad y^2-y-6=0$

$\Rightarrow (y+2)(y-3)=0 \Rightarrow y=3$ or $-2$

$\Rightarrow \quad y=3$ (as $y=2^x>0$)

$\Rightarrow 2^x=3$

$\Rightarrow \quad x=\log_2 3>0$

So, $x=\log_2 3$ is not a solution.

Therefore, number of real roots is one.

**65.** *(a)* Given quadratic equation is

$x^2+x\sin\theta-2\sin\theta=0,\ \theta\in\left(0,\dfrac{\pi}{2}\right)$

and its roots are $\alpha$ and $\beta$.

So, sum of roots $=\alpha+\beta=-\sin\theta$

and product of roots $=\alpha\beta=-2\sin\theta$

$\Rightarrow \quad \alpha\beta=2(\alpha+\beta) \quad \ldots$(i)

Now, the given expression is

$$\frac{\alpha^{12}+\beta^{12}}{(\alpha^{-12}+\beta^{-12})(\alpha-\beta)^{24}}$$

$$=\frac{\alpha^{12}+\beta^{12}}{\left(\dfrac{1}{\alpha^{12}}+\dfrac{1}{\beta^{12}}\right)(\alpha-\beta)^{24}}$$

$$=\frac{\alpha^{12}+\beta^{12}}{\left(\dfrac{\beta^{12}+\alpha^{12}}{\alpha^{12}\beta^{12}}\right)(\alpha-\beta)^{24}}$$

$$=\left[\frac{\alpha\beta}{(\alpha-\beta)^2}\right]^{12}=\left(\frac{\alpha\beta}{(\alpha+\beta)^2-4\alpha\beta}\right)^{12}$$

$$=\left[\frac{2(\alpha+\beta)}{(\alpha+\beta)^2-8(\alpha+\beta)}\right]^{12} \quad \text{[from Eq. (i)]}$$

$$=\left(\frac{2}{(\alpha+\beta)-8}\right)^{12}=\left(\frac{2}{-\sin\theta-8}\right)^{12}$$

$[\because \alpha+\beta=-\sin\theta]$

$$=\frac{2^{12}}{(\sin\theta+8)^{12}}$$

**66.** *(b)*

**key idea** (i) Use, Discriminant, $D=b^2-4ac<0$

Given quadratic equation is

$(1+m^2)x^2-2(1+3m)x$

$+(1+8m)=0 \ldots$(i)

Now, discriminant

$D=[-2(1+3m)]^2-4(1+m^2)(1+8m)$

$=4[(1+3m)^2-(1+m^2)(1+8m)]$

$=4[1+9m^2+6m$

$-(1+8m+m^2+8m^3)]$

$=4[-8m^3+8m^2-2m]$

$=-8m(4m^2-4m+1)=-8m(2m-1)^2$

According to the question there is no solution of the quadratic Eq. (i), then

$D<0$

$\therefore \quad -8m(2m-1)^2<0 \Rightarrow m>0$

So, there are infinitely many values of $m$ for which, there is no solution of the given quadratic equation.

**67.** *(c)* Given, $\alpha$ and $\beta$ are the roots of the quadratic equation

$x^2-2x+2=0$

$\Rightarrow \quad (x-1)^2+1=0 \Rightarrow (x-1)^2=-1$

$\Rightarrow \quad x-1=\pm i \quad$ [where, $i=\sqrt{-1}$]

$\Rightarrow \quad x=(1+i)$ or $(1-i)$

Clearly, if $\alpha=1+i$, then $\beta=1-i$

According to question,

$$\left(\frac{\alpha}{\beta}\right)^n=1$$

$$\Rightarrow \quad \left(\frac{1+i}{1-i}\right)^n=1$$

$$\Rightarrow \quad \left(\frac{(1+i)(1+i)}{(1-i)(1+i)}\right)^n=1$$

[by rationalization]

$$\Rightarrow \quad \left(\frac{1+i^2+2i}{1-i^2}\right)^n=1$$

$$\Rightarrow \quad \left(\frac{2i}{2}\right)^n=1 \Rightarrow i^n=1$$

So, minimum value of $n$ is 4. $[\because i^4=1]$

**68.** *(b)* Given, $x^2\sin\theta-x\sin\theta\cos\theta-x$

$+\cos\theta=0,$

where $0<\theta<45°$

$\Rightarrow x\sin\theta(x-\cos\theta)-1(x-\cos\theta)=0$

$\Rightarrow (x-\cos\theta)(x\sin\theta-1)=0$

$\Rightarrow x=\cos\theta,\ x=\operatorname{cosec}\theta$

$\Rightarrow \alpha=\cos\theta$ and $\beta=\operatorname{cosec}\theta$

($\because$ For $0<\theta<45°$, $\dfrac{1}{\sqrt{2}}<\cos\theta<1$ and

$\sqrt{2}<\operatorname{cosec}\theta<\infty$

$\Rightarrow \cos\theta<\operatorname{cosec}\theta$)

Now, consider,

$$\sum_{n=0}^{\infty}\left(\alpha^n+\frac{(-1)^n}{\beta^n}\right)=\sum_{n=0}^{\infty}\alpha^n+\sum_{n=0}^{\infty}\frac{(-1)^n}{\beta^n}$$

$$=(1+\alpha+\alpha^2+\alpha^3+\ldots\infty)$$

$$+\left(1-\frac{1}{\beta}+\frac{1}{\beta^2}-\frac{1}{\beta^3}+\ldots\infty\right)$$

$$=\frac{1}{1-\alpha}+\frac{1}{1-\left(-\dfrac{1}{\beta}\right)}$$

$$=\frac{1}{1-\alpha}+\frac{1}{1+\dfrac{1}{\beta}}$$

$$=\frac{1}{1-\cos\theta}+\frac{1}{1+\sin\theta} \quad \left[\because \frac{1}{\beta}=\sin\theta\right]$$

**69.** *(b)* $\dfrac{1}{x+p}+\dfrac{1}{x+q}=\dfrac{1}{r}$

$$\Rightarrow \quad \frac{x+p+x+q}{(x+p)(x+q)}=\frac{1}{r}$$

$(2x+p+q)r=x^2+px+qx+pq$

$x^2+(p+q-2r)x+pq-pr-qr=0$

Let $\alpha$ and $\beta$ be the roots.

$\therefore \quad \alpha+\beta=-(p+q-2r) \quad \ldots$(i)

and $\quad \alpha\beta=pq-pr-qr \quad \ldots$(ii)

$\because \quad \alpha=-\beta$ (given)

$\therefore$ In Eq. (i), we get

$\alpha+\beta=-(p+q-2r)=0$

$\Rightarrow \quad p+q=2r \quad \ldots$(iii)

Now, $\alpha^2+\beta^2=(\alpha+\beta)^2-2\alpha\beta$

From Eqs. (ii) and (iii), we get

$(0)^2-2(pq-pr-qr)$

$=-2pq+2r(p+q)=p^2+q^2$

**70.** *(d)* Given, quadratic equation is

$x(x+1)+(x+1)(x+2)+\ldots+(x+\overline{n-1})$

$(x+n)=10n$

$\Rightarrow \quad (x^2+x^2+\ldots+x^2)$

$+[(1+3+5+\ldots+(2n-1)]x$

$+[(1\cdot 2+2\cdot 3+\ldots+(n-1)n]=10n$

$$\Rightarrow nx^2+n^2x+\frac{n(n^2-1)}{3}-10n=0$$

$$\Rightarrow x^2+nx+\frac{n^2-1}{3}-10=0$$

$\Rightarrow \ 3x^2+3nx+n^2-31=0$

Let $\alpha$ and $\beta$ be the roots. Since, $\alpha$ and $\beta$ are consecutive.

$\therefore \ |\alpha-\beta|=1 \Rightarrow (\alpha-\beta)^2=1$

Again, $(\alpha-\beta)^2=(\alpha+\beta)^2-4\alpha\beta$

$\Rightarrow \quad 1=\left(\frac{-3n}{3}\right)^2-4\left(\frac{n^2-31}{3}\right)$

$\Rightarrow \quad 1=n^2-\frac{4}{3}(n^2-31)$

$\Rightarrow \quad 3=3n^2-4n^2+124$

$\Rightarrow n^2=121 \quad \Rightarrow n=\pm 11$

$\therefore \quad n=11 \quad [\because n>0]$

**71.** *(b)* $(a-1)(x^4+x^2+1)+(a+1)(x^2+x+1)^2=0$

$\Rightarrow (a-1)(x^2+x+1)(x^2-x+1)+(a+1)(x^2+x+1)^2=0$

$\Rightarrow (x^2+x+1)[(a-1)(x^2-x+1)+(a+1)(x^2+x+1)]=0$

$\Rightarrow (x^2+x+1)(ax^2+x+a)=0$

For roots to be distinct and real, $a\neq 0$ and $1-4a^2>0$

$\Rightarrow a\neq 0$ and $a^2<\frac{1}{4}$

$\Rightarrow a\in\left(-\frac{1}{2},0\right)\cup\left(0,\frac{1}{2}\right)$

**72.** *(a)* Let $\frac{1}{\sqrt{\alpha}}$ and $\frac{1}{\sqrt{\beta}}$ be the roots of $ax^2+bx+1=0$

$\frac{1}{\sqrt{\alpha}}+\frac{1}{\sqrt{\beta}}=\left(\frac{\sqrt{\alpha}+\sqrt{\beta}}{\sqrt{\alpha\beta}}\right)=-\frac{b}{a}$

$\left[\because \text{Sum of roots}=-\frac{b}{a}\right.$ and product of roots $\left.=\frac{c}{a}\right]$

$\frac{1}{\sqrt{\alpha}\sqrt{\beta}}=\frac{1}{a}\Rightarrow a=\sqrt{\alpha\beta}$

$b=-(\sqrt{\alpha}+\sqrt{\beta})$

$x(x+b^3)+(a^3-3abx)=0$

$\Rightarrow x^2+(b^3-3ab)x+a^3=0$

Putting values of $a$ and $b$, we get

$x^2+[-(\sqrt{\alpha}+\sqrt{\beta})]^3-3(\sqrt{\alpha\beta})(\sqrt{\alpha}+\sqrt{\beta})]x+(\alpha\beta)^{3/2}=0$

$\Rightarrow x^2-[\alpha^{3/2}+\beta^{3/2}+3\sqrt{\alpha\beta}(\sqrt{\alpha}+\sqrt{\beta})-3\sqrt{\alpha\beta}(\sqrt{\alpha}+\sqrt{\beta})]x+(\alpha\beta)^{3/2}=0$

$\Rightarrow x^2-(\alpha^{3/2}+\beta^{3/2})x+\alpha^{3/2}\beta^{3/2}=0$

Hence, roots of this equation are $\alpha^{3/2},\beta^{3/2}$.

**73.** *(b)* Given, $\alpha^3+\beta^3=-p$ and $\alpha\beta=q$

Let $\frac{\alpha^2}{\beta}$ and $\frac{\beta^2}{\alpha}$ be the root of required quadratic equation.

So, $\frac{\alpha^2}{\beta}+\frac{\beta^2}{\alpha}=\frac{\alpha^3+\beta^3}{\alpha\beta}=\frac{-p}{q}$

and $\frac{\alpha^2}{\beta}\times\frac{\beta^2}{\alpha}=\alpha\beta=q$

Hence, required quadratic equation is

$x^2-\left(\frac{-p}{q}\right)x+q=0$

$\Rightarrow \quad x^2+\frac{p}{q}x+q=0$

$\Rightarrow \quad qx^2+px+q^2=0$

**74.** *(d)* Given equation is

$x^2-(\sin\alpha-2)x-(1+\sin\alpha)=0$

Let $x_1$ and $x_2$ be two roots of quadratic equation.

$\therefore x_1+x_2=\sin\alpha-2$

and $x_1x_2=-(1+\sin\alpha)$

$(x_1+x_2)^2=(\sin\alpha-2)^2$

$=\sin^2\alpha+4-4\sin\alpha$

$\Rightarrow x_1^2+x_2^2=\sin^2\alpha+4-4\sin\alpha-2x_1x_2$

$=\sin^2\alpha+4-4\sin\alpha+2(1+\sin\alpha)$

$=\sin^2\alpha-2\sin\alpha+6 \quad \ldots(\text{i})$

Now, by putting

$\alpha=\frac{\pi}{6},\ \alpha=\frac{\pi}{4},\ \alpha=\frac{\pi}{3}$ and $\alpha=\frac{\pi}{2}$ in

Eq. (i) one by one, we get least value of $x_1^2+x_2^2$ at $\frac{\pi}{2}$.

Hence, $\alpha=\frac{\pi}{2}$

**75.** *(c)* Given $bx^2+cx+a=0$ has imaginary roots.

$\Rightarrow \quad c^2-4ab<0 \Rightarrow c^2<4ab$

$\Rightarrow \quad -c^2>-4ab \quad \ldots(\text{i})$

Let $f(x)=3b^2x^2+6bcx+2c^2$

Here, $\quad 3b^2>0$

So, the given expression has a minimum value.

$\therefore$ Minimum value $=\frac{-D}{4a}$

$=\frac{4ac-b^2}{4a}=\frac{4(3b^2)(2c^2)-36b^2c^2}{4(3b^2)}$

$=-\frac{12b^2c^2}{12b^2}=-c^2>-4ab$

[from Eq. (i)]

**76.** *(b)* Let $f(x)=x^2-2kx+k^2+k-5$

Since, both roots are less than 5.

Then,

$D\geq 0,\quad -\frac{b}{2a}<5$ and $f(5)>0$

Now, $D=4k^2-4(k^2+k-5)$

$=-4k+20\geq 0\Rightarrow k\leq 5 \quad \ldots(\text{i})$

$-\frac{b}{2a}<5 \Rightarrow k<5 \quad \ldots(\text{ii})$

and $f(5)>0$

$\Rightarrow 25-10k+k^2+k-5>0$

$\Rightarrow k^2-9k+20>0\Rightarrow (k-5)(k-4)>0$

$\Rightarrow k<4$ and $k>5 \quad \ldots(\text{iii})$

From Eqs. (i), (ii) and (iii), we get $k<4$

**77.** *(c)* Let $f(x)=a_nx^n+a_{n-1}x^{n-1}+\ldots+a_1x=0 \quad \ldots(\text{i})$

$f(0)=0$ and $f(\alpha)=0$

According to the Rolle's theorem,

$f'(x)=0$

has atleast one root between $(0,\alpha)$.

So, $f'(x)=0$ has a positive root less than $\alpha$.

**78.** *(a)*

> **key idea** Express the given as quadratic form such as $ax^2+bx+c=0$, then if $b^2-4ac=0$, so roots are real and equal.

Since, one of the roots of equation $x^2+px+12=0$ is 4.

$\therefore \quad 16+4p+12=0$

$\Rightarrow \quad 4p=-28 \Rightarrow p=-7$

So, the other equation is

$x^2-7x+q=0$ whose roots are equal.

Let the roots be $\alpha$ and $\alpha$.

$\therefore$ Sum of roots $=\alpha+\alpha=7/1$

$\Rightarrow \quad \alpha=7/2$

And product of roots $=\alpha\cdot\alpha=q$

$\Rightarrow \quad \left(\frac{7}{2}\right)^2=q$

$\therefore \quad q=\frac{49}{4}$

**79.** *(c)* Given equation is

$ax^2+bx+c=0.$

Let $\alpha$ and $\beta$ be the roots of the equation.

Then, $\alpha+\beta=-b/a$ and $\alpha\beta=c/a$

Also given, $\alpha+\beta=\frac{1}{\alpha^2}+\frac{1}{\beta^2}=\frac{\alpha^2+\beta^2}{\alpha^2\beta^2}$

$\Rightarrow \alpha+\beta=\left(\frac{\alpha+\beta}{\alpha\beta}\right)^2-\frac{2}{\alpha\beta}$

$$\Rightarrow \left(-\frac{b}{a}\right)=\left(\frac{-b/a}{c/a}\right)^2-\frac{2}{c/a}$$

$$\Rightarrow -\frac{b}{a}=\left(\frac{b}{c}\right)^2-\frac{2a}{c}$$

$$\Rightarrow \frac{2a}{b}=\frac{b}{c}+\frac{c}{a}$$

Since, $\frac{c}{a}, \frac{a}{b}$ and $\frac{b}{c}$ are in AP.

Hence, $\frac{a}{c}, \frac{b}{a}$ and $\frac{c}{b}$ are in HP.

**80.** *(a)* Since, one root of the quadratic equation $(a^2-5a+3)x^2+(3a-1)x+2=0$ is twice as large as the other, then let their roots be $\alpha$ and $2\alpha$.

$$\therefore \quad \alpha+2\alpha=-\frac{(3a-1)}{(a^2-5a+3)}$$

$$\Rightarrow \quad 3\alpha=-\frac{(3a-1)}{(a^2-5a+3)} \quad \text{...(i)}$$

and $\alpha\cdot 2\alpha=\dfrac{2}{(a^2-5a+3)}$

$$\Rightarrow \quad 2\alpha^2=\frac{2}{(a^2-5a+3)} \quad \text{...(ii)}$$

From Eqs. (i) and (ii), we get

$$\frac{(3a-1)^2}{9(a^2-5a+3)^2}=\frac{1}{(a^2-5a+3)}$$

$$\Rightarrow (3a-1)^2=9(a^2-5a+3)$$

$$\Rightarrow 9a^2-6a+1=9a^2-45a+27$$

$$\Rightarrow 45a-6a=27-1$$

$$\Rightarrow a=\frac{26}{39}=\frac{2}{3}$$

**81.** *(b)* Given $\alpha^2=5\alpha-3$

$\Rightarrow \alpha^2-5\alpha+3=0$ and $\beta^2=5\beta-3$

$\Rightarrow \beta^2-5\beta+3=0$

These two equations shows that $\alpha$ and $\beta$ are the roots of the equation $x^2-5x+3=0$.

$\therefore \alpha+\beta=5$ and $\alpha\beta=3$

Now, $\dfrac{\alpha}{\beta}+\dfrac{\beta}{\alpha}=\dfrac{\alpha^2+\beta^2}{\alpha\beta}$

$$=\frac{(\alpha+\beta)^2-2\alpha\beta}{\alpha\beta}$$

$=\dfrac{25-6}{3}=\dfrac{19}{3}$ and $\dfrac{\alpha}{\beta}\cdot\dfrac{\beta}{\alpha}=1$

Thus, the equation having roots $\alpha/\beta$ and $\beta/\alpha$ is given by

$$x^2-\left(\frac{\alpha}{\beta}+\frac{\beta}{\alpha}\right)x+\frac{\alpha}{\beta}\cdot\frac{\beta}{\alpha}=0$$

$$\Rightarrow x^2-19/3\,x+1=0$$

$$\Rightarrow 3x^2-19x+3=0$$

**82.** *(a)* By using Newton's formula in given quadratic equation

$\Rightarrow x^2-(t^2-5t+6)x+1=0$ can be expressed as

$$\Rightarrow a_{2025}-(t^2-5t+6)\,a_{2024}+a_{2023}=0$$

On rearranging

$$\Rightarrow \frac{a_{2025}+a_{2023}}{a_{2024}}=t^2-5t+6$$

$$\Rightarrow \frac{a_{2025}+a_{2023}}{a_{2024}}=t^2-5t+\frac{25}{4}-\frac{25}{4}+6$$

$$\Rightarrow \frac{a_{2025}+a_{2023}}{a_{2024}}=\left(t-\frac{5}{2}\right)^2-\frac{1}{4}$$

Minimum value of any square is 0

$\therefore$ Minimum value of

$$\frac{a_{2025}+a_{2023}}{a_{2024}}=0-\frac{1}{4}=-\frac{1}{4}$$

> **• Concept Enhancer**
>
> Here we use Newton's formula i.e. if $ax^2+bx+c=0$ be any quadratic eqation and $\alpha, \beta$ be its roots, then
>
> $\therefore \quad aS_n+bS_{n-1}+cS_{n-2}=0$
>
> where $\quad S_n=\alpha^n\pm\beta^n; n\in N$

**83.** *(d)* We have,

$$px^2+qx-r=0$$

Given that, $p$, $q$ and $r$ be the consecutive terms of a non-constant GP, then let

$$p=\frac{a}{r_1}, q=a, r=ar_1$$

and given, $\dfrac{1}{\alpha}+\dfrac{1}{\beta}=\dfrac{3}{4}$

$$\Rightarrow \quad \frac{\alpha+\beta}{\alpha\beta}=\frac{3}{4} \Rightarrow \frac{-\frac{q}{p}}{-\frac{r}{p}}=\frac{3}{4}$$

$$\Rightarrow \quad \frac{q}{r}=\frac{3}{4} \Rightarrow \frac{1}{r_1}=\frac{3}{4}$$

$$\Rightarrow \quad r_1=\frac{4}{3} \quad \text{...(i)}$$

Now, $(\alpha-\beta)^2=(\alpha+\beta)^2-4\alpha\beta$

$$=\left(\frac{-q}{p}\right)^2-4\left(\frac{-r}{p}\right)$$

$$=(r_1)^2+4(r_1^2)=5r_1^2$$

$$=5\left(\frac{4}{3}\right)^2 \quad \text{[from Eq. (i)]}$$

$$=\frac{80}{9}$$

> ***Alternate Solution***
>
> $\because$ $p, q, r$ in GP.
>
> $\therefore \quad q^2=pr$
>
> $\because \quad \dfrac{1}{\alpha}+\dfrac{1}{\beta}=\dfrac{3}{4}$
>
> $\Rightarrow \quad \dfrac{\alpha+\beta}{\alpha\beta}=\dfrac{3}{4}$
>
> $\therefore \quad \dfrac{q}{r}=\dfrac{3}{4}$
>
> Now, $|\alpha-\beta|=\dfrac{\sqrt{q^2+4pr}}{|p|}$
>
> $(\alpha-\beta)^2=\dfrac{q^2+4pr}{p^2}$
>
> $=\dfrac{5q^2}{p^2}=\dfrac{5q^2r^2}{q^4}=\dfrac{5r^2}{q^2}=5\times\dfrac{16}{9}=\dfrac{80}{9}$

**84.** *(d)* Given, $(a+b-2c)x^2+(b+c-2a)x+(c+a-2b)=0$

$\because$ 1 satisfies the equation.

$\alpha$ and 1 are roots of equation.

Product of roots

$$\alpha\cdot 1=\frac{c+a-2b}{a+b-2c}$$

$$\alpha=\frac{c+a-2b}{a+b-2c}$$

**Statement I**

$$\alpha\in(-1,0)$$

$$-1<\alpha<0$$

$$-1<\frac{c+a-2b}{a+b-2c}<0; \because a>b>c>0$$

$$0<\frac{c+a-2b+a+b-2c}{a+b-2c}<1$$

$$\because \quad a+b>c+c$$

$$a+b>2c$$

$$a+b-2c>0$$

$$0<\frac{2a-b-c}{a+b-2c}<1$$

$$\therefore \quad 2a-b-c>0$$

$$2a>b+c$$

$$a > \frac{b+c}{2}$$

$\therefore$ $b$ can't be GM between $a$ and $c$.

**Statement II**

$$\alpha \in (0, 1)$$
$$0 < \alpha < 1$$
$$0 < \frac{c+a-2b}{a+b-2c} < 1$$

Here, $b > c$ and $b < \frac{a+c}{2}$

$\therefore b$ may be GM between $a$ and $c$.

$\therefore$Statement I and statement II both are true.

**85.** *(b)* Given quadratic equation is

$$x^2 - x - 1 = 0 \quad \text{...(i)}$$

Also, given $\alpha$ and $\beta$ are roots of Eq. (i)

and $\quad S_n = 2023\alpha^n + 2024\beta^n$

Since, $\alpha$ and $\beta$ are roots of Eq. (i).

So, $\alpha^2 - \alpha - 1 = 0$ and $\beta^2 - \beta - 1 = 0$

$\Rightarrow \quad \alpha^2 = \alpha + 1$ and $\beta^2 = \beta + 1$

$\Rightarrow \alpha^{n-2}(\alpha^2) = \alpha^{n-2}(\alpha + 1)$ and

$\beta^{n-2}(\beta^2) = \beta^{n-2}(\beta + 1)$

$\Rightarrow \quad \alpha^n = \alpha^{n-1} + \alpha^{n-2} \quad$ ...(ii)

and $\quad \beta^n = \beta^{n-1} + \beta^{n-2} \quad$ ...(iii)

Given, $\quad S_n = 2023\alpha^n + 2024\beta^n$

$\Rightarrow \quad S_n = 2023(\alpha^{n-1} + \alpha^{n-2}) + 2024(\beta^{n-1} + \beta^{n-2})$

[from Eqs. (ii) and (iii)]

$\Rightarrow \quad S_n = 2023\alpha^{n-1} + 2023\alpha^{n-2} + 2024\beta^{n-1} + 2024\beta^{n-2}$

$\Rightarrow S_n = (2023\alpha^{n-1} + 2024\beta^{n-1}) + (2023\alpha^{n-2} + 2024\beta^{n-2})$

$\Rightarrow \quad S_n = S_{n-1} + S_{n-2} \quad$ ...(iv)

On putting $n = 12$ in Eq. (iv), we have

$$S_{12} = S_{11} + S_{10}$$

**86.** *(b)* We have, $x|x| - 5|x+2| + 6 = 0$

For, $x < -2$

$$x(-x) + 5(x+2) + 6 = 0$$
$$-x^2 + 5x + 10 + 6 = 0$$
$$x^2 - 5x - 16 = 0$$
$$x = \frac{5+\sqrt{89}}{2}, \frac{5-\sqrt{89}}{2}$$

For, $-2 \le x < 0$

$$x(-x) - 5(x+2) + 6 = 0$$
$$-x^2 - 5x - 10 + 6 = 0$$
$$-x^2 - 5x - 4 = 0$$
$$\Rightarrow \quad x^2 + 5x + 4 = 0$$
$$(x+4)(x+1) = 0 \Rightarrow x = -4, -1$$

$x = -4$ rejected

For, $x \ge 0$

$$x \times x - 5(x+2) + 6 = 0$$
$$x^2 - 5x - 10 + 6 = 0$$
$$x^2 - 5x - 4 = 0$$
$$x = \frac{-(-5) \pm \sqrt{(-5)^2 - 4(1)(-4)}}{2(1)}$$
$$= \frac{5 \pm \sqrt{25+16}}{2} = \frac{5 \pm \sqrt{41}}{2}$$
$$x = \frac{5+\sqrt{41}}{2}, \frac{5-\sqrt{41}}{2}$$
$$x = \frac{5-\sqrt{41}}{2} \text{ rejected}$$
$$\Rightarrow \quad x = -1, \frac{5+\sqrt{41}}{2}, \frac{5-\sqrt{89}}{2}$$

3 real roots we get.

**87.** *(a)* Given equation is

$$x^2 - \sqrt{2}x + 2 = 0 \quad \text{...(i)}$$

Apply the quadratic formula,

$$x = \frac{\sqrt{2} \pm \sqrt{2-8}}{2} = \frac{\sqrt{2} \pm \sqrt{6}i}{2}$$

Since, $\alpha, \beta$ are roots of Eq. (i), we get

$$\alpha = \frac{\sqrt{2}+\sqrt{6}i}{2}, \quad \beta = \frac{\sqrt{2}-\sqrt{6}i}{2}$$

$\Rightarrow \alpha = \sqrt{2}\, e^{i\pi/3} \quad \Rightarrow \beta = \sqrt{2} e^{-\pi/3\, i}$

$(\because e^{i\theta} = \cos\theta + i\sin\theta)$

Now, $\alpha^{14} = (\sqrt{2})^{14} e^{\frac{14\pi}{3}i}$

$$= 2^7 \cdot e^{\frac{2\pi i}{3}} = 128\left\{\cos\frac{2\pi}{3} + i\sin\frac{2\pi}{3}\right\}$$

Similarly,

$$\beta^{14} = 128\left\{\cos\frac{2\pi}{3} - i\sin\frac{2\pi}{3}\right\}$$

$$\therefore \alpha^{14} + \beta^{14} = 2 \times 128 \times \cos\frac{2\pi}{3}$$
$$= 2 \times 128 \times \left(-\frac{1}{2}\right) = -128$$

**88.** *(a)* We have,

$$|x^2 - 8x + 15| - 2x + 7 = 0$$
$$\Rightarrow \quad |x^2 - 5x - 3x + 15| - 2x + 7 = 0$$
$$\Rightarrow \quad |x(x-5) - 3(x-5)| - 2x + 7 = 0$$
$$\Rightarrow \quad |(x-3)(x-5)| - 2x + 7 = 0$$

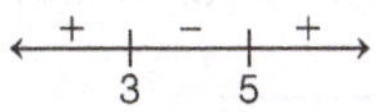


Sign for $|x^2 - 8x + 15|$

Now, when $x \le 3$ or $x \ge 5$

then, $x^2 - 8x + 15 - 2x + 7 = 0$

$$\Rightarrow \quad x^2 - 10x + 22 = 0$$
$$\Rightarrow \quad x^2 - 10x + 25 - 3 = 0$$
$$\Rightarrow \quad (x-5)^2 - 3 = 0 \Rightarrow x - 5 = \pm\sqrt{3}$$
$$\Rightarrow x = 5 + \sqrt{3}, 5 - \sqrt{3}$$

Since, $x \le 3$ or $x \ge 5$

$\therefore x = 5 + \sqrt{3}$

When, $3 < x < 5$, then

$$-(x^2 - 8x + 15) - 2x + 7 = 0$$
$$\Rightarrow \quad -x^2 + 8x - 15 - 2x + 7 = 0$$
$$\Rightarrow \quad -x^2 + 6x - 8 = 0$$
$$\Rightarrow \quad x^2 - 6x + 8 = 0$$
$$\Rightarrow \quad x^2 - 4x - 2x + 8 = 0$$
$$\Rightarrow \quad x(x-4) - 2(x-4) = 0$$
$$\Rightarrow (x-4)(x-2) = 0 \Rightarrow x = 2, 4$$

Since, $3 < x < 5$

$\therefore x = 4$

$\therefore$ Sum of roots $= (5 + \sqrt{3}) + 4 = 9 + \sqrt{3}$

**89.** *(a)* $x^2 + \sqrt{6}x + 3 = 0$

$$\Rightarrow x = \frac{-\sqrt{6} \pm \sqrt{6 - 4 \cdot 3}}{2} = \frac{-\sqrt{6} \pm \sqrt{-6}}{2}$$
$$= \frac{-\sqrt{6} \pm \sqrt{6}i}{2}$$
$$= \frac{\sqrt{6}}{2}(-1 \pm i) = \sqrt{3}\left(\frac{-1}{\sqrt{2}} \pm \frac{i}{\sqrt{2}}\right)$$

Let $\alpha = \sqrt{3}\left(-\frac{1}{\sqrt{2}} + \frac{i}{\sqrt{2}}\right)$,

$$\beta = \sqrt{3}\left(-\frac{1}{\sqrt{2}} - \frac{i}{\sqrt{2}}\right)$$

$$\Rightarrow \alpha = \sqrt{3}\left(\cos\left(\frac{3\pi}{4}\right) + i\sin\left(\frac{3\pi}{4}\right)\right),$$
$$\beta = \sqrt{3}\left(\cos\left(\frac{5\pi}{4}\right) + i\sin\left(\frac{5\pi}{4}\right)\right)$$

$$\Rightarrow \quad \alpha = \sqrt{3}\, e^{i\left(\frac{3\pi}{4}\right)}, \beta = \sqrt{3} e^{i\left(\frac{5\pi}{4}\right)}$$

$$\Rightarrow \alpha^4 = (\sqrt{3})^4\, e^{i(3\pi)}$$
$$= 9(\cos(3\pi) + i\sin(3\pi)) = -9$$

and $p^4 = (\sqrt{3})^4\, e^{i(5\pi)}$

$$= 9(\cos(5\pi) + i\sin(5\pi)) = -9$$

$$\therefore \frac{\alpha^{23} + \beta^{23} + \alpha^{14} + \beta^{14}}{\alpha^{15} + \beta^{15} + \alpha^{10} + \beta^{10}}$$

$$= \frac{(\alpha^4)^5 \cdot \alpha^3 + (\beta^4)^5 \cdot \beta^3 + (\alpha^4)^3 \cdot \alpha^2 + (\beta^4)^3 \cdot \beta^2}{(\alpha^4)^3 \cdot \alpha^3 + (\beta^4)^3 \cdot \beta^3 + (\alpha^4)^2 \cdot \alpha^2 + (\beta^4)^2 \cdot \beta^2}$$

$$= \frac{(-9)^5(\alpha^3 + \beta^3) + (-9)^3(\alpha^2 + \beta^2)}{(-9)^3(\alpha^3 + \beta^3) + (-9)^2(\alpha^2 + \beta^2)} \quad \text{...(i)}$$

Now, $\alpha+\beta=-\sqrt{6}, \alpha\beta=3$

$\Rightarrow \alpha^2+\beta^2=(\alpha+\beta)^2-2\alpha\beta$

$=(-\sqrt{6})^2-2\cdot3=0$

$\therefore$Eq. (i) becomes $=\dfrac{(-9)^5(\alpha^3+\beta^3)}{(-9)^3(\alpha^3+\beta^3)}$

$=(-9)^2=81$

**90.** *(a)* We have,

$(\sqrt{3}+\sqrt{2})^{x^2-4}+(\sqrt{3}-\sqrt{2})^{x^2-4}=10$

Let $(\sqrt{3}+\sqrt{2})^{x^2-4}=y$,

then $(\sqrt{3}-\sqrt{2})^{x^2-4}$

$=\left[\dfrac{(\sqrt{3}-\sqrt{2})(\sqrt{3}+\sqrt{2})}{(\sqrt{3}+\sqrt{2})}\right]^{x^2-4}$

$=\left[\dfrac{3-2}{\sqrt{3}+\sqrt{2}}\right]^{x^2-4}=\dfrac{1}{(\sqrt{3}+\sqrt{2})^{x^2-4}}$

$=\dfrac{1}{y}$

$\therefore\ y+\dfrac{1}{y}=10 \Rightarrow y^2-10y+1=0$

$\Rightarrow y=\dfrac{10\pm\sqrt{100-4}}{2}$

$=\dfrac{10\pm4\sqrt{6}}{2}=5\pm2\sqrt{6}$

Now, $y=5+2\sqrt{6}$

$\Rightarrow(\sqrt{3}+\sqrt{2})^{x^2-4}=5+2\sqrt{6}$

$\Rightarrow(\sqrt{3}+\sqrt{2})^{x^2-4}=(\sqrt{3}+\sqrt{2})^2$

$\Rightarrow x^2-4=2 \Rightarrow x=\pm\sqrt{6}$

Again, $y=5-2\sqrt{6}$

$\Rightarrow(\sqrt{3}+\sqrt{2})^{x^2-4}=5-2\sqrt{6}$

$\Rightarrow(\sqrt{3}+\sqrt{2})^{x^2-4}=(\sqrt{3}-\sqrt{2})^2$

$\Rightarrow(\sqrt{3}+\sqrt{2})^{x^2-4}=(\sqrt{3}+\sqrt{2})^{-2}$

$\Rightarrow x^2-4=-2 \Rightarrow x=\pm\sqrt{2}$

$\therefore S=\{-\sqrt{6},-\sqrt{2},\sqrt{2},\sqrt{6}\}$

So, $n(S)=4$

**91.** *(b)* We have, $e^{4x}+8e^{3x}+13e^{2x}-8e^x+1=0, x\in R$

Let $e^x=t$, then

$t^4+8t^3+13t^2-8t+1=0$

$\Rightarrow t^2+8t+13-\dfrac{8}{t}+\dfrac{1}{t^2}=0$

$\Rightarrow\left(t^2+\dfrac{1}{t^2}\right)+8\left(t-\dfrac{1}{t}\right)+13=0$

$\Rightarrow\left(t-\dfrac{1}{t}\right)^2+2+8\left(t-\dfrac{1}{t}\right)+13=0$

$\Rightarrow\left(t-\dfrac{1}{t}\right)^2+8\left(t-\dfrac{1}{t}\right)+15=0$

Again, let $t-\dfrac{1}{t}=z$, we get

$z^2+8z+15=0$

$\Rightarrow(z+3)(z+5)=0$

$\Rightarrow z=-3,-5$

$\therefore t-\dfrac{1}{t}=-3$ or $t-\dfrac{1}{t}=-5$

$\Rightarrow t^2+3t-1=0$ or $t^2+5t-1=0$

$\Rightarrow t=\dfrac{-3\pm\sqrt{13}}{2}$ or $\dfrac{-5\pm\sqrt{29}}{2}$

$\Rightarrow t=\dfrac{-3+\sqrt{13}}{2},\dfrac{-3-\sqrt{13}}{2}$

or $\dfrac{-5+\sqrt{29}}{2},\dfrac{-5-\sqrt{29}}{2}$

$\Rightarrow e^x=\dfrac{-3+\sqrt{13}}{2},\dfrac{-5+\sqrt{29}}{2}$

$[\because e^x>0]$

$\Rightarrow x=\log\left(\dfrac{-3+\sqrt{13}}{2}\right),\log\left(\dfrac{-5+\sqrt{29}}{2}\right)$

$\therefore$Equation has two solutions and both are negative as $\dfrac{-3+\sqrt{13}}{2}$ and $\dfrac{-5+\sqrt{29}}{2}$ both belongs to (0, 1).

**92.** *(d)* We have,

$\sqrt{x^2-4x+3}+\sqrt{x^2-9}$

$=\sqrt{4x^2-14x+6}$

$\Rightarrow\sqrt{(x-1)(x-3)}+\sqrt{(x+3)(x-3)}$

$=\sqrt{2(x-3)(2x-1)}$

$\Rightarrow\sqrt{x-3}\,[\sqrt{x-1}+\sqrt{x+3}$

$-\sqrt{2(2x-1)}]=0$

$\Rightarrow\sqrt{x-3}=0$ or $\sqrt{x-1}+\sqrt{x+3}$

$-\sqrt{2(2x-1)}=0$

Now, $\sqrt{x-3}=0\Rightarrow x=3$

Again,

$\sqrt{x-1}+\sqrt{x+3}-\sqrt{2(2x-1)}=0$

$\Rightarrow\quad\sqrt{x-1}+\sqrt{x+3}=\sqrt{2(2x-1)}$

Squaring both the sides, we get

$x-1+x+3+2\sqrt{(x-1)(x+3)}$

$=2(2x-1)$

$\Rightarrow\sqrt{(x-1)(x+3)}=x-2$ ... (i)

Squaring both the sides, we get

$(x-1)(x+3)=(x-2)^2$

$\Rightarrow 6x=7\Rightarrow x=\dfrac{7}{6}$

When $x=\dfrac{7}{6}$, from Eq. (i), we get

$\sqrt{\left(\dfrac{7}{6}-1\right)\left(\dfrac{7}{6}+3\right)}=\left(\dfrac{7}{6}-2\right)$

$\Rightarrow\dfrac{5}{6}=-\dfrac{5}{6}$, which is not possible.

So, there is only one root i.e. $x=3$

**93.** *(d)* Given equation is

$\log_{\cos x}\cot x+4\log_{\sin x}\tan x=1$, where $x\in\left(0,\dfrac{\pi}{2}\right)$

$\Rightarrow\dfrac{\log_e\cot x}{\log_e\cos x}+4\dfrac{\log_e\tan x}{\log_e\sin x}=1$

(using base change formula)

$\Rightarrow\dfrac{\log\cos x-\log\sin x}{\log\cos x}$

$+4\left(\dfrac{\log\sin x-\log\cos x}{\log\sin x}\right)=1$

$\Rightarrow(\log\cos x)(\log\sin x)-(\log(\sin x))^2$

$+4(\log\sin x)$

$(\log\cos x)-4(\log(\cos x))^2$

$=\log\sin x\cdot\log\cos x$

$\Rightarrow(\log\sin x)^2-4(\log\sin x)(\log\cos x)$

$+4(\log\cos x)^2=0$

$\Rightarrow(\log\sin x-2\log\cos x)^2=0$

$\Rightarrow\log\sin x=2\log\cos x$

$\Rightarrow\log\sin x=\log(\cos x)^2$

$\Rightarrow\sin x=\cos^2x\Rightarrow1-\sin^2x=\sin x$

$\Rightarrow\sin^2x+\sin x-1=0$

$\Rightarrow\sin x=\dfrac{-1\pm\sqrt{5}}{2}$

But, $x\in\left(0,\dfrac{\pi}{2}\right)$

So, $\sin x>0$

$\therefore\sin x=\dfrac{-1-\sqrt{5}}{2}<0$ (rejected)

Hence, $\sin x=\dfrac{-1+\sqrt{5}}{2}=\dfrac{\alpha+\sqrt{\beta}}{2}$

(given)

Therefore, $\alpha+\beta=-1+5=4$

**94.** *(c)* We have, $14x^2-31x+3\lambda=0$ ...(i)

and $\quad 35x^2-53x+4\lambda=0$ ...(ii)

$5\times$ Eq. (i) $-2\times$ Eq. (ii), we get

$\lambda=7x$

Since, common root of Eqs. (i) and (ii) is $\alpha$

$\therefore \quad \lambda = 7\alpha \quad \ldots$(iii)

$\alpha, \beta$ are roots of $14x^2 - 31x + 3\lambda = 0$

$\Rightarrow \quad \alpha\beta = \frac{3\lambda}{14}$

$\Rightarrow \quad \frac{\lambda}{7} \times \beta = \frac{3\lambda}{14} \Rightarrow \beta = \frac{3}{2} \quad \ldots$(iv)

Also, $\alpha + \beta = \frac{31}{14}$

$\Rightarrow \alpha + \frac{3}{2} = \frac{31}{14} \Rightarrow \alpha = \frac{31}{14} - \frac{3}{2}$

$\Rightarrow \quad \alpha = 5/7 \quad \ldots$(v)

$\therefore \quad \lambda = 7\alpha = 5 \quad \ldots$(vi)

Now, $\alpha, \gamma$ are roots of $35x^2 - 53x + 4\lambda = 0$

$\Rightarrow \alpha\gamma = \frac{4\lambda}{35}$

$\Rightarrow \frac{5}{7} \times \gamma = \frac{4 \times 5}{35} \Rightarrow \gamma = \frac{4}{5} \quad \ldots$(vii)

$\therefore \quad \frac{3\alpha}{\beta} = \frac{3 \times \left(\frac{5}{7}\right)}{\left(\frac{3}{2}\right)} = \frac{10}{7}$

and $\frac{4\alpha}{\gamma} = \frac{4 \times \left(\frac{5}{7}\right)}{\left(\frac{4}{5}\right)} = \frac{25}{7}$

So, the equation with roots $\frac{3\alpha}{\beta}$ and $\frac{4\alpha}{\gamma}$ is

$x^2 - \left(\frac{10}{7} + \frac{25}{7}\right)x + \frac{10}{7} \times \frac{25}{7} = 0$

i.e. $49x^2 - 245x + 250 = 0$

[$\because$ Equation with roots '$l$' and '$m$' is $x^2 - (l + m)x + lm = 0$]

**95.** *(b)* Given, $\alpha$ and $\beta$ are the roots of the equation $x^2 - \sqrt{2}x + \sqrt{6} = 0$

$\alpha + \beta = \sqrt{2}$ and $\alpha\beta = \sqrt{6}$

Now, $\frac{1}{\alpha^2} + 1 + \frac{1}{\beta^2} + 1 = \frac{\alpha^2 + \beta^2}{(\alpha\beta)^2} + 2$

$= \frac{(\alpha + \beta)^2 - 2\alpha\beta}{(\alpha\beta)^2} + 2$

$= \frac{2 - 2\sqrt{6}}{6} + 2 = \frac{14 - 2\sqrt{6}}{6}$

$\therefore \left(\frac{1}{\alpha^2} + 1\right)\left(\frac{1}{\beta^2} + 1\right) = \frac{1}{(\alpha\beta)^2} + \frac{\alpha^2 + \beta^2}{(\alpha\beta)^2} + 1$

$= \frac{1}{6} + \frac{2 - 2\sqrt{6}}{6} + 1 = \frac{9 - 2\sqrt{6}}{6}$

The equation whose roots are $\left(\frac{1}{\alpha^2} + 1, \frac{1}{\beta^2} + 1\right)$ is

$x^2 - \left(\frac{14 - 2\sqrt{6}}{6}\right)x + \frac{9 - 2\sqrt{6}}{6} = 0$

Here, $a = \frac{2\sqrt{6} - 14}{6}, b = \frac{9 - 2\sqrt{6}}{6}$

$a + b = \frac{2\sqrt{6} - 14 + 9 - 2\sqrt{6}}{6} = -\frac{5}{6}$

$\therefore x^2 - (a + b - 2)x + (a + b + 2) = 0$

$\Rightarrow x^2 - \left(-\frac{5}{6} - 2\right)x + \left(-\frac{5}{6} + 2\right) = 0$

$\Rightarrow x^2 + \frac{17}{6}x + \frac{7}{6} = 0$

$\Rightarrow 6x^2 + 17x + 7 = 0 \Rightarrow x = -\frac{1}{2}, -\frac{7}{3}$

$\therefore$ Both roots are real and negative.

**96.** *(b)* Given that, the equation is $x^4 + x^3 + x^2 + x + 1 = 0$ and $\alpha, \beta, \gamma$ and $\delta$ are roots of the above equation.

Since, $x^4 + x^3 + x^2 + x + 1 = 0$

$\Rightarrow \quad \frac{x^5 - 1}{x - 1} = 0 \; (x \neq 1)$

[$\because (x - 1)(x^4 + x^3 + x^2 + x + 1) = x^5 - 1$]

So, the roots of the above equation $\frac{x^5 - 1}{x - 1} = 0 \; (x \neq 1)$ are

$e^{2\pi i/5}, e^{4\pi i/5}, e^{6\pi i/5}, e^{8\pi i/5}$,

i.e. $\alpha, \beta, \gamma$ and $\delta$.

We know that, property of $n$th root of unity

$1^p + \alpha^p + (\alpha^2)^p + (\alpha^3)^p + \ldots\ldots + (\alpha^{n-1})^p$

$= \begin{cases} 0, & p \text{ is not a multiple of } n \\ n, & p \text{ is a multiple of } n \end{cases}$

Thus, from the above property, we get

$1^{2021} + \alpha^{2021} + \beta^{2021} + \gamma^{2021} + \delta^{2021} = 0$

$\Rightarrow \quad 1 + \alpha^{2021} + \beta^{2021} + \gamma^{2021} + \delta^{2021} = 0$

$\Rightarrow \quad \alpha^{2021} + \beta^{2021} + \gamma^{2021} + \delta^{2021} = -1$

**97.** *(b)* Given that, the equation is $3x^2 + \lambda x - 1 = 0$

and also given that $\alpha$ and $\beta$ are roots of the above equation.

According to the question,

$\frac{1}{\alpha^2} + \frac{1}{\beta^2} = 15 \quad \ldots$ (i)

Since, $\alpha$ and $\beta$ are roots of $3x^2 + \lambda x - 1 = 0$

Thus, $\alpha + \beta = -\frac{\lambda}{3}$ and $\alpha\beta = -\frac{1}{3}$

Hence, from Eq. (i), $\frac{\alpha^2 + \beta^2}{\alpha^2\beta^2} = 15$

$\Rightarrow \frac{(\alpha + \beta)^2 - 2\alpha\beta}{\alpha^2\beta^2} = 15$

$\Rightarrow \frac{\frac{\lambda^2}{9} + \frac{2}{3}}{1/9} = 15$

$\Rightarrow \lambda^2/9 = 1 \Rightarrow \lambda^2 = 9$

Now, $\alpha^3 + \beta^3 = (\alpha + \beta)(\alpha^2 + \beta^2 - \alpha\beta)$

$= (\alpha + \beta)((\alpha + \beta)^2 - 3\alpha\beta)$

$= \left(-\frac{\lambda}{3}\right)\left\{\left(-\frac{\lambda}{3}\right)^2 - 3\left(-\frac{1}{3}\right)\right\}$

$= \left(-\frac{\lambda}{3}\right)\left\{\frac{\lambda^2}{9} + 1\right\} = -\frac{\lambda}{3}\left(\frac{9}{9} + 1\right)$

[since, $\lambda^2 = 9$]

$= -\frac{\lambda}{3} \cdot 2 = -\frac{2\lambda}{3}$

Thus, $6(\alpha^3 + \beta^3)^2 = 6\left\{-\frac{2\lambda}{3}\right\}^2$

$= 6 \times \frac{4\lambda^2}{9}$

$= \frac{24\lambda^2}{9} = \frac{24 \times 9}{9} \quad [\because \lambda^2 = 9]$

$= 24$

**98.** *(a)* Given equation $x^2 + ax + b = 0$

It has two roots (not necessarily real $\alpha$ and $\beta$)

$\Rightarrow$ Either $\alpha = \beta$ or $\alpha \neq \beta$

I. If $\alpha = \beta \Rightarrow \alpha = \alpha^2 - 2 \Rightarrow \alpha = -1, 2$

When $\alpha = -1$, then $(a, b) = (2, 1)$

When $\alpha = 2$, then $(a, b) = (-4, 4)$

II. If $\alpha \neq \beta$, then

(a) $\alpha = \alpha^2 - 2$ and $\beta = \beta^2 - 2$

Here, $(\alpha, \beta) = (2, -1)$ or $(-1, 2)$

Hence $(a, b) = (-\alpha - \beta, \alpha\beta) = (-1, -2)$

(b) $\alpha = \beta^2 - 2$ and $\beta = \alpha^2 - 2$

Then $\alpha - \beta = \beta^2 - \alpha^2 = (\beta - \alpha)(\beta + \alpha)$

$\because \quad \alpha \neq \beta \Rightarrow \alpha + \beta = \beta^2 + \alpha^2 - 4$

or $\alpha + \beta = (\alpha + \beta)^2 - 2\alpha\beta - 4$

$\Rightarrow -1 = 1 - 2\alpha\beta - 4 \Rightarrow \alpha\beta = -1$

$\Rightarrow (a, b) = (-\alpha - \beta, \alpha\beta) = (1, -1)$

(c) $\alpha = \alpha^2 - 2 = \beta^2 - 2$ and $\alpha \neq \beta$

$\Rightarrow \alpha = -\beta$

Thus, $\alpha = 2, \beta = -2$ or $\alpha = -1, \beta = 1$

$\therefore \quad (a, b) = (0, -4)$ and $(0, -1)$

(d) $\beta = \alpha^2 - 2 = \beta^2 - 2$ and $\alpha \neq \beta$ (as in (c))

$\Rightarrow$ We get 6 pairs of $(a, b)$.

They are $(2, 1), (-4, 4), (-1, -2), (1, -1), (0, -4)$, and $(0, -1)$.

**99.** *(b)* $x + 1 - 2\log_2(3 + 2^x) + 2\log_4(10 - 2^{-x}) = 0$

$\Rightarrow x + 1 - 2\log_2(3 + 2^x) + \log_2\left(\dfrac{10\cdot 2^x - 1}{2^x}\right) = 0$

$\Rightarrow x + 1 - 2\log_2(3 + 2^x) + \log_2(10\cdot 2^x - 1) - x = 0$

$\Rightarrow 1 + \log_2\left(\dfrac{10\cdot 2^x - 1}{(3 + 2^x)^2}\right) = 0$

$\Rightarrow \dfrac{10\cdot 2^x - 1}{9 + (2^x)^2 + 6\cdot 2^x} = \dfrac{1}{2}$

$\Rightarrow (2^x)^2 - 14\cdot 2^x + 11 = 0$

Let $2^x = y$

$\Rightarrow y^2 - 14y + 11 = 0$

$y = \dfrac{14 \pm \sqrt{152}}{2} = 7 \pm \dfrac{\sqrt{152}}{2}$

$y_1 = 7 + \dfrac{\sqrt{152}}{2},\ y_2 = 7 - \dfrac{\sqrt{152}}{2}$

$\Rightarrow 2^{x_1} = 7 + \dfrac{\sqrt{152}}{2},\ 2^{x_2} = 7 - \dfrac{\sqrt{152}}{2}$

$\Rightarrow x_1 = \log_2\left(7 + \dfrac{\sqrt{152}}{2}\right),$

$x_2 = \log_2\left(7 - \dfrac{\sqrt{152}}{2}\right)$

$\therefore$ Sum of roots $= x_1 + x_2$

$= \log_2\left(49 - \dfrac{152}{4}\right) = \log_2 11$

**100.** *(a)* Given, $(3x^2 + 4x + 3)^2 - (k+1)(3x^2 + 4x + 3)$

$(3x^2 + 4x + 2) + k(3x^2 + 4x + 2)^2 = 0$

Let, $y = 3x^2 + 4x + 2$

Then, given equation becomes

$(y+1)^2 - (k+1)y(y+1) + ky^2 = 0$

$\Rightarrow \quad y^2 + 2y + 1 - ky^2 - ky - y^2 - y + ky^2 = 0$

$\Rightarrow y + 1 - ky = 0 \Rightarrow y(1 - k) = -1$

$\Rightarrow y = \dfrac{1}{k-1} \Rightarrow 3x^2 + 4x + 2 - \dfrac{1}{k-1} = 0$

For real roots, $D \geq 0$

$\Rightarrow 16 - 4\cdot 3\cdot\left(2 - \dfrac{1}{k-1}\right) \geq 0$

$\Rightarrow -8 + \dfrac{12}{k-1} \geq 0 \Rightarrow \dfrac{3}{k-1} \geq 2$

$\Rightarrow \dfrac{3 - 2k + 2}{k-1} \geq 0 \Rightarrow \dfrac{2k-5}{k-1} \leq 0$

$\Rightarrow k \in \left(1, \dfrac{5}{2}\right]$ $[\because k \neq 1]$

**101.** *(c)* $P(x) = x^2 + bx + c$

$\int_0^1 P(x)\,dx = 1$

$\Rightarrow \quad \int_0^1 (x^2 + bx + c)\,dx = 1$

$\Rightarrow \quad \left[\dfrac{x^3}{3} + \dfrac{bx^2}{2} + cx\right]_0^1 = 1$

$\Rightarrow \quad \dfrac{1}{3} + \dfrac{b}{2} + c = 1$

$\Rightarrow \quad b + 2c = 4/3$ ... (i)

And $P(x) = (x-2)\cdot Q(x) + 5$

When $x = 2,\quad P(2) = 5$

$4 + 2b + c = 5$

$c = 1 - 2b$ ... (ii)

Putting $c = 1 - 2b$ in Eq. (i),

$b + 2(1 - 2b) = 4/3$

$\Rightarrow \quad -3b + 2 = 4/3 \Rightarrow b = 2/9$

$\therefore \quad c = 1 - 4/9 = 5/9$

$9(b + c) = 9\left(\dfrac{2}{9} + \dfrac{5}{9}\right) = 7$

**102.** *(b)* Given, set $= \{x \in R : (|x| - 3)|x + 4| = 6\}$

As, we already know

$|x| = \begin{cases} x, & x \geq 0 \\ -x, & x < 0 \end{cases}$

and $|x + 4| = \begin{cases} x + 4, & x \geq -4 \\ -(x+4), & x < -4 \end{cases}$

Case I $x < -4$

$(-x - 3)(-x - 4) = 6$

$(x + 3)(x + 4) = 6$

$x = -6$ or $x = -1$

We will reject $x = -1$ as, $-1 > -4$

$\therefore$ When $x < -4$, $x = -6$ is the solution.

Case II $-4 \leq x < 0$

$(-x - 3)(x + 4) = 6$

$\Rightarrow \quad -(x+3)(x+4) = 6$

$\Rightarrow \quad -(x^2 + 7x + 12) = 6$

$\Rightarrow \quad x^2 + 7x + 18 = 0$

As, the discriminant of this quadratic equation is

$D = 7^2 - 4\cdot 18 = 49 - 72 = -23$

$\because \quad D = -23$ and $D < 0$

So, no real roots and as per the question, $x \in R$.

No solution when $-4 \leq x < 0$.

Case III $x \geq 0$

$(|x| - 3)|x + 4| = 6$

$\Rightarrow \quad (x - 3)(x + 4) = 6$

$x = \dfrac{-1 \pm \sqrt{1 + 72}}{2} = \dfrac{-1 \pm \sqrt{73}}{2}$

We will reject $x = \dfrac{-1 - \sqrt{73}}{2}$

as $\dfrac{-1 - \sqrt{73}}{2} < 0$ and here, $x \geq 0$.

So, $x = \dfrac{-1 + \sqrt{73}}{2}$, when $x \geq 0$.

$\therefore\ x = -6$ and $x = \dfrac{-1 + \sqrt{73}}{2}$

are the two solutions which belong to the set.

Hence, number of solutions = 2

**103.** *(d)* Centroid $= \left(\dfrac{10}{3}, \dfrac{7}{3}\right)$

$\alpha$ and $\beta$ are the roots of equation $ax^2 + bx + 1 = 0$

$\because a, b, c$ are in AP.

$\therefore \quad 2b = a + c$ ... (i)

Centroid $= \left(\dfrac{a + 2 + a}{3}, \dfrac{c + b + b}{3}\right)$

$= \left(\dfrac{2a + 2}{3}, \dfrac{c + 2b}{3}\right) = \left(\dfrac{10}{3}, \dfrac{7}{3}\right)$

$\Rightarrow \dfrac{2a + 2}{3} = \dfrac{10}{3}$ and $\dfrac{c + 2b}{3} = \dfrac{7}{3} \Rightarrow a = 4$

$\Rightarrow \quad c + a + c = 7$ $[\because 2b = a + c]$

$\Rightarrow \quad 2c = 7 - 4$ $[\because a = 4]$

$c = 3/2$

Also, $2b = a + c = 4 + \dfrac{3}{2}$

$\Rightarrow \quad b = 11/4$

Now, $\alpha$ and $\beta$ are roots of $ax^2 + bx + 1 = 0$

$\therefore \quad \alpha + \beta = \dfrac{-b}{a} = \dfrac{-11/4}{4}$

$\Rightarrow \quad \alpha + \beta = \dfrac{-11}{16} \Rightarrow \alpha\beta = \dfrac{1}{a} = \dfrac{1}{4}$

$\Rightarrow \quad \alpha\beta = \dfrac{1}{4}$

Now, $\alpha^2 + \beta^2 - \alpha\beta$

$= (\alpha + \beta)^2 - 3\alpha\beta = \left(\dfrac{-11}{16}\right)^2 - 3 \times \dfrac{1}{4}$

$= \dfrac{121 - 192}{256} = \dfrac{-71}{256}$

**104.** *(a)* Given, $t^2 - 9t + 8 = 0$, is satisfied by

$$e^{(\cos^2 x + \cos^4 x + \cos^6 x + \ldots \infty)\log_e 2}$$

Now, $a^{\log_c b} = b^{\log_c a}$

$$\therefore e^{(\cos^2 x + \cos^4 x + \ldots + \infty)\log_e 2}$$
$$= 2^{(\cos^2 x + \cos^4 x + \ldots + \infty)\log_e e}$$
$$= 2^{\cos^2 x + \cos^4 x + \ldots + \infty}$$

Here, $\cos^2 x + \cos^4 x + \ldots\infty$ are in GP, where $a = \cos^2 x$

$$r = \cos^2 x < 1 \quad \left(\because 0 < x < \frac{\pi}{2}\right)$$

$$\therefore \quad S_\infty = \frac{a}{1-r}$$

$$\therefore \quad S_\infty = \frac{\cos^2 x}{1-\cos^2 x} = \frac{\cos^2 x}{\sin^2 x} = \cot^2 x$$

$\therefore$ $\cos^2 x + \cos^4 x + \ldots + \infty = \cot^2 x$

Now, $2^{\cos^2 x + \cos^4 x + \ldots + \infty} = 2^{\cot^2 x}$

Now, roots of equation $t^2 - 9t + 8 = 0$, are

$$(t-1)(t-8) = 0$$
$$t = 1, 8$$

$\therefore$ $2^{\cot^2 x} = 1$ or $8$

$\Rightarrow$ $2^{\cot^2 x} = 1 = 2^0$ or $2^{\cot^2 x} = 8 = 2^3$

$\Rightarrow$ $\cot^2 x = 0$ or $\cot^2 x = 3$

$\Rightarrow$ $\cot x = 0$ or $\cot x = \sqrt{3}$

But here, $0 < x < \frac{\pi}{2}$.

$\therefore \cot x = 0$ not possible.

Hence, $\cot x = \sqrt{3}$ is the only possible value.

Now, $\dfrac{2\sin x}{\sin x + \sqrt{3}\cos x}$

Dividing numerator and denominator by $\sin x$, we get

$$\frac{2}{1+\sqrt{3}\cot x} = \frac{2}{1+\sqrt{3}\times\sqrt{3}} \quad [\because \cot x = \sqrt{3}]$$

$$= \frac{2}{1+3} = \frac{2}{4} = \frac{1}{2}$$

**105.** *(d)* Here, $u = \dfrac{2z+i}{z-ki}$, $z = x + iy$

$$u = \frac{2(x+iy)+i}{(x+iy)-ki} = \frac{2x+i(2y+1)}{x+i(y-k)} \times \frac{x-i(y-k)}{x-i(y-k)}$$

$$\Rightarrow \text{Re}(u) = \frac{2x^2 + (y-k)(2y+1)}{x^2+(y-k)^2}$$

$$\Rightarrow \text{Im}(u) = \frac{-2x(y-k)+x(2y+1)}{x^2+(y-k)^2}$$

Now, $\text{Re}(u) + \text{Im}(u) = 1$

$\Rightarrow 2x^2 + 2y^2 - 2ky + y - k - 2xy + 2kx + 2xy + x$

$= x^2 + y^2 + k^2 - 2ky$

$\Rightarrow x^2 + y^2 + x(2k+1) + y - k^2 - k = 0$

Since, the curve intersect at $Y$-axis

$\therefore x = 0$

$\Rightarrow y^2 + y - k(k+1) = 0$

Let $y_1$ and $y_2$ be roots of equations if $x = 0$

$\because y_1 + y_2 = -1; y_1 y_2 = -(k^2 + k)$

$\therefore (y_1 - y_2)^2 = (1 + 4k^2 + 4k)$

Given, $PQ = 5 \Rightarrow |y_1 - y_2| = 5$

$\Rightarrow 4k^2 + 4k - 24 = 0$

$\Rightarrow k = 2$ or $-3$

as $k > 0$, $\therefore$ $k = 2$

**106.** *(d)* It is given that for $a, b \in R$, $a \neq 0$, the quadratic equation $ax^2 - 2bx + 5 = 0$ has a repeated root $\alpha$.

So, $\quad \alpha = \dfrac{b}{a}$ and $D = 0 \quad \ldots$(i)

$\Rightarrow \quad 4b^2 - 20a = 0$

$\Rightarrow \quad b^2 = 5a \quad \ldots$(ii)

and $\quad \alpha^2 = \dfrac{5}{a} \quad \ldots$(iii)

and $\alpha, \beta$ are roots of another quadratic equation $x^2 - 2bx - 10 = 0$.

So, $\alpha + \beta = 2b$, $\alpha\beta = -10$

and $\alpha^2 - 2b\alpha - 10 = 0$

$$\Rightarrow \quad \frac{b^2}{a^2} - 2\frac{b^2}{a} - 10 = 0 \quad \left(\because \alpha = \frac{b}{a}\right)$$

$$\Rightarrow \quad \frac{5a}{a^2} - \frac{2(5a)}{a} - 10 = 0 \quad (\because b^2 = 5a)$$

$$\Rightarrow \quad \frac{5}{a} - 10 - 10 = 0$$

$\Rightarrow$ $20a = 5 \Rightarrow a = 1/4$

So, $b^2 = \dfrac{5}{4}$

$$\therefore \quad \alpha^2 = \frac{b^2}{a^2} = \frac{5/4}{1/16} = 20$$

Since, $\quad \alpha\beta = -10$

$\Rightarrow \quad \alpha^2\beta^2 = 100 \Rightarrow \beta^2 = 5$

$\therefore \quad \alpha^2 + \beta^2 = 20 + 5 = 25$

**107.** *(c)* Given equation is

$e^{4x} + e^{3x} - 4e^{2x} + e^x + 1 = 0$

Let $e^x = y > 0$, so

$y^4 + y^3 - 4y^2 + y + 1 = 0$

$$\Rightarrow \left(y^2 + \frac{1}{y^2}\right) + \left(y + \frac{1}{y}\right) - 4 = 0$$

[on dividing by $y^2$]

$$\Rightarrow \quad \left(y + \frac{1}{y}\right)^2 + \left(y + \frac{1}{y}\right) - 6 = 0$$

Again, let $y + \dfrac{1}{y} = t$, then

$$t^2 + t - 6 = 0$$

$\Rightarrow \quad (t-2)(t+3) = 0$

$\Rightarrow \quad t = -3, 2$

$\therefore \quad y + \dfrac{1}{y} = -3$ or $2$

$$\Rightarrow e^x + \frac{1}{e^x} = -3 \text{ or } 2 \Rightarrow e^x + \frac{1}{e^x} = 2$$

$$\left[\because e^x > 0 \Rightarrow e^x + \frac{1}{e^x} \neq -3 \;\forall x \in R\right]$$

$\Rightarrow$ $(e^x - 1)^2 = 0 \Rightarrow e^x = 1 \Rightarrow x = 0$

$\therefore$ Number of real roots of given equation is 1.

**108.** *(a)* Given equation, $3^x(3^x - 1) + 2$

$= |3^x - 1| + |3^x - 2|$

Let $3^x = t > 0$, $\forall x \in R$, so equation is

$$t^2 - t + 2 = |t-1| + |t-2| \quad \ldots\text{(i)}$$

$\because |t-1| + |t-2|$

$$= \begin{cases} 3-2t, & 0 < t \le 1 \\ 1 \quad , & 1 < t \le 2 \\ 2t-3, & t > 2 \end{cases}$$

Now, Case I

If $|t-1| + |t-2| = 3 - 2t$, for $0 < t \le 1$

then $t^2 - t + 2 = 3 - 2t$

$\Rightarrow t^2 + t - 1 = 0$

$$\Rightarrow \quad t = \frac{-1 \pm \sqrt{5}}{2} = \frac{\pm\sqrt{5}-1}{2}$$

$$\because \quad \frac{\sqrt{5}-1}{2} \in (0, 1]$$

$\therefore$ $t = \dfrac{\sqrt{5}-1}{2}$ is one of the solution.

**Case II**

If $|t-1| + |t-2| = 1$,

for $1 < t \le 2$

then $\quad t^2 - t + 2 = 1$

$\Rightarrow t^2 - t + 1 = 0$ have no real solution.

Case III

If $|t-1|+|t-2|=2t-3$, for $t>2$

then $t^2-t+2=2t-3$

$\Rightarrow \quad t^2-3t+5=0$ have no real solution.

Since, $S$ be the set of all real roots of the given equation, then $S$ is a singleton set.

**109.** *(c)* Given $\alpha$ and $\beta$ are roots of quadratic equation

$$375x^2-25x-2=0$$

$$\therefore \quad \alpha+\beta=\frac{25}{375}=\frac{1}{15} \quad \dots (i)$$

and $$\alpha\beta=-\frac{2}{375} \quad \dots (ii)$$

Now, $\lim_{n\to\infty}\sum_{r=1}^{n}\alpha^r+\lim_{n\to\infty}\sum_{r=1}^{n}\beta^r$

$=(\alpha+\alpha^2+\alpha^3+\dots+$ upto infinite terms$)+(\beta+\beta^2+\beta^3+\dots+$ upto infinite terms$)$

$$=\frac{\alpha}{1-\alpha}+\frac{\beta}{1-\beta} \quad \left[\because S_\infty=\frac{a}{1-r} \text{ for GP}\right]$$

$$=\frac{(\alpha+\beta)-2\alpha\beta}{1-(\alpha+\beta)+\alpha\beta}$$

On substituting the value of $\alpha+\beta=\frac{1}{15}$ and $\alpha\beta=\frac{-2}{375}$ from Eqs. (i) and (ii) respectively, we get

$$\frac{\frac{1}{15}+\frac{4}{375}}{1-\frac{1}{15}-\frac{2}{375}}=\frac{29}{375-25-2}=\frac{29}{348}$$

$$=\frac{1}{12}$$

**110.** *(a)* According to given information, we have the following graph

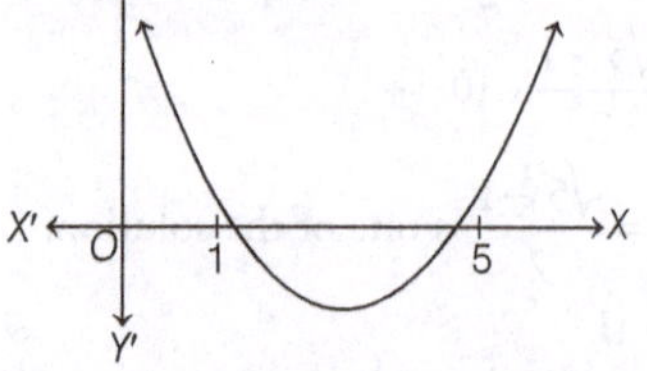


Now, the following conditions should satisfy

(i) $D>0 \Rightarrow b^2-4ac>0$

$$\Rightarrow \quad m^2-4\times1\times4>0$$

$$\Rightarrow \quad m^2-16>0$$

$$(m-4)(m+4)>0$$

$$\Rightarrow \quad m\in(-\infty,-4)\cup(4,\infty)$$

(ii) The vertex of the parabola should lie between $x=1$ and $x=5$

$$\therefore -\frac{b}{2a}\in(1,5)\Rightarrow 1<\frac{m}{2}<5$$

$$\Rightarrow \quad m\in(2,10)$$

(iii) $f(1)>0 \Rightarrow 1-m+4>0$

$\Rightarrow m<5$

$\Rightarrow m\in(-\infty,5)$

(iv) $f(5)>0 \Rightarrow 25-5m+4>0$

$\Rightarrow 5m<29$

$\Rightarrow m\in\left(-\infty,\frac{29}{5}\right)$

From the values of $m$ obtained in conditions (i), (ii), (iii) and (iv), we get $m\in(4,5)$.

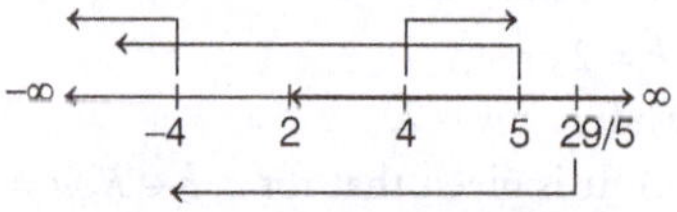


**111.** *(d)* For the roots of quadratic equation $ax^2+bx+c=0$ to be rational $D=(b^2-4ac)$ should be perfect square.

In the equation $6x^2-11x+\alpha=0$

$$a=6, b=-11 \text{ and } c=\alpha$$

$\therefore$ For roots to be rational

$D=(-11)^2-4(6)(\alpha)$ should be a perfect square.

$\Rightarrow D(\alpha)=121-24\alpha$ should be a perfect square

Now, $D(1)=121-24=97$ is not a perfect square.

$D(2)=121-24\times2=73$ is not a perfect square.

$D(3)=121-24\times3=49$ is a perfect square.

$D(4)=121-24\times4=25$ is a perfect square.

$D(5)=121-24\times5=1$ is a perfect square.

and for $\alpha\geq6$, $D(\alpha)<0$, hence imaginary roots.

$\therefore$ For 3 values of $\alpha$ $(\alpha=3,4,5)$, the roots are rational.

**112.** *(a)* Given, $(x^2-5x+5)^{x^2+4x-60}=1$

Clearly, this is possible when

I. $x^2+4x-60=0$ and $x^2-5x+5\neq0$

II. $x^2-5x+5=1$

III. $x^2-5x+5=-1$ and $x^2+4x-60$ = Even integer

**Case I** When $x^2+4x-60=0$

$$\Rightarrow \quad (x+10)(x-6)=0$$

$$\Rightarrow \quad x=-10 \text{ or } x=6$$

Note that, for these two values of $x$, $x^2-5x+5\neq0$

**Case II** When $x^2-5x+5=1$

$$\Rightarrow \quad x^2-5x+4=0$$

$$\Rightarrow \quad (x-4)(x-1)=0$$

$$\Rightarrow \quad x=4 \text{ or } x=1$$

**Case III** When $x^2-5x+5=-1$

$$\Rightarrow \quad x^2-5x+6=0$$

$$\Rightarrow \quad (x-2)(x-3)=0$$

$$\Rightarrow \quad x=2 \text{ or } x=3$$

Now, when $x=2$, $x^2+4x-60=4+8-60=-48$, which is an even integer.

When $x=3$, $x^2+4x-60=9+12-60=-39$, which is not an even integer.

Thus, in this case, we get $x=2$.

Hence, the sum of all real values of $x=-10+6+4+1+2=3$

**113.** *(c)* **Method I** Given, $\alpha$ and $\beta$ are the roots of the equation $x^2-6x-2=0$.

$\because \quad a_n=\alpha^n-\beta^n$ for $n\geq1$

$\therefore \quad a_{10}=\alpha^{10}-\beta^{10}$

$a_8=\alpha^8-\beta^8 \Rightarrow a_9=\alpha^9-\beta^9$

Now, consider

$$\frac{a_{10}-2a_8}{2a_9}=\frac{\alpha^{10}-\beta^{10}-2(\alpha^8-\beta^8)}{2(\alpha^9-\beta^9)}$$

$$=\frac{\alpha^8(\alpha^2-2)-\beta^8(\beta^2-2)}{2(\alpha^9-\beta^9)}$$

$$=\frac{\alpha^8\cdot6\alpha-\beta^8\cdot6\beta}{2(\alpha^9-\beta^9)}=\frac{6\alpha^9-6\beta^9}{2(\alpha^9-\beta^9)}$$

$$=\frac{6}{2}=3$$

**Method II** Since, $\alpha$ and $\beta$ are the roots of the equation

$$x^2-6x-2=0 \text{ or } x^2=6x+2$$

$$\therefore \quad \alpha^2=6\alpha+2$$

$$\Rightarrow \quad \alpha^{10}=6\alpha^9+2\alpha^8 \quad \dots(i)$$

Similarly, $\beta^{10}=6\beta^9+2\beta^8 \quad \dots(ii)$

On subtracting Eq. (ii) from Eq. (i), we get

$\alpha^{10}-\beta^{10}=6(\alpha^9-\beta^9)+2(\alpha^8-\beta^8)$

$\Rightarrow a_{10}=6a_9+2a_8 \quad (\because a_n=\alpha^n-\beta^n)$

$\Rightarrow a_{10}-2a_8=6a_9$

$\Rightarrow \dfrac{a_{10} - 2a_8}{2a_9} = 3$

**114.** *(d)*

> **key idea** If $ax^2 + bx + c = 0$ has roots $\alpha$ and $\beta$, then $\alpha + \beta = \dfrac{-b}{a}$ and $\alpha\beta = \dfrac{c}{a}$.

Since, $\alpha$ and $\beta$ are roots of $px^2 + qx + r = 0, p \neq 0$.

$\therefore \quad \alpha + \beta = \dfrac{-q}{p}, \alpha\beta = \dfrac{r}{p}$

Since, $p$, $q$ and $r$ are in AP.

$\therefore \quad 2q = p + r$

Also, $\dfrac{1}{\alpha} + \dfrac{1}{\beta} = 4 \Rightarrow \dfrac{\alpha + \beta}{\alpha\beta} = 4$

$\Rightarrow \alpha + \beta = 4\alpha\beta \Rightarrow \dfrac{-q}{p} = \dfrac{4r}{p}$

$\Rightarrow \quad q = -4r; \because 2q = p + r$

$\Rightarrow 2(-4r) = p + r \Rightarrow p = -9r$

$\because \alpha + \beta = \dfrac{-q}{p} = \dfrac{4r}{p} = \dfrac{4r}{-9r} = -\dfrac{4}{9}$

and $\alpha\beta = \dfrac{r}{p} = \dfrac{r}{-9r} = \dfrac{1}{-9}$

$\therefore \quad (\alpha - \beta)^2 = (\alpha + \beta)^2 - 4\alpha\beta$

$= \dfrac{16}{81} + \dfrac{4}{9} = \dfrac{16 + 36}{81}$

$\Rightarrow (\alpha - \beta)^2 = \dfrac{52}{81}$

$\Rightarrow |\alpha - \beta| = \dfrac{2}{9}\sqrt{13}$

**115.** *(d)* $x^2 - 4\sqrt{2}kx + 2e^{4 \ln k} - 1 = 0$

or, $x^2 - 4\sqrt{2}kx + 2k^4 - 1 = 0$

$\alpha + \beta = 4\sqrt{2}k$ ...(i)

and $\alpha \cdot \beta = 2k^4 - 1$ ...(ii)

$\left[\because \alpha + \beta = \dfrac{-b}{a} \text{ and } \alpha \cdot \beta = \dfrac{c}{a}\right]$

Squaring both sides of Eq. (i), we get

$(\alpha + \beta)^2 = (4\sqrt{2}k)^2$

$\Rightarrow \alpha^2 + \beta^2 + 2\alpha\beta = 32k^2$

$\Rightarrow 66 + 2\alpha\beta = 32k^2$

$\Rightarrow 66 + 2(2k^4 - 1) = 32k^2$

$\Rightarrow 66 + 4k^4 - 2 = 32k^2$

$\Rightarrow 4k^4 - 32k^2 + 64 = 0$

or, $k^4 - 8k^2 + 16 = 0$

$\Rightarrow (k^2)^2 - 8k^2 + 16 = 0$

$\Rightarrow (k^2 - 4)(k^2 - 4) = 0, k^2 = 4$

$\Rightarrow k = \pm 2$

Now,

$\alpha^3 + \beta^3 = (\alpha + \beta)(\alpha^2 + \beta^2 - \alpha\beta)$

$\therefore \alpha^3 + \beta^3 = (4\sqrt{2}k)[66 - (2k^4 - 1)]$

Putting, $k = -2$, ($k = +2$ cannot be taken because it does not satisfy the given equation)

$\therefore \alpha^3 + \beta^3 = (4\sqrt{2}(-2))$

$[66 - 2(-2)^4 + 1]$

$\alpha^3 + \beta^3 = (-8\sqrt{2})(66 - 32 + 1)$

$= (-8\sqrt{2})(35)$

$\therefore \alpha^3 + \beta^3 = -280\sqrt{2}$

**116.** *(b)* Given equation is

$e^{\sin x} - e^{-\sin x} = 4$

$\Rightarrow \quad e^{\sin x} - \dfrac{1}{e^{\sin x}} = 4$

Now, let $y = e^{\sin x}$

Hence, we get

$y - \dfrac{1}{y} = 4 \Rightarrow y^2 - 4y - 1 = 0$

$\Rightarrow y = \dfrac{4 \pm \sqrt{16 + 4}}{2} \Rightarrow y = 2 \pm \sqrt{5}$

On substituting the value of $y$, we get

$e^{\sin x} = 2 \pm \sqrt{5}$

Now, since sine is a bounded function, i.e., $-1 \leq \sin x \leq 1$. Hence, we get

$e^{-1} \leq e^{\sin x} \leq e \Rightarrow e^{\sin x} \in \left[\dfrac{1}{e}, e\right]$

Also, it is obvious that $2 + \sqrt{5} > e$

and $2 - \sqrt{5} < \dfrac{1}{e} \cdot \Rightarrow 2 \pm \sqrt{5} \notin \left[\dfrac{1}{e}, e\right]$

So, $e^{\sin x} = 2 + \sqrt{5}$ is not possible for any $x \in R$ and $e^{\sin x} = 2 - \sqrt{5}$ is also not possible for any $x \in R$.

Hence, we can say that the given equation has no solution.

**117.** *(a)* Given that $p(x) = f(x) - g(x)$ has only one root $-1$.

$\Rightarrow p(x) = (a - a_1)x^2 + (b - b_1)x + (c - c_1)$

has one root, that is, $-1$ only.

$\Rightarrow p'(x)$ will also has root as $-1$.

$\Rightarrow p'(x) = 0$ at $x = -1$

$\Rightarrow \quad 2(a - a_1)x + (b - b_1) = 0$ at $x = -1$

$\Rightarrow \quad -2(a - a_1) + (b - b_1) = 0$

$\Rightarrow \quad \dfrac{-(b - b_1)}{(a - a_1)} = -2$ ...(i)

Now, $p(x) = (a - a_1)x^2 + (b - b_1)x + (c - c_1)$

$\Rightarrow \dfrac{p(x)}{a - a_1} = x^2 + \dfrac{b - b_1}{a - a_1}x + \dfrac{(c - c_1)}{a - a_1}$

$\because \quad p(-1) = 0$

$\therefore \; 0 = (-1)^2 - \dfrac{(b - b_1)}{a - a_1} + \dfrac{(c - c_1)}{a - a_1}$

$\Rightarrow 0 = 1 - 2 + \dfrac{(c - c_1)}{a - a_1}$ [using Eq. (i)]

$\Rightarrow \quad \dfrac{c - c_1}{a - a_1} = 1$ ...(ii)

Also, given that $p(-2) = 2$

$\Rightarrow 4(a - a_1) - 2(b - b_1) + (c - c_1) = 2$ ...(iii)

From Eqs. (i), (ii) and (iii), we have

$4(a - a_1) - 4(a - a_1) + (a - a_1) = 2$

$\Rightarrow \quad a - a_1 = 2$

On substituting $a - a_1 = 2$ in Eq. (ii), we get

$c - c_1 = 2$

On substituting $a - a_1 = 2$ in Eq. (i), we get

$b - b_1 = 4$

Now, $p(2) = 4(a - a_1) + 2(b - b_1) + (c - c_1)$

$= 8 + 8 + 2 = 18$

**118.** *(b)* Since, both roots of equation $x^2 - 2mx + m^2 - 1 = 0$ are greater than $-2$ but less than 4.

$\therefore \quad D \geq 0, \quad -2 < -\dfrac{b}{2a} < 4,$

$f(4) > 0$ and $f(-2) > 0$

Now, $\quad D \geq 0;$

$4m^2 - 4m^2 + 4 \geq 0$

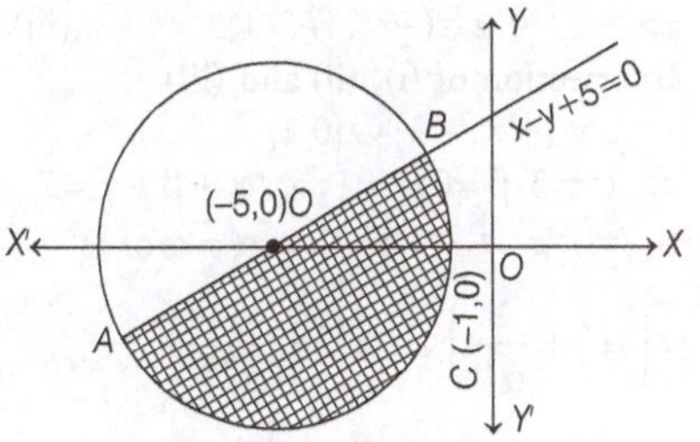


$\Rightarrow \quad 4 > 0, \forall m \in R$ ...(i)

$-2 < -\dfrac{b}{2a} < 4 \Rightarrow -2 < \left(\dfrac{2m}{2 \cdot 1}\right) < 4$

$\Rightarrow \quad -2 < m < 4$ ...(ii)

$f(4) > 0$

$\Rightarrow \quad 16 - 8m + m^2 - 1 > 0$

$\Rightarrow \quad (m - 3)(m - 5) > 0$

$\Rightarrow -\infty < m < 3$ and $5 < m < \infty$ ...(iii)

and $\quad f(-2) > 0$

$\Rightarrow \quad 4 + 4m + m^2 - 1 > 0$

$\Rightarrow \quad (m+3)(m+1) > 0$

$\Rightarrow \quad -\infty < m < -3$

and $\quad -1 < m < \infty \quad \ldots\text{(iv)}$

From Eqs. (i), (ii), (iii) and (iv), we get $m$ lie between $-1$ and 3.

**Alternate Solution**

The given equation is

$x^2 - 2mx + m^2 - 1 = 0$

$\Rightarrow x^2 - 2mx + m^2 = 1 \Rightarrow (x-m)^2 = 1$

$\Rightarrow \quad x - m = \pm 1 \Rightarrow x = m \pm 1$

Since, it is given that the roots of the equation are greater than $-2$ but less than 4.

$\therefore \quad -2 < x < 4 \Rightarrow -2 < m \pm 1 < 4$

$\Rightarrow -2 < m+1 < 4$ and $-2 < m-1 < 4$

$\Rightarrow -3 < m < 3$ and $-1 < m < 5 \Rightarrow -1 < m < 3$

**119.** (7) $f(x) = (1-a)x^2 + 2(a-3)x + 9$

$f(0) = 9 > 0$

$0 \geq 0 \Rightarrow 4(a-3)^2 \geq 4(1-a) \cdot 9$

$\Rightarrow \quad a \in (-\infty, 3] \cup [0, \infty) \quad \ldots\text{(i)}$

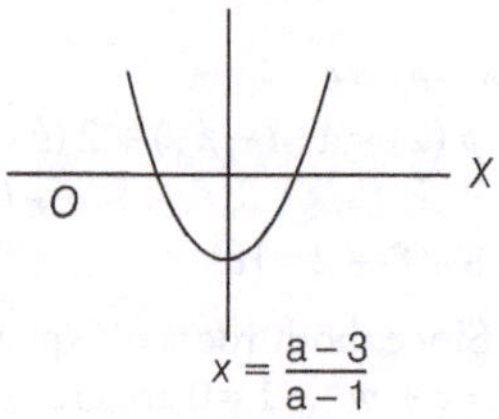


$x_1 + x_2 = \dfrac{-2(a-3)}{1-a}, \; x_1 x_2 = \dfrac{9}{1-a}$

$x_1 + x_2 > 0 \Rightarrow \dfrac{a-3}{a-1} > 0$

$\Rightarrow \quad a \in (-\infty, 1) \cup (3, \infty) \quad \ldots\text{(ii)}$

$x_1 x_2 > 0 \Rightarrow 1 - a > 0$

$\Rightarrow \quad a \in (-\infty, 1) \quad \ldots\text{(iii)}$

Intersection of (i), (ii) and (iii)

$a \in (-\infty, -3] \cup [0, 1)$

$\Rightarrow \alpha = 3, \beta = 0, \gamma = 1 \Rightarrow 2\alpha + \beta + \gamma = 7$

**120.** (11) $x^2 + x + 1 = 0 \Rightarrow \alpha = \omega$ or $\omega^2$

$\because \left(\alpha^k + \dfrac{1}{\alpha^k}\right)^2 = \alpha^{2k} + \dfrac{1}{\alpha^{2k}} + 2$

$= \omega^k + \dfrac{1}{\omega^k} + 2$

$\omega^k + \dfrac{1}{\omega^k} + 2 \begin{cases} 4 & k \text{ is multiple of } 3 \\ 1 & k \text{ is not multiple of } 3 \end{cases}$

$\therefore \displaystyle\sum_{k=1}^{n} \left(\alpha^k + \dfrac{1}{\alpha^k}\right)^2 = 20$

$(1+1+4) + (1+1+4) + (1+1+4) + (1+1) = 20$

$\therefore n = 11$

**121.** (31) $x^2 - ax - b = 0 <^{\alpha}_{\beta}$

$\alpha + \beta = a$ and $\alpha\beta = -b$

$\because \quad P_n = \alpha^n - \beta^n$

$\therefore \quad P_n - aP_{n-1} - bP_{n-2} = 0$

$\Rightarrow \quad P_n = aP_{n-1} + bP_{n-2}$

$P_6 = aP_5 + bP_4$

$45\sqrt{7}i = a(11\sqrt{7}i) + b(-3\sqrt{7}i)$

$45 = 11a - 3b \quad \ldots\text{(i)}$

and $\quad P_5 = aP_4 + bP_3$

$(11\sqrt{7}i) = a(-3\sqrt{7}i) + b(-5\sqrt{7}i)$

$11 = -3a - 5b \quad \ldots\text{(ii)}$

Solving Eqs. (i) and (ii), we get

$a = 3$ and $b = -4$

$\therefore \quad \alpha + \beta = 3$ and $\alpha\beta = 4$

$\alpha^2 + \beta^2 = (\alpha + \beta)^2 - 2\alpha\beta = 1$

$\alpha^4 + \beta^4 = (\alpha^2 + \beta^2)^2 - 2\alpha^2\beta^2$

$= 1 - 32 = -31$

$\therefore |\alpha^4 + \beta^4| = 31$

**122.** (117) We have,

$a(b-c)x^2 + b((c-a)x + c(a-b) = 0 \quad \ldots\text{(i)}$

$\therefore x = 1$ is a root of Eq. (i), we get

$\therefore$ Other root is sum of root

$\alpha + \beta = \dfrac{-b(c-a)}{a(b-c)} = 2$

$\Rightarrow \quad -bc + ab = 2ab - 2ac$

$\Rightarrow \quad 2ac = 15b \quad \ldots\text{(i)}$

$\Rightarrow \quad 2ac = 15\left(\dfrac{36}{5}\right)$

$\Rightarrow \quad ac = 54 \quad \ldots\text{(ii)}$

$a + c = 15$

$\Rightarrow a^2 + c^2 + 2ac = 225$

$\Rightarrow \quad a^2 + c^2 = 225 - 108 = 117$

**123.** (30) We have, $5x^3 - 15x - a = 0$

Let $f(x) = 5x^3 - 15x$

$f'(x) = 15x^2 - 15 = 15(x-1)(x+1)$

$\therefore \quad f(1) f(-1) < 0$

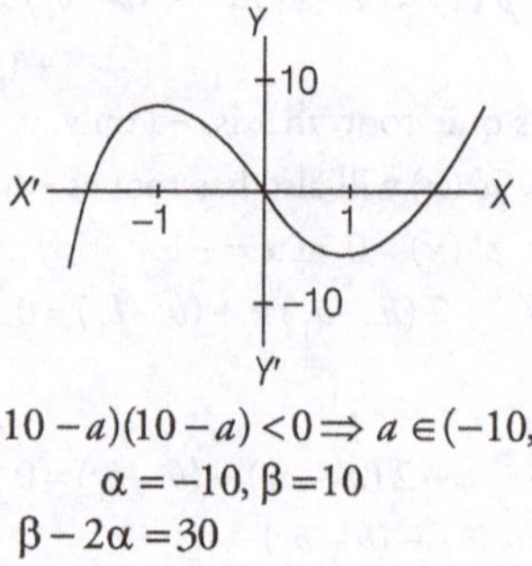


$\Rightarrow (-10 - a)(10 - a) < 0 \Rightarrow a \in (-10, 10)$

$\therefore \quad \alpha = -10, \beta = 10$

$\therefore \quad \beta - 2\alpha = 30$

**124.** (2) $|x+1||x+3| - 4|x+2| + 5 = 0$

**Case I** $x \leq -3$

$(x+1)(x+3) + 4(x+2) + 5 = 0$

$\Rightarrow x^2 + 4x + 3 + 4x + 13 = 0$

$\Rightarrow x^2 + 8x + 16 = 0$

$\Rightarrow (x+4)^2 = 0 \Rightarrow x = -4$

**Case II** $-3 \leq x \leq -2$

$-x^2 - 4x - 3 + 4x + 8 + 5 = 0$

$\Rightarrow -x^2 + 10 = 0 \Rightarrow x = \pm\sqrt{10}$

**Case III** $-2 \leq x \leq -1$

$-x^2 - 4x - 3 - 4x - 8 + 5 = 0$

$\Rightarrow \quad -x^2 - 8x - 6 = 0$

$\Rightarrow \quad x = -4 \pm \sqrt{10}$

**Case IV** $x \geq -1$

$x^2 + 4x + 3 - 4x - 8 + 5 = 0$

$\Rightarrow x^2 = 0 \Rightarrow x = 0$

Hence, number of solutions = 2

**125.** (4) **Method I**

Given quadratic $x^2 + \sqrt{2}x - 8 = 0$ has two roots $\alpha$ and $\beta$

Now, according to Newton's formula, above equation can be written as

$\Rightarrow U_{10} + \sqrt{2}U_9 - 8U_8 = 0$

$\Rightarrow \quad \dfrac{U_{10} + \sqrt{2}U_9}{2U_8} = 4$

**Method II**

Given quadratic $x^2 + \sqrt{2}x - 8 = 0$ has two roots $\alpha$ and $\beta$

$\Rightarrow \quad \alpha^2 + \sqrt{2}\alpha - 8 = 0$

and $\beta^2 + \sqrt{2}\beta - 8 = 0 \quad \ldots\text{(i)}$

According to the question,

$U_n = \alpha^n + \beta^n$

So, $\quad U_{10} = \alpha^{10} + \beta^{10} \quad \ldots\text{(ii)}$

$U_9 = \alpha^9 + \beta^9 \quad \ldots\text{(iii)}$

$U_8 = \alpha^8 + \beta^8 \quad \ldots\text{(iv)}$

Putting Eqs. (ii), (iii) and (iv) in $\dfrac{U_{10} + \sqrt{2}U_9}{2U_8}$,

$\dfrac{\alpha^{10} + \beta^{10} + \sqrt{2}\alpha^9 + \sqrt{2}\beta^9}{2(\alpha^8 + \beta^8)}$

$= \dfrac{\alpha^8(\alpha^2 + \sqrt{2}\alpha) + \beta^8(\beta^2 + \sqrt{2}\beta)}{2(\alpha^8 + \beta^8)}$

$= \dfrac{\alpha^8 \cdot 8 + \beta^8 \cdot 8}{2(\alpha^8 + \beta^8)}$

$[\because \alpha^2 + \sqrt{2}\alpha = 8$ and $\beta^2 + \sqrt{2}\beta = 8$, by Eq. (i)$]$

$= 4$

**126.** *(221)* $4x^4 + 8x^3 - 17x^2 - 12x + 9$

$= 4(x - x_1)(x - x_2)(x - x_3)(x - x_4)$

Now, lets substitute $x = 2i$ in above equation

$\Rightarrow 64 - 64i + 68 - 24i + 9$

$= 4(2i - x_1)(2i - x_2)(2i - x_3)(2i - x_4)$

$\Rightarrow 141 - 88i = 4(2i - x_1)(2i - x_2)(2i - x_3)(2i - x_4)$ ...(i)

On putting $x = -2i$, we get

$141 + 88i = 4(-2i - x_1)(-2i - x_2)(-2i - x_3)(-2i - x_4)$ ...(ii)

On multiple Eq. (i) by Eq. (ii), we get

$(141)^2 + (88)^2 = 16(4 + x_1^2)(4 + x_2^2)(4 + x_3^2)(4 + x_4^2)$

$\Rightarrow (4 + x_1^2)(4 + x_2^2)(4 + x_3^2)(4 + x_4^2) = \dfrac{27625}{16}$

According to the question,

$\dfrac{27625}{16} = \dfrac{125}{16} m \Rightarrow m = 221$

**127.** *(3)* $x|x+5| + 2|x+7| - 2 = 0$

Case I $x \le -7$

We get $-x^2 - 5x - 2x - 14 - 2 = 0$

$x^2 + 7x + 16 = 0$

$D = 49 - 64 < 0$

No real roots

$-7 < x < -5$

$-x^2 - 5x + 2x + 14 - 2 = 0$

$-x^2 - 3x + 12 = 0$

$x = \dfrac{-3 \pm \sqrt{57}}{2}$

$+$ : $\dfrac{-3 + \sqrt{57}}{2}$ $\quad$ $-$ : $\dfrac{-3 - \sqrt{57}}{2}$

Rejected

$x \ge -5$

$x^2 + 7x + 12 = 0$

$x = -3, -4$

Number of real solution $= 3$

**128.** *(3)* Given equation

$|x||x+2| - 5|x+1| - 1 = 0$

Clearly,

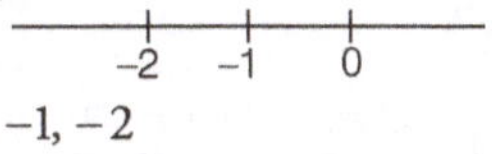

$x = 0, -1, -2$

When, $x \ge 0$, then

$x(x+2) - 5(x+1) - 1 = 0$

$\Rightarrow \quad x^2 + 2x - 5x - 5 - 1 = 0$

$\Rightarrow \quad x^2 - 3x - 6 = 0$

$\Rightarrow \quad x = \dfrac{3 \pm \sqrt{9 + 24}}{2} = \dfrac{3 \pm \sqrt{33}}{2}$

$\therefore$ One positive root.

When $-1 \le x < 0$, then

$(-x)(x+2) - 5(x+1) - 1 = 0$

$\Rightarrow \quad -x^2 - 2x - 5x - 5 - 1 = 0$

$\Rightarrow \quad x^2 + 7x + 6 = 0$

$\Rightarrow \quad (x+6)(x+1) = 0$

$x = -6, -1$

$\therefore$ One root in the range.

When $-2 \le x < -1$, then

$(-x)(x+2) + 5(x+1) - 1 = 0$

$\Rightarrow \quad -x^2 - 2x + 5x + 5 - 1 = 0$

$\Rightarrow \quad -x^2 + 3x + 4 = 0$

$\Rightarrow \quad x^2 - 3x - 4 = 0$

$\Rightarrow \quad (x-4)(x+1) = 0$

$x = -1, 4$

$\therefore$ No root in the range.

When $x < -2$, then

$(-x)(-1)(x+2) + 5(x+1) - 1 = 0$

$\Rightarrow \quad x(x+2) + 5x + 5 - 1 = 0$

$\Rightarrow \quad x^2 + 2x + 5x + 4 = 0$

$\Rightarrow \quad x^2 + 7x + 4 = 0$

$\Rightarrow \quad x = \dfrac{-7 \pm \sqrt{49 - 16}}{2} = \dfrac{7 \pm \sqrt{33}}{2}$

$\Rightarrow$ One root in the range.

$\therefore$ Total number of distinct roots $= 3$

**129.** *(1)* $x(x^2 + 3|x| + 5|x-1| + 6|x-2|) = 0$

$\Rightarrow \quad x = 0$ and

$x^2 + 3|x| + 5|x-1| + 6|x-2| = 0$

all terms are + ve.

$\therefore \quad$ LHS $\neq 0$

$\therefore$ Only one solution.

**130.** *(60)* We have,

$x^2 - 70x + \lambda = 0$

Now, sum of roots i.e. $\alpha + \beta = 70$ and product of roots $\alpha\beta = \lambda$ where $\lambda$ is not multiple of 2 and 3 and minimum value of $\lambda$.

$\because$ Possible value of $\alpha, \beta$

$\alpha = 5, \beta = 65$

$5 + 65 = 70$

and $\quad 5 \times 65 = 325$

$\therefore \quad \lambda = 325$

Now, $\dfrac{(\sqrt{\alpha - 1} + \sqrt{\beta - 1})(\lambda + 35)}{|\alpha - \beta|}$

$= \dfrac{(2 + 8)(360)}{60} = 60$

**131.** *(13)* We have,

$\alpha^6 + \alpha^4 + \beta^4 - 5\alpha^2$

$= \alpha^4(\alpha - 2) + \alpha^4 - 5\alpha^2 + (\beta - 2)^2$

$= \alpha^5 - \alpha^4 - 5\alpha^2 + \beta^2 - 4\beta + 4$

$= \alpha^3(\alpha - 2) - \alpha^4 - 5\alpha^2 + \beta - 2 - 4\beta + 4$

$= -2\alpha^3 - 5\alpha^2 - 3\beta + 2$

$= -2\alpha(\alpha - 2) - 5\alpha^2 - 3\beta + 2$

$= -7\alpha^2 + 4\alpha - 3\beta + 2$

$= -7(\alpha - 2) + 4\alpha - 3\beta + 2$

$= -3\alpha - 3\beta + 16 = -3(1) + 16 = 13$

**132.** *(5)* Since, $\alpha$ satisfies the equation $x^2 + x + 1 = 0$.

So, $\quad \alpha = \omega, \omega^2$

[which are roots of $x^2 + x + 1 = 0$]

Let $\alpha = \omega$ Then, $(1 + \alpha)^7 = -\omega^{14}$

$= -\omega^2 = 1 + \omega$

$= A + B\alpha + C\alpha^2$

$= A + B\omega + C\omega^2$

So, $A = 1$, $B = 1$ and $C = 0$

$\therefore 5(3A - 2B - C) = 5(3 - 2 - 0) = 5$

**133.** *(2)* Given,

$f(x) = e^{8x} - e^{6x} - 3e^{4x} - e^{2x} + 1, x \in R$

$\therefore f(x)$ cuts $X$-axis

$\Rightarrow f(x) = 0$

$\Rightarrow e^{8x} - e^{6x} - 3e^{4x} - e^{2x} + 1 = 0$

Let $e^{2x} = t$ $\quad$ [$\therefore t > 0$]

$\Rightarrow t^4 - t^3 - 3t^2 - t + 1 \backslash 0$

Divide it by $t^2$,

$t^2 - t = -3 - \dfrac{1}{t} + 1 = 0$

$\Rightarrow \left(t^2 + \dfrac{1}{t^2}\right) - \left(t + \dfrac{1}{t}\right) - 3 = 0$

$\Rightarrow \left(t - \dfrac{1}{t}\right)^2 - \left(t + \dfrac{1}{t}\right) - 5 = 0$

Now, let $y = t + \dfrac{1}{t} \Rightarrow$ [$y \ge 2$ and $t > 2$]

$\Rightarrow y^2 - y + 5 = 0$

$\Rightarrow y = \dfrac{1 + \sqrt{21}}{2}, \dfrac{1 - \sqrt{21}}{2}$

$\Rightarrow y = \dfrac{1 + \sqrt{21}}{2}$ $\quad$ [$\therefore y$ can't be negative]

$\Rightarrow e^{2x} + \dfrac{1}{e^{2x}} = \dfrac{1 + \sqrt{21}}{2}$

Adding 2 on both sides,

$\left(e^x + \dfrac{1}{e^x}\right)^2 = \dfrac{5 + \sqrt{21}}{2}$

$$\Rightarrow e^x + \frac{1}{e^x} = \sqrt{\frac{5+\sqrt{21}}{2}} \left[\because e^x + \frac{1}{e^x} \geq 2\right]$$

By using graph, we can easily decide it's solution.

$$\therefore \quad e^x + \frac{1}{e^x} = e^x + e^{-x}$$

$= \text{even function } [\because f(-x) = f(x)]$

Also, $e^x + e^{-x} \geq 2$

**134.** *(51)* We have, $a$ and $b$ are the roots of the equation $x^2 - 7x - 1 = 0$

$\Rightarrow \quad a^2 - 7a - 1 = 0$

$\Rightarrow \quad a^2 - 1 = 7a$ ...(i)

Squaring both sides, we get

$a^4 + 1 = 51a^2$

Similarly, $b^4 + 1 = 51b^2$ ...(ii)

Now, $\dfrac{a^{21} + b^{21} + a^{17} + b^{17}}{a^{19} + b^{19}}$

$$= \frac{a^{17}(a^4+1) + b^{17}(b^4+1)}{a^{19}+b^{19}}$$

$$= \frac{a^{17}(51a^2) + b^{17}(51b^2)}{a^{19}+b^{19}}$$

[∵ from Eqs. (i) and (iii)]

$$= \frac{51[a^{19} + b^{19}]}{a^{19} + b^{19}} = 51$$

**135.** *(8)* Given, $(2a)^{\log_e a} = (bc)^{\log_e b}$, when $2a > 0, bc > 0$

$\Rightarrow \log_e a(\log_e a + \log_e a)$

$= \log_e b(lnb + lnc)$ ... (i)

and $(b)^{1n2} = (a)^{1nc}$

$\Rightarrow \log_e 2 \cdot \log_e b = \log_e c \cdot \log_e a$ ...(ii)

Now, $\log_e a = x, \log_e b = y$

$\log_e 2 = p, \log_e c = z$

Now, from Eqs. (i) and (ii), we get

$p \cdot y = xz$ and $x(p + x) = y(y + z)$

$\therefore p = \dfrac{xz}{y}$

$\therefore x\left(\dfrac{x^2}{y} + x\right) = y(y+z)$

$\Rightarrow x^2 = \pm x^2 y = y^2(y+z)$

$\Rightarrow (x^2 - y^2)(y+z) = 0$

So, $x^2 - y^2 = 0, \quad x + z = 0$

$x = \pm y \quad\quad bc = 1$

if $x = y \Rightarrow \dfrac{a}{b} = 1$

$x = -y \Rightarrow ab = 1$

If $bc = 1 \Rightarrow (2a)^{\log_e a} = 1$

$\Rightarrow a = 1/2$

So, $6a + 5bc = 5 = 3 + 5 = 8$

**136.** *(9)* $x^2 - 12x + [x] + 31 = 0$

$\Rightarrow x^2 - 11x + 31 + [x] - x = 0$

$\Rightarrow x^2 - 11x + 31 = \{x\}$

$0 \leq x^2 - 11x + 31 < 1$

$x^2 - 11x + 30 < 0$ and $x^2 - 11x + 31 \geq 0$

$\Rightarrow x \in R \Rightarrow 5 < x < 6 \Rightarrow [x] = 5$

∴ The given equation be comes

$x^2 - 12x + 5 + 31 = 0$

$\Rightarrow x^2 - 12x + 36 = 0 \Rightarrow x = 6$

Which is not possible $\quad (\because 5 < x < 6)$

∴ There is no value of $x$

i.e., $m = 0$

Now, $x^2 - 5|x+2| - 4 = 0$

| Case-I | Case -II |
|---|---|
| $x \geq -2$ | $x < -2$ |
| $x^2 - 5x - 14 = 0$ | $x^2 + 5x + 6 = 0$ |
| $\Rightarrow x = 7, -2 \Rightarrow$ | $x = -3, -2$ |

$\therefore x = 7, -2, -3 \Rightarrow n = 3$

$\therefore m^2 + mn + n^2 = 0 + 0 + 3^2 = 9$

**137.** *(13)* We have,

$\because x^2 - 5ax + 1 = 0$ and $x^2 - ax - 5 = 0$

Let $\alpha$ be the common root.

Then, $\alpha^2 - 5a\alpha + 1 = 0 = \alpha^2 - a\alpha - 5$

$\Rightarrow 4a\alpha = 6 \Rightarrow \alpha = \dfrac{3}{2a}$

So, putting $\alpha = \dfrac{3}{2a}$ in equation

$\alpha^2 - 5a\alpha + 1 = 0$, we have

$$\Rightarrow \left(\frac{3}{2a}\right)^2 - 5a\left(\frac{3}{2a}\right) + 1 = 0$$

$$\Rightarrow \frac{9}{4a^2} - \frac{15}{2} + 1 = 0$$

$$\Rightarrow \frac{9}{4a^2} = \frac{15}{2} - 1 = \frac{13}{2} \Rightarrow a^2 = \frac{18}{52} = \frac{9}{26}$$

$$\Rightarrow \quad a = \frac{3}{\sqrt{26}} = \frac{3}{\sqrt{2\beta}} \Rightarrow 2\beta = 26$$

$\Rightarrow \quad \beta = 13$

**138.** *(9)* $x^7 + 3x^5 - 13x^3 - 15x = 0$

$\Rightarrow x(x^6 + 3x^4 - 13x^2 - 15) = 0$

$\Rightarrow x(x^6 + 2x^4 - 15x^2 + x^4 + 2x^2 - 15) = 0$

$\Rightarrow x[x^2(x^4 + 2x^2 - 15) + (x^4 + 2x^2 - 15)] = 0$

$\Rightarrow x(x^2 + 1)(x^4 + 2x^2 - 15) = 0$

$\Rightarrow x(x^2 + 1)(x^2 - 3)(x^2 + 5) = 0$

So, the roots are $0, \pm i, \pm\sqrt{3}, \pm\sqrt{5}i$

But, given $|\alpha_1| \geq |\alpha_2| \geq .... \geq |\alpha_7|$

Hence, $\alpha_1 = \sqrt{5}i, \alpha_3 = \sqrt{3}, \alpha_5 = i,$

$\alpha_7 = 0,$

$\alpha_2 = -\sqrt{5}i, \alpha_4 = -\sqrt{3}, \alpha_6 = -i,$

$\therefore \alpha_1\alpha_2 - \alpha_3\alpha_4 + \alpha_5\alpha_6 = 5 + 3 + 1 = 9$

**139.** *(45)* Given, $a \in R$, and $\alpha, \beta$ are root's of $x^2 + 60^{1/4}x + a = 0$

Sum of root's $= \alpha + \beta = -(60)^{1/4}$

Product of root's $= \alpha\beta = a$

And $\quad \alpha^4 + \beta^4 = -30$

So, $(\alpha^2 + \beta^2)^2 - 2\alpha^2\beta^2 = -30$

Now,

$[(\alpha+\beta)^2 - 2\alpha\beta]^2 - 2(\alpha\beta)^2 = -30$

$\Rightarrow [\{-(60)^{1/4}\}^2 - 2(a)]^2 - 2(a)^2 = -30$

$\Rightarrow [\sqrt{60} - 2a]^2 - 2a^2 = -30$

$\Rightarrow 60 + 4a^2 - 4a\sqrt{60} - 2a^2 = -30$

$\Rightarrow 2a^2 - 4a\sqrt{60} + 90 = 0$

Or $a^2 - 2a\sqrt{60} + 45 = 0$

It's a quadratic equation, so we will get 2 values of $a$ or 2 roots of the equation.

So, product of root's $= \dfrac{45}{1} = 45.$

**140.** *(16)* Given, $\alpha$ and $\beta$ are roots of the equation $x^2 - x - 4 = 0$ and $\alpha > \beta$.

$\alpha^2 - \alpha = 4$ and $\beta^2 - \beta = 4$

$P_n = \alpha^n - \beta^n, n \in N$

then, $\dfrac{P_{15}P_{16} - P_{14}P_{16} - P_{15}^2 + P_{14}P_{15}}{P_{13}P_{14}}$

$$= \frac{P_{15}P_{16} - P_{15}^2 - P_{14}P_{16} + P_{14}P_{15}}{P_{13}P_{14}}$$

$$= \frac{P_{15}(P_{16} - P_{15}) - P_{14}(P_{16} - P_{15})}{P_{13}P_{14}}$$

$$= \frac{(P_{16} - P_{15})(P_{15} - P_{14})}{P_{13}P_{14}}$$

$$= \frac{[(\alpha^{16} - \beta^{16}) - (\alpha^{15} - \beta^{15})][(\alpha^{15} - \beta^{15}) - (\alpha^{14} - \beta^{14})]}{(\alpha^{13} - \beta^{13})(\alpha^{14} - \beta^{14})}$$

$$= \frac{[\alpha^{14}(\alpha^2 - \alpha) - \beta^{14}(\beta^2 - \beta)][\alpha^{13}(\alpha^2 - \alpha) - \beta^{13}(\beta^2 - \beta)]}{(\alpha^{13} - \beta^{13})(\alpha^{14} - \beta^{14})}$$

$$= \frac{4(\alpha^{14} - \beta^{14}) \cdot 4(\alpha^{13} - \beta^{13})}{(\alpha^{13} - \beta^{13})(\alpha^{14} - \beta^{14})} = 16$$

**141.** *(272)*

**key idea** Let $ax^2 + bx + c = 0$ and $dx^2 + ex + f = 0$ have a common roots $\alpha$ (say).

Then, $a\alpha^2 + b\alpha + c = 0$ and $d\alpha^2 + e\alpha + f = 0$

Given, the equations
$(p^2+q^2)x^2-2q(p+r)x+q^2+r^2=0$
and $x^2+2x-8=0$ has a common root.

$x^2+2x-8=0$
$\Rightarrow x^2+4x-2x-8=0 \Rightarrow x=-4, 2$

Now, $(p^2+q^2)x^2-2pqx-2qrx+q^2+r^2=0$
$\Rightarrow (p^2x^2-2pqx+q^2)+(q^2x^2-2qrx+r^2)=0$
$\Rightarrow (px-q)^2+(qx-r)^2=0$
$\Rightarrow px-q=0$ and $qx-r=0$
$\Rightarrow x=\frac{q}{p}$ and $x=\frac{r}{q}$

As $p$, $q$ and $r$ are not all having same sign,
$x\neq 2,\ x=\frac{q}{p}=\frac{r}{q}=-4,$
$q=-4p$ and $r=-4q$
Also, $r=-4(-4p)=16p$
Now, $\frac{q^2+r^2}{p^2}=\frac{16p^2+256p^2}{p^2}$
$=\frac{272p^2}{p^2}=272$

**142.** *(2)* Given, equation dividing by $e^{2x}$,
$e^{2x}+4e^x-58+4e^{-x}+e^{-2x}=0$
$\Rightarrow (e^x+e^{-x})^2+4(e^x+e^{-x})-60=0$
Let $e^x+e^{-x}=t\in[2,\infty)$
$\Rightarrow t^2+4t-60=0$
$\Rightarrow t=6$ is only possible solution
$e^x+e^{-x}=6$
$\Rightarrow e^{2x}-6e^x+1=0$
Let $e^x=p$,
$p^2-6p+1=0$
$p=3\pm 2\sqrt{2}$
So, $x=\ln(3+2\sqrt{2})$ or $\ln(3-2\sqrt{2})$
∴Number of real roots is 2.

**143.** *(36)* $x^4-3x^3-2x^2+3x+1=0$
Dividing by $x^2$ both side, we get (where $x\neq 0$)
$x^2-3x-2+\frac{3}{x}+\frac{1}{x^2}=0$
$\Rightarrow x^2+\frac{1}{x^2}-2-3\left(x-\frac{1}{x}\right)=0$
$\Rightarrow \left(x-\frac{1}{x}\right)^2-3\left(x-\frac{1}{x}\right)=0$
$\left[\because \left(x-\frac{1}{x}\right)^2=x^2+\frac{1}{x^2}-2\right]$
$x-\frac{1}{x}=0,\ x-\frac{1}{x}=3$
$x^2-1=0,\ x^2-3x-1=0$
$x=\pm 1,\ \alpha+\beta=3$
$\gamma=1,\ \delta=-1,\ \alpha\beta=-1$
$\gamma^3+\delta^3+\alpha^3+\beta^3$
$=1-1+(\alpha+\beta)((\alpha+\beta)^2-3\alpha\beta)$
$=0+3(9-3(-1))=36$

**144.** *(4)* Let
$f(x)=3x^4+4x^3-12x^2+4=0$
Differentiating w.r.t. $x$,
$f'(x)=12x^3+12x^2-24x=0$
$\Rightarrow 12x(x^2+x-2)=0$
$\Rightarrow x(x+2)(x-1)=0$
Critical point $x=0, 1, -2$

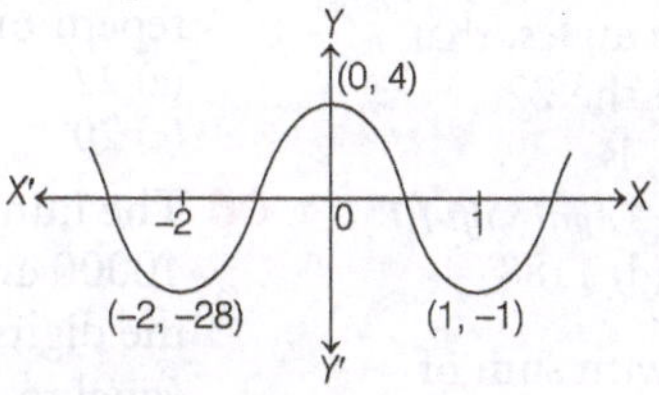


Graph of $y=f(x)$
Number of real roots = 4

**145.** *(66)* $\frac{2}{x-1}-\frac{1}{x-2}=\frac{2}{k}$
$\Rightarrow\ x\in R-\{1, 2\}$
$k(2x-4-x+1)=2(x^2-3x+2)$
$k(x-3)=2(x^2-3x+2)$
$2x^2-(6+k)x+3k+4=0$
For no real roots $b^2-4ac<0$
$\therefore (k+6)^2-8\cdot(3k+4)<0$
$\Rightarrow k^2-12k+4<0 \Rightarrow (k-6)^2-32<0$
$\Rightarrow (k-6)^2<32 \Rightarrow -4\sqrt{2}<k-6<4\sqrt{2}$
$\Rightarrow 6-4\sqrt{2}<k<6+4\sqrt{2}$
Integral $k\in\{1, 2, 3, 4, \ldots., 1\}$
Sum = 66

**146.** *(2)* Given equation,
$e^{4x}-e^{3x}-4e^{2x}-e^x+1=0$
Let $e^x=t>0$
$t^4-t^3-4t^2-t+1=0$
$\Rightarrow\ t^2-t-4-\frac{1}{t}+\frac{1}{t^2}=0$
$\Rightarrow\ t^2+\frac{1}{t^2}+2-\left(t+\frac{1}{t}\right)-6=0$
$\Rightarrow \left(t+\frac{1}{t}\right)^2-\left(t+\frac{1}{t}\right)-6=0$
Let $\alpha=t+1/t\geq 2$
$\Rightarrow\ \alpha^2-\alpha-6=0$
$\Rightarrow (\alpha-3)(\alpha+2)=0$
$\Rightarrow\ \alpha=3$ or $\alpha=-2$ (not possible)
$\Rightarrow\ t+1/t=3 \Rightarrow t^2-3t+1=0$
∴The number of real roots = 2

**147.** *(8)* Since the quadratic equation $2x^2+(a-10)x+\frac{33}{2}=2a$ has real roots.
So, $D\geq 0$
$\Rightarrow\ (a-10)^2-4(2)\left(\frac{33}{2}-2a\right)\geq 0$
$\Rightarrow a^2-20a+100-4(33-4a)\geq 0$
$\Rightarrow a^2-4a-32\geq 0$
$\Rightarrow (a+4)(a-8)\geq 0$
$\Rightarrow\ a\in(-\infty,-4]\cup[8,\infty)$
∴The least positive value of '$a$' for which the given quadratic equation has real roots is 8.

# C-05

JEE Main Chapterwise Topicwise Mathematics

In JEE Main, there are usually 1-2 questions per paper are visible from this chapter. In past 3 years, 36 conceptual and 24 numerical value type questions of moderate to difficulty level were asked from this chapter.

**For detailed analysis scan QR code**

# PERMUTATIONS AND COMBINATIONS

## Topic 1 Fundamental Principal of Counting, Permutations, Conditional and Circular Permutations

### Level 1

**Imp.** Fundamental Principal of Counting

**01** Line $L_1$ of slope 2 and line $L_2$ of slope $\frac{1}{2}$ intersect at the origin $O$. In the first quadrant, $P_1, P_2, \ldots, P_{12}$ are 12 points on line $L_1$ and $Q_1, Q_2, \ldots, Q_9$ are 9 points on line $L_2$. Then, the total number of triangles, that can be formed having vertices at three of the 22 points $O, P_1, P_2, \ldots, P_{12}, Q_1, Q_2, \ldots, Q_9$, is *[2025, 3 April Shift-II]*

(a) 1134 (b) 1026 (c) 1080 (d) 1188

**02** Let $P$ be the set of seven-digit numbers with sum of their digits equal to 11. If the numbers in $P$ are formed by using the digits 1, 2 and 3 only, then the number of elements in the set $P$ is *[2025, 29 Jan Shift-I]*

(a) 164 (b) 158
(c) 161 (d) 173

**03** The number of different 5-digit numbers greater than 50000 that can be formed using the digits 0, 1, 2, 3, 4, 5, 6, 7, such that the sum of their first and last digits should not be more than 8, is *[2025, 28 Jan Shift-I]*

(a) 4607 (b) 5720
(c) 5719 (d) 4608

**04** The number of words, which can be formed using all the letters of the word "DAUGHTER", so that all the vowels never come together is *[2025, 23 Jan Shift-I]*

(a) 34000 (b) 37000
(c) 36000 (d) 35000

**05** Number of ways of arranging 8 identical books into 4 identical shelves, where any number of shelves may remain empty is equal to *[2024, 29 Jan Shift-II]*

(a) 12 (b) 16
(c) 15 (d) 18

**06** Let $\alpha = \frac{(4!)!}{(4!)^{3!}}$ and $\beta = \frac{(5!)!}{(5!)^{4!}}$. Then, *[2024, 27 Jan Shift-II]*

(a) $\alpha \in N$ and $\beta \in N$ (b) $\alpha \notin N$ and $\beta \notin N$
(c) $\alpha \in N$ and $\beta \notin N$ (d) $\alpha \notin N$ and $\beta \in N$

**07** The total number of three-digit numbers, divisible by 3, which can be formed using the digits 1, 3, 5, 8, if repetition of digits is allowed, is *[2023, 15 April Shift-I]*

(a) 22 (b) 18
(c) 20 (d) 21

**08** The number of five digit numbers, greater than 40000 and divisible by 5, which can be formed using the digits 0, 1, 3, 5, 7 and 9 without repetition, is equal to *[2023, 12 April Shift-I]*

(a) 72 (b) 132 (c) 96 (d) 120

**09** The number of arrangements of the letters of the word "INDEPENDENCE" in which all the vowels always occur together is *[2023, 8 April Shift-I]*

(a) 16800 (b) 33600 (c) 18000 (d) 14800

**10** The number of integers, greater than 7000 that can be formed, using the digits 3, 5, 6, 7, 8 without repetition, is *[2023, 24 Jan Shift-II]*

(a) 120 (b) 220 (c) 48 (d) 168

**11** Let $R$ be a relation from the set {1, 2, 3, ..., 60} to itself such that $R = \{(a, b) : b = pq$, where $p, q \geq 3$ are prime numbers$\}$. Then, the number of elements in $R$, is *[2022, 29 July Shift-I]*

(a) 600 (b) 660
(c) 540 (d) 720

**12** The total number of 5-digit numbers, formed by using the digits 1, 2, 3, 5, 6, 7 without repetition, which are multiple of 6, is *[2022, 28 June Shift-I]*
(a) 36 (b) 48
(c) 60 (d) 72

**13** Team $A$ consists of 7 boys and $n$ girls and Team $B$ has 4 boys and 6 girls. If a total of 52 single matches can be arranged between these two teams when a boy plays against a boy and a girl plays against a girl, then $n$ is equal to *[2021, 17 March Shift-I]*
(a) 6 (b) 5
(c) 4 (d) 2

**14** A natural number has prime factorization given by $n=2^x 3^y 5^z$, where $y$ and $z$ are such that $y+z=5$ and $y^{-1}+z^{-1}=\frac{5}{6}$, $y>z$. Then, the number of odd divisors of $n$, including 1, is *[2021, 26 Feb Shift-II]*
(a) 11 (b) 6 (c) 12 (d) 14

**15** The value of $(2\cdot{}^1P_0 - 3\cdot{}^2P_1 + 4\cdot{}^3P_2 - ....$ up to 51th term$) + (1! - 2! + 3! - ...$ up to 51 th term$)$ is equal to *[2020, 3 Sep Shift-I]*
(a) $1-51(51)!$ (b) $1+(51)!$
(c) $1+(52)!$ (d) 1

**16** If the number of five-digit numbers with distinct digits and 2 at the 10th place is $336\,k$, then $k$ is equal to *[2020, 9 Jan Shift-I]*
(a) 8 (b) 7 (c) 4 (d) 6

**17** $n$-digit numbers are formed using only three digits 2, 5 and 7. The smallest value of $n$ for which 900 such distinct numbers can be formed, is *[JEE Main 2018]*
(a) 9 (b) 6 (c) 8 (d) 7

**18** The number of numbers between 2000 and 5000 that can be formed with the digits 0, 1, 2, 3, 4 (repetition of digits is not allowed) and are multiple of 3 is *[JEE Main 2018]*
(a) 36 (b) 48 (c) 24 (d) 30

**19** The number of ways in which 5 boys and 3 girls can be seated on a round table if a particular boy $B_1$ and a particular girl $G_1$ never sit adjacent to each other, is *[JEE Main 2017]*
(a) 7! (b) $5\times 6!$ (c) $6\times 6!$ (d) $5\times 7!$

**20** If the four letter words (need not be meaningful) are to be formed using the letters from the word "MEDITERRANEAN" such that the first letter is R and the fourth letters is E, then the total number of all such words is *[JEE Main 2016]*
(a) 110 (b) 59 (c) $\frac{11!}{(2!)^3}$ (d) 596

**21** The number of ways of selecting 15 teams from 15 men and 15 women, such that each team consists of a man and a woman, is *[JEE Main 2015]*
(a) 1120 (b) 1880 (c) 1960 (d) 1240

**22** The sum of the digits in the unit's place of all the 4-digit numbers formed by using the numbers 3, 4, 5 and 6, without repetition, is *[JEE Main 2014]*
(a) 432 (b) 108
(c) 36 (d) 18

**23** 5-digit numbers are to be formed using 2, 3, 5, 7, 9 without repeating the digits. If $p$ is the number of such numbers that exceed 20000 and $q$ be the number of those that lie between 30000 and 90000, then $p:q$ is *[JEE Main 2013]*
(a) 6 : 5 (b 3 : 2
(c) 4 : 3 (d) 5 : 3

**24** If seven women and seven men are to be seated around a circular table such that there is a man on either side of every woman, then the number of seating arrangements is *[AIEEE 2012]*
(a) 6! 7! (b) $(6!)^2$
(c) $(7!)^2$ (d) 7!

**25** How many ways are there to arrange the letters in the word 'GARDEN' with the vowels in alphabetical order? *[AIEEE 2004]*
(a) 120 (b) 240
(c) 360 (d) 480

**26** The number of ways in which 6 men and 5 women can dine at a round table, if no two women are to sit together, is given by *[AIEEE 2003]*
(a) $6!\times 5!$ (b) 30
(c) $5!\times 4!$ (d) $7!\times 5!$

**27** Number greater than 1000 but less than 4000 is formed using the digits 0, 2, 3, 4 repetition allowed is *[AIEEE 2002]*
(a) 125 (b) 105 (c) 128 (d) 625

**28** Total number of four digit odd numbers that can be formed using 0, 1, 2, 3, 5, 7 (using repetition allowed) are *[AIEEE 2002]*
(a) 216 (b) 375 (c) 400 (d) 720

## Level 2

**29** The number of ways, in which the letters $A$, $B$, $C$, $D$ and $E$ can be placed in the 8 boxes of the figure below so that no row remains empty and at most one letter can be placed in a box, is

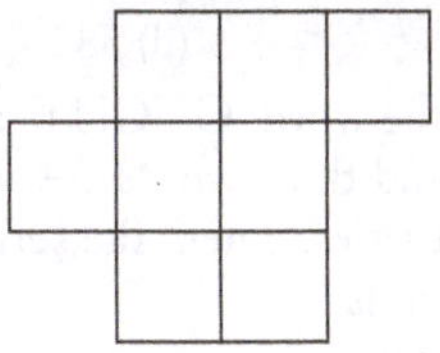

*[2025, 2 April Shift-II]*
(a) 5760 (b) 840 (c) 5880 (d) 960

**30** The number of sequences of ten terms, whose terms are either 0 or 1 or 2, that contain exactly five 1s and exactly three 2$s$, is equal to *[2025, 2 April Shift-I]*
(a) 2520 (b) 45 (c) 1820 (d) 360

**31** If all the words with or without meaning made using all the letters of the word "KANPUR" are arranged as in a dictionary, then the word at 440th position in this arrangement is *[2025, 29 Jan Shift-II]*

(a) PRNAUK (b) PRKANU
(c) PRKAUN (d) PRNAKU

**32** In a group of 3 girls and 4 boys, there are two boys $B_1$ and $B_2$. The number of ways, in which these girls and boys can stand in a queue such that all the girls stand together, all the boys stand together but $B_1$ and $B_2$ are not adjacent to each other is *[2025, 29 Jan Shift-II]*

(a) 144 (b) 96
(c) 72 (d) 120

**33** If all the words with or without meaning made using all the letters of the word 'NAGPUR' are arranged as in a dictionary, then the word at 315th position in this arrangement is *[2024, 6 April Shift-II]*

(a) NRAGUP (b) NRAPUG
(c) NRAPGU (d) NRAGPU

**34** All words, with or without meaning, are made using all the letters of the word MONDAY. These words are written as in a dictionary with serial numbers. The serial number of the word MONDAY is *[2023, 13 April Shift-II]*

(a) 327 (b) 328
(c) 324 (d) 326

**35** If the letters of the word MATHS are permuted and all possible words so formed are arranged as in a dictionary with serial numbers, then the serial number of the word THAMS is *[2023, 11 April Shift-II]*

(a) 104 (b) 102
(c) 103 (d) 101

**36** If the number of words, with or without meaning, which can be made using all the letters of the word MATHEMATICS in which C and S do not come together, is $(6!)k$, then $k$ is equal to *[2023, 8 April Shift-II]*

(a) 5670 (b) 1890
(c) 945 (d) 2835

**37** All the letters of the word PUBLIC are written in all possible orders and these words are written as in a dictionary with serial numbers. Then, the serial number of the word PUBLIC is *[2023, 6 April Shift-II]*

(a) 576 (b) 578
(c) 580 (d) 582

**38** The letters of the word 'OUGHT' are written in all possible ways and these words are arranged as in a dictionary in a series. Then, the serial number of the word 'TOUGH' is *[2023, 29 Jan Shift-II]*

(a) 84 (b) 89 (c) 86 (d) 79

**39** Five digit number divisible by 3 is formed using 0, 1, 2, 3, 4, 6 and 7 without repetition. Total number of such numbers are *[AIEEE 2002]*

(a) 312 (c) 3125 (b) 120 (d) 216

**40** Eight persons are to be transported from city $A$ to city $B$ in three cars of different makes. If each car can accommodate at most three persons, then the number of ways, in which they can be transported, is *[2023, 10 April Shift-II]*

(a) 1680 (b) 3360 (c) 1120 (d) 560

**41** The sum of all the 4-digit distinct numbers that can be formed with the digits 1, 2, 2 and 3 is *[2021, 18 March Shift-I]*

(a) 26664 (b) 122664
(c) 122234 (d) 22264

**42** Consider three boxes, each containing 10 balls labelled 1, 2, ..., 10. Suppose, one ball is randomly drawn from each of the boxes. Denote by $n_i$, the label of the ball drawn from the $i^{\text{th}}$ box, $(i = 1, 2, 3)$. Then, the number of ways in which the balls can be chosen such that $n_1 < n_2 < n_3$ is *[2019, 12 Jan Shift-I]*

(a) 120 (b) 164 (c) 240 (d) 82

**43** If all the words, with or without meaning, are written using the letters of the word QUEEN and are arranged as in English dictionary, then the position of the word QUEEN is *[JEE Main 2017]*

(a) 47th (b) 44th
(c) 45th (d) 46th

**44** If all the words (with or without meaning) having five letters, formed using the letters of the word SMALL and arranged as in a dictionary, then the position of the word SMALL is *[JEE Main 2016]*

(a) 46th (b) 59th
(c) 52nd (d) 58th

**45** 8-digit numbers are formed using the digits 1, 1, 2, 2, 2, 3, 4, 4. The number of such numbers in which the odd digits do no occupy odd places, is *[JEE Main 2014]*

(a) 160 (b) 120
(c) 60 (d) 48

**46** If the letters of the word 'SACHIN' are arranged in all possible ways and these words are written out as in dictionary, then the word 'SACHIN' appears at serial number *[AIEEE 2005]*

(a) 602 (b) 603
(c) 600 (d) 601

## Level 3

**47** The number of ways of selecting two numbers $a$ and $b$, $a \in \{2, 4, 6, ..., 100\}$ and $b \in \{1, 3, 5, ..., 99\}$ such that 2 is the remainder when $a + b$ is divided by 23 is *[2023, 30 Jan Shift-II]*

(a) 186 (b) 54
(c) 108 (d) 268

**48** The number of 6-digits numbers that can be formed using the digits 0, 1, 2, 5, 7 and 9 which are divisible by 11 and no digit is repeated, is *[2019, 10 April Shift-I]*

(a) 60 (b) 72 (c) 48 (d) 36

**49** The number of integers greater than 6000 that can be formed, using the digits 3, 5, 6, 7 and 8 without repetition, is *[JEE Main 2015]*
(a) 216 (b) 192 (c) 120 (d) 72

## Numerical Value Type Questions

**Imp.** Arrangement of Different Things

**50** All five letter words are made using all the letters $A$, $B$, $C$, $D$, $E$ and arranged as in an English dictionary with serial numbers. Let the word at serial number $n$ be denoted by $W_n$. Let the probability $P(W_n)$ of choosing the word $W_n$ satisfy $P(W_n) = 2P(W_{n-1})$, $n > 1$. If $P(CDBEA) = \dfrac{2^\alpha}{2^\beta - 1}$, $\alpha, \beta \in N$, then $\alpha + \beta$ is equal to ............ *[2025, 3 April Shift-I]*

**51** Let $m$ and $n$, $(m < n)$, be two 2-digit numbers. Then, the total numbers of pairs $(m, n)$ such that gcd $(m, n) = 6$, is ............... . *[2025, 4 April Shift-II]*

**52** If the number of seven-digit numbers, such that the sum of their digits is even is $m \cdot n \cdot 10^n : m, n \in \{1, 2, 3, ...., 9\}$, then $m + n$ is equal to ......... *[2025, 3 April Shift-I]*

**53** The number of ways, 5 boys and 4 girls can sit in a row so that either all the boys sit together or no two boys sit together, is......... *[2024, 23 Jan Shift-II]*

**54** All the letters of the word "GTWENTY" are written in all possible ways with or without meaning and these words are written as in a dictionary. The serial number of the word "GTWENTY" is ......... . *[2024, 29 Jan Shift-I]*

**55** Total numbers of 3-digit numbers that are divisible by 6 and can be formed by using the digits 1, 2, 3, 4, 5 with repetition, is ......... . *[2023, 13 April Shift-II]*

**56** The sum of all the four-digit numbers that can be formed using all the digits 2, 1, 2, 3 is equal to ......... *[2023, 10 April Shift-II]*

**57** The number of permutations of the digits 1, 2, 3, ..., 7 without repetition, which neither contain the string 153 nor the string 2467, is ........ *[2023, 10 April Shift-I]*

**58** The total number of six digit numbers, formed using the digits 4, 5, 9 only and divisible by 6, is ............ . *[2023, 1 Feb Shift-II]*

**59** The number of words, with or without meaning that can be formed using all the letters of the word 'ASSASSINATION', so that the vowels occur together, is ............ . *[2023, 1 Feb Shift-I]*

**60** Let 5-digit numbers be constructed using the digits 0, 2, 3, 4, 7, 9 with repetition allowed and are arranged in ascending order with serial numbers. Then, the serial number of the number 42923 is ........ *[2023, 31 Jan Shift-I]*

**61** Number of 4-digit numbers (the repetition of digits is allowed) which are made using the digits 1, 2, 3 and 5 and are divisible by 15, is equal to .......... *[2023, 30 Jan Shift-I]*

**62** Five digit numbers are formed using the digits 1, 2, 3, 5, 7 with repetitions and are written in descending order with serial numbers. e.g. the number 77777 has serial number 1. Then, the serial number of 35337 is ......... *[2023, 29 Jan Shift-I]*

**63** The number of 9-digit numbers, that can be formed using all the digits of the number 123412341, so that the even digits occupy only even places, is ........ *[2023, 24 Jan Shift-I]*

**64** The number of natural numbers lying between 1012 and 23421 that can be formed using the digits 2, 3, 4, 5, 6 (repetition of digits is not allowed) and divisible by 55 is ............... . *[2022, 29 July Shift-II]*

**65** Numbers are to be formed between 1000 and 3000, which are divisible by 4, using the digits 1, 2, 3, 4, 5 and 6 without repetition of digits. Then, the total number of such numbers is ............... . *[2022, 26 July Shift-II]*

**66** The number of 5- digit natural numbers, such that the product of their digits is 36, is ............ . *[2022, 26 July Shift-I]*

**67** The letters of the word 'MANKIND' are written in all possible orders and arranged in serial order as in an English dictionary. Then, the serial number of the word 'MANKIND' is .............. . *[2022, 25 July Shift-I]*

**68** The number of 6-digit numbers made by using the digits 1, 2, 3, 4, 5, 6, 7 without repetition and which are multiple of 15 is ........ *[2022, 30 June Shift-I]*

**69** The total number of four-digit numbers such that each of the first three digits is divisible by the last digit is equal to ........... . *[2022, 29 June Shift-II]*

**Imp.** Formula of Sets

**70** Let $b_1 b_2 b_3 b_4$ be a 4-element permutation with $b_i \in \{1, 2, 3, ......, 100\}$ for $1 \le i \le 4$ and $b_i \ne b_j$ for $i \ne j$, such that either $b_1, b_2, b_3$ are consecutive integers or $b_2, b_3, b_4$ are consecutive integers. Then, the number of such permutations $b_1 b_2 b_3 b_4$ is equal to ...... *[2022, 29 June Shift-I]*

**71** The total number of 3-digit numbers, whose greatest common divisor with 36 is 2, is ...... . *[2022, 26 June Shift-II]*

**72** The number of six letter words (with or without meaning), formed using all the letters of the word 'VOWELS', so that all the consonants never come together, is ...... . *[2021, 31 Aug Shift-I]*

**73** A number is called a palindrome if it reads the same backward as well as forward. For example 285582 is a six digit palindrome. The number of six digit palindromes, which are divisible by 55, is ...... . *[2021, 27 Aug Shift-I]*

**74** The sum of all 3-digit numbers less than or equal to 500, that are formed without using the digit 1 and they all are multiple of 11, is ...... *[2021, 26 Aug Shift-II]*

**75** The number of 3-digit even numbers, formed by the digits 0, 1, 3, 4, 6, 7, if the repetition of digits is not allowed, is ...... . *[2021, 26 Aug Shift-I]*

**76** If the digits are not allowed to repeat in any number formed by using the digits 0, 2, 4, 6, 8, then the number of all numbers greater than 10000 is equal to ......... . *[2021, 22 July Shift-II]*

**77** The missing value in the following figure is *[2021, 18 March Shift-I]*

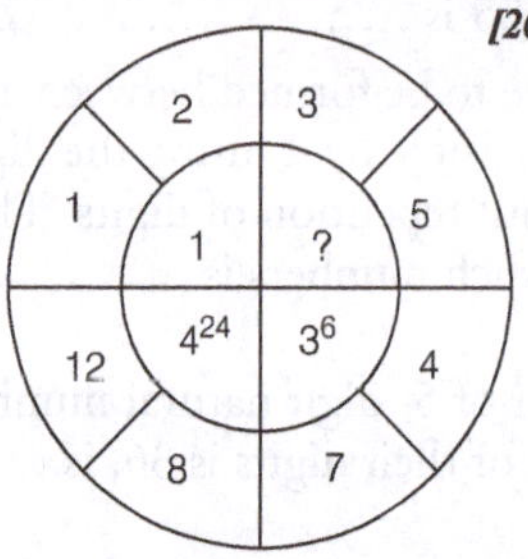


**78** The total number of numbers, lying between 100 and 1000 that can be formed with the digits 1, 2, 3, 4, 5, if the repetition of digits is not allowed and numbers are divisible by either 3 or 5, is ........... . *[2021, 25 Feb Shift-I]*

## Topic 2 Combinations, Conditional Combinations and Dearrangement

### Level 1

**Imp.** Combination

**01** From a group of 7 batsmen and 6 bowlers, 10 players are to be chosen for a team, which should include atleast 4 batsmen and atleast 4 bowlers. One batsmen and one bowler who are captain and vice-captain respectively of the team should be included. Then, the total number of ways such a selection can be made, is *[2025, 7 April Shift-I]*

(a) 155 (b) 145 (c) 165 (d) 135

**02** There are 12 points in a plane, no three of which are in the same straight line, except 5 points which are collinear. Then the total number of triangles that can be formed with the vertices at any three of these 12 points is *[2025, 8 April Shift-II]*

(a) 230 (b) 210
(c) 220 (d) 200

**03** From all the English alphabets, five letters are chosen and are arranged in alphabetical order. The total number of ways, in which the middle letter is '$M$' is *[2025, 22 Jan Shift-I]*

(a) 6084 (b) 14950
(c) 4356 (d) 5148

**04** Group $A$ consists of 7 boys and 3 girls, while group $B$ consists of 6 boys and 5 girls. The number of ways, 4 boys and 4 girls can be invited for a picnic if 5 of them must be from group $A$ and the remaining 3 from group $B$, is equal to *[2025, 24 Jan Shift-II]*

(a) 8575 (b) 8750
(c) 8925 (d) 9100

**05** The number of ways five alphabets can be chosen from the alphabets of the word MATHEMATICS, where the chosen alphabets are not necessarily distinct, is equal to *[2024, 8 April Shift-II]*

(a) 177 (b) 181
(c) 179 (d) 175

**06** If $n$ is the number of ways five different employees can sit into four indistinguishable offices where any office may have any number of persons including zero, then $n$ is equal to *[2024, 1 Feb Shift-I]*

(a) 51 (b) 43
(c) 53 (d) 47

**07** The number of ways in which 21 identical apples can be distributed among three children such that each child gets at least 2 apples, is *[2024, 31 Jan Shift-II]*

(a) 130 (b) 406
(c) 136 (d) 142

**08** The number of triplets $(x, y, z)$, where $x, y, z$ are distinct non-negative integers satisfying $x + y + z = 15$, is *[2023, 11 April Shift-I]*

(a) 136 (b) 80
(c) 114 (d) 92

**09** Let the number of elements in sets $A$ and $B$ be five and two respectively. Then, the number of subsets of $A \times B$ each having at least 3 and at most 6 elements is *[2023, 8 April Shift-I]*

(a) 782 (b) 792
(c) 752 (d) 772

**10** The number of ways, in which 5 girls and 7 boys can be seated at a round table so that no two girls sit together, is *[2023, 8 April Shift-I]*

(a) $7(360)^2$ (b) $7(720)^2$
(c) 720 (d) $126(5!)^2$

**11** A scientific committee is to be formed from 6 Indians and 8 foreigners, which includes at least 2 Indians and double the number of foreigners as Indians. Then, the number of ways, the committee can be formed, is *[2021, 24 Feb Shift-I]*

(a) 1050 (b) 1625 (c) 560 (d) 575

**12** Total number of 6-digit numbers in which only and all the five digits 1, 3, 5, 7 and 9 appear, is *[2020, 7 Jan Shift-I]*

(a) $6!$ (b) $\frac{5}{2}(6!)$ (c) $\frac{1}{2}(6!)$ (d) $5^6$

**13** From 6 different novels and 3 different dictionaries, 4 novels and 1 dictionary are to be selected and arranged in a row on a shelf, so that the dictionary is always in the middle. *[JEE Main 2018]*

The number of such arrangements is

(a) atleast 1000
(b) less than 500
(c) atleast 500 but less than 750
(d) atleast 750 but less than 1000

**14** How many different words can be formed by jumbling the letters in the word 'MISSISSIPPI' in which no two S are adjacent? *[AIEEE 2008]*

(a) $7 \cdot {}^6C_4 \cdot {}^8C_4$ (b) $8 \cdot {}^6C_4 \cdot {}^7C_4$
(c) $6 \cdot 7 \cdot {}^8C_4$ (d) $6 \cdot 8 \cdot {}^7C_4$

**15** The set $S = \{1, 2, 3, \ldots, 12\}$ is to be partitioned into three sets $A, B$ and $C$ of equal size.

Thus, $A \cup B \cup C = S$,
$A \cap B = B \cap C = A \cap C = \phi$

The number of ways to partition $S$ is *[AIEEE 2007]*

(a) $12!/3!(4!)^3$ (b) $12!/3!(3!)^4$
(c) $12!/(4!)^3$ (d) $12!/(3!)^4$

**16** At an election, a voter may vote for any number of candidates not greater than the number to be elected. There are 10 candidates and 4 are to be elected. If a voter votes for atleast one candidate, then the number of ways in which he can vote, is *[AIEEE 2006]*

(a) 6210 (b) 385
(c) 1110 (d) 5040

**17** The number of ways of distributing 8 identical balls in 3 distinct boxes, so that none of the boxes is empty, is *[AIEEE 2004]*

(a) 5 (b) 21
(c) $3^8$ (d) ${}^8C_3$

**18** A student is to answer 10 out of 13 questions in an examination such that he must choose atleast 4 from the first five questions. The number of choices available to him is *[AIEEE 2003]*

(a) 140 (b) 196
(c) 280 (d) 346

## Level 2

**19** Let $0 \le r \le n$. If ${}^{n+1}C_{r+1} : {}^nC_r : {}^{n-1}C_{r-1} = 55:35:21$, then $2n + 5r$ is equal to *[2024, 6 April Shift-II]*

(a) 62 (b) 60
(c) 50 (d) 55

**20** If for some $m, n$: ${}^6C_m + 2({}^6C_{m+1}) + {}^6C_{m+2} > {}^8C_3$ and ${}^{n-1}P_3 : {}^nP_4 = 1:8$, then ${}^nP_{m+1} + {}^{n+1}C_m$ is equal to *[2024, 31 Jan Shift-II]*

(a) 372 (b) 376 (c) 380 (d) 384

**21** ${}^{n-1}C_r = (k^2 - 8)\,{}^nC_{r+1}$ if and only if *[2024, 27 Jan Shift-I]*

(a) $2\sqrt{2} < k \le \sqrt{3}$ (b) $2\sqrt{3} < k \le 3\sqrt{2}$
(c) $2\sqrt{2} < k < 2\sqrt{3}$ (d) $2\sqrt{3} < k < 3\sqrt{3}$

**22** If ${}^{2n}C_3 : {}^nC_3 = 10:1$, then the ratio $(n^2 + 3n) : (n^2 - 3n + 4)$ is *[2023, 6 April Shift-II]*

(a) 27:11 (b) 2:1 (c) 65:37 (d) 35:16

**23** The value of

$$\frac{1}{1!50!} + \frac{1}{3!48!} + \frac{1}{5!46!} + \ldots + \frac{1}{49!2!} + \frac{1}{51!1!}$$ is *[2023, 1 Feb Shift-I]*

(a) $\frac{2^{51}}{50!}$ (b) $\frac{2^{51}}{51!}$ (c) $\frac{2^{50}}{51!}$ (d) $\frac{2^{50}}{50!}$

**24** The total number of positive integral solutions $(x, y, z)$, such that $xyz = 24$ is *[2021, 25 Feb Shift-I]*

(a) 36 (b) 24 (c) 45 (d) 30

**25** There are 3 sections in a question paper and each section contains 5 questions. A candidate has to answer a total of 5 questions, choosing at least one question from each section. Then the number of ways, in which the candidate can choose the questions, is *[2020, 5 Sep Shift-II]*

(a) 3000 (b) 1500 (c) 2255 (d) 2250

**26** A group of students comprises of 5 boys and $n$ girls. If the number of ways, in which a team of 3 students can randomly be selected from this group such that there is at least one boy and at least one girl in each team, is 1750, then $n$ is equal to *[2019, 12 April Shift-II]*

(a) 28 (b) 27
(c) 25 (d) 24

**27** A committee of 11 members is to be formed from 8 males and 5 females. If $m$ is the number of ways the committee is formed with at least 6 males and $n$ is the number of ways the committee is formed with atleast 3 females, then *[2019, 9 April Shift-I]*

(a) $m = n = 68$ (b) $m + n = 68$
(c) $m = n = 78$ (d) $n = m - 8$

**28** **Statement I** The number of ways of distributing 10 identical balls in 4 distinct boxes such that no box is empty, is ${}^9C_3$.

**Statement II** The number of ways of choosing any 3 places from 9 different places is ${}^9C_3$. *[AIEEE 2011]*

(a) Statement I is true, Statement II is true; Statement II is not a correct explanation of Statement I
(b) Statement I is true, Statement II is false
(c) Statement I is false, Statement II is true
(d) Statement I is true, Statement II is true; Statement II is a correct explanation of Statement I

## Level 3

**29** If $({}^{30}C_1)^2 + 2({}^{30}C_2)^2 + 3({}^{30}C_3)^2 + \ldots + ({}^{30}C_{30})^2 = \frac{\alpha 60!}{(30!)^2}$, then $\alpha$ is equal to **[2023, 24 Jan Shift-II]**

(a) 30 (b) 60
(c) 10 (d) 15

**30** The value of $\sum_{r=0}^{20} {}^{50-r}C_6$ is equal to **[2020, 4 Sep Shift-I]**

(a) ${}^{50}C_7 - {}^{30}C_7$
(b) ${}^{51}C_7 - {}^{30}C_7$
(c) ${}^{51}C_7 + {}^{30}C_7$
(d) ${}^{50}C_6 - {}^{30}C_6$

## Numerical Value Type Questions

**Imp.** Conditional Arrangement

**31** The number of 6-letter words with or without meaning that can be formed using the letters of the word MATHS such that any letter that appears in the word must appear at least twice, is ......... . **[2025, 29 Jan Shift-I]**

**32** The number of 3-digit numbers, formed using the digits 2, 3, 4, 5 and 7. When the repetition of digits is not allowed and which are not divisible by 3 is equal to ........ . **[2024, 8 April Shift-I]**

**33** The total number of words (with or without meaning) that can be formed out of the letters of the word 'DISTRIBUTION' taken four at a time, is equal to ........... . **[2024, 31 Jan Shift-I]**

**34** Let the digits $a, b, c$ be in AP. Nine-digit numbers are to be formed using each of these three digits thrice such that three consecutive digits are in AP at least once. How many such numbers can be formed? **[2023, 13 April Shift-I]**

**35** In an examination, 5 students have been allotted their seats as per their roll numbers. The number of ways, in which none of the students sits on the allotted seat, is ......... . **[2023, 11 April Shift-I]**

**36** The number of 4-letter words, with or without meaning, each consisting of 2 vowels and 2 consonants, which can be formed from the letters of the word UNIVERSE without repetition is....... . **[2023, 6 April Shift-II]**

**37** The number of ways of giving 20 distinct oranges to 3 children such that each child gets at least one orange is .......... **[2023, 6 April Shift-I]**

**38** If all the six digit numbers $x_1\ x_2\ x_3\ x_4\ x_5\ x_6$ with $0 < x_1 < x_2 < x_3 < x_4 < x_5 < x_6$ are arranged in the increasing order, then the sum of the digits in the 72th number is ........ . **[2023, 30 Jan Shift-II]**

**39** A class contains $b$ boys and $g$ girls. If the number of ways of selecting 3 boys and 2 girls from the class is 168, then $b + 3g$ is equal to ............ . **[2022, 28 July Shift-II]**

**40** The total number of three-digit numbers, with one digit repeated exactly two times, is ......... . **[2022, 25 June Shift-II]**

**41** In an examination, there are 5 multiple choice questions with 3 choices, out of which exactly one is correct. There are 3 marks for each correct answer, −2 marks for each wrong answer and 0 mark if the question is not attempted. Then, the number of ways a student appearing in the examination gets 5 marks is ...... . **[2022, 24 June Shift-I]**

**42** There are 5 students in class 10, 6 students in class 11 and 8 students in class 12. If the number of ways in which 10 students can be selected from them so as to include at least 2 students from each class and at most 5 students from the total 11 students of class 10 and 11 is 100 $k$, then $k$ is equal to ...... **[2021, 25 July Shift-I]**

**43** The number of times the digit 3 will be written when listing the integers from 1 to 1000 is ......... . **[2021, 18 March Shift-I]**

**44** The students $S_1, S_2, \ldots, S_{10}$ are to be divided into 3 groups $A$, $B$ and $C$ such that each group has at least one student and the group $C$ has at most 3 students. Then, the total number of possibilities of forming such groups is ......... . **[2021, 24 Feb Shift-II]**

# Topic 3 Application of Permutations and Combination in Geometry

## Level 1

**01** Let $n > 2$ be an integer. Suppose that there are $n$ Metro stations in a city located along a circular path. Each pair of stations is connected by a straight track only. Further, each pair of nearest stations is connected by blue line, whereas all remaining pairs of stations are connected by red line. If the number of red lines in 99 times the number of blue lines, then the value of $n$ is **[2020, 2 Sep Shift-II]**

(a) 201 (b) 200 (c) 101 (d) 199

**02** Suppose that 20 pillars of the same height have been erected along the boundary of a circular stadium. If the top of each pillar has been connected by beams with the top of all its non-adjacent pillars, then the total number of beams is **[2019, 10 April Shift-II]**

(a) 180 (b) 210
(c) 170 (d) 190

**03** Let $T_n$ be the number of all possible triangles formed by joining vertices of an $n$-sided regular polygon. If $T_{n+1} - T_n = 10$, then the value of $n$ is ***[JEE Main 2013]***

(a) 7 (b) 5
(c) 10 (d) 8

**04** The number of triangles whose vertices are at the vertices of a regular octagon but none of whose sides is a side of the octagon is ***[2024, 6 April Shift-I]***

(a) 16 (b) 56 (c) 24 (d) 48

## Level 2

**05** If the sides $AB$, $BC$ and $CA$ of a $\Delta ABC$ have 3, 5 and 6 interior points respectively, then the total number of triangles that can be constructed using these points as vertices, is equal to ***[2021, 17 March Shift-II]***

(a) 364 (b) 240 (c) 333 (d) 360

**06** Consider a rectangle $ABCD$ having 5, 7, 6, 9 points in the interior of the line segments $AB$, $CD$, $BC$, $DA$, respectively. Let $\alpha$ be the number of triangles having these points from different sides as vertices and $\beta$ be the number of quadrilaterals having these points from different sides as vertices. Then, $(\beta - \alpha)$ is equal to ***[2021, 16 March Shift-II]***

(a) 795 (b) 1173 (c) 1890 (d) 717

### Numerical Value Type Questions

**07** A person forgets his 4-digit ATM pin code. But he remembers that in the code all the digits are different, the greatest digit is 7 and the sum of the first two digits is equal to the sum of the last two digits. Then, the maximum number of trials necessary to obtain the correct code is ............ . ***[2023, 15 April Shift-I]***

**08** Some couples participated in a mixed doubles badminton tournament. If the number of matches played, so that no couple played in a match is 840, then the total number of persons, who participated in the tournament, is ........ ***[2023, 10 April Shift-I]***

## Key Index

### Topic 1

| | | | | | | | | | | | | | | | | | | | |
|---|---|---|---|---|---|---|---|---|---|---|---|---|---|---|---|---|---|---|---|
| 1 | (a) | 2 | (c) | 3 | (a) | 4 | (c) | 5 | (c) | 6 | (a) | 7 | (a) | 8 | (d) | 9 | (a) | 10 | (d) |
| 11 | (b) | 12 | (d) | 13 | (c) | 14 | (c) | 15 | (c) | 16 | (a) | 17 | (d) | 18 | (d) | 19 | (b) | 20 | (b) |
| 21 | (d) | 22 | (b) | 23 | (d) | 24 | (a) | 25 | (c) | 26 | (a) | 27 | (c) | 28 | (d) | 29 | (a) | 30 | (a) |
| 31 | (c) | 32 | (a) | 33 | (c) | 34 | (a) | 35 | (c) | 36 | (a) | 37 | (d) | 38 | (b) | 39 | (d) | 40 | (a) |
| 41 | (a) | 42 | (a) | 43 | (d) | 44 | (d) | 45 | (b) | 46 | (d) | 47 | (c) | 48 | (a) | 49 | (b) | 50 | (182) |
| 51 | (64) | 52 | (14) | 53 | (17280) | 54 | (553) | 55 | (16) | 56 | (26664) | 57 | (4898) | 58 | (81) | 59 | (50400) | 60 | (2997) |
| 61 | (21) | 62 | (1436) | 63 | (60) | 64 | (6) | 65 | (30) | 66 | (180) | 67 | (1492) | 68 | (360) | 69 | (1086) | 70 | (18915) |
| 71 | (150) | 72 | (576) | 73 | (100) | 74 | (7744) | 75 | (52) | 76 | (96) | 77 | (4) | 78 | (32) | | | | |

### Topic 2

| | | | | | | | | | | | | | | | | | | | |
|---|---|---|---|---|---|---|---|---|---|---|---|---|---|---|---|---|---|---|---|
| 1 | (a) | 2 | (b) | 3 | (d) | 4 | (c) | 5 | (c) | 6 | (a) | 7 | (c) | 8 | (c) | 9 | (b) | 10 | (d) |
| 11 | (b) | 12 | (b) | 13 | (a) | 14 | (a) | 15 | (c) | 16 | (b) | 17 | (b) | 18 | (b) | 19 | (c) | 20 | (a) |
| 21 | (a) | 22 | (b) | 23 | (c) | 24 | (d) | 25 | (d) | 26 | (c) | 27 | (c) | 28 | (a) | 29 | (d) | 30 | (b) |
| 31 | (1405) | 32 | (36) | 33 | (3734) | 34 | (1260) | 35 | (44) | 36 | (432) | 37 | (3483638676) | 38 | (32) | 39 | (17) | 40 | (243) |
| 41 | (40) | 42 | (238) | 43 | (300) | 44 | (31650) | | | | | | | | | | | | |

### Topic 3

| | | | | | | | | | | | | | | | |
|---|---|---|---|---|---|---|---|---|---|---|---|---|---|---|---|
| 1 | (a) | 2 | (c) | 3 | (b) | 4 | (a) | 5 | (c) | 6 | (d) | 7 | (72) | 8 | (16) |

# Detailed **Explanations**

## Topic 1

**01.** *(a)* Total number of ways to choose 3 points out of 22 is $^{22}C_3 = 1540$

The number of ways to choose 3 points on line $L_1$ (including 0) is $^{13}C_3 = 286$

The number of ways to choose 3 point, on line $L_2$ (including points 0) is $^{10}C_3 = 120$

$\therefore$ Number of triangles
$= 1540 - 286 - 120 = 1134$

**02.** *(c)* Case (I) Number of numbers created using 3221111 is

$$n_1 = \frac{7!}{4!2!} = \frac{7 \times 6 \times 5 \times 4!}{4! \times 2 \times 1} = 105$$

Case (II) Number of numbers created using 2222111 is

$$n_2 = \frac{7!}{4!3!} = \frac{7 \times 6 \times 5 \times 4!}{4! \times 3 \times 2} = 35$$

Case (III) Number of numbers created using 3311111 is

$$n_3 = \frac{7!}{5!2!} = \frac{7 \times 6 \times 5!}{5! \times 2 \times 1} = 21$$

Hence, total number of elements in the set

$P = n_1 + n_2 + n_3 = 105 + 35 + 21 = 161$

> • Concept **Enhancer**
>
> Number of permutation of $n$ things of which $P_1$ are alike of the one kind, $P_2$ are alike of second kind.....$P_r$ are of $r$th kind such that $P_1 + P_2 + \ldots + P_r = n$ is $\dfrac{n!}{P_1!P_2!\ldots P_r!}$

**03.** *(a)* 0, 1, 2, 3, 4, 5, 6, 7,

$\Rightarrow$ Total digits $= 8$

□□□□□ Number of digits

5 $\rightarrow$ 0, 1, 2, 3 $\rightarrow$ 8×8×8×4

6 $\rightarrow$ 0, 1, 2 $\rightarrow$ 8×8×8×3

7 $\rightarrow$ 0, 1 $\rightarrow$ 8×8×8×2

$$\text{Total} = 8 \times 8 \times 8 \times (4 + 3 + 2) - 1$$
$$= 4608 - 1 = 4607$$

**04.** *(c)* Given word is DAUGHTER

Total number of words $= 8!$

Total number of words all the vowels come together

AUE, DGHTR

$$1 + 5 = 6 \Rightarrow 6! \times 3!$$

$\therefore$ Required number of words when all the vowels never come together

$$= 8! - 6! \times 3!$$
$$= 6!(8 \times 7 - 6)$$
$$= 6! \times 50 = 720 \times 50 = 36000$$

**05.** *(c)* 8 identical objects into 4 identical shelves, where any number of shelves may remain empty is given by

0008, 0017, 0026, 0035, 0044 = 5

$$\left.\begin{matrix} 0 & 1 & 1 & 6 \\ 0 & 1 & 2 & 5 \\ 0 & 1 & 3 & 4 \\ 0 & 2 & 2 & 4 \\ 0 & 2 & 3 & 3 \end{matrix}\right] = 5$$

$$\left.\begin{matrix} 1 & 1 & 1 & 5 \\ 1 & 1 & 2 & 4 \\ 1 & 1 & 3 & 3 \\ 1 & 2 & 2 & 3 \end{matrix}\right] = 4$$

$$\left.\begin{matrix} 2 & 2 & 2 & 2 \end{matrix}\right] = 1$$

Total $= 15$

**06.** *(a)* Let 24 distinct objects are divided into 6 groups of 4 objects on each group.

Number of ways of formation of group

$$= \frac{24!}{(4!)^6 \cdot 6!} \in N$$

Similarly,

let 120 distinct objects are divided into 24 groups of 5 objects in each group.

$\therefore$ Number of ways of formation of groups

$$= \frac{(120)!}{(5!)^{24} \cdot 24!} \in N$$

Hence, $\alpha \in N$ and $\beta \in N$.

**07.** *(a)* I All three digits are alike

111, 333, 555, 888 $\rightarrow$ 4

II 2 digits are alike

$$558 \rightarrow \frac{3!}{2!} = 3;\ 885 \rightarrow \frac{3!}{2!} = 3$$

III All three digits are different

$$1, 3, 5 \rightarrow 6;\ 1, 3, 8 \rightarrow 6$$

$\therefore$ Total numbers = 22

**08.** *(d)* Since number is divisible by 5, so units digit can be 0 or 5.

Case 1 Units digit $= 0$

Since number is greater than 40000 and digits can't be repeated, the leading digit has only 3 options (5 or 7 or 9) Remaining digits can be filled in 4 and 3 and 2 ways.

Number of ways $= 4! \times 3$

Case 2 Units digit $= 5$

Here, leading digit has 2 options (7 or 9). Remaining digits can be filled in 4 and 3 and 2 ways. So, number of ways $= 4! \times 2$

$$\text{Total numbers} = 4! \times 3 + 4! \times 2$$
$$= 120$$

**09.** *(a)* Vowels – I, EEEE

Consonants – N, D, P, N, D, N, C

Number of ways in which vowels are always occur together

$$= \frac{8!}{3! \times 2!} \times \frac{5!}{4!} = 16800$$

**10.** *(d)* Given, digits are 3, 5, 6, 7, 8.

Since, there is no any repetition and number greater than 7000 will be of 4 and 5 digits.

For 4-digit number, the first place will contain 2 ways and the remaining will be arranged in $^4P_3$ ways.

$\therefore$ Number of ways

$$= 2 \times {}^4P_3 = 2 \times 24 = 48$$

Also, 5-digit number will be arranged in $^5P_5$ ways $= 5! = 120$ ways

$\therefore$ Total number of ways

$$= 120 + 48 = 168 \text{ ways}$$

**11.** *(b)* Now, number of possible values of $a = 60$, for $b = pq$,

If $p = 3$, then $q = 3, 5, 7, 11, 13, 17, 19$.

So, total cases $= 7$

If $p = 5$, then $q = 5, 7, 11$.

So, total cases $= 3$

If $p = 7$, then $q = 7$. So, total case $= 1$

Therefore, total cases

$$= 60(7 + 3 + 1)$$
$$= 660$$

**12.** *(d)* Number should be multiple of 6 and it should be even.

Total sum $= 1 + 2 + 3 + 5 + 6 + 7 = 24$

The number removed should be in 2 types.

Excluding 3 $\rightarrow$ 2 ways $= 4! \times 2 = 48$

Excluding 6 $\rightarrow$ 1 way $= 4! = 24$

Total cases $= 48 + 24 = 72$

**13.** *(c)* Number of matches when a girl plays against a girl $= n \times 6 = 6n$

Number of matches when a boy plays against a boy $= 7 \times 4 = 28$

According to the question, $28 + 6n = 52$

$\Rightarrow n = 4$

**14.** *(c)* Given, $y+z=5$ ...(i)

and $\frac{1}{y}+\frac{1}{z}=\frac{5}{6}$, $y>z$ ...(ii)

On solving Eqs. (i) and (ii), we get

$y=3$, $z=2$

$\Rightarrow \quad n=2^x \cdot 3^3 \cdot 5^2$

$=(2\cdot 2\cdot 2\ldots)(3\cdot 3\cdot 3)(5\cdot 5)$

$\therefore$ Number of odd divisors

$=(3+1)(2+1)=4\times 3=12$

**15.** *(c)* We have,

$(2\cdot {}^1P_0 - 3\cdot {}^2P_1 + 4\cdot {}^3P_2 - \ldots$ upto 51th terms) + $(1!-2!+3!-4!+\ldots$ upto 51th terms)

$=2!-3!+4!-\ldots-51!+52!+1!-2!+3!+\ldots+51!$

$=1+52!$

**16.** *(a)* It is given that the 10th place of 5 digit number with distinct digits is '2', then the ten thousand place, thousand place, hundred place and unit place we can fill in 8, 8, 7 and 6 ways only.

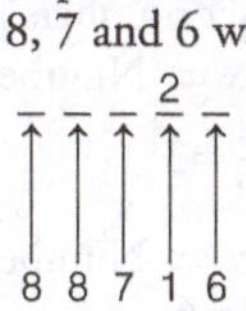


So, required number of five digit numbers is

$8\times 8\times 7\times 6=336\,k$ (given)

$\Rightarrow k=\frac{8\times 8\times 7\times 6}{336}=\frac{8\times 8\times 7}{56}=8$

**17.** *(d)* Since, $n$-digit numbers are formed using 2, 5 and 7 digits.

$\therefore$ The number of ways in which it can be done $=3^n$

Now, $3^6<900$ and $3^7>900$

$\therefore n=7$

**18.** *(d)* Number of 4-digit numbers between 2000 and 5000 which are multiple of $3=2\times 3!+2\times 3!+3!=30$

[$\because$ Numbers which are multiple of 3 can be formed by using (0, 1, 2, 3) or (0, 2, 3, 4)]

**19.** *(b)* Required number of ways

$=7!-6!\times 2$

$=6!(7-2)=5\times 6!$

**20.** *(b)* M, EEE, D, I, T, RR, AA, NN

R _ _ E

Two empty places can be filled with identical letters [EE, AA, NN] in 3 ways.

Two empty places can be filled with distinct letters [M, E, D, I, T, R, A, N] in ${}^8P_2$ ways.

$\therefore$ Number of words formed

$=3+{}^8P_2=59$

**21.** *(d)* Number of ways of selecting a man and a woman for a team from 15 men and 15 women $=15\times 15=(15)^2$

Number of ways of selecting a man and a woman for next team out of the remaining 14 men and 14 women $=14\times 14=(14)^2$

Similarly, for other teams.

Required number of ways

$=(15)^2+(14)^2+\ldots+(1)^2$

$=\frac{15\times 16\times 31}{6}$

$=1240$

**22.** *(b)* With 3 at unit place, total possible four-digit number (without repetition) will be $3!=6$

With 4 at unit place, total possible four-digit numbers will be $3!=6$

With 5 at unit place, total possible four-digit numbers will be $3!=6$

With 6 at unit place, total possible four-digit numbers will be $3!=6$

Sum of unit digit of all possible numbers

$=6\times 3+6\times 4+6\times 5+6\times 6$

$=108$

**23.** *(d)* $p$ : * * * * * place

5 4 3 2 1 ways

Total number of ways $=5!=120$

Since all numbers are $>20{,}000$

$\therefore$ All numbers 2, 3, 5, 7, 9 can come at first place.

$q$ : * * * * * place

3 4 3 2 1 ways

Total number of ways $=3\times 4!=72$

[$\because$ 2 and 9 cannot be put at first place]

So, $p:q=120:72=5:3$

**24.** *(a)* 7 women can be arranged around a circular table in 6! ways.

Among these 7 men can sit in 7! ways.

Hence, number of seating arrangement $=7!\times 6!$

**25.** *(c)* Total number of ways in which all letters can be arranged $=6!$.

There are two vowels (A, E) in the word 'GARDEN'. Total number of ways in which these two vowels can be arranged $=2!$

$\therefore$ Total number of required ways $=\frac{6!}{2!}$

$=360$

**26.** *(a)* First, we fix the position of men, the number of ways to sit men $=5!$ and the number of ways to sit women $={}^6P_5$

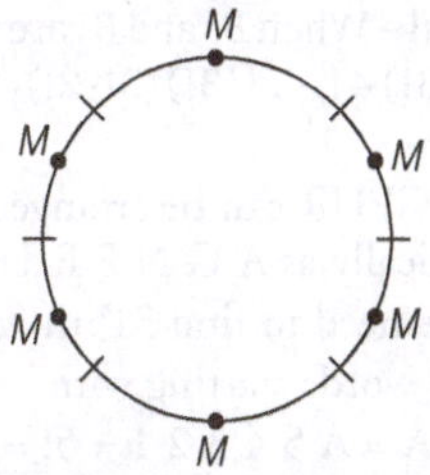


$\therefore$ Total number of ways

$=5!\times {}^6P_5=5!\times 6!$

**27.** *(c)* Let number formed be $x$.

$\therefore 1000<x<4000$, which means left extreme digit will be either 2 or 3.

$\therefore$ Required numbers

$=2\times 4\times 4\times 4=128$

**28.** *(d)* Total number of numbers formed using 0, 1, 2, 3, 5, 7 $=5\times 6\times 6\times 4$

$=36\times 20=720$

**29.** *(a)* $A, B, C, D$ and $E$ can be placed as per given condition, first of all we will find coefficient of $X^5$ in

$({}^3C_1X+{}^3C_2X^2+{}^3C_3X^3)^2$

$({}^2C_1X+{}^2C_2X^2)$

$=18+18+12=48$

$\therefore$ Total ways $=48\times 5!$ ($\because A, B, C, D$ and $E$ are different)

$=48\times 120$

$=5760$

**30.** *(a)* Exactly five 1s and exactly three 2s

$\Rightarrow$ 2 zeros must be added

$\Rightarrow$ 10 places to arrange 5 ones, 3 twos and 2 zeroes

$\Rightarrow \frac{10!}{2!3!5!}=2520$

Option (a) is correct.

**31.** *(c)* We have,

A, K, N, P, R, U

A .................. = 5! = 120

K .................. = 5! = 120

N .................. = 5! = 120

P A ............ = 4! = 24

P K ............ = 4! = 24

P N ............ = 4! = 24

P R A ...... = 3! = 6

P R K A N U ...... = 1

P R K A U N ...... = 1

Total $=440$

$\Rightarrow$ 440th word = PRKAUN

**32.** (a) Group : 3 Girls + 4 boys $<^{B_1}_{B_2}$

Number of required arrangements
$=$ Total $-$ When $B_1$ and $B_2$ are together
$= 2!(3!)4! - 2!(3!)(3!\cdot 2!)$
$= 144$

**33.** (c) NAGPUR can be arranged alphabetically as A G N P R U
Now, we need to find 315th word
So, total words starting with
A = A 5 4 3 2 1 = 5! = 120
Total words starting with
G = G 5 4 3 2 1 = 5! = 120
Total words starting with
N = N 5 4 3 2 1 = 5! = 120
Total words $= 120 + 120 + 120 = 360$
So, 315th words starts with N.
$\Rightarrow$ Now, words with N as first letter, and A as second letter
= N A 4 3 2 1 = 4! = 24
G as second letter
= N G 4 3 2 1 = 4! = 24
P as second letter
= N P 4 3 2 1 = 4! = 24
Total words so far
$= 240 + 24 \times 3 = 312$
So, we have to find next 3 words
$\Rightarrow$ N R A G P U
$\Rightarrow$ N R A G U P
$\Rightarrow$ N R A P G U $\rightarrow$ 315th words

**34.** (a) First arrange the letters of the word MONDAY in alphabetical order i.e. ADMNOY
$A$-----$=5!$
$D$-----$=5!$
Ⓜ $A$----$=4!$
Ⓜ $D$----$=4!$
ⓂⓃ----$=4!$
ⓂⓄ$A$---$=3!$
Ⓜ Ⓞ $D$---$=3!$
ⓂⓄⓃ $A$--$=2!$
ⓂⓄⓃⒹⒶⓎ $=327$
Rank $=327$

**35.** (c) To find the rank of word "THAMS"
We know that according to dictionary the letters would appear as A, H, M, S, T.
Words starting with letter "A"----
$= 4! = 24$
Words starting with letter "H"----
$= 4! = 24$
Words starting with letter "M" ----
$= 4! = 24$
Words starting with letter "S" ----
$= 4! = 24$
Words starting with letter TA ---
$= 3! = 6$
Words starting with letter THAMS $= 1$
Rank $= 103$
Hence, rank of word "THAMS" $= 103$

**36.** (a) Total number of words $= \dfrac{11!}{2!2!2!}$

$C$ and $S$ come together in $\dfrac{10!}{2!2!2!} \times 2!$ words

$\therefore$ Required number of words

$$= \frac{11!}{(2!)^3} - \frac{10!}{(2!)^2}$$

$$= \frac{10!}{(2!)^2}\left(\frac{11}{2} - 1\right) = \frac{10! \times 9}{8}$$

$$= 10 \times 9 \times 8 \times 7 \times 6! \times \frac{9}{8}$$

$$= 6! \times 5670 \Rightarrow k = 5670$$

**37.** (d) Given, word is PUBLIC and alphabetical order of letter is BCILPU, so number of words start with letter,
B is $5! = 120$
C -----is $5! = 120$
I -----is $5! = 120$
L -----is $5! = 120$
PB ----is $4! = 24$
PC ----is $4! = 24$
PI ----is $4! = 24$
PL ----is $4! = 24$
PUBC --is $2! = 2$
PUBI --is $2! = 2$
PUBLC - is $1! = 1$
PUBLIC - is $1! = 1$
$\therefore$ Serial number of the word PUBLIC
$= 4 \times 120 + 4 \times 24 + 2 \times 2 + 2 = 582$

**38.** (b) GHOTU : Alphabets involved are G, H, O, T, U
In dictionary order, these can be arranged as
G ----$\rightarrow 4! = 24$
H ----$\rightarrow 4! = 24$
O ----$\rightarrow 4! = 24$
TG ---$\rightarrow 3! = 6$
TH ----$\rightarrow 3! = 6$
TOG --$\rightarrow 2! = 2$
TOH --$\rightarrow 2! = 2$
TOUGH -$\rightarrow 1! = 1$
Total $= 89$

**39.** (d) Since a five-digit number is formed by using digits 0, 1, 2, 3, 4 and 5 divisible by 3, i.e. only possible when the sum of digits is multiple of three which gives two cases.
Case I Using digits 0, 1, 2, 4, 5 the number of ways $= 4 \times 4 \times 3 \times 2 \times 1 = 96$
Case II Using digits 1, 2, 3, 4, 5 the number of ways $= 5 \times 4 \times 3 \times 2 \times 1 = 120$
Therefore, total number formed
$= 120 + 96 = 216$

**40.** (a) Let $C_1$, $C_2$ and $C_3$ be the three cars in which 8 person are to be transported from city $A$ to city $B$.

$$\begin{matrix} C_1 & C_2 & C_3 \\ \downarrow & \downarrow & \downarrow \\ 3 & 3 & 2 \end{matrix}$$

Let the number of person transported in cars $C_1$, $C_2$ and $C_3$ are 3, 3 and 2 respectively.

There are total $\dfrac{8!}{3!\,3!\,2!}$ group.

$\therefore$ They can travel $\dfrac{8!}{3!3!2!} \times 3$ ways or 1680 ways

**41.** (a) Given, digits are $= 1, 2, 2, 3$
$\therefore$ Total distinct numbers $= 4!/2! = 12$
1 at unit place $\Rightarrow$ Number of such numbers $= \dfrac{3!}{2!} = 3$
2 at unit place $\Rightarrow$ Number of such numbers $= 3! = 6$
3 at unit place $\Rightarrow$ Number of such numbers $= 3!/2! = 3$
$\therefore$ Sum of digits at unit place is
$3 \times 1 + 6 \times 2 + 3 \times 3 = 24$
Hence, sum of all 4 digit such numbers
$= (3 + 12 + 9)(10^3 + 10^2 + 10 + 1)$
$= 26664$

**42.** (a) Let $n_1 = 1$, then $n_2$ can be 2, 3, ..., 9 and $n_3$ can be 3, ..., 10
$\therefore$ Number of ways
$= 8 + 7 + 6 + 5 + 4 + 3 + 2 + 1 = \dfrac{8 \times 9}{2}$
Similarly, when $n_1 = 2$, then $n_2$ can be 3, ..., 9 and $n_3$ can be 4, ..., 10.
$\therefore$ Number of ways
$= 7 + 6 + 5 + 4 + 3 + 2 + 1 = \dfrac{7 \times 8}{2}$
and so on.
$\therefore$ Total required ways

$$= \frac{8 \times 9}{2} + \frac{7 \times 8}{2} + \ldots + \frac{2 \times 3}{2} + \frac{1 \times 2}{2}$$

$$= \frac{240}{2} = 120$$

**43.** (d) We have E, E, N, Q, U
According to the English dictionary,
(i) Words starts with E
E _ _ _ _ $= 4! = 24$

(ii) Words starts with N

$N\_\_\_\_ = \frac{4!}{2!} = 12$

(iii) Words starts with QE

$QE\_\_\_ = 3! = 6$

(iv) Word starts with QN

$QN\_\_\_ = \frac{3!}{2!} = 3$

(v) Words QUEEN = 1

So, according to the English dictionary, the word QUEEN will be at $(24 + 12 + 6 + 3 + 1)$th position = 46th position

**44.** *(d)* Clearly, number of words start with $A = 4!/2! = 12$

Number of words start with

$L---- = 4! = 24$

Number of words start with

$M---- = \frac{4!}{2!} = 12$

Number of words start with

$SA--- = \frac{3!}{2!} = 3$

Number of words start with

$SL--- = 3! = 6$

Note that, next word will be 'SMALL'. Hence, the position of word 'SMALL' is 58th.

**45.** *(b)* In 8-digit numbers, 4-digit are odd places. Also, in the given 8 digits, there are three odd digits 1, 1 and 3. Number of ways in which three odd digits arranged at four even places

$= \frac{^4P_3}{2!} = \frac{4!}{2!}$

Number of ways the remaining five digits 2, 2, 2, 4 and 4 arranged at remaining five places $= \frac{5!}{3!2!}$

Hence, required number of 8-digit number $= \frac{4!}{2!} \times \frac{5!}{3!2!} = 120$

**46.** *(d)* In the word 'SACHIN', order of alphabets is A, C, H, I, N and S. Number of words start with $A = 5!$, so with C, H, I, N.

Now, words start with S and after that ACHIN are in ascending order of position, so $5 \cdot 5! = 600$ words are in dictionary before words start with S and position of this word is 601.

**47.** *(c)* Given that

$a \in \{2, 4, 6, 8, \ldots, 100\} \in$ even and

$b \in \{1, 3, 5, 7, \ldots, 99\} \in$ odd

$(a + b)_{max} = 100 + 99 = 199$

$(a + b)$ is divided by 23, remainder is 2.

$\Rightarrow \frac{a+b}{23} = \alpha + \frac{2}{23}, \ \alpha \in N$

$a + b = 23\alpha + 2$

$a + b = 25$

$a + b = 48$

$a + b = 71$

$a + b = 186$

$a + b = 94$

$a + b = 117$

$a + b = 140$

$a + b = 163$

On putting $\alpha = 1, 2, 3, \ldots, 8$, we get The sum of an even and an odd number cannot be even.

So, $a + b$ can be 25, 71, 117, 163.

When, $a + b = 25$

$\Rightarrow$ Number of ways = 12

$a + b = 71 \Rightarrow$ Number of ways = 35

$a + b = 117 \Rightarrow$ Number of ways = 42

$a + b = 163 \Rightarrow$ Number of ways = 19

So, total number of ways

$= 12 + 35 + 42 + 19 = 108$

**48.** *(a)*

**key idea** Use divisibility test of 11 and consider different situation according to given condition.

Since, the sum of given digits

$0 + 1 + 2 + 5 + 7 + 9 = 24$

Let the six-digit number be *abcdef* and to be divisible by 11, the difference of sum of odd placed digits and sum of even placed digits should be either 0 or a multiple of 11.

Hence, possible case is

$a + c + e = 12 = b + d + f$ (only)

Now, **Case I** set $\{a, c, e\} = \{0, 5, 7\}$ and set $\{b, d, f\} = \{1, 2, 9\}$

So, number of 6-digits numbers $= (2 \times 2!) \times (3!) = 24$ [$\because a$ can be selected in ways only either 5 or 7].

**Case II** Set $\{a, c, e\} = \{1, 2, 9\}$ and set $\{b, d, f\} = \{0, 5, 7\}$

So, number of 6-digits numbers

$= 3! \times 3! = 36$

So, total number of 6-digits numbers $= 24 + 36 = 60$

**49.** *(b)* The integer greater than 6000 may be of 4-digit or 5-digit. So, here two cases arise.

**Case I** When number is of 4 digit.

4-digit number can starts from 6, 7 or 8

6, 7 or 8 ↓

| 3 | 4 | 3 | 2 |

Thus, total number of 4-digit number, which are greater than 6000

$= 3 \times 4 \times 3 \times 2 = 72$

**Case II** When number is of 5-digit.

Total number of five digit number which are greater than $6000 = 5! = 120$

$\therefore$ Total number of integers $= 72 + 120 = 192$

**50.** *(182)* Alphabetical order = *A, B, C, D, E*

Words starting with $A = 4! = 24$

Words starting with $B = 4! = 24$

Words starting with $C$

Words starting with $CA = 3! = 6$

$CB = 3! = 6$

$CDA = 2! = 2$

$CDB = 2! = 2$

Serial number of $CDBEA = 63$

Let $P(W_1) = P \Rightarrow P(W_2) = 2P$,

$P(W_3) = 4P$ and so on

$\Rightarrow P(W_n) = 2^n P$

Total number of words $= 5! = 120$

Sum of probabilities is

$S_{120} = \frac{P(2^{120} - 1)}{2 - 1} = (2^{120} - 1)P$

Since, the sum of all probabilities is 1.

$P(2^{120} - 1) = 1$

$\Rightarrow \quad P = \frac{1}{2^{120} - 1}$

$(CDBEA) = P(w_{63}) = 2^{63-1} P = 2^{62} P$

$P(CDBEA) = 2^{62} \cdot \frac{1}{2^{120} - 1} = \frac{2^{62}}{2^{120} - 1}$

$\alpha = 62, \beta = 120$

$\therefore \quad \alpha + \beta = 182$

**51.** *(64)* 2 digit numbers divisible by 6

12, 18, 24 ......... 96 $\Rightarrow$ 15 terms.

$\because \gcd(m, n) = 6, m < n$

$\therefore$ Total number of possible ordered pair $(m, n)$

**Case I**

$m = 6k_1$

$n = 6k_2$

| $k_1$ | $k_2$ | |
|---|---|---|
| 2 | 3, 5, 7, 9, 11, 13, 15 | 7 |
| 4 | 5, 7, 9, 11, 13, 15 | 6 |
| 6 | 7, 11, 13, | 3 |
| 8 | 9, 11, 13, 15 | 4 |
| 10 | 11, 13, | 2 |
| 12 | 13 | 1 |
| 14 | 15 | 1 |

Total pairs = 24

Case II

$m = 6k_1$

$n = 6k_2$

| $k_1$ | $k_2$ | |
|---|---|---|
| 3 | 4, 5, 7, 8, 10, 11, 13, 14, 16 | 9 |
| 5 | 6, 7, 8, 9, 11, 12, 13, 14, 16 | 9 |
| 7 | 8, 9, 10, 11, 12, 13, 15, 16 | 8 |
| 9 | 10, 11, 13, 14, 16 | 5 |
| 11 | 12, 13, 14, 15, 16 | 5 |
| 13 | 14, 15, 16 | 3 |
| 15 | Total pairs =1/40 | |

$\therefore$ Total number of pairs = 64

**52.** *(14)* The first digit can be any number from 1 to 9 (9 options)

The other 6 digits can be any number from 0 to 9 (10 options each)

Total number of 7 digits numbers $= 9 \times 10^6$

Half of all number have even sum of digits

The number of 9-digit number with an even sum of digits

$$= \frac{9 \cdot 10^6}{2} = (4 \cdot 5) \times 10^6$$

$$4 \cdot 5 \times 10^6 = 9 \cdot 5 \cdot 10^5 = m \cdot n \cdot 10^n$$

$$m = 9, n = 5$$

$$\therefore \quad m + n = 14$$

**53.** *(17280)* 5 boys and 4 girls

$E_1$ = All boys sit together.

$E_2$ = No 2 boys sit together

$E_1 \cap E_2 = \phi$

$n(E_1) = 5!5!$

$n(E_2) = 4!\ 5!$

$n(E_1 \cap E_2) = 0$

$\therefore n(E_1 \cup E_2) = n(E_1) + n(E_2)$

$= 5!5! + 5!4!$

$= 6!\ 4! = 17280$

**54.** *(553)* Given, word GTWENTY

Words start with E $= \frac{6!}{2!} = 360$

Words start with GE $= \frac{5!}{2!} = 60$

Words start with GN $= \frac{5!}{2!} = 60$

Words start with GTE = 4! = 24

Words start with GTN = 4! = 24

Words start with GTT = 4! = 24

$\therefore$ Serial number of the word GTWENTY

$= 360 + 60 + 60 + 24 + 24 + 24 + 1$

$= 553$

**Alternate Solution**

| Increasing Position of letters in the word | G | T | W | E | N | T | Y |
|---|---|---|---|---|---|---|---|
| | 2 | 4 | 5 | 1 | 3 | 4 | 6 |
| | 6! | 5! | 4! | 3! | 2! | 1! | 0! |

$$= \frac{1\times 6!}{2!} + \frac{2\times 5!}{2!} + \frac{3\times 4!}{1!} + 0 + 0 + 0 + 0 + 1$$

$$= 360 + 120 + 72 + 1 = 553$$

**55.** *(16)* For 3-digit number to be divisible by 6, unit digit should be even and sum of digit is divisible by 3.

(2, 1, 3), (2, 3, 4), (2, 5, 5), (2, 2, 5), (2, 2, 2), (4, 1, 1), (4, 4, 1), (4, 4, 4), (4, 3, 5)

(2, 1, 3) ⇒ 132, 312

(2, 3, 4) ⇒ 342, 432, 324, 234

(2, 5, 5) ⇒ 552

(2, 2, 5) ⇒ 522, 252

(2, 2, 2) ⇒ 222

(4, 1, 1) ⇒ 114

(4, 4, 1) ⇒ 414, 144

(4, 4, 4) ⇒ 444

(4, 3, 5) ⇒ 354, 534

Required number is 16.

**56.** *(26664)* We have, four digits are 2, 1, 2, 3

Total numbers when 1 is at unit digit $= \frac{3!}{2!} = 3$

Total number when 2 is at unit digit $= 3! = 6$

Total numbers when 3 is at unit digit $= \frac{3!}{2!} = 3$

Sum of digits at unit place $= 3 \times 1 + 6 \times 2 + 3 \times 3 = 24$

$\therefore$ Required sum

$= 24 \times 1000 + 24 \times 100 + 24 \times 10 + 24 \times 1$

$= 24 \times 1111 = 26664$

**57.** *(4898)* Given that, digits are 1, 2, 3, 4, 5, 6, 7

Total permutations = 7!

Let $p$ = Number which can taking string 153

$q$ = Number which containing string 2467

$\therefore \quad n(p) = 5! \times 1 \Rightarrow n(q) = 4! \times 1$

$n(p \cap q) = 2!$

$\therefore \ n(p \cup q) = n(p) + n(q) - n(p \cap q)$

$= 5! + 4! - 2! = 120 + 24 - 2 = 142$

$\therefore n$ (neither string 143 nor string 2467)

$= 7! - 142 = 5040 - 142 = 4898$

**58.** *(81)* Taking single digit only one number can be formed, i.e. 444444.

Taking two digits, we get

(i) If (4, 5) are used, then

Total numbers formed $= \frac{5!}{3!\,2!} \times 1 = 10$

(ii) If (4, 9) are used, then

Total numbers formed $= \frac{5!}{3!\,2!} \times 1 = 10$

Taking three digits, we get

(i) 4, 5, 9, 4, 4, 4 are used, then

Total numbers formed = 5!/3! = 20

(ii) 4, 5, 9, 5, 5,5, are used, then

Total numbers formed = 5!/4! = 5

(iii) 4, 5, 9, 9, 9, 9 are used, then

Total numbers formed $= \frac{5!}{4!} = 5$

(iv) 4, 5, 9, 4, 5, 9 are used, then

Total numbers formed $= \frac{5!}{2!\,2!} = 30$

$\therefore$ Total numbers formed

$= 1 + 10 + 10 + 20 + 5 + 5 + 30 = 81$

**59.** *(50400)* In 'ASSASSINATION', vowels are AAIAIO.

$\therefore$ Number of words in which all vowels are together $= \frac{8!}{4!\,2!} \times \frac{6!}{3!\,2!}$

$$= \frac{8 \times 7 \times 6 \times 5}{2} \times \frac{6 \times 5 \times 4}{2}$$

$$= 840 \times 60 = 50400$$

**60.** *(2997)* Number starting with 2 and 3 $= 2 \times 6^4$

Number starting with 40 $= 6^3$

Number starting with 420, 422, 423, 424, 427 $= 5 \times 6^2$

Number starting with 4290 = 6

Next numbers are 42920, 42922, 42923

So, the serial number of 42923

$= 2 \times 6^4 + 6^3 + 5 \times 6^2 + 6 + 3 = 2997$

**61.** *(21)* If last place is fixed by 5.

So, possible combinations are shown below,

| 1 2 1 5 | 2 3 5 5 | 3 5 5 5 |
|---|---|---|
| Total=3 | Total=6 | Total=3 |
| 2 2 3 5 | 1 1 5 5 | 3 3 1 5 |
| Total=3 | Total=3 | Total=3 |

Hence, total number of 4-digits numbers which are also divisible by $15 = 3 + 6 + 3 + 3 + 3 + 3 = 21$

**Concept Enhancer**

For number to be divisible by 15, last digit of number should be 5 and the sum of digits must be divisible by 3.

**62.** *(1436)* We have, 5-digit numbers formed using 1, 2, 3, 5, 7 with repetitions and written in decreasing order.

Numbers starting with 7 .... $=5^4$

Numbers starting with 5 .... $=5^4$

Numbers starting with $3\cdot 7$ ... $=5^3$

Numbers starting with $3\cdot 5\cdot 7$ .. $=5^2$

Numbers starting with $3\cdot 5\cdot 5$ .. $=5^2$

Numbers starting with $3\cdot 5\cdot 3\cdot 7$ . $=5^1$

Numbers starting with $3\cdot 5\cdot 3\cdot 5$ . $=5^1$

Number starting with $3\cdot 5\cdot 3\cdot 3\cdot 7=5^0$

$\therefore$ Serial number of 3 5 3 3 7

$=2\cdot 5^4+5^3+2\cdot 5^2+2\cdot 5^1+5^0=1436$

**63.** *(60)*

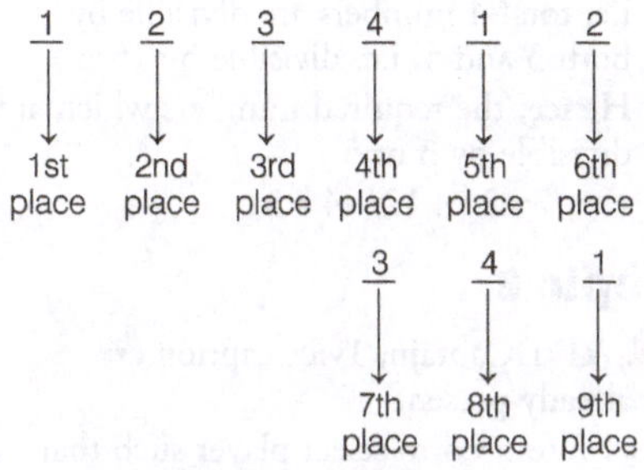


Even digits occupy at even places

$=\dfrac{4!}{2!\,2!}\times\dfrac{5!}{3!\,2!}=6\times 10=60$

**64.** *(6)* To find the numbers lying between 1012 and 23421 using digits 2, 3, 4, 5, 6 and divisible by 55. (repetition of digits is not allowed)

For divisibility by 55, number should be divisible by 5 and 11 both.

Finding 4 digit numbers

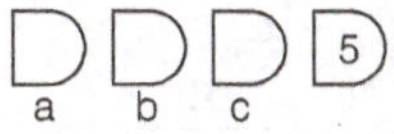


For divisibility by 11, sum of digits at odd places = Sum of digits at even places

$a+c=b+5$

For $b=2$, $a=3, c=4$

$a=4, c=3$

For $b=3$, $a=6, c=2$

$a=2, c=6$

For $b=4$, $a=6, c=3$

$a=3, c=6$

Also, 5-digit number is not possible.

So, total 6 possible four digit numbers are possible.

**65.** *(30)* The digit at thousand place can be either 1 or 2.

**Case I** When the digit at thousand place is 1, then

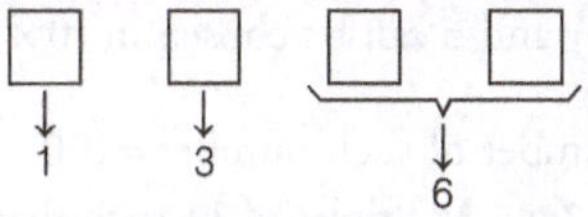


The last two digit can be 24, 32, 36, 52, 56, 64

**Case II** When the digit at thousand place is 2, then

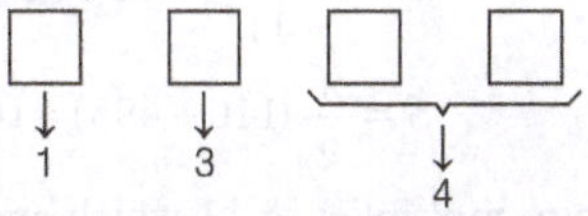


The last two digit can be 16, 36, 56, 64

$\therefore$ The total number of such numbers

$=1\times 3\times 6+1\times 3\times 4=18+12=30$

**66.** *(180)* Factors of 36 is 1, 2, 3, 4, 6, 9, 12, 18, 36 or $1\times 2^2\times 3^2$

**Case I** When digits are (1, 2, 2, 3, 3).

The number of such combinations is

$\dfrac{5!}{2!\,2!}=\dfrac{5\times 4\times 3}{1\times 2}=30$

**Case II** When digits are (1, 4, 3, 3, 1).

The number of such combination is

$\dfrac{5!}{2!\,2!}=\dfrac{5\times 4\times 3}{1\times 2}=30$

**Case III** When digits are (1, 9, 2, 2, 1).

The number of such combinations

$=\dfrac{5!}{2!\,2!}=30$

**Case IV** When digits are (1, 4, 9, 1, 1).

The number of such combination is

$5!/3!=20$

**Case V** When digits are (1, 2, 3, 6, 1).

The number of such combination is

$\dfrac{5!}{2!}=5\times 4\times 3=60$

**Case VI** When digits are (1, 6, 6, 1, 1).

The number of such combination is

$\dfrac{5!}{3!\,2!}=10$

$\therefore$ The total number of combinations are

$30+30+30+20+60+10=180$

**67.** *(1492)* The letters of the word 'MANKIND' are written in all possible order and arranged in serial order as in an english dictionary.

| Represent serial order in dictionary | ⑤ M | ① A | ⑥ N | ④ K | ③ I | ⑥ N | ② D |
|---|---|---|---|---|---|---|---|
| | 4/2! | 0 | 3/2! | 2 | 1 | 1 | 0 |
| | × | × | × | × | × | × | × |
| | 6! | 5! | 4! | 3! | 2! | 1! | 0! |
| Outcomes | 1440 | 0 | 36 | 12 | 2 | 1 | 0 |

Thus, the serial number of the word MANKIND is

$(1440+0+36+12+2+1)+1$

$=1491+1=1492$

**68.** *(360)* Given digits are 1, 2, 3, 4, 5, 6, 7.

Total positions are 6 with fixed unit digit.

The sum of digits 1, 2, 3, 4, 5, 6, 7 is 28. So, only 1, 4 and 7 can be removed. If 1 is removed, then the total number of numbers are 5!. Similarly, for 4 and 7.

Total number of ways $=5!\times 3=360$

**69.** *(1086)*

> **key idea** Assume 4-digit number is *abcd* and make cases for $d=1$ to 9 and solve further using multiplication principle of counting.

Let the four digit number be *abcd*.

It is given that *a*, *b* and *c* should be divisible by *d*.

Numbers *a*, *b*, *c* and *d* are chosen from 0, 1, 2, 3, 4, 5, 6, 7, 8, 9.

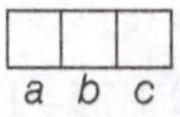


Number of such numbers

$d=1$ | 9 | 10 | 10 | = 900 [$\because$ first entry cannot be zero]

$d=2$ | 4 | 5 | 5 | = 100

$d=3$ | 3 | 4 | 4 | = 48

$d=4$ | 2 | 3 | 3 | = 18

$d=5$ | 1 | 2 | 2 | = 4

$d=6, 7, 8, 9$ | 1 | 2 | 2 | $\times 4=16$

Required total number

$=900+100+48+18+4+16$

$=1086$

**70.** *(18915)* $b_i\in\{1, 2, 3, \ldots, 100\}$

Let $P=$ Set when $b_1, b_2, b_3$ are consecutive

$n(P)=97+97+97+\ldots$ (98 times)

$=97\times 98$

Similarly, $Q=$ Set when $b_2, b_3, b_4$ are consecutive

$n(Q)=97\times 98$

$n(P\cap Q)=97$

$n(P\cup Q)=n(P)+n(Q)-n(P\cap Q)$

$=97\times 98+97\times 98-97$

$=97(98+98-1)$

$=97(195)=18915$

Hence, the number of required permutations is 18915.

**71.** *(150)* Total 3-digit number $=900$

Required number of numbers (a number whose GCD with 36 is 2)
$=$ (3- digit even number) $-$ (3-digit number that are multiples of 4) $-$ (3-digit number that are multiples of 6) $+$ (3-digit number that are multiples of 12)

3-digit even number $=\frac{900}{2}=450$

3- digit number divisible by 4,
100, 104, 108, ……, 996

$\therefore \quad 996=100+(n-1)4$

$\Rightarrow \quad \frac{896}{4}=n-1 \Rightarrow n=225$

3-digit number divisible by 6,
102, 108, 114, ……, 996

$\therefore \quad 996=102+(n-1)6$

$\frac{894}{6}=n-1 \Rightarrow n=150$

3-digit number divisible by 12
108, 120, ……996

$\therefore \quad 996=108+(n-1)12$

$\frac{888}{12}=n-1 \Rightarrow n=75$

Thus, required number
$=450-225-150+75=150$

> **Shortcut Method**
> Total 3-digit numbers $=900$
> Required number of numbers
> $=$ (3-digit even number) $-$ (multiple of 4) $-$ (multiple of 6) $+$ (multiple of 12)
> $=450-225-150+75=150$

**72.** *(576)* VOWELS (2 Vowel + 4 consonant)
All consonants must not be together
Total possibility of formation of 6 letter word $=6!$
The number of arrangement when all the consonant comes together $=3!\times 4!$
Number of arrangement when all the consonant never come together
$=$ Total $-$ All consonant together
$=6!-3!\,4!=576$

**73.** *(100)* Form of six digit palindrome number

*xyzzyx*

This will be divisible by 55.
Hence, $x=5$
and $5yzzy5$ will be divisible by 11.
$\Rightarrow (5+z+y)-(y+z+5)$ is divisible by 11
which is true for all values of $y$ and $z$
$\Rightarrow y$ and $z$ can be chosen in $10\times 10$ ways
Number of such number $=100$

**74.** *(7744)* Multiples of 11 such that they are of 3-digit and less than 500.
110, 121, 132, ..., 495

$$n=\frac{495-110}{11}+1=36$$

$$S=\frac{36}{2}(110+495)=10890$$

Again, multiples of 11 which are 3-digits, less than 500 and having 1 at hundred's place are 110, 121, 132, ..., 198

$$n_1=\left(\frac{198-110}{11}\right)+1=9$$

$$S_1=\frac{9}{2}(110+198)=1386$$

The multiply of 11 which are of 3-digits, less than 500 and having 1 at ten's place are 319, 418
$\therefore S_2=319+418=737$
The multiple of 11 which are 3-digits, less than 500 and having 1 at unit place are 231, 341, 451
$\therefore\ S_3=231+341+451=1023$
$\therefore$ Required sum $=S-S_1-S_2-S_3$
$=7744$

**75.** *(52)* **Case I** When 0 is at unit place
Total number $=5\times 4=20$
**Case II** When 4 or 6 are at unit place
Total numbers $=4\times 4\times 2=32$
[0 cannot come at place]
$\therefore$ Total number of required
$=20+32=52$

**76.** *(96)* 0, 2, 4, 6, 8

| 2/4/6/8 | | | | |
|---|---|---|---|---|
| 4 options × | 4 options × | 3 | ×2 × | 1 |

$\therefore$ Total $=4\times 4\times 3\times 2\times 1=96$

**77.** *(4)* As, we observe the pattern
Inside number $=(\text{difference})^{(\text{difference})!}$
i.e. $1=(2-1)^{(2-1)!},\ 4^{24}=(12-8)^{(12-8)!}$,
$3^6=(7-4)^{(7-4)!}$
$\therefore$ Required number $=2^{2!}=4$

**78.** *(32)* Given, digits $=\{1, 2, 3, 4, 5\}$
Numbers divisible by 3 (sum of digits divisible by 3).

| | | |
|---|---|---|
| | | |

**Case I** When sum is
$12\rightarrow 3, 4, 5\rightarrow 3!=6$
**Case II** When sum is
$9\rightarrow 2, 3, 4\rightarrow 3!=6$
**Case III** When sum is
$9\rightarrow 1, 3, 5\rightarrow 3!=6$
**Case IV** When sum is
$6\rightarrow 1, 2, 3\rightarrow 3!=6$
So, total numbers divisible by $3=6\times 4=24$
Numbers divisible by 5 (ending with 5)

| | | 5 |
|---|---|---|
| 4×3 | | 1 |

$=4\times 3=12$

So, total numbers divisible by $5=12$
Numbers divisible by 15, are 135, 315, 345, 435
i.e. total 4 numbers are divisible by both 3 and 5. i.e. divisible by 15.
Hence, the required numbers which are divisible by 3 or 5
$=24+12-4=32$

## Topic 2

**01.** *(a)* 1 Captain, 1vice caption are already present
$\Rightarrow$ We need to select player such that atleast 3 batsman and bowler must be there.

| Batsman | Bowler | Number of ways |
|---|---|---|
| 3 | 5 | ${}^6C_3\cdot{}^5C_5=20$ |
| 4 | 4 | ${}^6C_4\cdot{}^5C_4=75$ |
| 5 | 3 | ${}^6C_5\cdot{}^5C_3=60$ |

Total $=155$

**02.** *(b)* Total number of triangles
$$={}^{12}C_3-{}^5C_3$$
$$=\frac{12!}{3!\times 9!}-\frac{5!}{3!\times 2!}$$
$$=\frac{12\times 11\times 10}{6}-\frac{5\cdot 4}{2}$$
$$=20\times 11-10$$
$$=220-10=210$$
Option (b) is correct.

**03.** *(d)* Total alphabet $=26$

(12 alphabets) M (13 alphabets)

Remaining alphabet $=25$
Now, arrangement in alphabetical order
$$={}^{12}C_2\times{}^{13}C_2\times 1$$
$$=\frac{12\times 11}{2\times 1}\times\frac{13\times 12}{2\times 1}=5148$$

**04.** (c)

Group A: 7B, 3G → 5 taken from group A

Group B: 6B, 5G → 3 taken from group B

**Case I** (3$G$ and 2$B$) from group $A$ and (1$G$ and 2$B$) from group $B$.

**Case II** (2$G$ and 3$B$) from group $A$ and (2$G$ and 1$B$) from group $B$

**Case III** (1$G$ and 4$B$) from group $A$ and (3$G$ and 0$B$) from group $B$.

So, total = Case I + Case II + Case III

$= {}^7C_2 \times {}^3C_3 \times {}^5C_1 \times {}^6C_2 + {}^7C_3 \times {}^3C_2 \times {}^6C_1 \times {}^5C_2 + {}^7C_4 \times {}^3C_1 \times {}^6C_0 \times {}^5C_3$

$= 8925$

**05.** (c) AA, MM, TT, H, I, C, S, E

(a) All distinct $= {}^8C_5 = \frac{8\cdot7\cdot6}{3\cdot2\cdot1} = 56$

(b) 2 same 1st kind, 2 same 2nd kind, 1 different

$= {}^3C_2 \times {}^6C_1 = 3 \times 6 = 18$

(c) 2 same, 3 different

$= {}^3C_1 \times {}^7C_4 = 3 \times \frac{7\cdot6\cdot5}{3\cdot2\cdot1} = 105$

Hence, the total $= 56 + 18 + 105 = 179$

**06.** (a) Since, office are identical so we can distributes in following ways.

| I | II | III | IV | |
|---|---|---|---|---|
| 0 | 0 | 0 | 5 | $1 = 1$ |
| 0 | 0 | 1 | 4 | $\frac{5!}{4!1!} = 5$ |
| 0 | 1 | 1 | 3 | $\frac{5!}{3!2!} = 10$ |
| 0 | 1 | 2 | 2 | $\frac{5!}{1!2!2!} \times \frac{1}{2!} = 15$ |
| 1 | 1 | 1 | 2 | $\frac{5!}{1!1!1!2!} \times \frac{1}{3!} = 10$ |
| 0 | 0 | 2 | 3 | $\frac{5!}{3!2!} = 10$ |

$\therefore$ Total $= 1 + 5 + 10 + 15 + 10 + 10 = 51$

**07.** (c) Remaining apples after distributing 2 apples to each of three child

$= 21 - 6 = 15$

Number of ways $= {}^{15+3-1}C_{3-1}$

$= {}^{17}C_2 = \frac{17 \times 16}{2}$

$= 17 \times 8 = 136$

**08.** (c) We have, $x + y + z = 15$

Total number of solution $= {}^{15+3-1}C_{3-1}$

$= {}^{17}C_2 = \frac{17 \times 16}{2} = 136$

Now consider the case when $x = y \neq z$

| x | y | z |
|---|---|---|
| 1 | 1 | 13 |

i.e. $\frac{3!}{2!} = 3$

Other cases are

| | | |
|---|---|---|
| 2 | 2 | 11 |
| 3 | 3 | 9 |
| 7 | 7 | 1 |
| 4 | 4 | 7 |
| 6 | 6 | 3 |
| 0 | 0 | 15 |

$\therefore$ Total number of solution when $x = y \neq z$

$= 7 \times \frac{3!}{2!} = 21$

Now consider the case when $x = y = z$

i.e.

| | | |
|---|---|---|
| 5 | 5 | 5 |

$= 1$

$\therefore$ Total number of solution in which $x = y = z = 1$

$\therefore$ Required answer $= 136 - 21 - 1 = 114$

**09.** (b) Given, $n(A) = 5$ and $n(B) = 2$

Then, $n(A \times B) = 5 \times 2 = 10$

$\therefore$ Number of subsets

$= {}^{10}C_3 + {}^{10}C_4 + {}^{10}C_5 + {}^{10}C_6$

$= 120 + 210 + 252 + 210 = 792$

**10.** (d) Given that, number of boys = 7

Number of girls = 5

Number of ways boys can sit

$= (7-1)! = 6! = 720$

There are 7 gaps after sitting of boys in which 5 girls have to sit.

Total number of ways

$= 720 \times {}^7C_5 \cdot 5!$

$= 720 \times \frac{7 \times 6}{1 \times 2} \times 120$

$= 7 \times 6 \times 3 \times 120 \times 120$

$= 126\,(5!)^2$

**11.** (b)

> **key idea** (i) Number of ways to select $r$ persons out of $n$ persons $= {}^nC_r$
>
> (ii) Add the number of ways possible in each of the cases.

Given,

Number of Indians = 6

Number of foreigners = 8

Committee of at least 2 Indians and double number of foreigners is to be formed. Hence, the required cases are

$(2I, 4F) + (3I, 6F) + (4I, 8F)$

$= {}^6C_2 \times {}^8C_4 + {}^6C_3 \times {}^8C_6 + {}^6C_4 \times {}^8C_8$

$= (15 \times 70) + (20 \times 28) + (15 \times 1)$

$= 1050 + 560 + 15 = 1625$

**12.** (b) To make 6-digit numbers from given digits 1, 3, 5, 7 and 9, we must repeat a digit and we can done the same in ${}^5C_1$ ways.

Now, the arrangement of these 6-digits in which two are identical is $\frac{6!}{2!}$.

So, required numbers of 6-digit numbers

$= {}^5C_1 \frac{6!}{2!} = \frac{5}{2}(6!)$

**13.** (a) Given 6 different novels and 3 different dictionaries.

Number of ways of selecting 4 novels from 6 novels is ${}^6C_4 = \frac{6!}{2!4!} = 15$

Number of ways of selecting 1 dictionary from 3 dictionaries is

${}^3C_1 = \frac{3!}{1!2!} = 3$

$\therefore$ Total number of arrangement of 4 novels and 1 dictionary where dictionary is always in the middle, is

$15 \times 3 \times 4! = 45 \times 24 = 1080$

**14.** (a) Given word is MISSISSIPPI.

Here, I = 4 times, S = 4 times,

P = 2 times  M = 1time

_M_I_I_I_I_P_P_

$\therefore$ Required number of words

$= {}^8C_4 \times \frac{7!}{4!\,2!}$

$= {}^8C_4 \times \frac{7 \times 6!}{4!\,2!} = 7 \cdot {}^8C_4 \cdot {}^6C_4$

**15.** (c) Required number of ways

$= {}^{12}C_4 \times {}^8C_4 \times {}^4C_4$

$= \frac{12!}{8! \times 4!} \times \frac{8!}{4! \times 4!} \times 1 = \frac{12!}{(4!)^3}$

**16.** (b) Total number of ways

$= {}^{10}C_1 + {}^{10}C_2 + {}^{10}C_3 + {}^{10}C_4$

$= 10 + 45 + 120 + 210 = 385$

**17.** (b) The required number of ways

$= {}^{8-1}C_{3-1} = {}^7C_2$

$= \frac{7!}{2!5!} = \frac{7\cdot6}{2\cdot1} = 21$

**18.** (b) The number of choices available to him

$= {}^5C_4 \times {}^8C_6 + {}^5C_5 \times {}^8C_5$

$= 5 \times 28 + 1 \times 56$
$= 140 + 56 = 196$

**19.** *(c)* Given, $\dfrac{^{n+1}C_{r+1}}{^{n}C_r} = \dfrac{55}{35}$ ...(i)

and $\dfrac{^{n}C_r}{^{n-1}C_{r-1}} = \dfrac{35}{21}$ ...(ii)

Now, by using property

$$^{n}C_r = \frac{n}{r}\,^{n-1}C_{r-1}$$

By Eq. (i), $\dfrac{n+1}{r+1} \times \dfrac{^{n}C_r}{^{n}C_r} = \dfrac{55}{35}$

$\Rightarrow \dfrac{n+1}{r+1} = \dfrac{11}{7}$

$\Rightarrow 7n + 7 = 11r + 11$

$\Rightarrow 7n = 11r + 4$ ...(iii)

By Eq. (ii),

$\dfrac{n}{r}\dfrac{^{n-1}C_{r-1}}{^{n-1}C_{r-1}} = \dfrac{35}{21}$

$\Rightarrow \dfrac{n}{r} = \dfrac{5}{3}$

$\Rightarrow 3n = 5r$ ...(iv)

By Eqs. (iii) and (iv),

$n = 10$ and $r = 6$

$\Rightarrow 2n + 5r = 2 \times 10 + 5 \times 6 = 50$

**20.** *(a)*

**key idea** Use, $^{n}C_r + {^{n}C_{r-1}} = {^{n+1}C_r}$

Given,

$^{6}C_m + 2(^{6}C_{m+1}) + {^{6}C_{m+2}} > {^{8}C_3}$

$\Rightarrow (^{6}C_m + {^{6}C_{m+1}}) + (^{6}C_{m+1} + {^{6}C_{m+2}}) > {^{8}C_3}$

$\Rightarrow {^{7}C_{m+1}} + {^{7}C_{m+2}} > {^{8}C_3}$

${^{8}C_{m+2}} > {^{8}C_3}$

$m = 2$ and $\dfrac{^{n-1}P_3}{^{n}P_4} = \dfrac{1}{8}$

$$\frac{\dfrac{(n-1)!}{(n-4)!}}{\dfrac{n!}{(n-4)!}} = \frac{1}{8}$$

$\Rightarrow \dfrac{1}{n} = \dfrac{1}{8} \Rightarrow n = 8$

$^{n}P_{m+1} + {^{n+1}C_m} = {^{8}P_3} + {^{9}C_2}$

$= 8 \times 7 \times 6 + \dfrac{9 \times 8}{2}$

$= 336 + 36 = 372$

**21.** *(a)* Given, $^{n-1}C_r = (k^2 - 8)\,^{n}C_{r+1}$ ...(i)

$\Rightarrow {^{n-1}C_r} = (k^2 - 8)\dfrac{n}{r+1} \cdot {^{n-1}C_r}$

$\left[\because {^{n}C_r} = \dfrac{n}{r}\,{^{n-1}C_{r-1}}\right]$

$\Rightarrow \dfrac{1}{k^2 - 8} = \dfrac{n}{r+1}$ ...(ii)

From Eq. (i) to defined $^{n}C_{r+1}$,

$n \ge r + 1 \Rightarrow \dfrac{n}{r+1} \ge 1$

$\Rightarrow \dfrac{1}{k^2 - 8} \ge 1$

$\Rightarrow \dfrac{1 - (k^2 - 8)}{k^2 - 8} \ge 0$

$\Rightarrow \dfrac{k^2 - 9}{k^2 - 8} \le 0$

$\Rightarrow \dfrac{(k-3)(k+3)}{(k - 2\sqrt{2})(k + 2\sqrt{2})} \le 0$

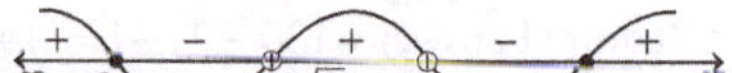


$\Rightarrow k \in [-3, -2\sqrt{2}) \cup (2\sqrt{2}, 3]$

So, $k \in (2\sqrt{2}, \sqrt{3}]$ [by given option]

**22.** *(b)* We have, $^{2n}C_3 : {^{n}C_3} = 10 : 1$

$\Rightarrow \dfrac{^{2n}C_3}{^{n}C_3} = \dfrac{10}{1}$

$\Rightarrow \dfrac{(2n)!}{3!(2n-3)!} \times \dfrac{3!(n-3)!}{n!} = \dfrac{10}{1}$

$\Rightarrow \dfrac{(2n)(2n-1)(2n-2)}{(n)(n-1)(n-2)} = \dfrac{10}{1}$

$\Rightarrow 4n^2 - 6n + 2 = 5(n^2 - 3n + 2)$

$\Rightarrow n^2 - 9n + 8 = 0$

$\Rightarrow (n-8)(n-1) = 0$

$\Rightarrow n = 8$ ($n = 1$ not valid)

$\therefore \dfrac{n^2 + 3n}{n^2 - 3n + 4} = \dfrac{88}{44} = \dfrac{2}{1}$

$= 2 : 1$

**23.** *(c)* We have,

$\dfrac{1}{1!50!} + \dfrac{1}{3!48!} + \dfrac{1}{5!46!} + \ldots + \dfrac{1}{49!2!} + \dfrac{1}{51!1!}$

$= \dfrac{1}{1!50!} + \dfrac{1}{3!48!} + \dfrac{1}{5!46!} + \ldots + \dfrac{1}{49!2!} + \dfrac{1}{51!0!}$ $[\because 0! = 1! = 1]$

$= \dfrac{1}{51!}\left[\dfrac{51!}{1!50!} + \dfrac{51!}{3!48!} + \dfrac{51!}{5!46!} + \ldots + \dfrac{51!}{49!2!} + \dfrac{51!}{51!0!}\right]$

$= \dfrac{1}{51!}(^{51}C_1 + {^{51}C_3} + \ldots + {^{51}C_{51}})$

$= \dfrac{1}{51!}\left(\dfrac{2^{51}}{2}\right)$

$\left[\because {^{n}C_1} + {^{n}C_3} + \ldots + {^{n}C_n} = \dfrac{2^n}{2}\right] = \dfrac{2^{50}}{51!}$

**Shortcut Method**

$\dfrac{1}{1!50!} + \dfrac{1}{3!48!} + \dfrac{1}{5!46!} + \ldots + \dfrac{1}{5!1!}$

$= \dfrac{1}{51!}\left(\dfrac{2^{51}}{2}\right) = \dfrac{2^{50}}{51!}$

**24.** *(d)* Given, $xyz = 24$

$\Rightarrow xyz = 2^3 \cdot 3^1$

Let $x = 2^{a_1} \cdot 3^{b_1}$,

$y = 2^{a_2} \cdot 3^{b_2}$,

$z = 2^{a_3} \cdot 3^{b_3}$

where, $a_1, a_2, a_3 \in \{0, 1, 2, 3\}$

$b_1, b_2, b_3 \in \{0, 1\}$

**Case I** $a_1 + a_2 + a_3 = 3$

$\therefore$ Non-negative solution

$= {^{3+3-1}C_{3-1}} = {^{5}C_2} = 10$

**Case II** $b_1 + b_2 + b_3 = 1$

$\therefore$ Non-negative solution

$= {^{1+3-1}C_{3-1}} = {^{3}C_2} = 3$

$\therefore$ Total solutions $= 10 \times 3 = 30$

**25.** *(d)* As each section has 5 questions, so number of ways to select 5 questions are $S_1 \quad S_2 \quad S_3$

| | | | |
|---|---|---|---|
| $S_1 \rightarrow$ | 1 | 1 | 3 |
| $S_2 \rightarrow$ | 1 | 3 | 1 |
| $S_3 \rightarrow$ | 3 | 1 | 1 |

and

| | | | |
|---|---|---|---|
| $S_1 \rightarrow$ | 1 | 2 | 2 |
| $S_2 \rightarrow$ | 2 | 1 | 2 |
| $S_3 \rightarrow$ | 2 | 2 | 1 |

$\therefore$ Total number of selection of 5 questions

$= 3 \times (^{5}C_1 \times {^{5}C_1} \times {^{5}C_3})$

$+ 3 \times (^{5}C_1 \times {^{5}C_2} \times {^{5}C_2})$

$= 3(5 \times 5 \times 10) + 3(5 \times 10 \times 10)$

$= 750 + 1500 = 2250$

**26.** *(c)* It is given that a group of students comprises of 5 boys and $n$ girls. The number of ways, in which a team of 3 students can be selected from this group such that each team consists of at least one boy and at least one girls, is
= (number of ways selecting one boy and 2 girls) + (number of ways selecting two boys and 1 girl)

$$=({}^5C_1 \times {}^nC_2)+({}^5C_2 \times {}^nC_1)=1750 \quad \text{[given]}$$

$$\Rightarrow \left(5\times \frac{n(n-1)}{2}\right)+\left(\frac{5\times 4}{2}\times n\right)=1750$$

$$\Rightarrow \quad n(n-1)+4n=\frac{2}{5}\times 1750$$

$$\Rightarrow \quad n^2+3n-700=0$$

$$\Rightarrow \quad (n-25)(n+28)=0$$

$$\Rightarrow \quad n=25 \quad [\because n \in N]$$

**27.** *(c)* Since there are 8 males and 5 females. Out of these 13 members committee of 11 members is to be formed. According to the question, $m$ = number of ways when there is at least 6 males

$$=({}^8C_6 \times {}^5C_5)+({}^8C_7 \times {}^5C_4)+({}^8C_8 \times {}^5C_4)$$

$$=(28\times 1)+(8\times 5)+(1\times 10)$$

$$=28+40+10=78$$

and $n$ = number of ways when there is at least 3 females

$$=({}^5C_3 \times {}^8C_8)+({}^5C_4 \times {}^8C_7)+({}^5C_5 \times {}^8C_6)$$

$$=10\times 1+5\times 8+1\times 28=78$$

So, $m=n=78$

**28.** *(a)* Let the number of ways of distributing $n$ identical objects among $r$ persons such that each person gets atleast one object is same as the number of ways of selecting $(r-1)$ places out of $(n-1)$ different places, *i.e.*, ${}^{n-1}C_{r-1}$.

**Statement I** Here, $n=10$ and $r=4$

$\therefore$ Required number of ways $={}^{10-1}C_{4-1}$
$={}^9C_3$

**Statement II** Required number of ways $={}^9C_3$

Hence, both the statements are true but Statement II is not a correct explanation of Statement I.

**29.** *(d)* Given,

$$({}^{30}C_1)^2+2({}^{30}C_2)^2+3({}^{30}C_3)^2+\ldots+30({}^{30}C_{30})^2$$

Let, $S=0({}^{30}C_0)^2+1({}^{30}C_1)^2+2({}^{30}C_2)^2+\ldots+30({}^{30}C_{30})^2 \quad \ldots\text{(i)}$

or $S=30({}^{30}C_{30})^2+29({}^{30}C_{29})^2+\ldots+0({}^{30}C_0)^2 \quad \ldots\text{(ii)}$

On adding Eqs. (i) and (ii), we get

$$2S=30\{({}^{30}C_0)^2+({}^{30}C_1)^2+\ldots+({}^{30}C_{30})^2\}$$

$$\therefore S=15\{{}^{30}C_0\cdot{}^{30}C_{30}+{}^{30}C_1\cdot{}^{30}C_{29}+\ldots+{}^{30}C_{30}\,{}^{30}C_0\}$$

$$[\because {}^nC_r={}^nC_{n-r}]$$

$$S=15\{{}^{60}C_{30}\}$$

$$S=15\times\frac{60!}{(30!)(30!)}=\alpha.\frac{60!}{(30!)^2}$$

$\therefore \alpha=15$

**30.** *(b)* The value of $\sum_{r=0}^{20} {}^{50-r}C_6$

$$={}^{50}C_6+{}^{49}C_6+{}^{48}C_6+\ldots+{}^{30}C_6$$

$$\Rightarrow \sum_{r=0}^{20}{}^{50-r}C_6+{}^{30}C_7={}^{30}C_7+{}^{30}C_6+{}^{31}C_6+{}^{32}C_6+\ldots+{}^{50}C_6$$

$$\Rightarrow \sum_{r=0}^{20}{}^{50-r}C_6+{}^{30}C_7$$

$$={}^{31}C_7+{}^{31}C_6+{}^{32}C_6+\ldots+{}^{50}C_6$$

$$[\text{as } {}^nC_r+{}^nC_{r+1}={}^{n+1}C_{r+1}]$$

Similarly, $\sum_{r=0}^{20}{}^{50-r}C_6+{}^{30}C_7={}^{51}C_7$

$$\Rightarrow \sum_{r=0}^{20}{}^{50-r}C_6={}^{51}C_7-{}^{30}C_7$$

**31.** *(1405)* **Case (I)** When single letter is used, then number of words = 5

**Case (II)** When two distinct letters are used, then number of words =

$${}^5C_2\left(\frac{6!}{2!4!}\times 2+\frac{6!}{3!3!}\right)$$

$$=10\left(\frac{6\times5\times4!}{2\times4!}\times 2+\frac{6\times5\times4\times3!}{3\times2\times1\times3!}\right)$$

$$=10\,(30+20)=500$$

**Case (III)** When three distinct letters are used, then number of words =

$${}^5C_3\times\frac{6!}{2!\,2!\,2!}=\frac{5!}{3!\,2!}\times\frac{6!}{2!2!2!}$$

$$=\frac{5\times4\times3!}{3!\times2}\times\frac{6\times5\times4\times3\times2!}{2\times2\times2}$$

$$=10\times 90=900$$

Hence, total number of words
$$=5+500+900=1405$$

**32.** *(36)* Given digits = 2, 3, 4, 5, 7

Total 3-digit numbers $={}^5C_3\times 3!=60$

Now, 3-digit numbers that are divisible by 3

| 234, | 237, | 345, | 357 |
|---|---|---|---|
| ↓ | ↓ | ↓ | ↓ |
| 3! | + 3! | + 3! | + 3! = 24 |

Hence, three digit numbers that are not divisible by 3 = Total − Divisible by 3
$=60-24=36$

**33.** *(3734)* DISTRIBUTION

3I, 2T, 1D, 1S, 1R, 1B, 1U, 1O, 1N

Number of word taken 4 at time.

| | Case | Number of words |
|---|---|---|
| 1. | All are different | ${}^9C_4\times 4!=3024$ |
| 2. | 1 Pair + 2 different | ${}^2C_1\times{}^8C_2\times\frac{4!}{2!}=672$ |
| 3. | 2 Pairs | ${}^2C_2\times\frac{4!}{2!2!}=6$ |
| 4. | 1 triplet + 1 Different | ${}^1C_1\times{}^8C_1\times\frac{4!}{3!}=32$ |
| | Total | 3734 |

**34.** *(1260)* Taking a consecutive 3-place as a group out of nine, we have 7 places to fill now which can be chosen in
${}^7C_1=7$ ways ...(i)

In the group, digits $a, b, c$ can be placed as $a\,b\,c$ or $c\,b\,a$ i.e. in 2 ways ...(ii)

In remaining 6 places, 2$a$'s, 2$b$'s and 2$c$'s have to be permuted which can be done in $\frac{6!}{2!\,2!\,2!}$ ways (Permutation of alike objects) ...(iii)

So, total numbers that can be formed

$$=7\times 2\times\frac{6!}{2!\,2!\,2!}=1260$$

**35.** *(44)* Number of way in which none of the students sits on the alloted seat

$$=5!\left(1-\frac{1}{1!}+\frac{1}{2!}-\frac{1}{3!}+\frac{1}{4!}-\frac{1}{5!}\right)$$

[use Dearrangement]

$$=120\left[1-1+\frac{1}{2}-\frac{1}{6}+\frac{1}{24}-\frac{1}{120}\right]$$

$$=44$$

**36.** *(432)* Given word is UNIVERSE

Here vowels are E, I, U and consonants are N, R, S, V

$\therefore$ Required number of 4-letters words, with or without meaning, each consisting of 2 vowels and 2 consonants

$= {}^3C_2 \times {}^4C_2 \times 4!$

$= {}^3C_1 \times {}^4C_2 \times 24$

$[\because {}^nC_r = {}^nC_{n-r}]$

$= 3 \times 6 \times 24 = 432$

**37.** *(3483638676)* Number of ways that 20 distinct oranges distributed among 3 children so that each child gets at least one orange

$= 3^{20} - {}^3C_1 (3-1)^{20} + {}^3C_2 (3-2)^{20} - {}^3C_3 (3-3)^{20}$

$= 3^{20} - {}^3C_1 2^{20} + {}^3C_2 1^{20}$

$= 3483638676$

**38.** *(32)* Let numbers are

1, 2, 3, 4, 5, 6, 7, 8, 9

Now, total number of numbers starting with 1 (fix on first place) $= {}^8C_5$

$= \dfrac{8!}{5!3!} = 56$

Now, fix 2, 3 on first and second place. So, total number of numbers starting with 2, 3 $= {}^6C_4 = 15$

Total number $= 56 + 15 = 71$

Now, 72 number will be 245678

So, the sum of 72 numbers digits

$= 2 + 4 + 5 + 6 + 7 + 8 = 32$

**39.** *(17)* Number of ways of selecting 3 boys from $b$ boys $= {}^bC_3$

Number of ways of selecting 2 girls from $g$ girls $= {}^gC_2$

Number of ways of selecting 3 boys and 2 girls from the class $= 168$

${}^bC_3 \times {}^gC_2 = 168$

$\Rightarrow \dfrac{b(b-1)(b-2)}{6} \times \dfrac{g(g-1)}{2} = 168$

$\Rightarrow b(b-1)(b-2) \times g(g-1) = 168 \times 6 \times 2$

$\Rightarrow b(b-1)(b-2) \times g(g-1) = 2016$

As $b$ and $g$ are natural number, so make factor of 206

$b(b-1)(b-2) \times g(g-1) = 6 \times 7 \times 8 \times 3 \times 2$

$\Rightarrow b = 8$ and $g = 3$

$\therefore b + 3g = 8 + 3 \times 3 = 8 + 9 = 17$

**40.** *(243)* Case I When all the digits are non-zero. Let the three digit be $x$, $x$ and $y$, where $x, y \neq 0$, then the number of formation of three digit number would be

${}^9C_2 \times 2! \times \dfrac{3!}{2!} = 216$

Case II When non-repeated digit would be zero, then total number of such three-digit numbers are ${}^9C_1 \times 2! = 18$

Case III When the repeated digit would be zero.

The total number of such three digit number are ${}^9C_1 = 9$

$\therefore$ The required number of three digit numbers are $216 + 18 + 9 = 243$.

**41.** *(40)* Let the student marks '$x$' correct answers and '$y$' incorrect. So, $3x - 2y = 5$ and $x + y \leq 5$ where $x, y \in W$.

Thus, only possible solution is $(x, y) = (3, 2)$

The student can mark correct answer by only one choice but for incorrect answers there are two choices. So, total number of ways a student appearing in the examination gets 5 marks is

${}^5C_3 (1)^3 \cdot (2)^2 = \dfrac{5!}{3! \times 2!} \times 1 \times 4 = 40$

**42.** *(238)*

| 10(5) | 11(6) | 12(8) |
|---|---|---|
| 2 + 1 | 2 | 2 + 3 Case I |
| 2 | 2 + 1 | 2 + 3 Case II |
| 2 | 2 | 2 + 4 Case III |

Case I ${}^5C_3 \times {}^6C_2 \times {}^8C_5 = 8400$

Case II ${}^5C_2 \times {}^6C_3 \times {}^8C_5 = 11200$

Case III ${}^5C_2 \times {}^6C_2 \times {}^8C_6 = 4200$

Total $= 8400 + 11200 + 4200 = 23800$

According to the question,

$100k = 23800 \Rightarrow k = 238$

**Shortcut Solution**

$\because {}^5C_3 + {}^6C_2 + {}^8C_5 = 8400$

${}^5C_2 \times {}^6C_3 \times {}^8C_5 = 11200$

${}^5C_2 \times {}^6C_2 \times {}^8C_6 = 4200$

Total $= 8400 + 11200 + 4200 = 23800$

$\Rightarrow 100k = 23800$

$\Rightarrow k = 238$

**43.** *(300)* Let the number be $xyz$, $0 \leq x, y, z \leq 9$

Case I '3' appears only one time

$\Rightarrow {}^3C_1 \times 9 \times 9 = 243$

Case II '3' appears two times

$\Rightarrow {}^3C_2 \times 2 \times 9 = 54$

Case III '3' appears three times

$\Rightarrow {}^3C_3 \times 3 = 3$

$\therefore$ Total $= 243 + 54 + 3 = 300$

**44.** *(31650)* Given, total students $= 10$

number of groups $= 3$ (i.e. $A$, $B$ and $C$)

Each group has atleast one student but group $C$ has atmost 3 students.

$\therefore$ There are 3 cases depending on number of students in group $C$.

Case I $C$ has 1 student, then $\left.\begin{matrix}A\\B\end{matrix}\right] \leftarrow 9$ students

$\therefore$ Number of ways $= {}^{10}C_1 \times [2^9 - 2]$

Case II $C$ has 2 students, then

$\left.\begin{matrix}A\\B\end{matrix}\right] \leftarrow 8$ Students.

$\therefore$ Number of ways $= {}^{10}C_2 \times [2^8 - 2]$

Case III $C$ has 3 students, then $\left.\begin{matrix}A\\B\end{matrix}\right] \leftarrow 7$ Students.

$\therefore$ Number of ways

$= {}^{10}C_3 \times [2^7 - 2]$

$\therefore$ Required number of possibilities

$= {}^{10}C_1 (2^9 - 2) + {}^{10}C_2 (2^8 - 2) + {}^{10}C_3 (2^7 - 2)$

$= 2^7 [{}^{10}C_1 \times 4 + {}^{10}C_2 \times 2 + {}^{10}C_3] - 20 - 90 - 240$

$= (128 \times 250) - 350 = 31650$

## Topic 3

**01.** *(a)* According to the given question,

Number of blue lines $= n$

and number of red lines $= {}^nC_2 - n$

$\because {}^nC_2 - n = 99n$

$\Rightarrow \dfrac{n(n-1)}{2} = 100n$

$\Rightarrow n - 1 = 200 \Rightarrow n = 201$

**02.** *(c)* It is given that, there are 20 pillars of the same height have been erected along the boundary of a circular stadium. Now, the top of each pillar has been connected by beams with the top of all its non-adjacent pillars, then total number of beams = number of diagonals of 20-sided polygon.

$\because {}^{20}C_2$ is selection of any two vertices of 20-sided polygon which included the sides as well.

So, required number of total beams

$= {}^{20}C_2 - 20$

[$\because$ the number of diagonals in a $n$-sided closed polygon $= {}^nC_2 - n$]

$= \dfrac{20 \times 19}{2} - 20 = 190 - 20 = 170$

**03.** *(b)* Given, $T_n = {}^nC_3$

$T_{n+1} = {}^{n+1}C_3$

$\therefore \quad T_{n+1} - T_n = {}^{n+1}C_3 - {}^nC_3 = 10$ [given]

$\Rightarrow {}^nC_2 + {}^nC_3 - {}^nC_3 = 10$

$\Rightarrow \quad {}^nC_2 = 10$

$\therefore \quad n = 5$

**04.** *(a)* Total triangle formed with vertices of octagon $= {}^8C_3$

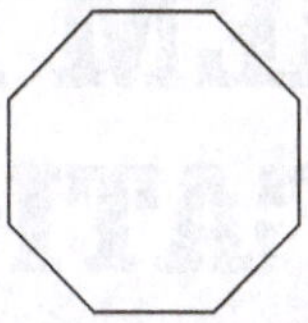

Now, one condition is given that no side of triangle coincides with side of octagon.

**Case I** Number of triangle with all sides coincides with octagon $= 0$

**Case II** Number of triangle with 2 sides coincides with octagon $= {}^8C_1$

($\because$ from one vertex two adjacent sides are originating)

**Case III** Number of triangle with 1 side coincides with octagon $= {}^8C_1 \times {}^4C_1$

($\because$ when we select one side, we have to leave adjacent sides also of selected side, so remaining vertices will be 4)

$\therefore$Total triangles with no sides coinciding with octagon is

$= {}^8C_3 - 0 - {}^8C_1 - {}^8C_1 \times {}^4C_1$

$= 56 - (0 + 8 + 32) = 16$

**05.** *(c)* **Method (I)** (Proper Method)

Whenever we construct a triangle, we must require three non-collinear points.

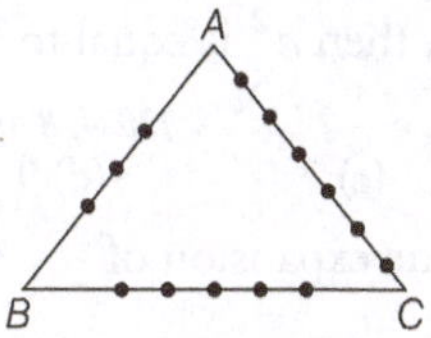

$\therefore$Total number of triangles using the points 3, 5 and 6 which are on the sides *AB*, *BC* and *CA*

= Either taking (one point from *AB*, *BC* and *CA*) or (one point from *AB* and two points from *BC*)or

(one point from *BC* and two points from *AB*) or

(one point from *AB* and two points from *AC*) or

(one point from *AC* and two points from *AB*) or

(one point from *BC* and two points from *AC*) or

(one point from *AC* and two points from *BC*)

$\Rightarrow$ Total number of triangles

$= ({}^3C_1 \times {}^5C_1 \times {}^6C_1) + ({}^3C_1 \times {}^5C_2) + ({}^5C_1 \times {}^3C_2) + ({}^3C_1 \times {}^6C_2) + ({}^6C_1 \times {}^3C_2) + ({}^5C_1 \times {}^6C_2) + ({}^6C_1 \times {}^5C_2)$

$= 90 + 30 + 15 + 45 + 18 + 75 + 60$

$= 333$

**Method (II) (Direct Method)**

Total number of points

$= 3 + 5 + 6 = 14$

Then, when we construct a triangle, we must select 3 points out of 14 but these points never be collinear.

$\therefore$Total number of triangles formed

$= {}^{14}C_3 - {}^3C_3 - {}^5C_3 - {}^6C_3$

$= 333$

**06.** *(d)*

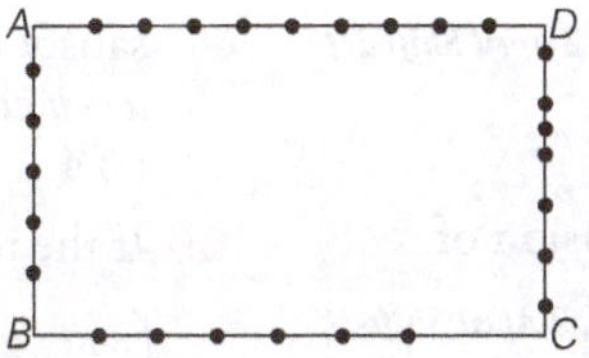

Number of triangles that can be formed from the points on 3 of the sides.

${}^5C_1 {}^7C_1 {}^6C_1 + {}^5C_1 {}^7C_1 {}^9C_1 + {}^5C_1 {}^6C_1 {}^9C_1 + {}^6C_1 {}^7C_1 {}^9C_1$

$= 210 + 315 + 270 + 378$

$\Rightarrow \alpha = 1173$

Number of quadrilaterals that can be formed by taking one point from each of the four vertex

${}^5C_1 {}^7C_1 {}^6C_1 {}^9C_1 = 5 \times 6 \times 7 \times 9$

$= 1890$

$\Rightarrow \quad \beta = 1890$

$\therefore \quad \beta - \alpha = 1890 - 1173$

$= 717$

> **Concept Enhancer**
>
> In a rectangle of $n \times p (n < p)$, numbers of rectangles of any size is ${}^{n+1}C_2$ or $\frac{np}{4}(n+1)(p+1)$ and number of square of any size is $\sum_{r=1}^{n}(n+1-r)(p+1-r)$.

**07.** *(72)* $\underline{a}\ \underline{b}\ \underline{c}\ \underline{d}$

According to condition $a + b = c + d$

If sum is 3 $\rightarrow$ (0, 3) (1, 2) [Not posible]

If sum is 4 $\rightarrow$ (0, 4) (1, 3) [Not posible]

If sum is 5 $\rightarrow$ (0, 5) (1, 4) (2, 3) [Not posible]

If sum is 6 $\rightarrow$ (0, 6) (1, 5) (2, 4) [Not posible]

If sum is 7 $\rightarrow$ (0, 7) (1, 6) (2, 5) (3, 4)

If sum is 8 $\rightarrow$ (1, 7) (2, 6) (3, 5)

If sum is 9 $\rightarrow$ (2, 7) (3, 6) (4, 5)

If sum is 10 $\rightarrow$ (3, 7) (4, 6)

If sum is 11 $\rightarrow$ (4, 7), (5, 6)

Now total trials

$4 \times {}^3C_1 \times 2! + 4 \times {}^2C_1 \times 2! + 4 \times {}^2C_1 \times 2! + 4 \times {}^1C_1 \times 2! + 4 \times {}^1C_1 \times 2! = 24 + 16 + 16 + 8 + 8$

$= 72$

**08.** *(16)* Let '*n*' be the total number of couples who participated in the tournament *t*.

According to question,

$2 \times {}^nC_2 \times {}^{n-2}C_2 = 840$

$\Rightarrow \quad {}^nC_2 \times {}^{n-2}C_2 = 420$

$\Rightarrow \quad \frac{n!}{2(n-2)!} \times \frac{n-2!}{(n-4)!\,2!} = 420$

$\Rightarrow \quad \frac{n(n-1)(n-2)(n-3)}{4} = 420$

Putting $n = 8$ satisfies the equation.

So, $n = 8$

$\therefore$ Total players $= 2n = 2 \times 8 = 16$

C-06

JEE Main Chapterwise Topicwise Mathematics

In JEE Main, there are usually 1-2 questions per paper are visible from this chapter. In past 3 years, 36 conceptual and 26 numericals questions of moderate to hard difficulty level were asked from this chapter.

**For detailed analysis scan QR code**

# BINOMIAL THEOREM AND ITS SIMPLE APPLICATIONS

## Topic 1 Binomial Theorem for a Positive Integer Index, General Term, Free from Radical Terms, Middle Term and Greatest Term

### Level 1

**01** The largest $n \in N$ such that $3^n$ divides 50! is *[2025, 2 April Shift-I]*

(a) 22 (b) 20 (c) 21 (d) 23

**02** If $\sum_{r=0}^{10}\left(\frac{10^{r+1}-1}{10^r}\right)\cdot {}^{11}C_{r+1} = \frac{\alpha^{11}-11^{11}}{10^{10}}$, then $\alpha$ is equal to *[2025, 2 April Shift-II]*

(a) 11 (b) 20
(c) 15 (d) 24

**03** The sum of all rational terms in the expansion of $(2+\sqrt{3})^8$ is *[2025, 3 April Shift-I]*

(a) 33845 (b) 16923
(c) 18817 (d) 3763

**04** The remainder when $((64)^{(64)})^{(64)}$ is divided by 7 is equal to *[2025, 7 April Shift-I]*

(a) 3 (b) 1 (c) 4 (d) 6

**05** The remainder, when $7^{103}$ is divided by 23 is equal to *[2025, 29 Jan Shift-II]*

(a) 6 (b) 17
(c) 9 (d) 14

**06** Suppose $A$ and $B$ are the coefficients of 30th and 12th terms, respectively in the binomial expansion of $(1+x)^{2n-1}$. If $2A = 5B$, then $n$ is equal to *[2025, 24 Jan Shift-II]*

(a) 21 (b) 20
(c) 19 (d) 22

**Imp.** General term of Binomial Expansion

**07** For some $n \neq 10$, let the coefficients of the $5^{th}$, $6^{th}$ and $7^{th}$ terms in the binomial expansion of $(1+x)^{n+4}$ be in AP. Then, the largest coefficient in the expansion of $(1+x)^{n+4}$ is *[2025, 24 Jan Shift-I]*

(a) 70 (b) 35 (c) 20 (d) 10

**08** Let $\alpha, \beta, \gamma$ and $\delta$ be the coefficients of $x^7, x^5, x^3$ and $x$ respectively in the expansion of $(x+\sqrt{x^3-1})^5 + (x-\sqrt{x^3-1})^5$, $x > 1$. If $u$ and $v$ satisfy the equations $\alpha u + \beta v = 18$, $\gamma u + \delta v = 20$, then $u + v$ equals *[2025, 22 Jan Shift-II]*

(a) 4 (b) 8 (c) 3 (d) 5

**09** If the term independent of $x$ in the expansion of $\left(\sqrt{a}x^2 + \frac{1}{2x^3}\right)^{10}$ is 105, then $a^2$ is equal to *[2024, 8 April Shift-II]*

(a) 6 (b) 4 (c) 2 (d) 9

**10** If the constant term in the expansion of $\left(\frac{\sqrt[5]{3}}{x} + \frac{2x}{\sqrt[3]{5}}\right)^{12}$, $x \neq 0$, is $\alpha \times 2^8 \times \sqrt[5]{3}$, then $25\alpha$ is equal to *[2024, 5 April Shift-II]*

(a) 724 (b) 742 (c) 639 (d) 693

**11** Let $m$ and $n$ be the coefficients of seventh and thirteenth terms respectively in the expansion of $\left\{\frac{1}{3}x^{1/3} + \frac{1}{2x^{2/3}}\right\}^{18}$. Then, $\left(\frac{n}{m}\right)^{1/3}$ is *[2024, 1 Feb Shift-II]*

(a) $\frac{9}{4}$ (b) $\frac{1}{9}$ (c) $\frac{4}{9}$ (d) $\frac{1}{4}$

**12** If $A$ denotes the sum of all the coefficients in the expansion of $(1-3x+10x^2)^n$ and $B$ denotes the sum of all the coefficient in the expansion of $(1+x^2)^n$, then **[2024, 27 Jan Shift-I]**
(a) $B=A^3$ (b) $A=3B$ (c) $3A=B$ (d) $A=B^3$

**13** Fractional part of the number $\frac{4^{2022}}{15}$ is equal to **[2023, 13 April Shift-I]**
(a) $\frac{8}{15}$ (b) $\frac{1}{15}$ (c) $\frac{14}{15}$ (d) $\frac{4}{15}$

**14** The remainder when $7^{2022}+3^{2022}$ is divided by 5 is **[2022, 28 July Shift-I]**
(a) 0 (b) 2 (c) 3 (d) 4

**15** The remainder, when $(2021)^{2022}+(2022)^{2021}$ is divided by 7 is **[2022, 27 July Shift-I]**
(a) 0 (b) 1 (c) 2 (d) 6

**16** Let $n \geq 5$ be an integer. If $9^n - 8n - 1 = 64\alpha$ and $6^n - 5n - 1 = 25\beta$, then $\alpha - \beta$ is equal to **[2022, 29 June Shift-II]**
(a) $1+{}^nC_2(8-5)+{}^nC_3(8^2-5^2)+\ldots+{}^nC_n(8^{n-1}-5^{n-1})$
(b) $1+{}^nC_3(8-5)+{}^nC_4(8^2-5^2)+\ldots+{}^nC_n(8^{n-2}-5^{n-2})$
(c) ${}^nC_3(8-5)+{}^nC_4(8^2-5^2)+\ldots+{}^nC_n(8^{n-2}-5^{n-2})$
(d) ${}^nC_4(8-5)+{}^nC_5(8^2-5^2)+\ldots+{}^nC_n(8^{n-3}-5^{n-3})$

**17** The remainder when $3^{2022}$ is divided by 5 is **[2022, 24 June Shift-I]**
(a) 1 (b) 2
(c) 3 (d) 4

**18** If the fourth term in the expansion of $(x+x^{\log_2 x})^7$ is 4480, then the value of $x$, where $x \in N$ is equal to **[2021, 17 March Shift-I]**
(a) 2 (b) 4
(c) 3 (d) 1

**19** The coefficient of $t^4$ in the expansion of $\left(\frac{1-t^6}{1-t}\right)^3$ is **[2019, 9 Jan Shift-II]**
(a) 12 (b) 10 (c) 15 (d) 14

**Imp.** Binomial Expansion

**20** The sum of the coefficients of all odd degree terms in the expansion of $(x+\sqrt{x^3-1})^5+(x-\sqrt{x^3-1})^5$ $(x>1)$ is **[JEE Main 2018]**
(a) $-1$ (b) 0 (c) 1 (d) 2

**21** The remainder left out when $8^{2n}-(62)^{2n+1}$ is divided by 9, is **[AIEEE 2009]**
(a) 0 (b) 2 (c) 7 (d) 8

**22** The coefficient of the middle term in the binomial expansion in powers of $x$ of $(1+\alpha x)^4$ and of $(1-\alpha x)^6$ is the same, if $\alpha$ is equal to **[AIEEE 2004]**
(a) $-\frac{5}{3}$ (b) $\frac{10}{3}$ (c) $-\frac{3}{10}$ (d) $\frac{3}{5}$

## Level 2

**23** The term independent of $x$ in the expansion of $\left(\frac{(x+1)}{(x^{2/3}+1-x^{1/3})}-\frac{(x-1)}{(x-x^{1/2})}\right)^{10}$, $x>1$, is **[2025, 2 April Shift-I]**
(a) 150 (b) 210 (c) 120 (d) 240

**24** If $\sum_{r=1}^{9}\left(\frac{r+3}{2^r}\right)\cdot {}^9C_r=\alpha\left(\frac{3}{2}\right)^9-\beta$, $\alpha, \beta \in N$, then $(\alpha+\beta)^2$ is equal to **[2025, 3 April Shift-I]**
(a) 81 (b) 9 (c) 18 (d) 27

**25** In the expansion of $\left(\sqrt[3]{2}+\frac{1}{\sqrt[3]{3}}\right)^n$, $n \in N$, if the ratio of 15th term from the beginning to the 15th term from the end is $\frac{1}{6}$, then the value of ${}^nC_3$ is **[2025, 4 April Shift-I]**
(a) 2300 (b) 4060 (c) 1040 (d) 4960

**26** If $1^2\cdot({}^{15}C_1)+2^2\cdot({}^{15}C_2)+3^2\cdot({}^{15}C_3)+\ldots+15^2\cdot({}^{15}C_{15})=2^m\cdot 3^n\cdot 5^k$, where $m, n, k \in N$, then $m+n+k$ is equal to **[2025, 4 April Shift-II]**
(a) 19 (b) 20 (c) 21 (d) 18

**27** The number of integral terms in the expansion of $\left(5^{\frac{1}{2}}+7^{\frac{1}{8}}\right)^{1016}$ is **[2025, 8 April Shift-II]**
(a) 128 (b) 129 (c) 127 (d) 130

**28** The least value of $n$ for which the number of integral terms in the binomial expansion of $(\sqrt[3]{7}+\sqrt[12]{11})^n$ is 183, is **[2025, 29 Jan Shift-I]**
(a) 2148 (b) 2184 (c) 2172 (d) 2196

**29** Let the coefficients of three consecutive terms $T_r$, $T_{r+1}$ and $T_{r+2}$ in the binomial expansion of $(a+b)^{12}$ be in a GP and let $p$ be the number of all possible values of $r$. Let $q$ be the sum of all rational terms in the binomial expansion of $(\sqrt[4]{3}+\sqrt[3]{4})^{12}$. Then, $p+q$ is equal to **[2025, 28 Jan Shift-II]**
(a) 295 (b) 283 (c) 287 (d) 299

**30** If in the expansion of $(1+x)^p(1-x)^q$, then the coefficients of $x$ and $x^2$ are 1 and $-2$, respectively, then $p^2+q^2$ is equal to *[2025, 23 Jan Shift-II]*

(a) 8 (b) 20
(c) 13 (d) 18

**31** The coefficient of $x^{70}$ in $x^2(1+x)^{98}+x^3(1+x)^{97}+x^4(1+x)^{96}+\ldots+x^{54}(1+x)^{46}$ is ${}^{99}C_p - {}^{46}C_q$. Then, a possible value of $p+q$ is *[2024, 9 April Shift-I]*

(a) 61 (b) 55
(c) 83 (d) 68

**32** The sum of all rational terms in the expansion of $(2^{1/5}+5^{1/3})^{15}$ is equal to *[2024, 4 April Shift-I]*

(a) 931 (b) 633
(c) 3133 (d) 6131

**33** Suppose $2-p, p, 2-\alpha, \alpha$ are the coefficients of four consecutive terms in the expansion of $(1+x)^n$. Then, the value of $p^2-\alpha^2+6\alpha+2p$ equals *[2024, 30 Jan Shift-II]*

(a) 8 (b) 6
(c) 10 (d) 4

**34** Let the number $(22)^{2022}+(2022)^{22}$ leave the remainder $\alpha$ when divided by 3 and $\beta$ when divided by 7. Then, $(\alpha^2+\beta^2)$ is equal to *[2023, 10 April Shift-II]*

(a) 13 (b) 5
(c) 10 (d) 20

**35** $25^{190}-19^{190}-8^{190}+2^{190}$ is divisible by *[2023, 8 April Shift-II]*

(a) 14 but not by 34 (b) 34 but not by 14
(c) neither 14 nor 34 (d) both 14 and 34

**36** Let $x=(8\sqrt{3}+13)^{13}$ and $y=(7\sqrt{2}+9)^9$. If $[t]$ denotes the greatest integer $\le t$, then *[2023, 30 Jan Shift-II]*

(a) $[x]$ is odd but $[y]$ is even
(b) $[x]+[y]$ is even
(c) $[x]$ is even but $[y]$ is odd
(d) $[x]$ and $[y]$ are both odd

**37** Let $K$ be the sum of the coefficient of the odd powers of $x$ in the expansion of $(1+x)^{99}$. Let $a$ be the middle term in the expansion of $\left(2+\frac{1}{\sqrt{2}}\right)^{200}$.

If $\frac{{}^{200}C_{99}K}{a}=\frac{2^l m}{n}$, where $m$ and $n$ are odd numbers, then the ordered pair $(l, n)$ is equal to *[2023, 29 Jan Shift II]*

(a) (51,99) (b) (50,51)
(c) (50,101) (d) (51,101)

**38** If $a_r$ is the coefficient of $x^{10-r}$ in the Binomial expansion of $(1+x)^{10}$, then $\sum_{r=1}^{10} r^3\left(\frac{a_r}{a_{r-1}}\right)^2$ is equal to *[2023, 25 Jan Shift-1]*

(a) 1210 (b) 4895
(c) 3025 (d) 5445

**39** The coefficient of $x^{101}$ in the expression $(5+x)^{500}+x(5+x)^{499}+x^2(5+x)^{498}+\ldots+x^{500}$, $x>0$, is *[2022, 25 June Shift-I]*

(a) ${}^{501}C_{101}(5)^{399}$ (b) ${}^{501}C_{101}(5)^{400}$
(c) ${}^{501}C_{100}(5)^{400}$ (d) ${}^{500}C_{101}(5)^{399}$

**40** If the coefficients of $x^7$ in $\left(x^2+\frac{1}{bx}\right)^{11}$ and $x^{-7}$ in $\left(x-\frac{1}{bx^2}\right)^{11}$, $b\ne 0$, are equal, then the value of $b$ is equal to *[2021, 27 July Shift-I]*

(a) 2 (b) $-1$
(c) 1 (d) $-2$

**41** The sum of all those terms which are rational numbers in the expansion of $(2^{1/3}+3^{1/4})^{12}$ is *[2021, 25 July Shift-II]*

(a) 89 (b) 27
(c) 35 (d) 43

**42** If the greatest value of the term independent of $x$ in the expansion of $\left(x\sin\alpha+a\frac{\cos\alpha}{x}\right)^{10}$ is $\frac{10!}{(5!)^2}$, then the value of $a$ is equal to *[2021, 25 July Shift-II]*

(a) $-1$ (b) 1
(c) $-2$ (d) 2

**43** For the natural numbers $m, n$, if $(1-y)^m(1+y)^n = 1+a_1y+a_2y^2+\ldots\ldots+a_{m+n}y^{m+n}$ and $a_1=a_2=10$, then the value of $(m+n)$ is equal to *[2021, 20 July Shift-II]*

(a) 88 (b) 64 (c) 100 (d) 80

**44** If $n$ is the number of irrational terms in the expansion of $(3^{1/4}+5^{1/8})^{60}$, then $(n-1)$ is divisible by *[2021, 16 March Shift-I]*

(a) 26 (b) 30 (c) 8 (d) 7

**45** Let $[x]$ denote greatest integer less than or equal to $x$. If for $n\in N$, $(1-x+x^3)^n=\sum_{j=0}^{3n} a_j x^j$, then $\sum_{j=0}^{\left[\frac{3n}{2}\right]} a_{2j}+4\sum_{j=0}^{\left[\frac{3n-1}{2}\right]} a_{2j}+1$ is equal to *[2021, 16 March Shift-I]*

(a) 2 (b) $2^{n-1}$ (c) 1 (d) $n$

**46** Let $\alpha > 0, \beta > 0$ be such that $\alpha^3 + \beta^2 = 4$. If the maximum value of the term independent of $x$ in the binomial expansion of $\left(\alpha x^{\frac{1}{9}} + \beta x^{-\frac{1}{6}}\right)^{10}$ is $10k$, then $k$ is equal to **[2020, 2 Sep Shift-I]**

(a) 84 (b) 176 (c) 352 (d) 336

**47** The term independent of $x$ in the expansion of $\left(\frac{1}{60} - \frac{x^8}{81}\right) \cdot \left(2x^2 - \frac{3}{x^2}\right)^6$ is equal to **[2019, 12 April Shift-II]**

(a) $-72$ (b) 36 (c) $-36$ (d) $-108$

**48** If the number of terms in the expansion of $\left(1 - \frac{2}{x} + \frac{4}{x^2}\right)^n$, $x \neq 0$, is 28, then the sum of the coefficients of all the terms in this expansion, is **[JEE Main 2016]**

(a) 64 (b) 2187 (c) 243 (d) 729

**Imp.** Properties of Union of Two Sets

**49** If $X = \{4^n - 3n - 1, n \in N\}$ and $Y = \{9(n-1) : n \in N\}$, where $N$ is the set of natural numbers, then $X \cup Y$ is equal to **[JEE Main 2014]**

(a) $Y - X$ (b) $X$ (c) $Y$ (d) $N$

**50** If $n$ is a positive integer, then $(\sqrt{3} + 1)^{2n} - (\sqrt{3} - 1)^{2n}$ is **[AIEEE 2012]**

(a) an irrational number
(b) an odd positive integer
(c) an even positive integer
(d) a rational number other than positive integers

**51** **Statement I** $\sum_{r=0}^{n} (r+1) \cdot {}^nC_r = (n+2)2^{n-1}$

**Statement II** $\sum_{r=0}^{n} (r+1)\, {}^nC_r \cdot x^r = (1+x)^n + nx(1+x)^{n-1}$ **[AIEEE 2008]**

(a) Statement I is false, Statement II is true
(b) Statement I is true, Statement II is true; Statement II is a correct explanation of Statement I
(c) Statement I is true, Statement II is true; Statement II is not a correct explanation of Statement I
(d) Statement I is true, Statement II is false

## Level 3

**52** The sum of the coefficient of $x^{2/3}$ and $x^{-2/5}$ in the binomial expansion of $\left(x^{2/3} + \frac{1}{2}x^{-2/5}\right)^9$ is **[2024, 9 April Shift-II]**

(a) 19/4 (b) 21/4
(c) 69/16 (d) 63/16

**Imp.** Properties of Arithmetic Progression

**53** If the coefficients of $x^4, x^5$ and $x^6$ in the expansion of $(1+x)^n$ are in the arithmetic progression, then the maximum value of $n$ is **[2024, 4 April Shift-II]**

(a) 21 (b) 14 (c) 7 (d) 28

**54** If the 1011$^{\text{th}}$ term from the end in the binomial expansion of $\left(\frac{4x}{5} - \frac{5}{2x}\right)^{2022}$ is 1024 times 1011$^{\text{th}}$ term from the beginning, then $|x|$ is equal to **[2023, 11 April Shift-II]**

(a) 10 (b) 12 (c) 8 (d) 15

**55** Among the statements : **[2023, 6 April Shift-II]**

(S1): $2023^{2022} - 1999^{2022}$ is divisible by 8.

(S2): $13(13)^n - 12n - 13$ is divisible by 144 for infinitely many $n \in N$.

(a) both (S1) and (S2) are incorrect
(b) only (S2) is correct
(c) both (S1) and (S2) are correct
(d) only (S1) is correct

**56** If the ratio of the fifth term from the beginning to the fifth term from the end in the expansion of $\left(\sqrt[4]{2} + \frac{1}{\sqrt[4]{3}}\right)^n$ is $\sqrt{6} : 1$, then the third term from the beginning is **[2023, 6 April Shift-I]**

(a) $30\sqrt{2}$ (b) $60\sqrt{2}$ (c) $60\sqrt{3}$ (d) $30\sqrt{3}$

**57** If the coefficient of $x^{15}$ in the expansion of $\left(ax^3 + \frac{1}{bx^{\frac{1}{3}}}\right)^{15}$ is equal to the coefficient of $x^{-15}$ in the expansion of $\left(ax^{\frac{1}{3}} - \frac{1}{bx^3}\right)^{15}$, where $a$ and $b$ are positive real numbers, then for each such ordered pair $(a, b)$ **[2023, 30 Jan Shift-I]**

(a) $ab = 1$ (b) $a = b$ (c) $a = 3b$ (d) $ab = 3$

**58** Consider the sequence $a_1, a_2, a_3, \ldots$ such that $a_1 = 1$, $a_2 = 2$ and $a_{n+2} = \frac{2}{a_{n+1}} + a_n$ for $n = 1, 2, 3, \ldots$ If

$$\left(\frac{a_1 + \frac{1}{a_2}}{a_3}\right) \cdot \left(\frac{a_2 + \frac{1}{a_3}}{a_4}\right) \cdot \left(\frac{a_3 + \frac{1}{a_4}}{a_5}\right) \cdots \left(\frac{a_{30} + \frac{1}{a_{31}}}{a_{32}}\right)$$

$= 2^{\alpha} \left({}^{61}C_{31}\right)$, then $\alpha$ is equal to **[2022, 28 July Shift-I]**

(a) $-30$ (b) $-31$ (c) $-60$ (d) $-61$

**59** A possible value of '$x$', for which the ninth term in the expansion of

$$\left\{3^{\log_3\sqrt{25^{x-1}+7}}+3^{\left(-\frac{1}{8}\right)\log_3(5^{x-1}+1)}\right\}^{10}$$ is equal to 180, is **[2021, 27 July Shift-II]**

(a) 0 (b) $-1$ (c) 2 (d) 1

**60** The lowest integer which is greater than $\left(1+\frac{1}{10^{100}}\right)^{10^{100}}$ is **[2021, 25 July Shift-II]**

(a) 3 (b) 4 (c) 2 (d) 1

**61** If $b$ is very small as compared to the value of $a$, so that the cube and other higher powers of $b/a$ can be neglected in the identity $\frac{1}{a-b}+\frac{1}{a-2b}+\frac{1}{a-3b}+\ldots+\frac{1}{a-nb}=\alpha n+\beta n^2+\gamma n^3$, then the value of $\gamma$ is **[2021, 25 July Shift-I]**

(a) $\frac{a^2+b}{3a^3}$ (b) $\frac{b^2}{3a^3}$

(c) $\frac{a+b^2}{3a^3}$ (d) $\frac{a+b}{3a^2}$

**62** The maximum value of the term independent of '$t$' in the expansion of $\left(tx^{1/5}+\frac{(1-x)^{1/10}}{t}\right)^{10}$, where $x\in(0,1)$ is **[2021, 26 Feb Shift-I]**

(a) $\frac{10!}{\sqrt{3}(5!)^2}$ (b) $\frac{2\cdot10!}{3\sqrt{3}(5!)^2}$

(c) $\frac{2\cdot10!}{3(5!)^2}$ (d) $\frac{10!}{3(5!)^2}$

**63** If for some positive integer $n$, the coefficients of three consecutive terms in the binomial expansion of $(1+x)^{n+5}$ are in the ratio 5 : 10 : 14, then the largest coefficient in this expansion is **[2020, 4 Sep Shift-II]**

(a) 330 (b) 462 (c) 792 (d) 252

**64** If some three consecutive coefficients in the binomial expansion of $(x+1)^n$ in powers of $x$ are in the ratio 2 : 15 : 70, then the average of these three coefficients is **[2019, 9 April Shift-II]**

(a) 964 (b) 227 (c) 232 (d) 625

**65** The coefficient of $x^{-5}$ in the binomial expansion of $\left(\frac{x+1}{x^{2/3}-x^{1/3}+1}-\frac{x-1}{x-x^{1/2}}\right)^{10}$, where $x\neq 0, 1$, is **[JEE Main 2017]**

(a) 1 (b) $-4$ (c) $-1$ (d) 4

**66** The coefficient of $x^7$ in the expansion of $(1-x-x^2+x^3)^6$ is **[AIEEE 2011]**

(a) $-132$ (b) $-144$

(c) 132 (d) 144

**67** In the expansion, powers of $x$ in the function $\frac{1}{(1-ax)(1-bx)}$ is $a_0+a_1x+a_2x^2+a_3x^3+\ldots$, then $a_n$ is equal to **[AIEEE 2006]**

(a) $\frac{a^n-b^n}{b-a}$ (b) $\frac{a^{n+1}-b^{n+1}}{b-a}$

(c) $\frac{b^{n+1}-a^{n+1}}{b-a}$ (d) $\frac{b^n-a^n}{b-a}$

## Numerical Value Type Questions

**68** The product of the last two digits of $(1919)^{1919}$ is ………… **[2025, 8 April Shift-II]**

**69** If $\sum_{r=1}^{30}\frac{r^2({}^{30}C_r)^2}{{}^{30}C_{r-1}}=\alpha\times 2^{29}$, then $\alpha$ is equal to …… . **[2025, 22 Jan Shift-II]**

**70** If $\sum_{r=0}^{5}\frac{{}^{11}C_{2r+1}}{2r+2}=\frac{m}{n}$, gcd $(m, n)=1$, then $m-n$ is equal to …… . **[2025, 22 Jan Shift-I]**

**71** If the second, third and fourth terms in the expansion of $(x+y)^n$ are 135, 30 and $\frac{10}{3}$ respectively, then $6(n^3+x^2+y)$ is equal to …… . **[2024, 6 April Shift-I]**

**72** If the constant term in the expansion of $(1+2x-3x^3)\left(\frac{3}{2}x^2-\frac{1}{3x}\right)^9$ is $p$, then $108p$ is equal to ………… . **[2024, 5 April Shift-I]**

**73** Let $a=1+\frac{{}^2C_2}{3!}+\frac{{}^3C_2}{4!}+\frac{{}^4C_2}{5!}+\ldots,$

$b=1+\frac{{}^1C_0+{}^1C_1}{1!}+\frac{{}^2C_0+{}^2C_1+{}^2C_2}{2!}+\frac{{}^3C_0+{}^3C_1+{}^3C_2+{}^3C_3}{3!}\ldots,$

Then, $\frac{2b}{a^2}$ is equal to …… . **[2024, 4 April Shift-I]**

**74** If the coefficient of $x^{30}$ in the expansion of $\left(1+\frac{1}{x}\right)^6(1+x^2)^7(1-x^3)^8; x\neq 0$ is $\alpha$, then $|\alpha|$ equals ……. . **[2024, 1 Feb Shift-I]**

**75** Let the coefficient of $x^r$ in the expansion of $(x+3)^{n-1}+(x+3)^{n-2}(x+2)+(x+3)^{n-3}(x+2)^2+\ldots+(x+2)^{n-1}$ be $\alpha_r$. If $\sum_{r=0}^{n}\alpha_r=\beta^n-\gamma^n$, $\beta,\gamma\in N$, then the value of $\beta^2+\gamma^2$ equals ......... . *[2024, 31 Jan Shift-II]*

**76** In the expansion of $(1+x)(1-x^2)\left(1+\frac{3}{x}+\frac{3}{x^2}+\frac{1}{x^3}\right)^5$, $x\neq 0$, the sum of the coefficients of $x^3$ and $x^{-13}$ is equal to .......... . *[2024, 31 Jan Shift-I]*

**77** Number of integral terms in the expansion of $\left\{7^{\left(\frac{1}{2}\right)}+11^{\left(\frac{1}{6}\right)}\right\}^{824}$ is equal to ......... . *[2024, 30 Jan Shift-I]*

**78** Remainder when $64^{32^{32}}$ is divided by 9 is equal to .... . *[2024, 29 Jan Shift-I]*

**79** The coefficient of $x^{2012}$ in the expansion of $(1-x)^{2008}(1+x+x^2)^{2007}$ is equal to ......... . *[2024, 27 Jan Shift-II]*

**80** The remainder, when $7^{103}$ is divided by 17, is ........... . *[2023, 13 April Shift-II]*

**81** The number of seven digit positive integers formed using the digits 1, 2, 3 and 4 only and sum of the digits equal to 12 is ......... . *[2023, 13 April Shift-I]*

**82** The number of integral terms in the expansion of $\left(3^{\frac{1}{2}}+5^{\frac{1}{4}}\right)^{680}$ is equal to ......... . *[2023, 11April Shift-II]*

**83** Let $[t]$ denote the greatest integer $\leq t$. If the constant term in the expansion of $\left(3x^2-\frac{1}{2x^5}\right)^7$ is $\alpha$, then $[\alpha]$ is equal to .......... . (where, $[\cdot]=$ GIF) *[2023, 8 April Shift-I]*

**84** If the term without $x$ in the expansion of $\left(x^{\frac{2}{3}}+\frac{\alpha}{x^3}\right)^{22}$, is 7315, then $|\alpha|$ is equal to ......... . *[2023, 1 Feb Shift-II]*

**85** The coefficient of $x^{-6}$ in the expansion of $\left(\frac{4x}{5}+\frac{5}{2x^2}\right)^9$ is ......... . *[2023, 31 Jan Shift-II]*

**86** Let $\alpha>0$, be the smallest number such that the expansion of $\left(x^{\frac{2}{3}}+\frac{2}{x^3}\right)^{30}$ has a term $\beta x^{-\alpha}$, $\beta\in N$. Then, $\alpha$ is equal to .......... *[2023, 31 Jan Shift-I]*

**87** 50th root of a number $x$ is 12 and 50th root of another number $y$ is 18. Then, the remainder obtained on dividing $(x+y)$ by 25 is ........ . *[2023, 30 Jan Shift-II]*

**88** The remainder when $(2023)^{2023}$ is divided by 35 is ......... . *[2023, 25 Jan Shift-II]*

**Imp.** $n^{th}$ term from end in the expansion of $(a+b)^n$

**89** Let the ratio of the fifth term from the beginning to the fifth term from the end in the binomial expansion of $\left(\sqrt[4]{2}+\frac{1}{\sqrt[4]{3}}\right)^n$, in the increasing powers of $1/\sqrt[4]{3}$ be $\sqrt[4]{6}:1$. If the sixth term from the beginning is $\alpha/\sqrt[4]{3}$, then $\alpha$ is equal to ......... . *[2022, 29 July Shift-I]*

**90** Let the coefficients of the middle terms in the expansion of $\left(\frac{1}{\sqrt{6}}+\beta x\right)^4$, $(1-3\beta x)^2$ and $\left(1-\frac{\beta}{2}x\right)^6$, $\beta>0$, respectively form the first three terms of an AP. If $d$ is the common difference of this AP, then $50-\frac{2d}{\beta^2}$ is equal to ...... . *[2022, 28 July Shift-II]*

**91** Let the 9th term in the binomial expansion of $(3+6x)^n$, in the increasing powers of $6x$, to be the greatest for $x=3/2$, the least value of $n$ is $n_0$. If $k$ is the ratio of the coefficient of $x^6$ to the coefficient of $x^3$, then $k+n_0$ is equal to ...... . *[2022, 27 July Shift-II]*

**92** If the coefficients of $x$ and $x^2$ in the expansion of $(1+x)^p(1-x)^q$, $p,q\leq 15$, are $-3$ and $-5$ respectively, then coefficient of $x^3$ is equal to ... *[2022, 26 July Shift-I]*

**93** If the maximum value of the term independent of $t$ in the expansion of $\left(t^2x^{\frac{1}{5}}+\frac{(1-x)^{\frac{1}{10}}}{t}\right)^{15}$, $x\geq 0$ is $k$, then $8k$ is equal to ......... . *[2022, 25 July Shift-I]*

**94** Let the coefficients of $x^{-1}$ and $x^{-3}$ in the expansion of $\left(2x^{1/5}-\dfrac{1}{x^{1/5}}\right)^{15}$, $x>0$, be $m$ and $n$ respectively. If $r$ is a positive integer such that $mn^2={}^{15}C_r\cdot 2^r$, then the value of $r$ is equal to ......... **[2022, 29 June Shift-II]**

**95** The number of positive integers $k$ such that the constant term in the binomial expansion of $\left(2x^3+\dfrac{3}{x^k}\right)^{12}$, $x\neq 0$ is $2^8\cdot l$, where $l$ is an odd integer, is ........... . **[2022, 28 June Shift-I]**

**96** If $({}^{40}C_0)+({}^{41}C_1)+({}^{42}C_2)+\ldots+({}^{60}C_{20})$
$$=m/n\ {}^{60}C_{20},$$
$m$ and $n$ are coprime, then $m+n$ is equal to ...... . **[2022, 26 June Shift-II]**

**97** If the sum of the coefficients of all the positive even powers of $x$ in the binomial expansion of $\left(2x^3+\dfrac{3}{x}\right)^{10}$ is $5^{10}-\beta\cdot 3^9$, then $\beta$ is equal to ......... . **[2022, 25 June Shift-II]**

**98** The remainder on dividing $1+3+3^2+3^3+\ldots+3^{2021}$ by 50 is ...... . **[2022, 24 June Shift-II]**

**99** If $\left(\dfrac{3^6}{4^4}\right)k$ is the term, independent of $x$, in the binomial expansion of $\left(\dfrac{x}{4}-\dfrac{12}{x^2}\right)^{12}$, then $k$ is equal to ......... **[2021, 31 Aug. Shift-I]**

**100** $3\times 7^{22}+2\times 10^{22}-44$ when divided by 18 leaves the remainder ......... . **[2021, 27 Aug. Shift-II]**

**101** If the coefficient of $x^7$ and $x^8$ in the expansion of $\left(2+\dfrac{x}{3}\right)^n$ are equal, then the value of $n$ is equal to ......... . **[2021, 25 July Shift-II]**

**102** The ratio of the coefficient of the middle term in the expansion of $(1+x)^{20}$ and the sum of the coefficients of two middle terms in expansion of $(1+x)^{19}$ is ......... . **[2021, 25 July Shift-I]**

**103** If the constant term, in Binomial expansion of $\left(2x^r+\dfrac{1}{x^2}\right)^{10}$ is 180, then $r$ is equal to ......... . **[2021, 22 July Shift-II]**

**104** The number of elements in the set $\{n\in\{1,2,3,\ldots,100\}:(11)^n>(10)^n+(9)^n\}$ is ......... . **[2021, 22 July Shift-II]**

**105** The number of rational terms in the binomial expansion of $(4^{1/4}+5^{1/6})^{120}$ is ...... . **[2021, 20 July Shift-I]**

**106** The term independent of $x$ in the expansion of $\left[\dfrac{x+1}{x^{2/3}-x^{1/3}+1}-\dfrac{x-1}{x-x^{1/2}}\right]^{10}$, $x\neq 1$, is equal to ............. . **[2021, 18 March Shift-II]**

**107** Let ${}^nC_r$ denote the binomial coefficient of $x^r$ in the expansion of $(1+x)^n$.
If $\sum_{k=0}^{10}(2^2+3k)\ {}^nC_k=\alpha\cdot 3^{10}+\beta\cdot 2^{10}$, $\alpha,\beta\in R$, then $\alpha+\beta$ is equal to ........... . **[2021, 18 March Shift-II]**

**108** Let the coefficients of third, fourth and fifth terms in the expansion of $\left(x+\dfrac{a}{x^2}\right)^n$, $x\neq 0$, be in the ratio $12:8:3$. Then, the term independent of $x$ in the expansion, is equal to ......... . **[2021, 17 March Shift-II]**

**109** If the remainder when $x$ is divided by 4 is 3, then the remainder when $(2020+x)^{2022}$ is divided by 8 is ............ . **[2021, 25 Feb Shift-II]**

**110** The natural number $m$, for which the coefficient of $x$ in the binomial expansion of $\left(x^m+\dfrac{1}{x^2}\right)^{22}$ is 1540, is ...... . **[2020, 5 Sep Shift-I]**

**111** If sum of the coefficients of all even powers of $x$ in the product $(1+x+x^2+\ldots+x^{2n})$ $(1-x+x^2-x^3+\ldots+x^{2n})$ is 61, then $n$ is equal to ......... . **[2020, 7 Jan Shift-I]**

## Topic 2 Multinomial Theorem

### Level 1

**01** Let $(a+bx+cx^2)^{10}=\sum_{i=0}^{20}p_ix^i$, $a,b,c\in N$. If $p_1=20$ and $p_2=210$, then $2(a+b+c)$ is equal to **[2023, 15 April Shift-I]**

(a) 15 (b) 12
(c) 8 (d) 6

### Level 2

**02** If the constant term in the expansion of $\left(3x^3-2x^2+\dfrac{5}{x^5}\right)^{10}$ is $2^k\cdot l$, where $l$ is an odd integer, then the value of $k$ is equal to **[2022, 29 June Shift-I]**

(a) 6 (b) 7
(c) 8 (d) 9

## Numerical Value Type Questions

Imp. Differentiation of Binomial Expansion

**03** Let $(1+x+x^2)^{10}=a_0-a_1x+a_2x^2+\ldots+a_{20}x^{20}$. If $(a_1+a_3+a_5+\ldots+a_{19})-11a_2=121k$, then $k$ is equal to ......... . *[2025, 3 April Shift-II]*

**04** The number of natural number between 212 and 999 such that the sum of their digits is 15, is ...... . *[2025, 28 Jan Shift-II]*

**05** The sum of all rational terms in the expansion of $(1+2^{\frac{1}{3}}+3^{\frac{1}{2}})^6$ is equal to ......... . *[2025, 23 Jan Shift-I]*

**06** The coefficient of $x^7$ in $(1-x+2x^3)^{10}$ is ..... *[2023, 10 April Shift-I]*

**07** The constant term in the expansion of $\left(2x+\frac{1}{x^7}+3x^2\right)^5$ ...... . *[2023, 25 Jan Shift-I]*

**08** If the coefficient of $a^7b^8$ in the expansion of $(a+2b+4ab)^{10}$ is $k\cdot 2^{16}$, then $k$ is equal to ......... . *[2021, 31 Aug Shift-II]*

**09** Let $(2x^2+3x+4)^{10}=\sum_{r=0}^{20} a_rx^r$. Then, $\frac{a_7}{a_{13}}$ is equal to ......... . *[2020, 4 Sep Shift-I]*

## • Key Index

### Topic 1

| | | | | | | | | | |
|---|---|---|---|---|---|---|---|---|---|
| **1** (a) | **2** (b) | **3** (c) | **4** (b) | **5** (d) | **6** (a) | **7** (b) | **8** (d) | **9** (b) | **10** (d) |
| **11** (c) | **12** (d) | **13** (b) | **14** (c) | **15** (a) | **16** (c) | **17** (d) | **18** (a) | **19** (c) | **20** (d) |
| **21** (b) | **22** (c) | **23** (b) | **24** (a) | **25** (a) | **26** (a) | **27** (a) | **28** (d) | **29** (b) | **30** (c) |
| **31** (c) | **32** (c) | **33** (*) | **34** (b) | **35** (b) | **36** (b) | **37** (c) | **38** (a) | **39** (a) | **40** (c) |
| **41** (d) | **42** (d) | **43** (d) | **44** (a) | **45** (a) | **46** (d) | **47** (c) | **48** (d) | **49** (c) | **50** (a) |
| **51** (b) | **52** (b) | **53** (b) | **54** (*) | **55** (c) | **56** (c) | **57** (a) | **58** (c) | **59** (d) | **60** (a) |
| **61** (b) | **62** (b) | **63** (b) | **64** (c) | **65** (a) | **66** (b) | **67** (c) | **68** (63) | **69** (465) | **70** (2035) |
| **71** (806) | **72** (54) | **73** (8) | **74** (678) | **75** (25) | **76** (118) | **77** (138) | **78** (1) | **79** (0) | **80** (12) |
| **81** (413) | **82** (171) | **83** (1275) | **84** (1) | **85** (5040) | **86** (2) | **87** (23) | **88** (7) | **89** (84) | **90** (57) |
| **91** (24) | **92** (23) | **93** (6006) | **94** (5) | **95** (2) | **96** (102) | **97** (83) | **98** (4) | **99** (55) | **100** (15) |
| **101** (55) | **102** (1) | **103** (8) | **104** (96) | **105** (21) | **106** (210) | **107** (19) | **108** (4) | **109** (1) | **110** (13) |
| **111** (30) | | | | | | | | | |

### Topic 2

| | | | | | | | | |
|---|---|---|---|---|---|---|---|---|
| **1** (b) | **2** (d) | **3** (239) | **4** (64) | **5** (612) | **6** (960) | **7** (1080) | **8** (315) | **9** (8) |

# Detailed **Explanations**

## Topic 1

**01.** (a) $\left[\frac{50}{3}\right]+\left[\frac{50}{3^2}\right]+\left[\frac{50}{3^3}\right]+\left[\frac{50}{3^4}\right]+\ldots$

$=16+5+1$

$=22$

Largest $n$ is 22

**02.** (b) $\sum_{r=0}^{10}\frac{10^{r+1}-1}{10^r}\cdot{}^{11}C_{r+1}$

$=\sum_{r=0}^{10}10\cdot{}^{11}C_{r+1}-\sum_{r=0}^{10}\left(\frac{1}{10}\right)^r\cdot{}^{11}C_{r+1}$

$=10\cdot(2^{11}-1)-10\left[\left(1+\frac{1}{10}\right)^{11}-1\right]$

$=10\cdot2^{11}-\frac{11^{11}}{10^{10}}$

$=\frac{20^{11}-11^{11}}{10^{10}}$

$\therefore\ \alpha=20$

**03.** (c) $S={}^8C_0(2)^8+{}^8C_1 2^7(\sqrt{3})+\ldots$

$+{}^8C_8(\sqrt{3})^8$

Sum of rational terms

$={}^8C_0(2)^8+{}^8C_2 2^6(\sqrt{3})^2$

$+{}^8C_4(2)^4(\sqrt{3})^4$

$+{}^8C_6(2)^2(\sqrt{3})^2+{}^8C_8(\sqrt{3})^8$

$=18817$

**04.** (b) $(64)^{64}=(63+1)^{64}$

$=63\lambda+1$

$(64)^{(64)^{(64)}}=(63+1)^{64}$

$63\lambda_1+1$

$\therefore$Required remainder when divided by 7 is 1

**05.** (d) $7^{103}=7(7)^{102}=7(343)^{34}$

$=7(345-2)^{34}$

$\Rightarrow\ 7^{103}=23k_1+7\cdot2^{34}$

Now, $7\cdot2^{34}=7\cdot2^2\cdot2^{32}=28\cdot(256)^4$

$=28(253+3)^4$

$\therefore 28\times81\Rightarrow(23+5)(69+12)$

$\equiv23k_2+60$

$\therefore$Remainder $=14$

**06.** (a)

> **key idea** Coefficient of $r$th term in $(1+x)^{2n}$ is ${}^{2n}C_{r-1}$

Given, $A=$ Coefficient of 30th term in the expansion of $(1+x)^{2n-1}$

and $B=$ Coefficient of 12th term in the expansion of $(1+x)^{2n-1}$

$\therefore\quad A={}^{2n-1}C_{29}$ and $B={}^{2n-1}C_{11}$

Given, $\quad 2A=5B$

$\Rightarrow\quad 2\cdot{}^{2n-1}C_{29}=5\ {}^{2n-1}C_{11}$

$\Rightarrow 2\cdot\frac{(2n-1)!}{29!(2n-30)!}=5\cdot\frac{(2n-1)!}{11!(2n-12)!}$

$\Rightarrow\quad\frac{1}{29\times28\times27\times\ldots\times12\times5}$

$=\frac{1}{(2n-12)(2n-13)(2n-14)\ldots(2n-29)\cdot2}$

$\Rightarrow\quad\frac{1}{30\times29\times\ldots\times12}$

$=\frac{1}{(2n-12)(2n-13)\ldots(2n-29)\cdot12}$

$\Rightarrow\quad 2n-12=30\Rightarrow n=21$

**07.** (b) Given coefficient of 5th, 6th and 7th terms in the binomial expansion of $(1+x)^{n+4}$ are in AP

i.e. ${}^{n+4}C_4,\ {}^{n+4}C_5,\ {}^{n+4}C_6$ are in AP

$\Rightarrow 2\times{}^{n+4}C_5={}^{n+4}C_4+{}^{n+4}C_6$

$\Rightarrow 4\times{}^{n+4}C_5=({}^{n+4}C_4+{}^{n+4}C_5)$

$+({}^{n+4}C_5+{}^{n+4}C_6)$

$\Rightarrow 4\times{}^{n+4}C_5={}^{n+5}C_5+{}^{n+5}C_6$

$\Rightarrow 4\times\frac{(n+4)!}{5!(n-1)!}=\frac{(n+6)!}{6!n!}$

$\Rightarrow 4=\frac{(n+6)(n+5)}{6n}$

$\Rightarrow n^2+11n+30=24n$

$\Rightarrow n^2-13n+30=0$

$\Rightarrow n=3, 10$ (rejected) ($\because n\neq10$ given)

$\therefore n+4=3+4=7$

$\therefore$Largest binomial coefficient in expansion of $(1+x)^7$ is the coefficient of middle term i.e. 5th term.

$={}^7C_4={}^7C_3=35$

**08.** (d) $\because (a+b)^n+(a-b)^n$

$=2$ (sum of all odd terms)

$\therefore\quad (x+\sqrt{x^3-1})^5+(x-\sqrt{x^3-1})^5$

$=2[{}^5C_0x^5+{}^5C_2x^3(\sqrt{x^3-1})^2$

$+{}^5C_4x^1(\sqrt{x^3-1})^4]$

$=2[x^5+10x^3(x^3-1)$

$+5x(x^6-2x^3+1)]$

$\alpha=$ Coefficient of $x^7=2\times5=10$

$\beta=$ Coefficient of $x^5=2\times1=2$

$\gamma=$ Coefficient of $x^3$

$=2(-10)=-20$

$\delta=$ Coefficient of $x=2\times5=10$

$\because\quad \alpha u+\beta v=18\Rightarrow10u+2v=18$

$\Rightarrow\quad 5u+v=9$ ...(i)

and $\quad\gamma u+\delta v=20$

$\Rightarrow\quad -20u+10v=20$

$\Rightarrow\quad -2u+v=2$ ...(ii)

By solving Eqs. (i) and (ii) we get

$7u=7\Rightarrow u=1$

and $\quad v=2+2=4$

$\therefore\quad u+v=5$

**09.** (b) $\left(\sqrt{a}x^2+\frac{1}{2x^3}\right)^{10}$

General term

$={}^{10}C_r(\sqrt{a}x^2)^{10-r}\left(\frac{1}{2x^3}\right)^r$

$\therefore$For the term independent of $x$

$\Rightarrow\quad 2(10-r)-3r=0$

$\Rightarrow\quad r=4$

Put $r=4$,

${}^{10}C_4(\sqrt{a}x^2)^6\left(\frac{1}{2x^3}\right)^4$

$=\frac{10\cdot9\cdot8\cdot7}{4\cdot3\cdot2\cdot1}\times a^3\times\frac{1}{2^4}$

$=210a^3\times\frac{1}{16}$

Now, $210a^3\times\frac{1}{16}=105$

$\Rightarrow\quad a^3=8$

$\Rightarrow\quad a^2=4$

**10.** (d) For $\left(\frac{\sqrt[5]{3}}{x}+\frac{2x}{\sqrt[3]{5}}\right)^{12}$

$T_{r+1}={}^{12}C_r\left(\frac{3^{1/5}}{x}\right)^{12-r}\left(\frac{2x}{5^{1/3}}\right)^r$

$={}^{12}C_r\ 3^{\frac{12-r}{5}}\cdot5^{-\frac{r}{3}}\cdot2^r\cdot x^{r-12+r}$

For constant term

$\Rightarrow r-12+r=0 \Rightarrow r=6$

$T_7 = {}^{12}C_6 \cdot 3^{6/5} \cdot 5^{-2} \cdot 2^6$

$= \frac{9\times11\times7}{25} \times 2^8 \times 3^{1/5}$

On comparing, $\alpha = \frac{9\times11\times7}{25}$

$\Rightarrow \quad 25\alpha = 693$

**11.** *(c)* We have,

$\left(\frac{1}{3}x^{1/3} + \frac{1}{2x^{2/3}}\right)^{18}$

Given, coefficients of $T_7 = m$

and coefficients of $T_{13} = n$

Now, $\frac{m}{n} = \frac{{}^{18}C_6\left(\frac{1}{3}\right)^{12}\cdot\left(\frac{1}{2}\right)^6}{{}^{18}C_{12}\left(\frac{1}{3}\right)^6\cdot\left(\frac{1}{2}\right)^{12}}$

$[\because {}^nC_r = {}^nC_{n-r}]$

$\Rightarrow \left(\frac{m}{n}\right)^{1/3} = \left(\frac{1}{3^{12}}\times 3^6 \times \frac{1}{2^6}\times 2^{12}\right)^{1/3}$

$= \left(\frac{2^6}{3^6}\right)^{1/3} = \frac{4}{9}$

$\therefore \left(\frac{m}{n}\right)^{1/3} = \frac{4}{9}$

**12.** *(d)* Given, $A$ = Sum of all the coefficients in the expansion of $(1-3x+10x^2)^n$

$B$ = Sum of all the coefficient in the expansion of $(1+x^2)^n$.

Put $x=1$ in $(1-3x+10x^2)^n$ to get $A$ and in similar way you can find $B$ also.

So, $A=(8)^n$ and $B=2^n$

Hence, $A=(2^n)^3 = B^3$

**13.** *(b)* $\frac{4^{2022}}{15} = \frac{4^{2(1011)}}{15}$

$= \frac{(16)^{1011}}{15} = \frac{(15+1)^{1011}}{15} = \frac{15K+1}{15}$

$\therefore \left\{\frac{4^{2022}}{15}\right\} = \left\{\frac{15K+1}{15}\right\} = \frac{1}{15}$

**14.** *(c)* Given, $7^{2022} + 3^{2022}$

$=(49)^{1011} + (9)^{1011}$

$=(50-1)^{1011} + (10-1)^{1011}$

$=5.\lambda - 1 + 5\cdot\mu - 1 = 5k-2$,

where $\lambda, \mu, k \in I$

$\therefore$ Remainder $=5-2=3$

**15.** *(a)* Let $N=(2021)^{2022}+(2022)^{2021}$

Nearest multiple of 7 is 2023.

$\therefore N=(2023-2)^{2022}+(2023-1)^{2021}$

$=(7\lambda-2)^{2022}+(7\lambda_1-1)^{2021}$

$=[(7\lambda-2)^3]^{674}+(7\lambda_1)^{2021}-2021$

$(7\lambda_1)^{2020}+\ldots-1$

$=(7\lambda_2-1)^{674}+(7\lambda_3-1]$

$=(7\lambda_4+1)+(7\lambda_5-1)$

$=7(\lambda_4+\lambda_5)=7\lambda_6$

$\therefore N$ is multiple of 7, hence remainder is 0.

**16.** *(c)* $9^n-8n-1=64\alpha$

and $6^n-5n-1=25\beta$

$9^n=(1+8)^n={}^nC_0+{}^nC_1 8^1$

$+{}^nC_2 8^2+{}^nC_3 8^3+{}^nC_4 8^4+\ldots$

$\Rightarrow \frac{9^n-8n-1}{64}={}^nC_2$

$+{}^nC_3 8+{}^nC_4 8^2+\ldots$

$\Rightarrow \alpha={}^nC_2+{}^nC_3 8+{}^nC_4 8^2+\ldots$

Similarly,

$\beta={}^nC_2+{}^nC_3 5+{}^nC_4 5^2+\ldots$

$\therefore \alpha-\beta={}^nC_3(8-5)+{}^nC_4(8^2-5^2)$

$+\ldots+{}^nC_n(8^{n-2}-5^{n-2})$

**17.** *(d)* Given, $3^{2022}$

$=(3^2)^{1011}=(9)^{1011}=(10-1)^{1011}$

Now applying binomial theorem,

$3^{2022}=10\{{}^{1011}C_0(10)^{1010}-$

${}^{1011}C_1(10)^{1009}+\ldots\ldots+{}^{1011}C_{1010}\}-1$

$\Rightarrow 3^{2022}=10k-1$, where $k\in Z$

$\Rightarrow 3^{2022}=10k+5-5-1$

[adding and subtracting by '5']

$\Rightarrow 3^{2022}=(10k-5)+(5-1)$

$\Rightarrow 3^{2022}=5(2k-1)+4$

$\Rightarrow 3^{2022}=5\lambda+4; \lambda\in Z$

Now, $3^{2022}$ i.e. $5\lambda+4$ is divided by 5, then the remainder is 4

Thus, the remainder is 4.

**18.** *(a)* $T_{r+1}={}^nC_r(x^{\log_2 x})^{n-r}x^r$

$T_4={}^7C_3(x^{\log_2 x})^4 x^3$

$=35x^{4\log_2 x}x^3$

$35x^{3+4\log_2 x}=4480$

$x^{3+4\log_2 x}=128=2^7$

Taking log on both sides, we get

$(3+4\log_2 x)(\log_2 x)=7$

$\Rightarrow 4(\log_2 x)^2+3\log_2 x-7=0$

$\Rightarrow (\log_2 x-1)(4\log_2 x+7)=0$

$\Rightarrow \log_2 x=1 \quad [\because x\in N]$

$\therefore x=2$

**19.** *(c)* Clearly,

$\left(\frac{1-t^6}{1-t}\right)^3=(1-t^6)^3(1-t)^{-3}$

$\therefore$ Coefficient of $t^4$ in $(1-t^6)^3(1-t)^{-3}$

= Coefficient of $t^4$ in $(1-t^{18}$

$-3t^6+3t^{12})(1-t)^{-3}$

= Coefficient of $t^4$ in $(1-t)^{-3}$

$={}^{3+4-1}C_4={}^6C_4=15$

[$\because$ coefficient of $x^r$ in $(1-x)^{-n}$

$={}^{n+r-1}C_r$]

**20.** *(d)*

> **key idea** Use, $(a+b)^n+(a-b)^n$
> $=2({}^nC_0a^n+{}^nC_2a^{n-2}b^2$
> $+{}^nC_4a^{n-4}b^4\ldots)$

We have,

$(x+\sqrt{x^3-1})^5+(x-\sqrt{x^3-1})^5, x>1$

$=2(x^5+10x^3(x^3-1)+5x(x^3-1)^4)$

$=2(x^5+10x^6-10x^3+5x^7-10x^4+5x)$

$\therefore$ Sum of coefficients of all odd degree terms is $2(1-10+5+5)=2$

**21.** *(b)* $8^{2n}-(62)^{2n+1}$

$=(1+63)^n-(63-1)^{2n+1}$

$=(1+63)^n+(1-63)^{2n+1}$

$=[(1+{}^nC_1 63+{}^nC_2(63)^2+\ldots$

$+(63)^n]+[1-{}^{(2n+1)}C_1 63+{}^{(2n+1)}$

$C_2(63)^2+\ldots+(-1)(63)^{(2n+1)}]$

$=2+63[{}^nC_1+{}^nC_2(63)+\ldots+$

$(63)^{n-1}+{}^{(2n+1)}C_2(63)-\ldots-(63)^{(2n)}]$

Hence, the remainder is 2.

**22.** *(c)* The coefficient of the middle term in powers of $x$ of $(1+\alpha x)^4={}^4C_2\alpha^2$.

The coefficient of the middle term in powers of $x$ of $(1-\alpha x)^6={}^6C_3(-\alpha)^3$.

According to given condition,

${}^4C_2\alpha^2={}^6C_3(-\alpha)^3$

$\Rightarrow \frac{4!}{2!2!}\alpha^2=-\frac{6!}{3!3!}\alpha^3$

$\Rightarrow 6\alpha^2=-20\alpha^3 \Rightarrow \alpha=-6/20$

$\therefore \alpha=-3/10$

**23.** *(b)* $x+1=(x^{1/3}+1)(x^{2/3}-x^{1/3}+1)$

and $x-1=(\sqrt{x}-1)(\sqrt{x}+1)$

$\therefore$ Given expression is $\left(x^{1/3}+1-1-\frac{1}{\sqrt{x}}\right)^{10}$

$=(x^{1/3}-x^{-1/2})^{10}$

General Term

$T_{r+1} = {}^{10}C_r\,(x^{1/3})^{10-r}(-x^{-1/2})^r$

$=(-1)^r\,{}^{10}C_r\,x^{\frac{10}{3}-\frac{r}{3}-\frac{r}{2}}$

Put $\frac{10}{3}-\frac{r}{3}-\frac{r}{2}=0 \Rightarrow r=4$

∴ Independent term of $x={}^{10}C_4$

$=\frac{10!}{4!6!}=\frac{10\times9\times8\times7}{4\times3\times2\times1}=210$

**24.** (a)

$\sum_{r=1}^{9}\left(\frac{r+3}{2^r}\right)\cdot{}^9C_r=\sum_{r=1}^{9}\frac{r}{2^r}\cdot\frac{9}{r}\,{}^8C_{r-1}$

$+\sum_{r=1}^{9}3\cdot{}^9C_r\left(\frac{1}{2}\right)^r$

$=\frac{9}{2}\sum_{r=1}^{9}{}^8C_{r-1}\left(\frac{1}{2}\right)^{r-1}+3\sum_{r=1}^{9}{}^9C_r\left(\frac{1}{2}\right)^r$

$=\frac{9}{2}\sum_{r=0}^{8}{}^8C_{r-1}\left(\frac{1}{2}\right)^r+3\sum_{r=1}^{9}{}^9C_r\left(\frac{1}{2}\right)^r$

$=\frac{9}{2}\sum_{r=0}^{8}{}^8C_{r-1}\left(\frac{1}{2}\right)^r+3\sum_{r=1}^{9}{}^9C_r\left(\frac{1}{2}\right)^r$

$=\frac{9}{2}\left(1+\frac{1}{2}\right)^8=3\left[\left(1+\frac{1}{2}\right)^7-{}^9C_6\left(\frac{1}{2}\right)^0\right]$

$=\frac{9}{2}\cdot\frac{3^8}{2^8}+3\left[\frac{3^9}{2^9}-1\right]$

$=\frac{3^{10}}{2^9}+\frac{3^{10}}{2^9}-3=4\cdot\frac{3^{10}}{2^{10}}-3$

$=4\left(\frac{3}{2}\right)^{10}-3=6\left(\frac{3}{2}\right)^9-3$

∴ $\alpha=6,\ \beta=3$

⇒ $(\alpha+\beta)^2=81$

**25.** (a) Expansion of $\left(2^{\frac{1}{3}}+\frac{1}{3^{\frac{1}{3}}}\right)^n$

$\frac{T_{15}\text{ from the beginning}}{T_{15}\text{ from the end}}=\frac{1}{6}$

$\frac{{}^nC_{14}\left(2^{\frac{1}{3}}\right)^{n-14}\cdot\left(\frac{1}{3^{\frac{1}{3}}}\right)^{14}}{{}^nC_{14}\left(\frac{1}{3^{\frac{1}{3}}}\right)^{n-14}\left(2^{\frac{1}{3}}\right)^{14}}=\frac{1}{6}$

$6^{\frac{n-28}{3}}=6^{-1}$

$\frac{n-28}{3}=-1$

$n=25$

∴ ${}^{25}C_3=\frac{25\times24\times23}{6}=2300$

**26.** (a)

$1^2\cdot{}^{15}C_1+2^2\cdot{}^{15}C_2+3^2\cdot{}^{15}C_3+\ldots.\ 15^2\cdot{}^{15}C_{15}$

$T_{r+1}=\sum_{r=1}^{15}r^2\cdot{}^{15}C_r$

$=\sum_{r=1}^{15}r^2\cdot\left(\frac{15}{r}\right){}^{14}C_{r-1}$

$=15\sum_{r=1}^{15}r\,{}^{14}C_{r-1}$

let $r-1=t$

$=15\sum_{t=0}^{14}(t+1)\cdot{}^{14}C_t$

$=15\left[\sum_{t=0}^{14}t\,{}^{14}C_t+\sum_{t=0}^{14}{}^{14}C_t\right]$

$=15\,(14\cdot2^{14-1}+2^{14})$

$=15\,(8\cdot2^{14})=3^1\times5^1\times2^{17}$

$m=17,\ n=1$ and $k=1$

∴ $m+n+k=19$

**27.** (a)

$T_{r+1}={}^{1016}C_r\,(5^{1/2})^{1016-r}\times(7^{1/8})^r$

$1016-r=2k$

$r=8\lambda$ [$k$, $\lambda$ belongs to integrals]

⇒ $1016-8\lambda=2k$

⇒ $2k+8\lambda=1016$

⇒ $k+4\lambda=508$

Also, $0\le r\le1016$

⇒ $0\le8\lambda\le1016$

⇒ $0\le\lambda\le127$

Also for every integral values of $\lambda\in[0,127)$ we are getting integral values of $k$

⇒ Number of terms $=127+1$ $=128$

**28.** (d)

> **key idea** $(r+1)^{th}$ term in expansion of $(a+b)^n$ is $T_{r+1}={}^nC_r\,a^{n-r}b^r$

General term

$=T_{r+1}={}^nC_r\,(7^{\frac{1}{3}})^{n-r}(11^{\frac{1}{12}})^r$

$={}^nC_r\,(7)^{\frac{n-r}{3}}(11)^{\frac{r}{12}}$

For integral terms, $r$ must be multiple of 12

∴ $r=12\lambda,\ \lambda\in W$

So, $r=0, 12, 24, 36, 48,$ ……

∵ Number of terms $=183$

∴ $r_{max}=12\times183=2196$

∴ Minimum value of $n$ will be 2196 as 3/2196 (i.e. 3 divides 2196).

**29.** (b)

> **key idea** General term in the expansion of $(x+a)^n$ is $T_{r+1}={}^nC_r\,x^{n-r}a^r$

Given, $T_r, T_{r+1}, T_{r+2}\ldots$ are in GP

⇒ $(T_{r+1})^2=T_r\cdot T_{r+2}$

⇒ $({}^{12}C_r\,a^{12-r}b^r)^2={}^{12}C_{r-1}\,a^{13-r}\cdot b^{r-1}\times{}^{12}C_{r+1}\,a^{11-r}\,b^{r+1}$

⇒ $({}^{12}C_r)^2={}^{12}C_{r-1}\times{}^{12}C_{r+1}$

⇒ $\frac{{}^{12}C_{r-1}}{{}^{12}C_r}=\frac{{}^{12}C_r}{{}^{12}C_{r+1}}$

⇒ $\frac{\frac{12!}{(r-1)!(13-r)!}}{\frac{12!}{r!(12-r)!}}=\frac{\frac{12!}{r!(12-r)!}}{\frac{12!}{(r+1)!(11-r)!}}$

⇒ $-r=13-r$ ⇒ $13=0$

Not possible

(no three consecutive binomial coefficients are in GP or HP but AP is possible) ⇒ $p=0$

Now, $T_{k+1}={}^{12}C_k\,(3^{1/4})^{12-k}(4^{1/3})^k$

$={}^{12}C_k\,3^{3-\frac{k}{4}}4^{\frac{k}{3}}$

For terms to be rational (4, 3) divides $k$

⇒ 12 divides $k$

⇒ 0, 12 → Two terms

∴ Sum of rational terms

$={}^{12}C_0 3^3 4^0+{}^{12}C_{12}4^4 3^0=3^3+4^4$

$=27+256=283=q$

∴ $p+q=0+283=283$

**30.** (c) General term of $(1+x)^p\,(1-x)^q$ is

$={}^pC_{r_1}\,x^{r_1}\cdot{}^qC_{r_2}\,(-1)^{r_2}x^{r_2}$

$={}^pC_{r_1}{}^qC_{r_2}\,(-1)^{r_2}\,x^{r_1+r_2}$

For coefficient of $x$,

$r_1+r_2=1$

∴ $r_1=1, r_2=0$ and $r_1=0, r_2=1$

∴ Coefficient

⇒ ${}^pC_1{}^qC_0\,(-1)^0+{}^pC_0{}^qC_1\,(-1)^1=7$

$\Rightarrow \quad p-q=1$ ...(i)

For coefficient of $x^2$, $r_1+r_2=2$

| $r_1$ | $r_2$ |
|---|---|
| 0 | 2 |
| 1 | 1 |
| 2 | 0 |

$\therefore$ Coefficient

$= {}^pC_0{}^qC_2(-1)^2+{}^pC_1{}^qC_1(-1)^1 + {}^pC_2{}^qC_0(-1)^0$

$\Rightarrow -2=\dfrac{q(q-1)}{2}-pq+\dfrac{p(p-1)}{2}$

$\Rightarrow -4=q^2-q-2pq+p^2-p$

$\Rightarrow -4=(p-q)^2-p-q$

$\Rightarrow -4=1-p-q$

$\Rightarrow p+q=5$ ...(ii)

On solving Eqs. (i) and (ii), we get

$p=3, q=2$

$\Rightarrow p^2+q^2=9+4=13$

**31.** *(c)* Coefficient of $x^{70}$ in

$x^2(1+x)^{98}+x^3(1+x)^{97}+x^4(1+x)^{96} +\ldots+x^{54}(1+x)^{46}$ is

$= {}^{98}C_{68}+{}^{97}C_{67}+{}^{96}C_{66}+\ldots + {}^{47}C_{17}+{}^{46}C_{16}$

$= {}^{46}C_{30}+{}^{47}C_{30}+\ldots+{}^{98}C_{30}$

$=({}^{46}C_{31}+{}^{46}C_{30})+{}^{47}C_{30}+\ldots + {}^{98}C_{30}-{}^{46}C_{31}$

$= {}^{47}C_{31}+{}^{47}C_{30}+\ldots\ldots {}^{98}C_{30}-{}^{46}C_{31}$

$= {}^{99}C_{31}-{}^{46}C_{31}={}^{99}C_p-{}^{46}C_q$

$\therefore$Possible values of $(p+q)$ are 62, 83.

$\therefore \quad p+q=83$

**32.** *(c)* General term

$T_{r+1}={}^{15}C_r\left(5^{\frac{1}{3}}\right)^r\left(2^{\frac{1}{5}}\right)^{15-r}$

$={}^{15}C_r 5^{\frac{r}{3}}\cdot 2^{\frac{15-r}{5}}, r\in\{0, 1, 2, \ldots, 15\}$

For rational terms,

$\dfrac{r}{3}\in$ integer and $\dfrac{r}{5}\in$ integer

$\Rightarrow$ 3 and 5 divides $r$

$\Rightarrow$ 15 divides $r$

$\Rightarrow r=0$ and 15

Hence, only two rational terms i.e.

${}^{15}C_0 2^3+{}^{15}C_{15}(5)^5=8+3125=3133$

**33.** (*) Binomial coefficients of 4 consecutive terms of $(1+x)^n$ are ${}^nC_r$, ${}^nC_{r+1}$, ${}^nC_{r+2}$, ${}^nC_{r+3}$, respectively.

Given,

${}^nC_r=2-p, {}^nC_{r+2}=2-\alpha,$

${}^nC_{r+1}=p, \quad {}^nC_{r+3}=\alpha$

$\therefore {}^nC_r+{}^nC_{r+1}=2$

and ${}^nC_{r+2}+{}^nC_{r+3}=2$

$\Rightarrow \quad {}^{n+1}C_{r+1}=2$

and $\quad {}^{n+1}C_{r+3}=2$

$\therefore \quad {}^{n+1}C_{r+1}={}^{n+1}C_{r+3}$

$\Rightarrow \quad r+1+r+3=n+1$

$\Rightarrow \quad 2r+4=n+1$

$\Rightarrow \quad n=2r+3$

Now, ${}^{n+1}C_{r+1}=2$

${}^{2r+4}C_{r+1}=2$

Not such value possible.

**34.** *(b)* We have, $(22)^{2022}+(2022)^{22}$

As 2022 is completely divisible by 3, so $(2022)^{22}$ is also divisible by 3

$(22)^{2022}=(21+1)^{2022}$

$=(3\times 7+1)^{2022}=3m+1$

$\Rightarrow (22)^{2022}$ leaves a remainder 1, when divided by 3.

$\therefore (22)^{2022}+(2022)^{22}$ leaves a remainder 1 when divided by 3.

$\therefore \quad \alpha=1$

$(22)^{2022}+(2022)^{22}=(21+1)^{2022} +(2023-1)^{22}$

$=7K+1+7\mu+1=7(K+\mu)+2$

$\Rightarrow (22)^{2022}+(2022)^{22}$ leaves a remainder 2 when divided by 7.

$\therefore \quad \beta=2$

Hence, $\alpha^2+\beta^2=1^2+2^2=5$

**35.** *(b)* $\because (x^n-a^n)$ is divisible by $(x-a)$.

$\therefore (25^{190}-19^{190})$ is divisible by $(25-19)$ i.e., 6

and $(8^{190}-2^{190})$ is divisible by $(8-2)=6$

$\Rightarrow (25^{190}-19^{190})-(8^{190}-2^{190})$ is divisible by 6

Again, $(25^{190}-8^{190})$ is divisible by $(25-8)=17$

and $(19^{190}-2^{190})$ is divisible by $(19-2)=17$

$\Rightarrow (25^{190}-8^{190})-(19^{190}-2^{190})$ is divisible by 17.

$\Rightarrow$ The given value is divisible by 6 as well as 17.

Hence, this is divisible by 2, 3, 17, 34 but not divisible by 14.

**36.** *(b)* $\because x=(8\sqrt{3}+13)^{13}$

$={}^{13}C_0(8\sqrt{3})^{13}(13)^0 + {}^{13}C_1(8\sqrt{3})^{12}(13)^1 +\ldots\ldots+{}^{13}C_{13}(8\sqrt{3})^0(13)^{13}$

and let $x_1=(8\sqrt{3}-13)^{13}$

$={}^{13}C_0(8\sqrt{3})^{13}(13)^0 - {}^{13}C_1(8\sqrt{3})^{12}(13)^1+\ldots - {}^{13}C_{13}(8\sqrt{3})^0(13)^{13}$

Therefore,

$x-x_1=2[{}^{13}C_1(8\sqrt{3})^{12}(13)^1+{}^{13}C_3 (8\sqrt{3})^{10}(13)^3\ldots\ldots {}^{13}C_{13}(8\sqrt{3})^0(13)^{13}]$

$\because (x-x_1)$ is multiple of 2 that means $(x-x_1)$ is even integer. Hence, $[x]$ is even.

Similarly,

$(y-y_1)=2[{}^9C_1(7\sqrt{2})^8(9)^1+{}^9C_3 (7\sqrt{2})^6(9)^3+\ldots+{}^9C_9(7\sqrt{2})^0(9)^9]$

$\because (y-y_1)$ is multiple of 2 means $(y-y_1)$ is even integer. Hence, $[y]$ is even.

**37.** *(c)*

> **key idea** For the solution firstly find the middle term using formula $T_{r+1}={}^nC_rx^{n-r}y^r$, then solve it.

Sum of coefficient of odd power of $x$ in $(1+x)^{99}$ is $K$.

So there are total 100 terms.

We know that

Sum of coefficient of even terms = Sum of coefficient of odd terms

$\Rightarrow$ Sum of coefficient of odd terms

$=\dfrac{\text{Sum of coefficient of all terms}}{2}$

$=\dfrac{2^{99}}{2}=2^{98}$

So, $K=2^{98}$

Since, $a$ is the middle term of $\left(2+\dfrac{1}{\sqrt{2}}\right)^{200}$

Total terms $=200+1=201$

So, $\left(\frac{201+1}{2}\right)$th term will be middle term ⇒ 101th term

So, $T_{101} = a$

$= {}^{200}C_{100}\, 2^{200-100}\left(\frac{1}{\sqrt{2}}\right)^{100}$

$= {}^{200}C_{100}\cdot 2^{50}$

Now, $\frac{{}^{200}C_{99}\cdot K}{a} = \frac{{}^{200}C_{99}\cdot 2^{98}}{{}^{200}C_{100}\cdot 2^{50}}$

$= \frac{25}{101}\cdot 2^{50}$

Compare with $\frac{2^{l} m}{n}$

So, $(l, n) = (50, 101)$

**38.** *(a)* In the expansion of $(1+x)^{10}$

$T_{r+1} = {}^{10}C_{10-r}\, x^{10-r} = {}^{10}C_r\, x^{10-r}$

Coefficient of $x^{10-r} = a_r = {}^{10}C_r$

We know that $\frac{{}^{n}C_r}{{}^{n}C_{r-1}} = \frac{n-r+1}{r}$

$\Rightarrow \frac{a_r}{a_{r-1}} = \frac{{}^{10}C_r}{{}^{10}C_{r-1}} = \frac{10-r+1}{r} = \frac{11-r}{r}$

Now, $\sum_{r=1}^{10} r^3\left(\frac{a_r}{a_{r-1}}\right)^2 = \sum_{r=1}^{10} r^3\left(\frac{11-r}{r}\right)^2$

$= \sum_{r=1}^{10} r(r^2-22r+121)$

$= \sum_{r=1}^{10} r^3 - 22\sum_{r=1}^{10} r^2 + 121\sum_{r=1}^{10} r$

$= \left[\frac{10(10+1)}{2}\right]^2 - 22\left[\frac{10\times(10+1)(20+1)}{6}\right]$

$+ 121\times\frac{10\times(10+1)}{2} = 1210$

**Alternate Solution**

$\sum_{r=1}^{10} r^3\left(\frac{a_r}{a_{r-1}}\right)^2 = \sum_{r=1}^{10} r^3\left(\frac{{}^{10}C_r}{{}^{10}C_{r-1}}\right)^2$

$= \sum_{r=1}^{10} r^3\times\frac{(11-r)^2}{r^2} = \sum_{r=1}^{10} r(11-r)^2$

$= \sum_{r=1}^{10}(r^2-22r+121)r$

$= \sum_{r=1}^{10}(r^3-22r^2+121r)$

we will get 1210.

**39.** *(a)* The given expression forms a GP with common ratio $\left(\frac{x}{5+x}\right)$ $[\because r > 1]$

$$\frac{x^{500}\dfrac{[(x+5)^{501}-x^{501}]}{x^{501}\left(\dfrac{5+x}{x}\right)}}{\dfrac{5}{x}}$$

⇒ Coefficient of $x^{101}$ in expression $\frac{1}{5}[(x+5)^{501}-x^{501}]$

$\Rightarrow \frac{1}{5}\times {}^{501}C_{101}\times 5^{501-101}$

$\Rightarrow {}^{501}C_{101}\times 5^{399}$

**40.** *(c)* Coefficient of $x^7$ in $\left(x^2+\frac{1}{bx}\right)^{11}$

General term $= {}^{11}C_r x^{3r-11}\, b^{r-11}$

$\Rightarrow 3r-11 = 7 \Rightarrow r = 6$

$\therefore$ Coefficient of $x^7 = {}^{11}C_6 b^{-5}$

Now, coefficient of $x^{-7}$ in $\left(x-\frac{1}{bx^2}\right)^{11}$.

General term $= {}^{11}C_r\left(\frac{-1}{b}\right)^{11-r} x^{3r-22}$

$\Rightarrow 3r-22 = -7$

$\Rightarrow r = 5$

Coefficient $= {}^{11}C_5\left(\frac{-1}{b}\right)^6 = {}^{11}C_5 b^{-6}$

Now, according to the question,

${}^{11}C_6 b^{-5} = {}^{11}C_5 b^{-6} \Rightarrow b = 1$

**41.** *(d)* In the given expansion of $\left(2^{\frac{1}{3}}+3^{\frac{1}{4}}\right)^{12}$

General term,

$T_{r+1} = {}^{12}C_r (2^{1/3})^r (3^{1/4})^{12-r}$

$(2^{1/3})^r$ will be a rational number when $r = 0, 3, 6, 9, 12$ and $(3^{1/4})^{12-r}$ will be rational number when $r = 0, 4, 8, 12$

Common $r = 0, 12$

If $r = 0$, then

$T_1 = {}^{12}C_0 (2^{1/3})^0 (3^{1/4})^{12} = 27$

If $r = 12$, then

$T_{13} = {}^{12}C_{12}\left(2^{\frac{1}{3}}\right)^{12}\left(3^{\frac{1}{4}}\right)^0 = 16$

So, $T_1 + T_{13} = 27 + 16 = 43$

**42.** *(d)* In the expansion of $\left(x\sin\alpha + a\frac{\cos\alpha}{x}\right)^{10}$

$T_{r+1} = {}^{10}C_r (x)^{10-2r}(\sin\alpha)^{10-r}(a\cos\alpha)^r$

$T_{r+1}$ is independent of $x$, when $10-2r = 0$

$\therefore r = 5$

$\Rightarrow T_6 = {}^{10}C_5 a^5\cdot\frac{1}{2^5}(\sin 2\alpha)^5$

For greatest value, $\sin 2\alpha = 1$

$= {}^{10}C_5 (a)^5\cdot\frac{1}{2^5}$

Given that the greatest value is $\frac{10!}{(5!)^2}$.

So, $\frac{10!}{(5!)^2}\cdot\frac{a^5}{2^5} = \frac{10!}{(5!)^2}$

$\Rightarrow a = 2$

**43.** *(d)* Now, $(1-y)^m(1+y)^n$

$= (1 - my + {}^{m}C_2 y^2 - \ldots + (-1)^m\cdot {}^{m}C_m y^m)$

$\times (1 + ny + {}^{n}C_2 y^2 + \ldots + {}^{n}C_n y^n)$

On expanding,

$(1-y)^m(1+y)^n = 1 + (n-m)y + ({}^{n}C_2 - mn + {}^{m}C_2)y^2 + \ldots$

$\therefore$ Coefficient of $y = n-m$

Coefficient of $y^2 = {}^{n}C_2 - mn + {}^{m}C_2$

Given expression have

Coefficient of $y = a_1 = 10$

Coefficient of $y^2 = a_2 = 10$

$\therefore n - m = 10$ ...(i)

and ${}^{n}C_2 + {}^{m}C_2 - mn = 10$ ...(ii)

From Eq. (ii), we get

$\frac{n!}{2!(n-2)!} + \frac{m!}{2!(m-2)!} - mn = 10$

$\Rightarrow n(n-1) + m(m-1) - 2mn = 20$

$\Rightarrow (m^2+n^2-2mn) - (m+n) = 20$

$\Rightarrow (-10)^2 - (m+n) = 20$ [using Eq. (i)]

$\Rightarrow m + n = 100 - 20 = 80$

**44.** *(a)* By using Binomial expansion, $(r+1)$th term,

$T_{r+1} = {}^{60}C_r 3^{r/4}\, 5^{\frac{60-r}{8}}$

Let $A$ be a set when $r$ is the multiple of 4.

$A = \{4, 8, 12, \ldots, 56, 60\}$

$n(A) = 15$

Let $B$ be a set of $r$, when $(60-r)$ is the multiple of 8.

$B = \{4, 12, 20, 28, 36, 44, 52, 60\}$

$n(B) = 8$

Now, $n(A\cap B) = 8$

So, there are only 8 terms out of 61 terms which will be rational numbers.

53 terms will be irrational.
So, $n = 53$
and $n - 1 = 52$ which is divisible only by 26 among the given options.

**45.** (a) Given, $(1 - x + x^3)^n = \sum_{j=0}^{3n} a_j x^j$

$\Rightarrow (1 - x + x^3)^n$
$= a_0 + a_1 x + a_2 x^2 + \ldots + a_{3n} x^{3n}$

Putting $x = 1$,
$$1 = a_0 + a_1 + a_2 + \ldots + a_{3n} \quad \ldots(i)$$
Putting $x = -1$,
$$1 = a_0 - a_1 + a_2 - a_3 + \ldots (-1)^{3n} a_{3n} \quad \ldots(ii)$$
Adding Eqs. (i) and (ii), we get
$$a_0 + a_2 + a_4 + \ldots = 1$$
On subtracting Eq. (ii) from Eq. (i), we get
$$a_1 + a_3 + a_5 + \ldots = 0$$
Now, $\sum_{j=0}^{\left[\frac{3n}{2}\right]} a_{2j} + 4 \sum_{j=0}^{\left[\frac{3n-1}{2}\right]} a_{2j} + 1$
$= [a_0 + a_2 + a_4 + \ldots] + 4[a_1 + a_3 + a_5 + \ldots] + 1$
$= 1 + 4 \times 0 + 1 = 2$

**46.** (d)

> **key idea** Use the inequality AM ≥ GM
> $$\frac{\alpha^3 + \beta^2}{2} \geq \sqrt{\alpha^3 \beta^2}$$
> $$\Rightarrow \alpha^6 \beta^4 \leq 16$$

∵ The general term in the expansion of binomial is
$$T_{r+1} = {}^{10}C_r \, \alpha^{10-r} \beta^r x^{\left(\frac{20 - 2r - 3r}{18}\right)}$$
For the term independent of $x$,
$$20 - 2r - 3r = 0 \Rightarrow r = 4$$
So, the value of the term independent of $x$ is
$${}^{10}C_4 \, \alpha^6 \beta^4 = 210 \, \alpha^6 \beta^4$$
Now, it is given that $\alpha > 0, \beta > 0$, such that
$\alpha^3 + \beta^2 = 4$, so according to AM-GM inequality, we have
$$\frac{\alpha^3 + \beta^2}{2} \geq \sqrt{\alpha^3 \beta^2}$$
$$\Rightarrow \quad \alpha^6 \beta^4 \leq 16$$
It is given that, the maximum value of $210 \, \alpha^6 \beta^4 = 10k$
$$\Rightarrow \quad 210 \times 16 = 10k$$
$$\Rightarrow \quad k = 336$$

**47.** (c) $T_{r+1} = {}^6C_r \, (-3)^r \, (2)^{6-r} \, x^{12 - 2r - 2r}$
$= {}^6C_r \, (-3)^r \, (2)^{6-r} \, x^{12 - 4r} \quad \ldots(i)$

Now, the term independent of $x$ in the expansion of $\left(\frac{1}{60} - \frac{x^8}{81}\right)\left(2x^2 - \frac{3}{x^2}\right)^6$
= the term independent of $x$ in the expansion of $\frac{1}{60}\left(2x^2 - \frac{3}{x^2}\right)^6$ + the term independent of $x$ in the expansion of $-\frac{x^8}{81}\left(2x^2 - \frac{3}{x^2}\right)^6$

[put $r = 5$ for first sum and $r = 5$ for second sum]
$$= \frac{1}{3}(-3)^3 \, 2^3 + \frac{3^5 \times 2(6)}{81}$$
$$= 36 - 72 = -36$$

**48.** (d) Clearly, number of terms in the expansion of
$\left(1 - \frac{2}{x} + \frac{4}{x^2}\right)^n$ is $\frac{(n+2)(n+1)}{2}$
or ${}^{n+2}C_2$.
[assuming $\frac{1}{x}$ and $\frac{1}{x^2}$ distinct]
$$\therefore \quad \frac{(n+2)(n+1)}{2} = 28$$
$$\Rightarrow \quad n = 6$$
Hence, sum of coefficients
$$= (1 - 2 + 4)^6 = 3^6 = 729$$

**49.** (c) $X = \{4^n - 3n - 1 : n \in N\}$,
$Y = \{9(n-1) : n \in N\}$
Now, $4^n - 3n - 1 = (1 + 3)^n - 3n - 1$
$= (1 + {}^nC_1 3 + {}^nC_2 3^2 + \ldots + 3^n) - 3n - 1$
$= (1 + 3n) + 9({}^nC_2 + \ldots + 3^{n-2}) - 3n - 1$
$= 9({}^nC_2 + \ldots + 3^{n-2}) = 9k$, where $k$ is an integer.
Also for $n = 1$, $4^n - 3n - 1 = 0$, a multiple of 9.
Every element in $X$ is a multiple of 9.
But $Y$ contains all multiple of 9.
Hence, $X \cup Y = Y$.

**50.** (a) $(\sqrt{3} + 1)^{2n} = {}^{2n}C_0 (\sqrt{3})^{2n} + {}^{2n}C_1 (\sqrt{3})^{2n-1} + {}^{2n}C_2 (\sqrt{3})^{2n-2} + \ldots + {}^{2n}C_{2n} (\sqrt{3})^{2n-2n}$

$(\sqrt{3} - 1)^{2n} = {}^{2n}C_0 (\sqrt{3})^{2n} (-1)^0 + {}^{2n}C_1 (\sqrt{3})^{2n-1} (-1)^1 + {}^{2n}C_2 (\sqrt{3})^{2n-2} (-1)^2 + \ldots + {}^{2n}C_{2n} (\sqrt{3})^{2n-2n} (-1)^{2n}$

Subtracting both the binomial, we get expansions, we get
$(\sqrt{3} + 1)^{2n} - (\sqrt{3} - 1)^{2n}$
$= 2[{}^{2n}C_1 (\sqrt{3})^{2n-1} + {}^{2n}C_3 (\sqrt{3})^{2n-3} + {}^{2n}C_5 (\sqrt{3})^{2n-5} + \ldots + {}^{2n}C_{2n-1} (\sqrt{3})^{2n-(2n-1)}]$

which is most certainly an irrational number because of odd powers of $\sqrt{3}$ in each of the terms.

**51.** (b) Since, $\sum_{r=0}^{n} {}^nC_r \cdot x^r = (1 + x)^n$

On multiplying by $x$, we get
$$\sum_{r=0}^{n} {}^nC_r \cdot x^{r+1} = x(1 + x)^n$$
On differentiating w.r.t. $x$, we get
$$\sum_{r=0}^{n} (r+1) \cdot {}^nC_r \cdot x^r = (1 + x)^n + nx(1 + x)^{n-1}$$
Hence, Statement II is true. If $x = 1$, then
$$\sum_{r=0}^{n} (r+1) \cdot {}^nC_r = 2^n + n(2)^{n-1} = (n+2)\,2^{n-1}$$
Hence, Statements I and II are true and Statement II is a correct explanation of Statement I.]

**52.** (b) In the expansion of $\left(x^{\frac{2}{3}} + \frac{x^{-\frac{2}{5}}}{2}\right)^9$
$$T_{r+1} = {}^9C_r \left(x^{\frac{2}{3}}\right)^{9-r} \left(\frac{x^{-\frac{2}{5}}}{2}\right)^r$$
$$= {}^9C_r \cdot \frac{1}{2^r} \cdot x^{\frac{18-2r}{3} - \frac{2r}{5}}$$
For coefficient of $x^{\frac{2}{3}}$
$$\Rightarrow \frac{18 - 2r}{3} - \frac{2r}{5} = \frac{2}{3}$$
$$\Rightarrow 5(9 - r) - 3r = 5$$
$$\Rightarrow \quad 45 - 8r = 5$$
$$r = 5$$

Coefficient of $x^{\frac{2}{3}} = \frac{{}^9C_5}{2^5}$

For coefficient of $x^{-\frac{2}{5}}$

$\Rightarrow \frac{18-2r}{3} - \frac{2r}{5} = \frac{-2}{5}$

$\Rightarrow 5(9-r)-3r=-3$

$45-8r=-3$

$r=6$

Coefficient of $x^{-\frac{2}{5}} = \frac{{}^9C_6}{2^6}$

$\therefore \text{Sum} = \frac{{}^9C_5}{2^5} + \frac{{}^9C_6}{2^6} = \frac{21}{4}$

**Concept Enhancer**

General term of binomial expansion

$(a+b)^n$ is $T_{r+1} = {}^nC_r a^{n-r} b^r$

**53.** (b) For $(1+x)^n$

Coefficient of $x^4 \Rightarrow {}^nC_4$

Coefficient of $x^5 = {}^nC_5$

Coefficient of $x^6 = {}^nC_6$

$\because$ ${}^nC_4, {}^nC_5, {}^nC_6$ are in AP

$\Rightarrow 2\cdot{}^nC_5 = {}^nC_4 + {}^nC_6$

$\Rightarrow$

$2 \times \frac{n!}{(n-5)!5!} = \frac{n!}{(n-4)!4!} + \frac{n!}{(n-6)!6!}$

$\Rightarrow \frac{2}{(n-5)\cdot 5} = \frac{1}{(n-4)(n-5)} + \frac{1}{6\times 5}$

$\Rightarrow n^2-21n+98=0 \Rightarrow n=14, 7$

$\therefore$Maximum value of $n=14$

**54.** (*) Given expression $\left(\frac{4x}{5} - \frac{5}{2x}\right)^{2022}$

So, 1011th term from the end

$= T_{1010+1} = {}^{2022}C_{1010}\left(\frac{-5}{2x}\right)^{1012}\left(\frac{4x}{5}\right)^{1010}$

and 1011th term from the beginning

$= T_{1010+1} = {}^{2022}C_{1010}\left(\frac{4x}{5}\right)^{1012}\left(\frac{-5}{2x}\right)^{1010}$

Given that,

1011th term from the end in above binomial expression is 1024 times 1011th term from the beginning.

Thus, ${}^{2022}C_{1010}\left(\frac{-5}{2x}\right)^{1012}\left(\frac{4x}{5}\right)^{1010}$

$= 1024\ {}^{2022}C_{1010}\left(\frac{4x}{5}\right)^{1012}\left(\frac{-5}{2x}\right)^{1010}$

$\Rightarrow \left(\frac{-5}{2x}\right)^2 = 1024\left(\frac{4x}{5}\right)^2$

$\Rightarrow x^4 = \frac{5^4}{64\times 1024}$

$\Rightarrow x^4 = \frac{5^4}{16^4}$

$\Rightarrow |x| = \frac{5}{16} = \frac{5}{2^4}$ (Bonus question)

None of the given option correct.

**55.** (c) We have,

$S1: (2023)^{2022} - (1999)^{2022}$

$= (1999+24)^{2022} - (1999)^{2022}$

$= {}^{2022}C_0\ (1999)^{2022}\ (24)^0 + {}^{2022}C_1\ (1999)^{2021}\ (24)^1 + {}^{2022}C_2(1999)^{2020}(24)^2 + \ldots - (1999)^{2022}$

$= {}^{2022}C_1\ (1999)^{2021}(24) + {}^{2022}C_2(1999)^{2022}(24)^2 + \ldots$

$= 24\,({}^{2022}C_1(1999)^{2021} + {}^{2022}C_2(1999)^{2022}(24) + \ldots)$

$\Rightarrow$ $S1$ is divisible by 24.

Now, $S2: 13(13)^n - 12n - 13$

Here, $13^n = (1+12)^n$

$= 1 + {}^nC_1 12 + {}^nC_2(12)^2 + {}^nC_3(12)^3 + \ldots$ upto $n$ terms.

$\therefore S2: 13\,(1 + {}^nC_1(12) + {}^nC_2(12)^2 + {}^nC_3(12)^3 + \ldots) - 12n - 13$

$= 13 + 156n + 13\,({}^nC_2(12)^2 + {}^nC_3(12)^3 + \ldots) - 12n - 13$

$= 144n + 144 \times 13\,({}^nC_2 + {}^nC_3(12) + \ldots)$

$144(n + 13({}^nC_2 + {}^nC_3(12)\ldots)$

$\Rightarrow$ $S2$ is divisible by 144 for infinitely many $n \in N$.

**56.** (c) We have,

$$\frac{\text{The fifth terms from the beginning in } \left(\sqrt[4]{2}\ \frac{1}{\sqrt[4]{3}}\right)^n}{\text{The fifth term from the end in } \left(\sqrt[4]{2}\ \frac{1}{\sqrt[4]{3}}\right)^n} = \frac{\sqrt{6}}{1}$$

$$\Rightarrow \frac{{}^nC_4\,(\sqrt[4]{2})^{n-4}\left(\frac{1}{\sqrt[4]{3}}\right)^4}{{}^nC_4\left(\frac{1}{\sqrt[4]{3}}\right)^{n-4}(\sqrt[4]{2})^4} = \frac{\sqrt{6}}{1}$$

[$\because$ $r$th term from end in the expansion of $(x+y)^n$ = $r$th terms from beginning in the expansion of $(y+x)^n$]

$$\Rightarrow \frac{{}^nC_4\,(2)^{\frac{n-4}{4}}\left(\frac{1}{3}\right)^{4/4}}{{}^nC_4\left(\frac{1}{3}\right)^{\frac{n-4}{4}}(2)^{4/4}} = \frac{\sqrt{6}}{1}$$

$\Rightarrow (2)^{\frac{n-8}{4}}(3)^{\frac{n-8}{4}} = 6^{1/2}$

$\Rightarrow 6^{\frac{n-8}{4}} = 6^{1/2} \Rightarrow n=10$

$\therefore T_3 = {}^{10}C_2\,(\sqrt[4]{2})^8\left(\frac{1}{\sqrt[4]{3}}\right)^2$

$= 45\,(2)^{\frac{8}{4}}\frac{1}{3^{1/2}} = 60\sqrt{3}$

**57.** (a) Now, general term in the expansion of $\left(ax^3 + \frac{1}{bx^{1/3}}\right)^{15}$ is

$T_{r+1} = {}^{15}C_r\,.(ax^3)^{15-r}\left(\frac{1}{bx^{1/3}}\right)^r$

$= {}^{15}C_r\,.(a)^{15-r}.(x)^{45-3r}\,.(b)^{-r}\,(x)^{-r/3}$

For coefficient of $x^{15}$, we put

$45 - 3r - \frac{r}{3} = 15 \Rightarrow 30 = \frac{10r}{3}$

$\Rightarrow r=9$

So, coefficient of $x^{15}$ in

$\left(ax^3 + \frac{1}{bx^{1/3}}\right)^{15} = {}^{15}C_9\,a^6 b^{-9}$ ...(i)

Again, we find general term in $\left(ax^{1/3} - \frac{1}{bx^3}\right)^{15}$ is

$T_{r+1} = {}^{15}C_r\,.(ax^{\frac{1}{3}})^{15-r}.\left(-\frac{1}{bx^3}\right)^r$

$\Rightarrow T_{r+1} = {}^{15}C_r\,.(a)^{15-r}.(x)^{\frac{15-r}{3}}.(-1)^r.(b)^{-r}.(x)^{-3r}$

For coefficient of $x^{-15}$, we get

$5 - \frac{r}{3} - 3r = -15$

$\Rightarrow \frac{10r}{3} = 20$

$\Rightarrow \quad r = 6$

$\therefore$ Coefficient of $x^{-15}$ in $\left(ax^{\frac{1}{3}} - \frac{1}{bx^3}\right)^{15}$

$= {}^{15}C_6 \cdot a^9 \times b^{-6} \times (-1)^6$

$= {}^{15}C_6 \cdot a^9 b^{-6}$ ...(ii)

According to the question,
Eq. (i) = Eq. (ii)

$\Rightarrow {}^{15}C_9\, a^6 . b^{-9} = {}^{15}C_6 \cdot a^9\, b^{-6}$

$\Rightarrow a^6 b^6 = a^9 b^9 \Rightarrow a^3 b^3 = 1,$

where $a, b > 0 \Rightarrow ab = 1$

**58.** *(c)* $a_1 = 1, a_2 = 2, a_{n+2} = \frac{2}{a_{n+1}} + a_n$

$\Rightarrow a_{n+2} \cdot a_{n+1} - a_{n+1} \cdot a_n = 2$

$a_3 \cdot a_2 - a_2 \cdot a_1 = 2$

$a_4 \cdot a_3 - a_3 \cdot a_2 = 2$

$a_5 \cdot a_4 - a_4 \cdot a_3 = 2$

..............................

..............................

$\therefore$ The series $a_1 \cdot a_2, a_2 \cdot a_3, a_3 \cdot a_4, a_4 \cdot a_5, \ldots$

$\Rightarrow 1 \cdot 2, 2 \cdot 2, 2 \cdot 3, 2 \cdot 4, \ldots.$

Now, $\prod_{n=1}^{n=30} \left(\frac{a_n + \frac{1}{a_{n+1}}}{a_{n+2}}\right)$

$= \prod_{n=1}^{n=30} \left(\frac{a_{n+2} - \frac{1}{a_{n+1}}}{a_{n+2}}\right)$

$= \prod_{n=1}^{30} \left(1 - \frac{1}{a_{n+2} \cdot a_{n+1}}\right)$

$= \left(1 - \frac{1}{2 \cdot 2}\right) \cdot \left(1 - \frac{1}{2 \cdot 3}\right) \cdot \left(1 - \frac{1}{2 \cdot 4}\right) \ldots \left(1 - \frac{1}{2 \cdot 31}\right)$

$\Rightarrow \frac{3 \cdot 5 \cdot 7 \cdot \ldots 61}{2^{30}\,(2 \cdot 3 \cdot 4 \cdot \ldots 31)}$

$= \frac{2^{-60} \cdot 61!}{31!\, 30!}$

$= 2^{-60} \cdot {}^{61}C_{30}$

$\therefore \quad \alpha = -60$

**59.** *(d)*

> **key idea** Use property of logarithm $a^{\log_a 6} = 6$

We have,

$\left\{3^{\log_3 \sqrt{25^{x-1} + 7}} + 3^{-\frac{1}{8} \cdot \log_3 (5^{x-1} + 1)}\right\}^{10}$

$= \left\{\sqrt{25^{x-1} + 7} + (5^{x-1} + 1)^{\frac{-1}{8}}\right\}^{10}$

Ninth term in the expansion is 180.

So, ${}^{10}C_8 (\sqrt{25^{x-1} + 7})^{10-8} \cdot \left[(5^{x-1} + 1)^{\frac{-1}{8}}\right]^8 = 180$

$\Rightarrow {}^{10}C_8 (25^{x-1} + 7)(5^{x-1} + 1)^{-1} = 180$

$\Rightarrow 45(25^{x-1} + 7)(5^{x-1} + 1)^{-1} = 180$

$\Rightarrow \frac{25^{x-1} + 7}{5^{x-1} + 1} = 4$

Let $\quad 5^{x-1} = t \Rightarrow \frac{t^2 + 7}{t + 1} = 4$

$\Rightarrow \quad t^2 - 4t + 3 = 0 \Rightarrow t = 3$ or $t = 1$

When $t = 3$

$5^{x-1} = 3 \Rightarrow x = \log_5 15$

When $\quad t = 1$

$5^{x-1} = 1 \quad [t = 5^{x-1}]$

$\Rightarrow \quad x = 1$

**60.** *(a)* Let $P = \left(1 + \frac{1}{10^{100}}\right)^{10^{100}}$

Let $x = 10^{100}$

$\Rightarrow P = \left(1 + \frac{1}{x}\right)^x$

$P = {}^xC_0 + {}^xC_1\left(\frac{1}{x}\right) + {}^xC_2\left(\frac{1}{x}\right)^2 + {}^xC_3\left(\frac{1}{x}\right)^3 + \ldots$ up to $(10^{100} + 1)$ term

$\Rightarrow \quad P = 1 + x\left(\frac{1}{x}\right) + \frac{x(x-1)}{2!} \cdot \frac{1}{x^2} + \frac{x(x-1)(x-2)}{3!} \cdot \frac{1}{x^3} + \ldots$ upto $(10^{100} + 1)$ terms

$\Rightarrow P = 1 + 1 + \left[\left(\frac{1}{2!} - \frac{1}{2!x}\right) + \left(\frac{1}{3!} + \ldots\right) + \ldots\right]$

$\Rightarrow P = 2 + \left(\text{Positive value less than } \frac{1}{2!} + \frac{1}{3!} + \frac{1}{4!} + \ldots\right)$

Now,

$\Rightarrow \quad \frac{1}{2!} + \frac{1}{3!} + \ldots = e - 2$

$\Rightarrow \ P = 2 +$ [Positive value less than $(e - 2)$]

$\Rightarrow \quad P \in (2, 3)$

So, lowest integer which is greater than $P$ is 3.

**61.** *(b)* $\frac{1}{a-b} + \frac{1}{a-2b} + \frac{1}{a-3b} + \ldots + \frac{1}{a-nb} = \alpha n + \beta n^2 + \gamma n^3$

$\Rightarrow \frac{1}{a}\left(\frac{1}{1 - \frac{b}{a}} + \frac{1}{1 - \frac{2b}{a}} + \frac{1}{1 - \frac{3b}{a}} + \ldots + \frac{1}{1 - \frac{nb}{a}}\right)$

$= \alpha n + \beta n^2 + \gamma n^3$

Let $b / a = x$

$\frac{1}{a}[(1-x)^{-1} + (1-2x)^{-1} + (1-3x)^{-1} + \ldots + (1-nx)^{-1}]$

$= \alpha n + \beta n^2 + \gamma n^3$

$\Rightarrow \frac{1}{a}(1 + x + x^2) + [1 + 2x + (2x)^2 + \ldots + [1 + nx + (nx)^2]$

$= \alpha n + \beta n^2 + \gamma n^3$

$\Rightarrow \frac{1}{a}\left(n + \frac{n(n+1)}{2}x + \frac{n(n+1)(2n+1)}{6}x^2\right)$

$= \alpha n + \beta n^2 + \gamma n^3$

Back substituting $x = \frac{b}{a}$, we get

$\frac{n}{a} + \frac{(n^2 + n)b}{2a^2} + \frac{n(n+1)(2n+1)b^2}{6a^3} = \alpha n + \beta n^2 + \gamma n^3$

Comparing coefficient of $n^3$,

we get $\frac{b^2}{3a^3} = \gamma$

**62.** *(b)* Using Binomial expansion, its $(r+1)$th term be,

$T_{r+1} = {}^{10}C_r\, (t)^{10-2r}\, (x)^{\frac{10-r}{5}}\, (1-x)^{r/10}$

If this term is independent of '$t$', then we have $10 - 2r = 0$ gives, $r = 5$

$\therefore \quad T_6 = {}^{10}C_5\, (x)^1 (1-x)^{1/2}$

Let $f(x)=x(1-x)^{1/2}$, to obtain its maximum value, we have to differentiate it and equate it to 0.
i.e. $f'(x)=0$

$\Rightarrow \dfrac{x}{2\sqrt{1-x}}(-1)+\sqrt{1-x}=0$

$\Rightarrow -3x+2=0 \Rightarrow x=2/3$ (Maximum value)

Thus, greatest term will be

$T_6={}^{10}C_5\left(\dfrac{2}{3}\right)\left(1-\dfrac{2}{3}\right)^{1/2}=\dfrac{10!\cdot 2}{(5!)^2(3\sqrt{3})}$

**63.** (b) Let the three consecutive terms in the binomial expansion of $(1+x)^{n+5}$ are ${}^{n+5}C_{r-1}$, ${}^{n+5}C_r$ and ${}^{n+5}C_{r+1}$

Now, according to the given information

${}^{n+5}C_{r-1}:{}^{n+5}C_r:{}^{n+5}C_{r+1}=5:10:14$

$\Rightarrow \dfrac{1}{(n-r+6)(n-r+5)}:\dfrac{1}{r(n-r+5)}:\dfrac{1}{(r+1)r}=5:10:14$

So, $\dfrac{r}{n-r+6}=\dfrac{5}{10}$

$\Rightarrow n+6=3r$ ...(i)

and $\dfrac{r+1}{n-r+5}=\dfrac{5}{7}$

$\Rightarrow 5n+18=12r$ ...(ii)

From Eqs. (i) and (ii), we have
$n=6, r=4$.
So, the largest coefficient in the expansion is same as the greatest binomial coefficient

$={}^{11}C_5$ or ${}^{11}C_6=\dfrac{11!}{5!6!}=462$

**64.** (c) Given binomial is $(x+1)^n$, whose general term, is $T_{r+1}={}^nC_r x^r$

According to the question, we have

${}^nC_{r-1}:{}^nC_r:{}^nC_{r+1}=2:15:70$

Now, $\dfrac{{}^nC_{r-1}}{{}^nC_r}=\dfrac{2}{15}\Rightarrow \dfrac{r}{n-r+1}=\dfrac{2}{15}$

$\Rightarrow 2n-17r+2=0$ ...(i)

Similarly, $\dfrac{{}^nC_r}{{}^nC_{r+1}}=\dfrac{15}{70}\Rightarrow \dfrac{r+1}{n-r}=\dfrac{3}{14}$

$\Rightarrow 3n-17r-14=0$ ...(ii)

On solving Eqs. (i) and (ii), we get

$n-16=0$

$\Rightarrow n=16$ and $r=2$

Now, the average

$=\dfrac{{}^{16}C_1+{}^{16}C_2+{}^{16}C_3}{3}=\dfrac{696}{3}=232$

**65.** (a) $\left(\dfrac{x+1}{x^{\frac{2}{3}}-x^{\frac{1}{3}}+1}-\dfrac{x-1}{x-x^{\frac{1}{2}}}\right)^{10}$

$=\left[\dfrac{(x^{1/3}+1)(x^{2/3}-x^{1/3}+1)}{(x^{2/3}-x^{1/3}+1)}-\dfrac{(\sqrt{x}-1)(\sqrt{x}+1)}{\sqrt{x}(\sqrt{x}-1)}\right]^{10}$

$=(x^{1/3}-1/x^{1/2})^{10}$

General term in given expansion

$={}^{10}C_r x^{\frac{10}{3}-\frac{r}{3}-\frac{r}{2}}(-1)^r$

Now for $x^{-5}$, $\dfrac{10}{3}-\dfrac{r}{3}-\dfrac{r}{2}=-5$

$\Rightarrow \dfrac{10}{3}+5=r\left(\dfrac{1}{3}+\dfrac{1}{2}\right)$

$\Rightarrow r=\dfrac{25}{3}\times\dfrac{6}{5}=10$

$\therefore$ Coefficient of $x^{-5}={}^{10}C_{10}(1)=1$

**66.** (b) Here, $(1-x-x^2+x^3)^6$

$=\{(1-x)-x^2(1-x)\}^6$

$=\{(1-x)(1-x^2)\}^6$

$=(1-x)^6\cdot(1-x^2)^6$

$=\left\{\sum_{r=0}^{6}(-1)^r\,{}^6C_r\cdot x^r\right\}\left\{\sum_{s=0}^{6}(-1)^s\,{}^6C_s\cdot x^{2s}\right\}$

$=\sum_{r=0}^{6}\sum_{s=0}^{6}(-1)^{r+s}\cdot{}^6C_r\cdot{}^6C_s\cdot x^{r+2s}$

For coefficient of $x^7$, we have

$r+2s=7$ i.e., $s=1, r=5$
or $s=2, r=3$ or $s=3, r=1$

$\therefore$ Coefficient of $x^7$ is

$\{(-1)^{5+1}\cdot{}^6C_5\cdot{}^6C_1\}+\{(-1)^{3+2}\cdot{}^6C_3\cdot{}^6C_2\}+\{(-1)^{1+3}\cdot{}^6C_1\cdot{}^6C_3\}$

$\Rightarrow (36)-(20)(15)+6(20)$

$\Rightarrow 36-300+120=-144$

**67.** (c) Now, $(1-ax)^{-1}(1-bx)^{-1}$

$=(1+ax+a^2x^2+\ldots)(1+bx+b^2x^2+\ldots)$

Hence, $a_n=$ Coefficient of $x^n$ in $(1-ax)^{-1}(1-bx)^{-1}$

$=a^0b^n+ab^{n-1}+\ldots+a^nb^0$

$=a^0b^n\left(1+\dfrac{a}{b}+\left(\dfrac{a}{b}\right)^2+\ldots\right)$

$=a^0b^n\left[\dfrac{\left(\dfrac{a}{b}\right)^{n+1}-1}{\dfrac{a}{b}-1}\right]$

[sum of $n$th term of a GP with common ratio $\dfrac{a}{b}$]

$=\dfrac{b^n(a^{n+1}-b^{n+1})}{a-b}\cdot\dfrac{b}{b^{n+1}}$ $[\because a_0=1=b_0]$

$=\dfrac{a^{n+1}-b^{n+1}}{a-b}$

**68.** (63) Let $l=(1919)^{1919}$

$=(2000-81)^{1919}$

$l={}^{1919}C_0(2000)^{1919}+\ldots{}^{1919}C_{1918}\times 2000\times(81)^{1918}-{}^{1919}C_{1919}\times(81)^{1919}$

Last 2 digit in $(1919)^{1919}$ is equivalent to remainder obtained by dividing $-(81)^{1919}$ with 100

Let $p=-(81)^{1919}$

$p=-(100-19)^{1919}$

$p=-[{}^{1919}C_0(100)^{1919}+\ldots{}^{1919}C_{1918}\times 100\times(19)^{1918}-{}^{1919}C_{1919}\times(19)^{1919}]$

Last two digit in $-(81)^{1919}$ is equivalent to remainder obtained by dividing $(19)^{1919}$ with 100

Let $q=(20-1)^{1919}$

$q={}^{1919}C_0(20)^{1919}+\ldots+{}^{1919}C_{1918}(20)-{}^{1919}C_{1919}$

Last 2 digit in $(1919)^{1919}$ is equivalent to remainder obtained.

When ${}^{1919}C_{1918}\times 20-1$ is divided by 100

$=1919\times 20-1=38380-1=38379$

Last 2 digit in $(1919)^{1919}$ is remainder when 38379 is divided by 100 is 79

Product of last 2 digit $=63$

**69.** *(465)* Given, $\sum_{r=1}^{30} \frac{r^2\,({}^{30}C_r)^2}{{}^{30}C_{r-1}} = \alpha\,(2^{29})$

$\because \quad \frac{{}^{30}C_r}{{}^{30}C_{r-1}} = \frac{30-r+1}{r} = \frac{31-r}{r}$

$\therefore \quad \text{LHS} = \sum_{r=1}^{30} r^2 \cdot {}^{30}C_r \cdot \frac{31-r}{r}$

$= \sum_{r=1}^{30} r \cdot {}^{30}C_r\,(31-r)$

$\Rightarrow \sum_{r=1}^{30} \frac{30!\,(31-r)}{(r-1)!\,(30-r)!}$

$\Rightarrow 30 \sum_{r=1}^{30} \frac{(31-r)\,29!}{(r-1)!\,(30-r)!}$

$= 30 \sum_{r=1}^{30} (31-r)\,{}^{29}C_{r-1}$

$= 30\left(31 \sum_{r=1}^{30} {}^{29}C_{r-1} - \sum_{r=1}^{30} r \cdot {}^{29}C_{r-1}\right)$

$= 30\left((31)(2^{29}) - \sum_{r=1}^{30} (r-1)\,{}^{29}C_{r-1} - \sum_{r=1}^{30} {}^{29}C_{r-1}\right)$

$= 30\left(31 \times 2^{29} - \sum_{r'=0}^{29} r' \cdot {}^{29}C_{r'} - 2^{29}\right)$

$= 30\,(31 \times 2^{29} - 29 \cdot 2^{28} - 2^{29})$

$= 30 \times 2^{28}\,(31 \times 2 - 29 - 2)$

$= 30 \times 2^{28} \times 31 = 15 \times 31 \times 2^{29}$

$\alpha = 15 \times 31 = 465$

> **Concept Enhancer**
>
> ${}^nC_r = \frac{n}{r} \cdot {}^{n-1}C_{r-1}$
>
> $\sum_{r=0}^{n} {}^nC_r = 2^n, \sum_{r=1}^{n} r \cdot {}^nC_r = n \cdot 2^{n-1}$

**70.** *(2035)* $\sum_{r=0}^{5} \frac{{}^{11}C_{2r+1}}{2r+2} = \frac{1}{12} \sum_{r=0}^{5} {}^{12}C_{2r+2}$

$= \frac{1}{12}\,({}^{12}C_2 + {}^{12}C_4 + {}^{12}C_6 + {}^{12}C_8 + {}^{12}C_{10} + {}^{12}C_{12})$

$= \frac{1}{12}\,(2^{11} - 1) = \frac{2^{11} - 1}{12}$

$= \frac{2047}{12} = \frac{m}{n}$

$\Rightarrow \quad m - n = 2047 - 12 = 2035$

> **Concept Enhancer**
>
> $\frac{{}^{n+1}C_{r+1}}{n+1} = \frac{{}^nC_r}{r+1}$

**71.** *(806)* We know that,

$(x+y)^n = {}^nC_0 x^n y^0 + {}^nC_1 x^{n-1} y^1 + {}^nC_2 x^{n-2} y^2 + \ldots + {}^nC_n x^0 y^n$

Now, second term,

$T_2 = {}^nC_1 x^{n-1} y = 135 \quad \ldots\text{(i)}$

Third term,

$T_3 = {}^nC_2 x^{n-2} y^2 = 30 \quad \ldots\text{(ii)}$

Fourth term,

$T_4 = {}^nC_3 x^{n-3} y^3 = \frac{10}{3} \quad \ldots\text{(iii)}$

By Eqs. (i) and (ii), we get

$\frac{{}^nC_1 x^{n-1} y}{{}^nC_2 x^{n-2} y^2} = \frac{135}{30}$

$\Rightarrow \quad \frac{2n}{n(n-1)} \cdot \frac{x}{y} = \frac{9}{2}$

$\Rightarrow \quad \frac{1}{n-1} \cdot \frac{x}{y} = \frac{9}{4} \quad \ldots\text{(iv)}$

By Eqs. (ii) and (iii), we get

$\frac{{}^nC_2 x^{n-2} y^2}{{}^nC_3 x^{n-3} y^3} = \frac{30 \times 3}{10}$

$\Rightarrow \quad \frac{n(n-1)}{2 \times 1} \times \frac{3 \times 2 \times 1}{n(n-1)(n-2)} \times \frac{x}{y} = 9$

$\Rightarrow \quad \frac{1}{(n-2)} \frac{x}{y} = 3 \quad \ldots\text{(v)}$

By Eqs. (iv) and (v), we get

$\frac{\frac{1}{n-1} \cdot \frac{x}{y}}{\frac{1}{(n-2)} \frac{x}{y}} = \frac{\frac{9}{4}}{\frac{3}{1}}$

$\Rightarrow \quad \frac{n-2}{n-1} = \frac{3}{4}$

$\Rightarrow \quad 4n - 8 = 3n - 3$

$\Rightarrow \quad n = 5$

Now, if $n = 5$, then

Eq. (i) will become ${}^5C_1 x^4 y = 135$

$\Rightarrow \quad x^4 y = 27$

and Eq. (v) will become $x = 9y$

Using both $\Rightarrow (9y)^4 \cdot y = 27$

$\Rightarrow \quad y^5 = \frac{3^3}{3^8} = \frac{1}{3^5}$

$\Rightarrow \quad y = \frac{1}{3}$

and $\quad x = 9 \times \frac{1}{3} = 3$

$\therefore 6(n^3 + x^2 + y) = 6\left(125 + 9 + \frac{1}{3}\right)$

$= 806$

**72.** *(54)* We have,

$(1 + 2x - 3x^2)\left(\frac{3}{2}x^2 - \frac{1}{3x}\right)^9$

Here, general term of $\left(\frac{3}{2}x^2 - \frac{1}{3x}\right)^9$

$= {}^9C_r \left(\frac{3}{2}x^2\right)^{9-r} (-1)^r \left(\frac{1}{3x}\right)^r$

$= {}^9C_r \frac{(3)^{9-r}}{(2)^{9-r}} (x^2)^{9-r} (-1)^r (3x)^{-r}$

$= {}^9C_r \frac{3^{9-2r}}{2^{9-r}} (-1)^r x^{18-3r}$

For constant term, put $18 - 3r = 0$

$\Rightarrow \quad r = 6$

On putting $r = 6$, we get

Coefficient of $x^0 = {}^9C_6 \frac{1}{216} x^0 = \frac{7}{18} x^0$

On putting $r = 7$, we get

Coefficient of

$x^{-3} = {}^9C_7 \frac{3^{-5}}{2^2} (-1)^7 x^{-3}$

$= -{}^9C_7 \frac{1}{3^5 \cdot 2^2} x^{-3} = -\frac{1}{27} x^{-3}$

$\therefore (1 + 2x - 3x^3)\left(\frac{7}{18} x^0 - \frac{1}{27} x^{-3}\right)$

Constant term $= \frac{7}{18} + \frac{3}{27} = \frac{7+2}{18} = \frac{1}{2}$

$\therefore \quad p = \frac{1}{2}$

$\therefore \; 108p = 108 \times \frac{1}{2} = 54$

> **Concept Enhancer**
>
> General term of Binomial expansion $(a+b)^n$ is
>
> $T_{r+1} = {}^nC_r a^{n-r} b^r.$

**73.** *(8)* Since,

$e^{1+x} = 1 + \frac{(1+x)}{1!} + \frac{(1+x)^2}{2!} + \frac{(1+x)^3}{3!} + \ldots$

$\Rightarrow \quad \frac{e^{1+x}}{1+x} = \frac{1}{1+x} + 1 + \frac{(1+x)}{2!} + \frac{(1+x)^2}{3!} + \frac{(1+x)^3}{4!} + \ldots$

Now, coefficient of $x^2$ in RHS is

$$1+\frac{^2C_2}{3}+\frac{^3C_2}{4}+\ldots=a$$

and coefficient of $x^2$ in LHS

$$e\left(1+x+\frac{x^2}{2!}\right)\ldots\left(1-x+\frac{x^2}{2!}+\ldots\right) \text{ is}$$

$$e-e+\frac{e}{2!}=a$$

$$b=1+\frac{2}{1!}+\frac{2^2}{2!}+\frac{2^3}{3!}+\ldots=e^2$$

$\therefore \quad 2b/a^2=8$

**74.** *(678)* Coefficient of $x^{30}$ in

$$\frac{(x+1)^6(1+x^2)^7(1-x^3)^8}{x^6}$$

and Coefficient of $x^{36}$ in

$(1+x)^6(1+x^2)^7(1-x^3)^8$

General term,

$$^6C_{r_1}\,^7C_{r_2}\,^8C_{r_3}(-1)^{r_3}x^{r_1+2r_2+3r_3}$$

$$r_1+2r_2+3r_3=36$$

**Case I**

| $r_1$ | $r_2$ | $r_3$ |
|---|---|---|
| 0 | 6 | 8 |
| 2 | 5 | 8 |
| 4 | 4 | 8 |
| 4 | 3 | 8 |

$\Rightarrow \quad r_1+2r_2=12$[taking $r_3=8$]

**Case II**

| $r_1$ | $r_2$ | $r_3$ |
|---|---|---|
| 1 | 7 | 7 |
| 3 | 6 | 7 |
| 5 | 5 | 7 |

$\Rightarrow \quad r_1+2r_2=15 \quad$ [taking $r_3=7$]

**Case III**

| $r_1$ | $r_2$ | $r_3$ |
|---|---|---|
| 4 | 7 | 6 |
| 6 | 6 | 6 |

$\Rightarrow \quad r_1+2r_2=18$ [taking $r_3=6$]

Coefficient

$=7+(25\times21)+(15\times35)+(35)$

$\quad-(6\times8)-(20\times7\times8)-(6\times21\times8)$

$\quad+(15\times28)+(7\times28)$

$=-678=\alpha$

Hence, $|\alpha|=678$

**75.** *(25)* $(x+3)^{n-1}+(x+3)^{n-2}(x+2)$

$+(x+3)^{n-3}(x+2)^2+\ldots+(x+2)^{n-1}$

$$=(x+3)^{n-1}\left[\frac{\left(\frac{x+2}{x+3}\right)^n-1}{\frac{x+2}{x+3}-1}\right]$$

$$=(x+3)^{n-1}\left[\frac{\frac{(x+2)^n}{(x+3)^n}-1}{\frac{-1}{x+3}}\right]$$

$=(x+3)^n-(x+2)^n$

Coefficient of

$x^r={}^nC_r3^r-{}^nC_r2^r=\alpha_r$

$$\sum_{r=0}^{n}\alpha_r=\sum_{r=0}^{n}{}^nC_r3^r-\sum_{r=0}^{n}{}^nC_r2^r$$

$=(1+3)^n-(1+2)^n$

$=4^n-3^n=\beta^n-\gamma^n$

$\beta=4,\gamma=3\Rightarrow\beta^2+\gamma^2=25$

**76.** *(118)* Expansion of $(1+x)(1-x^2)$

$$\left(1+\frac{3}{x}+\frac{3}{x^2}+\frac{1}{x^3}\right)^5$$

$$=(1+x)(1-x^2)\left[\left(1+\frac{1}{x}\right)^3\right]^5$$

$$=(1+x)(1-x^2)\frac{(1+x)^{15}}{x^{15}}$$

$$=\frac{(1+x)^{16}(1+x)(1-x)}{x^{15}}$$

$$=\frac{(1+x)^{17}(1-x)}{x^{15}}$$

$$\text{Expansion}=\frac{(1+x)^{17}}{x^{15}}-\frac{x(1+x)^{17}}{x^{15}}$$

Coefficient of $x^3$ in expansion is equal to coefficient of $x^{18}$ in the expansion.

$=0-1=-1$

and coefficient of $x^{-13}$ in the expansion is equal to coefficient of $x^2$ in the expansion

$={}^{17}C_2-{}^{17}C_1=17\times8-17$

$=17\times7=119$

$\therefore$Sum $=-1+119=118$

**77.** *(138)* We have,

$$T_{r+1}={}^nC_r11^{r/6}\cdot7^{\frac{824-r}{2}}$$

For integral term 6 should divided $r$

and $\dfrac{824-r}{2}$ must be integer

$\Rightarrow$ 2 must divide $r$

$\Rightarrow r$ divisible by 6

$\Rightarrow$ Possible values of $r\in\{0,1,2,\ldots,824\}$

$\Rightarrow$ For integer terms

$r\in\{0,6,12,\ldots,822\}$

$\therefore \quad 822=0+(n-1)6$

$\Rightarrow \quad n=138$

Hence, number of integral term is 138 term.

**78.** *(1)* $(64)^P=(9\times7+1)^P=(9\lambda+1)^P$;

$P=32^{32}$

${}^PC_0(9\lambda)^P+{}^PC_1(9\lambda)^{P-1}$

$+\ldots\ldots+{}^PC_{P-1}(9\lambda)+{}^PC_P1.$

$=M(a)+{}^PC_P1=M(9)+1$

$\because$ Remainder $=1$

**79.** *(0)* Since,

$(1-x^3)=(1-x)(1+x+x^2)$

$\Rightarrow(1-x^3)^{2007}=(1-x)^{2007}$

$(1+x+x^2)^{2007}$

Now, coefficient of $x^{2012}$ in

$(1-x)^{2008}(1+x+x^2)^{2007}$

$=$ Coefficient of $x^{2012}$ in

$(1-x^3)^{2007}(1-x)$

$=$ Coefficient of $x^{2012}$ in

${}^{2007}C_{r_1}(-x^3)^{r_1}$

$+{}^{2007}C_{r_2}(-1)^{r_2}\cdot x^{3r_2+1}$

$\because 3r_1=2012$, which is not possible for any $r_1\in W$

and $3r_2+1=2012$ also not possible for any $r_2\in W$

Hence, no any term containing $x^{2012}$.

Thus, coefficient of $x^{2012}$ is zero.

**80.** *(12)* $\quad 7\equiv7\pmod{17}$

$\Rightarrow \quad 7^2\equiv-2\pmod{17}$

$\Rightarrow \quad 7^6\equiv-8\pmod{17}$

$\Rightarrow \quad 7^8\equiv-1\pmod{17}$

$\Rightarrow \quad 7^{96}\equiv1\pmod{17}$

$\Rightarrow \quad 7^{103}\equiv12\pmod{17}$

Remainder = 12

**81.** *(413)* Given, $x_1+x_2+x_3+x_4+x_5+x_6+x_7=12$, where $x_i\in\{1,2,3,4\}$

Number of solution = Coefficient of $x^{12}$ in $(x^1+x^2+x^3+x^4)^7$

$$=\text{Coefficient of } x^{12} \text{ in}\left\{\frac{1-x^4}{1-x}\right\}^7\cdot x^7$$

$=$ Coefficient of $x^5$ in $(1-x^4)^7\,(1-x)^{-7}$ = Coefficient of $x^5$ in $(1-7x^4)\sum_{n=0}^{\infty}{}^{7+r-1}C_r\,x^r$

$$= {}^7C_0\,{}^{11}C_5 - {}^7C_1 \times {}^7C_1$$

$$= 462 - 49 = 413$$

**Alternate Solution**

$x_1+x_2+x_3+x_4+x_5+x_6+x_7=12$

| | | | | | | | |
|---|---|---|---|---|---|---|---|
| **Case 1** | 1 | 1 | 1 | 1 | 1 | 3 | 4 |
| **Case 2** | 1 | 1 | 1 | 1 | 2 | 2 | 4 |
| **Case 3** | 1 | 1 | 2 | 2 | 2 | 2 | 2 |
| **Case 4** | 1 | 1 | 1 | 2 | 2 | 2 | 3 |
| **Case 5** | 1 | 1 | 1 | 1 | 2 | 3 | 3 |

Total ways $=\dfrac{7!}{5!}+\dfrac{7!}{4!\cdot 2!}+\dfrac{7!}{5!\cdot 2!}+\dfrac{7!}{3!\cdot 3!}+\dfrac{7!}{4!\cdot 2!}=42+105+21+140+105=413$

**82.** *(171)* We have, $\left(3^{\frac{1}{2}}+5^{\frac{1}{4}}\right)^{680}$

General term of the expansion

$$\left(3^{\frac{1}{2}}+5^{\frac{1}{4}}\right)^{680} = {}^{680}C_r\,(3^{1/2})^{680-r}\,(5^{1/4})^r$$

$$= {}^{680}C_r \times 3^{\frac{680-r}{2}} \times 5^{\frac{r}{4}}$$

The term will be integral if $r$ is a multiple of 4.

$\therefore\ r = 0, 4, 8, 12, \ldots, 680$ (which is an AP)

$680 = 0 + (n-1)4$

$$n = \frac{680}{4} + 1 = 171$$

**83.** *(1275)* General term of expansion $\left(3x^2 - \dfrac{1}{2x^5}\right)^7$ is

$$T_{r+1} = {}^7C_r\,(3x^2)^{7-r}\left(-\frac{1}{2x^5}\right)^r$$

For term indendent of $x$

$$14 - 2r - 5r = 0 \Rightarrow r = 2$$

Required coefficient

$$= {}^7C_2\,3^{7-2}\left(-\frac{1}{2}\right)^2$$

$$= \frac{7\cdot 6}{2}\cdot 3^5\cdot\frac{1}{4} = 1275.75 = \alpha$$

$\therefore [a] = [1275.75] = 1275$

**84.** *(1)* We have, expansion is $\left(x^{\frac{2}{3}} + \dfrac{\alpha}{x^3}\right)^{22}$

$$\therefore T_{r+1} = {}^{22}C_r\left(x^{\frac{2}{3}}\right)^{22-r}\left(\frac{\alpha}{x^3}\right)^r$$

$$= {}^{22}C_r\,\alpha^r\,x^{\frac{2(22-r)}{3}-3r}$$

For term without $x$,

$$\frac{2(22-r)}{3} - 3r = 0$$

$\Rightarrow\quad 44 - 11r = 0 \Rightarrow r = 4$

Now, it is given that

Term without $x = 7315$

$\Rightarrow T_5 = 7315 \Rightarrow {}^{22}C_4\,\alpha^4 = 7315$

$\Rightarrow \dfrac{22\times 21\times 20\times 19}{4\times 3\times 2\times 1}\times\alpha^4 = 7315$

$\Rightarrow 7315\,\alpha^4 = 7315$

$\Rightarrow\quad \alpha^4 = 1 \Rightarrow \alpha = \pm 1 \Rightarrow |\alpha| = 1$

**85.** *(5040)* We have, $\left(\dfrac{4x}{5}+\dfrac{5}{2x^2}\right)^9$

$$\therefore T_{r+1} = {}^9C_r\left(\frac{4x}{5}\right)^{9-r}\left(\frac{5}{2x^2}\right)^r$$

$$= {}^9C_r\,2^{18-3r}\,5^{2r-9}\,x^{9-3r}$$

For coefficient of $x^{-6}$

$9 - 3r = -6 \Rightarrow 3r = 15 \Rightarrow r = 5$

$\therefore$ Coefficient of

$$x^{-6} = {}^9C_5\,2^{18-15}\,5^{10-9}$$

$$= \frac{9\times 8\times 7\times 6}{4\times 3\times 2\times 1}\times 8\times 5 = 5040$$

**86.** *(2)* The general term of the expansion $\left(x^{2/3}+\dfrac{2}{x^3}\right)^{30}$ is given by

$$T_{r+1} = {}^{30}C_r\,(x^{\frac{2}{3}})^{(30-r)}\left(\frac{2}{x^3}\right)^r$$

$$= {}^{30}C_r\,2^r\,x^{20-\frac{2}{3}r-3r}$$

Now, for term $\beta x^{-\alpha}$, $\beta \in N$ and $\alpha$ is smallest positive number, we have

$$20 - \frac{2}{3}r - 3r < 0 \Rightarrow 60 - 11r < 0$$

$\Rightarrow\quad r > 5.5 \Rightarrow r = 6$

Now, when $r = 6$

$$T_7 = {}^{30}C_6\,2^6\,.\,x^{-2}$$

On comparing with $\beta x^{-\alpha}$, $\alpha = 2$

**Concept Enhancer**

We know that the general term in $(x+y)^n$ $T_{r+1} = {}^nC_r x^{n-r} y^r$, then find general term of $\left(x^{2/3}+\dfrac{2}{x^3}\right)^{30}$.

**87.** *(23)* We have,

$x^{1/50} = 12 \Rightarrow x = 12^{50}$

And $y^{1/50} = 18 \Rightarrow y = 18^{50}$

So, $x + y = 12^{50} + 18^{50}$

$$= (12^2)^{25} + (18^2)^{25}$$

$$= (144)^{25} + (324)^{25}$$

$$= (150-6)^{25} + (325-1)^{25}$$

$\because$ 150 and 325 is divisible by 25.

$$= 25K - (6^{25} + 1)$$

$$= 25K - [(5+1)^{25} + 1]$$

$$= 25K - 25K_1 - (1+1) = 25K_2 - 2$$

So, when $x + y$ is dividing by 25

Then, remainder $= 25 - 2 = 23$

**88.** *(7)* $(2023)^{2023} = (1995+28)^{2023}$

$\Rightarrow (2023)^{2023} = {}^{2023}C_0(1995)^{2023} + {}^{2023}C_1(1995)^{2022}(28)^1 + \ldots + {}^{2023}C_{2022}(1995)(28)^{2022} + {}^{2023}C_{2023}(28)^{2023}$

So, $(2023)^{2023} = 1995$ [some integer] $+ (28)^{2023}$

And 1995 is divisible by 35, if we divide $(2023)^{2023}$ by 35, then remainder will be

$(28)^{2023} = (35-7)^{2023} = 35l - (7)^{2023}$

Again to get remainder, divide $-(7)^{2023}$ by 35.

So, $\quad (-7)^{2023} = (-7)\,(7)^{2022}$

$$= (-7)\,(7^3)^{674}$$

$$= (-7)\,(343)^{674} = (-7)\,(350-7)^{674}$$

$$= (-7)\,[35m - (7)^{674}]$$

Again to get remainder, divide $(-7)(-7)^{674}$ by 35.

So, $(-7)(+7)^{674} = -(7)^{675} = -(7^3)^{225}$

$$= -(343)^{225} = -(350-7)^{225}$$

Again $-(350-7)^{225}$ is equivalent to $35n + (7)^{225}$.

If we have to find remainder,

$$(7)^{225} = (7^3)^{75} = (343)^{75}$$

$$= (350-7)^{75}$$

and $\quad (350-7)^{75} \equiv -(7)^{75} = -(7^3)^{25}$

$(-7)^{75} = -(343)^{25} = -(350-7)^{25}$

Again $-(350-7)^{25} \equiv -(-7)^{25} = 7(7)^{24}$

$\because\ 7(7)^{24} = 7(7^3)^8 = 7(343)^8$

$$= 7(350-7)^8$$

Again $7(350-7)^8=7(-7)^8=(7^3)^3$

$\because \quad (7^3)^3=(343)^3=(350-7)^3$

Again $(350-7)^3 \equiv -(7)^3=-343$

$\because \quad -343=7-350$

So, remainder $=7$ of $\dfrac{(2023)^{2023}}{35}$

**89.** *(84)* $\left(\sqrt[4]{2}+\dfrac{1}{\sqrt[4]{3}}\right)^n=\left(2^{1/4}+\dfrac{1}{3^{1/4}}\right)^n$

Given, $\dfrac{T_5 \text{ from beginning}}{T_5 \text{ from end}}=\dfrac{\sqrt[4]{6}}{1}=\dfrac{6^{1/4}}{1}$

$T_5$ from end $=T_{n-5+2}=T_{n-3}$ from beginning

$$\Rightarrow \frac{T_5}{T_{n-3}}=\frac{T_{4+1}}{T_{n-4+1}}$$

$$=\frac{{}^nC_4(2^{1/4})^{n-4}\left(\frac{1}{3^{1/4}}\right)^4}{{}^nC_{n-4}(2^{1/4})^4\left(\frac{1}{3^{1/4}}\right)^{n-4}}=\frac{6^{1/4}}{1}$$

$$\Rightarrow 2^{\frac{n}{4}-2}\cdot 3^{\frac{n}{4}-2}=6^{1/4}$$

$$6^{\frac{n-8}{4}}=6^{1/4}$$

$\Rightarrow 6^{n-8}=6^1 \Rightarrow n-8=1 \Rightarrow n=9$

$T_6$ from beginning $T_6=T_{5+1}$

$$={}^9C_5(2^{1/4})^{9-5}\left(\frac{1}{3^{1/4}}\right)^5$$

$$={}^9C_5(2^{1/4})^{9-5}\,3^{-\frac{5}{4}}$$

$$\Rightarrow \frac{9\times 8\times 7\times 6}{4\times 3\times 2\times 1}\times 2\times 3^{-5/4}=\frac{\alpha}{3^{1/4}}$$

$$\Rightarrow \frac{3\times 42\times 2}{3}=\alpha \Rightarrow \alpha=84$$

**90.** *(57)* There are total 5 terms in the expansion of $\left(\dfrac{1}{\sqrt{6}}+\beta x\right)^4$.

Middle terms is 3rd term,

$$T_3={}^4C_2\left(\frac{1}{\sqrt{6}}\right)^2(\beta x)^2$$

Coefficient of the middle terms of this expansion $=\beta^2$

There are 3 terms in the expansion of $(1-3\beta x)^2$.

Middle term $(T_2)={}^2C_1(-3\beta x)=-6\beta x$

Coefficient of second terms of this expansion $=-6\beta$

There are 7 terms in the expansion $\left(1-\dfrac{\beta x}{2}\right)^6$.

Middle terms is 4th term

$$T_4={}^6C_3\left(-\frac{\beta}{2}x\right)^3$$

Coefficient of 4th term of this expansion

$$=-20\frac{\beta^3}{8}=-\frac{5\beta^3}{2}$$

The three terms of AP are $\beta^2, -6\beta, \dfrac{-5}{2}\beta^3$.

$$\therefore \quad 2(-6\beta)=\beta^2-\frac{5\beta^3}{2}$$

$$\Rightarrow \quad -12\beta=\beta^2-\frac{5\beta^3}{2}$$

$$\Rightarrow \quad -24\beta=2\beta^2-5\beta^3$$

$$\Rightarrow \quad 5\beta^2-2\beta-24=0 \qquad [\because \beta\neq 0]$$

$$\Rightarrow \quad 5\beta^2-12\beta+10\beta-24=0$$

$$\Rightarrow \quad \beta=\frac{12}{5}, \beta\neq -2$$

$$\therefore \quad d=-6\beta-\beta^2=-\beta(6+\beta)$$

$$=-\frac{12}{5}\left[6+\frac{12}{5}\right]$$

$$=-\frac{12}{5}\left(\frac{42}{5}\right)$$

$$=-\frac{504}{25}$$

$$\Rightarrow \quad 50-\frac{2d}{\beta^2}=50-\frac{2\left(-\frac{504}{25}\right)}{\frac{144}{25}}$$

$$=50+7=57$$

**91.** *(24)* If $T_{r+1}$ is the greatest term

Then, $\left[\dfrac{n+1}{1+\dfrac{3}{6x}}\right]=8$

where, $[x]$ greatest integer function

$$\left[\frac{3}{4}(n+1)\right]=8$$

$$\therefore \frac{3}{4}(n+1)\in[8,9) \Rightarrow n\in\left[\frac{29}{3},11\right).$$

$\therefore$ Least value of $n=n_0=10$

Now, binomial expansion is $(3+6x)^{10}$

$$\frac{\text{Coefficient of } x^6}{\text{Coefficient of } x^3}$$

$$=\frac{{}^{10}C_6\times 3^4\times 6^6}{{}^{10}C_3\times 3^7\times 6^3}=14=k$$

$\therefore k+n_0=14+10=24$

**92.** *(23)* $(1+x)^p(1-x)^q=[{}^pC_0+{}^pC_1x+{}^pC_2x^2+\ldots]$
$[{}^qC_0-{}^qC_1x+{}^qC_2x^2+\ldots]$

Coefficient of $x$ in $(1+x)^p(1-x)^q=-3$

$$\Rightarrow \quad -{}^pC_0{}^qC_1+{}^pC_1{}^qC_0=-3$$

$$\Rightarrow \quad -q+p=-3 \Rightarrow p=q-3$$

Coefficient of $x^2$ in $(1+x)^p(1-x)^q=-5$

$${}^pC_0{}^qC_2-{}^pC_1{}^qC_1+{}^pC_2{}^qC_0=-5$$

$$\Rightarrow \quad \frac{q(q-1)}{2}-pq+\frac{p(p-1)}{2}=-5$$

$$\Rightarrow q^2-q-2pq+p^2-p=-10$$

$$\Rightarrow q^2-q-2(q-3)q+(q-3)^2-(q-3)=-10$$

$$\Rightarrow \quad q^2-q-2q^2+6q+q^2+9-6q-q+3=-10$$

$$\Rightarrow \quad -2q=-22 \Rightarrow q=11$$

$$\Rightarrow \quad p=q-3=11-3=8$$

Now, coefficient of $x^3$ in $(1+x)^8(1-x)^{11}$

$$=-{}^8C_0{}^{11}C_3+{}^8C_1{}^{11}C_2-{}^8C_2{}^{11}C_1+{}^8C_3{}^{11}C_0$$

$$=\frac{-11\times 10\times 9}{1\times 2\times 3}+\frac{8\times 11\times 10}{1\times 2}-\frac{8\times 7}{1\times 2}\times 11+\frac{8\times 7\times 6}{1\times 2\times 3}$$

$$=-165+440-308+56=23$$

**93.** *(6006)* The given expression is

$$\left(t^2x^{1/5}+\frac{(1-x)^{1/10}}{t}\right)^{15}, x\geq 0$$

the above expression for term independent on $t$, then power of $t$ is equal to zero. Thus, general term of the above expression

$$={}^{15}C_r(t^2x^{1/5})^{15-r}\left(\frac{(1-x)^{1/10}}{t}\right)^r$$

For term independent of $t$,

$$2(15-r)-r=0 \Rightarrow r=10$$

Thus, 11th term of above expression is independent of $t$.

$$\therefore \quad t_{11}={}^{15}C_{10}\,x(1-x)$$

Now, maximum value of $x(1-x)$ occur at $x=1/2$

i.e. $[x(1-x)]_{\max}=\frac{1}{2}\left(1-\frac{1}{2}\right)=\frac{1}{4}$

Since, the maximum value of the term independent of $t$ in the expression is '$k$' [given]

Thus, $k={}^{15}C_{10}\times\frac{1}{4}=\frac{15!}{10!\cdot5!}\cdot\frac{1}{4}$

$$\Rightarrow 8k=\frac{2(15)!}{10!5!}=\frac{2\cdot15\cdot14\cdot13\cdot12\cdot11\cdot10!}{10!.5!}=6006$$

**94.** *(5)* Given, binomial expression is

$$\left(2x^{1/5}-\frac{1}{x^{1/5}}\right)^{15}$$

$$T_{r+1}={}^{15}C_r\,(2x^{1/5})^{15-r}.\left(\frac{1}{x^{1/5}}\right)^r(-1)^r$$

$$={}^{15}C_r\,2^{15-r}x^{\frac{15-r}{5}}\cdot x^{-\frac{r}{5}}\cdot(-1)^r$$

$$=(-1)^r\cdot{}^{15}C_r.2^{15-r}\cdot x^{\frac{15-2r}{5}}$$

Put $r=10$

$T_{11}={}^{15}C_{10}2^5=m$

Put $r=15$

$T_{16}=-{}^{15}C_{15}2^0=-1=n$

$\therefore\ mn^2={}^{15}C_{10}2^5\cdot(-1)^2={}^{15}C_5 2^5$

Therefore, required value of $r=5$.

> **Shortcut Method**
>
> $T_{r+1}=(-1)^r\,{}^{15}C_r\cdot2^{15-r}x^{\frac{15-2r}{5}}$
>
> For $x^{-1}$, $r=10$
>
> $m={}^{15}C_{10}2^5={}^{15}C_5\cdot2^5$
>
> For $x^{-3}, r=15\Rightarrow n=-1$
>
> $mn^2={}^{15}C_5\,2^5$
>
> $\Rightarrow\quad r=5$

**95.** *(2)* $T_{r+1}={}^{12}C_r(2x^3)^r\left(\frac{3}{x^k}\right)^{12-r}$

$x^{3r-(12-r)k}$ is constant

$\therefore\quad 3r-12k+rk=0$

$\Rightarrow\quad k=\frac{3r}{12-r}$

$\therefore$Possible values of $r$ are 3, 6, 8, 9, 10 and corresponding values of $k$ are 1, 3, 6, 9, 15, respectively.

${}^{12}C_r=220, 924, 495, 220, 66$

$\therefore$Possible values of $k$ for which we will get $2^8$ are 3, 6.

**96.** *(102)* We know,

${}^nC_{r-1}+{}^nC_r={}^{n+1}C_r$

Also $\quad{}^{40}C_0={}^{41}C_0=1$

So,

$${}^{40}C_0+{}^{41}C_1+{}^{42}C_2+.....+{}^{60}C_{20}$$

$$={}^{42}C_1+{}^{42}C_2+{}^{43}C_3+....+{}^{60}C_{20}$$

$$={}^{43}C_2+{}^{43}C_3+.....+{}^{60}C_{20}$$

$$={}^{60}C_{19}+{}^{60}C_{20}={}^{61}C_{20}$$

$$\therefore\ {}^{61}C_{20}=m/n\ {}^{60}C_{20}$$

$$\frac{61!}{20!(61-20)!}=\frac{m}{n}\frac{60!}{20!(60-20)!}$$

$$\Rightarrow\quad\frac{61!}{20!41!}=\frac{m}{n}\frac{60!}{20!40!}$$

$$\Rightarrow\quad\frac{61}{41}=\frac{m}{n}$$

$$\therefore\quad m+n=61+41=102$$

**97.** *(83)* General Term

$$T_{r+1}={}^{10}C_r\,(2x^3)^{10-r}\left(\frac{3}{x}\right)^r$$

$$={}^{10}C_r\,2^{10-r}\times3^r\times x^{30-3r-r}$$

$$={}^{10}C_r\times2^{10-r}\times3^r\times x^{30-4r}$$

$\Rightarrow\quad r\neq8, 9, 10$

Sum of the required coefficient

$$=\left(2(1)^3+\frac{3}{1}\right)^{10}-[{}^{10}C_8\times2^2\times3^8+{}^{10}C_9\times2\times3^9+{}^{10}C_{10}\times3^{10}]$$

$$=5^{10}-[{}^{10}C_2\times2^2\times3^8+{}^{10}C_1\times2\times3^9+{}^{10}C_0\times3^{10}]$$

$[\because{}^nC_r={}^nC_{n-r}]$

$$=5^{10}-3^9\left[\frac{10\times9}{1\times2}\times\frac{2^2}{3}+10\times2+3\right]$$

$$=5^{10}-3^9(83)\Rightarrow\beta=83$$

**98.** *(4)* Given, $1+3+3^2+...+3^{2021}$

$$=\frac{3^{2022}-1}{2}=\frac{(3^2)^{1011}-1}{2}$$

$$=\frac{(9)^{1011}-1}{2}=\frac{(10-1)^{1011}-1}{2}$$

$$=\frac{({}^{1011}C_0\cdot(10)^{1011}-{}^{1011}C_1\cdot(10)^{1010}+...+{}^{1011}C_{1010}10^1-C_{1011})-1}{2}$$

$$=\frac{100}{2}[\text{Integer}]+\frac{10110-2}{2}$$

$$=50\ (\text{Integer})+\frac{10108}{2}$$

$$=50\ \ (\text{Integer})+5054$$

Now, 50 (Integer)+5054 divided by 50, then we get a remainder 4.

**99.** *(55)* General term of $\left(\frac{x}{4}-\frac{12}{x^2}\right)^{12}$

$$T_{r+1}=(-1)^r\,{}^{12}C_r\left(\frac{x}{4}\right)^{12-r}\left(\frac{12}{x^2}\right)^r$$

$$=(-1)^r\left(\frac{{}^{12}C_r.12^r}{4^{12-r}}\right)x^{(12-r-2r)}$$

Term independent of $r$

$12-3r=0\Rightarrow\ r=4$

$$T_5=(-1)^4\left(\frac{{}^{12}C_4 12^4}{4^8}\right)=\frac{3^6}{4^4}k$$

$k=55$

**100.** *(15)* $3\times7^{22}+2\times10^{22}-44$

$=3\times(6+1)^{22}+2\times(9+1)^{22}-44$

Now, $(1+6)^{22}=1+{}^{22}C_1 6+{}^{22}C_2\cdot6^2+...+{}^{22}C_{22}6^{22}$

$(1+9)^{22}=1+{}^{22}C_1\,9+{}^{22}C_2\cdot9^2+...+{}^{22}C_{22}\,9^{22}$

$$\therefore 3\times(6+1)^{22}+2\times(9+1)^{22}-44$$

$$=3(1+6\lambda)+2(1+9\mu)-44$$

$$=18\lambda+3+18\mu+2-44$$

$$=18\delta-39=18\,\alpha+15$$

$3\times7^{22}+2\times10^{22}-44$, when divided by 18 leaves remainder 15.

**101.** *(55)* The coefficient of $x^7$ in the expansion of $\left(2+\frac{x}{3}\right)^n={}^nC_7 2^{n-7}\frac{1}{3^7}$

and the coefficient of

$x^8={}^nC_8 2^{n-8}\frac{1}{3^8}$

According to the question,

$${}^nC_7 2^{n-7}\cdot\frac{1}{3^7}={}^nC_8 2^{n-8}\frac{1}{3^8}$$

$$\Rightarrow\frac{n!}{7!(n-7)!}\cdot2\cdot2^{n-8}\frac{1}{3^7}=\frac{n!}{8!(n-8)!}2^{n-8}\times\frac{1}{3^8}$$

$$\Rightarrow\quad\frac{1}{7!(n-7)(n-8)!}\times2=\frac{1}{8\cdot7!(n-8)!}\times\frac{1}{3}\Rightarrow\frac{2}{n-7}=\frac{1}{24}$$

$\Rightarrow\quad n-7=48$

$\Rightarrow\quad n=55$

**102.** *(1)* Coefficient of middle term of $(1+x)^{20}$ is ${}^{20}C_{10}$.

Coefficient of middle term of $(1+x)^{19}$ is ${}^{19}C_9$ and ${}^{19}C_{10}$

According to the question,

$$\left(\frac{{}^{20}C_{10}}{{}^{19}C_9 + {}^{19}C_{10}}\right) = \frac{{}^{20}C_{10}}{{}^{20}C_{10}} = 1$$

**103.** *(8)* General term of $\left(2x^r + \frac{1}{x^2}\right)^{10}$

$$= {}^{10}C_k (2x^r)^{10-k} x^{-2k},$$

$$= {}^{10}C_k (2^{10-k})(x^{10r - rk - 2k})$$

Term independent of $r$

$10r - rk - 2k = 0$

$$\Rightarrow \quad k = \frac{10r}{r+2} \Rightarrow r = \frac{2k}{10-k}$$

$\because \quad k < 10$

If $k = 8$ and $r = -2 + 10 = 8$,

$k = 5$ and $r = 2$

$k = 6$ and $r = 3$

$\because \quad {}^{10}C_k 2^{10-k} = 180$

$\therefore r = 8$ is the possible value.

**104.** *(96)* $11^n > 10^n + 9^n$

$\Rightarrow \quad 11^n - 9^n > 10^n$

$\Rightarrow (10+1)^n - (10-1)^n > 10^n$

$$\{(10+1)^n - (10-1)^n = {}^nC_0 10^n + {}^nC_1 10^{n-1} + {}^nC_2 10^{n-2} + \ldots - \{{}^nC_0 10^n - {}^nC_1 10^{n-1} \ldots\}$$

$\Rightarrow 2({}^nC_1 10^{n-1} + {}^nC_3 10^{n-3} \ldots) > 10^n$

For $n = 1 \Rightarrow 2.1 \not> 10$

For $n = 2 \Rightarrow 2(2) \not> 100$

For $n = 3 \Rightarrow 2(3 \cdot 10^2 + 1) \not> 1000$

For $n = 4$

$2({}^4C_1 10^4 + {}^4C_3 10^1) = 8080 \not> 10^4$

For, $n = 5$

$\Rightarrow 2({}^5C_1 10^4 + {}^5C_3 10^2 + {}^5C_5) > 10^5$

Hence, $n \in \{5, 6, 7, \ldots\ldots 100\}$

$\therefore$ Number of elements $= 96$

**105.** *(21)* General term of $\left(4^{\frac{1}{4}} + 5^{\frac{1}{6}}\right)^{120}$

$$= {}^{120}C_r \left(4^{\frac{1}{4}}\right)^r \left(5^{\frac{1}{6}}\right)^{120-r}$$

$$= {}^{120}C_r 4^{\frac{r}{4}} 5^{20 - \frac{r}{6}}$$

$$= {}^{120}C_r 2^{\frac{r}{2}} 5^{20 - \frac{r}{6}}$$

For this term to be rational, $r$ should be a multiple of 2 and 6 i.e. $r$ should be a multiple of 6.

$r = \{0, 6, 12, 18, \ldots, 120\}$

Number of terms $= 21$

**106.** *(210)* $\left[\frac{x+1}{x^{2/3} - x^{1/3} + 1} - \frac{x-1}{x - x^{1/2}}\right]^{10}$

$$= \left[\frac{(x^{1/3})^3 + (1)^3}{x^{2/3} - x^{1/3} + 1} - \frac{(\sqrt{x}-1)(\sqrt{x}+1)}{\sqrt{x}(\sqrt{x}-1)}\right]^{10}$$

$$= \left[(x^{1/3} + 1) - \left(\frac{\sqrt{x}+1}{\sqrt{x}}\right)\right]^{10}$$

[use $(a^3 + b^3) = (a+b)(a^2 - ab + b^2)$]

$$= (x^{1/3} - x^{-1/2})^{10}$$

General term,

$$T_{r+1} = {}^{10}C_r (x^{1/3})^{10-r} (-x^{1/2})^r$$

$$= {}^{10}C_r (x)^{\frac{10-r}{3}} \cdot (-x)^{-\frac{r}{2}}$$

For term independent of $x$, we must put

$$\frac{10-r}{3} - \frac{r}{2} = 0$$

$\Rightarrow \quad 20 - 2r - 3r = 0$

$\Rightarrow \quad r = 4$

$$\therefore T_{4+1} = T_5 = {}^{10}C_4 = \frac{10!}{4!6!}$$

$$= \frac{10 \times 9 \times 8 \times 7 \times 6!}{4 \times 3 \times 2 \times 1 \times 6!} = 210$$

**107.** *(19)* $\sum_{k=0}^{10} (2^2 + 3k) \cdot {}^nC_k$

$$= 4 \sum_{k=0}^{10} {}^nC_k + 3 \cdot \sum_{k=0}^{10} k \cdot {}^nC_k.$$

$$= 4 \times 2^n + 3 \times n \sum_{k=1}^{10} {}^{n-1}C_{k-1}$$

$$\left[\because {}^nC_r = \frac{n}{r} {}^{n-1}C_{r-1}\right]$$

$$= 4 \times 2^n + 3n \times 2^{n-1}$$

$$= 2^n\left(4 + \frac{3n}{2}\right) = \left(4 + \frac{3n}{2}\right) \cdot 2^n + 0 \times 3^n$$

On comparing,

$$\left[0 \times 3^n + \left(4 + \frac{3n}{2}\right) \cdot 2^n\right]$$

with $[\alpha \cdot 3^{10} + \beta \cdot 2^{10}]$, we get

$n = 10, \alpha = 0, \beta = 19$

$\therefore \quad \alpha + \beta = 0 + 19 = 19$

**108.** *(4)* General term,

$$T_{r+1} = {}^nC_r \cdot a^r \cdot x^{n-3r}$$

$\therefore \quad T_3 = {}^nC_2 \cdot a^2 \cdot x^{n-6}$

$T_4 = {}^nC_3 a^3 \cdot x^{n-9}$

$T_5 = {}^nC_4 \cdot a^4 \cdot x^{n-12}$

Now, $\frac{\text{coefficient of } T_3}{\text{coefficient of } T_4}$

$$= \frac{{}^nC_2 \cdot a^2}{{}^nC_3 \cdot a^3} = \frac{3}{a(n-2)} = \frac{3}{2}$$

$\Rightarrow \quad a(n-2) = 2 \quad \ldots\text{(i)}$

Also, $\frac{\text{coefficient of } T_4}{\text{coefficient of } T_5} = \frac{{}^nC_3 \cdot a^3}{{}^nC_4 \cdot a^4}$

$$= \frac{4}{a(n-3)} = \frac{8}{3}$$

$\Rightarrow \quad a(n-3) = \frac{3}{2} \quad \ldots\text{(ii)}$

From Eqs. (i) and (ii), $n = 6, a = \frac{1}{2}$

For the term independent of '$x$',

$$r = \frac{n}{3} \Rightarrow r = \frac{6}{3} \Rightarrow r = 2$$

$\therefore$ Independent term is $T_3$.

Now,

$$T_3 = {}^6C_2 \cdot \left(\frac{1}{2}\right)^2 = \frac{15}{4} = 3.75 \approx 4$$

**109.** *(1)* Given, when $x$ is divided by 4, the remainder is 3.

Let $x = 4p + 3$, then

$(2020 + x)^{2022} = (2020 + 4p + 3)^{2022}$

$= (2024 + 4p - 1)^{2022} = (4k - 1)^{2022}$

($\because$ 2024 is divisible by 4)

$$= {}^{2022}C_0 (4K)^{2022} (-1)^0 + {}^{2022}C_1 (4K)^{2021} (-1)^1 + \ldots + {}^{2022}C_{2022} (4K)^0 (-1)^{2022}$$

On expansion $(2020 + x)^{2022}$, we get the form of $8\lambda + 1$. Since, each terms have 2022 and $4k$, so if we take 2 common from 2022 we get 8. Thus, each term have 8 in common. Hence, remainder is 1.

**110.** *(13)* The general term (i.e. $(r+1)$th term) in the expansion of binomial $\left(x^m + \frac{1}{x^2}\right)^{22}$ is

$$T_{r+1} = {}^{22}C_r (x^m)^{22-r} x^{-2r}$$

$\because$ The coefficient ${}^{22}C_r = 1540$ [given]

$\Rightarrow \quad r = 3$ or $19$

at $\quad r = 3,$

$$\Rightarrow \quad 19m = 7 \Rightarrow m = \frac{7}{19} \notin N$$

Now, at $r = 19$,

$\Rightarrow \quad 3m = 39 \Rightarrow m = 13 \in N$

So, the natural value of $m = 13$.

**111.** *(30)* Let $(1 + x + x^2 + \ldots + x^{2n})(1 - x + x^2 - x^3 + \ldots + x^{2n})$

$$= a_0 + a_1 x + a_2 x^2 + \ldots + a_{4n} x^{4n}$$

On putting $x = 1$, we get

$(1+1+1+...+1)(1-1+1-1+...+1)$
$(2n+1)$ terms $\quad (2n+1)$ terms

$$= a_0 + a_1 + a_2 + ... + a_{4n}$$

$\Rightarrow (2n+1)$

$$= a_0 + a_1 + a_2 + ... + a_{4n} \quad ...(i)$$

Similarly, on putting $x = -1$, we get

$(1-1+1-1+...+1)(1+1+1+...+1)$
$(2n+1)$ terms $\quad (2n+1)$terms

$$= a_0 - a_1 + a_2 - a_3 + ... + a_{4n}$$

$\Rightarrow (2n+1)$

$$= a_0 - a_1 + a_2 - a_3 ... + a_{4n} \quad ...(ii)$$

On adding Eqs. (i) and (ii), we get

$2(a_0 + a_2 + a_4 + ... + a_{4n}) = 2(2n+1)$

$\Rightarrow a_0 + a_2 + a_4 + ... + a_{4n} = 2n+1$

$\Rightarrow$ Sum of coefficients of all even powers of $x$ in the product

$(1 + x + x^2 + ... + x^{2n})$

$$(1 - x + x^2 - x^3 + ... + x^{2n})$$
$$= 2n + 1 = 61 \text{ [given]}$$

$\Rightarrow \quad 2n = 60 \Rightarrow n = 30$

## Topic 2

**01.** *(b)* $(a + bx + cx^2)^{10} = p_0 + p_1 x + p_2 x^2 + ... + p_{20} x^{20} = \sum_{i=0}^{20} p_i x^i$

Coefficient of $x$

$$\Rightarrow 20 = \frac{10!\, a^9 (bx)^1 (cx^2)^0}{9!1!}$$

$20 = 10a^9 b^1 \Rightarrow a = 1, b = 2$

$$\text{Coefficient of } x^2 = \frac{10!\, a^9 \times (cx^2)^1}{9!1!} + \frac{10!(a^8)(bx)^2(cx^2)^0}{8!2!}$$

$$= 10c + 180 = 210$$

$$10c = 30 \Rightarrow c = 3$$

$2(a + b + c) = 2(1 + 2 + 3)$
$= 2(6) = 12$

**02.** *(d)* Given, constant term in the expansion of $\left(3x^3 - 2x^2 + \frac{5}{x^5}\right)^{10}$ is $2^k \cdot l$, where $l$ is odd integer.

$$\text{General term, } T_{r+1} = \frac{10!}{r_1! r_2! r_3!}$$

$$(3x^3)^{r_1}(-2x^2)^{r_2}\left(\frac{5}{x^5}\right)^{r_3} = \frac{10!}{r_1! r_2! r_3!}(3)^{r_1} (-2)^{r_2} (5)^{r_3} (x)^{3r_1 + 2r_2 - 5r_3}$$

For constant term,

$$3r_1 + 2r_2 - 5r_3 = 0 \quad ...(i)$$

and $\quad r_1 + r_2 + r_3 = 10 \quad ...(ii)$

From Eq. (i) – Eq (ii) × 2, we get

$$r_1 - (5+2)r_3 = -20$$

$\Rightarrow \quad r_1 + 20 = 7r_3$

$\Rightarrow \quad r_1 = 1, r_3 = 3$ and $r_2 = 6$

$\therefore$Constant term

$$= \frac{10!}{(1)!(6)!(3)!}(3)^1(-2)^6(5)^3$$

$$= \frac{10\cdot 9\cdot 8\cdot 7}{3\cdot 2} \times 3 \times 2^6 \times 5^3$$

$$= 2^9 \times 3^2 \times 5^4 \times 7^1 = 2^k \times l$$

Hence, $k = 9$

**03.** *(239)* Given,

$$(1 + x + x^2)^{10} = 1 + a_1 x + a_2 x^2 + ... + a_{20} x^{20} \quad ...(i)$$

put $x = 1 \Rightarrow 3^{10} = 1 + a_1 + a_2 + ... + a_{20} \quad ...(ii)$

put $x = -1 \Rightarrow 1 = 1 - a_1 + a_2 - ... + a_{20} \quad ...(iii)$

From Eqs. (ii) and (iii), we get

$3^{10} - 1 = 2(a_1 + a_3 + a_5 + ... + a_{19})$

Differentiation of Eq. (i) w.r.t. $x$, we get

$10(1 + x + x^2)^9 (1 + 2x) =$
$\quad a_1 + 2a_2 x + ... + 20a_{20} x^{19}$

Again, differentiation w.r.t $x$ and put $x = 0$, we get

$$10[9(1 + x + x^2)^8(1+2x)^2 + (1 + x + x^2)^9(2)]$$
$$= 2a_2 + ... + 20 \times 19\, a_{20} x^{18}$$

$\Rightarrow \quad 10[9 + 2] = 2a_2$

$\Rightarrow \quad a_2 = 55$

$\Rightarrow (a_1 + a_2 + ... + a_{19}) - 11a_2$

$$= \frac{3^{10} - 1}{2} - 55 \times 11$$

$$= 28919 = 121x \quad \text{[given]}$$

$\Rightarrow \quad x = 239$

**04.** *(64)* Coefficient of $x^{15}$ in $(1 + x + x^2 + ... + x^9)^3$

$$= \text{Coefficient of } x^{15} \text{ in } \left(\frac{1 - x^{10}}{1 - x}\right)^3$$

$$= \text{Coefficient of } x^{15} \text{ in } (1 - x^{10})^3 (1 - x)^{-3}$$

$= \text{Coefficient of } x^{15}$ in $(1 - 3x^{10} + {}^3C_2\, x^{20} + ...)(1 - x)^{-3}$

$$= {}^{3+15-1}C_{15} - 3\; {}^{3+5-1}C_5$$

$$= {}^{17}C_{15} - 3\; {}^7C_5$$

$$= \frac{17 \times 16}{2} - 3 \times \frac{7 \times 6}{2}$$

$$= 73$$

$\therefore$ Number of less than 212 whose sum of digits is (078, 087, 069, 096, 195, 159, 186, 177, 168)

$\therefore$ Required number $= 73 - 9 = 64$

**05.** *(612)* We have, $(1 + 2^{\frac{1}{3}} + 3^{\frac{1}{2}})^6$

$$\Rightarrow \frac{6!}{r_1! r_2! r_3!}(1)^{r_1}(2)^{\frac{r_2}{3}} 3^{\frac{r_3}{2}}$$

| $r_1$ | $r_2$ | $r_3$ |
|---|---|---|
| 6 | 0 | 0 |
| 4 | 0 | 2 |
| 2 | 0 | 4 |
| 0 | 0 | 6 |
| 3 | 3 | 0 |
| 1 | 3 | 2 |
| 0 | 6 | 0 |

$\therefore$Required sum

$$= \frac{6!}{6!0!0!} + \frac{6!(3)}{4!0!2!} + \frac{6!}{2!0!4!}(3)^2 + \frac{6!}{0!0!6!}(3)^3 + \frac{6!}{3!3!0!}(2) + \frac{6!}{1!3!2!}(2^1)(3^1) + \frac{6!}{0!6!0!}2^2$$

$= 1 + 45 + 135 + 27 + 40 + 360 + 4$
$= 612$

**06.** *(960)* Given expression is $(1 - x + 2x^3)^{10}$

So, general term is

$$\frac{10!}{r_1! r_2! r_3!}(1)^{r_1}(-1)^{r_2} \cdot (2)^{r_3} \cdot (x)^{r_2 + 3r_3}$$

Where, $r_1 + r_2 + r_3 = 10$
and $\quad r_2 + 3r_3 = 7$

Now, for possibility,

| $r_1$ | $r_2$ | $r_3$ |
|---|---|---|
| 3 | 7 | 0 |
| 7 | 1 | 2 |
| 5 | 4 | 1 |

Thus, required coefficient

$$= \frac{10!}{3!7!}(-1)^7 + \frac{10!}{5!4!}(-1)^4(2) + \frac{10!}{7!2!}(-1)^1(2)^2$$

$= -120 + 2520 - 1440$
$= 960$

**07.** *(1080)* General term of

$$\left(2x+\frac{1}{x^7}+3x^2\right)^5$$

$$=\frac{5!(2x)^{n_1}(x^{-7})^{n_2}(3x^2)^{n_3}}{n_1!n_2!n_3!}$$

$$=\frac{5!2^{n_1}3^{n_3}}{n_1!n_2!n_3!}x^{n_1-7n_2+2n_3}$$

For constant term, $n_1-7n_2+2n_3=0$

$\Rightarrow n_1+2n_3=7n_2$ and $n_1+n_2+n_3=5$

Only possible, when $n_1=1, n_2=1,$ $n_3=3$

Constant term $=\dfrac{5!\cdot 2^1\cdot 3^3}{1!\,1!\,3!}$

$=5\times 4\times 2\times 27=40\times 27=1080$

**08.** *(315)* General term of

$(a+2b+4ab)^{10}$

$$=\frac{10!}{r_1!r_2!r_3!}(a)^{r_1}(2b)^{r_2}(4ab)^{r_3}$$

$$=\frac{10!}{r_1!r_2!r_3!}(a)^{r_1+r_3}2^{r_2}b^{r_2+r_3}.4^{r_3}$$

where, $r_1+r_2+r_3=10$ ...(i)

$r_1+r_3=7$ ...(ii)

$r_2+r_3=8$ ...(iii)

On adding Eqs. (ii) and (iii) and subtract Eq. (i), we get

$r_1+r_2+2r_3-r_1-r_2-r_3=15-10$

$\Rightarrow r_3=5$

From Eq. (iii), $r_2=8-5=3$

From Eq. (ii), $r_1=7-5=2$

$\therefore$ Required coefficient

$$=\frac{10!2^3}{2!3!5!}.4^5$$

$$=\frac{10!}{2!3!5!}\cdot 2^3\cdot 2^{10}$$

$$=\frac{10\times 9\times 8\times 7\times 6}{2\times 3\times 2}\cdot 2^{13}$$

$=315\times 2^{16}=k\cdot 2^{16}$ [given]

$\therefore\quad k=315$

**09.** *(8)* The general term in the expansion of trinomial $(2x^2+3x+4)^{10}$ is

$$\frac{10!}{n_1!n_2!n_3!}2^{n_1}3^{n_2}4^{n_3}x^{2n_1+n_2}$$

such that $n_1+n_2+n_3=10$

and $n_1, n_2, n_3\geq 0$

For $a_7$, $2n_1+n_2=7$

$$\therefore\quad a_7=\frac{10!}{0!7!3!}2^0 3^7 4^3+\frac{10!}{1!5!4!}2^1 3^5 4^4+\frac{10!}{2!3!5!}2^2 3^3 4^5+\frac{10!}{3!1!6!}2^3 3^1 4^6$$

and, for $a_{13}$, $2n_1+n_2=13$

$$\therefore a_{13}=\frac{10!}{3!7!0!}2^3 3^7+\frac{10!}{4!5!1!}2^4 3^5 4^1+\frac{10!}{5!3!2!}2^5 3^3 4^2+\frac{10!}{6!1!3!}2^6 3^1 4^3$$

$$\therefore\frac{a_7}{a_{13}}=\frac{\frac{10!}{7!3!}3^7 4^3+\frac{10!}{5!4!}2^1 3^5 4^4+\frac{10!}{2!3!5!}2^2 3^3 4^5+\frac{10!}{3!6!}2^3 3^1 4^6}{\frac{10!}{3!7!}2^3 3^7+\frac{10!}{4!5!}2^4 3^5 4^1+\frac{10!}{5!3!2!}2^5 3^3 4^2+\frac{10!}{6!3!}2^6 3^1 4^3}$$

$$=2^3\left[\frac{\frac{10!}{7!3!}2^3 3^7+\frac{10!}{4!5!}2^6 3^5+\frac{10!}{2!3!5!}2^9 3^3+\frac{10!}{6!3!}2^{12}3^1}{\frac{10!}{3!7!}2^3 3^7+\frac{10!}{4!5!}2^6 3^5+\frac{10!}{2!3!5!}2^9 3^3+\frac{10!}{6!3!}2^{12}3^1}\right]$$

$=2^3=8$

**Concept Enhancer**

The multinomial theorem is used to expand the sum of two or more terms raised to an integer power.

$$(x_1+x_2+\ldots+x_{12})^n=\sum_{n_1,n_2,n_3\ldots\ldots n_k\geq 0}\frac{n!}{n_1!n_2!\ldots n_k!}$$

where $n_1+n_2+\ldots+n_K=n$

# C-07

JEE Main Chapterwise Topicwise Mathematics

In JEE Main, there are usually 1-2 questions per paper are visible from this chapter. In past 3 years, 59 conceptual and 30 numerical questions of moderate difficulty level were asked from this chapter.

**For detailed analysis scan QR code**

# SEQUENCE AND SERIES

## Topic 1 Arithmetic Progression and Its Mean

### Level 1

**01** Let $a_n$ be the $n$th term of an AP. If $S_n = a_1 + a_2 + a_3 + \ldots + a_n = 700$, $a_6 = 7$ and $S_7 = 7$, then $a_n$ is equal to **[2025, 7 April Shift-II]**

(a) 56 (b) 64 (c) 70 (d) 65

**02** Let $a_1, a_2, a_3, \ldots$ be in an AP such that $\sum_{k=1}^{12} a_{2k-1} = -\frac{72}{5} a_1$, $a_1 \neq 0$. If $\sum_{k=1}^{n} a_k = 0$, then $n$ is **[2025, 2 April Shift-I]**

(a) 17 (b) 10 (c) 11 (d) 18

**03** $1 + 3 + 5^2 + 7 + 9^2 + \ldots$ upto 40 terms is equal to **[2025, 4 April Shift-I]**

(a) 33980 (b) 41880 (c) 43890 (d) 40870

**Imp.** Arithmetic Progression

**04** Consider an AP of positive integers, whose sum of the first three terms is 54 and the sum of the first twenty terms lies between 1600 and 1800. Then, its 11th term is **[2025, 29 Jan Shift-I]**

(a) 108 (b) 122 (c) 84 (d) 90

**05** In an arithmetic progression, if $S_{40} = 1030$ and $S_{12} = 57$, then $S_{30} - S_{10}$ is equal to **[2025, 24 Jan Shift-II]**

(a) 505 (b) 515
(c) 510 (d) 525

**06** If the first term of an AP is 3 and the sum of its first four terms is equal to one-fifth of the sum of the next four terms, then the sum of the first 20 terms is equal to **[2025, 23 Jan Shift-I]**

(a) $-120$ (b) $-1080$
(c) $-1020$ (d) $-1200$

**07** A software company sets up $m$ number of computer systems to finish an assignment in 17 days. If 4 computer systems crashed on the start of the second day, 4 more computer systems crashed on the start of the third day and so on, then it took 8 more days to finish the assignment. The value of $m$ is equal to **[2024, 6 April Shift-II]**

(a) 125 (b) 180 (c) 160 (d) 150

**08** For $x \geq 0$, the least value of $K$, for which $4^{1+x} + 4^{1-x}$, $\frac{K}{2}$, $16^x + 16^{-x}$ are three consecutive terms of an AP, is equal to **[2024, 5 April Shift-II]**

(a) 10 (b) 16 (c) 4 (d) 8

**09** Let $S_n$ denote the sum of the first $n$ terms of an arithmetic progression. If $S_{10} = 390$ and the ratio of the tenth and the fifth terms is $15:7$, then $S_{15} - S_5$ is equal to **[2024, 1 Feb Shift-II]**

(a) 690 (b) 790
(c) 890 (d) 800

**10** Let $S_n$ denote the sum of first $n$ terms of an arithmetic progression. If $S_{20} = 790$ and $S_{10} = 145$, then $S_{15} - S_5$ is **[2024, 30 Jan Shift-I]**

(a) 403 (b) 390
(c) 410 (d) 395

**11** If $\log_e a$, $\log_e b$, $\log_e c$ are in AP and $\log_e a - \log_e 2b$, $\log_e 2b - \log_e 3c$, $\log_e 3c - \log_e a$ are also in an AP, then $a:b:c$ is equal to **[2024, 29 Jan Shift-II]**

(a) $16:4:1$ (b) $9:6:4$
(c) $25:10:4$ (d) $6:3:2$

**12** The 20th term from the end of the progression $20, 19\frac{1}{4}, 18\frac{1}{2}, 17\frac{3}{4}, \ldots, -129\frac{1}{4}$ is **[2024, 27 Jan Shift-II]**

(a) $-110$ (b) $-115$
(c) $-118$ (d) $-100$

**13** The number of common terms in the progressions 4, 9, 14, 19, ..., up to 25th term and 3, 6, 9, 12, ..., up to 37th term is **[2024, 27 Jan Shift-I]**

(a) 5 (b) 8 (c) 9 (d) 7

**14** Let $s_1, s_2, s_3, \ldots, s_{10}$ respectively be the sum to 12 terms of 10 AP's whose first terms are 1, 2, 3, ..., 10 and the common differences are 1, 3, 5, ..., 19 respectively. Then, $\sum_{i=1}^{10} S_i$ is equal to **[2023, 6 April Shift-I]**

(a) 7380 (b) 7360 (c) 7220 (d) 7260

**15** Let α, β and γ be three positive real numbers. Let $f(x) = \alpha x^5 + \beta x^3 + \gamma x$, $x \in R$ and $g: R \to R$ be such that $g(f(x)) = x$ for all $x \in R$. If $a_1, a_2, a_3, \ldots, a_n$ be in arithmetic progression with mean zero, then the value of $f\left(g\left(\frac{1}{n}\sum_{i=1}^{n} f(a_i)\right)\right)$ is equal to **[2022, 28 July Shift-I]**

(a) 0 (b) 3 (c) 9 (d) 27

**16** Suppose, $a_1, a_2, \ldots, a_n, \ldots$ be an arithmetic progression of natural numbers. If the ratio of the sum of first five terms to the sum of first nine terms of the progression is 5 : 17 and $110 < a_{15} < 120$, then the sum of the first ten terms of the progression is equal to **[2022, 27 July Shift-I]**

(a) 290 (b) 380 (c) 460 (d) 510

**17** Let $a_1, a_2, a_3 \ldots$ be an AP. If $\frac{a_1 + a_2 + \ldots + a_{10}}{a_1 + a_2 + \ldots + a_p} = \frac{100}{p^2}$, $p \neq 10$, then $\frac{a_{11}}{a_{10}}$ is equal to **[2021, 31 Aug Shift-II]**

(a) 19/21 (b) 100/121 (c) 21/19 (d) 121/100

**18** Let $S_n$ be the sum of the first $n$ terms of an arithmetic progression, If $S_{3n} = 3S_{2n}$, then the value of $\frac{S_{4n}}{S_{2n}}$ is **[2021, 25 July Shift-I]**

(a) 6 (b) 4 (c) 2 (d) 8

**19** If the sum of first 11 terms of an AP, $a_1, a_2, a_3, \ldots$ is 0 $(a_1 \neq 0)$, then the sum of the AP $a_1, a_3, a_5, \ldots, a_{23}$ is $ka_1$, where $k$ is equal to **[2020, 2 Sep Shift-II]**

(a) $-\frac{121}{10}$ (b) $\frac{121}{10}$ (c) $\frac{72}{5}$ (d) $-\frac{72}{5}$

**20** If $a_1, a_2, a_3, \ldots, a_n$ are in AP and $a_1 + a_4 + a_7 + \ldots + a_{16} = 114$, then $a_1 + a_6 + a_{11} + a_{16}$ is equal to **[2019, 10 April Shift-I]**

(a) 64 (b) 76 (c) 98 (d) 38

**21** Let the sum of the first $n$ terms of a non-constant AP $a_1, a_2, a_3 \ldots$ be $50n + \frac{n(n-7)}{2}A$, where $A$ is a constant. If $d$ is the common difference of this AP, then the ordered pair $(d, a_{50})$ is equal to **[2019, 9 April Shift-I]**

(a) $(A, 50 + 46A)$ (b) $(50, 50 + 45A)$
(c) $(50, 50 + 46A)$ (d) $(A, 50 + 45A)$

**22** For any three positive real numbers $a, b$ and $c$, if $9(25a^2 + b^2) + 25(c^2 - 3ac) = 15b(3a + c)$, then **[JEE Main 2017]**

(a) $b, c$ and a are in GP (b) $b, c$ and $a$ are in AP
(c) $a, b$ and $c$ are in AP (d) $a, b$ and $c$ are in GP

**23** If 100 times the 100th term of an AP with non-zero common difference equal to the 50 times its 50th term, then the 150th term of this AP is **[AIEEE 2012]**

(a) −150 (b) 150 times its 50th term
(c) 150 (d) zero

**24** Let $T_r$ be the $r$th term of an AP whose first term is $a$ and common difference is $d$. If for some positive integers $m, n\ (m \neq n)$, $T_m = \frac{1}{n}$, and $T_n = \frac{1}{m}$, then $a - d$ is the equal to **[AIEEE 2004]**

(a) 0 (b) 1
(c) $\frac{1}{mn}$ (d) $\frac{1}{m} + \frac{1}{n}$

**25** Let $f(x)$ be a polynomial function of second degree. If $f(1) = f(-1)$ and $a, b, c$ are in AP, then $f'(a), f'(b)$ and $f'(c)$ are in **[AIEEE 2003]**

(a) AP
(b) GP
(c) HP
(d) Arithmetic Geometric progression

**Imp.** Properties of Logarithm

**26** If 1, $\log_3 \sqrt{(3^{1-x} + 2)}$, $\log_3(4 \cdot 3^x - 1)$ are in AP. Then, $x$ is equal to **[AIEEE 2002]**

(a) $\log_3 4$ (b) $1 - \log_3 4$
(c) $1 - \log_4 3$ (d) $\log_4 3$

## Level 2

**27** Consider two sets $A$ and $B$ each containing three numbers in AP. Let the sum and the product of the elements of $A$ be 36 and $p$ respectively and the sum and the product of the elements of $B$ be 36 and $q$ respectively. Let $d$ and $D$ be the common differences of AP's in $A$ and $B$ respectively such that $D = d + 3$, $d > 0$. If $\frac{p+q}{p-q} = \frac{19}{5}$, then $p - q$ is equal to **[2025, 4 April Shift-II]**

(a) 540 (b) 600
(c) 630 (d) 450

**28** The sum $1 + 3 + 11 + 25 + 45 + 71 + \ldots$ upto 20 terms, is equal to **[2025, 3 April Shift-I]**

(a) 6982 (b) 7130
(c) 7240 (d) 8124

**29** The number of terms of an AP is even, the sum of all the odd terms is 24, the sum of all the even terms is 30 and the last term exceeds the first by 21/2. Then, the number of terms which are integers in the AP is **[2025, 2 April Shift-II]**

(a) 10 (b) 6 (c) 4 (d) 8

**Imp.**

## Sum of General Term

**30** If the sum of the first 20 terms of the series $\frac{4\cdot1}{4+3\cdot1^2+1^4}+\frac{4\cdot2}{4+3\cdot2^2+2^4}+\frac{4\cdot3}{4+3\cdot3^2+3^4}+\frac{4\cdot4}{4+3\cdot4^2+4^4}+\ldots$ is $\frac{m}{n}$, where $m$ and $n$ are coprime, then $m+n$ is equal to **[2025, 4 April Shift-II]**

(a) 421 (b) 422 (c) 423 (d) 420

**31** Let $T_r$ be the $r$th term of an AP. If for some $m$, $T_m=\frac{1}{25}\cdot T_{25}=\frac{1}{20}$ and $20\sum_{r=1}^{25}T_r=13$, then $5m\sum_{r=m}^{2m}T_r$ is equal to **[2025, 28 Jan Shift-I]**

(a) 112 (b) 126 (c) 98 (d) 142

**32** Let $S_n=\frac{1}{2}+\frac{1}{6}+\frac{1}{12}+\frac{1}{20}+\ldots+$ upto $n$ terms. If the sum of the first six terms of an AP with first term $-p$ and common difference $p$ is $\sqrt{2026S_{2025}}$, then the absolute difference between 20th and 15th terms of the AP is **[2025, 24 Jan Shift-I]**

(a) 45 (b) 20 (c) 25 (d) 90

**33** Suppose that the number of terms in an AP is $2k$, $k\in N$. If the sum of all odd terms of the AP is 40, then sum of all even terms is 55 and the last term of the AP exceeds the first term by 27, then $k$ is equal to **[2025, 22 Jan Shift-II]**

(a) 4 (b) 6 (c) 8 (d) 5

**34** For positive integers $n$, if $4a_n=(n^2+5n+6)$ and $S_n=\sum_{k=1}^{n}\left(\frac{1}{a_k}\right)$, then the value of $507\,S_{2025}$ is **[2025, 28 Jan Shift-II]**

(a) 675 (b) 1350 (c) 540 (d) 135

**35** If $\sum_{r=1}^{n}T_r=\frac{(2n-1)(2n+1)(2n+3)(2n+5)}{64}$, then $\lim_{n\to\infty}\sum_{r=1}^{n}\left(\frac{1}{T_r}\right)$ is equal to **[2025, 22 Jan Shift-I]**

(a) 0 (b) 1/3
(c) 1 (d) 2/3

**36** If the sum of the series $\frac{1}{1\cdot(1+d)}+\frac{1}{(1+d)(1+2d)}+\ldots+\frac{1}{(1+9d)(1+10d)}$ is equal to 5, then $50d$ is equal to **[2024, 9 April Shift-I]**

(a) 15 (b) 20 (c) 5 (d) 10

**37** If $\frac{1}{\sqrt{1}+\sqrt{2}}+\frac{1}{\sqrt{2}+\sqrt{3}}+\ldots+\frac{1}{\sqrt{99}+\sqrt{100}}=m$ and $\frac{1}{1\cdot2}+\frac{1}{2\cdot3}+\ldots+\frac{1}{99\cdot100}=n$, then the point $(m,n)$ lies on the line **[2024, 5 April Shift-I]**

(a) $11x-100y=0$ (b) $11(x-2)-100(y-1)=0$
(c) $11(x-1)-100(y-2)=0$ (d) $11(x-1)-100y=0$

**38** The sum of the series $\frac{1}{1-3\cdot1^2+1^4}+\frac{2}{1-3\cdot2^2+2^4}+\frac{3}{1-3\cdot3^2+3^4}+\ldots$ up to 10terms is **[2024, 31 Jan Shift-I]**

(a) $\frac{45}{109}$ (b) $-\frac{55}{109}$ (c) $\frac{55}{109}$ (d) $-\frac{45}{109}$

**39** Let $a_1,a_2,a_3,\ldots,a_n$ be $n$ positive consecutive terms of an arithmetic progression. If $d>0$ is its common difference, then $\lim_{n\to\infty}\sqrt{\frac{d}{n}}\left(\frac{1}{\sqrt{a_1}+\sqrt{a_2}}+\frac{1}{\sqrt{a_2}+\sqrt{a_3}}+\ldots+\frac{1}{\sqrt{a_{n-1}}+\sqrt{a_n}}\right)$ is **[2023, 6 April Shift-I]**

(a) $\sqrt{d}$ (b) 0 (c) 1 (d) $\frac{1}{\sqrt{d}}$

**40** Let $a_1,a_2,a_3,\ldots$ be an AP. If $a_7=3$, the product $a_1a_4$ is minimum and the sum of its first $n$ terms is zero, then $n!-4a_{n(n+2)}$ is equal to **[2023, 31 Jan Shift-II]**

(a) $\frac{33}{4}$ (b) $\frac{381}{4}$ (c) 9 (d) 24

**41** If $n$ arithmetic means are inserted between $a$ and 100 such that the ratio of the first mean to the last mean is $1:7$ and $a+n=33$, then the value of $n$ is **[2022, 28 June Shift-II]**

(a) 21 (b) 22
(c) 23 (d) 24

**42** If $a_1,a_2,a_3\ldots$ and $b_1,b_1,b_3\ldots$ are in AP and $a_1=2$, $a_{10}=3$, $a_1b_1=1=a_{10}b_{10}$, then $a_4b_4$ is equal to **[2022, 27 June Shift-II]**

(a) $\frac{35}{27}$ (b) 1 (c) $\frac{27}{28}$ (d) $\frac{28}{27}$

**43** Let $a_1, a_2, \ldots\ldots, a_{21}$ be an AP such that $\sum_{n=1}^{20} \frac{1}{a_n a_{n+1}} = \frac{4}{9}$. If the sum of this AP is 189, then $a_6 a_{16}$ is equal to **[2021, 1 Sep Shift-II]**

(a) 57 (b) 72 (c) 48 (d) 36

**44** If $\tan\left(\frac{\pi}{9}\right), x, \tan\left(\frac{7\pi}{18}\right)$ are in arithmetic progression and $\tan\left(\frac{\pi}{9}\right), y, \tan\left(\frac{5\pi}{18}\right)$ are also in arithmetic progression, then $|x - 2y|$ is equal to **[2021, 27 July Shift-II]**

(a) 4 (b) 3
(c) 0 (d) 1

**45** Let $a_1, a_2, \ldots, a_n$ be a given AP. Whose common difference is an integer and $S_n = a_1 + a_2 + \ldots + a_n$. If $a_1 = 1$, $a_n = 300$ and $15 \le n \le 50$, then the ordered pair $(S_{n-4}, a_{n-4})$ is equal to **[2020, 4 Sep Shift-II]**

(a) (2490, 249) (b) (2480, 249)
(c) (2480, 248) (d) (2490, 248)

**46** The number of terms common to the two AP's 3, 7, 11, ..., 407 and 2, 9, 16, ....., 709 is ..... . **[2020, 9 Jan Shift-II]**

(a) 16 (b) 14 (c) 17 (d) 20

**Imp.** $n^{th}$ Term Formula of AP

**47** A man saves ₹ 200 in each of the first three months of his service. In each of the subsequent months, his saving increases by ₹ 40 more than the saving of immediately previous month. His total saving from the start of service will be ₹ 11040 after **[AIEEE 2011]**

(a) 19 months (b) 20 months
(c) 21 months (d) 18 months

## Numerical Value Type Questions

**48** Let $a_1, a_2, \ldots\ldots, a_{2024}$ be an arithmetic progression such that $a_1 + (a_5 + a_{10} + a_{15} + \ldots\ldots\ldots\ldots + a_{2020}) + a_{2024} = 2233$. Then, $a_1 + a_2 + a_3 + \ldots\ldots\ldots + a_{2024}$ is equal to.......... **[2025, 29 Jan Shift-II]**

**49** The interior angles of a polygon with $n$ sides are in an AP with common difference $6°$. If the largest interior angle of the polygon is $219°$, then $n$ is equal to....... **[2025, 28 Jan Shift-II]**

**50** The number of 3-digit numbers that are divisible by 2 and 3 but not divisible by 4 and 9, is **[2025, 24 Jan Shift-I]**

**51** The roots of the quadratic equation $3x^2 - px + q = 0$ are 10th and 11th terms of an arithmetic progression with common difference $\frac{3}{2}$. If the sum of the first 11 terms of this arithmetic progression is 88, then $q - 2p$ is equal to..... **[2025, 23 Jan Shift-II]**

**52** If $\left(\frac{1}{\alpha+1} + \frac{1}{\alpha+2} + \ldots + \frac{1}{\alpha+1012}\right) - \left(\frac{1}{2\cdot1} + \frac{1}{4\cdot3} + \frac{1}{6\cdot5} + \ldots + \frac{1}{2024\cdot2023}\right) = \frac{1}{2024}$, then $\alpha$ is equal to ............ . **[2024, 9 April Shift-II]**

**53** If a function $f$ satisfies $f(m+n) = f(m) + f(n)$ for all $m, n \in N$ and $f(1) = 1$, then the largest natural number $\lambda$ such that $\sum_{k=1}^{2022} f(\lambda + k) \le (2022)^2$ is equal to ............ **[2024, 9 April Shift-I]**

**54** An arithmetic progression is written in the following way

2
5 8
11 14 17
20 23 26 29
--------------------------------
--------------------------------------

The sum of all the terms of the 10th row is ________ . **[2024, 8 April Shift-II]**

**55** Let the positive integers be written in the form :

1
2 3
4 5 6
7 8 9 10
. .
. .

If the $k$th row contain exactly $k$ numbers for every natural number $k$, then the row in which the number 5310 will be, is .......... . **[2024, 8 April Shift-I]**

**56** Let $a_1, a_2, a_3, \ldots\ldots$ be in an arithmetic progression of positive terms. Let $A_k = a_1^2 - a_2^2 + a_3^2 - a_4^2 + \ldots\ldots + a_{2k-1}^2 - a_{2k}^2$.

If $A_3 = -153$, $A_5 = -435$ and $a_1^2 + a_2^2 + a_3^2 = 66$, then $a_{17} - A_7$ is equal to ........... . **[2024, 5 April Shift-I]**

**57** Let 3, 7, 11, 15, ..., 403 and 2, 5, 8, 11, ..., 404 be two arithmetic progressions. Then, the sum of the common terms in them, is equal to ......... . **[2024, 1 Feb Shift-I]**

**58** The sum of all those terms of the arithmetic progression 3, 8, 13, ..., 373, which are not divisible by 3, is equal to .......... **[2023, 10 April Shift-I]**

**59** The sum of the common terms of the following three arithmetic progressions. 3, 7, 11, 15, ......, 399, 2, 5, 8, 11, ......, 359 and 2, 7, 12, 17, ........., 197 is equal to ................ . **[2023, 1 Feb Shift-II]**

**60** The number of 3-digit numbers, that are divisible by either 2 or 3 but not divisible by 7, is ................ . **[2023, 1 Feb Shift-I]**

**61** Let $a_1 = 8, a_2, a_3, \ldots, a_n$ be an AP. If the sum of its first four terms is 50 and the sum of its last four terms is 170, then the product of its middle two terms is ...... . **[2023, 1 Feb Shift 1]**

**62** Let $a_1, a_2, \ldots, a_n$ be in AP. If $a_5 = 2a_7$ and $a_{11} = 18$, then $12\left(\frac{1}{\sqrt{a_{10}}+\sqrt{a_{11}}}+\frac{1}{\sqrt{a_{11}}+\sqrt{a_{12}}}+\ldots+\frac{1}{\sqrt{a_{17}}+\sqrt{a_{18}}}\right)$ is equal to ...... . **[2023, 31 Jan Shift-I]**

**63** For $p$ and $q \in R$, consider the real valued function $f(x) = (x-p)^2 - q, x \in R$ and $q > 0$, let $a_1, a_2, a_3$ and $a_4$ be in an arithmetic progression with mean $p$ and positive common difference. If $|f(a_i)| = 500\ i = 1, 2, 3$ and 4, then the absolute difference between the roots of $f(x) = 0$ is ......... . **[2022, 28 July Shift-I]**

**64** Different AP's are constructed with the first term 100, the last term 199, and integral common differences. The sum of the common differences of all such AP's having at least 3 terms and at most 33 terms is ...... . **[2022, 26 July Shift-II]**

**65** The number of 4-digit numbers which are neither multiple of 7 nor multiple of 3 is......... . **[2021, 31 Aug Shift-II]**

**66** If $\log_3 2, \log_3(2^x - 5), \log_3\left(2^x - \frac{7}{2}\right)$ are in an arithmetic progression, then the value of $x$ is equal to ............ . **[2021, 27 July Shift-I]**

## Topic 2 Geometric Progression and Its Mean

### Level 1

**01** Let $a_1, a_2, a_3$ ..... be a GP of increasing positive numbers. If $a_3 a_5 = 729$ and $a_2 + a_4 = \frac{111}{4}$, then $24(a_1 + a_2 + a_3)$ is equal to **[2025, 3 April Shift-I]**

(a) 128 (b) 131 (c) 130 (d) 129

**02** Let $a_1, a_2, a_3, \ldots\ldots$ be *a* GP of increasing positive terms.
If $a_1 a_5 = 28$ and $a_2 + a_4 = 29$, then $a_6$ is equal to **[2025, 22 Jan Shift-I]**

(a) 526 (b) 628 (c) 812 (d) 784

**03** Let $3, a, b, c$ be in AP and $3, a-1, b+1, c+9$ be in GP. Then, the arithmetic mean of $a, b$ and $c$ is **[2024, 1 Feb Shift-I]**

(a) 13 (b) 11 (c) $-4$ (d) $-1$

**04** Let 2nd, 8th and 44th terms of a non-constant AP be respectively the 1st, 2nd and 3rd terms of a GP If the first term of the AP is 1, then the sum of its first 20 terms is equal to **[2024, 31 Jan Shift-II]**

(a) 980 (b) 970 (c) 990 (d) 960

**05** Let $a$ and $b$ be two distinct positive real numbers. Let 11th term of a GP, whose first term is $a$ and third term is $b$, is equal to $p$th term of another GP, whose first term is $a$ and fifth term is $b$. Then, $p$ is equal to **[2024, 30 Jan Shift-II]**

(a) 20 (b) 21 (c) 24 (d) 25

**06** If in a GP of 64 terms, the sum of all the terms is 7times the sum of the odd terms of the GP, then the common ratio of the GP is equal to **[2024, 29 Jan Shift-I]**

(a) 7 (b) 6 (c) 5 (d) 4

**07** Let the sum of an infinite GP, whose first term is a and the common ratio is $r$ be 5. Let the sum of its first five terms be $\frac{98}{25}$. Then, the sum of the first 21 terms of an AP, whose first term is $10ar$, $n$th term is $a_n$ and the common difference is $10ar^2$ is equal to **[2022, 27 July Shift-II]**

(a) $21\,a_{11}$ (b) $22\,a_{11}$ (c) $15\,a_{16}$ (d) $14\,a_{16}$

**08** Let $A_1, A_2, A_3, \ldots$ be an increasing geometric progression of positive real numbers. If $A_1 A_3 A_5 A_7 = \frac{1}{1296}$ and $A_2 + A_4 = \frac{7}{36}$, then the value of $A_6 + A_8 + A_{10}$ is equal to **[2022, 28 June Shift-I]**

(a) 33 (b) 37 (c) 43 (d) 47

**09** Three numbers are in an increasing geometric progression with common ratio $r$. If the middle number is doubled, then the new numbers are in an arithmetic progression with common difference $d$. If the fourth term of GP is $3r^2$, then $r^2 - d$ is equal to **[2021, 31 Aug Shift-I]**

(a) $7 - 7\sqrt{3}$ (b) $7 + \sqrt{3}$
(c) $7 - \sqrt{3}$ (d) $7 + 3\sqrt{3}$

**10** If $0 < x < 1$ and $y = \frac{1}{2}x^2 + \frac{2}{3}x^3 + \frac{3}{4}x^4 + \ldots$, then the value of $e^{1+y}$ at $x = \frac{1}{2}$ is **[2021, 27 Aug Shift-II]**

(a) $\frac{1}{2}e^2$ (b) $2e$ (c) $\frac{1}{2}\sqrt{e}$ (d) $2e^2$

**Imp.** Sum of Infinite GP

**11** The product $2^{\frac{1}{4}} \cdot 4^{\frac{1}{16}} \cdot 8^{\frac{1}{48}} \cdot 16^{\frac{1}{128}} \cdot \ldots$ to $\infty$ is equal to **[2020, 9 Jan Shift-I]**

(a) $2^{1/4}$ (b) 2 (c) $2^{1/2}$ (d) 1

**12** The greatest positive integer $k$, for which $49^k + 1$ is a factor of the sum $49^{125} + 49^{124} + \ldots + 49^2 + 49 + 1$, is *[2020, 7 Jan Shift-I]*

(a) 32 (b) 63 (c) 65 (d) 60

**13** The sum $\sum_{k=1}^{20} k \frac{1}{2^k}$ is equal to *[2019, 8 April Shift-II]*

(a) $2 - \frac{11}{2^{19}}$ (b) $1 - \frac{11}{2^{20}}$ (c) $2 - \frac{3}{2^{17}}$ (d) $2 - \frac{21}{2^{20}}$

**14** Let $a$, $b$ and $c$ be the 7th, 11th and 13th terms respectively of a non-constant AP. If these are also the three consecutive terms of a GP, then $\frac{a}{c}$ is equal to *[2019, 9 Jan Shift-II]*

(a) 2 (b) 7/13 (c) 4 (d) 1/2

**15** Three positive numbers form an increasing GP. If the middle term in this GP is doubled, then new numbers are in AP. Then, the common ratio of the GP is *[JEE Main 2014]*

(a) $\sqrt{2} + \sqrt{3}$ (b) $3 + \sqrt{2}$ (c) $2 - \sqrt{3}$ (d) $2 + \sqrt{3}$

**16** The first two terms of a geometric progression add up to 12. The sum of the third and the fourth terms is 48. If the terms of the geometric progression are alternately positive and negative, then the first term is *[AIEEE 2008]*

(a) 4 (b) −4 (c) −12 (d) 12

**17** In a geometric progression consisting of positive terms, each term equals the sum of the next two terms. Then, the common ratio of this progression is equal to *[AIEEE 2007]*

(a) $\frac{1}{2}(1 - \sqrt{5})$ (b) $\frac{1}{2}\sqrt{5}$ (c) $\sqrt{5}$ (d) $\frac{1}{2}(\sqrt{5} - 1)$

**18** 5th term of a GP is 2, then the product of its 9 terms is *[AIEEE 2002]*

(a) 256 (b) 512
(c) 1024 (d) None of these

## Level 2

**19** If the sum of the second, fourth and sixth terms of a GP of positive terms is 21 and the sum of its eighth, tenth and twelfth terms is 15309, then the sum of its first nine terms is *[2025, 7 April Shift-II]*

(a) 755 (b) 760
(c) 750 (d) 757

**20** Let $x_1, x_2, x_3, x_4$ be in a geometric progression. If 2, 7, 9, 5 are subtracted respectively from $x_1, x_2, x_3, x_4$, then the resulting numbers are in an arithmetic progression. Then, the value of $\frac{1}{24}(x_1 x_2 x_3 x_4)$ is *[2025, 7 April Shift-I]*

(a) 216 (b) 36 (c) 72 (d) 18

**21** If $S = 5 + \frac{1}{7}(5 + \alpha) + \frac{1}{7^2}(5 + 2\alpha) + \frac{1}{7^3}(5 + 3\alpha) + \ldots + \infty$, then the value of $\alpha$ is *[2025, 24 Jan Shift-II]*

(a) 1 (b) 1/7 (c) 6 (d) 6/7

**22** Let $ABC$ be an equilateral triangle. A new triangle is formed by joining the middle points of all sides of the $\Delta ABC$ and the same process is repeated infinitely many times. If $P$ is the sum of perimeters and $Q$ is the sum of areas of all the triangles formed in this process, then *[2024, 6 April Shift-II]*

(a) $P^2 = 36\sqrt{3}Q$ (b) $P^2 = 6\sqrt{3}Q$
(c) $P = 36\sqrt{3}Q^2$ (d) $P^2 = 72\sqrt{3}Q$

**23** The value of $\frac{1 \times 2^2 + 2 \times 3^2 + \ldots + 100 \times (101)^2}{1^2 \times 2 + 2^2 \times 3 + \ldots + 100^2 \times 101}$ is *[2024, 4 April Shift-II]*

(a) $\frac{306}{305}$ (b) $\frac{31}{30}$ (c) $\frac{305}{301}$ (d) $\frac{32}{31}$

**24** Let the first three terms 2, $p$ and $q$, with $q \neq 2$, of a GP be respectively the 7th, 8th and 13th terms of an AP. If the 5th term of the GP in the $n$th term of the AP, then $n$ is equal to *[2024, 4 April Shift-I]*

(a) 151 (b) 177 (c) 169 (d) 163

**25** Let $\alpha$ and $\beta$ be the roots of the equation $px^2 + qx - r = 0$, where $p \neq 0$. If $p$, $q$ and $r$ are the consecutive terms of a non-constant GP and $\frac{1}{\alpha} + \frac{1}{\beta} = \frac{3}{4}$, then the value of $(\alpha - \beta)^2$ is *[2024, 1 Feb Shift-II]*

(a) 9 (b) 20/3 (c) 8 (d) 80/9

**26** If each term of a geometric progression $a_1, a_2, a_3, \ldots$ with $a_1 = \frac{1}{8}$ and $a_2 \neq a_1$ is the arithmetic mean of the next two terms and $S_n = a_1 + a_2 + \ldots + a_n$, then $S_{20} - S_{18}$ is equal to *[2024, 29 Jan Shift-II]*

(a) $2^{15}$ (b) $-2^{15}$ (c) $-2^{18}$ (d) $2^{18}$

**27** Let $a_1, a_2, a_3, \ldots$ be a GP of increasing positive numbers. Let the sum of its 6th and 8th terms be 2 and the product of its 3rd and 5th terms be 1/9. Then, $6(a_2 + a_4)(a_4 + a_6)$ is equal to *[2023, 13 April Shift-II]*

(a) $2\sqrt{2}$ (b) 3 (c) $3\sqrt{3}$ (d) 2

**28** Let the first term $a$ and the common ratio $r$ of a geometric progression be positive integers. If the sum of squares of its first three terms is 33033, then the sum of these three terms is equal to *[2023, 13 April Shift-I]*

(a) 210 (b) 220 (c) 231 (d) 241

**29** The parabolas : $ax^2+2bx+cy=0$ and $dx^2+2ex+fy=0$ intersect on the line $y=1$. If $a, b, c, d, e, f$ are positive real numbers and $a, b, c$ are in GP, then **[2023, 30 Jan Shift-II]**

(a) $d, e, f$ are in AP (b) $d, e, f$ are in GP

(c) $\frac{d}{a}, \frac{e}{b}, \frac{f}{c}$ are in GP (d) $\frac{d}{a}, \frac{e}{b}, \frac{f}{c}$ are in AP

**30** Consider two GP's $2, 2^2, 2^3, \ldots$ and $4, 4^2, 4^3 \ldots$ of 60 and $n$ terms, respectively. If the geometric mean of all the $60+n$ terms is $(2)^{\frac{225}{8}}$, then $\sum_{k=1}^{n} k(n-k)$ is equal to **[2022, 26 July Shift-I]**

(a) 560 (b) 1540 (c) 1330 (d) 2600

**31** Let $f: N \to R$ be a function such that $f(x+y)=2f(x)f(y)$ for natural numbers $x$ and $y$. If $f(1)=2$, then the value of $\alpha$ for which $\sum_{k=1}^{10} f(\alpha+k)=\frac{512}{3}(2^{20}-1)$ holds, is **[2022, 25 June Shift-I]**

(a) 2 (b) 3 (c) 4 (d) 6

**32** If $0<x<1$, then $\frac{3}{2}x^2+\frac{5}{3}x^3+\frac{7}{4}x^4+\ldots\ldots,$ is equal to **[2021, 27 Aug Shift-I]**

(a) $x\left(\frac{1+x}{1-x}\right)+\log_e(1-x)$ (b) $x\left(\frac{1-x}{1+x}\right)+\log_e(1-x)$

(c) $\frac{1-x}{1+x}+\log_e(1-x)$ (d) $\frac{1+x}{1-x}+\log_e(1-x)$

**33** If $x, y \in R, x>0$ $y=\log_{10} x+\log_{10} x^{1/3}+\log_{10} x^{1/9}+\ldots$ upto $\infty$ terms and $\frac{2+4+6+\ldots+2y}{3+6+9+\ldots+3y}=\frac{4}{\log_{10} x}$, then the ordered pair $(x, y)$ is equal to **[2021, 27 Aug Shift-I]**

(a) $(10^6, 6)$ (b) $(10^4, 6)$ (c) $(10^2, 3)$ (d) $(10^6, 9)$

**34** In an increasing geometric series, the sum of the second and the sixth term is $25/2$ and the product of the third and fifth term is 25. Then, the sum of 4th, 6th and 8th terms is equal to **[2021, 26 Feb Shift-I]**

(a) 30 (b) 26 (c) 35 (d) 32

**35** If $|x|<1, |y|<1$ and $x \neq y$, then the sum to infinity of the following series $(x+y)+(x^2+xy+y^2)+(x^3+x^2y+xy^2+y^3)+\ldots$ is **[2020, 2 Sep Shift-I]**

(a) $\frac{x+y+xy}{(1+x)(1+y)}$ (b) $\frac{x+y-xy}{(1-x)(1-y)}$

(c) $\frac{x+y+xy}{(1-x)(1-y)}$ (d) $\frac{x+y-xy}{(1+x)(1+y)}$

**36** Let $a, b$ and $c$ be in GP with common ratio $r$, where $a \neq 0$ and $0<r\leq\frac{1}{2}$. If $3a, 7b$ and $15c$ are the first three terms of an AP, then the 4th term of this AP is **[2019, 10 April Shift-II]**

(a) $5a$ (b) $\frac{2}{3}a$ (c) $a$ (d) $\frac{7}{3}a$

**37** If $a, b$ and $c$ be three distinct real numbers in GP and $a+b+c=xb$, then $x$ cannot be **[2019, 9 Jan Shift-I]**

(a) 4 (b) 2 (c) $-2$ (d) $-3$

**38** If the 2nd, 5th and 9th terms of a non-constant AP are in GP, then the common ratio of this GP is **[JEE Main 2016]**

(a) 8/5 (b) 4/3 (c) 1 (d) 7/4

**39** If $m$ is the AM of two distinct real numbers $l$ and $n\,(l, n>1)$ and $G_1, G_2$ and $G_3$ are three geometric means between $l$ and $n$, then $G_1^4+2G_2^4+G_3^4$ equals **[JEE Main 2015]**

(a) $4l^2mn$ (b) $4lm^2n$ (c) $4lmn^2$ (d) $4l^2m^2n^2$

## Level 3

**40** If $\frac{1}{1^4}+\frac{1}{2^4}+\frac{1}{3^4}+\ldots+\infty=\frac{\pi^4}{90}$.

$\frac{1}{1^4}+\frac{1}{3^4}+\frac{1}{5^4}+\ldots+\infty=\alpha,$

$\frac{1}{2^4}+\frac{1}{4^4}+\frac{1}{6^4}+\ldots+\infty=\beta,$

then $\frac{\alpha}{\beta}$ is equal to **[2025, 8 April Shift-II]**

(a) 18 (b) 23 (c) 15 (d) 14

**41** The sum

$1+\frac{1+3}{2!}+\frac{1+3+5}{3!}+\frac{1+3+5+7}{4!}+\ldots$ upto $\infty$ terms, is equal to **[2025, 3 April Shift-II]**

(a) $4e$ (b) $3e$ (c) $6e$ (d) $2e$

**42** Let $<a_n>$ be a sequence such that $a_0=0, a_1=\frac{1}{2}$ and $2a_{n+2}=5a_{n+1}-3a_n, n=0, 1, 2, 3 \ldots\ldots$

Then, $\sum_{k=1}^{100} a_k$ is equal to **[2025, 28 Jan Shift-I]**

(a) $3a_{99}+100$ (b) $3a_{100}-100$

(c) $3a_{100}+100$ (d) $3a_{99}-100$

**43** Let $a, ar, ar^2, \ldots\ldots\ldots$ be an infinite G P. If $\sum_{n=0}^{\infty} ar^n=57$ and $\sum_{n=0}^{\infty} a^3r^{3n}=9747$, then $a+18r$ is equal to **[2024, 9 April Shift-II]**

(a) 31 (b) 27 (c) 46 (d) 38

**44** In an increasing geometric progression of positive terms, the sum of the second and sixth terms is 70/3 and the product of the third and fifth terms is 49. Then, the sum of the 4th, 6th and 8th terms is equal to *[2024, 8 April Shift-II]*

(a) 84 (b) 78 (c) 96 (d) 91

**45** For $0 < c < b < a$, let $(a+b-2c)x^2 + (b+c-2a)x + (c+a-2b) = 0$ and $\alpha \neq 1$ be one of its root. Then, among the two statements. *[2024, 31 Jan Shift-II]*

I. If $\alpha \in (-1, 0)$, then $b$ cannot be the geometric mean of $a$ and $c$.

II. If $\alpha \in (0, 1)$, then $b$ may be the geometric mean of $a$ and $c$.

(a) Neither I nor II is true (b) Only II is true
(c) Only I is true (d) Both I and II are true

**46** If $2\tan^2\theta - 5\sec\theta = 1$ has exactly 7 solutions in the interval $\left[0, \frac{n\pi}{2}\right]$, for the least value of $n \in N$, then $\sum_{k=1}^{n} \frac{k}{2^k}$ is equal to *[2024, 27 Jan Shift-II]*

(a) $\frac{1}{2^{13}}(2^{14}-15)$ (b) $\frac{1}{2^{15}}(2^{14}-14)$

(c) $\frac{1}{2^{14}}(2^{15}-15)$ (d) $1-\frac{15}{2^{13}}$

**47** Let $a, b, c > 1$, $a^3, b^3$ and $c^3$ be in AP and $\log_a b$, $\log_c a$ and $\log_b c$ be in GP. If the sum of first 20 terms of an AP, whose first term is $\frac{a+4b+c}{3}$ and the common difference is $\frac{a-8b+c}{10}$ is $-444$, then $abc$ is equal to *[2023, 30 Jan Shift-II]*

(a) 216 (b) 343 (c) 343/8 (d) 125/8

**48** For three positive integers $p, q, r$, $x^{pq^2} = y^{qr} = z^{p^2r}$ and $r = pq + 1$ such that $3, 3\log_y x, 3\log_z y, 7\log_x z$ are in AP with common difference 1/2. Then, $r - p - q$ is equal to *[2023, 24 Jan Shift-I]*

(a) 12 (b) 2 (c) 6 (d) $-6$

## Numerical Value Type Questions

**49** The sum of the series $2 \times 1 \times {}^{20}C_4 - 3 \times 2 \times {}^{20}C_5 + 4 \times 3 \times {}^{20}C_6 - 5 \times 4 \times {}^{20}C_7 + \ldots + 18 \times 17 \times {}^{20}C_{20}$, is equal to ............... *[2025, 7 April Shift-II]*

**50** If the sum of the first 10 terms of the series $\frac{4 \cdot 1}{1+4 \cdot 1^4} + \frac{4 \cdot 2}{1+4 \cdot 2^4} + \frac{4 \cdot 3}{1+4 \cdot 3^4} + \ldots$ is $\frac{m}{n}$, where $\gcd(m, n) = 1$, then $m + n$ is equal to ......... . *[2025, 2 April Shift-II]*

**51** If the range of $f(\theta) = \frac{\sin^4\theta + 3\cos^2\theta}{\sin^4\theta + \cos^2\theta}$, $\theta \in R$ is $[\alpha, \beta]$. Then, the sum of the infinite GP, whose first term is 64 and the common ratio is $\alpha/\beta$, is equal to ......... . *[2024, 8 April Shift-I]*

**52** Let $\alpha = \sum_{r=0} (4r^2 + 2r + 1)\, {}^nC_r$ and $\beta = \left(\sum_{r=0}^{n} \frac{{}^nC_r}{r+1}\right) + \frac{1}{n+1}$. If $140 < \frac{2\alpha}{\beta} < 281$. Then, the value of $n$ is ........ . *[2024, 8 April Shift-I]*

**53** If $S(x) = (1+x) + 2(1+x)^2 + 3(1+x)^3 + \ldots + 60(1+x)^{60}$, $x \neq 0$ and $(60)^2 S(60) = a(b)^2 + b$, where $a$, be $N$, then $(a+b)$ equal to ......... . *[2024, 6 April Shift-II]*

**54** Let the first term of a series be $T_1 = 6$ and its $r$th term $T_r = 3T_{r-1} + 6^r$, $r = 2, 3, \ldots, n$. If the sum of the first $n$ terms of this series is $\frac{1}{5}(n^2 - 12n + 39)(4 \cdot 6^n - 5 \cdot 3^n + 1)$, then $n$ is equal to ............ *[2024, 6 April Shift-I]*

**55** If $1 + \frac{\sqrt{3}-\sqrt{2}}{2\sqrt{3}} + \frac{5-2\sqrt{6}}{18} + \frac{9\sqrt{3}-11\sqrt{2}}{36\sqrt{3}} + \frac{49-20\sqrt{6}}{180} + \ldots$ upto $\infty = 2 + \left(\sqrt{\frac{b}{a}} + 1\right)\log_e\left(\frac{a}{b}\right)$, where $a$ and $b$ are integers with $\gcd(a, b) = 1$, then $11a + 18b$ is equal to ...... . *[2024, 5 April Shift-II]*

**56** Let $S_n$ be the sum to $n$-terms of an arithmetic progression 3, 7, 11, ......... . If $40 < \left(\frac{6}{n(n+1)} \sum_{k=1}^{n} S_k\right) < 42$, then $n$ equals ......... . *[2024, 30 Jan Shift-II]*

**57** Let $\alpha = 1^2 + 4^2 + 8^2 + 13^2 + 19^2 + 26^2 + \ldots$ upto 10 terms and $\beta = \sum_{n=1}^{10} n^4$. If $4\alpha - \beta = 55k + 40$, then $k$ is equal to ......... . *[2024, 30 Jan Shift-I]*

**58** If $8 = 3 + \frac{1}{4}(3+p) + \frac{1}{4^2}(3+2p) + \frac{1}{4^3}(3+3p) + \ldots \infty$, then the value of $p$ is .............. . *[2024, 27 Jan Shift-I]*

**59** Let $0 < z < y < x$ be three real numbers such that $\frac{1}{x}, \frac{1}{y}, \frac{1}{z}$ are in an arithmetic progression and $x, \sqrt{2}y, z$ are in a geometric progression. If $xy + yz + zx = \frac{3}{\sqrt{2}} xyz$, then $3(x+y+z)^2$ is equal to ...... *[2023, 8 April Shift-I]*

**60** If $(20)^{19}+2(21)(20)^{18}+3(21)^2(20)^{17}+\ldots+20(21)^{19}$ $=k(20)^{19}$, then $k$ is equal to …… .

*[2023, 6 April Shift-II]*

**61** Let $\{a_k\}$ and $\{b_k\}$, $k\in N$ be two GPs with common ratios $r_1$ and $r_2$ respectively such that $a_1=b_1=4$ and $r_1<r_2$. Let $c_k=a_k+b_k, k\in N$. If $c_2=5$ and $c_3=\frac{13}{4}$, then $\sum_{k=1}^{\infty}c_k-(12a_6+8b_4)$ is equal to …… .

*[2023, 29 Jan Shift-II]*

**62** Let $a_1, a_2, a_3, \ldots$ be a GP of increasing positive numbers. If the product of fourth and sixth terms is 9 and the sum of fifth and seventh terms is 24, then $a_1a_9+a_2a_4a_9+a_5+a_7$ is equal to ……..

*[2023, 29 Jan Shift-I]*

**63** For the two positive numbers $a$, $b$, if $a$, $b$ and $\frac{1}{18}$ are in a geometric progression, while $\frac{1}{a}$, 10 and $\frac{1}{b}$ are in an arithmetic progression, then $16a+12b$ is equal to …… .

*[2023, 25 Jan Shift-II]*

**64** The 4th term of GP is 500 and its common ratio is $\frac{1}{m}$, $m\in N$. Let $S_n$ denote the sum of the first $n$ terms of this GP. If $S_6>S_5+1$ and $S_7<S_6+\frac{1}{2}$, then the number of possible values of $m$ is …… .

*[2023, 24 Jan Shift-I]*

**65** Let $x_1, x_2, x_3, \ldots, x_{20}$ be geometric progression with $x_1=3$ and the common ratio $\frac{1}{2}$. A new data is constructed replacing each $x_i$ by $(x_i-i)^2$. If $\bar{x}$ is the mean of new data, then the greatest integer less than or equal to $\bar{x}$ is ……… .

*[2022, 28 July Shift-I]*

**66** If $a_1(>0), a_2, a_3, a_4, a_5$ are in a GP $a_2+a_4=2a_3+1$ and $3a_2+a_3=2a_4$, then $a_2+a_4+2a_5$ is equal to …… .

*[2022, 26 June Shift-II]*

**67** If the value of

$$\left(1+\frac{2}{3}+\frac{6}{3^2}+\frac{10}{3^3}+\ldots\text{upto }\infty\right)^{\log_{(0.25)}\left(\frac{1}{3}+\frac{1}{3^2}+\frac{1}{3^3}+\ldots\text{upto }\infty\right)}$$

is $l$, then $l^2$ is equal to …… .

*[2021, 25 July Shift-I]*

**68** Let $\frac{1}{16}$, $a$ and $b$ be in G P and $\frac{1}{a},\frac{1}{b},6$ be in AP, where $a, b>0$. Then, $72(a+b)$ is equal to ……… .

*[2021, 16 March Shift-II]*

**69** The sum of first four terms of a geometric progression (GP) is $\frac{65}{12}$ and the sum of their respective reciprocals is $\frac{65}{18}$. If the product of first three terms of the GP is 1 and the third term is $\alpha$, then $2\alpha$ is ……… .

*[2021, 24 Feb Shift-II]*

**70** The value of $(0.16)^{\log_{2.5}\left(\frac{1}{3}+\frac{1}{3^2}+\frac{1}{3^3}+\ldots\text{ to }\infty\right)}$ is equal to …… .

*[2020, 3 Sep Shift-I]*

## Topic 3 Relation between AM and GM

### Level 1

**01** Let the range of the function $f(x)=\frac{1}{2+\sin 3x+\cos 3x}$, $x\in R$ be $[a, b]$. If $\alpha$ and $\beta$ are respectively the A M and the GM of $a$ and $b$, then $\frac{\alpha}{\beta}$ is equal to

*[2024, 9 April Shift-II]*

(a) $\pi$ (b) $\sqrt{2}$ (c) $\sqrt{\pi}$ (d) 2

**02** If the minimum value of $f(x)=\frac{5x^2}{2}+\frac{\alpha}{x^5}$, $x>0$ is 14, then the value of $\alpha$ is equal to

*[2022, 28 July Shift-I]*

(a) 32 (b) 64 (c) 128 (d) 256

**03** Let $x, y>0$, if $x^3y^2=2^{15}$, then the least value of $3x+2y$ is

*[2022, 24 June Shift-II]*

(a) 30 (b) 32 (c) 36 (d) 40

**Imp.** Maximum and Minimum Value of Expression

**04** The minimum value of $f(x)=a^{a^x}+a^{1-a^x}$, where $a$, $x\in R$ and $a>0$, is equal to

*[2021, 25 Feb Shift-II]*

(a) $a+1$ (b) $a+\frac{1}{a}$

(c) $2\sqrt{a}$ (d) $2a$

**05** Let $f:R\to R$ be such that for all $x\in R$, $(2^{1+x}+2^{1-x})$, $f(x)$ and $(3^x+3^{-x})$ are in AP, then the minimum value of $f(x)$ is

*[2020, 8 Jan Shift-I]*

(a) 0 (b) 4 (c) 3 (d) 2

**06** Let $x$, $y$ be positive real numbers and $m$, $n$ positive integers. The maximum value of the expression $\frac{x^m y^n}{(1+x^{2m})(1+y^{2n})}$ is

*[2019, 11 Jan Shift-II]*

(a) $\frac{1}{2}$ (b) 1 (c) $\frac{1}{4}$ (d) $\frac{m+n}{6mn}$

**07** If $a, b$ and $c$ are three distinct and real numbers in GP and $a+b+c=xb$, then $x$ cannot be *[2019, 9 Jan Shift-I]*

(a) 4 (b) −3 (c) −2 (d) 2

**08** If the arithmetic mean of two numbers $a$ and $b$, $a>b>0$, is five times their geometric mean, then $\frac{a+b}{a-b}$ is equal to *[JEE Main 2017]*

(a) $\frac{\sqrt{6}}{2}$ (b) $\frac{3\sqrt{2}}{4}$ (c) $\frac{7\sqrt{3}}{12}$ (d) $\frac{5\sqrt{6}}{12}$

**09** If $A>0, B>0$ and $A+B=\frac{\pi}{6}$, then the minimum value of $\tan A+\tan B$ is *[JEE Main 2016]*

(a) $\sqrt{3}-\sqrt{2}$ (b) $4-2\sqrt{3}$ (c) $\frac{2}{\sqrt{3}}$ (d) $2-\sqrt{3}$

**10** Let $G$ be the geometric mean of two positive numbers $a$ and $b$ and $M$ be the arithmetic mean of $\frac{1}{a}$ and $\frac{1}{b}$. If $\frac{1}{M}:G$ is 4 : 5, then $a:b$ can be *[JEE Main 2014]*

(a) 1 : 4 (b) 1 : 2
(c) 2 : 3 (d) 3 : 4

**11** Let $f:R\rightarrow R$ be a continuous function defined by $f(x)=\frac{1}{e^x+2e^{-x}}$.

**Statement I** $f(c)=\frac{1}{3}$, for some $c\in R$.

**Statement II** $0<f(x)\le\frac{1}{2\sqrt{2}}, \forall x\in R$. *[AIEEE 2010]*

(a) Statement I is false, Statement II is true
(b) Statement I is true, Statement II is true; Statement II is a correct explanation of Statement I
(c) Statement I is true, Statement II is true; Statement II is not a correct explanation for Statement I
(d) Statement I is true, Statement II is false

**12** If $p$ and $q$ are positive real numbers such that $p^2+q^2=1$, then the maximum value of $(p+q)$ is *[AIEEE 2007]*

(a) 2 (b) 1/2
(c) $1/\sqrt{2}$ (d) $\sqrt{2}$

## Level 2

**13** Let three real numbers $a, b, c$ be in arithmetic progression and $a+1, b, c+3$ be in geometric progression. If $a>10$ and the arithmetic mean of $a, b$ and $c$ is 8, then the cube of the geometric mean of $a, b$ and $c$ is *[2024, 4 April Shift-II]*

(a) 128 (b) 120
(c) 312 (d) 316

**14** Let $A_1$ and $A_2$ be two arithmetic means and $G_1, G_2, G_3$ be three geometric means of two distinct positive numbers. Then, $G_1^4+G_2^4+G_3^4+G_1^2G_3^2$ is equal to *[2023, 15 April Shift-II]*

(a) $(A_1+A_2)^2G_1G_3$ (b) $2(A_1+A_2)G_1^2G_3^2$
(c) $(A_1+A_2)G_1^2G_3^2$ (d) $2(A_1+A_2)G_1G_3$

**15** Let $a, b, c$ and $d$ be positive real numbers such that $a+b+c+d=11$. If the maximum value of $a^5b^3c^2d$ is $3750\beta$, then the value of $\beta$ is *[2023, 11 April Shift-II]*

(a) 55 (b) 90
(c) 108 (d) 110

**16** Let $x_1, x_2, \ldots, x_{100}$ be in an arithmetic progression, with $x_1=2$ and their mean equal to 200. If $y_i=i(x_i-i), 1\le i\le 100$, then the mean of $y_1, y_2, \ldots, y_{100}$ is *[2023, 11 April Shift-I]*

(a) 10049.50 (b) 10101.50
(c) 10051.50 (d) 10100

**17** Let $x, y$ and $z$ be positive real numbers such that $x+y+z=12$ and $x^3y^4z^5=(0.1)(600)^3$. Then, $x^3+y^3+z^3$ is equal to *[JEE Main 2016]*

(a) 342 (b) 216
(c) 258 (d) 270

## Numerical Value Type Questions

**18** If the arithmetic mean and geometric mean of the $p$th and $q$th terms of the sequence −16, 8, −4, 2, ... satisfy the equation $4x^2-9x+5=0$, then $p+q$ is equal to ……… *[2021, 26 Feb Shift-II]*

**19** If the arithmetic means (AMs) and three geometric means (GMs) are inserted between 3 and 243 such that 4th AM is equal to 2nd GM, then $m$ is equal to ………… . *[2020, 3 Sep Shift-II]*

# Key Index

## Topic 1

| | | | | | | | | | | | | | | | | | | | |
|---|---|---|---|---|---|---|---|---|---|---|---|---|---|---|---|---|---|---|---|
| **1** | (b) | **2** | (c) | **3** | (b) | **4** | (d) | **5** | (b) | **6** | (b) | **7** | (d) | **8** | (a) | **9** | (b) | **10** | (d) |
| **11** | (b) | **12** | (b) | **13** | (d) | **14** | (d) | **15** | (a) | **16** | (b) | **17** | (c) | **18** | (a) | **19** | (d) | **20** | (b) |
| **21** | (a) | **22** | (b) | **23** | (d) | **24** | (a) | **25** | (a) | **26** | (b) | **27** | (a) | **28** | (c) | **29** | (c) | **30** | (a) |
| **31** | (b) | **32** | (c) | **33** | (d) | **34** | (a) | **35** | (d) | **36** | (c) | **37** | (a) | **38** | (b) | **39** | (c) | **40** | (d) |
| **41** | (c) | **42** | (d) | **43** | (b) | **44** | (c) | **45** | (d) | **46** | (b) | **47** | (c) | **48** | (11132) | **49** | (20) | **50** | (125) |
| **51** | (474) | **52** | (1011) | **53** | (1010) | **54** | (1505) | **55** | (103) | **56** | (910) | **57** | (6699) | **58** | (9525) | **59** | (321) | **60** | (514) |
| **61** | (754) | **62** | (8) | **63** | (50) | **64** | (53) | **65** | (5143 | **66** | (3) | | | | | | | | |

## Topic 2

| | | | | | | | | | | | | | | | | | | | |
|---|---|---|---|---|---|---|---|---|---|---|---|---|---|---|---|---|---|---|---|
| **1** | (d) | **2** | (d) | **3** | (b) | **4** | (b) | **5** | (b) | **6** | (b) | **7** | (a) | **8** | (c) | **9** | (b) | **10** | (a) |
| **11** | (c) | **12** | (b) | **13** | (a) | **14** | (c) | **15** | (d) | **16** | (c) | **17** | (d) | **18** | (b) | **19** | (d) | **20** | (a) |
| **21** | (c) | **22** | (a) | **23** | (c) | **24** | (d) | **25** | (d) | **26** | (b) | **27** | (b) | **28** | (c) | **29** | (d) | **30** | (c) |
| **31** | (c) | **32** | (a) | **33** | (d) | **34** | (c) | **35** | (b) | **36** | (c) | **37** | (b) | **38** | (b) | **39** | (b) | **40** | (c) |
| **41** | (d) | **42** | (b) | **43** | (a) | **44** | (d) | **45** | (d) | **46** | (a) | **47** | (a) | **48** | (b) | **49** | (6460) | **50** | (441) |
| **51** | (96) | **52** | (5) | **53** | (3660) | **54** | (6) | **55** | (76) | **56** | (9) | **57** | (353) | **58** | (9) | **59** | (150) | **60** | (400) |
| **61** | (9) | **62** | (60) | **63** | (3) | **64** | (12) | **65** | (142) | **66** | (40) | **67** | (3) | **68** | (14) | **69** | (3) | **70** | (4) |

## Topic 3

| | | | | | | | | | | | | | | | | | | | |
|---|---|---|---|---|---|---|---|---|---|---|---|---|---|---|---|---|---|---|---|
| **1** | (b) | **2** | (c) | **3** | (d) | **4** | (c) | **5** | (c) | **6** | (c) | **7** | (d) | **8** | (d) | **9** | (b) | **10** | (a) |
| **11** | (b) | **12** | (d) | **13** | (b) | **14** | (a) | **15** | (b) | **16** | (a) | **17** | (b) | **18** | (10) | **19** | (39) | | |

# Detailed **Explanations**

## Topic 1

**01.** (b) Use $S_n = \frac{n}{2}[2a+(n-1)d]$ and

$a_n = a+(n-1)d$

$S_7 = 7 \Rightarrow \frac{7}{2}[2a+6d] = 7$

$\Rightarrow 2a+6d = 2 \Rightarrow a+3d = 1$ ...(i)

And $a_6 = 7 \Rightarrow a+5d = 7$ ...(ii)

Solving Eqs. (i) and (ii), we get $d = 3$, $a = -8$

Also, $S_n = a_1 + a_2 + a_3 + \ldots + a_n = 700$

$\frac{n}{2}[2a+(n-1)d] = 700$

$\frac{n}{2}[-16+(n-1)3] = 700$

$n(-16+3n-3) = 1400$

$3n^2 - 19n - 1400 = 0$

$3n^2 - 75n + 56n - 1400 = 0$

$3n(n-25)+56(n-25) = 0$

$(3n+56)(n-25) = 0$

Thus, $n = 25$

Now, $a_n = a_{25} = a+24d$
$= -8 + 24 \times 3 = 64$

**02.** (c) $\sum_{k=1}^{12} a_{2k-1} = -\frac{72}{5} \cdot a_1$

$\Rightarrow a_1 + a_3 + a_5 + \ldots + a_{23} = -\frac{72}{5}a$

$\Rightarrow a + a + 2d + \ldots + a + 22d = -\frac{72}{5}a$

$\Rightarrow 12a + 2d(1+2+\ldots+11) = -\frac{72}{5}a$

$\Rightarrow a = -5d$ ...(i)

Also, $\sum_{k=1}^{n} a_k = 0 \Rightarrow S_n = 0$

$\Rightarrow \quad 2a = -(n-1)d$

$\Rightarrow \quad n = 11 \quad (\because a = -5d)$

**03.** (b) $1+3+5^2+7+9^2+\ldots+40$ terms

$(1^2+5^2+9^2+\ldots+20 \text{ term})+$
$(3+7+11+\ldots+20 \text{ terms in AP})$

$\sum_{r=1}^{20}(4r-3)^2 + \frac{20}{2}[6+(20-1)4]$

$16\left(\frac{20}{6}(21)(41)\right) - 24\left(\frac{20}{2}(21)\right) + 9(20)$
$+10(6+19\times 4)$

$= 160 \times 7 \times 41 - 24 \times 210 + 180 + 820$
$= 41880$

**04.** (d) Consider AP as $a, a+d, a+2d, \ldots$ where $a, d \in N$.

Given sum of first three terms of AP
$= S_3 = 3a+3d = 54.$

$\Rightarrow \quad a+d = 18$ ...(i)

and $S_{20} = 10(2a+19d)$
$= 10(36+17d)$

According to the question,
$1600 < S_{20} < 1800$

$\Rightarrow \quad 1600 < 10(36+17d) < 1800$

$\Rightarrow \quad 160 < 36+17d < 180$

$\Rightarrow \quad 124 < 17d < 144$

$\Rightarrow \quad \frac{124}{17} < d < \frac{144}{17}$

Since, common difference will be natural number

$\therefore \quad d = 8$

$\Rightarrow \quad a+d = 18$

$\Rightarrow \quad a+8 = 18$

$\Rightarrow \quad a = 10$

Hence,

$a_{11} = 10 + (10 \times 8) = 10 + 80 = 90$

**05.** (b)

> **key idea** Sum of first $n$ terms of an AP is $\frac{n}{2}[2a+(n-1)d]$

Let $a$ = first term

$d$ = common difference of an AP

Also, $S_{40} = 1030$

$\Rightarrow \quad \frac{40}{2}(2a+39d) = 1030$ ...(i)

and $S_{12} = 57$

$\Rightarrow \quad \frac{12}{2}(2a+11d) = 57$ ...(ii)

On solving Eqs. (i) and (ii), we get

$a = -\frac{7}{2}$

and $\quad d = \frac{3}{2}$

$\therefore S_{30} - S_{10}$

$= \frac{30}{2}(2a+29d) - \frac{10}{2}(2a+9d)$

$= 20a + 390d$

$= 20 \times (-7/2) + 390(3/2) = 515$

**06.** (b)

> **key idea** Sum of '$n$' term in AP is $S_n = \frac{n}{2}[2a+(n-1)d]$

Given that $a = 3$

and $\quad S_4 = \frac{1}{5}(S_8 - S_4)$

$6S_4 = S_8$

$\Rightarrow \quad 6 \times \frac{4}{2}[2\times 3+3d]$

$= \frac{8}{2}[2\times 3+7d]$

$\Rightarrow \quad 3[6+3d] = 6+7d$

$\Rightarrow \quad 2d = -12$

$\Rightarrow \quad d = -6$

$S_{20} = 10[6+19\times(-6)]$
$= 10[6-114] = -1080$

**07.** (d) $17m = m+(m-4)+(m-4\times 2)$
$+\ldots+(m-4\times 24)$

$\Rightarrow \quad 17m = 25m - 4(1+2+\ldots+24)$

$\Rightarrow \quad 8m = \frac{4\cdot 24\cdot 25}{2} = 150$

**08.** (a) $4^{1+x}+4^{1-x}, \frac{K}{2}, 16^x+16^{-x}$ are in AP

$\Rightarrow \quad K = 4^{1+x}+4^{1-x}+16^x+16^{-x}$

$= 4\left(4^x+\frac{1}{4^x}\right)+\left(4^{2x}+\frac{1}{4^{2x}}\right)$

$\because \; P+\frac{1}{P} \geq 2$ for all $P > 0$

$\therefore \quad K_{\min} = 4\times 2+2 = 10$

**09.** (b) Given that

$S_{10} = 390$ and $\frac{a_{10}}{a_5} = \frac{15}{7}$

$S_{10} = \frac{10}{2}[2a+9d] = 390$

$\Rightarrow 2a+9d = 78$ ...(i)

$\frac{a_{10}}{a_5} = \frac{a+9d}{a+4d} = \frac{15}{7}$

$\Rightarrow 7a+63d = 15a+60d$

$\Rightarrow \quad 8a = 3d$ ...(ii)

On solving Eqs. (i) and (ii), we get

$a = 3, d = 8$

Now,

$S_{15} - S_5 = \frac{15}{2}[2a+14d] - \frac{5}{2}[2a+4d]$

$$=\frac{5}{2}[6a+42d-2a-4d]$$

$$=\frac{5}{2}[4a+38d]=5(2a+19d)$$

$$=5(6+19\times 8)=790$$

**10.** *(d)* Given that

$S_{20}=790$ and $S_{10}=145$

$$S_{20}=\frac{20}{2}[2a+19d]=790$$

$\Rightarrow 2a+19d=79$ ...(i)

$$\left[\because S_n=\frac{n}{2}[2a+(n-1)d]\right]$$

and $S_{10}=145$

$$\Rightarrow \frac{10}{2}[2a+9d]=145$$

$2a+9d=29$ ...(ii)

On solving Eqs. (i) and (ii),

$a=-8, d=5$

Now, $S_{15}-S_5$

$$=\frac{15}{2}[2a+14d]-\frac{5}{2}[2a+4d]$$

$$=\frac{15}{2}[-16+70]-\frac{5}{2}[-16+20]$$

$$=405-10=395$$

**11.** *(b)* $\log_e a, \log_e b, \log_e c$ are in AP ...(i)

$\Rightarrow 2\log b=\log a+\log c \Rightarrow b^2=ac$

$\because \log_e a-\log_e 2b$, $\log_e 2b-\log_e 3c$, $\log_e 3c-\log_e a$ are in AP

$\Rightarrow \log_e \frac{a}{2b}, \log_e \frac{2b}{3c}, \log_e \frac{3c}{a}$ are in AP

$$\Rightarrow 2\log_e \frac{2b}{3c}=\log_e \frac{a}{2b}+\log_e \frac{3c}{a}$$

$$\Rightarrow \left(\frac{2b}{3c}\right)^2=\frac{a}{2b}\times\frac{3c}{a}$$

$$\Rightarrow \frac{4b^2}{9c^2}=\frac{3c}{2b}\Rightarrow 8b^3=27c^3$$

$\Rightarrow 2b=3c\Rightarrow 6b=9c$ ...(ii)

$\because b^2 : ac$

$$\left(\frac{3c}{2}\right)^2 : ac$$

$\Rightarrow 9c=4a$ ...(iii)

From Eqs. (ii) and (iii), we get

$\therefore a:b:c=9:6:4$

**12.** *(b)* Given progression is

$$20, 19\frac{1}{4}, 18\frac{1}{2}, 17\frac{3}{4}, ...., -129\frac{1}{4}$$

i.e. $20, \frac{77}{4}, \frac{37}{2}, \frac{71}{3}, ..., -\frac{517}{4}$

i.e. last term $(t_n)$ or $l=-\frac{517}{4}$

As, we know that $a+(n-1)d=l$

$$\Rightarrow -\frac{517}{4}=20+(n-1)\left(-\frac{3}{4}\right)$$

$\Rightarrow -517=80+(-3n+3)$

$\Rightarrow -600=-3n\Rightarrow n=200$

Since, $r$th term from end

$=(n-r+1)$th term from beginning

$\Rightarrow$ 20th term from end

$=(200-20+1)$th term from beginning

$\Rightarrow$ 20th term from end $=181$th term from beginning

$$\text{Hence, } t_{181}=20+(181-1)\left(-\frac{3}{4}\right)$$

$$=20-135=-115$$

> **Alternate Solution**
>
> If we consider last term of given progression as first term.
>
> i.e. $a=-129\frac{1}{4}=-\frac{517}{4}$, then
>
> $d=$ common difference $=\frac{3}{4}$
>
> Hence, $T_{20}=-\frac{517}{4}+19\left(\frac{3}{4}\right)=-\frac{460}{4}=-115$

**13.** *(d)* First progression is 4, 9, 14, 19, … upto 25th term.

Second progression is 3, 6, 9, 12, … upto 37th term.

Now, $T_{25}$ (for first) $=4+(25-1)\times 5$
$=4+120=124$

and $T_{37}$ (for second)
$=3+(37-1)\times 3=3+108=111$

Let $d_1=$ common difference of 1st sequence $=5$

$d_2=$ common difference of 2nd sequence $=3$

and first common term $=9$

So, first common difference $=15$ (which is LCM of $d_1$ and $d_2$)

Hence, common terms are 9, 24, 39, 54, 69, 84, 99.

Thus, total number of common terms $=7$

**14.** *(d)* Given that $s_1, s_2, ..., s_{10}$ respectively be the sum of 12 terms of 10 APs, whose first terms are 1, 2, 3, … 10 and the common difference are 1, 3, 5, … 19, respectively.

The first terms are $a_i=i$ and the common differences are $d_i=2i-1$.

$$\text{Thus, } S_i=\frac{12}{2}[2i+11\times(2i-1)]$$

$$=6[24i-11]$$

$$\sum_{i=1}^{10} S_i=6\sum_{i=1}^{10}(24i-11)$$

$$=6[24\sum_{i=1}^{10} i-110]$$

$$=6\left[24\frac{(10)(10+1)}{2}-110\right]$$

$$=6[12\times 10\times 11-110]$$

$$=6[1320-110]$$

$$=6\times 1210=7260$$

**15.** *(a)* It is given that, AP with mean zero

$$\Rightarrow \frac{a_1+a_2+....a_n}{n}=0 \text{ i.e. } \frac{1}{n}\sum_{i=1}^{n} a_i=0$$

$$\text{Now, } \sum_{i=1}^{n} f(a_i)=\alpha(a_1^5+a_2^5+....a_n^5)+\beta(a_1^3+a_2^3+...+a_n^3)+\gamma(a_1+a_2+...a_n)$$

$$=0\,\alpha+0\beta+0\gamma=0$$

According to the question, $g(f(x))=x$

$\Rightarrow g$ is inverse of $f$ and $f$ is inverse of $g$.

$$\therefore f\left(g\left(\frac{1}{n}\sum_{i=1}^{n} f(a_i)\right)\right)=\frac{1}{n}\sum_{i=1}^{n} f(a_i)=0$$

**16.** *(b)* Let first term $=a$, common difference $=d$

$$\therefore \frac{S_5}{S_9}=\frac{(5/2)(2a+4d)}{(9/2)(2a+8d)}=\frac{5}{17}$$

$\Rightarrow 17a+34d=9a+36d$

$\Rightarrow 4a=d$

$a_{15}=a+14d=a+56a=57a, a\in N$

As, $110<a_{15}<120$

$\Rightarrow a_{15}=114$ $[110<57a<120]$

$$\therefore a=\frac{114}{57}=2$$

(Since $a_1, a_2, ..., a_n$ is natural number)

$\therefore d=8$

$$\therefore S_{10}=\frac{10}{2}[4+9\times 8]=380$$

**17.** *(c)* $$\frac{a_1+a_2+...+a_{10}}{a_1+a_2+...+a_p}=\frac{100}{p^2}$$

$$\Rightarrow \frac{S_{10}}{S_p}=\frac{100}{p^2}$$

$$\Rightarrow S_p=\frac{S_{10}\cdot p^2}{100}$$

$$\frac{a_{11}}{a_{10}}=\frac{S_{11}-S_{10}}{S_{10}-S_9}$$

$$=\frac{S_{10}\cdot\frac{121}{100}-S_{10}}{S_{10}-S_{10}\cdot\frac{81}{100}}$$

$$=\frac{\frac{121}{100}-1}{1-\frac{81}{100}}=\frac{21}{19}$$

**18.** (a) Let $S_n=An^2+Bn=n(An+B)$

$S_{3n}=3S_{2n}$

$\Rightarrow 3n[A(3n)+B]=3\cdot 2n\cdot[A(2n)+B]$

$\Rightarrow \quad 3An+B=4An+2B$

$\Rightarrow \quad An+B=0$

$\therefore \quad \frac{S_{4n}}{S_{2n}}=\frac{4n[A(4n)+B]}{2n[A(2n)+B]}$

$=2\left(\frac{4An-An}{2An-An}\right)=2\times 3=6$

**19.** (d) Let '$d$' be the common difference of the AP $a_1, a_2, a_3, \ldots\ldots (a_1\neq 0)$, so sum of first eleven terms is

$S_{11}=0$ [given]

$\Rightarrow \quad \frac{11}{2}[2a_1+(11-1)d]=0$

$\Rightarrow \quad a_1+5d=0$ ...(i)

Now, it is given that

$a_1+a_3+a_5+\ldots+a_{23}=ka_1$

$\Rightarrow \frac{12}{2}[2a_1+(12-1)(2d)]=ka_1$

$\Rightarrow \quad 6\left[2a_1+22\left(-\frac{a_1}{5}\right)\right]=ka_1$ [from Eq. (i)]

$\Rightarrow 6\left[\frac{10a_1-22a_1}{5}\right]=ka_1\Rightarrow k=-\frac{72}{5}$

**20.** (b) Given AP is $a_1, a_2, a_3, \ldots, a_n$

Let the above AP has common difference '$d$', then

$a_1+a_4+a_7+\ldots+a_{16}$

$=a_1+(a_1+3d)+(a_1+6d)+\ldots+(a_1+15d)$

$=6a_1+(3+6+9+12+15)d$

$=6a_1+45d=114$ [given]

$\Rightarrow \; 2a_1+15d=38$ ...(i)

Now, $a_1+a_6+a_{11}+a_{16}$

$=a_1+(a_1+5d)+(a_1+10d)+(a_1+15d)$

$=2(2a_1+15d)=2\times 38=76$ [from Eq. (i)]

**21.** (a)

> **key idea** Use the formula of sum of first $n$ terms of AP,
> i.e. $S_n=\frac{n}{2}[2a+(n-1)d]$

Given AP, is $a_1, a_2, a_3, \ldots$ having sum of first $n$-terms

$=\frac{n}{2}[2a_1+(n-1)d]$

$=50n+\frac{n(n-7)}{2}A$ [given]

$\Rightarrow \frac{1}{2}[2a_1+(n-1)d]=50+\frac{n-7}{2}A$

$\Rightarrow \frac{1}{2}[2a_1+nd-d]=\left(50-\frac{7}{2}A\right)+\frac{n}{2}A$

$\Rightarrow \left(a_1-\frac{d}{2}\right)+\frac{nd}{2}=\left(50-\frac{7}{2}A\right)+\frac{n}{2}A$

On comparing corresponding term, we get

$d=A$ and $a_1-\frac{d}{2}=50-\frac{7}{2}A$

$\Rightarrow a_1-\frac{A}{2}=50-\frac{7}{2}A$ [$\because d=A$]

$\Rightarrow \; a_1=50-3A$

So $a_{50}=a_1+49d$

$=(50-3A)+49A$ [$\because d=A$]

$=50+46A$

Therefore, $(d, a_{50})=(A, 50+46A)$

**22.** (b) We have,

$225a^2+9b^2+25c^2-75ac-45ab-15bc=0$

$\Rightarrow (15a)^2+(3b)^2+(5c)^2-(15a)(5c)-(15a)(3b)-(3b)(5c)=0$

$\Rightarrow \frac{1}{2}[(15a-3b)^2+(3b-5c)^2+(5c-15a)^2]=0$

$\Rightarrow \quad 15a=3b, 3b=5c$ and $5c=15a$

$\therefore \quad 15a=3b=5c$

$\Rightarrow \quad \frac{a}{1}=\frac{b}{5}=\frac{c}{3}=\lambda$ (say)

$\Rightarrow \; a=\lambda, b=5\lambda, c=3\lambda$

Hence, $b, c, a$ and are in AP.

**23.** (d) Given, 100 times the 100th term of an AP $=50$ times its 50th term.

To find the 150th term of the same AP.

Let $a$ be the first term and $d$ $(d\neq 0)$ be the common difference of the given AP, then

$T_{100}=a+(100-1)\,d=a+99\,d,$

$T_{50}=a+(50-1)\,d=a+49\,d,$

$T_{150}=a+(150-1)\,d=a+149\,d$

Now, according to the given data,

$100\times T_{100}=50\times T_{50}$

$\Rightarrow \quad 100(a+99\,d)=50(a+49\,d)$

$\Rightarrow \quad 2(a+99\,d)=(a+49\,d)$

$\Rightarrow \quad a+149\,d=0; \therefore T_{150}=0$

**24.** (a) Given that, $T_m=\frac{1}{n}$

$\Rightarrow \; a+(m-1)\,d=\frac{1}{n}$ ...(i)

and $T_n=\frac{1}{m}$

$\Rightarrow \; a+(n-1)\,d=\frac{1}{m}$ ...(ii)

On solving Eqs. (i) and (ii), we get

$a=d=\frac{1}{mn}$

$\therefore \quad a-d=0$

**25.** (a) Let $f(x)=Ax^2+Bx+C$

$\therefore \quad f(1)=A+B+C$

and $\quad f(-1)=A-B+C$

$\Rightarrow A+B+C=A-B+C$

$\Rightarrow \quad 2B=0$

$\Rightarrow \quad B=0$

$\therefore \quad f(x)=Ax^2+C$

$\Rightarrow \quad f'(x)=2Ax$

$\therefore \quad f'(a)=2Aa,$

$f'(b)=2Ab$

and $\quad f'(c)=2Ac$

Also, $a, b, c$ are in AP.

So, $2Aa$, $2Ab$ and $2Ac$ are in AP.

Hence, $f'(a)$, $f'(b)$ and $f'(c)$ are also in AP.

**26.** (b) Since, 1, $\log_3\sqrt{3^{1-x}+2}$, $\log_3(4\cdot 3^x-1)$ are in AP.

$\therefore 2\log_3(3^{1-x}+2)^{1/2}$

$=\log_3 3+\log_3(4\cdot 3^x-1)$

$\Rightarrow \log_3(3^{1-x}+2)=\log_3[3(4\cdot 3^x-1)]$

$\Rightarrow \quad 3^{1-x}+2=12\cdot 3^x-3$

$\therefore \quad \frac{3}{t}+2=12t-3$ [let $3^x=t$]

$\Rightarrow \; 12t^2-5t-3=0$

$\Rightarrow (3t+1)(4t-3)=0$

$\Rightarrow \quad t=-\frac{1}{3}, \frac{3}{4}$

$\Rightarrow 3^x = \frac{3}{4}$ [since, $3^x$ cannot be negative]

$\Rightarrow \quad \log_3\left(\frac{3}{4}\right) = x$

$\therefore \quad x = 1 - \log_3 4$

**27.** (a) Let

$A = \{a-d, a, a+d\}$

$B = \{b-D, b, b+D\}$

According to the question,

$3a = 36$ and $(a-d)\,a\,(a+d) = p$

$a = 12$ and $(12)(144 - d^2) = p$ ...(i)

and similarly $b = 12$ and

$12(144 - D^2) = q$ ...(ii)

$\Rightarrow \frac{144-d^2}{144-D^2} = \frac{p}{q}$

$\Rightarrow \frac{144-d^2}{144-(d+3)^2} = \frac{12}{7}$ [$\because D = d+3$]

$\left\{\because \frac{p+q}{p-q} = \frac{19}{5}, \frac{2p}{2q} = \frac{24}{14}, \frac{p}{q} = \frac{12}{7}\right\}$

$\Rightarrow 144 \times 7 - 7d^2 = 12(144) - 12(d+3)^2$

$\Rightarrow 12(d+3)^2 - 7d^2 + 144(7-12) = 0$

$\Rightarrow 5d^2 + 72d - 612 = 0$

$\Rightarrow d = \frac{-72 + \sqrt{(72)^2 + 4(5)(612)}}{10}$

$= \frac{-72+132}{10} = 6$

$\therefore \; d = 6, D = 9$

$\therefore \; p - q = 12(D^2 - d^2) = 12(81-36)$

$= 12 \times 45 = 540$

**28.** (c) $S = 1 + 3 + 11 + 25 + \ldots + T_n$

$S = 1 + 3 + 11 + \ldots + T_{n-1} + T_n$

$T_n = 1 + 2 + 8 + 14 + \ldots + (T_n - T_{n-1})$

$T_n = 1 + \frac{n-1}{2}[4 + (n-2)6]$

$= 1 + \left(\frac{n-1}{2}\right)[6n - 8]$

$= 1 + (n-1)(3n-4)$

$= 1 + 3n^2 - 4n - 3n + 4$

$T_n = 3n^2 - 7n + 5$

$S_n = \Sigma T_n = 3\Sigma n^2 - 7\Sigma n + \Sigma 5$

$= \frac{3n(n+1)(2n+1)}{6} - \frac{7n(n+1)}{2} + 5n$

$n = 20$

$\because S_{20}$

$= \frac{3 \times 20 \times 21 \times 41}{6} - \frac{7 \times 20 \times 21}{2} + 5 \times 20$

$= 7240$

**29.** (c) Let the number of terms $= 2n$, first term of A.P. is a and common difference is $d$.

Sum of odd term

$= \frac{n}{2}[2a + (n-1)2d] = 24$

$\Rightarrow \quad 2a + (n-1)2d = \frac{24 \times 2}{n}$

$\Rightarrow \quad a + (n-1)d = \frac{24}{n}$ ...(i)

Sum of even term

$\frac{n}{2}[2(a+d) + (n-1)2d] = 30$

$\Rightarrow \quad a + d + (n-1)d = \frac{30}{n}$ ...(ii)

From Eqs. (i) and (ii), we get

$d = \frac{30}{n} - \frac{24}{n}$

$\Rightarrow \quad nd = 6$ ...(iii)

Now, again given

$a + (2n-1)d - a = \frac{21}{2}$

$\Rightarrow \quad 2nd - d = \frac{21}{2} \Rightarrow d = 12 - \frac{21}{2}$

$d = 3/2$

Now, from Eq. (iii)

$n = 4$

From Eq. (i)

$a = \frac{24}{4} - nd + d = 6 - 6 + \frac{3}{2}$

$a = \frac{3}{2}$

Terms are $\frac{3}{2}, 3, \frac{9}{2}, 6, \frac{15}{2}, 9, \frac{21}{2}, 12$

$\therefore$Number of integers terms $= 4$

**30.** (a) $T_n = \frac{4n}{4 + 3n^2 + n^4} = \frac{4n}{n^4 + 3n^2 + 4}$

$= \frac{4n}{(n^4 + 4n^2 + 4) - n^2} = \frac{4n}{(n^2+2)^2 - n^2}$

$= \frac{4n}{(n^2 + n + 2)(n^2 - n + 2)}$

$= 2\left[\frac{1}{n^2 - n + 2} - \frac{1}{n^2 + n + 2}\right]$

$T_1 = 2\left(\frac{1}{2} - \frac{1}{4}\right)$

$T_2 = 2\left(\frac{1}{4} - \frac{1}{8}\right)$

$T_3 = 2\left(\frac{1}{8} - \frac{1}{14}\right)$

$\vdots$

$T_{20} = 2\left(\frac{1}{382} - \frac{1}{422}\right)$

$S_{20} = 2\left(\frac{1}{2} - \frac{1}{422}\right)$

$= \left(\frac{211-1}{422}\right)^2 = \frac{210}{211} = \frac{m}{n}$

$\therefore m + n = 210 + 211 = 421$

**31.** (b) $T_{25} = a + 24d = \frac{1}{20}$ ...(i)

$20 \sum_{r=1}^{25} T_r = 13$

$20 \times \frac{25}{2}[2a + 24d] = 13$

$a + a + 24d = \frac{13}{250}$

$a + \frac{1}{20} = \frac{13}{250}$

$\Rightarrow \quad a = \frac{13}{250} - \frac{1}{20} = \frac{26-25}{500}$

$a = \frac{1}{500}$

$\Rightarrow \quad \frac{1}{500} + 24d = \frac{1}{20}$

$\Rightarrow \quad 24d = \frac{1}{20} - \frac{1}{500}$

$24d = \frac{25-1}{500} \Rightarrow d = \frac{1}{500}$

$T_m = a + (m-1)\,d = \frac{1}{25}$

$\frac{1 + m - 1}{500} = \frac{1}{25}$

$\Rightarrow \quad m = 20$

$5m \sum_{r=m}^{2m} T_r = 5 \times 20 \sum_{r=20}^{40} T_r$

$= 100[T_{20} + T_{21} + T_{22} + \ldots + T_{40}]$

$= 100[S_{40} - S_{19}]$

$= 100\left[\frac{40}{2}\left(\frac{2}{500} + \frac{39}{500}\right) - \frac{19}{2}\left(\frac{2}{500} + \frac{18}{500}\right)\right]$

$= \frac{100}{2 \times 500}[40 \times 41 - 19 \times 20]$

$= \frac{1}{10} \times 20[82 - 19] = 126$

**32.** *(c)* Given,

$$S_n = \frac{1}{2}+\frac{1}{6}+\frac{1}{12}+\frac{1}{20}+\ldots+n \text{ terms}$$

$$S_{2025} = \sum_{n=1}^{2025}\frac{1}{n(n+1)} = \sum_{n=1}^{2025}\left(\frac{1}{n}-\frac{1}{n+1}\right)$$

$$=\left(\frac{1}{1}-\frac{1}{2}\right)+\left(\frac{1}{2}-\frac{1}{3}\right)+\left(\frac{1}{3}-\frac{1}{4}\right)+\ldots\ldots+\left(\frac{1}{2025}-\frac{1}{2026}\right)$$

$$=1-\frac{1}{2026}=\frac{2025}{2026}$$

Now, $\sqrt{2026\cdot S_{2025}} = \sqrt{2026\times\frac{2025}{2026}} = \sqrt{2025}=45$

According to the question,

$$\frac{6}{2}[-2p+(6-1)p]=45$$

$\Rightarrow \quad 9p=45 \Rightarrow p=5$

Hence, $|A_{20}-A_{15}|$

$$|[-5+19\times 5]-[-5+14\times 5]|$$
$$=|90-65|=25$$

**33.** *(d)*

> **key idea** For AP $S_n = \frac{n}{2}[2a+(n-1)d]$
>
> $a_n = a+(n-1)d$

Let AP $a_1, a_2, a_3 \ldots, a_{2k}$

Given, $\sum_{r=1}^{k} a_{2r-1} = 40$

and $\sum_{r=1}^{k} a_{2r} = 55$

$$\frac{k}{2}[2a_1+(k-1)2d]=40 \quad \ldots\text{(i)}$$

and $$\frac{k}{2}[2a_2+(k-1)2d]=55 \quad \ldots\text{(ii)}$$

Dividing Eq. (i) by Eq. (ii) we get

$$\frac{2a_1+(k-1)2d}{2a_2+(k-1)2d}=\frac{40}{55}$$

$$\Rightarrow \quad \frac{a_1+(k-1)d}{a_1+d+(k-1)d}=\frac{8}{11}$$

$$\Rightarrow \quad 11a_1+11(k-1)d=8a_1+8d+8(k-1)d$$

$$3a_1=8d-3(k-1)d \quad \ldots\text{(iii)}$$

$\because \quad a_{2k}=a_1+27$

$$a_1+(2k-1)d=a_1+27$$

$$d=\frac{27}{2k-1} \quad \ldots\text{(iv)}$$

From Eq. (i), we get

$$a_1+(k-1)d=\frac{40}{k}$$

$$\Rightarrow \quad a_1=\frac{40}{k}-(k-1)d$$

From Eq. (ii), we get

$$a_2+(k-1)d=\frac{55}{k}$$

$$\Rightarrow \quad a_1=\frac{55}{k}-(k-1)d-d$$

We get,

$$\frac{40}{k}-(k-1)d=\frac{55}{k}-(k-1)d-d$$

$$\Rightarrow \quad d=\frac{15}{k}$$

$$\Rightarrow \quad \frac{27}{2k-1}=\frac{15}{k} \Rightarrow 27k=30k-15$$

$$\Rightarrow \quad 3k=15$$

$$\therefore \quad k=5$$

**34.** *(a)* Given,

$$S_n=\sum_{k=1}^{n}\frac{1}{a_k}=\sum_{k=1}^{n}\frac{4}{k^2+5k+6}$$

$$=\sum_{k=1}^{n}\frac{4}{(k+2)(k+3)}$$

$$=4\sum_{k=1}^{n}\left(\frac{1}{k+2}-\frac{1}{k+3}\right)$$

$$=4\left[\left(\frac{1}{3}-\frac{1}{4}\right)+\left(\frac{1}{4}-\frac{1}{5}\right)+\ldots+\left(\frac{1}{n+2}-\frac{1}{n+3}\right)\right]$$

$$=4\left[\frac{1}{3}-\frac{1}{n+3}\right]=4\left(\frac{n+3-3}{3(n+3)}\right)$$

$$=\frac{4n}{3(n+3)}$$

$$\therefore S_{2025}=\frac{2025\times 4}{3\times 2028}=\frac{675\times 4}{2028}$$

$$\Rightarrow 507S_{2025}=\frac{507\times 675\times 4}{2028}$$

$$=\frac{675}{4}\times 4=675$$

**35.** *(d)*

$$\sum_{r=1}^{n}T_r=\frac{(2n-1)(2n+1)(2n+3)(2n+5)}{64}$$

$$T_n=S_n-S_{n-1}=\frac{(2n-1)(2n+1)(2n+3)(2n+5)-(2n-3)(2n-1)(2n+1)(2n+3)}{64}$$

$$=\frac{(2n-1)(2n+1)(2n+3)(2n+5-2n+3)}{64}$$

$$=\frac{(2n-1)(2n+1)(2n+3)}{8}$$

$$\therefore \sum_{r=1}^{n}\frac{1}{T_r}=8\sum_{r=1}^{n}\frac{1}{(2r-1)(2r+1)(2r+3)}$$

$$=\frac{8}{4}\left(\sum_{r=1}^{n}\frac{1}{(2r-1)(2r+1)}-\frac{1}{(2r+1)(2r+3)}\right)$$

$$=2\left[\left(\frac{1}{1\cdot 3}-\frac{1}{3\cdot 5}\right)+\left(\frac{1}{3\cdot 5}-\frac{1}{5\cdot 7}\right)\ldots\left(\frac{1}{(2n-1)(2n+1)}-\frac{1}{(2n+1)(2n+3)}\right)\right]$$

$$=2\left(\frac{1}{3}-\frac{1}{(2n+1)(2n+3)}\right)$$

At $n\to\infty$ required value $=2\times\frac{1}{3}=\frac{2}{3}$

**36.** *(c)* Given,

$$\frac{1}{1\cdot(1+d)}+\frac{1}{(1+d)(1+2d)}+\ldots+\frac{1}{(1+9d))(1+10d)}=5$$

$$\Rightarrow \frac{1}{d}\left[\frac{(1+d)-1}{1\cdot(1+d)}+\frac{(1+2d)-(1+d)}{(1+d)(1+2d)}+\ldots+\frac{(1+10d)-(1+9d)}{(1+9d)(1+10d)}\right]=5$$

$$\Rightarrow \frac{1}{d}\left[\left(1-\frac{1}{1+d}\right)+\left(\frac{1}{1+d}-\frac{1}{1+2d}\right)+\ldots+\left(\frac{1}{1+9d}-\frac{1}{1+10d}\right)\right]=5$$

$$\Rightarrow \frac{1}{d}\left[1-\frac{1}{1+10d}\right]=5$$

$$\Rightarrow \quad \frac{10d}{d(1+10d)}=5$$

$$\Rightarrow \quad 50d=5$$

**37.** *(a)* We have,

$$m=\frac{1}{\sqrt{1}+\sqrt{2}}+\frac{1}{\sqrt{2}+\sqrt{3}}+\ldots+\frac{1}{\sqrt{99}+\sqrt{100}}$$

$$\Rightarrow m=\frac{\sqrt{2}-\sqrt{1}}{(\sqrt{1}+\sqrt{2})(\sqrt{2}-\sqrt{1})}+\frac{\sqrt{3}-\sqrt{2}}{(\sqrt{2}+\sqrt{3})(\sqrt{3}-\sqrt{2})}+\ldots+\frac{\sqrt{100}-\sqrt{99}}{(\sqrt{99}+\sqrt{100})(\sqrt{100}-\sqrt{99})}$$

$$\Rightarrow m=(\sqrt{2}-\sqrt{1})+(\sqrt{3}-\sqrt{2})+\ldots..+(\sqrt{100}-\sqrt{99})$$

$$=\sqrt{100}-\sqrt{1}=10-1=9$$

Also, $n=\frac{1}{1\cdot 2}+\frac{1}{2\cdot 3}+\ldots+\frac{1}{99\cdot 100}$

$$n=\frac{2-1}{1\cdot 2}+\frac{3-2}{2\cdot 3}+\ldots+\frac{100-99}{99\cdot 100}$$

$$\Rightarrow \quad n=1-\frac{1}{2}+\frac{1}{2}-\frac{1}{3}+\ldots+\frac{1}{99}-\frac{1}{100}$$

$$\Rightarrow \quad n=1-\frac{1}{100}=\frac{99}{100}$$

$$\therefore \ (m, n)=\left(9, \frac{99}{100}\right)$$

which lies on the line $11x-100y=0$

**38.** *(b)* $S=\frac{1}{1-3\cdot 1^2+1^4}+\frac{2}{1-3\cdot 2^2+2^4}+\frac{3}{1-3\cdot 3^2+3^4}+\ldots 10$ term

$$T_n=\frac{n}{1-3n^2+n^4}$$

$$=\frac{n}{(n^2-n-1)(n^2+n-1)}$$

$$=\frac{1}{2}\left[\frac{1}{n^2-n-1}-\frac{1}{n^2+n-1}\right]$$

$$S=\sum_{n=1}^{10} T_n$$

$$=\frac{1}{2}\left[\sum_{n=1}^{10}\frac{1}{n^2-n-1}-\frac{1}{n^2+n-1}\right]$$

$$=\frac{1}{2}\left[\left(\frac{1}{-1}-\frac{1}{1}\right)+\left(\frac{1}{1}-\frac{1}{5}\right)+\left(\frac{1}{5}-\left(-\frac{1}{11}\right)\right)+\ldots\left(\frac{1}{89}-\frac{1}{109}\right)\right]$$

$$=\frac{1}{2}\left(-1-\frac{1}{109}\right)=\frac{-55}{109}$$

**39.** *(c)* To find

$$\lim_{n\to\infty}\sqrt{\frac{d}{n}}\left(\frac{1}{\sqrt{a_1}+\sqrt{a_2}}+\frac{1}{\sqrt{a_2}+\sqrt{a_3}}+\ldots+\frac{1}{\sqrt{a_{n-1}}+\sqrt{a_n}}\right)$$

Now, $\frac{1}{\sqrt{a_1}+\sqrt{a_2}}+\frac{1}{\sqrt{a_2}+\sqrt{a_3}}+\ldots+\frac{1}{\sqrt{a_{n-1}}+\sqrt{a_n}}$

$$=\frac{\sqrt{a_2}-\sqrt{a_1}}{a_2-a_1}+\frac{\sqrt{a_3}-\sqrt{a_2}}{a_3-a_2}+\ldots+\frac{\sqrt{a_n}-\sqrt{a_{n-1}}}{a_n-a_{n-1}}$$

$$=\frac{\sqrt{a_2}-\sqrt{a_1}+\sqrt{a_3}-\sqrt{a_2}+\ldots+\sqrt{a_n}-\sqrt{a_{n-1}}}{d}$$

$(\because a_2-a_1=a_3-a_2=\ldots a_n-a_{n-1}=d)$

$$=\frac{\sqrt{a_n}-\sqrt{a_1}}{d}$$

$$=\frac{\sqrt{a_1+(n-1)d}-\sqrt{a_1}}{d}$$

$$\therefore \lim_{n\to\infty}\sqrt{\frac{d}{n}}\left(\frac{\sqrt{a_1+(n-1)d}-\sqrt{a_1}}{d}\right)$$

$$=\lim_{n\to\infty}\left[\frac{1}{\sqrt{d}}\left(\frac{\sqrt{a_1+(n-1)d}-\sqrt{a_1}}{\sqrt{n}}\right)\right]$$

$$=\lim_{n\to\infty}\left[\frac{1}{\sqrt{d}}\left(\sqrt{\frac{a_1}{n}+\left(d-\frac{d}{n}\right)}-\sqrt{\frac{a_1}{n}}\right)\right]$$

$$=\frac{1}{\sqrt{d}}(\sqrt{0+d-0}-\sqrt{0})=\frac{\sqrt{d}}{\sqrt{d}}=1$$

**40.** *(d)*

> **key idea** (i) Find the value of $a_1 a_4 = p$ (let) in terms of $d$.
> (ii) $n! = 1\times 2\times 3\times 4\times \ldots \times n$

We have, $a_7=3$

$$\Rightarrow \quad a+6d=3\Rightarrow a=3-6d$$

Let $P=a_1 a_4=a(a+3d)$

$$=(3-6d)(3-6d+3d)$$

$$=9-27d+18d^2$$

$$P=18d^2-27d+9$$

$$=18\left[d^2-\frac{3}{2}d+\frac{1}{2}\right]$$

$$=18\left[\left(d-\frac{3}{4}\right)^2-\frac{9}{16}+\frac{1}{2}\right]$$

$$=18\left[\left(d-\frac{3}{4}\right)^2-\frac{1}{16}\right]$$

$P$ is minimum when

$$\left(d-\frac{3}{4}\right)^2=0\Rightarrow d=\frac{3}{4}$$

So, $P$ is minimum at $d=\frac{3}{4}$

Now, $a=3-6d$

$$\Rightarrow \quad a=3-6\times\frac{3}{4}=\frac{-3}{2}$$

Again, $S_n=0$

$$\Rightarrow \quad \frac{n}{2}[2a+(n-1)d]=0$$

$$\Rightarrow \quad 2a+(n-1)d=0$$

$$\Rightarrow 2\times\left(-\frac{3}{2}\right)+(n-1)\left(\frac{3}{4}\right)=0$$

$$\Rightarrow \quad n=5$$

Now, $n!-4a_{n(n+2)}=5!-4a_{35}$

$$=120-4(a+34d)$$

$$=120-4\left[\frac{-3}{2}+34\times\frac{3}{4}\right]$$

$$=120+6-102=24$$

**41.** *(c)* $a, A_1, A_2, \ldots, A_n, 100.$

Let $d$ be the common difference of above AP.

According to the question,

$$\frac{a+d}{100-d}=\frac{1}{7}$$

$$\Rightarrow \quad 7a+8d=100 \quad \ldots (i)$$

and $\quad a+n=33 \quad \ldots$ (ii)

and $\quad 100=a+(n+1)d$

$$\Rightarrow \quad 100=a+(34-a)\left(\frac{100-7a}{8}\right)$$

[using Eqs. (i) and (ii)]

$$\Rightarrow \quad 800=8a+7a^2-338a+3400$$

$$\Rightarrow \quad 7a^2-330a+2600=0$$

$$\Rightarrow \quad a=10, \frac{260}{7}, \text{ but } a\neq\frac{260}{7}$$

From Eq. (ii), we get

$$n=23$$

**42.** *(d)* We have, $a_1 = 2, a_{10} = 3,$

$$a_1 b_1 = a_{10}b_{10} = 1$$

Let $d_1$ be the common difference of an AP $a_1, a_2, a_3, \ldots\ldots$ and $d_2$ be the common difference of an AP $b_1, b_2, b_3, \ldots\ldots$

$$a_{10} = a_1 + 9d_1$$

$$\Rightarrow \quad 3 = 2 + 9d_1$$

$$\Rightarrow \quad d_1 = \frac{1}{9}$$

$$b_{10} = b_1 + 9d_2 = \frac{1}{3} = \frac{1}{2} + 9d_2$$

$$\Rightarrow \quad d_2 = -\frac{1}{54}$$

$$a_4 = a_1 + 3d_1 = 2 + 3\left(\frac{1}{9}\right) = \frac{7}{3}$$

$$b_4 = b_1 + 3d_2 = \frac{1}{2} + 3\left(-\frac{1}{54}\right) = \frac{8}{18}$$

$$\therefore \quad a_4 b_4 = 28/27$$

**43.** *(b)*

> **key idea** (i) $\sum \frac{1}{a_n a_{n+1}} = \frac{1}{d}\sum\left(\frac{1}{a_n} - \frac{1}{a_n + d}\right)$
>
> where, $d$ = common difference
>
> (ii) Sum of first $n$-terms of an A P
>
> $= \frac{n}{2}[2a + (n-1)d]$

Let $d$ be the common difference of an AP $a_1, a_2, \ldots, a_{21}$ and $\sum_{n=1}^{20} \frac{1}{a_n a_{n+1}} = \frac{4}{9}$

$$\Rightarrow \quad \sum_{n=1}^{20} \frac{1}{a_n(a_n + d)} = \frac{4}{9}$$

$$\Rightarrow \quad \frac{1}{d}\sum_{n=1}^{20}\left(\frac{1}{a_n} - \frac{1}{a_n + d}\right) = \frac{4}{9}$$

$$\Rightarrow \frac{1}{d}\left[\frac{1}{a_1} - \frac{1}{a_2} + \frac{1}{a_2} - \frac{1}{a_3} + \ldots + \frac{1}{a_{20}} - \frac{1}{a_{21}}\right] = \frac{4}{9}$$

$$\Rightarrow \quad \frac{1}{d}\left(\frac{1}{a_1} - \frac{1}{a_{21}}\right) = \frac{4}{9}$$

$$\Rightarrow \quad \frac{1}{d}\left(\frac{a_{21} - a_1}{a_1 a_{21}}\right) = \frac{4}{9}$$

$$\Rightarrow \quad \frac{20}{a_1 a_{21}} = \frac{4}{9}$$

$$\Rightarrow \quad a_1 a_{21} = 45 \quad [\because a_{21} = a_1 + 20d]$$

$$\Rightarrow a_1(a_1 + 20d) = 45 \qquad \ldots(i)$$

Also sum of first 21 terms $= 189$

$$\Rightarrow \quad \frac{21}{2}(2a_1 + 20d) = 189$$

$$\Rightarrow \quad a_1 + 10d = 9 \qquad \ldots(ii)$$

By Eqs. (i) and (ii), we get $a_1 = 3, d = \frac{3}{5}$

or $\quad a_1 = 15, d = -\frac{3}{5}$

So, $a_6 a_{16} = (a_1 + 5d)(a_1 + 15d)$
$= 72$

**44.** *(c)* If $\tan\left(\frac{\pi}{9}\right)$, $x$, $\tan\left(\frac{7\pi}{18}\right)$ are in AP.

So, $x = \frac{1}{2}\left[\tan\frac{\pi}{9} + \tan\left(\frac{7\pi}{18}\right)\right]$

($\because$ if $a, b, c$ are in AP, so, $b = \frac{a+c}{2}$)

And $\tan\left(\frac{\pi}{9}\right)$, $y$, $\tan\left(\frac{5\pi}{18}\right)$ are in AP.

Now, $x - 2y = \frac{1}{2}\left[\tan\frac{\pi}{9} + \tan\frac{7\pi}{18}\right] - \left(\tan\frac{\pi}{9} + \tan\frac{5\pi}{18}\right)$

$$\Rightarrow |x - 2y|$$

$$= \left|\frac{\cot\frac{\pi}{9} - \tan\frac{\pi}{9}}{2} - \tan\frac{5\pi}{18}\right|$$

$$\left\{\because \tan\frac{5\pi}{18} = \cot\frac{2\pi}{9} \text{ and } \tan\frac{7\pi}{18} = \cot\frac{\pi}{9}\right\}$$

$$= \left|\cot\frac{2\pi}{9} - \cot\frac{2\pi}{9}\right| = 0$$

$$\left[\because \cot 2A = \frac{\cot^2 A - 1}{2\cot a}\right]$$

**45.** *(d)* For the given AP,

$$S_n = \frac{n}{2}[a_1 + a_n] = \frac{n}{2}[1 + 300] \quad \ldots(i)$$

As, $\quad a_1 = 1$ and $a_n = 300$

and $300 = 1 + (n-1)d$, where $d$ is the common difference of AP.

$$\Rightarrow \quad d = \frac{299}{n-1} = \frac{23 \times 13}{n-1}$$

$\because \quad d$ is an integer so $n - 1 = 23$, or $13$

$$\Rightarrow \quad n = 24 \text{ or } 14$$

As it is given that, $15 \le n \le 50$

So, $\quad n = 24 \quad \Rightarrow \quad d = 13$

$$\therefore \quad S_{n-4} = S_{20} = \frac{20}{2}[2 + 19 \times 13]$$
$$= 10 \times 249 = 2490$$

and $a_{n-4} = a_{20} = 1 + (19 \times 13) = 248$

$\therefore$ The ordered pair

$(S_{n-4}, a_{n-4}) = (S_{20}, a_{20}) = (2490, 248)$

**46.** *(b)* Given AP's are 3, 7, 11, ..., 407 and 2, 9, 16, ..., 709.

The terms, which are common in above AP's are also in an AP have common difference = LCM $\{d_1, d_2\}$

(where $d_1$ and $d_2$ are common differences of given two AP.s)

and LCM $\{4, 7\} = 28 = d$ (let)

Now, let $m$th term of first AP is first common term of second AP, which is $n$th term of second AP.

So, $\quad 3 + (m-1)4 = 2 + (n-1)7$

$$\Rightarrow \quad 4m - 1 = 7n - 5$$

$$\Rightarrow \quad 7n - 4m = 4$$

So, $\quad m = 6$ and $n = 4$

So, first term which is common, is $3 + (6-1)4 = 23$

Now, let there are $p$ terms common to the two given AP's, so

$23 + (p-1)28 \le 407$

$$\Rightarrow \quad 28p \le 412$$

$$\Rightarrow \quad p \le 14.7$$

So, $p = 14$, therefore 14 terms are common to the two given AP's.

Hence, answer is 14.

> **Shortcut Method**
>
> From the observation, first common term is 23.
>
> and LCM of $(d_1, d_2) = 4 \times 7 = 28$
>
> Where, $d_1, d_2$ = common difference of respective APs
>
> So, total number of common terms
>
> $= \left[\frac{407 - 23}{28}\right] + 1$
>
> ([.] = Greatest integer function = 13 + 1 = 14)

**47.** *(c)* Let the time taken to save ₹ 11040 be $(n+3)$ months. For first three months, he saves ₹ 200 each month. In $(n+3)$ months,

$$3 \times 200 + \frac{n}{2}\{2(240) + (n-1) \times 40\} = 11040$$

$$\Rightarrow \quad 600 + \frac{n}{2}\{40(12 + n - 1)\} = 11040$$

$$\Rightarrow \quad 600 + 20n(n + 11) = 11040$$

$\Rightarrow \quad n^2 + 11n - 522 = 0$

$\Rightarrow \quad n(n+29) - 18(n+29) = 0$

$\Rightarrow \quad (n-18)(n+29) = 0$

$\therefore \quad n = 18, n = -29, \quad$ (neglecting)

$\therefore$ Total time $= (n+3) = 21$ months

**48.** *(11132)* Given,

$a_1, a_2, \ldots\ldots, a_{2024}$ be arithmetic progression such that

$$a_1 + (a_5 + a_{10} + \ldots\ldots + a_{2020}) + a_{2024} = 2233$$

In an AP the sum of terms equidistant from ends is equal

$\Rightarrow a_1 + a_{2024} = a_5 + a_{2020} = a_{10} + a_{2015}$

$\Rightarrow$ 203 pairs

$\Rightarrow 203(a_1 + a_{2024}) = 2233$

Hence, $S_{2024} = \dfrac{2024}{2}(a_1 + a_{2024})$

$= 1012 \times 11 = 11132$

**49.** *(20)* $a + (n-1)d = 219°$

$\Rightarrow \quad a + (n-1)6 = 219° \quad \ldots$(i)

$\Rightarrow \quad \dfrac{n}{2}[2a + (n-1)6] = (n-2)180°$

$\Rightarrow \quad \dfrac{n}{2}[a + 219] = (n-2)180°$

$\Rightarrow \quad \dfrac{n}{2}[219 + 219 - (n-1)6] = (n-2)180°$

$\Rightarrow \quad n[219 - 3(n-1)] = (n-2)180°$

$\Rightarrow \quad n[222 - 3n] = 180n - 360$

$\Rightarrow \quad 3n^2 - 222n + 180n - 360 = 0$

$\Rightarrow \quad 3n^2 - 42n - 360 = 0$

$\Rightarrow n^2 - 14n - 120 = 0$

$\Rightarrow (n+6)(n-20) = 0 \Rightarrow n = 20$

**50.** *(125)* number of 3-digit numbers which are divisible by 6 is 102, 108, 114, … , 996

$\therefore$Total number $= 150$

And number divisible by 36 is 108, 144, … 972

$\therefore$Total number $= 25$

Hence, required number = (divisible by 6) − (divisible by 36)

$= 150 - 25 = 125$

Method II

$\because$ Total number of 3-digit numbers

$= 999 - 99 = 900$

Now, total number of 3-digit numbers which are divisible by 2 and 3 i.e., 6

$= \dfrac{900}{6} = 150$

and total number of 2-digit numbers which are divisible by 4 and 9 i.e., by

$36 = \dfrac{900}{36} = 25$

Hence, total number of 3-digit number which are divisible by 2 and 3 but not by 4 and 9 $= 150 - 25 = 125$

**51.** *(474)*

$$3x^2 - px + q = 0 \;\; \begin{matrix} \nearrow a_{10} \\ \searrow a_{11} \end{matrix}$$

$a_{10}$ = 10th term of AP

$a_{11}$ = 11th term of AP

$d = \dfrac{3}{2}$

$\Rightarrow \quad S_{11} = 88$

$\Rightarrow \quad \dfrac{11}{2}(2a + 10d) = 88$

$\Rightarrow \quad a + 5d = 8$

$\Rightarrow \quad a = 8 - 5\left(\dfrac{3}{2}\right) = \dfrac{1}{2}$

$\therefore \quad a_{10} = a + 9d = \dfrac{1}{2} + \dfrac{27}{2} = 14$

$\therefore \quad a_{11} = a + 10d = \dfrac{1}{2} + \dfrac{30}{2} = \dfrac{31}{2}$

$\because \quad a_{10} + a_{11} = \dfrac{p}{3}$

$\Rightarrow \quad 14 + \dfrac{31}{2} = \dfrac{p}{3} \Rightarrow \dfrac{59}{2} = \dfrac{p}{3}$

$\Rightarrow \quad 2p = 177$

and $\quad a_{10}a_{11} = \dfrac{q}{3}$

$\Rightarrow \quad 14 \times \dfrac{31}{2} = \dfrac{q}{3}$

$\Rightarrow \quad q = 21 \times 31 = 651$

$\therefore \quad q - 2p = 651 - 177 = 474$

**52.** *(1011)* $\left(\dfrac{1}{\alpha+1} + \dfrac{1}{\alpha+2} + \ldots + \dfrac{1}{\alpha+1012}\right)$

$$-\left(\frac{1}{2\cdot1} + \frac{1}{4\cdot3} + \frac{1}{6\cdot5} + \ldots - \frac{1}{2024\cdot2023}\right) = \frac{1}{2024}$$

$$\Rightarrow \left(\frac{1}{\alpha+1} + \frac{1}{\alpha+2} + \ldots + \frac{1}{\alpha+1012}\right) - \left\{\left(\frac{1}{1} - \frac{1}{2}\right) + \left(\frac{1}{3} - \frac{1}{4}\right) + \ldots \left(\frac{1}{2023} - \frac{1}{2024}\right)\right\} = \frac{1}{2024}$$

$$\Rightarrow \left(\frac{1}{\alpha+1} + \frac{1}{\alpha+2} + \ldots \frac{1}{\alpha+1012}\right) - \left\{\left(1 + \frac{1}{2} + \frac{1}{3} + \ldots \frac{1}{2023}\right) - \frac{1}{2024} - 2\left(\frac{1}{2} + \frac{1}{4} + \ldots \frac{1}{2022}\right)\right\} = \frac{1}{2024}$$

$$\Rightarrow \frac{1}{\alpha+1} + \frac{1}{\alpha+2} + \ldots + \frac{1}{\alpha+1012} - \left(\frac{1}{1} + \frac{1}{2} + \ldots + \frac{1}{2023}\right) + \frac{1}{2024} + 2\left(\frac{1}{2} + \frac{1}{4} + \ldots + \frac{1}{2022}\right) = \frac{1}{2024}$$

$$\Rightarrow \frac{1}{\alpha+1} + \frac{1}{\alpha+2} + \ldots + \frac{1}{\alpha+1012} = \frac{1}{1012} + \frac{1}{1013} + \ldots + \frac{1}{2023}$$

$\therefore \quad \alpha = 1011$

**53.** *(1010)*

Given, $f(m+n) = f(m) + f(n)$

So, $f(x) = mx$

$\because \quad f(1) = 1 \Rightarrow 1 = m \times 1 \Rightarrow m = 1$

$\therefore \quad f(x) = x$

Now, $\displaystyle\sum_{k=1}^{2022} f(\lambda + k) \le (2022)^2$

$\Rightarrow \displaystyle\sum_{k=1}^{2022} (\lambda + k) \le (2022)^2$

$\Rightarrow (\lambda + \lambda + \lambda + \ldots + \lambda) + (1 + 2 + 3 + \ldots + 2022) \le (2022)^2$

(2022 terms)

$\Rightarrow 2022\lambda + \dfrac{2022 \times 2023}{2} \le (2022)^2$

$\Rightarrow \lambda \le 2022 - \dfrac{2023}{2}$

$\Rightarrow \lambda \le \dfrac{2021}{2} = 1010.5$

$\therefore$ Largest natural number of $\lambda$ is 1010.

**54.** *(1505)*

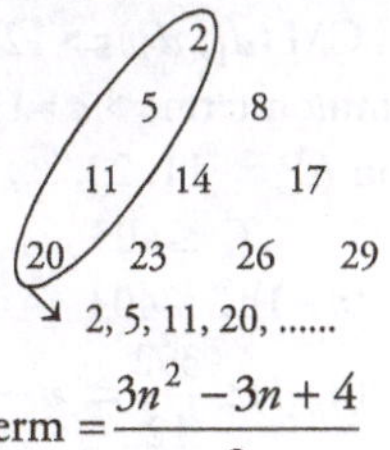


General term $= \dfrac{3n^2 - 3n + 4}{2}$

$T_{10} = \dfrac{3 \times 100 - 3 \times 10 + 4}{2}$

$= \dfrac{300 - 26}{2} = 137$

$d = 3$

So, sum $= \dfrac{10}{2}[2 \times 137 + 9 \times 3]$

$= 5(274 + 27) = 1505$

**55.** *(103)*

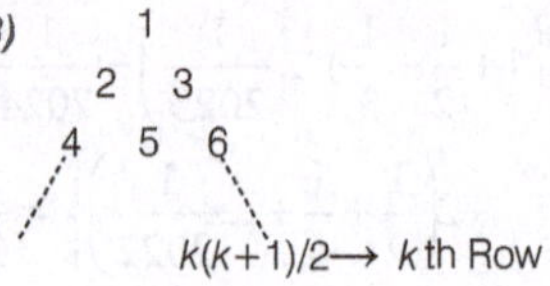


For 5310

$$\frac{k(k+1)}{2} < 5310 \Rightarrow k(k+1) < 10620$$

For $k=102$ (maximum value of $k$)

$$k(k+1) < 10620$$

That means, 5310 will contain in next row.

So, required row $=102+1=103$

**56.** *(910)* We have,

$$A_k = a_1^2 - a_2^2 + a_3^2 - a_4^2 + a_{2k-1}^2 - a_{2k}^2$$
$$=(-d)(a_1 + a_2 + \ldots + a_{2k})$$
$$=(-d)\left(\frac{2k}{2}\right)(2a_1 + (2k-1)d)$$

$\Rightarrow \quad A_3 = (-3d)(2a_1 + 5d) = -153$

$\Rightarrow \quad (d)(2a_1 + 5d) = 51$ ...(i)

Also, $A_5 = (-5d)(2a_1 + 9d) = -435$

$= (d)(2a_1 + 9d) = 87$ ...(ii)

From Eqs. (i) and (ii), we get

$4d^2 = 36 \Rightarrow d = \pm 3$

$\therefore \quad 3(2a_1 + 9d) = 87$

$\Rightarrow \quad 2a_1 + 9d = 29$

$\Rightarrow \quad 2a_1 = 29 - 27 = 2 \Rightarrow a_1 = 1$

$\therefore a_{17} - A_7 = (a + 16d) - (-7d)(2 + 13d)$

$= (1 + 16 \times 3) - (-7 \times 3)(2 + 13 \times 3)$

$= 49 + 7 \times 3(2 + 39)$

$= 49 + 21 \times 41 = 910$

**57.** *(6699)* Ist AP $\Rightarrow$ 3, 7, 11, ..., 403

$\Rightarrow \quad d_1 = 4$

IInd AP $\Rightarrow$ 2, 5, 8, ..., 404 $\Rightarrow d_2 = 3$

Common difference of common AP

$\Rightarrow D = \text{LCM}(d_1, d_2) \Rightarrow 12$

First common term $\Rightarrow a = 11$

Common AP $\Rightarrow$ 11, 23, ..., $T_n$

$$T_n \le 403$$

$\Rightarrow \quad 11 + (n-1)12 \le 403$

$\Rightarrow \quad (n-1) \le \frac{392}{12} \Rightarrow n - 1 \le 32.66$

$\Rightarrow \quad n \le 33.66 \Rightarrow n = 33$

Sum of common terms

$$S = \frac{33}{2}[2\cdot 11 + (33-1)12]$$
$$S = 6699$$

**58.** *(9525)* The sum of all those term of the arithmetic progression 3, 8, 13, ..., 373, which are not divisible by 3 is

$= (3 + 8 + 13 + 18 + \ldots + 373)$

$-(3 + 18 + 33 \ldots + 363)$

$= \frac{75}{2}(3 + 373) - \frac{25}{2}(3 + 363)$

$= \frac{75}{2} \times 376 - \frac{25}{2} \times 366$

$= 75 \times 188 - 25 \times 183$

$= 14100 - 4575 = 9525$

**59.** *(321)* We have,

$S_1$ : 3, 7, 11, 15, ......, 399

$S_2$ : 2, 5, 8, 11, ......, 359

$S_3$ : 2, 7, 12, 17, ......, 197

The common terms in the three AP's are 47, 107, 167

$\therefore$ Required sum $= 47 + 107 + 167 = 321$

**60.** *(514)* Numbers which are divisible by 2 are 100, 102, 104, ......, 998

$\therefore \quad a_n = a + (n-1)d$

$\Rightarrow \quad 998 = 100 + (n-1)\,2$

$\Rightarrow \quad n = 450$

Numbers which are divisible by 3 are 102, 105, 108, ...., 999

$\therefore \quad a_n = a + (n-1)d$

$\Rightarrow 999 = 102 + (n-1) \times 3 \Rightarrow n = 300$

Numbers which are divisible by 2 and 3 are 102, 108,114, ... 996

$\therefore \quad a_n = a + (n-1)d$

$\Rightarrow \quad 996 = 102 + (n-1) \times 6 \Rightarrow n = 150$

Numbers which are divisible by 2 and 7 are 112, 126, 140, ...., 994

$\therefore \quad a_n = a + (n-1)\,d$

$\Rightarrow \quad 994 = 112 + (n-1)14 \Rightarrow n = 64$

Numbers which are divisible by 3 and 7 are 105, 126, 147, ...., 987

$\therefore \quad a_n = a + (n-1)d$

$\Rightarrow \quad 987 = 105 + (n-1)\,21$

$\Rightarrow \quad n = 43$

Numbers which are divisible by 2, 3, 7 are 126, 168, 252, ...., 966

$\therefore \quad a_n = a + (n-1)\,d$

$\Rightarrow \quad 966 = 126 + (n-1)\,42 \Rightarrow n = 21$

Let $A$ : Numbers which are divisible by 2

$B$ : Numbers which are divisible by 3

$C$ : Numbers which are divisible by 7

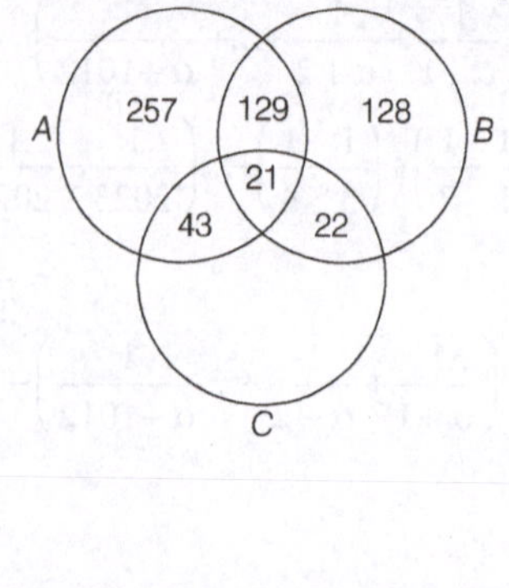


$\therefore$ Total required number

$= 257 + 129 + 128 = 514$

**61.** *(754)* We have, $a_1 = 8, a_2, a_3, \ldots, a_n$ are in AP.

Now, $S_4 = 50$

$\Rightarrow \frac{4}{2}[2a_1 + (4-1)d] = 50 \Rightarrow d = 3$

Again, sum of last four terms $= 170$

$\Rightarrow (a + (n-1)d) + (a + (n-2)d)$
$+ (a + (n-3)d) + (a + (n-4)d) = 170$

$\Rightarrow \quad 4a + (4n - 10)d = 170$

$\Rightarrow \quad n = 14$

$\therefore$ Middle terms are $a_7$ and $a_8$.

Required product $= a_7 \times a_8$

$= (a + 6d)(a + 7d)$

$= 26 \times 29 = 754$

**62.** *(8)* Given, $a_5 = 2a_7$ and $a_{11} = 18$

Now, $\quad a_5 = 2a_7$

$\Rightarrow \quad a + 4d = 2(a + 6d)$

$\Rightarrow \quad a + 4d = 2a + 12d$

$\Rightarrow \quad a = -8d$ ...(i)

and $\quad a_{11} = 18$

$\Rightarrow \quad a + 10d = 18$ ...(ii)

From Eqs. (i) and (ii), we get

$a = -72, d = 9$

Now,

$$12\left(\frac{1}{\sqrt{a_{10}} + \sqrt{a_{11}}} + \frac{1}{\sqrt{a_{11}} + \sqrt{a_{12}}} + \ldots + \frac{1}{\sqrt{a_{17}} + \sqrt{a_{18}}}\right)$$

$$= 12\left(\frac{\sqrt{a_{11}} - \sqrt{a_{10}}}{a_{11} - a_{10}} + \frac{\sqrt{a_{12}} - \sqrt{a_{11}}}{a_{12} - a_{11}} + \ldots + \frac{\sqrt{a_{18}} - \sqrt{a_{17}}}{a_{18} - a_{17}}\right)$$

$$= 12\left(\frac{\sqrt{a_{11}} - \sqrt{a_{10}}}{d} + \frac{\sqrt{a_{12}} - \sqrt{a_{11}}}{d} + \ldots + \frac{\sqrt{a_{18}} - \sqrt{a_{17}}}{d}\right)$$

$$= \frac{12}{d}(\sqrt{a_{18}} - \sqrt{a_{10}})$$

$$= \frac{12}{9}(\sqrt{a + 17d} - \sqrt{a + 9d})$$

$$= \frac{12}{9}[\sqrt{-72 + 153} - \sqrt{-72 + 81}]$$

$$= \frac{12}{9}[9 - 3] = 8$$

**63.** *(50)* $\because a_1, a_2, a_3$ and $a_4$ are in AP and its mean is $p$.

$\therefore a_1 = p - 3d, a_2 = p - d,$

$a_3 = p + d, a_4 = p + 3d$

where, $d > 0$

$\because \quad |f(a_i)| = 500$ [given]

$|9d^2 - q| = 500$

and $|d^2 - q| = 500$ ... (i)

Either, $9d^2 - q = d^2 - q$

or $2q = 8d^2$

$\Rightarrow d = 0$ not acceptable because $d > 0$

$\therefore 9d^2 - q = q - d^2$

$\Rightarrow 10d^2 = 2q$

$\Rightarrow 5d^2 - q = 0$ $[\because 2q = 8d^2]$

Roots of $f(x) = 0$ are $p + \sqrt{q}$ and $p - \sqrt{q}$.

$\therefore$Absolute difference between roots $= |2\sqrt{q}| = 50$

**64.** *(53)* First term, $a = 100$

Last term, $l = 199$

Let 3 terms are $a, a + d, a + 2d$

$l = a + (n-1)d$

$n =$ Number of terms $= 3$

Then, $d_1 = \frac{199 - 100}{2} = \frac{99}{2} \notin$ Integer

If $n = 4$, $d_2 = \frac{99}{3} = 33 \in$ Integer

If $n = 10$, $d_2 = \frac{99}{9} = 11 \in$ Integer

If $n = 12$, $d_4 = \frac{99}{11} = 9 \in$ Integer

$\therefore$ Sum of differences $= 33 + 11 + 9 = 53$

**65.** *(5143)* Total 4-digit number $= 9 \times 10 \times 10 \times 10 = 9000$

4 -digit number divisible by 7

1001, 1008, ..., 9996

Number of 4-digit number divisible by 7

$= \frac{9996 - 1001}{7} + 1 = 1286$

4 -digit number divisible by 3

1002, 1005, ..., 9999

Number of 4-digit number divisible by 3

$= \frac{9999 - 1002}{3} + 1 = 3000$

4-digit number divisible by 21

1008, 1031, ..., 9996

Number of 4- digit number divisible by 21

$= \frac{9996 - 1008}{21} + 1 = 429$

$\therefore$ Number of 4- digit numbers neither divisible by 7 nor 3

$= 9000 - 1286 - 3000 + 429 = 5143$

**66.** *(3)* $\log_3 2, \log_3 (2^x - 5),$

$\log_3\left(2^x - \frac{7}{2}\right)$ are in AP

$\Rightarrow 2\log_3(2^x - 5) = \log_3 2 + \log_3\left(2^x - \frac{7}{2}\right)$

$\Rightarrow \log_3(2^x - 5)^2 = \log_3\left[2 \cdot \left(2^x - \frac{7}{2}\right)\right]$

$\Rightarrow (2^x - 5)^2 = 2 \cdot 2^x - 7$

$\Rightarrow (2^x)^2 + 25 - 10 \cdot 2^x - 2 \cdot 2^x + 7 = 0$

$\Rightarrow (2^x - 4)(2^x - 8) = 0$

$\Rightarrow \quad 2^x = 4$ or $8 \Rightarrow x = 2$ or $3$

If $x = 2$, then

$\log_3(2^x - 5) = \log_3(2^2 - 5)$

Here, argument is negative, so, $x \neq 2$.

Hence, $x = 3$

## Topic 2

**01.** *(d)* $a_3 a_5 = 729$

$(a_1 r^2)(a_1 r^4) = 729$

$(a_1 r^3)^2 = (27)^2$

$\Rightarrow a_1 r^3 = 27$

$\Rightarrow a_4 = a_1 r^3 = 27$

Also, $a_2 + a_4 = \frac{111}{4}$

$\Rightarrow a_2 = \frac{111}{4} - 27$

$\Rightarrow a_2 = \frac{3}{4}$

$a_1 r = \frac{3}{4}$

$\therefore \frac{a_1 r^3}{a_1 r} = r^2 = \frac{27}{3} \times 4 = 36 \Rightarrow r = 6$

$a_1 \times 6 = \frac{3}{4} \Rightarrow a_1 = \frac{1}{8}$

$\therefore 24(a_1 + a_2 + a_3) = 24a_1(1 + r + r^2)$

$= 24 \times \frac{1}{8}(1 + 6 + 36)$

$= 3 \times 43 = 129$

**02.** *(d)*

> **key idea** General term of GP.
> $T_n = ar^{n-1}$

$a_1, a_2, a_3, \ldots$ are in GP.

$a_1 a_5 = 28$ and $a_2 + a_4 = 29$

Let $a$ be the first term and $r$ be the common ratio.

$a(ar^4) = 28$ and $ar + ar^3 = 29$

$a^2 r^4 = 28$

$\Rightarrow ar^2 = 2\sqrt{7}$

$\Rightarrow \frac{2\sqrt{7}}{r} + 2\sqrt{7}r = 29$

$\Rightarrow 2\sqrt{7}r^2 - 29r + 2\sqrt{7} = 0$

$\Rightarrow 2\sqrt{7}r^2 - 28r - r + 2\sqrt{7} = 0$

$\Rightarrow 2\sqrt{7}r(r - 2\sqrt{7}) - 1(r - 2\sqrt{7}) = 0$

$\Rightarrow (r - 2\sqrt{7})(2\sqrt{7}r - 1) = 0$

$\Rightarrow r = 2\sqrt{7}, r = \frac{1}{2\sqrt{7}}$

$\because$ Increasing GP $\Rightarrow r = 2\sqrt{7}$

$\therefore \quad a = \frac{1}{2\sqrt{7}}$

$\therefore a_6 = ar^5 = (2\sqrt{7})^4$

$= (28)^2 = 784$

**03.** *(b)* Let AP be $3, 3 + d, 3 + 2d, 3 + 3d$

So, $3 + d = a, 3 + 2d = b, 3 + 3d = c$

Also, let GP be $3, 3r, 3r^2, 3r^3$

So, $3r = a - 1, 3r^2 = b + 1, 3r^3 = c + 9$

On comparing, we get

$3r = 3 + d - 1$

$\Rightarrow \quad d = 3r - 2$ ...(i)

and $\quad 3r^2 = 3 + 2d + 1$

$\Rightarrow \quad 3r^2 = 2d + 4$ ...(ii)

On solving Eqs. (i) and (ii), we get

$r = 2, a = 7, b = 11, c = 15$

Hence, AM of $a, b$ and $c$

$= \frac{7 + 11 + 15}{3} = 11$

**04.** *(b)* Let first term of AP be $a$ and common difference be $d$.

Given, $\quad a = 1$

2nd term of AP, $T_2 = a + d = 1 + d$

8th term of AP $T_8 = a + 7d = 1 + 7d$

44th term of AP,

$T_{44} = a + 43d = 1 + 43d$

Given, $1 + d, 1 + 7d, 1 + 43d$ are in GP.

$(1 + 7d)^2 = (1 + d)(1 + 43d)$

$\Rightarrow 1+49d^2+14d=1+44d+43d^2$

$\Rightarrow \quad 6d^2-30d=0$

$\Rightarrow \quad d=5$ [given, non-constant AP]

So, $S_{20}=\dfrac{20}{2}[2\times1+(20-1)\times5]$

$=970$

**05.** *(b)* For Ist GP,

$T_1=a,$

$T_3=b=ar_1^2$ ...(i)

For 2nd GP,

$T_1'=a,$

$T_5'=b=ar_2^4$ ...(ii)

According to the question,

$T_{11}=T_p'$

[$\because$ General term of GP $=ar^{n-1}$]

$\Rightarrow ar_1^{10}=a\cdot r_2^{p-1}$

$\Rightarrow a\left(\dfrac{b}{a}\right)^5=a\left(\dfrac{b}{a}\right)^{\frac{p-1}{4}}$

[by Eqs. (i) and (ii)]

$\therefore \quad 5=\dfrac{p-1}{4}\Rightarrow p=21$

**06.** *(b)* Let $a$ be the first term and $r$ be the common ratio of the GP

$\therefore (a+ar+ar^2+\ldots+ar^{63})$

$=7(a+ar^2+\ldots+ar^{62})$

$\Rightarrow \dfrac{a(1-r^{64})}{1-r}=\dfrac{7a(1-r^{64})}{1-r^2}$

$\Rightarrow \quad 1=\dfrac{7}{1+r}\Rightarrow 1+r=7$

$\Rightarrow \quad r=6$

**07.** *(a)* $n$ th term of the AP is $a_n$

Now, $S_{21}=\dfrac{21}{2}[a_1+a_{21}]$

$=21\times\left(\dfrac{a_1+a_{21}}{2}\right)$

$\because a_1=10ar,\ a_{21}=10ar+200ar^2,$

$\because a_{11}=10ar+100ar^2,$

$S_{21}=\dfrac{21}{2}[20ar+200ar^2]$

$=21[10ar+100ar^2]=21\times a_{11}$

**08.** *(c)* Given, $A_1A_3A_5A_7=\dfrac{1}{1296}$

$\Rightarrow \dfrac{A_4}{r^3}\cdot\dfrac{A_4}{r}\cdot A_4r\cdot A_4r^3=\dfrac{1}{1296}$

$\Rightarrow (A_4)^4=\dfrac{1}{1296}\Rightarrow A_4=\dfrac{1}{6}$

$\because A_2+A_4=\dfrac{7}{36}\Rightarrow A_2=\dfrac{7}{36}-\dfrac{1}{6}=\dfrac{1}{36}$

$\therefore \quad A_6=1, A_8=6$ and $A_{10}=36$

So, $A_6+A_8+A_{10}=1+6+36=43$

**09.** *(b)* Let three numbers be $\dfrac{a}{r}$, $a$, $ar$.

According to the question,

$\dfrac{a}{r}$, $2a$, $ar$ are in AP

$4a=ar+\dfrac{a}{r}\Rightarrow r+\dfrac{1}{r}=4$

$\Rightarrow r^2-4r+1=0\Rightarrow r=2\pm\sqrt{3}$

$T_4$ of GP $=3r^2$

$3r^2=ar^2$

$a=3$

$r=2+\sqrt{3}$

$d=2a-\dfrac{a}{r}=3\sqrt{3}$

$\Rightarrow r^2-d=(2+\sqrt{3})^2-3\sqrt{3}$

$=7+4\sqrt{3}-3\sqrt{3}=7+\sqrt{3}$

**10.** *(a)* $y=\dfrac{1}{2}x^2+\dfrac{2}{3}x^3+\dfrac{3}{4}x^4+\ldots$

$\Rightarrow y=\left(1-\dfrac{1}{2}\right)x^2+\left(1-\dfrac{1}{3}\right)x^3+\left(1-\dfrac{1}{4}\right)x^4+.$

$=(x^2+x^3+x^4+\ldots)-\left(\dfrac{x^2}{2}+\dfrac{x^3}{3}+\dfrac{x^4}{4}+\ldots\right)$

$=\dfrac{x^2}{1-x}+x-\left(x+\dfrac{x^2}{2}+\dfrac{x^3}{3}+\dfrac{x^4}{4}\ldots.\right)$

$\therefore \quad y=\dfrac{x}{1-x}+\ln(1-x)$

Put $x=\dfrac{1}{2}$, we get

$y=1-\ln 2$

Then, $e^{1+y}=e^{1+1-\ln 2}=e^{2-\ln 2}$

$=e^2\cdot e^{\ln 2^{-1}}=\dfrac{1}{2}e^2$

**11.** *(c)* Given product,

$2^{\frac{1}{4}}\cdot4^{\frac{1}{16}}\cdot8^{\frac{1}{48}}\cdot16^{\frac{1}{128}}\ldots$ to $\infty$

$2^{\frac{1}{4}}\cdot2^{\frac{2}{16}}\cdot2^{\frac{3}{48}}\cdot2^{\frac{4}{128}}\ldots$ to $\infty$

$=2^{\left(\frac{1}{4}+\frac{1}{8}+\frac{1}{16}+\frac{1}{32}+\ldots\text{to}\,\infty\right)}$

$\because \dfrac{1}{4}+\dfrac{1}{8}+\dfrac{1}{16}+\dfrac{1}{32}+\ldots$ to $\infty$ is a GP of infinite terms having first term 1/4 and common ratio 1/2.

So, the product is $2^{\left(\frac{1/4}{1-1/2}\right)}$

$=2^{\left(\frac{1/4}{1/2}\right)}=2^{1/2}$

**12.** *(b)* The sum of series

$49^{125}+49^{124}+\ldots+49^2+49+1$

$=49^0+49^1+49^2+\ldots+49^{124}+49^{125}$

$=\dfrac{1(49^{126}-1)}{49-1}.$

$=\dfrac{(49^{63})^2-1^2}{48}$

$=\dfrac{(49^{63}-1)(49^{63}+1)}{48}$

$\because 49^{63}+1$ is a factor of the above sum.

$\therefore$ The greatest positive integer $k$ is 63.

**13.** *(a)* Let $S=\displaystyle\sum_{k=1}^{20}k\left(\frac{1}{2^k}\right)$

$\Rightarrow S=\dfrac{1}{2}+\dfrac{2}{2^2}+\dfrac{3}{2^3}+\dfrac{4}{2^4}+\ldots+\dfrac{20}{2^{20}}$ ...(i)

On multiplying by $\left(\dfrac{1}{2}\right)$ both sides, we get

$\dfrac{S}{2}=\dfrac{1}{2^2}+\dfrac{2}{2^3}+\dfrac{3}{2^4}+\ldots+\dfrac{19}{2^{20}}+\dfrac{20}{2^{21}}$ ...(ii)

On subtracting Eq. (ii) from Eq. (i), we get

$S-\dfrac{S}{2}=\dfrac{1}{2}+\dfrac{1}{2^2}+\dfrac{1}{2^3}+\ldots+\dfrac{1}{2^{20}}-\dfrac{20}{2^{21}}$

$\Rightarrow \dfrac{S}{2}=\dfrac{\frac{1}{2}\left(1-\frac{1}{2^{20}}\right)}{1-1/2}-\dfrac{20}{2^{21}}$

$\Rightarrow \dfrac{S}{2}=1-\dfrac{11}{2^{20}}\Rightarrow S=2-\dfrac{11}{2^{19}}$

**14.** *(c)* Let $A$ be the 1st term of AP and $d$ be the common difference.

$\therefore$ 7th term $=a=A+6d$

[$\because n$th term $=A+(n-1)d$]

11th term $=b=A+10d$

13th term $=c=A+12d$

$\because a, b, c$ are also in GP

$$\therefore \quad b^2 = ac$$

$$\Rightarrow (A+10d)^2 = (A+6d)(A+12d)$$

$$\Rightarrow A^2 + 20Ad + 100d^2 = A^2 + 18Ad + 72d^2$$

$$\Rightarrow \quad 2Ad + 28d^2 = 0$$

$$\Rightarrow \quad 2d(A+14d) = 0$$

$$\Rightarrow d = 0 \text{ or } A + 14d = 0$$

But $\quad d \neq 0$

[∵ The series is non constant AP]

$$\Rightarrow \quad A = -14d$$

$$\Rightarrow \quad \frac{a}{c} = \frac{-8d}{-2d} = 4$$

**15.** (d) Let $a, ar, ar^2$ be in GP $(r > 1)$.

According to the question,

$a, 2ar, ar^2$ in AP.

$$\Rightarrow 4ar = a + ar^2 \Rightarrow r^2 - 4r + 1 = 0$$

$$\Rightarrow \quad r = \frac{4 \pm \sqrt{16-4}}{2} = 2 \pm \sqrt{3}$$

$$\Rightarrow \quad r = 2 + \sqrt{3} \quad [\because \text{AP is increasing}]$$

**16.** (c) Since,

$$a + ar = a(1+r) = 12 \quad \ldots\text{(i)}$$

and $ar^2 + ar^3 = ar^2(1+r) = 48 \quad \ldots\text{(ii)}$

From Eqs. (i) and (ii), we get

$$r^2 = 4 \Rightarrow r = -2$$

[Since, the series is alternately sign, so we take negative values]

On putting the value of $r$ in Eq. (i), we get

$$a = -12$$

**17.** (d) Since, each term is equal to the sum of two preceding terms.

$$\therefore \quad ar^{n-1} = ar^n + ar^{n+1}$$

$$\Rightarrow \quad \frac{1}{r} = 1 + r \Rightarrow r^2 + r - 1 = 0$$

$$\therefore \quad r = \frac{\sqrt{5}-1}{2} \quad \left[\because r \neq \frac{-\sqrt{5}-1}{2}\right]$$

**18.** (b) Since, 5th term of a GP = 2

$$\therefore \quad ar^4 = 2 \quad \ldots\text{(i)}$$

Where, $a$ and $r$ are the first term and common ratio of a GP. Now, required product

$$= a \times ar \times ar^2 \times ar^3 \times ar^4 \times ar^5 \times ar^6 \times ar^7 \times ar^8$$

$$= (ar^4)^9 = 2^9 = 512 \quad \text{[from Eq. (i)]}$$

**19.** (d) The $n$th term of a GP is $a_n = ar^{n-1}$

According to the question,

$$a_2 + a_4 + a_6 = 21$$

$$\Rightarrow \quad ar + ar^3 + ar^5 = 21$$

$$\Rightarrow \quad ar(1 + r^2 + r^4) = 21 \quad \ldots\text{(i)}$$

And $a_8 + a_{10} + a_{12} = 15309$

$$\Rightarrow ar^7 + ar^9 + ar^{11} = 15309$$

$$ar^7(1 + r^2 + r^4) = 15309 \quad \ldots\text{(ii)}$$

Divide Eqs. (ii) by (i)

$$\frac{ar^7(1+r^2+r^4)}{ar(1+r^2+r^4)} = \frac{15309}{21} \Rightarrow r^6 = 729$$

$$\Rightarrow \quad r = 3$$

From Eq. (i), $ar(1 + r^2 + r^4) = 21$

$$a \times 3(1 + 3^2 + 3^4) = 21$$

$$a = \frac{21}{273} = \frac{1}{13}$$

The sum is given by the formula

$$S_n = \frac{a(r^n - 1)}{r - 1}$$

$$\Rightarrow \quad S_9 = \frac{\frac{1}{13}(3^9 - 1)}{3 - 1} = \frac{1}{26}(19683 - 1)$$

$$= 757$$

**20.** (a) $x_1 = a, \quad x_2 = ar$

$$x_3 = ar^2, \; x_4 = ar^3$$

$a - 2, ar - 7, ar^2 - 9, ar^3 - 5$ are in AP

$$a_2 - a_1 = a_3 - a_2$$

$$ar - 7 - a + 2 = ar^2 - 9 - ar + 7$$

$$\Rightarrow \quad a(r-1)^2 = -3 \quad \ldots\text{(i)}$$

Similarly, $a_2 - a_1 = a_4 - a_3$

$$ar - 7 - a + 2 = ar^3 - 5 - ar^2 + 9$$

$$\Rightarrow \quad a(r-1) - 5 = ar^2(r-1) + 4$$

$$\Rightarrow a(r-1)(r^2 - 1) = -9 \quad \ldots\text{(ii)}$$

On dividng in Eqs. (i) and (ii), we get

$$\Rightarrow \quad r + 1 = 3$$

$$\Rightarrow \quad r = 2$$

Using Eq. (i), we get

$$a(1)(1) = -3$$

$$a = -3$$

$$x_1 = -3, x_2 = -6, x_3 = -12, x_4 = -24$$

$$\frac{1}{24}(x_1 \cdot x_2 \cdot x_3 \cdot x_4) = 216$$

**21.** (c) Let

$$S = 5 + \frac{1}{7}(5 + \alpha) + \frac{1}{7^2}(5 + 2\alpha) + \ldots + \infty \quad \ldots\text{(i)}$$

$$\frac{1}{7}S = \frac{1}{7}(5) + \frac{1}{7^2}(5 + \alpha) + \ldots + \infty \quad \ldots\text{(ii)}$$

On subtracting Eq. (ii) from Eq. (i), we get

$$S - \frac{S}{7} = 5 + \frac{1}{7}\alpha\left(\frac{1}{1 - 1/7}\right)$$

$$\Rightarrow \quad \frac{6}{7}S = 5 + \frac{\alpha}{7} \times \frac{7}{6}$$

$$\Rightarrow \quad \frac{6}{7}S = 5 + \frac{\alpha}{6} \Rightarrow \frac{6}{7} \times 7 = 5 + \frac{\alpha}{6}$$

$$[\because s = 7 \text{ given}]$$

$$\Rightarrow \quad \alpha = 6$$

**22.** (a) According to the question,

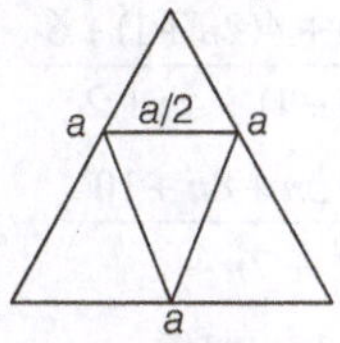


$$\text{Area of first } \Delta = \frac{\sqrt{3}a^2}{4}$$

$$\text{Area of second } \Delta = \frac{\sqrt{3}}{4}\left(\frac{a}{2}\right)^2 = \frac{\sqrt{3}a^2}{16}$$

$$\text{Area of third } \Delta = \frac{\sqrt{3}a^2}{64}$$

Sum of area

$$= \frac{\sqrt{3}a^2}{4}\left(1 + \frac{1}{4} + \frac{1}{16} + \ldots\right)$$

$$Q = \frac{\sqrt{3}a^2}{4}\left(\frac{4}{3}\right) = \frac{a^2}{\sqrt{3}} \quad \ldots\text{(i)}$$

Perimeter of first triangle $= 3a$

Perimeter of second triangle $= \dfrac{3a}{2}$

Perimeter of third triangle $= \dfrac{3a}{4}$

$$P = 3a\left(1 + \frac{1}{2} + \frac{1}{4} + \ldots\right)$$

$$P = 3a \cdot 2 = 6a$$

$$a = \frac{P}{6} \quad \ldots\text{(ii)}$$

By Eqs. (i) and (ii), we get

$$Q = \frac{1}{\sqrt{3}}\frac{P^2}{36} \Rightarrow P^2 = 36\sqrt{3}Q$$

**23.** (c) $\dfrac{1 \times 2^2 + 2 \times 3^2 + \ldots 100 \times (101)^2}{1^2 \times 2 + 2^2 \times 3 + \ldots 100^2 \times 101}$

$$\Rightarrow \quad \frac{\sum_{n=1}^{100} r(r+1)^2}{\sum_{n=1}^{100} r^2(r+1)} = \frac{\Sigma(r^3 + 2r^2 + r)}{\Sigma r^3 + r^2}$$

$$=\frac{\left[\frac{n}{2}(n+1)\right]^2+\frac{2n}{6}(n+1)(2n+1)+\frac{n}{2}(n+1)}{\left[\frac{n}{2}(n+1)\right]^2+\frac{n}{6}(n+1)(2n+1)}$$

$$=\frac{\frac{n}{2}(n+1)+\frac{2(2n+1)}{3}+1}{\frac{n}{2}(n+1)+\frac{2n+1}{3}}$$

$$=\frac{3n(n+1)+4(2n+1)+6}{3(n)(n+1)+4n+2}$$

$$\Rightarrow \frac{3n^2+3n+8n+10}{3n^2+7n+2}$$

$$\Rightarrow \frac{3n^2+11n+10}{3n^2+7n+2}=1+\frac{4n+8}{3n^2+7n+2}$$

Put $n=100$

$$\Rightarrow 1+\frac{408}{30702}=1+\frac{4}{301}=\frac{305}{301}$$

**Concept Enhancer**

$$\Sigma n^3=\left(\frac{n(n+1)}{2}\right)^2$$

and $\Sigma n^2=\frac{n(n+1)(2n+1)}{6}$

**24.** *(d)* Since, 2, $p$, $q$ are in GP.

So, $p^2=2q$ ...(i)

According to the question,

$a+6d=2$ ...(ii)

$a+7d=p$ ...(iii)

$a+12d=q$ ...(iv)

where $d$ = common difference

Subtracting Eq. (iii) from Eq. (ii),

$p-2=d$ ...(v)

Subtracting Eq. (iv) from Eq. (iii),

$q-p=5d=5(p-2)$ (from Eq. (v))

$\Rightarrow q=6p-10$

From Eq. (i), $p^2=2(6p-10)$

$\Rightarrow p^2-12p+20=0 \Rightarrow p=10, 2$

When $p=10$, then $q=50$ and $d=8$ and $a=-46$

So, GP is 2, 10, 50, 250, 1250

Now, $ar^4=a+(n-1)d$

$\Rightarrow 1250=-46+(n-1)8$

$\therefore n=163$

**25.** *(d)* We have, $px^2+qx-r=0$

Given that, $p$, $q$ and $r$ be the consecutive terms of a non-constant GP, then let

$$p=\frac{a}{r_1}, q=a, r=ar_1$$

and given, $\frac{1}{\alpha}+\frac{1}{\beta}=\frac{3}{4}$

$$\Rightarrow \frac{\alpha+\beta}{\alpha\beta}=\frac{3}{4} \Rightarrow \frac{\frac{q}{p}}{-\frac{r}{p}}=\frac{3}{4}$$

$$\Rightarrow \frac{q}{r}=\frac{3}{4} \Rightarrow \frac{1}{r_1}=\frac{3}{4}$$

$$\Rightarrow r_1=\frac{4}{3} \quad ...(i)$$

Now, $(\alpha-\beta)^2=(\alpha+\beta)^2-4\alpha\beta$

$$=\left(\frac{-q}{p}\right)^2-4\left(\frac{-r}{p}\right)$$

$$=(r_1)^2+4(r_1^2)=5r_1^2$$

$$=5\left(\frac{4}{3}\right)^2 \quad \text{[from Eq. (i)]}$$

$$=\frac{80}{9}$$

**26.** *(b)* $a_1, a_2, a_3, \ldots$ are in GP

$$a_1=\frac{1}{8}$$

$$a_2=\frac{a_3+a_4}{2} \quad ...(i)$$

$\therefore a_1=a$ and $r$ are the common ratio

$\therefore$ Eq. (i) becomes,

$$ar=\frac{ar^2+ar^3}{2}$$

$\Rightarrow 2=r+r^2$

$\Rightarrow r^2+r-2=0 \Rightarrow r^2+2r-r-2=0$

$\Rightarrow (r+2)(r-1)=0$

$\therefore r=-2, 1$

$a_2 \neq a_1 \Rightarrow r \neq 1$

$\therefore r=-2$

Now, $S_{20}-S_{18}=a_{19}+a_{20}$

$$=ar^{18}+ar^{19}=ar^{18}(1+r)$$

$$=\frac{1}{8}(-2)^{18}(1-2)$$

$$=\frac{-2^{18}}{2^3}=-2^{15}$$

**27.** *(b)* Let $a_1, a_2, a_3, \ldots$ be a GP of increasing positive numbers and given that sum of its 6th and 8th terms be 2.

So, $ar^5+ar^7=2$ ...(i)

and also given that, product of its 3rd and 5th terms be $\frac{1}{9}$.

So, $(ar^2)(ar^4)=\frac{1}{9} \Rightarrow a^2r^6=\frac{1}{9}$ ...(ii)

Since, $r>0$

So, from Eq. (i), we get

$ar^5(1+r^2)=2$ ...(iii)

and from Eq. (ii), we get $(ar^3)^2=\frac{1}{9}$

$\Rightarrow ar^3=\pm\frac{1}{3}$

Since, $r>0$ and $a>0$

$\Rightarrow ar>0$, so $ar^3=\frac{1}{3}$

From Eq. (iii), we get

$ar^3\cdot r^2(1+r^2)=2$

$\Rightarrow \frac{1}{3}(r^2+r^4)=2$

$\Rightarrow r^4+r^2-6=0$

$\Rightarrow r^2=2$, $r^2=-3$ (not possible)

$\Rightarrow r=\sqrt{2} \Rightarrow \therefore a=\frac{1}{6\sqrt{2}}$

Now, $6(a_2+a_4)(a_4+a_6)$

$$=6(ar+ar^3)(ar^3+ar^5)$$

$$=6a^2r^4(1+r^2)^2$$

$$=6\times\frac{1}{36\times 2}\times 4\times(1+2)^2$$

$$=\frac{1}{12}\times 4\times 9=3$$

**28.** *(c)* Given that, the first term $a$ and common ratio $r$ of a geometric progression be positive integer. So their first three terms are $a$, $ar$, $ar^2$

$a^2+a^2r^2+a^2r^4=33033$

$\Rightarrow a^2(1+r^2+r^4)$

$=3\times 7\times 11\times 11\times 13$

$=3\times 7\times 13\times 11^2$

$\therefore a^2=11^2$

$\Rightarrow a=11$ and $1+r^2+r^4=273$

$\Rightarrow r^2+r^4=272 \Rightarrow r^4+r^2-272=0$

$\Rightarrow (r^2+17)(r^2-16)=0$

$\Rightarrow r^2=-17$ and $r^2-16=0$

[$\because r^2=-17$ not possible]

$\Rightarrow \quad r = \pm 4 \Rightarrow r = 4\,(r > 0)$

So, sum of these first three terms are $a + ar + ar^2$

$$11 + 44 + 176 = 231$$

**29.** *(d)* Given that $ax^2 + 2bx + cy = 0$ and $dx^2 + 2ex + fy = 0$ intersect on the line $y = 1$

$$ax^2 + 2bx + c = 0 \quad \ldots(i)$$

and $$dx^2 + 2ex + f = 0 \quad \ldots(ii)$$

$a, b, c$ are in GP.

So, $b^2 = ac \Rightarrow b = \sqrt{a}\sqrt{c}$

Put $b = \sqrt{ac}$ in Eq. (i), we get

$$ax^2 + 2\sqrt{a}\sqrt{c}\,x + c = 0$$

$$(\sqrt{a}x + \sqrt{c})^2 = 0 \Rightarrow x = -\frac{\sqrt{c}}{\sqrt{a}}$$

Put this value in Eq. (ii), we get

$$\frac{cd}{a} + f = 2e\sqrt{\frac{c}{a}}$$

$$\Rightarrow \quad cd + af = 2e\sqrt{ac}$$

$$\Rightarrow \quad cd + af = 2eb \quad [\because b = \sqrt{ac}]$$

Divided by $ac$,

$$\frac{d}{a} + \frac{f}{c} = \frac{2eb}{b^2} \quad [\because b^2 = ac]$$

$$\frac{d}{a} + \frac{f}{c} = \frac{2e}{b}$$

$\Rightarrow \quad \frac{d}{a}, \frac{e}{b}, \frac{f}{c}$ are in AP.

**30.** *(c)* There are total $60 + n$ terms in both the series.

GM of both the series $= (2)^{\frac{225}{8}}$

$$[2 \times 2^2 \times 2^3 \times \ldots 2^{60} \times 4 \times 4^2 \times 4^3 \times \ldots 4^n]^{\frac{1}{60+n}} = (2)^{\frac{225}{8}}$$

$$\Rightarrow \quad [2^{1+2+3+\ldots+60} \times 4^{1+2+3+\ldots+n}]^{\frac{1}{60+n}} = 2^{\frac{225}{8}}$$

$$\Rightarrow \left[2^{\frac{60\times 61}{2}} \times 4^{\frac{n(n+1)}{2}}\right]^{\frac{1}{60+n}} = (2)^{\frac{225}{8}}$$

$$\Rightarrow \quad [2^{30\times 61 + n(n+1)}]^{\frac{1}{60+n}} = (2)^{\frac{225}{8}}$$

$$\Rightarrow \quad 2^{\frac{n^2+n+1830}{60+n}} = (2)^{\frac{225}{8}}$$

$$\Rightarrow \quad \frac{n^2+n+1830}{60+n} = \frac{225}{8}$$

$$\Rightarrow \quad 8n^2 - 217n + 1140 = 0$$

$$\Rightarrow \quad (8n - 57)(n - 20) = 0$$

$$\Rightarrow \quad n = 20 \quad \left[\because n \neq \frac{57}{8}\right]$$

$$\sum_{k=1}^{20} k(20 - k) = 20\sum_{k=1}^{20} k - \sum_{k=1}^{20} k^2$$

$$= 20 \times \frac{20 \times 21}{2} - \frac{20(21) \times 41}{6}$$

$$= 4200 - 2870$$

$$= 1330$$

**31.** *(c)* Given, $f(x+y) = 2f(x)f(y)$

and $f(1) = 2$

Then, $f(2) = 2f(1)f(1) = 2^3$

$$f(3) = 2f(1)f(2) = 2^5$$

$$f(4) = 2f(1)f(3) = 2^7$$

and $f(10) = 2f(1)f(9) = 2^{19}$

$$\sum_{k=1}^{10} f(\alpha + k) = \frac{512}{3}(2^{20} - 1)$$

$$\Rightarrow 2f(\alpha)[f(1) + f(2) + \ldots.. + f(10)] = \frac{512}{3}(2^{20} - 1)$$

$$\Rightarrow 2f(\alpha)[2 + 2^3 + \ldots. + 2^{19}] = \frac{512}{3}(2^{20} - 1)$$

$$= 2f(\alpha)\frac{2(2^{20} - 1)}{2^2 - 1} = \frac{512}{3}(2^{20} - 1)$$

$$\Rightarrow \quad f(\alpha) = 128 = 2^7$$

$$\Rightarrow \quad f(\alpha) = f(4) \Rightarrow \alpha = 4$$

**32.** *(a)*

> **key idea** Use the formula of sum of infinite GP $= \frac{a}{1-r}, |r| < 1$ and logarithmic series and solve further.

We have,

$$\frac{3}{2}x^2 + \frac{5}{3}x^3 + \frac{7}{4}x^4 + \ldots.$$

$$= \left(2 - \frac{1}{2}\right)x^2 + \left(2 - \frac{1}{3}\right)x^3 + \left(2 - \frac{1}{4}\right)x^4 + \ldots$$

$$= 2(x^2 + x^3 + x^4 + \ldots) - \left(\frac{x^2}{2} + \frac{x^3}{3} + \frac{x^4}{4} + \ldots\right)$$

$$= 2 \cdot \frac{x^2}{1-x} - [-\log_e(1-x) - x]$$

[Using sum of infinite GP $= \frac{a}{1-r}$ and logarithmic series]

$$= \frac{2x^2}{1-x} + x + \log_e(1-x)$$

$$= \frac{2x^2 + x - x^2}{1-x} + \log_e(1-x)$$

$$= \frac{x^2 + x}{1-x} + \log_e(1-x)$$

$$= x\left(\frac{1+x}{1-x}\right) + \log_e(1-x)$$

**33.** *(d)* Given,

$$\frac{2 + 4 + 6 + \ldots + 2y}{3 + 6 + 9 + \ldots + 3y} = \frac{4}{\log_{10} x}$$

$$\Rightarrow \quad \frac{2(1 + 2 + 3 + \ldots + y)}{3(1 + 2 + 3 + \ldots + y)} = \frac{4}{\log_{10} x}$$

$$\Rightarrow \quad \frac{2}{3} = \frac{4}{\log_{10} x} \Rightarrow \log_{10} x = 6$$

$$\Rightarrow \quad x = 10^6$$

Now, $y = \log_{10} x + \log_{10} x^{1/3} + \log_{10} x^{\frac{1}{9}} + \ldots$ upto $\infty$ terms.

$$= \log_{10}\left(x \cdot x^{\frac{1}{3}} \cdot x^{\frac{1}{9}} \ldots \infty \text{ terms}\right)$$

$$= \log_{10} x^{1 + \frac{1}{3} + \frac{1}{9} + \ldots \infty \text{ terms}}$$

$$= \log_{10} x^{\frac{1}{1 - \frac{1}{3}}} = \log_{10} x^{3/2}$$

$$= \log_{10}(10^6)^{3/2} \quad [\because x = 10^6]$$

$$\Rightarrow \quad y = 6 \times \frac{3}{2} = 9$$

$$\therefore x = 10^6, y = 9, (x, y) = (10^6, 9)$$

> **Shortcut Method**
>
> $$\frac{2(1+2+3+\ldots+y)}{3(1+2+3+\ldots+y)} = \frac{4}{\log_{10} x}$$
>
> $\log_{10} x = 6 \Rightarrow x = 10^6$
>
> Now $y = \log_{10} x + \frac{1}{3}\log_{10} x + \frac{1}{3^2}\log_{10} x + \ldots \infty$ term
>
> $y = \log_{10} x\left(1 + \frac{1}{3} + \frac{1}{3^2} + \frac{1}{3^3} + \ldots \infty \text{ term}\right)$
>
> $y = 6 \times \frac{1}{1 - \frac{1}{3}} = 6 \times \frac{3}{2} \quad y = 9$
>
> $\therefore (x, y) = (10^6, 9)$

**34.** *(c)* Let the first term of geometric series be $a$ and common ratio be $r$.

Given that $T_2 + T_6 = \frac{25}{2}$

$\Rightarrow \quad ar(1+r^4) = 25/2 \quad \ldots \text{(i)}$

Also, $(T_3)(T_5) = 25$

$\Rightarrow \quad a^2r^6 = 25 \quad \ldots \text{(ii)}$

On squaring Eq. (i), we get

$$a^2r^2(1+r^4)^2 = \left(\frac{25}{2}\right)^2 \quad \ldots \text{(iii)}$$

On dividing Eq. (ii) by Eq. (iii), we get

$$\frac{a^2r^2(1+r^4)^2}{a^2r^6} = \frac{(25)^2}{4(25)}$$

$\Rightarrow \quad \frac{(1+r^4)^2}{r^4} = \frac{25}{4}$

$\Rightarrow \quad 4(1+r^4)^2 = 25r^4$

$\Rightarrow \quad 4(1+r^8+2r^4) = 25r^4$

$\Rightarrow \quad 4r^8 - 17r^4 + 4 = 0$

$\Rightarrow \quad (r^4-4)(4r^4-1) = 0$

Gives, $r^4 = 4$ or $r^4 = \frac{1}{4}$

We have to find sum of 4th, 6th and 8th term, i.e.

$T_4 + T_6 + T_8 = ar^3 + ar^5 + ar^7$

$= ar^3(1+r^2+r^4) \quad \ldots \text{(iv)}$

Using Eq. (ii), $ar^3 = 5$

Also, we take $r^4 = 4$ because given series is increasing and $r^2 = 2$.

$\therefore \quad T_4 + T_6 + T_8 = 5(1+2+4)$
$= 5(7) = 35$

**35.** *(b)* Given series of infinite terms, if $|x|<1, |y|<1$ and $x \neq y$ is

$S = (x+y) + (x^2+xy+y^2) + (x^3+x^2y+xy^2+y^3) + \ldots$

On multiplying $(x-y)$ both sides, we get

$(x-y)S = (x^2-y^2) + (x^3-y^3) + (x^4-y^4) + \ldots$

$= [x^2+x^3+x^4+\ldots] - [y^2+y^3+y^4+\ldots]$

$= \frac{x^2}{1-x} - \frac{y^2}{1-y}$ as $|x|<1$ and $|y|<1$

$= \frac{(x^2-y^2)-(x^2y-y^2x)}{(1-x)(1-y)}$

$\Rightarrow \quad (x-y)S = \frac{(x-y)[(x+y)-xy]}{(1-x)(1-y)}$

$\Rightarrow \quad S = \frac{x+y-xy}{(1-x)(1-y)}$

**36.** *(c)*

> **key idea** Use $n$th term of AP i.e., $a_n = a + (n-1)d$, If $a, A$ and $b$ are in AP, then $2A = a+b$ and $n$th term of GP i.e., $a_n = ar^{n-1}$.

It is given that, the terms $a, b, c$ are in GP with common ratio $r$, where $a \neq 0$ and $0 < r \le 1/2$.

So, let, $b = ar$ and $c = ar^2$

Now, the terms $3a, 7b$ and $15c$ are the first three terms of an AP, then

$2(7b) = 3a + 15c \Rightarrow 14ar = 3a + 15ar^2$

$\Rightarrow \quad 5r(3r-1) - 3(3r-1) = 0$

$\Rightarrow \quad (3r-1)(5r-3) = 0 \Rightarrow r = \frac{1}{3}$ or $\frac{3}{5}$

As, $r \in \left(0, \frac{1}{2}\right]$, so $r = \frac{1}{3}$

Now, the common difference of AP

$= 7b - 3a = -\frac{2a}{3}$

So, 4th term of AP

$= 3a + 3\left(\frac{-2a}{3}\right) = a$

**37.** *(b)* Let $b = ar$ and $c = ar^2$, where $r$ is the common ratio.

Then, $\quad a + b + c = xb$

$\Rightarrow \quad a + ar + ar^2 = xar$

$\Rightarrow \quad 1 + r + r^2 = xr \quad \ldots \text{(i)}$ $\quad [\because a \neq 0]$

We know that $r + \frac{1}{r} \ge 2$ (for $r > 0$)

and $r + \frac{1}{r} \le -2$ (for $r < 0$)

[using AM $\ge$ GM]

$\therefore \quad 1 + r + \frac{1}{r} \ge 3$ or $1 + r + \frac{1}{r} \le -1$

$\Rightarrow \quad x \ge 3$ or $x \le -1$

$\Rightarrow \quad x \in (-\infty, -1] \cup [3, \infty)$

Hence, $x$ cannot be 2.

**38.** *(b)* Let $a$ be the first term and $d$ be the common difference. Then, we have $a+d$, $a+4d$, $a+8d$ in GP,

i.e. $\quad (a+4d)^2 = (a+d)(a+8d)$

$\Rightarrow a^2 + 16d^2 + 8ad = a^2 + 8ad + ad + 8d^2$

$\Rightarrow \quad 8d^2 = ad$

$\Rightarrow \quad 8d = a \quad [\because d \neq 0]$

Now, common ratio,

$$r = \frac{a+4d}{a+d} = \frac{4}{3}$$

**39.** *(b)* Given, $m$ is the AM of $l$ and $n$.

$\therefore \quad l + n = 2m \quad \ldots \text{(i)}$

$\therefore l, G_1, G_2, G_3, n$ are in GP.

Let $r$ be the common ratio of this GP.

$\therefore \quad G_1 = lr \Rightarrow G_2 = lr^2$

$\Rightarrow \quad G_3 = lr^3 \Rightarrow n = lr^4$

$\Rightarrow \quad r = \left(\frac{n}{l}\right)^{1/4}$

Now, $G_1^4 + 2G_2^4 + G_3^4$

$= (lr)^4 + 2(lr^2)^4 + (lr^3)^4$

$= l^4 \times r^4(1 + 2r^4 + r^8)$

$= l^4 \times r^4(r^4+1)^2$

$= l^4 \times \frac{n}{l}\left(\frac{n+l}{l}\right)^2 = ln \times 4m^2$

$= 4lm^2n$

**40.** *(c)* Given $\frac{1}{1^4} + \frac{1}{2^4} + \frac{1}{3^4} + \ldots + \infty = \frac{\pi^4}{90}$

$\frac{1}{1^4} + \frac{1}{3^4} + \frac{1}{5^4} + \ldots + \infty = \alpha$

$\frac{1}{2^4} + \frac{1}{4^4} + \frac{1}{6^4} + \ldots + \infty = \beta$

$\frac{1}{2^4}\left(\frac{1}{1^4} + \frac{1}{2^4} + \frac{1}{3^4} + \ldots + \infty\right) = \beta$

$\frac{1}{16} \times \frac{\pi^4}{90} = \beta$

Also $\frac{1}{1^4} + \frac{1}{2^4} + \frac{1}{3^4} + \frac{1}{4^4} + \ldots + \infty = \frac{\pi^4}{90}$

$\left(\frac{1}{1^4} + \frac{1}{3^4} + \frac{1}{5^4} + \ldots + \infty\right) + \left(\frac{1}{2^4} + \frac{1}{4^4} + \frac{1}{6^4} + \ldots + \infty\right) = \frac{\pi^4}{90}$

$\alpha + \beta = \frac{\pi^4}{90}$

$\Rightarrow \quad \frac{\alpha}{\beta} + 1 = \frac{\pi^4}{90\beta}$

$\Rightarrow \quad \frac{\alpha}{\beta} + 1 = \frac{\pi^4}{90} \times \frac{16 \times 90}{\pi^4}$

$\Rightarrow \quad \frac{\alpha}{\beta} = 15$

**41.** *(d)* $S = 1 + \frac{1+3}{2!} + \frac{1+3+5}{3!} + \ldots + \infty$

We know that,

$$e^x = \sum_{n=0}^{\infty} \frac{x^n}{n!} = 1 + x + \frac{x^2}{2!} + \frac{x^3}{3!} + \ldots$$

Now, sum of $n$ odd terms :

$1 + 3 + 5 + \ldots + (2n-1) = n^2$

$n$ th term , $a_n = \frac{n^2}{n!}$

$$a_n = \frac{n^2}{n!} = \frac{n}{(n-1)!}$$

$$a_n = \frac{n-1+1}{(n-1)!} = \frac{n-1}{(n-1)!} + \frac{1}{(n-1)!}$$

$$a_n = \frac{1}{(n-2)!} + \frac{1}{(n-1)!}$$

sum of series,

$$S = 1 + \sum_{n=r}^{\infty}\left(\frac{1}{(n-2)!} + \frac{1}{(n-1)!}\right)$$

Separate the sum

$$S = 1 + \sum_{n=2}^{\infty} \frac{1}{(n-2)!} + \sum_{n=2}^{\infty} \frac{1}{(n-1)!}$$

Let $n-2 = k$ and $n-1 = i$

$$S = 1 + \sum_{k=0}^{\infty} \frac{1}{k!} + \sum_{j=1}^{\infty} \frac{1}{j!}$$

$$\sum_{k=0}^{\infty} \frac{1}{k!} = e^1 = e$$

$$\sum_{j=1}^{\infty} \frac{1}{j!} = e^1 - 1 = e - 1$$

substitutions the values

$$S = 1 + e + (e-1) = 2e$$

**42.** *(b)*

**key idea** Sum of first $n$ terms of a GP,

$$S_n = \frac{a(1-r^n)}{1-r}$$

$$2a_{n+2} = 5a_{n+1} - 3a_n$$

$$\Rightarrow \quad 2a_{n+2} - 5a_{n+1} + 3a_n = 0$$

$$\Rightarrow \quad 2x^2 - 5x + 3 = 0$$

have roots $\alpha$ and $\beta$.

Then, $a_n = p\alpha^n + q\beta^n$

and $\quad 2x^2 - 5x + 3 = 0$

$$\Rightarrow \quad 2x^2 - 2x - 3x + 3 = 0$$

$$\Rightarrow \quad 2x(x-1) - 3(x-1) = 0$$

$$S = \sum_{n=2}^{\infty} \frac{1}{(n-2)!} + \sum_{n=1}^{\infty} \frac{1}{(n-1)!}$$

$$S = e + e = 2e$$

$$\Rightarrow \quad (2x-3)(x-1) = 0$$

$$x = 1, 3/2$$

$$\therefore \quad a_n = p(1)^n + q(3/2)^n$$

Let $\alpha = 1$ and $\beta = \frac{3}{2}$

Put $n = 0$

$\Rightarrow a_0 = p + q \Rightarrow p + q = 0 \quad \ldots$(i)

Put $n = 1 \Rightarrow a_1 = p + q\left(\frac{3}{2}\right) = \frac{1}{2}$

$$\Rightarrow -q + \frac{3}{2}q = \frac{1}{2} \Rightarrow \frac{q}{2} = \frac{1}{2} \Rightarrow q = 1$$

$$\Rightarrow \quad p = -1$$

$$\therefore \quad a_n = -4 + \left(\frac{3}{2}\right)^n$$

$$\Rightarrow \quad S_n = \Sigma a_n$$

$$\therefore \quad \sum_{k=1}^{100} a_k = a_1 + a_2 + \ldots + a_{100}$$

$$= -100 + \frac{3}{2} + \left(\frac{3}{2}\right)^2 + \ldots + \left(\frac{3}{2}\right)^{100}$$

$$= -100 + \frac{3}{2}\left[1 \cdot \frac{(1-(3/2)^{100})}{1-3/2}\right]$$

$$= -100 - \frac{3/2}{1/2}(1-(3/2)^{100})$$

$$= -100 - 3(1-(3/2)^{100})$$

$$= -100 + 3a_{100}$$

$$= 3a_{100} - 100$$

**43.** *(a)* $\sum_{n=0}^{\infty} ar^n = 57$

$$\Rightarrow \quad \frac{a}{1-r} = 57$$

$\Rightarrow \quad a = 57(1-r) \quad \ldots$(i)

$\because \quad \sum_{n=0}^{\infty} a^3 r^{3n} = 9747$

$\frac{a^3}{1-r^3} = 9747 \quad \ldots$(ii)

By Eqs. (i) and (ii), we get

$$\frac{(57)^3(1-r)^3}{(1-r)(1+r+r^2)} = 9747$$

$$\Rightarrow \quad \frac{1+r^2-2r}{1+r^2+r} = \frac{9747}{57 \times 57 \times 57} = \frac{1}{19}$$

$$\Rightarrow \quad 19 + 19r^2 - 38r = 1 + r^2 + r$$

$$\Rightarrow \quad 18r^2 - 39r + 18 = 0$$

$$\Rightarrow \quad 6r^2 - 13r + 6 = 0$$

$$\Rightarrow \quad 6r^2 - 9r - 4r + 6 = 0$$

$$\Rightarrow \quad 3r(2r-3) - 2(2r-3) = 0$$

$$r = \frac{3}{2}, \frac{2}{3} \quad \left(\because r = \frac{3}{2} \text{ reject}\right)$$

$$\therefore \quad r = \frac{2}{3} \Rightarrow a = 19$$

$$\therefore a + 18r = 19 + 18 \times \frac{2}{3} = 31$$

**Concept Enhancer**

Here, we use

$$\sum_{n=0}^{\infty} ar^n = a + ar + ar^2 + ar^3 + \ldots \infty$$

$$= \frac{a}{1-r}$$

and $\sum_{n=0}^{\infty} a^3 r^{3n} = a^3 + a^3r^3 + a^3r^6 + \ldots \infty$

$$= \frac{a^3}{1-r^3}$$

**44.** *(d)* Given, $T_2 + T_6 = \frac{70}{3}$

$$\Rightarrow \quad ar + ar^5 = \frac{70}{3}$$

$\Rightarrow ar(1+r^4) = \frac{70}{3} \quad \ldots$(i)

and $T_3 \cdot T_5 = 49 \Rightarrow ar^2 \cdot ar^4 = 49$

$$\Rightarrow \quad a^2 r^6 = 49 \Rightarrow ar^3 = 7 \Rightarrow a = \frac{7}{r^3}$$

Put the value of $a$ in Eq. (i),

$$\frac{7}{r^3} \times r(1+r^4) = \frac{70}{3}$$

$$\Rightarrow \quad \frac{1+r^4}{r^2} = \frac{10}{3} \Rightarrow 3 + 3r^4 = 10r^2$$

$$\Rightarrow \quad 3(r^2)^2 - 10r^2 + 3 = 0$$

Let $r^2 = t$

$$3t^2 - 10t + 3 = 0$$

$$\Rightarrow \quad 3t^2 - 9t - t + 3 = 0$$

$$\Rightarrow \quad 3t(t-3) - 1(t-3) = 0$$

$$\Rightarrow (t-3)(3t-1) = 0 \Rightarrow t = 3, 1/3$$

So, $r^2 = 3$ or $r^2 = 1/3$

$\Rightarrow r \neq \frac{1}{\sqrt{3}} \quad$ (GP is increasing)

$$\Rightarrow r = \sqrt{3}$$

$$T_4 + T_6 + T_8 = ar^3 + ar^5 + ar^7$$

$$= ar^3(1 + r^2 + r^4)$$

$$= 7(1 + 3 + 9) = 91$$

**45.** *(d)* Given, $(a + b - 2c)x^2 + (b + c - 2a)x + (c + a - 2b) = 0$

$\because$ 1 satisfies the equation.

$\alpha$ and 1 are roots of equation.

Product of roots

$$\alpha \cdot 1 = \frac{c+a-2b}{a+b-2c}$$

$$\alpha = \frac{c+a-2b}{a+b-2c}$$

**Statement I**

$\alpha \in (-1, 0)$

$-1 < \alpha < 0$

$-1 < \dfrac{c+a-2b}{a+b-2c} < 0; \because \; a > b > c > 0$

$0 < \dfrac{c+a-2b+a+b-2c}{a+b-2c} < 1$

$\because \quad a+b > c+c$

$a+b > 2c$

$a+b-2c > 0$

$0 < \dfrac{2a-b-c}{a+b-2c} < 1$

$\therefore \quad 2a-b-c > 0$

$2a > b+c$

$a > \dfrac{b+c}{2}$

$\therefore b$ can't be GM between $a$ and $c$.

**Statement II** $\alpha \in (0, 1)$

$0 < \alpha < 1$

$0 < \dfrac{c+a-2b}{a+b-2c} < 1$

Here, $b > c$ and $b < \dfrac{a+c}{2}$

$\therefore b$ may be GM between $a$ and $c$.

$\therefore$Statement I and statement II both are true.

**46.** *(a)* Given, $2\tan^2\theta - 5\sec\theta = 1$

$\Rightarrow \quad 2(\sec^2\theta - 1) - 5\sec\theta - 1 = 0$

$\Rightarrow \quad 2\sec^2\theta - 5\sec\theta - 3 = 0$

$\Rightarrow 2\sec^2\theta - 6\sec\theta + \sec\theta - 3 = 0$

$\Rightarrow \quad (2\sec\theta + 1)(\sec\theta - 3) = 0$

$\because \quad \sec\theta = -\dfrac{1}{2}$ (not possible)

$\therefore \quad \sec\theta = 3 \Rightarrow \cos\theta = 1/3$

From graph,

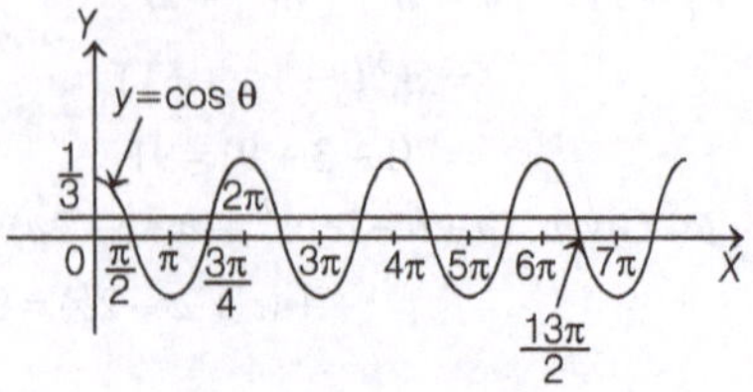


2 solutions in $[0, 2\pi]$,

2 solutions in $[2\pi, 4\pi]$,

2 solutions in $[4\pi, 6\pi]$,

and 1 solution in $\left[6\pi, \dfrac{13\pi}{2}\right]$.

Hence, $\quad n = 13$

Now, $\displaystyle\sum_{k=1}^{13} \frac{k}{2^k} = S$ [say]

So, $\quad S = \dfrac{1}{2} + \dfrac{2}{2^2} + \dfrac{3}{2^3} + \ldots + \dfrac{13}{2^{13}}$

$\dfrac{1}{2}S = \dfrac{1}{2^2} + \dfrac{2}{2^3} + \ldots + \dfrac{13}{2^{14}}$

$\dfrac{1}{2}S = \dfrac{1}{2} + \dfrac{1}{2^2} + \dfrac{1}{2^3} + \ldots + \dfrac{1}{2^{13}} - \dfrac{13}{2^{14}}$

$$\Rightarrow \frac{1}{2}S = \frac{1}{2}\left[\frac{\left(1-\frac{1}{2^{13}}\right)}{\left(1-\frac{1}{2}\right)} - \frac{13}{2^{14}}\right]$$

$$\Rightarrow S = 2\left(\frac{2^{13}-1}{2^{13}}\right) - \frac{13}{2^{13}} = \frac{2^{14}-15}{2^{13}}$$

**47.** *(a)* Given that $a^3$, $b^3$ and $c^3$ be in AP, then

$2b^3 = a^3 + c^3 \quad \ldots(i)$

and $\log_a b$, $\log_c a$ and $\log_b c$ in GP

$(\log_c a)^2 = (\log_a b)(\log_b c)$

$= \log_a b \times \dfrac{\log_a c}{\log_a b} \quad \left[\because \log_b x = \dfrac{\log_a x}{\log_a b}\right]$

$\Rightarrow \quad (\log_c a)^3 = 1$

$\Rightarrow \quad a = c$

Put in Eq. (i), we get

$2b^3 = a^3 + a^3$

$\Rightarrow \quad b = a$

$\therefore \quad a = b = c$

Also given,

$a_1 = \dfrac{a+4b+c}{3}$ and $d = \dfrac{a-8b+c}{10}$

and $S_{20} = -444$

$\Rightarrow \quad -444 = \dfrac{20}{2}[2 \times a_1 + 19d]$

$= 10\left[\dfrac{2(a+4b+c)}{3} + 19\dfrac{[a-8b+c]}{10}\right]$

$= 10\left[4a + 19\left(-\dfrac{3a}{5}\right)\right]$

$= 10\left(\dfrac{20a-57a}{5}\right) = 2(-37a)$

$\Rightarrow \quad a = \dfrac{222}{37} = 6$

Now, $abc = a^3 \quad [\because a = b = c]$

$= 6^3 = 216$

**48.** *(b)* Given,

$x^{pq^2} = y^{qr} = z^{p^2r} = \lambda$ (say)

Since, $x^{pq^2} = \lambda$ gives $pq^2 = \log_x \lambda$

$\Rightarrow \log_\lambda x = \dfrac{1}{pq^2} \quad \ldots(i)$

$qr = \log_y \lambda$

$\Rightarrow \log_\lambda y = \dfrac{1}{qr} \quad \ldots(ii)$

$\Rightarrow \log_\lambda z = \dfrac{1}{p^2r} \quad \ldots(iii)$

Now, $\log_y x = \dfrac{\log_\lambda x}{\log_\lambda y} = \dfrac{qr}{pq^2} \quad \ldots(iv)$

[Using Eqs. (i) and (iii)]

$\log_z y = \dfrac{\log_\lambda y}{\log_\lambda z} = \dfrac{\frac{1}{qr}}{\frac{1}{p^2r}} = \dfrac{p^2r}{qr} = \dfrac{p^2}{q}$

$\ldots(v)$

[Using Eqs. (ii) and (iii)]

and $\log_x z = \dfrac{\log_\lambda z}{\log_\lambda x} = \dfrac{\frac{1}{p^2r}}{\frac{1}{pq^2}} = \dfrac{q^2}{pr} \ldots(vi)$

[Using Eqs. (i) and (iii)]

Given, 3, $\dfrac{3r}{pq}, \dfrac{3p^2}{q}, \dfrac{7q^2}{pr}$ are in AP.

Then, $\dfrac{3r}{pq} - 3 = \dfrac{1}{2} \Rightarrow r = \dfrac{7}{6}pq \quad \ldots(vii)$

Also given, $r = pq + 1$

$\Rightarrow \quad pq = 6 \quad \ldots(viii)$

From Eq. (vii),

$r = 7 \quad \ldots(ix)$

Now, $\dfrac{3p^2}{q} - \dfrac{3r}{pq} = \dfrac{1}{2}$

$\Rightarrow \quad \dfrac{3p^2}{q} = \dfrac{1}{2} + \dfrac{3r}{pq} = \dfrac{8}{2} = 4 \quad \ldots(x)$

From Eq. (viii), $q = \dfrac{6}{p} \Rightarrow \dfrac{1}{q} = \dfrac{p}{6}$

From Eq. (x), $3p^2 \times \dfrac{p}{6} = 4 \Rightarrow p = 2$

$\therefore pq = 6$ gives us $q = 3$

Hence, $r - p - q = 7 - 2 - 3 = 2$

**49.** *(6460)* The general term of the series is

$T_n = (-1)^r\ (r+2)(r+1)\ 20_{C_{r+4}}$

$$S = \sum_{r=0}^{16} (-1)^r\ (r+2)(r+1)\ 20_{C_{r+4}}$$

Also, $(r+2)(r+1) = (r+2)(r+1)$

$$\frac{(20-r-2)(20-r-3)}{(20-r-2)(20-r-3)} =$$

We also known that $k(k-1)n_{C_k}$

$= n(n-1) \times n - 2_{C_{k-2}}$

So, $(r+2)(r+1)\ {}^{20}C_{r+4} = (r+2)(r+1)$

$$\frac{20!}{(16-r)!(r+4)!} =$$

$$\frac{20!}{(r+2)!(16-r)!} = 20 \times 19 \times 18_{C_{r+2}}$$

$$S = \sum_{r=0}^{16} (-1)^r\ 20 \times 19 \times 18_{C_{r+2}}$$

$$= 380 \sum_{r=0}^{16} (-1)^r\ 18_{C_{r+2}}$$

Let $k = r + 2$ then $r = k - 2$, when $r = 0$, $k = 2$, when $r = 16$, $k = 18$

$$S = 380 \sum_{k=2}^{18} (-1)^{k-2}\ 18_{C_k}$$

$$= 380 \sum_{k=2}^{18} (-1)^k 18_{C_k} =$$

We know that $\sum_{k=0}^{n} (-1)^k n_{C_k} = 0$

So, $\sum_{k=0}^{18} (-1)^k 18_{C_k} = 18_{C_0} - 18_{C_1} + \sum_{k=2}^{18} (-1)^k 18_{C_k} = 0$

Thus,

$\sum_{k=2}^{18} (-1)^k 18_{C_k} = -18_{C_0} + 18_{C_1}$

$= -1 + 18 = 17$

Therefore, $S = 380 \times 17 = 6460$

**50.** *(441)*

$$\frac{4 \cdot 1}{1 + 4 \cdot 1^4} + \frac{4 \cdot 2}{1 + 4 \cdot 2^4} + \frac{4 \cdot 3}{1 + 4 \cdot 3^4} + \ldots$$

$$T_r = \frac{4r}{1 + 4r^2}$$

$$= \frac{4r}{(2r^2+1)^2 - (2r)^2}$$

$$T_r = \frac{(2r^2 + 2r + 1) - (2r^2 - 2r + 1)}{(2r^2 + 2r + 1)(2r^2 - 2r + 1)}$$

$$T_r = \left(\frac{1}{r^2 + (r-1)^2} - \frac{1}{r^2 + (r+1)^2}\right)$$

$$\sum_{r=1}^{10} T_r = 1 - \frac{1}{221} = \frac{220}{221}$$

$m + n = 220 + 221 = 441$

**51.** *(96)* $f(\theta) = \dfrac{\sin^4\theta + 3\cos^2\theta}{\sin^4\theta + \cos^2\theta}$

$$= \frac{\sin^4\theta + \cos^2\theta + 2\cos^2\theta}{\sin^4\theta + \cos^2\theta}$$

$$= 1 + \frac{2\cos^2\theta}{\sin^4\theta + \cos^2\theta}$$

$$= 1 + \frac{2\cos^2\theta}{1 + \cos^4\theta - 2\cos^2\theta + \cos^2\theta}$$

$$= 1 + \frac{2\cos^2\theta}{1 + \cos^4\theta - \cos^2\theta}$$

$$= 1 + \frac{2}{\dfrac{1}{\cos^2\theta} + \cos^2\theta - 1}$$

[divide and multiply by $\cos^2\theta$]

Let $g(\theta) = \dfrac{2}{\dfrac{1}{\cos^2\theta} + \cos^2\theta - 1}$

Maximum of $g(\theta) \Rightarrow$ when denominator is minimum and

minimum of $\left\{\dfrac{1}{\cos^2\theta} + \cos^2\theta - 1\right\} = 1$

So, maximum of $g(\theta) = 2$

Similarly, minimum of $g(\theta) = \dfrac{2}{\infty} = 0$

So, range of

$f(\theta) = 1 + g(\theta) = [0+1, 2+1] = [1, 3]$

$\qquad\qquad\qquad\qquad\qquad\downarrow\ \downarrow$

$\qquad\qquad\qquad\qquad\qquad\alpha\ \ \beta$

Required sum of GP,

$$S_\infty = \frac{a}{1-r} = \frac{64}{\left(1 - \dfrac{1}{3}\right)} = \frac{64 \times 3}{2} = 96$$

**52.** *(5)* $\alpha = \sum_{r=0}^{n} (4r^2 + 2r + 1) \cdot {}^nC_r$

$$= \sum_{r=0}^{n} 4r^2 \cdot {}^nC_r + \sum_{r=0}^{n} 2r \cdot {}^nC_r + \sum_{r=0}^{n} {}^nC_r$$

$= 4n(n+1) \times 2^{n-2} + 2n \cdot 2^{n-1} + 2^n$

$= 2^n[n(n+1) + n + 1] = 2^n(n+1)^2$

and $\beta = \left(\sum_{r=0}^{n} \dfrac{{}^nC_r}{n+1}\right) + \dfrac{1}{n+1}$

$$= \frac{2^{n+1} - 1}{n+1} + \frac{1}{n+1} = \frac{2^{n+1}}{n+1}$$

So, $\dfrac{2\alpha}{\beta}$

$$= \frac{2^{n+1}(n+1)^2}{2^{n+1}} \times (n+1) = (n+1)^3$$

According to the question,

$140 < (n+1)^3 < 281$

So, $\quad n + 1 = 6 \Rightarrow n = 5$

> **Concept Enhancer**
>
> Here we use $\sum r^2\ {}^nC_r = n(n+1)2^{n-2}$
>
> and $\quad \sum r^2\ {}^nC_r = n \cdot 2^{n-1}$

**53.** *(3660)* $S(x) = (1+x) + 2(1+x)^2 + 3(1+x)^3 + \ldots + 60(1+x)^{60}$

$(1+x)\ S(x) = (1+x)^2 + 2(1+x)^3 + \ldots + 59(1+x)^{60} + 60(1+x)^{61}$

$\Rightarrow$ Subtracting both

$$\Rightarrow -xS = \frac{(1+x)(1+x)^{60} - 1}{x} - 60(1+x)^{61}$$

$x^2 S(y) = 60x(1+x)^{61} + (1+x) - (1+x)^{61}$

Put $\ x = 60$

$(60)^2 S(60) = (60)^2 \times (61)^{61} + 61 - (61)^{61}$

$= (61 \times 59)(61)^2 + 61$

$\therefore\ a = 61 \times 59$ and $b = 61$

So, $a + b = 3660$

**54.** *(6)* $T_r = 3T_{r-1} + 6^r$; $r = 2, 3, 4, \ldots, n$

$T_2 = 3T_1 + 6^2 = 3 \cdot 6 + 6^2 \qquad$ ...(i)

$T_3 = 3T_2 + 6^3 = 3(3 \cdot 6 + 6^2) + 6^3$

$= 3^2 \cdot 6 + 3 \cdot 6^2 + 6^3 \qquad$ ...(ii)

$\Rightarrow\ T_r = 3^{r-1} \cdot 6 + 3^{r-2} \cdot 6^2 + \ldots + 6^r$

$$\Rightarrow\ T_r = 3^{r-1} \cdot 6\left[1 + \frac{6}{3} + \left(\frac{6}{3}\right)^2 + \ldots + \left(\frac{6}{3}\right)^{r-1}\right]$$

$\Rightarrow T_r = 3^{r-1} \cdot 6\,[1 + 2 + 2^2 + \ldots + 2^{r-1}]$

$$T_r = 6 \cdot 3^{r-1} \cdot 1 \cdot \frac{(1 - 2^r)}{(-1)}$$

$$T_r = \frac{6 \cdot 3^r}{3}(2^r - 1)$$

$T_r = 2(6^r - 3^r)$

$\Rightarrow S_n = 2\Sigma(6^r - 3^r)$

$\Rightarrow S_n = 2\left[\frac{6}{5}(6^n - 1) - \frac{3}{2}(3^n - 1)\right]$

$S_n = 2\left[\frac{12(6^n - 1) - 15(3^n - 1)}{10}\right]$

$= \frac{3}{5}[4 \cdot 6^4 - 5 \cdot 3^n + 1]$

$\therefore n^2 - 12n + 39 = 3$

$\Rightarrow \quad n = 6$

**55.** *(76)* Let

$S = 1 + \frac{x}{2\sqrt{3}} + \frac{x^2}{18} + \frac{x^3}{36\sqrt{3}} + \frac{x^4}{180} + \ldots + \infty$

Put $\frac{x}{\sqrt{3}} = t$, where $x = \sqrt{3} - \sqrt{2}$

$\Rightarrow S = 1 + \frac{t}{2} + \frac{t^2}{6} + \frac{t^3}{12} + \frac{t^4}{20} + \ldots$

$= 1 + \left(1 - \frac{1}{2}\right)t + \left(\frac{1}{2} - \frac{1}{3}\right)t^2 + \left(\frac{1}{3} - \frac{1}{4}\right)t^3 + \ldots$

$\Rightarrow S = \left(1 + t + \frac{t^2}{2} + \frac{t^3}{3} + \ldots\right) - \left(\frac{t}{2} + \frac{t^2}{3} + \frac{t^3}{4} + \ldots\right)$

$\Rightarrow S = \left(t + \frac{t^2}{2} + \frac{t^3}{3} + \ldots\right) - \frac{1}{t}\left(t + \frac{t^2}{2} + \frac{t^3}{3} + \ldots\right) + 2$

$\Rightarrow S = 2 - \log(1 - t)\left(1 - \frac{1}{t}\right)$

$= 2 + \log(1 - t)\left(\frac{1}{t} - 1\right)$

$= 2 + \left(\frac{1}{t} - 1\right)\log(1 - t)$

$= 2 + \left(\frac{\sqrt{3}}{\sqrt{3} - \sqrt{2}} - 1\right)\log\left(1 - \frac{\sqrt{3} - \sqrt{2}}{\sqrt{3}}\right)$

$= 2 + \left(\frac{\sqrt{2}}{\sqrt{3} - \sqrt{2}}\right)\log\left(\frac{\sqrt{2}}{\sqrt{3}}\right)$

$= 2 + \sqrt{2}(\sqrt{3} + \sqrt{2})\log\left(\frac{2}{3}\right)^{1/2}$

$= 2 + \left(\frac{\sqrt{6} + 2}{2}\right)\log\frac{2}{3}$

$= 2 + \left(\sqrt{\frac{3}{2}} + 1\right)\ln\frac{2}{3}$

$\therefore \; a = 2, b = 3 \Rightarrow 11a + 18b = 76$

**56.** *(9)* Given, 3, 7, 11, .... are in AP

$\therefore \quad S_k = \frac{k}{2}[6 + (k - 1)4]$

$= k(3 + 2k - 2)$

$= k(2k + 1) = 2k^2 + k$

$\Rightarrow \sum_{k=1}^{n} S_k = \Sigma 2k^2 + \Sigma k$

$= 2\left[\frac{n}{6}(n + 1)(2n + 1)\right] + \frac{n}{2}(n + 1)$

$= n(n + 1)\left[\frac{2n + 1}{3} + \frac{1}{2}\right]$

$= \frac{n(n + 1)}{6}(4n + 5)$

$\Rightarrow \frac{6\Sigma S_k}{n(n + 1)} = 4n + 5$

$\because \quad 40 < 4n + 5 < 42$

$35 < 4n < 37 \Rightarrow \frac{35}{4} < n < \frac{37}{4}$

$\therefore \quad n = 9$

**57.** *(353)*

> **key idea** Use the formula $\Sigma n^2 = \frac{n(n+1)(2n+1)}{6}$ and $\Sigma n^3 = \left[\frac{n(n+1)}{2}\right]^2$

We have,

$\alpha = 1^2 + 4^2 + 8^2 + 13^2 + \ldots$up to 10

$\beta = \sum_{n=1}^{10} n^4$

Let $S = 1 + 4 + 8 + 13 + \ldots + t_n$

$S = 1 + 4 + 8 + 13 \ldots t_{n-1} + t_n$

Subtracting both, we get

$\Rightarrow t_n = 1 + |3 + 4 + 5 + \ldots (n - 1)\text{ term}|$

$\Rightarrow t_n = 1 + \frac{n - 1}{2}(6 + (n - 2)1)$

$\Rightarrow t_n = 1 + \frac{(n - 1)}{2}(n + 4)$

$= \frac{(n - 1)(n + 4) + 2}{2}$

$= \frac{n^2 + 3n - 4 + 2}{2} = \frac{n^2 + 3n - 2}{2}$

$\alpha = \sum_{n=1}^{10}\left(\frac{n^2 + 3n - 2}{2}\right)^2$

$4\alpha = \sum_{n=1}^{10}(n^2 + 3n - 2)^2$

$= \sum_{n=1}^{10}(n^4 + 9n^2 + 4 + 6n^3 - 12n - 4n^2)$

$4\alpha - \beta = \sum_{n=1}^{10}(6n^3 + n^2 - 12n + 4)$

$= 6\sum_{n=1}^{10} n^3 + \sum_{n=1}^{10} n^2 - 12\sum_{n=1}^{10} n + 4\sum_{n=1}^{10} 1$

$= 6\left[\frac{10(11)}{2}\right]^2 + \frac{10 \times 11 \times 21 \times 5}{6} - 12 \times \frac{10 \times 11}{2} + 40$

$= 55(330 + 35 - 12) + 40$

$= 55 \times 353 + 40$

$k = 353$

**58.** *(9)* Given,

$8 = 3 + \frac{1}{4}(3 + p) + \frac{1}{4^2}(3 + 2p) + \frac{1}{4^3}(3 + 3p) + \ldots\infty$

$\Rightarrow \quad 8 = \frac{3}{1 - \frac{1}{4}} + \frac{p \times \left(\frac{1}{4}\right)}{\left(1 - \frac{1}{4}\right)^2}$

[as we know, sum of infinite terms of AGP $= \frac{a}{1 - r} + \frac{r \cdot d}{(1 - r)^2}$]

$\Rightarrow \quad 8 = \frac{3}{1} \times \frac{4}{3} + \frac{p}{4} \times \frac{16}{9} \Rightarrow 4 = \frac{16p}{36}$

$\Rightarrow \quad p = 9$

> **Concept Enhancer**
> The sum of an infinite arithmetic-geometric series of the form $a, (a + d)r, (a + 2d)r^2 \ldots\ldots \infty$ is given by $\frac{a}{1 - r} + \frac{r \cdot d}{(1 - r)^2}$

**59.** *(150)* Given, $\frac{1}{x}, \frac{1}{y}, \frac{1}{z}$ are in AP

$\Rightarrow \quad \frac{2}{y} = \frac{1}{x} + \frac{1}{z} \quad \ldots(i)$

$x, \sqrt{2}y, z$ are in GP

$\Rightarrow \quad (\sqrt{2}y)^2 = xz$

$\Rightarrow \quad 2y^2 = xz \quad \ldots(ii)$

$xy + yz + zx = \frac{3}{\sqrt{2}}xyz$

$\Rightarrow \quad \frac{1}{z} + \frac{1}{x} + \frac{1}{y} = \frac{3}{\sqrt{2}} \quad \ldots(iii)$

From Eqs. (i) and (iii), we get

$$\frac{2}{y}+\frac{1}{y}=\frac{3}{\sqrt{2}}$$

$$\Rightarrow \quad y=\sqrt{2}$$

From Eq. (ii), we get

$$2(\sqrt{2})^2 = xz \Rightarrow xz=4 \quad \ldots(iv)$$

From Eq. (iii), we get

$$\frac{1}{z}+\frac{1}{x}+\frac{1}{\sqrt{2}}=\frac{3}{\sqrt{2}} \Rightarrow \frac{x+z}{xz}=\sqrt{2}$$

$$\Rightarrow \quad x+z=4\sqrt{2} \quad (\because xz=4)$$

$$\therefore \quad 3(x+y+z)^2 = 3(4\sqrt{2}+\sqrt{2})^2$$

$$=3\times 50 = 150$$

**60.** *(400)* Given, $(20)^{19}+2(21)(20)^{18}$

$$+3(21)^2(20)^{17}+\ldots\ldots+20(21)^{19}$$

$$=k(20)^{19}$$

$$\Rightarrow (20)^{19}\left[1+2\left(\frac{21}{20}\right)+3\left(\frac{21}{20}\right)^2 +\ldots+20\left(\frac{21}{20}\right)^{19}\right]=k(20)^{19}$$

$$\Rightarrow k=1+2\left(\frac{21}{20}\right)+3\left(\frac{21}{20}\right)^2+\ldots+20\left(\frac{21}{20}\right)^{19} \quad \ldots(i)$$

$$\text{Now, } k\left(\frac{21}{20}\right)=\left(\frac{21}{20}\right)+2\left(\frac{21}{20}\right)^2+3\left(\frac{21}{20}\right)^3+\ldots+20\left(\frac{21}{20}\right)^{20} \quad \ldots(ii)$$

From Eq. (i) – Eq. (ii), we get

$$k\left(\frac{-1}{20}\right)=1+\frac{21}{20}+\left(\frac{21}{20}\right)^2+\ldots+\left(\frac{21}{20}\right)^{19}-20\left(\frac{21}{20}\right)^{20}$$

$$\Rightarrow k\left(\frac{-1}{20}\right)=\frac{1\left(\left(\frac{21}{20}\right)^{20}-1\right)}{\frac{21}{20}-1}-20\left(\frac{21}{20}\right)^{20}$$

$$\Rightarrow k\left(\frac{-1}{20}\right)=20\left(\left(\frac{21}{20}\right)^{20}-1\right)-20\left(\frac{21}{20}\right)^{20}$$

$$=20\left(\frac{21}{20}\right)^{20}-20-20\left(\frac{21}{20}\right)^{20}$$

$$\Rightarrow k\left(\frac{-1}{20}\right)=-20 \Rightarrow k=400$$

**61.** *(9)* We know that $k$th term of GP of

$$a_k=a_1r_1^{k-1}$$

$$b_k=b_1r_2^{k-1} \quad [\because \{a_k\} \text{ and } \{b_k\} \text{ in GP}]$$

Given that $c_k=a_k+b_k$

$$\Rightarrow \quad c_2=a_2+b_2$$

$$\Rightarrow \quad 5=a_1r_1^1+b_1r_2^1$$

$$\Rightarrow \quad r_1+r_2=\frac{5}{4} \quad \ldots(i)$$

and $\quad c_3=a_3+b_3$

$$\Rightarrow \quad \frac{13}{4}=a_1r_1^2+b_1r_2^2$$

$$\Rightarrow \quad r_1^2+r_2^2=\frac{13}{16} \quad \ldots(ii)$$

From Eq. (i), squaring both sides, we get

$$r_1^2+r_2^2+2r_1r_2=\frac{25}{16} \Rightarrow r_1r_2=\frac{3}{8}$$

$$\text{Now, } r_1-r_2=\sqrt{(r_1+r_2)^2-4r_1r_2}$$

$$=\sqrt{\frac{25}{16}-\frac{3}{2}}=\frac{1}{4} \quad \ldots(iii)$$

From Eqs. (i) and (iii), we get

$r_1=\frac{3}{4}$ and $r_2=\frac{2}{4}$ but, here $r_1>r_2$

But, we need as per question, $r_1<r_2$, so lets assume

$$r_1=\frac{2}{4}=\frac{1}{2} \text{ and } r_2=\frac{3}{4}$$

$$\text{Now, } a_6=4\left(\frac{1}{2}\right)^5 \Rightarrow 12a_6=\frac{3}{2}$$

$$\text{and } b_4=4\left(\frac{3}{4}\right)^3 \Rightarrow 8b_4=\frac{27}{2}$$

So, $12a_6+8b_4=15$

$$\text{and } \sum_{k=1}^{\infty}c_k=\sum_{k=1}^{\infty}(a_k+b_k)$$

$$=\frac{a_1}{1-r_1}+\frac{b_1}{1-r_2}$$

$$=\frac{4}{1-\frac{1}{2}}+\frac{4}{1-\frac{3}{4}}=24$$

$$\text{So, } \sum_{k=1}^{\infty}c_k-(12a_6+8b_4)=24-15=9$$

> **Concept Enhancer**
>
> $S=\frac{a}{1-r}$, where $S$ is equal sum of infinite terms of GP, whose first term is $a$ and the common ratio is $r$, where $|r|<1$.

**62.** *(60)* We have,

$a_1, a_2, a_3, \ldots$ an increasing GP of positive numbers.

Let common ratio be '$r$'.

Given that $a_4 \cdot a_6=9$

$$\Rightarrow a_1r^3\cdot a_1r^5=9 \Rightarrow (a_1r^4)^2=9$$

$$\Rightarrow \quad a_1r^4=3 \quad \ldots(i)$$

Now, $a_1a_9+a_2a_4a_9+(a_5+a_7)$

$$=a_1\cdot a_1r^8+(a_1r)(a_1r^3)(a_1r^8)+(24)$$

$$=(a_1r^4)^2+(a_1r^4)^3+24$$

$$=3^2+3^3+24 \quad \text{[from Eq. (i)]}$$

$$=60$$

**63.** *(3)* If $a$ and $b$ are positive numbers and $a, b, \frac{1}{18}$ are in GP.

$$\text{So,} \quad b^2=a.\frac{1}{18}$$

$$\text{or} \quad a=18b^2 \quad \ldots(i)$$

$\because \frac{1}{a}, 10, \frac{1}{b}$ are in AP.

$$\text{So, } 2(10)=\frac{1}{a}+\frac{1}{b}$$

$$\Rightarrow \quad 20=\frac{a+b}{ab} \quad \text{or} \quad a+b=20ab$$

From Eq. (i), we get

$$18b^2+b=20(18b^2)b$$

$$\text{or} \quad 360b^3-18b^2-b=0$$

$$b(12b-1)(30b+1)=0$$

$$b=0 \text{ or } b=\frac{1}{12} \text{ or } b=-\frac{1}{30}$$

$$b=\frac{1}{12}, \text{ as } b>0 \Rightarrow a=\frac{1}{8}$$

Now,

$$16a+12b=16\left(\frac{1}{8}\right)+12\left(\frac{1}{12}\right)$$

$$\text{or} \quad 16a+12b=3$$

**64.** *(12)* Let $a$ = first term of GP

common ratio of GP $=\frac{1}{m}, m\in N$ [given] ...(i)

Also, $T_4=500$ [given]

$\Rightarrow \dfrac{a}{m^3} = 500$ [using Eq. (i)]

Since, $S_n - S_{n-1} = ar^{n-1}$ ...(ii)

Given, $S_6 > S_5 + 1$ and $S_7 - S_6 < \dfrac{1}{2}$

$\Rightarrow S_6 - S_5 > 1$ and $\dfrac{a}{m^6} < \dfrac{1}{2}$ [Using Eq. (ii)]

$\Rightarrow ar^5 > 1$ and $\dfrac{500}{m^3} < \dfrac{1}{2}$

$\Rightarrow \dfrac{500}{m^2} > 1$ and $m^3 > 1000$

$\Rightarrow m^2 < 500$ ...(iii)

and $m^3 > 10^3$ ...(iv)

From Eqs. (iii) and (iv), we get $m = 11, 12, 13, ..., 22$ because $m \in N$. Hence, number of possible values of $m$ is 12.

**65.** *(142)* Since, $\sum_{i=1}^{20} x_i = \dfrac{3\left[1-\left(\frac{1}{2}\right)^{20}\right]}{1-\frac{1}{2}}$

$= 6\left(1 - \dfrac{1}{2^{20}}\right)$

Now, $\sum_{i=1}^{20}(x_i - i)^2 = \sum_{i=1}^{20} x_i^2 - 2\sum_{i=1}^{20} x_i i + \sum_{i=1}^{20} i^2$ ...(i)

$\therefore \sum_{i=1}^{20} x_i^2 = \dfrac{9\left[1-\left(\frac{1}{4}\right)^{20}\right]}{1-\frac{1}{4}} = 12\left(1 - \dfrac{1}{2^{40}}\right)$

$\sum_{i=1}^{20} i^2 = \dfrac{1}{6} \times 20 \times 21 \times 41 = 2870$

and $\sum_{i=1}^{20} x_i i = 3 + 2\left(\dfrac{3}{2}\right) + 3\left(\dfrac{3}{2^2}\right) + 4\left(\dfrac{3}{2^3}\right) + .....$ [AGP]

$= 6\left(2 - \dfrac{22}{2^{20}}\right)$

$\therefore \bar{x} = \dfrac{\sum_{i=1}^{20}(x_i - i)^2}{20}$

$= \dfrac{12 - \frac{12}{2^{40}} + 2870 - 12\left(2 - \frac{22}{2^{20}}\right)}{20}$

$= \dfrac{2858}{20} + \left(\dfrac{-12}{2^{40}} + \dfrac{22}{2^{20}}\right) \times \dfrac{1}{20}$

$\Rightarrow \bar{x} = 142$

**66.** *(40)* Given, $a_1 > 0$

and $a_2, a_3, a_4, a_5$ are in GP

$3a_2 + a_3 = 2a_4$

$\Rightarrow 3ar + ar^2 = 2ar^3$

$\Rightarrow 2r^2 - r - 3 = 0$

$\Rightarrow r = -1$ and $r = \dfrac{3}{2}$

and $a_2 + a_4 = 2a_3 + 1$

$\Rightarrow ar + ar^3 = 2ar^2 + 1$

$\Rightarrow a(r + r^3 - 2r^2) = 1 \Rightarrow a = \dfrac{8}{3}$

when $r = \dfrac{3}{2}$ and $a = -\dfrac{1}{4}$ when $r = -1$

[rejected as $a_1 > 0$]

Now, $a_2 + a_4 + 2a_5 = \dfrac{8}{3} \times \dfrac{3}{2} + \dfrac{8}{3} \times \dfrac{27}{8} + 2 \times \dfrac{8}{3} \times \dfrac{81}{16}$

$= 4 + 9 + 27 = 40$

**67.** *(3)* Let $l = \alpha^\beta$

$\alpha = 1 + \dfrac{2}{3} + \dfrac{6}{3^2} + \dfrac{10}{3^3} + .... \infty$ ...(i)

$\dfrac{\alpha}{3} = \dfrac{1}{3} + \dfrac{2}{3^2} + \dfrac{6}{3^3} + .....$ ...(ii)

On subtracting Eq. (ii) from Eq. (i), we get

$\dfrac{2\alpha}{3} = 1 + \dfrac{1}{3} + \dfrac{4}{3^2} + \dfrac{4}{3^3} + .....$

$= 1 + \dfrac{1}{3} + \dfrac{4}{3}\left(\dfrac{1}{1-\frac{1}{3}}\right) = 2$

$\therefore \alpha = 3$

$\beta = \log_{0.25}\left(\dfrac{1}{3} + \dfrac{1}{3^2} + ....\right)$

$= \log_{0.25}\left(\dfrac{\frac{1}{3}}{1-\frac{1}{3}}\right) = \log_{\frac{1}{4}}\dfrac{1}{2} = \dfrac{1}{2}$

$\therefore l = 3^{1/2}$

$\Rightarrow l^2 = 3$

**68.** *(14)* Given, GP $= \dfrac{1}{16}, a, b$

$\Rightarrow a^2 = \dfrac{b}{16}$

and given, AP $= \dfrac{1}{a}, \dfrac{1}{b}, 6$

$\Rightarrow \dfrac{2}{b} = \dfrac{1}{a} + 6 \Rightarrow \dfrac{2}{16a^2} = \dfrac{1}{a} + 6$

$\Rightarrow 48a^2 + 8a - 1 = 0$

$\Rightarrow (4a+1)(12a-1) = 0$

$a = -1/4$ or $1/12$

As per the question, $a > 0$

$\therefore a = 1/12$

$b = 16a^2 = 16 \cdot \dfrac{1}{144} = \dfrac{1}{9}$

$\therefore 72(a+b) = 6 + 8 = 14$

**69.** *(3)* Let four numbers in GP be $a, ar, ar^2, ar^3$.

According to the question,

$a + ar + ar^2 + ar^3 = \dfrac{65}{12}$ ... (i)

and $\dfrac{1}{a} + \dfrac{1}{ar} + \dfrac{1}{ar^2} + \dfrac{1}{ar^3} = \dfrac{65}{18}$

$\Rightarrow \dfrac{1}{a}\left(\dfrac{1 + r + r^2 + r^3}{r^3}\right) = \dfrac{65}{18}$ ... (ii)

On dividing Eq. (i) by Eq. (ii), we get

$a^2r^3 = \dfrac{18}{12} = \dfrac{3}{2}$

Also, product of first three terms $= 1$

$a \times ar \times ar^2 = 1$

$\Rightarrow a^3r^3 = 1$

$\Rightarrow a^3 \times \dfrac{3}{2a^2} = 1 \left[\because r^3 = \dfrac{3/2}{a^2}\right]$

$\Rightarrow a = \dfrac{2}{3}$

and $r^3 = \dfrac{3/2}{(2/3)^2} = \left(\dfrac{3}{2}\right)^3 \Rightarrow r = \dfrac{3}{2}$

According to the question,

third term $= \alpha = ar^2 = \dfrac{2}{3} \times \dfrac{3}{2} \times \dfrac{3}{2} = \dfrac{3}{2}$

$\therefore 2\alpha = 2 \times \dfrac{3}{2} = 3$

**70.** *(4)* Since,

$(0.16)^{\log_{2.5}\left(\frac{1}{3} + \frac{1}{3^2} + \frac{1}{3^3} + ... \text{to } \infty\right)}$

$= (0.16)^{\log_{2.5}\left(\frac{1/3}{1-1/3}\right)} = (0.16)^{\log_{2.5}\left(\frac{1}{2}\right)}$

$$=\left(\frac{16}{100}\right)^{\log_{2.5}(2)^{-1}}=\left(\frac{100}{16}\right)^{\log_{2.5}(2)}$$

$$=\left(\frac{10}{4}\right)^{2\log_{2.5}2}=(2.5)^{\log_{2.5}2^2}$$

$$=2^2=4$$

## Topic 3

**01.** *(b)*

> **key idea** First use $-\sqrt{2}\le\sin\theta+\cos\theta\le\sqrt{2}$ and get the value of [a, b], then use $AM=\frac{a+b}{2}$ and $GM=\sqrt{ab}$.

We have, $f(x)=\dfrac{1}{2+\sin 3x+\cos 3x}$

$$-\sqrt{2}\le\sin 3x+\cos 3x\le\sqrt{2}$$

or $2-\sqrt{2}\le 2+\sin 3x+\cos 3x\le 2+\sqrt{2}$

or $\dfrac{1}{2+\sqrt{2}}\le\dfrac{1}{2+\sin 3x+\cos 3x}\le\dfrac{1}{2-\sqrt{2}}$

$\therefore a=\dfrac{1}{2+\sqrt{2}}$ and $b=\dfrac{1}{2-\sqrt{2}}$

Now, $\alpha=\left(\dfrac{\dfrac{1}{2+\sqrt{2}}+\dfrac{1}{2-\sqrt{2}}}{2}\right)$

$$=\frac{2-\sqrt{2}+2+\sqrt{2}}{(4-2)\times 2}=\frac{4}{2\times 2}=1$$

and $\beta=\sqrt{\dfrac{1}{(2+\sqrt{2})}\times\dfrac{1}{(2-\sqrt{2})}}$

$$=\sqrt{\frac{1}{(4-2)}}=\frac{1}{\sqrt{2}}$$

$\therefore\quad\dfrac{\alpha}{\beta}=\dfrac{1}{\left(\dfrac{1}{\sqrt{2}}\right)}=\sqrt{2}$

**02.** *(c)* Given, $f(x)=\dfrac{5x^2}{2}+\dfrac{\alpha}{x^5}$

$$=\frac{x^2}{2}+\frac{x^2}{2}+\frac{x^2}{2}+\frac{x^2}{2}+\frac{x^2}{2}+\frac{\alpha}{2x^5}+\frac{\alpha}{2x^5}\ge 7\left(\frac{\alpha^2}{2^7}\right)^{1/7}$$

Now, minimum value of $f(x)=14$

$\Rightarrow\ \dfrac{7\alpha^{2/7}}{2}=14\Rightarrow\quad\alpha^{2/7}=2^2$

$\Rightarrow\ \alpha=2^7=128$

**03.** *(d)* Given $x^3y^2=2^{15}$

$\Rightarrow x\cdot x\cdot x\cdot y\cdot y=2^{15}$

Using $AM\ge GM$

$\Rightarrow\ \dfrac{x+x+x+y+y}{5}\ge(x^3y^2)^{1/5}$

$\Rightarrow\ \dfrac{3x+2y}{5}\ge(2^{15})^{1/5}=2^3$

$\Rightarrow\ 3x+2y\ge 40$

Least value of $3x+2y$ is 40.

**04.** *(c)* We already know, Arithmetic mean $\ge$ Geometric mean, Let us take AM and GM of two terms $a^{a^x}$ and $a^{1-a^x}$,

$\Rightarrow\quad AM=\dfrac{a^{a^x}+a^{1-a^x}}{2}$ and GM

$$=\sqrt{a^{a^x}.a^{1-a^x}}$$

$\Rightarrow\ \dfrac{a^{a^x}+a^{1-a^x}}{2}\ge\sqrt{a^{a^x}\cdot a^{1-a^x}}$

$\Rightarrow\ a^{a^x}+a^{1-a^x}\ge 2\sqrt{a^1}$

$\therefore$ Minimum value of $f(x)=a^{a^x}+a^{1-a^x}$ is $2\sqrt{a}$.

**05.** *(c)* Since, $(2^{1+x}+2^{1-x}),\ f(x),\ (3^x+3^{-x})$ are in AP.

$$\therefore f(x)=\left(\frac{2^{1-x}+2^{1+x}+3^x+3^{-x}}{2}\right)$$

$$\ge\sqrt{2^2}+\sqrt{3^0}$$

[using $AM\ge GM$]

$\Rightarrow f(x)\ge 3$

**06.** *(c)* Consider $\dfrac{x^my^n}{(1+x^{2m})(1+y^{2n})}$

$$=\frac{1}{(x^m+x^{-m})(y^n+y^{-n})}$$

By using $AM\ge GM$ (because $x,y\in R^+$), we get

$(x^m+x^{-m})\ge 2$ and $(y^n+y^{-n})\ge 2$

$\Rightarrow\quad(x^m+x^{-m})(y^n+y^{-n})\ge 4$

$\Rightarrow\quad\dfrac{1}{(x^m+x^{-m})(y^n+y^{-n})}\le\dfrac{1}{4}$

$\therefore$ Maximum value $=\dfrac{1}{4}$.

**07.** *(d)* Since, $a$, $b$ and $c$ are in GP so

$a=\dfrac{b}{r}$ and $c=br$.

We have, $a+b+c=xb$

$\Rightarrow\quad\dfrac{b}{r}+b+br=xb$

[$\because a,b,c$ are in GP where, $|r|\ne 1$]

$\Rightarrow b\left(\dfrac{1}{r}+1+r\right)=xb$

$\Rightarrow\ \dfrac{1}{r}+r+1=x\ \Rightarrow\ x-1=r+\dfrac{1}{r}$

$\Rightarrow\ x-1>2$ or $x-1<-2$

[using AM > GM]

$\Rightarrow x>3$ or $x<-1$.

Hence, $x$ cannot be 2.

**08.** *(d)* We have, AM $=5$ GM

$\Rightarrow\ \dfrac{a+b}{2}=5\sqrt{ab}\Rightarrow\dfrac{(a+b)^2}{4ab}=25$

By Dividendo,

$$\frac{(a+b)^2}{(a+b)^2-4ab}=\frac{25}{25-1}$$

$\Rightarrow\ \dfrac{(a+b)^2}{(a-b)^2}=\dfrac{25}{24}\Rightarrow\dfrac{a+b}{a-b}=\dfrac{5\sqrt{6}}{12}$

**09.** *(b)* We have,

$$\tan(A+B)=\frac{\tan A+\tan B}{1-\tan A\tan B}$$

$\Rightarrow\dfrac{1}{\sqrt{3}}=\dfrac{y}{1-\tan A\tan B}$

Where, $y=\tan A+\tan B$

$\Rightarrow\tan A\tan B=1-\sqrt{3}y$

Also $AM\ge GM$

$\Rightarrow\dfrac{\tan A+\tan B}{2}\ge\sqrt{\tan A\tan B}$

$\Rightarrow y\ge 2\sqrt{1-\sqrt{3}y}\ \Rightarrow\ y^2\ge 4-4\sqrt{3}y$

$\Rightarrow y^2+4\sqrt{3}y-4\ge 0$

$\Rightarrow y\le-2\sqrt{3}-4$ or $y\ge-2\sqrt{3}+4$

($y\le-2\sqrt{3}-4$ is not possible as $\tan A\tan B>0$)

**10.** *(a)* $G=\sqrt{ab}$ and $M=\dfrac{\frac{1}{a}+\frac{1}{b}}{2}=\dfrac{a+b}{2ab}$

Given that $\dfrac{1}{M}:G=4:5$

$\Rightarrow\ \dfrac{2ab}{(a+b)\sqrt{ab}}=\dfrac{4}{5}\Rightarrow\dfrac{a+b}{2\sqrt{ab}}=\dfrac{5}{4}$

$\Rightarrow\ \dfrac{a+b+2\sqrt{ab}}{a+b-2\sqrt{ab}}=\dfrac{5+4}{5-4}$

{Using Componendo and Dividendo}

$\Rightarrow\ \left(\dfrac{\sqrt{b}+\sqrt{a}}{\sqrt{b}-\sqrt{a}}\right)^2=\dfrac{9}{1}\Rightarrow\dfrac{\sqrt{b}+\sqrt{a}}{\sqrt{b}-\sqrt{a}}=\dfrac{3}{1}$

$\Rightarrow\ \dfrac{\sqrt{b}+\sqrt{a}+\sqrt{b}-\sqrt{a}}{\sqrt{b}+\sqrt{a}-\sqrt{b}+\sqrt{a}}=\dfrac{3+1}{3-1}$

{Using Componendo and Dividendo}

$\Rightarrow \sqrt{\dfrac{b}{a}} = \dfrac{4}{2} = 2 \Rightarrow \dfrac{b}{a} = \dfrac{4}{1}$

$\dfrac{a}{b} = \dfrac{1}{4} \Rightarrow a : b = 1 : 4$

**11.** *(b)* $\because f(x) = \dfrac{1}{e^x + \dfrac{2}{e^x}}$

Using, AM $\geq$ GM

$$\frac{e^x + \dfrac{2}{e^x}}{2} \geq \left(e^x \cdot \frac{2}{e^x}\right)^{1/2}, \text{ as } e^x > 0$$

$$\Rightarrow \quad e^x + \frac{2}{e^x} \geq 2\sqrt{2}$$

$$\Rightarrow 0 < \frac{1}{e^x + \dfrac{2}{e^x}} \leq \frac{1}{2\sqrt{2}}$$

$$\therefore \quad 0 < f(x) \leq \frac{1}{2\sqrt{2}}, \forall x \in R$$

Statement II is true and Statement I is also true as for some $c$.

$\Rightarrow f(c) = \dfrac{1}{3}$ [for $c = 0$]

which lies between 0 and $\dfrac{1}{2\sqrt{2}}$.

So, statement II is correct explanation of statement I.

> **Shortcut Method**
>
> $f(x) = \dfrac{1}{e^x + 2e^{-x}} = \dfrac{e^x}{e^{2x} + 2}$
>
> $\Rightarrow f'(x) = \dfrac{(e^{2x} + 2)e^x - 2e^{2x} \cdot e^x}{(e^{2x+2})^2}$
>
> $\Rightarrow f'(x) = 0 \Rightarrow e^{2x} + 2 = 2e^{2x}$
>
> $\Rightarrow e^{2x} = 2 \Rightarrow e^x = \sqrt{2}$
>
> Maximum value of
>
> $f(x) = \dfrac{\sqrt{2}}{4} = \dfrac{1}{2\sqrt{2}} \Rightarrow 0 < f(x) \leq \dfrac{1}{2\sqrt{2}}, \forall x \in R$
>
> Since, $0 < \dfrac{1}{3} < \dfrac{1}{2\sqrt{2}}$
>
> $\Rightarrow$ For some $c \in R, f(c) = 1/3$

**12.** *(d)* Applying AM-GM inequality in the positive real numbers $p^2$ and $q^2$,

$$\frac{p^2 + q^2}{2} \geq \sqrt{p^2 q^2} = pq \quad [\because \text{AM} \geq \text{GM}]$$

$$\Rightarrow \quad pq \leq \frac{1}{2} \quad [\because p^2 + q^2 = 1] \quad \ldots\text{(i)}$$

Now, $(p+q)^2 = p^2 + q^2 + 2pq$

$$\Rightarrow \quad (p+q)^2 = 1 + 2pq$$

$$\Rightarrow \quad (p+q)^2 \leq 1 + 1 \quad \text{[from Eq. (i)]}$$

$$\Rightarrow \quad p + q \leq \sqrt{2}$$

**13.** *(b)* Given, $a, b, c$ are in AP

$\Rightarrow 2b = a + c$

and $(a+1), b, c+3$ are in GP

$\Rightarrow b^2 = (a+1)(c+3)$

AM of $a, b, c = 8$

$$\frac{a+b+c}{3} = 8 \Rightarrow a + b + c = 24$$

$$\Rightarrow \quad 3b = 24 \Rightarrow b = 8$$

$a + c = 16$ and $(a+1)(c+3) = 64$

$\Rightarrow c = 16 - a$

Then, $(a+1)(16 - a + 3) = 64$

$$\Rightarrow \quad (a+1)(19-a) = 64$$

$$\Rightarrow \quad 19a - a^2 + 19 - a = 64$$

$$\Rightarrow \quad a^2 - 18a + 45 = 0$$

$$\Rightarrow \quad (a-15)(a-3) = 0$$

$$a = 15, 3 \quad \because a > 10$$

$$\Rightarrow \quad a = 15 \text{ and } c = 1$$

$\therefore a = 15, b = 8, c = 1$

$\therefore$GM of $a, b, c = (abc)^{1/3}$

$$= (15 \times 8 \times 1)^{1/3} = (120)^{1/3}$$

Cube of GM $= 120$

**14.** *(a)* Let $a, b$ be two positive number

$a, A_1, A_2, b$ in AP

$$\Rightarrow \quad \frac{a+b}{2} = \frac{A_1 + A_2}{2}$$

$$\Rightarrow \quad a + b = A_1 + A_2 \quad \ldots\text{(i)}$$

also, $a, G_1, G_2, G_3, b$

$$\Rightarrow \quad b = ar^4 \Rightarrow \sqrt[4]{\frac{b}{a}} = r$$

$$\Rightarrow \quad G_1 = ar = a\sqrt[4]{\frac{b}{a}},$$

$$\Rightarrow \quad G_2 = ar^2 = a \cdot \left(\sqrt[4]{\frac{b}{a}}\right)^2 = a\sqrt{\frac{b}{a}}$$

$$\Rightarrow \quad G_3 = ar^3 = a\left(\sqrt[4]{\frac{b}{a}}\right)^3 = a\left(\frac{b}{a}\right)^{3/4}$$

$\therefore G_1^4 + G_2^4 + G_3^4 + G_1^2 G_3^2$

$$= \left(a\left(\frac{b}{a}\right)^{1/4}\right)^4 + \left(a\left(\frac{b}{a}\right)^{1/2}\right)^4 + \left[a\left(\frac{b}{a}\right)^{3/4}\right]^4 + \left[a\left(\frac{b}{a}\right)^{1/4}\right]^2\left[a\left(\frac{b}{a}\right)^{3/4}\right]^2$$

$$= a^4 \times \frac{b}{a} + a^4 \times \frac{b^2}{a^2} + a^4 \times \frac{b^3}{a^3} + a^2\frac{b^{1/2}}{a^{1/2}} \times \frac{a^2 \times b \times b^{1/2}}{a \times a^{1/2}}$$

$$= a^3 b + a^2 b^2 + ab^3 + a^2 b^2$$

$$= a^2 b(a+b) + ab^2(a+b)$$

$$= (a+b)(a^2 b + ab^2)$$

$$= (A_1 + A_2)[ab(a+b)]$$

$$= (A_1 + A_2)(A_1 + A_2)ab$$

$$= (A_1 + A_2)^2 G_1 G_3$$

**15.** *(b)* Given, $a, b, c, d > 0$ and $a, b, c, d \in R$

and $a + b + c + d = 11$

Also, $(a^5 b^3 c^2 d)_{\max} = 3750\beta$

Clearly, we know that for all the real numbers if min/max is given or asked then use AM $\geq$ GM

So, $\left(\dfrac{a}{5} + \dfrac{a}{5} + \dfrac{a}{5} + \dfrac{a}{5} + \dfrac{a}{5}\right) + \left(\dfrac{b}{3} + \dfrac{b}{3} + \dfrac{b}{3}\right) + \left(\dfrac{c}{2} + \dfrac{c}{2}\right) + d = 11$

$\Rightarrow$ AM $\geq$ GM

$$\Rightarrow \left(\frac{\dfrac{a}{5} \times 5 + \dfrac{b}{3} \times 3 + \dfrac{c}{2} \times 2 + d}{5 + 3 + 2 + 1}\right) \geq \left[\left(\frac{a}{5}\right)^5\left(\frac{b}{3}\right)^3\left(\frac{c}{2}\right)^2 d\right]^{\frac{1}{11}}$$

$$\Rightarrow \quad \left(\frac{11}{11}\right)^{11} \geq \frac{a^5 b^3 c^2 d}{5^5 \times 3^3 \times 2^2}$$

$$\Rightarrow (a^5 b^3 c^2 d)_{\max} = 1^{11} \times 5^5 \times 3^3 \times 2^2$$

$$\Rightarrow \ 3750\beta = 5^5 \times 3^3 \times 2^2$$

$$\Rightarrow \quad \beta = 5 \times 3^2 \times 2 \Rightarrow \beta = 90$$

Hence, value of $\beta$ is 90.

**16.** *(a)* We have, where, $x_1, x_2, \ldots, x_{100}$ are in AP with first term as 2.

$$\text{Mean} = 200 \Rightarrow \frac{\sum_{i=1}^{100} x_i}{100} = 200$$

$$\Rightarrow \frac{100}{2} \times [2 \times 2 + 99d] = 20000$$

$$4 + 99d = 400$$

$d = 4$

Also, $y_i = i(x_i - i) = i[2+(i-1)4-i]$

$= i[3i-2] = 3i^2 - 2i$

$$\text{Required mean} = \frac{\sum_{i=1}^{n} y_i}{100}$$

$$= \frac{1}{100}\left[\sum_{i=1}^{n}(3i^2-2i)\right]$$

$$= \frac{1}{100}\left[\frac{3\times100\times101\times201}{6} - 2\times\frac{100\times101}{2}\right]$$

$$= \frac{20301}{2} - 101 = \frac{20099}{2} = 10049.5$$

**17.** *(b)* Given, $x+y+z=12$

Now, AM $\geq$ GM

$$\Rightarrow \frac{3\left(\frac{x}{3}\right)+4\left(\frac{y}{4}\right)+5\left(\frac{z}{5}\right)}{12} \geq \left[\left(\frac{x}{3}\right)^3\left(\frac{y}{4}\right)^4\left(\frac{z}{5}\right)^5\right]^{\frac{1}{12}}$$

$$\Rightarrow \frac{x^3y^4z^5}{3^3 4^4 5^5} \leq 1 \Rightarrow x^3y^4z^5 \leq 3^3\cdot4^4\cdot5^5$$

But, given $x^3y^4z^5 = (0.1)(600)^3$

$\therefore$ All the numbers are equal.

$$\therefore \quad \frac{x}{3} = \frac{y}{4} = \frac{z}{5} = k \text{ (say)}$$

$\Rightarrow \quad x = 3k,\ y = 4k,\ z = 5k$

But $\quad x+y+z = 12$

$\Rightarrow \quad 3k+4k+5k = 12$

$\Rightarrow \quad k = 1$

$\therefore\ x = 3,\ y = 4,\ z = 5$

So, $x^3+y^3+z^3 = 216$

**18.** *(10)* If AM and GM satisfy the equation $4x^2-9x+5=0$, then AM and GM are nothing but roots of this quadratic equation,

$$4x^2-9x+5=0$$

$\Rightarrow \quad (x-1)(4x-5)=0 \Rightarrow \quad x = 1,\ 5/4$

Then, AM $= 5/4$ and GM $= 1$

[$\because$ AM $\geq$ GM]

Again, the given series is

$-16, 8, -4, 2 \ldots\ldots$

which is a geometric progression series with common ratio $-1/2$, then

$$p\text{th term} = -16\left(\frac{-1}{2}\right)^{p-1} = t_p$$

$$q\text{th term} = -16\left(\frac{-1}{2}\right)^{q-1} = t_q$$

$\because$ Geometric mean $= 1 \Rightarrow \sqrt{t_p t_q} = 1$

$$\Rightarrow (-16)\left(\frac{-1}{2}\right)^{p-1}\cdot(-16)\left(\frac{-1}{2}\right)^{q-1} = 1$$

$$\Rightarrow (-16)^2\left(\frac{-1}{2}\right)^{p+q-2} = 1$$

$$\Rightarrow (-2)^8\frac{(+1)^{p+q-2}}{(-2)^{p+q-2}} = 1$$

$$\Rightarrow (-2)^8 = (-2)^{p+q-2}$$

$\Rightarrow p+q-2 = 8$

$\Rightarrow \quad p+q = 10$

**19.** *(39)* The 4th AM out of $m$ AMs inserted between 3 and 243 is

$$A_4 = 3+4\frac{243-3}{m+1} \quad \ldots\text{(i)}$$

and the 2nd GM out of three GMs inserted between 3 and 243 is

$$G_2 = 3\left(\left(\frac{243}{3}\right)^{\frac{1}{3+1}}\right)^2 = 3(81)^{1/2} = 27$$

$\because \quad A_4 = G_2$

$$\Rightarrow \quad 3+4\left(\frac{240}{m+1}\right) = 27$$

$\Rightarrow \quad m+1 = 40$

$\Rightarrow \quad m = 39$

# C-08

JEE Main Chapterwise Topicwise Mathematics

In JEE Main, there is usually 0-1 question per paper is visible from this chapter. In past 34 years, 34 conceptual and 6 numerical questions of moderate difficulty level were asked from this chapter.

**For detailed analysis scan QR code**

# STRAIGHT LINES AND PAIR OF STRAIGHT LINES

## Topic 1 Rectangular Cartesian Coordinates and Straight Lines

### Level 1

**01** Let $ABCD$ be a tetrahedron such that the edges $AB$, $AC$ and $AD$ are mutually perpendicular. Let the areas of the triangles $ABC$, $ACD$ and $ADB$ be 5, 6 and 7 square units respectively. Then, the area (in square units) of the $\Delta BCD$ is equal to *[2025, 2 April Shift-I]*

(a) $\sqrt{340}$ (b) 12 (c) $7\sqrt{3}$ (d) $\sqrt{110}$

**02** Let the area of the triangle formed by a straight line $L: x+by+c=0$ with coordinate axes be 48 square units. If the perpendicular drawn from the origin to the line $L$ makes an angle of 45° with the positive $X$-axis, then the value of $b^2+c^2$ is *[2025, 2 April Shift-II]*

(a) 97 (b) 90 (c) 93 (d) 83

**03** A line passes through the origin and makes equal angles with the positive coordinate axes. It intersects the lines $L_1: 2x+y+6=0$ and $L_2: 4x+2y-p=0$, $p>0$, at the points $A$ and $B$, respectively. If $AB=\dfrac{9}{\sqrt{2}}$ and the foot of the perpendicular from the point $A$ on the line $L_2$ is $M$, then $AM/BM$ is equal to *[2025, 3 April Shift-I]*

(a) 5 (b) 3 (c) 4 (d) 2

**Imp.** Differentiability of Function

**04** Let the function $f(x)=(x^2-1)|x^2-ax+2|+\cos|x|$ be not differentiable at the two points $x=\alpha=2$ and $x=\beta$. Then, the distance of the point $(\alpha,\beta)$ from the line $12x+5y+10=0$ is equal to *[2025, 29 Jan Shift-II]*

(a) 5 (b) 4 (c) 3 (d) 2

**05** Two equal sides of an isosceles triangle are along $-x+2y=4$ and $x+y=4$. If $m$ is the slope of its third side, then the sum of all possible distinct values of $m$ is *[2025, 28 Jan Shift-II]*

(a) 6 (b) $-2\sqrt{10}$ (c) 12 (d) $-6$

**06** Let $A$, $B$, $C$ be three points in $XY$-plane, whose position vector are given by $\sqrt{3}\hat{i}+\hat{j}$, $\hat{i}+\sqrt{3}\hat{j}$ and $a\hat{i}+(1-a)\hat{j}$, respectively with respect to the origin O. If the distance of the point $C$ from the line bisecting the angle between the vectors **OA** and **OB** is $\dfrac{9}{\sqrt{2}}$, then the sum of all the possible values of a is *[2025, 28 Jan Shift-II]*

(a) 1 (b) $\dfrac{9}{2}$ (c) 0 (d) 2

**07** Let the area of $\Delta PQR$ with vertices $P(5,4)$, $Q(-2,4)$ and $R(a,b)$ be 35 square units. If its orthocenter and centroid are $O\left(2,\dfrac{14}{5}\right)$ and $C(c,d)$ respectively, then $c+2d$ is equal to *[2025, 23 Jan Shift-I]*

(a) $\dfrac{7}{3}$ (b) $\dfrac{8}{3}$ (c) 2 (d) 3

**08** Let the $\Delta PQR$ be the image of the triangle with vertices (1, 3), (3, 1) and (2, 4) in the line $x+2y=2$. If the centroid of $\Delta PQR$ is the point $(\alpha,\beta)$, then $15(\alpha-\beta)$ is equal to *[2025, 22 Jan Shift-I]*

(a) 21 (b) 22 (c) 24 (d) 19

**09** If the locus of the point whose distances from the points (2, 1) and (1, 3) are in the ratio 5 : 4, is $ax^2+by^2+cxy+dx+ey+170=0$, then the value of $a^2+2b+3c+4d+e$ is equal to *[2024, 6 April Shift-II]*

(a) 5 (b) 37 (c) 437 (d) $-27$

**10** Let $\left(5, \frac{a}{4}\right)$ be the circumcenter of a triangle with vertices $A(a, -2)$, $B(a, 6)$ and $C\left(\frac{a}{4}, -2\right)$. Let $\alpha$ denote the circumradius, $\beta$ denote the area and $\gamma$ denote the perimeter of the triangle. Then, $\alpha + \beta + \gamma$ is **[2024, 29 Jan Shift-I]**

(a) 30 (b) 60 (c) 53 (d) 62

**11** Let $\alpha_1, \alpha_2$ $(\alpha_1 < \alpha_2)$ be the values of $\alpha$ for the points $(\alpha, -3)$, $(2, 0)$ and $(1, \alpha)$ to be collinear. Then the equation of the line, passing through $(\alpha_1, \alpha_2)$ and making an angle of $\pi/3$ with the positive direction of the $X$-axis, is **[2022, 30 July Shift-I]**

(a) $x - \sqrt{3}y - 3\sqrt{3} - 1 = 0$ (b) $\sqrt{3}x - y + \sqrt{3} + 3 = 0$
(c) $x - \sqrt{3}y + 3\sqrt{3} + 1 = 0$ (d) $\sqrt{3}x - y + \sqrt{3} - 3 = 0$

**12** Let $A(1, 1)$, $B(-4, 3)$, $C(-2, -5)$ be vertices of a $\Delta ABC$, $P$ be a point on side $BC$, and $\Delta_1$ and $\Delta_2$ be the areas of $\Delta APB$ and $\Delta ABC$, respectively. If $\Delta_1 : \Delta_2 = 4 : 7$, then the area enclosed by the lines $AP$, $AC$ and the $X$-axis is **[2022, 27 July Shift-I]**

(a) 1/4 (b) 3/4 (c) 1/2 (d) 1

**13** If $p$ and $q$ are the lengths of the perpendiculars from the origin on the lines, $x \operatorname{cosec} \alpha - y \sec \alpha = k \cot 2\alpha$ and $x \sin \alpha + y \cos \alpha = k \sin 2\alpha$ respectively, then $k^2$ is equal to **[2021, 31 Aug Shift-I]**

(a) $4p^2 + q^2$ (b) $2p^2 + q^2$
(c) $p^2 + 2q^2$ (d) $p^2 + 4q^2$

**14** Let $A$ be a fixed point $(0, 6)$ and $B$ be a moving point $(2t, 0)$. Let $M$ be the mid-point of $AB$ and the perpendicular bisector of $AB$ meets the $Y$-axis at $C$. The locus of the mid-point $P$ of $MC$ is **[2021, 27 Aug Shift-I]**

(a) $3x^2 - 2y - 6 = 0$ (b) $3x^2 + 2y - 6 = 0$
(c) $2x^2 + 3y - 9 = 0$ (d) $2x^2 - 3y + 9 = 0$

**Imp.** Equation of Angle Bisector

**15** Let $ABC$ be a triangle with $A(-3, 1)$ and $\angle ACB = \theta$, $0 < \theta < \frac{\pi}{2}$. If the equation of the median through $B$ is $2x + y - 3 = 0$ and the equation of angle bisector of $C$ is $7x - 4y - 1 = 0$, then $\tan \theta$ is equal to **[2021, 26 Aug Shift-I]**

(a) 1/2 (b) 3/4 (c) 4/3 (d) 2

**16** In a $\Delta PQR$, the coordinates of the points $P$ and $Q$ are $(-2, 4)$ and $(4, -2)$, respectively. If the equation of the perpendicular bisector of $PR$ is $2x - y + 2 = 0$, then the centre of the circumcircle of the $\Delta PQR$ is **[2021, 17 March Shift-I]**

(a) $(-1, 0)$ (b) $(-2, -2)$
(c) $(0, 2)$ (d) $(1, 4)$

**17** The intersection of three lines $x - y = 0$, $x + 2y = 3$ and $2x + y = 6$ is a **[2021, 26 Feb Shift-I]**

(a) right angled triangle (b) equilateral triangle
(c) isosceles triangle (d) None of these

**18** The image of the point $(3, 5)$ on the line $x - y + 1 = 0$, lies on **[2021, 25 Feb Shift-I]**

(a) $(x-2)^2 + (y-2)^2 = 12$ (b) $(x-4)^2 + (y+2)^2 = 16$
(c) $(x-4)^2 + (y-4)^2 = 8$ (d) $(x-2)^2 + (y-4)^2 = 4$

**19** A man is walking on a straight line. The arithmetic mean of the reciprocals of the intercepts of this line on the coordinate axes is $1/4$. Three stones $A$, $B$ and $C$ are placed at the points (1,1), (2, 2) and (4, 4), respectively. Then, which of these stones is / are on the path of the man ? **[2021, 24 Feb Shift-I]**

(a) $A$ only (b) $B$ only
(d) All the three (c) $C$ only

**20** If the perpendicular bisector of the line segment joining the points $P(1, 4)$ and $Q(k, 3)$ has $y$-intercept equal to $-4$, then a value of $k$ is **[2020, 4 Sep Shift-II, 2008]**

(a) $\sqrt{15}$ (b) $-4$ (c) $\sqrt{14}$ (d) $-2$

**21** The locus of the mid-points of the perpendiculars drawn from points on the line, $x = 2y$ to the line $x = y$ is **[2020, 7 Jan Shift-II]**

(a) $5x - 7y = 0$ (b) $2x - 3y = 0$
(c) $3x - 2y = 0$ (d) $7x - 5y = 0$

**22** If the two lines $x + (a-1)y = 1$ and $2x + a^2 y = 1$, $(a \in R - \{0, 1\})$ are perpendicular, then the distance of their point of intersection from the origin is **[2019, 9 April Shift-II]**

(a) $\frac{2}{5}$ (b) $\frac{\sqrt{2}}{5}$
(c) $\frac{2}{\sqrt{5}}$ (d) $\sqrt{\frac{2}{5}}$

**23** Let $O(0, 0)$ and $A(0, 1)$ be two fixed points, then the locus of a point $P$ such that the perimeter of $\Delta AOP$ is 4, is **[2019, 08 April Shift-II]**

(a) $8x^2 - 9y^2 + 9y = 18$ (b) $9x^2 - 8y^2 + 8y = 16$
(c) $9x^2 + 8y^2 - 8y = 16$ (d) $8x^2 + 9y^2 - 9y = 18$

**24** If the straight line, $2x - 3y + 17 = 0$ is perpendicular to the line passing through the points (7, 17) and $(15, \beta)$, then $\beta$ equals **[2019, 12 Jan Shift-I]**

(a) $\frac{35}{3}$ (b) $-5$
(c) $-\frac{35}{3}$ (d) 5

**25** If in a parallelogram $ABDC$, the coordinates of $A$, $B$ and $C$ are respectively (1, 2), (3, 4) and (2, 5), then the equation of the diagonal $AD$ is **[2019, 11 Jan Shift-II]**

(a) $3x + 5y - 13 = 0$ (b) $3x - 5y + 7 = 0$
(c) $5x - 3y + 1 = 0$ (d) $5x + 3y - 11 = 0$

**26** Consider the set of all lines $px + qy + r = 0$ such that $3p + 2q + 4r = 0$. Which one of the following statements is true? *[2019, 9 Jan Shift-I]*
(a) Each line passes through the origin.
(b) The lines are concurrent at the point $\left(\frac{3}{4}, \frac{1}{2}\right)$
(c) The lines are all parallel
(d) The lines are not concurrent

**27** Let the orthocentre and centroid of a triangle be $A(-3, 5)$ and $B(3, 3)$, respectively. If $C$ is the circumcentre of this triangle, then the radius of the circle having line segment $AC$ as diameter, is *[JEE Main 2018]*
(a) $\sqrt{10}$ (b) $2\sqrt{10}$ (c) $3\sqrt{\frac{5}{2}}$ (d) $\frac{3\sqrt{5}}{2}$

**28** In a $\Delta ABC$, coordinates of $A$ are (1, 2) and the equations of the medians through $B$ and $C$ are $x + y = 5$ and $x = 4$ respectively. Then, area of $\Delta ABC$ (in sq. units) is *[JEE Main 2018, 15]*
(a) 5 (b) 9 (c) 12 (d) 4

**29** A square, of each side 2 lies above the $X$-axis and has one vertex at the origin. If one of the sides passing through the origin makes an angle 30° with the positive direction of the $X$-axis, then the sum of the $X$-coordinates of the vertices of the square is *[JEE Main 2017]*
(a) $2\sqrt{3} - 1$ (b) $2\sqrt{3} - 2$
(c) $\sqrt{3} - 2$ (d) $\sqrt{3} - 1$

**30** A straight line through origin $O$ meets the lines $3y = 10 - 4x$ and $8x + 6y + 5 = 0$ at points $A$ and $B$ respectively. Then $O$ divides the segment $AB$ in the ratio *[JEE Main 2016]*
(a) 2 : 3 (b) 1 : 2 (c) 4 : 1 (d) 3 : 4

**31** The number of points having both coordinates as integers that lie in the interior of the triangle with vertices (0, 0), (0, 41) and (41, 0) is *[JEE Main 2015]*
(a) 901 (b) 861 (c) 820 (d) 780

**32** The points $(0, 8/3)$, (1, 3) and (82, 30) *[JEE Main 2015]*
(a) form an obtuse angled triangle
(b) form an acute angled triangle
(c) form a right angled triangle
(d) lie on a straight line

**33** If the three distinct lines $x + 2ay + a = 0$, $x + 3by + b = 0$ and $x + 4ay + a = 0$ are concurrent, then the point $(a, b)$ lies on a *[JEE Main 2014]*
(a) circle (b) hyperbola
(c) straight line (d) parabola

**34** If a line $L$ is perpendicular to the line $5x - y = 1$, and the area of the triangle formed by the line $L$ and the coordinate axes is 5, then the distance of line $L$ from the line $x + 5y = 0$ is *[JEE Main 2014]*
(a) $7/\sqrt{5}$ (b) $5/\sqrt{13}$
(c) $7/\sqrt{13}$ (d) $5/\sqrt{7}$

**35** A ray of light along $x + \sqrt{3}y = \sqrt{3}$ gets reflected upon reaching $X$-axis, the equation of the reflected ray is *[JEE Main 2013]*
(a) $y = x + \sqrt{3}$ (b) $\sqrt{3}y = x - \sqrt{3}$
(c) $y = \sqrt{3}x - \sqrt{3}$ (d) $\sqrt{3}y = x - 1$

**36** The $x$-coordinate of the incentre of the triangle that has the coordinates of mid-points of its sides as (0, 1), (1, 1) and (1, 0) is *[JEE Main 2013]*
(a) $2 + \sqrt{2}$ (b) $2 - \sqrt{2}$ (c) $1 + \sqrt{2}$ (d) $1 - \sqrt{2}$

**37** If the three lines $x - 3y = p$, $ax + 2y = q$ and $ax + y = r$ form a right-angled triangle then *[JEE Main 2013]*
(a) $a^2 - 9a + 18 = 0$ (b) $a^2 - 6a - 12 = 0$
(c) $a^2 - 6a - 18 = 0$ (d) $a^2 - 9a + 12 = 0$

**38** If the $x$-intercept of some line $L$ is double as that of the line, $3x + 4y = 12$ and the $y$-intercept of $L$ is half as that of the same line, then the slope of $L$ is *[JEE Main 2013]*
(a) $-3$ (b) $-\frac{3}{8}$ (c) $-\frac{3}{2}$ (d) $-\frac{3}{16}$

**39** If the line $2x + y = k$ passes through the point which divides the line segment joining the points (1, 1) and (2, 4) in the ratio 3 : 2, then $k$ is equal to *[AIEEE 2012]*
(a) 29/5 (b) 5 (c) 6 (d) 11/5

**40** The point of intersection of the lines $(a^3 + 3)x + ay + a - 3 = 0$ and $(a^5 + 2)x + (a + 2)y + 2a + 3 = 0$ ($a$ real) lies on the $Y$-axis for *[AIEEE 2012]*
(a) no value of $a$ (b) more than two values of $a$
(c) exactly one value of $a$ (d) exactly two values of $a$

**41** If the point $(1, a)$ lies between the straight lines $x + y = 1$ and $2(x + y) = 3$ then $a$ lies in interval *[AIEEE 2012]*
(a) $(3/2, \infty)$ (b) $(1, 3/2)$ (c) $(-\infty, 0)$ (d) $(0, 1/2)$

**42** The line parallel to $X$-axis and passing through the point of intersection of lines $ax + 2by + 3b = 0$ and $bx - 2ay - 3a = 0$, where $(a, b) \neq (0, 0)$ is
(a) above $X$-axis at a distance 2/3 from it *[AIEEE 2012, 2005]*
(b) above $X$-axis at a distance 3/2 from it
(c) below $X$-axis at a distance 3/2 from it
(d) below $X$-axis at a distance 2/3 from it

**43** Consider the straight lines
$L_1 : x - y = 1$; $L_2 : x + y = 1$
$L_3 : 2x + 2y = 5$; $L_4 : 2x - 2y = 7$
The correct statement is *[AIEEE 2012]*
(a) $L_1 \parallel L_4$, $L_2 \parallel L_3$, $L_1$ intersect $L_4$.
(b) $L_1 \perp L_2$, $L_1 \parallel L_3$, $L_1$ intersect $L_2$.
(c) $L_1 \perp L_2$, $L_2 \parallel L_3$, $L_1$ intersect $L_4$.
(d) $L_1 \perp L_2$, $L_1 \perp L_3$, $L_2$ intersect $L_4$.

**44** If $a, b, c \in R$ and 1 is a root of equation $ax^2 + bx + c = 0$, then the curve $y = 4ax^2 + 3bx + 2c$, $a \neq 0$ intersect $X$-axis at **[AIEEE, 2012]**
(a) two distinct points whose coordinates are always rational numbers
(b) no point
(c) exactly two distinct points
(d) exactly one point

**45** The lines $x + y = |a|$ and $ax - y = 1$ intersect each other in the first quadrant. Then, the set of all possible values of $a$ in the interval **[AIEEE 2011]**
(a) $(-1, 1]$ (b) $(0, \infty)$ (c) $[1, \infty)$ (d) $(-1, \infty)$

**46** The lines $p(p^2 + 1)x - y + q = 0$ and $(p^2 + 1)^2 x + (p^2 + 1)y + 2q = 0$ are perpendicular to a common line for **[AIEEE 2009]**
(a) exactly one value of $p$ (b) exactly two values of $p$
(c) more than two values of $p$ (d) no value of $p$

**Imp.** Shortest Distance between Line and Curve

**47** The shortest distance between the line $y - x = 1$ and the curve $x = y^2$ is **[AIEEE 2009]**
(a) $\frac{2\sqrt{3}}{8}$ (b) $\frac{3\sqrt{2}}{5}$ (c) $\frac{\sqrt{3}}{4}$ (d) $\frac{3\sqrt{2}}{8}$

**48** A straight line through the point $A(3, 4)$ is such that its intercept between the axes is bisected at $A$. Its equation is **[AIEEE 2006]**
(a) $3x - 4y + 7 = 0$ (b) $4x + 3y = 24$
(c) $3x + 4y = 25$ (d) $x + y = 7$

**49** If $(a, a^2)$ falls inside the angle made by the lines $y = \frac{x}{2}, x > 0$ and $y = 3x, x > 0$, then $a$ belongs to **[AIEEE 2006]**
(a) $(3, \infty)$ (b) $\left(\frac{1}{2}, 3\right)$ (c) $\left(-3, -\frac{1}{2}\right)$ (d) $\left(0, \frac{1}{2}\right)$

**50** The normal to the curve $x = a(1 + \cos\theta)$, $y = a\sin\theta$ at $\theta$ always passes through the fixed point **[AIEEE 2004]**
(a) $(0, 0)$ (b) $(0, a)$
(c) $(a, 0)$ (d) $(a, a)$

**51** If $x_1, x_2, x_3$ and $y_1, y_2, y_3$ are both in GP with the same common ratio, then the points $(x_1, y_1), (x_2, y_2)$ and $(x_3, y_3)$ **[AIEEE 2003]**
(a) lie on a straight line (b) lie on an ellipse
(c) lie on a circle (d) are vertices of a triangle

**52** If the equation of the locus of a point equidistant from the points $(a_1, b_1)$ and $(a_2, b_2)$ is $(a_1 - a_2)x + (b_1 - b_2)y + c = 0$, then the value of $c$ is **[AIEEE 2003]**
(a) $\frac{1}{2}(a_2^2 + b_2^2 - a_1^2 - b_1^2)$ (b) $a_1^2 - a_2^2 + b_1^2 - b_2^2$
(c) $\frac{1}{2}(a_1^2 + a_2^2 + b_1^2 + b_2^2)$ (d) $\sqrt{a_1^2 + b_1^2 - a_2^2 - b_2^2}$

**53** A triangle with vertices $(4, 0)$, $(-1, -1)$, $(3, 5)$ is **[AIEEE 2002]**
(a) isosceles and right angled
(b) isosceles but not right angled
(c) right angled but not isosceles
(d) neither right angled nor isosceles

**54** Locus of mid-point of the portion between the axes of $x\cos\alpha + y\sin\alpha = p$ where $p$ is constant is **[AIEEE 2002]**
(a) $x^2 + y^2 = \frac{4}{p^2}$ (b) $x^2 + y^2 = 4p^2$
(c) $\frac{1}{x^2} + \frac{1}{y^2} = \frac{2}{p^2}$ (d) $\frac{1}{x^2} + \frac{1}{y^2} = \frac{4}{p^2}$

**55** A straight line through the point $(2, 2)$ intersects the lines $\sqrt{3}x + y = 0$ and $\sqrt{3}x - y = 0$ at the points $A$ and $B$. The equation to the line $AB$ so that the $\Delta OAB$ is equilateral, is **[AIEEE 2002]**
(a) $x - 2 = 0$ (b) $y - 2 = 0$
(c) $x + y - 4 = 0$ (d) None of the above

## Level 2

**56** Let the equation $x(x + 2)(12 - k) = 2$ have equal roots. Then, the distance of the point $\left(k, \frac{k}{2}\right)$ from the line $3x + 4y + 5 = 0$ is **[2025, 3 April Shift-II]**
(a) 15 (b) $5\sqrt{3}$
(c) $15\sqrt{5}$ (d) 12

**57** Consider the lines $x(3\lambda + 1) + y(7\lambda + 2) = 17\lambda + 5$, $\lambda$ being a parameter, all passing through a point $P$. One of these lines (say $L$) is farthest from the origin. If the distance of $L$ from the point $(3, 6)$ is $d$, then the value of $d^2$ is **[2025, 3 April Shift-II]**
(a) 20 (b) 30 (c) 10 (d) 15

**Imp.** Number of terms in a Binomial Expansion

**58** For an integer $n \geq 2$, if the arithmetic mean of all coefficients in the binomial expansion of $(x + y)^{2n-3}$ is 16, then the distance of the point $P$ $(2n - 1, n^2 - 4n)$ from the line $x + y = 8$ is **[2025, 4 April Shift-I]**
(a) $5\sqrt{2}$ (b) $3\sqrt{2}$ (c) $\sqrt{2}$ (d) $2\sqrt{2}$

**59** Let $ABC$ be the triangle such that the equations of lines $AB$ and $AC$ be $3y - x = 2$ and $x + y = 2$, respectively, and the points $B$ and $C$ lie on $X$-axis. If $P$ is the orthocentre of the $\Delta ABC$, then the area of the $\Delta PBC$ is equal to **[2025, 7 April Shift-I]**
(a) 4 (b) 6 (c) 10 (d) 8

**60** If the orthocenter of the triangle formed by the lines $y = x + 1$, $y = 4x - 8$ and $y = mx + c$ is at $(3, -1)$, then $m - c$ is **2025, 7 April Shift-II]**
(a) $-2$ (b) 4 (c) 2 (d) 0

**61** Let $a$ be the length of a side of a square $OABC$ with $O$ being the origin. Its side $OA$ makes an acute angle $\alpha$ with the positive $X$-axis and the equations of its diagonals are $(\sqrt{3}+1)x+(\sqrt{3}-1)y=0$ and $(\sqrt{3}-1)x-(\sqrt{3}+1)y+8\sqrt{3}=0$. Then, $a^2$ is equal to *[2025, 8 April Shift-II]*

(a) 16 (b) 48 (c) 32 (d) 24

**62** Let the point $\left(\frac{11}{2}, \alpha\right)$ lie on or inside the triangle with sides $x+y=11$, $x+2y=16$ and $2x+3y=29$. Then, the product of the smallest and the largest values of $\alpha$ is equal to *[2025, 24 Jan Shift-II]*

(a) 44 (b) 22
(c) 55 (d) 33

**Imp.** Image of a Point on the Line

**63** Let the lines $3x-4y-\alpha=0$, $8x-11y-33=0$ and $2x-3y+\lambda=0$ be concurrent. If the image of the point (1, 2) in the line $2x-3y+\lambda=0$ is $\left(\frac{57}{13}, \frac{-40}{13}\right)$, then $|\alpha\lambda|$ is equal to *[2025, 24 Jan Shift-I]*

(a) 84 (b) 113 (c) 101 (d) 91

**64** Let the range of the function

$$f(x)=6+16\cos x\cdot\cos\left(\frac{\pi}{3}-x\right)\cdot\cos\left(\frac{\pi}{3}+x\right)\cdot\sin 3x\cdot\cos 6x,\ x\in R$$

be $[\alpha,\beta]$. Then, the distance of the point $(\alpha,\beta)$ from the line $3x+4y+12=0$ is *[2025, 23 Jan Shift-II]*

(a) 9 (b) 10
(c) 8 (d) 11

**65** A rod of length eight units moves such that its ends $A$ and $B$ always lie on the lines $x-y+2=0$ and $y+2=0$, respectively. If the locus of the point $P$ that divides the rod $AB$ internally in the ratio 2:1 is $9(x^2+\alpha y^2+\beta xy+\gamma x+28y)-76=0$, then $\alpha-\beta-\gamma$ is equal to *[2025, 23 Jan Shift-II]*

(a) 22 (b) 24
(c) 23 (d) 21

**66** Let two straight lines drawn from the origin $O$ intersect the line $3x+4y=12$ at the points $P$ and $Q$ such that $\Delta OPQ$ is an isosceles triangle and $\angle POQ=90°$. If $l=OP^2+PQ^2+QO^2$, then the greatest integer less than or equal to $l$ is *[2024, 5 April Shift-I]*

(a) 44 (b) 42 (c) 46 (d) 48

**67** The distance of the point (2, 3) from the line $2x-3y+28=0$, measured parallel to the line $\sqrt{3}x-y+1=0$ is equal to *[2024, 29 Jan Shift-II]*

(a) $6\sqrt{3}$ (b) $3+4\sqrt{2}$ (c) $4\sqrt{2}$ (d) $4+6\sqrt{3}$

**68** Let $R$ be the interior region between the lines $3x-y+1=0$ and $x+2y-5=0$ containing the origin. The set of all values of $a$, for which the points $(a^2, a+1)$ lie in $R$, is *[2024, 27 Jan Shift-II]*

(a) $(-3,-1)\cup\left(\frac{1}{3},1\right)$ (b) $(-3,0)\cup\left(\frac{2}{3},1\right)$

(c) $(-3,-1)\cup\left(-\frac{1}{3},1\right)$ (d) $(-3,0)\cup\left(\frac{1}{3},1\right)$

**69** The portion of the line $4x+5y=20$ in the first quadrant is trisected by the lines $L_1$ and $L_2$ passing through the origin. The tangent of an angle between the lines $L_1$ and $L_2$ is *[2024, 27 Jan Shift-I]*

(a) $\frac{25}{41}$ (b) $\frac{30}{41}$ (c) $\frac{2}{5}$ (d) $\frac{8}{5}$

**70** If $(\alpha,\beta)$ is the orthocenter of the $\Delta ABC$ with vertices $A(3,-7)$, $B(-1,2)$ and $C(4,5)$, then $9\alpha-6\beta+60$ is equal to *[2023, 15 April Shift-I]*

(a) 30 (b) 35
(c) 40 (d) 25

**71** If the point $\left(\alpha, \frac{7\sqrt{3}}{3}\right)$ lies on the curve traced by the mid-points of the line segments of the lines $x\cos\theta+y\sin\theta=7$, $\theta\in\left(0,\frac{\pi}{2}\right)$ between the coordinate axes, then $\alpha$ is equal to *[2023, 12 April Shift-I]*

(a) $7\sqrt{3}$ (b) 7
(c) $-7\sqrt{3}$ (d) $-7$

**72** The equations of the sides $AB$, $BC$ and $CA$ of a $\Delta ABC$ are $2x+y=0$, $x+py=39$ and $x-y=3$ respectively and $P(2,3)$ is its circumcentre. Then, which of the following is not true? *[2022, 27 July Shift-II]*

(a) $(AC)^2=9p$ (b) $(AC)^2+p^2=136$
(c) $32<\text{area}(\Delta ABC)<36$ (d) $34<\text{area}(\Delta ABC)<38$

**73** A point $P$ moves so that the sum of squares of its distances from the points (1, 2) and (–2, 1) is 14. Let $f(x,y)=0$ be the locus of $P$, which intersects the $X$-axis at the points $A$ and $B$ and the $Y$-axis at the points $C$ and $D$. Then, the area of the quadrilateral $ABCD$ is equal to *[2022, 26 July Shift-I]*

(a) $\frac{9}{2}$ (b) $\frac{3\sqrt{17}}{2}$ (c) $\frac{3\sqrt{17}}{4}$ (d) 9

**74** The distance between the two points $A$ and $A'$ which lie on $y=2$ such that both the line segments $AB$ and $A'B$ (where $B$ is the point (2, 3)) subtend angle $\frac{\pi}{4}$ at the origin, is equal to *[2022, 29 June Shift-I]*

(a) 10 (b) $\frac{48}{5}$ (c) $\frac{52}{5}$ (d) 3

**75** Let $a, b$ and $c$ be the length of sides of a $\Delta ABC$ such that $\frac{a+b}{7}=\frac{b+c}{8}=\frac{c+a}{9}$. If $r$ and $R$ are the radius of incircle and radius of circumcircle of the $\Delta ABC$, respectively, then the value of $R/r$ is equal to **[2022, 25 June Shift-I]**

(a) $\frac{5}{2}$ (b) 2 (c) $\frac{3}{2}$ (d) 1

**76** Two sides of a parallelogram are along the lines $4x+5y=0$ and $7x+2y=0$. If the equation of one of the diagonals of the parallelogram is $11x+7y=9$, then other diagonal passes through the point **[2021, 27 July Shift-II]**

(a) (1, 2) (b) (2, 2) (c) (2, 1) (d) (1, 3)

**77** The equation of one of the straight lines which passes through the point (1,3) and makes an angle $\tan^{-1}(\sqrt{2})$ with the straight line, $y+1=3\sqrt{2}x$ is **[2021, 18 March Shift-I]**

(a) $4\sqrt{2}x+5y-(15+4\sqrt{2})=0$

(b) $5\sqrt{2}x+4y-(15+4\sqrt{2})=0$

(c) $4\sqrt{2}x+5y-4\sqrt{2}=0$

(d) $4\sqrt{2}x-5y-(5+4\sqrt{2})=0$

**78** Let $L$ denote the line in the $XY$-plane with $x$ and $y$-intercepts as 3 and 1 respectively. Then, the image of the point $(-1,-4)$ in this line is **[2020, 6 Sep Shift-II]**

(a) $\left(\frac{11}{5},\frac{28}{5}\right)$ (b) $\left(\frac{29}{5},\frac{8}{5}\right)$

(c) $\left(\frac{8}{5},\frac{29}{5}\right)$ (d) $\left(\frac{29}{5},\frac{11}{5}\right)$

**79** A $\Delta ABC$ lying in the first quadrant has two vertices as $A(1,2)$ and $B(3,1)$. If $\angle BAC=90°$, and ar$(\Delta ABC)=5\sqrt{5}$ sq units, then the abscissa of the vertex $C$ is **[2020, 4 Sep Shift-I]**

(a) $2+\sqrt{5}$ (b) $1+2\sqrt{5}$ (c) $2\sqrt{5}-1$ (d) $1+\sqrt{5}$

**80** If a $\Delta ABC$ has vertices $A(-1,7)$, $B(-7,1)$ and $C(5,-5)$, then its orthocentre has coordinates **[2020, 3 Sep Shift-II]**

(a) $(3,-3)$ (b) $(-3,3)$

(c) $\left(-\frac{3}{5},\frac{3}{5}\right)$ (d) $\left(\frac{3}{5},-\frac{3}{5}\right)$

**81** A rectangle is inscribed in a circle with a diameter lying along the line $3y=x+7$. If the two adjacent vertices of the rectangle are $(-8,5)$ and $(6,5)$, then the area of the rectangle (in sq. units) is : **[2019, 09 April Shift-II]**

(a) 84 (b) 98 (c) 72 (d) 56

**82** The foot of the perpendicular drawn from the origin, on the line, $3x+y=\lambda\ (\lambda\neq 0)$ is $P$. If the line meets $X$-axis at $A$ and $Y$-axis at $B$, then the ratio $BP:PA$ is **[JEE Main 2018, 15]**

(a) 9 : 1 (b) 1 : 3 (c) 1 : 9 (d) 3 : 1

**83** Let $k$ be an integer such that triangle with vertices $(k,-3k)$, $(5,k)$ and $(-k,2)$ has area 56 sq units. Then the orthocentre of this triangle is at the point **[JEE Main 2017]**

(a) $\left(1,\frac{3}{4}\right)$ (b) $\left(1,-\frac{3}{4}\right)$ (c) $\left(2,\frac{1}{2}\right)$ (d) $\left(2,-\frac{1}{2}\right)$

**84** Two sides of a rhombus are along the lines, $x-y+1=0$ and $7x-y-5=0$. If its diagonals intersect at $(-1,-2)$, then which one of the following is a vertex of this rhombus? **[JEE Main 2016]**

(a) $(-3,-9)$ (b) $(-3,-8)$ (c) $\left(\frac{1}{3},-\frac{8}{3}\right)$ (d) $\left(-\frac{10}{3},-\frac{7}{3}\right)$

**85** A ray of light is incident along line which meets another line, $7x-y+1=0$, at the point (0, 1). The ray is then reflected from this point along the line, $y+2x=1$. Then the equation of the line of incidence of the ray of light is **[JEE Main 2016]**

(a) $41x-25y+25=0$ (b) $41x+25y-25=0$

(c) $41x-38y+38=0$ (d) $41x+38y-38=0$

**86** The base of an equilateral triangle is along the line given by $3x+4y=9$. If a vertex of the triangle is (1, 2), then the length of a side of the triangle is **[JEE Main 2014]**

(a) $\frac{2\sqrt{3}}{15}$ (b) $\frac{4\sqrt{3}}{15}$ (c) $\frac{4\sqrt{3}}{5}$ (d) $\frac{2\sqrt{3}}{5}$

**87** If a line intercepted between the coordinate axes is trisected at point $A(4,3)$, which is nearer to $X$-axis, then its equation is **[JEE Main 2014]**

(a) $4x-3y=7$ (b) $3x+2y=18$

(c) $3x+8y=36$ (d) $x+3y=13$

**88** If the extremities of the base of an isosceles triangle are the points $(2a,0)$ and $(0,a)$ and the equation of one of the sides is $x=2a$, then the area of the triangle, in square units is **[JEE Main 2013]**

(a) $\frac{5}{4}a^2$ (b) $\frac{5}{2}a^2$ (c) $\frac{25a^2}{4}$ (d) $5a^2$

**89** If the image of point $P(2,3)$ in a line $L$ is $Q(4,5)$, then the image of point $R(0,0)$ in the same line is **[JEE Main 2013]**

(a) (2, 2) (b) (4, 5) (c) (3, 4) (d) (7, 7)

**90** The lines $L_1: y-x=0$ and $L_2: 2x+y=0$ intersect the line $L_3: y+2=0$ at $P$ and $Q$, respectively. The bisector of the acute angle between $L_1$ and $L_2$ intersects $L_3$ at $R$.

**Statement I** The ratio $PR:RQ$ equals $2\sqrt{2}:\sqrt{5}$.

**Statement II** In any triangle, bisector of an angle divides the triangle into two similar triangles. **[AIEEE 2011]**

(a) Statement I is true, Statement II is true; Statement II is not a correct explanation of Statement I

(b) Statement I is true, Statement II is false

(c) Statement I is false, Statement II is true

(d) Statement I is true, Statement II is true; Statement II is a correct explanation of Statement I

**91** The curve that passes through the point (2, 3) and has the property that the segment of any tangent to it lying between the coordinate axes is bisected by the point of contact, is given by **[AIEEE 2011]**

(a) $\left(\frac{x}{2}\right)^2+\left(\frac{y}{3}\right)^2=2$ (b) $2y-3x=0$

(c) $y=6/x$ (d) $x^2+y^2=13$

**92** The line $L$ given by $\frac{x}{5}+\frac{y}{b}=1$ passes through the point (13, 32). The line $K$ is parallel to $L$ and has the equation $\frac{x}{c}+\frac{y}{3}=1$. Then, the distance between $L$ and $K$ is **[AIEEE 2010]**

(a) $\frac{23}{\sqrt{15}}$ (b) $\sqrt{17}$ (c) $\frac{17}{\sqrt{15}}$ (d) $\frac{23}{\sqrt{17}}$

**93** Let $P=(-1,0), Q=(0,0)$ and $R=(3,3\sqrt{3})$ be three points. The equation of the bisector of the $\angle PQR$ is **[AIEEE 2007]**

(a) $\sqrt{3}x+y=0$ (b) $x+\frac{\sqrt{3}}{2}y=0$

(c) $\frac{\sqrt{3}}{2}x+y=0$ (d) $x+\sqrt{3}y=0$

**94** A square of side $a$ lies above the $X$-axis and has one vertex at the origin. The side passing through the origin makes an angle $\alpha\left(0<\alpha<\frac{\pi}{4}\right)$ with the positive direction of $X$-axis. The equation of its diagonal not passing through the origin is **[AIEEE 2003]**

(a) $y(\cos\alpha-\sin\alpha)-x(\sin\alpha-\cos\alpha)=a$

(b) $y(\cos\alpha+\sin\alpha)+x(\sin\alpha-\cos\alpha)=a$

(c) $y(\cos\alpha+\sin\alpha)+x(\sin\alpha+\cos\alpha)=a$

(d) $y(\cos\alpha+\sin\alpha)+x(\cos\alpha-\sin\alpha)=a$

## Level 3

**95** A line passing through the point $P(a,0)$ makes an acute angle $\alpha$ with the positive $X$-axis. Let this line be rotated about the point $P$ through an angle $\frac{\alpha}{2}$ in the clock-wise direction. If in the new position, the slope of the line is $2-\sqrt{3}$ and its distance from the origin is $\frac{1}{\sqrt{2}}$, then the value of $3a^2\tan^2\alpha-2\sqrt{3}$ is **[2025, 8 April Shift-II]**

(a) 4 (b) 5 (c) 6 (d) 8

**96** Let the line $x+y=1$ meet the axes of $x$ and $y$ at $A$ and $B$, respectively. A right angled $\Delta AMN$ is inscribed in the $\Delta OAB$, where $O$ is the origin and the points $M$ and $N$ lie on the where $OB$ and $AB$, respectively. If the area of the $\Delta AMN$ is $\frac{4}{9}$ of the area of the $\Delta OAB$ and $AN:NB=\lambda:1$, then the sum of all possible values(s) of is $\lambda$. **[2025, 29 Jan Shift-II]**

(a) 2 (b) $\frac{5}{2}$ (c) $\frac{1}{2}$ (d) $\frac{13}{6}$

**Imp.** Image of the Centroid of Triangle in the Line

**97** Let $ABC$ be a triangle formed by the lines $7x-6y+3=0, x+2y-31=0$ and $9x-2y-19=0$. Let the point $(h, k)$ be the image of the centroid of $\Delta ABC$ in the line $3x+6y-53=0$. Then, $h^2+k^2+hk$ is equal to **[2025, 29 Jan Shift-I]**

(a) 37 (b) 36 (c) 47 (d) 40

**98** In a $\Delta ABC$, suppose $y=x$ is the equation of the bisector of the $\angle B$ and the equation of the side $AC$ is $2x-y=2$. If $2AB=BC$ and the points $A$ and $B$ are respectively (4, 6) and $(\alpha,\beta)$, then $\alpha+2\beta$ is equal to **[2024, 29 Jan Shift-I]**

(a) 42 (b) 45 (c) 48 (d) 39

**99** Let $C(\alpha,\beta)$ be the circumcenter of the triangle formed by the lines $4x+3y=69, 4y-3x=17$ and $x+7y=61$ Then, $(\alpha-\beta)^2+\alpha+\beta$ is equal to **[2023, 8 April Shift-I]**

(a) 16 (b) 17 (c) 18 (d) 15

**100** If the orthocentre of the triangle, whose vertices are (1, 2), (2, 3) and (3, 1) is $(\alpha,\beta)$, then the quadratic equation whose roots are $\alpha+4\beta$ and $4\alpha+\beta$, is **[2023, 31 JanShift-I]**

(a) $x^2-20x+99=0$ (b) $x^2-19x+90=0$

(c) $x^2-18x+80=0$ (d) $x^2-22x+120=0$

**101** A straight line cuts off the intercepts $OA=a$ and $OB=b$ on the positive directions of $X$-axis and $Y$-axis, respectively. If the perpendicular from origin $O$ to this line makes an angle of $\pi/6$ with positive direction of $Y$-axis and the area of $\Delta OAB$ is $\frac{98}{3}\sqrt{3}$, then $a^2-b^2$ is equal to **[2023, 30 Jan Shift-I]**

(a) $\frac{392}{3}$ (b) $\frac{196}{3}$ (c) 196 (d) 98

**102** Let $m_1, m_2$ be the slopes of two adjacent sides of a square of side $a$ such that $a^2+11a+3(m_1^2+m_2^2)=220$. If one vertex of the square is $10(\cos\alpha-\sin\alpha)$, $10(\sin\alpha+\cos\alpha)$, where $\alpha\in\left(0,\frac{\pi}{2}\right)$ and the equation of one diagonal is $(\cos\alpha-\sin\alpha)x+(\sin\alpha+\cos\alpha)y=10$, then $72(\sin^4\alpha+\cos^4 x)+a^2-3a+13$ is equal to **[2022, 29 July Shift-II]**

(a) 119 (b) 128 (c) 145 (d) 155

**103** Let $A(\alpha, -2)$, $B(\alpha, 6)$ and $C\left(\frac{\alpha}{4}, -2\right)$ be vertices of a $\Delta ABC$. If $\left(5, \frac{\alpha}{4}\right)$ is the circumcentre of $\Delta ABC$, then which of the following is not correct about $\Delta ABC$ *[2022, 29 July Shift-II]*

(a) area is 24 units (b) perimeter is 25 units
(c) circumradius is 5 units (d) inradius is 2 units

**104** Let a triangle be bounded by the lines $L_1: 2x + 5y = 10$; $L_2: -4x + 3y = 12$ and the line $L_3$, which passes through the point $P(2, 3)$, intersects $L_2$ at $A$ and $L_1$ at $B$. If the point $P$ divides the line segment $AB$, internally in the ratio $1:3$, then the area of the triangle is equal to *[2022, 28 June Shift-II]*

(a) 110/13 (b) 132/13 (c) 142/13 (d) 151/13

**105** Let $A(-1, 1)$, $B(3, 4)$ and $C(2, 0)$ be given three points. A line $y = mx$, $m > 0$ intersects lines $AC$ and $BC$ at point $P$ and $Q$, respectively. Let $A_1$ and $A_2$ be the areas of $\Delta ABC$ and $\Delta PQC$, respectively, such that $A_1 = 3A_2$, then the value of $m$ is equal to *[2021, 16 March Shift-II]*

(a) 4/15 (b) 1 (c) 2 (d) 3

**106** Two sides of a parallelogram are along the lines, $x + y = 3$ and $x - y + 3 = 0$. If its diagonals intersect at (2, 4), then one of its vertex is *[2019, 10 Jan Shift-II]*

(a) (3, 6) (b) (2, 6) (c) (2, 1) (d) (3, 5)

**107** Let $A(h, k)$, $B(1, 1)$ and $C(2, 1)$ be the vertices of a right angled triangle with $AC$ as its hypotenuse. If the area of the triangle is 1, then the set of values which $k$ can take is given by *[AIEEE 2007]*

(a) {1, 3} (b) {0, 2} (c) {–1, 3} (d) {–3, –2}

## Numerical Value Type Questions

**108** If $\alpha = 1 + \sum_{r=1}^{6} (-3)^{r-1}\, {}^{12}C_{2r-1}$, then the distance of the point $(12, \sqrt{3})$ from the line $\alpha x - \sqrt{3}y + 1 = 0$ is ...... *[2025, 28 Jan Shift-I]*

**109** Let the distance between two parallel lines be 5 units and a point $P$ lie between the lines at a unit distance from one of them. An equilateral $\Delta PQR$ is formed such that $Q$ lies on one of the parallel lines, while $R$ lies on the other. Then, $(QR)^2$ is equal to ......... . *[2025, 22 Jan Shift-II]*

**Imp.** Orthocentre and Centroid of a Triangle

**110** Let $A(6, 8)$, $B(10\cos\alpha, -10\sin\alpha)$ and $C(-10\sin\alpha, 10\cos\alpha)$ be the vertices of a triangle. If $L(a, 9)$ and $G(h, k)$ be its orthocenter and centroid respectively, then $(5a - 3h + 6k + 100\sin 2\alpha)$ is equal to ......... . *[2025, 22 Jan Shift-II]*

**111** Let a ray of light passing through the point (3, 10) reflects on the line $2x + y = 6$ and the reflected ray passes through the point (7, 2). If the equation of the incident ray is $ax + by + 1 = 0$, then $a^2 + b^2 + 3ab$ is equal to ........... . *[2024, 8 April Shift-II]*

**112** Let $ABC$ be an isosceles triangle in which $A$ is at $(-1, 0)$, $\angle A = \frac{2\pi}{3}$, $AB = AC$ and $B$ is on the positive $X$-axis. If $BC = 4\sqrt{3}$ and the line $BC$ intersects the line $y = x + 3$ at $(\alpha, \beta)$, then $\frac{\beta^4}{\alpha^2}$ is ......... . *[2024, 1 Feb Shift-II]*

**113** Let the equations of two adjacent sides of a parallelogram $ABCD$ be $2x - 3y = -23$ and $5x + 4y = 23$. If the equation of its one diagonal $AC$ is $3x + 7y = 23$ and the distance of $A$ from the other diagonal is $d$, then $50\,d^2$ is equal to ........ . *[2023, 10 April Shift-II]*

**114** The equation of the sides $AB$, $BC$ and $CA$ of a $\Delta ABC$ are $2x + y = 0$, $x + py = 15a$ and $x - y = 3$, respectively. If its orthocentre is $(2, a)$, $-\frac{1}{2} < a < 2$, then $p$ is equal to ...... . *[2022, 26 July Shift-I]*

**115** Let $A\left(\frac{3}{\sqrt{a}}, \sqrt{a}\right)$, $a > 0$, be a fixed point in the $XY$-plane. The image of $A$ in $Y$-axis be $B$ and the image of $B$ in $X$-axis be $C$. If $D(3\cos\theta, a\sin\theta)$ is a point in the fourth quadrant such that the maximum area of $\Delta ACD$ is 12 sq units, then $a$ is equal to ...... . *[2022, 24 June Shift-I]*

**116** A man starts walking from the point $P(-3, 4)$, touches the $X$-axis at $R$, and then turns to reach at the point $Q(0, 2)$. The man is walking at a constant speed. If the man reaches the point $Q$ in the minimum time, then $50((PR)^2 + (RQ)^2)$ is equal to ......... . *[2021, 1 Sep Shift-II]*

**117** Let the points of intersections of the lines $x - y + 1 = 0$, $x - 2y + 3 = 0$ and $2x - 5y + 11 = 0$ are the mid points of the sides of a $\Delta ABC$. Then, the area of the $\Delta ABC$ is ......... . *[2021, 1 Sep Shift-II]*

**118** If the line, $2x - y + 3 = 0$ is at a distance $\frac{1}{\sqrt{5}}$ and $\frac{2}{\sqrt{5}}$ from the lines $4x - 2y + \alpha = 0$ and $6x - 3y + \beta = 0$, respectively, then the sum of all possible values of $\alpha$ and $\beta$ is ...... . *[2020, 5 Sep Shift-I]*

## Topic 2 Pair of Straight Lines

### Level 1

**01** Let the equation of the pair of lines, $y = px$ and $y = qx$ can be written as $(y - px)(y - qx) = 0$. Then, the equation of the pair of the angle bisectors of the line $x^2 - 4xy - 5y^2 = 0$ is **[2021, 25 July Shift-II]**

(a) $x^2 - 3xy + y^2 = 0$ (b) $x^2 + 4xy - y^2 = 0$
(c) $x^2 + 3xy - y^2 = 0$ (d) $x^2 - 3xy - y^2 = 0$

**02** If $x^2 - y^2 + 2hxy + 2gx + 2fy + c = 0$ is the locus of a point, which moves such that it is always equidistant from the lines $x + 2y + 7 = 0$ and $2x - y + 8 = 0$, then the value of $g + c + h - f$ equals **[2024, 30 Jan Shift-II]**

(a) 29 (b) 14 (c) 8 (d) 6

**03** If the curve $x^2 + 2y^2 = 2$ intersects the line $x + y = 1$ at two points $P$ and $Q$, then the angle subtended by the line segment $PQ$ at the origin is **[2021, 25 Feb Shift-II]**

(a) $\frac{\pi}{2} + \tan^{-1}\left(\frac{1}{4}\right)$ (b) $\frac{\pi}{2} - \tan^{-1}\left(\frac{1}{4}\right)$
(c) $\frac{\pi}{2} + \tan^{-1}\left(\frac{1}{3}\right)$ (d) $\frac{\pi}{2} - \tan^{-1}\left(\frac{1}{3}\right)$

**04** If one of the line of $my^2 + (1 - m^2)xy - mx^2 = 0$ is a bisector of the angle between lines $xy = 0$, then $m$ is **[AIEEE 2007]**

(a) $-\frac{1}{2}$ (b) $-2$ (c) $\pm 1$ (d) 2

**05** If the sum of the slopes of the lines given by $x^2 - 2cxy - 7y^2 = 0$ is four times their product, then $c$ has the value **[AIEEE 2004]**

(a) 1 (b) $-1$
(c) 2 (d) $-2$

**06** If one of the lines given by $6x^2 - xy + 4cy^2 = 0$ is $3x + 4y = 0$, then $c$ is equal to **[AIEEE 2004]**

(a) 1 (b) $-1$
(c) 3 (d) $-3$

**07** If the pair of straight lines $x^2 - 2pxy - y^2 = 0$ and $x^2 - 2qxy - y^2 = 0$ be such that each pair bisects the angle between the other pair, then **[AIEEE 2003]**

(a) $p = q$ (b) $p = -q$
(c) $pq = 1$ (d) $pq = -1$

**08** If the pair of lines $ax^2 + 2hxy + by^2 + 2gx + 2fy + c = 0$ intersect on the $Y$-axis, then **[AIEEE 2002]**

(a) $2fgh = bg^2 + ch^2$ (b) $bg^2 \neq ch^2$
(c) $abc = 2fgh$ (d) None of these

### Numerical Value Type Question

**09** If the line $l_1 : 3y - 2x = 3$ is the angular bisector of the lines $l_2 : x - y + 1 = 0$ and $l_3 : \alpha x + \beta y + 17 = 0$, then $\alpha^2 + \beta^2 - \alpha - \beta$ is equal to ......... . **[2023, 11 April Shift-II]**

## Key Index

### Topic 1

| | | | | | | | | | |
|---|---|---|---|---|---|---|---|---|---|
| 1 (d) | 2 (a) | 3 (b) | 4 (c) | 5 (a) | 6 (a) | 7 (d) | 8 (b) | 9 (b) | 10 (c) |
| 11 (b) | 12 (c) | 13 (a) | 14 (c) | 15 (c) | 16 (b) | 17 (c) | 18 (d) | 19 (b) | 20 (b) |
| 21 (a) | 22 (d) | 23 (c) | 24 (d) | 25 (c) | 26 (b) | 27 (c) | 28 (b) | 29 (b) | 30 (c) |
| 31 (d) | 32 (d) | 33 (c) | 34 (b) | 35 (b) | 36 (b) | 37 (a) | 38 (d) | 39 (c) | 40 (a) |
| 41 (d) | 42 (c) | 43 (d) | 44 (d) | 45 (c) | 46 (a) | 47 (d) | 48 (b) | 49 (b) | 50 (c) |
| 51 (a) | 52 (a) | 53 (a) | 54 (d) | 55 (b) | 56 (a) | 57 (a) | 58 (b) | 59 (b) | 60 (d) |
| 61 (b) | 62 (d) | 63 (d) | 64 (d) | 65 (c) | 66 (c) | 67 (d) | 68 (d) | 69 (b) | 70 (d) |
| 71 (b) | 72 (d) | 73 (b) | 74 (c) | 75 (a) | 76 (b) | 77 (a) | 78 (a) | 79 (b) | 80 (b) |
| 81 (a) | 82 (d) | 83 (c) | 84 (c) | 85 (c) | 86 (b) | 87 (b) | 88 (b) | 89 (d) | 90 (b) |
| 91 (c) | 92 (d) | 93 (a) | 94 (d) | 95 (a) | 96 (a) | 97 (a) | 98 (a) | 99 (b) | 100 (a) |
| 101 (a) | 102 (b) | 103 (b) | 104 (b) | 105 (b) | 106 (a) | 107 (c) | 108 (5) | 109 (28) | 110 (145) |
| 111 (1) | 112 (36) | 113 (529) | 114 (3) | 115 (8) | 116 (1250) | 117 (6) | 118 (30) | | |

### Topic 2

| | | | | | | | | |
|---|---|---|---|---|---|---|---|---|
| 1 (c) | 2 (b) | 3 (a) | 4 (c) | 5 (c) | 6 (d) | 7 (d) | 8 (a) | 9 (348) |

# Detailed **Explanations**

## Topic 1

**01.** *(d)* $ar(\Delta ABC)=5$

$$\frac{1}{2}\times bc=5$$

$$\Rightarrow \quad bc=10$$

$$ar(\Delta ACD)=6$$

$$=\frac{1}{2}\times cd=6\Rightarrow cd=12$$

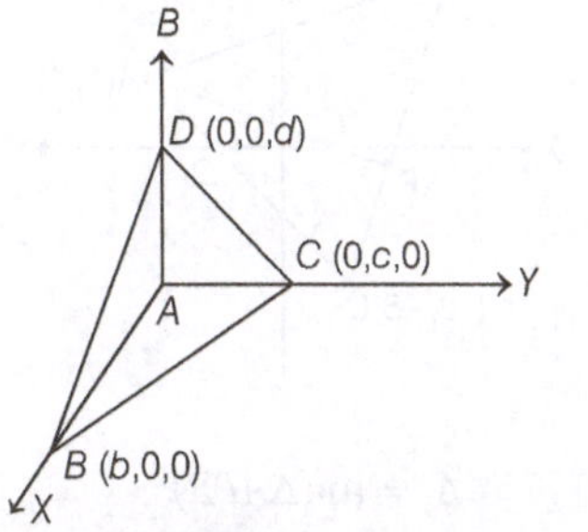


$$ar(\Delta ABD)=7$$

$$\Rightarrow \quad bd=14$$

$$\text{Area }(\Delta BCD)=\frac{1}{2}(\mathbf{BC}\times \mathbf{BD})$$

$$\mathbf{BC}=<-b,c,0>$$

$$\mathbf{BD}=<-b,0,d>$$

$$|\mathbf{BC}\times \mathbf{BD}|=\sqrt{c^2d^2+b^2d^2+b^2c^2}$$

$$=\sqrt{440}$$

$$ar(\Delta BCD)=\sqrt{110}$$

**02.** *(a)* $L:x+by=c$

Slope of this equation is $=-\frac{1}{b}$

Given, $-\frac{1}{b}=\tan 135°$

$$\Rightarrow \quad b=1$$

$$\therefore \quad x+y=c$$

Now, area of the $\Delta$ formed by line $L$ and coordinate axes.

$$=\frac{c^2}{2}=48$$

$$\Rightarrow \quad c^2=96$$

$$\Rightarrow \quad b^2+c^2=97$$

**03.** *(b)* The line passing through origin and making equal angles with the axes is

$$y=x\Rightarrow 2x+x+6$$

$$\Rightarrow \quad x=-2,\ y=-2$$

$$\therefore \quad A\equiv(-2,-2)$$

Also, $4x+2x-p=0\Rightarrow x=p/6$

$$\Rightarrow \quad B=\left(\frac{p}{6},\frac{p}{6}\right)$$

Given, $AB=\frac{9}{\sqrt{2}}$

$$\Rightarrow \quad \sqrt{\left(\frac{p}{6}+2\right)^2+\left(\frac{p}{6}+2\right)^2}=\frac{9}{\sqrt{2}}$$

$$\Rightarrow \quad \left(\frac{p}{6}+2\right)\sqrt{2}=\frac{9}{\sqrt{2}}\Rightarrow\frac{p+12}{6}=\frac{9}{2}$$

$$p=15\Rightarrow B\equiv\left(\frac{5}{2},\frac{5}{2}\right)$$

$$\text{AM}=\frac{|4(-2)+2(-2)-15|}{\sqrt{4^2+2^2}}=\frac{27}{2\sqrt{5}}$$

$$\because BM^2=AB^2-AM^2$$

$$=\left(\frac{9\sqrt{2}}{2}\right)^2-\left(\frac{27}{2\sqrt{5}}\right)^2=\frac{81}{20}$$

$$\Rightarrow \quad BM=\frac{9}{2\sqrt{5}}$$

$$\frac{AM}{BM}=\frac{27}{2\sqrt{5}}\times\frac{2\sqrt{5}}{9}=3$$

**04.** *(c)* $\because \cos|x|$ is always differentiable

$\therefore$ We have to check only for $|x^2-ax+2|$

$\therefore$ Not differentiable at $x^2-ax+2=0$

One root is given $\alpha=2$

$\therefore \quad 4-2a+2=0\Rightarrow a=3$

$\Rightarrow$ Other root $\beta=1$

But for $x=1$, $f(x)$ is differentiable

So, it is bonus

$\therefore$ Distance from $(\alpha,\beta)$ i.e. $(2,1)$ to line $12x+5y+10=0$

$$=\left|\frac{12(2)+5(1)+10}{\sqrt{144+25}}\right|=\frac{39}{13}=3$$

**05.** *(a)* Given, lines $L_1:-x+2y=4$ and $L_2:x+y=4$

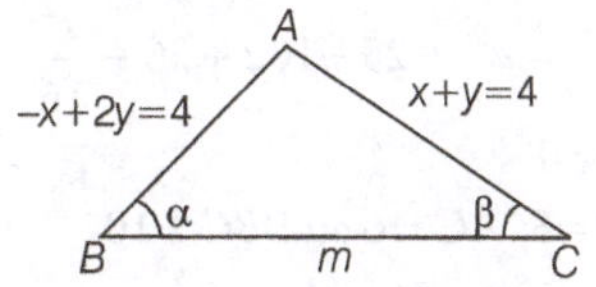


Slope, $m_1=\frac{1}{2}$ and $m_2=-1$

In $\Delta ABC$, $\tan\alpha=\tan\beta$

$$\frac{m-m_1}{1+mm_1}=\frac{m_2-m}{1+mm_2}$$

$$\Rightarrow \quad \frac{m-\frac{1}{2}}{1+\frac{1}{2}m}=\frac{-1-m}{1-m}$$

$$\Rightarrow \quad \frac{2m-1}{2+m}=\frac{-m-1}{1-m}$$

$$\Rightarrow \quad m^2-6m-1=0$$

$$\Rightarrow \quad m=\frac{6\pm\sqrt{40}}{2}\Rightarrow m=3\pm\sqrt{10}$$

$\therefore$ Sum $=6$

**06.** *(a)*

> **key idea** Distance of point $P(x_1,y_1)$ from $ax+by+c=0$ is
>
> $$\left|\frac{ax_1+by_1+c}{\sqrt{a^2+b^2}}\right|$$

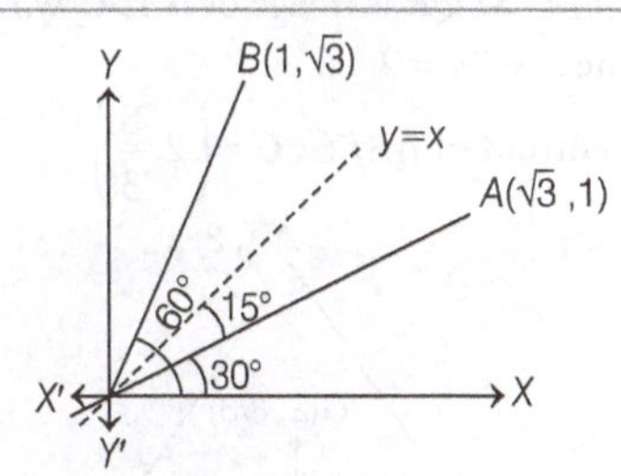


Distance of point $(a,1-a,0)$ from line $y-x=0$ is $\frac{9}{\sqrt{2}}$

$$\Rightarrow \left|\frac{1-a-a}{\sqrt{2}}\right|=\frac{9}{\sqrt{2}}$$

$$\Rightarrow 1-2a=\pm 9\Rightarrow 1\pm 9=2a$$

$$\Rightarrow \quad -8,10=2a\Rightarrow a=-4,5$$

$$\therefore \quad \text{Sum}=-4+5=1$$

**07.** *(d)* Given,

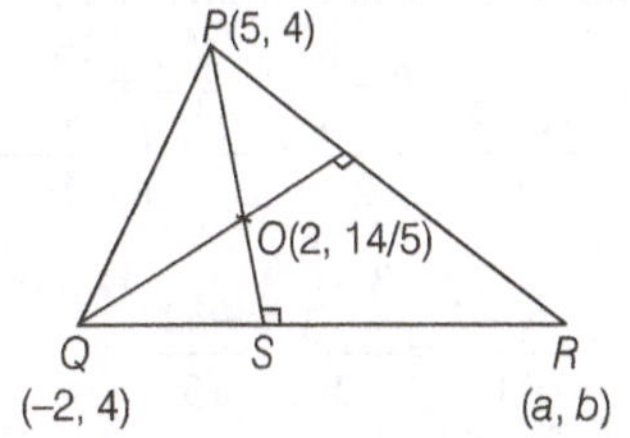

$\Rightarrow$ Slope of $PS = \dfrac{\frac{14}{5}-4}{2-5} = \dfrac{-6}{5\times(-3)} = \dfrac{2}{5}$

and Slope of $QR = \dfrac{-5}{2}$

$\Rightarrow$ Equation of $QR$ is $y-4 = \dfrac{-5}{2}(x+2)$

$$5x+2y+2=0 \quad \ldots\text{(i)}$$

Similarly, equation of line $PR$

$$10x-3y-38=0 \quad \ldots\text{(ii)}$$

From, (i) and (ii), we get

$$10x+4y+4=0$$
$$\underline{\begin{matrix}10x-3y-38=0\\ -\quad+\quad+\end{matrix}}$$
$$7y+42 \Rightarrow y=-6$$
$$y=-6$$
$$\Rightarrow \quad x=2$$

Point $R \equiv (2,-6) = (a,b)$ [given]

$\Rightarrow a=2, b=-6$

Centroid

$$= \left(\frac{5-2+2}{3}, \frac{4+4-6}{3}\right) = \left(\frac{5}{3}, \frac{2}{3}\right)$$

$= (c, d)$ [given]

$\Rightarrow c = \dfrac{5}{3}, d = \dfrac{2}{3}$

$\therefore \quad c+2d = \dfrac{5}{3}+\dfrac{4}{3} = 3$

**08.** *(b)* $\because \Delta PQR$ is image of $\Delta ABC$ w.r.t. line $x+2y=2$

Centroid of $\Delta ABC$, $G = \left(2, \dfrac{8}{3}\right)$

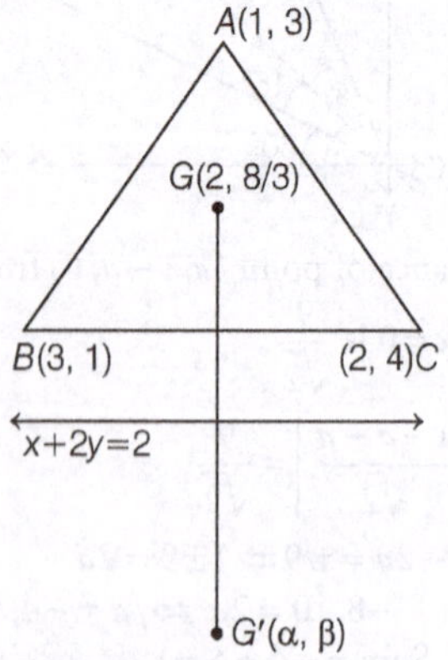


$\therefore$Centroid of $\Delta PQR$ is $G'(\alpha,\beta)$

That is image of $G$ w.r.t. $x+2y-2=0$

$$\frac{\alpha-2}{1} = \frac{\beta-\frac{8}{3}}{2} = \frac{-2\left(2+\frac{16}{3}-2\right)}{1+4}$$

$$\Rightarrow \quad \frac{\alpha-2}{1} = \frac{\beta-\frac{8}{3}}{2} = \frac{-32}{15}$$

$$\Rightarrow \quad \alpha = 2-\frac{32}{15} = -\frac{2}{15}$$

$$\Rightarrow \quad \beta = \frac{8}{3}-\frac{64}{15} = -\frac{24}{15}$$

$\therefore\ 15(\alpha-\beta) = 22$

> **Time Saver Tech**
>
> Let the image of a point $(x_1, y_1)$ with respect to $ax+by+c=0$ be $(x_2, y_2)$, then
>
> $$\frac{x_2-x_1}{a} = \frac{y_2-y_1}{b} = \frac{-2(ax_1+by_1+c)}{a^2+b^2}$$

**09.** *(b)* Let the point be $(h, k)$

According to the question,

$$\frac{\sqrt{(h-2)^2+(k-1)^2}}{\sqrt{(h-1)^2+(k-3)^2}} = \frac{5}{4}$$

By squaring and simplification,

$$16[(h-2)^2+(k-1)^2] = 25[(h-1)^2+(k-3)^2]$$

$$\Rightarrow 16[h^2+4-4h+k^2+1-2k] = 25[h^2+1-2h+k^2+9-6k]$$

$$\Rightarrow 9h^2+9k^2+14h-118k+170=0$$

Now, replace $(h, k)$ with $(x, y)$, so the locus is

$$9x^2+9y^2+14x-118y+170=0$$

Now, comparing it with

$$ax^2+by^2+cxy+dx+ey+170=0,$$

we have

$a=9, b=9, c=0, d=14, e=-118$

So, the value of

$$a^2+2b+3c+4d+e$$
$$=81+18+0+56-118=37$$

**10.** *(c)* We have,

$A(a,-2)$, $B(a,6)$ and $C\left(\dfrac{a}{4},-2\right)$

and $O\left(5, \dfrac{a}{4}\right)$

$$\therefore\ AO = BO \Rightarrow (a-5)^2 + \left(-2-\frac{a}{4}\right)^2$$
$$= (a-5)^2 + \left(6-\frac{a}{4}\right)^2$$
$$\Rightarrow a^2+25-10a+4+\frac{a^2}{16}+a$$
$$= a^2+25-10a+36+\frac{a^2}{16}-3a$$

$\Rightarrow a=8$

$AB=8$, $AC=6$ and $BC=10$

$\therefore\ \alpha=5, \beta=24$ and $\gamma=24$

$$\therefore \alpha+\beta+\gamma = 5+24+24$$
$$=53$$

**11.** *(b)* Given point $(\alpha,-3)$, $(2,0)$ and $(1,\alpha)$.

For collinear,

$$\left(\frac{\alpha-0}{-1}\right) = \frac{\alpha+3}{1-\alpha}$$

$$\Rightarrow \quad \alpha = -1, 3$$

$$\Rightarrow \quad \alpha_1 = -1, \alpha_2 = 3$$

Equation is $(y-y_1) = m(x-x_1)$

$$\Rightarrow \quad y-3 = \sqrt{3}(x+1)$$

**12.** *(c)*

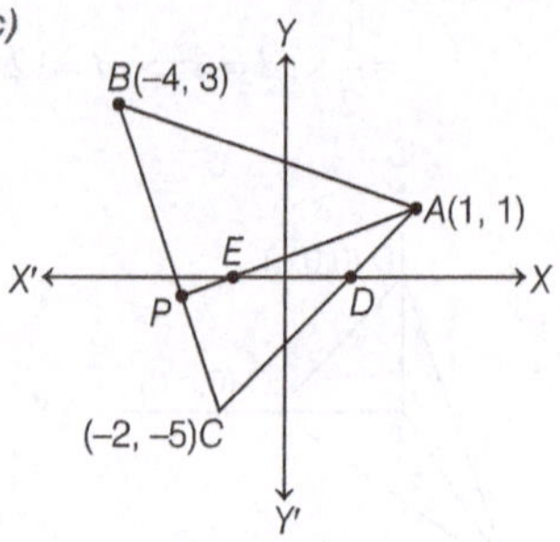


$$\Delta_1 = \text{ar}(\Delta APB),$$
$$\Delta_2 = \text{ar}(\Delta ABC)$$

Now, $\dfrac{\Delta_1}{\Delta_2} = \dfrac{4}{7}$

$$\Rightarrow \quad \frac{\text{ar }(\Delta PAB)}{\text{ar }(\Delta APC)} = \frac{4}{3}$$

$$\therefore \quad \frac{BP}{PC} = \frac{4}{3}$$

$$\therefore\ P = \left(\frac{(-4)\times 3+(-2)\times 4}{3+4}, \frac{3\times 3+(-5)\times 4}{3+4}\right)$$

$$\Rightarrow \quad P = \left(\frac{-20}{7}, \frac{-11}{7}\right)$$

Let $D$ be $(h, 0)$

$$\Rightarrow \quad \frac{1-0}{1-h} = \frac{-5-0}{-2-h}$$

$$\Rightarrow \quad D = \left(\frac{1}{2}, 0\right)$$

Let $E$ be $(k, 0)$

$$\Rightarrow \quad \frac{1-0}{1-k} = \frac{\frac{-11}{7}-0}{\frac{-20}{7}-k}$$

$$\Rightarrow \quad E = \left(\frac{-1}{2}, 0\right)$$

$$\therefore\ \text{ar}(\Delta AED) = \frac{1}{2}\times 1\times 1 = \frac{1}{2}$$

**13.** *(a)* $p = \dfrac{k\cot 2\alpha}{\sqrt{\operatorname{cosec}^2\alpha + \sec^2\alpha}}$

and $q = \dfrac{k\sin 2\alpha}{\sqrt{\sin^2\alpha + \cos^2\alpha}}$

$\Rightarrow \quad p = \dfrac{k\left(\dfrac{\cos 2\alpha}{\sin 2\alpha}\right)}{\sqrt{\dfrac{\sin^2\alpha + \cos^2\alpha}{\sin^2\alpha\cos^2\alpha}}}$

$= \dfrac{k\cos 2\alpha}{\sin 2\alpha} \times \sin\alpha\cos\alpha$

$\Rightarrow \quad p = \left(\dfrac{k}{2}\right)\cos 2\alpha$ and $q = k\sin 2\alpha$

$\Rightarrow \cos 2\alpha = (2p/k)$ and $\sin 2\alpha = (q/k)$

$\Rightarrow \sin^2 2\alpha + \cos^2 2\alpha = 1$

$\Rightarrow \quad \dfrac{4p^2}{k^2} + \dfrac{q^2}{k^2} = 1 \Rightarrow 4p^2 + q^2 = k^2$

**14.** *(c)* Given, $A(0, 6)$ and $B(2t, 0)$

Mid-point of $AB = M(t, 3)$

Equation of perpendicular bisector of $AB$ passes through $M$.

$\therefore \quad y - 3 = \dfrac{t}{3}(x - t) \quad \ldots \text{(i)}$

Intersection of Eq. (i) on $Y$-axis

$C\left(0, 3 - \dfrac{t^2}{3}\right)$

Let mid-point of $MC$ is $(h, k)$.

Then, $\quad (h, k) = \left(\dfrac{t}{2}, 3 - \dfrac{t^2}{6}\right)$

$\Rightarrow \quad h = \dfrac{t}{2}, k = 3 - \dfrac{t^2}{6}$

Eliminating $t$, we get

$2h^2 = 3(3 - k)$

Locus of $(h, k)$

$2x^2 = 3(3 - y) \Rightarrow 2x^2 + 3y - 9 = 0$

**15.** *(c)*

**key idea** (i) Mid points of $(x_1, y_1)$ and $(x_2, y_2)$ is
$\left[\left(\dfrac{x_1 + x_2}{2}\right), \left(\dfrac{y_1 + y_2}{2}\right)\right]$

(ii) Slope of line joining $(x_1, y_1)$ and $(x_2, y_2) = \dfrac{(y_2 - y_1)}{(x_2 - x_1)}$

(iii) $\tan\theta = \left[\dfrac{m_2 - m_1}{1 + m_2 m_1}\right]$

Given, the equation of median through $B$ i.e. $BE: 2x + y - 3 = 0$

Equation, of angle bisector of $C$ i.e. $CD: 7x - 4y = 1$

Since, $E$ satisfies the equation of $BE$.

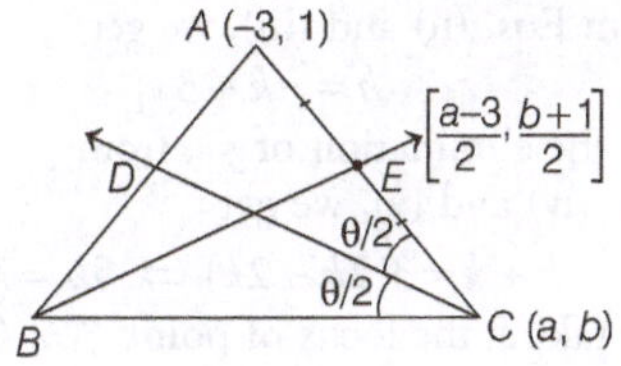


$\therefore \quad 2\left(\dfrac{a-3}{2}\right) + \left(\dfrac{b+1}{2}\right) - 3 = 0$

$2a - 6 + b + 1 - 6 = 0$

$2a + b = 11 \quad \ldots \text{(i)}$

Since, $C$ satisfies $CD$.

$\therefore \quad 7a - 4b = 1 \quad \ldots \text{(ii)}$

On solving Eqs. (i) and (ii), we get

$a = 3, b = 5$

Slope of $AC = 2/3$, Slope of $CD = 7/4$

$\therefore \quad \tan\left(\dfrac{\theta}{2}\right) = \left|\dfrac{\dfrac{2}{3} - \dfrac{7}{4}}{1 + \dfrac{14}{12}}\right| = \dfrac{1}{2}$

Now, $\tan\theta = \dfrac{2\tan\left(\dfrac{\theta}{2}\right)}{1 - \tan^2\left(\dfrac{\theta}{2}\right)} = \dfrac{2\left(\dfrac{1}{2}\right)}{1 - \dfrac{1}{4}} = \dfrac{4}{3}$

**16.** *(b)* Let $O$ be the centre of the circumcircle.

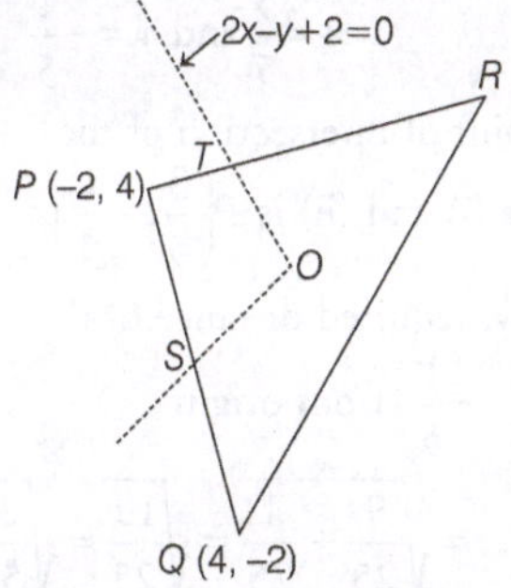


And $T$ be the mid-point of $PR$.

So, equation of $OT$ is given as

$2x - y + 2 = 0$

Let $S$ be the mid-point of $PQ$.

Now, $S$ will be $\left(\dfrac{-2 + 4}{2}, \dfrac{4 - 2}{2}\right) = (1, 1)$

Slope of the line $PQ = \dfrac{-2 - 4}{4 + 2} = -1$

Equation of $OS \ : y - 1 = 1(x - 1)$

$y = x$

Now, coordinates of $O$ will be the intersection of lines $OS$ and $OT$.

$\begin{cases} y = x \\ 2x - y + 2 = 0 \end{cases}$

$\Rightarrow \quad 2x - x + 2 = 0 \Rightarrow x = -2$

$\Rightarrow \quad y = -2 \Rightarrow O = (-2, -2)$

**17.** *(c)* Given lines, $x - y = 0$, $x + 2y = 3$, $2x + y = 6$

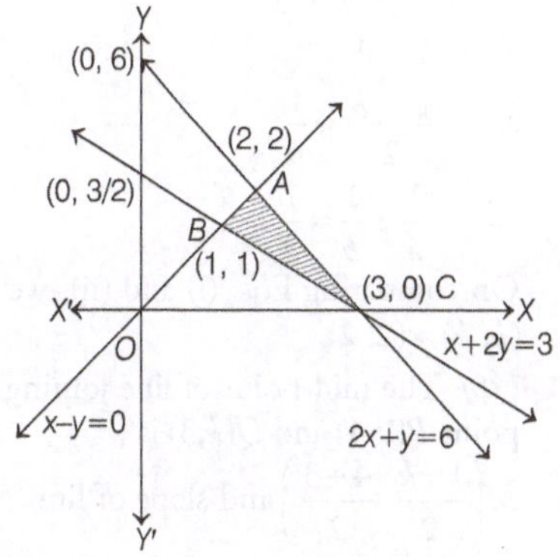


The only triangle which include all three lines is $\Delta ABC$.

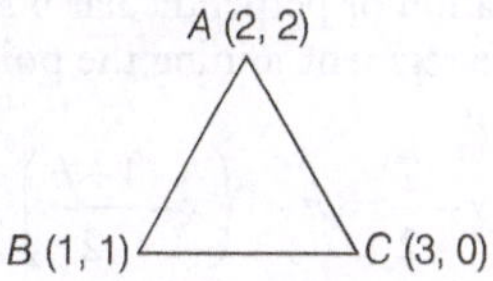


Now, $AB = \sqrt{(2-1)^2 + (2-1)^2} = \sqrt{2}$

$AC = \sqrt{(2-3)^2 + (2-0)^2} = \sqrt{5}$

$BC = \sqrt{(3-1)^2 + (0-1)^2} = \sqrt{5}$

$\Rightarrow AC = BC$ (two sides are equal)

$\Rightarrow \Delta ABC$ is isosceles triangle.

**18.** *(d)* Image of $P(3, 5)$ on the line $x - y + 1 = 0$ is

$\dfrac{x-3}{1} = \dfrac{y-5}{-1} = \dfrac{-2(3-5+1)}{2}$

$\Rightarrow \quad \dfrac{x-3}{1} = \dfrac{y-5}{-1} = 1$

$\Rightarrow \quad \dfrac{x-3}{1} = 1$ and $\dfrac{y-5}{-1} = 1$

$x = 4, y = 4$

$\therefore$ Required image is at $(4, 4)$.

Clearly, this point lies on $(x-2)^2 + (y-4)^2 = 4$ as $(4, 4)$ satisfies this equation.

**19.** *(b)* Given, position of $A = (1, 1)$

Position of $B = (2, 2)$

Position of $C = (4, 4)$

Let $x$-intercept be $a$ and $y$-intercept be $b$.

Equation of line traced is

$\dfrac{x}{a} + \dfrac{y}{b} = 1$

This is the equation of path, let a point $(h, k)$ lie on this path, then

$\dfrac{h}{a} + \dfrac{k}{b} = 1 \quad \ldots \text{(i)}$

Also, arithmetic mean of reciprocal of $a$ and $b = \frac{1}{4}$

$$\therefore \quad \frac{\frac{1}{a}+\frac{1}{b}}{2} = \frac{1}{4}$$

$$\Rightarrow \quad \frac{1}{a}+\frac{1}{b} = \frac{1}{2} \quad \ldots(\text{ii})$$

On comparing Eqs. (i) and (ii), we get $(h, k) = (2, 2)$.

**20.** *(b)* The mid-point of line joining point $P(1, 4)$ and $Q(k,3)$ is $M\left(\frac{1+k}{2}, \frac{4+3}{2}\right)$ and slope of line perpendicular to line $PQ$ is $-\frac{1-k}{4-3} = k-1$.

$\therefore$ Equation of perpendicular bisector of the line segment joining the points $P$ and $Q$ is

$$y - \frac{7}{2} = (k-1)\left(x - \frac{1+k}{2}\right) \quad \ldots(\text{i})$$

and line (i) passes through $(0, -4)$, so

$$-4 - \frac{7}{2} = (k-1)\left(-\frac{1+k}{2}\right)$$

$$\Rightarrow \quad 15 = k^2 - 1 \Rightarrow k = \pm 4$$

**21.** *(a)* Let a point $A(x_1, y_1)$ on the line $x = 2y$ and the middle point of the perpendicular drawn from point $A(x_1, y_1)$ to the line $x = y$ is $P(h, k)$.

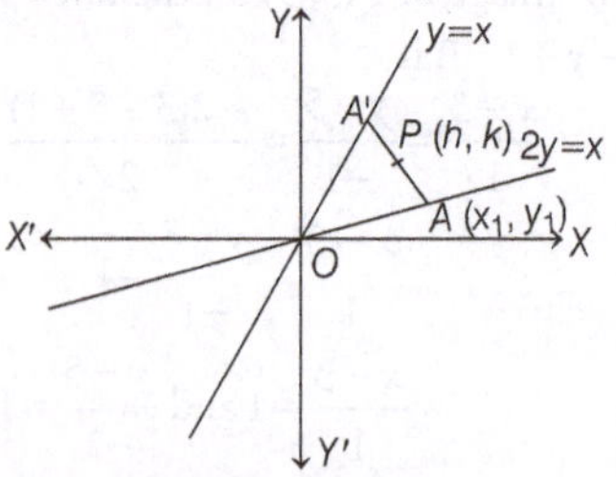


So, the coordinate of foot of perpendicular to line $y = x$ from point $A(x_1, y_1)$ is $A'(2h - x_1, 2k - y_1)$.

Since, the point $A'$ lies on line $y = x$, so we have

$$2h - x_1 = 2k - y_1 \quad \ldots(\text{i})$$

and as point $A$ lies on line $2y = x$, so we have

$$x_1 = 2y_1 \quad \ldots(\text{ii})$$

and as $PA$ is perpendicular to line $y = x$, so

$$\frac{y_1 - k}{x_1 - h} = -1$$

$$\Rightarrow \quad y_1 - k = -x_1 + h \quad \ldots(\text{iii})$$

From Eqs. (i) and (ii), we get

$$2h = 2k + y_1 \quad \ldots(\text{iv})$$

From Eqs. (ii) and (iii), we get

$$h = -k + 3y_1 \quad \ldots(\text{v})$$

On the elimination of $y_1$, from Eqs. (iv) and (v), we get

$$h + k = 3(2h - 2k) \Rightarrow 5h = 7k$$

On taking the locus of point $P(h, k)$, we get

$$5x - 7y = 0$$

**22.** *(d)* Given, lines $x + (a-1)y = 1$ and $2x + a^2 y = 1$, where $a \in R - \{0, 1\}$ are perpendicular to each other.

$$\therefore \quad \left(-\frac{1}{a-1}\right) \times \left(-\frac{2}{a^2}\right) = -1$$

[$\because$ If lines are perpendicular, then product of their slopes is $-1$]

$$\Rightarrow \quad a^2(a-1) = -2$$

$$\Rightarrow \quad a^3 - a^2 + 2 = 0$$

$$\Rightarrow \quad (a+1)(a^2 - 2a + 2) = 0$$

$$\Rightarrow \quad a = -1$$

$\therefore$ Equation of lines are

$$x - 2y = 1 \quad \ldots(\text{i})$$

and

$$2x + y = 1 \quad \ldots(\text{ii})$$

On solving Eqs. (i) and (ii), we get

$$x = \frac{3}{5} \text{ and } y = -\frac{1}{5}$$

$\therefore$ Point of intersection of the lines (i) and (ii) is $P\left(\frac{3}{5}, -\frac{1}{5}\right)$.

Now, required distance of the point $P\left(\frac{3}{5}, -\frac{1}{5}\right)$ from origin

$$= \sqrt{\frac{9}{25} + \frac{1}{25}} = \sqrt{\frac{10}{25}} = \sqrt{\frac{2}{5}}$$

**23.** *(c)* Given vertices of $\Delta AOP$ are $O(0, 0)$ and $A(0, 1)$.

Let the coordinates of point $P$ be $(x, y)$.

Clearly, perimeter

$$= OA + AP + OP = 4 \text{ (given)}$$

$$\Rightarrow \quad \sqrt{(0-0)^2 + (0-1)^2} + \sqrt{(0-x)^2 + (1-y)^2} + \sqrt{x^2 + y^2} = 4$$

$$\Rightarrow 1 + \sqrt{x^2 + (y-1)^2} + \sqrt{x^2 + y^2} = 4$$

$$\Rightarrow \quad \sqrt{x^2 + y^2 - 2y + 1} = 3 - \sqrt{x^2 + y^2}$$

$$\Rightarrow x^2 + y^2 - 2y + 1 = 9 + x^2 + y^2 - 6\sqrt{x^2 + y^2}$$

[squaring both sides]

$$\Rightarrow \quad 1 - 2y = 9 - 6\sqrt{x^2 + y^2}$$

$$\Rightarrow \quad 6\sqrt{x^2 + y^2} = 2y + 8$$

$$\Rightarrow \quad 3\sqrt{x^2 + y^2} = y + 4$$

$$\Rightarrow \quad 9(x^2 + y^2) = (y+4)^2$$

[squaring both sides]

$$\Rightarrow \quad 9x^2 + 9y^2 = y^2 + 8y + 16$$

$$\Rightarrow \quad 9x^2 + 8y^2 - 8y = 16$$

Thus, the locus of point $P(x, y)$ is

$$9x^2 + 8y^2 - 8y = 16$$

**24.** *(d)*

> **key idea** (i) Slope of line passing through two given point $(x_1, y_1)$ and $(x_2, y_2)$ is defined as $\frac{y_2 - y_1}{x_2 - x_1}$.
>
> (ii) If two lines are perpendicular then product of their slope is equal to $-1$.

Slope of the line $2x - 3y + 17 = 0$ is $\frac{2}{3} = m_1$ (let) and the slope of line joining the points $(7, 17)$ and $(15, \beta)$ is

$$\frac{\beta - 17}{15 - 7} = \frac{\beta - 17}{8} = m_2 \text{ (let)}$$

According to the question, $m_1 m_2 = -1$

$$\Rightarrow \quad \frac{2}{3} \times \frac{\beta - 17}{8} = -1 \Rightarrow \beta - 17 = -12$$

$$\Rightarrow \quad \beta = 5$$

**25.** *(c)* According to given information, we have the following figure

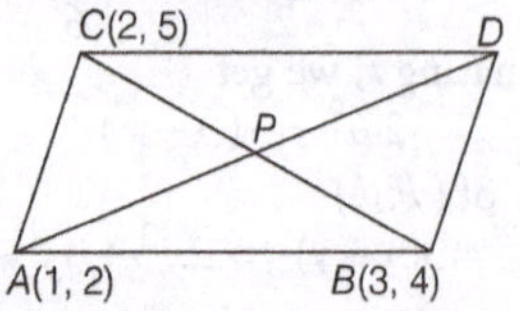


We know that, diagonals of a parallelogram intersect at mid-point.

$\therefore$ $P$ = Mid-point of $BC$ and so,

$$P \equiv \left(\frac{5}{2}, \frac{9}{2}\right)$$

Now, equation of $AD$ is

$$(y - 2) = \frac{\frac{9}{2} - 2}{\frac{5}{2} - 1}(x - 1)$$

$$\Rightarrow \quad y - 2 = \frac{5}{3}(x - 1)$$

$$\Rightarrow \quad 3y - 6 = 5x - 5$$

$$\Rightarrow \quad 5x - 3y + 1 = 0$$

**26.** *(b)* Given, $px+qy+r=0$ is the equation of line such that

$$3p+2q+4r=0$$

Consider, $3p+2q+4r=0$

$$\Rightarrow \quad \frac{3p}{4}+\frac{2q}{4}+r=0$$

(dividing the equation by 4)

$$\Rightarrow \quad p\left(\frac{3}{4}\right)+q\left(\frac{1}{2}\right)+r=0$$

$\Rightarrow \left(\frac{3}{4},\frac{1}{2}\right)$ satisfy $px+qy+r=0$

So, the lines always passes through the point $\left(\frac{3}{4},\frac{1}{2}\right)$.

**27.** *(c)*

> **key idea** Othrocentre, centroid and circumcentre are collinear and centroid divide orthocentre and circumcentre in 2 : 1 ratio.

We have orthocentre and centroid of a triangle be $A(-3, 5)$ and $B(3, 3)$ respectively and $C$ circumcentre.

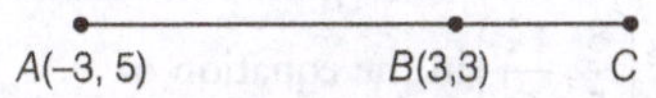


Clearly,

$$AB=\sqrt{(3+3)^2+(3-5)^2}$$
$$=\sqrt{36+4}=2\sqrt{10}$$

We know that, $AB:BC=2:1$

$$\Rightarrow \quad BC=\sqrt{10}$$

Now, $AC=AB+BC=2\sqrt{10}+\sqrt{10}$

$$=3\sqrt{10}$$

Since, $AC$ is a diameter of circle.

$$\therefore \quad r=\frac{AC}{2}$$

$$\Rightarrow \quad r=\frac{3\sqrt{10}}{2}=3\sqrt{\frac{5}{2}}$$

**28.** *(b)* Median through $C$ is $x=4$

$\Delta ABC$ has vertex $A(1, 2)$.

So, the $x$-coordinate of $C$ is 4. Let $C\equiv(4, y)$, then the mid-point of $A(1, 2)$ and $C(4, y)$ is $D$ which lies on the median through $B$.

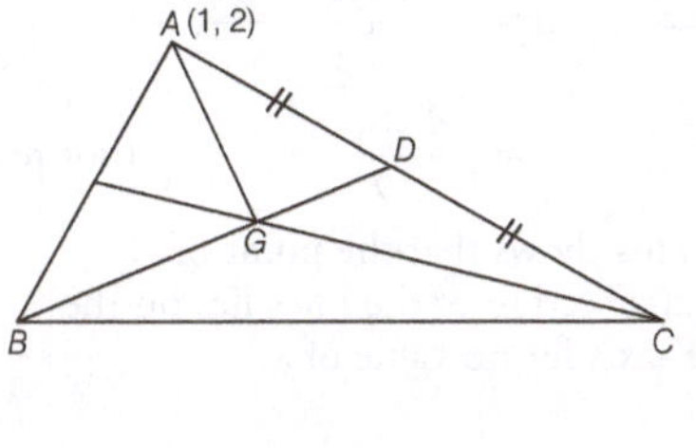


$$\therefore D\equiv\left(\frac{1+4}{2},\frac{2+y}{2}\right)$$

Now, $\dfrac{1+4+2+y}{2}=5$

$\left[\because D \text{ lies on median through } B\ (x+y=5)\right]$

$$\Rightarrow \quad y=3$$

So, $\quad C\equiv(4, 3)$

The centroid of the triangle is the intersection of the medians. Here the medians $x=4$ and $x+y=5$ intersect at $G(4, 1)$.

The area of triangle,

$$\Delta ABC=3\times\Delta AGC$$
$$=3\times\frac{1}{2}[1(1-3)+4(3-2)+4(2-1)]=9$$

**29.** *(b)*

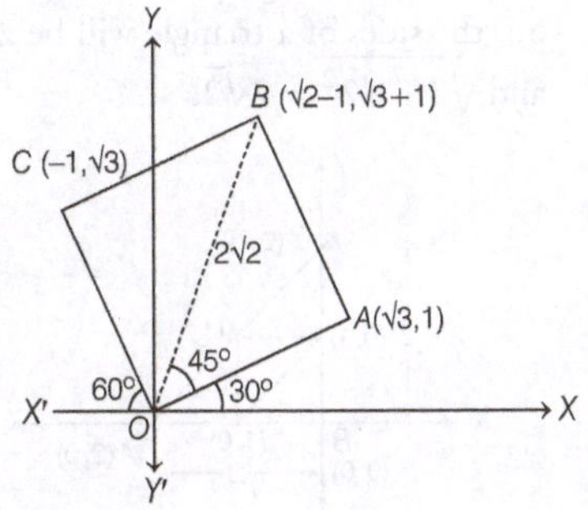


Coordinate of $A$,

$$\frac{x_1}{\cos 30^\circ}=\frac{y_1}{\sin 30^\circ}=2$$

$$\Rightarrow \quad x_1=\sqrt{3},\ y_1=1 \quad \text{...(i)}$$

Coordinate of $C$,

$$\frac{x_2}{\cos 120^\circ}=\frac{y_2}{\sin 120^\circ}=2$$

$$\Rightarrow \quad x_2=-1$$
$$\Rightarrow \quad y_2=\sqrt{3} \quad \text{...(ii)}$$

Coordinate of $B$,

$$\frac{x_3}{\cos 75^\circ}=\frac{y_3}{\sin 75^\circ}=2\sqrt{2}$$

$$\Rightarrow \quad x_3=\sqrt{3}-1$$

and $\quad y_3=\sqrt{3}+1$

$\therefore$ Required sum $x_1+x_2+x_3+0$

$$=\sqrt{3}+(-1)+(\sqrt{3}-1)+0$$
$$=2\sqrt{3}-2$$

**30.** *(c)* Length of perpendicular to $4x+3y=10$ from origin $(0, 0)$

$$P_1=\left|\frac{4\times0+3\times0-10}{\sqrt{4^2+3^2}}\right|$$

$$=\frac{10}{5}=2$$

Length of perpendicular to $8x+6y+5=0$ from origin $(0, 0)$

$$P_2=\left|\frac{8\times0+6\times0+5}{\sqrt{6^2+8^2}}\right|$$

$$=\frac{5}{10}=\frac{1}{2}$$

$\because$ Lines are parallel to each other

$\therefore$ Ratio will be $4:1$ or $1:4$.

**31.** *(d)* Required points $(x, y)$ are such that it satisfy

$$x+y<41$$

and $\quad x>0,\ y>0$

Number of positive integral solution of the equation $x+y+k=41$ will be number of integral coordinates in the bounded region.

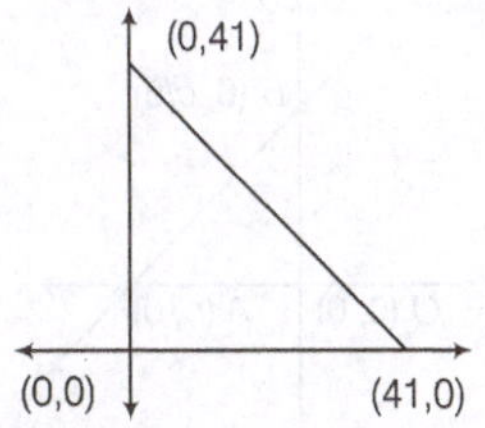


$\therefore$ Total number of integral coordinates

$$={}^{41-1}C_{3-1}={}^{40}C_2$$

$$=\frac{40!}{2!\,38!}=780$$

> ***Alternate Solution***
>
> Consider the following figure :
>
> 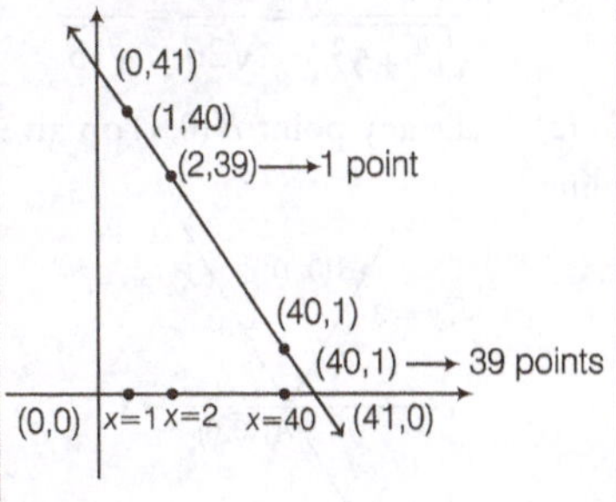
> 
>
> Clearly, the number of required points
>
> $=1+2+3+\ldots+39$
>
> $=\frac{39}{2}(39+1)=780$

**32.** *(d)* The points are $A\left(0,\frac{8}{3}\right)$, $B(1, 3)$ and $C(82, 30)$.

Slope of $AB=\frac{1}{3}$, Slope of $BC=\frac{1}{3}$

So, the given points lies on the same line.

Hence, $A$, $B$, $C$ are collinear.

**33.** *(c)* Given the equation,

$$x+2ay+a=0 \quad \text{...(i)}$$
$$x+3by+b=0 \quad \text{...(ii)}$$
$$x+4ay+a=0 \quad \text{...(iii)}$$

On subtracting Eq. (iii) from Eq. (i), we get

$$-2ay=0\,;\; ay=0 \Rightarrow y=0$$

On putting the value of $y$ in Eq. (i), we get

$$x+0+a=0$$
$$x=-a$$

Putting the value of $x$ and $y$ in Eq. (ii), we get

$$-a+b=0 \Rightarrow a=b$$

Thus, $(a, b)$ lies on a straight line.

**34.** *(b)* Given equation of line $L$, perpendicular to line $5x-y=1$ be

$$x+5y=c$$

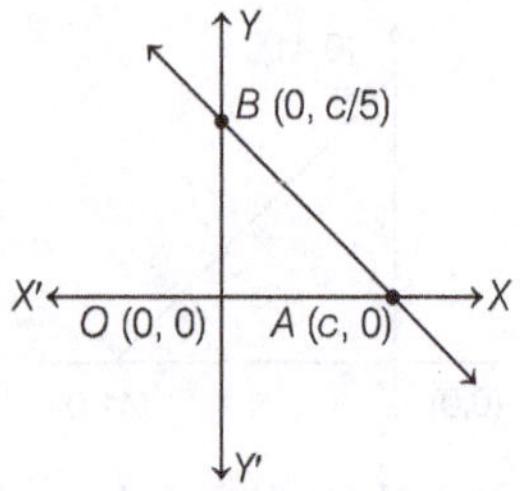


Given that, area of $\Delta AOB$ is 5.

So $\quad 5=\dfrac{1}{2}\left[c\left(\dfrac{c}{5}\right)\right] \Rightarrow c=\pm 5\sqrt{2}$

The distance between the lines $x+5y=0$ and $x+5y=\pm 5\sqrt{2}$ is

$$\frac{|\pm 5\sqrt{2}-0|}{\sqrt{1^2+5^2}}=\frac{5\sqrt{2}}{\sqrt{26}}=\frac{5}{\sqrt{13}}$$

**35.** *(b)* Take any point $B\,(0, 1)$ on given line.

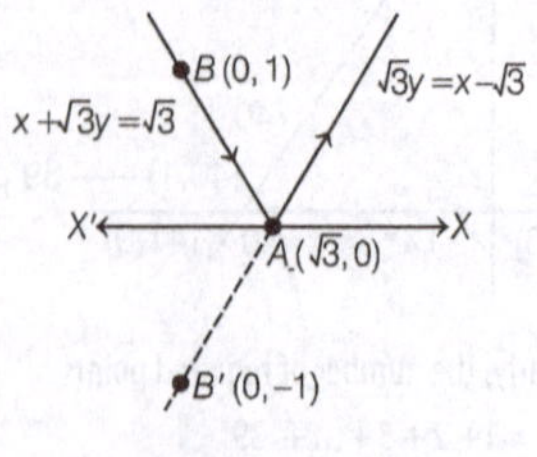


Equation of $AB'$,

$$y-0=\frac{-1-0}{0-\sqrt{3}}(x-\sqrt{3})$$

$$\Rightarrow \quad -\sqrt{3}y=-x+\sqrt{3}$$

$$\Rightarrow \quad x-\sqrt{3}y=\sqrt{3}$$

$$\Rightarrow \quad \sqrt{3}y=x-\sqrt{3}$$

**Alternate Solution**

Slope of given line is $-\dfrac{1}{\sqrt{3}}$.

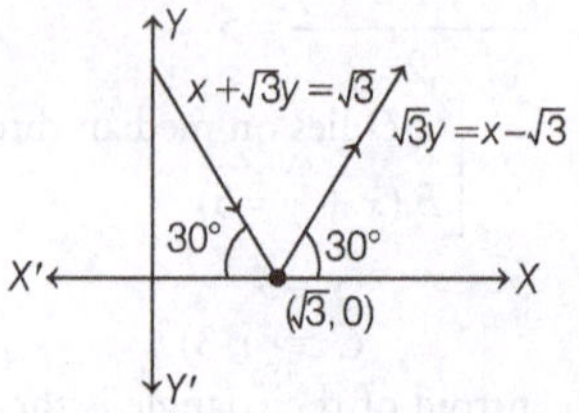


From figure, equation of reflected rays is

$$y=\frac{1}{\sqrt{3}}(x-\sqrt{3})$$

$$\Rightarrow \quad \sqrt{3}y=x-\sqrt{3}$$

**36.** *(b)* Given, mid-points of side of a triangle are (0, 1), (1, 1) and (1, 0). Plotting these points on a graph paper and make a triangle.

So, the sides of a triangle will be 2, 2 and $\sqrt{2^2+2^2}=2\sqrt{2}$.

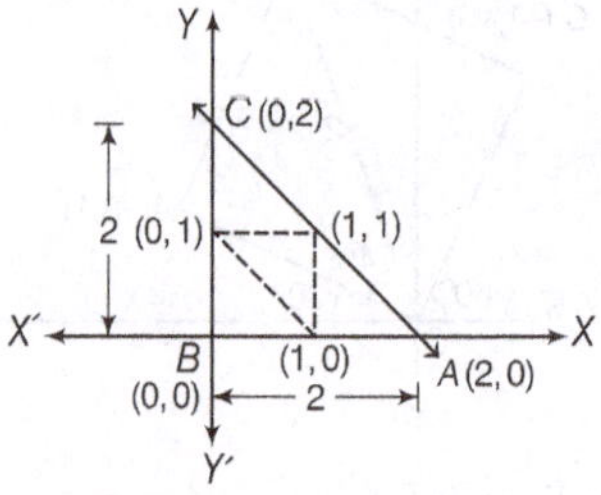


$x$-coordinate of incentre

$$=\frac{2\times 0+2\sqrt{2}\cdot 0+2\cdot 2}{2+2+2\sqrt{2}}$$

$$=\frac{2}{2+\sqrt{2}}\times\frac{2-\sqrt{2}}{2-\sqrt{2}}$$

$$=2-\sqrt{2}$$

**37.** *(a)* Given, three lines $x-3y=p$, $ax+2y=q$ and $ax+y=r$ form a right angled triangle

$\therefore$ Product of slopes of any two lines $=-1$

Suppose, $ax+2y=q$ and $x-3y=p$ are perpendicular to each other.

$$\therefore \quad \frac{-a}{2}\times\frac{1}{3}=-1$$

$$\Rightarrow \quad a=6$$

Now, consider option one-by-one $a=6$ satisfies only option (a).

$\therefore$ Required answer is

$$a^2-9a+18=0$$

**38.** *(d)* Given line $3x+4y=12$

$$\frac{3x}{12}+\frac{4y}{12}=1$$

$$\Rightarrow \quad \frac{x}{4}+\frac{y}{3}=1$$

$\Rightarrow$ $x$-intercept $=4$

and $y$-intercept $=3$

$\therefore$ Required line is $\dfrac{x}{8}+\dfrac{y}{3/2}=1$

$$\Rightarrow \quad 3x+16y=24$$

$$\Rightarrow \quad y=\frac{-3}{16}x+\frac{24}{16}$$

Hence, required slope is $\dfrac{-3}{16}$.

**39.** *(c)* Given line $L: 2x+y=k$ passes through point (say $P$) which divides a line segment (say $AB$) in ratio 3 : 2, where $A(1, 1)$ and $B(2, 4)$.

Using section formula, the coordinates of the point $P$ which divides $AB$ internally in the ratio 3 : 2 are

$$P\left(\frac{3\times 2+2\times 1}{3+2}, \frac{3\times 4+2\times 1}{3+2}\right)$$

$$\equiv P\left(\frac{8}{5}, \frac{14}{5}\right)$$

Also, since the line $L$ passes through $P$, hence substituting the coordinates of $P\left(\dfrac{8}{5}, \dfrac{14}{5}\right)$ in the equation of $L: 2x+y=k$, we get

$$2\left(\frac{8}{5}\right)+\left(\frac{14}{5}\right)=k$$

$$\Rightarrow \quad k=6$$

**40.** *(a)* Given equation of lines are

$(a^3+3)x+ay+a-3=0$ and

$(a^5+2)x+(a+2)y+2a+3=0$

where, $a$ is a real number.

Since, point of intersection of lines lies on $Y$-axis.

$\therefore$ Putting $x=0$ in each equation, we get

$$ay+a-3=0 \quad \text{...(i)}$$
$$(a+2)y+2a+3=0 \quad \text{...(ii)}$$

On solving Eqs. (i) and (ii), we get

$$(a+2)(a-3)-a(2a+3)=0$$

$$\Rightarrow \quad a^2-a-6-2a^2-3a=0$$

$$\Rightarrow \quad -a^2-4a-6=0$$

$$\Rightarrow \quad a^2+4a+6=0$$

$$\Rightarrow \quad a=\frac{-4\pm\sqrt{16-24}}{2}$$

$$=\frac{-4\pm\sqrt{-8}}{2} \quad \text{(not real)}$$

This shows that the point of intersection of the lines lies on the $Y$-axis for no value of $a$.

**41.** (d)

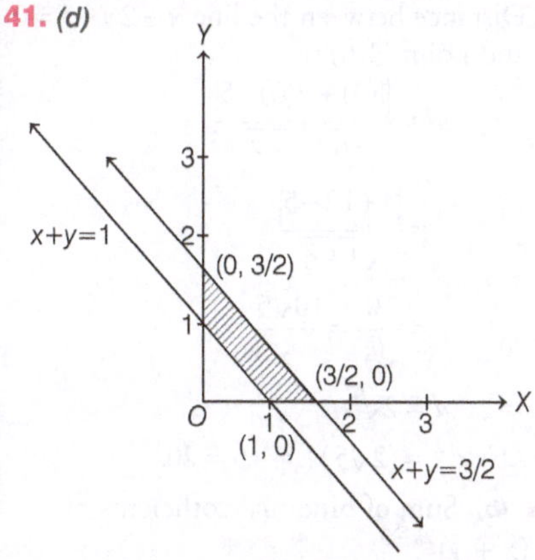


Given, the point (1, $a$) lies between straight line $x + y = 1$ and $2(x + y) = 3$

$\therefore$ Putting $x = 1$ in $2(x + y) = 3$, we get

$$2(1 + y) = 3 \Rightarrow y = \frac{3}{2} - 1 = \frac{1}{2}$$

Thus $a$ lies in $\left(0, \frac{1}{2}\right)$.

**42.** (c) Given lines are $ax + 2by + 3b = 0$ and $bx - 2ay - 3a = 0$

Hence, required line is parallel to $X$-axis.

$\therefore \quad x = 0$

On putting $x = 0$ in given equation, we get

$$2by = -3b \Rightarrow y = -\frac{3}{2}$$

The required line is below $X$-axis at a distance of $\frac{3}{2}$ from it.

**43.** (d) Given straight lines are

$L_1 : x - y = 1$ ...(i)

$L_2 : x + y = 1$ ...(ii)

$L_3 : 2x + 2y = 5$ ...(iii)

$L_4 : 2x - 2y = 7$ ...(iv)

$L_1 \perp L_2$ is correct statement

($\because$ Product of their slopes $= -1$)

$L_1 \perp L_3$ is also correct statements

($\because$ Product of their slopes $= -1$)

Now, $L_2 : x + y = 1$

$L_4 : 2x - 2y = 7$

On solving Eqs. (iv) and (ii), we get

$$2x - 2(1 - x) = 7$$

$$\Rightarrow \quad 2x - 2 + 2x = 7$$

$$\Rightarrow \quad x = \frac{9}{4} \text{ and } y = \frac{-5}{4}$$

Hence, $L_2$ intersects $L_4$.

**44.** (d) Given $ax^2 + bx + c = 0$

$$\Rightarrow \quad ax^2 = -bx - c$$

Given curve

$$y = 4ax^2 + 3bx + 2c$$

$$= 4[-bx - c] + 3bx + 2c$$

$$= -4bx - 4c + 3bx + 2c = -bx - 2c$$

$\because$ This curve intersects $X$-axis.

$\therefore$ Putting $y = 0$, we get

$$-bx - 2c = 0 \Rightarrow -bx = 2c \Rightarrow x = \frac{-2c}{b}$$

Thus, given curve intersects $X$-axis at exactly one point.

**45.** (c) As, $x + y = |a|$ and $ax - y = 1$, intersect in I quadrant.

Therefore, $x$ and $y$ intercept are positive.

$$\therefore \quad x = \frac{1 + |a|}{1 + a} \geq 0 \text{ and } y = \frac{a|a| - 1}{a + 1} \geq 0$$

$\Rightarrow 1 + a \geq 0$ and $a|a| - 1 \geq 0$

$\Rightarrow \quad a \geq -1$ and $a|a| \geq 1$ ...(i)

Case I If $\quad -1 < a < 0$

$\Rightarrow \quad -a^2 > 1$ [not possible]

Case II If $\quad a \geq 0 \Rightarrow a^2 \geq 1$

$\Rightarrow \quad a \geq 1$ ...(ii)

$\therefore \quad a \geq 1$ or $a \in [1, \infty)$

**46.** (a) Lines perpendicular to same line are parallel to each other.

$$\therefore \quad \frac{p(p^2 + 1)}{(p^2 + 1)^2} = \frac{-1}{p^2 + 1} \Rightarrow p = -1$$

Hence, there is exactly one value of $p$.

**47.** (d) Let $(b^2, b)$ be the point of shortest distance on $x = y^2$.

Then, distance between $(b^2, b)$ and line $x - y + 1 = 0$ is given by

$$D = \left| \frac{ax_1 + by_1 + c}{\sqrt{a^2 + b^2}} \right| = \frac{|b^2 - b + 1|}{\sqrt{2}}$$

$$= \frac{1}{\sqrt{2}} \left| \left(b - \frac{1}{2}\right)^2 + \frac{3}{4} \right|$$

It is minimum when $b = \frac{1}{2}$

and $D_{\min} = \frac{3}{4\sqrt{2}} = \frac{3\sqrt{2}}{8}$

**48.** (b) Since, $A$ is the mid-point of line $PQ$.

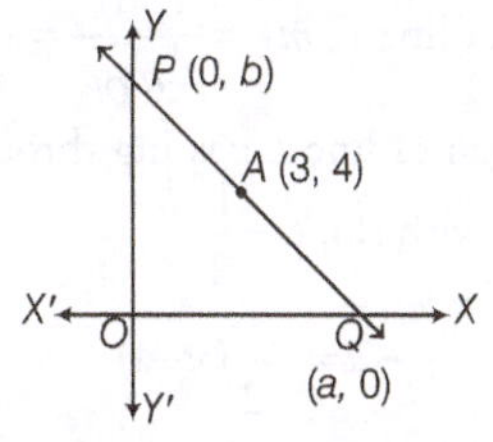


$$\therefore \quad 3 = \frac{a + 0}{2} \Rightarrow a = 6$$

and $\quad 4 = \frac{0 + b}{2} \Rightarrow b = 8$

Hence, the equation of line is

$$\frac{x}{6} + \frac{y}{8} = 1 \text{ or } 4x + 3y = 24$$

**49.** (b) The graph of equations $x - 2y = 0$ and $3x - y = 0$ is as shown in the figure. Since, given point $(a, a^2)$ lies in the shaded region.

Then, $a^2 - \frac{a}{2} > 0$ and $a^2 - 3a < 0$

$\Rightarrow \quad a(2a - 1) > 0$ and $a(a - 3) < 0$

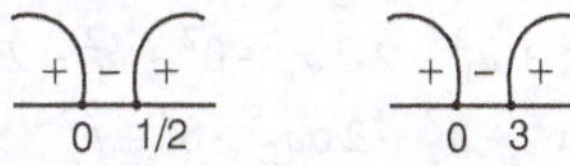


$$\Rightarrow a \in (-\infty, 0) \cup \left(\frac{1}{2}, \infty\right) \text{ and } 0 < a < 3$$

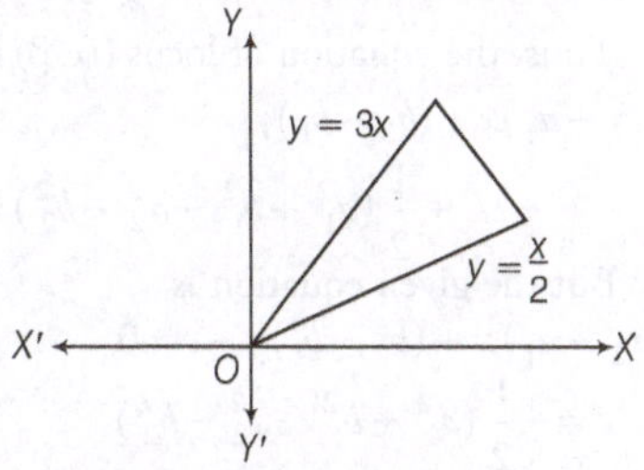


i.e., $a \in (0, 3) \Rightarrow a \in \left(\frac{1}{2}, 3\right)$

**50.** (c) The equation of normal at $\theta$ is

$$y - y_1 = -\frac{1}{dy/dx}(x - x_1)$$

Point $(a(1 + \cos\theta), a\sin\theta)$

$$\Rightarrow y - a\sin\theta = \frac{\sin\theta}{\cos\theta}(x - a(1 + \cos\theta)),$$

$$y = \tan\theta(x - a)$$

Comparing with it $(y - y_1) = m(x - x_1)$

We find that the normal of the curve passes through fixed point $(a, 0)$.

**51.** (a) Since, $x_1, x_2, x_3$ and $y_1, y_2, y_3$ are in GP.

Then, $\quad x_2 = rx_1, \quad x_3 = r^2 x_1$

and $\quad y_2 = ry_1, \quad y_3 = r^2 y_1,$

where, $r$ is a common ratio.

The points become $(x_1, y_1)$, $(rx_1, ry_1)$ and $(r^2x_1, r^2y_1)$.

$$\text{Now, } \begin{vmatrix} x_1 & y_1 & 1 \\ x_2 & y_2 & 1 \\ x_3 & y_3 & 1 \end{vmatrix} = \begin{vmatrix} x_1 & y_1 & 1 \\ rx_1 & ry_1 & 1 \\ r^2x_1 & r^2y_1 & 1 \end{vmatrix}$$

Taking $x_1$ common from $C_1$ and $y_1$ from $C_2$

$$= x_1 y_1 \begin{vmatrix} 1 & 1 & 1 \\ r & r & 1 \\ r^2 & r^2 & 1 \end{vmatrix} = x_1 y_1 (0) = 0$$

[since, two columns are identical]

Hence, these points lie on a straight line.

**52.** (a) Let $P(\alpha, \beta)$ be the point which is equidistant from $A(a_1, b_1)$ and $B(a_2, b_2)$.

$\therefore \quad PA = PB$

$\Rightarrow (\alpha - a_1)^2 + (\beta - b_1)^2 = (\alpha - a_2)^2 + (\beta - b_2)^2$

$\Rightarrow \alpha^2 + a_1^2 - 2\alpha a_1 + \beta^2 + b_1^2 - 2\beta b_1 = \alpha^2 + a_2^2 - 2\alpha a_2 + \beta^2 + b_2^2 - 2\beta b_2$

$\Rightarrow 2(a_2 - a_1)\alpha + 2(b_2 - b_1)\beta + (a_1^2 + b_1^2 - a_2^2 - b_2^2) = 0$

Thus, the equation of locus $(\alpha, \beta)$ is

$(a_2 - a_1)x + (b_2 - b_1)y + \frac{1}{2}(a_1^2 + b_1^2 - a_2^2 - b_2^2) = 0$

But the given equation is

$(a_2 - a_1)x + (b_2 - b_1)y - c = 0$

$\therefore \; c = -\frac{1}{2}(a_1^2 + b_1^2 - a_2^2 - b_2^2)$

$= \frac{1}{2}(a_2^2 + b_2^2 - a_1^2 - b_1^2)$

**53.** (a) Let the vertices of $\Delta ABC$ be $A(4, 0)$, $B(-1, -1)$ and $C(3, 5)$.

Now, $AB = \sqrt{(-1-4)^2 + (-1-0)^2}$

$= \sqrt{25+1} = \sqrt{26}$

$BC = \sqrt{(3+1)^2 + (5+1)^2}$

$= \sqrt{16+36} = \sqrt{52}$

and $CA = \sqrt{(4-3)^2 + (0-5)^2}$

$= \sqrt{1+25} = \sqrt{26}$

$\therefore \; CA^2 + AB^2 = (\sqrt{26})^2 + (\sqrt{26})^2$

$= 26 + 26 = 52 = BC^2$

$\Rightarrow CA^2 + AB^2 = BC^2$

Thus, the triangle is isosceles and right angled triangle.

**54.** (d) Given, $x \cos\alpha + y \sin\alpha = p$

$$\Rightarrow \frac{x}{\frac{p}{\cos\alpha}} + \frac{y}{\frac{p}{\sin\alpha}} = 1$$

This line touches the axes at $\left(\frac{p}{\cos\alpha}, 0\right)$ and $\left(0, \frac{p}{\sin\alpha}\right)$.

So, coordinates of mid-point are

mid-point $(x_1, y_1) = \left(\frac{p}{2\cos\alpha}, \frac{p}{2\sin\alpha}\right)$

$x_1 = \frac{p}{2\cos\alpha}$ and $y_1 = \frac{p}{2\sin\alpha}$;

$\Rightarrow \cos\alpha = \frac{p}{2x_1}$ and $\sin\alpha = \frac{p}{2y_1}$

$\because \quad \cos^2\alpha + \sin^2\alpha = 1$

Locus of $(x_1, y_1)$ is $\frac{1}{x^2} + \frac{1}{y^2} = \frac{4}{p^2}$.

**55.** (b) Given, lines $\sqrt{3}x + y = 0$ makes an angle of 120° with $OX$ and $\sqrt{3}x - y = 0$ makes an angle of 60° with $OX$. So, the required line is $y - 2 = 0$.

**56.** (a) $x(x+2)(12-k) = 2$

$\Rightarrow \quad x^2 + 2x - \frac{2}{12-k} = 0$

$D = 0$

$\Rightarrow \quad 4 - 4\left(\frac{-2}{12-k}\right) = 0$

$\Rightarrow \quad 1 + \frac{2}{12-k} = 0 \Rightarrow k = 14$

$\left(k, \frac{k}{2}\right) \equiv (14, 7)$

$\Rightarrow \quad d = \left|\frac{3\times14 + 4\times7 + 5}{5}\right| = 15$

**57.** (a) $x(3\lambda+1) + y(7\lambda+2) = 17\lambda + 5$

$\lambda(3x + 7y - 17) + (x + 2y - 5) = 0$

$3x + 7y - 17 = 0 \quad \ldots(i)$

$(x + 2y - 5 = 0) \times (-3) \quad \ldots(ii)$

$-3x - 6y + 15 = 0 \quad \ldots(iii)$

Adding Eqs (i) and (iii), we get

$\Rightarrow \quad y - 2 = 0$

$\Rightarrow \quad y = 2$

Put $y = 2$ in Eq. (ii), we get

$x + (2)(2) - 5 = 0$

$x = 1$

Point $P \equiv (1, 2)$

Slope of line OP, $m_{OP} = \frac{2-0}{1-0} = 2$

Slope of line $L$, $m_L = -\frac{1}{m_{OP}} = -\frac{1}{2}$

Equation of line $L$ passing through $P(1,2)$ with slope $-\frac{1}{2}$

$y - 2 = -\frac{1}{2}(x-1)$

$\Rightarrow \quad 2y - 4 = -x + 1$

$\Rightarrow \quad x + 2y - 5 = 0$

Distance between the line $x + 2y - 5 = 0$ and point $(3, 6)$ is

$d = \frac{|1(3) + 2(6) - 5|}{\sqrt{1^2 + 2^2}}$

$= \frac{|3 + 12 - 5|}{\sqrt{1+4}}$

$= \frac{10}{\sqrt{5}} = \frac{10\sqrt{5}}{5}$

$\Rightarrow \quad d = 2\sqrt{5}$

$\therefore \quad d^2 = (2\sqrt{5})^2 = 4.5 = 20.$

**58.** (b) Sum of binomial cofficients of $(x+y)^{2n-3}$

$= 2^{2n-3}$ (Put $x = y = 1$)

Number of terms in $(x+y)^{2n-3} = 2n - 3 + 1$

$= 2n - 2$

$\therefore \quad AM = \frac{2^{2n-3}}{2n-2} = \frac{2^{2n-4}}{n-1}$

$16 = \frac{2^{2n-4}}{n-1}$ [$\because$ AM = 16]

$\therefore \quad n = 5$

$\therefore P(2n-1, n^2 - 4n) = P(9, 5)$

Distance between $P(9, 5)$ and line $x + y - 8 = 0$

$D = \left|\frac{9+5-8}{\sqrt{1+1}}\right| = \frac{6}{\sqrt{2}} = 3\sqrt{2}$

**59.** (b) Equation of line $AB$ is $3y - x = 2$ and $AC$ is $x + y = 2$

In line $AB$,

When $y = 0$, $x = -2$

$\therefore B(-2, 0)$

In line $AC$,

When $y = 0$, $x = 2$

$\therefore \; C(2, 0)$

Equation of altitude of $AC$,

$x - y = \mu$

$x - y = -2$

($\because$ it is passing through point $(-2, 0)$)

Equation of altitude of $AB$ is

$y + 3x = \lambda$

$\Rightarrow y + 3x = 6$

($\because$ it is passing through $(2, 0)$)

Solving both altitudes

orthocenter $P \equiv (1, 3)$

$\therefore$ Area of $\Delta PBC = \frac{1}{2} \times 3 \times 4 = 6$

**60.** (d) The orthocenter is the intersection point of the altitudes of a triangle. Find the equations of two altitudes of the triangle and use the orthocenter coordinates to solve for $m-c$.

For $y=x+1$, $y=4x-8$

$\Rightarrow x+1=4x-8$

$\Rightarrow x=3$, then $y=4$

$\therefore$ Point $B(3, 4)$

For $y=x+1$, $y=mx+c$

$\Rightarrow x+1=mx+c$

$\Rightarrow x=\frac{c-1}{1-m}$, then $y=\frac{c-1}{1-m}+1=\frac{c-m}{1-m}$

Point $A\left(\frac{c-1}{1-m}, \frac{c-m}{1-m}\right)$

For $y=4x-8$, $y=mx+c$

$\Rightarrow 4x-8=mx+c$

$\Rightarrow x=\frac{c+8}{4-m}$, then

$y=4\left(\frac{c+8}{4-m}\right)-8=\frac{4c+8m}{4-m}$

Point $C\left(\frac{c+8}{4-m}, \frac{4c+8m}{4-m}\right)$

Altitude from $A$ to the line $y=4x-8$ has slope $-\frac{1}{4}$

Altitude from $B$ to the line $y=mx+c$ has slope $-\frac{1}{m}$

Altitude from $C$ to the line $y=x+1$ has slope $-1$.

Equation of altitude from $A$ to $BC$ is

$y-y_1=m(x-x_1)$

$y-\frac{c-m}{1-m}=-\frac{1}{4}\left(x-\frac{c-1}{1-m}\right)$

$\Rightarrow \frac{y(1-m)-(c-m)}{1-m}$

$=-\frac{1}{4}\left[\frac{x(1-m)-(c-1)}{1-m}\right]$

$\Rightarrow 4y(1-m)-4(c-m)=(c-1)-x(1-m)$

This equation passes through orthocentre $(3, -1)$,

$\Rightarrow -4(1-m)-4(c-m)=(c-1)-3(1-m)$

$\Rightarrow -4+4m-4c+4m=c-1-3+3m$

$\Rightarrow 8m-4c-c-3m=0$

$5m=5c \Rightarrow m=c$

Equation of altitude from $B$ to $AC$ is

$y-y_1=m(x-x_1)$

$y-4=-\frac{1}{m}(x-3)$

$my-4m=-x+3$

This equation passes through orthocentre $(3, -1)$,

$-m-4m=-3+3$

$\Rightarrow \quad m=0$

Equation of altitude from $C$ to $AB$ is

$y-y_1=m(x-x_1)$

$\Rightarrow y-\frac{4c+8m}{4-m}=-1\left(x-\frac{c+8}{4-m}\right)$

$\Rightarrow \frac{y(4-m)-(4c+8m)}{4-m}$

$=\frac{c+8-(4-m)x}{4-m}$

$\Rightarrow y(4-m)-(4c+8m)=c+8-(4-m)x$

This equation passes through orthocentre $(3, -1)$,

$-(4-m)-(4c+8m)=c+8-(4-m)3$

$\Rightarrow -4+m-4c-8m=c+8-12+3m$

$\Rightarrow -7m-4c=3m+c$

$\Rightarrow -10m=5c \Rightarrow c=-2m$

But $m=c$

so, $c=-2c$ not valid

Thus, $m-c=0-0=0$

**61.** *(b)*

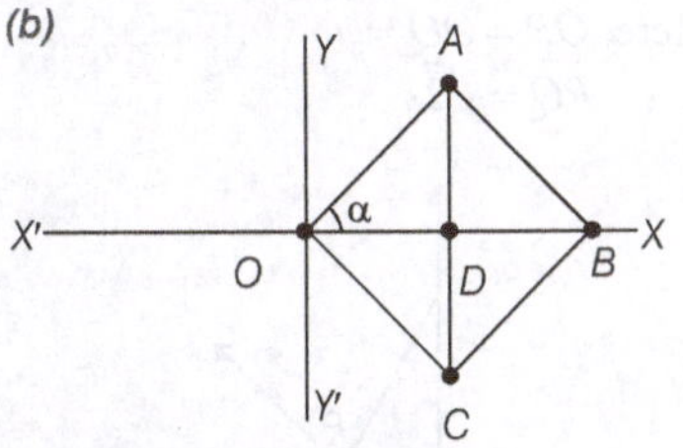


Given equation of $OB$

$(\sqrt{3}+1)x+(\sqrt{3}-1)y=0$ ...(i)

Equation of $AC$

$(\sqrt{3}-1)x-(\sqrt{3}+1)y+8\sqrt{3}=0$ ...(ii)

Solving Eqs. (i) and (ii)

we get point $D$ which is point of intersection of Diagonals of square

$(\sqrt{3}-1)x+\frac{(\sqrt{3}+1)\times(\sqrt{3}+1)x}{\sqrt{3}-1}+8\sqrt{3}=0$

$\Rightarrow x[(\sqrt{3}-1)^2+(\sqrt{3}+1)^2]=8\sqrt{3}(1-\sqrt{3})$

$\Rightarrow \quad x(8)=8\sqrt{3}(1-\sqrt{3})$

$\Rightarrow \quad x=\sqrt{3}-3$

$\Rightarrow \quad y=\frac{(\sqrt{3}+1)x}{1-\sqrt{3}}$

$=\frac{(\sqrt{3}+1)\times\sqrt{3}(1-\sqrt{3})}{1-\sqrt{3}}$

$y=3+\sqrt{3}$

Let side of square be a

$\Rightarrow \quad OD=\frac{a}{\sqrt{2}} \Rightarrow OD^2=\frac{a^2}{2}$

$\Rightarrow \quad \frac{a^2}{2}=(\sqrt{3}-3)^2+(3+\sqrt{3})^2$

$a^2=2(3+9)\times 2=48$

**62.** *(d)* Clearly, $x=\frac{11}{2}$ intersect

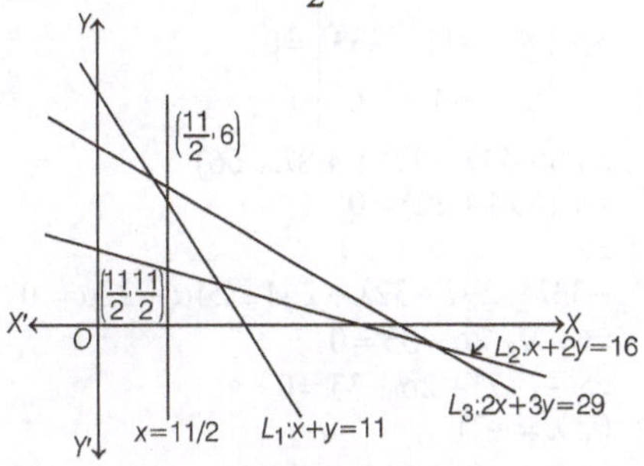


$L_1: x+y=11$ at $\left(\frac{11}{2}, \frac{11}{2}\right)$

and $L_3: 2x+3y-29=0$ at $\left(\frac{11}{2}, 6\right)$

So, $\quad \alpha \in\left[\frac{11}{2}, 6\right]$

i.e., $\alpha_{\min}=\frac{11}{2}$ and $\alpha_{\max}=6$

So, $\quad \alpha_{\min}\times\alpha_{\max}=\frac{11}{2}\times 6=33$

**63.** *(d)* $A\equiv(1, 2)$, $B=\left(\frac{57}{13}, -\frac{40}{13}\right)$

According to the question $R$ is the mid-point of $AB$

$\therefore \quad R\equiv\left(\frac{\frac{57}{13}+1}{2}, \frac{\frac{-40}{13}+2}{2}\right)$

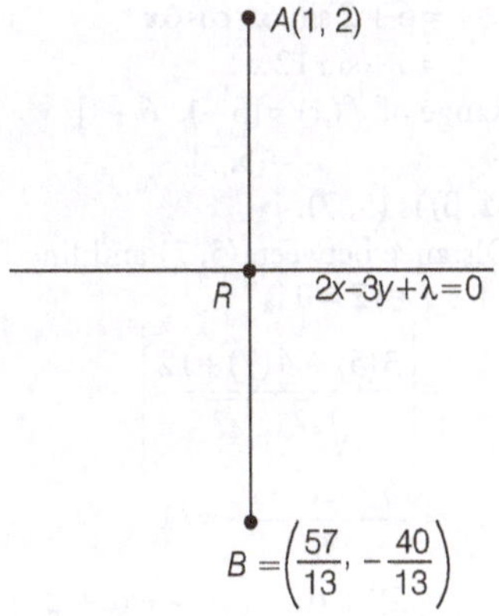


$=\left(\frac{35}{13}, \frac{-7}{13}\right)$

$\because R$ lies on the line $2x-3y+\lambda=0$

So, $\quad 2\left(\frac{35}{13}\right)-3\left(\frac{-7}{13}\right)+\lambda=0$

$\Rightarrow \quad \lambda=\frac{-91}{13}=-7$

Also, $3x-4y-\alpha=0$, $8x-11y-33=0$

and $2x-3y+\lambda=0$ are concurrent.

So, $\begin{vmatrix} 3 & -4 & -\alpha \\ 8 & -11 & -33 \\ 2 & -3 & \lambda \end{vmatrix}=0$

$\Rightarrow 3(-11\lambda-99)+4(8\lambda+66)$
$-\alpha(-24+22)=0$

$\Rightarrow$
$-33\lambda-297+32\lambda+264+24\alpha-22\alpha=0$

$\Rightarrow -\lambda+2\alpha-33=0$

$\Rightarrow -(-7)+2\alpha-33=0$

$(\because \lambda=-7)$

$\Rightarrow 2\alpha=26 \;\Rightarrow\; \alpha=13$

$\therefore\; |\alpha\lambda|=|13\times(-7)|=91$

**• Concept Enhancer**

Lines $a_1x+b_1y+c_1=0, a_2x+b_2y+c_2=0$ and $a_3x+b_3y+c_3=0$ are concurrent, if

$\begin{vmatrix} a_1 & b_1 & c_1 \\ a_2 & b_2 & c_2 \\ a_3 & b_3 & c_3 \end{vmatrix}=0$

**64.** *(d)* $f(x)=6+16\cos x\cos\left(\frac{\pi}{3}-x\right)$

$\cos\left(\frac{\pi}{3}+x\right)\times\sin 3x\cos 6x$

$\because \cos A\cos(60°-A)\cos(60°+A)$
$=\frac{1}{4}\cos 3A$

$\therefore f(x)=6+16\left(\frac{1}{4}\cos 3x\right)\sin 3x\cos 6x$

$=6+4\cos 3x\sin 3x\cos 6x$
$=6+2\sin 6x\cos 6x$
$=6+\sin 12x$

$\therefore$ Range of $f(x)=[6-1,\;6+1]$
$=[5,7]$

$\therefore(\alpha,\beta)$ is $(5, 7)$.

$\therefore$ Distance between $(5, 7)$ and line $3x+4y+12=0$ is

$=\left|\frac{3(5)+4(7)+12}{\sqrt{3^2+4^2}}\right|$

$=\frac{15+28+12}{5}=11$

**65.** *(c)* $h=\frac{2\beta+\alpha}{3},\; k=\frac{-4+\alpha+2}{3}$

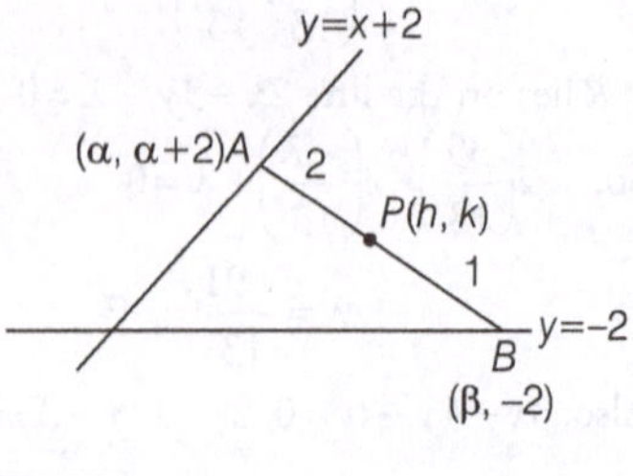


$\Rightarrow\quad 2\beta=3h-\alpha,\;\alpha=3k+2$

$\Rightarrow\quad 2\beta=3h-3k-2$

$\because\quad AB=8$

$(\alpha-\beta)^2+(\alpha+4)^2=(8)^2$

$\left(3k+2-\left(\frac{3h-3k-2}{2}\right)\right)^2$
$+(3k+2+4)^2=64$

$\Rightarrow\frac{(9k-3h+6)^2}{4}+(3k+6)^2=64$

$\Rightarrow 9\,[(3k-h+2)^2+4\,(k+2)^2]=64\times 4$

Put $h=x$ and $k=y$

$9\,(x^2+13y^2-6xy-4x+28y)-76$

$\therefore\qquad \alpha=13, \beta=-6$

and $\qquad \gamma=-4$

$\Rightarrow \alpha-\beta-\gamma=13+6+4=23$

**66.** *(c)* Given, $\Delta OPQ$ be isosceles triangle and $\angle POQ=90°$.

Here, $OP=OQ=a$

$\therefore\qquad PQ=\sqrt{2}a$

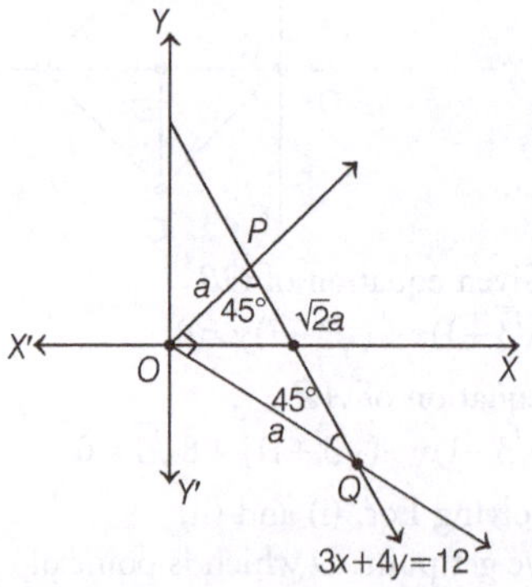


Let slope of $OP$ be $m_1$

$\therefore$ Equation of line $OP$ is $y=m_1x$

and slope of $3x+4y=12$ is $-\frac{3}{4}$

i.e. $m_2=-\frac{3}{4}$

Now, $\tan 45°=\left|\frac{m_1-m_2}{1+m_1m_2}\right|$

$\Rightarrow\qquad 1=\left|\frac{m_1+\frac{3}{4}}{1-\frac{3}{4}m_1}\right|$

$\Rightarrow 1-\frac{3}{4}m_1=m_1+\frac{3}{4}$

$\Rightarrow\quad \frac{7}{4}m_1=\frac{1}{4}$

$\Rightarrow\qquad m_1=\frac{1}{7}$

$\therefore$ Equation of line $OP$ is $y=\frac{1}{7}x$

Intersecting point of lines $y=\frac{1}{7}x$ and

$3x+4y=12$ is $P\left(\frac{84}{25},\frac{12}{25}\right)$

and $OP^2=a^2=\left(\frac{84}{25}-0\right)^2+\left(\frac{12}{25}-0\right)^2$

$\Rightarrow\quad a^2=\frac{7056}{625}+\frac{144}{625}=\frac{7200}{625}=\frac{288}{25}$

$\therefore\qquad l=OP^2+PQ^2+OQ^2$
$=a^2+2a^2+a^2=4a^2$
$=4\times\frac{288}{25}=46.08$

$\therefore\qquad [l]=46$

**67.** *(d)*

**key idea** The angle between two lines whose slopes $m_1$ and $m_2$ is given by formula

$\tan\theta=\left|\frac{m_1-m_2}{1+m_1m_2}\right|$

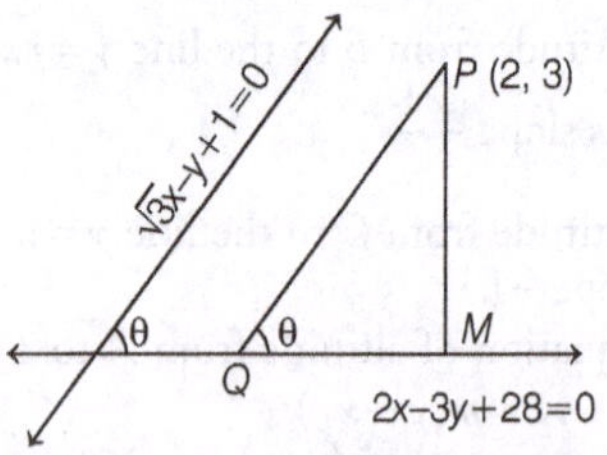


The distance of $P(2, 3)$ from the line $2x-3y+28=0$ measured parallel to the line $\sqrt{3}x-y+1=0$ is $PQ$.

Let $\theta$ be the angle between two given lines

$\tan\theta=\left|\frac{\frac{2}{3}-\sqrt{3}}{1+\frac{2}{3}\sqrt{3}}\right|$

$\left(\because m_1=\frac{2}{3}\text{ and } m_2=-\sqrt{3}\right)$

$=\tan\theta=\left|\frac{2-3\sqrt{3}}{3+2\sqrt{3}}\right|$

$\tan\theta=\frac{3\sqrt{3}-2}{2\sqrt{3}+3}$

Now, $PM=\left|\frac{2(2)-3(3)+28}{\sqrt{4+9}}\right|=\frac{23}{\sqrt{13}}$

$\because\qquad PQ=\frac{PM}{\sin\theta}=PM\operatorname{cosec}\theta$

$$= \frac{23}{\sqrt{13}} \left[ \frac{\sqrt{(2\sqrt{3}+3)^2 + (3\sqrt{3}-2)^2}}{3\sqrt{3}-2} \right]$$

$$= \frac{23}{\sqrt{13}} \left( \frac{\sqrt{12+9+12\sqrt{3}+27 + 4 - 12\sqrt{3}}}{3\sqrt{3}-2} \right)$$

$$= \frac{23}{\sqrt{13}} \times \frac{\sqrt{52}}{(3\sqrt{3}-2)}$$

$$= \frac{46}{3\sqrt{3}-2} \times \frac{3\sqrt{3}+2}{3\sqrt{3}+2}$$

$$= \frac{46(3\sqrt{3}+2)}{23} = 4 + 6\sqrt{3}$$

**68.** *(d)*

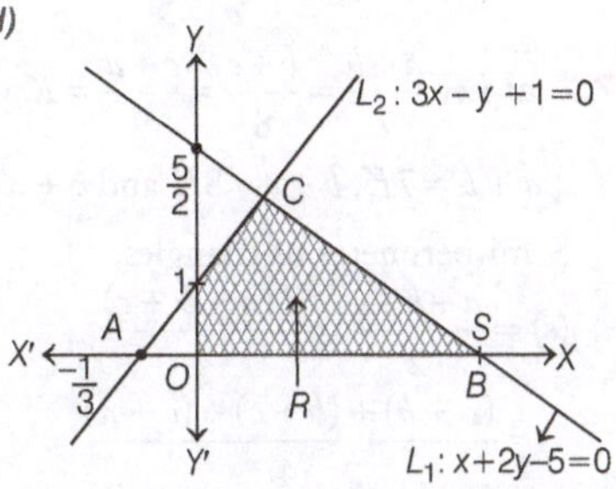


Let $L_1: x + 2y - 5 = 0$

$L_2: 3x - y + 1 = 0$

$R$ = Shaded region, where $P(a^2, a+1)$ should lie.

**For line $L_1$** : Origin and $P$ lies same side w.r.t. $L_1$

$$L_1(O) \cdot L_1(P) < 0$$

$\Rightarrow a^2 + 2(a+1) - 5 < 0$

$\Rightarrow a^2 + 2a - 3 < 0$

$\Rightarrow (a-1)(a+3) < 0$

$\Rightarrow -3 < a < 1$ ...(i)

**For line $L_2$**

$$L_2(O) \cdot L_2(P) > 0$$

$\Rightarrow 3a^2 - a - 1 + 1 > 0$

$\Rightarrow 3a^2 - a > 0$

$\Rightarrow a(3a-1) > 0$

$\Rightarrow a < 0$ or $a > \frac{1}{3}$ ...(ii)

From Eqs. (i) and (ii), we have

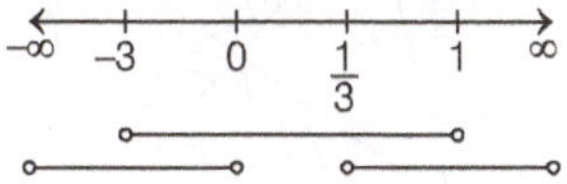


$\Rightarrow a \in (-3, 0) \cup \left(\frac{1}{3}, 1\right)$

**69.** *(b)* Let $\theta$ = Angle between $L_1$ and $L_2$

According to the question,

$$PA = AB = BQ$$

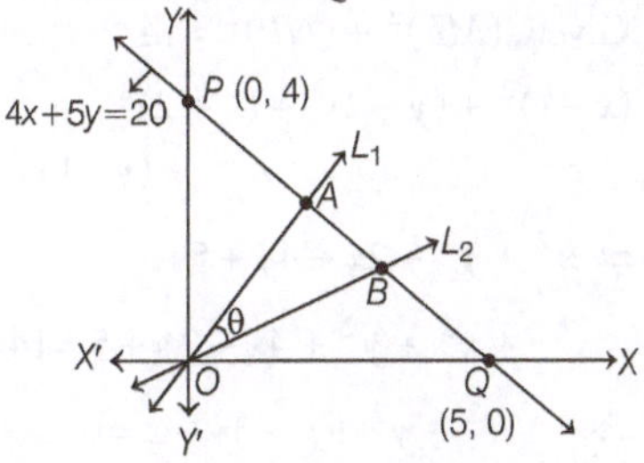


So, coordinates of $A \equiv \left(\frac{5}{3}, \frac{8}{3}\right)$

and coordinates of $B \equiv \left(\frac{10}{3}, \frac{4}{3}\right)$

$$\therefore m_{OA} = \text{Slope of } OA = \frac{\frac{8}{3} - 0}{\frac{5}{3} - 0} = \frac{8}{5}$$

$$m_{OB} = \text{Slope of } OB = \frac{\frac{4}{3} - 0}{\frac{10}{3} - 0} = \frac{2}{5}$$

$$\text{Now, } \tan\theta = \left| \frac{m_{OA} - m_{OB}}{1 + m_{OA} \times m_{OB}} \right|$$

$$= \left| \frac{\frac{8}{5} - \frac{2}{5}}{1 + \frac{8}{5} \times \frac{2}{5}} \right| = \left| \frac{\frac{6}{5}}{\frac{25+16}{25}} \right| = \frac{30}{41}$$

**70.** *(d)*

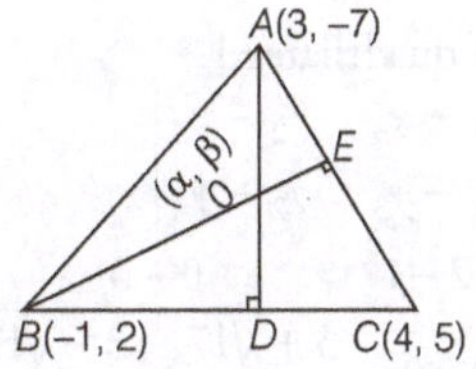


$$m_{AD} = \frac{-1}{m_{BC}} = \frac{-5}{3} \left[ \because m_{BC} = \frac{5-2}{4+1} = \frac{3}{5} \right]$$

$$m_{AC} = \frac{5+7}{4-3} = \frac{12}{1}; \; m_{BE} = -\frac{1}{12}$$

$$AD: (y+7) = \frac{-5}{3}(x-3)$$

$$3y + 21 = -5x + 15$$

$$5x + 3y + 21 - 15 = 0$$

$\Rightarrow 5x + 3y + 6 = 0$ ...(i)

$$BE: (y-2) = \frac{-1}{12}(x+1)$$

$$12y - 24 = -x - 1$$

$$x + 12y - 24 + 1 = 0$$

$\Rightarrow x + 12y - 23 = 0$ ...(ii)

$\Rightarrow x = 23 - 12y$

On putting the value of $x$ in Eq. (i), we get

$$5(23 - 12y) + 3y + 6 = 0$$

$\Rightarrow 115 - 60y + 3y + 6 = 0$

$$y = \frac{121}{57}$$

$\Rightarrow x = 23 - \frac{12 \times 121}{57} = -\frac{141}{57}$

Orthocentre $O(\alpha, \beta) = \left(\frac{-141}{57}, \frac{121}{57}\right)$

$$= 9\alpha - 6\beta + 60$$

$$= 9 \times \left(\frac{-141}{57}\right) - 6\left(\frac{121}{57}\right) + 60$$

$$= \frac{3 \times (-141)}{19} - \frac{2 \times 121}{19} + 60$$

$$= \frac{-423}{19} - \frac{242}{19} + 60 = 25$$

**71.** *(b)* We have the line $x \cos\theta + y \sin\theta = 7$, $\theta \in \left(0, \frac{\pi}{2}\right)$

$$\Rightarrow \frac{x}{7\sec\theta} + \frac{y}{7\operatorname{cosec}\theta} = 1$$

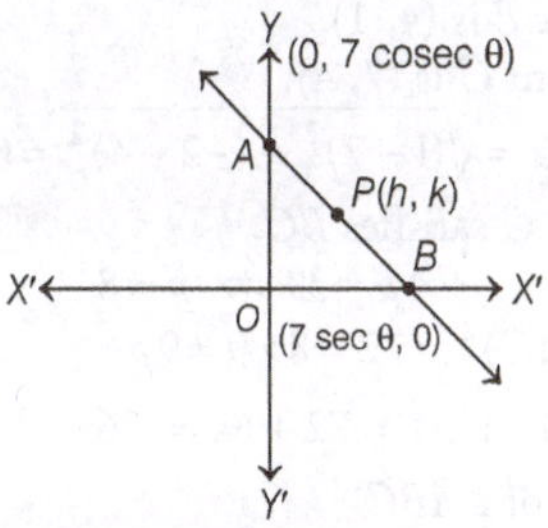


Let $P(h, k)$ be mid-point of the line segment $AB$.

$$\therefore h = \frac{7\sec\theta}{2}, \; k = \frac{7\operatorname{cosec}\theta}{2}$$

Since, $\theta \in \left(0, \frac{\pi}{2}\right) \Rightarrow h$ and $k$ are positive. Now, $\cos\theta = \frac{7}{2h}$, $\sin\theta = \frac{7}{2k}$

$$\Rightarrow \left(\frac{7}{2h}\right)^2 + \left(\frac{7}{2k}\right)^2 = 1$$

Also, $\left(\alpha, \frac{7\sqrt{3}}{3}\right)$ lies on this curve

$$\Rightarrow \left(\frac{7}{2\alpha}\right)^2 + \left(\frac{7 \cdot 3}{2 \cdot 7\sqrt{3}}\right)^2 = 1$$

$$\Rightarrow \frac{49}{4\alpha^2} + \frac{3}{4} = 1 \Rightarrow \alpha^2 = 49$$

$\Rightarrow \alpha = \pm 7$

Since $h > 0$ i.e. $x$-coordinate of points on curve are positive.

**72.** (d) $AB \to 2x + y = 0$

$AC \to x - y = 3$

$BC \to x + py = 39$

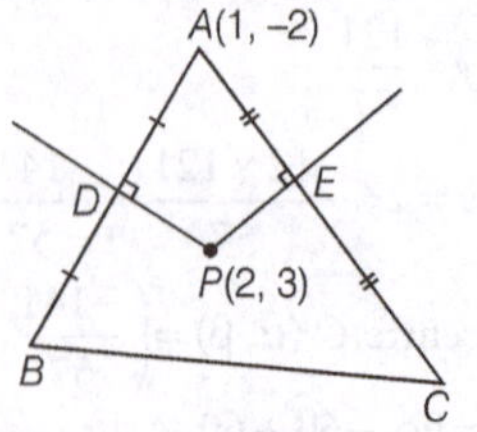


Point of intersection of $AB$, $AC$ is vertex $A = (1, -2)$

Let $PD$ be perpendicular bisector of $AB$.

Slope of $AB = -2$

$\therefore$ Slope of $PD = \frac{1}{2}$

$\therefore$ Equation of $PD$ is $x - 2y = -4$

$\therefore$ Point $D$ is $\left(\frac{-4}{5}, \frac{8}{5}\right)$

$\therefore$ Point $B$ is $\left(\frac{-13}{5}, \frac{26}{5}\right)$.

Similarly, equation of $PE$ is

$x + y = 5$

Point $E$ is (4, 1).

$\therefore$ Point $C$ is (7, 4).

$\therefore AC = \sqrt{(1-7)^2 + (-2-4)^2} = 6\sqrt{2}$

Also, $C$ satisfies $BC$.

$\therefore \quad 7 + 4p = 39 \Rightarrow p = 8$

$\Rightarrow (AC)^2 = 72 = 8 \times 9 = 9p$

$(AC)^2 + p^2 = 72 + 64 = 136$

Area of $\Delta ABC$

$$= \frac{1}{2}\begin{vmatrix} 1 & -2 & 1 \\ \frac{-13}{5} & \frac{26}{5} & 1 \\ 7 & 4 & 1 \end{vmatrix}$$

$$= \left|\frac{1}{2}\left[1\left(\frac{26}{5} - 4\right) + 2\left(\frac{-13}{5} - 7\right) + 1\left(\frac{-52}{5} - \frac{182}{5}\right)\right]\right|$$

$$= \left|\frac{1}{2}\left[\frac{6}{5} - \frac{96}{5} - \frac{234}{5}\right]\right| = 32.4$$

**73.** (b)

**key idea** (i) The orthocentre of a triangle is the point where the perpendicular drawn from the vertices to the opposite sides of the triangle intersect each other.

(ii) For perpendicular lines, product of their slopes is equal to – 1.

Let the coordinate of point $P$ be $(x, y)$.

Let the point $M$ and $N$ be (1, 2) and (– 2, 1) respectively.

Given, $(MP)^2 + (NP)^2 = 14$

$$(x-1)^2 + (y-2)^2 + (x+2)^2 + (y-1)^2 = 14$$

$$\Rightarrow x^2 + y^2 - 2x - 4y + 5 + x^2 + y^2 + 4x - 2y + 5 = 14$$

$$\Rightarrow \quad x^2 + y^2 + x - 3y - 2 = 0$$

$\therefore$ Locus of point $P$ is

$x^2 + y^2 + x - 3y - 2 = 0$

For the intersection of locus of point $P$ with $X$-axis, substitute $y = 0$

$$\Rightarrow \quad x^2 + x - 2 = 0$$

$$\Rightarrow x(x+2) - 1(x+2) = 0$$

$$\Rightarrow \quad x = -2, 1$$

Let point $A(-2, 0)$ and $B(1, 0)$.

For the intersection of locus of point $P$ with $Y$-axis, substitute $x = 0$.

$$y^2 - 3y - 2 = 0$$

$$\Rightarrow \quad y = \frac{3 \pm \sqrt{9+8}}{2} = \frac{3 \pm \sqrt{17}}{2}$$

$\therefore$ Point $C\left(0, \frac{3+\sqrt{17}}{2}\right)$ and point $D\left(0, \frac{3-\sqrt{17}}{2}\right)$.

Area of quadrilateral

$$= \frac{1}{2}\begin{vmatrix} x_1 - x_3 & x_2 - x_4 \\ y_1 - y_3 & y_2 - y_4 \end{vmatrix}$$

$$= \frac{1}{2}\begin{vmatrix} -2-1 & 0-0 \\ 0-0 & \frac{3+\sqrt{17}}{2} - \frac{3-\sqrt{17}}{2} \end{vmatrix}$$

$$= \left|\frac{1}{2}\right| - 3(\sqrt{17}) \left|\right| = \frac{3\sqrt{17}}{2} \text{ sq. units}$$

**74.** (c) $A$ and $A'$ lie on the line $y = 2$

And point $B$ is (2, 3).

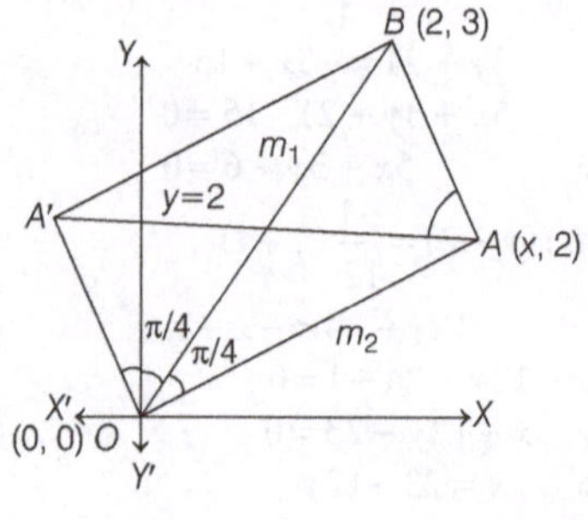


$$m_1 = \frac{3-0}{2-0} = \frac{3}{2}, \; m_2 = \frac{2-0}{x-0} = \frac{2}{x}$$

$$\tan\theta = \left|\frac{m_1 - m_2}{1 + m_1 m_2}\right|$$

$$\Rightarrow \quad \tan\frac{\pi}{4} = \left|\frac{\frac{3}{2} - \frac{2}{x}}{1 + \frac{3}{2}\cdot\frac{2}{x}}\right| \Rightarrow 1 = \left|\frac{3x-4}{2x+6}\right|$$

$$\Rightarrow \quad \pm(2x+6) = 3x - 4$$

It gives $\quad x = -\frac{2}{5}, 10; \; x_1 = -\frac{2}{5}$

and $\quad x_2 = 10$

$$\therefore \quad AA' = \sqrt{\left(10 + \frac{2}{5}\right)^2 + (2-2)^2}$$

$$= 10 + \frac{2}{5} = \frac{52}{5}$$

**75.** (a) Let $\frac{a+b}{7} = \frac{b+c}{8} = \frac{c+a}{9} = K$

$\therefore a + b = 7K, b + c = 8K$ and $c + a = 9K$

Semi-perimeter of triangles,

$$(s) = \frac{a+b+c}{2} = \frac{2(a+b+c)}{4}$$

$$= \frac{(a+b)+(b+c)+(c+a)}{4}$$

$$= \frac{7K + 8K + 9K}{4} = 6K$$

$a = 2s - (b+c) = 12K - 8K = 4K$

Similarly, $b = 3K$ and $c = 5K$

Now, $\Delta = \sqrt{s(s-a)(s-b)(s-c)}$

$= \sqrt{6K(2K)(3K)(K)} = 6K^2$

Radius of incircle $(r) = \frac{\Delta}{S} = \frac{6K^2}{6K} = K$

Radius of circumcircle $(R) = \frac{abc}{4\Delta}$

$$= \frac{4K \times 3K \times 5K}{4 \times 6K^2} = \frac{5K}{2}$$

$$\therefore \quad \frac{R}{r} = \frac{5K}{2} \div K = \frac{5}{2}$$

**76.** (b) Given, two sides of parallelogram are

$$4x + 5y = 0 \text{ and } 7x + 2y = 0$$

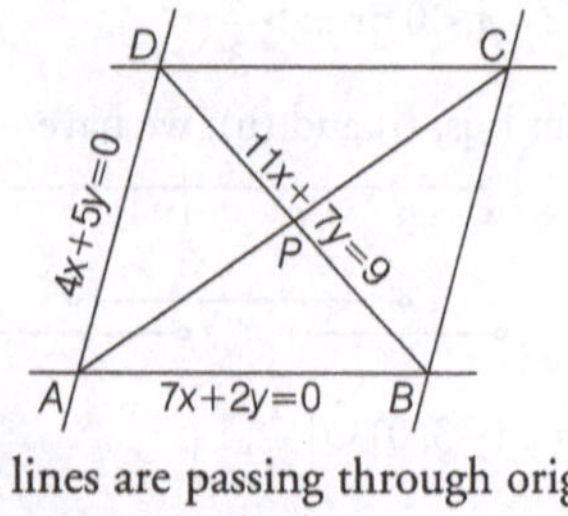


Both lines are passing through origin.

Thus, point $A = (0,0)$

The equation of one diagonal is
$11x+7y=9$

Point $D$ is the point of intersection of
$4x+5y=0$
and $\quad 11x+7y=9$

So, coordinate of point $D=\left(\frac{5}{3},-\frac{4}{3}\right)$

Also, point $B$ is the point of intersection of $7x+2y=0$ and $11x+7y=9$

So, coordinate of point $B=\left(-\frac{2}{3},\frac{7}{3}\right)$

We know that, diagonals of parallelogram bisects each other.
Let $P$ is the middle point of $BD$.
So, coordinate of

$$P=\left(\frac{\frac{5}{3}+\left(-\frac{2}{3}\right)}{2},\frac{-\frac{4}{3}+\frac{7}{3}}{2}\right)=\left(\frac{1}{2},\frac{1}{2}\right)$$

Now, equation of diagonal $AC$

$$y-0=\frac{\frac{1}{2}-0}{\frac{1}{2}-0}(x-0)\Rightarrow y=\frac{\frac{1}{2}}{\frac{1}{2}}x$$

$\Rightarrow y=x$

$\therefore$ Diagonal $AC$ passes through (2, 2).

**77.** *(a)* **Method I** Let $m=$ Slope of required line

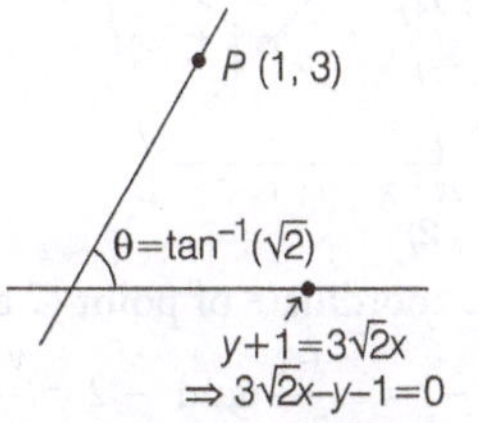


$\therefore$ Equation of required line

$$y-3=m(x-1) \quad \ldots\text{(i)}$$

Given, equation of line is

$$3\sqrt{2}x-y-1=0 \quad \ldots\text{(ii)}$$

Since, angle $\theta$ between Eqs. (i) and (ii) is $\tan^{-1}(\sqrt{2})$

i.e. $\quad \tan\theta=\sqrt{2}$

$$\Rightarrow \left|\frac{m-3\sqrt{2}}{1+3\sqrt{2}m}\right|=\sqrt{2} \quad (m=3\sqrt{2})$$

($\because$ Slope of Eq. (i) $=m$ and slope of Eq. (ii) $=3\sqrt{2}$)

Squaring on both sides,

$m^2-6\sqrt{2}m+18=2(1+18m^2+6\sqrt{2}m)$

$\Rightarrow 35m^2+18\sqrt{2}m-16=0$

$$\therefore \quad m=\frac{-18\sqrt{2}\pm\sqrt{648+2240}}{70}$$

$$=\frac{-18\sqrt{2}\pm 38\sqrt{2}}{70}$$

$$\Rightarrow \quad m=\frac{2\sqrt{2}}{7},-\frac{4}{5}\sqrt{2}$$

For $m=\frac{2\sqrt{2}}{7}$ and point (1, 3) equation of required line will be

$$y-3=\frac{2\sqrt{2}}{7}(x-1)$$

$\Rightarrow 2\sqrt{2}x-7y+21-2\sqrt{2}=0$

(options are not matching so, neglect this)

For $m=\frac{-4\sqrt{2}}{5}$, equation of required line will be

$$y-3=\frac{-4\sqrt{2}}{5}(x-1)$$

$\Rightarrow \quad 5y-15=-4\sqrt{2}x+4\sqrt{2}$

$\Rightarrow \quad 4\sqrt{2}x+5y-15-4\sqrt{2}=0$

$\Rightarrow \quad 4\sqrt{2}x+5y-(15+4\sqrt{2})=0$

**Method II** Let equation of required line be $y=mx+c$ $\quad\ldots$(i)

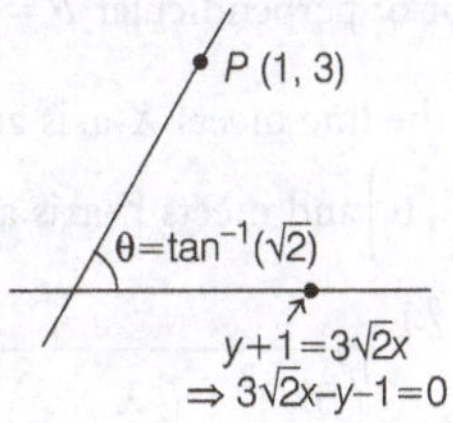


$\because$ Eq (i) passes through (1, 3).

$\therefore \quad 3=m+c\Rightarrow c=3-m \quad \ldots$(ii)

Let $\theta=$ angle between Eq. (i) and the given line

$$y+1=3\sqrt{2}x=\tan^{-1}(\sqrt{2})$$

$\therefore \quad \tan\theta=\sqrt{2}$

$$\Rightarrow \left|\frac{m-3\sqrt{2}}{1+3\sqrt{2}m}\right|=\sqrt{2}\Rightarrow \frac{m-3\sqrt{2}}{1+3\sqrt{2}m}=\pm\sqrt{2}$$

(using $|x|=a,\ a>0;\ \therefore x=\pm a$)

$$\Rightarrow \frac{m-3\sqrt{2}}{1+3\sqrt{2}m}=\sqrt{2} \text{ or } \frac{m-3\sqrt{2}}{1+3\sqrt{2}m}=-\sqrt{2}$$

$\Rightarrow \quad \sqrt{2}+6m=m-3\sqrt{2}$

or $\quad -\sqrt{2}-6m=m-3\sqrt{2}$

$\Rightarrow \quad 5m=-4\sqrt{2}$ or $-7m=-2\sqrt{2}$

$$\Rightarrow \quad m=\frac{-4\sqrt{2}}{5} \text{ or } m=\frac{2\sqrt{2}}{7}$$

For the value of $m=\frac{2\sqrt{2}}{7}$ equation of required line will not match to the given options. So, neglect this value of $m$.

From Eq. (ii), we get

$c=3-m$

$$=3+\frac{4\sqrt{2}}{5} \quad \left(\because m=\frac{-4\sqrt{2}}{5}\right)$$

$$=\frac{15+4\sqrt{2}}{5}$$

From Eq. (i), we get

$$y=\frac{-4\sqrt{2}}{5}x+\frac{15+4\sqrt{2}}{5}$$

$\Rightarrow \quad 4\sqrt{2}x+5y-(15+4\sqrt{2})=0$

**78.** *(a)* It is given that the line '$L$' made $x$ and $y$ intercepts 3 and 1 respectively, so equation of line $L$ is

$$\frac{x}{3}+\frac{y}{1}=1\Rightarrow x+3y=3 \quad \ldots\text{(i)}$$

Now, image of point $(-1,-4)$ with respect to line (i) is $(x_1,y_1)$, then

$$\frac{x_1+1}{1}=\frac{y_1+4}{3}=-2\left(\frac{-1-12-3}{1+9}\right)=\frac{16}{5}$$

Since, image $(h,k)$ or $(x_1,y_1)$ w.r.t. the line mirror $ax+by+c=0$ is given by

$$\left[\frac{h-x_1}{a}=\frac{k-y_1}{b}=\frac{-2(ax_1+by_1+c)}{a^2+b^2}\right]$$

$$\Rightarrow \quad x_1=\frac{16}{5}-1=\frac{11}{5}$$

and $\quad y_1=\left(\frac{16}{5}\times 3\right)-4=\frac{28}{5}$

$$\therefore \quad (x_1,y_1)=\left(\frac{11}{5},\frac{28}{5}\right)$$

**79.** *(b)* According to given information,
Let $AC=l$ and

$AB=\sqrt{(3-1)^2+(1-2)^2}=\sqrt{5}$

$\because$The area of right angled $\Delta ABC$

$$=\frac{1}{2}(AC)(AB)=5\sqrt{5} \quad \text{[given]}$$

$$\Rightarrow \quad \frac{1}{2}l(\sqrt{5})=5\sqrt{5}\Rightarrow l=10 \text{ units}$$

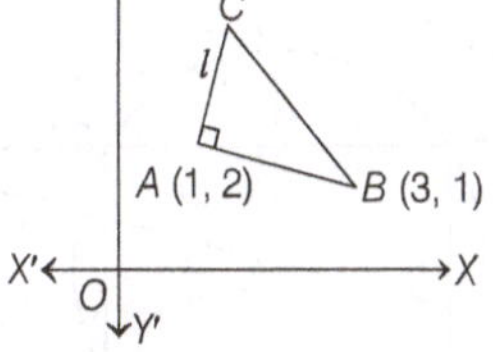


Now according to concept of rotation abscissa of the vertex $C$ we will get easily if we write the coordinate of the vertices in complex form, like $A(1+2i)$, $B(3+i)$ and let $C(x_1+iy_1)$.

Then, $\frac{x_1+iy_1-1-2i}{3+i-1-2i}=\frac{AC}{AB}e^{i\frac{\pi}{2}}$

$\Rightarrow \frac{(x_1-1)+i(y_1-2)}{2-i}=\frac{10}{\sqrt{5}}i$

$\Rightarrow (x_1-1)+i(y_1-2)=2\sqrt{5}i(2-i)$

$\Rightarrow (x_1-1)+i(y_1-2)=2\sqrt{5}+4\sqrt{5}i$

$\therefore x_1=1+2\sqrt{5}$ is abscissa of the vertex $C$.

**80.** *(b)* The vertices of given $\Delta ABC$ are $A(-1, 7)$, $B(-7, 1)$ and $C(5, -5)$.

Since, slope of line $BC$ is

$$m_1=\frac{-5-1}{5+7}=\frac{-6}{12}=-\frac{1}{2}$$

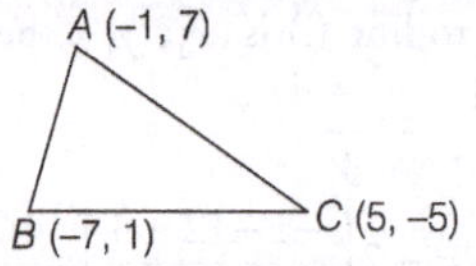


So, slope of altitude through vertex $A$ is

$$m_1'=-\frac{1}{m}=2$$

$\therefore$ Equation of altitude through vertex $A$ is $y-7=2(x+1)$

$\Rightarrow 2x-y+9=0$ ...(i)

Similarly, the slope of line $CA$ is

$$m_2=\frac{7-(-5)}{-1-5}=\frac{12}{-6}=-2$$

So, slope of altitude through vertex $B$ is

$$m_2'=\frac{1}{2}$$

$\therefore$ Equation of altitude through vertex $B$ is

$y-1=\frac{1}{2}(x+7)$

$\Rightarrow x-2y+9=0$ ...(ii)

As, the point of intersection of altitudes Eqs. (i) and (ii) is the orthocentre.

Then, the orthocentre is $(-3, 3)$.

**81.** *(a)* Given situation

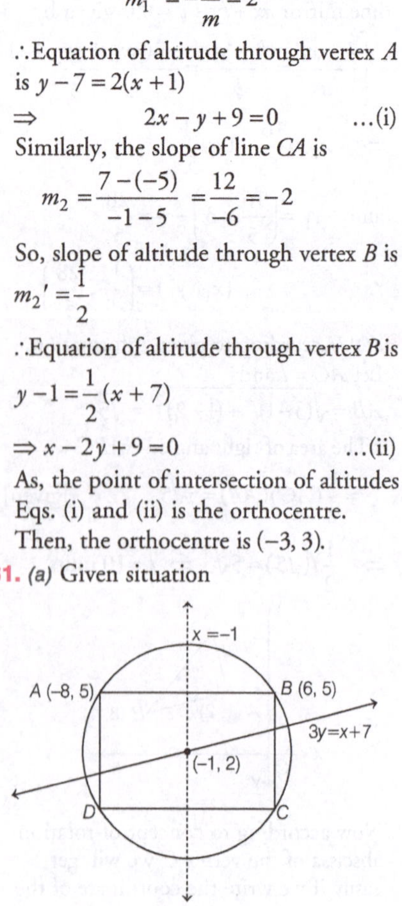


$\because$ Perpendicular bisect of $AB$ will pass from centre.

$\therefore$ Equation of perpendicular bisector $x=-1$

Hence, centre of the circle is $(-1, 2)$.

Let coordinate of $D$ is $(x, y)$.

$\Rightarrow \frac{x+6}{2}=-1$ and $\frac{y+5}{2}=2$

$\Rightarrow x=-8$ and $y=-1$,

$\therefore D\equiv(-8,-1)$

$|AD|=6$ and $|AB|=14$

Area of the rectangle $=6\times 14=84$

**82.** *(d)* Equation of the line, which is perpendicular to the line, $3x+y=\lambda$ $(\lambda\neq 0)$ and passing through origin, is given by

$$\frac{x-0}{3}=\frac{y-0}{1}$$

For foot of perpendicular

$$r=\frac{|(3\times 0)+(1\times 0)-\lambda|}{3^2+1^2}=\frac{\lambda}{10}$$

So, foot of perpendicular $P=\left(\frac{3\lambda}{10},\frac{\lambda}{10}\right)$

Given the line meets $X$-axis at

$A=\left(\frac{\lambda}{3},0\right)$ and meets $Y$-axis at

$B=(0,\lambda)$

So, $BP=\sqrt{\left(\frac{3\lambda}{10}\right)^2+\left(\frac{\lambda}{10}-\lambda\right)^2}$

$\Rightarrow BP=\sqrt{\frac{9\lambda^2}{100}+\frac{81\lambda^2}{100}} \Rightarrow BP=\sqrt{\frac{90\lambda^2}{100}}$

Now,

$PA=\sqrt{\left(\frac{\lambda}{3}-\frac{3\lambda}{10}\right)^2+\left(0+\frac{\lambda}{10}\right)^2}$

$\Rightarrow PA=\sqrt{\frac{\lambda^2}{900}+\frac{\lambda^2}{100}} \Rightarrow PA=\sqrt{\frac{10\lambda^2}{900}}$

Therefore, $BP:PA=3:1$

**83.** *(c)* Given, area $=56$ sq units

$$\Delta=\begin{vmatrix} x_1 & y_1 & 1\\ x_2 & y_2 & 1\\ x_3 & y_3 & 1\end{vmatrix}$$

$$\begin{vmatrix} k & -3k & 1\\ 5 & k & 1\\ -k & 2 & 1\end{vmatrix}=\pm 56$$

Expanding, we get

$k(k-2)-5(-3k-2)-k(-3k-k)=\pm 56$

$\Rightarrow k^2-2k+15k+10+3k^2+k^2=\pm 56$

$\Rightarrow 5k^2+13k+10=\pm 56$

Taking the positive sign

$5k^2+13k-46=0$

$\Rightarrow (5k+23)(k-2)=0$

$\therefore k=2$ is an integer

Taking the negative sign

$5k^2+13k+66=0$

Here, $D=(13^2-4\times 5\times 66)<0$

$\Rightarrow$ There is no solution in this case.

So, the vertices are $A(2,-6)$, $B(5, 2)$ and $C(-2, 2)$.

The equation of altitude from $A$ is $x=2$ and the equation of altitude from $C$ is $y-2=-\frac{3}{8}(x+2)$

$\because 3x+8y-10=0$

Solving the two we get the orthocentre as $\left(2,\frac{1}{2}\right)$.

**84.** *(c)* As the given lines $x-y+1=0$ and $7x-y-5=0$ are not parallel, therefore they represent the adjacent sides of the rhombus.

On solving $x-y+1=0$ and $7x-y-5=0$, we get $x=1$ and $y=2$.

Thus, one of the vertex is $A(1, 2)$.

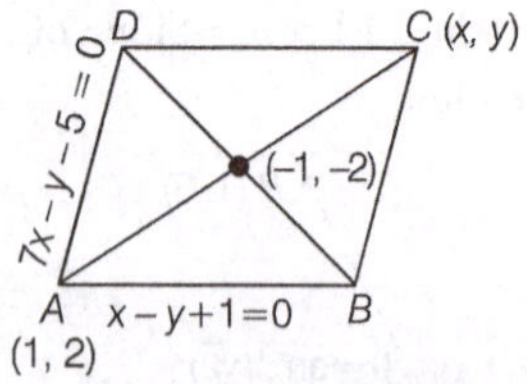


Let the coordinate of point $C$ be $(x, y)$.

Then, $-1=\frac{x+1}{2}$ and $-2=\frac{y+2}{2}$

$\Rightarrow x+1=-2$ and $y=-4-2$

$\Rightarrow x=-3$ and $y=-6$

Hence, coordinates of $C=(-3,-6)$

Note that, vertices $B$ and $D$ will satisfy $x-y+1=0$ and $7x-y-5=0$, respectively.

Since, option (c) satisfies $7x-y-5=0$, therefore coordinate of vertex $D$ is $\left(\frac{1}{3},\frac{-8}{3}\right)$.

**85.** *(c)* Let slope of incident ray be $m$.

$\therefore$ Angle of incidence = angle of reflection

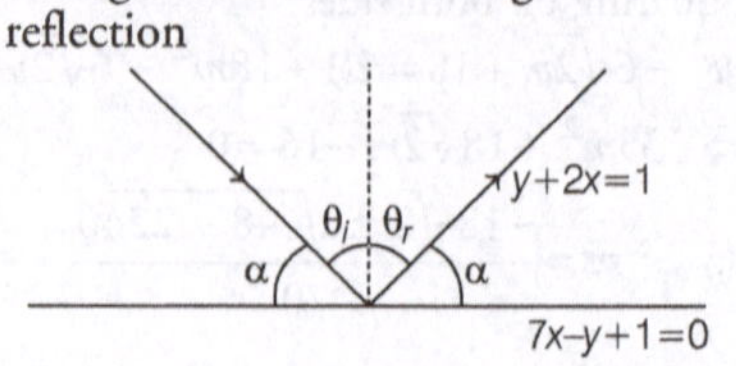

$\therefore \quad \left|\dfrac{m-7}{1+7m}\right| = \left|\dfrac{-2-7}{1-14}\right| = \dfrac{9}{13}$

$\Rightarrow \quad \dfrac{m-7}{1+7m} = \dfrac{9}{13}$

or $\quad \dfrac{m-7}{1+7m} = -\dfrac{9}{13}$

$\Rightarrow \quad 13m - 91 = 9 + 63m$

or $\quad 13m - 91 = -9 - 63m$

$\Rightarrow \quad 50m = -100$

or $\quad 76m = 82$

$\Rightarrow \quad m = -2$ or $m = \dfrac{41}{38}$

$\Rightarrow y - 1 = -2(x-0)$

or $y - 1 = \dfrac{41}{38}(x - 0)$

i.e, $\quad 2x + y - 1 = 0$

or $\quad 38y - 38 - 41x = 0$

$\Rightarrow \quad 41x - 38y + 38 = 0$

**86.** *(b)* Given, $3x + 4y = 9$

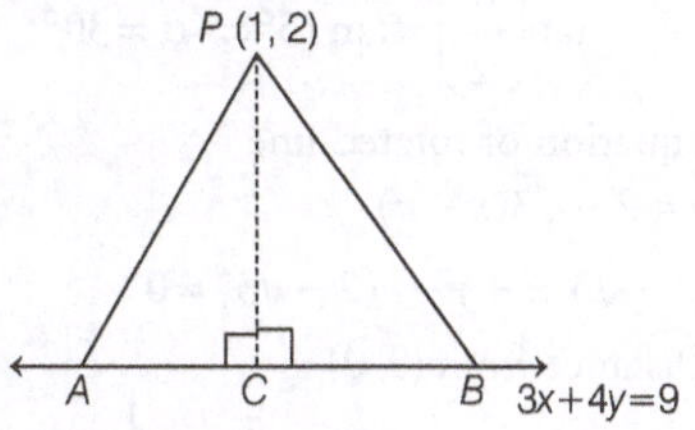


Shortest distance of a point $(a_1, b_1)$ from line $ax + by = c$ is

$$d = \left|\frac{aa_1 + bb_1 - c}{\sqrt{a^2 + b^2}}\right|$$

Now shortest distance of $P(1, 2)$ from $3x + 4y = 9$ is

$$PC = d = \left|\frac{3(1) + 4(2) - 9}{\sqrt{3^2 + 4^2}}\right| = \frac{2}{5}$$

We know that $\Delta APB$ is an equilateral triangle.

Let '$a$' be its side, then $PB = a$, $CB = \dfrac{a}{2}$

Now, in $\Delta PCB$,

$(PB)^2 = (PC)^2 + (CB)^2$

(by Pythagoras theorem)

$$a^2 = \left(\frac{2}{5}\right)^2 + \frac{a^2}{4}$$

$$a^2 - \frac{a^2}{4} = \frac{4}{25}$$

$$\Rightarrow \frac{3a^2}{4} = \frac{4}{25};\ a^2 = \frac{16}{75}$$

$$\Rightarrow \quad a = \sqrt{\frac{16}{75}} = \frac{4}{5\sqrt{3}} \times \frac{\sqrt{3}}{\sqrt{3}} = \frac{4\sqrt{3}}{15}$$

$\therefore$ Length of side of equilateral triangle

$$a = \frac{4\sqrt{3}}{15}$$

**87.** *(b)* Let $a$ be $x$-intercept and $b$ is $y$-intercept of the required line.

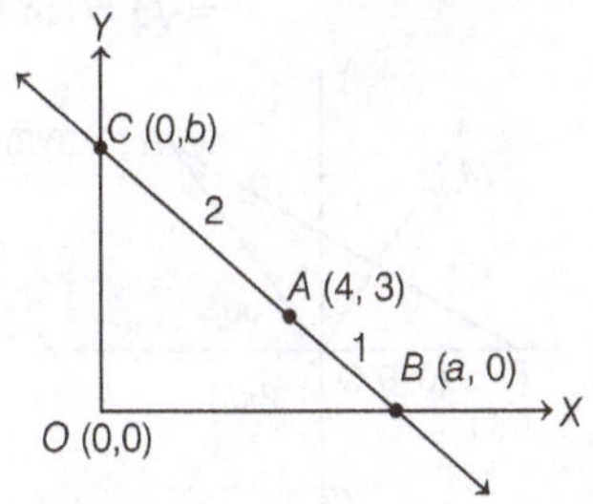


$A$ divides $CB$ in $2:1$

$$\Rightarrow 4 = \left(\frac{1 \times 0 + 2 \times a}{1+2}\right) = \frac{2a}{3}$$

$\Rightarrow a = 6 \Rightarrow$ coordinate of $B$ is $(6, 0)$.

$$3 = \left(\frac{1 \times b + 2 \times 0}{1+2}\right) = \frac{b}{3}$$

$\Rightarrow b = 9 \Rightarrow C = (0, 9)$

Slope of line passing through $(6, 0)$ and $(0, 9)$, $m = \dfrac{9}{-6} = -\dfrac{3}{2}$

Equation of line $y - 0 = \dfrac{-3}{2}(x - 6)$

$$2y = -3x + 18$$

$$\Rightarrow \quad 3x + 2y = 18$$

**88.** *(b)* Given, points $(2a, 0)$ and $(0, a)$.

Let $y$-coordinate of $C = b$

$\therefore C = (2a, b)$

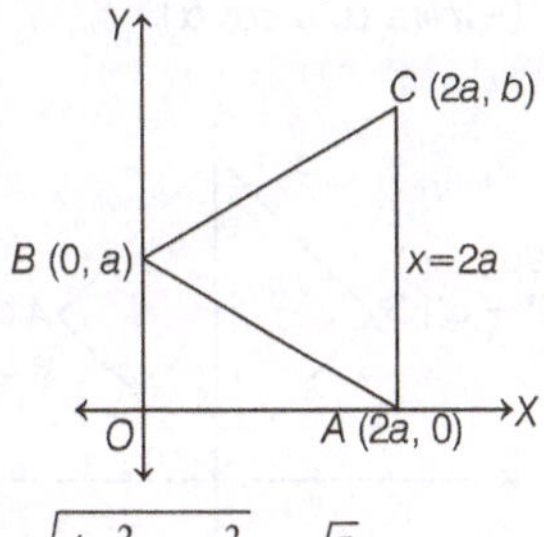


$AB = \sqrt{4a^2 + a^2} = \sqrt{5}a$

Now, $AC = BC$

$\Rightarrow \quad b = \sqrt{4a^2 + (b-a)^2}$

On squaring both sides, we get

$\Rightarrow b^2 = 4a^2 + b^2 + a^2 - 2ab$

$\Rightarrow 2ab = 5a^2 \Rightarrow b = \dfrac{5a}{2}$

$\therefore \quad C = \left(2a, \dfrac{5a}{2}\right)$

Hence, area of the triangle

$$= \frac{1}{2}\left\|\begin{matrix} 2a & 0 & 1 \\ 0 & a & 1 \\ 2a & \frac{5a}{2} & 1 \end{matrix}\right\| = \frac{1}{2}\left\|\begin{matrix} 2a & 0 & 1 \\ 0 & a & 1 \\ 0 & \frac{5a}{2} & 0 \end{matrix}\right\|$$

$$= \frac{1}{2} \times \left|2a\left(-\frac{5a}{2}\right)\right|$$

$$= \frac{5a^2}{2} \text{ sq units}$$

**89.** *(d)* Mid-point of $P(2, 3)$ and $Q(4, 5)$ $= (3, 4)$

Slope of $PQ = 1$

Slope of the line $L = -1$

Mid-point $(3, 4)$ lies on the line $L$.

Equation of line $L$,

$$y - 4 = -1(x - 3)$$

$\Rightarrow \quad x + y - 7 = 0 \quad$ ...(i)

Let image of point $R(0, 0)$ be $S(a_1, b_1)$

Mid-point of $RS = \left(\dfrac{a_1}{2}, \dfrac{b_1}{2}\right)$

Mid-point $\left(\dfrac{a_1}{2}, \dfrac{b_1}{2}\right)$ lies on the line (i)

$\therefore \quad a_1 + b_1 = 14 \quad$ ...(ii)

Slope of $RS = \dfrac{b_1}{a_1}$

Since $RS \perp$ line $L$.

$\therefore \quad \dfrac{b_1}{a_1} \times (-1) = -1$

$a_1 = b_1 \quad$ ...(iii)

From Eqs. (ii) and (iii), we get

$$a_1 = b_1 = 7$$

Hence, the image of $R = (7, 7)$

**90.** *(b)* Here, $L_1 : y - x = 0$,

$L_2 : 2x + y = 0$

and $\quad L_3 : y + 2 = 0$ shown as

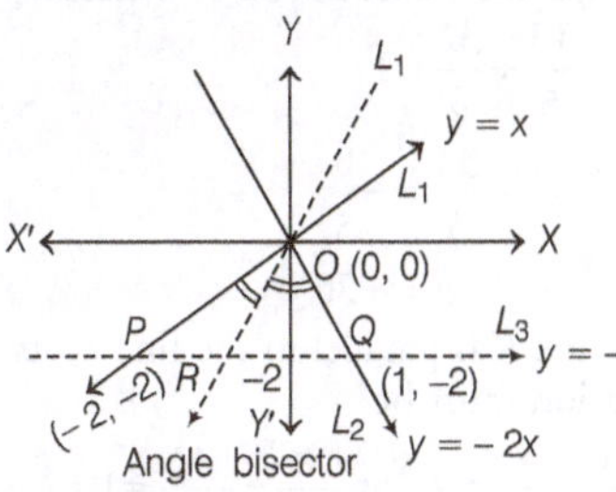


$$|PO| = \sqrt{4+4} = 2\sqrt{2}$$

$$|OQ| = \sqrt{1+4} = \sqrt{5}$$

Since, $OR$ is angle bisector.

$$\therefore \quad \frac{OP}{OQ} = \frac{PR}{RQ}$$

In a triangle, angle bisector divides the opposite sides in the ratio of side containing the angle.

$\Rightarrow \quad \dfrac{PR}{RQ}=\dfrac{2\sqrt{2}}{\sqrt{5}}$

So Statement I is true.

But, it does not divide the triangle in two similar triangles. So, Statement II is false.

**91.** *(c)*

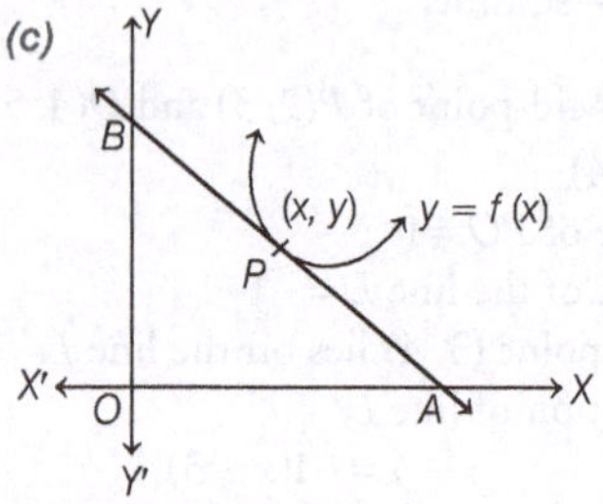


Equation of $AB$ is

$$\frac{Y-y}{X-x}=\left(\frac{dy}{dx}\right)_{(x,\,y)} \quad \ldots\text{(i)}$$

$x$-intercept is $\left(x-y\cdot\left(\dfrac{dx}{dy}\right),0\right)$

$y$-intercept is $\left(0,\ y-x\dfrac{dy}{dx}\right)$

As $P$ is mid-point of $AB$.

$\therefore 2x=x-y\cdot\left(\dfrac{dx}{dy}\right) \Rightarrow y\dfrac{dx}{dy}=-x$

$\therefore \quad \dfrac{dx}{x}+\dfrac{dy}{y}=0$

Integrating both sides, we get

$\log x+\log y=\log c$

$\Rightarrow xy=c$, as it passes through (2, 3).

$\therefore \quad c=6 \Rightarrow xy=6$

**92.** *(d)* Since, the line $L$ is passing through the point (13, 32). Therefore,

$$\frac{13}{5}+\frac{32}{b}=1$$

$\Rightarrow \quad \dfrac{32}{b}=-\dfrac{8}{5}$

$\Rightarrow \quad b=-20$

The line $K$ is parallel to the line $L$, its equation must be

$$\frac{x}{5}-\frac{y}{20}=a \quad \text{or} \quad \frac{x}{5a}-\frac{y}{20a}=1$$

On comparing with $\dfrac{x}{c}+\dfrac{y}{3}=1$,

we get

$$20a=-3,\ c=5a=5\left(-\frac{3}{20}\right)=-\frac{3}{4}$$

Hence, the distance between lines

$$=\frac{|a-1|}{\sqrt{\dfrac{1}{25}+\dfrac{1}{400}}}=\frac{\left|\dfrac{-3}{20}-1\right|}{\sqrt{\dfrac{17}{400}}}=\frac{23}{\sqrt{17}}$$

**93.** *(a)* Now, slope of $QR=\dfrac{3\sqrt{3}-0}{3-0}$

$=\sqrt{3}=\tan\theta$

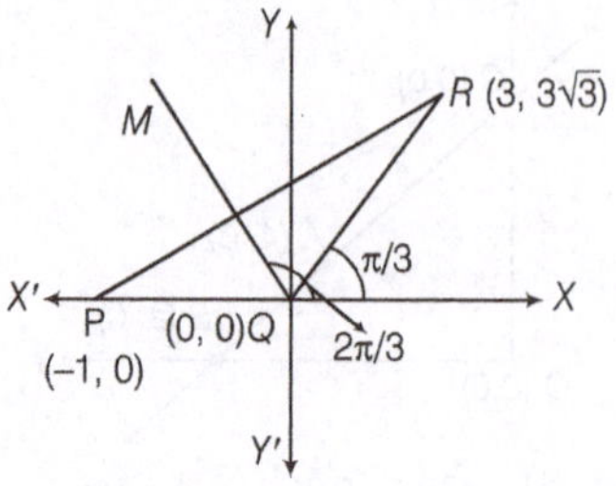


$\Rightarrow \quad \theta=\pi/3$

The angle $PQR$ is $\dfrac{2\pi}{3}$, so the line $QM$ makes an angle $\dfrac{2\pi}{3}$ from positive direction of $X$-axis.

Slope of the line $QM=\tan\dfrac{2\pi}{3}=-\sqrt{3}$

Hence, equation of line $QM$ is $y=-\sqrt{3}x$

or $\quad \sqrt{3}x+y=0$

**94.** *(d)* Since, line $OA$ makes an angle $\alpha$ with $X$-axis and given $OA=a$, then coordinates of $A$ are $(a\cos\alpha,\ a\sin\alpha)$.

Also, $OB\perp OA$, then $OB$ makes an angle $(90°+\alpha)$ with $X$-axis, then coordinates of $B$ are

$[a\cos(90°+\alpha),\ a\sin(90°+\alpha)]$.

i.e., $\quad(-a\sin\alpha,\ a\cos\alpha)$

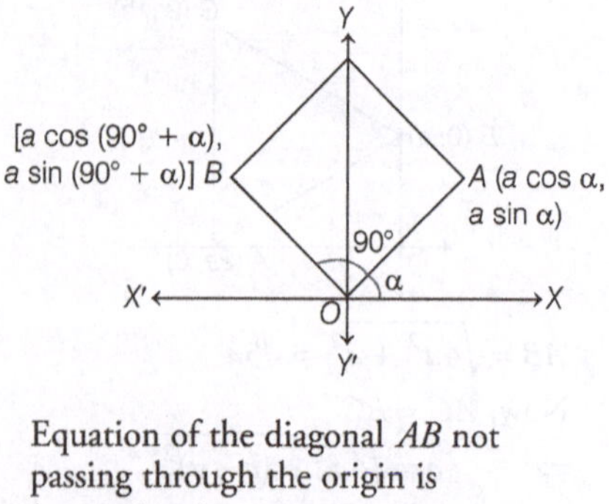


Equation of the diagonal $AB$ not passing through the origin is

$$(y-a\sin\alpha)=\frac{a\cos\alpha-a\sin\alpha}{-a\sin\alpha-a\cos\alpha}(x-a\cos\alpha)$$

$\Rightarrow (\sin\alpha+\cos\alpha)(y-a\sin\alpha)$

$=(\sin\alpha-\cos\alpha)(x-a\cos\alpha)$

$\Rightarrow y(\sin\alpha+\cos\alpha)+x(\cos\alpha-\sin\alpha)$

$=a\sin\alpha(\sin\alpha+\cos\alpha)$

$\quad -a\cos\alpha(\sin\alpha-\cos\alpha)$

$=a(\sin^2\alpha+\sin\alpha\cos\alpha$

$\quad -\cos\alpha\sin\alpha+\cos^2\alpha)$

$\therefore y(\sin\alpha+\cos\alpha)+x(\cos\alpha-\sin\alpha)=a$

**95.** *(a)*

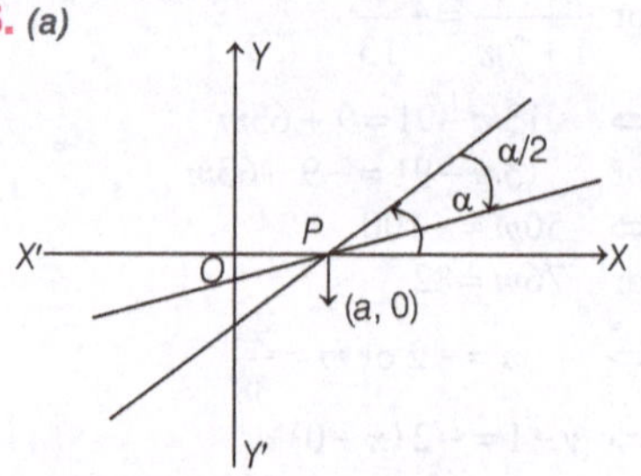


Inclination of rotated line $=\alpha/2$

Slope of rotated line $=\tan\left(\dfrac{\alpha}{2}\right)$

$\Rightarrow \quad \tan\left(\dfrac{\alpha}{2}\right)=2-\sqrt{3}$

$\tan\left(\dfrac{\alpha}{2}\right)=\tan 15° \Rightarrow \alpha=30°$

Equation of rotated line

$y=2-\sqrt{3}(x-a)$

$(2-\sqrt{3})x-y-a(2-\sqrt{3})=0$

Distance from (0, 0) to

$(2-\sqrt{3})x-y-a(2-\sqrt{3})=\dfrac{1}{\sqrt{2}}$

$$\frac{1}{\sqrt{2}}=\frac{(2-\sqrt{3})|a|}{\sqrt{(2-\sqrt{3})^2+1^2}}$$

$\Rightarrow \quad a^2=\dfrac{(2-\sqrt{3})^2+1^2}{2(2-\sqrt{3})^2}$

$\therefore \quad 3a^2\tan^2\alpha-2\sqrt{3}$

$$=\frac{3\times(8-4\sqrt{3})}{2(2-\sqrt{3})^2}\times\tan^2 30°-2\sqrt{3}$$

$$=\frac{12(2-\sqrt{3})}{2(2-\sqrt{3})^2}\times\frac{1}{3}-2\sqrt{3}$$

$$=\frac{2}{2-\sqrt{3}}-2\sqrt{3}$$

$$=\frac{2-4\sqrt{3}+6}{2-\sqrt{3}}=4$$

**96.** *(a)*

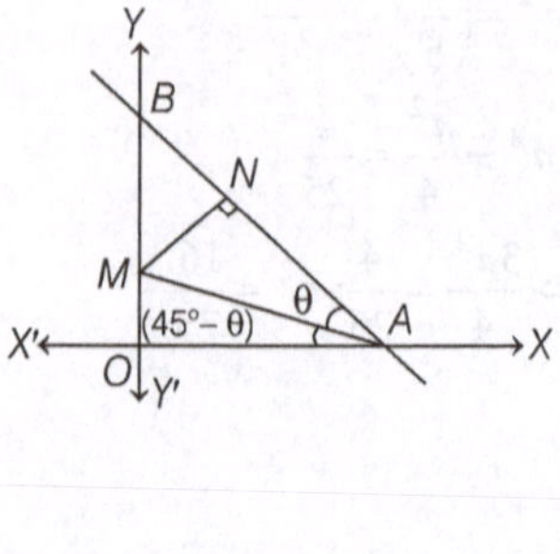

$$\text{ar}(AOB) = \frac{1}{2}$$

$$\text{ar}(\Delta AMN) = \frac{4}{9} \times \frac{1}{2} = \frac{2}{9}$$

Equation of $AB$ is $x + y = 1$

$OA = 1$, $AM = \sec(45° - \theta)$

$AN = \sec(45° - \theta) \cos\theta$

$MN = \sec(45° - \theta) \sin\theta$

Now, ar $(\Delta AMN)$

$$= \frac{1}{2}\sec^2(45° - \theta)\sin\theta \cdot \cos\theta = \frac{2}{9}$$

$$\Rightarrow \tan\theta = 2, \frac{1}{2}$$

$\tan\theta = 2$ is rejected

So, $\dfrac{AN}{NB} = \dfrac{\lambda}{1} = \cot\theta = 2$

**97.** *(a)* Given equation of sides of $\Delta ABC$ are

$AB: 7x - 6y + 3 = 0$ ...(i)

$BC: 9x - 2y - 19 = 0$ ...(ii)

$CA: x + 2y - 31 = 0$ ...(iii)

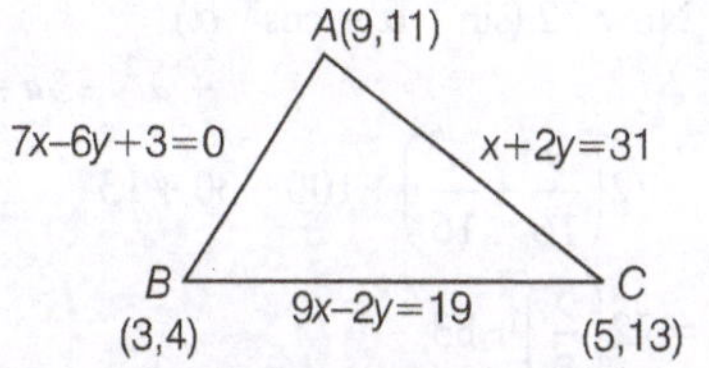


On solving these equations taken two at a time we get the coordinates of $A$, $B$ and $C$ such that

$A \equiv (9, 11)$, $B \equiv (3, 4)$ and $C \equiv (5, 13)$

Now, centroid of

$$\Delta ABC = \left(\frac{9+3+5}{3}, \frac{11+4+13}{3}\right)$$

$$= \left(\frac{17}{3}, \frac{28}{3}\right)$$

Let image of the centroid $P\left(\dfrac{17}{3}, \dfrac{28}{3}\right)$ in the line mirror $3x + 6y - 53 = 0$ is $Q(h, k)$.

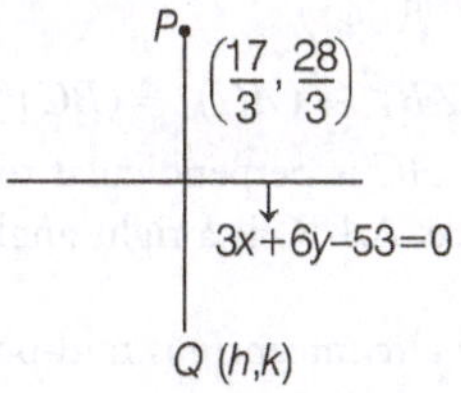


So, $m_{PQ} \times m_{\text{line}} = -1$

$$\Rightarrow \left(\frac{k - \frac{28}{3}}{h - \frac{17}{3}}\right) \times \left(-\frac{1}{2}\right) = -1 \quad \text{...(iv)}$$

$$\text{and } 3\left(\frac{h + \frac{17}{3}}{2}\right) + 6\left(\frac{k + \frac{28}{3}}{2}\right) = 53 \quad \text{...(v)}$$

[$\because$ mid-point of $PQ$ lies on $3x + 6y - 53 = 0$]

On solving Eqs. (iv) and (v), we get

$h = 3$, $k = 4$

$\therefore h^2 + k^2 + hk = 9 + 16 + 12 = 37$

**98.** *(a)*

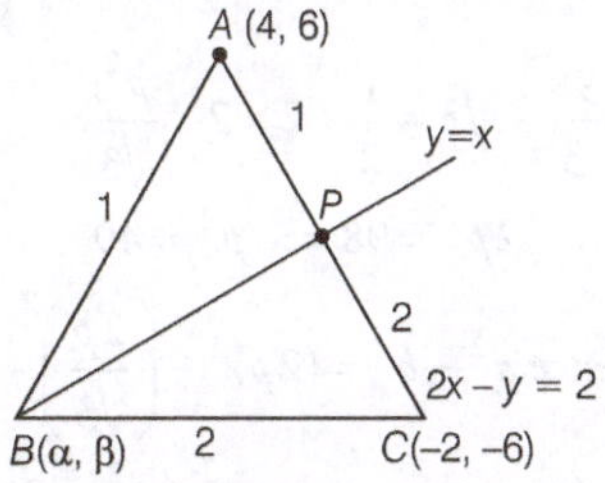


Given, $2AB = BC$

$$\Rightarrow \frac{AB}{BC} = \frac{1}{2}$$

$\Rightarrow AP : PC = 1 : 2$

We have,

$y = x$ ...(i)

$2x - y = 2$ ...(ii)

On solving Eqs. (i) and (ii), we get point $P(2, 2)$

Let, coordinates of $C$ be $(x, y)$

$\therefore$ by section formula, we have

$$2 = \frac{1 \times x + 2 \times 4}{1+2} \Rightarrow x = -2$$

$$\text{and} \quad 2 = \frac{1 \times y + 2 \times 6}{1+2}$$

$\Rightarrow y = -6$

$\therefore C \equiv (-2, -6)$

$\because$ Point $B$ lies on $y = x$

$\therefore (\alpha = \beta)$

$$\because \frac{AB}{BC} = \frac{1}{2} \Rightarrow |BC| = 2|AB|$$

$$\Rightarrow \sqrt{(\alpha+2)^2 + (\alpha+6)^2} = 2\sqrt{(\alpha-4)^2 + (\alpha-6)^2}$$

$$\Rightarrow (\alpha+2)^2 + (\alpha+6)^2 = 4[(\alpha-4)^2 + (\alpha-6)^2]$$

$$\Rightarrow \alpha^2 - 16\alpha + 28 = 0 \Rightarrow \alpha = 2, 14$$

$\alpha = 2$ is given point $P$, $\Rightarrow \alpha = 14$

$\therefore \alpha + 2\beta = \alpha + 2\alpha = 3\alpha = 3 \times 14 = 42$

**99.** *(b)* Given, sides of triangle

$4x + 3y = 69$ ...(i)

$4y - 3x = 17$

$\Rightarrow 3x - 4y = -17$ ...(ii)

$x + 7y = 61$ ...(iii)

Lines (i) and (ii) are perpendicular to each other

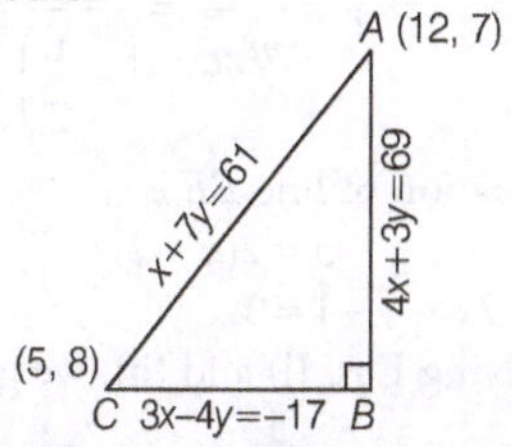


$\therefore$ Circumcentre of a right angled triangle is the mid-point of the hypotenuse.

Solving Eqs. (i) and (iii) after multiply by 4 in Eq (iii),

$4x + 3y = 69$

$4x + 28y = 244$

$-\quad -\quad -$

$-25y = -175$

$\Rightarrow y = 7$ and $x = 12$

$A(12, 7)$

Solving Eqs. (ii) and (iii) for $C$

$3x - 4y = -17$

$3x + 21y = 183$

$-\quad -\quad -$

$-25y = -200$

$\Rightarrow y = 8$ and $x = 5$

$C(5, 8)$

Coordianates of circumcentre

= mid-point of $AC$

$$= \left(\frac{5+12}{2}, \frac{7+8}{2}\right)$$

$$\therefore \alpha = \frac{17}{2}, \beta = \frac{15}{2}$$

$$\therefore (\alpha - \beta)^2 + \alpha + \beta = 1 + 16 = 17$$

**100.** *(a)*

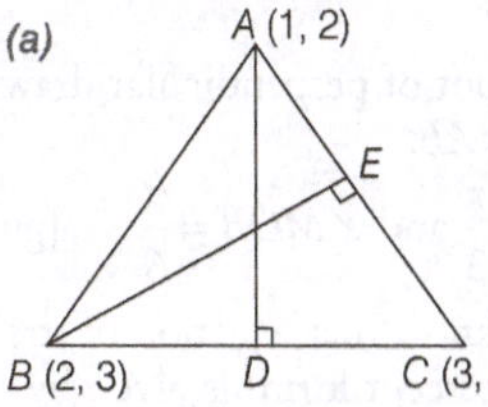


$$\text{Slope of } BC, m_{BC} = \frac{1-3}{3-2} = \frac{-2}{1} = -2$$

$$\therefore \text{Slope of } AD, m_{AD} = \frac{-1}{m_{BC}} = \frac{-1}{-2} = \frac{1}{2}$$

So, equation of line $AD$ is

$$y - 2 = \frac{1}{2}(x - 1)$$

$\Rightarrow x - 2y + 3 = 0$ ...(i)

$$\text{Slope of } AC, m_{AC} = \frac{2-1}{1-3} = \frac{1}{-2} = \frac{-1}{2}$$

$\therefore$ Slope of $BE$,

$$m_{BE} = \frac{-1}{m_{AC}} = \frac{-1}{\left(-\frac{1}{2}\right)} = 2$$

So, equation of line $BE$ is

$$y - 3 = 2(x - 2)$$

$\Rightarrow \quad 2x - y - 1 = 0 \qquad \text{...(ii)}$

On solving Eqs. (i) and (ii), we get

$$x = \frac{5}{3}, y = \frac{7}{3}$$

$\therefore$ Orthocentre of $\Delta ABC$ is $\left(\frac{5}{3}, \frac{7}{3}\right)$.

$\therefore \quad \alpha = \frac{5}{3}$ and $\beta = \frac{7}{3}$

Now, Let $\alpha_1 = \alpha + 4\beta$
and $\beta_1 = 4\alpha + \beta$.
Then,

$$\alpha_1 = \alpha + 4\beta = \frac{5}{3} + 4 \times \frac{7}{3} = \frac{33}{3} = 11$$

and $\beta_1 = 4\alpha + \beta = 4 \times \frac{5}{3} + \frac{7}{3} = \frac{27}{3} = 9$

Again, $\alpha_1 + \beta_1 = 11 + 9 = 20$
and $\alpha_1\beta_1 = 11 \times 9 = 99$

So, quadratic equation, whose roots are $\alpha_1$ and $\beta_1$ is

$$x^2 - (\alpha_1 + \beta_1)x + \alpha_1\beta_1 = 0$$

$\Rightarrow \quad x^2 - 20x + 99 = 0$

**101.** *(a)* Given, $OA = a$, $OB = b$

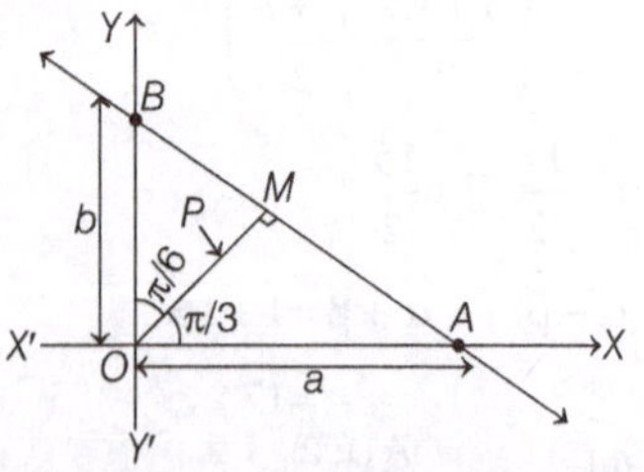


Let $M$ = foot of perpendicular drawn from $O$ to $AB$.

$\angle MOA = \frac{\pi}{3}$ and $\angle MOB = \frac{\pi}{6}$ [given]

Since, we know that equation of straight line (in intercept form) is given by

$$\frac{x}{a} + \frac{y}{b} = 1 \qquad \text{...(i)}$$

Also, we can write equation of straight line in normal form as

$$x \cos\frac{\pi}{3} + y \sin\frac{\pi}{3} = p$$

where $p$ = length of perpendicular drawn from origin to the given line $AB$.

$\Rightarrow \quad \frac{x}{2} + \frac{y\sqrt{3}}{2} = p \Rightarrow \frac{x}{2p} + \frac{\sqrt{3}y}{2p} = 1$

$\Rightarrow \quad \frac{x}{2p} + \frac{y}{\left(\frac{2p}{\sqrt{3}}\right)} = 1 \qquad \text{...(ii)}$

On comparing Eqs. (i) and (ii), we get

$$a = 2p \text{ and } b = \frac{2p}{\sqrt{3}}$$

Now, area of $\Delta OAB = \frac{1}{2} \times a \times b$

$$= \frac{1}{2} \times (2p) \times \left(\frac{2p}{\sqrt{3}}\right)$$

$\Rightarrow \quad \frac{98}{3} \times \sqrt{3} = \frac{1}{2} \times 2 \times 2 \times \frac{p^2}{\sqrt{3}}$

$\Rightarrow \quad 2p^2 = 98 \Rightarrow p^2 = 49$

Hence, $a^2 - b^2 = (2p)^2 - \left(\frac{2p}{\sqrt{3}}\right)^2$

$$= 4p^2 - \frac{4p^2}{3} = \frac{8}{3}p^2$$

$\Rightarrow a^2 - b^2 = \frac{8}{3} \times 49 = \frac{392}{3}$

**102.** *(b)* Given equation,

$$a^2 + 11a + 3(m_1^2 + m_2^2) = 220 \qquad \text{...(i)}$$

$$m_1m_2 = -1, m_2 = -\frac{1}{m_1}$$

Put the value of $m_2$ in Eq. (i)

$$a^2 + 11a + 3\left(m_1^2 + \frac{1}{m_1^2}\right) = 220$$

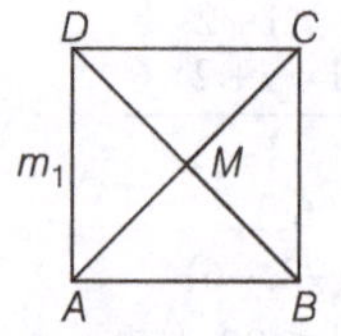


Equation of diagonal $AC$

$$(\cos\alpha - \sin\alpha)x + (\sin\alpha + \cos\alpha)y = 10$$

$$BD = (\sin\alpha - \cos\alpha)x + (\sin\alpha - \cos\alpha)y = 0$$

One vertex of square

$$= [10(\cos\alpha - \sin\alpha), 10(\sin\alpha - \cos\alpha)]$$

Slope of $AC = \left(\frac{\sin\alpha - \cos\alpha}{\sin\alpha + \cos\alpha}\right)$

Equation of line making an angle $\frac{\pi}{4}$ with $AC$

$$(m_1, m_2) = \frac{\frac{m \pm \tan\frac{\pi}{4}}{4}}{1 \pm m\tan\frac{\pi}{4}} = \frac{m+1}{1-m}$$

or $\frac{m-1}{1+m}$

$$\frac{\frac{\sin\alpha - \cos\alpha}{\sin\alpha + \cos\alpha} + 1}{1 - \left(\frac{\sin\alpha - \cos\alpha}{\sin\alpha + \cos\alpha}\right)} = \frac{\frac{\sin\alpha - \cos\alpha}{\sin\alpha + \cos\alpha} - 1}{1 + \frac{\sin\alpha - \cos\alpha}{\sin\alpha + \cos\alpha}}$$

$m_1, m_2 = \tan\alpha, -\cot\alpha$

Mid-point of $AC$ and $BD$

$= M(5(\cos\alpha - \sin\alpha), 5(\cos\alpha + \sin\alpha))$

$B(10(\cos\alpha - \sin\alpha), 10(\cos\alpha + \sin\alpha))$

$a = AB = \sqrt{2}\ BM = \sqrt{2}\,(5\sqrt{2})$

$\Rightarrow \quad a = 10$

$\because \quad a^2 + 11a + 3\left(m_1^2 + \frac{1}{m_1^2}\right) = 220$

Hence $\tan^2\alpha = 3$, $\tan^2\alpha = \frac{1}{3}$,

$\alpha = \frac{\pi}{3}$ or $\frac{\pi}{6}$.

Now $72(\sin^4\alpha + \cos^4\alpha) + a^2 - 3a + 13$

$= 72\left(\frac{9}{16} + \frac{1}{16}\right) + 100 - 30 + 13$

$= 72\left(\frac{5}{8}\right) + 83$

$= 45 + 83 = 128$

**103.** *(b)* Given,

$$A(\alpha, -2), B(\alpha, 6), C\left(\frac{\alpha}{4}, -2\right)$$

$$AB = \sqrt{0^2 + 8^2} = 8$$

$$AC = \sqrt{\left(\alpha - \frac{\alpha}{4}\right)^2 + 0} = \frac{3\alpha}{4}$$

$$BC = \sqrt{\left(\alpha - \frac{\alpha}{4}\right)^2 + (6+2)^2}$$

$$= \sqrt{\left(\frac{3\alpha}{4}\right)^2 + 64}$$

So, $(AB)^2 + (AC)^2 = (BC)^2$

Also, $AC$ is perpendicular to $AB$.
Hence, $\Delta ABC$ is a right angled triangle at $A$.

Now circumcentre is mid-point of $BC$.

$$\left(5, \frac{\alpha}{4}\right) \equiv \left(\frac{\alpha + \frac{\alpha}{4}}{2}, \frac{6-2}{2}\right)$$

So, $\frac{\alpha}{4} = 2$

$\Rightarrow \quad \alpha = 8$

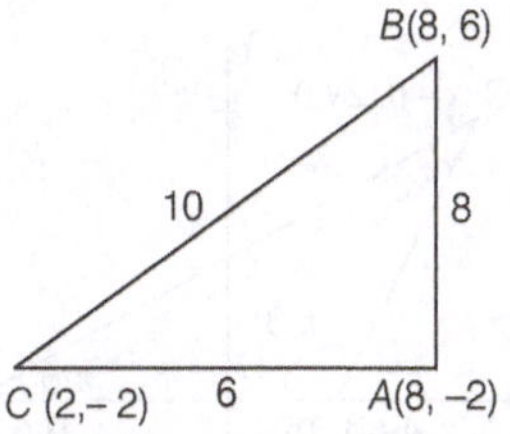


So, area of $\Delta ABC = \frac{1}{2} \times 6 \times 8 = 24$

Perimeter $= 6 + 8 + 10 = 24$

Circumradius $= \frac{BC}{2} = \frac{10}{2} = 5$

Inradius $= \frac{\Delta}{S} = \frac{24}{12} = 2 \quad \left[\because s = \frac{a+b+c}{2}\right]$

**104.** *(b)* Given, $L_1 : 2x + 5y = 10$

$L_2 : -4x + 3y = 12$

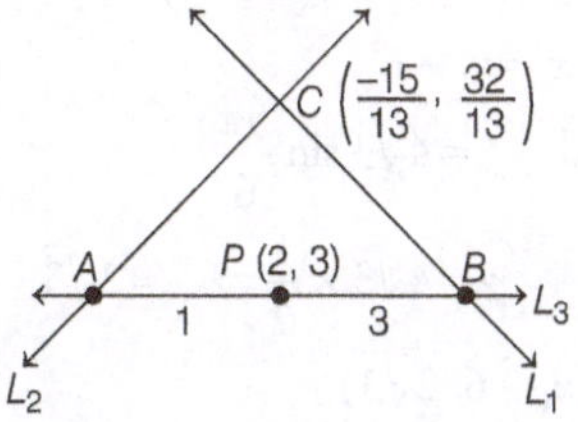


On solving $L_1$ and $L_2$, we get

$$C \equiv \left(\frac{-15}{13}, \frac{32}{13}\right)$$

Now, let $A\left(l_1, \frac{1}{3}(12 + 4l_1)\right)$

and $B\left(l_2, \frac{1}{5}(10 - 2l_2)\right)$

$$\therefore \quad \frac{3l_1 + l_2}{4} = 2$$

and $$\frac{(12 + 4l_1) + \frac{10 - 2l_2}{5}}{4} = 3$$

So, $\quad 3l_1 + l_2 = 8 \quad \ldots(\text{i})$

and $\quad 10l_1 - l_2 = -5 \quad \ldots(\text{ii})$

Solving Eqs. (i) and (ii), we get

$$(l_1, l_2) = \left(\frac{3}{13}, \frac{95}{13}\right)$$

$\therefore A = \left(\frac{3}{13}, \frac{56}{13}\right)$ and $B = \left(\frac{95}{13}, -\frac{12}{13}\right)$

Area of $\Delta ABC$

$$= \left|\frac{1}{2}\left[\frac{3}{13}\left(-\frac{44}{13}\right) - \frac{56}{13}\left(\frac{110}{13}\right) + 1\left(\frac{2860}{169}\right)\right]\right|$$

$= \frac{132}{13}$ sq. units

**105.** *(b)* Given, points $A(-1, 1)$, $B(3, 4)$, $C(2, 0)$.

Equation of $AC$ is $\frac{y-1}{x+1} = \frac{0-1}{2+1}$

$\Rightarrow \quad 3y - 3 = -x - 1$

$\Rightarrow \quad x + 3y = 2 \quad \ldots(\text{i})$

On solving Eq. (i) and $y = mx$,

we get $P\left(\frac{2}{3m+1}, \frac{2m}{3m+1}\right)$

Equation of $BC = \frac{y-0}{x-2} = \frac{4-0}{3-2}$

$\Rightarrow \quad y = 4x - 8 \quad \ldots(\text{ii})$

Similarly, on solving Eq. (ii) and $y = mx$, we get

$Q\left(\frac{8}{4-m}, \frac{8m}{4-m}\right)$

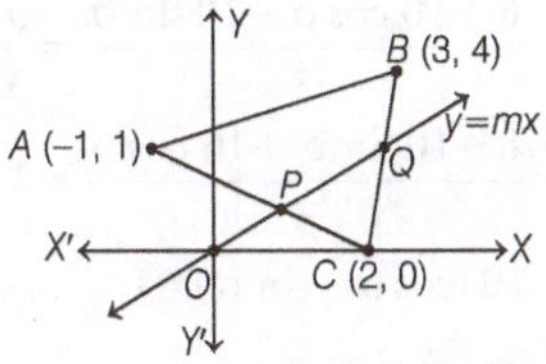


Area of $\Delta ABC = 3$ Area of $\Delta PQC$ [given]

$$\frac{1}{2}\begin{vmatrix} -1 & 1 & 1 \\ 2 & 0 & 1 \\ 3 & 4 & 1 \end{vmatrix} = 3 \times \frac{1}{2}\begin{vmatrix} 2 & 0 & 1 \\ \frac{8}{4-m} & \frac{8m}{4-m} & 1 \\ \frac{2}{3m+1} & \frac{2m}{3m+1} & 1 \end{vmatrix}$$

$$\Rightarrow 13 = 3\left(\frac{1}{4-m}\right)\left(\frac{1}{3m+1}\right)\begin{vmatrix} 2 & 0 & 1 \\ 8 & 8m & 4-m \\ 2 & 2m & 3m+1 \end{vmatrix}$$

$$\Rightarrow 13 = \frac{3}{4 + 11m - 3m^2} \times (52m^2)$$

$$\Rightarrow 15m^2 - 11m - 4 = 0$$

$\Rightarrow \quad m = 1, \frac{-4}{15} \quad$ [but $m > 0$]

$\Rightarrow \quad m = 1$

**106.** *(a)* According to given information, we have the following figure

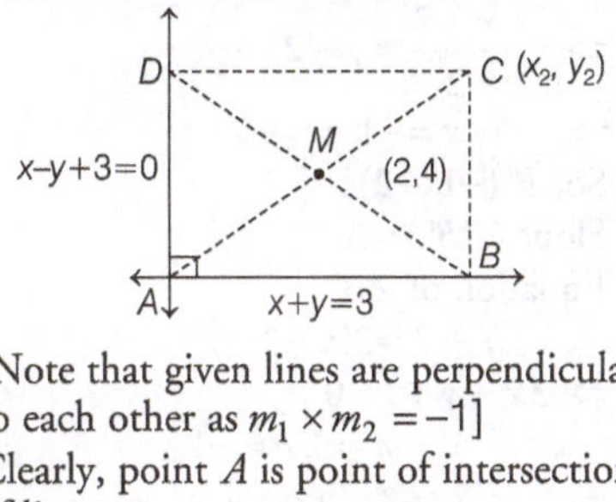


[Note that given lines are perpendicular to each other as $m_1 \times m_2 = -1$]

Clearly, point $A$ is point of intersection of lines

$$x + y = 3 \quad \ldots(\text{i})$$

and $\quad x - y = -3 \quad \ldots(\text{ii})$

So, $\quad A = (0, 3)$

[solving Eqs. (i) and (ii)]

Now, as point $M$ (2, 4) is mid-point of line joining the points $A$ and $C$, so

$$(2, 4) = \left(\frac{0 + x_2}{2}, \frac{3 + y_2}{2}\right)$$

$$\left[\because \text{mid-point} = \left(\frac{x_1 + x_2}{2}, \frac{y_1 + y_2}{2}\right)\right]$$

$\Rightarrow \quad 2 = \frac{0 + x_2}{2}; 4 = \frac{3 + y_2}{2}$

$\Rightarrow \quad x_2 = 4$ and $y_2 = 5$

$\therefore$ Thus, $C \equiv (4, 5)$

Now, equation of line $BC$ is given by

$$(y - y_1) = m(x - x_1)$$

$$y - 5 = 1(x - 4)$$

[line $BC$ is parallel to $x - y + 3 = 0$ and slope of $x - y + 3 = 0$ is $\frac{(-1)}{(-1)} = 1$]

$\Rightarrow \quad y = x + 1 \quad \ldots(\text{iii})$

and equation of line $DC$ is

$$y - 5 = -1(x - 4)$$

[line $DC$ is parallel to $x + y = 3$ and slope of $x + y = 3$ is $\frac{-1}{1} = -1$]

$\Rightarrow \quad x + y = 9 \quad \ldots(\text{iv})$

On solving Eqs. (i) and (iii), we get $B$ (1, 2) and on solving Eqs. (ii) and (iv), we get $D$ (3, 6).

**107.** *(c)* Since, $A(h, k)$, $B(1, 1)$ and $C(2, 1)$ are the vertices of a right angled $\Delta ABC$.

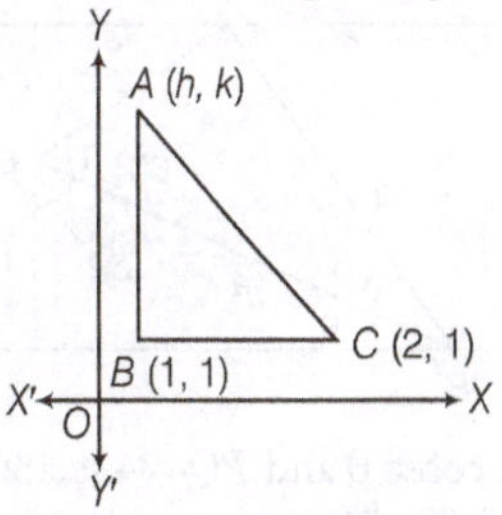


Now, $AB = \sqrt{(1-h)^2 + (1-k)^2}$

$BC = \sqrt{(2-1)^2 + (1-1)^2} = 1$

and $CA = \sqrt{(h-2)^2 + (k-1)^2}$

From Pythagoras theorem,

$$AC^2 = AB^2 + BC^2$$

$\Rightarrow 4 + h^2 - 4h + k^2 + 1 - 2k$

$= h^2 + 1 - 2h + k^2 + 1 - 2k + 1$

$\Rightarrow \quad 5 - 4h = 3 - 2h \Rightarrow h = 1 \quad \ldots(\text{i})$

Now, given that area of triangle is 1.

Then, area $(\Delta ABC) = \frac{1}{2} \times AB \times BC$

$\Rightarrow 1 = \frac{1}{2} \times \sqrt{(1-h)^2 + (1-k)^2} \times 1$

$\Rightarrow 2 = \sqrt{(1-h)^2 + (1-k)^2}$ ...(ii)

$\Rightarrow 2 = \sqrt{(k-1)^2}$ [from Eq. (i)]

$\Rightarrow 4 = k^2 + 1 - 2k$

$\Rightarrow \quad k^2 - 2k - 3 = 0$

$\Rightarrow \quad (k-3)(k+1) = 0$

$\Rightarrow \quad k = -1, 3$

Thus, the set of values of $k$ is $\{-1, 3\}$.

**108.** *(5)* $\alpha = 1 + \sum_{r=1}^{6} (-3)^{r-1} \, {}^{12}C_{2r-1}$

$\because C_1 - C_3 + C_5 - \ldots$

$= \frac{(1+i)^n - (1-i)^n}{2} = 2^{n/2} \sin \frac{n\pi}{4}$

$C_1\sqrt{3} - C_3(\sqrt{3})^3 + C_5(\sqrt{3})^5 - \ldots$

$= \frac{(1+i\sqrt{3})^{12} - (1-i\sqrt{3})^{12}}{2}$

$= \frac{2^{12}\,[(e^{i\pi/3})^{12} - (e^{-i\pi/3})^{12}]}{2}$

$= 2^{11}\,(e^{i4\pi} - e^{-i4\pi})$

$= 2^{11}\,(1-1) = 0 \Rightarrow \alpha = 1$

Line : $\quad \alpha x - \sqrt{3}y + 1 = 0$

Distance $= \left|\frac{12 - 3 + 1}{\sqrt{1+3}}\right| = \left|\frac{10}{2}\right| = 5$

**109.** *(28)* Let side of equilateral $\Delta PQR$ be $a$.

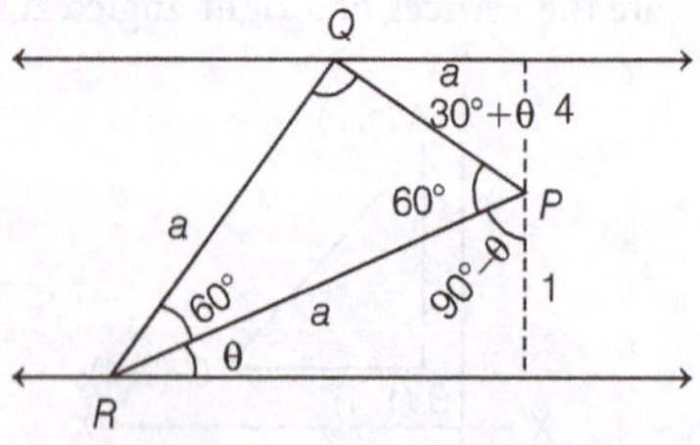


$PR = 1 \operatorname{cosec} \theta$ and $PQ = 4 \sec(30° + \theta)$

$\because \quad PR = PQ$

$\operatorname{cosec} \theta = 4 \sec(30° + \theta)$

$\Rightarrow \quad 4 \sin \theta = \cos(30° + \theta)$

$\Rightarrow 4 \sin \theta = \cos 30° \cos \theta - \sin 30° \sin \theta$

$\Rightarrow \quad 4 \sin \theta = \frac{\sqrt{3}}{2} \cos \theta - \frac{1}{2} \sin \theta$

$\Rightarrow \quad \frac{9}{2} \sin \theta = \sqrt{3} \cos \theta$

$\Rightarrow \quad \tan \theta = \frac{2\sqrt{3}}{9} = \frac{2}{3\sqrt{3}}$

$\therefore \quad (QR)^2 = (PR)^2$

$= \frac{(3\sqrt{3})^2 + 1}{1} = 28$

**110.** *(145)* Given $A(6, 8)$, $B(10 \cos \alpha, -10 \sin \alpha)$ and $C(-10 \sin \alpha, 10 \cos \alpha)$

$\because$ All three points $A$, $B$ and $C$ lies on the circle

$x^2 + y^2 = 100$

$\therefore$ Circumcentre of $\Delta ABC$ is $O(0, 0)$ orthocenter $H(a, 9)$ centroid $G(h, k)$

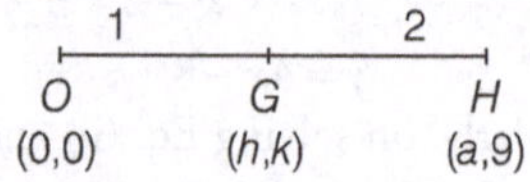


$h = \frac{a}{3}$ and $k = \frac{9}{3} = 3$

$\frac{6 + 10 \cos \alpha - 10 \sin \alpha}{3} = \frac{a}{3}$

and $\frac{8 - 10 \sin \alpha + 10 \cos \alpha}{3} = 3$

$10(\cos \alpha - \sin \alpha) = 1$

$\cos \alpha - \sin \alpha = \frac{1}{10}$

$\therefore \quad \frac{a}{3} = \frac{6 + 10\left(\frac{1}{10}\right)}{3} \Rightarrow a = 7$

Now, $5a - 3h + 6k + 100 \sin 2\alpha$

$= 5(7) - 3\left(\frac{7}{3}\right) + 6(3)$

$+ 100[1 - (\cos \alpha - \sin \alpha)^2]$

$= 35 - 7 + 18 + 100\left(1 - \frac{1}{100}\right)$

$= 28 + 18 + 99$

$= 145$

**111.** *(1)*

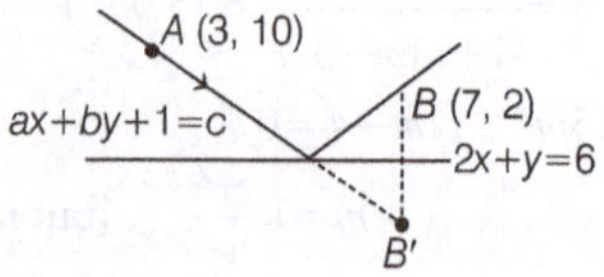


For $B'$,

$\frac{x-7}{2} = \frac{y-2}{1} = -2\left(\frac{14+2-6}{5}\right)$

$\Rightarrow \quad \frac{x-7}{2} = y - 2 = -4$

$\Rightarrow \quad x = -1, y = -2$

So, $B'(-1, -2)$

Slope $(AB') = 3$

Equation of $AB'$

$y + 2 = 3(x+1)$

$\Rightarrow 3x - y + 1 = 0$

$a = 3, b = -1$

Therefore, $a^2 + b^2 + 3ab = 9 + 1 - 9 = 1$

**112.** *(36)* We have,

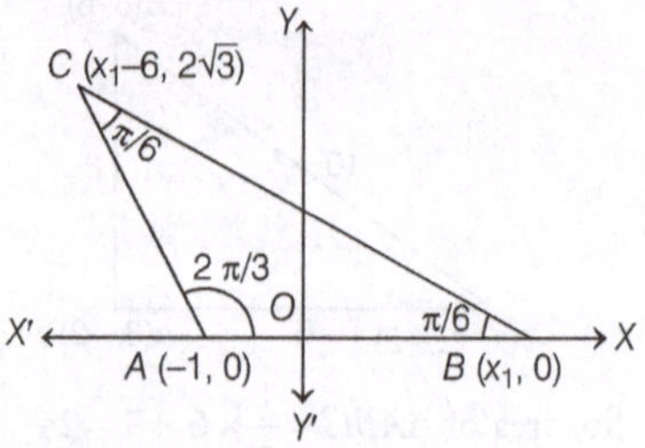


Coordinates of '$C$' are

$\frac{x - x_1}{\cos \frac{5\pi}{6}} = \frac{y - 0}{\sin \frac{5\pi}{6}} = 4\sqrt{3}$

$\therefore \quad x = 4\sqrt{3} \cos \frac{5\pi}{6} + x_1$

$= 4\sqrt{3} \times \left(\frac{-\sqrt{3}}{2}\right) + x_1$

$\Rightarrow x = x_1 - 6$

and $\quad y = 4\sqrt{3} \sin \frac{5\pi}{6}$

$\Rightarrow \quad y = 4\sqrt{3} \times \frac{1}{2} \Rightarrow y = 2\sqrt{3}$

$C(x_1 - 6, 2\sqrt{3})$

$M_{AC} = -\sqrt{3}$

$\therefore \frac{0 - 2\sqrt{3}}{-1 - x_1 + 6} = -\sqrt{3} \Rightarrow x_1 = 3$

$BC \Rightarrow x + \sqrt{3}y = 3, \; y = x + 3$

$\Rightarrow \quad y - 3 + \sqrt{3}y = 3$

$\Rightarrow \quad y = \frac{6}{\sqrt{3}+1},$

$x = \frac{6}{\sqrt{3}+1} - 3 = \frac{6 - 3\sqrt{3} - 3}{\sqrt{3}+1}$

$x = \frac{3 - 3\sqrt{3}}{\sqrt{3}+1} = \frac{-6}{(1+\sqrt{3})^2}$

So, $\frac{\beta^4}{\alpha^2} = 36$

**113.** *(529)*

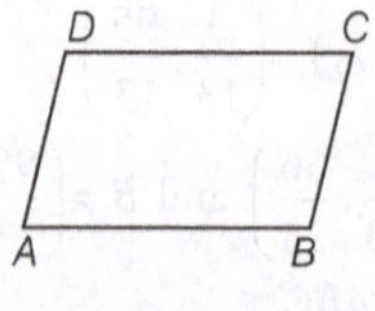


We have, $ABCD$ is a parallelogram.

Let equation of $AB$ be

$2x - 3y = -23$ ...(i)

and equation of $BC$ be

$5x + 4y = 23$ ...(ii)

Equation of $AC$ is

$3x + 7y = 23$ ...(iii)

Solving Eqs. (i) and (ii), we get

$x=-1$ and $y=7$

∴ Coordinate of $B$ is $(-1, 7)$

Solving Eqs. (ii) and (iii), we get

$x=3, y=2$

∴ Coordinate of $C$ is $(3, 2)$.

Solving Eqs. (i) and (iii), we get

$x=-4$ and $y=5$

∴ Coordinate of $A$ is $(-4, 5)$.

Let $M$ be the intersection point of diagonal.

Coordinate of $M$ is $\left(\frac{-4+3}{2}, \frac{5+2}{2}\right)$ or $\left(-\frac{1}{2}, \frac{7}{2}\right)$.

Equation of $BD$ is $y-7=\left(\frac{7-\frac{7}{2}}{-1+\frac{1}{2}}\right)(x+1)$

$\Rightarrow \quad 7x+y=0$

Distance of $A$ from diagonal

$BD=\frac{|7\times(-4)+5|}{\sqrt{7^2+1^2}}$

$\therefore \quad d=\frac{23}{\sqrt{50}}$

Hence, $\quad 50d^2=(23)^2=529$

**114.** *(3)*

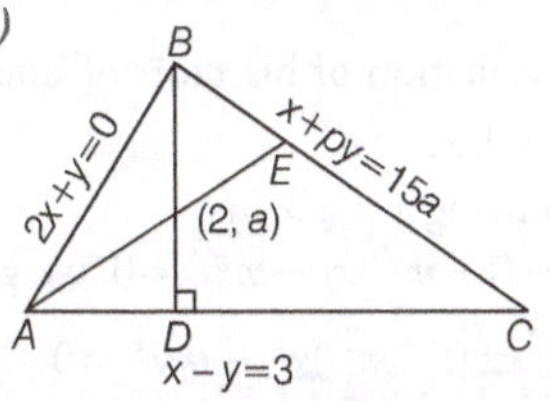


For determining coordinate of point $A$,

$2x+y=0$

$\Rightarrow \quad y=-2x$

$x-y=3$

$\Rightarrow x-(-2x)=3$

$\Rightarrow \quad x=1$

Coordinates of $A$ is $(1, -2)$.

For determining coordinate of point $B$.

On putting $y=-2x$ in equation $x+py=15a$, we get

$x-2px=15a$

$\Rightarrow \quad x=\frac{15a}{1-2p}$

$\therefore$ Coordinate $B$ is $\left(\frac{15a}{1-2p}, \frac{-30a}{1-2p}\right)$.

Slope of $AE=\frac{a+2}{2-1}=a+2$

Slope of $BD=\frac{a+\frac{30a}{1-2p}}{2-\frac{15a}{1-2p}}$

$=\frac{31a-2ap}{2-4p-15a}$

Slope of $BC=-\frac{1}{p}$

Slope of $AC=1$

As $AE$ is perpendicular to $BC$,

Slope of $AE\times$ Slope of $BC=-1$

$\Rightarrow \quad (a+2)\left(-\frac{1}{p}\right)=-1$

$\Rightarrow \quad a+2=p$

$\Rightarrow \quad a=p-2 \quad$ ...(i)

As $BD$ is perpendicular to $AC$,

Slope of $BD\times$ Slope of $AC=-1$

$\Rightarrow \quad \frac{31a-2ap}{2-4p-15a}\times 1=(-1)$

$\Rightarrow \quad 16a-2ap-4p+2=0$

$\Rightarrow \quad 16(p-2)-2p(p-2)$
$-4p+2=0$

$\Rightarrow \quad p^2-8p+15=0$

$\Rightarrow \quad p=3$ or $5$

If $p=3$, then $a=1$ and if $p=5$, then $a=3$

As, $-\frac{1}{2}<a<2, a\neq 3$

$\Rightarrow \quad p\neq 5$

Hence, $p=3$

**115.** *(8)* Given that $A\left(\frac{3}{\sqrt{a}}, \sqrt{a}\right), a>0$ be a fixed point in the $XY$-plane, the image of $A$ in $Y$-axis is $B\left(-\frac{3}{\sqrt{a}}, \sqrt{a}\right)$ and the image of $B$ in $X$-axis is $C\left(\frac{-3}{\sqrt{a}}, -\sqrt{a}\right)$ and also given $D(3\cos\theta, a\sin\theta)$ in the fourth quadrant.

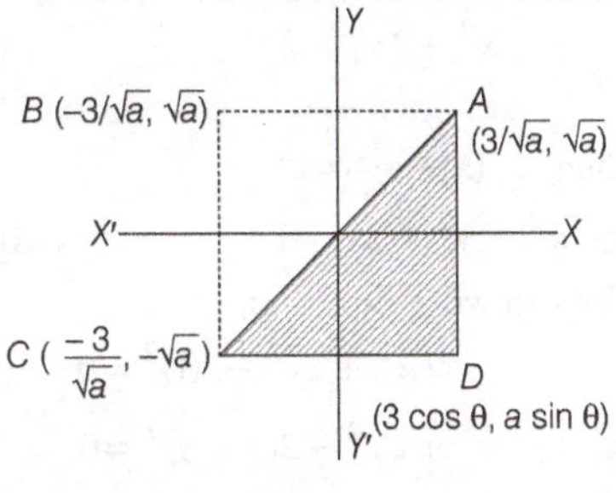


Thus, area of

$$\Delta ACD=\frac{1}{2}\begin{vmatrix}\frac{3}{\sqrt{a}} & \sqrt{a} & 1\\ -\frac{3}{\sqrt{a}} & -\sqrt{a} & 1\\ 3\cos\theta & a\sin\theta & 1\end{vmatrix}$$

Applying $R_1\rightarrow R_1+R_2$,

$$\Delta=\frac{1}{2}\begin{vmatrix}0 & 0 & 2\\ -\frac{3}{\sqrt{a}} & -\sqrt{a} & 1\\ 3\cos\theta & a\sin\theta & 1\end{vmatrix}$$

$\Rightarrow \quad \Delta=\left|\frac{1}{2}2\{-3\sqrt{a}\sin\theta+3\sqrt{a}\cos\theta\}\right|$

$\Rightarrow \quad \Delta=|3\sqrt{a}(-\sin\theta+\cos\theta)|$

$\Rightarrow \quad \Delta=3\sqrt{a}\,|-\sin\theta+\cos\theta|$

$\therefore \Delta_{max}=3\sqrt{a}\cdot\sqrt{2}=12 \quad$ [given]

$\Rightarrow \sqrt{a}=2\sqrt{2}$

$\Rightarrow \quad a=(2\sqrt{2})^2=8$

**116.** *(1250)* Let $P'$ be the mirror image of $P$ about $X$-axis.

$\therefore \quad PR=P'R$

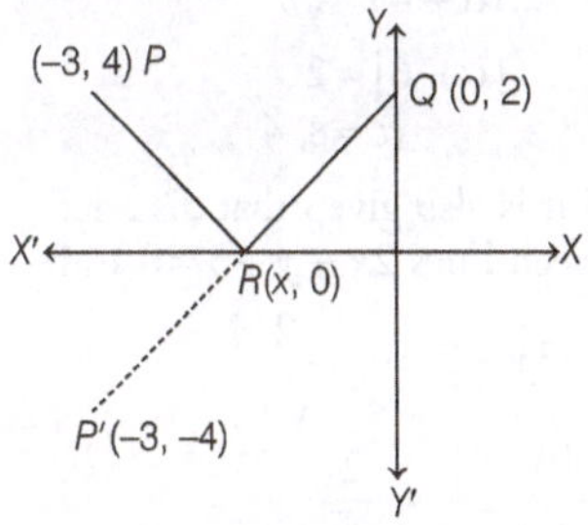


For minimum distance,

$PR+RQ=P'R+RQ$

$P', R, Q$ are collinear.

$\Rightarrow$ Slope of $P'R=$ Slope of $RQ$

$\Rightarrow \quad \frac{-4-0}{-3-x}=\frac{2-0}{0-x}$

$\Rightarrow \quad 4x=-6-2x$

$\Rightarrow \quad x=-1$

$\Rightarrow \quad R=(-1, 0)$

$\therefore PR^2+RQ^2=P'R^2+RQ^2$

$=(-3+1)^2+(-4)^2+1^2+2^2=25$

So, $50\times(PR^2+RQ^2)$

$=50\times 25=1250$

**117.** *(6)* Let point of intersection of lines be $A, B, C$.

Solving system of equations, we get point of intersections $(1, 2), (7, 5), (2, 3)$

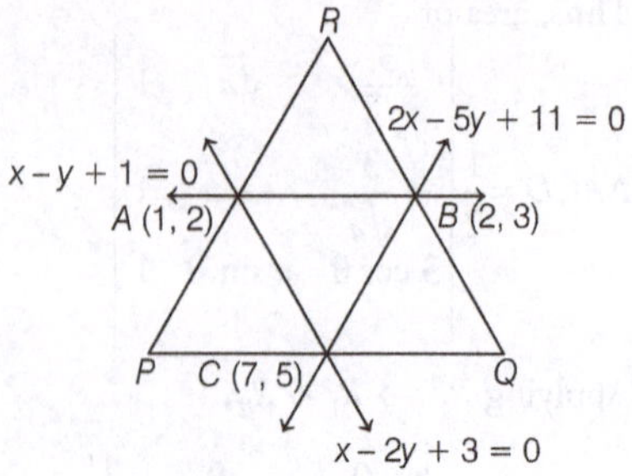


$$\text{Area of } \Delta ABC = \frac{1}{2}\begin{vmatrix} 1 & 2 & 1 \\ 7 & 5 & 1 \\ 2 & 3 & 1 \end{vmatrix}$$

$$= \frac{1}{2}(3) = \frac{3}{2}$$

Area $\Delta PQR = 4 \times \Delta ABC$

$$= 4\left(\frac{3}{2}\right) = 6 \text{ sq units}$$

**118.** *(30)* It is given that distance between lines $2x - y + 3 = 0$ and $4x - 2y + \alpha = 0$ is $\frac{1}{\sqrt{5}}$

$$\therefore \quad \frac{|6-\alpha|}{\sqrt{16+4}} = \frac{1}{\sqrt{5}}$$

$$\Rightarrow \quad |\alpha - 6| = 2$$

$$\Rightarrow \quad \alpha = 8, 4$$

And it is also given that distance between lines $2x - y + 3 = 0$ and $6x - 3y + \beta = 0$ is $\frac{2}{\sqrt{5}}$.

$$\therefore \quad \frac{|\beta - 9|}{\sqrt{36+9}} = \frac{2}{\sqrt{5}}$$

$$\Rightarrow \quad |\beta - 9| = 6$$

$$\Rightarrow \quad \beta = 15, 3$$

So, sum of all possible values of $\alpha$ and $\beta$ $= 4 + 8 + 3 + 15 = 30$.

## Topic 2

**01.** *(c)* Equation of angle bisector of homogeneous equation of pair of straight line $ax^2 + 2hxy + by^2 = 0$ is

$$\frac{x^2 - y^2}{a - b} = \frac{xy}{h}$$

For $x^2 - 4xy - 5y^2 = 0$

$$a = 1, h = -2, b = -5$$

So, equation of angle bisector is

$$\frac{x^2 - y^2}{1-(-5)} = \frac{xy}{-2}$$

$$\Rightarrow \quad x^2 - y^2 = -3xy$$

$$\Rightarrow \quad x^2 + 3xy - y^2 = 0$$

So, combined equation of angle bisector is $x^2 + 3xy - y^2 = 0$.

- **Concept Enhancer**

The joint equation of bisector of the angles between the lines represented by the equation $ax^2 + 2hxy + by^2 = 0$ is

$$\frac{x^2 - y^2}{a - b} = \frac{xy}{h}$$

$$\Rightarrow hx^2 - (a-b)xy - hy^2 = 0$$

**02.** *(b)* Given, $x^2 - y^2 + 2hxy + 2gx + 2fy + c = 0$ ...(i)

locus of point $(x, y)$ which is equidistant from $x + 2y + 7 = 0$ and $2x - y + 8 = 0$

$\therefore$ Equation will be

$$\frac{x + 2y + 7}{\sqrt{1+4}} = \pm \frac{2x - y + 8}{\sqrt{4+1}}$$

$$x + 2y + 7 = \pm(2x - y + 8)$$

Squaring both sides, we get

$$x^2 + 4y^2 + 49 + 4xy + 28y + 14x$$

$$= 4x^2 + y^2 + 64 - 4xy - 16y + 32x$$

$$\Rightarrow 3x^2 - 3y^2 - 8xy + 18x - 44y + 15 = 0$$

$$\Rightarrow x^2 - y^2 - \frac{8}{3}xy + 6x - \frac{44}{3}y + 5 = 0 \quad \text{...(ii)}$$

Comparing Eqs. (i) and (ii), we get

| | | | |
|---|---|---|---|
| $2h = -\frac{8}{3}$ | $2g = 6$ | $2f = -\frac{44}{3}$ | $c = 5$ |
| $\Rightarrow h = -\frac{4}{3}$ | $\Rightarrow g = 3$ | $\Rightarrow f = -\frac{22}{3}$ | |

$$\therefore g + c + h - f = 3 + 5 - \frac{4}{3} + \frac{22}{3}$$

$$= 8 + 6 = 14$$

**03.** *(a)* Curve $x^2 + 2y^2 = 2$ intersect the line $x + y = 1$ at points $P$ and $Q$. First we have to find any common relation between these two curves. Use substitution for the same as follows,

$$x^2 + 2y^2 = 2 \quad \text{...(i)}$$

$$x + y = 1$$

then $\quad (x + y)^2 = 1^2$

$$\Rightarrow x^2 + y^2 + 2xy = 1 \quad \text{...(ii)}$$

We can write Eq. (i) as,

$$x^2 + 2y^2 - 2(1)^2 = 0$$

$$\Rightarrow \quad x^2 + 2y^2 - 2(x + y)^2 = 0$$

[using Eq. (ii) in Eq. (i)]

$$\Rightarrow x^2 + 2y^2 - 2x^2 - 2y^2 - 4xy = 0$$

$$\Rightarrow \quad -x^2 - 4xy = 0 \Rightarrow -x(x + 4y) = 0$$

Given, $x = 0$ and $x + 4y = 0$ or $y = \frac{-1}{4}x$

Draw the line $y = \frac{-1}{4}x$ on graph and take arbitrary point (any one) as follows, From given graph,

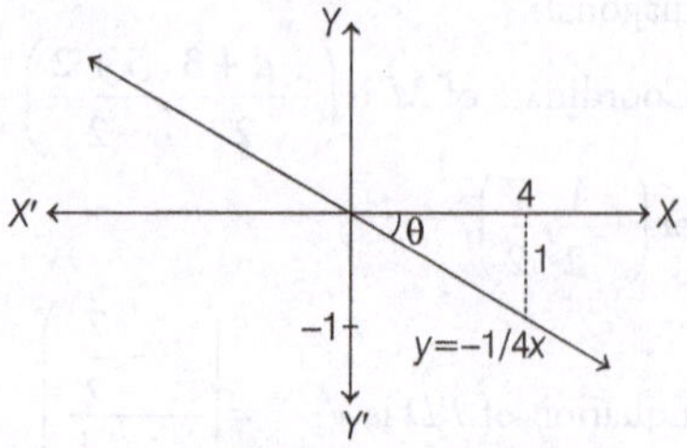


$$\tan\theta = \frac{1}{4} \Rightarrow \theta = \tan^{-1}\left(\frac{1}{4}\right)$$

We have two lines, $y = -\frac{1}{4}x$ and $x = 0$ (i.e. $Y$ - axis). Thus, any line joining these two curves makes an angle $\left(\frac{\pi}{2} + \theta\right)$ at origin.

$\therefore$ Required answer is $\frac{\pi}{2} + \tan^{-1}\left(\frac{1}{4}\right)$.

**04.** *(c)* Equation of bisectors of line $xy = 0$ are $y = \pm x$.

On putting $y = \pm x$ in $my^2 + (1 - m^2)xy - mx^2 = 0$, we get

$$mx^2 \pm (1 - m^2)x^2 - mx^2 = 0$$

$$\Rightarrow \quad (1 - m^2)x^2 = 0$$

$$\Rightarrow \quad m = \pm 1$$

**05.** *(c)* The given pair of lines is

$$x^2 - 2cxy - 7y^2 = 0.$$

On comparing the standard equation $ax^2 + 2hxy + by^2 = 0$,

we get $a = 1, \quad 2h = -2c, \quad b = -7$

Now, $\quad m_1 + m_2 = -\frac{2h}{b} = -\frac{2c}{7}$

and $\quad m_1 m_2 = \frac{a}{b} = -\frac{1}{7}$

According to the given condition,

$$m_1 + m_2 = 4m_1 m_2$$

$$\therefore \quad -\frac{2c}{7} = -\frac{4}{7}$$

$$\Rightarrow \quad c = 2$$

**• Concept Enhancer**

If $ax^2 + 2hxy + by^2 = 0$ be an equation of pair of straight lines.

(i) Slope of first line,

$$m_1 = \frac{-h+\sqrt{h^2-ab}}{b}$$

and slope of the second line,

$$m_2 = \frac{-h-\sqrt{h^2-ab}}{b}$$

$$\therefore m_1 + m_2 = -\frac{2h}{b} = -\frac{\text{Coefficient of } xy}{\text{Coefficient of } y^2}$$

$$\text{and } m_1m_2 = \frac{a}{b} = \frac{\text{Coefficient of } x^2}{\text{Coefficient of } y^2}$$

**06.** *(d)* Since, one of the two lines is $3x + 4y = 0$. Then, $y = -\frac{3x}{4}$ will satisfy the equation

$$6x^2 - xy + 4cy^2 = 0$$

$$\therefore 6x^2 - x\left(-\frac{3x}{4}\right) + 4c\left(-\frac{3x}{4}\right)^2 = 0$$

$$\Rightarrow \quad 6x^2 + \frac{3x^2}{4} + 4c \cdot \frac{9x^2}{16} = 0$$

$$\Rightarrow \quad x^2(27 + 9c) = 0$$

$$\therefore \quad c = -3$$

**07.** *(d)* The given equation is

$$x^2 - 2pxy - y^2 = 0$$

On comparing with standard equation $ax^2 + 2hxy + by^2 = 0$, we get $a = 1$, $b = -1$, $h = -p$

Equation of the bisector of angles is

$$\frac{x^2 - y^2}{a - b} = \frac{xy}{h}$$

$$\therefore \quad \frac{x^2 - y^2}{1+1} = \frac{xy}{-p}$$

$$\Rightarrow \quad x^2 - y^2 = -\frac{2xy}{p}$$

$$\Rightarrow \quad x^2 + \frac{2xy}{p} - y^2 = 0 \quad \ldots\text{(i)}$$

But given bisector equation of angles is

$$x^2 - 2qxy - y^2 = 0 \quad \ldots\text{(ii)}$$

On comparing Eqs. (i) and (ii), we get

$$\frac{2}{p} = -2q \Rightarrow pq = -1$$

**08.** *(a)* We have,

$$ax^2 + 2hxy + by^2 + 2gx + 2fy + c = 0$$

On putting $x = 0$ in the given equation

$$by^2 + 2fy + c = 0$$

For unique point of intersection,

$$f^2 - bc = 0$$

$$\Rightarrow \quad af^2 - abc = 0$$

We know that for pair of straight line

$$abc + 2fgh - af^2 - bg^2 - ch^2 = 0$$

$$\Rightarrow \quad 2fgh - bg^2 - ch^2 = 0$$

**09.** *(348)* Given : $l_1$ is angular bisector of $l_2$ and $l_3$.

Let angle between $l_1$ and $l_2$ is $\theta$.

$$\tan\theta = \left|\frac{m_1 - m_2}{1 + m_1m_2}\right| \quad \left[\because m_1 = \frac{2}{3}, m_2 = 1\right]$$

$$\tan\theta = \left|\frac{\frac{2}{3} - 1}{1 + \frac{2}{3}}\right| = \frac{1}{5}$$

$$\text{Then, } \tan\theta = \frac{1}{5} = \left|\frac{m_1 - m_3}{1 + m_1m_3}\right|$$

$$\Rightarrow \quad \left|\frac{m - \frac{2}{3}}{1 + \frac{2m}{3}}\right| = \frac{1}{5}$$

$$\Rightarrow \quad 5m - \frac{10}{3} = \pm\left(1 + \frac{2m}{3}\right)$$

$$\text{and} \quad 5m - \frac{10}{3} = -\left(1 + \frac{2m}{3}\right)$$

$$m = 1; \quad m = \frac{7}{17}$$

$\therefore$ Intersection point of $l_1$ and $l_2$ passes through

$\Rightarrow l_3$ passes through (0, +1)

$$\Rightarrow \quad y - 1 = \frac{7x}{17}$$

$$\Rightarrow \quad 7x - 17y + 17 = 0$$

$$\alpha = 7$$

$$\text{and} \quad \beta = -17$$

Hence, $\alpha^2 + \beta^2 - \alpha - \beta$

$$= 49 + 289 - 7 + 17$$

$$= 49 + 289 + 10$$

$$= 338 + 10 = 348$$

# C-09

JEE Main Chapterwise Topicwise Mathematics

In JEE Main, there are usually 1-2 questions per paper are visible from this chapter. In past 3 years, 90 conceptual and 47 numerical questions of moderate difficulty level were asked.

**For detailed analysis scan QR code**

# CONIC SECTIONS

## Topic 1 Circle and System of Circle

### Level 1

**01** If the four distinct points (4, 6), (–1, 5), (0, 0) and $(k, 3k)$ lie on a circle of radius $r$, then $10k + r^2$ is equal to *[2025, 3 April Shift-II]*

(a) 32 (b) 35 (c) 33 (d) 34

**Imp.** Condition of two Circles Touch Externally

**02** Let $C_1$ be the circle in the third quadrant of radius 3, that touches both coordinate axes. Let $C_2$ be the circle with centre (1, 3) that touches $C_1$ externally at the point $(\alpha, \beta)$. If $(\beta - \alpha)^2 = \frac{m}{n} \cdot \gcd(m, n) = 1$, then $m + n$ is equal to *[2025, 7 April Shift-I]*

(a) 9 (b) 13 (c) 22 (d) 31

**03** Let a circle $C$ pass through the points (4, 2) and (0, 2) and its centre lie on $3x + 2y + 2 = 0$. Then, the length of the chord of the circle $C$, whose mid-point is (1, 2) is *[2025, 29 Jan Shift-II]*

(a) $\sqrt{3}$ (b) $2\sqrt{2}$ (c) $2\sqrt{3}$ (d) $4\sqrt{2}$

**04** If $P(6, 1)$ be the orthocentre of the triangle whose vertices are $A(5, -2)$, $B(8, 3)$ and $C(h, k)$, then the point $C$ lies on the circle *[2024, 6 April Shift-II]*

(a) $x^2 + y^2 - 52 = 0$ (b) $x^2 + y^2 - 61 = 0$
(c) $x^2 + y^2 - 65 = 0$ (d) $x^2 + y^2 - 74 = 0$

**05** Let the circle $C_1 : x^2 + y^2 - 2(x + y) + 1 = 0$ and $C_2$ be a circle having centre at $(-1, 0)$ and radius 2. If the line of the common chord of $C_1$ and $C_2$ intersects the $Y$-axis at the point $P$, then the square of the distance of $P$ from the centre of $C_1$ is *[2024, 5 April Shift-II]*

(a) 2 (b) 4 (c) 6 (d) 1

**06** Let $C$ be a circle with radius $\sqrt{10}$ units and centre at the origin. Let the line $x + y = 2$ intersects the circle $C$ at the points $P$ and $Q$. Let $MN$ be a chord of $C$ of length 2 units and slope $-1$. Then, a distance (in units) between the chord $PQ$ and the chord $MN$ is *[2024, 5 April Shift-II]*

(a) $3 - \sqrt{2}$ (b) $\sqrt{2} - 1$ (c) $2 - \sqrt{3}$ (d) $\sqrt{2} + 1$

**07** Let a circle $C$ of radius 1 and closer to the origin be such that the lines passing through the point (3, 2) and parallel to the coordinate axes touch it. Then, the shortest distance of the circle $C$ from the point (5,5) is *[2024, 5 April Shift-I]*

(a) $4\sqrt{2}$ (b) 4 (c) $2\sqrt{2}$ (d) 5

**08** Let a variable line passing through the centre of the circle $x^2 + y^2 - 16x - 4y = 0$, meet the positive coordinate axes at the points $A$ and $B$. Then, the minimum value of $OA + OB$, where $O$ is the origin is equal to *[2024, 31 Jan Shift-II]*

(a) 18 (b) 24 (c) 20 (d) 12

**09** If the circle $x^2 + y^2 - 2gx + 6y - 19c = 0$, $c \in R$ passes through the point (6, 1) and its centre lies on the line $x - 2cy = 8$, then the length of intercept made by the circle on $X$-axis is *[2022, 27 July Shift-I]*

(a) $\sqrt{11}$ (b) 4
(c) 3 (d) $2\sqrt{3}$

**10** Let the abscissae of the two points $P$ and $Q$ on a circle be the roots $x^2 - 4x - 6 = 0$ and the ordinates of $P$ and $Q$ be the roots of $y^2 + 2y - 7 = 0$. If $PQ$ is a diameter of the circle $x^2 + y^2 + 2ax + 2by + c = 0$, then the value of $(a + b - c)$ is *[2022, 26 July Shift-II]*

(a) 12 (b) 13
(c) 14 (d) 16

**11** Consider a circle $C$ which touches the $Y$-axis at (0, 6) and cuts off an intercept $6\sqrt{5}$ on the $X$-axis. Then, the radius of the circle $C$ is equal to **[2021, 27 July Shift-II]**

(a) $\sqrt{53}$ (b) 9
(c) 8 (d) $\sqrt{82}$

**12** The line $2x - y + 1 = 0$ is a tangent to the circle at the point (2, 5) and the centre of the circle lies on $x - 2y = 4$. Then, the radius of the circle is **[2021, 17 March Shift-I]**

(a) $3\sqrt{5}$ (b) $5\sqrt{3}$
(c) $5\sqrt{4}$ (d) $4\sqrt{5}$

**13** Choose the incorrect statement about the two circles whose equations are given below
$x^2 + y^2 - 10x - 10y + 41 = 0$
and $x^2 + y^2 - 16x - 10y + 80 = 0$ **[2021, 17 March Shift-I]**

(a) Distance between two centres is the average of radii of both the circles.
(b) Both circle's centres lie inside region of one another.
(c) Both circles pass through the centre of each other.
(d) Circles have two intersection points.

**14** If the length of the chord of the circle, $x^2 + y^2 = r^2 (r > 0)$ along the line, $y - 2x = 3$ is $r$, then $r^2$ is equal to **[2020, 5 Sep Shift-II]**

(a) 9/5 (b) 12 (c) 24/5 (d) 12/5

**15** A circle passes through the points (2, 3) and (4, 5). If its centre lies on the line, $y - 4x + 3 = 0$, then its radius is equal to **[JEE Main 2018]**

(a) $\sqrt{5}$ (b) 1 (c) $\sqrt{2}$ (d) 2

**16** The tangent to the circle $C_1 : x^2 + y^2 - 2x - 1 = 0$ at the point (2, 1) cuts off a chord of length 4 from a circle $C_2$ whose centre is (3, −2). Then radius of $C_2$ is **[JEE Main 2018]**

(a) $\sqrt{6}$ (b) 2 (c) $\sqrt{2}$ (d) 3

**17** Let the orhtocentre and centroid of a triangle be $A(-3, 5)$ and $B(3, 3)$, respectively. If $C$ is the circumcentre of this triangle, then the radius of the circle having line segment $AC$ as diameter, is **[JEE Main 2018]**

(a) $\frac{3\sqrt{5}}{2}$ (b) $\sqrt{10}$
(c) $2\sqrt{10}$ (d) $3\sqrt{\frac{5}{2}}$

**18** If $y + 3x = 0$ is the equation of a chord of the circle, $x^2 + y^2 - 30x = 0$, then the equation of the circle with this chord as diameter is **[JEE Main 2015]**

(a) $x^2 + y^2 + 3x + 9y = 0$ (b) $x^2 + y^2 - 3x + 9y = 0$
(c) $x^2 + y^2 - 3x - 9y = 0$ (d) $x^2 + y^2 + 3x - 9y = 0$

**19** If the circle $x^2 + y^2 - 6x - 8y + (25 - a^2) = 0$ touches the axis of $x$, then $a$ equals **[JEE Main 2013]**

(a) 0 (b) ± 4
(c) ± 2 (d) ± 3

**20** The equation of the circle passing through the point (1, 0) and (0, 1) and having the smallest radius is **[AIEEE 2011]**

(a) $x^2 + y^2 + x + y - 2 = 0$
(b) $x^2 + y^2 - 2x - 2y + 1 = 0$
(c) $x^2 + y^2 - x - y = 0$
(d) $x^2 + y^2 + 2x + 2y - 7 = 0$

**21** The circle $x^2 + y^2 = 4x + 8y + 5$ intersects the line $3x - 4y = m$ at two distinct points, if **[AIEEE 2010]**

(a) $-85 < m < -35$ (b) $-35 < m < 15$
(c) $15 < m < 65$ (d) $35 < m < 85$

**22** If $P$ and $Q$ are the points of intersection of the circles $x^2 + y^2 + 3x + 7y + 2p - 5 = 0$ and $x^2 + y^2 + 2x + 2y - p^2 = 0$, then there is a circle passing through $P, Q$ and (1, 1) for **[AIEEE 2009]**

(a) all values of $p$ (b) all except one value of $p$
(c) all except two values of $p$ (d) exactly one value of $p$

**23** The point diametrically opposite to the point $P(1, 0)$ on the circle $x^2 + y^2 + 2x + 4y - 3 = 0$ is **[AIEEE 2008]**

(a) (3, 4) (b) (3, − 4)
(c) (− 3, 4) (d) (−3, − 4)

**24** If the lines $2x + 3y + 1 = 0$ and $3x - y - 4 = 0$ lie along diameters of a circle of circumference $10\pi$, then the equation of the circle is **[AIEEE 2004]**

(a) $x^2 + y^2 - 2x + 2y - 23 = 0$
(b) $x^2 + y^2 - 2x - 2y - 23 = 0$
(c) $x^2 + y^2 + 2x + 2y - 23 = 0$
(d) $x^2 + y^2 + 2x - 2y - 23 = 0$

**25** The greatest distance of the point $P(10, 7)$ from the circle $x^2 + y^2 - 4x - 2y - 20 = 0$ is **[AIEEE 2002]**

(a) 10 units (b) 15 units
(c) 5 units (d) None of these

**26** The centres of a set of circles, each of radius 3, lie on the circle $x^2 + y^2 = 25$. The locus of any point in the set is **[AIEEE 2002]**

(a) $4 \le x^2 + y^2 \le 64$ (b) $x^2 + y^2 \le 25$
(c) $x^2 + y^2 \ge 25$ (d) $3 \le x^2 + y^2 \le 0$

**27** The equation of the tangent to the circle $x^2 + y^2 + 4x - 4y + 4 = 0$ which make equal intercepts on the positive co-ordinate axes, is **[AIEEE 2002]**

(a) $x + y = 2$ (b) $x + y = 2\sqrt{2}$
(c) $x + y = 4$ (d) $x + y = 8$

## Level 2

**28** Let the equation of the circle, which touches $X$-axis at the point $(a, 0)$, $a>0$ and cuts off an intercept of length $b$ on $Y$-axis be $x^2+y^2-\alpha x+\beta y+\gamma=0$. If the circle lies below $X$-axis, then the ordered pair $(2a, b^2)$ is equal to [2025, 28 Jan Shift-I]

(a) $(\alpha, \beta^2-4\gamma)$ (b) $(\gamma, \beta^2+4\alpha)$
(c) $(\gamma, \beta^2-4\alpha)$ (d) $(\alpha, \beta^2+4\gamma)$

**29** A circle $C$ of radius 2 lies in the second quadrant and touches both the coordinate axes. Let $r$ be the radius of a circle that has centre at the point $(2, 5)$ and intersects the circle $C$ at exactly two points. If the set of all possible values of $r$ is the interval $(\alpha, \beta)$, then $3\beta-2\alpha$ is equal to [2025, 22 Jan Shift-I]

(a) 14 (b) 12 (c) 15 (d) 10

**30** Let the parabola $y=x^2+px-3$ meet the coordinate axes at the points $P$, $Q$ and $R$. If the circle $C$ with centre at $(-1,-1)$ passes through the points $P$, $Q$ and $R$, then the area of $\Delta PQR$ is [2025, 22 Jan Shift-I]

(a) 5 (b) 6 (c) 7 (d) 4

**Imp.** Section Formula

**31** Let the circles $C_1:(x-\alpha)^2+(y-\beta)^2=r_1^2$ and $C_2:(x-8)^2+\left(y-\frac{15}{2}\right)^2=r_2^2$ touch each other externally at the point (6, 6). If the point (6, 6) divides the line segment joining the centres of the circles $C_1$ and $C_2$ internally in the ratio 2 : 1, then $(\alpha+\beta)+4(r_1^2+r_2^2)$ equals [2024, 8 April Shift-I]

(a) 125 (b) 145 (c) 110 (d) 130

**32** The number of common tangents, to the circles $x^2+y^2-18x-15y+131=0$ and $x^2+y^2-6x-6y-7=0$, are [2023, 15 April Shift-I]

(a) 2 (b) 3
(c) 1 (d) 4

**33** Let $A$ be the point (1, 2) and $B$ be any point on the curve $x^2+y^2=16$. If the centre of the locus of the point $P$, which divides the line segment $AB$ in the ratio 3 : 2 is the point $C(\alpha, \beta)$, then the length of the line segment $AC$ is [2023, 10 April Shift-II]

(a) $\frac{3\sqrt{5}}{5}$ (b) $\frac{6\sqrt{5}}{5}$
(c) $\frac{2\sqrt{5}}{5}$ (d) $\frac{4\sqrt{5}}{5}$

**34** A line segment $AB$ of length $\lambda$ moves such that the points $A$ and $B$ remain on the periphery of a circle of radius $\lambda$. Then, the locus of the point, that divides the line segment $AB$ in the ratio 2 : 3, is a circle of radius [2023, 10 April Shift-I]

(a) $\frac{\sqrt{19}}{7}\lambda$ (b) $\frac{\sqrt{19}}{5}\lambda$ (c) $\frac{3}{5}\lambda$ (d) $\frac{2}{3}\lambda$

**35** Let $O$ be the origin and $OP$ and $OQ$ be the tangents to the circle $x^2+y^2-6x+4y+8=0$ at the points $P$ and $Q$ on it. If the circumcircle of the $\Delta OPQ$ passes through the point $\left(\alpha, \frac{1}{2}\right)$, then a value of $\alpha$ is [2023, 8 April Shift-II]

(a) 5/2 (b) 1 (c) $-1/2$ (d) 3/2

**36** Let $y=x+2$, $4y=3x+6$ and $3y=4x+1$ be three tangent lines to the circle $(x-h)^2+(y-k)^2=r^2$. Then, $h+k$ is equal to [2023, 30 Jan Shift-I]

(a) $5(1+\sqrt{2})$ (b) $5\sqrt{2}$
(c) 5 (d) 6

**37** Let $C$ be the centre of the circle $x^2+y^2-x+2y=\frac{11}{4}$ and $P$ be a point on the circle. A line passes through the point $C$, makes an angle of $\frac{\pi}{4}$ with the line $CP$ and intersects the circle at $Q$ and $R$. Then, area of the $\Delta PQR$ (in unit$^2$) is [2022, 28 July Shift-I]

(a) 2 (b) $2\sqrt{2}$
(c) $8\sin\left(\frac{\pi}{8}\right)$ (d) $8\cos\left(\frac{\pi}{8}\right)$

**38** For $t\in(0, 2\pi)$, if $ABC$ is an equilateral triangle with vertices $A(\sin t, -\cos t)$, $B(\cos t, \sin t)$ and $C(a, b)$ such that its orthocentre lies on a circle with centre $\left(1, \frac{1}{3}\right)$, then $(a^2-b^2)$ is equal to [2022, 28 July Shift-I]

(a) 8/3 (b) 8
(c) 77/9 (d) 80/9

**39** Let a $\Delta ABC$ be inscribed in the circle $x^2-\sqrt{2}(x+y)+y^2=0$ such that $\angle BAC=\frac{\pi}{2}$. If the length of side $AB$ is $\sqrt{2}$, then the area of the $\Delta ABC$ is equal to [2022, 29 June Shift-II]

(a) $\frac{(\sqrt{2}+\sqrt{6})}{3}$ (b) $\frac{(\sqrt{6}+\sqrt{3})}{2}$
(c) $\frac{(3+\sqrt{3})}{4}$ (d) $\frac{(\sqrt{6}+2\sqrt{3})}{4}$

**40** Let the tangent to the circle $C_1 : x^2 + y^2 = 2$ at the point $M(-1, 1)$ intersect the circle $C_2 : (x-3)^2 + (y-2)^2 = 5$, at two distinct points $A$ and $B$. If the tangents to $C_2$ at the points $A$ and $B$ intersect at $N$, then the area of the $\Delta ANB$ is equal to **[2022, 29 June Shift-I]**

(a) 1/2 (b) 2/3 (c) 1/6 (d) 5/3

**41** If the tangents drawn at the points $O(0, 0)$ and $P(1+\sqrt{5}, 2)$ on the circle $x^2 + y^2 - 2x - 4y = 0$ intersect at the point $Q$, then the area of the $\Delta OPQ$ is equal to **[2022, 28 June Shift-I]**

(a) $\frac{3+\sqrt{5}}{2}$ (b) $\frac{4+2\sqrt{5}}{2}$ (c) $\frac{5+3\sqrt{5}}{2}$ (d) $\frac{7+3\sqrt{5}}{2}$

**42** The set of values of $k$ for which the circle $C : 4x^2 + 4y^2 - 12x + 8y + k = 0$ lies inside the fourth quadrant and the point $\left(1, -\frac{1}{3}\right)$ lies on or inside the circle $C$ is **[2022, 27 June Shift-II]**

(a) an empty set (b) $\left(6, \frac{65}{9}\right]$

(c) $\left[\frac{80}{9}, 10\right)$ (d) $\left(9, \frac{92}{9}\right]$

**43** A circle $C$ touches the line $x = 2y$ at the point (2,1) and intersects the circle $C_1 : x^2 + y^2 + 2y - 5 = 0$ at two points $P$ and $Q$ such that $PQ$ is a diameter of $C_1$. Then, the diameter of $C$ is **[2021, 26 Aug Shift-II]**

(a) $7\sqrt{5}$ (b) 15 (c) 285 (d) $4\sqrt{15}$

**44** Two tangents are drawn from the point $P(-1, 1)$ to the circle $x^2 + y^2 - 2x - 6y + 6 = 0$. If these tangents touch the circle at points $A$ and $B$, and if $D$ is a point on the circle such that length of the segments $AB$ and $AD$ are equal, then the area of the $\Delta ABD$ is equal to **[2021, 27 July Shift-I]**

(a) 2 (b) $(3\sqrt{2} + 2)$

(c) 4 (d) $3(\sqrt{2} - 1)$

**45** Let $P$ and $Q$ be two distinct points on a circle which has center at $C(2,3)$ and which passes through origin $O$. If $OC$ is perpendicular to both the line segments $CP$ and $CQ$, then the set $\{P, Q\}$ is equal to **[2021, 27 July Shift-I]**

(a) $\{(4, 0), (0, 6)\}$

(b) $\{(2+2\sqrt{2}, 3-\sqrt{3}), (2-2\sqrt{2}, 3+\sqrt{5})\}$

(c) $\{(2+2\sqrt{2}, 3+\sqrt{5}), (2-2\sqrt{2}, 3-\sqrt{5})$

(d) $\{(-1, 5), (5, 1)\}$

**Imp.** Nature of Roots

**46** Let the circle $S : 36x^2 + 36y^2 - 108x + 120y + C = 0$ be such that it neither intersects nor touches the coordinate axes. If the point of intersection of the lines $x - 2y = 4$ and $2x - y = 5$ lies inside the circle $S$, then **[2021, 22 July Shift-II]**

(a) $\frac{25}{9} < C < \frac{13}{3}$ (b) $100 < C < 165$

(c) $81 < C < 156$ (d) $100 < C < 156$

**47** For the four circles $M$, $N$, $O$ and $P$ following four equations are given

Circle $M : x^2 + y^2 = 1$

Circle $N : x^2 + y^2 - 2x = 0$

Circle $O : x^2 + y^2 - 2x - 2y + 1 = 0$

Circle $P : x^2 + y^2 - 2y = 0$

If the centre of circle $M$ is joined with centre of the circle $N$, further centre of circle $N$ is joined with centre of the circle $O$, centre of circle $O$ is joined with the centre of circle $P$ and lastly, centre of circle $P$ is joined with centre of circle $M$, then these lines form the sides of a **[2021, 18 March Shift-I]**

(a) rhombus (b) square

(c) rectangle (d) parallelogram

**48** Let the tangent to the circle $x^2 + y^2 = 25$ at the point $R(3, 4)$ meet $X$-axis and $Y$-axis at points $P$ and $Q$, respectively. If $r$ is the radius of the circle passing through the origin $O$ and having centre at the incentre of the $\Delta OPQ$, then $r^2$ is equal to **[2021, 17 March Shift-II]**

(a) $\frac{529}{64}$ (b) $\frac{625}{72}$ (c) $\frac{585}{66}$ (d) $\frac{125}{72}$

**49** Let $A(1, 4)$ and $B(1, -5)$ be two points. Let $P$ be a point on the circle $(x-1)^2 + (y-1)^2 = 1$, such that $(PA)^2 + (PB)^2$ have maximum value, then the points $P$, $A$ and $B$ lie on **[2021, 26 Feb Shift-II]**

(a) a straight line (b) a hyperbola

(c) an ellipse (d) a parabola

**50** In the circle given below, let $OA = 1$ unit, $OB = 13$ unit and $PQ \perp OB$. Then, the area of the $\Delta PQB$ (in square units) is **[2021, 26 Feb Shift-I]**

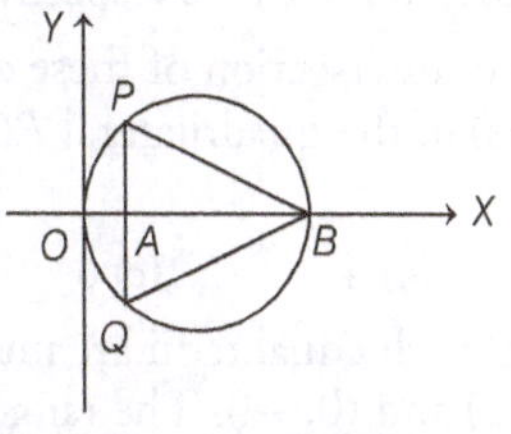


(a) $24\sqrt{2}$ (b) $24\sqrt{3}$ (c) $26\sqrt{3}$ (d) $26\sqrt{2}$

**51** For which of the following curves, the line $x+\sqrt{3}y=2\sqrt{3}$ is the tangent at the point $\left(\frac{3\sqrt{3}}{2},\frac{1}{2}\right)$? **[2021, 24 Feb Shift-II]**

(a) $x^2+y^2=7$ (b) $y^2=\frac{1}{6\sqrt{3}}x$

(c) $2x^2-18y^2=9$ (d) $x^2+9y^2=9$

**52** If a line, $y=mx+c$ is a tangent to the circle, $(x-3)^2+y^2=1$ and it is perpendicular to a line $L_1$, where $L_1$ is the tangent to the circle, $x^2+y^2=1$ at the point $\left(\frac{1}{\sqrt{2}},\frac{1}{\sqrt{2}}\right)$, then **[2020, 8 Jan Shift-II]**

(a) $c^2+7c+6=0$ (b) $c^2-6c+7=0$

(c) $c^2-7c+6=0$ (d) $c^2+6c+7=0$

**53** Let the tangents drawn from the origin to the circle, $x^2+y^2-8x-4y+16=0$ touch it at the points $A$ and $B$. Then, $(AB)^2$ is equal to **[2020, 7 Jan Shift-II]**

(a) 56/5 (b) 52/2 (c) 64/5 (d) 32/5

**54** The locus of the centres of the circles, which touch the circle, $x^2+y^2=1$ externally, also touch the $Y$-axis and lie in the first quadrant, is **[2019, 10 April Shift-II]**

(a) $y=\sqrt{1+2x}, x\geq 0$ (b) $y=\sqrt{1+4x}, x\geq 0$

(c) $x=\sqrt{1+2y}, y\geq 0$ (d) $x=\sqrt{1+4y}, y\geq 0$

**55** The common tangent to the circles $x^2+y^2=4$ and $x^2+y^2+6x+8y-24=0$ also passes through the point **[2019, 9 April Shift-II]**

(a) $(6,-2)$ (b) $(4,-2)$ (c) $(-6,4)$ (d) $(-4,6)$

**56** A rectangle is inscribed in a circle with a diameter lying along the line $3y=x+7$. If the two adjacent vertices of the rectangle are $(-8, 5)$ and $(6, 5)$, then the area of the rectangle (in sq units) is **[2019, 9 April Shift-II]**

(a) 72 (b) 84 (c) 98 (d) 56

**57** Let $C_1$ and $C_2$ be the centres of the circles $x^2+y^2-2x-2y-2=0$ and $x^2+y^2-6x-6y+14=0$ respectively. If $P$ and $Q$ are the points of intersection of these circles, then the area (in sq units) of the quadrilateral $PC_1QC_2$ is **[2019, 12 Jan Shift-I]**

(a) 8 (b) 4 (c) 6 (d) 9

**58** Two circles with equal radii are intersecting at the points (0, 1) and (0, −1). The tangent at the point (0,1) to one of the circles passes through the centre of the other circle. Then, the distance between the centres of these circles is **[2019, 11 Jan Shift-I]**

(a) $\sqrt{2}$ (b) $2\sqrt{2}$ (c) 1 (d) 2

**59** If the area of an equilateral triangle inscribed in the circle, $x^2+y^2+10x+12y+c=0$ is $27\sqrt{3}$ sq units, then $c$ is equal to **[2019, 10 Jan Shift-II]**

(a) 13 (b) 20 (c) −25 (d) 25

**60** If the tangent at (1, 7) to the curve $x^2=y-6$ touches the circle $x^2+y^2+16x+12y+c=0$, then the value of $c$ is **[JEE Main 2018]**

(a) 95 (b) 195 (c) 185 (d) 18

**61** If a circle $C$, whose radius is 3, touches externally the circle, $x^2+y^2+2x-4y-4=0$ at the point (2, 2), then the length of the intercept cut by this circle $C$, on the $X$-axis is equal to **[JEE Main 2018]**

(a) $2\sqrt{3}$ (b) $3\sqrt{2}$ (c) $\sqrt{5}$ (d) $2\sqrt{5}$

**62** The two adjacent sides of a cyclic quadrilateral are 2 and 5 and the angle between them is 60°. If the area of the quadrilateral is $4\sqrt{3}$, then the perimeter of the quadrilateral is **[JEE Main 2017]**

(a) 12.5 (b) 13.2

(c) 12 (d) 13

**63** If a point $P$ has co-ordinates (0, −2) and $Q$ is any point on the circle, $x^2+y^2-5x-y+5=0$, then the maximum value of $(PQ)^2$ is **[JEE Main 2017]**

(a) $\frac{25+\sqrt{6}}{2}$ (b) $14+5\sqrt{3}$ (c) $\frac{47+10\sqrt{6}}{2}$ (d) $8+5\sqrt{3}$

**64** If two parallel chords of a circle, having diameter 4 units, lie on the opposite sides of the centre and subtend angles $\cos^{-1}\left(\frac{1}{7}\right)$ and $\sec^{-1}(7)$ at the centre respectively, then the distance between these chords, is **[JEE Main 2017]**

(a) $\frac{4}{\sqrt{7}}$ (b) $\frac{8}{\sqrt{7}}$ (c) $\frac{8}{7}$ (d) $\frac{16}{7}$

**65** If one of the diameters of the circle, given by the equation, $x^2+y^2-4x+6y-12=0$, is a chord of a circle $S$, whose centre is at $(-3, 2)$, then the radius of $S$ is **[JEE Main 2016]**

(a) $5\sqrt{2}$ (b) $5\sqrt{3}$ (c) 5 (d) 10

**66** Equation of the tangent to the circle, at the point (1, −1) whose centre is the point of intersection of the straight lines $x-y=1$ and $2x+y=3$ is **[JEE Main 2016]**

(a) $x+4y+3=0$ (b) $3x-y-4=0$

(c) $x-3y-4=0$ (d) $4x+y-3=0$

**67** A circle passes through $(-2, 4)$ and touches the $Y$-axis at $(0, 2)$. Which one of the following equations can represent a diameter of this circle? *[JEE Main 2016]*

(a) $2x - 3y + 10 = 0$ (b) $3x + 4y - 3 = 0$
(c) $4x + 5y - 6 = 0$ (d) $5x + 2y + 4 = 0$

**68** Locus of the image of the point $(2, 3)$ in the line $(2x - 3y + 4) + k(x - 2y + 3) = 0, k \in R$, is a *[JEE Main 2015]*

(a) circle of radius $\sqrt{2}$.
(b) circle of radius $\sqrt{3}$.
(c) straight line parallel to $X$-axis
(d) straight line parallel to $Y$-axis

**69** If the incentre of an equilateral triangle is $(1, 1)$ and the equation of its cone side is $3x + 4y + 3 = 0$, then the equation of the circumcircle of this triangle is *[JEE Main 2015]*

(a) $x^2 + y^2 - 2x - 2y - 14 = 0$
(b) $x^2 + y^2 - 2x - 2y - 2 = 0$
(c) $x^2 + y^2 - 2x - 2y + 2 = 0$
(d) $x^2 + y^2 - 2x - 2y - 7 = 0$

**70** Let the tangents drawn to the circle, $x^2 + y^2 = 16$ from the point $P(0, h)$ meet the $X$-axis at point $A$ and $B$. If the area of $\Delta APB$ is minimum, then $h$ is equal to *[JEE Main 2015]*

(a) $4\sqrt{2}$ (b) $3\sqrt{3}$ (c) $3\sqrt{2}$ (d) $4\sqrt{3}$

**71** The number of common tangents to the circles $x^2 + y^2 - 4x - 6y - 12 = 0$ and $x^2 + y^2 + 6x + 18y + 26 = 0$ is *[JEE Main 2015]*

(a) 1 (b) 2 (c) 3 (d) 4

**72** Let $C$ be the circle with centre at $(1, 1)$ and radius 1. If $T$ is the circle centred at $(0, y)$ passing through origin and touching the circle $C$ externally, then the radius of $T$ is equal to *[JEE Main 2014]*

(a) $\frac{\sqrt{3}}{\sqrt{2}}$ (b) $\frac{\sqrt{3}}{2}$
(c) $\frac{1}{2}$ (d) $\frac{1}{4}$

**73** For the two circles $x^2 + y^2 = 16$ and $x^2 + y^2 - 2y = 0$, there is/are *[JEE Main 2014]*

(a) one pair of common tangents
(b) two pair of common tangents
(c) three pair of common tangent
(d) no common tangent

**74** The circle passing through $(1, -2)$ and touching the axis of $x$ at $(3, 0)$ also passes through the point *[JEE Main 2013]*

(a) $(-5, 2)$ (b) $(2, -5)$
(c) $(5, -2)$ (d) $(-2, 5)$

**75** Given A circle, $2x^2 + 2y^2 = 5$ and a parabola, $3y^2 = 4\sqrt{5}x$.

**Statement I** An equation of a common tangent to these curves is $y = x + \sqrt{5}$.

**Statement II** If the line, $y = mx + \sqrt{5}/m, (m \neq 0)$ is the common tangent, then $m$ satisfies $m^4 - 3m^2 + 2 = 0$. *[JEE Main 2013]*

(a) Statement I is true, Statement II is true; Statement II is a correct explanation of Statement I
(b) Statement I is true, Statement II is true; Statement II is not a correct explanation of Statement I
(c) Statement I is true, Statement II is false
(d) Statement I is false, Statement II is true

**76** If each of the lines $5x + 8y = 13$ and $4x - y = 3$ contains a diameter of the circle $x^2 + y^2 - 2(a^2 - 7a + 11)x - 2(a^2 - 6a + 6)y + b^3 + 1 = 0$, then *[JEE Main 2013]*

(a) $a = 5$ and $b \notin (-1, 0)$ (b) $a = 1$ and $b \notin (-1, 1)$
(c) $a = 2$ and $b \notin (-\infty, 1)$ (d) $a = 5$ and $b \notin (-\infty, 1)$

**77** If the line $y = mx + 1$ meets the circle $x^2 + y^2 + 3x = 0$ in two points equidistant from and on opposite sides of $X$-axis, then *[AIEEE 2012]*

(a) $3m + 2 = 0$ (b) $3m - 2 = 0$
(c) $2m + 3 = 0$ (d) $2m - 3 = 0$

**78** If three distinct points $A, B, C$ are given in the 2-dimensional co-ordinate plane such that the ratio of the distance of each one of them from the point $(1, 0)$ to the distance from $(-1, 0)$ is equal to $1/2$, then the circumcentre of the $\Delta ABC$ is at the point *[AIEEE 2012]*

(a) $\left(\frac{5}{3}, 0\right)$ (b) $(0, 0)$ (c) $\left(\frac{1}{3}, 0\right)$ (d) $(3, 0)$

**79** The length of the diameter of the circle which touches the $X$-axis at the point $(1, 0)$ and passes through the point $(2, 3)$ is *[AIEEE 2012]*

(a) 10/3 (b) 3/5 (c) 6/5 (d) 5/3

**80** The two circles $x^2 + y^2 = ax$ and $x^2 + y^2 = c^2, (c > 0)$ touch each other, if *[AIEEE 2011]*

(a) $|a| = c$ (b) $a = 2c$ (c) $|a| = 2c$ (d) $2|a| = c$

**81** Three distinct points $A$, $B$ and $C$ are given in the 2-dimensional coordinates plane such that the ratio of the distance of any one of them from the point $(1, 0)$ to the distance from the point $(-1, 0)$ is equal to $1/3$. Then, the circumcentre of the triangle $ABC$ is at the point *[AIEEE 2009]*

(a) $\left(\frac{5}{4}, 0\right)$ (b) $\left(\frac{5}{2}, 0\right)$ (c) $\left(\frac{5}{3}, 0\right)$ (d) $(0, 0)$

**82** Consider a family of circles which are passing through the point $(-1, 1)$ and are tangent to $X$-axis. If $(h, k)$ are the coordinate of the centre of the circle, then the set of values of $k$ is given by the interval **[AIEEE 2007]**

(a) $-\frac{1}{2} \le k \le \frac{1}{2}$ (b) $k \le \frac{1}{2}$

(c) $0 \le k \le \frac{1}{2}$ (d) $k \ge \frac{1}{2}$

**83** If the lines $3x - 4y - 7 = 0$ and $2x - 3y - 5 = 0$ are two diameters of a circle of area $49\pi$ sq units, then the equation of the circle is **[AIEEE 2006]**

(a) $x^2 + y^2 + 2x - 2y - 62 = 0$

(b) $x^2 + y^2 - 2x + 2y - 62 = 0$

(c) $x^2 + y^2 - 2x + 2y - 47 = 0$

(d) $x^2 + y^2 + 2x - 2y - 47 = 0$

**84** Let $C$ be the circle with centre $(0, 0)$ and radius 3 units. The equation of the locus of the mid-points of the chords of the circle $C$ that subtend an angle of $\frac{2\pi}{3}$ at its centre, is **[AIEEE 2006]**

(a) $x^2 + y^2 = 1$ (b) $x^2 + y^2 = \frac{27}{4}$

(c) $x^2 + y^2 = \frac{9}{4}$ (d) $x^2 + y^2 = \frac{3}{2}$

**85** A circle touches the $X$-axis and also touches the circle with centre at $(0, 3)$ and radius 2. The locus of the centre of the circle is **[AIEEE 2005]**

(a) a parabola (b) a hyperbola (c) a circle (d) an ellipse

**86** If a circle passes through the point $(a, b)$ and cuts the circle $x^2 + y^2 = p^2$ orthogonally, then the equation of the locus of its centre is **[AIEEE 2005]**

(a) $2ax + 2by - (a^2 + b^2 + p^2) = 0$

(b) $x^2 + y^2 - 2ax - 3by + (a^2 - b^2 - p^2) = 0$

(c) $2ax + 2by - (a^2 - b^2 + p^2) = 0$

(d) $x^2 + y^2 - 3ax - 4by + (a^2 + b^2 - p^2) = 0$

**87** If the pair of lines $ax^2 + 2(a + b)xy + by^2 = 0$ lie along diameters of a circle and divide the circle into four sectors such that the area of one of the sectors is thrice the area of another sector, then **[AIEEE 2005]**

(a) $3a^2 + 2ab + 3b^2 = 0$ (b) $3a^2 + 10ab + 3b^2 = 0$

(c) $3a^2 - 2ab + 3b^2 = 0$ (d) $3a^2 - 10ab + 3b^2 = 0$

**88** If a circle passes through the point $(a, b)$ and cuts the circle $x^2 + y^2 = 4$ orthogonally, then the locus of its centre is **[AIEEE 2004]**

(a) $2ax + 2by + (a^2 + b^2 + 4) = 0$

(b) $2ax + 2by - (a^2 + b^2 + 4) = 0$

(c) $2ax - 2by + (a^2 + b^2 + 4) = 0$

(d) $2ax - 2by - (a^2 + b^2 + 4) = 0$

**89** The intercept on the line $y = x$ by the circle $x^2 + y^2 - 2x = 0$ is $AB$. Equation of the circle on $AB$ as a diameter is **[AIEEE 2004]**

(a) $x^2 + y^2 - x - y = 0$ (b) $x^2 + y^2 - x + y = 0$

(c) $x^2 + y^2 + x + y = 0$ (d) $x^2 + y^2 + x - y = 0$

**90** If the two circles $(x - 1)^2 + (y - 3)^2 = r^2$ and $x^2 + y^2 - 8x + 2y + 8 = 0$ intersect in two distinct points, then **[AIEEE 2003]**

(a) $2 < r < 8$ (b) $r < 2$ (c) $r = 2$ (d) $r > 2$

**91** The equation of a circle with origin as a centre and passing through equilateral triangle whose median is of length $3a$ is **[AIEEE 2002]**

(a) $x^2 + y^2 = 9a^2$ (b) $x^2 + y^2 = 16a^2$

(c) $x^2 + y^2 = 4a^2$ (d) $x^2 + y^2 = a^2$

**92** The centre of the circle passing through $(0, 0)$ and $(1, 0)$ and touching the circle $x^2 + y^2 = 9$ is **[AIEEE 2002]**

(a) $\left(\frac{1}{2}, \frac{1}{2}\right)$ (b) $\left(\frac{1}{2}, -\sqrt{2}\right)$ (c) $\left(\frac{3}{2}, \frac{1}{2}\right)$ (d) $\left(\frac{1}{2}, \frac{3}{2}\right)$

## Level 3

**93** Let the line $x + y = 1$ meet the circle $x^2 + y^2 = 4$ at the points $A$ and $B$. If the line perpendicular to $AB$ and passing through the mid-point of the chord $AB$ intersects the circle at $C$ and $D$, then the area of the quadrilateral $ADBC$ is equal to **[2025, 29 Jan Shift-I]**

(a) $\sqrt{14}$ (b) $5\sqrt{7}$ (c) $3\sqrt{7}$ (d) $2\sqrt{14}$

**94** Let circle $C$ be the image of $x^2 + y^2 - 2x + 4y - 4 = 0$ in the line $2x - 3y + 5 = 0$ and $A$ be the point on $C$ such that $OA$ is parallel to $X$-axis and $A$ lies on the right hand side of the centre $O$ of $C$. If $B(\alpha, \beta)$ with $\beta < 4$, lies on $C$ such that the length of the arc $AB$ is $\left(\frac{1}{6}\right)$th of the perimeter of $C$, then $\beta - \sqrt{3}\alpha$ is equal to **[2025, 24 Jan Shift-I]**

(a) $3 + \sqrt{3}$ (b) $4 - \sqrt{3}$

(c) 3 (d) 4

**95** Let a circle passing through $(2, 0)$ have its centre at the point $(h, k)$. Let $(x_c, y_c)$ be the point of intersection of the lines $3x + 5y = 1$ and $(2 + c)x + 5c^2 y = 1$. If $h = \lim_{c \to 1} x_c$ and $k = \lim_{c \to 1} y_c$, then the equation of the circle is **[2024, 9 April Shift-I]**

(a) $5x^2 + 5y^2 - 4x - 2y - 12 = 0$

(b) $25x^2 + 25y^2 - 2x + 2y - 60 = 0$

(c) $25x^2 + 25y^2 - 20x + 2y - 60 = 0$

(d) $5x^2 + 5y^2 - 4x + 2y - 12 = 0$

**96** A square is inscribed in the circle $x^2+y^2-10x-6y+30=0$. One side of this square is parallel to $y=x+3$. If $(x_i, y_i)$ are the vertices of the square, then $\Sigma(x_i^2+y_i^2)$ is equal to *[2024, 4 April Shift-I]*

(a) 156 (b) 152 (c) 148 (d) 160

**97** Let the centre of a circle $C$ be $(\alpha, \beta)$ and its radius $r<8$. Let $3x+4y=24$ and $3x-4y=32$ be two tangents and $4x+3y=1$ be a normal to $C$. Then, $(\alpha-\beta+r)$ is equal to *[2023, 13 April Shift-II]*

(a) 6 (b) 9 (c) 7 (d) 5

**98** If the tangents at the points $P$ and $Q$ on the circle $x^2+y^2-2x+y=5$ meet at the point $R\left(\frac{9}{4}, 2\right)$, then the area of the $\Delta PQR$ is *[2023, 6 April Shift-II]*

(a) 5/4 (b) 13/4 (c) 13/8 (d) 5/8

**99** The set of all values of $a^2$ for which the line $x+y=0$ bisects two distinct chords drawn from a point $P\left(\frac{1+a}{2}, \frac{1-a}{2}\right)$ on the circle $2x^2+2y^2-(1+a)x-(1-a)y=0$, is equal to *[2023, 31 Jan Shift-II]*

(a) (0, 4] (b) $(4, \infty)$ (c) $(8, \infty)$ (d) (2, 12]

**100** Let a circle $C_1$ be obtained on rolling the circle $x^2+y^2-4x-6y+11=0$ upwards 4 units on the tangent $T$ to it at the point (3, 2). Let $C_2$ be the image of $C_1$ in $T$. Let $A$ and $B$ be the centres of circles $C_1$ and $C_2$ respectively, $M$ and $N$ be respectively the feet of perpendiculars drawn from $A$ and $B$ on the $X$-axis. Then, the area of the trapezium $AMNB$ is *[2023, 31 Jan Shift-I]*

(a) $3+2\sqrt{2}$ (b) $2(2+\sqrt{2})$
(c) $4(1+\sqrt{2})$ (d) $2(1+\sqrt{2})$

**101** Let the tangents at the points $A(4, -11)$ and $B(8, -5)$ on the circle $x^2+y^2-3x+10y-15=0$, intersect at the point $C$. Then, the radius of the circle, whose centre is $C$ and the line joining $A$ and $B$ is its tangent, is equal to *[2023, 29 Jan Shift-I]*

(a) $2\sqrt{13}$ (b) $\frac{2\sqrt{13}}{3}$
(c) $\frac{3\sqrt{3}}{4}$ (d) $\sqrt{13}$

**102** The points of intersection of the line $ax+by=0$, $(a \neq b)$ and the circle $x^2+y^2-2x=0$ are $A(\alpha, 0)$ and $B(1, \beta)$. The image of the circle with $AB$ as a diameter in the line $x+y+2=0$ is *[2023, 25 Jan Shift-I]*

(a) $x^2+y^2-5x-5y+12=0$
(b) $x^2+y^2+3x+5y+8=0$
(c) $x^2+y^2+5x+5y+12=0$
(d) $x^2+y^2+3x+3y+4=0$

**103** The locus of the mid-points of the chords of the circle $C_1:(x-4)^2+(y-5)^2=4$ which subtend an angle $\theta_i$ at the centre of the circle $C_1$ is a circle of radius $r_i$. If $\theta_1=\frac{\pi}{3}, \theta_3=\frac{2\pi}{3}$ and $r_1^2=r_2^2+r_3^2$, then $\theta_2$ is equal to *[2023, 24 Jan Shift-II]*

(a) $\frac{3\pi}{4}$ (b) $\frac{\pi}{4}$ (c) $\frac{\pi}{6}$ (d) $\frac{\pi}{2}$

**104** Let the tangents at two points $A$ and $B$ on the circle $x^2+y^2-4x+3=0$ meet at origin $O(0, 0)$. Then, the area of the $\Delta OAB$ is *[2022, 28 July Shift-II]*

(a) $\frac{3\sqrt{3}}{2}$ (b) $\frac{3\sqrt{3}}{4}$ (c) $\frac{3}{2\sqrt{3}}$ (d) $\frac{3}{4\sqrt{3}}$

**105** A circle $C_1$ passes through the origin $O$ and has diameter 4 on the positive $X$-axis. The line $y=2x$ gives a chord $OA$ of circle $C_1$. Let $C_2$ be the circle with $OA$ as a diameter. If the tangent to $C_2$ at the point $A$ meets the $X$-axis at $P$ and $Y$-axis at $Q$, then $QA:AP$ is equal to *[2022, 27 July Shift-II]*

(a) 1 : 4 (b) 1 : 5 (c) 2 : 5 (d) 1 : 3

**106** Let $C$ be a circle passing through the points $A(2, -1)$ and $B(3, 4)$. The line segment $AB$ is not a diameter of $C$. If $r$ is the radius of $C$ and its centre lies on the circle $(x-5)^2+(y-1)^2=\frac{13}{2}$, then $r^2$ is equal to *[2022, 26 June Shift-I]*

(a) 32 (b) $\frac{65}{2}$ (c) $\frac{61}{2}$ (d) 30

**107** Let $r_1$ and $r_2$ be the radii of the largest and smallest circles respectively, which pass through the point (–4, 1) and having their centres on the circumference of the circle $x^2+y^2+2x+4y-4=0$. If $\frac{r_1}{r_2}=a+b\sqrt{2}$, then $a+b$ is equal to *[2021, 20 July Shift-II]*

(a) 3 (b) 11 (c) 5 (d) 7

**108** Let the lengths of intercepts on $X$-axis and $Y$-axis made by the circle $x^2+y^2+ax+2ay+c=0$, $(a<0)$ be $2\sqrt{2}$ and $2\sqrt{5}$, respectively. Then, the shortest distance from origin to a tangent to this circle which is perpendicular to the line $x+2y=0$, is equal to *[2021, 16 March Shift-II]*

(a) $\sqrt{11}$ (b) $\sqrt{7}$ (c) $\sqrt{6}$ (d) $\sqrt{10}$

**109** A square is inscribed in the circle $x^2+y^2-6x+8y-103=0$ with its sides parallel to the coordinate axes. Then, the distance of the vertex of this square which is nearest to the origin is *[2019, 11 Jan Shift-I]*

(a) 6 (b) 13
(c) $\sqrt{41}$ (d) $\sqrt{137}$

**110** If a circle $C$ passing through the point (4, 0) touches the circle $x^2+y^2+4x-6y=12$ externally at the point (1, – 1), then the radius of $C$ is *[2019, 10 Jan Shift-I]*

(a) 5 (b) $2\sqrt{5}$
(c) $\sqrt{57}$ (d) 4

## Numerical Value Type Questions

**Imp.** Condition of Tangency

**111** The absolute difference between the squares of the radii of the two circles passing through the point (– 9, 4) and touching the lines $x+y=3$ and $x-y=3$, is equal to ........... *[2025, 2 April Shift-I]*

**112** Let the centre of a circle, passing through the points (0, 0), (1, 0) and touching the circle $x^2+y^2=9$, be $(h,k)$. Then, for all possible values of the coordinates of the centre $(h,k)$, $4(h^2+k^2)$ is equal to .......... *[2024, 9 April Shift-I]*

**113** Equations of two diameters of a circle are $2x-3y=5$ and $3x-4y=7$. The line joining the points $\left(-\frac{22}{7},-4\right)$ and $\left(-\frac{1}{7},3\right)$ intersects the circle at only one point $P(\alpha,\beta)$. Then, $17\beta-\alpha$ is equal to … . *[2024, 29 Jan Shift-I]*

**114** Let the mirror image of a circle $c_1: x^2+y^2-2x-6y+\alpha=0$ in line $y=x+1$ be $c_2: 5x^2+5y^2+10gx+10fy+38=0$. If $r$ is the radius of circle $c_1$, then $\alpha+6r^2$ is equal to *[2024, 29 Jan Shift-I]*

**115** Two circles in the first quadrant of radii $r_1$ and $r_2$ touch the coordinate axes. Each of them cuts off an intercept of 2 units with the line $x+y=2$. Then, $r_1^2+r_2^2-r_1r_2$ is equal to ......... . *[2023, 12 April Shift-I]*

**116** Consider a circle $C_1: x^2+y^2-4x-2y=\alpha-5$. Let its mirror image in the line $y=2x+1$ be another circle $C_2: 5x^2+5y^2-10fx-10gy+36=0$. Let $r$ be the radius of $C_2$. Then, $\alpha+r$ is equal to ..... *[2023, 8 April Shift-I]*

**117** A circle passing through the point $P(\alpha,\beta)$ in the first quadrant touches the two coordinate axes at the points $A$ and $B$. The point $P$ is above the line $AB$. The point $Q$ on the line segment $AB$ is the foot of perpendicular from $P$ on $AB$. If $PQ$ is equal to 11 units, then the value of $\alpha\beta$ is ........ *[2023, 6 April Shift-I]*

**118** Let $P(a_1,b_1)$ and $Q(a_2,b_2)$ be two distinct points on a circle with center $C(\sqrt{2},\sqrt{3})$. Let $O$ be the origin and $OC$ be perpendicular to both $CP$ and $CQ$. If the area of $\Delta OCP$ is $\frac{\sqrt{35}}{2}$, then $a_1^2+a_2^2+b_1^2+b_2^2$ is equal to ....... *[2023, 30 Jan Shift-II]*

**119** A circle with centre (2, 3) and radius 4 intersects the line $x+y=3$ at the points $P$ and $Q$. If the tangents at $P$ and $Q$ intersect at the point $S(\alpha,\beta)$, then $4\alpha-7\beta$ is equal to ........ *[2023, 29 Jan Shift-II]*

**120** Points $P(-3,2)$, $Q(9,10)$ and $R(\alpha,4)$ lie on a circle $C$ with $PR$ as its diameter. The tangents to $C$ at the points $Q$ and $R$ intersect at the point $S$. If $S$ lies on the line $2x-ky=1$, then $k$ is equal to ........ *[2023, 25 Jan Shift-II]*

**121** Let $AB$ be a chord of length 12 of the circle $(x-2)^2+(y+1)^2=169/4$. If tangents drawn to the circle at points $A$ and $B$ intersect at the point $P$, then five times the distance of point $P$ from chord $AB$ is equal to ............... . *[2022, 29 July Shift-II]*

**122** If the circles $x^2+y^2+6x+8y+16=0$ and $x^2+y^2+2(3-\sqrt{3})x+x+2(4-\sqrt{6})y=k+6\sqrt{3}+8\sqrt{6}, k>0$, touch internally at the point $P(\alpha,\beta)$, then $(\alpha+\sqrt{3})^2+(\beta+\sqrt{6})^2$ is equal to ........ *[2022, 25 July Shift-II]*

**123** If one of the diameters of the circle $x^2+y^2-2\sqrt{2}x-6\sqrt{2}y+14=0$ is a chord of the circle $(x-2\sqrt{2})^2+(y-2\sqrt{2})^2=r^2$, then the value of $r^2$ is equal to......... *[2022, 28 June Shift-II]*

**124** Let the lines $y+2x=\sqrt{11}+7\sqrt{7}$ and $2y+x=2\sqrt{11}+6\sqrt{7}$ be normal to a circle $C: (x-h)^2+(y-k)^2=r^2$. If the line $\sqrt{11}y-3x=\frac{5\sqrt{77}}{3}+11$ is tangent to the circle $C$, then the value of $(5h-8k)^2+5r^2$ is equal to ...... . *[2022, 28 June Shift-I]*

**125** A rectangle $R$ with end points of the one of its sides as (1, 2) and (3, 6) is inscribed in a circle. If the equation of a diameter of the circle is $2x - y + 4 = 0$, then the area of $R$ is ...... . *[2022, 27 June Shift-I]*

**126** Let the abscissae of the two points $P$ and $Q$ be the roots of $2x^2 - rx + p = 0$ and the ordinates of $P$ and $Q$ be the roots of $x^2 - sx - q = 0$. If the equation of the circle described on $PQ$ as diameter is $2(x^2 + y^2) - 11x - 14y - 22 = 0$, then $2r + s - 2q + p$ is equal to ........... . *[2022, 25 June Shift-I]*

**127** Let a circle $C : (x - h)^2 + (y - k)^2 = r^2, k > 0$, touch the $X$-axis at (1, 0). If the line $x + y = 0$ intersects the circle $C$ at $P$ and $Q$ such that the length of the chord $PQ$ is 2, then the value of $h + k + r$ is equal to ...... . *[2022, 24 June Shift-I]*

**128** Let $B$ be the centre of the circle $x^2 + y^2 - 2x + 4y + 1 = 0$. Let the tangents at two points $P$ and $Q$ on the circle intersect at the point $A$ (3, 1). Then $8\left(\frac{\text{area } \Delta APQ}{\text{area } \Delta BPQ}\right)$ is equal to ........... . *[2021, 31 Aug Shift-II]*

**129** If the variable line $3x + 4y = \alpha$ lies between the two circles $(x - 1)^2 + (y - 1)^2 = 1$ and $(x - 9)^2 + (y - 1)^2 = 4$, without intercepting a chord on either circle, then the sum of all the integral values of $\alpha$ is ........... . *[2021, 31 Aug Shift-I]*

**130** Two circles each of radius 5 units touch each other at the point (1, 2). If the equation of their common tangent is $4x + 3y = 10$ and $C_1(\alpha, \beta)$ and $C_2(\gamma, \delta)$, $C_1 \neq C_2$ are their centres, then $|(\alpha + \beta)(\gamma + \delta)|$ is equal to ........... . *[2021, 27 Aug Shift-II]*

**131** Let the equation $x^2 + y^2 + px + (1 - p)y + 5 = 0$ represent circles of varying radius $r \in (0, 5]$. Then, the number of elements in the set $S = \{q : q = p^2$ and $q$ is an integer$\}$ is ........... . *[2021, 27 Aug Shift-I]*

**132** The locus of a point, which moves such that the sum of squares of its distances from the points (0, 0), (1, 0), (0, 1) (1, 1) is 18 units, is a circle of diameter $(d)$. Then, $d^2$ is equal to ........... . *[2021, 26 Aug Shift-I]*

**133** Let a point $P$ be such that its distance from the point (5, 0) is thrice the distance of $P$ from the point (− 5, 0). If the locus of the point $P$ is a circle of radius $r$, then $4r^2$ is equal to ......... . *[2021, 24 Feb Shift-II]*

**134** If the area of the triangle formed by the positive $X$-axis, the normal and the tangent to the circle $(x - 2)^2 + (y - 3)^2 = 25$ at the point (5, 7) is $A$, then $24A$ is equal to ......... . *[2021, 24 Feb Shift-II]*

**135** If one of the diameters of the circle $x^2 + y^2 - 2x - 6y + 6 = 0$ is a chord of another circle '$C$', whose centre is at (2, 1), then its radius is .......... . *[2021, 24 Feb Shift-I]*

**136** The diameter of the circle, whose centre lies on the line $x + y = 2$ in the first quadrant and which touches both the lines $x = 3$ and $y = 2$, is ...... . *[2020, 3 Sep Shift-I]*

**137** The number of integral values of $k$ for which the line, $3x + 4y = k$ intersects the circle, $x^2 + y^2 - 2x - 4y + 4 = 0$ at two distinct points is ...... . *[2020, 2 Sep Shift-I]*

## Topic 2 Parabola

### Level 1

**Imp.** AM-GM inequalites

**01** Let the focal chord $PQ$ of the parabola $y^2 = 4x$ make an angle of 60° with the positive $X$-axis, where $P$ lies in the first quadrant. If the circle, whose one diameter is $PS$, $S$ being the focus of the parabola, touches the $Y$-axis at the point $(0, \alpha)$, then $5\alpha^2$ is equal to *[2025, 2 April Shift-I]*

(a) 20 (b) 25 (c) 15 (d) 30

**02** Let the point $P$ of the focal chord $PQ$ of the parabola $y^2 = 16x$ be (1, − 4). If the focus of the parabola divides the chord $PQ$ in the ratio $m : n$, gcd $(m, n) = 1$, then $m^2 + n^2$ is equal to *[2025, 2 April Shift-II]*

(a) 37 (b) 10
(c) 17 (d) 26

**03** The axis of a parabola is the line $y = x$ and its vertex and focus are in the first quadrant at distances $\sqrt{2}$ and $2\sqrt{2}$ units from the origin, respectively. If the point k is: $(1, k)$ lies on the parabola, then a possible value of $k$ is *[2025, 4 April Shift-II]*

(a) 8 (b) 4 (c) 3 (d) 9

**04** Let $P$ be the parabola, whose focus is (−2, 1) and directrix is $2x + y + 2 = 0$. Then, the sum of the ordinates of the points on $P$, whose abscissa is −2, is *[2025, 7 April Shift-I]*

(a) $\frac{3}{4}$ (b) $\frac{5}{2}$ (c) $\frac{1}{4}$ (d) $\frac{3}{2}$

**05** Let $PQ$ be a chord of the parabola $y^2 = 12x$ and the mid-point of $PQ$ be at (4, 1).

Then, which of the following point lies on the line passing through the points $P$ and $Q$? **[2024, 4 April Shift-II]**

(a) $\left(\frac{3}{2}, -16\right)$ (b) $\left(\frac{1}{2}, -20\right)$
(c) $(3, -3)$ (d) $(2, -9)$

**06** If the shortest distance of the parabola $y^2 = 4x$ from the centre of the circle $x^2 + y^2 - 4x - 16y + 64 = 0$ is $d$, then $d^2$ is equal to **[2024, 27 Jan Shift-I]**

(a) 24 (b) 16
(c) 20 (d) 36

**07** If the length of the latus rectum of a parabola, whose focus is $(a, a)$ and the tangent at its vertex is $x + y = a$, is 16, then $|a|$ is equal to **[2022, 27 July Shift-II]**

(a) $2\sqrt{2}$ (b) $2\sqrt{3}$ (c) $4\sqrt{2}$ (d) 4

**08** The length of the latus rectum of a parabola, whose vertex and focus are on the positive $X$-axis at a distance $R$ and $S(> R)$ respectively from the origin, is **[2021, 31 Aug Shift-I]**

(a) $4(S + R)$ (b) $2(S - R)$ (c) $4(S - R)$ (d) $2(S + R)$

**09** Let the latus rectum of the parabola $y^2 = 4x$ be the common chord to the circles $C_1$ and $C_2$ each of them having radius $2\sqrt{5}$. Then, the distance between the centres of the circles $C_1$ and $C_2$ is **[2020, 3 Sep Shift-II]**

(a) 8 (b) $4\sqrt{5}$ (c) $8\sqrt{5}$ (d) 12

**10** If two tangents drawn from a point $P$ to the parabola $y^2 = 4x$ are at right angles, then the locus of $P$ is **[AIEEE 2010]**

(a) $x = 1$ (b) $2x + 1 = 0$ (c) $x = -1$ (d) $2x - 1 = 0$

**11** Let $P$ be the point (1, 0) and $Q$ a point on the the locus $y^2 = 8x$. The locus of mid point of $PQ$ is **[AIEEE 2005]**

(a) $x^2 - 4y + 2 = 0$ (b) $x^2 + 4y + 2 = 0$
(c) $y^2 + 4x + 2 = 0$ (d) $y^2 - 4x + 2 = 0$

**12** The normal at the point $(bt_1^2, 2bt_1)$ on a parabola meets the parabola again in the point $(bt_2^2, 2bt_2)$, then **[AIEEE 2003]**

(a) $t_2 = -t_1 - \frac{2}{t_1}$ (b) $t_2 = -t_1 + \frac{2}{t_1}$
(c) $t_2 = t_1 - \frac{2}{t_1}$ (d) $t_2 = t_1 + \frac{2}{t_1}$

**13** The equation of the directrix of the parabola $y^2 + 4y + 4x + 2 = 0$ is **[AIEEE 2002]**

(a) $x = -1$ (b) $x = 1$
(c) $x = -3/2$ (d) $x = 3/2$

## Level 2

**14** The radius of the smallest circle which touches the parabolas $y = x^2 + 2$ and $x = y^2 + 2$ is **[2025, 3 April Shift-I]**

(a) $\frac{7\sqrt{2}}{16}$ (b) $\frac{7\sqrt{2}}{8}$ (c) $\frac{7\sqrt{2}}{4}$ (d) $\frac{7\sqrt{2}}{2}$

**15** Two parabolas have the same focus (4, 3) and their directrices are the $X$-axis and the $Y$-axis, respectively. If these parabolas intersects the points $A$ and $B$, then $(AB)^2$ is equal to **[2025, 29 Jan Shift-I]**

(a) 392 (b) 384 (c) 96 (d) 192

**16** Let $ABCD$ be a trapezium whose vertices lie on the parabola $y^2 = 4x$. Let the sides $AD$ and $BC$ of the trapezium be parallel to $Y$-axis. If the diagonal $AC$ is of length 25/4 and it passes through the point (1, 0), then the area of $ABCD$ is **[2025, 28 Jan Shift-I]**

(a) 25/2 (b) 125/8 (c) 75/4 (d) 75/8

**17** If the equation of the parabola with vertex $V\left(\frac{3}{2}, 3\right)$ and the directrix $x + 2y = 0$ is $\alpha x^2 + \beta y^2 - \gamma xy - 30x - 60y + 225 = 0$, then $\alpha + \beta + \gamma$ is equal to **[2025, 24 Jan Shift-II]**

(a) 7 (b) 6 (c) 9 (d) 8

**18** Let the shortest distance from $(a, 0)$ $a > 0$, to the parabola $y^2 = 4x$ be 4. Then, the equation of the circle passing through the point $(a, 0)$ and the focus of the parabola and having its centre on the axis of the parabola is **[2025, 23 Jan Shift-II]**

(a) $x^2 + y^2 - 8x + 7 = 0$ (b) $x^2 + y^2 - 10x + 9 = 0$
(c) $x^2 + y^2 - 6x + 5 = 0$ (d) $x^2 + y^2 - 4x + 3 = 0$

**19** If the line $3x - 2y + 12 = 0$ intersects the parabola $4y = 3x^2$ at the points $A$ and $B$, then at the vertex of the parabola, the line segment $AB$ subtends an angle equal to **[2025, 23 Jan Shift-I]**

(a) $\frac{\pi}{2} - \tan^{-1}\left(\frac{3}{2}\right)$ (b) $\tan^{-1}\left(\frac{11}{9}\right)$
(c) $\tan^{-1}\left(\frac{4}{5}\right)$ (d) $\tan^{-1}\left(\frac{9}{7}\right)$

**20** Let $P(4, 4\sqrt{3})$ be a point on the parabola $y^2 = 4ax$ and $PQ$ be a focal chord of the parabola. If $M$ and $N$ are the foot of perpendiculars drawn from $P$ and $Q$, respectively on the directrix of the parabola, then the area of the quadrilateral $PQMN$ is equal to **[2025, 22 Jan Shift-II]**

(a) $\frac{343\sqrt{3}}{8}$ (b) $17\sqrt{3}$ (c) $\frac{263\sqrt{3}}{8}$ (d) $\frac{34\sqrt{3}}{3}$

**21** Let $PQ$ be a focal chord of the parabola $y^2 = 36x$ of length 100, making an acute angle with the positive $X$-axis. Let the ordinate of $P$ be positive and $M$ be the point on the line segment $PQ$ such that $PM : MQ = 3 : 1$. Then, which of the following points does not lie on the line passing through $M$ and perpendicular to the line $PQ$? **[2023, 13 April Shift-I]**

(a) (3, 33) (b) (6, 29) (c) (−3, 43) (d) (−6, 45)

**22** The parabolas : $ax^2 + 2bx + cy = 0$ and $dx^2 + 2ex + fy = 0$ intersect on the line $y = 1$. If $a, b, c, d, e, f$ are positive real numbers and $a, b, c$ are in GP, then **[2023, 30 Jan Shift-II]**

(a) $d, e, f$ are in AP (b) $d, e, f$ are in GP
(c) $\frac{d}{a}, \frac{e}{b}, \frac{f}{c}$ are in GP (d) $\frac{d}{a}, \frac{e}{b}, \frac{f}{c}$ are in AP

**23** Let $P(a, b)$ be a point on the parabola $y^2 = 8x$ such that the tangent at $P$ passes through the centre of the circle $x^2 + y^2 - 10x - 14y + 65 = 0$. Let $A$ be the product of all possible values of $a$ and $B$ be the product of all possible values of $b$. Then, the value of $A + B$ is equal to **[2022, 27 July Shift-I]**

(a) 0 (b) 25 (c) 40 (d) 65

**24** The equation of a common tangent to the parabolas $y = x^2$ and $y = -(x-2)^2$ is **[2022, 26 July Shift-II]**

(a) $y = 4(x-2)$ (b) $y = 4(x-1)$
(c) $y = 4(x+1)$ (d) $y = 4(x+2)$

**25** A particle is moving in the $XY$-plane along a curve $C$ passing through the point (3, 3). The tangent to the curve $C$ at the point $P$ meets the $X$-axis at $Q$. If the $Y$-axis bisects the segment $PQ$, then $C$ is a parabola with **[2022, 24 June Shift-II]**

(a) length of latus rectum 3 (b) length of latus rectum 6
(c) focus $\left(\frac{4}{3}, 0\right)$ (d) focus $\left(0, \frac{3}{4}\right)$

**26** Consider the parabola with vertex $\left(\frac{1}{2}, \frac{3}{4}\right)$ and the directrix $y = 1/2$. Let $P$ be the point where the parabola meets the line $x = -\frac{1}{2}$. If the normal to the parabola at $P$ intersects the parabola again at the point $Q$, then $(PQ)^2$ is equal to **[2021, 01 Sep Shift-II]**

(a) 75/8 (b) 125/16 (c) 25/2 (d) 15/2

**27** A tangent and a normal are drawn at the point $P(2, -4)$ on the parabola $y^2 = 8x$, which meet the directrix of the parabola at the points $A$ and $B$ respectively. If $Q(a, b)$ is a point such that $AQBP$ is a square, then $2a + b$ is equal to **[2021, 27 Aug Shift-I]**

(a) −16 (b) −18
(c) −12 (d) −20

**28** If a line along a chord of the circle $4x^2 + 4y^2 + 120x + 675 = 0$, passes through the point $(-30, 0)$ and is tangent to the parabola $y^2 = 30x$, then the length of this chord is **[2021, 26 Aug Shift-I]**

(a) 5 (b) 7
(c) $5\sqrt{3}$ (d) $3\sqrt{5}$

**29** Let $C$ be the locus of the mirror image of a point on the parabola $y^2 = 4x$ with respect to the line $y = x$. Then, the equation of tangent to $C$ at $P(2, 1)$ is **[2021, 16 March Shift-II]**

(a) $x - y = 1$ (b) $2x + y = 5$
(c) $x + 3y = 5$ (d) $x + 2y = 5$

**Imp.** **Equation of Normal to the Parabola**

**30** If the three normals drawn to the parabola, $y^2 = 2x$ pass through the point $(a, 0)$, $a \neq 0$, then $a$ must be greater than **[2021, 16 March Shift-I]**

(a) $\frac{1}{2}$ (b) $-\frac{1}{2}$ (c) −1 (d) 1

**31** If the curve $y = ax^2 + bx + c$, $x \in R$, passes through the point (1,2) and the tangent line to this curve at origin is $y = x$, then the possible values of $a, b, c$ are **[2021, 24 Feb Shift-II]**

(a) $a = \frac{1}{2}, b = \frac{1}{2}, c = 1$ (b) $a = 1, b = 0, c = 1$
(c) $a = 1, b = 1, c = 0$ (d) $a = -1, b = 1, c = 1$

**32** If $P$ is a point on the parabola $y = x^2 + 4$ which is closest to the straight line $y = 4x - 1$, then the coordinates of $P$ are **[2021, 24 Feb Shift-II]**

(a) (3, 13) (b) (1, 5) (c) (−2, 8) (d) (2, 8)

**33** The centre of the circle passing through the point (0, 1) and touching the parabola $y = x^2$ at the point (2, 4) is **[2020, 6 Sep Shift-II]**

(a) $\left(\frac{-53}{10}, \frac{16}{5}\right)$ (b) $\left(\frac{6}{5}, \frac{53}{10}\right)$ (c) $\left(\frac{3}{10}, \frac{16}{5}\right)$ (d) $\left(-\frac{16}{5}, \frac{53}{10}\right)$

**34** If the common tangent of the parabolas, $y^2 = 4x$ and $x^2 = 4y$ also touches the circle, $x^2 + y^2 = c^2$, then $c$ is equal to **[2020, 5 Sep Shift-I]**

(a) $\frac{1}{\sqrt{2}}$ (b) $\frac{1}{4}$ (c) $\frac{1}{2}$ (d) $\frac{1}{2\sqrt{2}}$

**35** The area (in sq. units) of an equilateral triangle inscribed in the parabola $y^2 = 8x$, with one of its vertices on the vertex of this parabola, is **[2020, 2 Sep Shift-II]**

(a) $64\sqrt{3}$ (b) $256\sqrt{3}$ (c) $192\sqrt{3}$ (d) $128\sqrt{3}$

**36** The locus of a point which divides the line segment joining the point (0, – 1) and a point on the parabola, $x^2 = 4y$, internally in the ratio 1 : 2, is *[2020, 8 Jan Shift-I]*
(a) $4x^2 - 3y = 2$ (b) $x^2 - 3y = 2$
(c) $9x^2 - 12y = 8$ (d) $9x^2 - 3y = 2$

**37** If $y = mx + 4$ is a tangent to both the parabolas, $y^2 = 4x$ and $x^2 = 2by$, then $b$ is equal to *[2020, 7 Jan Shift-I]*
(a) –32 (b) –128 (c) –64 (d) 128

**38** The tangents to the curve $y = (x-2)^2 - 1$ at its points of intersection with the line $x - y = 3$, intersect at the point *[2019, 12 April Shift-II]*
(a) $\left(\frac{5}{2}, 1\right)$ (b) $\left(-\frac{5}{2}, -1\right)$
(c) $\left(\frac{5}{2}, -1\right)$ (d) $\left(-\frac{5}{2}, 1\right)$

**39** If the line $ax + y = c$, touches both the curves $x^2 + y^2 = 1$ and $y^2 = 4\sqrt{2}x$, then $|c|$ is equal to *[2019, 10 April Shift-II]*
(a) $1/\sqrt{2}$ (b) 2 (c) $\sqrt{2}$ (d) 1/2

**40** If one end of a focal chord of the parabola, $y^2 = 16x$ is at (1, 4), then the length of this focal chord is *[2019, 9 April Shift-I]*
(a) 22 (b) 25 (c) 24 (d) 20

**41** If the parabolas $y^2 = 4b(x-c)$ and $y^2 = 8ax$ have a common normal, then which one of the following is a valid choice for the ordered triad $(a, b, c)$? *[2019, 10 Jan Shift-I]*
(a) $\left(\frac{1}{2}, 2, 0\right)$ (b) (1, 1, 0) (c) (1, 1, 3) (d) $\left(\frac{1}{2}, 2, 3\right)$

**42** Axis of a parabola lies along $X$-axis. If its vertex and focus are at distances 2 and 4 respectively from the origin, on the positive $X$-axis, then which of the following points does not lie on it? *[2019, 9 Jan Shift-I]*
(a) (4, –4) (b) $(6, 4\sqrt{2})$ (c) (8, 6) (d) $(5, 2\sqrt{6})$

**43** If the tangent at (1, 7) to the curve $x^2 = y - 6$ touches the circle $x^2 + y^2 + 16x + 12y + c = 0$, then the value of $c$ is *[JEE Main 2018]*
(a) 195 (b) 185 (c) 85 (d) 95

**44** Tangent and normal are drawn at $P(16, 16)$ on the parabola $y^2 = 16x$, which intersect the axis of the parabola at $A$ and $B$, respectively. If $C$ is the centre of the circle through the points $P$, $A$ and $B$ and $\angle CPB = \theta$, then a value of $\tan\theta$ is *[JEE Main 2018]*
(a) 1/2 (b) 2 (c) 3 (d) 4/3

**45** Let $O$ be the vertex and $Q$ be any point on the parabola $x^2 = 8y$. If the point $P$ divides the line segment $OQ$ internally in the ratio 1:3, then the locus of $P$ is *[JEE Main 2015]*
(a) $x^2 = y$ (b) $y^2 = x$ (c) $y^2 = 2x$ (d) $x^2 = 2y$

**46** The slope of the line touching both the parabolas $y^2 = 4x$ and $x^2 = -32y$ is *[JEE Main 2014]*
(a) 1/2 (b) 3/2 (c) 1/8 (d) 2/3

**47** The shortest distance between line $y - x = 1$ and curve $x = y^2$ is *[AIEEE 2011, 2009]*
(a) $\frac{3\sqrt{2}}{8}$ (b) $\frac{8}{3\sqrt{2}}$ (c) $\frac{4}{\sqrt{3}}$ (d) $\frac{\sqrt{3}}{4}$

**48** The locus of the vertices of the family of parabolas $y = \frac{a^3x^2}{3} + \frac{a^2x}{2} - 2a$ is *[AIEEE 2006]*
(a) $xy = \frac{3}{4}$ (b) $xy = \frac{35}{16}$ (c) $xy = \frac{64}{105}$ (d) $xy = \frac{105}{64}$

## Level 3

**49** A line passing through the point $A(-2, 0)$, touches the parabola $P: y^2 = x - 2$ at the point $B$ in the first quadrant. The area, region bounded by the line $AB$, parabola $P$ and the $X$-axis, is *[2025, 4 April Shift-II]*
(a) 8/3 (b) 2 (c) 3 (d) 7/3

**50** Let $R$ be the focus of the parabola $y^2 = 20x$ and the line $y = mx + c$ intersect the parabola at two points $P$ and $Q$. Let the point $G(10, 10)$ be the centroid of the $\Delta PQR$. If $c - m = 6$, then $(PQ)$ is *[2023, 8 April Shift-I]*
(a) 325 (b) 317 (c) 296 (d) 346

**51** If $P(h, k)$ be a point on the parabola $x = 4y^2$, which is nearest to the point $Q(0, 33)$, then the distance of $P$ from the directrix of the parabola $y^2 = 4(x + y)$ is equal to *[2023, 30 Jan Shift-I]*
(a) 2 (b) 8 (c) 4 (d) 6

**52** The equations of two sides of a variable triangle are $x = 0$ and $y = 3$ and its third side is a tangent to the parabola $y^2 = 6x$. The locus of its circumcentre is *[2023, 25 Jan Shift-II]*
(a) $4y^2 - 18y - 3x + 18 = 0$ (b) $4y^2 - 18y + 3x + 18 = 0$
(c) $4y^2 - 18y - 3x - 18 = 0$ (d) $4y^2 + 18y + 3x + 18 = 0$

**53** The distance of the point $(6, -2\sqrt{2})$ from the common tangent $y = mx + c, m > 0$, of the curves $x = 2y^2$ and $x = 1 + y^2$ is *[2023, 25 Jan Shift-I]*
(a) 5 (b) $\frac{1}{3}$ (c) $\frac{14}{3}$ (d) $5\sqrt{3}$

**54** The equations of the sides $AB$ and $AC$ of a triangle $ABC$ are $(\lambda+1)x+\lambda y=4$ and $\lambda x+(1-\lambda)y+\lambda=0$, respectively. Its vertex $A$ is on the $Y$-axis and its orthocentre is (1, 2). The length of the tangent from the point $C$ to the part of the parabola $y^2=6x$ in the first quadrant is **[2023, 24 Jan Shift-II]**

(a) $2\sqrt{2}$ (b) 2 (c) $\sqrt{6}$ (d) 4

**55** Let a tangent to the curve $y^2=24x$ meet the curve $xy=2$ at the points $A$ and $B$. Then, the mid points of such line segment $AB$ lies on a parabola with the **[2023, 24 Jan Shift-I]**

(a) length of latus rectum $=2$ (b) directrix $4x=-3$

(c) length of latus rectum $=\frac{3}{2}$ (d) directrix $4x=3$

**56** If the tangents drawn at the points $P$ and $Q$ on the parabola $y^2=2x-3$ intersect at the point $R(0, 1)$, then the orthocentre of the $\Delta PQR$ is **[2022, 28 July Shift-I]**

(a) (0, 1) (b) (2, –1) (c) (6, 3) (d) (2, 1)

**57** The tangents at the points $A(1, 3)$ and $B(1, -1)$ on the parabola $y^2-2x-2y=1$ meet at the point $P$. Then, the area (in sq. unit) of the $\Delta PAB$ is **[2022, 25 July Shift-II]**

(a) 4 (b) 6 (c) 7 (d) 8

**58** Let $P: y^2=4ax$, $a>0$ be a parabola with focus $S$. Let the tangents to the parabola $P$ make an angle of $\pi/4$ with the line $y=3x+5$ touch the parabola $P$ at $A$ and $B$ Then, the value of $a$ for which $A$, $B$ and $S$ are collinear is **[2022, 29 June Shift-II]**

(a) 8 only (b) 2 only (c) 1/4 only (d) any $a>0$

**59** If the equation of the parabola, whose vertex is at (5, 4) and the directrix is $3x+y-29=0$, is $x^2+ay^2+bxy+cx+dy+e=0$, then $a+b+c+d+e$ is equal to **[2022, 27 June Shift-II]**

(a) 575 (b) –575 (c) 576 (d) –576

**60** Let the normal at the point $P$ on the parabola $y^2=6x$ pass through the point (5, –8). If the tangent at $P$ to the parabola intersects its directrix at the point $Q$, then the ordinate of the point $Q$ is **[2022, 26 June Shift-I]**

(a) –3 (b) $-\frac{9}{4}$ (c) $-\frac{5}{2}$ (d) –2

**61** If the line $y=4+kx$, $k>0$, is the tangent to the parabola $y=x-x^2$ at the point $P$ and $V$ is the vertex of the parabola, then the slope of the line through $P$ and $V$ is **[2022, 25 June Shift-II]**

(a) $\frac{3}{2}$ (b) $\frac{26}{9}$ (c) $\frac{5}{2}$ (d) $\frac{23}{6}$

**62** Let $x=2t$, $y=\frac{t^2}{3}$ be a conic. Let $S$ be the focus and $B$ be the point on the axis of the conic such that $SA\perp BA$, where $A$ is any point on the conic. If $k$ is the ordinate of the centroid of $\Delta SAB$, then $\lim_{t\to 1} k$ is equal to **[2022, 25 June Shift-I]**

(a) 17/18 (b) 19/18 (c) 11/18 (d) 13/18

**63** If $y=m_1x+c_1$ and $y=m_2x+c_2$, $m_1\neq m_2$ are two common tangents of circle $x^2+y^2=2$ and parabola $y^2=x$, then the value of $8|m_1 m_2|$ is equal to **[2022, 25 June Shift-I]**

(a) $3+4\sqrt{2}$ (b) $-5+6\sqrt{2}$

(c) $-4+3\sqrt{2}$ (d) $7+6\sqrt{2}$

**64** Let $x^2+y^2+Ax+By+C=0$ be a circle passing through (0, 6) and touching the parabola $y=x^2$ at (2, 4). Then, $A+C$ is equal to … . **[2022, 24 June Shift-I]**

(a) 16 (b) 88/5 (c) 72 (d) –8

**65** Let $P$ be a variable point on the parabola $y=4x^2+1$. Then, the locus of the mid point of the point $P$ and the foot of the perpendicular drawn from the point $P$ to the line $y=x$ is **[2021, 20 July Shift-II]**

(a) $(3x-y)^2+(x-3y)+2=0$

(b) $2(3x-y)^2+(x-3y)+2=0$

(c) $(3x-y)^2+2(x-3y)+2=0$

(d) $2(x-3y)^2+(3x-y)+2=0$

**66** Let the tangent to the parabola $S: y^2=2x$ at the point $P(2, 2)$ meet the $X$-axis at $Q$ and normal at it meet the parabola $S$ at the point $R$. Then, the area (in sq units) of $\Delta PQR$ is equal to **[2021, 20 July Shift-I]**

(a) 25/2 (b) 35/2 (c) 15/2 (d) 25

**67** The locus of the mid-point of the line segment joining the focus of the parabola $y^2=4ax$ to a moving point of the parabola, is another parabola whose directrix is **[2021, 24 Feb Shift-I]**

(a) $x=a$ (b) $x=-\frac{a}{2}$ (c) $x=0$ (d) $x=\frac{a}{2}$

**68** Let $P$ be a point on the parabola, $y^2=12x$ and $N$ be the foot of the perpendicular drawn from $P$ on the axis of the parabola. A line is now drawn through the mid-point $M$ and $PN$, parallel to its axis which meets the parabola at $Q$. If the $y$-intercept of the line $NQ$ is 4./3, then **[2020, 3 Sep Shift-I]**

(a) $PN=4$ (b) $MQ=\frac{1}{3}$

(c) $MQ=\frac{1}{4}$ (d) $PN=3$

**69** The area (in sq units) of the smaller of the two circles that touch the parabola, $y^2 = 4x$ at the point (1, 2) and the $X$-axis is *[2019, 9 April Shift-II]*

(a) $8\pi(3-2\sqrt{2})$ (b) $4\pi(3+\sqrt{2})$

(c) $8\pi(2-\sqrt{2})$ (d) $4\pi(2-\sqrt{2})$

**70** Let $A$ (4, −4) and $B$(9, 6) be points on the parabola, $y^2 = 4x$. Let $C$ be chosen on the arc $AOB$ of the parabola, where $O$ is the origin, such that the area of $\Delta ACB$ is maximum. Then, the area (in sq. units) of $\Delta ACB$, is *[2019, 9 Jan Shift-II]*

(a) $31\frac{1}{4}$ (b) 32 (c) $31\frac{3}{4}$ (d) $30\frac{1}{2}$

**71** The centres of those circles which touch the circle, $x^2 + y^2 - 8x - 8y - 4 = 0$, externally and also touch the $X$-axis, lie on *[JEE Main 2016]*

(a) a circle

(b) an ellipse which is not a circle

(c) a hyperbola

(d) a parabola

**72** Let $P$ be the point on the parabola, $y^2 = 8x$, which is at a minimum distance from the centre $C$ of the circle, $x^2 + (y+6)^2 = 1$. Then, the equation of the circle, passing through $C$ and having its centre at $P$ is *[JEE Main 2016]*

(a) $x^2 + y^2 - 4x + 8y + 12 = 0$

(b) $x^2 + y^2 - x + 4y - 12 = 0$

(c) $x^2 + y^2 - \frac{x}{4} + 2y - 24 = 0$

(d) $x^2 + y^2 - 4x + 9y + 18 = 0$

## Numerical Value Type Questions

**73** Let $r$ be the radius of the circle, which touches $X$- axis at point $(a, 0)$, $a < 0$ and the parabola $y^2 = 9x$ at the point (4, 6). Then, $r$ is equal to ........... *[2025, 8 April Shift-II]*

**74** Let $y^2 = 12x$ be the parabola and $S$ be its focus. Let $PQ$ be a focal chord of the parabola such that $(SP)(SQ) = 147/4$.
Let $C$ be the circle described taking $PQ$ as a diameter. If the equation of a circle $C$ is $64x^2 + 64y^2 - \alpha x - 64\sqrt{3}y = \beta$, then $\beta - \alpha$ is equal to......... *[2025, 29 Jan Shift-II]*

**75** Let $A$ and $B$ be the two points of intersection of the line $y + 5 = 0$ and the mirror image of the parabola $y^2 = 4x$ with respect to the line $x + y + 4 = 0$. If $d$ denotes the distance between $A$ and $B$ and $a$ denotes the area of $\Delta SAB$, where $S$ is the focus of the parabola $y^2 = 4x$, then the value of $(a + d)$ is *[2025, 28 Jan Shift-II]*

**76** The focus of the parabola $y^2 = 4x + 16$ is the centre of the circle $C$ of radius 5. If the value of $\lambda$, for which $C$ passes through the point of intersection of the lines $3x - y = 0$ and $x + \lambda y = 4$ are $\lambda_1$ and $\lambda_2, \lambda_1 < \lambda_2$, then $12\lambda_1 + 29\lambda_2$ is equal to....... *[2025, 23 Jan Shift-II]*

**77** Consider the circle $C: x^2 + y^2 = 4$ and the parabola $P: y^2 = 8x$. If the set of all value of $\alpha$, for which three chords of the circle $C$ on three distinct lines passing through the point $(\alpha, 0)$ are bisected by the parabola $P$ is the interval $(p, q)$, then $(2q - p)^2$ is equal to ...... . *[2024, 9 April Shift-II]*

**78** Let a line perpendicular to the line $2x - y = 10$ touch the parabola $y^2 = 4(x-9)$ at the point $P$. The distance of the point $P$ from the centre of the circle $x^2 + y^2 - 14x - 8y + 56 = 0$ is ...... *[2024, 5 April Shift-II]*

**79** Suppose $AB$ is a focal chord of the parabola $y^2 = 12x$ of length $l$ and slople $m < \sqrt{3}$. If the distance of the chord $AB$ from the origin is $d$, then $ld^2$ is equal to ......... . *[2024, 5 April Shift-I]*

**80** Let the length of the focal chord $PQ$ of the parabola $y^2 = 12x$ be 15 units. If the distance of $PQ$ from the origin is $p$, then $10p^2$ is equal to ...... . *[2024, 4 April Shift-I]*

**81** Let $P(\alpha, \beta)$ be a point on the parabola $y^2 = 4x$. If $P$ also lies on the chord of the parabola $x^2 = 8y$, whose mid-point is $\left(1, \frac{5}{4}\right)$, then $(\alpha - 28)(\beta - 8)$ is equal to *[2024, 29 Jan Shift-II]*

**82** Let a common tangent to the curves $y^2 = 4x$ and $(x-4)^2 + y^2 = 16$ touch the curves at the points $P$ and $Q$. Then, $(PQ)^2$ is equal to ...... *[2023, 10 April Shift-I]*

**83** The ordinates of the points $P$ and $Q$ on the parabola with focus (3, 0) and directrix $x = -3$ are in the ratio 3 : 1. If $R(\alpha, \beta)$ is the point of intersection of the tangents to the parabola at $P$ and $Q$, then $\left(\frac{\beta}{\alpha}\right)^2$ is equal to .......... . *[2023, 8 April Shift-II]*

**84** Let the tangent to the curve $x^2 + 2x - 4y + 9 = 0$ at the point $P(1, 3)$ on it meet the $Y$-axis at $A$. Let the line passing through $P$ and parallel to the line $x - 3y = 6$ meet the parabola $y^2 = 4x$ at $B$. If $B$ lies on the line $2x - 3y = 8$, then $(AB)^2$ is equal to ......... *[2023, 6 April Shift-I]*

**85** If the $x$-intercept of a focal chord of the parabola, $y^2 = 8x + 4y + 4$ is 3, then the length of this chord is equal to ........... . **[2023, 1 Feb Shift-II]**

**86** Let $S$ be the set of all $a \in N$ such that the area of the triangle formed by the tangent at the point $P(b, c)$, $b, c \in N$, on the parabola $y^2 = 2ax$ and the lines $x = b$, $y = 0$ is 16 unit$^2$, then $\sum_{a \in S} a$ is equal to .............. **[2023, 31 Jan Shift-II]**

**87** A triangle is formed by the tangents at the point (2, 2) on the curves $y^2 = 2x$ and $x^2 + y^2 = 4x$ and the line $x + y + 2 = 0$. If $r$ is the radius of its circumcircle, then $r^2$ is equal to ....... **[2023, 29 Jan Shift-II]**

**88** Let the function $f(x) = 2x^2 - \log_e x, x > 0$, be decreasing in $(0, a)$ and increasing in $(a, 4)$. A tangent to the parabola $y^2 = 4ax$ at a point $p$ on it passes through the point $(8a, 8a - 1)$ but does not pass through the point $\left(-\frac{1}{a}, 0\right)$. If the equation of the normal at $P$ is $\frac{x}{\alpha} + \frac{y}{\beta} = 1$, then $\alpha + \beta$ is equal to ...... . **[2022, 26 July Shift-I]**

**89** A circle of radius 2 units passes through the vertex and the focus of the parabola $y^2 = 2x$ and touches the parabola $y = \left(x - \frac{1}{4}\right)^2 + \alpha$, where $\alpha > 0$. Then, $(4\alpha - 8)^2$ is equal to ...... . **[2022, 27 June Shift-I]**

**90** Let $P_1$ be a parabola with vertex (3, 2) and focus (4, 4) and $P_2$ be its mirror image with respect to the line $x + 2y = 6$. Then, the directrix of $P_2$ is $x + 2y =$ ...... . **[2022, 24 June Shift-II]**

**91** The equation of a circle is $\text{Re}(z^2) + 2[\text{Im}(z)]^2 + 2\text{Re}(z) = 0$. where, $z = x + iy$. A line which passes through the center of the given circle and the vertex of the parabola, $x^2 - 6x - y + 13 = 0$, has $y$-intercept equal to ...... . **[2021, 25 July Shift-II]**

**92** If the point on the curve $y^2 = 6x$, nearest to the point $(3, 3/2)$ is $(\alpha, \beta)$, then $2(\alpha + \beta)$ is equal to **[2021, 20 July Shift-II]**

**93** A line is a common tangent to the circle $(x - 3)^2 + y^2 = 9$ and the parabola $y^2 = 4x$. If the two points of contact $(a, b)$ and $(c, d)$ are distinct and lie in the first quadrant, then $2(a + c)$ is equal to ......... . **[2021, 25 Feb Shift-II]**

# Topic 3 Ellipse

## Level 1

**01** If the length of the minor axis of an ellipse is equal to one fourth of the distance between the foci, then the eccentricity of the ellipse is **[2025, 2 April Shift-II]**

(a) $\frac{\sqrt{3}}{16}$ (b) $\frac{3}{\sqrt{19}}$ (c) $\frac{\sqrt{5}}{7}$ (d) $\frac{4}{\sqrt{17}}$

**02** If $S$ and $S'$ are the foci of the ellipse $\frac{x^2}{18} + \frac{y^2}{9} = 1$ and $P$ be a point on the ellipse, then $\min(SP \cdot S'P) + \max(SP \cdot S'P)$ is equal to **[2025, 2 April Shift-I]**

(a) $3(6 + \sqrt{2})$ (b) $3(1 + \sqrt{2})$
(c) 9 (d) 27

**03** The centre of a circle $C$ is at the centre of the ellipse $E: \frac{x^2}{a^2} + \frac{y^2}{b^2} = 1, a > b$. Let $C$ pass through the foci $F_1$ and $F_2$ of $E$ such that the circle $C$ and the ellipse $E$ intersect at four points. Let $P$ be one of these four points. If the area of the $\Delta PF_1F_2$ is 30 and the length of the major axis of $E$ is 17, then the distance between the foci of $E$ is **[2025, 4 April Shift-II]**

(a) 12 (b) $\frac{13}{2}$ (c) 26 (d) 13

**04** The length of the latus-rectum of the ellipse, whose foci are (2, 5) and (2, – 3) and eccentricity is $\frac{4}{5}$ is **[2025, 4 April Shift-I]**

(a) $\frac{6}{5}$ (b) $\frac{50}{3}$ (c) $\frac{10}{3}$ (d) $\frac{18}{5}$

**05** Let the length of a latus rectum of an ellipse $\frac{x^2}{a^2} + \frac{y^2}{b^2} = 1$ be 10. If its eccentricity is the minimum value of the function $f(t) = t^2 + t + \frac{11}{12}, t \in R$, then $a^2 + b^2$ is equal to **[2025, 7 April Shift-II]**

(a) 115 (b) 126
(c) 120 (d) 125

**06** If $\alpha x + \beta y = 109$ is the equation of the chord of the ellipse $\frac{x^2}{9} + \frac{y^2}{4} = 1$, whose mid-point is $\left(\frac{5}{2}, \frac{1}{2}\right)$, then $\alpha + \beta$ is equal to **[2025, 29 Jan Shift-II]**

(a) 58 (b) 46
(c) 37 (d) 72

**07** If the mid-point of a chord of the ellipse $\frac{x^2}{9}+\frac{y^2}{4}=1$ is $\left(\sqrt{2},\frac{4}{3}\right)$ and the length of the chord is $\frac{2\sqrt{\alpha}}{3}$, then $\alpha$ is **[2025, 28 Jan Shift-II]**

(a) 26 (b) 18 (c) 20 (d) 22

**08** The equation of the chord of ellipse $\frac{x^2}{25}+\frac{y^2}{16}=1$, whose mid-point is (3, 1) is **[2025, 24 Jan Shift-II]**

(a) $5x+16y=31$ (b) $4x+122y=134$
(c) $48x+25y=169$ (d) $25x+101y=176$

**09** If the length of the minor axis of an ellipse is equal to half of the distance between the foci, then the eccentricity of the ellipse is **[2024, 30 Jan Shift-I]**

(a) $\frac{\sqrt{3}}{2}$ (b) $\frac{2}{\sqrt{5}}$ (c) $\frac{1}{\sqrt{3}}$ (d) $\frac{\sqrt{5}}{3}$

**10** Let $a$ and $L$ pass through the point of intersection of the lines $bx+10y-8=0$ and $2x-3y=0$, $b\in R-\left\{\frac{4}{3}\right\}$. If the line $L$ also passes through the point (1, 1) and touches the circle $17(x^2+y^2)=16$, then the eccentricity of the ellipse $\frac{x^2}{5}+\frac{y^2}{b^2}=1$ is **[2022, 29 July Shift-I]**

(a) $\frac{2}{\sqrt{5}}$ (b) $\sqrt{\frac{3}{5}}$ (c) $\frac{1}{\sqrt{5}}$ (d) $\sqrt{\frac{2}{5}}$

**11** If $m$ is the slope of a common tangent to the curves $\frac{x^2}{16}+\frac{y^2}{9}=1$ and $x^2+y^2=12$, then $12m^2$ is equal to **[2022, 26 June Shift-II]**

(a) 6 (b) 9 (c) 10 (d) 12

**12** The line $12x\cos\theta+5y\sin\theta=60$ is tangent to which of the following curves? **[2021, 31 Aug Shift-I]**

(a) $x^2+y^2=169$ (b) $144x^2+25y^2=3600$
(c) $25x^2+12y^2=3600$ (d) $x^2+y^2=60$

**13** Let $E_1:\frac{x^2}{a^2}+\frac{y^2}{b^2}=1$, $a>b$. Let $E_2$ be another ellipse such that it touches the end points of major axis of $E_1$ and the foci of $E_2$ are the end points of minor axis of $E_1$. If $E_1$ and $E_2$ have same eccentricities, then its value is **[2021, 22 July Shift-II]**

(a) $\frac{-1+\sqrt{5}}{2}$ (b) $\frac{-1+\sqrt{8}}{2}$
(c) $\frac{-1+\sqrt{3}}{2}$ (d) $\frac{-1+\sqrt{6}}{2}$

**14** Which of the following points lies on the locus of the foot of perpendicular drawn upon any tangent to the ellipse, $\frac{x^2}{4}+\frac{y^2}{2}=1$ from any of its foci? **[2020, 6 Sep Shift-I]**

(a) $(-2,\sqrt{3})$ (b) $(-1,\sqrt{2})$ (c) $(-1,\sqrt{3})$ (d) (1, 2)

**15** If the point $P$ on the curve, $4x^2+5y^2=20$ is farthest from the point $Q(0,-4)$, then $PQ^2$ is equal to **[2020, 5 Sep Shift-I]**

(a) 29 (b) 21 (c) 48 (d) 36

**16** A focus of an ellipse is at the origin. The directrix is the line $x=4$ and the eccentricity is $1/2$, then the length of semi-major axis is **[AIEEE 2008]**

(a) $\frac{5}{3}$ (b) $\frac{8}{3}$ (c) $\frac{2}{3}$ (d) $\frac{4}{3}$

**17** An ellipse has $OB$ as semi-minor axis, $F$ and $F'$ its foci and the $\angle FBF$ is a right angle. Then, the eccentricity of the ellipse is **[AIEEE 2005]**

(a) $\frac{1}{\sqrt{3}}$ (b) $\frac{1}{4}$
(c) $\frac{1}{2}$ (d) $\frac{1}{\sqrt{2}}$

**18** The eccentricity of an ellipse with its centre at the origin is $\frac{1}{2}$. If one of the directrices is $x=4$, then the equation of the ellipse is **[AIEEE 2004]**

(a) $3x^2+4y^2=1$ (b) $3x^2+4y^2=12$
(c) $4x^2+3y^2=12$ (d) $4x^2+3y^2=1$

**19** The equation of the ellipse whose foci are $(\pm 2,0)$ and eccentricity is 1/2, is **[AIEEE 2002]**

(a) $\frac{x^2}{12}+\frac{y^2}{16}=1$ (b) $\frac{x^2}{16}+\frac{y^2}{12}=1$
(c) $\frac{x^2}{16}+\frac{y^2}{8}=1$ (d) None of these

## Level 2

**20** Let $C$ be the circle of minimum area enclosing the ellipse $E:\frac{x^2}{a^2}+\frac{y^2}{b^2}=1$ with eccentricity $\frac{1}{2}$ and foci $(\pm 2,0)$. Let $PQR$ be a variable triangle, whose vertex $P$ is on the circle $C$ and the side $QR$ of length $2a$ is parallel to the major axis of $E$ and contains the point of intersection of $E$ with the negative $Y$-axis. Then, the maximum area of the $\Delta PQR$ is **[2025, 3 April Shift-II]**

(a) $8(3+\sqrt{2})$ (b) $8(2+\sqrt{3})$
(c) $6(2+\sqrt{3})$ (d) $6(3+\sqrt{2})$

**21** Let $p$ be the number of all triangles that can be formed by joining the vertices of a regular polygon $P$ of $n$ sides and $q$ be the number of all quadrilaterals that can be formed by joining the vertices of $P$. If $p+q=126$, then the eccentricity of the ellipse $\frac{x^2}{16}+\frac{y^2}{n}=1$ is **[2025, 7 April Shift-II]**

(a) $\frac{1}{2}$ (b) $\frac{\sqrt{7}}{4}$

(c) $\frac{3}{4}$ (d) $\frac{1}{\sqrt{2}}$

**22** Let the product of the focal distances of the point $\left(\sqrt{3}, \frac{1}{2}\right)$ on the ellipse $\frac{x^2}{a^2}+\frac{y^2}{b^2}=1, (a>b)$ be $\frac{7}{4}$. Then, the absolute difference of the eccentricities of two such ellipses is **[2025, 24 Jan Shift-I]**

(a) $\frac{3-2\sqrt{2}}{2\sqrt{3}}$ (b) $\frac{1-2\sqrt{2}}{\sqrt{3}}$

(c) $\frac{1-\sqrt{3}}{\sqrt{2}}$ (d) $\frac{3-2\sqrt{2}}{3\sqrt{2}}$

**23** The length of the chord of the ellipse $\frac{x^2}{4}+\frac{y^2}{2}=1$, whose mid-point is $\left(1, \frac{1}{2}\right)$ is **[2025, 24 Jan Shift-I]**

(a) $\sqrt{15}$ (b) $\frac{5}{3}\sqrt{15}$

(c) $\frac{2}{3}\sqrt{15}$ (d) $\frac{1}{3}\sqrt{15}$

**24** Let $P$ be a point on the ellipse $\frac{x^2}{9}+\frac{y^2}{4}=1$. Let the line passing through $P$ and parallel to $Y$-axis meet the circle $x^2+y^2=9$ at point $Q$ such that $P$ and $Q$ are on the same side of the $X$-axis. Then, the eccentricity of the locus of the point $R$ on $PQ$ such that $PR:RQ=4:3$ as $P$ moves on the ellipse is **[2024, 1 Feb Shift-II]**

(a) $\frac{11}{19}$ (b) $\frac{13}{21}$

(c) $\frac{\sqrt{139}}{23}$ (d) $\frac{\sqrt{13}}{7}$

**25** If the radius of the largest circle with centre (2, 0) inscribed in the ellipse $x^2+4y^2=36$ is $r$, then $12r^2$ is equal to **[2023, 11 April Shift-II]**

(a) 69 (b) 72

(c) 115 (d) 92

**26** Consider ellipses $E_k : kx^2+k^2y^2=1, k=1, 2, \ldots 20$. Let $C_k$ be the circle which touches the four chords joining the end points (one minor axis and another on major axis) of the ellipse $E_k$. If $r_k$ is the radius of the circle $C_k$, then the value of $\sum_{k=1}^{20}\frac{1}{r_k^2}$ is **[2023, 11 April Shift-I]**

(a) 3320 (b) 2870 (c) 3080 (d) 3210

**27** Let a circle of radius 4 be concentric to the ellipse $15x^2+19y^2=285$. Then, the common tangents are inclined to the minor axis of the ellipse at the angle **[2023, 10 April Shift-II]**

(a) $\frac{\pi}{4}$ (b) $\frac{\pi}{3}$ (c) $\frac{\pi}{12}$ (d) $\frac{\pi}{16}$

**28** In a group of 100 persons 75 speak English and 40 speak Hindi. Each person speaks at least one of the two languages. If the number of persons, who speak only English is $\alpha$ and the number of persons who speak only Hindi is $\beta$, then the eccentricity of the ellipse $25(\beta^2x^2+\alpha^2y^2)=\alpha^2\beta^2$ is **[2023, 6 April Shift-II]**

(a) $\frac{3\sqrt{15}}{12}$ (b) $\frac{\sqrt{129}}{12}$ (c) $\frac{\sqrt{117}}{12}$ (d) $\frac{\sqrt{119}}{12}$

**39** If the maximum distance of normal to the ellipse $\frac{x^2}{4}+\frac{y^2}{b^2}=1, b<2$, from the origin is 1, then the eccentricity of the ellipse is **[2023, 31 Jan Shift-I]**

(a) $\frac{1}{\sqrt{2}}$ (b) $\frac{1}{2}$ (c) $\frac{\sqrt{3}}{4}$ (d) $\frac{\sqrt{3}}{2}$

**30** Let a line $L$ passes through the point of intersection of the lines $bx+10y-8=0$ and $2x-3y=0$, $b\in R-\left\{\frac{4}{3}\right\}$. If the line $L$ also passes through the point (1, 1) and touches the circle $17(x^2+y^2)=16$, then the eccentricity of the ellipse $\frac{x^2}{5}+\frac{y^2}{b^2}=1$ is **[2022, 29 July Shift-I]**

(a) $\frac{2}{\sqrt{5}}$ (b) $\sqrt{\frac{3}{5}}$ (c) $\frac{1}{\sqrt{5}}$ (d) $\sqrt{\frac{2}{5}}$

**31** Let the eccentricity of an ellipse $\frac{x^2}{a^2}+\frac{y^2}{b^2}=1, a>b$, be 1/4. If this ellipse passes through the point $\left(-4\sqrt{\frac{2}{5}}, 3\right)$, then $a^2+b^2$ is equal to **[2022, 27 June Shift-I]**

(a) 29 (b) 31

(c) 32 (d) 34

**32** The locus of the mid-point of the line segment joining the point (4, 3) and the points on the ellipse $x^2 + 2y^2 = 4$ is an ellipse with eccentricity *[2022, 26 June Shift-II]*

(a) $\frac{\sqrt{3}}{2}$ (b) $\frac{1}{2\sqrt{2}}$ (c) $\frac{1}{\sqrt{2}}$ (d) $\frac{1}{2}$

**33** The locus of mid-points of the line segments joining (– 3, –5) and the points on the ellipse $\frac{x^2}{4} + \frac{y^2}{9} = 1$ is *[2021, 31 Aug Shift-II]*

(a) $9x^2 + 4y^2 + 18x + 8y + 145 = 0$
(b) $36x^2 + 16y^2 + 90x + 56y + 145 = 0$
(c) $36x^2 + 16y^2 + 108x + 80y + 145 = 0$
(d) $36x^2 + 16y^2 + 72x + 32y + 145 = 0$

**34** If a tangent to the ellipse $x^2 + 4y^2 = 4$ meets the tangents at the extremities of its major axis at $B$ and $C$, then the circle with $BC$ as diameter passes through the point *[2021, 25 July Shift-II]*

(a) $(\sqrt{3}, 0)$ (b) $(\sqrt{2}, 0)$ (c) (1, 1) (d) (–1, 1)

**35** Let an ellipse $E \Rightarrow \frac{x^2}{a^2} + \frac{y^2}{b^2} = 1, a^2 > b^2$, passes through $\left(\sqrt{\frac{3}{2}}, 1\right)$ and has eccentricity $1/\sqrt{3}$. If a circle, centered at focus $F(\alpha, 0), \alpha > 0$, of $E$ and radius $2/\sqrt{3}$, intersects $E$ at two points $P$ and $Q$, then $PQ^2$ is equal to *[2021, 25 July Shift-I]*

(a) 8/3 (b) 4/3 (c) 16/3 (d) 3

**36** Let a tangent be drawn to the ellipse $\frac{x^2}{27} + y^2 = 1$ at $(3\sqrt{3}\cos\theta, \sin\theta)$, where $\theta \in \left(0, \frac{\pi}{2}\right)$. Then the value of $\theta$, such that the sum of intercepts on axes made by this tangent is minimum is equal to *[2021, 18 March Shift-II]*

(a) $\pi/8$ (b) $\pi/4$ (c) $\pi/6$ (d) $\pi/3$

**37** Let $L$ be a tangent line to the parabola $y^2 = 4x - 20$ at (6, 2). If $L$ is also a tangent to the ellipse $\frac{x^2}{2} + \frac{y}{b} = 1$, then the value of $b$ is equal to *[2021, 17 March Shift-II]*

(a) 11 (b) 14 (c) 16 (d) 20

**38** The length of the minor axis (along $Y$ - axis) of an ellipse in the standard form is $\frac{4}{\sqrt{3}}$. If this ellipse touches the line, $x + 6y = 8$; then its eccentricity is *[2020, 9 Jan Shift-II]*

(a) $\sqrt{\frac{5}{6}}$ (b) $\frac{1}{2}\sqrt{\frac{11}{3}}$ (c) $\frac{1}{3}\sqrt{\frac{11}{3}}$ (d) $\frac{1}{2}\sqrt{\frac{5}{3}}$

**39** If the line $x - 2y = 12$ is tangent to the ellipse $\frac{x^2}{a^2} + \frac{y^2}{b^2} = 1$ at the point $\left(3, \frac{-9}{2}\right)$, then the length of the latusrectum of the ellipse is *[2019, 10 April Shift-I]*

(a) $8\sqrt{3}$ (b) 9
(c) 5 (d) $12\sqrt{2}$

**40** Let $S$ and $S'$ be the foci of an ellipse and $B$ be any one of the extremities of its minor axis. If $\Delta S'BS$ is a right angled triangle with right angle at $B$ and area $(\Delta S'BS) = 8$ sq units, then the length of a latus rectum of the ellipse is *[2019, 12 Jan Shift-II]*

(a) $2\sqrt{2}$ (b) $4\sqrt{2}$
(c) 2 (d) 4

**41** Let the length of the latus rectum of an ellipse with its major-axis along $X$-axis and centre at the origin, be 8. If the distance between the foci of this ellipse is equal to the length of its minor axis, then which one of the following points lies on it? *[2019, 11 Jan Shift-II]*

(a) $(4\sqrt{2}, 2\sqrt{3})$ (b) $(4\sqrt{3}, 2\sqrt{2})$
(c) $(4\sqrt{2}, 2\sqrt{2})$ (d) $(4\sqrt{3}, 2\sqrt{3})$

**42** The eccentricity of an ellipse whose centre is at the origin is 1/2. If one of its directrices is $x = -4$, then the equation of the normal to it at $\left(1, \frac{3}{2}\right)$ is *[JEE Main 2017]*

(a) $2y - x = 2$ (b) $4x - 2y = 1$
(c) $4x + 2y = 7$ (d) $x + 2y = 4$

**43** The area (in sq units) of the quadrilateral formed by the tangents at the end points of the latus rectum to the ellipse $\frac{x^2}{9} + \frac{y^2}{5} = 1$ is *[JEE Main 2015]*

(a) 27/4 (b) 18
(c) 27/2 (d) 27

**44** The equation of the circle passing through the foci of the ellipse $\frac{x^2}{16} + \frac{y^2}{9} = 1$ and having centre at (0, 3) is *[JEE Main 2013]*

(a) $x^2 + y^2 - 6y - 7 = 0$ (b) $x^2 + y^2 - 6y + 7 = 0$
(c) $x^2 + y^2 - 6y - 5 = 0$ (d) $x^2 + y^2 - 6y + 5 = 0$

**45** An ellipse is drawn by taking a diameter of the circle $(x-1)^2 + y^2 = 1$ as its semi-minor axis and a diameter of the circle $x^2 + (y-2)^2 = 4$ is semi-major axis. If the centre of the ellipse is at the origin and its axes are the coordinate axes, then the equation of the ellipse is *[AIEEE 2012]*

(a) $4x^2 + y^2 = 4$ (b) $x^2 + 4y^2 = 8$
(c) $4x^2 + y^2 = 8$ (d) $x^2 + 4y^2 = 16$

**46** Equation of the ellipse whose axes are the axes of coordinates and which passes through the point (–3, 1) and has eccentricity $\sqrt{2/5}$ is **[AIEEE 2011]**

(a) $5x^2+3y^2-48=0$ (b) $3x^2+5y^2-15=0$
(c) $5x^2+3y^2-32=0$ (d) $3x^2+5y^2-32=0$

**47** The ellipse $x^2+4y^2=4$ is inscribed in a rectangle aligned with the coordinate axes, which in turn is inscribed in another ellipse that passes through the point (4, 0). Then, the equation of the ellipse is **[AIEEE 2009]**

(a) $x^2+12y^2=16$ (b) $4x^2+48y^2=48$
(c) $4x^2+64y^2=48$ (d) $x^2+16y^2=16$

## Level 3

**48** Let for two distinct values of $p$ the lines $y=x+p$ touch the ellipse $E:\frac{x^2}{4^2}+\frac{y^2}{3^2}=1$ at the points $A$ and $B$. Let the line $y=x$ intersect $E$ at the points $C$ and $D$. Then, the area of the quadrilateral $ABCD$ is equal to **[2025, 4 April Shift-II]**

(a) 48 (b) 36 (c) 24 (d) 20

**59** Let the ellipse $3x^2+py^2=4$ pass through the centre $C$ of the circle $x^2+y^2-2x-4y-11=0$ of radius $r$. Let $f_1, f_2$ be the focal distances of the point C on the ellipse. Then, $6f_1f_2-r$ is equal to **[2025, 8 April Shift-II]**

(a) 78 (b) 74 (c) 70 (d) 68

**50** Let the ellipse $E_1:\frac{x^2}{a^2}+\frac{y^2}{b^2}=1$, $a>b$ and $E_2:\frac{x^2}{A^2}+\frac{y^2}{B^2}=1$, $A<B$ have same eccentricity $1/\sqrt{3}$. Let the product of their lengths of latus rectums be $32/\sqrt{3}$ and the distance between the foci of $E_1$ be 4. If $E_1$ and $E_2$ meets at $A$, $B$, $C$ and $D$, then the area of the quadrilateral $ABCD$ equals **[2025, 29 Jan Shift-I]**

(a) $\frac{24\sqrt{6}}{5}$ (b) $\frac{12\sqrt{6}}{5}$ (c) $\frac{18\sqrt{6}}{5}$ (d) $6\sqrt{6}$

**51** Let the line $2x+3y-k=0, k>0$, intersect the $X$-axis and $Y$-axis at the points $A$ and $B$, respectively. If the equation of the circle having the line segment $AB$ as a diameter is $x^2+y^2-3x-2y=0$ and the length of the latus rectum of the ellipse $x^2+9y^2=k^2$ is $\frac{m}{n}$, where $m$ and $n$ are coprime, then $2m+n$ is equal to **[2024, 5 April Shift-I]**

(a) 13 (b) 12
(c) 11 (d) 10

**52** Let $P$ be a parabola with vertex (2, 3) and directrix $2x+y=6$. Let an ellipse $E:\frac{x^2}{a^2}+\frac{y^2}{b^2}=1$, $a>b$, of eccentricity $\frac{1}{\sqrt{2}}$ pass through the focus of the parabola $P$. Then, the square of the length of the latus rectum of $E$, is **[2024, 31 Jan Shift-II]**

(a) $\frac{385}{8}$ (b) $\frac{512}{25}$ (c) $\frac{656}{25}$ (d) $\frac{347}{8}$

**53** Let the tangent and normal at the point $(3\sqrt{3}, 1)$ on the ellipse $\frac{x^2}{36}+\frac{y^2}{4}=1$ meet the $Y$-axis at the points $A$ and $B$, respectively. Let the circle $C$ be drawn taking $AB$ as a diameter and the line $x=2\sqrt{5}$ intersect $C$ at the points $P$ and $Q$. If the tangents at the points $P$ and $Q$ on the circle intersect at the point $(\alpha, \beta)$, then $\alpha^2-\beta^2$ is equal to **[2023, 13 April Shift-I]**

(a) 60 (b) $\frac{314}{5}$
(c) 61 (d) $\frac{304}{5}$

**54** Let $P\left(\frac{2\sqrt{3}}{\sqrt{7}}, \frac{6}{\sqrt{7}}\right)$, $Q$, $R$ and $S$ be four points on the ellipse $9x^2+4y^2=36$. Let $PQ$ and $RS$ be mutually perpendicular and pass through the origin. If $\frac{1}{(PQ)^2}+\frac{1}{(RS)^2}=\frac{p}{q}$, where $p$ and $q$ are coprime, then $p+q$ is equal to **[2023, 12 April Shift-II]**

(a) 137 (b) 157
(c) 147 (d) 143

**55** Let $A$ be a point on the $X$-axis. Common tangents are drawn from $A$ to the curves $x^2+y^2=8$ and $y^2=16x$. If one of these tangents touches the two curves at $Q$ and $R$, then $(QR)^2$ is equal to **[2023, 30 Jan Shift-II]**

(a) 76 (b) 72
(c) 64 (d) 81

**56** The acute angle between the pair of tangents drawn to the ellipse $2x^2+3y^2=5$ from the point (1, 3) is **[2022, 26 July Shift-II]**

(a) $\tan^{-1}\left(\frac{16}{7\sqrt{5}}\right)$ (b) $\tan^{-1}\left(\frac{24}{7\sqrt{5}}\right)$
(c) $\tan^{-1}\left(\frac{32}{7\sqrt{5}}\right)$ (d) $\tan^{-1}\left(\frac{3+8\sqrt{5}}{35}\right)$

**57** Let $PQ$ be a focal chord of the parabola $y^2 = 4x$ such that it subtends an angle of $\frac{\pi}{2}$ at the point (3, 0). Let the line segment $PQ$ be also a focal chord of the ellipse $E: \frac{x^2}{a^2} + \frac{y^2}{b^2} = 1, a^2 > b^2$. If $e$ is the eccentricity of the ellipse $E$, then the value of $\frac{1}{e^2}$ is equal to *[2022, 28 June Shift-I]*

(a) $1+\sqrt{2}$ (b) $3+2\sqrt{2}$ (c) $1+2\sqrt{3}$ (d) $4+5\sqrt{3}$

**58** On the ellipse $\frac{x^2}{8} + \frac{y^2}{4} = 1$. Let $P$ be a point in the second quadrant such that the tangent at $P$ to the ellipse is perpendicular to the line $x + 2y = 0$. Let $S$ and $S'$ be the foci of the ellipse and $e$ be its eccentricity. If $A$ is the area of the $\Delta SPS$, then the value of $(5 - e^2) \cdot A$ is *[2021, 26 Aug Shift-I]*

(a) 6 (b) 12 (c) 14 (d) 24

**59** A ray of light through (2, 1) is reflected at a point $P$ on the $Y$-axis and then passes through the point (5, 3). If this reflected ray is the directrix of an ellipse with eccentricity $1/3$ and the distance of the nearer focus from this directrix is $8/\sqrt{53}$, then the equation of the other directrix can be *[2021, 27 July Shift-I]*

(a) $11x + 7y + 8 = 0$ or $11x + 7y - 15 = 0$
(b) $11x - 7y - 8 = 0$ or $11x + 7y + 15 = 0$
(c) $2x - 7y + 29 = 0$ or $2x - 7y - 7 = 0$
(d) $2x - 7y - 39 = 0$ or $2x - 7y - 7 = 0$

**60** Let $S_1 \Rightarrow x^2 + y^2 = 9$ and $S_2 \Rightarrow (x-2)^2 + y^2 = 1$. Then, the locus of center of a variable circle $S$ which touches $S_1$ internally and $S_2$ externally always passes through the points *[2021, 18 March Shift-II]*

(a) $(0, \pm\sqrt{3})$ (b) $\left(\frac{1}{2}, \pm\frac{\sqrt{5}}{2}\right)$ (c) $\left(2, \pm\frac{3}{2}\right)$ (d) $(1, \pm 2)$

**61** Let the line $y = mx$ and the ellipse $2x^2 + y^2 = 1$ intersect at a point $P$ in the first quadrant. If the normal to this ellipse at $P$ meets the coordinate axes at $\left(-\frac{1}{3\sqrt{2}}, 0\right)$ and $(0, \beta)$, then $\beta$ is equal to *[2020, 8 Jan Shift-I]*

(a) $\frac{2\sqrt{2}}{3}$ (b) $\frac{\sqrt{2}}{3}$ (c) $\frac{2}{\sqrt{3}}$ (d) $\frac{2}{3}$

**62** If the normal to the ellipse $3x^2 + 4y^2 = 12$ at a point $P$ on it is parallel to the line, $2x + y = 4$ and the tangent to the ellipse at $P$ passes through $Q(4, 4)$ then $PQ$ is equal to *[2019, 12 April Shift-I]*

(a) $\frac{5\sqrt{5}}{2}$ (b) $\frac{\sqrt{61}}{2}$
(c) $\frac{\sqrt{221}}{2}$ (d) $\frac{\sqrt{157}}{2}$

**63** The locus of the foot of perpendicular drawn from the centre of the ellipse $x^2 + 3y^2 = 6$ on any tangent to it, is *[JEE Main 2014]*

(a) $(x^2 - y^2)^2 = 6x^2 + 2y^2$ (b) $(x^2 - y^2)^2 = 6x^2 - 2y^2$
(c) $(x^2 + y^2)^2 = 6x^2 + 2y^2$ (d) $(x^2 + y^2)^2 = 6x^2 - 2y^2$

**64** Area of the greatest rectangle that can be inscribed in the ellipse $\frac{x^2}{a^2} + \frac{y^2}{b^2} = 1$ is *[AIEEE 2005]*

(a) $a/b$ sq unit (b) $\sqrt{ab}$ sq unit
(c) $ab$ sq unit (d) $2ab$ sq unit

## Numerical Value Type Questions

**65** Let $C$ be the circle $x^2 + (y-1)^2 = 2$, $E_1$ and $E_2$ be two ellipses whose centres lie at the origin and major axes lie on $X$-axis and $Y$-axis respectively. Let the straight line $x + y = 3$ touch the curves $C$, $E_1$ and $E_2$ at $P(x_1, y_1)$, $Q(x_2, y_2)$ and $R(x_3, y_3)$ respectively. Given that $P$ is the mid-point of the line segment $QR$ and $PQ = \frac{2\sqrt{2}}{3}$, the value of $9(x_1 y_1 + x_2 y_2 + x_3 y_3)$ is equal to ........ *[2025, 4 April Shift-I]*

**66** Let $E_1 : \frac{x^2}{9} + \frac{y^2}{4} = 1$ be an ellipse. Ellipses $E_i$s are constructed such that their centres and eccentricities are same as that of $E_1$ and the length of minor axis of $E_i$ is the length of major axis of $E_{i+1}$ $(i \geq 1)$. If $A_i$ is the area of the ellipse $E_i$, then $\frac{5}{\pi}\left(\sum_{i=1}^{\infty} A_i\right)$ is equal to ...... *[2025, 28 Jan Shift-I]*

**67** Let an ellipse with centre (1, 0) and latus rectum of length 1/2 have its major axis along $X$-axis. If its minor axis subtends an angle 60° at the foci, then the square of the sum of the lengths of its minor and major axes is equal to ............ *[2023, 15 April Shift-I]*

**68** The line $x = 8$ is the directirix of the ellipse $E: \frac{x^2}{a^2} + \frac{y^2}{b^2} = 1$ with the corresponding focus (2, 0). If the tangent to $E$ at the point $P$ in the first quadrant passes through the point $(0, 4\sqrt{3})$ and intersects the $X$-axis at $Q$, then $(3PQ)^2$ is equal to ......... . *[2023, 1 Feb Shift-II]*

**69** Let $C$ be the largest circle centred at (2, 0) and inscribed in the ellipse $\frac{x^2}{36}+\frac{y^2}{16}=1$. If $(1, \alpha)$ lies on $C$, then $10\alpha^2$ is equal to ......... *[2023, 24 Jan Shift-I]*

**70** Let a tangent to the curve $9x^2+16y^2=144$ intersect the coordinate axes at the points $A$ and $B$. Then, the minimum length of the line segment $AB$ is ........ *[2023, 24 Jan Shift-I]*

**71** Let the tangents at the points $P$ and $Q$ on the ellipse $x^2/2+y^2/4=1$ meet at the point $R(\sqrt{2}, 2\sqrt{2}-2)$. If $S$ is the focus of the ellipse on its negative major axis, then $SP^2+SQ^2$ is equal to ............ *[2022, 28 July Shift-II]*

**72** If the length of the latus rectum of the ellipse $x^2+4y^2+2x+8y-\lambda=0$ is 4 and $l$ is the length of its major axis, then $\lambda+l$ is equal to ......... . *[2022, 27 July Shift-I]*

**73** Let the common tangents to the curves $4(x^2+y^2)=9$ and $y^2=4x$ intersect at the point $Q$. Let an ellipse, centered at the origin $O$, has lengths of semi-minor and semi-major axes equal to $OQ$ and 6, respectively. If $e$ and $l$ respectively denote the eccentricity and the length of the latusrectum of this ellipse, then $l/e^2$ is equal to...... . *[2022, 26 June Shift-I]*

**74** If two tangents drawn from a point $(\alpha, \beta)$ lying on the ellipse $25x^2+4y^2=1$ to the parabola $y^2=4x$ are such that the slope of one tangent is four times the other, then the value of $(10\alpha+5)^2+(16\beta^2+50)^2$ equals ...... . *[2022, 24 June Shift-I]*

**75** If the minimum area of the triangle formed by a tangent to the ellipse $\frac{x^2}{b^2}+\frac{y^2}{4a^2}=1$ and the coordinate axis is $kab$, then $k$ is equal to *[2021, 27 Aug Shift-I]*

**76** Let $E$ be an ellipse whose axes are parallel to the co-ordinates axes, having its center at (3, −4), one focus at (4, −4) and one vertex at (5, −4). If $mx-y=4, m>0$ is a tangent to the ellipse $E$, then the value of $5m^2$ is equal to ......... . *[2021, 27 July Shift-II]*

# Topic 4 Hyperbola

## Level 1

**Imp.** Eccentricity of Ellipse and Hyperbola

**01** Let $e_1$ and $e_2$ be the eccentricities of the ellipse $\frac{x^2}{b^2}+\frac{y^2}{25}=1$ and the hyperbola $\frac{x^2}{16}-\frac{y^2}{b^2}=1$, respectively. If $b<5$ and $e_1e_2=1$, then the eccentricity of the ellipse having its axes along the coordinate axes and passing through all four foci (two of the ellipse and two of the hyperbola) is *[2025, 7 April Shift-II]*

(a) $\frac{\sqrt{7}}{4}$ (b) $\frac{4}{5}$ (c) $\frac{\sqrt{3}}{2}$ (d) $\frac{3}{5}$

**02** Let $E: \frac{x^2}{a^2}+\frac{y^2}{b^2}=1, a>b$ and $H: \frac{x^2}{A^2}-\frac{y^2}{B^2}=1$. Let the distance between the foci of $E$ and the foci of $H$ be $2\sqrt{3}$. If $a-A=2$ and the ratio of the eccentricities of $E$ and $H$ is $\frac{1}{3}$, then the sum of the lengths of their latus rectums is equal to *[2025, 22 Jan Shift-II]*

(a) 8 (b) 10 (c) 7 (d) 9

**03** Let the foci of a hyperbola be (1, 14) and (1, −12). If it passes through the point (1, 6), then the length of its latus-rectum is *[2025, 22 Jan Shift-I]*

(a) $\frac{144}{5}$ (b) $\frac{288}{5}$ (c) $\frac{24}{5}$ (d) $\frac{25}{6}$

**04** Let $P$ be a point on the hyperbola $H: \frac{x^2}{9}-\frac{y^2}{4}=1$ in the first quadrant such that the area of triangle formed by $P$ and the two foci of $H$ is $2\sqrt{13}$. Then, the square of the distance of $P$ from the origin is *[2024, 30 Jan Shift-II]*

(a) 20 (b) 26 (c) 22 (d) 18

**05** Let $H$ be the hyperbola, whose foci are $(1\pm\sqrt{2}, 0)$ and eccentricity is $\sqrt{2}$. Then, the length of its latus rectum is *[2023, 31 Jan Shift-II]*

(a) 3 (b) 2 (c) 5/2 (d) 3/2

**06** Let $a > 0$, $b > 0$. Let $e$ and $l$ respectively be the eccentricity and length of the latus rectum of the hyperbola $\frac{x^2}{a^2} - \frac{y^2}{b^2} = 1$. Let $e'$ and $l'$ respectively be the eccentricity and length of the latus rectum of its conjugate hyperbola. If $e^2 = \frac{11}{14} l$ and $(e')^2 = \frac{11}{8} l'$, then the value of $77a + 44b$ is equal to *[2022, 28 June Shift-II]*

(a) 100 (b) 110 (c) 120 (d) 130

**07** Let the eccentricity of the hyperbola $H: \frac{x^2}{a^2} - \frac{y^2}{b^2} = 1$ be $\sqrt{5/2}$ and length of its latus rectum be $6\sqrt{2}$. If $y = 2x + c$ is a tangent to the hyperbola $H$, then the value of $c^2$ is equal to *[2022, 28 June Shift-I]*

(a) 18 (b) 20 (c) 24 (d) 32

**08** The normal to the hyperbola $\frac{x^2}{a^2} - \frac{y^2}{9} = 1$ at the point $(8, 3\sqrt{3})$ on it passes through the point. *[2022, 26 June Shift-II]*

(a) $(15, -2\sqrt{3})$ (b) $(9, 2\sqrt{3})$
(c) $(-1, 9\sqrt{3})$ (d) $(-1, 6\sqrt{3})$

**09** The locus of the mid-points of the chords of the hyperbola $x^2 - y^2 = 4$, which touch the parabola $y^2 = 8x$, is *[2021, 26 Aug Shift-II]*

(a) $y^3(x-2) = x^2$ (b) $x^3(x-2) = y^2$
(c) $y^2(x-2) = x^3$ (d) $x^2(x-2) = y^3$

**10** Let a line $L: 2x + y = k$, $k > 0$ be a tangent to the hyperbola $x^2 - y^2 = 3$. If $L$ is also a tangent to the parabola $y^2 = \alpha x$, then $\alpha$ is equal to *[2021, 22 July Shift-II]*

(a) 12 (b) −12 (c) 24 (d) −24

**11** If a hyperbola passes through the point $P(10, 16)$ and it has vertices at $(\pm 6, 0)$, then the equation of the normal to it at $P$ is *[2020, 8 Jan Shift-II]*

(a) $3x + 4y = 94$ (b) $x + 2y = 42$
(c) $2x + 5y = 100$ (d) $x + 3y = 58$

**12** If the vertices of a hyperbola be at $(-2, 0)$ and $(2, 0)$ and one of its foci be at $(-3, 0)$, then which one of the following points does not lie on this hyperbola? *[2019, 12 Jan Shift-I]*

(a) $(2\sqrt{6}, 5)$ (b) $(6, 5\sqrt{2})$ (c) $(4, \sqrt{15})$ (d) $(-6, 2\sqrt{10})$

**13** If a hyperbola has length of its conjugate axis equal to 5 and the distance between its foci is 13, then the eccentricity of the hyperbola is *[2019, 11 Jan Shift-II]*

(a) 13/12 (b) 2 (c) 13/8 (d) 13/6

**14** The eccentricity of the hyperbola whose length of the latusrectum is equal to 8 and the length of its conjugate axis is equal to half of the distance between its foci, is *[JEE Main 2016]*

(a) $4/3$ (b) $4/\sqrt{3}$ (c) $2/\sqrt{3}$ (d) $\sqrt{3}$

**15** The equation of the hyperbola whose foci are $(-2, 0)$ and $(2, 0)$ and eccentricity is 2, is given by *[AIEEE 2011]*

(a) $-3x^2 + y^2 = 3$ (b) $x^2 - 3y^2 = 3$
(c) $3x^2 - y^2 = 3$ (d) $-x^2 + 3y^2 = 3$

**16** For the hyperbola $\frac{x^2}{\cos^2 \alpha} - \frac{y^2}{\sin^2 \alpha} = 1$, which of the following remains constant when $\alpha$ varies? *[AIEEE 2007]*

(a) Eccentricity (b) Directrix
(c) Abscissae of vertices (d) Abscissae of foci

**17** The locus of a point $P(\alpha, \beta)$ moving under the condition that the line $y = \alpha x + \beta$ is a tangent to the hyperbola $\frac{x^2}{a^2} - \frac{y^2}{b^2} = 1$, is *[AIEEE 2005]*

(a) a hyperbola (b) a parabola
(c) a circle (d) an ellipse

**18** The equation of the chord joining two points $(x_1, y_1)$ and $(x_2, y_2)$ on the rectangular hyperbola $xy = c^2$ is *[AIEEE 2002]*

(a) $\frac{x}{x_1 + x_2} + \frac{y}{y_1 + y_2} = 1$ (b) $\frac{x}{x_1 - x_2} + \frac{y}{y_1 - y_2} = 1$
(c) $\frac{x}{y_1 + y_2} + \frac{y}{x_1 + x_2} = 1$ (d) $\frac{x}{y_1 - y_2} + \frac{y}{x_1 - x_2} = 1$

## Level 2

**19** If $A$ and $B$ are points of intersection of the circle $x^2 + y^2 - 8x = 0$ and the hyperbola $\frac{x^2}{9} - \frac{y^2}{4} = 1$ and a point $P$ moves on the lines $2x - 3y + 4 = 0$, then the centroid of $\Delta PAB$ lies on the line : *[2025, 28 Jan Shift-I]*

(a) $6x - 9y = 20$ (b) $9x - 9y = 32$
(c) $x + 9y = 36$ (d) $4x - 9y = 12$

**20** Let the foci of a hyperbola $H$ coincide with the foci of the ellipse $E: \frac{(x-1)^2}{100} + \frac{(y-1)^2}{75} = 1$ and the eccentricity of the hyperbola $H$ be the reciprocal of the eccentricity of the ellipse $E$. If the length of the transverse axis of $H$ is $\alpha$ and the length of its conjugate axis is $\beta$, then $3\alpha^2 + 2\beta^2$ is equal to *[2024, 9 April Shift-II]*

(a) 242 (b) 205
(c) 237 (d) 225

**21** Let $H: \frac{-x^2}{a^2} + \frac{y^2}{b^2} = 1$ be the hyperbola, whose eccentricity is $\sqrt{3}$ and the length of the latus rectum is $4\sqrt{3}$. Suppose the point $(\alpha, 6)$, $\alpha > 0$ lies on $H$. If $\beta$ is the product of the focal distances of the point $(\alpha, 6)$, then $\alpha^2 + \beta$ is equal to *[2024, 8 April Shift-I]*

(a) 172 (b) 169
(c) 171 (d) 170

**22** For $0 < \theta < \frac{\pi}{2}$, if the eccentricity of the hyperbola $x^2 - y^2 \operatorname{cosec}^2 \theta = 5$ is $\sqrt{7}$ times eccentricity of the ellipse $x^2 \operatorname{cosec}^2 \theta + y^2 = 5$, then the value of $\theta$ is *[2024, 1 Feb Shift-I]*

(a) $\frac{\pi}{6}$ (b) $\frac{\pi}{3}$
(c) $\frac{\pi}{4}$ (d) $\frac{5\pi}{12}$

**23** Let $e_1$ be the eccentricity of the hyperbola $\frac{x^2}{16} - \frac{y^2}{9} = 1$ and $e_2$ be the eccentricity of the ellipse $\frac{x^2}{a^2} + \frac{y^2}{b^2} = 1$, $a > b$, which passes through the foci of the hyperbola. If $e_1 e_2 = 1$, then the length of the chord of the ellipse parallel to the $X$-axis and passing through $(0, 5)$ is *[2024, 27 JanShift-II]*

(a) $8\sqrt{5}/3$ (b) $4\sqrt{5}$
(c) $3\sqrt{5}$ (d) $10\sqrt{5}/3$

**24** Let $P(x_0, y_0)$ be the point on the hyperbola $3x^2 - 4y^2 = 36$, which is nearest to the line $3x + 2y = 1$. Then, $\sqrt{2}(y_0 - x_0)$ is equal to *[2023, 1 Feb Shift-II]*

(a) 9 (b) $-9$
(c) $-3$ (d) 3

**25** If the line $x - 1 = 0$ is directrix of the hyperbola $kx^2 - y^2 = 6$, then the hyperbola passes through the point *[2022, 26 July Shift-I]*

(a) $(-2\sqrt{5}, 6)$ (b) $(-\sqrt{5}, 3)$
(c) $(\sqrt{5}, -2)$ (d) $(2\sqrt{5}, 3\sqrt{6})$

**26** Let $\lambda x - 2y = \mu$ be a tangent to the hyperbola $a^2x^2 - y^2 = b^2$. Then, $\left(\frac{\lambda}{a}\right)^2 - \left(\frac{\mu}{b}\right)^2$ is equal to *[2022, 24 June Shift-I]*

(a) $-2$ (b) $-4$
(c) 2 (d) 4

**27** The point $P(-2\sqrt{6}, \sqrt{3})$ lies on the hyperbola $\frac{x^2}{a^2} - \frac{y^2}{b^2} = 1$ having eccentricity $\sqrt{5}/2$. If the tangent and normal at $P$ to the hyperbola intersect its conjugate axis at the point $Q$ and $R$ respectively, then $QR$ is equal to *[2021, 26 Aug Shift-II]*

(a) $4\sqrt{3}$ (b) 6
(c) $6\sqrt{3}$ (d) $3\sqrt{6}$

**28** The locus of the centroid of the triangle formed by any point $P$ on the hyperbola $16x^2 - 9y^2 + 32x + 36y - 164 = 0$, and its foci is *[2021, 25 July Shift-I]*

(a) $16x^2 - 9y^2 + 32x + 36y - 36 = 0$
(b) $9x^2 - 16y^2 + 36x - 32y - 144 = 0$
(c) $16x^2 - 9y^2 + 32x - 36y - 144 = 0$
(d) $9x^2 - 16y^2 + 36x - 32y - 36 = 0$

**29** A hyperbola passes through the foci of the ellipse $\frac{x^2}{25} + \frac{y^2}{16} = 1$ and its transverse and conjugate axes coincide with major and minor axes of the ellipse, respectively. If the product of their eccentricities is one, then the equation of the hyperbola is *[2021, 25 Feb Shift-II]*

(a) $\frac{x^2}{9} - \frac{y^2}{16} = 1$ (b) $\frac{x^2}{9} - \frac{y^2}{4} = 1$
(c) $\frac{x^2}{9} - \frac{y^2}{25} = 1$ (d) $x^2 - y^2 = 9$

**30** The equation of a common tangent to the curves, $y^2 = 16x$ and $xy = -4$, is *[2019, 12 April Shift-II]*

(a) $x - y + 4 = 0$ (b) $x + y + 4 = 0$
(c) $x - 2y + 16 = 0$ (d) $2x - y + 2 = 0$

**31** If the line $y = mx + 7\sqrt{3}$ is normal to the hyperbola $\frac{x^2}{24} - \frac{y^2}{18} = 1$, then a value of $m$ is *[2019, 9 April Shift-I]*

(a) $\frac{3}{\sqrt{5}}$ (b) $\frac{\sqrt{15}}{2}$
(c) $\frac{2}{\sqrt{5}}$ (d) $\frac{\sqrt{5}}{2}$

**32** If the eccentricity of the standard hyperbola passing through the point $(4, 6)$ is 2, then the equation of the tangent to the hyperbola at $(4, 6)$ is *[2019, 8 April Shift-II]*

(a) $3x - 2y = 0$
(b) $x - 2y + 8 = 0$
(c) $2x - y - 2 = 0$
(d) $2x - 3y + 10 = 0$

**33** Tangents are drawn to the hyperbola $4x^2 - y^2 = 36$ at the points $P$ and $Q$. If these tangents intersect at the point $T(0, 3)$, then the area (in sq units) of $\Delta PTQ$ is **[JEE Main 2018]**

(a) $45\sqrt{5}$ (b) $54\sqrt{3}$
(c) $60\sqrt{3}$ (d) $36\sqrt{5}$

**34** If a hyperbola passes through the point $P(\sqrt{2}, \sqrt{3})$ and has foci at $(\pm 2, 0)$, then the tangent to this hyperbola at $P$ also passes through the point **[JEE Main 2017]**

(a) $(3\sqrt{2}, 2\sqrt{3})$ (b) $(2\sqrt{2}, 3\sqrt{3})$
(c) $(\sqrt{3}, \sqrt{2})$ (d) $(-\sqrt{2}, -\sqrt{3})$

**35** The foci of the ellipse $\dfrac{x^2}{16} + \dfrac{y^2}{b^2} = 1$ and the hyperbola $\dfrac{x^2}{144} - \dfrac{y^2}{81} = \dfrac{1}{25}$ coincide. Then, the value of $b^2$ is **[AIEEE 2003]**

(a) 1 (b) 5
(c) 7 (d) 6

## Level 3

**36** Consider a hyperbola $H$ having centre at the origin and foci on the $X$-axis. Let $C_1$ be the circle touching the hyperbola $H$ and having the centre at the origin. Let $C_2$ be the circle touching the hyperbola $H$ at its vertex and having the centre at one of its foci. If areas (in sq units) of $C_1$ and $C_2$ are $36\pi$ and $4\pi$, respectively, then the length (in units) of latus rectum of $H$ is **[2024, 4 April Shift-II]**

(a) $\dfrac{14}{3}$ (b) $\dfrac{28}{3}$
(c) $\dfrac{11}{3}$ (d) $\dfrac{10}{3}$

**37** If the foci of a hyperbola are same as that of the ellipse $\dfrac{x^2}{9} + \dfrac{y^2}{25} = 1$ and the eccentricity of the hyperbola is $\dfrac{15}{8}$ times the eccentricity of the ellipse, then the smaller focal distance of the point $\left(\sqrt{2}, \dfrac{14}{3}\sqrt{\dfrac{2}{5}}\right)$ on the hyperbola, is equal to **[2024, 31 Jan Shift-I]**

(a) $7\sqrt{\dfrac{2}{5}} - \dfrac{8}{3}$ (b) $14\sqrt{\dfrac{2}{5}} - \dfrac{4}{3}$
(c) $14\sqrt{\dfrac{2}{5}} - \dfrac{16}{3}$ (d) $7\sqrt{\dfrac{2}{5}} + \dfrac{8}{3}$

**38** Let the focal chord of the parabola $P: y^2 = 4x$ along the line $L: y = mx + c, m > 0$ meet the parabola at the points $M$ and $N$. Let the line $L$ be a tangent to the hyperbola $H: x^2 - y^2 = 4$. If $O$ is the vertex of $P$ and $F$ is the focus of $H$ on the positive $X$-axis, then the area of the quadrilateral $OMFN$ is **[2022, 29 July Shift-I]**

(a) $2\sqrt{6}$ (b) $2\sqrt{14}$
(c) $4\sqrt{6}$ (d) $4\sqrt{14}$

**39** Let the hyperbola $H: \dfrac{x^2}{a^2} - \dfrac{y^2}{b^2} = 1$ passes through the point $(2\sqrt{2}, -2\sqrt{2})$. A parabola is drawn whose focus is same as the focus of $H$ with positive abscissa and the directrix of the parabola passes through the other focus of $H$. If the length of the latusrectum of the parabola is $e$ times the length of the latusrectum of $H$, where $e$ is the eccentricity of $H$, then which of the following points lies on the parabola? **[2022, 28 July Shift-II]**

(a) $(2\sqrt{3}, 3\sqrt{2})$ (b) $(3\sqrt{3}, -6\sqrt{2})$
(c) $(\sqrt{3}, -\sqrt{6})$ (d) $(3\sqrt{6}, 6\sqrt{2})$

**40** Let the foci of the ellipse $\dfrac{x^2}{16} + \dfrac{y^2}{7} = 1$ and the hyperbola $\dfrac{x^2}{144} - \dfrac{y^2}{\alpha} = \dfrac{1}{25}$ coincide. Then, the length of the latus rectum of the hyperbola is **[2022, 25 July Shift-II]**

(a) $\dfrac{32}{9}$ (b) $\dfrac{18}{5}$
(c) $\dfrac{27}{4}$ (d) $\dfrac{27}{10}$

**41** The locus of the mid-points of the chord of the circle, $x^2 + y^2 = 25$ which is tangent to the hyperbola, $\dfrac{x^2}{9} - \dfrac{y^2}{16} = 1$ is **[2021, 16 March Shift-I]**

(a) $(x^2 + y^2)^2 - 16x^2 + 9y^2 = 0$
(b) $(x^2 + y^2)^2 - 9x^2 + 144y^2 = 0$
(c) $(x^2 + y^2)^2 - 9x^2 - 16y^2 = 0$
(d) $(x^2 + y^2)^2 - 9x^2 + 16y^2 = 0$

### Numerical Value Type Questions

**42** Let the product of the focal distances of the point $P(4, 2\sqrt{3})$ on the hyperbola $H: \dfrac{x^2}{a^2} - \dfrac{y^2}{b^2} = 1$ be 32. Let the length of the conjugate axis of $H$ be $p$ and the length of its latus rectum be $q$. Then, $p^2 + q^2$ is equal to ........ **[2025, 3 April Shift-II]**

**43** If the equation of the hyperbola with foci (4, 2) and (8, 2) is $3x^2 - y^2 - \alpha x + \beta y + \gamma = 0$, then $\alpha + \beta + \gamma$ is equal to ......... . **[2025, 3 April Shift-II]**

**44** Consider the hyperbola $\frac{x^2}{a^2}-\frac{y^2}{b^2}=1$ having one of its focus at $P(-3,0)$. If the latus rectum through its other focus subtends a right angle at $P$ and $a^2b^2=\alpha\sqrt{2}-\beta, \alpha, \beta \in N$, then $\alpha+\beta$ is ......... .

**[2025, 7 April Shift-I]**

**45** Let the lengths of the transverse and conjugate axes of a hyperbola in standard form be $2a$ and $2b$ respectively and one focus and the corresponding directrix of this hyperbola be $(-5,0)$ and $5x+9=0$, respectively. If the product of the focal distances of a point $(\alpha, 2\sqrt{5})$ on the hyperbola is $p$, then $4p$ is equal to ......... .

**[2025, 7 April Shift-II]**

**46** Let $H_1:\frac{x^2}{a^2}-\frac{y^2}{b^2}=1$ and $H_2:-\frac{x^2}{A^2}+\frac{y^2}{B^2}=1$ be two hyperbolas having length of latus-rectums $15\sqrt{2}$ and $12\sqrt{5}$, respectively. Let their eccentricities be $e_1=\sqrt{5/2}$ and $e_2$ respectively. If the product of the lengths of their transverse axes is $100\sqrt{10}$, then $25\,e_2^2$ is equal to ......... .

**[2025, 24 Jan Shift-II]**

**47** Let the circle $C$ touches the line $x-y+1=0$ have the centre on the positive $X$-axis and cut off a chord of length $\frac{4}{\sqrt{13}}$ along the line $-3x+2y=1$. Let $H$ be the hyperbola $\frac{x^2}{\alpha^2}-\frac{y^2}{\beta^2}=1$, whose one of the foci is the centre of $C$ and the length of the transverse axis is the diameter of $C$. Then, $2\alpha^2+3\beta^2$ is equal to ......... .

**[2025, 23 Jan Shift-I]**

**48** Let $S$ be the focus of the hyperbola $\frac{x^2}{3}-\frac{y^2}{5}=1$, on the positive $X$-axis. Let $C$ be the circle with its centre at $A(\sqrt{6},\sqrt{5})$ and passing through the point $S$. If $O$ is the origin and $SAB$ is a diameter of $C$, then the square of the area of the $\Delta OSB$ is equal to ________ .

**[2024, 8 April Shift-II]**

**49** The length of the latus rectum and directrices of a hyperbola with eccentricity $e$ are 9 and $x=\pm\frac{4}{\sqrt{3}}$, respectively. Let the line $y-\sqrt{3}x+\sqrt{3}=0$ touch this hyperbola at $(x_0, y_0)$. If $m$ is the product of the focal distances of the point $(x_0, y_0)$, then $4e^2+m$ is equal to

**[2024, 6 April Shift-II]**

**50** Let the foci and length of the latus rectum of an ellipse $\frac{x^2}{a^2}+\frac{y^2}{b^2}=1, a>b$ be $(\pm 5,0)$ and $\sqrt{50}$, respectively. Then, the square of the eccentricity of the hyperbola $\frac{x^2}{b^2}-\frac{y^2}{a^2b^2}=1$ equals

**[2024, 31 Jan Shift-I]**

**51** Let the latus rectum of the hyperbola $\frac{x^2}{9}-\frac{y^2}{b^2}=1$ subtend an angle of $\frac{\pi}{3}$ at the centre of the hyperbola. If $b^2$ is equal to $\frac{l}{m}(1+\sqrt{n})$, where $l$ and $m$ are coprime numbers, then $l^2+m^2+n^2$ is equal to ......... .

**[2024, 30 Jan Shift-I]**

**52** The foci of a hyperbola are $(\pm 2, 0)$ and its eccentricity is $\frac{3}{2}$. A tangent, perpendicular to the line $2x+3y=6$, is drawn at a point in the first quadrant on the hyperbola. If the intercepts made by the tangent on the $X$-and $Y$-axes are $a$ and $b$ respectively, then $|6a|+|5b|$ is equal to ...... .

**[2023, 13 April Shift-II]**

**53** Let $m_1$ and $m_2$ be the slopes of the tangents drawn from the point $P(4, 1)$ to the hyperbola $H=\frac{y^2}{25}-\frac{x^2}{16}=1$. If $Q$ is the point from which the tangents drawn to $H$ have slopes $|m_1|$ and $|m_2|$ and they make positive intercepts $\alpha$ and $\beta$ on the $X$-axis, then $\frac{(PQ)^2}{\alpha\beta}$ is equal to ......... .

**[2023, 13 April Shift-I]**

**54** Let $H_n=\frac{x^2}{1+x}-\frac{y^2}{3+x}=1, n\in N$. Let $k$ be the smallest even value of $n$ such that the eccentricity of $H_k$ is a rational number. If $l$ is the length of the latus rectum of $H_k$, then $21l$ is equal to ............ .

**[2023, 11 April Shift-I]**

**55** Let the eccentricity of an ellipse $\frac{x^2}{a^2}+\frac{y^2}{b^2}=1$ is reciprocal to that of the hyperbola $2x^2-2y^2=1$. If the ellipse intersects the hyperbola at right angles, then square of length of the latus rectum of the ellipse is .........

**[2023, 6 April Shift-II]**

**56** Let the tangent to the parabola $y^2 = 12x$ at the point $(3, \alpha)$ be perpendicular to the line $2x + 2y = 3$. Then, the square of distance of the point $(6, -4)$ from the normal to the hyperbola $\alpha^2 x^2 - 9y^2 = 9\alpha^2$ at its point $(\alpha - 1, \alpha + 2)$ is equal to ........... .

**[2023, 11 April Shift-I]**

**57** The vertices of a hyperbola $H$ are $(\pm 6, 0)$ and its eccentricity is $\frac{\sqrt{5}}{2}$. Let $N$ be the normal to $H$ at a point in the first quadrant and parallel to the line $\sqrt{2}x + y = 2\sqrt{2}$. If $d$ is the length of the line segment of $N$ between $H$ and the $Y$-axis, then $d^2$ equal to .............. .

**[2023, 25 Jan Shift-I]**

**58** A common tangent $T$ to the curves $C_1 : \frac{x^2}{4} + \frac{y^2}{9} = 1$ and $C_2 : \frac{x^2}{42} - \frac{y^2}{143} = 1$ does not pass through the fourth quadrant. If $T$ touches $C_1$ at $(x_1, y_1)$ and $C_2$ at $(x_2, y_2)$, then $|2x_1 + x_2|$ is equal to ........... .

**[2022, 27 July Shift-II]**

**59** Let $H : \frac{x^2}{a^2} - \frac{y^2}{b^2} = 1$, $a > 0, b > 0$, be a hyperbola such that the sum of lengths of the transverse and the conjugate axes is $4(2\sqrt{2} + \sqrt{14})$. If the eccentricity $H$ is $\frac{\sqrt{11}}{2}$, then value of $a^2 + b^2$ is equal to ...... .

**[2022, 29 June Shift-I]**

**60** Let a line $L_1$ be tangent to the hyperbola $\frac{x^2}{16} - \frac{y^2}{4} = 1$ and let $L_2$ be the line passing through the origin and perpendicular to $L_1$. If the locus of the point of intersection of $L_1$ and $L_2$ is $(x^2 + y^2)^2 = \alpha x^2 + \beta y^2$, then $\alpha + \beta$ is equal to ...... .

**[2022, 26 June Shift-II]**

**61** Let the eccentricity of the hyperbola $\frac{x^2}{a^2} - \frac{y^2}{b^2} = 1$ be $\frac{5}{4}$. If the equation of the normal at the point $\left(\frac{8}{\sqrt{5}}, \frac{12}{5}\right)$ on the hyperbola is $8\sqrt{5}x + \beta y = \lambda$, then $\lambda - \beta$ is equal to ......... .

**[2022, 24 June Shift-II]**

**62** Let the hyperbola $H : \frac{x^2}{a^2} - y^2 = 1$ and the ellipse $E : 3x^2 + 4y^2 = 12$ be such that the length of latus rectum of $H$ is equal to the length of latus rectum of $E$. If $e_H$ and $e_E$ are the eccentricities of $H$ and $E$ respectively, then the value of $12(e_H^2 + e_E^2)$ is equal to ...... .

**[2022, 24 June Shift-II]**

**63** Let $A(\sec\theta, 2\tan\theta)$ and $B(\sec\phi, 2\tan\phi)$, where $\theta + \phi = \pi/2$, be two points on the hyperbola $2x^2 - y^2 = 2$. If $(\alpha, \beta)$ is the point of the intersection of the normals to the hyperbola at $A$ and $B$, then $(2\beta)^2$ is equal to ......... .

**[2021, 27 Aug Shift-II]**

**64** The locus of the point of intersection of the lines $(\sqrt{3})kx + ky - 4\sqrt{3} = 0$ and $\sqrt{3}x - y - 4(\sqrt{3})k = 0$ is a conic, whose eccentricity is ............ .

**[2021, 25 Feb Shift-I]**

## Key Index

### Topic 1

| | | | | | | | | | | | | | | | | | | | |
|---|---|---|---|---|---|---|---|---|---|---|---|---|---|---|---|---|---|---|---|
| 1 | (b) | 2 | (c) | 3 | (c) | 4 | (c) | 5 | (a) | 6 | (a) | 7 | (b) | 8 | (a) | 9 | (d) | 10 | (a) |
| 11 | (b) | 12 | (a) | 13 | (b) | 14 | (d) | 15 | (d) | 16 | (a) | 17 | (d) | 18 | (b) | 19 | (b) | 20 | (c) |
| 21 | (b) | 22 | (c) | 23 | (d) | 24 | (a) | 25 | (b) | 26 | (a) | 27 | (b) | 28 | (a) | 29 | (c) | 30 | (b) |
| 31 | (d) | 32 | (b) | 33 | (a) | 34 | (b) | 35 | (a) | 36 | (c) | 37 | (b) | 38 | (b) | 39 | (*) | 40 | (c) |
| 41 | (c) | 42 | (d) | 43 | (a) | 44 | (c) | 45 | (d) | 46 | (d) | 47 | (b) | 48 | (b) | 49 | (a) | 50 | (b) |
| 51 | (d) | 52 | (d) | 53 | (c) | 54 | (a) | 55 | (a) | 56 | (b) | 57 | (b) | 58 | (d) | 59 | (d) | 60 | (a) |
| 61 | (d) | 62 | (c) | 63 | (b) | 64 | (b) | 65 | (b) | 66 | (a) | 67 | (a) | 68 | (a) | 69 | (a) | 70 | (a) |
| 71 | (c) | 72 | (d) | 73 | (d) | 74 | (c) | 75 | (b) | 76 | (d) | 77 | (b) | 78 | (a) | 79 | (a) | 80 | (a) |
| 81 | (a) | 82 | (d) | 83 | (c) | 84 | (c) | 85 | (a) | 86 | (a) | 87 | (a) | 88 | (b) | 89 | (a) | 90 | (a) |
| 91 | (c) | 92 | (b) | 93 | (d) | 94 | (d) | 95 | (c) | 96 | (b) | 97 | (c) | 98 | (d) | 99 | (c) | 100 | (c) |
| 101 | (b) | 102 | (c) | 103 | (d) | 104 | (b) | 105 | (a) | 106 | (b) | 107 | (c) | 108 | (c) | 109 | (c) | 110 | (a) |
| 111 | (768) | 112 | (9) | 113 | (2) | 114 | (12) | 115 | (7) | 116 | (2) | 117 | (121) | 118 | (24) | 119 | (11) | 120 | (3) |
| 121 | (72) | 122 | (25) | 123 | (10) | 124 | (816) | 125 | (16) | 126 | (7) | 127 | (7) | 128 | (18) | 129 | (165) | 130 | (40) |
| 131 | (61) | 132 | (16) | 133 | (56.25) | 134 | (1225) | 135 | (3) | 136 | (3) | 137 | (9) | | | | | | |

### Topic 2

| | | | | | | | | | | | | | | | | | | | |
|---|---|---|---|---|---|---|---|---|---|---|---|---|---|---|---|---|---|---|---|
| 1 | (c) | 2 | (c) | 3 | (d) | 4 | (d) | 5 | (b) | 6 | (c) | 7 | (c) | 8 | (c) | 9 | (a) | 10 | (c) |
| 11 | (d) | 12 | (a) | 13 | (d) | 14 | (c) | 15 | (d) | 16 | (c) | 17 | (c) | 18 | (c) | 19 | (d) | 20 | (a) |
| 21 | (c) | 22 | (d) | 23 | (d) | 24 | (b) | 25 | (a) | 26 | (b) | 27 | (a) | 28 | (d) | 29 | (a) | 30 | (d) |
| 31 | (c) | 32 | (d) | 33 | (d) | 34 | (a) | 35 | (c) | 36 | (c) | 37 | (b) | 38 | (c) | 39 | (c) | 40 | (b) |
| 41 | (c) | 42 | (c) | 43 | (d) | 44 | (b) | 45 | (d) | 46 | (a) | 47 | (a) | 48 | (d) | 49 | (a) | 50 | (a) |
| 51 | (d) | 52 | (b) | 53 | (a) | 54 | (a) | 55 | (d) | 56 | (b) | 57 | (d) | 58 | (d) | 59 | (d) | 60 | (b) |
| 61 | (c) | 62 | (d) | 63 | (c) | 64 | (a) | 65 | (b) | 66 | (a) | 67 | (c) | 68 | (c) | 69 | (a) | 70 | (a) |
| 71 | (d) | 72 | (a) | 73 | (30) | 74 | (1328) | 75 | (14) | 76 | (15) | 77 | (80) | 78 | (10) | 79 | (108) | 80 | (72) |
| 81 | (–192) | 82 | (32) | 83 | (16) | 84 | (292) | 85 | (16) | 86 | (146) | 87 | (10) | 88 | (45) | 89 | (63) | 90 | (10) |
| 91 | (1) | 92 | (9) | 93 | (9) | | | | | | | | | | | | | | |

### Topic 3

| | | | | | | | | | | | | | | | | | | | |
|---|---|---|---|---|---|---|---|---|---|---|---|---|---|---|---|---|---|---|---|
| 1 | (d) | 2 | (d) | 3 | (d) | 4 | (d) | 5 | (b) | 6 | (a) | 7 | (d) | 8 | (c) | 9 | (b) | 10 | (b) |
| 11 | (b) | 12 | (b) | 13 | (a) | 14 | (c) | 15 | (d) | 16 | (b) | 17 | (d) | 18 | (b) | 19 | (b) | 20 | (b) |
| 21 | (d) | 22 | (a) | 23 | (c) | 24 | (d) | 25 | (d) | 26 | (c) | 27 | (b) | 28 | (d) | 29 | (d) | 30 | (b) |
| 31 | (b) | 32 | (c) | 33 | (c) | 34 | (a) | 35 | (c) | 36 | (c) | 37 | (b) | 38 | (b) | 39 | (b) | 40 | (d) |
| 41 | (b) | 42 | (b) | 43 | (d) | 44 | (a) | 45 | (d) | 46 | (d) | 47 | (a) | 48 | (c) | 49 | (c) | 50 | (a) |
| 51 | (c) | 52 | (c) | 53 | (d) | 54 | (b) | 55 | (b) | 56 | (b) | 57 | (b) | 58 | (a) | 59 | (c) | 60 | (c) |
| 61 | (b) | 62 | (a) | 63 | (c) | 64 | (d) | 65 | (46) | 66 | (54) | 67 | (5) | 68 | (39) | 69 | (118) | 70 | (7) |
| 71 | (13) | 72 | (75) | 73 | (4) | 74 | (2929) | 75 | (2) | 76 | (3) | | | | | | | | |

### Topic 4

| | | | | | | | | | | | | | | | | | | | |
|---|---|---|---|---|---|---|---|---|---|---|---|---|---|---|---|---|---|---|---|
| 1 | (d) | 2 | (a) | 3 | (b) | 4 | (c) | 5 | (b) | 6 | (d) | 7 | (b) | 8 | (c) | 9 | (c) | 10 | (d) |
| 11 | (c) | 12 | (b) | 13 | (a) | 14 | (c) | 15 | (c) | 16 | (d) | 17 | (a) | 18 | (a) | 19 | (a) | 20 | (d) |
| 21 | (c) | 22 | (b) | 23 | (d) | 24 | (b) | 25 | (c) | 26 | (d) | 27 | (c) | 28 | (a) | 29 | (a) | 30 | (a) |
| 31 | (c) | 32 | (c) | 33 | (a) | 34 | (b) | 35 | (c) | 36 | (b) | 37 | (a) | 38 | (b) | 39 | (b) | 40 | (d) |
| 41 | (d) | 42 | (120) | 43 | (141) | 44 | (1994) | 45 | (189) | 46 | (55) | 47 | (19) | 48 | (40) | 49 | (61) | 50 | (51) |
| 51 | (182) | 52 | (12) | 53 | (8) | 54 | (306) | 55 | (2) | 56 | (116) | 57 | (216) | 58 | (20) | 59 | (88) | 60 | (12) |
| 61 | (85) | 62 | (42) | 63 | (9) | 64 | (2) | | | | | | | | | | | | |

# Detailed **Explanations**

## Topic 1

**01.** *(b)*

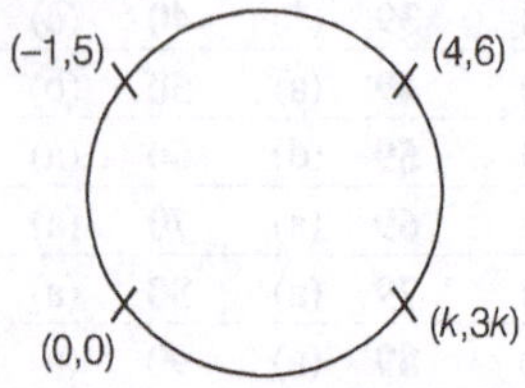


Let equation of circle be
$(x-a)^2+(y-b)^2=r^2$

Circle passing through the points (0, 0), (–1, 5),(4, 6) and (*k*, 3*k*)

$$\Rightarrow \quad (4-a)^2+(6-b)^2=r^2 \quad \ldots(\text{i})$$

$$a^2+b^2=r^2 \quad \ldots(\text{ii})$$

$$(-1-a)^2+(5-b)^2=r^2 \quad \ldots(\text{iii})$$

$$(k-a)^2+(3k-b)^2=r^2 \quad \ldots(\text{iv})$$

From Eqs (i), (ii), (iii) and (iv), we get

$$a=2,\ b=3,\ r=\sqrt{13} \text{ and } k=\frac{11}{5}$$

$$\therefore \quad 10k+r^2=10\times\frac{11}{5}+13$$

$$=22+13=35$$

**02.** *(c)*

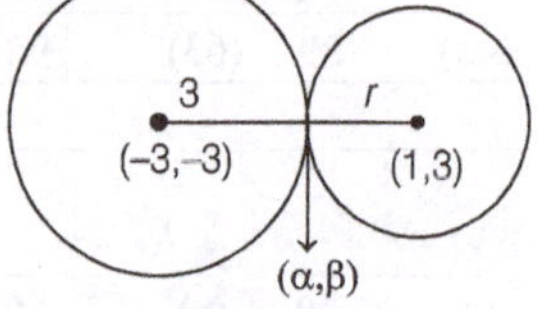


$$\sqrt{16+36}=r+3$$

$$r=\sqrt{52}-3$$

$$\Rightarrow \quad \frac{3-3r}{r+3}=\alpha$$

$$\frac{9-3r}{3+r}=\beta$$

$$\Rightarrow \quad (\beta-\alpha)^2=\frac{m}{n} \quad \text{[given]}$$

$$\Rightarrow \quad \frac{(9-3r-3+3r)^2}{(r+3)^2}=\frac{m}{n}$$

$$\Rightarrow \quad \frac{36}{52}=\frac{m}{n}$$

$$\Rightarrow \quad \frac{m}{n}=\frac{9}{13}$$

$$\Rightarrow \quad m+n=22$$

**03.** *(c)*

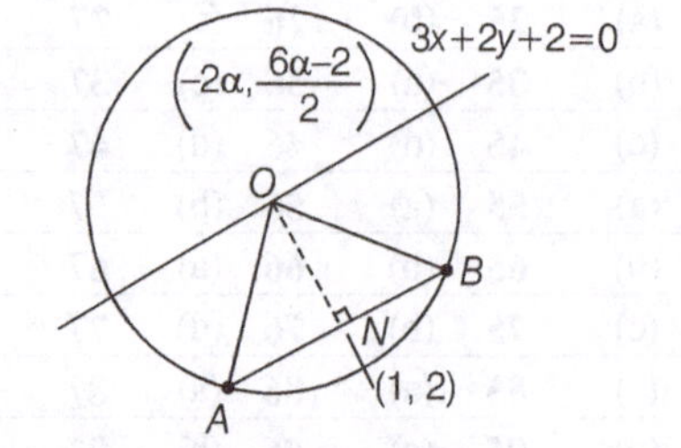


Let the centre be

$$\left(-2\alpha, \frac{6\alpha-2}{2}\right)=(-2\alpha, 3\alpha-1)$$

Centre is at equal distance from (4, 2) and (0, 2)

$$=\sqrt{(4+2\alpha)^2+(3\alpha-3)^2}$$

$$=\sqrt{(-2\alpha-0)^2+(3\alpha-3)^2}$$

$$=4\alpha^2+16+16\alpha=4\alpha^2$$

$$\Rightarrow \quad \alpha=-1$$

$\therefore$ Centre $O=(2,-4)$

$$r=OA=\sqrt{(2-4)^2+(2+4)^2}=\sqrt{40}$$

Mid-point of chord is $N\equiv(1, 2)$

$$\therefore \quad ON=\sqrt{37}$$

$$\therefore \text{Length of chord}=2\sqrt{r^2-(ON)^2}$$

$$=2\sqrt{40-37}=2\sqrt{3}$$

**04.** *(c)* Orthocentre is the intersection of altitudes.

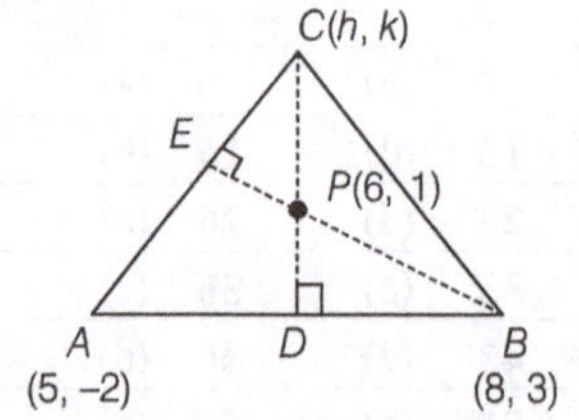


So, Slope of $BP\times$ Slope of $AC=-1$

$$\Rightarrow \quad \left(\frac{3-1}{8-6}\right)\times\left(\frac{k+2}{h-5}\right)=-1$$

$$\Rightarrow \quad k+2=-h+5$$

$$\Rightarrow \quad h+k=3 \quad \ldots(\text{i})$$

and slope of $CP\times$ slope of $AB=-1$

$$\Rightarrow \quad \left(\frac{k-1}{h-6}\right)\times\left(\frac{3+2}{8-5}\right)=-1$$

$$\Rightarrow \quad 5k-5=-3h+18$$

$$\Rightarrow \quad 3h+5k=23 \quad \ldots(\text{ii})$$

On solving Eqs. (i) and (ii), we get

$$h=-4,\ k=7$$

On checking, option (c) is correct

$$x^2+y^2-65=0$$

**05.** *(a)*

> **key idea** Common chord of two circles $S_1$ and $S_2$ can be found by $S_1-S_2=0$.

$$S_1: x^2+y^2-2x-2y+1=0$$

$$S_2: x^2+y^2+2x-3=0$$

$\because$ Given $\{C_2(-1, 0)$ and $r=2\}$

Equation of common chord

$$\Rightarrow \quad S_1-S_2=0$$

$$\Rightarrow \quad -4x-2y+4=0$$

$$\Rightarrow \quad 2x+y=2$$

At *Y*-axis $\Rightarrow$ $P(0, 2)$

Distance between centre (–1, 0) and

$$P=\sqrt{(-1-0)^2+(2-0)^2}=\sqrt{2}$$

Its square = 2

**06.** *(a)* $C: x^2+y^2=10$

$$AN=\frac{MN}{2}=1$$

$\because$ In $\Delta OAN$,

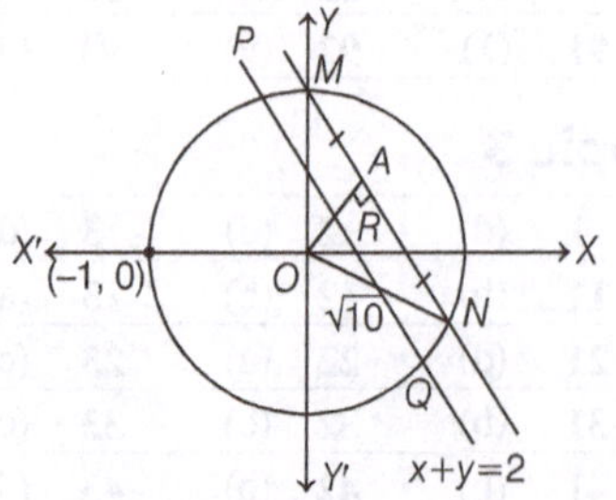


$$\Rightarrow \quad ON^2=OA^2+AN^2$$

$$\Rightarrow \quad 10=OA^2+1$$

$$\Rightarrow \quad OA=3$$

Perpendicular distance of centre from

$$PQ=\left|\frac{0+0-2}{\sqrt{2}}\right|=\sqrt{2}=OR$$

Perpendicular distance between *MN* and $PQ=AR=OA-OR$

$$=3-\sqrt{2}$$

**07.** *(b)* We have, circle with radius 1 cm and lines passing through the point (3, 2) and parallel to the coordinate axes.

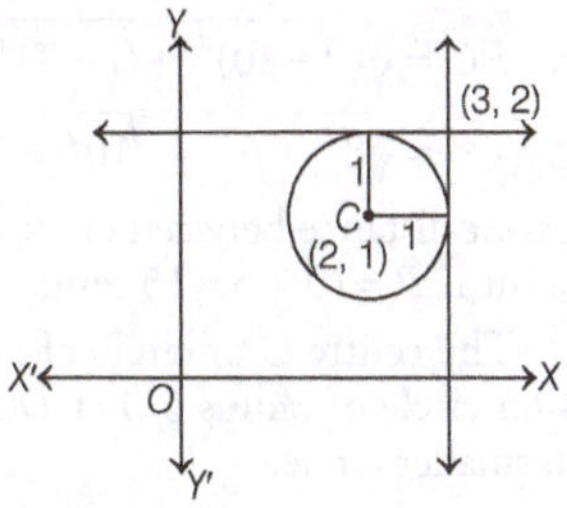


$\therefore$ Shortest distance of the circle $C$ from the point (5, 5)

$=\sqrt{(5-2)^2+(5-1)^2}-$ (radius of circle)

$=\sqrt{9+16}-1=5-1=4$

**08.** *(a)* Circle : $x^2+y^2-16x-4y=0$

Let the equation of line passing through centre of circle $C(8, 2)$ with slope $m$ is $y-2=m(x-8)$

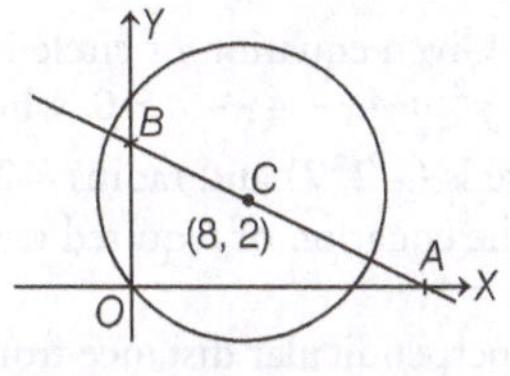


$$A=\left(8-\frac{2}{m}, 0\right)$$

$$B=(0, 2-8m)$$

$$OA+OB=8-\frac{2}{m}+2-8m$$

$$=10-\left(\frac{2}{m}+8m\right)$$

AM $\geq$ GM

$$\frac{\frac{2}{m}+8m}{2}\geq\sqrt{\frac{2}{m}\cdot 8m}$$

$$\frac{2}{m}+8m\geq 8$$

Since, $m<0, \left(\frac{2}{m}+8m\right)_{\max}=-8$

Minimum value of $OA+OB=10-(-8)=18$

**09.** *(d)* Given, equation of circle is $x^2+y^2-2gx+6y-19c=0$ passes through the point (6, 1)

$$12g+19c=43 \quad \ldots(i)$$

Centre $(g, -3)$ lies on given line.

So, $\quad g+6c=8 \quad \ldots(ii)$

On solving Eqs. (i) and (ii), we get

$$c=1 \text{ and } g=2$$

Equation of circle is

$x^2+y^2-4x+6y-19=0$

$x$-intercept $=2\sqrt{g^2-c}=2\sqrt{3}$

**10.** *(a)* Given, abscissae and ordinates of points $P$ and $Q$ are the roots of $x^2-4x-6=0$ and $y^2+2y-7=0$, respectively.

Diametric form of circle is $(x-x_1)(x-x_2)+(y-y_1)(y-y_2)=0$

$\Rightarrow$ Equation of circle,

$x^2-4x-6+y^2+2y-7=0$

$\Rightarrow \quad x^2+y^2-4x+2y-13=0$

Compare it with the given equation of circle, we get

$$a=-2, b=1, c=-13$$

$$\therefore \quad a+b-c=-2+1-(-13)$$

$$=-1+13$$

$$=12$$

**11.** *(b)* Let a circle $C$ which touches the $Y$-axis at (0, 6) and cuts off an intercept $6\sqrt{5}$ on the $X$-axis.

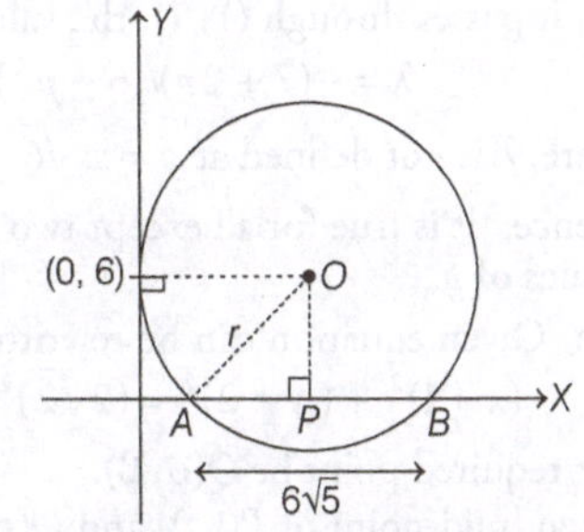


Chord $AB=6\sqrt{5}$

So, $AP=3\sqrt{5}$ (perpendicular from centre bisects the chord)

and $\quad OP=6$

In $\Delta APO$, $\quad r^2=(6)^2+(3\sqrt{5})^2$

$\Rightarrow \quad r=9$

**12.** *(a)* Given, $2x-y+1=0$ is a tangent to the circle at (2, 5).

So, normal at (2, 5) will be

$$\frac{(y-5)}{(x-2)}=\frac{-1}{2}$$

$\Rightarrow \quad x+2y=12 \quad \ldots(i)$

Now, it is also given that centre lies on $x-2y=4$.

On solving Eqs. (i) and (ii), we get

$\Rightarrow \quad x=8, y=2$

Radius will be the distance between (8, 2) and (2, 5)

$$r^2=(8-2)^2+(2-5)^2$$

$\Rightarrow \quad r^2=36+9 \Rightarrow r=\sqrt{45}=3\sqrt{5}$

**13.** *(b)* $C_1 \Rightarrow x^2+y^2-10x-10y+41=0$

$\Rightarrow \quad (x-5)^2+(y-5)^2=3^2$

$\Rightarrow$ Centre $=(5, 5)$ and Radius $=3$

$C_2 \Rightarrow x^2+y^2-16x-10y+80=0$

$\Rightarrow (x-8)^2+(y-5)^2=3^2$

$\Rightarrow$ Centre $=(8, 5)$ and Radius $=3$

Now, distance between centres

$$=\sqrt{(8-5)^2+(5-5)^2}=3$$

Average radii $=\frac{3+3}{2}=3$

$\therefore$ Option (a) is correct.

$C_1(8, 5)=(8-5)^2+(5-5)^2-9=0$

$C_2(5, 5)=(5-8)^2+(5-5)^2-9=0$

Centres of each other lies on circumference of each other.

**14.** *(d)* Equation of given circle

$$x^2+y^2=r^2 \quad \ldots(i)$$

and the chord $\quad y=2x+3 \quad \ldots(ii)$

From the figure,

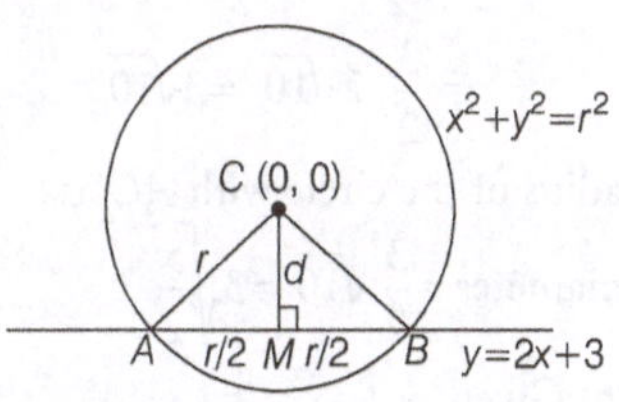


$$AB=r$$

$$d=\sqrt{r^2-\frac{r^2}{4}}=\frac{\sqrt{3}}{2}r$$

and $\quad d=\frac{3}{\sqrt{1+4}}$

$$\therefore \quad \frac{3}{\sqrt{5}}=\frac{\sqrt{3}}{2}r \Rightarrow r^2=\frac{12}{5}$$

**15.** *(d)* Equation of the line passing through the points (2, 3) and (4, 5) is

$$y-3=\left(\frac{5-3}{4-2}\right)(x-2)$$

$\Rightarrow \quad x-y+1=0 \ldots(i)$

Equation of the perpendicular line passing through the mid-point (3, 4) is

$$x+y-7=0 \quad \ldots(ii)$$

Since, centre lies on

$$y-4x+3=0 \quad \ldots(iii)$$

The intersects of the given line at the centre of the circle is found to be (2, 5).

Therefore, the radius

$=\sqrt{(2-2)^2+(5-3)^2}=2$ units

**16.** *(a)* The equation of tangent on $C_1$ at (2, 1) is

$$2x+y-(x+2)-1=0$$

or $\quad x+y=3$

$AB$ is the tangent to circle $C_1$ and chord of circle $C_2$

Length, $C_2M = \left|\dfrac{3-2-3}{\sqrt{2}}\right| = \sqrt{2}$

Also, length of the chord is $l = 4$

$\therefore$ Radius of $C_2 = r = \sqrt{\left(\dfrac{l}{2}\right)^2 + d^2}$

$= \sqrt{(2)^2 + (\sqrt{2})^2} = \sqrt{6}$

**17.** *(d)* Given, $A(-3, 5)$, $B(3, 3)$.

We know that the centroid divides orthocentre and circumcentre in the ratio $2:1$.

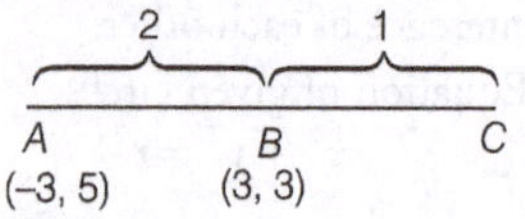


$AC = \dfrac{3}{2}AB = \dfrac{3}{2}\cdot\sqrt{6^2 + 2^2}$

$= \dfrac{3}{2}\cdot 2\sqrt{10} = 3\sqrt{10}$

Radius of the circle with $AC$ as

$\therefore$diameter $= \dfrac{3}{2}\sqrt{10} = 3\sqrt{\dfrac{5}{2}}$

**18.** *(b)* Given, $y + 3x = 0$

Equation of a chord of the circle is

$y = -3x$ ...(i)

$\Rightarrow x^2 + (-3x)^2 - 30x = 0$

$\Rightarrow 10x^2 - 30x = 0$

$\Rightarrow 10x(x-3) = 0 \Rightarrow x = 0, y = 0$

and $x = 3$, $y = -9$ are end points of diameter.

Hence, the equation of the circle is

$(x-3)(x-0) + (y+9)(y-0) = 0$

$\Rightarrow x^2 + y^2 - 3x + 9y = 0$

**19.** *(b)*

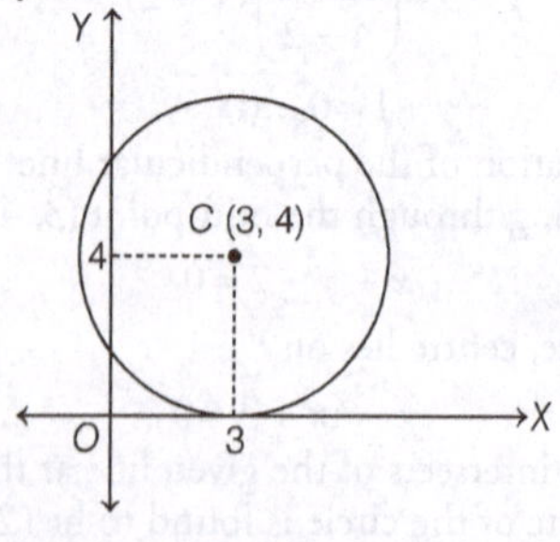


Given, the circle

$x^2 + y^2 - 6x - 8y + (25 - a^2) = 0$

Radius $= 4 = \sqrt{9 + 16 - (25 - a^2)}$

$\Rightarrow \quad a = \pm 4$

**20.** *(c)* Circle whose diametric end points are (1, 0) and (0, 1) will be of smallest radius.

$\Rightarrow (x-1)(x-0) + (y-0)(y-1) = 0$

$\Rightarrow x^2 + y^2 - x - y = 0$

**21.** *(b)* Since, the coordinates of the centre of the circle are (2, 4).

Also, $r^2 = 4 + 16 + 5 = 25$

The line will intersect the circle at two distinct points, if the distance of (2, 4) from $3x - 4y = m$ is less than radius of the circle.

$\therefore \quad \dfrac{|6 - 16 - m|}{5} < 5$

$\Rightarrow -25 < 10 + m < 25$

$\therefore \quad -35 < m < 15$

**22.** *(c)* Let

$S = x^2 + y^2 + 3x + 7y + 2p - 5 = 0$

and $S' = x^2 + y^2 + 2x + 2y - p^2 = 0$

Equation of the required circle is

$S + \lambda S' = 0$.

As, it passes through (1, 1), the value of

$\lambda = -(7 + 2p)/(6 - p^2)$

Here, $\lambda$ is not defined at $p = \pm\sqrt{6}$

Hence, it is true for all except two values of $p$.

**23.** *(d)* Given equation can be rewritten as

$(x+1)^2 + (y+2)^2 = (2\sqrt{2})^2$

Let required point be $Q(\alpha, \beta)$.

Then, mid-point of $P(1, 0)$ and $Q(\alpha, \beta)$ is the centre of the circle.

i.e., $\dfrac{\alpha + 1}{2} = -1$ and $\dfrac{\beta + 0}{2} = -2$

$\Rightarrow \quad \alpha = -3$

and $\quad \beta = -4$

So, required point is $(-3, -4)$.

**24.** *(a)* Given lines $2x + 3y + 1 = 0$ and $3x - y - 4 = 0$ are the diameters of circle.

The intersection of two lines is the centre of circle $(1, -1)$.

Circumference of circle $= 10\pi$ [given]

$\Rightarrow \quad 2\pi r = 10\pi \Rightarrow r = 5$

$\therefore$Equation of circle having centre $(1, -1)$ and radius 5 is

$(x-1)^2 + (y+1)^2 = 5^2$

$\Rightarrow \quad x^2 + y^2 - 2x + 2y - 23 = 0$

**25.** *(b)* Given, equation of circle

$x^2 + y^2 - 4x - 2y - 20 = 0$

whose centre is $C(2, 1)$ and radius is 5.

At point (10, 7),

$S_1 = 10^2 + 7^2 - 4\times 10 - 2\times 7 - 20$

$= 75 > 0$

So, $P$ lies outside the circle.

Now, $PC = \sqrt{(2-10)^2 + (1-7)^2}$

$= \sqrt{8^2 + 6^2} = \sqrt{10^2} = 10$

$\therefore$Greatest distance between circle and the point is $P = 10 + 5 = 15$ units.

**26.** *(a)* $\because$ The centre $C$ of circle of radius 3 lies on circle of radius 5. Let $D(x, y)$ in the smaller circle.

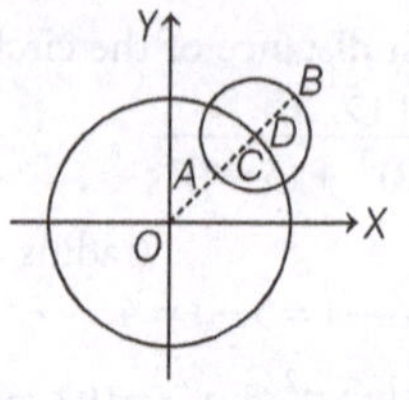


We should have, $OA \le OD \le OB$

$\Rightarrow (5-3) \le \sqrt{x^2 + y^2} \le 5 + 3$

$\Rightarrow 4 \le x^2 + y^2 \le 64$

**27.** *(b)* Given equation of circle is $x^2 + y^2 + 4x - 4y + 4 = 0$, whose centre is $(-2, 2)$ and radius $= 2$.

Let the equation of required tangent be $x + y = a$.

The perpendicular distance from centre to the circle is equal to radius of circle.

$\therefore \quad \left|\dfrac{-2 + 2 - a}{\sqrt{2}}\right| = 2$

$\Rightarrow \quad a = 2\sqrt{2}$

Hence, the equation of tangent is $x + y = 2\sqrt{2}$.

**28.** *(a)* $x^2 + y^2 - \alpha x + \beta y + \gamma = 0$

Centre $= \left(\dfrac{\alpha}{2}, -\dfrac{\beta}{2}\right)$

Radius $= \sqrt{\left(\dfrac{\alpha}{2}\right)^2 + \left(\dfrac{\beta}{2}\right)^2 - \gamma}$

Clearly, radius $= \dfrac{\beta}{2}$

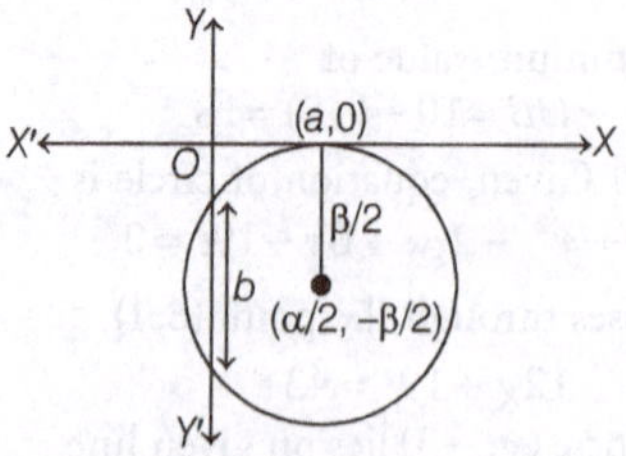


$\Rightarrow \quad \sqrt{\dfrac{\alpha^2}{4} + \dfrac{\beta^2}{4} - \gamma} = \dfrac{\beta}{2}$

On squaring both sides, we get

$\dfrac{\alpha^2}{4} + \dfrac{\beta^2}{4} - \gamma = \dfrac{\beta^2}{4}$

$$\frac{\alpha^2}{4} = \gamma \quad \ldots(i)$$

∵ $(a, 0)$ satisfies the given equation of circle

$$\Rightarrow \quad a^2 - a\alpha + \gamma = 0$$

$$a^2 - a\alpha + \frac{\alpha^2}{4} = 0 \Rightarrow \left(a - \frac{\alpha}{2}\right)^2 = 0$$

$$\Rightarrow \quad a = \frac{\alpha}{2} \Rightarrow 2a = \alpha$$

Length of $y$-intercept $= 2\sqrt{g^2 - c}$

$$= 2\sqrt{\left(\frac{\beta}{2}\right)^2 - \gamma} = b$$

$$\Rightarrow \quad 4\left(\frac{\beta^2}{4} - \gamma\right) = b^2$$

$$\beta^2 - 4\gamma = b^2$$

$$\Rightarrow \quad (2a, b^2) = (\alpha, \beta^2 - 4\gamma)$$

**29.** *(c)*

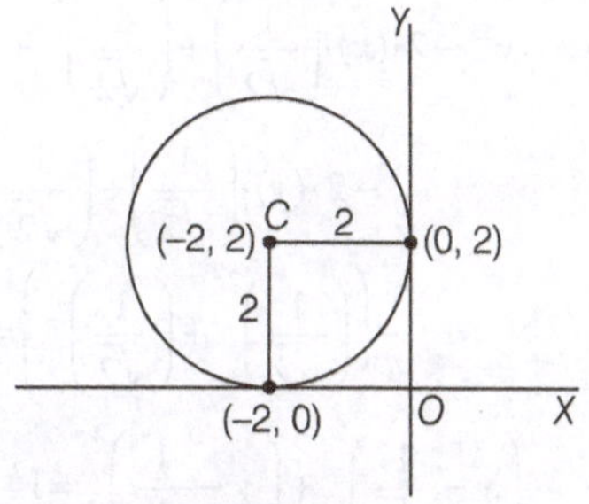


Equation of circle $C$:

$$(x+2)^2 + (y-2)^2 = 2^2$$

Centre $C_1(-2, 2)$, $r_1 = 2$

∵ An another circle having centre $C_2(2, 5)$ and radius $r$ intersects the circle $C$ at exactly 2 points.

$$\therefore \quad |r_1 - r_2| < C_1C_2 < r_1 + r_2$$

$$|r - 2| < 5 < r + 2$$

$$-5 < r - 2 < 5 \text{ and } r + 2 > 5$$

$$-3 < r < 7 \text{ and } r > 3$$

$$\therefore \quad r \in (3, 7)$$

$$\Rightarrow \alpha = 3, \beta = 7$$

$$\therefore \; 3\beta - 2\alpha = 21 - 6 = 15$$

**30.** *(b)* $y = x^2 + px - 3$

∵ $x = 0, y = -3$

∴ $(0, -3)$ be point where parabola cuts $Y$-axis.

∴ Equation of circle $C$,

$$(x+1)^2 + (y+1)^2 = (-1-0)^2 + (-1+3)^2$$

$$\Rightarrow \quad (x+1)^2 + (y+1)^2 = 5$$

Circle cuts $X$-axis at $y = 0$.

$$\therefore \quad (x+1)^2 = 4 \Rightarrow x + 1 = \pm 2$$

$$\Rightarrow \quad x = 1, -3$$

∴ Points are $(1, 0)$, $(-3, 0)$ and $(0, -3)$.

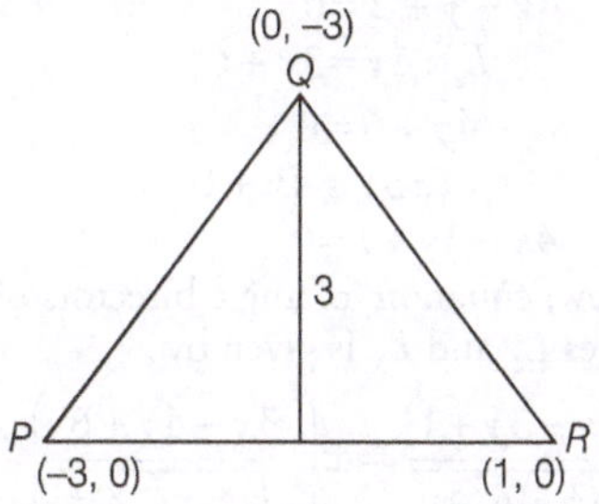


Area of $\Delta PQR = \frac{1}{2} \times 4 \times 3 = 6$

**31.** *(d)* $C_1 : (x - \alpha)^2 + (y - \beta)^2 = r_1^2$

$$C_2 : (x-8)^2 + \left(y - \frac{15}{2}\right)^2 = r_2^2$$

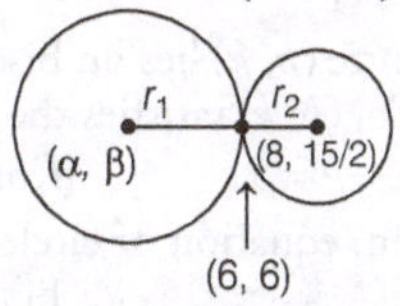


$$r_1 : r_2 = 2 : 1$$

So, $\frac{2 \times 8 + \alpha}{2 + 1} = 6$

$$\Rightarrow \quad \alpha = 2$$

and $\frac{2 \times \frac{15}{2} + \beta}{2+1} = 6 \Rightarrow \beta = 3$

So, $(\alpha, \beta) = (2, 3)$

Now, $r_1 = \sqrt{(6-\alpha)^2 + (6-\beta)^2}$

$$= \sqrt{(4)^2 + (3)^2} = 5$$

$$\Rightarrow \quad r_1^2 = 25$$

and $r_2^2 = \left(\sqrt{(8-6)^2 + \left(\frac{15}{2} - 6\right)^2}\right)^2$

$$= 4 + \frac{9}{4} = \frac{25}{4}$$

Therefore, $(\alpha + \beta) + 4(r_1^2 + r_2^2)$

$$= (2+3) + 4\left(25 + \frac{25}{4}\right)$$

$$= 5 + 4 \times \frac{125}{4} = 130$$

**32.** *(b)* $S_1 = x^2 + y^2 - 18x - 15y + 131 = 0$

$$C_1 : (-g_1, -f_1) = \left(9, \frac{15}{2}\right)$$

$$S_2 = x^2 + y^2 - 6x - 6y - 7 = 0$$

$$C_2 = (-g_2, -f_2) = (3, 3)$$

$$r_1 = \sqrt{g_1^2 + f_1^2 - c_1}$$

$$= \sqrt{(9)^2 + \left(\frac{15}{2}\right)^2 - 131}$$

$$= \sqrt{81 + \frac{225}{4} - 131} = \frac{5}{2}$$

$$r_2 = \sqrt{(3)^2 + (3)^2 + 7} = \sqrt{9 + 9 + 7}$$

$$= \sqrt{25} = 5$$

$$C_1C_2 = \sqrt{(9-3)^2 + \left(\frac{15}{2} - 3\right)^2}$$

$$= \sqrt{(6)^2 + \left(\frac{9}{2}\right)^2}$$

$$= \sqrt{36 + \frac{81}{4}} = \sqrt{\frac{144 + 81}{4}}$$

$$= \sqrt{\frac{225}{4}} = \frac{15}{2}$$

$$|C_1C_2| = \left|\frac{15}{2}\right| = \frac{5}{2} + \frac{5}{1}$$

$$= |C_1C_2| = r_1 + r_2$$

⇒ Number of common tangents are 3.

**33.** *(a)* We have, equation of circle is

$$x^2 + y^2 = 16$$

Let any point on the circle be $B(4\cos\theta, 4\sin\theta)$

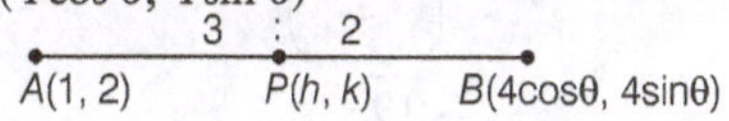


$h = \frac{12\cos\theta + 2}{5}$ and $k = \frac{12\sin\theta + 4}{5}$

$$\Rightarrow \quad \cos\theta = \frac{5h - 2}{12}$$

and $\sin\theta = \frac{5k - 4}{12}$

As, $\cos^2\theta + \sin^2\theta = 1$,

$$\left(\frac{5h-2}{12}\right)^2 + \left(\frac{5k-4}{12}\right)^2 = 1$$

$$\Rightarrow \left(h - \frac{2}{5}\right)^2 + \left(k - \frac{4}{5}\right)^2 = \left(\frac{12}{5}\right)^2$$

∴ Locus of point $P$ is

$$\left(x - \frac{2}{5}\right)^2 + \left(y - \frac{4}{5}\right)^2 = \left(\frac{12}{5}\right)^2$$

Which is equation of circle with center $\left(\frac{2}{5}, \frac{4}{5}\right)$.

Hence, $AC = \sqrt{\left(1 - \frac{2}{5}\right)^2 + \left(2 - \frac{4}{5}\right)^2}$

$$= \frac{\sqrt{45}}{5} = \frac{3\sqrt{5}}{5}$$

**34.** *(b)* Given, length of $AB = \lambda$

So, $AC = \dfrac{\lambda}{2}$ and $AM = \dfrac{2\lambda}{5}$

$CM = AC - AM = \dfrac{\lambda}{2} - \dfrac{2\lambda}{5} = \dfrac{\lambda}{10}$

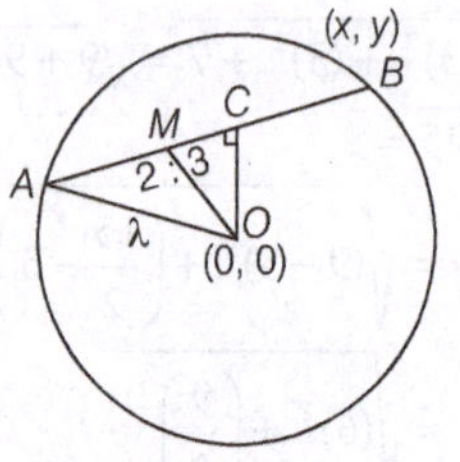


Now, from $\Delta OCM$, we get

$OM^2 = CM^2 + OC^2$

$\Rightarrow x^2 + y^2 = \left(\dfrac{\lambda}{10}\right)^2 + \left(\lambda^2 - \dfrac{\lambda^2}{4}\right)$

$\Rightarrow x^2 + y^2 = \dfrac{\lambda^2}{100} + \dfrac{3\lambda^2}{4}$

$= \dfrac{\lambda^2 + 75\lambda^2}{100}$

$\Rightarrow x^2 + y^2 = \dfrac{76\lambda^2}{100} = \dfrac{19\lambda^2}{25}$

$\therefore x^2 + y^2 = \left(\dfrac{\sqrt{19}\lambda}{5}\right)^2$

So, radius is $\dfrac{\sqrt{19}\lambda}{5}$.

**35.** *(a)* Equation of circle

$x^2 + y^2 - 6x + 4y + 8 = 0$

Centre $= (3, -2)$, Radius $= \sqrt{5}$

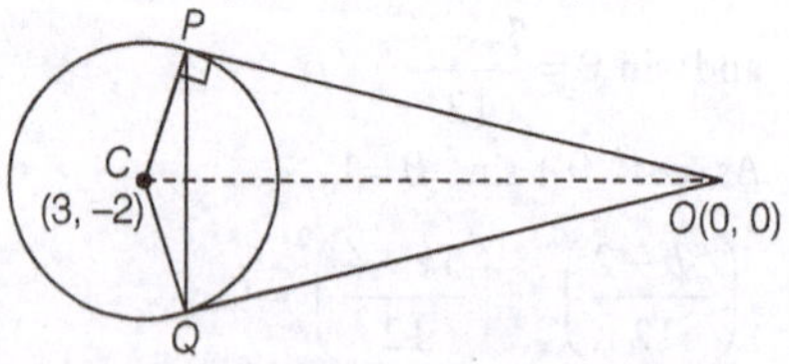


Equation of circumcircle of $\Delta OPQ$

$(x-0)(x-3) + (y-0)(y+2) = 0$

($\because \angle OPC = 90^\circ$, $\therefore$ $OC$ is diameter)

$\Rightarrow x^2 + y^2 - 3x + 2y = 0$

$\because$ it passes through the point $(\alpha, 1/2)$,

$\therefore \alpha^2 + \dfrac{1}{4} - 3\alpha + 2 \times \dfrac{1}{2} = 0$

$\Rightarrow \quad \alpha^2 - 3\alpha + \dfrac{5}{4} = 0$

$\Rightarrow \quad 4\alpha^2 - 12\alpha + 5 = 0$

$\Rightarrow \quad \alpha = \dfrac{1}{2}, \dfrac{5}{2}$

**36.** *(c)* Given equation of circle is

$(x-h)^2 + (y-k)^2 = r^2 \quad \ldots \text{(i)}$

Let $L_1 : y = x + 2$

$\Rightarrow \quad x - y + 2 = 0$

$L_2 : 4y = 3x + 6$

$\Rightarrow 3x - 4y + 6 = 0$

$L_3 : 3y = 4x + 1$

$\Rightarrow 4x - 3y + 1 = 0$

Now, equation of angle bisectors of lines $L_2$ and $L_3$ is given by

$\dfrac{4x - 3y + 1}{\sqrt{4^2 + (-3)^2}} = \pm\left(\dfrac{3x - 4y + 6}{\sqrt{3^2 + (-4)^2}}\right)$

$\Rightarrow \dfrac{4x - 3y + 1}{5} = \pm\left(\dfrac{3x - 4y + 6}{5}\right)$

$\Rightarrow 4x - 3y + 1 = 3x - 4y + 6$

(taking + ve sign)

$\Rightarrow x + y = 5 \quad \ldots \text{(ii)}$

Since, centre $(h, k)$ lies on bisector of $L_2$ and $L_3$. So, $(h, k)$ satisfies the Eq. (ii).

Hence, $h + k = 5$ [from Eq. (ii)]

**37.** *(b)* Given, equation of circle is

$x^2 + y^2 - x + 2y - \dfrac{11}{4} = 0$

$\because$ Centre $= \left(\dfrac{1}{2}, -1\right)$

and radius $= \sqrt{\left(\dfrac{1}{2}\right)^2 + (-1)^2 - \left(\dfrac{-11}{4}\right)} = 2$

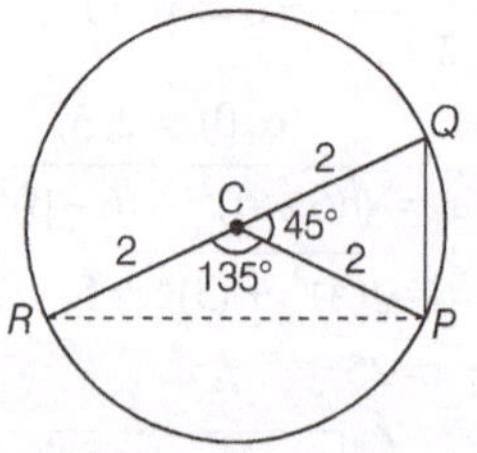


$QR =$ Diameter $= 2 \times$ Radius $= 4$

$\therefore$ Area of $\Delta PQR = \dfrac{1}{2}(2)(2\sin 45^\circ)$

$+ \dfrac{1}{2}(2)(2\sin 135^\circ)$

$= \dfrac{2}{\sqrt{2}} + \dfrac{2}{\sqrt{2}} = 2\sqrt{2}$

**38.** *(b)* Let $p \equiv \sin t$, $q \equiv \cos t$

Let orthocentre be $(h, k)$.

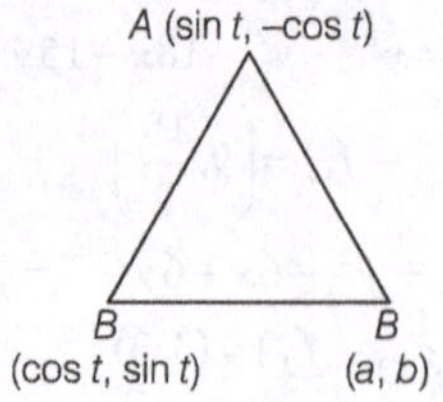


Since, if it is an equilateral triangle, hence orthocentre coincides with centroid.

$\Rightarrow \quad a + p + q = 3h, b + p - q = 3k$

$\Rightarrow \quad (3h - a)^2 + (3k - b)^2 = (p+q)^2 + (p-q)^2 = 2(p^2 + q^2) = 2$

$\therefore \quad \left(h - \dfrac{a}{3}\right)^2 + \left(k - \dfrac{b}{3}\right)^2 = \dfrac{2}{9}$

Now, circle centre at $\left(\dfrac{a}{3}, \dfrac{b}{3}\right)$.

We are given that $\dfrac{a}{3} = 1, \dfrac{b}{3} = \dfrac{1}{3}$

$\Rightarrow \quad a = 3, b = 1$

$\therefore \quad a^2 - b^2 = 8$

**39.** *(*)* $x^2 - \sqrt{2}(x + y) + y^2 = 0$,

$\angle BAC = \dfrac{\pi}{2}$ and $AB = \sqrt{2}$

$\Rightarrow \quad x^2 - \sqrt{2}x + y^2 - \sqrt{2}y = 0$

$\Rightarrow \quad x^2 - 2\cdot(x)\cdot\left(\dfrac{1}{\sqrt{2}}\right) + \left(\dfrac{1}{\sqrt{2}}\right)^2 + y^2 - 2\cdot(y)\cdot\left(\dfrac{1}{\sqrt{2}}\right) + \left(\dfrac{1}{\sqrt{2}}\right)^2 - \left(\left(\dfrac{1}{\sqrt{2}}\right)^2 + \left(\dfrac{1}{\sqrt{2}}\right)^2\right) = 0$

$\Rightarrow \quad \left(x - \dfrac{1}{\sqrt{2}}\right)^2 + \left(y - \dfrac{1}{\sqrt{2}}\right)^2 = 1^2$

$\therefore$ Radius $= 1$ unit

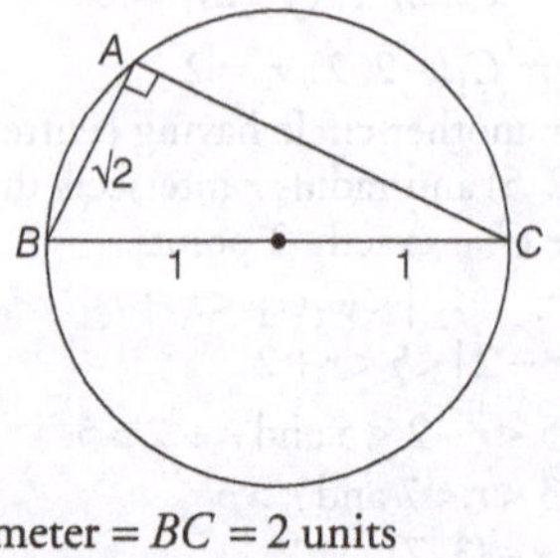


Diameter $= BC = 2$ units

$AC = \sqrt{BC^2 - AB^2} = \sqrt{4 - 2} = \sqrt{2}$

$\text{ar}(\Delta ABC) = \dfrac{1}{2} \times AB \times AC$

$= \dfrac{1}{2} \times \sqrt{2} \times \sqrt{2} = \dfrac{2}{2} = 1$

**40.** *(c)*

> **key idea** (i) Equation of the tangent to the circle $x^2 + y^2 = r^2$ at point $(x_1, y_1)$ is given by $xx_1 + yy_1 = r^2$
>
> (ii) Draw the diagram as per given question and solve further.

$C_1 : x^2 + y^2 = 2$

$C_2 : (x-3)^2 + (y-2)^2 = 5$

Line $AB : x - y + 2 = 0$

$$CP = \left|\frac{3-2+2}{\sqrt{(1)^2+(-1)^2}}\right| = \frac{3}{\sqrt{2}}$$

$$AP = \sqrt{(CA)^2 - (CP)^2}$$

$$= \sqrt{5 - \frac{9}{2}} = \frac{1}{\sqrt{2}}$$

$$\tan\theta = \frac{CP}{AP} = \frac{\frac{3}{\sqrt{2}}}{\frac{1}{\sqrt{2}}} = 3$$

$$\therefore \sin\theta = \frac{3}{\sqrt{1+9}} = \frac{3}{\sqrt{10}} = \frac{AP}{AN}$$

$$\therefore AN = \frac{\sqrt{10}}{3} \times \frac{1}{\sqrt{2}} = \frac{\sqrt{5}}{3} = BN$$

Area of $\Delta ANB = \frac{1}{2} \times AB \times PN$

$$= \frac{1}{2} \times 2AP \times PN$$

$$= \frac{1}{\sqrt{2}} \times [AN(\cos\theta)]$$

$$= \frac{1}{\sqrt{2}} \times \frac{\sqrt{5}}{3} \times \sqrt{1 - \frac{9}{10}}$$

$$= \frac{1}{\sqrt{2}} \times \frac{\sqrt{5}}{3} \times \frac{1}{\sqrt{10}} = \frac{1}{6}$$

**41.** *(c)*

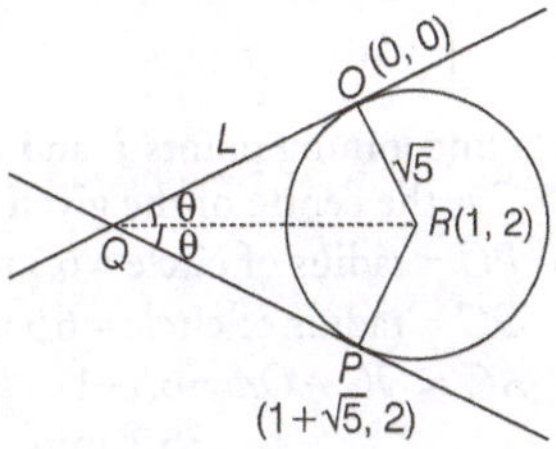


$$\tan\theta = \frac{\sqrt{5}}{L}$$

$$\tan 2\theta = 2$$

$$\Rightarrow \frac{2\tan\theta}{1-\tan^2\theta} = 2$$

$$\Rightarrow \tan\theta = \frac{\sqrt{5}-1}{2} \text{ (as } \theta \text{ is acute)}$$

$$\text{Area} = \frac{1}{2}L^2 \sin 2\theta$$

$$= \frac{1}{2} \cdot \frac{5}{\tan^2\theta} \cdot 2\sin\theta\cos\theta$$

$$= \frac{5\sin\theta\cos\theta}{\sin^2\theta} . \cos^2\theta$$

$$= 5\cot\theta \cdot \cos^2\theta$$

$$= 5 \cdot \frac{2}{\sqrt{5}-1} \cdot \frac{1}{1+\left(\frac{\sqrt{5}-1}{2}\right)^2}$$

$$= \frac{10}{\sqrt{5}-1} \cdot \frac{4}{4+6-2\sqrt{5}}$$

$$= \frac{40}{2\sqrt{5}(\sqrt{5}-1)^2} = \frac{\sqrt{5}(3+\sqrt{5})}{2}$$

**42.** *(d)* Given,

$$C : 4x^2 + 4y^2 - 12x + 8y + k = 0$$

$$C : x^2 + y^2 - 3x + 2y + \frac{k}{4} = 0$$

[Divide by 4]

$$\text{Centre} = \left(\frac{3}{2}, -1\right)$$

$$\text{radius} = \sqrt{\frac{9}{4} + 1 - \frac{k}{4}}$$

According to question, radius < 1

$$\Rightarrow \sqrt{\frac{9}{4} + 1 - \frac{k}{4}} < 1$$

$$\Rightarrow \sqrt{\frac{13-k}{4}} < 1$$

$$\Rightarrow \quad k > 9 \quad \ldots\text{(i)}$$

Since, $\left(1, -\frac{1}{3}\right)$ lies on or inside the circle.

$$\therefore \quad 4(1)^2 + 4\left(\frac{-1}{3}\right)^2 - 12(1) + 8\left(-\frac{1}{3}\right) + k \le 0$$

$$\Rightarrow \quad k \le \frac{92}{9} \quad \ldots\text{(ii)}$$

From Eqs. (i) and (ii), we get

$$9 < k \le \frac{92}{9} \text{ i.e. } k \in \left(9, \frac{92}{9}\right]$$

**43.** *(a)* Circle $C$ touches $x = 2y$ at (2, 1), then equation of circle $C$ is given by

$$(x-2)^2 + (y-1)^2 + \lambda(x-2y) = 0$$

$$\Rightarrow x^2 - 4x + 4 + y^2 - 2y + 1 + \lambda x - 2\lambda y = 0$$

$$\Rightarrow x^2 + y^2 + x(\lambda - 4) + y(-2-2\lambda) + 5 = 0 \quad \ldots\text{(i)}$$

Since, circles $C$ and $C_1$ intersect each other at $PQ$, so $PQ$ is common chord.

$\therefore$ Equation of $PQ$ will be $C - C_1 = 0$

$\Rightarrow \quad x(\lambda-4) + y(-4-2\lambda) + 10 = 0$

Again $PQ$ is diameter of circle $C_1$.

So, $PQ$ passes through centre of $C_1$.

Now, equation of $C_1$ is

$$x^2 + y^2 + 2y - 5 = 0$$

$\therefore$ Centre of $C_1$ is (0, −1).

So, $PQ$ passes through (0, −1)

$$\therefore \quad 0(\lambda-4) - 1(-4-2\lambda) + 10 = 0$$

$$\Rightarrow \quad \lambda = -7$$

On putting $\lambda = -7$ in Eq. (i), we get equation of circle $C$ as

$$x^2 + y^2 - 11x + 12y + 5 = 0$$

$$\therefore \text{Radius of } C = \sqrt{\left(\frac{11}{2}\right)^2 + (-6)^2 - 5}$$

$$= \sqrt{\frac{121}{4} + 36 - 5} = \sqrt{\frac{245}{4}} = \frac{\sqrt{245}}{2}$$

$\therefore$ Diameter of $C = 2 \times$ Radius

$$= \sqrt{245} = 7\sqrt{5}$$

**44.** *(c)* $C : x^2 + y^2 - 2x - 6y + 6 = 0$

$$(x-1)^2 + (y-3)^2 = 2^2$$

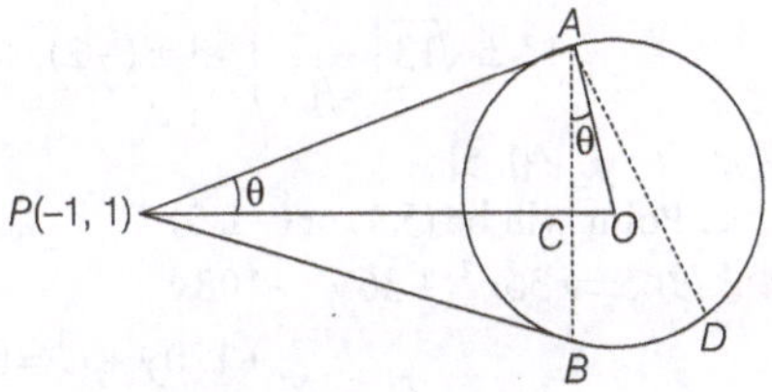


$AP =$ Length of tangent $= \sqrt{S_1}$

$$\Rightarrow \quad AP^2 = (-1-1)^2 + (1-3)^2 - 4$$

$$\therefore \quad AP = 2$$

$$\tan\theta = \frac{AO}{AP} = \frac{2}{2} = 1 \quad (\therefore \theta = 45^\circ)$$

In $\Delta ACO$,

$$\frac{AC}{AO} = \cos\theta$$

$$\therefore \quad AC = 2 \times \frac{1}{\sqrt{2}} = \sqrt{2}$$

$\therefore \quad \Delta OAC \cong \Delta OAM$

$\Delta OAC \cong \Delta OBC$

$\Delta OBC \cong \Delta ODM$

$\therefore$ Area of $\Delta ABD = 4$ Area of $\Delta ACO$

$$= 4 \times \frac{1}{2}(AO \times AC \times \sin\theta)$$

$$= 2 \times 2 \times \sqrt{2} \times \frac{1}{\sqrt{2}} = 4$$

**45.** *(d)*

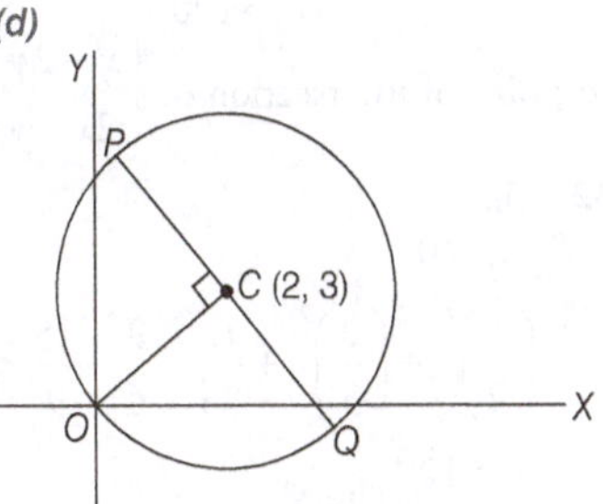

Equation of circle is

$(x-2)^2+(y-3)^2=(2-0)^2+(3-0)^2$

$\Rightarrow (x-2)^2+(y-3)^2=13$

Slope of $OC=\dfrac{3-0}{2-0}=\dfrac{3}{2}$

$\therefore$ Slope of $PQ=-\dfrac{2}{3}=\tan\theta$

Points at a distance of $\sqrt{13}$ units from $O$ with slope $=-\dfrac{2}{3}$

Using, $\begin{cases} x_2=x_1\pm r\cos\theta \\ y_2=y_1\pm r\sin\theta \end{cases}$

So, $\quad x_2=2\pm\sqrt{13}\left(\dfrac{3}{\sqrt{13}}\right)=2\pm3$

$\therefore \quad x_2:\{5,-1\}$

$y_2:3\pm\sqrt{13}\left(\dfrac{-2}{\sqrt{13}}\right)=3\pm(-2)$

$\therefore \quad y_2:\{1,5\}$

$\therefore$ Point will be $(5,1)$ or $(-1,5)$.

**46.** *(d)* $S\Rightarrow 36x^2+36y^2-108x+120y+C=0$

$x^2+y^2-3x+\left(\dfrac{10}{3}\right)y+\dfrac{C}{36}=0$

$\left(x-\dfrac{3}{2}\right)^2+\left(y+\dfrac{5}{3}\right)^2+\dfrac{C}{36}-\dfrac{9}{4}-\dfrac{25}{9}=0$

When $x=0$, $36y^2+120y+C=0$

Since, it neither touch the coordinate axis nor intersects. So, the roots of this equation must be imaginary.

$D<0$

$\Rightarrow 120^2-4\times36C<0$

$\Rightarrow \quad C>\dfrac{120\times120}{4\times36}$

$C>100 \quad \ldots(i)$

Similarly, when $y=0$,

$36x^2-108x+C=0$

$108^2-4\times36\,C<0$

$\Rightarrow \quad C>81 \quad \ldots(ii)$

From Eqs. (i) and (ii), we get

$\Rightarrow \quad C>100$

and point of intersection of $\begin{cases} x-2y=4 \\ 2x-y=5 \end{cases}$

is $(2,-1)$.

So, $S_{(2,1)}<0$

$\left(\dfrac{1}{2}\right)^2+\left(\dfrac{2}{3}\right)^2+\dfrac{C}{36}-\dfrac{9}{4}-\dfrac{25}{9}<0$

$C<156$

$C\in(100,156)$

**47.** *(b)* Given,

$M\Rightarrow x^2+y^2=1$, its centre $O_1\equiv(0,0)$

$N\Rightarrow x^2+y^2-2x=0$, its centre $O_2\equiv(1,0)$

$O\Rightarrow x^2+y^2-2x-2y+1=0$, its centre $O_3\equiv(1,1)$

$P\Rightarrow x^2+y^2-2y=0$, its centre $O_4\equiv(0,1)$

We see that

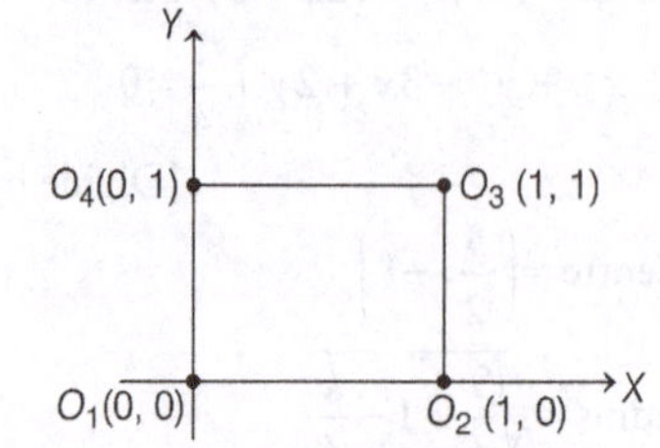


$O_1O_2=O_2O_3=O_3O_4=O_4O_1$

and $O_1O_3=O_2O_4$

Hence, $O_1O_2O_3O_4$ form a square.

**48.** *(b)* Given equation of tangent to the circle $x^2+y^2=25$ at the point $R(3,4)$.

Now, equation $3x+4y=25$.

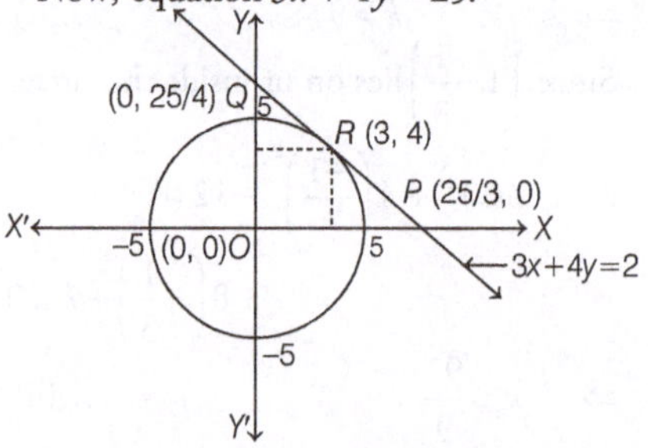


Now, point $P\left(\dfrac{25}{3},0\right)$ and $Q\left(0,\dfrac{25}{4}\right)$.

Now,

$OP+OQ+PQ=\dfrac{25}{3}+\dfrac{25}{4}+\dfrac{125}{12}=25$

Incentre of $\Delta OPQ$

$=\left(\dfrac{\dfrac{25}{3}\times0+\dfrac{125}{12}\times0+\dfrac{25}{4}\times\dfrac{25}{3}}{25},\ \dfrac{\dfrac{25}{3}\times\dfrac{25}{4}+\dfrac{125}{12}\times0+\dfrac{25}{4}\times0}{25}\right)$

$=\left(\dfrac{25}{12},\dfrac{25}{12}\right)$

$\therefore r^2=\left(\dfrac{25}{12}\right)^2+\left(\dfrac{25}{12}\right)^2$

[$\because$ Distance from origin to incentre is $r$]

$=\dfrac{625}{72}$

**49.** *(a)* Given, $P$ be a point on circle

$(x-1)^2+(y-1)^2=1$

So, co-ordinate of $P$ will be of form $P(1+\cos\theta, 1+\sin\theta)$.

Given, $\quad A(1,4)$ and $B(1,-5)$

Then, $\quad (PA)^2+(PB)^2$

$=\cos^2\theta+(\sin\theta-3)^2+\cos^2\theta+(\sin\theta+6)^2$

$=47+6\sin\theta$

Now, $(PA)^2+(PB)^2=47+6\sin\theta$ is Maximum, when

$\cos\theta=0$

$P(1,2),\ A(1,4),\ B(1,-5)$

$\therefore P$, $A$ and $B$ are collinear points on the line $x=1$.

**50.** *(b)* Given, $OA=1$ unit, $OB=13$ units

Since, $OB$ is diameter of circle.

Then, radius $(r)=13/2=6.5$ unit

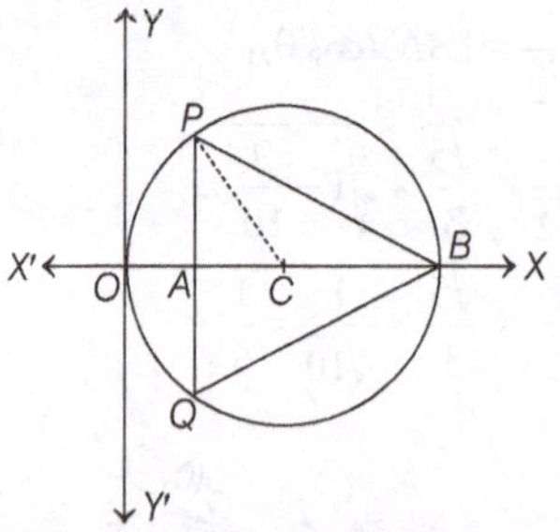


Draw a line joining points $P$ and $C$, where $C$ is the centre of the given circle.

Then, $PC=$ radius of circle $=6.5$ units

$OC=$ radius of circle $=6.5$ units

Now, $AC=OC-OA=6.5-1$

$=5.5$ units

Then, using Pythagoras theorem,

$(PA)^2=(PC)^2-(AC)^2$

$=(6.5)^2-(5.5)^2=(1)\ (12)=12$

$\therefore \quad PA=\sqrt{12}$

Then, $PQ=2PA=2\sqrt{12}$

Hence, area of $\Delta PQB=\dfrac{1}{2}\times$ (Base) $\times$ (Height)

$=\dfrac{1}{2}\times(PQ)\times(AB)$

$=\dfrac{1}{2}\times(PQ)\times(OB-OA)$

$=\dfrac{1}{2}\times(2\sqrt{12})\times(13-1)$

$=24\sqrt{3}$ sq units

**51.** *(d)* Given, line $x+\sqrt{3}y=2\sqrt{3}$

and point $\left(\frac{3\sqrt{3}}{2},\frac{1}{2}\right)$

From options, we take the conic
$x^2+9y^2=9$

Equation of any tangent at $(x_1, y_1)$ is

$$xx_1+9\cdot yy_1=9$$

$$\Rightarrow \quad \frac{3\sqrt{3}}{2}x+9\times\frac{1}{2}\times y=9$$

$$\left[\because (x_1, y_1)=\left(\frac{3\sqrt{3}}{2},\frac{1}{2}\right)\right]$$

$$\Rightarrow \frac{3\sqrt{3}}{2}(x+\sqrt{3}y)=9$$

$$\Rightarrow x+\sqrt{3}y=2\sqrt{3}$$

Clearly, $x+\sqrt{3}y=2\sqrt{3}$ is a tangent to the curve $x^2+9y^2=9$.

**52.** *(d)* Since, equation of tangent to the circle $x^2+y^2=1$ at point $\left(\frac{1}{\sqrt{2}},\frac{1}{\sqrt{2}}\right)$ is

$T=0$.

$$\Rightarrow \quad \frac{x}{\sqrt{2}}+\frac{y}{\sqrt{2}}=1$$

$$\Rightarrow \quad x+y=\sqrt{2} \quad \ldots\text{(i)}$$

$\because$ Line $L_1: x+y=\sqrt{2}$ is perpendicular to line $y=mx+c$, so $m=1$, since line $y=mx+c$ is tangent to the circle $(x-3)^2+y^2=1$, where $m=1$, then

$y=x+c$ is equivalent to

$y=1(x-3)\pm 1\sqrt{1+1^2}$.

So, $\quad c=-3\pm\sqrt{2}\Rightarrow c+3=\pm\sqrt{2}$

$\Rightarrow (c+3)^2=2\Rightarrow c^2+6c+7=0$

**53.** *(c)* The equation of chord of contact $AB$ to circle. $x^2+y^2-8x-4y+16=0$ w.r.t point origin (0, 0) is $T=0$

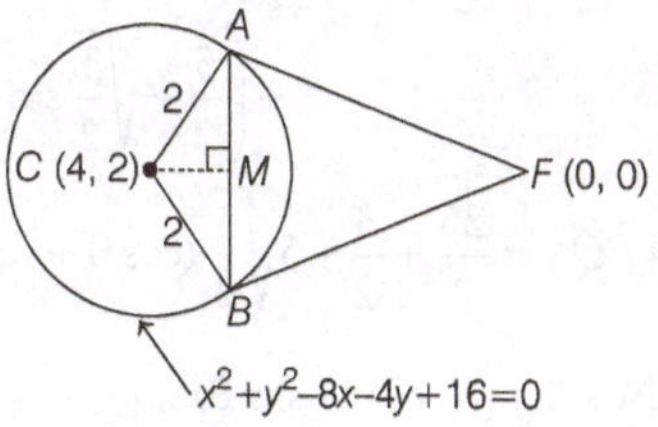


$\Rightarrow 0x+0y-4(x+0)-2(y+0)+16=0$

$\Rightarrow 2x+y-8=0 \quad \ldots\text{(i)}$

In right angled $\Delta AMC$,

$$CM=\frac{|2(4)+2-8|}{\sqrt{2^2+1^2}}=\frac{2}{\sqrt{5}}$$

$$\because AM^2=CA^2-CM^2=2^2-\left(\frac{2}{\sqrt{5}}\right)^2$$

$$\Rightarrow AM=4/\sqrt{5}$$

$$\because AB^2=(2AM)^2=\left(\frac{2\times 4}{\sqrt{5}}\right)^2=\frac{64}{5}$$

**54.** *(a)* Let $(h, k)$ be the centre of the circle and radius $r=h$, as circle touch the $Y$-axis and other circle $x^2+y^2=1$ whose centre (0, 0) and radius is 1.

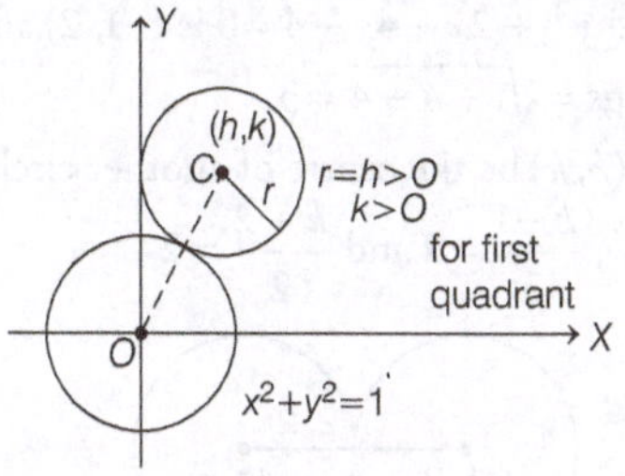


$\therefore \quad OC=r+1$

[$\because$ If circles touch each other externally, then $C_1C_2=r_1+r_2$]

$\Rightarrow \quad \sqrt{h^2+k^2}=h+1, h>0$

and $k>0$, for first quadrant.

$\Rightarrow \quad h^2+k^2=h^2+2h+1$

$\Rightarrow \quad k^2=2h+1$

$\Rightarrow \quad k=\sqrt{1+2h},$

as $\quad k>0$

Now, on taking locus of centre $(h, k)$, we get

$$y=\sqrt{1+2x}, x\geq 0$$

**55.** *(a)* Given circles are $x^2+y^2=4$, centre $c_1(0, 0)$ and radius, $r_1=2$ and $x^2+y^2+6x+8y-24=0$, centre $c_2(-3, -4)$ and radius $r_2=7$

$\because \quad c_1c_2=\sqrt{9+16}=5$

and $|r_1-r_2|=5$

$\because \quad c_1c_2=|r_1-r_2|=5$

$\therefore$circle $x^2+y^2=4$ touches the circle $x^2+y^2+6x+8y-24=0$ internally.

So, equation of common tangent is

$$S_1-S_2=0$$

$\Rightarrow \quad 6x+8y-20=0$

$\Rightarrow \quad 3x+4y=10 \quad \ldots\text{(i)}$

The common tangent passes through the point $(6, -2)$, from the given options.

**56.** *(b)* Given points are $(-8, 5)$ and $(6, 5)$ in which $y$-coordinate is same, i.e. these points lie on horizontal line $y=5$.

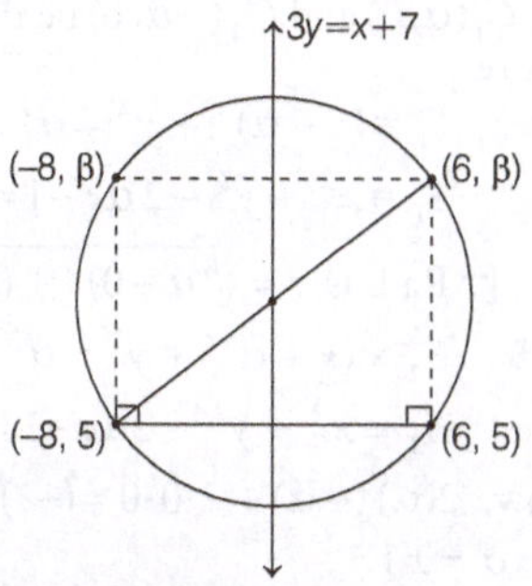


Let $(-8, \beta)$ and $(6, \beta)$ are the coordinates of the other vertices of rectangle as shown in the figure.

Since, the mid-point of line joining points $(-8, 5)$ and $(6, \beta)$ lies on the line $3y=x+7$.

$$\therefore \quad 3\left(\frac{5+\beta}{2}\right)=\frac{-8+6}{2}+7$$

$$\Rightarrow \quad 15+3\beta=-2+14\Rightarrow\beta=-1$$

Now, area of rectangle
$=|-8-6|\times|\beta-5|=14\times 6=84$

**57.** *(b)* Given circles,

$x^2+y^2-2x-2y-2=0 \quad \ldots\text{(i)}$

and $x^2+y^2-6x-6y+14=0 \ldots\text{(ii)}$

are intersecting each other orthogonally, because

$$2(1)(3)+2(1)(3)=14-2$$

[$\because$ Two circles are intersected orthogonally, if $2g_1g_2+2f_1f_2=c_1+c_2$]

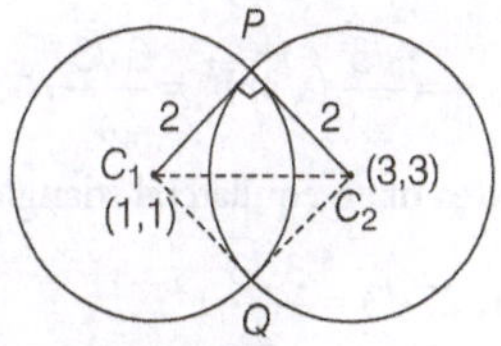


So, area of quadrilateral
$PC_1QC_2=2\times\text{ar}(\Delta PC_1C_2)$.

$$=2\times\left(\frac{1}{2}\times 2\times 2\right)=4 \text{ sq units}$$

**58.** *(d)* Clearly, circles are orthogonal because tangent at one point of intersection is passing through centre of the other.

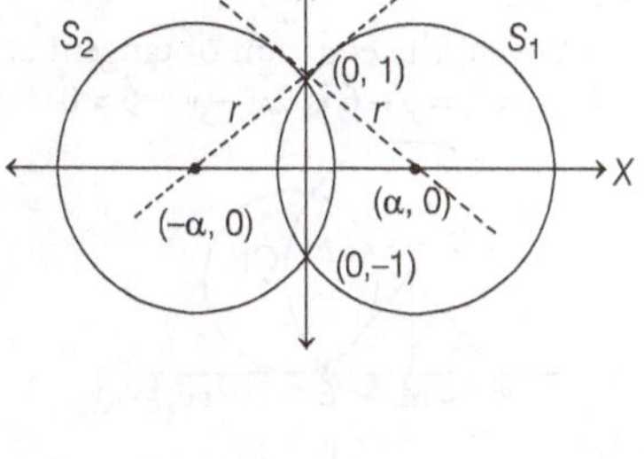

Let $C_1(\alpha, 0)$ and $C_2(-\alpha, 0)$ be the centres.

Then, $S_1 \equiv (x-\alpha)^2 + y^2 = \alpha^2 + 1$

$\Rightarrow \quad S_1 \equiv x^2 + y^2 - 2\alpha x - 1 = 0$

[∵ Radius, $r = \sqrt{(\alpha - 0)^2 + (0-1)^2}$ ]

and $\quad S_2 \equiv (x+\alpha)^2 + y^2 = \alpha^2 + 1$

$\Rightarrow \quad S_2 \equiv x^2 + y^2 + 2\alpha x - 1 = 0$

Now, $2(\alpha)(-\alpha) + 2\cdot 0 \cdot 0 = (-1) + (-1)$

$\Rightarrow \alpha = \pm 1$

[∵ Condition of orthogonality is $2g_1g_2 + 2f_1f_2 = c_1 + c_2$]

$\therefore \; C_1(1, 0)$ and $C_2(-1, 0) \Rightarrow C_1C_2 = 2$

**59.** *(d)*

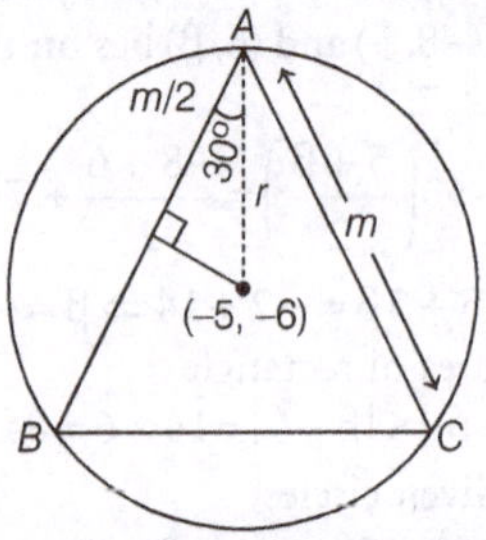


Let the sides of equilateral triangle inscribed in the circle be $a$, then

$$\cos 30° = \frac{m}{2r} \Rightarrow \frac{\sqrt{3}}{2} = \frac{m}{2r}$$

$$\Rightarrow \quad m = \sqrt{3}r$$

Then, area of the equilateral triangle

$$= \frac{\sqrt{3}}{4} m^2$$

$$= \frac{\sqrt{3}}{4}(\sqrt{3}r)^2 = \frac{3\sqrt{3}}{4} r^2$$

The area of an equilateral triangle $= 27\sqrt{3}$

Then, $27\sqrt{3} = \dfrac{3\sqrt{3}}{4} r^2$

$$\Rightarrow \quad r^2 = 36 \Rightarrow r = 6$$

But $\left(-\dfrac{1}{2}\text{ coefficient of } x\right)^2 + \left(-\dfrac{1}{2}\text{ coefficient of } y\right)^2 - \text{constant term} = r^2$

$\Rightarrow (-5)^2 + (-6)^2 - c = 36$

$\Rightarrow c = 25$

**60.** *(a)* Given, the equation of tangent at $(1, 7)$ to $x^2 = y - 6$ is $2x - y + 5 = 0$

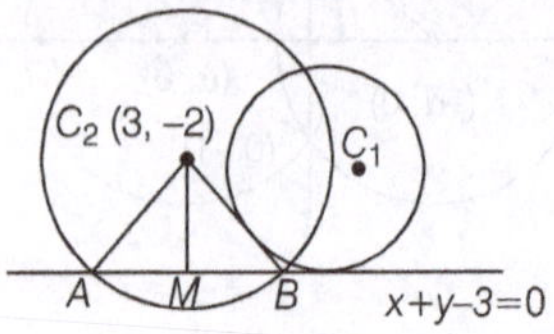


$$x^2 + y^2 + 2gx + 2fy + c = 0$$

On comparing equation, then

The perpendicular distance of centre $(-8, -6)$ to the line $2x - y + 5 = 0$ should be equal to the radius of the circle.

$$\therefore \; \sqrt{64 + 36 - c} = \frac{|-16 + 6 + 5|}{\sqrt{5}}$$

$$\Rightarrow \quad \sqrt{5} = \sqrt{100 - c}$$

$$\Rightarrow \quad c = 95$$

**61.** *(d)* Given, centre of circle $x^2 + y^2 + 2x - 4y - 4 = 0$ is $(-1, 2)$ and radius $= \sqrt{1+4+4} = 3$

Let $(h, k)$ be the centre of another circle.

Now, $\dfrac{h-1}{2} = 2$ and $\dfrac{k+2}{2} = 2$

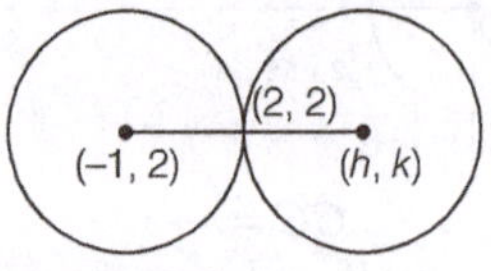


$\Rightarrow h = 4 + 1$ and $k = 4 - 2$

$\Rightarrow h = 5$ and $k = 2$

So, centre of required circle is $(5, 2)$ and radius $= 3$

∴Equation of circle becomes

$(x-5)^2 + (y-2)^2 = (3)^2$

$\Rightarrow x^2 + y^2 - 10x - 4y + 20 = 0 \quad$ ...(i)

Length of intercept made by Eq. (i) on $X$-axis

$$= 2\sqrt{g^2 - c}$$

$$= 2\sqrt{25 - 20} \quad [\because g = -5, c = 20]$$

$$= 2\sqrt{5}$$

**62.** *(c)* Given, $AB = 5$, $AC = 2$

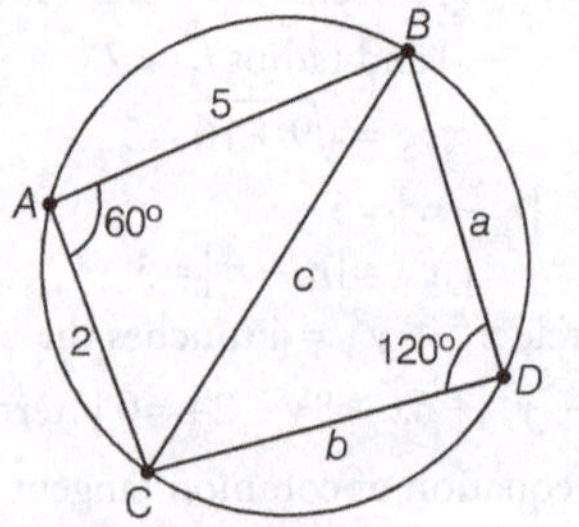


On applying cosine rule,

$\cos\theta = \dfrac{a^2 + b^2 - c^2}{2ab}$ and $\theta = 60°$

$$\Rightarrow \cos 60° = \frac{4 + 25 - c^2}{2 \cdot 2 \cdot 5}$$

$$\Rightarrow \quad c^2 = 19 \Rightarrow c = \sqrt{19}$$

Also, $\cos\theta = \dfrac{a^2 + b^2 - c^2}{2ab}$ and $\theta = 120°$

$$\Rightarrow \quad -\frac{1}{2} = \frac{a^2 + b^2 - 19}{2ab}$$

$$\Rightarrow \quad a^2 + b^2 + ab = 19$$

∴ Area

$$= \frac{1}{2} \times 2 \times 5 \sin 60° + \frac{1}{2} ab \sin 120°$$

$$= 4\sqrt{3}$$

$$\frac{ab}{4} = 4 - \frac{5}{2} = \frac{3}{2}$$

$$\Rightarrow \quad ab = 6$$

$\therefore \; a^2 + b^2 = 13$

$\Rightarrow \; a = 2, b = 3$

Perimeter = Sum of all sides

$= 2 + 5 + 2 + 3 = 12$

**63.** *(b)* Given the equation and coordinates $(0, -2)$

$$x^2 + y^2 - 5x - y + 5 = 0$$

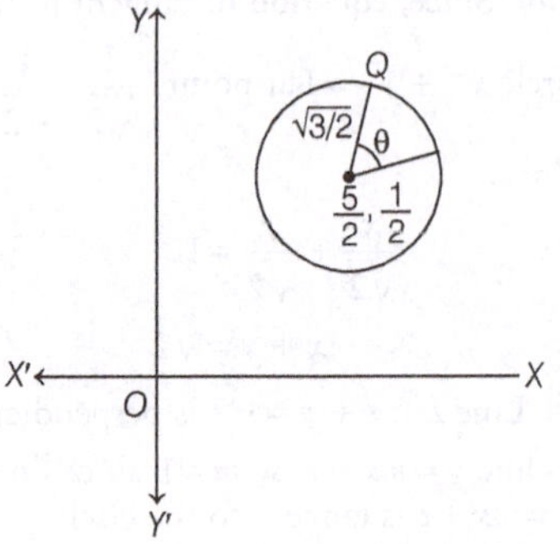


$$\Rightarrow \left(x - \frac{5}{2}\right)^2 + \left(y - \frac{1}{2}\right)^2 = \frac{3}{2}$$

On circle

$$\left[Q \equiv \left(\frac{5}{2} + \sqrt{\frac{3}{2}} \cos\theta, \frac{1}{2} + \sqrt{\frac{3}{2}} \sin\theta\right)\right]$$

$$\Rightarrow PQ^2 = \left(\frac{5}{2} + \sqrt{\frac{3}{2}} \cos\theta\right)^2 + \left(\frac{5}{2} + \sqrt{\frac{3}{2}} \sin\theta\right)^2$$

$$\Rightarrow PQ^2 = \frac{25}{2} + \frac{3}{2} + 5\sqrt{\frac{3}{2}}(\cos\theta + \sin\theta)$$

$$= 14 + 5\sqrt{\frac{3}{2}}(\cos\theta + \sin\theta)$$

∴ Maximum value of $PQ^2$

$[\because \text{Max}(\sin\theta + \cos\theta) = \sqrt{2}]$

$$= 14 + 5\sqrt{\frac{3}{2}} \times \sqrt{2}$$

$$= 14 + 5\sqrt{3}$$

**64.** *(b)* Given, $d=4$

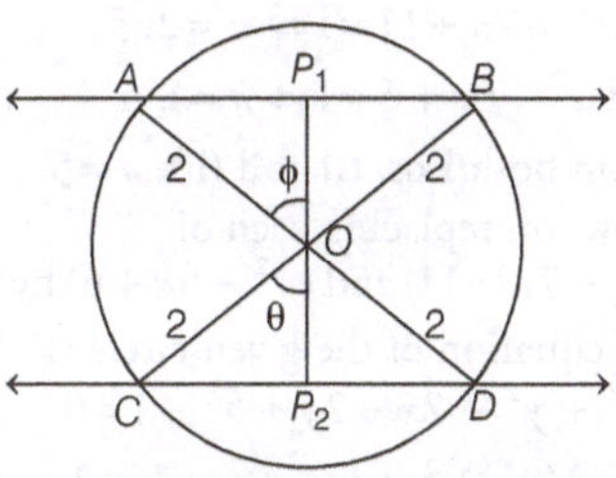


Hence, $\cos 2\theta=\frac{1}{7} \Rightarrow 2\cos^2\theta-1=\frac{1}{7}$

$\Rightarrow \quad 2\cos^2\theta=\frac{8}{7} \Rightarrow \cos^2\theta=\frac{4}{7}$

$\Rightarrow \quad \cos\theta=2/\sqrt{7}$

Also, $\sec^2 2\phi=7 \Rightarrow \frac{1}{2\cos^2\phi-1}=7$

$\Rightarrow 2\cos^2\phi-1=\frac{1}{7} \Rightarrow 2\cos^2\phi=\frac{8}{7}$

$\Rightarrow \cos\phi=2/\sqrt{7}$

$$P_1P_2=r\cos\theta+r\cos\phi$$
$$=\frac{4}{\sqrt{7}}+\frac{4}{\sqrt{7}}=\frac{8}{\sqrt{7}}$$

**65.** *(b)* Given equation of circle is $x^2+y^2-4x+6y-12=0$, whose centre is $(2,-3)$ and radius

$$=\sqrt{2^2+(-3)^2+12}$$
$$=\sqrt{4+9+12}=5$$

Now, according to given information, we have the following figure.

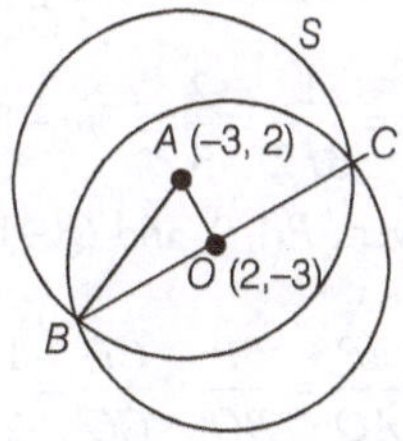


$$x^2+y^2-4x+6y-12=0$$

Clearly, $AO \perp BC$, as $O$ is mid-point of the chord.

Now, in $\Delta AOB$, we have

$$OA=\sqrt{(-3-2)^2+(2+3)^2}$$
$$=\sqrt{25+25}=\sqrt{50}=5\sqrt{2}$$

and $\quad OB=5$

$\therefore \quad AB=\sqrt{OA^2+OB^2}$

$$=\sqrt{50+25}=\sqrt{75}=5\sqrt{3}$$

**66.** *(a)* Given, point of intersection of straight lines $x-y=1$ and $2x+y=3$ is $(4/3, 1/3)$

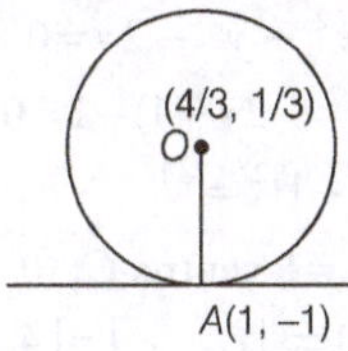


$$\text{Slope of } OA=\frac{\frac{1}{3}+1}{\frac{4}{3}-1}=\frac{\frac{4}{3}}{\frac{1}{3}}=4$$

Now, slope of tangent $=-\frac{1}{4}$

Equation of tangent, $y+1=-\frac{1}{4}(x-1)$

$4y+4=-x+1 \Rightarrow x+4y+3=0$

**67.** *(a)*

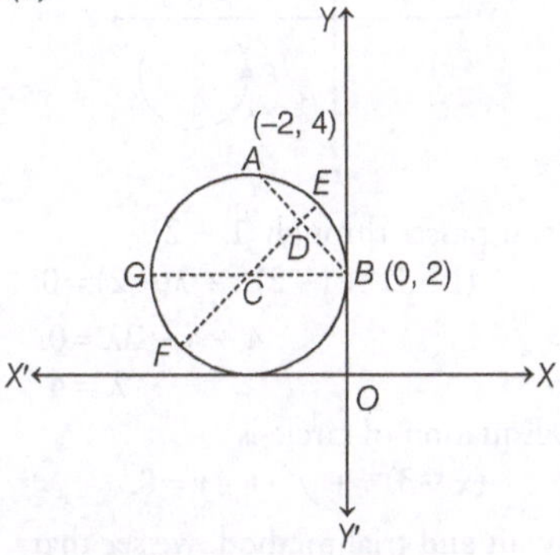


We know that, $EF \perp AB$ and $BG \perp Y$-axis.

Mid-point of chord $AB=(-1, 3)$

Slope of $AB=\frac{4-2}{-2-0}=-1$

$\because \quad EF \perp AB$

$\therefore \quad$ Slope of $EF=1$

Equation of $EF$, $y-3=1(x+1)$

$\Rightarrow \quad y=x+4 \quad$ ...(i)

Equation of $BG$, $y=2 \quad$ ...(ii)

On solving Eqs. (i) and (ii), we get $x=-2$, $y=2$

Since, $C$ be the point of intersection of $EF$ and $BG$, so centre, $C=(-2, 2)$

Hence, $2x-3y+10=0$ is the equation of the diameter.

**68.** *(a)* Intersection point of $2x-3y+4=0$ and $x-2y+3=0$ is $(1, 2)$.

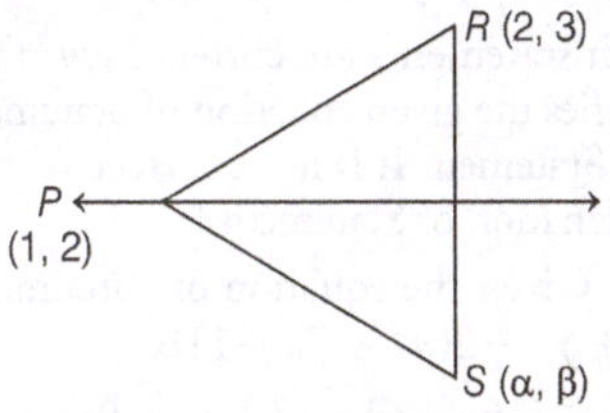


Locus of the image of $R(2, 3)$ is $S(\alpha, \beta)$.

Since, $P$ is the fixed point for given family of lines

So, $PR=PS$

$(\alpha-1)^2+(\beta-2)^2=(2-1)^2+(3-2)^2$

$\Rightarrow (\alpha-1)^2+(\beta-2)^2=1+1=2$

$(x-1)^2+(y-2)^2=(\sqrt{2})^2$

On comparing with the equation $(x-a)^2+(y-b)^2=r^2$

The locus of circle with centre $(1, 2)$ and $r=\sqrt{2}$

**69.** *(a)* $\because r=$ Perpendicular distance from $(1, 1)$ to $3x+4y+3=0$

$$=\left|\frac{3+4+3}{\sqrt{9+16}}\right|=2$$

As, for equilateral triangle incentre, circumcentre and centroid coincide. Therefore, circumcentre is $(1, 1)$.

$\therefore \quad R/r=2/1 \quad$ [$\because XD$ median]

$\therefore \quad R=2r=4$

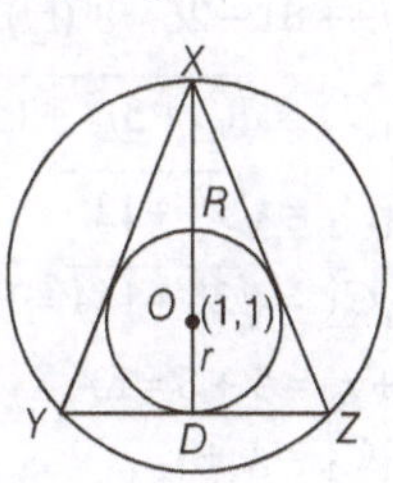


The equation of circle $(x-1)^2+(y-1)^2=16$

$\Rightarrow x^2+y^2-2x-2y-14=0$

**70.** *(a)* Given, tangent of circle $x^2+y^2=16$ and point $P(0, h)$

$$OP=\frac{4}{\sin\theta},\ OB=\frac{4}{\cos\theta}$$

Area of $\Delta APB$

$$=OC\times OB=\frac{16}{\sin\theta\cos\theta}=\frac{32}{\sin 2\theta}$$

For minimum area, maximum value of $\sin 2\theta=1 \Rightarrow \theta=45°$

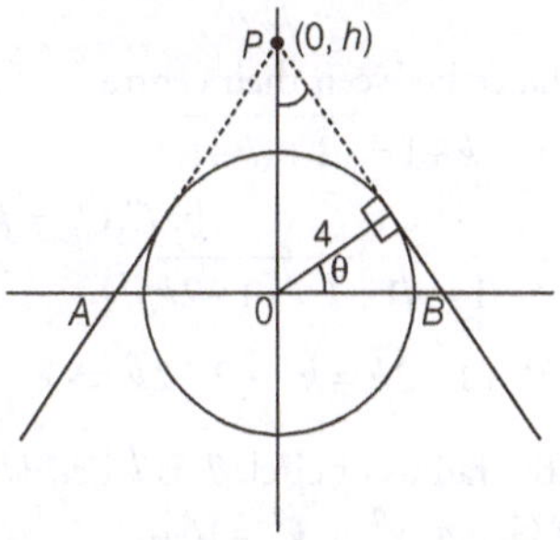


So, $h=\frac{4}{\sin 45°}=4\sqrt{2}$

**71.** *(c)*

**key idea** Number of common tangents depend on the position of the circle with respect to each other.

(i) If circles touch externally

$\Rightarrow$ $C_1C_2 = r_1 + r_2$, 3 common tangents

(ii) If circles touch internally

$\Rightarrow$ $C_1C_2 = r_2 - r_1$, 1 common tangent

(iii) If circles do not touch each other, 4 common tangents

(iv) If one circle is inside other circle

$\Rightarrow$ $|C_1C_2| < |r_2 - r_1|$, No common tangent

Given equations of circles are

$$x^2 + y^2 - 4x - 6y - 12 = 0 \quad \ldots(i)$$
$$x^2 + y^2 + 6x + 18y + 26 = 0 \quad \ldots(ii)$$

Centre of circle (i) is $C_1(2, 3)$ and radius

$$= \sqrt{4+9+12} = 5\,(r_1) \quad \text{(say)}$$

Centre of circle (ii) is $C_2(-3, -9)$ and radius $= \sqrt{9+81-26} = 8\ (r_2)$ (say)

Now, $C_1C_2 = \sqrt{(2+3)^2 + (3+9)^2}$

$\Rightarrow$ $C_1C_2 = \sqrt{5^2 + 12^2}$

$\Rightarrow$ $C_1C_2 = \sqrt{25+144} = 13$

$\therefore$ $r_1 + r_2 = 5 + 8 = 13$

Also, $C_1C_2 = r_1 + r_2$

Thus, both circles touch each other externally. Hence, there are three common tangents.

**72.** *(d)* Let the coordinate of the centre of $T$ be $(0, k)$.

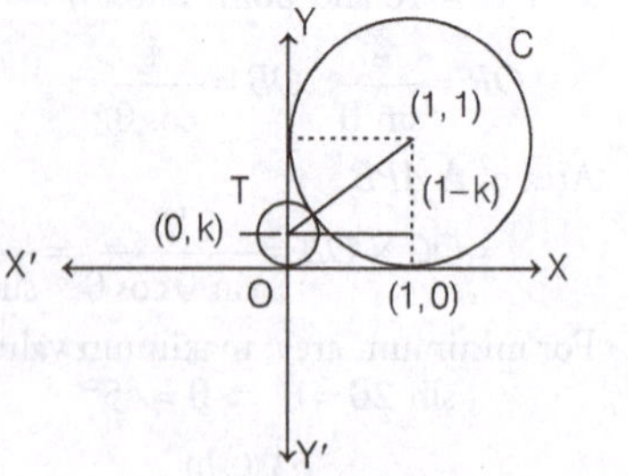


Distance between their centre

$$k+1 = \sqrt{1+(k-1)^2}$$
$$[\because C_1C_2 = k+1]$$

$\Rightarrow$ $k+1 = \sqrt{1+k^2+1-2k}$

$\Rightarrow$ $k^2 + 1 + 2k = k^2 + 2 - 2k \Rightarrow k = \dfrac{1}{4}$

So, the radius of circle $T$ is $k$ i.e., 1/4.

**73.** *(d)* Given, $x^2 + y^2 = 16$ or $x^2 + y^2 = 4^2$

Radius of circle, $r_1 = 4$, centre $C_1(0, 0)$

We have, $x^2 + y^2 - 2y = 0$

$\Rightarrow x^2 + (y^2 - 2y + 1) - 1 = 0$

or $x^2 + (y-1)^2 = 1^2$

Radius, $r^2 = 1$, centre, $C_2(0, 1)$

$|C_1C_2| = 1 \Rightarrow |r_2 - r_1| = |4-1| = 3$

$|C_1C_2| < |r_2 - r_1|$

Thus, one circle is inside other circle.Hence, there is no common tangent.

**74.** *(c)* Let the equation of circle be

$$(x-3)^2 + (y-0)^2 + \lambda y = 0$$

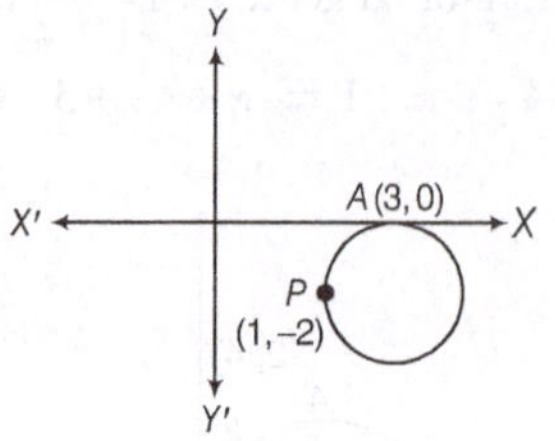


As, it passes through (1, – 2).

$\therefore$ $(1-3)^2 + (-2)^2 + \lambda(-2) = 0$

$\Rightarrow$ $4 + 4 - 2\lambda = 0$

$\Rightarrow$ $\lambda = 4$

$\therefore$ Equation of circle is

$$(x-3)^2 + y^2 + 4y = 0.$$

By hit and trial method, we see that point (5, – 2) satisfies equation of circle.

**75.** *(b)* Equation of circle can be rewritten as $x^2 + y^2 = 5/2$

Here, centre is (0, 0) and radius is $\sqrt{5/2}$.

Let common tangent be

$$y = mx + \sqrt{5}/m \quad \ldots(i)$$

The perpendicular from centre to the tangent is equal to radius.

$\therefore$ $\dfrac{\sqrt{5}/m}{\sqrt{1+m^2}} = \sqrt{\dfrac{5}{2}}$

$\Rightarrow$ $m\sqrt{1+m^2} = \sqrt{2}$

$\Rightarrow$ $m^2(1+m^2) = 2$

$\Rightarrow m^4 + m^2 - 2 = 0$

$\Rightarrow$ $(m^2+2)(m^2-1) = 0 \Rightarrow m = \pm 1$

$$[\because m^2 + 2 \neq 0, \text{ as } m \in R]$$

Thus, from Eq. (i),

$\therefore$ $y = \pm(x + \sqrt{5})$

Both statements are correct as $m = \pm 1$ satisfies the given equation of Statement II but Statement II is not a correct explanation of Statement I.

**76.** *(d)* Given the equation of circumcircle

$$x^2 + y^2 - 2(a^2 - 7a + 11)x$$
$$- 2(a^2 - 6a + 6)y + b^3 + 1 = 0$$

Hence, centre = (1, 1)

$\therefore$ $a^2 - 7a + 11 = 1 \Rightarrow a = 2, 5 \quad \ldots(i)$

and $a^2 - 6a + 6 = 1 \Rightarrow a = 1, 5 \quad \ldots(ii)$

From both Eqs. (i) and (ii), $a = 5$

Now, on replacing each of $(a^2 - 7a + 11)$ and $(a^2 - 6a + 6)$ by 1, the equation of the given circle is

$$x^2 + y^2 - 2x - 2y + b^3 + 1 = 0$$

$\Rightarrow$ $(x-1)^2 + (y-1)^2 + b^3 = 1$

$\Rightarrow$ $b^3 = 1 - [(x-1)^2 + (y-1)^2]$

$\therefore$ $b \in (-\infty, 1)$

**77.** *(b)* Given the equation of circle

$x^2 + y^2 + 3x = 0$, centre $B = \left(-\dfrac{3}{2}, 0\right)$

Radius = 3/2 units

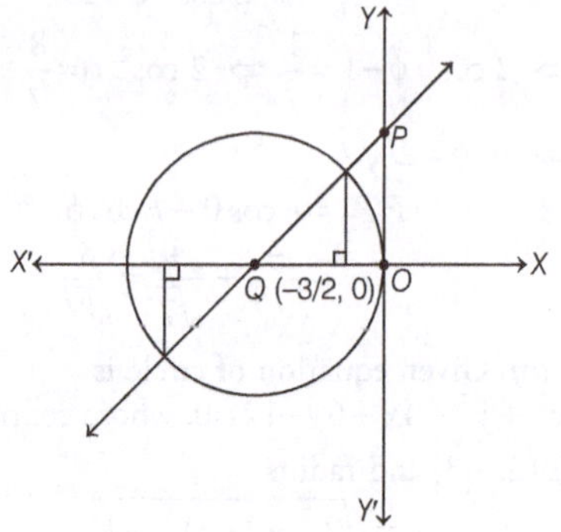


Line : $y = mx + 1$, $y$-intercept of the line $= 1$

$\therefore$ $P = (0, 1)$

Slope of line, $m = \tan\theta = \dfrac{OP}{OQ}$

$\Rightarrow$ $m = \dfrac{1}{3/2} = \dfrac{2}{3} \Rightarrow 3m - 2 = 0$

**78.** *(a)* Given, $P(1, 0)$ and $Q(-1, 0)$, $A(x, y)$.

Given: $\dfrac{AP}{AQ} = \dfrac{BP}{BQ} = \dfrac{CP}{CQ} = \dfrac{1}{2}$

$\Rightarrow$ $2AP = AQ \Rightarrow 4(AP)^2 = AQ^2$

$\Rightarrow$ $4[(x-1)^2 + y^2] = (x+1)^2 + y^2$

$\Rightarrow$ $4(x^2 + 1 - 2x) + 4y^2$

$$= x^2 + 1 + 2x + y^2$$

$\Rightarrow$ $3x^2 + 3y^2 - 8x - 2x + 4 - 1 = 0$

$\Rightarrow$ $x^2 + y^2 - \dfrac{10}{3}x + 1 = 0 \quad \ldots(i)$

$A$ lies on the circle given by Eq. (i). As $B$ and $C$ also follow the same condition they must lie on the same circle.

$\therefore$ Centre of circumcircle of $\Delta ABC$= centre of circle given by (i) $= \left(\dfrac{5}{3}, 0\right)$

**79.** (a) Given

(i) A circle which touches X-axis at the point (1, 0).

(ii) The circle also passes through the point (2, 3).

To find the length of the diameter of the circle.

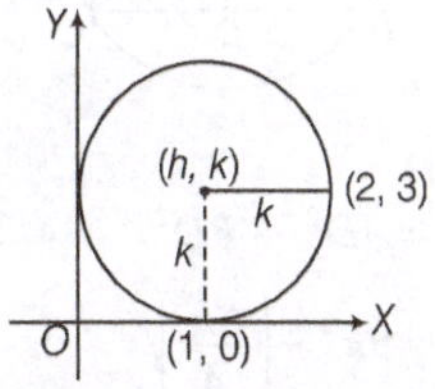


Let us assume that the coordinates of the centre of the circle are $C(h, k)$ and its radius is $r$.

Now, since the circle touches $X$-axis at (1, 0), hence its radius should be equal to ordinate of centre.

$\Rightarrow \quad r = k$

Hence, the equation of the circle is

$$(x-h)^2 + (y-k)^2 = k^2$$

Also, given that the circle passes through points (1, 0) and (2, 3). Hence, substituting them in the equation of the circle, we get

$$(1-h)^2 + (0-k)^2 = k^2 \quad \ldots(i)$$

$$(2-h)^2 + (3-k)^2 = k^2 \quad \ldots(ii)$$

From Eq. (i), we get $h = 1$

On substituting in Eq. (ii), we get

$$(2-1)^2 + (3-k)^2 = k^2$$

$\Rightarrow \quad k = 5/3$

The diameter of the circle is $2k = 10/3$

**80.** (a) $x^2 + y^2 - ax = 0$ and $x^2 + y^2 = c^2$ touch each other.

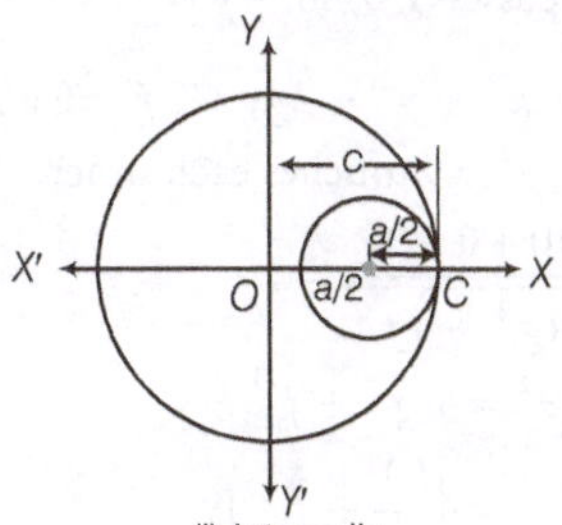


(i) Internally

(i) If circles touch internally,

$$\left| c - \frac{a}{2} \right| = \frac{a}{2}$$

$$\Rightarrow c - \frac{a}{2} = \frac{a}{2}$$

$\Rightarrow \quad c = a,\ c > 0$

$\therefore \quad |a| = c$

(ii) If circles touch externally,

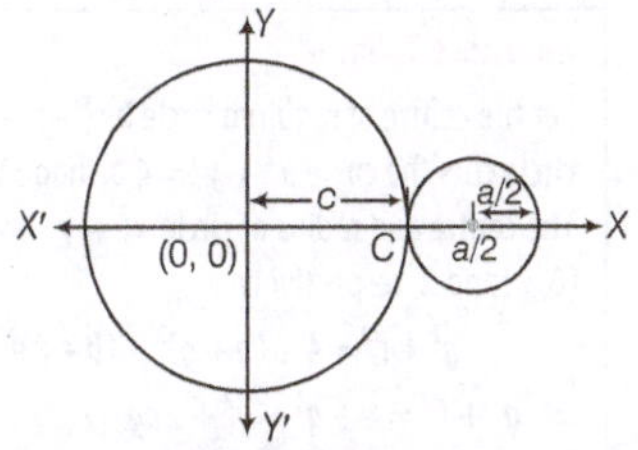


(ii) Externally

$$\left| c + \frac{a}{2} \right| = \frac{a}{2} \Rightarrow c + \frac{a}{2} = \frac{a}{2}$$

$\therefore c = 0$, i.e., not possible as $c > 0$.

Hence, the circles should touch internally and $|a| = c$.

**81.** (a) Given that $P(1, 0)$, $Q(-1, 0)$

and $\dfrac{AP}{AQ} = \dfrac{BP}{BQ} = \dfrac{CP}{CQ} = \dfrac{1}{3}$

$\Rightarrow \ 3AP = AQ$

Let $A = (x, y)$, then $3AP = AQ$

$\Rightarrow \quad 9AP^2 = AQ^2$

$\Rightarrow \quad 9(x-1)^2 + 9y^2 = (x+1)^2 + y^2$

$$\Rightarrow \quad x^2 + y^2 - \frac{20}{8}x + 1 = 0 \quad \ldots(i)$$

$\therefore$ $A$ lies on the circle given by Eq. (i).

As, $B$ and $C$ also follow the same condition.

$\therefore$Centre of circumcircle of $\Delta ABC$.

= Centre of circle given by Eq. (i)

$$= \left(\frac{5}{4}, 0\right)$$

**82.** (d) If $(h, k)$ are the co-ordinates of the centre of the circles and it touch the $X$-axis, then the equation of circle is

$(x-h)^2 + (y-k)^2 = k^2$

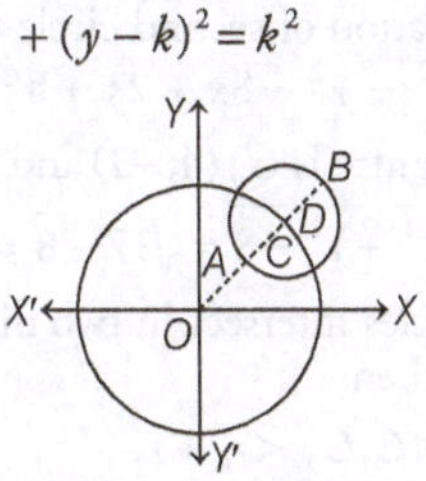


($\because$ Radius of circle $= k$ because circle is tangent to $X$-axis)

$\because$ Equation of circle passing through $(-1, 1)$

$\therefore \quad (-1-h)^2 + (1-k)^2 = k^2$

$\Rightarrow \quad h^2 + 2h - 2k + 2 = 0$

$D \geq 0$

$\therefore \ (2)^2 - 4 \times 1 \cdot (-2k + 2) \geq 0$

$\Rightarrow \quad 1 + 2k - 2 \geq 0$

$\Rightarrow \quad k \geq \dfrac{1}{2}$

**83.** (c) The given equations of diameters are

$$3x - 4y - 7 = 0 \quad \ldots(i)$$

and $\quad 2x - 3y - 5 = 0 \quad \ldots(ii)$

On solving Eqs. (i) and (ii), we get

$$x = 1$$

and $\quad y = -1$

So, the intersection of two diameters is the centre of circle, is $(1, -1)$.

$\Rightarrow \quad \pi r^2 = 49\pi$

$\Rightarrow \quad r = 7$ units

$\therefore$ Equation of required circle is

$(x-1)^2 + (y+1)^2 = 49$

$\Rightarrow \quad x^2 + y^2 - 2x + 2y - 47 = 0$

**84.** (c) Let the coordinates of a point $P$ be $(h, k)$ which is mid-point of the chord $AB$.

Now, $OP = \sqrt{(h-0)^2 + (k-0)^2}$

$= \sqrt{h^2 + k^2}$

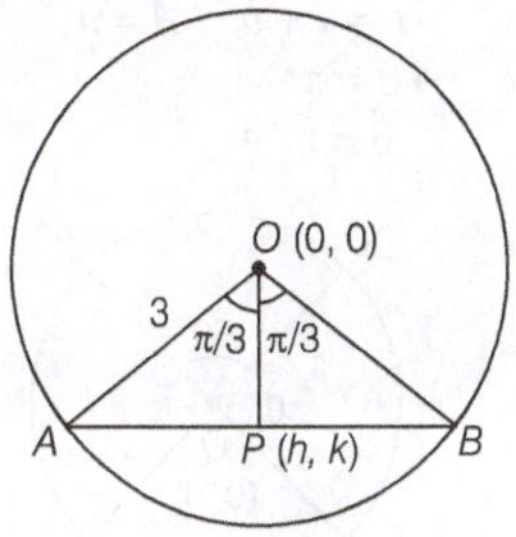


In $\Delta AOP$, $\cos\dfrac{\pi}{3} = \dfrac{OP}{OA}$

$$\Rightarrow \frac{1}{2} = \frac{\sqrt{h^2 + k^2}}{3} \Rightarrow h^2 + k^2 = \frac{9}{4}$$

Hence, the required locus is

$$x^2 + y^2 = \frac{9}{4}$$

**85.** (a) Since, circle touches the $X$-axis and also touches the circle with the centre at (0, 3) and radius 2, then

$C_1C_2 = r_1 + r_2$

$\therefore \quad h^2 + (k-3)^2 = (|k| + 2)^2$

$h^2 + k^2 + 9 - 6k = k^2 + 4 + 4|k|$

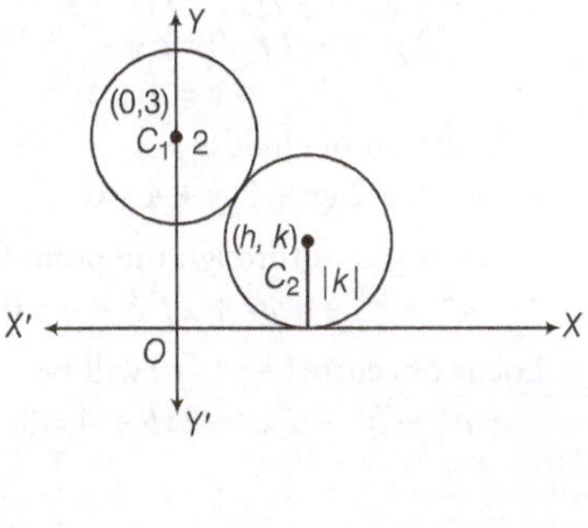

$\therefore$ Locus of centre of circle is

$$x^2 = -5 + 6y + 4|y|$$

$\Rightarrow \quad x^2 = 10y - 5 \quad [\because y > 0]$

This equation represents a parabol *a*. Thus, locus of the centre of the circle is a parabola.

**86.** *(a)* Let the equation of circle be $x^2 + y^2 + 2gx + 2fy + c = 0$. It cuts the circle $x^2 + y^2 = p^2$ orthogonally.

$\because \quad 2g_1g_2 + 2f_1f_2 = c_1 + c_2$

$\therefore \quad 2g(0) + 2f(0) = c - p^2$

$\Rightarrow \quad c = p^2$

Also, it passes through $(a, b)$.

$\therefore \quad a^2 + b^2 + 2ga + 2fb + p^2 = 0$

So, locus of $(-g, -f)$ is

$$a^2 + b^2 - 2ax - 2by + p^2 = 0$$

$\Rightarrow \quad 2ax + 2by - (a^2 + b^2 + p^2) = 0$

**87.** *(a)* Given equation of pair of lines is

$$ax^2 + 2(a+b)xy + by^2 = 0$$

Here, $H = a + b, \quad A = a, \quad B = b$

Since, $4\theta = \pi$

$\Rightarrow \quad \theta = \pi/4$

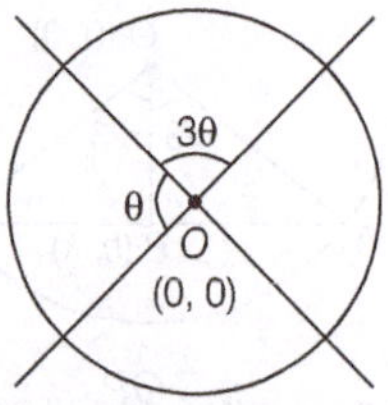


Angle between lines is given by

$$\tan\theta = \frac{2\sqrt{H^2 - AB}}{A + B}$$

$$\Rightarrow \tan\frac{\pi}{4} = 1 = \frac{2\sqrt{(a+b)^2 - ab}}{a + b}$$

$\Rightarrow a^2 + b^2 + 2ab = 4(a^2 + b^2 + ab)$

$\Rightarrow 3a^2 + 3b^2 + 2ab = 0$

**88.** *(b)* Let the equation of circle be

$$x^2 + y^2 + 2gx + 2fy + c = 0$$

It cuts the circle $x^2 + y^2 = 4$ orthogonally, if

$$2g_1g_2 + 2f_1f_2 = c_1 + c_2$$

$\therefore \quad 2g \cdot 0 + 2f \cdot 0 = c - 4$

$\Rightarrow \quad c = 4$

$\therefore$ Equation of circle is

$x^2 + y^2 + 2gx + 2fy + 4 = 0$

Since, it passes through the point $(a, b)$.

$\therefore \quad a^2 + b^2 + 2ag + 2fb + 4 = 0$

Locus of centre $(-g, -f)$ will be

$$a^2 + b^2 - 2xa - 2yb + 4 = 0$$

$\Rightarrow 2ax + 2by - (a^2 + b^2 + 4) = 0$

**Alternate Solution**

Let the centre of required circle be $(-g, -f)$. This circle cuts the circle $x^2 + y^2 = 4$ orthogonally. The centre and radius of circle $x^2 + y^2 = 4$ are $(0, 0)$ and 2, respectively.

$\therefore \quad g^2 + f^2 = 4 + (a+g)^2 + (b+f)^2$

$\Rightarrow g^2 + f^2 = 4 + a^2 + g^2 + 2ag + 2bf + b^2 + f^2$

$\Rightarrow \quad 4 + a^2 + b^2 + 2ag + 2bf = 0$

Hence, the locus of centre is

$$2ax + 2by - (a^2 + b^2 + 4) = 0$$

**89.** *(a)* Given, equation of line is

$$y = x \quad \ldots(i)$$

And equation of circle is

$$x^2 + y^2 - 2x = 0 \quad \ldots(ii)$$

From Eqs. (i) and (ii), we get

$$x^2 + x^2 - 2x = 0$$

$\Rightarrow \quad x = 0, \quad x = 1$

On putting the values of $x$ in Eq. (i) respectively, we get

$$y = 0, \quad y = 1$$

Let coordinates of $A$ be (0, 0) and coordinates of $B$ be (1, 1).

$\therefore$ Equation of circle when $AB$ as a diameter, is

$(x - 0)(x - 1) + (y - 0)(y - 1) = 0$

$\Rightarrow \quad x^2 + y^2 - x - y = 0$

**90.** *(a)* The equation of first circle

$$(x-1)^2 + (y-3)^2 = r^2$$

whose centre is $C_1$ (1, 3) and radius $r_1 = r$.

and equation of second circle

$$x^2 + y^2 - 8x + 2y + 8 = 0$$

whose centre is $C_2$ (4, –1) and radius

$$r_2 = \sqrt{4^2 + 1^2 - 8} = \sqrt{17 - 8} = \sqrt{9} = 3$$

Two circles intersect in two distinct points, then

$r_1 - r_2 < C_1C_2 < r_1 + r_2$

$\Rightarrow \quad r - 3 < \sqrt{(4-1)^2 + (-1-3)^2} < r + 3$

$\Rightarrow \quad r - 3 < \sqrt{9 + 16} < r + 3$

$\Rightarrow \quad r - 3 < 5 < r + 3$

$\Rightarrow \quad r - 3 < 5 \quad \text{and} \quad 5 < r + 3$

$\Rightarrow \quad r < 8 \quad \text{and} \quad 2 < r$

$\therefore \quad 2 < r < 8$

**91.** *(c)* Given, $ABC$ be an equilateral triangle, whose median is $AD$.

So, $AD \perp BC$

Given, $AD = 3a$

Let $AB = BC = AC = x$

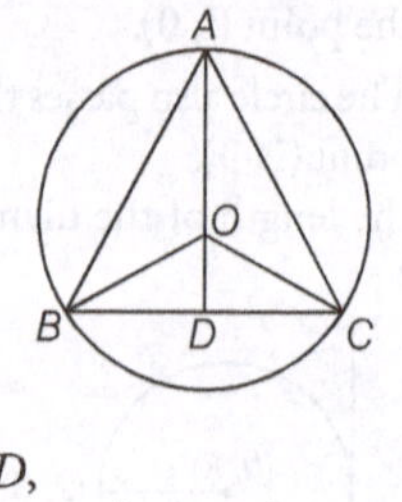


In $\Delta ABD$,

$$AB^2 = AD^2 + BD^2$$

$\Rightarrow \quad x^2 = 9a^2 + \left(\frac{x^2}{4}\right)$

$\Rightarrow \frac{3}{4}x^2 = 9a^2$

$\Rightarrow \quad x^2 = 12a^2$

In $\Delta OBD$,

$$r^2 = (3a - r)^2 + \frac{x^2}{4}$$

$\Rightarrow \quad 6ar = 12a^2$

$\Rightarrow \quad r = 2a$

So, equation of circle is $x^2 + y^2 = 4a^2$

**92.** *(b)* The equation of circle be $x^2 + y^2 + 2gx + 2fy + c = 0$.

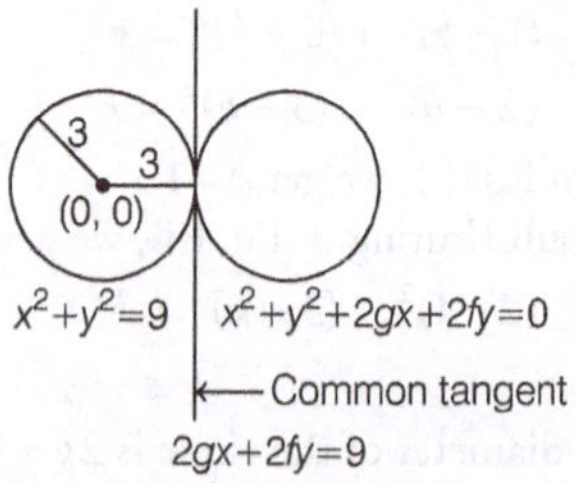


As, it passes through (0, 0), so $c = 0$ and as it passes (1, 0) so, $-g = \frac{1}{2}$.

Now, $x^2 + y^2 + 2gx + 2fy = 0$ and $x^2 + y^2 = 9$ touches each other.

$$\therefore \frac{|0 + 0 - 9|}{\sqrt{4g^2 + 4f^2}} = 3$$

$\Rightarrow \quad 9^2 = 4(g^2 + f^2)9$

$$= 4\left(\frac{1}{4} + f^2\right)9$$

$\Rightarrow \quad 9 = 1 + 4f^2$

$\Rightarrow \quad f^2 = 2$

$\therefore \quad f = \pm\sqrt{2}$

$\therefore$ Centre of the required circle be $\left(\frac{1}{2}, \sqrt{2}\right)\left(\frac{1}{2}, -\sqrt{2}\right)$

**93.** *(d)* First of all we have to write equation of line $CD$.

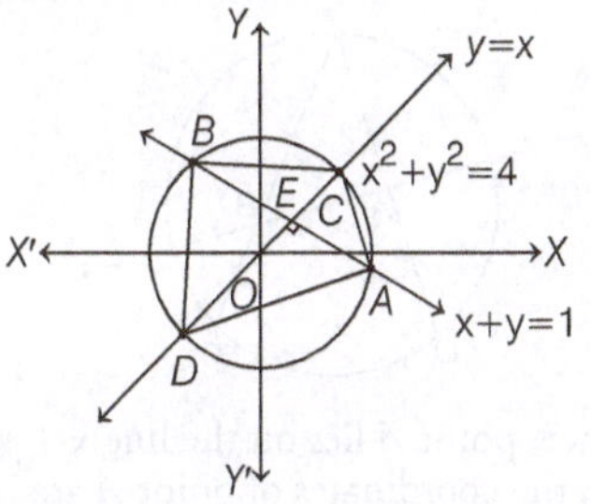


As $\quad CD \perp AB$

$$\therefore \quad M_{CD} = \frac{-1}{m_{AB}} = \frac{-1}{-1} = 1$$

and $CD$ passes through mid-point of $AB$.
when we solve $x + y = 1$ with $x^2 + y^2 = 4$.
then we get coordinats of $A$ and $B$

i.e. $A \equiv \left(\frac{(1+\sqrt{7})}{2}, \frac{1-\sqrt{7}}{2}\right)$ and

$B \equiv \left(\frac{1-\sqrt{7}}{2}, \frac{1+\sqrt{7}}{2}\right)$ and thus, we find that mid-point of $AB$ is $\left(\frac{1}{2}, \frac{1}{2}\right)$.

So, equation of line $CD$ will be

$$y - \frac{1}{2} = 1\left(x - \frac{1}{2}\right)$$

$$\Rightarrow \quad y = x \quad \ldots \text{(i)}$$

Now, solving $y = x$ with $x^2 + y^2 = 4$ to get coordinates of $C$ and $D$. Then,

$$x^2 + x^2 = 4$$

$$\Rightarrow \quad x^2 = 2$$

$$\Rightarrow \quad x = \pm\sqrt{2}$$

$$\therefore \quad y = \pm\sqrt{2}$$

So, $C \equiv (\sqrt{2}, \sqrt{2})$ and $D \equiv (-\sqrt{2}, -\sqrt{2})$ [$\because C$ lies in first quadrant and $D$ lies in third quadrant]

Now, area of quadrilateral $ADBC$

$$= 2 \times (\text{area of } \Delta BCD)$$

$$= 2 \times \frac{1}{2}\begin{vmatrix} \sqrt{2} & \sqrt{2} & 1 \\ \frac{1-\sqrt{7}}{2} & \frac{1+\sqrt{7}}{2} & 1 \\ -\sqrt{2} & -\sqrt{2} & 1 \end{vmatrix}$$

$$= 2\sqrt{14} \text{ sq unit}$$

**94.** *(d)* Centre and radius of given circle are $(1, -2)$ and 3, respectively.

Now, image of $(1, -2)$ in $2x - 3y + 5 = 0$ is

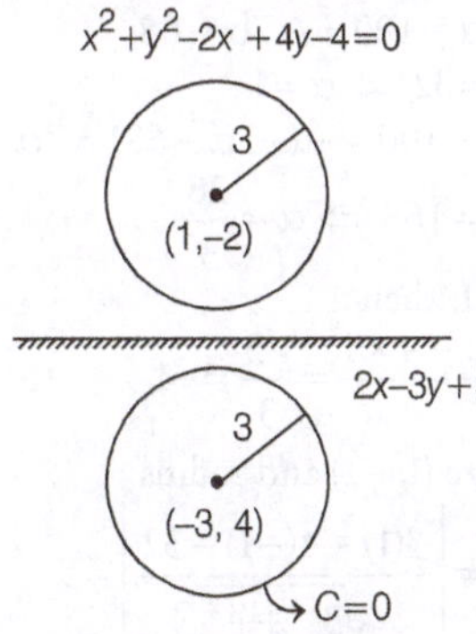


$$\frac{x-1}{2} = \frac{y+2}{-3} = \frac{-2(2+6+5)}{13} = -2$$

$\Rightarrow x = -3, y = 4$

$\therefore$ Centre of $C = 0$ is $(-3, 4)$ and radius of $C = 0$ will be 3 (as given circle)

So, equation of $C_2$ is

$(x+3)^2 + (y-4)^2 = (3)^2$

According to the question,

$$l(\text{arc } AB) = \frac{1}{6} \times 2\pi r$$

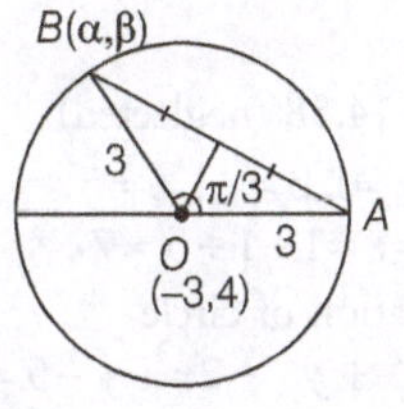


$$\Rightarrow r\theta = \frac{\pi r}{3} \quad (\because l = r\theta)$$

$$\Rightarrow \quad \theta = \pi/3$$

$\because (\alpha, \beta)$ lies on $C = 0$,

$$\therefore (\alpha+3)^2 + (\beta-4)^2 = 9 \quad \ldots \text{(i)}$$

$$(\alpha+6)^2 + (\beta-4)^2 = 27 \quad \ldots \text{(ii)}$$

$$\text{(ii)} - \text{(i)} \Rightarrow (\alpha+6)^2 - (\alpha+3)^2 = 18$$

$$\Rightarrow 6\alpha = -9 \Rightarrow \alpha = \frac{-3}{2}$$

$$\therefore \quad \beta = 4 - \frac{3\sqrt{3}}{2}$$

Hence, $\beta - \sqrt{3}\alpha = 4 - \frac{3\sqrt{3}}{2} - \sqrt{3}\left(\frac{-3}{2}\right)$

$$= 4 - \frac{3\sqrt{3}}{2} + \frac{3\sqrt{3}}{2} = 4$$

**95.** *(c)* Since, $(x_c, y_c)$ be the point of intersection of the lines.

$$3x + 5y = 1 \quad \ldots \text{(i)}$$

and $(2+c)x + 5c^2 y = 1 \quad \ldots$(ii)

On solving Eqs. (i) and (ii), we get

$$(2+c)x + 5c^2\left(\frac{1-3x}{5}\right) = 1$$

$$\Rightarrow x = \frac{1-c^2}{2+c-3c^2} = \frac{(1-c)(1+c)}{(1-c)(2+3c)}$$

$$\Rightarrow x = \frac{1+c}{2+3c} \quad \left(\because x_c = \frac{1+c}{2+3c}\right)$$

and $y = \frac{1-3x}{5} = \frac{-1}{5(3c+2)} = y_c$

Now, $h = \lim_{c \to 1} x_c = \lim_{c \to 1} \frac{(1+c)}{2+3c} = \frac{2}{5}$

$$k = \lim_{c \to 1} y_c = \lim_{c \to 1}\left[\frac{-1}{5(3c+2)}\right] = \frac{-1}{25}$$

$$\therefore \text{Centre} = \left(\frac{2}{5}, \frac{-1}{25}\right),$$

$$\text{Radius} = \sqrt{\left(2 - \frac{2}{5}\right)^2 + \left(0 - \frac{1}{25}\right)^2}$$

$$= \sqrt{\frac{64}{25} + \frac{1}{625}} = \frac{\sqrt{1601}}{25}$$

So, equation of the circle is

$$\left(x - \frac{2}{5}\right)^2 + \left(y + \frac{1}{25}\right)^2 = \frac{1601}{(25)^2}$$

$$\Rightarrow 25x^2 + 25y^2 - 20x + 2y - 60 = 0$$

**96.** *(b)*

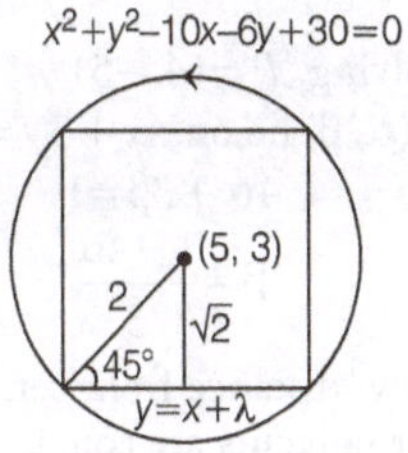


Given equation of circle is

$$x^2 + y^2 - 10x - 6y + 30 = 0$$

$\therefore$ Centre of the circle

$$= \left(-\frac{1}{2} \times (-10), -\frac{1}{2} \times (-6)\right) = (5, 3)$$

Also, one side of square is parallel to the line $y = x + 3$.

So, equation of that side of square is given by, $y = x + \lambda$.

Now, distance of $(5, 3)$ to the line $y = x + \lambda$ is $\sqrt{2}$.

$$\Rightarrow \left|\frac{3-5-\lambda}{\sqrt{2}}\right| = \sqrt{2} \Rightarrow |\lambda + 2| = 2$$

$$\Rightarrow \quad \lambda + 2 = \pm 2 \Rightarrow \lambda = 0, -4$$

So, the lines are $y = x$ and $y = x - 4$

Now, solving these lines with the circles as

$y = x$ and $x^2 + y^2 - 10x - 6y + 30 = 0$,

we get

$x=3, y=3$ and $x=5, y=5$

(by putting $y=x$ in the equation of circle)

and $y=x-4$

and $x^2+y^2-10x-6y+30=0$,

we get $x=7, y=3$ and $x=5, y=1$

So, vertices are (5, 5), (3, 3), (7, 3), (5, 1).

Hence, $\Sigma x_i^2+y_i^2=9+9+25+25+25+1+49+9=152$

**97.** *(c)*

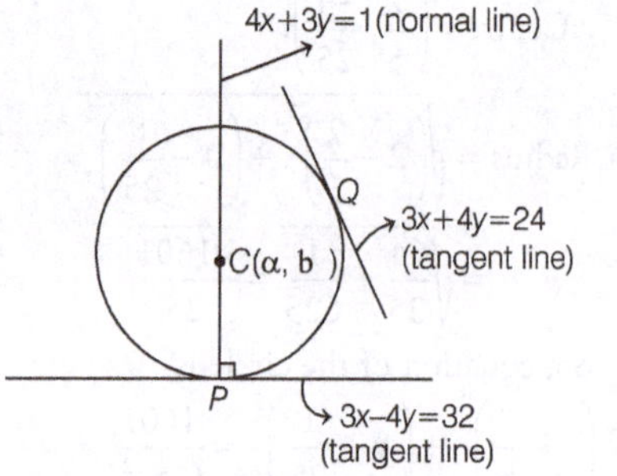


First solving equations $4x+3y=1$ and $3x-4y=32$ to get interchanging point $P$.

After solving, $P=(4,-5)$

Centre $(\alpha, \beta)$ lie on $4x+3y=1$, so that

$$4\alpha+3\beta=1$$

$$\Rightarrow \quad \beta=\frac{1-4\alpha}{3}$$

We know, distance from centre of two different tangents are equal.

So,distance from the centre to the tangent line $3x-4y=32$ and $3x+4y=24$ are equal.

Thus, $$\left|\frac{3\alpha-4\left(\frac{1-4\alpha}{3}\right)-32}{5}\right|=\left|\frac{3\alpha+4\left(\frac{1-4\alpha}{3}\right)-24}{5}\right|$$

$$\Rightarrow \left|\frac{\frac{9\alpha-4+16\alpha-96}{3}}{5}\right|$$

$$\Rightarrow \left|\frac{25\alpha-100}{15}\right|=\left|\frac{-7\alpha-68}{15}\right|$$

Either

$$\Rightarrow |25\alpha-100|=|-7\alpha-68|$$

$$25\alpha-100=-7\alpha-68$$

$$\Rightarrow 32\alpha=32 \Rightarrow \alpha=1$$

Or $25\alpha-100=-(-7\alpha-68)=7\alpha+68$

$$\Rightarrow 18\alpha=168 \Rightarrow \alpha=\frac{28}{3}$$

For $\alpha=1$, then

$$\beta=\frac{1-4\times 1}{3}=\frac{-3}{3}=-1$$

So, centre $(1,-1)$ and radius

$$=\left|\frac{3(1)-4(-1)-32}{\sqrt{3^2+4^2}}\right|=5$$

For, $\alpha=28/3$, then

$$\beta=\frac{1-4\times\frac{28}{3}}{3}=\frac{-109}{9}$$

So, centre $\left(\frac{28}{3}, \frac{-109}{9}\right)$ and radius

$$=\left|\frac{3\left(\frac{28}{3}\right)-4\left(\frac{-109}{9}\right)-32}{5}\right|=\frac{656}{45}$$

Radius = 14.58 (neglected)

Hence, $\alpha=1, \beta=-1, n=5$

$\therefore \alpha-\beta=r=1+1+5=7$

**98.** *(d)* Equation of circle

$$x^2+y^2-2x+y-5=0 \quad \ldots(i)$$

On comparing with

$x^2+y^2+2gx+2fy+c=0$

$\therefore \quad 2g=-2, 2f=1, c=-5$

$\Rightarrow \quad g=-1, f=1/2, c=-5$

$\therefore$ Radius of the circle

$$r=\sqrt{(-1)^2+\left(\frac{1}{2}\right)^2+5}=\frac{5}{2}$$

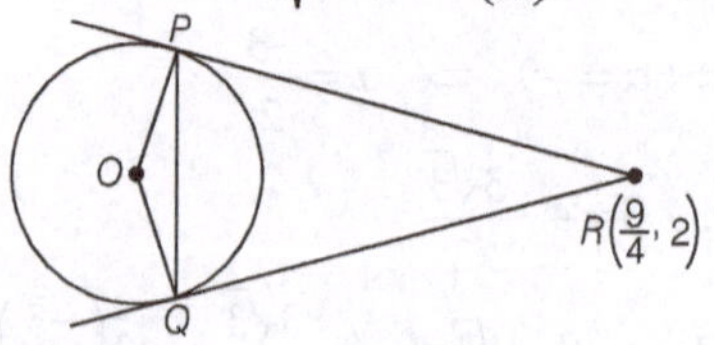


Length of tangent $PR=\sqrt{S_1}$

$$l=\sqrt{\left(\frac{9}{4}\right)^2+(2)^2-2\left(\frac{9}{4}\right)+2-5}=\frac{5}{4}$$

$$\therefore \text{Area of } \Delta PQR=\frac{rl^3}{r^2+l^2}$$

$$=\frac{\left(\frac{5}{2}\right)\left(\frac{5}{4}\right)^3}{\left(\frac{5}{2}\right)^2+\left(\frac{5}{4}\right)^2}=\frac{\frac{5}{2}\times\frac{125}{64}\times 16}{125}=\frac{5}{8}$$

**99.** *(c)* $P\left(\frac{1+a}{2}, \frac{1-a}{2}\right)$

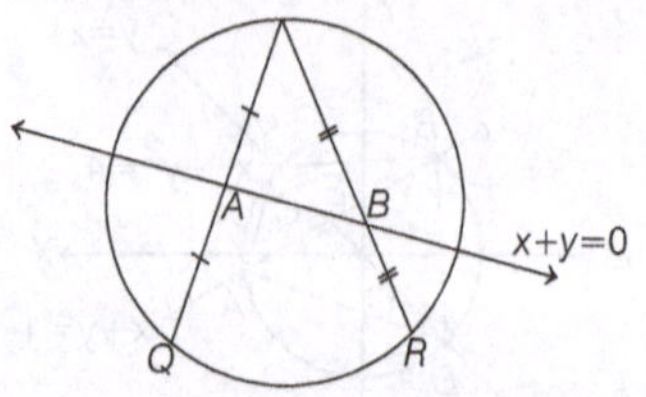


Since, point $A$ lies on the line $x+y=0$. So, the coordinates of point $A$ are $(k,-k)$. Let the coordinates of $Q$ are $(\alpha, \beta)$.

Since, $A$ is mid-point of $PQ$, then

$$\left(\frac{\frac{1+a}{2}+\alpha}{2}, \frac{\frac{1-a}{2}+\beta}{2}\right)=(k,-k)$$

$$\Rightarrow \alpha=2k-\frac{(1+a)}{2} \text{ and } \beta=-2k-\frac{(1-a)}{2}$$

Now, $Q$ lies on circle

$2x^2+2y^2-(1+a)x-(1-a)y=0$,

then $2\alpha^2+2\beta^2-(1+a)\alpha-(1-a)\beta=0$

$$\Rightarrow 2\left[2k-\frac{(1+a)}{2}\right]^2+2\left[-2k-\frac{(1-a)}{2}\right]^2-(1+a)\left[2k-\frac{(1+a)}{2}\right]-(1-a)\left[-2k-\frac{(1-a)}{2}\right]=0$$

$$\Rightarrow 2\left[4k^2+\frac{(1+a)^2}{4}-2k(1+a)\right]+2\left[4k^2+\frac{(1-a)^2}{4}+2k(1-a)\right]-2k(1+a)+\frac{(1+a)^2}{2}+2k(1-a)+\frac{(1-a)^2}{2}=0$$

$$\Rightarrow 16k^2-6k(1+a-1+a)+(1+a)^2+(1-a)^2=0$$

$$\Rightarrow 16k^2-12ka+2(1+a^2)=0$$

$$\Rightarrow \quad 8k^2-6ka+(1+a^2)=0$$

Since, $x+y=0$ bisects two distinct chords, so

$$D>0$$

$$\Rightarrow \quad 36a^2-32(1+a^2)>0$$

$$\Rightarrow \quad 36a^2-32-32a^2>0$$

$$\Rightarrow \quad 4a^2>32 \Rightarrow a^2>8$$

$$\therefore \quad a^2\in(8,\infty)$$

**100.** *(c)* Given, equation of circle is

$x^2+y^2-4x-6y+11=0$

Equation of tangent to above circle at (3, 2) is

$$3x + 2y - 2(x+3) - 3(y+2) + 11 = 0$$

$$\Rightarrow x - y - 1 = 0$$

When the circle $x^2 + y^2 - 4x - 6y + 11 = 0$ is rolled 4 units upward at the tangent $x - y - 1 = 0$, the centre of new circle $c_1$ becomes $(2 + 2\sqrt{2}, 3 + 2\sqrt{2})$ and radius remains same $\sqrt{21}$.

Since, $c_2$ is image of $c_1$ in tangent $x - y - 1 = 0$.

So, centre of $c_2$ becomes $(4 + 2\sqrt{2}, 1 + 2\sqrt{2})$.

$\therefore$ $A$ and $B$ are $(2 + 2\sqrt{2}, 3 + 2\sqrt{2})$ and $(4 + 2\sqrt{2}, 1 + 2\sqrt{2})$, respectively.

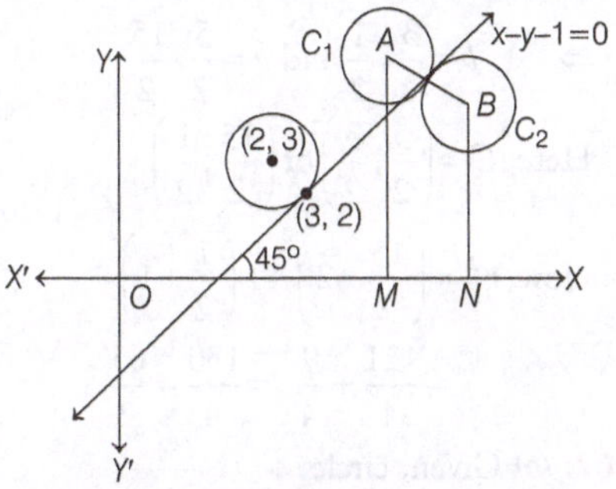


Again foot of perpendicular from $A$ and $B$ on $X$-axis are $M(2 + 2\sqrt{2}, 0)$ and $N(4 + 2\sqrt{2}, 0)$.

$\therefore$ Area of trapezium $AMNB = \frac{1}{2} \times$ (Sum of parallel sides) $\times$ (Perpendicular distance between them)

$$= \frac{1}{2} \times (AM + BN) \times MN$$

$$= \frac{1}{2} \times (3 + 2\sqrt{2} + 1 + 2\sqrt{2}) \times 2$$

$$= 4 + 4\sqrt{2} = 4(1 + \sqrt{2})$$

**101.** *(b)*

**key idea** The equation of tangent at point $p(x_1, y_1)$ to a circle $x^2 + y^2 + 2gx + 2fy + c = 0$ is $xx_1 + yy_1 + g(x + x_1) + f(y + y_1) + c = 0$

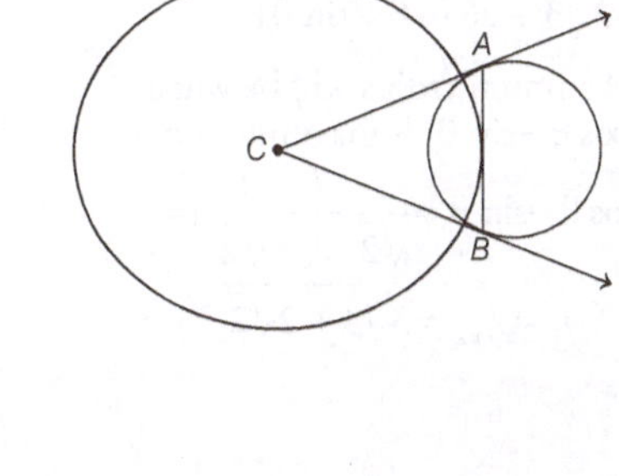


We have, circle

$$x^2 + y^2 - 3x + 10y - 15 = 0$$

Tangent at $A(4, -11)$ will be $T = 0$

i.e. $x(4) + y(-11) - \frac{3}{2}(x + 4) + 5(y - 11) - 15 = 0$

i.e. $5x - 12y - 152 = 0$ ... (i)

Tangent at $B(8, -5)$ will be $T = 0$

i.e. $x(8) + y(-5) - \frac{3}{2}(x + 8) + 5(y - 5) - 15 = 0$

i.e. $x = 8$ ...(ii)

Since, these tangents intersect at $C$.

On solving Eqs. (i) and (ii), we get

$$x = 8, \; y = \frac{-28}{3}$$

i.e. $C\left(8, -\frac{28}{3}\right)$

Using two points of the line, equation of $AB$ is

$$(y + 5) = \frac{(-11) + 5}{4 - 8} \times (x - 8)$$

i.e. $2(y + 5) = 3(x - 8)$

i.e. $3x - 2y - 34 = 0$ ...(iii)

Since, this line is tangent to the circle with centre $C$.

So, radius of this circle = Perpendicular distance of $C$ from $AB$

$$\Rightarrow \text{Radius} = \frac{\left|3 \cdot 8 + 2 \cdot \frac{28}{3} - 34\right|}{\sqrt{3^2 + (-2)^2}} \text{ units}$$

$\because$ Distance of $(x_1, y_1)$ from $ax + by + c = 0$ is $\frac{|ax_1 + by_1 + c|}{\sqrt{a^2 + b^2}}$ units

$\left(\because \text{ Distance of } (x_1, y_1) \text{ from } ax + by + c = 0 \text{ is units}\right)$

$$\Rightarrow \text{Radius} = \frac{26}{3\sqrt{13}} \text{ units}$$

$$= \frac{2\sqrt{13}}{3} \text{ units}$$

**102.** *(c)* Given line, $ax + by = 0$ ...(i)

$(a \neq b)$ and circle

$$x^2 + y^2 - 2x = 0 \quad \text{...(ii)}$$

From Eq. (i), $y = -\frac{ax}{b}$ puting in Eq. (ii), we get

$$x^2 + \frac{a^2}{b^2}x^2 - 2x = 0$$

$$\Rightarrow x\left(\frac{a^2 + b^2}{b^2}x - 2\right) = 0$$

$$x = 0 \text{ or } x = \frac{2b^2}{a^2 + b^2}$$

Now, circle and line cuts at $A(\alpha, 0)$ and $B(1, \beta)$.

So, only possible value is $\alpha = 0$ and

$$\frac{2b^2}{a^2 + b^2} = 1$$

$$a^2 = b^2 \Rightarrow a = -b \quad [\because a \neq b]$$

Line $ax + by = 0$

$ax - ay = 0 \Rightarrow x = y$

So, points $A(0, 0)$ and $B(1, 1)$.

Now, equation of circle with $AB$ as diameter

$$(x - 0)(x - 1) + (y - 0)(y - 1) = 0$$

$$x^2 + y^2 - x - y = 0$$

Centre $\left(\frac{1}{2}, \frac{1}{2}\right)$; $r = \sqrt{\frac{1}{4} + \frac{1}{4}} = \frac{1}{\sqrt{2}}$

Let image of $\left(\frac{1}{2}, \frac{1}{2}\right)$ in the line $x + y + 2 = 0$ is $(x, y)$.

$$\frac{x - \frac{1}{2}}{1} = \frac{y - \frac{1}{2}}{1} = \frac{-2\left(\frac{1}{2} + \frac{1}{2} + 2\right)}{1^2 + 1^2}$$

$$x - \frac{1}{2} = y - \frac{1}{2} = -3$$

$$x = y = -3 + \frac{1}{2} = \frac{-5}{2}$$

So, equation of image circle

$$\left(x + \frac{5}{2}\right)^2 + \left(y + \frac{5}{2}\right)^2 = \left(\frac{1}{\sqrt{2}}\right)^2$$

$$\Rightarrow x^2 + y^2 + 5x + 5y + 12 = 0$$

**103.** *(d)* Given equation is

$$(x - 4)^2 + (y - 5)^2 = 2^2$$

$$\therefore \quad r = 2$$

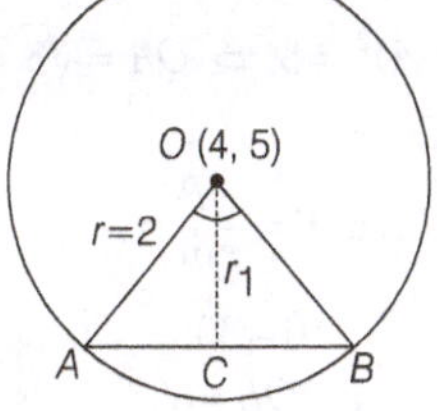


and $O(4, 5)$

$$\because \quad \theta_1 = \frac{\pi}{3}$$

$$\therefore \quad \frac{\theta_1}{2} = \frac{\pi}{6}$$

$\therefore$ From $\Delta OAC$,

$$2\cos\left(\frac{\theta_1}{2}\right) = r_1$$

$$2\cos\frac{\pi}{6} = 2\times\frac{\sqrt{3}}{2} = \sqrt{3}$$

$$\therefore \quad r_1 = \sqrt{3}$$

If $\theta_3 = \dfrac{2\pi}{3}$

$$\therefore 2\cos\left(\frac{\theta_3}{2}\right) = 2\cos\frac{\pi}{3} = 2\times\frac{1}{2} = 1 = r_3$$

$$\because \quad r_1^2 = r_2^2 + r_3^2$$

$$3 = r_2^2 + 1^2$$

$$\Rightarrow r_2^2 = 2 \Rightarrow r_2 = \sqrt{2}$$

$\therefore$ From $\Delta OAC$,

$$2\cos\left(\frac{\theta_2}{2}\right) = \sqrt{2}$$

$$\cos\left(\frac{\theta_2}{2}\right) = \frac{1}{\sqrt{2}}$$

$$\therefore \quad \frac{\theta_2}{2} = \frac{\pi}{4}$$

$$\therefore \quad \theta_2 = \frac{\pi}{2}$$

**104.** *(b)* Given, equation of circle is

$$x^2 + y^2 - 4x + 3 = 0$$

$$\Rightarrow \quad x^2 - 4x + 4 + y^2 = 4 - 3$$

$$\Rightarrow \quad (x-2)^2 + (y-0)^2 = (1)^2$$

Centre of circle is $P(2, 0)$ and radius $= 1$ unit

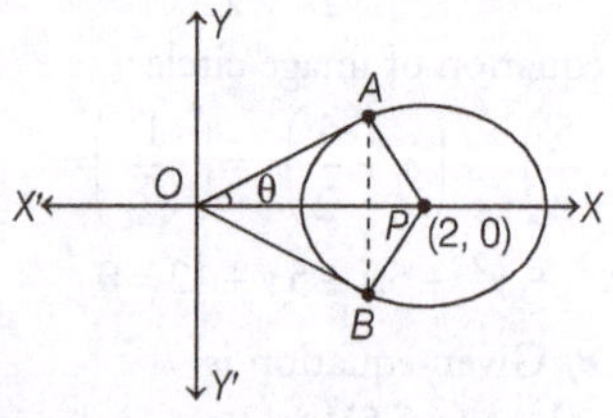


$OP = 2$ units and $AP = 1$ unit

In $\Delta OAP$,

$$(OP)^2 = (OA)^2 + (AP)^2$$

$$\Rightarrow \quad (OA)^2 = 3 \Rightarrow OA = \sqrt{3} \text{ units}$$

Let $\angle AOP = \theta$

In $\Delta AOP$, $\sin\theta = \dfrac{AP}{OP} = \dfrac{1}{2}$

$$\Rightarrow \quad \theta = 30^\circ$$

$$\angle AOB = 60^\circ$$

In $\Delta OAB$, $OA = OB$

$$\therefore \quad \angle ABO = \angle BAO = \frac{180^\circ - 60^\circ}{2}$$

$$= 60^\circ$$

$\therefore \Delta OAB$ is an equilateral triangle.

$$\text{Area of } \Delta OAB = \frac{\sqrt{3}}{4}\times S^2$$

$$= \frac{\sqrt{3}}{4}\times(OA)^2 = \frac{3\sqrt{3}}{4} \text{ sq units}$$

**105.** *(a)* Centre of $C_1$ is (2, 0).

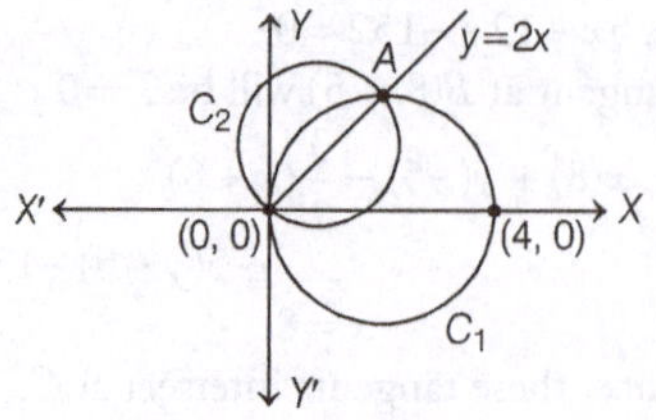


Equation of $C_1$ is $(x-2)^2 + y^2 = 4$

$$x^2 + y^2 - 4x = 0 \quad \text{...(i)}$$

To find $A$, on putting $y = 2x$

$$x^2 + 4x^2 - 4x = 0 \Rightarrow 5x^2 - 4x = 0$$

$$\Rightarrow x = 0 \text{ or } x = \frac{4}{5} \Rightarrow A = \left(\frac{4}{5}, \frac{8}{5}\right)$$

Equation of $C_2$ is

$$(x)\left(x - \frac{4}{5}\right) + (y)\left(y - \frac{8}{5}\right) = 0$$

$$\Rightarrow \quad x^2 + y^2 - \frac{4x}{5} - \frac{8y}{5} = 0$$

Equation of tangent at $A\left(\dfrac{4}{5}, \dfrac{8}{5}\right)$ is

$$T = 0$$

$$\Rightarrow \frac{4}{5}x + \frac{8}{5}y - \frac{2}{5}\left(x + \frac{4}{5}\right) - \frac{4}{5}\left(y + \frac{8}{5}\right) = 0$$

$$\Rightarrow \quad \frac{2x}{5} + \frac{4y}{5} - \frac{8}{5} = 0$$

$$\Rightarrow \quad x + 2y - 4 = 0$$

$$P \equiv (4, 0),\ Q \equiv (0, 2)$$

$$\therefore \quad \frac{QA}{AP} = \frac{\sqrt{\dfrac{16}{25} + \dfrac{4}{25}}}{\sqrt{\dfrac{256}{25} + \dfrac{64}{25}}} = \frac{\sqrt{20}}{\sqrt{320}} = \frac{1}{4}$$

**106.** *(b)* Let $C = (a, b)$

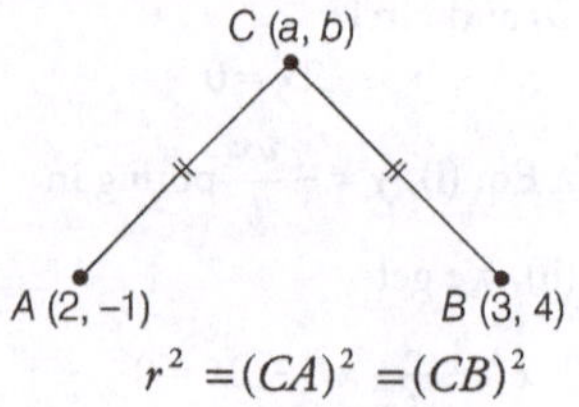


$$r^2 = (CA)^2 = (CB)^2$$

$$\Rightarrow \quad (CA)^2 = (CB)^2$$

$$\Rightarrow \quad (a-2)^2 + (b+1)^2 = (a-3)^2 + (b-4)^2$$

$$\Rightarrow \quad a = 10 - 5b \quad \text{...(i)}$$

Now, $c(a, b)$ lies on the circle

$$(x-5)^2 + (y-1)^2 = \frac{13}{2}$$

Then, $(a-5)^2 + (b-1)^2 = \dfrac{13}{2}$

$$\Rightarrow \quad (10 - 5b - 5)^2 + (b-1)^2 = \frac{13}{2}$$

$$\Rightarrow \quad (5 - 5b)^2 + (b-1)^2 = \frac{13}{2}$$

$$\Rightarrow 25 + 25b^2 - 50b + b^2 + 1 - 2b = \frac{13}{2}$$

$$\Rightarrow \quad 2 + 2b^2 - 4b = \frac{1}{2}$$

$$\Rightarrow \quad 4b^2 - 8b + 3 = 0$$

$$\Rightarrow \quad (2b-3)(2b-1) = 0$$

$$\Rightarrow \quad b = \frac{3}{2}, \frac{1}{2} \text{ and } a = \frac{5}{2}, \frac{15}{2}$$

Here, $C = \left(\dfrac{5}{2}, \dfrac{3}{2}\right)$ or $\left(\dfrac{15}{2}, \dfrac{1}{2}\right)$

Now, $r^2 = \left(\dfrac{15}{2} - 2\right)^2 + \left(\dfrac{1}{2} + 1\right)^2$

$$= \frac{121}{4} + \frac{9}{4} = \frac{130}{4} = \frac{65}{2}$$

**107.** *(c)* Given, circle

$$\Rightarrow \quad x^2 + y^2 + 2x + 4y - 4 = 0$$

$$\Rightarrow (x^2 + 2x + 1) + (y^2 + 4y + 4) - 1 - 4 - 4 = 0$$

$$\Rightarrow \quad (x+1)^2 + (y+2)^2 = 9$$

Any point on the given circle be $[(3\cos\theta - 1), (3\sin\theta - 2)]$

[to find the point coordinates, take $x + 1 = 3\cos\theta$ and $y + 2 = 3\sin\theta$]

Now, circle passes through $(-4, 1)$ and their centres lie on the given circle. So, the centre coordinate of that circle be $(3\cos\theta - 1, 3\sin\theta - 2)$. Since, it passes through $(-4, 1)$, then radius of this circle be

$$r = \sqrt{(3\cos\theta - 1 + 4)^2 + (3\sin\theta - 2 - 1)^2}$$

$$= \sqrt{9\cos^2\theta + 9 + 18\cos\theta + 9\sin^2\theta + 9 - 18\sin\theta}$$

$$= \sqrt{27 + 18(\cos\theta - \sin\theta)}$$

$$= 3\sqrt{3 + 2(\cos\theta - \sin\theta)}$$

Maximum radius will be when $(\cos\theta - \sin\theta)$ is maximum i.e.

$$\cos\theta - \sin\theta = \frac{1}{\sqrt{2}} - \left(-\frac{1}{\sqrt{2}}\right) = \frac{2}{\sqrt{2}} = \sqrt{2}$$

$$\therefore \quad r_1 = r_{\max} = 3\sqrt{3 + 2\sqrt{2}}$$

Minimum radius will be when $(\cos\theta - \sin\theta)$ is minimum i.e. $-\sqrt{2}$.

$\therefore \quad r_2 = r_{\min} = 3\sqrt{3 - 2\sqrt{2}}$

Given, $\dfrac{r_1}{r_2} = a + b\sqrt{2}$,

then $\dfrac{r_1^2}{r_2^2} = (a + b\sqrt{2})^2$

$\Rightarrow \quad \dfrac{9(3+2\sqrt{2})}{9(3-2\sqrt{2})} = (a + b\sqrt{2})^2$

$\Rightarrow \quad \dfrac{(3+2\sqrt{2})^2}{1} = (a + b\sqrt{2})^2$

On comparing coefficients, $a = 3$, $b = 2$

$\therefore \quad a + b = 5$

**108.** *(c)* $x^2 + y^2 + ax + 2ay + c = 0$

$(a < 0)$

For a circle in standard form

$x^2 + y^2 + 2gx + 2fy + c = 0$

$x$-intercept $= 2\sqrt{g^2 - c}$

$y$-intercept $= 2\sqrt{f^2 - c}$

Here, $2g = a \Rightarrow g = \dfrac{a}{2}$

$2f = 2a \Rightarrow f = a$

$x$-intercept $\Rightarrow 2\sqrt{\dfrac{a^2}{4} - c} = 2\sqrt{2}$

$\Rightarrow \quad \dfrac{a^2}{4} - c = 2 \quad \ldots \text{(i)}$

$y$-intercept $\Rightarrow 2\sqrt{a^2 - c} = 2\sqrt{5}$

$a^2 - c = 5 \quad \ldots \text{(ii)}$

On subtracting Eq. (ii) from Eq. (i), we get

$\Rightarrow \quad a = \pm 2$

As, $a < 0$, so, $a = -2$, $c = -1$

So, $x^2 + y^2 - 2x - 4y - 1 = 0$

$\Rightarrow \quad (x-1)^2 + (y-2)^2 = 6$

Centre (1, 2), Radius $= \sqrt{6}$

Equation of tangent is perpendicular to $x + 2y = 0$.

So, equation of tangent will be $2x - y + \lambda = 0$.

Now, perpendicular distance of this line from (1, 2) will be $\sqrt{6}$ units.

$$\frac{|2\cdot 1 - 2 + \lambda|}{\sqrt{5}} = \sqrt{6}$$

$\Rightarrow \quad \lambda = \pm\sqrt{30}$

So, $T \Rightarrow 2x - y \pm \sqrt{30} = 0$

Now, shortest distance from origin (0, 0) to tangent $T$ will be

$$\frac{|2\cdot 0 - 0 \pm \sqrt{30}|}{\sqrt{2^2 + 1^2}} = \frac{\sqrt{30}}{\sqrt{5}}$$

$$= \sqrt{6}$$

**109.** *(c)* Given equation of circle is $x^2 + y^2 - 6x + 8y - 103 = 0$, which can be written as

$(x-3)^2 + (y+4)^2 = 128 = (8\sqrt{2})^2$

$\therefore$ Centre $= (3, -4)$ and radius $= 8\sqrt{2}$

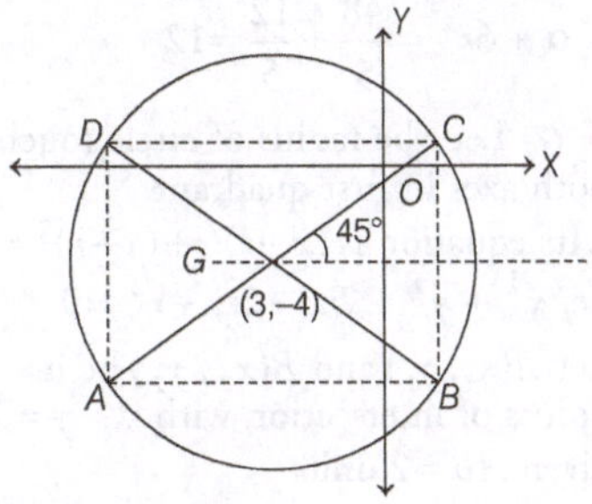


For the coordinates of $A$ and $C$.

Consider, $\dfrac{x-3}{1/\sqrt{2}} = \dfrac{y+4}{1/\sqrt{2}} = \pm 8\sqrt{2}$

[Using distance (parametric) form of line, $\dfrac{x - x_1}{\cos\theta} = \dfrac{y - y_1}{\sin\theta} = \pm r$]

$\Rightarrow \quad x = 3 \pm 8,\ y = -4 \pm 8$

$\therefore \quad A(-5, -12)$ and $C(11, 4)$

Similarly, for the coordinates of $B$ and $D$, consider

$$\frac{x-3}{-\frac{1}{\sqrt{2}}} = \frac{y+4}{\frac{1}{\sqrt{2}}} = \pm 8\sqrt{2}$$

[in this case, $\theta = 135°$]

$\Rightarrow \quad x = 3 \mp 8,\ y = -4 \pm 8$

$\therefore B(11, -12)$ and $D(-5, 4)$

Now, $OA = \sqrt{25 + 144} = \sqrt{169} = 13$;

$OB = \sqrt{121 + 144} = \sqrt{265}$

$OC = \sqrt{121 + 16} = \sqrt{137}$

and $\quad OD = \sqrt{25 + 16} = \sqrt{41}$

**110.** *(a)* Equation of tangent to the circle $x^2 + y^2 + 4x - 6y - 12 = 0$ at $(1, -1)$ is given by

$xx_1 + yy_1 + 2(x + x_1) - 3(y + y_1) - 12 = 0$, where $x_1 = 1$ and $y_1 = -1$

$\Rightarrow \ x - y + 2(x+1) - 3(y-1) - 12 = 0$

$\Rightarrow 3x - 4y - 7 = 0$

Now, equation of family of circles touching the line $3x - 4y - 7 = 0$ at point $(1, -1)$ is given by

$(x-1)^2 + (y+1)^2 + \lambda(3x - 4y - 7) = 0$

So, the equation of required circle will be $(x-1)^2 + (y+1)^2$

$+ \lambda(3x - 4y - 7) = 0$,

for some $\lambda \in R \quad \ldots \text{(i)}$

$\because$ The required circle passes through (4, 0).

$\therefore \ (4-1)^2 + (0+1)^2$

$+ \lambda(3 \times 4 - 4 \times 0 - 7) = 0$

$\Rightarrow \lambda = -2$

On substituting $\lambda = -2$ in Eq. (i), we get

$(x-1)^2 + (y+1)^2 - 2(3x - 4y - 7) = 0$

$\Rightarrow \ x^2 + y^2 - 8x + 10y + 16 = 0$

On comparing it with

$x^2 + y^2 + 2gx + 2fy + c = 0$, we get

$g = -4$, $f = 5$, $c = 16$

$\therefore \quad$ Radius $= \sqrt{g^2 + f^2 - c}$

$= \sqrt{16 + 25 - 16} = 5$

**111.** *(768)*

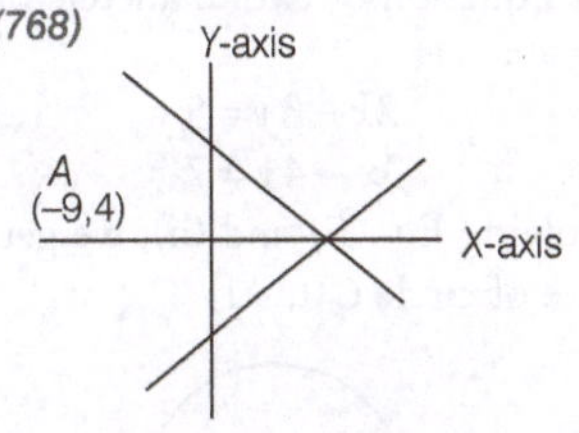


$\because \quad x + y = 3$

and $\quad x - y = 3$ are tangents

$\therefore$ Both circles centre will lies on $X$-axis

$\therefore \quad (x - \alpha)^2 + y^2 = r^2$

Hence, centre is $C(\alpha, 0)$

$\Rightarrow \quad r = \sqrt{(\alpha + a)^2 + 16} \quad \ldots \text{(i)}$

Also, $\left|\dfrac{\alpha - 3}{\sqrt{2}}\right| = k \quad \ldots \text{(ii)}$

From Eqs. (i) and (ii), we get

$\sqrt{(\alpha+9)^2 + 16} = \left|\dfrac{\alpha - 3}{\sqrt{2}}\right| \Rightarrow \alpha = -5, -37$

$r_1$ (when $\alpha = -5$) $= 4\sqrt{2}$

$r_2$ (when $\alpha = -37$) $= 20\sqrt{2}$

$|r_1^2 - r_2^2| = |32 - 800| = 768$

**112.** *(9)* Let $C_1 : x^2 + y^2 = 9$

and $C_2 : (x-h)^2 + (y-k)^2 = h^2 + k^2$

$\Rightarrow x^2 + y^2 - 2hx - 2ky = 0 \quad \ldots \text{(i)}$

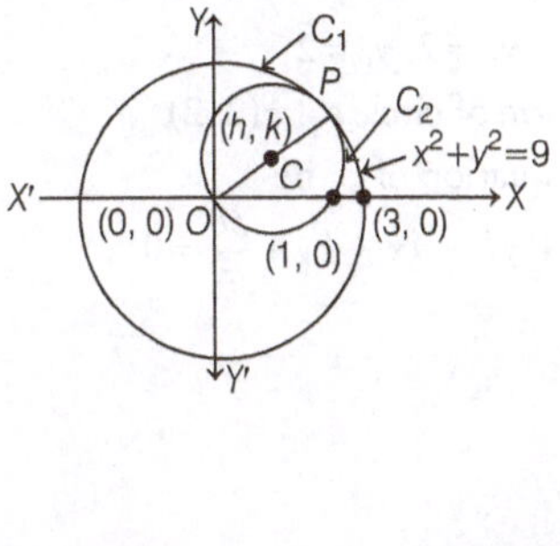

$\because$ Eq. (i) passes through (1, 0), then

$1+0-2h=0 \Rightarrow h=\dfrac{1}{2}$

$\because \quad OC=\dfrac{OP}{2}$

$\sqrt{\left(\dfrac{1}{2}\right)^2+k^2}=\dfrac{3}{2}$

$\Rightarrow \quad \dfrac{1}{4}+k^2=\dfrac{9}{4}$

$\Rightarrow \quad k^2=2 \Rightarrow k=\pm\sqrt{2}$

$\therefore$ Possible values of $(h, k)$ are $\left(\dfrac{1}{2}, \sqrt{2}\right)$, $\left(\dfrac{1}{2}, -\sqrt{2}\right)$

Thus, $4(h^2+k^2)=4\left(\dfrac{1}{4}+2\right)=4\left(\dfrac{9}{4}\right)=9$

**113.** *(2)* Equation of two diameters of a circle are

$2x-3y=5 \quad \ldots$(i)

and $\quad 3x-4y=7 \quad \ldots$(ii)

On solving Eqs. (i) and (ii), we get

Centre of circle $C(1, -1)$

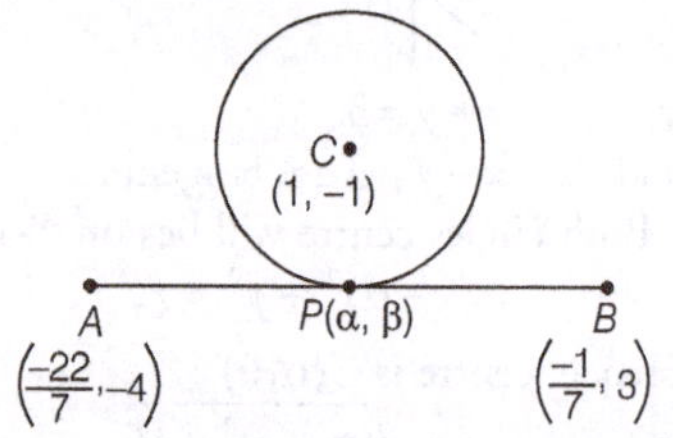


Equation of $AB$ is $7x-3y+10=0 \quad \ldots$(i)

and equation of $CP$ is $3x+7y+4=0 \quad \ldots$(ii)

On solving Eqs. (i) and (ii), we get

$\alpha=\dfrac{-41}{29}, \beta=\dfrac{1}{29}$

$\therefore 17\beta-\alpha=2$

**114.** *(12)* Given, the equation

$c_1 = x^2+y^2-2x-6y+\alpha=0$

Image of centre $c_1 \equiv (1, 3)$ in $x-y+1=0$ is given by

$\dfrac{x_1-1}{1}=\dfrac{y_1-3}{-1}=\dfrac{-2(1-3+1)}{1^2+1^2}$

$\Rightarrow \quad x_1=2, y_1=2$

Centre of circle $c_2=(2, 2)$

$\therefore$ Equation of $c_2$ be

$x^2+y^2-4x-4y+\dfrac{38}{5}=0$

On comparing the equation $x^2+y^2+2gx+2fy+c=0$.

Now, radius of $c_2$ is $\sqrt{4+4-\dfrac{38}{5}}$

$\Rightarrow \quad r_2=\sqrt{2/5}$

$(\text{radius of } c_1)^2=(\text{radius of } c_2)^2$

$\Rightarrow \quad 10-\alpha=\dfrac{2}{5} \Rightarrow \alpha=\dfrac{48}{5}$

$\therefore \alpha+6r^2=\dfrac{48}{5}+\dfrac{12}{5}=12$

**115.** *(7)* Let $r$ be radius of circle touching both axes in first quadrant.

$\therefore$ Its equation is $(x-r)^2+(y-r)^2=r^2$

i.e. $x^2+y^2-2rx-2ry+r^2=0 \quad \ldots$(i)

Let $A(x_1, y_1)$ and $B(x_2, y_2)$ be its points of intersection with $x+y=2$ given $AB=2$ units

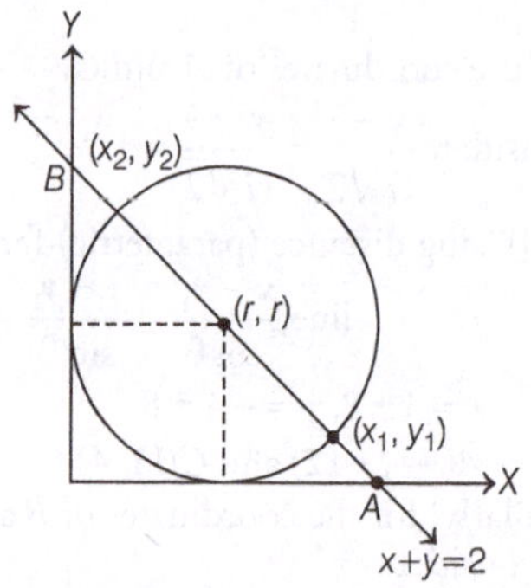


$\Rightarrow AB^2=4$

$\Rightarrow (x_1-x_2)^2+(y_1-y_2)^2=4$

$\Rightarrow (x_1+x_2)^2-4(x_1x_2)+(y_1+y_2)^2-4(y_1y_2)=4 \ldots$(ii)

Here, equation of line $x+y=2$

$\Rightarrow x=2-y$

Putting the value of $x=2-y$ in Eq. (i), we get

$(2-y)^2+y^2-2r(2-y)-2ry+r^2=0$

$2y^2-4y+r^2-4r+4=0$, whose roots are $y_1, y_2$

$\Rightarrow y_1+y_2=-\dfrac{(-4)}{2}=2,$

$y_1y_2=\dfrac{r^2-4r+4}{2} \quad \ldots$(iii)

Now, $x_1+x_2=(2-y_1)+(2-y_2)$

$=4-(y_1+y_2)=2,$

$x_1x_2=(2-y_1)(2-y_2)$

$=4-2(y_1+y_2)+y_1y_2$

$=\dfrac{r^2-4r+4}{2} \quad \ldots$(iv)

Putting in Eq. (ii), we get

$2^2-4\left(\dfrac{r^2-4r+4}{2}\right)+2^2-4\left(\dfrac{r^2-4r+4}{2}\right)=4$

$\Rightarrow (r-2)^2=1, \Rightarrow r-2=\pm1 \Rightarrow r=2\pm1$

$\Rightarrow r=1, 3$

Let $r_1=1, r_2=3$

$\therefore r_1^2+r_2^2-r_1r_2=1^2+3^2-1\times3=7$

**116.** *(2)* Given circles,

$C_1: x^2+y^2-4x-2y=\alpha-5$

Centre (2, 1)

$C_2: 5x^2+5y^2-10fx-10gy+36=0$

Centre $(f, g)$

Line $L: y=2x+1$

Image of (2, 1) in line $L$ in $(f, g)$

$\dfrac{x-2}{2}=\dfrac{y-1}{-1}=\dfrac{-2(4-1+1)}{5}=-\dfrac{8}{5}$

$\Rightarrow \quad x=-\dfrac{6}{5}=f \Rightarrow y=\dfrac{13}{5}=g$

Radius of both the circles is same

$\therefore 4+1-(5-\alpha)=f^2+g^2-\dfrac{36}{5}$

$\Rightarrow \alpha=1$

$\therefore$ radius of $C_2$, $r=\sqrt{\dfrac{36}{25}+\dfrac{169}{25}-\dfrac{36}{5}}=1$

$\alpha+r=1+1=2$

**117.** *(121)* Let equation of circle be

$(x-a)^2+(y-a)^2=a^2 \quad \ldots$(i)

Since, Eq. (i) passes through $P(\alpha, \beta)$

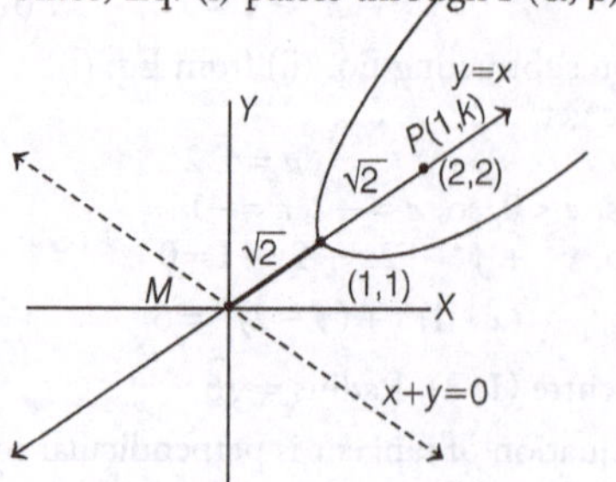


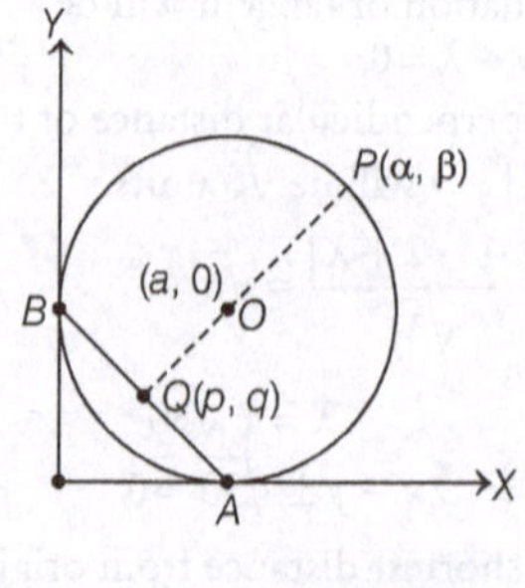


$\therefore (\alpha-a)^2+(\beta-a)^2=a^2$

$\therefore (\alpha - a)^2 + (\beta - a)^2 = a^2$

$\Rightarrow \alpha^2 + \beta^2 - 2\alpha a - 2\beta a + a^2 = 0$ ...(ii)

Equation of $AB = \frac{x}{a} + \frac{y}{a} = 1$

$\Rightarrow \quad x + y = a$ ...(iii)

Let $Q(p, q)$ be the foot of the perpendicular from $P$ to line Eq. (iii)

$\therefore \frac{p - \alpha}{1} = \frac{q - \beta}{1} = \frac{-(\alpha + \beta - a)}{(1)^2 + (1)^2}$

$\Rightarrow \frac{p - \alpha}{1} = \frac{q - \beta}{1} = \frac{-(\alpha + \beta - a)}{2}$

$\Rightarrow p - \alpha = \frac{-(\alpha + \beta - a)}{2}$

and $q - \beta = \frac{-(\alpha + \beta - a)}{2}$

Now, $PQ^2 = (p - \alpha)^2 + (q - \beta)^2$

$= \frac{1}{4}(\alpha + \beta - a)^2 + \frac{1}{4}(\alpha + \beta - a)^2$

$\Rightarrow (11)^2 = \frac{1}{2}(\alpha + \beta - a)^2$

$\Rightarrow \alpha^2 + \beta^2 + a^2 + 2\alpha\beta - 2\beta a - 2\alpha a = 242$

$\Rightarrow 2\alpha\beta = 242$ [using Eq. (ii)]

$\Rightarrow \alpha\beta = 121$

**118.** *(24)* We have,

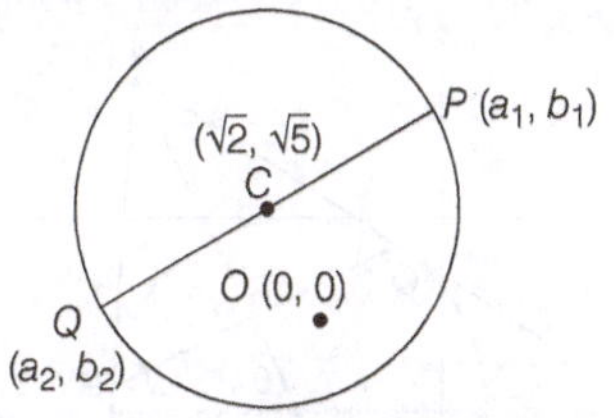


$\because OC \perp CP$ and $CQ$

and $OC = \sqrt{(\sqrt{2} - 0)^2 + (\sqrt{3} - 0)^2}$

$= \sqrt{2 + 3} = \sqrt{5}$

and $PQ =$ Diameter

and Area of $\Delta OCP$

$= \frac{1}{2}(OC) \times (CP) = \frac{\sqrt{35}}{2}$

$\Rightarrow \sqrt{5} \times CP = \sqrt{35}$

$\Rightarrow CP = \sqrt{7} =$ Radius of circle

Now, $\quad CP^2 + OC^2 = OP^2$

$\Rightarrow a_1^2 + b_1^2 = 7 + 5 = 12$ ...(i)

and $OQ^2 = CQ^2 + OC^2 \quad [\because CQ = CP]$

$\Rightarrow \quad a_2^2 + b_2^2 = 12$ ...(ii)

adding Eqs. (i) and (ii), we get

$a_1^2 + a_2^2 + b_1^2 + b_2^2 = 24$

**119.** *(11)* Given, a circle with centre (2, 3) and radius 4.

Equation of the circle

$(x - 2)^2 + (y - 3)^2 = 16$

$\Rightarrow x^2 + y^2 - 4x - 6y - 3 = 0$ ...(i)

Chord of contact : $T = 0$

$\Rightarrow x\alpha + y\beta - 2(x + \alpha) - 3(y + \beta) - 3 = 0$

$\Rightarrow x(\alpha - 2) + y(\beta - 3) = 3 + 2\alpha + 3\beta$

$\therefore \frac{\alpha - 2}{1} = \frac{\beta - 3}{1} = \frac{2\alpha + 3\beta + 3}{3}$

$\therefore \alpha - 2 = \beta - 3 \Rightarrow \alpha - \beta = -1$

$\therefore \alpha - 2 = \frac{2\alpha + 3\alpha + 6}{3} \Rightarrow \alpha = -6$

$\therefore \quad \beta = -5$

So, $4\alpha - 7\beta = 4(-6) - 7(-5)$

$= -24 + 35 = 11$

**120.** *(3)* Given, $PR$ is a diameter.

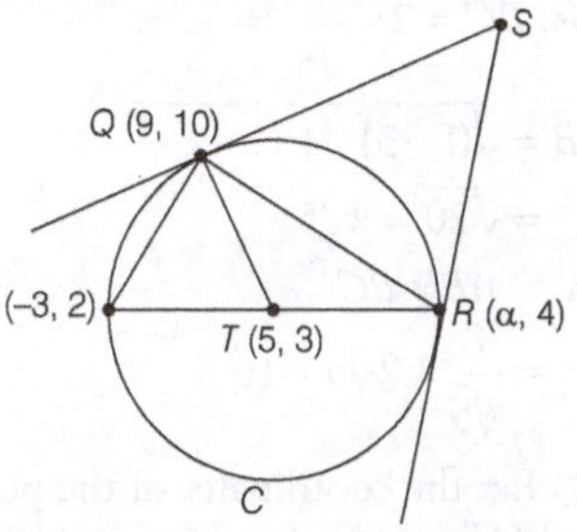


$\angle PQR = 90°$

$\Rightarrow$ Slope of $PQ \times$ slope of $RQ = -1$

$\Rightarrow \left(\frac{10 - 2}{9 + 3}\right) \times \left(\frac{10 - 4}{9 - \alpha}\right) = -1$

$\Rightarrow \frac{8}{12} \times \frac{6}{9 - \alpha} = -1$

$\Rightarrow \quad \alpha = 13$

Now, centre of circle,

$T = T\left(\frac{13 - 3}{2}, \frac{4 + 2}{2}\right) = T(5, 3)$

Since, tangents to $C$ at the points $Q$ and $R$ intersect at the point $S(x, y)$

$\therefore$ Slope of $QT = \frac{10 - 3}{9 - 5} = \frac{7}{4}$

Slope of tangent, $QS = -\frac{4}{7}$

Slope of $RT = \frac{4 - 3}{13 - 5} = \frac{1}{8}$

Slope of tangent, $RS = -8$

Now, equation of $QS$,

$\Rightarrow \quad y - 10 = -\frac{4}{7}(x - 9)$

$\Rightarrow \quad 4x + 7y = 106$ ...(i)

And equation of $RS$ is

$y - 4 = (-8)(x - 13)$

$\Rightarrow \quad y - 4 = -8x + 104$

$\Rightarrow \quad 8x + y - 108 = 0$ ...(ii)

On solving Eqs. (i) and (ii), we get

$x = \frac{25}{2}$ and $y = 8$

$\therefore$ Since, $S\left(\frac{25}{2}, 8\right)$ lies on $2x - ky = 1$

$\Rightarrow 2\left(\frac{25}{2}\right) - k(8) = 1$

$\therefore \quad k = 3$

**121.** *(72)* Circle $(x - 2)^2 + (y + 1)^2 = \frac{169}{4}$,

Length of $AB = 12$

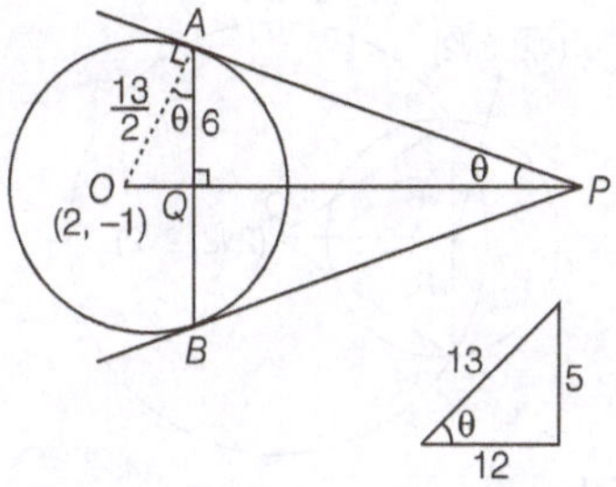


So, $\cos\theta = \frac{6}{\left(\frac{13}{2}\right)} = \frac{12}{13}$

$\Rightarrow \tan\theta = \frac{5}{12}$

In $\Delta AQP$, $\tan\theta = \frac{6}{QP} = \frac{5}{12}$, $QP = \frac{72}{5}$

$\Rightarrow \quad 5QP = 72$

**122.** *(25)* Given circle

$x^2 + y^2 + 6x + 8y + 16 = 0$ with centre $(-3, -4)$ and radius 3 units.

The circle

$x^2 + y^2 + 2(3 - \sqrt{3})x + 2(4 - \sqrt{6})y$

$= k + 6\sqrt{3} + 8\sqrt{6}, k > 0$ has centre

$(\sqrt{3} - 3, \sqrt{6} - 4)$ and radius $\sqrt{k + 34}$

So, these two circles touch internally hence $\sqrt{3 + 6} = |\sqrt{k + 34} - 3|$.

Here, $k = 2$ is only possible $(\because k > 0)$

Equation of common tangent to two circles is

$2\sqrt{3}x + 2\sqrt{6}y + 16 + 6\sqrt{3} + 8\sqrt{6} + k = 0$

So, $k = 2$, then equation is

$x + \sqrt{2}y + 3 + 4\sqrt{2} + 3\sqrt{3} = 0$ ...(i)

So, $(\alpha, \beta)$ are foot of perpendicular from $(-3, -4)$ to line (i), then

$$\frac{\alpha+3}{1}=\frac{\beta+4}{\sqrt{2}}=\frac{-(3-4\sqrt{2}+3+4\sqrt{2}+3\sqrt{3})}{1+2}$$

$$\therefore \alpha+3=\frac{\beta+4}{\sqrt{2}}=-\sqrt{3}$$

$\Rightarrow (\alpha+\sqrt{3})^2=9$ and $(\beta+\sqrt{6})^2=16$

Therefore, $(\alpha+\sqrt{3})^2+(\beta+\sqrt{6})^2=25$

**123.** *(10)* Given, equation of circle

$x^2+y^2-2\sqrt{2}x-6\sqrt{2}y+14=0$

Centre $=(\sqrt{2},3\sqrt{2})$

Radius $=\sqrt{(\sqrt{2})^2+(3\sqrt{2})^2-14}=\sqrt{6}$

$\Rightarrow$ Diameter $=2\sqrt{6}$

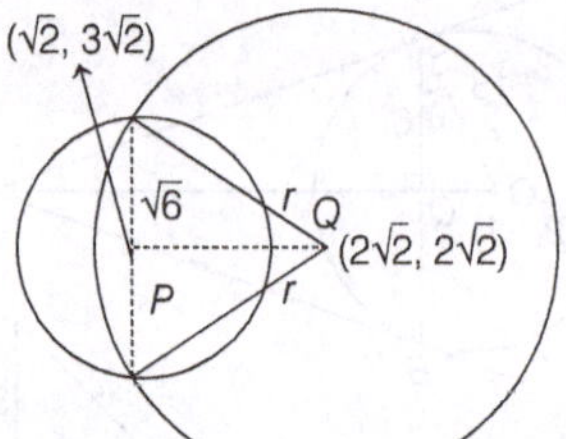


If this diameter is chord to

$(x-2\sqrt{2})^2+(y-2\sqrt{2})^2=r^2$, then

$$r^2=6+\left(\sqrt{(\sqrt{2})^2+(\sqrt{2})^2}\right)^2$$

$\Rightarrow \quad r^2=6+4=10$

$\therefore \quad r^2=10$

**124.** *(816)* Let $L_1: y+2x=\sqrt{11}+7\sqrt{7}$

and $L_2: 2y+x=2\sqrt{11}+6\sqrt{7}$

Point of intersection of these two lines is centre of circle

i.e. $\left(\frac{8}{3}\sqrt{7},\sqrt{11}+\frac{5}{3}\sqrt{7}\right)$

Perpendicular from centre to line

$3x-\sqrt{11}y+\left(\frac{5\sqrt{77}}{3}+11\right)=0$

is radius of circle

$$\Rightarrow r=\left|\frac{8\sqrt{7}-11-\frac{5}{3}\sqrt{77}+\frac{5\sqrt{77}}{3}+11}{\sqrt{20}}\right|$$

$$=\left|4\sqrt{\frac{7}{5}}\right|=4\sqrt{\frac{7}{5}} \text{ units}$$

So, $(5h-8k)^2+5r^2$

$$=\left(\frac{40}{3}\sqrt{7}-8\sqrt{11}-\frac{40}{3}\sqrt{7}\right)^2+5\cdot16\cdot\frac{7}{5}$$

$=64\times11+112=816$

**125.** *(16)*

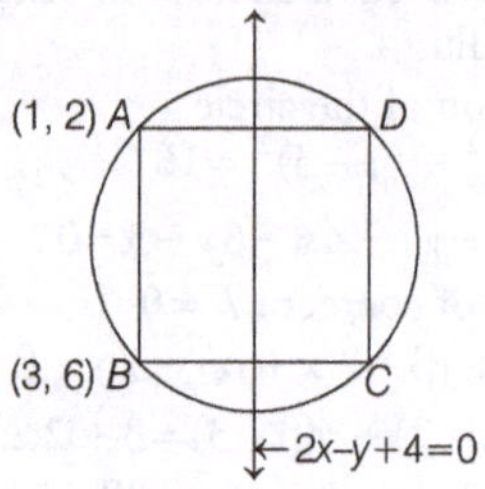


Equation of line $AB$

$y=2x$

Slope of $AB=2$

Slope of given diameter $=2$

So, the diameter is parallel to $AB$.

Distance between diameter and line $AB$

$$=\frac{4}{\sqrt{4+1}}=\frac{4}{\sqrt{5}}$$

Thus, $BC=2\times\frac{4}{\sqrt{5}}=\frac{8}{\sqrt{5}}$

$AB=\sqrt{(1-3)^2+(2-6)^2}$

$=\sqrt{20}=2\sqrt{5}$

Area $=AB\times BC$

$$=\frac{8}{\sqrt{5}}\times2\sqrt{5}=16$$

**126.** *(7)* Let the coordinates of the points $P$ and $Q$ be $(x_1,y_1)$ and $(x_2,y_2)$, respectively.

According to the question,

$$x_1+x_2=\frac{r}{2}, x_1x_2=\frac{p}{2}$$

Also, $y_1+y_2=s,\ y_1y_2=-q$

Equation of circle with diameter $PQ$ is

$$(x-x_1)(x-x_2)+(y-y_1)(y-y_2)=0$$

$$\Rightarrow \quad x^2-(x_1+x_2)x+y^2-(y_1+y_2)y+x_1x_2+y_1y_2=0$$

Compare it with given equation of circle

$$x^2+y^2-\frac{11}{2}x-7y-11=0$$

$$\Rightarrow \quad x_1+x_2=\frac{11}{2}, y_1+y_2=7,$$

$$x_1x_2+y_1y_2=-11$$

$$\Rightarrow \quad \frac{r}{2}=\frac{11}{2}, s=7,$$

$$\frac{p}{2}-q=-11$$

$$\Rightarrow \quad r=11, s=7,$$

$$p-2q=-22$$

$$\therefore\ 2r+s-2q+p=2\times11+7-22=7$$

**127.** *(7)* From the shown figure, the length of chord $PQ$ is 2 units.

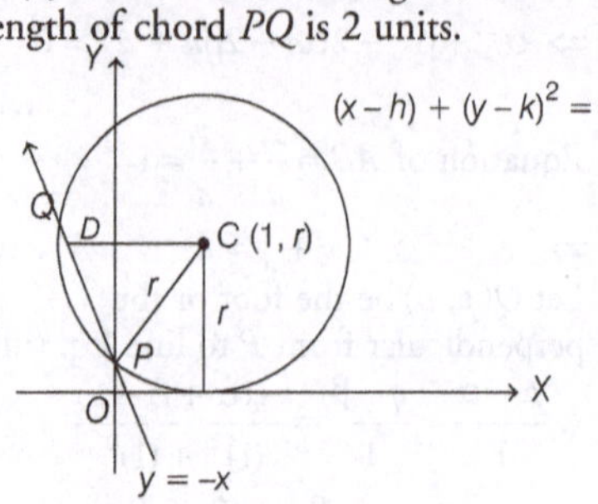


$$CD=\frac{1+r}{\sqrt{2}}$$

$$\therefore PQ=2\sqrt{r^2-\left(\frac{1+r}{\sqrt{2}}\right)^2}$$

$$\Rightarrow \quad 2=2\sqrt{r^2-\left(\frac{1+r}{\sqrt{2}}\right)^2}$$

$$\Rightarrow \quad 1=\sqrt{r^2-\left(\frac{1+r}{\sqrt{2}}\right)^2}$$

$\Rightarrow \quad 2=2r^2-1-r^2-2r$

$\Rightarrow \quad r^2-2r-3=0 \Rightarrow (r-3)\cdot(r+1)=0$

$\Rightarrow \quad r=-1$ or $r=3$

So, $k=r=3$ and $h=1$ [given]

Hence, $h+k+r=1+3+3=7$

**128.** *(18)*

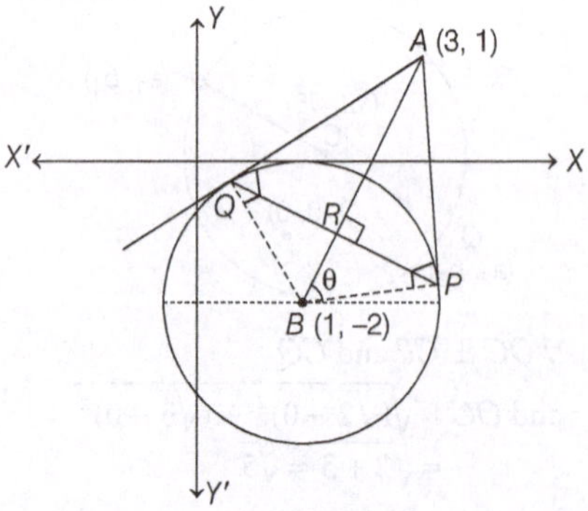


Radius $=\sqrt{1+4-1}=2$

$AB=\sqrt{3^2+2^2}=\sqrt{13}$

In $\Delta ABP$

$AP^2=AB^2-BP^2=13-4=9$

$AP=3$

$\Rightarrow AQ=AP=3$

Let $\angle ABP=\theta$, $\angle BAP=90-\theta$

In $\Delta ABP$, $\tan\theta=3/2$

$$\sin\theta=\frac{3}{\sqrt{13}}, \cos\theta=\frac{2}{\sqrt{13}}$$

In $\Delta ARP$,

$$\cos(90-\theta)=\frac{AR}{AP}$$

$\Rightarrow \quad AR = 3\sin\theta$

In $\Delta BRP$

$\cos\theta = \dfrac{BR}{BP}$

$\Rightarrow BR = 2\cos\theta$

$$= \frac{\text{Area}(\Delta APQ)}{\text{Area}(\Delta BPQ)} = \frac{\frac{1}{2}\times PQ\times AR}{\frac{1}{2}\times PQ\times BR}$$

$$= \frac{AR}{RB} = \frac{3\sin\theta}{2\cos\theta} = \frac{9}{4}$$

$$\Rightarrow \quad 8\left(\frac{\text{Area}(\Delta APQ)}{\text{Area}(\Delta BPQ)}\right) = 18$$

**129.** *(165)*

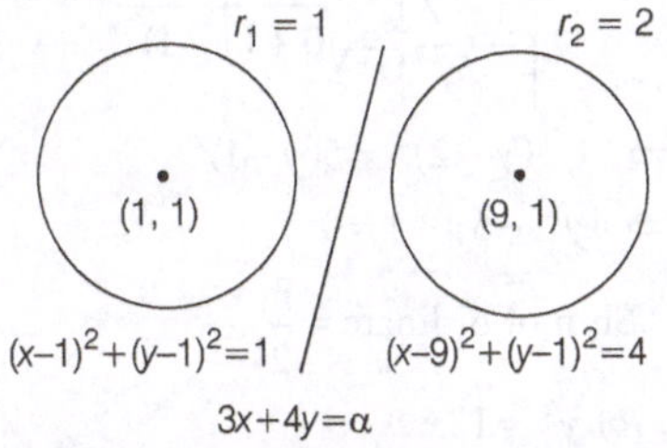


There are 2 possible cases

1. Both centres should lie on different sides of line.
2. Line can be tangent also

Centre of both the circles are (1, 1) and (9, 1), respectively.

$(3+4-\alpha)(27+4-\alpha) < 0$

$(7-\alpha)(31-\alpha) < 0$

$\alpha \in (7, 31)$ ... (i)

$P_1$ = Distance of line from centre (1, 1)

$P_1 \geq r_1$

$\dfrac{|3+4-\alpha|}{5} \geq 1 \quad |7-\alpha| \geq 5$

$\alpha \in (-\infty, 2] \cup [12, \infty)$ ... (ii)

Similarly, $P_2$ = Distance of line from (9, 1)

$\dfrac{|27+4-\alpha|}{5} \geq 2$

$\alpha \in (-\infty, 21] \cup [41, \infty)$ ... (iii)

Now, $(A) \cap (B) \cap (C)$

$\Rightarrow \quad \alpha \in [12, 21]$

Sum $= (12 + 13 + 14 + \ldots + 21)$

$$= \left(\frac{21\times 22}{2}\right) - \left(\frac{11\times 12}{2}\right)$$

$= 231 - 66 = 165$

**130.** *(40)* Equation of circles

$(x-1)^2 + (y-2)^2$

$+ \lambda(4x + 3y - 10) = 0$

$x^2 + y^2 + (2\lambda - 1)\,2x + \left(\frac{3}{2}\lambda - 2\right)$

$2y + 5 - 10\lambda = 0$

$$r = \sqrt{(2\lambda-1)^2 + \left(\frac{3}{2}\lambda - 2\right)^2 - 5 + 10\lambda}$$

$= 5$

$$\Rightarrow 4\lambda^2 + 1 - 4\lambda + \frac{9}{4}\lambda^2 + 4$$
$$- 6\lambda - 5 + 10\lambda = 25$$

$$\Rightarrow \quad \frac{25}{4}\lambda^2 - 25 = 0 \Rightarrow \lambda = \pm 2$$

For $\lambda = 2$

$x^2 + y^2 + 6x + 2y - 15 = 0,\ C_1(-3, -1)$

For $\lambda = -2$

$x^2 + y^2 - 10x - 10y + 25 = 0,\ C_2(5, 5)$

$|(\alpha+\beta)(\gamma+\delta)| = |-4\times 10| = 40$

**131.** *(61)* Given equation,

$x^2 + y^2 + px + (1-p)y + 5 = 0$

$$\text{Radius, } r = \sqrt{\left(\frac{p}{2}\right)^2 + \left(\frac{1-p}{2}\right)^2 - 5}$$

$$= \sqrt{\frac{2p^2 - 2p - 19}{4}}$$

Now, $\quad r \in (0, 5] \Rightarrow 0 < r \leq 5$

$$\Rightarrow \quad 0 < \sqrt{\frac{2p^2 - 2p - 19}{4}} \leq 5$$

$$\Rightarrow \quad 0 < \frac{2p^2 - 2p - 19}{4} \leq 25$$

$\Rightarrow \quad 2p^2 - 2p - 19 > 0$

and $2p^2 - 2p - 19 \leq 100$

$\Rightarrow \quad 2p^2 - 2p - 19 > 0$

and $2p^2 - 2p - 19 \leq 100$

$$\Rightarrow p \in \left(-\infty, \frac{1-\sqrt{39}}{2}\right) \cup \left(\frac{1+\sqrt{39}}{2}, \infty\right)$$

$$\text{and} \quad p \in \left[\frac{1-\sqrt{239}}{2}, \frac{1+\sqrt{239}}{2}\right]$$

$$p \in \left[\frac{1-\sqrt{239}}{2}, \frac{1-\sqrt{39}}{2}\right] \cup \left[\frac{1+\sqrt{39}}{2}, \frac{1+\sqrt{239}}{2}\right]$$

$$\Rightarrow p^2 \in \left[\left(\frac{1-\sqrt{39}}{2}\right)^2, \left(\frac{1+\sqrt{239}}{2}\right)^2\right]$$

$\because \quad q = p^2,\ q \in Z$

$\Rightarrow \quad q = 7, 8, 9, 10, \ldots, 67$

$\Rightarrow \quad$ 61 integers in $S$

$\therefore \quad n(S) = 61$

**132.** *(16)* Let $P(x, y)$ be the required point.

Then, $(x^2 + y^2) + [x^2 + (y-1)^2]$

$+ [(x-1)^2 + y^2]$

$+ [(x-1)^2 + (y-1)^2] = 18$

$\Rightarrow \quad 4(x^2 + y^2) - 4y - 4x = 14$

$\Rightarrow \quad x^2 + y^2 - x - y - \dfrac{7}{2} = 0$

$$\text{Radius} = \sqrt{\left(\frac{1}{2}\right)^2 + \left(\frac{1}{2}\right)^2 + \frac{7}{2}} = 2$$

Diameter $= 4 \quad \Rightarrow \quad d^2 = 16$

**133.** *(56.25)* Let $P$ be $(h, k)$, $A(5, 0)$ and $B(-5, 0)$.

Given $\quad PA = 3PB$

$\Rightarrow \quad PA^2 = 9PB^2$

$\Rightarrow (h-5)^2 + k^2 = 9[(h+5)^2 + k^2]$

$\Rightarrow 8h^2 + 8k^2 + 100h + 200 = 0$

$\therefore$ Locus of $P$ is $x^2 + y^2 + \dfrac{25}{2}x + 25 = 0$

$$\text{Centre} = \left(\frac{-25}{4}, 0\right)$$

$$\text{and } r^2 = \left(\frac{-25}{4}\right)^2 + 0^2 - 25$$

$$= \frac{625}{16} - 25 = \frac{225}{16}$$

$$\therefore\ 4r^2 = 4\times\frac{225}{16} = \frac{225}{4} = 56.25$$

**134.** *(1225)*

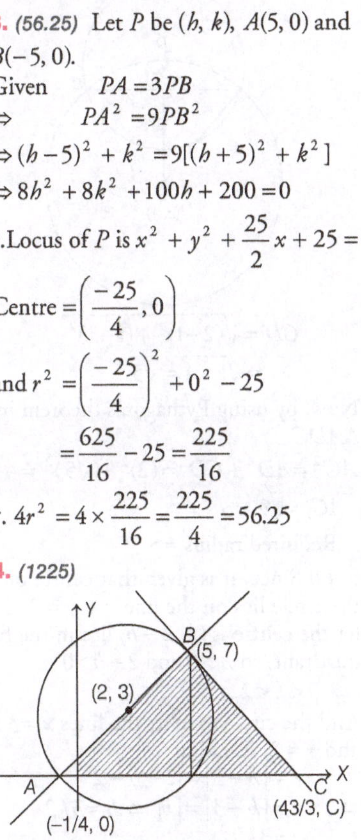


Given, equation of circle

$x^2 + y^2 - 4x - 6y - 12 = 0$

$$\text{Equation of normal, } \frac{y - y_1}{y_1 + f} = \frac{x - x_1}{x_1 + g}$$

$\Rightarrow \quad 4x - 3y + 1 = 0$

Equation of tangents

$xx_1 + yy_1 + g(x + x_1) + f(y + y_1) + c = 0$

$3x + 4y - 43 = 0$

$$\text{Area of triangle} = \frac{1}{2}\left(\frac{43}{3} + \frac{1}{4}\right)\times(7)$$

$$= \frac{1}{2}\left(\frac{172+3}{12}\right)\times 7$$

$$\therefore \quad A = \frac{1225}{24} \Rightarrow 24A = 1225$$

**135.** (3)

> **key idea** (i) Perpendicular from centre to chord bisects the chord
> (ii) Draw the figure and apply Pythagoras theorem to find the radius of circle c.

Given, circle

$\Rightarrow x^2 + y^2 - 2x - 6y + 6 = 0$

Coordinate of $C = (1, 3)$

Radius $= r = \sqrt{1^2 + 3^2 - 6}$

$= \sqrt{4} = 2$ units

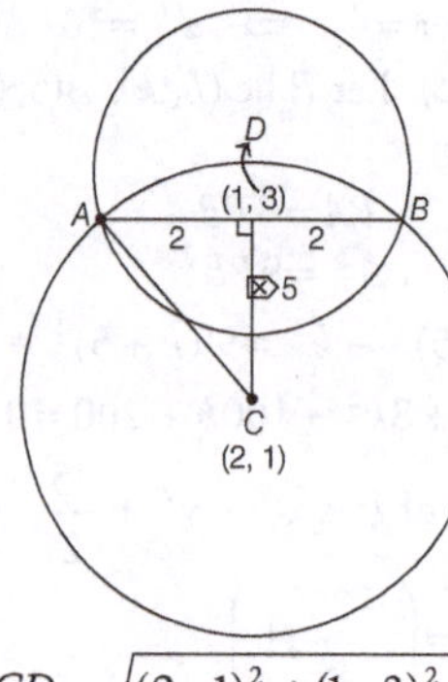


$CD = \sqrt{(2-1)^2 + (1-3)^2}$

$= \sqrt{1+4} = \sqrt{5}$

Now, by using Pythagoras theorem in $\Delta ADC$,

$AC^2 = AD^2 + CD^2 = (2)^2 + (\sqrt{5})^2 = 4 + 5$

$AC = \sqrt{9} = 3$

$\therefore$ Required radius $= 3$

**136.** (3) Since, it is given that centre of the circle lies on the line $x + y = 2$, so let the centre is $(h, 2 - h)$ lies in the first quadrant, so $h > 0$ and $2 - h > 0$

$\Rightarrow 0 < h < 2$

And the circle touches the lines $x = 3$ and $y = 2$, so radius

$r = |h - 3| = |2 - h - 2|$

$\Rightarrow r = |h - 3| = |h| \Rightarrow h = 3/2$

$\therefore \quad r = 3/2$

$\Rightarrow$ Diameter of the circle $= 3$

**137.** (9) Since, the line $3x + 4y = k$ intersects the circle

$x^2 + y^2 - 2x - 4y + 4 = 0$

$\Rightarrow (x-1)^2 + (y-2)^2 = 1$ at two distinct points, so perpendicular distance from centre (1, 2) to the line $3x + 4y - k = 0$ must be less than the radius 1.

$\therefore \left|\dfrac{3 + 8 - k}{\sqrt{9 + 16}}\right| < 1$

$\Rightarrow |11 - k| < 5 \Rightarrow (k - 11) \in (-5, 5)$

$\Rightarrow k \in (6, 16)$

$\Rightarrow k = 7, 8, 9, 10, 11, 12, 13, 14, 15$

$\therefore$ Number of integral values of $k$ is 9.

## Topic 2

**01.** (c) $PT: ty = x + at^2$

$PS = PT$

Slope of tangent $PT$

$= \dfrac{1}{t} = \tan 30° = \dfrac{1}{\sqrt{3}}$

$t = \sqrt{3}$

$\alpha = at = \sqrt{3}\ (a = 1)$

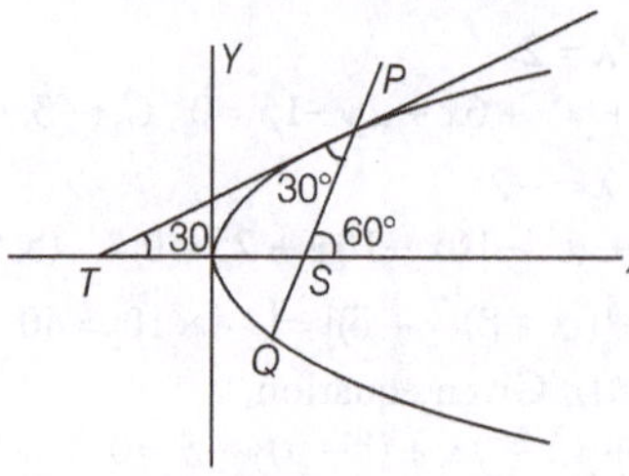


$\therefore \quad 5\alpha^2 = 15$

**02.** (c) $P(at^2, 2at) \Rightarrow P(4t^2, 8t) \equiv (1, -4)$

$\Rightarrow \quad t = -\dfrac{1}{2}$

$Q\left(\dfrac{a}{t^2}, -\dfrac{2a}{t}\right)$

$S \equiv (4, 0)$ is the focus

$PS = a + at^2$

$QS = a + \dfrac{a}{t^2}$

$\Rightarrow \quad \dfrac{PS}{QS} = t^2 = \dfrac{1}{4} = \dfrac{m}{n}$

$\Rightarrow m^2 + n^2 = 1 + 16 = 17$

**03.** (d) focus $= (2, 2)$

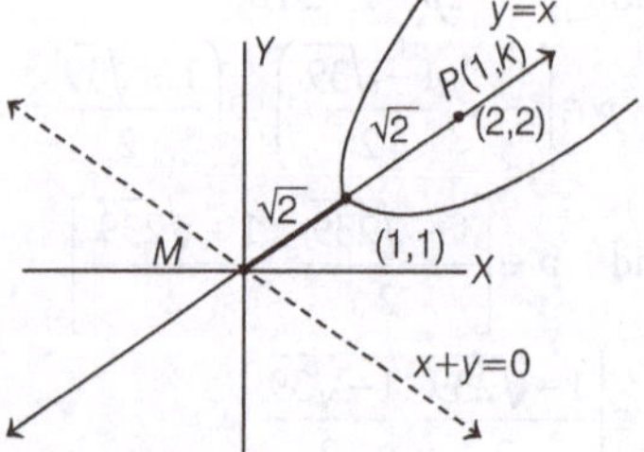


Equation of direction

$x + y = 0$

Equation of parabola

$SP = PM$

$\sqrt{(1-2)^2 + (k-2)^2}$

$= \left|\dfrac{1 + k}{\sqrt{1+1}}\right|$

$(x-2)^2 + (y-2)^2 = \dfrac{(x+y)^2}{2}$

$\therefore \quad (1, k)$

$1 + (k-2)^2 = \dfrac{(k+1)^2}{2}$

$\therefore \quad k = 9$

**04.** (d)

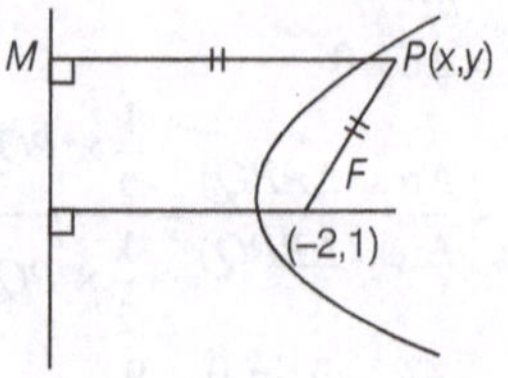


$PM = PF$

$\Rightarrow \left|\dfrac{2x + y + 2}{\sqrt{5}}\right| = \sqrt{(x+2)^2 + (y-1)^2}$

Now, obscissa of $P$ is $-2 \Rightarrow x = -2$

$\left|\dfrac{y - 2}{\sqrt{5}}\right| = \sqrt{0 + (y-1)^2}$

$\Rightarrow \quad (y-2)^2 = 5(y-1)^2$

$\Rightarrow 4y^2 - 6y + 1 = 0$

$\therefore$ Sum of ordinate $= \dfrac{3}{2}$

**05.** (b) $y^2 = 12x$

Equation of chord $PQ$, whose mid-point (4, 1) is $T = S_1$

$\Rightarrow yy_1 - 12\left(\dfrac{x_1 + x}{2}\right) = y_1^2 - 12x_1$

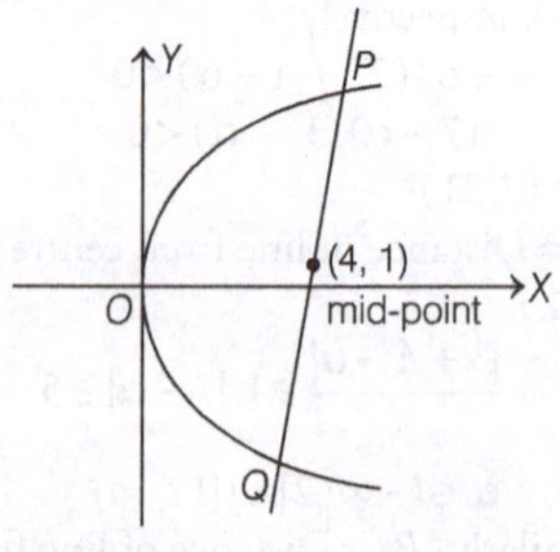


$y(1) - \dfrac{12(x+4)}{2} = (1)^2 - 12(4)$

$\Rightarrow \quad y - 6(x + 4) = 1 - 48$

$\Rightarrow \quad y - 6x - 24 = -47$

$\Rightarrow \quad 6x - y = 23$

$\because \left(\dfrac{1}{2}, -20\right)$ satisfy above equation.

$\therefore$ Lines passes through $\left(\dfrac{1}{2}, -20\right)$.

> **Concept Enhancer**
> Equation of chord of parabola $y^2 = 4ax$ whose mid-point is $(x_1, y_1)$,
> $yy_1 - \dfrac{4a(x + x_1)}{2} = y_1^2 - 4ax_1$

**06.** (c) Given equation of parabola is
$y^2 = 4x$

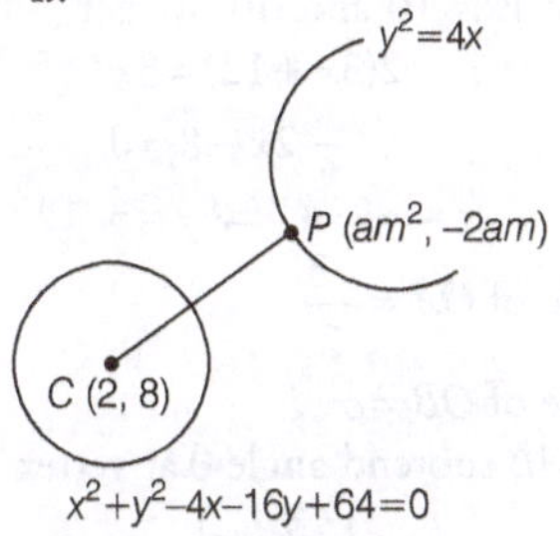


and equation of circle is
$$x^2 + y^2 - 4x - 16y + 64 = 0$$
Now, equation of normal to parabola is given by
$y = mx - 2m - m^3$, where $m$ = slope
This normal passing through centre of the circle which is (2, 8).
So, $8 = 2m - 2m - m^3$
$\Rightarrow m^3 = -8 = (-2)^3 \Rightarrow m = -2$
Coordinates of any point $P$ taken on given parabola is of the form $(am^2, -2am)$ i.e (4, 4)
Now, $CP = \sqrt{(4-2)^2 + (4-8)^2}$
$= \sqrt{4+16} = \sqrt{20} = d$ [given]
$\therefore \quad d^2 = 20$

**07.** (c) Given, equation of tangent at vertex is $x + y - a = 0$
Focus is $(a, a)$.
$\therefore$ Latus rectum $= 4\left|\dfrac{a + a - a}{\sqrt{2}}\right| = 16$
$\Rightarrow \left|\dfrac{a}{\sqrt{2}}\right| = 4 \Rightarrow \dfrac{a^2}{2} = 16$
$\Rightarrow a^2 = 32 \Rightarrow |a| = 4\sqrt{2}$

**08.** (c)

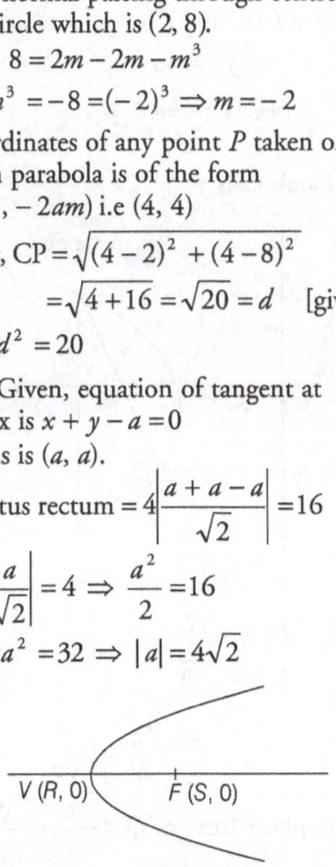


Length of $LR$ of $y^2 = 4ax$ is $4a$.
Or Length of $LR = 4 \times$ [Distance between focus and vertex]
Similarly here, Length of $LR = 4(S - R)$

**09.** (a) Let $AB$ be the latus rectum of parabola $y^2 = 4x$ having length 4.

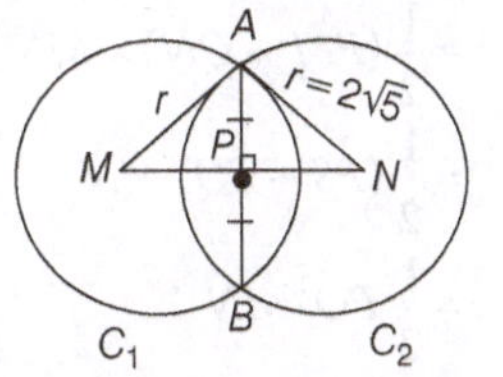


From the figure as $AB$ divides $MN$ equally.
So, $MN = 2MP = 2\sqrt{AM^2 - PA^2}$ as
$AB = 2PA = 2\sqrt{(2\sqrt{5})^2 - 2^2} = 8$

**10.** (c) We know that the locus of point $P$ from which two perpendicular tangents are drawn to the parabola, is the directrix of the parabola.
Hence, the required locus is $x = -1$.

**11.** (d) Let $M(h, k)$ be point of locus of mid point of $PQ$
$\Rightarrow \dfrac{x_1 + 1}{2} = h, \dfrac{y_1 + 0}{2} = k,$

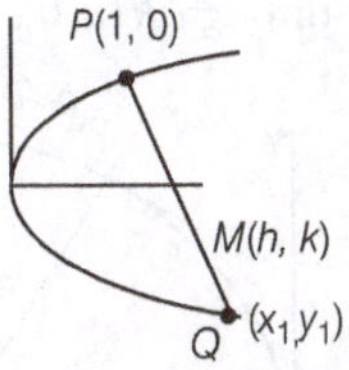


$\therefore \quad x_1 = 2h - 1, \; y_1 = 2k$
Now, $(x_1, y_1)$ lines on $y^2 = 8x$
$\Rightarrow \quad (2k)^2 = 8(2h - 1)$
$\therefore$ locus is
$\Rightarrow \quad y^2 = (2x - 1)$
$\Rightarrow \; y^2 - 4x + 2 = 0$

**12.** (a) Equation of the normal at point $(bt_1^2, 2bt_1)$ on parabola is
$$y = -t_1 x + 2bt_1 + bt_1^3$$
It also passes through $(bt_2^2, 2bt_2)$, then
$$2bt_2 = -t_1 \cdot bt_2^2 + 2bt_1 + bt_1^3$$
$\Rightarrow \; 2t_2 - 2t_1 = -t_1(t_2^2 - t_1^2)$
$= -t_1(t_2 + t_1)(t_2 - t_1)$
$\Rightarrow \quad 2 = -t_1(t_2 + t_1)$
$\therefore \quad t_2 = -t_1 - \dfrac{2}{t_1}$

**13.** (d) Given equation of parabola can be rewritten as
$$(y + 2)^2 = -4\left(x - \frac{1}{2}\right)$$
Let $\quad y + 2 = Y$ and $\quad x - \dfrac{1}{2} = X$
$\therefore \quad Y^2 = -4X$
Here, $\quad a = 1$
$\therefore$ Equation of directrix is $X = a$.
$\therefore \quad x - 1/2 = 1 \quad \Rightarrow \quad x = 3/2$

**14.** (c) $y^2 = x - 2 \Rightarrow 2y\dfrac{dy}{dx} = 1 \Rightarrow y\dfrac{dy}{dx} = \dfrac{1}{2}$
$$x^2 = y - 2 \Rightarrow 2x = \frac{dy}{dx} = 1$$
$$x = 1/2$$
Point on $y^2 = x - 2$ will be $\left(\dfrac{9}{4}, \dfrac{1}{2}\right)$
Point on $x^2 = y - 2$ will be $\left(\dfrac{1}{2}, \dfrac{9}{4}\right)$

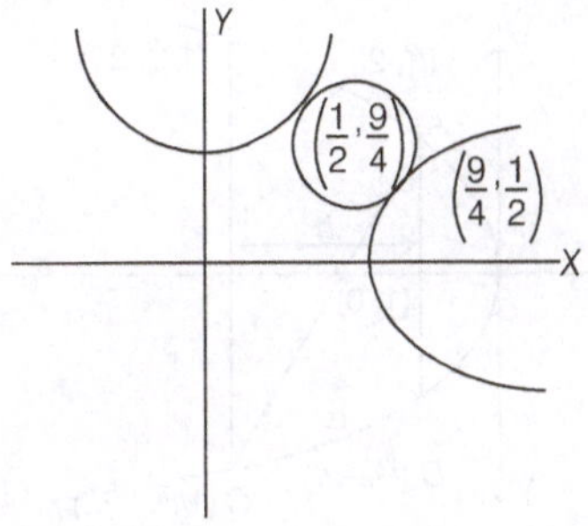


Diameter $= 2r$
= Shortest distance btween two points.
$= \sqrt{\left(\dfrac{9}{4} - \dfrac{1}{2}\right)^2 + \left(\dfrac{1}{2} - \dfrac{9}{4}\right)^2}$
$= \sqrt{\dfrac{2 \times 49}{16}} = \dfrac{7\sqrt{2}}{4}$

**15.** (d) According to the question, scenario, diagram is created as below-
Now, let us consider that coordinates of $A$ and $B$ are $(x_1, y_1)$ and $(x_2, y_2)$, respectively.

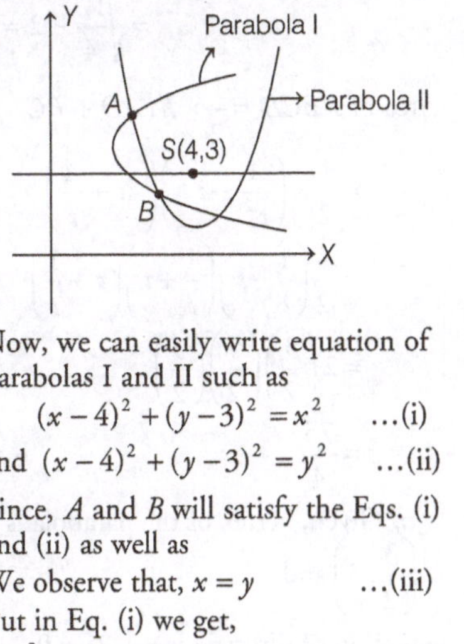


Now, we can easily write equation of parabolas I and II such as
$$(x - 4)^2 + (y - 3)^2 = x^2 \quad \ldots\text{(i)}$$
and $(x - 4)^2 + (y - 3)^2 = y^2 \quad \ldots$(ii)
Since, $A$ and $B$ will satisfy the Eqs. (i) and (ii) as well as
We observe that, $x = y \quad \ldots$(iii)
Put in Eq. (i) we get,
$x^2 - 14x + 25 = 0 \Rightarrow x_1 + x_2 = 14$
and $\quad x_1 x_2 = 25$
$\therefore \quad AB^2 = (x_1 - x_2)^2 + (y_1 - y_2)^2$
$= 2(x_1 - x_2)^2$
$[\because x = y \therefore x_1 = y_1 \text{ and } x_2 = y_2]$
$= 2[(x_1 + x_2)^2 - 4x_1 x_2]$
$= 2[(14)^2 - 4 \times 25]$
$= 2 \times [196 - 100]$
$= 2 \times 96 = 192$

**16.** *(c)* $y^2 = 4x \Rightarrow a = 1$

$\Rightarrow$ Focus $= (1, 0)$

Hence, $AC$ is a focal chord.

Length of focal chord

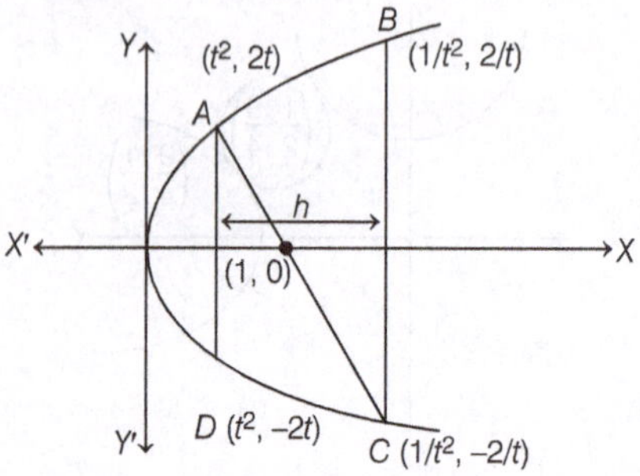


$$= a\left(t + \frac{1}{t}\right)^2 = \frac{25}{4}$$

$$\left(t + \frac{1}{t}\right)^2 = \frac{25}{4} \Rightarrow t + \frac{1}{t} = \frac{5}{2}$$

$$\Rightarrow \quad t^2 + \frac{1}{t^2} + 2 = \frac{25}{4}$$

$$\Rightarrow \quad t^2 + \frac{1}{t^2} = \frac{17}{4}$$

$$\Rightarrow \quad \left|t - \frac{1}{t}\right| = \sqrt{t^2 + \frac{1}{t^2} - 2}$$

$$= \sqrt{\frac{17}{4} - 2} = \frac{3}{2}$$

$$\text{Area of } ABCD = \frac{1}{2} \times h(AD + BC)$$

$$= \frac{1}{2} \times \left(\frac{1}{t^2} - t^2\right)\left(4t + \frac{4}{t}\right)$$

$$= \frac{4}{2}\left(\frac{1}{t} - t\right)\left(\frac{1}{t} + t\right)\left(t + \frac{1}{t}\right)$$

$$= 2\left(\frac{3}{2}\right)\left(\frac{5}{2}\right)\left(\frac{5}{2}\right)$$

$$= \frac{75}{4}$$

**17.** *(c)* Given, vertex of the parabola is $V\left(\frac{3}{2}, 3\right)$ and

equation of directrix is $x + 2y = 0$ ...(i)

So, equation of axis is

$$y - 3 = 2\left(x - \frac{3}{2}\right)$$

[$\because$ axis is $\perp$ to directrix and passes through focus]

$$\Rightarrow \quad y - 2x = 0 \quad \text{...(ii)}$$

On solving Eqs. (i) and (ii) we, get $x = 0$ and $y = 0$ which is coordinates of foot of directrix, i.e. (0, 0).

and thus focus $\equiv (3, 6)$

Now, $PS^2 = PM^2$

$$\Rightarrow (x - 3)^2 + (y - 6)^2 = \left(\frac{x + 2y}{\sqrt{5}}\right)^2$$

$$\Rightarrow 4x^2 + y^2 - 4xy - 30x - 60y + 225 = 0$$

Comparing this with

$$\alpha x^2 + \beta y^2 - \gamma xy - 30x - 60y + 225 = 0,$$

we get $\alpha = 4, \beta = 1$ and $\gamma = 4$

So, $\quad \alpha + \beta + \gamma = 9$

**18.** *(c)* Parabola $y^2 = 4x$

Normal at $P(t)$ to the parabola

$$y + tx = 2t + t^3$$

$\because$ It passes through $R(a, 0)$.

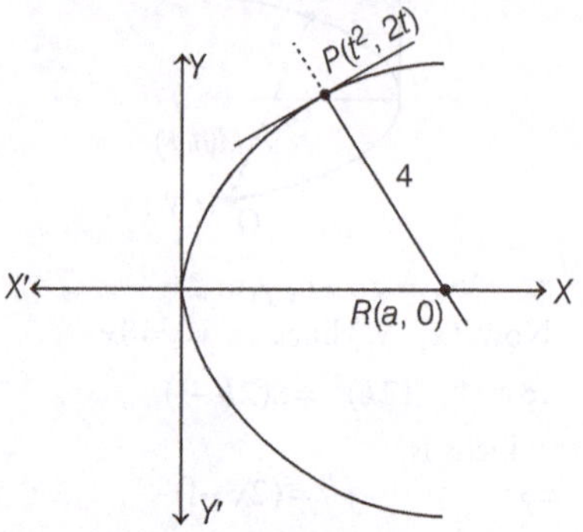


$$\Rightarrow \quad at = 2t + t^3$$

$$\Rightarrow \quad a = 2 + t^2 \Rightarrow a - t^2 = 2$$

$$\because \quad PR = 4$$

$$PR^2 = 16$$

$$\Rightarrow \quad (a - t^2)^2 + (2t)^2 = 16$$

$$\Rightarrow \quad (2)^2 + 4t^2 = 16$$

$$\Rightarrow \quad t^2 = 3$$

$$\therefore \quad t = \pm\sqrt{3}$$

$$a = 2 + 3 = 5 \Rightarrow R(5, 0)$$

$\because$ Focus of parabola $= (1, 0)$

$\because$ (1, 0) and (5, 0) are end points of diameter.

$\therefore$Equation of circle

$$(x - 1)(x - 5) + (y - 0)(y - 0) = 0$$

$$x^2 + y^2 - 6x + 5 = 0$$

**19.** *(d)* Given lines

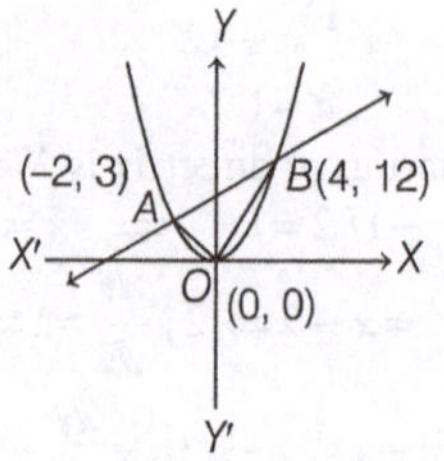


$$3x - 2y + 12 = 0 \quad \text{...(i)}$$

$$4y = 3x^2 \quad \text{...(ii)}$$

Form Eqs. (i) and (ii), we get

$$2(3x + 12) = 3x^2$$

$$x^2 - 2x - 8 = 0$$

$$x = -2, 4 \Rightarrow y = 3, 12$$

$$\text{Slope of } OA = \frac{-3}{2}$$

$$\text{Slope of } OB = 3$$

Let $AB$ subtend angle $\theta$ at vertex

$$\tan\theta = \left|\frac{\frac{3}{2} - 3}{1 - \frac{9}{2}}\right|$$

$$\Rightarrow \quad \tan\theta = \frac{9}{7} \Rightarrow \theta = \tan^{-1}\left(\frac{9}{7}\right)$$

**20.** *(a)* $\because P(4, 4\sqrt{3})$ lies on parabola

$$y^2 = 4ax$$

$$\therefore \quad (4\sqrt{3})^2 = 4a \times 4$$

$$\Rightarrow \quad a = 3$$

$\therefore$ Parabola is $y^2 = 12x$

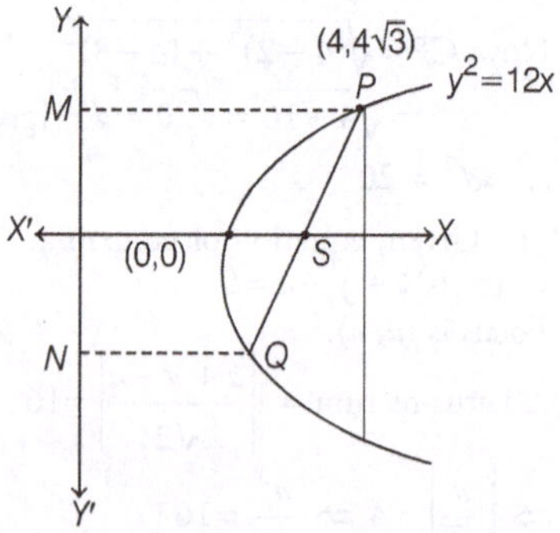


$$P(4, 4\sqrt{3})$$

$$2at = 4\sqrt{3}$$

$$t = \frac{4\sqrt{3}}{6} = \frac{2}{\sqrt{3}}$$

$$\text{Parametric form of } Q = -\frac{1}{t} = -\frac{\sqrt{3}}{2}$$

$$\therefore \quad Q = \left(3\left(-\frac{\sqrt{3}}{2}\right)^2, 2(3)\left(-\frac{\sqrt{3}}{2}\right)\right)$$

$$= \left(\frac{9}{4}, -3\sqrt{3}\right)$$

$\therefore$ Area of trapezium $PQNM$

$$= \frac{1}{2}(PM + QN) \times MN$$

$$= \frac{1}{2}(PS + SQ)\, MN$$

$$= \frac{1}{2}\, PQ \cdot MN$$

$$=\frac{1}{2}\sqrt{\left(\frac{9}{4}-4\right)^2+(-3\sqrt{3}-4\sqrt{3})^2}\cdot 7\sqrt{3}$$

$$=\frac{1}{2}\times 7\times\frac{7}{4}\times 7\sqrt{3}=\frac{343\sqrt{3}}{8}$$

**21.** (c) Given that $PQ$ be a focal chord of the parabola $y^2=36x$ of length 100.

Thus, length $=100 \Rightarrow 9\left(t+\frac{1}{t}\right)^2=100$

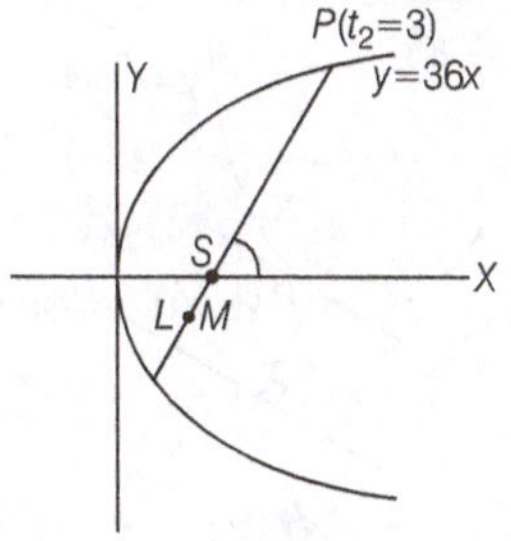


$\Rightarrow \left(t+\frac{1}{t}\right)^2=\frac{100}{9} \Rightarrow t+\frac{1}{t}=\pm\frac{10}{3}$

$\therefore \quad t=3$ or $1/3$

Thus, $P=(81, 54)$ and $Q=(1, -6)$ and $M=(21, 9)$

Thus, required line $y-9=-\frac{4}{3}(x-21)$

$\Rightarrow 3y-27=-4x+84 \Rightarrow 4x+3y=111$

Clearly, (3, 33), (6, 29), and (–6, 45) all the points satisfied the given line which passing through $M(21, 9)$ and perpendicualr to $PQ$.

So, only (–3, 43) not hold$4x+3y=111$.

**22.** (d) $\because$ at $y=1$

$$ax^2+2bx+c=0 \quad \ldots\text{(i)}$$

and $dx^2+2ex+f=0 \quad \ldots\text{(ii)}$

$\because a, b, c$ are in GP.

So, $b^2=ac \Rightarrow b=\sqrt{a}\sqrt{c}$

Put $b=\sqrt{ac}$ in Eq. (i), we get

$ax^2+2\sqrt{a}\sqrt{c}\,x+c=0$

$\Rightarrow (\sqrt{a}x+\sqrt{c})^2=0 \therefore x=-\frac{\sqrt{c}}{\sqrt{a}}$

Put this value in Eq. (ii), we get

$$d\frac{c}{a}-2e\cdot\frac{\sqrt{c}}{\sqrt{a}}+f=0$$

$\Rightarrow \frac{cd}{a}+f=2e\sqrt{\frac{c}{a}} \Rightarrow cd+af=2e\sqrt{ac}$

$\Rightarrow cd+af=2e\cdot b \quad [\because b=\sqrt{ac}]$

Divided by $ac$, $\frac{cd}{ac}+\frac{af}{ac}=\frac{2eb}{ac}$

$$\frac{d}{a}+\frac{f}{c}=\frac{2eb}{b^2} \quad [\because b^2=ac]$$

$$\frac{d}{a}+\frac{f}{c}=\frac{2e}{b}$$

$\Rightarrow \frac{d}{a}, \frac{e}{b}, \frac{f}{c}$ are in AP.

**23.** (d) $y^2=8x$

Let $p(t)=(a, b) \Rightarrow a=2t^2, b=4t$

Tangent at $P$ is $x-ty+2t^2=0$

This passes through (5, 7).

$\therefore 5-7t+2t^2=0$

$\Rightarrow$ Two possible points are $p(t_1), p(t_2)$, where, $t_1+t_2=\frac{7}{2}, t_1\cdot t_2=\frac{5}{2}$

Now, $A=(2t_1)^2\times(2t_2)^2=4(t_1t_2)^2$

$$=4\times\frac{25}{4}=25$$

and $B=(4t_1)\times(4t_2)=16\times\frac{5}{2}=40$

$\therefore A+B=65$

**24.** (b) Let the equation of the tangent of the parabola $y=x^2$ be

$$y=mx-\frac{m^2}{4}$$

Now, the tangent also touches the parabola $y=-(x-2)^2$,

so $mx-\frac{m^2}{4}=-(x-2)^2$

$\Rightarrow \quad 4mx-m^2=-4(x^2-4x+4)$

$\Rightarrow \quad 4mx-m^2+4x^2-16x+16=0$

$\Rightarrow \quad 4x^2+4(m-4)x+16-m^2=0$

As the tangent touches parabola at only one point, $D=0$

$$[4(m-4)]^2-4\times 4(16-m^2)=0$$

$\Rightarrow \quad 16[m^2-8m+16-16+m^2]=0$

$\Rightarrow \quad 2m^2-8m=0 \Rightarrow m=0, 4$

Equation of tangent is $y=0$ or $y=4x-4=4(x-1)$

**25.** (a) Let $P(h, k)$ be a point on the curve.

Equation of line passing $(h, k)$ is $y-k=m(x-h)$ where $m=dy/dx$

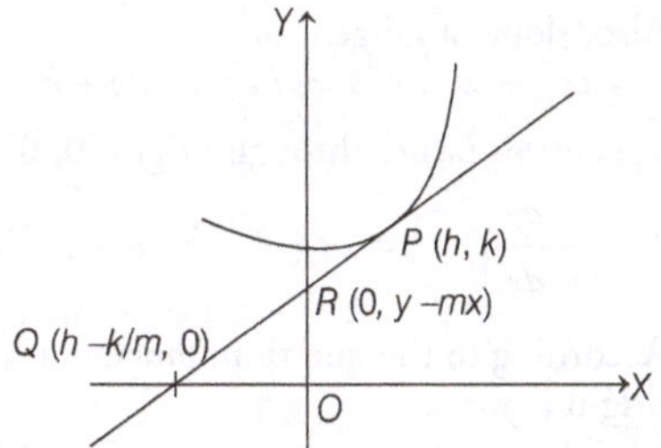


At $Q$, $y=0 \Rightarrow x=h-\frac{k}{m}$

As $Y$-axis bisects $PQ$,

$$\frac{h+h-\frac{k}{m}}{2}=0 \Rightarrow h+h-\frac{k}{m}=0$$

$$\Rightarrow \quad \frac{dy}{y}=\frac{dx}{2x}$$

[Replacing $h$ by $x$ and $k$ by $y$]

$\Rightarrow \ln y=\frac{1}{2}\ln x+\ln C$, where $C$ is a constant

$\Rightarrow \quad y^2=Cx$

It passes through the point (3,3)

$\Rightarrow \quad C=3$

Hence, $y^2=3x$

Then, length of latus rectum is 3.

**26.** (b) Vertex $\left(\frac{1}{2}, \frac{3}{4}\right)$

Equation of directrix $y=\frac{1}{2}$

Equation of parabola is

$$\left(x-\frac{1}{2}\right)^2=y-\frac{3}{4}$$

Point on parabola $P\left(\frac{-1}{2}, \frac{7}{4}\right)$

Equation of normal at $P\left(\frac{-1}{2}, \frac{7}{4}\right)$ is

$x=2y-4$

This normal cuts the parabola at $Q(2, 3)$

$$(PQ)^2=\left(2+\frac{1}{2}\right)^2+\left(3-\frac{7}{4}\right)^2=\frac{125}{16}$$

**27.** (a) Given, parabola $y^2=8x \quad \ldots\text{(i)}$

Equation of tangent at $P(2, -4)$ is

$$-4y=4(x+2)$$

Or $\quad x+y+2=0 \quad \ldots\text{(ii)}$

and Equation of normal to the parabola is

$$x-y+C=0$$

$\therefore$Normal passes through $(2, -4)$

$\therefore \quad C=-6$

Normal: $\quad x-y=6 \quad \ldots\text{(iii)}$

Equation of directrix of parabola

$$x=-2 \quad \ldots\text{(iv)}$$

Point of intersection of tangent and normal with directrix are $x=-2$ at $A(-2, 0)$ and $B(-2, -8)$ respectively.

$Q(a, b)$ and $P(2, -4)$ are given and $AQBP$ is a square.
Mid-point of $AB$ = Mid-point of $PQ$

$\Rightarrow \quad (-2, -4) = \left(\frac{a+2}{2}, \frac{b-4}{2}\right)$

$\Rightarrow \quad a = -6, b = -4$

$\Rightarrow \quad 2a + b = -16$

**28.** *(d)* Equation of tangent to $y^2 = 30x$ is

$$y = mx + \frac{30}{4m}$$

Now, this tangent passes through $(-30, 0)$.

$\therefore \quad 0 = -30m + \frac{30}{4m}$

$\Rightarrow \quad \frac{30}{4m} = 30\,m \Rightarrow m^2 = \frac{1}{4}$

$\Rightarrow \quad m = \pm\frac{1}{2}$

$\therefore$ Equation of tangent is

$y = \frac{x}{2} + 15$ or $y = -\frac{x}{2} - 15$

Now equation of circle is

$x^2 + y^2 + 30x + \frac{675}{4} = 0$

Let perpendicular distance of the tangent from the centre $(-15, 0)$ of the circle $= p$

$$\therefore \quad p = \frac{\left|\frac{-15}{2} + 15\right|}{\sqrt{1 + \frac{1}{4}}} = 3\sqrt{5}$$

$\therefore$ Length of chord $= 2\sqrt{r^2 - p^2}$

$$= 2\sqrt{\left(15^2 + 0 - \frac{675}{4}\right) - 45} = 3\sqrt{5}$$

where, $r$ is radius of the given circle.

**29.** *(a)* The mirror image of any point $(\alpha, \beta)$ with respect to line $y = x$ is simply $(\beta, \alpha)$.

Let $(h, k)$ be the mirror image of a point on parabola $y^2 = 4ax$

Then, $(k, h)$ will be the mirror image of $(h, k)$ and it will lie on parabola.

So, $\quad y^2 = 4x$

$\quad h^2 = 4k$

$\Rightarrow \quad x^2 = 4y$

Hence, Locus is $x^2 = 4y$ ...(i)

For finding equation of tangent differentiate Eq. (i) w.r.t. $x$

$$2x = 4\frac{dy}{dx}$$

$\Rightarrow \quad \left(\frac{dy}{dx}\right)_{(2,1)} = \left(\frac{2}{2}\right) = 1 \Rightarrow \frac{y-1}{x-2} = 1$

$\quad y = x - 1$

$\therefore$ Equation of tangent $\Rightarrow y = x - 1$

$\Rightarrow \quad x - y = 1$

**30.** *(d)* Given, equation of parabola

$\Rightarrow y^2 = 2x$

Equation of normal of parabola, $y^2 = 4ax$ is $tx + y = 2at + at^3$

Here, $4a = 2$

So, equation of normal

$\Rightarrow \quad tx + y = t + t^3/2$

$\quad t^3 + (2 - 2x)t - 2y = 0$

As, there are three normals which are passing through $(a, 0)$, so there must be three roots of this equation.

$t^3 + (2 - 2a)t - 2.0 = 0$

$\Rightarrow \quad t^3 + (2 - 2a)t = 0$ (roots $t_1$, $t_2$, $t_3$)

$\therefore \quad t_1 + t_2 + t_3 = 0$

and $\quad t_1t_2 + t_2t_3 + t_3t_1 = (2 - 2a)$

$\quad t_1, t_2, t_3 \in R$

So, $t_1^2 + t_2^2 + t_3^2 > 0$

$(t_1 + t_2 + t_3)^2 - 2(t_1t_2 + t_2t_3 + t_3t_1) > 0$

$\Rightarrow \quad 0 - 2(2 - 2a) > 0 \Rightarrow a - 1 > 0$

$\Rightarrow \quad a > 1$

> ***Shortcut Method***
>
> For standard parabola $y^2 = 4ax$, for more than 3 normals (on axis) $x > \frac{L}{2}$; $L$ = length of latus rectum
>
> Now, for $y^2 = 2x$; $\left(a = \frac{1}{2}\right)$
>
> $LR = 2$
>
> for point $(a, 0)$
>
> $a > \frac{L}{2} \Rightarrow a > 1$

**31.** *(c)* Given curve,

$\Rightarrow y = ax^2 + bx + c, x \in R$ and point $(1, 2)$

$\because$ The given curve passes through $(1, 2)$.

$\therefore \quad 2 = a + b + c$

Also, slope of tangent of $y = ax^2 + bx + c$ is $dy/dx = 2ax + b$

$\because$ Tangent passes through origin $(0, 0)$.

$\therefore \quad \left.\frac{dy}{dx}\right|_{(0,0)} = 2a \times 0 + b = b$ ... (i)

According to the question, tangent at origin is $y = x$

$\therefore$ Its slope is 1. ... (ii)

From Eqs. (i) and (ii),

$\quad b = 1$

Also, $a + b + c = 2 \Rightarrow a + c = 1$

From the option look for $b = 1$ and $a + c = 1$

The only correct order triplet is $a = 1$, $b = 1$, $c = 0$.

**32.** *(d)* Given, curve $y = x^2 + 4$ and, line $y = 4x - 1$

Here, $\quad y = x^2 + 4$

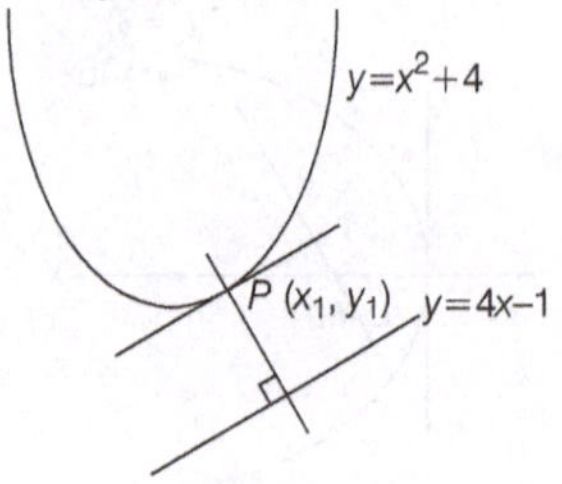


$\therefore \quad \frac{dy}{dx} = 2x$ ... (i)

and $\quad y = 4x - 1$

$\quad \frac{dy}{dx} = 4$ ... (ii)

Let the required point be $P(x_1, y_1)$.

$\therefore \quad \left.\frac{dy}{dx}\right|_P = 2x_1$ ... (iii)

$\because$ Slopes will be equal.

$\therefore\ 2x_1 = 4$ [From Eqs. (ii) and (iii)]

$\Rightarrow \quad x_1 = \frac{4}{2} = 2$

Now, the given point $P(x_1, y_1)$ lies on curve $y = x^2 + 4$,

$\therefore \quad y_1 = x_1^2 + 4$

$\Rightarrow \quad y_1 = 2^2 + 4 = 8$

Hence, required coordinate of $P = (2, 8)$

**33.** *(d)* It is given that circle passes through the points $A(0, 1)$ and $B(2, 4)$, so the centre of the circle is point of intersection of perpendicular bisector of $AB$ and the normal to the parabola $y = x^2$ at point $B(2, 4)$, as circle touches the parabola at point $B(2, 4)$ itself.

Now, equation of normal to the parabola $y = x^2$ at point $B(2, 4)$ is

$\quad y - 4 = -1/4(x - 2)$

$\Rightarrow \quad x + 4y = 18$ ...(i)

and equation of perpendicular bisector of points $A(0, 1)$ and $B(2, 4)$ is

$\quad y - 5/2 = -2/3(x - 1)$

$\Rightarrow \quad 4x + 6y = 19$ ...(ii)

From Eqs. (i) and (ii), we get

$\Rightarrow \quad y = 53/10$

$\therefore$ Coordinate of centre is $\left(-\frac{16}{5}, \frac{53}{10}\right)$.

**34.** *(a)* Equation of tangent to the parabola $y^2 = 4x$ having slope '$m$' is

$$y = mx + \frac{1}{m} \quad \ldots(i)$$

$\because$ Line (i) is also tangent to the parabola $x^2 = 4y$

So, the discriminant of

$x^2 - 4mx - \frac{4}{m} = 0$ must be zero, so

$16m^2 + \frac{16}{m} = 0 \Rightarrow m = -1$

Therefore, the equation of common tangent to given parabolas is

$$x + y + 1 = 0 \quad \ldots(ii)$$

It is given that line (ii) is tangent to the circle $x^2 + y^2 = c^2$

$\therefore \quad c = 1/\sqrt{2}$

**35.** *(c)* Equation of given parabola is $y^2 = 8x$ and the equilateral triangle inscribed in the given parabola having one of its vertices on the vertex of the parabola, so according to symmetry other two vertices of the equilateral triangle, we can take as, $A(2t^2, 4t)$ and $B(2t^2, -4t)$

$\therefore$ Area of $\Delta OAB$ is

$$\Delta = \frac{1}{2}(2t^2)(8t) = \frac{\sqrt{3}}{4}(8t)^2$$

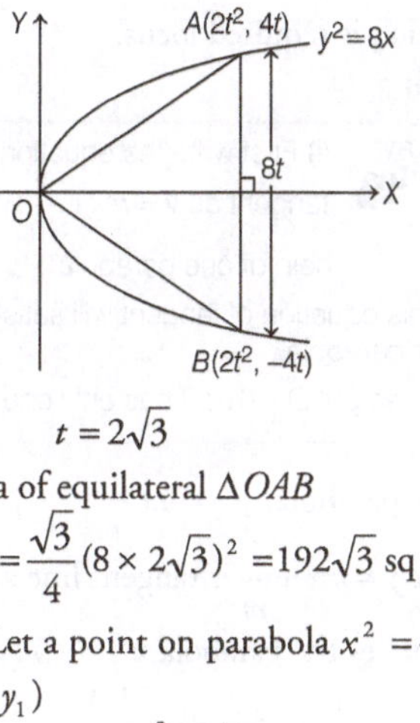


$\Rightarrow \quad t = 2\sqrt{3}$

$\therefore$ Area of equilateral $\Delta OAB$

$$= \frac{\sqrt{3}}{4}(8 \times 2\sqrt{3})^2 = 192\sqrt{3} \text{ sq unit}$$

**36.** *(c)* Let a point on parabola $x^2 = 4y$ is $A(x_1, y_1)$

So $\quad x_1^2 = 4y_1 \quad \ldots(i)$

and an another point $P(h, k)$ which divides the line segment joining the point $B(0, -1)$ and $A(x_1, y_1)$ internally in the ratio 1 : 2, so

$$h = \frac{(2 \times 0) + (1 \times x_1)}{2 + 1}$$

and $\quad k = \frac{2 \times (-1) + (1 \times y_1)}{2 + 1}$

$\Rightarrow x_1 = 3h$ and $y_1 = 3k + 2 \quad \ldots(ii)$

Now, on the elimination of $(x_1, y_1)$ with the help of Eq. (i) and relation (ii), we get $9h^2 = 4(3k + 2)$

On taking locus of point $P(h, k)$, we have

$$9x^2 = 12y + 8 \Rightarrow 9x^2 - 12y = 8$$

**37.** *(b)* As we know, equation of tangent to the parabola $y^2 = 4x$, having slope '$m$' is

$$y = mx + \frac{1}{m} \quad \ldots(i)$$

On comparing the Eq. (i) with the equation of given tangent $y = mx + 4$,

we get $\frac{1}{m} = 4 \Rightarrow m = \frac{1}{4}$

$\therefore$ Equation of the tangent is

$y = \frac{1}{4}x + 4$, now it is tangent to the parabola $x^2 = 2by$, so on solving the equation of parabola $x^2 = 2by$ and tangent $y = \frac{1}{4}x + 4$, we must get only a common point, so $x^2 = 2b\left(\frac{1}{4}x + 4\right)$

$\Rightarrow 2x^2 - bx - 16b = 0$ is a quadratic equation having one solution.

So, $\quad D = 0 \Rightarrow b^2 + 4(2)(16b) = 0$

$\Rightarrow \quad b = -128 \quad [\because b$ cannot be zero$]$

**38.** *(c)* Given equation of parabola is

$$y = (x - 2)^2 - 1$$

$\Rightarrow \quad y = x^2 - 4x + 3 \quad \ldots(i)$

Now, let $(x_1, y_1)$ be the point of intersection of tangents of parabola (i) and line.

$$x - y = 3, \text{ then}$$

Equation of chord of contact of point $(x_1, y_1)$ w.r.t. parabola (i) is

$$T = 0$$

$\Rightarrow \quad \frac{1}{2}(y + y_1) = xx_1 - 2(x + x_1) + 3$

$\Rightarrow \quad y + y_1 = 2x(x_1 - 2) - 4x_1 + 6$

$\Rightarrow \quad 2x(x_1 - 2) - y = 4x_1 + y_1 - 6,$

this equation represent the line $x - y = 3$ only, so on comparing, we get

$$\frac{2(x_1 - 2)}{1} = \frac{-1}{-1} = \frac{4x_1 + y_1 - 6}{3}$$

$\Rightarrow \quad x_1 = \frac{5}{2}$ and $y_1 = -1$

So, the required point is $\left(\frac{5}{2}, -1\right)$.

**39.** *(c)*

**key idea** Use the equation of tangent of slope $m$ to the parabola $y^2 = 4ax$ is $y = mx + \frac{a}{m}$ and a line $ax + by + c = 0$ touches the circle $x^2 + y^2 = r^2$, if $\frac{|c|}{\sqrt{a^2 + b^2}} = r$.

Since, equation of given parabola is $y^2 = 4\sqrt{2}x$ and equation of tangent line is $ax + y = c$ or $y = -ax + c$,

then $c = \frac{\sqrt{2}}{m} = \frac{\sqrt{2}}{-a}$

$[\because m = \text{slope of line} = -a]$

$[\because$ Line $y = mx + c$ touches the parabola $y^2 = 4ax$ iff $c = a/m]$.

Then, equation of tangent line becomes

$$y = -ax - \frac{\sqrt{2}}{a} \quad \ldots(i)$$

$\because$ Line (i) is also tangent to the circle $x^2 + y^2 = 1$.

$\therefore \quad$ Radius $= 1 = \frac{\left|-\frac{\sqrt{2}}{a}\right|}{\sqrt{1 + a^2}}$

$\Rightarrow \quad \sqrt{1 + a^2} = \left|-\frac{\sqrt{2}}{a}\right| \Rightarrow 1 + a^2 = \frac{2}{a^2}$

$\Rightarrow \quad a^4 + a^2 - 2 = 0$

$\Rightarrow \quad (a^2 + 2)(a^2 - 1) = 0$

$\Rightarrow \quad a^2 = 1 \quad [\because a^2 > 0, \forall a \in R]$

$\therefore \quad |c| = \frac{\sqrt{2}}{|a|} = \sqrt{2}$

**40.** *(b)*

**key idea** (i) First find the focus of the given parabola

(ii) Then, find the slope of the focal chord by using $m = \frac{y_2 - y_1}{x_2 - x_1}$

(iii) Now, find the length of the focal chord by using the formula $4a \operatorname{cosec}^2 \alpha$

Equation of given parabola is $y^2 = 16x$, its focus is (4, 0).

$\therefore$ Slope of focal chord having one end point is (1, 4) is $m = \tan \alpha = \frac{4 - 0}{1 - 4} = -\frac{4}{3}$

[Where, $\alpha$ is the inclination of focal chord with $X$-axis]

Since, the length of focal chord
$$=4a\,\text{cosec}^2\alpha$$
$\therefore$ The required length of the focal chord $=16\,[1+\cot^2\alpha]$
$$[\because a=4 \text{ and } \text{cosec}^2\alpha=1+\cot^2\alpha]$$
$$=16\left[1+\frac{9}{16}\right]=25 \text{ units}$$
$$\left[\because \cot\alpha=\frac{1}{\tan\alpha}=-\frac{3}{4}\right]$$

**41.** *(c)* Normal to parabola $y^2=4ax$ is given by
$$y=mx-2am-am^3$$
$\therefore$ Normal to parabola $y^2=4b(x-c)$ is
$$y=m(x-c)-2bm-bm^3$$
[replacing $a$ by $b$ and $x$ by $x-c$]
$$=mx-(2b+c)m-bm^3 \quad \ldots \text{(i)}$$
and normal to parabola $y^2=8ax$ is
$$y=mx-4am-2am^3 \quad \ldots\text{(ii)}$$
[replacing a by $2a$]

For common normal, we should have
$$mx-4am-2am^3$$
$$=mx-(2b+c)m-bm^3$$
[Using Eqs. (i) and (ii)]
$$\Rightarrow \quad 4am+2am^3=(2b+c)m+bm^3$$
$$\Rightarrow \quad (2a-b)m^3+(4a-2b-c)m=0$$
$$\Rightarrow \quad m((2a-b)m^2+(4a-2b-c))=0$$
$$\Rightarrow \quad m=0 \text{ or } m^2=\frac{2b+c-4a}{2a-b}$$
$$=\frac{c}{2a-b}-2$$
As, $m^2>0$, therefore $\dfrac{c}{2a-b}>2$

Note that if $m=0$, then all options satisfy.
($\because y=0$ is a common normal) and if common normal is other than the axis, then only option (c) satisfies.
$$\left[\because \text{for option (c), } \frac{c}{2a-b}=\frac{3}{2-1}=3>2\right]$$

**42.** *(c)* According to given information, we have the following figure.

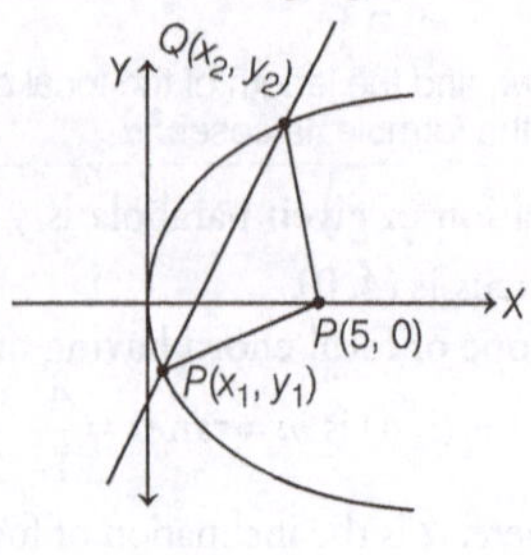


Now, if the origin is shifted to (2, 0) and $(X, Y)$ are the coordinates with respect to new origin, then equation of parabola is $Y^2=4aX$,
where, $X=x-2$ and $Y=y$
and $a=4-2=2$
$$\therefore \quad y^2=8(x-2)$$
Note that (8, 6) is the only point which does not satisfy the equation.

**43.** *(d)*

> **key idea** Equation of tangent to the curve $x^2=4ay$ at $(x_1, y_1)$ is
> $$xx_1=4a\left(\frac{y+y_1}{2}\right)$$

Tangent to the curve $x^2=y-6$ at (1, 7) is
$$x=\frac{y+7}{2}-6 \Rightarrow 2x-y+5=0 \quad \ldots\text{(i)}$$
Equation of circle is
$$x^2+y^2+16x+12y+c=0$$
Centre $(-8, -6)$
$$r=\sqrt{8^2+6^2-c}=\sqrt{100-c}$$
Since, line $2x-y+5=0$ also touches the circle.
$$\therefore \quad \sqrt{100-c}=\left|\frac{2(-8)-(-6)+5}{\sqrt{2^2+1^2}}\right|$$
$$\Rightarrow \quad \sqrt{100-c}=|-\sqrt{5}|$$
$$\Rightarrow \quad 100-c=5$$
$$\Rightarrow \quad c=95$$

**44.** *(b)* Equation of tangent and normal to the curve $y^2=16x$ at (16, 16) is $x-2y+16=0$ and $2x+y-48=0$, respectively.

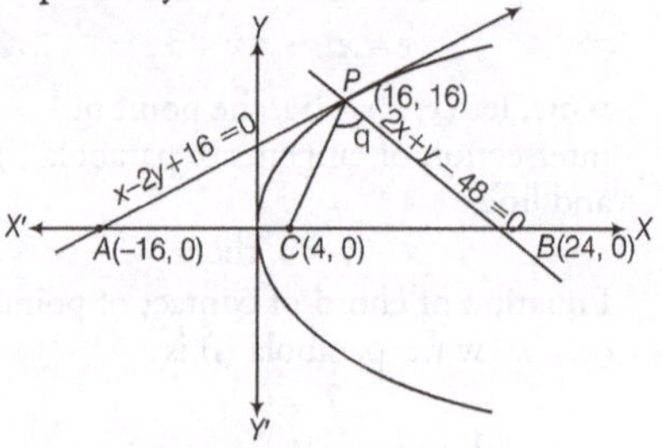


$$A=(-16, 0); \quad B=(24, 0)$$
$C$ is the centre of circle passing through $PAB$
i.e., $\quad C=(4, 0)$
$$\text{Slope of } PC=\frac{16-0}{16-4}=\frac{16}{12}=\frac{4}{3}=m_1$$
$$\text{Slope of } PB=\frac{16-0}{16-24}=\frac{16}{-8}=-2=m_2$$
$$\tan\theta=\left|\frac{m_1-m_2}{1+m_1m_2}\right|$$
$$\Rightarrow \quad \tan\theta=\left|\frac{\frac{4}{3}+2}{1-\left(\frac{4}{3}\right)(2)}\right|$$
$$\Rightarrow \quad \tan\theta=2$$

**45.** *(d)* Equation of parabola is
$$x^2=8y \quad \ldots\text{(i)}$$
Let any point $Q$ on the parabola (i) is $(4t, 2t^2)$.

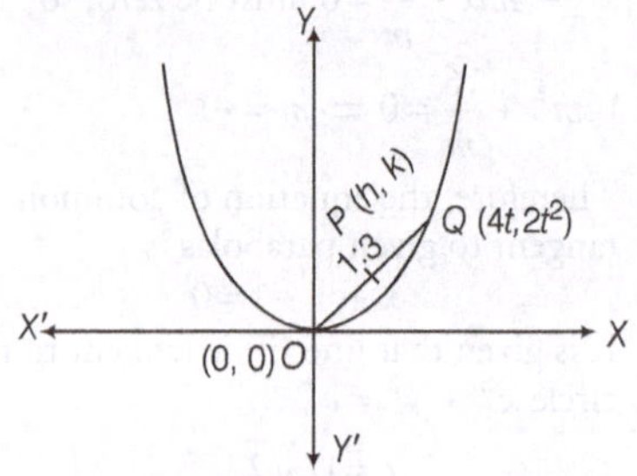


Let $P(h, k)$ be the point which divides the line segment joining (0,0) and $(4t, 2t^2)$ in the ratio 1 : 3.
$$\therefore \quad h=\frac{1\times 4t+3\times 0}{4} \Rightarrow h=t$$
$$\text{and} \quad k=\frac{1\times 2t^2+3\times 0}{4} \Rightarrow k=\frac{t^2}{2}$$
$$\Rightarrow \quad k=\frac{1}{2}h^2 \quad (\because t=h)$$
$$\Rightarrow \quad 2k=h^2$$
$$\Rightarrow \quad 2y=x^2$$
which is required locus.

**46.** *(a)*

> **key idea** (i) First write the equation of tangent as $y=mx+\frac{a}{m}$ with the help of one parabola.
> (ii) This equation of tangent will satisfy the other parabola.
> (iii) Then put $D=0$ as it has only one value.

For parabola, $y^2=4x$

Let $y=mx+\dfrac{1}{m}$ is tangent line and it touches the parabola $x^2=-32y$.
$$\therefore \quad x^2=-32\left(mx+\frac{1}{m}\right)$$
$$\Rightarrow \quad x^2+32mx+\frac{32}{m}=0$$
$$\because \quad D=0$$

$\therefore\ (32m)^2 - 4\cdot\left(\frac{32}{m}\right)=0 \Rightarrow m^3 = \frac{1}{8}$

$\Rightarrow \quad m = 1/2$

**47.** *(a)* **To find** The shortest distance between $y - x = 1$ and $x = y^2$ along the common normal.

$\therefore$ Tangent at $P$ is parallel to

$$y = x + 1 \quad \ldots\text{(i)}$$

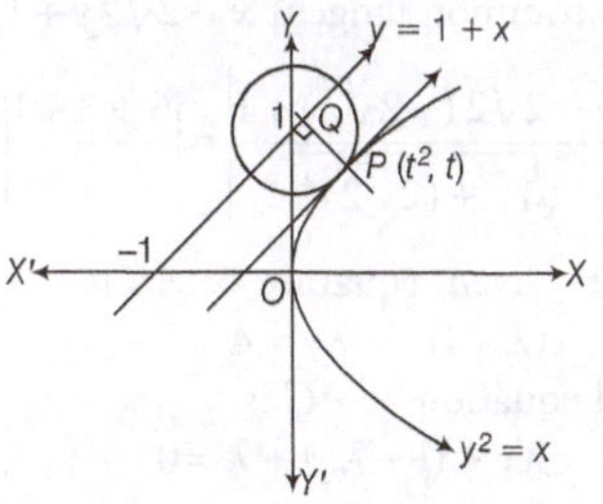


$\therefore$ Slope of tangent at $P(t^2, t)$,

$$\Rightarrow \frac{dy}{dx} = \left(\frac{1}{2y}\right)_{(t^2,\, t)} = \frac{1}{2t} \quad \ldots\text{(ii)}$$

$\Rightarrow 1/2t = 1$ [Eqs. (i) and (ii) are parallel]

$\Rightarrow \quad t = 1/2 \quad \therefore \quad P(1/4, 1/2)$

Shortest distance $= |PQ|$

$$= \frac{\left|\frac{1}{4} - \frac{1}{2} + 1\right|}{\sqrt{1+1}} = \frac{3}{4\sqrt{2}}$$

Hence, shortest distance is $\frac{3\sqrt{2}}{8}$.

**48.** *(d)* The given family of parabolas is

$$y = \frac{a^3x^2}{3} + \frac{a^2x}{2} - 2a$$

$$\Rightarrow y + 2a = \frac{a^3}{3}\left(x^2 + \frac{3}{2a}x\right)$$

$$\Rightarrow y + 2a = \frac{a^3}{3}\left(x^2 + \frac{3}{2a}x + \frac{9}{16a^2} - \frac{9}{16a^2}\right)$$

$$\Rightarrow y + 2a = \frac{a^3}{3}\left(x + \frac{3}{4a}\right)^2 - \frac{9}{16a^2} \times \frac{a^3}{3}$$

$$\Rightarrow \quad y + 2a + \frac{3a}{16} = \frac{a^3}{3}\left(x + \frac{3}{4a}\right)^2$$

$$\Rightarrow \left(y + \frac{35a}{16}\right) = \frac{a^3}{3}\left(x + \frac{3}{4a}\right)^2$$

Thus, the vertices of parabola is $\left(-\frac{3}{4a}, -\frac{35a}{16}\right)$.

Let $\quad h = -\frac{3}{4a}$ and $k = -\frac{35a}{16}$

Now, $hk = 105/64$

Thus, the locus of vertices of the parabola is

$$xy = \frac{105}{64}$$

**49.** *(a)* Let equation of line $AB$

$$y - 0 = m(x + 2)$$
$$y = mx + 2m$$

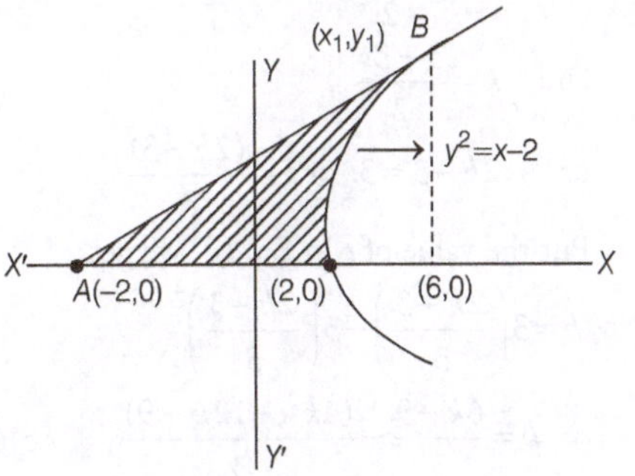


It is tangent to parabola

$(mx + 2m)^2 = x - 2$

$m^2x^2 + 4m^2x - x + 4m^2 + 2 = 0$

$D = 0$

$\Rightarrow (4m^2 - 1)^2 - 4(m^2)(4m^2 + 2) = 0$

$\Rightarrow 16m^4 - 8m^2 + 1 - 16m^4 - 8m^2 = 0$

$\Rightarrow 16m^2 = 1$

$\Rightarrow \quad m^2 = \frac{1}{16}$

$$m = \pm\frac{1}{4} \Rightarrow m > 0 \Rightarrow m = \frac{1}{4}$$

Equation of tangent $y = \frac{x}{4} + \frac{1}{2}$

$$\frac{x}{4} - y + \frac{1}{2} = 0$$

Let $B(x_1 y_1) \Rightarrow \left(\frac{x + x_1}{2}\right) - yy_1 - 2 = 0$

$$\frac{x}{2} - yy_1 - 2 + \frac{x_1}{2} = 0$$

$$\therefore \quad \frac{\frac{1}{4}}{\frac{1}{2}} = \frac{-1}{-y_1} = \frac{\frac{1}{2}}{-2 + \frac{x_1}{2}}$$

$\Rightarrow \quad y_1 = 2, x_1 = 6$

$\because$ Area required

$$= \frac{1}{2} \times 8 \times 2 - \int_2^6 y\, dx$$

$$= 8 - \int_2^6 (\sqrt{x - 2})\, dx$$

$$= 8 - \frac{[(x-2)^{\frac{3}{2}}]_2^6}{\frac{3}{2}}$$

$$= 8 - \frac{2}{3}(4^{\frac{3}{2}} - 0)$$

$$= 8 - \frac{16}{3} = \frac{8}{3}$$

**50.** *(a)* Focus of $y^2 = 20x$ is $R(5, 0)$

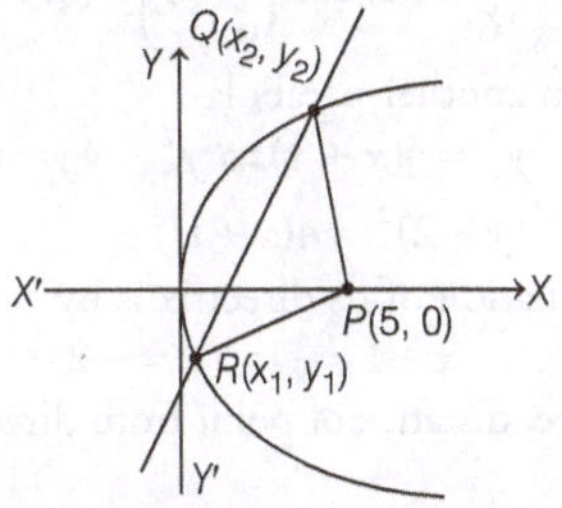


Centroid of $PQR$

$G(10, 10)$

$$\equiv G\left(\frac{x_1 + x_2 + 5}{3}, \frac{y_1 + y_2 + 0}{3}\right)$$

$\Rightarrow x_1 + x_2 = 25, y_1 + y_2 = 30$

$C - m = 60 \quad \Rightarrow C = m + 6$

line $y = mx + m + 6 \Rightarrow x = \frac{y - (m+6)}{m}$

Eliminating $x$ in parabola $y^2 = 20x$, we get

$$y^2 = 20\left(\frac{y - (m+6)}{m}\right)$$

$\Rightarrow my^2 - 20y + 20(m + 6) = 0$

whose roots are $y_1$ and $y_2$, then sum of roots are

$y_1 + y_2 = 20/m = 30 \Rightarrow m = 2/3$

$\therefore C = \frac{2}{3} + 6 = \frac{20}{3}$

for $m = 2/3$

$$\Rightarrow \frac{2}{3}y^2 - 20y + 20\left(\frac{2}{3} + 6\right) = 0$$

$\Rightarrow y^2 - 30y + 200 = 0 \Rightarrow y = 20, 10$

for $y = 20, x = 20$; for $y = 10, x = 5$

$PQ^2 = (x_2 - x_1)^2 + (y_2 - y_1)^2$

$= (20 - 5)^2 + (20 - 10)^2 = 325$

**51.** *(d)* Given equation of parabola is $x = 4y^2$.

Now, comparing it with $y^2 = 4ax$, we get $a = 1/16$

Equation of normal is

$y = -tx + 2at + at^3$

$$\Rightarrow y = -tx + \frac{2}{16}t + \frac{1}{16}t^3 \quad \ldots \text{(i)}$$

Since, Eq. (i) passes through (0, 33).

$$\Rightarrow 33 = \frac{t}{8} + \frac{t^3}{16} \Rightarrow t^3 + 2t - 528 = 0$$

$$\Rightarrow (t-8)(t^2 + 8t + 66) = 0$$

$$\Rightarrow \; t = 8 \; (\because D < 0 \text{ for } t^2 + 8t + 66)$$

$$\therefore P \equiv (at^2, 2at)$$

$$= \left(\frac{1}{16} \times 64, 2 \times \frac{1}{16} \times 8\right) = (4, 1)$$

Given another parabola :

$$y^2 = 4(x+y) \Rightarrow y^2 - 4y = 4x$$

$$\Rightarrow \quad (y-2)^2 = 4(x+1)$$

$\therefore$ Equation of its directrix is by

$$x + 1 = -1 \Rightarrow x = -2$$

Hence, distance of point from directrix $= 6$

**52.** *(b)* Two sides of a triangle are $x = 0$ and $y = 3$

Third side is tangent to $y^2 = 6x$

Parametric co-ordinates of $y^2 = 6x$ are $(6t^2, 6t)$

Let's point of tangency is $p(6t^2, 6t)$

$\because$ Tangent at any point $(x_1, y_1)$ on $y^2 = 4ax$ is

$$yy_1 = 49\left(\frac{x + x_1}{2}\right)$$

So, equation of tangent at point $P(6t^2, 6t)$ on parabola is

$$y(6t) = 6\left(\frac{x + 6t^2}{2}\right)$$

$$\Rightarrow \quad 2yt = x + 6t^2$$

as $\quad x - 2yt + 6t^2 = 0$

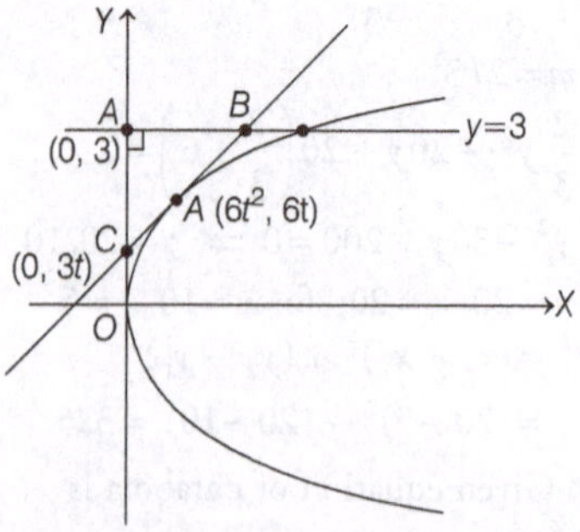


Intersection of tangent and $y = 3 \equiv$ vertex $B$, so put $y = 3$ in the equation of tangent

$$x - 2(3)t + 6t^2 = 0 \Rightarrow x = 6t - 6t^2$$

and $y = 3 \;\; B(6t - 6t^2, 3)$

Vertex $C \equiv$ Intersection of $x = 0$ and tangent

So, put $x = 0$ in the equation of tangent

$$0 - 2yt + 6t^2 = 0 \Rightarrow y = 3t$$

Vertex $C(0, 3t)$

$\Delta ABC$ is right angled triangle, right angled at vertex $A$ as $x = 0$ and $y = 3$ are perpendicular to each other.

So, Let its circumcentre is $(h, k)$

$$\text{So, } h = \frac{0 + 6t - 6t^2}{2} = 3t - 3t^2 \quad \ldots \text{(i)}$$

$$\text{and} \quad k = \frac{3 + 3t}{2}$$

$$\text{or} \quad 2k - 3 = 3t \text{ or } t = \frac{(2k-3)}{3}$$

Put the value of $t$ in Eq. (i), we get

$$h = 3\left(\frac{2k-3}{3}\right) - 3\left(\frac{2k-3}{3}\right)^2$$

$$\Rightarrow h = \frac{6k-9}{3} - \frac{(4k^2 - 12k + 9)}{3}$$

$$\Rightarrow 3h = 6k - 9 - (4k^2 - 12k + 9)$$

$$\Rightarrow 3h = -4k^2 + 18k - 18$$

$$\Rightarrow 4k^2 - 18k + 3h + 18 = 0$$

Now, replace $(h, k)$ with $(x, y)$

$$4y^2 - 18y + 3x + 18 = 0$$

**53.** *(a)* For parabola $x = 2y^2$

$$\Rightarrow \quad y^2 = 4 \times \frac{1}{8}x \Rightarrow a = \frac{1}{8}$$

Equation of tangent for parabola, $x = 2y^2$ is

$$y = mx + \frac{a}{m} \Rightarrow y = mx + \frac{1}{8m} \quad \ldots \text{(i)}$$

And for parabola, $x = 1 + y^2$

$$\Rightarrow \quad y^2 = 4 \times \frac{1}{4}(x - 1) \Rightarrow a = \frac{1}{4}$$

Equation of tangent for parabola, $x = 1 + y^2$ is

$$Y = mX + \frac{a}{m} \Rightarrow y = m(x-1) + \frac{1}{4m}$$

$$y = mx + \frac{1}{4m} - m \quad \ldots \text{(ii)}$$

Since, Eqs. (i) and (ii) represent same tangent

$$\text{So, } \frac{1}{1} = \frac{m}{m} = \frac{\frac{1}{8m}}{\frac{1}{4m} - m}$$

$$\frac{1}{4m} - m = \frac{1}{8m} \Rightarrow \frac{1}{8m} = m$$

$$m^2 = \frac{1}{8} \Rightarrow m = \frac{1}{2\sqrt{2}} \quad (\text{Given, } m > 0)$$

Put in Eq. (i).

So, common tangent is,

$$y = \frac{x}{2\sqrt{2}} + \frac{1}{2\sqrt{2}} \Rightarrow x - 2\sqrt{2}\,y + 1 = 0$$

Distance of the point $(6, -2\sqrt{2})$ from the common tangent $x - 2\sqrt{2}y + 1 = 0$

$$= \left|\frac{6 - 2\sqrt{2}(-2\sqrt{2}) + 1}{\sqrt{1^2 + (2\sqrt{2})^2}}\right| = \left|\frac{6 + 8 + 1}{3}\right| = 5$$

**54.** *(a)* Given, equation of $AB$ is

$$(\lambda + 1)x + \lambda y = 4 \quad \ldots \text{(i)}$$

and equation of $AC$ is

$$\lambda x + (1 - \lambda)y + \lambda = 0 \quad \ldots \text{(ii)}$$

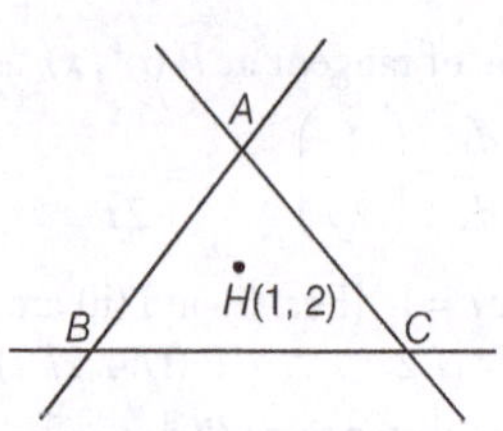


When, $x = 0$, then $\lambda = 2$

Equation of $AB$ is $3x + 2y = 4$

and Equation of $AC$ is $2x - y + 2 = 0$

$\therefore$Co-ordinates of $A$ are (0, 2)

Since, $CH \perp AB$ and let co-ordinates of $C$ are $(a, b)$.

$$\therefore \left(\frac{b-2}{a-1}\right) \times \left(\frac{-3}{2}\right) = -1$$

$$\Rightarrow \quad 3b - 6 = 2a - 2$$

$$\Rightarrow \quad 3b - 2a = 4 \quad \ldots \text{(iii)}$$

Since, $C$ lies on $AC$

$$\therefore \quad 2a - b + 2 = 0 \quad \ldots \text{(iv)}$$

On solving Eqs. (iii) and (iv), we get

$$a = -\frac{1}{2} \text{ and } b = 1$$

$$\text{So, } C\left(-\frac{1}{2}, 1\right)$$

Let the equation of tangent be

$$y = mx + \frac{3}{2m}$$

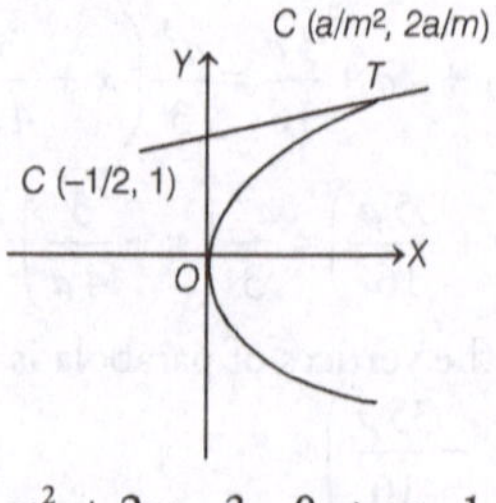


$$m^2 + 2m - 3 = 0 \Rightarrow m = 1, -3$$

So, tangent which touches in first quadrant at $T$ is

$$T=\left(\frac{a}{m^2},\frac{2a}{m}\right)=\left(\frac{3}{2},3\right)$$

$\Rightarrow CT=\sqrt{4+4}=2\sqrt{2}$

**55.** *(d)* Given, equation of parabola is

$$y^2=24x \quad \text{...(i)}$$

and equation of curve (hyperbola) is

$$xy=2 \quad \text{...(ii)}$$

On comparing Eq. (i) with $y^2=4ax$, we get

$$4a=24 \Rightarrow a=6$$

Any point taken on the parabola (i) is of the form $(at^2, 2at)$ i.e., $(6t^2, 12t)$.

$\therefore$ Equation of tangent $AB$ at $(6t^2, 12t)$ is given by $T=0$.

i.e., $y\times(12t)=24\times\frac{1}{2}\times(x+6t^2)$

($\because$ Equation of tangent to the parabola $y^2=4ax$ at $(x_1, y_1)$ is $T=0$

i.e., $yy_1=4a\times\frac{1}{2}\times(x+x_1) \Rightarrow ty=x+6t^2$

$$\Rightarrow \quad x-ty+6t^2=0 \quad \text{...(iii)}$$

Also, equation of chord $AB$ using $M(h, k)$ as mid-point is given by $T=S_1$

i.e., $\frac{xk+yh}{2}-2=hk-2$

$$\Rightarrow xk+yh-2hk=0 \quad \text{...(iv)}$$

Now, Eqs. (iii) and (iv) represents the equation of same line.

So, $\frac{k}{1}=\frac{h}{-t}=\frac{-2hk}{6t^2} \Rightarrow k^2=-3h$

$\Rightarrow$ Locus of M is $y^2=-3x$.

Hence, directrix is $4x=3$ ($\because$ Equation of directrix for the parabola $y^2=-4ax$ is $x=a$)

**56.** *(b)*

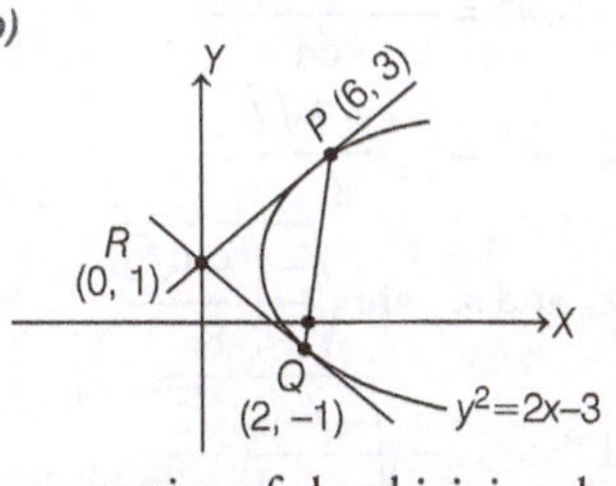


Now, equation of chord joining the points $P$ and $Q$ is given by

$$y+1=x-2 \Rightarrow x-y=3 \quad \text{... (i)}$$

and equation of parabola is $y^2=2x-3$ ... (ii)

For finding the coordinates of $P$ and $Q$, we solve the Eqs. (i) and (ii), we get $P\equiv(6, 3)$ and $Q\equiv(2, -1)$.

Now, slope of

$$PQ=M_{PQ}=\frac{3-(-1)}{6-2}=\frac{4}{4}=1$$

$$M_{RQ}=\frac{1-(-1)}{0-2}=\frac{2}{-2}=-1$$

$\because M_{PQ}\times M_{RQ}=-1$

$\therefore PQ\perp RQ.$

Hence, $\angle PQR=90°$

As, we know that orthocentre of a right angled triangle is that vertex where right angle is formed.

$\therefore$ Orthocentre $=Q\equiv(2, -1)$.

**57.** *(d)* The given parabola is

$$y^2-2x-2y=1$$

This can be written as,

$$(y-1)^2=2(x+1)$$

Now, from the above given information the tangent at the point $A(1, 3)$ on the parabola $y^2-2x-2y=1$ is

$$y(3)-(x+1)-(y+3)=1$$

$$\Rightarrow \quad 3y-x-1-y-3=1$$

$$\Rightarrow \quad 2y-x=5 \quad \text{...(i)}$$

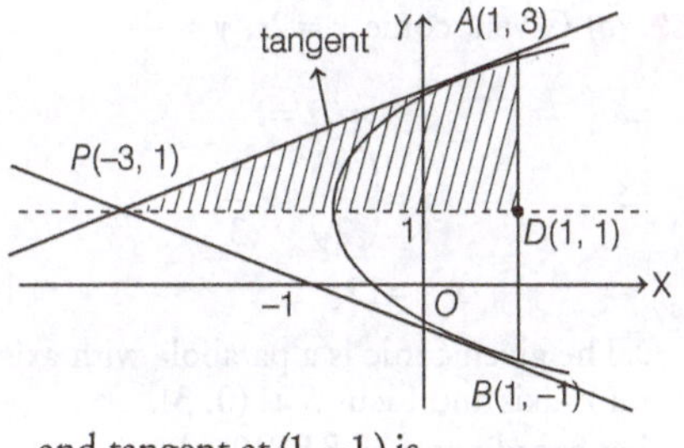


and tangent at $(1, -1)$ is

$$y(-1)-(x+1)-(y-1)=1$$

$$\Rightarrow \quad -y-x-1-y+1=1$$

$$\Rightarrow \quad x+2y=-1 \quad \text{...(ii)}$$

Now, solving Eqs. (i) and (ii), we get

$$x=-3 \text{ and } y=1$$

Hence, the point $P(-3, 1)$

Thus, area of $\Delta PAB$ is

$$2\times\frac{1}{2}\times(1-(-3))\times(3-1)=8 \text{ sq units}$$

**58.** *(d)*

**key idea** (i) The angle between two lines with slopes $m_1$ and $m_2$ is

$$\tan\theta=\left|\frac{m_1-m_2}{1+m_1m_2}\right|$$

(ii) The equation of tangent to parabola $y^2=4ax$ is $y=mx+\frac{a}{m}$ and the point of contact is $\left(\frac{a}{m^2},\frac{2a}{m}\right)$

(iii) If $(x_1, y_1)$, $(x_2, y_2)$ and $(x_3, y_3)$ are collinear, then $\begin{vmatrix} x_1 & y_2 & 1 \\ x_2 & y_2 & 1 \\ x_3 & y_3 & 1 \end{vmatrix}=0$

Given, $y^2=4ax, a>0$

and $y=3x+5$

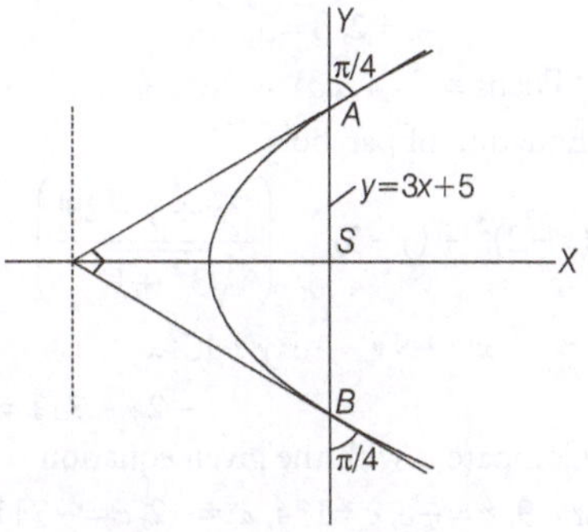


Let slope of tangent be $m$,

$$\tan\theta=\left|\frac{3-m}{1+3m}\right|$$

$$\Rightarrow \quad \tan\frac{\pi}{4}=\left|\frac{3-m}{1+3m}\right|$$

$$\Rightarrow \quad \pm1=\frac{3-m}{1+3m}$$

Taking positive sign, we get

$$1\times(1+3m)=3-m$$

$$\Rightarrow \quad 4m=2$$

$$\Rightarrow \quad m=\frac{1}{2}$$

Taking negative sign, we get

$$-1(1+3m)=3-m$$

$$\Rightarrow \quad -2m=4$$

$$\Rightarrow \quad m=-2$$

$$\because \quad m=\frac{1}{2}, -2$$

It means tangents are perpendicular to each other. As we know that the tangent at extremities of a focal chord of a parabola intersect at right angle on directrix.

Therefore, $A$, $S$ and $B$ are collinear for all $a>0$.

**59.** *(d)*

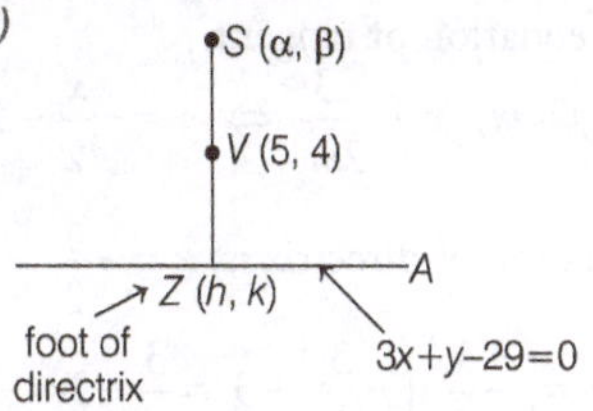


$$\frac{h-5}{3}=\frac{k-4}{1}=\frac{-(3\times5+4-29)}{10}$$

$$\Rightarrow \quad \frac{h-5}{3}=\frac{k-4}{1}=1$$

$$\Rightarrow \quad h=8, k=5$$

$$\therefore \quad z=(8, 5)$$

Now, $\frac{\alpha+8}{2}=5$ and $\frac{\beta+5}{2}=4$

$\alpha=2, \beta=3$

$\therefore$ Focus $\equiv S=(2,3)$

Equation of parabola,

$$(x-2)^2+(y-3)^2=\left(\frac{|3x+y-29|}{\sqrt{3^2+1^2}}\right)^2$$

$$\Rightarrow \quad x^2+9y^2-6xy+134x-2y-711=0$$

Compare it with the given equation

$a=9, b=-6, c=134, d=-2, e=-711$

$\therefore a+b+c+d+e=9$
$-6+134-2-711=-576$

**60.** *(b)* $y^2=6x \Rightarrow 4a=6 \Rightarrow a=\frac{3}{2}$

Equation of normal

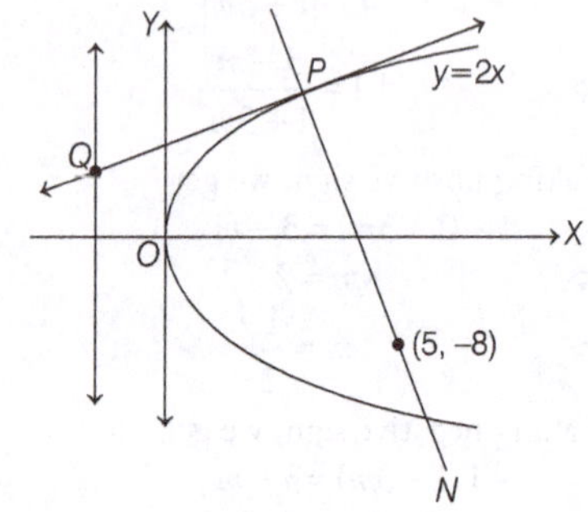


$$y=mx-3m-\frac{3}{2}m^3$$

Point (5, –8) satisfies it.

$$\Rightarrow \quad -8=5m-3m-\frac{3}{2}m^3$$

$$\Rightarrow 2m-\frac{3}{2}m^3=-8$$

$$\Rightarrow \quad m=2=m_N$$

$\Rightarrow$ Slope of tangent at $P$,

$$m_T=-\frac{1}{2}$$

So, equation of tangent,

$$y=m_T\,x+\frac{3}{2m_T} \Rightarrow y=-\frac{x}{2}-3$$

Equation of directrix is, $x=-\frac{3}{2}$

$$\Rightarrow y=\left(-\frac{1}{2}\right)\left(-\frac{3}{2}\right)-3=\frac{3}{4}-3=-\frac{9}{4}$$

$\therefore$ So ordinate $=-\frac{9}{4}$

**61.** *(c)* Line $y=4+kx$ touches the parabola $y=x-x^2$ at point $P$.

$\Rightarrow \quad 4+kx=x-x^2$ has only one root

$\Rightarrow \quad x^2+(k-1)x+4=0$ has only one root

$\Rightarrow \quad (k-1)^2-4\times 4=0$

$\Rightarrow \quad (k-1-4)(k-1+4)=0$

$\Rightarrow \quad k=5, k\neq -3$, as $k>0$

$\therefore \quad x^2+4x+4=0 \Rightarrow x=-2$

If $x=-2$, then $y=-6$

$\therefore$ Co-ordinate of point $P$ is $(-2,-6)$.

The given parabola is $y=x-x^2$

$$y-\frac{1}{4}=-\left(x^2-x+\frac{1}{4}\right)$$

$$\Rightarrow \quad y-\frac{1}{4}=-\left(x-\frac{1}{2}\right)^2$$

Vertex of the parabola is $V=\left(\frac{1}{2},\frac{1}{4}\right)$.

$$\text{Slope of } PV=\frac{\frac{1}{4}+6}{\frac{1}{2}+2}=\frac{\frac{25}{4}}{\frac{5}{2}}=\frac{5}{2}$$

**62.** *(d)* Given, conic $x=2t$, $y=\frac{t^2}{3}$

$$\Rightarrow \quad x/2=t$$

$$\Rightarrow \quad y=\left(\frac{x}{2}\right)^2\times\frac{1}{3}$$

$$x^2=12y$$

$\therefore$ The given conic is a parabola with axis on $Y$-axis and focus $S$ as (0, 3).

Let coordinates of $B$ be $(0, y)$.

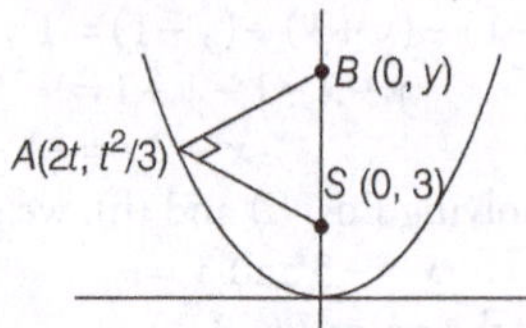


As $AB\perp AS$, $M_{AB}\times M_{AS}=-1$

$$\left(\frac{y-\frac{t^2}{3}}{0-2t}\right)\times\left(\frac{3-\frac{t^2}{3}}{0-2t}\right)=-1$$

$$\Rightarrow \quad (3y-t^2)(9-t^2)=-36t^2$$

$$\Rightarrow \quad 27y-9t^2-3yt^2+t^4+36t^2=0$$

$$\Rightarrow \quad y=\frac{t^4+27t^2}{3t^2-27}=\frac{t^4+27t^2}{3(t^2-9)}$$

$k$ is ordinate of centroid of $\Delta SAB$

$$k=\frac{3+y+\frac{t^2}{3}}{3}$$

$$\Rightarrow \quad 9k=9+3y+t^2$$

$$\Rightarrow \quad 9k=9+\frac{t^4+27t^2}{t^2-9}+t^2$$

$$\Rightarrow 9k=\frac{9(t^2-9)+t^4+27t^2+t^2(t^2-9)}{t^2-9}$$

$$\Rightarrow k=\frac{9t^2-81+t^4+27t^2+t^4-9t^2}{9(t^2-9)}$$

$$\Rightarrow \; k=\frac{2t^4+27t^2-81}{9(t^2-9)}$$

$$\lim_{t\to 1}k=\lim_{t\to 1}\frac{2t^4+27t^2-81}{9(t^2-9)}$$

$$=\frac{2+27-81}{9(1-9)}$$

$$=\frac{52}{72}=\frac{13}{18}$$

**63.** *(c)* Let the equation of tangent of parabola $y^2=x$ is

$$y=mx+\frac{1}{4m}$$

$y=mx+\frac{1}{4m}$ is also the equation of tangent of circle $x^2+y^2=2$

$$\Rightarrow \quad \left(\frac{1}{4m}\right)^2=2(1+m^2)$$

$$\Rightarrow \quad \frac{1}{16m^2}=2+2m^2$$

$$\Rightarrow \quad 32m^4+32m^2-1=0$$

$$\Rightarrow \quad m^2=\frac{-32\pm\sqrt{(32)^2-4(32)(-1)}}{2\times 32}$$

$$\Rightarrow \quad m^2=\frac{-32\pm\sqrt{32\times 36}}{2\times 32}$$

$$\Rightarrow \quad m^2=\frac{-32+24\sqrt{2}}{64}$$

$$\Rightarrow \quad m^2=\frac{-4+3\sqrt{2}}{8}$$

$\therefore m_1$ and $m_2$ are $\sqrt{\frac{-4+3\sqrt{2}}{8}}$

and $-\sqrt{\frac{-4+3\sqrt{2}}{8}}$

$$m_1\,m_2=\sqrt{\frac{(-4+3\sqrt{2})^2}{(8)^2}}$$

$$\therefore 8|m_1\,m_2|=8\left[\frac{-4+3\sqrt{2}}{8}\right]$$

$$=-4+3\sqrt{2}$$

**64.** (a) Given that, circle is

$$x^2 + y^2 + Ax + By + C = 0 \quad \dots \text{(i)}$$

Parabola is $\quad y = x^2 \quad \dots$ (ii)

Since, Eqs. (i) and (ii) touching at (2, 4).

$\therefore$ (2, 4) lies on Eq. (i),

$\Rightarrow 2A + 4B + C + 20 = 0$

$\Rightarrow \quad 2A + 4B + C = -20 \quad \dots$ (iii)

Now, differentiating Eq. (i) w.r.t. '$x$', we get

$$2x + 2y\frac{dy}{dx} + A + B\frac{dy}{dx} = 0 \quad \dots \text{(iv)}$$

Also differentiating Eq. (ii) w.r.t. $x$,

we get $\frac{dy}{dx} = 2x \Rightarrow \left.\frac{dy}{dx}\right|_{(2,4)} = 4$

Now, from Eq. (iv), we get

$$4 + 32 + A + 4B = 0 \quad \left[\text{put } \left.\frac{dy}{dx}\right|_{(2,4)} = 4\right]$$

$\Rightarrow \quad A + 4B = -36 \quad \dots$ (v)

Now, subtracting Eq. (v) from Eq. (iii), we get $(2A + 4B + C) - (A + 4B)$

$= (-20) - (-36)$

$\Rightarrow \quad A + C = 16$

Thus, $A + C$ is equal to 16.

**65.** (b) Given, parabola $y = 4x^2 + 1$

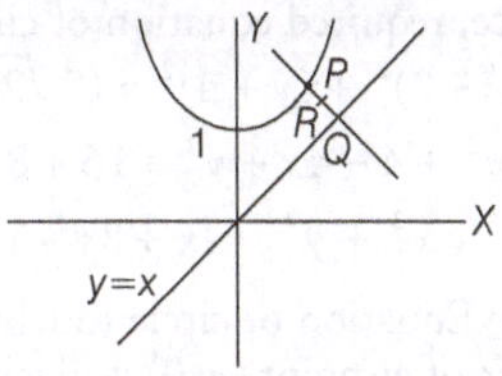


Let $R(a, b)$ be mid-point of line joining point $P$ and $Q$ where $PQ$ is perpendicular to line $y = x$.

Let coordinates of $P$ be $P(x, y)$, $Q(q, q)$ and $R(a, b)$ then, $a = \frac{x + q}{2}$

and $b = \frac{y + q}{2}$

Now, slope of line $y = x$ is $m_1 = 1$

Slope of line $PQ$ be

$$\frac{b - q}{a - q} = m_2 \quad \text{(say)}$$

$\because$ Line $y = x$ and $PQ$ are perpendicular to each other, $m_1 \cdot m_2 = -1$

$$\Rightarrow \quad \frac{b - q}{a - q} = -1$$

$$\Rightarrow \quad b - q = q - a \Rightarrow q = \frac{b + a}{2}$$

$$\therefore \quad a = \frac{x + q}{2} = \frac{x + \left(\frac{b + a}{2}\right)}{2}$$

$$= \frac{2x + b + a}{4}$$

$$\Rightarrow \quad x = \frac{4a - b - a}{2} = \frac{3a - b}{2}$$

and $\quad b = \frac{y + q}{2} = \frac{y + \left(\frac{b + a}{2}\right)}{2}$

$$= \frac{2y + b + a}{4} \Rightarrow y = \frac{3b - a}{2}$$

Put $(x, y)$ in equation of parabola as $P(x, y)$ is variable point on parabola

$$\frac{3b - a}{2} = 4\left(\frac{3a - b}{2}\right)^2 + 1$$

$\Rightarrow \quad (3b - a) = 2(3a - b)^2 + 2$

Replace $(a, b)$ as $(x, y)$

$\Rightarrow (3y - x) = 2(3x - y)^2 + 2$

or $\quad 2(3x - y)^2 + (x - 3y) + 2 = 0$

**66.** (a) $y^2 = 2x$

At (2, 2), equation of tangent

$\Rightarrow \quad y(2) = (x + 2)$

$T_i \Rightarrow \quad 2y = x + 2$

Normal at (2, 2)

$\Rightarrow \quad y - 2 = -2(x - 2)$

$N_i \Rightarrow y = -2x + 6$

$R$ is point of intersection of $T$ and line $y = 0$

$R(-2, 0)$

$$(2, 2) = \left(\frac{t^2}{2}, t\right) \Leftrightarrow t = 2$$

If normal at $t_1$ meets the parabola again at $t_2$,

$$t_2 = -t_1 - \frac{2}{t_1}$$

$$-t_2 = -2 - \frac{2}{2} = -3$$

$\therefore \quad t_2 = 3$

$$R: (at_2^2, 2at_2) \Rightarrow \left(\frac{1}{2} \cdot 9, 2 \cdot \frac{1}{2} \cdot 3\right)$$

$$\Rightarrow \left(\frac{9}{2}, 3\right)$$

$\therefore P(2, 2), Q(-2, 1), R\left(\frac{9}{2}, 3\right)$

$$\text{Area} = \frac{1}{2}\begin{vmatrix} 1 & 1 & 1 \\ 2 & -2 & \frac{9}{2} \\ 2 & 1 & 3 \end{vmatrix} = \frac{25}{2}$$

**67.** (c) Given, equation of parabola

$\Rightarrow \quad y^2 = 4ax$

Focus $= S(a, 0)$

Let any point on the parabola be $P(at^2, 2at)$.

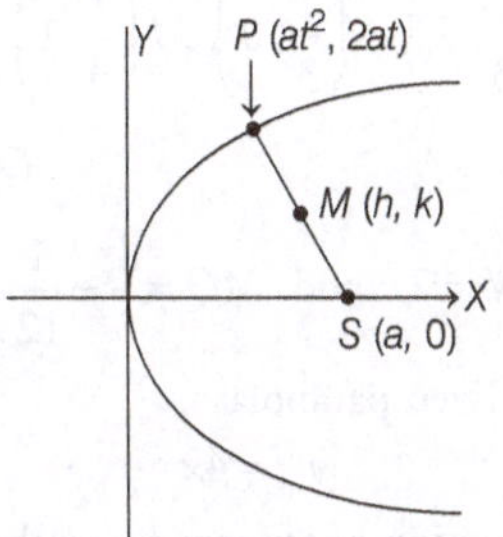


and let the mid-point of $PS$ be $M(h, k)$.

$$\therefore \quad h = \frac{at^2 + a}{2};\ k = \frac{2at + 0}{2}$$

$$\Rightarrow \quad t^2 = \frac{2h - a}{a};\ t = \frac{k}{a} \Rightarrow t^2 = \frac{k^2}{a^2}$$

Now, $\frac{2h - a}{a} = \frac{k^2}{a^2}$

$$\Rightarrow \quad 2h - a = \frac{k^2}{a}$$

$$\Rightarrow \quad k^2 = a(2h - a)$$

$\therefore$ Locus of $(h, k)$ is

$$y^2 = a(2x - a)$$

$$y^2 = 2a\left(x - \frac{a}{2}\right)$$

$\therefore$ The directrix of this parabola is

$$x - \frac{a}{2} = -\frac{a}{2} \Rightarrow x = 0.$$

**68.** (c) Equation of given parabola is

$$y^2 = 12x \quad \dots \text{(i)}$$

Now, let a point $P(3t^2, 6t)$ on the parabola, since point $N$ is the foot of perpendicular drawn from $P$ on the axis of the parabola (i.e., $X$-axis) so coordinate of $N$ is $(3t^2, 0)$ and $M$ is mid-point of $PN$, so coordinate of $M$ is $(3t^2, 3t)$. Now, equation of the parallel to the axis of parabola through point $M$ is

$$y = 3t \quad \dots \text{(ii)}$$

So, point of intersection of parabola (i) and line (ii) is $Q\left(\frac{3t^2}{4}, 3t\right)$, so

equation of line $NQ$ is

$$y - 0 = \frac{3t}{\frac{3t^2}{4} - 3t^2}(x - 3t^2) \quad \dots \text{(iii)}$$

For $y$-intercept made by line (iii) put $x=0$ then

$$y=\frac{1}{(1/4)-1}(-3t)=\frac{12t}{3}=\frac{4}{3} \quad \text{[given]}$$

$\Rightarrow \quad t=1/3$

$$\therefore P\left(\frac{1}{3},2\right), N\left(\frac{1}{3},0\right), M\left(\frac{1}{3},1\right), Q\left(\frac{1}{12},1\right)$$

$\therefore \quad PN=2,$ and $MQ=\frac{1}{3}-\frac{1}{12}=\frac{1}{4}$

**69.** *(a)* Given parabola

$$y^2=4x \quad \ldots\text{(i)}$$

So, equation of tangent to parabola (i) at point (1, 2) is $2y=2(x+1)$

[$\because$ equation of the tangent to the parabola $y^2=4ax$ at a point $(x_1, y_1)$ is given by $yy_1=2a(x+x_1)$]

$\Rightarrow \quad y=x+1 \quad \ldots$(ii)

Now, equation of circle, touch the parabola at point (1, 2) is

$(x-1)^2+(y-2)^2+\lambda(x-y+1)=0$

$\Rightarrow x^2+y^2+(\lambda-2)x+(-4-\lambda)y+(5+\lambda)=0 \ldots$(iii)

Also, circle (iii) touches the $X$-axis, so $g^2=c$

$$\Rightarrow \left(\frac{\lambda-2}{2}\right)^2=5+\lambda \Rightarrow \lambda^2-8\lambda-16=0$$

$$\Rightarrow \quad \lambda=\frac{8\pm\sqrt{64+64}}{2}$$

$$\Rightarrow \quad \lambda=4\pm\sqrt{32}=4\pm4\sqrt{2}$$

Now, radius of circle is

$$r=\sqrt{g^2+f^2-c}$$

$$\Rightarrow \quad r=|f| \quad [\because g^2=c]$$

$$=\left|\frac{\lambda+4}{2}\right|=\frac{8+4\sqrt{2}}{2} \text{ or } \frac{8-4\sqrt{2}}{2}$$

For least area $r=\frac{8-4\sqrt{2}}{2}$

$$=4-2\sqrt{2} \text{ units}$$

So, area $=\pi r^2=\pi(16+8-16\sqrt{2})$

$$=8\pi(3-2\sqrt{2}) \text{ sq units}$$

**70.** *(a)* According to given information, we have the following figure.

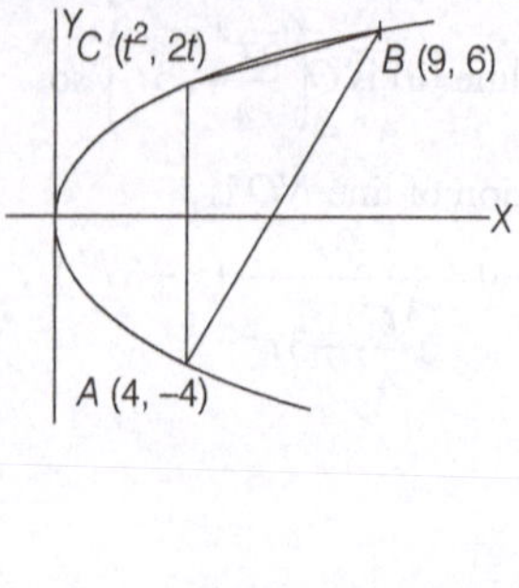


[For $y^2=4ax$, parametric coordinates of a point is $(at^2, 2at)$.]

$\therefore$For $y^2=4x$, let coordinates of $C$ be $(t^2, 2t)$.

Then, area of

$$\Delta ABC=\frac{1}{2}\left\|\begin{matrix} t^2 & 2t & 1 \\ 9 & 6 & 1 \\ 4 & -4 & 1 \end{matrix}\right\|$$

$$=5|t^2-t-6|$$

Let $\quad A(t)=5|t^2-t-6| \quad \ldots$(i)

Clearly, $A(4,-4)\equiv A(t_1^2, 2t_1)$

$\Rightarrow 2t_1=-4$

$\Rightarrow t_1=-2$

and $B(9, 6)\equiv B(t_2^2, 2t_2)$

$\Rightarrow 2t_2=6 \Rightarrow t_2=3$

Since, $C$ is on the arc $AOB$, the parameter $t$ for point $C\in(-2, 3)$.

Let $f(t)=t^2-t-6$

$\Rightarrow f'(t)=2t-1$

Now, $\quad f'(t)=0 \Rightarrow t=\frac{1}{2}$

$\therefore$For $A(t)$, critical point is at $t=\frac{1}{2}$

$$\text{Now, } A\left(\frac{1}{2}\right)=5\left|\left(\frac{1}{2}\right)^2-\frac{1}{2}-6\right|=\frac{125}{4}$$

$$=31\frac{1}{4} \quad \text{[using Eq. (i)]}$$

**71.** *(d)* Given equation of circle is

$$x^2+y^2-8x-8y-4=0,$$

whose centre is $C(4, 4)$ and radius

$$=\sqrt{4^2+4^2+4}=\sqrt{36}=6$$

Let the centre of required circle be $C_1(x, y)$. Now, as it touch the $X$-axis, therefore its radius $=|y|$. Also, it touch the circle

$$x^2+y^2-8x-8y-4=0,$$

therefore $CC_1=6+|y|$

$$\Rightarrow \sqrt{(x-4)^2+(y-4)^2}=6+|y|$$

$$\Rightarrow \quad x^2+16-8x+y^2+16-8y=36+y^2+12|y|$$

$$\Rightarrow \quad x^2-8x-8y-4=12|y|$$

**Case I** If $y>0$, then we have

$$x^2-8x-8y-4=12y$$

$$\Rightarrow \quad x^2-8x-4=20y$$

$$\Rightarrow \quad (x-4)^2-20=20y$$

$\Rightarrow (x-4)^2=20(y+1)$, which is a parabola

**Case II** If $y<0$, then we have

$$x^2-8x-8y-4=-12y$$

$$\Rightarrow \quad x^2-8x-8y-4+12y=0$$

$$\Rightarrow \quad x^2-8x-4=-4y$$

$$\Rightarrow \quad (x-4)^2=20-4y$$

$$\Rightarrow \quad (x-4)^2=-4(y-5)$$

which is again a parabola.

**72.** *(a)* Centre of circle $x^2+(y+6)^2=1$ is $C(0,-6)$.

Let the parametric point $P$ be $(2t^2, 4t)$.

Now, let

$$D=CP=\sqrt{(2t^2)^2+(4t+6)^2}$$

$$\Rightarrow \quad D=\sqrt{4t^4+16t^2+36+48t}$$

On squaring on both side, we get

$$\Rightarrow \quad D^2=4t^4+16t^2+48t+36$$

Let $F(t)=4t^4+16t^2+48t+36$

For minimum, $F'(t)=0$

$$\Rightarrow \quad 16t^3+32t+48=0$$

$$\Rightarrow \quad t^3+2t+3=0$$

$$\Rightarrow (t+1)(t^2-t+3)=0 \Rightarrow t=-1$$

Thus, coordinates of point $P$ are $(2,-4)$.

Now, $CP=\sqrt{2^2+(-4+6)^2}$

$$=\sqrt{4+4}=2\sqrt{2}$$

Hence, required equation of circle is

$$(x-2)^2+(y+4)^2=(2\sqrt{2})^2$$

$$\Rightarrow \quad x^2+4-4x+y^2+16+8y=8$$

$$\Rightarrow \quad x^2+y^2-4x+8y+12=0$$

**73.** *(30)* Equation of circle touching $X$ axis of at point $(a, 0)$ is given by

$S_{\text{point}}+\lambda L=0$

$(x-a)^2+y^2+\lambda y=0$

$\Rightarrow \quad$ (4, 6) satisfies Eq. of circle

$$(4-a)^2+36+6\lambda=0 \quad \ldots\text{(i)}$$

Equation of tangent at (4, 6) to $y^2=9x$

$$y\times6=\frac{9}{2}(x+4)$$

$$12y=9(x+4)$$

$$3x-4y+12=0$$

Equation of circle is

$$x^2+y^2-2xa+\lambda y+a^2=0$$

$$\text{Centre}=\left(a, -\frac{\lambda}{2}\right)$$

$$\text{radius}=\sqrt{a^2+\frac{\lambda^2}{4}-a^2}=\frac{|\lambda|}{2}$$

$$\frac{|3a+2\lambda+12|}{\sqrt{3^2+4^2}}=\frac{|\lambda|}{2}$$

$$|3a+2\lambda+12|=\frac{5}{2}|\lambda|$$

$$\Rightarrow 3a+2\lambda+12=\pm\frac{5\lambda}{2}$$

$$\Rightarrow 6a+4\lambda+24=\pm 5\lambda$$

$$\Rightarrow \quad 6a+24=\lambda \text{ or } 6a+24=-9\lambda$$

$$\Rightarrow \quad 36a+144=6\lambda$$

or $\quad 2a+8=-3\lambda$

$$\Rightarrow \quad 36a+144=-36-(4-a)^2$$

or $\quad 4a+16=-6\lambda$

$$36a+144=-36-16-a^2+8a$$

$$a^2+28a+196=0$$

$$(a+14)^2=0$$

$$\Rightarrow \quad a=-14$$

Also, $\quad 4a+16=-6\lambda$

$$4a+16=36+(4-a)^2$$

$$4a+16=36+16-8a+a^2$$

$$\Rightarrow a^2-12a+36=0$$

$$\Rightarrow \quad (a-6)^2=0$$

$$a=6$$

Put $a<0$

$$\Rightarrow \quad a=-14$$

$$6\lambda=-36-(4-a)^2$$

$$6\lambda=-36-18^2=-360$$

$$\Rightarrow \quad \lambda=-60$$

$$r=\frac{|-60|}{2}=30$$

$$r=30$$

**74.** *(1328)*

> **key idea** Equation of the circle, when the coordinates of end points of a diameter are $(x_1, y_1)$ and $(x_2, y_2)$ is $(x-x_1)(x-x_2)+(y-y_1)(y-y_2)=0$

We have, $y^2=12x$

$$\Rightarrow \quad a=3$$

and $SP\times SQ=\frac{147}{4}$

Let $P(3t^2, 6t)$ and $t_1t_2=-1$

(end of focal chord)

So, $Q\left(\frac{3}{t^2}, \frac{-6}{t}\right)$

$S(3, 0)$

$\therefore \quad SP\times SQ=PM_1\times QM_2$

(distance from directrix)

$$(3+3t^2)\left(3+\frac{3}{t^2}\right)=\frac{147}{4}$$

$$\Rightarrow \quad \frac{(1+t^2)^2}{t^2}=\frac{49}{12}$$

$$\Rightarrow \quad t^2=\frac{3}{4}, \frac{4}{3} \Rightarrow t=\pm\frac{\sqrt{3}}{2}, \pm\frac{2}{\sqrt{3}}$$

Considering $t=\frac{\sqrt{3}}{2}$

$P\left(\frac{9}{4}, -3\sqrt{3}\right)$ and $Q(4, 4\sqrt{3})$

Hence, diametric circle

$$(x-4)\left(x-\frac{9}{4}\right)+(y+3\sqrt{3})(y-4\sqrt{3})=0$$

$$\Rightarrow x^2+y^2-\frac{25}{4}x-\sqrt{3}y-27=0$$

$$\Rightarrow \alpha=400, \beta=1728$$

$$\beta-\alpha=1328$$

**75.** *(14)*

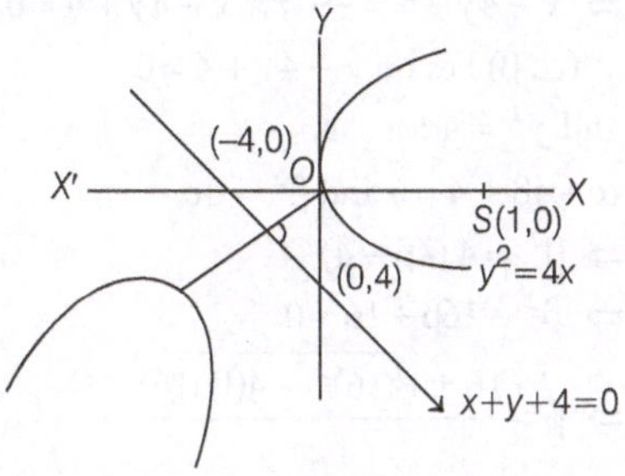


Let $L_1\equiv x+y+4=0$

image of point $(t^2, 2t)$ in $x+y+4=0$ is

$$\frac{x-t^2}{1}=\frac{y-2t}{1}$$

$$=\frac{-2(t^2+2t+4)}{2}$$

$$=-(t^2+2t+4)$$

$$\Rightarrow x=-2t-4, y=-t^2-4$$

$\therefore$ Image of point is $y=-\left(\frac{x+4}{-2}\right)^2-4$

$$\Rightarrow \quad (x+4)^2=-4(y+4)$$

Given, $\quad y+5=0$

$$\Rightarrow \quad y=-5$$

$$\therefore (x+4)^2=-4(-5+4)=4$$

$$x+4=\pm 2 \Rightarrow x=\pm 2-4=-2, -6$$

$\therefore$ Points are $(-2, -5)$ and $(-6, -5)$

$\Rightarrow d=4$

$\therefore$ Area

$$\Delta SAB=\frac{1}{2}\left\|\begin{matrix}1 & 0 & 1\\ -2 & -5 & 1\\ -6 & -5 & 1\end{matrix}\right\|=10=a$$

$$\therefore a+d=4+10=14$$

> **Concept Enhancer**
> Let the image of a point $(x_1, y_1)$ with respect to $ax+by+c=0$ be $(x_2, y_2)$, then
> $$\frac{x_2-x_1}{a}=\frac{y_2-y_1}{b}=-\frac{2(ax_1+by_1+c)}{a^2+b^2}$$

**76.** *(15)* Equation of parabola

$$y^2=4(x+4)$$

$$\therefore \quad 4a=4 \Rightarrow a=1$$

Focus $(-4+1, 0)=(-3, 0)$

$\therefore$ Equation of circle $C$,

$$(x+3)^2+y^2=5^2$$

Point of intersection $P$ of $3x-y=0$ and $x+\lambda y=4$

$$=x(1+3\lambda)=4$$

$$\Rightarrow \quad x=\frac{4}{3\lambda+1} \text{ and } y=\frac{12}{3\lambda+1}$$

$\because P$ lies on circle $C$.

$$\Rightarrow \left(\frac{4}{3\lambda+1}+3\right)^2+\left(\frac{12}{3\lambda+1}\right)^2=25$$

$$\Rightarrow (9\lambda+7)^2+144=25(9\lambda^2+6\lambda+1)$$

$$\Rightarrow 81\lambda^2+49+126\lambda+144$$

$$=225\lambda^2+150\lambda+25$$

$$\Rightarrow \quad 144\lambda^2+24\lambda-168=0$$

$$\Rightarrow \quad 36\lambda^2+6\lambda-42=0$$

$$\Rightarrow 6\lambda^2+\lambda-7=0$$

$$\lambda=-\frac{7}{6}, 1$$

$$\therefore \quad \lambda_1=-\frac{7}{6}, \lambda_2=1$$

$[\because \lambda_2>\lambda_1]$

$$\Rightarrow \quad 12\lambda_1+29\lambda_2=-14+29=15$$

**77.** *(80)* $C: x^2+y^2=4$

$$P: y^2=8x$$

Equation of chord bisected at $(2t^2, 4t)$

$$T=S_1$$

$$x(2t^2)+y(4t)-4=(2t^2)^2+(4t)^2-4$$

$\because$ Line passes through $(\alpha, 0)$

$$\alpha(2t^2)=4t^4+16t^2$$

$$\alpha=2t^2+8$$

$$\frac{\alpha-8}{2}=t^2$$

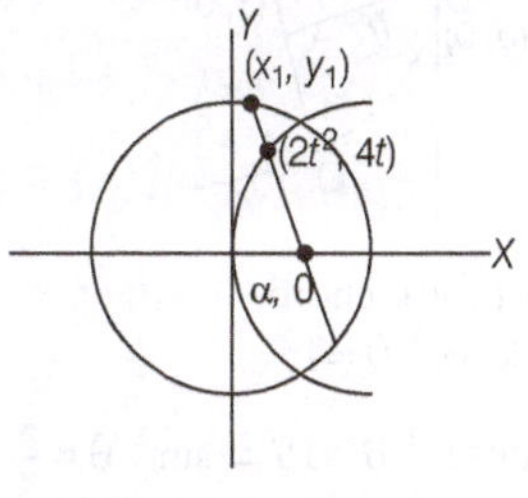


Also, $4t^4+16t^2-4<0$

$$\therefore \quad t^2=-2+\sqrt{5}$$

$$\alpha=4+2\sqrt{5}$$

$$\therefore \quad \alpha\in(8, 4+2\sqrt{5})$$

$$\therefore \quad (2q-p)^2=80$$

**78.** *(10)* $y^2 = 4(x-9)$

Slope of tangent $= -\frac{1}{2}$

Point of contact $P\left[9 + \frac{1}{\left(-\frac{1}{2}\right)^2}, \frac{2\times 1}{-1/2}\right]$

$= (13, -4)$

Centre of circle $C(7, 4)$

Distance

$CP = \sqrt{(13-7)^2 + (-4-4)^2} = 10$

**79.** *(108)* Given, equation of parabola $y^2 = 12x$

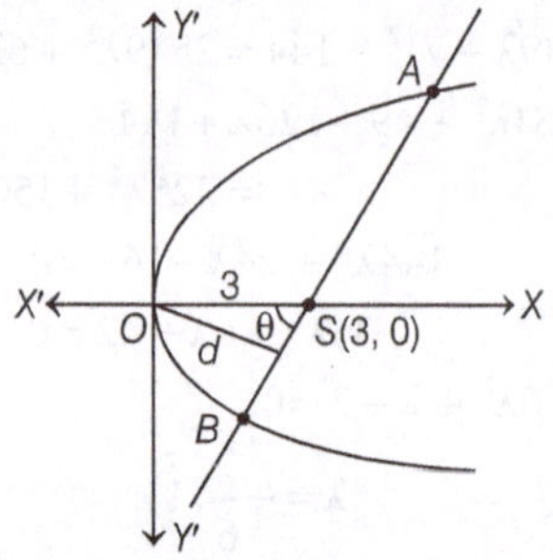


Length of focal chord *AB*,

$$l = 4a \operatorname{cosec}^2\theta$$

$$= 4 \times 3 \times \left(\frac{3}{d}\right)^2 = 12 \times \frac{9}{d^2}$$

$\therefore\ ld^2 = 12 \times 9 = 108$

**80.** *(72)* Given equation of parabola is $y^2 = 12x$

and length of focal chord $= 15$

As we know that,

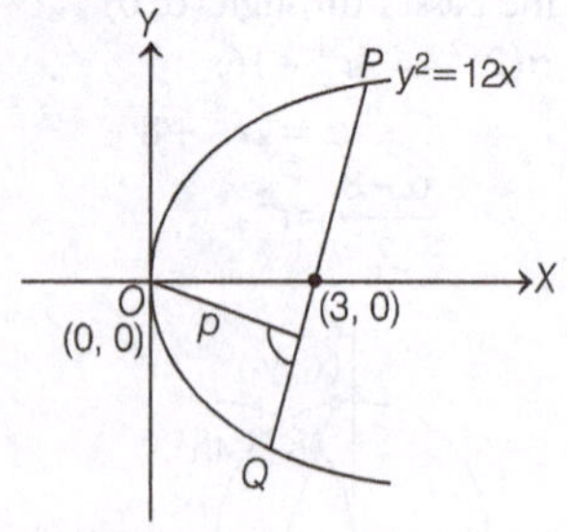


Length of focal chord

$4a \operatorname{cosec}^2\theta = 15$

$\Rightarrow\ 12 \operatorname{cosec}^2\theta = 15 \Rightarrow \sin^2\theta = \frac{4}{5}$

$\Rightarrow\ \tan^2\theta = 4 \Rightarrow \tan\theta = 2$

Hence, equation of *PQ* is $\frac{y-0}{x-3} = 2$

$(\because m = \tan\theta = 2 = \text{slope})$

$\Rightarrow\ y = 2x - 6 \Rightarrow 2x - y - 6 = 0$

$\therefore\ p = \left|\frac{0-0-6}{\sqrt{4+1}}\right| = \frac{6}{\sqrt{5}}$

Hence, $10p^2 = 10 \times \frac{36}{5} = 2 \times 36 = 72$

**81.** *(–192)* Equation of chord bisected at point $\left(1, \frac{5}{4}\right)$ to parabola $x^2 = 8y$

$$T = S_1$$

$xx_1 - 4(y + y_1) = x_1^2 - 8y_1$

$\because\ (x_1, y_1) = \left(1, \frac{5}{4}\right)$

$\Rightarrow x - 4y - 5 = -9 \Rightarrow x - 4y + 4 = 0$

$\because$ (α, β) lies on $x - 4y + 4 = 0$

and $y^2 = 4x$

$\alpha - 4\beta + 4 = 0$ and $\beta^2 = 4\alpha$

$\Rightarrow \beta^2 = 4(4\beta - 4)$

$\Rightarrow \beta^2 - 16\beta + 16 = 0$

$\Rightarrow \beta = \frac{16 \pm \sqrt{(16)^2 - 4(1)(16)}}{2}$

$\Rightarrow \beta = 8 \pm 4\sqrt{3}$

$8 - \beta = \mp 4\sqrt{3}$

Now, $\alpha = 4\beta - 4 = 28 \pm 16\sqrt{3}$

$\therefore (\alpha - 28)(\beta - 8) = (\pm 16\sqrt{3})(\mp 4\sqrt{3})$

$= -192$

**82.** *(32)* Tangent of slope in to the parabola $y^2 = 4x$ is given by

$y = mx + (1/m)$

and tangent of slope '*m*' to the line $x(x-4)^2 + y^2 = 16$ is given by

$$y = m(x-4) \pm 4\sqrt{1+m^2}$$

For common tangent

$$1/m = -4m \pm 4\sqrt{1+m^2}$$

$\Rightarrow \left(\left(\frac{1}{m} + 4m\right)^2 = (\pm 4\sqrt{1+m^2})^2\right)$

[squaring both sides]

$\Rightarrow\ \frac{1}{m^2} + 16m^2 + 8 = 16 + 16m^2$

$\Rightarrow\ \frac{1}{m^2} = 8 \ \Rightarrow m = \pm\frac{1}{2\sqrt{2}}$

Their the point of contact on parabola is $(8, 4\sqrt{2})$.

Length of tangent *PQ* from $(8, 4\sqrt{2})$ on the circle is

$\Rightarrow\ PQ = \sqrt{(8-4)^2 + (4\sqrt{2})^2 - 16}$

$\Rightarrow\ PQ = \sqrt{16 + 32 - 16}$

$\Rightarrow\ PQ = \sqrt{32} \Rightarrow (PQ)^2 = 32$

**83.** *(16)* Focus $= (3, 0)$

Directrix, $x = -3 \Rightarrow 2a = 6 \Rightarrow a - 3$

$\Rightarrow$ Equation of parabola $y^2 = 4 \times 3 \times x$

i.e., $y^2 = 12x$

We know that,

$\alpha = at_1t_2$ and $\beta = a(t_1 + t_2)$

or $\alpha = 3t \cdot 3t,\ \beta = 3(t + 3t)$

$\alpha = 9t^2,\ \beta = 12t$

$\therefore\ \left(\frac{\beta}{\alpha}\right)^2 = \frac{144t^2}{9t^2} = 16$

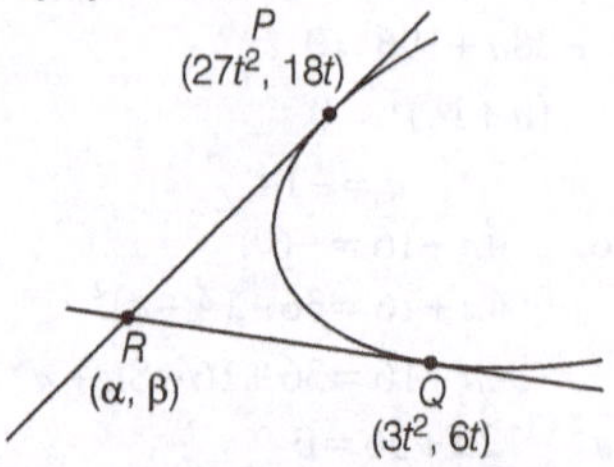


**84.** *(292)* Given, equation of curve is

$x^2 + 2x - 4y + 9 = 0$ ...(i)

equation of tangent at *P*(1, 3) to the given curve Eq. (i)

$x(1) + 2\left(\frac{x+1}{2}\right) - 4\left(\frac{y+3}{2}\right) + 9 = 0$

$\Rightarrow 2x + 2x + 2 - 4y - 12 + 18 = 0$

$\Rightarrow 4x - 4y + 8 = 0 \Rightarrow x - y + 2 = 0$

which is meet the *Y*-axis at *A*

$\therefore A \equiv (0, 2)$

Equation of line passing through *P* and parallel to $x - 3y = 6$ is

$x - 3y + 8 = 0$ ...(ii)

Since, line (ii) meet the parabola $y^2 = 4x$ at *B*

$\therefore y^2 = 4(3y - 8) \Rightarrow y^2 - 12y + 32 = 0$

$\Rightarrow (y-4)(y-8) = 0$

$\therefore$ Possible coordinates of *B* are (4, 4) and (16, 8).

Since, (4, 4) does not satisfies line $2x - 3y = 8$.

Thus, *B* is (16, 8)

$\therefore\ AB^2 = (16-0)^2 + (8-2)^2$

$= 256 + 36 = 292$

**85.** *(16)* We have, $y^2 = 8x + 4y + 4$

$\Rightarrow\ (y-2)^2 = 8(x+1)$

$\therefore\ a = 2$

Let $y - 2 = Y,\ x + 1 = X$

So, equation of parabola is $Y^2 = 8X$

$\therefore$ Focus $= (a, 0) = (2, 0)$

$\therefore\ X = 2, Y = 0$

$\Rightarrow\ x + 1 = 2,\ y - 2 = 0$

$\Rightarrow\ x = 1,\ y = 2$

So, focus of given parabola is (1, 2).

$\therefore$ Equation of focal chord of slope $m$ is $y-2=m(x-1)$.

But, $x$-intercept is 3

$\therefore \quad 0-2=m(3-1) \Rightarrow m=-1$

$\therefore$ Equation of focal chord is

$$y-2=-1(x-1)$$

$\Rightarrow \quad x+y-3=0$

Now, $\quad m=-1$

$\Rightarrow \quad \tan\theta=-1 \Rightarrow \theta=135°$

Again, length of focal chord

$=4a\operatorname{cosec}^2\theta$

$=4\times 2\operatorname{cosec}^2(135°)=8(\sqrt{2})^2=16$

**86.** *(146)* Equation of tangent at $P(b, c)$ on the parabola $y^2=2ax$ is $cy=a(x+b)$

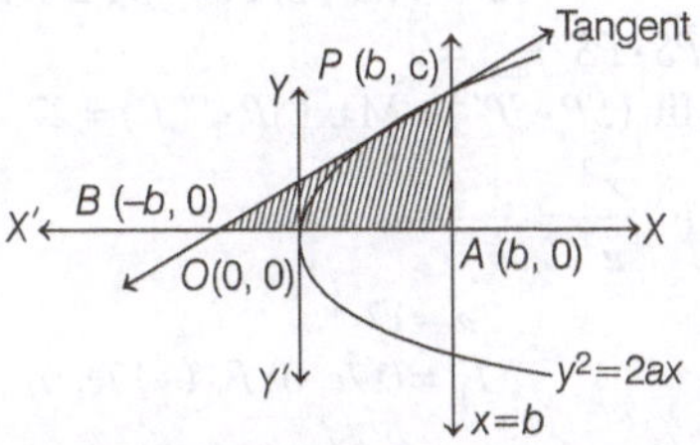


$\Rightarrow \quad ax-cy+ab=0$

Tangent meet $X$-axis at $(-b, 0)$

Now, $ar(\Delta ABP)=\frac{1}{2}AB\times AP$

$$=\frac{1}{2}\times(2b)\times(c)=bc$$

But, $ar(\Delta ABP)=16$

$\therefore \quad bc=16$

Since, $b, c\in N$, so possible values of $(b, c)=(1,16), (2,8), (4,4), (8,2), (16,1)$

Now, $(b, c)$ lies on parabola, so

$$c^2=2ab \Rightarrow a=\frac{c^2}{2b}$$

On putting values of $b$ and $c$, we get

$a=128, 16, 2, \frac{1}{4}, \frac{1}{32}$

But, $a\in N \therefore a=128, 16, 2$

$\therefore \sum_{a\in S} a=128+16+2=146$

**87.** *(10)* Tangent at (2, 2) on $y^2=2x$ is

$x-2y+2=0$

Tangent at (2, 2) on $x^2+y^2=4x$ is

$y=2\rightarrow L_1$

Equation of line is $x+y+2=0\rightarrow L_2$

We have to find intersection point of the line and tangent on the circle at (2,2)

The intersection point of $L_1$ and $L_2$ is $A(-6, 2)$.

$\tan(\angle A)=\left|\frac{m_1-m_2}{1+m_1m_2}\right|$ where,

$m_1=$ slope of line $L_1=0$

$\Rightarrow \quad \tan A=\frac{1}{1}=1$

$m_2=$ slope of $L_2=-1$

$\Rightarrow \quad \angle A=45°$

Now, circumradius (radius of circumcircle) of $\Delta ABC$ is

$r=\frac{a}{2\sin A}$ $\quad$ [where, $a=BC=\sqrt{20}$]

$=\frac{\sqrt{20}}{2\sin 45°}=\frac{\sqrt{20}}{2.\frac{1}{\sqrt{2}}}=\sqrt{10}$

Hence, $r^2=(\sqrt{10})^2=10$

**88.** *(45)* Given, $f(x)=2x^2-\log_e x$

$$f'(x)=4x-\frac{1}{x}$$

$$f'(x)=0 \Rightarrow 4x^2-1=0$$

$\Rightarrow \quad x=\frac{1}{2}\left[x\neq-\frac{1}{2}\right]$

$\therefore f(x)$ is decreasing in $\left(0, \frac{1}{2}\right)$ and increasing in $\left(\frac{1}{2}, \infty\right)$.

$\therefore$ Equation of parabola is $y^2=2x$

Equation of tangent of parabola $y^2=2x$ is $y=mx+\frac{1}{2m}$

The tangent passes through $(8a, 8a-1)$, i.e. (4, 3)

$$3=4m+\frac{1}{2m}$$

$\Rightarrow 8m^2-6m+1=1 \Rightarrow m=\frac{1}{2}$ or $\frac{1}{4}$

Equation of tangent can be $y=\frac{1}{2}x+1$ or $y=\frac{1}{4}x+2$

But the tangent $y=\frac{1}{2}x+1$ does pass through $\left(-\frac{1}{a}, 0\right)$ i.e., $(-2, 0)$, so rejected.

$\therefore$The required tangent of parabola is $y=\frac{1}{4}x+2$

$\therefore$Slope of normal is $-4$.

Equation of normal of parabola is

$$y=mx-2am-am^3$$

$\Rightarrow y=-4x-2\left(\frac{1}{2}\right)(-4)-\frac{1}{2}(-4)^3$

$\Rightarrow 4x+y=36 \Rightarrow \frac{x}{9}+\frac{y}{36}=1$

Here, $\alpha=9, \beta=36$

$\Rightarrow \quad \alpha+\beta=9+36=45$

**89.** *(63)* Vertex and focus of parabola $y^2=2x$ are $V(0, 0)$ and $S\left(\frac{1}{2}, 0\right)$ respectively.

Let equation of circle be

$$(x-h)^2+(y-k)^2=4$$

$\because$ Circle passes through (0, 0).

$\Rightarrow \quad h^2+k^2=4 \quad \ldots(i)$

$\because$ Circle passes through $\left(\frac{1}{2}, 0\right)$

$$\left(\frac{1}{2}-h\right)^2+k^2=4$$

$\Rightarrow h^2+k^2-h=\frac{15}{4} \quad \ldots(ii)$

On solving Eqs. (i) and (ii), we get

$4-h=\frac{15}{4} \quad \Rightarrow h=\frac{1}{4}$ and $k=\frac{\sqrt{63}}{4}$

$k=\frac{-\sqrt{63}}{4}$ is rejected as circle with centre $\left(\frac{1}{4}, \frac{-\sqrt{63}}{4}\right)$.

can not touch given parabola.

Equation of circle is

$$\left(x-\frac{1}{4}\right)^2+\left(y-\frac{\sqrt{63}}{4}\right)^2=4$$

$\Rightarrow \quad \alpha=2+\frac{\sqrt{63}}{4}=\frac{8+\sqrt{63}}{4}$

$\Rightarrow \quad 4\alpha-8=\sqrt{63}$

$\Rightarrow (4\alpha-8)^2=63$

**90.** *(10)* Given that $P_1$ be a parabola with vertex (3, 2) and focus (4, 4).

The point of intersection of directrix with axis of parabola $=A=(2, 0)$

Image of $A(2, 0)$ w.r.t. line $x+2y=6$ is $B(x_2, y_2)$.

$\therefore \quad \frac{x_2-2}{1}=\frac{y_2-0}{2}=\frac{-2(2+0-6)}{5}$

$\therefore B(x_2, y_2)=\left(\frac{18}{5}, \frac{16}{5}\right)$.

The point $B$ is the point of intersection of directrix with axis of parabola $P_2$.

Thus, $x + 2y = k$ must have point $\left(\frac{18}{5}, \frac{16}{5}\right)$

$\Rightarrow \frac{18}{5} + \frac{32}{5} = k \Rightarrow k = 10$

$\Rightarrow x + 2y = 10$

**91.** *(1)* The equation of circle

$\text{Re}(z^2) + 2[\text{Im}(z)]^2 + 2\text{Re}(z) = 0$,

where $z = x + iy$

$\Rightarrow (x^2 - y^2) + 2y^2 + 2x = 0$

$[\because z^2 = (x^2 - y^2) + 2xyi]$

$\Rightarrow x^2 + y^2 + 2x = 0$

$\Rightarrow (x+1)^2 + y^2 = 1$

Centre $= (-1, 0)$

Now, parabola $\Rightarrow x^2 - 6x - y + 13 = 0$

$\Rightarrow x^2 - 6x + 9 = y - 4$

$\Rightarrow (x-3)^2 = y - 4$

Vertex = Equation of line passing through centre $(-1, 0)$ and vertex $(3, 4)$ is

$y - 0 = \frac{4-0}{3+1}(x+1)$

$\Rightarrow y = x + 1$

On comparing, $y = mx + c$

$\Rightarrow$ $y$-intercept $= 1$

**92.** *(9)* Shortest distance is along the normal.

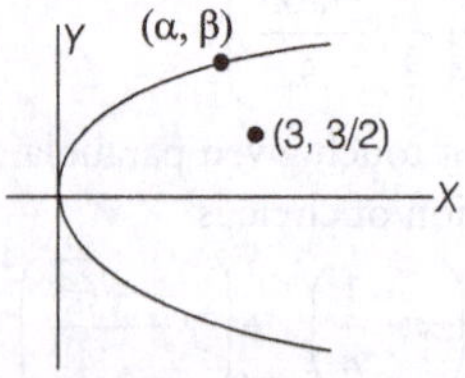


Equation of normal at $(\alpha, \beta)$

$y - \beta = -\frac{\beta}{2a}(x - \alpha)$

$(\because y^2 = 4ax \Rightarrow 4a = 6 \Rightarrow 2a = 3)$

$\Rightarrow 3y - 3\beta = -\beta(x - \alpha)$

This normal pass through $\left(3, \frac{3}{2}\right)$.

$\frac{9}{2} - 3\beta = -3\beta + \alpha\beta \Rightarrow \alpha\beta = \frac{9}{2}$ ...(i)

Now, from $y^2 = 6\beta^2 = 6\alpha$ ...(ii)

From Eqs. (i) and (ii), we get

$\alpha = 3/2, \beta = 3$

$\therefore 2(\alpha + \beta) = 2\left(\frac{3}{2} + 3\right) = 9$

**93.** *(9)* Given, circle $\Rightarrow (x-3)^2 + y^2 = 9$

Parabola $\Rightarrow y^2 = 4x$

Let equation of common tangent be

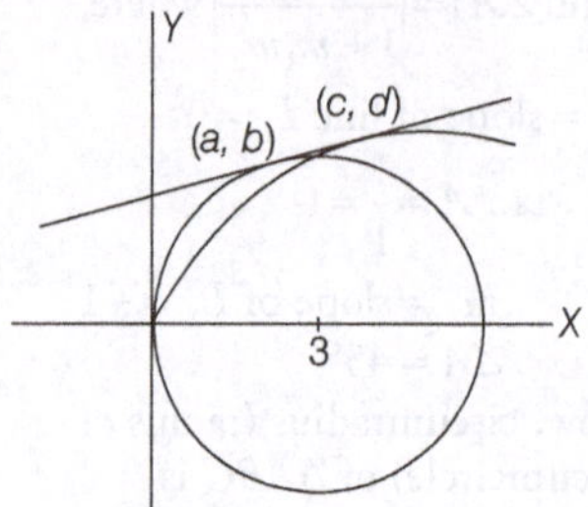


$y = mx + \frac{a}{m}$

$\Rightarrow y = mx + \frac{1}{m}$ $(\because y^2 = 4x)$

$\Rightarrow m^2x - my + 1 = 0$

The above line is tangent to circle.

$\therefore$ Perpendicular from (3, 0) to line = 3

$\Rightarrow \left|\frac{(3m^2 - 0 + 1)}{\sqrt{m^2 + m^4}}\right| = 3$

$\Rightarrow (3m^2 + 1)^2 = 9(m^2 + m^4)$

$\Rightarrow m = \pm\frac{1}{\sqrt{3}}$

Tangent is $y = \frac{1}{\sqrt{3}}x + \sqrt{3} \Rightarrow m = \frac{1}{\sqrt{3}}$

or $y = \frac{(-1)}{\sqrt{3}} \times (-\sqrt{3})$ (rejected)

For parabola, point of contact is

$(c, d) = \left(\frac{a}{m^2}, \frac{2a}{m}\right)$

$\therefore (c, d) = (3, 2\sqrt{3})$

Again, solve circle and line equation, we get

$(x-3)^2 + \left[\left(\frac{1}{\sqrt{3}}x + \sqrt{3}\right)\right]^2 = 9$

$\Rightarrow x^2 + 9 - 6x + \frac{1}{3}x^2 + 3 + 2x = 9$

$\Rightarrow \frac{4}{3}x^2 - 4x + 3 = 0 \Rightarrow x = \frac{3}{2} = a$

$\therefore 2(a + c) = 2\left(\frac{3}{2} + 3\right) = 9$

## Topic 3

**01.** *(d)* $2b = \frac{1}{4}(2ae)$ [given]

$b/a = e/4$

but we know, $e^2 = 1 - \frac{b^2}{a^2}$

$\Rightarrow e^2 = 1 - \frac{e^2}{16} \Rightarrow \frac{17e^2}{16} = 1 \Rightarrow e = \frac{4}{\sqrt{17}}$

**02.** *(d)*

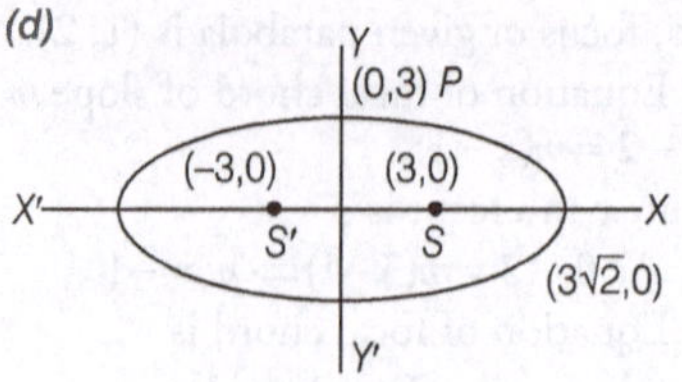


$a = 3\sqrt{2},\ b = 3$ and $e = 1/\sqrt{2}$

$\Rightarrow PS + PS' = 2a = 6\sqrt{2}$

$\frac{PS + PS'}{2} \geq \sqrt{(PS)\cdot(PS')}$

$\Rightarrow (PS \times PS')_{\max} = 18$

Minima happens when $P$ lies on major axis

$\Rightarrow P \equiv (3\sqrt{2}, 0)$

$PS = 3\sqrt{2} - 3,\ PS' = 3\sqrt{2} + 3$

$(PS \cdot PS')_{\min} = 9$

$\text{Min}(SP \cdot SP') + \text{Max}(SP \cdot S'P) = 27$

**03.** *(d)* $\frac{x^2}{a^2} + \frac{y^2}{b^2} = 1,\ a > b$

$2a = 17$

$F_1 = (17e, 0)\ F_2\ (-17e, 0)$

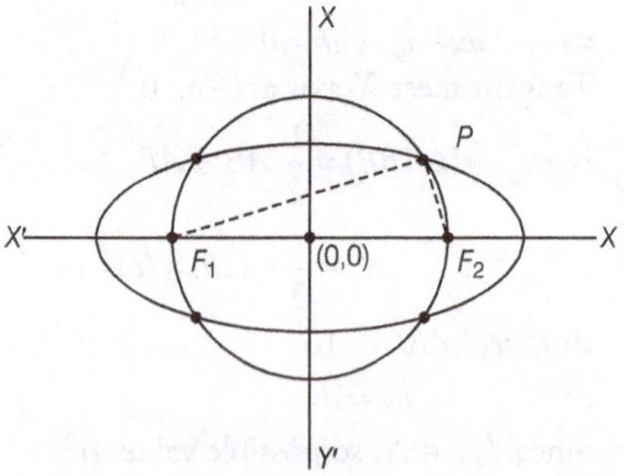


Area of $\Delta PF_1F_2 = 30$

$\frac{1}{2} \times PF_1 \cdot PF_2 = 30$

$(PF_1)(17 - PF_1) = 60$

$PF_1 = 5$ or $12$

$PF_2 = 12$ or $5$

$\therefore F_1F_2 = \sqrt{(PF_1)^2 + (PF_2)^2} = 13$

**04.** *(d)* $F_1\ (2, 5);\ F_2\ (2, -3)$

$e = 4/5$

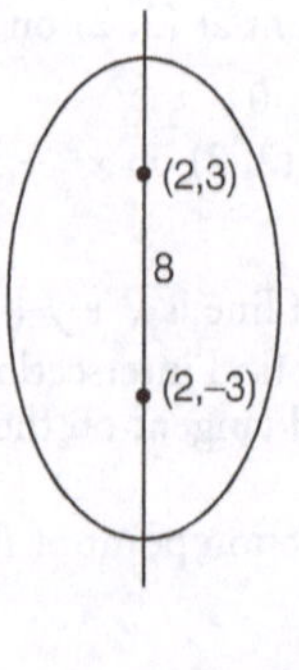

$$\therefore \quad 2be = 8$$
$$be = 4$$
$$b = 5$$
$$e = \sqrt{1 - \frac{a^2}{b^2}}$$
$$\frac{16}{25} = 1 - \frac{9^2}{25}$$
$$a^2 = 9 \Rightarrow a = 3$$
$$\text{LR} = \frac{2a^2}{b} = \frac{2 \times 9}{5} = \frac{18}{5}$$

**05.** *(b)* Length of latus rectum is $\frac{2b^2}{a} = 10$

and eccentricity $e = \sqrt{1 - \frac{b^2}{a^2}}$

$$f(t) = t^2 + t + \frac{11}{2} \Rightarrow f'(t) = 2t + 1 = 0$$
$$\Rightarrow t = -\frac{1}{2}$$

Minima occur at $t = -\frac{1}{2}$

Minimum value $f\left(-\frac{1}{2}\right) = \left(-\frac{1}{2}\right)^2$

$$-\frac{1}{2} + \frac{11}{2} = \frac{1}{4} - \frac{1}{2} + \frac{11}{12} = \frac{3 - 6 + 11}{12}$$
$$= \frac{2}{3}$$
$$\Rightarrow \quad e = \sqrt{1 - \frac{b^2}{a^2}} = \frac{2}{3} \text{ and } \frac{2b^2}{a} = 10$$
$$\Rightarrow b^2 = 5a$$
$$\frac{4}{9} = 1 - \frac{5a}{a^2} \Rightarrow \frac{4}{9} = 1 - \frac{5}{a}$$
$$\Rightarrow -\frac{5}{9} = -\frac{5}{a} \Rightarrow a = 9$$

Then $b^2 = 5a = 45$

Thus, $a^2 + b^2 = 81 + 45 = 126$

**06.** *(a)* Given, equation of ellipase

$$\frac{x^2}{9} + \frac{y^2}{4} = 1$$

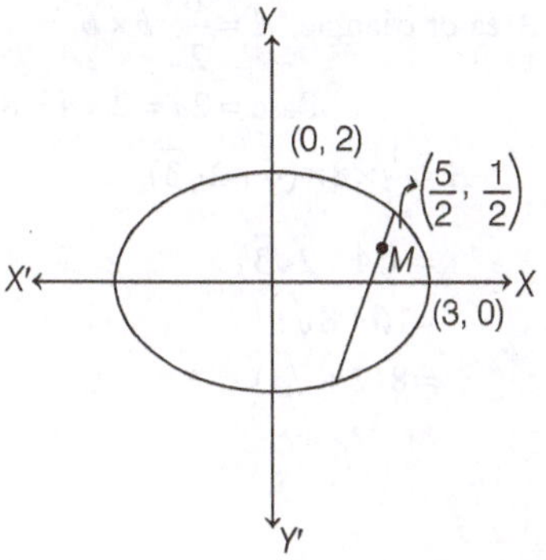


Equation of chord

$$\frac{5}{2}\left(\frac{x}{9}\right) + \frac{1}{2}\left(\frac{y}{4}\right) = \frac{25}{36} + \frac{1}{16}$$
$$\Rightarrow \frac{5x}{18} + \frac{y}{8} = \frac{100 + 9}{144} = \frac{109}{144}$$
$$\Rightarrow 40x + 18y = 109$$

Compare with $\alpha x + \beta y = 109$, we get

$$\Rightarrow \quad \alpha = 40, \beta = 18$$
$$\Rightarrow \quad \alpha + \beta = 58$$

> **Concept Enhancer**
>
> The equation of the chord of the ellipse $\frac{x^2}{a^2} + \frac{y^2}{b^2} = 1$ bisected at the point $(x_1, y_1)$ is $T = S_1$

**07.** *(d)* Equation of chord is $T = S_1$

$$\Rightarrow \quad \frac{x\sqrt{2}}{9} + \frac{y}{3} = \frac{2}{3} \Rightarrow \sqrt{2}\,x + 3y = 6$$

Now, $\dfrac{\left(\dfrac{6 - 3y}{\sqrt{2}}\right)^2}{9} + \dfrac{y^2}{4} = 1$

$$\Rightarrow \quad (y - 2)(3y - 2) = 0$$
$$\Rightarrow \quad y = 2, \frac{2}{3}$$

If $y = 2$, then $x = 0$

If $y = \frac{2}{3}$, then $x = 2\sqrt{2}$

Length of chord is $\sqrt{8 + \frac{16}{9}} = \frac{2\sqrt{22}}{3}$

$$\frac{2\sqrt{\alpha}}{3} = \frac{2\sqrt{22}}{3}$$
$$\Rightarrow \quad \alpha = 22$$

**08.** *(c)* Given equation of ellipse,

$$\frac{x^2}{25} + \frac{y^2}{16} = 1$$

∴ Equation of chord with given middle point (3, 1) is $T = S_1$

$$\Rightarrow \frac{x \cdot x_1}{25} + \frac{y \cdot y_1}{16} - 1 = \frac{x_1^2}{25} + \frac{y_1^2}{16} - 1$$
$$\Rightarrow \quad \frac{3x}{25} + \frac{y}{16} = \frac{9}{25} + \frac{1}{16}$$
$$\Rightarrow \quad 48x + 25y = 144 + 25$$
$$\Rightarrow \quad 48x + 25y = 169$$

**09.** *(b)* If the minor axis is $a$ and $e$ is eccentricity. According to the question,

$$ae = 2b$$
$$\Rightarrow \quad \frac{4b^2}{a^2} = e^2$$
$$\Rightarrow 4(1 - e^2) = e^2 \qquad \left[\because e = \sqrt{1 - \frac{b^2}{a^2}}\right]$$
$$\Rightarrow \quad 4 = 5e^2 \Rightarrow e^2 = \frac{4}{5} \Rightarrow e = \frac{2}{\sqrt{5}}$$

Hence, the eccentricity $2/\sqrt{5}$.

**10.** *(b)* Given, line $L$ is passing through the point of intersection of $bx + 10y - 8 = 0$ and $2x - 3y = 0$ is $(bx + 10y - 8) + \mu(2x - 3y) = 0$.

∵ Line passing through (1, 1), so $\mu = b + 2$.

Now, line $(3b + 4)x - (3b - 4)y - 8 = 0$ is tangent of circle $17(x^2 + y^2) = 16$

So, $\dfrac{8}{\sqrt{(3b + 4)^2 + (3b - 4)^2}} = \dfrac{4}{\sqrt{17}}$

$$\Rightarrow \quad b^2 = 2 \Rightarrow e = \sqrt{\frac{3}{5}}$$
$$\{\therefore b^2 = a^2(1 - e^2)\}$$

**11.** *(b)* Curves $\frac{x^2}{16} + \frac{y^2}{9} = 1; x^2 + y^2 = 12$

To find $= 12m^2$

Let the slope of common tangent be $m$.

Equation of tangent to $x^2 + y^2 = 12$

$$y = mx \pm \sqrt{12 + 12m^2}$$

Equation of tangent to ellipse

$$\frac{x^2}{16} + \frac{y^2}{9} = 1$$
$$y = mx \pm \sqrt{16m^2 + 9}$$

Both equations are identical,

$$mx \pm \sqrt{12 + 12m^2} = mx \pm \sqrt{16m^2 + 9}$$
$$\Rightarrow \quad 12 + 12m^2 = 16m^2 + 9$$
$$3 = 4m^2 \Rightarrow m^2 = 3/4$$
$$\therefore \quad 12m^2 = 12 \times \frac{3}{4} = 9$$

**12.** *(b)* $12x\cos\theta + 5y\sin\theta = 60$

$$\Rightarrow \quad \frac{x\cos\theta}{5} + \frac{y\sin\theta}{12} = 1$$

The given equation is the equation of the tangent which passes through $(5\cos\theta, 12\sin\theta)$.

Hence, the given equation is tangent to $\frac{x^2}{25} + \frac{y^2}{144} = 1$ at $(5\cos\theta, 12\sin\theta)$.

**13.** *(a)* $e^2 = 1 - \frac{b^2}{a^2} \Rightarrow e^2 = 1 - \frac{a^2}{c^2}$

$$\Rightarrow \quad \frac{b^2}{a^2} = \frac{a^2}{c^2} \Rightarrow c = \frac{a^2}{b}$$

Also, $b = ce \Rightarrow c = b/e$

$\Rightarrow \quad \frac{b}{e} = \frac{a^2}{b} \quad \Rightarrow \quad e = \frac{b^2}{a^2} = 1 - e^2$

$\Rightarrow \quad e^2 + e - 1 = 0 \Rightarrow e = \frac{-1+\sqrt{5}}{2}$

**14.** *(c)* As we know that locus of the foot of perpendicular drawn upon any tangent to the ellipse $\frac{x^2}{4} + \frac{y^2}{2} = 1$ from any of its foci is the auxilliary circle and its equation is

$$x^2 + y^2 = 4 \qquad \ldots(i)$$

So, from the option the auxilliary circle (i) passes through the point $(-1, \sqrt{3})$.

**15.** *(d)* Equation of given curve

$$4x^2 + 5y^2 = 20$$

$\Rightarrow \quad \frac{x^2}{5} + \frac{y^2}{4} = 1 \qquad \ldots(i)$

Let a point $P(\sqrt{5}\cos\theta, 2\sin\theta)$ on the curve (i).

$PQ^2 = (\sqrt{5}\cos\theta - 0)^2 + (2\sin\theta + 4)^2$

$[\because Q = (0, -4)]$

$\therefore PQ^2 = 5\cos^2\theta + 16 + 4\sin^2\theta + 16\sin\theta$

$= 21 - \sin^2\theta + 16\sin\theta$

$= 21 - (8 - \sin\theta)^2 + 64$

$= 85 - (8 - \sin\theta)^2$

As, the point $P$ is farthest from the point $Q$,

so $\quad \sin\theta = 1$

And, so $\quad PQ^2 = 85 - 49 = 36$

**16.** *(b)* $\because \frac{a}{e} - ae = 4$ and $e = 1/2$

$\therefore \quad 2a - \frac{a}{2} = 4$

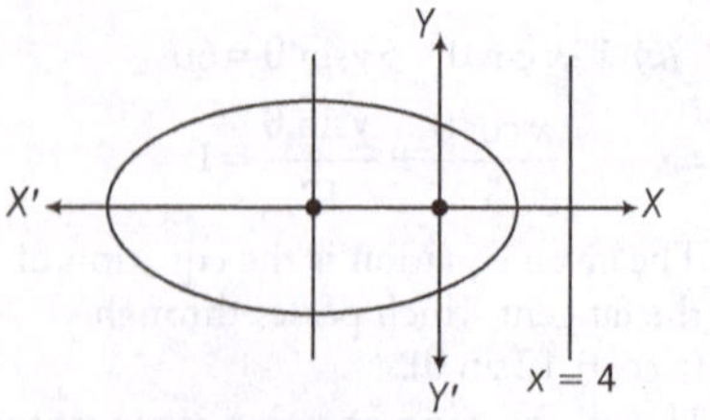


$\Rightarrow \quad \frac{3a}{2} = 4$

$\Rightarrow \quad a = \frac{8}{3}$

**17.** *(d)* Since, $\angle FBF' = 90°$, then

$\angle OBF' = 45°$ and $\angle BF'O = 45°$

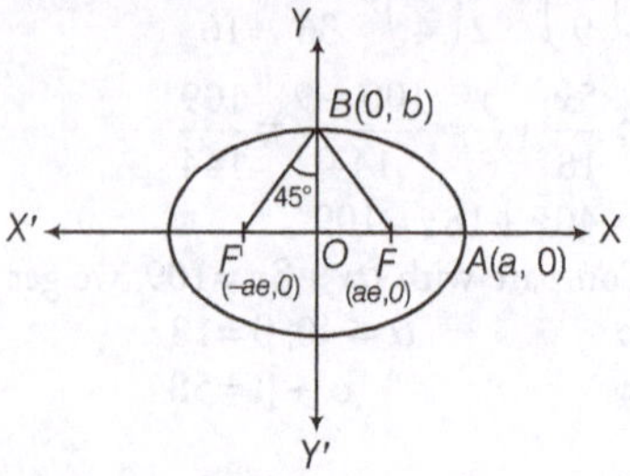


$\Rightarrow \quad ae = b$

[Since, $\Delta BOF'$ is an isosceles triangle]

and $e^2 = 1 - \frac{b^2}{a^2} \Rightarrow e^2 = 1 - \frac{a^2e^2}{a^2}$

$\Rightarrow \quad e^2 = 1 - e^2 \Rightarrow 2e^2 = 1$

$\therefore \quad e = \frac{1}{\sqrt{2}}$

[Since, $e$ cannot be negative]

**Alternate Solution**

Since, $F$ and $F'$ are foci of an ellipse, whose coordinates are $(ae, 0)$ and $(-ae, 0)$ respectively and coordiantes of $B$ are $(0, b)$.

$\therefore$ Slope of $BF = \frac{b}{-ae}$ and slope of

$BF' = \frac{b}{ae}; \quad \because \angle FBF' = 90°$

$\therefore \quad -\frac{b}{ae} \cdot \left(\frac{b}{ae}\right) = -1$

$\Rightarrow \quad b^2 = a^2e^2$

$\therefore \quad e^2 = 1 - \frac{b^2}{a^2} = 1 - \frac{a^2e^2}{a^2}$

$\Rightarrow \quad 2e^2 = 1; \therefore e = \frac{1}{\sqrt{2}}$

**18.** *(b)* Since, equation of directrix is $x = 4$, then major axis of an ellipse is along $X$-axis.

$\therefore \quad \frac{a}{e} = 4 \Rightarrow a = 4 \times \frac{1}{2} \quad \left[\because e = \frac{1}{2}\right]$

$\Rightarrow \quad a = 2$

Now, $b^2 = a^2(1 - e^2)$

$\therefore \quad b^2 = 4\left(1 - \frac{1}{4}\right) = 4 \times \frac{3}{4}$

$\Rightarrow \quad b^2 = 3$

Hence, equation of ellipse is

$$\frac{x^2}{4} + \frac{y^2}{3} = 1$$

or $\quad 3x^2 + 4y^2 = 12$

**19.** *(b)* Given that,

$e = \frac{1}{2}$,

$ae = 2$

$\Rightarrow \quad a = 4$

$\therefore \quad b^2 = a^2(1 - e^2)$

$= 16\left(1 - \frac{1}{4}\right) = 12$

Hence, equation of an ellipse is

$$\frac{x^2}{16} + \frac{y^2}{12} = 1$$

**20.** *(b)* foci $\equiv (\pm 2, 0)$ [given]

$ae = 2$

$a \cdot \frac{1}{2} = 2 \qquad \left[\because e = \frac{1}{2} \text{ (given)}\right]$

$a = 4$

using $b^2 = a^2 - (ae)^2$

$b^2 = 4^2 - 2^2$

$= 16 - 4 = 12$

$b = \sqrt{12} = 2\sqrt{3}$

intersection point occurs when $x = 0$

$\frac{x^2}{4^2} + \frac{y^2}{(2\sqrt{3})^2} = 1$

$\Rightarrow \quad \frac{0}{4^2} + \frac{y^2}{(2\sqrt{3})^2} = 1$

$\Rightarrow \quad \frac{y^2}{12} = 1$

$\Rightarrow \quad y^2 = 12$

$y = -2\sqrt{3}$ [considering $y$ to be negative]

intersection point is $(0, -2\sqrt{3})$

The circle has minimum area when it circumscribes the ellipse.

The radius of this circle $= a$

$r = a = 4$

height $h$ = distance from point $(0, -2\sqrt{3})$ to the circles top

$h = r + 2\sqrt{3} = 4 + 2\sqrt{3}$

$\therefore$Area of triangle, $A = \frac{1}{2} \times b \times h$

Base $= 2a = 2 \times 4 = 8$

$A = \frac{1}{2} \times 8 \times (4 + 2\sqrt{3})$

$= 4(4 + 2\sqrt{3})$

$= 16 + 8\sqrt{3}$

$= 8(2 + \sqrt{3})$

**21.** (*d*) It is given that $p$ be the number of all triangles that can be formed by joining the vertices of a regular polygon $P$ and $q$ be the number of all quadrilaterals that can be formed by joining the vertices of $P$.

A triangle can be formed by joining three vertices and a quadrilateral formed by joining four vertices.

The number of ways to choose $k$ items from $n$ is given by ${}^nC_r = \frac{n!}{(n-r)!}$

The number of triangles $p$ is the number of ways to choose three vertices from $n$

$$p = {}^nC_3 = \frac{n!}{(n-3)!}$$

The number of quadrilaterals $q$ is the number of ways to choose four vertices from $n$.

$$q = {}^nC_4 = \frac{n!}{(n-4)!}$$

Also, given that $p+q=126$

$${}^nC_3 + {}^nC_4 = 126 \Rightarrow {}^{n+1}C_4 = 126$$

$$[\because {}^nC_{r-1} + {}^nC_r = {}^{n+1}C_4]$$

$${}^{n+1}C_4 = 126$$

$\Rightarrow$ Use ${}^9C_4 = 126$ this gives $n+1=9$

$\Rightarrow n = 8$

The equation of ellipse $\frac{x^2}{16} + \frac{y^2}{n} = 1$

becomes $\frac{x^2}{16} + \frac{y^2}{8} = 1$

The eccentricity

$$e = \sqrt{1 - \frac{b^2}{a^2}} = \sqrt{1 - \frac{8}{16}} = \frac{1}{\sqrt{2}}$$

**22.** (*a*) As, we know that focal distance of an ellipse $= a \pm ex$

$\therefore$ Its product

$$= (a+ex)(a-ex)$$
$$= a^2 - e^2x^2$$
$$= a^2 - e^2(\sqrt{3})^2$$
$$= a^2 - 3e^2 = \frac{7}{4} \quad \text{[given]}$$

$$\Rightarrow \quad 4a^2 = 7 + 12e^2 \quad \ldots\text{(i)}$$

Also, $(\sqrt{3}, 1/2)$ lies on ellipse

$$\frac{x^2}{a^2} + \frac{y^2}{b^2} = 1$$

So, $$\frac{3}{a^2} + \frac{1}{4b^2} = 1$$

$$\Rightarrow \quad \frac{3}{a^2} + \frac{1}{4a^2(1-e^2)} = 1$$

$$[\because b^2 = a^2(1-e^2)]$$

$$\Rightarrow 12(1-e^2) + 1 = 4a^2(1-e^2)$$

$$\Rightarrow 13 - 12e^2 = (7+12e^2)(1-e^2) \quad \text{(using Eq. (i))}$$

$$\Rightarrow 13 - 12e^2 = 7 - 7e^2 + 12e^2 - 12e^4$$

$$\Rightarrow 12e^4 - 17e^2 + 6 = 0$$

$$\Rightarrow e^2 = \frac{17 \pm \sqrt{289-288}}{24} = \frac{17 \pm 1}{24}$$

$$\Rightarrow e^2 = \frac{3}{4}, \frac{2}{3}$$

$$\Rightarrow \quad e = \frac{\sqrt{3}}{2} \text{ and } \sqrt{2/3}$$

$[\because 0 < e < 1$ always for an ellipse$]$

$\therefore$ Absolute difference

$$= \left| \frac{\sqrt{3}}{2} - \sqrt{2/3} \right|$$

$$= \left| \sqrt{2/3} - \frac{\sqrt{3}}{2} \right|$$

$$= \frac{3 - 2\sqrt{2}}{2\sqrt{3}}$$

**23.** (*c*) Ellipse : $\frac{x^2}{4} + \frac{y^2}{2} = 1$

Equation of chord whose mid-point is $\left(1, \frac{1}{2}\right)$ is

$$T = S_1$$

$$\Rightarrow \quad \frac{x(1)}{4} + \frac{y(1/2)}{2} - 1 = \frac{1}{4} + \frac{1}{8} - 1$$

$$\Rightarrow \quad \frac{x}{4} + \frac{y}{4} = \frac{3}{8}$$

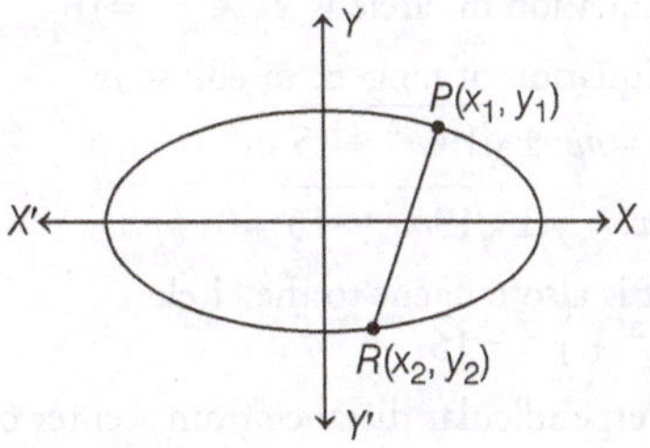


$$\Rightarrow \quad x + y = \frac{3}{2} \quad \ldots\text{(i)}$$

Eq. (i) represents equation of chord $PR$.

$$PR = \sqrt{(x_2 - x_1)^2 + (y_2 - y_1)^2}$$
$$= |x_2 - x_1| \sqrt{1+1}$$
$$= \sqrt{2}\,|x_2 - x_1|$$

Put $y = \frac{3}{2} - x$ in equation of ellipse

$$\Rightarrow \quad \frac{x^2}{4} + \frac{\left(\frac{3}{2} - x\right)^2}{2} = 1$$

$$\Rightarrow \quad x^2 + 2\left(\frac{9}{4} + x^2 - 3x\right) = 4$$

$$\Rightarrow \quad x^2 + \frac{9}{2} + 2x^2 - 6x = 4$$

$$\Rightarrow \quad 3x^2 - 6x + \frac{1}{2} = 0$$

$$\Rightarrow \quad 6x^2 - 12x + 1 = 0$$

$$\Rightarrow \quad |x_2 - x_1| = \frac{\sqrt{D}}{a}$$

$$= \frac{\sqrt{144 - 24}}{6}$$

$$= \frac{\sqrt{120}}{6}$$

$$\therefore \quad PR = \sqrt{2}\,\frac{\sqrt{120}}{6}$$

$$= \frac{2}{6}\sqrt{60}$$

$$= \frac{2}{3}\sqrt{15}$$

**24.** (*d*) We have,

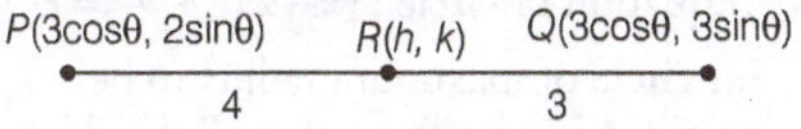


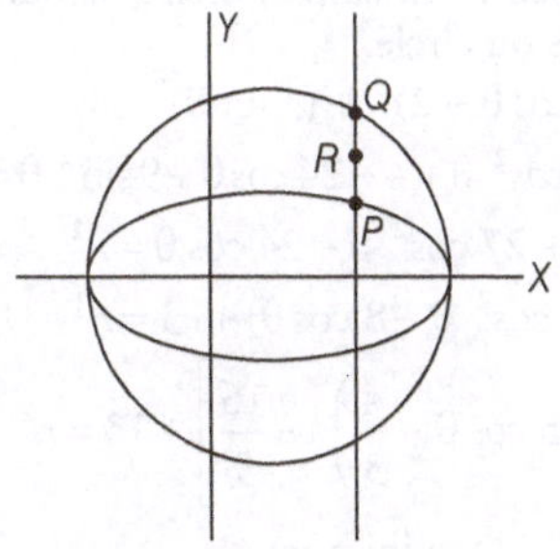


$$\therefore \quad h = \frac{12\cos\theta + 9\cos\theta}{7}$$

$$\Rightarrow \quad h = 3\cos\theta$$

and $k = \frac{6\sin\theta + 12\sin\theta}{7}$

$$\Rightarrow \quad k = \frac{18}{7}\sin\theta$$

Hence, locus of the point $R$ on $PQ$.

$$\frac{h^2}{3^2} + \frac{k^2}{\left(\frac{18}{7}\right)^2} = 1$$

here, $a^2 = 3^2, b^2 = \left(\frac{18}{7}\right)^2$

Now, eccentricity,

$$e^2 = \sqrt{1 - \frac{b^2}{a^2}}$$

$$= \sqrt{1 - \frac{\left(\frac{18}{7}\right)^2}{3^2}}$$

$$= \sqrt{1 - \frac{36}{49}} = \sqrt{\frac{13}{49}}$$

So, $e = \frac{\sqrt{13}}{7}$

**25.** *(d)* Given, equation on ellipse

$$\frac{x^2}{6^2} + \frac{y^2}{3^2} = 1$$

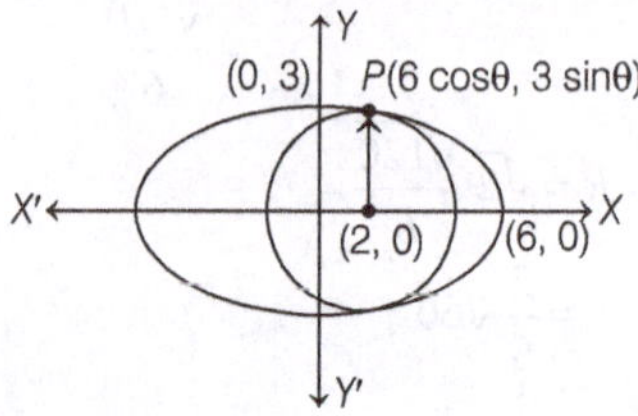


Let $P$ be a point on ellipse.
Then, $P(6 \cos \theta, 3 \sin \theta)$
$\therefore$Equation of circle : $(x-2)^2 + y^2 = r^2$
For circle of maximum radius to be inscribed in an ellipse, then $P$ should also lie on circle.

$\Rightarrow (6 \cos \theta - 2)^2 + (3 \sin \theta)^2 = r^2$

$\Rightarrow 36 \cos^2 \theta + 4 - 24 \cos\theta + 9 \sin^2 \theta = r^2$

$\Rightarrow 13 + 27 \cos^2 \theta - 24 \cos \theta = r^2$

$\Rightarrow 3(9 \cos^2 \theta - 8 \cos \theta) + 13 = r^2$

$\Rightarrow 3\left[\left(3 \cos \theta - \frac{4}{3}\right)^2 - \frac{16}{9}\right] + 13 = r^2$

For, $r \longrightarrow$ minimum

$\Rightarrow \left(3 \cos \theta - \frac{4}{3}\right)^2 = 0$

Then, $r^2 = 13 - \frac{16}{3} = \frac{23}{3}$

$\Rightarrow\ 12r^2 = 23 \times 4 = 92$

**26.** *(c)* We have,

$E_k = kx^2 + k^2y^2 = 1, k = 1, 2, \ldots, 20$

$$\Rightarrow \quad \frac{x^2}{\frac{1}{k}} + \frac{y^2}{\frac{1}{k^2}} = 1$$

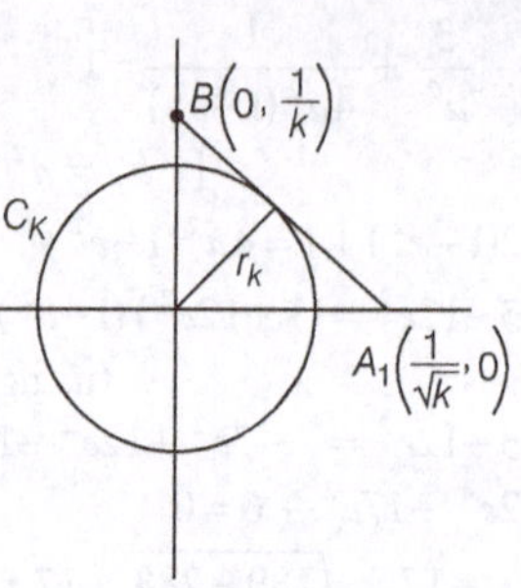


Equation of $AB$ is $\frac{x}{\frac{1}{\sqrt{k}}} + \frac{y}{\frac{1}{k}} = 1$

or $\qquad \sqrt{k}x + ky = 1$

$r_k$ is the radius of circle $C_k$.
So, $r_k$ = Perpendicular distance from (0, 0) to $AB$

$$r_k = \frac{|0+0-1|}{\sqrt{(\sqrt{k})^2 + k^2}} = \frac{1}{\sqrt{k+k^2}}$$

$$\frac{1}{r_k^2} = k + k^2$$

$$\sum_{k=1}^{20} \frac{1}{r_k^2} = \sum_{k=1}^{20} k + \sum_{k=1}^{20} k^2$$

$$= \frac{20 \times 21}{2} + \frac{20 \times 21 \times 41}{6}$$

$$= 210 + 2870$$

$$= 3080$$

**27.** *(b)* We have, equation of ellipse
$15x^2 + 19y^2 = 285$

or $\quad \frac{x^2}{19} + \frac{y^2}{15} = 1$

Let the coordinate of center of circle be (0, 0).
Equation of circle is $x^2 + y^2 = 16$
Equation of tangent of ellipse is
$y = mx \pm \sqrt{19m^2 + 15}$ or
$mx - y \pm \sqrt{19m^2 + 15} = 0$
It is also tangent to the circle
$x^2 + y^2 = 16$
Perpendicular distance from center of circle to tangent = 4

$$\frac{\left|0 - 0 \pm \sqrt{19m^2 + 15}\right|}{\sqrt{m^2 + 1}} = 4$$

Squaring both side, we get
$19m^2 + 15 = 16m^2 + 16$

$3m^2 = 1 \quad \Rightarrow m^2 = \frac{1}{3} \Rightarrow \quad m = \pm \frac{1}{\sqrt{3}}$

$\tan \theta = \frac{1}{\sqrt{3}} \Rightarrow \theta = \frac{\pi}{6}$ with $X$-axis or $\pi/3$ with $Y$-axis

Hence, the common tangents are inclined to the minor axis of the ellipse at an angle of $\pi/3$.

**28.** *(d)* Let $E$ be the person speak English
$\therefore n(E) = 75$
and $H$ be the person speak Hindi
$\therefore n(H) = 40$
Let number of persons who speak both English and Hindi are $t$

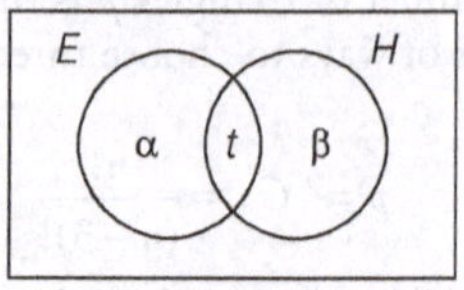


$\therefore \qquad \alpha + t + \beta = 100 \qquad \ldots$(i)

$\qquad \alpha + t = 75 \qquad \ldots$(ii)

and $\qquad \beta + t = 40 \qquad \ldots$(iii)

From Eqs. (i) and (ii), $\beta = 25$
From Eqs. (i) and (iii), $\alpha = 60$
and from Eq. (i), $t = 15$
We have, equation of ellipse
$25(\beta^2x^2 + \alpha^2y^2) = \alpha^2\beta^2$

$\Rightarrow \quad 25\left(\frac{x^2}{\alpha^2} + \frac{y^2}{\beta^2}\right) = 1$

$\Rightarrow \quad 25\left(\frac{x^2}{3600} + \frac{y^2}{625}\right) = 1$

$\Rightarrow \qquad \frac{x^2}{144} + \frac{y^2}{25} = 1$

$\therefore$ Eccentricity, $e = \sqrt{1 - \frac{b^2}{a^2}}$

$$= \sqrt{1 - \frac{25}{144}} = \frac{\sqrt{119}}{12}$$

**29.** *(d)* We have, equation of ellipse is

$\frac{x^2}{4} + \frac{y^2}{b^2} = 1, b < 2$

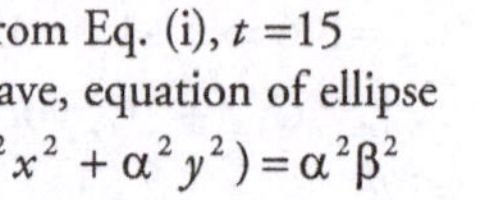


Equation of normal on ellipse at $p$
$2x \sec \theta - by \operatorname{cosec} \theta = 4 - b^2$
Now, distance from origin (0, 0)

$$= \left|\frac{4 - b^2}{\sqrt{4\sec^2 \theta + b^2 \operatorname{cosec}^2 \theta}}\right|_{\max} = 1$$

Now,

$4\sec^2\theta + b^2\operatorname{cosec}^2\theta = 4(1+\tan^2\theta) + b^2(1+\cot^2\theta)$

$= 4 + b^2 + 4\tan^2\theta + b^2\cot^2\theta$

$\ge 4 + b^2 + 4b$ $[\because 4\tan^2\theta + b^2\cot^2\theta \ge 4b]$

$\ge (2+b)^2$

$\therefore \quad \left|\dfrac{4-b^2}{2+b}\right| = 1 \Rightarrow |2-b| = 1$

$\Rightarrow \quad 2-b = \pm 1 \Rightarrow b = 3, 1$

$\Rightarrow \quad b = 1 \quad [\because b < 2]$

Now, eccentricity of ellipse $= \sqrt{1 - \dfrac{b^2}{a^2}}$

$= \sqrt{1 - \dfrac{1}{4}} = \dfrac{\sqrt{3}}{2}$

**30.** *(b)* $L$ is line passing through the intersection of the lines $bx + 10y - 8 = 0$ and $2x - 3y = 0$.

$L : (bx + 10y - 8) + \lambda(2x - 3y) = 0$

Also passes through (1, 1)

$(b + 10 - 8) + \lambda(2 - 3) = 0$

$\Rightarrow \quad \lambda = b + 2$

So, equation of line $L$ :

$(bx + 10y - 8) + (b+2)(2x - 3y) = 0$

$L : (3b+4)x - (3b-4)y - 8 = 0$

This line $L$ is tangent to circle

$x^2 + y^2 = \dfrac{16}{17}$

So, distance from centre to tangent = Radius

$\left|\dfrac{-8}{\sqrt{(3b+4)^2 + (3b-4)^2}}\right| = \dfrac{4}{\sqrt{17}}$

$\Rightarrow 9b^2 + 16 + 24b + 9b^2 + 16 - 24b = 68$

$\Rightarrow \quad 18b^2 = 36 \Rightarrow b^2 = 2$

Equation of ellipse, $\dfrac{x^2}{5} + \dfrac{y^2}{2} = 1$

$\Rightarrow \quad e = \sqrt{1 - \dfrac{b^2}{a^2}} = \sqrt{1 - \dfrac{2}{5}} = \sqrt{\dfrac{3}{5}}$

**31.** *(b)* $\dfrac{x^2}{a^2} + \dfrac{y^2}{b^2} = 1; \; a > b$

$e^2 = 1 - \dfrac{b^2}{a^2} \Rightarrow \dfrac{b^2}{a^2} = 1 - \dfrac{1}{16} = \dfrac{15}{16}$

$\Rightarrow \quad b^2 = \dfrac{15}{16}a^2 \quad \dfrac{x^2}{a^2} + \dfrac{y^2}{b^2} = 1$

Passing through $\left(-4\sqrt{\dfrac{2}{5}}, 3\right)$

$\dfrac{16 \times \dfrac{2}{5}}{a^2} + \dfrac{9}{b^2} = 1 \Rightarrow \dfrac{32}{5a^2} + \dfrac{9}{b^2} = 1$

$\Rightarrow \dfrac{32}{5a^2} + \dfrac{9}{\dfrac{15}{16}a^2} = 1 \Rightarrow \dfrac{80}{5a^2} = 1$

$\Rightarrow \quad a^2 = 16 \text{ and } b^2 = 15$

$\therefore \quad a^2 + b^2 = 16 + 15 = 31$

**32.** *(c)*

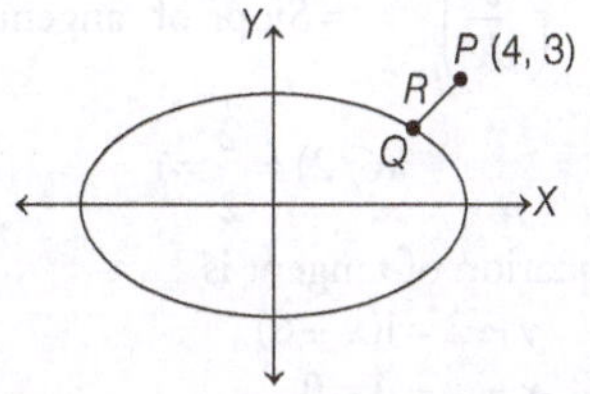


Equation of ellipse is $x^2 + 2y^2 = 4$

$\Rightarrow \quad \dfrac{x^2}{4} + \dfrac{y^2}{2} = 1$

Let the point $Q(2\cos\theta, \sqrt{2}\sin\theta)$.

Let the mid-point of $PQ$ be $R = (h, k)$

$\Rightarrow 2h = 4 + 2\cos\theta$ and $2k = 3 + \sqrt{2}\sin\theta$

$\Rightarrow \quad (h-2)^2 + \left(\dfrac{2k-3}{\sqrt{2}}\right)^2 = 1$

$\Rightarrow \quad \dfrac{(x-2)^2}{1} + \dfrac{4\left(y - \dfrac{3}{2}\right)^2}{2} = 1$

$\therefore$ Locus is $\dfrac{(x-2)^2}{1} + \dfrac{\left(y - \dfrac{3}{2}\right)^2}{1/2} = 1$

$\Rightarrow \quad e^2 = 1 - \dfrac{1}{2} = \dfrac{1}{2}$

$\Rightarrow \quad e = \dfrac{1}{\sqrt{2}}$

**33.** *(c)* Let $(2\sin\theta, 3\cos\theta)$ be the point on ellipse $\dfrac{x^2}{4} + \dfrac{y^2}{9} = 1$ and let mid-point of the line segments joining $(-3, -5)$ and $(2\sin\theta, 3\cos\theta)$ will be $(h, k)$.

Then, $\dfrac{2\sin\theta - 3}{2} = h, \; \dfrac{3\cos\theta - 5}{2} = k$

$\Rightarrow \sin\theta = \dfrac{2h+3}{2}, \; \cos\theta = \dfrac{2k+5}{3}$

$\therefore \quad \sin^2\theta + \cos^2\theta = 1$

$\Rightarrow \left(\dfrac{2h+3}{2}\right)^2 + \left(\dfrac{2k+5}{3}\right)^2 = 1$

$\Rightarrow \quad \dfrac{(4h^2 + 9 + 12h)}{4} + \dfrac{(4k^2 + 25 + 20k)}{9} = 1$

$\Rightarrow 36h^2 + 16k^2 + 108h + 80k + 145 = 0$

Locus of $(h, k)$

$36x^2 + 16y^2 + 108x + 80y + 145 = 0$

**34.** *(a)* Equation of ellipse

$\Rightarrow x^2 + 4y^2 = 4$

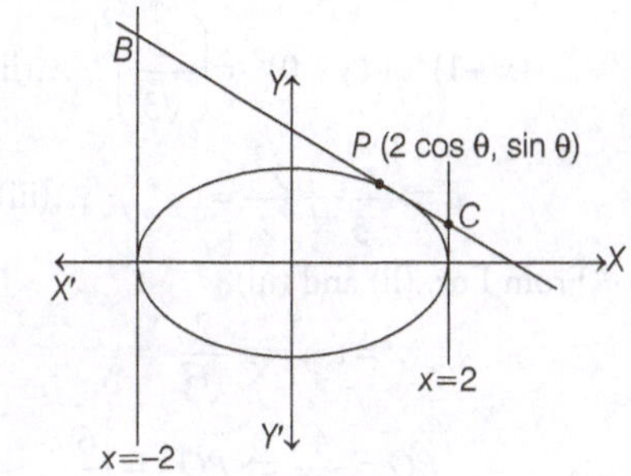


$\Rightarrow \quad \dfrac{x^2}{4} + \dfrac{y^2}{1} = 1$

Extremities end of major axis $= (\pm 2, 0)$

Let any point $P$ on ellipse$(2\cos\theta, \sin\theta)$

Equation of tangent to the ellipse at the point $P$ is $\dfrac{2\cos\theta x}{4} + \dfrac{y\sin\theta}{1} = 1$

Coordinate of $B\left(-2, \cot\dfrac{\theta}{2}\right)$ and $C\left(2, \tan\dfrac{\theta}{2}\right)$.

Equation of circle whose end points of diameter are $B$ and $C$ is

$(x-2)(x+2) + \left(y - \cot\dfrac{\theta}{2}\right)\left(y - \tan\dfrac{\theta}{2}\right) = 0$

$x^2 + y^2 - \left(\tan\dfrac{\theta}{2} + \cot\dfrac{\theta}{2}\right)y - 3 = 0$

At $\quad y = 0, x = \pm\sqrt{3}$

Hence, circle is passing through the point $(\pm\sqrt{3}, 0)$.

**35.** *(c)* $E \Rightarrow \dfrac{x^2}{a^2} + \dfrac{y^2}{b^2} = 1$ is passes through $\left(\sqrt{\dfrac{3}{2}}, 1\right)$.

$\therefore \quad \dfrac{3}{2a^2} + \dfrac{1}{b^2} = 1 \quad \ldots\text{(i)}$

$a^2 - b^2 = a^2e^2$

$\Rightarrow \; a^2 - b^2 = \dfrac{a^2}{3} \Rightarrow 2a^2 = 3b^2$

On putting $2a^2 = 3b^2$ in Eq. (i), we get

$\dfrac{3}{3b^2} + \dfrac{1}{b^2} = 1 \Rightarrow 2 = b^2 \Rightarrow a^2 = 3$

$\text{Focus}\left(\sqrt{3}.\frac{1}{\sqrt{3}},0\right) \Rightarrow \text{Focus} = (1, 0)$

$\text{Centre} = (1, 0), \text{Radius} = \frac{2}{\sqrt{3}}$

$$(x-1)^2 + (y-0)^2 = \left(\frac{2}{\sqrt{3}}\right)^2 \quad \text{...(ii)}$$

$$E \Rightarrow \frac{x^2}{3} + \frac{y^2}{2} = 1 \quad \text{...(iii)}$$

From Eqs. (ii) and (iii),

$$x = 1, y = \pm\frac{2}{\sqrt{3}}$$

$$\therefore \quad PQ = \frac{4}{\sqrt{3}} \Rightarrow PQ^2 = \frac{16}{3}$$

**36.** *(c)* Given, equation of ellipse is

$$\frac{x^2}{27} + y^2 = 1$$

Now, equation of tangent to the ellipse at $P(3\sqrt{3}\cos\theta, \sin\theta)$ is given by

$$\frac{3\sqrt{3}\cos\theta \cdot x}{27} + \sin\theta \cdot y = 1$$

$$\Rightarrow \quad \frac{x\cos\theta}{3\sqrt{3}} + y\sin\theta = 1 \quad \text{... (i)}$$

$X$-intercept of Eq. (i) is

$x = 3\sqrt{3}\sec\theta = OA$ (say)

$Y$-intercept of Eq. (i) is

$y = \text{cosec}\,\theta = OB$ (say)

$\therefore$ Sum of intercepts

$= 3\sqrt{3}\sec\theta + \text{cosec}\,\theta = f(\theta)$ (say)

$$\Rightarrow f'(\theta) = 3\sqrt{3}\sec\theta\tan\theta - \text{cosec}\,\theta \cdot \cot\theta$$

Put $\quad f'(\theta) = 0$

$$\Rightarrow \quad \frac{3\sqrt{3}\sin\theta}{\cos^2\theta} = \frac{\cos\theta}{\sin^2\theta}$$

$$\Rightarrow \quad \tan^3\theta = \frac{1}{3\sqrt{3}} \Rightarrow \theta = \pi/6$$

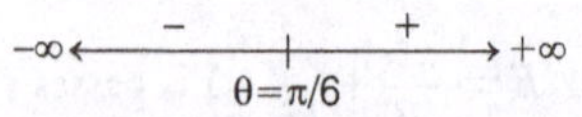


$\because f'(\theta)$ changes sign from negative to positive when moving from left to right.

$\therefore f(\theta)$ will attain minima at $\theta = \pi/6$

**37.** *(b)*

> **key idea** (i) First find the slope of line $L$ of (6, 2) then write the equation of tangent.
>
> (ii) Then use the condition of tangency to the ellipse as $c^2 = a^2m^2 + b^2$
>
> if $E: \frac{x^2}{a^2} + \frac{y^2}{b^2} = 1 \; L: y = mx + c$

Given, equation of parabola is

$$y^2 = 4x - 20 \quad \text{... (i)}$$

Differentiating Eq. (i) w.r.t. $x$, we get

$$2y\frac{dy}{dx} = 4 \Rightarrow \frac{dy}{dx} = \frac{2}{y}$$

$$\therefore \quad \left(\frac{dy}{dx}\right)_{(6,\,2)} = \text{Slope of tangent at}$$

$$(6, 2) = \frac{2}{2} = 1$$

$\therefore$Equation of tangent is

$$y - 2 = 1(x - 6)$$

$$\Rightarrow \quad x - y - 4 = 0 \quad \text{... (ii)}$$

As, we know the condition of tangency to the ellipse,

A straight line $y = mx + c$ will be tangent to the ellipse

$x^2/a^2 + y^2/b^2 = 1$

$$c^2 = a^2m^2 + b^2$$

Then, $\quad 16 = 2(1)^2 + b$

[here, $c = -4, m = 1, b = \sqrt{b}$]

$$\Rightarrow \quad b = 14$$

**38.** *(b)* According to given information,

Let the equation of ellipse

$$\frac{x^2}{a^2} + \frac{3y^2}{4} = 1 \quad \text{...(i)}$$

$\because$ Ellipse (i) touches the line $x + 6y = 8$, then

$$\left(\frac{8}{6}\right)^2 = a^2\left(\frac{1}{6}\right)^2 + \frac{4}{3} \Rightarrow \frac{16}{9} = \frac{a^2}{36} + \frac{4}{3}$$

$$\Rightarrow \frac{a^2}{36} = \frac{4}{9} \Rightarrow a^2 = 16$$

Now, eccentricity

$$e = \sqrt{1 - \frac{b^2}{a^2}} = \sqrt{1 - \frac{4}{3\times16}} = \frac{1}{2}\sqrt{\frac{11}{3}}$$

**39.** *(b)* Equation of given ellipse is

$$\frac{x^2}{a^2} + \frac{y^2}{b^2} = 1 \quad \text{...(i)}$$

Now, equation of tangent at the point $\left(3, -\frac{9}{2}\right)$ on the ellipse (i) is

$$\Rightarrow \quad \frac{3x}{a^2} - \frac{9y}{2b^2} = 1 \quad \text{...(ii)}$$

[$\because$ the equation of the tangent to the ellipse $\frac{x^2}{a^2} + \frac{y^2}{b^2} = 1$ at the point $(x_1, y_1)$ is $\frac{xx_1}{a^2} + \frac{yy_1}{b^2} = 1$]

$\because$ Tangent (ii) represent the line $x - 2y = 12$, so

$$\frac{\frac{1}{3}}{a^2} = \frac{\frac{2}{9}}{2b^2} = \frac{12}{1}$$

$$\Rightarrow \quad a^2 = 36 \text{ and } b^2 = 27$$

Now, Length of latus rectum

$$= \frac{2b^2}{a} = \frac{2\times27}{6} = 9 \text{ units}$$

**40.** *(d)* Let the ellipse be $\frac{x^2}{a^2} + \frac{y^2}{b^2} = 1$.

Then, according to given information, we have the following figure.

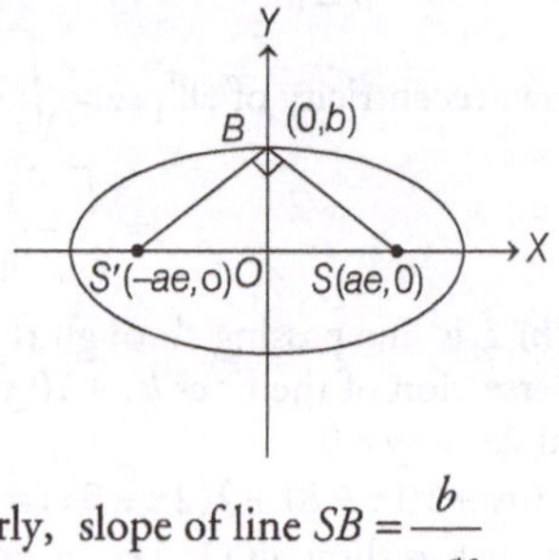


Clearly, slope of line $SB = \frac{b}{-ae}$

and slope of line $S'B = \frac{b}{ae}$

$\because$ Lines $SB$ and $S'B$ are perpendicular, so

$$\left(\frac{b}{-ae}\right)\cdot\left(\frac{b}{ae}\right) = -1$$

[$\because$ Product of slopes of two perpendicular lines is $(-1)$]

$$\Rightarrow \quad b^2 = a^2e^2 \quad \text{...(i)}$$

Also, it is given that area of $\Delta S'BS = 8$

$$\therefore \quad \frac{1}{2}a^2 = 8$$

[$\because S'B = SB = a$ because $S'B + SB = 2a$ and $S'B = SB$]

$$\Rightarrow \quad a^2 = 16 \Rightarrow a = 4 \quad \text{...(ii)}$$

$$\because \quad e^2 = 1 - \frac{b^2}{a^2} = 1 - e^2 \quad \text{[from Eq. (i)]}$$

$$\Rightarrow 2e^2 = 1 \Rightarrow e^2 = \frac{1}{2} \quad \text{...(iii)}$$

From Eqs. (i) and (iii), we get

$$b^2 = a^2\left(\frac{1}{2}\right) = 16\left(\frac{1}{2}\right)$$

[using Eq. (ii)]

$$\Rightarrow \quad b^2 = 8$$

Now, length of latus rectum

$$= \frac{2b^2}{a} = \frac{2\times8}{4} = 4 \text{ units}$$

**41.** *(b)* Let the equation of ellipse be

$$\frac{x^2}{a^2} + \frac{y^2}{b^2} = 1$$

According the problem, we have

$$\frac{2b^2}{a}=8 \text{ and } 2ae=2b$$

[length of latus rectum $=\frac{2b^2}{a}$ and length of minor axis $=2b$]

$$\Rightarrow \quad b\left(\frac{b}{a}\right)=4 \text{ and } \frac{b}{a}=e$$

$$\Rightarrow b(e)=4 \Rightarrow b=4\cdot\frac{1}{e} \quad \ldots\text{(i)}$$

Also, we know that $b^2=a^2(1-e^2)$

$$\Rightarrow \quad \frac{b^2}{a^2}=1-e^2$$

$$\Rightarrow \quad e^2=1-e^2 \quad \left[\because \frac{b}{a}=e\right]$$

$$\Rightarrow \quad 2e^2=1 \Rightarrow e=\frac{1}{\sqrt{2}} \quad \ldots\text{(ii)}$$

From Eqs. (i) and (ii), we get

$$b=4\sqrt{2}$$

Now, $a^2=\dfrac{b^2}{1-e^2}=\dfrac{32}{1-\frac{1}{2}}=64$

$\therefore$ Equation of ellipse be $\dfrac{x^2}{64}+\dfrac{y^2}{32}=1$

Now, check all the options.

Only $(4\sqrt{3}, 2\sqrt{2})$, satisfy the above equation.

**42.** *(b)* We have, $e=\dfrac{1}{2}$

and $\quad \dfrac{-a}{e}=-4$

$\therefore \quad a=2$

Now, $b^2=a^2(1-e^2)=(2)^2\left[1-\left(\frac{1}{2}\right)^2\right]$

$$=4\left(1-\frac{1}{4}\right)=3 \Rightarrow b=\sqrt{3}$$

$\therefore$ Equation of the ellipse is

$$\frac{x^2}{(2)^2}+\frac{y^2}{(\sqrt{3})^2}=1 \Rightarrow \frac{x^2}{4}+\frac{y^2}{3}=1$$

Now, the equation of normal at $\left(1, \frac{3}{2}\right)$ is

$$\frac{a^2x}{x_1}-\frac{b^2y}{y_1}=a^2-b^2$$

$$\Rightarrow \frac{4x}{1}-\frac{3y}{(3/2)}=4-3 \Rightarrow 4x-2y=1$$

**43.** *(d)* Given equation of ellipse is

$$\frac{x^2}{9}+\frac{y^2}{5}=1 \quad \ldots\text{(i)}$$

$\therefore \ a^2=9, b^2=5 \Rightarrow a=3, b=\sqrt{5}$

Now, $\quad e=\sqrt{1-\dfrac{b^2}{a^2}}=\sqrt{1-\dfrac{5}{9}}=\dfrac{2}{3}$

foci $=(\pm ae, 0)=(\pm 2, 0)$

and $\quad \dfrac{b^2}{a}=\dfrac{5}{3}$

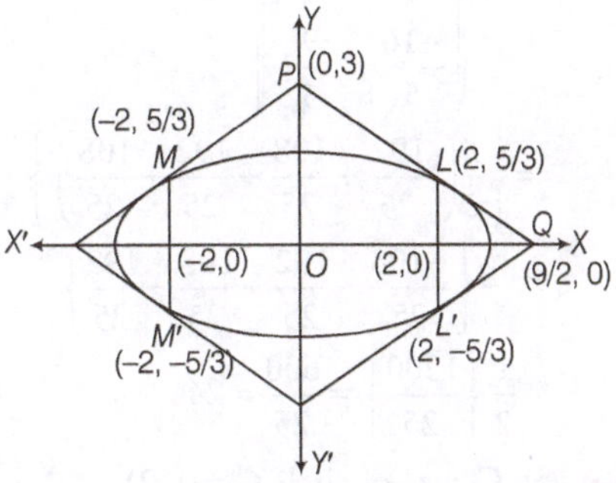


$\therefore$ Extremities of one of latusrectum are $\left(2, \frac{5}{3}\right)$ and $\left(2, \frac{-5}{3}\right)$

$\therefore$ Equation of tangent at $\left(2, \frac{5}{3}\right)$ is,

$$\frac{x(2)}{9}+\frac{y(5/3)}{5}=1$$

or $\quad 2x+3y=9 \quad \ldots\text{(ii)}$

Eq.(ii) intersects $X$ and $Y$-axes at $\left(\frac{9}{2}, 0\right)$ and (0, 3), respectively.

$\therefore$ Area of quadrilateral $=4\times$ Area of $\Delta POQ$

$$=4\times\left(\frac{1}{2}\times\frac{9}{2}\times 3\right)=27 \text{ sq units}$$

**44.** *(a)* Given equation of ellipse is

$$\frac{x^2}{16}+\frac{y^2}{9}=1$$

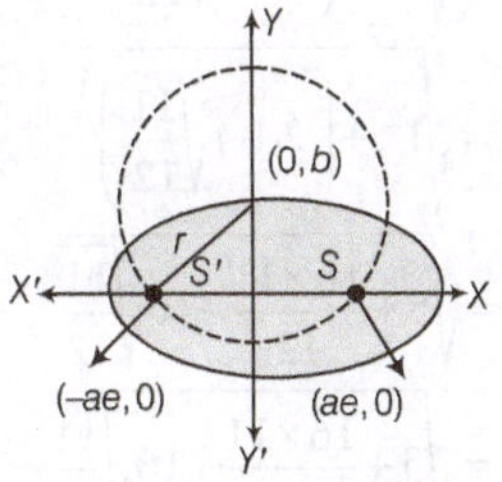


Here, $a=4, b=3, e=\sqrt{1-\dfrac{9}{16}} \Rightarrow \dfrac{\sqrt{7}}{4}$

$\therefore$ Foci is $\quad (\pm ae, 0)=\left(\pm 4\times\dfrac{\sqrt{7}}{4}, 0\right)$

$$=(\pm\sqrt{7}, 0)$$

$\therefore$ Radius of the circle, $r=\sqrt{(ae)^2+b^2}$

$$=\sqrt{7+9}=\sqrt{16}=4$$

Now, equation of circle is

$$(x-0)^2+(y-3)^2=16$$

$$\therefore \quad x^2+y^2-6y-7=0$$

**45.** *(d)* Given

(i) An ellipse whose semi-minor axis coincides with one of the diameters of the circle $(x-1)^2+y^2=1$.

(ii) The semi-major axis of the ellipse coincides with one of the diameters of circle $x^2+(y-2)^2=4$.

(iii) The centre of the ellipse is at origin.

(iv) The axes of the ellipse are coordinate axes.

To find The equation of the ellipse.

Diameter of circle $(x-1)^2+y^2=1$ is 2 units and that of circle $x^2+(y-2)^2=4$ is 4 units.

$\Rightarrow$ Semi-minor axis of ellipse, $b=2$ units and semi-major axis of ellipse, $a=4$ units.

Hence, the equation of the ellipse is

$$\frac{x^2}{a^2}+\frac{y^2}{b^2}=1 \Rightarrow \frac{x^2}{16}+\frac{y^2}{4}=1$$

$$\therefore \quad x^2+4y^2=16$$

**46.** *(d)* Given, $\dfrac{x^2}{a^2}+\dfrac{y^2}{b^2}=1$

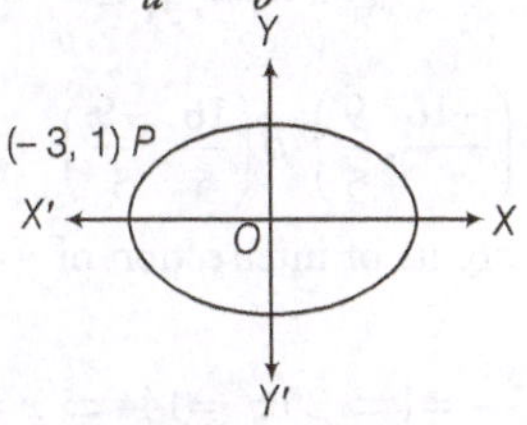


Passes through $P(-3, 1)$ and $e=\sqrt{2/5}$.

$$\therefore \quad \frac{9}{a^2}+\frac{1}{b^2}=1 \text{ and } e^2=1-\frac{b^2}{a^2}$$

$$\Rightarrow \quad \frac{9}{a^2}+\frac{5}{3a^2}=1 \text{ and } \frac{2}{5}=1-\frac{b^2}{a^2}$$

$$\Rightarrow \quad \frac{27+5}{3a^2}=1 \text{ and } \frac{b^2}{a^2}=\frac{3}{5}$$

$$\Rightarrow \quad a^2=\frac{32}{3} \text{ and } b^2=\frac{32}{5}$$

$\therefore$ Equation of ellipse

$$\Rightarrow \quad \frac{3x^2}{32}+\frac{5y^2}{32}=1 \Rightarrow 3x^2+5y^2=32$$

**47.** *(a)* Let the equation of the required ellipse be $\frac{x^2}{16}+\frac{y^2}{b^2}=1$.

But the ellipse passes through the point (2, 1).

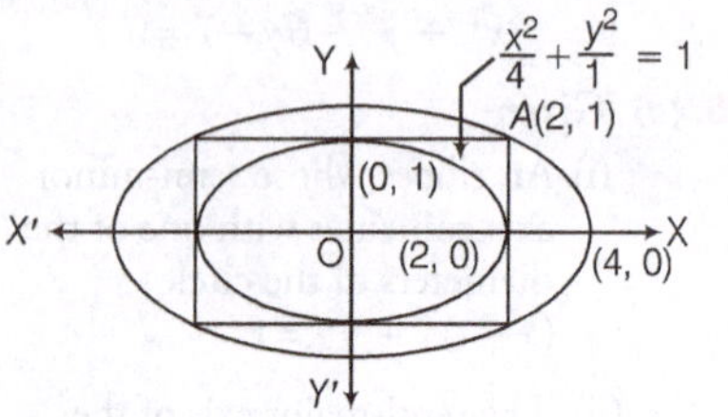


$\Rightarrow \quad \frac{1}{4}+\frac{1}{b^2}=1 \Rightarrow \frac{1}{b^2}=\frac{3}{4}$

$\therefore \quad b^2=4/3$

Hence, equation is $\frac{x^2}{16}+\frac{3y^2}{4}=1$

$\Rightarrow \quad x^2+12y^2=16$

**48.** *(c)* $E:\frac{x^2}{4^2}+\frac{y^2}{3^2}=1$

$y=x+p$ is tangent of ellipse $E$

let point contact $(x_1, y_1)$

$\therefore \quad c^2=a^2m^2+b^2$

$\Rightarrow \quad p^2=4^2(1)^2+3^2$

$\Rightarrow \quad p^2=25 \Rightarrow p=\pm 5$

$\frac{xx_1}{16}+\frac{yy_1}{9}-1=0 \Rightarrow x-y+p=0$

$\frac{x_1}{16}=\frac{-y_1}{9}=\frac{-1}{p}$

$x_1=\frac{-16}{p}, y_1=\frac{9}{p}$

$\therefore \quad A\left(\frac{-16}{5},\frac{9}{5}\right), B\left(\frac{16}{5},\frac{-9}{5}\right)$

Now, points of intersection of $y=x$ and $E$

$\frac{x^2}{16}+\frac{x^2}{9}=1 \Rightarrow 25x^2=144 \Rightarrow x=\pm\frac{12}{5}$

$\therefore C\left(\frac{12}{5},\frac{12}{5}\right)$ and $D=\left(\frac{-12}{5},\frac{-12}{5}\right)$

Area of quadratic $ABCD$

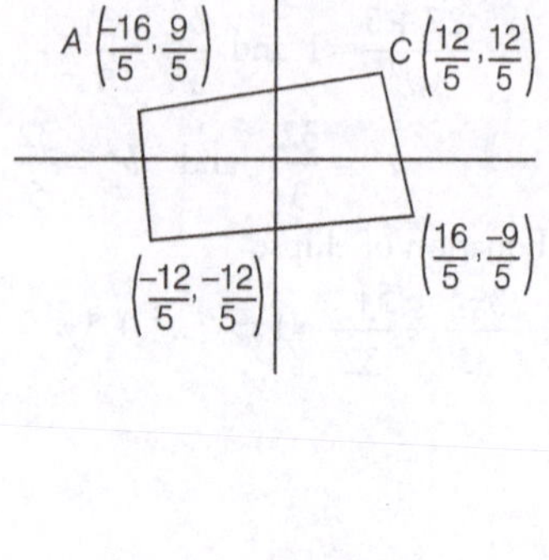


$$=\frac{1}{2}\begin{Vmatrix} \frac{-16}{5} & \frac{9}{5} \\ \frac{-12}{5} & \frac{-12}{5} \\ \frac{16}{5} & \frac{-9}{5} \\ \frac{12}{5} & \frac{12}{5} \\ \frac{-16}{5} & \frac{9}{5} \end{Vmatrix}$$

$$=\left\{\frac{1}{2}\left(\frac{192}{25}+\frac{108}{25}+\frac{192}{25}+\frac{108}{25}\right)\right\}-\left\{\frac{-108}{25}-\frac{192}{25}-\frac{108}{25}-\frac{192}{25}\right\}$$

$$=\frac{1}{2}\left\{\frac{1200}{25}\right\}=\frac{600}{25}=24$$

**49.** *(c)* Centre of circle $C=(1, 2)$

$r=\sqrt{1^2+2^2+11}=4$

As ellipse satisfies (1, 2)

$\Rightarrow \quad 3+4p=4$

$\Rightarrow \quad p=\frac{1}{4}$

So, equation of ellipse is $3x^2+\frac{y^2}{4}=4$

$\Rightarrow \quad \frac{x^2}{4/3}+\frac{y^2}{16}=1$

So, $e^2=1-\frac{4}{3\times 16}=\frac{11}{12}$

So, $f_1=\left(0, 4\sqrt{\frac{11}{12}}\right)$

[$\because$ Foci of ellipse are $(0, \pm be)$]

$f_2=\left(0, -4\sqrt{\frac{11}{12}}\right)$

So, $f_1=\sqrt{1^2+\left(2-4\sqrt{\frac{11}{12}}\right)^2}$

$f_2=\sqrt{1^2+\left(2+4\sqrt{\frac{11}{12}}\right)^2}$

$f_1=\sqrt{5+\frac{16\times 11}{12}-16\sqrt{\frac{11}{12}}}$

$f_2=\sqrt{5+\frac{16\times 11}{12}+16\sqrt{\frac{11}{12}}}$

$6f_1f_2-r=3(2f_1f_2)-r$

$=3[(f_1+f_2)^2-f_1^2-f_2^2]-r$

$=3[(2\times 4)^2-f_1^2-f_2^2]-4$

[$\because f_1+f_2=28$]

$=3\left[64-\left(5+\frac{16\times 11}{12}\right)\times 2\right]-4$

$=3\left[64-10-\frac{32\times 11}{12}\right]-4$

$=3\times 54-88-4$

$=162-92=70$

Option (c) is correct.

**50.** *(a)* Given $E_1:\frac{x^2}{a^2}+\frac{y^2}{b^2}=1, a>b$

(horizontal ellipse)

$E_2:\frac{x^2}{A^2}+\frac{y^2}{B^2}=1, A<B$

(vertical ellipse)

both are having eccentricity $=\frac{1}{\sqrt{3}}=e$

Also, given, distance between focal of $E_1=4 \Rightarrow 2ae=4$

$\Rightarrow 2a\left(\frac{1}{\sqrt{3}}\right)=4 \Rightarrow a=2\sqrt{3}$

$\because \quad b^2=a^2(1-e^2)$

$=(2\sqrt{3})^2\left(1-\left(\frac{1}{\sqrt{3}}\right)^2\right)$

$=12\left(1-\frac{1}{3}\right)=12\times\frac{2}{3}=8$

According to the question, product of LR's of both ellipse $=\frac{32}{\sqrt{3}}$

i.e. $\frac{2b^2}{a}\times\frac{2A^2}{B}=\frac{32}{\sqrt{3}}$

$\Rightarrow 2\left(\frac{8}{2\sqrt{3}}\right)\cdot\frac{2A^2}{B}=\frac{32}{\sqrt{3}}$

$\Rightarrow \quad A^2=2B$

Also, $A^2=B^2(1-e^2)$

$\Rightarrow 2B=B^2\left(1-\frac{1}{3}\right) \Rightarrow 2B=B^2\times\frac{2}{3}$

$\Rightarrow B^2-3B=0 \Rightarrow B=0, 3$ but $B\neq 0$

$\therefore \quad B=3$

$\therefore \quad A^2=6$

So, $\quad E_1:\frac{x^2}{12}+\frac{y^2}{8}=1 \quad$ ...(i)

$E_2:\frac{x^2}{6}+\frac{y^2}{9}=1 \quad$ ...(ii)

On solving Eqs. (i) and (ii), we get

$(x, y)\equiv\left(\frac{\sqrt{6}}{\sqrt{5}},\frac{6}{\sqrt{5}}\right),\left(\frac{-\sqrt{6}}{\sqrt{5}},\frac{6}{\sqrt{5}}\right),$

$\left(\frac{\sqrt{6}}{\sqrt{5}}, \frac{-6}{\sqrt{5}}\right)\left(\frac{-\sqrt{6}}{\sqrt{5}}, \frac{-6}{\sqrt{5}}\right)$

These coordinates shows that this forms a rectangle.

So, ar (quadrant $\cdot ABCD$)

$= \text{ar (rectangle } ABCD)$

$= \frac{2\sqrt{6}}{\sqrt{5}} \times \frac{12}{\sqrt{5}}$

$= \frac{24\sqrt{6}}{5}$

**51.** *(c)* Given equation of line $2x + 3y - k = 0$ which intersect the $X$-axis and $Y$-axis at the points $A$ and $B$.

When $y = 0$, then coordinate of $A$ is $\left(\frac{k}{2}, 0\right)$ and when $x = 0$, then coordinate of $B$ is $\left(0, \frac{k}{3}\right)$.

Equation of circle with diameter $AB$ is

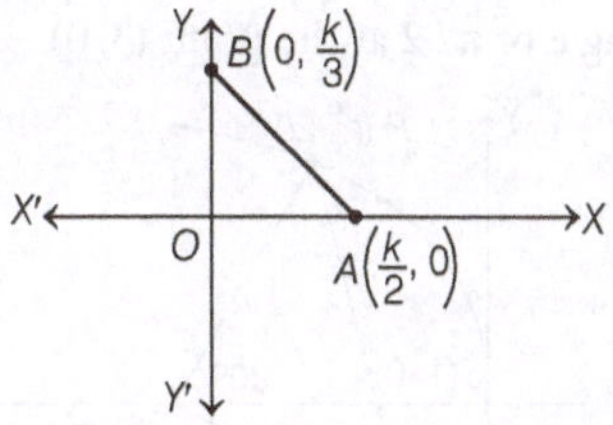


$$(x - 0)\left(x - \frac{k}{2}\right) + \left(y - \frac{k}{3}\right)(y - 0) = 0$$

$$\Rightarrow x^2 + y^2 - \frac{kx}{2} - \frac{ky}{3} = 0$$

On comparing with

$$x^2 + y^2 - 3x - 2y = 0,$$

we get $\frac{k}{2} = 3 \Rightarrow k = 6$

Now, equation of ellipse

$$x^2 + 9y^2 = k^2 = 36$$

$$\Rightarrow \frac{x^2}{36} + \frac{y^2}{4} = 1$$

Latus-rectum $= \frac{2b^2}{a} = \frac{2 \times 4}{6} = \frac{4}{3}$

$\Rightarrow \quad m = 4, n = 3$

$\therefore \quad 2m + n = 8 + 3 = 11$

**52.** *(c)*

> **key idea** First use
>
> $e = \sqrt{1 - \frac{b^2}{a^2}}$, then
>
> Length of latus rectum $= \frac{2b^2}{a}$

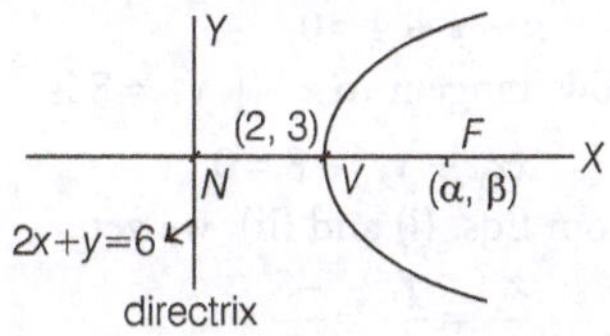


So, equation of axis is

$$y - 3 = \frac{1}{2}(x - 2) \Rightarrow x - 2y + 4 = 0$$

Solving $2x + y = 6$ and $x - 2y + 4 = 0$

$N\left(\frac{8}{5}, \frac{14}{5}\right)$

Since, $V$ is the mid-point of $N$ and $F$

$$\frac{\frac{8}{5} + \alpha}{2} = 2, \ \alpha = 4 - \frac{8}{5} = \frac{12}{5}$$

$$\Rightarrow \quad \frac{\frac{14}{5} + \beta}{2} = 3$$

$$\beta = 6 - \frac{14}{5} = \frac{16}{5}$$

Given that ellipse passes through $(\alpha, \beta)$.

$$\therefore \quad \frac{\alpha^2}{a^2} + \frac{\beta^2}{b^2} = 1$$

$$\frac{(12/5)^2}{a^2} + \frac{(16/5)^2}{b^2} = 1 \quad \text{...(i)}$$

Also, eccentricity of ellipse is $\frac{1}{\sqrt{2}}$

$$\sqrt{1 - \frac{b^2}{a^2}} = \frac{1}{\sqrt{2}} \Rightarrow 1 - \frac{b^2}{a^2} = \frac{1}{2}$$

$$\Rightarrow \quad a^2 = 2b^2$$

Put in Eq. (i),

$$\frac{144/25}{2b^2} + \frac{256/25}{b^2} = 1$$

$$b^2 = \frac{72}{25} + \frac{256}{25} = \frac{328}{25}$$

Square of length of latus rectum

$$= \left(\frac{2b^2}{a}\right)^2 = \frac{4 \times b^4}{a^2}$$

$$= \frac{4b^4}{2b^2} = 2b^2 = \frac{656}{25}$$

**53.** *(d)* Given, equation on ellipse

$$\frac{x^2}{36} + \frac{y^2}{4} = 1$$

Tangent equation is $\frac{x \cdot (3\sqrt{3})}{36} + \frac{y \cdot 1}{4} = 1$

$$\Rightarrow \quad \frac{x\sqrt{3}}{12} + \frac{y}{4} = 1$$

$$\Rightarrow \quad x\sqrt{3} + 3y = 12 \quad \text{...(i)}$$

Since, Eq. (i) meet $Y$-axis at $A$, (0, 4).

Now, normal equation is

$$\frac{x \cdot 36}{3\sqrt{3}} - \frac{y \cdot 4}{1} = 36 - 4 = 32$$

$$\Rightarrow \quad \frac{12x}{\sqrt{3}} - 4y = 32 \quad \text{...(ii)}$$

Since, Eq. (ii) meet $Y$-axis at $B$,

$\therefore B = (0, -8)$

Now, equation of circle $AB$ is

$$(x - 0)(x - 0) + (y - 4)(y + 8) = 0$$

$$\Rightarrow x^2 + y^2 + 4y - 32 = 0$$

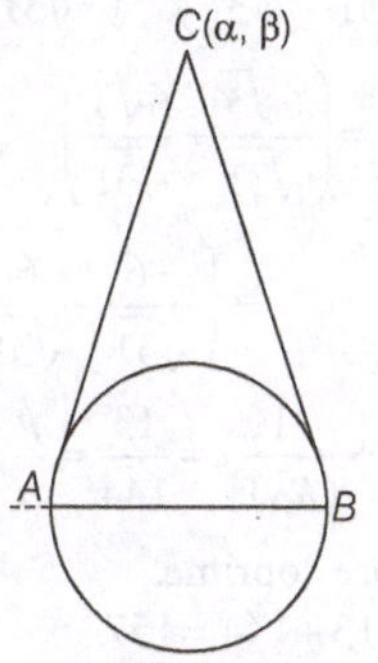


Now, equation of chord of contact from $(\alpha, \beta)$ with respect to circle $\alpha x + \beta y + 2(y + \beta) - 32 = 0$, which is same as $x = 2\sqrt{5}$

Thus, $\beta + 2 = 0$ and $\frac{32 - 2\beta}{\alpha} = 2\sqrt{5}$

$$\therefore \beta = -2 \Rightarrow \frac{36}{2} = 2\sqrt{5} \Rightarrow \alpha = \frac{18}{\sqrt{5}}$$

Thus, $(\alpha^2 - \beta^2) = \left(\frac{324}{5} - 4\right) = \frac{304}{5}$

**54.** *(b)* Ellipse $9x^2 + 4y^2 = 36$,

$$P\left[\frac{2\sqrt{3}}{\sqrt{7}}, \frac{6}{\sqrt{7}}\right]$$

Since, $PQ$ passes through origin,

so slope of $PQ = \frac{6/\sqrt{7} - 0}{2\sqrt{3}/\sqrt{7} - 0} = \sqrt{3}$

$\therefore$ Equation of $PQ$ is $y = \sqrt{3}x$

solving with ellipse, $9x^2 + 4(\sqrt{3}x)^2 = 36$

$$\Rightarrow 21x^2 = 36 \Rightarrow x = \frac{\pm 2\sqrt{3}}{\sqrt{7}} \Rightarrow y = \frac{\pm 6}{\sqrt{7}}$$

$$\therefore Q = \left(\frac{-2\sqrt{3}}{\sqrt{7}}, \frac{-6}{\sqrt{7}}\right)$$

$$\Rightarrow (PQ)^2 = \left(\frac{2\sqrt{3}}{\sqrt{7}} + \frac{2\sqrt{3}}{\sqrt{7}}\right)^2 + \left(\frac{6}{\sqrt{7}} + \frac{6}{\sqrt{7}}\right)^2$$

$$= \frac{48}{7} + \frac{144}{7} = \frac{192}{7}$$

Since, $RS \perp PQ$,

$\therefore$ Slope of $RS = \frac{-1}{\sqrt{3}}$

So, equation of $RS$ is $y = \frac{-x}{\sqrt{3}}$

Solving with ellipse,

$$9x^2 + 4\left(\frac{-x}{\sqrt{3}}\right)^2 = 36$$

Ler $R\left(\frac{6\sqrt{3}}{\sqrt{31}}, \frac{-6}{\sqrt{31}}\right), S\left(\frac{-6\sqrt{3}}{\sqrt{31}}, \frac{6}{\sqrt{31}}\right)$

$$\Rightarrow (RS)^2 = \left(\frac{6\sqrt{3}}{\sqrt{31}} + \frac{6\sqrt{3}}{\sqrt{31}}\right)^2 + \left(\frac{-6}{\sqrt{31}} - \frac{6}{\sqrt{31}}\right)^2 = \frac{576}{31}$$

$$\therefore \frac{1}{(PQ)^2} + \frac{1}{(RS)^2} = \frac{13}{144} = \frac{p}{q},$$

$p$ and $q$ are coprime.

$\therefore p + q = 13 + 144 = 157$

**55.** *(b)* We know that, tangent at parabola $y^2 = 4ax$ is given by

$$y = mx + \frac{a}{m} \qquad [m = \text{slope}]$$

And tangent at ellipse, $x^2 + y^2 = a^2$ is given by

$$y = mx \pm \sqrt{a^2m^2 + b^2}$$

So, in this case

Tangent at parabola $y^2 = 16x$

$$y = mx + \frac{4}{m}$$

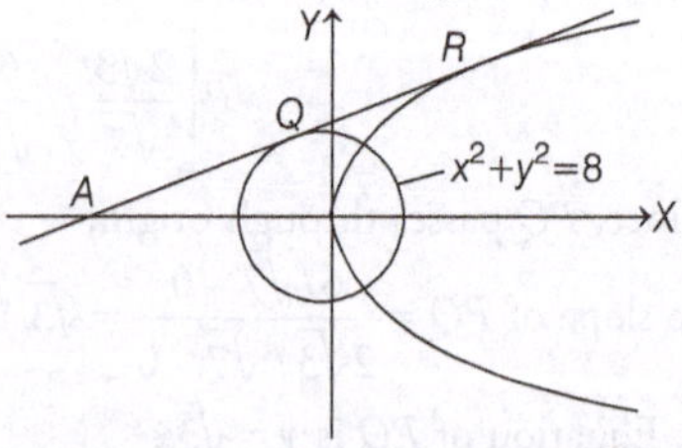


And tangent at curve $x^2 + y^2 = 8$

$$y = mx \pm \sqrt{8m^2 + 8}$$

Both tangent are same, so

$$\frac{4}{m} = \pm\sqrt{8m^2 + 8}$$

$$\Rightarrow \frac{16}{m^2} = 8m^2 + 8 \Rightarrow \frac{2}{m^2} = m^2 + 1$$

$$\Rightarrow (m^2 + 2)(m^2 - 1) = 0$$

Take, $m = \pm 1$

So, common tangent

$$\Rightarrow \quad y = x + 4$$

$$\Rightarrow \quad x - y + 4 = 0 \qquad \ldots\text{(i)}$$

Now, tangent to $x^2 + y^2 = 8$ is

$$xx_1 + yy_1 - 8 = 0 \qquad \ldots\text{(ii)}$$

From Eqs. (i) and (ii), we get

$$\frac{x_1}{1} = \frac{y_1}{-1} = \frac{-8}{4}$$

$$\Rightarrow \quad x_1 = -2, y_1 = 2$$

Point $Q(-2, 2)$, tangent at $y^2 = 16x$ is

$$\Rightarrow \quad yy_1 = 16\left(\frac{x + x_1}{2}\right)$$

$$yy_1 = 8x + 8x_1$$

$$8x - yy_1 + 8x_1 = 0 \qquad \ldots\text{(iii)}$$

From Eqs. (i) and (iii), we get

$$\frac{8}{1} = \frac{-y_1}{-1} = \frac{8x_1}{4} \Rightarrow y_1 = 8, x_1 = 4$$

So, $R(4, 8)$

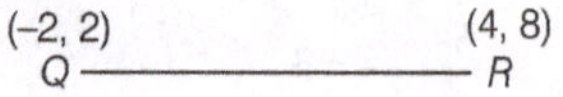


Now, $QR^2 = (-2-4)^2 + (2-8)^2$

$= 36 + 36 = 72$

**56.** *(b)* Given, equation of ellipse is $2x^2 + 3y^2 = 5$

$$\Rightarrow \quad \frac{x^2}{5/2} + \frac{y^2}{5/3} = 1$$

Equation of tangent to the ellipse is

$$y = mx \pm \sqrt{a^2m^2 + b^2}$$

Here, $a^2 = \frac{5}{2}, b^2 = \frac{5}{3}$

$$y = mx \pm \sqrt{\frac{5}{2}m^2 + \frac{5}{3}}$$

The ellipse passes through (1, 3), so

$$3 = m \pm \sqrt{\frac{5}{2}m^2 + \frac{5}{3}}$$

$$\Rightarrow \quad 3 - m = \pm\sqrt{\frac{5}{2}m^2 + \frac{5}{3}}$$

On squaring both sides,

$$9 + m^2 - 6m = \frac{5}{2}m^2 + \frac{5}{3}$$

$$\Rightarrow \quad 54 + 6m^2 - 36m = 15m^2 + 10$$

$$\Rightarrow \quad 9m^2 + 36m - 44 = 0$$

Let the pair of lines are $y = m_1x + c_1$ and $y = m_2x + c_2$

Here, $m_1 + m_2 = -\frac{36}{9} = -4$ and

$$m_1m_2 = -\frac{44}{9}$$

$$|m_1 - m_2| = \sqrt{(m_1 + m_2)^2 - 4m_1m_2}$$

$$= \sqrt{16 - 4 \times \left(-\frac{44}{9}\right)}$$

$$= \sqrt{\frac{144 + 176}{9}} = \sqrt{\frac{320}{9}} = \frac{8\sqrt{5}}{3}$$

$\because \quad \tan\theta = \left|\frac{m_1 - m_2}{1 + m_1m_2}\right|$, where $\theta$ is the angle between the tangents

$$= \left|\frac{\frac{8\sqrt{5}}{3}}{1 + \left(-\frac{44}{9}\right)}\right|$$

$$= \left|\frac{8\sqrt{5} \times 3}{-35}\right| = \frac{24}{7\sqrt{5}}$$

$$\therefore \quad \theta = \tan^{-1}\left(\frac{24}{7\sqrt{5}}\right)$$

**57.** *(b)* Given, that $PQ$ is a focal chord of the parabola $y^2 = 4x$ which subtends an angle of $\pi/2$ at the point (3, 0).

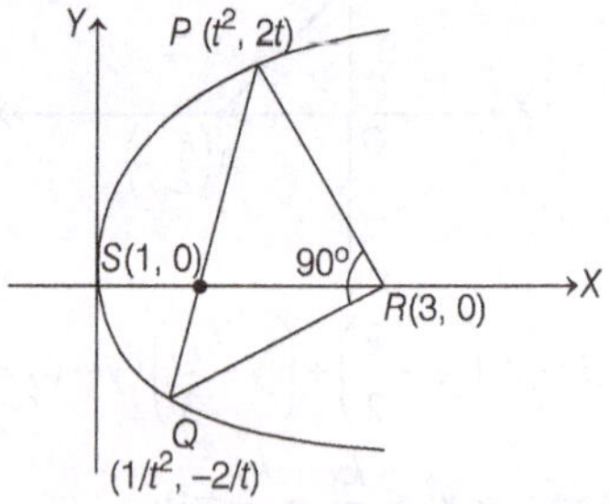


$$m_{PR} \cdot m_{RQ} = -1$$

$$\Rightarrow \quad \frac{(0 - 2t)}{(3 - t^2)} \times \frac{\left(0 - \left(-\frac{2}{t}\right)\right)}{\left(3 - \frac{1}{t^2}\right)} = -1$$

$$\Rightarrow \quad \frac{4t^2}{9t^2 - 3 - 3t^4 + t^2} = 1$$

$$\Rightarrow \quad 4t^2 = 10t^2 - 3 - 3t^4$$

$$\Rightarrow \quad 3t^4 - 6t^2 + 3 = 0$$

$$\Rightarrow \quad (t^2)^2 - 2(t)^2(1) + (1)^2 = 0$$

$$\Rightarrow \quad (t^2 - 1)^2 = 0 \Rightarrow t = 1$$

$\Rightarrow P$ and $Q$ must be end points of latus rectum.

$P(1, 2)$ and $Q(1, -2)$

$$\therefore \quad \frac{2b^2}{a} = 4 \text{ and } ae = 1$$

$$\because \quad b^2 = a^2(1 - e^2) \Rightarrow 2a = a^2 - 1$$

$$\Rightarrow \quad a^2 - 2a - 1 = 0 \Rightarrow a = 1 + \sqrt{2}$$

Now, $e^2 = 1 - \dfrac{b^2}{a^2}$

$\Rightarrow e^2 = 1 - \dfrac{2a}{a^2} = 1 - \dfrac{2}{a} = \dfrac{a-2}{a} = \dfrac{\sqrt{2}-1}{\sqrt{2}+1}$

$\Rightarrow \quad \dfrac{1}{e^2} = \dfrac{\sqrt{2}+1}{\sqrt{2}-1} \times \dfrac{\sqrt{2}+1}{\sqrt{2}+1}$

$= 3 + 2\sqrt{2}$

$\Rightarrow \quad \dfrac{1}{e^2} = 3 + 2\sqrt{2}$

**58.** *(a)* Given, equation of ellipse

$$\frac{x^2}{8} + \frac{y^2}{4} = 1$$

Then, equation of tangent at $(x_1, y_1)$ will be $\dfrac{xx_1}{8} + \dfrac{yy_1}{4} = 1$

Since, tangent is perpendicular to the line $x + 2y = 0$, then

$$\left(\frac{-x_1}{2y_1}\right)\left(\frac{-1}{2}\right) = -1$$

$\Rightarrow \quad x_1 = -4y_1$

Also, $\quad \dfrac{x_1^2}{8} + \dfrac{y_1^2}{4} = 1$

$\Rightarrow \quad \dfrac{16y_1^2}{8} + \dfrac{y_1^2}{4} = 1 \quad [\because x_1 = -4y_1]$

$\Rightarrow \quad y_1^2 = \dfrac{4}{9}$

$\Rightarrow \quad y_1 = \pm\dfrac{2}{3} \quad \Rightarrow \quad y_1 = \dfrac{2}{3}$

$[\because (x_1, y_1)$ lies in second quadrant$]$

And $\quad x_1 = -4y_1 = -4 \times \dfrac{2}{3} = \dfrac{-8}{3}$

$\therefore \quad P\left(\dfrac{-8}{3}, \dfrac{2}{3}\right)$

Again, $\quad a^2 - b^2 = a^2e^2$

$8 - 4 = 8e^2 \Rightarrow e = 1/\sqrt{2}$

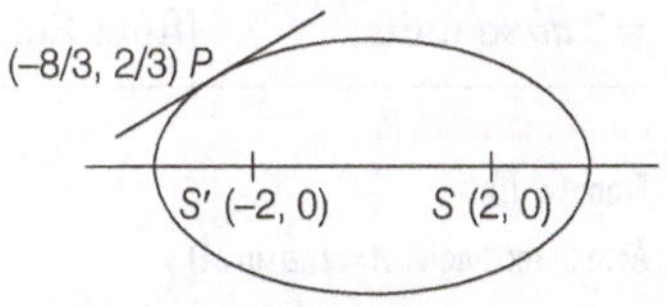


Now, $S$ and $S'$ are the foci of the ellipse,

So, $S : (ae, 0)$

$$= \left(2\sqrt{2}\cdot\frac{1}{\sqrt{2}}, 0\right) = (2, 0)$$

and $S' : (-ae, 0) = \left(-2\sqrt{2}.\dfrac{1}{\sqrt{2}}, 0\right)$

$= (-2, 0)$

Area of $\Delta SPS' = \dfrac{1}{2} \times 4 \times \dfrac{2}{3} = \dfrac{4}{3}$

$\left[\because \text{base} = 4, \text{height} = \dfrac{2}{3}\right]$

So, $(5 - e^2)A = \left(5 - \dfrac{1}{2}\right)\dfrac{4}{3} = \dfrac{9}{2}.\dfrac{4}{3} = 6$

**59.** *(c)* Equation of the reflected ray will be

$$L \Rightarrow \frac{y-3}{x-5} = m$$

Now, the image of (2, 1) w.r.t. line $x = 0$ should lie on the reflected line.

Image of $(2, 1) = (-2, 1)$

So, $\quad \dfrac{1-3}{-2-5} = m \Rightarrow m = \dfrac{2}{7}$

So, equation of reflected ray

$\Rightarrow \quad \dfrac{y-3}{x-5} = \dfrac{2}{7}$

$7y - 21 = 2x - 10$

$\Rightarrow 2x - 7y + 11 = 0$

(equation of directrix)

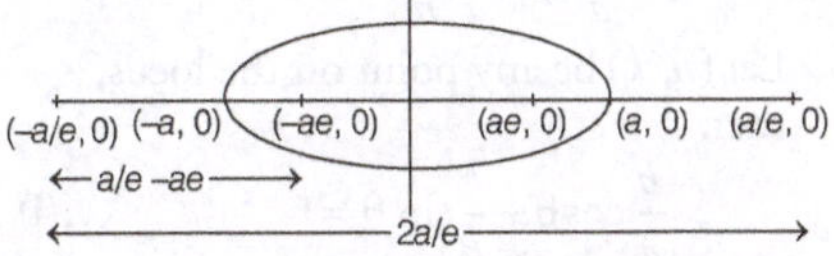


Now, $\quad e = \dfrac{1}{3}$ [given]

So, $\quad \dfrac{a}{e} - ae = \dfrac{a}{1/3} - \dfrac{1}{3}a = \dfrac{8a}{3}$

$\Rightarrow \quad \dfrac{8a}{3} = \dfrac{8}{\sqrt{53}}$ [given]

So, $\quad a = \dfrac{3}{\sqrt{53}}$

and $\quad \dfrac{2a}{e} = \left(2\cdot\dfrac{3}{\sqrt{53}}\right) \times \left(\dfrac{3}{1}\right) = \dfrac{18}{\sqrt{53}}$

Now, another directrix will be parallel to the first directrix and lie at a distance of $\dfrac{18}{\sqrt{53}}$ units.

So, let the equation of another directrix be $2x - 7y + \lambda = 0$

Accordingly, $\dfrac{|\lambda - 11|}{\sqrt{2^2 + 7^2}} = \dfrac{18}{\sqrt{53}}$

$|\lambda - 11| = 18 \Rightarrow \lambda = 11 \pm 18$

$\Rightarrow \quad \lambda = 29$ or $-7$

So, equation of another directrix will be $2x - 7y + 29 = 0$ or $2x - 7y - 7 = 0$

**60.** *(c)* Given,

$S_1 \Rightarrow x^2 + y^2 = 9$

$S_2 \Rightarrow (x-2)^2 + y^2 = 1$

$\therefore \quad C_1 \equiv (0, 0)$ and $r_1 = 3$

$C_2 \equiv (2, 0), r_2 = 1$

Let centre of variable circle be $P(h, k)$ and radius be $r$.

From the given data, it is clear that given circles touch each other internally. Hence, variable circle touches $S_1$ internally and $S_2$ externally as shown in figure below.

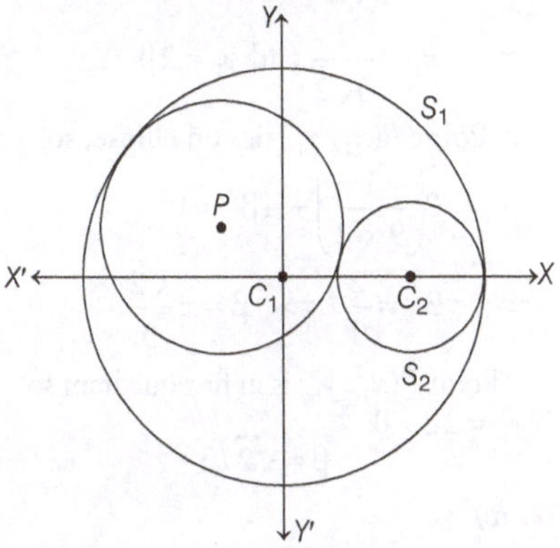


Clearly, $\quad PC_1 = 3 - r$

and $\quad PC_2 = 1 + r$

and $\quad C_1C_2 = 2$

$\therefore \quad PC_1 + PC_2 = 4$

which is greater than $C_1C_2$.

Hence, locus of $P$ is an ellipse whose foci are $C_1$ and $C_2$ and major axis is $2a = 4$.

and $\quad 2ae = C_1C_2 = 2 \Rightarrow e = 1/2$

$\Rightarrow \quad b^2 = 4\left(1 - \dfrac{1}{4}\right) = 3$

Centre is mid-point of $C_1$ (0, 0) and $C_2$ (2, 0).

$\therefore$Centre of ellipse is (1, 0).

$\dfrac{(x-1)^2}{4} + \dfrac{y^2}{3} = 1$ is the equation of locus of centre of variable circle which is satisfied by $\left(2, \pm\dfrac{3}{2}\right)$.

**61.** *(b)* Since equation of normal to the ellipse at $P$ meets the coordinate axes at $\left(-\dfrac{1}{3\sqrt{2}}, 0\right)$ and $(0, \beta)$ is

$$\frac{x}{-\frac{1}{3\sqrt{2}}} + \frac{y}{\beta} = 1 \qquad \ldots(i)$$

Now, let point $P(x_1, y_1)$, so equation of normal to ellipse $\frac{x^2}{1/2}+\frac{y^2}{1}=1$ at point $P$ is

$$\Rightarrow \frac{x}{2x_1}-\frac{y}{y_1}=\frac{1}{2}-1$$

$$\Rightarrow \frac{x}{-2x_1}+\frac{y}{y_1}=\frac{1}{2} \quad \dots\text{(ii)}$$

$\because$ Eqs. (i) and (ii) represents same line.

$$\therefore \frac{-2x_1}{-\frac{1}{3\sqrt{2}}}=\frac{y_1}{\beta}=\frac{2}{1}$$

$$\Rightarrow x_1=\frac{1}{3\sqrt{2}} \text{ and } y_1=2\beta$$

$\because$ Point $P(x_1, y_1)$ lies on ellipse, so

$$2\left(\frac{1}{9\times 2}\right)+4\beta^2=1$$

$$\Rightarrow \beta^2=\frac{2}{9} \Rightarrow \beta=\pm\frac{\sqrt{2}}{3}$$

$\because$ Point $P(x_1, y_1)$ is in first quadrant so $y_1=2\beta>0$

$$\therefore \beta=\sqrt{2}/3$$

**62.** *(a)*

> **key idea** Equation of tangent and normal to the ellipse $\frac{x^2}{a^2}+\frac{y^2}{b^2}=1$ at point $P(x_1, y_1)$ is
> $$T=0 \Rightarrow \frac{xx_1}{a^2}+\frac{yy_1}{b^2}=1$$
> and $\frac{a^2x}{x_1}-\frac{b^2y}{y_1}=a^2-b^2$ respectively.

Equation of given ellipse is
$3x^2+4y^2=12$

$$\Rightarrow \frac{x^2}{4}+\frac{y^2}{3}=1 \quad \dots\text{(i)}$$

Now, let point $P(2\cos\theta, \sqrt{3}\sin\theta)$, so equation of tangent to ellipse (i) at point $P$ is

$$\frac{x\cos\theta}{2}+\frac{y\sin\theta}{\sqrt{3}}=1 \quad \dots\text{(ii)}$$

Since, tangent (ii) passes through point $Q(4, 4)$

$$\therefore 2\cos\theta+\frac{4}{\sqrt{3}}\sin\theta=1 \quad \dots\text{(iii)}$$

and equation of normal to ellipse (i) at point $P$ is

$$\frac{4x}{2\cos\theta}-\frac{3y}{\sqrt{3}\sin\theta}=4-3$$

$$\Rightarrow 2x\sin\theta-\sqrt{3}\cos\theta y=\sin\theta\cos\theta \quad \dots\text{(iv)}$$

Since, normal Eq. (iv) is parallel to line, $2x+y=4$

$\therefore$ Slope of normal Eq. (iv) = slope of line,

$$\frac{2}{\sqrt{3}}\tan\theta=-2$$

$$\Rightarrow \tan\theta=-\sqrt{3} \Rightarrow \theta=120^\circ$$

$$\Rightarrow (\sin\theta, \cos\theta)=\left(\frac{\sqrt{3}}{2}, -\frac{1}{2}\right)$$

Hence, point $P\left(-1, \frac{3}{2}\right)$

$$\text{Now, } PQ=\sqrt{(4+1)^2+\left(4-\frac{3}{2}\right)^2}$$

[given cordinates of $Q \equiv (4, 4)$]

$$=\sqrt{25+\frac{25}{4}}=\frac{5\sqrt{5}}{2}$$

**63.** *(c)* Equation of ellipse is

$$x^2+3y^2=6.$$

Equation of the tangent is

$$\frac{x\cos\theta}{a}+\frac{y\sin\theta}{b}=1.$$

Let $(h, k)$ be any point on the locus, then

$$\frac{h}{a}\cos\theta+\frac{k}{b}\sin\theta=1 \quad \dots\text{(i)}$$

Slope of the tangent line is $\frac{-b}{a}\cot\theta$.

Slope of perpendicular drawn from centre (0, 0) to $(h, k)$ is $\frac{k}{h}$.

Since, both the lines are perpendicular.

$$\therefore \left(\frac{k}{h}\right)\times\left(-\frac{b}{a}\cot\theta\right)=-1$$

$$\Rightarrow \frac{\cos\theta}{ha}=\frac{\sin\theta}{kb}=\alpha \quad \text{[say]}$$

$$\Rightarrow \cos\theta=\alpha ha, \sin\theta=\alpha kb$$

From Eq. (i),

$$\frac{h}{a}(\alpha ha)+\frac{k}{b}(\alpha kb)=1$$

$$\Rightarrow h^2\alpha+k^2\alpha=1$$

$$\Rightarrow \alpha=\frac{1}{h^2+k^2} \quad \dots\text{(ii)}$$

Also, $\sin^2\theta+\cos^2\theta=1$

$$\Rightarrow (\alpha kb)^2+(\alpha ha)^2=1$$

$$\Rightarrow \alpha^2k^2b^2+\alpha^2h^2a^2=1$$

$$\Rightarrow \frac{k^2b^2}{(h^2+k^2)^2}+\frac{h^2a^2}{(h^2+k^2)^2}=1$$

[Using Eq. (ii)]

$$\Rightarrow \frac{2k^2}{(h^2+k^2)^2}+\frac{6h^2}{(h^2+k^2)^2}=1$$

$[\because a^2=6 \text{ and } b^2=2]$

$$\Rightarrow 6x^2+2y^2=(x^2+y^2)^2$$

[replacing $k$ by $y$ and $h$ by $x$]

**64.** *(d)* Let the coordinates of the vertices of rectangle $ABCD$ be
$A(a\cos\theta, b\sin\theta)$, $B(-a\cos\theta, b\sin\theta)$, $C(-a\cos\theta, -b\sin\theta)$ and $D(a\cos\theta, -b\sin\theta)$

Then length of rectangle,
$AB=2a\cos\theta$
and breadth of rectangle, $AD=2b\sin\theta$

$\therefore$ Area of rectangle, $A=AB\times AD$
$=2a\cos\theta\times 2b\sin\theta$

$$\Rightarrow A=2ab\sin 2\theta \quad \dots\text{(i)}$$

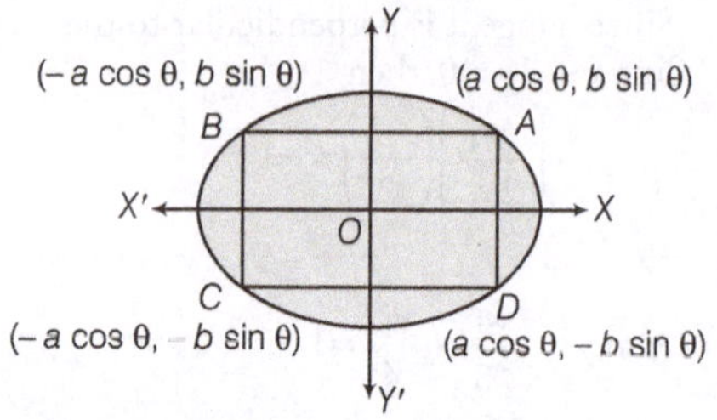


On differentiating Eq. (i) w.r.t. $\theta$, we get

$$dA/d\theta=2\times 2ab\cos 2\theta$$

For maxima or minima, put $dA/d\theta=0$

$$\Rightarrow 4ab\cos 2\theta=0$$

$$\therefore 2\theta=\frac{\pi}{2} \Rightarrow \theta=\frac{\pi}{4}$$

$$\text{Now, } \frac{d^2A}{d\theta^2}=-8ab\sin 2\theta$$

$$\text{At } \theta=\frac{\pi}{4}, \quad \left(\frac{d^2A}{d\theta^2}\right)<0$$

$\therefore$ Area is maximum at $\theta=\pi/4$.

$\Rightarrow$ Maximum area of rectangle
$=2ab$ sq units [from Eq. (i)]

> **Alternate Solution**
> From Eq. (i),
> Area of rectangle, $A=2ab\sin 2\theta$
> $\therefore$ $A$ is maximum when $\sin 2\theta=1$
> $\Rightarrow$ Maximum area of rectangle $=2ab$ sq units

**65.** *(46)* $E_1: \frac{x^2}{a^2}+\frac{y^2}{b^2}=1 \quad a>b$

$$E_2: \frac{x^2}{c^2}+\frac{y^2}{d^2}=1 \quad c<d$$

$$C: x^2+(y-1)^2=2$$

Equation of tangent at $P(x_1, y_1)$

$xx_1 + y(y_1 - 1) = y_1 + 1$

Comparing with $x + y = 3$

We get $x_1 = 1$

$y_1 = 2$

$\therefore \quad P(1, 2)$

$\because$ Parametric form of $x + y = 3$ is

$$\frac{x-1}{-\frac{1}{\sqrt{2}}} = \frac{y-2}{\frac{1}{\sqrt{2}}} = \pm\frac{2\sqrt{2}}{3}\left(\because PQ = \frac{2\sqrt{2}}{3}\right)$$

by solving we get $Q\left(\frac{5}{3}, \frac{4}{3}\right)$ and $R\left(\frac{1}{3}, \frac{8}{3}\right)$

$$\therefore \quad 9(x_1y_1 + x_2y_2 + x_3y_3)$$
$$= 9\left(2 + \frac{5}{3}\left(\frac{4}{3}\right) + \left(\frac{1}{3}\right)\left(\frac{8}{3}\right)\right) = 46$$

**66.** *(54)* Given,

$$e_1 = e_i$$

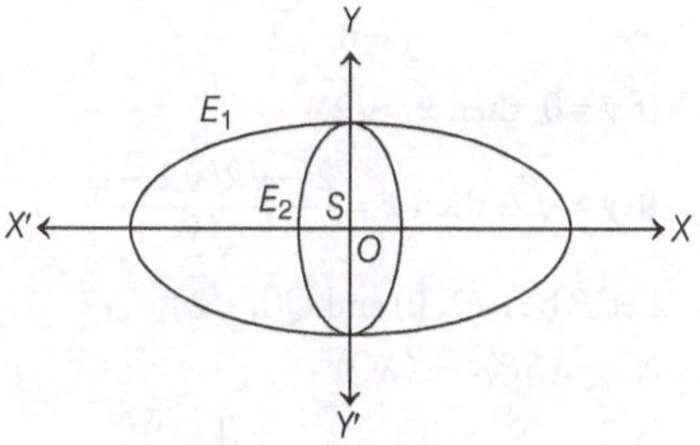


$$b_i = a_{i+1}$$
$$e_1 = \sqrt{1 - 4/9} = \sqrt{5}/3$$

Area of $E_i = A_i = \pi a_i b_i$

and area of $E_{i+1} = A_{i+1}$

$$= \pi a_{i+1} + b_{i+1}$$
$$e_i^2 = e_{i+1}^2$$
$$\Rightarrow \quad 1 - \frac{b_i^2}{a_i^2} = 1 - \frac{b_{i+1}^2}{e_{i+1}^2} \Rightarrow \frac{b_i^4}{a_i^2} = b_{i+1}^2$$
$$\Rightarrow \quad b_{i+1} = \frac{b_i^2}{a_i}$$
$$\Rightarrow \quad A_{i+1} = \pi b_i\left(\frac{b_i^2}{a_i}\right) = \pi a_i b_i\left(\frac{b_i}{a_i}\right)^2$$
$$= A_i[1 - e_i^2]$$
$$= A_i(1 - 5/9) = A_i.4/9$$
$$\Rightarrow \quad \frac{5}{\pi}\sum_{i=1}^{\infty} A_i = \frac{5}{\pi}\cdot S_1 \times \left[1 - \frac{4}{9}\right]$$
$$= \frac{5}{\pi} \times \pi \times 3 \times 2 \times \frac{9}{5}$$

$(S_1 = \pi \times 2 \times 3 = 6\pi)$

$$= 54$$

**67.** *(5)*

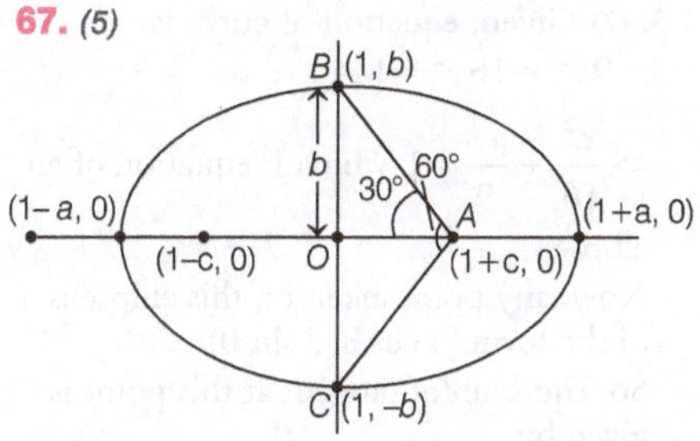


$$LR = \frac{1}{2} \Rightarrow \frac{2b^2}{a} = \frac{1}{2} \Rightarrow b^2 = \frac{a}{4}$$
$$\Rightarrow b = \frac{\sqrt{a}}{2}$$

In $\Delta AOB$, $\tan 30° = \frac{b}{OA}$

$$\Rightarrow \frac{1}{\sqrt{3}} = \frac{\sqrt{a}}{2 \times OA} \Rightarrow OA = \frac{\sqrt{3a}}{2}$$

$A(1 + OA, 0) \equiv A(1 + c, 0)$

$1 + OA = 1 + c \quad \Rightarrow \quad OA = c$

$$c = OA \Rightarrow c^2 = OA^2 \Rightarrow a^2 - b^2 = \frac{3a}{4}$$
$$a^2 - \frac{a}{4} = \frac{3a}{4} \quad \Rightarrow \quad 4a^2 - a = 3a$$
$$4a^2 - 4a = 0 \quad \Rightarrow \quad 4a(a-1) = 0$$
$$a = 1, a = 0 \quad \Rightarrow \quad a = 1$$
$$\Rightarrow b = \frac{\sqrt{a}}{2} = \frac{1}{2}$$

Sum of square of minor axis and major axis is

$$(2a)^2 + (2b)^2 = (2 \times 1)^2 + \left(2 \times \frac{1}{2}\right)^2$$
$$= 4 + 1 = 5$$

**68.** *(39)* Given, ellipse $E : \frac{x^2}{a^2} + \frac{y^2}{b^2} = 1$

Since, focus of ellipse is (2, 0) and directrix is $x = 8$, then

$$ae = 2 \text{ and } \frac{a}{e} = 8$$
$$\therefore \quad (ae)\left(\frac{a}{e}\right) = 2 \times 8$$
$$\Rightarrow \quad a^2 = 16 \Rightarrow \quad a = 4$$
$$\therefore \quad e = \frac{2}{a} = \frac{2}{4} = \frac{1}{2}$$

Again $\quad e = \sqrt{1 - \frac{b^2}{a^2}}$

$$\Rightarrow \quad \frac{1}{2} = \sqrt{1 - \frac{b^2}{16}}$$
$$\Rightarrow \quad \frac{1}{4} = \frac{16 - b^2}{16}$$
$$\Rightarrow \quad b^2 = 12$$

$\therefore$ Equation of ellipse is $\frac{x^2}{16} + \frac{y^2}{12} = 1$

Let $P(4\cos\theta, 2\sqrt{3}\sin\theta)$. So, equation of tangent at $P$ is

$$\frac{4\cos\theta\, x}{16} + \frac{2\sqrt{3}\sin\theta\, y}{12} = 1$$
$$\text{or } \frac{x\cos\theta}{4} + \frac{y\sin\theta}{2\sqrt{3}} = 1 \quad \ldots\text{(i)}$$

Since, tangent passes through $(0, 4\sqrt{3})$, then

$$0 + \frac{4\sqrt{3}\sin\theta}{2\sqrt{3}} = 1$$
$$\Rightarrow \quad \sin\theta = \frac{1}{2}$$
$$\Rightarrow \quad \theta = 30°$$

$\therefore$Equation of tangent is

$$\frac{x}{4} \times \frac{\sqrt{3}}{2} + \frac{y}{2\sqrt{3}} \times \frac{1}{2} = 1$$
$$\Rightarrow \quad \frac{\sqrt{3}x}{8} + \frac{y}{4\sqrt{3}} = 1$$

Again, point $P$ is $(4\cos 30°, 2\sqrt{3}\sin 30°)$ i.e. $(2\sqrt{3}, \sqrt{3})$

Also, tangent intersect $X$-axis at $\left(\frac{8}{\sqrt{3}}, 0\right)$.

So, $Q$ is $\left(\frac{8}{\sqrt{3}}, 0\right)$

$$\therefore \quad PQ^2 = \left(\frac{8}{\sqrt{3}} - 2\sqrt{3}\right)^2 + (0 - \sqrt{3})^2$$
$$= \left(\frac{8-6}{\sqrt{3}}\right)^2 + 3 = \frac{4}{3} + 3 = \frac{13}{3}$$
$$\therefore (3PQ)^2 = 9\,PQ^2 = 9 \times \frac{13}{3} = 39$$

**69.** *(118)* Given equation of ellipse is

$$\frac{x^2}{36} + \frac{y^2}{16} = 1 \Rightarrow \frac{x^2}{(6)^2} + \frac{y^2}{(4)^2} = 1$$

Let $R \equiv (1, \alpha) \Rightarrow P \equiv (2, 0)$

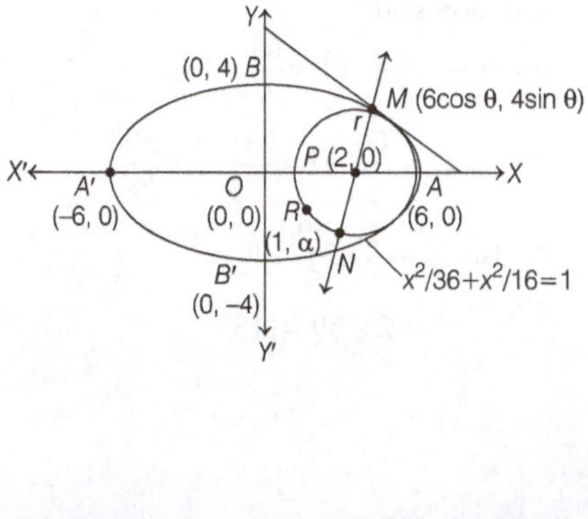

Also, given $C$ be the largest circle having centre (2, 0) inscribed in the given ellipse.

Now, co-ordinates of any point $M$ taken on the ellipse is $(6\cos\theta, 4\sin\theta)$.

So, equation of normal $MN$ to the ellipse at $M(6\cos\theta, 4\sin\theta)$ is given by

$$\frac{36x}{6\cos\theta} - \frac{16y}{4\sin\theta} = 36 - 16$$

$$\Rightarrow \frac{6x}{\cos\theta} - \frac{4y}{\sin\theta} = 20$$

$$\Rightarrow \frac{3x}{\cos\theta} - \frac{2y}{\sin\theta} = 10 \quad \text{...(i)}$$

Since, Eq. (i) is also normal to the given circle $C$ and passing through the point $P(2, 0)$.

So, $\frac{3\times 2}{\cos\theta} - \frac{2\times 0}{\sin\theta} = 10$

$$\Rightarrow 6 = 10\cos\theta \Rightarrow \cos\theta = \frac{3}{5}$$

$$\therefore \sin\theta = \frac{4}{5}$$

[Using $\sin^2\theta = 1 - \cos^2\theta$]

So, co-ordinates of $M$ is $\left(6\times\frac{3}{5}, 4\times\frac{4}{5}\right)$

i.e. $\left(\frac{18}{5}, \frac{16}{5}\right)$

$\therefore$ Largest radius of the circle

$$= r = \sqrt{\left(\frac{18}{5} - 2\right)^2 + \left(\frac{16}{5} - 0\right)^2}$$

$$= \sqrt{\left(\frac{8}{5}\right)^2 + \left(\frac{16}{5}\right)^2}$$

$$\Rightarrow r = \sqrt{\frac{320}{25}} = \frac{\sqrt{320}}{5}$$

Hence, equation of the required circle having centre at $P(2, 0)$ and radius $r$ is

$$(x-2)^2 + (y-0)^2 = \left(\frac{\sqrt{320}}{5}\right)^2$$

$$\Rightarrow (x-2)^2 + y^2 = \frac{320}{25} = \frac{64}{5} \quad \text{...(ii)}$$

Since, Eq. (ii) passes through $R(1, \alpha)$ as question said.

So, $(1-2)^2 + \alpha^2 = \frac{64}{5}$

$$\Rightarrow \alpha^2 = \frac{64}{5} - 1 = \frac{64-5}{5} = \frac{59}{5}$$

$$\therefore 10\alpha^2 = 10 \times \frac{59}{5}$$

$$= 2 \times 59 = 118$$

**70.** (7) Given, equation of curve is

$$9x^2 + 16y^2 = 144$$

$\Rightarrow \frac{x^2}{16} + \frac{y^2}{9} = 1$, which is equation of an ellipse.

Now, any point taken on this ellipse is of the form $(4\cos\theta, 3\sin\theta)$.

So, equation of tangent at this point is given by

$$\frac{x(4\cos\theta)}{16} + \frac{y(3\sin\theta)}{9} = 1$$

$$\Rightarrow \frac{x\cos\theta}{4} + \frac{y\sin\theta}{3} = 1 \quad \text{...(i)}$$

According to the question, Eq. (i) intersect the coordinate axes at $A$ and $B$.

So, $A = (4\sec\theta, 0)$

and $B = (0, 3\,\text{cosec}\,\theta)$

Now, length of line segment

$$AB = \sqrt{(4\sec\theta - 0)^2 + (0 - 3\,\text{cosec}\,\theta)^2}$$

$$\Rightarrow AB = \sqrt{16\sec^2\theta + 9\,\text{cosec}^2\theta}$$

$$= \sqrt{16(1+\tan^2\theta) + 9(1+\cot^2\theta)}$$

$$\Rightarrow AB = \sqrt{25 + 16\tan^2\theta + 9\cot^2\theta}$$

Now, $16\tan^2\theta$ and $9\cot^2\theta$ are two positive real numbers using AM $\geq$ GM

$$\Rightarrow \frac{16\tan^2\theta + 9\cot^2\theta}{2} \geq \sqrt{16\tan^2\theta \times 9\cot^2\theta}$$

$$\Rightarrow 16\tan^2\theta + 9\cot^2\theta \geq 2\sqrt{16\tan^2\theta \times 9 \times \frac{1}{\tan^2\theta}}$$

$$\Rightarrow 16\tan^2\theta + 9\cot^2\theta = 2\times 4\times 3 = 24$$

$$\Rightarrow 25 + 16\tan^2\theta + 9\cot^2\theta \geq 2\sqrt{16\times 9\times 1} \geq 24 + 25$$

$$\Rightarrow \sqrt{25 + 16\tan^2\theta + 9\cot^2\theta} \geq \sqrt{49}$$

$$\Rightarrow AB \geq 7$$

$\therefore$ Minimum length of line segment $AB = 7$

**71.** (13) Given, equation of ellipse

$$\frac{x^2}{2} + \frac{y^2}{4} = 1$$

Here, $a^2 = 2, b^2 = 4$

$$e = \sqrt{1 - \frac{a^2}{b^2}} = \sqrt{1 - \frac{1}{2}}$$

$$= \sqrt{\frac{1}{2}} = \frac{1}{\sqrt{2}}$$

$\therefore$ Co-ordinates of $S$ is $(0, -\sqrt{2})$.

Equation of chord of contact from the point $(\sqrt{2}, 2\sqrt{2} - 2)$ is

$$\frac{x\sqrt{2}}{2} + \frac{y(2\sqrt{2}-2)}{4} = 1$$

$$\Rightarrow \frac{x}{\sqrt{2}} = 1 - \frac{y(\sqrt{2}-1)}{2}$$

$$\Rightarrow x = \frac{2 - y(\sqrt{2}-1)}{2}$$

Substituting value of $x$ in equation of ellipse

$$\frac{\left(\frac{2 - y(\sqrt{2}-1)}{\sqrt{2}}\right)^2}{2} + \frac{y^2}{4} = 1$$

$$\Rightarrow 4 + y^2(\sqrt{2}-1)^2 - 4y(\sqrt{2}-1) + y^2 = 4$$

$$\Rightarrow y^2[2 + 1 - 2\sqrt{2} + 1] - 4y(\sqrt{2}-1) = 0$$

$$\Rightarrow y^2(4 - 2\sqrt{2}) - 4y(\sqrt{2}-1) = 0$$

$$\Rightarrow 2\sqrt{2}y^2(\sqrt{2}-1) - 4y(\sqrt{2}-1) = 0$$

$$\Rightarrow y = 0, \sqrt{2}$$

If $y = 0$, then $x = \sqrt{2}$

If $y = \sqrt{2}$, then $x = \frac{2 - \sqrt{2}(\sqrt{2}-1)}{\sqrt{2}} = 1$

Let $P$ be $(\sqrt{2}, 0)$ and $Q(1, \sqrt{2})$

$$\therefore (SP)^2 + (SQ)^2$$

$$(\sqrt{2}-0)^2 + (0+\sqrt{2})^2 + (1-0)^2 + (\sqrt{2}+\sqrt{2})^2$$

$$= 2 + 2 + 1 + 8 = 13$$

**72.** (75) Ellipse is

$$(x+1)^2 + 4(y+1)^2 = \lambda + 5$$

$$\Rightarrow \frac{(x+1)^2}{\lambda+5} + \frac{(y+1)^2}{\left(\frac{\lambda+5}{4}\right)} = 1$$

$\therefore$ Now, Length of latusrectum,

$$4 = \frac{\frac{2(\lambda+5)}{4}}{\sqrt{\lambda+5}}$$

$$\Rightarrow 8 = \sqrt{\lambda+5}$$

$$\Rightarrow \lambda = 59$$

$\therefore$ Length of major axis $l$

$$= 2\sqrt{\lambda+5} = 2\times 8 = 16$$

$$\therefore \lambda + l = 59 + 16 = 75$$

**73.** (4) Let $y = mx + c$ is the common tangent.

So, $c = \frac{1}{m} = \pm\frac{3}{2}\sqrt{1+m^2}$

$$\Rightarrow m^2 = \frac{1}{3} \Rightarrow m = \pm\frac{1}{\sqrt{3}}$$

So, equation of common tangents will be

$y = \pm\frac{1}{\sqrt{3}}x \pm \sqrt{3}$, which intersects at $Q(-3, 0)$.

Here, for ellipse, $a = 3, b = 6$

Then, $e = \sqrt{1 - \frac{a^2}{b^2}} = \sqrt{1 - \frac{9}{36}}$

$$e = \frac{\sqrt{3}}{2}$$

then, $l = \frac{2a^2}{b} = \frac{2\times 9}{6} = 3$

Now, $\frac{l}{e^2} = \frac{3}{3} \times 4 = 4$

**74.** *(2929)* The given ellipse is $25x^2 + 4y^2 = 1$

Since, $(\alpha, \beta)$ lies on the given ellipse,

$$25\alpha^2 + 4\beta^2 = 1 \quad \dots \text{(i)}$$

Tangent to the parabola $y^2 = 4x$ is

$y = mx + \frac{1}{m}$ passing through $(\alpha, \beta)$.

So, $\alpha m^2 - \beta m + 1 = 0$ has roots $m_1$ and $4m_1$

$$m_1 + 4m_1 = \frac{\beta}{\alpha} \text{ and } m_1 \cdot 4m_1 = \frac{1}{\alpha}$$

$\Rightarrow \quad 4\beta^2 = 25\alpha \quad \dots \text{(ii)}$

From Eqs. (i) and (ii), we get

$$25(\alpha^2 + \alpha) = 1 \quad \dots \text{(iii)}$$

Now, $(10\alpha + 5)^2 + (16\beta^2 + 50)^2$

$= 25(2\alpha + 1)^2 + 2500(2\alpha + 1)^2$

$= 2525(4\alpha^2 + 4\alpha + 1)$

$= 2525\left(\frac{4}{25} + 1\right) \quad \left[\because \alpha^2 + \alpha = \frac{1}{25}\right]$

$= 101(29) = 2929$

**75.** *(2)* Equation of an ellipse

$B(0, 2a \operatorname{cosec} \theta)$

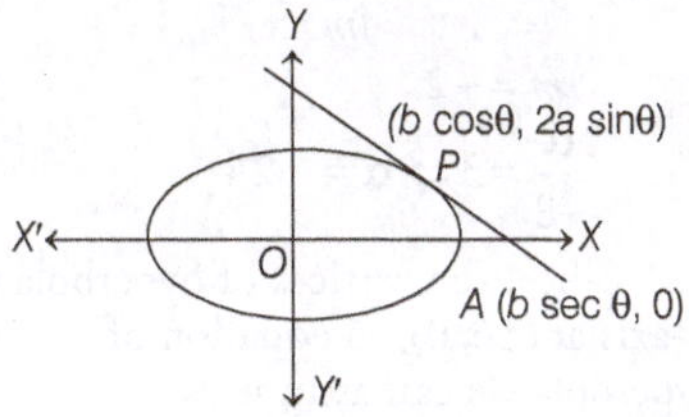


$$\frac{x^2}{b^2} + \frac{y^2}{4a^2} = 1$$

Equation of tangent at $P(b\cos\theta, 2a\sin\theta)$ is

$$\frac{x\cos\theta}{b} + \frac{y\sin\theta}{2a} = 1$$

Point of intersection of tangent and coordinate axis are

$A(b\sec\theta, 0)$ and $B(0, 2a\operatorname{cosec}\theta)$

Area of $\Delta OAB$ d

$= \frac{1}{2}(b\sec\theta)(2a\operatorname{cosec}\theta)$

$= \frac{ab}{\sin\theta\cos\theta} = \frac{2ab}{\sin 2\theta}$

Area$_{\min} = 2ab$, when $\sin 2\theta \pm 1$

$\therefore \quad k = 2$

**76.** *(3)* $E$ be an ellipse whose centre $C(3, -4)$ and focus $F(4, -4)$ and one vertex $A(5, -4)$.

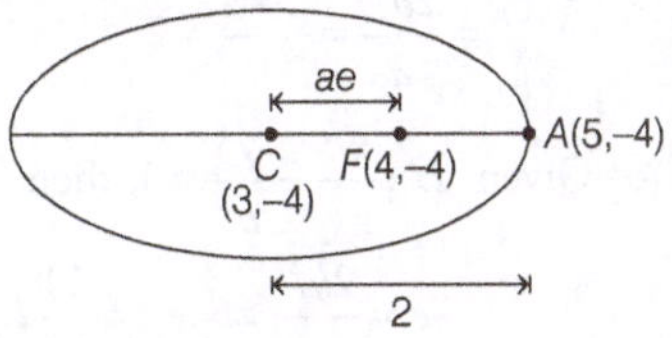


Hence, $a = 2$ and $ae = 1$

$\Rightarrow \quad e = 1/2$

Now, $b^2 = a^2 - (ae)^2 = 4 - 1 \Rightarrow b^2 = 3$

So, equation of ellipse

$$E \Rightarrow \frac{(x-3)^2}{4} + \frac{(y+4)^2}{3} = 1$$

Intersecting with given tangent

$mx - y = 4$

$\Rightarrow y + 4 = mx$

$$\frac{x^2 - 6x + 9}{4} + \frac{m^2x^2}{3} = 1$$

$\Rightarrow \quad (3 + 4m^2)x^2 - 18x + 15 = 0$

Now, $D = 0$

$\Rightarrow 324 - 4(3 + 4m^2)(15) = 0$

$\Rightarrow 144 - 240m^2 = 0 \Rightarrow 5m^2 = 3$

## Topic 4

**01.** *(d)* $\frac{x^2}{b^2} + \frac{y^2}{25} = 1$, Since $b < 5$, the major axis is along the $Y$-axis.

Eccentricity, $e_1 = \sqrt{1 - \frac{b^2}{a^2}} = \sqrt{1 - \frac{b^2}{25}}$

Eccentricity of the hyperbola,

$$e_2 = \sqrt{1 + \frac{b^2}{a^2}} = \sqrt{1 + \frac{b^2}{16}}$$

Also, $e_1e_2 = 1 \Rightarrow \sqrt{1 - \frac{b^2}{25}}\sqrt{1 + \frac{b^2}{16}} = 1$

$$1 + \frac{b^2}{16} - \frac{b^2}{25} - \frac{b^4}{400} = 1$$

$$\frac{b^2}{16} - \frac{b^2}{25} - \frac{b^4}{400} = 0 \Rightarrow \frac{b^2}{16} - \frac{b^2}{25} = \frac{b^4}{400}$$

$$\frac{9b^2}{400} = \frac{b^4}{400} \Rightarrow 9b^2 = b^4 \Rightarrow b^2 = 9, b \neq 0$$

Foci of the ellipse $(0, \pm\sqrt{25 - b^2})$

$= (0, \pm\sqrt{25 - 9}) = (0, \pm 4)$

Foci of hyperbola $(\pm\sqrt{a^2 + b^2}, 0)$

$= (\pm\sqrt{16 + 9}, 0) = (\pm 5, 0)$

The equation of the new ellipse passes through $(0, \pm 4)$, $(\pm 5, 0)$

$$\frac{x^2}{25} + \frac{y^2}{16} = 1$$

Eccentricity of new ellipse,

$$e = \sqrt{1 - \frac{16}{25}} = \frac{3}{5}$$

**02.** *(a)* $E: \frac{x^2}{a^2} + \frac{y^2}{b^2} = 1$

$a > b \Rightarrow$ focii $(ae, 0), (-ae, 0)$

$H: \frac{x^2}{A^2} - \frac{y^2}{B^2} = 1$

$\Rightarrow \quad$ focii $(Ae', 0), (-Ae', 0)$

$\because$ Given $2ae = 2\sqrt{3}$ and $2Ae' = 2\sqrt{3}$

$ae = \sqrt{3} \quad Ae' = \sqrt{3}$

$ae = Ae' \Rightarrow \frac{a}{A} = \frac{e'}{e} \Rightarrow 3$

$\Rightarrow \quad a = 3A$ and $a - A = 2$

$2A = 2 \Rightarrow A = 1 \Rightarrow a = 3$

$Ae' = \sqrt{3}$ and $ae = \sqrt{3}$

$e' = \sqrt{3}, \quad e = 1/\sqrt{3}$

Sum of lengths of their latus-rectum

$= \frac{2b^2}{a} + \frac{2B^2}{A}$

$= \frac{2a^2(1 - e^2)}{a} + \frac{2A^2[(e')^2 - 1]}{A}$

$= 2(3)\left(1 - \frac{1}{3}\right) + 2(1)(3 - 1) = 4 + 4 = 8$

**03.** *(b)* Foci of hyperbola $(1, 14)$, $(1, -12)$

$\therefore \ x = 1$ be the transverse axis.

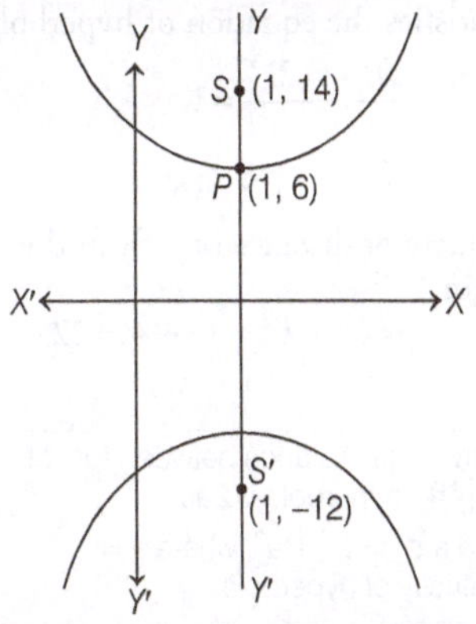

$|SP - S'P| = 2b$

$\Rightarrow 2b = |8 - 18|$

$\Rightarrow b = 5$ and $SS' = 2be$

$26 = 2be$

$be = 13 \Rightarrow e = 13/5$

$\therefore$ Length of latus-rectum

$$= \frac{2a^2}{b} = \frac{2b^2(e^2 - 1)}{b}$$

$$= 2(5)\left(\frac{144}{25}\right) = \frac{288}{5}$$

**04.** *(c)* $H : \frac{x^2}{9} - \frac{y^2}{4} = 1$

Here $a^2 = 9, b^2 = 4$

$\therefore \quad a = 3, b = 2$

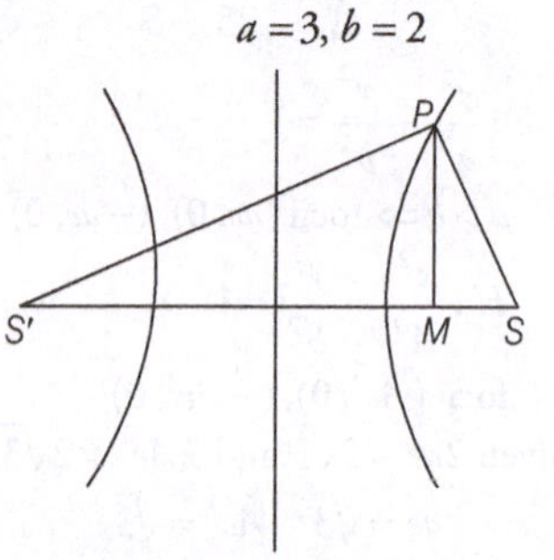


Let $S$ and $S'$ are focii of hyperbola.

$$\therefore \quad e = \sqrt{1 + \frac{b^2}{a^2}} = \sqrt{1 + \frac{4}{9}} = \frac{\sqrt{13}}{3}$$

$\because \quad S = (ae, 0)$ and $S'(-ae, 0)$

$= (\sqrt{13}, 0)$ and $(-\sqrt{13}, 0)$

$\therefore \quad SS' = 2\sqrt{13}$

$\therefore \quad$ Area of $\Delta PSS' = 2\sqrt{13}$

$\frac{1}{2} \times SS' \times PM = 2\sqrt{13}$

$\Rightarrow \frac{1}{2} \times 2\sqrt{13} \times PM = 2\sqrt{13}$

$\Rightarrow \quad PM = 2$

Let coordinate of $P(h, k)$

$\therefore \quad k = 2 \quad [\because PM = 2]$

$P$ satisfies the equation of hyperbola.

$$\therefore \quad \frac{h^2}{9} - \frac{(2)^2}{4} = 1$$

$\Rightarrow \quad h^2 = 18$

$\therefore$Square of distance of $p$ from the origin

$= h^2 + k^2 = 18 + 4 = 22$

**05.** *(b)*

> **key idea** (i) Distance between foci of hyperbola is $2ae$.
> (ii) And $a^2e^2 = b^2 + a^2$, where $e$ = eccentricity of hyperbola.

We have,

Foci of hyperbola are $S(1 + \sqrt{2}, 0)$ and $S'(1 - \sqrt{2}, 0)$.

Now, we know that

$SS' = 2ae \quad \Rightarrow 2\sqrt{2} = 2ae$

$\Rightarrow a = 1 \quad [\because e = \sqrt{2}]$

Again, $e = \sqrt{1 + \frac{b^2}{a^2}}$

$\Rightarrow \quad \sqrt{2} = \sqrt{1 + \frac{b^2}{(1)^2}} \Rightarrow b^2 = 1$

Now, length of latus rectum

$$= \frac{2b^2}{a} = \frac{2 \times 1}{1} = 2$$

**06.** *(d)* Given, $H : \frac{x^2}{a^2} - \frac{y^2}{b^2} = 1$, then

$$e = \frac{2b^2}{a} \Rightarrow e^2 = \frac{11}{14} l$$

$\Rightarrow \quad a^2 + b^2 = \frac{11}{7} b^2 a \quad \ldots \text{(i)}$

and $\quad e'^2 = \frac{11}{8} l'$

$\Rightarrow \quad a^2 + b^2 = \frac{11}{4} a^2 b \quad \ldots \text{(ii)}$

From Eqs. (i) and (ii), $7a = 4b$

Then, from Eq. (i), we get

$$\frac{16}{49} b^2 + b^2 = \frac{11}{7} b^2 . \frac{4b}{7}$$

$\Rightarrow \quad 44b = 65$ and $77a = 65$

$\therefore \quad 77a + 44b = 130$

**07.** *(b)* Given, $e = \sqrt{5/2}$

$\Rightarrow \quad 1 + \frac{b^2}{a^2} = \frac{5}{2} \Rightarrow \frac{b^2}{a^2} = \frac{3}{2}$

and $\frac{2b^2}{a} = 6\sqrt{2} \Rightarrow 2 \cdot \frac{3}{2} \cdot a = 6\sqrt{2}$

$a = 2\sqrt{2}, b^2 = 12$

Now,

$c^2 = a^2m^2 - b^2 = 8 \cdot 4 - 12 = 20$

**08.** *(c)* Given $\frac{x^2}{a^2} - \frac{y^2}{9} = 1$; point $(8, 3\sqrt{3})$

$\frac{64}{a^2} - \frac{9 \times 3}{9} = 1 \quad \Rightarrow \quad a^2 = 16$

$\therefore$ Equation of normal at point $(8, 3\sqrt{3})$

$$(y - y_1) = \frac{-y_1 a^2}{x_1 b^2}(x - x_1)$$

$$\Rightarrow \quad (y - 3\sqrt{3}) = \frac{-3\sqrt{3}}{8} \times \frac{16}{9}(x - 8)$$

$$y - 3\sqrt{3} = \frac{-2}{\sqrt{3}}(x - 8)$$

Verify $(-1, 9\sqrt{3})$ in the equation of normal

$$y - 3\sqrt{3} = -\frac{2}{\sqrt{3}}(-1 - 8) \Rightarrow y = 9\sqrt{3}$$

Thus, $(-1, 9\sqrt{3})$ is the required point.

**09.** *(c)* Let the locus of the mid-point of the chord is $(h, k)$. Then, chord bisected at mid-point $(h, k)$ is

$T = S_1$

$xh - yk = h^2 - k^2 \quad \ldots \text{(i)}$

Now, this is also a tangent of $y^2 = 8x$

The equation of the tangent of slope $m$ to the parabola $y^2 = 8x$ is given by

Tangent : $\quad y = mx + \frac{2}{m}$

$\Rightarrow \quad m^2 x - my = -2 \quad \ldots \text{(ii)}$

Eqs. (i) and (ii) are coincide.

$$\therefore \quad \frac{h}{m^2} = \frac{-k}{-m} = \frac{h^2 - k^2}{-2}$$

$\Rightarrow \quad h = km \Rightarrow m = \frac{h}{k}$

$$\therefore \quad \frac{k^2}{h} = \frac{h^2 - k^2}{-2}$$

$-2k^2 = h^3 - hk^2 \Rightarrow h^3 = k^2(h - 2)$

Therefore, locus of mid-point of the chords, $x^3 = y^2(x - 2)$

**10.** *(d)* $T : 2x + y = k$

$H : x^2 - y^2 = 3$

$P : y^2 = \alpha x$

If $T$ is tangent, then

$y = mx \pm \sqrt{a^2m^2 - b^2}$

$\Rightarrow \quad y = -2x - k$

On comparing both sides we get

$m = -2$ and $k = 3$

If $2x + y = 3$ is also tangent to $y^2 = \alpha x$,

Then, $\quad y = mx + \frac{\alpha}{4m}$

$\because \quad m = -2$

$\therefore \quad \frac{\alpha}{-8} = 3 \Rightarrow \alpha = -24$

**11.** *(c)* Since, the vertices of hyperbola on $X$-axis at $(\pm 6, 0)$, so equation of hyperbola we can assume as

$$\frac{x^2}{a^2} - \frac{y^2}{b^2} = 1 \text{ and } |a| = 6$$

and hyperbola passes through point $P(10, 16)$ so

$$\frac{(10)^2}{6^2} - \frac{(16)^2}{b^2} = 1$$

⇒ $b^2 = \dfrac{36 \times 256}{64} = 144$

So, equation of hyperbola is

$\dfrac{x^2}{36} - \dfrac{y^2}{144} = 0$, and the equation of normal to hyperbola at point $P$ is

$$\frac{36x}{10} + \frac{144y}{16} = 36 + 144$$

⇒ $2x + 5y = 100$

**12.** *(b)* The vertices of hyperbola are given as $(\pm 2, 0)$ and one of its foci is at $(-3, 0)$.

$\therefore (a, 0) = (2, 0)$ and $(-ae, 0), = (-3, 0)$

$a = 2$ and $-ae = -3$

⇒ $e = \dfrac{3}{2}$

Also, $\dfrac{9}{4} = 1 + \dfrac{b^2}{4} \Rightarrow b^2 = 5$

So, equation of the hyperbola is

$$\frac{x^2}{4} - \frac{y^2}{5} = 1 \quad \ldots \text{(i)}$$

The point $(6, 5\sqrt{2})$ from the given options does not satisfy the above equation of hyperbola.

**13.** *(a)* We know that in $\dfrac{x^2}{a^2} - \dfrac{y^2}{b^2} = 1$, where $b^2 = a^2(e^2 - 1)$, the length of conjugate axis is $2b$ and distance between the foci is $2ae$.

$\therefore$ According the problem, $2b = 5$ and $2ae = 13$

Now, $\left(\dfrac{5}{2}\right)^2 = a^2e^2 - a^2$

⇒ $\dfrac{25}{4} = \dfrac{(2ae)^2}{4} - a^2$

⇒ $a^2 = \dfrac{169 - 25}{4}$ ⇒ $a = 6$

Now, $2ae = 13$

⇒ $2 \times 6 \times e = 13$

⇒ $e = \dfrac{13}{12}$

**14.** *(c)* We have, $\dfrac{2b^2}{a} = 8$ and $2b = ae$

⇒ $b^2 = 4a$ and $2b = ae$

⇒ $4b^2 = a^2e^2$

⇒ $4a^2(e^2 - 1) = a^2e^2$ $\quad [\because a \neq 0]$

⇒ $3e^2 = 4$

⇒ $e = \dfrac{2}{\sqrt{3}}$ $\quad [\because e > 0]$

**15.** *(c)*

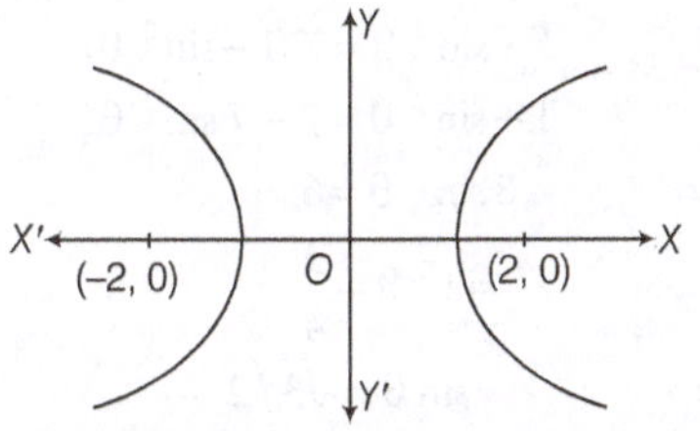


Let equation of hyperbola be

$$\frac{x^2}{a^2} - \frac{y^2}{b^2} = 1$$

where, $2ae = 4$

and $e = 2$

⇒ $a = 1$

$\because a^2e^2 = a^2 + b^2 \Rightarrow 4 = 1 + b^2$

$\therefore b^2 = 3$

Thus, equation of hyperbola is

$\dfrac{x^2}{1} - \dfrac{y^2}{3} = 1$ or $3x^2 - y^2 = 3$

**16.** *(d)* The given equation of hyperbola is

$$\frac{x^2}{\cos^2 \alpha} - \frac{y^2}{\sin^2 \alpha} = 1$$

Here, $a^2 = \cos^2 \alpha$ and $b^2 = \sin^2 \alpha$

Now, $e = \sqrt{1 + \dfrac{b^2}{a^2}}$

⇒ $e = \sqrt{1 + \dfrac{\sin^2 \alpha}{\cos^2 \alpha}} \Rightarrow e = \sec \alpha$

Coordinates of foci are $(\pm ae, 0)$ *i.e.*, $(\pm 1, 0)$.

Hence, abscissae of foci remains constant when $\alpha$ varies.

**17.** *(a)* Since, the line $y = \alpha x + \beta$ is tangent to the hyperbola $\dfrac{x^2}{a^2} - \dfrac{y^2}{b^2} = 1$.

$\therefore \beta^2 = a^2\alpha^2 - b^2$.

So, locus of $(\alpha, \beta)$ is

$y^2 = a^2x^2 - b^2$

⇒ $a^2x^2 - y^2 - b^2 = 0$

Since, this equation represents a hyperbola, so locus of a point $P(\alpha, \beta)$ is a hyperbola.

**18.** *(a)* The mid-point of the chord is

$$\left(\frac{x_1 + x_2}{2}, \frac{y_1 + y_2}{2}\right).$$

The equation of the chord $T = S_1$.

$$\therefore x\left(\frac{y_1 + y_2}{2}\right) + y\left(\frac{x_1 + x_2}{2}\right) = 2\left(\frac{x_1 + x_2}{2}\right)\left(\frac{y_1 + y_2}{2}\right)$$

⇒ $x(y_1 + y_2) + y(x_1 + x_2) = (x_1 + x_2)(y_1 + y_2)$

⇒ $\dfrac{x}{x_1 + x_2} + \dfrac{y}{y_1 + y_2} = 1$

**19.** *(a)* Put $y^2 = 8x - x^2$

$$\frac{x^2}{9} - \frac{8x - x^2}{4} = 1$$

⇒ $4x^2 - 72x + 9x^2 = 36$

⇒ $13x^2 - 72x - 36 = 0$

⇒ $13x^2 - 78x + 6x - 36 = 0$

⇒ $x = 6$ or $x = \dfrac{-6}{13}$ (not possible)

If $x = 6$, then $y = \pm 2\sqrt{3}$

⇒ $A \equiv (6, 2\sqrt{3})$ and $B \equiv (6, -2\sqrt{3})$

and $P \equiv (\alpha, \beta)$, then $2\alpha - 3\beta + 4 = 0$

Centroid of $\Delta PAB$ is

$$\left(\frac{\alpha + 12}{3}, \frac{\beta}{3}\right) \equiv (h, k)$$

$\alpha = 3h - 12, \beta = 3k$.

⇒ $6h - 24 - 9k + 4 = 0$

⇒ $6h - 9k = 20$

Locus is $6x - 9y = 20$

**20.** *(d)*

> **key idea** For ellipse use eccentricity, $e = \sqrt{1 - \dfrac{a^2}{b^2}}$ and for hyperbola use eccentricity, $e = \sqrt{1 + \dfrac{a^2}{b^2}}$

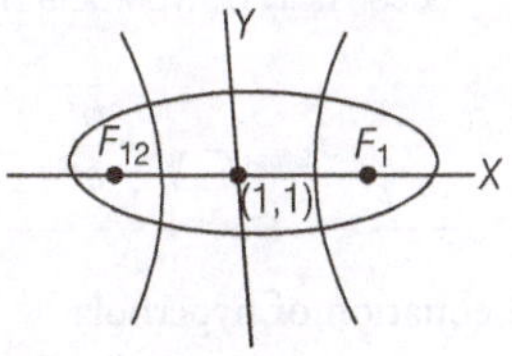


$$E: \frac{(x-1)^2}{100} + \frac{(y-1)^2}{75} = 1$$

$e_1 = \sqrt{1 - \dfrac{75}{100}} = \dfrac{1}{2}$

$e_2 = 2$

$F_1(6, 1)$ and $F_2(-4, 1)$

$2ae_2 = 10 \Rightarrow a = \dfrac{5}{2} \Rightarrow 2a = 5$

$\therefore \alpha = 5$

Now, $e_2^2 = 1 + \dfrac{b^2}{a^2} \Rightarrow b^2 = 3a^2$

⇒ $b = \dfrac{5\sqrt{3}}{2}$

$\Rightarrow \quad 2b = 5\sqrt{3} = \beta$

$\therefore 3\alpha^2 + 2\beta^2 = 3 \times 25 + 2 \times 25 \times 3 = 225$

**21.** *(c)* $\dfrac{-x^2}{a^2} + \dfrac{y^2}{b^2} = +1$

$$e = \sqrt{1 + \frac{a^2}{b^2}} = \sqrt{3}$$

$\Rightarrow \quad 1 + \dfrac{a^2}{b^2} = 3 \Rightarrow \dfrac{a^2}{b^2} = 2 \quad \text{...(i)}$

and latus rectum $= \dfrac{2a^2}{b} = 4\sqrt{3}$

By Eq. (i),

$\dfrac{4b^2}{b} = 4\sqrt{3} \Rightarrow b = \sqrt{3}$

$\Rightarrow \quad b^2 = 3$ and $a^2 = 2 \times 3 = 6$

Now, $\dfrac{-x^2}{6} + \dfrac{y^2}{3} = 1$

Put $(\alpha, 6)$ on it

$\dfrac{-\alpha^2}{6} + \dfrac{36}{3} = 1$

$\Rightarrow \quad -\alpha^2 = -66 \Rightarrow \alpha^2 = 66$

For the given equation,

focal distances $= |ey_1 - a|, |ey_1 + a|$

So, $\beta = (ey_1 + b)(ey_1 - b) = e^2 y_1^2 - b^2$

$= 3 \times 6^2 - 3 = 105$

So, $\alpha^2 + \beta = 66 + 105 = 171$

**22.** *(b)*

> **key idea** Use following step wise approach
>
> **Step 1** Apply formula
>
> $e_{\text{Hyperbola}} = \sqrt{1 + \dfrac{b^2}{a^2}}$ to get eccentricity of hyperbola. Then, apply
>
> $e_{\text{Ellipse}} = \sqrt{1 - \dfrac{b^2}{a^2}}$

Given equation of hyperbola

$x^2 - y^2 \operatorname{cosec}^2 \theta = 5$

$\Rightarrow \quad \dfrac{x^2}{5} - \dfrac{y^2}{5 \sin^2 \theta} = 1$

$e_1 = \sqrt{1 + \dfrac{5 \sin^2 \theta}{5}} = \sqrt{1 + \sin^2 \theta}$

Equation of ellipse

$x^2 \operatorname{cosec}^2 \theta + y^2 = 5$

$\Rightarrow \quad \dfrac{x^2}{5 \sin^2 \theta} + \dfrac{y^2}{5} = 1$

$e = \sqrt{1 - \dfrac{5 \sin^2 \theta}{5}}$

$e_2 = \sqrt{1 - \sin^2 \theta}$

Given, $\quad e_1 = \sqrt{7} e_2$

$1 + \sin^2 \theta = 7(1 - \sin^2 \theta)$

$1 + \sin^2 \theta = 7 - 7 \sin^2 \theta$

$\Rightarrow \quad 8 \sin^2 \theta = 6$

$\Rightarrow \quad \sin^2 \theta = \dfrac{3}{4}$

$\Rightarrow \quad \sin \theta = \sqrt{3}/2$

$\theta = \pi/3$

**23.** *(d)* Given, $e_1$ = eccentricity of the hyperbola $\dfrac{x^2}{16} - \dfrac{y^2}{9} = 1$

$e_2$ = eccentricity of the ellipse $\dfrac{x^2}{a^2} + \dfrac{y^2}{b^2} = 1, a > b.$

and $e_1 e_2 = 1$

Now, $e_1 = \dfrac{5}{4} \quad \left[\because e_1 = \sqrt{1 + \dfrac{b^2}{a^2}}\right]$

$\therefore \quad e_2 = \dfrac{4}{5}$

Also, given ellipse is passing through $(\pm 5, 0)$ which is foci of the hyperbola $(\pm ae, 0)$.

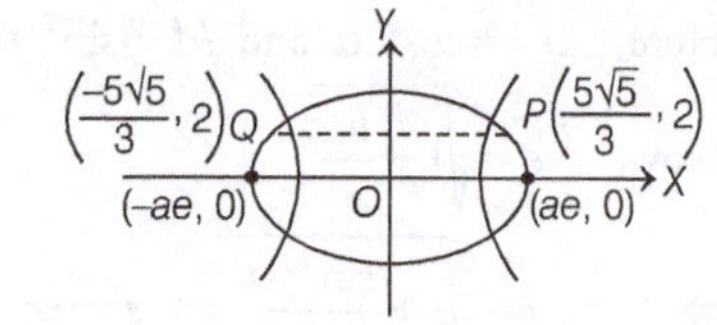


Also, $1 - \dfrac{b^2}{a^2} = \dfrac{16}{25} \Rightarrow b = 3$

So, equation of ellipse $\dfrac{x^2}{25} + \dfrac{y^2}{9} = 1$

Hence, coordinates of points of chords are $\left(\pm \dfrac{5\sqrt{5}}{3}, 2\right)$.

Thus, length of chord $(L_{PQ}) = \dfrac{10\sqrt{5}}{3}$

**24.** *(b)* We have, equation of the hyperbola is

$3x^2 - 4y^2 = 36$

$T : 3x_0 x - 4y_0 y = 36$

For $(x_0, y_0)$ to be nearest to line $3x + 2y = 1$, tangent $T$ must be parallel to $3x + 2y = 1$

$\therefore$ Slope of $T$ = Slope of line $(3x + 2y = 1)$

$\Rightarrow \quad \dfrac{-3x_0}{-4y_0} = -\dfrac{3}{2} \Rightarrow x_0 = -2y_0$

Now, $(x_0, y_0)$ lies on hyperbola

$\therefore \quad 3x_0^2 - 4y_0^2 = 36$

$\Rightarrow \quad 12y_0^2 - 4y_0^2 = 36 \quad [\because x_0 = -2y_0]$

$\Rightarrow \quad y_0 = \pm \dfrac{3}{\sqrt{2}}$

$\therefore \quad x_0 = -2y_0 = \mp 3\sqrt{2}$

So, points are $\left(3\sqrt{2}, -\dfrac{3}{\sqrt{2}}\right)$

and $\left(-3\sqrt{2}, \dfrac{3}{\sqrt{2}}\right)$

But, $\left(3\sqrt{2}, -\dfrac{3}{\sqrt{2}}\right)$ is nearest to the line $3x + 2y = 1$

$\therefore P(x_0, y_0) = \left(3\sqrt{2}, \dfrac{-3}{\sqrt{2}}\right)$

So, $\sqrt{2}(y_0 - x_0) = \sqrt{2}\left(\dfrac{-3}{\sqrt{2}} - 3\sqrt{2}\right)$

$= -3 - 6 = -9$

**25.** *(c)* Given, equation of hyperbola is $kx^2 - y^2 = 6$

$\dfrac{x^2}{\frac{6}{k}} - \dfrac{y^2}{6} = 1 \Rightarrow e = \sqrt{1 + \dfrac{b^2}{a^2}} = \sqrt{1 + k}$

Equation of directrix; $x = \pm \dfrac{a}{e}$

$\Rightarrow \quad x = \pm \sqrt{\dfrac{6}{k}} \times \dfrac{1}{\sqrt{1+k}}$

$\Rightarrow \quad \dfrac{\sqrt{6}}{\sqrt{k}\sqrt{1+k}} = 1 \Rightarrow 6 = k^2 + k$

$\Rightarrow \quad k(k+3) - 2(k+3) = 0$

$\Rightarrow \quad k = 2 \quad [\because k \neq -3]$

$\therefore$ The equation of hyperbola is

$\dfrac{x^2}{3} - \dfrac{y^2}{6} = 1$

From the options,

(a), $(-2\sqrt{5}, 6); \dfrac{20}{3} - \dfrac{36}{6} \neq 1$

(b), $(-\sqrt{5}, 3); \dfrac{5}{3} - \dfrac{9}{6} \neq 1$

(c), $(\sqrt{5}, -2); \dfrac{5}{3} - \dfrac{4}{6} = 1$

(d), $(2\sqrt{5}, 3\sqrt{6}) = \dfrac{20}{3} - 9 \neq 1$

$\therefore$ The hyperbola passes through the point $(\sqrt{5}, -2)$.

**26.** *(d)* The given equation of hyperbola

$a^2 x^2 - y^2 = b^2$

i.e. $\frac{x^2}{\left(\frac{b^2}{a^2}\right)} - \frac{y^2}{b^2} = 1$ ... (i)

Tangent of Eq. (i) is

$$y = mx \pm \sqrt{\frac{b^2m^2}{a^2} - b^2} \quad \text{... (ii)}$$

Given that $\lambda x - 2y = \mu$ be a tangent to the hyperbola.

So, $y = \frac{\lambda x}{2} - \frac{\mu}{2}$ ... (iii)

Now comparing Eqs. (ii) and (iii), we get

$$m = \frac{\lambda}{2}, \frac{b^2m^2}{a^2} - b^2 = \frac{\mu^2}{4}$$

Substituting $m = \frac{\lambda}{2}$ in $\frac{b^2m^2}{a^2} - b^2 = \frac{\mu^2}{4}$,

we get

$$\frac{b^2}{a^2} \times \frac{\lambda^2}{4} - b^2 = \frac{\mu^2}{4}$$

$\Rightarrow \frac{\lambda^2}{a^2} - \frac{\mu^2}{b^2} = 4$

$\Rightarrow \left(\frac{\lambda}{a}\right)^2 - \left(\frac{\mu}{b}\right)^2 = 4$

**27.** *(c)* $a^2 + b^2 = a^2e^2$

$a^2 + b^2 = a^2(5/4) \quad 4b^2 = a^2$

$\frac{x^2}{a^2} - \frac{y^2}{b^2} = 1$

∵ Point $P(-2\sqrt{6}, \sqrt{3})$ lies on given ellipse.

∴ $\frac{24}{4b^2} - \frac{3}{b^2} = 1 \Rightarrow b^2 = 3 \Rightarrow a^2 = 12$

Equation of tangent at $P$

$$\frac{x(-2\sqrt{6})}{12} - \frac{y(\sqrt{3})}{3} = 1$$

For conjugate axis, put $x = 0$

∴ $Q(0, -\sqrt{3})$

Equation of normal at $P$

$$\frac{y - \sqrt{3}}{x + 2\sqrt{6}} = \left(\frac{-1}{\sqrt{3}}\right) \times \left(\frac{12}{-2\sqrt{6}}\right)$$

$\Rightarrow y - \sqrt{3} = 2\sqrt{12} = 4\sqrt{3}$

$\Rightarrow y = 5\sqrt{3}$

∴ $R(0, 5\sqrt{3}) \Rightarrow QR = 6\sqrt{3}$

**28.** *(a)* Given, hyperbola is

$16x^2 - 9y^2 + 32x + 36y - 164 = 0$

$\Rightarrow 16(x+1)^2 - 9(y-2)^2$

$= 164 + 16 - 36 = 144$

$\Rightarrow \frac{(x+1)^2}{9} - \frac{(y-2)^2}{16} = 1$

Eccentricity, $e = \sqrt{1 + \frac{16}{9}} = \frac{5}{3}$

$\Rightarrow$ Foci are (4, 2) and (−6, 2).

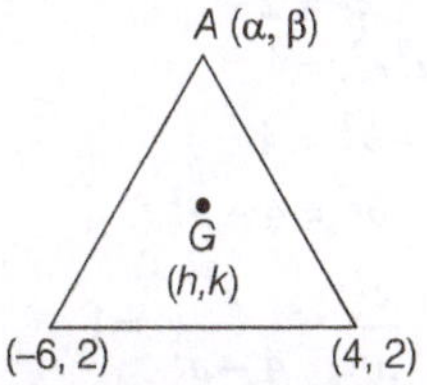


Let the centroid be $(h, k)$ and $A(\alpha, \beta)$ be point on hyperbol*(a)*

So, $h = \frac{\alpha - 6 + 4}{3}, k = \frac{\beta + 2 + 2}{3}$

$\Rightarrow \alpha = 3h + 2, \beta = 3k - 4$

$(\alpha, \beta)$ lies on hyperbola, so

$16(3h + 2 + 1)^2 - 9(3k - 4 - 2)^2 = 144$

$\Rightarrow 16(h^2 + 2h + 1) - 9$

$(k^2 - 4k + 4) = 16$

$\Rightarrow 16x^2 - 9y^2 + 32x + 36y - 36 = 0$

**29.** *(a)* Given, ellipse is $\frac{x^2}{25} + \frac{y^2}{16} = 1$

or $\frac{x^2}{(5)^2} + \frac{y^2}{(4)^2} = 1, \ a = 5, b = 4$

Now, focci of ellipse $= (\pm ae, 0)$ where

$e = \sqrt{a^2 - b^2}$

Put the values of $a$ and $b$, we get

$\Rightarrow ae = \sqrt{5^2 - 4^2} = \sqrt{25 - 16} = \sqrt{9}$

∴ Focus $= (\pm 3, 0)$

According to question, hyperbola passes through the focus of ellipse.

Let equation of hyperbola be

$$\frac{x^2}{a^2} - \frac{y^2}{b^2} = 1$$

Since, it passes through $(\pm 3, 0)$, we get

$\frac{(\pm 3)^2}{a^2} - \frac{0}{b^2} = 1, a^2 = 9$

Also, given that product of eccentricities is 1.

$\Rightarrow \left(\sqrt{1 - \frac{16}{25}}\right)\left(\sqrt{1 + \frac{b^2}{9}}\right) = 1$

(using formula of eccentricity of ellipse and hyperbola)

Squaring on both sides,

$$1 + \frac{b^2}{9} = \frac{25}{9}$$

$\Rightarrow b^2 = 16$

Thus, equation of hyperbola is

$$\frac{x^2}{9} - \frac{y^2}{16} = 1.$$

**30.** *(a)* Given equation of curves are

$y^2 = 16x$ (parabola) ...(i)

and $xy = -4$ (rectangular hyperbola) ...(ii)

Clearly, equation of tangent having slope '$m$' to parabola (i) is

$$y = mx + \frac{4}{m} \quad \text{...(iii)}$$

Now, eliminating $y$ from Eqs. (ii) and (iii), we get

$\Rightarrow mx^2 + \frac{4}{m}x + 4 = 0,$

which will give the points of intersection of tangent and rectangular hyperbola.

Since, line $y = mx + \frac{4}{m}$ is also a tangent to the rectangular hyperbola.

∴Discriminant of quadratic equation $mx^2 + \frac{4}{m}x + 4 = 0$, should be zero.

[∵ there will be only one point of intersection]

$\Rightarrow D = \left(\frac{4}{m}\right)^2 - 4(m)(4) = 0$

$\Rightarrow m^3 = 1$

$\Rightarrow m = 1$

So, equation of required tangent is

$y = x + 4.$

**31.** *(c)*

**key idea** The equation of the normals of slope $m$ to the hyperbola $\frac{x^2}{a^2} - \frac{y^2}{b^2} = 1$ are $y = mx \pm \frac{m(a^2 + b^2)}{\sqrt{a^2 - b^2m^2}}$

Given equation of hyperbola, is

$$\frac{x^2}{24} - \frac{y^2}{18} = 1 \quad \text{...(i)}$$

Since, the equation of the normals of slope m to the hyperbola $\frac{x^2}{a^2} - \frac{y^2}{b^2} = 1$, are given by

$$y = mx \pm \frac{m(a^2 + b^2)}{\sqrt{a^2 - b^2m^2}}$$

∴ Equation of normals of slope $m$, to the hyperbola (i), are

$$y = mx \pm \frac{m(24 + 18)}{\sqrt{24 - m^2(18)}} \quad \text{...(ii)}$$

∵ Line $y = mx + 7\sqrt{3}$ is normal to hyperbola (i).

$\therefore$ On comparing with Eq. (ii), we get

$$\pm\frac{m(42)}{\sqrt{24-18m^2}}=7\sqrt{3}$$

$\Rightarrow \quad 12m^2=24-18m^2$

$\Rightarrow \quad 5m^2=4 \Rightarrow m=\pm\frac{2}{\sqrt{5}}$

**32.** *(c)* Let the equation of standard hyperbola is

$$\frac{x^2}{a^2}-\frac{y^2}{b^2}=1 \quad \ldots\text{(i)}$$

Now, eccentricity of hyperbola is

$$\sqrt{1+\frac{b^2}{a^2}}=2 \quad \text{(given)}$$

$\Rightarrow \quad b^2=3a^2 \quad \ldots\text{(ii)}$

Since, hyperbola (i) passes through the point (4, 6)

$$\therefore \quad \frac{16}{a^2}-\frac{36}{b^2}=1 \quad \ldots\text{(iii)}$$

On solving Eqs. (ii) and (iii), we get

$$a^2=4 \text{ and } b^2=12 \quad \ldots\text{(iv)}$$

Now, equation of tangent to hyperbola (i) at point (4, 6), is

$$\frac{4x}{a^2}-\frac{6y}{b^2}=1$$

$\Rightarrow \quad \frac{4x}{4}-\frac{6y}{12}=1$

$\Rightarrow \quad 2x-y-2=0$

**33.** *(a)* Tangents are drawn to the hyperbola $4x^2-y^2=36$ at the point $P$ and $Q$.

Tangent intersects at point $T(0, 3)$

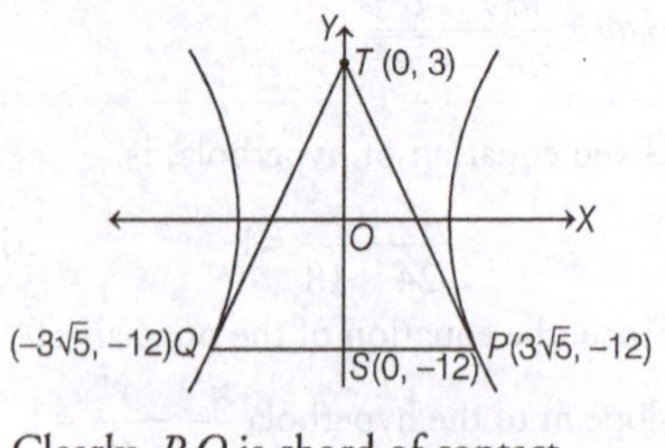


Clearly, $PQ$ is chord of contact.

$\therefore$ Equation of $PQ$ is $-3y=36$

$\Rightarrow \quad y=-12$

Solving the curve $4x^2-y^2=36$ and $y=-12$, we get $x=\pm 3\sqrt{5}$

Area of $\Delta PQT=\frac{1}{2}\times PQ\times ST$

$$=\frac{1}{2}(6\sqrt{5}\times 15)$$

$$=45\sqrt{5}$$

**34.** *(b)* Let the equation of hyperbola be

$$\frac{x^2}{a^2}-\frac{y^2}{b^2}=1.$$

$\therefore \quad ae=2$

$\Rightarrow \quad a^2e^2=4$

$\Rightarrow \quad a^2+b^2=4$

$\Rightarrow \quad b^2=4-a^2$

$$\therefore \quad \frac{x^2}{a^2}-\frac{y^2}{4-a^2}=1$$

Since, $(\sqrt{2}, \sqrt{3})$ lie on hyperbola.

$$\therefore \quad \frac{2}{a^2}-\frac{3}{4-a^2}=1$$

$\Rightarrow \quad a^4-9a^2+8=0$

$\Rightarrow \quad (a^4-8)(a^4-1)=0$

$\Rightarrow \quad a^4=8, a^4=1$

$\therefore \quad a=1$

Now, equation of hyperbola is

$$\frac{x^2}{1}-\frac{y^2}{3}=1.$$

$\therefore$ Equation of tangent at $(\sqrt{2}, \sqrt{3})$ is given by

$$\sqrt{2}x-\frac{\sqrt{3}y}{3}=1 \Rightarrow \sqrt{2}x-\frac{y}{\sqrt{3}}=1$$

which passes through the point $(2\sqrt{2}, 3\sqrt{3})$.

**35.** *(c)* Given equation of hyperbola can be rewritten as

$$\frac{x^2}{\left(\frac{12}{5}\right)^2}-\frac{y^2}{\left(\frac{9}{5}\right)^2}=1$$

$\therefore$ Eccentricity, $e'^2=1+\frac{b'^2}{a'^2}$

$\Rightarrow \quad e'^2=1+\frac{9}{16}=\frac{25}{16}$

$\therefore \quad e'=5/4$

The foci of a hyperbola are

$$(\pm a'e', 0)=\left(\pm\frac{12}{5}\times\frac{5}{4}, 0\right)$$

$$=(\pm 3, 0)$$

Given equation of ellipse is

$$\frac{x^2}{16}+\frac{y^2}{b^2}=1$$

Foci of an ellipse are $(\pm ae, 0)=(\pm 4e, 0)$ but given focus of ellipse and hyperbola coincide, then

$$4e=3$$

$\Rightarrow \quad e=\frac{3}{4}$

Also, $b^2=a^2(1-e^2)=16\left(1-\frac{9}{16}\right)$

$$=16-9=7$$

**36.** *(b)* Let $H:\frac{x^2}{a^2}-\frac{y^2}{b^2}=1$

$$b^2=a^2(e^2-1)$$

$\therefore$ Equation of circle $\Rightarrow x^2+y^2=a^2$

Area $=36\pi$

$\Rightarrow \quad \pi a^2=36\pi \Rightarrow a=6$

Now radius of $c_2$ can be $a(e-1)$ or $a(e+1)$.

For $r=a(e-1)$

Area $=4\pi$

$\Rightarrow \quad \pi a^2(e-1)^2=4\pi$

$\Rightarrow \quad (e-1)^2=\frac{4}{36}=\frac{1}{9}$

$\Rightarrow \quad e-1=\frac{1}{3} \Rightarrow e=\frac{4}{3}$

For $r=a(e+1)$

Area $=4\pi$

$\Rightarrow \quad \pi r^2=4\pi$

$\Rightarrow \quad a^2(e+1)^2=4$

$\Rightarrow \quad (e+1)^2=\frac{1}{9}$

$\Rightarrow \quad e+1=\frac{1}{3}$

$\Rightarrow \quad e=-\frac{2}{3}$ (rejected)

$\therefore \quad b^2=a^2(e^2-1)=36\left(\frac{16}{9}-1\right)$

$$=4\times 7=28$$

Length of latusrectum

$$=\frac{2b^2}{a}=\frac{2\times 28}{6}=\frac{28}{3}$$

**Alternate Solution**

Area of $C_1=\pi a^2=36\pi \Rightarrow a^2=36$

Area of $C_2=\pi(c-a)=4\pi$

$(c-6)^2=4, c=8$

$\therefore b^2=c^2-a^2=64-36=28$

Length of latus rectum

$$=\frac{2b^2}{a}=\frac{2\times 28}{6}=\frac{28}{3}$$

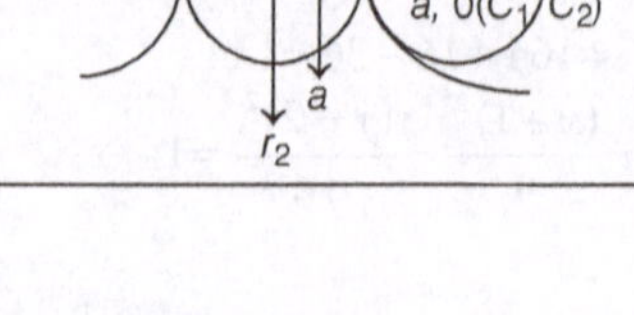

**37.** *(a)* Ellipse

$$\frac{x^2}{9}+\frac{y^2}{25}=1 \quad \ldots(i)$$

$a=3,\ b=5$

$$e=\sqrt{1-\frac{a^2}{b^2}}=\sqrt{1-\frac{9}{25}}=\frac{4}{5} \quad \ldots(ii)$$

For Eq. (ii), $(0, \pm be)=(0, \pm 4)$

$\therefore\ e_H=\frac{15}{8}\times e_{\text{ellipse}}=\frac{15}{8}\times\frac{4}{5}=\frac{3}{2}$

Let equation of hyperbola

$$\frac{x^2}{a^2}-\frac{y^2}{b^2}=-1$$

focii $(0, \pm 4)$, $be_H=4$

$$b\times\frac{3}{2}=4 \Rightarrow b=\frac{8}{3}$$

$\Rightarrow\ a^2=b^2(e_H^2-1)=\frac{64}{9}\left(\frac{9}{4}-1\right)$

$\Rightarrow\ a^2=\frac{80}{9}$

$$\therefore\ \frac{x^2}{\frac{80}{9}}-\frac{y^2}{\frac{64}{9}}=-1$$

Directrix, $y=\pm\frac{b}{e}=\pm\frac{16}{9}$

$PS=ePM$

$M=$ Distance between $P\left(\sqrt{2}, \frac{14}{3}\sqrt{\frac{2}{5}}\right)$ and directrix

$$=\frac{3}{2}\left|\frac{14}{3}\sqrt{\frac{2}{5}}-\frac{16}{9}\right|$$

$$=7\sqrt{\frac{2}{5}}-\frac{8}{3}$$

**38.** *(b)*

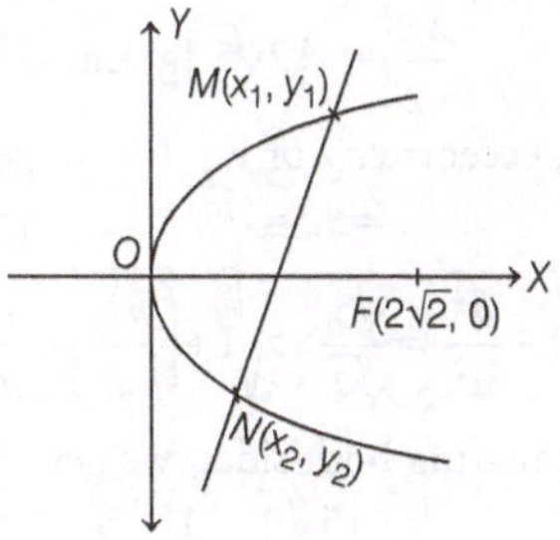


$H: x^2-y^2=4$

$$\Rightarrow\ \frac{x^2}{4}-\frac{y^2}{4}=1$$

Focus $(ae, 0)=F(2\sqrt{2}, 0)$

$y^2=4x$, $a=1$, focus $(a, 0)=(1, 0)$

Focal chord line $L: y=mx+c\,; m>0$

Line $L$ passes through $(1, 0)$

$$0=m+c \quad \ldots(i)$$

$\Rightarrow\ m=-c$

Also line $L$ is tangent to hyperbola

$$\frac{x^2}{4}-\frac{y^2}{4}=1$$

$c=\pm\sqrt{a^2m^2-b^2}=\pm\sqrt{4m^2-4}$

$\Rightarrow\ 4m^2-4=m^2 \quad [\because m=-c]$

$\Rightarrow\ m=2/\sqrt{3}\,(m>0)$

Equation of line $L: y=\frac{2}{\sqrt{3}}x-\frac{2}{\sqrt{3}}$

Now, this line $L$ cuts the parabola $y^2=4x$ at two points $M$ and $N$

$$\left(\frac{2x-2}{\sqrt{3}}\right)^2=4x$$

$\Rightarrow\ (x-1)^2=3x$

$\Rightarrow\ x^2-5x+1=0$

Now, $y^2=\frac{4(\sqrt{3}y+2)}{2}$

$\Rightarrow\ y^2-2\sqrt{3}y-4=0$

$\Rightarrow\ y_1+y_2=2\sqrt{3}$

$y_1y_2=-4$

Area of quadrilateral $OMFN$

$$=\frac{1}{2}\left|\begin{vmatrix}0 & x_1 & 2\sqrt{2} & x_2 & 0\\ 0 & y_1 & 0 & y_2 & 0\end{vmatrix}\right|$$

$$=\frac{1}{2}|0-2\sqrt{2}y_1+2\sqrt{2}y_2-0+0|$$

$$=\frac{2\sqrt{2}}{2}|y_2-y_1|$$

$$=\sqrt{2}\sqrt{(y_1+y_2)^2-4y_1y_2}$$

$$=\sqrt{2}\sqrt{12+16}=2\sqrt{14}$$

**39.** *(b)* Given, equation of hyperbola

$$\frac{x^2}{a^2}-\frac{y^2}{b^2}=1$$

Hyperbola passes through

$(2\sqrt{2}, -2\sqrt{2})$, $\frac{8}{a^2}-\frac{8}{b^2}=1$

$$8b^2-8a^2=a^2b^2 \quad \ldots(i)$$

Focus of parabola $=$ Focus of hyperbola with positive abscissa

$$A=ae. \quad \ldots(ii)$$

Equation of directrix of parabola is

$$x=-ae$$

Length of latusrectum of hyperbola is $e$ times the length of latusrectum of hyperbola.

$$4A=\frac{2b^2}{a}\times e \Rightarrow 4a=\frac{2b^2}{a}$$

$[\because$ From Eq. (ii)$]$

$\Rightarrow 2a^2=b^2$

Substitute it in Eq. (i), we get

$$8(2a^2)-8a^2=a^2(2a^2)$$

$\Rightarrow\ a^2=4=8 \quad [\because a\neq 0]$

$b^2=2a^2=8$

$\therefore\ e=\sqrt{1+\frac{8}{4}}=\sqrt{3}$

$\Rightarrow\ A=ae=2\sqrt{3}$

Equation of parabola is

$$y^2=4Ax=4\times 2\sqrt{3}x$$
$$=8\sqrt{3}x$$

From options, $=8\sqrt{3}\times 3\sqrt{3}$

$\Rightarrow\ 72=72$

Hence, $(3\sqrt{3}, -6\sqrt{2})$ lies on the parabola.

**40.** *(d)* The given equation of ellipse is

$$\frac{x^2}{16}+\frac{y^2}{7}=1$$

Thus, eccentricity is $\sqrt{1-\frac{7}{16}}=\frac{3}{4}$

So, foci is $(\pm ae, 0)=(\pm 3, 0)$

Now, equation of hyperbola is

$$\frac{x^2}{\left(\frac{144}{25}\right)}-\frac{y^2}{\left(\frac{\alpha}{25}\right)}=1$$

Now, eccentricity

$$(e)=\sqrt{1+\frac{\left(\frac{\alpha}{25}\right)}{\left(\frac{144}{25}\right)}}=\frac{\sqrt{\alpha+144}}{12}$$

Then, foci is $(\pm ae, 0)$

$$=\left(\pm\frac{12}{5}\cdot\frac{1}{12}\sqrt{144+\alpha}, 0\right)$$

$$=\left(\pm\frac{1}{5}\sqrt{144+\alpha}, 0\right)$$

Given that, the foci of the ellipse and hyperbola are coincide, then

$$3=\frac{1}{5}\sqrt{144+\alpha}$$

$\Rightarrow\ \alpha=81$

Thus, the hyperbola is

$$\frac{x^2}{\left(\frac{12}{5}\right)^2}-\frac{y^2}{\left(\frac{9}{5}\right)^2}=1$$

Hence, the length of latusrectum

$$=2\left(\frac{81/25}{12/5}\right)$$

$$=2\times\frac{81}{25}\times\frac{5}{12}=\frac{27}{10}$$

**41.** *(d)*

> **key idea** Equation of tangent having slope '$m$' to hyperbola $\frac{x^2}{a^2}-\frac{y^2}{b^2}=1$ is
> $y=mx\pm\sqrt{a^2m-b^2}$

Let the mid-point of the chord of the circle be $(h, k)$.
The equation of a chord through its mid-point is simply $T=S_1$

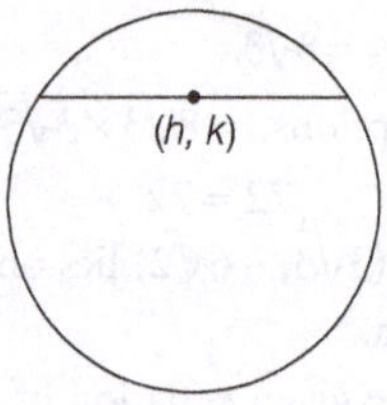


Here, $hx+ky=h^2+k^2$

The equation of tangent to hyperbola $\frac{x^2}{a^2}-\frac{y^2}{b^2}=1$ is

$$y=mx\pm\sqrt{a^2m-b^2}$$

Here, $y=mx\pm\sqrt{9m^2-16}$

So, $y=\frac{-hx}{k}+\frac{h^2+k^2}{k}$ ... (i)

and $y=mx\pm\sqrt{9\,m^2-16}$ ... (ii)

Eqs. (i) and (ii) are identical, so

$$\frac{1}{1}=\frac{-h/k}{m}=\frac{h^2+k^2}{k\sqrt{9m^2-16}}\Rightarrow m=\frac{-h}{k}$$

$$\Rightarrow\ 9m^2-16=\frac{(h^2+k^2)^2}{k^2}$$

$$\Rightarrow\ 9\left(\frac{h^2}{k^2}\right)-16=\frac{(h^2+k^2)^2}{k^2}$$

$$\Rightarrow\ 9h^2-16k^2=(h^2+k^2)^2\ \begin{cases}h\to x\\ k\to y\end{cases}$$

So, $(x^2+y^2)^2-9x^2+16y^2=0$

**42.** *(120)* Given, the product of focal distance is 32

$$\Rightarrow\quad a^2+e^2x^2-x^2=32$$

$$\Rightarrow\quad a^2+\frac{16b^2}{a^2}=32 \quad ...(ty)$$

Put coordinate of point $P$ is ellipse.

$$\Rightarrow\quad \frac{16}{a^2}-\frac{12}{b^2}=1 \quad ...(ii)$$

From eqs. (i) and (ii), we get $a^2=40$,

$$b^2=20$$

$$\Rightarrow\quad p=2b=4\sqrt{5},\ q=\frac{2b^2}{a}=2\sqrt{10}$$

$p^2+q^2=120$

**43.** *(141)* Centre of the hyperbola is the mid-point of the foci $\equiv(6, 2)$
The distance between the centre and focus $=c=-\sqrt{(8-c)^2+(2-2)^2}=2$
Since, the foci have the same y-coordinate, the transverse axis is horizontal
$\Rightarrow$ Equation of hyperbola is

$$\frac{(x-6)^2}{a^2}-\frac{(y-2)^2}{b^2}=1 \quad ...(i)$$

But given equation of hyperbola is
$3x^2-y^2-\alpha x+\beta y+r=0$
Compare with Eqs. (i), we get

$$\Rightarrow\quad b^2=3$$

$$\Rightarrow\quad a^2=c^2-b^2=4-3=1$$

putting this value in Eqs., we get
$3(x-6)^2-(y-2)^2=3$
$3x^2-36x+108-y^2+4y-4-3=0$

$$3x^2-y^2-36x+4y+101=0$$

$$\alpha=36,\ \beta=4 \text{ and } Y=101$$

$\therefore\quad \alpha+\beta+\gamma=36+4+101=141$

**44.** *(1994)*

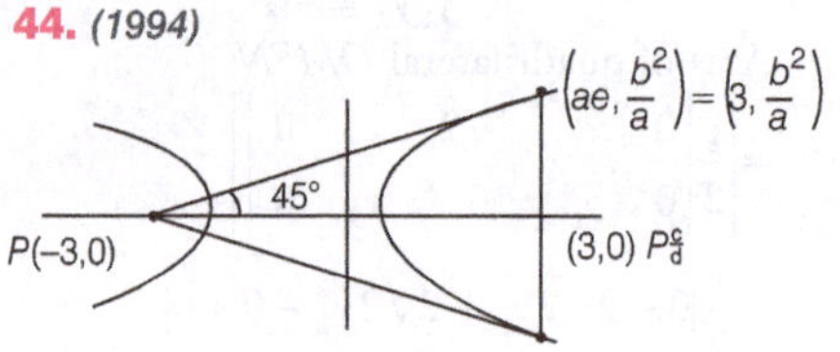


$$AP^1=PP^1$$

$$\frac{b^2}{a}=6 \quad ...(i)$$

$$ae=3$$

$$\Rightarrow\quad a\cdot\sqrt{1+\frac{b^2}{a^2}}=3$$

$$\Rightarrow\quad a^2+b^2=9 \quad ...(ii)$$

$$\Rightarrow\quad a^2+6a-9=0 \quad (\because b^2=6a)$$

$$\Rightarrow\quad a=3(\sqrt{2}-1)$$

$$\Rightarrow\quad a^2b^2=9(3-2\sqrt{2})6\cdot 3(\sqrt{2}-1)$$

$$=162\,(5\sqrt{2}-7)$$

$$\alpha=162\times 5=810$$

$$\beta=162\times 7=1134$$

$$\Rightarrow\ \alpha+\beta=1944$$

**45.** *(189)* The distance from the centre to the focus is $ae=5$
Now, the directrix is $5x+9=0$

$$\Rightarrow\quad x=-\frac{9}{5}$$

Distance from the centre of the directrix is $\frac{a}{3}\cdot$

$$\frac{a}{e}=\frac{9}{5}\Rightarrow\frac{a}{5/a}=\frac{9}{5}\Rightarrow a^2=9\Rightarrow a=3$$

Then $e=c/a=5/3$
Use formula

$$b^2=a^2(e^2-1)=9\left(\frac{25}{9}-1\right)=16$$

$$b=4$$

For a hyperbola $\frac{x^2}{a^2}-\frac{y^2}{b^2}=1$, the focal distance from $(x, y)$ are $a+ex$ and $a-ex$.
The product $=(a+ex)(a-ex)$
$=a^2-e^2x^2$
Use $p=a^2-e^2x^2$

$$p=3^2-\left(\frac{5}{3}\right)^2\alpha^2\Rightarrow p=9-\frac{25}{9}\alpha^2$$

Also, $(\alpha, 2\sqrt{5})$ on the hyperbola is

$$\frac{\alpha^2}{9}-\frac{20}{16}=1$$

$$\alpha^2=9\left(1+\frac{20}{16}\right)=9\times\frac{36}{16}=\frac{81}{4}$$

$$p=e^2\alpha^2-a^2=\frac{25}{9}\times\frac{81}{4}-9=\frac{189}{4}$$

Calculate $4p=4\times\frac{189}{4}=189$

**46.** *(55)* Given, $H_1:\frac{x^2}{a^2}-\frac{y^2}{b^2}=1$ ...(i)

and $\quad H_2:\frac{-x^2}{A^2}+\frac{y^2}{B^2}=1$ ...(ii)

Now, length of latus-rectum of Eq. (i), we get

$$\frac{2b^2}{a}=15\sqrt{2}\ \text{[given]} \quad ...(iii)$$

and, length of latus-rectum of Eq. (ii)

$$\frac{2A^2}{B}=12\sqrt{5}\ \text{[given]} \quad ...(iv)$$

Also, eccentricity of Eq. (i), we get

$$=e_1=\sqrt{5/2} \quad \text{[given]}$$

$$\Rightarrow\sqrt{1+\frac{b^2}{a^2}}=\frac{\sqrt{5}}{\sqrt{2}}\Rightarrow\sqrt{1+\left(\frac{b^2}{a}\right)\times\frac{1}{a}}=\frac{\sqrt{5}}{\sqrt{2}}$$

On squaring both sides, we get

$$\Rightarrow\quad 1+\frac{15\sqrt{2}}{2}\times\frac{1}{a}=\frac{5}{2}$$

$$\Rightarrow\quad \frac{1}{a}=\left(\frac{5}{2}-1\right)\times\frac{2}{15\sqrt{2}}=\frac{1}{5\sqrt{2}}$$

$$\therefore\quad a=5\sqrt{2}$$

and hence $b=5\sqrt{3}$ Also, given product of the lengths of transverse axis of Eqs. (i) and (ii) $=100\sqrt{10}$

i.e., $\quad(2a)(2B)=100\sqrt{10}$

⇒ $2\times 5\sqrt{2}\times 2\times B = 100\sqrt{10}$

⇒ $B=5\sqrt{5}$

∴ $\frac{2A^2}{5\sqrt{5}}=12\sqrt{5}$

⇒ $A^2 = 6\sqrt{5}\times 5\sqrt{5}=150$

⇒ $A=5\sqrt{6}$ [∵ $A>0$]

Now, $e_2^2 = 1+\frac{A^2}{B^2}=1+\frac{150}{125}$

$=1+\frac{30}{25}=\frac{55}{25}$

⇒ $25e_2^2=55$

**47.** *(19)* Given that

$x-y+1=0$ ...(i)

$CM=\left|\frac{-3\alpha-1}{\sqrt{13}}\right|$

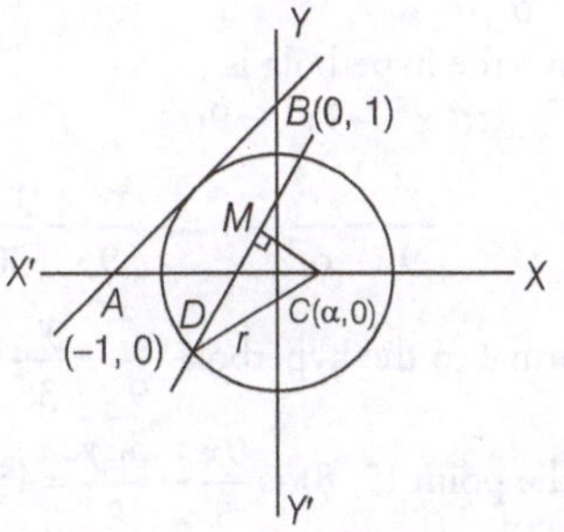


$r=\left|\frac{\alpha+1}{\sqrt{2}}\right|$

⇒ $(\alpha+1)^2=2r^2$ ...(i)

Now, in $\Delta CMD$

$CD^2=DM^2+CM^2$

$r^2=\frac{(3\alpha+1)^2}{13}+\frac{4}{13}$

$(3\alpha+1)^2+4=13r^2$ ...(ii)

From Eqs. (i) and (ii), we get

$(3\alpha+1)^2+4=13\frac{(\alpha+1)^2}{2}$

⇒ $18\alpha^2+12\alpha+10=13\alpha^2+26\alpha+13$

⇒ $5\alpha^2-14\alpha-3=0$

⇒ $5\alpha^2-15\alpha+\alpha-3=0$

$\alpha=\frac{-1}{5}, 3$ [∵ $\alpha>0$ ⇒ $\alpha=3$]

From Eq. (i), we get $r=2\sqrt{2}$

Now, $\alpha e=3$, $2\alpha=4\sqrt{2}$

⇒ $\alpha=2\sqrt{2}$

$\alpha^2e^2=9$

⇒ $\alpha^2\left(1+\frac{\beta^2}{\alpha^2}\right)=9$

⇒ $\alpha^2+\beta^2=9 ⇒ \beta^2=1$

∴ $2\alpha^2+3\beta^2=2(8)+3=19$

**48.** *(40)* $\frac{x^2}{3}-\frac{y^2}{5}=1$

Focus of hyperbola $=(ae, 0)$

Here, $a=\sqrt{3}$ and $b^2=a^2(e^2-1)$

⇒ $5=3(e^2-1)$

⇒ $e^2=\frac{5}{3}+1=\frac{8}{3}$

⇒ $e=\sqrt{8/3}$

So, $(ae, 0)=(\sqrt{8}, 0)=S$

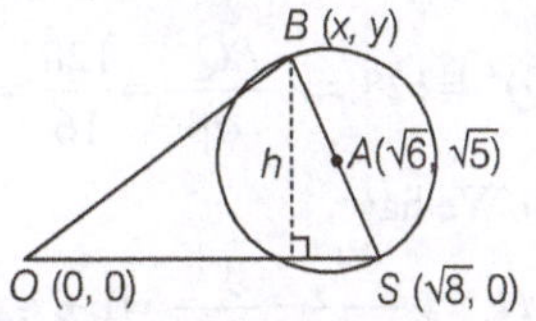


$\frac{x+\sqrt{8}}{2}=\sqrt{6} ⇒ x=2\sqrt{6}-\sqrt{8}$

and $\frac{y}{2}=\sqrt{5} ⇒ y=2\sqrt{5}$

$B(x, y)=(2\sqrt{6}-\sqrt{8}, 2\sqrt{5})$

$h=2\sqrt{5} ⇒ OS=\sqrt{8}$

Therefore, area $(\Delta OSB)=\frac{1}{2}(OS)\times h$

$=\frac{1}{2}\times\sqrt{8}\times 2\sqrt{5}=\sqrt{40}$

$A^2=40$

**49.** *(61)* Given, $\frac{2b^2}{a}=9$ and $\frac{a}{e}=\pm\frac{4}{\sqrt{3}}$

Equation of tangent

$y-\sqrt{3}x+\sqrt{3}=0$

By equation of tangent

Let slope ⇒ $S=\sqrt{3}$

Constant $=-\sqrt{3}$

By condition of tangency

⇒ $6=6a^2-9a ⇒ a=2, b^2=9$

Equation of hyperbola is

$\frac{x^2}{4}-\frac{y^2}{9}=1$ and for tangent

Point of contact is $(4, 3\sqrt{3})=(x_0, y_0)$

Now, $e=\sqrt{1+\frac{9}{4}}=\frac{\sqrt{13}}{2}$

Again, product of focal distances

$m=(x_0e+a)(x_0e-a)$

$m+4e^2=20e^2-a^2$

$=20\times\frac{13}{4}-4=61$

**50.** *(51)* $\frac{x^2}{a^2}+\frac{y^2}{b^2}=1$

Foci $=(\pm 5, 0)=(\pm ae, 0)$

Length of Latus rectum $=\sqrt{50}=\frac{2b^2}{a}$

⇒ $ae=5$, $\frac{2b^2}{a}=5\sqrt{2}$

⇒ $a=\frac{5}{e}$, $b^2=\frac{5\sqrt{2}a}{2}$

⇒ $b^2=a^2(1-e^2)$

⇒ $\frac{5\sqrt{2}a}{2}=a^2(1-e^2)$

⇒ $\frac{5\sqrt{2}}{2}=a(1-e^2)$

⇒ $\frac{5\sqrt{2}}{2}=\frac{5}{e}(1-e^2)$

⇒ $\frac{e}{\sqrt{2}}=1-e^2$

⇒ $\sqrt{2}e^2+e-\sqrt{2}=0$

⇒ $(e+\sqrt{2})(\sqrt{2}e-1)=0$

⇒ $e=\frac{1}{\sqrt{2}} ⇒ a=5\sqrt{2}$

⇒ $b=5$

Now, hyperbola

⇒ $\frac{x^2}{b^2}-\frac{y^2}{a^2b^2}=1$

⇒ $a^2b^2=b^2(e^2-1)$

⇒ $a^2=e^2-1$

⇒ $e^2=1+a^2$

$=1+50$

$e^2=51$

**51.** *(182)* We have, $\frac{x^2}{9}-\frac{y^2}{b^2}=1$

From figure in $\Delta AOB$,

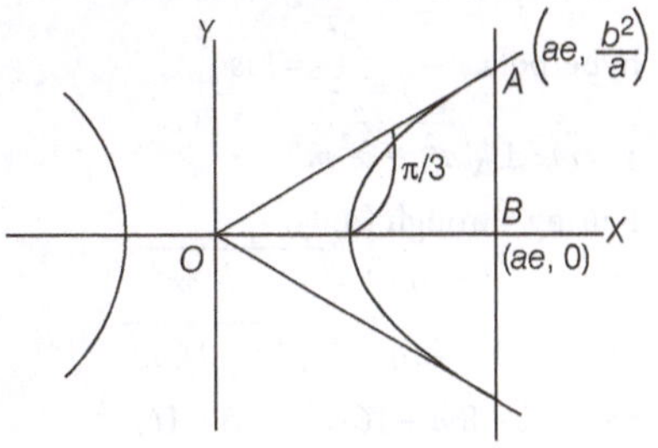


$\tan\pi/6=\frac{b^2/a}{ae} ⇒ 1/\sqrt{3}=\frac{b^2}{a^2e}$

⇒ $e=\frac{\sqrt{3}b^2}{a^2}$ or $e^2=1+\frac{b^2}{a}$

⇒ $1+\frac{b^2}{9}=\frac{3b^4}{81} ⇒ b^4=3b^2+27$

$b^4 - 3b^2 - 27 \Rightarrow b^2 = \frac{2}{3}(1+\sqrt{13})$

$\therefore \quad l = 3, m = 2$ and $n = 13$

$\therefore \quad l^2 + m^2 + n^2 = 182$

**52.** *(12)* Let the equation of hyperbola $\frac{x^2}{a^2} - \frac{y^2}{b^2} = 1$, where $ae = 2$ and $e = \frac{3}{2}$ [Given]

$\Rightarrow \quad a = \frac{4}{3}$ also $b^2 = a^2e^2 - a^2$

$\Rightarrow b^2 = \frac{16}{9} \times \frac{9}{4} - \frac{16}{9} \Rightarrow b^2 = 4 - \frac{16}{9}$

$\Rightarrow b^2 = \frac{20}{9}$

Given, tangent perpendicular to the line $2x + 3y = 6$.

So, slope of tangent $= 3/2$

The equation of tangent of hyperbola is

$y = mx \pm \sqrt{a^2m^2 - b^2}$

$= \frac{3x}{2} \pm \sqrt{\frac{16}{9} \times \frac{9}{4} - \frac{20}{9}}$

$= \frac{3x}{2} \pm \sqrt{4 - \frac{20}{9}} = \frac{3x}{2} \pm \frac{4}{3}$

$\Rightarrow \quad y = \frac{3x}{2} \pm \frac{4}{3}$

Intercept on $X$-axis $(y = 0)$

$\Rightarrow \quad a = \frac{8}{9}$

Intercept on $Y$-axis (b)

$\therefore \quad b = \frac{4}{3}$

So, $\quad |6a| + |5b| = \left|6 \times \frac{8}{9}\right| + \left|5 \times \frac{4}{3}\right|$

$= \frac{16}{3} + \frac{20}{3} = \frac{36}{3} = 12$

**53.** (8) Equation of tangent to the hyperbola, $\frac{y^2}{a^2} - \frac{x^2}{b^2} = 1$ is

$y = mx \pm \sqrt{a^2 - b^2m^2}$

Pairing through (4, 1)

$\Rightarrow \quad 1 = 4m \pm \sqrt{a^2 - b^2m^2}$

$\Rightarrow \quad (1 - 4m) = \pm\sqrt{25 - 16m^2}$

$\Rightarrow \quad 1 - 8m + 16m^2 = 25 - 16m^2$

$\Rightarrow \quad 32m^2 - 8m - 24 = 0$

$\Rightarrow \quad 4m^2 - m - 3 = 0 \Rightarrow m = 1, \frac{-3}{4}$

Given that tangent drawn to given hyperbola to have positive slopes.

So, equation of tangent with positive slopes 1 and 3/4 are

$(y - 1) = 1 \cdot (x - 4) = y = x - 3 \quad \ldots(i)$

and $(y + 1) = \frac{3}{4}(x - 4) = 4y = 3x - 16 \quad \ldots(ii)$

With intercept on $X$-axis.

Thus, $\alpha = \frac{16}{3}$ and $\beta = 3$

On solving Eqs. (i) and (ii), we get

$x = -4, y = -7$

Thus, $p = (4, 1)$ and $Q = (-4, -7)$

$\therefore \quad PQ = \sqrt{(4+4)^2 + (1+7)^2}$

$= \sqrt{64 + 64} = \sqrt{128}$

$(PQ)^2 = 128 \Rightarrow \frac{PQ^2}{\alpha\beta} = \frac{128}{16} = 8$

**54.** *(306)* We have,

$H_n \Rightarrow \frac{x^2}{1+x} - \frac{y^2}{3+x} = 1, n \in N$

Here, $a^2 = 1 + x$ and $b^2 = 3 + x$

Eccentiricity $(e) = \sqrt{1 + \frac{b^2}{a^2}}$

$= \sqrt{1 + \left(\frac{3+x}{1+x}\right)} = \sqrt{\frac{2x+4}{x+1}} = \sqrt{\frac{2(x+2)}{x+1}}$

The smallest even value for which $c \in \theta$ is 48

$\therefore \quad K = 48$

$e = \sqrt{\frac{2(48+2)}{48+1}} = \frac{10}{7}$

$\Rightarrow \quad a^2 = n + 1 = 49, b^2 = n + 3 = 51$

$\therefore 21l = 21 \times \left(\frac{2b^2}{a}\right) = 21 \times 2 \times \frac{51}{7} = 306$

**55.** *(2)* Equation of hyperbola is

$2x^2 - 2y^2 = 1$

$\Rightarrow \quad \frac{x^2}{1/2} - \frac{y^2}{1/2} = 1$

Here, $\quad a = b$

$\therefore$Eccentricity of rectangular hyperbola is $\sqrt{2}$.

$\therefore$Eccentricity of ellipse is $1/\sqrt{2}$.

Since, ellipse intersects the hyperbola at right angles.

$\therefore$Ellipse and the hyperbola are co-focal

$\therefore$Foci of hyperbola = (1, 0)

$\Rightarrow$ Foci of ellipse, $ae = 1$

$\Rightarrow a\left(\frac{1}{\sqrt{2}}\right) = 1 \Rightarrow a = \sqrt{2}$

$\Rightarrow \quad e^2 = 1 - \frac{b^2}{a^2} \Rightarrow \frac{1}{2} = 1 - \frac{b^2}{2}$

$\Rightarrow \quad \frac{b^2}{2} = \frac{1}{2}$

$\Rightarrow \quad b^2 = 1$

$\therefore$Length of the latus rectum $= \frac{2b^2}{a}$

$= \frac{2}{\sqrt{2}} = \sqrt{2}$

$\therefore$Square of length of the latus rectum $= 2$

**56.** *(116)* Given, equation of parabola $y^2 = 12x$

Since point $(3, \alpha)$ lies on $y^2 = 12x$

$\Rightarrow \quad \alpha^2 = 12 \times 3 = 36 \Rightarrow \alpha = \pm 6$

Since, tangent to the parabola $y^2 = 12x$ at the point $(3, \alpha)$ be $\perp^r$ to the line $2x + 2y = 3$

So, $\quad m = -1$

$\Rightarrow \quad \frac{6}{\alpha} = 17\alpha = 6 \quad (\alpha = -6$ rejected)

Now, the hyperbola is

$\alpha^2x^2 - 9y^2 = 9\alpha^2$

$\Rightarrow \quad \frac{x^2}{9} - \frac{y^2}{\alpha^2} = 1 \Rightarrow \frac{x^2}{9} - \frac{y^2}{30} = 1$

Normal to the hyperbola $\frac{x^2}{9} - \frac{y^2}{36} = 1$ at the point (5, 8) is $\frac{9x}{5} + \frac{36y}{8} = 45$

$\Rightarrow 72x + 180y = 1800 \Rightarrow 2x + 5y = 50$

Now, distance of $(6, -4)$ from the normal to the given hyperbola is equal to $\left|\frac{2(6) + 5(-4) + 50}{\sqrt{2^2 + 5^2}}\right|$

$= \left|\frac{12 - 20 - 50}{\sqrt{29}}\right| = \left|\frac{-56}{\sqrt{29}}\right| = \frac{58}{\sqrt{29}}$

Distance $= 2\sqrt{29}$

$\therefore(\text{Distance})^2 = (2\sqrt{29})^2 = 4 \times 29 = 116$

**57.** *(216)* Vertices of hyperbola $H$ are $(\pm 6, 0) \equiv (\pm a, 0)$

So, $a = 6$

Eccentricity, $e = \frac{\sqrt{5}}{2}$

$\sqrt{1 + \frac{b^2}{a^2}} = \frac{\sqrt{5}}{2} \Rightarrow b^2 = 9, b = 3$

So, equation of hyperbola,

$H: \frac{x^2}{36} - \frac{y^2}{9} = 1$

Now, equation of normal for hyperbola $H$

$\frac{ax}{\sec\theta} + \frac{by}{\tan\theta} = a^2 + b^2$

$6\cos\theta\, x + 3\cot\theta\, y = 45$

This normal is parallel to
$\sqrt{2}x + y = 2\sqrt{2}$

$$\frac{-\sqrt{2}}{1} = \frac{-6\cos\theta}{3\cot\theta}$$

$$\sin\theta = \frac{1}{\sqrt{2}} \Rightarrow \theta = \frac{\pi}{4}$$

So, equation of normal is :

$$6\cos\frac{\pi}{4}x + 3\cot\frac{\pi}{4}y = 45$$

$$6\cdot\frac{x}{\sqrt{2}} + 3\cdot1\cdot y = 45 \Rightarrow \sqrt{2}x + y = 15$$

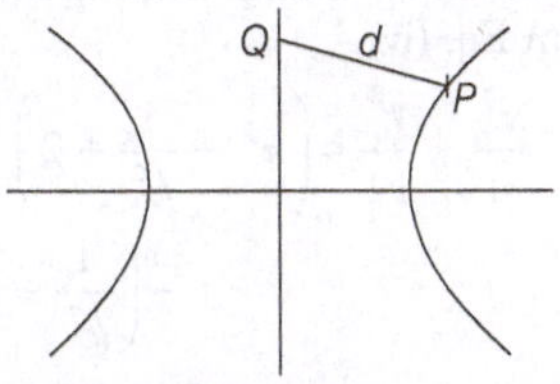


Let $P = (a\sec\theta, b\tan\theta)$

$$\equiv\left(6\sec\frac{\pi}{4}, 3\tan\frac{\pi}{4}\right), P(6\sqrt{2}, 3)$$

$Q$ lies on $Y$-axis, put $x = 0$ in equation of normal

$$0 + y = 15 \Rightarrow y = 15$$

$Q(0, 15)$

$$d^2 = (6\sqrt{2} - 0)^2 + (3 - 15)^2$$
$$= 72 + 144 = 216$$

**58.** *(20)* Given, $C_1 : \frac{x^2}{4} + \frac{y^2}{9} = 1$

and $C_2 : \frac{x^2}{42} - \frac{y^2}{143} = 1$

Tangent to $C_1$ is $y = mx \pm \sqrt{4m^2 + 9}$

Tangent to $C_2$ is

$$y = mx \pm \sqrt{42m^2 - 143}$$

For common tangent,
$4m^2 + 9 = 42m^2 - 143 \Rightarrow m = \pm 2$

If tangent does not pass through 4th quadrant, then $m > 0$ and $y$-intercept is positive.

∴Tangent is $y = 2x + 5$

This tangent touches $C_1$ at $(x_1\ y_1)$.

∴Tangent is $\frac{x_1x}{4} + \frac{yy_1}{9} = 1$

Comparing, we get $\frac{x_1/4}{2} = \frac{1}{-5}$

$\Rightarrow x_1 = -8/5$

Also point of contact with $C_2$ is $(x_2, y_2)$.

Tangent is $\frac{xx_2}{42} - \frac{yy_2}{143} = 1$

$$\therefore \frac{x_2/42}{2} = \frac{1}{-5}$$

$$\Rightarrow x_2 = \frac{-84}{5}$$

$$\therefore |2x_1 + x_2| = \frac{100}{5} = 20$$

**59.** *(88)*

> **key idea** Sum of length of transverse axis and conjugate axis of hyperbola = 2a + 2b

$$H: \frac{x^2}{a^2} - \frac{y^2}{b^2} = 1$$

$$e = \frac{\sqrt{11}}{2} \Rightarrow e^2 = \frac{11}{4} \Rightarrow 1 + \frac{b^2}{a^2} = \frac{11}{4}$$

$$\Rightarrow b^2 = \frac{7}{4}a^2 \Rightarrow b^2 = \left(\frac{\sqrt{7}}{2}a\right)^2$$

$$\therefore \frac{x^2}{a^2} - \frac{y^2}{\left(\frac{\sqrt{7}}{2}a\right)^2} = 1$$

Length of transverse axis $= 2a$
Length of conjugate axis $= 2b$

$$= 2\left(\frac{\sqrt{7}}{2}a\right) = \sqrt{7}a$$

Sum of length of transverse axis and conjugate axis.

$$2a + \sqrt{7}a = 4(2\sqrt{2} + \sqrt{14})$$

$$\Rightarrow a(2 + \sqrt{7}) = 4\sqrt{2}(2 + \sqrt{7})$$

$$\Rightarrow a = 4\sqrt{2}$$

$$\Rightarrow b^2 = \left(\frac{\sqrt{7}}{2}\times 4\sqrt{2}\right)^2 = 4\times14 = 56$$

$$\therefore a^2 + b^2 = 32 + 56 = 88$$

**60.** *(12)* Equation of hyperbola

$$\frac{x^2}{16} - \frac{y^2}{4} = 1$$

$$L_1 : y = mx \pm \sqrt{16m^2 - 4} \quad \ldots(i)$$

$$L_2 : y = \frac{-1}{m}x$$

$$\Rightarrow m = \frac{-x}{y} \quad \ldots(ii)$$

Using Eq. (ii) in Eq. (i)

$$y = -\left(\frac{x}{y}\right)x \pm \sqrt{16\left(\frac{-x}{y}\right)^2 - 4}$$

$$\Rightarrow y = \frac{-x^2}{y} \pm \frac{\sqrt{16x^2 - 4y^2}}{y}$$

$$\Rightarrow (y^2 + x^2)^2 = 16x^2 - 4y^2 \quad \ldots(iii)$$

Given, equation

$$(x^2 + y^2)^2 = \alpha x^2 + \beta y^2 \quad \ldots(iv)$$

on comparing Eqs. (iii) and (iv), we get

$$\alpha = 16, \beta = -4$$

$$\therefore \alpha + \beta = 16 - 4 = 12$$

**61.** *(85)* $\because e = 5/4$

$$\Rightarrow 1 + \frac{b^2}{a^2} = \left(\frac{5}{4}\right)^2 \Rightarrow 9a^2 = 16b^2$$

The point $\left(\frac{8}{\sqrt{5}}, \frac{12}{5}\right)$ satisfy the hyperbola.

$$\frac{64}{5a^2} - \frac{144}{25b^2} = 1$$

$$\Rightarrow \frac{64\times9}{5\times16b^2} - \frac{144}{25b^2} = 1$$

$$\Rightarrow b = \frac{6}{5} \text{ and } a = \frac{8}{5}$$

Equation of normal at the point $\left(\frac{8}{\sqrt{5}}, \frac{12}{5}\right)$.

$$= \frac{\sqrt{5}a^2x}{8} + \frac{5b^2y}{12} = a^2 + b^2$$

$$\Rightarrow \sqrt{5}\times\frac{64}{25\times8}x + \frac{5}{12}\times\frac{36}{25}y = 4$$

$$\Rightarrow 8\sqrt{5}x + 15y = 100$$

Compare it with given equation
$8\sqrt{5}x + \beta y = \lambda \Rightarrow \beta = 15$ and $\lambda = 100$

Hence, $\lambda - \beta = 100 - 15 = 85$

**62.** *(42)* The given equation of hyperbola

$$H: \frac{x^2}{a^2} - y^2 = 1$$

and ellipse $E: 3x^2 + 4y^2 = 12$

$$\Rightarrow \frac{x^2}{4} + \frac{y^2}{3} = 1$$

Latusrectum of hyperbola

$$(l) = \frac{2\times(1)^2}{a} = \frac{2}{a}$$

and latusrectum of ellipse

$$(l) = \frac{2\times\sqrt{(3)^2}}{2} = \frac{2\times3}{2} = 3$$

Since, given that, length of latusrectum of hyperbola = length of latusrectum of parabola

$$\Rightarrow \frac{2}{a} = 3 \Rightarrow a = \frac{2}{3}$$

$e_H$ is eccentricity of

$$H = \sqrt{1 + \frac{b^2}{a^2}} = \sqrt{1 + \frac{9}{4}} = \sqrt{\frac{13}{4}}$$

and $e_E$ is eccentricity of

$$E=\sqrt{1-\frac{b^2}{a^2}}=\sqrt{1-\frac{3}{4}}=\sqrt{\frac{1}{4}}$$

Thus $12(e_E^2+e_H^2)=12\left(\frac{1}{4}+\frac{13}{4}\right)$

$$=12\left(\frac{14}{4}\right)$$

$$=\frac{168}{4}=42$$

**63.** *(9)* Let $A(\sec\theta, 2\tan\theta)$ and $B(\sec\phi, 2\tan\phi)$

Equation of normal of $\sec\theta, 2\tan\theta$ of hyperble $\frac{x^2}{1}-\frac{y^2}{2}=1$ is

$$x\cos\theta+2y\cot\theta=3 \quad \ldots\text{(i)}$$

and $AB$ $(\sec\phi, 2\tan\phi)$ is

$$x\cos\phi+2y\cot\phi=3 \quad \ldots\text{(ii)}$$

Solving Eqs. (i) and (ii), we get

Multiply $\cos\phi$ by Eq. (i) and $\cos\theta$ by Eq. (ii) and subtracting we get

$$2y(\cot\theta\cos\phi-\cot\phi\cos\theta)=3(\cos\phi-\cos\theta)$$

$$\Rightarrow 2y\left(\frac{\cos\phi\cos\theta}{\sin\theta}-\frac{\cos\phi\cos\theta}{\sin\phi}\right)=3(\cos\phi-\cos\theta)$$

$$\Rightarrow 2y\cos\phi\cos\theta\left(\frac{\sin\phi-\sin\theta}{\sin\theta\sin\phi}\right)=3(\cos\phi-\cos\theta)$$

$$=\frac{3(\cos\phi-\cos\theta)\tan\theta\tan\phi}{\sin\phi-\sin\theta}$$

$$=\frac{3\left(2\sin\left(\frac{\phi+\theta}{2}\right)\sin\left(\frac{\theta-\phi}{2}\right)\right)\tan\theta\tan\phi}{2\sin\left(\frac{\phi-\theta}{2}\right)\cos\left(\frac{\theta+\phi}{2}\right)}$$

$$=-3\tan\left(\frac{\theta+\phi}{2}\right)\tan\theta\tan\phi$$

$$=-3\tan\frac{\pi}{4}\tan\theta\tan\left(\frac{\pi}{2}-\theta\right)$$

$$[\because \phi+\theta=2]$$

$$\Rightarrow 2y=-3\tan\theta\cot\theta=-3$$

$$\therefore (2y)^2=9\Rightarrow(2\beta)^2=9$$

**64.** *(2)* Given, lines are

$$\sqrt{3}kx+ky-4\sqrt{3}=0 \quad \ldots\text{(i)}$$

$$\sqrt{3}x-y-4\sqrt{3}\,k=0 \quad \ldots\text{(ii)}$$

Multiply Eq. (ii) by $k$ and then adding Eqs. (i) and (ii),

$$(2\sqrt{3}x)k=4\sqrt{3}+4\sqrt{3}k^2$$

$$\therefore \quad x=\frac{4\sqrt{3}(1+k^2)}{2\sqrt{3}k}=2\left(k+\frac{1}{k}\right)$$

On subtracting Eq. (i) from Eq. (ii) $\times k$

$$2ky=4\sqrt{3}-4\sqrt{3}k^2$$

$$y=\frac{4\sqrt{3}(1-k^2)}{2k}=2\sqrt{3}\left(\frac{1}{k}-k\right)$$

We have,

$$\frac{x}{2}=\left(k+\frac{1}{k}\right) \quad \ldots\text{(iii)}$$

$$\frac{y}{2\sqrt{3}}=\left(\frac{1}{k}-k\right) \quad \ldots\text{(iv)}$$

On squaring and subtracting Eq. (iii) from Eq. (iv),

$$\frac{x^2}{4}-\frac{y^2}{12}=\left(k^2+\frac{1}{k^2}+2\right)-\left(\frac{1}{k^2}+k^2-2\right)$$

$$\frac{x^2}{16}-\frac{y^2}{48}=1$$

Clearly, this is a hyperbola

$$\therefore \quad e^2=1+\frac{b^2}{a^2}=1+\frac{48}{16}=1+3$$

$$e^2=4$$

$$\therefore \quad e=\sqrt{4}=2 \quad (\because e \text{ is positive})$$

# C-10

JEE Main Chapterwise Topicwise Mathematics

In JEE Main, there is usually 0-1 question per paper is visible from this chapter. In past 3 years, 27 Conceptual and 10 numerical questions of moderate difficulty level were asked from this chapter.

**For detailed analysis scan QR code**

# LIMITS AND DERIVATIVES

## Topic 1 Limits of Algebraic, Exponential and Logarithmic Function

### Level 1

**Imp.** Limits of Algebraic Function

**01** $\lim_{x\to\infty} \dfrac{(2x^2-3x+5)(3x-1)^{x/2}}{(3x^2+5x+4)\sqrt{(3x+2)^x}}$ is equal to **[2025, 23 Jan Shift-II]**

(a) $\dfrac{2}{3\sqrt{e}}$ (b) $\dfrac{2e}{\sqrt{3}}$ (c) $\dfrac{2e}{3}$ (d) $\dfrac{2}{\sqrt{3e}}$

**02** $\lim_{x\to 0} \dfrac{e-(1+2x)^{\frac{1}{2x}}}{x}$ is equal to **[2024, 9 April Shift-II]**

(a) 0 (b) $e$ (c) $e-e^2$ (d) $\dfrac{-2}{e}$

**03** If $\alpha, \beta$ are the distinct roots of $x^2+bx+c=0$, then

$\lim_{x\to\beta} \dfrac{e^{2(x^2+bx+c)}-1-2(x^2+bx+c)}{(x-\beta)^2}$ is equal to

**[2021, 27 Aug Shift-I]**

(a) $b^2+4c$ (b) $2(b^2+4c)$

(c) $2(b^2-4c)$ (d) $b^2-4c$

**04** Let $f: R \to R$ be a function such that $f(2)=4$ and $f'(2)=1$. Then, the value of $\lim_{x\to 2}\dfrac{x^2f(2)-4f(x)}{x-2}$ is equal to **[2021, 27 July Shift-I]**

(a) 4 (b) 8

(c) 16 (d) 12

**05** $\lim_{x\to 0}\dfrac{x\left(e^{(\sqrt{1+x^2+x^4}-1)/x}-1\right)}{\sqrt{1+x^2+x^4}-1}$ **[2020, 5 Sep Shift-II]**

(a) is equal to $\sqrt{e}$ (b) is equal to 1

(c) is equal to 0 (d) does not exist

**06** $\lim_{x\to 0}\left(\dfrac{3x^2+2}{7x^2+2}\right)^{1/x^2}$ is equal to **[2020, 8 Jan Shift-I]**

(a) $e^2$ (b) $e$ (c) $1/e^2$ (d) $1/e$

**07** Let $f(x)=5-|x-2|$ and $g(x)=|x+1|$, $x\in R$. If $f(x)$ attains maximum value at $\alpha$ and $g(x)$ attains minimum value at $\beta$, then

$\lim_{x\to -\alpha\beta}\dfrac{(x-1)(x^2-5x+6)}{x^2-6x+8}$ is equal to **[2019, 12 April Shift-II]**

(a) $1/2$ (b) $-3/2$

(c) $-1/2$ (d) $3/2$

**08** $\lim_{x\to 0}\dfrac{(27+x)^{1/3}-3}{9-(27+x)^{2/3}}$ equals **[JEE Main 2018]**

(a) $-\dfrac{1}{3}$ (b) $\dfrac{1}{6}$ (c) $-\dfrac{1}{6}$ (d) $\dfrac{1}{3}$

**09** $\lim_{x\to 3}\dfrac{\sqrt{3x}-3}{\sqrt{2x-4}-\sqrt{2}}$ is equal to **[JEE Main 2017]**

(a) $\sqrt{3}$ (b) $\dfrac{\sqrt{3}}{2}$ (c) $\dfrac{1}{2\sqrt{2}}$ (d) $\dfrac{1}{\sqrt{2}}$

**10** If $\lim_{x\to\infty}\left(1+\dfrac{a}{x}-\dfrac{4}{x^2}\right)^{2x}=e^3$, then $a$ is equal to **[JEE Main 2016]**

(a) 2 (b) 3/2 (c) 1/2 (d) 2/3

**11** Let $f: R\to[0,\infty)$ be such that $\lim_{x\to 5} f(x)$ exists and

$\lim_{x\to 5}\dfrac{[f(x)]^2-9}{\sqrt{|x-5|}}=0$. Then, $\lim_{x\to 5} f(x)$ is equal to **[AIEEE 2011]**

(a) 3 (b) 0

(c) 1 (d) 2

**12** If $\lim_{x\to 0}\dfrac{\log(3+x)-\log(3-x)}{x}=k$, then the value of $k$ is **[AIEEE 2003]**

(a) 0 (b) $-1/3$

(c) 2/3 (d) $-2/3$

**13** Let $f(a)=g(a)=k$ and their $n$th derivatives $f^n(a), g^n(a)$ exist and are not equal for some $n$. Further, if

$$\lim_{x\to a}\frac{f(a)g(x)-f(a)-g(a)f(x)+g(a)}{g(x)-f(x)}=4,$$

then the value of $k$ is **[AIEEE 2003]**

(a) 4 (b) 2 (c) 1 (d) 0

**14** $\lim_{x\to 0}\frac{\log x^n-[x]}{[x]}$, $n\in N$, ($[x]$ denotes greatest integer less than or equal to $x$) **[AIEEE 2002]**

(a) has value $-1$ (b) has value 0
(c) has value 1 (d) does not exist

## Level 2

**Imp.** Functional Equation

**15** Let $f: R-\{0\}\to R$ be a function such that $f(x)-6f\left(\frac{1}{x}\right)=\frac{35}{3x}-\frac{5}{2}$.

If the $\lim_{x\to 0}\left(\frac{1}{\alpha x}+f(x)\right)=\beta;\alpha,\beta\in R$, then $\alpha+2\beta$ is equal to **[2025, 24 Jan Shift-I]**

(a) 6 (b) 3 (c) 4 (d) 5

**16** $\lim_{n\to\infty}\left\{\left(2^{\frac{1}{2}}-2^{\frac{1}{3}}\right)\left(2^{\frac{1}{2}}-2^{\frac{1}{5}}\right)\ldots\left(2^{\frac{1}{2}}-2^{\frac{1}{2n-1}}\right)\right\}$ is equal to **[2023, 6 April Shift-II]**

(a) $1/\sqrt{2}$ (b) $\sqrt{2}$ (c) 0 (d) 1

**17** The set of all values of $a$ for which $\lim_{x\to\infty}[x-5]-[2x+2]=0$, where $[\alpha]$ denotes the greatest integer less than or equal to $\alpha$ is equal to **[2023, 25 Jan Shift-I]**

(a) $[-7.5,-6.5)$ (b) $(-7.5,-6.5)$
(c) $(-7.5,-6.5]$ (d) $[-7.5,-6.5]$

**18** Let $\beta=\lim_{x\to 0}\frac{\alpha x-(e^{3x}-1)}{\alpha x(e^{3x}-1)}$ for some $\alpha\in R$. Then, the value of $\alpha+\beta$ is **[2022, 26 July Shift-II]**

(a) 14/5 (b) 3/2 (c) 5/2 (d) 7/2

**19** $\lim_{n\to\infty}\left(1+\frac{1+\frac{1}{2}+\ldots\ldots+\frac{1}{n}}{n^2}\right)^n$ is equal to **[2021, 25 Feb Shift-I]**

(a) 1/2 (b) 0 (c) $1/e$ (d) 1

**20** $\lim_{x\to a}\frac{(a+2x)^{1/3}-(3x)^{1/3}}{(3a+x)^{1/3}-(4x)^{1/3}}$ $(a\neq 0)$ is equal to **[2020, 3 Sep Shift-II]**

(a) $\left(\frac{2}{3}\right)^{4/3}$ (b) $\left(\frac{2}{9}\right)^{4/3}$ (c) $\left(\frac{2}{9}\right)\left(\frac{2}{3}\right)^{1/3}$ (d) $\left(\frac{2}{3}\right)\left(\frac{2}{9}\right)^{1/3}$

**21** $\lim_{y\to 0}\frac{\sqrt{1+\sqrt{1+y^4}}-\sqrt{2}}{y^4}$ **[2019, 9 Jan Shift-I]**

(a) exists and equals $\frac{1}{4\sqrt{2}}$
(b) does not exist
(c) exists and equals $\frac{1}{2\sqrt{2}}$
(d) exists and equals $\frac{1}{2\sqrt{2}(\sqrt{2}+1)}$

**22** For each $t\in R$, let $[t]$ be the greatest integer less than or equal to $t$. Then, $\lim_{x\to 0^+}x\left(\left[\frac{1}{x}\right]+\left[\frac{2}{x}\right]+\ldots+\left[\frac{15}{x}\right]\right)$ **[JEE Main 2018]**

(a) is equal to 0 (b) is equal to 15
(c) is equal to 120 (d) does not exist (in R)

**23** $\lim_{x\to\infty}\left(\frac{x^2+5x+3}{x^2+x+2}\right)^x$ is equal to **[AIEEE 2002]**

(a) $e^4$ (b) $e^2$ (c) $e^3$ (d) $e$

## Level 3

**24** The value of $\lim_{n\to\infty}\left(\sum_{k=1}^{n}\frac{k^3+6k^2+11k+5}{(k+3)!}\right)$ is **[2025, 29 Jan Shift-I]**

(a) 4/3 (b) 5/3 (c) 7/3 (d) 2

**25** If $\lim_{x\to\infty}\left(\left(\frac{e}{1-e}\right)\left(\frac{1}{e}-\frac{x}{1+x}\right)\right)^x=\alpha$, then the value of $\frac{\log_e\alpha}{1+\log_e\alpha}$ equals **[2025, 22 Jan Shift-II]**

(a) $e$ (b) $e^{-2}$ (c) $e^2$ (d) $e^{-1}$

**26** $\lim_{n\to\infty}\frac{(1^2-1)(n-1)+(2^2-2)(n-2)+\ldots((n-1)^2-(n-1))\cdot 1}{(1^3+2^3+\ldots+n^3)-(1^2+2^2+\ldots+n^2)}$ is equal to **[2024, 6 April Shift-II]**

(a) 3/4 (b) 1/3 (c) 1/2 (d) 2/3

**27** The value of

$$\lim_{n\to\infty}\frac{(1+2-3)+(4+5-6)+\ldots+(3n-2)+(3n-1)-3n}{\sqrt{2n^4+4n+3}-\sqrt{n^4+5n+4}}$$ is **[2023, 25 Jan Shift-I]**

(a) $\frac{3}{2}(\sqrt{2}+1)$ (b) $\frac{3}{2\sqrt{2}}$ (c) $3(\sqrt{2}+1)$ (d) $\frac{\sqrt{2}+1}{2}$

**28** $\lim_{x\to\infty}\frac{(\sqrt{3x+1}+\sqrt{3x-1})^6+(\sqrt{3x+1}-\sqrt{3x-1})^6}{(x+\sqrt{x^2-1})^6+(x-\sqrt{x^2-1})^6}x^3$ **[2023, 24 Jan Shift-II]**

(a) is equal to 27/2 (b) does not exist
(c) is equal to 27 (d) is equal to 9

**29** If $\lim_{n\to\infty}(\sqrt{n^2-n-1}+n\alpha+\beta)=0$, then $8(\alpha+\beta)$ is equal to *[2022, 25 July Shift-I]*
(a) 4 (b) −8 (c) −4 (d) 8

**30** Let $a$ be an integer such that $\lim_{x\to 7}\frac{18-[1-x]}{[x]-3a}$ exists, where $[t]$ is greatest integer $\le t$. Then, $a$ is equal to *[2022, 27 June Shift-I]*
(a) −6 (b) −2 (c) 2 (d) 6

**31** The value of $\lim_{n\to\infty}\frac{[r]+[2r]+\ldots\ldots+[nr]}{n^2}$, where $r$ is non-zero real number and $[r]$ denotes the greatest integer less than or equal to $r$, is equal to *[2021, 17 March Shift-II]*
(a) $r/2$ (b) $r$ (c) $2r$ (d) 0

## Numerical Value Type Questions

**Imp.** Sum of Roots of Quadratic Equation

**32** For $t>-1$, let $\alpha_t$ and $\beta_t$ be the roots of the equation
$$\left((t+2)^{\frac{1}{7}}-1\right)x^2+\left((t+2)^{1/6}-1\right)x+\left((t+2)^{\frac{1}{21}}-1\right)=0.$$
If $\lim_{t\to 1^+}\alpha_t=a$ and $\lim_{t\to -1^+}\beta_t=b$, then $72(a+b)^2$ is equal to ………… . *[2025, 7 April Shift-II]*

**33** Let $[t]$ be the greatest integer less than or equal to t. Then, the least value of $p\in N$ for which
$$\lim_{x\to 0^+}\left(x\left(\left[\frac{1}{x}\right]+\left[\frac{2}{x}\right]+\ldots+\left[\frac{p}{x}\right]\right)-x^2\left(\left[\frac{1}{x^2}\right]+\left[\frac{2^2}{x^2}\right]+\ldots+\left[\frac{9^2}{x^2}\right]\right)\right)\ge 1$$
is equal to ………… . *[2025, 29 Jan Shift-I]*

**34** If $\lim_{t\to 0}\left(\int_0^1(3x+5)^t\,dx\right)^{\frac{1}{t}}=\frac{\alpha}{5e}\left(\frac{8}{5}\right)^{\frac{2}{3}}$, then $\alpha$ is equal to……. *[2025, 29 Jan Shift-II]*

**35** If $\lim_{x\to 1}\frac{(5x+1)^{\frac{1}{3}}-(x+5)^{\frac{1}{3}}}{(2x+3)^{\frac{1}{2}}-(x+4)^{\frac{1}{2}}}=\frac{m\sqrt{5}}{n(2n)^{\frac{2}{3}}}$, where $\gcd(m,n)=1$, then $8m+12n$ is equal to…… . *[2024, 4 April Shift-I]*

**36** If $\lim_{x\to 0}\frac{ax^2e^x-b\log_e(1+x)+cxe^{-x}}{x^2\sin x}=1$, then $16(a^2+b^2+c^2)$ is equal to ……… . *[2024, 31 Jan Shift-II]*

**37** Let $\sum_{n=0}^{\infty}\frac{n^3((2n)!)+(2n-1)(n!)}{(n!)((2n)!)}=ae+\frac{b}{e}+c$, where $a, b, c\in Z$ and $e=\sum_{n=0}^{\infty}\frac{1}{n!}$. Then, $a^2-b+c$ is equal to ………. *[2023, 30 Jan Shift-I]*

**38** Let $[f]$ denote the greatest integer $\le t$ and $\{t\}$ denote the fractional part of $t$. Then, integral value of $\alpha$ for which the left hand limit of the function
$$f(x)=[1+x]+\frac{\alpha^{2[x]+\{x\}}+[x]-1}{2[x]+\{x\}}$$
at $x=0$ is equal to $\alpha-\frac{4}{3}$ is ……… . *[2022, 27 June Shift-II]*

**39** Let $f(x)=x^6+2x^4+x^3+2x+3$, $x\in R$. Then, the natural number $n$ for which $\lim_{x\to 1}\frac{x^n f(1)-f(x)}{x-1}=44$ is ……… . *[2021, 1 Sep Shift-II]*

**40** If $\lim_{x\to 0}\frac{ax-(e^{4x}-1)}{ax(e^{4x}-1)}$ exists and is equal to $b$, then the value of $a-2b$ is ……… . *[2021, 25 Feb Shift-II]*

**41** If $\lim_{x\to 1}\frac{x+x^2+x^3+\ldots+x^n-n}{x-1}=820$, $(n\in N)$ then the value of $n$ is equal to ……… *[2020, 2 Sep Shift-I]*

# Topic 2 Limits of Trigonometric Functions

## Level 1

**01** $\lim_{x\to 0^+}\frac{\tan\left(5(x)^{\frac{1}{3}}\right)\log_e(1+3x^2)}{(\tan^{-1}(3\sqrt{x}))^2\left(e^{5x^{4/3}}-1\right)}$ is equal to *[2025, 7 April Shift-I]*
(a) $\frac{1}{15}$ (b) $\frac{5}{3}$ (c) $\frac{1}{3}$ (d) 1

**02** If $\lim_{x\to 0}\frac{\cos(2x)+a\cos(4x)-b}{x^4}$ is finite, then $(a+b)$ is equal to *[2025, 2 April Shift-II]*
(a) −1 (b) 0 (c) 1/2 (d) 3/4

**03** For $\alpha,\beta,\gamma\in R$, if $\lim_{x\to 0}\frac{x^2\sin\alpha x+(\gamma-1)e^{x^2}}{\sin 2x-\beta x}=3$, then $\beta+\gamma-\alpha$ is equal to *[2025, 2 April Shift-I]*
(a) 7 (b) −1 (c) 4 (d) 6

**04** $\lim_{x \to 0} \operatorname{cosec} x\left(\sqrt{2\cos^2 x + 3\cos x} - \sqrt{\cos^2 x + \sin x + 4}\right)$ is **[2025, 24 Jan Shift-I]**

(a) $\frac{1}{\sqrt{15}}$ (b) $-\frac{1}{2\sqrt{5}}$ (c) 0 (d) $\frac{1}{2\sqrt{5}}$

**05** $\lim_{x \to 0} \frac{e^{2|\sin x|} - 2|\sin x| - 1}{x^2}$ **[2024, 31 Jan Shift-I]**

(a) does not exist (b) is equal to −1
(c) is equal to 1 (d) is equal to 2

**06** $\lim_{x \to \infty} \left(\left(\frac{1-\cos^2(3x)}{\cos^3(4x)}\right)\left(\frac{\sin^3(4x)}{(\log_e(2x+1))^5}\right)\right)$ is equal to ....... **[2023, 8 April Shift-I]**

(a) 24 (b) 9 (c) 18 (d) 15

**07** $\lim_{t \to 0}\left(1^{\frac{1}{\sin^2 t}} + 2^{\frac{1}{\sin^2 t}} + \ldots + n^{\frac{1}{\sin^2 t}}\right)^{\sin^2 t}$ is equal to **[2023, 24 Jan Shift-I]**

(a) $\frac{n(n+1)}{2}$ (b) $n$ (c) $n^2 + n$ (d) $n^2$

**08** The value of $\lim_{x \to 0}\left(\frac{x}{\sqrt[8]{1-\sin x} - \sqrt[8]{1+\sin x}}\right)$ is equal to **[2021, 27 July Shift-II]**

(a) 0 (b) 4 (c) −4 (d) −1

**09** The value of

$\lim_{x \to 0^+} \frac{\cos^{-1}(x-[x]^2)\cdot\sin^{-1}(x-[x]^2)}{x-x^3}$, where $[x]$ denotes the greatest integer $\leq x$ is **[2021, 17 March Shift-I]**

(a) $\pi$ (b) 0 (c) $\pi/4$ (d) $\pi/2$

**10** $\lim_{x \to 0}\left(\tan\left(\frac{\pi}{4}+x\right)\right)^{1/x}$ is equal to **[2020, 2 Sep Shift-II]**

(a) $e$ (b) 2 (c) 1 (d) $e^2$

**11** $\lim_{x \to 0} \frac{\sin^2 x}{\sqrt{2} - \sqrt{1+\cos x}}$ equals **[2019, 8 April Shift-I]**

(a) $4\sqrt{2}$ (b) $\sqrt{2}$ (c) $2\sqrt{2}$ (d) 4

**12** $\lim_{x \to 0} \frac{(1-\cos 2x)(3+\cos x)}{x \tan 4x}$ is equal to **[JEE Main 2015, 2013]**

(a) 4 (b) 3 (c) 2 (d) 1/2

**13** $\lim_{x \to 0} \frac{\sin(\pi \cos^2 x)}{x^2}$ is equal to **[JEE Main 2014]**

(a) $\pi/2$ (b) 1 (c) $-\pi$ (d) $\pi$

## Level 2

**14** If $\lim_{x \to 1^+} \frac{(x-1)(6+\lambda\cos(x-1)) + \mu\sin(1-x)}{(x-1)^3} = -1$ where $\lambda, \mu \in R$, then $\lambda + \mu$ is equal to **[2025, 4 April Shift-I]**

(a) 18 (b) 19 (c) 17 (d) 20

**15** For some $a, b$ let

$$f(x) = \begin{vmatrix} a + \frac{\sin x}{x} & 1 & b \\ a & 1 + \frac{\sin x}{x} & b \\ a & 1 & b + \frac{\sin x}{x} \end{vmatrix},$$

$x \neq 0$, $\lim_{x \to 0} f(x) = \lambda + \mu a + vb$.

Then, $(\lambda + \mu + v)^2$ is equal to **[2025, 24 Jan Shift-II]**

(a) 25 (b) 16 (c) 36 (d) 9

**16** If $\lim_{x \to 0} \frac{3 + \alpha\sin x + \beta\cos x + \log_e(1-x)}{3\tan^2 x} = \frac{1}{3}$, then $2\alpha - \beta$ is equal to **[2024, 27 Jan Shift-II]**

(a) 2 (b) 5 (c) 7 (d) 1

**17** If $\lim_{x \to 0} \frac{e^{ax} - \cos(bx) - \frac{cxe^{-cx}}{2}}{1-\cos(2x)} = 17$, then $5a^2 + b^2$ is equal to **[2023, 13 April Shift-II]**

(a) 76 (b) 72 (c) 64 (d) 68

**18** Let $x = 2$ be a root of the equation $x^2 + px + q = 0$ and

$$f(x) = \begin{cases} \frac{1-\cos(x^2 - 4px + q^2 + 8q + 16)}{(x-2p)^4}, & x \neq 2p \\ 0, & x = 2p \end{cases}$$

Then, $\lim_{x \to 2p^+} [f(x)]$, where $[\,.\,]$ denotes greatest integer function, is **[2023, 29 Jan Shift-I]**

(a) 1 (b) −1 (c) 2 (d) 0

**19** $\lim_{x \to \frac{\pi}{4}} \frac{8\sqrt{2} - (\cos x + \sin x)^7}{\sqrt{2} - \sqrt{2}\sin 2x}$ is equal to **[2022, 25 July Shift-II]**

(a) 14 (b) 7 (c) $14\sqrt{2}$ (d) $7\sqrt{2}$

**20** $\lim_{x \to \frac{\pi}{2}} \tan^2 x\left((2\sin^2 x + 3\sin x + 4)^{1/2} - (\sin^2 x + 6\sin x + 2)^{\frac{1}{2}}\right)$ is equal to **[2022, 25 June Shift-II]**

(a) $\frac{1}{12}$ (b) $-\frac{1}{18}$ (c) $-\frac{1}{12}$ (d) $\frac{1}{6}$

**21** If $\lim_{x \to 0} \frac{\sin^{-1} x - \tan^{-1} x}{3x^3}$ is equal to $L$, then the value of $(6L+1)$ is **[2021, 18 March Shift-I]**

(a) 1/6 (b) 1/2 (c) 6 (d) 2

**22** For each $t \in R$, let $[t]$ be the greatest integer less than or equal to $t$. Then,

$$\lim_{x \to 1+} \frac{(1-|x|+\sin|1-x|)\sin\left(\frac{\pi}{2}[1-x]\right)}{|1-x|[1-x]}$$ **[2019, 10 Jan Shift-I]**

(a) equals 0 (b) does not exist
(c) equals −1 (d) equals 1

**23** $\lim_{x \to 0} \frac{x \tan 2x - 2x \tan x}{(1-\cos 2x)^2}$ equals **[JEE Main 2018]**

(a) $-\frac{1}{2}$ (b) $\frac{1}{4}$ (c) $\frac{1}{2}$ (d) 1

**24** $\lim_{x \to \pi/2} \frac{\cot x - \cos x}{(\pi - 2x)^3}$ equals **[JEE Main 2017]**

(a) $\frac{1}{24}$ (b) $\frac{1}{16}$ (c) $\frac{1}{8}$ (d) $\frac{1}{4}$

**25** $\lim_{x \to 0} \frac{(1-\cos 2x)^2}{2x \tan x - x \tan 2x}$ is **[JEE Main 2016]**

(a) 2 (b) 1/2 (c) −2 (d) 1/2

**26** If α and β are the distinct roots of $ax^2 + bx + c = 0$, then $\lim_{x \to \alpha} \frac{1-\cos(ax^2+bx+c)}{(x-\alpha)^2}$ is equal to **[AIEEE 2005]**

(a) $\frac{1}{2}(\alpha-\beta)^2$ (b) $-\frac{a^2}{2}(\alpha-\beta)^2$

(c) 0 (d) $\frac{a^2}{2}(\alpha-\beta)^2$

**27** $\lim_{x \to \frac{\pi}{2}} \frac{\left[1-\tan\left(\frac{x}{2}\right)\right](1-\sin x)}{\left[1+\tan\left(\frac{x}{2}\right)\right](\pi-2x)^3}$ is equal to **[AIEEE 2003]**

(a) 1/8 (b) 0 (c) 1/32 (d) ∞

## Level 3

**Imp.** Expansion of Trigonometric and Logarithm Series

**28** Given below are two statements

**Statement I**

$$\lim_{x \to 0}\left(\frac{\tan^{-1} x + \log_e \sqrt{\frac{1+x}{1-x}} - 2x}{x^5}\right) = \frac{2}{5}$$

**Statement II** $\lim_{x \to 1}\left(x^{\frac{2}{1-x}}\right) = \frac{1}{e^2}$

In the light of the above statements, choose the correct answer from the options given below **[2025, 8 April Shift-II]**

(a) Both Statement I and Statement II are false
(b) Statement I is true but Statement II is false
(c) Both Statement I and Statement II are true
(d) Statement I is false but Statement II is true

**29** If $\alpha > \beta > 0$ are the roots of the equation $ax^2 + bx + 1 = 0$, and $\lim_{x \to \frac{1}{\alpha}}\left(\frac{1-\cos(x^2+bx+a)}{2(1-\alpha x)^2}\right)^{1/2} = \frac{1}{k}\left(\frac{1}{\beta}-\frac{1}{\alpha}\right)$, then $k$ is equal to **[2023, 8 April Shift-II]**

(a) α (b) 2α (c) 2β (d) β

**30** Let $f$, $g$ and $h$ be the real valued functions defined on $R$ as

$$f(x) = \begin{cases} \frac{x}{|x|}, & x \neq 0 \\ 1, & x = 0 \end{cases}, \quad g(x) = \begin{cases} \frac{\sin(x+1)}{(x+1)}, & x \neq -1 \\ 1, & x = -1 \end{cases}$$

and $h(x) = 2[x] - f(x)$, where $[x]$ is the greatest integer $\le x$. Then, the value of $\lim_{x \to 1} g(h(x-1))$ is **[2023, 30 Jan Shift-II]**

(a) 1 (b) sin (1) (c) 0 (d) −1

**31** If $\lim_{x \to 0} \frac{\alpha e^x + \beta e^{-x} + \gamma \sin x}{x \sin^2 x} = \frac{2}{3}$, where α, β and $\gamma \in R$, then which of the following is not correct? **[2022, 29 July Shift-I]**

(a) $\alpha^2 + \beta^2 + \gamma^2 = 6$ (b) $\alpha\beta + \beta\gamma + \gamma\alpha + 1 = 0$
(c) $\alpha\beta^2 + \beta\gamma^2 + \gamma\alpha^2 + 3 = 0$ (d) $\alpha^2 - \beta^2 + \gamma^2 = 4$

**32** $\lim_{x \to 0} \frac{\cos(\sin x) - \cos x}{x^4}$ is equal to **[2022, 26 June Shift-II]**

(a) 1/3 (b) 1/4 (c) 1/6 (d) 1/12

**33** The value of $\lim_{h \to 0} 2\left\{\frac{\sqrt{3}\sin\left(\frac{\pi}{6}+h\right) - \cos\left(\frac{\pi}{6}+h\right)}{\sqrt{3}h(\sqrt{3}\cos h - \sin h)}\right\}$ is **[2021, 26 Feb Shift-I]**

(a) 4/3 (b) $2/\sqrt{3}$ (c) 3/4 (d) 2/3

**34** Let $[x]$ denote the greatest integer less than or equal to $x$. Then, $\lim_{x \to 0} \frac{\tan(\pi \sin^2 x) + (|x| - \sin(x[x]))^2}{x^2}$ **[2019, 11 Jan Shift-I]**

(a) equals π (b) equals π + 1
(c) equals 0 (d) does not exist

## Numerical Value Type Questions

**35** If $\lim_{x \to 0}\left(\frac{\tan x}{x}\right)^{\frac{1}{x^2}} = p$, then $96 \log_e p$ is equal to ......... . **[2025, 3 April Shift-II]**

**36** Let $f(x)=\lim_{n\to\infty}\sum_{r=0}^{n}\left(\frac{\tan(x/2^{r+1})+\tan^3(x/2^{r+1})}{1-\tan^2(x/2^{r+1})}\right)$, then $\lim_{x\to 0}\frac{e^x-e^{f(x)}}{(x-f(x))}$ is equal to **[2025, 28 Jan Shift-II]**

**37** Let $\{x\}$ denote the fractional part of $x$ and $f(x)=\frac{\cos^{-1}(1-\{x\}^2)\sin^{-1}(1-\{x\})}{\{x\}-\{x\}^3}$, $x\neq 0$. If $L$ and $R$ respectively denotes the left hand limit and the right hand limit of $f(x)$ at $x=0$, then $\frac{32}{\pi^2}(L^2+R^2)$ is equal to ....... . **[2024, 1 Feb Shift-I]**

**38** $\lim_{x\to 0}\left(\frac{(x+2\cos x)^3+2(x+2\cos x)^2+3\sin(x+2\cos x)}{(x+2)^3+2(x+2)^2+3\sin(x+2)}\right)^{100/x}$ is equal to ......... . **[2022, 28 July Shift-I]**

**39** If $\lim_{x\to 0}\frac{\alpha xe^x-\beta\log_e(1+x)+\gamma x^2e^{-x}}{x\sin^2 x}=10$, $\alpha,\beta,\gamma\in R$, then the value of $\alpha+\beta+\gamma$ is .... . **[2021, 20 July Shift-II]**

**40** If $\lim_{x\to 0}\left\{\frac{1}{x^8}\left(1-\cos\frac{x^2}{2}-\cos\frac{x^2}{4}+\cos\frac{x^2}{2}\cos\frac{x^2}{4}\right)\right\}=2^{-k}$, then the value of $k$ is .......... . **[2020, 3 Sep Shift-I]**

## Topic 3 Methods of Differentiation

### Level 1

**01** Let $f:(-\infty,\infty)-\{0\}\to R$ be a differentiable function such that $f'(1)=\lim_{a\to\infty}a^2 f\left(\frac{1}{a}\right)$. Then,

$$\lim_{a\to\infty}\frac{a(a+1)}{2}\tan^{-1}\left(\frac{1}{a}\right)+a^2-2\log_e a$$

is equal to **[2024, 6 April Shift-I]**

(a) $\frac{5}{2}+\frac{\pi}{8}$ (b) $\frac{3}{2}+\frac{\pi}{4}$ (c) $\frac{3}{4}+\frac{\pi}{8}$ (d) $\frac{3}{8}+\frac{\pi}{4}$

**02** Let $f(x)$ be a differentiable function at $x=a$ with $f'(a)=2$ and $f(a)=4$.
Then, $\lim_{x\to a}\frac{xf(a)-af(x)}{x-a}$ equals **[2021, 26 Feb Shift-II]**

(a) $2a+4$ (b) $4-2a$
(c) $2a-4$ (d) $a+4$

**03** Let $f(1)=-2$ and $f'(x)\geq 4.2$ for $1\leq x\leq 6$. The possible value of $f(6)$ lies in the interval. **[JEE Main 2013]**

(a) $[15, 19)$ (b) $(-\infty, 12)$
(c) $[12, 15)$ (d) $[19, \infty)$

**04** If $f(x)=3x^{10}-7x^8+5x^6-21x^3+3x^2-7$, then $\lim_{\alpha\to 0}\frac{f(1-\alpha)-f(1)}{\alpha^3+3\alpha}$ is **[AIEEE 2012]**

(a) $-\frac{53}{3}$ (b) $\frac{53}{3}$ (c) $-\frac{55}{3}$ (d) $\frac{55}{3}$

**05** If $f(x+y)=f(x)\cdot f(y)$, $\forall\, x, y$ and $f(5)=2$, $f'(0)=3$, then $f'(5)$ is **[AIEEE 2002]**

(a) 0 (b) 1 (c) 6 (d) 2

**06** If $f(1)=1$, $f'(1)=2$, then $\lim_{x\to 1}\frac{\sqrt{f(x)}-1}{\sqrt{x}-1}$ is **[AIEEE 2002]**

(a) 2 (b) 4 (c) 1 (d) 1/2

### Level 2

**07** If $f(x)=x^3-x^2f'(1)+xf''(2)-f'''(3)$, $x\in R$, then **[2023, 24 Jan Shift-II]**

(a) $f(1)+f(2)+f(3)=f(0)$
(b) $3f(1)+f(2)=f(3)$
(c) $f(3)-f(2)=f(1)$
(d) $2f(0)-f(1)+f(3)=f(2)$

### Level 3

**08** Let $f(x)$ be a polynomial of degree 4 having extreme values at $x=1$ and $x=2$. If $\lim_{x\to 0}\left(\frac{f(x)}{x^2}+1\right)=3$, then $f(-1)$ is equal to **[JEE Main 2018, 2015]**

(a) $\frac{1}{2}$ (b) $\frac{3}{2}$ (c) $\frac{5}{2}$ (d) $\frac{9}{2}$

### Numerical Value Type Questions

**09** The value of $\lim_{x\to 0}2\left(\frac{1-\cos x\sqrt{\cos 2x}\sqrt[3]{\cos 3x}\ldots\sqrt[10]{\cos 10x}}{x^2}\right)$ is ....... . **[2024, 8 April Shift-I]**

**10** Let $f:R\to R$ satisfy the equation $f(x+y)=f(x)\cdot f(y)$ for all $x,y\in R$ and $f(x)\neq 0$ for any $x\in R$. If the function $f$ is differentiable at $x=0$ and $f'(0)=3$, then $\lim_{h\to 0}\frac{1}{h}(f(h)-1)$ is equal to ...... . **[2021, 18 March Shift-II]**

## Key Index

### Topic 1

| | | | | | | | | | |
|---|---|---|---|---|---|---|---|---|---|
| 1 (a) | 2 (b) | 3 (c) | 4 (d) | 5 (b) | 6 (c) | 7 (a) | 8 (c) | 9 (d) | 10 (b) |
| 11 (a) | 12 (c) | 13 (a) | 14 (d) | 15 (c) | 16 (c) | 17 (a) | 18 (c) | 19 (d) | 20 (d) |
| 21 (a) | 22 (c) | 23 (a) | 24 (b) | 25 (a) | 26 (b) | 27 (a) | 28 (c) | 29 (c) | 30 (a) |
| 31 (a) | 32 (98) | 33 (24) | 34 (64) | 35 (100) | 36 (81) | 37 (26) | 38 (3) | 39 (7) | 40 (5) |
| 41 (40) | | | | | | | | | |

### Topic 2

| | | | | | | | | | |
|---|---|---|---|---|---|---|---|---|---|
| 1 (c) | 2 (c) | 3 (a) | 4 (b) | 5 (d) | 6 (c) | 7 (b) | 8 (c) | 9 (d) | 10 (d) |
| 11 (a) | 12 (c) | 13 (d) | 14 (a) | 15 (b) | 16 (b) | 17 (d) | 18 (d) | 19 (a) | 20 (a) |
| 21 (d) | 22 (a) | 23 (c) | 24 (b) | 25 (c) | 26 (d) | 27 (c) | 28 (c) | 29 (b) | 30 (a) |
| 31 (c) | 32 (c) | 33 (a) | 34 (d) | 35 (32) | 36 (1) | 37 (18) | 38 (1) | 39 (3) | 40 (8) |

### Topic 3

| | | | | | | | | | |
|---|---|---|---|---|---|---|---|---|---|
| 1 (a) | 2 (b) | 3 (d) | 4 (b) | 5 (c) | 6 (a) | 7 (d) | 8 (d) | 9 (55) | 10 (3) |

# Detailed Explanations

### Topic 1

**01.** *(a)* $\lim_{x\to\infty} \frac{(2x^2-3x+5)(3x-1)^{x/2}}{(3x^2+5x+4)\sqrt{(3x+2)^x}}$

$$= \lim_{x\to\infty} \frac{2x^2-3x+5}{3x^2+5x+4} \cdot \lim_{x\to\infty}\left(\frac{3x-1}{3x+2}\right)^{x/2}$$

$$= \lim_{x\to\infty} \frac{x^2\left(2-\frac{3}{x}+\frac{5}{x^2}\right)}{x^2\left(3+\frac{5}{x}+\frac{4}{x^2}\right)} e^{\lim_{x\to\infty}\left(\frac{3x-1}{3x+2}-1\right)\cdot\frac{x}{2}}$$

$$= \frac{2}{3} e^{\lim_{x\to\infty}\left(\frac{-3}{3x+2}\right)\left(\frac{x}{2}\right)} = \frac{2}{3} e^{\lim_{x\to\infty}\frac{-3}{\left(3+\frac{2}{x}\right)}\left(\frac{1}{2}\right)}$$

$$= \frac{2}{3} e^{\frac{-3}{3\times 2}} = \frac{2}{3}\cdot e^{-\frac{1}{2}} = \frac{2}{3\sqrt{e}}$$

**02.** *(b)* $\lim_{x\to 0} = \frac{e-(1+2x)^{1/2x}}{x}$

$$= \lim_{x\to 0} = \frac{e-e^{\frac{1}{2x}\ln(1+2x)}}{x} = \lim_{x\to 0}(-e)\frac{(e^{\frac{\ln(1+2x)}{2x}-1}-1)}{x}$$

$$= \lim_{x\to 0}(-e)\frac{e^{\frac{\ln(1+2x)}{2x}-1}-1}{\frac{\ln(1+2x)}{2x}-1} \times \frac{\frac{\ln(1+2x)}{2x}-1}{x}$$

$$= \lim_{x\to 0}(-e)\frac{\ln(1+2x)-2x}{2x^2}$$

$$= \lim_{x\to 0}(-e)\frac{\left[\left(2x-\frac{(2x)^2}{2}+\ldots\right)-2x\right]}{2x^2} = e\times 1 = e$$

**03.** *(c)* $\because$ $\alpha$ and $\beta$ are distinct roots of $x^2+bx+c=0$

$\Rightarrow x^2+bx+c=(x-\alpha)(x-\beta)=0$

$$\text{Now, } \lim_{x\to\beta} \frac{e^{2(x^2+bx+c)}-1-2(x^2+bx+c)}{(x-\beta)^2}$$

$$= \lim_{x\to\beta} \frac{e^{2(x-\alpha)(x-\beta)}-1-2(x-\alpha)(x-\beta)}{(x-\beta)^2}$$

$$= \lim_{h\to 0} \frac{e^{2(\beta-\alpha+h)h}-1-2(\beta-\alpha+h)h}{h^2} \quad [\because \text{Put } x=\beta+h]$$

$$= \lim_{h\to 0} \frac{1+2(\beta-\alpha+h)h+\frac{|2(\beta-\alpha+h)h|^2}{2!}+\ldots-1-2h(\beta-\alpha+h)}{h^2}$$

$$= \lim_{h\to 0} \frac{2(\beta-\alpha+h)^2h^2+\ldots}{h^2}$$

$$= 2(\beta-\alpha)^2 = 2(b^2-4c)$$

**04.** (d) $f(2)=4,\ f'(2)=1$

Now, $\lim_{x\to 2}\dfrac{x^2 f(2)-4f(x)}{x-2}$

On applying L' Hospital rule as 0/0 form on putting $x=2$

So, $\lim_{x\to 2}\dfrac{2x\,f(2)-4f'(x)}{1}=2\cdot 2\cdot f(2)-4f'(2)$

$=4\cdot 4-4\cdot 1=12$

**05.** (b) $\lim_{x\to 0}\dfrac{x(e^{(\sqrt{1+x^2+x^4}-1)/x}-1)}{\sqrt{1+x^2+x^4}-1}=\lim_{x\to 0}\dfrac{e^{(\sqrt{1+x^2+x^4}-1)/x}-1}{(\sqrt{1+x^2+x^4}-1)/x}=1$

$\left[\because \lim_{y\to 0}\dfrac{e^y-1}{y}=1\right]$

**06.** (c) Given limit, $\lim_{x\to 0}\left(\dfrac{3x^2+2}{7x^2+2}\right)^{1/x^2}=P$ (let), having $1^\infty$ form.

$\therefore\quad P=e^{\lim_{x\to 0}\frac{1}{x^2}\left(\frac{3x^2+2}{7x^2+2}-1\right)}$

{$\because$ If $\lim_{x\to a}(f(x))^{g(x)}$ have indeterminant form $1^\infty$, then

$\lim_{x\to a}(f(x))^{g(x)}=e^{\lim_{x\to a}g(x)(f(x)-1)}$}

$\Rightarrow P=e^{\lim_{x\to 0}\frac{1}{x^2}\left(\frac{3x^2+2-7x^2-2}{7x^2+2}\right)}=e^{\lim_{x\to 0}\frac{1}{x^2}\left(-\frac{4x^2}{7x^2+2}\right)}$

$=e^{\lim_{x\to 0}\frac{-4}{7x^2+2}}$

On applying limit, we get $P=e^{-4/2}=e^{-2}=\dfrac{1}{e^2}$

**07.** (a) Given functions are $f(x)=5-|x-2|$

and $g(x)=|x+1|$, where $x\in R$.

Clearly, maximum of $f(x)$ occurred at $x=2$, so $\alpha=2$.

and minimum of $g(x)$ occurred at $x=-1$, so $\beta=-1$.

$\Rightarrow \alpha\beta=-2$

Now, $\lim_{x\to -\alpha\beta}\dfrac{(x-1)(x^2-5x+6)}{x^2-6x+8}$

$=\lim_{x\to 2}\dfrac{(x-1)(x-3)(x-2)}{(x-4)(x-2)}\qquad[\because \alpha\beta=-2]$

$=\lim_{x\to 2}\dfrac{(x-1)(x-3)}{(x-4)}$

$=\dfrac{(2-1)(2-3)}{(2-4)}=\dfrac{1\times(-1)}{(-2)}=\dfrac{1}{2}$

**08.** (c) We have, $\lim_{x\to 0}\dfrac{(27+x)^{1/3}-3}{9-(27+x)^{2/3}}$

$=\lim_{x\to 0}\dfrac{3\left[\left(1+\frac{x}{27}\right)^{1/3}-1\right]}{9\left[1-\left(1+\frac{x}{27}\right)^{2/3}\right]}=\lim_{x\to 0}\dfrac{\left[\left(1+\frac{x}{3\times 27}+\dots\right)-1\right]}{3\left[1-\left(1+\frac{2x}{3\times 27}+\dots\right)\right]}$

$=\lim_{x\to 0}\dfrac{1(x/81)}{3(-2x/81)}=-\dfrac{1}{6}$

**09.** (d) We have, $\lim_{x\to 3}\dfrac{\sqrt{3x}-3}{\sqrt{2x-4}-\sqrt{2}}$

$=\lim_{x\to 3}\dfrac{\sqrt{3x}-3}{\sqrt{2x-4}-\sqrt{2}}\times\dfrac{\sqrt{2x-4}+\sqrt{2}}{\sqrt{2x-4}+\sqrt{2}}$

$=\lim_{x\to 3}\dfrac{(\sqrt{3x}-3)(\sqrt{2x-4}+\sqrt{2})}{(2x-4-2)}$

$=\lim_{x\to 3}\dfrac{(\sqrt{3x}-3)(\sqrt{3x}+3)(\sqrt{2x-4}+\sqrt{2})}{(2x-6)(\sqrt{3x}+3)}$

$=\lim_{x\to 3}\dfrac{(3x-9)(\sqrt{2x-4}+\sqrt{2})}{2(x-3)(\sqrt{3x}+3)}$

$=\lim_{x\to 3}\dfrac{3(x-3)(\sqrt{2x-4}+\sqrt{2})}{2(x-3)(\sqrt{3x}+3)}=\dfrac{3}{2}\times\dfrac{(\sqrt{2}+\sqrt{2})}{(3+3)}=\dfrac{1}{\sqrt{2}}$

**10.** (b) We have, $\lim_{x\to\infty}\left(1+\dfrac{a}{x}-\dfrac{4}{x^2}\right)^{2x}=e^3$

Now, $\lim_{x\to\infty}\left(1+\dfrac{a}{x}-\dfrac{4}{x^2}\right)^{2x}$ ($1^\infty$ form)

$=e^{\lim_{x\to\infty}\left[\left(1+\frac{a}{x}-\frac{4}{x^2}-1\right)2x\right]}=e^{\lim_{x\to\infty}\left(2a-\frac{8}{x}\right)}=e^{2a}$

Hence, $e^{2a}=e^3$

$\therefore\quad 2a=3\Rightarrow a=\dfrac{3}{2}$

**11.** (a) Given, $\lim_{x\to 5}f(x)$ exists and $\lim_{x\to 5}\dfrac{[f(x)]^2-9}{\sqrt{|x-5|}}=0$

$\Rightarrow\ \lim_{x\to 5}[f(x)]^2-9=0\Rightarrow(\lim_{x\to 5}[f(x)])^2=9$

$\therefore\ \lim_{x\to 5}f(x)=3,-3$

But, $f:R\to[0,\infty)$

$\therefore$ Range of $f(x)\geq 0\Rightarrow \lim_{x\to 5}f(x)=3$

**12.** (c) $\because \lim_{x\to 0}\dfrac{\log(3+x)-\log(3-x)}{x}=k$

[Using L'Hospital rule]

$\Rightarrow \lim_{x\to 0}\dfrac{\left(\frac{1}{3+x}+\frac{1}{3-x}\right)}{1}=k\Rightarrow \dfrac{1}{3}+\dfrac{1}{3}=k$

$\therefore\ k=\dfrac{2}{3}$

**13.** (a) $\lim_{x\to a}\dfrac{f(a)g(x)-f(a)-g(a)f(x)+g(a)}{g(x)-f(x)}=4$

On applying L'Hospital rule, we get

$\lim_{x\to a}\dfrac{f(a)g'(x)-g(a)f'(x)}{g'(x)-f'(x)}=4$

$\Rightarrow \lim_{x\to a}\dfrac{kg'(x)-kf'(x)}{g'(x)-f'(x)}=4\ \Rightarrow k=4$

**14.** *(d)* $\lim_{x\to 0} \frac{\log x^n - [x]}{[x]} = \lim_{x\to 0} \frac{n \log x}{[x]} - 1$

which does not exist as $\lim_{x\to 0} \frac{\log x}{[x]}$ does not exist.

**15.** *(c)* Given $f(x) - 6f(1/x) = \frac{35}{3x} - \frac{5}{2}$ ...(i)

Replace $x$ by $1/x$ in Eq. (i), we get

$$f(1/x) - 6f(x) = \frac{35x}{3} - \frac{5}{2} \quad \text{...(ii)}$$

On solving Eqs. (i) and (ii), we get

$$f(x) = -2x - \frac{1}{3x} + \frac{1}{2}$$

Now, $\beta = \lim_{x\to 0}\left(\frac{1}{\alpha x} + f(x)\right)$

$\Rightarrow \quad \beta = \lim_{x\to 0}\left(\frac{1}{\alpha x} - 2x - \frac{1}{3x} + \frac{1}{2}\right)$

limits exists finitely iff $\alpha = 3 \Rightarrow \beta = 1/2$

Hence, $\alpha + 2\beta = 3 + 2 \times \frac{1}{2} = 4$

**16.** *(c)* Let $L = \lim_{n\to\infty} \{(2^{\frac{1}{2}} - 2^{\frac{1}{3}})(2^{\frac{1}{2}} - 2^{\frac{1}{5}}) \ldots (2^{\frac{1}{2}} - 2^{\frac{1}{2n+1}})\}$

Here, $(2^{\frac{1}{2}} - 2^{\frac{1}{3}})^n < (2^{\frac{1}{2}} - 2^{\frac{1}{3}})(2^{\frac{1}{2}} - 2^{\frac{1}{5}})$

$$\ldots (2^{\frac{1}{2}} - 2^{\frac{1}{2n+1}}) < (2^{\frac{1}{2}} - 2^{\frac{1}{2n+1}})^n$$

$\Rightarrow \lim_{n\to\infty} (2^{\frac{1}{2}} - 2^{\frac{1}{3}})^n < L < \lim_{n\to\infty} (2^{\frac{1}{2}} - 2^{\frac{1}{2n+1}})^n$

Since, $\lim_{n\to\infty} (2^{\frac{1}{2}} - 2^{\frac{1}{3}})^n = 0$

and $\lim_{n\to\infty} (2^{\frac{1}{2}} - 2^{\frac{1}{2n+1}})^n = 0$

$\therefore \quad L = 0$

**17.** *(a)* We have, $\lim_{x\to\alpha} ([x-5] - [2x+2]) = 0$

$$[\alpha - 5] - [2\alpha + 2] = 0$$
$$[\alpha - 5] = [2\alpha + 2] \quad \text{...(i)}$$

For algebraic calculation,

$$\alpha - 5 = 2\alpha + 2$$

$\Rightarrow \quad \alpha = -7$

So, its value will be around $-7$

For $\alpha = -7.5$ in Eq. (i),

$$[-7.5 - 5] = [-15 + 2]$$

$\Rightarrow \quad [-12.5] = [-13]$

$\Rightarrow \quad -13 = -13 \Rightarrow$ True

For $\alpha = -6.5$ in Eq. (i),

$$[-6.5 - 5] = [-13 + 2]$$

$\Rightarrow \quad [-11.5] = [-11]$

$\Rightarrow \quad -12 \neq -11$

$\Rightarrow$ Not satisfies.

$\therefore$ The required range will be $\alpha \in [-7.5, -6.5)$

**18.** *(c)* Given, $\beta = \lim_{x\to 0} \frac{\alpha x - (e^{3x} - 1)}{\alpha x(e^{3x} - 1)}$

$\Rightarrow \quad \beta = \lim_{x\to 0} \frac{1}{e^{3x} - 1} - \lim_{x\to 0} \frac{1}{\alpha x}$

$\Rightarrow \quad \beta = \lim_{x\to 0} \frac{3x}{e^{3x} - 1} \times \frac{1}{3x} - \lim_{x\to 0} \frac{1}{\alpha x}$

$\Rightarrow \quad \beta = \lim_{x\to 0}\left[\frac{1}{3x} - \frac{1}{\alpha x}\right] \quad \left[\because \lim_{x\to 0} \frac{e^{3x} - 1}{3x} = 1\right]$

For the existence of limits, $\alpha = 3$ as $\infty - \infty$ is an indeterminant for *m*.

Now, $\quad \beta = \lim_{x\to 0} \frac{3x - (e^{3x} - 1)}{3x \cdot (e^{3x} - 1)}$

On applying L' Hospital rule, we get

$$\lim_{x\to 0} \frac{(3 - 3e^{3x})}{3x \times 3e^{3x} + 3(e^{3x} - 1)}$$
$$= \lim_{x\to 0} \frac{3(1 - e^{3x})}{9xe^{3x} + 3(e^{3x} - 1)}$$

Again, applying L' Hospital rule, we get

$$\beta = \lim_{x\to 0} \frac{3(-3e^{3x})}{9x(3e^{3x}) + 9e^{3x} + 3(3e^{3x})} = -\frac{1}{2}$$

$\therefore \quad \alpha + \beta = 3 + \left(-\frac{1}{2}\right) = \frac{5}{2}$

**19.** *(d)* Given, limit is of the form $1^\infty$.

So, $l = \exp\left[\lim_{n\to\infty} \frac{1 + \frac{1}{2} + \frac{1}{3} + \ldots + \frac{1}{n}}{n}\right]$

$$\left[\because \lim_{x\to a} [f(x)]^{g(x)} = e^{\lim_{x\to a} g(x)[f(x) - 1]}\right]$$

Now, $0 \le 1 + \frac{1}{2} + \frac{1}{3} + \ldots\ldots + \frac{1}{n} \le 1 + \frac{1}{\sqrt{2}} + \frac{1}{\sqrt{3}} + \ldots\ldots + \frac{1}{\sqrt{n}}$

$$\le 2\sqrt{n} - 1$$

Now, $\lim_{n\to\infty} \frac{2\sqrt{n} - 1}{n} = \lim_{n\to\infty} \frac{2}{\sqrt{n}} - \frac{1}{n} = 0$

$\therefore$ Using Sandwich rule

$$l = \exp(0) = 1$$

**20.** *(d)* $\lim_{x\to a} \frac{(a + 2x)^{1/3} - (3x)^{1/3}}{(3a + x)^{1/3} - (4x)^{1/3}}, (a \neq 0) \quad \left[\frac{0}{0} \text{ form}\right]$

Put $x = a + h$

So, $\quad \lim_{h\to 0} \frac{(a + 2a + 2h)^{1/3} - (3a + 3h)^{1/3}}{(3a + a + h)^{1/3} - (4a + 4h)^{1/3}}$

$$= \lim_{h\to 0} \frac{(3a)^{1/3}\left[\left(1 + \frac{2h}{3a}\right)^{1/3} - \left(1 + \frac{3h}{3a}\right)^{1/3}\right]}{(4a)^{1/3}\left[\left(1 + \frac{h}{4a}\right)^{1/3} - \left(1 + \frac{4h}{4a}\right)^{1/3}\right]}$$

$$= \lim_{h\to 0}\left(\frac{3}{4}\right)^{1/3}\left[\frac{1+\frac{2h}{9a}-1-\frac{3h}{9a}+\text{higher degree terms}}{1+\frac{h}{12a}-1-\frac{4h}{12a}+\text{higher degree terms}}\right]$$

$$=\left(\frac{3}{4}\right)^{1/3}\left(\frac{\frac{2}{9}-\frac{3}{9}}{\frac{1}{12}-\frac{4}{12}}\right)=\left(\frac{3}{4}\right)^{1/3}\left(\frac{\frac{-1}{9}}{\frac{-3}{12}}\right)$$

$$=\left(\frac{3}{4}\right)^{1/3}\frac{4}{(3)^2}=\frac{4^{1-\frac{1}{3}}}{3^{2-\frac{1}{3}}}=\frac{4^{2/3}}{3^{5/3}}=\frac{2^{4/3}}{3^{5/3}}=\frac{2}{3}\left(\frac{2}{9}\right)^{1/3}$$

**21.** (a) Clearly, $\lim_{y\to 0}\frac{\sqrt{1+\sqrt{1+y^4}}-\sqrt{2}}{y^4}$

$$=\lim_{y\to 0}\frac{\sqrt{1+\sqrt{1+y^4}}-\sqrt{2}}{y^4}\times\frac{\sqrt{1+\sqrt{1+y^4}}+\sqrt{2}}{\sqrt{1+\sqrt{1+y^4}}+\sqrt{2}}$$

[Rationalising the numerator]

$$=\lim_{y\to 0}\frac{(1+\sqrt{1+y^4})-2}{y^4(\sqrt{1+\sqrt{1+y^4}}+\sqrt{2})}\qquad [\because (a+b)(a-b)=a^2-b^2]$$

$$=\lim_{y\to 0}\frac{\sqrt{1+y^4}-1}{y^4(\sqrt{1+\sqrt{1+y^4}}+\sqrt{2})}\times\frac{\sqrt{1+y^4}+1}{\sqrt{1+y^4}+1}$$

[Again, rationalising the numerator]

$$=\lim_{y\to 0}\frac{y^4}{y^4(\sqrt{1+\sqrt{1+y^4}}+\sqrt{2})(\sqrt{1+y^4}+1)}=\frac{1}{2\sqrt{2}\times 2}$$

(by cancelling $y^4$ and then by direct substitution).

$$=\frac{1}{4\sqrt{2}}$$

**22.** (c) We have, $\lim_{x\to 0^+} x\left(\left[\frac{1}{x}\right]+\left[\frac{2}{x}\right]+\ldots+\left[\frac{15}{x}\right]\right)$

We know, $[x]=x-\{x\}$

$\therefore\quad \left[\frac{1}{x}\right]=\frac{1}{x}-\left\{\frac{1}{x}\right\}$

Similarly, $\left[\frac{n}{x}\right]=\frac{n}{x}-\left\{\frac{n}{x}\right\}$

$\therefore$Given limit $=\lim_{x\to 0^+} x\left(\frac{1}{x}-\left\{\frac{1}{x}\right\}+\frac{2}{x}-\left\{\frac{2}{x}\right\}+\ldots\frac{15}{x}-\left\{\frac{15}{x}\right\}\right)$

$$=\lim_{x\to 0^+}(1+2+3+\ldots+15)-x\left(\left\{\frac{1}{x}\right\}+\left\{\frac{2}{x}\right\}+\ldots+\left\{\frac{15}{x}\right\}\right)$$

$=120-0=120$ $\quad\left[\because 0\le\left\{\frac{n}{x}\right\}<1\text{, therefore } 0\le x\left\{\frac{n}{x}\right\}<x\Rightarrow \lim_{x\to 0^+}x\left\{\frac{n}{x}\right\}=0\right]$

**23.** (a) Now, $\lim_{x\to\infty}\left(\frac{x^2+5x+3}{x^2+x+2}\right)^x=\lim_{x\to\infty}\left(1+\frac{4x+1}{x^2+x+2}\right)^x$

$$=\lim_{x\to\infty}\left[\left(1+\frac{4x+1}{x^2+x+2}\right)^{1/\frac{(4x+1)}{x^2+x+2}}\right]^{\frac{(4x+1)x}{x^2+x+2}}$$

$$=e^{\lim_{x\to\infty}\frac{\left(4+\frac{1}{x}\right)}{1+\frac{1}{x}+\frac{2}{x^2}}}=e^4\qquad\left[\because \lim_{x\to 0}\left(1+\frac{1}{x}\right)^x=e\right]$$

**24.** (b) $\lim_{n\to\infty}\sum_{k=1}^{n}\frac{k^3+6k^2+11k+5}{(k+3)!}$

$$=\lim_{n\to\infty}\sum_{k=1}^{n}\frac{k^3+6k^2+11k+6-1}{(k+3)!}$$

$$=\lim_{n\to\infty}\sum_{k=1}^{n}\frac{(k+1)(k+2)(k+3)-1}{(k+3)(k+2)(k+1)(k)!}$$

$$=\lim_{n\to\infty}\sum_{k=1}^{n}\left(\frac{1}{k!}-\frac{1}{(k+3)!}\right)$$

$$=\left(\frac{1}{1!}+\frac{1}{2!}+\frac{1}{3!}+\ldots..+\infty\right)-\left(\frac{1}{4!}+\frac{1}{5!}+\frac{1}{6!}+\ldots+\infty\right)$$

As $n\to\infty$

$$=(e-1)-\left(e-1-\frac{1}{1!}-\frac{1}{2!}-\frac{1}{3!}\right)$$

$$\left(\because e=1+1+\frac{1}{2!}+\frac{1}{3!}+\frac{1}{4!}+\ldots+\infty\right)$$

$$=1+\frac{1}{2}+\frac{1}{6}=\frac{10}{6}=\frac{5}{3}$$

**25.** (a) $\alpha=\lim_{x\to\infty}\left(\left(\frac{e}{1-e}\right)\left(\frac{1}{e}-\frac{x}{1+x}\right)\right)^x\qquad (1^\infty \text{ form})$

$$=e^{\lim_{x\to\infty}\left[\left(\frac{e}{1-e}\right)\left(\frac{1}{e}-\frac{x}{1+x}\right)-1\right]x}$$

$$=e^{\lim_{x\to\infty}x\left(\frac{e}{1-e}\right)\left(\frac{1}{e}-\frac{x}{1+x}-\frac{1-e}{e}\right)}$$

$$=e^{\lim_{x\to\infty}x\left(\frac{e}{1-e}\right)\left(1-\frac{x}{1+x}\right)}=e^{\lim_{x\to\infty}x\left(\frac{e}{1-e}\right)\left(\frac{1}{1+x}\right)}$$

$$=e^{\lim_{x\to\infty}\frac{e}{1-e}\left(\frac{1}{1+1/x}\right)}$$

$$\alpha=e^{\frac{e}{1-e}}$$

$$\Rightarrow\ \log_e\alpha=\frac{e}{1-e}$$

$$\therefore\ \frac{\log_e\alpha}{1+\log_e\alpha}=\frac{\frac{e}{1-e}}{1+\frac{e}{1-e}}=\frac{e}{1-e+e}=e$$

**26.** *(b)* Given expansion can be written as

$$= \lim_{n\to\infty} \frac{\sum_{r=1}^{n-1} (r^2 - r)(n-r)}{\sum_{r=1}^{n} r^3 - \sum_{r=1}^{n} r^2}$$

$$= \lim_{n\to\infty} \frac{\sum_{r=1}^{n-1} (-r^3 + r^2(n+1) - nr)}{\left(\frac{n(n+1)}{2}\right)^2 - \frac{n(n+1)(2n+1)}{6}}$$

$$= \lim_{n\to\infty} \frac{\frac{-n^2(n-1)^2}{4} + \frac{(n+1)(n-1)n(2n-1)}{6} - \frac{n^2}{2}(n-1)}{\frac{n(n+1)}{2}\left(\frac{n(n+1)}{2} - \frac{2n+1}{3}\right)}$$

$$= \lim_{n\to\infty} \frac{\frac{-3n(n-1)^2 + 2(n+1)(n-1)(2n-1) - 6n(n-1)}{12}}{\frac{(n+1)}{2}\frac{[3n(n+1) - 2(2n+1)]}{6}}$$

$$= \lim_{n\to\infty} \frac{(n-1)[-3n(n-1) + 2(2n^2 + 2n - n - 1) - 6n]}{(n+1)[3n^2 + 3n - 4n - 2]}$$

$$= \lim_{n\to\infty} \frac{(n-1)[-3n^2 + 3n + 4n^2 + 2n - 2 - 6n]}{(n+1)(3n^2 - n - 2)}$$

$$= \lim_{n\to\infty} \frac{(n-1)[n^2 - n - 2]}{(n+1)[3n^2 - n - 2]} = \frac{1}{3}$$

**27.** *(a)* $\lim_{n\to\infty} \frac{(1+2-3) + (4+5-6) + \ldots + (3n-2) + (3n-1) - 3n}{\sqrt{2n^4 + 4n + 3} - \sqrt{n^4 + 5n + 4}}$

$$= \lim_{n\to\infty} \frac{0 + 3 + 6 + 9 + 12 + \ldots + (3n-3)}{\sqrt{2n^4 + 4n + 3} - \sqrt{n^4 + 5n + 4}}$$

$$= \lim_{n\to\infty} \frac{\sum_{r=1}^{n} (3r-3)}{(\sqrt{2n^4 + 4n + 3} - \sqrt{n^4 + 5n + 4})}$$

$$= \lim_{n\to\infty} \frac{\frac{3\cdot n(n+1)}{2} - 3}{n^2\left(\sqrt{2 + \frac{4}{n^3} + \frac{3}{n^4}} - \sqrt{1 + \frac{5}{n^3} + \frac{4}{n^4}}\right)}$$

$$= \lim_{n\to\infty} \frac{\frac{3}{2}\left(1 + \frac{1}{n}\right) - \frac{3}{n^2}}{\left(\sqrt{2 + \frac{4}{n^3} + \frac{3}{n^4}} - \sqrt{1 + \frac{5}{n^3} + \frac{4}{n^4}}\right)}$$

Put $n = \infty$

$$= \frac{\frac{3}{2}(1+0) - 0}{(\sqrt{2+0+0} - \sqrt{1+0+0})} = \frac{3}{2(\sqrt{2}-1)} = \frac{3}{2}(\sqrt{2}+1)$$

**28.** *(c)* Let $L = \lim_{x\to\infty}$

$$\frac{(\sqrt{3x+1} + \sqrt{3x-1})^6 + (\sqrt{3x+1} - \sqrt{3x-1})^6}{(x + \sqrt{x^2-1})^6 + (x - \sqrt{x^2-1})^6} x^3$$

Now, we know that $(a+b)^6 + (a-b)^6$

$$= 2\,({}^6C_0 a^6 + {}^6C_2 a^4 b^2 + {}^6C_4 a^2 b^4 + {}^6C_6 b^6)$$

$\therefore L = \lim_{x\to\infty} x^3$

$$\left[\frac{2[{}^6C_0(3x+1)^3 + {}^6C_2(3x+1)^2(3x-1) + {}^6C_4(3x+1)(3x-1)^2 + {}^6C_6(3x-1)^3]}{2[{}^6C_0 x^6 + {}^6C_2 x^4(x^2-1) + {}^6C_4 x^2(x^2-1)^2 + {}^6C_6(x^2-1)^3]}\right]$$

$$= \lim_{x\to\infty} \frac{x^6\left[\left(3+\frac{1}{x}\right)^3 + 15\left(3+\frac{1}{x}\right)^2\left(3-\frac{1}{x}\right) + 15\left(3+\frac{1}{x}\right)\left(3-\frac{1}{x}\right)^2 + \left(3-\frac{1}{x}\right)^3\right]}{x^6\left[1 + 15\left(1-\frac{1}{x^2}\right) + 15\left(1-\frac{1}{x^2}\right)^2 + \left(1-\frac{1}{x^2}\right)^3\right]}$$

$$= \lim_{x\to\infty} \frac{\left[\left(3+\frac{1}{x}\right)^3 + 15\left(3+\frac{1}{x}\right)^2\left(3-\frac{1}{x}\right) + 15\left(3+\frac{1}{x}\right)\left(3-\frac{1}{x}\right)^2 + \left(3-\frac{1}{x}\right)^3\right]}{\left[1 + 15\left(1-\frac{1}{x^2}\right) + 15\left(1-\frac{1}{x^2}\right)^2 + \left(1-\frac{1}{x^2}\right)^3\right]}$$

$$= \frac{(3)^3 + 15(3)^2(3) + 15(3)(3)^2 + (3)^3}{1 + 15 + 15 + 1}$$

$$= \frac{27 + 405 + 405 + 27}{32} = \frac{864}{32} = 27$$

**29.** *(c)* $\lim_{n\to\infty}\left(\sqrt{n^2 - n - 1} + n\alpha + \beta\right) = 0$

Let $n = \frac{1}{t}$ as $n \to \infty \Rightarrow t \to 0$

$$\Rightarrow \quad \lim_{t\to 0}\left[\sqrt{\frac{1}{t^2} - \frac{1}{t} - 1} + \frac{\alpha}{t}\right] = -\beta$$

$$\Rightarrow \quad \lim_{t\to 0}\left(\frac{\sqrt{1 - t - t^2}}{t} + \frac{\alpha}{t}\right) = -\beta$$

$$\Rightarrow \quad \lim_{t\to 0}\left(\frac{\sqrt{1 - t - t^2} + \alpha}{t}\right) = -\beta$$

Since, $\beta \neq 0$, thus clearly from above $\alpha = -1$ otherwise the above explanation is invalid.

Now, $\lim_{t\to 0}\left(\frac{\sqrt{1-t-t^2}-1}{t}\right)=-\beta$

$$\lim_{t\to 0}\left(\frac{1-t-t^2-1}{t(\sqrt{1-t-t^2}+1)}\right)=-\beta \quad \text{[by Rationalising]}$$

$$\Rightarrow \lim_{t\to 0}\frac{-t(1+t)}{t(\sqrt{1-t-t^2}+1)}=-\beta$$

$$\Rightarrow \lim_{t\to 0}\frac{-(1+t)}{\sqrt{1-t-t^2}+1}=-\beta$$

$$\Rightarrow -1/2=-\beta \Rightarrow \beta=1/2$$

Thus, $8(\alpha+\beta)=8\left(-1+\frac{1}{2}\right)=8\left(-\frac{1}{2}\right)=-4$

**30.** *(a)* Given function of limit is

$$\lim_{x\to 7}\frac{18-[1-x]}{[x]-3a}$$

$$\text{LHL}=\lim_{x\to 7}\frac{18-[1-x]}{[x]-3a}=\lim_{h\to 0}\frac{18-[1-(7-h)]}{[7-h]-3a}$$

When, $-6\le x<-5$, then greatest integer function would be $-6$ and when $6\le x<7$, then greatest integer would be 6.

$$=\frac{18-(-6)}{6-3a}=\frac{24}{6-3a}$$

$$\text{RHL}=\lim_{x\to 7}\frac{18-[1-x]}{[x]-3a}$$

$$=\lim_{h\to 0}\frac{18-[1-(7+h)]}{[7+h]-3a}$$

When, $-7\le x<-6$, then greatest integer function would be $-7$ and when $7\le x<8$, then greatest integer function would be 7.

$$=\frac{18-(-7)}{7-3a}=\frac{25}{7-3a}$$

Take LHL = RHL

$$\frac{24}{6-3a}=\frac{25}{7-3a}$$

$$\Rightarrow a=-6$$

**31.** *(a)*

> **key idea** **Sandwich Theorem**
>
> $\Rightarrow$ Let $g(x)\le f(x)\le h(x)$ and $\lim_{x\to a} g(x)=\lim_{x\to a} h(x)=l$
>
> $\therefore \lim_{x\to a} f(x)=l$

As, we know that $r\le[r]<r+1$

$$2r\le[2r]<2r+1$$
$$3r\le[3r]<3r+1$$
$$\vdots \quad \vdots \quad \vdots$$
$$nr\le[nr]<nr+1$$

Adding $(r+2r+3r+4r+\ldots+nr)\le[r]+[2r]+[3r]+[4r]+\ldots[nr]<(r+1)+(2r+1)+(3r+1)+(4r+1)+\ldots+(nr+1)$

$$\Rightarrow r(1+2+3+4+\ldots+n)\le[r]+[2r]+[3r]+\ldots+[nr]<(r+2r+3r+\ldots+nr)+\underbrace{(1+1+1+\ldots+1)}_{n\text{-times}}$$

$$\Rightarrow r\cdot\frac{n(n+1)}{2}\le[r]+[2r]+[3r]+\ldots+[nr]<\frac{r\cdot(n(n+1))}{2}+n$$

$$\Rightarrow \frac{r\cdot\left(\frac{n(n+1)}{2}\right)}{n^2}\le\frac{[r]+[2r]+[3r]+\ldots+[nr]}{n^2}<\frac{r\cdot\left(\frac{n(n+1)}{2}\right)+n}{n^2}$$

Now, $\lim_{n\to\infty}\frac{n(n+1)\cdot r}{2\cdot n^2}=\lim_{n\to\infty}\frac{n\cdot n(1+1/n)\cdot r}{2n^2}=\frac{(1+0)\cdot r}{2}=\frac{r}{2}$ ...(i)

and $\lim_{n\to\infty}\frac{\frac{n(n+1)}{2}\cdot r+n}{n^2}=\lim_{n\to\infty}\frac{n(n+1)\cdot r+2n}{2n^2}$

$$=\lim_{n\to\infty}\frac{n^2\left\{\left(1+\frac{1}{n}\right)r+\frac{2}{n}\right\}}{2n^2}$$

$$=\frac{(1+0)\cdot r+0}{2}=\frac{r}{2} \quad \ldots\text{(ii)}$$

From Eqs. (i) and (ii), by Sandwich theorem, we conclude that,

$$\lim_{n\to\infty}\frac{[r]+[2r]+[3r]+\ldots+[nr]}{n^2}=\frac{r}{2}$$

**32.** *(98)* We know that sum of roots, $\alpha+\beta=-\frac{b}{a}$

$$a+b=\lim_{t\to -1^+}\alpha_t+\beta_t=\lim_{t\to -1^+}-\left[\frac{((t+2)^{1/6}-1)}{((t+2)^{1/7}-1)}\right]$$

$$a+b=\lim_{h\to 0}\alpha_t+\beta_t=\lim_{h\to 0}-\left[\frac{((1+h)^{1/6}-1)}{((1+h)^{1/7}-1)}\right]$$

$$=\lim_{h\to 0}-\left[\frac{1+\frac{1}{6}h+\frac{\frac{1}{6}\left(\frac{1}{6}-1\right)}{2!}h^2+\ldots}{1+\frac{1}{7}h+\frac{\frac{1}{7}\left(\frac{1}{7}-1\right)}{2!}h^2+\ldots}\right]$$

$$=\frac{-\frac{1}{6}}{\frac{1}{7}}=-\frac{7}{6}$$

Thus, $72(a+b)^2=72\times\left(-\frac{7}{6}\right)^2$

$$=72\times\frac{49}{36}=98$$

**33.** *(24)*

**key idea** For $x \in R$, $x = [x] + \{x\}$, where $[\cdot]$ is GIF and $\{x\}$ be the fractional part of $x$.

Given, $$\lim_{x\to 0^+}\left(x\left(\left[\frac{1}{x}\right]+\left[\frac{2}{x}\right]+\left[\frac{3}{x}\right]+\ldots+\left[\frac{p}{x}\right]\right)-x^2\left(\left[\frac{1}{x^2}\right]+\left[\frac{2^2}{x^2}\right]+\left[\frac{3^2}{x^2}\right]+\ldots+\left[\frac{9^2}{x^2}\right]\right)\right)\geq 1$$

We can write above limit in sigma ($\Sigma$) notation as

$$\lim_{x\to 0^+}\left(\sum_{r=1}^{p} x\left[\frac{r}{x}\right]-\sum_{k=1}^{9} x^2\left[\frac{k^2}{x^2}\right]\right)\geq 1$$

$$\Rightarrow \sum_{r=1}^{p}\lim_{x\to 0^+} x\left[\frac{r}{x}\right]-\sum_{k=1}^{9}\lim_{x\to 0^+} x^2\left[\frac{k^2}{x^2}\right]\geq 1 \quad \ldots(i)$$

Now, $\lim_{x\to 0^+} x\left[\frac{r}{x}\right]=\lim_{x\to 0^+} x\cdot\left(\frac{r}{x}-\left\{\frac{r}{x}\right\}\right)$ where $\{\cdot\}$ denotes fractional part function

$$=\lim_{x\to 0^+}\left(r-x\left\{\frac{r}{x}\right\}\right)$$

$$=\lim_{x\to 0^+} r-\lim_{x\to 0^+} x\left\{\frac{r}{x}\right\}=r-0\left(\because 0\leq\left\{\frac{r}{x}\right\}<1\right)$$

$$=r$$

Similarly, we can find

$$=\lim_{x\to 0^+} x^2\left[\frac{k^2}{x^2}\right]$$

$$=\lim_{x\to 0^+} x^2\left(\frac{k^2}{x^2}-\left\{\frac{k^2}{x^2}\right\}\right)$$

$$=k^2-0=k^2$$

From Eq. (i), we get

$$\sum_{r=1}^{p} r-\sum_{k=1}^{9} k^2\geq 1$$

$$\Rightarrow \frac{p(p+1)}{2}-\frac{9\times(9+1)(2\times 9+1)}{6}\geq 1$$

$$\Rightarrow \frac{p(p+1)}{2}-285\geq 1$$

$$\Rightarrow p(p+1)\geq 2\times 286$$

$$\Rightarrow p(p+1)\geq 572$$

$\therefore$ Least natural value of $p=24$ $\quad(\because 24^2=576>572.)$

**34.** *(64)* Given limit is $1^\infty$ form

Now, $$L=e^{\lim_{t\to 0}\frac{\left(\left(\frac{8x+5}{3(t+1)}\right)_0^1-1\right)}{t}}$$

$$=e^{\lim_{t\to 0}\frac{8^{t+1}-5^{t+1}-3t-3}{3t(t+1)}}$$

$$=e^{\frac{8\ln 8-5\ln 5-3}{3}}$$

$$=\left(\frac{8}{5}\right)^{\frac{2}{3}}\left(\frac{64}{5e}\right)=\frac{\alpha}{5e}\left(\frac{8}{5}\right)^{\frac{2}{3}}$$

On comparing, we get $\alpha=64$

**35.** *(100)* Let $L=\lim_{x\to 1}\frac{(5x+1)^{1/3}-(x+5)^{1/3}}{(2x+3)^{1/2}-(x+4)^{1/2}}$ $\quad\left(\text{form}\frac{0}{0}\right)$

$$\therefore \quad L=\lim_{x\to 1}\frac{\frac{1}{3}(5x+1)^{\frac{-2}{3}}\cdot 5-\frac{1}{3}(x+5)^{\frac{-2}{3}}}{\frac{1}{2}(2x+3)^{\frac{-1}{2}}\cdot 2-\frac{1}{2}(x+4)^{\frac{-1}{2}}}$$

(Using L' Hospital Rule)

$$=\frac{8\sqrt{5}}{3\cdot 6^{2/3}}=\frac{m\sqrt{5}}{n(2n)^{2/3}} \quad \text{(given)}$$

On comparing, we get

$$m=8 \text{ and } n=3$$

Thus, $\quad 8m+12n=100$

**36.** *(81)* $$\lim_{x\to 0}\frac{ax^2e^x-b\log_e(1+x)+cxe^{-x}}{x^2\sin x}=1$$

$$\lim_{x\to 0}\frac{ax^2\left(1+x+\frac{x^2}{2!}+\frac{x^3}{3!}+\ldots\right)-b\left(x-\frac{x^2}{2}+\frac{x^3}{3}-\ldots\right)+cx\left(1-x+\frac{x^2}{2!}-\frac{x^3}{3!}+\ldots\right)}{x^3\left(\frac{\sin x}{x}\right)}=1$$

$$\Rightarrow \lim_{x\to 0}\frac{(c-b)x+\left(\frac{b}{2}-c+a\right)x^2+\left(a-\frac{b}{3}+\frac{c}{2}\right)x^3+\ldots}{x^3}=1$$

$$c-b=0 \quad \ldots(i)$$

$$\frac{b}{2}-c+a=0 \quad \ldots(ii)$$

$$a-\frac{b}{3}+\frac{c}{2}=1 \quad \ldots(iii)$$

From Eqs. (i), (ii) and (iii), we get

$$a=\frac{3}{4},\ b=c=\frac{3}{2}$$

$$\therefore 16(a^2+b^2+c^2)=16\left(\frac{9}{16}+\frac{9}{4}+\frac{9}{4}\right)=9+72=81$$

**37.** *(26)* $$\sum_{n=0}^{\infty}\frac{n^3((2n)!)+(2n-1)(n!)}{(n!)((2n)!)}$$

$$=\sum_{n=0}^{\infty}\frac{1}{(n-3)!}+\sum_{n=0}^{\infty}\frac{3}{(n-2)!}+\sum_{n=0}^{\infty}\frac{1}{(n-1)!}+\sum_{n=0}^{\infty}\frac{1}{(2n-1)!}-\sum_{n=0}^{\infty}\frac{1}{(2n)!}$$

$$=e+3e+e+\frac{1}{2}\left(e-\frac{1}{e}\right)-\frac{1}{2}\left(e+\frac{1}{e}\right)$$

$$=5e-\frac{1}{e}=ae+\frac{b}{e}+c$$

where $a, b, c \in$
On comparing $a=5, b=-1, c=0$

$$a^2-b+c=25-(-1)+0=26$$

**38.** *(3)* Given function is $f(x)=[1+x]+\dfrac{\alpha^{2[x]+\{x\}}+[x]-1}{2[x]+\{x\}}$

$$\lim_{x\to 0^-} f(x)=\alpha-\frac{4}{3}$$

Here, $0\le x<1$, then $[x]$ is 0 and $-1\le x<0$, then $[x]$ is $-1$.

$$\Rightarrow \quad 0+\frac{\alpha^{-1}-2}{-1}=\alpha-\frac{4}{3}$$

$$\Rightarrow \quad 2-\frac{1}{\alpha}=\alpha-\frac{4}{3} \Rightarrow \alpha+\frac{1}{\alpha}=\frac{10}{3}$$

$$\Rightarrow \quad 3\alpha^2-10\alpha+3=0 \Rightarrow 3\alpha^2-9\alpha-\alpha+3=0$$

$$\Rightarrow \quad 3\alpha(\alpha-3)-1(\alpha-3)=0 \Rightarrow (\alpha-3)(3\alpha-1)=0,$$

So, $\alpha=3, \alpha\in I$

**39.** *(7)* Given that $f(x)=x^6+2x^4+x^3+2x+3$

and $\lim_{x\to 1}\dfrac{x^n f(1)-f(x)}{x-1}=44$

$$\Rightarrow \lim_{x\to 1}\frac{9x^n-(x^6+2x^4+x^3+2x+3)}{x-1}=44 \quad [\because f(1)=9]$$

Since, we get 0/0 form, therefore by L' Hospital rule, we have

$$\lim_{x\to 1}\frac{9nx^{n-1}-(6x^5+8x^3+3x^2+2)}{1}=44$$

$$\Rightarrow \quad 9n-(19)=44 \Rightarrow 9n=63 \Rightarrow n=7$$

**40.** *(5)*

> **key idea** (i) Firstly check $\lim_{x\to 0}\dfrac{ax-(e^{4x}-1)}{ax(e^{4x}-1)}$ in $\dfrac{0}{0}$ form
>
> (ii) Apply L-Hospital's rule

$$\lim_{x\to 0}\frac{ax-(e^{4x}-1)}{ax(e^{4x}-1)}=L \text{ (say)} \quad \left[\frac{0}{0}\text{ form}\right]$$

On applying L' Hospital rule, we get

$$L=\lim_{x\to 0}\frac{a-4e^{4x}}{a(e^{4x}-1)+ax(4e^{4x})}$$

[Limit exist everywhere except $a=4$]

Again, apply L' Hospital rule,

$$L=\lim_{x\to 0}\frac{-16e^{4x}}{a(4e^{4x})+a(4e^{4x})+ax(16e^{4x})}=\frac{-16}{4a+4a}=\frac{-2}{a}$$

$$=\frac{-2}{4}=\frac{-1}{2} \quad [\text{Use } a=4]$$

Given, $\quad L=b \Rightarrow \dfrac{-2}{a}=\dfrac{-1}{2}=b$

Then, $a-2b=4-2\left(\dfrac{-1}{2}\right)=4+1=5$

**41.** *(40)* Given, $\lim_{x\to 1}\dfrac{x+x^2+x^3+\ldots+x^n-n}{x-1}=820, (n\in N)$

$$\Rightarrow \lim_{x\to 1}\left\{\frac{x-1}{x-1}+\frac{x^2-1}{x-1}+\frac{x^3-1}{x-1}+\ldots+\frac{x^n-1}{x-1}\right\}=820$$

$$\Rightarrow \quad 1+2+3+\ldots+n=820 \Rightarrow \frac{n(n+1)}{2}=820$$

$$\Rightarrow \quad n(n+1)=1640=40\times 41 \Rightarrow n=40$$

## Topic 2

**01.** *(c)* $\lim_{x\to 0^+}\dfrac{\tan(5x^{1/3})\cdot\ln(1+3x^2)}{(\tan^{-1}(3\sqrt{x}))^2\cdot(e^{5x^{\frac{4}{3}}}-1)}$

$$\Rightarrow \lim_{x\to 0^+}\frac{\left(\dfrac{\tan(5x^{\frac{1}{3}})}{5x^{\frac{1}{3}}}\right)\cdot\left(\dfrac{\ln(1+3x^2)}{3x^2}\right)\cdot 15x^{\frac{7}{3}}}{\left(\dfrac{\tan^{-1}(3\sqrt{x})}{3\sqrt{x}}\right)^2\cdot\left(\dfrac{e^{5x^{\frac{4}{3}}}-1}{5x^{\frac{4}{3}}}\right)\cdot 9x\cdot 5x^{\frac{4}{3}}}$$

$$\Rightarrow \lim_{x\to 0^+}\frac{\left(\dfrac{\tan(5x^{\frac{1}{3}})}{5x^{\frac{1}{3}}}\right)\cdot\left(\dfrac{\ln(1+3x^2)}{3x^2}\right)}{\left(\dfrac{\tan^{-1}(3\sqrt{x})}{3\sqrt{x}}\right)^2\cdot\left(\dfrac{e^{5x^{\frac{4}{3}}}-1}{5x^{\frac{4}{3}}}\right)}\cdot\frac{15x^{\frac{7}{3}}}{45x^{\frac{7}{3}}}=\frac{1}{3}$$

**02.** *(c)* $\lim_{x\to 0}\dfrac{\cos(2x)+a\cos(4x)-b}{x^4}$

$$\lim_{x\to 0}\frac{\left(1-\dfrac{4x^2}{2!}+\dfrac{16x^4}{4!}-\ldots\right)+a\left(1-\dfrac{(4x)^2}{2!}+\dfrac{(4x)^4}{4!}-\ldots\right)-b}{x^4}$$

For limit to be finite

$$1+a-b=0 \text{ and } -2-8a=0$$

$$\Rightarrow \quad a=-\frac{1}{4} \text{ and } b=1-\frac{1}{4}=\frac{3}{4}$$

$$\Rightarrow \quad a+b=-\frac{1}{4}+\frac{3}{4}=\frac{1}{2}$$

**03.** *(a)* $\lim_{x\to 0}\dfrac{x^2\sin(\alpha x)+(\gamma-1)e^{x^2}}{\sin(2x)-\beta x}=3$

$$\Rightarrow \lim_{x\to 0}\frac{x^2\left[\alpha x-\dfrac{\alpha^3x^3}{3!}+\ldots\right]+(\gamma-1)\left[1+x^2+\dfrac{x^4}{2!}+\ldots\right]}{\left[2x-\dfrac{8x^3}{3!}+\dfrac{(2x)^5}{5!}-\ldots\right]-\beta x}=3$$

$$\Rightarrow \lim_{x\to 0}\frac{(\gamma-1)(1+x^2)+\alpha x^3+(\gamma-1)\dfrac{x^4}{2!}\ldots}{(2-\beta)x-\dfrac{8x^3}{3!}+\ldots}=3$$

Limit will be exist when $\gamma-1=0$ and $2-\beta=0$
$\Rightarrow \gamma=1, \beta=2$

$$\alpha x^3+(\gamma-1)\frac{x^4}{2!} \Rightarrow \alpha=4$$

$$\therefore \quad \beta+\gamma-\alpha=2+1+4=7$$

**04.** (b) $\lim_{x\to 0} \operatorname{cosec}x(\sqrt{2\cos^2 x+3\cos x}-\sqrt{\cos^2 x+\sin x+4})$

$$=\lim_{x\to 0}\left\{\frac{\operatorname{cosec}x(\cos^2 x+3\cos x-\sin x-4)}{\sqrt{2\cos^2 x+3\cos x}+\sqrt{\cos^2 x+\sin x+4}}\right\}$$

(on rationalisation)

$$=\frac{1}{2\sqrt{5}}\lim_{x\to 0}\frac{(\cos^2 x+3\cos x-\sin x-4)}{\sin x}$$

$$=\frac{1}{2\sqrt{5}}\lim_{x\to 0}\left[\frac{-2\cos x\cdot\sin x-3\sin x-\cos x}{\cos x}\right]$$

[Using L'Hospital Rule]

$$=\frac{-1}{2\sqrt{5}}$$

**05.** (d) $\lim_{x\to 0}\dfrac{e^{2|\sin x|}-2|\sin x|-1}{x^2}$

$$e^x=1+x+\frac{x^2}{2!}+\frac{x^3}{3!}+\ldots$$

$$e^{2|\sin x|}=1+2|\sin x|+\frac{(2|\sin x|)^2}{2!}+\frac{(2|\sin x|)^3}{3!}+\ldots$$

$\therefore$

$$\lim_{x\to 0}=\frac{1+2|\sin x|+\dfrac{4|\sin x|^2}{2!}+\dfrac{8|\sin x|^3}{3!}+\ldots-2|\sin x|-1}{x^2}$$

$$=\lim_{x\to 0}\frac{|\sin x|^2}{x^2}\left[\frac{4}{2!}+\frac{8}{3!}|\sin x|+\ldots\right]$$

$$=\lim_{x\to 0}\frac{\sin^2 x}{x^2}\left[\frac{4}{2!}+\frac{8}{3!}|\sin x|+\ldots\right]=\frac{4}{2!}=2$$

**06.** (c) $\lim_{x\to 0}\left[\left(\dfrac{1-\cos^2 3x}{\cos^3 4x}\right)\left(\dfrac{\sin^3 4x}{[\log_e(2x+1)]^5}\right)\right]$

$$=\lim_{x\to 0}\left[\frac{\sin^2 3x}{(3x)^2}\times\frac{\sin^3 4x}{(4x)^3}\times\frac{1}{\cos^3 4x}\times\frac{1}{\left[\dfrac{\log(1+2x)}{2x}\right]^5}\times\frac{3^2\cdot 4^3}{2^5}\right]$$

$$=\frac{3^2\cdot 4^3}{2^5}=9\times 2=18$$

**07.** (b) Let $L=\lim_{t\to 0}\left(1^{\frac{1}{\sin^2 t}}+2^{\frac{1}{\sin^2 t}}+\ldots+n^{\frac{1}{\sin^2 t}}\right)^{\sin^2 t}$

$$=\lim_{t\to 0} n\left[\left(\frac{1}{n}\right)^{\operatorname{cosec}^2 t}+\left(\frac{2}{n}\right)^{\operatorname{cosec}^2 t}+\left(\frac{3}{n}\right)^{\operatorname{cosec}^2 t}+\ldots+\left(\frac{n-1}{n}\right)^{\operatorname{cosec}^2 t}+\left(\frac{n}{n}\right)^{\operatorname{cosec}^2 t}\right]^{\sin^2 t}$$

$$=\lim_{t\to 0} n\cdot[0+0+0+0+\ldots+0+1]^{\sin^2 t}$$

($\because a^x\to 0$, when $x\to\infty$; $0<a<1$) $=n$

[$\because$ (exact 1)$^{\text{(tends to zero)}}$ = always one]

**08.** (c) $\lim_{x\to 0}\left(\dfrac{x}{\sqrt[8]{1-\sin x}-\sqrt[8]{1+\sin x}}\right)$

We get $\dfrac{0}{0}$ form, using L'Hospital rule, we get

$$\lim_{x\to 0}\frac{1}{\dfrac{-\cos x}{8(1-\sin x)^{7/8}}-\dfrac{\cos x}{8(1+\sin x)^{7/8}}}$$

$$=\frac{1}{\dfrac{-\cos 0}{8(1-\sin 0)^{7/8}}-\dfrac{\cos 0}{8(1+\sin 0)^{7/8}}}=-4$$

**09.** (d) $\lim_{x\to 0^+}\dfrac{\cos^{-1}(x-[x]^2)\cdot\sin^{-1}(x-[x]^2)}{x-x^3}$

Put the value $x\to 0+h$.

$$\lim_{h\to 0}\frac{\cos^{-1}(h-0)\cdot\sin^{-1}(h-0)}{h-h^3}=\lim_{h\to 0}\frac{\cos^{-1}h\cdot\sin^{-1}h}{h(1-h)(1+h)}$$

$$=\lim_{h\to 0}\left(\frac{\sin^{-1}h}{h}\right)\left[\frac{\cos^{-1}h}{(1-h)(1+h)}\right]=1\cdot\frac{\pi}{2}=\frac{\pi}{2}$$

**10.** (d) Given, $\lim_{x\to 0}\left(\tan\left(\dfrac{\pi}{4}+x\right)\right)^{1/x}$ [Having $1^\infty$ form]

$$=e^{\lim_{x\to 0}\left\{\frac{1}{x}\left[\tan\left(\frac{\pi}{4}+x\right)-1\right]\right\}}=e^{\lim_{x\to 0}\left\{\frac{1}{x}\left(\frac{1+\tan x}{1-\tan x}-1\right)\right\}}$$

$$=e^{\lim_{x\to 0}\left(\frac{1}{x}\times\frac{2\tan x}{1-\tan x}\right)}=e^{\lim_{x\to 0}2\times\frac{\tan x}{x}\times\frac{1}{1-\tan x}}$$

$$=e^{2(1)(1)}=e^2 \qquad \left[\because \lim_{x\to 0}\frac{\tan x}{x}=1\right]$$

**11.** (a) Given limit is $\lim_{x\to 0}\dfrac{\sin^2 x}{\sqrt{2}-\sqrt{1+\cos x}}$ $\left[\dfrac{0}{0}\text{ form}\right]$

$$=\lim_{x\to 0}\frac{\sin^2 x}{\sqrt{2}-\sqrt{2}\cos\frac{x}{2}}=\lim_{x\to 0}\frac{\sin^2 x}{\sqrt{2}\left(1-\cos\frac{x}{2}\right)}$$

$$=\lim_{x\to 0}\frac{\sin^2 x}{\sqrt{2}\times 2\sin^2\left(\frac{x}{4}\right)}=\frac{16}{2\sqrt{2}}=4\sqrt{2}$$

**12.** (c) We have, $\lim_{x\to 0}\dfrac{(1-\cos 2x)(3+\cos x)}{x\tan 4x}$

$$=\lim_{x\to 0}\frac{2\sin^2 x(3+\cos x)}{x\times\dfrac{\tan 4x}{4x}\times 4x}$$

$$=\lim_{x\to 0}\frac{2\sin^2 x}{x^2}\times\lim_{x\to 0}\frac{(3+\cos x)}{4}\times\frac{1}{\lim_{x\to 0}\dfrac{\tan 4x}{4x}}$$

$$=2\times\frac{4}{4}\times 1 \qquad \left(\because \lim_{\theta\to 0}\frac{\sin\theta}{\theta}=1\text{ and }\lim_{\theta\to 0}\frac{\tan\theta}{\theta}=1\right)$$

$$=2$$

**13.** *(d)* $\lim_{x\to 0}\dfrac{\sin(\pi\cos^2 x)}{x^2}$

$$=\lim_{x\to 0}\frac{\sin(\pi(1-\sin^2 x))}{x^2}=\lim_{x\to 0}\frac{\sin(\pi-\pi\sin^2 x)}{x^2}$$

$$=\lim_{x\to 0}\frac{\sin(\pi\sin^2 x)}{x^2}=\lim_{x\to 0}\frac{\sin(\pi\sin^2 x)}{\pi\sin^2 x}\times(\pi)\left(\frac{\sin^2 x}{x^2}\right)=\pi$$

$$\left[\because \lim_{x\to 0}\frac{\sin x}{x}=1\right]$$

**14.** *(a)* $\lim_{x\to 1^+}\dfrac{(x-1)(6+\lambda\cos(x-1))+\mu\sin(1-x)}{(x-1)^3}=-1$

Put $x=1+h$

$$\lim_{h\to 0}\frac{h(6+\lambda\cos h)-\mu\sin h}{h^3}=-1$$

Using L'Hospital Rule,

$$\lim_{h\to 0}\frac{(6+\lambda\cos h)+h(-\lambda\sin h)-\mu\cos h}{3h^2}=-1$$

Limit exist only when

$$6+\lambda-\mu=0 \quad \text{...(i)}$$

Again, using L'Hospital rule

$$\lim_{h\to 0}\frac{-\lambda\sin h-\lambda\sin h-h\lambda\cos h+\mu\sin h}{6h}=-1$$

$$\lim_{h\to 0}\frac{-\lambda\cos h-\lambda\cos h-\lambda\cos h+h\lambda\sin h+\mu\cos h}{6}=-1$$

$$\Rightarrow \frac{-3\lambda+\mu}{6}=-1$$

$$\Rightarrow -3\lambda+\mu=-6 \quad \text{...(ii)}$$

From Eqa. (i) and (ii), we get

$$-2\lambda=-12 \Rightarrow \lambda=6 \text{ and } \mu=12$$

$$\therefore \quad \lambda+\mu=18$$

**15.** *(b)* Since, we know that $\lim_{x\to 0}\dfrac{\sin x}{x}=1$

So, $\lim_{x\to 0^+}f(x)=\lim_{x\to 0}\begin{vmatrix} a+\dfrac{\sin x}{x} & 1 & b \\ a & 1+\dfrac{\sin x}{x} & b \\ a & 1 & b+\dfrac{\sin x}{x}\end{vmatrix}$

$$=\begin{vmatrix} a+1 & 1 & b \\ a & 1+1 & b \\ a & 1 & b+1\end{vmatrix}$$

$$=(a+1)(2(b+1)-b)+1(ab-a(b+1))-ba$$

$$=(a+1)(b+2)+(-a)-ab$$

$$=ab+2a+b+2-a-ab$$

$$=a+b+2=\lambda+\mu a+vb \quad \text{[given]}$$

$$\therefore \quad \lambda=2,\ \mu=1,\ v=1$$

Hence, $(\lambda+\mu+\gamma)^2=(2+1+1)^2=4^2=16$

**16.** *(b)* Given, $\lim_{x\to 0}\dfrac{3+\alpha\sin x+\beta\cos x+\log_e(1-x)}{3\tan^2 x}=\dfrac{1}{3}$

$$\Rightarrow \lim_{x\to 0}\frac{\dfrac{3+\alpha\sin x+\beta\cos x+\log_e(1-x)}{x^2}}{\dfrac{3\tan^2 x}{x^2}\times x^2}=\frac{1}{3}$$

For limit to be exist, $\beta+3=0\Rightarrow\beta=-3$

So, $\lim_{x\to 0}\dfrac{\alpha\cos x-\beta\sin x-\dfrac{1}{1-x}}{(2x)}=1$

$$\left[\text{using L' Hospital rule and } \lim_{x\to 0}\left(\frac{\tan x}{x}\right)^2=1\right]$$

For limit to be exist, Numerator $\to 0$

$\Rightarrow \alpha-0-1=0\Rightarrow\alpha=1$ [for Method (2) to solve this question is "using expansion technique"]

Hence, $2\alpha-\beta=2\times 1-(-3)=5$

**17.** *(d)* Given, $\lim_{x\to 0}\dfrac{e^{ax}-\cos(bx)-\dfrac{cxe^{-cx}}{2}}{1-\cos 2x}=17$

On expansion above, we get

$$\Rightarrow \lim_{x\to 0}\frac{\left(1+ax+\dfrac{a^2x^2}{2!}+\ldots\right)-\left(1-\dfrac{x^2b^2}{2!}+\ldots\right)-\dfrac{cx}{2}\left(1-cx+\dfrac{c^2x^2}{2!}\ldots\right)}{1-1+\dfrac{4x^2}{2!}-\ldots}=17$$

(All the neighbour terms are negligible)

$$\Rightarrow \lim_{x\to 0}\frac{\left(a-\dfrac{c}{2}\right)x+x^2\left(\dfrac{a^2}{2}+\dfrac{b^2}{2}+\dfrac{c^2}{2}\right)}{2x^2}=17$$

If above limit exist, then $a-\dfrac{c}{2}=0$

$$a=\frac{c}{2}\Rightarrow c=2a \text{ and } \frac{\dfrac{a^2}{2}+\dfrac{b^2}{2}+\dfrac{c^2}{2}}{2}=17$$

$$\Rightarrow \frac{a^2+b^2+c^2}{4}=17\Rightarrow a^2+b^2+4a^2=68 \quad \{\because c=2a\}$$

$$\Rightarrow 5a^2+b^2=68$$

**18.** *(d)* $x=2$ is a root of $x^2+px+q=0$

$$\Rightarrow 2^2+2p+q=0\Rightarrow q+4=-2p \quad \text{...(i)}$$

$$\therefore \quad q^2+8q+16=(q+4)^2=4p^2 \quad \text{...(ii)}$$

and $x^2-4px+(q^2+8q+16)$

$$=x^2-4px+4p^2=(x-2p)^2 \quad \text{...(iii)}$$

We have, $f(x)=\begin{cases}\dfrac{1-\cos(x^2-4px+q^2+8q+16)}{(x-2p)^4}, & x\neq 2p\\ 0 & x=2p\end{cases}$

$$\Rightarrow f(x)=\begin{cases}\dfrac{1-\cos(x-2p)^2}{(x-2p)^4}, & x\neq 2p\\ 0, & x=2p\end{cases}$$

Now, as $x\to 2p^+$, $x-2p\to 0^+$

$\therefore \lim_{x\to 2p^+} f(x)=\frac{1}{2}$ $\left[\because \lim_{y\to 0}\frac{1-\cos y}{y^2}=\frac{1}{2}\right]$

Hence, $\lim_{x\to 2p^+}[f(x)]=0$, where [.] is the greatest integer function.

**19.** (a) $\lim_{x\to\frac{\pi}{4}}\frac{8\sqrt{2}-(\cos x+\sin x)^7}{\sqrt{2}-\sqrt{2}\sin 2x}$ $\left[\frac{0}{0}\text{ form}\right]$

Using L' Hospital rule,

$$=\lim_{x\to\frac{\pi}{4}}\frac{-7(\sin x+\cos x)^6(-\sin x+\cos x)}{-2\sqrt{2}\cos 2x}\quad\left[\frac{0}{0}\text{ form}\right]$$

$$=\frac{-7\left(\frac{1}{\sqrt{2}}+\frac{1}{\sqrt{2}}\right)^6}{-2\sqrt{2}}\cdot\lim_{x\to\frac{\pi}{4}}\frac{(\cos x-\sin x)}{\cos 2x}\quad\left[\frac{0}{0}\text{ form}\right]$$

$$=\frac{7(\sqrt{2})^6}{2\sqrt{2}}\lim_{x\to\frac{\pi}{4}}\frac{(-\sin x-\cos x)}{-2\sin 2x}$$

$$=\frac{56}{2\sqrt{2}}\cdot\frac{1}{2}\lim_{x\to\frac{\pi}{4}}\frac{(\sin x+\cos x)}{\sin 2x}=\frac{56}{4\sqrt{2}}\cdot\frac{\sqrt{2}}{1}=14$$

**20.** (a) $\lim_{x\to\frac{\pi}{2}}\left[\tan^2 x\left((2\sin^2 x+3\sin x+4)^{\frac{1}{2}}-(\sin^2 x+6\sin x+2)^{\frac{1}{2}}\right)\right]$

$$=\lim_{x\to\pi/2}[\tan^2 x[\sqrt{2\sin^2 x+3\sin x+4}-\sqrt{\sin^2 x+6\sin x+2}]$$

$$=\lim_{x\to\frac{\pi}{2}}\frac{\tan^2 x[(2\sin^2 x+3\sin x+4)-(\sin^2 x+6\sin x+2)]}{\sqrt{2\sin^2 x+3\sin x+4}+\sqrt{\sin^2 x+6\sin x+2}}$$

$$=\lim_{x\to\frac{\pi}{2}}\frac{\sin^2 x[\sin^2 x-3\sin x+2]}{\cos^2 x[\sqrt{2\sin^2 x+3\sin x+4}+\sqrt{\sin^2 x+6\sin x+2}]}$$

$$=\lim_{x\to\frac{\pi}{2}}\frac{\sin^2 x(\sin x-1)(\sin x-2)}{(1-\sin^2 x)[\sqrt{2\sin^2 x+3\sin x+4}+\sqrt{\sin^2 x+6\sin x+2}]}$$

$$=\lim_{x\to\frac{\pi}{2}}\frac{\sin^2 x(-1)(\sin x-2)}{(1+\sin x)[\sqrt{2\sin^2 x+3\sin x+4}+\sqrt{\sin^2 x+6\sin x+2}]}$$

Substitute $x=\pi/2$

$$=\frac{1(-1)[1-2]}{(2)[\sqrt{9}+\sqrt{9}]}=\frac{1}{2(3+3)}=\frac{1}{12}$$

**21.** (d) Given, $L=\lim_{x\to 0}\frac{\sin^{-1}x-\tan^{-1}x}{3x^3}$

$$\Rightarrow L=\lim_{x\to 0}\frac{\left(x+\frac{1^2\cdot x^3}{3!}+\frac{1^2\cdot 3^2\cdot x^5}{5!}+\dots\right)-\left(x-\frac{x^3}{3}+\frac{x^5}{5}\dots\right)}{3x^3}$$

[Using expansion of $\sin^{-1}x$ and $\tan^{-1}x$]

$$\Rightarrow L=\lim_{x\to 0}\frac{\left(\frac{x^3}{3!}+\frac{9x^5}{5!}+\dots\right)-\left(-\frac{x^3}{3}+\frac{x^5}{5}\dots\right)}{3x^3}$$

$$\Rightarrow L=\lim_{x\to 0}\frac{x^3\left[\left(\frac{1}{6}+\frac{9x^2}{120}+\dots\right)+\left(\frac{1}{3}-\frac{x^2}{5}+\dots\right)\right]}{3x^3}$$

$$\Rightarrow L=\frac{1/6+1/3}{3}=\frac{1}{6}$$

$$\therefore 6L+1=6\times\frac{1}{6}+1=2$$

**22.** (a) Given, $\lim_{x\to 1^+}\frac{(1-|x|+\sin|1-x|)\sin\left(\frac{\pi}{2}[1-x]\right)}{|1-x|[1-x]}$

Put $x=1+h$, then $x\to 1^+\Rightarrow h\to 0^+$

$$\therefore \lim_{x\to 1^+}\frac{(1-|x|+\sin|1-x|)\sin\left(\frac{\pi}{2}[1-x]\right)}{|1-x|[1-x]}$$

$$=\lim_{h\to 0^+}\frac{(1-|h+1|+\sin|-h|)\sin\left(\frac{\pi}{2}[-h]\right)}{|-h|[-h]}$$

$$=\lim_{h\to 0^+}\frac{(1-(h+1)+\sin h)\sin\left(\frac{\pi}{2}[-h]\right)}{h[-h]}$$

$(\because |-h|=h \text{ and } |h+1|=h+1 \text{ as } h>0)$

$$=\lim_{h\to 0^+}\frac{(-h+\sin h)\sin(\pi/2(-1))}{h(-1)}$$

$(\because [x]=-1 \text{ for } -1\le x<0 \text{ and } h\to 0^+\Rightarrow -h\to 0^-)$

$$=\lim_{h\to 0^+}\frac{(-h+\sinh)}{-h}\sin\left(\frac{-\pi}{2}\right)$$

$$=\lim_{h\to 0^+}\frac{(-h+\sin h)}{-h}(-1)=\lim_{h\to 0^+}\frac{\sin h-h}{h}$$

$$=\lim_{h\to 0^+}\left(\frac{\sin h}{h}\right)-\lim_{h\to 0^+}\left(\frac{h}{h}\right)=1-1=0\quad\left[\because \lim_{h\to 0^+}\frac{\sin h}{h}=1\right]$$

**23.** (c) Let $f(x)=\lim_{x\to 0}\frac{x\tan 2x-2x\tan x}{(1-\cos 2x)^2}$

$$=\lim_{x\to 0}\frac{\frac{x\sin 2x}{\cos 2x}-\frac{2x\sin x}{\cos x}}{4\sin^4 x}$$

$$=\lim_{x\to 0}\frac{x\cos x\cdot\sin 2x-2x\sin x\cos 2x}{4\sin^4 x}\times\lim_{x\to 0}\frac{1}{\cos 2x\cdot\cos x}$$

$$=\lim_{x\to 0}\frac{2x\cos^2 x\sin x-2x\sin x\cos 2x}{4\sin^4 x}\times 1$$

$$=\lim_{x\to 0}\frac{x(\cos^2 x-\cos 2x)}{2\sin^3 x}=\lim_{x\to 0}\frac{x(\cos^2 x-\cos^2 x+\sin^2 x)}{2\sin^3 x}$$

$$=\lim_{x\to 0}\frac{x}{2\sin x}=\lim_{x\to 0}\frac{1}{2\frac{\sin x}{x}}=\frac{1}{2}\lim_{x\to 0}\frac{1}{\frac{\sin x}{x}}=\frac{1}{2}$$

**24.** *(b)* $\lim_{x\to\pi/2}\dfrac{\cot x-\cos x}{(\pi-2x)^3}=\lim_{x\to\pi/2}\dfrac{1}{8}\cdot\dfrac{\cos x(1-\sin x)}{\sin x\left(\dfrac{\pi}{2}-x\right)^3}$

$$=\lim_{h\to0}\frac{1}{8}\cdot\frac{\cos\left(\frac{\pi}{2}-h\right)\left[1-\sin\left(\frac{\pi}{2}-h\right)\right]}{\sin\left(\frac{\pi}{2}-h\right)\left(\frac{\pi}{2}-\frac{\pi}{2}+h\right)^3}\quad\left[\text{Put } x=\frac{\pi}{2}-h\right]$$

$$=\frac{1}{8}\lim_{h\to0}\frac{\sin h(1-\cos h)}{\cos h\cdot h^3}=\frac{1}{8}\lim_{h\to0}\frac{\sin h\left(2\sin^2\frac{h}{2}\right)}{\cos h\cdot h^3}$$

$$=\frac{1}{4}\lim_{h\to0}\frac{\sin h\cdot\sin^2\left(\frac{h}{2}\right)}{h^3\cos h}=\frac{1}{4}\lim_{h\to0}\left(\frac{\sin h}{h}\right)\left(\frac{\sin\frac{h}{2}}{\frac{h}{2}}\right)^2\cdot\frac{1}{\cos h}\cdot\frac{1}{4}$$

$$=\frac{1}{4}\times\frac{1}{4}=\frac{1}{16}$$

**25.** *(c)* $\lim_{x\to0}\dfrac{(1-\cos 2x)^2}{2x\tan x-x\tan 2x}$

$$=\lim_{x\to0}\frac{(2\sin^2 x)^2}{2x\left(x+\frac{x^3}{3}+\frac{2x^5}{15}+\ldots\right)-x\left(2x+\frac{2^3x^3}{3}+2\frac{2^5x^5}{15}+\ldots\right)}$$

$$=\lim_{x\to0}\frac{4\left(x-\frac{x^3}{3!}+\frac{x^5}{5!}-\ldots\right)^4}{x^4\left(\frac{2}{3}-\frac{8}{3}\right)+x^6\left(\frac{4}{15}-\frac{64}{15}\right)+\ldots}$$

$$=\lim_{x\to0}\frac{4\left(1-\frac{x^2}{3!}+\frac{x^4}{5!}-\ldots\right)^4}{-2+x^2\left(-\frac{60}{15}\right)+\ldots}=-2$$

***Alternate Solution***

$$\lim_{x\to0}\frac{(1-\cos 2x)^2}{2x\tan x-x\tan 2x}=\lim_{x\to0}\frac{4\sin^4 x}{2x\tan x-\frac{x\cdot2\tan 2x}{1-\tan^2 x}}$$

$$\left[\because \tan 2x=\frac{2\tan x}{1-\tan^2 x}\text{ and }1-\cos 2x=2\sin^2 x\right]$$

$$=\lim_{x\to0}\frac{4\sin^4 x(1-\tan^2 x)}{2x\tan x(1-\tan^2 x-1)}$$

$$=\lim_{x\to0}\frac{4\sin^4 x(1-\tan^2 x)}{-2x^4\frac{\tan^3 x}{x^3}}=-2$$

**26.** *(d)* Now, $\lim_{x\to\alpha}\dfrac{1-\cos(ax^2+bx+c)}{(x-\alpha)^2}$

$$=\lim_{x\to\alpha}\frac{2\sin^2\left(\frac{ax^2+bx+c}{2}\right)}{(x-\alpha)^2}$$

$$=\lim_{x\to\alpha}\frac{2\sin^2\left(\frac{a}{2}(x-\alpha)(x-\beta)\right)}{\left(\frac{a}{2}\right)^2(x-\alpha)^2(x-\beta)^2}\cdot\left(\frac{a}{2}\right)^2(x-\beta)^2$$

$$=\lim_{x\to\alpha}\frac{a^2}{2}(x-\beta)^2\quad\left[\because \lim_{x\to0}\frac{\sin x}{x}=1\right]$$

$$=\frac{a^2}{2}(\alpha-\beta)^2$$

**27.** *(c)* $\lim_{x\to\frac{\pi}{2}}\dfrac{\left(1-\tan\frac{x}{2}\right)(1-\sin x)}{\left(1+\tan\frac{x}{2}\right)(\pi-2x)^3}$

Let $x=\dfrac{\pi}{2}-h$ as $x\to\dfrac{\pi}{2}, h\to0$

$$\therefore \lim_{h\to0}\frac{1-\tan\left(\frac{\pi}{4}-\frac{h}{2}\right)}{1+\tan\left(\frac{\pi}{4}-\frac{h}{2}\right)}\cdot\frac{(1-\cos h)}{(2h)^3}$$

$$=\lim_{h\to0}\tan\left[\frac{\pi}{4}-\left(\frac{\pi}{4}-\frac{h}{2}\right)\right]\cdot\frac{2\sin^2\frac{h}{2}}{8h^3}$$

$$=\lim_{h\to0}\frac{1}{4}\cdot\frac{\tan h/2}{h/2\times2}\times\left(\frac{\sin h/2}{h/2}\right)^2\times\frac{1}{4}$$

$$=\frac{1}{32}\lim_{h\to0}\left(\frac{\tan h/2}{h/2}\right)\lim_{h\to0}\left(\frac{\sin h/2}{h/2}\right)^2=\frac{1}{32}$$

$$\left[\because \lim_{\theta\to0}\frac{\sin\theta}{\theta}=1\text{ and }\lim_{\theta\to0}\frac{\tan\theta}{\theta}=1\right]$$

**28.** *(c)* $\lim_{x\to0}\left(\dfrac{\tan^{-1}x+\ln\sqrt{\dfrac{1+x}{1-x}}-2x}{x^5}\right)$

$$\lim_{x\to0}\left(\frac{\tan^{-1}x+\frac{1}{2}[\ln(1+x)-\ln(1-x)]-2x}{x^5}\right)$$

Using Expansion

$$\lim_{x\to0}\frac{\left[\left(x-\frac{x^3}{3}+\frac{x^5}{5}-\ldots\right)+\frac{1}{2}\left[\left(x-\frac{x^2}{2}+\frac{x^3}{3}-\frac{x^4}{4}+\frac{x^5}{5}-\ldots\right)-\left(-x-\frac{x^2}{2}-\frac{x^3}{3}-\frac{x^4}{4}-\frac{x^5}{5}-\ldots\right)\right]-2x\right]}{x^5}$$

$$\lim_{x\to 0}\frac{\left[2x-\frac{x^3}{3}+\frac{x^3}{3}+\frac{x^5}{5}+\frac{x^5}{5}+(\text{terms containing higher power of } x \text{ greater than } 5)-2x\right]}{x^5}$$

$$\lim_{x\to 0}\frac{x^5\left(\frac{2}{5}+\text{terms containing } x\right)}{x^5}=\frac{2}{5}$$

Statement II

$$\lim_{x\to 1}(x^{2/1-x})$$

$$=\lim_{x\to 1}\frac{1}{x^{2/x-1}}$$

$$=\frac{1}{e^{\lim_{x\to 1}\frac{2}{x-1}(x-1)}} \qquad [\because \lim_{x\to 0}(1+x)^{1/x}=e]$$

$$=\frac{1}{e^2}$$

**29.** *(b)* $\alpha, \beta$ are the roots of the equation $ax^2+bx+1=0$

$\Rightarrow$ The roots of the equation $x^2+bx+a=0$ are $\frac{1}{\alpha}$ and $\frac{1}{\beta}$.

$$\Rightarrow \quad x^2+bx+a=\left(x-\frac{1}{\alpha}\right)\left(x-\frac{1}{\beta}\right)$$

$\therefore$ The given limit

$$=\lim_{x\to\frac{1}{\alpha}}\left[\frac{1-\cos\left(x-\frac{1}{\alpha}\right)\left(x-\frac{1}{\beta}\right)}{2\alpha^2\left(x-\frac{1}{\alpha}\right)^2}\right]^{1/2}$$

$$=\lim_{x\to\frac{1}{\alpha}}\left[\frac{1-\cos\left(x-\frac{1}{\alpha}\right)\left(x-\frac{1}{\beta}\right)}{2\alpha^2\left(x-\frac{1}{\alpha}\right)^2\left(x-\frac{1}{\beta}\right)^2}\right]^{1/2}\times\left(x-\frac{1}{\beta}\right)$$

$$=\lim_{x\to\frac{1}{\alpha}}\left[\frac{2\sin^2\left(\frac{\left(x-\frac{1}{\alpha}\right)\left(x-\frac{1}{\beta}\right)}{2}\right)}{2\alpha^2\left\{\frac{\left(x-\frac{1}{\alpha}\right)\left(x-\frac{1}{\beta}\right)}{2}\right\}^2\times 4}\right]^{1/2}\left(x-\frac{1}{\beta}\right)$$

$$=\frac{1}{2\alpha}\left(\frac{1}{\beta}-\frac{1}{\alpha}\right)=\frac{1}{2\alpha}\left(\frac{1}{\beta}-\frac{1}{\alpha}\right)$$

$\therefore \quad k=2\alpha$

**30.** *(a)* Given that

$$f(x)=\begin{cases}\frac{x}{|x|} & x\neq 0\\ 1 & x=0\end{cases},\quad g(x)=\begin{cases}\frac{\sin(x+1)}{(x+1)} & x\neq -1\\ 1 & x=-1\end{cases}$$

and $h(x)=2[x]-f(x)$

We have to find, $\lim_{x\to 1} g(h(x-1))$

Now, LHS $=\lim_{x\to 1^-}=\lim_{x\to 1^-} g(h(0^-))=\lim_{x\to 1^-} g(2[0^-]-f(0^-))$

$$=\lim_{x\to 1^-} g\left(2(-1)-\frac{0}{|0^-|}\right)=\lim_{x\to 1^-} g(-2+1)=\lim_{x\to 1^-} g(-1)=1$$

RHS $=\lim_{x\to 1^+} g(h(x-1))=\lim_{x\to 1^+} g(h(1^+-1))=\lim_{x\to 1^+} g(h(0^+))$

$$=\lim_{x\to 1^+} g(2[0^+]-f(0^+))=\lim_{x\to 1^+} g\left(2\times 0-\frac{0^+}{|0^+|}\right)$$

$$=\lim_{x\to 1^+} g(-1)=1$$

LHS = RHS So, $\lim_{x\to 1} g(h(x-1))=1$

**31.** *(c)* $\lim_{x\to 0}\frac{\alpha e^x+\beta e^{-x}+\gamma\sin x}{x\sin^2 x}=\frac{2}{3}$

$$\lim_{x\to 0}\frac{\alpha\left(1+x+\frac{x^2}{2!}+\frac{x^3}{3!}+\ldots\right)+\beta\left(1-x+\frac{x^2}{2!}-\frac{x^3}{3!}+\ldots\right)+\gamma\left(x-\frac{x^3}{3!}+\frac{x^5}{5!}-\ldots\right)}{x\left(x-\frac{x^3}{3!}+\frac{x^5}{5!}-\ldots\right)^2}$$

Now, here constant term $=0$

$$\alpha+\beta=0 \qquad \ldots\text{(i)}$$

Also coefficient of $x$ should be zero

i.e. $\quad \alpha-\beta+\gamma=0 \qquad \ldots$(ii)

and coefficent of $x^2$ should be zero

$$\frac{\alpha}{2!}+\frac{\beta}{2!}=0 \quad\Rightarrow\quad \alpha+\beta=0$$

So, $$\lim_{x\to 0}\frac{x^3\left(\frac{\alpha}{3!}-\frac{\beta}{3!}-\frac{\gamma}{3!}\right)+x^4\left(\frac{\alpha}{4!}+\frac{\beta}{4!}\right)}{x^3}=\frac{2}{3}$$

$$\frac{\alpha}{3!}-\frac{\beta}{3!}-\frac{\gamma}{3!}=\frac{2}{3}\Rightarrow \alpha-\beta-\gamma=4 \qquad \ldots\text{(iii)}$$

On adding Eqs. (ii) and (iii), we get

$$\alpha-\beta=2 \qquad \ldots\text{(iv)}$$

From Eqs. (i) and (iv), we get

$$\alpha=1, \beta=-1, \gamma=-2$$

From option (c), $\alpha\beta^2+\beta\gamma^2+\gamma\alpha^2+3$

$$=1-4-2+3=-2\neq 0$$

**32.** (c) $\lim_{x\to 0}\frac{\cos(\sin x)-\cos x}{x^4}$

$$=\lim_{x\to 0}\frac{2\sin\left(\frac{\sin x+x}{2}\right)\cdot\sin\left(\frac{x-\sin x}{2}\right)}{x^4}\qquad\left[\frac{0}{0}\text{ form}\right]$$

$$=\lim_{x\to 0}\frac{2\sin\left(\frac{\sin x+x}{2}\right)}{\left(\frac{\sin x+x}{2}\right)}\cdot\frac{\sin\left(\frac{x-\sin x}{2}\right)}{\left(\frac{x-\sin x}{2}\right)}\times\left(\frac{x+\sin x}{2}\right)\times\left(\frac{x-\sin x}{2}\right)\frac{1}{x^4}$$

$$=\lim_{x\to 0}2\left[\frac{\sin\left(\frac{\sin x+x}{2}\right)}{\left(\frac{\sin x+x}{2}\right)}\right]\left[\frac{\sin\left(\frac{x-\sin x}{2}\right)}{\left(\frac{x-\sin x}{2}\right)}\right]\left(\frac{x^2-\sin^2 x}{4}\right)\cdot\frac{1}{x^4}$$

We know, $\lim_{t\to 0}\frac{\sin t}{t}=1$, thus we get

$$\lim_{x\to 0}2\left(\frac{x^2-\sin^2 x}{4x^4}\right)\qquad\left[\frac{0}{0}\text{ form}\right]$$

Using L' Hospital rule, we get

$$\lim_{x\to 0}2\left(\frac{2x-2\sin x\cos x}{16x^3}\right)=\lim_{x\to 0}\frac{2x-\sin 2x}{8x^3}\qquad\left[\frac{0}{0}\text{ form}\right]$$

Using L' Hospital rule, we get

$$\lim_{x\to 0}\frac{2-2\cos 2x}{24x^2}\qquad\left[\frac{0}{0}\text{ form}\right]$$

$$=\lim_{x\to 0}\left(\frac{4\sin 2x}{48x}\right)=\frac{1}{6}\lim_{x\to 0}\cdot\frac{\sin 2x}{2x}\qquad\left[\lim_{t\to 0}\frac{\sin t}{t}=1\right]$$

$$=\frac{1}{6}$$

**33.** (a) $\lim_{h\to 0}2\left\{\frac{\sqrt{3}\sin\left(\frac{\pi}{6}+h\right)-\cos\left(\frac{\pi}{6}+h\right)}{\sqrt{3}h(\sqrt{3}\cos h-\sin h)}\right\}$

$$=\lim_{h\to 0}2\left\{\frac{2\left(\frac{\sqrt{3}}{2}\sin\left(\frac{\pi}{6}+h\right)-\frac{1}{2}\cos\left(\frac{\pi}{6}+h\right)\right)}{2\times\sqrt{3}h\left(\frac{\sqrt{3}}{2}\cos h-\frac{1}{2}\sin h\right)}\right\}$$

$$=\lim_{h\to 0}2\left\{\frac{\cos\left(\frac{\pi}{6}\right)\sin\left(\frac{\pi}{6}+h\right)-\sin\left(\frac{\pi}{6}\right)\cos\left(\frac{\pi}{6}+h\right)}{\sqrt{3}h\left(\cos\frac{\pi}{6}\cos h-\sin\frac{\pi}{6}\sin h\right)}\right\}$$

$$=\lim_{h\to 0}2\left\{\frac{\sin\left(\frac{\pi}{6}+h-\frac{\pi}{6}\right)}{\sqrt{3}h\cos\left(h+\frac{\pi}{6}\right)}\right\}$$

$$=\lim_{h\to 0}\frac{2}{\sqrt{3}}\left\{\frac{\sin h}{h\cos(h+\pi/6)}\right\}$$

$$=\frac{2}{\sqrt{3}}\cdot\lim_{h\to 0}\frac{\sin h}{h}\cdot\frac{1}{\cos(h+\pi/6)}=\frac{2}{\sqrt{3}}\cdot(1)\cdot\frac{1}{\cos(\pi/6)}$$

$$=\frac{2}{\sqrt{3}}\cdot 1\cdot\frac{2}{\sqrt{3}}=\frac{4}{3}$$

**34.** (d) At $x=0$,

$$\text{RHL}=\lim_{x\to 0^+}\frac{\tan(\pi\sin^2 x)+(|x|-\sin(x[x]))^2}{x^2}$$

$$=\lim_{x\to 0^+}\frac{\tan(\pi\sin^2 x)+(x-\sin(x\cdot 0))^2}{x^2}\qquad\left[\begin{array}{l}\because |x|=x\text{ for }x>0\\ \text{and }[x]=0\text{ for }0\le x<1\end{array}\right]$$

$$=\lim_{x\to 0^+}\frac{\tan(\pi\sin^2 x)+x^2}{x^2}$$

$$=\lim_{x\to 0^+}\left(\frac{\tan(\pi\sin^2 x)}{\pi\sin^2 x}\cdot\frac{\pi\sin^2 x}{x^2}+1\right)$$

$$=\pi\lim_{x\to 0^+}\frac{\tan(\pi\sin^2 x)}{\pi\sin^2 x}\cdot\lim_{x\to 0^+}\frac{\sin^2 x}{x^2}+1$$

$$=\pi+1\qquad\left[\because\lim_{x\to 0}\frac{\tan x}{x}=1\text{ and }\lim_{x\to 0}\frac{\sin x}{x}=1\right]$$

$$\text{and LHL}=\lim_{x\to 0^-}\frac{\tan(\pi\sin^2 x)+(|x|-\sin(x[x]))^2}{x^2}$$

$$=\lim_{x\to 0^-}\frac{\tan(\pi\sin^2 x)+(-x-\sin(x(-1)))^2}{x^2}$$

$$\left[\begin{array}{l}\because |x|=-x\text{ for }x<0\\ \text{and }[x]=-1\text{ for }-1\le x<0\end{array}\right]$$

$$=\lim_{x\to 0^-}\frac{\tan(\pi\sin^2 x)+(x+\sin(-x))^2}{x^2}$$

$$=\lim_{x\to 0^-}\frac{\tan(\pi\sin^2 x)+(x-\sin x)^2}{x^2}\qquad[\because\sin(-\theta)=-\sin\theta]$$

$$=\lim_{x\to 0^-}\left(\frac{\tan(\pi\sin^2 x)+x^2+\sin^2 x-2x\sin x}{x^2}\right)$$

$$=\lim_{x\to 0^-}\left(\frac{\tan(\pi\sin^2 x)}{x^2}+1+\frac{\sin^2 x}{x^2}-\frac{2x\sin x}{x^2}\right)$$

$$=\lim_{x\to 0^-}\left(\frac{\tan(\pi\sin^2 x)}{\pi\sin^2 x}\cdot\frac{\pi\sin^2 x}{x^2}+1+\frac{\sin^2 x}{x^2}-2\frac{\sin x}{x}\right)$$

$$=\lim_{x\to 0^-}\frac{\tan(\pi\sin^2 x)}{\pi\sin^2 x}\cdot\lim_{x\to 0^-}\frac{\pi\sin^2 x}{x^2}+1+\lim_{x\to 0^-}\frac{\sin^2 x}{x^2}-2\lim_{x\to 0^-}\frac{\sin x}{x}$$

$$=\pi+1+1-2=\pi$$

$\Rightarrow$ RHL $\neq$ LHL

$\therefore$ Limit does not exist.

**35.** *(32)* We know that $\tan x = x + \frac{x^3}{3} + \frac{2x^5}{15} + \ldots$

Now, $\lim_{x\to 0}\left(\frac{\tan x}{x}\right)^{\frac{1}{x^2}} = p$

$\Rightarrow e^{\lim_{x\to 0}\left[\frac{\tan x - x}{x}\right]\times\frac{1}{x^2}} = p$

$\Rightarrow e^{\lim_{x\to 0}\left[\frac{x+\frac{x^3}{3}+\frac{2x^5}{15}+\ldots-x}{x^3}\right]} = p$

$\Rightarrow e^{1/3} = p$

$\therefore 96\log_e p = 96\log_e(e^{1/3}) = 96\times\frac{1}{3} = 32$

**36.** *(1)* Since, $\frac{\tan\theta+\tan^3\theta}{1-\tan^2\theta} = \frac{2\tan\theta-\tan\theta+\tan^3\theta}{1-\tan^2\theta}$

$$= \frac{2\tan\theta}{1-\tan^2\theta} - \frac{\tan\theta(1-\tan^2\theta)}{1-\tan^2\theta}$$

$$= \tan 2\theta - \tan\theta$$

$$\therefore \quad f(x) = \lim_{n\to\infty}\sum_{r=0}^{n}\left(\frac{\tan\left(\frac{x}{2^{r+1}}\right)+\tan^3\left(\frac{x}{2^{r+1}}\right)}{1-\tan^2\left(\frac{x}{2^{r+1}}\right)}\right)$$

$$= \lim_{n\to\infty}\sum_{r\to 0}^{n}\tan\left(\frac{2x}{2^{r+1}}\right)-\tan\left(\frac{x}{2^{r+1}}\right)$$

$$= \lim_{n\to\infty}\left[\left(\tan\frac{x}{1}-\tan\frac{x}{2}\right)+\left(\tan\frac{x}{2}-\tan\frac{x}{2^2}\right)+\ldots+\left(\tan\frac{x}{2^n}-\tan\frac{x}{2^{n+1}}\right)\right]$$

$$= \lim_{n\to\infty}\left(\tan x - \tan\frac{x}{2^{n+1}}\right) = \tan x - 0 = \tan x$$

$$\therefore \quad \lim_{x\to 0}\frac{e^x - e^{f(x)}}{(x - f(x))} = \lim_{x\to 0}\frac{e^x - e^{\tan x}}{x - \tan x}$$

$$= \lim_{x\to 0} e^{\tan x}\lim_{x\to 0}\frac{e^{x-\tan x}-1}{x-\tan x} = e^0\cdot 1 = 1$$

**37.** *(18)* Finding right hand limit

$$\lim_{x\to 0^+} f(x) = \lim_{h\to 0} f(0+h) = \lim_{h\to 0} f(h)$$

$$= \lim_{h\to 0}\frac{\cos^{-1}(1-h^2)\sin^{-1}(1-h)}{h(1-h^2)}$$

$$= \lim_{h\to 0}\frac{\cos^{-1}(1-h^2)}{h}\left(\frac{\sin^{-1}1}{1}\right)$$

Let $\cos^{-1}(1-h^2) = \theta \Rightarrow \cos\theta = 1-h^2$

$$= \frac{\pi}{2}\lim_{\theta\to 0}\frac{\theta}{\sqrt{1-\cos\theta}} = \frac{\pi}{2}\lim_{\theta\to 0}\frac{1}{\sqrt{\frac{1-\cos\theta}{\theta^2}}} = \frac{\pi}{2}\frac{1}{\sqrt{\frac{1}{2}}}$$

$$\therefore \quad R = \frac{\pi}{\sqrt{2}}$$

$$\therefore \quad L = \lim_{x\to 0^-} f(x) = \lim_{h\to 0} f(-h)$$

$$= \lim_{h\to 0}\frac{\cos^{-1}(1-\{-h\}^2)\sin^{-1}(1-\{-h\})}{\{-h\}-\{-h\}^3}$$

$$= \lim_{h\to 0}\frac{\cos^{-1}(1-(-h+1)^2)\sin^{-1}(1-(-h+1))}{(-h+1)-(-h+1)^3}$$

$$= \lim_{h\to 0}\frac{\cos^{-1}(-h^2+2h)\sin^{-1}h}{(1-h)(1-(1-h)^2)}$$

$$= \lim_{h\to 0}\left(\frac{\pi}{2}\right)\frac{\sin^{-1}h}{(1-(1-h)^2)} = \frac{\pi}{2}\lim_{h\to 0}\left(\frac{\sin^{-1}h}{-h^2+2h}\right)$$

$$= \frac{\pi}{2}\lim_{h\to 0}\left(\frac{\sin^{-1}h}{h}\right)\left(\frac{1}{-h+2}\right)$$

$$\therefore \quad L = \frac{\pi}{4}$$

$$\Rightarrow \quad \frac{32}{\pi^2}(L^2+R^2) = \frac{32}{\pi^2}\left(\frac{\pi^2}{2}+\frac{\pi^2}{16}\right) = 16+2 = 18$$

**38.** *(1)* Given, $\lim_{x\to 0}\left(\frac{(x+2\cos x)^3+2(x+2\cos x)^2+3\sin(x+2\cos x)}{(x+2)^3+2(x+2)^2+3\sin(x+2)}\right)^{\frac{100}{x}}$

Let $x + 2\cos x = p$ and $x + 2 = q$

As $x\to 0$, then $p\to 2$ and $q\to 2$

$$\Rightarrow \quad \lim_{x\to 0}\left(\frac{p^3+2p^2+3\sin p}{q^3+2q^2+3\sin q}\right)^{\frac{100}{x}}$$

$$= e^{\lim_{x\to 0}\frac{100}{x}\frac{(p^3-q^3)+2(p^2-q^2)+3(\sin p-\sin q)}{q^3+2q^2+3\sin q}} \quad [\because \text{we get } 1^\infty \text{ form}]$$

$$= e^0 = 1 \quad \left[\because \lim_{x\to 0}\frac{p-q}{x} - \lim_{x\to 0}\frac{2(\cos x-1)}{x} = 0\right]$$

**39.** *(3)* Given, $\lim_{x\to 0}\frac{\alpha x e^x - \beta\log_e(1+x)+\gamma x^2 e^{-x}}{x\sin^2 x} = 10$

Now, $\lim_{x\to 0}\frac{\alpha x e^x - \beta\log_e(1+x)+\gamma x^2 e^{-x}}{x\sin^2 x}$

$$= \lim_{x\to 0}\frac{\alpha x\left(1+\frac{x}{1!}+x^2/2!+\ldots\right)-\beta\left(x-\frac{x^2}{2}+\frac{x^3}{3}\ldots\right)+\gamma x^2\left(1-\frac{x}{1!}+x^2/2!\ldots\right)}{x\sin^2 x}$$

$$= \lim_{x\to 0}\frac{x(\alpha-\beta)+x^2(\alpha+\beta/2+\gamma)+x^3\left(\frac{\alpha}{2}-\frac{\beta}{3}-\gamma\right)\ldots}{x\sin^2 x}$$

To exists limit, the numerator must have degree greater than or equal to denominator.

Degree of denominator $=3$

$\therefore$ For limit to exist, $\alpha-\beta=0$ ...(i)

and $\alpha+\frac{\beta}{2}+\gamma=0$ ...(ii)

Also, for terms greater than degree 3, gives 0 as $x\to 0$

$\therefore \frac{\alpha}{2}-\frac{\beta}{3}-\gamma=10$ ...(iii)

From Eq. (i), $\beta=\alpha$

From Eq. (ii), $\gamma=-\left(\frac{\alpha}{2}+\alpha\right)=-\frac{3\alpha}{2}$

On putting these in Eq. (iii), we get

$\frac{\alpha}{2}-\frac{\alpha}{3}+\frac{3\alpha}{2}=10\Rightarrow\alpha=6$

Now, $\alpha=\beta\Rightarrow\beta=6$

Again, $\gamma=-3\alpha/2\Rightarrow\gamma=-9$

$\therefore \alpha+\beta+\gamma=6+6-9=12-9=3$

**40.** *(8)* Given, $\lim_{x\to 0}\left\{\frac{1}{x^8}\left(1-\cos\frac{x^2}{2}-\cos\frac{x^2}{4}+\cos\frac{x^2}{2}\cos\frac{x^2}{4}\right)\right\}=2^{-k}$

$\Rightarrow \lim_{x\to 0}\frac{1}{x^8}\left(1-\cos\frac{x^2}{2}\right)\left(1-\cos\frac{x^2}{4}\right)=2^{-k}$

$\Rightarrow \lim_{x\to 0}\frac{2\sin^2\left(\frac{x^2}{4}\right)\cdot 2\sin^2\frac{x^2}{8}}{x^8}=2^{-k}$

$\Rightarrow \lim_{x\to 0} 4\left\{\left[\frac{\sin\frac{x^2}{4}}{x^2/4}\times\frac{\sin\frac{x^2}{8}}{\frac{x^2}{8}}\right]^2\frac{\frac{x^8}{(32)^2}}{x^8}\right\}=2^{-k}$

$\Rightarrow 4/2^{10}=2^{-k}\Rightarrow 2^{-k}=2^{-8}\Rightarrow k=8$

## Topic 3

**01.** *(a)* Given, $f:(-\infty,\infty)-\{0\}\to R$

$f'(1)=\lim_{a\to\infty}a^2 f(1/a)$

Now, we have to find

$\lim_{a\to\infty}\frac{a(a+1)}{2}\tan^{-1}\left(\frac{1}{a}\right)+a^2-2\ln(a)$

$=\lim_{a\to\infty}a^2\left(\frac{\left(1+\frac{1}{a}\right)}{2}\tan^{-1}\left(\frac{1}{a}\right)+1+\frac{2}{a^2}\ln\frac{1}{a}\right)$

$\Rightarrow f(x)=\frac{1}{x^2}(1+x)\tan^{-1}x+1+2x^2\ln(x)$

$f'(x)=\frac{1}{2}\left(\frac{1+x}{1+x^2}+\tan^{-1}x\right)+4x\ln x+2x$

$f'(1)=\frac{1}{2}\left(1+\frac{\pi}{4}\right)+2=\frac{5}{2}+\frac{\pi}{8}$

**02.** *(b)* $\lim_{x\to a}\frac{xf(a)-af(x)}{x-a}$

$=\lim_{x\to a}\frac{xf(a)-af(x)+af(a)-af(a)}{x-a}$

$=\lim_{x\to a}\frac{(x-a)f(a)-a[f(x)-f(a)]}{x-a}$

$=\lim_{x\to a}\frac{(x-a)f(a)}{x-a}-a\lim_{x\to a}\frac{f(x)-f(a)}{x-a}$

$=f(a)-af'(a)=4-a(2)$ [Given, $f(a)=4$, $f'(a)=2$]

$=4-2a$

**03.** *(d)* Given, $f(1)=-2$ and $f'(x)\geq 4.2$ for $1\leq x\leq 6$

Consider $f'(x)=\frac{f(x+h)-f(x)}{h}$

$\Rightarrow f(x+h)-f(x)=f'(x)\,h\geq(4.2)h$

So, $f(x+h)\geq f(x)+(4.2)h$

Put $x=1$ and $h=5$, we get

$f(6)\geq f(1)+5(4.2)\Rightarrow f(6)\geq 19$

Hence, $f(6)$ lies in $[19,\infty)$.

**04.** *(b)* Let $f(x)=3x^{10}-7x^8+5x^6-21x^3+3x^2-7$

$f'(x)=30x^9-56x^7+30x^5-63x^2+6x$

$f'(1)=30-56+30-63+6=66-63-56=-53$

Consider $\lim_{\alpha\to 0}\frac{f(1-\alpha)-f(1)}{\alpha^3+3\alpha}=\lim_{\alpha\to 0}\frac{f'(1-\alpha)(-1)-0}{3\alpha^2+3}$

(By using L' Hospital rule)

$=\frac{f'(1-0)(-1)}{3(0)^2+3}=\frac{-f'(1)}{3}=\frac{53}{3}$

**05.** *(c)* Given, $f(x+y)=f(x)\cdot f(y)$

$\therefore f(0+0)=(f(0))^2\Rightarrow f(0)=0$ or $f(0)=1$ but $f(0)\neq 0$

Now,

$f'(x)=\lim_{h\to 0}\frac{f(x+h)-f(x)}{h}=\lim_{h\to 0}\frac{f(x)f(h)-f(x)}{h}$

$f'(x)=f(x)\lim_{h\to 0}\frac{f(h)-1}{h}$

$\Rightarrow \therefore f'(0)=f(0)\lim_{h\to 0}\frac{f(h)-1}{h}$

$\Rightarrow 3=\lim_{h\to 0}\frac{f(h)-1}{h}$ [$\because f(0)=1$]

Now, $f'(x)=f(x)\lim_{h\to 0}\frac{f(h)-1}{h}$

$\therefore f'(5)=f(5)\times 3=2\times 3=6$

**06.** *(a)* $\lim_{x\to 1}\frac{\sqrt{f(x)}-1}{\sqrt{x}-1}$ $\left[\frac{0}{0}\text{ form}\right]$

$=\lim_{x\to 1}\frac{1}{2\sqrt{f(x)}}\times\frac{2\sqrt{x}}{1}\times f'(x)=\frac{2\times 1\times 2}{2}=2$

**07.** *(d)* Given, $f(x)=x^3-x^2f'(1)+xf''(2)-f'''(3)$

Let $f'(1)=a$, $f''(2)=b$ and $f'''(3)=c$

$\therefore \quad f(x) = x^3 - ax^2 + bx - c$

$f'(x) = 3x^2 - 2ax + b$

$\Rightarrow \quad f'(1) = 3 - 2a + b = a \quad \ldots(i)$

$f''(x) = 6x - 2a$

$\Rightarrow \quad f''(2) = 12 - 2a = b \quad \ldots(ii)$

$f'''(x) = 6 \Rightarrow f'''(3) = 6 = c$

$\Rightarrow \quad c = 6$

From Eqs. (i) and (ii), we get

$3 - 2a + (12 - 2a) = a$

$15 = 5a \Rightarrow a = 3$

$\therefore$ From Eqs. (ii), $b = 12 - 2 \times 3 = 6$

$b = 6$

$\therefore \quad f(0) = -6,\ f(1) = 1 - 3 + 6 - 6 = -2$

$f(2) = 8 - 12 + 12 - 6 = 2$

$f(3) = 27 - 27 + 18 - 6 = 12$

$\therefore$ Option (d) satisfies.

**Alternate Solution**

Here, $f(x) = x^3 - x^2 f'(1) + x f''(2) - f'''(3) \quad \ldots(i)$

$f'(x) = 3x^2 - 2x f'(1) + f''(2) \quad \ldots(ii)$

$f''(x) = 6x - 2f'(1) \quad \ldots(iii)$

$\Rightarrow \quad f'''(x) = 6 \Rightarrow f'''(3) = 6$

From Eq. (iii),

$f''(2) = 6 \times 2 - 2f'(1) = 12 - 2f'(1) \quad \ldots(iv)$

From Eqs. (ii) and (iv),

$f'(1) = 3 - 2f'(1) + 12 - 2f'(1) = 15 - 4f'(1) \Rightarrow f'(1) = 3$

$\therefore \quad f''(2) = 12 - 2 \times 3 = 6$ [from Eq. (iv)]

and $f(0) = -6$

$\therefore \quad f(0) = -6, f(1) = -2, f(2) = 2$ and $f(3) = 12$

**08.** *(d)* $\because f(x)$ has extreme values at $x = 1$ and $x = 2$.

$\because \quad f'(1) = 0$ and $f'(2) = 0$

As, $f(x)$ is a polynomial of degree 4.

Suppose $f(x) = Ax^4 + Bx^3 + Cx^2 + Dx + E$

$$\because \lim_{x \to 0}\left(\frac{f(x)}{x^2} + 1\right) = 3 \Rightarrow \lim_{x \to 0}\left(\frac{Ax^4 + Bx^3 + Cx^2 + Dx + E}{x^2} + 1\right) = 3$$

$$\Rightarrow \lim_{x \to 0}\left(Ax^2 + Bx + C + \frac{D}{x} + \frac{E}{x^2} + 1\right) = 3$$

As limit has finite value, so $D = 0$ and $E = 0$

Now, $A(0)^2 + B(0) + C + 0 + 0 + 1 = 3$

$\Rightarrow C + 1 = 3 \Rightarrow C = 2,\ f'(x) = 4Ax^3 + 3Bx^2 + 2Cx + D$

$f'(1) = 0 \quad \Rightarrow 4A(1) + 3B(1) + 2C(1) + D = 0$

$4A + 3B = -4 \quad \ldots(i)$

$f'(2) = 0 \Rightarrow 4A(8) + 3B(4) + 2C(2) + D = 0$

$\Rightarrow \quad 8A + 3B = -2 \quad \ldots(ii)$

From Eqs. (i) and (ii), we get $A = 1/2$ and $B = -2$

So, $\quad f(x) = \dfrac{x^4}{2} - 2x^3 + 2x^2$

Therefore, $f(-1) = \dfrac{(-1)^4}{2} - 2(-1)^3 + 2(-1)^2 = \dfrac{1}{2} + 2 + 2 = \dfrac{9}{2}$

Hence, $\quad f(-1) = 9/2$

**09.** *(55)* $$L = \lim_{x \to 0} 2\left(\frac{1 - \cos x\,(\cos 2x)^{1/2}\,(\cos 3x)^{1/3} \ldots (\cos 10x)^{1/10}}{x^2}\right)$$

At $x = 0$, $L$ makes $= \dfrac{0}{0}$ form

So, we will use L-Hospital's rule

$\because \quad \dfrac{d}{dx}(x^2) = 2x$ and $\dfrac{d}{dx}(1) = 0$

and for $\dfrac{d}{dx}(1 + \cos x(\cos 2x)^{1/2} \ldots. (\cos 10x)^{1/10})$

Let $y = \cos x\,(\cos 2x)^{\frac{1}{2}} \ldots (\cos 10x)^{\frac{1}{10}}$

Let $\log(y) = \log(\cos x) + \dfrac{1}{2}\log\cos(2x) + \ldots + \dfrac{1}{10}\log(\cos 10x)$

$$\frac{1}{y}\left(\frac{dy}{dx}\right) = \tan x + \frac{1}{2} \times 2\tan 2x + \ldots + \frac{1}{10} \times 10\tan 10x$$

$$\frac{dy}{dx} = y(\tan x + \tan 2x + \ldots + \tan 10x)$$

$$\text{So, } L = \lim_{x \to 0} 2\left(\frac{y(\tan x + \tan 2x + \ldots + \tan 10x)}{2x}\right)$$

$\because \lim_{x \to 0} \cos x\,(\cos 2x)^{\frac{1}{2}} \ldots. (\cos 10x)^{\frac{1}{10}} = 1$

$$= \lim_{x \to 0} 1 \times \left(\frac{\tan x}{x} + \frac{2\tan 2x}{2x} + \ldots + \frac{10\tan 10x}{10x}\right)$$

$$= 1 + 2 + 3 + \ldots + 10 = \frac{10 \times 11}{2} = 55$$

**10.** *(3)* **Method I** Given, $f(x + y) = f(x) \cdot f(y)\ \forall\ x, y \in R$

$\therefore \quad f(x) = a^x \Rightarrow f'(x) = a^x \cdot \log(a)$

Now, $f'(0) = \log(a) \Rightarrow 3 = \log(a)$

$\Rightarrow \quad a = e^3$

$\therefore \quad f(x) = (e^3)^x = e^{3x}$

$\Rightarrow \quad f(h) = e^{3h}$

$$\text{Now, } \lim_{h \to 0}\left(\frac{f(h) - 1}{h}\right) = \lim_{h \to 0}\left(\frac{e^{3h} - 1}{h}\right)$$

$$= \lim_{h \to 0}\left(\frac{e^{3h} - 1}{3h} \times 3\right) = 3 \times 1 = 3$$

**Method II** Let $L = \lim_{h \to 0} \dfrac{1}{h}(f(h) - 1) \quad \left[\dfrac{0}{0} \text{ form}\right]$

and $\quad f(x + y) = f(x) \cdot f(y)$

Put $x = y = 0$

$\therefore f(0) = f(0) \cdot f(0) \Rightarrow [f(0)]^2 = f(0)$

$\Rightarrow \quad [f(0)]^2 - f(0) = 0 \Rightarrow f(0)[f(0) - 1] = 0$

$\Rightarrow \quad f(0) = 0,\ f(0) = 1$

$f(0) = 0$ rejected because $f(x) \neq 0,\ \forall\ x \in R \Rightarrow f(0) = 1$

Using L' Hospital rule, we get

$$L = \lim_{h \to 0} \frac{f'(h) - 0}{1} = f'(0) = 3$$

# C-11

JEE Main Chapterwise Topicwise Mathematics

In JEE Main, there is usually 0-1 question per paper is visible from this chapter. In past 3 years, 30 conceptual and 10 numericals questions of moderate difficulty level were asked from this chapter.

**For detailed analysis scan QR code**

# STATISTICS

## Topic 1 Calculation of Mean, Median & Mode of Grouped and Ungrouped Data

### Level 1

**01** Marks obtained by all the students of class 12 are presented in a frequency distribution with classes of equal width. Let the median of this grouped data be 14 with median class interval 12-18 and median class frequency 12. If the number of students whose marks are less than 12 is 18, then the total number of students is *[2025, 23 Jan Shift-I]*

(a) 52 (b) 44 (c) 48 (d) 40

**02** The mean of the data set comprising of 16 observations is 16. If one of the observation valued 16 is deleted and three new observations valued 3, 4 and 5 are added to the data, then the mean of the resultant data is *[JEE Main 2015]*

(a) 16.8 (b) 16 (c) 15 (d) 14

**03** The average marks of boys in a class is 52 and that of girls is 42. The average marks of boys and girls combined is 50. The percentage of boys in the class is *[AIEEE 2007]*

(a) 40% (b) 20% (c) 80% (d) 60%

**Imp.** Mean, Median & Mode

**04** If in a frequency distribution, the mean and median are 21 and 22 respectively, then its mode is approximately *[AIEEE 2005]*

(a) 24.0 (b) 25.5 (c) 20.5 (d) 22

**05** The median of a set of 9 distinct observations is 20.5. If each of the largest 4 observations of the set is increased by 2, then the median of the new set *[AIEEE 2003]*

(a) is increased by 2
(b) is decreased by 2
(c) is two times the original median
(d) remains the same as that of the original set

### Level 2

**06** Let $M$ denote the median of the following frequency distribution. *[2024, 30 Jan Shift-I]*

| Class | 0-4 | 4-8 | 8-12 | 12-16 | 16-20 |
|---|---|---|---|---|---|
| Frequency | 3 | 9 | 10 | 8 | 6 |

Then, $20M$ is equal to

(a) 32 (b) 416 (c) 104 (d) 208

**07** If for some $x \in R$, the frequency distribution of the marks obtained by 20 students in a test is

| Marks | 2 | 3 | 5 | 7 |
|---|---|---|---|---|
| Frequency | $(x+1)^2$ | $2x-5$ | $x^2-3x$ | $x$ |

Then, the mean of the marks is *[2019, 10 April Shift-I]*

(a) 3.0 (b) 2.8 (c) 2.5 (d) 3.2

**08** In a class of 100 students, there are 70 boys whose average marks in a subject are 75. If the average marks of the complete class is 72, then what is the average of the girls? *[AIEEE 2002]*

(a) 73 (b) 65 (c) 68 (d) 74

#### Numerical Value Type Questions

**09** Let $x_1, x_2, x_3, \ldots, x_{20}$ be in geometric progression with $x_1 = 3$ and the common ratio 1/2. A new data is constructed replacing each $x_i$ by $(x_i - i)^2$. If $\bar{x}$ is the mean of new data, then the greatest integer less than or equal to $\bar{x}$ is ......... . *[2022, 28 July Shift-I]*

**10** The mean of 10 numbers $7 \times 8, 10 \times 10, 13 \times 12, 16 \times 14, \ldots$ is. *[2021, 31 Aug Shift-I]*

**11** Consider the following frequency distribution

| Class | 10-20 | 20-30 | 30-40 | 40-50 | 50-60 |
|---|---|---|---|---|---|
| Frequency | $\alpha$ | 110 | 54 | 30 | $\beta$ |

If the sum of all frequencies is 584 and median is 45, then $|\alpha - \beta|$ is equal to ...... . *[2021, 25 July Shift-I]*

**12** Consider the following frequency distribution

| Class | 0-6 | 6-12 | 12-18 | 18-24 | 24-30 |
|---|---|---|---|---|---|
| Frequency | $a$ | $b$ | 12 | 9 | 5 |

If mean $= \dfrac{309}{22}$ and median = 14, then the value $(a-b)^2$ is equal to ....... **[2021, 22 July Shift-II]**

**13** The mean age of 25 teachers in a school is 40 yr. A teacher retires at the age of 60 yr and a new teacher is appointed in his place. If the mean age of the teachers in this school now is 39 yr, then the age (in years) of the newly appointed teacher is .... **[2021, 18 March Shift-I]**

## Topic 2 Calculation of Standard Deviation, Variance & Mean Deviation for Grouped and Ungrouped Data

### Level 1

**01** If the mean and the variance of 6, 4, $a$, 8, $b$, 12, 10, 13 are 9 and 9.25 respectively, then $a+b+ab$ is equal to **[2025, 2 April Shift-II]**

(a) 106 (b) 100
(c) 103 (d) 105

**02** For a statistical data $x_1, x_2, \ldots\ldots, x_{10}$ statistical of 10 values a student obtained the mean as 5.5 and $\sum_{i=1}^{10} x_i^2 = 371$. He later found that he had noted two values in the data incorrectly as 4 and 5, instead of the correct values 6 and 8, respectively. The variance of the corrected data is **[2025, 24 Jan Shift-I]**

(a) 5 (b) 9
(c) 4 (d) 7

**03** If the variance of the frequency distribution

| $x$ | $c$ | $2c$ | $3c$ | $4c$ | $5c$ | $6c$ |
|---|---|---|---|---|---|---|
| $f$ | 2 | 1 | 1 | 1 | 1 | 1 |

is 160, then the value of $c \in N$ is **[2024, 9 April Shift-II]**

(a) 8 (b) 5
(c) 6 (d) 7

**04** Two dice $A$ and $B$ are rolled. Let the numbers obtained on $A$ and $B$ be $\alpha$ and $\beta$, respectively. If the variance of $\alpha - \beta$ is $\dfrac{p}{q}$, where $p$ and $q$ are coprime, then the sum of the positive divisors of $p$ is equal to **[2023, 12 April Shift-I]**

(a) 36 (b) 72
(c) 48 (d) 31

**05** Let the mean and variance of 12 observations be 9/2 and 4 respectively. Later on, it was observed that two observations were considered as 9 and 10 instead of 7 and 14 respectively. If the correct variance is $\dfrac{m}{n}$, where $m$ and $n$ are coprime, then $m+n$ is equal to **[2023, 8 April Shift-II]**

(a) 316 (b) 315
(c) 314 (d) 317

**06** The mean and variance of a set of 15 numbers are 12 and 14 respectively. The mean and variance of another set of 15 numbers are 14 and $\sigma^2$ respectively. If the variance of all the 30 numbers in the two sets is 13, then $\sigma^2$ is equal to **[2023, 6 April Shift-I]**

(a) 9 (b) 11
(c) 10 (d) 12

**07** The mean and variance of 5 observations are 5 and 8, respectively. If 3 observations are 1, 3, 5, then the sum of cubes of the remaining two observations is **[2023, 1 Feb Shift-I]**

(a) 1216 (b) 1072
(c) 1792 (d) 1456

**08** The number of values of $a \in N$ such that the variance of 3, 7, 12, $a$, $43-a$ is a natural number is **[2022, 29 June Shift-II]**

(a) 0 (b) 2
(c) 5 (d) infinite

**09** If $\sum_{i=1}^{n}(x_i - a) = n$ and $\sum_{i=1}^{n}(x_i - a)^2 = na$, $(n, a > 1)$, then the standard deviation of $n$ observations $x_1, x_2, \ldots, x_n$ is **[2020, 6 Sep Shift-I]**

(a) $a-1$ (b) $n\sqrt{a-1}$
(c) $\sqrt{n(a-1)}$ (d) $\sqrt{a-1}$

**Imp.** Standard Deviation & Variance

**10** If $\sum_{i=1}^{9}(x_i - 5) = 9$ and $\sum_{i=1}^{9}(x_i - 5)^2 = 45$, then the standard deviation of the 9 items $x_1, x_2, \ldots, x_9$ is **[JEE Main 2018]**

(a) 9 (b) 4 (c) 2 (d) 3

**11** If the mean of the data : 7, 8, 9, 7, 8, 7, $\lambda$, 8 is 8, then the variance of this data is **[JEE Main 2018]**

(a) 2 (b) 7/8
(c) 9/8 (d) 1

**12** If the standard deviation of the numbers 2, 3, $a$ and 11 is 3.5, then which of the following is true? **[JEE Main 2016]**

(a) $3a^2 - 26a + 55 = 0$ (b) $3a^2 - 32a + 84 = 0$
(c) $3a^2 - 34a + 91 = 0$ (d) $3a^2 - 23a + 44 = 0$

**13** All the students of a class performed poorly in Mathematics. The teacher decided to give grace marks of 10 to each of the students. Which of the following statistical measure will not change even after the grace marks were given? ***[JEE Main 2013]***

(a) Mean (b) Median (c) Mode (d) Variance

**14** Let $x_1, x_2, \ldots, x_n$ be $n$ observations $\bar{x}$ be their arithmetic mean and $\sigma^2$ be the variance.

**Statement I** Variance of $2x_1, 2x_2, \ldots, 2x_n$ is $4\sigma^2$.

**Statement II** Arithmetic mean of $2x_1, 2x_2, \ldots, 2x_n$ is $4\bar{x}$. ***[AIEEE 2012]***

(a) Statement I is false, Statement II is true
(b) Statement I is true, Statement II is true; Statement II is a correct explanation for Statement I
(c) Statement I is true, Statement II is true; Statement II is not a correct explanation for Statement I
(d) Statement I is true, Statement II is false

**Imp.** Mean Deviation About the Median

**15** If the mean deviations about the median of the numbers $a, 2a, \ldots, 50a$ is 50, then $|a|$ is equal to ***[AIEEE 2011]***

(a) 3 (b) 4 (c) 5 (d) 2

**16 Statement I** The variance of first $n$ even natural numbers is $\frac{n^2-1}{4}$.

**Statement II** The sum of first $n$ natural numbers is $\frac{n(n+1)}{2}$ and the sum of squares of first $n$ natural numbers is $\frac{n(n+1)(2n+1)}{6}$. ***[AIEEE 2009]***

(a) Statement I is true, Statement II is true; Statement II is a correct explanation for Statement I
(b) Statement I is true, Statement II is true; Statement II is not a correct explanation for Statement I
(c) Statement I is true, Statement II is false
(d) Statement I is false, Statement II is true

**17** The mean of the numbers $a, b, 8, 5, 10$ is 6 and the variance is 6.80. Then, which one of the following gives possible values of $a$ and $b$? ***[AIEEE 2008]***

(a) $a = 3$ and $b = 4$ (b) $a = 0$ and $b = 7$
(c) $a = 5$ and $b = 2$ (d) $a = 1$ and $b = 6$

**18** Suppose, a population $A$ has 100 observations 101, 102, ..., 200 and another population $B$ has 100 observations 151, 152, ... , 250. If $V_A$ and $V_B$ represent the variances of the two populations respectively, then $V_A / V_B$ is ***[AIEEE 2006]***

(a) $\frac{9}{4}$ (b) $\frac{4}{9}$
(c) $\frac{2}{3}$ (d) 1

**19** Let $x_1, x_2, \ldots, x_n$ be $n$ observations such that $\Sigma x_i^2 = 400$ and $\Sigma x_i = 80$. Then, a possible value of $n$ among the following is ***[AIEEE 2005]***

(a) 12 (b) 9 (c) 18 (d) 15

**20** Consider the following statements

I. Mode can be computed from histogram
II. Median is not independent of change of scale
III. Variance is independent of change of origin and scale

Which of these is/are correct? ***[AIEEE 2004]***

(a) only I (b) only II (c) I and II (d) I, II and III

## Level 2

**21** Let the Mean and Variance of five observations $x_1 = 1, x_2 = 3, x_3 = a, x_4 = 7$ and $x_5 = b, a > b$, be 5 and 10 respectively. Then, the variance of the observations $n + x_n, n = 1, 2, \ldots, 5$ is ***[2025, 3 April Shift-II]***

(a) 17.4 (b) 16 (c) 16.4 (d) 17

**22** Let the mean and the standard deviation of the observation 2, 3, 3, 4, 5, 7, $a$, $b$ be 4 and $\sqrt{2}$ respectively. Then, the mean deviation about the mode of these observations is ***[2025, 4 April Shift-II]***

(a) 1 (b) 3/4 (c) 1/2 (d) 2

**23** The mean and standard deviation of 100 observations are 40 and 5.1, respectively. By mistake one observation is taken as 50 instead of 40. If the correct mean and the correct standard deviation are $\mu$ and $\sigma$ respectively, then $10(\mu + \sigma)$ is equal to ***[2025, 7 April Shift-I]***

(a) 449 (b) 451 (c) 447 (d) 445

**24** The frequency distribution of the age of students in a class of 40 students is given below

| Age | 15 | 16 | 17 | 18 | 19 | 20 |
|---|---|---|---|---|---|---|
| Number of Students | 5 | 8 | 5 | 12 | $x$ | $y$ |

If the mean deviation about the median is 1.25, then $4x + 5y$ is equal to ***[2024, 9 April Shift-I]***

(a) 44 (b) 43 (c) 46 (d) 47

**25** The mean and standard deviation of 20 observations are found to be 10 and 2, respectively. On rechecking, it was found that an observation by mistake was taken 8 instead of 12. The correct standard deviation is ***[2024, 6 April Shift-I]***

(a) 1.8 (b) $\sqrt{3.96}$ (c) $\sqrt{3.86}$ (d) 1.94

**26** Let $\alpha, \beta \in R$. Let the mean and the variance of 6 observations $-3, 4, 7, -6, \alpha, \beta$ be 2 and 23, respectively. The mean deviation about the mean of these 6 observations is ***[2024, 5 April Shift-I]***

(a) $\frac{16}{3}$ (b) $\frac{14}{3}$ (c) $\frac{11}{3}$ (d) $\frac{13}{3}$

**27** Let the mean and the variance of 6 observations $a, b, 68, 44, 48, 60$ be 55 and 194, respectively. If $a > b$, then $a + 3b$ is **[2024, 31 Jan Shift-II]**

(a) 180 (b) 190
(c) 200 (d) 210

**28** Let $a_1, a_2, ..., a_{10}$ be 10 observations such that $\sum_{k=1}^{10} a_k = 50$ and $\sum_{\forall k<j} a_k \cdot a_j = 1100$. Then, the standard deviation of $a_1, a_2, ...., a_{10}$ is equal to **[2024, 27 Jan Shift-I]**

(a) 10 (b) $\sqrt{5}$
(c) 5 (d) $\sqrt{115}$

**29** The mean and standard deviation of 10 observations are 20 and 8 respectively. Later on, it was observed that one observation was recorded as 50 instead of 40. Then, the correct variance is **[2023, 15 April Shift-I]**

(a) 11 (b) 14
(c) 13 (d) 12

**30** Let the mean of 6 observations 1, 2, 4, 5, $x$ and $y$ be 5 and their variance be 10. Then, their mean deviation about the mean is equal to **[2023, 11 April Shift-II]**

(a) 10/3 (b) 3
(c) 7/3 (d) 8/3

**31** Let $\mu$ be the mean and $\sigma$ be the standard deviation of the distribution

| $x_i$ | 0 | 1 | 2 | 3 | 4 | 5 |
|---|---|---|---|---|---|---|
| $f_i$ | $k+2$ | $2k$ | $k^2-1$ | $k^2-1$ | $k^2+1$ | $k-3$ |

where $\Sigma f_i = 62$. If $[x]$ denotes the greatest integer $\leq x$, then $[\mu^2 + \sigma^2]$ is equal to **[2023, 10 April Shift-II]**

(a) 8 (b) 9 (c) 7 (d) 6

**32** Let $S$ be the set of all values of $a_1$ for which the mean deviation about the mean of 100 consecutive positive integers $a_1, a_2, a_3, ..., a_{100}$ is 25. Then, $S$ is **[2023, 30 Jan Shift-II]**

(a) $N$ (b) {9} (c) $\phi$ (d) {99}

**33** The mean and variance of 7 observations are 8 and 16, respectively. If one observation 14 is omitted and $a$ and $b$ are respectively mean and variance of remaining 6 observations, then $a + 3b - 5$ is equal to **[2023, 30 Jan Shift-I]**

(a) 37 (b) 40 (c) 45 (d) 42

**34** Let $X = \{11, 12, 13, ..., 40, 41\}$ and $Y = \{61, 62, 63, ..., 90, 91\}$ be the two sets of observations. If $\bar{x}$ and $\bar{y}$ are their respective means and $\sigma^2$ is the variance of all the observations in $X \cup Y$, then $|\bar{x} + \bar{y} - \sigma^2|$ is equal to **[2023, 29 Jan Shift-II]**

(a) 603 (b) 605 (c) 601 (d) 608

**35** The mean and variance of the marks obtained by the students in a test are 10 and 4, respectively. Later, the marks of one of the students in increased from 8 to 12. If the new mean of the marks is 10.2, then their new variance is equal to **[2023, 25 Jan Shift-I]**

(a) 4.04 (b) 4.08
(c) 3.92 (d) 3.96

**36** Let the mean and the variance of 5 observations $x_1, x_2, x_3, x_4, x_5$ be $\frac{24}{5}$ and $\frac{194}{25}$ respectively. If the mean and variance of the first 4 observation are 7/2 and $a$ respectively, then $(4a + x_5)$ is equal to **[2022, 29 June Shift-I]**

(a) 13 (b) 15
(c) 17 (d) 18

**37** Suppose, a class has 7 students. The average marks of these students in the mathematics examination is 62, and their variance is 20. A student fails in the examination if he/she gets less than 50 marks, then in worst case, the number of students can fail is **[2022, 28 June Shift-II]**

(a) 4 (b) 5
(c) 0 (d) 2

**38** The mean and standard deviation of 15 observations are found to be 8 and 3 respectively. On rechecking it was found that, in the observations, 20 was misread as 5. Then, the correct variance is equal to **[2022, 28 June Shift-I]**

(a) 17 (b) 15
(c) 19 (d) 14

**39** The mean and variance of the data 4, 5, 6, 6, 7, 8, $x$, $y$ where $x < y$ are 6 and 9/4 respectively. Then, $x^4 + y^2$ is equal to **[2022, 27 June Shift-II]**

(a) 162 (b) 320
(c) 674 (d) 420

**40** The mean and standard deviation of 50 observations are 15 and 2 respectively. It was found that one incorrect observation was taken such that the sum of correct and incorrect observations is 70. If the correct mean is 16, then the correct variance is equal to **[2022, 26 June Shift-II]**

(a) 10 (b) 36
(c) 43 (d) 60

**41** The mean and variance of 7 observations are 8 and 16 respectively. If two observations are 6 and 8, then the variance of the remaining 5 observations is **[2021, 31 Aug Shift-II]**

(a) $\frac{92}{5}$ (b) $\frac{134}{5}$
(c) $\frac{536}{25}$ (d) $\frac{112}{5}$

**42** The mean and standard deviation of 20 observations were calculated as 10 and 2.5 respectively. It was found that by mistake one data value was taken as 25 instead of 35. If $\alpha$ and $\sqrt{\beta}$ are the mean and standard deviation respectively for correct data, then $(\alpha, \beta)$ is **[2021, 26 Aug Shift-I]**

(a) (11, 26) (b) (10.5, 25)
(c) (11, 25) (d) (10.5, 26)

**43** Let the mean and variance of the frequency distribution be 6 and 6.8, respectively.

| $x$ | $x_1 = 2$ | $x_2 = 6$ | $x_3 = 8$ | $x_4 = 9$ |
|---|---|---|---|---|
| $f$ | 4 | 4 | $\alpha$ | $\beta$ |

If $x_3$ is changed from 8 to 7, then the mean for the new data will be **[2021, 27 July Shift-II]**

(a) 4 (b) 5
(c) 17/3 (d) 16/3

**44** If the mean and variance of the following data :

6, 10, 7, 13, $a$, 12, $b$, 12 are 9 and $\frac{37}{4}$ respectively,

then $(a-b)^2$ is equal to **[2021, 27 July Shift-I]**

(a) 24 (b) 12
(c) 32 (d) 16

**45** If the mean and variance of six observations 7, 10, 11, 15, $a$, $b$ are 10 and $\frac{20}{3}$, respectively, then the value of $|a-b|$ is equal to **[2021, 20 July Shift-II]**

(a) 9 (b) 11
(c) 7 (d) 1

**46** The mean of 6 distinct observations is 6.5 and their variance is 10.25. If 4 out of 6 observations are 2, 4, 5 and 7, then the remaining two observations are **[2021, 20 July Shift-I]**

(a) 10, 11 (b) 3, 18
(c) 8, 13 (d) 1, 20

**47** Consider three observations $a$, $b$ and $c$, such that $b = a + c$. If the standard deviation of $a+2$, $b+2$, $c+2$ is $d$, then which of the following is true? **[2021, 16 March Shift-I]**

(a) $b^2 = 3(a^2 + c^2) + 9d^2$
(b) $b^2 = a^2 + c^2 + 3d^2$
(c) $b^2 = 3(a^2 + c^2 + d^2)$
(d) $b^2 = 3(a^2 + c^2) - 9d^2$

**48** The mean and variance of 8 observations are 10 and 13.5 respectively. If 6 of these observations are 5,7,10, 12, 14, 15, then the absolute difference of the remaining two observations is **[2020, 4 Sep Shift-I]**

(a) 9 (b) 3
(c) 7 (d) 5

**49** Let the observations $x_i (1 \le i \le 10)$ satisfy the equations, $\sum_{i=1}^{10} (x_i - 5) = 10$ and $\sum_{i=1}^{10} (x_i - 5)^2 = 40$. If $\mu$ and $\lambda$ are the mean and the variance of the observations, $x_1 - 3, x_2 - 3, \ldots, x_{10} - 3$, then the ordered pair $(\mu, \lambda)$ is equal to **[2020, 9 Jan Shift-I]**

(a) (6, 3) (b) (3, 6) (c) (3, 3) (d) (6, 6)

**50** If the data $x_1, x_2, \ldots, x_{10}$ is such that the mean of first four of these is 11, the mean of the remaining six is 16 and the sum of squares of all of these is 2000, then the standard deviation of this data is **[2019, 12 April Shift-I]**

(a) $2\sqrt{2}$ (b) 2
(c) 4 (d) $\sqrt{2}$

**51** If both the mean and the standard deviation of 50 observations $x_1, x_2, \ldots, x_{50}$ are equal to 16, then the mean of $(x_1 - 4)^2, (x_2 - 4)^2, \ldots, (x_{50} - 4)^2$ is **[2019, 10 April Shift-II]**

(a) 480 (b) 400
(c) 380 (d) 525

**52** The mean and variance of seven observations are 8 and 16 respectively. If 5 of the observations are 2, 4, 10, 12, 14, then the product of the remaining two observations is **[2019, 8 April Shift-I]**

(a) 45 (b) 49
(c) 48 (d) 40

**53** The mean and the variance of five observations are 4 and 5.20, respectively. If three of the observations are 3, 4 and 4, then the absolute value of the difference of the other two observations is **[2019, 12 Jan Shift-II]**

(a) 1 (b) 7
(c) 5 (d) 3

**54** The mean of five observations is 5 and their variance is 9.20. If three of the given five observations are 1, 3 and 8, then a ratio of other two observations is **[2019, 10 Jan Shift-I]**

(a) 4 : 9 (b) 6 : 7
(c) 10 : 3 (d) 5 : 8

**55** If the mean deviation of the numbers 1, $1+d$, ..., $1+100d$ from their mean is 255, then a value of $d$ is **[JEE Main 2016, AIEEE 2009]**

(a) 10.1 (b) 5.05
(c) 20.2 (d) 10

**56** For two data sets, each of size 5, the variances are given to be 4 and 5 and the corresponding means are given to be 2 and 4, respectively. The variance of the combined data set is **[AIEEE 2010]**

(a) $\frac{5}{2}$ (b) $\frac{11}{2}$
(c) 6 (d) $\frac{13}{2}$

**57** In a series of $2n$ observations, half of them equal $a$ and remaining half equal $-a$. If the standard deviation of the observations is 2, then $|a|$ equal to *[AIEEE 2004]*

(a) $\frac{1}{n}$ (b) $\sqrt{2}$

(c) 2 (d) $\frac{\sqrt{2}}{n}$

**58** In an experiment with 15 observations on $x$, the following results were available

$$\Sigma x^2 = 2830, \quad \Sigma x = 170$$

One observation that was 20, was found to be wrong and was replaced by the correct value 30. Then, the corrected variance is *[AIEEE 2003]*

(a) 78 (b) 188.66

(c) 177.33 (d) 8.33

## Level 3

**Imp.** Mean and Variance

**59** Let $x_1, x_2 ..., x_{10}$ be ten observations such that $\sum_{i=1}^{10} (x_i - 2) = 30$, $\sum_{i=1}^{10} (x_i - \beta)^2 = 98$, $\beta > 2$ and their variance is $\frac{4}{5}$. If $\mu$ and $\sigma^2$ are respectively the mean and the variance of $2(x_1 - 1) + 4\beta$, $2(x_2 - 1) + 4\beta, ..., 2(x_{10} - 1) + 4\beta$, then $\frac{\beta\mu}{\sigma^2}$ is equal to *[2025, 29 Jan Shift-I]*

(a) 120 (b) 110 (c) 100 (d) 90

**60** Consider 10 observations $x_1, x_2, ..., x_{10}$ such that $\sum_{i=1}^{10} (x_i - \alpha) = 2$ and $\sum_{i=1}^{10} (x_i - \beta)^2 = 40$, where $\alpha, \beta$ are positive integers. Let the mean and the variance of the observations be $\frac{6}{5}$ and $\frac{84}{25}$, respectively. Then, $\frac{\beta}{\alpha}$ is equal to *[2024, 1 Feb Shift-II]*

(a) 2 (b) 1

(c) 3/2 (d) 5/2

**61** Let the median and the mean deviation about the median of 7 observations 170, 125, 230, 190, 210, $a$, $b$ be 170 and $\frac{205}{7}$, respectively. Then, the mean deviation about the mean of these 7 observations is *[2024, 1 Feb Shift-I]*

(a) 32 (b) 28

(c) 30 (d) 31

**62** If the mean and variance of five observations are $\frac{24}{5}$ and $\frac{194}{25}$ respectively and the mean of the first four observations is $\frac{7}{2}$, then the variance of the first four observations is equal to *[2024, 29 Jan Shift-II]*

(a) 4/5 (b) 77/12

(c) 5/4 (d) 105/4

**63** Let sets $A$ and $B$ have 5 elements each. Let the mean of the elements in sets $A$ and $B$ be 5 and 8 respectively and the variance of the elements in sets $A$ and $B$ be 12 and 20 respectively. $A$ new set $C$ of 10 elements is formed by subtracting 3 from each element of $A$ and adding 2 to each element of $B$. Then, the sum of the mean and variance of the elements of $C$ is ........... . *[2023, 11 April Shift-I]*

(a) 36 (b) 38

(c) 40 (d) 32

**64** Let $9 = x_1 < x_2 < ... < x_7$ be in an AP with common difference $d$. If the standard deviation of $x_1, x_2 ..., x_7$ is 4 and the mean is $\bar{x}$, then $\bar{x} + x_6$ is equal to *[2023, 1 Feb Shift-II]*

(a) $2\left(9 + \frac{8}{\sqrt{7}}\right)$ (b) $18\left(1 + \frac{1}{\sqrt{3}}\right)$

(c) 34 (d) 25

**65** Let the mean and standard deviation of marks of class A of 100 students be respectively 40 and $\alpha$ (> 0) and the mean and standard deviation of marks of class B of $n$ students be respectively 55 and $30 - \alpha$. If the mean and variance of the marks of the combined class of $100 + n$ students are respectively 50 and 350, then the sum of variances of classes A and B is *[2023, 31 Jan Shift-II]*

(a) 650 (b) 450

(c) 500 (d) 900

**66** The mean of the numbers $a$, $b$, 8, 5, 10 is 6 and their variance is 6.8. If $M$ is the mean deviation of the numbers about the mean, then 25 $M$ is equal to *[2022, 26 June Shift-I]*

(a) 60 (b) 55

(c) 50 (d) 45

**67** The first of the two samples in a group has 100 items with mean 15 and standard deviation 3. If the whole group has 250 items with mean 15.6 and standard deviation $\sqrt{13.44}$, then the standard deviation of the second sample is *[2021, 25 July Shift-II]*

(a) 8 (b) 6

(c) 4 (d) 5

**68** Let in a series of $2n$ observations, half of them are equal to $a$ and remaining half are equal to $-a$. Also, by adding a constant $b$ in each of these observations, the mean and standard deviation of new set become 5 and 20, respectively. Then, the value of $a^2 + b^2$ is equal to **[2021, 18 March Shift-II]**

(a) 425 (b) 650
(c) 250 (d) 925

**69** The outcome of each of 30 items was observed, 10 items gave an outcome $\frac{1}{2} - d$ each, 10 items gave outcome $\frac{1}{2}$ each and the remaining 10 items gave outcome $\frac{1}{2} + d$ each. If the variance of this outcome data is $\frac{4}{3}$, then $|d|$ equals **[2019, 11 Jan Shift-I]**

(a) 2/3 (b) $\sqrt{5}/2$
(c) $\sqrt{2}$ (d) 2

**70** In a group of data, there are $n$ observations, $x, x_2, ...., x_n$. If $\sum_{i=1}^{n}(x_i + 1)^2 = 9n$ and $\sum_{i=1}^{n}(x_i - 1)^2 = 5n$, the standard deviation of the data is **[2019, 9 Jan Shift-II]**

(a) 2 (b) $\sqrt{7}$
(c) 5 (d) $\sqrt{5}$

**71** The mean of 5 observations is 5 and their variance is 124. If three of the observations are 1, 2 and 6, then the mean deviation from the mean of the data is **[JEE Main 2016]**

(a) 2.5 (b) 2.6
(c) 2.8 (d) 2.4

## Numerical Value Type Questions

**72** The variance of the number 8, 21, 34, 47, ........., 320 is ......... . **[2025, 23 Jan Shift-II]**

**73** Let $a, b, c \in N$ and $a < b < c$. Let the mean, the mean deviation about the mean and the variance of the 5 observations 9, 25, $a$, $b$, $c$ be 18, 4 and $\frac{136}{5}$, respectively. Then, $2a + b - c$ is equal to ......... . **[2024, 8 April Shift-II]**

**74** The variance $\sigma^2$ of the data

| $x_i$ | 0 | 1 | 5 | 6 | 10 | 12 | 17 |
|---|---|---|---|---|---|---|---|
| $f_i$ | 3 | 2 | 3 | 2 | 6 | 3 | 3 |

is ......... . **[2024, 30 Jan Shift-II]**

**75** If the mean and variance of the data 65, 68, 58, 44, 48, 45, 60, $\alpha$, $\beta$, 60 where, $\alpha > \beta$, are 56 and 66.2 respectively, then $\alpha^2 + \beta^2$ is equal to ....... **[2024, 29 Jan Shift-I]**

**76** The mean and standard deviation of 15 observations were found to be 12 and 3, respectively. On rechecking it was found that an observation was read as 10 in place of 12. If $\mu$ and $\sigma^2$ denote the mean and variance of the correct observations respectively, then $15(\mu + \mu^2 + \sigma^2)$ is equal to ......... . **[2024, 27 Jan Shift-II]**

**77** The mean and standard deviation of the marks of 10 students were found to be 50 and 12, respectively. Later, it was observed that two marks 20 and 25 were wrongly read as 45 and 50 respectively. Then, the correct variance is ........... . **[2023, 13 April Shift-II]**

**78** Let the mean of the data

| $x$ | 1 | 3 | 5 | 7 | 9 |
|---|---|---|---|---|---|
| Frequency ($f$) | 4 | 24 | 28 | $\alpha$ | 8 |

be 5. If $m$ and $\sigma^2$ are respectively the mean deviation about the mean and the variance of the data, then $\frac{3\alpha}{m + \sigma^2}$ is equal to ........... . **[2023, 13 April Shift-I]**

**79** If the mean of the frequency distribution

| Class | 0-10 | 10-20 | 20-30 | 30-40 | 40-50 |
|---|---|---|---|---|---|
| Frequency | 2 | 3 | $x$ | 5 | 4 |

is 28, then its variance is ....... **[2023, 10 April Shift-I]**

**80** Let the mean and variance of 8 numbers $x$, $y$, 10, 12, 6, 12, 4, 8 be 9 and 9.25 respectively. If $x > y$, then $3x - 2y$ is equal to ........... **[2023, 8 April Shift-I]**

**81** If the mean and variance of the frequency distribution

| $x_i$ | 2 | 4 | 6 | 8 | 10 | 12 | 14 | 16 |
|---|---|---|---|---|---|---|---|---|
| $f_i$ | 4 | 4 | $\alpha$ | 15 | 8 | $\beta$ | 4 | 5 |

are 9 and 15.08 respectively, then the value of $\alpha^2 + \beta^2 - \alpha\beta$ is ...... **[2023, 6 April Shift-II]**

**82** Let the mean and the variance of 20 observations $x_1, x_2, .... x_{20}$ be 15 and 9, respectively. For $\alpha \in R$, if the mean of $(x_1 + \alpha)^2, (x_2 + \alpha)^2, ......, (x_{20} + \alpha)^2$ is 178, then the square of the maximum value of $\alpha$ is equal to ......... . **[2022, 29 July Shift-I]**

**83** The mean and variance of 10 observation were calculated as 15 and 15 respectively by a student who took by mistake 25 instead of 15 for one observation. Then, the correct standard deviation is ........... . **[2022, 27 July Shift-I]**

**84** An online exam is attempted by 50 candidates out of which 20 are boys. The average marks obtained by boys is 12 with a variance 2. The variance of marks obtained by 30 girls is also 2. The average marks of all 50 candidates is 15. If $\mu$ is the average marks of girls and $\sigma^2$ is the variance of marks of 50 candidates, then $\mu + \sigma^2$ is equal to ........... . *[2022, 27 Aug Shift-II]*

**85** Let $n$ be an odd natural number such that the variance of 1, 2, 3, 4, ..., $n$ is 14. Then, $n$ is equal to ........... . *[2022, 27 Aug Shift-I]*

**86** Let the mean and variance of four numbers 3, 7, $x$ and $y$ $(x > y)$ be 5 and 10 respectively. Then, the mean of four numbers $3 + 2x$, $7 + 2y$, $x + y$ and $x - y$ is ........... . *[2022, 26 Aug Shift-II]*

**87** The mean and standard deviation of 40 observations are 30 and 5 respectively. It was noticed that two of these observations 12 and 10 were wrongly recorded. If $\sigma$ is the standard deviation of the data after omitting the two wrong observations from the data, then $38\sigma^2$ is equal to ............... . *[2022, 26 July Shift-II]*

**88** Consider a set of $3n$ numbers having variance 4. In this set, the mean of first $2n$ numbers is 6 and the mean of the remaining $n$ numbers is 3. A new set is constructed by adding 1 into each of first $2n$ numbers and subtracting 1 from each of the remaining $n$ numbers. If the variance of the new set is $k$, then $9k$ is equal to ......... . *[2021, 17 March Shift-II]*

**89** Consider the statistics of two sets of observations as follows

| | Size | Mean | Variance |
|---|---|---|---|
| Observation I | 10 | 2 | 2 |
| Observation II | $n$ | 3 | 1 |

If the variance of the combined set of these two observations is 17/9, then the value of $n$ is equal to ......... . *[2021, 16 March Shift-II]*

**90** Let $x_1, x_2, \ldots, x_{18}$ be eighteen observations, such that $\sum_{i=1}^{18}(x_i - \alpha) = 36$ and $\sum_{i=1}^{18}(x_i - \beta)^2 = 90$, where $\alpha$ and $\beta$ are distinct real numbers. If the standard deviation of these observations is 1, then the value of $|\alpha - \beta|$ is ......... . *[2021, 26 Feb Shift-II]*

**91** If the variance of 10 natural numbers 1, 1, 1,...., 1, $k$ is less than 10, then the maximum possible value of $k$ is ......... . *[2021, 24 Feb Shift-II]*

## Key Index

### Topic 1

| | | | | | | | | | |
|---|---|---|---|---|---|---|---|---|---|
| **1** (b) | **2** (d) | **3** (c) | **4** (a) | **5** (d) | **6** (d) | **7** (b) | **8** (b) | **9** (142) | **10** (398) |
| **11** (164) | **12** (4) | **13** (35) | | | | | | | |

### Topic 2

| | | | | | | | | | |
|---|---|---|---|---|---|---|---|---|---|
| **1** (c) | **2** (d) | **3** (d) | **4** (c) | **5** (d) | **6** (c) | **7** (b) | **8** (a) | **9** (d) | **10** (c) |
| **11** (d) | **12** (d) | **13** (d) | **14** (d) | **15** (b) | **16** (d) | **17** (a) | **18** (d) | **19** (c) | **20** (c) |
| **21** (b) | **22** (a) | **23** (a) | **24** (a) | **25** (b) | **26** (d) | **27** (a) | **28** (b) | **29** (c) | **30** (d) |
| **31** (a) | **32** (a) | **33** (a) | **34** (a) | **35** (d) | **36** (b) | **37** (c) | **38** (a) | **39** (b) | **40** (c) |
| **41** (c) | **42** (d) | **43** (c) | **44** (d) | **45** (d) | **46** (a) | **47** (d) | **48** (c) | **49** (c) | **50** (b) |
| **51** (b) | **52** (c) | **53** (b) | **54** (a) | **55** (a) | **56** (b) | **57** (c) | **58** (a) | **59** (c) | **60** (a) |
| **61** (c) | **62** (c) | **63** (b) | **64** (c) | **65** (c) | **66** (a) | **67** (c) | **68** (a) | **69** (c) | **70** (d) |
| **71** (*) | **72** (8788) | **73** (33) | **74** (29) | **75** (6344) | **76** (2521) | **77** (269) | **78** (8) | **79** (151) | **80** (25) |
| **81** (25) | **82** (4) | **83** (2) | **84** (25) | **85** (13) | **86** (12) | **87** (238) | **88** (68) | **89** (5) | **90** (4) |
| **91** (11) | | | | | | | | | |

# Detailed **Explanations**

## Topic 1

**01.** *(b)* We know that

$$\text{Median} = l + \left(\frac{\frac{N}{2} - cf}{f}\right) \times h$$

$$\Rightarrow \quad 14 = 12 + \left(\frac{\frac{N}{2} - 18}{12}\right) \times 6$$

$$\Rightarrow \quad 2 \times 2 = \frac{N}{2} - 18$$

$$\Rightarrow \quad N = 44$$

**02.** *(d)* Given, $\dfrac{x_1 + x_2 + x_3 + \ldots + x_{16}}{16} = 16$

$$\Rightarrow \quad \sum_{i=1}^{16} x_i = 16 \times 16$$

Sum of new observations

$$= \sum_{i=1}^{18} y_i = (16 \times 16 - 16) + (3 + 4 + 5)$$

$= 252$

Number of observations $= 18$

$$\therefore \quad \text{New mean} = \frac{\sum_{i=1}^{18} y_i}{18} = \frac{252}{18} = 14$$

**03.** *(c)* Let the number of boys and girls be $x$ and $y$, respectively.

$$\therefore \quad 52x + 42y = 50(x + y)$$

$$\Rightarrow \quad 52x + 42y = 50x + 50y$$

$$\Rightarrow \quad 2x = 8y \Rightarrow x = 4y$$

$\therefore$ Total number of students in the class

$$= x + y = 4y + y = 5y$$

$\therefore$ Required percentage of boys

$$= \frac{4y}{5y} \times 100\% = 80\%$$

**04.** *(a)* Given that mean = 21 and median = 22

Using the Emperical relation,

$$\text{Mode} = 3 \ \text{Median} - 2 \ \text{Mean}$$

$$\therefore \quad \text{Mode} = 3(22) - 2(21)$$

$$= 66 - 42 = 24$$

**05.** *(d)* Number of terms $(n) = 9$

$$\text{Median} = \left(\frac{n+1}{2}\right)\text{th term} \quad [\because \text{As } n \text{ is odd}]$$

$\therefore$ Median is 5th term

If each of the largest 4 observations of the set is increased by 2, then it doesn't affect the 5th term or order of the terms.

The median remains same as that of the original set.

**06.** *(d)* We have,

| $x_i$ | 0-4 | 4-8 | 8-12 | 12-16 | 16-20 |
|---|---|---|---|---|---|
| $f_i$ | 3 | 9 | 10 | 8 | 6 |
| $cf$ | 3 | 12 | 22 | 30 | 36 |

$$\therefore \quad N = \sum f = 36$$

$$\frac{N}{2} = \frac{36}{2} = 18$$

So, we have median lies in the class 8-12

$$l_1 = 8, f = 10, h = 4, cf = 12$$

So, here we apply formula

$$M = l_1 + \frac{N/2 - cf}{f} \times h$$

$$= 8 + \frac{18 - 12}{10} \times 4$$

$$\Rightarrow \quad M = 10.4$$

So, $\quad 20M = 208$

**07.** *(b)*

> **key idea** Use $\sum_{i=1}^{n} f_i$ = Number of students
>
> and Mean $(\bar{x}) = \dfrac{\Sigma f_i x_i}{\Sigma f_i}$

The given frequency distribution, for some $x \in R$ of marks obtained by 20 students is

| Marks | 2 | 3 | 5 | 7 |
|---|---|---|---|---|
| Frequency | $(x+1)^2$ | $2x-5$ | $x^2-3x$ | $x$ |

$\because$ Number of students $= 20 = \Sigma f_i$

$$\Rightarrow \quad (x+1)^2 + (2x-5) + (x^2 - 3x) + x = 20$$

$$\Rightarrow \quad (x^2 + 2x + 1) + (2x - 5) + (x^2 - 3x) + x = 20$$

$$\Rightarrow \quad 2x^2 + 2x - 24 = 0$$

$$\Rightarrow \quad x^2 + x - 12 = 0$$

$$\Rightarrow \quad (x + 4)(x - 3) = 0$$

$$\Rightarrow \quad x = 3 \qquad [\text{As } x > 0]$$

$$\text{Now, mean } (\bar{x}) = \frac{\Sigma f_i x_i}{\Sigma f_i}$$

$$= \frac{2(x+1)^2 + 3(2x-5) + 5(x^2 - 3x) + 7x}{20}$$

$$= \frac{2(4)^2 + 3(1) + 5(0) + 7(3)}{20}$$

$$= \frac{32 + 3 + 21}{20} = \frac{56}{20} = 2.8$$

**08.** *(b)* Since, total number of students $= 100$

and number of boys $= 70$

$\therefore$ Number of girls $= (100 - 70) = 30$

Now, the total marks of 100 students

$$= 100 \times 72 = 7200$$

And total marks of 70 boys

$$= 70 \times 75 = 5250$$

Total marks of 30 girls

$$= 7200 - 5250 = 1950$$

$$\therefore \text{ Average marks of 30 girls} = \frac{1950}{30} = 65$$

**09.** *(142)* Since, $x_1, x_2, x_3, \ldots, x_{20}$ are in GP, where $x_1 = 3$ and $r$ = common ratio $= 1/2$

So, $\quad x_1 = 3 \Rightarrow x_2 = \dfrac{3}{2}$

$$x_3 = \frac{3}{2^2} \Rightarrow x_4 = \frac{3}{2^3}$$

$$\ldots\ldots\ldots \qquad \ldots\ldots\ldots$$

$$x_{20} = \frac{3}{2^{19}}$$

$$\text{Now, } \sum_{i=1}^{20} x_i = x_1 + x_2 + x_3 + \ldots + x_{20}$$

$$= 3 + \frac{3}{2} + \frac{3}{2^2} + \frac{3}{2^3} + \ldots + \frac{3}{2^{19}}$$

$$= 3\left(1 + \frac{1}{2} + \frac{1}{2^2} + \frac{1}{2^3} + \ldots + \frac{1}{2^{19}}\right)$$

$$= 3\left[\frac{1\left(1 - \left(\frac{1}{2}\right)^{20}\right)}{1 - \frac{1}{2}}\right] = \frac{3}{1/2}\left(1 - \frac{1}{2^{20}}\right)$$

$$\Rightarrow \quad \sum_{i=1}^{20} x_i = 6\left(1 - \frac{1}{2^{20}}\right) \qquad \ldots \text{(i)}$$

According to the question,

$$\bar{x} = \frac{\sum_{i=1}^{20}(x_i - i)^2}{n}$$

$$= \frac{\sum_{i=1}^{20} x_i^2 + \sum_{i=1}^{20} i^2 - 2\sum_{i=1}^{20} x_i i}{n} \quad \dots\text{(ii)}$$

From Eq. (i), we get

$$\sum_{i=1}^{20}(x_i)^2 = x_1^2 + x_2^2 + x_3^2 + \dots x_{20}^2$$

$$= 9 + \frac{9}{2^2} + \frac{9}{2^4} + \frac{9}{2^6} + \dots + \frac{9}{2^{38}}$$

$$= 9\left[\frac{1-\left(\frac{1}{4}\right)^{20}}{1-\frac{1}{4}}\right] = 9 \times \frac{4}{3}\left[1 - \frac{1}{4^{20}}\right]$$

$$\Rightarrow \sum_{i=1}^{20}(x_i)^2 = 12\left[1 - \left(\frac{1}{2^{40}}\right)\right] \dots\text{(iii)}$$

and $\sum_{i=1}^{20} i^2 = (1)^2 + (2)^2 + (3)^2 + \dots + (20)^2$

$$= \frac{20 \times (20+1)(2 \times 20 + 1)}{6}$$

$$= 2870 \quad \dots\text{(iv)}$$

Also, $\sum_{i=1}^{20} x_i i = 3 + 2\cdot\frac{3}{2} + 3\cdot\frac{3}{2^2} + 4\cdot\frac{3}{2^3} + 5\left(\frac{3}{2^4}\right) + \dots + 20\left(\frac{3}{2^{19}}\right)$

which are in arithmetic co-geometric (AG) series.

Let $z = \sum_{i=1}^{20} x_i i = 3 + 2\cdot\frac{3}{2} + 3\cdot\frac{3}{2^2} + 4\cdot\frac{3}{2^3} + \dots + 20.\frac{3}{2^{19}} \dots\text{(v)}$

$$\frac{1}{2}z = \frac{3}{2} + 2\cdot\frac{3}{2^2} + 3\cdot\frac{3}{2^3} + \dots + 19\cdot\frac{3}{2^{19}} + 20\cdot\frac{3}{2^{20}} \dots\text{(vi)}$$

On subtracting Eq. (vi) from Eq. (v), we get

$$\left(z - \frac{z}{2}\right) = \left(3 + \frac{3}{2} + \frac{3}{2^2} + \frac{3}{2^3} + \frac{3}{2^4} + \dots + \frac{3}{2^{19}} - 20\cdot\frac{3}{2^{20}}\right)$$

$$\Rightarrow \frac{z}{2} = 3\left(1 + \frac{1}{2} + \frac{1}{2^2} + \frac{1}{2^3} + \dots + \frac{1}{2^{19}}\right) - 20\cdot\frac{3}{2^{20}}$$

$$\Rightarrow \frac{z}{2} = 3\left[\frac{1-\left(\frac{1}{2}\right)^{20}}{1-\frac{1}{2}}\right] - \frac{60}{2^{20}}$$

$$= 6\left(1 - \frac{1}{2^{20}}\right) - \frac{60}{2^{20}}$$

$$\Rightarrow z = 6\left(2 - \frac{22}{2^{20}}\right)$$

$$\therefore \sum_{i=1}^{20} x_i i = 6\left(2 - \frac{22}{2^{20}}\right) \dots\text{(vii)}$$

On putting the values from Eqs. (iii), (iv) and (vii) in Eq. (ii), we get

$$\bar{x} = \frac{12 - \frac{12}{2^{40}} + 2870 - 12\left(2 - \frac{22}{2^{20}}\right)}{20}$$

$$= \frac{2858}{20} + \left(\frac{-12}{2^{40}} + \frac{264}{2^{20}}\right) \times \frac{1}{20}$$

$$\therefore [\bar{x}] = 142$$

**10.** *(398)* For AP : $(7 \times 8), (10 \times 10), (13 \times 12), \dots$

7, 10, 13, …

$$a_n = 7 + (n-1)3 = 3n + 4$$

For AP : 8, 10, 12, …

$$b_n = 8 + (n-1)2 = 2n + 6$$

So, $T_n = (3n+4)(2n+6)$

$$= 6n^2 + 26n + 24$$

Sum $S_n = \Sigma T_n$

$$= 6\cdot\left[\frac{n(n+1)(2n+1)}{6}\right] + 26\left[\frac{n(n+1)}{2}\right] + 24n$$

$$\text{Mean} = \frac{\text{Sum}}{10} = \frac{10\cdot 11\cdot 21}{10} + \frac{13\cdot 10\cdot 11}{10} + \frac{24\cdot 10}{10} = 398$$

**11.** *(164)* Sum of frequencies = 584

$$\alpha + \beta + 110 + 54 + 30 = 584$$

$$\alpha + \beta = 390$$

Median is at $\frac{584}{2}$ = 292th term

$\because$ Median = 45 (lies in class 40 – 50)

$$\Rightarrow \alpha + 110 + 54 + 15 = 292 \Rightarrow \alpha = 113$$

$$\Rightarrow \beta = 390 - 113 = 277$$

$$\Rightarrow \alpha - \beta = 113 - 277 = -164$$

$$\therefore |\alpha - \beta| = 164$$

**12.** *(4)*

| Class | Frequency | $x_i$ | $f_i x_i$ |
|---|---|---|---|
| 0-6 | $a$ | 3 | $3a$ |
| 6-12 | $b$ | 9 | $9b$ |
| 12-18 | 12 | 15 | 180 |
| 18-24 | 9 | 21 | 189 |
| 24-30 | 5 | 27 | 135 |
| | $\Sigma n = a + b + 26$ | | $\Sigma f_i x_i = 3a + 9b + 504$ |

$$\text{Mean} = \frac{\Sigma f_i x_i}{\Sigma n} = \frac{3a + 9b + 504}{a + b + 26} = \frac{309}{22}$$

$$\Rightarrow 66a + 198b + 11088 = 309a + 309b + 8034$$

$$\Rightarrow 243a + 111b = 3054$$

$$\Rightarrow 81a + 37b = 1018$$

Median

$$= 12 + \frac{\frac{a+b+26}{2} - (a+b)}{12} \times 6 = 14$$

$$\Rightarrow \frac{a+b}{4} = \frac{9}{2} \Rightarrow a + b = 18$$

So, $81a + 37b = 1018$ and $a + b = 18$

$\Rightarrow a = 8$ and $b = 10$

$\therefore (a-b)^2 = 4$

**13.** *(35)* Given, $n = 25, \bar{x} = 40$

$$\therefore \bar{x} = \frac{\sum x}{n}$$

$$\Rightarrow \sum x = 25 \times 40 = 1000$$

Let $x$ be the age of newly appointed teacher.

New sum $= 1000 - 60 + x = 940 + x$

New average = 39

$$\therefore 39 = \frac{940 + x}{25}$$

$$\Rightarrow 975 = 940 + x \Rightarrow x = 35$$

## Topic 2

**01.** *(c)* $\frac{a+b+8+12+10+6+4+13}{8} = 9$

$$\Rightarrow a + b = 19 \quad \dots\text{(i)}$$

$$\frac{\Sigma x_i^2}{8} - 9^2 = 9.25$$

$$\Rightarrow a^2 + b^2 + 8^2 + 12^2 + 10^2 + 6^2 + 4^2 + 13^2 = (81 + 9.25) \times 8$$

$$\Rightarrow a^2 + b^2 = 193 \quad \dots\text{(ii)}$$

$$\Rightarrow \quad ab = \frac{(a+b)^2 - (a^2+b^2)}{2}$$

$$= \frac{361-193}{2} = 84$$

[using Eqs. (i) and (ii)]

$$\Rightarrow a + b + ab = 19 + 84 = 103$$

**02.** *(d)* Given, mean $= \bar{x} = 5.5$

$$\Rightarrow \sum_{i=1}^{10} x_i = 5.5 \times 10 = 55 \text{ and } \sum_{i=1}^{10} x_i^2 = 371$$

$$(\Sigma x_i)_{\text{new}} = 55 - (4+5) + (6+8) = 60$$

$$(\Sigma x_i^2)_{\text{new}} = 371 - (4^2+5^2) + (6^2+8^2) = 430$$

$$\text{Variance } (\sigma^2) = \frac{\Sigma x_i^2}{10} - \left(\frac{\Sigma x_i}{10}\right)^2$$

$$\Rightarrow \sigma^2 = \frac{430}{10} - \left(\frac{60}{10}\right)^2 = 43 - 36 = 7$$

**03.** *(d)*

| $x$ | $c$ | $2c$ | $3c$ | $4c$ | $5c$ | $6c$ |
|---|---|---|---|---|---|---|
| $f$ | 2 | 1 | 1 | 1 | 1 | 1 |

$$\text{Var }(X) = \frac{\Sigma f_i x_i^2}{\Sigma f_i} - \left(\frac{\Sigma f_i x_i}{\Sigma f_i}\right)^2$$

$$160 = \frac{2c^2 + 4c^2 + 9c^2 + 16c^2 + 25c^2 + 36c^2}{7} - \left(\frac{2c + 2c + 3c + 4c + 5c + 6c}{7}\right)^2$$

$$\Rightarrow 160 = \frac{92c^2}{7} - \frac{484c^2}{49} = \frac{160c^2}{49}$$

$$\therefore \quad c^2 = 49$$

$$c = 7 \quad \text{(taking only positive sign)}$$

**04.** *(c)* Since, $\alpha, \beta \in \{1, 2, 3, 4, 5, 6\}$

$\therefore \alpha - \beta \in \{0, 1, 2, 3, 4, 5\}$

$$\text{Mean, } \bar{X} = \frac{0+1+2+3+4+5}{6} = \frac{5}{2}$$

$$\text{Variance} = \frac{\sum_{i=1}^{n}(x_i - \bar{X})^2}{n}$$

$$= \frac{\left(0-\frac{5}{2}\right)^2 + \left(1-\frac{5}{2}\right)^2 + \ldots + \left(5-\frac{5}{2}\right)^2}{6}$$

$$= \frac{35}{12} = \frac{p}{q} \quad \text{[given]}$$

i.e. $p = 35$ with divisors $= 1, 5, 7, 35$

Required sum $= 1 + 5 + 7 + 35 = 48$

**05.** *(d)* Given,

$$\frac{\Sigma x}{12} = \frac{9}{2} \text{ and } \frac{\Sigma x^2}{12} - \left(\frac{9}{2}\right)^2 = 4$$

$$\Rightarrow \quad \Sigma x = 54 \text{ and } \Sigma x^2 = 291$$

After correction

$$\Sigma x_{\text{corr.}} = 54 - (9+10) + (7+14) = 56$$

$$\Sigma x^2_{\text{corr.}} = 291 - (81+100) + 49 + 196 = 355$$

$$\sigma^2_{\text{corr.}} = \frac{355}{12} - \left(\frac{56}{12}\right)^2 = \frac{281}{36} = \frac{m}{n}$$

$\therefore m + n = 317$

**06.** *(c)* We know that, if $n_1, n_2$ are the sizes, $\bar{X}_1, \bar{X}_2$ are the means and $\sigma_1, \sigma_2$ are the standard deriation of the series, then the combine variance of the series is

$$\sigma^2 = \frac{n_1\sigma_1^2 + n_2\sigma_2^2}{n_1 + n_2} + \frac{n_1 \cdot n_2}{(n_1+n_2)^2}(\bar{X}_1 - \bar{X}_2)^2$$

$$\Rightarrow \quad 13 = \frac{15 \times 14 + 15 \times \sigma^2}{15+15} + \frac{15 \times 15}{(15+15)^2}(12-14)^2$$

$$\Rightarrow \quad 13 = \frac{14 + \sigma^2}{2} + \frac{1}{4} \times 4$$

$$\Rightarrow \quad 14 + \sigma^2 = 2 \times 12 \Rightarrow \sigma^2 = 10$$

**07.** *(b)* Let the remaining two observations are $x$ and $y$.

$$\text{Then, mean} = \frac{1+3+5+x+y}{5} = 5$$

$$\Rightarrow \quad x + y = 16 \quad \ldots\text{(i)}$$

$$\text{Also, variance} = \frac{\Sigma x_i^2}{N} - (\bar{X})^2$$

$$\Rightarrow 8 = \frac{1^2 + 3^2 + 5^2 + x^2 + y^2}{5} - (5)^2$$

$$\Rightarrow \quad x^2 + y^2 = 130 \quad \ldots\text{(ii)}$$

Now, we know that

$$(x+y)^2 = x^2 + y^2 + 2xy$$

From Eqs. (i) and (ii),

$$(16)^2 = 130 + 2xy$$

$$\Rightarrow \quad xy = 63$$

$$\text{Required sum} = x^3 + y^3$$

$$= (x+y)^3 - 3xy(x+y)$$

$$= (16)^3 - 3(63)(16)$$

$$= 4096 - 3024 = 1072$$

**08.** *(a)* Given, numbers are 3, 7, 12, $a$, $43 - a$

Mean = 13

$$\text{Variance} = \frac{\Sigma x_i^2}{n} - (\bar{x})^2$$

$$= \frac{9 + 49 + 144 + a^2 + (43-a)^2}{5} - 13^2 \in N$$

$$= \frac{202 + 2a^2 - 86a + 1849}{5} - 169$$

$$= \frac{1206 + 2a^2 - 86a}{5}$$

$$= \frac{1205 - 85a}{5} + \frac{2a^2 - a + 1}{5} \in N$$

$$\Rightarrow \quad \frac{2a^2 - a + 1}{5} \in N$$

$\Rightarrow 2a^2 - a + 1 - 5N = 0$ must have solution as natural numbers.

Its $D = 40N - 7$ always has 3 at unit place.

$\Rightarrow$ $D$ cannot be a perfect square.

So, $a$ cannot be integer.

> **Shortcut Method**
>
> $$\text{Mean } (\bar{x}) = \frac{3+7+12+a+43-a}{5} = 13$$
>
> $$\text{Variance} = \frac{9+49+144+a^2+(43-a)^2}{5} - (13)^2 \in N$$
>
> $$\Rightarrow \frac{2a^2 - a + 1}{5} \in N$$
>
> $\Rightarrow a^2 - a + 1 - 5n = 0$ must have solution as natural number.
>
> Now, its discriminant $D = 40n - 7$
>
> $\Rightarrow D$ can not be perfect square as all perfect square will be form of $4P$ or $4P+1$ for $P \in N$
>
> $\Rightarrow a$ can not be integer.

**09.** *(d)* It is given that $\sum_{i=1}^{n}(x_i - a) = n$

and $\sum_{i=1}^{n}(x_i - a)^2 = na, \ (n, a > 1)$

$\therefore$ Standard deviation of '$n$' observations

$$= \sqrt{\frac{\sum_{i=1}^{n}(x_i - a)^2}{n} - \left(\frac{\sum_{i=1}^{n}(x_i - a)}{n}\right)^2}$$

$$= \sqrt{\frac{na}{n} - \left(\frac{n}{n}\right)^2} = \sqrt{a-1}$$

**10.** (c)

> **key idea** Standard deviation is remain unchanged, if observations are added or subtracted by a fixed number.

We have, $\sum_{i=1}^{9}(x_i - 5) = 9$

and $\sum_{i=1}^{9}(x_1 - 5)^2 = 45$

$$SD = \sqrt{\frac{\sum_{i=1}^{9}(x_i-5)^2}{9} - \left(\frac{\sum_{i=1}^{9}(x_i-5)}{9}\right)^2}$$

$$= \sqrt{\frac{45}{9} - \left(\frac{9}{9}\right)^2} = \sqrt{5-1} = \sqrt{4} = 2$$

**11.** (d) Let the mean of the given observations is

$$\bar{x} = \frac{7+8+9+7+8+7+\lambda+8}{8} = 8$$

$[\because \bar{x} = 8 \text{ (given)}]$

$\Rightarrow 54 + \lambda = 64 \Rightarrow \lambda = 64 - 54 = 10$

Now, $\Sigma x_i^2 = (7)^2 + (8)^2 + (9)^2 + (7)^2 + (8)^2 + (7)^2 + (10)^2 + (8)^2$

$= 49 + 64 + 81 + 49 + 64 + 49 + 100 + 64 = 520$

Variance, $\sigma^2 = \frac{\Sigma x_i^2}{n} - (\bar{x})^2$

$\therefore$ Variance $= \frac{520}{8} - (8)^2 = 65 - 64 = 1$

**12.** (d) Let $a, b, c, d, e$ be five observations.

$\therefore \quad SD = \sqrt{\frac{\Sigma(a-\bar{x})^2}{5}} = 0$

$\Rightarrow \quad \Sigma(a-\bar{x})^2 = 0$

$\Rightarrow (a-\bar{x})^2 + (b-\bar{x})^2 + (c-\bar{x})^2 + (d-\bar{x})^2 + (e-\bar{x})^2 = 0$

$\Rightarrow a - \bar{x} = b - \bar{x} = c - \bar{x} = d - \bar{x} = e - \bar{x} = 0$

$\Rightarrow a = b = c = d = e = \bar{x}$

$\Rightarrow a = b = c = d = e = 9 \qquad [\because \bar{x} = 9]$

$\therefore a + b + c + d + e = 9 \times 5 = 45$

Let $x$ be the change in term $d$. Then,

New sum $= a + b + c + (d + x) + e = 5 \times 10 = 50 \Rightarrow x = 5$

$\therefore$ New observations becomes 9, 9, 9, 14, 9

New SD $= \sqrt{\frac{\Sigma(a-\bar{x})^2}{5}}$

$= \sqrt{\frac{1^2+1^2+1^2+4^2+1^2}{5}}$

$= \sqrt{\frac{20}{5}} = 2$

**13.** (d)

> **key idea** The variance is independent of change of origin.

If initially all marks were $x_i$, then

$$\sigma_i^2 = \frac{\Sigma(x_i - \bar{x})^2}{N}$$

Now, each is increased by 10.

$$\therefore \text{ New } \sigma_i^2 = \frac{\Sigma[(x_i+10)-(\bar{x}+10)]^2}{N} = \sigma_i^2$$

So, variance will not change whereas mean, median and mode will increase by 10.

**14.** (d) Given, $\bar{x}$ is the AM and $\sigma^2$ is the variance of $n$ observations $x_1, x_2, x_3, \ldots, x_n$. AM of $2x_1, 2x_2, 2x_3, \ldots, 2x_n$

$$= \frac{2x_1 + 2x_2 + 2x_3 + \ldots + 2x_n}{n}$$

$$= 2\left(\frac{x_1 + x_2 + x_3 + \ldots + x_n}{n}\right) = 2\bar{x}$$

Hence, it is obvious that Statement II is false.

Variance of $2x_1, 2x_2, 2x_3, \ldots, 2x_n$ = Variance $(2x_i) = 2^2$ Variance $(x_i) = 4\sigma^2$

Hence, Statement I is correct.

Finally, Statement I is true and Statement II is false.

**15.** (b) Median of $a, 2a, 3a, 4a, \ldots, 50a$ is

$$\frac{25a + 26a}{2} = (25.5)\,a$$

$\left\{\because \text{ If total term is even, then Median } = \frac{1}{2}\left[\left(\frac{n}{2}\right)\text{th term} + \left(\frac{n}{2}+1\right)\text{th term}\right]\right\}$

MD (about median)

$$= \frac{\sum_{i=1}^{50}|x_i - \text{Median}|}{N}$$

$\Rightarrow \quad 50 = \frac{1}{50}\{2|a| \cdot (0.5 + 1.5 + 2.5 + \ldots + 24.5)\}$

$\Rightarrow 2500 = 2|a| \cdot \frac{25}{2}(25)$

$\Rightarrow |a| = 4$

**16.** (d) Sum of $n$ even natural numbers $= n(n+1)$

$\Rightarrow \quad \text{Mean } (\bar{x}) = \frac{n(n+1)}{n} = n + 1$

$\therefore \text{ Variance} = \left[\frac{1}{n}\sum (x_i)^2\right] - (\bar{x})^2$

$= \frac{1}{n}[2^2 + 4^2 + \ldots + (2n)^2] - (n+1)^2$

$= \frac{1}{n}2^2[1^2 + 2^2 + \ldots + n^2] - (n+1)^2$

$= \frac{4}{n}\left(\frac{n(n+1)(2n+1)}{6}\right) - (n+1)^2$

$= \frac{(n+1)[2(2n+1) - 3(n+1)]}{3}$

$= \frac{(n+1)(n-1)}{3} = \frac{n^2-1}{3}$

Hence, Statement I is false.

Statement II is true because they are standard formulae.

**17.** (a) According to the given condition,

$$6.80 = \frac{\left[(6-a)^2 + (6-b)^2 + (6-8)^2 + (6-5)^2 + (6-10)^2\right]}{5}$$

$\Rightarrow 34 = (6-a)^2 + (6-b)^2 + 4 + 1 + 16$

$\Rightarrow \quad (6-a)^2 + (6-b)^2 = 13 = 9 + 4$

$\Rightarrow \quad (6-a)^2 + (6-b)^2 = 3^2 + 2^2$

$\Rightarrow \quad a = 3, \quad b = 4$

**18.** (d) Since, variance is independent of change of origin. Therefore, variance of observations 101, 102, ... , 200 is same as variance of 151, 152, ..., 250.

$\therefore \quad V_A = V_B \Rightarrow \frac{V_A}{V_B} = 1$

**19.** (c) Given that $\Sigma x_i^2 = 400$ and $\Sigma x_i = 80$

$\because \quad \sigma^2 \geq 0$

$\therefore \quad \frac{\Sigma x_i^2}{n} - \left(\frac{\Sigma x_i}{n}\right)^2 \geq 0$

$\Rightarrow \quad \frac{400}{n} - \frac{6400}{n^2} \geq 0$

$\therefore \quad n \geq 16$

**20.** (c)

> **key idea** Mode and Median are dependent of change of origin but variance does not.

It is true that mode can be computed from histogram and median is not independent of change of scale.

But variance is independent of change of origin and not of scale. Hence, option (c) is correct.

**21.** (b) $5 = \dfrac{1+3+a+7+b}{5}$

$\Rightarrow \quad 25 = 11 + a + b$

$\Rightarrow a + b = 14 \quad \ldots\text{(i)}$

and $10 = \dfrac{1^2+3^2+a^2+7^2+b^2}{5} - 5^2$

$\Rightarrow (10+25)\times 5 = a^2+b^2+59$

$\Rightarrow \quad 175 - 59 = a^2 + b^2$

$\Rightarrow \quad a^2 + b^2 = 116 \quad \ldots\text{(ii)}$

$(a+b)^2 - (a^2+b^2) = 2ab$

$\Rightarrow \quad 2ab = 14^2 - 116 = 80 \Rightarrow ab = 40$

$\Rightarrow (a-b)^2 = (a+b)^2 - 4ab = 14^2$

$-2\times 80 = 196 - 160 = 36$

$a - b = \pm 6$ and $a + b = 14$

$\Rightarrow \quad a + b = 14$

and $\quad a - b = 6$

$2a = 20$

$\Rightarrow \quad a = 10$

and $\quad b = 4$

Now, digits $= n + x_n$

$= 1 + x_1, 2 + x_2, 3 + x_3, 4 + x_4, 5 + x_5$

$= 2, 5, 13, 11, 9$

New Mean

$= \dfrac{2+5+13+11+9}{5} = \dfrac{40}{5} = 8$

New Variance

$= \dfrac{(2-8)^2+(5-8)^2+(13-8)^2+(11-8)^2+(9-8)^2}{5}$

$= \dfrac{36+9+25+9+1}{5} = \dfrac{80}{5} = 16$

**22.** (a) $x_i \to 2, 3, 3, 4, 5, 7, a, b$

Mean $= 4 \Rightarrow \dfrac{\sum x_i}{8} = 4$

$2 + 3 + 3 + 4 + 5 + 7 + a + b = 32$

$\Rightarrow \quad a + b = 8 \quad \ldots\text{(i)}$

and standard deviation $= \sqrt{2}$

$\therefore$ Variance $= (\sqrt{2})^2 = 2$

$\dfrac{\sum x_i^2}{8} - (\bar{x})^2 = 2$

$\Rightarrow \quad 4+9+9+16+25+49 + a^2 + b^2 = 8(2 + (4)^2)]$

$\Rightarrow \quad a^2 + b^2 = 144 - 112$

$\Rightarrow \quad a^2 + b^2 = 32$

$\Rightarrow \; a^2 + (8-a)^2 = 32$ [using Eq. (i)]

$\Rightarrow \; 2a^2 - 16a + 32 = 0$

$\Rightarrow \quad a^2 - 8a + 16 = 0$

$\Rightarrow \quad (a-4)^2 = 0$

$\Rightarrow a = 4$ and $b = 4$

$\therefore$ Observations are 2, 3, 3, 4, 5, 7, 4, 4

$\therefore$ Mode = 4 (occur maximum time)

$\therefore$ Mean deviation about mode

$\dfrac{\sum|x_i - m|}{n}$

$= \dfrac{2+1+1+0+1+3+0+0}{8}$

$= \dfrac{8}{8} = 1$

**23.** (a) Let the observations be $x_1 x_2, \ldots, x_{99}, 50$

Mean

$= \dfrac{x_1 + x_2 + x_3 + \ldots + x_{99} + 50}{100} = 40$

$\Rightarrow x_1 + x_2 + x_3 + \ldots + x_{99} = 3950$

Correct Mean $= \dfrac{3950 + 40}{100}$

$\mu = \dfrac{399}{10} = 39.9$

$(\text{SD})^2 = \dfrac{\sum\limits_{i=1}^{99} x_i^2 + 2500}{100} - (40)^2$

$\sum\limits_{i=1}^{99} x_i^2 = 160101$

$(\text{Correct SD})^2 = \dfrac{160101 + 1600}{100} - \left(\dfrac{399}{10}\right)^2$

$\text{SD} = 5$

$10(\mu + \sigma) = 10(39.9 + 5)$

$= 10(44.9) = 449$

**24.** (a) From given information,

$x + y = 10 \quad \ldots\text{(i)}$

| Age ($x_i$) | $f$ | $\lvert x_i - M\rvert$ | $f_i \lvert x_i - M\rvert$ |
|---|---|---|---|
| 15 | 5 | 3 | 15 |
| 16 | 8 | 2 | 16 |
| 17 | 5 | 1 | 5 |
| 18 | 12 | 0 | 0 |
| 19 | $x$ | 1 | $x$ |
| 20 | $y$ | 2 | $2y$ |

Median $= 18 = M$ (say)

$\therefore \quad \text{MD} = \dfrac{\Sigma f_i \lvert x_i - M\rvert}{\Sigma f_i}$

$\Rightarrow \quad 1.25 = \dfrac{36 + x + 2y}{40}$

$\Rightarrow x + 2y = 14 \quad \ldots\text{(ii)}$

On solving Eqs. (i) and (ii), we get

$x = 6,\; y = 4$

Hence, $4x + 5y = 24 + 20 = 44$

**25.** (b) According to the question,

Mean $(\bar{x}) = 10$ and standard deviation $(\sigma) = 2$

Now,

$\bar{x} = \dfrac{x_1 + x_2 + x_3 + \ldots + x_{20}}{20} = 10$

$\Rightarrow \sum\limits_{i=1}^{20} x_i = 200$

Now, one entry was wrong, taken 8 instead of 12

so, correct sum $= 200 - 8 + 12 = 204$

Correct mean $= \dfrac{204}{20} = 10.2$

We know,

Variance, $\quad \sigma^2 = \dfrac{\Sigma x_i^2}{n} - (\bar{x})^2 = 4$

$\Rightarrow \dfrac{\Sigma x_i^2}{20} - (10)^2 = 4$

$\Rightarrow \quad \Sigma x_i^2 = 2080$

Now, the correct summation of square of observations

$\Rightarrow \quad \Sigma x_i^2 = 2080 - 64 + 144 = 2160$

Correct variance,

$\sigma_n^2 = \dfrac{2160}{20} - (10.2)^2$

$\Rightarrow \quad \sigma_n^2 = 3.96$

Correct standard deviation, $\sigma_n = \sqrt{3.96}$

**26.** (d) Given, observations are $-3, 4, 7, -6, \alpha, \beta$

Mean = 2 and variance = 23

$\Rightarrow \dfrac{-3 + 4 + 7 + (-6) + \alpha + \beta}{6} = 2$

and $\dfrac{\Sigma x_i^2}{n} - (\mu)^2 = 23, \mu = \text{Mean}$

$\Rightarrow \alpha + \beta = 10 \quad \ldots\text{(i)}$

and $\Sigma x_i^2 = n(23 + \mu^2) = 6 \times (23 + 4)$

$= 27 \times 6$

$\Rightarrow 9 + 16 + 49 + 36 + \alpha^2 + \beta^2 = 162$

$\Rightarrow \alpha^2 + \beta^2 = 52 \quad \ldots\text{(ii)}$

On solving, Eqs. (i) and (ii), we get

$\alpha = 4$ and $\beta = 6$

So, mean deviation about mean

$$= \frac{\Sigma|x_i - \bar{x}|}{n}$$

$$= \frac{|-3-2|+|4-2|+|7-2| + |-6-2|+|4-2|+|6-2|}{6}$$

$$= \frac{5+2+5+8+2+4}{6} = \frac{26}{6} = \frac{13}{3}$$

**27.** (a) Mean = 55

$$\frac{a+b+68+44+48+60}{6} = 55$$

$$\Rightarrow \quad a+b+220 = 330$$

$$\Rightarrow \quad a+b = 110 \quad ...(i)$$

Also, variance = 194

$$\Sigma\frac{(x_i-\bar{x})^2}{n} = 194$$

$$\Rightarrow \Sigma\frac{(x_i-55)^2}{6} = 194$$

$$\Rightarrow \Sigma(x_i-55)^2 = 1164$$

$$\Rightarrow (a-55)^2+(b-55)^2+(68-55)^2 + (44-55)^2+(48-55)^2 + (60-55)^2 = 1164$$

$$\Rightarrow a^2-110a+3025+b^2-110b+3025 + 169+121+49+25 = 1164$$

$$a^2+b^2-110(a+b) = -5250$$

$$a^2+b^2-110\times 110 = -5250$$

$$\Rightarrow \quad a^2+b^2 = 6850 \quad ...(ii)$$

From Eq. (i), $b = 110 - a$ put in Eq. (ii),

$$a^2+(110-a)^2 = 6850$$

$$\Rightarrow \quad 2a^2-220a+5250 = 0$$

$$\Rightarrow \quad a^2-110a+2625 = 0$$

$$a = 35, 75$$

If $a = 75, b = 35$ $\quad (\because a > b)$

$$a+3b = 75+105 = 180$$

**28.** (b) Given, $\sum_{k=1}^{10} a_k = 50$

$$\Rightarrow a_1+a_2+a_3+...+a_{10} = 50 \quad ...(i)$$

and $\sum_{\forall k<j} a_k \cdot a_j = 1100$ $\quad ...(ii)$

On squaring Eq. (i), we get

$$(a_1+a_2+a_3+...+a_{10})^2 = (50)^2 = 2500$$

$$\Rightarrow \quad \sum_{i=1}^{10} a_i^2 + 2\sum_{k<j} a_k a_j = 2500$$

[when open whole square of LHS and written in $\Sigma$ notation]

$$\Rightarrow \sum_{i=1}^{10} a_i^2 = 2500 - 2(1100)$$

[using Eq. (ii)]

$$\Rightarrow \sum_{i=1}^{10} a_i^2 = 300$$

$$\text{Now, SD} = \sigma = \sqrt{\frac{\Sigma a_i^2}{10} - \left(\frac{\Sigma a_i}{10}\right)^2}$$

$$= \sqrt{\frac{300}{10} - \left(\frac{50}{10}\right)^2}$$

$$= \sqrt{30-25}$$

$$= \sqrt{5}$$

**29.** (c) Given, $n = 10, \bar{x} = 20, \sigma_x = 8$

Correct value = 40, Wrong value = 50

$$\bar{x} = \frac{\Sigma x_i}{n}$$

$$20 = \frac{\Sigma x_i}{10} \Rightarrow \Sigma x_i = 200$$

Corrected $\Sigma x_i$ = 200 − (Wrong value) + Correct value

$$= 200 - 50 + 40 = 190$$

$$\text{Corrected mean} = \frac{\text{Corrected } \Sigma x_i}{n}$$

$$= \frac{190}{10} = 19$$

$$\sigma_x^2 = \frac{\Sigma x_i^2}{n} - (\bar{x})^2$$

$$\Rightarrow \quad (8)^2 = \frac{\Sigma x_i^2}{10} - (20)^2$$

$$\Rightarrow \quad 64+400 = \frac{\Sigma x_i^2}{10}$$

$$464 = \frac{\Sigma x_i^2}{10} \Rightarrow 4640 = \Sigma x_i^2$$

$$\text{Corrected } \Sigma x_i^2 = 4640 - (\text{Wrong value})^2 + (\text{Corrected value})^2$$

$$= 4640 - (50)^2 + (40)^2$$

$$= 4640 - 2500 + 1600$$

$$= 3740$$

$$\text{Corrected variance} = \frac{\text{Corrected } \Sigma x_i^2}{n} - (\text{Corrected mean})^2$$

$$= \frac{3740}{10} - (19)^2 = 374 - 361 = 13$$

**30.** (d) Given observations are 1, 2, 4, 5, $x$ and $y$

and mean $(\bar{X}) = 5$

and variance $(\sigma^2) = 10$

$$\therefore \bar{X} = \text{Mean} = \frac{\text{Sum of observations}}{\text{Total observations}}$$

$$\Rightarrow \quad 5 = \frac{1+2+4+5+x+y}{6}$$

$$\Rightarrow x+y = 18 \quad ...(i)$$

$$\therefore \text{Variance} = \sigma^2 = \frac{\Sigma x_i^2}{n} - (\bar{X})^2$$

$$10 = \frac{1^2+2^2+4^2+5^2+x^2+y^2}{6} - (5)^2$$

$$\Rightarrow \quad 210 = x^2+y^2+46$$

$$\Rightarrow x^2+y^2 = 164 \quad ...(ii)$$

On squaring Eq. (i), we get

$$x^2+y^2+2xy = 324$$

From Eq. (ii),

$$2xy+164 = 324$$

$$\Rightarrow \quad 2xy = 160$$

$$xy = 80 \quad ..(iii)$$

From Eqs. (ii) and (iii), we get

$$x = 10, y = 8 \text{ or } x = 8, y = 10$$

$$\text{Mean deviation} = \frac{\sum|x_i - \bar{X}|}{n}$$

$$= \frac{|1-5|+|2-5|+|4-5|+|5-5| + |10-5|+|8-5|}{6}$$

$$= \frac{4+3+1+0+5+3}{6} = \frac{16}{6} = \frac{8}{3}$$

Hence, mean deviation about the mean is 8/3.

**31.** (a) We have, $\Sigma f_i = 62$

$$(k+2)+2k+(k^2-1)) + (k^2-1)+(k^2+1)+(k-3) = 62$$

$$\Rightarrow \quad 3k^2+4k-64 = 0$$

$$\Rightarrow (3k+16)(k-4) = 0$$

$\Rightarrow k = 4$ [Rejecting negative value of $k$]

| $x_i$ | 0 | 1 | 2 | 3 | 4 | 5 |
|---|---|---|---|---|---|---|
| $f_i$ | 6 | 8 | 15 | 15 | 17 | 1 |

$$\mu = \frac{\Sigma f_i x_i}{\Sigma f_i} = \frac{0+8+30+45+68+5}{62}$$

$$= \frac{156}{62}$$

$$\text{and } \sigma^2 = \frac{\Sigma f_i x_i^2}{\Sigma f_i} - \left(\frac{\Sigma f_i x_i}{\Sigma f_i}\right)^2$$

$$= \frac{0\times 6+1\times 8+4\times 15 +9\times 15+16\times 17+25\times 1}{62} - \left(\frac{156}{62}\right)^2$$

$$= \frac{500}{62} - \left(\frac{156}{62}\right)^2 = \frac{500}{62} - \mu^2$$

$$\therefore \quad \sigma^2 + \mu^2 = \frac{500}{62}$$

Hence, $[\sigma^2 + \mu^2] = 8$

**32.** (a)

> **key idea** Use Formula,
> Mean deviation about mean $= \frac{\Sigma|x_i - \bar{x}|}{n}$

Given that mean deviation about mean of 100 consecutive positive integers is 25.

Let first integer is $a_1$.

Then, $a_2 = a_1 + 1$

$a_3 = a_1 + 2$

$a_4 = a_1 + 3$

..............

..............

$a_{100} = a_1 + 99$

Now, mean of 100 positive consecutive integers

$$\bar{x} = \frac{a_1 + (a_1+1) + (a_1+2) + (a_1+3) + \ldots (a_1+99)}{100}$$

$$= \frac{100\, a_1 + (1+2+\ldots.99)}{100} = a_1 + 49.5$$

$$\therefore \quad \text{MD} = \frac{\Sigma|x_i - \bar{x}|}{n} = 25$$

$$\Rightarrow \frac{2\,(49.5 + 48.5 + \ldots .0.5)}{100} = 25$$

$\Rightarrow 2 \times 1250 = 2500$

It is true, $\forall\, a_i \in N$

So, set $S = N$

**33.** (a) Given, mean of 7 observations = 8 and variance of 7 observations = 16

After one observation 14 is omitted, then new mean = $a$ and new variance = $b$

Let $x_7 = 14$

Now, Mean

$$= \frac{x_1 + x_2 + x_3 + \ldots + x_7}{7} = 8$$

$$\Rightarrow \frac{x_1 + x_2 + x_3 + \ldots + 14}{7} = 8$$

$$\Rightarrow x_1 + x_2 + x_3 + \ldots + x_6 = 42$$

So, $\frac{x_1 + x_2 + x_3 + \ldots . + x_6}{6}$

$$= \frac{42}{6} = 7 = a \quad \text{[given]}$$

Also, variance $= \frac{\Sigma x_i^2}{7} - (8)^2$

Old variance = 16 [given]

$\Sigma x_i^2 = (16 + 64) \times 7 = 80 \times 7 = 560$

$$\Rightarrow x_1^2 + x_2^2 + x_3^2 + \ldots + x_6^2 + x_7^2 = 560$$

$$\Rightarrow x_1^2 + x_2^2 + x_3^2 + \ldots . + x_6^2 = 560 - (14)^2 = 364$$

$$\therefore b = \frac{x_1^2 + x_2^2 + x_3^2 + \ldots + x_6^2}{6} - (7)^2$$

$$\text{(new variance)} = \frac{364}{6} - 49 = \frac{70}{6}$$

$$\text{Hence, } a + 3b - 5 = 7 + 3 \times \frac{70}{6} - 5 = 7 + 35 - 5 = 37$$

**34.** (a)

> **key idea** To find the solution of this type of problems. Firstly find the mean, then use the value of mean, find the variance of the given data with the help of $\sigma^2 = \frac{\Sigma x_i^2 + \Sigma y_i^2}{n} - \left(\frac{\Sigma x_i + \Sigma y_i}{n}\right)^2$.

Let $X = \{11, 12, \ldots., 40, 41\}$ and $Y = \{61, 62, \ldots\ldots, 90, 91\}$ be the two sets of observations.

$$\therefore \bar{X} = \frac{\frac{31}{2}(11+41)}{31} = 26 \text{ (Mean of } X)$$

$$\bar{Y} = \frac{\frac{31}{2}(61+91)}{31} = 76 \text{ (Mean of } Y)$$

Now, variance $(\sigma^2)$ of $X \cup Y$ is,

$$\sigma^2 = \frac{\Sigma x_i^2 + \Sigma y_i^2}{62} - \left(\frac{\Sigma x_i + \Sigma y_i}{62}\right)^2$$

$$\sigma^2 = \frac{(11^2 + 12^2 + \ldots + 41^2) + (61^2 + 62^2 + \ldots + 91^2)}{62} - (51)^2$$

$= 3306 - 2601 = 705$

$\therefore |\bar{x} + \bar{y} - \sigma^2| = |26 + 76 - 705| = 603$

**35.** (d) Given, Mean = 10

$$\frac{\sum_{i=1}^{n} x_i}{n} = 10 \Rightarrow \sum_{i=1}^{n} x_i = 10n \quad \ldots\text{(i)}$$

And Variance = 4

$$\frac{\Sigma x_i^2}{n} - \left(\frac{\Sigma x_i}{n}\right)^2 = 4 \Rightarrow \Sigma x_i^2 = 104n$$

Now, the marks of one of the students is increased from 8 to 12.

Now, mean = 10.2

$$\frac{\Sigma x_i - 8 + 12}{n} = 10.2$$

$$\Sigma x_i = (10.2)n - 4 \quad \ldots\text{(ii)}$$

From Eqs. (i) and (ii), we get

$10n = (10.2)n - 4$

$0.2n = 4 \Rightarrow n = 20$

Now, variance

$$= \frac{\Sigma x_i^2 - 8^2 + 12^2}{20} - (10.2)^2$$

$= 104 + 4 - 104.04$

New, variance = 3.96

**36.** (b)

> **key idea** Use formula, mean $= \frac{\Sigma x}{n}$
> Variance $= \frac{\Sigma x^2}{n} - \left(\frac{\Sigma x}{n}\right)^2$

$$\text{Given, } \bar{x} = \frac{\Sigma x_i}{5} = \frac{24}{5} \quad \ldots\text{(i)}$$

$$\text{and} \quad \sigma^2 = \frac{\Sigma x_1^2}{5} - (\bar{x})^2 = \frac{194}{25} \quad \ldots\text{(ii)}$$

$\therefore \quad \Sigma x_i = 24$ [from Eq. (i)]

$$\Rightarrow x_1 + x_2 + x_3 + x_4 + x_5 = 24 \quad \ldots\text{(iii)}$$

$$\text{and } \frac{\Sigma x_i^2}{5} - \left(\frac{24}{5}\right)^2 = \frac{194}{25} \quad \text{[from Eq. (ii)]}$$

$$\Rightarrow \Sigma x_i^2 = \frac{194}{5} + \frac{576}{5}$$

$$\Rightarrow \Sigma x_i^2 = \frac{770}{5} = 154$$

$$\Rightarrow x_1^2 + x_2^2 + x_3^2 + x_4^2 + x_5^2 = 154 \quad \ldots\text{(iv)}$$

$$\because \frac{x_1 + x_2 + x_3 + x_4}{4} = \frac{7}{2}$$

$\therefore \quad x_1 + x_2 + x_3 + x_4 = 14$

$\Rightarrow \quad x_5 = 10$ [from Eq. (iii)]

$$\text{and } \frac{x_1^2 + x_2^2 + x_3^2 + x_4^2}{4} - \left(\frac{7}{2}\right)^2 = a$$

$\Rightarrow \quad x_1^2 + x_2^2 + x_3^2 + x_4^2 = 4a + 49$

$\Rightarrow \quad x_5^2 = 154 - (4a + 49)$ [from Eq. (iv)]

$\Rightarrow \quad 10^2 = 154 - 4a - 49$

$\Rightarrow \quad 100 = 105 - 4a \Rightarrow 4a = 5$

$\therefore \quad 4a + x_5 = 5 + 10 = 15$

**37.** *(c)* According to given data

$$\frac{\sum_{i=1}^{7}(x_i-62)^2}{7}=20$$

$$\Rightarrow \sum_{i=1}^{7}(x_i-62)^2=140$$

$$\Rightarrow (x_1-62)^2+(x_2-62)^2+\ldots+(x_7-62)^2=140$$

So, for any $x_i$, $(x_i-62)^2\le 140$

$\Rightarrow x_i>50, \forall\, i=1,2,3,\ldots,7$

So, no student is going to score less than 50.

**38.** *(a)* We have, $\frac{\Sigma x_i^2}{15}-8^2=9$

$\Rightarrow \Sigma x_i^2=15\times 73=1095$

Let $\bar{x}_c$ be corrected mean, $\bar{x}_c=9$

$\Sigma x_c^2=1095-25+400=1470$

Corrected variance

$=\frac{1470}{15}-(9)^2=98-81=17$

**39.** *(b)* According to the question, we have

$\bar{x}=6$

$$\Rightarrow \frac{4+5+6+6+7+8+x+y}{8}=6$$

$\Rightarrow x+y=12 \qquad \ldots(i)$

and variance $=9/4$

$$\Rightarrow \frac{9}{4}=\frac{\Sigma x_j^2}{n}-(\bar{x})^2$$

$$\Rightarrow \frac{9}{4}=\frac{16+25+36+36+49+64+x^2+y^2}{8}-36$$

$\Rightarrow x^2+y^2=80$

$\Rightarrow (x+y)^2-2xy=80$

$\Rightarrow (12)^2-80=2xy \qquad$ [from Eq. (i)]

$\Rightarrow xy=32$

Now, $(x-y)^2=x^2+y^2-2xy$

$=80-2\times 32=16$

$\Rightarrow x-y=\pm 4 \qquad \ldots(ii)$

From Eqs. (i) and (ii), we get

$x+y=12$

$x-y=4$ or $x-y=-4$

$\Rightarrow x=8$ or $x=4$ or $y=4$, $y=8$

So, $x=4$ and $y=8 \qquad (\because x<y)$

$\therefore x^4+y^2=256+64=320$

**40.** *(c)* Mean $=15$

Standard deviation $=2$

Number of observation $=50$

Sum of correct and incorrect observation $=70$

$(\text{Mean})_c=16$

Variance = ?

$$(\text{Mean})_{\text{incorrect}}=\frac{\Sigma x_i}{50}=15$$

$\Rightarrow (\Sigma x_i)_{\text{incorrect}}=750$

$$(\text{Mean})_c=\frac{(\Sigma x_i)_c}{50}=16$$

$\Rightarrow (\Sigma x_i)_c=800$

$\therefore (\Sigma x_i)_c-(\Sigma x_i)_{\text{incorrect}}$

$=800-750=50$

If wrong observation was $x$, then corrected one is $x+50$

$\Rightarrow x+(x+50)=70 \Rightarrow x=10$

Thus, correct observation $=60$

$$\text{Variance}=\frac{\Sigma x_i^2}{50}-(15)^2=2^2$$

$$\Rightarrow \frac{\Sigma x_i^2}{50}=229$$

$\Rightarrow (\Sigma x_i^2)_{\text{incorrect}}=11450$

Now,

$(\Sigma x_i^2)_{\text{correct}}=11450-(10)^2+(60)^2$

$=14950$

$$\text{Variance of correct set}=\frac{14950}{50}-(16)^2$$

$=299-256=43$

**41.** *(c)* Let $a$, $b$, $c$, $d$ and $e$ are 5 remaining observations, $n=7$, mean $=8$ and variance $=16$

Sum of observations $=7\times 8=56$

Mean of remaining observations

$$=\frac{56-8-6}{5}=\frac{42}{5}$$

$$\text{And variance}=\frac{\Sigma x^2}{n}-(\bar{x})^2$$

$$\Rightarrow 16=\frac{\Sigma x^2}{7}-64 \Rightarrow \Sigma x^2=560$$

$\Rightarrow a^2+b^2+c^2+d^2+e^2$

$=560-8^2-6^2=460$

Variance of 5 remaining observations

$$=\frac{460}{5}-\left(\frac{42}{5}\right)^2=\frac{536}{25}$$

**42.** *(d)* $\bar{x}=\frac{\Sigma x_i}{20}=10$

$\Rightarrow \Sigma x_i=200$

$$\sigma^2=\frac{\Sigma x_i^2}{20}-100=6.25$$

$\Sigma x_i^2=20\times 106.25=2125$

Now, replacing 25 with 35, then

$\Sigma x_i-25+35=210$

$\Sigma x_i^2-25^2+35^2=2725$

$$\text{New mean}=\frac{210}{20}=10.5=(\alpha)$$

$$\text{New SD}=\sqrt{\frac{2725}{20}-(10.5)^2}$$

$\text{SD}=\sqrt{136.25-110.25}=\sqrt{26}=\sqrt{\beta}$

$(\alpha,\beta)=(10.5, 26)$

**43.** *(c)* Given,

| $x$ | $x_1=2$ | $x_2=6$ | $x_3=8$ | $x_4=9$ |
|---|---|---|---|---|
| $f$ | 4 | 4 | $\alpha$ | $\beta$ |

$$\text{Mean}=6 \Rightarrow \frac{8+24+8\alpha+9\beta}{8+\alpha+\beta}=6$$

$$\left(\because \text{Mean}=\frac{\sum f_i x_i}{\sum f_i}\right)$$

$\Rightarrow 32+8\alpha+9\beta=48+6\alpha+6\beta$

$\Rightarrow 2\alpha+3\beta=16 \qquad \ldots(i)$

Also, variance $=6.8$

$$\Rightarrow \frac{4\times 16+4\times\alpha+9\times\beta}{8+\alpha+\beta}=6.8$$

$$\left(\because \text{Variance}=\frac{\sum f_i(x_i-\bar{x})^2}{\sum f_i}\right)$$

$\Rightarrow 64+4\alpha+9\beta=(8+\alpha+\beta)\,6.8$

$\Rightarrow 28\alpha-22\beta=96$

$\Rightarrow 14\alpha-11\beta=48 \qquad \ldots(ii)$

From Eqs. (i) and (ii),

$\alpha=5$ and $\beta=2$

If $x_3$ is changed from 8 to 7 then,

| $x$ | $x_1=2$ | $x_2=6$ | $x_3=7$ | $x_4=9$ |
|---|---|---|---|---|
| $f$ | 4 | 4 | 5 | 2 |

$$\text{New mean}=\frac{8+24+35+18}{15}=\frac{17}{3}$$

**44.** *(d)* Given data

$=\{6, 10, 7, 13, a, 12, b, 12\}$

Mean $=9$ and variance $=37/4$

Now, mean

$$=\frac{\Sigma x_i}{n}=\frac{6+10+7+13+a+12+b+12}{8}$$

or $9=\frac{60+a+b}{8}$

$\Rightarrow a+b=12 \qquad \ldots(i)$

$$\text{Variance} = \frac{\sum x_i^2}{n} - \left(\frac{\sum x_i}{n}\right)^2$$

$$\Rightarrow \frac{37}{4} = \frac{6^2+10^2+7^2+13^2+a^2+12^2+b^2+12^2}{8} - 9^2$$

$$\Rightarrow \frac{37}{4} = \frac{642+a^2+b^2}{8} - 81$$

$$\Rightarrow 74 = a^2+b^2-6$$

$$\Rightarrow a^2+b^2 = 80 \quad \ldots(ii)$$

$$\therefore (a+b)^2 = a^2+b^2+2ab$$

$$2ab = (a+b)^2-(a^2+b^2)$$

On putting the values from Eqs. (i) and (ii), we get

$$2ab = 12^2 - 80 = 64$$

Now, $(a-b)^2 = a^2+b^2-2ab$

$$= 80-64 = 16$$

**45.** *(d)* Given, mean = 10

i.e. $\frac{7+10+11+15+a+b}{6} = 10$

$$\Rightarrow a+b = 17 \quad \ldots(i)$$

Also, variance = 20/3

i.e. $\frac{(7)^2+(10)^2+(11)^2+(15)^2+a^2+b^2}{6} - (\text{Mean})^2 = \frac{20}{3}$

$$\Rightarrow \frac{495+a^2+b^2}{6} - 100 = \frac{20}{3}$$

$$\Rightarrow a^2+b^2 = 145 \quad \ldots(ii)$$

Using Eqs. (i) and (ii), put $b = 17-a$ in Eq. (ii),

$$a^2+(17-a)^2 = 145$$

$$\Rightarrow a^2+289+a^2-34a = 145$$

$$\Rightarrow a^2-17a+72 = 0$$

$$\Rightarrow (a-9)(a-8) = 0$$

$$\therefore a = 9 \text{ or } a = 8$$

Using $b = 17-a$

$$\therefore b = 8 \text{ or } b = 9$$

$$\therefore |a-b| = 1$$

**46.** *(a)*

> **key idea** (i) Mean of $[x_1, x_2, \ldots, x_n] = \frac{(x_1+x_2+\ldots+x_n)}{n}$
>
> (ii) Variance
>
> $$\sigma^2 = \sum_{i=1}^{n} \frac{(x_i-\bar{x})^2}{n} = \frac{\sum_{i=1}^{n} x_i^2}{n} - (\bar{x})^2$$

Mean = 6.5

$$\therefore \text{Total} = 6 \times 6.5 = 39$$

Let the remaining two observations be $x_1$ and $x_2$

$$\sum x_i = 39$$

$$\Rightarrow 2+4+5+7+x_1+x_2 = 39$$

$$\Rightarrow x_1+x_2 = 21$$

$$\text{Variance} = \frac{\sum x_i^2}{n} - \left(\frac{\sum x_i}{n}\right)^2$$

$$\Rightarrow 10.25 = \frac{\sum x_i^2}{6} - 6.5^2$$

$$\therefore \sum x_i^2 = 6(10.25+42.25) = 52.5 \times 6$$

$$\sum x_i^2 = 52.5 \times 6$$

$$4+16+25+49+x_1^2+x_2^2 = 315$$

$$\Rightarrow x_1^2+x_2^2 = 221$$

$$\Rightarrow x_1^2+(21-x_1)^2 = 221$$

$$\Rightarrow 2x_1^2-42x_1+441 = 221$$

$$\Rightarrow x_1^2-21x_1+110 = 0$$

$$\Rightarrow (x_1-11)(x_1-10) = 0$$

$$\Rightarrow x_1 = 10 \text{ or } 11$$

$$x_2 = 11 \text{ or } 10$$

Remaining two observations are 10 and 11.

**47.** *(d)*

> **key idea** (i) Mean of $[x_1, x_3, \ldots, x_n] = \frac{x_1+x_2+\ldots+x_n}{n}$
>
> (ii) $\text{SD} = \sqrt{\frac{\sum x^2}{n} - (\bar{x})^2}$

Given, three observations = $a, b, c$

$$b = a+c$$

Standard deviation = $\sigma$

$$\sigma = \sqrt{\frac{\Sigma(x_i-\mu)^2}{N}}$$

When, $x_1 = a+2$

$$x_2 = b+2, \quad x_3 = c+2$$

$$\mu = \frac{(a+2)+(b+2)+(c+2)}{3} = \frac{2b}{3}+2$$

$$x_1-\mu = (a+2)-(2b/3+2) = a-2b/3$$

$$x_2-\mu = (b+2)-(2b/3+2) = b-2b/3$$

$$x_3-\mu = (c+2)-(2b/3+2) = c-2b/3$$

$$d = \sqrt{\frac{(a-2b/3)^2+(b-2b/3)^2+(c-2b/3)^2}{3}}$$

$$3d^2 = a^2+b^2+c^2+\frac{12b^2}{9} - \frac{4}{3}(ab+b^2+bc)$$

$$\Rightarrow 3d^2 = a^2+b^2+c^2+\frac{4}{3}b^2-\frac{4}{3}[b(a+c)+b^2]$$

$$\Rightarrow 3d^2 = a^2+b^2+c^2+\frac{4}{3}b^2-\frac{4}{3}[b \cdot b+b^2]$$

$$\Rightarrow 3d^2 = a^2+b^2+c^2+\frac{4}{3}b^2-\frac{4}{3}\cdot 2b^2$$

$$\Rightarrow 3d^2 = a^2+c^2-\frac{b^2}{3}$$

$$\Rightarrow b^2 = 3a^2+3c^2-9d^2$$

**48.** *(c)* Let remaining two observations are $x$ and $y$, so mean

$$= \frac{5+7+10+12+14+15+x+y}{8} = 10$$

$$\Rightarrow x+y = 80-63$$

$$\Rightarrow x+y = 17 \quad \ldots(i)$$

and Variance

$$= \frac{25+49+100+144+196+225+x^2+y^2}{8} - 100 = 13.5$$

$$\Rightarrow \frac{739+x^2+y^2}{8} = 113.5$$

$$\Rightarrow x^2+y^2 = 169 \quad \ldots(ii)$$

From Eqs. (i) and (ii), we get

$x = 12$ and $y = 5$

$$\therefore |x-y| = |12-5| = 7$$

**49.** *(c)* For the observations $x_i\,(1 \le i \le 10)$, we have

$$\sum_{i=1}^{10}(x_i-5) = 10$$

$$\Rightarrow \sum_{i=1}^{10} x_i - 50 = 10 \Rightarrow \sum_{i=1}^{10} x_i = 60 \quad \ldots(i)$$

and $\sum_{i=1}^{10}(x_i-5)^2 = 40$

$$\Rightarrow \sum_{i=1}^{10} x_i^2 - 10\sum_{i=1}^{10} x_i + 10(5^2) = 40$$

$$\Rightarrow \sum_{i=1}^{10} x_i^2 - 10(60) + 250 = 40,$$

[from Eq. (i)]

$$\Rightarrow \sum_{i=1}^{10} x_i^2 = 390 \quad \ldots\text{(ii)}$$

Now, mean

$$\mu = \frac{\sum_{i=1}^{10}(x_i - 3)}{10} = \frac{\sum_{i=1}^{10} x_i - 30}{10}$$

$$= \frac{60-30}{10} = \frac{30}{10} = 3$$

and variance $\lambda = \dfrac{\sum_{i=1}^{10}(x_i - 3)^2}{10} - (\mu)^2$

$$= \frac{\sum_{i=1}^{10} x_i^2 - 6\sum_{i=1}^{10} x_i + 10(3^2)}{10} - 9$$

$$= \frac{390 - 6(60) + 90 - 90}{10} = \frac{30}{10} = 3$$

So, the ordered pair $(\mu, \lambda) = (3, 3)$

**50.** *(b)* Given 10 observations are $x_1, x_2, x_3, \ldots, x_{10}$.

$$\therefore \frac{x_1 + x_2 + x_3 + x_4}{4} = 11$$

$$\Rightarrow x_1 + x_2 + x_3 + x_4 = 44 \quad \ldots\text{(i)}$$

and $\dfrac{x_5 + x_6 + x_7 + x_8 + x_9 + x_{10}}{6} = 16$

$$\Rightarrow x_5 + x_6 + x_7 + x_8 + x_9 + x_{10} = 96 \quad \ldots\text{(ii)}$$

So, mean of given 10 observations

$$= \frac{44+96}{10} = \frac{140}{10} = 14$$

Since, the sum of squares of all the observations = 2000

$$\therefore x_1^2 + x_2^2 + x_3^2 + \ldots + x_{10}^2 = 2000 \quad \ldots\text{(iii)}$$

Now, $\sigma^2 = (\text{standard deviation})^2$

$$= \frac{\Sigma x_i^2}{10} - \left(\frac{\Sigma x_i}{10}\right)^2 = \frac{2000}{10} - (14)^2$$

$$= 200 - 196 = 4$$

So, $\sigma = 2$

**51.** *(b)* It is given that both mean and standard deviation of 50 observations $x_1, x_2, x_3, \ldots, x_{50}$ are equal to 16,

So, mean $= \dfrac{\Sigma x_i}{50} = 16 \quad \ldots\text{(i)}$

and standard deviation

$$= \sqrt{\frac{\Sigma x_i^2}{50} - \left(\frac{\Sigma x_i}{50}\right)^2} = 16$$

$$\Rightarrow \frac{\Sigma x_i^2}{50} - (16)^2 = (16)^2$$

$$\Rightarrow \frac{\Sigma x_i^2}{50} = 2 \times 256 = 512 \quad \ldots\text{(ii)}$$

Now, mean of $(x_1 - 4)^2, (x_2 - 4)^2, \ldots, (x_{50} - 4)^2$

$$= \frac{\Sigma(x_i - 4)^2}{50} = \frac{\Sigma(x_i^2 - 8x_i + 16)}{50}$$

$$= \frac{\Sigma x_i^2}{50} - 8\left(\frac{\Sigma x_i}{50}\right) + \frac{16}{50}\Sigma 1$$

$$= 512 - (8 \times 16) + \left(\frac{16}{50} \times 50\right)$$

[from Eqs. (i) and (ii)]

$$= 512 - 128 + 16 = 400$$

**52.** *(c)* Let the remaining two observations are $a$ and $b$. According to the question,

$$\text{Mean} = \frac{2+4+10+12+14+a+b}{7} = 8$$

$$\Rightarrow a + b = 14 \quad \ldots\text{(i)}$$

and variance

$$= \frac{a^2 + b^2 + 4 + 16 + 100 + 144 + 196}{7} - 8^2 = 16$$

$$\Rightarrow a^2 + b^2 = 100 \quad \ldots\text{(ii)}$$

We know that

$$(a+b)^2 = (a^2 + b^2) + 2ab$$

$$\Rightarrow (14)^2 = 100 + 2ab$$

[From Eqs. (i) and (ii)]

$$\Rightarrow ab = 48$$

**53.** *(b)* Given, mean $\bar{x} = 4$ and variance $\sigma^2 = 5.20$ and numbers of observation $n = 5$

Let $x_1 = 3, x_2 = 4, x_3 = 4$ and $x_4, x_5$ be the five observations

So, $\sum_{i=1}^{5} x_i = 5 \cdot \bar{x} = 5 \times 4 = 20$

$$\Rightarrow x_1 + x_2 + x_3 + x_4 + x_5 = 20$$

$$\Rightarrow 3 + 4 + 4 + x_4 + x_5 = 20$$

$$\Rightarrow x_4 + x_5 = 9 \quad \ldots\text{(i)}$$

Now, variance $\sigma^2 = \dfrac{\sum_{i=1}^{5} x_i^2}{5} - (\bar{x})^2$

$$\Rightarrow \frac{x_1^2 + x_2^2 + x_3^2 + x_4^2 + x_5^2}{5} - (4)^2 = 5.20$$

$$\Rightarrow \frac{9 + 16 + 16 + x_4^2 + x_5^2}{5} = 16 + 5.20$$

$$\Rightarrow x_4^2 + x_5^2 = 65 \quad \ldots\text{(ii)}$$

$$\because (x_4 + x_5)^2 = x_4^2 + x_5^2 + 2x_4x_5$$

$$\therefore 81 = 65 + 2x_4x_5$$

[from Eqs. (i) and (ii)]

$$\Rightarrow 16 = 2x_4x_5 \quad \ldots\text{(iii)}$$

Now, $(|x_4 - x_5|)^2 = x_4^2 + x_5^2 - 2x_4x_5$

$$= 65 - 16$$

[from Eqs. (ii) and (iii)]

$$= 49$$

$$\Rightarrow |x_4 - x_5| = 7$$

**54.** *(a)* Let 1, 3, 8, $x$ and $y$ be the five observations.

Then, mean $\bar{x} = \dfrac{\Sigma x_i}{n}$

$$\Rightarrow \bar{x} = \frac{1+3+8+x+y}{5} = 5 \quad \text{[given]}$$

$$\Rightarrow x + y = 25 - 12 = 13$$

$$\Rightarrow x + y = 13 \quad \ldots\text{(i)}$$

and variance $= \sigma^2 = \dfrac{\Sigma(x_i - \bar{x})^2}{n}$

$$= \frac{\left[\begin{array}{l}(1-5)^2 + (3-5)^2 + (8-5)^2 \\ \quad + (x-5)^2 + (y-5)^2\end{array}\right]}{5} = 9.2$$

[given]

$$\Rightarrow 16 + 4 + 9 + (x^2 - 10x + 25) + (y^2 - 10y + 25) = 46$$

$$\Rightarrow x^2 + y^2 - 10(x + y) = 46 - 79$$

$$\Rightarrow x^2 + y^2 - 10 \times 13 = -33$$

$$(\because x + y = 13)$$

$$\Rightarrow x^2 + y^2 = 97 \quad \ldots\text{(ii)}$$

Let $\dfrac{y}{x} = t \Rightarrow y = xt$

On putting $y = xt$ in Eq. (i), we get

$$x(1+t) = 13$$

$$\Rightarrow x^2(1+t)^2 = 169 \quad \ldots\text{(iii)}$$

On putting $y = xt$ in Eq. (ii), we get

$$x^2(1+t^2) = 97 \quad \ldots\text{(iv)}$$

On dividing Eq (iii) by Eq. (iv), we get

$$\frac{x^2(1+t)^2}{x^2(1+t^2)} = \frac{169}{97}$$

$$\Rightarrow 97(t^2 + 2t + 1) = 169(1 + t^2)$$

$$\Rightarrow (4t - 9)(9t - 4) = 0$$

$$\Rightarrow t = 9/4 \text{ or } t = \frac{4}{9}$$

**55.** *(a)* Given numbers are
$1, 1+d, 1+2d, ..., 1+100d.$

$\therefore n = 101$

$\therefore$ Mean, $\bar{x} = \frac{1}{101}[1+(1+d)+(1+2d)\ldots(1+100d)]$

$= \frac{1}{101} \times \frac{101}{2}[2+100d] = 1+50d$

$\therefore$ Mean deviation from mean

$= \frac{1}{101}[|1-(1+50d)|+|(1+d)-(1+50d)|+\ldots+|(1+100d)-(1+50d)|]$

$= \frac{2d}{101}(1+2+3+\ldots+50)$

$= \frac{2d}{101} \times \frac{50 \times 51}{2} = \frac{2550}{101}d$

Now, $\frac{2250}{101}d = 255$ [given]

$\Rightarrow d = 10.1$

**56.** *(b)* $\because \sigma_x^2 = 4$ and $\sigma_y^2 = 5$

Also, $\bar{x} = 2$ and $\bar{y} = 4$

Now, $\frac{\Sigma x_i}{5} = 2 \Rightarrow \Sigma x_i = 10$, $\Sigma y_i = 20$

and $\sigma_x^2 = \left(\frac{1}{5}\Sigma x_i^2\right) - (\bar{x})^2$,

$\sigma_x^2 = \frac{1}{5}\Sigma x_1^2 - 4$

$\sigma_y^2 = \frac{1}{5}(\Sigma y_i^2) - 16$

$\Rightarrow \Sigma x_i^2 = 40$ and $\Sigma y_i^2 = 105$

$\therefore$ Variance of the combined data set is

$\sigma_z^2 = \frac{1}{10}(\Sigma x_i^2 + \Sigma y_i^2) - \left(\frac{\bar{x}+\bar{y}}{2}\right)^2$

$= \frac{1}{10}(40+105) - 9$

$= \frac{145-90}{10} = \frac{55}{10} = \frac{11}{2}$

**57.** *(c)* In the $2n$ observations, half of them equal to $a$ and remaining half equal to $-a$. Then, the mean of total $2n$ observations is equal to zero.

$\therefore \text{SD} = \sqrt{\frac{\Sigma(x-\bar{x})^2}{N}} \Rightarrow 2 = \sqrt{\frac{\Sigma x^2}{2n}}$

$\Rightarrow 4 = \frac{\Sigma x^2}{2n} \Rightarrow 4 = \frac{2na^2}{2n} \Rightarrow a^2 = 4$

$\therefore |a| = 2$

**58.** *(a)* Given, $N = 15$, $\Sigma x^2 = 2830$, $\Sigma x = 170$

Since, one observation 20 was replaced by 30, then

Corrected $\Sigma x^2 = 2830 - 400 + 900 = 3330$

and corrected $\Sigma x = 170 - 20 + 30 = 180$

Corrected variance,

$\sigma^2 = \frac{\Sigma x^2}{N} - \left(\frac{\Sigma x}{N}\right)^2 = \frac{3330}{15} - \left(\frac{180}{15}\right)^2$

$= 78$

**59.** *(c)*

> **key idea** Mean and variance of $n$ observations $x_1, x_2, ..., x_n$ are $\bar{x} = \frac{\Sigma x_i}{n}$, $\sigma^2 = \frac{1}{n}\sum_{i=1}^{n} x_i^2 - (\bar{x})^2$, respectively.

Given, variance $= 4/5$

$\Rightarrow \frac{\Sigma x_i^2}{10} - \left(\frac{\Sigma x_i}{10}\right)^2 = \frac{4}{5}$

$\because \sum_{i=1}^{10}(x_i - 2) = 30$

$\therefore \sum_{i=1}^{10} x_i = 50$

$\therefore \left(\frac{\Sigma x_i^2}{10}\right)^2 = (5)^2 = 25$

$\Rightarrow \frac{\Sigma x_i^2}{10} - 25 = \frac{4}{5}$

$\Rightarrow \Sigma x_i^2 = 258$

Now, $\sum_{i=1}^{10}(x_i - \beta)^2 = 98$

$\Rightarrow \sum_{i=1}^{10}(x_i^2 - 2\beta \cdot x_i + \beta^2) = 98$

$\Rightarrow 258 - 2\beta(50) + 10\beta^2 = 98$

$\Rightarrow (\beta - 8)(\beta - 2) = 0$

$\Rightarrow \beta = 2, 8$

but $\beta > 2$ [given]

$\therefore \beta = 8$ (only)

Now, $2(x_1 - 1) + 4\beta, 2(x_2 - 1) + 4\beta, \ldots 2(x_{10} - 1) + 4\beta$

$= 2x_1 + 30, 2x_2 + 30, \ldots\ldots, 2x_{10} + 30$

So, mean of above observation

$\mu = 2(5) + 30 = 40$

[$\because$ mean of $x_1, x_2, x_3 \ldots, x_{10}$ is 5]

and variance $= \sigma^2 = 2^2\left(\frac{4}{5}\right) = \frac{16}{5}$

Hence, $\frac{\beta\mu}{\sigma^2} = \frac{8 \times 40}{\frac{16}{5}} = 100$

**60.** *(a)* Given that $\bar{x} = \frac{\sum_{i=1}^{10} x_i}{10} = \frac{6}{5}$

$\Rightarrow \sum_{i=1}^{10} x_i = 12$

and $\frac{\Sigma x_i^2}{10} - (\bar{x})^2 = \frac{84}{25}$

$\Rightarrow \frac{\sum_{i=1}^{10} x_i^2}{10} - \frac{36}{25} = \frac{84}{25}$

$\Rightarrow \sum_{i=1}^{10} x_i^2 = 48$

and $\sum_{i=1}^{10}(x_i - \alpha) = 2$,

$\sum_{i=1}^{10} x_i - 10\alpha = 2$,

$\Rightarrow 12 - 10\alpha = 2$

$\Rightarrow \alpha = 1$

Now, $\sum_{i=1}^{10}(x_i - \beta)^2 = 40$

$\Rightarrow \sum_{i=1}^{10} x_i^2 + 10\beta^2 - 2\beta\sum_{i=1}^{10} x_i = 40$

$\Rightarrow 48 + 10\beta^2 - 24\beta = 40$

$\Rightarrow 10\beta^2 - 24\beta + 8 = 0$

$\Rightarrow 5\beta^2 - 12\beta + 4 = 0$

$\beta = 2$ or $\beta = \frac{2}{5}$ [reject]

$\therefore \frac{\beta}{\alpha} = \frac{2}{1} = 2$

**61.** *(c)*

> **key idea** Use formula of mean deviation for simple distribution
>
> $\text{MD} = \frac{\sum_{i=1}^{n}|x_i - \bar{x}|}{n}$
>
> Where, $n$ = Number of terms
>
> $\bar{x}$ = Mean or median

Given data,
170, 125, 230, 190, 210, $a$, $b$

$\therefore$ MD about median

$$= \frac{|170-170|+|125-170|+|230-170|+|190-170|+|210-170|+|a-170|+|b-170|}{7}$$

$$\Rightarrow \frac{205}{7} = \frac{0+45+60+20+40+170-a+170-b}{7}$$

$\therefore a+b=300$

Mean

$$= \frac{170+125+230+190+210+a+b}{7}$$

$$= \frac{170+125+230+190+210+300}{7}$$

$$= \frac{1225}{7} = 175$$

Mean deviation about mean

$$= \frac{|170-175|+|125-175|+|230-175|+|190-175|+|210-175|+|a-175|+|b-175|}{7}$$

$$= \frac{5+50+55+15+35+350-300}{7}$$

$$= \frac{5+50+55+15+35+50}{7} = \frac{210}{7} = 30$$

**62.** (c) $\bar{X} = \frac{24}{5}$

$$\Rightarrow \frac{x_1+x_2+x_3+x_4+x_5}{5} = \frac{24}{5}$$

$$= \Sigma x_i = 24 \quad \dots(i)$$

$$\text{Var}(X) = \frac{194}{25}$$

Mean of first four observation $= \frac{7}{2}$

$$\frac{x_1+x_2+x_3+x_4}{4} = \frac{7}{2}$$

$$x_1+x_2+x_3+x_4 = 14 \quad \dots(ii)$$

By Eqs. (i) and (ii), $x_5 = 10$

Now, $\text{Var}(X) = \frac{194}{25}$

$$\Rightarrow \frac{\Sigma x_i^2}{5} - (\bar{X})^2 = \frac{194}{25}$$

$$\Rightarrow \frac{x_1^2+x_2^2+x_3^2+x_4^2+x_5^2}{5} - \left(\frac{24}{5}\right)^2 = \frac{194}{25}$$

$$\Rightarrow \frac{x_1^2+x_2^2+x_3^2+x_4^2+x_5^2}{5} = \frac{194}{25} + \frac{576}{25}$$

$$\Rightarrow x_1^2+x_2^2+x_3^2+x_4^2 = \frac{770}{5} - 100 = \frac{270}{5} = 54$$

New variance

$$\frac{\sum_{i=1}^{4} x_i^2}{4} - (\bar{X})^2 = \frac{54}{4} - \left(\frac{7}{2}\right)^2 = \frac{5}{4}$$

**63.** (b) We have, $n(A) = n(B) = 5$

Let $A = \{a_1, a_2, a_3, a_4, a_5\}$

and $B = \{b_1, b_2, b_3, b_4, b_5\}$

Mean of element is set $A(\text{a}) = 5$

$$\text{a} = \frac{\sum_{i=1}^{5} a_i}{5} = 5 \Rightarrow \sum_{i=1}^{5} a_i = 25$$

Mean of element in set $B(\text{b}) = 8$

$$\text{b} = \frac{\sum_{i=1}^{5} b_i}{5} = 8 \Rightarrow \sum_{i=1}^{5} b_i = 40$$

Variance of element in set $A = 12$

$$\frac{\sum_{i=1}^{5} (a_i)^2}{5} - (\text{a})^2 = 12$$

$$\sum_{i=1}^{5} a_i^2 = (12+25)\times 5 = 185$$

Variance of element in set $B = 20$

$$\frac{\sum_{i=1}^{5} b_i^2}{5} - (b)^2 = 20$$

$$\sum_{i=1}^{5} b_i^2 = (20+64)\times 5 = 420$$

Let $C = \{C_1, C_2, C_3, C_4, C_5, \dots C_{10}\}$

Mean of $C$

$$(\text{c}) = \frac{\left(\sum_{i=1}^{5} a_i - 15\right) + \left(\sum_{i=1}^{5} b_i + 10\right)}{10}$$

$$= \frac{(25-15)+(40+10)}{10} = 6$$

$$\text{Variance of } C = \frac{\sum_{i=1}^{5} C_i^2}{10} - (\text{c})^2$$

$$= \frac{\sum_{i=1}^{5} (a_i-3)^2 + \sum_{i=1}^{5} (b_i+2)^2}{10} - (6)^2$$

$$= \frac{\sum_{i=1}^{5} a_i^2 + \sum_{i=1}^{5} b_i^2 - 6\sum_{i=1}^{5} a_i + 4\sum_{i=1}^{5} b_i + 65}{10} - 36$$

$$= \frac{185+420-150+160+65}{10} - 36 = 32$$

Mean + Variance of element of $C$

$$= 6+32 = 38$$

**64.** (c) We have, $x_1, x_2, x_3, \dots, x_7$ are in AP with $x_1 = 9$ and common difference $= d$

$$\text{Now, Mean, } \bar{x} = \frac{\sum_{i=1}^{7} x_i}{7}$$

$$= \frac{x_1+x_2+x_3+\dots+x_7}{7}$$

$$= \frac{\frac{7}{2}[2x_1+(7-1)d]}{7} = \frac{1}{2}[2x_1+6d]$$

$$= x_1 + 3d = x_4$$

$$\text{Again, variance} = \frac{\sum_{i=1}^{7} (x_i-\bar{x})^2}{7}$$

$$\Rightarrow (4)^2 = \frac{\sum_{i=1}^{7} (x_i-x_4)^2}{7}$$

$$\Rightarrow 16 = \frac{(3d)^2+(2d)^2+d^2+0+d^2+(2d)^2+(3d)^2}{7}$$

$$\Rightarrow 16 = \frac{28d^2}{7}$$

$$\Rightarrow d = 2$$

$$[\because x_1 < x_2 < x_3 \dots < x_7]$$

$\therefore \quad \bar{x} = x_1 + 3d = 9 + 3\times 2 = 15$

and $\quad x_6 = x_1 + 5d = 9 + 5\times 2 = 19$

So, $\bar{x} + x_6 = 15 + 19 = 34$

**65.** (c) We have,

| | $A$ | $B$ | Combined $(A+B)$ |
|---|---|---|---|
| Mean | $\bar{X}_1 = 40$ | $\bar{X}_2 = 55$ | $\bar{X}_{12} = 50$ |
| SD | $\sigma_1 = \alpha$ | $\sigma_2 = 30 - \alpha$ | $\sigma_{12} = 350$ |
| Total number of observation | $n_1 = 100$ | $n_2 = n$ | $n_{12} = 100 + n$ |

$$\text{Now, } \bar{X}_{12} = \frac{n_1\bar{X}_1 + n_2\bar{X}_2}{n_1+n_2}$$

$$\Rightarrow 50 = \frac{100\times 40 + n\times 55}{100+n}$$

$$\Rightarrow 5000 + 50n = 4000 + 55n$$

$$\Rightarrow n = 200$$

Again, $\sigma_1^2 = \frac{\Sigma X_{i1}^2}{n_1} - (\bar{X}_1)^2$

$\Rightarrow \alpha^2 = \frac{\Sigma X_{i1}^2}{100} - (40)^2$

$\Rightarrow \Sigma X_{i1}^2 = (\alpha^2 + 1600) \times 100$

and $\sigma_2^2 = \frac{\Sigma X_{i2}^2}{n_2} - (\bar{X}_2)^2$

$\Rightarrow (30-\alpha)^2 = \frac{\Sigma X_{i2}^2}{n} - (55)^2$

$\Rightarrow \Sigma X_{i2}^2 = ((30-\alpha)^2 + 3025) \times 200$ $[\because n = 200]$

Now, $\sigma_{12}^2 = \frac{\Sigma X_{i1}^2 + \Sigma X_{i2}^2}{n_1 + n_2} - (\bar{X}_{12})^2$

$\Rightarrow 350 = \frac{(\alpha^2+1600)\times 100 + ((30-\alpha)^2 + 3025)\times 200}{300} - (50)^2$

$\Rightarrow \alpha^2 - 40\alpha + 300 = 0$

$\Rightarrow (\alpha - 10)(\alpha - 30) = 0$

$\Rightarrow \alpha = 10, 30$

$\Rightarrow \alpha = 10$ [$\alpha = 30$ is not possible]

$\therefore \sigma_1^2 + \sigma_2^2 = \alpha^2 + (30-\alpha)^2 = (10)^2 + (20)^2 = 500$

**66.** *(a)* Observations are $a, b, 8, 5, 10$

Mean, $\mu = \frac{a+b+8+5+10}{5} = 6$

$\therefore a + b = 7$

Variance, $\sigma^2 = \frac{\Sigma(x_i - \mu)^2}{N}$

$\Rightarrow \sigma^2 = \frac{(a-6)^2 + (b-6)^2 + 2^2 + 1^2 + 4^2}{5}$

$\Rightarrow 6.8 \times 5 = a^2 + b^2 - 12a - 12b + 93$

$\Rightarrow a^2 + b^2 = 25$

$\Rightarrow ab = 12$

$\therefore a = 3, b = 4$

Mean deviation about mean $= \frac{\Sigma|x_i - \mu|}{N}$

$M = \frac{|3-6| + |4-6| + |8-6| + |5-6| + |10-6|}{5}$

$= \frac{3+2+2+1+4}{5}$

$= \frac{12}{5}$

$\therefore 25M = \frac{12}{5} \times 25 = 60$

**67.** *(c)* Combined mean = 15.6

$\Rightarrow \frac{100 \times 15 + 150 \times \bar{x}_B}{250} = 15.6$

$\Rightarrow 1500 + 150\bar{x}_B = 3900$

$\Rightarrow 150\bar{x}_B = 2400$

$\Rightarrow \bar{x}_B = \frac{2400}{150} = 16$

Combined SD $= \sqrt{13.44}$

$\Rightarrow$ Combined variance $(\sigma^2) = 13.44$

$\sigma^2 = \frac{\sum x_i^2}{n} - (\bar{x})^2$

$\Rightarrow 13.44 = \frac{\sum x_i^2}{250} - (15.6)^2$

$\Rightarrow \sum x_i^2 = 64200$ ...(i)

For sample $A$,

$9 = \frac{\sum (x_i)_A^2}{100} - 225$

$\Rightarrow \sum (x_i)_A^2 = 23400$

Now, $\sum (x_i)_B^2 = 64200 - 23400 = 40800$

SD of sample $B = \sqrt{\frac{\sum (x_i)_B^2}{n_B} - (\bar{x}_B)^2}$

$= \sqrt{\frac{40800}{150} - 256} = 4$

> **Concept Enhancer**
>
> Suppose we have two sets of data containing $n_1$ and $n_2$ observations with means $\bar{X}_1$ and $\bar{X}_2$ and variances $S_1^2$ and $S_2^2$.
>
> If $\bar{X}_C$ is the combined mean and if $S_C^2$ is the combined variance of $n_1 + n_2$ observation.
>
> Combined mean $(\bar{X}_C) = \frac{n_1\bar{X}_1 + n_2\bar{X}_2}{n_1 + n_2}$
>
> Combined standard deviation $= \sqrt{\text{Combined variance}}$

**68.** *(a)* Let observations be denoted by $x_i$ for $1 \le i < 2n$.

$\therefore$ Mean $(\bar{x}) = \frac{\Sigma x_i}{2n}$

$= \frac{(a + a + a + \ldots + a) - (a + a + a + \ldots + a)}{2n}$

$\Rightarrow \bar{x} = 0$ and $\sigma_x^2 = \frac{\Sigma x_1^2}{2n} - (\bar{x})^2$

$= \frac{\overbrace{a^2 + a^2 + a^2 + \ldots + a^2}^{2n \text{ times}}}{2n} - 0 = a^2$

$\Rightarrow \sigma_x^2 = a^2 \Rightarrow \sigma_x = a$

According to the question, adding a constant $b$, then new mean

$(\bar{y}) = \bar{x} + b = 5$

$\Rightarrow 0 + b = 5 \Rightarrow b = 5$

and new SD $(\sigma_y) = \sigma_x$ (no change in SD)

$\Rightarrow \sigma_y = \sigma_x = 20$

$\Rightarrow a = 20$

$\therefore a^2 + b^2 = 400 + 25 = 425$

**69.** *(c)* We know that, $\sigma^2 = \frac{\Sigma x^2}{n} - \mu^2$

$= \frac{10\left(\frac{1}{2} - d\right)^2 + 10 \times \frac{1}{4} + 10\left(\frac{1}{2} + d\right)^2}{30} - \left(\frac{10\left(\frac{1}{2} - d\right) + 10 \times \frac{1}{2} + 10\left(\frac{1}{2} + d\right)}{30}\right)^2$ $\left[\because \mu = \frac{\Sigma x_i}{n}\right]$

$= \frac{20\left(\frac{1}{4} + d^2\right) + 5/2}{30} - \left(\frac{1}{4}\right)$

$= \frac{\frac{15}{2} + 20d^2}{30} - \frac{1}{4} = \frac{1}{4} + \frac{2d^2}{3} - \frac{1}{4} = \frac{2}{3}d^2$

$\therefore \frac{2}{3}d^2 = \frac{4}{3}$

$\Rightarrow d^2 = 2$

$\Rightarrow |d| = \sqrt{2}$

**70.** *(d)* We have, $\sum_{i=1}^{n} (x_i + 1)^2 = 9n$ ...(i)

and $\sum_{i=1}^{n} (x_i - 1)^2 = 5n$ ...(ii)

On subtracting Eq. (ii) from Eq. (i), we get

$$\sum_{i=1}^{n}\{(x_i+1)^2-(x_i-1)^2\}=4n$$

$$\Rightarrow \sum_{i=1}^{n}4x_i=4n \Rightarrow \sum_{i=1}^{n}x_i=n$$

$$\Rightarrow \frac{\sum_{i=1}^{n}x_i}{n}=1$$

$\therefore$ Mean $(\bar{x})=1$

Now, standard deviation

$$=\sqrt{\frac{\sum_{i=1}^{n}(x_i-\bar{x})^2}{n}}=\sqrt{\frac{\sum_{i=1}^{n}(x_i-1)^2}{n}}$$

$$=\sqrt{\frac{5n}{n}}=\sqrt{5}$$

**71.** (*) Let the other two observations be $a$ and $b$.

According to the question, mean = 5

$a+b+9=25 \Rightarrow a+b=16$ ...(i)

Also, variance = 124

$$\Rightarrow \frac{\Sigma x_i^2}{n}-(\bar{x})^2=124$$

$$\Rightarrow \frac{1+4+36+a^2+b^2}{5}-25=124$$

$\Rightarrow a^2+b^2=704$ ...(ii)

$\Rightarrow (a+b)^2-2ab=704$

$\Rightarrow -2ab=448$ ...(iii)

Now,

$(a-b)^2=a^2+b^2-2ab=704+448$ [Using Eqs. (ii) and (iii)]

$\Rightarrow (a-b)^2=1152$

$\Rightarrow a-b=\pm 24\sqrt{2}$ ...(iv)

From Eqs. (i) and (iv), we get

$a=8+12\sqrt{2}, b=8-12\sqrt{2}$

or $a=8-12\sqrt{2}, b=8+12\sqrt{2}$

Now, mean deviation about mean

$$=\frac{\Sigma|x_i-\bar{x}|}{n}$$

$$=\frac{|1-5|+|2-5|+|6-5|+|8+12\sqrt{2}-5|+|8-12\sqrt{2}-5|}{5}$$

$$=\frac{8+24\sqrt{2}}{5}\approx 8.4$$

**72.** (8788) 8, 21, 34, 47, ..., 320

This series is in AP.

For $n$, $n=\frac{320-8}{13}+1=\frac{312}{13}+1$

$=24+1=25$

$$\Rightarrow \bar{X}=\frac{\frac{n}{2}(8+320)}{n}=4+160=164$$

$\Rightarrow a_n=8+(n-1)13=13n-5$

$a_n^2=169n^2+25-130n$

$$\sum a_i^2=169\left(\frac{n}{6}\right)(n+1)(2n+1)+25n-130\frac{n}{2}(n+1)$$

$$=n\left(\frac{169}{6}(n+1)(2n+1)+25-65(n+1)\right)$$

$\therefore n=25$

$$\sum x_i^2=25\left(\frac{169}{6}(26)(51)+25-65\times 26\right)$$

$$\frac{\sum x_i^2}{25}=35684$$

$\therefore$ Var $(X)$

$$=\frac{\sum x_i^2}{25}-(\bar{X})^2=35684-(164)^2$$

$=8788$

**Time Saver Tech**

Variance of $a, a+d, a+2d, \ldots\ldots, a+(n-1)d$ is $d^2\left(\frac{n^2-1}{12}\right)$

$\therefore$ Given, 8, 21, 34, ......, 320

Here, $d=13, n=25$

$\therefore$ Variance

$$=13^2\left(\frac{25^2-1}{12}\right)=169\times\left(\frac{625-1}{12}\right)=8788$$

**73.** (33) $a, b, c \in N, a<b<c$

$$\bar{X}=\text{Mean}=\frac{9+25+a+b+c}{5}=18$$

$a+b+c=56$

$$\text{Mean deviation}=\frac{\Sigma|X_i-\bar{X}|}{n}=4$$

$\Rightarrow 9+7+|18-a|+|18-b|+|18-c|=20$

$\Rightarrow |18-a|+|18-b|+|18-c|=4$

$$\text{Variance}=\frac{\Sigma|X_i-\bar{X}|^2}{n}=\frac{136}{5}$$

$=81+49+|18-a|^2+|18-b|^2+|18-c|^2=136$

$=(18-a)^2+(18-b)^2+(18-c)^2=6$

Possible values

$(18-a)^2=1, (18-b)^2=1, (18-c)^2=4$ $(\because a<b<c)$

So $18-a=1, 18-b=-1, 18-c=-2$

$a=17, b=19, c=20$

$\therefore 2a+b-c=2\times 17+19-20$

$=34+19-20=33$

**74.** (29)

| $x_i$ | $f_i$ | $f_ix_i$ | $f_ix_i^2$ |
|---|---|---|---|
| 0 | 3 | 0 | 0 |
| 1 | 2 | 2 | 2 |
| 5 | 3 | 15 | 75 |
| 6 | 2 | 12 | 72 |
| 10 | 6 | 60 | 600 |
| 12 | 3 | 36 | 432 |
| 17 | 3 | 51 | 867 |
| Total $\Sigma f_i=22$ | | $\Sigma f_ix_i=176$ | $\Sigma f_ix_i^2=2048$ |

$$\because \text{Var}(\sigma^2)=\left(\frac{\Sigma f_ix_i^2}{\Sigma f_i}\right)-\left(\frac{\Sigma f_ix_i}{\Sigma f_i}\right)^2$$

$$=\frac{2048}{22}-\left(\frac{176}{22}\right)^2=\frac{1024}{11}-64$$

$=93.0909-64=29.0909$

**75.** (6344) Given, $\bar{x}=56$ and $\sigma^2=66.2$

$$\text{Variance, } \sigma^2=\frac{\Sigma x_i^2}{n}-(\bar{x}_i)^2$$

$$\Rightarrow 66.2=\frac{65^2+68^2+58^2+44^2+48^2+45^2+60^2+\alpha^2+\beta^2+60^2}{10}-(56)^2$$

$$\Rightarrow 66.2=\frac{\alpha^2+\beta^2+25678}{10}-(56)^2$$

$\Rightarrow 662+3136\times 10=\alpha^2+\beta^2+25678$

$\Rightarrow \alpha^2+\beta^2=6344$

**76.** (2521) Let $\mu'$ = incorrect mean,

$\mu$ = correct mean [given]

and $\sigma'$ = standard deviation,

$\sigma$ = standard deviation (corrected) [given]

Now we have,

$$\mu'=\frac{\Sigma x_i}{15}=12 \text{ [given]} \Rightarrow \Sigma x_i=180$$

According to the question,

corrected $\Sigma x_i=180-10+12$

$\Rightarrow \quad \mu \text{ (correct mean)} = \dfrac{182}{15}$

Also, $\sigma' = \sqrt{\dfrac{\Sigma x_i^2}{15} - 144} = 3$

$\Rightarrow \Sigma x_i^2 = 2295$

Correct $\Sigma x_i^2 = 2295 - 100 + 144$
$= 2339$ [correct variance]
$= \dfrac{2339}{15} - \dfrac{182 \times 182}{15 \times 15}$

Hence, required value
$= 15(\mu + \mu^2 + \sigma^2)$
$= 15\left(\dfrac{182}{15} + \dfrac{182 \times 182}{15 \times 15} + \dfrac{2339}{15} - \dfrac{182 \times 182}{15 \times 15}\right)$
$= 15\left(\dfrac{182}{15} + \dfrac{2339}{15}\right) = 2521$

**77.** *(269)* Given that the mean and standard deviation of the marks of 10 students were found to be 50 and 12

$\bar{x} = 50$ and $\sigma = 12$

$\Rightarrow \quad \sigma^2 = 144$

Now, $\Sigma x = 10 \times (\bar{x}) = 10 \times 50 = 500$

$\Sigma x_{i(\text{correct})} = 500 + 20 + 25 - 45 - 50$
$= 450$

Now, $\dfrac{\Sigma x_i^2}{10} - (50)^2 = 144$ (given)

$\Sigma x^2_{i(\text{correct})} = \{144 + (50)^2\} \times 10 - (45)^2 - (50)^2 + (20)^2 + (25)^2$
$= 26440 - 4525 + 1025$
$= 26440 - 3500$

$\Sigma x^2_{i(\text{correct})} = 22940$

Correct variance
$= \dfrac{\Sigma (x_{i_{\text{correct}}})^2}{10} - \left(\dfrac{\Sigma x_{i_{\text{correct}}}}{10}\right)^2$
$= \dfrac{22940}{10} - (45)^2 = 2294 - 2025 = 269$

**78.** *(8)* Given, mean = 5

$\bar{x} = \dfrac{\sum x_i f_i}{\sum f_i} = 5$

$\Rightarrow \dfrac{4 + 72 + 140 + 7\alpha + 72}{64 + \alpha} = 5$

$\Rightarrow \quad 288 + 7\alpha = 320 + 5\alpha$

$\Rightarrow \quad 2\alpha = 32 \Rightarrow \alpha = 16$

$m = \text{MD about } \bar{x} = \dfrac{\sum f_i |x_i - \bar{x}|}{\sum f_i}$

$= \dfrac{4 \times 4 + 24 \times 2 + 28 \times 0 + 16 \times 2 + 8 \times 4}{64 + 16}$

$m = \dfrac{128}{80} = \dfrac{8}{5}$

$\sigma^2 = \text{Variance} = \dfrac{\sum f_i (x_i - \bar{x}_i)^2}{\sum f_i}$

$= \dfrac{4 \times 16 + 24 \times 4 + 28 \times 0 + 16 \times 4 + 8 \times 16}{80}$

$= \dfrac{352}{80}$

$\therefore \quad \dfrac{3\alpha}{m + \sigma^2} = \dfrac{3 \times 16}{\dfrac{8}{5} + \dfrac{352}{80}} = 8$

**79.** *(151)* Given, mean is 28

So, $\dfrac{2 \times 5 + 3 \times 15 + x \times 25 + 5 \times 35 + 4 \times 45}{14 + x}$
$= 28$

$\Rightarrow \dfrac{10 + 45 + 25x + 175 + 180}{14 + x} = 28$

$\Rightarrow 410 + 25x = 392 + 28x$

$\Rightarrow \quad 3x = 18$

$\Rightarrow \quad x = 6$

$\therefore \text{Variance} = \left(\dfrac{\sum x_i^2 f_i}{\sum f_i}\right) - (\text{Mean})^2$

$= \left(\dfrac{2 \times 5^2 + 3 \times 15^2 + 6 \times 25^2 + 5 \times 35^2 + 4 \times 45^2}{20}\right) - (28)^2$

$= \left(\dfrac{50 + 675 + 3750 + 6125 + 8100}{20}\right) - (28)^2$

$= \left(\dfrac{18720}{20}\right) - (28)^2$

$= 935 - 784 = 151$

**80.** *(25)* Mean = 9

$\Rightarrow \dfrac{x + y + 12 + 6 + 4 + 8 + 12 + 10}{8} = 9$

$\Rightarrow x + y = 20 \quad \ldots\text{(i)}$

Variance = 9.25

$\Rightarrow \dfrac{x^2 + y^2 + 12^2 + 6^2 + 4^2 + 8^2 + 12^2 + 10^2}{8} - 9^2$

$= 9.25$

$\Rightarrow \quad x^2 + y^2 = 218 \quad \ldots\text{(ii)}$

Eliminating $y$ from Eqs. (i) and (ii), we get

$x^2 + (20 - x)^2 = 218$

$\Rightarrow \quad x^2 - 20x + 91 = 0$

$\Rightarrow x = 7, 13$

$\Rightarrow x = 13, y = 7 \ (x > y)$

$\therefore 3x - 2y = 39 - 14 = 25$

**81.** *(25)*

| $x_i$ | $f_i$ | $x_i^2$ | $f_i x_i$ | $f_i x_i^2$ |
|---|---|---|---|---|
| 2 | 4 | 4 | 8 | 16 |
| 4 | 4 | 16 | 16 | 64 |
| 6 | $\alpha$ | 36 | $6\alpha$ | $36\alpha$ |
| 8 | 15 | 64 | 120 | 960 |
| 10 | 8 | 100 | 80 | 800 |
| 12 | $\beta$ | 144 | $12\beta$ | $144\beta$ |
| 14 | 4 | 196 | 56 | 784 |
| 16 | 5 | 256 | 80 | 1280 |
| | $\Sigma f_i = 40 + \alpha + \beta$ | | $\Sigma f_i x_i = 360 + 6\alpha + 12\beta$ | $\Sigma f_i x_i^2 = 3904 + 36\alpha + 144\beta$ |

Given, mean = 9

$\Rightarrow \dfrac{\Sigma f_i x_i}{\Sigma f_i} = 9 \Rightarrow \dfrac{360 + 6\alpha + 12\beta}{40 + \alpha + \beta} = 9$

$\Rightarrow 360 + 6\alpha + 12\beta = 9(40 + \alpha + \beta)$

$\Rightarrow \quad 3\beta = 3\alpha \Rightarrow \alpha = \beta \quad \ldots\text{(i)}$

variance = 15.08

$\Rightarrow \dfrac{\Sigma f_i x_i^2}{\Sigma f_i} - \left(\dfrac{\Sigma f_i x_i}{\Sigma f_i}\right)^2 = 15.08$

$\Rightarrow \dfrac{3904 + 36\alpha + 144\beta}{40 + \alpha + \beta} - (9)^2 = 15.08$

$\Rightarrow \dfrac{3904 + 180\alpha}{40 + 2\alpha} = 81 + 15.08$

$\Rightarrow 3904 + 180\alpha = 96.08\,(40 + 2\alpha)$

$\Rightarrow 3904 + 180\alpha = 3843.2 + 192.16\alpha$

$\Rightarrow 60.8 = 12.16\alpha \Rightarrow \alpha = 5 = \beta$

$\therefore \quad \alpha^2 + \beta^2 - \alpha\beta = 25 + 25 - 25 = 25$

**82.** *(4)* Observations : $x_1, x_2, x_3, \ldots, x_{20}$

Mean, $\dfrac{\sum_{n=1}^{20} x_n}{20} = 15 \Rightarrow \sum_{n=1}^{20} x_n = 300$

Variance, $\frac{\Sigma x_i^2}{20} - \left(\frac{\Sigma x_i}{20}\right)^2 = 9$

$\Rightarrow \frac{\Sigma x_i^2}{20} - (15)^2 = 9$

$\Rightarrow \Sigma x_i^2 = 20(225+9) = 4680$

Mean of $(x_1+\alpha)^2, (x_2+\alpha)^2 + \ldots (x_{20}+\alpha)^2$ is 178

$\Rightarrow \frac{\Sigma(x_i+\alpha)^2}{20} = 178$

$\Rightarrow \Sigma x_i^2 + 2\alpha\Sigma x_i + \alpha^2\Sigma 1 = 3560$

$\Rightarrow 4680 + 2\alpha(300) + \alpha^2 \cdot 20 = 3560$

$\Rightarrow \alpha^2 + 30\alpha + 56 = 0$

$\Rightarrow \alpha = -2, -28$

Maximum of $\alpha = -2$

Square of maximum value of $\alpha = 4$

**83.** *(2)* Mean $= \frac{\Sigma x_i}{10} = 15,$

variance $= \frac{\Sigma x_i^2}{10} - 225 = 15$

$\Rightarrow \Sigma x_i = 150$ and $\Sigma x_i^2 = 2400$

Actual mean $= \frac{(\Sigma x_i) - 25 + 15}{10} = 14$

Actual variance

$= \frac{(\Sigma x_i^2) - 625 + 225}{10} - (14)^2$

$= 200 - 196 = 4$

$\therefore$ Actual standard deviation $= \sqrt{4} = 2$

**84.** *(25)* Number of boys = 20

Number of girls = 30

$\sigma_B^2 = 2, \bar{X}_B = 12, \sigma_G^2 = 2$

$\bar{X}_G = \frac{50 \times 15 - 20 \times 12}{30}$

$= \frac{510}{30} = 17 = \mu$

Variance of 50 candidates

$\sigma^2 = \frac{20\sigma_B^2 + 30\sigma_G^2}{50} + \frac{20 \cdot 30}{(20+30)^2}(\bar{X}_B - \bar{X}_G)^2$

$= \frac{20 \times 2 + 30 \times 2}{50} + \frac{600}{2500} \times 25 = 8$

$\therefore \mu + \sigma^2 = 17 + 8 = 25$

**85.** *(13)* Variance $= \frac{\Sigma x^2}{n} - (\bar{x})^2$ $[\because \bar{x} = \text{means}]$

$\Rightarrow 14 = \frac{1^2 + 2^2 + 3^3 + \ldots + n^2}{n} - \left(\frac{1+2+3+\ldots+n}{n}\right)^2$

$\Rightarrow 14 = \frac{n(n+1)(2n+1)}{6n} - \left(\frac{n(n+1)}{2n}\right)^2$

$\Rightarrow 14 = \frac{(n+1)(2n+1)}{6} - \frac{(n+1)^2}{4}$

$\Rightarrow 14 = \left(\frac{n+1}{12}\right)[2(2n+1) - 3(n+1)]$

$\Rightarrow 168 = (n+1)(n-1)$

$\Rightarrow n^2 - 1 = 168$

$\Rightarrow n^2 = 169$

$\Rightarrow n = 13$

**86.** *(12)* Given, mean of 3, 7, $x$ and $y$ is 5.

So, $3 + 7 + x + y = 20$

$\Rightarrow x + y = 10$ ...(i)

and Variance of 3, 7, $x$ and $y$ is 10

$\Rightarrow \frac{3^2 + 7^2 + x^2 + y^2}{4} - (5)^2 = 10$

$\Rightarrow x^2 + y^2 = 82$ ...(ii)

$\Rightarrow x^2 + (10-x)^2 = 82$ [Using Eq. (i)]

$\Rightarrow x^2 - 10x + 9 = 0$

$\Rightarrow (x-9)(x-1) = 0$

$\Rightarrow x = 1, 9$

$\Rightarrow (x, y) = (9, 1)$ [as $x > y$]

Mean $= \frac{3 + 2x + 7 + 2y + x + y + x - y}{4}$

$= \frac{10 + 4x + 2y}{4}$

$= \frac{10 + 4(9) + 2(1)}{4} = \frac{48}{4} = 12$

**87.** *(238)* Here, $n = 40$, Mean = 30 and $\sigma = 5$

$\because$ Mean = 30 $\Rightarrow \frac{\Sigma x_i}{40} = 30$

$\Rightarrow \Sigma x_i = 1200$

New, $\Sigma x_i = 1200 - 12 - 10$

$= 1200 - 22 = 1178$

$\therefore \sigma = \sqrt{\frac{\Sigma x_i^2}{n} - \left(\frac{\Sigma x_i}{n}\right)^2}$

$\Rightarrow 5 = \sqrt{\frac{\Sigma x_i^2}{40} - 900}$

$\Rightarrow 40(25 + 900) = \Sigma x_i^2$

$\Rightarrow \Sigma x_i^2 = 37000$

New, $\Sigma x_i^2 = 37000 - (12)^2 - (10)^2$

$= 37000 - 144 - 100$

$= 37000 - 244 = 36756$

Now, $\sigma = \sqrt{\frac{\text{New } \Sigma x_i^2}{n} - \left(\frac{\text{New } \Sigma x_i}{n}\right)^2}$

$\Rightarrow \sigma = \sqrt{\frac{36756}{38} - \left(\frac{1178}{38}\right)^2}$

$\Rightarrow \sigma^2 = \frac{36756}{38} - \frac{(1178)^2}{(38)^2}$

$\Rightarrow 38\sigma^2 = 36756 - 36518$

$\Rightarrow 38\sigma^2 = 238$

**88.** *(68)* $x : x_1, x_2, x_3, \ldots, x_{n+1}, x_{n+2}, \ldots x_{2n}, x_{2n+1}, \ldots, x_{3n}$

$\frac{1}{3n}\sum_{i=1}^{3n} x_i^2 - (\bar{x})^2 = 4$

$\left[\text{where, } \bar{x} = \frac{n_1\bar{x}_1 + n_2\bar{x}_2}{n_1 + n_2} = \frac{2n(6) + n(3)}{3n} = 5\right]$

$\frac{1}{3n}\sum_{i=1}^{3n} x_i^2 = 25 + 4 = 29$

$\Rightarrow \sum_{i=1}^{3n} (x_i)^2 = 87n$

$x' : x_1 + 1, x_2 + 1, \ldots, x_{2n} + 1, x_{(2n+1)} - 1, \ldots, x_{3n} - 1$

$\bar{x}' = \frac{2n(6+1) + n(3-1)}{3n} = \frac{16}{3}$

$k = \frac{1}{3n}\sum_{i=1}^{3n} x_i'^2 - \frac{256}{9}$

$= \frac{1}{3n}\left[\sum_{i=1}^{3n} x_i'^2 + 2\sum_{i=1}^{3n} x_i' + 3n - \sum_{i=2n+1}^{3n} x_i' + 3n\right] - \frac{256}{9}$

$= \frac{1}{3n}[87n + 2(12n) - 2 \cdot 3n + 3n] - \frac{256}{9}$

$\Rightarrow k = 36 - \frac{256}{9} \Rightarrow 9k = 324 - 256 = 68$

**89.** *(5)*

| | Size | Mean | Variance |
|---|---|---|---|
| Observation I | 10 | 2 | 2 |
| Observation II | $n$ | 3 | 1 |

Variance of combined set is 17/9.

Let the observations for group I be set $A : \{x_1, x_2, x_3, \ldots, x_{10}\}$

$$\frac{\Sigma x_i}{10} = 2$$

$$\Rightarrow \Sigma x_i = 20$$

and $\Sigma\left(\frac{x_i^2}{10}\right) - \left(\frac{\Sigma x_i}{10}\right)^2 = 2$

$$\Rightarrow \frac{\Sigma x_i^2}{10} - \frac{400}{100} = 2 \Rightarrow \Sigma x_i^2 = 60$$

Let the observation for group II be set $B:\{y_1, y_2, y_3, \ldots, y_{10}\}$

$$\frac{\Sigma y_i}{n} = 3 \Rightarrow \Sigma y_i = 3n$$

and $\frac{\Sigma y_i^2}{n} - \left(\frac{\Sigma y_i}{n}\right)^2 = 1$

$$\frac{\Sigma y_i^2}{n} - 9 = 1 \Rightarrow \Sigma y_i^2 = 10n$$

Combined variance $= 17/9$

$$\sigma^2 = \frac{\Sigma(x_i^2 + y_i^2)}{n+10} - \left[\Sigma\left(\frac{x_i + y_i}{n+10}\right)\right]^2$$

$$\Rightarrow \frac{17}{9} = \frac{60 + 10n}{n+10} - \left[\frac{20 + 3n}{n+10}\right]^2$$

$$\Rightarrow \frac{17}{9} = \frac{(10n+60)(n+10) - (3n+20)^2}{(n+10)^2}$$

$$\Rightarrow \frac{17}{9} = \frac{n^2 + 40n + 200}{n^2 + 20n + 100}$$

$$\Rightarrow 8n^2 - 20n - 100 = 0$$

$$\Rightarrow 4(2n+5)(n-5) = 0$$

$$\Rightarrow n = 5$$

**90.** *(4)* Given, $\sum_{i=1}^{18} (x_i - \alpha) = 36$

and $\sum_{i=1}^{18} |x_i - \beta|^2 = 90$

Now, $\sum_{i=1}^{18} (x_i - \alpha) = 36$

$$\sum_{i=1}^{18} x_i - 18\alpha = 36$$

$$\sum_{i=1}^{18} x_i = 36 + 18\alpha \quad \ldots\text{(i)}$$

Again, $\sum_{i=1}^{18} x_i^2 + 18\beta^2 - 2\beta \sum_{i=1}^{18} x_i = 90$

$$\Rightarrow \sum_{i=1}^{18} x_i^2 + 18\beta^2 - 2\beta(36 + 18\alpha) = 90$$

[Using Eq. (i)]

$$\Rightarrow \sum_{i=1}^{18} x_i^2 = 90 - 18\beta^2 + 2\beta(36 + 18\alpha) \quad \ldots\text{(ii)}$$

Given, standard deviation $= 1$

i.e., $\sigma^2 = 1$

$$\Rightarrow \frac{\sum_{i=1}^{18} x_i^2}{18} - \left(\frac{\sum_{i}^{18} x_i}{18}\right)^2 = 1$$

$$\Rightarrow \frac{1}{18}(90 - 18\beta^2 + 36\alpha\beta + 72\beta) - \left(\frac{36 + 18\alpha}{18}\right)^2 = 1$$

$$\Rightarrow -\beta^2 + 2\alpha\beta + 4\beta - \alpha^2 - 4\alpha = 0$$

$$\Rightarrow (\alpha - \beta)(\alpha - \beta + 4) = 0$$

$$\Rightarrow \alpha - \beta = -4 \quad (\because \alpha \neq \beta)$$

$$|\alpha - \beta| = |-4| = 4$$

**91.** *(11)* Given, 10 natural numbers $= 1, 1, 1, \ldots 1, k$

According to the question, variance $(\sigma^2) < 10$

$$\because \sigma^2 = \frac{\Sigma x^2}{n} - \left(\frac{\Sigma x}{n}\right)^2$$

$$\sigma^2 = \frac{(9 + k^2)}{10} - \left(\frac{9 + k}{10}\right)^2 < 10$$

$$\Rightarrow 10(9 + k^2) - (81 + k^2 + 18k) < 1000$$

$$\Rightarrow 90 + 10k^2 - 81 - k^2 - 18k < 1000$$

$$\Rightarrow 9k^2 - 18k + 9 < 1000$$

$$\Rightarrow (k-1)^2 < \frac{1000}{9}$$

$$\Rightarrow k < \frac{10\sqrt{10}}{3} + 1 \approx 11.54$$

$\therefore$ Maximum possible value of $k$ is 11.

# C-12

JEE Main Chapterwise Topicwise Mathematics

In JEE Main, there is usually 0-1 question per paper is visible from this chapter. In past 3 years, 1 conceptual and 5 numericals value type questions of moderate difficulty level were asked.

**For detailed analysis scan QR code**

# PROBABILITY

## Topic 1 Probability of an Event, Mutually Exclusive & Exhaustive Events, Equally Likely Events

### Level 1

**01** Let $S$ be the set of all the words that can be formed by arranging all the letters of the words GARDEN. From the set $S$, one word is selected at random. The probability that the selected word will not have vowels in alphabetical order is. *[2025, 28 Jan Shift-II]*

(a) 1/3 (b) 1/2 (c) 1/4 (d) 2/3

**02** Let $A = [a_{ij}]$ be a square matrix of order 2 with entries either 0 or 1. Let $E$ be the event that $A$ is an invertible matrix. Then, the probability $P(E)$ is *[2025, 24 Jan Shift-II]*

(a) $\frac{1}{8}$ (b) $\frac{3}{16}$ (c) $\frac{3}{8}$ (d) $\frac{5}{8}$

**03** If an unbiased dice is rolled thrice, then the probability of getting a greater number in the $i$th roll than the number obtained in the $(i-1)$th roll, $i = 2, 3$, is equal to *[2024, 9 April Shift-II]*

(a) 3/54 (b) 1/54 (c) 2/54 (d) 5/54

**Imp.** Nature of Roots of Quadratic Equation

**04** The coefficients $a$, $b$ and $c$ in the quadratic equation $ax^2 + bx + c = 0$ are from the set (1, 2, 3, 4,5, 6). If the probability of this equation having one real root bigger than the other is $p$, then $216p$ equals *[2024, 5 April Shift-II]*

(a) 57 (b) 38
(c) 76 (d) 19

**05** A coin is biased so that a head is twice as likely to occur as a tail. If the coin is tossed 3 times, then the probability of getting two tails and one head is *[2024, 31 Jan Shift-II]*

(a) $\frac{1}{27}$ (b) $\frac{2}{9}$ (c) $\frac{2}{27}$ (d) $\frac{1}{9}$

**06** Two marbles are drawn in succession from a box containing 10 red, 30 white, 20 blue and 15 orange marbles with replacement being made after each drawing. Then, the probability that first drawn marble is red and second drawn marble is white, is *[2024, 31 Jan Shift-I]*

(a) $\frac{4}{75}$ (b) $\frac{2}{3}$ (c) $\frac{4}{25}$ (d) $\frac{2}{25}$

**07** A fair die is thrown until 2 appears. Then, the probability that 2 appears in even number of throws, is *[2024, 29 Jan Shift-I]*

(a) $\frac{5}{11}$ (b) $\frac{1}{6}$ (c) $\frac{5}{6}$ (d) $\frac{6}{11}$

**08** An urn contains 6 white and 9 black balls. Two successive draws of 4 balls are made without replacement. The probability, that the first draw gives all white balls and the second draw gives all black balls, is *[2024, 27 Jan Shift-II]*

(a) $\frac{5}{256}$ (b) $\frac{3}{715}$
(c) $\frac{5}{715}$ (d) $\frac{3}{256}$

**09** Three dice are rolled. If the probability of getting different numbers on the three dice is $\frac{p}{q}$, where $p$ and $q$ are co-prime, then $q - p$ is equal to *[2023, 6 April Shift-II]*

(a) 4 (b) 3
(c) 1 (d) 2

**10** Let $\Omega$ be the sample space and $A \subseteq \Omega$ be an event. Given below are two statements

(S1) : If $P(A) = 0$, then $A = \phi$

(S2) : If $P(A) = 1$, then $A = \Omega$.

Then, *[2023, 24 Jan Shift-I]*

(a) only (S2) is true
(b) Both (S1) and (S2) are true
(c) only (S1) is true
(d) Both (S1) and (S2) are false

**11** The probability, that in a randomly selected 3-digit number at least two digits are odd, is *[2022, 28 June Shift-I]*

(a) 19/36 (b) 15/36
(c) 13/36 (d) 23/36

**12** Five numbers $x_1, x_2, x_3, x_4, x_5$ are randomly selected from the numbers 1, 2, 3, ..., 18 and are arranged in the increasing order $(x_1 < x_2 < x_3 < x_4 < x_5)$. The probability that $x_2 = 7$ and $x_4 = 11$ is *[2022, 27 June Shift-I]*

(a) $\frac{1}{136}$ (b) $\frac{1}{72}$
(c) $\frac{1}{68}$ (d) $\frac{1}{34}$

**Imp.** Mappings of Function

**13** Let $S = \{1, 2, 3, 4, 5, 6\}$. Then, the probability that a randomly chosen onto function $g$ from $S$ to $S$ satisfies $g(3) = 2g(1)$ is *[2021, 31 Aug Shift-II]*

(a) 1/10 (b) 1/15 (c) 1/5 (d) 1/30

**14** Let a computer program generate only the digits 0 and 1 to form a string of binary numbers with probability of occurrence of 0 at even places be 1/2 and probability of occurrence of 0 at the odd places be 1/3. Then, the probability that '10' is followed by '01' is equal to *[2021, 17 March Shift-II]*

(a) $\frac{1}{18}$ (b) $\frac{1}{3}$ (c) $\frac{1}{6}$ (d) $\frac{1}{9}$

**15** Two dices are rolled. If both dices have six faces numbered 1, 2, 3, 5, 7 and 11, then the probability that the sum of the numbers on the top faces is less than or equal to 8 is *[2021, 17 March Shift-I]*

(a) $\frac{4}{9}$ (b) $\frac{17}{36}$ (c) $\frac{5}{12}$ (d) $\frac{1}{2}$

**16** A seven digit number is formed using digits 3, 3, 4, 4, 4, 5, 5. The probability, that numbers formed is divisible by 2, is *[2021, 26 Feb Shift-II]*

(a) $\frac{6}{7}$ (b) $\frac{1}{7}$ (c) $\frac{3}{7}$ (d) $\frac{4}{7}$

**17** The coefficients $a, b$ and $c$ of the quadratic equation, $ax^2 + bx + c = 0$ are obtained by throwing a dice three times. The probability that this equation has equal roots is *[2021, 25 Feb Shift-I]*

(a) $\frac{1}{72}$ (b) $\frac{5}{216}$ (c) $\frac{1}{36}$ (d) $\frac{1}{54}$

**18** Out of 11 consecutive natural numbers if three numbers are selected at random (without repetition), then the probability that they are in AP with positive common difference, is *[2020, 6 Sep Shift-I]*

(a) $\frac{15}{101}$ (b) $\frac{5}{101}$ (c) $\frac{5}{33}$ (d) $\frac{10}{99}$

**19** If there of the six vertices of a regular hexagon are chosen at random, then the probability that the triangle formed with these chosen vertices is equilateral is *[2019, 12 April Shift-I]*

(a) $\frac{1}{10}$ (b) $\frac{1}{5}$
(c) $\frac{3}{10}$ (d) $\frac{3}{20}$

**20** Two different families $A$ and $B$ are blessed with equal number of children. There are 3 tickets to be distributed amongst the children of these families so that no child gets more than one ticket. If the probability that all the tickets go to the children of the family $B$ is 1/12, then the number of children in each family is *[JEE Main 2018]*

(a) 3 (b) 5
(c) 4 (d) 6

**21** Let $X$ be a set containing 10 elements and $P(X)$ be its power set. If $A$ and $B$ are picked up at random from $P(X)$, with replacement, then the probability that $A$ and $B$ have equal number of elements, is *[JEE Main 2015]*

(a) $\frac{^{20}C_{10}}{2^{10}}$ (b) $\frac{(2^{10}-1)}{2^{20}}$ (c) $\frac{(2^{10}-1)}{2^{10}}$ (d) $\frac{^{20}C_{10}}{2^{20}}$

**22** An urn contains nine balls out of which three are red, four are blue and two are green. Three balls are drawn at random without replacement from the urn. The probability that the three balls have different colours, is *[AIEEE 2010]*

(a) $\frac{1}{3}$ (b) $\frac{2}{7}$ (c) $\frac{1}{21}$ (d) $\frac{2}{23}$

**23** Three houses are available in a locality. Three persons apply for the houses. Each applies for one house without consulting others. The probability that all the three apply for the same house, is *[AIEEE 2005]*

(a) $\frac{7}{9}$ (b) $\frac{8}{9}$ (c) $\frac{1}{9}$ (d) $\frac{2}{9}$

**24** $A$ and $B$ play a game, where each is asked to select a number from 1 to 25. If the two numbers match, both of them win a prize. The probability that they will not win a prize in a single trial, is *[AIEEE 2002]*

(a) $\frac{1}{25}$ (b) $\frac{24}{25}$
(c) 2/25 (d) None of these

## Level 2

**25** The probability, of forming a 12 persons committee from 4 engineers, 2 doctors and 10 professors containing at least 3 engineers and at least 1 doctor, is *[2025, 4 April Shift-I]*

(a) $\frac{17}{26}$ (b) $\frac{103}{182}$ (c) $\frac{19}{26}$ (d) $\frac{129}{182}$

**26** Two numbers $k_1$ and $k_2$ are randomly chosen from the set of natural numbers. Then, the probability that the value of $i^{k_1} + i^{k_2}$, $(i = \sqrt{-1})$ is non-zero equals *[2025, 28 Jan Shift-I]*

(a) $\frac{3}{4}$ (b) $\frac{1}{4}$

(c) $\frac{2}{3}$ (d) $\frac{1}{2}$

**27** A board has 16 squares as shown in the figure

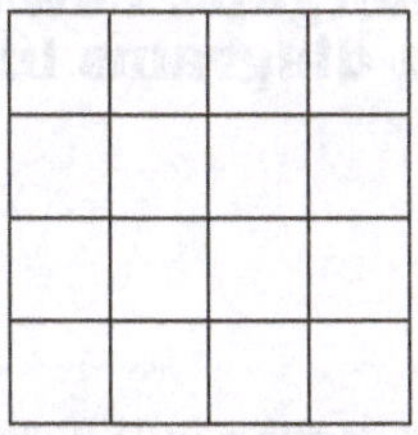

Out of these 16 squares, two squares are chosen at random. The probability that they have no side in common is *[2025, 23 Jan Shift-II]*

(a) $\frac{7}{10}$ (b) $\frac{23}{30}$

(c) $\frac{3}{5}$ (d) $\frac{4}{5}$

**28** One die has two faces marked 1, two faces marked 2, one face marked 3 and one face marked 4. Another die has one face marked 1, two faces marked 2, two faces marked 3 and one face marked 4. The probability of getting the sum of numbers to be 4 or 5, when both the dice are thrown together, is *[2025, 23 Jan Shift-I]*

(a) $\frac{1}{2}$ (b) $\frac{2}{3}$

(c) $\frac{4}{9}$ (d) $\frac{3}{5}$

**29** Let the sum of two positive integers be 24. If the probability, that their product is not less than 3/4 times their greatest possible product is $\frac{m}{n}$, where gcd $(m, n) = 1$, then $n - m$ equals *[2024, 8 April Shift-I]*

(a) 8 (b) 10

(c) 11 (d) 9

**30** Two integers $x$ and $y$ are chosen with replacement from the set $[0, 1, 2, 3, ..., 10]$. Then, the probability that $|x - y| \geq 5$, is *[2024, 30 Jan Shift-I]*

(a) $\frac{62}{121}$ (b) $\frac{30}{121}$

(c) $\frac{60}{121}$ (d) $\frac{31}{121}$

**31** Let $N$ be the sum of the numbers appeared when two fair dice are rolled and let the probability that $N-2$, $\sqrt{3N}$, $N+2$ are in geometric progression be $k/48$. Then, the value of $k$ is *[2023, 25 Jan Shift-II]*

(a) 2 (b) 8 (c) 16 (d) 4

**32** Let $S$ be the sample space of all five digit numbers. If $p$ is the probability that a randomly selected number from $S$, is multiple of 7 but not divisible by 5, then $9p$ is equal to *[2022, 27 July Shift-I]*

(a) 1.0146 (b) 1.2085

(c) 1.0285 (d) 1.1521

**33** If the numbers appeared on the two throws of a fair six faced die are $\alpha$ and $\beta$, then the probability that $x^2 + \alpha x + \beta > 0, \forall x \in R$, is *[2022, 25 July Shift-I]*

(a) $\frac{17}{36}$ (b) $\frac{4}{9}$ (c) $\frac{1}{2}$ (d) $\frac{19}{36}$

**34** When a certain biased die is rolled, a particular face occurs with probability $\frac{1}{6} - x$ and its opposite face occurs with probability $\frac{1}{6} + x$. All other faces occur with probability 1/6. Note that opposite faces sum to 7 in any die. If $0 < x < \frac{1}{6}$, and the probability of obtaining total sum $= 7$, when such a die is rolled twice is 13/96, then the value of $x$ is *[2021, 27 Aug Shift-I]*

(a) 1/16 (b) 1/8

(c) 1/9 (d) 1/12

**35** The probability that a randomly selected 2-digit number belongs to the set $\{n \in N : (2^n - 2)$ is a multiple of 3$\}$ is equal to *[2021, 27 July Shift-I]*

(a) $\frac{1}{6}$ (b) $\frac{2}{3}$ (c) $\frac{1}{2}$ (d) $\frac{1}{3}$

**36** Let 9 distinct balls be distributed among 4 boxes, $B_1, B_2, B_3$ and $B_4$. If the probability that $B_3$ contains exactly 3 balls is $k\left(\frac{3}{4}\right)^9$, then $k$ lies in the set *[2021, 25 July Shift-I]*

(a) $\{x \in R : |x-3| < 1\}$ (b) $\{x \in R : |x-2| \leq 1\}$

(c) $\{x \in R : |x-13| < 1\}$ (d) $\{x \in R : |x-5| \leq 1\}$

**37** The probability of selecting integers $a \in [-5, 30]$ such that $x^2 + 2(a+4)x - 5a + 64 > 0$, for all $x \in R$, is *[2021, 20 July Shift-I]*

(a) $\frac{7}{36}$ (b) $\frac{2}{9}$

(c) $\frac{1}{6}$ (d) $\frac{1}{4}$

**38** If two different numbers are taken from the set {0, 1, 2, 3, ..., 10}, then the probability that their sum as well as absolute difference are both multiple of 4, is *[JEE Main 2017]*

(a) $\frac{6}{55}$ (b) $\frac{12}{55}$ (c) $\frac{14}{45}$ (d) $\frac{7}{55}$

## Level 3

**39** Let $S = \{w_1, w_2, ....\}$ be the sample space associated to a random experiment. Let $P(w_n) = \frac{P(w_{n-1})}{2}, n \geq 2$. Let $A = \{2k + 3l : k, l \in N\}$ and $B = \{w_n : n \in A\}$. Then, $P(B)$ is equal to *[2023, 29 Jan Shift-II]*

(a) $\frac{1}{16}$ (b) $\frac{1}{32}$ (c) $\frac{3}{64}$ (d) $\frac{3}{32}$

### Numerical Value Type Questions

**40** Three distinct numbers are selected randomly from the set {1, 2, 3, ..., 40}. If the probability that the selected numbers are in an increasing GP, is $\frac{m}{n}$, $\gcd(m, n) = 1$, then $m + n$ is equal to ........ *[2025, 2 April Shift-I]*

**41** Let $a$, $b$ and $c$ denote the outcome of three independent rolls of a fair tetrahedral die, whose four faces are marked 1, 2, 3, 4. If the probability that $ax^2 + bx + c = 0$ has all real roots is $\frac{m}{n}$, $gcd(m, n) = 1$, then $m + n$ is equal to .......... *[2024, 9 April Shift-I]*

**42** A fair $n(n > 1)$ faces die is rolled repeatedly until a number less than $n$ appears. If the mean of the number of tosses required is $n/9$, then $n$ is equal to ......... . *[2023, 12 April Shift-I]*

**43** A bag contains six balls of different colours. Two balls are drawn in succession with replacement. The probability that both the balls are of the same colour is $p$. Next four balls are drawn in succession with replacement and the probability that exactly three balls are of the same colour is $q$. If $p : q = m : n$, where $m$ and $n$ are coprime, then $m + n$ is equal to ....... . *[2023, 30 Jan Shift-II]*

**44** Let $A$ be the event that the absolute difference between two randomly chosen real numbers in the sample space [0, 60] is less than or equal to $a$. If $P(A) = \frac{11}{36}$, then $a$ is equal to ......... . *[2023, 30 Jan Shift-II]*

**45** Let $S = \{E_1, E_2 .... E_8\}$ be a sample space of random experiment such that $P(E_n) = \frac{n}{36}$ for every $n = 1, 2, ...., 8$. Then, the number of elements in the set $\left\{A \subseteq S : P(A) \geq \frac{4}{5}\right\}$ is ......... . *[2022, 27 June Shift-II]*

**46** If the probability that a randomly chosen 6-digit number formed by using digits 1 and 8 only is a multiple of 21 is $p$, then 96 $p$ is equal to ...... *[2022, 26 June Shift-II]*

**47** An urn contains 5 red marbles, 4 black marbles and 3 white marbles. Then, the number of ways in which 4 marbles can be drawn so that at the most three of them are red is ......... . *[2020, 8 Jan Shift-I]*

# Topic 2 Use of Union, Intersection, De-Morgan's Law and Venn-diagrams in Probability

## Level 1

**Imp.** Independent Event

**01** Let Ajay will not appear in JEE exam with probability $p = \frac{2}{7}$, while both Ajay and Vijay will appear in the exam with probability $q = \frac{1}{5}$. Then, the probability, that Ajay will appear in the exam and Vijay will not appear is *[2024, 7 Feb Shift-II]*

(a) $\frac{18}{35}$ (b) $\frac{9}{35}$ (c) $\frac{24}{35}$ (d) $\frac{3}{35}$

**02** In a box, there are 20 cards, out of which 10 are labelled as $A$ and the remaining 10 are labelled as $B$. Cards are drawn at random, one after the other and with replacement, till a second $A$-card is obtained. The probability that the second $A$-card appears before the third $B$-card is *[2020, 9 Jan Shift-I]*

(a) 15/16 (b) 9/16
(c) 13/16 (d) 11/16

**03** Let $S = \{1, 2, \ldots, 20\}$. A subset $B$ of $S$ is said to be 'nice', if the sum of the elements of $B$ is 203. Then, the probability that a randomly chosen subset of $S$ is 'nice', is *[2019, 11 Jan Shift-II]*

(a) $\frac{6}{2^{20}}$ (b) $\frac{4}{2^{20}}$
(c) $\frac{7}{2^{20}}$ (d) $\frac{5}{2^{20}}$

## Level 2

**Imp.** Intersection of Events

**04** If $A$ and $B$ are two events such that $P(A) = 0.7$, $P(B) = 0.4$ and $P(A \cap \overline{B}) = 0.5$, where $\overline{B}$ denotes the complement of $B$, then $P(B \mid (A \cup \overline{B}))$ is equal to *[2025, 8 April Shift-II]*

(a) $\frac{1}{3}$ (b) $\frac{1}{6}$ (c) $\frac{1}{4}$ (d) $\frac{1}{2}$

**05** An integer is chosen at random from the integers 1, 2, 3, ..., 50. The probability that the chosen integer is a multiple of atleast one of 4, 6 and 7 is
**[2024, 29 Jan Shift-II]**

(a) $\frac{8}{25}$ (b) $\frac{9}{50}$
(c) $\frac{14}{25}$ (d) $\frac{21}{50}$

**06** Let $E_1$, $E_2$ and $E_3$ be three mutually exclusive events such that $P(E_1)=\frac{2+3p}{6}$, $P(E_2)=\frac{2-p}{8}$ and $P(E_3)=\frac{1-p}{2}$. If the maximum and minimum values of $p$ are $p_1$ and $p_2$ respectively, then $(p_1+p_2)$ is equal to
**[2022, 26 July Shift-I]**

(a) $\frac{2}{3}$ (b) $\frac{5}{3}$
(c) $\frac{5}{4}$ (d) 1

**07** Events $A$, $B$ and $C$ are mutually exclusive events such that $P(A)=\frac{3x+1}{3}$, $P(B)=\frac{1-x}{4}$ and $P(C)=\frac{1-2x}{2}$. The set of possible values of $x$ are in the interval
**[AIEEE 2003]**

(a) $\left[\frac{1}{3},\frac{1}{2}\right]$ (b) $\left[\frac{1}{3},\frac{2}{3}\right]$
(c) $\left[\frac{1}{3},\frac{13}{3}\right]$ (d) [0, 1]

## Level 3

**08** Let $S=\{1,2,3,\ldots,2022\}$. Then, the probability, that a randomly chosen number $n$ from the set $S$ such that HCF $(n, 2022)=1$, is
**[2022, 29 July Shift-I]**

(a) $\frac{128}{1011}$ (b) $\frac{166}{1011}$
(c) $\frac{127}{337}$ (d) $\frac{112}{337}$

# Key Index

### Topic 1

| | | | | | | | | | |
|---|---|---|---|---|---|---|---|---|---|
| 1 (b) | 2 (c) | 3 (d) | 4 (b) | 5 (b) | 6 (a) | 7 (a) | 8 (b) | 9 (a) | 10 (b) |
| 11 (a) | 12 (c) | 13 (a) | 14 (d) | 15 (b) | 16 (c) | 17 (b) | 18 (c) | 19 (a) | 20 (b) |
| 21 (d) | 22 (b) | 23 (c) | 24 (b) | 25 (d) | 26 (a) | 27 (d) | 28 (a) | 29 (b) | 30 (b) |
| 31 (d) | 32 (c) | 33 (a) | 34 (b) | 35 (c) | 36 (a) | 37 (b) | 38 (a) | 39 (c) | 40 (4949) |
| 41 (19) | 42 (10) | 43 (14) | 44 (10) | 45 (19) | 46 (33) | 47 (490) | | | |

### Topic 2

| | | | | | | | |
|---|---|---|---|---|---|---|---|
| 1 (a) | 2 (d) | 3 (d) | 4 (c) | 5 (d) | 6 (b) | 7 (a) | 8 (d) |

# Detailed **Explanations**

## Topic 1

**01.** *(b)* There are 2 vowels $A$ and $E$ only

So, probability the selected word will not have vowels in alphabetical order

$$1-\frac{\left(\frac{6!}{2!}\right)}{6!}=\frac{1}{2}$$

**02.** *(c)* Case I $A=\begin{bmatrix}1&1\\0&1\end{bmatrix}\rightarrow 4$ ways.

Case II $A=\begin{bmatrix}1&0\\0&1\end{bmatrix}$ and $\begin{bmatrix}0&1\\1&0\end{bmatrix}\rightarrow 2$ ways.

$\therefore P(E)$

$$=\frac{\text{Total number of favourable cases}}{\text{Total cases}}$$

$$=\frac{6}{16}=\frac{3}{8}$$

**03.** *(d)* Favourable cases $= {}^6C_3$

Total out comes $=6^3$

$\therefore$Probability of getting greater number than previous number are

$$=\frac{{}^6C_3}{6^3}=\frac{20}{216}=\frac{5}{54}$$

**04.** *(b)* $ax^2+bx+c=0$

One root bigger than other

$\therefore$Roots are real and distinct

$$D>0 \Rightarrow b^2-4ac>0$$

$b=3$, then $(a, c)=(1, 1), (1, 2), (2, 1)$

$b=4$, then $(a, c)=(1, 1), (1, 2), (2, 1)$
$(1, 3), (3, 1)$

$b=5$, then $(a, c)=(1, 1), (1, 2), (2, 1)$
$(1, 3), (3, 1), (1, 4), (4, 1), (1, 5),$
$(5, 1), (1, 6), (6, 1), (2, 3), (3, 2), (2, 2)$

$b=6$, then $(a, c)=(1, 1), (1, 2), (2, 1)$
$(1, 3), (3, 1), (1, 4), (4, 1), (1, 5),$
$(5, 1), (1, 6), (6, 1), (2, 3), (3, 2),$
$(2, 2), (2, 4), (4, 2)$.

Favourable cases = 38

$$\therefore \quad p=\frac{38}{6\times6\times6}\Rightarrow 216p=38$$

**05.** *(b)* Probability of tail $=\frac{1}{3}$

Probability of head $=\frac{2}{3}$

Two tails and one head
$=$TTH, THT, HTT

Probability of getting 2 tails and 1 head

$$=\left(\frac{1}{3}\times\frac{1}{3}\times\frac{2}{3}\right)+\left(\frac{1}{3}\times\frac{2}{3}\times\frac{1}{3}\right)+\left(\frac{2}{3}\times\frac{1}{3}\times\frac{1}{3}\right)=\frac{2}{9}$$

**06.** *(a)* 10 Red, 30 White, 20 Blue and 15 Marbles.

Probability of drawing first marble red and second marble white is

$$=P(R)\times P(W)$$

$$=\frac{10}{75}\times\frac{30}{75}=\frac{4}{75}$$

**07.** *(a)* We have,

$$P(2)=\frac{1}{6} \text{ and } P(\overline{2})=\frac{5}{6}$$

$\therefore P$[2 obtained on even numbered toss]

$$=\frac{5}{6}\times\frac{1}{6}+\left(\frac{5}{6}\right)^3\times\frac{1}{6}+\left(\frac{5}{6}\right)^5\times\frac{1}{6}+\ldots$$

$$=\frac{\frac{5}{6}\times\frac{1}{6}}{1-\left(\frac{5}{6}\right)^2}=\frac{5}{36}\times\frac{36}{11}=\frac{5}{11}$$

**08.** *(b)* Probability of required event

$$=\frac{{}^6C_4}{{}^{15}C_4}\times\frac{{}^9C_4}{{}^{11}C_4}$$

$$=\frac{15\times24}{15\times14\times13\times12}\times\frac{9\times8\times7\times6\times24}{24\times11\times10\times9\times8}$$

$$=\frac{1}{13\times7}\times\frac{7\times3}{11\times5}=\frac{3}{715}$$

**09.** *(a)* Total number of outcomes $=6\times6\times6=216$

Number of outcomes getting different numbers on the three dice are

$${}^6P_3=\frac{6!}{3!}=120$$

$$\therefore \text{ Required probability}=\frac{120}{216}=\frac{5}{9}$$

$\therefore \quad p=5$ and $q=9$

$\therefore \quad q-p=9-5=4$

**10.** *(b)* Given,

$\Omega=$sample space

$A=$ an event such that $A\subseteq\Omega$

Now, as $S1: P(A)=0$, then $A$ must be $\phi$.

So, $S1$ is true.

Also, as $S2: P(A)=1$, then $A$ must be equal to sample space $(\Omega)$.

So, $S2$ is also true.

**11.** *(a)* Required cases = Total ways – all digits even – exactly one digit odd

Total ways $=900$ ways

All digits even $=4\times5\times5=100$ ways

One digit odd $=\underset{\text{odd}}{5}\times5\times5=125$ ways

$4\times\underset{\text{odd}}{5}\times5=100$ ways

$4\times5\times\underset{\text{odd}}{5}=100$ ways

$$\therefore\text{Required probability}=\frac{900-425}{900}=\frac{19}{36}$$

**12.** *(c)* Number of ways to select and arrange $x_1, x_2, x_3, x_4, x_5$ in increasing order selected from 1, 2, 3, ….,18.

$$n(S)={}^{18}C_5$$

$$x_1 \; x_2 \; x_3 \; x_4 \; x_5$$

7 11

$$n(E)={}^6C_1\times{}^3C_1\times{}^7C_1$$

(for $x_1<7$) (for $x_3\in\{8, 9, 10\}$) (for $x_5>11$)

$$P(E)=\frac{6\times3\times7}{{}^{18}C_5}=\frac{126}{8568}=\frac{1}{68}$$

**13.** *(a)* $\because g: S\rightarrow S$

where, $S=\{1, 2, 3, 4, 5, 6\}$

Total number of onto function $=6!$

Given condition $g(3)=2g(1)$

If $g(1)=1$, then $g(3)=2$

$g(1)=2$, then $g(3)=4$

and $g(1)=3$, then $g(3)=6$

There are only 3 possible cases

Number of onto functions satisfying

$$g(3)=2g(1)=3\times4!$$

$$\therefore \text{ Required probability}=\frac{3\times4!}{6!}=\frac{1}{10}$$

**14.** *(d)* Probability of 0 at even place

$=P(0$ at even place$)=\frac{1}{2}$ [given]

and $P(0$ at odd places$)=\frac{1}{3}$ [given]

$\therefore P(1$ at even place$)=\frac{1}{2}$

and $P(1$ at odd place$)=1-\frac{1}{3}=\frac{2}{3}$

$\therefore P(10 \text{ is followed by } 01) = \text{Concepts} \begin{pmatrix} \ldots 1001 \\ EOEO \\ or\ (+) \\ OEOE \end{pmatrix},$

Where, $E = \text{Even}$

and $O = \text{odd}$

$$\left(\frac{1}{2}\times\frac{1}{3}\times\frac{1}{2}\times\frac{2}{3}\right)+\left(\frac{2}{3}\times\frac{1}{2}\times\frac{1}{3}\times\frac{1}{2}\right)$$

$$=\frac{1}{18}+\frac{1}{18}=\frac{1}{9}$$

**15.** *(b)* Six faces have numbers $\{1, 2, 3, 5, 7, 11\}$.

Sum of the numbers on top faces of both the sides is less than or equal to 8

$= (1, 1), (1, 2), (1, 3), (1, 5), (1, 7) \rightarrow 5$

$(2, 1), (2, 2), (2, 3), (2, 5) \rightarrow 4$

$(3, 1), (3, 2), (3, 3), (3, 5) \rightarrow 4$

$(5, 1), (5, 2), (5, 3) \rightarrow 3$

$(7, 1) \rightarrow 1$

$n(E) = 5 + 4 + 4 + 3 + 1 = 17$

$n(S) = 6 \times 6 = 36$

$\therefore$ Probability $= \dfrac{17}{36}$

**16.** *(c)* Given, digits 3, 3, 4, 4, 4, 5, 5

Total 7-digit number from given digits will be

$$\frac{7!}{3!\,2!\,2!}=\frac{7\times6\times5\times4}{2\times2}=210$$

Now, if the 7-digit number is divisible by 2, then last digit must be 4.

Now, 7-digit number divisible by 2 will be

$$=\frac{6!}{2!\,2!\,2!}=\frac{6\times5\times4\times3}{2\times2}=90$$

Required probability $=\dfrac{90}{210}=\dfrac{9}{21}=\dfrac{3}{7}$

**17.** *(b)* Given, $ax^2 + bx + c = 0$

According to the question,

$a, b, c \in \{1, 2, 3, 4, 5, 6\}$

$\therefore \quad n(S) = 6 \times 6 \times 6 = 216$

roots are equal, so

$D = 0 \Rightarrow b^2 - 4ac = 0$

$\Rightarrow \quad b^2 = 4ac \Rightarrow ac = \dfrac{b^2}{4}$

If $\quad b = 2, ac = 1 \Rightarrow (a = 1, c = 1)$

If $\quad b = 4, ac = 4 \Rightarrow (a = 1, c = 4)$ or $(a = 4, c = 1)$ or $(a = 2, c = 2)$

If $\quad b = 6, ac = 9 \Rightarrow (a = 3, c = 3)$

$\therefore$ Required probability $= 5/216$

**18.** *(c)* We have to select three numbers at random out of 11 consecutive natural numbers such that they are in AP, then two numbers must be either even or odd and third will be selected automatically, so we can do this in ${}^6C_2 + {}^5C_2$ ways and total number of ways to select 3 numbers out of 11 consecutive natural numbers is ${}^{11}C_3$.

So, required probability

$$=\frac{{}^6C_2+{}^5C_2}{{}^{11}C_3}=\frac{15+10}{165}=\frac{25}{165}=\frac{5}{33}$$

**19.** *(a)* Since, there is a regular hexagon, then the number of ways of choosing three vertices is ${}^6C_3$. And, there is only two ways i.e. choosing vertices of a regular hexagon alternate, here $A_1, A_3, A_5$ or $A_2, A_4, A_6$ will result in an equilateral triangle.

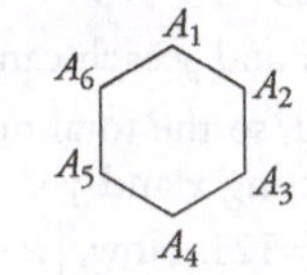


$\therefore$ Required probability

$$=\frac{2}{{}^6C_3}=\frac{2}{20}=\frac{1}{10}$$

**20.** *(b)* Let the number of children in each family be $x$.

So, total number of children in both families are $2x$.

According to the question,

$$\frac{1}{12}=\frac{{}^xC_3\cdot 3!}{{}^{2x}C_3\cdot 3!}$$

$$\Rightarrow \frac{x\cdot(x-1)\cdot(x-2)}{(2x)(2x-1)(2x-2)}=\frac{1}{12}$$

$$\Rightarrow \frac{(x-2)}{4(2x-1)}=\frac{1}{12}\Rightarrow 3(x-2)=2x-1$$

$$\Rightarrow \quad x=5$$

**21.** *(d)*

**key idea** Use Formula : $({}^nC_0)^2 + ({}^nC_1)^2 + \ldots + ({}^nC_n)^2 = \dfrac{(2n)!}{(n!)^2}$.

Required probability

$$=\frac{({}^{10}C_0)^2+({}^{10}C_1)^2+({}^{10}C_2)^2+\ldots+({}^{10}C_{10})^2}{(2^{10})^2}$$

$$=\frac{{}^{20}C_{10}}{2^{20}}$$

**22.** *(b)* Total number of cases $= {}^9C_3 = 84$

Number of favourable cases

$= {}^3C_1 \cdot {}^4C_1 \cdot {}^2C_1 = 24$

$\therefore \quad P = \dfrac{24}{84} = \dfrac{2}{7}$

**23.** *(c)* All the three persons has three options to apply a house.

$\therefore$ Total number of cases $= 3^3$

Now, favourable cases $= 3$ (as either all has applied for house 1 or 2 or 3)

$\therefore \quad$ Required probability $= \dfrac{3}{3^3} = \dfrac{1}{9}$

**24.** *(b)* The total number of ways in which numbers can be choosed

$= 25 \times 25 = 625$

The number of ways in which either players can choose same numbers $= 25$

$\therefore$ Probability that they win a prize

$= \dfrac{25}{625} = \dfrac{1}{25}$

Thus, the probability that they will not win a prize in a single trial

$= 1 - \dfrac{1}{25} = \dfrac{24}{25}$

**25.** *(d)*

Persons → 4 engineers, 2 Doctors, 10 Professors → 12 persons ↓ atleast 3E,1D

$\therefore$ Number of favourable cases

3 E 1 D 8F + 3 E 2 D 7 F + 4 E 1 D 7 F + 4 E 2 D 6 F

$= {}^4C_3 \cdot {}^2C_1 \cdot {}^{10}C_8 + {}^4C_3 \cdot {}^2C_2 \cdot {}^{10}C_7 + {}^4C_4 \cdot {}^2C_1 \cdot {}^{10}C_7 + {}^4C_4 \cdot {}^2C_2 \cdot {}^{10}C_6$

$= 1290$

$\therefore$ Required probability $= \dfrac{1290}{{}^{16}C_{12}}$

$= \dfrac{1290 \times 4 \times 3 \times 2 \times 1}{16 \times 15 \times 14 \times 13}$

$= \dfrac{129}{182}$

**26.** *(a)* $i^{k_1} + i^{k_2} \neq 0,\ k_1, k_2 \in N$

| $k_1$ | $k_2$ | $i^{k_1} + i^{k_2}$ |
|---|---|---|
| $4n$ | $4m + 2$ | 0 |
| $4n + 2$ | $4m$ | 0 |
| $4n + 1$ | $4m + 3$ | 0 |
| $4n + 3$ | $4m + 1$ | 0 |

$k_1 = 4n, 4n+1, 4n+2, 4n+3$
$k_2 = 4m, 4m+1, 4m+2, 4m+3$
$P$ (probability sum is zero)

$$= \frac{4}{16} = \frac{1}{4}$$

∴ Required probability

$$= 1 - \frac{1}{4} = \frac{3}{4}$$

**27.** *(d)* Total ways of selecting any 2 squares out of $16 = {}^{16}C_2 = 120$

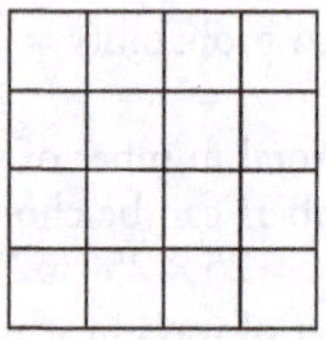

Total ways for selecting common side squares

$$= 3 \times 4 + 3 \times 4$$

↓ Horizontal ↓ Vertical

$$= 24$$

∴Required probability

$$= 1 - \frac{24}{120} = 1 - \frac{1}{5}$$

$$= \frac{4}{5}$$

**28.** *(a)* $(\text{Die})_1 = \{1, 1, 2, 2, 3, 4\}$

$(\text{Die})_2 = \{1, 2, 2, 3, 3, 4\}$

Probability (getting sum 4 or 5)
$= P(\text{Sum } 4) + P(\text{Sum } 5)$

$$= \frac{2}{6} \times \frac{2}{6} + \frac{2}{6} \times \frac{2}{6} + \frac{1}{6} \times \frac{1}{6} + \frac{2}{6} \times \frac{1}{6}$$

$$+ \frac{2}{6} \times \frac{2}{6} + \frac{1}{6} \times \frac{2}{6} + \frac{1}{6} \times \frac{1}{6}$$

$$= \frac{18}{36}$$

∴Probability $= \frac{1}{2}$

**29.** *(b)* Let the numbers be $x$ and $y$.

So, $x + y = 24$; $x, y \in N$

and $x \cdot y \geq \frac{3}{4}$ (maximum product)

Maximum product will be found.

When $x = y = \frac{24}{2} = 12$

So, $xy \geq \frac{3}{4}(12 \times 12)$

$\Rightarrow xy \geq \frac{3}{4} \times 144$

$\Rightarrow xy \geq 108$

Possible cases

| $x$ | $y$ | $xy$ |
|---|---|---|
| 1 | 23 | 23 |
| 2 | 22 | 44 |
| ⋮ | ⋮ | ⋮ |
| 6 | 18 | 108 |
| ⋮ | ⋮ | ⋮ |
| 11 | 13 | 143 |
| 12 | 12 | 144 |
| 13 | 11 | 143 |
| ⋮ | ⋮ | ⋮ |
| 18 | 6 | 108 |

∴ Total combination $= 13$

So, $P(xy \geq 108) = \frac{13}{23} = \frac{m}{n}$

So, $n - m = 23 - 13 = 10$

**30.** *(b)* Since, $x$ and $y$ each can take values from 0 to 10, so the total number of ways of selecting $x$ and $y$ is

$11 \times 11 = 121$. Now, $|x - y| > 5$

$\Rightarrow x - y < -5$ or $x - y > 5$

When $x - y > 5$, we have following cases

When $x = 0$, $y = 0 \rightarrow 1$ case
$x = 7$, $y = 0, 1 \rightarrow 2$ cases
$x = 8$, $y = 0, 1, 2, \rightarrow 3$ cases
$x = 9$, $y = 0, 1, 2, 3 \rightarrow 4$ cases
$x = 10, 0, 1, 2, 3, 4 \rightarrow 5$ cases

Hence, the total number of cases 15.
Similarly, we have 15 cases for $x - y < 5$.
There are 30 pairs values of $x$ and $y$ satisfying these two in equalities.
So, favourable number of ways is 30.

Hence, required probability is $\frac{30}{121}$.

**31.** *(d)*

> **key idea** If $a$, $b$ and $c$ are in GP, then $b^2 = ac$.

Let $N$ be the sum of numbers appeared when two fair dice are rolled.
Given, $N - 2, \sqrt{3N}, N + 2$ are in GP.

$$[\because a, b, c \text{ are in GP, then } b^2 = ac]$$

So, $(\sqrt{3N})^2 = (N-2)(N+2)$

$$3N = N^2 - 4$$

$$N^2 - 3N - 4 = 0$$

$$(N-4)(N+1) = 0$$

$$N = 4 \text{ and } N = -1$$

$N = -1$ is rejected as it cannot be negative.
So, sum $= N = 4$
So, probability of getting sum 4 when two fair dice are rolled

$$= \frac{\text{Favourable outcomes}}{\text{Total outcomes}}$$

$\because$ Total outcomes $= 6 \times 6 = 36$
$\because$ Favourable outcomes

$= \{\text{sum} = 4\}$
$= \{(1, 3), (3, 1), (2, 2)\}$

So, probability $= 3/36 = 1/12$

But, it's given that the probability $= \frac{k}{48}$

So, $\frac{1}{12} = \frac{k}{48}$

$\Rightarrow k = 4$

**32.** *(c)* $n(S) = 9000$ and $S$ contains numbers from 10000 to 99999.

Now, $\left[\frac{10000}{7}\right] = 1428$

$\left[\frac{99999}{7}\right] = 14285$

∴ Number of numbers divisible by 7
$= 14285 - 1428 = 12857$

Now, $\left[\frac{10000}{35}\right] = 285$

$\left[\frac{99999}{35}\right] = 2857$

∴ Numbers divisible by 35
$= 2857 - 285 = 2572$

∴ Numbers divisible by 7 but not 5
$= 12857 - 2572$
$= 10285$

∴ Probability $= \frac{10285}{90000} = p$

$\Rightarrow \quad 9p = \frac{10285}{10000} = 1.0285$

**33.** *(a)* Given the numbers appeared on the two throws of a fair six faced die are $\alpha$ and $\beta$.

Thus, $\alpha, \beta \in \{1, 2, 3, \ldots, 6\}$.
Total possible choice of $\alpha$ and $\beta = 6 \times 6 = 36$
Now, $x^2 + \alpha x + \beta > 0$, $\forall x \in R$ hold
If we have $\alpha^2 - 4\beta < 0$. i.e. $\alpha^2 < 4\beta$
If $\alpha = 1$, $\beta$ can be 1, 2, 3, 4,5, 6, i.e. 6 choices
It $\alpha = 2$, $\beta$ can be 2, 3, 4, 5, 6 i.e. 5 choices
If $\alpha = 3$, $\beta$ can be 3, 4, 5, 6 i.e. 4 choices
If $\alpha = 4$, $\beta$ can be 5, 6 i.e. 2 choices

If $\alpha = 5$, no choices
If $\alpha = 6$, no choices
Hence, total favourable outcomes
$= 6 + 5 + 4 + 2 + 0 + 0 = 17$
Hence, required probability
$$= \frac{\text{Total favourable outcomes}}{\text{Total possible choice}} = \frac{17}{36}$$

**Shortcut Method**

Total possible choice of $\alpha$ and $\beta$
$= 6 \times 6 = 36$
Total favourable outcomes
$= 6 + 5 + 4 + 2 + 0 = 17$
Required probability $= \frac{17}{36}$

**34.** (b) Let $P_1 = \frac{1}{6} - x$

$P_6 = \frac{1}{6} + x$ [sum of opposite faces is 7]

$P_2 = P_3 = P_4 = P_5 = 1/6$
$P(\text{sum} = 7)$
$= 2(P_6 \times P_1 + P_5 \times P_2 + P_4 \times P_3)$

$\Rightarrow \frac{13}{96} = 2 \left[\left(\frac{1}{6} + x\right)\left(\frac{1}{6} - x\right) + \left(\frac{1}{6}\right)^2 + \left(\frac{1}{6}\right)^2\right]$

$\Rightarrow \frac{13}{96} = 2\left[\frac{1}{36} - x^2 + \frac{1}{36} + \frac{1}{36}\right]$

$\Rightarrow \frac{1}{6} - 2x^2 = \frac{13}{96} \Rightarrow x^2 = \frac{1}{64}$

$\Rightarrow x = \frac{1}{8}$

**35.** (c) Sample set $S = \{10, 11, 12, \ldots, 99\}$
$n(S) = 99 - 10 + 1 = 90$

| $n$ | $2^n$ | $2^n - 2$ | Divisible by 3 |
|---|---|---|---|
| 1 | 2 | 0 | Yes |
| 2 | 4 | 2 | No |
| 3 | 8 | 6 | Yes |
| 4 | 16 | 14 | No |
| 5 | 32 | 30 | Yes |
| 6 | 64 | 62 | No |
| 7 | 128 | 126 | Yes |
| 8 | 256 | 254 | No |
| 9 | 512 | 510 | Yes |
| 10 | 1024 | 1022 | No |
| 11 | 2048 | 2046 | Yes |
| 12 | 4096 | 4094 | No |

As we can see, for all the odd values of $n$, $2^n - 2$ is divisible by 3. So, the event set will be all the two digit odd numbers.

$$E = \{11, 13, 15, \ldots, 99\}$$
$$n(E) = \left(\frac{99 - 11}{2} + 1\right) = 45$$
$$\therefore \text{Probability} = \frac{n(E)}{n(S)} = \frac{45}{90} = \frac{1}{2}$$

**36.** (a) $B_3$ contains exactly 3 balls among 9 balls. So, number of ways $= {}^9C_3$

And the rest of the 6 balls can be placed in $3^6$ ways.

If $E$ is the event and $S$ is the sample,
$$n(E) = {}^9C_3 \times 3^6 \text{ and } n(S) = 4^9$$
$$\text{Probability} = \frac{{}^9C_3 \times 3^6}{4^9}$$
According to the question,
$$\frac{{}^9C_3 \times 3^6}{4^9} = k\frac{3^9}{4^9}$$
$$k = \frac{{}^9C_3}{3^3} = \frac{9 \times 8 \times 7}{1 \times 2 \times 3 \times 27} = \frac{28}{9}$$
which satisfies $|x - 3| < 1$.

**37.** (b) $f(x) = x^2 + 2(a + 4)x - 5a + 64$
$f(x) > 0$,
When, $D < 0$ as coefficient of $x^2 > 0$
$4(a + 4)^2 - 4(64 - 5a) < 0$
$\Rightarrow a^2 + 16 + 8a - 64 + 5a < 0$
$\Rightarrow a^2 + 13a - 48 < 0$
$\Rightarrow (a + 16)(a - 3) < 0$
$a \in (-16, 3)$
$\Rightarrow a \in [-5, 2]$
According to the question, $a \in [-5, 30]$
$n(S) = 30 + 5 + 1 = 36$
$n(E) = 2 + 5 + 1 = 8$
$$\therefore \text{Probability} = \frac{n(E)}{n(S)} = \frac{8}{36} = \frac{2}{9}$$

**38.** (a) Total number of ways of selecting 2 different numbers from
$$\{0, 1, 2, \ldots, 10\} = {}^{11}C_2 = 55$$
Let two numbers selected be $x$ and $y$.
Then, $x + y = 4m$ ...(i)
and $|x - y| = 4n$ ...(ii)
$\Rightarrow x + y = 4m,\ x - y = 4n$
or $x + y = 4m,\ -x + y = 4n$
$\Rightarrow x = 2(m + n)$
or $x = 2(m - n)$
and $y = 2(m - n)$
and $y = 2(m + n)$
$\therefore$ $x$ and $y$ both are even numbers.

| $x$ | $y$ |
|---|---|
| 0 | 4, 8 |
| 2 | 6, 10 |
| 4 | 0, 8 |
| 6 | 2, 10 |
| 8 | 0, 4 |
| 10 | 2, 6 |

or

| $y$ | $x$ |
|---|---|
| 0 | 4, 8 |
| 2 | 6, 10 |
| 4 | 0, 8 |
| 6 | 2, 10 |
| 8 | 0, 4 |
| 10 | 2, 6 |

As both the numbers are choosen at once.
$$\therefore \text{Required probability} = \frac{6}{55}$$

**39.** (c)

**key idea** If $|r| < 1$, then $a + ar + ar^2 + \ldots = \frac{a}{1 - r}$

Let $S = \{w_1, w_2, \ldots\}$ be the sample space.

Let $P(w_n) = \frac{P(w_{n-1})}{2}$, $n \geq 2$.

Let $A = \{2k + 3l : k, l \in N\}$
$= \{5, 7, 8, 9, \ldots\}$
$= N - \{1, 2, 3, 4, 6\}$
and $B = \{w_n : n \in A\}$

Now, $P(w_2) = \frac{P(w_1)}{2}$

$P(w_3) = \frac{P(w_1)}{2^2}$

$P(w_4) = \frac{P}{2^3}$ [let $P(w_1) = P$]

$\vdots$

$P(w_n) = \frac{P}{2^{n-1}}$

$\therefore P(B) = 1 - P(B^c)$ ...(i)

Now, we need to find $P(w_1)$.
As we know total probability $= 1$
$\Rightarrow P(w_1) + P(w_2) + P(w_3) + \ldots + P(w_n) + \ldots = 1$

$\Rightarrow P + \frac{P}{2} + \frac{P}{2^2} + \ldots + \frac{P}{2^{n-1}} + \ldots = 1$

$\Rightarrow P\Big/\left(1 - \frac{1}{2}\right) = 1 \left[\because \text{Sum of GP} = \frac{a}{1 - r}\right]$

$\Rightarrow P = 1/2$

So, $P(B) = 1 - (P(w_1) + P(w_2) + P(w_3) + P(w_4) + P(w_6))$

$$= 1 - \left(\frac{1}{2} + \frac{1}{2^2} + \frac{1}{2^3} + \frac{1}{2^4} + \frac{1}{2^6}\right)$$
$$= 1 - \left(\frac{32 + 16 + 8 + 4 + 1}{64}\right)$$
$$= \frac{64 - 61}{64} = \frac{3}{64}$$

**40.** *(4949)*

| Common ratio | Last triplet | Total |
|---|---|---|
| $r=2$ | 10, 20, 40 | 10 |
| $r=3$ | 4, 12, 36 | 4 |
| $r=4$ | 2, 8, 32 | 2 |
| $r=5$ | 1, 5, 25 | 1 |
| $r=6$ | 1, 6, 36 | 1 |
| | Total | 18 |

Total choices $= {}^{40}C_3 = 9880$

Required probability

$$=\frac{18}{9880}=\frac{9}{4940}=\frac{m}{n}$$

$\therefore \quad m+n=4949$

**41.** *(19)* $a, b, c \in \{1, 2, 3, 4\}$

Now, $ax^2+bx+c=0$ has all real roots, if $D \geq 0$.

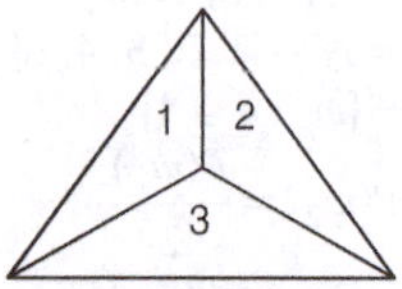


Tetrahedral dice

$\Rightarrow b^2-4ac \geq 0$

Let $b=1 \Rightarrow 1-4ac \geq 0$ (not feasible)

$b=2 \Rightarrow 4-4ac \geq 0 \Rightarrow 1 \geq ac$

$\Rightarrow a=1, c=1$

$b=3 \Rightarrow 9-4ac \geq 0 \Rightarrow \frac{9}{4} \geq ac$

$\Rightarrow (a=1, c=1), (a=1, c=2),$
$(a=2, c=1)$

$b=4 \Rightarrow 16-4ac \geq 0 \Rightarrow 4 \geq ac$

$\Rightarrow (a=1, c=2), (a=2, c=1),$
$(a=1, c=3)$
$(a=3, c=1), (a=1, c=4), (a=4, c=1),$
$(a=2, c=2)$

$\therefore$ Required probability $=\frac{12}{4\times4\times4}$

$$=\frac{3}{16}=\frac{m}{n} \quad \text{[given]}$$

$\therefore \quad m+n=19$

**42.** *(10)* Since, $A$ fair $n(n>1)$ faces die is rolled repeatedly until a number less than $n$ appears. Thus, mean of the number of tosses

$$=1\cdot\left(\frac{1}{n}\right)^0\left(\frac{n-1}{n}\right)+2\left(\frac{1}{n}\right)^1\left(\frac{n-1}{n}\right)$$
$$+3\left(\frac{1}{n}\right)^2\left(\frac{n-1}{n}\right)+\ldots$$
$$=1\left(\frac{n-1}{n}\right)\left(1+\frac{2}{n}+\frac{3}{n^2}+\ldots\right)$$
$$=\left(\frac{n-1}{n}\right)\left(1-\frac{1}{n}\right)^{-2}$$
$$\{\because (1-x)^{-2}=1+2x+3x^2+\ldots\}$$
$$=\left(\frac{n-1}{n}\right)\left(\frac{n-1}{n}\right)^{-2}=\frac{n}{n-1}$$

Again, the mean of number of tosses given $=\frac{n}{9}$

So, $\quad \frac{n}{9}=\frac{n}{n-1} \Rightarrow n^2-n-9n=0$

$\Rightarrow \quad n^2-10n=0 \Rightarrow n=10 \quad \{\because n>1\}$

Hence, the required value of $n$ is 10.

**43.** *(14)* $\because$ Bag contains balls of different colours.

$p=$ Probability of both drawn balls are of same colour $=\frac{{}^6C_1}{6\times6}=\frac{1}{6}$

$q=$ Probability that exactly three from four balls are of the same colour

$$=\frac{{}^6C_1\times{}^5C_1\times{}^4C_1}{6\times6\times6\times6}=\frac{6\times5\times4}{6\times6\times6\times6}$$
$$=\frac{5}{54}$$
$$\frac{p}{q}=\frac{(1/6)}{(5/54)}=\frac{1}{6}\times\frac{54}{5}=\frac{9}{5}=\frac{m}{n}$$

So, $\quad m=9$ and $n=5$

Therefore, $m+n=9+5=14$

**44.** *(10)* Let the two real numbers be $x$ and $y$. Then,

$$|x-y|<a$$
$\Rightarrow \quad -a<x-y<a$

$\Rightarrow \quad x-y<a$ and $x-y>-a$

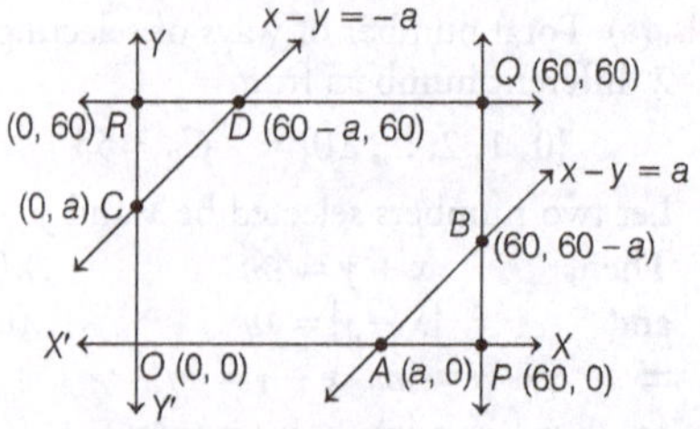


Required probability,

$$P(A)=\frac{ar(OABQDC)}{ar(OPQR)}$$
$$=\frac{ar(OPQR)-ar(\Delta APB)-ar(\Delta CDR)}{ar(OPQR)}$$
$$=\frac{(60)^2-\frac{1}{2}(60-a)^2-\frac{1}{2}(60-a)^2}{(60)^2}$$
$$=\frac{3600-(60-a)^2}{3600}$$
$$=\frac{3600-3600-a^2+120a}{3600}=\frac{120a-a^2}{3600}$$

It is given that $P(A)=\frac{11}{36}$

$\Rightarrow \quad \frac{120a-a^2}{3600}=\frac{11}{36}$

$\Rightarrow \quad 120a-a^2=1100$

$\Rightarrow \quad a^2-120a+1100=0$

$\Rightarrow \quad (a-10)(a-110)=0$

$\Rightarrow \quad a=10, 110$

$\Rightarrow \quad a=10 \quad [\because a\neq110]$

**45.** *(19)* Given, $S=\{E_1, E_2, \ldots, E_8\}$ is a sample space of a random experiment such that $P(E_n)=n/36$, for every $n=1, 2, 3, \ldots, 8$

Hence, $P(A)=\frac{\text{Any possible sum of } \{1, 2, 3, \ldots, 8\}}{36}$

Let '$a$' = any possible sum of $\{1,2,3,\ldots,8\}$

$\Rightarrow \quad P(A)=a/36$

Given that $P(A)\geq\frac{4}{5}, A\subseteq S$

$\Rightarrow \quad \frac{a}{36}\geq\frac{4}{5} \Rightarrow a\geq\frac{4}{5}\times36$

$\Rightarrow \quad a\geq28.8$

Thus, the maximum possible value of $a$ is 29, 30, 31, 32, 33 ..., 36

If one of the number from $\{1, 2, \ldots, 8\}$ is left, then $a\geq29$ by 3 ways.

Similarly, by leaving terms more 2 or 3 also, we get 16 more combinations.

$\therefore$ Total number of different set $A$ possible is $16+3=19$

**46.** *(33)* Total possibilities to make a 6-digit number using only 1 and 8.

Every position has 2 possibility either 1 or 8

2 2 2 2 2 2

$\therefore \quad 2^6$

Number divisible by 21 :

1. for divisibility by 3
   (sum of digits must be divisible by 3)
   (a) Number of 1's = Number of 8
   (b) Every digit is 1
   (c) Every digit is 8.

2. for divisiblity by 7.

$|2 \text{ (last digit)} - \text{remaining number}| = 7k, k \in z$

(a) all are $1s \Rightarrow 1$ number

(b) all are $8s \Rightarrow 1$ number

(c) 3, 1's and 3, 8's $\Rightarrow \dfrac{6!}{3!3!}$

Total number divisible by 21 = number divisible by 3 and 7

$$= \frac{6!}{3!3!} + 2 = \left(\frac{6 \times 5 \times 4 \times 3!}{3! \times 3 \times 2}\right) + 2 = 22$$

$$\therefore \text{Probability} = \frac{22}{2^6} = p$$

$$96p = 96 \times \frac{22}{2^6}$$

$$= 96 \times \frac{22}{2 \times 2 \times 2 \times 2 \times 2 \times 2} = 33$$

**47.** *(490)* It is given that an urn contains 5 red marbles, 4 blacks marbles and 3 white marbles, then the number of ways in which 4 marbles can be drawn so that at most three of them are red = (total number of ways to drawn 4 marbles) – (numbers of ways to drawn 4 marbles so that all are red)

$$= {}^{12}C_4 - {}^5C_4 = \frac{12 \times 11 \times 10 \times 9}{4 \times 3 \times 2} - 5$$

$$= (11 \times 5 \times 9) - 5$$

$$= 5(99 - 1) = 5 \times 98 = 490$$

## Topic 2

**01.** *(a)* Given, Ajay will not appear in exam $\quad P(\overline{A}) = \dfrac{2}{7}$

$$\Rightarrow \quad P(A) = \frac{5}{7}$$

and both Ajay and Vijay will appear in the exam with probability

$$P(A \cap V) = \frac{1}{5} = P(A) \cdot P(V)$$

[$\because P(A)$ and $P(V)$ are independent]

$$\therefore \quad P(A) \cdot P(V) = \frac{1}{5}$$

$$P(V) = \frac{7}{25} \quad \left[\because P(A) = \frac{5}{7}\right]$$

Now, probability that Ajay will appear in the exam and Vijay will not appear

$$P(A \cap \overline{V}) = P(A)\,P(\overline{V})$$

$$= \frac{5}{7} \times \left(1 - \frac{7}{25}\right) \quad [\because P(\overline{V}) = 1 - P(V)]$$

$$= \frac{18}{35}$$

**02.** *(d)* According to given information, the required probability is equal to

$$P(A \cap A) + P(A \cap B \cap A) + P(B \cap A \cap A) + P(B \cap B \cap A \cap A) + P(B \cap A \cap B \cap A) + P(A \cap B \cap B \cap A)$$

$$= \left(\frac{1}{2} \times \frac{1}{2}\right) + \left(\frac{1}{2} \times \frac{1}{2} \times \frac{1}{2}\right) + \left(\frac{1}{2} \times \frac{1}{2} \times \frac{1}{2}\right) + \left(\frac{1}{2} \times \frac{1}{2} \times \frac{1}{2} \times \frac{1}{2}\right) + \left(\frac{1}{2} \times \frac{1}{2} \times \frac{1}{2} \times \frac{1}{2}\right) + \left(\frac{1}{2} \times \frac{1}{2} \times \frac{1}{2} \times \frac{1}{2}\right)$$

$$= \frac{1}{4} + \frac{2}{8} + \frac{3}{16} = \frac{4 + 4 + 3}{16} = \frac{11}{16}$$

**03.** *(d)* Number of subset of $S = 2^{20}$

Sum of elements in $S$ is

$$1 + 2 + \ldots + 20 = \frac{20(21)}{2} = 210$$

$$\left[\because 1 + 2 + \ldots + n = \frac{n(n+1)}{2}\right]$$

Clearly, the sum of elements of a subset would be 203, if we consider it as follows

$S - \{7\}$, $S - \{1, 6\}$ $S - \{2, 5\}$, $S - \{3, 4\}$

$S - \{1, 2, 4\}$

$\therefore$ Number of favourable cases = 5

Hence, required probability $= \dfrac{5}{2^{20}}$

**04.** *(c)* Given $P(A) = 0.7$

$P(B) = 0.4$

$P(A \cap \overline{B}) = 0.5$

To find $P(B / (A \cup \overline{B})$

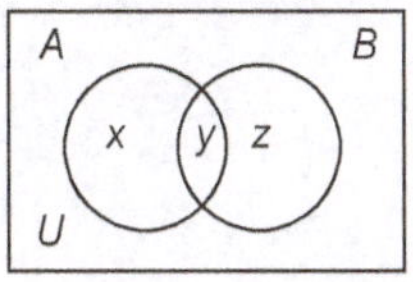


Since, $P(A \cap \overline{B}) = P(A) - P(A \cap B)$

$0.5 = 0.7 - P(A \cap B)$

$\Rightarrow \quad P(A \cap B) = 0.2$

$P(A \cup \overline{B}) = P(A) + 1 - P(A \cup B)$

$= P(A) + 1 - [P(A) + P(B) - P(A \cap B)]$

$= P(A \cap B) + 1 - P(B)$

$= 0.2 + 1 - 0.4 = 0.8$

$\therefore P(B \cap (A \cup \overline{B}) = P(A \cap B)$

$= 0.2$

$$\text{So, } P(B / (A \cup \overline{B}) = \frac{P(B \cap (A \cup \overline{B}))}{P(A \cup \overline{B})}$$

$$= \frac{0.2}{0.8} = \frac{1}{4}$$

**05.** *(d)*

> **key idea** Use the formula to get final result
> $n(A \cup B \cup C) = n(A) + n(B) + n(C) - n(A \cap B) - n(B \cap C) - n(C \cap A) + n(A \cap B \cap C)$

$U = \{1, 2, 3, \ldots, 50\}$

$A$ = Set of integers which is multiple of 4

$B$ = Set of integers which is multiple of 6

$C$ = Set of integers which is multiple of 7

$n(A) = 12$

$n(B) = 8$

$n(C) = 7$

$n(A \cap B) = 4$

$n(B \cap C) = 1$

$n(C \cap A) = 1$

$n(A \cap B \cap C) = 0$

$$\therefore n(A \cup B \cup C) = n(A) + n(B) + n(C) - n(A \cap B) - n(B \cap C) - n(C \cap A) + n(A \cap B \cap C)$$

$$= 12 + 8 + 7 - 4 - 1 - 1 + 0 = 21$$

$$\therefore \text{Required probability} = \frac{21}{50}$$

**06.** *(b)*

> **key idea** $E_1, E_2$ and $E_3$ are three mutually exclusive events.
> $P(E_1 \cup E_2 \cup E_3) = P(E_1) + P(E_2) + P(E_3)$

As we know that $0 \le P(E) \le 1$

So, $0 \le P(E_1) \le 1$

$$\Rightarrow \quad 0 \le \frac{2 + 3p}{6} \le 1 \Rightarrow 0 \le 2 + 3p \le 6$$

$$\Rightarrow -2 \le 3p \le 4 \;\Rightarrow\; -\frac{2}{3} \le p \le \frac{4}{3}$$

$$\Rightarrow \quad p \in \left[-\frac{2}{3}, \frac{4}{3}\right]$$

Similarly, for $P(E_2)$, $p \in [-6, 2]$ and for $P(E_3)$, $p \in [-1, 1]$.

$E_1, E_2$ and $E_3$ are three mutually exclusive events,

$$P(E_1 \cup E_2 \cup E_3) = P(E_1) + P(E_2) + P(E_3)$$

$$0 \le P(E_1 \cup E_2 \cup E_3) \le 1$$

$$\Rightarrow \quad 0 \le P(E_1) + P(E_2) + P(E_3) \le 1$$

$$\Rightarrow \quad 0 \le \frac{2 + 3p}{6} + \frac{2 - p}{8} + \frac{1 - p}{2} \le 1$$

$$\Rightarrow \quad 0 \le \frac{8 + 12p + 6 - 3p + 12 - 12p}{24} \le 1$$

$\Rightarrow \quad 0 \le 26 - 3p \le 24$

$\Rightarrow -26 \le -3p \le -2$

$\Rightarrow \quad 2 \le 3p \le 26$

$\Rightarrow \quad \frac{2}{3} \le p \le \frac{26}{3}$

$\therefore \quad p \in \left[\frac{2}{3}, \frac{26}{3}\right]$

Taking intersection of all the above value of $p$, we get

$$p \in \left[\frac{2}{3}, 1\right]$$

$\Rightarrow \quad p_1 = 1$ and $p_2 = \frac{2}{3}$

$\therefore p_1 + p_2 = 1 + \frac{2}{3} = \frac{5}{3}$

**07.** *(a)*

**key idea** For any event $E, 0 \le P(E) \le 1$.

Since, $0 \le P(A) \le 1, 0 \le P(B) \le 1$, $0 \le P(C) \le 1$

and $0 \le P(A) + P(B) + P(C) \le 1$

$\therefore \quad 0 \le \frac{3x+1}{3} \le 1$

$\Rightarrow \quad -\frac{1}{3} \le x \le \frac{2}{3}$ ...(i)

$0 \le \frac{1-x}{4} \le 1$

$\Rightarrow \quad -3 \le x \le 1$ ...(ii)

$0 \le \frac{1-2x}{2} \le 1$

$\Rightarrow \quad -\frac{1}{2} \le x \le \frac{1}{2}$ ...(iii)

and $\quad 0 \le \frac{3x+1}{3} + \frac{1-x}{4} + \frac{1-2x}{2} \le 1$

$\Rightarrow \quad 0 \le 13 - 3x \le 12$

$\Rightarrow \quad \frac{1}{3} \le x \le \frac{13}{3}$ ...(iv)

From Eqs. (i), (ii), (iii) and (iv), we get

$$\frac{1}{3} \le x \le \frac{1}{2}$$

**08.** *(d)* $2022 = 2 \times 3 \times 337$

Let $A$ = Number which are divisible by 2 from {1, 2, 3, ....., 2022}

$n(A) = 1011$

$B$ = Number which are divisible by 3 from {1, 2, 3, ....., 2022}

$n(B) = 674$

$n(A \cap B)$ = Number which are divisible by 6 from

{6, 12, 18, ...... , 2022}

$n(A \cap B) = 337$

$n(A \cup B) = n(A) + n(B) - n(A \cap B)$

$= 1011 + 674 - 337 = 1348$

$C$ = Number which are divisible by 337 from {1, 2, ....., 2022}

$C$ = {337, 674, 1011, 1348,1685, 2022}

But number 674, 1011, 1348, 2022 are already counted in $(A \cup B)$.

So, $n(C) = 2$

So, total number which are divisible by either 2 or 3 or 337 = 1348 + 2 = 1350

$n$ = Number for which HCF of $n$ and 2022 is 1

$= 2022 - 1350$

$\Rightarrow \quad n = 672$

$\therefore$ Probability $= \frac{672}{2022} = \frac{112}{337}$

# C-13

JEE Main Chapterwise Topicwise Mathematics

In JEE Mains, there is usually 0-1 question per paper is visible from this chapter. In past 3 years, 49 conceptual and 18 numerical questions of moderate difficulty level were asked.

**For detailed analysis scan QR code**

# RELATIONS AND FUNCTIONS

## Topic 1 Types of Relations, Equivalence Relations, Functions (or Mappings)

### Level 1

**Imp.** Reflective and Symmetric Relations

**01** Let $A = \{-2, -1, 0, 1, 2, 3\}$. Let $R$ be a relation on $A$ defined by $xRy$ if and only if $y = \max\{x, 1\}$. Let $l$ be the number of elements in $R$. Let $m$ and $n$ be the minimum number of elements required to be added in $R$ to make it reflexive and symmetric relations, respectively. Then, $l + m + n$ is equal to *[2025, 3 April Shift-II]*

(a) 12 (b) 13 (c) 14 (d) 11

**02** Define a relation $R$ on the interval $[0, \pi/2)$ by $x\,R\,y$ if and only if $\sec^2 x - \tan^2 y = 1$. Then, $R$ is *[2025, 29 Jan Shift-I]*

(a) both reflexive and transitive but not symmetric
(b) reflexive but neither symmetric not transitive
(c) both reflexive and symmetric but not transitive
(d) an equivalence relation

**03** The relation $R = \{(x, y) : x, y \in Z \text{ and } x + y \text{ is even}\}$ *[2025, 28 Jan Shift-I]*

(a) symmetric and transitive but not reflexive
(b) reflexive and transitive but not symmetric
(c) an equivalence relation
(d) reflexive and symmetric but not transitive

**04** The function $f : (-\infty, \infty) \to (-\infty, 1)$ defined by $f(x) = \dfrac{2^x - 2^{-x}}{2^x + 2^{-x}}$ is *[2025, 24 Jan Shift-II]*

(a) Neither one-one nor onto
(b) Both one-one and onto
(c) One-one but not onto
(d) Onto but not one-one

**05** Let $X = R \times R$, Define a relation $R$ on $X$ as $(a_1, b_1) R (a_2, b_2) \Leftrightarrow b_1 = b_2$

**Statement I** $R$ is an equivalence relation

**Statement II** For some $(a, b) \in X$, then set $S = \{(x, y) \in X : (x, y)\,R\,(a, b)\}$ represents a line parallel to $y = x$.

In the light of the above statements, choose the correct answer from the options given below *[2025, 23 Jan Shift-II]*

(a) Both statement I and Statement II are true.
(b) Statement I is true but Statement II is false.
(c) Both statement I and Statement II are false.
(d) Statement I is false but Statement II is true.

**06** The number of non-empty equivalence relations on the set $\{1, 2, 3\}$ is *[2025, 22 Jan Shift-I]*

(a) 5 (b) 7
(c) 4 (d) 6

**07** Let $[t]$ be the greatest integer less than or equal to $t$. Let $A$ be the set of all prime factors of 2310 and $f : A \to Z$ be the function $f(x) = \left[\log_2\left(x^2 + \left[\dfrac{x^3}{5}\right]\right)\right]$. The number of one-to-one functions from $A$ to the range of $f$ is *[2024, 8 April Shift-I]*

(a) 20 (b) 25 (c) 24 (d) 120

**08** Let $A = \{1, 3, 7, 9, 11\}$ and $B = \{2, 4, 5, 7, 8, 10, 12\}$. Then, the total number of one-one maps $f : A \to B$, such that $f(1) + f(3) = 14$, is *[2024, 5 April Shift-I]*

(a) 180 (b) 120 (c) 480 (d) 240

**09** Let a relation $R$ on $N \times N$ be defined as: $(x_1, y_1) R (x_2, y_2)$ if and only if $x_1 \le x_2$ or $y_1 \le y_2$. *[2024, 4 April Shift-II]*

Consider the two statements :

I : $R$ is reflexive but not symmetric.

II : $R$ is transitive

Then, which one of the following is true?

(a) Only I is correct.
(b) Only II is correct.
(c) Neither I nor II is correct.
(d) Both I and II are correct.

**Imp.** Equivalence Relation

**10** If $R$ is the smallest equivalence relation on the set $\{1, 2, 3, 4\}$ such that $\{(1, 2), (1, 3)\} \subset R$, then the number of elements in $R$ is *[2024, 29 Jan Shift-II]*

(a) 15 (b) 10
(c) 8 (d) 12

**11** Let $R$ be a relation on $Z \times Z$ defined by $(a, b) R (c, d)$ if and only if $ad - bc$ is divisible by 5. Then, $R$ is *[2024, 29 Jan Shift-I]*

(a) Reflexive but neither symmetric nor transitive
(b) Reflexive and transitive but not symmetric
(c) Reflexive, symmetric and transitive
(d) Reflexive and symmetric but not transitive

**12** The function $f : N - \{1\} \to N$; defined by $f(n) =$ the highest prime factor of $n$, is *[2024, 27 Jan Shift-I]*

(a) neither one-one nor onto (b) both one-one and onto
(c) onto only (d) one-one only

**13** Let $S = \{1, 2, 3, ..., 10\}$. Suppose $M$ is the set of all the subsets of $S$, then the relation $R = \{(A, B) : A \cap B \ne \phi;\ A, B \in M\}$ is *[2024, 27 Jan Shift-I]*

(a) symmetric only
(b) symmetric and reflexive only
(c) reflexive only
(d) symmetric and transitive only

**14** Let $A = \{1, 2, 3, 4, 5, 6, 7\}$. Then, the relation $R = \{(x, y) \in A \times A : x + y = 7\}$ is *[2023, 8 April Shift-II]*

(a) reflexive but neither symmetric nor transitive
(b) an equivalence relation
(c) transitive but neither symmetric nor reflexive
(d) symmetric but neither reflexive nor transitive

**15** Let $P(S)$ denote the power set of $S = \{1, 2, 3, \ldots, 10\}$. Define the relations $R_1$ and $R_2$ on $P(S)$ as $AR_1B$, if $(A \cap B^c) \cup (B \cap A^c) = \phi$ and $AR_2B$, if $A \cup B^c = B \cup A^c, \forall A, B \in P(S)$. Then, *[2023, 1 Feb Shift-II]*

(a) only $R_1$ is an equivalence relation
(b) Both $R_1$ and $R_2$ are not equivalence relations
(c) only $R_2$ is an equivalence relation
(d) Both $R_1$ and $R_2$ are equivalence relations

**16** Let $R$ be a relation defined on $N$ as $a R b$, if $2a + 3b$ is a multiple of 5, $a, b \in N$. Then, $R$ is *[2023, 29 Jan Shift-II]*

(a) symmetric but not transitive
(b) an equivalence relation
(c) transitive but not symmetric
(d) Not reflexive

**17** Let $R$ be a relation from the set $\{1, 2, 3, \ldots, 60\}$ to itself such that $R = \{(a, b) : b = pq$, where $p, q \ge 3$ are prime numbers$\}$. Then, the number of elements in $R$ is *[2022, 29 July Shift-I]*

(a) 600 (b) 660 (c) 540 (d) 720

**18** The number of bijective functions $f : \{1, 3, 5, 7, ..., 99\} \to \{2, 4, 6, 8, ..., 100\}$, such that $f(3) \ge f(9) \ge f(15) \ge f(21) \ge \ldots \ge f(99)$ is *[2022, 25 July Shift-II]*

(a) ${}^{50}P_{17}$ (b) ${}^{50}P_{33}$ (c) $33! \times 17!$ (d) $50!/2$

**19** Let a set $A = A_1 \cup A_2 \cup \ldots \cup A_k$, where $A_i \cap A_j = \phi$ for $i \ne j, 1 \le i, j \le k$. Define the relation $R$ from $A$ to $A$ by $R = \{(x, y) : y \in A_i$, if and only if $x \in A_i, 1 \le i \le k\}$. Then, $R$ is *[2022, 29 June Shift-I]*

(a) reflexive, symmetric but not transitive
(b) reflexive, transitive but not symmetric
(c) reflexive but not symmetric and transitive
(d) an equivalence relation

**Imp.** Type of Function

**20** Let a function $f : N \to N$ be defined by

$$f(n) = \begin{cases} 2n, & n = 2, 4, 6, 8, \ldots \\ n - 1, & n = 3, 7, 11, 15, \ldots \\ \dfrac{n+1}{2}, & n = 1, 5, 9, 13, \ldots \end{cases}$$

then $f$ is *[2022, 28 June Shift (I)]*

(a) one-one but not onto (b) onto but not one-one
(c) Neither one-one nor onto (d) one-one and onto

**21** Which of the following is not correct for relation $R$ on the set of real numbers ? *[2021, 31 Aug Shift-I]*

(a) $(x, y) \in R \Leftrightarrow 0 < |x| - |y| \le 1$ is neither transitive nor symmetric.
(b) $(x, y) \in R \Leftrightarrow 0 < |x - y| \le 1$ is symmetric and transitive.
(c) $(x, y) \in R \Leftrightarrow |x| - |y| \le 1$ is reflexive but not symmetric.
(d) $(x, y) \in R \Leftrightarrow |x - y| \le 1$ is reflexive and symmetric.

**22** Let $A = \{x \in R : x$ is not a positive integer$\}$. Define a function $f : A \to R$ as $f(x) = \dfrac{2x}{x-1}$, then $f$ is *[2021, 9 Jan Shift-II]*

(a) not injective
(b) Neither injective nor surjective
(c) surjective but not injective
(d) injective but not surjective

**23** If the function $f: R-\{1,-1\} \to A$ defined by $f(x)=\frac{x^2}{1-x^2}$, is surjective, then $A$ is equal to **[2019, 9 April Shift-I]**

(a) $R-\{-1\}$ (b) $[0, \infty)$ (c) $R-[-1, 0)$ (d) $R-(-1, 0)$

**24** For real $x$, let $f(x)=x^3+5x+1$, then **[AIEEE 2009]**

(a) $f$ is one-one but not onto $R$
(b) $f$ is onto $R$ but not one-one
(c) $f$ is one-one and onto $R$
(d) $f$ is neither one-one nor onto $R$

**25** Let $W$ denotes the words in the English dictionary define the relation $R$ by $R=\{(x, y) \in W \times W$: the words $x$ and $y$ have atleast one letter in common$\}$. Then, $R$ is **[AIEEE 2006]**

(a) reflexive, symmetric and not transitive
(b) reflexive, symmetric and transitive
(c) reflexive, not symmetric and transitive
(d) Not reflexive, symmetric and transitive

**26** Let $R=\{(3, 3), (6, 6), (9, 9), (12, 12), (6, 12), (3, 9), (3, 12), (3, 6)\}$ be a relation on the set $A=\{3, 6, 9, 12\}$. The relation is **[AIEEE 2005]**

(a) reflexive and symmetric only.
(b) an equivalence relation.
(c) reflexive only.
(d) reflexive and transitive only.

**27** Let $f:(-1, 1) \to B$ be a function defined by $f(x)=\tan^{-1}\left(\frac{2x}{1-x^2}\right)$, then $f$ is both one-one and onto when $B$ is in the interval **[AIEEE 2005]**

(a) $\left(-\frac{\pi}{2}, \frac{\pi}{2}\right)$ (b) $\left[-\frac{\pi}{2}, \frac{\pi}{2}\right]$

(c) $\left[0, \frac{\pi}{2}\right)$ (d) $\left(0, \frac{\pi}{2}\right)$

## Level 2

**28** Let $A=\{0, 1, 2, 3, 4, 5\}$. Let $R$ be a relation on $A$ defined by $(x, y) \in R$ if and only if $\max\{x, y\} \in \{3, 4\}$. Then, among the statements

$(S_1)$: The number of elements in $R$ is 18, and
$(S_2)$: The relation $R$ is symmetric but neither reflexive nor transitive **[2025, 8 April Shift-II]**

(a) only $(S_2)$ is true (b) both are true
(c) both are false (d) only $(S_1)$ is true

**29** Let $A=\{-3, -2, -1, 0, 1, 2, 3\}$ and $R$ be a relation on $A$ defined by $xRy$ if and only if $2x-y \in \{0, 1\}$. Let $l$ be the number of elements in $R$. Let $m$ and $n$ be the minimum number of elements required to be added in $R$ to make it reflexive and symmetric relations, respectively. Then, $l+m+n$ is equal to **[2025, 4 April Shift-II]**

(a) 15 (b) 18
(c) 16 (d) 17

**30** Let $A$ be the set of all functions $f: Z \to Z$ and $R$ be a relation on $A$ such that $R=\{(f, g): f(0)=g(1)$ and $f(1)=g(0)\}$. Then $R$ is: **[2025, 2 April Shift-I]**

(a) Reflexive but neither symmetric nor transitive
(b) Transitive but neither reflexive nor symmetric
(c) Symmetric but neither reflective nor transitive
(d) Symmetric and transitive but not reflective

**31** Let $R=\{(1, 2), (2, 3) (3, 3)\}$ be a relation defined on the set $\{1, 2, 3, 4\}$. Then, the minimum number of elements, needed to be added in $R$ so that $R$ becomes an equivalence relation, is **[2025, 23 Jan Shift-I]**

(a) 10 (b) 9 (c) 8 (d) 7

**32** Let $A=\{2, 3, 6, 8, 9, 11\}$ and $B=\{1, 4, 5, 10, 15\}$. Let $R$ be a relation on $A \times B$ defined by $(a, b)R(c, d)$ if and only if $3ad-7bc$ is an even integer. Then, the relation $R$ is **[2024, 8 April Shift-II]**

(a) an equivalence relation.
(b) reflexive but not symmetric.
(c) transitive but not symmetric.
(d) reflexive and symmetric but not transitive.

**33** Let $f(x)=\begin{cases} -a & \text{if } -a \le x \le 0 \\ x+a & \text{if } 0<x \le a \end{cases}$, where $a>0$ and $g(x)=(f(|x|)-|f(x)|)/2$. **[2024, 8 April Shift-II]**

Then, the function $g:[-a, a] \to [-a, a]$ is

(a) one-one. (b) onto.
(c) both one-one and onto. (d) neither one-one nor onto.

**34** Let $A=\{1,2,3,4,5\}$. Let $R$ be a relation on $A$ defined by $xRy$ if and only if $4x \le 5y$. Let $m$ be the number of elements in $R$ and $n$ be the minimum number of elements from $A \times A$ that are required to be added to $R$ to make it a symmetric relation. Then, $m+n$ is equal to **[2024, 6 April Shift-II]**

(a) 26 (b) 23 (c) 24 (d) 25

**35** Let the relations $R_1$ and $R_2$ on the set $X=\{1, 2, 3, \ldots, 20\}$ be given by $R_1=\{(x, y): 2x-3y=2\}$ and $R_2=\{(x, y): -5x+4y=0\}$. If $M$ and $N$ be the minimum number of elements required to be added in $R_1$ and $R_2$, respectively, in order to make the relations symmetric, then $M+N$ equals **[2024, 6 April Shift-I]**

(a) 16 (b) 10
(c) 12 (d) 8

**36** Let $A=\{1, 3, 4, 6, 9\}$ and $B=\{2, 4, 5, 8, 10\}$. Let $R$ be a relation defined on $A \times B$, such that $R=\{((a_1, b_1), (a_2, b_2)): a_1 \le b_2$ and $b_1 \le a_2\}$. Then, the number of elements in the set $R$ is **[2023, 11 April Shift-II]**

(a) 180 (b) 52
(c) 160 (d) 26

**37** Let $A = \{2, 3, 4\}$ and $B = \{8, 9, 12\}$. Then, the number of elements in the relation $R = \{((a_1, b_1), (a_2, b_2)) \in (A \times B, A \times B) : a_1$ divides $b_2$ and $a_2$ divides $b_1\}$ is *[2023, 10 April Shift-II]*

(a) 24 (b) 18
(c) 36 (d) 12

**38** Among the relations
$S = \{(a, b) : a, b \in R - \{0\}, 2 + \frac{a}{b} > 0\}$ and
$T = \{(a, b) : a, b \in R, a^2 - b^2 \in Z\}$, *[2023, 31 Jan Shift-II]*

(a) both $S$ and $T$ are symmetric
(b) $T$ is symmetric but $S$ is not
(c) $S$ is transitive but $T$ is not
(d) neither $S$ nor $T$ is transitive

**39** The minimum number of elements that must be added to the relation $R = \{(a, b), (b, c)\}$ on the set $\{a, b, c\}$, so that it becomes symmetric and transitive is *[2023, 30 Jan Shift-I]*

(a) 4 (b) 5
(c) 3 (d) 7

**40** The relation $R = \{(a, b) : \gcd(a, b) = 1, 2a \neq b, a, b \in Z\}$ is *[2023, 24 Jan Shift-I]*

(a) Neither symmetric nor transitive
(b) reflexive but not symmetric
(c) symmetric but not transitive
(d) transitive but not reflexive

**41** Let $R_1$ and $R_2$ be two relations defined on $R$ by $a R_1 b \Leftrightarrow ab \geq 0$ and $a R_2 b \Leftrightarrow a \geq b$. Then, *[2022, 27 July Shift-I]*

(a) $R_1$ is an equivalence relation but not $R_2$
(b) $R_2$ is an equivalence relation but not $R_1$
(c) Both $R_1$ and $R_2$ are equivalence relations
(d) Neither $R_1$ nor $R_2$ is an equivalence relation

**42** Let $R = \{(P, Q) \mid$, $P$ and $Q$ are at the same distance from the origin$\}$ be a relation, then the equivalence class of $(1, -1)$ is the set *[2021, 26 Feb Shift-I]*

(a) $S = \{(x, y) \mid x^2 + y^2 = 4\}$
(b) $S = \{(x, y) \mid x^2 + y^2 = 1\}$
(c) $S = \{(x, y) \mid x^2 + y^2 = \sqrt{2}\}$
(d) $S = \{(x, y) \mid x^2 + y^2 = 2\}$

**43** A function $f$ from the set of natural numbers to integers defined by

$$f(n) = \begin{cases} n - 1/2, \text{ when } n \text{ is odd} \\ -n/2, \text{ when } n \text{ is even} \end{cases} \text{ is}$$

*[AIEEE 2003]*

(a) one-one but not onto.
(b) onto but not one-one.
(c) one-one and onto both.
(d) Neither one-one nor onto.

## ■ Level 3

**44** Let $f : [0, 3] \to A$ be defined by $f(x) = 2x^3 - 15x^2 + 36x + 7$ and $g : [0, \infty) \to B$ be defined by $g(x) = \frac{x^{2025}}{x^{2025} + 1}$. If both the functions are onto and $S = \{x \in Z : x \in A \text{ or } x \in B\}$, then $n(S)$ is equal to *[2025, 28 Jan Shift-I]*

(a) 36 (b) 29 (c) 31 (d) 30

**45** Let $A = \{1, 2, 3, 4\}$ and $B = \{1, 4, 9, 16\}$. Then, the number of many-one functions $f : A \to B$ such that $1 \in f(A)$ is equal to *[2025, 22 Jan Shift-II]*

(a) 151 (b) 139
(c) 163 (d) 127

**46** The function $f(x) = \frac{x^2 + 2x - 15}{x^2 - 4x + 9}$, $x \in R$ is *[2024, 6 April Shift-I]*

(a) onto but not one-one. (b) one-one but not onto.
(c) neither one-one nor onto (d) both one-one and onto.

**47** Consider the relations $R_1$ and $R_2$ defined as $aR_1 b \Leftrightarrow a^2 + b^2 = 1$ for all $a, b \in R$ and $(a, b) R_2 (c, d) \Leftrightarrow a + d = b + c$ for all $(a, b)$, $(c, d) \in N \times N$. Then, *[2024, 1 Feb Shift-II]*

(a) only $R_1$ is an equivalence relation
(b) only $R_2$ is an equivalence relation
(c) $R_1$ and $R_2$ both are equivalence relations
(d) neither $R_1$ nor $R_2$ is an equivalence relation

**48** Let $R$ be a relation on $R$, given by $R = \{(a, b) : 3a - 3b + \sqrt{7}$ is an irrational number$\}$. Then, $R$ is *[2023, 1 Feb Shift-I]*

(a) an equivalence relation
(b) reflexive and transitive but not symmetric
(c) reflexive and symmetric but not transitive
(d) reflexive but neither symmetric nor transitive

**49** Let $R$ be a relation of $N \times N$ defined by $(a, b) R (c, d)$, if and only if $ad(b - c) = bc(a - d)$. Then, $R$ is *[2023, 31 Jan Shift-I]*

(a) transitive but neither reflexive nor symmetric
(b) symmetric and transitive but not reflexive
(c) symmetric but neither reflexive nor transitive
(d) reflexive and symmetric but not transitive

**50** Let $f : R \to R$ be a function such that $f(x) = \frac{x^2 + 2x + 1}{x^2 + 1}$. Then, *[2023, 29 Jan Shift-I]*

(a) $f(x)$ is one-one in $[1, \infty)$ but not in $(-\infty, \infty)$
(b) $f(x)$ is many-one in $(-\infty, -1)$
(c) $f(x)$ is one-one in $(-\infty, \infty)$
(d) $f(x)$ is many-one in $(1, \infty)$.

**51** For $\alpha \in N$, consider a relation $R$ on $N$ given by $R = \{(x, y) : 3x + \alpha y$ is a multiple of 7$\}$. The relation $R$ is an equivalence relation, if and only if
(a) $\alpha = 14$ **[2022, 28 July Shift-I]**
(b) $\alpha$ is a multiple of 4
(c) 4 is the remainder when $\alpha$ is divided by 10
(d) 4 is the remainder when $\alpha$ is divided by 7

**52** Let $R$ be the set of real numbers.
**Statement I** $A = \{(x, y) \in R \times R : y - x$ is an integer$\}$ is an equivalence relation on $R$.
**Statement II** $B = \{(x, y) \in R \times R : x = \alpha y$ for some rational number $\alpha\}$ is an equivalence relation on $R$. **[AIEEE 2011]**
(a) Statement I is true, Statement II is true; Statement II is not a correct explanation of Statement I
(b) Statement I is true, Statement II is false
(c) Statement I is false, Statement II is true
(d) Statement I is true, Statement II is true, Statement II is a correct explanation of Statement I

## Numerical Value Type Questions

**Imp.** Reflexive and Transitive Relations

**53** The number of relations on the set $A = \{1, 2, 3\}$, containing at most 6 elements including (1, 2). which are reflexive and transitive but not symmetric, is ………… **[2025, 7 April Shift-I]**

**54** Number of functions $f : \{1, 2, \ldots, 100\} \to \{0, 1\}$ that assign 1 to exactly one of the positive integers less than or equal to 98 is equal to ……… . **[2025, 24 Jan Shift-II]**

**55** Let $A = \{1, 2, 3\}$. The number of relations on $A$, containing (1, 2) and (2, 3), which are reflexive and transitive but not symmetric is ……… . **[2025, 22 Jan Shift-II]**

**56** Let $A = \{(x, y) : 2x + 3y = 23, x, y \in N\}$ and $B = \{x : (x, y) \in A\}$. Then, the number of one-one functions from $A$ to $B$ is equal to ……… . **[2024, 9 April Shift-II]**

**57** Let $A = \{1, 2, 3, \ldots, 100\}$. Let $R$ be a relation on $A$ defined by $(x, y) \in R$ if and only if $2x = 3y$. Let $R_1$ be a symmetric relation on $A$ such that $R \subset R_1$ and the number of elements in $R_1$ is $n$. Then, the minimum value of $n$ is ……… . **[2024, 31 Jan Shift-II]**

**58** Let $A = \{1, 2, 3, 4\}$ and $R = \{(1, 2), (2, 3), (1, 4)\}$ be a relation on $A$. Let $S$ be the equivalence relation on $A$ such that $R \subset S$ and the number of elements in $S$ is $n$. Then, the minimum value of $n$ is ……… . **[2024, 31 Jan Shift-I]**

**59** The number of symmetric relations defined on the set {1, 2, 3, 4} which are not reflexive is …… . **[2024, 30 Jan Shift-II]**

**60** Let $A = \{1, 2, 3, 4\}$ and $R$ be a relation on the set $A \times A$ defined by $R = \{((a, b), (c, d)) : 2a + 3b = 4c + 5d\}$. Then ,the number of elements in $R$ is ……… . **[2023, 15 April Shift-I]**

**61** Let $A = \{-4, -3, -2, 0, 1, 3, 4\}$ and $R = \{(a, b) \in A \times A : b = |a| \text{ or } b^2 = a + 1\}$ be a relation on $A$. Then, the minimum number of elements, that must be added to the relation $R$ so that it becomes reflexive and symmetric, is ………… . **[2023, 13 April Shift-II]**

**62** The number of relations, on the set {1, 2, 3} containing (1, 2) and (2, 3), which are reflexive and transitive but not symmetric, is ………… . **[2023, 12 April Shift-I]**

**63** Let $A = \{1, 2, 3, 4, 5\}$ and $B = \{1, 2, 3, 4, 5, 6\}$. Then, the number of functions $f : A \to B$ satisfying $f(1) + f(2) = f(4) - 1$ is equal to ………… . **[2023, 11 April Shift-II]**

**64** Let $R = \{a, b, c, d, e\}$ and $S = \{1, 2, 3, 4\}$. Total number of onto functions $f : R \to S$, such that $f(a) \neq 1$, is equal to ………. **[2023, 8 April Shift-II]**

**65** Let $A = \{0, 3, 4, 6, 7, 8, 9, 10\}$ and $R$ be the relation defined on $A$ such that $R = \{(x, y) \in A \times A : x - y$ is odd positive integer or $x - y = 2\}$. The minimum number of elements that must be added to the relation $R$, so that it is a symmetric relation, is equal to ………. **[2023, 8 April Shift-I]**

**66** Let $A = \{1, 2, 3, 4, \ldots, 10\}$ and $B = \{0, 1, 2, 3, 4\}$. The number of elements in the relation $R = \{(a, b) \in A \times A : 2(a - b)^2 + 3(a - b) \in B\}$ is ………. **[2023, 6 April Shift-I]**

**67** The minimum number of elements that must be added to the relation $R = \{(a, b), (b, c), (b, d)\}$ on the set $\{a, b, c, d\}$, so that it is an equivalence relation, is ……… . **[2023, 24 Jan Shift-II]**

**68** Let $S = \{1, 2, 3, 4\}$. Then, the number of elements in the set $\{f : S \times S \to S : f$ is onto and $f(a, b) = f(b, a) \geq a, \forall (a, b) \in S \times S\}$ is ……… . **[2022, 28 June Shift-II]**

**69** The number of one-one functions $f : \{a, b, c, d\} \to \{0, 1, 2, \ldots, 10\}$, such that $2f(a) - f(b) + 3f(c) + f(d) = 0$ is …… . **[2022, 24 June Shift-I]**

**70** Let $A = \{0, 1, 2, 3, 4, 5, 6, 7\}$. Then, the number of bijective functions $f : A \to A$, such that $f(1) + f(2) = 3 - f(3)$ is equal to ……… . **[2021, 22 July Shift-II]**

## Topic 2 Composition of Functions, Inverse of a Function

### Level 1

**01** Let $f: R-\{0\} \to (-\infty, 1)$ be a polynomial of degree 2, satisfying $f(x) f\left(\frac{1}{x}\right) = f(x) + f\left(\frac{1}{x}\right)$.

If $f(K) = -2K$, then the sum of squares of all possible values of $K$ is **[2025, 28 Jan Shift-II]**

(a) 7 (b) 6 (c) 1 (d) 9

**02** Let $f(x) = x^2 + 9$, $g(x) = \frac{x}{x-9}$ and $a = fog(10)$, $b = gof(3)$. If $e$ and $l$ denote the eccentricity and the length of the latus rectum of the ellipse $\frac{x^2}{a} + \frac{y^2}{b} = 1$, then $8e^2 + l^2$ is equal to **[2024, 9 April Shift-I]**

(a) 6 (b) 12 (c) 8 (d) 16

**03** If $f(x) = \frac{4x+3}{6x-4}$, $x \neq \frac{2}{3}$ and $(fof)(x) = g(x)$, where $g: R - \left\{\frac{2}{3}\right\} \to R - \left\{-\frac{2}{3}\right\}$, then $(gogog)(4)$ is equal to **[2024, 31 Jan Shift-I]**

(a) $\frac{19}{20}$ (b) $-4$ (c) $-\frac{19}{20}$ (d) 4

**04** Consider function $f: A \to B$ and $g: B \longrightarrow C(A, B, C \subseteq R)$ such that $(gof)^{-1}$ exists, then **[2021, 25 July Shift-II]**

(a) $f$ and $g$ both are one-one
(b) $f$ and $g$ both are onto
(c) $f$ is one-one and $g$ is onto
(d) $f$ is onto and $g$ is one-one

**05** Let $A = \{1, 2, 3, \ldots, 10\}$ and $f: A \to A$ be defined as
$$f(x) = \begin{cases} x+1, & \text{if } x \text{ is odd} \\ x, & \text{if } x \text{ is even} \end{cases}$$
Then, the number of possible functions $g: A \to A$, such that $gof = f$ is **[2021, 26 Feb Shift-II]**

(a) $10^5$ (b) ${}^{10}C_5$ (c) $5^5$ (d) $5!$

### Level 2

**Imp.** Composition of Function

**06** Let $f, g: (1, \infty) \to R$ be defined as $f(x) = \frac{2x+3}{5x+2}$ and $g(x) = \frac{2-3x}{1-x}$. If the range of the function $fog: [2, 4] \to R$ is $[\alpha, \beta]$, then $\frac{1}{\beta - \alpha}$ is equal to **[2025, 4 April Shift-II]**

(a) 2 (b) 29 (c) 56 (d) 68

**07** Let $f: R \to R$ be a function defined by $f(x) = (2+3a)x^2 + \left(\frac{a+2}{a-1}\right)x + b$, $a \neq 1$.

If $f(x+y) = f(x) + f(y) + 1 - \frac{2}{7}xy$, then the value of $28\sum_{i=1}^{5} |f(i)|$ is **[2025, 28 Jan Shift-I]**

(a) 735 (b) 675 (c) 545 (d) 715

**08** If $f(x) = \frac{2^x}{2^x + \sqrt{2}}$, $x \in R$, then $\sum_{k=1}^{81} f\left(\frac{k}{82}\right)$ is equal to **[2025, 28 Jan Shift-I]**

(a) 82 (b) $\frac{81}{2}$ (c) $81\sqrt{2}$ (d) 41

**09** Let $f, g: R \to R$ be defined as
$f(x) = |x-1|$ and $g(x) = \begin{cases} e^x, & x \geq 0 \\ x+1, & x \leq 0 \end{cases}$. Then, the function $f(g(x))$ is **[2024, 5 April Shift-II]**

(a) one-one but not onto. (b) neither one-one nor onto.
(c) onto but not one-one. (d) both one-one and onto.

**10** Let $f: R \to R$ and $g: R \to R$ be defined as
$f(x) = \begin{cases} \log_e x, & x > 0 \\ e^{-x}, & x \leq 0 \end{cases}$ and $g(x) = \begin{cases} x, & x \geq 0 \\ e^x, & x < 0 \end{cases}$.
Then, $gof: R \to R$ is **[2024, 1 Feb Shift-II]**

(a) onto but not one-one
(b) neither one-one nor onto
(c) one-one but not onto
(d) both one-one and onto

**11** If $f(x) = \begin{cases} 2+2x, & -1 \leq x < 0 \\ 1 - \frac{x}{3}, & 0 \leq x \leq 3 \end{cases}$;
$g(x) = \begin{cases} -x, & -3 \leq x \leq 0 \\ x, & 0 < x \leq 1 \end{cases}$, then range of $(fog)(x)$ is **[2024, 29 Jan Shift-I]**

(a) $[0, 3)$ (b) $(0, 1]$ (c) $[0, 1]$ (d) $[0, 1)$

**12** Let $f: R \to R$ be defined as $f(x) = x - 1$ and $g: R - \{1, -1\} \to R$ be defined as $g(x) = \frac{x^2}{x^2 - 1}$.
Then, the function $fog$ is **[2022, 26 June Shift-II]**

(a) one-one but not onto function
(b) onto but not one-one function
(c) Both one-one and onto function
(d) Neither one-one nor onto function

**13** Let $f: R \to R$ be defined as $f(x) = x^3 + x - 5$. If $g(x)$ is a function, such that $f[g(x)] = x, \forall x \in R$, then $g'(63)$ is equal to **[2022, 25 June Shift-I]**

(a) $\frac{1}{49}$ (b) $\frac{3}{49}$ (c) $\frac{43}{49}$ (d) $\frac{91}{49}$

**14** Let $f : R - \left\{\frac{\alpha}{6}\right\} \longrightarrow R$ be defined by $f(x) = \frac{5x+3}{6x-\alpha}$. Then, the value of $\alpha$ for which $(fof)(x) = x, \forall x \in R - \left\{\frac{\alpha}{6}\right\}$ is **[2021, 20 July Shift-II]**

(a) No such $\alpha$ exists (b) 5
(c) 8 (d) 6

**15** Let $f, g : N \to N$, such that $f(n+1) = f(n) + f(1) \forall n \in N$ and $g$ be any arbitrary function. Which of the following statements is not true? **[2021, 25 Feb Shift-I]**

(a) If $fog$ is one-one, then $g$ is one-one.
(b) If $f$ is onto, then $f(n) = n, \forall n \in N$.
(c) $f$ is one-one.
(d) If $g$ is onto, then $fog$ is one-one.

**16** Let $f : R \to R$ be defined as $f(x) = 2x - 1$ and $g : R - \{1\} \to R$ be defined as $g(x) = \frac{x - 1/2}{x-1}$

Then, the composition function $f(g(x))$ is **[2021, 24 Feb Shift-I]**

(a) one-one but not onto (b) onto but not one-one
(c) Neither one-one nor onto (d) Both one-one and onto

**17** The inverse function of $f(x) = \frac{8^{2x} - 8^{-2x}}{8^{2x} + 8^{-2x}}, x \in (-1, 1)$ is **[2020, 8 Jan Shift-I]**

(a) $\frac{1}{4}(\log_8 e)\log_e\left(\frac{1-x}{1+x}\right)$

(b) $\frac{1}{4}\log_e\left(\frac{1-x}{1+x}\right)$

(c) $\frac{1}{4}(\log_8 e)\log_e\left(\frac{1+x}{1-x}\right)$

(d) $\frac{1}{4}\log_e\left(\frac{1+x}{1-x}\right)$

**18** Let $f(x) = x^2, x \in R$. For any $A \subseteq R$, define $g(A) = \{x \in R : f(x) \in A\}$. If $S = [0, 4]$, then which one of the following statements is not true? **[2019, 10 April Shift-I]**

(a) $f(g(S)) = S$ (b) $g(f(S)) \neq S$
(c) $g(f(S)) = g(S)$ (d) $f(g(S)) \neq f(S)$

**19** The function $f : R \to \left[-\frac{1}{2}, \frac{1}{2}\right]$ defined as $f(x) = \frac{x}{1+x^2}$ is **[JEE Main 2017]**

(a) invertible.
(b) injective but not surjective.
(c) surjective but not injective.
(d) Neither injective nor surjective.

## Level 3

**20** Let $f : R - \left\{\frac{-1}{2}\right\} \to R$ and $g : R - \left\{\frac{-5}{2}\right\} \to R$ be defined as $f(x) = \frac{2x+3}{2x+1}$ and $g(x) = \frac{|x|+1}{2x+5}$. Then, the domain of the function $fog$ is **[2024, 27 Jan Shift-II]**

(a) $R - \left\{-\frac{7}{4}\right\}$ (b) $R - \left\{-\frac{5}{2}, -\frac{7}{4}\right\}$

(c) $R$ (d) $R - \left\{-\frac{5}{2}\right\}$

**Imp.** Domain & Range of the Function

**21** Let $D$ be the domain of the function $f(x) = \sin^{-1}\left(\log_{3x}\left(\frac{6+2\log_3 x}{-5x}\right)\right)$.

If the range of the function $g : D \to R$ defined by $g(x) = x - [x]$, ($[x]$ is the greatest integer function), is $(\alpha, \beta)$, then $\alpha^2 + \frac{5}{\beta}$ is equal to **[2023, 12 April Shift-I]**

(a) 135 (b) 46 (c) 45 (d) 136

**22** If $f(x) = \frac{(\tan 1°)x + \log_e(123)}{x \log_e(1234) - (\tan 1°)}, x > 0$, then the least value of $f(f(x)) + f\left(f\left(\frac{4}{x}\right)\right)$ is **[2023, 10 April Shift-I]**

(a) 2 (b) 8 (c) 4 (d) 0

**23** Suppose, $f : R \to (0, \infty)$ be a differentiable function such that $5f(x+y) = f(x) \cdot f(y), \forall x, y \in R$. If $f(3) = 320$, then $\sum_{n=0}^{5} f(n)$ is equal to **[2023, 30 Jan Shift-I]**

(a) 6525 (b) 6825 (c) 6575 (d) 6875

**24** Let $f(x) = \frac{x-1}{x+1}, x \in R - \{0, -1, 1\}$. If $f^{n+1}(x) = f(f^n(x)), \forall n \in N$, then $f^6(6) + f^7(7)$ is equal to **[2022, 26 June Shift-I]**

(a) 7/6 (b) $-3/2$ (c) 7/12 (d) $-11/12$

**25** Let $f : R \to R$ and $g : R \to R$ be two functions defined by $f(x) = \log_e(x^2+1) - e^{-x} + 1$ and $g(x) = \frac{1-2e^{2x}}{e^x}$. Then, for which of the following range of $\alpha$, the inequality $f\left(g\left(\frac{(\alpha-1)^2}{3}\right)\right) > f\left(g\left(\alpha - \frac{5}{3}\right)\right)$ holds? **[2022, 25 June Shift-I]**

(a) (2, 3) (b) (−2, −1) (c) (1, 2) (d) (−1, 1)

**26** Let $g : N \to N$ be defined as
$g(3n+1) = 3n+2,\ g(3n+2) = 3n+3,$
$g(3n+3) = 3n+1, \forall n \geq 0.$
Then, which of the following statements is true ? **[2021, 25 July Shift-I]**

(a) There exists an onto function $f : N \to N$ such that $fog = f$
(b) There exists a one-one function $f : N \to N$ such that $fog = f$
(c) $gogog = g$
(d) There exists a function $f : N \to N$ such that $gof = f$

**27** Let $f$ be a function defined by $f(x) = (x-1)^2 + 1,$ $(x \geq 1).$
**Statement I** The set $\{x : f(x) = f^{-1}(x)\} = \{1, 2\}$
**Statement II** $f$ is bijection and
$f^{-1}(x) = 1 + \sqrt{x-1}, x \geq 1$ **[AIEEE 2011]**

(a) Statement I is false, Statement II is true.
(b) Statement I is true, Statement II is true; Statement II is a correct explanation of Statement I.
(c) Statement I is true, Statement II is true; Statement II is not a correct explanation of Statement I.
(d) Statement I is true, Statement II is false.

**28** Let $f : N \to Y$ be a function defined as
$f(x) = 4x + 3,$ where $Y = \{y \in N : y = 4x + 3$ for some $x \in N\}.$
Show that $f$ is invertible and its inverse is **[AIEEE 2008]**

(a) $f^{-1}(y) = \dfrac{y-3}{4}$
(b) $f^{-1}(y) = \dfrac{3y+4}{3}$
(c) $f^{-1}(y) = 4 + \dfrac{y+3}{4}$
(d) $f^{-1}(y) = \dfrac{y+3}{4}$

## Numerical Value Type Questions

**29** Consider the function $f : R \to R$ defined by
$f(x) = \dfrac{2x}{\sqrt{1+9x^2}}.$ If the composition of
$\underbrace{fo\,(fo\,fo\,fo \dots fo)}_{10 \text{ times}}(x) = \dfrac{2^{10}x}{\sqrt{1+9\alpha x^2}},$ then the value of $\sqrt{3\alpha+1}$ is equal to...... . **[2024, 5 April Shift-I]**

**30** Let $f^1(x) = \dfrac{3x+2}{2x+3}, x \in R - \left\{\dfrac{-3}{2}\right\}$
For $n \geq 2$, defines $f^n(x) = f^1 o f^{n-1}(x).$
If $f^5(x) = \dfrac{ax+b}{bx+a},$ gcd $(a, b) = 1$, then $a + b$ is equal to ......... . **[2023, 30 Jan Shift-I]**

**31** For some $a, b, c \in N$, let $f(x) = ax - 3$ and $g(x) = x^b + c, x \in R.$ If $(fog)^{-1}(x) = \left[\dfrac{x-7}{2}\right]^{\frac{1}{3}}$, then $(fog)(ac) + (gof)(b)$ is equal to ...... **[2023, 25 Jan Shift-I]**

**32** Let $f(x)$ be a quadratic polynomial with leading coefficient 1 such that $f(0) = p,\ p \neq 0$ and $f(1) = \dfrac{1}{3}.$
If the equation $f(x) = 0$ and $fofofof(x) = 0$ have a common real root, then $f(-3)$ is equal to ......... **[2022, 25 July Shift-II]**

**33** Let $S = \{1, 2, 3, 4, 5, 6, 7, 8, 9, 10\}$. Define
$f : S \to S$ as $f(n) = \begin{cases} 2n, & \text{if } n = 1, 2, 3, 4, 5 \\ 2n - 11, & \text{if } 6, 7, 8, 9, 10 \end{cases}$
Let $g : S \to S$ be a function such that
$fog(n) = \begin{cases} n+1, & \text{if } n \text{ is odd} \\ n-1, & \text{if } n \text{ is even} \end{cases}$
$g(10)\,((g(1) + g(2) + g(3) + g(4) + g(5))$ is equal to .................. . **[2022, 27 June Shift-II]**

**34** Let $f : R \to R$ be a function defined by
$f(x) = \left(2\left(1 - \dfrac{x^{25}}{2}\right)(2 + x^{25})\right)^{1/50}$. If the function $g(x) = f(f(f(x))) + f(f(x))$, the greatest integer less than or equal to $g(1)$ is ......... . **[2022, 25 June Shift-I]**

## • Key Index

### Topic 1

| | | | | | | | | | |
|---|---|---|---|---|---|---|---|---|---|
| 1 (a) | 2 (d) | 3 (c) | 4 (c) | 5 (b) | 6 (a) | 7 (d) | 8 (d) | 9 (d) | 10 (b) |
| 11 (d) | 12 (a) | 13 (a) | 14 (d) | 15 (d) | 16 (b) | 17 (b) | 18 (b) | 19 (d) | 20 (d) |
| 21 (b) | 22 (d) | 23 (c) | 24 (c) | 25 (a) | 26 (d) | 27 (a) | 28 (a) | 29 (d) | 30 (c) |
| 31 (d) | 32 (d) | 33 (d) | 34 (d) | 35 (b) | 36 (c) | 37 (c) | 38 (b) | 39 (d) | 40 (a) |
| 41 (d) | 42 (d) | 43 (c) | 44 (d) | 45 (a) | 46 (c) | 47 (b) | 48 (d) | 49 (c) | 50 (a) |
| 51 (d) | 52 (b) | 53 (5) | 54 (392) | 55 (3) | 56 (24) | 57 (66) | 58 (16) | 59 (960) | 60 (5) |
| 61 (7) | 62 (4) | 63 (360) | 64 (180) | 65 (19) | 66 (18) | 67 (9) | 68 (37) | 69 (31) | 70 (720) |

### Topic 2

| | | | | | | | | | |
|---|---|---|---|---|---|---|---|---|---|
| 1 (b) | 2 (c) | 3 (d) | 4 (c) | 5 (a) | 6 (c) | 7 (b) | 8 (b) | 9 (b) | 10 (b) |
| 11 (c) | 12 (d) | 13 (a) | 14 (b) | 15 (d) | 16 (a) | 17 (c) | 18 (c) | 19 (c) | 20 (d) |
| 21 (a) | 22 (c) | 23 (b) | 24 (b) | 25 (a) | 26 (a) | 27 (b) | 28 (a) | 29 (1024) | 30 (3125) |
| 31 (2039) | 32 (25) | 33 (190) | 34 (2) | | | | | | |

# Detailed **Explanations**

## Topic 1

**01.** (a) $R=\{(-2,1),(-1,1),(0,1),$
$(1,1),(2,2),(3,3)\}$

$l=6$ (number of elements)

$R$ needs $(-2,-2),(-1,-1)$ and $(0,0)$ to make it reflexive

$\Rightarrow m=3$

$R$ needs $(1,-1),(1,-2),(1,0)$

To make it symmetric

$\Rightarrow n=3$

$\therefore l+m+n=6+3+3=12$

**02.** (d) $xRy \Rightarrow \sec^2 x-\tan^2 y=1$

**To check Reflexivity**

$xRx \Rightarrow \sec^2 x-\tan^2 x=1$ which is true.

So, $R$ is reflexive.

**To check symmetricity**

$xRy \Rightarrow yRx$

i.e. $\quad \sec^2 x-\tan^2 y=1$

$\Rightarrow \tan^2 x+1-\sec^2 y+1=1$

$\Rightarrow \quad \tan^2 x-\sec^2 y=-1$

$\Rightarrow \quad \sec^2 y-\tan^2 x=1 \Rightarrow y\,R\,x.$

So, $R$ is symmetric.

**To check Transitivity**

$x\,R\,y,\ y\,R\,z \Rightarrow x\,R\,z$

i.e. $\quad \sec^2 x-\tan^2 y=1 \quad \ldots(i)$

$\Rightarrow \quad \sec^2 y-\tan^2 z=1 \quad \ldots(ii)$

On adding Eqs. (i) and (ii), we get

$\sec^2 x+\sec^2 y-\tan^2 y-\tan^2 z=2$

$\Rightarrow \quad \sec^2 x+1-\tan^2 z=2$

$\Rightarrow \quad \sec^2 x-\tan^2 z=1 \Rightarrow x\,R\,z$

$\therefore$ $R$ is transitive.

Hence, $R$ is an equivalence relation.

> **Time Saver Tech**
>
> Given relation $xRy \Rightarrow \sec^2 x-\tan^2 y=1$
>
> $\sec^2 x=1+\tan^2 y$
>
> $\sec^2 x=\sec^2 y$
>
> $x=y;\ x,y \in \left(0,\dfrac{\pi}{2}\right)$
>
> $\therefore x=y$ is reflexive, symmetric and Transitive
>
> $\therefore$ It is equivalence relation.

**03.** (c) $R=\{(x,y): x, y \in Z$ and $x+y$ is even$\}$

**Reflexive** Let $x \in Z$.

Then, $\quad x+x=2x$, which is given, $\forall\, x \in Z$.

$\Rightarrow R$ is reflexive.

**Symmetric** Let $x, y \in Z$, Then,

$(x,y) \in R$

$\Rightarrow \quad x+y$ is an even number.

Therefore, $y+x$ will also be an even number because $x+y=y+x$, $\forall\, x, y \in Z$

$\Rightarrow \quad R$ is symmetric.

**Transitive** Let $x, y, z \in Z$ and $(x,y) \in R, (y,z) \in R$

$\Rightarrow \quad x+y=2n$ for some $n \in Z$

and $\quad y+z=2m$ for some $m \in Z$,

then $(x+y)+(y+z)=2n+2m$

$x+2y+z=2n+2m$

$x+z=2(n+m-y)$

$\Rightarrow x+z$ is an even number $\forall\, n, m, y \in Z$.

Therefore, $(x,z) \in R$.

$\Rightarrow R$ is transitive.

Hence, $R$ is an equivalence relation.

**04.** (c) Given,

$$f(x)=\frac{2^{2x}-1}{2^{2x}+1}=\frac{2^{2x}+1-2}{2^{2x}+1}$$

$$= \frac{2^{2x}+1}{2^{2x}+1} - \frac{2}{2^{2x}+1}$$

$$\Rightarrow \quad f(x) = 1 - \frac{2}{2^{2x}+1} \quad \ldots \text{(i)}$$

On differentiating Eq. (i) w.r.t $x$, we get

$$f'(x) = 0 + \frac{2}{(2^{2x}+1)^2} \times 2^{2x} \cdot \ln 2 \cdot 2$$

as $2^{2x} > 0, \forall x \in R$ and $\ln 2 > 0$ always

So, $\quad f'(x) > 0, \forall x \in R$

So, $f(x)$ is strictly increasing function, i.e. $f(x)$ is one-one.

Now, $f(-\infty) = -1$ and $f(\infty) = 1$

$\therefore f(x) \in (-1, 1) \neq$ codomain of $f$ (as $c_f : (-\infty, 1)$)

i.e. $f(x)$ is not onto.

Hence, $f(x)$ is one-one but not onto.

**05.** *(b)* **Statement I**

**Reflexive** $(a_1, b_1)\, R\, (a_1, b_1) \Rightarrow b_1 = b_1$

It is always true.

$\therefore R$ is reflexive relation.

**Symmetric**

$(a_1, b_1)\, R\, (a_2, b_2) \Rightarrow b_1 = b_2 \Rightarrow b_2 = b_1$

$\Rightarrow \quad (a_2, b_2)\, R\, (a_1, b_1)$

$\therefore R$ is symmetric.

**Transitive**

$(a_1, b_1)\, R\, (a_2, b_2) \Rightarrow b_1 = b_2$

and $\quad (a_2, b_2)\, R\, (a_3, b_3) \Rightarrow b_2 = b_3$

$\therefore \quad b_1 = b_3 \Rightarrow (a_1, b_1)\, R\, (a_3, b_3)$

$\therefore R$ is transitive.

$\therefore R$ is equivalence relation.

$\therefore$ Statement I is true.

**Statement II**

$S = \{(x, y) \in X : (x, y)\, R\, (a, b)\}$

$\because \quad (a, b) \in X$

$\because \quad (x, y)\, R\, (a, b) \Rightarrow y = b$

It does not represent a line parallel to $y = x$

$\therefore$ Statement II is not true.

**Equivalence relation** Any relation on a set A which is reflexive, symmetric and transitive is called an equivalence relation.

**06.** *(a)*

**key idea** A relation is equivalence when
(i) $(a, a) \in R \ \forall a \in A$
(ii) $(a, b) \in R \Rightarrow (b, a) \in R \ \forall a, b \in A$
(iii) $(a, b) \in R$,
$(b, c) \in R \Rightarrow (a, c) \in R \ \forall a, b, c \in A$

$A = \{1, 2, 3\}$

$\therefore$ Non-empty equivalence relation

$R_1 = \{(1, 1), (2, 2), (3, 3)\}$

$R_2 = \{(1, 1), (2, 2), (3, 3), (1, 2), (2, 1)\}$

$R_3 = \{(1, 1), (2, 2), (3, 3), (2, 3), (3, 2)\}$

$R_4 = \{(1, 1), (2, 2), (3, 3), (1, 3), (3, 1)\}$

$R_5 = \{(1, 1), (2, 2), (3, 3), (1, 2), (2, 1), (1, 3), (3, 1), (2, 3), (3, 2)\}$

$\therefore$ Number of relation $= 5$

**07.** *(d)* Prime factors of 2310

$= 2 \times 3 \times 5 \times 7 \times 11$

$= 2, 3, 5, 7, 11$

and $f(x) = \left[\log_2\left(x^2 + \left[\frac{x^3}{5}\right]\right)\right]$

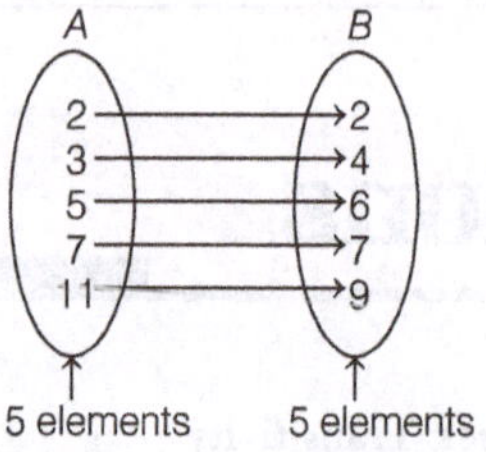


(put the values of $x$ and find the values of $f(x)$)

$\{\because [\ ]$ is a greatest function$\}$

So, number of one-to-one function

$= {}^5P_5 = 5! = 120$

**08.** *(d)* Given, $A = \{1, 3, 7, 9, 11\}$

$B = \{2, 4, 5, 7, 8, 10, 12\}$

and $f(1) + f(3) = 14$

Consider when

$f(1) = 2, f(3) = 12$

or $\ f(1) = 12, f(3) = 2$

$\therefore$Number of one-one function

$= 2 \times {}^5P_3$

Again consider when

$f(1) = 4, f(3) = 10$

or $f(1) = 10, f(3) = 4$

$\therefore$Number of one-one function $= 2 \times {}^5P_3$

$\therefore$Total number of one-one function

$= 4 \times {}^5P_3$

$= 4 \times 5 \times 4 \times 3 = 240$

**09.** *(d)* $(x_1, y_1)\, R\, (x_2, y_2)$

$\Rightarrow x_1 \leq x_2$ or $y_1 \leq y_2$

For $\quad (x_1, y_1), (x_1, y_1)$

$\Rightarrow x_1 \leq x_1$ or $y_1 \leq y_1$ which is true for all natural number.

$\therefore R$ is reflexive.

$(1, 1) R (2, 3)$ but $((2, 3), (1, 1)) \notin R$

$\therefore R$ is not symmetric.

For transitive $x_1 < x_2, y_1 < y_2$

$x_2 < x_3, y_2 < y_3$

$\therefore x_1 < x_3$ and $y_1 < y_3$

$R$ is transitive.

**10.** *(b)* $A = \{1, 2, 3, 4\}$

$R = \{(1, 2), (1, 3)\}$

For reflexive

We need to have

$(1, 1), (2, 2), (3, 3), (4, 4)$

For symmetric

$(1, 2) \in R$

$\Rightarrow \quad (2, 1) \in R$

$\Rightarrow \quad (1, 3) \in R$

$\Rightarrow \quad (3, 1) \in R$

$\therefore R$ becomes $\{(1, 1), (2, 2), (3, 3), (4, 4), (1, 2), (2, 1), (1, 3), (3, 1)\}$

Now, $(3, 1)$ and $(1, 2)$ both belongs to $R$

So, for transitive $(3, 2)$ must be on $R$

$\therefore$ $(3, 2)$ must be added

Again, for symmetric $(2, 3)$ also be an element

$\therefore R = \{(1, 1)\ (2, 2)\ (3, 3)\ (4, 4)\ (1, 2)\ (2, 1)\ (1, 3)\ (3, 1)\ (3, 2)\ (2, 3)\}$

$\therefore$ Number of elements in $R = 10$

**11.** *(d)* Given, $(a, b)\, R\, (c, d)$

Such that $ad - bc$ is divisible by 5

For reflexive, $(a, b)\, R\, (a, b)$

$\Rightarrow ab - ba = 0$ which is divisible by 5

$\Rightarrow R$ is reflexive

For symmetric, $(a, b)\, R\, (c, d)$

$\Rightarrow \ ad - bc$ is divisible by 5

$\Rightarrow -(ad - bc)$ is divisible by 5

$\Rightarrow \ cb - da$ is divisible by 5

$\Rightarrow \ (c, d)\, R\, (a, b)$

$\Rightarrow \ R$ is symmetric

For transitive, $(a, b)\, R\, (c, d)$ and $(c, d)\, R\, (b, a)$

but $(a, b)$ is not related to $(b, a)$

So, $R$ is not transitive.

**12.** *(a)* Given, $f : N - \{1\} \rightarrow N$

$f(n) =$ The highest prime factor of $n$.

We take an example to check its injectivity and surjectivity.

$f(2) = 2 = f(4)$

So, $f$ is many one.
Also, 4 is not image of any element.
Hence, $f$ is into.
Thus, $f$ is many one into i.e '$f$' is neither one-one nor onto.

**13.** (a) Given, $S = \{1, 2, 3, 4, ..., 10\}$
and $R = \{(A, B) : A \cap B \neq \phi; A, B \in M\}$

**To check Reflexivity**

$M$ is subset of '$S$' as question said.

So, $\phi \in M$

for $\phi \cap \phi = \phi$

$\Rightarrow$ but relation is $A \cap B \neq \phi$

So, it is not reflexive.

**To check Symmetricity**

$ARB \Rightarrow A \cap B \neq \phi$

$BRA \Rightarrow B \cap A \neq \phi$

So, it is symmetric.

**To check Transitivity**

If $A = \{1, 2\}$
$B = \{2, 3\ 4)\}$
$C = \{4, 5, 6)\}$

$ARB$ and $BRC$ but $A$ does not related to $C$ i.e. $A\not R C$.

So, it is not transitive.

**14.** (d) $x + y = 7$

$y = 7 - x$

$R = \{(1, 6), (2, 5), (3, 4), (4, 3), (5, 2), (6, 1)\}$

$\because (a, b) \in R \Rightarrow (b, a) \in R$

$\therefore$ Given relation is symmetric but neither reflexive nor transitive.

**15.** (d) We have, $AR_1B$

$\Rightarrow (A \cap B^C) \cup (B \cap A^C) = \phi$

$\Rightarrow A \cap B^C = B \cap A^C = \phi$

$\Rightarrow A = B$

So, $R_1$ is an equivalence relation.

Again, $AR_2B \Rightarrow A \cup B^C = B \cup A^C$

$\Rightarrow A^C \cap B = B^C \cap A$

$\Rightarrow A = B$

So, $R_2$ is an equivalence relation.

**16.** (b) For reflexive, $aRa \Rightarrow 2a + 3a = 5a$,
It is a multiple of 5, so $R$ is reflexive.
For symmetric, $aRb \Rightarrow 2a + 3b$ multiple of 5

$bRa \Rightarrow 2b + 3a$

$= \underbrace{(5a + 5b)}_{\text{Multiple of 5}} - \underbrace{(2a + 3b)}_{\text{Multiple of 5}}$

So, $R$ is symmetric

For transitive,

$aRb \Rightarrow 2a + 3b$ Multiple of 5

$bRc \Rightarrow 2b + 3c$ Multiple of 5

$\Rightarrow 2a + 5b + 3c$ Multiple of 5

$\underbrace{(2a + 3c)}_{\text{Multiple of 5}} + \underbrace{5b}_{\text{Multiple of 5}}$

So, $aRc$ is transitive.

$\therefore$ An equivalence relation.

**17.** (b) Given, $R = \{(a, b) : b = pq$, where $p, q \geq 3$ are prime numbers$\}$
$a$ can be anything from $\{1, 2, 3, \dots, 60\}$.
So, total possible values of $a$ are 60.

For $b = pq$

If $p = 3, q = 3, 5, 7, 11, 13, 17, 19$

If $p = 5, q = 5, 7, 11$

If $p = 7, q = 7$

Total possible values of $b = 11$

Total cases $= 60 \times 11 = 660$

**18.** (b) An function is one-one and onto out of 50 elements of domain set 17 elements are following restriction

$f(3) \geq f(9) \geq f(15) \geq f(21) \geq \dots \geq f(99)$

So, all elements 3, 9, 15, ..., 99 i.e. 17 elements are 1 choice.
Remaining $50 - 17 = 33$ elements has taken from 50 elements.

$\therefore$ Number of ways $= {}^{50}P_{33}$

**19.** (d)

**key idea** (i) Recall the definition of symmetric, reflextive and transitive relation.
(ii) A relation is said to be equivalence relation it if is symmetric, reflexive and transitive.

Let $A = \{1, 2, 3\}$

Then, $R = \{(1, 1)\ (1, 2), (1, 3)\ (2, 1), (2, 2), (2, 3), (3, 1), (3, 2), (3, 3)\}$

(i) $R$ is reflexive. $[\because (a, a) \in R]$

(ii) $R$ is symmetric.
$[\because (a, b) \in R \Rightarrow (b, a) \in R]$

(iii) $R$ is transitive.
$[\because (a, b) \in R$ and $(b, c) \in R \Rightarrow (a, c) \in R]$

$\therefore R$ is an equivalence relation.

**20.** (d) When, $n = 2, 4, 6, 8$, then $2n$ is the multiple of 4.
When, $n = 3, 7, 11, 15$ then $(n - 1)$ is not multiple of 4.
When, $n = 1, 5, 9, 13$, then $\left(\frac{n+1}{2}\right)$ is the odd number.
Every number gives exactly one value.
Thus, $f$ is one-one and onto.

**21.** (b) According to the question, let's consider option (b), (2, 3) and (3, 4) satisfy $0 < |x - y| \leq 1$ but (2, 4) does not satisfy it.

**22.** (d) A function $f : A \to R$ given by

$f(x) = \frac{2x}{x - 1}$

$\Rightarrow f(x_1) = f(x_2)$

$\Leftrightarrow \frac{2x_1}{x_1 - 1} = \frac{2x_2}{x_2 - 1}$

So, $f$ is one-one.
As $f(x) \neq 2$ for any $x \in A \Rightarrow f$ is not into.
Hence, $f$ is injective but not surjective.

**23.** (c)

**key idea** For surjective function, range of $f(x)$ = codomain

$f(x) = \frac{x^2}{1 - x^2} = y$ (let)

$\Rightarrow x^2 = y(1 - x^2)$ $[\because x^2 \neq 1]$

$\Rightarrow x^2 = \frac{y}{1 + y}$ [provided $y \neq -1$]

$\because x^2 \geq 0 \Rightarrow \frac{y}{1 + y} \geq 0$

$\Rightarrow y \in (-\infty, -1) \cup [0, \infty)$

$\therefore$ Set $A$ should be $R - [-1, 0)$.

**24.** (c) Given $f(x) = x^3 + 5x + 1$

Now, $f'(x) = 3x^2 + 5 > 0, \ \forall\ x \in R$

Thus, $f(x)$ is strictly increasing function.
So, $f(x)$ is one-one function.
Clearly, $f(x)$ is a continuous function and also increasing on $R$.

$\therefore \lim_{x \to -\infty} f(x) = -\infty$

and $\lim_{x \to \infty} = \infty$

Hence, $f(x)$ takes every value between $-\infty$ and $\infty$.
Thus, $f(x)$ is onto function.

**25.** (a) Let $W = \{CAT, TOY, YOU, \dots\}$

Clearly, $R$ is reflexive and symmetric but not transitive.

$[\because CAT\ R\ TOY, TOY\ R\ YOU \not\Rightarrow CAT\ R\ YOU]$

**26.** (d) Since, for every elements of $A$, there exists elements (3, 3), (6, 6), (9, 9), (12, 12) $\in R$

$\Rightarrow R$ is reflexive relation.

Now, $(6, 12) \in R$ but $(12, 6) \notin R$, so it is not a symmetric relation.

Also, $(3, 6), (6, 12) \in R \Rightarrow (3, 12) \in R$

$\therefore R$ is transitive relation.

**27.** *(a)* Given, $x \in (-1, 1)$

$\Rightarrow \quad \tan^{-1} x \in \left(-\frac{\pi}{4}, \frac{\pi}{4}\right)$

Given that $f(x) = \tan^{-1}\left(\frac{2x}{1-x^2}\right)$

$= 2\tan^{-1} x \quad [\because x^2 < 1]$

So, $f(x) \in \left(-\frac{\pi}{2}, \frac{\pi}{2}\right)$

Hence, function is one-one onto.

**28.** *(a)* Let $\max(x, y) = 3$

$\Rightarrow \{(0, 3), (1, 3), (2, 3), (3, 3), (3, 0), (3, 1), (3, 2)\}$

Let $\max(x, y) = 4$

$\Rightarrow \{(0, 4), (1, 4), (2, 4), (3, 4), (4, 4), (4, 0), (4, 1), (4, 2), (4, 3)\}$

Total 16 elements.

$R$ is not reflexive : Max $(x, y) \in \{3, 4\}$

$R$ is symmetric

$\Rightarrow$ Max $(y, x) \in (3, 4)$

Max $(x, y) \in \{3, 4\}$

and Max $(y, z) \in \{3, 4\}$

$\Rightarrow$ Max $= \{x, z\}$

$\Rightarrow$ Not transitive

$(0, 4) \in R_1$ and $(4, 0) \in R$

$\Rightarrow \quad (0, 0) \notin R$

**29.** *(d)* $A = \{-3, -2, -1, 0, 1, 2, 3\}$

$\because 2x - y \in \{0, 1\}$

$\Rightarrow 2x - y = 0$ or $2x - y = 1$

$y = 2x, y = 2x - 1$

$x = -1, y = -2$

$x = -1, y = -3$

$x = 0, y = 0$

$x = 0, y = -1$

$x = 1, y = 2$

$x = 1, y = 1$

$x = 2, y = 3$

$\therefore R = \{(-1, -2)(0, 0)(1, 2)(-1, -3)(0, -1)(1, 1)(2, 3)\}$

for reflexive needs to be added elements

$\Rightarrow$

$(-3, -3), (-2, -2), (-1, -1), (2, 2), (3, 3)$

For symmetric neds to be added elements

$\Rightarrow (-2, -1), (-3, -1), (-1, 0)(2, 1), (3, 2)$

$\therefore l = 7$

$m = 5$

$n = 5$

$\therefore l + m + n = 17$

**30.** *(c)* For $R$ to be reflexive $(f, f)$ must be in $R$

This means $f(0) = f(1)$ and $f(1) = f(0)$ must be true $\forall f$

But $f(0) \neq f(1)$ always

Therefore, $R$ is not reflexive

If $(f, g) \in R$, then $f(0) = g(1)$ and $f(1) = g(0)$

$\because f(0) = g(1) \Rightarrow g(1) = f(0)$

and $f(1) = g(0)$

$\Rightarrow g(0) = f(1)$

$R$ is symmetric

If $(f, g) \in R$ and $(g, h) \in R$, then

$f(0) = g(1)$

$f(1) = g(0)$, $g(0) = h(1)$ and $g(1) = h(0)$

Since, $f(0) = g(1)$ and $g(1) = h(0)$, then $f(0)$ is not necessarily equal to $h(0)$

Therefore, $R$ is not transitive.

$\therefore$ The relation $R$ is symmetric but not reflexive or transitive.

**31.** *(d)* $A = \{1, 2, 3, 4\}$

$R = \{(1, 2), (2, 3), (3, 3)\}$

Minimum elements will be added in $R$ $(1, 1), (2, 2), (4, 4)$ $(2, 1), (3, 2), (3, 1), (1, 3)$, then $R$ become an equivalence relation

$\therefore$ Minimum number of elements $= 7$

**32.** *(d)* $A = \{2, 3, 6, 8, 9, 11\}$,

$B = \{1, 4, 5, 10, 15\}$

**Reflexive** $(a, b)\ R(a, b)$

So, $3ad - 7bc$

$\Rightarrow 3ab - 7ba = -4ab$ always even

So, it is reflexive

**Symmetric** If $3ad - 7bc =$ even

**Case I** odd odd

**Case II** even even

$(c, d)\ R(a, b)$

$\Rightarrow \quad 3bc - 7ad$

**Case I** odd odd

**Case II** even even

So, symmetric relation.

**Transitive** Set $(3, 4)\ R(6, 4)$ satisfy relation

Set $(6, 4)\ R(3, 1)$ satisfy relation

But $(3, 4)\ R(3, 1)$ does not satisfy relation

So, it is not transitive. Hence, reflexive and symmetric but not transitive.

**33.** *(d)* $f(x) = \begin{cases} -a, & -a \le x \le 0 \\ x + a, & 0 < x \le a \end{cases}$

So, $y = f(x) \Rightarrow$

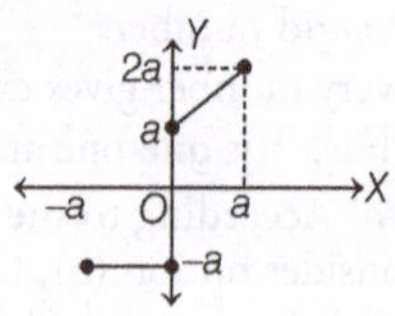


$y = f(|x|) \Rightarrow$

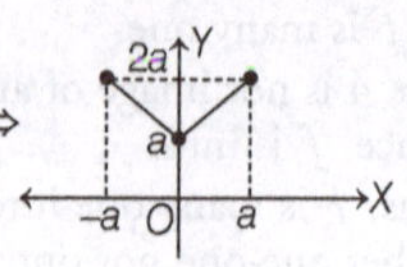


$y = |f(x)| \Rightarrow$

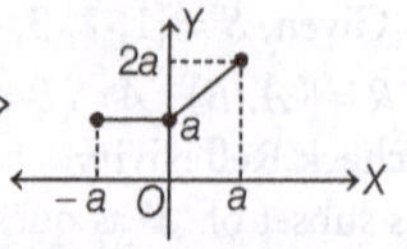


Now, $g(x) = \frac{f|x| - |f(x)|}{2}$

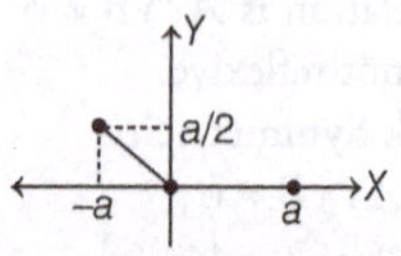


Hence, $g(x)$ is neither one-one nor onto.

**34.** *(d)* Given, $A = \{1, 2, 3, 4, 5\}$

Relation $R$ on $A$ is defined as $xRy$ iff $4x \le 5y$

Now, for $x = 1$, $y \ge \frac{4}{5}$

$x = 2$, $y \ge \frac{8}{5}$ ... so on.

$\therefore R = \{(1, 1), (1, 2), (1, 3), (1, 4), (1, 5), (2, 2), (2, 3), (2, 4), (2, 5), (3, 3), (3, 4), (3, 5), (4, 4), (4, 5), (5, 4), (5, 5)\}$

According to the question, $m = 16$

Now, for $R$ to be symmetric, we will have to add

For $(1, 2) \in R$, $(2, 1)$ has to be added

For $(1, 3) \in R$, $(3, 1)$ has to be added

For $(1, 4) \in R$, $(4, 1)$ has to be added

For $(1, 5) \in R$, $(5, 1)$ has to be added

For $(2, 3) \in R$, $(3, 2)$ has to be added

For $(2, 4) \in R$, $(4, 2)$ has to be added

For $(2, 5) \in R$, $(5, 2)$ has to be added

For $(3, 4) \in R$, $(4, 3)$ has to be added

For $(3, 5) \in R$, $(5, 3)$ has to be added

$n = 9$

$\Rightarrow m + n = 25$

**35.** *(b)* According to the question,

Set $X = \{1, 2, 3, \ldots, 20\}$

Now, $R_1 = \{(x, y) : 2x - 3y = 2\}$

$\Rightarrow \quad y = \frac{2x - 2}{3}$,

Here, numerator is always even and denominator is 3, so numerator have to be multiple of 3 to get natural number.

So, $R_1 = \{(4, 2), (7, 4), (10, 6), (13, 8), (16, 10), (19, 12)\}$

Minimum number of elements needs to be added to make $R_1$ symmetric $=6$

So, $M=6$

Now, $R_2=\{(x, y): -5x+4y=0\}$

$\Rightarrow y=\frac{5}{4}x \Rightarrow x$ must be multiple of 4

$\therefore R_2=\{(4, 5), (8, 10), (12, 15), (16, 20)\}$

Minimum number of elements needs to be added to make $R_2$ symmetric $=4$

So, $N=4$

$\therefore M+N=6+4=10$

**36.** *(c)* Given, $A=\{1, 3, 4, 6, 9\}$ and $B=\{2, 4, 5, 8, 10\}$

$R$ is defined on $A\times B$

i.e., $R=\{(a_1, b_1), (a_2, b_2): a_1 \le b_2$ and $b_1 \le a_2\}$

If $a_1=1$, then $b_2$ can be 2/4/5/8/10

$a_1=3$, then $b_2$ can be 4/5/8/10

$a_1=4$, then $b_2$ can be 4/5/8/10

$a_1=6$, then $b_2$ can be 8/10

$a_1=9$, then $b_2$ can be 10 only

So, number of ways $=5+4+4+2+1$
$=16$

Similarly, for $b_1=2 \Rightarrow a_2$ can be 3/4/6/9

$b_1=4 \Rightarrow a_2$ can be 4/6/9

$b_1=5 \Rightarrow a_2$ can be 6/9

$b_1=8 \Rightarrow a_2$ can be 9 only

and $b_1 \ne 10$

Thus, number of ways $=4+3+2+1$
$=10$

Finally, number of elements in set $R$ is $16\times 10=160$

**37.** *(c)* We have,

$A=\{2, 3, 4\}, B=\{8, 9, 12\}$

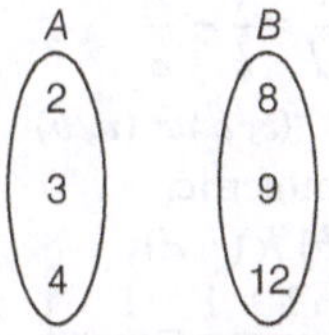


$R=\{((a_1, b_1), (a_2, b_2))\in(A\times B, A\times B)$
$a_1$ divides $b_2$ and $a_2$ divides $b_1\}$

2 divides 8 and 12, 3 divides 9 and 12, and 4 divides 8 and 12.

$\therefore$ Each element in $A$ divides exactly 2 element is $B$.

The number of cases that $a_1$ divides $b_2$ $=3\times 2=6$

[$\because$ Each element has 2 choices]

Similarly, the number of cases that $a_2$ divides $b_1=3\times 2=6$

Hence, the required number of cases $=6\times 6=36$

**38.** *(b)* Let $a=-1, b=10$. Then,

$$2+\frac{a}{b}=2-\frac{1}{10}=\frac{19}{10}>0$$

So, $(-1, 10)\in S$

Now, $a=10, b=-1$.

Then, $2+\frac{a}{b}=2-10=-8<0$

So, $(10, -1)\notin S$

Hence, $S$ is not symmetric.

We know that

If $a^2-b^2\in Z$, then $b^2-a^2\in Z$

$\therefore \quad (a, b)\in T \Rightarrow (b, a)\in T$

So, $T$ is symmetric.

Again, $a^2-b^2\in Z$ and $b^2-c^2\in Z$.

$\Rightarrow \quad a^2-b^2+b^2-c^2\in Z$

$\Rightarrow \quad a^2-c^2\in Z$

$\therefore \quad (a, b)\in T, (b, c)\in T$

$\Rightarrow \quad (a, c)\in T$

So, $T$ is transitive.

**39.** *(d)* For symmetric

$(a, b), (b, c)\in R$

$\Rightarrow \quad (b, a), (c, b)\in R$

For transitive

$(a, b), (b, c)\in R \Rightarrow (a, c)\in R$

Now, symmetric

$\therefore \quad (a, c)\in R \Rightarrow (c, a)\in R$

Transitive

$\therefore (a, b), (b, a)\in R \Rightarrow (a, a)\in R$

$(b, c), (c, b)\in R \Rightarrow (b, b)$ and $(c, c)\in R$

$\therefore$ Elements to be added so that $R$ becomes symmetric and transitive is $=\{(b, a), (c, b), (a, c), (c, a), (a, a), (b, b), (c, c)\}$

Hence, total number of elements to be added $=7$

**40.** *(a)* **To check reflexivity**

$(a, a)\Rightarrow$ gcd of $(a, a)=1$ and $b=a\ne 2a$, $\forall\, a\in z$. which is true.

**To check symmetric**

Take, $a=2$ and $b=1$

$\Rightarrow$ gcd$(2, 1)=1$

Also, $2a=4\ne b$

Now, when $a=1, b=2$, then

gcd$(1, 2)=1$

Also, $2a=2=b$

Hence, $2a=b$

So, $R$ is not symmetric.

**To check transitivity**

consider $a=14, b=19$ and $c=21$

gcd$(a, b)=1$

gcd$(b, c)=1$

gcd$(a, c)=7$

Hence, $R$ is not transitive.

Finally, $R$ is neither symmetric nor transitive.

**41.** *(d)* $R_1=\{(a, b): a, b\in R, ab\ge 0\}$

(i) $a\cdot a=a^2\ge 0 \Rightarrow aR_1a, \forall\, a\in R$

$\therefore R_1$ is reflexive.

(ii) If $aR_1b \Rightarrow ab\ge 0 \Rightarrow ba\ge 0$;

$\therefore bR_1a$

$\therefore R_1$ is symmetric.

(iii) If $aR_1b$ and $bR_1c \Rightarrow ab\ge 0$ and $bc\ge 0$

$\therefore (ab)(bc)\ge 0$

$\Rightarrow (ac)b^2\ge 0$

Now, if $b\ne 0$, $ac\ge 0$

But, when $b=0$, $ac$ can be negative.

Also, it is not always true that $ac\ge 0$.

$\therefore \quad R_1$ is not transitive.

$\Rightarrow \quad R_1$ is not equivalence relation.

Now, $R_2=\{(a, b): a, b\in R, a\ge b\}$

(i) $a=a \Rightarrow aR_2a \Rightarrow R_2$ is reflexive.

(ii) If $aR_2b$, then $a\ge b$, but $b\ge a$ is not true for all $a, b\in R$

e.g. $a=2, b=1, 2>1$ but $1<2$

$\therefore R_2$ is not symmetric.

$\therefore R_2$ is not an equivalence relation.

**42.** *(d)* Let $P(a, b)$ and $Q(c, d)$ are any two points.

Given, $OP=OQ$

i.e. $\sqrt{a^2+b^2}=\sqrt{c^2+d^2}$

On squaring on both sides, we get

$$a^2+b^2=c^2+d^2 \quad \ldots(i)$$

$R(x, y), S(1, -1), OR=OS$ (equivalence class)

This gives $OR=\sqrt{x^2+y^2}$ and $OS=\sqrt{2}$

$\Rightarrow \sqrt{x^2+y^2}=\sqrt{2} \Rightarrow x^2+y^2=2$

(Squaring on both sides)

$\therefore \quad S=\{(x, y): x^2+y^2=2\}$

**43.** *(c)* Let $x, y\in N$ and both be even.

Then, $f(x)=f(y)$

$\Rightarrow \quad -\frac{x}{2}=-\frac{y}{2} \Rightarrow x=y$

Again, $x, y\in N$ and both are odd.

Then, $f(x)=f(y)$

$\Rightarrow \quad \frac{x-1}{2}=\frac{y-1}{2} \Rightarrow x=y$

So, mapping is one-one.

Since, each negative integer is an image of even natural number and positive integer is an image of odd natural number. So, mapping is onto. Hence, mapping is one-one and onto.

**44.** (d) $f(x) = 2x^3 - 15x^2 + 36x + 7$

$\Rightarrow f'(x) = 6x^2 - 30x + 36$

Put $f'(x) = 0 \Rightarrow 6(x^2 - 5x + 6) = 0$

$\Rightarrow 6(x-2)(x-3) = 0 \Rightarrow x = 2, 3$

$\therefore f(2) = 35, f(3) = 34$ and $f(0) = 7$

$\Rightarrow \quad A = [7, 35]$

$\Rightarrow$ Total number of integer in $A = 29$

Now, $g(x) = \dfrac{x^{2025}}{x^{2025}+1} = 1 - \dfrac{1}{x^{2025}+1}$

$\Rightarrow B = [0, 1)$

$\Rightarrow$ Number of integer in $B = 1$

So, total number of integers in the range $A \cup B = 30$

(because no element is common in $A$ and $B$)

**45.** (a) Total function $= 4^4 = 256$

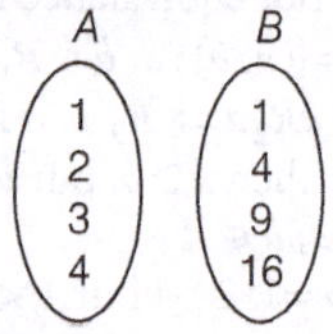


Total possible one-one function $= 4! = 24$

$\therefore$Many one function $= 256 - 24 = 232$

Many one function where $1 \notin f(A)$

$= 3 \cdot 3 \cdot 3 \cdot 3 = 81$

$\therefore$Required number of many one function

$= 232 - 81 = 151$

**46.** (c) Let $y = \dfrac{x^2 + 2x - 15}{x^2 - 4x + 9}$

Let find value of $y$ at $\infty$ and $-\infty$

$\Rightarrow \quad y|_{\infty} = \lim\limits_{x \to \infty} \dfrac{x^2 + 2x - 15}{x^2 - 4x + 9}$

$= \lim\limits_{x \to \infty} \dfrac{1 + \frac{2}{x} - \frac{15}{x^2}}{1 - \frac{4}{x} + \frac{9}{x^2}} = 1$

and $y|_{-\infty} = \lim\limits_{x \to -\infty} \dfrac{1 + \frac{2}{x} - \frac{15}{x^2}}{1 - \frac{4}{x} + \frac{9}{x^2}} = 1$

For 2 values of $x$, we are getting same $y$.

$\therefore$ Many one.

Now lets find range, $y = \dfrac{x^2 + 2x - 15}{x^2 - 4x + 9}$

$\Rightarrow yx^2 - 4yx + 9y = x^2 + 2x - 15$

$\Rightarrow (y-1)x^2 - (4y+2)x + (9y+15) = 0$ (where $y \neq 1$)

If above equation hold for all $x$, then its discriminant should be greater than or equal to zero.

$\Rightarrow (4y+2)^2 - 4 \times (y-1)(9y+15) \geq 0$

$\Rightarrow 16y^2 + 4 + 16y - (36y^2 + 60y - 36y - 60) \geq 0$

$\Rightarrow -20y^2 - 8y + 64 \geq 0$

$\Rightarrow \quad 5y^2 + 2y - 16 \leq 0$

$\Rightarrow \quad (y+2)(5y-8) \leq 0$

$\Rightarrow y \notin R \Rightarrow$ Into

**47.** (b) We have,

$a\, R_1\, b \Leftrightarrow a^2 + b^2 = 1, \forall\, a, b \in R$

For reflexive $\quad a = b \Rightarrow a^2 + a^2 = 1$

$a = \pm \dfrac{1}{\sqrt{2}}$

$\therefore$Only $\left(\dfrac{1}{\sqrt{2}}, \dfrac{1}{\sqrt{2}}\right)$ and $\left(-\dfrac{1}{\sqrt{2}}, \dfrac{-1}{\sqrt{2}}\right)$ will be present in $R_1$.

So, $R_1$ is not reflexive and not equivalence.

Now, $(a, b)\, R_2\, (c, d) \Leftrightarrow a + d = b + c, \forall\, (a, b), (c, d) \in N \times N$

Checking reflexive, put $a = b = c = d$

$\Rightarrow a + a = a + a$

Which is satisfied $\forall$ ordered pair $\in N \times N$

$\therefore R_2$ is reflexive.

Checking for symmetric.

Let suppose $((a, b), (c, d))$ be in $R_2$

So, $\quad a + d = b + c \quad$ ...(i)

To check $((c, d), (a, b))$ lies in $R_2$

Eq. (i) can be written as

$c + b = d + a$

$\therefore R_2$ is symmetric

Checking for transitive

Let suppose $((a, b), (c, d))$ and $((c, d), (e, f))$ are in $R_2$

So, $\quad a + d = b + c \quad$ ...(ii)

and $\quad c + f = d + e \quad$ ...(iii)

Adding Eqs. (ii) and (iii), we get

$a + f = b + e$

$\therefore ((a, b), (e, f))$ also in $R_2$

$\therefore R_2$ is also transitive.

So, $R_2$ is equivalence.

**48.** (d) Now, $3a - 3a + \sqrt{7} = \sqrt{7}$, which is an irrational.

So, $(a, a) \in R$. Hence, $R$ is reflexive.

Now, take $a = \dfrac{\sqrt{7}}{3}, b = 0$.

Then, $3a - 3b + \sqrt{7}$

$= 2\sqrt{7}$ which is irrational.

$\therefore \left(\dfrac{\sqrt{7}}{3}, 0\right) \in R$

Now, if $a = 0, b = \dfrac{\sqrt{7}}{3}$, then

$3a - 3b + \sqrt{7} = 0$, which is not irrational

$\therefore \left(0, \dfrac{\sqrt{7}}{3}\right) \notin R$

So, $R$ is not symmetric.

Similarly, $(0, \sqrt{3}) \in R, \left(\sqrt{3}, \dfrac{\sqrt{7}}{3}\right) \in R$

but $\left(0, \dfrac{\sqrt{7}}{3}\right) \notin R$.

So, $R$ is not transitive.

**49.** (c) Now, $ad(b-c) = bc(a-d)$

$\Rightarrow \quad \dfrac{b-c}{bc} = \dfrac{a-d}{ad}$

$\Rightarrow \quad \dfrac{1}{a} - \dfrac{1}{b} = \dfrac{1}{d} - \dfrac{1}{c}$

Now, $(a, b)\, R\, (a, b)$

$\Rightarrow \quad \dfrac{1}{a} - \dfrac{1}{b} = \dfrac{1}{b} - \dfrac{1}{a}$

which is not true.

So, $R$ is not reflexive.

Again, $(a, b)\, R\, (c, d)$

$\Rightarrow \quad \dfrac{1}{a} - \dfrac{1}{b} = \dfrac{1}{d} - \dfrac{1}{c}$

$\Rightarrow \quad \dfrac{1}{c} - \dfrac{1}{d} = \dfrac{1}{b} - \dfrac{1}{a}$

$\Rightarrow \quad (c, d)\, R\, (a, b)$

So, $R$ is symmetric.

Again, $(a, b)\, R\, (c, d)$

$\Rightarrow \quad \dfrac{1}{a} - \dfrac{1}{b} = \dfrac{1}{d} - \dfrac{1}{c} \quad$ ...(i)

and $(c, d)\, R\, (e, f)$

$\Rightarrow \quad \dfrac{1}{c} - \dfrac{1}{d} = \dfrac{1}{f} - \dfrac{1}{e} \quad$ ...(ii)

From Eqs. (i) and (ii), we get

$\dfrac{1}{a} - \dfrac{1}{b} = \dfrac{1}{e} - \dfrac{1}{f}$

$\neq (a, b)\, R\, (e, f)$

So, $R$ is not transitive.

**50.** *(a)*

> **key idea** A function $f: A \to B$ is a one-one, if for every $a_1, a_2 \in A$ such that $a_1 \neq a_2$, then $f(a_1) \neq f(a_2)$.
> And $f: A \to B$ is many-one, if for every $a_1, a_2 \in A$ such that $a_1 \neq a_2$, then $f(a_1) = f(a_2)$.

We have,

$$f: R \to R,\ f(x) = \frac{x^2+2x+1}{x^2+1} = 1 + \frac{2x}{x^2+1}$$

For $x_1, x_2 \in R$, let $f(x_1) = f(x_2)$

$\Rightarrow \quad \dfrac{x_1}{1+x_1^2} = \dfrac{x_2}{1+x_2^2}$

$\Rightarrow \quad x_1 + x_1 x_2^2 - x_2 - x_2 x_1^2 = 0$

$\Rightarrow \quad (x_1 - x_2)(1 - x_1 x_2) = 0$

$\Rightarrow \quad x_1 = x_2$

or, $\quad x_1 x_2 = 1$

For $x_1, x_2 \in R$; $\exists\, x_2 \in R$, $\forall\, x_1 \in R$, such that $x_2 x_1 = 1$

$\therefore f(x)$ is many- one in $R$ i.e. $(-\infty, \infty)$.

For $x_1, x_2 \in [1, \infty)$, $\exists\, x_2 \in [1, \infty)$ for any $x_1 \in [1, \infty)$, such that $x_1 x_2 = 1$

$\therefore f(x)$ is one-one in $[1, \infty)$.

For $x_1, x_2 \in (-\infty, -1]$, $\exists\, x_2 \in (-\infty, -1]$ for any $x_1 \in (-\infty, -1]$, such that $x_1 x_2 = 1$

$\therefore f(x)$ is one-one in $(-\infty, -1]$.

**51.** *(d)* For $R$ to be reflexive $\Rightarrow x\,R\,x$

$\Rightarrow \quad 3x + \alpha x = 7K$

$\Rightarrow \quad (3+\alpha)x = 7K$

$\Rightarrow \quad 3 + \alpha = 7\lambda$

$\Rightarrow \quad \alpha = 7\lambda - 3 = 7N + 4,\ K, \lambda, N \in I$

$\therefore$ When, $\alpha$ divided by 7 its leaves remainder as 4.

$R$ to be symmetric $xRy \Rightarrow yRx$

$3x + \alpha y = 7N_1,\ 3y + \alpha x = 7N_2$

$\Rightarrow \quad (3+\alpha)(x+y) = 7(N_1 + N_2) = 7N_3$

which holds when $3 + \alpha$ is multiple of 7.

$\therefore \quad \alpha = 7N + 4$

$R$ is be transitive $xRy$ and $yRz$ $\Rightarrow xRz$.

$3x + \alpha y = 7N_1$

and $\quad 3y + \alpha z = 7N_2$

and $\quad 3x + \alpha z = 7N_3$

$\therefore \quad 3x + 7N_2 - 3y = 7N_3$

$\therefore \quad 7N_1 - \alpha y + 7N_2 - 3y = 7N_3$

$\therefore \quad 7(N_1 + N_2) - (3+\alpha)y = 7N_4$

$\therefore \quad (3+\alpha)y = 7N$

which is true again when $3 + \alpha$ divisible by 7, i.e. when $\alpha$ divided by 7, the remainder is 4.

**52.** *(b)* **Statement I**

(a) Reflexive

$xRx : (x - x)$ is an integer. i.e., true

$\therefore$ Reflexive

(b) Symmetric

$xRy : (x - y)$ is an integer.

$\Rightarrow (y - x)$ is an integer.

$\Rightarrow yRx$

$\therefore$ Symmetric

(c) Transitive

$xRy$ and $yRz$

$\Rightarrow (x - y)$ is an integer and $(y - z)$ is an integer.

$\Rightarrow (x - y) + (y - z)$ is an integer.

$\Rightarrow (x - z)$ is an integer.

$\Rightarrow xRz$

$\therefore$ Transitive

Hence, $A$ is an equivalence relation.

**Statement II**

If $\alpha = \dfrac{1}{2}$, then for reflexive, we have

$xRx \Rightarrow x = \dfrac{1}{2}x$, which is not true, $\forall\, x \in R - \{0\}$.

$\therefore B$ is not reflexive on $R$.

Hence, $B$ is not an equivalence relation on $R$.

Hence, Statement I is true, Statement II is false.

**53.** *(5)* Since, relation needs to be reflexive the ordered pairs (1, 1), (2, 2), (3, 3) need to the there and (1, 2) is also to be included.

$R_0 = \{(1, 1), (2, 2), (3, 3), (1, 2)\}$ the base relation.

$\because\ A \times A$ contain $3 \times 3 = 9$ ordered pairs, remaining 5 ordered pairs are (2, 1), (1, 3), (3, 1), (2, 3), (3, 2)

we have to add atmost two ordered pairs to $R_0$ such that resulting relation in reflexive, transitive but not symmetric.

Following are the only possibilities.

$R = R_0 \cup \{(1, 3)\}$

or $\ R_0 \cup \{(3, 2)\}$

or $R_0 \cup \{(1, 3), (3, 1)\}$

or $R_0 \cup \{(1, 3), (3, 2)\}$

or $R_0 \cup \{(3, 1), (3, 2)\}$

only five cases.

$\therefore$Number of relations = 5

**54.** *(392)*

(ways to connect one of {1,2,3,---, 98}to 1) $\times$ (99 can have image 1 or 0)

$\downarrow \qquad\qquad \downarrow$

98 $\times$ 2

$\times$ (100 can have image 0 or 1)

$\downarrow$

$\times$ 2

$= 392$

**55.** *(3)* $A = \{1, 2, 3\}$

$R = \{(1, 2), (2, 3)\}$

for reflexive $(1, 1), (2, 2), (3, 3) \in R$

for transitive $(1, 2), (2, 3) \in R$

$\Rightarrow \quad (1, 3) \in R$

Relation which are reflexive, transitive but not symmetric

$R_1 = \{(1, 1)\,(2, 2)\,(3, 3)\,(1, 2)\,(2, 3)\,(1, 3)\}$

$R_2 = \{(1, 1)\,(2, 2)\,(3, 3)\,(1, 2)\,(2, 1)\,(2, 3)\,(1, 3)\}$

$R_3 = \{(1, 1)\,(2, 2)\,(3, 3)\,(1, 2)\,(2, 3)\,(3, 2)\,(1, 3)\}$

**56.** *(24)* $2x + 3y = 23$

$x = 1 \quad y = 7$

$x = 4 \quad y = 5$

$x = 7 \quad y = 3$

$x = 10 \quad y = 1$

| $A$ | $B$ |
|---|---|
| (1, 7) | 1 |
| (4, 5) | 4 |
| (7, 3) | 7 |
| (10, 1) | 10 |

$\therefore$Number of one-one function $= 4! = 24$

**57.** *(66)* $A = \{1, 2, 3, \ldots, 100\}$

$(x, y) \in R$ and $2x = 3y$

Then, $R = \{(3, 2), (6, 4), (9, 6), (12, 8), \ldots, (99, 66)\}$

Number of elements in $R = 33$

$R_1$ be a symmetric relation on $A$ such that $R \subset R_1$

$\therefore R_1 = \{(3, 2), (2, 3), (6, 4), (4, 6), \ldots (99, 66), (66, 99)\}$

So minimum number of elements in $R_1 = 66$

So, $\quad n = 66$

**58.** *(16)* $A = \{1, 2, 3, 4\}$

$R = \{(1, 2), (2, 3), (1, 4)\}$

Now, for making $R$ to be equivalence we have to make it reflexive, symmetric and transitive.
For reflexive add

$$(1, 1)\ (2, 2)\ (3, 3)\ (4, 4)$$

For symmetric add

$$(2, 1)\ (3, 2)\ (4, 1)$$

$R$ becomes
$\{(1, 1), (2, 2), (3, 3), (4, 4), (1, 2), (2, 1), (2, 3), (3, 2), (1, 4), (4, 1)\}$
For transitive
$(1, 2), (2, 3) \Rightarrow (1, 3)$ so add it.
$(3, 1) \rightarrow$ for symmetric
$(4, 1), (1, 2) = (4, 2)$ so add it
$(2, 4) \rightarrow$ for symmetric
$\therefore$ To make $R$ an equivalence, we have to add all possible doublet
$n(R) = 4 \times 4 = 16$

**59.** *(960)* Total number of symmetric relation which are no reflective

$$= 2^{\frac{n^2+n}{2}} - 2^{\frac{n^2-n}{2}}$$

$$= 2^{\frac{16+4}{2}} - 2^{\frac{16-4}{2}} = 2^{10} - 2^6$$

$$= 1024 - 64 = 960$$

**60.** *(5)* $A = \{1, 2, 3, 4\}$
$R = \{((a, b), (c, d)) : 2a + 3b = 4c + 5d\}$
Let $\alpha = 2a + 3b = 4c + 5d$
Now, $2a = \{2, 4, 6, 8\}, 3b = \{3, 6, 9, 12\}$
$4c = \{4, 8, 12, 16\}, 5d = \{5, 10, 15, 20\}$
$\therefore\ 2a + 3b = \{5, 8, 11, 14, 7, 10, 13, 16, 9, 12, 15, 18, 11, 14, 17, 20\}$
$= \{5, 7, 8, 9, 10, 11, 12, 13, 14, 15, 16, 17, 18, 20\}$
and $4c + 5d = \{9, 14, 19, 24, 13, 18, 17, 23, 26, 27, 31, 32, 36, 22, 21, 26\}$
$= \{9, 13, 14, 17, 18, 19, 21, 22, 23, 24, 26, 27, 28, 31, 32, 36\}$
Since, $2a + 3b = 4c + 5d$
$\therefore$ Possible values are $\{9, 13, 14, 17, 18\}$
So, pairs of $\{(a, b), (c, d)\} = 5$
$\therefore$ Total number of elements in $R$ is 5.

**61.** *(7)* Let $A = \{-4, -3, -2, 0, 1, 3, 4\}$
and $R = \{(a, b) \in A \times A, b = |a|$ or $b^2 = a + 1\}$ be a relation on $A$
So, $R = \{(-4, 4), (-3, 3), (3, 3), (4, 4), (0, 0), (3, -2), (0, 1), (1, 1)\}$
This relation $R$ becomes reflexive, if we add only $(-2, -2), (-4, -4), (-3, -3)$ elements and this relation $R$ becomes symmetric, if we add only $(4, -4), (3, -3), (-2, 3), (1, 0)$ elements.
Thus, the required number be 7.

**62.** *(4)* Given, set $A = \{1, 2, 3\}$
For relation $R$ to be reflexive,
$\Rightarrow (1, 1), (2, 2), (3, 3) \in R$
$\therefore R = \{(1, 1), (2, 2), (3, 3), (1, 2), (2, 3), \ldots\}$
Since, $(1, 2), (2, 3) \in R \Rightarrow (1, 3) \in R$ to be transitive
$\therefore R_1 = \{(1, 1), (2, 2), (3, 3), (1, 2), (2, 3), (1, 3)\}$ is a required relation.
Now, $R_2 = \{R_1, (2, 1)\}$ is reflexive, transitive but not symmetric
$R_3 = \{R_1, (3, 2)\}$ and $R_4 = \{R_1, (3, 1), (2, 1)\}$ are total 4 such possible relations

**63.** *(360)* Given,
$A = \{1, 2, 3, 4, 5\}$ and
$B = \{1, 2, 3, 4, 5, 6\}$
and $f : A \rightarrow B$ such that
$f(1) + f(2) + 1 = f(4)$
Possible cases are
1. $f(1) = 1, f(2) = 1, f(4) = 3$
2. $f(1) = 1, f(2) = 2, f(4) = 4$
3. $f(1) = 2, f(2) = 1, f(4) = 4$
4. $f(1) = 2, f(2) = 2, f(4) = 5$
5. $f(1) = 3, f(2) = 1, f(4) = 5$
6. $f(1) = 1, f(2) = 3, f(4) = 5$
7. $f(1) = 1, f(2) = 4, f(4) = 6$
8. $f(1) = 4, f(2) = 1, f(4) = 6$
9. $f(1) = 2, f(2) = 3, f(4) = 6$
10. $f(1) = 3, f(2) = 2, f(4) = 6$

But along with that the remaining elements from set $A$ are 3 and 5, which should also be associated with the elements of set $B$ to form a function.
Thus, total ways will be $10 \times 6 \times 6 = 360$ functions.

**64.** *(180)* Total number of onto function

$$= \frac{5!}{3!\,2!} \times 4! = 240$$

When, $f(a) = 1$
Number of onto function

$$= 4! + \frac{4!}{2!\,2!} \times 3! = 24 + 36 = 60$$

$\therefore$ Required number of onto functions
$= 240 - 60 = 180$

**65.** *(19)* $A = \{0, 3, 4, 6, 7, 8, 9, 10\}$
$R = \{(x, y) : x, y \in A, x - y = 2$ or odd positive integer$\}$
$R = \{(6, 4), (8, 6), (9, 7), (10, 8), (3, 0), (7, 0), (9, 0), (4, 3), (6, 3), (8, 3), (10, 3), (7, 4), (9, 4), (7, 6), (9, 6), (8, 7), (10, 7), (9, 8), (10, 9)\}$
To make it symmetric 19 more elements should be added.

**66.** *(18)* Given,
$A = \{1, 2, 3, 4, \ldots, 10\}$
and $B = \{0, 1, 2, 3, 4\}$
Now,
$2(a - b)^2 + 3(a - b) \in B$
$= (a - b)\{2(a - b) + 3\} \in B$
Since, $(a, b) \in A \times A$
$\therefore a - b = 0 \Rightarrow a = b$
$\Rightarrow$ 10 order pairs
(Since, $A$ has 10 elements)
and $a - b = -2$
$\Rightarrow$ 8 order pairs
$\{(1, 3), (2, 4), (3, 5), (4, 6), (5, 7), (6, 8), (7, 9), (8, 10)\}$
$\therefore$ Total number of elements in the relation $R = 10 + 8 = 18$

**67.** *(9)* Given relation,
$R = \{(a, b), (b, c), (b, d)\}$
and let $A = \{a, b, c, d\}$
For reflexive,
The elements should be $(a, a), (b, b), (c, c)$ and $(d, d)$
For symmetric,
If $(a, b) \in R \Rightarrow (b, a) \in R$
If $(b, c) \in R \Rightarrow (c, b) \in R$
If $(b, d) \in R \Rightarrow (d, b) \in R$
For transitive,
If $(a, b) \in R$ and $(b, c) \in R$
$\Rightarrow (a, c) \in R$ ...(i)
If $(a, c) \in R \Rightarrow (c, a) \in R$
$\therefore$ Total number $= 4 + 4 + 1 = 9$

**68.** *(37)* There are total 16 ordered pairs in $S \times S$. We can write all these ordered pairs as follows.
$A = \{(1, 1)\}$
$B = \{(1, 4), (2, 4), (3, 4), (4, 4), (4, 3), (4, 2), (4, 1)\}$
$C = \{(1, 3), (2, 3), (3, 3), (3, 2), (3, 1)\}$
$D = \{(1, 2), (2, 2), (2, 1)\}$
All elements of set $B$ have image 4 and only element of $A$ has image 1. All elements of set $C$ have image 3 or 4 and all elements of set $D$ have image 2 or 3 or 4.
We will solve this question in two cases.
**Case I** When no element of set $C$ has image 3.
Number of onto functions $= 2$
(When elements of Set $D$ have images 2 or 3).
**Case II** When atleast one element of set $C$ has image 3.

Number of onto functions
$=(2^3-1)(1+2+2)=35$

Total number of function = 37

**69.** *(31)* Given, $3f(c)+2f(a)+f(d)=f(b)$

| Value of $f(c)$ | Value of $f(a)$ | Number of functions |
|---|---|---|
| 0 | 1 | 7 |
| | 2 | 5 |
| | 3 | 3 |
| | 4 | 2 |
| 1 | 0 | 6 |
| | 2 | 2 |
| | 3 | 1 |
| 2 | 0 | 3 |
| | 1 | 1 |
| 3 | 0 | 1 |

Thus, total number of functions is 31.

**70.** *(720)* $f(1)+f(2)=3-f(3)$

$A=\{0,1,2,3,4,5,6,7\}$

$f:A\to A$

So, $f(1)+f(2)+f(3)=3$

$0+1+2=3$ is the only possibility.

So, $f(1)$ can be either 0 or 1 or 2.

Similarly, $f(2)$ and $f(3)$ can be 0, 1 and 2. and $\underbrace{\{0,4,5,6,7\}}\longrightarrow\{3,4,5,6,7\}$

They have 5! choices. And $\underbrace{\{0,1,2\}}$

They have 3! choices.

Number of bijective functions
$=3!\times5!=720$

## Topic 2

**01.** *(b)* $f(x)\,f\left(\frac{1}{x}\right)=f(x)+f\left(\frac{1}{x}\right)$

$\Rightarrow f(x)=1\pm x^2$ [degree 2 is given]

$\Rightarrow f(x)=1+x^2$ [not possible]

Since, range of $f(x)$ is $(-\infty,1)$ [given]

$\Rightarrow f(x)=1-x^2\Rightarrow1-K^2=-2K$

$\Rightarrow K^2-2K-1=0$

$\Rightarrow K_1+K_2=2$

and $K_1K_2=-1$

$\Rightarrow K_1^2+K_2^2=(K_1+K_2)^2-2K_1K_2$
$=4+2=6$

**02.** *(c)* Given, $f(x)=x^2+9$

and $g(x)=\frac{x}{x-9}$

and $a=f(g(10))=f\left(\frac{10}{10-9}\right)$
$=f(10)=109$

and $b=g(f(3))=g(9+9)=g(18)$
$=\frac{18}{9}=2$

Now, equation of ellipse is
$$\frac{x^2}{109}+\frac{y^2}{2}=1$$

$\therefore$ Eccentricity $(e)=\sqrt{1-\frac{2}{109}}=\sqrt{\frac{107}{109}}$

$\Rightarrow e^2=\frac{107}{109}$

and $l=\frac{2(2)}{\sqrt{109}}=\frac{4}{\sqrt{109}}$

$\therefore 8e^2+l^2=\frac{8(107)}{109}+\frac{16}{109}=8$

**03.** *(d)* $f(x)=\frac{4x+3}{6x-4}\;x\neq\frac{2}{3}$

$fof(x)=g(x)$

$\Rightarrow g(x)=fof(x)$
$$=\frac{4\left(\frac{4x+3}{6x-4}\right)+3}{6\left(\frac{4x+3}{6x-4}\right)-4}$$
$$=\frac{34x}{34}=x$$

$g(x)=x\Rightarrow gogog(x)=x$

$gogog(4)=4$

**04.** *(c)* Given functions, $f:A\to B$ and $g:B\to C$ $(A,B,C\subseteq R)$

$\therefore (gof)^{-1}$ exists

$\Rightarrow gof$ is a bijective function.

$\Rightarrow$ '$f$' must be 'one-one' and '$g$' must be 'onto' function.

**05.** *(a)* $f(x)=\begin{cases}x+1, & x\text{ is odd.}\\ x, & x\text{ is even.}\end{cases}$

Given, $g:A\to A$ such that
$g(f(x))=f(x)$

When, $x$ is even, then
$g(x)=x$ ...(i)

When, $x$ is odd, then
$g(x+1)=x+1$ ...(ii)

This implies, $g(x)=x$, $x$ is even.

If $x$ is odd, then $g(x)$ can take any value in set $A$.

So, number of $g(x)=10^5$

**06.** *(c)* $f:(1,\infty)\to R,\quad f(x)=\frac{2x+3}{5x+2}$

$g:(1,\infty)\to R,\quad g(x)=\frac{2-3x}{1-x}$

$fog(x)=f\{g(x)\}$
$$=f\left(\frac{2-3x}{1-x}\right)$$
$$=\frac{2\left(\frac{2-3x}{1-x}\right)+3}{5\left(\frac{2-3x}{1-x}\right)+2}=\frac{7-9x}{12-17x}$$

$fog$ define when $12-17x\neq0\Rightarrow x\neq\frac{12}{17}$

and $fog'(x)>0\Rightarrow fog$ is increasing function

$fog(2)=\frac{7-18}{12-34}=\frac{1}{2}$

$fog(4)=\frac{7-36}{12-68}=\frac{29}{56}$

$\therefore$ Range $=\left[\frac{1}{2},\frac{29}{56}\right]$

$\beta-\alpha=\frac{29}{56}-\frac{1}{2}=\frac{1}{56}$

$\frac{1}{\beta-\alpha}=56$

**07.** *(b)* $f(x)=(2+3a)x^2+\left(\frac{a+2}{a-1}\right)x+b$

$f(x+y)=f(x)+f(y)+1-\frac{2}{7}xy$

$\Rightarrow(2+3a)(x+y)^2+\left(\frac{a+2}{a-1}\right)(x+y)+b$
$=(2+3a)(x^2+y^2)+\left(\frac{a+2}{a-1}\right)(x+y)$
$+2b+1-\frac{2}{7}xy$

$\Rightarrow (2+3a)(2xy)=b+1-\frac{2}{7}xy$

$\left(2+3a+\frac{1}{7}\right)2xy-(b+1)=0$

$\Rightarrow 3a=-\frac{15}{7}$ and $b=-1$

$\therefore a=-\frac{5}{7}$ and $b=-1$

$$f(x)=-\frac{1}{7}x^2+\left(\frac{-\frac{5}{7}+2}{-\frac{5}{7}-1}\right)x-1$$
$$=-\frac{1}{7}x^2-\frac{9}{12}x-1$$
$$=-\frac{1}{7}x^2-\frac{3}{4}x-1$$

$28 \sum_{i=1}^{5} |f(i)|$

$= 28\left[\sum_{i=1}^{5} \frac{1}{7} i^2 + \frac{3}{4} \sum_{i=1}^{5} i + 5\right]$

$= 28\left[\frac{1}{7} \times \frac{5 \times 6 \times 11}{6} + \frac{3}{4} \times \frac{5 \times 6}{2} + 5\right]$

$= 4 \times 5 \times 11 + 7 \times 45 + 140$

$= 220 + 315 + 140 = 675$

**08.** (b) $f(x) = \dfrac{2^x}{2^x + \sqrt{2}}$

$\Rightarrow f(1-x) = \dfrac{2^{1-x}}{2^{1-x} + \sqrt{2}} = \dfrac{2 \cdot 2^{-x}}{2 \cdot 2^{-x} + \sqrt{2}}$

$\Rightarrow f(1-x) = \dfrac{2}{2 + \sqrt{2} \cdot 2^x} = \dfrac{\sqrt{2}}{\sqrt{2} + 2^x}$

Then,

$f(x) + f(1-x) = \dfrac{2^x}{2^x + \sqrt{2}} + \dfrac{\sqrt{2}}{\sqrt{2} + 2^x}$

$= \dfrac{2^x + \sqrt{2}}{2^x + \sqrt{2}} = 1 \Rightarrow \sum_{k=1}^{81} f\left(\frac{k}{82}\right)$

$= f\left(\frac{1}{82}\right) + f\left(\frac{2}{82}\right) + f\left(\frac{3}{82}\right)$

$+ \ldots + f\left(\frac{41}{82}\right) + \ldots + f\left(\frac{80}{82}\right) + f\left(\frac{81}{82}\right)$

$= f\left(\frac{1}{81}\right) + f\left(\frac{81}{82}\right) + f\left(\frac{2}{82}\right) + f\left(\frac{80}{82}\right)$

$+ \ldots. + f\left(\frac{41}{82}\right)$

$= (1 + 1 + \ldots + 1)_{40 \text{ times}} + f\left(\frac{1}{2}\right)$

$= 40 + \dfrac{\sqrt{2}}{\sqrt{2} + \sqrt{2}} = 40 + \dfrac{1}{2}$

$= \dfrac{81}{2}$

**09.** (b) $f : R \to R$

$f(x) = |x - 1|$

and

$g(x) = \begin{cases} e^x & x \ge 0 \quad e^x \ge 1 \quad g \ge 1 \\ x + 1 & x \le 0 \quad x + 1 \le 1 \quad g \le 1 \end{cases}$

$= \begin{cases} -(x-1) & x < 1 \\ x - 1 & x \ge 1 \end{cases}$

$fog = f\{g(x)\} = \begin{cases} -(g-1) & g < 1 \\ g - 1 & g \ge 1 \end{cases}$

$fog = \begin{cases} -(x + 1 - 1) & x < 0 \\ e^x - 1 & x \ge 0 \end{cases}$

$= \begin{cases} -x & x < 0 \\ e^x - 1 & x \ge 0 \end{cases}$

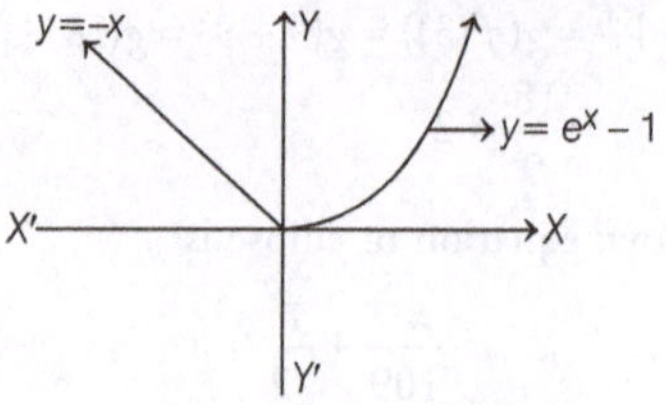


$\because$ It is neither one-one nor onto.

**10.** (b) $g(f(x)) = \begin{cases} f(x), & f(x) \ge 0 \\ e^{f(x)}, & f(x) < 0 \end{cases}$

$g(f(x)) = \begin{cases} \ln x, & x \ge 1 \\ e^{\ln x}, & 0 < x < 1 \\ e^{-x}, & x \le 0 \end{cases}$

$g(f(x)) = \begin{cases} e^{-x}, & x \le 0 \\ x, & 0 < x < 1 \\ \ln x, & x \ge 1 \end{cases}$

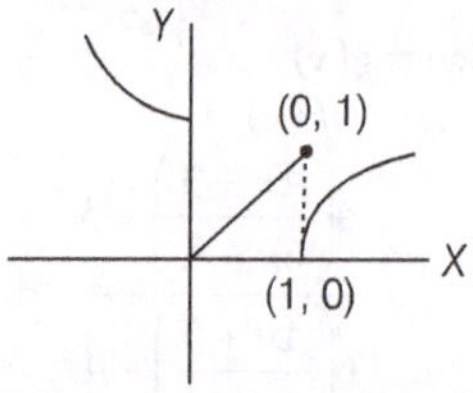


$g(f(x))$ is clearly many-one and into function.

**11.** (c) $f(x) = \begin{cases} 2 + 2x, & x \in [-1, 0) \\ 1 - \dfrac{x}{3}, & x \in [0, 3] \end{cases}$

$g(x) = \begin{cases} x, & x \in (0, 1] \\ -x, & x \in [-3, 0] \end{cases}$

$\therefore f(g(x))$

$= \begin{cases} 2 + 2|g(x)|, & |x| \in [-1, 0) \Rightarrow x \in \phi \\ 1 - \dfrac{|g(x)|}{3}, & 0 \le g(x) \le 3 \end{cases}$

$\Rightarrow f(g(x)) = 1 - \dfrac{x}{3}, x \in [0, 1]$

$= 1 + \dfrac{x}{3}, x \in [-3, 0]$

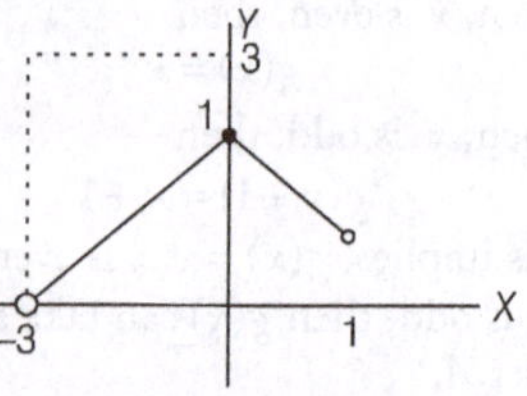


$\therefore$ Range of $fog(x)$ is [0, 1]

**12.** (d) $f(x) = x - 1$ and $g(x) = \dfrac{x^2}{x^2 - 1}$

$fog(x) = f(g(x)) = \dfrac{x^2}{x^2 - 1} - 1 = \dfrac{1}{x^2 - 1}$

$fog(x) = \dfrac{1}{x^2 - 1}$ is an even function,

$x \ne \pm 1$

$\Rightarrow$ It is many one function.

Now, $y = \dfrac{1}{x^2 - 1} \Rightarrow x^2 = \dfrac{1}{y} + 1$

$\dfrac{y+1}{y} \ge 0 \qquad (\because x^2 \ge 0)$

Here, $y > 0$ or $y + 1 \ge 0 \Rightarrow y \ge -1$

$y \in (0, \infty)$

Range $\ne$ codomain $\Rightarrow fog(x)$ is into function.

> **Concept Enhancer**
>
> A function $f$ is even, if the following equations hold for all $x$ and $-x$ in the domain of $f : f(x) = f(-x)$.

**13.** (a) $f(x) = x^3 + x - 5$

$f[g(x)] = x, \forall x \in R$

$g(x) = f^{-1}(x) \Rightarrow g(f(x)) = x$

On differentiating both sides w.r.t. $x$, we get

$g'[f(x)] f'(x) = 1$

$g'[f(x)] = \dfrac{1}{f'(x)} \qquad \ldots$(i)

$g'(63)$ can be determined when

$f(x) = 63$

$\Rightarrow \quad x^3 + x - 68 = 0 \therefore x = 4$

$\Rightarrow \quad f'(x) = 3x^2 + 1$

$\Rightarrow \quad f'(4) = 49$

$\therefore \quad g'(63) = \dfrac{1}{49}$

**14.** (b) $f(x) = \dfrac{5x + 3}{6x - \alpha}$

Now, $fof(x) = f\left(\dfrac{5x+3}{6x-\alpha}\right)$

$= \dfrac{5\left(\dfrac{5x+3}{6x-\alpha}\right) + 3}{6\left(\dfrac{5x+3}{6x-\alpha}\right) - \alpha}$

$= \dfrac{5(5x+3) + 3(6x - \alpha)}{6(5x+3) - \alpha(6x - \alpha)}$

$\because$ Given, $fof(x) = x$

$\Rightarrow \quad 25x + 15 + 18x - 3\alpha$

$= 30x^2 + 18x - 6\alpha x^2 + \alpha^2 x$

$\Rightarrow x^2(30 - 6\alpha) + x(\alpha^2 - 25)$

$+ 3\alpha - 15 = 0$

On comparing coefficients, we get

$30 - 6\alpha = 0$

$\Rightarrow \quad \alpha = 5$

**15.** (d) Given,
$f(n+1)=f(n)+f(1), \forall n \in N$
$\Rightarrow f(n+1)-f(n)=f(1)$
It is an AP with common difference $=f(1)$
Also, general term
$\Rightarrow T_n = f(1)+(n-1)f(1)=nf(1)$
Clearly, $f(n)$ is one-one.
For $fog$ to be one-one, $g$ must be one-one.
For $f$ to be onto, $f(n)$ should take all the values of natural numbers.
As, $f(x)$ is increasing, $f(1)=1$
$\Rightarrow \quad f(n)=n$
If $g$ is many-one, then $fog$ is many one.
So, if $g$ is onto, then $fog$ is one-one is incorrect.

**16.** (a)

**key idea** (i) If a function is strictly increasing or strictly decreasing in entire domain, the function is one-one.
(ii) If range of a function is equal to co-domain then function is onto.
(iii) If range of a function is completely a subset of co-domain, then function is into.

$$f[g(x)]=2g(x)-1$$
$$=2\times\left(\frac{x-\frac{1}{2}}{x-1}\right)-1=\frac{2x-1}{x-1}-1=\frac{x}{x-1}$$
$$\therefore \quad f[g(x)]=1+\frac{1}{x-1}$$

Now, draw the graph of $1+\dfrac{1}{x-1}$,

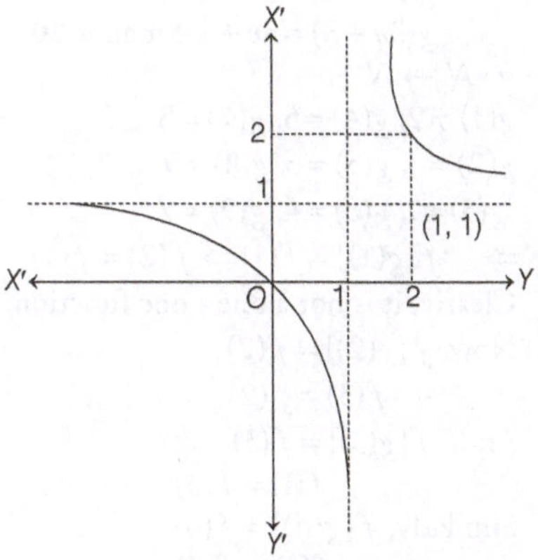


$\because$ Any horizontal line does not cut the graph at more than one points, so it is one-one and here, co-domain and range are not equal, so it is into.
Hence, the required function is one-one into.

**17.** (c) Given function,
$$f(x)=\frac{8^{2x}-8^{-2x}}{8^{2x}+8^{-2x}}, x\in(-1,1)$$
$$=\frac{8^{4x}-1}{8^{4x}+1}=y \text{ (let)}$$
On applying componendo and dividendo law, we get
$$8^{4x}=\frac{1+y}{1-y}$$
On applying logarithm having base '8' both sides, we get
$$4x=\log_8\left(\frac{1+y}{1-y}\right)$$
$$\Rightarrow \quad x=\frac{1}{4}\log_8\left(\frac{1+y}{1-y}\right)$$
$$=\frac{1}{4}(\log_8 e)\log_e\left(\frac{1+y}{1-y}\right)$$
$$f^{-1}(x)=(1/4)(\log_8 e)\log_e\left(\frac{1+x}{1-x}\right).$$

**18.** (c) Now, for $S=[0,4]$
$g(S)=\{x\in R: f(x)\in S=[0,4]\}$
$=\{x\in R: x^2\in[0,4]\}$
$=\{x\in R: x\in[-2,2]\}$
$\Rightarrow g(S)=[-2,2]$
So, $f(g(S))=[0,4]=S$
Now,
$f(S)=\{x^2: x\in S=[0,4]\}=[0,16]$
and $g(f(S))=\{x\in R: f(x)\in f(S)$
$=[0,16]\}$
$=\{x\in R: f(x)\in[0,16]\}$
$=\{x\in R: x^2\in[0,16]\}$
$=\{x\in R: x\in[-4,4]\}$
$=[-4,4]$
From above, it is clear that $g(f(S))\neq g(S)$.

**19.** (c) We have, $f(x)=\dfrac{x}{1+x^2}$
$$\therefore \quad f\left(\frac{1}{x}\right)=\frac{\frac{1}{x}}{1+\frac{1}{x^2}}=\frac{x}{1+x^2}=f(x)$$
$$\therefore \quad f\left(\frac{1}{2}\right)=f(2)$$
$$\text{or} \quad f\left(\frac{1}{3}\right)=f(3) \text{ and so on.}$$
So, $f(x)$ is many-one function.
Again, let $\quad y=f(x)$
$\Rightarrow \quad y+x^2y=x$
$\Rightarrow \quad yx^2-x+y=0$
As, $\quad x\in R$ and $D\geq 0$
$\therefore \quad (-1)^2-4(y)(y)\geq 0$
$\Rightarrow \quad 1-4y^2\geq 0 \Rightarrow y\in\left[\frac{-1}{2},\frac{1}{2}\right]$
$\therefore$ Range $=$ Codomain $=[-1/2, 1/2]$
So, $f(x)$ is surjective.
Hence, $f(x)$ is surjective but not injective.

**20.** (d) Given, $f(x)=\dfrac{2x+3}{2x+1}$
and $g(x)=\dfrac{|x|+1}{2x+5}$
For $fog(x)$; $g(x)\neq -\dfrac{1}{2}$
$$\Rightarrow \quad \frac{|x|+1}{2x+5}\neq -\frac{1}{2} \quad \text{...(i)}$$
**Case I** When $x\geq 0$
$$\frac{x+1}{2x+5}=\frac{-1}{2}$$
$\Rightarrow \quad 2x+2=-2x-5$
$\Rightarrow \quad 4x=-7$
$\Rightarrow \quad x=-\dfrac{7}{4}$ but $x\geq 0$
$\Rightarrow \quad x\in\phi$
**Case II** When $x<0$
$$\Rightarrow \quad \frac{-x+1}{2x+5}=-\frac{1}{2}$$
$\Rightarrow \quad -2x+2=-2x-5$
$\Rightarrow \quad 2=-5$ (not possible)
Hence, domain of
$f(g(x))=$ domain of $g(x)$
i.e. $\quad D_{fog(x)}=R-\left\{-\dfrac{5}{2}\right\}$

**21.** (a) $f(x)=\sin^{-1}\left(\log_{3x}\left(\dfrac{6+2\log_3 x}{-5x}\right)\right)$
From domain, $x>0$ ...(i)
$$\frac{6+2\log_3 x}{-5x}>0$$
$\Rightarrow 6+2\log_3 x<0 \quad (\because x>0)$
$\log_3 x<-3$
$\Rightarrow \quad x<(3)^{-3}$ ...(ii)
From Eqs. (i) and (ii) $x\in\left(0,\dfrac{1}{27}\right)$
for this $\log_{3x}\left(\dfrac{6+2\log_3 x}{-5x}\right)\in(-1,1)$
$\therefore$ Domain of $f(x)$ is $\left(0,\dfrac{1}{27}\right)$

Now, $g(x)=x-[x]$, for $x\in\left(0,\frac{1}{27}\right)$,

$$[x]=0$$

$\therefore \quad g(x)=x$

So, range of $g(x)$ is $\left(0,\frac{1}{27}\right)$

Comparing with $(\alpha,\beta)$, $\alpha=0,\beta=\frac{1}{27}$

$\therefore \alpha^2+\frac{5}{\beta}=0+5\times 27=135$

**22.** *(c)* Given, $f(x)=\dfrac{(\tan 1^\circ)x+\log_e(123)}{x\log_e(1234)-(\tan 1^\circ)}$

Let us consider a similar function of $f(x)$,

$\therefore f(x)=\dfrac{Ax+B}{Cx-A}$

Now, $f(f(x))=\dfrac{A\left(\frac{Ax+B}{Cx-A}\right)+B}{C\left(\frac{Ax+B}{Cx-A}\right)-A}$

$$=\frac{A^2x+AB+BCx-AB}{ACx+BC-ACx+A^2}$$

$$=\frac{x(A^2+BC)}{(BC+A^2)}=x$$

$\therefore \quad f(f(x))=x$

Similarly, $f\left(f\left(\frac{4}{x}\right)\right)=\frac{4}{x}$

Apply AM $\geq$ GM

$$\frac{\left(x+\frac{4}{x}\right)}{2}\geq\sqrt{x\cdot\frac{4}{x}}=2 \quad [\because x>0]$$

$\Rightarrow \quad x+\frac{4}{x}=4$

$\Rightarrow \quad f(f(x))+f\left(f\left(\frac{4}{x}\right)\right)\geq 4$

Hence, the least value is 4.

**23.** *(b)*

> **key idea** Sum of Geometric progression is equal to
> $\dfrac{a(r^n-1)}{r-1}$ (where '$a$' is first term and '$r$' is common ratio and also $|r|>1$ and '$n$' is number of terms

Given functional equation is

$5f(x+y)=f(x)\cdot f(y);\ \forall\ x,y\in R \quad \ldots$(i)

On putting $x=y=0$ in Eq. (i), we get

$$5f(0+0)=f(0)\cdot f(0)$$

$\Rightarrow \quad f(0)(f(0)-5)=0$

$\Rightarrow \quad f(0)=5$

($\because f(0)\neq 0$ because $f:R\to(0,\infty)$ i.e. 0 is excluded from co-domain of $f$ and we know that range of $f\leq$ co-domain of $f$)

Again putting $y=1$ in Eq. (i), we get

$$5f(x+1)=f(x)\cdot f(1) \quad \ldots\text{(ii)}$$

$\Rightarrow \quad \dfrac{f(x+1)}{f(x)}=\left[\dfrac{f(1)}{5}\right]$

$\Rightarrow \quad \dfrac{f(1)}{f(0)}\times\dfrac{f(2)}{f(1)}\times\dfrac{f(3)}{f(2)}=\left(\dfrac{f(1)}{5}\right)^3$

$\Rightarrow \quad f(1)=20$

From Eq. (ii),

$$f(x+1)=4f(x)$$

Now, $\sum_{n=0}^{5}f(n)=f(0)+f(1)+f(2)+f(3)+f(4)+f(5)$

$=5+5\times 4+5\times 4^2+5\times 4^3+5\times 4^4+5\times 4^5$

$=5\left[\dfrac{1(4^6-1)}{(4-1)}\right]=\dfrac{5\times(4^6-1)}{3}$

$=6825$

**24.** *(b)* Given, $f(x)=\dfrac{x-1}{x+1}$

$$f(f(x))=f\left(\frac{x-1}{x+1}\right)=\frac{\frac{x-1}{x+1}-1}{\frac{x-1}{x+1}+1}$$

$\therefore \quad f^2(x)=-\dfrac{1}{x}$

Then, $f^3(x)=f(f^2(x))=f\left(-\dfrac{1}{x}\right)$

$\Rightarrow \quad f^3(x)=\dfrac{-\frac{1}{x}-1}{-\frac{1}{x}+1}$

$\therefore \quad f^3(x)=\dfrac{x+1}{1-x}$

So, $\quad f^6(x)=f^3(f^3(x))$

$$=\frac{\frac{x+1}{1-x}+1}{1-\frac{x+1}{1-x}}=-\frac{1}{x}$$

and $\quad f^7(x)=f(f^6(x))$

$$=f\left(-\frac{1}{x}\right)=\frac{x+1}{1-x}$$

Then, $f^6(6)+f^7(7)$

$$=-\frac{1}{6}+\left(\frac{8}{-6}\right)=-\frac{3}{2}$$

$\therefore \quad f^6(6)+f^7(7)=-\dfrac{3}{2}$

**25.** *(a)* $f(x)=\log_e(x^2+1)-e^{-x}+1$

$$f'(x)=\frac{2x}{x^2+1}+e^{-x}$$

$f'(x)>0, \forall x\in R$

$\therefore f(x)$ is increasing function.

$$f\left(g\left(\frac{(\alpha-1)^2}{3}\right)\right)>f\left(g\left(\alpha-\frac{5}{3}\right)\right)$$

$\Rightarrow \quad g\left(\dfrac{(\alpha-1)^2}{3}\right)>g\left(\alpha-\dfrac{5}{3}\right)$

$$g(x)=\frac{1-2e^{2x}}{e^x}$$

$$g'(x)=-e^{-x}-2e^x<0$$

$\therefore g(x)$ is an decreasing function.

$\therefore \quad$ If $g\left(\dfrac{(\alpha-1)^2}{3}\right)>g\left(\alpha-\dfrac{5}{3}\right)$, then

$$\frac{(\alpha-1)^2}{3}<\alpha-\frac{5}{3}$$

$\Rightarrow \quad (\alpha-1)^2<3\alpha-5$

$\Rightarrow \quad \alpha^2-5\alpha+6<0\Rightarrow\alpha\in(2,3)$

**26.** *(a)* Given, $g(3n+1)=3n+2$

$g(3n+2)=3n+3$

$g(3n+3)=3n+1$, for all $n\geq 0$

$g:N\to N$

$g(1)=2,\ g(4)=5,\ g(7)=8$

$g(2)=3,\ g(5)=6,\ g(8)=9$

$g(3)=1,\ g(6)=4,\ g(9)=7$

$\Rightarrow \quad f[g(1)]=f(1)\Rightarrow f(2)=f(1)$

Clearly, it is not a one - one function.

Now, $f[g(2)]=f(2)$

$\quad f(3)=f(2)$

And, $f[g(3)]=f(3)$

$\quad f(1)=f(3)$

Similarly, $f[g(4)]=f(4)$

$\quad f(5)=f(4)$

And, so on

$\quad f(1)=f(2)=f(3)$

$$f(4)=f(5)=f(6)$$

Now, there can be a possibility such that

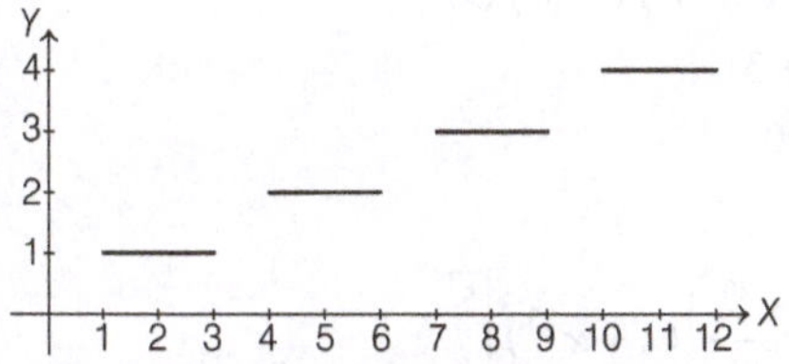


So, $f(x)$ can be onto function.

When $f(1)=f(2)=f(3)=1$

$f(4)=f(5)=f(6)=2$

and so on.

**27.** *(b)* $f(x)=(x-1)^2+1$,

as $x\ge 1$

When, $f(x)=f^{-1}(x)$

$\Rightarrow\quad f(x)=x$

$\Rightarrow (x-1)^2+1=x$

$\Rightarrow\quad (x-1)^2-(x-1)=0$

$\Rightarrow\quad x=1, 2$

$\therefore\quad \{x: f(x)=f^{-1}(x)\}=\{1, 2\}$

Also, let $f(x)=y$

$\therefore\quad y=(x-1)^2+1$

$\Rightarrow\quad (x-1)=\pm\sqrt{y-1}$

$\therefore\quad x=1\pm\sqrt{y-1}$

$\because\quad x\ge 1$

Neglecting $1-\sqrt{y-1}$

$\Rightarrow\quad x=1+\sqrt{y-1}$

$\therefore\quad f^{-1}(x)=1+\sqrt{x-1}$

So, both statements are correct and Statement II is correct explanation of Statement I.

**28.** *(a)*

> **key idea** For finding inverse of any function, first isolate $x$ in terms of $y$ and then by replacing $x$ to $y$ and $y$ to $x$. We get the inverse of any function.

The given function is

$f: N\to Y$ defined as

$f(x)=4x+3$,

where $Y=\{y\in N: y=4x+3$ for some $x\in N\}$

Now, $f(x)=4x+3\Rightarrow f'(x)=4>0$

$\Rightarrow$ $f$ is a strictly increasing function.

$\Rightarrow$ $f$ is one-one.

Also, here $y=\{7, 8, 9, \dots, \infty\}$

and range of $f=\{7, 8, 9, \dots, \infty\}$

$\because$ Range of $f=Y=$ Codomain of $f$

$\Rightarrow$ $f$ is onto. $\Rightarrow$ $f$ is invertible.

Now, let $y\in Y$ such that $f(x)=y$

$\Rightarrow\quad y=4x+3\Rightarrow x=\dfrac{y-3}{4}$

$\therefore$ Inverse of $f$ is $f^{-1}(y)=\dfrac{y-3}{4}$.

**29.** *(1024)* $f\{f(x)\}=\dfrac{2f(x)}{\sqrt{1+9f^2(x)}}$

$$=\frac{4x}{\sqrt{1+9x^2+9\cdot 2^2x^2}}$$

$$f(f(f(x)))=\frac{2^3x/\sqrt{1+9x^2}}{\sqrt{1+9(1+2^2)\cdot\dfrac{2^2x^2}{1+9x^2}}}$$

$$=\frac{2^3x}{\sqrt{1+9x^2(1+2^2+2^4)}}$$

$\therefore$ By observation

$9\alpha=9(1+2^2+2^4+\dots+2^{18})$

$\Rightarrow\quad \alpha=1+2^2+2^4+\dots 2^{18}$ (GP)

$$=\frac{1((2^2)^{10}-1)}{2^2-1}=\frac{2^{20}-1}{3}$$

$\Rightarrow\quad 3\alpha+1=2^{20}$

$\Rightarrow\quad \sqrt{3\alpha+1}=2^{10}=1024$

**30.** *(3125)* Given, $f^1(x)=\dfrac{3x+2}{2x+3}$,

where $x\in R-\left\{\dfrac{-3}{2}\right\}$

Also given, $f^n(x)=f^1 of^{n-1}(x)$

For $n=2$, $f^2(x)=f^1(of^1(x))$

$=f^1(f^1(x))$

$$=f^1\left(\frac{3x+2}{2x+3}\right)$$

$$=\frac{3\left(\dfrac{3x+2}{2x+3}\right)+2}{2\left(\dfrac{3x+2}{2x+3}\right)+3}$$

$$f^2(x)=\frac{13x+12}{12x+13}$$

Similarly, for $n=3$

$$f^3(x)=\frac{63x+62}{62x+63}$$

for $n=5$

$$\therefore\quad f^5(x)=\frac{1563x+1562}{1562x+1563}$$

$$=\frac{ax+b}{bx+a}$$

Hence, $a+b=1563+1562=3125$

**31.** *(2039)* $(fog)^{-1}x=\left(\dfrac{x-7}{2}\right)^{1/3}=y$ (say)

$(fog)^{-1}x=y$ and $\left(\dfrac{x-7}{2}\right)^{1/3}=y$

$x=fog(y)$ and $x=2y^3+7$

$fog(y)=2y^3+7$

$fog(x)=2x^3+7$ ... (i)

Now, $f(x)=ax-3$ and $g(x)=x^b+c$

$(fog)x=a(x^b+c)-3$

$=ax^b+ac-3$ ... (ii)

On comparing Eqs. (i) and (ii), we get

$a=2, b=3, ac-3=7$

$2c=10\Rightarrow c=5$

$gof(x)=(2x-3)^3+5$

Hence,

$fog(ac)+gof(b)=fog(10)+gof(3)$

$=(2\times 10^3+7)+[(6-3)^3+5]$

$=2039$

**32.** *(25)* Let the roots of quadratic polynomial be $t$ and $s$.

Then, $f(x)=(x-t)(x-s)$

$f(0)=p$

$\Rightarrow\quad st=p$ [given]

and $f(1)=1/3$

$\Rightarrow\quad (1-t)(1-s)=\dfrac{1}{3}$

Now, assume that $t$ is the common root of $f(x)=0$ and $fofofo(x)=0$

Take, $fofofof(x)=0$

$\Rightarrow\quad fofof(0)=0$

$\Rightarrow\quad fof(p)=0$

So, $f(p)$ is either $t$ or $s$

$(p-t)(p-s)=t$

$(ts-t)(ts-s)=t$

$\Rightarrow (s-1)(t-1)s=1$ $(\because t\ne 0)$

So, $s=3$

$$(1-t)(1-3)=\frac{1}{3}$$

$$\Rightarrow \quad t=\frac{7}{6}$$

$$f(x)=\left(x-\frac{7}{6}\right)(x-3)$$

$$f(-3)=\left(-3-\frac{7}{6}\right)(-3-3)=25$$

**33.** *(190)* From given $f(n)=\begin{cases} 2n, & \text{if } n=1, 2, 3, 4, 5 \\ 2n-11, & \text{if } n=6, 7, \ldots, 10 \end{cases}$

Put $n=1, 2, 3, 4, \ldots, 10$

$f(1)=2, f(2)=4, f(3)=6,$
$f(4)=8, \ldots f(6)=1, f(7)=3,$
$f(8)=5 \ldots f(10)=9.$

Take, $fog(n)=\begin{cases} (n+1), & \text{if } n \text{ is odd} \\ (n-1), & \text{if } n \text{ is even} \end{cases}$

Put $n=1, 2, 3, 4, \ldots, 10$

$f(g(1))=2, f(g(2))=1, f(g(3))=4, f(g(4))=3, f(g(5))=6,$
$f(g(10))=9$

As, $f(g(10))=9$, and $f(10)=9$, then $g(10)=10$

Similarly, $g(1)=1, g(2)=6, g(3)=2, g(4)=7, g(5)=3$

Put the values in the expression,

$g(10)\,(g(1)+g(2)+g(3)+g(4)+g(5))$

$\Rightarrow 10(1+6+2+7+3)$

$\Rightarrow \quad 10\times(19)=190$

**34.** *(2)* From given,

$$f(x)=\left[2\left(1-\frac{x^{25}}{2}\right)(2+x^{25})\right]^{\frac{1}{50}}$$

$$=(4-x^{50})^{\frac{1}{50}}$$

$$\text{Let } f(f(x))=\left(4-\left((4-x^{50})^{\frac{1}{50}}\right)^{50}\right)^{\frac{1}{50}}$$

$$=x$$

Now, $g(x)=f(f(f(x)))+f(f(x))$
$\qquad =f(x)+x$

Put $x=1$ in above equation, we get

$g(1)=f(1)+1$

$[g(1)]=[3^{1/50}+1]=2$

# C-14

JEE Main Chapterwise Topicwise Mathematics

In JEE Mains, there is usually 0-1 question per paper is visible from this chapter. In past 3 years, 22 conceptual and 9 numerical questions of easy to moderate difficulty level were asked from this chapter.

**For detailed analysis scan QR code**

# INVERSE TRIGONOMETRIC FUNCTIONS

## Topic 1 Domain and Range of Inverse Trigonometric Functions, Principal Value of Inverse Trigonometric Functions

### Level 1

**Imp.** Principle Value of Inverse Trigonometric Function

**01** Considering the principal values of the inverse trigonometric functions, $\sin^{-1}\left(\frac{\sqrt{3}}{2}x+\frac{1}{2}\sqrt{1-x^2}\right)$, $-\frac{1}{2}<x<\frac{1}{\sqrt{2}}$, is equal to *[2025, 4 April shift - I]*

(a) $\frac{5\pi}{6}-\sin^{-1}x$ (b) $\frac{-5\pi}{6}-\sin^{-1}x$

(c) $\frac{\pi}{6}+\sin^{-1}x$ (d) $\frac{\pi}{4}+\sin^{-1}x$

**02** Let $[x]$ denote the greatest integer less than or equal to $x$. Then, the domain of $f(x)=\sec^{-1}(2[x]+1)$ is *[2025, 28 Jan shift - II]*

(a) $(-\infty,-1]\cup[1,\infty)$ (b) $(-\infty,-1]\cup[0,\infty)$

(c) $(-\infty,\infty)-\{0\}$ (d) $(-\infty,\infty)$

**03** If $a=\sin^{-1}(\sin(5))$ and $b=\cos^{-1}(\cos(5))$, then $a^2+b^2$ is equal to *[2024, 31 Jan shift - II]*

(a) $4\pi^2-25$ (b) $4\pi^2-20\pi+50$

(c) 25 (d) $8\pi^2-40\pi+50$

**04** The range of $f(x)=4\sin^{-1}\left(\frac{x^2}{x^2+1}\right)$ is *[2023, 13 April Shift-II]*

(a) $[0,\pi)$ (b) $[0,\pi]$ (c) $[0,2\pi)$ (d) $[0,2\pi]$

**05** The domain of the function $f(x)=\sin^{-1}\left(\frac{x^2-3x+2}{x^2+2x+7}\right)$ is *[2022, 29 July Shift-II]*

(a) $[1,\infty)$ (b) $[-1,2]$

(c) $[-1,\infty)$ (d) $(-\infty,2]$

**06** Considering the principal values of the inverse trigonometric functions, the sum of all the solutions of the equation $\cos^{-1}(x)-2\sin^{-1}(x)=\cos^{-1}(2x)$ is equal to *[2022, 28 July Shift-I]*

(a) 0 (b) 1 (c) $\frac{1}{2}$ (d) $-\frac{1}{2}$

**07** $\cos^{-1}(\cos(-5))+\sin^{-1}(\sin(6))-\tan^{-1}(\tan(12))$ is equal to (The inverse trigonometric functions take the principal values) *[2021, 01 Sep Shift-II]*

(a) $3\pi-11$ (b) $4\pi-9$

(c) $4\pi-11$ (d) $3\pi+1$

**08** The largest interval lying in $\left(\frac{-\pi}{2},\frac{\pi}{2}\right)$ for which the function, $f(x)=4^{-x^2}+\cos^{-1}\left(\frac{x}{2}-1\right)+\log(\cos x)$ is defined, is *(AIEEE 2007)*

(a) $\left[-\frac{\pi}{4},\frac{\pi}{2}\right)$ (b) $\left[0,\frac{\pi}{2}\right)$

(c) $[0,\pi]$ (d) $\left(-\frac{\pi}{2},\frac{\pi}{2}\right)$

**09** The trigonometric equation $\sin^{-1}x=2\sin^{-1}a$, has a solution for *[AIEEE 2003]*

(a) $\frac{1}{2}<|a|<\frac{1}{\sqrt{2}}$ (b) all real values of $a$

(c) $|a|\le\frac{1}{\sqrt{2}}$ (d) $|a|\ge\frac{1}{\sqrt{2}}$

## Level 2

**10** If the domain of the function $f(x)=\sin^{-1}\left(\frac{x-1}{2x+3}\right)$ is $R-(\alpha, \beta)$, then $12\alpha\beta$ is equal to *[2024, 9 April Shift-I]*
(a) 36 (b) 32 (c) 40 (d) 24

**Imp.** Domain of Inverse Trigonometric Functions

**11** If the domain of the function
$\sin^{-1}\left(\frac{3x-22}{2x-19}\right)+\log_e\left(\frac{3x^2-8x+5}{x^2-3x-10}\right)$ is $(\alpha, \beta]$, then $3\alpha+10\beta$ is equal to *[2024, 4 April Shift-I]*
(a) 100 (b) 98 (c) 97 (d) 95

**12** For $\alpha, \beta, \gamma = \theta$, if $\sin^{-1}\alpha+\sin^{-1}\beta+\sin^{-1}\gamma=\pi$ and $(\alpha+\beta+\gamma)(\alpha-\gamma+\beta)=3\alpha\beta$, then $\gamma$ equals *[2024, 31 Jan Shift-I]*
(a) $\frac{\sqrt{3}-1}{2\sqrt{2}}$ (b) $\sqrt{3}$ (c) $\frac{\sqrt{3}}{2}$ (d) $\frac{1}{\sqrt{2}}$

**13** If the domain of the function
$f(x)=\log_e\left(\frac{2x+3}{4x^2+x-3}\right)+\cos^{-1}\left(\frac{2x-1}{x+2}\right)$ is $(\alpha, \beta]$, then the value of $5\beta-4\alpha$ is equal to *[2024, 30 Jan Shift-II]*
(a) 11 (b) 9 (c) 12 (d) 10

**14** If the domain of the function
$f(x)=\cos^{-1}\left(\frac{2-|x|}{4}\right)+\{\log_e(3-x)\}^{-1}$ is $[-\alpha, \beta)-[\gamma]$, then $\alpha+\beta+\gamma$ is equal to *[2024, 30 Jan Shift-I]*
(a) 8 (b) 12 (c) 9 (d) 11

**15** Considering only the principal values of inverse trigonometric function, the number of positive real values of $x$ satisfying $\tan^{-1}(x)+\tan^{-1}(2x)=\frac{\pi}{4}$ is *[2024, 27 Jan Shift-II]*
(a) 2 (b) 0
(c) More than 2 (d) 1

**16** Let $S=\left\{x\in R: 0<x<1 \text{ and } 2\tan^{-1}\left(\frac{1-x}{1+x}\right)=\cos^{-1}\left(\frac{1-x^2}{1+x^2}\right)\right\}$.
If $n(S)$ denotes the number of elements in $S$, then *[2023, 1 Feb Shift-II]*
(a) $n(S)=1$ and the element in $S$ is less than $1/2$.
(b) $n(S)=2$ and only one element in $S$ is less than $1/2$.
(c) $n(S)=0$.
(d) $n(S)=1$ and the elements in $S$ is more than $1/2$.

**17** Considering only the principal values of the inverse trigonometric functions, the domain of the function $f(x)=\cos^{-1}\left(\frac{x^2-4x+2}{x^2+3}\right)$ is *[2022, 28 July Shift-I]*
(a) $\left(-\infty, \frac{1}{4}\right]$ (b) $\left[-\frac{1}{4}, \infty\right)$
(c) $\left(-\frac{1}{3}, \infty\right)$ (d) $\left(-\infty, \frac{1}{3}\right]$

**18** If $0<x<\frac{1}{\sqrt{2}}$ and $\frac{\sin^{-1}x}{\alpha}=\frac{\cos^{-1}x}{\beta}$, then the value of $\sin\left(\frac{2\pi\alpha}{\alpha+\beta}\right)$ is *[2022, 26 July Shift-II]*
(a) $4\sqrt{(1-x^2)(1-2x^2)}$ (b) $4x\sqrt{(1-x^2)(1-2x^2)}$
(c) $2x\sqrt{(1-x^2)(1-4x^2)}$ (d) $4\sqrt{(1-x^2)(1-4x^2)}$

**19** If the inverse trigonometric functions take principal values, then
$\cos^{-1}\left(\frac{3}{10}\cos\left(\tan^{-1}\frac{4}{3}\right)+\frac{2}{5}\sin\left(\tan^{-1}\left(\frac{4}{3}\right)\right)\right)$ is equal to *[2022, 26 June Shift-II]*
(a) 0 (b) $\pi/4$
(c) $\pi/3$ (d) $\pi/6$

**20** A possible value of $\tan\left(\frac{1}{4}\sin^{-1}\frac{\sqrt{63}}{8}\right)$ is *[2021, 24 Feb Shift-II]*
(a) $\frac{1}{\sqrt{7}}$ (b) $2\sqrt{2}-1$ (c) $\sqrt{7}-1$ (d) $\frac{1}{2\sqrt{2}}$

**21** The domain of the function $f(x)=\sin^{-1}\left(\frac{|x|+5}{x^2+1}\right)$ is $(-\infty, -a]\cup[a, \infty)$. Then a is equal to *[2020, 2 Sep Shift-I]*
(a) $\frac{\sqrt{17}}{2}+1$ (b) $\frac{\sqrt{17}-1}{2}$
(c) $\frac{1+\sqrt{17}}{2}$ (d) $\frac{\sqrt{17}}{2}$

## Numerical Value Type Questions

**22** If $y=\cos\left(\frac{\pi}{3}+\cos^{-1}\frac{x}{2}\right)$, then $(x-y)^2+3y^2$ is equal to ......... *[2025, 2 April Shift-II]*

**23** If the domain of the function $f(x)=\sec^{-1}\left(\frac{2x}{5x+3}\right)$ is $[\alpha, \beta)\cup(\gamma, \delta]$, then $|3\alpha+10(\beta+\gamma)+21\delta|$ is equal to ........... *[2023, 10 April Shift-II]*

**24** For $k \in R$, let the solution of the equation $\cos(\sin^{-1}(x \cot(\tan^{-1}(\cos(\sin^{-1} x))))) = k, 0 < |x| < \frac{1}{\sqrt{2}}$ be $\alpha$ and $\beta$, where the inverse trigonometric functions take only principal values. If the solutions of the equation $x^2 - bx - 5 = 0$ are $\frac{1}{\alpha^2} + \frac{1}{\beta^2}$ and $\frac{\alpha}{\beta}$, then $\frac{b}{k^2}$ is equal to ......... . *[2022, 27 July Shift-I]*

## Topic 2 Properties of Inverse Trigonometric Functions, Sum upto $n$-terms & Infinite Terms of Inverse Series

### Level 1

**Imp.** Sum of Infinite Terms of Inverse Series

**01** The sum of the infinite series $\cot^{-1}\left(\frac{7}{4}\right) + \cot^{-1}\left(\frac{19}{4}\right) + \cot^{-1}\left(\frac{39}{4}\right) + \cot^{-1}\left(\frac{67}{4}\right) + \ldots$ is *[2025, 4 April Shift-II]*

(a) $\frac{\pi}{2} + \cot^{-1}\left(\frac{1}{2}\right)$ (b) $\frac{\pi}{2} - \tan^{-1}\left(\frac{1}{2}\right)$

(c) $\frac{\pi}{2} - \cot^{-1}\left(\frac{1}{2}\right)$ (d) $\frac{\pi}{2} + \tan^{-1}\left(\frac{1}{2}\right)$

**02** $\cos\left(\sin^{-1}\frac{3}{5} + \sin^{-1}\frac{5}{13} + \sin^{-1}\frac{33}{65}\right)$ is equal to *[2025, 28 Jan Shift-I]*

(a) 1 (b) 32/65 (c) 0 (d) 33/65

**03** If $\alpha > \beta > \gamma > 0$, then the expression

$$\cot^{-1}\left\{\beta + \frac{(1+\beta^2)}{(\alpha-\beta)}\right\} + \cot^{-1}\left\{\gamma + \frac{(1+\gamma^2)}{(\beta-\gamma)}\right\} + \cot^{-1}\left\{\alpha + \frac{(1+\alpha^2)}{(\gamma-\alpha)}\right\}$$ is equal to *[2025, 24 Jan Shift-II]*

(a) 0 (b) $3\pi$

(c) $\frac{\pi}{2} - (\alpha + \beta + \gamma)$ (d) $\pi$

**04** If $\frac{\pi}{2} \le x \le \frac{3\pi}{4}$, then $\cos^{-1}\left(\frac{12}{13}\cos x + \frac{5}{13}\sin x\right)$ is equal to *[2025, 23 Jan Shift-I]*

(a) $x + \tan^{-1}\frac{5}{12}$ (b) $x - \tan^{-1}\frac{5}{12}$

(c) $x - \tan^{-1}\frac{4}{3}$ (d) $x + \tan^{-1}\frac{4}{5}$

**05** $\tan^{-1}\left(\frac{1+\sqrt{3}}{3+\sqrt{3}}\right) + \sec^{-1}\left(\sqrt{\frac{8+4\sqrt{3}}{6+3\sqrt{3}}}\right)$ is equal to *[2023, 24 Jan Shift-I]*

(a) $\pi/4$ (b) $\pi/6$ (c) $\pi/2$ (d) $\pi/3$

**06** If $(\sin^{-1} x)^2 - (\cos^{-1} x)^2 = a, 0 < x < 1, a \ne 0$, then the value of $2x^2 - 1$ is *[2021, 27 Aug Shift-I]*

(a) $\cos\left(\frac{4a}{\pi}\right)$ (b) $\sin\left(\frac{2a}{\pi}\right)$

(c) $\cos\left(\frac{2a}{\pi}\right)$ (d) $\sin\left(\frac{4a}{\pi}\right)$

**07** $2\pi - \left(\sin^{-1}\frac{4}{5} + \sin^{-1}\frac{5}{13} + \sin^{-1}\frac{16}{65}\right)$ is equal to *[2020, 3 Sep Shift-I]*

(a) $\pi/2$ (b) $5\pi/4$ (c) $3\pi/2$ (d) $7\pi/4$

**08** The value of $\sin^{-1}\left(\frac{12}{13}\right) - \sin^{-1}\left(\frac{3}{5}\right)$ is equal to *[2019, 12 April Shift-I]*

(a) $\pi - \sin^{-1}\left(\frac{63}{65}\right)$ (b) $\frac{\pi}{2} - \sin^{-1}\left(\frac{56}{65}\right)$

(c) $\frac{\pi}{2} - \cos^{-1}\left(\frac{9}{65}\right)$ (d) $\pi - \cos^{-1}\left(\frac{33}{65}\right)$

**09** For $x \in \left(0, \frac{1}{4}\right)$, if the derivative of $\tan^{-1}\left(\frac{6x\sqrt{x}}{1-9x^3}\right)$ is $\sqrt{x} \cdot g(x)$, then $g(x)$ equals *[JEE Main 2017]*

(a) $\frac{9}{1+9x^3}$ (b) $\frac{3x\sqrt{x}}{1-9x^3}$ (c) $\frac{3x}{1-9x^3}$ (d) $\frac{3}{1+9x^3}$

**10** Let $\tan^{-1} y = \tan^{-1} x + \tan^{-1}\left(\frac{2x}{1-x^2}\right)$, where $|x| < \frac{1}{\sqrt{3}}$. Then, a value of $y$ is *[JEE Main 2015]*

(a) $\frac{3x - x^3}{1 - 3x^2}$ (b) $\frac{3x + x^3}{1 - 3x^2}$ (c) $\frac{3x - x^3}{1 + 3x^2}$ (d) $\frac{3x + x^3}{1 + 3x^2}$

**11** If $f(x) = 2\tan^{-1} x + \sin^{-1}\left(\frac{2x}{1+x^2}\right), x > 1$, then $f(5)$ is equal to *[JEE Main 2015]*

(a) $\frac{\pi}{2}$ (b) $\pi$

(c) $4\tan^{-1}(5)$ (d) $\tan^{-1}\left(\frac{65}{156}\right)$

**12** If $x, y, z$ are in AP and $\tan^{-1} x$, $\tan^{-1} y$ and $\tan^{-1} z$ are also in AP, then *(JEE Main 2013)*

(a) $2x = 3y = 6z$ (b) $6x = 3y = 2z$

(c) $6x = 4y = 3z$ (d) $x = y = z$

**13** The value of $\cot\left(\text{cosec}^{-1}\frac{5}{3} + \tan^{-1}\frac{2}{3}\right)$ is *[AIEEE 2008]*

(a) $\frac{5}{17}$ (b) $\frac{6}{17}$ (c) $\frac{3}{17}$ (d) $\frac{4}{17}$

**14** If $\sin^{-1}\left(\frac{x}{5}\right)+\operatorname{cosec}^{-1}\left(\frac{5}{4}\right)=\frac{\pi}{2}$, then the value of $x$ is *[AIEEE 2007]*

(a) 1 (b) 3
(c) 4 (d) 5

**15** $\tan^{-1}\left(\frac{1}{4}\right)+\tan^{-1}\left(\frac{2}{9}\right)$ is equal to *[AIEEE 2002]*

(a) $\frac{1}{2}\cos^{-1}\left(\frac{3}{5}\right)$ (b) $\frac{1}{2}\sin^{-1}\left(\frac{3}{5}\right)$
(c) $\frac{1}{2}\tan^{-1}\left(\frac{3}{5}\right)$ (d) $\tan^{-1}\left(\frac{1}{2}\right)$

## Level 2

**16** The value of $\cot^{-1}\frac{\sqrt{1+\tan^2(2)}-1}{\tan(2)}-\cot^{-1}\left(\frac{\sqrt{1+\tan^2\left(\frac{1}{2}\right)}+1}{\tan\left(\frac{1}{2}\right)}\right)$ is equal to *[2025, 8 April Shift-II]*

(a) $\pi+\frac{5}{2}$ (b) $\pi-\frac{3}{2}$ (c) $\pi-\frac{5}{4}$ (d) $\pi+\frac{3}{2}$

**17** Using the principal values of the inverse trigonometric functions, the sum of the maximum and the minimum values of $16[(\sec^{-1}x)^2+(\operatorname{cosec}^{-1}x)^2]$ is *[2025, 22 Jan Shift-I]*

(a) $24\pi^2$ (b) $22\pi^2$ (c) $18\pi^2$ (d) $31\pi^2$

**Imp.** Minimum & Maximum Value of Function

**18** Given that the inverse trigonometric function assumes principal values only. Let $x, y$ be any two real numbers in $[-1, 1]$ such that $\cos^{-1}x-\sin^{-1}y=\alpha, \frac{-\pi}{2}\le\alpha\le\pi$

Then, the minimum value of $x^2+y^2+2xy\sin\alpha$ is *[2024, 4 April Shift-II]*

(a) 0 (b) $\frac{-1}{2}$ (c) $\frac{1}{2}$ (d) $-1$

**19** If $\sin^{-1}\frac{\alpha}{17}+\cos^{-1}\frac{4}{5}-\tan^{-1}\frac{77}{36}=0, 0<\alpha<13$, then $\sin^{-1}(\sin\alpha)+\cos^{-1}(\cos\alpha)$ is equal to *[2023, 31 Jan Shift-I]*

(a) 0 (b) $\pi$ (c) 16 (d) $16-5\pi$

**20** Let $a_1=1, a_2, a_3, a_4, \ldots\ldots$ be consecutive natural numbers. Then, $\tan^{-1}\left(\frac{1}{1+a_1a_2}\right)+\tan^{-1}\left(\frac{1}{1+a_2a_3}\right)+\ldots+\tan^{-1}\left(\frac{1}{1+a_{2021}a_{2022}}\right)$ is equal to *[2023, 30 Jan Shift-II]*

(a) $\frac{\pi}{4}-\cot^{-1}(2022)$
(b) $\tan^{-1}(2022)-\frac{\pi}{4}$
(c) $\frac{\pi}{4}-\tan^{-1}(2022)$
(d) $\cot^{-1}(2022)-\frac{\pi}{4}$

**21** $\tan\left(2\tan^{-1}\frac{1}{5}+\sec^{-1}\frac{\sqrt{5}}{2}+2\tan^{-1}\frac{1}{8}\right)$ is equal to *[2022, 26 July Shift-I]*

(a) 1 (b) 2
(c) 1/4 (d) 5/4

**22** The value of $\cot\left(\sum_{n=1}^{50}\tan^{-1}\left(\frac{1}{1+n+n^2}\right)\right)$ is *[2022, 27 June Shift-II]*

(a) $\frac{26}{25}$ (b) $\frac{25}{26}$
(c) $\frac{50}{51}$ (d) $\frac{52}{51}$

**23** $\sin^{-1}\left(\sin\frac{2\pi}{3}\right)+\cos^{-1}\left(\cos\frac{7\pi}{6}\right)+\tan^{-1}\left(\tan\frac{3\pi}{4}\right)$ is equal to *[2022, 27 June Shift-I]*

(a) $\frac{11\pi}{12}$ (b) $\frac{17\pi}{12}$
(c) $\frac{31\pi}{12}$ (d) $-\frac{3\pi}{4}$

**24** Let $M$ and $m$ respectively be the maximum and minimum values of the function $f(x)=\tan^{-1}(\sin x+\cos x)$ in $\left[0,\frac{\pi}{2}\right]$, then the value of $\tan(M-m)$ is equal to *[2021, 27 Aug Shift-II]*

(a) $2+\sqrt{3}$ (b) $2-\sqrt{3}$
(c) $3+2\sqrt{2}$ (d) $3-2\sqrt{2}$

**25** $A$ 10 inches long pencil $AB$ with mid-point $C$ and a small eraser $P$ are placed on the horizontal top of a table such that $PC = \sqrt{5}$ inches and $\angle PCB = \tan^{-1}(2)$. The acute angle through which the pencil must be rotated about $C$ so that the perpendicular distance between eraser and pencil becomes exactly 1 inch is **[2021, 26 Aug. Shift-II]**

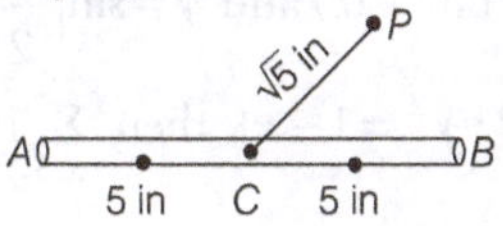


(a) $\tan^{-1}\left(\frac{3}{4}\right)$ (b) $\tan^{-1}(1)$
(c) $\tan^{-1}\left(\frac{4}{3}\right)$ (d) $\tan^{-1}\left(\frac{1}{2}\right)$

**26** The value of
$\tan\left(2\tan^{-1}\left(\frac{3}{5}\right)+\sin^{-1}\left(\frac{5}{13}\right)\right)$ is equal to **[2021, 20 July Shift-II]**

(a) $\frac{-181}{69}$ (b) $\frac{220}{21}$
(c) $\frac{-291}{76}$ (d) $\frac{151}{63}$

**27** The number of solutions of the equation
$\sin^{-1}\left[x^2+\frac{1}{3}\right]+\cos^{-1}\left[x^2-\frac{2}{3}\right]=x^2$, for $x \in [-1, 1]$,
and $[x]$ denotes the greatest integer less than or equal to $x$, is **[2021, 17 March Shift-II]**

(a) 2 (b) 0
(c) 4 (d) infinite

**28** If $\cot^{-1}(\alpha)=\cot^{-1}2+\cot^{-1}8+\cot^{-1}18$ $+\cot^{-1}32+\dots$ upto 100 terms, then $\alpha$ is **[2021, 17 March Shift-I]**

(a) 1.01 (b) 1.00
(c) 1.02 (d) 1.03

**29** The sum of possible values of $x$ for
$\tan^{-1}(x+1)+\cot^{-1}\left(\frac{1}{x-1}\right)=\tan^{-1}\left(\frac{8}{31}\right)$ is **[2021, 17 March Shift-I]**

(a) $\frac{-32}{4}$ (b) $-\frac{31}{4}$
(c) $-\frac{30}{4}$ (d) $-\frac{33}{4}$

**30** Let $S_k = \sum_{r=1}^{k}\tan^{-1}\left(\frac{6^r}{2^{2r+1}+3^{2r+1}}\right)$. Then, $\lim_{k\to\infty} S_k$ is equal to **[2021, 16 March Shift-I]**

(a) $\tan^{-1}\left(\frac{3}{2}\right)$ (b) $\frac{\pi}{2}$
(c) $\cot^{-1}\left(\frac{3}{2}\right)$ (d) $\tan^{-1}(3)$

**31** $\operatorname{cosec}\left[2\cot^{-1}(5)+\cos^{-1}\left(\frac{4}{5}\right)\right]$ is equal to **[2021, 25 Feb. Shift-II]**

(a) 56/33 (b) 65/33
(c) 65/56 (d) 75/56

**32** If $\cos^{-1}\left(\frac{2}{3x}\right)+\cos^{-1}\left(\frac{3}{4x}\right)=\frac{\pi}{2}, x>\frac{3}{4}$, then $x$ is equal to **[2019, 9 Jan Shift-I]**

(a) $\frac{\sqrt{145}}{10}$ (b) $\frac{\sqrt{146}}{12}$ (c) $\frac{\sqrt{145}}{12}$ (d) $\frac{\sqrt{145}}{11}$

**33** $\cot^{-1}(\sqrt{\cos\alpha})-\tan^{-1}(\sqrt{\cos\alpha})=x$, then $\sin x$ is equal to **[AIEEE 2002]**

(a) $\tan^2\left(\frac{\alpha}{2}\right)$ (b) $\cot^2\left(\frac{\alpha}{2}\right)$
(c) $\tan\alpha$ (d) $\cot\left(\frac{\alpha}{2}\right)$

## Level 3

**34** Let $S$ be the set of all solutions of the equation
$\cos^{-1}(2x)-2\cos^{-1}(\sqrt{1-x^2})=\pi, x\in\left[-\frac{1}{2},\frac{1}{2}\right]$.
Then, $\sum_{x\in S} 2\sin^{-1}(x^2-1)$ is equal to **[2023, 1 Feb Shift-I]**

(a) $\pi-2\sin^{-1}\left(\frac{\sqrt{3}}{4}\right)$ (b) $\frac{-2\pi}{3}$
(c) 0 (d) $\pi-\sin^{-1}\left(\frac{\sqrt{3}}{4}\right)$

**35** The value of $\cot\left(\sum_{n=1}^{19}\cot^{-1}\left(1+\sum_{p=1}^{n}2p\right)\right)$ is **[2019, 10 Jan Shift-II]**

(a) 23/22 (b) 21/19 (c) 19/21 (d) 22/23

## Numerical Value Type Questions

**Imp.** Roots of Quadratic Equation

**36** If for some $\alpha, \beta : \alpha \le \beta, \alpha+\beta=8$ and $\sec^2(\tan^{-1}\alpha)+\operatorname{cosec}^2(\cot^{-1}\beta)=36$, then $\alpha^2+\beta$ is ...... . **[2025, 24 Jan Shift-I]**

**37** Let $S=\{x:\cos^{-1}x=\pi+\sin^{-1}x+\sin^{-1}(2x+1)\}$. Then, $\sum_{x\in S}(2x-1)^2$ is equal to ........... . **[2025, 29 Jan Shift-I]**

**38** Let the inverse trigonometric function take principal values. The number of real solution of the equation $2\sin^{-1}x+3\cos^{-1}x=\frac{2\pi}{5}$, is ......... . **[2024, 9 April Shift-II]**

**39** For $n \in N$, if $\cot^{-1} 3 + \cot^{-1} 4 + \cot^{-1} 5 + \cot^{-1} n = \frac{\pi}{4}$, then $n$ is equal to .......... .

[2024, 6 April Shift-I]

**40** For $x \in (-1, 1]$, the number of solutions of the equation $\sin^{-1} x = 2\tan^{-1} x$ is equal to ........... .

[2023, 13 April Shift-II]

**41** If $S = \left\{ x \in R : \sin^{-1}\left(\frac{x+1}{\sqrt{x^2+2x+2}}\right) - \sin^{-1}\left(\frac{x}{\sqrt{x^2+1}}\right) = \frac{\pi}{4} \right\}$, then $\sum_{x \in S}\left(\sin\left((x^2+x+5)\frac{\pi}{2}\right) - \cos((x^2+x+5)\pi)\right)$ is equal to ......... .

[2023, 13 April Shift-I]

**42** If the sum of all the solutions of $\tan^{-1}\left(\frac{2x}{1-x^2}\right) + \cot^{-1}\left(\frac{1-x^2}{2x}\right) = \frac{\pi}{3}, -1 < x < 1, x \neq 0$, is $\alpha - \frac{4}{\sqrt{3}}$, then $\alpha$ is equal to ........

[2023, 25 Jan Shift-I]

**43** Let $x = \sin(2\tan^{-1}\alpha)$ and $y = \sin\left(\frac{1}{2}\tan^{-1}\frac{4}{3}\right)$. If $S = \{\alpha \in R : y^2 = 1 - x\}$, then $\sum_{\alpha \in S} 16\alpha^3$ is equal to ............ .

[2022, 25 July Shift-II]

**44** $50\tan\left(3\tan^{-1}\left(\frac{1}{2}\right) + 2\cos^{-1}\left(\frac{1}{\sqrt{5}}\right)\right) + 4\sqrt{2}\tan\left(\frac{1}{2}\tan^{-1}(2\sqrt{2})\right)$ is equal to ...... .

[2022, 29 June Shift-I]

## Key Index

### Topic 1

| | | | | | | | | | |
|---|---|---|---|---|---|---|---|---|---|
| 1 (c) | 2 (d) | 3 (d) | 4 (c) | 5 (c) | 6 (a) | 7 (c) | 8 (b) | 9 (c) | 10 (b) |
| 11 (c) | 12 (c) | 13 (c) | 14 (d) | 15 (d) | 16 (a) | 17 (b) | 18 (b) | 19 (c) | 20 (a) |
| 21 (c) | 22 (3) | 23 (24) | 24 (12) | | | | | | |

### Topic 2

| | | | | | | | | | |
|---|---|---|---|---|---|---|---|---|---|
| 1 (b) | 2 (c) | 3 (d) | 4 (b) | 5 (d) | 6 (b) | 7 (c) | 8 (b) | 9 (a) | 10 (a) |
| 11 (b) | 12 (d) | 13 (b) | 14 (b) | 15 (d) | 16 (c) | 17 (b) | 18 (a) | 19 (b) | 20 (b) |
| 21 (b) | 22 (a) | 23 (a) | 24 (d) | 25 (a) | 26 (b) | 27 (b) | 28 (a) | 29 (b) | 30 (c) |
| 31 (c) | 32 (c) | 33 (a) | 34 (b) | 35 (b) | 36 (14) | 37 (5) | 38 (0) | 39 (47) | 40 (2) |
| 41 (4) | 42 (2) | 43 (130) | 44 (29) | | | | | | |

# Detailed **Explanations**

## Topic 1

**01.** *(c)* $y = \sin^{-1}\left(\frac{\sqrt{3}}{2}x + \frac{1}{2}\sqrt{1-x^2}\right)$

$\frac{-1}{2} < x < \frac{1}{\sqrt{2}}$

Let $\theta = \sin^{-1} x,\ \theta \in \left(\frac{-\pi}{6}, \frac{\pi}{4}\right)$

$\Rightarrow x = \sin\theta \Rightarrow$ We get

$y = \sin^{-1}\left(\frac{\sqrt{3}}{2}\sin\theta + \frac{1}{2}\cos\theta\right)$

$= \sin^{-1}\left(\sin\left(\theta + \frac{\pi}{6}\right)\right)$

$\because\ \theta + \frac{\pi}{6} \in \left(0, \frac{\sqrt{5}\pi}{12}\right)$

$y = \theta + \frac{\pi}{6} = \sin^{-1} x + \frac{\pi}{6}$

**02.** *(d)* Given, $f(x) = \sec^{-1}(2[x]+1)$

$2[x]+1 \geq 1$ or $2[x]+1 \leq -1$

$\Rightarrow\ [x] \geq 0$ or $[x] \leq -1$

$\Rightarrow\ x \in [0, \infty)$ or $x \in (-\infty, 0]$

$\Rightarrow\ x \in (-\infty, \infty)$

**03.** *(d)* We have,

$a = \sin^{-1}(\sin 5) = 5 - 2\pi$

and $b = \cos^{-1}(\cos 5) = 2\pi - 5$

So, $a^2 + b^2 = (5-2\pi)^2 + (2\pi-5)^2$

$= 50 + 8\pi^2 - 40\pi$

**04.** *(c)* Given that

$f(x) = 4\sin^{-1}\left(\frac{x^2}{x^2+1}\right)$

$\frac{x^2}{x^2+1} = \frac{x^2+1-1}{x^2+1} = 1 - \frac{1}{x^2+1}$

So, $0 \leq \frac{x^2}{x^2+1} < 1$

Thus,

$\sin^{-1}(0) \leq \sin^{-1}\left(\frac{x^2}{x^2+1}\right) < \sin^{-1}(1)$

$\Rightarrow\ 0 \leq f(x) < 2\pi$

Thus, range of $f(x) = [0, 2\pi)$

**05.** *(c)* Consider

$f(x) = \sin^{-1}\left(\frac{x^2-3x+2}{x^2+2x+7}\right)$

where, $-1 \leq \frac{x^2-3x+2}{x^2+2x+7} \leq 1$

$\Rightarrow\ -x^2 - 2x - 7 \leq x^2 - 3x + 2 \leq x^2 + 2x + 7$

$\Rightarrow\ 0 \leq 2x^2 - x + 9$ and $5x + 5 \geq 0$

$\Rightarrow\ 2x^2 - x + 9 \geq 0,\ x \in R$

and $x + 1 \geq 0$

$\therefore$ Domain of $f(x) = [-1, \infty)$

**06.** *(a)* We have,

$\cos^{-1} x - 2\sin^{-1} x = \cos^{-1} 2x$

$\Rightarrow\ \frac{\pi}{2} - 3\sin^{-1} x = \frac{\pi}{2} - \sin^{-1} 2x$

$\Rightarrow\ 3\sin^{-1} x = \sin^{-1} 2x$

$\Rightarrow\ \sin^{-1}(3x - 4x^3) = \sin^{-1} 2x$

$\Rightarrow\ 3x - 4x^3 = 2x$

$\Rightarrow\ x(1-4x^2) = 0 \Rightarrow x = 0, \frac{-1}{2}, \frac{1}{2}$

$\therefore$ Required sum $= 0 + \frac{1}{2} + \left(\frac{-1}{2}\right) = 0$

**07.** *(c)* $\cos^{-1}(\cos(-5)) + \sin^{-1}(\sin(6)) - \tan^{-1}(\tan(12))$

$= 2\pi - 5 + (-2\pi + 6) - (12 - 4\pi)$

$= 4\pi - 11$

**08.** *(b)* $f(x)$ is defined if

$-1 \leq \frac{x}{2} - 1 \leq 1$ and $\cos x > 0$

or $0 \leq x \leq 4$ and $-\frac{\pi}{2} < x < \frac{\pi}{2}$

$\therefore\ 0 \leq x < \frac{\pi}{2}.$

**09.** *(c)*

**key idea** The range of $\sin^{-1} x$ is $\left[-\frac{\pi}{2}, \frac{\pi}{2}\right]$.

Given that, $\sin^{-1} x = 2\sin^{-1} a$

$\because\ -\frac{\pi}{2} \leq \sin^{-1} x \leq \frac{\pi}{2}$

$\Rightarrow\ -\frac{\pi}{2} \leq 2\sin^{-1} a \leq \frac{\pi}{2}$

$\Rightarrow\ -\frac{\pi}{4} \leq \sin^{-1} a \leq \frac{\pi}{4}$

$\Rightarrow\ \sin\left(-\frac{\pi}{4}\right) \leq a \leq \sin\frac{\pi}{4}$

$\Rightarrow\ -\frac{1}{\sqrt{2}} \leq a \leq \frac{1}{\sqrt{2}}$

$\Rightarrow\ |a| \leq \frac{1}{\sqrt{2}}$

**10.** *(b)* Given, $f(x) = \sin^{-1}\left(\frac{x-1}{2x+3}\right)$

For $f(x)$ to be defined

$2x + 3 \neq 0$ and $\left|\frac{x-1}{2x+3}\right| \leq 1$

$\Rightarrow x \neq -\frac{3}{2}$ and $|x-1| \leq |2x+3|$

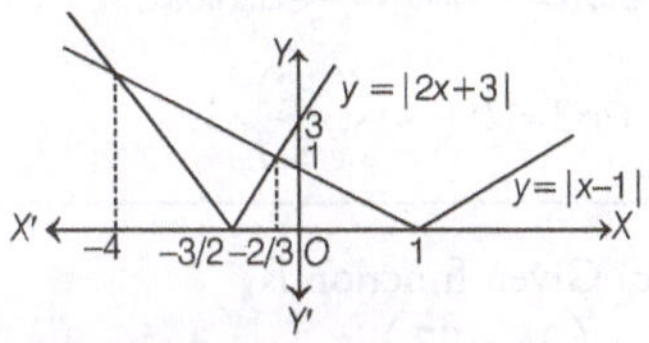


From above graph i.e. graph of $y = |2x+3|$ and $y = |x-1|$, we observe that, for $|2x+3| \geq |x-1|$

$\Rightarrow x \in (-\infty, -4] \cup \left[-\frac{2}{3}, \infty\right)$ i.e.

$R - \left(-4, -\frac{2}{3}\right)$

So, $\alpha = -4$ and $\beta = -\frac{2}{3}$

Hence, $12\alpha\beta = 32$

**Alternate Solution**

Given, $f(x) = \sin^{-1}\left(\frac{x-1}{2x+3}\right)$

For $f(x)$ to be defined

$2x+3 \neq 0$ and $-1 \leq \frac{x-1}{2x+3} \leq 1$

$\Rightarrow x \neq -\frac{3}{2}$ and $\begin{cases} \frac{x-1}{2x+3} \geq -1 \\ \frac{x-1}{2x+3} \leq 1 \end{cases}$

$\Rightarrow \begin{cases} \frac{x-1+2x+3}{2x+3} \geq 0 \\ \frac{x-1-2x-3}{2x+3} \leq 0 \end{cases}$

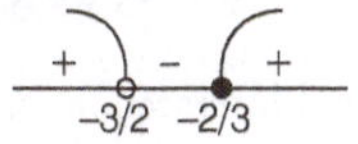


$\Rightarrow \begin{cases} \frac{3x+2}{2x+3} \geq 0 \\ \frac{-x-4}{2x+3} \leq 0 \end{cases} \Rightarrow \begin{cases} \frac{3x+2}{2x+3} \geq 0 \\ \frac{x+4}{2x+3} \geq 0 \end{cases}$

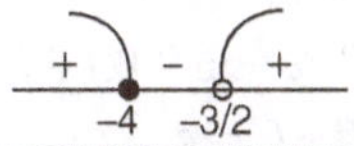

$$\Rightarrow \begin{cases} x\in\left(-\infty,-\frac{3}{2}\right)\cup\left[-\frac{2}{3},\infty\right) \\ x\in(-\infty,-4]\cup\left(-\frac{3}{2},\infty\right) \end{cases}$$

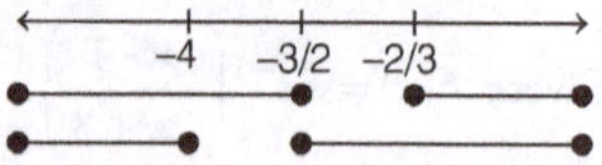


$\Rightarrow x\in(-\infty,-4]\cup\left[-\frac{2}{3},\infty\right)$

i.e. $x\in R-\left(-4,-\frac{2}{3}\right)$

So, $\alpha=-4$ and $\beta=\frac{-2}{3}$ and hence,

$12\alpha\beta=12\times(-4)\times\left(\frac{-2}{3}\right)=32$

**11.** *(c)* Given function is

$$\sin^{-1}\left(\frac{3x-22}{2x-19}\right)+\log_e\left(\frac{3x^2-8x+5}{x^2-3x-10}\right)$$

and it is defined if

$$-1\le\frac{3x-22}{2x-19}\le 1 \text{ and } \frac{3x^2-8x+5}{x^2-3x-10}>0$$

$$\Rightarrow \begin{cases} \frac{3x-22-2x+19}{2x-19}\le 0 \\ \frac{3x-22+2x-19}{2x-19}\ge 0 \end{cases}$$

and $\frac{(3x-5)(x-1)}{(x-5)(x+2)}>0$

$$\Rightarrow \begin{cases} \frac{x-3}{2x-19}\le 0 \\ \frac{5x-41}{2x-19}\ge 0 \end{cases}$$

and $x\in(-\infty,-2)\cup\left(1,\frac{5}{3}\right)\cup(5,\infty)$

$$\Rightarrow \begin{cases} x\in\left[3,\frac{19}{2}\right) \\ x\in\left(-\infty,\frac{41}{5}\right]\cup\left(\frac{19}{2},\infty\right) \end{cases}$$

and $x\in(-\infty,-2)\cup\left(1,\frac{5}{3}\right)\cup(5,\infty)$

Thus, final intersection is, $x\in\left(5,\frac{41}{5}\right]$

So, $\alpha=5$ and $\beta=41/5$

Hence, $3\alpha+10\beta=97$

**12.** *(c)* Let $\sin^{-1}\alpha=A, \sin^{-1}\beta=B$ and $\sin^{-1}\gamma=C$

$\because \quad \sin^{-1}\alpha+\sin^{-1}\beta+\sin^{-1}\gamma=\pi$

$\Rightarrow \quad A+B+C=\pi$

Given, $(\alpha+\beta+\gamma)(\alpha-\gamma+\beta)=3\alpha\beta$

$$(\alpha+\beta)^2-\gamma^2=3\alpha\beta$$

$$\alpha^2+\beta^2-\gamma^2=\alpha\beta$$

$$\frac{\alpha^2+\beta^2-\gamma^2}{2\alpha\beta}=\frac{1}{2}$$

$\therefore \quad \cos C=\frac{1}{2}$

$\Rightarrow \quad \sqrt{1-\sin^2 C}=\frac{1}{2}$

$\Rightarrow \quad \sqrt{1-\gamma^2}=\frac{1}{2}$

$\Rightarrow \quad \gamma=\frac{\sqrt{3}}{2}$

• Concept **Enhancer**

$\cos A=\frac{b^2+c^2-a^2}{2bc}$

$\cos B=\frac{a^2+c^2-b^2}{2ac}$

$\cos C=\frac{a^2+b^2-c^2}{2ab}$

**13.** *(c)* $f(x)=\log_e\left(\frac{2x+3}{4x^2+x-3}\right)+\cos^{-1}\left(\frac{2x-1}{x+2}\right)$

Domain of $f(x)$

$$\frac{2x+3}{4x^2+x-3}>0 \text{ and } -1\le\frac{2x-1}{x+2}\le 1$$

$$\frac{2x+3}{(4x-3)(x+1)}>0 \text{ and } \frac{2x-1}{x+2}+1\ge 0$$

and $\frac{2x-1}{x+2}-1\le 0$

| | | |
|---|---|---|
| $\frac{2x+3}{(4x-3)(x+1)}>0$ 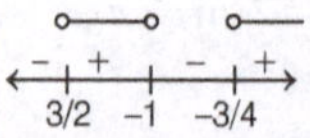  | $\frac{3x+1}{x+2}\ge 0$ 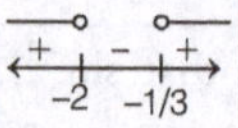  | $\frac{x-3}{x+2}\le 0$ 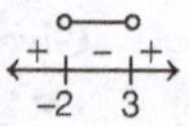  |

Common of all solutions,

we get $x\in\left(\frac{3}{4},3\right]$

$\therefore \quad \alpha=\frac{3}{4}, \beta=3$

$\therefore 5\beta-4\alpha=15-3=12$

**14.** *(d)* We have,

$$y=\cos^{-1}\left(\frac{2-|x|}{4}\right)+\{\log_e(3-x)\}^{-1}$$

$\therefore \quad -1\le\left(\frac{2-|x|}{4}\right)\le 1$

$\Rightarrow \quad -4\le 2-|x|\le 4$

$\Rightarrow \quad -6\le -|x|\le 2$

$\Rightarrow \quad -2\le |x|\le 6$

$\Rightarrow \quad |x|\le 6$

$\Rightarrow \quad x\in[-6,6]$ ... (i)

Now, $\quad 3-x\ne 1$

and $\quad x\ne 2$ ... (ii)

and $\quad 3-x>0$

$\quad x<3$ ... (iii)

From Eqs. (i) , (ii) and (iii) we get

$x\in[-6,3)-[2]$

$\alpha=6, \beta=3, \gamma=2$

$\therefore \quad \alpha+\beta+\gamma=11$

**15.** *(d)* Given, $\tan^{-1}x+\tan^{-1}(2x)=\frac{\pi}{4}$

$\Rightarrow \tan^{-1}\left(\frac{3x}{1-2x^2}\right)=\frac{\pi}{4}$

$\Rightarrow \quad \frac{3x}{1-2x^2}=1 \Rightarrow 3x=1-2x^2$

$\Rightarrow \quad 2x^2+3x-1=0$

$\Rightarrow \quad x=\frac{-3\pm\sqrt{9+8}}{4}$

$\Rightarrow \quad x=\frac{-3\pm\sqrt{17}}{4}$

$\Rightarrow \quad x=\frac{-3-\sqrt{17}}{4}<0$

or $\quad x=\frac{-3+\sqrt{17}}{4}>0$

Hence, total number of positive real values of $x=1$.

**16.** *(a)* We have,

$$2\tan^{-1}\left(\frac{1-x}{1+x}\right)=\cos^{-1}\left(\frac{1-x^2}{1+x^2}\right)$$

Puting $x=\tan\theta$. Also $0<x<1$

$\Rightarrow 0<\theta<\frac{\pi}{4}$

Now,

$$2\tan^{-1}\left(\frac{1-\tan\theta}{1+\tan\theta}\right)=\cos^{-1}\left(\frac{1-\tan^2\theta}{1+\tan^2\theta}\right)$$

$\Rightarrow \quad 2\tan^{-1}\tan\left(\frac{\pi}{4}-\theta\right)=\cos^{-1}\cos 2\theta$

$\Rightarrow \quad 2\left(\frac{\pi}{4}-\theta\right)=2\theta \Rightarrow \frac{\pi}{4}-\theta=\theta$

$\Rightarrow \quad \theta=\frac{\pi}{8} \Rightarrow \tan^{-1}x=\frac{\pi}{8}$

$\Rightarrow \quad x=\tan\frac{\pi}{8}=\sqrt{2}-1\approx 0.414$

$\therefore \quad x < \frac{1}{2}$

So, $n(S) = 1$ and the element in $S$ is less than 1/2.

**17.** *(b)* Here, $f(x) = \cos^{-1}\left(\frac{x^2 - 4x + 2}{x^2 + 3}\right)$

For domain, $-1 \le \frac{x^2 - 4x + 2}{x^2 + 3} \le 1$

Case I $x^2 - 4x + 2 \ge -x^2 - 3$

$\Rightarrow 2x^2 - 4x + 5 \ge 0 \Rightarrow x \in (-\infty, \infty)$ ...(i)

Case II $x^2 - 4x + 2 \le x^2 + 3$

$\Rightarrow -4x + 2 \le 3 \Rightarrow x \ge \frac{-1}{4}$ ...(ii)

From Eqs. (i) and (ii), we get

$x \in \left[\frac{-1}{4}, \infty\right)$

So, $D_f = \left[\frac{-1}{4}, \infty\right)$.

**18.** *(b)*

**key idea** Use inverse trigonometric properties

(i) $\sin^{-1} x + \cos^{-1} x = \frac{\pi}{2}$

(ii) $2 \sin^{-1} x = \sin^{-1} (2x\sqrt{1 - x^2})$

Let $\frac{\sin^{-1} x}{\alpha} = \frac{\cos^{-1} x}{\beta} = k$

$\sin^{-1} x = \alpha k$ and $\cos^{-1} x = k\beta$

As we know that, $\sin^{-1} x + \cos^{-1} x = \frac{\pi}{2}$

$\Rightarrow \quad k\alpha + k\beta = \frac{\pi}{2} \Rightarrow k(\alpha + \beta) = \frac{\pi}{2}$

Now, $\sin\left(\frac{2\pi\alpha}{\alpha + \beta}\right) = \sin\left(\frac{2\pi\alpha}{\frac{\pi}{2k}}\right)$ $\left[\because \alpha + \beta = \frac{\pi}{2k}\right]$

$= \sin(4\alpha k) = \sin(4 \sin^{-1} x)$ $[\because \alpha k = \sin^{-1} x]$

$= \sin[2 \sin^{-1} 2x\sqrt{1 - x^2}]$

$[\because 2 \sin^{-1} x = \sin^{-1} (2x\sqrt{1 - x^2})]$

$= \sin(\sin^{-1} 2(2x\sqrt{1 - x^2})$

$\sqrt{1 - (2x\sqrt{1 - x^2})^2}$

$= 4x\sqrt{1 - x^2}\ \sqrt{1 - 4x^2(1 - x^2)}$

$= 4x\sqrt{1 - x^2}\ \sqrt{1 - 4x^2 + 4x^4}$

$= 4x\sqrt{1 - x^2}\ \sqrt{(1 - 2x^2)^2}$

$= 4x(1 - 2x^2)\sqrt{1 - x^2}$

**19.** *(c)* $\cos^{-1}\left\{\frac{3}{10}\cos\left(\tan^{-1}\frac{4}{3}\right) + \frac{2}{5}\sin\left(\tan^{-1}\frac{4}{3}\right)\right\}$

Let $\tan^{-1}\left(\frac{4}{3}\right) = \theta \Rightarrow \tan\theta = \frac{4}{3}$

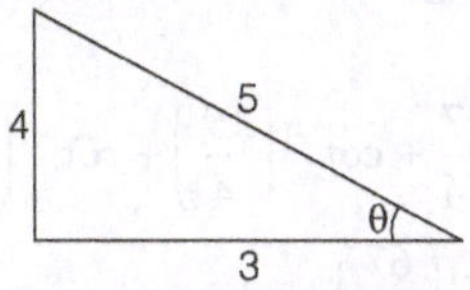


$\therefore \quad \cos\theta = \frac{3}{5}$ and $\sin\theta = \frac{4}{5}$

So, $\cos^{-1}\left\{\frac{3}{10}\cos\left(\tan^{-1}\frac{4}{3}\right) + \frac{2}{5}\sin\left(\tan^{-1}\frac{4}{3}\right)\right\}$

$= \cos^{-1}\left\{\frac{3}{10}\cos\theta + \frac{2}{5}\sin\theta\right\}$

$= \cos^{-1}\left\{\frac{3}{10} \times \frac{3}{5} + \frac{2}{5} \times \frac{4}{5}\right\}$

$= \cos^{-1}\left\{\frac{9}{50} + \frac{8}{25}\right\} = \cos^{-1}\left\{\frac{1}{2}\right\}$

$= \frac{\pi}{3}$

**20.** *(a)* Given, $\tan\left(\frac{1}{4}\sin^{-1}\frac{\sqrt{63}}{8}\right)$

Let $\sin^{-1}\frac{\sqrt{63}}{8} = \theta$

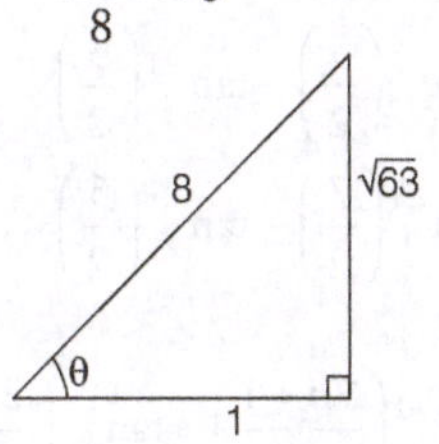


$\Rightarrow \quad \sin\theta = \frac{\sqrt{63}}{8} \Rightarrow \cos\theta = \frac{1}{8}$

Also, $\cos\frac{\theta}{2} = \sqrt{\frac{1 + \cos\theta}{2}}$

$= \sqrt{\frac{1 + \frac{1}{8}}{2}} = \sqrt{\frac{9}{16}} = \frac{3}{4}$

$\therefore \quad \tan\left(\frac{1}{4}\sin^{-1}\frac{\sqrt{63}}{8}\right) = \tan\left(\frac{\theta}{4}\right)$

$= \sqrt{\frac{1 - \cos\frac{\theta}{2}}{1 + \cos\frac{\theta}{2}}} = \sqrt{\frac{1 - \frac{3}{4}}{1 + \frac{3}{4}}} = \frac{1}{\sqrt{7}}$

**21.** *(c)* Given function,

$f(x) = \sin^{-1}\left(\frac{|x| + 5}{x^2 + 1}\right)$

For domain of '$f$'

$\frac{|x| + 5}{x^2 + 1} \le 1$, as $\frac{|x| + 5}{x^2 + 1} > 0, \forall x \in R$

$\Rightarrow |x| + 5 \le x^2 + 1$ as $x^2 + 1$ is positive, $\forall x \in R$

$\Rightarrow \quad x^2 - |x| - 4 \ge 0$

$\Rightarrow \quad |x|^2 - |x| - 4 \ge 0$

$[\because \forall x \in R, x^2 = |x|^2]$

$\because$ For $|x|^2 - |x| - 4 = 0$

$\Rightarrow |x| = \frac{1 \pm \sqrt{1 + 16}}{2} \Rightarrow |x| = \frac{1 \pm \sqrt{17}}{2}$

$\Rightarrow \quad |x| = \frac{1 + \sqrt{17}}{2}$ as $|x| \ge 0$

$\therefore \quad |x| \ge \frac{\sqrt{17} + 1}{2}$

$\Rightarrow x \in \left(-\infty, -\frac{\sqrt{17} + 1}{2}\right] \cup \left[\frac{\sqrt{17} + 1}{2}, \infty\right)$

$\therefore \quad a = \frac{\sqrt{17} + 1}{2}$

**22.** *(3)* $y = \cos\left(\frac{\pi}{3} + \cos^{-1}\frac{x}{2}\right)$

$y = \cos\frac{\pi}{3} \cdot \cos\left(\cos^{-1}\frac{x}{2}\right) - \sin\frac{\pi}{3} \cdot \sin\left(\cos^{-1}\frac{x}{2}\right)$

$\Rightarrow \quad y = \frac{1}{2} \cdot \frac{x}{2} - \frac{\sqrt{3}}{2}\sqrt{1 - \frac{x^2}{4}}$

$\Rightarrow \quad 4y = x - \sqrt{3}\ \sqrt{4 - x^2}$

$\Rightarrow \quad (x - y)^2 + 3y^2 = 3$

**23.** *(24)* We have $f(x) = \sec^{-1}\left(\frac{2x}{5x + 3}\right)$

$\frac{2x}{5x + 3} \le -1$ or $\frac{2x}{5x + 3} \ge 1$

$\frac{2x + 5x + 3}{5x + 3} \le 0$ or $\frac{2x - 5x - 3}{5x + 3} \ge 0$

$\frac{7x+3}{5x+3} \le 0$ or $\frac{-3(x+1)}{5x+3} \ge 0$

Case I $7x+3 \le 0$ and $5x+3>0$

$$x \le -\frac{3}{7} \text{ or } x > -\frac{3}{5}$$

$$\Rightarrow \quad -\frac{3}{5} < x \le -\frac{3}{7}$$

Case II $7x+3 \ge 0$ and $5x+3<0$

$$x \ge -\frac{3}{7} \text{ and } x < -\frac{3}{5}$$

which is not possible.

Case III $x+1 \ge 0$ and $5x+3<0$

$$x \ge -1 \text{ and } x < -\frac{3}{5} \Rightarrow -1 \le x < -\frac{3}{5}$$

Case IV $x+1 \le 0$ and $5x+3 \ge 0$

$$x \le -1 \text{ and } x \ge -\frac{3}{5}$$

which is not possible.

$\therefore$ Domain is $\left[-1, -\frac{3}{5}\right) \cup \left(-\frac{3}{5}, -\frac{3}{7}\right]$

$\therefore \alpha = -1, \beta = -\frac{3}{5}, \gamma = -\frac{3}{5}, \delta = -\frac{3}{7}$

$$|3\alpha + 10(\beta+\gamma) + 21\delta|$$

$$= \left|-3 + 10\left(-\frac{3}{5} - \frac{3}{5}\right) + 21\left(-\frac{3}{7}\right)\right| = 24$$

**24.** *(12)*

$$\cos(\sin^{-1}(x\cot(\tan^{-1}(\cos(\sin^{-1}x))))) = k, \; 0<|x|<\frac{1}{\sqrt{2}}$$

$$\Rightarrow \cos(\sin^{-1}(x\cot(\tan^{-1}(\sqrt{1-x^2})))) = k$$

$$\Rightarrow \quad \cos\left(\sin^{-1}\left(\frac{x}{\sqrt{1-x^2}}\right)\right) = k$$

$$\Rightarrow \quad \sqrt{1 - \frac{x^2}{1-x^2}} = k \Rightarrow \frac{1-2x^2}{1-x^2} = k^2$$

$$\therefore \; (k^2-2)x^2 = k^2 - 1$$

$$\Rightarrow \quad x^2 = \frac{k^2-1}{k^2-2}$$

$$\therefore \quad \alpha^2 = \beta^2 = \frac{k^2-1}{k^2-2}$$

$$\Rightarrow \frac{1}{\alpha^2} + \frac{1}{\beta^2} = 2\left(\frac{k^2-2}{k^2-1}\right) \text{ and } \frac{\alpha}{\beta} = -1$$

As $\frac{1}{\alpha^2} + \frac{1}{\beta^2}$ and $\frac{\alpha}{\beta}$ are roots of

$$x^2 - bx - 5 = 0$$

$$b = \frac{1}{\alpha^2} + \frac{1}{\beta^2} + \frac{\alpha}{\beta} = \frac{2k^2-4}{k^2-1} - 1 = \frac{k^2-3}{k^2-1}$$

Also $\left(\frac{1}{\alpha^2} + \frac{1}{\beta^2}\right) \times \left(\frac{\alpha}{\beta}\right) = -5$

$$\Rightarrow \quad 2\left(\frac{k^2-2}{k^2-1}\right) = 5 \Rightarrow k^2 = \frac{1}{3}$$

$$\therefore \quad b = \frac{\frac{1}{3}-3}{\frac{1}{3}-1} = 4 \Rightarrow \frac{b}{k^2} = \frac{4}{1/3} = 12$$

## Topic 2

**01.** *(b)*

$$\cot^{-1}\frac{7}{4} + \cot^{-1}\left(\frac{19}{4}\right) + \cot^{-1}\left(\frac{39}{4}\right) + \cot^{-1}\left(\frac{67}{4}\right) + \ldots$$

$$T_n = \cot^{-1}\left(\frac{4n^2+3}{4}\right) = \tan^{-1}\left(\frac{4}{4n^2+3}\right)$$

$$= \tan^{-1}\left(\frac{4}{4+(4n^2-1)}\right)$$

$$= \tan^{-1}\left(\frac{1}{1+\frac{(2n-1)(2n+1)}{4}}\right)$$

$$= \tan^{-1}\left(\frac{\frac{2n+1}{2} - \frac{2n-1}{2}}{1+\left(\frac{2n+1}{2}\right)\left(\frac{2n-1}{2}\right)}\right)$$

$$= \tan^{-1}\left(\frac{2n+1}{2}\right) - \tan^{-1}\left(\frac{2n-1}{2}\right)$$

$$T_1 = \tan^{-1}\left(\frac{3}{2}\right) - \tan^{-1}\left(\frac{1}{2}\right)$$

$$T_2 = \tan^{-1}\left(\frac{5}{2}\right) - \tan^{-1}\left(\frac{3}{2}\right)$$

$$T_3 = \tan^{-1}\left(\frac{7}{2}\right) - \tan^{-1}\left(\frac{5}{2}\right)$$

$$\vdots$$

$$T_n = \tan^{-1}\left(\frac{2n+1}{2}\right) - \tan^{-1}\left(\frac{2n-1}{2}\right)$$

$$S_n = \tan^{-1}\left(\frac{2n+1}{2}\right) - \tan^{-1}\frac{1}{2}$$

$$S_\infty = \tan^{-1}(\infty) - \tan^{-1}\frac{1}{2}$$

$$= \frac{\pi}{2} - \tan^{-1}\frac{1}{2}$$

**02.** *(c)* Given,

$$\cos\left(\sin^{-1}\frac{3}{5} + \sin^{-1}\frac{5}{13} + \sin^{-1}\frac{33}{65}\right)$$

$$= \cos\left(\sin^{-1}\left(\frac{3}{5}\times\frac{12}{13} + \frac{4}{5}\times\frac{5}{13}\right) + \sin^{-1}\frac{33}{65}\right)$$

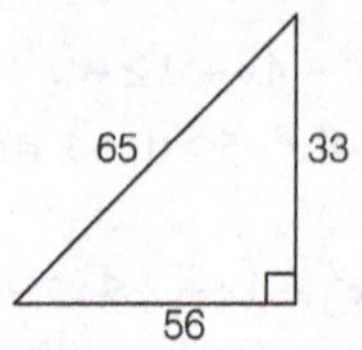


$$= \cos\left(\sin^{-1}\left(\frac{56}{65}\right) + \sin^{-1}\left(\frac{33}{65}\right)\right)$$

$$= \cos\left(\cos^{-1}\left(\frac{33}{65}\right) + \sin^{-1}\left(\frac{33}{65}\right)\right)$$

$$= \cos\left(\frac{\pi}{2}\right) = 0$$

> **Concept Enhancer**
>
> $\sin^{-1}x + \sin^{-1}y$
>
> $= \sin^{-1}[x\sqrt{1-y^2} + y\sqrt{1-x^2}]$,
>
> if $x, y \in [-1, 1]$ and $x^2 + y^2 \le 1$

**03.** *(d)*

$$\cot^{-1}\left(\beta + \frac{1+\beta^2}{\alpha-\beta}\right) + \cot^{-1}\left(\gamma + \frac{1+\gamma^2}{\beta-\gamma}\right) + \cot^{-1}\left(\alpha + \frac{1+\alpha^2}{\gamma-\alpha}\right)$$

$$= \cot^{-1}\left(\frac{\alpha\beta+1}{\alpha-\beta}\right) + \cot^{-1}\left(\frac{\beta\gamma+1}{\beta-\gamma}\right) + \cot^{-1}\left(\frac{\alpha\gamma+1}{\gamma-\alpha}\right)$$

$$= \tan^{-1}\left(\frac{\alpha-\beta}{1+\alpha\beta}\right) + \tan^{-1}\left(\frac{\beta-\gamma}{1+\beta\gamma}\right) + \tan^{-1}\left(\frac{\gamma-\alpha}{1+\gamma\alpha}\right)$$

$$= \tan^{-1}\alpha - \tan^{-1}\beta + \tan^{-1}\beta - \tan^{-1}\gamma + (\pi + \tan^{-1}\gamma - \tan^{-1}\alpha)$$

$$[\because \alpha > \beta > \gamma > 0 \therefore \gamma - \alpha < 0]$$

$$= \pi$$

**04.** *(b)*

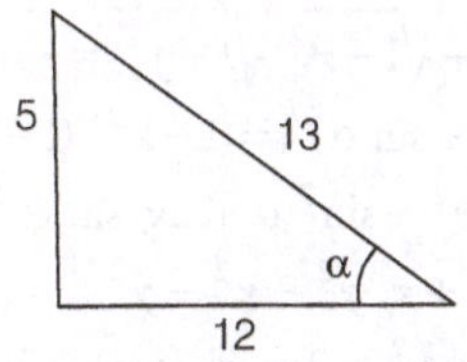


Let $\cos\alpha = \frac{12}{13} \Rightarrow \sin\alpha = \frac{5}{13}$ and

$\tan\alpha = \frac{5}{12}$

$$\because \cos^{-1}\left[\frac{12}{13}\cos x + \frac{5}{13}\sin x\right]$$

$$= \cos^{-1}[\cos x\cos\alpha + \sin x \sin\alpha]$$

$$= \cos^{-1}(\cos(x-\alpha)) = x - \alpha$$

$$\left[\because x - \alpha \in \left(\frac{-\pi}{2}, \frac{\pi}{2}\right)\right]$$

$$= x - \tan^{-1}\frac{5}{12}$$

$$\left[\because \cos\alpha = \frac{12}{13} \Rightarrow \tan\alpha = \frac{5}{12}\right]$$

**05.** *(d)* Given,

$$\tan^{-1}\left(\frac{1+\sqrt{3}}{3+\sqrt{3}}\right) + \sec^{-1}\left(\sqrt{\frac{8+4\sqrt{3}}{6+3\sqrt{3}}}\right)$$

$$= \tan^{-1}\left(\frac{(1+\sqrt{3})}{\sqrt{3}(\sqrt{3}+1)}\right) + \sec^{-1}\left(\sqrt{\frac{4(2+\sqrt{3})}{3(2+\sqrt{3})}}\right)$$

$$= \tan^{-1}(1/\sqrt{3}) + \sec^{-1}(2/\sqrt{3})$$

$$= \frac{\pi}{6} + \frac{\pi}{6} = \frac{2\pi}{6} = \frac{\pi}{3}$$

**06.** *(b)* Given, $(\sin^{-1}x)^2 - (\cos^{-1}x)^2 = a$

$$\Rightarrow (\sin^{-1}x + \cos^{-1}x)(\sin^{-1}x - \cos^{-1}x) = a$$

$$\Rightarrow \frac{\pi}{2}(\sin^{-1}x - \cos^{-1}x) = a$$

$$\Rightarrow \frac{\pi}{2} - 2\cos^{-1}x = \frac{2a}{\pi}$$

$$\Rightarrow 2\cos^{-1}x = \frac{\pi}{2} - \frac{2a}{\pi}$$

$$\Rightarrow \cos^{-1}(2x^2 - 1) = \frac{\pi}{2} - \frac{2a}{\pi}$$

$$\Rightarrow 2x^2 - 1 = \cos\left(\frac{\pi}{2} - \frac{2a}{\pi}\right)$$

$$\Rightarrow 2x^2 - 1 = \sin(2a/\pi)$$

**07.** *(c)* We have,

$$2\pi - \left(\sin^{-1}\frac{4}{5} + \sin^{-1}\frac{5}{13} + \sin^{-1}\frac{16}{65}\right)$$

$$= 2\pi - \left(\tan^{-1}\frac{4}{3} + \tan^{-1}\frac{5}{12} + \tan^{-1}\frac{16}{63}\right)$$

$$= 2\pi - \left(\tan^{-1}\frac{\frac{4}{3}+\frac{5}{12}}{1-\frac{4}{3}\times\frac{5}{12}} + \tan^{-1}\frac{16}{63}\right)$$

$$= 2\pi - \left(\tan^{-1}\frac{63}{16} + \tan^{-1}\frac{16}{63}\right)$$

$$= 2\pi - \left(\cot^{-1}\frac{16}{63} + \tan^{-1}\frac{16}{63}\right)$$

$$= 2\pi - \frac{\pi}{2} = \frac{3\pi}{2}$$

**08.** *(b)*

> **key idea** *Use formulae*
>
> (i) $\sin^{-1}x - \sin^{-1}y = \sin^{-1}(x\sqrt{1-y^2} - y\sqrt{1-x^2})$
>
> (ii) $\sin^{-1}x = \cos^{-1}\sqrt{1-x^2}$ and
>
> (iii) $\sin^{-1}\theta + \cos^{-1}\theta = \frac{\pi}{2}$

We have, $\sin^{-1}\left(\frac{12}{13}\right) - \sin^{-1}\left(\frac{3}{5}\right)$

$$= \sin^{-1}\left(\frac{12}{13}\sqrt{1-\left(\frac{3}{5}\right)^2} - \frac{3}{5}\sqrt{1-\left(\frac{12}{13}\right)^2}\right)$$

$$= \sin^{-1}\left(\frac{33}{65}\right) = \cos^{-1}\sqrt{1-\left(\frac{33}{65}\right)^2}$$

$$[\because \sin^{-1}x = \cos^{-1}\sqrt{1-x^2}]$$

$$= \cos^{-1}\sqrt{\frac{3136}{4225}} = \cos^{-1}\left(\frac{56}{65}\right)$$

$$= \frac{\pi}{2} - \sin^{-1}\left(\frac{56}{65}\right)$$

$$\left[\because \sin^{-1}\theta + \cos^{-1}\theta = \frac{\pi}{2}\right]$$

**09.** *(a)* Let $y = \tan^{-1}\left(\frac{6x\sqrt{x}}{1-9x^3}\right)$

$$= \tan^{-1}\left[\frac{2\cdot(3x^{3/2})}{1-(3x^{3/2})^2}\right]$$

$$= 2\tan^{-1}(3x^{3/2})$$

$$\left[\because 2\tan^{-1}x = \tan^{-1}\frac{2x}{1-x^2}\right]$$

$$\therefore \frac{dy}{dx} = 2\cdot\frac{1}{1+(3x^{3/2})^2}\cdot 3\times\frac{3}{2}(x)^{1/2}$$

$$= \frac{9}{1+9x^3}\cdot\sqrt{x}$$

$$\therefore g(x) = \frac{9}{1+9x^3}$$

**10.** *(a)* Given,

$$\tan^{-1}y = \tan^{-1}x + \tan^{-1}\left(\frac{2x}{1-x^2}\right),$$

where $|x| < \frac{1}{\sqrt{3}}$

$$\Rightarrow \tan^{-1}y = \tan^{-1}\left\{\frac{x + \frac{2x}{1-x^2}}{1 - x\left(\frac{2x}{1-x^2}\right)}\right\}$$

$$\left[\because \tan^{-1}x + \tan^{-1}y = \tan^{-1}\left(\frac{x+y}{1-xy}\right),\ x>0, y>0, xy<1\right]$$

$$\Rightarrow \tan^{-1}y = \tan^{-1}\left(\frac{3x-x^3}{1-3x^2}\right)$$

$$\Rightarrow y = \frac{3x-x^3}{1-3x^2}$$

**11.** *(b)* We have,

$$f(x) = 2\tan^{-1}x + \sin^{-1}\left(\frac{2x}{1+x^2}\right)$$

$$= 2\tan^{-1}x + \pi - 2\tan^{-1}x$$

$$\Rightarrow f(x) = \pi$$

So, $f(5) = \pi$

**12.** *(d)* As $x, y, z$ are in AP

$$\Rightarrow 2y = x + z \quad \text{...(i)}$$

$\tan^{-1}x$, $\tan^{-1}y$ and $\tan^{-1}z$ are in AP, then

$$2\tan^{-1}y = \tan^{-1}x + \tan^{-1}z$$

$$2\tan^{-1}y = \tan^{-1}\left(\frac{x+z}{1-xz}\right)$$

Thus, $y^2 = xz$ ...(ii)

From Eqs. (i) and (ii), we get $x = y = z$.

**13.** *(b)* Since, $\text{cosec}^{-1}\left(\frac{5}{3}\right) = \tan^{-1}\left(\frac{3}{4}\right)$

$$\therefore \cot\left(\tan^{-1}\frac{3}{4} + \tan^{-1}\frac{2}{3}\right)$$

$$= \cot\left(\tan^{-1}\left[\frac{\frac{3}{4}+\frac{2}{3}}{1-\frac{1}{2}}\right]\right)$$

$$= \cot\left[\tan^{-1}\left(\frac{17}{6}\right)\right]$$

$$= \frac{6}{17}$$

**Alternate Solution**

Since $\text{cosec}^{-1}\left(\frac{5}{3}\right)=\tan^{-1}\left(\frac{3}{4}\right)$

$\therefore \cot\left(\tan^{-1}\left(\frac{3}{4}\right)+\tan^{-1}\left(\frac{2}{3}\right)\right)$

$=\cot\left(\cot^{-1}\left(\frac{4}{3}\right)+\cot^{-1}\left(\frac{3}{2}\right)\right)$ $\left[\because \tan^{-1}\frac{1}{x}=\cot^{-1}x\right]$

$=\cot\left(\cot^{-1}\left(\frac{\frac{4}{3}\times\frac{3}{2}-1}{\frac{4}{3}+\frac{3}{2}}\right)\right)$ $\left[\because \cot^{-1}(A)+\cot^{-1}(B)=\cot^{-1}\left(\frac{AB-1}{A+B}\right)\right]$

$=\cot\left(\cot^{-1}\left(\frac{6}{17}\right)\right)=\frac{6}{17}$

**14.** *(b)*

$\because \quad \sin^{-1}\left(\frac{x}{5}\right)+\text{cosec}^{-1}\left(\frac{5}{4}\right)=\frac{\pi}{2}$

$\Rightarrow \quad \sin^{-1}\left(\frac{x}{5}\right)+\sin^{-1}\left(\frac{4}{5}\right)=\frac{\pi}{2}$

$\Rightarrow \quad \sin^{-1}\left(\frac{x}{5}\right)=\frac{\pi}{2}-\sin^{-1}\left(\frac{4}{5}\right)$

$\Rightarrow \quad \sin^{-1}\left(\frac{x}{5}\right)=\cos^{-1}\left(\frac{4}{5}\right)$

$\Rightarrow \quad \sin^{-1}\left(\frac{x}{5}\right)=\sin^{-1}\left(\frac{3}{5}\right)$

$\therefore \quad x=3$

**15.** *(d)* $\tan^{-1}\left(\frac{1}{4}\right)+\tan^{-1}\left(\frac{2}{9}\right)$

$=\tan^{-1}\left(\frac{\frac{1}{4}+\frac{2}{9}}{1-\frac{1}{4}\times\frac{2}{9}}\right)=\tan^{-1}\left(\frac{1}{2}\right)$

**16.** *(c)* $\cot^{-1}\left(\frac{\sqrt{1+\tan^2(2)}-1}{\tan 2}\right)$

$=\cot^{-1}\left(\frac{|\sec 2|-1}{\tan 2}\right)=\cot^{-1}\left(\frac{-\sec 2-1}{\tan 2}\right)$

$=\pi-\cot^{-1}\left(\frac{1+\cos 2}{\sin 2}\right)$

$=\pi-\cot^{-1}\left(\frac{2\cos^2 1}{2\sin 1\cdot\cos 1}\right)$

$=\pi-\cot^{-1}(\cot 1)=\pi-1$

$\cot^{-1}\left(\frac{\sqrt{1+\tan^2\left(\frac{1}{2}\right)}+1}{\tan\left(\frac{1}{2}\right)}\right)$

$\cot^{-1}\left(\frac{1+\cos\left(\frac{1}{2}\right)}{\sin\left(\frac{1}{2}\right)}\right)$

$=\cot^{-1}\left(\frac{2\cos^2\left(\frac{1}{4}\right)}{2\sin\left(\frac{1}{4}\right)\cos\left(\frac{1}{4}\right)}\right)$

$=\cot^{-1}\left(\cot\frac{1}{4}\right)=\frac{1}{4}$

So, $\pi-1-\frac{1}{4}\Rightarrow\pi-\frac{5}{4}$

**17.** *(b)* Let $y=(\sec^{-1}x)^2+(\text{cosec}^{-1}x)^2$

$=(\sec^{-1}x+\text{cosec}^{-1}x)^2-2\sec^{-1}x\,\text{cosec}^{-1}x$

$=\left(\frac{\pi}{2}\right)^2-2(\sec^{-1}x)\left(\frac{\pi}{2}-\sec^{-1}x\right)$

$=\frac{\pi^2}{4}+2(\sec^{-1}x)^2-\pi\sec^{-1}x$

$\Rightarrow 2\left((\sec^{-1}x)^2-\frac{\pi}{2}(\sec^{-1}x)+\frac{\pi^2}{8}\right)$

$\Rightarrow 2\left[\left(\sec^{-1}x-\frac{\pi}{4}\right)^2+\frac{\pi^2}{8}-\frac{\pi^2}{16}\right]$

$\Rightarrow 2\left(\sec^{-1}x-\frac{\pi}{4}\right)^2+\frac{\pi^2}{8}$

$\because \quad \sec^{-1}x\in[0,\pi]-\left\{\frac{\pi}{2}\right\}$

$\therefore$Minimum value of $y=0+\frac{\pi^2}{8}$

Maximum value of

$y=2\left(\frac{9\pi^2}{16}\right)+\frac{\pi^2}{8}=\frac{10\pi^2}{8}$

$\therefore$Required sum

$=16\left(\frac{\pi^2}{8}+\frac{10\pi^2}{8}\right)=22\pi^2$

**18.** *(a)* $\cos^{-1}x-\sin^{-1}y=\alpha;$

$-\frac{\pi}{2}\le\alpha\le\pi$

$\Rightarrow \quad \cos^{-1}x-\frac{\pi}{2}+\cos^{-1}y=\alpha$

$\Rightarrow \quad \cos^{-1}x+\cos^{-1}y=\frac{\pi}{2}+\alpha$

$\Rightarrow \quad \cos^{-1}(xy-\sqrt{1-x^2}\sqrt{1-y^2})=\frac{\pi}{2}+\alpha$

$\Rightarrow \quad xy-\sqrt{1-x^2}\sqrt{1-y^2}=-\sin\alpha$

$(xy+\sin\alpha)^2=(1-x^2)(1-y^2)$

$x^2y^2+\sin^2\alpha+2xy\sin\alpha$

$=1+x^2y^2-x^2-y^2$

$\Rightarrow \quad x^2+y^2+2xy\sin\alpha=1-\sin^2\alpha$

$=\cos^2\alpha$

$\therefore$Minimum value of

$x^2+y^2+2xy\sin\alpha=0$

**19.** *(b)* We have,

$\sin^{-1}\frac{\alpha}{17}+\cos^{-1}\frac{4}{5}-\tan^{-1}\frac{77}{36}=0$

$\Rightarrow \quad \sin^{-1}\frac{\alpha}{17}=\tan^{-1}\frac{77}{36}-\cos^{-1}\frac{4}{5}$

$=\sin^{-1}\frac{77}{85}-\sin^{-1}\frac{3}{5}$

$=\sin^{-1}\left[\frac{77}{85}\sqrt{1-\left(\frac{3}{5}\right)^2}-\frac{3}{5}\sqrt{1-\left(\frac{77}{85}\right)^2}\right]$

$=\sin^{-1}\left(\frac{200}{425}\right)$

$\therefore \quad \sin^{-1}\frac{\alpha}{17}=\sin^{-1}\frac{8}{17}\Rightarrow\frac{\alpha}{17}=\frac{8}{17}$

$\Rightarrow \quad \alpha=8$

Now, $\sin^{-1}(\sin\alpha)+\cos^{-1}(\cos\alpha)$

$=\sin^{-1}(\sin 8)+\cos^{-1}(\cos 8)$

$=\sin^{-1}\sin(3\pi-8)+\cos^{-1}\cos(8-2\pi)$

$=3\pi-8+8-2\pi=\pi$

**Alternate Solution**

Given, $\sin^{-1}\frac{\alpha}{17}+\cos^{-1}\frac{4}{5}-\tan^{-1}\frac{77}{36}=0$

$\Rightarrow \sin^{-1}\frac{\alpha}{17}=\tan^{-1}\frac{77}{36}-\cos^{-1}\frac{4}{5}$

We can write

$\cos^{-1}\frac{4}{5}=\tan^{-1}\frac{3}{4}$

$\therefore \quad \sin^{-1}\frac{\alpha}{17}=\tan^{-1}\frac{77}{36}-\tan^{-1}\frac{3}{4}$

$=\tan^{-1}\left[\frac{\frac{77}{36}-\frac{3}{4}}{1+\frac{77}{36}\cdot\frac{3}{4}}\right]$

$\Rightarrow \sin^{-1}\frac{\alpha}{17}=\tan^{-1}\frac{8}{15}\Rightarrow\sin^{-1}\frac{\alpha}{17}=\sin^{-1}\frac{8}{17}$

$\therefore \quad \alpha=8$

Now, $\sin^{-1}(\sin\alpha)+\cos^{-1}(\cos\alpha)$

$=\sin^{-1}(\sin 8)+\cos^{-1}(\cos 8)$

$=\sin^{-1}\sin(3\pi-8)+\cos^{-1}\cos(8-2\pi)$

$=3\pi-8+8-2\pi=\pi$

**20.** *(b)* Given that

$a_1=1, a_2=2, \ldots, a_{2022}=2022$,

Now, $\tan^{-1}\left(\dfrac{1}{1+a_1a_2}\right)+\tan^{-1}\left(\dfrac{1}{1+a_2a_3}\right)+\ldots+\tan^{-1}\left(\dfrac{1}{1+a_{2021}a_{2022}}\right)$

On putting $a_1=1, a_2=2, \ldots\ldots, a_{2021}=2021, a_{2022}=2022$, we get

$$\tan^{-1}\left[\frac{1}{1+1\cdot2}\right]+\tan^{-1}\left[\frac{1}{1+2\cdot3}\right]+\ldots+\tan^{-1}\left[\frac{1}{1+2021\cdot2022}\right]$$

$$\Rightarrow \tan^{-1}\left[\frac{2-1}{1+2\cdot1}\right]+\tan^{-1}\left[\frac{3-2}{1+3\cdot2}\right]+\ldots+\tan^{-1}\left[\frac{2022-2021}{1+2022\cdot2021}\right]$$

$$\Rightarrow \tan^{-1}(2)-\tan^{-1}(1)+\tan^{-1}(3)-\tan^{-1}(2)+\ldots\ldots+\tan^{-1}(2022)-\tan^{-1}(2021)$$

$$\left[\because \tan^{-1}x-\tan^{-1}y=\tan^{-1}\left(\frac{x-y}{1+x\cdot y}\right)\right]$$

$$\Rightarrow \tan^{-1}(2022)-\tan^{-1}(1)$$

$$\Rightarrow \tan^{-1}(2022)-\frac{\pi}{4}$$

**21.** *(b)* Given,

$$\tan\left(2\tan^{-1}\frac{1}{5}+\sec^{-1}\frac{\sqrt5}{2}+2\tan^{-1}\frac{1}{8}\right)$$

$$=\tan\left[2\tan^{-1}\frac{1}{5}+2\tan^{-1}\frac{1}{8}+\sec^{-1}\left(\frac{\sqrt5}{2}\right)\right]$$

$$=\tan\left[2\tan^{-1}\left[\frac{\frac{1}{5}+\frac{1}{8}}{1-\frac{1}{5}\times\frac{1}{8}}\right]+\sec^{-1}\frac{\sqrt5}{2}\right]$$

$$=\tan\left[2\tan^{-1}\left(\frac{1}{3}\right)+\tan^{-1}\frac{1}{2}\right]$$

$$\left[\because \sec^{-1}\frac{\sqrt5}{2}=\tan^{-1}\frac{1}{2}\right]$$

$$=\tan\left[\tan^{-1}\left[\frac{2\times\frac{1}{3}}{1-\left(\frac{1}{3}\right)^2}\right]+\tan^{-1}\frac{1}{2}\right]$$

$$=\tan\left[\tan^{-1}\frac{3}{4}+\tan^{-1}\frac{1}{2}\right]$$

$$=\tan\left[\tan^{-1}\left(\frac{\frac{3}{4}+\frac{1}{2}}{1-\frac{3}{4}\times\frac{1}{2}}\right)\right]$$

$$=\tan[\tan^{-1}(2)]=2$$

**22.** *(a)* Let $E=$

$$\cot\left[\sum_{n=1}^{50}\tan^{-1}\frac{1}{1+n(n+1)}\right]$$

$$=\cot\left[\sum_{n=1}^{50}\tan^{-1}\frac{(n+1)-n}{1+(n+1)n}\right]$$

$$=\cot\left[\sum_{n=1}^{50}(\tan^{-1}(n+1)-\tan^{-1}n)\right] \quad \ldots\text{(i)}$$

Now, $\left[\sum_{n=1}^{50}(\tan^{-1}(n+1)-\tan^{-1}n\right]$

$$=(\tan^{-1}2-\tan^{-1}1)+(\tan^{-1}3-\tan^{-1}2)+(\tan^{-1}4-\tan^{-1}3)+\ldots\ldots+(\tan^{-1}51-\tan^{-1}50)$$

$$=\tan^{-1}51-\tan^{-1}1$$

$$=\tan^{-1}\left(\frac{51-1}{1+51\times1}\right)$$

$$=\tan^{-1}\left(\frac{25}{26}\right)=\cot^{-1}\left(\frac{26}{25}\right)$$

$$\therefore E=\cot\left(\cot^{-1}\frac{26}{25}\right)=\frac{26}{25}$$

[using Eq.(i)]

**23.** *(a)* Given,

$$\sin^{-1}\left(\sin\frac{2\pi}{3}\right)+\cos^{-1}\left(\cos\frac{7\pi}{6}\right)+\tan^{-1}\left(\tan\frac{3\pi}{4}\right)$$

Now,

$$\sin^{-1}\left(\sin\frac{2\pi}{3}\right)=\sin^{-1}\left(\sin\left(\pi-\frac{\pi}{3}\right)\right)$$

$$=\sin^{-1}\left(\sin\frac{\pi}{3}\right)=\frac{\pi}{3}$$

$$\cos^{-1}\left(\cos\frac{7\pi}{6}\right)=\cos^{-1}\left(\cos\left(\pi+\frac{\pi}{6}\right)\right)$$

$$=\cos^{-1}\left(-\cos\frac{\pi}{6}\right)$$

$$=\pi-\cos^{-1}\left(\cos\frac{\pi}{6}\right)$$

$$=\pi-\frac{\pi}{6}=\frac{5\pi}{6}$$

$$\tan^{-1}\left(\tan\frac{3\pi}{4}\right)=\tan^{-1}\left(\tan\left(\pi-\frac{\pi}{4}\right)\right)$$

$$=\tan^{-1}\left(-\tan\frac{\pi}{4}\right)$$

$$=-\tan^{-1}\left(\tan\frac{\pi}{4}\right)=\frac{-\pi}{4}$$

So, $\sin^{-1}\left(\sin\dfrac{2\pi}{3}\right)+\cos^{-1}\left(\cos\dfrac{7\pi}{6}\right)+\tan^{-1}\left(\tan\dfrac{3\pi}{4}\right)$

$$=\frac{\pi}{3}+\frac{5\pi}{6}-\frac{\pi}{4}=\frac{11\pi}{12}$$

**24.** *(d)* We have,

$f(x)=\tan^{-1}(\sin x+\cos x)$

where, $x\in[0, \pi/2]$

$$\Rightarrow \quad 1\le\sin x+\cos x\le\sqrt2$$

$$[\because -\sqrt{A^2+B^2}\le A\sin x+B\cos x\le\sqrt{A^2+B^2}]$$

$$\Rightarrow \quad \tan^{-1}(1)\le\tan^{-1}(\sin x+\cos x)\le\tan^{-1}(\sqrt2)$$

$\therefore \quad m=\tan^{-1}(1)$ and $M=\tan^{-1}(\sqrt2)$

$$\therefore \quad M-m=\tan^{-1}\sqrt2-\tan^{-1}(\sqrt1)$$

$$=\tan^{-1}\left(\frac{\sqrt2-1}{1+\sqrt2}\right)$$

$$=\tan^{-1}(3-2\sqrt2)$$

$$\Rightarrow \quad \tan(M-m)=3-2\sqrt2$$

**25.** *(a)* Let the angle before rotation be $\theta$, then

$$\theta=\angle PCB=\tan^{-1}(2)$$

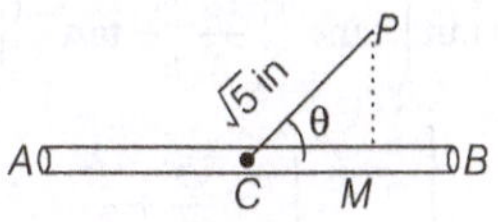


Now, after rotation let angle become $\alpha$.

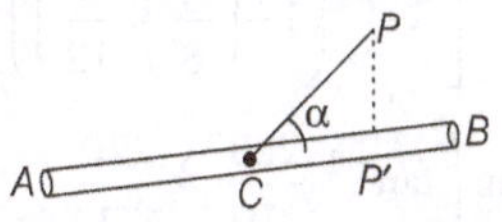

In $\Delta PCM$, $PC=\sqrt{5}$

$$PM=\sqrt{5}\sin\theta=\sqrt{5}\left(\frac{2}{\sqrt{5}}\right)=2$$

After rotation perpendicular distance becomes $PP'=1$

$$PC\sin\alpha=1\Rightarrow\sqrt{5}\sin\alpha=1$$

$$\Rightarrow \quad \sin\alpha=1/\sqrt{5}$$

$$\Rightarrow \quad \tan\alpha=\frac{\sin\alpha}{\sqrt{1-\sin^2\alpha}}=\frac{1}{2}$$

$\therefore$ Rotated angle

$$=\theta-\alpha=\tan^{-1}(2)-\tan^{-1}\left(\frac{1}{2}\right)$$

$$=\tan^{-1}\left(\frac{2-1/2}{1+2\times1/2}\right)=\tan^{-1}\left(\frac{3}{4}\right)$$

**26.** *(b)* Let

$$A=\tan\left(2\tan^{-1}\left(\frac{3}{5}\right)+\sin^{-1}\left(\frac{5}{13}\right)\right)\quad\ldots\text{(i)}$$

Now, using

$2\tan^{-1}x=\tan^{-1}\dfrac{2x}{1-x^2}$, we get

$$2\tan^{-1}\left(\frac{3}{5}\right)=\tan^{-1}\left(\frac{2\left(\frac{3}{5}\right)}{1-\left(\frac{3}{5}\right)^2}\right)$$

$$=\tan^{-1}\left(\frac{15}{8}\right)$$

Let $\sin^{-1}\left(\frac{5}{13}\right)=\theta$, then $\sin\theta=\frac{5}{13}$

$$\therefore\ \tan\theta=\frac{\sin\theta}{\sqrt{1-\sin^2\theta}}=\frac{\frac{5}{13}}{\sqrt{1-\left(\frac{5}{13}\right)^2}}$$

$$=\frac{5}{12}$$

$$\Rightarrow \quad \theta=\tan^{-1}\left(\frac{5}{12}\right)$$

From Eq. (i), we get

$$A=\tan\left(\tan^{-1}\left(\frac{15}{8}\right)+\tan^{-1}\left(\frac{5}{12}\right)\right)$$

$$=\tan\left[\tan^{-1}\left(\frac{\frac{15}{8}+\frac{5}{12}}{1-\left(\frac{15}{8}\right)\left(\frac{5}{12}\right)}\right)\right]$$

$$=\tan\left(\tan^{-1}\frac{220}{21}\right)=\frac{220}{21}$$

**27.** *(b)* Given,

$$\sin^{-1}\left[x^2+\frac{1}{3}\right]+\cos^{-1}\left[x^2-\frac{2}{3}\right]=x^2$$

$$\Rightarrow \sin^{-1}\left[\left(x^2-\frac{2}{3}+1\right)\right]+\cos^{-1}\left[x^2-\frac{2}{3}\right]=x^2$$

$$\Rightarrow\ \sin^{-1}\left[\left(x^2-\frac{2}{3}\right)+1\right]+\cos^{-1}\left[x^2-\frac{2}{3}\right]=x^2\ldots\text{(i)}$$

$$(\because [x+n]=[x]+n,\ n\in I)$$

$\because \left[x^2-\frac{2}{3}\right]$ gives always integral values.

$$\because \quad -1\le x\le 1$$

$$\therefore \quad 0\le x^2\le 1$$

$$\Rightarrow \quad -\frac{2}{3}\le x^2-\frac{2}{3}\le 1-\frac{2}{3}$$

$$\Rightarrow \quad -\frac{2}{3}\le x^2-\frac{2}{3}\le\frac{1}{3}$$

$$\Rightarrow \quad -0.66\le x^2-\frac{2}{3}\le 0.33$$

$\therefore \left[x^2-\frac{2}{3}\right]=-1, 0$ are possibilities.

**Case I** If $\left[x^2-\frac{2}{3}\right]=0$

Then, Eq. (i) becomes,

$$\sin^{-1}(1)+\cos^{-1}(0)=x^2$$

$$\Rightarrow \quad x^2=\pi\Rightarrow x=\pm\sqrt{\pi}$$

But at this value of

$$x^2,\left[x^2-\frac{2}{3}\right]\ne 0$$

$$\therefore \quad x^2=\pi \quad \text{(Rejected)}$$

**Case II** If $\left[x^2-\frac{2}{3}\right]=-1$

Then, Eq. (i) becomes,

$$\sin^{-1}(0)+\cos^{-1}(-1)=x^2$$

$$\Rightarrow \quad x^2=\pi$$

$$\Rightarrow \quad x=\pm\sqrt{\pi}$$

But at this value of $x^2$, $\left[x^2-\frac{2}{3}\right]\ne -1$

$$\therefore \quad x^2=\pi \quad \text{(Rejected also)}$$

Hence, there is no solution for Eq. (i).

$\therefore$ Total number of solution (s) $=0$

**28.** *(a)* Given,

$\cot^{-1}\alpha=\cot^{-1}2+\cot^{-1}8+\cot^{-1}18$ $+\cot^{-1}32\ldots$ upto 100 terms

Let $T_n$ be the $n$th term of $\cot^{-1}\alpha$.

$$T_n=\cot^{-1}(2n^2)=\tan^{-1}\left(\frac{1}{2n^2}\right)$$

$$=\tan^{-1}\left[\frac{(2n+1)-(2n-1)}{1+(2n+1)(2n-1)}\right]$$

$$\because\ \tan^{-1}x-\tan^{-1}y=\tan^{-1}\left(\frac{x-y}{1+xy}\right)$$

$$\therefore\ \tan^{-1}\left\{\frac{(2n+1)-(2n-1)}{1+(2n+1)(2n-1)}\right\}$$

$$=\tan^{-1}(2n+1)-\tan^{-1}(2n-1)$$

$$T_1=\tan^{-1}3-\tan^{-1}1$$

$$T_2=\tan^{-1}5-\tan^{-1}3$$

$$T_3=\tan^{-1}7-\tan^{-1}5$$

$$\vdots$$

$$T_{99}=\tan^{-1}199-\tan^{-1}197$$

$$T_{100}=\tan^{-1}201-\tan^{-1}199$$

$$\sum T_r=\tan^{-1}201-\tan^{-1}1$$

$$=\tan^{-1}\left(\frac{201-1}{1+201\cdot1}\right)$$

$$=\tan^{-1}\left(\frac{200}{202}\right)$$

$$\Rightarrow\ \cot^{-1}(\alpha)=\tan^{-1}\left(\frac{200}{202}\right)$$

$$\Rightarrow\ \cot^{-1}(\alpha)=\cot^{-1}\left(\frac{202}{200}\right)$$

$$\therefore\ \alpha=\frac{202}{200}=\frac{101}{100}=1.01$$

**29.** *(b)* Given, $\tan^{-1}(x+1)+\cot^{-1}\left(\frac{1}{x-1}\right)$

$$=\tan^{-1}\left(\frac{8}{31}\right)$$

$$\Rightarrow\ \tan^{-1}(x+1)+\tan^{-1}(x-1)=\tan^{-1}(8/31)$$

$$\Rightarrow \tan^{-1}\left[\frac{(x+1)+(x-1)}{1-(x+1)(x-1)}\right]=\tan^{-1}\left(\frac{8}{31}\right)$$

$$\Rightarrow\ \tan^{-1}\left[\frac{2x}{1-(x^2-1)}\right]=\tan^{-1}\left(\frac{8}{31}\right)$$

$$\Rightarrow \quad \frac{2x}{2-x^2}=\frac{8}{31}$$

$$\Rightarrow \quad 62x=16-8x^2$$

$$\Rightarrow \quad 8x^2+62x-16=0$$

$$\Rightarrow 2(x+8)(4x-1)=0\Rightarrow x=-8, 1/4$$

So, the sum of possible value of $x$ is

$$=\frac{1}{4}+(-8)=-\frac{31}{4}$$

**30.** (c)

> **key idea** Write $6^r$ as $2^r 3^{r+1} - 3^r 2^{r+1}$. Then factor out the common factor $2^{2r+1}$ from numerator and denominator.
>
> After that, evaluate the value of the obtain expression by using following properties.
>
> $\tan^{-1} x - \tan^{-1} y = \tan^{-1}\left(\frac{x-y}{1+xy}\right)$
>
> $\lim_{k\to\infty} (x)^k = 0;\ 0 < x < 1$
>
> $\tan^{-1}(x) = \cot^{-1}\left(\frac{1}{x}\right);\ 0 < x < 1$

Given, $S_k = \sum_{r=1}^{k} \tan^{-1}\left(\frac{6^r}{2^{2r+1}+3^{2r+1}}\right)$

$$= \sum_{r=1}^{k} \tan^{-1}\left(\frac{2^r 3^{r+1} - 3^r 2^{r+1}}{2^{2r+1}\left[1+\left(\frac{3}{2}\right)^{2r+1}\right]}\right)$$

$$= \sum_{r=1}^{k} \tan^{-1}\left(\frac{2^{2r+1}\left[\left(\frac{3}{2}\right)^{r+1} - \left(\frac{3}{2}\right)^{r}\right]}{2^{2r+1}\left[1+\left(\frac{3}{2}\right)^{2r+1}\right]}\right)$$

$$= \sum_{r=1}^{k} \tan^{-1}\left(\frac{\left(\frac{3}{2}\right)^{r+1} - \left(\frac{3}{2}\right)^{r}}{1+\left(\frac{3}{2}\right)^{2r+1}}\right)$$

$$= \sum_{r=1}^{k}\left[\tan^{-1}\left(\frac{3}{2}\right)^{r+1} - \tan^{-1}\left(\frac{3}{2}\right)^{r}\right]$$

$$= \tan^{-1}\left(\frac{3}{2}\right)^{2} - \tan^{-1}\left(\frac{3}{2}\right) +$$

$$\tan^{-1}\left(\frac{3}{2}\right)^{3} - \tan^{-1}\left(\frac{3}{2}\right)^{2} +$$

$$\vdots$$

$$\tan^{-1}\left(\frac{3}{2}\right)^{k+1} - \tan^{-1}\left(\frac{3}{2}\right)^{k}$$

$$\Rightarrow S_k = \tan^{-1}\left(\frac{3}{2}\right)^{k+1} - \tan^{-1}\left(\frac{3}{2}\right)$$

When $k \to \infty$, $\tan^{-1}\left(\frac{3}{2}\right)^{k+1} \to \pi/2$

$$\lim_{k\to\infty} S_k = \frac{\pi}{2} - \tan^{-1}\left(\frac{3}{2}\right) = \cot^{-1}\left(\frac{3}{2}\right)$$

**31.** (c) $\operatorname{cosec}[2\cot^{-1}(5) + \cos^{-1}(4/5)]$

$$= \operatorname{cosec}\left[2\tan^{-1}\left(\frac{1}{5}\right) + \cos^{-1}\left(\frac{4}{5}\right)\right]$$

$$\left[\because \tan^{-1} x = \cot^{-1}\left(\frac{1}{x}\right)\right]$$

$$= \operatorname{cosec}\left[\tan^{-1}\left(\frac{2\left(\frac{1}{5}\right)}{1-\left(\frac{1}{5}\right)^2}\right) + \cos^{-1}\left(\frac{4}{5}\right)\right]$$

$$\left[\therefore 2\tan^{-1}\theta = \tan^{-1}\left(\frac{2\theta}{1-\theta^2}\right)\right]$$

$$= \operatorname{cosec}\left(\tan^{-1}\frac{5}{12} + \cos^{-1}\frac{4}{5}\right)$$

Let $\tan^{-1}\left(\frac{5}{12}\right) = x$, then $\tan x = \frac{5}{12}$

gives $\sin x = \frac{5}{13}$, $\cos x = \frac{12}{13}$

Let $\cos^{-1}\left(\frac{4}{5}\right) = y$, then $\cos y = \frac{4}{5}$ gives, $\sin y = 3/5$

Now,

$$\operatorname{cosec}(x+y) = \frac{1}{\sin(x+y)}$$

$$= \frac{1}{\sin x \cos y + \cos x \sin y}$$

$$= \frac{1}{\left(\frac{5}{13}\right)\left(\frac{4}{5}\right) + \left(\frac{12}{13}\right)\left(\frac{3}{5}\right)} = \frac{65}{56}$$

**32.** (c)

> **key idea** *Use the formula,*
>
> $\cos^{-1} x + \cos^{-1} y = \cos^{-1}(xy - \sqrt{1-x^2}\sqrt{1-y^2})$

We have,

$$\cos^{-1}\left(\frac{2}{3x}\right) + \cos^{-1}\left(\frac{3}{4x}\right) = \frac{\pi}{2}$$

$$\Rightarrow \cos^{-1}\left(\frac{2}{3x}\cdot\frac{3}{4x} - \sqrt{1-\frac{4}{9x^2}}\sqrt{1-\frac{9}{16x^2}}\right) = \frac{\pi}{2}$$

$$[\because \cos^{-1} x + \cos^{-1} y = \cos^{-1}(xy - \sqrt{1-x^2}\sqrt{1-y^2})]$$

$$\Rightarrow \cos^{-1}\left(\frac{1}{2x^2} - \frac{\sqrt{9x^2-4}\sqrt{16x^2-9}}{12x^2}\right) = \frac{\pi}{2}$$

$$\Rightarrow \frac{6-\sqrt{9x^2-4}\sqrt{16x^2-9}}{12x^2} = \cos\frac{\pi}{2} = 0$$

$$\Rightarrow \sqrt{9x^2-4}\sqrt{16x^2-9} = 6$$

On squaring both sides, we get

$$(9x^2-4)(16x^2-9) = 36$$

$$\Rightarrow 144x^4 - 145x^2 = 0$$

$$\Rightarrow x^2(144x^2 - 145) = 0$$

$$\Rightarrow x = 0$$

or $x = \pm\sqrt{\frac{145}{144}}$

$$= \pm\frac{\sqrt{145}}{12}$$

But $x > \frac{3}{4}$

$\therefore$ $x = \sqrt{145}/12$

**33.** (a) Given that,

$$\cot^{-1}(\sqrt{\cos\alpha}) - \tan^{-1}(\sqrt{\cos\alpha}) = x \quad \ldots(i)$$

We know that,

$$\cot^{-1}(\sqrt{\cos\alpha}) + \tan^{-1}(\sqrt{\cos\alpha}) = \pi/2 \quad \ldots(ii)$$

$$\left[\because \cot^{-1} x + \tan^{-1} x = \frac{\pi}{2}\right]$$

On adding Eqs. (i) and (ii), we get

$$2\cot^{-1}(\sqrt{\cos\alpha}) = \frac{\pi}{2} + x$$

$$\Rightarrow \sqrt{\cos\alpha} = \cot\left(\frac{\pi}{4} + \frac{x}{2}\right)$$

$$\Rightarrow \sqrt{\cos\alpha} = \frac{\cot\frac{x}{2} - 1}{1 + \cot\frac{x}{2}}$$

$$\Rightarrow \sqrt{\cos\alpha} = \frac{\cos\frac{x}{2} - \sin\frac{x}{2}}{\cos\frac{x}{2} + \sin\frac{x}{2}}$$

On squaring both sides, we get

$$\cos\alpha = \frac{\cos^2\frac{x}{2} + \sin^2\frac{x}{2} - 2\sin\frac{x}{2}\cos\frac{x}{2}}{\cos^2\frac{x}{2} + \sin^2\frac{x}{2} + 2\sin\frac{x}{2}\cos\frac{x}{2}}$$

$$\Rightarrow \cos\alpha = \frac{1-\sin x}{1+\sin x}$$

$$\Rightarrow \frac{1-\tan^2\frac{\alpha}{2}}{1+\tan^2\frac{\alpha}{2}}=\frac{1-\sin x}{1+\sin x}$$

Applying componendo and dividendo rule, we get

$$\sin x=\tan^2(\alpha/2)$$

**34.** *(b)* Given,

$$\cos^{-1}2x-2\cos^{-1}(\sqrt{1-x^2})=\pi$$

$$\Rightarrow \cos^{-1}2x=\pi+2\cos^{-1}(\sqrt{1-x^2})$$

$$\Rightarrow 2x=\cos(\pi+2\cos^{-1}(\sqrt{1-x^2}))$$

$$\Rightarrow 2x=-\cos(2\cos^{-1}(\sqrt{1-x^2}))$$

$$\Rightarrow 2x=-\cos[\cos^{-1}(2(\sqrt{1-x^2})^2-1)]$$

$$[\because 2\cos^{-1}x=\cos^{-1}(2x^2-1)]$$

$$\Rightarrow 2x=-\cos\cos^{-1}(2-2x^2-1)$$

$$\Rightarrow 2x=-(1-2x^2)\Rightarrow 2x=-1+2x^2$$

$$\Rightarrow 2x^2-2x-1=0$$

$$\Rightarrow x=\frac{-(-2)\pm\sqrt{(-2)^2-4(2)(-1)}}{2\times 2}$$

$$=\frac{2\pm 2\sqrt{3}}{4}=\frac{1}{2}\pm\frac{\sqrt{3}}{2}$$

$$\therefore x=\frac{1}{2}-\frac{\sqrt{3}}{2}\in\left[\frac{-1}{2},\frac{1}{2}\right]$$

$$\therefore S=\left\{x:x=\frac{1-\sqrt{3}}{2}\right\}$$

Now, $\sum_{x\in S}=2\sin^{-1}(x^2-1)$

$$=2\sin^{-1}\left(\left(\frac{1-\sqrt{3}}{2}\right)^2-1\right)$$

$$=2\sin^{-1}\left(-\frac{\sqrt{3}}{2}\right)$$

$$=-2\sin^{-1}\frac{\sqrt{3}}{2}=-2\times\frac{\pi}{3}=-\frac{2\pi}{3}$$

**35.** *(b)* Consider,

$$\cot\left(\sum_{n=1}^{19}\cot^{-1}\left(1+\sum_{p=1}^{n}2p\right)\right)$$

$$=\cot\left(\sum_{n=1}^{19}\cot^{-1}(1+n(n+1))\right)$$

$$\left[\because \sum_{p=1}^{n}P=\frac{n(n+1)}{2}\right]$$

$$=\cot\left(\sum_{n=1}^{19}\cot^{-1}(1+n+n^2)\right)$$

$$=\cot\left(\sum_{n=1}^{19}\tan^{-1}\frac{1}{1+n(n+1)}\right)$$

$$\left[\because \cot^{-1}x=\tan^{-1}\frac{1}{x},\text{ if } x>0\right]$$

$$=\cot\left(\sum_{n=1}^{19}\tan^{-1}\left(\frac{(n+1)-n}{1+n(n+1)}\right)\right)$$

$$[\text{putting } 1=(n+1)-n]$$

$$=\cot\sum_{n=1}^{19}(\tan^{-1}(n+1)-\tan^{-1}n)$$

$$\left[\because \tan^{-1}\frac{x-y}{1+xy}=\tan^{-1}x-\tan^{-1}y\right]$$

$$=\cot[(\tan^{-1}2-\tan^{-1}1)+(\tan^{-1}3-\tan^{-1}2)+\ldots\ldots+(\tan^{-1}20-\tan^{-1}19)]$$

$$=\cot(\tan^{-1}20-\tan^{-1}1)$$

$$=\cot\left[\left(\frac{\pi}{2}-\cot^{-1}20\right)-\left(\frac{\pi}{2}-\cot^{-1}1\right)\right]$$

$$[\because \tan^{-1}x+\cot^{-1}x=\pi/2]$$

$$=\cot(\cot^{-1}1-\cot^{-1}20)$$

$$=\frac{\cot(\cot^{-1}1)\cot(\cot^{-1}20)+1}{\cot(\cot^{-1}20)-\cot(\cot^{-1}1)}$$

$$\left[\because \cot(A-B)]=\frac{\cot A\cot B+1}{\cot B-\cot A}\right]$$

$$=\frac{(1\times 20)+1}{20-1}=\frac{21}{19}$$

$$[\because \cot(\cot^{-1}x)=x]$$

**36.** *(14)* Given,

$$\sec^2(\tan^{-1}\alpha)+\operatorname{cosec}^2(\cot^{-1}\beta)=36 \quad \ldots(i)$$

let $\tan^{-1}\alpha=x$ and $\cot^{-1}\beta=y$

So, that $\alpha=\tan x$ and $\beta=\cot y$

From Eq. (i), we get

$$\sec^2x+\operatorname{cosec}^2y=36$$

$$\Rightarrow (1+\tan^2x)+(1+\cot^2y)=36$$

$$\Rightarrow 2+\tan^2x+\cot^2y=36$$

$$\Rightarrow 2+\alpha^2+\beta^2=36$$

$$\Rightarrow \alpha^2+\beta^2=34 \quad \ldots(ii)$$

Also, given $\alpha+\beta=8$ ...(iii)

On solving Eqs. (ii) and (iii), we get

$$\because (\alpha+\beta)^2=\alpha^2+\beta^2+2\alpha\beta$$

$$\Rightarrow (8)^2=34+2\alpha\beta$$

$$\Rightarrow \frac{64-34}{2}=\alpha\beta$$

$$\Rightarrow \alpha\beta=15 \quad \ldots(iv)$$

So, we observe that $\alpha,\beta$ are the roots of the equation

$$x^2-8x+15=0$$

$$[\because \alpha+\beta=8 \text{ and } \alpha\beta=15]$$

$$\Rightarrow (x-3)(x-5)=0$$

$$\Rightarrow x=3,5$$

$$\therefore \alpha=3,\beta=5 \quad [\because \alpha\le\beta \text{ given}]$$

$\alpha=3,\beta=5$ because $\alpha\le\beta$ [given]

$$\therefore \alpha^2+\beta=(3)^2+5=14$$

**37.** *(5)* Given,

$$\cos^{-1}x=\pi+\sin^{-1}x+\sin^{-1}(2x+1) \quad \ldots(i)$$

$$\Rightarrow \cos^{-1}x+\cos^{-1}x=\pi+\cos^{-1}x+\sin^{-1}x+\sin^{-1}(2x+1)$$

$$\Rightarrow 2\cos^{-1}x=\pi+\frac{\pi}{2}+\sin^{-1}(2x+1)$$

$$=\frac{3\pi}{2}+\sin^{-1}(2x+1)$$

$$\Rightarrow 2m=\frac{3\pi}{2}+n$$

where $m=\cos^{-1}x$

and $n=\sin^{-1}(2x+1)$

$$\Rightarrow \cos(2m)=\cos\left(\frac{3\pi}{2}+n\right)=\sin n$$

$$\Rightarrow 2\cos^2m-1=\sin n$$

$$\Rightarrow 2x^2-1=2x+1$$

$$(\because m=\cos^{-1}x,\ n=\sin^{-1}(2x+1))$$

$$\therefore x^2-x-1=0$$

$$\Rightarrow x=\frac{1\pm\sqrt{5}}{2}$$

But $x=\frac{1+\sqrt{5}}{2}$ (rejected) because at this value of $x$, Eq. (i) will be undefined.

$$\therefore x=\frac{1-\sqrt{5}}{2}$$

$$\text{So, }\sum_{x\in s}(2x-1)^2=\left[2\times\left(\frac{1-\sqrt{5}}{2}\right)-1\right]^2$$

$$=[1-\sqrt{5}-1]^2$$

$$=5$$

**38.** *(0)* $2\sin^{-1}x+3\cos^{-1}x=\frac{2\pi}{5}$

$$\pi+\cos^{-1}x=\frac{2\pi}{5}$$

$$\Rightarrow \cos^{-1}x=\frac{-3\pi}{5}$$

Not possible.

Thus, there are zero solution.

**39.** *(47)*

> **key idea** Firstly convert the given equation in the terms of $\tan^{-1} x$, then use the formula
> $$\tan^{-1} x + \tan^{-1} y = \tan^{-1}\left(\frac{x+y}{1-xy}\right)$$

According to the question,

$$\cot^{-1} 3 + \cot^{-1} 4 + \cot^{-1} 5 + \cot^{-1} n = \frac{\pi}{4}$$

By using $\cot^{-1} x = \tan^{-1}\left(\frac{1}{x}\right)$

and $\tan^{-1} x + \tan^{-1} y = \tan^{-1}\left(\frac{x+y}{1-xy}\right)$

$$\Rightarrow \tan^{-1}\left(\frac{1}{3}\right) + \tan^{-1}\left(\frac{1}{4}\right) + \tan^{-1}\left(\frac{1}{5}\right) + \tan^{-1}\left(\frac{1}{n}\right) = \frac{\pi}{4}$$

$$\Rightarrow \tan^{-1}\left(\frac{\frac{1}{3}+\frac{1}{4}}{1-\frac{1}{3}\times\frac{1}{4}}\right) + \tan^{-1}\left(\frac{\frac{1}{5}+\frac{1}{n}}{1-\frac{1}{5}\times\frac{1}{n}}\right) = \frac{\pi}{4}$$

$$\Rightarrow \tan^{-1}\left(\frac{7}{11}\right) + \tan^{-1}\left(\frac{n+5}{5n-1}\right) = \frac{\pi}{4}$$

$$\Rightarrow \tan^{-1}\left(\frac{\frac{7}{11}+\frac{(n+5)}{5n-1}}{1-\left(\frac{7}{11}\right)\left(\frac{n+5}{5n-1}\right)}\right) = \frac{\pi}{4}$$

$$\Rightarrow \frac{35n-7+11n+55}{11(5n-1)-7(n+5)} = 1$$

$$\Rightarrow n = 47$$

**40.** *(2)* Given equation is

$\sin^{-1} x = 2\tan^{-1} x$

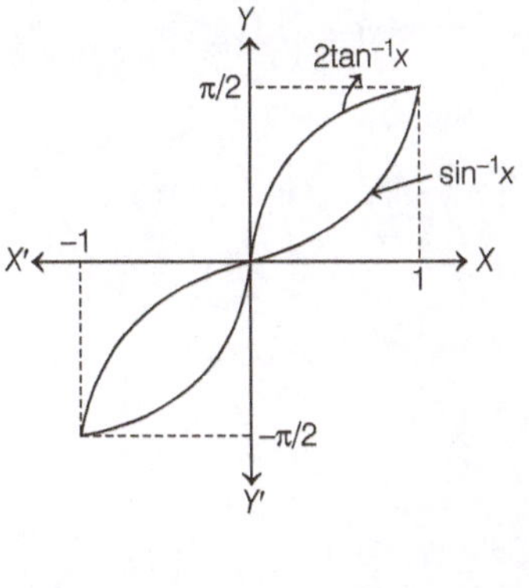


Clearly, from the above figure, the two graphs cut at two points in $-1 < x \le 1$. So, required number of solution is 2.

> **Alternate Solution**
>
> Given equation is $\sin^{-1} x = 2\tan^{-1} x$
>
> $\Rightarrow x = \sin(2\tan^{-1} x)$ ...(i)
>
> Let $\tan^{-1} x = \theta \Rightarrow \tan\theta = x$
>
> $\therefore \sin\theta = \frac{x}{\sqrt{x^2+1}}, \cos\theta = \frac{1}{\sqrt{x^2+1}}$
>
> From Eq. (i), we have
>
> $x = \sin 2\theta$
>
> $= 2\sin\theta\cos\theta$
>
> $= 2\times\frac{x}{\sqrt{x^2+1}}\times\frac{1}{\sqrt{x^2+1}}$
>
> $\Rightarrow x = \frac{2x}{x^2+1}$
>
> $\Rightarrow x^3 + x - 2x = 0$
>
> $\Rightarrow x^3 - x = 0$
>
> $\Rightarrow x = 0, -1, 1$
>
> But $x \ne -1$
>
> $[\because x \in (-1, 1)]$
>
> $\therefore x = 0, 1$
>
> Hence, the required number of solutions is 2.

**41.** *(4)* Given,

$$\sin^{-1}\left(\frac{x+1}{\sqrt{x^2+2x+2}}\right) - \sin^{-1}\left(\frac{x}{\sqrt{x^2+1}}\right) = \frac{\pi}{4}$$

$$\Rightarrow \tan^{-1}(x+1) - \tan^{-1} x = \frac{\pi}{4}$$

$$\Rightarrow \tan^{-1}\left(\frac{(x+1)-x}{1+(x+1)x}\right) = \frac{\pi}{4}$$

$$\Rightarrow \tan^{-1}\left(\frac{1}{1+x^2+x}\right) = \frac{\pi}{4}$$

$$\Rightarrow \tan^{-1}\left(\frac{1}{1+x^2+x}\right) = \frac{\pi}{4}$$

$$\Rightarrow \frac{1}{1+x+x^2} = \tan\left(\frac{\pi}{4}\right) = 1$$

$$\Rightarrow \frac{1}{1+x+x^2} = 1$$

$$\Rightarrow 1 + x + x^2 = 1$$

$$\Rightarrow x(x+1) = 0$$

$$\Rightarrow x = 0 \text{ or } x = -1$$

$$\therefore S = \{-1, 0\}$$

Now, $\sum_{x \in S}\left(\sin(x^2+x+5)\frac{\pi}{2} - \cos((x^2+x+5)\pi)\right)$

$$= \left[\sin\left(\frac{5\pi}{2}\right) - \cos 5\pi\right] + \left[\sin\left(\frac{5\pi}{2}\right) - \cos 5\pi\right]$$

$$= 2\left[\sin\left(2\pi + \frac{\pi}{2}\right) - (-1)\right]$$

$$= 2[1+1] = 4$$

**42.** *(2)*

> **key idea** We use property,
> $\cot^{-1} x = \tan^{-1}\frac{1}{x}, x > 0$ and
> $\cot^{-1} x = \pi + \tan^{-1}\frac{1}{x}, x < 0$ to solve
> $$\tan^{-1}\left(\frac{2x}{1-x^2}\right) + \cot^{-1}\left(\frac{1-x^2}{2x}\right) = \frac{\pi}{3}$$

$$\tan^{-1}\left(\frac{2x}{1-x^2}\right) + \cot^{-1}\left(\frac{1-x^2}{2x}\right) = \frac{\pi}{3},$$

$-1 < x < 1, x \ne 0$

We know that

$$\cot^{-1} x = \tan^{-1}\frac{1}{x} : x > 0$$

$$\cot^{-1} x = \pi + \tan^{-1}\frac{1}{x} : x < 0$$

Now, **Case I** $x > 0$

$$\tan^{-1}\frac{2x}{1-x^2} + \tan^{-1}\frac{2x}{1-x^2} = \frac{\pi}{3}$$

$$\Rightarrow 2\tan^{-1}\left(\frac{2x}{1-x^2}\right) = \frac{\pi}{3}$$

$$\tan^{-1}\frac{2x}{1-x^2} = \frac{\pi}{6}$$

$$\Rightarrow \frac{2x}{1-x^2} = \frac{1}{\sqrt{3}}$$

$$x^2 + 2\sqrt{3}x - 1 = 0$$

$$x = \frac{-2\sqrt{3} \pm \sqrt{12+4}}{2}$$

$$= -\sqrt{3} \pm 2$$

But $x > 0 \Rightarrow x = 2 - \sqrt{3}$

**Case II** $x < 0$

$$\tan^{-1}\frac{2x}{1-x^2} + \tan^{-1}\frac{2x}{1-x^2} + \pi = \frac{\pi}{3}$$

$$\tan^{-1}\left(\frac{2x}{1-x^2}\right) = -\frac{\pi}{3}$$

$$\frac{2x}{1-x^2} = -\sqrt{3}$$

$\Rightarrow \sqrt{3}\,x^2 - 2x - \sqrt{3} = 0$

$$x = \frac{2 \pm \sqrt{4+12}}{2\sqrt{3}} = \frac{2 \pm 4}{2\sqrt{3}} = \frac{-1}{\sqrt{3}}, \sqrt{3}$$

But $x < 0 \Rightarrow x = \dfrac{-1}{\sqrt{3}}$

Given, sum of all solutions $= \alpha - \dfrac{4}{\sqrt{3}}$

$$2 - \sqrt{3} - \frac{1}{\sqrt{3}} = \alpha - \frac{4}{\sqrt{3}} \Rightarrow \alpha = 2$$

**43.** *(130)* Given, $x = \sin(2\tan^{-1}\alpha)$

$$= \sin\left(\sin^{-1}\left(\frac{2\alpha}{1+\alpha^2}\right)\right)$$

$$= \frac{2\alpha}{1+\alpha^2} \quad \text{...(i)}$$

and $\quad y = \sin\left(\dfrac{1}{2}\tan^{-1}\dfrac{4}{3}\right)$

Let $\dfrac{1}{2}\tan^{-1}\dfrac{4}{3} = \theta$

$\Rightarrow \quad \tan 2\theta = \dfrac{4}{3} \Rightarrow \dfrac{2\tan\theta}{1-\tan^2\theta} = \dfrac{4}{3}$

$\Rightarrow \quad 4 - 4\tan^2\theta = 6\tan\theta$

$\Rightarrow \quad 4\tan^2\theta + 6\tan\theta - 4 = 0$

$\Rightarrow \quad 2(2\tan\theta - 1)(\tan\theta + 2) = 0$

$\Rightarrow \quad \tan\theta = \dfrac{1}{2} \Rightarrow \sin\theta = \dfrac{1}{\sqrt{5}}$

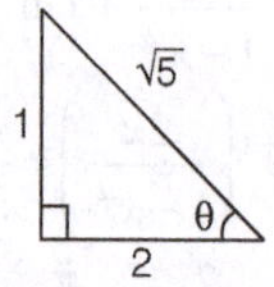


$\Rightarrow \quad \theta = \sin^{-1}(1/\sqrt{5})$

Hence, $y = \sin\left(\sin^{-1}\dfrac{1}{\sqrt{5}}\right) = \dfrac{1}{\sqrt{5}}$

Now, $y^2 = 1 - x$

$\Rightarrow \quad \dfrac{1}{5} = 1 - \dfrac{2\alpha}{1+\alpha^2}$

$\Rightarrow \quad 2\alpha^2 - 5\alpha + 2 = 0 \Rightarrow \alpha = 2, \dfrac{1}{2}$

Now, $\displaystyle\sum_{\alpha \in S} 16\alpha^3 = 16(2)^3 + 16\left(\frac{1}{2}\right)^3$

$$= 16(8) + 16\left(\frac{1}{8}\right) = 128 + 2 = 130$$

**44.** *(29)*

**key idea** Convert $\cos^{-1}\left(\dfrac{1}{\sqrt{5}}\right)$ in terms of $\tan^{-1}$ and simplify further using $\tan^{-1}x + \tan^{-1}y = \tan^{-1}\left(\dfrac{x+y}{1-xy}\right)$

$$50\tan\left(3\tan^{-1}\left(\frac{1}{2}\right) + 2\cos^{-1}\left(\frac{1}{\sqrt{5}}\right)\right) + 4\sqrt{2}\tan\left(\frac{1}{2}\tan^{-1}(2\sqrt{2})\right)$$

$$= 50\tan\left(3\tan^{-1}\left(\frac{1}{2}\right) + 2\tan^{-1}\left(\frac{2}{1}\right)\right) + 4\sqrt{2}\tan\left(\frac{1}{2}\tan^{-1}2\sqrt{2}\right)$$

$$= 50\tan\left(\tan^{-1}\left(\frac{1}{2}\right) + 2\left(\tan^{-1}\frac{1}{2} + \tan^{-1}2\right)\right) + 4\sqrt{2}\tan(\theta) \quad \text{... (i)}$$

Let $\theta = \dfrac{1}{2}\tan^{-1}2\sqrt{2}$ in Eq. (i) where $\theta$ is in I-quadrant

$\Rightarrow \quad \tan 2\theta = 2\sqrt{2}$

$\Rightarrow \quad \dfrac{2\tan\theta}{1-\tan^2\theta} = 2\sqrt{2}$

$\Rightarrow \quad \tan\theta = \sqrt{2}(1-\tan^2\theta)$

$\Rightarrow \quad \sqrt{2}\tan^2\theta + \tan\theta - \sqrt{2} = 0$

$\Rightarrow \quad (\tan\theta + \sqrt{2})(\sqrt{2}\tan\theta - 1) = 0$

$\Rightarrow \quad \tan\theta = \dfrac{1}{\sqrt{2}}$

Therefore, the expression in Eq. (i) is

$$50\tan\left(\tan^{-1}\frac{1}{2} + 2(\cot^{-1}2 + \tan^{-1}2)\right) + 4\sqrt{2}\tan(\theta)$$

$$= 50\tan\left(\tan^{-1}\frac{1}{2} + 2 \times \frac{\pi}{2}\right) + 4\sqrt{2} \times \frac{1}{\sqrt{2}}$$

$$= 50\tan\left(\pi + \tan^{-1}\frac{1}{2}\right) + 4$$

$$= 50\tan\left(\tan^{-1}\frac{1}{2}\right) + 4$$

$$= 50 \times \frac{1}{2} + 4$$

$$= 25 + 4 = 29$$

**Concept Enhancer**

$$\tan^{-1}x_1 + \tan^{-1}x_2 + \ldots + \tan^{-1}x_n = \tan^{-1}\frac{(S_1 - S_3 + S_5 - \ldots)}{1 - S_2 + S_4 - S_6 + \ldots}$$

where $S_k$ denotes the sum of the products of $x_1, x_2, \ldots, x_n$ taken $k$ at a time.

# C-15

JEE Main Chapterwise Topicwise Mathematics

In JEE Mains, there is usually 0-1 question per paper is visible from this chapter. In past 3 years, 17 conceptual and 5 numerical questions of moderate difficulty level were asked from this chapter.

# MATRICES

## Topic 1 Order of Matrices, Type of Matrices, Operations on Matrices

### Level 1

**01** Let $\alpha$ be a solution of $x^2 + x + 1 = 0$ and for some $a$ and $b$ in $R$, $[4\ a\ b]\begin{bmatrix} 1 & 16 & 13 \\ -1 & -1 & 2 \\ -2 & -14 & -8 \end{bmatrix} = [0\ 0\ 0]$. If $\dfrac{4}{\alpha^4} + \dfrac{m}{\alpha^a} + \dfrac{n}{\alpha^b} = 3$, then $m + n$ is equal to ........... *[2025, 8 April Shift-II]*

(a) 11 (b) 7 (c) 3 (d) 8

**02** Let $A$ be a $3 \times 3$ real matrix such that $A^2(A - 2I) - 4(A - I) = O$, where $I$ and $O$ are the identity and null matrices, respectively. If $A^5 = \alpha A^2 + \beta A + \gamma I$, where $\alpha$, $\beta$ and $\gamma$ are real constants, then $\alpha + \beta + \gamma$ is equal to *[2025, 2 April Shift-II]*

(a) 76 (b) 12 (c) 4 (d) 20

**03** Let $A = [a_{ij}]$ be a matrix of order $3 \times 3$ with $a_{ij} = (\sqrt{2})^{i+j}$. If the sum of all the elements in the third row of $A^2$ is $\alpha + \beta\sqrt{2}$, $\alpha, \beta \in Z$, then $\alpha + \beta$ is equal to *[2025, 29 Jan Shift-II]*

(a) 280 (b) 224 (c) 210 (d) 168

**04** Let $\alpha$ and $\beta$ be real numbers. Consider a $3 \times 3$ matrix $A$ such that $A^2 = 3A + \alpha I$. If $A^4 = 21A + \beta I$, then *[2023, 29 Jan Shift-I]*

(a) $\alpha = 1$ (b) $\alpha = 4$ (c) $\beta = 8$ (d) $\beta = -8$

**05** The number of square matrices of order 5 with entries from the set {0, 1}, such that the sum of all the elements in each row is 1 and the sum of all the elements in each column is also 1, is *[2023, 24 Jan Shift-II]*

(a) 120 (b) 125 (c) 225 (d) 150

**06** If $A$ and $B$ are two non-zero $n \times n$ matrices such that $A^2 + B = A^2 B$, then *[2023, 24 Jan Shift-I]*

(a) $A^2 B = BA^2$ (b) $A^2 B = I$

(c) $A^2 = I$ or $B = I$ (d) $AB = I$

**Imp.** Characteristic Equation of Matrix

**07** Let $A = \begin{bmatrix} 1 & 2 \\ -2 & -5 \end{bmatrix}$. Let $\alpha, \beta, \in R$ be such that $\alpha A^2 + \beta A = 2I$. Then, $\alpha + \beta$ is equal to *[2022, 27 July Shift-I]*

(a) $-10$ (b) $-6$ (c) 6 (d) 10

**08** If the matrix $A = \begin{bmatrix} 0 & 2 \\ K & -1 \end{bmatrix}$ satisfies $A(A^3 + 3I) = 2I$, then the value of $K$ is *[2021, 27 Aug Shift-I]*

(a) $\dfrac{1}{2}$ (b) $-\dfrac{1}{2}$

(c) $-1$ (d) 1

**09** Let $A = \begin{bmatrix} 1 & 2 \\ -1 & 4 \end{bmatrix}$. If $A^{-1} = \alpha I + \beta A$, $\alpha, \beta \in R$, $I$ is a $2 \times 2$ identity matrix, then $4(\alpha - \beta)$ is equal to *[2021, 27 July Shift-I]*

(a) 8/3 (b) 4

(c) 2 (d) 5

**10** Let $A = \begin{bmatrix} 1 & 0 & 0 \\ 1 & 1 & 0 \\ 1 & 1 & 1 \end{bmatrix}$ and $B = A^{20}$. Then, the sum of the elements of the first column of $B$ is *[JEE Main 2018]*

(a) 211 (b) 251

(c) 231 (d) 210

**11** Let $A$ be a $3 \times 3$ matrix such that $A^2 - 5A + 7I = 0$

**Statement I** $A^{-1} = \dfrac{1}{7}(5I - A)$

**Statement II** The polynomial $A^3 - 2A^2 - 3A + I$ can be reduced to $5(A - 4I)$. Then, *[JEE Main 2016]*

(a) Both the statements are true.

(b) Both the statements are false.

(c) Statement I is true, but Statement II is false.

(d) Statement I is false, but Statement II is true.

**12** If $A=\begin{bmatrix}1 & 0\\ 1 & 1\end{bmatrix}$ and $I=\begin{bmatrix}1 & 0\\ 0 & 1\end{bmatrix}$, then which one of the following holds, $\forall n \geq 1$, by the principle of mathematical induction? *[AIEEE 2005]*

(a) $A^n = 2^{n-1}A + (n-1)I$
(b) $A^n = nA + (n-1)I$
(c) $A^n = 2^{n-1}A - (n-1)I$
(d) $A^n = nA - (n-1)I$

## Level 2

**13** Let the matrix $A=\begin{bmatrix}1 & 0 & 0\\ 1 & 0 & 1\\ 0 & 1 & 0\end{bmatrix}$ satisfy $A^n = A^{n-2} + A^2 - I$ for $n \geq 3$. Then, the sum of all the elements of $A^{50}$ is *[2025, 4 April Shift-II]*

(a) 44 (b) 53
(c) 52 (d) 39

**14** Let $A=\begin{bmatrix}2 & a & 0\\ 1 & 3 & 1\\ 0 & 5 & b\end{bmatrix}$. If $A^3 = 4A^2 - A - 21I$, where $I$ is the identity matrix of order $3\times 3$. Then, $2a+3b$ is equal to *[2024, 8 April Shift-I]*

(a) −10 (b) −9
(c) −13 (d) −12

**15** If $A=\dfrac{1}{2}\begin{bmatrix}1 & \sqrt{3}\\ -\sqrt{3} & 1\end{bmatrix}$, then *[2023, 1 Feb Shift-II]*

(a) $A^{30} - A^{25} = 2I$ (b) $A^{30} + A^{25} - A = I$
(c) $A^{30} + A^{25} + A = I$ (d) $A^{30} = A^{25}$

**Imp.** nth Term of GP

**16** Let $A=[a_{ij}]$ be a square matrix of order 3 such that $a_{ij} = 2^{j-i}$, $\forall\, i, j = 1, 2, 3$. Then, the matrix $A^2 + A^3 + \ldots + A^{10}$ is equal to *[2022, 29 June Shift-I]*

(a) $\left(\dfrac{3^{10}-3}{2}\right)A$ (b) $\left(\dfrac{3^{10}-1}{2}\right)A$
(c) $\left(\dfrac{3^{10}+1}{2}\right)A$ (d) $\left(\dfrac{3^{10}+3}{2}\right)A$

**17** Let $\alpha$ be a root of the equation $x^2+x+1=0$ and the matrix $A=\dfrac{1}{\sqrt{3}}\begin{bmatrix}1 & 1 & 1\\ 1 & \alpha & \alpha^2\\ 1 & \alpha^2 & \alpha^4\end{bmatrix}$, then the matrix $A^{31}$ is equal to *[2020, 7 Jan Shift-I]*

(a) $A^3$ (b) $I_3$ (c) $A^2$ (d) $A$

## Level 3

**18** Let $A=[a_{ij}]$ be a $3\times 3$ matrix such that
$$A\begin{bmatrix}0\\1\\0\end{bmatrix}=\begin{bmatrix}0\\0\\1\end{bmatrix}, A\begin{bmatrix}4\\1\\3\end{bmatrix}=\begin{bmatrix}0\\1\\0\end{bmatrix} \text{ and } A\begin{bmatrix}2\\1\\2\end{bmatrix}=\begin{bmatrix}1\\0\\0\end{bmatrix}$$, then $a_{23}$ equals *[2025, 23 Jan Shift-II]*

(a) 2 (b) −1
(c) 1 (d) 0

**19** Let $A=\begin{bmatrix}1 & \frac{1}{51}\\ 0 & 1\end{bmatrix}$. If $B=\begin{bmatrix}1 & 2\\ -1 & -1\end{bmatrix}A\begin{bmatrix}-1 & -2\\ 1 & 1\end{bmatrix}$, then the sum of all the elements of the matrix $\sum_{n=1}^{50} B^n$ is equal to *[2023, 12 April Shift-I]*

(a) 100 (b) 75
(c) 125 (d) 50

**20** Let $A=\begin{bmatrix}1 & 0 & 0\\ 0 & 4 & -1\\ 0 & 12 & -3\end{bmatrix}$. Then, the sum of the diagonal elements of the matrix $(A+I)^{11}$ is equal to *[2023, 31 Jan Shift-I]*

(a) 2050 (b) 4094
(c) 6144 (d) 4097

**21** Let $A=\begin{bmatrix}1 & 0 & 0\\ 0 & 1 & 1\\ 1 & 0 & 0\end{bmatrix}$. Then, $A^{2025} - A^{2020}$ is equal to *[2021, 26 Aug Shift-II]*

(a) $A^6 - A$ (b) $A^5$
(c) $A^5 - A$ (d) $A^6$

**22** Let $P=\begin{bmatrix}1 & 0 & 0\\ 3 & 1 & 0\\ 9 & 3 & 1\end{bmatrix}$ and $Q=[q_{ij}]$ be two $3\times 3$ matrices such that $Q - P^5 = I_3$. Then, $\dfrac{q_{21}+q_{31}}{q_{32}}$ is equal to *[2019, 12 Jan Shift-I]*

(a) 10 (b) 135
(c) 9 (d) 15

### Numerical Value Type Questions

**23** Let $S=\{m \in Z : A^{m^2} + A^m = 3I - A^{-6}\}$, where $A=\begin{bmatrix}2 & -1\\ 1 & 0\end{bmatrix}$. Then, $n(S)$ is equal to ........... . *[2025, 29 Jan Shift-I]*

**Imp.** Integral Power of Matrix

**24** Let $A = \begin{bmatrix} 2 & -1 \\ 1 & 1 \end{bmatrix}$. If the sum of the diagonal elements of $A^{13}$ is $3^n$, then $n$ is equal to ......... . *[2019, 8 April Shift-I]*

**25** Let $A = \begin{bmatrix} 0 & 1 & 2 \\ a & 0 & 3 \\ 1 & c & 0 \end{bmatrix}$, where $a, c \in R$. If $A^3 = A$ and the positive value of $a$ belongs to the interval $(n-1, n]$, where $n \in N$, then $n$ is equal to ............ . *[2023, 11 April Shift-I]*

**26** Let $A = [a_{ij}]$, $a_{ij} \in Z \cap [0, 4], 1 \le i, j \le 2$.
The number of matrices $A$ such that the sum of all entries is a prime number $p \in (2, 13)$ is ............ . *[2023, 31 Jan Shift-II]*

**27** Let $X = \begin{bmatrix} 1 \\ 1 \\ 1 \end{bmatrix}$ and $A = \begin{bmatrix} -1 & 2 & 3 \\ 0 & 1 & 6 \\ 0 & 0 & -1 \end{bmatrix}$. For $k \in N$, if $X' A^k X = [33]$, then $k$ is equal to ....... . *[2022, 29 July Shift-II]*

**28** Let $A = \begin{bmatrix} 1 & -1 \\ 2 & \alpha \end{bmatrix}$ and $B = \begin{bmatrix} \beta & 1 \\ 1 & 0 \end{bmatrix}$, $\alpha, \beta \in R$. Let $\alpha_1$ be the value of $\alpha$ which satisfies $(A+B)^2 = A^2 + \begin{bmatrix} 2 & 2 \\ 2 & 2 \end{bmatrix}$ and $\alpha_2$ be the value of $\alpha$ which satisfies $(A+B)^2 = B^2$, then the value of $|\alpha_1 - \alpha_2|$ ............ . *[2022, 28 July Shift-I]*

**29** Let $A = \begin{bmatrix} 1 & a & a \\ 0 & 1 & b \\ 0 & 0 & 1 \end{bmatrix}$, $a, b \in R$.

If for some $n \in N$, $A^n = \begin{bmatrix} 1 & 48 & 2160 \\ 0 & 1 & 96 \\ 0 & 0 & 1 \end{bmatrix}$, then $n + a + b$ is equal to ............ . *[2022, 25 July Shift-II]*

**30** Let $M = \begin{bmatrix} 0 & -\alpha \\ \alpha & 0 \end{bmatrix}$, where $\alpha$ is a non-zero real number and $N = \sum_{k=1}^{49} M^{2k}$. If $(I - M^2)N = -2I$, then the positive integral value of $\alpha$ is ............ . *[2022, 29 June Shift-II]*

**31** Let $A = \begin{bmatrix} 1+i & 1 \\ -i & 0 \end{bmatrix}$, where $i = \sqrt{-1}$.
Then, the number of elements in the set $\{n \in \{1, 2, \ldots, 100\} : A^n = A\}$ is ...... . *[2022, 28 June Shift-II]*

**32** Let $X = \begin{bmatrix} 0 & 1 & 0 \\ 0 & 0 & 1 \\ 0 & 0 & 0 \end{bmatrix}$, $Y = \alpha I + \beta X + \gamma X^2$ and $Z = \alpha^2 I - \alpha\beta X + (\beta^2 - \alpha\gamma)X^2$, $\alpha, \beta, \gamma \in R$. If $Y^{-1} = \begin{bmatrix} 1/5 & -2/5 & 1/5 \\ 0 & 1/5 & -2/5 \\ 0 & 0 & 1/5 \end{bmatrix}$, then $(\alpha - \beta + \gamma)^2$ is equal to .......... . *[2022, 26 June Shift-II]*

**33** Let $A = \begin{bmatrix} 2 & -2 \\ 1 & -1 \end{bmatrix}$ and $B = \begin{bmatrix} -1 & 2 \\ -1 & 2 \end{bmatrix}$. Then, the number of elements in the set $\{(n, m) : n, m \in \{1, 2, \ldots, 10\}$ and $nA^n + mB^m = I\}$ is ...... . *[2022, 25 June Shift-II]*

**34** Let $S = \left\{ \begin{bmatrix} -1 & a \\ 0 & b \end{bmatrix}, a, b \in \{1, 2, 3, \ldots 100\} \right\}$ and let $T_n = \{A \in S : A^{n(n+1)} = I\}$. Then, the number of elements in $\bigcap_{n=1}^{100} T_n$ is ...... . *[2022, 24 June Shift-II]*

**35** If $A = \begin{bmatrix} 1 & 1 & 1 \\ 0 & 1 & 1 \\ 0 & 0 & 1 \end{bmatrix}$ and $M = A + A^2 + A^3 + \ldots + A^{20}$, then the sum of all the elements of the matrix $M$ is equal to ............... . *[2021, 27 July Shift-II]*

**36** Let $A = \begin{bmatrix} 0 & 1 & 0 \\ 1 & 0 & 0 \\ 0 & 0 & 1 \end{bmatrix}$. Then, the number of $3 \times 3$ matrices $B$ with entries from the set $\{1, 2, 3, 4, 5\}$ and satisfying $AB = BA$ is ......... . *[2021, 22 July Shift-II]*

**37** Let $A = \begin{bmatrix} 1 & -1 & 0 \\ 0 & 1 & -1 \\ 0 & 0 & 1 \end{bmatrix}$ and $B = 7A^{20} - 20A^7 + 2I$, where $I$ is an identity matrix of order $3 \times 3$. If $B = [b_{ij}]$, then $b_{13}$ is equal to ......... . *[2021, 20 July Shift-I]*

**38** If the matrix $A = \begin{bmatrix} 1 & 0 & 0 \\ 0 & 2 & 0 \\ 3 & 0 & -1 \end{bmatrix}$ satisfies the equation $A^{20} + \alpha A^{19} + \beta A = \begin{bmatrix} 1 & 0 & 0 \\ 0 & 4 & 0 \\ 0 & 0 & 1 \end{bmatrix}$ for some real numbers $\alpha$ and $\beta$, then $\beta - \alpha$ is equal to ______. *[2021, 26 Feb Shift-II]*

**39** Let $A = \begin{bmatrix} x & 1 \\ 1 & 0 \end{bmatrix}$, $x \in R$ and $A^4 = [a_{ij}]$. If $a_{11} = 109$, then $a_{22}$ is equal to .......... . [2020, 3 Sep Shift-I]

## Topic 2 Transpose of a Matrix, Symmetric, Skew-symmetric & Orthogonal Matrices

### Level 1

**01** Let $A$ be a square matrix such that $AA^T = I$. Then, $\frac{1}{2} A[(A + A^T)^2 + (A - A^T)^2]$ is equal to [2024, 29 Jan Shift - I]

(a) $A^2 + A^T$ (b) $A^3 + A^T$
(c) $A^2 + I$ (d) $A^3 + I$

**02** The number of symmetric matrices of order 3, with all the entries from the set {0, 1, 2, 3, 4, 5, 6, 7, 8, 9}, is [2023, 13 April Shift-I]

(a) $9^{10}$ (b) $10^6$
(c) $10^9$ (d) $6^{10}$

**03** Let $P$ be a square matrix such that $P^2 = I - P$. For $\alpha, \beta, \gamma, \delta \in N$. If $P^\alpha + P^\beta = \gamma I - 29P$ and $P^\alpha - P^\beta = \delta I - 13P$, then $\alpha + \beta + \gamma - \delta$ is equal to [2023, 6 April Shift-II]

(a) 22 (b) 18
(c) 24 (d) 40

**04** Let $A = \begin{bmatrix} 1 \\ 1 \\ 1 \end{bmatrix}$ and $B = \begin{bmatrix} 9^2 & -10^2 & 11^2 \\ 12^2 & 13^2 & -14^2 \\ -15^2 & 16^2 & 17^2 \end{bmatrix}$, then the value of $A'BA$ is [2022, 26 July Shift-II]

(a) 1224 (b) 1042
(c) 540 (d) 539

**05** The total number of matrices

$A = \begin{bmatrix} 0 & 2y & 1 \\ 2x & y & -1 \\ 2x & -y & 1 \end{bmatrix}$, $(x, y \in R, x \neq y)$ for which $A^T A = 3I_3$ is [2019, 9 April Shift-II]

(a) 2 (b) 4
(c) 3 (d) 6

**06** For two $3 \times 3$ matrices $A$ and $B$, let $A + B = 2B'$ and $3A + 2B = I_3$, where $B'$ is the transpose of $B$ and $I_3$ is $3 \times 3$ identity matrix. Then, [JEE Main 2017]

(a) $10A + 5B = 3I_3$ (b) $5A + 10B = 2I_3$
(c) $3A + 6B = 2I_3$ (d) $B + 2A = I_3$

### Level 2

**07** Let $A = \begin{bmatrix} \frac{1}{\sqrt{2}} & -2 \\ 0 & 1 \end{bmatrix}$ and $P = \begin{bmatrix} \cos\theta & -\sin\theta \\ \sin\theta & \cos\theta \end{bmatrix}$, $\theta > 0$. If $B = PAP^T$, $C = P^T B^{10} P$ and the sum of the diagonal elements of $C$ is $\frac{m}{n}$, where $\gcd(m, n) = 1$, then $m + n$ is [2025, 28 Jan Shift - I I]

(a) 127 (b) 65
(c) 258 (d) 2049

**08** Let $P = \begin{bmatrix} \frac{\sqrt{3}}{2} & \frac{1}{2} \\ -\frac{1}{2} & \frac{\sqrt{3}}{2} \end{bmatrix}$, $A = \begin{bmatrix} 1 & 1 \\ 0 & 1 \end{bmatrix}$ and $Q = PAP^T$. If $P^T Q^{2007} P = \begin{bmatrix} a & b \\ c & d \end{bmatrix}$, then $2a + b - 3c - 4d$ is equal to [2023, 8 April Shift-I]

(a) 2007 (b) 2004 (c) 2005 (d) 2006

**09** Let $A$ and $B$ be any two $3 \times 3$ symmetric and skew-symmetric matrices, respectively. Then, which of the following is not true? [2022, 28 July Shift-II]

(a) $A^4 - B^4$ is a symmetric matrix
(b) $AB - BA$ is a symmetric matrix
(c) $B^5 - A^5$ is a skew-symmetric matrix
(d) $AB + BA$ is a skew-symmetric matrix

**10** Let $A = \begin{bmatrix} 0 & -2 \\ 2 & 0 \end{bmatrix}$. If $M$ and $N$ are two matrices given by $M = \sum_{k=1}^{10} A^{2k}$ and $N = \sum_{k=1}^{10} A^{2k-1}$, then $MN^2$ is [2022, 25 June Shift-I]

(a) a non-identity symmetric matrix
(b) a skew-symmetric matrix
(c) neither symmetric nor skew-symmetric matrix
(d) an identity matrix

Imp. Trace of Matrix

**11** Let $A + 2B = \begin{bmatrix} 1 & 2 & 0 \\ 6 & -3 & 3 \\ -5 & 3 & 1 \end{bmatrix}$ and $2A - B = \begin{bmatrix} 2 & -1 & 5 \\ 2 & -1 & 6 \\ 0 & 1 & 2 \end{bmatrix}$. If $Tr(A)$ denotes the sum of all diagonal elements of the matrix $A$, then $Tr(A) - Tr(B)$ has value equal to [2021, 18 March Shift-I]

(a) 1 (b) 2 (c) 0 (d) 3

**12** Let $a, b, c \in R$ be all non-zero and satisfy $a^3 + b^3 + c^3 = 2$. If the matrix $A = \begin{bmatrix} a & b & c \\ b & c & a \\ c & a & b \end{bmatrix}$ satisfies $A^T A = I$, then a value of $abc$ can be **[2020, 2 Sep Shift-II]**

(a) $-\frac{1}{3}$ (b) $\frac{1}{3}$

(c) 3 (d) $\frac{2}{3}$

**13** Let $A = \begin{bmatrix} 0 & 2q & r \\ p & q & -r \\ p & -q & r \end{bmatrix}$. If $AA^T = I_3$, then $|p|$ is **[2019, 11 Jan Shift-I]**

(a) $\frac{1}{\sqrt{5}}$ (b) $\frac{1}{\sqrt{2}}$

(c) $\frac{1}{\sqrt{3}}$ (d) $\frac{1}{\sqrt{6}}$

**14** If $P = \begin{bmatrix} \frac{\sqrt{3}}{2} & \frac{1}{2} \\ -\frac{1}{2} & \frac{\sqrt{3}}{2} \end{bmatrix}$, $A = \begin{bmatrix} 1 & 1 \\ 0 & 1 \end{bmatrix}$ and $Q = PAP^T$, then $P^T Q^{2015} P$ is **[JEE Main 2016]**

(a) $\begin{bmatrix} 0 & 2015 \\ 0 & 0 \end{bmatrix}$ (b) $\begin{bmatrix} 2015 & 0 \\ 1 & 2015 \end{bmatrix}$

(c) $\begin{bmatrix} 1 & 2015 \\ 0 & 1 \end{bmatrix}$ (d) $\begin{bmatrix} 2015 & 1 \\ 0 & 2015 \end{bmatrix}$

**15** If $A = \begin{bmatrix} 1 & 2 & 2 \\ 2 & 1 & -2 \\ a & 2 & b \end{bmatrix}$ is a matrix satisfying the equation $AA^T = 9I$, where $I$ is a $3 \times 3$ identity matrix, then the ordered pair $(a, b)$ is equal to **[JEE Main 2015]**

(a) (2, 1) (b) (−2, −1)
(c) (2, −1) (d) (−2, 1)

**16** Let $A$ and $B$ be two symmetric matrices of order 3.
**Statement I** $A(BA)$ and $(AB)A$ are symmetric matrices.
**Statement II** $AB$ is symmetric matrix, if matrix multiplication of $A$ with $B$ is commutative. **[AIEEE 2011]**

(a) Statement I is true, Statement II is true; Statement II is not a correct explanation of Statement I
(b) Statement I is true, Statement II is false
(c) Statement I is false, Statement II is true
(d) Statement I is true, Statement II is true; Statement II is a correct explanation of Statement I

## Level 3

**17** If $A = \begin{bmatrix} 1 & 5 \\ \lambda & 10 \end{bmatrix}$, $A^{-1} = \alpha A + \beta I$ and $\alpha + \beta = -2$, then $4\alpha^2 + \beta^2 + \lambda^2$ is equal to **[2023, 8 April Shift-II]**

(a) 14 (b) 12 (c) 19 (d) 10

**18** Let $A$, $B$, $C$ be $3 \times 3$ matrices such that $A$ is symmetric and $B$ and $C$ are skew-symmetric. Consider the statements
(S1) $A^{13} B^{26} - B^{26} A^{13}$ is symmetric.
(S2) $A^{26} C^{13} - C^{13} A^{26}$ is symmetric.
Then, **[2023, 25 Jan Shift-II]**

(a) Both S1 and S2 are false (b) only S1 is true
(c) only S2 is true (d) Both S1 and S2 are true

**19** Let $A$ be a symmetric matrix of order 2 with integer entries. If the sum of the diagonal elements of $A^2$ is 1, then the possible number of such matrices is **[2021, 26 Feb Shift-I]**

(a) 4 (b) 1 (c) 6 (d) 12

**Imp.** Symmetric and Skew-Symmetric Matrix

**20** Let $M$ denote the set of all real matrices of order $3 \times 3$ and let $S = \{-3, -2, -1, 1, 2\}$. Let
$S_1 = \{A = [a_{ij}] \in M : A = A^T \text{ and } a_{ij} \in S, \forall i, j\}$,
$S_2 = \{A = [a_{ij}] \in M : A = -A^T \text{ and } a_{ij} \in S, \forall i, j\}$,
$S_3 = \{A = [a_{ij}] \in M : a_{11} + a_{22} + a_{33} = 0 \text{ and } a_{ij} \in S, \forall i, j\}$.
If $n(S_1 \cup S_2 \cup S_3) = 125\,\alpha$, then $\alpha$ equals ...... **[2025, 28 Jan Shift-I]**

## Numerical Value Type Questions

**21** Let $A = \begin{bmatrix} \cos\theta & 0 & -\sin\theta \\ 0 & 1 & 0 \\ \sin\theta & 0 & \cos\theta \end{bmatrix}$. If for some $\theta \in (0, \pi)$. $A^2 = A^T$, then the sum of the diagonal elements of the matrix $(A + I)^3 + (A - I)^3 - 6A$ is equal to .......... **[2025, 4 April Shift-I]**

**22** Let $A$ be a symmetric matrix such that $|A| = 2$ and $\begin{bmatrix} 2 & 1 \\ 3 & 3/2 \end{bmatrix} A = \begin{bmatrix} 1 & 2 \\ \alpha & \beta \end{bmatrix}$. If the sum of the diagonal elements of $A$ is $S$, then $\frac{\beta \cdot S}{\alpha^2}$ is equal to .......... . **[2023, 29 Jan Shift-II]**

**23** Let $S$ be the set containing all $3 \times 3$ matrices with entries from $\{-1, 0, 1\}$. The total number of matrices $A \in S$ such that the sum of all the diagonal elements of $A^T A$ is 6, is ......... . **[2022, 27 July Shift-I]**

**24** The total number of $3\times 3$ matrices $A$ having entries from the set {0, 1, 2, 3}, such that the sum of all the diagonal entries of $AA^T$ is 9, is equal to ......... . *[2021, 16 March Shift-I]*

**25** Let $A=\begin{bmatrix} x & y & z \\ y & z & x \\ z & x & y \end{bmatrix}$, where $x$, $y$ and $z$ are real numbers such that $x+y+z>0$ and $xyz=2$. If $A^2=I_3$, then the value of $x^3+y^3+z^3$ is ............. . *[2021, 25 Feb Shift I]*

**26** Let $M$ be any $3\times 3$ matrix with entries from the set {0, 1, 2}. The maximum number of such matrices, for which the sum of diagonal elements of $M^T M$ is seven, is ........ *[2021, 25 Feb Shift-I]*

**27** The number of all $3\times 3$ matrices $A$, with enteries from the set $\{-1, 0, 1\}$ such that the sum of the diagonal elements of $AA^T$ is 3, is ......... . *[2020, 8 Jan Shift-I]*

## Key Index

### Topic 1

| | | | | | | | | | | | | | | | | | | | |
|---|---|---|---|---|---|---|---|---|---|---|---|---|---|---|---|---|---|---|---|
| 1 | (a) | 2 | (b) | 3 | (b) | 4 | (d) | 5 | (a) | 6 | (a) | 7 | (d) | 8 | (a) | 9 | (b) | 10 | (c) |
| 11 | (a) | 12 | (d) | 13 | (b) | 14 | (c) | 15 | (b) | 16 | (a) | 17 | (a) | 18 | (b) | 19 | (a) | 20 | (d) |
| 21 | (a) | 22 | (a) | 23 | (2) | 24 | (7) | 25 | (2) | 26 | (204) | 27 | (10) | 28 | (2) | 29 | (24) | 30 | (1) |
| 31 | (25) | 32 | (100) | 33 | (1) | 34 | (100) | 35 | (2020) | 36 | (3125) | 37 | (910) | 38 | (4) | 39 | (10) | | |

### Topic 2

| | | | | | | | | | | | | | | | | | | | |
|---|---|---|---|---|---|---|---|---|---|---|---|---|---|---|---|---|---|---|---|
| 1 | (b) | 2 | (b) | 3 | (c) | 4 | (d) | 5 | (b) | 6 | (a) | 7 | (b) | 8 | (c) | 9 | (c) | 10 | (a) |
| 11 | (b) | 12 | (b) | 13 | (b) | 14 | (c) | 15 | (b) | 16 | (a) | 17 | (a) | 18 | (c) | 19 | (a) | 20 | (1613) |
| 21 | (6) | 22 | (5) | 23 | (5376) | 24 | (766) | 25 | (7) | 26 | (540) | 27 | (672) | | | | | | |

# Detailed **Explanations**

## Topic 1

**01.** (a) $[4-a-2b, \quad 64-a-14b, \quad 52+2a-8b] = [0 \quad 0 \quad 0]$

$\Rightarrow \quad a+2b=4 \quad \ldots(i)$

$a+14b=64 \quad \ldots(ii)$

$2a-8b=-52 \quad \ldots(iii)$

From Eqs. (i) and (ii), we get

$b=5 \quad a=-6$

Given $\dfrac{4}{\alpha^4}+\dfrac{m}{\alpha^{-6}}+\dfrac{n}{\alpha^5}=3$

Since, $\alpha$ is solution of $x^2+x+1=0$

$\alpha=\omega$ or $\omega^2$

$$\frac{4}{\omega^4}+m\omega^6+\frac{n}{\omega^5}=3$$

$$\frac{4}{\omega}+m+\frac{n}{\omega^2}=3$$

$$4\omega^2+m+n\omega=3$$

$$4\left(\frac{-1+\sqrt{3}i}{2}\right)+m+n\left(\frac{-1-\sqrt{3}i}{2}\right)=3$$

$$-2+2\sqrt{3}i+m-\frac{n}{2}-i\frac{\sqrt{3}n}{2}=3$$

Equating real and imaginary parts

$$m-\frac{n}{2}=5$$

$$2\sqrt{3}=\frac{\sqrt{3}n}{2}\Rightarrow n=4$$

$$m-2=5$$

$$m=7$$

$\therefore \quad m+4=11$

**02.** (b) Given, $A^2(A-2I)-4(A-I)=0$

$\Rightarrow A^3=2A^2+4A-4I$

$\Rightarrow A^4=2(2A^2+4A-4I)+4A^2-4A$

$\Rightarrow A^4=8A^2+4A-8I$

$\Rightarrow A^5=8(2A^2+4A-4I)+4A^2-8A$

$=20A^2+24A-32I$

Given, $\quad A^5=\alpha A^2+\beta A+\gamma I$

$\Rightarrow \alpha+\beta+\gamma=20+24-32=12$

**03.** (b) We have,

$A=[a_{ij}]$, a $3\times 3$ matrix,

where $a_{ij}=(\sqrt{2})^{i+j}$

$$\Rightarrow A=\begin{bmatrix}(\sqrt{2})^2 & (\sqrt{2})^3 & (\sqrt{2})^4\\ (\sqrt{2})^3 & (\sqrt{2})^4 & (\sqrt{2})^5\\ (\sqrt{2})^4 & (\sqrt{2})^5 & (\sqrt{2})^6\end{bmatrix}$$

$$\Rightarrow A=\begin{bmatrix}2 & 2\sqrt{2} & 4\\ 2\sqrt{2} & 4 & 4\sqrt{2}\\ 4 & 4\sqrt{2} & 8\end{bmatrix}$$

$$A^2=2^2\begin{bmatrix}1 & \sqrt{2} & 2\\ \sqrt{2} & 2 & 2\sqrt{2}\\ 2 & 2\sqrt{2} & 4\end{bmatrix}\begin{bmatrix}1 & \sqrt{2} & 2\\ \sqrt{2} & 2 & 2\sqrt{2}\\ 2 & 2\sqrt{2} & 4\end{bmatrix}$$

$$=4\begin{bmatrix}- & - & -\\ - & - & -\\ (2+4+8) & (2\sqrt{2}+4\sqrt{2}+8\sqrt{2}) & (4+8+16)\end{bmatrix}$$

Sum of elements of 3rd row

$=4(14+14\sqrt{2}+28)=4(42+14\sqrt{2})$

$=168+56\sqrt{2}=\alpha+\beta\sqrt{2}$

$\therefore \alpha+\beta=168+56=224$

**04.** (d) We have, $\alpha, \beta\in R$ and $3\times 3$ matrix $A$ such that $A^2=3A+\alpha I \quad \ldots$ (i)

and $\quad A^4=21A+\beta I \quad \ldots$ (ii)

Consider $A^4=A^2\cdot A^2$

$=(3A+\alpha I)(3A+\alpha I)$ [From Eq. (i)]

$=9A^2+6\alpha A+\alpha^2 I$

From Eq. (i),

$A^4=9(3A+\alpha I)+6\alpha A+\alpha^2 I$

$=(27+6\alpha)A+(\alpha^2+9\alpha)I \quad \ldots$(iii)

From Eqs. (ii) and (iii), we get

$21A+\beta I=(27+6\alpha)A+(\alpha^2+9\alpha)I$

$\Rightarrow \quad (\beta-\alpha^2-9\alpha)I=(6+6\alpha)A$

For a general $3\times 3$ matrix $A$, we get

$6+6\alpha=0\Rightarrow 6\alpha=-6$

$\Rightarrow \quad \alpha=-1 \quad \ldots$(iv)

and $\quad \beta-\alpha^2-9\alpha=0$

$\Rightarrow \quad \beta=\alpha^2+9\alpha=(-1)^2+9(-1)$

$=1-9=-8$

$\therefore \quad \beta=-8$

**05.** (a) Given the order of the matrix is 5.

$\because$ Sum of each row as well as each column $=1$. It means, if the first position of each row or column is 1, then remaining position contains 0. Also, the position of 1 in each row or column is as per the permutation law.

Hence, number of ways

$=5\times 4\times 3\times 2\times 1$

$=120$ ways

**06.** (a) Given, $A^2+B=A^2B \quad \ldots$(i)

$(A^2-I)(B-I)=I \quad \ldots$(ii)

Again, $A^2+B=A^2B$

$\Rightarrow \quad B=A^2B-A^2=A^2(B-I)$

$\Rightarrow \quad A^2=B(B-I)^{-1}$

$\Rightarrow \quad A^2=B(A^2-I)$ [from Eq. (ii)]

$\Rightarrow \quad A^2=BA^2-B$

$\Rightarrow \quad A^2+B=BA^2$

So, $A^2B=BA^2$ [using Eq. (i)]

**07.** (d) $A=\begin{bmatrix}1 & 2\\ -2 & -5\end{bmatrix}$

Characteristic equation of $A$ is

$|A-\lambda I|=0$

$$\Rightarrow \begin{vmatrix}1-\lambda & 2\\ -2 & -5-\lambda\end{vmatrix}=0$$

$\Rightarrow \quad \lambda^2+4\lambda-1=0$

$\therefore \quad A^2+4A-I=0$

$\Rightarrow \quad 2A^2+8A=2I$

$\Rightarrow \quad \alpha=2, \beta=8$

$\therefore \quad \alpha+\beta=10$

**08.** (a) Given matrix, $A=\begin{bmatrix}0 & 2\\ K & -1\end{bmatrix}$

Characteristic equation of $A$ is

$|A-\lambda I|=0$

$$\Rightarrow \begin{vmatrix}-\lambda & 2\\ K & -1-\lambda\end{vmatrix}=0$$

$\Rightarrow \quad \lambda^2+\lambda-2k=0$

$\because$ Every square matrix satisfied its own characteristic equation.

$\therefore \quad A^2+A-2KI=0$

$\Rightarrow \quad A^2=2KI-A$

$\Rightarrow \quad A^4=4K^2I+A^2-2(2KI)(A)$

$\Rightarrow \quad A^4=4K^2I+2KI-A-4KA$

$\Rightarrow \quad A^4=(4K^2+2K)I-(1+4K)A \quad \ldots$(i)

Now, $A(A^3+3I)=2I$

$\Rightarrow \quad A^4=2I-3A \quad \ldots$ (ii)

On comparing Eqs. (i) and (ii), we get

$1+4K=3$

$\Rightarrow \quad K=\dfrac{1}{2}$

**Alternate Solution**

Given, $A=\begin{bmatrix}0 & 2\\ K & -1\end{bmatrix}$

$$A^2=\begin{bmatrix}0 & 2\\ K & -1\end{bmatrix}\begin{bmatrix}0 & 2\\ K & -1\end{bmatrix}=\begin{bmatrix}2K & -2\\ -K & 2K+1\end{bmatrix}$$

$$A^4=A^2\cdot A^2=\begin{bmatrix}2K & -2\\ -K & 2K+1\end{bmatrix}\begin{bmatrix}2K & -2\\ -K & 2K+1\end{bmatrix}$$

$$=\begin{bmatrix}4K^2+2K & -8K-2\\ -4K^2-K & 4K^2+6K+1\end{bmatrix}$$

$\because A(A^3+3I)=2I$

$\Rightarrow A^4+3A=2I$

$$\Rightarrow \begin{bmatrix}4K^2+2K & -8K-2\\ -4K^2-K & 4K^2+6K+1\end{bmatrix}+3\begin{bmatrix}0 & 2\\ K & -1\end{bmatrix}=2\begin{bmatrix}1 & 0\\ 0 & 1\end{bmatrix}$$

$$\Rightarrow \begin{bmatrix}4K^2+2K & -8K+4\\ -4K^2+2K & 4K^2+6K-2\end{bmatrix}=\begin{bmatrix}2 & 0\\ 0 & 2\end{bmatrix}$$

On comparing both sides, we get

$$-8K+4=0 \Rightarrow K=\frac{1}{2}$$

**09.** *(b)* Given, $A=\begin{bmatrix}1 & 2\\ -1 & 4\end{bmatrix}$

$$\Rightarrow A^2=\begin{bmatrix}1 & 2\\ -1 & 4\end{bmatrix}\begin{bmatrix}1 & 2\\ -1 & 4\end{bmatrix}=\begin{bmatrix}-1 & 10\\ -5 & 14\end{bmatrix}$$

Now, $A^{-1}=\alpha I+\beta A$ ...(i)

On multiplying $A$ in Eq. (i), we get

$$A^{-1}A=\alpha IA+\beta AA \Rightarrow I=\alpha A+\beta A^2$$

$$\Rightarrow \begin{bmatrix}1 & 0\\ 0 & 1\end{bmatrix}=\begin{bmatrix}\alpha & 2\alpha\\ -\alpha & 4\alpha\end{bmatrix}+\begin{bmatrix}-\beta & 10\beta\\ -5\beta & 14\beta\end{bmatrix}$$

$$\Rightarrow \begin{bmatrix}1 & 0\\ 0 & 1\end{bmatrix}=\begin{bmatrix}\alpha-\beta & 2\alpha+10\beta\\ -\alpha-5\beta & 4\alpha+14\beta\end{bmatrix}$$

On comparing, we get

$\alpha-\beta=1$ ...(ii)

$\alpha+5\beta=0$ ...(iii)

On putting $\alpha-\beta=1$, we get

$4(\alpha-\beta)=4\times 1=4$

**Alternate Solution**

$$A^2=\begin{bmatrix}-1 & 10\\ -5 & 14\end{bmatrix}$$

$$A^{-1}=\alpha I+\beta A \Rightarrow I=\alpha A+\beta A^2$$

$$\begin{bmatrix}1 & 0\\ 0 & 1\end{bmatrix}=\begin{bmatrix}\alpha-\beta & 2\alpha+10\beta\\ -\alpha-5\beta & 4\alpha+14\beta\end{bmatrix}$$

On comparing, we get

$\alpha-\beta=1$

$\therefore$ $4(\alpha-\beta)=4$

**10.** *(c)* $A=\begin{bmatrix}1 & 0 & 0\\ 1 & 1 & 0\\ 1 & 1 & 1\end{bmatrix}$, $A^2=\begin{bmatrix}1 & 0 & 0\\ 2 & 1 & 0\\ 3 & 2 & 1\end{bmatrix}$;

$$A^3=\begin{bmatrix}1 & 0 & 0\\ 3 & 1 & 0\\ 6 & 3 & 1\end{bmatrix};\ A^4=\begin{bmatrix}1 & 0 & 0\\ 4 & 1 & 0\\ 10 & 4 & 1\end{bmatrix}$$

$$A^{20}=\begin{bmatrix}1 & 0 & 0\\ 20 & 1 & 0\\ 210 & 20 & 1\end{bmatrix}$$

We have,

$B=A^{20}$

$\therefore$ Sum of the elements of the first column of $B=1+20+210=231$

**11.** *(a)* Given, $A^2-5A=-7I$

$\Rightarrow AAA^{-1}-5AA^{-1}=-7IA^{-1}$

$\Rightarrow A^{-1}=\frac{1}{7}(5I-A)$ $\quad[\because AA^{-1}=I]$

Also, $A^3-2A^2-3A+I$

$=A(5A-7I)-2A^2-3A+I$

$=5A^2-7A-2A^2-3A+I$

$=3A^2-10A+I$

$=3(5A-7I)-10A+I$

$=5A-20I$

$=5(A-4I)$

**12.** *(d)* Given that $A=\begin{bmatrix}1 & 0\\ 1 & 1\end{bmatrix}$

$$\therefore\ A^2=\begin{bmatrix}1 & 0\\ 1 & 1\end{bmatrix}\begin{bmatrix}1 & 0\\ 1 & 1\end{bmatrix}=\begin{bmatrix}1 & 0\\ 2 & 1\end{bmatrix}$$

$$A^3=\begin{bmatrix}1 & 0\\ 2 & 1\end{bmatrix}\begin{bmatrix}1 & 0\\ 1 & 1\end{bmatrix}=\begin{bmatrix}1 & 0\\ 3 & 1\end{bmatrix}$$

$\vdots\ \vdots\ \vdots\ \vdots\ \vdots\ \vdots$

$A^n=\begin{bmatrix}1 & 0\\ n & 1\end{bmatrix}$ can be verified by induction.

Now, taking option (d).

$$nA-(n-1)I=\begin{bmatrix}n & 0\\ n & n\end{bmatrix}-\begin{bmatrix}n-1 & 0\\ 0 & n-1\end{bmatrix}$$

$$=\begin{bmatrix}1 & 0\\ n & 1\end{bmatrix}=A^n$$

**13.** *(b)* $A=\begin{bmatrix}1 & 0 & 0\\ 1 & 0 & 1\\ 0 & 1 & 0\end{bmatrix}$

$A^n=A^{n-2}+A^2-I$

$A^n-A^{n-2}=A^2-I$

Put $n=4 \Rightarrow A^4-A^2=A^2-I$

$n=6 \Rightarrow A^6-A^4=A^2-I$

$n=8 \Rightarrow A^8-A^6=A^2-I$

$\vdots$

$n=50,\ A^{50}-A^{48}=A^2-I$

$\therefore A^{50}-A^2=24(A^2-I)$

$A^{50}=25A^2-24I$ ...(i)

$$\text{Now, } A^2=\begin{bmatrix}1 & 0 & 0\\ 1 & 0 & 1\\ 0 & 1 & 0\end{bmatrix}\begin{bmatrix}1 & 0 & 0\\ 1 & 0 & 1\\ 0 & 1 & 0\end{bmatrix}$$

$$=\begin{bmatrix}1 & 0 & 0\\ 1 & 1 & 0\\ 1 & 0 & 1\end{bmatrix}$$

$$\therefore A^{50}=\begin{bmatrix}25-24 & 0 & 0\\ 25 & 25-24 & 0\\ 25 & 0 & 25-24\end{bmatrix}$$

Sum of all elements $=1+1+1+25+25$

$=53$

**14.** *(c)*

> **key idea** Use, if $A$ is $3\times 3$ matrices, then $A^3-\text{Tr}(A)(A)^2+\Sigma A_{ii}\ (A)+|A|\ I=0$
> where $A_{ij}$ = cofactor of $a_{ij}$ to solve the questions.

$$A=\begin{bmatrix}2 & a & 0\\ 1 & 3 & 1\\ 0 & 5 & b\end{bmatrix}$$

and $A^3=4A^2-A-21I$

So, $x^3-4x^2+x+21=0$ is the characteristic equation of matrix $A$.

So, sum of roots $=T_r(A)$

$\Rightarrow\quad -(-4)=2+3+b$

$\Rightarrow\quad 5+b=4$

$\Rightarrow\quad b=-1$ ...(i)

Product of roots $=|A|$

$\Rightarrow\quad -21=2(3b-5)-a(b)$

Put $b=-1$ from Eq. (i),

$-21=2(-8)+a$

$\Rightarrow\quad a=-21+16=-5$

Therefore,

$2a+3b=2(-5)+3(-1)$

$=-13$

**15.** *(b)*

> **key idea** By converting number into $\sin\theta$ and $\cos\theta$, it is easy to determine the value of higher power of matrix $A$ (such as $A^{10}, A^{30}, \ldots, A^n$)

We have,

$$A=\frac{1}{2}\begin{bmatrix}1 & \sqrt{3}\\ -\sqrt{3} & 1\end{bmatrix}$$

$$= \begin{bmatrix} \frac{1}{2} & \frac{\sqrt{3}}{2} \\ -\frac{\sqrt{3}}{2} & \frac{1}{2} \end{bmatrix} = \begin{bmatrix} \cos 60° & \sin 60° \\ -\sin 60° & \cos 60° \end{bmatrix}$$

We know that if $A_\theta = \begin{bmatrix} \cos\theta & \sin\theta \\ -\sin\theta & \cos\theta \end{bmatrix}$

Then, $A_\theta^n = \begin{bmatrix} \cos n\theta & \sin n\theta \\ -\sin n\theta & \cos n\theta \end{bmatrix}$

Now, $A_{\frac{\pi}{3}} = \begin{bmatrix} \cos\frac{\pi}{3} & \sin\frac{\pi}{3} \\ -\sin\frac{\pi}{3} & \cos\frac{\pi}{3} \end{bmatrix}$

$$\therefore A_{\frac{\pi}{3}}^{30} = \begin{bmatrix} \cos\left(30\cdot\frac{\pi}{3}\right) & \sin\left(30\cdot\frac{\pi}{3}\right) \\ -\sin\left(30\cdot\frac{\pi}{3}\right) & \cos\left(30\cdot\frac{\pi}{3}\right) \end{bmatrix}$$

$$= \begin{bmatrix} \cos 10\pi & \sin 10\pi \\ -\sin 10\pi & \cos 10\pi \end{bmatrix} = \begin{bmatrix} 1 & 0 \\ 0 & 1 \end{bmatrix} = I$$

Again, $A_{\frac{\pi}{3}}^{25} = \begin{bmatrix} \cos\left(25\cdot\frac{\pi}{3}\right) & \sin\left(25\cdot\frac{\pi}{3}\right) \\ -\sin\left(25\cdot\frac{\pi}{3}\right) & \cos\left(25\cdot\frac{\pi}{3}\right) \end{bmatrix}$

$$= \begin{bmatrix} \frac{1}{2} & \frac{\sqrt{3}}{2} \\ -\frac{\sqrt{3}}{2} & \frac{1}{2} \end{bmatrix} = A$$

$$\therefore A^{30} + A^{25} - A = I + A - A = I$$

**16.** *(a)*

> **key idea** The sum of $n$ terms of GP is given by the formula $S_n = \frac{a(r^n - 1)}{r - 1}$, where $a$ is first term, $r$ is common ratio and $n$ is the number of terms.

Given, $A = [a_{ij}]$

$a_{ij} = 2^{j-i}$ for all $i, j = 1, 2, 3$.

$$A = \begin{bmatrix} 1 & 2 & 2^2 \\ 2^{-1} & 1 & 2 \\ 2^{-2} & 2^{-1} & 1 \end{bmatrix} = \begin{bmatrix} 1 & 2 & 4 \\ 1/2 & 1 & 2 \\ 1/4 & 1/2 & 1 \end{bmatrix}$$

$$A^2 = \begin{bmatrix} 1 & 2 & 4 \\ 1/2 & 1 & 2 \\ 1/4 & 1/2 & 1 \end{bmatrix}\begin{bmatrix} 1 & 2 & 4 \\ 1/2 & 1 & 2 \\ 1/4 & 1/2 & 1 \end{bmatrix}$$

$$= \begin{bmatrix} 3 & 6 & 12 \\ 3/2 & 3 & 6 \\ 3/4 & 3/2 & 3 \end{bmatrix} = 3A$$

$$\therefore A^3 = A^2 \cdot A = (3A)A = 3A^2$$

$$= 3(3A) = 3^2 A$$

$$\therefore A^2 + A^3 + A^4 + \dots + A^{10}$$
$$= 3A + 3^2 A + 3^3 A + \dots + 3^9 A$$
$$= 3\left[\frac{3^9 - 1}{3 - 1}\right]A$$
$$= \left(\frac{3^{10} - 3}{2}\right)A$$

**17.** *(a)* It is given that $\alpha$ is the root of equation $x^2 + x + 1 = 0$, so $\alpha^2 + \alpha + 1 = 0$ and $\alpha^3 = 1$.

Now, matrix $A = \frac{1}{\sqrt{3}}\begin{bmatrix} 1 & 1 & 1 \\ 1 & \alpha & \alpha^2 \\ 1 & \alpha^2 & \alpha^4 \end{bmatrix}$

$$= \frac{1}{\sqrt{3}}\begin{bmatrix} 1 & 1 & 1 \\ 1 & \alpha & \alpha^2 \\ 1 & \alpha^2 & \alpha \end{bmatrix} \quad [\because \alpha^3 = 1]$$

$$\therefore A^2 = \left(\frac{1}{\sqrt{3}}\right)^2 \begin{bmatrix} 1 & 1 & 1 \\ 1 & \alpha & \alpha^2 \\ 1 & \alpha^2 & \alpha \end{bmatrix}\begin{bmatrix} 1 & 1 & 1 \\ 1 & \alpha & \alpha^2 \\ 1 & \alpha^2 & \alpha \end{bmatrix}$$

$$= \frac{1}{3}\begin{bmatrix} 1+1+1 & 1+\alpha+\alpha^2 & 1+\alpha^2+\alpha \\ 1+\alpha+\alpha^2 & 1+\alpha^2+\alpha^4 & 1+\alpha^3+\alpha^3 \\ 1+\alpha^2+\alpha & 1+\alpha^3+\alpha^3 & 1+\alpha^4+\alpha^2 \end{bmatrix}$$

$$= \frac{1}{3}\begin{bmatrix} 3 & 0 & 0 \\ 0 & 0 & 3 \\ 0 & 3 & 0 \end{bmatrix} = \begin{bmatrix} 1 & 0 & 0 \\ 0 & 0 & 1 \\ 0 & 1 & 0 \end{bmatrix}$$

$[\because \alpha^3 = 1$ and $\alpha^2 + \alpha + 1 = 0]$

$$\therefore A^4 = A^2 \cdot A^2$$

$$= \begin{bmatrix} 1 & 0 & 0 \\ 0 & 0 & 1 \\ 0 & 1 & 0 \end{bmatrix}\begin{bmatrix} 1 & 0 & 0 \\ 0 & 0 & 1 \\ 0 & 1 & 0 \end{bmatrix} = \begin{bmatrix} 1 & 0 & 0 \\ 0 & 1 & 0 \\ 0 & 0 & 1 \end{bmatrix} = I$$

$$\therefore A^8 = I = A^{16} = A^{20} = A^{24} = A^{28}$$
$$\because A^{31} = A^{28} \cdot A^3 = I \cdot A^3 = A^3$$
$$\Rightarrow A^{31} = A^3$$

**18.** *(b)* Let $A = \begin{bmatrix} a_{11} & a_{12} & a_{13} \\ a_{21} & a_{22} & a_{23} \\ a_{31} & a_{32} & a_{33} \end{bmatrix}$

$$\because A\begin{bmatrix} 0 \\ 1 \\ 0 \end{bmatrix} = \begin{bmatrix} 0 \\ 0 \\ 1 \end{bmatrix}$$

$$\Rightarrow \begin{bmatrix} a_{12} \\ a_{22} \\ a_{32} \end{bmatrix} = \begin{bmatrix} 0 \\ 0 \\ 1 \end{bmatrix}$$

$$\Rightarrow a_{12} = 0, a_{22} = 0, a_{32} = 1$$

and $A\begin{bmatrix} 4 \\ 1 \\ 3 \end{bmatrix} = \begin{bmatrix} 0 \\ 1 \\ 0 \end{bmatrix}$

$$4a_{11} + a_{12} + 3a_{13} = 0$$
$$\Rightarrow 4a_{11} + 3a_{13} = 0 \quad \dots(i)$$
$$4a_{21} + a_{22} + 3a_{23} = 1$$
$$\Rightarrow 4a_{21} + 3a_{23} = 1 \quad \dots(ii)$$
$$4a_{31} + a_{32} + 3a_{33} = 0$$
$$\Rightarrow 4a_{31} + 3a_{33} = 0 \quad \dots(iii)$$

and $A\begin{bmatrix} 2 \\ 1 \\ 2 \end{bmatrix} = \begin{bmatrix} 1 \\ 0 \\ 0 \end{bmatrix}$

$$\Rightarrow 2a_{11} + a_{12} + 2a_{13} = 1$$
$$\Rightarrow 2a_{11} + 2a_{13} = 1 \quad \dots(iv)$$
$$2a_{21} + a_{22} + 2a_{23} = 0$$
$$\Rightarrow a_{21} + a_{23} = 0 \quad \dots(v)$$
$$2a_{31} + a_{32} + 2a_{33} = 0$$
$$\Rightarrow a_{31} + a_{33} = 0 \quad \dots(vi)$$

Solving Eqs. (ii) and (v), we get

$$-a_{23} = 1 \Rightarrow a_{23} = -1$$

**19.** *(a)*

> **key idea** Matrix $A$ is said to be symmetric matrix, if $A^T = A$.
> Matrix $A$ is said to be skew-symmetric matrix, if $A^T = -A$.

Given, $A = \begin{bmatrix} 1 & \frac{1}{51} \\ 0 & 1 \end{bmatrix}$

$$\Rightarrow A^2 = \begin{bmatrix} 1 & \frac{1}{51} \\ 0 & 1 \end{bmatrix}\begin{bmatrix} 1 & \frac{1}{51} \\ 0 & 1 \end{bmatrix}$$

$$= \begin{bmatrix} 1 & \frac{2}{51} \\ 0 & 1 \end{bmatrix} \text{ and so on}$$

$$\Rightarrow A^n = \begin{bmatrix} 1 & \frac{n}{51} \\ 0 & 1 \end{bmatrix} \quad \dots(i)$$

$$B = \begin{bmatrix} 1 & 2 \\ -1 & -1 \end{bmatrix} A \begin{bmatrix} -1 & -2 \\ 1 & 1 \end{bmatrix}$$

$$B^2 = \begin{bmatrix} 1 & 2 \\ -1 & -1 \end{bmatrix} A \begin{bmatrix} -1 & -2 \\ 1 & 1 \end{bmatrix}\begin{bmatrix} 1 & 2 \\ -1 & -1 \end{bmatrix} A \begin{bmatrix} -1 & -2 \\ 1 & 1 \end{bmatrix}$$

$$= \begin{bmatrix} 1 & 2 \\ -1 & -1 \end{bmatrix} A \begin{bmatrix} 1 & 0 \\ 0 & 1 \end{bmatrix} A \begin{bmatrix} -1 & -2 \\ 1 & 1 \end{bmatrix}$$

$$= \begin{bmatrix} 1 & 2 \\ -1 & -1 \end{bmatrix} A^2 \begin{bmatrix} -1 & -2 \\ 1 & 1 \end{bmatrix} \text{ and so on}$$

$$\Rightarrow B^n = \begin{bmatrix}1 & 2\\ -1 & -1\end{bmatrix} A^n \begin{bmatrix}-1 & -2\\ 1 & 1\end{bmatrix}$$

$$\therefore \sum_{n=1}^{50} B^n = \begin{bmatrix}1 & 2\\ -1 & -1\end{bmatrix}(A + A^2 + \ldots + A^{50})\begin{bmatrix}-1 & -2\\ 1 & 1\end{bmatrix}$$

$$= \begin{bmatrix}1 & 2\\ -1 & -1\end{bmatrix}\left(\begin{bmatrix}1 & \frac{1}{51}\\ 0 & 1\end{bmatrix} + \begin{bmatrix}1 & \frac{2}{51}\\ 0 & 1\end{bmatrix} + \ldots + \begin{bmatrix}1 & \frac{50}{51}\\ 0 & 1\end{bmatrix}\right)\begin{bmatrix}-1 & -2\\ 1 & 1\end{bmatrix}$$

$$= \begin{bmatrix}1 & 2\\ -1 & -1\end{bmatrix}\begin{bmatrix}50 & \frac{1+2+\ldots+50}{51}\\ 0 & 50\end{bmatrix}\begin{bmatrix}-1 & -2\\ 1 & 1\end{bmatrix}$$

$$= \begin{bmatrix}1 & 2\\ -1 & -1\end{bmatrix}\begin{bmatrix}50 & 25\\ 0 & 50\end{bmatrix}\begin{bmatrix}-1 & -2\\ 1 & 1\end{bmatrix}$$

$$= \begin{bmatrix}1 & 2\\ -1 & -1\end{bmatrix}\begin{bmatrix}-25 & -75\\ 50 & 50\end{bmatrix}$$

$$= \begin{bmatrix}75 & 25\\ -25 & 25\end{bmatrix}$$

$\therefore$ Sum of elements $=100$

**20.** *(d)* We have,

$$A = \begin{bmatrix}1 & 0 & 0\\ 0 & 4 & -1\\ 0 & 12 & -3\end{bmatrix}$$

Then,

$$A^2 = A\cdot A = \begin{bmatrix}1 & 0 & 0\\ 0 & 4 & -1\\ 0 & 12 & -3\end{bmatrix}\begin{bmatrix}1 & 0 & 0\\ 0 & 4 & -1\\ 0 & 12 & -3\end{bmatrix}$$

$$= \begin{bmatrix}1 & 0 & 0\\ 0 & 4 & -1\\ 0 & 12 & -3\end{bmatrix} = A$$

$\therefore \quad A^3 = A^2\cdot A = A\cdot A = A^2 = A$

Similarly, $A^4 = A^5 = \ldots = A^{11} = A$.

Now, $(A+I)^{11} = {}^{11}C_0 A^{11} + {}^{11}C_1 A^{10} I + {}^{11}C_2 A^9 I^2 + \ldots + {}^{11}C_{11} A^0 I^{11}$

$= {}^{11}C_0 A + {}^{11}C_1 A + {}^{11}C_2 A + \ldots + {}^{11}C_{10} A + {}^{11}C_{11} I$

$= A({}^{11}C_0 + {}^{11}C_1 + {}^{11}C_2 + \ldots + {}^{11}C_{10}) + I$

$= A(2^{11} - 1) + I$

$$= (2^{11}-1)\begin{bmatrix}1 & 0 & 0\\ 0 & 4 & -1\\ 0 & 12 & -3\end{bmatrix} + \begin{bmatrix}1 & 0 & 0\\ 0 & 1 & 0\\ 0 & 0 & 1\end{bmatrix}$$

$$= \begin{bmatrix}2^{11}-1+1 & 0 & 0\\ 0 & 4(2^{11}-1)+1 & -(2^{11}-1)\\ 0 & 12(2^{11}-1) & -3(2^{11}-1)+1\end{bmatrix}$$

$\therefore$ Sum of all diagonal elements

$= 2^{11} + 4(2^{11}-1) + 1 - 3(2^{11}-1) + 1$

$= 2\times 2^{11} + 1 = 2^{12} + 1$

$= 4097$

**21.** *(a)* $A = \begin{bmatrix}1 & 0 & 0\\ 0 & 1 & 1\\ 1 & 0 & 0\end{bmatrix}$

Now,

$$A^2 = \begin{bmatrix}1 & 0 & 0\\ 0 & 1 & 1\\ 1 & 0 & 0\end{bmatrix}\begin{bmatrix}1 & 0 & 0\\ 0 & 1 & 1\\ 1 & 0 & 0\end{bmatrix} = \begin{bmatrix}1 & 0 & 0\\ 1 & 1 & 1\\ 1 & 0 & 0\end{bmatrix}$$

$$A^3 = \begin{bmatrix}1 & 0 & 0\\ 1 & 1 & 1\\ 1 & 0 & 0\end{bmatrix}\begin{bmatrix}1 & 0 & 0\\ 0 & 1 & 1\\ 1 & 0 & 0\end{bmatrix} = \begin{bmatrix}1 & 0 & 0\\ 2 & 1 & 1\\ 1 & 0 & 0\end{bmatrix}$$

$$A^4 = \begin{bmatrix}1 & 0 & 0\\ 2 & 1 & 1\\ 1 & 0 & 0\end{bmatrix}\begin{bmatrix}1 & 0 & 0\\ 0 & 1 & 1\\ 1 & 0 & 0\end{bmatrix} = \begin{bmatrix}1 & 0 & 0\\ 3 & 1 & 1\\ 1 & 0 & 0\end{bmatrix}$$

$\vdots \qquad \vdots \qquad \vdots$

$$A^n = \begin{bmatrix}1 & 0 & 0\\ n-2 & 1 & 1\\ 1 & 0 & 0\end{bmatrix}\begin{bmatrix}1 & 0 & 0\\ 0 & 1 & 1\\ 1 & 0 & 0\end{bmatrix} = \begin{bmatrix}1 & 0 & 0\\ n-1 & 1 & 1\\ 1 & 0 & 0\end{bmatrix}$$

$$A^{2025} - A^{2020} = \begin{bmatrix}1 & 0 & 0\\ 2024 & 1 & 1\\ 1 & 0 & 0\end{bmatrix} - \begin{bmatrix}1 & 0 & 0\\ 2019 & 1 & 1\\ 1 & 0 & 0\end{bmatrix} = \begin{bmatrix}0 & 0 & 0\\ 5 & 0 & 0\\ 0 & 0 & 0\end{bmatrix}$$

Now, $A^6 - A = \begin{bmatrix}1 & 0 & 0\\ 5 & 1 & 1\\ 1 & 0 & 0\end{bmatrix} - \begin{bmatrix}1 & 0 & 0\\ 0 & 1 & 1\\ 1 & 0 & 0\end{bmatrix} = \begin{bmatrix}0 & 0 & 0\\ 5 & 0 & 0\\ 0 & 0 & 0\end{bmatrix}$

$\therefore \quad A^{2025} - A^{2020} = A^6 - A$

**22.** *(a)* Given, matrix

$$P = \begin{bmatrix}1 & 0 & 0\\ 3 & 1 & 0\\ 9 & 3 & 1\end{bmatrix} = \begin{bmatrix}0 & 0 & 0\\ 3 & 0 & 0\\ 9 & 3 & 0\end{bmatrix} + \begin{bmatrix}1 & 0 & 0\\ 0 & 1 & 0\\ 0 & 0 & 1\end{bmatrix}$$

$\Rightarrow \quad P = X + I$ (let)

Now, $P^5 = (I + X)^5$

$= I + {}^5C_1(X) + {}^5C_2(X^2) + {}^5C_3(X^3) + \ldots + {}^5C_5(X^5)$

$[\because I^n = I,\ I\cdot A = A$ and $(a+x)^n = {}^nC_0 a^n + {}^nC_1 a^{n-1}x + \ldots + {}^nC_n x^n]$

Here,

$$X^2 = \begin{bmatrix}0 & 0 & 0\\ 3 & 0 & 0\\ 9 & 3 & 0\end{bmatrix}\begin{bmatrix}0 & 0 & 0\\ 3 & 0 & 0\\ 9 & 3 & 0\end{bmatrix} = \begin{bmatrix}0 & 0 & 0\\ 0 & 0 & 0\\ 9 & 0 & 0\end{bmatrix}$$

and $X^3 = X^2\cdot X = \begin{bmatrix}0 & 0 & 0\\ 0 & 0 & 0\\ 9 & 0 & 0\end{bmatrix}\begin{bmatrix}0 & 0 & 0\\ 3 & 0 & 0\\ 9 & 3 & 0\end{bmatrix}$

$$= \begin{bmatrix}0 & 0 & 0\\ 0 & 0 & 0\\ 0 & 0 & 0\end{bmatrix}$$

$$\Rightarrow \quad X^4 = X^5 = \begin{bmatrix}0 & 0 & 0\\ 0 & 0 & 0\\ 0 & 0 & 0\end{bmatrix}$$

So, $P^5 = I + 5\begin{bmatrix}0 & 0 & 0\\ 3 & 0 & 0\\ 9 & 3 & 0\end{bmatrix} + 10\begin{bmatrix}0 & 0 & 0\\ 0 & 0 & 0\\ 9 & 0 & 0\end{bmatrix}$

$$= \begin{bmatrix}1 & 0 & 0\\ 15 & 1 & 0\\ 135 & 15 & 1\end{bmatrix}$$

and $Q = I + P^5 = \begin{bmatrix}2 & 0 & 0\\ 15 & 2 & 0\\ 135 & 15 & 2\end{bmatrix} = [q_{ij}]$

$\Rightarrow \quad q_{21} = 15,\ q_{31} = 135$ and $q_{32} = 15$

Hence, $\dfrac{q_{21} + q_{31}}{q_{32}} = \dfrac{15 + 135}{15}$

$= \dfrac{150}{15} = 10$

**23.** *(2)* Given, $A = \begin{bmatrix}2 & -1\\ 1 & 0\end{bmatrix}$

$$\Rightarrow A^2 = A\cdot A = \begin{bmatrix}2 & -1\\ 1 & 0\end{bmatrix}\begin{bmatrix}2 & -1\\ 1 & 0\end{bmatrix}$$

$$= \begin{bmatrix}3 & -2\\ 2 & -1\end{bmatrix}$$

$$\Rightarrow A^3 = A^2\cdot A = \begin{bmatrix}3 & -2\\ 2 & -1\end{bmatrix}\begin{bmatrix}2 & -1\\ 1 & 0\end{bmatrix}$$

$$= \begin{bmatrix} 4 & -3 \\ 3 & -2 \end{bmatrix}$$

$$\Rightarrow A^4 = A^3 \cdot A = \begin{bmatrix} 4 & -3 \\ 3 & -2 \end{bmatrix}\begin{bmatrix} 2 & -1 \\ 1 & 0 \end{bmatrix}$$

$$= \begin{bmatrix} 5 & -4 \\ 4 & -3 \end{bmatrix}$$

$$\Rightarrow A^5 = \begin{bmatrix} 6 & -5 \\ 5 & -4 \end{bmatrix}, A^6 = \begin{bmatrix} 7 & -6 \\ 6 & -5 \end{bmatrix}$$

and so on

$$\Rightarrow A^m = \begin{bmatrix} m+1 & -m \\ m & -(m-1) \end{bmatrix}$$

and $A^{m^2} = \begin{bmatrix} m^2+1 & -m^2 \\ m^2 & -(m^2-1) \end{bmatrix}$

Now, $A^{m^2} + A^m = 3I - A^{-6}$

$$\Rightarrow \begin{bmatrix} m^2+1 & -m^2 \\ m^2 & -(m^2-1) \end{bmatrix} + \begin{bmatrix} m+1 & -m \\ m & -(m-1) \end{bmatrix}$$

$$= 3\begin{bmatrix} 1 & 0 \\ 0 & 1 \end{bmatrix} - \begin{bmatrix} -5 & 6 \\ -6 & 7 \end{bmatrix} \quad [\because A^{-6} = (A^6)^{-1}]$$

$$= \begin{bmatrix} 8 & -6 \\ 6 & -4 \end{bmatrix}$$

$\Rightarrow m^2 + 1 + m + 1 = 8 \Rightarrow m^2 + m - 6 = 0$

$\Rightarrow m = -3, 2$

$\therefore n(S) = 2$

**24.** *(7)* $A = \begin{bmatrix} 2 & -1 \\ 1 & 1 \end{bmatrix}$

$$A^2 = \begin{bmatrix} 2 & -1 \\ 1 & 1 \end{bmatrix} \cdot \begin{bmatrix} 2 & -1 \\ 1 & 1 \end{bmatrix}$$

$$= \begin{bmatrix} 3 & -3 \\ 3 & 0 \end{bmatrix} = 3\begin{bmatrix} 1 & -1 \\ 1 & 0 \end{bmatrix}$$

$$A^4 = 3^2\begin{bmatrix} 1 & -1 \\ 1 & 0 \end{bmatrix}\begin{bmatrix} 1 & -1 \\ 1 & 0 \end{bmatrix} = 3^2\begin{bmatrix} 0 & -1 \\ 1 & -1 \end{bmatrix}$$

$$A^8 = 3^4\begin{bmatrix} 0 & -1 \\ 1 & -1 \end{bmatrix}\begin{bmatrix} 0 & -1 \\ 1 & -1 \end{bmatrix}$$

$$= 3^4\begin{bmatrix} -1 & 1 \\ -1 & 0 \end{bmatrix}$$

$$A^{12} = A^4 \times A^8$$

$$= 3^2 \times 3^4 \begin{bmatrix} 0 & -1 \\ 1 & -1 \end{bmatrix}\begin{bmatrix} -1 & 1 \\ -1 & 0 \end{bmatrix}$$

$$= 3^6\begin{bmatrix} 1 & 0 \\ 0 & 1 \end{bmatrix}$$

$$A^{13} = A \times A^{12}$$

$$= 3^6\begin{bmatrix} 1 & 0 \\ 0 & 1 \end{bmatrix}\begin{bmatrix} 2 & -1 \\ 1 & 1 \end{bmatrix}$$

$$= 3^6\begin{bmatrix} 2 & -1 \\ 1 & 1 \end{bmatrix} = \begin{bmatrix} 2\times 3^6 & -3^6 \\ 3^6 & 3^6 \end{bmatrix}$$

Sum of diagonal elements

$$= 2 \times 3^6 + 3^6$$

$$= 3^6(2+1) \Rightarrow 3^7 = 3^n$$

Hence, $n = 7$

**25.** *(2)* We have, $A = \begin{bmatrix} 0 & 1 & 2 \\ a & 0 & 3 \\ 1 & c & 0 \end{bmatrix}$,

where $a, c \in R$

$$A^2 = \begin{bmatrix} 0 & 1 & 2 \\ a & 0 & 3 \\ 1 & c & 0 \end{bmatrix}\begin{bmatrix} 0 & 1 & 2 \\ a & 0 & 3 \\ 1 & c & 0 \end{bmatrix}$$

$$= \begin{bmatrix} a+2 & 2c & 3 \\ 3 & a+3c & 2a \\ ac & 1 & 2+3c \end{bmatrix}$$

$$A^3 = \begin{bmatrix} a+2 & 2c & 3 \\ 3 & a+3c & 2a \\ ac & 1 & 2+3c \end{bmatrix}\begin{bmatrix} 0 & 1 & 2 \\ a & 0 & 3 \\ 1 & c & 0 \end{bmatrix}$$

$$= \begin{bmatrix} 2ac+3 & a+2+3c & 2a+4+6c \\ a(a+3c)+2a & 3+2ac & 6+3a+9c \\ a+2+3c & ac+2c+3c^2 & 2ac+3 \end{bmatrix}$$

$A^3 = A$ (given)

$\therefore \ 2ac + 3 = 0,\ a + 2 + 3c = 1$

and $a^2 + 2a + 3ac = a$

$$\Rightarrow a^2 + 2a + 3\left(-\frac{3}{2}\right) = a$$

$\Rightarrow \ 2a^2 + 2a - 9 = 0$

When $a = 1$, $2a^2 + 2a - 9 < 0$ and

When $a = 2$, $2a^2 + 2a - 9 > 0$

$\therefore$ Positive value of $a \in (1, 2]$

Hence, $n = 2$

**26.** *(204)* Let $A = \begin{bmatrix} a_{11} & a_{12} \\ a_{21} & a_{22} \end{bmatrix}$, such that

$a_{ij} \in \{0, 1, 2, 3, 4\}$

Now, $a_{11} + a_{12} + a_{21} + a_{22} = p$, where $p$ is prime number $p \in (2, 13)$.

$\therefore \ a_{11} + a_{12} + a_{21} + a_{22} = 3$ or $5$ or $7$ or $11$

Now, $(x^0 + x^1 + x^2 + x^3 + x^4)^4$

$$= \left(\frac{1-x^5}{1-x}\right)^4 = (1-x^5)^4(1-x)^{-4}$$

$$= {}^4C_{r_1}(-x^5)^{r_1} \times {}^{4+r_2-1}C_{r_2}x^{r_2}$$

$$= {}^4C_{r_1} \times {}^{4+r_2-1}C_{r_2}(-1)^{r_1}x^{5r_1+r_2}$$

$\therefore \ 5r_1 + r_2 = 3$ or $5$ or $7$ or $11$

Now, when, $5r_1 + r_2 = 3 \Rightarrow r_1 = 0, r_2 = 3$

When, $5r_1 + r_2 = 5 \Rightarrow r_1 = 0, r_2 = 5$ or $r_1 = 1, r_2 = 0$

When, $5r_1 + r_2 = 7 \Rightarrow r_1 = 0, r_2 = 7$ or $r_1 = 1, r_2 = 2$,

When, $5r_1 + r_2 = 11 \Rightarrow r_1 = 0, r_2 = 11$ or $r_1 = 1, r_2 = 6$ or $r_1 = 2, r_2 = 1$

Sum of all coefficients

$$= {}^4C_0 \times {}^6C_3 + {}^4C_0 \times {}^8C_5 - {}^4C_1 \times {}^3C_0 + {}^4C_0 \times {}^{10}C_7 - {}^4C_1 \times {}^5C_2 + {}^4C_0 \times {}^{14}C_{11} - {}^4C_1 \times {}^9C_6 + {}^4C_2 \times {}^4C_1 = 204$$

**27.** *(10)* Given, $X = \begin{bmatrix} 1 \\ 1 \\ 1 \end{bmatrix}, A = \begin{bmatrix} -1 & 2 & 3 \\ 0 & 1 & 6 \\ 0 & 0 & -1 \end{bmatrix}$

$$A^2 = \begin{bmatrix} 1 & 0 & 6 \\ 0 & 1 & 0 \\ 0 & 0 & 1 \end{bmatrix}, A^3 = \begin{bmatrix} -1 & 2 & -3 \\ 0 & 1 & 6 \\ 0 & 0 & -1 \end{bmatrix}$$

$$A^4 = \begin{bmatrix} 1 & 0 & 12 \\ 0 & 1 & 0 \\ 0 & 0 & 1 \end{bmatrix}, A^5 = \begin{bmatrix} -1 & 2 & -9 \\ 0 & 1 & 6 \\ 0 & 0 & -1 \end{bmatrix}$$

$$A^6 = \begin{bmatrix} 1 & 0 & 18 \\ 0 & 1 & 0 \\ 0 & 0 & 1 \end{bmatrix}$$

When, $k$ is even, $A^k = \begin{bmatrix} 1 & 0 & 3k \\ 0 & 1 & 0 \\ 0 & 0 & 1 \end{bmatrix}$

$X' A^k X = 33$

$$\Rightarrow [1 \ 1 \ 1]\begin{bmatrix} 1 & 0 & 3k \\ 0 & 1 & 0 \\ 0 & 0 & 1 \end{bmatrix}\begin{bmatrix} 1 \\ 1 \\ 1 \end{bmatrix} = [33]$$

$$\Rightarrow [1 \ 1 \ 1]\begin{bmatrix} 1+3k \\ 1 \\ 1 \end{bmatrix} = [33]$$

$\Rightarrow \ 1 + 3k + 1 + 1 = 33$

$\Rightarrow \ k = 10$

When, $k$ is odd.

$X' A^k X = [33]$

$\Rightarrow X' A \cdot A^{k-1} X = [33]$

$$[1\ \ 1\ \ 1]\begin{bmatrix}-1 & 2 & 3\\ 0 & 1 & 6\\ 0 & 0 & -1\end{bmatrix}$$

$$\begin{bmatrix}1 & 0 & 3(k-1)\\ 0 & 1 & 0\\ 0 & 0 & 1\end{bmatrix}\begin{bmatrix}1\\ 1\\ 1\end{bmatrix}=[33]$$

$\Rightarrow -3k+2+3+8=33$

$\Rightarrow k=\dfrac{-20}{3}\notin N$

Hence, $k=10,\ k\in N$

**28.** *(2)* Given, $A=\begin{bmatrix}1 & -1\\ 2 & \alpha\end{bmatrix}$,

$B=\begin{bmatrix}\beta & 1\\ 1 & 0\end{bmatrix}\quad A+B=\begin{bmatrix}1+\beta & 0\\ 3 & \alpha\end{bmatrix}$

$$(A+B)^2=\begin{bmatrix}(1+\beta)^2 & 0\\ 3(1+\beta)+3\alpha & \alpha^2\end{bmatrix},$$

$$B^2=\begin{bmatrix}\beta^2+1 & \beta\\ \beta & 1\end{bmatrix}$$

and $\quad A^2=\begin{bmatrix}-1 & -1-\alpha\\ 2+2\alpha & \alpha^2-2\end{bmatrix}$

$$(A+B)^2=A^2+\begin{bmatrix}2 & 2\\ 2 & 2\end{bmatrix}$$

$$\Rightarrow \begin{bmatrix}2 & 2\\ 2 & 2\end{bmatrix}+\begin{bmatrix}-1 & -1-\alpha\\ 2+2\alpha & \alpha^2-2\end{bmatrix}$$

$$=\begin{bmatrix}(1+\beta)^2 & 0\\ 3(1+\beta)+3\alpha & \alpha^2\end{bmatrix}$$

$\Rightarrow \quad \alpha=1=\alpha_1$

Now, $(A+B)^2=B^2$

$$\Rightarrow \begin{bmatrix}(1+\beta)^2 & 0\\ 3(1+\beta)+3\alpha & \alpha^2\end{bmatrix}=\begin{bmatrix}\beta^2+1 & \beta\\ \beta & 1\end{bmatrix}$$

$\Rightarrow \quad \beta=0,\ \alpha=-1=\alpha_2$

$\therefore \quad |\alpha_1-\alpha_2|=2$

**29.** *(24)* Given, $A=\begin{bmatrix}1 & a & a\\ 0 & 1 & b\\ 0 & 0 & 1\end{bmatrix}$, $a,b\in R$

$$=\begin{bmatrix}1 & 0 & 0\\ 0 & 1 & 0\\ 0 & 0 & 1\end{bmatrix}+\begin{bmatrix}0 & a & a\\ 0 & 0 & b\\ 0 & 0 & 0\end{bmatrix}=I+B$$

where, $B=\begin{bmatrix}0 & a & a\\ 0 & 0 & b\\ 0 & 0 & 0\end{bmatrix}$

$$B^2=\begin{bmatrix}0 & a & a\\ 0 & 0 & b\\ 0 & 0 & 0\end{bmatrix}\begin{bmatrix}0 & a & a\\ 0 & 0 & b\\ 0 & 0 & 0\end{bmatrix}$$

$$=\begin{bmatrix}0 & 0 & ab\\ 0 & 0 & 0\\ 0 & 0 & 0\end{bmatrix}$$

and $B^3=\begin{bmatrix}0 & 0 & ab\\ 0 & 0 & 0\\ 0 & 0 & 0\end{bmatrix}\begin{bmatrix}0 & a & a\\ 0 & 0 & b\\ 0 & 0 & 0\end{bmatrix}$

$$=\begin{bmatrix}0 & 0 & 0\\ 0 & 0 & 0\\ 0 & 0 & 0\end{bmatrix}$$

Now, $A^n=(I+B)^n={}^nC_0I+{}^nC_1B$
$+{}^nC_2B^2+{}^nC_3B^3+\ldots+{}^nC_nB^n$

$$=\begin{bmatrix}1 & 0 & 0\\ 0 & 1 & 0\\ 0 & 0 & 1\end{bmatrix}+\begin{bmatrix}0 & na & na\\ 0 & 0 & nb\\ 0 & 0 & 0\end{bmatrix}+\begin{bmatrix}0 & 0 & \frac{n(n-1)}{2}ab\\ 0 & 0 & 0\\ 0 & 0 & 0\end{bmatrix}$$

$$=\begin{bmatrix}1 & na & na+\frac{n(n-1)}{2}ab\\ 0 & 1 & nb\\ 0 & 0 & 1\end{bmatrix}$$

$$=\begin{bmatrix}1 & 48 & 2160\\ 0 & 1 & 96\\ 0 & 0 & 1\end{bmatrix}\qquad \text{[given]}$$

Now, comparing above, we get

$na=48$ and $nb=96$ and

$na+\dfrac{n(n-1)ab}{2}=2160$

$\Rightarrow \quad 48+48(n-1)a=2160$

$\Rightarrow \quad 1+na-a=45\Rightarrow a=4$

$\therefore n=12$, then $nb=96\Rightarrow b=8$

$\therefore n+a+b=12+4+8=24$

**30.** *(1)* $M=\begin{bmatrix}0 & -\alpha\\ \alpha & 0\end{bmatrix}$

$$N=\sum_{k=1}^{49}M^{2k}\text{ and }(I-M^2)N=-2I$$

$$M^2=\begin{bmatrix}0 & -\alpha\\ \alpha & 0\end{bmatrix}\begin{bmatrix}0 & -\alpha\\ \alpha & 0\end{bmatrix}$$

$$=\begin{bmatrix}-\alpha^2 & 0\\ 0 & -\alpha^2\end{bmatrix}=-\alpha^2I$$

$$N=\sum_{k=1}^{49}M^{2k}=\sum_{k=1}^{49}(M^2)^k$$

$=(M^2)^1+(M^2)^2+(M^2)^3+\ldots+(M^2)^{49}$

$=[-\alpha^2+\alpha^4-\alpha^6+\ldots-\alpha^{98}]\,I$

$=-\dfrac{\alpha^2[1-(-\alpha^2)^{49}]}{1-(-\alpha^2)}I=\dfrac{-\alpha^2[1+\alpha^{98}]}{1+\alpha^2}I$

$\Rightarrow \ (I-M^2)\,N=(1+\alpha^2)$

$$\left[\frac{-\alpha^2(1+\alpha^{98})}{(1+\alpha^2)}\right]I=-2I$$

$\Rightarrow \quad \alpha^2(1+\alpha^{98})=2$

$\Rightarrow \quad \alpha^2=1\Rightarrow \alpha=\pm 1$

$\therefore \quad \alpha=1$

> **Alternate Solution**
>
> $M=\begin{bmatrix}0 & -\alpha\\ \alpha & 0\end{bmatrix}$
>
> $M^2=\begin{bmatrix}-\alpha^2 & 0\\ 0 & -\alpha^2\end{bmatrix}=-\alpha^2I$
>
> Now, $N=[-\alpha^2+\alpha^4-\alpha^6+\ldots]I$
>
> $=\dfrac{-\alpha^2(1+\alpha^{98})}{1+\alpha^2}I$
>
> $\therefore (I-M^2)N=-2I$
>
> $\Rightarrow -\alpha^2(1+\alpha^{98})=-2\Rightarrow \alpha=1$

**31.** *(25)* Given, $A=\begin{bmatrix}1+i & 1\\ -i & 0\end{bmatrix}$

$$\therefore A^2=\begin{bmatrix}1+i & 1\\ -i & 0\end{bmatrix}\begin{bmatrix}1+i & 1\\ -i & 0\end{bmatrix}$$

$$=\begin{bmatrix}i & 1+i\\ 1-i & -i\end{bmatrix}$$

$$A^4=\begin{bmatrix}i & 1+i\\ 1-i & -i\end{bmatrix}\begin{bmatrix}i & 1+i\\ 1-i & -i\end{bmatrix}=I$$

So, $A^5=A,\ A^9=A$ and so on.

Clearly, $n=1,5,9,\ldots,97$

Number of values of $n=25$

**32.** *(100)* $X=\begin{bmatrix}0 & 1 & 0\\ 0 & 0 & 1\\ 0 & 0 & 0\end{bmatrix};\ X^2=\begin{bmatrix}0 & 0 & 1\\ 0 & 0 & 0\\ 0 & 0 & 0\end{bmatrix}$

$Y=\alpha I+\beta X+\gamma X^2$

$$Y=\begin{bmatrix}\alpha & 0 & 0\\ 0 & \alpha & 0\\ 0 & 0 & \alpha\end{bmatrix}+\begin{bmatrix}0 & \beta & 0\\ 0 & 0 & \beta\\ 0 & 0 & 0\end{bmatrix}+\begin{bmatrix}0 & 0 & \gamma\\ 0 & 0 & 0\\ 0 & 0 & 0\end{bmatrix}$$

$$Y=\begin{bmatrix}\alpha & \beta & \gamma\\ 0 & \alpha & \beta\\ 0 & 0 & \alpha\end{bmatrix}$$

$YY^{-1}=I$

$$\Rightarrow \begin{bmatrix}\alpha & \beta & \gamma\\ 0 & \alpha & \beta\\ 0 & 0 & \alpha\end{bmatrix}\begin{bmatrix}1/5 & -2/5 & 1/5\\ 0 & 1/5 & -2/5\\ 0 & 0 & 1/5\end{bmatrix}$$

$$=\begin{bmatrix}1 & 0 & 0\\ 0 & 1 & 0\\ 0 & 0 & 1\end{bmatrix}$$

$$\Rightarrow \begin{bmatrix} \frac{\alpha}{5} & \frac{-2\alpha+\beta}{5} & \frac{\alpha-2\beta+\gamma}{5} \\ 0 & \frac{\alpha}{5} & \frac{-2\alpha+\beta}{5} \\ 0 & 0 & \frac{\alpha}{5} \end{bmatrix} = \begin{bmatrix} 1 & 0 & 0 \\ 0 & 1 & 0 \\ 0 & 0 & 1 \end{bmatrix}$$

$\alpha = 5, \beta = 10$ and $\gamma = 15$

So, $(\alpha - \beta + \gamma)^2 = (5 - 10 + 15)^2 = 100$

**33.** *(1)* Given, $A = \begin{bmatrix} 2 & -2 \\ 1 & -1 \end{bmatrix}$

$$\Rightarrow \quad A^2 = \begin{bmatrix} 2 & -2 \\ 1 & -1 \end{bmatrix}\begin{bmatrix} 2 & -2 \\ 1 & -1 \end{bmatrix} = \begin{bmatrix} 2 & -2 \\ 1 & -1 \end{bmatrix}$$

$\therefore \quad A^n = A, \forall n \in N$

$$B^2 = \begin{bmatrix} -1 & 2 \\ -1 & 2 \end{bmatrix}\begin{bmatrix} -1 & 2 \\ -1 & 2 \end{bmatrix} = \begin{bmatrix} -1 & 2 \\ -1 & 2 \end{bmatrix}$$

$\therefore \quad B^m = B, \forall m \in N$

Now, $nA^n + mB^m = I$

$$n\begin{bmatrix} 2 & -2 \\ 1 & -1 \end{bmatrix} + m\begin{bmatrix} -1 & 2 \\ -1 & 2 \end{bmatrix} = \begin{bmatrix} 1 & 0 \\ 0 & 1 \end{bmatrix}$$

[$\because A^n = A$ and $B^m = B$]

$$= \begin{bmatrix} 2n-m & -2n+2m \\ n-m & -n+2m \end{bmatrix} = \begin{bmatrix} 1 & 0 \\ 0 & 1 \end{bmatrix}$$

$\Rightarrow \quad 2n - m = 1, -n + 2m = 1$ and $n - m = 0 \Rightarrow m = n$

On putting $m = n$ in equation $2n - m = 1$, we get

$2n - n = 1 \Rightarrow n = 1$

$\therefore \quad (m, n) = (1, 1)$

Hence, the number of element in set is 1.

**34.** *(100)* Given,

$$S = \left\{ \begin{bmatrix} -1 & a \\ 0 & b \end{bmatrix}; a, b \in \{1, 2, 3, \dots, 100\} \right\}$$

Since, $A = \begin{bmatrix} -1 & a \\ 0 & b \end{bmatrix}$, then all even power of $A$ or $\begin{bmatrix} 1 & 0 \\ 0 & 1 \end{bmatrix}$, so we observe that $b = 1$

but $a \in \{1, 2, 3 \dots 100\}$

Since, $T_n = \{A \in S : A^{n(n+1)} = I\}$

$T_1 = \{A \in S : A^2 = I\} = I$

$T_2 = \{A \in S : A^6 = I\} = I$

$T_{100} = \{A \in S : A^{100 \times (101)} = I\} = I$

Thus, $T_1, T_2, \dots, T_n$ all are $I$ for $b = 1$ and each value of $a$.

Thus, $\bigcap_{n=1}^{100} T_n = 100$

**35.** *(2020)* We have, $A = \begin{bmatrix} 1 & 1 & 1 \\ 0 & 1 & 1 \\ 0 & 0 & 1 \end{bmatrix}$

$$A^2 = \begin{bmatrix} 1 & 1 & 1 \\ 0 & 1 & 1 \\ 0 & 0 & 1 \end{bmatrix}\begin{bmatrix} 1 & 1 & 1 \\ 0 & 1 & 1 \\ 0 & 0 & 1 \end{bmatrix} = \begin{bmatrix} 1 & 2 & 3 \\ 0 & 1 & 2 \\ 0 & 0 & 1 \end{bmatrix}$$

$$A^3 = \begin{bmatrix} 1 & 2 & 3 \\ 0 & 1 & 2 \\ 0 & 0 & 1 \end{bmatrix}\begin{bmatrix} 1 & 1 & 1 \\ 0 & 1 & 1 \\ 0 & 0 & 1 \end{bmatrix} = \begin{bmatrix} 1 & 3 & 6 \\ 0 & 1 & 3 \\ 0 & 0 & 1 \end{bmatrix}$$

$\vdots$

$$A^n = \begin{bmatrix} 1 & n & \sum n \\ 0 & 1 & n \\ 0 & 0 & 1 \end{bmatrix}$$

Now, $M = A + A^2 + A^3 + \dots + A^{20}$

$$= \begin{bmatrix} 1 & 1 & 1 \\ 0 & 1 & 1 \\ 0 & 0 & 1 \end{bmatrix} + \begin{bmatrix} 1 & 2 & 3 \\ 0 & 1 & 2 \\ 0 & 0 & 1 \end{bmatrix} + \dots + \begin{bmatrix} 1 & n & \sum n \\ 0 & 1 & n \\ 0 & 0 & 1 \end{bmatrix}$$

$$= \begin{bmatrix} 20 & \sum n & \sum(\sum n) \\ 0 & 20 & \sum n \\ 0 & 0 & 20 \end{bmatrix}$$

Now, $\sum_{r=1}^{20} n = 1 + 2 + \dots + 20$

$$= \frac{20 \times 21}{2} = 210$$

$$\sum_{r=1}^{20}(n) = \sum_{r=1}^{20} \frac{n(n+1)}{2}$$

$$= \frac{1}{2}\sum_{r=1}^{20}(n^2 + n)$$

$$= \frac{1}{2}\left[\frac{20 \times 21 \times 41}{6} + \frac{20 \times 21}{2}\right]$$

$$= \frac{1}{2}[2870 + 210] = 1540$$

$$\text{Hence, } M = \begin{bmatrix} 20 & 210 & 1540 \\ 0 & 20 & 210 \\ 0 & 0 & 20 \end{bmatrix}$$

Sum of all elements = 2020.

**36.** *(3125)*

$$A = \begin{bmatrix} 0 & 1 & 0 \\ 1 & 0 & 0 \\ 0 & 0 & 1 \end{bmatrix}, B = \begin{bmatrix} a_1 & a_2 & a_3 \\ b_1 & b_2 & b_3 \\ c_1 & c_2 & c_3 \end{bmatrix}$$

$\because AB = BA$ [given]

$$\Rightarrow \begin{bmatrix} b_1 & b_2 & b_3 \\ a_1 & a_2 & a_3 \\ c_1 & c_2 & c_3 \end{bmatrix} = \begin{bmatrix} a_2 & a_1 & a_3 \\ b_2 & b_1 & b_3 \\ c_2 & c_1 & c_3 \end{bmatrix}$$

$\Rightarrow b_1 = a_2,\ b_2 = a_1,\ b_3 = a_3$ and $c_1 = c_2$

$$\therefore \quad B = \begin{bmatrix} a_1 & a_2 & a_3 \\ a_2 & a_1 & a_3 \\ c_1 & c_1 & c_3 \end{bmatrix}$$

Number of distinct elements in $B$ is 5 and according to question,

$a_{ij} \in \{1, 2, 3, 4, 5\}$.

So, number of matrices

$= 5 \times 5 \times 5 \times 5 \times 5$

$= 3125$

**37.** *(910)* Given, $A = \begin{bmatrix} 1 & -1 & 0 \\ 0 & 1 & -1 \\ 0 & 0 & 1 \end{bmatrix}$

$B = 7A^{20} - 20A^7 + 2I$

$$\therefore \quad A^2 = \begin{bmatrix} 1 & -2 & 1 \\ 0 & 1 & -2 \\ 0 & 0 & 1 \end{bmatrix}$$

$$\Rightarrow \quad A^3 = \begin{bmatrix} 1 & -3 & 3 \\ 0 & 1 & -3 \\ 0 & 0 & 1 \end{bmatrix}$$

$$\Rightarrow \quad A^4 = \begin{bmatrix} 1 & -4 & 6 \\ 0 & 1 & -4 \\ 0 & 0 & 1 \end{bmatrix}$$

$$\Rightarrow \quad A^n = \begin{bmatrix} 1 & -n & \frac{n(n-1)}{2} \\ 0 & 1 & -n \\ 0 & 0 & 1 \end{bmatrix}$$

$$\therefore \quad b_{13} = 7\left(\frac{20 \times 19}{2}\right) - 20\left(\frac{7 \times 6}{2}\right) + 2(0)$$

$\Rightarrow \quad b_{13} = 1330 - 420 = 910$

**38.** *(4)* Given, $A = \begin{bmatrix} 1 & 0 & 0 \\ 0 & 2 & 0 \\ 3 & 0 & -1 \end{bmatrix}$

$$\therefore A^2 = \begin{bmatrix} 1 & 0 & 0 \\ 0 & 4 & 0 \\ 0 & 0 & 1 \end{bmatrix} \Rightarrow A^3 = \begin{bmatrix} 1 & 0 & 0 \\ 0 & 8 & 0 \\ 3 & 0 & -1 \end{bmatrix}$$

$$\Rightarrow A^4 = \begin{bmatrix} 1 & 0 & 0 \\ 0 & 16 & 0 \\ 0 & 0 & 1 \end{bmatrix}$$

$$\therefore A^{19}=\begin{bmatrix}1&0&0\\0&2^{19}&0\\3&0&-1\end{bmatrix}$$

$$\Rightarrow A^{20}=\begin{bmatrix}1&0&0\\0&2^{20}&0\\0&0&1\end{bmatrix}$$

So, $A^{20}+\alpha A^{19}+\beta A$

$$=\begin{bmatrix}1+\alpha+\beta&0&0\\0&2^{20}+\alpha 2^{19}+2\beta&0\\3\alpha+3\beta&0&1-\alpha-\beta\end{bmatrix}=\begin{bmatrix}1&0&0\\0&4&0\\0&0&1\end{bmatrix}$$

On comparing, we get

$1-\alpha-\beta=1 \Rightarrow \alpha+\beta=0$

and $2^{20}+\alpha 2^{19}+2\beta=4$

$\Rightarrow 2^{20}+\alpha(2^{19}-2)=4$ [Using $\alpha+\beta=0$]

$\Rightarrow \alpha=\dfrac{4-2^{20}}{2^{19}-2}=-2$

$\Rightarrow \beta=2$

$\therefore \beta-\alpha=2-(-2)=2+2=4$

**39.** *(10)* Given matrix, $A=\begin{bmatrix}x&1\\1&0\end{bmatrix}$

$$\Rightarrow A^2=\begin{bmatrix}x&1\\1&0\end{bmatrix}\begin{bmatrix}x&1\\1&0\end{bmatrix}=\begin{bmatrix}x^2+1&x\\x&1\end{bmatrix}$$

$$\Rightarrow A^4=\begin{bmatrix}x^2+1&x\\x&1\end{bmatrix}\begin{bmatrix}x^2+1&x\\x&1\end{bmatrix}=\begin{bmatrix}(x^2+1)^2+x^2&x(x^2+2)\\x(x^2+2)&x^2+1\end{bmatrix}$$

$\because A^4=[a_{ij}]$ and $a_{11}=109$

$\Rightarrow (x^2+1)^2+x^2=109 \Rightarrow x^2=9$

$\therefore a_{22}=x^2+1=9+1=10$

## Topic 2

**01.** *(b)* We have, $AA^T=I$

$$\frac{1}{2}A[(A+A^T)^2+(A-A^T)^2]$$

$$=\frac{1}{2}A[A^2+(A^T)^2+2AA^T+A^2+(A^T)^2-2AA^T]$$

$$=\frac{1}{2}A[2A^2+2(A^T)^2]=A^3+A^T$$

**02.** *(b)* Let symmetric matrix of order $3\times 3$ is $\begin{bmatrix}a&x&y\\x&b&z\\y&z&c\end{bmatrix}$

Thus, $x$ can be filled in 10 ways.

$y$ can be filled in 10 ways.

$z$ can be filled in 10 ways.

$\therefore$ Elements of diagonal can be filled in $10^3$ ways. Thus, total number of matrix formed which are symmetric also is $10^3\times 10^3=10^6$

**03.** *(c)* We have, $P^2=I-P$ ...(i)

$\Rightarrow P^4=(I-P)^2=I+P^2-2P$

$=2I-3P$ [using Eq. (i)]

$\Rightarrow P^8=(2I-3P)^2=4I+9P^2-12P$

$=13I-21P$ [using Eq. (i)]

and $P^6=(I-P)(2I-3P)$

$=2I-3P-2P+3P^2$

$=5I-8P$ [using Eq. (i)]

Now, $P^8+P^6=18I-29P$

and $P^8-P^6=8I-13P$

$\therefore \alpha=8, \beta=6, \gamma=18, \delta=8$

$\therefore \alpha+\beta+\gamma-\delta=8+6+18-8=24$

**04.** *(d)* Given, $A=\begin{bmatrix}1\\1\\1\end{bmatrix}$

and $B=\begin{bmatrix}9^2&-10^2&11^2\\12^2&13^2&-14^2\\-15^2&16^2&17^2\end{bmatrix}$

Now, $A'BA=$

$$[1\ 1\ 1]\begin{bmatrix}81&-100&121\\144&169&-196\\-225&256&289\end{bmatrix}\begin{bmatrix}1\\1\\1\end{bmatrix}$$

$$=[1\ 1\ 1]\begin{bmatrix}81-100+121\\144+169-196\\-225+256+289\end{bmatrix}$$

$$=[1\ 1\ 1]\begin{bmatrix}102\\117\\320\end{bmatrix}=539$$

**05.** *(b)* Given, matrix $A=\begin{bmatrix}0&2y&1\\2x&y&-1\\2x&-y&1\end{bmatrix}$, $(x, y\in R, x\neq y)$ for which $A^TA=3I_3$

$$\Rightarrow \begin{bmatrix}0&2x&2x\\2y&y&-y\\1&-1&1\end{bmatrix}\begin{bmatrix}0&2y&1\\2x&y&-1\\2x&-y&1\end{bmatrix}=\begin{bmatrix}3&0&0\\0&3&0\\0&0&3\end{bmatrix}$$

$$\Rightarrow \begin{bmatrix}8x^2&0&0\\0&6y^2&0\\0&0&3\end{bmatrix}=\begin{bmatrix}3&0&0\\0&3&0\\0&0&3\end{bmatrix}$$

Here, two matrices are equal, therefore equating the corresponding elements, we get

$8x^2=3$ and $6y^2=3 \Rightarrow x=\pm\sqrt{3/8}$ and $y=\pm\dfrac{1}{\sqrt{2}}$

$\because$ There are 2 different values of $x$ and $y$ each.

So, 4 matrices are possible such that $A^TA=3I_3$.

**06.** *(a)* Given, $A+B=2B'$ ...(i)

$\Rightarrow A'+B'=2B \Rightarrow B=\dfrac{A'+B'}{2}$

On putting the value in Eq. (i), we get

$A+\left(\dfrac{B'+A'}{2}\right)=2B'$

$\Rightarrow 2A+B'+A'=4B' \Rightarrow A=\dfrac{3B'-A'}{2}$

Now, $3A+2B=I_3$ ...(ii)

$\Rightarrow 3\left(\dfrac{3B'-A'}{2}\right)+2\left(\dfrac{A'+B'}{2}\right)=I_3$

$\Rightarrow 11B'-A'=2I_3 \Rightarrow (11B'-A')'=(2I_3)'$

$\Rightarrow 11B-A=2I_3$ ...(iii)

On multiplying Eq. (iii) by 3 and then adding Eq. (ii), we get

$35B=7I_3 \Rightarrow B=I_3/5$

From Eq. (iii), we get

$11\dfrac{I_3}{5}-A=2I_3 \Rightarrow A=\dfrac{I_3}{5}$

$\therefore 5A=5B=I_3 \Rightarrow 10A+5B=3I_3$

**07.** *(b)* $P=\begin{bmatrix}\cos\theta&-\sin\theta\\\sin\theta&\cos\theta\end{bmatrix}$

$P^T=\begin{bmatrix}\cos\theta&\sin\theta\\-\sin\theta&\cos\theta\end{bmatrix}$

$PP^T=P^TP=I$

Now, $C=P^TB^{10}P$

$=P^T(PAP^T)(PAP^T)\ldots\ldots(PAP^T)P$

$=A^{10}$

Now, $A^2=\begin{bmatrix}\left(\dfrac{1}{\sqrt{2}}\right)^2&-\sqrt{2}-2\\0&1\end{bmatrix}$

$A^3=\begin{bmatrix}\left(\dfrac{1}{\sqrt{2}}\right)^3&-\sqrt{2}-3\\0&1\end{bmatrix}$

$A^n=\begin{bmatrix}\left(\dfrac{1}{\sqrt{2}}\right)^n&-\sqrt{2}-n\\0&1\end{bmatrix}$

$$tr(A^n) = \left(\frac{1}{\sqrt{2}}\right)^n + 1$$

$$tr(A^{10}) = \left(\frac{1}{\sqrt{2}}\right)^{10} + 1 = \frac{33}{32}$$

$m = 33$ and $n = 32$

$\therefore \quad m + n = 65$

**08.** *(c)* Given,

$$A = \begin{bmatrix} 1 & 1 \\ 0 & 1 \end{bmatrix}, \quad P = \begin{bmatrix} \frac{\sqrt{3}}{2} & \frac{1}{2} \\ \frac{-1}{2} & \frac{\sqrt{3}}{2} \end{bmatrix}$$

$$\Rightarrow A^2 = \begin{bmatrix} 1 & 2 \\ 0 & 1 \end{bmatrix}, \quad P^T = \begin{bmatrix} \frac{\sqrt{3}}{2} & \frac{-1}{2} \\ \frac{1}{2} & \frac{\sqrt{3}}{2} \end{bmatrix}$$

$$\Rightarrow A^3 = \begin{bmatrix} 1 & 2 \\ 0 & 1 \end{bmatrix}\begin{bmatrix} 1 & 1 \\ 0 & 1 \end{bmatrix} = \begin{bmatrix} 1 & 3 \\ 0 & 1 \end{bmatrix}$$

$$\therefore \quad A^n = \begin{bmatrix} 1 & n \\ 0 & 1 \end{bmatrix}$$

$$PP^T = \begin{bmatrix} \frac{\sqrt{3}}{2} & \frac{1}{2} \\ \frac{-1}{2} & \frac{\sqrt{3}}{2} \end{bmatrix}\begin{bmatrix} \frac{\sqrt{3}}{2} & -\frac{1}{2} \\ \frac{1}{2} & \frac{\sqrt{3}}{2} \end{bmatrix}$$

$$= \begin{bmatrix} 1 & 0 \\ 0 & 1 \end{bmatrix} = I$$

Now, $\begin{bmatrix} a & b \\ c & d \end{bmatrix} = P^T Q^{2007} P$

$= P^T(PAP^T)(PAP^T)(PAP^T) \ldots (PAP^T)P$

$= IAIA \ldots IAI = A^{2007}$

$$\Rightarrow \begin{bmatrix} a & b \\ c & d \end{bmatrix} = \begin{bmatrix} 1 & 2007 \\ 0 & 1 \end{bmatrix}$$

$$\left[\because A^n = \begin{bmatrix} 1 & n \\ 0 & 1 \end{bmatrix}\right]$$

On comparing both sides,

$a = 1, b = 2007, c = 0, d = 1$

$\therefore 2a + b - 3c - 4d = 2005$

**09.** *(c)* Given, $A$ and $B$ are any two $3 \times 3$ symmetric and skew-symmetric matrix, respectively.

$\Rightarrow A^T = A, B^T = -B$

Let $C = A^4 - B^4$

$C^T = (A^4 - B^4)^T = (A^4)^T - (B^4)^T$

$= (A^T)^4 - (B^T)^4 = A^4 - (-B)^4$

$[\because A^T = A \text{ and } B^T = -B]$

$= A^4 - B^4 \quad [\because (-B)^4 = B^4]$

$= C$

$\therefore C$ is a symmetric matrix.

Let $D = AB - BA$

$D^T = (AB - BA)^T$

$= (AB)^T - (BA)^T$

$= B^T A^T - A^T B^T \quad [\because (AB)^T = B^T A^T]$

$= -BA + AB$

$[\because A^T = A \text{ and } B^T = -B]$

$= D$

$\therefore D$ is a symmetric matrix.

Let $E = B^5 - A^5$

$E^T = (B^5 - A^5)^T = (B^5)^T - (A^5)^T$

$= (B^T)^5 - (A^T)^5 = (-B)^5 - A^5$

$[\because A^T = A, B^T = -B]$

$= -B^5 - A^5$

$= -[B^5 + A^5] \neq -E$

$\therefore E$ is neither symmetric nor skew-symmetric.

Let $F = AB + BA$

$F^T = (AB + BA)^T = (AB)^T + (BA)^T$

$= B^T A^T + A^T B^T = -BA - AB$

$= -(AB + BA) = -F$

$\therefore F$ is a skew-symmetric.

**10.** *(a)* $\because A = \begin{bmatrix} 0 & -2 \\ 2 & 0 \end{bmatrix}$

$$\therefore \quad A^2 = \begin{bmatrix} -4 & 0 \\ 0 & -4 \end{bmatrix} = -4I$$

$$A^3 = A^2 \cdot A$$

$$= \begin{bmatrix} -4 & 0 \\ 0 & -4 \end{bmatrix}\begin{bmatrix} 0 & -2 \\ 2 & 0 \end{bmatrix}$$

$$= \begin{bmatrix} 0 & 8 \\ -8 & 0 \end{bmatrix}$$

$$= -4A$$

$$M = \sum_{k=1}^{10} (A^2)^k$$

$= [(-4) + (-4)^2 + \ldots + (-4)^{10}]I$

$= -4\lambda I$, which is symmetric.

where $\lambda = [1 + (-4) + (-4)^2 + \ldots + (-4)^9]$

$$N = \sum_{k=1}^{10} A^{2k-1}$$

$= A[1 + (-4) + (-4)^3 + \ldots + (-4)^9]$

$= \lambda A$ which is skew-symmetric.

Now, $N^2$ is symmetric.

$MN = (-4\lambda I)(\lambda A) = -4\lambda^2 A = NM$

$MN^2 = (MN)N$

$= (NM)N \quad [\because MN = NM]$

$= N(MN) = N(NM)$

$= N^2 M$

Hence, $MN^2$ is non-identity symmetric matrix.

**11.** *(b)* Given,

$$A + 2B = \begin{bmatrix} 1 & 2 & 0 \\ 6 & -3 & 3 \\ -5 & 3 & 1 \end{bmatrix} \quad \ldots\text{(i)}$$

$$\text{and} \quad 2A - B = \begin{bmatrix} 2 & -1 & 5 \\ 2 & -1 & 6 \\ 0 & 1 & 2 \end{bmatrix} \quad \ldots\text{(ii)}$$

On multiplying Eq. (ii) by 2, we get

$$4A - 2B = \begin{bmatrix} 4 & -2 & 10 \\ 4 & -2 & 12 \\ 0 & 2 & 4 \end{bmatrix} \quad \ldots\text{(iii)}$$

On adding Eqs. (i) and (iii), we get

$$5A = \begin{bmatrix} 5 & 0 & 10 \\ 10 & -5 & 15 \\ -5 & 5 & 5 \end{bmatrix}$$

$$\therefore \quad A = \begin{bmatrix} 1 & 0 & 2 \\ 2 & -1 & 3 \\ -1 & 1 & 1 \end{bmatrix}$$

$\therefore Tr(A) = 1 - 1 + 1 = 1$

From Eq. (i),

$$B = \frac{1}{2}\left\{\begin{bmatrix} 1 & 2 & 0 \\ 6 & -3 & 3 \\ -5 & 3 & 1 \end{bmatrix} - \begin{bmatrix} 1 & 0 & 2 \\ 2 & -1 & 3 \\ -1 & 1 & 1 \end{bmatrix}\right\}$$

$$= \begin{bmatrix} 0 & 1 & -1 \\ 2 & -1 & 0 \\ -2 & 1 & 0 \end{bmatrix}$$

$\therefore \text{Tr}(B) = 0 - 1 + 0 = -1$

Hence, $Tr(A) - \text{Tr}(B) = 1 - (-1) = 2$

**12.** *(b)* Here, $A^T A = \begin{bmatrix} a & b & c \\ b & c & a \\ c & a & b \end{bmatrix}\begin{bmatrix} a & b & c \\ b & c & a \\ c & a & b \end{bmatrix}$

$$= \begin{bmatrix} a^2 + b^2 + c^2 & ab + bc + ca & ac + ba + cb \\ ba + cb + ac & b^2 + c^2 + a^2 & bc + ca + ab \\ ca + ab + bc & cb + ac + ba & c^2 + a^2 + b^2 \end{bmatrix}$$

$\because \quad A^T A = I$ [given]

$\therefore a^2 + b^2 + c^2 = 1$ and $ab + bc + ca = 0$ ...(i)

Now, as $(a + b + c)^2 = a^2 + b^2 + c^2 + 2(ab + bc + ca)$

$\Rightarrow \quad a + b + c = \pm 1 \quad \ldots\text{(ii)}$

As we know,

$a^3 + b^3 + c^3 - 3abc = (a + b + c)(a^2 + b^2 + c^2 - ab - bc - ca)$

$\Rightarrow 2 - 3abc = (a + b + c)(1 - 0)$

[From Eq. (i) and $a^3 + b^3 + c^3 = 2$ [given]

$\Rightarrow \quad abc = \dfrac{2-(a+b+c)}{3}$

If $\quad a+b+c=1 \Rightarrow abc = \dfrac{1}{3}$, or

If $\quad a+b+c=-1 \Rightarrow abc = 1$

Now, from the options $abc = 1/3$

**13.** *(b)* Given, $AA^T = I$

$$\Rightarrow \begin{bmatrix} 0 & 2q & r \\ p & q & -r \\ p & -q & r \end{bmatrix}\begin{bmatrix} 0 & p & p \\ 2q & q & -q \\ r & -r & r \end{bmatrix} = \begin{bmatrix} 1 & 0 & 0 \\ 0 & 1 & 0 \\ 0 & 0 & 1 \end{bmatrix}$$

$$\Rightarrow \begin{bmatrix} 0+4q^2+r^2 & 0+2q^2-r^2 & 0-2q^2+r^2 \\ 0+2q^2-r^2 & p^2+q^2+r^2 & p^2-q^2-r^2 \\ 0-2q^2+r^2 & p^2-q^2-r^2 & p^2+q^2+r^2 \end{bmatrix} = \begin{bmatrix} 1 & 0 & 0 \\ 0 & 1 & 0 \\ 0 & 0 & 1 \end{bmatrix}$$

We know that, if two matrices are equal, then corresponding elements are also equal, so

$$4q^2 + r^2 = 1 = p^2 + q^2 + r^2, \quad \ldots(i)$$

$$2q^2 - r^2 = 0 \Rightarrow r^2 = 2q^2 \quad \ldots(ii)$$

and $\quad p^2 - q^2 - r^2 = 0 \quad \ldots(iii)$

Using Eqs. (ii) and (iii), we get

$$p^2 = 3q^2 \quad \ldots(iv)$$

Using Eqs. (ii) and (iv) in Eq. (i), we get

$4q^2 + 2q^2 = 1$

$\Rightarrow \quad 6q^2 = 1 \Rightarrow 2p^2 = 1$ [using Eq. (iv)]

$p^2 = \dfrac{1}{2} \Rightarrow |p| = \dfrac{1}{\sqrt{2}}$

**14.** *(c)* $P = \begin{bmatrix} \frac{\sqrt{3}}{2} & \frac{1}{2} \\ -\frac{1}{2} & \frac{\sqrt{3}}{2} \end{bmatrix}, P^T = \begin{bmatrix} \frac{\sqrt{3}}{2} & \frac{-1}{2} \\ \frac{1}{2} & \frac{\sqrt{3}}{2} \end{bmatrix}$

Now, $PP^T = P^TP = I$

Since,

$Q^{2015} = (PAP^T)(PAP^T)\ldots$ (2015 terms)

$= PA^{2015}P^T$

$P^TQ^{2015}P = A^{2015}$

$A^2 = \begin{bmatrix} 1 & 1 \\ 0 & 1 \end{bmatrix}\begin{bmatrix} 1 & 1 \\ 0 & 1 \end{bmatrix} = \begin{bmatrix} 1 & 2 \\ 0 & 1 \end{bmatrix}$

$\Rightarrow A^3 = \begin{bmatrix} 1 & 2 \\ 0 & 1 \end{bmatrix}\begin{bmatrix} 1 & 1 \\ 0 & 1 \end{bmatrix} = \begin{bmatrix} 1 & 3 \\ 0 & 1 \end{bmatrix}$

Again, applying equation ultimately, we get

$A^{2015} = \begin{bmatrix} 1 & 2015 \\ 0 & 1 \end{bmatrix}$

So, $P^TQ^{2015}P = \begin{bmatrix} 1 & 2015 \\ 0 & 1 \end{bmatrix}$

**15.** *(b)* Given, $A = \begin{bmatrix} 1 & 2 & 2 \\ 2 & 1 & -2 \\ a & 2 & b \end{bmatrix}$

$\Rightarrow A^T = \begin{bmatrix} 1 & 2 & a \\ 2 & 1 & 2 \\ 2 & -2 & b \end{bmatrix}$

We have, $AA^T = 9I$

$$\Rightarrow \begin{bmatrix} 1 & 2 & 2 \\ 2 & 1 & -2 \\ a & 2 & b \end{bmatrix}\begin{bmatrix} 1 & 2 & a \\ 2 & 1 & 2 \\ 2 & -2 & b \end{bmatrix} = \begin{bmatrix} 9 & 0 & 0 \\ 0 & 9 & 0 \\ 0 & 0 & 9 \end{bmatrix}$$

$$\Rightarrow \begin{bmatrix} 9 & 0 & a+2b+4 \\ 0 & 9 & 2a+2-2b \\ a+2b+4 & 2a+2-2b & a^2+4+b^2 \end{bmatrix} = \begin{bmatrix} 9 & 0 & 0 \\ 0 & 9 & 0 \\ 0 & 0 & 9 \end{bmatrix}$$

On comparing both sides, we get

$$a + 2b + 4 = 0 \quad \ldots(i)$$

$$2a + 2 - 2b = 0 \quad \ldots(ii)$$

$$a^2 + 4 + b^2 = 9 \quad \ldots(iii)$$

From Eqs. (i) and (ii), we get

Ordered pair $(a, b) = (-2, -1)$

**16.** *(a)* Since, $A$ and $B$ are symmetric matrices.

$\therefore \quad A^T = A$ and $B^T = B$

Now, to check $A(BA)$ is symmetric.

Consider $[A(BA)]^T = (BA)^T \cdot A^T$

$= (A^TB^T)A^T = (AB)A = A(BA)$

So, $\quad [A(BA)]^T = A(BA)$

$\Rightarrow A(BA)$ is symmetric.

Similarly, $(AB)A$ is symmetric.

So, Statement I is true. Also, Statement II is true, as if $A$ and $B$ are symmetric.

$\Rightarrow (AB)$ is symmetric, iff $AB = BA$.

*i.e.*, $AB$ commutative.

Hence, both the statements are true but Statement II is not a correct explanation of Statement I.

**17.** *(a)* Characteristic equation is

$|A - xI| = 0$

$\Rightarrow \quad \begin{vmatrix} 1-x & 5 \\ \lambda & 10-x \end{vmatrix} = 0$

$\Rightarrow \quad x^2 - 11x + 10 - 5\lambda = 0$

$\Rightarrow \quad A^2 - 11A + (10-5\lambda)I = 0$

$\Rightarrow (10-5\lambda)I = -A^2 + 11A$

$\Rightarrow (10-5\lambda)A^{-1} = -A\cdot(AA^{-1}) + 11A\cdot A^{-1}$

$\Rightarrow (10-5\lambda)A^{-1} = -A + 11I$

$\Rightarrow A^{-1} = \left(\dfrac{-1}{10-5\lambda}\right)A + \left(\dfrac{11}{10-5\lambda}\right)I$

$\Rightarrow \alpha = -\dfrac{1}{10-5\lambda}$ and $\beta = \dfrac{11}{10-5\lambda}$

$\alpha + \beta = -2$ [given]

$\Rightarrow \dfrac{-1}{10-5\lambda} + \dfrac{11}{10-5\lambda} = -2$

$\Rightarrow \quad \dfrac{10}{10-5\lambda} = -2$

$\Rightarrow \quad \lambda = 3$

$\therefore \quad \alpha = \dfrac{1}{5}$ and $\beta = -\dfrac{11}{5}$

$\therefore 4\alpha^2 + \beta^2 + \lambda^2 = \dfrac{4}{25} + \dfrac{121}{25} + 3^2$

$= \dfrac{4+121+225}{25} = \dfrac{350}{25} = 14$

**Alternate Solution**

Given, $\quad A = \begin{bmatrix} 1 & 5 \\ \lambda & 10 \end{bmatrix}$

and $A^{-1} = \alpha A + \beta I \quad \ldots(i)$

Post multiply by $A$ in Eq. (i) both sides, we get

$A^{-1}A = \alpha AA + \beta I - A$

$I = \alpha A^2 + \beta A$

Now,

$A^2 = \begin{bmatrix} 1 & 5 \\ \lambda & 10 \end{bmatrix}\begin{bmatrix} 1 & 5 \\ \lambda & 10 \end{bmatrix} = \begin{bmatrix} 1+5\lambda & 55 \\ 11\lambda & 5\lambda+100 \end{bmatrix}$

$\alpha\begin{bmatrix} 1+5\lambda & 55 \\ 11\lambda & 5\lambda+100 \end{bmatrix} + \beta\begin{bmatrix} 1 & 5 \\ \lambda & 10 \end{bmatrix} = \begin{bmatrix} 1 & 0 \\ 0 & 1 \end{bmatrix}$

$\Rightarrow \begin{bmatrix} \alpha+5\lambda\alpha+\beta & 55\alpha+5\beta \\ 11\alpha\lambda+\beta\lambda & 5\alpha\lambda+100\alpha+10\beta \end{bmatrix} = \begin{bmatrix} 1 & 0 \\ 0 & 1 \end{bmatrix}$

On comparing both sides, we get

$\alpha + \beta + 5\alpha\lambda = 1, \ 55\alpha + 5\beta = 0$

$\Rightarrow \alpha + \beta + 5\alpha\lambda = 1, \ \beta = -11\alpha$

$\Rightarrow \alpha - 11\alpha + 5\alpha\lambda = 1 \Rightarrow \alpha(5\lambda - 10) = 1$

$\Rightarrow \alpha = \dfrac{1}{5\lambda-10}$ and $\beta = -\dfrac{11}{5\lambda-10}$

$\because$ Given, $\alpha + \beta = -2$

$\dfrac{1}{5\lambda-10} + \left(-\dfrac{11}{5\lambda-10}\right) = -2$

$\Rightarrow -\dfrac{10}{5\lambda-10} = -2 \Rightarrow \lambda = 3$

$\therefore \alpha = \dfrac{1}{5}$ and $\beta = -\dfrac{11}{5}$

$\therefore 4\alpha^2 + \beta^2 + \lambda^2 = \dfrac{4}{25} + \dfrac{121}{25} + 9$

$= \dfrac{4+121+225}{25} = \dfrac{350}{25} = 14$

**18.** (c) Let $A, B, C$ be $3 \times 3$ matrices, where $A$ is symmetric, $B$ and $C$ are skew-symmetric matrices.

$(S1): A^{13}B^{26} - B^{26}A^{13}$ is symmetric

Let $\quad P = A^{13}B^{26} - B^{26}A^{13}$

Now, taking transpose both the sides, we get

$$P^T = (A^{13}B^{26} - B^{26}A^{13})^T$$

$$\left[\begin{array}{l} \because (M+N)^T = M^T N^T \\ \text{and } (AB)^T = B^T A^T \\ \text{and } (M^n)^T = (M^T)^n \end{array}\right]$$

So, $\quad P^T = (A^{13}B^{26})^T - (B^{26}A^{13})^T$

$$P^T = (B^{26})^T(A^{13})^T - (A^{13})^T(B^{26})^T$$

$$P^T = (B^T)^{26}(A^T)^{13} - (A^T)^{13}(B^T)^{26}$$

[$\because$ If $A$ is a symmetric matrix, so $A^T = A$ but $B$ is a skew-symmetric matrix, so $B^T = -B$]

Now, $P^T = (-B)^{26}(A)^{13} - (A^{13})(-B)^{26}$

$\Rightarrow P^T = B^{26}A^{13} - A^{13}B^{26}$

$\Rightarrow P^T = -(A^{13}B^{26} - A^{13}B^{26})$

$\Rightarrow P^T = -P$ skew-symmetric matrix

So, $(S1)$ is a wrong statement.

Now, $(S2): A^{26}C^{13} - C^{13}A^{26}$ is symmetric.

Let's $\quad Q = A^{26}C^{13} - C^{13}A^{26}$

Now, taking transpose both the sides, we get

$$Q^T = (A^{26}C^{13} - C^{13}A^{26})^T$$

$\Rightarrow Q^T = (A^{26}C^{13})^T - (C^{13}A^{26})^T$

$\Rightarrow Q^T = (C^{13})^T(A^{26})^T - (A^{26})^T(C^{13})^T$

$\Rightarrow Q^T = (C^T)^{13}(A^T)^{26} - (A^T)^{26}(C^T)^{13}$

$\Rightarrow Q^T = (-C)^{13}(A)^{26} - (A)^{26}(-C)^{13}$

$\Rightarrow Q^T = -C^{13}A^{26} + A^{26}C^{13}$

$$Q^T = A^{26}C^{13} - C^{13}A^{26}$$

$Q^T = Q$ symmetric matrix

So, S2 is a correct statement.

So, only S2 is true.

**19.** (a) Let $A$ be the matrix as follows,

$A = \begin{bmatrix} a & b \\ b & c \end{bmatrix}$, since $A$ is symmetric matrix.

Now, $A^2 = \begin{bmatrix} a & b \\ b & c \end{bmatrix}\begin{bmatrix} a & b \\ b & c \end{bmatrix}$

$$= \begin{bmatrix} a^2+b^2 & ab+bc \\ ab+bc & b^2+c^2 \end{bmatrix}$$

Given that the sum of diagonal entries of $A^2$ is 1.

i.e. $a^2 + b^2 + b^2 + c^2 = 1$

or $\qquad a^2 + 2b^2 + c^2 = 1$

**Case I** $a = 0$

Then, $2b^2 + c^2 = 1$

(a) $c = 0$ gives, $b^2 = \frac{1}{2}$ or $b = \pm\frac{1}{\sqrt{2}}$

$\therefore a = 0, b = 1/\sqrt{2}, c = 0$

$a = 0, b = -1/\sqrt{2}, c = 0 \qquad$ (2 matrices)

(b) $\quad b = 0$, gives $c^2 = 1$ or $c = \pm 1$

$\therefore a = 0, b = 0, c = 1$

and $a = 0, b = 0, c = -1$ (2 matrices)

**Case II** $b = 0$, then $a^2 + c^2 = 1$

(a) $\quad a = 0$, then $c = \pm 1$

$a = 0, b = 0, c = 1$ and $a = 0, b = 0, c = -1$

This is repeated case.

(b) $c = 0$, then $a = \pm 1$

$a = 1, b = 0, c = 0$ and $a = -1, b = 0, c = 0$

Again 2 matrices.

Thus, only acceptable matrices are as follows

$$A = \begin{bmatrix} 0 & 0 \\ 0 & 1 \end{bmatrix}, \begin{bmatrix} 0 & 0 \\ 0 & -1 \end{bmatrix}, \begin{bmatrix} 1 & 0 \\ 0 & 0 \end{bmatrix}, \begin{bmatrix} -1 & 0 \\ 0 & 0 \end{bmatrix}$$

Then, possible number of such matrices are 4.

**20.** (1613) $S_1 = \{A = [a_{ij}] \in M : A = A^T$ and $a_{ij} \in S, \forall\, i, j\}$

$$n(S_1) = 5^6$$

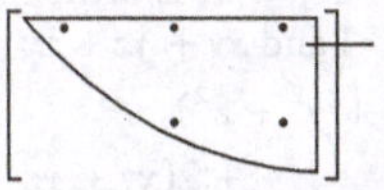

6 free elements and each element has 5 choices.

$S_2 = \{A = [a_{ij}] \in M : A = -A^T$ and $a_{ij} \in S, \forall\, i, j\}$

Skew-symmetric matrix has diagonal entries as zero.

But $\quad 0 \notin S$

Then, $n(S_2) = 0$

$S_3 = \{A = [a_{ij}] \in M, :$
$a_{11} + a_{22} + a_{33} = 0$ and $a_{ij} \in S, \forall\, i, j]$

$$\begin{bmatrix} a_{11} & \cdot & \cdot \\ \cdot & a_{22} & \cdot \\ \cdot & \cdot & a_{33} \end{bmatrix} \text{ 6 free elements}$$

and each has 5 choices.

$a_{11} + a_{22} + a_{33} = 0$

$(1, 2, -3) \rightarrow 3!$ arrangements

$(1, 1, -2) \rightarrow 3$ arrangements.

$(-1, -1, 2) \rightarrow 3$ arrangements

Total $= 12$

$n(S_3) = 5^6 \times 12$

$n(S_1 \cap S_2) = 0, n(S_2 \cap S_3) = 0$

and $\quad n(S_1 \cap S_3) = 5^3 \cdot 12$

$$\begin{bmatrix} a_{11} & & \\ & a_{22} & \\ & & a_{33} \end{bmatrix}$$

$a_{11} + a_{22} + a_{33} = 0$

Clearly, $\quad n(S_1 \cap S_2 \cap S_3) = 0$

$\Rightarrow n(S_1 \cup S_2 \cup S_3) = n(S_1) + n(S_2) + n(S_3) - n(S_1 \cap S_2) - n(S_2 \cap S_3) - n(S_1 \cap S_3) + n(S_1 \cap S_2 \cap S_3) = 0$

$= 5^6 + 5^6 \cdot 12 - 5^3 \cdot 12$

$= 5^3\,[125 + 125 \times 12 - 12] = 5^3\,(1613)$

$\Rightarrow \quad \alpha = 1613$

**21.** (6) $A = \begin{bmatrix} \cos\theta & 0 & -\sin\theta \\ 0 & 1 & 0 \\ \sin\theta & 0 & \cos\theta \end{bmatrix}$

$A^2 = A^T$

$\therefore A$ is orthogonal matrix

$\therefore \qquad A^T = A^{-1}$

$\Rightarrow \qquad A^2 = A^{-1}$

$\Rightarrow \qquad A^3 = I$

Let $B = (A+I)^3 + (A-I)^3 - 6A$

$= 2(A^3 + 3A) - 6A$

$= 2(3A + I) - 6A$

$= 2I$

$$B = \begin{bmatrix} 2 & 0 & 0 \\ 0 & 2 & 0 \\ 0 & 0 & 2 \end{bmatrix}$$

$\therefore$ Sum of diagonal elements $= 6$

**22.** (5) Let $A$ be a symmetric matrix say

$A = \begin{bmatrix} a & c \\ c & b \end{bmatrix}$

such that $|A| = 2 \Rightarrow ab - c^2 = 2$

and $\qquad \begin{bmatrix} 2 & 1 \\ 3 & 3/2 \end{bmatrix} A = \begin{bmatrix} 1 & 2 \\ \alpha & \beta \end{bmatrix}$

$\Rightarrow \qquad \begin{bmatrix} 2 & 1 \\ 3 & 3/2 \end{bmatrix}\begin{bmatrix} a & c \\ c & b \end{bmatrix} = \begin{bmatrix} 1 & 2 \\ \alpha & \beta \end{bmatrix}$

$\Rightarrow \qquad \begin{bmatrix} 2a+c & 2c+b \\ 3a+\frac{3}{2}c & 3c+\frac{3}{2}b \end{bmatrix} = \begin{bmatrix} 1 & 2 \\ \alpha & \beta \end{bmatrix}$

$\Rightarrow\; 2a + c = 1, \;\; 2c + b = 2$

$3a + \frac{3}{2}c = \alpha$ and $3c + \frac{3}{2}b = \beta$

Now, solving $2a + c = 1$ and $2c + b = 2$, we get

$$4a = b$$

Now, $ab - c^2 = 2 \Rightarrow \quad 4a^2 - c^2 = 2$

$\Rightarrow \qquad 4a^2 - (1-2a)^2 = 2$

$\Rightarrow \qquad 4a - 1 = 2 \Rightarrow a = \frac{3}{4}$

$\therefore \qquad b = 4a = 3$

$$c = 1 - 2a = -\frac{1}{2}$$

$\therefore \qquad A = \begin{bmatrix} 3/4 & -1/2 \\ -1/2 & 3 \end{bmatrix}$,

Given, $a + b = 5a \Rightarrow S = \frac{15}{4}$

Now, $\alpha = 3\left(\frac{3}{4}\right) + \frac{3}{2}\left(-\frac{1}{2}\right) = \frac{3}{2}$

$\beta = 3\left(-\frac{1}{2}\right) + \frac{3}{2}(3) = \frac{6}{2} = 3$

$\therefore \frac{\beta \cdot s}{\alpha^2} = \frac{3 \cdot \left(\frac{15}{4}\right)}{\frac{9}{4}} = \frac{45}{9} = 5$

**23.** *(5376)* Let $A = \begin{bmatrix} a_1 & b_1 & c_1 \\ a_2 & b_2 & c_2 \\ a_3 & b_3 & c_3 \end{bmatrix}$,

$A^T = \begin{bmatrix} a_1 & a_2 & a_3 \\ b_1 & b_2 & b_3 \\ c_1 & c_2 & c_3 \end{bmatrix}$

$\therefore A^T A = \begin{bmatrix} a_1 & a_2 & a_3 \\ b_1 & b_2 & b_3 \\ c_1 & c_2 & c_3 \end{bmatrix}\begin{bmatrix} a_1 & b_1 & c_1 \\ a_2 & b_2 & c_2 \\ a_3 & b_3 & c_3 \end{bmatrix}$

Trace $(A^T A) = a_1^2 + b_1^2 + c_1^2 + a_2^2 + b_2^2 + c_2^2 + a_3^2 + b_3^2 + c_3^2$

Given, trace $(A^T A) = 6$

So, total number of such matrices is ${}^9C_6 \times 2^6 = 5376$

**24.** *(766)* Let $A = \begin{bmatrix} a_1 & a_2 & a_3 \\ b_1 & b_2 & b_3 \\ c_1 & c_2 & c_3 \end{bmatrix}$

$\Rightarrow A^T = \begin{bmatrix} a_1 & b_1 & c_1 \\ a_2 & b_2 & c_2 \\ a_3 & b_3 & c_3 \end{bmatrix}$

$AA^T = \begin{bmatrix} a_1 & a_2 & a_3 \\ b_1 & b_2 & b_3 \\ c_1 & c_2 & c_3 \end{bmatrix}\begin{bmatrix} a_1 & b_1 & c_1 \\ a_2 & b_2 & c_2 \\ a_3 & b_3 & c_3 \end{bmatrix}$

$= \begin{bmatrix} a_1^2 + a_2^2 + a_3^2 & a_1b_1 + a_2b_2 + a_3b_3 & a_1c_1 + a_2c_2 + a_3c_3 \\ a_1b_1 + a_2b_2 + a_3b_3 & b_1^2 + b_2^2 + b_3^2 & b_1c_1 + b_2c_2 + b_3c_3 \\ a_1c_1 + a_2c_2 + a_3c_3 & b_1c_1 + b_2c_2 + b_3c_3 & c_1^2 + c_2^2 + c_3^2 \end{bmatrix}$

Now, sum of all the diagonal entries of $AA^T = 9$

$\Rightarrow a_1^2 + a_2^2 + a_3^2 + b_1^2 + b_2^2 + b_3^2 + c_1^2 + c_2^2 + c_3^2 = 9$

Given, $S : \{0, 1, 2, 3\}$

Hence, $a_1, a_2, a_3, b_1, b_2, b_3, c_1, c_2, c_3 \in \{0, 1, 2, 3\}$

**Case I** When All 1*s*

Number of matirces $= {}^9C_9 = 1$

**Case II** When one 3*s*, remaining 0*s*

Number of matrices $= {}^9C_1 = 9$

**Case III** When one 2*s* five 1*s* three 0*s*

Number of matrices $= {}^9C_1 \times {}^8C_5 = 504$

**Case IV** When two 2*s* one 1*s* six 0*s*

Number of matrices $= {}^9C_2 \times {}^7C_1 = 252$

So, total number of ways

$= 1 + 9 + 504 + 252 = 766$

**25.** *(7)* Given, $A^2 = I$

$\Rightarrow AA' = I$ $[\because A' = A] \Rightarrow A$ is orthogonal.

So, $x^2 + y^2 + z^2 = 1$ and $xy + yz + zx = 0$

$(x + y + z)^2 = (x^2 + y^2 + z^2) + 2(xy + yz + zx)$

$= 1 + 2(0)$

$\Rightarrow x + y + z = 1$ $\quad [\because x + y + z > 0]$

Thus, $x^3 + y^3 + z^3 - 3xyz = (x + y + z)(x^2 + y^2 + z^2 - xy - yz - zx)$

$\Rightarrow x^3 + y^3 + z^3 = (x + y + z)(x^2 + y^2 + z^2 - (xy + yz + zx)) + 3xyz$

$= (1)(1 - 0) + 3(2) = 7$

**26.** *(540)* Given, $M$ is a $3 \times 3$ matrix.

Let $M = \begin{bmatrix} a & b & c \\ d & e & f \\ g & h & i \end{bmatrix}$, then

Now, $M^T M = \begin{bmatrix} a & d & g \\ b & e & h \\ c & f & i \end{bmatrix} \times \begin{bmatrix} a & b & c \\ d & e & f \\ g & h & i \end{bmatrix}$

Sum of diagonal elements

$= a^2 + b^2 + c^2 + d^2 + e^2 + f^2 + g^2 + h^2 + i^2 = 7$ ...(i)

According to the question, the entries are $\{0, 1, 2\}$. $[\because M^T M = 7]$

i.e. $\{a, b, c, \ldots, h, i\} = \{0, 1, 2\}$

So, for Eq. (i) to be true, there are two cases.

**Case I** When 7-1's are there and 2-0's are there.

$\Rightarrow {}^9C_7 \times {}^2C_2 = 36$ ways of arrangements.

**Case II** When 1-2's is there, 3-1's and 5-0's are there.

${}^9C_1 \times {}^8C_3 \times {}^5C_5 = 9 \times \frac{8!}{3!5!} \times 1$

$= 504$ ways of arrangements.

$\therefore$Total possible arrangements

$= 36 + 504 = 540$

**27.** *(672)* Let a $3 \times 3$ matrix $A = [a_{ij}]_{3\times 3}$, $i, j \in \{1, 2, 3\}$ So, sum of diagonal elements of $AA^T$

$=$ trace of $AA^T = tr([a_{ij}][a_{ji}])$

$= a_{11}^2 + a_{12}^2 + a_{13}^2 + a_{21}^2 + a_{22}^2 + a_{23}^2 + a_{31}^2 + a_{32}^2 + a_{33}^2 = 3$ [given]

$\because a_{ij} \in \{-1, 0, 1\}$

So, possible combinations are

0, 0, 0, 0, 0, 0, 1, 1, 1

0, 0, 0, 0, 0, 0, −1, 1, 1

0, 0, 0, 0, 0, 0, −1, −1, 1

0, 0, 0, 0, 0, 0, −1, −1, −1

$\therefore$Number of such matrices such that $tr(AA^T) = 3$ is

$= {}^9C_6 \times 2 \times 2 \times 2$

$= \frac{9 \times 8 \times 7}{3 \times 2} \times 2 \times 2 \times 2$

$= 84 \times 8 = 672$

# C-16

JEE Main Chapterwise Topicwise Mathematics

In JEE Main, there are usually 1-2 questions per paper are visible from this chapter. In past 3 years, 78 conceptual and 20 numerical questions of moderate difficulty level were asked from this chapter.
**For detailed analysis scan QR code**

# DETERMINANTS

## Topic 1 Determinants of Matrices of Order Two and Three, Singular and Non-singular Matrices, Multiplication of Two Determinants of Same Order

### Level 1

**01** Let $M$ and $m$, respectively be the maximum and the minimum values of

$$f(x)=\begin{vmatrix}1+\sin^2 x & \cos^2 x & 4\sin 4x\\ \sin^2 x & 1+\cos^2 x & 4\sin 4x\\ \sin^2 x & \cos^2 x & 1+4\sin 4x\end{vmatrix},$$

$x\in R$. Then, $M^4-m^4$ is equal to **[2025, 29 Jan Shift-I]**

(a) 1040 (b) 1215
(c) 1295 (d) 1280

**Imp.** Co-factors of Matrix

**02** Let $\alpha\beta\neq 0$ and $A=\begin{bmatrix}\beta & \alpha & 3\\ \alpha & \alpha & \beta\\ -\beta & \alpha & 2\alpha\end{bmatrix}$.

If $B=\begin{bmatrix}3\alpha & -9 & 3\alpha\\ -\alpha & 7 & -2\alpha\\ -2\alpha & 5 & -2\beta\end{bmatrix}$ is the matrix of co-factors of the elements of $A$, then $\det(AB)$ is equal to **[2024, 5 April Shift-II]**

(a) 64 (b) 343 (c) 125 (d) 216

**03** Let $A=\begin{bmatrix}1 & 0 & 0\\ 0 & \alpha & \beta\\ 0 & \beta & \alpha\end{bmatrix}$ and $|2A|^3=2^{21}$ where, $\alpha,\beta\in Z$.

Then, a value of $\alpha$ is **[2024, 29 Jan Shift-I]**

(a) 3 (b) 5 (c) 9 (d) 17

**04** Let $A=\begin{bmatrix}4 & -2\\ \alpha & \beta\end{bmatrix}$.

If $A^2+4A+18I=0$, then $\det(A)$ is equal to **[2022, 27 July Shift-II]**

(a) −18 (b) 18 (c) −50 (d) 50

***05** Let $A=\begin{bmatrix}[x+1] & [x+2] & [x+3]\\ [x] & [x+3] & [x+3]\\ [x] & [x+2] & [x+4]\end{bmatrix}$, where $[t]$ denotes the greatest integer less than or equal to $t$. If $\det(A)=192$, then the set of values of $x$ is the interval **[2021, 27 Aug Shift-II]**

(a) [68, 69) (b) [62, 63) (c) [65, 66) (d) [60, 61)

***06** The number of distinct real roots of

$$\begin{vmatrix}\sin x & \cos x & \cos x\\ \cos x & \sin x & \cos x\\ \cos x & \cos x & \sin x\end{vmatrix}=0 \text{ in the interval } -\frac{\pi}{4}\leq x\leq\frac{\pi}{4} \text{ is}$$

**[2021, 25 July Shift-II]**

(a) 4 (b) 1
(c) 2 (d) 3

**07** If $x, y, z$ are in arithmetic progression with common difference $d$, $x\neq 3d$, and the determinant of the matrix $\begin{bmatrix}3 & 4\sqrt{2} & x\\ 4 & 5\sqrt{2} & y\\ 5 & k & z\end{bmatrix}$ is zero, then the value of $k^2$ is **[2021, 17 March Shift-II]**

(a) 72 (b) 12
(c) 36 (d) 6

**08** The value of $\begin{vmatrix}(a+1)(a+2) & a+2 & 1\\ (a+2)(a+3) & a+3 & 1\\ (a+3)(a+4) & a+4 & 1\end{vmatrix}$ is **[2021, 26 Feb Shift-I]**

(a) $(a+2)(a+3)(a+4)$ (b) −2
(c) $(a+1)(a+2)(a+3)$ (d) 0

**(Marked Qns. are from topics, not included in latest JEE Main syllabus.)*

***09** Let $A$ be a $3 \times 3$ matrix with $\det(A) = 4$. Let $R_i$ denote the ith row of $A$. If a matrix B is obtained by performing the operation $R_2 \to 2R_2 + 5R_3$ on $2A$, then $\det(B)$ is equal to **[2021, 25 Feb Shift-II]**

(a) 16 (b) 80 (c) 64 (d) 128

**10** Let $A$ be a $2 \times 2$ real matrix with entries from $\{0,1\}$ and $|A| \neq 0$. Consider the following two statements

(P) If $A \neq I_2$, then $|A| = -1$

(Q) If $|A| = 1$, then $\text{tr}(A) = 2$,

where $I_2$ denotes $2 \times 2$ identity matrix and $\text{tr}(A)$ denotes the sum of the diagonal entries of $A$. Then, **[2020, 2 Sep Shift-I]**

(a) (P) is false and (Q) is true

(b) Both (P) and (Q) are false

(c) (P) is true and (Q) is false

(d) Both (P) and (Q) are true

**11** Let $a - 2b + c = 1$.

If $f(x) = \begin{vmatrix} x+a & x+2 & x+1 \\ x+b & x+3 & x+2 \\ x+c & x+4 & x+3 \end{vmatrix}$, then **[2020, 9 Jan Shift-II]**

(a) $f(-50) = 501$ (b) $f(50) = 1$

(c) $f(-50) = -1$ (d) $f(50) = -501$

**Imp.** Sum of Roots of Cubic Equation

**12** The sum of the real roots of the equation

$\begin{vmatrix} x & -6 & -1 \\ 2 & -3x & x-3 \\ -3 & 2x & x+2 \end{vmatrix} = 0$, is equal to **[2019, 10 April Shift-II]**

(a) 0 (b) $-4$ (c) 6 (d) 1

**13** If $\Delta_1 = \begin{vmatrix} x & \sin\theta & \cos\theta \\ -\sin\theta & -x & 1 \\ \cos\theta & 1 & x \end{vmatrix}$

and $\Delta_2 = \begin{vmatrix} x & \sin 2\theta & \cos 2\theta \\ -\sin 2\theta & -x & 1 \\ \cos 2\theta & 1 & x \end{vmatrix}$, $x \neq 0$,

then for all $\theta \in \left(0, \frac{\pi}{2}\right)$ **[2019, 10 April Shift-I]**

(a) $\Delta_1 + \Delta_2 = -2(x^3 + x - 1)$

(b) $\Delta_1 - \Delta_2 = -2x^3$

(c) $\Delta_1 + \Delta_2 = -2x^3$

(d) $\Delta_1 - \Delta_2 = x(\cos 2\theta - \cos 4\theta)$

**14** Let $\alpha$ and $\beta$ be the roots of the equation $x^2 + x + 1 = 0$. Then, for $y \neq 0$ in $R$,

$\begin{vmatrix} y+1 & \alpha & \beta \\ \alpha & y+\beta & 1 \\ \beta & 1 & y+\alpha \end{vmatrix}$ is equal to **[2019, 9 April Shift-I]**

(a) $y(y^2-1)$ (b) $y(y^2-3)$ (c) $y^3-1$ (d) $y^3$

**15** If $\begin{vmatrix} x-4 & 2x & 2x \\ 2x & x-4 & 2x \\ 2x & 2x & x-4 \end{vmatrix} = (A + Bx)(x - A)^2$,

then the ordered pair $(A, B)$ is equal to **[JEE Main 2018]**

(a) $(-4, -5)$ (b) $(-4, 3)$ (c) $(-4, 5)$ (d) $(4, 5)$

**16** The least value of the product $xyz$ for which the determinant $\begin{vmatrix} x & 1 & 1 \\ 1 & y & 1 \\ 1 & 1 & z \end{vmatrix}$ is non-negative is **[JEE Main 2015]**

(a) $-2\sqrt{2}$ (b) $-16\sqrt{2}$ (c) $-8$ (d) $-1$

**17** If $\alpha, \beta \neq 0$, $f(n) = \alpha^n + \beta^n$ and

$\begin{vmatrix} 3 & 1+f(1) & 1+f(2) \\ 1+f(1) & 1+f(2) & 1+f(3) \\ 1+f(2) & 1+f(3) & 1+f(4) \end{vmatrix}$

$= K(1-\alpha)^2 (1-\beta)^2 (\alpha-\beta)^2$, then $K$ is equal to **[JEE Main 2014]**

(a) $\alpha\beta$ (b) $1/\alpha\beta$ (c) 1 (d) $-1$

**18** Let $P$ and $Q$ be $3 \times 3$ matrices, $P \neq Q$. If $P^3 = Q^3$ and $P^2Q = Q^2P$, then determinant of $(P^2 + Q^2)$ is equal to **[AIEEE 2012]**

(a) $-2$ (b) 1 (c) 0 (d) $-1$

**19 Statement I** Determinant of a skew- symmetric matrix of order 3 is zero.

**Statement II** For any matrix $A$, $\det(A^T) = \det(A)$ and $\det(-A) = -\det(A)$.

Then, **[AIEEE 2011]**

(a) Statement I is true and Statement II is false

(b) Both statements are true

(c) Both statements are false

(d) Statement I is false and Statement II is true

**20** Let $A = \begin{bmatrix} 5 & 5\alpha & \alpha \\ 0 & \alpha & 5\alpha \\ 0 & 0 & 5 \end{bmatrix}$. If $|A^2| = 25$, then $|\alpha|$ is equal to **[AIEEE 2007]**

(a) $5^2$ (b) 1 (c) 1/5 (d) 5

**21** If $D=\begin{vmatrix}1 & 1 & 1\\ 1 & 1+x & 1\\ 1 & 1 & 1+y\end{vmatrix}$ for $x\neq 0,\ y\neq 0$, then $D$ is **[AIEEE 2007]**

(a) divisible by neither $x$ nor $y$
(b) divisible by both $x$ and $y$
(c) divisible by $x$ but not $y$
(d) divisible by $y$ but not $x$

***22** If $a_1, a_2, \ldots, a_n, \ldots$ are in GP, then the determinant

$$\Delta=\begin{vmatrix}\log a_n & \log a_{n+1} & \log a_{n+2}\\ \log a_{n+3} & \log a_{n+4} & \log a_{n+5}\\ \log a_{n+6} & \log a_{n+7} & \log a_{n+8}\end{vmatrix}$$

is equal to **[AIEEE 2005, 2004]**

(a) 2 (b) 4 (c) 0 (d) 1

**Imp.** Cube Root of Unity

**23** If 1, $\omega$ and $\omega^2$ are the cube roots of unity, then

$\Delta=\begin{vmatrix}1 & \omega^n & \omega^{2n}\\ \omega^n & \omega^{2n} & 1\\ \omega^{2n} & 1 & \omega^n\end{vmatrix}$ is equal to **[AIEEE 2003]**

(a) 0 (b) 1 (c) $\omega$ (d) $\omega^2$

***24** If $l, m$ and $n$ are the $p$th, $q$th and $r$th terms of a GP and all positive, then $\begin{vmatrix}\log l & p & 1\\ \log m & q & 1\\ \log n & r & 1\end{vmatrix}$ is equal to **[AIEEE 2002]**

(a) 3 (b) 2 (c) 1 (d) 0

**25** If $\begin{vmatrix}6i & -3i & 1\\ 4 & 3i & -1\\ 20 & 3 & i\end{vmatrix}=x+iy$, then **[AIEEE 2002]**

(a) $x=3, y=1$ (b) $x=1, y=3$
(c) $x=0, y=3$ (d) $x=0, y=0$

## Level 2

**26** If $\alpha\neq a, \beta\neq b, \gamma\neq c$ and $\begin{vmatrix}\alpha & b & c\\ a & \beta & c\\ a & b & \gamma\end{vmatrix}=0$, then

$\dfrac{a}{\alpha-a}+\dfrac{b}{\beta-b}+\dfrac{\gamma}{\gamma-c}$ is equal to **[2024, 8 April Shift-II]**

(a) 0 (b) 3 (c) 2 (d) 1

**27** Let $A=\begin{bmatrix}2 & a & 0\\ 1 & 3 & 1\\ 0 & 5 & b\end{bmatrix}$. If $A^3=4A^2-A-21I$, where $I$ is the identity matrix of order $3\times 3$. Then, $2a+3b$ is equal to **[2024, 8 April Shift-I]**

(a) $-10$ (b) $-9$ (c) $-13$ (d) $-12$

***28** For $\alpha, \beta\in R$ and a natural number $n$. Let

$A_r=\begin{vmatrix}r & 1 & \dfrac{n^2}{2}+\alpha\\ 2r & 2 & n^2-\beta\\ 3r-2 & 3 & \dfrac{n(3n-1)}{2}\end{vmatrix}$. Then, $2A_{10}-A_8$ is **[2024, 6 April Shift-I]**

(a) $4\alpha+2\beta$ (b) $2\alpha+4\beta$
(c) 0 (d) $2n$

**29** If $A=\begin{bmatrix}\sqrt{2} & 1\\ -1 & \sqrt{2}\end{bmatrix}$, $B=\begin{bmatrix}1 & 0\\ 1 & 1\end{bmatrix}$, $C=ABA^T$ and $X=A^TC^2A$, then det $X$ is equal to **[2024, 1 Feb Shift-I]**

(a) 243 (b) 729 (c) 891 (d) 27

***30** The values of $\alpha$, for which $\begin{vmatrix}1 & \dfrac{3}{2} & \alpha+\dfrac{3}{2}\\ 1 & \dfrac{1}{3} & \alpha+\dfrac{1}{3}\\ 2\alpha+3 & 3\alpha+1 & 0\end{vmatrix}=0$, lie in the interval **[2024, 27 Jan Shift-II]**

(a) $(-2, 1)$ (b) $(-3, 0)$
(c) $\left(-\dfrac{3}{2}, \dfrac{3}{2}\right)$ (d) $(0, 3)$

**31** If $\begin{vmatrix}x+1 & x & x\\ x & x+\lambda & x\\ x & x & x+\lambda^2\end{vmatrix}=\dfrac{9}{8}(103x+81)$, then $\lambda, \dfrac{\lambda}{3}$ are the roots of the equation **[2023, 11 April Shift-II]**

(a) $4x^2+24x+27=0$
(b) $4x^2-24x+27=0$
(c) $4x^2+24x-27=0$
(d) $4x^2-24x-27=0$

**32** Let $f(x)=\begin{vmatrix}1+\sin^2 x & \cos^2 x & \sin 2x\\ \sin^2 x & 1+\cos^2 x & \sin 2x\\ \sin^2 x & \cos^2 x & 1+\sin 2x\end{vmatrix}$, $x\in\left[\dfrac{\pi}{6}, \dfrac{\pi}{3}\right]$. If $\alpha$ and $\beta$ respectively are the maximum and the minimum values of $f$, then **[2023, 1 Feb Shift-I]**

(a) $\beta^2+2\sqrt{\alpha}=\dfrac{19}{4}$

(b) $\alpha^2+\beta^2=\dfrac{9}{2}$

(c) $\alpha^2-\beta^2=4\sqrt{3}$

(d) $\beta^2-2\sqrt{\alpha}=\dfrac{19}{4}$

**33** Let $\alpha$ be a root of the equation $(a-c)x^2+(b-a)x+(c-b)=0$, where $a, b, c$ are distinct real numbers such that the matrix $\begin{bmatrix} \alpha^2 & \alpha & 1 \\ 1 & 1 & 1 \\ a & b & c \end{bmatrix}$ is singular. Then, the value of

$$\frac{(a-c)^2}{(b-a)(c-b)}+\frac{(b-a)^2}{(a-c)(c-b)}+\frac{(c-b)^2}{(a-c)(b-a)} \text{ is}$$

*[2023, 24 Jan Shift-I]*

(a) 6 (b) 9 (c) 12 (d) 3

**34** Let $A=\begin{bmatrix} 2 & 3 \\ a & 0 \end{bmatrix}$, $a \in R$ be written as $P+Q$, where $P$ is a symmetric matrix and $Q$ is skew-symmetric matrix. If det $(Q)=9$, then the modulus of the sum of all possible values of determinant of $P$ is equal to *[2021, 20 July Shift-I]*

(a) 36 (b) 24 (c) 45 (d) 18

**35** Let $A=[a_{ij}]$ be a $3 \times 3$ matrix, where

$$a_{ij}=\begin{cases} 1 & , \quad \text{if } i=j \\ -x & , \quad \text{if } |i-j|=1 \\ 2x+1 & , \quad \text{otherwise} \end{cases}$$

Let a function $f: R \to R$ be defined as $f(x)=\det(A)$. Then, the sum of maximum and minimum values of $f$ on $R$ is equal to *[2021, 20 July Shift-I]*

(a) $-\frac{20}{27}$ (b) $\frac{88}{27}$ (c) $\frac{20}{27}$ (d) $-\frac{88}{27}$

**36** The solutions of the equation

$$\begin{vmatrix} 1+\sin^2 x & \sin^2 x & \sin^2 x \\ \cos^2 x & 1+\cos^2 x & \cos^2 x \\ 4\sin 2x & 4\sin 2x & 1+4\sin 2x \end{vmatrix}$$

$=0, (0<x<\pi)$, are *[2021, 18 March Shift-I]*

(a) $\frac{\pi}{12}, \frac{\pi}{6}$ (b) $\frac{\pi}{6}, \frac{5\pi}{6}$ (c) $\frac{5\pi}{12}, \frac{7\pi}{12}$ (d) $\frac{7\pi}{12}, \frac{11\pi}{12}$

**37** If $A=\begin{bmatrix} 0 & \sin\alpha \\ \sin\alpha & 0 \end{bmatrix}$ and $\det\left(A^2-\frac{1}{2}I\right)=0$, then a possible value of $\alpha$ is *[2021, 17 March Shift-I]*

(a) $\frac{\pi}{2}$ (b) $\frac{\pi}{3}$ (c) $\frac{\pi}{4}$ (d) $\frac{\pi}{6}$

**38** The maximum value of

$$f(x)=\begin{vmatrix} \sin^2 x & 1+\cos^2 x & \cos 2x \\ 1+\sin^2 x & \cos^2 x & \cos 2x \\ \sin^2 x & \cos^2 x & \sin 2x \end{vmatrix}, x \in R \text{ is}$$

*[2021, 16 March Shift-II]*

(a) $\sqrt{7}$ (b) 3/4
(c) $\sqrt{5}$ (d) 5

**39** Let $\theta=\frac{\pi}{5}$ and $A=\begin{bmatrix} \cos\theta & \sin\theta \\ -\sin\theta & \cos\theta \end{bmatrix}$.

If $B=A+A^4$, then det $(B)$ is equal to *[2020, 6 Sep Shift-II]*

(a) is one (b) line in (2, 3)
(c) is zero (d) lies in (1, 2)

**Imp.** **Maximum & Minimum Value of Function**

**40** Let $m$ and $M$ be respectively the minimum and maximum values of

$$\begin{vmatrix} \cos^2 x & 1+\sin^2 x & \sin 2x \\ 1+\cos^2 x & \sin^2 x & \sin 2x \\ \cos^2 x & \sin^2 x & 1+\sin 2x \end{vmatrix}.$$

Then, the ordered pair $(m, M)$ is equal to *[2020, 6 Sep Shift-I]*

(a) $(-3, 3)$ (b) $(-3, -1)$ (c) $(-4, -1)$ (d) $(1, 3)$

**41** If $a+x=b+y=c+z+1$, where $a, b, c, x, y, z$ are non-zero distinct real numbers, then $\begin{vmatrix} x & a+y & x+a \\ y & b+y & y+b \\ z & c+y & z+c \end{vmatrix}$ is equal to *[2020, 5 Sep Shift-II]*

(a) $y(b-a)$ (b) $y(a-b)$ (c) 0 (d) $y(a-c)$

**42** If $\Delta=\begin{vmatrix} x-2 & 2x-3 & 3x-4 \\ 2x-3 & 3x-4 & 4x-5 \\ 3x-5 & 5x-8 & 10x-17 \end{vmatrix}=$

$Ax^3+Bx^2+Cx+D$, then $B+C$ is equal to *[2020, 3 Sep Shift-I]*

(a) $-1$ (b) 1 (c) $-3$ (d) 9

**43** If $A=\begin{bmatrix} 1 & \sin\theta & 1 \\ -\sin\theta & 1 & \sin\theta \\ -1 & -\sin\theta & 1 \end{bmatrix}$, then for all $\theta \in \left(\frac{3\pi}{4}, \frac{5\pi}{4}\right)$, det$(A)$ lies in the interval *[2019, 12 Jan Shift-II]*

(a) $\left(\frac{3}{2}, 3\right]$ (b) $\left[\frac{5}{2}, 4\right)$ (c) $\left(0, \frac{3}{2}\right]$ (d) $\left(1, \frac{5}{2}\right]$

***44** If $\begin{vmatrix} a-b-c & 2a & 2a \\ 2b & b-c-a & 2b \\ 2c & 2c & c-a-b \end{vmatrix}$

$=(a+b+c)(x+a+b+c)^2, x \neq 0$ and $a+b+c \neq 0$, then $x$ is equal to *[2019, 11 Jan Shift-II]*

(a) $-(a+b+c)$ (b) $-2(a+b+c)$
(c) $2(a+b+c)$ (d) $abc$

**45** Let $d \in R$, and $A = \begin{bmatrix} -2 & 4+d & (\sin\theta)-2 \\ 1 & (\sin\theta)+2 & d \\ 5 & (2\sin\theta)-d & (-\sin\theta)+2+2d \end{bmatrix}$, $\theta \in [0, 2\pi]$. If the minimum value of det($A$) is 8, then the value of $d$ is **[2019, 10 Jan Shift-I]**

(a) $-5$ (b) $-7$
(c) $2(\sqrt{2}+1)$ (d) $2(\sqrt{2}+2)$

**46** If $S = \left\{ x \in [0, 2\pi] : \begin{vmatrix} 0 & \cos x & -\sin x \\ \sin x & 0 & \cos x \\ \cos x & \sin x & 0 \end{vmatrix} = 0 \right\}$, then $\sum_{x \in S} \tan\left(\frac{\pi}{3}+x\right)$ is equal to **[JEE Main 2017]**

(a) $-2+\sqrt{3}$ (b) $4+2\sqrt{3}$ (c) $-4-2\sqrt{3}$ (d) $-2-2\sqrt{3}$

**47** If $A = \begin{bmatrix} -4 & -1 \\ 3 & 1 \end{bmatrix}$, then the determinant of the matrix $(A^{2016} - 2A^{2015} - A^{2014})$ is **[JEE Main 2016]**

(a) $-175$ (b) 2014 (c) 2016 (d) $-25$

**48** Let $a$, $b$ and $c$ be such that $(b+c) \neq 0$.

If $\begin{vmatrix} a & a+1 & a-1 \\ -b & b+1 & b-1 \\ c & c-1 & c+1 \end{vmatrix} + \begin{vmatrix} a+1 & b+1 & c-1 \\ a-1 & b-1 & c+1 \\ (-1)^{n+2}a & (-1)^{n+1}b & (-1)^n c \end{vmatrix} = 0$,

then the value of $n$ is **[AIEEE 2009]**

(a) zero (b) any even integer
(c) any odd integer (d) any integer

## Numerical Value Types Questions

**49** The number of singular matrices of order 2, whose elements are from the set {2, 3, 6, 9}, is ………… . **[2025, 7 April Shift-I]**

***50** Let $D_k = \begin{vmatrix} 1 & 2k & 2k-1 \\ n & n^2+n+2 & n^2 \\ n & n^2+n & n^2+n+2 \end{vmatrix}$. If $\sum_{k=1}^{n} D_k = 96$, then $n$ is equal to ………… . **[2023, 12 April Shift-I]**

**51** Let $A_1$, $A_2$, $A_3$ be the three AP with the same common difference $d$ and having their first terms as $A$, $A+1$, $A+2$, respectively. Let $a$, $b$, $c$ be the 7th, 9th, 17th terms of $A_1$, $A_2$, $A_3$, respectively such that $\begin{vmatrix} a & 7 & 1 \\ 2b & 17 & 1 \\ c & 17 & 1 \end{vmatrix} + 70 = 0$.

If $a = 29$, then the sum of first 20 terms of an AP whose first term is $c-a-b$ and common difference is $d/12$ is equal to ….. **[2023, 25 Jan Shift-I]**

**52** Let $P$ and $P+2$ be prime numbers and let

$$\Delta = \begin{vmatrix} P! & (P+1)! & (P+2)! \\ (P+1)! & (P+2)! & (P+3)! \\ (P+2)! & (P+3)! & (P+4)! \end{vmatrix}$$

Then, the sum of the maximum values of $\alpha$ and $\beta$ such that $P^\alpha$ and $(P+2)^\beta$ divide $\Delta$, is … . **[2022, 29 July Shift-I]**

**53** Let $f(x) = \begin{vmatrix} \sin^2 x & -2+\cos^2 x & \cos 2x \\ 2+\sin^2 x & \cos^2 x & \cos 2x \\ \sin^2 x & \cos^2 x & 1+\cos 2x \end{vmatrix}$, $x \in [0, \pi]$. Then, the maximum value of $f(x)$ is equal to ………… . **[2021, 27 July Shift-I]**

**54** Let $M = \left[ A = \begin{bmatrix} a & b \\ c & d \end{bmatrix} : a, b, c, d \in \{\pm 3, \pm 2, \pm 1, 0\} \right]$.

Define $f : M \to Z$, as $f(A) = \det(A)$, for all $A \in M$, where $Z$ is set of all integers. Then the number of $A \in M$ such that $f(A) = 15$ is equal to …… . **[2021, 25 July Shift-I]**

***55** Let $a, b, c, d$ be in arithmetic progression with common difference $\lambda$.

If $\begin{vmatrix} x+a-c & x+b & x+a \\ x-1 & x+c & x+b \\ x-b+d & x+d & x+c \end{vmatrix} = 2$,

then value of $\lambda^2$ is equal to …… . **[2021, 20 July Shift-I]**

**Imp.** Arithmetic Mean

**56** If 1, $\log_{10}(4^x - 2)$ and $\log_{10}\left(4^x + \frac{18}{5}\right)$ are in arithmetic progression for a real number $x$, then the value of the determinant $\begin{vmatrix} 2\left(x-\frac{1}{2}\right) & x-1 & x^2 \\ 1 & 0 & x \\ x & 1 & 0 \end{vmatrix}$ is equal to ……… . **[2021, 17 March Shift-II]**

**57** Let $A = \begin{bmatrix} a & b \\ c & d \end{bmatrix}$ and $B = \begin{bmatrix} \alpha \\ \beta \end{bmatrix} \neq \begin{bmatrix} 0 \\ 0 \end{bmatrix}$, such that $AB = B$ and $a+d = 2021$, then the value of $ad - bc$ is equal to ……………. **[2021, 17 March Shift-II]**

## Topic 2 Area of Triangle using Determinants

### Level 1

**01** Let $A(a, 0)$, $B(b, 2b+1)$ and $C(0, b)$, $b \neq 0, |b| \neq 1$, be points such that the area of $\Delta ABC$ is 1 sq. unit, then the sum of all possible values of a is **[2021, 27 Aug Shift-II]**

(a) $\frac{-2b}{b+1}$ (b) $\frac{2b}{b+1}$ (c) $\frac{2b^2}{b+1}$ (d) $\frac{-2b^2}{b+1}$

### Level 2

**02** Let the area of the triangle with vertices $A(1, \alpha)$, $B(\alpha, 0)$ and $C(0, \alpha)$ be 4 sq units. If the points $(\alpha, -\alpha)$, $(-\alpha, \alpha)$ and $(\alpha^2, \beta)$ are collinear, then $\beta$ is equal to **[2021, 24 June Shift-II]**

(a) 64 (b) −8 (c) −64 (d) 512

### Level 3

**03** Let $k$ be an integer such that the triangle with vertices $(k, -3k)$, $(5, k)$ and $(-k, 2)$ has area 28 sq units. Then, the orthocentre of this triangle is at the point **[JEE Main 2017]**

(a) $\left(2, -\frac{1}{2}\right)$ (b) $\left(1, \frac{3}{4}\right)$ (c) $\left(1, -\frac{3}{4}\right)$ (d) $\left(2, \frac{1}{2}\right)$

## Topic 3 Minors and Cofactors, Adjoint & Evaluation of Inverse of a Square Matrix using Determinants

### Level 1

**01** Let $A$ be a $3 \times 3$ matrix such that $|\text{adj}(\text{adj}(\text{adj}A))| = 81$. If

$$S = \left\{ n \in Z : (|\text{adj}(\text{adj}A)|)^{\frac{(n-1)^2}{2}} = |A|^{(3n^2 - 5n - 4)} \right\}, \text{ then}$$

$\sum_{n \in S} \left| A^{(n^2+n)} \right|$ is equal to **[2025, 7 April Shift-I]**

(a) 750 (b) 866 (c) 820 (d) 732

**02** If $A$, $B$ and $(\text{adj}(A^{-1}) + \text{adj}(B^{-1}))$ are non-singular matrices of same order, then the inverse of $A(\text{adj}(A^{-1}) + \text{adj}(B^{-1}))^{-1} B$, is equal to **[2025, 23 Jan Shift-I]**

(a) $\frac{1}{|AB|}(\text{adj}(B) + \text{adj}(A))$ (b) $\text{adj}(B^{-1}) + \text{adj}(A^{-1})$

(c) $AB^{-1} + A^{-1}B$ (d) $\frac{AB^{-1}}{|A|} + \frac{BA^{-1}}{|B|}$

**03** For a $3 \times 3$ matrix $M$, let trace $(M)$ denote the sum of all the diagonal elements of $M$. Let $A$ be a $3 \times 3$ matrix such that $|A| = \frac{1}{2}$ and trace $(A) = 3$. If $B = \text{adj}(\text{adj}(2A))$, then the value of $|B| + \text{trace}(B)$ equals **[2025, 22 Jan Shift-II]**

(a) 174 (b) 280 (c) 132 (d) 56

**04** Let $B = \begin{bmatrix} 1 & 3 \\ 1 & 5 \end{bmatrix}$ and $A$ be a $2 \times 2$ matrix such that $AB^{-1} = A^{-1}$. If $BCB^{-1} = A$ and $C^4 + \alpha C^2 + \beta I = O$, then $2\beta - \alpha$ is equal to **[2024, 9 April Shift-II]**

(a) 8 (b) 10 (c) 16 (d) 2

**05** Let $A = \begin{bmatrix} 1 & 2 \\ 0 & 1 \end{bmatrix}$ and

$B = I + \text{adj}(A) + (\text{adj } A)^2 + \ldots + (\text{adj } A)^{10}$.

Then, the sum of all the elements of the matrix $B$ is **[2024, 4 April Shift-II]**

(a) −88 (b) −124 (c) −110 (d) 22

**Imp.** Determinant of Adjoint of Matrix

***06** If $A = \frac{1}{5!6!7!} \begin{bmatrix} 5! & 6! & 7! \\ 6! & 7! & 8! \\ 7! & 8! & 9! \end{bmatrix}$, then $|\text{adj}(\text{adj}(2A))|$ is equal to **[2023, 10 April Shift-II]**

(a) $2^{12}$ (b) $2^{20}$ (c) $2^{16}$ (d) $2^8$

**07** If $A$ is a $3 \times 3$ matrix and $|A| = 2$, then $|3 \text{ adj}(|3A| A^2)|$ is equal to **[2023, 10 April Shift-I]**

(a) $3^{12} \cdot 6^{10}$ (b) $3^{12} \cdot 6^{11}$ (c) $3^{11} \cdot 6^{10}$ (d) $3^{10} \cdot 6^{11}$

**08** Let $A = \begin{bmatrix} 2 & 1 & 0 \\ 1 & 2 & -1 \\ 0 & -1 & 2 \end{bmatrix}$. If $|\text{adj}(\text{adj}(\text{adj} 2A))| = (16)^n$, then $n$ is equal to **[2023, 8 April Shift-I]**

(a) 8 (b) 10 (c) 12 (d) 9

**09** If $P$ is a $3 \times 3$ real matrix such that $P^T = aP + (a-1)I$, where $a > 1$, then **[2023, 30 Jan Shift-II]**

(a) $P$ is a singular matrix (b) $|\text{Adj } P| = 1$

(c) $|\text{Adj } P| > 1$ (d) $|\text{Adj } P| = 1/2$

**10** Let $x, y, z > 1$ and $A = \begin{bmatrix} 1 & \log_x y & \log_x z \\ \log_y x & 2 & \log_y z \\ \log_z x & \log_z y & 3 \end{bmatrix}$.

Then, $|\text{adj}(\text{adj } A^2)|$ is equal to **[2023, 25 Jan Shift-I]**

(a) $6^4$ (b) $4^8$ (c) $2^8$ (d) $2^4$

**11** Let $A$ be a $3 \times 3$ matrix such that $|\text{adj}(\text{adj}(\text{adj } A))| = 12^4$. Then, $|A^{-1} \text{ adj } A|$ is equal to **[2023, 24 Jan Shift-II]**

(a) 12 (b) 1 (c) $2\sqrt{3}$ (d) $\sqrt{6}$

**12** Let $A$ be a matrix of order $3 \times 3$ and det $(A) = 2$. Then, det (det $(A)$ adj (5 adj $(A^3)$)) is equal to **[2022, 28 June Shift-I]**

(a) $512 \times 10^6$ (b) $256 \times 10^6$ (c) $1024 \times 10^6$ (d) $256 \times 10^{11}$

**13** Let $A$ and $B$ be two $3 \times 3$ real matrices such that $(A^2 - B^2)$ is invertible matrix. If $A^5 = B^5$ and $A^3 B^2 = A^2 B^3$, then the value of the determinant of the matrix $A^3 + B^3$ is equal to **[2021, 27 July Shift-II]**

(a) 2 (b) 4 (c) 1 (d) 0

**14** If $B = \begin{bmatrix} 5 & 2\alpha & 1 \\ 0 & 2 & 1 \\ \alpha & 3 & -1 \end{bmatrix}$ is the inverse of a $3 \times 3$ matrix $A$, then the sum of all values of $\alpha$ for which det $(A) + 1 = 0$, is **[2019, 12 April Shift-I]**

(a) 0 (b) $-1$ (c) 1 (d) 2

**15** If $\begin{bmatrix} 1 & 1 \\ 0 & 1 \end{bmatrix} \cdot \begin{bmatrix} 1 & 2 \\ 0 & 1 \end{bmatrix} \cdot \begin{bmatrix} 1 & 3 \\ 0 & 1 \end{bmatrix} \cdots \begin{bmatrix} 1 & n-1 \\ 0 & 1 \end{bmatrix} = \begin{bmatrix} 1 & 78 \\ 0 & 1 \end{bmatrix}$, then the inverse of $\begin{bmatrix} 1 & n \\ 0 & 1 \end{bmatrix}$ is **[2019, 9 April Shift-I]**

(a) $\begin{bmatrix} 1 & 0 \\ 12 & 1 \end{bmatrix}$ (b) $\begin{bmatrix} 1 & -13 \\ 0 & 1 \end{bmatrix}$ (c) $\begin{bmatrix} 1 & 0 \\ 13 & 1 \end{bmatrix}$ (d) $\begin{bmatrix} 1 & -12 \\ 0 & 1 \end{bmatrix}$

**16** If $A = \begin{bmatrix} 2 & -3 \\ -4 & 1 \end{bmatrix}$, then adj $(3A^2 + 12A)$ is equal to **[JEE Main 2016]**

(a) $\begin{bmatrix} 72 & -84 \\ -63 & 51 \end{bmatrix}$ (b) $\begin{bmatrix} 51 & 63 \\ 84 & 72 \end{bmatrix}$

(c) $\begin{bmatrix} 51 & 84 \\ 63 & 72 \end{bmatrix}$ (d) $\begin{bmatrix} 72 & -63 \\ -84 & 51 \end{bmatrix}$

**17** If $A$ is a $3 \times 3$ non-singular matrix such that $AA' = A'A$ and $B = A^{-1}A'$, then $BB'$ is equal to **[JEE Main 2014]**

(a) $I + B$ (b) $I$ (c) $B^{-1}$ (d) $(B^{-1})'$

$[(AB)^T = B^T A^T$ and $A^{-1}A = I$

(where, $A^T = A' =$ Transpose of $A$)]

**18** If $P = \begin{bmatrix} 1 & \alpha & 3 \\ 1 & 3 & 3 \\ 2 & 4 & 4 \end{bmatrix}$ is the adjoint of a $3 \times 3$ matrix $A$ and $|A| = 4$, then $\alpha$ is equal to **[JEE Main 2013]**

(a) 4 (b) 11 (c) 5 (d) 0

**19** Let $A$ be $2 \times 2$ matrix.

**Statement I** adj (adj A) = $A$

**Statement II** $|\text{adj } A| = |A|$ **[AIEEE 2009]**

(a) Statement I is false, Statement II is true
(b) Statement I is true, Statement II is true; Statement II is a correct explanation of Statement I
(c) Statement I is true, Statement II is true; Statement II is not a correct explanation of Statement I
(d) Statement I is true, Statement II is false

**20** Let $A$ be a square matrix all of whose entries are integers. Then, which one of the following is true? **[AIEEE 2008]**

(a) If det $(A) = \pm 1$, then $A^{-1}$ need not exist
(b) If det $(A) = \pm 1$, then $A^{-1}$ exists but all its entries are not necessarily integers
(c) If det $(A) \neq \pm 1$, then $A^{-1}$ exists and all its entries are non-integers
(d) If det $(A) = \pm 1$, then $A^{-1}$ exists and all its entries are integers

**21** If $A^2 - A + I = O$, then the inverse of $A$ is **[AIEEE 2005]**

(a) $I - A$ (b) $A - I$
(c) $A$ (d) $A + I$

**22** Let $A = \begin{bmatrix} 0 & 0 & -1 \\ 0 & -1 & 0 \\ -1 & 0 & 0 \end{bmatrix}$. The only correct statement about the matrix $A$ is **[AIEEE 2004]**

(a) $A$ is a zero matrix
(b) $A = (-1)I$, where $I$ is a unit matrix
(c) $A^{-1}$ does not exist
(d) $A^2 = I$

## Level 2

**23** Let $A = \begin{bmatrix} 2 & 2+p & 2+p+q \\ 4 & 6+2p & 8+3p+2q \\ 6 & 12+3p & 20+6p+3q \end{bmatrix}$. If det (adj(adj)(3$A$))) $= 2^m \cdot 3^n$, $m, n \in N$, then $m + n$ is equal to **[2025, 8 April Shift-I]**

(a) 24 (b) 26
(c) 20 (d) 22

**24** Let $a \in R$ and $A$ be a matrix of order $3 \times 3$ such that det$(A) = -4$ and $A + I = \begin{bmatrix} 1 & a & 1 \\ 2 & 1 & 0 \\ a & 1 & 2 \end{bmatrix}$, where $I$ is the identity matrix of order $3 \times 3$. If det$((a+1)\,\text{adj}((a-1)A))$ is $2^m 3^n$, $m, n \in \{0, 1, 2, \ldots, 20\}$, then $m + n$ is equal to **[2025, 2 April Shift-I]**

(a) 17 (b) 15
(c) 14 (d) 16

**Imp.** Minor and Cofactor of Matrices

**25** Let $A=[a_{ij}]=\begin{bmatrix}\log_5 128 & \log_4 5\\ \log_5 8 & \log_4 25\end{bmatrix}$

If $A_{ij}$ is the cofactor of $a_{ij}$, $C_{ij}=\sum_{k=1}^{2} a_{ik}A_{jk}, 1\le i,j\le 2$ and $C=|C_{ij}|$, then $8|C|$ is equal to [2025, 9 Jan Shift-I]

(a) 262 (b) 222 (c) 288 (d) 242

**26** If $A$ is a square matrix of order 3 such that $\det(A)=3$ and $\det(\text{adj}(-4\,\text{adj}(-3\,\text{adj}(3\text{adj}(2A)^{-1}))))=2^m 3^n$, then $m+2n$ is equal to [2024, 6 April Shift-II]

(a) 3 (b) 4 (c) 6 (d) 2

**Imp.** Properties of Adjoint of Matrices

**27** Let $A$ and $B$ be two square matrices of order 3 such that $|A|=3$ and $|B|=2$. Then, $|A^T A(\text{adj}(2A))^{-1}(\text{adj}(4B))(\text{adj}(AB))^{-1}AA^T|$ is equal to [2024, 5 April Shift-I]

(a) 32 (b) 108
(c) 81 (d) 64

**28** Let $R=\begin{bmatrix}x&0&0\\0&y&0\\0&0&z\end{bmatrix}$ be a non-zero $3\times 3$ matrix, where

$x\sin\theta=y\sin\left(\theta+\frac{2\pi}{3}\right)=z\sin\left(\theta+\frac{4\pi}{3}\right)\neq 0,$

$\theta\in(0.2\pi)$. For a square matrix $M$, let trace $(M)$ denote the sum of all the diagonal entries of $M$. Then, among the statements :

**Statement I** Trace $(R)=0$

**Statement II** If trace (adj (adj $(R)$)) $=0$, then $R$ has exactly one non-zero entry. [2024, 30 Jan Shift-II]

(a) Only Statement I is true
(b) Neither Statement I nor II is true
(c) Both I and II are true
(d) Only Statement II is true

**29** Let $A=\begin{bmatrix}2&1&2\\6&2&11\\3&3&2\end{bmatrix}$ and $P=\begin{bmatrix}1&2&0\\5&0&2\\7&1&5\end{bmatrix}$. The sum of the prime factors of $|P^{-1}AP-2I|$ is equal to [2024, 29 Jan Shift-II]

(a) 27 (b) 26 (c) 23 (d) 66

**30** Let the determinant of a square matrix $A$ of order $m$ be $m-n$, where $m$ and $n$ satisfy $4m+n=22$ and $17m+4n=93$. If $\det(n\,\text{adj}(\text{adj}(mA)))=3^a5^b6^c$, then $a+b+c$ is equal to [2023, 15 April Shift-I]

(a) 101 (b) 96 (c) 109 (d) 84

**31** Let for $A=\begin{bmatrix}1&2&3\\\alpha&3&1\\1&1&2\end{bmatrix}$, $|A|=2$. If $|2\,\text{adj}(2\,\text{adj}(2\,\text{adj}(2A)))|=32^n$. Then, $3n+\alpha$ is equal to [2023, 13 April Shift-II]

(a) 10 (b) 11
(c) 12 (d) 9

**32** Let $B=\begin{bmatrix}1&3&\alpha\\1&2&3\\\alpha&\alpha&4\end{bmatrix}$, $\alpha>2$ be the adjoint of a matrix $A$ and $|A|=2$. Then, $[\alpha\ -2\alpha\ \alpha]\,B\begin{bmatrix}\alpha\\-2\alpha\\\alpha\end{bmatrix}$ is equal to [2023, 13 April Shift-I]

(a) 32 (b) 16
(c) −16 (d) 0

**33** Let $A=\begin{bmatrix}m&n\\p&q\end{bmatrix}$, $d=|A|\neq 0$ and $|A-d(\text{Adj }A)|=0$. Then, [2023, 30 Jan Shift-I]

(a) $1+d^2=(m+q)^2$
(b) $1+d^2=m^2+q^2$
(c) $(1+d)^2=m^2+q^2$
(d) $(1+d)^2=(m+q)^2$

**34** The set of all values of $t\in R$, for which the matrix $\begin{bmatrix}e^t & e^{-t}(\sin t-2\cos t) & e^{-t}(-2\sin t-\cos t)\\ e^t & e^{-t}(2\sin t+\cos t) & e^{-t}(\sin t-2\cos t)\\ e^t & e^{-1}\cos t & e^{-t}\sin t\end{bmatrix}$ is invertible, is [2023, 29 Jan Shift-II]

(a) $\{k\pi, k\in Z\}$ (b) $R$
(c) $\left\{k\pi+\frac{\pi}{4}, k\in Z\right\}$ (d) $\left\{(2k+1)\frac{\pi}{2}, k\in Z\right\}$

***35** Let the matrix $A=\begin{bmatrix}0&1&0\\0&0&1\\1&0&0\end{bmatrix}$ and the matrix $B_0=A^{49}+2A^{98}$. If $B_n=\text{adj}(B_{n-1})$, $n\ge 1$, then det $(B_4)$ is equal to [2022, 28 July Shift-I]

(a) $3^{28}$ (b) $3^{30}$
(c) $3^{32}$ (d) $3^{36}$

**36** Let $A$ be a $2\times 2$ matrix with det $(A)=-1$ and det $((A+I)(\text{Adj }(A)+I))=4$. Then, the sum of the diagonal elements of $A$ can be [2022, 26 July Shift-I]

(a) −1 (b) 2
(c) 1 (d) $-\sqrt{2}$

**37** If $A=\begin{bmatrix}\frac{1}{\sqrt{5}} & \frac{2}{\sqrt{5}}\\ \frac{-2}{\sqrt{5}} & \frac{1}{\sqrt{5}}\end{bmatrix}$, $B=\begin{bmatrix}1 & 0\\ i & 1\end{bmatrix}$, $i=\sqrt{-1}$, and $Q=A^T BA$, then the inverse of the matrix $AQ^{2021}A^T$ is equal to **[2021, 26 Aug Shift-I]**

(a) $\begin{bmatrix}\frac{1}{\sqrt{5}} & -2021\\ 2021 & \frac{1}{\sqrt{5}}\end{bmatrix}$ (b) $\begin{bmatrix}1 & 0\\ -2021i & 1\end{bmatrix}$

(c) $\begin{bmatrix}1 & 0\\ 2021i & 1\end{bmatrix}$ (d) $\begin{bmatrix}1 & -2021i\\ 0 & 1\end{bmatrix}$

**38** Let $A=\begin{bmatrix}1 & 2\\ -1 & 4\end{bmatrix}$. If $A^{-1}=\alpha I+\beta A$, $\alpha, \beta \in R$, $I$ is a $2\times 2$ identity matrix, then $4(\alpha-\beta)$ is equal to **[2021, 27 July Shift-I]**

(a) 5 (b) 8/3
(c) 2 (d) 4

**39** Let $A$ be a $3\times 3$ matrix such that $\text{adj}\, A=\begin{bmatrix}2 & -1 & 1\\ -1 & 0 & 2\\ 1 & -2 & -1\end{bmatrix}$ and $B=\text{adj (adj A)}$. If $|A|=\lambda$ and $|(B^{-1})^T|=\mu$, then the ordered pair, $(|\lambda|, \mu)$ is equal to **[2020, 3 Sep Shift-II]**

(a) (3, 81) (b) $\left(9, \frac{1}{81}\right)$ (c) $\left(9, \frac{1}{9}\right)$ (d) $\left(3, \frac{1}{81}\right)$

**40** If the matrices $A=\begin{bmatrix}1 & 1 & 2\\ 1 & 3 & 4\\ 1 & -1 & 3\end{bmatrix}$, $B=\text{adj}\, A$ and $C=3A$, then $\frac{|\text{adj}\, B|}{|C|}$ is equal to **[2020, 9 Jan Shift-I]**

(a) 16 (b) 2 (c) 8 (d) 72

**41** Let $A$ and $B$ be two invertible matrices of order $3\times 3$. If $\det(ABA^T)=8$ and $\det(AB^{-1})=8$, then $\det(BA^{-1}B^T)$ is equal to **[2019, 11 Jan Shift-II]**

(a) 1 (b) 1/4 (c) 1/16 (d) 16

**42** If $A=\begin{bmatrix}1 & 2 & 2\\ 2 & 1 & -2\\ a & 2 & b\end{bmatrix}$ is a matrix satisfying the equation $AA^T=9I$, where $I$ is $3\times 3$ identity matrix, then the ordered pair $(a, b)$ is equal to **[JEE Main 2015]**

(a) (2, –1) (b) (–2, 1) (c) (2, 1) (d) (–2, –1)

**43** Let $A=\begin{bmatrix}1 & -1 & 1\\ 2 & 1 & -3\\ 1 & 1 & 1\end{bmatrix}$ and $10B=\begin{bmatrix}4 & 2 & 2\\ -5 & 0 & \alpha\\ 1 & -2 & 3\end{bmatrix}$

If $B$ is the inverse of matrix $A$, then $\alpha$ is equal to **[AIEEE 2004]**

(a) –2 (b) 1 (c) 2 (d) 5

## Level 3

**44** Let $A=\begin{bmatrix}\frac{1}{\sqrt{10}} & \frac{3}{\sqrt{10}}\\ \frac{-3}{\sqrt{10}} & \frac{1}{\sqrt{10}}\end{bmatrix}$ and $B=\begin{bmatrix}1 & -i\\ 0 & 1\end{bmatrix}$, where $i=\sqrt{-1}$.

If $M=A^T BA$, then the inverse of the matrix $AM^{2023}A^T$ is **[2023, 25 Jan Shift-II]**

(a) $\begin{bmatrix}1 & 0\\ 2023i & 1\end{bmatrix}$ (b) $\begin{bmatrix}1 & -2023i\\ 0 & 1\end{bmatrix}$

(c) $\begin{bmatrix}1 & 0\\ -2023i & 1\end{bmatrix}$ (d) $\begin{bmatrix}1 & 2023i\\ 0 & 1\end{bmatrix}$

**45** Let $\alpha \in (0, \infty)$ and $A=\begin{bmatrix}1 & 2 & \alpha\\ 1 & 0 & 1\\ 0 & 1 & 2\end{bmatrix}$. If det $|\text{adj}\,(2A-2A^T)\cdot \text{adj}\,(A-2A^T)|=2^8$, then $(\det(A))^2$ is equal to **[2022, 4 April Shift-I]**

(a) 36 (b) 49 (c) 16 (d) 1

**46** If $A=\begin{bmatrix}\cos\theta & -\sin\theta\\ \sin\theta & \cos\theta\end{bmatrix}$, then the matrix $A^{-50}$ when $\theta=\frac{\pi}{12}$, is equal to **[2019, 9 Jan Shift-I]**

(a) $\begin{bmatrix}\frac{1}{2} & \frac{\sqrt{3}}{2}\\ -\frac{\sqrt{3}}{2} & \frac{1}{2}\end{bmatrix}$ (b) $\begin{bmatrix}\frac{\sqrt{3}}{2} & -\frac{1}{2}\\ \frac{1}{2} & \frac{\sqrt{3}}{2}\end{bmatrix}$

(c) $\begin{bmatrix}\frac{\sqrt{3}}{2} & \frac{1}{2}\\ -\frac{1}{2} & \frac{\sqrt{3}}{2}\end{bmatrix}$ (d) $\begin{bmatrix}\frac{1}{2} & -\frac{\sqrt{3}}{2}\\ \frac{\sqrt{3}}{2} & \frac{1}{2}\end{bmatrix}$

## Numerical Value Type Questions

**47** Let $I$ be the identity matrix of order $3\times 3$ and for the matrix $A=\begin{bmatrix}\lambda & 2 & 3\\ 4 & 5 & 6\\ 7 & -1 & 2\end{bmatrix}$, $|A|=-1$. Let $B$ be the inverse of the matrix adj $(A\,\text{adj}\,(A^2))$. Then, $|(\lambda_B+I)|$ is equal to ……… . **[2025, 3 April Shift-II]**

**Imp.** Properties of Adjoint of a Matrix

**48** Let $A$ be a square matric of order 3 such that $\det(A) = -2$ and $\det(3\text{adj}(-6\text{adj}(3A))) = 2^{m+n}$, $m > n$. Then, $4m + 2n$ is equal to ..... [2025, 22 Jan Shift-I]

**49** Let $A$ be a non-singular matrix of order 3. If $\det(3\text{adj}(2\text{adj}((\det A)A))) = 3^{-13} \cdot 2^{-10}$ and $\det(3\,\text{adj}(2A)) = 2^m \cdot 3^n$, then $|3m + 2n|$ is equal to ........... [2024, 9 April Shift-I]

**50** Let $A$ be a $2 \times 2$ symmetric matrix such that $A\begin{bmatrix}1\\1\end{bmatrix} = \begin{bmatrix}3\\7\end{bmatrix}$ and the determinant of $A$ be 1.

If $A^{-1} = \alpha A + \beta I$, where $I$ is an identity matrix of order $2 \times 2$, then $\alpha + \beta$ equals....... . [2024, 4 April Shift-II]

**51** Let $A$ be a $3 \times 3$ matrix and $\det(A) = 2$. If $n = \det(\underbrace{\text{adj}(\text{adj}(\ldots(\text{adj}}_{2024\text{-times}}\,A))))$, then the remainder when $n$ is divided by 9 is equal to ......... . [2024, 31 Jan Shift-II]

**52** Let $A$ be a $n \times n$ matrix such that $|A| = 2$. If the determinant of the matrix $\text{Adj}(2 \cdot \text{Adj}(2A^{-1}))$ is $2^{84}$, then $n$ is equal to ....... . [2023, 31 Jan Shift-II]

**53** The number of matrices $A = \begin{bmatrix} a & b \\ c & d \end{bmatrix}$, where $a, b, c$ and $d \in \{-1, 0, 1, 2, 3, \ldots, 10\}$, such that $A = A^{-1}$, is ............... . [2022, 26 July Shift-II]

**54** Let $X = \begin{bmatrix} 0 & 1 & 0 \\ 0 & 0 & 1 \\ 0 & 0 & 0 \end{bmatrix}$, $Y = \alpha I + \beta X + \gamma X^2$ and $Z = \alpha^2 I - \alpha\beta X + (\beta^2 - \alpha\gamma) X^2$, $\alpha, \beta, \gamma \in R$. If $Y^{-1} = \begin{bmatrix} 1/5 & -2/5 & 1/5 \\ 0 & 1/5 & -2/5 \\ 0 & 0 & 1/5 \end{bmatrix}$, then $(\alpha - \beta + \gamma)^2$ is equal to ......... . [2022, 26 June Shift-II]

**55** Let $A$ be a $3 \times 3$ real matrix.
If $\det(2\text{Adj}(2\,\text{Adj}(\text{Adj}(2A)))) = 2^{41}$, then the value of $\det(A^2)$ equal ........... . [2021, 26 Aug Shift-II]

**56** Let $A = \{a_{ij}\}$ be a $3 \times 3$ matrix, where

$$a_{ij} = \begin{cases} (-1)^{j-i}, & \text{if } i < j \\ 2, & \text{if } i = j, \\ (-1)^{i+j}, & \text{if } i > j \end{cases}$$

then $\det[3\,\text{Adj}(2A^{-1})]$ is equal to ......... . [2021, 20 July Shift-II]

**57** If $A = \begin{bmatrix} 0 & -\tan\left(\dfrac{\theta}{2}\right) \\ \tan\left(\dfrac{\theta}{2}\right) & 0 \end{bmatrix}$ and $(I_2 + A)(I_2 - A)^{-1} = \begin{bmatrix} a & -b \\ b & a \end{bmatrix}$, then $13(a^2 + b^2)$ is equal to ...... . [2021, 25 Feb Shift-I]

**58** Let $P = \begin{bmatrix} 3 & -1 & -2 \\ 2 & 0 & \alpha \\ 3 & -5 & 0 \end{bmatrix}$, where $\alpha \in R$. Suppose $Q = [q_{ij}]$ is a matrix satisfying $PQ = kI_3$ for some non-zero $k \in R$. If $q_{23} = -\dfrac{k}{8}$ and $|Q| = \dfrac{k^2}{2}$, then $\alpha^2 + k^2$ is equal to ......... . [2021, 24 Feb Shift-I]

## Topic 4 Solutions of System of Linear Equations in Two or Three Variables Matrices

### Level 1

**01** Let the system of equations
$2x + 3y + 5z = 9,$
$7x + 3y - 2z = 8,$
$12x + 3y - (4 + \lambda)z = 16 - \mu,$
have infinitely many solutions. Then, the radius of the circle centred at $(\lambda, \mu)$ and touching the line $4x = 3y$ is [2025, 7 April Shift-I]

(a) $\dfrac{17}{5}$ (b) 7 (c) $\dfrac{21}{5}$ (d) $\dfrac{7}{5}$

**02** If the system of equations
$$2x + \lambda y + 3z = 5$$
$$3x + 2y - z = 7$$
$$4x + 5y + \mu z = 9$$
has infinitely many solutions, then $(\lambda^2 + \mu^2)$ is equal to [2025, 2 April Shift-II]

(a) 18 (b) 26 (c) 22 (d) 30

**03** If the system of linear equations

$$3x + y + \beta z = 3$$
$$2x + \alpha y - z = -3$$
$$x + 2y + z = 4$$

has infinitely many solutions, then the value of $22\beta - 9\alpha$ is *[2025, 2 April Shift-I]*

(a) 49 (b) 43 (c) 31 (d) 37

**04** Let $\alpha, \beta (\alpha \neq \beta)$ be the values of $m$ for which the equations $x + y + z = 1$; $x + 2y + 4z = m$ and $x + 4y + 10z = m^2$ have infinitely many solutions.

Then, the value of $\sum_{n=1}^{10} (n^{\alpha} + n^{\beta})$ is equal to *[2025, 29 Jan Shift-II]*

(a) 3080 (b) 560 (c) 3410 (d) 440

**05** If the system of equations

$$x + 2y - 3z = 2$$
$$2x + \lambda y + 5z = 5$$
$$14x + 3y + \mu z = 33$$

has infinitely many solutions, then $\lambda + \mu$ is equal to *[2025, 24 Jan Shift-II]*

(a) 12 (b) 13 (c) 10 (d) 11

**06** If the system of equations

$$2x - y + z = 4$$
$$5x + \lambda y + 3z = 12$$
$$100x - 47y + \mu z = 212.$$

has infinitely many solutions, then $\mu - 2\lambda$ is equal to *[2025, 24 Jan Shift-I]*

(a) 59 (b) 57 (c) 55 (d) 56

**07** The system of equations
$x + y + z = 6$, $x + 2y + 5z = 9$, $x + 5y + \lambda z = \mu$,
has no solution, if *[2025, 23 Jan Shift-II]*

(a) $\lambda = 15, \mu \neq 17$ (b) $\lambda \neq 17, \mu \neq 18$
(c) $\lambda = 17, \mu = 18$ (d) $\lambda = 17, \mu \neq 18$

**08** If the system of equations

$$(\lambda - 1)x + (\lambda - 4)y + \lambda z = 5$$
$$\lambda x + (\lambda - 1)y + (\lambda - 4)z = 7$$
$$(\lambda + 1)x + (\lambda + 2)y - (\lambda + 2)z = 9$$

has infinitely many solutions, then $\lambda^2 + \lambda$ is equal to *[2025, 23 Jan Shift-I]*

(a) 20 (b) 10 (c) 12 (d) 6

**Imp.** Solution of System of Linear Equations

**09** If the system of linear equations

$$x + y + 2z = 6,$$
$$2x + 3y + az = a + 1,$$
$$-x - 3y + bz = 2b,$$

where $a, b \in R$ has infinitely many solutions, then $7a + 3b$ is equal to *[2025, 22 Jan Shift-II]*

(a) 9 (b) 22 (c) 16 (d) 12

**10** If the system of equation $x + 4y - z = \lambda$, $7x + 9y + \mu z = -3$, $5x + y + 2z = -1$ has infinitely many solutions, then $(2\mu + 3\lambda)$ is equal to *[2024, 8 April Shift-II]*

(a) $-2$ (b) $-3$ (c) 2 (d) 3

**11** If the system of equations

$$x + (\sqrt{2}\sin\alpha)y + (\sqrt{2}\cos\alpha)z = 0,$$
$$x + (\cos\alpha)y + (\sin\alpha)z = 0$$

and $x + (\sin\alpha)y - (\cos\alpha)z = 0$

has a non-trivial solution, then $\alpha \in \left(0, \frac{\pi}{2}\right)$ is equal to *[2024, 4 April Shift-I]*

(a) $\frac{3\pi}{4}$ (b) $\frac{11\pi}{24}$
(c) $\frac{7\pi}{24}$ (d) $\frac{5\pi}{24}$

**12** If the system of equations
$2x + 3y - z = 5$,
$x + \alpha y + 3z = -4$, and $3x - y + \beta z = 7$ *[2024, 1 Feb Shift-I]*
has infinitely many solutions, then $13\alpha\beta$ is equal to

(a) 1220 (b) 1110
(c) 1120 (d) 1210

**13** For the system of linear equations $\alpha x + y + z = 1$, $x + \alpha y + z = 1$, $x + y + \alpha z = \beta$, which one of the following statements is not correct? *[2023, 1 Feb Shift-II]*

(a) It has no solution, if $\alpha = -2$ and $\beta = 1$
(b) $x + y + z = \frac{3}{4}$, if $\alpha = 2$ and $\beta = 1$
(c) It has infinitely many solutions, if $\alpha = 1$ and $\beta = 1$
(d) It has infinitely many solutions, if $\alpha = 2$ and $\beta = -1$

**14** Let $S$ denote the set of all real values of $\lambda$ such that the system of equations

$$\lambda x + y + z = 1, \; x + \lambda y + z = 1$$

and $x + y + \lambda z = 1$

is inconsistent, then $\sum_{\lambda \in S} (|\lambda|^2 + |\lambda|)$ is equal to *[2023, 1 Feb Shift-I]*

(a) 12 (b) 6
(c) 4 (d) 2

**15** For $\alpha, \beta \in R$, suppose the system of linear equations *[2023, 30 Jan Shift-II]*

$$x - y + z = 5, \; 2x + 2y + \alpha z = 8$$

and $3x - y + 4z = \beta$

has infinitely many solutions. Then, $\alpha$ and $\beta$ are the roots of

(a) $x^2 + 14x + 24 = 0$
(b) $x^2 + 18x + 56 = 0$
(c) $x^2 - 18x + 56 = 0$
(d) $x^2 - 10x + 16 = 0$

Imp.

## Existence of Solution

**16** Let $S_1$ and $S_2$ be respectively the sets of all $a \in R - \{0\}$ for which the system of linear equations

$$ax + 2ay - 3az = 1,$$
$$(2a+1)x + (2a+3)y + (a+1)z = 2$$
and $(3a+5)x + (a+5)y + (a+2)z = 3$

has unique solution and infinitely many solutions. Then, **[2023, 25 Jan Shift-I]**

(a) $S_1 = \phi$ and $S_2 = R - \{0\}$
(b) $S_1$ is an infinite set and $n(S_2) = 2$
(c) $n(S_1) = 2$ and $S_2$ is an infinite set
(d) $S_1 = R - \{0\}$ and $S_2 = \phi$

**17** If the system of linear equations $2x + y - z = 7$, $x - 3y + 2z = 1$ and $x + 4y + \delta z = k$, where $\delta, k \in R$ has infinitely many solutions, then $\delta + k$ is equal to **[2022, 29 June Shift-I]**

(a) $-3$ (b) 3 (c) 6 (d) 9

**18** If the system of linear equations

$2x + 3y - z = -2$, $x + y + z = 4$

and $x - y + |\lambda| z = 4\lambda - 4$

where $\lambda \in R$, has no solution, then **[2023, 28 June Shift-II]**

(a) $\lambda = 7$ (b) $\lambda = -7$
(c) $\lambda = 8$ (d) $\lambda^2 = 1$

**19** If the system of equations $\alpha x + y + z = 5$, $x + 2y + 3z = 4$, $x + 3y + 5z = \beta$, has infinitely many solutions, then the ordered pair $(\alpha, \beta)$ is equal to **[2022, 26 June Shift-II]**

(a) $(1, -3)$ (b) $(-1, 3)$ (c) $(1, 3)$ (d) $(-1, -3)$

**20** The system of equations

$-kx + 3y - 14z = 25$, $-15x + 4y - kz = 3$

and $-4x + y + 3z = 4$

is consistent for all $k$ in the set **[2022, 25 June Shift-II]**

(a) $R$ (b) $R - \{-11, 13\}$
(c) $R - \{13\}$ (d) $R - \{-11, 11\}$

**21** The number of values of $\alpha$ for which the system of equations **[2022, 24 June Shift-I]**

$x + y + z = \alpha$, $\alpha x + 2\alpha y + 3z = -1$

and $x + 3\alpha y + 5z = 4$

is inconsistent, is

(a) 0 (b) 1 (c) 2 (d) 3

**22** Consider the system of linear equations

$-x + y + 2z = 0$, $3x - ay + 5z = 1$,

$2x - 2y - az = 7$

Let $S_1$ be the set of all $a \in R$ for which the system is inconsistent and $S_2$ be the set of all $a \in R$ for which the system has infinitely many solutions. If $n(S_1)$ and $n(S_2)$ denote the number of elements in $S_1$ and $S_2$ respectively, then **[2021, 1 Sep Shift-II]**

(a) $n(S_1) = 2$ and $n(S_2) = 2$ (b) $n(S_1) = 1$ and $n(S_2) = 0$
(c) $n(S_1) = 2$ and $n(S_2) = 0$ (d) $n(S_1) = 0$ and $n(S_2) = 2$

**23** If $\alpha + \beta + \gamma = 2\pi$, then the system of equations

$$x + (\cos\gamma)y + (\cos\beta)z = 0,$$
$$(\cos\gamma)x + y + (\cos\alpha)z = 0$$
and $(\cos\beta)x + (\cos\alpha)y + z = 0$ has **[2021, 31 Aug Shift-II]**

(a) no solution
(b) infinitely many solution
(c) exactly two solutions
(d) a unique solution

**24** If the following system of linear equations

$2x + y + z = 5$, $x - y + z = 3$ and $x + y + az = b$

has no solution, then **[2021, 31 Aug Shift-I]**

(a) $a = -\frac{1}{3}$ and $b \neq \frac{7}{3}$ (b) $a \neq \frac{1}{3}$ and $b = \frac{7}{3}$
(c) $a \neq -\frac{1}{3}$ and $b = \frac{7}{3}$ (d) $a = \frac{1}{3}$ and $b \neq \frac{7}{3}$

**25** Let $|\lambda|$ be the greatest integer less than or equal to $\lambda$. The set of all values of $\lambda$ for which the system of linear equations $x + y + z = 4$, $3x + 2y + 5z = 3$, $9x + 4y + (28 + |\lambda|)z = |\lambda|$ has a solution is **[2021, 27 Aug Shift-II]**

(a) $R$ (b) $(-\infty, -9) \cup (-9, \infty)$
(c) $[-9, -8)$ (d) $(-\infty, -9) \cup [-8, \infty)$

**26** Let the system of linear equations

$4x + \lambda y + 2z = 0$, $2x - y + z = 0$

$\mu x + 2y + 3z = 0$, $\lambda, \mu \in R$.

has a non-trivial solution. Then which of the following is true? **[2021, 18 March Shift-II]**

(a) $\mu = 6, \lambda \in R$ (b) $\lambda = 2, \mu \in R$
(c) $\lambda = 3, \mu \in R$ (d) $\mu = -6, \lambda \in R$

**27** The following system of linear equations

$2x + 3y + 2z = 9$, $3x + 2y + 2z = 9$

and $x - y + 4z = 8$ **[2021, 25 Feb Shift-II]**

(a) does not have any solution
(b) has a unique solution
(c) has infinitely many solutions
(d) has a solution $(\alpha, \beta, \gamma)$ satisfying $\alpha + \beta^2 + \gamma^3 = 12$

**28** Let $A$ and $B$ be $3 \times 3$ real matrices, such that $A$ is symmetric matrix and $B$ is skew-symmetric matrix. Then, the system of linear equations $(A^2B^2 - B^2A^2)X = O$, where $X$ is a $3 \times 1$ column matrix of unknown variables and $O$ is a $3 \times 1$ null matrix, has **[2021, 24 Feb Shift-II]**

(a) no solution
(b) exactly two solutions
(c) infinitely many solutions
(d) a unique solution

**29** Let $\lambda \in R$. The system of linear equations $2x_1 - 4x_2 + \lambda x_3 = 1, x_1 - 6x_2 + x_3 = 2,$ $\lambda x_1 - 10x_2 + 4x_3 = 3$ is inconsistent for *[2020, 5 Sep Shift-I]*
(a) exactly two values of $\lambda$
(b) exactly one positive value of $\lambda$
(c) every value of $\lambda$
(d) exactly one negative value of $\lambda$

**30** If for some $\alpha$ and $\beta$ in $R$, the intersection of the following three planes $x + 4y - 2z = 1, x + 7y - 5z = \beta$ and $x + 5y + \alpha z = 5$ is a line in $R^3$, then $\alpha + \beta$ is equal to *[2020, 9 Jan Shift-I]*
(a) 0 (b) 10 (c) −10 (d) 2

**31** If the system of linear equations
$2x + 2ay + az = 0, 2x + 3by + bz = 0,$
$2x + 4cy + cz = 0$
where $a, b, c \in R$ are non-zero and distinct; has a non-zero solution, then *[2020, 7 Jan Shift-I]*
(a) $a + b + c = 0$ (b) $\frac{1}{a}, \frac{1}{b}, \frac{1}{c}$ are in AP
(c) $a, b, c$ are in AP (d) $a, b, c$ are in GP

**32** If the system of linear equations
$x - 2y + kz = 1, 2x + y + z = 2,$
$3x - y - kz = 3$ has a solution $(x, y, z)$, $z \neq 0$, then $(x, y)$ lies on the straight line whose equation is *[2019, 8 April Shift-II]*
(a) $3x - 4y - 4 = 0$ (b) $3x - 4y - 1 = 0$
(c) $4x - 3y - 4 = 0$ (d) $4x - 3y - 1 = 0$

**33** The greatest value of $c \in R$ for which the system of linear equations $x - cy - cz = 0, cx - y + cz = 0,$ $cx + cy - z = 0$ has a non-trivial solution, is *[2019, 8 April Shift-I]*
(a) −1 (b) 1/2
(c) 2 (d) 0

**34** An ordered pair $(\alpha, \beta)$ for which the system of linear equations $(1 + \alpha)x + \beta y + z = 2, \alpha x + (1 + \beta)y + z = 3$ and $ax + \beta y + 2z = 2$
has a unique solution, is *[2019, 12 Jan Shift-I]*
(a) (2, 4) (b) (− 4, 2) (c) (1, − 3) (d) (−3, 1)

**35** The system of linear equations
$x + \lambda y - z = 0, \lambda x - y - z = 0$ and $x + y - \lambda z = 0$
has a non-trivial solution for *[JEE Main 2016]*
(a) infinitely many values of $\lambda$ (b) exactly one value of $\lambda$
(c) exactly two values of $\lambda$ (d) exactly three values of $\lambda$

**36** If the trivial solution is the only solution of the system of equations
$x - ky + z = 0, kx + 3y - kz = 0$ and $3x + y - z = 0$
Then, the set of all values of $k$ is *[AIEEE 2011]*
(a) $\{2, -3\}$ (b) $R - \{2, -3\}$
(c) $R - \{2\}$ (d) $R - \{-3\}$

***37** Consider the system of linear equations
$x_1 + 2x_2 + x_3 = 3, 2x_1 + 3x_2 + x_3 = 3$
and $3x_1 + 5x_2 + 2x_3 = 1$
The system has *[AIEEE 2010]*
(a) infinite number of solutions
(b) exactly 3 solutions
(c) a unique solution
(d) no solution

**38** Let $a, b$ and $c$ be any real numbers. Suppose that there are real numbers $x, y, z$ not all zero such that $x = cy + bz, y = az + cx$ and $z = bx + ay$. Then, $a^2 + b^2 + c^2 + 2abc$ is equal to *[AIEEE 2008]*
(a) 1 (b) 2 (c) −1 (d) 0

**39** If the system of linear equations
$x + 2ay + az = 0, x + 3by + bz = 0$ and $x + 4cy + cz = 0$
has a non-zero solution, then $a, b$ and $c$ *[AIEEE 2003]*
(a) are in AP (b) are in GP
(c) are in HP (d) satisfy $a + 2b + 3c = 0$

## Level 2

**40** Let $\lambda, \mu \in R$. If the system of equations
$3x + 5y + \lambda z = 3,$
$7x + 11y - 9z = 2$
and $97x + 155y - 189z = \mu$ *[2024, 9 April Shift-I]*
has infinitely many solutions, then $\mu + 2\lambda$ is equal to
(a) 27 (b) 24 (c) 22 (d) 25

**41** Let the system of equations $x + 2y + 3z = 5,$ $2x + 3y + z = 9, 4x + 3y + \lambda z = \mu$ have infinite number of solutions. Then, $\lambda + 2\mu$ is equal to *[2024, 1 Feb Shift-II]*
(a) 17 (b) 28 (c) 22 (d) 15

**42** If the system of linear equations
$x - 2y + z = -4, 2x + \alpha y + 3z = 5, 3x - y + \beta z = 3$
has infinitely many solutions, then $12\alpha + 13\beta$ is equal to *[2024, 31 Jan Shift-I]*
(a) 58 (b) 64 (c) 60 (d) 54

***43** Consider the system of linear equations
$x + y + z = 4\mu, x + 2y + 2\lambda z = 10\mu,$
$x + 3y + 4\lambda^2 z = \mu^2 + 15$, where $\lambda, \mu \in R$. Which one of the following statements is not correct? *[2024, 30 Jan Shift-I]*
(a) The system has infinite number of solutions if $\lambda = 1/2$ and $\mu = 15$
(b) The system is inconsistent if $\lambda = \frac{1}{2}$ and $\mu \neq 1$
(c) The system has unique solution if $\lambda \neq \frac{1}{2}$ and $\mu \neq 1, 15$
(d) The system is consistent if $\lambda \neq \frac{1}{2}$

**44** If the system of equations
$2x+y-z=5, 2x-5y+\lambda z=\mu, x+2y-5z=7$
has infinitely many solutions, then
$(\lambda+\mu)^2+(\lambda-\mu)^2$ is equal to *[2023, 13 April Shift-II]*
(a) 920 (b) 904 (c) 916 (d) 912

**45** If the system of linear equations
$7x+11y+\alpha z=13, 5x+4y+7z=\beta,$
$175x+194y+57z=361$ has infinitely many solutions, then $\alpha+\beta+2$ is equal to *[2023, 11 April Shift-II]*
(a) 4 (b) 3 (c) 6 (d) 5

**46** For the system of linear equations
$2x-y+3z=5, 3x+2y-z=7, 4x+5y+\alpha z=\beta$
which of the following is not correct? *[2023, 10 April Shift-I]*
(a) The system has infinitely many solutions for $\alpha=-5$ and $\beta=9$
(b) The system has a unique solution for $\alpha\neq-5$ and $\beta=8$
(c) The system has infinitely many solutions for $\alpha=-6$ and $\beta=9$
(d) The system is inconsistent for $\alpha=-5$ and $\beta=8$

**47** For the system of equations
$$x+y+z=6,\ x+2y+\alpha z=10$$
and $x+3y+5z=\beta$, which one of the following is not true? *[2023, 6 April Shift-II]*
(a) System has infinitely many solutions for $\alpha=3, \beta=14$.
(b) System has no solution for $\alpha=3, \beta=24$.
(c) System has a unique solution for $\alpha=3, \beta\neq14$.
(d) System has a unique solution for $\alpha=-3, \beta=14$.

**48** If the system of equations
$$x+y+az=b,$$
$$2x+5y+2z=6$$
and $x+2y+3z=3$
has infinitely many solutions, then $2a+3b$ is equal to *[2023, 6 April Shift-I]*
(a) 23 (b) 25 (c) 28 (d) 20

**49** For the system of linear equations
$2x+4y+2az=b, x+2y+3z=4$ and
$2x-5y+2z=8$ which of the following is not correct? *[2023, 13 April Shift-II]*
(a) It has infinitely many solutions if $a=3, b=6$
(b) It has unique solution if $a=b=8$
(c) It has infinitely many solutions if $a=3, b=8$
(d) It has unique solution if $a=b=6$

**50** Let $S$ be the set of all values of $\theta\in[-\pi,\pi]$ for which the system of linear equations
$x+y+\sqrt{3}z=0, -x+(\tan\theta)y+\sqrt{7}z=0$ and
$x+y+(\tan\theta)z=0$ has non-trivial solution. Then $\frac{120}{\pi}\sum_{\theta\in S}\theta$ is equal to *[2023, 8 April Shift-II]*
(a) 40 (b) 10 (c) 30 (d) 20

**51** For the system of linear equations
$$x+y+z=6,\ \alpha x+\beta y+7z=3$$
and $x+2y+3z=14,$
which of the following is not true? *[2023, 31 Jan Shift-I]*
(a) If $\alpha=\beta=7$, then the system has no solution
(b) There is a unique point $(\alpha, \beta)$ on the line $x+2y+18=0$ for which the system has infinitely many solutions
(c) If $\alpha=\beta$ and $\alpha\neq7$, then the system has a unique solution
(d) For every point $(\alpha, \beta)\neq(7, 7)$ on the line $x-2y+7=0$, the system has infinitely many solutions

**52** Let the system of linear equations
$$x+y+kz=2, 2x+3y-z=1$$
and $3x+4y+2z=k$
have infinitely many solutions. Then, the system
$$(k+1)x+(2k-1)y=7$$
and $(2k+1)x+(k+5)y=10$ has *[2023, 30 Jan Shift-I]*
(a) no solution
(b) unique solution satisfying $x-y=1$
(c) unique solution satisfying $x+y=1$
(d) infinitely many solutions

**53** If the system of equations
$$x+2y+3z=3,\ 4x+3y-4z=4$$
and $8x+4y-\lambda z=9+\mu$
has infinitely many solutions, then the ordered pair $(\lambda, \mu)$ is equal to *[2023, 24 Jan Shift-II]*
(a) $\left(-\frac{72}{5}, -\frac{21}{5}\right)$ (b) $\left(\frac{72}{5}, -\frac{21}{5}\right)$
(c) $\left(-\frac{72}{5}, \frac{21}{5}\right)$ (d) $\left(\frac{72}{5}, \frac{21}{5}\right)$

**54** The number of $q\in(0, 4\pi)$ for which the system of linear equations $3(\sin 3\theta)x-y+z=2$; $3(\cos 2\theta)x+4y+3z=3$ and $6x+7y+7z=9$ has no solution, is *[2022, 25 July Shift-I]*
(a) 6 (b) 7 (c) 8 (d) 9

**55** The ordered pair $(a, b)$, for which the system of linear equations $3x-2y+z=b, 5x-8y+9z=3,$ $2x+y+az=-1$ has no solution, is *[2022, 26 June Shift-I]*
(a) $\left(3, \frac{1}{3}\right)$ (b) $\left(-3, \frac{1}{3}\right)$ (c) $\left(-3, -\frac{1}{3}\right)$ (d) $\left(3, -\frac{1}{3}\right)$

**56** Two fair dice are thrown. The numbers on them are taken as $\lambda$ and $\mu$, and a system of linear equations $x+y+z=5, x+2y+3z=\mu$ and $x+3y+\lambda z=1$ is constructed. If $p$ is the probability that the system has a unique solution and $q$ is the probability that the system has no solution, then *[2021, 26 Aug Shift-II]*
(a) $p=\frac{1}{6}$ and $q=\frac{1}{36}$ (b) $p=\frac{5}{6}$ and $q=\frac{5}{36}$
(c) $p=\frac{5}{6}$ and $q=\frac{1}{36}$ (d) $p=\frac{1}{6}$ and $q=\frac{5}{36}$

**57** Let $\theta \in \left(0, \frac{\pi}{2}\right)$. If the system of linear equations

$(1+\cos^2\theta)x+\sin^2\theta y+4\sin 3\theta z=0$

$\cos^2\theta x+(1+\sin^2\theta)y+4\sin 3\theta z=0$

$\cos^2\theta x+\sin^2\theta\, y+(1+4\sin 3\theta)z=0$

has a non-trivial solution, then the value of $\theta$ is **[2021, 26 Aug Shift-I]**

(a) $\frac{4\pi}{9}$ (b) $\frac{7\pi}{18}$ (c) $\frac{\pi}{18}$ (d) $\frac{5\pi}{18}$

**58** The value of $k \in R$, for which the following system of liner equations $3x-y+4z=3$, $x+2y-3z=-2$ and $6x+5y+kz=-3$ has infinitely many solutions, is **[2021, 20 July Shift-II]**

(a) 3 (b) −5 (c) 5 (d) −3

**59** Let $A=\begin{bmatrix} i & -i \\ -i & i \end{bmatrix}$, $i=\sqrt{-1}$. Then, the system of linear equations $A^8\begin{bmatrix} x \\ y \end{bmatrix}=\begin{bmatrix} 8 \\ 64 \end{bmatrix}$ has **[2021, 16 March Shift-I]**

(a) a unique solution
(b) infinitely many solutions
(c) no solution
(d) exactly two solutions

**60** Consider the following system of equations $x+2y-3z=a$, $2x+6y-11z=b$ and $x-2y+7z=c$ where, $a$, $b$ and $c$ are real constants. Then, the system of equations **[2021, 26 Feb Shift-II]**

(a) has a unique solution, when $5a=2b+c$
(b) has infinite number of solutions when $5a=2b+c$
(c) has no solution for all $a$, $b$ and $c$
(d) has a unique solution for all $a$, $b$ and $c$

**61** The system of linear equations $3x-2y-kz=10$, $2x-4y-2z=6$, $x+2y-z=5m$ is inconsistent, if **[2021, 24 Feb Shift-I]**

(a) $k\neq 3, m\neq 4/5$ (b) $k=3, m=4/5$
(c) $k=3, m\neq 4/5$ (d) $k\neq 3, m\in R$

**62** Let $S$ be the set of all $\lambda \in R$ for which the system of linear equations

$2x-y+2z=2$, $x-2y+\lambda z=-4$, $x+\lambda y+z=4$

has no solution. Then, the set $S$ **[2020, 2 Sep Shift-I]**

(a) contains more than two elements
(b) contains exactly two elements
(c) is an empty set
(d) is a singleton

**63** For which of the following ordered pairs $(\mu, \delta)$, the system of liner equations $x+2y+3z=1$, $3x+4y+5z=\mu$, $4x+4y+4z=\delta$ is inconsistent? **[2020, 8 Jan Shift-I]**

(a) (4, 6) (b) (4, 3)
(c) (1, 0) (d) (3, 4)

**64** If the system of equations $2x+3y-z=0$, $x+ky-2z=0$ and $2x-y+z=0$ has a non-trivial solution $(x, y, z)$, then $\frac{x}{y}+\frac{y}{z}+\frac{z}{x}+k$ is equal to **[2019, 9 April Shift-II]**

(a) −4 (b) 1/2
(c) −1/4 (d) 3/4

**65** If the system of linear equations

$x-4y+7z=g$, $3y-5z=h$,
$-2x+5y-9z=k$

is consistent, then **[2019, 9 Jan Shift-II]**

(a) $2g+h+k=0$ (b) $g+2h+k=0$
(c) $g+h+k=0$ (d) $g+h+2k=0$

**66** If the system of linear equations $2x+2y+3z=a$, $3x-y+5z=b$ and $x-3y+2z=c$ where $a, b, c$ are non-zero real numbers, has more than one solution, then **[2019, 11 Jan Shift-I]**

(a) $b-c-a=0$ (b) $a+b+c=0$
(c) $b-c+a=0$ (d) $b+c-a=0$

**67** If the system of linear equations $x+ky+3z=0$, $3x+ky-2z=0$ and $2x+4y-3z=0$ has a non-zero solution $(x, y, z)$, then $\frac{xz}{y^2}$ is equal to **[JEE Main 2018]**

(a) −10 (b) 10 (c) −30 (d) 30

**68** The set of all values of $\lambda$ for which the system of linear equations $2x_1-2x_2+x_3=\lambda x_1$, $2x_1-3x_2+2x_3=\lambda x_2$ and $-x_1+2x_2=\lambda x_3$ has a non-trivial solution, **[JEE Main 2015]**

(a) is an empty set
(b) is a singleton set
(c) contains two elements
(d) contains more than two elements

**69** The number of values of $k$, for which the system of equations $(k+1)x+8y=4k$ and $kx+(k+3)y=3k-1$ has no solution, is **[JEE Main 2013]**

(a) ∞ (b) 1 (c) 2 (d) 3

## Level 3

**70** Let the system of equations

$x+5y-z=1$
$4x+3y-3z=7$
$24x+y+\lambda z=\mu$

$\lambda, \mu \in R$ have infinitely many solutions. Then, the number of the solutions of this system, if $x$, $y$, $z$ are integers and satisfy $7\le x+y+z\le 77$, is **[2025, 7 April Shift-II]**

(a) 5 (b) 4 (c) 6 (d) 3

**71** Let $A$ be a $3 \times 3$ real matrix such that

$$A\begin{bmatrix}1\\0\\1\end{bmatrix}=2\begin{bmatrix}1\\0\\1\end{bmatrix}, A\begin{bmatrix}-1\\0\\1\end{bmatrix}=4\begin{bmatrix}-1\\0\\1\end{bmatrix}, A\begin{bmatrix}0\\1\\0\end{bmatrix}=2\begin{bmatrix}0\\1\\0\end{bmatrix}$$

Then, the system $(A-3I)\begin{bmatrix}x\\y\\z\end{bmatrix}=\begin{bmatrix}1\\2\\3\end{bmatrix}$ has **[2024, 31 Jan Shift-II]**

(a) exactly two solutions
(b) unique solution
(c) infinitely many solutions
(d) no solution

**72** Consider the system of linear equations $x+y+z=5$, $x+2y+\lambda^2 z=9$, $x+3y+\lambda z=\mu$, where $\lambda, \mu \in R$. Then, which of the following statement is not correct? **[2024, 30 Jan Shift-II]**

(a) System has unique solution, if $\lambda \neq 1$ and $\mu \neq 13$
(b) System is inconsistent, if $\lambda = 1$ and $\mu \neq 13$
(c) System has infinite number of solutions, if $\lambda = 1$ and $\mu = 13$
(d) System is consistent, if $\lambda \neq 1$ and $\mu = 13$

**73** Consider the following system of equations $\alpha x + 2y + z = 1$, $2\alpha x + 3y + z = 1$, $3x + \alpha y + 2z = \beta$ for some $\alpha, \beta \in R$. Then, which of the following is not correct. **[2023, 29 Jan Shift-I]**

(a) It has a solution, $\forall \alpha \neq 1$ and $\beta = 2$
(b) It has no solution, if $\alpha = -1$ and $\beta \neq 2$
(c) It has no solution, for $\alpha = 3$ and $\beta \neq 2$
(d) It has no solution, for $\alpha = -1$ and $\beta \in R$

**74** Let $\alpha, \beta, \gamma$ be the real roots of the equation, $x^3 + ax^2 + bx + c = 0$, $(a, b, c \in R$ and $a, b \neq 0)$. If the system of equations (in $u, v, w$) given by $\alpha u + \beta v + \gamma w = 0$, $\beta u + \gamma v + \alpha w = 0$; $\gamma u + \alpha v + \beta w = 0$ has non-trivial solution, then the value of $a^2 / b$ is **[2021, 18 March Shift-I]**

(a) 5
(b) 3
(c) 1
(d) 0

## Numerical Value Type Questions

**75** Let $A$ be a $3 \times 3$ matrix such that $X^T AX = O$ for all non-zero $3 \times 1$ matrices $X = \begin{bmatrix}x\\y\\z\end{bmatrix}$.

If $A\begin{bmatrix}1\\1\\1\end{bmatrix}=\begin{bmatrix}1\\4\\-5\end{bmatrix}, A\begin{bmatrix}1\\2\\1\end{bmatrix}=\begin{bmatrix}0\\4\\-8\end{bmatrix}$ and $\det(\text{adj}(2(A+I))) = 2^\alpha 3^\beta 5^\gamma$, $\alpha, \beta, \gamma \in N$, then $\alpha^2 + \beta^2 + \gamma^2$ is

**76** If the system of equations $2x + 7y + \lambda z = 3$, $3x + 2y + 5z = 4$ and $x + \mu y + 32z = -1$ has infinitely many solutions, then $(\lambda - \mu)$ is equal to ........... . **[2024, 6 April Shift-II]**

**77** Let $\alpha\beta\gamma = 45$, $\alpha, \beta, \gamma \in R$. If $x(\alpha, 1, 2) + y(1, \beta, 2) + z(2, 3, \gamma) = (0, 0, 0)$ for some $x, y, z \in R$, $xyz \neq 0$, then $6\alpha + 4\beta + \gamma$ is equal to ......... . **[2024, 6 April Shift-I]**

**78** Let $A=\begin{bmatrix}2&0&1\\1&1&0\\1&0&1\end{bmatrix}$, $B=[B_1, B_2, B_3]$, where $B_1, B_2, B_3$ be column matrics and $AB_1=\begin{bmatrix}1\\0\\0\end{bmatrix}$, $AB_2=\begin{bmatrix}2\\3\\0\end{bmatrix}$, $AB_3=\begin{bmatrix}3\\2\\1\end{bmatrix}$.

If $\alpha = |B|$ and $\beta$ is the sum of all the diagonal elements of $B$, then $\alpha^3 + \beta^3$ is equal to ...... . **[2024, 27 Jan Shift-I]**

**79** Let $S$ be the set of values of $\lambda$, for which the system of equations $6\lambda x - 3y + 3z = 4\lambda^2$, $2x + 6\lambda y + 4z = 1$ and $3x + 2y + 3\lambda z = \lambda$ has no solution. Then, $12\sum_{\lambda \in S}|\lambda|$ is equal to ...... . **[2023, 13 April Shift-II]**

**80** If the system of linear equations $2x + y - z = 3$, $x - y - z = \alpha$ and $3x + 3y + \beta z = 3$ has infinitely many solution, then $\alpha + \beta - \alpha\beta$ is equal to ............ . **[2021, 27 Aug Shift-I]**

**81** For real numbers $\alpha$ and $\beta$, consider the following system of linear equations $x + y - z = 2$, $x + 2y + \alpha z = 1$, $2x - y + z = \beta$. If the system has infinite solutions, then $\alpha + \beta$ is equal to ......... . **[2021, 27 July Shift-I]**

**82** The sum of distinct values of $\lambda$ for which the system of equations

$(\lambda - 1)x + (3\lambda + 1)y + 2\lambda z = 0$
$(\lambda - 1)x + (4\lambda - 2)y + (\lambda + 3)z = 0$
$2x + (3\lambda + 1)y + 3(\lambda - 1) = 0$

has non-zero solutions, is ........ **[2020, 6 Sep Shift-II]**

# Key Index

## Topic 1

| | | | | | | | | | |
|---|---|---|---|---|---|---|---|---|---|
| **1** (d) | **2** (d) | **3** (b) | **4** (b) | **5** (b) | **6** (b) | **7** (a) | **8** (b) | **9** (c) | **10** (a) |
| **11** (b) | **12** (a) | **13** (c) | **14** (d) | **15** (c) | **16** (c) | **17** (c) | **18** (c) | **19** (a) | **20** (c) |
| **21** (b) | **22** (c) | **23** (a) | **24** (d) | **25** (d) | **26** (a) | **27** (c) | **28** (a) | **29** (b) | **30** (b) |
| **31** (b) | **32** (d) | **33** (d) | **34** (a) | **35** (d) | **36** (d) | **37** (c) | **38** (c) | **39** (d) | **40** (b) |
| **41** (b) | **42** (c) | **43** (a) | **44** (b) | **45** (a) | **46** (d) | **47** (d) | **48** (c) | **49** (36) | **50** (6) |
| **51** (495) | **52** (4) | **53** (6) | **54** (16) | **55** (1) | **56** (2) | **57** (2020) | | | |

## Topic 2

| | | |
|---|---|---|
| **1** (d) | **2** (c) | **3** (d) |

## Topic 3

| | | | | | | | | | |
|---|---|---|---|---|---|---|---|---|---|
| **1** (d) | **2** (a) | **3** (b) | **4** (b) | **5** (a) | **6** (c) | **7** (c) | **8** (b) | **9** (b) | **10** (c) |
| **11** (c) | **12** (a) | **13** (d) | **14** (c) | **15** (b) | **16** (b) | **17** (b) | **18** (b) | **19** (c) | **20** (d) |
| **21** (a) | **22** (d) | **23** (a) | **24** (d) | **25** (d) | **26** (b) | **27** (d) | **28** (b) | **29** (b) | **30** (b) |
| **31** (b) | **32** (c) | **33** (d) | **34** (b) | **35** (c) | **36** (b) | **37** (b) | **38** (d) | **39** (d) | **40** (c) |
| **41** (c) | **42** (d) | **43** (d) | **44** (d) | **45** (c) | **46** (c) | **47** (38) | **48** (34) | **49** (14) | **50** (5) |
| **51** (7) | **52** (5) | **53** (50) | **54** (100) | **55** (4) | **56** (108) | **57** (13) | **58** (17) | | |

## Topic 4

| | | | | | | | | | |
|---|---|---|---|---|---|---|---|---|---|
| **1** (d) | **2** (b) | **3** (c) | **4** (d) | **5** (a) | **6** (b) | **7** (d) | **8** (c) | **9** (c) | **10** (b) |
| **11** (d) | **12** (c) | **13** (b) | **14** (b) | **15** (c) | **16** (d) | **17** (b) | **18** (b) | **19** (c) | **20** (d) |
| **21** (b) | **22** (c) | **23** (b) | **24** (d) | **25** (a) | **26** (a) | **27** (b) | **28** (c) | **29** (d) | **30** (b) |
| **31** (b) | **32** (c) | **33** (b) | **34** (a) | **35** (d) | **36** (b) | **37** (d) | **38** (a) | **39** (c) | **40** (d) |
| **41** (a) | **42** (a) | **43** (b) | **44** (c) | **45** (a) | **46** (c) | **47** (c) | **48** (a) | **49** (a) | **50** (d) |
| **51** (d) | **52** (c) | **53** (b) | **54** (b) | **55** (c) | **56** (b) | **57** (b) | **58** (b) | **59** (c) | **60** (b) |
| **61** (c) | **62** (b) | **63** (b) | **64** (b) | **65** (a) | **66** (a) | **67** (b) | **68** (d) | **69** (b) | **70** (d) |
| **71** (b) | **72** (a) | **73** (d) | **74** (b) | **75** (44) | **76** (38) | **77** (55) | **78** (28) | **79** (24) | **80** (5) |
| **81** (5) | **82** (3) | | | | | | | | |

# Detailed **Explanations**

## Topic 1

**01.** *(d)* Given,

$$f(x)=\begin{vmatrix}1+\sin^2 x & \cos^2 x & 4\sin 4x\\ \sin^2 x & 1+\cos^2 x & 4\sin 4x\\ \sin^2 x & \cos^2 x & 1+4\sin 4x\end{vmatrix}$$

$$\Rightarrow f(x)=\begin{vmatrix}1+\sin^2 x & \cos^2 x & 4\sin 4x\\ -1 & 1 & 0\\ -1 & 0 & 1\end{vmatrix}$$

(applying $R_2\to R_2-R_1$ and $R_3\to R_3-R_1$)

$\Rightarrow f(x)=2+4\sin 4x$ (expanding along $R_1$)

$[\because -1\le\sin 4x\le 1\Rightarrow -4\le 4\sin 4x\le 4]$

$\Rightarrow -4+2\le 2+4\sin 4x\le 4+2$

$\Rightarrow -2\le f(x)\le 6$

$\therefore \quad f(x)|_{\min}=m=-2$

and $f(x)|_{\max}=M=6$

Hence, $M^4-m^4=(6)^4-(-2)^4=1280$

**02.** *(d)* $A=\begin{bmatrix}\beta & \alpha & 3\\ \alpha & \alpha & \beta\\ -\beta & \alpha & 2\alpha\end{bmatrix}$

$$B=\begin{bmatrix}3\alpha & -9 & 3\alpha\\ -\alpha & 7 & -2\alpha\\ -2\alpha & 5 & -2\beta\end{bmatrix}$$

= cofactor matrix of $A$

$A_{21}$ = cofactor of $a_{21}$

$2\alpha^2-3\alpha=\alpha$

$\alpha=0,\ 2$ (accept)

Now, $2\alpha^2-\alpha\beta=3\alpha$

$\alpha=2,\ \beta=1$

$\therefore \quad |AB|=|A||B|=|A||A|^2=|A|^3$

Now, $|A|=\begin{vmatrix}1 & 2 & 3\\ 2 & 2 & 1\\ -1 & 2 & 7\end{vmatrix}=6$

$\therefore$ Required value $=6^3=216$

**03.** *(b)* We have,

$|2A|^3=2^{21}$

$\Rightarrow \quad |2A|=2^7$

$\Rightarrow \quad 8|A|=2^7$

$\Rightarrow \quad |A|=2^4$

Also, $\quad A=\begin{bmatrix}1 & 0 & 0\\ 0 & \alpha & \beta\\ 0 & \beta & \alpha\end{bmatrix}$

$\Rightarrow \quad |A|=\alpha^2-\beta^2=2^4$

$\Rightarrow (\alpha-\beta)(\alpha+\beta)=16$

$\Rightarrow \quad \alpha-\beta=2$ and $\alpha+\beta=8$

$\Rightarrow \quad \alpha=5$ and $\beta=3$

**04.** *(b)* $A=\begin{bmatrix}4 & -2\\ \alpha & \beta\end{bmatrix}$ satisfies

$A^2+4A+18I=0$

Characteristic equation of $A$ is

$A^2-(T_r(A))A+|A|I=0$

Comparing this with given equation, we get $|A|=18$

**05.** *(b)* Given,

$$A=\begin{bmatrix}[x+1] & [x+2] & [x+3]\\ [x] & [x+3] & [x+3]\\ [x] & [x+2] & [x+4]\end{bmatrix}$$

$$A=\begin{bmatrix}[x]+1 & [x]+2 & [x]+3\\ [x] & [x]+3 & [x]+3\\ [x] & [x]+2 & [x]+4\end{bmatrix}$$

$(\because [x+n]=n+[x],\ n\in I)$

Applying $R_1\to R_1-R_3$, $R_2\to R_2-R_3$,

$$A=\begin{bmatrix}1 & 0 & -1\\ 0 & 1 & -1\\ [x] & [x]+2 & [x]+4\end{bmatrix}$$

$\det(A)=1([x]+4+[x]+2)-1\,(-[x])$

$=3[x]+6$

$\because \quad \det(A)=192$

$\Rightarrow \quad 3[x]+6=192$

$\Rightarrow \quad [x]=62$

$\Rightarrow \quad 62\le x<63$

$\Rightarrow \quad x\in[62,63)$

**06.** *(b)* We have,

$$\begin{vmatrix}\sin x & \cos x & \cos x\\ \cos x & \sin x & \cos x\\ \cos x & \cos x & \sin x\end{vmatrix}=0,\ -\frac{\pi}{4}\le x\le\frac{\pi}{4}$$

Apply $R_1\to R_1-R_2$ and $R_2\to R_2-R_3$,

$$\begin{vmatrix}\sin x-\cos x & \cos x-\sin x & 0\\ 0 & \sin x-\cos x & \cos x-\sin x\\ \cos x & \cos x & \sin x\end{vmatrix}=0$$

$$\Rightarrow (\sin x-\cos x)^2\begin{vmatrix}1 & -1 & 0\\ 0 & 1 & -1\\ \cos x & \cos x & \sin x\end{vmatrix}=0$$

On expanding,

$(\sin x-\cos x)^2\,[1(\sin x+\cos x)$

$+1(\cos x)]=0$

$\Rightarrow \ (\sin x-\cos x)^2\ (\sin x+2\cos x)=0$

$\therefore \quad \sin x-\cos x=0$

$\Rightarrow \quad \sin x=\cos x\Rightarrow x=\pi/4$

and $\sin x+2\cos x=0$

$\Rightarrow \quad \sin x=2\cos x$

$\Rightarrow \quad x\notin\left[\frac{\pi}{4},\frac{\pi}{4}\right]$

Hence, the number of distinct real roots $=1$

**07.** *(a)* **Method I**

Given, $x$, $y$ and $z$ are in AP with common difference $=d$

$\therefore x=$ First term

$y=$ Second term of AP

$=$ First term $+$ Common difference

$\Rightarrow \quad y=x+d \quad$ ...(i)

and $z=$ Third term of AP $=$ Second term $+$ Common difference

$\Rightarrow \ z=(x+d)+d=x+2d \quad$ ...(ii)

Also, given $x\ne 3d \quad$ ...(iii)

$$\begin{vmatrix}3 & 4\sqrt{2} & x\\ 4 & 5\sqrt{2} & y\\ 5 & k & z\end{vmatrix}=0$$

Expanding along $C_1$

$x(4k-25\sqrt{2})-y(3k-20\sqrt{2})$

$+z(15\sqrt{2}-16\sqrt{2})=0$

$\Rightarrow x(4k-25\sqrt{2})-\dfrac{x+z}{2}(3k-20\sqrt{2})$

$-z\sqrt{2}=0$

$\Rightarrow x(8k-50\sqrt{2})-x(3k-20\sqrt{2})$

$-z(3k-20\sqrt{2})-z2\sqrt{2}=0$

$\Rightarrow x(5k-30\sqrt{2})-z(3k-18\sqrt{2})=0$

$\Rightarrow \quad (5x-3z)(k-6\sqrt{2})=0$

Either $k-6\sqrt{2}=0$ or $3z-5x=0$

$\Rightarrow k=6\sqrt{2}$ or $3(x+2d)-5x=0$ [from Eq. (ii)]

$\Rightarrow x=3d$ which is not possible as in Eq. (iii).

$\therefore k=6\sqrt{2}$ is only one solution.

Hence, $k^2=(6\sqrt{2})^2 \Rightarrow k^2=72$

**Method II**

Here, $x=$ First term

$y=$ Second term $=x+d$, where

$d=$ Common difference

$z=$ Third term $=x+2d$

Now, given matrices will be of the form

$$\begin{bmatrix} 3 & 4\sqrt{2} & x \\ 4 & 5\sqrt{2} & x+d \\ 5 & k & x+2d \end{bmatrix}$$

According to the question,

$$\begin{vmatrix} 3 & 4\sqrt{2} & x \\ 4 & 5\sqrt{2} & x+d \\ 5 & k & x+2d \end{vmatrix}=0$$

Applying $R_2 \to R_2-R_1$ and $R_3 \to R_3-R_1$, we get

$$\begin{vmatrix} 3 & 4\sqrt{2} & x \\ 1 & \sqrt{2} & d \\ 2 & k-4\sqrt{2} & 2d \end{vmatrix}=0$$

Again, applying $R_3 \to R_3-2R_2$, we have

$$\begin{vmatrix} 3 & 4\sqrt{2} & x \\ 1 & \sqrt{2} & d \\ 0 & k-6\sqrt{2} & 0 \end{vmatrix}=0$$

$\Rightarrow \quad -(k-6\sqrt{2})(3d-x)=0$ (Expanding along $R_3$)

$\Rightarrow \quad k-6\sqrt{2}=0$ or $3d-x=0$

$\Rightarrow \quad k=6\sqrt{2}$ or $x=3d$ which is not acceptable.

According to given problem condition.

$\therefore \ k^2=(6\sqrt{2})^2$

$\Rightarrow k^2=72$

**08.** *(b)* Let $A=\begin{vmatrix} (a+1) & (a+2) & a+2 & 1 \\ (a+2) & (a+3) & a+3 & 1 \\ (a+3) & (a+4) & a+4 & 1 \end{vmatrix}$

Expanding along $C_3$

$A=[(a+4)(a+2)(a+3)$
$\quad -(a+3)(a+3)(a+4)]$
$\quad -1[(a+1)(a+2)(a+4)$
$\quad -(a+2)(a+3)(a+4)]$
$\quad +1[(a+1)(a+2)(a+3)$
$\quad -(a+2)(a+2)(a+3)]$

$A=-(a+4)(a+3)+2(a+2)(a+4)-(a+2)(a+3)$

$A=(a+4)[-a-3+2a+4]-(a+2)(a+3)$

$A=(a+4)(a+1)-(a+2)(a+3)$

$A=a^2+5a+4-a^2-5a-6=-2$

**09.** *(c)* Let $A=\begin{bmatrix} a & b & c \\ d & e & f \\ g & h & i \end{bmatrix}$

Then, $2A=\begin{bmatrix} 2a & 2b & 2c \\ 2d & 2e & 2f \\ 2g & 2h & 2i \end{bmatrix}$

Now, perform the operation $R_2 \to 2R_2+5R_3$ on $2A$, we get

$$B=\begin{bmatrix} 2a & 2b & 2c \\ 4d+10g & 4e+10h & 4f+10i \\ 2g & 2h & 2i \end{bmatrix}$$

$$|B|=\begin{vmatrix} 2a & 2b & 2c \\ 4d+10g & 4e+10h & 4f+10i \\ 2g & 2h & 2i \end{vmatrix}$$

Using property of invariance to calculate $|B|$, apply $R_2 \to R_2-5R_3$

$$|B|=\begin{vmatrix} 2a & 2b & 2c \\ 4d & 4e & 4f \\ 2g & 2h & 2i \end{vmatrix}$$

$$=2\times 4\times 2\begin{vmatrix} a & b & c \\ d & e & f \\ g & h & i \end{vmatrix}=16\times \det(A)$$

$=16\times 4=64 \quad [\because \det(A)=4]$

**10.** *(a)* It is given that, $A$ is a $2\times 2$ real matrix with entries from {0, 1} and $|A|\neq 0$, then following matrices are possible cases

(i) $A=\begin{bmatrix} 1 & 0 \\ 0 & 1 \end{bmatrix}$, here $|A|=1$ and $\text{tr}(A)=2$

(ii) $A=\begin{bmatrix} 1 & 1 \\ 0 & 1 \end{bmatrix}$ or $\begin{bmatrix} 1 & 0 \\ 1 & 1 \end{bmatrix}$, here, $|A|=1$ and $\text{tr}(A)=2$

(iii) $A=\begin{bmatrix} 0 & 1 \\ 1 & 0 \end{bmatrix}$, here $A\neq I_2$ and $|A|=-1$

(iv) $A=\begin{bmatrix} 1 & 1 \\ 1 & 0 \end{bmatrix}$ or $\begin{bmatrix} 0 & 1 \\ 1 & 1 \end{bmatrix}$, here $A\neq I_2$ and $|A|=-1$

So, the given statements,

$(P)$ If $A\neq I_2$, then $|A|=-1$, is true as in cases (iii) and (iv). But in case (ii) $A\neq I_2$ and $|A|=1$

$(Q)$ If $|A|=1$, then $tr(A)=2$, is true as cases (i) and (ii).

**11.** *(b)* $f(x)=\begin{vmatrix} x+a & x+2 & x+1 \\ x+b & x+3 & x+2 \\ x+c & x+4 & x+3 \end{vmatrix}$

Expanding along $C_3$

$f(x)=(x+1)[x^2+4x+bx+4b-x^2-3x-cx-3c]$
$\quad -(x+2)[x^2+4x+ax+4a-x^2-2x-cx-2c]$
$\quad +(x+3)[x^2+3x+ax+3a-x^2-2x-bx-2b]$

$=(x+1)[x+(b-c)x+(4b-3c)]$
$\quad -(x+2)[2x+(a-c)x+(4a-2c)]$
$\quad +(x+3)[x+(a-b)x+(3a-2b)]$

$=x^2(0)+x(0)+(-2b+a+c)$

$f(x)=1 \quad [\because a-2b+c=1]$

$f(50)=1$

**12.** *(a)* Given equation

$$\begin{vmatrix} x & -6 & -1 \\ 2 & -3x & x-3 \\ -3 & 2x & x+2 \end{vmatrix}=0$$

On expansion of determinant along $R_1$, we get

$x[(-3x)(x+2)-2x(x-3)]+6[2(x+2)+3(x-3)]-1[2(2x)-(-3x)(-3)]=0$

$\Rightarrow x[-3x^2-6x-2x^2+6x]+6[2x+4+3x-9]-1[4x-9x]=0$

$\Rightarrow x(-5x^2)+6(5x-5)-1(-5x)=0$

$\Rightarrow \quad -5x^3+30x-30+5x=0$

$\Rightarrow \quad 5x^3-35x+30=0$

$\Rightarrow \quad x^3-7x+6=0.$

Since all roots are real

$\therefore$Sum of roots $=-\dfrac{\text{Coefficient of } x^2}{\text{Coefficient of } x^3}=0$

**13.** *(c)* Given determinants are

$$\Delta_1=\begin{vmatrix} x & \sin\theta & \cos\theta \\ -\sin\theta & -x & 1 \\ \cos\theta & 1 & x \end{vmatrix}$$

$= x(-x^2-1) - \sin\theta(-x\sin\theta - \cos\theta)$
$\quad + \cos\theta(-\sin\theta + x\cos\theta)$

$= -x^3 - x + x(\sin^2\theta + \cos^2\theta)$

$= -x^3 - x + x(1) = -x^3$

and
$$\Delta_2 = \begin{vmatrix} x & \sin 2\theta & \cos 2\theta \\ -\sin 2\theta & -x & 1 \\ \cos 2\theta & 1 & x \end{vmatrix}, x \neq 0$$

$= -x^3$ (similarly as $\Delta_1$)

So, according to options, we get

$\Delta_1 + \Delta_2 = -2x^3$

**14.** *(d)* Given, quadratic equation is $x^2 + x + 1 = 0$ having roots $\alpha, \beta$.

Then, $\alpha + \beta = -1$ and $\alpha\beta = 1$

Now, given determinant
$$\Delta = \begin{vmatrix} y+1 & \alpha & \beta \\ \alpha & y+\beta & 1 \\ \beta & 1 & y+\alpha \end{vmatrix}$$

Expanding along $R_1$

$\Delta = (y+1)[y^2 + (\alpha+\beta)y + \alpha\beta - 1]$
$\quad - \alpha[y\alpha + \alpha^2 - \beta] + \beta[\alpha - \beta y - \beta^2]$

$\Delta = (y+1)[y^2 + (\alpha+\beta)y + \alpha\beta - 1]$
$\quad + \alpha\beta - y\alpha^2 - \alpha^3 + \beta\alpha - \beta^2 y - \beta^3$

$\Delta = (y+1)[y^2 + (\alpha+\beta)y + \alpha\beta - 1]$
$\quad + 2\alpha\beta - y(\alpha^2 + \beta^2) - (\alpha^3 + \beta^3)$

[$\because \alpha + \beta = -1$, $\alpha\beta = 1$,
$\alpha^2 + \beta^2 = (\alpha+\beta)^2 - 2\alpha\beta = 1 - 2 = -1$
and $\alpha^3 + \beta^3 = (\alpha+\beta)^3 - 3\alpha\beta(\alpha+\beta)$
$= -1 + 3 = 2$

$\Delta = (y+1)[y^2 - y + 1 - 1] + 2 + y - 2$

$\Delta = (y+1)(y^2 - y) + 2 + y - 2$

$= y^3 - y^2 + y^2 - y + y = y^3$

**15.** *(c)* Given, $\begin{vmatrix} x-4 & 2x & 2x \\ 2x & x-4 & 2x \\ 2x & 2x & x-4 \end{vmatrix}$

$= (A + Bx)(x - A)^2$

Expanding the determinant in LHS along $R_1$,

$(x-4)[(x-4)^2 - 4x^2]$
$\quad -2x[2x^2 - 8x - 4x^2]$
$\quad +2x[4x^2 - 2x^2 + 8x]$

$\Rightarrow (x-4)[x^2 + 16 - 8x - 4x^2]$
$\quad -2x[-2x^2 - 8x - 2x^2 - 8x]$

$\Rightarrow (x-4)[-3x^2 - 8x + 16]$
$\quad -2x[-4x^2 - 16x]$

$\Rightarrow (x-4)[-3x^2 - 8x + 16] + 8x^2[x+4]$

$\Rightarrow -(x-4)(3x-4)(x+4) + 8x^2(x+4)$

$\Rightarrow (x+4)[-(x-4)(3x-4) + 8x^2]$

$\Rightarrow (x+4)[-3x^2 + 16x - 16 + 8x^2]$

$\Rightarrow (x+4)(5x-4)(x+4)$

$\Rightarrow (x+4)^2(5x-4)$

$\therefore (5x-4)(x+4)^2 = (A + Bx)(x - A)^2$

Equating, we get $A = -4$ and $B = 5$

**16.** *(c)* $\begin{vmatrix} x & 1 & 1 \\ 1 & y & 1 \\ 1 & 1 & z \end{vmatrix} = xyz - (x + y + z) + 2$

$\because$ AM $\geq$ GM

$\Rightarrow \dfrac{x+y+z}{3} \geq (xyz)^{\frac{1}{3}}$

$\Rightarrow x + y + z \geq 3(xyz)^{\frac{1}{3}}$

For least value of $xyz$, $xyz$-3

$(xyz)^{\frac{1}{3}} + 2 \geq 0$

$\Rightarrow t^3 - 3t + 2 \geq 0$ [Putting $t = (xyz)^{\frac{1}{3}}$]

$\Rightarrow (t+2)(t^2 - 2t + 1) \geq 0 \Rightarrow t = -2, 1$

So, least value of $t^3 = xyz$ is $-8$.

**17.** *(c)* Let
$$\Delta = \begin{vmatrix} 3 & 1 + f(1) & 1 + f(2) \\ 1 + f(1) & 1 + f(2) & 1 + f(3) \\ 1 + f(2) & 1 + f(3) & 1 + f(4) \end{vmatrix}$$

$$\Rightarrow \Delta = \begin{vmatrix} 3 & 1+\alpha+\beta & 1+\alpha^2+\beta^2 \\ 1+\alpha+\beta & 1+\alpha^2+\beta^2 & 1+\alpha^3+\beta^3 \\ 1+\alpha^2+\beta^2 & 1+\alpha^3+\beta^3 & 1+\alpha^4+\beta^4 \end{vmatrix}$$

$$= \begin{vmatrix} 1\cdot1+1\cdot1+1\cdot1 & 1\cdot1+1\cdot\alpha+1\cdot\beta & 1\cdot1+1\cdot\alpha^2+1\cdot\beta^2 \\ 1\cdot1+\alpha\cdot1+\beta\cdot1 & 1\cdot1+\alpha\cdot\alpha+\beta\cdot\beta & 1\cdot1+\alpha\cdot\alpha^2+\beta\cdot\beta^2 \\ 1\cdot1+\alpha^2\cdot1+\beta^2\cdot1 & 1\cdot1+\alpha^2\cdot\alpha+\beta^2\cdot\beta & 1\cdot1+\alpha^2\cdot\alpha^2+\beta^2\cdot\beta^2 \end{vmatrix}$$

$$= \begin{vmatrix} 1 & 1 & 1 \\ 1 & \alpha & \beta \\ 1 & \alpha^2 & \beta^2 \end{vmatrix} \begin{vmatrix} 1 & 1 & 1 \\ 1 & \alpha & \beta \\ 1 & \alpha^2 & \beta^2 \end{vmatrix} = \begin{vmatrix} 1 & 1 & 1 \\ 1 & \alpha & \beta \\ 1 & \alpha^2 & \beta^2 \end{vmatrix}^2$$

On expanding, we get

$\Delta = (1-\alpha)^2(1-\beta)^2(\alpha-\beta)^2$

Hence, $K(1-\alpha)^2(1-\beta)^2(\alpha-\beta)^2$
$= (1-\alpha)^2(1-\beta)^2(\alpha-\beta)^2$

$\therefore \quad K = 1$

**18.** *(c)* Given,

(i) Two matrices $P$ and $Q$ of order $3 \times 3$ such that $P \neq Q$.

(ii) $P^3 = Q^3$ and $P^2Q = Q^2P$

To find the value of determinant of $P^2 + Q^2$.

On subtracting the given equations, we get

$P^3 - P^2Q = Q^3 - Q^2P$

$\Rightarrow \quad P^2(P - Q) = Q^2(Q - P)$

$\Rightarrow \quad (P - Q)(P^2 + Q^2) = 0$

$\Rightarrow \quad P^2 + Q^2 = 0 \quad [\because P \neq Q]$

$\Rightarrow \quad |P^2 + Q^2| = 0$

**19.** *(a)* Determinant of skew-symmetric matrix of odd order is zero and of even order is perfect square.

So, Statement I is true.

Now, $\det(A^T) = \det(A)$

and $\det(-A) = (-1)^n \det(A)$

So, Statement II is false.

**20.** *(c)* Since, $A = \begin{bmatrix} 5 & 5\alpha & \alpha \\ 0 & \alpha & 5\alpha \\ 0 & 0 & 5 \end{bmatrix}$

$$\therefore A^2 = \begin{bmatrix} 5 & 5\alpha & \alpha \\ 0 & \alpha & 5\alpha \\ 0 & 0 & 5 \end{bmatrix} \begin{bmatrix} 5 & 5\alpha & \alpha \\ 0 & \alpha & 5\alpha \\ 0 & 0 & 5 \end{bmatrix}$$

$$= \begin{bmatrix} 25 & 25\alpha + 5\alpha^2 & 10\alpha + 25\alpha^2 \\ 0 & \alpha^2 & 5\alpha^2 + 25\alpha \\ 0 & 0 & 25 \end{bmatrix}$$

$$\Rightarrow |A^2| = \begin{vmatrix} 25 & 25\alpha + 5\alpha^2 & 10\alpha + 25\alpha^2 \\ 0 & \alpha^2 & 5\alpha^2 + 25\alpha \\ 0 & 0 & 25 \end{vmatrix}$$

$$= 25 \begin{vmatrix} \alpha^2 & 25\alpha + 5\alpha^2 \\ 0 & 25 \end{vmatrix}$$

$= 625\alpha^2$

$\Rightarrow 625\alpha^2 = 25$ [given]

$\therefore \quad |\alpha| = \dfrac{1}{5}$

**21.** *(b)* Given that $D = \begin{vmatrix} 1 & 1 & 1 \\ 1 & 1+x & 1 \\ 1 & 1 & 1+y \end{vmatrix}$

Expanding along $C_1$

$D = 1(1 + x + y + xy - 1) - 1(1 + y - 1)$
$\quad + 1(1 - 1 - x)$

$D = x + y + xy - y - x = xy$

**22.** (c) Since, $a_1, a_2, \ldots, a_n, \ldots$ are in GP, then

$\log a_n, \log a_{n+1}, \log a_{n+2}, \ldots,$
$\log a_{n+8}, \ldots$ are in AP.

Given that

$$\Delta = \begin{vmatrix} \log a_n & \log a_{n+1} & \log a_{n+2} \\ \log a_{n+3} & \log a_{n+4} & \log a_{n+5} \\ \log a_{n+6} & \log a_{n+7} & \log a_{n+8} \end{vmatrix}$$

$$\therefore \Delta = \begin{vmatrix} a & a+d & a+2d \\ a+3d & a+4d & a+5d \\ a+6d & a+7d & a+8d \end{vmatrix} = 0$$

Where, $a$ and $d$ are the first term and common difference of an AP.

Applying $C_2 \to C_2 - C_1$,
$C_3 \to C_3 - C_1$,

$$\Delta = \begin{vmatrix} a & d & 2d \\ a+3d & d & 2d \\ a+6d & d & 2d \end{vmatrix} = 2\begin{vmatrix} a & d & d \\ a+3d & d & d \\ a+6d & d & d \end{vmatrix}$$

$= 0$ [Since, two columns are identical]

**23.** (a) Given that, $\Delta = \begin{vmatrix} 1 & \omega^n & \omega^{2n} \\ \omega^n & \omega^{2n} & 1 \\ \omega^{2n} & 1 & \omega^n \end{vmatrix}$

$$= 1(\omega^{3n} - 1) - \omega^n(\omega^{2n} - \omega^{2n}) + \omega^{2n}(\omega^n - \omega^{4n})$$

$$= 1(1-1) - 0 + \omega^{2n}(\omega^n - \omega^n)$$

$[\because \omega^3 = 1]$

$= 0$

**24.** (d) Since, $l$, $m$ and $n$ are the $p$th, $q$th and $r$th terms of a GP whose first term is $A$ and common ratio is $R$.

$\therefore \quad l = AR^{p-1}$

$\Rightarrow \log l = \log A + (p-1)\log R$

Similarly, $\log m = \log A + (q-1)\log R$

and $\log n = \log A + (r-1)\log R$

Now, $\begin{vmatrix} \log l & p & 1 \\ \log m & q & 1 \\ \log n & r & 1 \end{vmatrix}$

$$= \begin{vmatrix} \log A + (p-1)\log R & p & 1 \\ \log A + (q-1)\log R & q & 1 \\ \log A + (r-1)\log R & r & 1 \end{vmatrix}$$

Applying

$C_1 \to C_1 - [C_3 \log A + (C_2 - C_3)\log R]$,

$$= \begin{vmatrix} 0 & p & 1 \\ 0 & q & 1 \\ 0 & r & 1 \end{vmatrix} = 0$$

**25.** (d) Given that,

$$\begin{vmatrix} 6i & -3i & 1 \\ 4 & 3i & -1 \\ 20 & 3 & i \end{vmatrix} = x + iy]$$

Expanding along $C_3$

$$1(12 - 60i) + 1(18i + 60i) + i(18i^2 + 12)i = x + iy$$

$$12 - 60i + 78i - 18i - 12 = x + iy$$

$$0 = x + iy \Rightarrow 0 + 0i = x + iy$$

$\therefore \quad x = 0$ and $y = 0$

**26.** (a) $\begin{vmatrix} \alpha & b & c \\ a & \beta & c \\ a & b & \gamma \end{vmatrix} = 0$

$\Rightarrow \alpha(\beta\gamma - bc) - b(a\gamma - ac) + c(ab - a\beta) = 0$

$\Rightarrow \alpha\beta\gamma - bc\alpha - ba\gamma + abc + abc - ca\beta = 0$

$\Rightarrow \gamma(\alpha - a)(\beta - b) - b(\alpha - a)(c - \gamma) + a(b - \beta)(c - \gamma) = 0$

Divide and multiply by
$(\alpha - a)(\beta - b)(\gamma - c)$

$$\Rightarrow \frac{\gamma}{(\gamma - c)} + \frac{b}{(\beta - b)} + \frac{a}{(\alpha - a)} = 0$$

***Alternate Solution***

Applying $R_1 \to R_1 - R_2, R_2 \to R_2 - R_3$,

$$\begin{vmatrix} \alpha - a & b - \beta & 0 \\ 0 & \beta - b & c - \gamma \\ a & b & \gamma \end{vmatrix} = 0$$

$\Rightarrow (\alpha - a)[\gamma(\beta - b) - b(c - \gamma)] - (b - \beta)[-a(c - \gamma)] = 0$

$\Rightarrow \gamma(\alpha - a)(\beta - b) - b(\alpha - a)(c - \gamma) + a(b - \beta)(c - \gamma) = 0$

Divide and multiply by $(\alpha - a)(\beta - b)(\gamma - c)$,

$$\frac{\gamma}{(\gamma - c)} + \frac{b}{(\beta - b)} + \frac{a}{(\alpha - a)} = 0$$

**27.** (c)

**key idea** Use, if $A$ is $3 \times 3$ matrices, then $A^3 - \text{Tr}(A)(A)^2 + \Sigma A_{ii}(A) + |A|I = 0$ where $A_{ii}$ = cofactor of $a_{ii}$ to solve the questions.

$$A = \begin{bmatrix} 2 & a & 0 \\ 1 & 3 & 1 \\ 0 & 5 & b \end{bmatrix}$$

and $A^3 = 4A^2 - A - 21I$

So, $x^3 - 4x^2 + x + 21 = 0$ is the characteristic equation of matrix $A$.

So, sum of roots $= T_r(A)$

$\Rightarrow \; -(-4) = 2 + 3 + b$

$\Rightarrow \; 5 + b = 4 \Rightarrow b = -1$ ...(i)

Product of roots $= |A|$

$\Rightarrow -21 = 2(3b - 5) - a(b)$

Put $b = -1$ from Eq. (i),

$-21 = 2(-8) + a$

$\Rightarrow \quad a = -21 + 16 = -5$

Therefore,

$2a + 3b = 2(-5) + 3(-1) = -13$

**28.** (a) According to the question,

$$A_r = \begin{vmatrix} r & 1 & \frac{n^2}{2} + \alpha \\ 2r & 2 & n^2 - \beta \\ 3r - 2 & 3 & \frac{n}{2}(3n - 1) \end{vmatrix}$$

Applying $C_1 \to C_1 - rC_2$

$$A_r = \begin{vmatrix} 0 & 1 & \frac{n^2}{2} + \alpha \\ 0 & 2 & n^2 - \beta \\ -2 & 3 & \frac{n}{2}(3n - 1) \end{vmatrix}$$

$$A_r = -2\left[(n^2 - \beta) - 2\left(\frac{n^2}{2} + \alpha\right)\right]$$

$$A_r = +2[2\alpha + \beta]$$

$A_r$ is independent of $r$

So, $2A_{10} - A_8 = 4[2\alpha + \beta] - 2[2\alpha + \beta]$
$= 2[2\alpha + \beta] = 4\alpha + 2\beta$

**29.** (b) Given, $A = \begin{bmatrix} \sqrt{2} & 1 \\ -1 & \sqrt{2} \end{bmatrix}$,

$$B = \begin{bmatrix} 1 & 0 \\ 1 & 1 \end{bmatrix}$$

$$A^T = \begin{bmatrix} \sqrt{2} & -1 \\ 1 & \sqrt{2} \end{bmatrix}$$

$$AA^T = \begin{bmatrix} \sqrt{2} & 1 \\ -1 & \sqrt{2} \end{bmatrix}\begin{bmatrix} \sqrt{2} & -1 \\ 1 & \sqrt{2} \end{bmatrix}$$

$$= \begin{bmatrix} 3 & 0 \\ 0 & 3 \end{bmatrix} = 3I$$

$C = ABA^T$

$C^2 = (ABA^T)(ABA^T)$

$C^2 = ABA^T ABA^T$

$C^2 = AB3IBA^T = 3AB^2A^T$

$X = A^T C^2 A$

$= 3A^T AB^2 A^T A = 27B^2$

$X = 27I \left[ B^2 = \begin{bmatrix} 1 & 0 \\ 1 & 1 \end{bmatrix}\begin{bmatrix} 1 & 0 \\ 1 & 1 \end{bmatrix} = I \right]$

$|X| = 27 \times 27 = 729$

**30.** *(b)* Given determinant is

$$\begin{vmatrix} 1 & \frac{3}{2} & \alpha+\frac{3}{2} \\ 1 & \frac{1}{3} & \alpha+\frac{1}{3} \\ 2\alpha+3 & 3\alpha+1 & 0 \end{vmatrix}=0$$

Applying $C_3 \to C_3 - (\alpha C_1 + C_2)$, we get

$$\begin{vmatrix} 1 & 3/2 & 0 \\ 1 & 1/3 & 0 \\ 2\alpha+3 & 3\alpha+1 & -(2\alpha^2+6\alpha+1) \end{vmatrix}=0$$

$$\Rightarrow [-(2\alpha^2+6\alpha+1)]\left(\frac{1}{3}-\frac{3}{2}\right)=0$$

$$\Rightarrow \frac{7}{6}(2\alpha^2+6\alpha+1)=0$$

$$\Rightarrow 2\alpha^2+6\alpha+1=0$$

$$\Rightarrow \alpha=\frac{-6\pm\sqrt{36-4\times2\times1}}{2\times2}$$

$$\Rightarrow \alpha=\frac{-6\pm\sqrt{28}}{4}$$

$$\Rightarrow \alpha=\frac{-3\pm\sqrt{7}}{2}=\frac{-3\pm2.65}{2} \quad [\because \sqrt{7}\cong2.645]$$

$$\Rightarrow \alpha=(-0.175) \text{ or } (-2.82)\in(-3,0)$$

**Alternate Solution**

Given, determinant is

$$\begin{vmatrix} 1 & \frac{3}{2} & \alpha+\frac{3}{2} \\ 1 & \frac{1}{3} & \alpha+\frac{1}{3} \\ 2\alpha+3 & 3\alpha+1 & 0 \end{vmatrix}=0$$

$$\Rightarrow \left(\alpha+\frac{3}{2}\right)\left[3\alpha+1-\frac{2\alpha}{3}-1\right]-\left(\alpha+\frac{1}{3}\right)\left[3\alpha+1-3\alpha-\frac{9}{2}\right]=0$$

$$\Rightarrow \left(\alpha+\frac{3}{2}\right)\left[\frac{7\alpha}{3}\right]-\left(\alpha+\frac{1}{3}\right)\left[-\frac{7}{2}\right]=0$$

$$\Rightarrow \frac{7\alpha^2}{3}+\frac{7\alpha}{2}+\frac{7}{2}\alpha+\frac{7}{6}=0$$

$$\Rightarrow \frac{7\alpha^2}{3}+14\alpha+\frac{7}{6}=0$$

$$\Rightarrow 14\alpha^2+84\alpha+7=0 \Rightarrow 2\alpha^2+6\alpha+1=0$$

$$\Rightarrow \alpha=\frac{-6\pm\sqrt{36-4\times2\times1}}{2\times2} \Rightarrow \alpha=\frac{-6\pm\sqrt{28}}{4}$$

$$\Rightarrow \alpha=\frac{-3\pm\sqrt{7}}{2}=\frac{-3\pm2.65}{2} \quad [\because\sqrt{7}\cong2.645]$$

$$\Rightarrow \alpha=(-0.175) \text{ or } (-2.82)\in(-3,0)$$

**31.** *(b)* Given, $\begin{vmatrix} x+1 & x & x \\ x & x+\lambda & x \\ x & x & x+\lambda^2 \end{vmatrix}$

$$=\frac{9}{8}(103x+81)$$

Here, we can substitute any value of $x$.

Hence, put $x=0$

Then, $\begin{vmatrix} 1 & 0 & 0 \\ 0 & \lambda & 0 \\ 0 & 0 & \lambda^2 \end{vmatrix}=\frac{9}{8}\times81 \Rightarrow \lambda^3=\frac{3^6}{2^3}$

$$\Rightarrow \lambda=\frac{3^2}{2}=\frac{9}{2}$$

$$\Rightarrow \lambda=\frac{9}{2} \Rightarrow \frac{\lambda}{3}=\frac{3}{2}$$

If $\lambda$, $\lambda/3$ are roots then, quadratic equation will be

$$x^2-\left(\lambda+\frac{\lambda}{3}\right)x+\frac{\lambda^2}{3}=0$$

$$\Rightarrow x^2-\left(\frac{12}{2}\right)x+\frac{27}{4}=0$$

$$\Rightarrow x^2-6x+\frac{27}{4}=0$$

or $4x^2-24x+27=0$

**32.** *(d)*

> **key idea** When it's given, that determinants is a function, apply elementary operations to reduce it into simple form and then find the final solution with the help of the given data.

We have,

$$f(x)=\begin{vmatrix} 1+\sin^2 x & \cos^2 x & \sin 2x \\ \sin^2 x & 1+\cos^2 x & \sin 2x \\ \sin^2 x & \cos^2 x & 1+\sin 2x \end{vmatrix}$$

Expanding along $R_1$

$f(x)=(1+\sin^2 x)$

$[(1+\cos^2 x)(1+\sin 2x)-\cos^2 x\sin 2x]$

$-\cos^2 x[\sin^2 x(1+\sin 2x)-\sin^2 x\sin 2x]$

$+\sin 2x[\cos^2 x\sin^2 x-\sin^2 x(1+\cos^2 x)]$

$f(x)=(1+\sin^2 x)[1+\sin 2x+\cos^2 x$

$+\cos^2 x\sin 2x-\cos^2 x\sin 2x]$

$-\cos^2 x[\sin^2 x+\sin^2 x\sin 2x$

$-\sin^2 x\sin 2x]$

$+\sin 2x[\cos^2 x\sin^2 x-\sin^2 x$

$-\sin^2 x\cos^2 x]$

$=(1+\sin^2 x)[1+\sin 2x+\cos^2 x]$

$-\cos^2 x[\sin^2 x]$

$+\sin 2x[-\sin^2 x]$

$=1+\sin 2x+\cos^2 x$

$+\sin^2 x+\sin^2 x\sin 2x+\sin^2 x\cos^2 x$

$-\cos^2 x\sin^2 x-\sin 2x\sin^2 x$

$=2+\sin 2x$

$f(x)=2+\sin 2x$

Now, $x\in\left[\frac{\pi}{6},\frac{\pi}{3}\right] \Rightarrow 2x\in\left[\frac{\pi}{3},\frac{2\pi}{3}\right]$

$\therefore \sin 2x\in\left[\frac{\sqrt{3}}{2},1\right]$

Now, $f_{min}=2+\frac{\sqrt{3}}{2}$ and $f_{max}=2+1=3$

$\therefore \alpha=3$ and $\beta=2+\frac{\sqrt{3}}{2}$

Again, $\beta^2=\left(2+\frac{\sqrt{3}}{2}\right)^2$

$$=4+\frac{3}{4}+2\sqrt{3}=\frac{19}{4}+2\sqrt{3}$$

$$\therefore \beta^2-2\sqrt{\alpha}=\frac{19}{4}+2\sqrt{3}-2\sqrt{3}=\frac{19}{4}$$

**33.** *(d)* Given quadratic equation is $(a-c)x^2+(b-a)x+(c-b)=0$ and the matrix $\begin{bmatrix} \alpha^2 & \alpha & 1 \\ 1 & 1 & 1 \\ a & b & c \end{bmatrix}$ is singular.

So, its determinant value will be zero.

i.e. $\Delta=0 \Rightarrow \begin{vmatrix} \alpha^2 & \alpha & 1 \\ 1 & 1 & 1 \\ a & b & c \end{vmatrix}=0$

$\Rightarrow \alpha^2(c-b)-\alpha(c-a)+(b-a)=0$

i.e. when we put $\alpha=1$, then the given matrix will be singular.

Now, $\frac{(a-c)^2}{(b-a)(c-b)}+\frac{(b-a)^2}{(a-c)(c-b)}$

$+\frac{(c-b)^2}{(a-c)(b-a)}=\frac{(c-a)^2}{(a-b)(b-c)}$

$+\frac{(a-b)^2}{(b-c)(c-a)}+\frac{(b-c)^2}{(a-b)(c-a)}$

$$=\frac{(c-a)^3+(a-b)^3+(b-c)^3}{(a-b)(b-c)(c-a)}$$

Here, we use the formula $x^3+y^3+z^3=3xyz$, when $x+y+z=0$ i.e. here $(c-a)+(a-b)+(b-c)=0$.

So, $(c-a)^3+(a-b)^3+(b-c)^3=3(c-a)(a-b)(b-c)$

Hence, required expression will become

$$\frac{3(a-b)(b-c)(c-a)}{(a-b)(b-c)(c-a)}=3$$

**Alternate Solution**

$$\frac{(a-c)^2}{(b-a)(c-b)}+\frac{(b-a)^2}{(a-c)(c-b)}+\frac{(c-b)^2}{(a-c)(b-a)}$$

$$=\frac{(a-c)^3+(b-a)^3+(c-b)^3}{(-a+b)(c-b)(a-c)}$$

Here, $(a-c)+(b-a)+(c-b)=0$.

Now, using the result $x^3+y^3+z^3=3xyz$ when $(x+y+z)=0$, we get

$$\Rightarrow (a-c)^3+(b-a)^3+(c-b)^3 = 3(a-c)(b-a)(c-b)$$

Hence, required expression we become

$$\frac{3(a-c)(b-a)(c-b)}{(b-a)(c-b)(a-c)}=3$$

**34.** (a) $A=\begin{bmatrix}2 & 3\\ a & 0\end{bmatrix}$, $A^T=\begin{bmatrix}2 & a\\ 3 & 0\end{bmatrix}$

$$A=\frac{A+A^T}{2}+\frac{A-A^T}{2}$$

Let $P=\dfrac{A+A^T}{2}$ and $Q=\dfrac{A-A^T}{2}$

$$Q=\begin{pmatrix}0 & \dfrac{3-a}{2}\\ \dfrac{a-3}{2} & 0\end{pmatrix}$$

$$\text{Det }(Q)=9,\ 0-\left(\frac{3-a}{2}\right)\left(\frac{a-3}{2}\right)=9$$

$$\Rightarrow \left(\frac{a-3}{2}\right)^2=9$$

$$\Rightarrow (a-3)^2=36$$

$$a-3=\pm 6$$

$$\Rightarrow a=9,-3$$

$$P=\begin{bmatrix}2 & \dfrac{a+3}{2}\\ \dfrac{a+3}{2} & 0\end{bmatrix}$$

$$P=\begin{bmatrix}2 & 6\\ 6 & 0\end{bmatrix} \text{ or } \begin{bmatrix}2 & 0\\ 0 & 0\end{bmatrix}$$

$$|P|=-36 \text{ or } 0$$

$$\therefore\ |-36+0|=36$$

**35.** (d)

> **key idea** For a function $f(x)$ to be minimum or maximum. We put $f'(x)=0$ to find the critical points.

Given, $a_{ij}=\begin{cases}1 & ,i=j\\ -x & ,|i-j|=1\\ 2x+1, & \text{otherwise}\end{cases}$

$A=[a_{ij}]$ be $3\times 3$ matrix.

$$\therefore\quad A=\begin{bmatrix}1 & -x & 2x+1\\ -x & 1 & -x\\ 2x+1 & -x & 1\end{bmatrix}$$

$$|A|=4x^3-4x^2-4x$$

Let $f(x)=4x^3-4x^2-4x$

$\Rightarrow f'(x)=12x^2-8x-4$

Now, putting $f'(x)=0$

$\Rightarrow\ 4(3x^2-2x-1)=0$

$\Rightarrow\ (x-1)(3x+1)=0$

$$\underset{}{+}\quad -1/3 \quad - \quad 1 \quad +$$

$\Rightarrow\ x=1,-1/3$

At $f(1)\rightarrow$ minimum

$f(1)=4(1^3-1^2-1)=-4$

$f\left(-\dfrac{1}{3}\right)\rightarrow$ maximum

$f\left(-\dfrac{1}{3}\right)=\dfrac{20}{27}$

$\therefore\quad \text{Sum}=-4+\dfrac{20}{27}=\dfrac{-88}{27}$

**36.** (d) Given,

$$\begin{vmatrix}1+\sin^2 x & \sin^2 x & \sin^2 x\\ \cos^2 x & 1+\cos^2 x & \cos^2 x\\ 4\sin 2x & 4\sin 2x & 1+4\sin 2x\end{vmatrix}=0$$

$(0<x<\pi)$

Expanding along $R_1$,

$(1+\sin^2 x)[(1+\cos^2 x)(1+4\sin 2x)$
$-4\cos^2 x\sin 2x]-\sin^2 x$
$[\cos^2 x(1+4\sin 2x)-4\cos^2 x\sin 2x]$
$+\sin^2 x[4\cos^2 x\sin 2x$
$-4\sin 2x(1+\cos^2 x)]=0$

$\Rightarrow (1+\sin^2 x)$
$[1+4\sin 2x+\cos^2 x+4\sin 2x\cos^2 x$
$-4\cos^2 x\sin 2x]$
$-\sin^2 x[\cos^2 x+4\sin 2x\cos^2 x$
$-4\cos^2 x\sin 2x]$
$+\sin^2 x[4\cos^2 x\sin 2x-4\sin 2x$
$-4\sin 2x\cos^2 x]=0$

$\Rightarrow (1+\sin^2 x)[1+4\sin 2x+\cos^2 x]$
$-\sin^2 x[\cos^2 x]-4\sin^2 x\sin 2x=0$

$\Rightarrow 1+4\sin 2x+\cos^2 x+\sin^2 x$
$+4\sin^2 x\sin 2x$
$+\sin^2 x\cos^2 x-\sin^2 x\cos^2 x$
$-4\sin^2 x\sin 2x=0$

$\Rightarrow 2+4\sin 2x=0$

$\Rightarrow \sin 2x=-\dfrac{1}{2}$

$\Rightarrow 2x=\pi+\dfrac{\pi}{6},2\pi-\dfrac{\pi}{6}$

$\Rightarrow 2x=\dfrac{7\pi}{6},\dfrac{11\pi}{6}\Rightarrow x=\dfrac{7\pi}{12},\dfrac{11\pi}{12}$

**37.** (c) $A=\begin{bmatrix}0 & \sin\alpha\\ \sin\alpha & 0\end{bmatrix}$

and $\det\left(A^2-\dfrac{1}{2}I\right)=0$

$$A^2=\begin{bmatrix}0 & \sin\alpha\\ \sin\alpha & 0\end{bmatrix}\begin{bmatrix}0 & \sin\alpha\\ \sin\alpha & 0\end{bmatrix}$$

$$A^2=\begin{bmatrix}\sin^2\alpha & 0\\ 0 & \sin^2\alpha\end{bmatrix}$$

$$\frac{I}{2}=\begin{bmatrix}1/2 & 0\\ 0 & 1/2\end{bmatrix}$$

Now, $\left[A^2-\dfrac{I}{2}\right]$

$$=\begin{bmatrix}\sin^2\alpha-1/2 & 0\\ 0 & \sin^2\alpha-1/2\end{bmatrix}$$

$\because\ \det\left[A^2-\dfrac{1}{2}I\right]=0$

$$\Rightarrow \begin{vmatrix}\sin^2\alpha-1/2 & 0\\ 0 & \sin^2\alpha-1/2\end{vmatrix}=0$$

$$\Rightarrow \left[\sin^2\alpha-\frac{1}{2}\right]^2=0$$

$$\sin\alpha=\pm\frac{1}{\sqrt{2}}$$

$\Rightarrow\quad \alpha=\pi/4$ is one possibility.

**38.** (c)

> **key idea** If $f(\theta)=A\sin\theta+B\cos\theta$, thenthe range of $f(\theta)$ will be $-\sqrt{A^2+B^2}\le f(\theta)\le\sqrt{A^2+B^2}$

Given,

$$f(x)=\begin{vmatrix}\sin^2 x & 1+\cos^2 x & \cos 2x\\ 1+\sin^2 x & \cos^2 x & \cos 2x\\ \sin^2 x & \cos^2 x & \sin 2x\end{vmatrix}$$

Expanding along $R_1$

$f(x)=\sin^2 x(\cos^2 x\sin 2x$
$-\cos^2 x\cos 2x)-(1+\cos^2 x)$
$[(1+\sin^2 x)\sin 2x-\sin^2 x\cos 2x]$
$+\cos 2x[\cos^2 x+\cos^2 x\sin^2 x$
$-\cos^2 x\sin^2 x]$

$f(x) = \sin^2 x \cos^2 x$
$(\sin 2x - \cos 2x) - (1+\cos^2 x)$
$[\sin 2x + \sin^2 x(\sin 2x - \cos 2x)]$
$+ \cos 2x \cos^2 x$

$f(x) = \sin^2 x \cos^2 x(\sin 2x - \cos 2x)$
$- \sin 2x - \sin^2 x(\sin 2x - \cos 2x)$
$- \cos^2 x \sin 2x - \cos^2 x \sin^2 x(\sin 2x - \cos 2x) + \cos 2x \cos^2 x$

$f(x) = -\sin 2x - \sin^2 x(\sin 2x - \cos 2x) - \cos^2 x(\sin 2x - \cos 2x)$

$f(x) = -\sin 2x - (\sin^2 x + \cos^2 x)(\sin 2x - \cos 2x)$

$f(x) = -\sin 2x - \sin 2x + \cos 2x$
$f(x) = -(2\sin 2x - \cos 2x)$

As, we know that, if
$f(\theta) = A\sin\theta + B\cos\theta$
Then,

$$-\sqrt{A^2+B^2} \le f(\theta) \le \sqrt{A^2+B^2}$$

Here, we have, $f(x) = 2\sin 2x - \cos 2x$

$$-\sqrt{2^2+1^2} \le f(x) \le \sqrt{2^2+1^2}$$
$$-\sqrt{5} \le f(x) \le \sqrt{5}$$

So, maximum value of $f(x)$ is $\sqrt{5}$.

**39.** *(d)* Given matrix, $A = \begin{bmatrix} \cos\theta & \sin\theta \\ -\sin\theta & \cos\theta \end{bmatrix}$

So, $A^2 = \begin{bmatrix} \cos\theta & \sin\theta \\ -\sin\theta & \cos\theta \end{bmatrix} \begin{bmatrix} \cos\theta & \sin\theta \\ -\sin\theta & \cos\theta \end{bmatrix}$

$= \begin{bmatrix} \cos 2\theta & \sin 2\theta \\ -\sin 2\theta & \cos 2\theta \end{bmatrix}$

$\therefore \quad A^4 = \begin{bmatrix} \cos 4\theta & \sin 4\theta \\ -\sin 4\theta & \cos 4\theta \end{bmatrix}$

Since, it is given, matrix $B = A + A^4$

$\Rightarrow$

$B = \begin{bmatrix} \cos\theta + \cos 4\theta & \sin\theta + \sin 4\theta \\ -(\sin\theta + \sin 4\theta) & \cos\theta + \cos 4\theta \end{bmatrix}$

$\therefore \det(B) = (\cos\theta + \cos 4\theta)^2 + (\sin\theta + \sin 4\theta)^2$

$= 2 + 2\cos(3\theta)$

So, at $\theta = \pi/5$

$$\det(B) = 2 + 2\cos\frac{3\pi}{5} = 4\cos^2\left(\frac{3\pi}{10}\right)$$

$$= 4\left(\frac{\sqrt{10-2\sqrt{5}}}{4}\right)^2 = \frac{1}{4}(10-2\sqrt{5})$$

$$= \frac{10 - 2(2.236)}{4} \qquad [\because \sqrt{5} \approx 2.236]$$

$$= \frac{10 - 4.472}{4} = \frac{5.528}{4} \in (1, 2)$$

**40.** *(b)*

$$D = \begin{vmatrix} \cos^2 x & 1+\sin^2 x & \sin 2x \\ 1+\cos^2 x & \sin^2 x & \sin 2x \\ \cos^2 x & \sin^2 x & 1+\sin 2x \end{vmatrix}$$

Expanding along $R_1$

$D = \cos^2 x(\sin^2 x + \sin^2 x \sin 2x - \sin^2 x \sin 2x)$
$-(1+\sin^2 x)(1+\cos^2 x + \sin 2x + \sin 2x \cos^2 x - \cos^2 x \sin 2x)$
$+ \sin 2x(\sin^2 x + \sin^2 x \cos^2 x - \cos^2 x \sin^2 x)$

$D = \cos^2 x \sin^2 x - (1+\sin^2 x)$
$(1+\cos^2 x + \sin 2x) + \sin 2x \sin^2 x$

$D = \cos^2 x \sin^2 x - 1 - \cos^2 x - \sin 2x$
$-\sin^2 x - \sin^2 x \cos^2 x - \sin^2 x \sin 2x$

$D = -\cos^2 x - 1 - \sin^2 x - \sin 2x$
$= -(2 + \sin 2x)$

$\therefore$ Minimum value of $D$ is '$m$'
$= -(2+1) = -3$

and maximum value of $D$ is '$M$'
$= -(2-1) = -1$

$\therefore$ The ordered pair $(m, M)$ is $(-3, -1)$.

**41.** *(b)* Given determinant

$$D = \begin{vmatrix} x & a+y & x+a \\ y & b+y & y+b \\ z & c+y & z+c \end{vmatrix}$$

$$= \begin{vmatrix} x & a+y & x \\ y & b+y & y \\ z & c+y & z \end{vmatrix} + \begin{vmatrix} x & a+y & a \\ y & b+y & b \\ z & c+y & c \end{vmatrix}$$

$$= 0 + \begin{vmatrix} x & a & a \\ y & b & b \\ z & c & c \end{vmatrix} + \begin{vmatrix} x & y & a \\ y & y & b \\ z & y & c \end{vmatrix} = y\begin{vmatrix} x & 1 & a \\ y & 1 & b \\ z & 1 & c \end{vmatrix}$$

Applying $C_1 \to C_1 + C_3$, we get

$$D = y\begin{vmatrix} x+a & 1 & a \\ y+b & 1 & b \\ z+c & 1 & c \end{vmatrix}$$

It is given that
$a + x = b + y = c + z + 1 = \lambda$ (let), so

$$D = y\begin{vmatrix} \lambda & 1 & a \\ \lambda & 1 & b \\ \lambda-1 & 1 & c \end{vmatrix}$$

Applying $R_1 \to R_1 - R_3$
and $R_2 \to R_2 - R_3$, we get

$$D = y\begin{vmatrix} 1 & 0 & a-c \\ 1 & 0 & b-c \\ \lambda-1 & 1 & c \end{vmatrix}$$

$= y(-1)[(b-c) - (a-c)] = y(a-b)$

**42.** *(c)* Given determinant is

$$\Delta = \begin{vmatrix} x-2 & 2x-3 & 3x-4 \\ 2x-3 & 3x-4 & 4x-5 \\ 3x-5 & 5x-8 & 10x-17 \end{vmatrix}$$

Expanding along $R_1$

$\Delta = (x-2)[(3x-4)(10x-17) - (4x-5)(5x-8)] - (2x-3)$
$[(2x-3)(10x-17) - (4x-5)(3x-5)]$
$+ (3x-4)[(2x-3)(5x-8) - (3x-4)(3x-5)]$

$\Delta = (x-2)[30x^2 - 91x + 68 - 20x^2 + 57x - 40]$
$- (2x-3)[20x^2 - 64x + 51 - 12x^2 + 35x - 25]$
$+ (3x-4)[10x^2 - 31x + 24 - 9x^2 + 27x - 20]$

$\Delta = (x-2)[10x^2 - 34x + 28] - (2x-3)$
$[8x^2 - 29x + 26] + (3x-4)[x^2 - 4x + 4]$

$\Delta = 2(x-2)[5x^2 - 17x + 14] - (2x-3)$
$(x-2)(8x-13) + (3x-4)(x-2)^2$

$\Delta = 2(x-2)^2(5x-7) - (2x-3)$
$(x-2)(8x-13) + (3x-4)(x-2)^2$

$\Delta = (x-2)^2[10x - 14 + 3x - 4] - (2x-3)$
$(x-2)(8x-13)$

$\Delta = (x-2)^2(13x-18) - (2x-3)$
$(x-2)(8x-13)$

$\Delta = (x-2)[(x-2)(13x-18)$
$-(2x-3)(8x-13)]$

$\Delta = (x-2)[-3x^2 + 6x - 3]$
$= -3x^3 + 12x^2 - 15x + 6$
$= Ax^3 + Bx^2 + Cx + D$ [given]

On comparing the coefficient of different terms, we get
$A = -3, B = 12, C = -15$ and $D = 6$
$\therefore B + C = -3$

**43.** *(a)* Given matrix

$$A = \begin{bmatrix} 1 & \sin\theta & 1 \\ -\sin\theta & 1 & \sin\theta \\ -1 & -\sin\theta & 1 \end{bmatrix}$$

$$\Rightarrow \det(A) = |A| = \begin{vmatrix} 1 & \sin\theta & 1 \\ -\sin\theta & 1 & \sin\theta \\ -1 & -\sin\theta & 1 \end{vmatrix}$$

$=1(1+\sin^2\theta)-\sin\theta(-\sin\theta+\sin\theta)+1(\sin^2\theta+1)$

$\Rightarrow |A|=2(1+\sin^2\theta)$ ...(i)

As we know that, for $\theta \in \left(\frac{3\pi}{4}, \frac{5\pi}{4}\right)$

$\sin\theta \in \left(-\frac{1}{\sqrt{2}}, \frac{1}{\sqrt{2}}\right)$

$\Rightarrow \sin^2\theta \in \left[0, \frac{1}{2}\right)$

$\Rightarrow 1+\sin^2\theta \in \left[0+1, \frac{1}{2}+1\right)$

$\Rightarrow 1+\sin^2\theta \in \left[1, \frac{3}{2}\right)$

$\Rightarrow 2(1+\sin^2\theta) \in [2, 3)$

$\Rightarrow |A| \in [2,3) \subset \left(\frac{3}{2}, 3\right]$

**44.** *(b)* Let

$$\Delta = \begin{vmatrix} a-b-c & 2a & 2a \\ 2b & b-c-a & 2b \\ 2c & 2c & c-a-b \end{vmatrix}$$

Applying $R_1 \to R_1+R_2+R_3$, we get

$$\Delta = \begin{vmatrix} a+b+c & a+b+c & a+b+c \\ 2b & b-c-a & 2b \\ 2c & 2c & c-a-b \end{vmatrix}$$

$$=(a+b+c)\begin{vmatrix} 1 & 1 & 1 \\ 2b & b-c-a & 2b \\ 2c & 2c & c-a-b \end{vmatrix}$$

(Taking common $(a+b+c)$ from $R_1$)

Applying $C_2 \to C_2 - C_1$ and $C_3 \to C_3 - C_1$, we get

$$\Delta=(a+b+c)\begin{vmatrix} 1 & 0 & 0 \\ 2b & -(a+b+c) & 0 \\ 2c & 0 & -(a+b+c) \end{vmatrix}$$

Now, expanding along $R_1$, we get

$\Delta=(a+b+c)\{(a+b+c)^2-0\}$

$=(a+b+c)^3=(a+b+c)(x+a+b+c)^2$ [Given]

$\Rightarrow (x+a+b+c)^2=(a+b+c)^2$

$\Rightarrow x+a+b+c=\pm(a+b+c)$

$\Rightarrow x=-2(a+b+c)$ $[\because x \neq 0]$

**45.** *(a)*

$$A=\begin{bmatrix} -2 & 4+d & \sin\theta-2 \\ 1 & (\sin\theta)+2 & d \\ 5 & (2\sin\theta)-d & (-\sin\theta)+2+2d \end{bmatrix}$$

Expanding along $R_1$

$A=-2[(\sin\theta+2)(-\sin\theta+2+2d)-(2\sin\theta-d)d]$
$-(4+d)[-\sin\theta+2+2d-5(d)]$
$+(\sin\theta-2)[2\sin\theta-d-5(\sin\theta+2)]$

$=-2[-\sin^2\theta+2\sin\theta+2d\sin\theta-2\sin\theta+4+4d-2d\sin\theta+d^2]$
$-(d+4)[-\sin\theta+2+2d-5d]$
$+(\sin\theta-2)[2\sin\theta-d-5\sin\theta-10]$

$=-2[-\sin^2\theta+4+4d+d^2]-(4+d)[-\sin\theta+2-3d]+(\sin\theta-2)[-3\sin\theta-10-d]$

$=2\sin^2\theta-8-8d-2d^2+4\sin\theta-8+12d+d\sin\theta-2d+3d^2-3\sin^2\theta-10\sin\theta-d\sin\theta+6\sin\theta+20+2d$

$=-\sin^2\theta+4+4d+d^2$

$=-\sin^2\theta+(d+2)^2$

$A=(d+2)^2-\sin^2\theta$

Note that $|A|$ will be minimum if $\sin^2\theta$ is maximum i.e. if $\sin^2\theta$ takes value 1.

$\because |A|_{\min}=8,$

therefore $(d+2)^2-1=8$

$\Rightarrow (d+2)^2=9 \Rightarrow d+2=\pm 3$

$\Rightarrow d=1, -5$

**46.** *(d)* If $A=\begin{vmatrix} 0 & \cos x & -\sin x \\ \sin x & 0 & \cos x \\ \cos x & \sin x & 0 \end{vmatrix}$

$\therefore |A|=0(0-\cos x\sin x)-\cos x(0-\cos^2 x)-\sin x(\sin^2 x-0)=0$

$\Rightarrow \cos^3 x-\sin^3 x=0$

$\Rightarrow \tan^3 x=1$

$\Rightarrow \tan x=1$

$\sum_{x\in S}\tan\left(\frac{\pi}{3}+x\right)=$

$\sum_{x\in I}\frac{\sqrt{3}+\tan x}{1-\sqrt{3}\tan x}=\frac{\sqrt{3}+1}{1-\sqrt{3}}\times\frac{1+\sqrt{3}}{1+\sqrt{3}}$

(by rationalising denominator) $[\because \tan x=1]$

$=\frac{1+3+2\sqrt{3}}{-2}=\frac{4}{-2}-\frac{2\sqrt{3}}{2}$

$=-2-2\sqrt{3}$

**47.** *(d)* Given, $A=\begin{bmatrix} -4 & -1 \\ 3 & 1 \end{bmatrix}$

$\Rightarrow A^2=\begin{bmatrix} -4 & -1 \\ 3 & 1 \end{bmatrix}\begin{bmatrix} -4 & -1 \\ 3 & 1 \end{bmatrix}$

$=\begin{bmatrix} 13 & 3 \\ -9 & -2 \end{bmatrix}$

and $|A|=-1$

So, $A^{2016}-2A^{2015}-A^{2014}$

$=A^{2014}(A^2-2A-I)$

$\therefore |A^{2016}-2A^{2015}-A^{2014}|$

$=|A^{2014}||A^2-2A-I|$

$=|A|^{2014}\begin{vmatrix} 20 & 5 \\ -15 & -5 \end{vmatrix}$

$=(-1)^{2014}(-100+75)=-25$

**48.** *(c)* $\begin{vmatrix} a & a+1 & a-1 \\ -b & b+1 & b-1 \\ c & c-1 & c+1 \end{vmatrix}+(-1)^n\begin{vmatrix} a+1 & b+1 & c-1 \\ a-1 & b-1 & c+1 \\ a & -b & c \end{vmatrix}$

$=\begin{vmatrix} a & a+1 & a-1 \\ -b & b+1 & b-1 \\ c & c-1 & c+1 \end{vmatrix}+(-1)^n\begin{vmatrix} a+1 & a-1 & a \\ b+1 & b-1 & -b \\ c-1 & c+1 & c \end{vmatrix}$ $[\because |A|=|A^T|]$

$=\begin{vmatrix} a & a+1 & a-1 \\ -b & b+1 & b-1 \\ c & c-1 & c+1 \end{vmatrix}+(-1)^{n+1}\begin{vmatrix} a+1 & a & a-1 \\ b+1 & -b & b-1 \\ c-1 & c & c+1 \end{vmatrix}$ $[C_2 \leftrightarrow C_3]$

$=\begin{vmatrix} a & a+1 & a-1 \\ -b & b+1 & b-1 \\ c & c-1 & c+1 \end{vmatrix}+(-1)^{n+2}\begin{vmatrix} a & a+1 & a-1 \\ -b & b+1 & b-1 \\ c & c-1 & c+1 \end{vmatrix}$ $[C_1 \leftrightarrow C_2]$

$=[1+(-1)^{n+2}]\begin{vmatrix} a & a+1 & a-1 \\ -b & b+1 & b-1 \\ c & c-1 & c+1 \end{vmatrix}$

This is equal to zero only, if $n+2$ is odd i.e., $n$ is an odd integer.

**49.** (36) Let $A = \begin{pmatrix} a & b \\ c & d \end{pmatrix}$

for $A$ to be singular matrix

$ad = bc$

Case I Exactly 1 number in used

$\Rightarrow {}^4C_1$ ways

Case II Exactly 2 numbers are used

$\Rightarrow {}^4C_2$ ways

Case III Exactly 3 numbers are used

$\Rightarrow$ none will be singular

Case IV Exactly 4 number are used

$\Rightarrow ab = cd \Rightarrow 2 \times 9 = 3 \times 6$

$\Rightarrow {}^4C_1 \times 2! = 8$ matrices

$\therefore$ Total ways $= 4 + 6 \times 4 + 8 = 36$ matrices.

**50.** (6) $D_k = \begin{vmatrix} 1 & 2k & 2k-1 \\ n & n^2+n+2 & n^2 \\ n & n^2+n & n^2+n+2 \end{vmatrix}$

$$\sum_{k=1}^{n} D_k = \begin{vmatrix} \underset{n-\text{times}}{1+1+\ldots+1} & 2(1+2+3+\ldots+n) & \underset{n-\text{times}}{1+3+5+\ldots} \\ n & n^2+n+2 & n^2 \\ n & n^2+n & n^2+n+2 \end{vmatrix}$$

$$\Rightarrow 96 = \begin{vmatrix} n & n(n+1) & n^2 \\ n & n^2+n+2 & n^2 \\ n & n^2+n & n^2+n+2 \end{vmatrix}$$

Applying $R_1 \to R_1 - R_3$, $R_2 \to R_2 - R_3$,

$$96 = \begin{vmatrix} 0 & 0 & -n-2 \\ 0 & 2 & -n-2 \\ n & n^2+n & n^2+n+2 \end{vmatrix}$$

$\Rightarrow \quad 96 = (-n-2)\,(0 \cdot (n^2+n) - 2n)$

$\Rightarrow \quad 96 = 2n(n+2)$

$\Rightarrow \quad n^2 + 2n - 48 = 0$

$\Rightarrow \quad (n+8)(n-6) = 0 \Rightarrow n = -8, 6$

But $n \in N$, $\therefore n = 6$

**51.** (495) $A_1, A_2, A_3 \to$ AP

$A_1 = A$, $A_2 = A+1$, $A_3 = A+2$,

common difference $= d$

$a =$ 7th term of $A_1 = A + 6d$

$b =$ 9th term of $A_2 = A + 1 + 8d$

$c =$ 17th term of $A_3 = A + 2 + 16d$

$$\Rightarrow \begin{vmatrix} a & 7 & 1 \\ 2b & 17 & 1 \\ c & 17 & 1 \end{vmatrix} + 70 = 0$$

$$\begin{vmatrix} A+6d & 7 & 1 \\ 2(A+1+8d) & 17 & 1 \\ A+2+16d & 17 & 1 \end{vmatrix} + 70 = 0$$

Expanding along $C_3$

$1[34(A+1+8d) - 17(A+2+16d)]$
$-1[17A + 102d - 7A - 14 - 112d]$
$+1[17A + 102d - 14A - 14 - 112d]$
$+70 = 0$

$[34A + 34 + 272d - 17A - 34 - 272d]$
$-[10A - 10d - 14]$
$+[3A - 10d - 14] + 70 = 0$

$17A - 10A + 10d + 14 + 3A - 10d$
$-14 + 70 = 0$

$\Rightarrow \quad 10A + 70 = 0$

$\Rightarrow \quad A = -7$

Given, $\quad a = 29$

$A + 6d = 29 \Rightarrow -7 + 6d = 29$

$\Rightarrow \quad d = 6$

New AP

First term $= A' = c - a - b$

$= (A + 2 + 16d) - (A + 6d) - (A + 1 + 8d)$

$= -A + 1 + 2d = 7 + 1 + 12 = 20$

Common difference, $D = \dfrac{d}{12} = \dfrac{6}{12} = \dfrac{1}{2}$

$\therefore$Sum of first 20 terms $= \dfrac{20}{2}[2A' + 19D]$

$= 10\left[40 + \dfrac{19}{2}\right] = 495$

**52.** (4) $\Delta = \begin{vmatrix} P! & (P+1)! & (P+2)! \\ (P+1)! & (P+2)! & (P+3)! \\ (P+2)! & (P+3)! & (P+4)! \end{vmatrix}$

Expanding along $R_1$

$\Delta = P![(P+2)!(P+4)!$
$-(P+3)!(P+3)!]$
$-(P+1)![(P+1)!(P+4)!$
$-(P+2)!(P+3)!]$
$+(P+2)![(P+1)!(P+3)!$
$-(P+2)!(P+2)!]$

$\Delta = P!(P+2)!(P+3)![P+4-P-3]$
$-(P+1)!(P+1)!(P+3)!$
$[P+4-P-2] + (P+2)!(P+1)!(P+2)!$
$[P+3-P-2]$

$\Delta = P!(P+2)!(P+3)! - 2(P+1)!$
$(P+1)!(P+3)! + (P+2)!(P+1)!(P+2)!$

$\Delta = P!(P+1)!(P+2)![(P+3)(P+2)$
$-2(P+1)(P+3) + (P+2)(P+1)]$

$\Delta = P!(P+1)!(P+2)![(P+3)$
$(P+2-2P-2) + (P+2)(P+1)]$

$\Delta = P!(P+1)!(P+2)![-P(P+3)$
$+(P+2)(P+1)]$

$\Delta = P!(P+1)!(P+2)![-P^2 - 3P + P^2$
$+3P + 2]$

$\Delta = 2P!(P+1)!(P+2)!$

$\Delta$ is divisible by $P^3$ and $(P+2)$

So, $\alpha = 3$, $\beta = 1$, $\alpha + \beta = 4$

**53.** (6)

$$f(x) = \begin{vmatrix} \sin^2 x & -2 + \cos^2 x & \cos 2x \\ 2 + \sin^2 x & \cos^2 x & \cos 2x \\ \sin^2 x & \cos^2 x & 1 + \cos 2x \end{vmatrix}$$

Expanding along $R_1$

$f(x) = \sin^2 x[\cos^2 x + \cos^2 x \cos 2x$
$-\cos^2 x \cos 2x] + (2 - \cos^2 x)$
$[2 + 2\cos 2x + \sin^2 x + \sin^2 x \cos 2x$
$-\sin^2 x \cos 2x] + \cos 2x[2\cos^2 x$
$+\cos^2 x \sin^2 x - \cos^2 x \sin^2 x]$

$f(x) = \sin^2 x[\cos^2 x] + (2 - \cos^2 x)$
$[2 + 2\cos 2x + \sin^2 x] + \cos 2x[2\cos^2 x]$

$f(x) = \sin^2 x \cos^2 x + 4 + 4\cos 2x$
$+2\sin^2 x - 2\cos^2 x$
$-2\cos^2 x \cos 2x - \cos^2 x \sin^2 x$
$+2\cos 2x \cos^2 x$

$f(x) = 4 + 4\cos 2x + 2\sin^2 x - 2\cos^2 x$

$f(x) = 4 + 4(1 - 2\sin^2 x) + 2\sin^2 x$
$-2(1 - \sin^2 x)$

$f(x) = 4 + 4 - 8\sin^2 x + 2\sin^2 x - 2$
$+2\sin^2 x$

$f(x) = 6 - 4\sin^2 x$

$f(x) = 2(3 - 2\sin^2 x)$

For $f(x)$ to be maximum, $\sin^2 x$ should be minimum. Minimum value of $\sin^2 x$ is 0.

$\therefore [f(x)]_{\max} = 2(3 - 0) = 6$

**54.** (16) $A = \begin{bmatrix} a & b \\ c & d \end{bmatrix}$, $|A| = ad - bc$

$a, b, c, d = \{-3, -2, -1, 0, 1, 2, 3\}$

Case I $ad = 9$ and $bc = -6$

$$\left\{\begin{matrix} a = 3, & d = 3 \\ a = -3, & d = -3 \end{matrix}\right. \text{ and } \left\{\begin{matrix} b = 2, & c = -3 \\ b = -2, & c = 3 \\ b = 3, & c = -2 \\ b = -3, & c = 2 \end{matrix}\right.$$

Total possible pairs $= 4 \times 2 = 8$

Case II $ad = 6$ and $bc = -9$

$$\begin{cases} a=2, & d=3 \\ a=-2, & d=-3 \\ a=3, & d=2 \\ a=-3, & d=-2 \end{cases}$$

and $\begin{cases} b=-3, & c=3 \\ b=3, & c=-3 \end{cases}$

Total possible pairs $= 4\times 2 = 8$

Case I + Case II $= 8+8=16$

**55.** *(1)* Given $a, b, c, d$ are in AP and $\lambda$ is common difference.

$\therefore b=a+\lambda, c=a+2\lambda, d=a+3\lambda$

$$\begin{vmatrix} x+a-c & x+b & x+a \\ x-1 & x+c & x+b \\ x-b+d & x+d & x+c \end{vmatrix}=2$$

Applying $C_2 \to C_2 - C_3$

$$\begin{vmatrix} x+a-c & b-a & x+a \\ x-1 & c-b & x+b \\ x-b+d & d-c & x+c \end{vmatrix}=2$$

$$\Rightarrow \begin{vmatrix} x-2\lambda & \lambda & x+a \\ x-1 & \lambda & x+b \\ x+2\lambda & \lambda & x+c \end{vmatrix}=2$$

$[\because b-a=c-b=d-c=\lambda;$
$c-a=d-b=2\lambda]$

$$\Rightarrow \lambda\begin{vmatrix} x-2\lambda & 1 & x+a \\ x-1 & 1 & x+b \\ x+2\lambda & 1 & x+c \end{vmatrix}=2$$

Applying

$R_2 \to R_2 - R_1, R_3 \to R_3 - R_1$

$$\lambda\begin{vmatrix} x-2\lambda & 1 & x+a \\ 2\lambda-1 & 0 & b-a \\ 4\lambda & 0 & c-a \end{vmatrix}=2$$

$$\Rightarrow \lambda\begin{vmatrix} x-2\lambda & 1 & x+a \\ 2\lambda-1 & 0 & \lambda \\ 4\lambda & 0 & 2\lambda \end{vmatrix}=2$$

$\Rightarrow \quad -\lambda[2\lambda(2\lambda-1)-4\lambda^2]=2$

$\Rightarrow \quad 2\lambda^2=2 \Rightarrow \lambda^2=1$

**56.** *(2)* Given 1, $\log_{10}(4^x-2)$, $\log_{10}\left(4^x+\frac{18}{5}\right)$ are in AP.

$\therefore 2\log_{10}(4^x-2)=1+\log_{10}\left(4^x+\frac{18}{5}\right)$

$=\log_{10}10+\log_{10}\left(4^x+\frac{18}{5}\right)$

$\Rightarrow \log_{10}(4^x-2)^2$

$=\log_{10}\left(10\times\left(4^x+\frac{18}{5}\right)\right)$

$\Rightarrow \quad (4^x-2)^2=10\times 4^x+36$

$\Rightarrow \quad (4^x)^2-4(4^x)+4=10\times 4^x+36$

$\Rightarrow (4^x)^2-14(4^x)-32=0$

$\Rightarrow \quad (4^x-16)(4^x+2)=0$

$\Rightarrow \quad 4^x=16$ or $4^x=-2$

(Rejected because $4^x>0, \forall x\in R$)

$\Rightarrow \quad 4^x=4^2 \Rightarrow x=2$

$$\therefore \begin{vmatrix} 2\left(x-\frac{1}{2}\right) & x-1 & x^2 \\ 1 & 0 & x \\ x & 1 & 0 \end{vmatrix} = \begin{vmatrix} 3 & 1 & 4 \\ 1 & 0 & 2 \\ 2 & 1 & 0 \end{vmatrix} = 2$$

**Alternate Solution**

$2\log_{10}(4^x-2)^2=\log_{10}\left(4^x+\frac{18}{5}\right)+l$

$\Rightarrow \log_{10}(4^x-2)^2=\log_{10}\left(10\times\left(4^x+\frac{18}{5}\right)\right)$

$\Rightarrow (4^x-16)(4^x+2)=0$

$\Rightarrow 4^x=16$ or $4^x=-2$ (rejected $4^x>0$)

$x=2$

$$\therefore \begin{vmatrix} 2\left(x-\frac{1}{2}\right) & x-1 & x^2 \\ 1 & 0 & x \\ x & 1 & 0 \end{vmatrix} = \begin{vmatrix} 3 & 1 & 4 \\ 1 & 0 & 2 \\ 2 & 1 & 0 \end{vmatrix}=2$$

**57.** *(2020)* **Method I**

Given, $A=\begin{bmatrix} a & b \\ c & d \end{bmatrix}, B=\begin{bmatrix} \alpha \\ \beta \end{bmatrix} \neq \begin{bmatrix} 0 \\ 0 \end{bmatrix}$ i.e.

$B\neq 0$ and $AB=B$

$\Rightarrow AB-B=0 \Rightarrow B(A-I)=0$

$\Rightarrow \quad |(A-I)B|=0$

$\because \quad B\neq 0$

$\therefore \quad |A-I|=0$

$$\Rightarrow \begin{vmatrix} (a-1) & b \\ c & (d-1) \end{vmatrix}=0$$

$\Rightarrow \quad (a-1)(d-1)-bc=0$

$\Rightarrow \quad ad-(a+d)+1-bc=0$

$\Rightarrow \quad ad-2021+1-bc=0$

$[\because a+d=2021]$

$\Rightarrow \quad ad-bc=2020$

**Method II**

Given, $\quad A=\begin{bmatrix} a & b \\ c & d \end{bmatrix}, B=\begin{bmatrix} \alpha \\ \beta \end{bmatrix}$

$AB=B$

$$\Rightarrow \begin{bmatrix} a & b \\ c & d \end{bmatrix}\begin{bmatrix} \alpha \\ \beta \end{bmatrix}=\begin{bmatrix} \alpha \\ \beta \end{bmatrix} \Rightarrow \begin{bmatrix} a\alpha+b\beta \\ c\alpha+d\beta \end{bmatrix}=\begin{bmatrix} \alpha \\ \beta \end{bmatrix}$$

$\Rightarrow \quad a\alpha+b\beta=\alpha$

$\Rightarrow \quad \alpha(a-1)=-b\beta$ ... (i)

and $c\alpha+d\beta=\beta \Rightarrow c\alpha=\beta(1-d)$ ... (ii)

On dividing Eq. (i) by Eq. (ii), we get

$$\frac{a-1}{c}=\frac{-b}{1-d}$$

$\Rightarrow \quad (a-1)(1-d)=-bc$

$\Rightarrow \quad a-1-ad+d=-bc$

$\Rightarrow \quad a+d-1=ad-bc$

$\Rightarrow \quad 2021-1=ad-bc$

$\Rightarrow \quad ad-bc=2020$

## Topic 2

**01.** *(d)* $A(a, 0), B(b, 2b+1), C(0, b)$

$$\text{Area of } \Delta ABC=\frac{1}{2}\begin{vmatrix} a & 0 & 1 \\ b & 2b+1 & 1 \\ 0 & b & 1 \end{vmatrix}=\pm 1$$

$\Rightarrow \frac{1}{2}[a(b+1)+b^2]=\pm 1$

$\Rightarrow a=\frac{2-b^2}{b+1}$ or $\frac{-2-b^2}{b+1}$

Sum of all possible values of

$a=\frac{-2b^2}{b+1}$

**02.** *(c)* Given that the area of triangle whose vertices are $A(1,\alpha)$ $B(\alpha, 0), C(0,\alpha)$ is 4 (unit)$^2$.

$$\text{Thus, } \frac{1}{2}\begin{vmatrix} 1 & \alpha & 1 \\ \alpha & 0 & 1 \\ 0 & \alpha & 1 \end{vmatrix}=4$$

$\Rightarrow |1(0-\alpha)-\alpha(\alpha)+\alpha^2|=8$

$\Rightarrow \quad \alpha=\pm 8$

Take, $\alpha=8$

Thus, $(8,-8), (-8,8)$ and $(64,\beta)$ are collinear.

$$\text{Then, } \begin{vmatrix} 8 & -8 & 1 \\ -8 & 8 & 1 \\ 64 & \beta & 1 \end{vmatrix}=0$$

$\Rightarrow 8(8-\beta)+8(-8-64)+1$

$(-8\beta-512)=0$

$\Rightarrow 8-\beta-72-\beta-64=0$

[Divided both side by 8]

$\Rightarrow \quad -2\beta=128 \quad \Rightarrow \quad \beta=-64$

Similarly, take $\alpha=-8$, then also $\beta=-64$

Hence, the value of $\beta$ is $-64$.

**03.** (d) Given, vertices of triangle are $(k, -3k)$, $(5, k)$ and $(-k, 2)$.

$$\therefore \frac{1}{2}\begin{vmatrix} k & -3k & 1 \\ 5 & k & 1 \\ -k & 2 & 1 \end{vmatrix} = \pm 28$$

$$\Rightarrow \begin{vmatrix} k & -3k & 1 \\ 5 & k & 1 \\ -k & 2 & 1 \end{vmatrix} = \pm 56$$

$$\Rightarrow k(k-2) + 3k(5+k) + 1(10+k^2) = \pm 56$$

$$\Rightarrow k^2 - 2k + 15k + 3k^2 + 10 + k^2 = \pm 56$$

$$\Rightarrow 5k^2 + 13k + 10 = \pm 56$$

$$\Rightarrow 5k^2 + 13k + 66 = 0$$

or $5k^2 + 13k - 46 = 0$

$$\Rightarrow k = 2 \quad [\because k \in I]$$

Thus, the coordinates of vertices of the triangle are $A(2, -6)$, $B(5, 2)$ and $C(-2, 2)$.

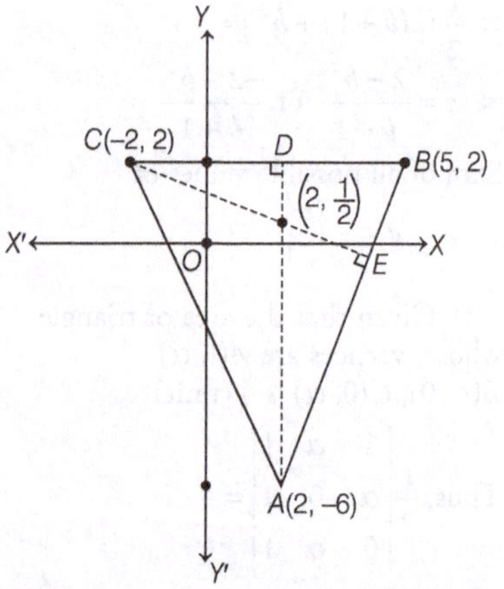


Now, equation of altitude from vertex $A$ is

$$y - (-6) = \frac{-1}{\left(\frac{2-2}{-2-5}\right)}(x-2)$$

$$\Rightarrow x = 2 \quad \ldots(i)$$

Equation of altitude from vertex $C$ is

$$y - 2 = \frac{-1}{\left[\frac{2-(-6)}{5-2}\right]}[x - (-2)]$$

$$\Rightarrow 3x + 8y - 10 = 0 \quad \ldots(ii)$$

On solving Eqs. (i) and (ii), we get $x = 2$ and $y = 1/2$.

$\therefore$ Orthocentre $= \left(2, \frac{1}{2}\right)$

## Topic 3

**01.** (d) $|\text{adj}(\text{adj}(\text{adj}\, A))| = 81$

$$\Rightarrow |A|^{(n-1)^3} = 3^4$$

$$\Rightarrow |A|^8 = 3^4$$

$$\Rightarrow |A| = 3^{\frac{1}{2}}$$

$$|\text{adj}(\text{adj}\, A)|^{\frac{(n-1)^2}{2}} = |A|^{3n^2-5n-4}$$

$$|A|^{2(n-1)^2} = |A|^{3n^2-5n-4}$$

$$\Rightarrow 2(n-1)^2 = 3n^2 - 5n - 4$$

$$\Rightarrow n^2 - n - 6 = 0$$

$$\Rightarrow n = -2, 3$$

$$\sum_{n \in S} |A^{n^2+n}| = |A^2| + |A^{12}|$$

$$= 3 + 3^6 = 732$$

**02.** (a) $[A(\text{adj}(A^{-1}) + \text{adj}(B^{-1}))^{-1} B]^{-1}$

$B^{-1}(\text{adj}(A^{-1}) + \text{adj}(B^{-1})A^{-1}$

$B^{-1}\text{adj}(A^{-1})A^{-1} + B^{-1}(\text{adj}(B^{-1}))A^{-1}$

$B^{-1}|A^{-1}|I + |B^{-1}|IA^{-1}$

$$\frac{B^{-1}}{|A|} + \frac{A^{-1}}{|B|} = \frac{\text{adj}B}{|B||A|} + \frac{\text{adj}A}{|A||B|}$$

$$= \frac{1}{|A||B|}(\text{adj}B + \text{adj}A)$$

> • Concept **Enhancer**
>
> (i) $(ABC)^{-1} = C^{-1}B^{-1}A^{-1}$
>
> (ii) $(A^{-1})^{-1} = A$
>
> (iii) $A\ \text{adj}\ A = |A|I$

**03.** (b) $|A| = \frac{1}{2} tr(A) = 3, n = 3$

$$B = \text{adj}\ (\text{adj}\ 2A)$$

$$= |2A|^{n-2}\ (2A)$$

$$= |2A|(2A) = 2^3\,|A| \cdot 2A$$

$$= 2^4 \times \frac{1}{2} \times A = 2^3 A$$

$$|B| = |2^3 A| = (2^3)^3\,|A| = 2^8 = 256$$

$$trace(B) = 2^3\ trace(A) = 2^3 \times 3 = 24$$

$$\therefore |B| + trace(B) = 256 + 24 = 280$$

**04.** (b) $B = \begin{bmatrix} 1 & 3 \\ 1 & 5 \end{bmatrix}$

$BCB^{-1} = A$ and $(BCB^{-1})(BCB^{-1}) = A^2$

$$BC^2B^{-1} = A^2 \quad \ldots(i)$$

and $AB^{-1} = A^{-1}$

$$B^{-1} = A^{-1} \cdot A^{-1}$$

(Pre multiply by $A^{-1}$)

In Eq. (i) Pre multiply of $B^{-1}$ and post multiply by $B$, we get

$$C^2 = B^{-1}AAB = A^{-1}A^{-1} \cdot AAB$$

$$C^2 = B$$

$$\therefore C^4 + \alpha C^2 + \beta I = 0$$

$$\Rightarrow B^2 + \alpha B + \beta I = 0$$

which is characteristics equation of $B$.

Now, $|B - \lambda I| = 0$

$$\Rightarrow \begin{vmatrix} 1-\lambda & 3 \\ 1 & 5-\lambda \end{vmatrix} = 0$$

$$\Rightarrow 5 + \lambda^2 - 6\lambda - 3 = 0$$

$$\lambda^2 - 6\lambda + 2 = 0$$

$$\therefore B^2 - 6B + 2I = 0$$

$$\alpha = -6, \beta = 2$$

$$\therefore 2\beta - \alpha = 4 + 6 = 10$$

**05.** (a) $A = \begin{bmatrix} 1 & 2 \\ 0 & 1 \end{bmatrix} \Rightarrow \text{Adj}\ A = \begin{bmatrix} 1 & -2 \\ 0 & 1 \end{bmatrix}$

$$(\text{Adj}\ A)^2 = \begin{bmatrix} 1 & -4 \\ 0 & 1 \end{bmatrix}$$

$$\vdots\ (\text{Adj}\ A)^{10} = \begin{bmatrix} 1 & -20 \\ 0 & 1 \end{bmatrix}$$

$$\therefore B = I + \text{adj}\,A + (\text{adj}\ A)^2 + \ldots + (\text{adj}\ A)^{10}$$

$$= \begin{bmatrix} 1 & 0 \\ 0 & 1 \end{bmatrix} + \begin{bmatrix} 1 & -2 \\ 0 & 1 \end{bmatrix} + \begin{bmatrix} 1 & -4 \\ 0 & 1 \end{bmatrix} + \ldots \begin{bmatrix} 1 & -20 \\ 0 & 1 \end{bmatrix}$$

$$= \begin{bmatrix} 11 & -110 \\ 0 & 11 \end{bmatrix}$$

Sum of elements of $B = -88$

**06.** (c) We have, $A = \frac{1}{5!6!7!}\begin{bmatrix} 5! & 6! & 7! \\ 6! & 7! & 8! \\ 7! & 8! & 9! \end{bmatrix}$

$$|A| = \frac{1}{5!6!7!} \times 5!6!7! \begin{vmatrix} 1 & 6 & 42 \\ 1 & 7 & 56 \\ 1 & 8 & 72 \end{vmatrix}$$

Applying $R_3 \to R_3 - R_2$ and $R_2 \to R_2 - R_1$,

$$|A| = \begin{vmatrix} 1 & 6 & 42 \\ 0 & 1 & 14 \\ 0 & 1 & 16 \end{vmatrix} = 2$$

$$|\text{adj adj}\ (2A)| = |2A|^{(n-1)^2}$$

$$= |2A|^4 = (2^3\,|A|)^4$$

$$= 2^{12}\,|A|^4 = 2^{16}$$

**07.** (c) Given that $A$ is $3\times 3$ matrix and $|A|=2$

Now, $|3\text{adj}(|3A|A^2)|$

$=3^3|\text{adj}(|3A|A^2)|$

$=3^3|\text{adj}(54A^2)|$ $\quad [\because |kA|=k^3|A|]$

$=3^3|54A^2|^2=3^3\times(54^3)^2\times|A|^4$

$=3^3\times(54)^6\times 2^4$ $\quad [\because |A|=2 \text{ given}]$

$=3^3\times(3^3\times 2)^6\times 2^4$

$=3^{11}\times 3^{10}\times 2^{10}=3^{11}\times 6^{10}$

**08.** (b) $|\text{adj}(\text{adj}(\text{adj}\,2A))|$

$=|2A|^{(n-1)^3}=|2A|^8=(2^3|A|)^8$

$=2^{24}|A|^8$

Now, $A=\begin{bmatrix}2&1&0\\1&2&-1\\0&-1&2\end{bmatrix}$

$|A|=4$

$\therefore |\text{adj}(\text{adj}(\text{adj}\,2A))|=2^{24}\cdot 4^8$

$=2^{24}\cdot 2^{16}$

$=2^{40}=(16)^{10}=(16)^n$ $\quad$ [given]

$\therefore \quad n=10$

**09.** (b) Given that $P^T=aP+(a-1)I$ $\quad$ ...(i)

$\Rightarrow \quad (P^T)^T=aP^T+(a-1)I^T$

$\Rightarrow \quad P=aP^T+(a-1)I$ $\quad$ ...(ii)

$[\because (A^T)^T=A \text{ and } I^T=I]$

From Eq. (i), putting the value of $P^T$ in Eq. (ii), we get

$P=a[aP+(a-1)I]+(a-1)I$

$\Rightarrow \quad P=a^2P+a(a-1)I+(a-1)I$

$\Rightarrow (1-a^2)P=(a-1)(a+1)I$

$\Rightarrow \quad (1-a^2)P=(a^2-1)I$

$\therefore \quad P=-I$

So, $\quad \text{Adj}\,P=I$

$\Rightarrow \quad |\text{Adj}\,P|=1$

**10.** (c) Given,

$$A=\begin{bmatrix}1&\log_x y&\log_x z\\ \log_y x&2&\log_y z\\ \log_z x&\log_z y&3\end{bmatrix}$$

$$|A|=\begin{vmatrix}1&\dfrac{\log y}{\log x}&\dfrac{\log z}{\log x}\\ \dfrac{\log x}{\log y}&2&\dfrac{\log z}{\log y}\\ \dfrac{\log x}{\log z}&\dfrac{\log y}{\log z}&3\end{vmatrix}$$

Expanding along $R_1$

$$|A|=1(6-1)-\frac{\log y}{\log x}\left(\frac{3\log x}{\log y}-\frac{\log x}{\log y}\right)+\frac{\log z}{\log x}\left(\frac{\log x}{\log z}-\frac{2\log x}{\log z}\right)$$

$|A|=5-2-1=2$

We know that, $|\text{adj}(\text{adj}(A))|=|A|^{(n-1)^2}$

So, $|\text{adj}(\text{adj}A^2)|=|A^2|^{(3-1)^2}$

$=|A^2|^4=|A|^8=2^8$

> **Concept Enhancer**
>
> By using the change of base formula, the logarithmic expression "$\log_a x$" can be written as,
>
> $\dfrac{\log_b x}{\log_b a}$, where $a, b, x>0$ and $a, b\neq 1$.

**11.** (c) For $3\times 3$ matrix, $n=3$

Given, $|\text{adj (adj (adj } A))|=12^4$

$|A|^{(n-1)^3}=12^4$

$|A|^{2^3}=12^4$

$\Rightarrow \quad |A|^8=12^4 \Rightarrow |A|=\sqrt{12}=2\sqrt{3}$

$\therefore \quad |A|=2\sqrt{3}$

$\therefore |A^{-1}\text{adj}\,A|=\left|\dfrac{1}{|A|}\right|\cdot|\text{adj}\,A|=\dfrac{|A|^2}{|A|}=|A|$

$=2\sqrt{3}$

**12.** (a) Given, $|A|=2$

Now, $||A|\,\text{adj}\,(5\,\text{adj}\,A^3)|$

$=|5^2\cdot|A|\cdot\text{adj (adj } A^3)|$

$=25^3\cdot|A|^3\cdot|\text{adj}(A)^3|^2$

$=25^3\cdot 2^3\cdot|A^3|^4$

$=25^3\cdot 2^3\cdot 2^{12}=(25\times 2\times 2)^3\times 2^9$

$=10^6\cdot 512=512\times 10^6$

**13.** (d) Let $C=A^2-B^2, |C|\neq 0$

$(\because C$ is invertible matrix, $|C|\neq 0)$

Now, $\quad A^5=B^5$ $\quad$ ...(i)

and $\quad A^3B^2=A^2B^3$ $\quad$ ...(ii)

Subtracting Eq. (ii) from Eq. (i),

$A^5-A^3B^2=B^5-A^2B^3$

$\Rightarrow \quad A^3(A^2-B^2)=B^3(B^2-A^2)$

$\Rightarrow \quad A^3(A^2-B^2)+B^3(A^2-B^2)=0$

$\Rightarrow \quad (A^3+B^3)(A^2-B^2)=0$

Multiplying inverse of $(A^2-B^2)$ on both sides

$A^3+B^3=0$

**14.** (c) Given matrix $B$ is the inverse matrix of $3\times 3$ matrix $A$,

where $B=\begin{bmatrix}5&2\alpha&1\\0&2&1\\\alpha&3&-1\end{bmatrix}$

We know that

$\det(A)=\dfrac{1}{\det(B)}$ $\quad \left[\because \det(A^{-1})=\dfrac{1}{\det(A)}\right]$

Since,

$\det(A)+1=0$ $\quad$ [Given]

$\dfrac{1}{\det(B)}+1=0$

$\Rightarrow \quad \det(B)=-1$

$\Rightarrow \quad 5(-2-3)-2\alpha(0-\alpha)+1(0-2\alpha)=-1$

$\Rightarrow -25+2\alpha^2-2\alpha=-1$

$\Rightarrow \quad 2\alpha^2-2\alpha-24=0$

$\Rightarrow \quad \alpha^2-\alpha-12=0$

$\Rightarrow \quad (\alpha-4)(\alpha+3)=0$

$\Rightarrow \quad \alpha=-3, 4$

So, required sum of all values of $\alpha$ is $4-3=1$

**15.** (b) Given,

$$\begin{bmatrix}1&1\\0&1\end{bmatrix}\begin{bmatrix}1&2\\0&1\end{bmatrix}\begin{bmatrix}1&3\\0&1\end{bmatrix}\cdots\begin{bmatrix}1&n-1\\0&1\end{bmatrix}=\begin{bmatrix}1&78\\0&1\end{bmatrix}$$

$\because \quad \begin{bmatrix}1&1\\0&1\end{bmatrix}\begin{bmatrix}1&2\\0&1\end{bmatrix}=\begin{bmatrix}1&2+1\\0&1\end{bmatrix},$

$\begin{bmatrix}1&2+1\\0&1\end{bmatrix}\begin{bmatrix}1&3\\0&1\end{bmatrix}=\begin{bmatrix}1&3+2+1\\0&1\end{bmatrix},$

$\vdots \qquad \vdots \qquad \vdots$

$\therefore \quad \begin{bmatrix}1&1\\0&1\end{bmatrix}\begin{bmatrix}1&2\\0&1\end{bmatrix}\begin{bmatrix}1&3\\0&1\end{bmatrix}\cdots\begin{bmatrix}1&n-1\\0&1\end{bmatrix}$

$=\begin{bmatrix}1&(n-1)+(n-2)+\ldots+3+2+1\\0&1\end{bmatrix}$

$=\begin{bmatrix}1&\dfrac{n(n-1)}{2}\\0&1\end{bmatrix}=\begin{bmatrix}1&78\\0&1\end{bmatrix}$

Since, both matrices are equal, so equating corresponding element, we get

$\dfrac{n(n-1)}{2}=78\Rightarrow n(n-1)=156$

$=13\times 12=13(13-1)\Rightarrow n=13$

So, $\quad A=\begin{bmatrix}1&13\\0&1\end{bmatrix}=A^{-1}=\begin{bmatrix}1&-13\\0&1\end{bmatrix}$

$\left[\because \text{If } |A|=1 \text{ and } A=\begin{bmatrix} a & b \\ c & d \end{bmatrix}, \text{ then } A^{-1}=\begin{bmatrix} d & -b \\ -c & a \end{bmatrix}\right]$

**16.** *(b)* We have, $A=\begin{bmatrix} 2 & -3 \\ -4 & 1 \end{bmatrix}$

$\therefore A^2 = A\cdot A = \begin{bmatrix} 2 & -3 \\ -4 & 1 \end{bmatrix}\begin{bmatrix} 2 & -3 \\ -4 & 1 \end{bmatrix}$

$=\begin{bmatrix} 16 & -9 \\ -12 & 13 \end{bmatrix}$

Now, $3A^2+12A = 3\begin{bmatrix} 16 & -9 \\ -12 & 13 \end{bmatrix} + 12\begin{bmatrix} 2 & -3 \\ -4 & 1 \end{bmatrix}$

$=\begin{bmatrix} 48 & -27 \\ -36 & 39 \end{bmatrix}+\begin{bmatrix} 24 & -36 \\ -48 & 12 \end{bmatrix}$

$=\begin{bmatrix} 72 & -63 \\ -84 & 51 \end{bmatrix}$

$\therefore$ Adj $(3A^2+12A)=\begin{bmatrix} 51 & 63 \\ 84 & 72 \end{bmatrix}$

**17.** *(b)* If $A$ is non-singular matrix, then $|A|\neq 0$.

Given, $AA^T = A^TA$ and $B=A^{-1}A^T$

$BB' = (A^{-1}A^T)(A^{-1}A^T)^T$

$= A^{-1}A^T A(A^{-1})^T \quad [\because (A')'=A]$

$= A^{-1}AA^T(A^{-1})^T \quad [\because AA^T = A^TA]$

$= A^T(A^{-1})^T = A^T(A^{-1})^T$

$=(A^{-1}A)^T = I^T = I \quad [\because (AB)^T = B^TA^T]$

**18.** *(b)* If $A$ is matrix of order $n$, then $|\text{adj } A| = |A|^{n-1}$.

Given, $P=\begin{bmatrix} 1 & \alpha & 3 \\ 1 & 3 & 3 \\ 2 & 4 & 4 \end{bmatrix}$

$\therefore |P| = 1(12-12)-\alpha(4-6)+3(4-6)$

$=2\alpha-6$

$\because \quad P=\text{adj}(A)$ [Given]

$\therefore \quad |P| = |\text{adj } A| = |A|^2 = 16$

$\Rightarrow \quad 2\alpha-6=16 \Rightarrow 2\alpha = 22$

$\therefore \quad \alpha = 11$

**19.** *(c)* Given that $A$ is $2\times 2$ matrix.

So, $n=2$.

$|\text{adj } A| = |A|^{n-1} = |A|^{2-1} = |A|$

$\text{adj (adj } A) = |A|^{n-2}A = |A|^0 A = A$

Hence, both the statements are true but Statement II is not a correct explanation of Statement I.

**20.** *(d)* As $\det(A)=\pm 1$, $A^{-1}$ exists.

And $A^{-1} = \dfrac{1}{\det(A)}(\text{adj } A) = \pm(\text{adj } A)$

All entries in adj $(A)$ are integers.

Hence, $A^{-1}$ has integer entries.

**21.** *(a)* $\because A^2 - A + I = O$

$\Rightarrow A^{-1}A^2 - A^{-1}A + A^{-1}I = O$

$\Rightarrow (A^{-1}A)A - (A^{-1}A) + A^{-1} = O$

$\Rightarrow A - I + A^{-1} = O \Rightarrow A^{-1} = I - A$

**22.** *(d)* The given matrix,

$A=\begin{bmatrix} 0 & 0 & -1 \\ 0 & -1 & 0 \\ -1 & 0 & 0 \end{bmatrix}$

(a) It is clear that $A$ is not a zero matrix.

(b) Now, $(-1)I = -1\begin{bmatrix} 1 & 0 & 0 \\ 0 & 1 & 0 \\ 0 & 0 & 1 \end{bmatrix}$

$=\begin{bmatrix} -1 & 0 & 0 \\ 0 & -1 & 0 \\ 0 & 0 & -1 \end{bmatrix} \neq A$

$\therefore \quad (-1)I \neq A$

(c) Now, $|A| = 0+0-1\begin{vmatrix} 0 & -1 \\ -1 & 0 \end{vmatrix}$

$=-1(-1)=1$

Since, $|A|\neq 0$, so $A^{-1}$ exists.

(d) Now,

$A^2=\begin{bmatrix} 0 & 0 & -1 \\ 0 & -1 & 0 \\ -1 & 0 & 0 \end{bmatrix}\begin{bmatrix} 0 & 0 & -1 \\ 0 & -1 & 0 \\ -1 & 0 & 0 \end{bmatrix}$

$\Rightarrow A^2 = \begin{bmatrix} 1 & 0 & 0 \\ 0 & 1 & 0 \\ 0 & 0 & 1 \end{bmatrix} = I$

**23.** *(a)* $\text{adj}(kA) = k^{n-1}\text{adj}(A)$

$\text{adj}(3A) = 3^2\,\text{adj}(A) = 9\text{adj}(A)$

$\text{adj}(9\,\text{adj}(A)) = 9^2\,\text{adj}(\text{adj}(A))$

$\text{adj}(\text{adj}(3A)) = 3^4\,\text{adj}(\text{adj}(A))$

$\det(\text{adj}(\text{adj}(3A))) = \det(3^4\,\text{adj}(\text{adj}(A)))$

$=(3^4)^3\,\det(\text{adj}(\text{adj}(A)))$

$=3^{12}\{\det(A)\}^4$

$\det(A)=\begin{vmatrix} 2 & 2+p & 2+p+q \\ 4 & 6+2p & 8+3p+2q \\ 6 & 12+3p & 20+6p+3q \end{vmatrix}=8$

$3^{12}\times(2^3)^4 = 2^m\times 3^n$

$\Rightarrow m+n = 12+12 = 24$

**24.** *(d)* $A+I=\begin{pmatrix} 1 & a & 1 \\ 2 & 1 & 0 \\ a & 1 & 2 \end{pmatrix}$

$A=\begin{pmatrix} 0 & a & 1 \\ 2 & 0 & 0 \\ a & 1 & 1 \end{pmatrix}$

$|A| = -4 \Rightarrow -2a+2 = -4$

$\Rightarrow \quad a=3$

$|4\,adj(2A)| = 4^3|adj(2A)|$

$=4^3|2A|^2 = 4^3\cdot 2^6|A|^2$

$=2^{12}(-4)^2 = 12^{16}\cdot 3^0$

$m=16, n=0 \Rightarrow m+n=16$

**25.** *(d)* Given $A=\begin{bmatrix} \log_5 128 & \log_4 5 \\ \log_5 8 & \log_4 25 \end{bmatrix}$

$\therefore |A| = (\log_5 128\times\log_4 25) - (\log_4 5\times\log_5 8)$

$=(\log_5 128\times 2\times\log_4 5) - \left(\log_5 8\times\dfrac{1}{\log_5 4}\right)$

$=2\times\log_4 128 - \log_4 8$

$=\log_4(128)^2 - \log_4 8$

$=\log_4\left(\dfrac{128\times 128}{8}\right)$

$=\log_4(4^2\times 4^3\times 2)$

$=\log_4(4^5)+\log_4 2$

$=5+\dfrac{1}{2}=\dfrac{11}{2}$

Now,

$C_{11} = \sum_{k=1}^{2} a_{1k}\cdot A_{1k} = a_{11}A_{11}+a_{12}A_{12}$

$=|A| = \dfrac{11}{2}$

$C_{12} = \sum_{k=1}^{2} a_{1k}A_{2k}$

$=a_{11}A_{21}+a_{12}A_{22} = 0$

$C_{21} = \sum_{k=1}^{2} a_{2k}\cdot A_{1k}$

$=a_{21}A_{11}+a_{22}A_{12} = 0$

$C_{22} = \sum_{k=1}^{2} a_{2k}A_{2k} = a_{21}A_{21}+a_{22}A_{22}$

$=|A| = \dfrac{11}{2}$

$\therefore C=\begin{bmatrix} 11/2 & 0 \\ 0 & 11/2 \end{bmatrix}$

$\Rightarrow |C| = \frac{11}{2} \times \frac{11}{2} - 0 = \frac{121}{4}$

Hence, $8|C| = 8 \times \frac{121}{4} = 242$

**26.** *(b)* According to the question,
$|A| = 3$, where $A$ is matrix of order 3
and given $|\text{adj}(-4\,\text{adj}(-3\,\text{adj}(3\,\text{adj}((2A)^{-1}))))| = 2^m 3^n$

Now, LHS $\Rightarrow$
$|-4\,\text{adj}(-3\,\text{adj}(3\,\text{adj}((2A)^{-1})))|^{3-1}$
$[\because |\text{adj}\,A| = |A|^{n-1}]$

$= (-4)^6 |\text{adj}(-3\,\text{adj}(3\,\text{adj}((2A)^{-1})))|^2$

$= 4^6 \cdot |-3\,\text{adj}(3\,\text{adj}((2A)^{-1}))|^4$

$= 4^6 \cdot (-3)^{3\cdot 4} |\text{adj}(3\,\text{adj}((2A)^{-1}))|^4$

$= 4^6 \cdot 3^{12} |3\,\text{adj}((2A)^{-1})|^8$

$= 4^6 \cdot 3^{12} \cdot 3^{24} |\text{adj}(2A^{-1})|^8$

$= 4^6 \cdot 3^{36} |2A^{-1}|^{16}$

$= 4^6 \cdot 3^{36} \frac{1}{|2A|^{16}} = 2^{12} \cdot 3^{36} \times \frac{1}{2^{48}|A|^{16}}$

$= 2^{12} \cdot 3^{36} \times \frac{1}{2^{48} \cdot 3^{16}} = \frac{3^{20}}{2^{36}} = 2^{-36} \cdot 3^{20}$

$= 2^m \cdot 3^n = \text{RHS}$

On comparing,
$m = -36, n = 20$

$\Rightarrow m + 2n = -36 + 2 \times 20 = 4$

**27.** *(d)* Given, $|A| = 3, |B| = 2$

$\therefore |A^T A (\text{adj}(2A))^{-1} (\text{adj}(4B))(\text{adj}(AB))^{-1} AA^T|$

$= |A^T||A||\text{adj}(2A)^{-1}||\text{adj}(4B)| |(\text{adj}(AB))^{-1}||A||A^T|$

$= 3 \times 3 \times \frac{1}{|\text{adj}(2A)|} |4^{3-1}\,\text{adj}(B)| \times \frac{1}{|\text{adj}(AB)|} \times 3 \times 3$

$= 3^4 \times \frac{1}{|2^{3-1}\,\text{adj}(A)|} \times 4^2 \times |\text{adj}(B)| \times \left|\frac{1}{\text{adj}(B) \cdot \text{adj}(A)}\right|$

$= 3^4 \times \frac{1}{(2^2)^3 |\text{adj}(A)|} \times (4^2)^3 \times |B|^{3-1} \times \frac{1}{|B|^{3-1}|A|^{3-1}}$

$= \frac{3^4}{2^6 \times 3^{3-1}} \times 4^6 \times 2^2 \times \frac{1}{2^2 \times 3^2}$

$= \frac{3^4}{2^6 \times 3^2} \times 2^{12} \times 2^2 \times \frac{1}{2^2 \times 3^2} = 2^6$

$= 64$

**28.** *(b)*

> **key idea** Consider $A$ is a non-singular matrix, then use adj (adj $A$) $= |A|^{n-2} A$

Given, $R = \begin{bmatrix} x & 0 & 0 \\ 0 & y & 0 \\ 0 & 0 & z \end{bmatrix}$,

$x \sin\theta = y \sin\left(\theta + \frac{2\pi}{3}\right) = z \sin\left(\theta + \frac{4\pi}{3}\right) \neq 0$

$\because \sin\theta + \sin\left(\theta + \frac{2\pi}{3}\right) + \sin\left(\theta + \frac{4\pi}{3}\right)$

$= \sin\theta + \sin\theta \cos\frac{2\pi}{3} + \cos\theta \sin\frac{2\pi}{3}$
$+ \sin\theta \cos\frac{4\pi}{3} + \cos\theta \sin\frac{4\pi}{3} = 0, \forall \theta \in R$

$\therefore \frac{1}{x} + \frac{1}{y} + \frac{1}{z} = 0 \Rightarrow xy + yz + zx = 0$

**Statement I** $\text{Tr}(R) = 0$

$x + y + z = 0$ and $xy + yz + zx = 0$
$\Rightarrow x = y = z = 0$ which is not possible.
$\therefore$ Statement I is false.

**Statement II** $Tr[\text{adj}(\text{adj}\,R)] = 0$

$\because \quad \text{adj}(\text{adj}\,R) = |R|R$
$\therefore \text{Tr}(\text{adj}(\text{adj}\,R)) = \text{Tr}(|R|R)$
$= |R|(x + y + z)$
$= xyz(x + y + z) \neq 0$

$\therefore$ Statement II is false.

**29.** *(b)* $A = \begin{bmatrix} 2 & 1 & 2 \\ 6 & 2 & 11 \\ 3 & 3 & 2 \end{bmatrix}$ and $P = \begin{bmatrix} 1 & 2 & 0 \\ 5 & 0 & 2 \\ 7 & 1 & 5 \end{bmatrix}$

Let $B = P^{-1}AP - 2I$

Pre multiply by $P$,
$PB = AP - 2P$
$[\because PP^{-1} = I \text{ and } PI = P]$

Post multiply by $P^{-1}$,
$PBP^{-1} = A - 2I$

Now,
$|PBP^{-1}| = |A - 2I|$

$|P| \cdot |B| \cdot |P^{-1}| = \begin{vmatrix} 2-2 & 1 & 2 \\ 6 & 2-2 & 11 \\ 3 & 3 & 2-2 \end{vmatrix}$

$|P| \cdot |B| \cdot \frac{1}{|P|} = \begin{vmatrix} 0 & 1 & 2 \\ 6 & 0 & 11 \\ 3 & 3 & 0 \end{vmatrix}$

$|B| = -1(-33) + 2(18)$
[expand w.r.t. $R_1$]

$\Rightarrow |P^{-1}AP - 2I| = 33 + 36 = 69$

Prime factors of 69 is 3 and 23
$\therefore$ Sum $= 3 + 23 = 26$

**30.** *(b)* $|A| = m - n$
and $4m + n = 22$
$\Rightarrow \quad n = 22 - 4m$
Also, $17m + 4n = 93$
$17m + 88 - 16m = 93 \Rightarrow m = 5$
$\therefore \quad n = 22 - 4 \times 5 = 2$
$|A| = m - n = 5 - 2 = 3$

$\det(n\,\text{adj}(\text{adj}(mA))) = 3^a 5^b 6^c$
$= n^m [\det(mA)]^{(m-1)^2}$
$= n^m (m^m \det(A))^{(m-1)^2}$
$= n^m \cdot m^{m(m-1)^2} (|A|)^{(m-1)^2}$

$\because n = 2, m = 5$ and $|A| = 3$
$\therefore 2^5 \cdot 5^{5(5-1)^2} (3)^{(5-1)^2} = 2^5 \cdot 5^{5 \times 4^2} (3)^{4^2}$
$= 2^5 \cdot 5^{80} (3)^{16} = 2^5 \times 3^5 \times 3^{11} \times 5^{80}$
$= (6)^5 \times (3)^{11} \times (5)^{80}$
$= 3^{11} \times 5^{80} \times 6^5 = 3^a 5^b 6^c$

$\Rightarrow a = 11, b = 80, c = 5$
$a + b + c = 11 + 80 + 5 = 96$

**31.** *(b)* Given, $A = \begin{bmatrix} 1 & 2 & 3 \\ \alpha & 3 & 1 \\ 1 & 1 & 2 \end{bmatrix}$

and $|A| = 2$

$\therefore \quad |A| = \begin{vmatrix} 1 & 2 & 3 \\ \alpha & 3 & 1 \\ 1 & 1 & 2 \end{vmatrix}$

$= 1(5) + 2(1 - 2\alpha) + 3(\alpha - 3)$
$= -\alpha - 2$

Given, $|A| = 2 \Rightarrow -\alpha - 2 = 2$
$\Rightarrow -\alpha = 4 \Rightarrow \alpha = -4$

Now, $|2\,\text{adj}(2\,\text{adj}(2A)|$
$= 2^3 |\text{adj}(2\,\text{adj}(2A)|$
$= 8|\text{adj}(2 \cdot 2^3\,\text{adj}(A)|$
$= 8|\text{adj}(2^3\,\text{adj}\,A)|$
$= 8\,|2^6\,\text{adj}(\text{adj}\,A)|$
$= 8 \cdot (2^6)^3 |\text{adj}(\text{adj}\,A)|$
$= 2^3 \cdot 2^{18} \cdot |A|^4$
$(\because |\text{adj}(\text{adj}\,A)| = |A|^{(n-1)^2})$

$= 2^3 \cdot 2^{18} \cdot 2^4$

$= 2^{25} = (2^5)^5 = (32)^5 = (32)^n$ [Given]

So, $n = 5$

$\therefore \alpha = -4$

Thus, $3n + \alpha = 3 \times 5 - 4 = 15 - 4 = 11$

**32.** *(c)* Given, $B = \begin{bmatrix} 1 & 3 & \alpha \\ 1 & 2 & 3 \\ \alpha & \alpha & 4 \end{bmatrix}$ $\alpha > 2$

Since, $B$ is adjoint of matrix $A$, then

$|B| = |\text{adj } A| = |A|^{3-1} = (2)^2 = 4$

So, $\begin{vmatrix} 1 & 3 & \alpha \\ 1 & 2 & 3 \\ \alpha & \alpha & 4 \end{vmatrix} = 4$

$\Rightarrow 1(8 - 3\alpha) - 3(4 - 3\alpha) + \alpha(\alpha - 2\alpha) = 4$

$\Rightarrow 8 - 3\alpha - 12 + 9\alpha + \alpha^2 - 2\alpha^2 = 4$

$\Rightarrow -\alpha^2 + 6\alpha - 8 = 0$

$\therefore \alpha = 2, 4$

Since, $\alpha > 2$ so only $\alpha = 4$

Thus, $[\alpha, -2\alpha \ \ \alpha] \, B \begin{bmatrix} \alpha \\ -2\alpha \\ \alpha \end{bmatrix}$

$= [4 \ -8 \ 4] \begin{bmatrix} 1 & 3 & 4 \\ 1 & 2 & 3 \\ 4 & 4 & 4 \end{bmatrix} \begin{bmatrix} 4 \\ -8 \\ 4 \end{bmatrix} = -16$

**33.** *(d)* Given,

$A = \begin{bmatrix} m & n \\ p & q \end{bmatrix}$

$d = |A| \neq 0$ ...(i)

and $|A - d(\text{adj } A)| = 0$ ...(ii)

Now,

$A - d(\text{adj } A) = \begin{bmatrix} m & n \\ p & q \end{bmatrix} - d \begin{bmatrix} q & -n \\ -p & m \end{bmatrix}$

$= \begin{bmatrix} m - qd & n(1+d) \\ p(1+d) & q - md \end{bmatrix}$

From Eq. (ii), $\begin{vmatrix} m - qd & n(1+d) \\ p(1+d) & q - md \end{vmatrix} = 0$

$\Rightarrow (m - qd)(q - md) - np(1+d)^2 = 0$

$\Rightarrow mq - m^2 d - q^2 d + mqd^2 - np(1+d)^2 = 0$

$\Rightarrow (mq - np) + d^2(mq - np) - d(m^2 + q^2 + 2np) = 0$

$\Rightarrow d + d^3 - d((m+q)^2 - 2d) = 0$ [$\because d = mq - np$]

$\Rightarrow 1 + d^2 = (m+q)^2 - 2d$

$\Rightarrow (1+d)^2 = (m+q)^2$

**34.** *(b)* If $|A| \neq 0$, then $A$ to be invertible

Let $A = \begin{bmatrix} e^t & e^{-t}(\sin t - 2\cos t) & e^{-t}(-2\sin t - \cos t) \\ e^t & e^{-t}(2\sin t + \cos t) & e^{-t}(\sin t - 2\cos t) \\ e^t & e^{-t}\cos t & e^{-t}\sin t \end{bmatrix}$

Expanding along $C_1$

$A = e^t[e^{-2t}(2\sin t + \cos t)\sin t - e^{-2t}\cos t(\sin t - 2\cos t)]$
$- e^t[e^{-2t}(\sin t - 2\cos t)\sin t - e^{-2t}(-2\sin t - \cos t)\cos t]$
$+ e^t[e^{-2t}(\sin t - 2\cos t)^2 - e^{-2t}(2\sin t + \cos t)^2]$

$A = e^{-t}[2\sin^2 t + \cos t \sin t - \cos t \sin t + 2\cos^2 t] - e^{-t}[\sin^2 t - 2\cos t \sin t + 2\sin t \cos t + \cos^2 t]$
$+ e^{-t}[\sin^2 t + 4\cos^2 t - 4\sin t \cos t + 4\sin^2 t + \cos^2 t + 4\sin t \cos t]$

$A = e^{-t}[2] - e^{-t}[1] + e^{-t}[1 + 4]$

$A = 6e^{-t} \neq 0$

$\Rightarrow e^{-t} \neq 0$

$\therefore$ The set of all values of $t \in R$, for which the matrix $A$ is invertible is $R$.

• Concept **Enhancer**

If the determinant of a given matrix is not equal to 0, then the matrix is invertible and we can find the inverse of such matrix.

**35.** *(c)* Given, $A = \begin{bmatrix} 0 & 1 & 0 \\ 0 & 0 & 1 \\ 1 & 0 & 0 \end{bmatrix}$

$\therefore A^2 = \begin{bmatrix} 0 & 1 & 0 \\ 0 & 0 & 1 \\ 1 & 0 & 0 \end{bmatrix} \begin{bmatrix} 0 & 1 & 0 \\ 0 & 0 & 1 \\ 1 & 0 & 0 \end{bmatrix}$

$= \begin{bmatrix} 0 & 0 & 1 \\ 1 & 0 & 0 \\ 0 & 1 & 0 \end{bmatrix}$

Applying $R_1 \leftrightarrow R_2$,

$A^2 = -\begin{bmatrix} 1 & 0 & 0 \\ 0 & 0 & 1 \\ 0 & 1 & 0 \end{bmatrix}$

Again applying $R_2 \leftrightarrow R_3$,

$A^2 = +\begin{bmatrix} 1 & 0 & 0 \\ 0 & 1 & 0 \\ 0 & 0 & 1 \end{bmatrix} = I$

Now,

$B_0 = A^{49} + 2A^{98} = (A^2)^{24} \cdot A + 2(A^2)^{49}$
$= (I)^{24} \cdot A + 2(I)^{49} = I \cdot A + 2I$

$B_0 = A + 2I = \begin{bmatrix} 0 & 1 & 0 \\ 0 & 0 & 1 \\ 1 & 0 & 0 \end{bmatrix} + 2\begin{bmatrix} 1 & 0 & 0 \\ 0 & 1 & 0 \\ 0 & 0 & 1 \end{bmatrix}$

$= \begin{bmatrix} 2 & 1 & 0 \\ 0 & 2 & 1 \\ 1 & 0 & 2 \end{bmatrix}$

$\Rightarrow |B_0| = 2(4 - 0) - 0 + 1(1 - 0) = 9$

[Expanding along $C_1$]

Since, $B_n = \text{adj}(B_{n-1})$

$\therefore B_4 = \text{adj}(B_3) = \text{adj}(\text{adj}(B_2))$
$= \text{adj}(\text{adj}(\text{adj}(B_1)))$
$= \text{adj}(\text{adj}(\text{adj}(\text{adj}(B_0))))$

$\Rightarrow B_4 = |B_0|^{(3-1)^4} = |B_0|^{16}$
$= (9)^{16} = (3^2)^{16} = 3^{32}$

**36.** *(b)* Given, $|A| = -1$

and $|(A + I)(\text{adj } A + I)| = 4$

Now, Let $A = \begin{bmatrix} a & b \\ c & d \end{bmatrix}$,

$\text{adj } A = \begin{bmatrix} d & -b \\ -c & a \end{bmatrix}$

As, $|A| = -1$, $ad - bc = -1$

$\Rightarrow |(A + I)(\text{adj } A + I)| = 4$

$\Rightarrow |A \text{ adj } A + \text{adj } A + A + I| = 4$

$\Rightarrow ||A| I + \text{adj } A + A + I| = 4$

[$\because A \text{ adj } A = |A| I_n$]

$\Rightarrow |-I + \text{adj } A + A + I| = 4$

[$\because |A| = -1$]

$\Rightarrow |\text{adj } A + A| = 4$

$\Rightarrow \begin{vmatrix} d + a & 0 \\ 0 & a + d \end{vmatrix} = 4$

$\Rightarrow (a + d)^2 = 4$

$\Rightarrow a + d = \pm 2$

**37.** *(b)* $AA^T = \begin{bmatrix} \frac{1}{\sqrt{5}} & \frac{2}{\sqrt{5}} \\ \frac{-2}{\sqrt{5}} & \frac{1}{\sqrt{5}} \end{bmatrix} \begin{bmatrix} \frac{1}{\sqrt{5}} & \frac{-2}{\sqrt{5}} \\ \frac{2}{\sqrt{5}} & \frac{1}{\sqrt{5}} \end{bmatrix}$

$AA^T = \begin{bmatrix} 1 & 0 \\ 0 & 1 \end{bmatrix} = I$

Now, $Q^2 = A^T BAA^T BA$

$\Rightarrow Q^2 = A^T B^2 A$

Again,

$Q^3 = (A^T BA)(A^T B^2 A) = A^T B^3 A$

Similarly, $Q^{2021} = A^T B^{2021} A$

$\Rightarrow \quad AQ^{2021}A^T = A(A^TB^{2021}A)A^T$
$= (AA^T)B^{2021}(AA^T) = B^{2021}$

As, $\quad B = \begin{bmatrix} 1 & 0 \\ i & 1 \end{bmatrix} \Rightarrow B^2 = \begin{bmatrix} 1 & 0 \\ 2i & 1 \end{bmatrix}$,

Similarly, $B^{2021} = \begin{bmatrix} 1 & 0 \\ 2021i & 1 \end{bmatrix}$

$(B^{2021})^{-1} = \dfrac{\text{adj}(B^{2021})}{|B^{2021}|} = \begin{bmatrix} 1 & 0 \\ -2021i & 1 \end{bmatrix}$

**38.** *(d)* Given, $A = \begin{bmatrix} 1 & 2 \\ -1 & 4 \end{bmatrix}$

$A^{-1} = \alpha I + \beta A,\ \alpha, \beta \in R$

Now, $\quad A^{-1} = \dfrac{\text{adj}\,A}{|A|}$

If $X = \begin{bmatrix} a & b \\ c & d \end{bmatrix}$, then adj $X = \begin{bmatrix} d & -b \\ -c & a \end{bmatrix}$

Similarly, adj A $= \begin{bmatrix} 4 & -2 \\ 1 & 1 \end{bmatrix}$

$|A| = 4 - (-2) = 6$

$\therefore \quad A^{-1} = \dfrac{1}{6}\begin{bmatrix} 4 & -2 \\ 1 & 1 \end{bmatrix}$

Now, $\quad A^{-1} = \alpha I + \beta A$

$= \begin{bmatrix} \alpha & 0 \\ 0 & \alpha \end{bmatrix} + \begin{bmatrix} \beta & 2\beta \\ -\beta & 4\beta \end{bmatrix}$

$= \begin{bmatrix} \alpha+\beta & 2\beta \\ -\beta & \alpha+4\beta \end{bmatrix}$

$\Rightarrow \begin{bmatrix} \alpha+\beta & 2\beta \\ -\beta & \alpha+4\beta \end{bmatrix} = \begin{bmatrix} \frac{2}{3} & -\frac{1}{3} \\ \frac{1}{6} & \frac{1}{6} \end{bmatrix}$

$\therefore\ \beta = -\dfrac{1}{6}$ and $\alpha + \beta = \dfrac{2}{3}$

$\Rightarrow\ \alpha = \dfrac{4}{6} + \dfrac{1}{6} = \dfrac{5}{6}$

$\therefore \quad 4(\alpha - \beta) = 4\left(\dfrac{5}{6} + \dfrac{1}{6}\right) = 4$

**39.** *(d)* It is given that

$\text{adj}(A) = \begin{bmatrix} 2 & -1 & 1 \\ -1 & 0 & 2 \\ 1 & -2 & -1 \end{bmatrix}$

$\Rightarrow |\text{adj}\,A| = 2(0+4) + 1(1-2) + 1(2) = 9$

$\because \quad |\text{adj}\,A| = |A|^{3-1}$

$\Rightarrow \quad |A|^2 = |\text{adj}\,A| \Rightarrow |A| = \pm 3$

$\Rightarrow \quad |\lambda| = 3$

$\because \quad B = \text{adj}(\text{adj}\,A)$

$\Rightarrow \quad |B| = |A|^{(3-1)^2} = |A|^4 = 81$

$\because \quad |(B^{-1})^T| = |B^{-1}| = \dfrac{1}{|B|} = \dfrac{1}{81}$

$\Rightarrow \quad \mu = \dfrac{1}{81}$

$\therefore\ (|\lambda|, \mu)$ is $\left(3, \dfrac{1}{81}\right)$

**40.** *(c)* Given matrices, $A = \begin{bmatrix} 1 & 1 & 2 \\ 1 & 3 & 4 \\ 1 & -1 & 3 \end{bmatrix}$

$B = \text{adj}(A)$ and $C = 3A$

Now, $|\text{adj}(B)| = |\text{adj}(\text{adj}(A))|$

$[\because B = \text{adj}(A)]$

and $|\text{adj}(\text{adj}(A))| = |A|^{(n-1)^2}$, where $n$ is the order of square matrix $A$.

$\therefore\ |\text{adj}(\text{adj}(A))| = |A|^{(3-1)^2} = |A|^4$

and $|C| = |3A| = 3^3|A| = 27|A|$

$\because |KA| = K^n|A|$, where $K$ is a scalar and $n$ is the order of square matrix $A$.

$\therefore \quad \dfrac{|\text{adj}\,B|}{|C|} = \dfrac{|A|^4}{27|A|} = \dfrac{|A|^3}{27}$

$\because \quad |A| = \begin{vmatrix} 1 & 1 & 2 \\ 1 & 3 & 4 \\ 1 & -1 & 3 \end{vmatrix}$

$= 1(9+4) - 1(3-4) + 2(-1-3)$

$= 13 + 1 - 8 = 6$

So, $\dfrac{|\text{adj}\,B|}{|C|} = \dfrac{|A|^3}{27} = \dfrac{6^3}{27} = 2^3 = 8$

**41.** *(c)* Given, $|ABA^T| = 8$

$\Rightarrow |A||B||A^T| = 8$

$[\because |XY| = |X||Y|]$

$\therefore \quad |A|^2|B| = 8 \quad \ldots(\text{i})$

$[\because |A^T| = |A|]$

Also, we have

$|AB^{-1}| = 8$

$\Rightarrow \quad |A||B^{-1}| = 8$

$\Rightarrow \quad \dfrac{|A|}{|B|} = 8 \quad \ldots(\text{ii})$

$\left[\because |A^{-1}| = |A|^{-1} = \dfrac{1}{|A|}\right]$

On multiplying Eqs. (i) and (ii), we get

$z|A|^3 = 8 \cdot 8 = 4^3$

$\Rightarrow \quad |A| = 4$

$\Rightarrow \quad |B| = \dfrac{|A|}{8} = \dfrac{4}{8} = \dfrac{1}{2}$

Now, $|BA^{-1}B^T| = |B|\dfrac{1}{|A|}|B|$

$= \left(\dfrac{1}{2}\right)\dfrac{1}{4}\left(\dfrac{1}{2}\right) = \dfrac{1}{16}$

**42.** *(d)* Given, $A = \begin{bmatrix} 1 & 2 & 2 \\ 2 & 1 & -2 \\ a & 2 & b \end{bmatrix}$

$\therefore\ A^T = \begin{bmatrix} 1 & 2 & a \\ 2 & 1 & 2 \\ 2 & -2 & b \end{bmatrix}$

$AA^T = \begin{bmatrix} 1 & 2 & 2 \\ 2 & 1 & -2 \\ a & 2 & b \end{bmatrix}\begin{bmatrix} 1 & 2 & a \\ 2 & 1 & 2 \\ 2 & -2 & b \end{bmatrix}$

$= \begin{bmatrix} 9 & 0 & a+4+2b \\ 0 & 9 & 2a+2-2b \\ a+4+2b & 2a+2-2b & a^2+4+b^2 \end{bmatrix}$

It is given that

$AA^T = 9I$

$\Rightarrow$

$\begin{bmatrix} 9 & 0 & a+4+2b \\ 0 & 9 & 2a+2-2b \\ a+4+2b & 2a+2-2b & a^2+4+b^2 \end{bmatrix}$

$= 9\begin{bmatrix} 1 & 0 & 0 \\ 0 & 1 & 0 \\ 0 & 0 & 1 \end{bmatrix} = \begin{bmatrix} 9 & 0 & 0 \\ 0 & 9 & 0 \\ 0 & 0 & 9 \end{bmatrix}$

On comparing, we get

$a + 4 + 2b = 0$

$\Rightarrow \quad a + 2b = -4 \quad \ldots(\text{i})$

$2a + 2 - 2b = 0$

$\Rightarrow \quad a - b = -1 \quad \ldots(\text{ii})$

and $a^2 + 4 + b^2 = 9 \quad \ldots(\text{iii})$

On solving Eqs. (i) and (ii), we get

$a = -2, b = -1$

This satisfies Eq. (iii).

Hence, $(a, b) \equiv (-2, -1)$.

**43.** *(d)* Since, $B$ is the inverse of matrix $A$, *i.e.*, $B = A^{-1}$.

$\therefore\ 10A^{-1} = \begin{bmatrix} 4 & 2 & 2 \\ -5 & 0 & \alpha \\ 1 & -2 & 3 \end{bmatrix}$

$\Rightarrow\ 10A^{-1}A = \begin{bmatrix} 4 & 2 & 2 \\ -5 & 0 & \alpha \\ 1 & -2 & 3 \end{bmatrix}A$

$$\Rightarrow 10I = \begin{bmatrix} 4 & 2 & 2 \\ -5 & 0 & \alpha \\ 1 & -2 & 3 \end{bmatrix}\begin{bmatrix} 1 & -1 & 1 \\ 2 & 1 & -3 \\ 1 & 1 & 1 \end{bmatrix} \quad [\because A^{-1}A = I]$$

$$\Rightarrow \begin{bmatrix} 10 & 0 & 0 \\ 0 & 10 & 0 \\ 0 & 0 & 10 \end{bmatrix} = \begin{bmatrix} 10 & 0 & 0 \\ -5+\alpha & 5+\alpha & -5+\alpha \\ 0 & 0 & 10 \end{bmatrix}$$

$$\Rightarrow -5+\alpha = 0 \text{ and } 5+\alpha = 10$$

$$\Rightarrow \alpha = 5$$

**44.** *(d)* Given, $A = \begin{bmatrix} \frac{1}{\sqrt{10}} & \frac{3}{\sqrt{10}} \\ -\frac{3}{\sqrt{10}} & \frac{1}{\sqrt{10}} \end{bmatrix}$

$B = \begin{bmatrix} 1 & -i \\ 0 & 1 \end{bmatrix}$, where $i = \sqrt{-1}$

So,

$$A \cdot A^T = \begin{bmatrix} \frac{1}{\sqrt{10}} & \frac{3}{\sqrt{10}} \\ -\frac{3}{\sqrt{10}} & \frac{1}{\sqrt{10}} \end{bmatrix}\begin{bmatrix} \frac{1}{\sqrt{10}} & -\frac{3}{\sqrt{10}} \\ \frac{3}{\sqrt{10}} & \frac{1}{\sqrt{10}} \end{bmatrix}$$

$$A \cdot A^T = \begin{bmatrix} 1 & 0 \\ 0 & 1 \end{bmatrix} = I$$

Given, $M = A^T BA$

So, $M^2 = (A^T \cdot B \cdot I \cdot B \cdot A)$

So, $M^2 = A^T B^2 A$

Now, $B^2 = \begin{bmatrix} 1 & -i \\ 0 & 1 \end{bmatrix}\begin{bmatrix} 1 & -i \\ 0 & 1 \end{bmatrix}$

$B^2 = \begin{bmatrix} 1 & -2i \\ 0 & 1 \end{bmatrix}$

Now,

$M^4 = (A^T BA)(A^T BA)(A^T BA)(A^T BA)$

So, $M^4 = A^T B^2 \cdot B^2 A$

Now, $B^2 \cdot B^2 = \begin{bmatrix} 1 & -2i \\ 0 & 1 \end{bmatrix}\begin{bmatrix} 1 & -2i \\ 0 & 1 \end{bmatrix}$

$B^4 = \begin{bmatrix} 1 & -4i \\ 0 & 1 \end{bmatrix}$

So, if we calculate $B^{2023}$,

it is $B^{2023} = \begin{bmatrix} 1 & -2023i \\ 0 & 1 \end{bmatrix}$

So, $M^{2023} = A^T \begin{bmatrix} 1 & -2023i \\ 0 & 1 \end{bmatrix} A$

$$AM^{2023}A^T = \begin{bmatrix} 1 & -2023i \\ 0 & 1 \end{bmatrix}$$

$$\therefore (AM^{2023}A^T)^{-1} = \begin{bmatrix} 1 & 2023i \\ 0 & 1 \end{bmatrix}$$

**45.** *(c)* Given, $A = \begin{bmatrix} 1 & 2 & \alpha \\ 1 & 0 & 1 \\ 0 & 1 & 2 \end{bmatrix}$

and

$|\text{adj}(2A - 2A^T) \cdot \text{adj}(A - 2A^T)| = 2^8$

$\Rightarrow |\text{adj}((A - 2A^T) . (2A - A^T))| = 2^8$

$[\because \text{adj}(AB) = \text{adj}(B) . \text{adj}(A)]$

$\Rightarrow |(A - 2A^T)(2A - A^T)|^2 = 2^8$

$[\because |\text{adj } A| = |A|^{(n-1)}]$

$\Rightarrow |A - 2A^T| \cdot |2A - A^T| = 2^4$

$[\because |AB|) = |A||B|]$

$\because (A - 2A^T)^T = A^T - (2A^T)^T$

$= A^T - (2A^T)^T$

$(A^T)^T = A^T - 2A$

$\Rightarrow |A - 2A^T| = |A^T - 2A|$

So, $|A - 2A^T||A - 2A^T| = 16$

$\Rightarrow |A - 2A^T|^2 = 16$

$\Rightarrow |A - 2A^T| = \pm 4$ ... (i)

Now,

$$A - 2A^T = \begin{bmatrix} 1 & 2 & \alpha \\ 1 & 0 & 1 \\ 0 & 1 & 2 \end{bmatrix} - 2\begin{bmatrix} 1 & 1 & 0 \\ 2 & 0 & 1 \\ \alpha & 1 & 2 \end{bmatrix}$$

$$= \begin{bmatrix} -1 & 0 & \alpha \\ -3 & 0 & -1 \\ -2\alpha & -1 & -2 \end{bmatrix}$$

$\therefore |A - 2A^T| = -(-1)(1 + 3\alpha)$

(Expanding along $C_2$)

$= 1 + 3\alpha = \pm 4$ [From Eq. (i)]

$\Rightarrow 1 + 3\alpha = 4$

or $1 + 3\alpha = -4$

$\Rightarrow \alpha = 1 \in (0, \infty)$

or $\alpha = -\frac{5}{3} \notin (0, \infty)$

$$\therefore |A| = \begin{vmatrix} 1 & 2 & 1 \\ 1 & 0 & 1 \\ 0 & 1 & 2 \end{vmatrix}$$

$= -1 - 3 = -4$

$\Rightarrow |A|^2 = 16$

**46.** *(c)* We have,

$$A = \begin{bmatrix} \cos\theta & -\sin\theta \\ \sin\theta & \cos\theta \end{bmatrix}$$

$\therefore |A| = \cos^2\theta + \sin^2\theta = 1$

and $\text{adj } A = \begin{bmatrix} \cos\theta & \sin\theta \\ -\sin\theta & \cos\theta \end{bmatrix}$

$\Rightarrow A^{-1} = \begin{bmatrix} \cos\theta & \sin\theta \\ -\sin\theta & \cos\theta \end{bmatrix}$

Note that $A^{-50} = (A^{-1})^{50}$

Now, $A^{-2} = (A^{-1})(A^{-1})$

$$\Rightarrow A^{-2} = \begin{bmatrix} \cos\theta & \sin\theta \\ -\sin\theta & \cos\theta \end{bmatrix}\begin{bmatrix} \cos\theta & \sin\theta \\ -\sin\theta & \cos\theta \end{bmatrix}$$

$$= \begin{bmatrix} \cos^2\theta - \sin^2\theta & \cos\theta\sin\theta + \sin\theta\cos\theta \\ -\cos\theta\sin\theta - \cos\theta\sin\theta & -\sin^2\theta + \cos^2\theta \end{bmatrix}$$

$$= \begin{bmatrix} \cos 2\theta & \sin 2\theta \\ -\sin 2\theta & \cos 2\theta \end{bmatrix}$$

Also, $A^{-3} = (A^{-2})(A^{-1})$

$$A^{-3} = \begin{bmatrix} \cos 2\theta & \sin 2\theta \\ -\sin 2\theta & \cos 2\theta \end{bmatrix}\begin{bmatrix} \cos\theta & \sin\theta \\ -\sin\theta & \cos\theta \end{bmatrix}$$

$$= \begin{bmatrix} \cos 3\theta & \sin 3\theta \\ -\sin 3\theta & \cos 3\theta \end{bmatrix}$$

Similarly, $A^{-50} = \begin{bmatrix} \cos 50\theta & \sin 50\theta \\ -\sin 50\theta & \cos 50\theta \end{bmatrix}$

$$= \begin{bmatrix} \cos\frac{25}{6}\pi & \sin\frac{25}{6}\pi \\ -\sin\frac{25}{6}\pi & \cos\frac{25}{6}\pi \end{bmatrix} \quad \left(\text{When, } \theta = \frac{\pi}{12}\right)$$

$$= \begin{bmatrix} \cos\frac{\pi}{6} & \sin\frac{\pi}{6} \\ -\sin\frac{\pi}{6} & \cos\frac{\pi}{6} \end{bmatrix}$$

$$\left[\because \cos\left(\frac{25\pi}{6}\right) = \cos\left(4\pi + \frac{\pi}{6}\right) = \cos\frac{\pi}{6} \text{ and } \sin\left(\frac{25\pi}{6}\right) = \sin\left(4\pi + \frac{\pi}{6}\right) = \sin\frac{\pi}{6}\right]$$

$$= \begin{bmatrix} \frac{\sqrt{3}}{2} & \frac{1}{2} \\ \frac{-1}{2} & \frac{\sqrt{3}}{2} \end{bmatrix}$$

**47.** (38) $B=(\text{adj}(A\text{ adj }A^2))^{-1}$

$=(\text{adj}(A|A|^2 A^{-2}))^{-1}$

$(\text{adj }A=|A|A^{-1})$

$=(\text{adj}(A\cdot(-1)^2 A^{-2}))^{-1}(|A|=-1)$

$=(\text{adj}(A\cdot A^{-2}))^{-1}$

$=(\text{adj }A^{-1})^{-1}=(\text{adj }A)$

$(\text{adj }A^{-1}=(\text{adj }A)^{-1})$

$=|A|A^{-1}=-1A^{-1}=-A^{-1}$

$(\text{adj }A=|A|A^{-1})$

Given, $|A|=1$

$\Rightarrow \lambda(16)-2(-34)+3(-39)=4$

$\Rightarrow \lambda=3$

$\therefore |\lambda_B+I|=|3(-A^{-1})+I|=\left|-3\dfrac{\text{adj }A}{|A|}+I\right|$

$=|3\text{ adj }A+I|$

$=3|A|A^{-1}+I|=|3(-1)A^{-1}+I|$

$=|-3A^{-1}+I|$

$=|-3A^{-1}+AA^{-1}|=|(A-3I)A^{-1}|$

$=|A-3I||A^{-1}|=38$

**48.** (34) $|A|=-2$

Order of matrix $A=3(n)$

$\therefore$ $|3\text{adj}(-6\text{ adj }3A)|$

$|3\text{ adj}(-6\cdot 3^{3-1}\text{ adj }A|$

$=|3(-6\cdot 3^2)^{3-1}\text{adj adj }A|$

$=|3\cdot 6^2\cdot 3^4\text{ adj adj }A|$

$=3^3\cdot(6^2)^3(3^4)^3\,|\text{adj adj }A|$

$=3^3\cdot 2^6\cdot 3^6\cdot 3^{12}\,|A|^{(n-1)^2}$

$=2^6\cdot 3^{21}\cdot(-2)^4=2^{10}\,3^{21}$

$\therefore$ $m+n=10$

$mn=21$ and $m>n$

$m=7$ and $n=3$

$\Rightarrow 4m+2n=28+6=34$

> **Concept Enhancer**
>
> $\text{adj}(KA)=k^{n-1}\text{adj}(A)$ and $|KA|=K^n|A|$

**49.** (14) $|3\text{ adj}(2\text{ adj}(|A|A)|$

$=|3\text{ adj}(2|A|^2\text{ adj }A|$

$=|3\cdot 2^2\,|A|^4\text{ adj}(\text{adj }A)|$

$=2^6\cdot 3^3\,|A|^{12}\,|A|^4=2^6\cdot 3^3\cdot|A|^{16}$

$=2^{-10}\cdot 3^{-13}$ [given]

So, $|A|^{16}=2^{-16}\cdot 3^{-16}$

$\Rightarrow |A|=2^{-1}\cdot 3^{-1}$ ...(i)

Now, $|3\text{ adj}(2A)|=|3\cdot 2^2\text{ adj}(A)|$

$=2^6\cdot 3^3\,|A|^2=2^m\cdot 3^n$ [given]

$\Rightarrow 2^6\cdot 3^3\cdot 2^{-2}\cdot 3^{-2}=2^m\cdot 3^n$

(from Eq. (i))

$\Rightarrow 2^m\cdot 3^n=2^4\,3^1$

$\therefore$ $m=4, n=1$

Hence, $|3m+2n|=|12+2|=|14|=14$

**50.** (5) Let $A=\begin{bmatrix}a & b\\ b & d\end{bmatrix}$

$\begin{bmatrix}a & b\\ b & d\end{bmatrix}\begin{bmatrix}1\\1\end{bmatrix}=\begin{bmatrix}3\\7\end{bmatrix}$ and $|A|=1$

$ad-b^2=1, a+b=3, b+d=7$

$a=3-b, d=7-b$

Now, $ad-b^2=1$

$(3-b)(7-b)-b^2=1$

$\Rightarrow$ $21-10b=1$

$\therefore$ $b=2, a=1$ and $d=5$

$\therefore$ $A=\begin{bmatrix}1 & 2\\ 2 & 5\end{bmatrix}$

$A^{-1}=\alpha A+\beta I$

$\Rightarrow \begin{bmatrix}5 & -2\\ -2 & 1\end{bmatrix}=\begin{bmatrix}\alpha & 2\alpha\\ 2\alpha & 5\alpha\end{bmatrix}+\begin{bmatrix}\beta & 0\\ 0 & \beta\end{bmatrix}$

$\alpha+\beta=5\Rightarrow 2\alpha=-2$

$\alpha=-1, \beta=6$

$\therefore$ $\alpha+\beta=5$

**51.** (7) For a $3\times 3$ matrix

$|\text{adj }A|=|A|^2$

$\dfrac{|\text{adj}(\text{adj}(\text{adj}\ldots(\text{adj }A)))|}{2024\text{ times}}=|A|^{2^{2024}}$

$=2^{2^{2024}}$

Now, $2^{2024}=4^{1012}=(3+1)^{1012}$

$=3k+1$ (form)

where $k$ is odd.

So $|\text{adj}(\text{adj}(\ldots(\text{adj }A)))|=2^{3k+1}$

$=2^{3k}\cdot 2=2\cdot 8^k$

$=2(9-1)^k=9\times 2M-2$

$=9P-2=9P+7$

Remainder $=7$

**52.** (5) We have, $|\text{adj}(2\text{ adj}(2A^{-1}))|=2^{84}$

$\Rightarrow$ $|2\text{ adj}(2A^{-1})|^{n-1}=2^{84}$

$\Rightarrow$ $(2^n\,|\text{adj}(2A^{-1})|^{n-1}=2^{84}$

$\Rightarrow$ $2^{n(n-1)}(|2A^{-1}|^{(n-1)})^{n-1}=2^{84}$

$\Rightarrow$ $2^{n(n-1)}(2^n|A^{-1}|)^{(n-1)^2}=2^{84}$

$\Rightarrow$ $2^{n(n-1)}\left(2^n\dfrac{1}{|A|}\right)^{(n-1)^2}=2^{84}$

$\Rightarrow$ $2^{n(n-1)}\left(\dfrac{2^n}{2}\right)^{(n-1)^2}=2^{84}$

$\Rightarrow$ $2^{n(n-1)+(n-1)^3}=2^{84}$

$\Rightarrow$ $n(n-1)+(n-1)^3=84$

$\Rightarrow$ $(n-1)[n+n^2-2n+1]=84$

$\Rightarrow$ $(n-1)(n^2-n+1)=84$

$\Rightarrow$ $n=5$

> **Concept Enhancer**
>
> (i) If A is a non-singular matrix of order $n$, then $|\text{adj }A|=|A|^{n-1}$
>
> (ii) $|A^{-1}|=|A|^{-1}$
>
> (iii) $|kA|=k^n|A|$, where A is a matrix of order $n$.

**53.** (50) Given, $A=\begin{bmatrix}a & b\\ c & d\end{bmatrix}$

$|A|=\begin{vmatrix}a & b\\ c & d\end{vmatrix}=ad-bc$

$A^{-1}$ exist if $ad-bc\neq 0$

If $A=A^{-1}$, then $A^2=I$

$\begin{bmatrix}a & b\\ c & d\end{bmatrix}\begin{bmatrix}a & b\\ c & d\end{bmatrix}=\begin{bmatrix}1 & 0\\ 0 & 1\end{bmatrix}$

$\Rightarrow \begin{bmatrix}a^2+bc & ab+bd\\ ac+cd & bc+d^2\end{bmatrix}=\begin{bmatrix}1 & 0\\ 0 & 1\end{bmatrix}$

$\Rightarrow \begin{bmatrix}a^2+bc & b(a+d)\\ c(a+d) & bc+d^2\end{bmatrix}=\begin{bmatrix}1 & 0\\ 0 & 1\end{bmatrix}$

$ab+bd=0$ ...(i)

$ac+cd=0$ ...(ii)

$a^2+bc=1$ ...(iii)

$bc+d^2=1$ ...(iv)

On subtracting Eq. (iv) from Eq. (iii), we get

$a^2-d^2=0$

$(a+d)(a-d)=0$

$\Rightarrow$ $a+d=0$ or $a-d=0$

**Case I** When, $a+d=0$ or $a=-d$

$(a, d)$ can be $(-1,1)$, $(0,0)$ and $(1,-1)$

(a) When, $(a,d)=(-1,1)$

$a^2+bc=1$

$1+bc=1$

$bc=0$

When, $b=0$, then possibility of $c$ is 12.

When, $c=0$, then possibility of $b$ is 12.

$\therefore$ Total number of possibility $= 24 - 1 = 23$

as (0, 0) is repeated.

(b) When, $(a, d) = (0, 0)$

$a^2 + bc = 1 \Rightarrow bc = 1$

Possible value of $(b, c)$ is $(1, 1), (-1, -1)$

(c) When, $(a, d) = (1, -1)$

There are 23 possible value

**Case II** When $a - d = 0$, then $a = d$

From Eq. (ii),

$ac + cd = 0 \Rightarrow 2ac = 0$

$a \neq 0, c = 0$

From Eq. (i),

$d \neq 0, b = 0$

From Eq. (iii),

$a^2 = 1 \Rightarrow a = \pm 1$

There are only 2 possible values.

Total number of possible values $= 23 + 23 + 2 + 2 = 50$

**54.** *(100)* Given,

$$X = \begin{bmatrix} 0 & 1 & 0 \\ 0 & 0 & 1 \\ 0 & 0 & 0 \end{bmatrix}, Y = \alpha I + \beta X + \gamma X^2$$

$$Z = \alpha^2 I - \alpha\beta X + (\beta^2 - \alpha\gamma) X^2$$

$$Y^{-1} = \begin{bmatrix} \frac{1}{5} & \frac{-2}{5} & \frac{1}{5} \\ 0 & \frac{1}{5} & \frac{-2}{5} \\ 0 & 0 & \frac{1}{5} \end{bmatrix}$$

To find $(\alpha - \beta + \gamma)^2$

From given value of $X$

$$X^2 = \begin{bmatrix} 0 & 0 & 1 \\ 0 & 0 & 0 \\ 0 & 0 & 0 \end{bmatrix}$$

Let $Y = \begin{bmatrix} \alpha & \beta & \gamma \\ 0 & \alpha & \beta \\ 0 & 0 & \alpha \end{bmatrix}$

Then, $Z = \alpha^2 I - \alpha\beta X + (\beta^2 - \alpha\gamma) X^2$

$$= \alpha^2 \begin{bmatrix} 1 & 0 & 0 \\ 0 & 1 & 0 \\ 0 & 0 & 1 \end{bmatrix} - \alpha\beta \begin{bmatrix} 0 & 1 & 0 \\ 0 & 0 & 1 \\ 0 & 0 & 0 \end{bmatrix} + (\beta^2 - \alpha\gamma) \begin{bmatrix} 0 & 0 & 1 \\ 0 & 0 & 0 \\ 0 & 0 & 0 \end{bmatrix}$$

$$= \begin{bmatrix} \alpha^2 & -\alpha\beta & \beta^2 - \alpha\gamma \\ 0 & \alpha^2 & -\alpha\beta \\ 0 & 0 & \alpha^2 \end{bmatrix}$$

Let $Y \cdot Y^{-1} = I$

$$\begin{bmatrix} \alpha & \beta & \gamma \\ 0 & \alpha & \beta \\ 0 & 0 & \alpha \end{bmatrix} \begin{bmatrix} \frac{1}{5} & \frac{-2}{5} & \frac{1}{5} \\ 0 & \frac{1}{5} & \frac{-2}{5} \\ 0 & 0 & \frac{1}{5} \end{bmatrix} = \begin{bmatrix} 1 & 0 & 0 \\ 0 & 1 & 0 \\ 0 & 0 & 1 \end{bmatrix}$$

$$\frac{\alpha}{5} = 1 \Rightarrow \alpha = 5$$

$$\therefore \quad -\frac{2}{5}\alpha + \frac{\beta}{5} = 0 \Rightarrow \frac{\beta}{5} = \frac{2}{5}\alpha$$

$$\beta = 10$$

and $\frac{\alpha}{5} - \frac{2}{5}\beta + \frac{\gamma}{5} = 0$

$$\Rightarrow \frac{5}{5} - \frac{2}{5} \times 10 + \frac{\gamma}{5} = 0 \Rightarrow 1 - 4 + \frac{\gamma}{5} = 0$$

$$\Rightarrow \gamma = 15$$

$$\therefore (\alpha - \beta + \gamma)^2 = (5 - 10 + 15)^2 = 100$$

**55.** *(4)* We have, $A$ is a square matrix of $3 \times 3$. Now, $2\text{Adj}(2\text{Adj}(\text{Adj}(2A)))$

$= 2\text{Adj}(2\text{Adj}(2^{3-1}\text{Adj}A))$

$[\because \text{adj}(KA) = K^{n-1} \text{ adj } A]$

$= 2\text{Adj}(2\text{Adj}(4\text{Adj}\,A))$

$= 2\text{Adj}(2 \times 4^{3-1} \text{ Adj Adj } A)$

$= 2\text{Adj}(32 \text{ Adj Adj } A)$

$= 2 \times 32^{3-1} \text{ Adj Adj Adj } A$

$= 2^{11} \text{ Adj Adj Adj } A$

$= 2^{11} \text{ Adj}(|A|^{3-2} A)$

$= 2^{11} \text{ Adj}(|A| A)$

$= 2^{11} \times |A|^{3-1} \text{ Adj } A$

$= 2^{11} \times |A|^2 \text{ Adj } A$

Now, $|2\text{Adj}(2\text{ Adj}(\text{Adj}(2A)))| = 2^{41}$

$\Rightarrow |2^{11} \times |A|^2 \text{ Adj } A| = 2^{41}$

$\Rightarrow (2^{11})^3 (|A|^2)^3 |\text{Adj } A| = 2^{41}$

$\Rightarrow 2^{33} |A|^6 |A|^{3-1} = 2^{41}$

$\Rightarrow |A|^8 = 2^8 \Rightarrow |A| = \pm 2$

Now, $|A^2| = |A|^2 = (\pm 2)^2 = 4$

**56.** *(108)* Given, $A = \begin{bmatrix} 2 & -1 & 1 \\ -1 & 2 & -1 \\ 1 & -1 & 2 \end{bmatrix}$

Now,

$|A| = 2(4-1) + 1(-2+1) + 1(1-2)$

$= 6 - 1 - 1 = 4$

Then, $|3\text{ adj}(2A^{-1})| = 3^3 |\text{adj}(2A^{-1})|$

$= 27|(2A^{-1})|^2$

$= 27|2A^{-1}| \cdot |2A^{-1}|$

$= 27 \cdot 2^3 |A^{-1}| \cdot 2^3 |A^{-1}|$

$= 27 \cdot 2^6 |A^{-1}|^2 = 27 \cdot 2^6 \cdot \frac{1}{|A|^2}$

$= 27 \cdot 2^6 \cdot \frac{1}{4^2} = 108$

**57.** *(13)* $A = \begin{bmatrix} 0 & -\tan\left(\frac{\theta}{2}\right) \\ \tan\left(\frac{\theta}{2}\right) & 0 \end{bmatrix}$

and $(I_2 + A)(I_2 - A)^{-1} = \begin{bmatrix} a & -b \\ b & a \end{bmatrix}$

$\Rightarrow |(I_2 + A)(I_2 - A)^{-1}| = a^2 + b^2$

$$\Rightarrow \quad a^2 + b^2 = \frac{|I_2 + A|}{|I_2 - A|} \quad \dots \text{(i)}$$

Now,

$$I_2 + A = \begin{bmatrix} 1 & 0 \\ 0 & 1 \end{bmatrix} + \begin{bmatrix} 0 & -\tan\left(\frac{\theta}{2}\right) \\ \tan\left(\frac{\theta}{2}\right) & 0 \end{bmatrix}$$

$$= \begin{bmatrix} 1 & -\tan\left(\frac{\theta}{2}\right) \\ \tan\left(\frac{\theta}{2}\right) & 1 \end{bmatrix}$$

Similarly,

$$I_2 - A = \begin{bmatrix} 1 & \tan\left(\frac{\theta}{2}\right) \\ -\tan\left(\frac{\theta}{2}\right) & 1 \end{bmatrix}$$

Here,

$$|I_2 + A| = |I_2 - A| = \left(1 + \tan^2\left(\frac{\theta}{2}\right)\right)$$

$$\Rightarrow \quad \frac{|I_2 + A|}{|I_2 - A|} = 1 \quad \dots \text{(ii)}$$

From Eqs. (i) and (ii), $a^2 + b^2 = 1$

Now, $13(a^2 + b^2) = 13 \times 1 = 13$

**58.** *(17)* Given, $P = \begin{bmatrix} 3 & -1 & -2 \\ 2 & 0 & \alpha \\ 3 & -5 & 0 \end{bmatrix}$

$$\Rightarrow |P| = \begin{vmatrix} 3 & -1 & -2 \\ 2 & 0 & \alpha \\ 3 & -5 & 0 \end{vmatrix} = (20 + 12\alpha)$$

According to the question,

$$PQ = kI_3 \Rightarrow Q = kP^{-1}I_3$$

Now, $Q = \dfrac{k}{|P|}(\text{adj } P)\, I_3$

$$= \frac{k}{(20+12\alpha)}\begin{bmatrix} 5\alpha & -10 & -\alpha \\ 3\alpha & -6 & -3\alpha-4 \\ -10 & 12 & -2 \end{bmatrix}\begin{bmatrix} 1 & 0 & 0 \\ 0 & 1 & 0 \\ 0 & 0 & 1 \end{bmatrix}$$

$\because$ Given, $q_{23} = -\dfrac{k}{8}$

$\therefore \quad \dfrac{k}{(20+12\alpha)}(-3\alpha-4) = -\dfrac{k}{8}$

$\Rightarrow \quad \alpha = -1$

Also, $|Q| = \dfrac{k^3|I|}{|P|} \Rightarrow \dfrac{k^2}{2} = \dfrac{k^3}{20+12\alpha}$

$\Rightarrow \; 20 + 12\alpha = 2k$

$\Rightarrow \quad k = 4$

$\therefore$Required value of

$k^2 + \alpha^2 = 4^2 + (-1)^2 = 17$

## Topic 4

**01.** *(d)* $\Delta = \begin{vmatrix} 2 & 3 & 5 \\ 7 & 3 & -2 \\ 12 & 3 & -(4+\lambda) \end{vmatrix} = 15\lambda - 75$

$\Rightarrow \quad 15\lambda - 75 = 0$

$\Rightarrow \quad \lambda = 5$

$$\Delta_1 = 0 = \begin{vmatrix} 9 & 3 & 5 \\ 8 & 3 & -2 \\ 16-\mu & 3 & -9 \end{vmatrix}$$

($\because$ infinite many solution, $\Delta = 0 = \Delta_1 = \Delta_2 = \Delta_3$)

$\Rightarrow \quad 21\mu - 189 = 0$

$\Rightarrow \quad \mu = 9$

$\therefore$Radius of circle $= \dfrac{|4\cdot 5 - 3\cdot 9|}{\sqrt{4^2+3^2}} = \dfrac{7}{5}$

**02.** *(b)* $2x + \lambda y + 3z = 5$

$3x + 2y - z = 7$

$4x + 5y + \mu z = 9$

For infinite solutions

$\Delta = 0 = \Delta_1 = \Delta_2 = \Delta_3$

$$\Delta = \begin{vmatrix} 2 & \lambda & 3 \\ 3 & 2 & -1 \\ 4 & 5 & \mu \end{vmatrix}$$

$\Rightarrow \quad -4\lambda - 3\lambda\mu + 4x + 31 = 0$

$$\Delta_1 = \begin{vmatrix} 5 & \lambda & 3 \\ 7 & 2 & -1 \\ 9 & 5 & \mu \end{vmatrix} = 0$$

$\Rightarrow \quad \mu + 5 = 0 \Rightarrow \mu = -5$

$$\Delta_2 = \begin{vmatrix} 2 & 5 & 3 \\ 3 & 7 & -1 \\ 4 & 9 & \mu \end{vmatrix} = 0$$

$\Rightarrow \quad \lambda + 1 = 0 \Rightarrow \lambda = -1$

$\therefore \lambda^2 + \mu^2 = 25 + 1 = 26$

**03.** *(c)* System of linear equations has infinite many solutions

$\Rightarrow \Delta = 0 = \Delta_1 = \Delta_2 = \Delta_3$

$$\Delta_2 = 0 \Rightarrow \begin{vmatrix} 3 & 3 & \beta \\ 2 & -3 & -1 \\ 1 & 4 & 1 \end{vmatrix} = 0 \Rightarrow \beta = \frac{6}{11}$$

$$\Delta_3 = 0 \Rightarrow \begin{vmatrix} 3 & 1 & 3 \\ 2 & \alpha & -3 \\ 1 & 2 & 4 \end{vmatrix} = 0$$

$\Rightarrow \; \alpha = -\dfrac{19}{9} \Rightarrow 22\beta - 9\alpha = 31$

**04.** *(d)* $\Delta = \begin{vmatrix} 1 & 1 & 1 \\ 1 & 2 & 4 \\ 1 & 4 & 10 \end{vmatrix}$

$= 1(20-16) - 1(10-4) + 1(4-2)$

$= 4 - 6 + 2 = 0$

For infinite solutions

$\Delta_x = \Delta_y = \Delta_z = 0$

$m^2 - 3m + 2 = 0$

$m = 1, 2$

$\alpha = 1, \beta = 2$

$$\therefore \sum_{n=1}^{10}(n^\alpha + n^\beta) = \sum_{n=1}^{10} n^1 + \sum_{n=1}^{10} n^2$$

$$= \frac{10(11)}{2} + \frac{10(11)(21)}{6}$$

$= 55 + 385 = 440$

**05.** *(a)* From given system of linear equation, we get

$$\Delta = \begin{vmatrix} 1 & 2 & -3 \\ 2 & \lambda & 5 \\ 14 & 3 & \mu \end{vmatrix},$$

$$\Delta_1 = \begin{vmatrix} 2 & 2 & -3 \\ 5 & \lambda & 5 \\ 33 & 3 & \mu \end{vmatrix},$$

$$\Delta_2 = \begin{vmatrix} 1 & 2 & -3 \\ 2 & 5 & 5 \\ 14 & 33 & \mu \end{vmatrix},$$

and $\Delta_3 = \begin{vmatrix} 1 & 2 & 2 \\ 2 & \lambda & 5 \\ 14 & 3 & 33 \end{vmatrix}$

For infinitely many solutions

Put $\Delta = 0$, $\Delta_1 = 0$, $\Delta_2 = 0$ and $\Delta_3 = 0$

For $\Delta = 0 \Rightarrow \lambda\mu + 42\lambda - 4\mu + 107 = 0$ ...(i)

$\Delta_1 = 0 \Rightarrow 2\lambda\mu + 99\lambda - 10\mu + 255 = 0$ ...(ii)

$\Delta_2 = 0 \Rightarrow 13 - \mu = 0 \Rightarrow \mu = 13$

$\Delta_3 = 0 \Rightarrow 5\lambda + 5 = 0 \Rightarrow \lambda = -1$

We observe that $\mu = 13$, $\lambda = -1$ also satisfies Eqs. (i) and (ii).

Hence, $\lambda + \mu = -1 + 13 = 12$

**06.** *(b)* According to the question,

$$\Delta = 0 \Rightarrow \begin{vmatrix} 2 & -1 & 1 \\ 5 & \lambda & 3 \\ 100 & -47 & \mu \end{vmatrix} = 0$$

$\Rightarrow 2(\lambda\mu + 141) + 1(5\mu - 300)$
$+ 1(-235 - 100\lambda) = 0$ ... (i)

Now, $\Delta_3 = 0 \Rightarrow \begin{vmatrix} 2 & -1 & 4 \\ 5 & \lambda & 12 \\ 100 & -47 & 212 \end{vmatrix} = 0$

$\Rightarrow 2(212\lambda + 564) + 1(212 \times 5 - 1200)$
$+ 4(-47 \times 5 - 100 - \lambda) = 0$

$\Rightarrow 6\lambda = -12 \Rightarrow \lambda = -2$

On putting in Eq. (i), we get

$2(-2\mu + 141) + 5\mu - 300 - 235 + 200 = 0$

$\Rightarrow \mu = 53$

Hence, $\mu - 2\lambda = 53 - 2 \times (-2) = 57$

> • Concept **Enhancer**
>
> The system of equations $a_1x + b_1y + c_1 = d_1$, $a_2x + b_2y + c_2 = d_2$ and $a_3x + b_3y + c_3 = d_3$ is consistent with many infinite solutions if $\Delta = 0 = \Delta_1 = \Delta_2 = \Delta_3$

**07.** *(d)* $\quad x + y + z = 6$

$x + 2y + 5z = 9$

$x + 5y + \lambda z = \mu$

For no solution,

$\Delta = 0$ and atleast one in $\Delta_x$, $\Delta_y$ or $\Delta_z$ is non-zero.

Now, $\Delta = \begin{vmatrix} 1 & 1 & 1 \\ 1 & 2 & 5 \\ 1 & 5 & \lambda \end{vmatrix} = 0$

$\Rightarrow 1(2\lambda - 25) - 1(\lambda - 5) + 1(5 - 2) = 0$

$\Rightarrow \quad 2\lambda - 25 - \lambda + 5 + 3 = 0$

$\Rightarrow \quad \lambda = 17$

$\Delta_z \neq 0$

$$\Rightarrow \begin{vmatrix} 1 & 1 & 6 \\ 1 & 2 & 9 \\ 1 & 5 & \mu \end{vmatrix} \neq 0$$

$$\Rightarrow 1(2\mu - 45) - 1(\mu - 9) + 6(3) \neq 0$$
$$\Rightarrow 2\mu - 45 - \mu + 9 + 18 \neq 0 \Rightarrow \mu \neq 18$$

**Concept Enhancer**

If a system of linear equations in 3 variables has no solution than $\Delta = 0$ and in $\Delta_x, \Delta_y, \Delta_z$ atleast one is non-zero.

**08.** *(c)*

**key idea** A system of linear equations in 3 variables has infinite solution, then $\Delta = \Delta_x = \Delta_y = \Delta_z = 0$

For infinitely many solution

$$D \equiv \begin{vmatrix} \lambda-1 & \lambda-4 & \lambda \\ \lambda & \lambda-1 & \lambda-4 \\ \lambda+1 & \lambda+2 & -(\lambda+2) \end{vmatrix} = 0$$

$$\Rightarrow (\lambda-3)(2\lambda+1) = 0 \quad \ldots(i)$$

$$D_x = \begin{vmatrix} 5 & \lambda-4 & \lambda \\ 7 & \lambda-1 & \lambda-4 \\ 9 & \lambda+2 & -(\lambda+2) \end{vmatrix} = 0$$

$$\Rightarrow 2(3-\lambda)(23-2\lambda) = 0$$
$$\Rightarrow \lambda = 3 \quad \ldots(ii)$$

[common solution of Eqs. (i) and (ii)]

$$\therefore \lambda^2 + \lambda = 9 + 3 = 12$$

**09.** *(c)* $x + y + 2z = 6$

$2x + 3y + az = a + 1$

$-x - 3y + bz = 2b$

For infinite many solution

$$\Delta = \Delta_x = \Delta_y = \Delta_z = 0$$

$$\Delta = \begin{vmatrix} 1 & 1 & 2 \\ 2 & 3 & a \\ -1 & -3 & b \end{vmatrix} = 0$$

$$\Rightarrow 1(3b + 3a) - 1(2b + a) - 2(-6 + 3) = 0$$
$$\Rightarrow 2a + b - 6 = 0 \quad \ldots(i)$$

$$\Delta_z = \begin{vmatrix} 1 & 1 & 6 \\ 2 & 3 & a+1 \\ -1 & -3 & 2b \end{vmatrix} = 0$$

$$= 1(6b + 3a + 3) - 1(4b + a + 1) + 6(-6 + 3) = 0$$
$$\Rightarrow a + b - 8 = 0 \quad \ldots(ii)$$

Solving Eqs. (i) and (ii), we get

$a = -2$ and $b = 10$

$7a + 3b = 7(-2) + 3(10) = 16$

**10.** *(b)* Given system of equations has infinitely many solutions so, $\Delta = 0$

$$\Rightarrow \begin{vmatrix} 1 & 4 & -1 \\ 7 & 9 & \mu \\ 5 & 1 & 2 \end{vmatrix} = 0$$

$$\Rightarrow (18 - \mu) - 4(14 - 5\mu) - 1(-38) = 0$$
$$\Rightarrow 18 - \mu - 56 + 20\mu + 38 = 0$$
$$\Rightarrow \mu = 0$$
$$\Delta_x = \Delta_y = \Delta_z = 0$$

(for infinite solutions)

$$\text{So, } \Delta_x = \begin{vmatrix} \lambda & 4 & -1 \\ -3 & 9 & 0 \\ -1 & 1 & 2 \end{vmatrix} = 0$$

$$\Rightarrow \lambda(18) - 4(-6) - 1(-3 + 9) = 0$$
$$\Rightarrow 18\lambda = -24 + 6 = -18$$
$$\Rightarrow \lambda = -1$$

Therefore, $2\mu + 3\lambda = -3$

**11.** *(d)* For non-trivial solution

$$\begin{vmatrix} 1 & \sqrt{2}\sin\alpha & \sqrt{2}\cos\alpha \\ 1 & \sin\alpha & -\cos\alpha \\ 1 & \cos\alpha & \sin\alpha \end{vmatrix} = 0$$

$$\Rightarrow 1 - \sqrt{2}\sin\alpha(\sin\alpha + \cos\alpha) + \sqrt{2}\cos\alpha(\cos\alpha - \sin\alpha) = 0$$

(Expanding along $R_1$)

$$\Rightarrow 1 + \sqrt{2}\cos 2\alpha - \sqrt{2}\sin 2\alpha = 0$$
$$\Rightarrow \cos 2\alpha - \sin 2\alpha = -1/\sqrt{2}$$
$$\Rightarrow \cos(2\alpha + \pi/4) = -1/2$$
$$\Rightarrow 2\alpha + \frac{\pi}{4} = 2n\pi \pm \frac{2\pi}{3}, n \in I$$
$$\Rightarrow \alpha + \frac{\pi}{8} = n\pi \pm \frac{\pi}{3}, n \in I$$

For $n = 0$, $\alpha = \frac{\pi}{3} - \frac{\pi}{8} = \frac{5\pi}{24}$

**12.** *(c)* Given system of equation

$2x + 3y - z = 5,$

$x + \alpha y + 3z = -4$

and $3x - y + \beta z = 7$

has infinitely many solution

$$\begin{vmatrix} 2 & 3 & -1 \\ 1 & \alpha & 3 \\ 3 & -1 & \beta \end{vmatrix} = 0$$

$$2(\alpha\beta + 3) - 3(\beta - 9) - 1(-1 - 3\alpha) = 0$$
$$2\alpha\beta + 6 - 3\beta + 27 + 1 + 3\alpha = 0$$
$$2\alpha\beta - 3\beta + 3\alpha + 34 = 0 \ldots(i)$$

$$\text{Also, } \begin{vmatrix} 2 & 3 & 5 \\ 1 & \alpha & -4 \\ 3 & -1 & 7 \end{vmatrix} = 0$$

$$2(7\alpha - 4) - 3(7 + 12) + 5(-1 - 3\alpha) = 0$$
$$14\alpha - 8 - 57 - 5 - 15\alpha = 0$$
$$\alpha = -70 \quad \ldots(ii)$$

Put, $\alpha = -70$ in Eq. (i), we get

$$-140\beta - 3\beta - 210 + 34 = 0$$
$$\beta = \frac{-176}{143} = -\frac{16}{13}$$
$$\therefore 13\alpha\beta = 13(-70)\left(-\frac{16}{13}\right) = 1120$$

**13.** *(b)* We have, $\alpha x + y + z = 1$

$x + \alpha y + z = 1$

$x + y + \alpha z = \beta$

$$\therefore D = \begin{vmatrix} \alpha & 1 & 1 \\ 1 & \alpha & 1 \\ 1 & 1 & \alpha \end{vmatrix}$$

$$= \alpha(\alpha^2 - 1) - 1(\alpha - 1) + 1(1 - \alpha)$$
$$= (\alpha - 1)^2(\alpha + 2)$$

Now, if $D = 0$

$\Rightarrow (\alpha - 1)^2(\alpha + 2) = 0 \Rightarrow \alpha = 1, -2$

So, at $\alpha = 1, \beta = 1,$

all equations become same.

Hence, system has infinite solution.

Again, at $\alpha = -2, \beta = 1, D = 0$

So, system has no solution.

On adding all equations, we have

$$(2 + \alpha)x + (2 + \alpha)y + (2 + \alpha)z = \beta + 2$$
$$\Rightarrow x + y + z = \frac{\beta + 2}{\alpha + 2}$$

When, $\alpha = 2, \beta = 1 \Rightarrow x + y + z = \frac{3}{4}$

**Concept Enhancer**

Let $AX = B$ be a system of $m$ linear equations in $n$ variables.

If $|A| \neq 0$, then the system is consistent and has unique solution.

If $|A| = 0$ and $(\text{adj } A)B = 0$, then system is consistent and has infinite number of solutions.

If $|A| = 0$ and $(\text{adj } A)B \neq 0$, then system is inconsistent and has no solutions.

**14.** *(b)* We have, $\lambda x + y + z = 1$

$x + \lambda y + z = 1$

$x + y + \lambda z = 1$

$$\therefore D = \begin{vmatrix} \lambda & 1 & 1 \\ 1 & \lambda & 1 \\ 1 & 1 & \lambda \end{vmatrix}$$

$$= \lambda(\lambda^2 - 1) - 1(\lambda - 1) + 1(1 - \lambda)$$
$$= (\lambda - 1)^2(\lambda + 2)$$

Now, when $D = 0$

$$\Rightarrow (\lambda - 1)^2(\lambda + 2) = 0 \Rightarrow \lambda = 1, -2$$

But when $\lambda=1$, all the equations becomes identical, so system will have infinite solution.

So, for inconsistency, $\lambda=-2$

$$\therefore \sum_{\lambda \in S}(|\lambda|^2+|\lambda|)=|-2|^2+|-2|$$

$$=4+2=6$$

**Concept Enhancer**

The system is inconsistent, if at least one of the determinants, $D_x, D_y$ or $D_z$ has a value not equal to zero and the denominator determinant has a value of zero.

**15.** *(c)* Given that system of linear equations,

$x-y+z=5,\ 2x+2y+\alpha z=8$

and $\quad 3x-y+4z=\beta$

has infinitely many solution, then

$$\Delta=\Delta x=\Delta y=\Delta z=0$$

$$\Delta=\begin{vmatrix}1 & -1 & 1\\ 2 & 2 & \alpha\\ 3 & -1 & 4\end{vmatrix}=0$$

$$8+\alpha+8-3\alpha-8=0 \Rightarrow \alpha=4$$

$$\Delta z=\begin{vmatrix}1 & -1 & 5\\ 2 & 2 & 8\\ 3 & -1 & \beta\end{vmatrix}=0$$

$$4\beta=56 \Rightarrow \beta=14$$

$$\Rightarrow \quad \alpha=4 \text{ and } \beta=14$$

So, equation is

$$x^2-(\alpha+\beta)+\alpha\cdot\beta=0$$

$$\Rightarrow \quad x^2-18x+56=0$$

**16.** *(d)* For unique and infinitely many solutions

$$\Delta \neq 0$$

$$\begin{vmatrix}a & 2a & -3a\\ 2a+1 & 2a+3 & a+1\\ 3a+5 & a+5 & a+2\end{vmatrix} \neq 0$$

$$a[(2a+3)(a+2)-(a+1)(a+5)]-2a[(2a+1)(a+2)-(3a+5)(a+1)]$$
$$-3a[(2a+1)(a+5)-(3a+5)(2a+3)] \neq 0$$

$$a[15a^2+31a+36] \neq 0$$

Now, $15a^2+31a+36$ is always positive

So, $\quad a \neq 0$

$$a \in R-\{0\}$$

Hence, $S_1=R-\{0\}$ and $S_2=\phi$

**17.** *(b)* System of linear equations,

$$2x+y-z=7$$
$$x-3y+2z=1$$
$$x+4y+\delta z=k,$$

where $\delta, k \in R$

For infinitely many solutions, $\Delta=0$

$$\begin{vmatrix}2 & 1 & -1\\ 1 & -3 & 2\\ 1 & 4 & \delta\end{vmatrix}=0$$

$$\Rightarrow -1(4+3)-2(8-1)+\delta(-6-1)=0$$

$$\Rightarrow \quad \delta=-3$$

and $\begin{vmatrix}2 & 1 & 7\\ 1 & -3 & 1\\ 1 & 4 & k\end{vmatrix}=0$

$$\Rightarrow 7(4+3)-1(8-1)+k(-6-1)=0$$

$$\Rightarrow k=6$$

$$\Rightarrow \quad \delta+k=-3+6=3$$

**18.** *(b)* The system of given equation has no solution,

$$\Delta=\begin{vmatrix}2 & 3 & -1\\ 1 & 1 & 1\\ 1 & -1 & |\lambda|\end{vmatrix}=0 \Rightarrow |\lambda|=7$$

But at $\lambda=7$, $D_x=D_y=D_z=0$

$P_1: 2x+3y-z=-2$

$P_2: x+y+z=4$

$P_3: x-y+|\lambda|z=4\lambda-4$

So, clearly $5P_2-2P_1=P_3$, so at $\lambda=7$, system of equation is having infinite solutions.

So, $\lambda=-7$ is correct answer.

**19.** *(c)* $\quad \alpha x+y+z=5$

$$x+2y+3z=4$$
$$x+3y+5z=\beta$$

For infinitely many solutions,

$$\Delta=\Delta_1=\Delta_2=\Delta_3=0$$

$$\Delta=\begin{vmatrix}\alpha & 1 & 1\\ 1 & 2 & 3\\ 1 & 3 & 5\end{vmatrix}$$

$$=\alpha(1)-(5-3)+(3-2)$$
$$=\alpha-2+1=\alpha-1$$

$$\therefore \quad \Delta=0$$

$$\Rightarrow \quad \alpha-1=0$$

$$\Rightarrow \quad \alpha=1$$

$$\Delta_1=\begin{vmatrix}5 & 1 & 1\\ 4 & 2 & 3\\ \beta & 3 & 5\end{vmatrix}=0$$

$$\Rightarrow \quad 5(10-9)-4(5-3)+\beta(3-2)=0$$

$$\Rightarrow \quad 5-8+\beta=0$$

$$\Rightarrow \quad -3+\beta=0$$

$$\Rightarrow \quad \beta=3$$

$$\therefore \quad (\alpha,\beta)=(1,3)$$

**20.** *(d)* The given system of equations is consistent if $\Delta \neq 0$

$$\begin{vmatrix}-k & 3 & -14\\ -15 & 4 & -k\\ -4 & 1 & 3\end{vmatrix} \neq 0$$

$$\Rightarrow -k(12+k)-3(-45-4k)$$
$$-14(-15+16) \neq 0$$

$$\Rightarrow -k^2+121 \neq 0 \Rightarrow k^2 \neq 121$$

$$\Rightarrow \quad k \neq -11 \text{ or } 11$$

Hence, the system is consistent if

$$k \in R-\{-11, 11\}$$

**21.** *(b)* The given system of equation is

$$x+y+z=\alpha$$
$$\alpha x+2\alpha y+3z=-1$$

$x+3\alpha y+5z=4$ is inconsistent if $\Delta=0$

Where, $\Delta=\begin{vmatrix}1 & 1 & 1\\ \alpha & 2\alpha & 3\\ 1 & 3\alpha & 5\end{vmatrix}=0$

$$=1(10\alpha-9\alpha)-1(5\alpha-3)+1(3\alpha^2-2\alpha)$$
$$=3\alpha^2-6\alpha+3=0$$
$$\Rightarrow (\alpha-1)^2=0$$

Hence, the value of '$\alpha$' is 1.

**22.** *(c)* For inconsistent system of equations $\Delta=0$ and atleast one is non-zero in $\Delta_1, \Delta_2$ and $\Delta_3$

$$\Delta=\begin{vmatrix}-1 & 1 & 2\\ 3 & -a & 5\\ 2 & -2 & -a\end{vmatrix}=0$$

$$\Rightarrow \quad a^2-7a+12=0$$

$$\Rightarrow \quad a=3, 4$$

$$\Delta_x=\begin{vmatrix}0 & 1 & 2\\ 1 & -a & 5\\ 7 & -2 & -a\end{vmatrix}$$

$$=15a+31$$

$$\Delta_x \neq 0 \text{ for } a=3, 4$$

$$\Rightarrow \quad n(S_1)=2$$

Now, for infinitely many solutions, if $\Delta=0$ also $\Delta_x=\Delta_y=\Delta_z=0$ which is not possible for any real value of $a$

$$\Rightarrow \quad n(S_2)=0$$

**23.** *(b)* Given $\alpha+\beta+\gamma=2\pi$

$$\Delta=\begin{vmatrix}1 & \cos\gamma & \cos\beta\\ \cos\gamma & 1 & \cos\alpha\\ \cos\beta & \cos\alpha & 1\end{vmatrix}$$

$$=1-\cos^2\alpha-\cos\gamma(\cos\gamma-\cos\alpha\cos\beta)$$
$$+\cos\beta(\cos\alpha\cos\gamma-\cos\beta)$$

$=1-\cos^2\alpha-\cos^2\beta-\cos^2\gamma$
$+2\cos\alpha\cos\beta\cos\gamma$
$=\sin^2\alpha-\cos^2\beta$
$-\cos\gamma(\cos\gamma-2\cos\alpha\cos\beta)$
$=-\cos(\alpha+\beta)\cos(\alpha-\beta)-\cos\gamma$
$(\cos(2\pi-(\alpha+\beta))-2\cos\alpha\cos\beta)$
$=-\cos(2\pi-\gamma)\cos(\alpha-\beta)-\cos\gamma$
$(\cos(\alpha+\beta)-2\cos\alpha\cos\beta)$
$=-\cos\gamma\cos(\alpha-\beta)+\cos\gamma$
$(\cos\alpha\cos\beta+\sin\alpha\sin\beta)$
$=-\cos\gamma\cos(\alpha-\beta)+\cos\gamma\cos(\alpha-\beta)$
$=0$

So, the system of equation has infinitely many solutions.

**24.** *(d)* $\Delta=\begin{vmatrix}2&1&1\\1&-1&1\\1&1&a\end{vmatrix}$
$=2(-a-1)-1(a-1)+(1+1)$
$=1-3a$

$\Delta_3=\begin{vmatrix}2&1&5\\1&-1&3\\1&1&b\end{vmatrix}=2(-b-3)-1$
$(b-3)+5(1+1)=7-3b$

Now, $z=\Delta_3/\Delta$

If $\Delta=0$ and $\Delta_3\neq0$, then no solution

$1-3a=0$

$\Rightarrow 7-3b\neq0 \Rightarrow a=\frac{1}{3}, b\neq\frac{7}{3}$

**25.** *(a)* Given, system of equations

$x+y+z=4$
$3x+2y+5z=3$
$9x+4y+(28+|\lambda|)z=|\lambda|$

$\Delta=\begin{vmatrix}1&1&1\\3&2&5\\9&4&28+|\lambda|\end{vmatrix}$

$=1(56+2|\lambda|-20)-1(84+3|\lambda|-45)$
$+1(-6)$
$=-(|\lambda|+9)$

If $\Delta\neq0$ i.e $|\lambda|+9\neq0$, then system of equation has unique solution.

If $|\lambda|+9=0$, then

$\Delta_1=\Delta_2=\Delta_3=0,$

the system of equation has infinite solution.

$\Rightarrow \lambda\in R$

**26.** *(a)* Given, system of linear equations

$4x+\lambda y+2z=0$
$2x-y+z=0$
$\mu x+2y+3z=0$

For non-trivial solution, $\Delta=0$

$\begin{vmatrix}4&\lambda&2\\2&-1&1\\\mu&2&3\end{vmatrix}=0$

$\Rightarrow 4(-3-2)-\lambda(6-\mu)+2(4+\mu)=0$
$\Rightarrow -\lambda(6-\mu)-2(6-\mu)=0$
$\Rightarrow (6-\mu)(\lambda+2)=0$
$\Rightarrow \lambda=-2$ and $\mu\in R$ or $\mu=6$
and $\lambda\in R$.

**27.** *(b)* The given system of equations is non-homogeneous and it can be written

as, $\begin{bmatrix}2&3&2\\3&2&2\\1&-1&4\end{bmatrix}\begin{bmatrix}x\\y\\z\end{bmatrix}=\begin{bmatrix}9\\9\\8\end{bmatrix}$

i.e,. $AX=B$

Now,

$|A|=2(8+2)-3(12-2)+2(-3-2)$
$=20-30-10=-20\neq0$

$\because |A|\neq0$, then this system have unique solution.

**28.** *(c)* Given, $A$ be a $3\times3$ matrix.

$A$ is symmetric and $B$ is skew-symmetric.

$\therefore \quad A^T=A, B^T=-B$

Let $A^2B^2-B^2A^2=P$
$P^T=(A^2B^2-B^2A^2)^T$
$=(A^2B^2)^T-(B^2A^2)^T$
$=(B^2)^T(A^2)^T-(A^2)^T(B^2)^T$
$=B^2A^2-A^2B^2=-(A^2B^2-B^2A^2)$
$=-P \Rightarrow P^T=-P$

$P$ is skew-symmetric.

$\therefore |P|=0$

Hence, $PX=0$ have infinite solutions.

**29.** *(d)* Given system of linear equations

$2x_1-4x_2+\lambda x_3=1$
$x_1-6x_2+x_3=2$
and $\lambda x_1-10x_2+4x_3=3$

is inconsistent, so $\Delta=0$

$\Rightarrow \begin{vmatrix}2&-4&\lambda\\1&-6&1\\\lambda&-10&4\end{vmatrix}=0$

$\Rightarrow 2(-24+10)+4(4-\lambda)$
$+\lambda(-10+6\lambda)=0$
$\Rightarrow 6\lambda^2-14\lambda-12=0$
$\Rightarrow 3\lambda^2-7\lambda-6=0$
$\Rightarrow 3\lambda(\lambda-3)+2(\lambda-3)=0$
$\Rightarrow \lambda=-\frac{2}{3}, 3$

On adding first and second given linear equations we have

$3x_1-10x_2+(\lambda+1)x_3=3$

And at $\lambda=3$, it is same as the third linear equation, so for $\lambda=3$, there are infinitly many solution, but for $\lambda=\frac{-2}{3}$ both are different, so for $\lambda=\frac{-2}{3}$, the system of linear equations is inconsistent.

**30.** *(b)* It is given that the three planes

$x+4y-2z=1,\ x+7y-5z=\beta$
and $x+5y+\alpha z=5$

having intersection is a line in $R^3$, so

$\Delta=0$

$\Rightarrow \begin{vmatrix}1&4&-2\\1&7&-5\\1&5&\alpha\end{vmatrix}=0$

$\Rightarrow 1(7\alpha+25)-4(\alpha+5)-2(5-7)=0$
$\Rightarrow 3\alpha+9=0$
$\Rightarrow \alpha=-3$ and $\Delta_3=0$

$\Rightarrow \begin{vmatrix}1&4&1\\1&7&\beta\\1&5&5\end{vmatrix}=0$

$\Rightarrow 1(35-5\beta)-4(5-\beta)+1(5-7)=0$
$\Rightarrow \beta=13$
$\therefore \alpha+\beta=-3+13=10$

**31.** *(b)* Given system of linear equations

$2x+2ay+az=0$ ...(i)
$2x+3by+bz=0$ ...(ii)
and $2x+4cy+cz=0$ ...(iii)

Where, $a, b, c\in R$ are non-zero and distinct has a non-zero solution, means

$\Delta=0$

$\begin{vmatrix}2&2a&a\\2&3b&b\\2&4c&c\end{vmatrix}=0$

Expanding along $C_1$

$2(3bc-4cb)-2(2ac-4ac)$
$+2(2ab-3ab)=0$
$\Rightarrow -2bc+4ac-2ab=0$
$\Rightarrow 2ac=ab+bc$
$\Rightarrow \frac{2}{b}=\frac{1}{c}+\frac{1}{a}$ [on dividing by $abc$]
$\Rightarrow \frac{1}{a},\frac{1}{b},\frac{1}{c}$ are in AP.

**32.** *(c)* Given system of linear equations

$x-2y+kz=1$ ...(i)
$2x+y+z=2$ ...(ii)
and $3x-y-kz=3$ ...(iii)

has a solution $(x, y, z), z \neq 0$.

On adding Eqs. (i) and (iii), we get

$$x - 2y + kz + 3x - y - kz = 1 + 3$$
$$4x - 3y = 4$$
$$\Rightarrow \quad 4x - 3y - 4 = 0$$

This is the required equation of the straight line in which point $(x, y)$ lies.

**33.** *(b)* Given system of linear equations is

$$x - cy - cz = 0,$$
$$cx - y + cz = 0$$
and $$cx + cy - z = 0$$

We know that a homogeneous system of linear equations have non-trivial solutions iff

$$\Delta = 0 \Rightarrow \begin{vmatrix} 1 & -c & -c \\ c & -1 & c \\ c & c & -1 \end{vmatrix} = 0$$

$$\Rightarrow 1(1 - c^2) + c(-c - c^2) - c(c^2 + c) = 0$$
$$\Rightarrow \quad 2c^3 + 3c^2 - 1 = 0$$
$$\Rightarrow \quad (c + 1)(2c - 1)(c + 1) = 0$$
$$\Rightarrow \quad c = -1 \text{ or } \frac{1}{2}$$

Clearly, the greatest value of $c$ is $\frac{1}{2}$.

> • Concept **Enhancer**
>
> Homogeneous system of equations let $AX = 0$ is a system of $n$ linear equations in $n$ variables.
>
> (a) If $|A| \neq 0$, then it has only one solution $x = 0$, is called the trival solution.
>
> (b) If $|A| = 0$, then the system has infinitely many solutions, called non-trivial solution.

**34.** *(a)* Given system of linear equations,

$$(1 + \alpha)x + \beta y + z = 2$$
$$\alpha x + (1 + \beta)y + z = 3$$
$$\alpha x + \beta y + 2z = 2$$

has a unique solution, if

$$\begin{vmatrix} 1+\alpha & \beta & 1 \\ \alpha & (1+\beta) & 1 \\ \alpha & \beta & 2 \end{vmatrix} \neq 0$$

Expanding along $C_3$,

$$1(\beta\alpha - \alpha - \alpha\beta) - 1(\beta + \beta\alpha - \beta\alpha)$$
$$\Rightarrow \quad +2(1 + \alpha + \beta + \beta\alpha - \alpha\beta) \neq 0$$
$$\Rightarrow \quad -\alpha - \beta + 2 + 2\alpha + 2\beta \neq 0$$
$$\Rightarrow \quad \alpha + \beta + 2 \neq 0$$

Note that, only (2, 4) satisfy the Eq. (i)

**35.** *(d)* Given, system of linear equations is $x + \lambda y - z = 0$;
$\lambda x - y - z = 0$; $x + y - \lambda z = 0$

Note that, given system will have a non-trivial solution only if determinant of coefficient matrix is zero, i.e.

$$\begin{vmatrix} 1 & \lambda & -1 \\ \lambda & -1 & -1 \\ 1 & 1 & -\lambda \end{vmatrix} = 0$$

$$\Rightarrow 1(\lambda + 1) - \lambda(-\lambda^2 + 1) - 1(\lambda + 1) = 0$$
$$\Rightarrow \quad \lambda + 1 + \lambda^3 - \lambda - \lambda - 1 = 0$$
$$\Rightarrow \quad \lambda^3 - \lambda = 0 \Rightarrow \lambda(\lambda^2 - 1) = 0$$
$$\Rightarrow \quad \lambda = 0 \text{ or } \lambda = \pm 1$$

Hence, given system of linear equation has a non-trivial solution for exactly three values of $\lambda$.

**36.** *(b)* $x - ky + z = 0$, $kx + 3y - kz = 0$ and $3x + y - z = 0$ has trivial solution.

$$\Rightarrow \begin{vmatrix} 1 & -k & 1 \\ k & 3 & -k \\ 3 & 1 & -1 \end{vmatrix} \neq 0$$

$$\Rightarrow 1(-3 + k) + k(-k + 3k) + 1(k - 9) \neq 0$$
$$\Rightarrow \quad k - 3 + 2k^2 + k - 9 \neq 0$$
$$\Rightarrow 2k^2 + 2k - 12 \neq 0$$
$$\Rightarrow k^2 + k - 6 \neq 0$$
$$\Rightarrow (k + 3)(k - 2) \neq 0$$
$$\therefore \quad k \neq 2, -3$$
$$\Rightarrow k \in R - \{2, -3\}$$

**37.** *(d)* The given system of linear equations can be put in the matrix form as

$$\begin{bmatrix} 1 & 2 & 1 \\ 2 & 3 & 1 \\ 3 & 5 & 2 \end{bmatrix} \begin{bmatrix} x_1 \\ x_2 \\ x_3 \end{bmatrix} = \begin{bmatrix} 3 \\ 3 \\ 1 \end{bmatrix}$$

$$\sim \begin{bmatrix} 1 & 2 & 1 \\ 0 & -1 & -1 \\ 0 & -1 & -1 \end{bmatrix} \begin{bmatrix} x_1 \\ x_2 \\ x_3 \end{bmatrix} = \begin{bmatrix} 3 \\ -3 \\ -8 \end{bmatrix}$$

$$\left[\text{by } R_2 \to R_2 - 2R_1, \; R_3 \to R_3 - 3R_1\right]$$

$$\sim \begin{bmatrix} 1 & 2 & 1 \\ 0 & -1 & -1 \\ 0 & 0 & 0 \end{bmatrix} \begin{bmatrix} x_1 \\ x_2 \\ x_3 \end{bmatrix} = \begin{bmatrix} 3 \\ -3 \\ -5 \end{bmatrix}$$

$$[\text{by } R_3 \to R_3 - R_2]$$

Clearly, the given system of equations has no solution.

> ***Alternate Solution***
>
> Subtracting the addition of first two equations from third equation, we get
> $0 = -5$, which is an absurd result.
>
> Hence, the given system of equations has no solution.

**38.** *(a)* Given equations are

$$x - cy - bz = 0, \; cx - y + az = 0$$
and $$bx + ay - z = 0$$

For non-zero solution, $\begin{vmatrix} 1 & -c & -b \\ c & -1 & a \\ b & a & -1 \end{vmatrix} = 0$

$$\Rightarrow 1(1 - a^2) + c(-c - ab) - b(ac + b) = 0$$
$$\Rightarrow 1 - a^2 - c^2 - abc - abc - b^2 = 0$$
$$\Rightarrow \quad a^2 + b^2 + c^2 + 2abc = 1$$

**39.** *(c)* Since, the system of linear equations has a non-zero solution, then

$$\begin{vmatrix} 1 & 2a & a \\ 1 & 3b & b \\ 1 & 4c & c \end{vmatrix} = 0$$

Expanding along $C_1$

$$1(3bc - 4bc) - 1(2ac - 4ac)$$
$$+1(2ab - 3ab) = 0$$
$$\Rightarrow -bc + 2ac - ab = 0$$
$$\Rightarrow \quad 2ac = bc + ab$$

On dividing by $abc$ both sides, we get

$$\frac{2}{b} = \frac{1}{a} + \frac{1}{c}$$

So, $a, b$ and $c$ are in HP.

**40.** *(d)* Given system of equations

$$3x + 5y + \lambda z = 3 \quad \text{...(i)}$$
$$7x + 11y - 9z = 2 \quad \text{...(ii)}$$
$$97x + 155y - 189z = \mu \quad \text{...(iii)}$$

On solving Eqs. (i), (iii), we get

$$-4x + (31\lambda + 189)z = 93 - \mu \quad \text{...(iv)}$$

On solving Eq. (ii) $\times 155$ – Eq.(iii) $\times 11$, we get,

$$1085x + 1705y - 1395z = 310$$
$$1067x + 1705y - 2079z = 11\mu$$
$$- \quad - \quad + \quad -$$
$$18x + 684z = 310 - 11\mu \quad \text{...(v)}$$

Again, Eq. (iv) $\times 9$ + Eq. (v) $\times 2$, we get

$$-36x + 9(31\lambda + 189)z = 9(93 - \mu)$$
$$36x + 1368z = 2(310 - 11\mu)$$
$$(279\lambda + 3069)z = 1457 - 31\mu$$

For infinite solution

$$\lambda = \frac{-3069}{279} = \frac{-341}{31}$$

and $$\mu = \frac{1457}{31}$$

Hence, $$\mu + 2\lambda = \frac{1457 - 682}{31}$$
$$= \frac{775}{31} = 25$$

**Alternate Solution**

Since, the system of equations has infinitely many solutions

So, $\Delta=0, \Delta x=0, \Delta y=0$ and $\Delta z=0$

$\therefore \begin{vmatrix} 3 & 5 & \lambda \\ 7 & 11 & -9 \\ 97 & 155 & -189 \end{vmatrix}$

$\Rightarrow 3(11\times(-189)+9\times155)-5(7\times-189)+9(97))+\lambda(7\times155-97\times11)=0$

$\Rightarrow -2052+2250+18\lambda=0$

$\Rightarrow \lambda=\frac{-198}{18}=-11$ and $\begin{vmatrix} 3 & 5 & 3 \\ 7 & 11 & 2 \\ 97 & 155 & -\mu \end{vmatrix}=0$

$\Rightarrow 3(11\mu-310)-5(7\mu-194)+3(7\times155-97\times11)=0$

$\Rightarrow 33\mu-930-35\mu+970+54=0$

$\Rightarrow -2\mu=-94 \Rightarrow \mu=47$

Hence, $\mu+2\lambda=47-22=25$

**41.** *(a)*

**key idea** For infinite numbers of solution we have to solve for $\Delta=0$ and $\Delta_x=\Delta_y=\Delta_z=0$

We have,

$x+2y+3z=5$ ...(i)

$2x+3y+z=9$ ...(ii)

$4x+3y+\lambda z=\mu$ ...(iii)

Now, $\Delta=\begin{vmatrix} 1 & 2 & 3 \\ 2 & 3 & 1 \\ 4 & 3 & \lambda \end{vmatrix}$

For infinite number of solutions, $\Delta=0$

$\Delta=1(3\lambda-3)-2(2\lambda-4)+3(6-12)=0$

$\Rightarrow 3\lambda-3-4\lambda+8-18=0$

$\Rightarrow -\lambda-13=0 \Rightarrow \lambda=-13$

Also, $\Delta_1=0$

$\therefore \begin{vmatrix} 5 & 2 & 3 \\ 9 & 3 & 1 \\ \mu & 3 & -13 \end{vmatrix}=0$ $[\because \lambda=-13]$

$\Rightarrow 5(-39-3)-2(-\mu-117)+3(27-3\mu)=0$

$\Rightarrow -210+2\mu+234+81-9\mu=0$

$\Rightarrow 105-7\mu=0 \Rightarrow \mu=15$

Now, $\lambda+2\mu=-13+2\times15=17$

**42.** *(a)*

**key idea** Use Cramer's rule, i.e. a system of equations is consistent with infinitely many solution, if $D=0$ and $D_1=D_2=D_3=0$

System

$x-2y+z=-4$

$2x+\alpha y+3z=5$

$3x-y+\beta z=3$ has infinite solution

$\Delta=\Delta_x=\Delta_y=\Delta_z=0$

$\Delta_y=0$

$\begin{vmatrix} 1 & -4 & 1 \\ 2 & 5 & 3 \\ 3 & 3 & \beta \end{vmatrix}=0$

$1(5\beta-9)+4(2\beta-9)+1(6-15)=0$

$\Rightarrow 13\beta-9-36-9=0$

$\Rightarrow 13\beta=54$

$\Delta_z=0$

$\begin{vmatrix} 1 & -2 & -4 \\ 2 & \alpha & 5 \\ 3 & -1 & 3 \end{vmatrix}=0$

$\Rightarrow 1(3\alpha+5)+2(6-15)-4(-2-3\alpha)=0$

$\Rightarrow 3\alpha+5-18+8+12\alpha=0$

$\Rightarrow 15\alpha=5 \Rightarrow \alpha=1/3$

$\therefore 12\alpha+13\beta=4+54=58$

**43.** *(b)* Given equations are

$x+y+z=4\mu,$

$x+2y+2\lambda z=10\mu$

and $x+3y+4\lambda^2 z=\mu^2+15$

We can write as $[A/B]$ form

$\left[\begin{array}{ccc|c} 1 & 1 & 1 & 4\mu \\ 1 & 2 & 2\lambda & 10\mu \\ 1 & 3 & 4\lambda^2 & \mu^2+15 \end{array}\right]$

Applying $R_2 \leftrightarrow R_2-R_1$ and $R_3 \leftrightarrow R_3-R_2$,

$\sim\left[\begin{array}{ccc|c} 1 & 1 & 1 & 4\mu \\ 0 & 1 & 2\lambda-1 & 6\mu \\ 0 & 1 & 4\lambda^2-2\lambda & \mu^2-10\mu+15 \end{array}\right]$

Applying $R_3 \leftrightarrow R_3-R_2$,

$\sim\left[\begin{array}{ccc|c} 1 & 1 & 1 & 4\mu \\ 0 & 1 & 2\lambda-1 & 6\mu \\ 0 & 0 & 4\lambda^2-4\lambda+1 & \mu^2-16\mu+15 \end{array}\right]$

For system has unique solution, if $4\lambda^2-4\lambda+1\neq0$

$\Rightarrow (2\lambda-1)^2\neq0 \Rightarrow \lambda\neq1/2$

and $\mu^2-16\mu+15\neq0$

$\Rightarrow (\mu-1)(\mu-15)\neq0 \Rightarrow \mu\neq1, 15$

Hence, system has unique solution, if $\lambda\neq1/2$ and $\mu\neq1, 15$.

For infinite solution $\lambda=\frac{1}{2}, \mu=1$ or 15.

**44.** *(c)* Given system of equations are

$2x+y-z=5$

$2x-5y+\lambda z=\mu$

$x+2y-5z=7$

has infinitely many solutions.

So, $\begin{vmatrix} 2 & 1 & -1 \\ 2 & -5 & \lambda \\ 1 & 2 & -5 \end{vmatrix}=0$

$\Rightarrow 2(25-2\lambda)+1(\lambda+10)-1(4+5)=0$

$\Rightarrow 50-4\lambda+\lambda+10-9=0$

$\Rightarrow 3\lambda=51 \Rightarrow \lambda=17$

and $\begin{vmatrix} 2 & 1 & 5 \\ 2 & -5 & \mu \\ 1 & 2 & 7 \end{vmatrix}=0$

$\Rightarrow 2(-35-2\mu)-1(14-\mu)+5(4+5)=0$

$\Rightarrow -70-4\mu-14+\mu+45=0$

$\Rightarrow 3\mu=39 \Rightarrow \mu=13$

Thus, $(\lambda+\mu)^2+(\lambda-\mu)^2$

$=2(\lambda^2+\mu^2)=2(17^2+13^2)$

$=2(289+169)=2(458)=916$

**45.** *(a)* The given system of linear equations is

$7x+11y+\alpha z=13$

$5x+4y+7z=\beta$

$175x+194y+57z=361$

has infinitely many solution,

So, $\begin{vmatrix} 7 & 11 & \alpha \\ 5 & 4 & 7 \\ 175 & 194 & 57 \end{vmatrix}=0$

$\Rightarrow 7(228-1358)+11(1225-285)+\alpha(970-700)=0$

$\Rightarrow 7(-1130)+11(940)+\alpha(270)=0$

$\Rightarrow 270\alpha+10340-7910=0$

$\Rightarrow 270\alpha+2430=0$

$\Rightarrow \alpha=-\frac{2430}{270}=-9$

and $\begin{vmatrix} 7 & 11 & 13 \\ 5 & 4 & \beta \\ 175 & 194 & 361 \end{vmatrix}=0$

$\Rightarrow 7(1444-194\beta)+11(175\beta-1805)+13(970-700)=0$

$\Rightarrow 10108-1358\beta+1925\beta-19855+3510=0$

$\Rightarrow 567\beta-6237=0$

$\Rightarrow \beta=\frac{6237}{567}=11$

Thus, $\alpha+\beta+2=-9+11+2=2+2=4$

**46.** (c) Given system of linear equation is

$$2x-y+3z=5$$
$$3x+2y-z=7$$
$$4x+5y+\alpha z=\beta$$

Now, $\Delta=\begin{vmatrix}2&-1&3\\3&2&-1\\4&5&\alpha\end{vmatrix}=7(\alpha+5)$

So, this system of equation has unique solution if $\alpha\neq-5$

and $\Delta_1=\begin{vmatrix}5&-1&3\\7&2&-1\\\beta&5&\alpha\end{vmatrix}=17\alpha-5\beta+130$

and $\Delta_2=\begin{vmatrix}2&5&3\\3&7&-1\\4&\beta&\alpha\end{vmatrix}=-11\beta+\alpha+104$

and $\Delta_3=\begin{vmatrix}2&-1&5\\3&2&7\\4&5&\beta\end{vmatrix}=7(\beta-9)$

For infinity many solutions,

$\Delta=\Delta_1=\Delta_2=\Delta_3=0$

So, $\Delta=0$ $\Delta_3=0$

$\Rightarrow$ $7(\alpha+5)=0\Rightarrow 7(\beta-9)=0$

$\Rightarrow$ $\alpha=-5\Rightarrow\beta=9$

and $\beta=8$ and $\Delta_3\neq0$

And, if $\alpha=-5$ then $\Delta=0$, then system of linear equation is in consistent.

So, only option (c) is correct.

**47.** (c) Given system of equation

$$x+y+z=6 \quad\ldots\text{(i)}$$
$$x+2y+\alpha z=10 \quad\ldots\text{(ii)}$$
$$x+3y+5z=\beta \quad\ldots\text{(iii)}$$

Here, $\Delta=\begin{vmatrix}1&1&1\\1&2&\alpha\\1&3&5\end{vmatrix}$

$=1(10-3\alpha)-1(5-\alpha)+1(3-2)$

$=10-3\alpha-5+\alpha+1=6-2\alpha$

For unique solution, $\Delta\neq0$

$\Rightarrow$ $6-2\alpha\neq0$

$\Rightarrow$ $\alpha\neq3$

When, $\alpha=3$

$\Delta_1=\begin{vmatrix}6&1&1\\10&2&3\\\beta&3&5\end{vmatrix}=\begin{vmatrix}0&1&0\\-2&2&1\\\beta-18&3&2\end{vmatrix}$

$=-1(-4-\beta+18)$

$=\beta-14$ and $\Delta_2=\begin{vmatrix}1&6&1\\1&10&3\\1&\beta&5\end{vmatrix}$

$=1(50-3\beta)-6(5-3)+1(\beta-10)$

$=50-3\beta-12+\beta-10$

$=28-2\beta=2(14-\beta)$

and $\Delta_3=\begin{vmatrix}1&1&6\\1&2&10\\1&3&\beta\end{vmatrix}$

$=1(2\beta-30)-1(\beta-10)+6(3-2)$

$=2\beta-30-\beta+10+6=\beta-14$

Thus, at $\beta=14$,

$\Delta_1=\Delta_2=\Delta_3=0$

$\Rightarrow$ $\alpha=3,\beta=14$ system has infinite solutions

**48.** (a) Given, system of equations

$$x+y+az=b$$
$$2x+5y+2z=6$$
and $$x+2y+3z=3$$

Since, given system of equation has infinite many solutions.

$\therefore D=0$ and $D_1=D_2=D_3=0$

Here, $D=\begin{vmatrix}1&1&a\\2&5&2\\1&2&3\end{vmatrix}=0$

$\Rightarrow 1(15-4)-1(6-2)+a(4-5)=0$

$\Rightarrow$ $-a+11-4=0$

$\Rightarrow$ $a=7$

and $D_1=\begin{vmatrix}b&1&a\\6&5&2\\3&2&3\end{vmatrix}=0$

$\Rightarrow b(15-4)-1(18-6)+a(12-15)=0$

$\Rightarrow$ $11b-12-3a=0$

$\Rightarrow$ $11b-12-21=0$ $(\because a=7)$

$\Rightarrow$ $11b-33=0\Rightarrow b=3$

$\therefore 2a+3b=2(7)+3(3)$

$=14+9=23$

**49.** (a) Given, system of linear equation

$$2x+4y+2az=b,$$
$$x+2y+3z=4$$
$$2x-5y+2z=8$$

Now, $D=\begin{vmatrix}2&4&2a\\1&2&3\\2&-5&2\end{vmatrix}$

$=2(4+15)-4(2-6)+2a(-5-4)$

$=38-4(-4)+2a(-9)=38+16-18a$

For non zero solution,

If $D=0$, then $54-18a=0$

$\Rightarrow a=3$

Thus, if $a\neq3$ then $\Delta\neq0$ so, unique solution.

Now, $D_1=\begin{vmatrix}b&4&6\\4&2&3\\8&-5&2\end{vmatrix}=0$

$\Rightarrow b(4+15)-4(8-24)+6(-20-16)$

$\Rightarrow 19b+64-216\Rightarrow 19b-152$

$\Rightarrow$ $b=\dfrac{152}{19}$

Thus, if $a=3$ and $b=8$, then it has infinitely many solution.

**50.** (d) For non trivial solutions

$\begin{vmatrix}1&1&\sqrt{3}\\-1&\tan\theta&\sqrt{7}\\1&1&\tan\theta\end{vmatrix}=0$

$\Rightarrow(\tan^2\theta-\sqrt{7})-(-\tan\theta-\sqrt{7})$
$+\sqrt{3}(-1-\tan\theta)=0$

$\Rightarrow\tan^2\theta-(\sqrt{3}-1)\tan\theta-\sqrt{3}=0$

$\tan\theta=\dfrac{(\sqrt{3}-1)\pm\sqrt{(\sqrt{3}-1)^2+4\sqrt{3}}}{2}$

$=\dfrac{(\sqrt{3}-1)\pm\sqrt{4+2\sqrt{3}}}{2}$

$=\dfrac{(\sqrt{3}-1)\pm\sqrt{(\sqrt{3}+1)^2}}{2}$

$=\dfrac{(\sqrt{3}-1)+(\sqrt{3}+1)}{2},\dfrac{(\sqrt{3}-1)-(\sqrt{3}+1)}{2}$

$\tan\theta=\sqrt{3},-1$

$\Rightarrow$ $\theta=\left\{\dfrac{\pi}{3},\dfrac{-2\pi}{3},\dfrac{3\pi}{4},\dfrac{-\pi}{4}\right\}$

$\therefore$ $\dfrac{120}{\pi}\Sigma\theta=\dfrac{120}{\pi}\left(\dfrac{\pi}{3}-\dfrac{2\pi}{3}+\dfrac{3\pi}{4}-\dfrac{\pi}{4}\right)$

$=\dfrac{120}{\pi}\cdot\dfrac{\pi}{6}=20$

**51.** (d) We have, $x+y+z=6$

$$\alpha x+\beta y+7z=3$$
$$x+2y+3z=14$$

$\therefore A=\begin{bmatrix}1&1&1\\\alpha&\beta&7\\1&2&3\end{bmatrix}$

$|A|=1(3\beta-14)-1(3\alpha-7)+1(2\alpha-\beta)$

$=2\beta-\alpha-7$

Now, if $\alpha=\beta$

and $\alpha\neq7$, then

$|A|=2\alpha-\alpha-7=\alpha-7\neq0$

So, system has unique solution.

If $\alpha=\beta=7$, then $|A|=0$. Also, system of equations are $x+y+z=6$,

$$7x+7y+7z=3$$
$$\Rightarrow \quad x+y+z=3/7$$
and, $\quad x+2y+3z=14$

Since, first and second equation contradict each other.

So, system has no solution.

Again, $(\alpha, \beta) \neq (7, 7)$ on the line $x-2y+7=0, |A| \neq 0$.

So, system has unique solution.

So, option (d) is false.

> **Alternate Solution**
>
> Given equations are
>
> $x+y+z=6 \quad \ldots$(i)
>
> $\alpha x+\beta y+7z=3 \quad \ldots$(ii)
>
> $x+2y+3z=14 \quad \ldots$(iii)
>
> On solving Eqs. (i) and (iii), we get
>
> $x=-2+z$ and $y=8-2z$
>
> Now, put the value of $x$ and $y$ in Eq. (ii), we get
>
> $\alpha(-2+z)+\beta(8-2z)+7z=3$
>
> $\Rightarrow (\alpha-2\beta+7)z=2\alpha-8\beta+3$
>
> So, equations have unique solution if
>
> $\alpha-2\beta+7 \neq 0$
>
> And equations have no solution, if
>
> $\alpha-2\beta+7=0$ and $2\alpha+8\beta+3 \neq 0$
>
> Hence, for every point $(\alpha, \beta) \neq (7, 7)$ on the line $x-2y+7=0$, the system has infinitely many solutions.

**52.** *(c)* Given system of linear equations

$$\left.\begin{aligned} x+y+kz&=2 \\ 2x+3y-z&=1 \\ 3x+4y+2z&=k \end{aligned}\right\} \quad \ldots \text{(i)}$$

Also, given another system of linear equations is

$$\left.\begin{aligned} (k+1)x+(2k-1)y&=7 \\ (2k+1)x+(k+5)y&=10 \end{aligned}\right\} \quad \ldots \text{(ii)}$$

Since, system (i) have infinitely many solutions.

So, $\quad \Delta=0$

$$\Rightarrow \begin{vmatrix} 1 & 1 & k \\ 2 & 3 & -1 \\ 3 & 4 & 2 \end{vmatrix}=0$$

$$\Rightarrow 1(6+4)-1(4+3)+k(8-9)=0$$
$$\Rightarrow \quad k=3$$

For this value of $k$, system (ii) will become

$$4x+5y=7 \quad \ldots \text{(iii)}$$
$$7x+8y=10 \quad \ldots \text{(iv)}$$

Clearly, they have a unique solution.

Subtract Eq. (iii) from Eq. (iv), we get

$$3x+3y=3$$
$$\Rightarrow \quad x+y=1$$

**53.** *(b)* We have,

$$x+2y+3z=3$$
$$4x+3y-4z=4$$
$$8x+4y-\lambda z=9+\mu$$

For infinitely many solutions,

$|A|=0$ and $(\text{adj } A)\cdot B=0$ in the form $AX=B$.

$$\therefore \begin{vmatrix} 1 & 2 & 3 \\ 4 & 3 & -4 \\ 8 & 4 & -\lambda \end{vmatrix}=0$$

$$1(-3\lambda+16)-2(-4\lambda+32)+3(16-24)=0$$
$$-3\lambda+16+8\lambda-64-24=0$$
$$5\lambda-72=0$$
$$\Rightarrow \quad \lambda=\frac{72}{5}$$

$$\text{adj } A=\begin{bmatrix} A_{11} & A_{12} & A_{13} \\ A_{21} & A_{22} & A_{23} \\ A_{31} & A_{32} & A_{33} \end{bmatrix}^T = \begin{bmatrix} A_{11} & A_{21} & A_{31} \\ A_{12} & A_{22} & A_{32} \\ A_{13} & A_{23} & A_{33} \end{bmatrix},$$

where $A_{ij}$ are cofactors.

$$\because \quad A=\begin{bmatrix} 1 & 2 & 3 \\ 4 & 3 & -4 \\ 8 & 4 & -\frac{72}{5} \end{bmatrix}$$

$\therefore \quad (\text{adj } A)B$

$$=\begin{bmatrix} -\frac{136}{5} & \frac{204}{5} & -17 \\ \frac{128}{5} & -\frac{192}{5} & 16 \\ -8 & 12 & -5 \end{bmatrix} \cdot \begin{bmatrix} 3 \\ 4 \\ 9+\mu \end{bmatrix}=0$$

From 3rd row expansion,

$$-24+48-45-5\mu=0$$
$$-21-5\mu=0$$
$$\Rightarrow \quad \mu=-\frac{21}{5}$$

> **Concept Enhancer**
>
> The system of an equation has infinitely many solutins when the lines are coincident and they have the same $y$-intercept.
>
> If $AX=B$ be a system of $m$ linear equation in $n$ variable.
>
> If $|A|=0$ and adj $A \cdot B=0$, then system is consistent and has infinite number of solutions.

**54.** *(b)* Given, system of equation is

$$\left.\begin{aligned} 3(\sin 3\theta)x-y+z&=2 \\ 3(\cos 2\theta)x+4y+3z&=3 \\ 6x+7y+7z&=9 \end{aligned}\right\} \quad \ldots \text{(i)}$$

$$\text{Thus, } \Delta=\begin{vmatrix} 3\sin 3\theta & -1 & 1 \\ 3\cos 2\theta & 4 & 3 \\ 6 & 7 & 7 \end{vmatrix}$$

$$=3\sin 3\theta(28-21)+1(21\cos 2\theta-18)+1(21\cos 2\theta-24)$$
$$=21\sin 3\theta+42\cos 2\theta-42$$

Now, for no solution of above system (i).

Then, $\Delta=0$

$$\Rightarrow \quad 21\sin 3\theta+42\cos 2\theta-42=0$$
$$\Rightarrow \quad \sin 3\theta+2\cos 2\theta-2=0$$
$$\Rightarrow \quad \sin 3\theta=2(2\sin^2\theta)$$
$$\Rightarrow \quad 3\sin\theta-4\sin^3\theta=4\sin^2\theta$$
$$\Rightarrow \sin\theta(3-4\sin^2\theta-4\sin\theta)=0$$
$$\sin\theta=0 \text{ or } \sin\theta=\frac{1}{2} \text{ and } -\frac{3}{2}$$

But given $\theta \in (0, 4\pi)$ thus, $\sin\theta \neq -\frac{3}{2}$

$$\therefore \theta=\pi, 2\pi, 3\pi, \frac{\pi}{6}, \frac{5\pi}{6}, \frac{13\pi}{6}, \frac{17\pi}{6}.$$

Thus, '7' value of $\theta \in (0, 4\pi)$.

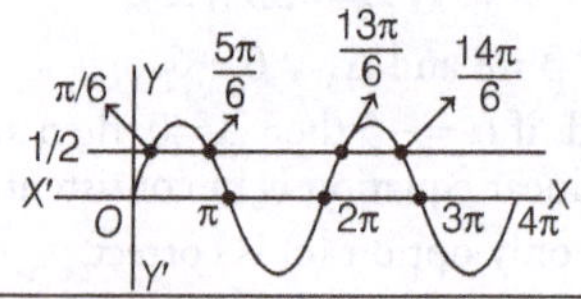


> **Alternate Solution**
>
> $\sin 3\theta+2\cos 2\theta-2=0$
>
> $\Rightarrow \quad \sin\theta(3-4\sin^2\theta-4\sin\theta)=0$
>
> $\Rightarrow \quad \sin\theta=0 \text{ or } \sin\theta=\frac{1}{2} \text{ and } -\frac{3}{2}$
>
> (But $\sin\theta \neq -\frac{3}{2}$ as $\theta \in (0, 4\pi)$)
>
> $\therefore \quad \theta=\pi, 2\pi, 3\pi, \frac{\pi}{6}, \frac{5\pi}{6}, \frac{13\pi}{6}, \frac{17\pi}{6}$

**55.** *(c)* Given, equations are

$$3x-2y+z=b$$
$$5x-8y+9z=3$$
$$2x+y+az=-1$$

For no solutions, $\Delta=0$ and atleast one of $\Delta_x, \Delta_y, \Delta_z \neq 0$.

$$\Delta=\begin{vmatrix} 3 & -2 & 1 \\ 5 & -8 & 9 \\ 2 & 1 & a \end{vmatrix}=0$$

$$\Rightarrow 3(-8a-9)+2(5a-18)+5+16=0$$
$$\Rightarrow \quad a=-3$$

If $\Delta_z = \begin{vmatrix} 3 & -2 & b \\ 5 & -8 & 3 \\ 2 & 1 & -1 \end{vmatrix} = 0$

$\Rightarrow 3(8-3) + 2(-5-6) + b(5+16) = 0$

$\Rightarrow \quad 15 - 22 + 21b = 0$

$\Rightarrow \qquad b = 1/3$

i.e. for $b \neq \frac{1}{3}$, there are no solution.

Thus, $(a, b) \equiv (-3, -1/3)$.

**56.** *(b)*

> **key idea** Probability of the event $= \dfrac{\text{Favourable events}}{\text{Total number of events}}$

Given, $x + y + z = 5, x + 2y + 3z = \mu$ and $x + 3y + \lambda z = 1$

$$\Delta = \begin{vmatrix} 1 & 1 & 1 \\ 1 & 2 & 3 \\ 1 & 3 & \lambda \end{vmatrix}$$

$1(2\lambda - 9) - 1(\lambda - 3) + 1(3 - 2)$

For unique solution $\Delta \neq 0$

$\Rightarrow \qquad \lambda \neq 5$

And $\Delta_1$ or $\Delta_2$ or $\Delta_3 \neq 0$

$\Delta_3 = \begin{vmatrix} 1 & 1 & 5 \\ 1 & 2 & \mu \\ 1 & 3 & 1 \end{vmatrix} = (2 - 3\mu) + (\mu - 1) + 5$

$= 6 - 2\mu$

If $\Delta_3 \neq 0$ and $\Delta = 0$, then no solution

$\mu \neq 3$ and $\lambda = 5$

$p =$ Probability of unique solution $= \frac{5}{6}$

$q =$ Probability of no solution

$= \frac{5}{6} \cdot \frac{1}{6} = \frac{5}{36}$

**57.** *(b)* For non-trivial solution $\Delta = 0$

$\Rightarrow$

$$\begin{vmatrix} 1 + \cos^2\theta & \sin^2\theta & 4\sin 3\theta \\ \cos^2\theta & 1 + \sin^2\theta & 4\sin 3\theta \\ \cos^2\theta & \sin^2\theta & 1 + 4\sin 3\theta \end{vmatrix} = 0$$

Expanding along $R_1$

$(1 + \cos^2\theta)$

$[1 + 4\sin 3\theta + \sin^2\theta + 4\sin^2\theta \sin 3\theta$

$\quad - 4\sin^2\theta \sin 3\theta] - \sin^2\theta[\cos^2\theta$

$\quad + 4\cos^2\theta \sin 3\theta - 4\cos^2\theta \sin 3\theta]$

$\quad + 4\sin 3\theta[\cos^2\theta \sin^2\theta - \cos^2\theta$

$\qquad - \cos^2\theta \sin^2\theta] = 0$

$\Rightarrow (1 + \cos^2\theta)[1 + 4\sin 3\theta + \sin^2\theta]$

$\quad - \sin^2\theta \cos^2\theta - \cos^2\theta \sin 3\theta = 0$

$\Rightarrow 1 + 4\sin 3\theta + \sin^2\theta + \cos^2\theta$

$\quad + 4\cos^2\theta \sin\theta + \cos^2\theta \sin^2\theta$

$\quad - \sin^2\theta \cos^2\theta - 4\cos^2\theta \sin 3\theta = 0$

$\Rightarrow 2 + 4\sin 3\theta = 0$

$\Rightarrow \sin 3\theta = -\frac{1}{2} \Rightarrow 3\theta = \left(\pi + \frac{\pi}{6}\right)$

$\Rightarrow \qquad \theta = 7\pi/18$

**58.** *(b)* Given system of equation can be written in form $AX = B$, where $A$ is $3 \times 3$, $X$ is $3 \times 1$ and $B$ is $3 \times 1$ matrix as follows.

$$\begin{bmatrix} 3 & -1 & 4 \\ 1 & 2 & -3 \\ 6 & 5 & k \end{bmatrix} \begin{bmatrix} x \\ y \\ z \end{bmatrix} = \begin{bmatrix} 3 \\ -2 \\ -3 \end{bmatrix}$$

This system of linear equation has infinitely many solutions when $|A| = 0$

i.e. $\begin{vmatrix} 3 & -1 & 4 \\ 1 & 2 & -3 \\ 6 & 5 & k \end{vmatrix} = 0$

$\Rightarrow 3(2k + 15) + (k + 18) + 4(5 - 12) = 0$

$\Rightarrow 6k + 45 + k + 18 + 20 - 48 = 0$

$\Rightarrow 7k + (35) = 0 \Rightarrow 7k = -35$

$\therefore \; k = -5$

**59.** *(c)* Let $= A \begin{bmatrix} i & -i \\ -i & i \end{bmatrix}$

$A^2 = A \cdot A = \begin{bmatrix} i & -i \\ -i & i \end{bmatrix} \begin{bmatrix} i & -i \\ -i & i \end{bmatrix}$

$= \begin{bmatrix} i^2 + i^2 & -i^2 - i^2 \\ -i^2 - i^2 & i^2 + i^2 \end{bmatrix}$

$= \begin{bmatrix} -2 & 2 \\ 2 & -2 \end{bmatrix} A^4 = \begin{bmatrix} -2 & 2 \\ 2 & -2 \end{bmatrix} \begin{bmatrix} -2 & 2 \\ 2 & -2 \end{bmatrix}$

$= \begin{bmatrix} 4 + 4 & -4 - 4 \\ -4 - 4 & 4 + 4 \end{bmatrix} = \begin{bmatrix} 8 & -8 \\ -8 & 8 \end{bmatrix}$

Similarly, $A^8 = A^4 \cdot A^4$

$= \begin{bmatrix} 8 & -8 \\ -8 & 8 \end{bmatrix} \begin{bmatrix} 8 & -8 \\ -8 & 8 \end{bmatrix}$

$= \begin{bmatrix} 64 + 64 & -64 - 64 \\ -64 - 64 & 64 + 64 \end{bmatrix}$

$= \begin{bmatrix} 128 & -128 \\ -128 & 128 \end{bmatrix}$

Now, $A^8 \begin{bmatrix} x \\ y \end{bmatrix} = \begin{bmatrix} 8 \\ 64 \end{bmatrix}$

$\Rightarrow \begin{bmatrix} 128 & -128 \\ -128 & 128 \end{bmatrix} \begin{bmatrix} x \\ y \end{bmatrix} = \begin{bmatrix} 8 \\ 64 \end{bmatrix}$

$\Rightarrow 128x - 128y = 8$

$x - y = \frac{1}{16}$ ...(i)

and $-128x + 128y = 64$

$\Rightarrow \quad x - y = -\frac{1}{2}$ ...(ii)

From Eqs. (i) and (ii), we get these two lines are parallel.

So, there will be no solution.

**60.** *(b)* Given, system of equation can be written as $AX = B$, where

$A = \begin{bmatrix} 1 & 2 & -3 \\ 2 & 6 & -11 \\ 1 & -2 & 7 \end{bmatrix}, X = \begin{bmatrix} x \\ y \\ z \end{bmatrix}, B = \begin{bmatrix} a \\ b \\ c \end{bmatrix}$

Then, $|A| = \begin{vmatrix} 1 & 2 & -3 \\ 2 & 6 & -11 \\ 1 & -2 & 7 \end{vmatrix}$

$= 1(42 - 22) - 2(14 + 11) - 3(-4 - 6)$

$= 20 - 50 + 30 = 0$

$|A_1| = \begin{vmatrix} a & 2 & -3 \\ b & 6 & -11 \\ c & -2 & 7 \end{vmatrix}$

$= a(42 - 22) - 2(7b + 11c) - 3(-2b - 6c)$

$= 20a - 8b - 4c = 4(5a - 2b - c)$

$|A_2| = \begin{vmatrix} 1 & a & -3 \\ 2 & b & -11 \\ 1 & c & 7 \end{vmatrix}$

$= 1(7b + 11c) - a(14 + 11) - 3(2c - b)$

$= -25a + 10b + 5c = -5(5a - 2b - c)$

$|A_3| = \begin{vmatrix} 1 & 2 & a \\ 2 & 6 & b \\ 1 & -2 & c \end{vmatrix} = -10a + 4b + 2c$

$= -2(5a - 2b - c)$

For infinite solution,

$|A| = |A_1| = |A_2| = |A_3| = 0$

$\Rightarrow \qquad 5a - 2b - c = 0$

$\Rightarrow \qquad 5a = 2b + c$

**61.** *(c)* Given, $3x - 2y - kz = 10$

$2x - 4y - 2z = 6$

$x + 2y - z = 5m$

Condition for inconsistency $\Rightarrow \Delta = 0$ and at least one of the $\Delta_x, \Delta_y, \Delta_z$ is non-zero.

Now, $\Delta = \begin{vmatrix} 3 & -2 & -k \\ 2 & -4 & -2 \\ 1 & 2 & -1 \end{vmatrix}$

$= 3(4 + 4) + 2(-2 + 2) - k(4 + 4)$

$= 24 - 8k$

Now, $\Delta = 0$

$\therefore \quad 24-8k=0 \Rightarrow k=\frac{24}{8}=3$

$\Delta_x=\begin{vmatrix}10 & -2 & -k\\ 6 & -4 & -2\\ 5m & 2 & -1\end{vmatrix}$

$=10(4+4)+2(-6+10m)-3(12+20m)$

$=32-40m=8(4-5m)$

$\Delta_y=\begin{vmatrix}3 & 10 & -k\\ 2 & 6 & -2\\ 1 & 5m & -1\end{vmatrix}$

$=3(-6+10m)-10(-2+2)-3(10m-6)$

$=0$

$\Delta_z=\begin{vmatrix}3 & -2 & 10\\ 2 & -4 & 6\\ 1 & 2 & 5m\end{vmatrix}$

$=3(-20m-12)+2(10m-6)+10(4+4)$

$=-40m+32=8(4-5m)$

Here, either $\Delta_x$ or $\Delta_z \neq 0$

$\Rightarrow \quad 8(4-5m)\neq 0$

$\Rightarrow \quad m\neq 4/5$

Hence, the required values are $k=3$, $m\neq 4/5$.

**62.** *(b)* Given system of linear equations is

$2x-y+2z=2$

$x-2y+\lambda z=-4$

and $x+\lambda y+z=4$

has no solution.

$\therefore \quad \Delta=0 \Rightarrow \begin{vmatrix}2 & -1 & 2\\ 1 & -2 & \lambda\\ 1 & \lambda & 1\end{vmatrix}=0$

$\Rightarrow 2(-2-\lambda^2)+1(1-\lambda)+2(\lambda+2)=0$

$\Rightarrow \quad -2\lambda^2+\lambda+1=0$

$\Rightarrow \quad 2\lambda^2-\lambda-1=0$

$\Rightarrow \quad (2\lambda+1)(\lambda-1)=0$

$\Rightarrow \lambda=-\frac{1}{2}$ or 1 and $\Delta_1=\begin{vmatrix}2 & -1 & 2\\ -4 & -2 & \lambda\\ 4 & \lambda & 1\end{vmatrix}$

$=2(-2-\lambda^2)+1(-4-4\lambda)+2(-4\lambda+8)$

$=-2\lambda^2-12\lambda+8=-2(\lambda^2+6\lambda-4)$

at $\lambda=-\frac{1}{2}$,

$\Delta_1=-2\left(\frac{1}{4}-3-4\right)=\frac{27}{2}\neq 0$

$\therefore$ System has no, solution for $\lambda=-1/2$ and, at $\lambda=1$,

$\Delta_1=-2(1+6-4)=-6\neq 0$

$\therefore$ System has no solution for $\lambda=1$.

Hence, set $S$ contains exactly two elements.

**63.** *(b)* Given system of linear equations

$x+2y+3z=1 \quad \ldots$(i)

$3x+4y+5z=\mu \quad \ldots$(ii)

$4x+4y+4z=\delta \quad \ldots$(iii)

$\because \quad \Delta=\begin{vmatrix}1 & 2 & 3\\ 3 & 4 & 5\\ 4 & 4 & 4\end{vmatrix}$

$=1(16-20)-2(12-20)+3(12-16)$

$=-4+16-12=0 \Rightarrow \Delta=0$

Now, $\Delta_1=\begin{vmatrix}1 & 2 & 3\\ \mu & 4 & 5\\ \delta & 4 & 4\end{vmatrix}$

$=1(16-20)-2(4\mu-5\delta)+3(4\mu-4\delta)$

$=4\mu-2\delta-4$

Similarly, $\Delta_2=\begin{vmatrix}1 & 1 & 3\\ 3 & \mu & 5\\ 4 & \delta & 4\end{vmatrix}$

$=1(4\mu-5\delta)-1(12-20)+3(3\delta-4\mu)$

$=4\delta-8\mu+8$

and $\Delta_3=\begin{vmatrix}1 & 2 & 1\\ 3 & 4 & \mu\\ 4 & 4 & \delta\end{vmatrix}$

$=1(4\delta-4\mu)-2(3\delta-4\mu)+1(12-16)$

$=4\mu-2\delta-4$

From the options if $(\mu, \delta)=(4, 3)$, then $4\delta-8\mu+8\neq 0$

Now according to Cramer's rule if $\Delta=0$ and any one of $\Delta_1$, $\Delta_2$ or $\Delta_3$ is not zero, then system is called inconsistent.

$\therefore$ The ordered pair $(\mu, \delta)=(4, 3)$, then the system of linear equations is inconsistent.

> **Concept Enhancer**
>
> Let the system of equations be
>
> $a_1x+b_1y+c_1z=d_1$
>
> $a_2x+b_2y+c_2z=d_2$
>
> $a_3x+b_3y+c_3z=d_3$
>
> Then, the solution of the system of equation is $x=\frac{D_1}{D}, y=\frac{D_2}{D}, z=\frac{D_3}{D}$, it is called cramer's rule.
>
> If $D=0$ and atleast one of the determinant $D_1, D_2, D_3$ is non-zero, then the given system of linear equation is inconsistent, i.e. having no solution.

**64.** *(b)* Given system of linear equations

$2x+3y-z=0,$

$x+ky-2z=0$

and $2x-y+z=0$ has a non-trivial solution $(x, y, z)$.

$\therefore \quad \Delta=0 \Rightarrow \begin{vmatrix}2 & 3 & -1\\ 1 & k & -2\\ 2 & -1 & 1\end{vmatrix}=0$

$2(k-2)-3(1+4)-1(-1-2k)=0$

$\Rightarrow \quad 4k=18$

$\Rightarrow \quad k=\frac{9}{2}$

So, system of linear equations is

$2x+3y-z=0 \quad \ldots$(i)

$2x+9y-4z=0 \quad \ldots$(ii)

and $\quad 2x-y+z=0 \quad \ldots$(iii)

From Eqs. (i) and (ii), we get

$6y-3z=0, \frac{y}{z}=\frac{1}{2}$

From Eqs. (i) and (iii), we get

$4x+2y=0$

$\Rightarrow \quad \frac{x}{y}=-\frac{1}{2}$

So, $\frac{x}{z}=\frac{x}{y}\times\frac{y}{z}=-\frac{1}{4}$

$\Rightarrow \quad \frac{z}{x}=-4 \quad \left[\because \frac{y}{z}=\frac{1}{2} \text{ and } \frac{x}{y}=\frac{1}{2}\right]$

$\therefore \frac{x}{y}+\frac{y}{z}+\frac{z}{x}+k=-\frac{1}{2}+\frac{1}{2}-4+\frac{9}{2}=\frac{1}{2}$

**65.** *(a)* Here, $D=\begin{vmatrix}1 & -4 & 7\\ 0 & 3 & -5\\ -2 & 5 & -9\end{vmatrix}$

$=1(-27+25)+4(0-10)+7(0+6)$

[expanding along $R_1$]

$=-2-40+42=0$

$\therefore$ The system of linear equations have infinite many solutions.

[$\because$ system is consistent and does not have unique solution as $D=0$]

$\Rightarrow \quad D_1=D_2=D_3=0$

Now, $D_1=0 \Rightarrow \begin{vmatrix}g & -4 & 7\\ h & 3 & -5\\ k & 5 & -9\end{vmatrix}=0$

$\Rightarrow g(-27+25)+4(-9h+5k)+7(5h-3k)=0$

$\Rightarrow \quad -2g-36h+20k+35h-21k=0$

$\Rightarrow \quad -2g-h-k=0 \Rightarrow 2g+h+k=0$

**66.** *(a)* We know that, if the system of equations

$a_1x+b_1y+c_1z=d_1$

$a_2x+b_2y+c_2z=d_2$

$$a_3x + b_3y + c_3z = d_3$$

has more than one solution, then $D = 0$ and $D_1 = D_2 = D_3 = 0$. In the given problem,

$$D_1 = 0 \Rightarrow \begin{vmatrix} a & 2 & 3 \\ b & -1 & 5 \\ c & -3 & 2 \end{vmatrix} = 0$$

$\Rightarrow a(-2+15) - 2(2b - 5c) + 3(-3b + c) = 0$

$\Rightarrow \quad 13a - 4b + 10c - 9b + 3c = 0$

$\Rightarrow \quad 13a - 13b + 13c = 0$

$\Rightarrow \quad a - b + c = 0$

$\Rightarrow \quad b - a - c = 0$

**67.** *(b)* We have, $x + ky + 3z = 0$; $3x + ky - 2z = 0$; $2x + 4y - 3z = 0$

System of equation has non-zero solution, if

$$\begin{vmatrix} 1 & k & 3 \\ 3 & k & -2 \\ 2 & 4 & -3 \end{vmatrix} = 0$$

$\Rightarrow (-3k + 8) - k(-9 + 4) + 3(12 - 2k) = 0$

$\Rightarrow \quad -3k + 8 + 9k - 4k + 36 - 6k = 0$

$\Rightarrow \quad -4k + 44 = 0 \Rightarrow k = 11$

Let $z = \lambda$, then we get

$x + 11y + 3\lambda = 0 \quad \ldots(i)$

$3x + 11y - 2\lambda = 0 \quad \ldots(ii)$

and $\quad 2x + 4y - 3\lambda = 0 \quad \ldots(iii)$

On solving Eqs. (i) and (ii), we get

$x = 5\lambda/2, \; y = -\lambda/2, \; z = \lambda$

$$\Rightarrow \frac{xz}{y^2} = \frac{5\lambda^2}{2 \times \left(-\frac{\lambda}{2}\right)^2} = 10$$

**68.** *(d)* Given system of linear equations

$2x_1 - 2x_2 + x_3 = \lambda x_1$

$\Rightarrow \quad (2 - \lambda)x_1 - 2x_2 + x_3 = 0 \quad \ldots(i)$

$2x_1 - 3x_2 + 2x_3 = \lambda x_2$

$\Rightarrow \quad 2x_1 - (3 + \lambda)x_2 + 2x_3 = 0 \quad \ldots(ii)$

$-x_1 + 2x_2 = \lambda x_3$

$\Rightarrow \quad -x_1 + 2x_2 - \lambda x_3 = 0 \quad \ldots(iii)$

Since, the system has non-trivial solution.

$$\therefore \begin{vmatrix} 2-\lambda & -2 & 1 \\ 2 & -(3+\lambda) & 2 \\ -1 & 2 & -\lambda \end{vmatrix} = 0$$

$\Rightarrow (2 - \lambda)(3\lambda + \lambda^2 - 4) + 2(-2\lambda + 2) + 1(4 - 3 - \lambda) = 0$

$\Rightarrow (2 - \lambda)(\lambda^2 + 3\lambda - 4) + 4(1 - \lambda) + (1 - \lambda) = 0$

$\Rightarrow (2 - \lambda)(\lambda + 4)(\lambda - 1) + 5(1 - \lambda) = 0$

$\Rightarrow (\lambda - 1)(\lambda^2 + 2\lambda - 3) = 0$

$\Rightarrow (\lambda - 1)^2(\lambda + 3) = 0 \Rightarrow \lambda = 1, 1, -3$

**69.** *(b)* Given equations can be written in matrix form, $\quad AX = B$

Where, $A = \begin{bmatrix} k+1 & 8 \\ k & k+3 \end{bmatrix}$,

$X = \begin{bmatrix} x \\ y \end{bmatrix}$ and $B = \begin{bmatrix} 4k \\ 3k - 1 \end{bmatrix}$

For no solution, $|A| = 0$ and $(\text{adj } A)B \neq 0$

Now, $|A| = \begin{vmatrix} k+1 & 8 \\ k & k+3 \end{vmatrix} = 0$

$\Rightarrow (k + 1)(k + 3) - 8k = 0$

$\Rightarrow k^2 - 4k + 3 = 0 \Rightarrow (k - 1)(k - 3) = 0$

$\Rightarrow k = 1$ or $= 3$

Now, $\quad \text{adj } A = \begin{bmatrix} k+3 & -8 \\ -k & k+1 \end{bmatrix}$

Now, $(\text{adj } A)B = \begin{bmatrix} k+3 & -8 \\ -k & k+1 \end{bmatrix}\begin{bmatrix} 4k \\ 3k-1 \end{bmatrix}$

$= \begin{bmatrix} (k+3)(4k) - 8(3k-1) \\ -4k^2 + (k+1)(3k-1) \end{bmatrix}$

$= \begin{bmatrix} 4k^2 - 12k + 8 \\ -k^2 + 2k - 1 \end{bmatrix}$

Putting $k = 1$

$(\text{adj } A)B = \begin{bmatrix} 4 - 12 + 8 \\ -1 + 2 - 1 \end{bmatrix} = \begin{bmatrix} 0 \\ 0 \end{bmatrix}$ (not true)

Putting $k = 3$

$(\text{adj } A)B = \begin{bmatrix} 36 - 36 + 8 \\ -9 + 6 - 1 \end{bmatrix} = \begin{bmatrix} 8 \\ -4 \end{bmatrix} \neq 0$ (true)

Hence, only one value of $k$ exists.

> ***Alternate Solution***
>
> Condition for the system of equations has no solution.
>
> $\frac{a_1}{a_2} = \frac{b_1}{b_2} \neq \frac{c_1}{c_2}$
>
> $\therefore \quad \frac{k+1}{k} = \frac{8}{k+3} \neq \frac{4k}{3k-1}$
>
> Take $\frac{k+1}{k} = \frac{8}{k+3}$
>
> $\Rightarrow \quad k^2 + 4k + 3 = 8k \Rightarrow k^2 - 4k + 3 = 0$
>
> $\Rightarrow \quad (k - 1)(k - 3) = 0$ and $k = 1, 3$
>
> If $k = 1$, then $\frac{8}{1+3} \neq \frac{4 \cdot 1}{2}$ (not true)
>
> If $k = 3$, then $\frac{8}{6} \neq \frac{4 \cdot 3}{9 - 1}$ (true)
>
> $k = 3$
>
> Hence, only one value of $k$ exists.

**70.** *(d)* $\Delta = 0 \Rightarrow \begin{vmatrix} 1 & 5 & -1 \\ 4 & 3 & -3 \\ 24 & 1 & \lambda \end{vmatrix} = 0$

This gives

$1(3\lambda + 9) - 5(4\lambda + 72) - 1(-68) = 0$

$\lambda = -17$

Also, $\Delta_1 = 0 \Rightarrow \begin{vmatrix} 1 & 5 & -1 \\ 7 & 3 & -3 \\ \mu & 1 & \lambda \end{vmatrix} = 0$

$\Rightarrow \quad \mu = 45$

Use first two equations $x + 5y - z = 1$, $4x + 3y - 3z = 7$

Multiply first equation by 3 and subtract second equation,

$-x + 12y = -4$

$x = 12y + 4$

From $x + 5y - z = 1$

$\Rightarrow 12y + 4 + 5y - z = 1$

$17y + 4 - z = 1 \Rightarrow z = 17y + 3$

From $7 \leq x + y + z \leq 77$

$7 \leq 12y + 4 + y + 17y + 3 \leq 77$

$7 \leq 30y + 7 \leq 77$

$0 \leq 30y \leq 70$

$0 \leq y \leq \frac{7}{3}$

$y$ is an integer, $y = 0, 1, 2$

For $y = 0$, $x = 4$, $z = 3$, then

$x + y + z = 7$

For $y = 1$, $x = 16$, $z = 20$, then

$x + y + z = 37$

For $y = 2$, $x = 28$, $z = 37$, then

$x + y + z = 67$

There are three solutions.

OR

A system of linear equations has infinite solutions if the determinant of the coefficient matrix is zero and the equations are consistent.

For a system to have infinite solutions, one equation must be a linear combination of the others.

Multiply Eq. (i) by $a$ and (ii) by $b$ and add them

$a(x + 5y - z) + b(4x + 3y - 3z) = a + 7b$

$(a + 4b)x + (5a + 3b)y + (-a - 3b)z = a + 7b$

Compare with $24x + y + \lambda z = \mu$

$a + 4b = 24$, $5a + 3b = 1$, $-a - 3b = \lambda$, $a + 7b = \mu$

Solve $a + 4b = 24$ and $5a + 3b = 1$

We get $a=-4, b=7$

Then, $-a-3b=\lambda \Rightarrow \lambda=-17$

$a+7b=\mu \Rightarrow \mu=45$

Use first two equations $x+5y-z=1$, $4x+3y-3z=7$

Multiply first equation by 3 and subtract second equation,

$-x+12y=-4$

$x=12y+4$

From $x+5y-z=1$

$\Rightarrow 12y+4+5y-z=1$

$17y+4-z=1 \Rightarrow z=17y+3$

From $7 \le x+y+z \le 77$

$7 \le 12y+4+y+17y+3 \le 77$

$7 \le 30y+7 \le 77$

$0 \le 30y \le 70 \Rightarrow 0 \le y \le \frac{7}{3}$

$y$ is an integer, $y=0, 1, 2$

For $y=0, x=4, z=3$, then $x+y+z=7$

For $y=1, x=16, z=20$, then $x+y+z=37$

For $y=2, x=28, z=37$, then $x+y+z=67$

There are three solutions.

**71.** *(b)* Let $A=\begin{bmatrix} a_1 & b_1 & c_1 \\ a_2 & b_2 & c_2 \\ a_3 & b_3 & c_3 \end{bmatrix}$

Given, $A\begin{bmatrix}1\\0\\1\end{bmatrix}=2\begin{bmatrix}1\\0\\1\end{bmatrix}$

$$\begin{bmatrix} a_1 & b_1 & c_1 \\ a_2 & b_2 & c_2 \\ a_3 & b_3 & c_3 \end{bmatrix}\begin{bmatrix}1\\0\\1\end{bmatrix}=\begin{bmatrix}2\\0\\2\end{bmatrix}$$

$$\Rightarrow \begin{bmatrix} a_1+c_1 \\ a_2+c_2 \\ a_3+c_3 \end{bmatrix}=\begin{bmatrix}2\\0\\2\end{bmatrix}$$

$a_1+c_1=2$ ...(i)

$a_2+c_2=0$ ...(ii)

$a_3+c_3=2$ ...(iii)

and $A\begin{bmatrix}-1\\0\\1\end{bmatrix}=4\begin{bmatrix}-1\\0\\1\end{bmatrix}$

$$\begin{bmatrix} a_1 & b_1 & c_1 \\ a_2 & b_2 & c_2 \\ a_3 & b_3 & c_3 \end{bmatrix}\begin{bmatrix}-1\\0\\1\end{bmatrix}=\begin{bmatrix}-4\\0\\4\end{bmatrix}$$

$$\Rightarrow \begin{bmatrix} -a_1+c_1 \\ -a_2+c_2 \\ -a_3+c_3 \end{bmatrix}=\begin{bmatrix}-4\\0\\4\end{bmatrix}$$

$-a_1+c_1=-4$ ...(iv)

$-a_2+c_2=0$ ...(v)

$-a_3+c_3=4$ ...(vi)

From Eqs. (i) and (iv), we get

$c_1=-1, a_1=3$

From Eqs. (ii) and (v), we get

$a_2=0, c_2=0$

From Eqs. (iii) and (vi), we get

$a_3=-1, c_3=3$

Also, $A\begin{bmatrix}0\\1\\0\end{bmatrix}=2\begin{bmatrix}0\\1\\0\end{bmatrix}$

$$\Rightarrow \begin{bmatrix} a_1 & b_1 & c_1 \\ a_2 & b_2 & c_2 \\ a_3 & b_3 & c_3 \end{bmatrix}\begin{bmatrix}0\\1\\0\end{bmatrix}=\begin{bmatrix}0\\2\\0\end{bmatrix}$$

$$\Rightarrow \begin{bmatrix} b_1 \\ b_2 \\ b_3 \end{bmatrix}=\begin{bmatrix}0\\2\\0\end{bmatrix}$$

$b_1=0, b_2=2, b_3=0$

$$A=\begin{bmatrix} 3 & 0 & -1 \\ 0 & 2 & 0 \\ -1 & 0 & 3 \end{bmatrix}$$

Now,

$$A-3I=\begin{bmatrix} 3 & 0 & -1 \\ 0 & 2 & 0 \\ -1 & 0 & 3 \end{bmatrix}-\begin{bmatrix} 3 & 0 & 0 \\ 0 & 3 & 0 \\ 0 & 0 & 3 \end{bmatrix}$$

$$=\begin{bmatrix} 0 & 0 & -1 \\ 0 & -1 & 0 \\ -1 & 0 & 0 \end{bmatrix}$$

$$(A-3I)\begin{bmatrix}x\\y\\z\end{bmatrix}=\begin{bmatrix}1\\2\\3\end{bmatrix}$$

$$\Rightarrow \begin{bmatrix} 0 & 0 & -1 \\ 0 & -1 & 0 \\ -1 & 0 & 0 \end{bmatrix}\begin{bmatrix}x\\y\\z\end{bmatrix}=\begin{bmatrix}1\\2\\3\end{bmatrix}$$

$$\begin{bmatrix}-z\\-y\\-x\end{bmatrix}=\begin{bmatrix}1\\2\\3\end{bmatrix}$$

$x=-3, y=-2, z=-1$

It is a unique solution.

**72.** *(a)* Given, system is

$x+y+z=5$,

$x+2y+\lambda^2 z=9$,

and $x+3y+\lambda z=\mu$

where $\lambda, \mu \in R$

For unique solution $\Rightarrow \Delta \ne 0$

$$\Delta=\begin{vmatrix} 1 & 1 & 1 \\ 1 & 2 & \lambda^2 \\ 1 & 3 & \lambda \end{vmatrix} \ne 0$$

$2\lambda^2-\lambda-1 \ne 0 \Rightarrow \lambda \ne 1, -\frac{1}{2}$

$$\Delta x=\begin{vmatrix} 5 & 1 & 1 \\ 9 & 2 & \lambda^2 \\ \mu & 3 & \lambda \end{vmatrix}$$

$=10\lambda-15\lambda^2-9\lambda+\mu\lambda^2+27-2\mu$

$=-15\lambda^2+\lambda+\mu\lambda^2-2\mu-27$

If $\lambda=1, \Delta x=-15+1+\mu-2\mu+27$

$=13-\mu$

$\Rightarrow \Delta x=0, \mu=13$

$$\Delta y=\begin{vmatrix} 1 & 5 & 1 \\ 1 & 9 & \lambda^2 \\ 1 & \mu & \lambda \end{vmatrix}$$

$=4\lambda+5\lambda^2-\lambda^2\mu+\mu-9$

If $\lambda=1 \Delta y=0$

$$\Delta z=\begin{vmatrix} 1 & 1 & 5 \\ 1 & 2 & 9 \\ 1 & 3 & \mu \end{vmatrix}$$

$=1(2\mu-27)-1(\mu-9)+5(1)$

$=\mu-13$

$\because \Delta z=0 \Rightarrow \mu=13$

$\therefore$ Infinite solution $\Rightarrow \lambda=1$ and $\mu=13$

For no solution $\Rightarrow \lambda=\frac{-1}{2}$, 1 and $\mu \ne 13$

For unique solution $\Rightarrow \lambda \ne 1$

**73.** *(d)* We have,

$$\left.\begin{aligned} \alpha x+2y+z&=1 \\ 2\alpha x+3y+z&=1 \\ 3x+\alpha y+2z&=\beta \end{aligned}\right\} \quad \text{... (i)}$$

Let $A=\begin{bmatrix} \alpha & 2 & 1 \\ 2\alpha & 3 & 1 \\ 3 & \alpha & 2 \end{bmatrix}$,

$X=\begin{bmatrix}x\\y\\z\end{bmatrix}, B=\begin{bmatrix}1\\1\\\beta\end{bmatrix}$

So, we have $AX=B$

For $|A| \neq 0$, we have a unique solution i.e.

$$\begin{vmatrix} \alpha & 2 & 1 \\ 2\alpha & 3 & 1 \\ 3 & \alpha & 2 \end{vmatrix} \neq 0$$

$$\alpha(3\cdot 2-\alpha\cdot 1)-2(2\alpha\cdot 2-3\cdot 1)+1(2\alpha\cdot\alpha-3\cdot 3)\neq 0$$

$\Rightarrow \quad \alpha^2-2\alpha-3\neq 0$

$\Rightarrow \quad (\alpha+1)(\alpha-3)\neq 0$

$\Rightarrow \quad \alpha \neq -1, 3$

i.e. the system of equations has a solution, $\forall\ \alpha \in R-\{-1, 3\}$.

For no solution, $|A|=0$ and $\text{adj}(A)\cdot B\neq 0$

i.e. $\alpha=-1, 3$ and $\text{adj}(A)\cdot B\neq 0$

For $\alpha=-1$,

$$A=\begin{bmatrix} -1 & 2 & 1 \\ -2 & 3 & 1 \\ 3 & -1 & 2 \end{bmatrix}$$

and $\text{adj}(A) = [\text{Co-factor matrix}]$

$$=\begin{bmatrix} 7 & 7 & -7 \\ -5 & -5 & 5 \\ -1 & -1 & 1 \end{bmatrix}$$

Now, $\text{adj}(A)\cdot B=$

$$\begin{bmatrix} 7 & 7 & -7 \\ -5 & -5 & 5 \\ -1 & -1 & 1 \end{bmatrix}\begin{bmatrix} 1 \\ 1 \\ \beta \end{bmatrix}=\begin{bmatrix} 14-7\beta \\ -10+5\beta \\ -2+\beta \end{bmatrix}$$

For $\beta=2$, $\text{adj}(A)\cdot B=0$

So, it will have no solution, $\forall \beta\neq 2$.

For $\alpha=3$

$$A=\begin{bmatrix} 3 & 2 & 1 \\ 6 & 3 & 1 \\ 3 & 3 & 2 \end{bmatrix}$$

$\Rightarrow \text{adj}(A) = [\text{Co-factor matrix}]$

$$=\begin{bmatrix} 3 & -9 & 9 \\ -1 & 3 & -3 \\ -1 & 3 & -3 \end{bmatrix}$$

Now, $\text{adj}(A)\cdot B$

$$=\begin{bmatrix} 3 & -9 & 9 \\ -1 & 3 & -3 \\ -1 & 3 & -3 \end{bmatrix}\begin{bmatrix} 1 \\ 1 \\ \beta \end{bmatrix}$$

$$=\begin{bmatrix} -6+9\beta \\ 2-3\beta \\ 2-3\beta \end{bmatrix}$$

For any value of $\beta$, $\text{adj}(A)\cdot B\neq 0$

$\therefore$ It will have no solution.

**• Concept Enhancer**

Let $AX=B$ be a system of $n$ linear equation in $n$ variables then

(i) If $|A|\neq 0$, then the system of equations is consistent and has a unique solution

(ii) If $|A|=0$ and $(\text{adj } A)B=0$, then the system of equations is consistent and has infinitely many solutions.

(iii) If $|A|=0$ and $(\text{adj } A)B\neq 0$, then the system of equations is inconsistent i.e. having no solution.

**74.** *(b)* Given, $\alpha, \beta, \gamma$ are the real roots of $x^3+ax^2+bx+c=0$, where $a, b, c\in R$ and $a, b\neq 0$

$\therefore \alpha+\beta+\gamma=$ Sum of roots taken one at a time $=-a$

$\alpha\beta+\beta\gamma+\gamma\alpha=$ Sum of roots taken two at a time $=b$

$\alpha\beta\gamma=$ Product of roots $=-c$

Also, given system of equations in $u, v, w$

$$\left.\begin{aligned} \alpha u+\beta v+\gamma w&=0 \\ \beta u+\gamma v+\alpha w&=0 \\ \gamma u+\alpha v+\beta w&=0 \end{aligned}\right\}$$

has non-trivial solution.

$$\therefore\ \Delta=0\Rightarrow\begin{vmatrix} \alpha & \beta & \gamma \\ \beta & \gamma & \alpha \\ \gamma & \alpha & \beta \end{vmatrix}=0$$

$$\Rightarrow \alpha(\beta\gamma-\alpha^2)-\beta(\beta^2-\gamma\alpha)+\gamma(\alpha\beta-\gamma^2)=0$$

[Expanding along $R_1$]

$\Rightarrow \alpha\beta\gamma-\alpha^3-\beta^3+\alpha\beta\gamma+\alpha\beta\gamma-\gamma^3=0$

$\Rightarrow \alpha^3+\beta^3+\gamma^3=3\alpha\beta\gamma$

Then, (using standard result),

Either $\quad \alpha+\beta+\gamma=0$ or $\alpha=\beta=\gamma$

If $\quad \alpha+\beta+\gamma=0$, then $-a=0$

$\Rightarrow a=0$ which is not possible according to given condition.

$\therefore \quad \alpha+\beta+\gamma=0$ (not possible)

Now, $\quad \alpha+\beta+\gamma=-a$

$\Rightarrow \quad \alpha+\alpha+\alpha=-a \quad (\because \alpha=\beta=\gamma)$

$\Rightarrow \quad a=-3\alpha \qquad \ldots$(i)

$\alpha\beta+\beta\gamma+\gamma\alpha=b$

$\Rightarrow \quad b=3\alpha^2 \qquad \ldots$(ii)

Using Eqs. (i) and (ii),

$$\frac{a^2}{b}=3$$

**75.** *(44)* Given, $X^T AX=0$

$$[x\ y\ z]\begin{bmatrix} a_1 & a_2 & a_3 \\ b_1 & b_2 & b_3 \\ c_1 & c_2 & c_3 \end{bmatrix}\begin{bmatrix} x \\ y \\ z \end{bmatrix}=0$$

$$\Rightarrow [x\ y\ z]\begin{bmatrix} a_1x+a_2y+a_3z \\ b_1x+b_2y+b_3z \\ c_1x+c_2y+c_3z \end{bmatrix}=0$$

$\Rightarrow x(a_1x+a_2y+a_3z)+y(b_1x+b_2y+b_3z)+z(c_1x+c_2y+c_3z)=0$

$\Rightarrow a_1=0, b_2=0, c_3=0, a_2+b_1=0, a_3+c_1=0, b_3=c_2=0$

So, $A$ will be skew-symmetric matrix.

$$A=\begin{bmatrix} 0 & x & y \\ -x & 0 & z \\ -y & -z & 0 \end{bmatrix}$$

Given $A\begin{bmatrix} 1 \\ 1 \\ 1 \end{bmatrix}=\begin{bmatrix} 1 \\ 4 \\ -5 \end{bmatrix}$

$$\Rightarrow \begin{bmatrix} 0 & x & y \\ -x & 0 & z \\ -y & -z & 0 \end{bmatrix}\begin{bmatrix} 1 \\ 1 \\ 1 \end{bmatrix}=\begin{bmatrix} 1 \\ 4 \\ -5 \end{bmatrix}$$

$\Rightarrow x+y=1, -x+z=4, y+z=5$

Also, given $A\begin{bmatrix} 1 \\ 2 \\ 1 \end{bmatrix}=\begin{bmatrix} 0 \\ 4 \\ -8 \end{bmatrix}$

$$\Rightarrow \begin{bmatrix} 0 & x & y \\ -x & 0 & z \\ -y & -z & 0 \end{bmatrix}\begin{bmatrix} 1 \\ 2 \\ 1 \end{bmatrix}=\begin{bmatrix} 0 \\ 4 \\ -8 \end{bmatrix}$$

$$\Rightarrow \left.\begin{aligned} 2x+y&=0 \\ -x+z&=4 \\ -y-2z&=-8 \end{aligned}\right\}$$ on solving these we

get $\quad x=-1, y=2, z=3$

$$\therefore A=\begin{bmatrix} 0 & -1 & 2 \\ 1 & 0 & 3 \\ -2 & -3 & 0 \end{bmatrix}$$

Now, $2(A+I)=\begin{bmatrix} 2 & -2 & 4 \\ 2 & 2 & 6 \\ -4 & -6 & 2 \end{bmatrix}$

$\Rightarrow |2(A+I)|=120$

$\because |\text{adj } 2(A+I)|=|2(A+I)|^{(n-1)}$

$(\because |\text{adj } A|=|A|^{n-1})$

$=|2(A+I)|^2=(120)^2=2^6\cdot 3^2\cdot 5^2$

By comparing with $2^\alpha\cdot 3^\beta\cdot 5^\gamma$, we get

$\alpha=6, \beta=2, \gamma=2$

$$\therefore \alpha^2+\beta^2+\gamma^2 = 6^2+2^2+2^2 = 36+4+4 = 44$$

**76.** *(38)* For infinitely many solution

$$D = D_x = D_y = D_z = 0$$

$$D_z = \begin{vmatrix} 2 & 7 & 3 \\ 3 & 2 & 4 \\ 1 & \mu & -1 \end{vmatrix} = 0$$

$$\Rightarrow \mu = -39$$

$$D = \begin{vmatrix} 2 & 7 & \lambda \\ 3 & 2 & 5 \\ 1 & -39 & 32 \end{vmatrix} = 0$$

$$\Rightarrow \lambda = -1 \Rightarrow \lambda - \mu = 38$$

**77.** *(55)* Given, $\alpha\beta\gamma = 45$ and $\alpha, \beta, \gamma \in R$

$$\Rightarrow x(\alpha, 1, 2) + y(1, \beta, 2) + z(2, 3, \gamma) = (0, 0, 0)$$

$x, y, z \in R, xyz \neq 0$

$$\Rightarrow \alpha x + y + 2z = 0 \Rightarrow x + \beta y + 3z = 0$$

$$2x + 2y + \gamma z = 0 \Rightarrow xyz \neq 0$$

$\Rightarrow$ non-trivial solution

$$\begin{vmatrix} \alpha & 1 & 2 \\ 1 & \beta & 3 \\ 2 & 2 & \gamma \end{vmatrix} = 0$$

$$\Rightarrow \alpha(\beta\gamma - 6) - 1(\gamma - 6) + 2(2 - 2\beta) = 0$$

$$\Rightarrow \alpha\beta\gamma - 6\alpha - \gamma + 6 + 4 - 4\beta = 0$$

$$\Rightarrow 6\alpha + 4\beta + \gamma = 55$$

**78.** *(28)* Given, $A = \begin{bmatrix} 2 & 0 & 1 \\ 1 & 1 & 0 \\ 1 & 0 & 1 \end{bmatrix}$

and $B = [B_1, B_2, B_3]$

Let $B_1 = \begin{bmatrix} x_1 \\ y_1 \\ z_1 \end{bmatrix}, B_2 = \begin{bmatrix} x_2 \\ y_2 \\ z_2 \end{bmatrix}, B_3 = \begin{bmatrix} x_3 \\ y_3 \\ z_3 \end{bmatrix}$

Since, $AB_1 = \begin{bmatrix} 1 \\ 0 \\ 0 \end{bmatrix}$ [given]

$$\Rightarrow \begin{bmatrix} 2 & 0 & 1 \\ 1 & 1 & 0 \\ 1 & 0 & 1 \end{bmatrix}\begin{bmatrix} x_1 \\ y_1 \\ z_1 \end{bmatrix} = \begin{bmatrix} 1 \\ 0 \\ 0 \end{bmatrix}$$

$$\Rightarrow x_1 = 1, y_1 = -1, z_1 = -1$$

and $AB_2 = \begin{bmatrix} 2 \\ 3 \\ 0 \end{bmatrix}$ [given]

$$\Rightarrow \begin{bmatrix} 2 & 0 & 1 \\ 1 & 1 & 0 \\ 1 & 0 & 1 \end{bmatrix}\begin{bmatrix} x_2 \\ y_2 \\ z_2 \end{bmatrix} = \begin{bmatrix} 2 \\ 3 \\ 0 \end{bmatrix}$$

$$\Rightarrow x_2 = 2, y_2 = 1, z_2 = -2$$

and $AB_3 = \begin{bmatrix} 3 \\ 2 \\ 1 \end{bmatrix}$

$$\Rightarrow \begin{bmatrix} 2 & 0 & 1 \\ 1 & 1 & 0 \\ 1 & 0 & 1 \end{bmatrix}\begin{bmatrix} x_3 \\ y_3 \\ z_3 \end{bmatrix} = \begin{bmatrix} 3 \\ 2 \\ 1 \end{bmatrix}$$

$$\Rightarrow x_3 = 2, y_3 = 0, z_3 = -1$$

Hence, $B = \begin{bmatrix} 1 & 2 & 2 \\ -1 & 1 & 0 \\ -1 & -2 & -1 \end{bmatrix} \Rightarrow |B| = 3$

According to the question,

$$\alpha = |B| = 3$$

and $\beta =$ Sum of all the diagonal elements of $B = 1$

Thus, $\alpha^3 + \beta^3 = 27 + 1 = 28$

**79.** *(24)* We have, the given system of equation has no solution.

$$\Rightarrow \Delta = 0$$

i.e. $\begin{vmatrix} 6\lambda & -3 & 3 \\ 2 & 6\lambda & 4 \\ 3 & 2 & 3\lambda \end{vmatrix} = 0$

$$6\lambda(18\lambda^2 - 8) + 3(6\lambda - 12) + 3(4 - 18\lambda) = 0$$

$$18\lambda^3 - 14\lambda - 4 = 0$$

$$(\lambda - 1)(3\lambda + 1)(3\lambda + 2) = 0$$

$$\Rightarrow \lambda = 1, -\frac{1}{3}, -\frac{2}{3}$$

$$12\sum_{\lambda \in S}|\lambda| = 12\left[|1| + \left|-\frac{1}{3}\right| + \left|-\frac{2}{3}\right|\right]$$

$$= 12\left[1 + \frac{1}{3} + \frac{2}{3}\right] = 24$$

**80.** *(5)* Given, system of equation

$$2x + y - z = 3,$$

$$x - y - z = \alpha$$

and $3x + 3y + \beta z = 3$

has infinitely many solutions,

if $\Delta = 0, \Delta_1 = \Delta_2 = \Delta_3 = 0$

Now, $\Delta = \begin{vmatrix} 2 & 1 & -1 \\ 1 & -1 & -1 \\ 3 & 3 & \beta \end{vmatrix} = 0$

$$\Rightarrow 2(-\beta + 3) - 1(\beta + 3) - 1(3 + 3) = 0$$

$$\Rightarrow \beta = -1$$

$$\Delta_1 = \begin{vmatrix} 3 & 1 & -1 \\ \alpha & -1 & -1 \\ 3 & 3 & -1 \end{vmatrix} = 0$$

$$\Rightarrow 3(1 + 3) - 1(-\alpha + 3) - 1(3\alpha + 3) = 0$$

$$\Rightarrow 12 + \alpha - 3 - 3\alpha - 3 = 0 \Rightarrow \alpha = 3$$

Also, $\Delta_2 = \begin{vmatrix} 2 & 3 & -1 \\ 1 & \alpha & -1 \\ 3 & 3 & -1 \end{vmatrix} = 0$

$$\Rightarrow 2(-\alpha + 3) - 3(-1 + 3) - 1(3 - 3\alpha) = 0$$

$$\Rightarrow \alpha = 3 \text{ and } \Delta_3 = \begin{vmatrix} 2 & 1 & 3 \\ 1 & -1 & \alpha \\ 3 & 3 & 3 \end{vmatrix} = 0$$

$$\Rightarrow 2(-3 - 3\alpha) - 1(3 - 3\alpha) + 3(3 + 3) = 0$$

$$\Rightarrow -3\alpha + 9 = 0 \Rightarrow \alpha = 3$$

$$\therefore \alpha = 3, \beta = -1$$

So, $\alpha + \beta - \alpha\beta = 3 - 1 - 3(-1) = 5$

**81.** *(5)* Given equation are $x + y - z = 2$

$$x + 2y + \alpha z = 1, \ 2x - y + z = \beta$$

It has infinite solutions, when

$$\Delta_1 = \Delta_2 = \Delta_3 = \Delta = 0$$

$$\begin{vmatrix} 1 & 1 & -1 \\ 1 & 2 & \alpha \\ 2 & -1 & 1 \end{vmatrix} = 0$$

$$\Rightarrow (2 + \alpha) + (2\alpha - 1) - (-1 - 4) = 0$$

$$\Rightarrow 3\alpha + 6 = 0 \ \therefore \ \alpha = -2$$

$$\Rightarrow \Delta_3 = 0 \Rightarrow \begin{vmatrix} 1 & 1 & 2 \\ 1 & 2 & 1 \\ 2 & -1 & \beta \end{vmatrix} = 0$$

$$\Rightarrow 2\beta + 1 + 2 - \beta - 10 = 0$$

$$\Rightarrow \beta - 7 = 0 \Rightarrow \beta = 7$$

$$\therefore \alpha + \beta = -2 + 7 = 5$$

**82.** *(3)* Given system of equations

$$(\lambda - 1)x + (3\lambda + 1)y + 2\lambda z = 0$$

$$(\lambda - 1)x + (4\lambda - 2)y + (\lambda + 3)z = 0$$

and $2x + (3\lambda + 1)y + 3(\lambda - 1)z = 0$ has non-zero solutions, so

$$\Delta = 0 \Rightarrow \begin{vmatrix} \lambda - 1 & 3\lambda + 1 & 2\lambda \\ \lambda - 1 & 4\lambda - 2 & \lambda + 3 \\ 2 & 3\lambda + 1 & 3(\lambda - 1) \end{vmatrix} = 0$$

$$\Rightarrow (\lambda - 1)[6(2\lambda - 1)(\lambda - 1) - (\lambda + 3)(3\lambda + 1)] - (3\lambda + 1)[3(\lambda - 1)^2 - 2(\lambda + 3)] + 2\lambda[(\lambda - 1)(3\lambda + 1) - 2(4\lambda - 2)] = 0$$

$$\Rightarrow (\lambda - 1)[12\lambda^2 - 18\lambda + 6 - 3\lambda^2 - 10\lambda - 3] - (3\lambda + 1)[3\lambda^2 - 6\lambda + 3 - 2\lambda - 6] + 2\lambda[3\lambda^2 - 2\lambda - 1 - 8\lambda + 4] = 0$$

$$\Rightarrow \lambda[9\lambda^2 - 28\lambda + 3 - 9\lambda^2 + 24\lambda + 9 + 6\lambda^2 - 20\lambda + 6] - 1[12\lambda^2 - 36\lambda] = 0$$

$$\Rightarrow \lambda[6\lambda^2 - 36\lambda + 54] = 0$$

$$\Rightarrow \lambda(\lambda - 3)^2 = 0 \Rightarrow \lambda = 0, 3, 3$$

$\therefore$ Sum of distinct values of $\lambda$ is 3.

# C-17

JEE Main Chapterwise Topicwise Mathematics

In JEE Main, there is usually 0-1 question per paper is visible from this chapter. In past 3 years, 42 conceptual and 18 numerical questions of easy to moderate difficulty level were asked from this chapter.

**For detailed analysis scan QR code**

# CONTINUITY AND DIFFERENTIABILITY

## Topic 1 Continuity

### Level 1

**Imp.** Continuity of Function

**01** Let $f(x)=\begin{cases}(1+ax)^{1/x}, & x<0\\ 1+b, & x=0\\ \dfrac{(x+4)^{1/2}-2}{(x+c)^{1/3}-2}, & x>0\end{cases}$

be continuous at $x=0$. Then, $e^a bc$ is equal to **[2025, 3 April Shift-I]**

(a) 72 (b) 36 (c) 64 (d) 48

**02** If the function $f(x)=\dfrac{\sin 3x+\alpha \sin x-\beta \cos 3x}{x^3}$, $x \in R$, is continuous at $x=0$, then $f(0)$ is equal to **[2024, 5 April Shift-I]**

(a) 2 (b) −2
(c) 4 (d) −4

**03** Let : $R \to R$ be a function given by

$$f(x)=\begin{cases}\dfrac{1-\cos 2x}{x^2}, & x<0\\ \alpha, & x=0\\ \dfrac{\beta\sqrt{1-\cos x}}{x}, & x>0\end{cases}$$

where $\alpha, \beta \in R$. If $f$ is continuous at $x=0$, then $\alpha^2+\beta^2$ is equal to **[2024, 4 April Shift-I]**

(a) 3 (b) 12
(c) 48 (d) 6

**Imp.** Continuity of Greatest Integer Functions & Modulus Function

**04** Let $f(x)=[x^2-x]+|-x+[x]|$, where $x \in R$ and $[t]$ denotes the greatest integer less than or equal to $t$. Then, $f$ is **[2023, 11 April Shift-I]**

(a) continuous at $x=0$ and $x=1$
(b) continuous at $x=1$, but not continuous at $x=0$
(c) not continuous at $x=0$ and $x=1$
(d) continuous at $x=0$, but not continuous at $x=1$

**05** Let the function

$$f(x)=\begin{cases}\dfrac{\log_e(1+5x)-\log_e(1+\alpha x)}{x}; & \text{if } x \neq 0\\ 10 & ; \text{if } x=0\end{cases}$$

be continuous at $x=0$. Then, $\alpha$ is equal to **[2022, 29 July Shift-II]**

(a) 10 (b) −10
(c) 5 (d) −5

**06** Let $f: R \to R$ be defined as

$$f(x)=\begin{cases}[e^x], & x<0\\ ae^x+[x-1], & 0 \le x<1\\ b+[\sin(\pi x)], & 1 \le x<2\\ [e^{-x}]-c, & x \ge 2\end{cases}$$

where $a, b, c \in R$ and $[t]$ denotes greatest integer less than or equal to $t$. Then, which of the following statement is true ? **[2022, 28 June Shift-I]**

(a) There exists $a, b, c \in R$ such that $f$ is continuous on $R$.
(b) If $f$ is discontinuous at exactly one point, then $a+b+c=1$.
(c) If $f$ is discontinuous at exactly one point, then $a+b+c \neq 1$.
(d) $f$ is discontinuous at atleast two points, for any values of $a$, $b$ and $c$.

**07** Let $a, b \in R, b \neq 0$. Define a function

$$f(x)=\begin{cases} a\sin\frac{\pi}{2}(x-1), & \text{for } x \le 0 \\ \frac{\tan 2x - \sin 2x}{bx^3}, & \text{for } x > 0 \end{cases}$$

If $f$ is continuous at $x=0$, then $10-ab$ is equal to **[2021, 26 Aug Shift-I]**

(a) 12 (b) 10 (c) 14 (d) 11

**08** Let a function $f: R \to R$ be defined as

$$f(x)=\begin{cases} \sin x - e^x, & \text{if } x \le 0 \\ a+[-x], & \text{if } 0<x<1 \\ 2x-b, & \text{if } x \ge 1 \end{cases}$$

where, $[x]$ is the greatest integer less than or equal to $x$. If $f$ is continuous on $R$, then $(a+b)$ is equal to **[2021, 20 July Shift-I]**

(a) 4 (b) 3 (c) 2 (d) 5

**09** Let $f: R \to R$ be defined as

$$f(x)=\begin{cases} 2\sin\left(-\frac{\pi x}{2}\right), & \text{if } x<-1 \\ |ax^2+x+b|, & \text{if } -1 \le x \le 1 \\ \sin(\pi x), & \text{if } x>1 \end{cases}$$

If $f(x)$ is continuous on $R$, then $a+b$ equals **[2021, 26 Feb Shift-II]**

(a) −3 (b) −1 (c) 3 (d) 1

**10** If $f: R \to R$ is a function defined by $f(x)=[x-1]\cos\left(\frac{2x-1}{2}\right)\pi$, where $[\cdot]$ denotes the greatest integer function, then $f$ is **[2021, 24 Feb Shift-I]**

(a) discontinuous only at $x=1$

(b) discontinuous at all integral values of $x$ except at $x=1$

(c) continuous only at $x=1$

(d) continuous for every real $x$

**11** Let $[t]$ denote the greatest integer $\le t$ and $\lim_{x\to 0} x\left[\frac{4}{x}\right]=A$. Then, the function, $f(x)=[x^2]\sin(\pi x)$ is discontinuous, when $x$ is equal to **[2020, 9 Jan Shift-II]**

(a) $\sqrt{A+1}$ (b) $\sqrt{A+21}$

(c) $\sqrt{A}$ (d) $\sqrt{A+5}$

**12** If the function $f(x)=\begin{cases} a|\pi-x|+1, & x \le 5 \\ b|x-\pi|+3, & x>5 \end{cases}$ is continuous at $x=5$, then the value of $a-b$ is **[2019, 9 April Shift-II]**

(a) $\frac{-2}{\pi+5}$ (b) $\frac{2}{\pi+5}$ (c) $\frac{2}{\pi-5}$ (d) $\frac{2}{5-\pi}$

**Imp.** Discontinuity of function

**13** Let $f:[-1,3] \to R$ be defined as

$$f(x)=\begin{cases} |x|+[x], & -1 \le x<1 \\ x+|x|, & 1 \le x<2 \\ x+[x], & 2 \le x \le 3, \end{cases}$$

where, $[t]$ denotes the greatest integer less than or equal to $t$. Then, $f$ is discontinuous at **[2019, 8 April Shift-II]**

(a) four or more points (b) only two points

(c) only three points (d) only one point

**14** Let $f(x)=\begin{cases} (x-1)^{\frac{1}{2-x}}, & x>1, x \neq 2 \\ k, & x=2 \end{cases}$

The value of $k$ for which $f$ is continuous at $x=2$ is **[JEE Main 2018]**

(a) $e^{-1}$ (b) $e$ (c) $e^{-2}$ (d) 1

**15** The value of $k$ for which the function

$$f(x)=\begin{cases} \left(\frac{4}{5}\right)^{\frac{\tan 4x}{\tan 5x}}, & 0<x<\frac{\pi}{2} \\ k+\frac{2}{5}, & x=\frac{\pi}{2} \end{cases}$$

is continuous at $x=\frac{\pi}{2}$, is **[JEE Main 2017]**

(a) $\frac{17}{20}$ (b) $\frac{3}{5}$ (c) $-\frac{2}{5}$ (d) $\frac{2}{5}$

**16** Let $a, b \in R, (a \neq 0)$. If the function $f$ defined as

$$f(x)=\begin{cases} \frac{2x^2}{a}, & 0 \le x<1 \\ a, & 1 \le x<\sqrt{2} \\ \frac{2b^2-4b}{x^3}, & \sqrt{2} \le x<\infty \end{cases}$$

is continuous in the interval $[0, \infty)$, then an ordered pair $(a, b)$ is **[JEE Main 2016]**

(a) $(-\sqrt{2}, 1-\sqrt{3})$ (b) $(\sqrt{2}, -1+\sqrt{3})$

(c) $(\sqrt{2}, 1-\sqrt{3})$ (d) $(-\sqrt{2}, 1+\sqrt{3})$

**17** Let $k$ be a non-zero real number. If

$$f(x)=\begin{cases} \frac{(e^x-1)^2}{\sin\left(\frac{x}{k}\right)\log\left(1+\frac{x}{4}\right)}, & x \neq 0 \\ 12, & x=0 \end{cases}$$

is a continuous function, then the value of $k$ is **[JEE Main 2015]**

(a) 1 (b) 2

(c) 3 (d) 4

**18** Define $f(x)$ as the product of two real functions $f_1(x) = x, x \in R$ and

$f_2(x) = \begin{cases} \sin\frac{1}{x}, & \text{if } x \neq 0 \\ 0, & \text{if } x = 0 \end{cases}$ as follows

$f(x) = \begin{cases} f_1(x) \cdot f_2(x), & \text{if } x \neq 0 \\ 0, & \text{if } x = 0 \end{cases}$

**Statement I** $f(x)$ is continuous on $R$.

**Statement II** $f_1(x)$ and $f_2(x)$ are continuous on $R$. ***[AIEEE 2011]***

(a) Statement I is false, Statement II is true
(b) Statement I is true, Statement II is true; Statement II is correct explanation of Statement I.
(c) Statement I is true, Statement II is true; Statement II is not a correct explanation of Statement I
(d) Statement I is true, Statement II is false

**19** The function $f : R/\{0\} \to R$ given by

$f(x) = \frac{1}{x} - \frac{2}{e^{2x} - 1}$ can be made continuous at $x = 0$ by defining $f(0)$ as ***[AIEEE 2007]***

(a) 2 (b) –1 (c) 0 (d) 1

**20** Let $f(x) = \frac{1 - \tan x}{4x - \pi}, x \neq \frac{\pi}{4}, x \in \left[0, \frac{\pi}{2}\right]$. If $f(x)$ is continuous in $\left[0, \frac{\pi}{2}\right]$, then $f\left(\frac{\pi}{4}\right)$ is equal to ***[AIEEE 2004]***

(a) 1 (b) 1/2 (c) –1/2 (d) –1

## Level 2

**21** Let $f : R \to R$ be a continuous function satisfying $f(0) = 1$ and $f(2x) - f(x) = x$ for all $x \in R$. If $\lim_{n \to \infty} \left\{ f(x) - f\left(\frac{x}{2^n}\right) \right\} = G(x)$, then $\sum_{r=1}^{10} G(r^2)$ is equal to ***[2025, 4 April Shift-I]***

(a) 540 (b) 420 (c) 215 (d) 385

**22** If the function

$$f(x) = \begin{cases} \frac{2}{x}\{\sin(k_1 + 1)x + \sin(k_2 - 1)x\}, & x < 0 \\ 4, & x = 0 \\ \frac{2}{x}\log_e\left(\frac{2 + k_1 x}{2 + k_2 x}\right), & x > 0 \end{cases}$$

is continuous at $x = 0$, then $k_1^2 + k_2^2$ is equal to ***[2025, 23 Jan Shift-I]***

(a) 5 (b) 10 (c) 8 (d) 20

**23** Let $f : [-1, 2] \to R$ be given by $f(x) = 2x^2 + x + [x^2] - [x]$, where $[t]$ denotes the greatest integer less than or equal to $t$. The number of points, where $f$ is not continuous, is ***[2024, 5 April Shift-II]***

(a) 3 (b) 6
(c) 4 (d) 5

**24** If the function $f(x) = \begin{cases} \frac{72^x - 9^x - 8^x + 1}{\sqrt{2} - \sqrt{1 + \cos x}}, & x \neq 0 \\ a \log_e 2 \log_e 3, & x = 0 \end{cases}$ is continuous at $x = 0$, then the value of $a^2$ is equal to ***[2024, 4 April Shift-II]***

(a) 1152 (b) 1250
(c) 968 (d) 746

**25** If for $p \neq q \neq 0$, the function

$f(x) = \frac{\sqrt[7]{p(729 + x)} - 3}{\sqrt[3]{729 + qx} - 9}$ is continuous at $x = 0$, then ***[2022, 27 July Shift-II]***

(a) $7pq\, f(0) - 1 = 0$ (b) $63q\, f(0) - p^2 = 0$
(c) $21q\, f(0) - p^2 = 0$ (d) $7pq\, f(0) - 9 = 0$

**26** If $f(x) = \begin{cases} x + a, & x \leq 0 \\ |x - 4|, & x > 0 \end{cases}$ and

$g(x) = \begin{cases} x + 1, & x < 0 \\ (x - 4)^2 + b, & x \geq 0 \end{cases}$ are continuous on $R$,

then $(gof)(2) + (fog)(-2)$ is equal to ***[2022, 26 July Shift-I]***

(a) –10 (b) 10
(c) 8 (d) –8

**27** Let $f, g : R \to R$ be functions defined by

$f(x) = \begin{cases} [x], & x < 0 \\ |1 - x|, & x \geq 0 \end{cases}$

and $g(x) = \begin{cases} e^x - x, & x < 0 \\ (x - 1)^2 - 1, & x \geq 0 \end{cases}$

where $[x]$ denote the greatest integer less than or equal to $x$. Then, the function $fog$ is discontinuous at exactly ***[2022, 28 June Shift-I]***

(a) one point (b) two points
(c) three points (d) four points

**28** If the function $f(x) = \begin{cases} \frac{1}{x}\log_e\left(\frac{1 + \frac{x}{a}}{1 - \frac{x}{b}}\right), & x < 0 \\ k, & x = 0 \\ \frac{\cos^2 x - \sin^2 x - 1}{\sqrt{x^2 + 1} - 1}, & x > 0 \end{cases}$

is continuous at $x = 0$, then $\frac{1}{a} + \frac{1}{b} + \frac{4}{k}$ is equal to ***[2021, 31 Aug Shift-I]***

(a) –5 (b) 5
(c) –4 (d) 4

**29** Let $f: R \to R$ be defined as

$$f(x)=\begin{cases}\dfrac{\lambda|x^2-5x+6|}{\mu(5x-x^2-6)}, & x<2\\ e^{\frac{\tan(x-2)}{x-[x]}}, & x>2\\ \mu, & x=2\end{cases}$$

where, $[x]$ is the greatest integer less than or equal to $x$. If $f$ is continuous at $x=2$, then $\lambda+\mu$ is equal to [2021, 25 July Shift-I]

(a) $e(-e+1)$ (b) $e(e-2)$
(c) 1 (d) $2e-1$

**30** Let $f: R \to R$ be defined as

$$f(x)=\begin{cases}\dfrac{x^3}{(1-\cos 2x)^2}\log_e\left(\dfrac{1+2xe^{-2x}}{(1-xe^{-x})^2}\right), & x\neq 0\\ \alpha, & x=0\end{cases}$$

If $f$ is continuous at $x=0$, then $\alpha$ is equal to [2021, 22 July Shift-II]

(a) 1 (b) 3
(c) 0 (d) 2

**31** Let $f: R \to R$ be a function defined as

$$f(x)=\begin{cases}\dfrac{\sin(a+1)x+\sin 2x}{2x}, & \text{if } x<0\\ b, & \text{if } x<0\\ \dfrac{\sqrt{x+bx^3}-\sqrt{x}}{bx^{5/2}}, & \text{if } x>0\end{cases}$$

If $f$ is continuous at $x=0$, then the value of $a+b$ is equal to [2021, 18 March Shift-II]

(a) $-\dfrac{5}{2}$ (b) $-2$ (c) $-3$ (d) $-\dfrac{3}{2}$

**32** If a function $f(x)$ defined by

$$f(x)=\begin{cases}ae^x+be^{-x}, & -1\le x<1\\ cx^2, & 1\le x\le 3\\ ax^2+2cx, & 3<x\le 4\end{cases}$$

be continuous for some $a, b, c \in R$ and $f'(0)+f'(2)=e$, then the value of $a$ is [2020, 2 Sep Shift-I]

(a) $\dfrac{e}{e^2+3e+13}$ (b) $\dfrac{e}{e^2-3e-13}$
(c) $\dfrac{1}{e^2-3e+13}$ (d) $\dfrac{e}{e^2-3e+13}$

**33** If $f(x)=\begin{cases}\dfrac{\sin(p+1)x+\sin x}{x}, & x<0\\ q, & x=0\\ \dfrac{\sqrt{x+x^2}-\sqrt{x}}{x^{3/2}}, & x>0\end{cases}$

is continuous at $x=0$, then the ordered pair $(p, q)$ is equal to [2019, 10 April Shift-I]

(a) $\left(-\dfrac{3}{2}, -\dfrac{1}{2}\right)$ (b) $\left(-\dfrac{1}{2}, \dfrac{3}{2}\right)$
(c) $\left(\dfrac{5}{2}, \dfrac{1}{2}\right)$ (d) $\left(-\dfrac{3}{2}, \dfrac{1}{2}\right)$

**34** If $f: R \to R$ is a function defined by $f(x)=[x]\cos\left(\dfrac{2x-1}{2}\right)\pi$, where $[x]$ denotes the greatest integer function, then $f$ is [AIEEE 2012]

(a) continuous for every real $x$
(b) discontinuous only at $x=0$
(c) discontinuous only at non-zero integral values of $x$
(d) continuous only at $x=0$

## Level 3

**35** For $a, b>0$, let

$$f(x)=\begin{cases}\dfrac{\tan((a+1)x)+b\tan x}{x}, & x<0\\ 3, & x=0\\ \dfrac{\sqrt{ax+b^2x^2}-\sqrt{ax}}{b\sqrt{a}\cdot x\sqrt{x}}, & x>0\end{cases}$$

be a continuous function at $x=0$. Then, $\dfrac{b}{a}$ is equal to [2024, 8 April Shift-II]

(a) 4 (b) 5 (c) 8 (d) 6

**36** Consider the function

$$f(x)=\begin{cases}\dfrac{a(7x-12-x^2)}{b|x^2-7x+12|}, & x<3\\ 2^{\frac{\sin(x-3)}{x-[x]}}, & x>3,\\ b, & x=3\end{cases}$$

where $[x]$ denotes the greatest integer less than or equal to $x$. If $S$ denotes the set of all ordered pairs $(a, b)$ such that $f(x)$ is continuous at $x=3$, then the number of elements in $S$, is [2024, 27 Jan Shift-I]

(a) 4 (b) 2
(c) 1 (d) Infinitely many

**37** If the function

$$f(x)=\begin{cases}(1+|\cos x|)^{\frac{\lambda}{|\cos x|}}, & 0<x<\dfrac{\pi}{2}\\ \mu, & x=\dfrac{\pi}{2}\\ e^{\frac{\cot 6x}{\cot 4x}}, & \dfrac{\pi}{2}<x<\pi\end{cases}$$

is continuous at $x = \pi/2$, then $9\lambda + 6\log_e \mu + \mu^6 - e^{6\lambda}$ is equal to [2023, 25 Jan Shift-II]

(a) 10 (b) $2e^4 + 8$
(c) 8 (d) 11

**38** The function $f : R \to R$ defined by

$$f(x) = \lim_{n\to\infty} \frac{\cos(2\pi x) - x^{2n}\sin(x-1)}{1 + x^{2n+1} - x^{2n}}$$ is continuous for all $x$ in [2022, 28 July Shift-II]

(a) $R - \{-1\}$ (b) $R - \{-1, 1\}$
(c) $R - \{1\}$ (d) $R - \{0\}$

**39** If the function

$$f(x) = \begin{cases} \dfrac{\log_e(1 - x + x^2) + \log_e(1 + x + x^2)}{\sec x - \cos x}, & x \in \left(\dfrac{-\pi}{2}, \dfrac{\pi}{2}\right) - \{0\} \\ k & x = 0 \end{cases}$$

is continuous at $x = 0$, then $k$ is equal to [2022, 26 July Shift-I]

(a) 1 (b) $-1$ (c) $e$ (d) 0

**40** Let $f : \left(-\dfrac{\pi}{4}, \dfrac{\pi}{4}\right) \to R$ be defined as

$$f(x) = \begin{cases} (1 + |\sin x|)^{\frac{3a}{|\sin x|}}, & \dfrac{-\pi}{4} < x < 0 \\ b, & x = 0 \\ e^{\frac{\cot 4x}{\cot 2x}}, & 0 < x < \dfrac{\pi}{4} \end{cases}$$

If $f$ is continuous at $x = 0$, then the value of $6a + b^2$ is equal to [2021, 27 July Shift I]

(a) $1 - e$ (b) $e - 1$ (c) $1 + e$ (d) $e$

**41** Let $\alpha \in R$ be such that the function

$$f(x) = \begin{cases} \dfrac{\cos^{-1}(1 - \{x\}^2)\sin^{-1}(1 - \{x\})}{\{x\} - \{x\}^3}, & x \neq 0 \\ \alpha, & x = 0 \end{cases}$$ is continuous at

$x = 0$, where $\{x\} = x - [x]$, $[x]$ is the greatest integer less than or equal to $x$. Then, [2021, 16 March Shift-II]

(a) $\alpha = \dfrac{\pi}{\sqrt{2}}$ (b) $\alpha = 0$
(c) No such $\alpha$ exists (d) $\alpha = \dfrac{\pi}{4}$

## Numerical Value Type Questions

**42** The number of points of discontinuity of the function $f(x) = \left[\dfrac{x^2}{2}\right] - [\sqrt{x}]$, $x \in [0, 4]$, where $[\cdot]$ denotes the greatest integer function, is ...... [2025, 7 April Shift-I]

**43** If the function $f(x) = \dfrac{\tan(\tan x) - \sin(\sin x)}{\tan x - \sin x}$ is continuous at $x = 0$, then $f(0)$ is equal to ......... . [2025, 7 April Shift-II]

**44** Let $f : (0, \pi) \to R$ be a function given by

$$f(x) = \begin{cases} \left(\dfrac{8}{7}\right)^{\frac{\tan 8x}{\tan 7x}}, & 0 < x < \dfrac{\pi}{2} \\ a - 8, & x = \dfrac{\pi}{2} \\ (1 + |\cot x|)^{\frac{b}{a}|\tan x|}, & \dfrac{\pi}{2} < x < \pi \end{cases}$$

where $a, b \in Z$. If $f$ is continuous at $x = \dfrac{\pi}{2}$, then $a^2 + b^2$ is equal to .......... . [2024, 9 April Shift-I]

**45** Let $[t]$ denote the greatest integer less than or equal to $t$. Let $f : [0, \infty] \to R$ be a function defined by $f(x) = \left[\dfrac{x}{2} + 3\right] - [\sqrt{x}]$. Let $S$ be the set of all points in the interval $[0, 8]$ at which $f$ is not continuous. Then, $\sum_{a \in S} a$ is equal to ......... . [2024, 6 April Shift-II]

**46** Let $[x]$ be the greatest integer $\le x$. Then, the number of points in the interval $(-2, 1)$, where the function $f(x) = |[x]| + \sqrt{x - [x]}$ is discontinuous is ......... . [2023, 12 April Shift-I]

**47** Let $f(x) = [2x^2 + 1]$ and $g(x) = \begin{cases} 2x - 3, & x < 0 \\ 2x + 3, & x \ge 0 \end{cases}$, where $[t]$ is the greatest integer $\le t$. Then, in the open interval $(-1, 1)$, the number of points where $fog$ is discontinuous is equal to ......... [2022, 25 June Shift-II]

**48** Let $[t]$ denote the greatest integer $\le t$. The number of points where the function

$$f(x) = [x]|x^2 - 1| + \sin\left(\frac{\pi}{[x] + 3}\right) - [x + 1], x \in (-2, 2)$$

is not continuous is ........ [2021, 1 Sep Shift-II]

**49** If the function $f(x) = \dfrac{\cos(\sin x) - \cos x}{x^4}$ is continuous at each point in its domain and $f(0) = \dfrac{1}{k}$, then $k$ is ......... . [2021, 17 March Shift-I]

**50** Let $f: R \to R$ and $g: R \to R$ be defined as

$$f(x)=\begin{cases} x+a, & x<0 \\ |x-1|, & x\geq 0 \end{cases} \text{ and } g(x)=\begin{cases} x+1, & x<0 \\ (x-1)^2+b, & x\geq 0 \end{cases}$$

where $a, b$ are non-negative real numbers. If $(gof)(x)$ is continuous for all $x \in R$, then $a+b$ is equal to......

*[2021, 16 March Shift-II]*

**51** Let $f(x)=x\left[\frac{x}{2}\right]$, for $-10<x<10$, where $[t]$ denotes the greatest integer function. Then, the number of points of discontinuity of $f$ is equal to ....... .

*[2020, 5 Sep Shift-I]*

## Topic 2 Differentiability

### Level 1

**01** Let $f$ be a differentiable function on R such that $f(2)=1$, $f'(2)=4$. Let $\lim_{x\to 0}(f(2+x))^{3/x}=e^{\alpha}$.

Then, the number of times the curve $y=4x^3-4x^2-4(\alpha-7)x-\alpha$ meets $X$-axis is

*[2025, 4 April Shift-II]*

(a) 2 (b) 0 (c) 3 (d) 1

**02** Let $[x]$ denote the greatest integer function and let $m$ and $n$, respectively be the numbers of the points, where the function $f(x)=[x]+|x-2|$, $-2<x<3$ is not continuous and not differentiable.

Then, $m+n$ is equal to *[2025, 24 Jan Shift-II]*

(a) 6 (b) 8 (c) 9 (d) 7

**03** Let $f:(-\infty,\infty)-\{0\}\to R$ be a differentiable function such that $f'(1)=\lim_{a\to\infty} a^2 f\left(\frac{1}{a}\right)$. Then,

$$\lim_{a\to\infty}\frac{a(a+1)}{2}\tan^{-1}\left(\frac{1}{a}\right)+a^2-2\log_e a$$

is equal to *[2024, 6 April Shift-I]*

(a) $\frac{5}{2}+\frac{\pi}{8}$ (b) $\frac{3}{2}+\frac{\pi}{4}$ (c) $\frac{3}{4}+\frac{\pi}{8}$ (d) $\frac{3}{8}+\frac{\pi}{4}$

**Imp.** Number of Points where Function is not Differentiable

**04** The number of points, where the function $f: R\to R$, $f(x)=|x-1|\cos|x-2|\sin|x-1|+(x-3)|x^2-5x+4|$, is not differentiable, is

*[2022, 29 July Shift-I]*

(a) 1 (b) 2 (c) 3 (d) 4

**05** Let $f(x)=\min\{1, 1+x\sin x\}$, $0\leq x\leq 2\pi$. If $m$ is the number of points, where $f$ is not differentiable and $n$ is the number of points, where $f$ is not continuous, then the ordered pair $(m, n)$ is equal to

*[2022, 26 June Shift-II]*

(a) (2, 0) (b) (1, 0) (c) (1, 1) (d) (2, 1)

**06** The function $f(x)=|x^2-2x-3|\cdot e^{|9x^2-12x+4|}$ is not differentiable at exactly *[2021, 31 Aug Shift-I]*

(a) four points (b) three points
(c) two points (d) one point

**07** Let $[t]$ denote the greatest integer less than or equal to $t$. Let $f(x)=x-[x]$, $g(x)=1-x+[x]$, and $h(x)=\min\{f(x), g(x)\}$, $x\in[-2,2]$. Then $h$ is

*[2021, 26 Aug Shift-II]*

(a) continuous in [– 2, 2] but not differentiable at more than four points in (–2, 2)
(b) not continuous at exactly three points in [– 2, 2]
(c) continuous in [– 2, 2] but not differentiable at exactly three points in (–2, 2)
(d) not continuous at exactly four points in [– 2, 2]

**08** Let $f:[0,\infty)\to[0,3]$ be a function defined by

$$f(x)=\begin{cases}\max\{\sin t: 0\leq t\leq x\}, & 0\leq x<\pi \\ 2+\cos x, & x>\pi\end{cases}$$

Then which of the following is true?

*[2021, 27 July Shift-II]*

(a) $f$ is continuous everywhere but not differentiable exactly at one point in $(0,\infty)$
(b) $f$ is differentiable everywhere in $(0,\infty)$
(c) $f$ is not continuous exactly at two points in $(0,\infty)$
(d) $f$ is continuous everywhere but not differentiable exactly at two points in $(0,\infty)$

**09** Let $f: R\to R$ be a function defined by $f(x)=\max\{x, x^2\}$. Let $S$ denote the set of all points in $R$, where $f$ is not differentiable.

Then, *[2020, 6 Sep Shift-II]*

(a) {0, 1} (b) {0}
(c) $\phi$ {an empty set} (d) {1}

**10** Let $f: R\to R$ be differentiable at $c\in R$ and $f(c)=0$. If $g(x)=|f(x)|$, then at $x=c$, $g$ is *[2019, 10 April Shift-I]*

(a) not differentiable
(b) differentiable if $f'(c)\neq 0$
(c) not differentiable if $f'(c)=0$
(d) differentiable if $f'(c)=0$

**11** Let $f(x)=15-|x-10|$; $x\in R$. Then, the set of all values of $x$, at which the function, $g(x)=f(f(x))$ is not differentiable, is *[2019, 9 April Shift-I]*

(a) {5, 10, 15, 20} (b) {5, 10, 15}
(c) {10} (d) {10, 15}

**12** Let $S$ be the set of all points in $(-\pi,\pi)$ at which the function, $f(x)=\min\{\sin x, \cos x\}$ is not differentiable. Then, $S$ is a subset of which of the following?

*[2019, 12 Jan Shift-I]*

(a) $\left\{-\frac{\pi}{4}, 0, \frac{\pi}{4}\right\}$ (b) $\left\{-\frac{\pi}{2}, -\frac{\pi}{4}, \frac{\pi}{4}, \frac{\pi}{2}\right\}$

(c) $\left\{-\frac{3\pi}{4}, -\frac{\pi}{4}, \frac{3\pi}{4}, \frac{\pi}{4}\right\}$ (d) $\left\{-\frac{3\pi}{4}, -\frac{\pi}{2}, \frac{\pi}{2}, \frac{3\pi}{4}\right\}$

**13** Let $S = \{(\lambda, \mu) \in R \times R : f(t) = (|\lambda| e^{|t|} - \mu) \sin(2|t|)$, $t \in R$, is differentiable function$\}$. Then, $S$ is a subset of **[JEE Main 2018]**

(a) $[0, \infty) \times R$ (b) $R \times (-\infty, 0)$
(c) $R \times [0, \infty)$ (d) $(-\infty, 0) \times R$

**14** For $x \in R$, $f(x) = |\log 2 - \sin x|$ and $g(x) = f(f(x))$, then **[JEE Main 2016]**

(a) $g$ is not differentiable at $x = 0$
(b) $g'(0) = \cos(\log 2)$
(c) $g'(0) = -\cos(\log 2)$
(d) $g$ is differentiable at $x = 0$ and $g'(0) = -\sin(\log 2)$

**15** If the function $g(x) = \begin{cases} k\sqrt{x+1}, & 0 \le x \le 3 \\ mx + 2, & 3 < x \le 5 \end{cases}$ is differentiable, then the value of $k + m$ is **[JEE Main 2015]**

(a) 2 (b) 16/5
(c) 10/3 (d) 4

**16** If function $f(x)$ is differentiable at $x = a$, then $\lim_{x \to a} \frac{x^2 f(a) - a^2 f(x)}{x - a}$ is equal to **[AIEEE 2011]**

(a) $2af(a) + a^2 f'(a)$ (b) $-a^2 f'(a)$
(c) $af(a) - a^2 f'(a)$ (d) $2af(a) - a^2 f'(a)$

**17** If $f : (-1, 1) \to R$ is a differentiable function with $f(0) = -1$ and $f'(0) = 1$. Let $g(x) = [f(2f(x) + 2)]^2$. Then, $g'(0)$ is equal to **[AIEEE 2010]**

(a) 4 (b) $-4$ (c) 0 (d) $-2$

**18** The set of points, where $f(x) = \frac{x}{1 + |x|}$ is differentiable, is **[AIEEE 2006]**

(a) $(-\infty, -1) \cup (-1, \infty)$ (b) $(-\infty, \infty)$
(c) $(0, \infty)$ (d) $(-\infty, 0) \cup (0, \infty)$

**19** If $f(x)$ is differentiable at $x = 1$ and $\lim_{h \to 0} \frac{1}{h} f(1 + h) = 5$, then $f'(1)$ is equal to **[AIEEE 2005]**

(a) 6 (b) 5 (c) 4 (d) 3

## Level 2

**20** Let $f(x) = \left|2x^2 + 5|x| - 3\right|$, $x \in R$. If $m$ and $n$ denote the number of points where $f$ is not continuous and not differentiable respectively, then $m + n$ is equal to **[2024, 1 Feb Shift-II]**

(a) 3 (b) 0 (c) 2 (d) 5

**21** Consider the function $f : (0, \infty) \to R$, defined by $f(x) = e^{-|\log_e x|}$. If $m$ and $n$ are respectively, the number of points at which $f$ is not continuous and $f$ is not differentiable, then $m + n$ is **[2024, 31 Jan Shift-II]**

(a) 1 (b) 2
(c) 3 (d) 0

**22** Consider the function $f : (0, 2) \to R$ defined by $f(x) = \frac{x}{2} + \frac{2}{x}$ and the function $g(x)$ defined by

$$g(x) = \begin{cases} \min\{f(t)\}, & 0 < t \le x \text{ and } 0 < x \le 1 \\ \frac{3}{2} + x, & 1 < x < 2 \end{cases}$$

. Then, **[2024, 27 Jan Shift-II]**

(a) $g$ is neither continuous nor differentiable at $x = 1$
(b) $g$ is not continuous for all $x \in (0, 2)$
(c) $g$ is continuous but not differentiable at $x = 1$
(d) $g$ is continuous and differentiable for all $x \in (0, 2)$

**23** Let $[x]$ denote the greatest integer function and $f(x) = \max\{1 + x + [x], 2 + x, x + 2[x]\}$, $0 \le x \le 2$. Let $m$ be the number of points in $[0, 2]$, where $f$ is not continuous and $n$ be the number of points in $(0, 2)$, where $f$ is not differentiable. Then $(m + n)^2 + 2$ is equal to **[2023, 15 April Shift-I]**

(a) 6 (b) 3 (c) 11 (d) 2

**24** For the differentiable function $f : R - \{0\} \to R$, let $3f(x) + 2f\left(\frac{1}{x}\right) = \frac{1}{x} - 10$, then $\left| f(3) + f'\left(\frac{1}{4}\right) \right|$ is equal to **[2023, 13 April Shift-I]**

(a) $\frac{33}{5}$ (b) $\frac{29}{5}$ (c) 7 (d) 13

**25** Let $f$ and $g$ be two functions defined by

$$f(x) = \begin{cases} x + 1, & x < 0 \\ |x - 1|, & x \ge 0 \end{cases} \text{ and } g(x) = \begin{cases} x + 1, & x < 0 \\ 1, & x \ge 0 \end{cases}$$

Then, $(gof)(x)$ is **[2023, 11 April Shift-II]**

(a) continuous everywhere but not differentiable exactly at one point
(b) continuous everywhere but not differentiable at $x = 1$
(c) differentiable everywhere
(d) not continuous at $x = -1$

**26** $f, g : R \to R$ be two real valued functions defined as

$$f(x) = \begin{cases} -|x + 3|, & x < 0 \\ e^x, & x \ge 0 \end{cases} \text{ and}$$

$$g(x) = \begin{cases} x^2 + k_1 x, & x < 0 \\ 4x + k_2, & x \ge 0 \end{cases}$$, where $k_1$ and $k_2$ are real constants. If $(gof)$ is differentiable at $x = 0$, then $(gof)(-4) + (gof)(4)$ is equal to **[2022, 26 June Shift-I]**

(a) $4(e^4 + 1)$ (b) $2(2e^4 + 1)$
(c) $4e^4$ (d) $2(2e^4 - 1)$

**27** Let $K$ be the set of all real values of $x$, where the function $f(x) = \sin|x| - |x| + 2(x - \pi)\cos|x|$ is not differentiable. Then, the set $K$ is equal to **[2019, 11 Jan Shift-II]**

(a) $\{0\}$ (b) $\phi$ (an empty set)
(c) $\{\pi\}$ (d) $\{0, \pi\}$

**28** Let $S = \{t \in R : f(x) = |x - \pi| \cdot (e^{|x|} - 1) \sin|x|$ is not differentiable at $t\}$. Then, the set $S$ is equal to **[JEE Main 2018]**

(a) $\phi$ (an empty set) (b) $\{0\}$
(c) $\{\pi\}$ (d) $\{0, \pi\}$

**29** If $f(x) = \begin{cases} xe^{-\left(\frac{1}{|x|}+\frac{1}{x}\right)}, & x \neq 0 \\ 0, & x = 0 \end{cases}$, then $f(x)$ is **[AIEEE 2003]**

(a) continuous as well as differentiable for all $x$
(b) continuous for all $x$ but not differentiable at $x = 0$
(c) Neither differentiable nor continuous at $x = 0$
(d) discontinuous everywhere

## Level 3

**30** Let $f : R \to R$ be defined as

$$f(x) = \begin{cases} \dfrac{a - b\cos 2x}{x^2}, & x < 0 \\ x^2 + cx + 2, & 0 \le x \le 1 \\ 2x + 1, & x > 1 \end{cases}$$

If $f$ is continuous everywhere in $R$ and $m$ is the number of points where $f$ is not differential, then $m + a + b + c$ equals **[2024, 1 Feb Shift-I]**

(a) 4 (b) 2 (c) 3 (d) 1

**31** Let $f(x) = \begin{cases} x^2 \sin\left(\dfrac{1}{x}\right), & x \neq 0 \\ 0, & x = 0 \end{cases}$

Then, at $x = 0$ **[2023, 24 Jan Shift-I]**

(a) $f$ and $f'$ both are continuous
(b) $f$ is continuous but $f'$ is not continuous
(c) $f$ is continuous but not differentiable
(d) $f'$ is continuous but not differentiable

**32** If $f(x) = \begin{cases} \dfrac{1}{|x|} & ; |x| \ge 1 \\ ax^2 + b & ; |x| < 1 \end{cases}$

is differentiable at every point of the domain, then the values of $a$ and $b$ are respectively **[2021, 18 March shift-I]**

(a) $\frac{1}{2}, \frac{1}{2}$ (b) $\frac{1}{2}, -\frac{3}{2}$ (c) $\frac{5}{2}, -\frac{3}{2}$ (d) $-\frac{1}{2}, \frac{3}{2}$

**33** Let the functions $f : R \to R$ and $g : R \to R$ be defined as $f(x) = \begin{cases} x + 2, & x < 0 \\ x^2, & x \ge 0 \end{cases}$

and $g(x) = \begin{cases} x^3, & x < 1 \\ 3x - 2, & x \ge 1 \end{cases}$

Then, the number of points in $R$, where $(fog)(x)$ is not differentiable is equal to **[2021, 16 March Shift-I]**

(a) 3 (b) 1 (c) 0 (d) 2

**34** Let $f(x) = \begin{cases} -1, & -2 \le x < 0 \\ x^2 - 1, & 0 \le x \le 2 \end{cases}$

and $g(x) = |f(x)| + f(|x|)$.

Then, in the interval $(-2, 2)$, $g$ is **[2019, 11 Jan Shift-I]**

(a) not differentiable at one point
(b) not differentiable at two points
(c) differentiable at all points
(d) not continuous

**35** Let $f : (-1, 1) \longrightarrow R$ be a function defined by $f(x) = \max\{-|x|, -\sqrt{1 - x^2}\}$. If $K$ is the set of all points at which $f$ is not differentiable, then $K$ has exactly **[2019, 10 Jan Shift-II]**

(a) three elements (b) five elements
(c) two elements (d) one element

**36** Let $f(x) = \begin{cases} \max\{|x|, x^2\}, & |x| \le 2 \\ 8 - 2|x|, & 2 < |x| \le 4 \end{cases}$

Let $S$ be the set of points in the interval $(-4, 4)$ at which $f$ is not differentiable. Then, $S$ **[2019, 10 Jan Shift-I]**

(a) equals $\{-2, -1, 0, 1, 2\}$ (b) equals $\{-2, 2\}$
(c) is an empty set (d) equals $\{-2, -1, 1, 2\}$

**37** Consider the function, $f(x) = |x - 2| + |x - 5|, x \in R$.

**Statement I** $f'(4) = 0$

**Statement II** $f$ is continuous in $[2, 5]$, differentiable in $(2, 5)$ and $f(2) = f(5)$.

(a) Statement I is false, Statement II is true
(b) Statement II is true, Statement II is true; Statement II is a correct explanation of Statement I
(c) Statement I is true, Statement II is true; Statement II is not a correct explanation of Statement I
(d) Statement I is true, Statement II is false **[AIEEE 2012]**

**38** Let $f(x) = \begin{cases} (x - 1)\sin\dfrac{1}{x - 1}, & \text{if } x \neq 1 \\ 0, & \text{if } x = 1 \end{cases}$

Then, which one of the following is true? **[AIEEE 2008]**

(a) $f$ is differentiable at $x = 1$ but not at $x = 0$
(b) $f$ is neither differentiable at $x = 0$ nor at $x = 1$
(c) $f$ is differentiable at $x = 0$ and at $x = 1$
(d) $f$ is differentiable at $x = 0$ but not at $x = 1$

### Numerical Value Type Questions

**39** Let $m$ and $n$ be the number of points at which the function $f(x) = \max\{x, x^3, x^5, \ldots, x^{21}\}, x \in R, x \in R$, is not differentiable and not continuous, respectively. Then, $m + n$ is equal to .......... **[2025, 4 April Shift-I]**

**40** Let

$$f(x)=\begin{cases}3x, & x<0\\ \min\{1+x+[x], x+2[x]\}, & 0\le x\le 2, \text{ where}\\ 5, & x>2\end{cases}$$

[.] denotes greatest integer function. If $\alpha$ and $\beta$ are the number of points, where $f$ is not continuous and is not differentiable, respectively, then $\alpha+\beta$ equals ……. **[2025, 28 Jan Shift-I]**

**41** If the function $f(x)=\begin{cases}\dfrac{1}{|x|}, & |x|\ge 2\\ ax^2+2b, & |x|<2\end{cases}$ is differentiable on $R$, then $48(a+b)$ is equal to ……… . **[2024, 24 Jan Shift-I]**

**42** Let for a differentiable function $f:(0,\infty)\to R$,

$$f(x)-f(y)\ge \log_e\left(\frac{x}{y}\right)+x-y, \forall\, x, y\in(0,\infty).$$

Then, $\sum_{x=1}^{20} f'\left(\dfrac{1}{x^2}\right)$ is equal to …. . **[2024, 27 Jan Shift-I]**

**43** Let $f:(-2,2)\to R$ be defined by

$$f(x)=\begin{cases}x[x], & -2<x<0\\ (x-1)[x], & 0\le x<2\end{cases}$$

where $[x]$ denotes the greatest integer function. If $m$ and $n$ respectively are the number of points in $(-2, 2)$ at which $y=|f(x)|$ is not continuous and not differentiable, then $m+n$ is equal to …… **[2023, 10 April Shift-I]**

**44** Let $k$ and $m$ be positive real numbers such that the function $f(x)=\begin{cases}3x^2+k\sqrt{x+1}, & 0<x<1\\ mx^2+k^2, & x\ge 1\end{cases}$ is differentiable for all $x>0$. Then, $\dfrac{8f'(8)}{f'\left(\dfrac{1}{8}\right)}$ is equal to

………… **[2023, 8 April Shift-II]**

**45** Let $a\in Z$ and $[t]$ be the greatest integer $\le t$. Then, the number of points, where the function $f(x)=[a+13\sin x], x\in(0,\pi)$ is not differentiable, is ………. **[2023, 6 April Shift-I]**

**46** If $[t]$ denotes the greatest integer $\le t$, then number of points, at which the function

$$f(x)=4|2x+3|+9\left[x+\frac{1}{2}\right]-12[x+20] \text{ is not}$$

differentiable in the open interval $(-20, 20)$, is ………… . **[2022, 29 July Shift-II]**

**47** Let $f$ and $g$ be twice differentiable even functions on $(-2,2)$ such that $f\left(\dfrac{1}{4}\right)=0, f\left(\dfrac{1}{2}\right)=0, f(1)=1$ and $g\left(\dfrac{3}{4}\right)=0, g(1)=2$. Then, the minimum number of solutions of $f(x)g''(x)+f'(x)g'(x)=0$ in $(-2,2)$ is equal to ……… . **[2022, 29 June Shift-II]**

**48** Let $f:[0,3]\to R$ be defined by

$f(x)=\min\{x-[x], 1+[x]-x\}$

where, $[x]$ is the greatest integer less than or equal to $x$. Let $P$ denote the set containing all $x\in(0,3)$, where $f$ is discontinuous and $Q$ denotes the set containing all $x\in(0,3)$, where $f$ is not differentiable. Then, the sum of number of elements in $P$ and $Q$ is equal to ………… . **[2021, 27 July Shift-I]**

**49** Let $f:R\to R$ be a function defined as

$$f(x)=\begin{cases}3\left(1-\dfrac{|x|}{2}\right), & \text{if } |x|\le 2\\ 0, & \text{if } |x|>2\end{cases}$$

Let $g:R\to R$ be given by $g(x)=f(x+2)-f(x-2)$. If $n$ and $m$ denote the number of points in $R$, where $g$ is not continuous and not differentiable respectively, then $n+m$ is equal to ……… . **[2021, 22 July Shift-II]**

**50** Let a function $g:[0,4]\to R$ be defined as

$$g(x)=\begin{cases}\max\limits_{0\le t\le x}(t^3-6t^2+9t-3), & 0\le x\le 3\\ 4-x, & 3<x\le 4\end{cases}$$ then the number of points in the interval $(0, 4)$ where $g(x)$ is not differentiable, is ………… . **[2021, 20 July Shift-II]**

**51** A function $f$ is defined on $[-3, 3]$ as

$$f(x)=\begin{cases}\min\{|x|, 2-x^2\}, & -2\le x\le 2\\ [|x|], & 2<|x|\le 3\end{cases}$$

where, $[x]$ denotes the greatest integer $\le x$. The number of points, where $f$ is not differentiable in $(-3, 3)$ is ………. **[2021, 25 Feb Shift-II]**

**52** The number of points at which the function $f(x)=|2x+1|-3|x+2|+|x^2+x-2|, x\in R$ is not differentiable, is ………. **[2021, 25 Feb Shift-I]**

**53** Let $S$ be the set of points where the function, $f(x)=|2-|x-3||, x\in R$, is not differentiable.

Then $\sum_{x\in S} f(f(x))$ is equal to …. . **[2020, 7 Jan Shift-I]**

## Topic 3 Differentiation of Inverse/Implicit/Parametric/Logarithmic/Composite Functions, A Function w.r.t. to Another Function

### Level 1

**01** If $y(x)=\begin{vmatrix} \sin x & \cos x & \sin x+\cos x+1 \\ 27 & 28 & 27 \\ 1 & 1 & 1 \end{vmatrix}$, $x\in R$, then $\frac{d^2y}{dx^2}+y$ is equal to **[2025, 3 April Shift-I]**

(a) 1 (b) $-1$ (c) 28 (d) 27

**02** Suppose for a differentiable function $h$, $h(0)=0$, $h(1)=1$ and $h'(0)=h'(1)=2$. If $g(x)=h(e^x)e^{h(x)}$, then $g'(0)$ is equal to **[2024, 6 April Shift-II]**

(a) 3 (b) 5 (c) 4 (d) 8

**03** Let $f:R-\{0\}\to R$ be a function satisfying $f\left(\frac{x}{y}\right)=\frac{f(x)}{f(y)}$ for all $x, y, f(y)\neq 0$. If $f'(1)=2024$, then **[2024, 30 Jan Shift-II]**

(a) $xf'(x)+f(x)=2024$ (b) $xf'(x)-2023f(x)=0$
(c) $xf'(x)+2024f(x)=0$ (d) $xf'(x)-2024f(x)=0$

**04** If $(a+\sqrt{2}b\cos x)(a-\sqrt{2}b\cos y)=a^2-b^2$, where $a>b>0$, then $\frac{dx}{dy}$ at $\left(\frac{\pi}{4},\frac{\pi}{4}\right)$ is **[2020, 4 Sep Shift-I]**

(a) $\frac{a-b}{a+b}$ (b) $\frac{a+b}{a-b}$ (c) $\frac{2a+b}{2a-b}$ (d) $\frac{a-2b}{a+2b}$

**05** Let $f$ and $g$ be differentiable functions on $R$ such that $fog$ is the identity function. If for some $a, b\in R$, $g'(a)=5$ and $g(a)=b$, then $f'(b)$ is equal to **[2020, 9 Jan Shift-II]**

(a) 1/5 (b) 5 (c) 2/5 (d) 1

**06** Let $x^k+y^k=a^k$, $(a, k>0)$ and $\frac{dy}{dx}+\left(\frac{y}{x}\right)^{\frac{1}{3}}=0$, then $k$ is **[2020, 7 Jan Shift-I]**

(a) 4/3 (b) 3/2 (c) 2/3 (d) 1/3

**07** Let $f(x)=\log_e(\sin x)$, $(0<x<\pi)$ and $g(x)=\sin^{-1}(e^{-x})$, $(x\geq 0)$. If $\alpha$ is a positive real number such that $a=(fog)'(\alpha)$ and $b=(fog)(\alpha)$, then **[2019, 9 April Shift-II]**

(a) $a\alpha^2-b\alpha-a=0$ (b) $a\alpha^2-b\alpha-a=1$
(c) $a\alpha^2+b\alpha-a=-2\alpha^2$ (d) $a\alpha^2+b\alpha+a=0$

**08** If $f(1)=1$, $f'(1)=3$, then the derivative of $f(f(f(x)))+(f(x))^2$ at $x=1$ is **[2019, 8 April Shift-II]**

(a) 12 (b) 9 (c) 15 (d) 33

**09** If $f(x)=\begin{vmatrix} \cos x & x & 1 \\ 2\sin x & x^2 & 2x \\ \tan x & x & 1 \end{vmatrix}$, then $\lim_{x\to 0}\frac{f'(x)}{x}$ **[JEE Main 2018]**

(a) exists and is equal to 0 (b) exists and is equal to $-2$
(c) exists and is equal to 2 (d) does not exist

**10** If $f(x)=\sin^{-1}\left(\frac{2\times 3^x}{1+9^x}\right)$, then $f'\left(-\frac{1}{2}\right)$ equals **[JEE Main 2018]**

(a) $\sqrt{3}\log_e\sqrt{3}$ (b) $-\sqrt{3}\log_e 3$
(c) $-\sqrt{3}\log_e\sqrt{3}$ (d) $\sqrt{3}\log_e 3$

**11** If for $x\in\left(0,\frac{1}{4}\right)$, the derivative of $\tan^{-1}\left(\frac{6x\sqrt{x}}{1-9x^3}\right)$ is $\sqrt{x}\cdot g(x)$, then $g(x)$ equals **[JEE Main 2017]**

(a) $\frac{3x\sqrt{x}}{1-9x^3}$ (b) $\frac{3x}{1-9x^3}$ (c) $\frac{3}{1+9x^3}$ (d) $\frac{9}{1+9x^3}$

**12** If $g$ is the inverse of a function $f$ and $f'(x)=\frac{1}{1+x^5}$, then $g'(x)$ is equal to **[JEE Main 2014]**

(a) $5x^4$ (b) $\frac{1}{1+\{g(x)\}^5}$
(c) $1+\{g(x)\}^5$ (d) $1+x^5$

**13** If $y=\sec(\tan^{-1}x)$, then $\frac{dy}{dx}$ at $x=1$ is equal to **[JEE Main 2013]**

(a) $\frac{1}{\sqrt{2}}$ (b) $\frac{1}{2}$
(c) 1 (d) $\sqrt{2}$

**14** If $x^m\cdot y^n=(x+y)^{m+n}$, then $\frac{dy}{dx}$ is **[AIEEE 2006]**

(a) $\frac{y}{x}$ (b) $\frac{x+y}{xy}$
(c) $xy$ (d) $\frac{x}{y}$

**15** If $x^y=e^{x-y}$, then $\frac{dy}{dx}$ is equal to **[AIEEE 2002]**

(a) $\frac{1+x}{1+\log x}$ (b) $\frac{1-\log x}{1+\log x}$
(c) not defined (d) $\frac{\log x}{(1+\log x)^2}$

### Level 2

**16** Let $f(x)=x^5+2x^3+3x+1$, $x\in R$ and $g(x)$ be a function such that $g(f(x))=x$, for all $x\in R$, then $\frac{g(7)}{g'(7)}$ is equal to **[2024, 5 Jan Shift-I]**

(a) 1 (b) 7
(c) 42 (d) 14

**17** Let $f(x) = x^5 + 2e^{x/4}$ for all $x \in R$. Consider a function $g(x)$ such that $(gof)(x) = x$ for all $x \in R$. Then, the value of $8g'(2)$ is *[2024, 4 April Shift-I]*
(a) 4 (b) 2
(c) 8 (d) 16

**18** If $f(x) = \begin{vmatrix} x^3 & 2x^2+1 & 1+3x \\ 3x^2+2 & 2x & x^3+6 \\ x^3-x & 4 & x^2-2 \end{vmatrix}$ for all $x \in R$, then $2f(0) + f'(0)$ is equal to *[2024, 31 Jan Shift-I]*
(a) 42 (b) 24 (c) 18 (d) 48

**19** Suppose $f(x) = \dfrac{(2^x + 2^{-x}) \tan x \sqrt{\tan^{-1}(x^2 - x + 1)}}{(7x^2 + 3x + 1)^3}$.
Then, the value of $f'(0)$ is equal to *[2024, 29 Jan Shift-I]*
(a) 0 (b) $\pi/2$ (c) $\pi$ (d) $\sqrt{\pi}$

**20** Consider the function $f : \left[\dfrac{1}{2}, 1\right] \to R$ defined by $f(x) = 4\sqrt{2}x^3 - 3\sqrt{2}x - 1$.
Consider the statements
I. The curve $y = f(x)$ intersects the $X$-axis exactly at one point.
II. The curve $y = f(x)$ intersects the $X$-axis at $x = \cos\dfrac{\pi}{12}$.
Then, *[2024, 29 Jan Shift-I]*
(a) Only I is correct (b) Both I and II are correct
(c) Both I and II are incorrect (d) Only II is correct

**21** The value of $\log_e 2 \dfrac{d}{dx}(\log_{\cos x} \operatorname{cosec} x)$ at $x = \dfrac{\pi}{4}$ is *[2022, 26 July Shift-II]*
(a) $-2\sqrt{2}$ (b) $2\sqrt{2}$
(c) $-4$ (d) 4

**22** If $y(x) = \cot^{-1}\left(\dfrac{\sqrt{1+\sin x} + \sqrt{1-\sin x}}{\sqrt{1+\sin x} - \sqrt{1-\sin x}}\right)$, $x \in \left(\dfrac{\pi}{2}, \pi\right)$, then $\dfrac{dy}{dx}$ at $x = \dfrac{5\pi}{6}$ is *[2021, 27 Aug Shift-II]*
(a) 0 (b) $-1$
(c) $\dfrac{-1}{2}$ (d) $\dfrac{1}{2}$

**23** Let $f(x) = \cos\left(2\tan^{-1}\sin\left(\cot^{-1}\sqrt{\dfrac{1-x}{x}}\right)\right)$, $0 < x < 1$.
Then, *[2021, 26 Aug Shift-I]*
(a) $(1-x)^2 f'(x) - 2(f(x))^2 = 0$
(b) $(1+x)^2 f'(x) + 2(f(x))^2 = 0$
(c) $(1-x)^2 f'(x) + 2(f(x))^2 = 0$
(d) $(1+x)^2 f'(x) - 2f(x))^2 = 0$

**24** If $y(\alpha) = \sqrt{2\left(\dfrac{\tan\alpha + \cot\alpha}{1+\tan^2\alpha}\right) + \dfrac{1}{\sin^2\alpha}}$, $\alpha \in \left(\dfrac{3\pi}{4}, \pi\right)$, then $\dfrac{dy}{d\alpha}$ at $\alpha = \dfrac{5\pi}{6}$ is *[2020, 7 Jan Shift-I]*
(a) $-\dfrac{1}{4}$ (b) $\dfrac{4}{3}$ (c) $-4$ (d) 4

**Imp.** **Logarithmic Differentiation**

**25** For $x > 1$, if $(2x)^{2y} = 4e^{2x-2y}$, then $(1 + \log_e 2x)^2 \dfrac{dy}{dx}$ is equal to *[2019, 12 Jan Shift-I]*
(a) $\dfrac{x\log_e 2x + \log_e 2}{x}$ (b) $\dfrac{x\log_e 2x - \log_e 2}{x}$
(c) $x\log_e 2x$ (d) $\log_e 2x$

**26** If $x\log_e(\log_e x) - x^2 + y^2 = 4 (y > 0)$, then $\dfrac{dy}{dx}$ at $x = e$ is equal to *[2019, 11 Jan Shift-I]*
(a) $\dfrac{e}{\sqrt{4+e^2}}$ (b) $\dfrac{(2e-1)}{2\sqrt{4+e^2}}$ (c) $\dfrac{(1+2e)}{\sqrt{4+e^2}}$ (d) $\dfrac{(1+2e)}{2\sqrt{4+e^2}}$

**27** If $\sin y = x\sin(a+y)$, then $\dfrac{dy}{dx}$ is equal to *[AIEEE 2002]*
(a) $\dfrac{\sin a}{\sin^2(a+y)}$ (b) $\dfrac{\sin^2(a+y)}{\sin a}$
(c) $\sin a \sin^2(a+y)$ (d) $\dfrac{\sin^2(a-y)}{\sin a}$

## Level 3

**28** Let $g(x)$ be a linear function and
$$f(x) = \begin{cases} g(x), & x \le 0 \\ \left(\dfrac{1+x}{2+x}\right)^{\frac{1}{x}}, & x > 0 \end{cases}$$, is continuous at $x = 0$.
If $f'(1) = f(-1)$, then the value $g(3)$ is *[2024, 31 Jan Shift-I]*
(a) $\log_e\left(\dfrac{4}{9}\right) - 1$ (b) $\log_e\left(\dfrac{4}{9e^{1/3}}\right)$
(c) $\dfrac{1}{3}\log_e\left(\dfrac{4}{9e^{1/3}}\right)$ (d) $\dfrac{1}{3}\log_e\left(\dfrac{4}{9}\right) + 1$

**29** Let $f(x) = \dfrac{\sin x + \cos x - \sqrt{2}}{\sin x - \cos x}$, $x \in [0, \pi] - \left\{\dfrac{\pi}{4}\right\}$. Then, $f\left(\dfrac{7\pi}{12}\right) f''\left(\dfrac{7\pi}{12}\right)$ is equal to *[2023, 8 April Shift-I]*
(a) $\dfrac{2}{9}$ (b) $\dfrac{-1}{3\sqrt{3}}$
(c) $\dfrac{-2}{3}$ (d) $\dfrac{2}{3\sqrt{3}}$

**30** Let $y = f(x) = \sin^3\left(\frac{\pi}{3}\left(\cos\left(\frac{\pi}{3\sqrt{2}}(-4x^3+5x^2+1)^{\frac{3}{2}}\right)\right)\right)$.

Then, at $x = 1$, [2023, 31 Jan Shift-I]

(a) $2y' + 3\pi^2 y = 0$ (b) $2y' + \sqrt{3}\pi^2 y = 0$
(c) $y' + 3\pi^2 y = 0$ (d) $\sqrt{2}y' - 3\pi^2 y = 0$

**31** Let $f(\theta) = 3\left(\sin^4\left(\frac{3\pi}{2} - \theta\right) + \sin^4(3\pi + \theta)\right) - 2(1 - \sin^2 2\theta)$ and $S = \left\{\theta \in [0, \pi] : f'(\theta) = -\frac{\sqrt{3}}{2}\right\}$.

If $4\beta = \sum_{\theta \in S} \theta$, then $f(\beta)$ is equal to [2023, 29 Jan Shift-I]

(a) $\frac{5}{4}$ (b) $\frac{3}{2}$
(c) $\frac{11}{8}$ (d) $\frac{9}{8}$

**32** Let $y = y(x)$ be a function of $x$ satisfying $y\sqrt{1-x^2} = k - x\sqrt{1-y^2}$ where $k$ is a constant and $y\left(\frac{1}{2}\right) = -\frac{1}{4}$. Then $\frac{dy}{dx}$ at $x = \frac{1}{2}$ is equal to [2020, 7 Jan Shift-II]

(a) $\frac{\sqrt{5}}{2}$ (b) $-\frac{\sqrt{5}}{2}$
(c) $\frac{2}{\sqrt{5}}$ (d) $-\frac{\sqrt{5}}{4}$

**33** The derivative of $\tan^{-1}\left(\frac{\sin x - \cos x}{\sin x + \cos x}\right)$, with respect to $\frac{x}{2}$, where $\left(x \in \left(0, \frac{\pi}{2}\right)\right)$ is [2019, 12 April Shift-II]

(a) 1 (b) 2/3
(c) 1/2 (d) 2

**34** If $2y = \left(\cot^{-1}\left(\frac{\sqrt{3}\cos x + \sin x}{\cos x - \sqrt{3}\sin x}\right)\right)^2$, $x \in \left(0, \frac{\pi}{2}\right)$, then $\frac{dy}{dx}$ is equal to [2019, 8 April Shift-I]

(a) $\frac{\pi}{6} - x$ (b) $x - \frac{\pi}{6}$
(c) $\frac{\pi}{3} - x$ (d) $2x - \frac{\pi}{3}$

**35** If $x = \sqrt{2^{\operatorname{cosec}^{-1} t}}$ and $y = \sqrt{2^{\sec^{-1} t}}$ $(|t| \geq 1)$, then $\frac{dy}{dx}$ is equal to [JEE Main 2018]

(a) $-y/x$ (b) $x/y$ (c) $-x/y$ (d) $y/x$

## Numerical Value Type Questions

**36** The value of $\lim_{x \to 0} 2\left(\frac{1 - \cos x\sqrt{\cos 2x}\sqrt[3]{\cos 3x} \ldots \sqrt[10]{\cos 10x}}{x^2}\right)$ is ....... [2024, 8 April Shift-I]

**37** If $y = \frac{(\sqrt{x}+1)(x^2-\sqrt{x})}{x\sqrt{x}+x+\sqrt{x}} + \frac{1}{15}(3\cos^2 x - 5)\cos^3 x$, then $96\, y'\left(\frac{\pi}{6}\right)$ is equal to ......... [2024, 1 Feb Shift-II]

**38** Let $f(x) = 2^x - x^2, x \in R$. If $m$ and $n$ are respectively the number of points at which the curves $y = f(x)$ and $y = f'(x)$ intersect the $X$-axis, then the value of $m + n$ is ........... [2024, 29 Jan Shift-I]

**39** Let $f : R \to R$ satisfy $f(x + y) = 2^x f(y) + 4^y f(x), \forall x, y \in R$. If $f(2) = 3$, then $14 \cdot \frac{f'(4)}{f'(2)}$ is equal to ......... [2022, 26 June Shift-II]

**40** If $f(x) = \sin\left(\cos^{-1}\left(\frac{1-2^{2x}}{1+2^{2x}}\right)\right)$ and its first derivative with respect to $x$ is $-\frac{b}{a}\log_e 2$ when $x = 1$, where $a$ and $b$ are integers, then the minimum value of $|a^2 - b^2|$ is ......... [2021, 17 March Shift-I]

## Topic 4 Infinite Series, Successive Differentiation

### Level 1

**01** If $\log_e y = 3\sin^{-1} x$, then $(1 - x^2)y'' - xy'$ at $x = \frac{1}{2}$ is equal to [2024, 9 April Shift-II]

(a) $3e^{\pi/6}$ (b) $9e^{\pi/6}$
(c) $9e^{\pi/2}$ (d) $3e^{\pi/2}$

**02** If $f(x) = \begin{cases} x^3 \sin\left(\frac{1}{x}\right), & x \neq 0 \\ 0, & x = 0 \end{cases}$, then [2024, 6 April Shift-I]

(a) $f''(0) = 1$ (b) $f''(0) = 0$
(c) $f''\left(\frac{2}{\pi}\right) = \frac{12 - \pi^2}{2\pi}$ (d) $f''\left(\frac{2}{\pi}\right) = \frac{24 - \pi^2}{2\pi}$

**03** If $y = \tan^{-1}(\sec x^3 - \tan x^3)$, $\frac{\pi}{2} < x^3 < \frac{3\pi}{2}$, then [2022, 24 June Shift-II]

(a) $xy'' + 2y' = 0$ (b) $x^2 y'' - 6y + \frac{3\pi}{2} = 0$
(c) $x^2 y'' - 6y + 3\pi = 0$ (d) $xy'' - 4y' = 0$

**04** If $2x = y^{\frac{1}{5}} + y^{\frac{1}{5}}$ and $(x^2-1)\dfrac{d^2y}{dx^2} + \lambda x\dfrac{dy}{dx} + ky = 0$, then $\lambda + k$ is equal to **[JEE Main 2017]**

(a) $-23$ (b) $-24$
(c) 26 (d) $-26$

**05** $\dfrac{d^2x}{dy^2}$ is equal to **[AIEEE 2011]**

(a) $-\left(\dfrac{d^2y}{dx^2}\right)^{-1}\left(\dfrac{dy}{dx}\right)^{-3}$ (b) $\left(\dfrac{d^2y}{dx^2}\right)\left(\dfrac{dy}{dx}\right)^{-2}$

(c) $-\left(\dfrac{d^2y}{dx^2}\right)\left(\dfrac{dy}{dx}\right)^{-3}$ (d) $\left(\dfrac{d^2y}{dx^2}\right)^{-1}$

**06** If the equation $a_nx^n + a_{n-1}x^{n-1} + \ldots + a_1x = 0$, $a_1 \neq 0, n \geq 2$, has a positive root $x = \alpha$, then the equation $na_nx^{n-1} + (n-1)a_{n-1}x^{n-2} + \ldots + a_1 = 0$ has a positive root, which is **[AIEEE 2005]**

(a) smaller than $\alpha$
(b) greater than $\alpha$
(c) equal to $\alpha$
(d) greater than or equal to $\alpha$

**07** If $2a + 3b + 6c = 0$, then at least one root of the equation $ax^2 + bx + c = 0$ lies in the interval **[AIEEE 2004, 2002]**

(a) (2, 3) (b) (1, 2) (c) (0, 1) (d) (1, 3)

## Level 2

**08** Let $f(x) = ax^3 + bx^2 + cx + 41$ be such that $f(1) = 40$, $f'(1) = 2$ and $f''(1) = 4$. Then, $a^2 + b^2 + c^2$ is equal to **[2024, 9 April Shift-I]**

(a) 73 (b) 54
(c) 62 (d) 51

**09** Let $g: R \to R$ be a non-constant twice differentiable function such that $g'\left(\dfrac{1}{2}\right) = g'\left(\dfrac{3}{2}\right)$. If a real valued function $f$ is defined as $f(x) = \dfrac{1}{2}[g(x) + g(2-x)]$, then **[2024, 30 Jan Shift-I]**

(a) $f''(x) = 0$ for exactly one $x$ in (0, 1)
(b) $f''(x) = 0$ for atleast two $x$ in (0, 2)
(c) $f''(x) = 0$ for no $x$ in (0, 1)
(d) $f'\left(\dfrac{3}{2}\right) + f'\left(\dfrac{1}{2}\right) = 1$

**10** If $y(x) = x^x, x > 0$, then $y''(2) - 2y'(2)$ is equal to **[2023, 1 Feb Shift-II]**

(a) $4(\log_e 2)^2 - 2$ (b) $4\log_e 2 + 2$
(c) $8\log_e 2 - 2$ (d) $4(\log_e 2)^2 + 2$

**11** Let $y(x) = (1+x)(1+x^2)(1+x^4)(1+x^8)(1+x^{16})$. Then, $y' - y''$ at $x = -1$ is equal to **[2023, 25 Jan Shift-I]**

(a) 496 (b) 464
(c) 976 (d) 944

**12** Let $f: S \to S$, where $S = (0, \infty)$ be a twice differentiable function, such that $f(x+1) = x\,f(x)$. If $g: S \to R$ be defined as $g(x) = \log_e f(x)$, then the value of $|g''(5) - g''(1)|$ is equal to **[2021, 16 March Shift-II]**

(a) $\dfrac{205}{144}$ (b) $\dfrac{197}{144}$ (c) $\dfrac{187}{144}$ (d) 1

**13** If the function $f(x) = \begin{cases} k_1(x-\pi)^2 - 1, & x \leq \pi \\ k_2 \cos x, & x > \pi \end{cases}$ is twice differentiable, then the ordered pair $(k_1, k_2)$ is equal to **[2020, 5 Sep Shift-I]**

(a) (1, 1) (b) $\left(\dfrac{1}{2}, 1\right)$ (c) (1, 0) (d) $\left(\dfrac{1}{2}, -1\right)$

**14** If $y^2 + \log_e(\cos^2 x) = y, x \in \left(-\dfrac{\pi}{2}, \dfrac{\pi}{2}\right)$, then **[2020, 3 Sep Shift-I]**

(a) $y''(0) = 0$ (b) $|y'(0)| + |y''(0)| = 1$
(c) $|y''(0)| = 2$ (d) $|y'(0)| + |y''(0)| = 3$

**15** Let $f: R \to R$ be a function such that $f(x) = x^3 + x^2f'(1) + xf''(2) + f'''(3), x \in R$. Then, $f(2)$ equals **[2019, 10 Jan Shift-I]**

(a) 30 (b) $-4$
(c) $-2$ (d) 8

**16** If $x = 3\tan t$ and $y = 3\sec t$, then the value of $\dfrac{d^2y}{dx^2}$ at $t = \dfrac{\pi}{4}$, is **[2019, 9 Jan Shift-II]**

(a) $\dfrac{1}{6}$ (b) $\dfrac{1}{6\sqrt{2}}$

(c) $\dfrac{1}{3\sqrt{2}}$ (d) $\dfrac{3}{2\sqrt{2}}$

**17** If $x^2 + y^2 + \sin y = 4$, then the value of $\dfrac{d^2y}{dx^2}$ at the point $(-2, 0)$ is **[JEE Main 2018]**

(a) $-34$ (b) $-32$
(c) $-2$ (d) 4

**18** Let $f$ be a polynomial function such that $f(3x) = f'(x)\,f''(x)$, for all $x \in R$. Then, **[JEE Main 2017]**

(a) $f(2) - f'(2) + f''(2) = 10$
(b) $f''(2) - f(2) = 4$
(c) $f''(2) - f'(2) = 0$
(d) $f(2) + f'(2) = 28$

**19** Let $f(x)=x|x|$ and $g(x)=\sin x$

**Statement I** $gof$ is differentiable at $x=0$ and its derivative is continuous at that point.

**Statement II** $gof$ is twice differentiable at $x=0$.

*[AIEEE 2009]*

(a) Statement I is false, Statement II is true
(b) Statement I is true, Statement II is true; Statement II is a correct explanation of Statement I
(c) Statement I is true, Statement II is true; Statement II is not a correct explanation of Statement I
(d) Statement I is true, Statement II is false

**20** If $f(x)=x^n$, then the value of

$$f(1)-\frac{f'(1)}{1!}+\frac{f''(1)}{2!}-\frac{f'''(1)}{3!}+\ldots+\frac{(-1)^n f^n(1)}{n!}$$ is

*[AIEEE 2003]*

(a) $2^n$ (b) $2^{n-1}$
(c) 0 (d) 1

**21** If $y=(x+\sqrt{1+x^2})^n$, then $(1+x^2)\frac{d^2y}{dx^2}+x\frac{dy}{dx}$ is equal to *[AIEEE 2002]*

(a) $n^2 y$ (b) $-n^2 y$
(c) $-y$ (d) $2x^2 y$

## Level 3

**22** If $y(\theta)=\dfrac{2\cos\theta+\cos 2\theta}{\cos 3\theta+4\cos 2\theta+5\cos\theta+2}$, then at $\theta=\dfrac{\pi}{2}$, $y''+y'+y$ is equal to *[2024, 5 April Shift-II]*

(a) $\frac{3}{2}$ (b) $\frac{1}{2}$ (c) 1 (d) 2

**23** Let $y=\log_e\left(\dfrac{1-x^2}{1+x^2}\right)$, $-1<x<1$. Then, at $x=\dfrac{1}{2}$, the value of $225(y'-y'')$ is equal to *[2024, 29 Jan Shift-II]*

(a) 736 (b) 742 (c) 732 (d) 746

**24** If $e^y+xy=e$, the ordered pair $\left(\dfrac{dy}{dx},\dfrac{d^2y}{dx^2}\right)$ at $x=0$ is equal to *[2019, 12 April Shift-I]*

(a) $\left(\frac{1}{e},-\frac{1}{e^2}\right)$ (b) $\left(-\frac{1}{e},\frac{1}{e^2}\right)$
(c) $\left(\frac{1}{e},\frac{1}{e^2}\right)$ (d) $\left(-\frac{1}{e},-\frac{1}{e^2}\right)$

**25** If $y=[x+\sqrt{x^2-1}]^{15}+[x-\sqrt{x^2-1}]^{15}$, then $(x^2-1)\dfrac{d^2y}{dx^2}+x\dfrac{dy}{dx}$ is equal to *[JEE Main 2017]*

(a) $225y^2$ (b) $224y^2$ (c) $125y$ (d) $225y$

## Numerical Value Type Questions

**Imp.** Successive Differentiation

**26** Let $f:R\to R$ be a thrice differentiable function such that $f(0)=0, f(1)=1, f(2)=-1, f(3)=2$ and $f(4)=-2$. Then, the minimum number of zeroes of $(3f'f''+ff''')(x)$ is ........... *[2024, 4 April Shift-II]*

**27** Let $f(x)=x^3+x^2 f'(1)+xf''(2)+f'''(3), x\in R$. Then, $f'(10)$ is equal to ......... *[2024, 27 Jan Shift-I]*

**28** Let $f(x)=\sum_{k=1}^{10} kx^k, x\in R$. If $2f(2)+f'(2)=119(2)^n+1$, then $n$ is equal to ......... *[2023, 13 April Shift-II]*

**29** For the curve $C:(x^2+y^2-3)+(x^2-y^2-1)^5=0$, the value of $3y'-y^3y''$, at the point $(\alpha,\alpha)$, $\alpha>0$, on $C$ is equal to ........... *[2022, 27 July Shift-II]*

**30** If $y^{1/4}+y^{-1/4}=2x$, and $(x^2-1)\dfrac{d^2y}{dx^2}+\alpha x\dfrac{dy}{dx}+\beta y=0$ then $|\alpha-\beta|$ is equal to ......... *[2021, 27 Aug Shift-I]*

**31** If $y=y(x)$ is an implicit function of $x$ such that $\log_e(x+y)=4xy$, then $\dfrac{d^2y}{dx^2}$ at $x=0$ is equal to ......... *[2021, 26 Aug Shift-I]*

## Key Index

### Topic 1

| | | | | | | | | | |
|---|---|---|---|---|---|---|---|---|---|
| 1 (d) | 2 (d) | 3 (b) | 4 (b) | 5 (d) | 6 (b) | 7 (c) | 8 (b) | 9 (b) | 10 (d) |
| 11 (a) | 12 (d) | 13 (c) | 14 (a) | 15 (b) | 16 (c) | 17 (c) | 18 (d) | 19 (d) | 20 (c) |
| 21 (d) | 22 (b) | 23 (c) | 24 (a) | 25 (b) | 26 (d) | 27 (b) | 28 (a) | 29 (a) | 30 (a) |
| 31 (d) | 32 (d) | 33 (d) | 34 (a) | 35 (d) | 36 (c) | 37 (a) | 38 (b) | 39 (a) | 40 (c) |
| 41 (c) | 42 (8) | 43 (2) | 44 (81) | 45 (17) | 46 (2) | 47 (62) | 48 (2) | 49 (6) | 50 (1) |
| 51 (8) | | | | | | | | | |

### Topic 2

| | | | | | | | | | |
|---|---|---|---|---|---|---|---|---|---|
| 1 (a) | 2 (b) | 3 (a) | 4 (b) | 5 (b) | 6 (c) | 7 (a) | 8 (b) | 9 (a) | 10 (d) |
| 11 (b) | 12 (c) | 13 (c) | 14 (b) | 15 (a) | 16 (d) | 17 (b) | 18 (b) | 19 (b) | 20 (a) |
| 21 (a) | 22 (c) | 23 (b) | 24 (d) | 25 (a) | 26 (d) | 27 (b) | 28 (a) | 29 (b) | 30 (b) |
| 31 (b) | 32 (d) | 33 (b) | 34 (a) | 35 (a) | 36 (a) | 37 (c) | 38 (d) | 39 (3) | 40 (5) |
| 41 (15) | 42 (2890) | 43 (4) | 44 (309) | 45 (25) | 46 (79) | 47 (4) | 48 (5) | 49 (4) | 50 (1) |
| 51 (5) | 52 (2) | 53 (3) | | | | | | | |

### Topic 3

| | | | | | | | | | |
|---|---|---|---|---|---|---|---|---|---|
| 1 (b) | 2 (c) | 3 (d) | 4 (b) | 5 (a) | 6 (c) | 7 (b) | 8 (d) | 9 (b) | 10 (a) |
| 11 (d) | 12 (c) | 13 (a) | 14 (a) | 15 (d) | 16 (d) | 17 (d) | 18 (a) | 19 (d) | 20 (b) |
| 21 (d) | 22 (c) | 23 (c) | 24 (d) | 25 (b) | 26 (b) | 27 (b) | 28 (b) | 29 (a) | 30 (a) |
| 31 (a) | 32 (b) | 33 (d) | 34 (b) | 35 (a) | 36 (55) | 37 (105) | 38 (5) | 39 (248) | 40 (481) |

### Topic 4

| | | | | | | | | | |
|---|---|---|---|---|---|---|---|---|---|
| 1 (c) | 2 (d) | 3 (b) | 4 (b) | 5 (c) | 6 (a) | 7 (c) | 8 (d) | 9 (b) | 10 (a) |
| 11 (a) | 12 (a) | 13 (b) | 14 (c) | 15 (c) | 16 (b) | 17 (a) | 18 (c) | 19 (d) | 20 (c) |
| 21 (a) | 22 (d) | 23 (a) | 24 (b) | 25 (d) | 26 (5) | 27 (202) | 28 (10) | 29 (16) | 30 (17) |
| 31 (40) | | | | | | | | | |

# Detailed **Explanations**

## Topic 1

**01.** *(d)* $f(x)=\begin{cases}(1+ax)^{\frac{1}{x}}, & x<0\\ 1+b, & x=0\\ \dfrac{(x+4)^{\frac{1}{2}}-2}{(x+c)^{\frac{1}{3}}-2}, & x>0\end{cases}$

$$\text{LHL}=\lim_{x\to 0^-} f(x)=\lim_{h\to 0}(1+ah)^{\frac{1}{h}}\ (1^{\infty})$$

$$=e^{\lim\limits_{h\to 0}(1-ah-1)\times\frac{-1}{h}}$$

$$=e^{\lim\limits_{h\to 0}-ah\times-\frac{1}{h}}=e^{a}$$

$$\text{RHL}=\lim_{x\to 0^+} f(x)=\lim_{h\to 0}\frac{(h+4)^{\frac{1}{2}}-2}{(h+c)^{\frac{1}{3}}-2}$$

$$=\lim_{h\to 0}\frac{\frac{1}{2}(h+4)^{\frac{-1}{2}}}{\frac{1}{3}(h+c)^{\frac{1}{3}-1}}\qquad[\therefore C=8]$$

$$\frac{3}{4}C^{2/3}=\frac{3}{4}\times(2^3)^{2/3}$$
$$=3$$

Given $f(x)$ is continuous at $x=0$.

$\Rightarrow\quad e^a=1+b=3$

$e^a bc=48$

**02.** *(d)*

> **key idea** Use expansion series of $\sin x=x-\dfrac{x^3}{3!}+\dfrac{x^5}{5!}-\ldots\infty, x\in R$
> and $\cos x=1-\dfrac{x^2}{2!}+\dfrac{x^4}{4!}-\ldots\infty,$ $x\in R$

Given,

$$f(x)=\frac{\sin 3x+\alpha\sin x-\beta\cos 3x}{x^3}, x\in R$$

is continuous at $x=0$

$\therefore \lim\limits_{x\to 0} f(x)=f(0)$

$$\Rightarrow \lim_{x\to 0}\frac{\left[3x-\frac{(3x)^3}{3!}+\frac{(3x)^5}{5!}+\ldots+\alpha\left(x-\frac{x^3}{3!}+\frac{x^5}{5!}\ldots\right)-\beta\left(1-\frac{(3x)^2}{2!}+\frac{(3x)^4}{4!}\ldots\right)\right]}{x^3}=f(0)$$

$$\Rightarrow \lim_{x\to 0}\frac{-\beta+x(3+\alpha)+\frac{9\beta x^2}{2!}+\left(-\frac{27}{3!}-\frac{\alpha}{3!}\right)x^3\ldots}{x^3}=f(0)$$

Limit will be exist, if

$\beta=0, 3+\alpha=0$ and $\dfrac{-27}{3!}-\dfrac{\alpha}{3!}=f(0)$

$\Rightarrow\quad \beta=0, \alpha=-3$

$\therefore\quad f(0)=\dfrac{-27}{6}+\dfrac{3}{6}=\dfrac{-27+3}{6}=-4$

**03.** *(b)* Given,

$$f(x)=\begin{cases}\dfrac{1-\cos 2x}{x^2}, & x<0\\ \alpha, & x=0\\ \dfrac{\beta\sqrt{1-\cos x}}{x}, & x>0\end{cases}\quad\text{where,}$$

$\alpha, \beta\in R$

Since, $f(x)$ is continuous at $x=0$.

So, LHL = RHL

= Value of function ... (i)

Now, $f(0^-)=\lim\limits_{x\to 0^-}\dfrac{2\sin^2 x}{x^2}=2$

(Using $1-\cos 2x=2\sin^2 x$)

$$f(0^+)=\lim_{x\to 0^+}\beta\times\sqrt{2}\,\frac{\sin\frac{x}{2}}{2.\frac{x}{2}}=\frac{\beta}{\sqrt{2}}$$

$[\because x\to 0^+;\ \therefore\ |\sin x/2|=\sin x/2]$

and $f(0)=\alpha$

From Eq. (i),

$\alpha=2$ and $\beta=2\sqrt{2}$

Hence, $\alpha^2+\beta^2=4+8=12$

**04.** *(b)* We have,

$f(x)=[x^2-x]+|-x+[x]|$

$=[x(x-1)]+\{x\}$

$$f(x)=\begin{cases}x+1, & -0.5<x<0\\ 0, & x=0\\ -1+x, & 0<x<1\\ 0, & x=1\\ x-1, & 1<x<1.5\end{cases}$$

At $x=0$,

$\text{LHL}=\lim\limits_{x\to 0^-} f(x)=1$

$\text{RHL}\ \lim\limits_{x\to 0^+} f(x)=-1$

$f(0)=0$

$\therefore f(x)$ is not continuous at $x=0$

At $x=1$,

$\text{LHL}=\lim\limits_{x\to 1^-} f(x)=-1+1=0$

$\text{RHL}=\lim\limits_{x\to 1^+} f(x)=1-1=0$

$f(1)=0$

$\therefore f(x)$ is continuous at $x=1$

Hence, $f(x)$ is continuous at $x=1$, but not continuous at $x=0$.

**05.** *(d)* $f(x)=\begin{cases}\dfrac{\log_e(1+5x)-\log_e(1+\alpha x)}{x}, & x\neq 0\\ 10, & x=0\end{cases}$

continuous at $x=0$

Then, $\lim\limits_{x\to 0} f(x)=10$

$$\lim_{x\to 0}\frac{\log_e(1+5x)-\log_e(1+\alpha x)}{x}=10$$

$$\lim_{x\to 0}\frac{\log(1+5x)}{5x}\times 5-\frac{\log(1+\alpha x)}{\alpha x}\times\alpha=10$$

$\Rightarrow\quad 5-\alpha=10\Rightarrow\alpha=-5$

**06.** *(b)* $f(x)=\begin{cases}0, & x<0\\ ae^x-1, & 0\le x<1\\ b, & x=1\\ b-1, & 1<x<2\\ -c, & x\ge 2\end{cases}$

To be continuous at $x=0$

$\Rightarrow\quad a-1=0\Leftrightarrow a=1$

To be continuous at $x=1$

$\Rightarrow \quad ae0-1=b \Rightarrow b=-1$ not possible

and to be continuous at $x=2$

$$b-1=-c \Rightarrow b+c=1$$

If $a=1$ and $b+c=1$, then $f(x)$ is discontinuous at exactly one point.

**07.** *(c)* For continuity

LHL at $0=f(0)=$ RHL at 0

$$\text{LHL}=\lim_{x\to 0^-} a\sin\frac{\pi}{2}(x-1)$$

$$=-a\sin\frac{\pi}{2}=-a \quad \ldots(\text{i})$$

$$\text{RHL}=\lim_{x\to 0^+}\frac{\tan 2x-\sin 2x}{bx^3}$$

$$=\lim_{x\to 0^+}\frac{\sin 2x(1-\cos 2x)}{bx^3.\cos 2x}$$

$$=\lim_{x\to 0^+}2\left(\frac{\sin 2x}{2x}\right)\frac{(2\sin^2 x)}{x^2}$$

$$.\frac{1}{b\cos 2x}=\frac{4}{b} \quad \ldots(\text{ii})$$

From Eqs. (i) and (ii), we get

$$-a=\frac{4}{b}\Rightarrow ab=-4\Rightarrow 10-ab=14$$

**08.** *(b)* $f(x)=\begin{cases}\sin x-e^x & ,-x\le 0\\ a+[-x] & ,0<x<1\\ 2x-b & ,x\ge 1\end{cases}$

$f(x)$ is continuous.

So, $\quad f(0^-)=0-e^0=-1$

$$f(0^+)=a-1$$

$\Rightarrow \quad a-1=-1\Rightarrow a=0$

$$f(1^-)=\lim_{h\to 0} a+[-1-h]=a-1$$

$$f(1^+)=2(1+h)-b=2-b$$

$\therefore \quad 2-b=a-1\Rightarrow b=2+1=3$

$\therefore \quad a+b=3$

**09.** *(b)* Given, $f(x)$ is continuous on $R$.

If $f(x)$ is continuous, then

$f$ is continuous at $x=1$

$$\Rightarrow \quad \lim_{x\to 1^-} f(x)=f(1)=\lim_{x\to 1^+} f(x)$$

$$\Rightarrow \quad |a+1+b|=\sin\pi=0$$

$$\Rightarrow \quad a+b=-1 \quad \ldots(\text{i})$$

Also, $f$ is continuous at $x=-1$

$$\Rightarrow \lim_{x\to -1^-} f(x)=f(-1)=\lim_{x\to -1^+} f(x)$$

$$\Rightarrow 2\sin\left(\frac{-\pi}{2}(-1)\right)=|a-1+b|$$

$$\Rightarrow \quad 2=|a+b-1| \quad \ldots(\text{ii})$$

Eq. (ii) is satisfied.

$\therefore \quad a+b=-1$

**10.** *(d)* Given,

$f(x)=[x-1]\cos\left(\frac{2x-1}{2}\right)\pi$, where $[\cdot]$ is greatest integer function and $f:R\to R$

$\because$ It is a greatest integer function then we need to check its continuity at $x\in I$ except these it is continuous.

Let $x=n$ where $n\in I$

$$\text{Then, LHL}=\lim_{x\to n^-}[x-1]\cos\left(\frac{2x-1}{2}\right)\pi$$

$$=0$$

$$\text{RHL}=\lim_{x\to n^+}[x-1]\cos\left(\frac{2x-1}{2}\right)\pi$$

and $\quad f(n)=0.$

Here, $\lim_{x\to n^-} f(x)=\lim_{x\to n^+} f(x)=f(n)$

$\therefore$ It is continuous at every integers. Therefore, the given function is continuous for all real $x$.

**11.** *(a)* As we know that $x-1<[x]\le x$, where $[x]$ denote the greatest integer $\le x$.

$$\text{So,}\quad \frac{4}{x}-1<\left[\frac{4}{x}\right]\le\frac{4}{x}$$

$$\therefore \lim_{x\to 0} x\left(\frac{4}{x}-1\right)<\lim_{x\to 0} x\left[\frac{4}{x}\right]\le\lim_{x\to 0} x\frac{4}{x}$$

$$\Rightarrow \lim_{x\to 0}(4-x)<\lim_{x\to 0} x\left[\frac{4}{x}\right]\le\lim_{x\to 0} 4$$

So, according to Squeeze theorem, we have

$$\lim_{x\to 0} x\left[\frac{4}{x}\right]=4=A \quad \text{[given]}$$

Now, the function $f(x)=[x^2]\sin(\pi x)$ is continuous at every integral value of $x$, so according to given options $\sqrt{A+1}=\sqrt{5}\notin$ Integer.

So $f(x)$ is discontinuous when $x=\sqrt{5}$ $=\sqrt{A+1}$.

**12.** *(d)* Given function

$$f(x)=\begin{cases}a|\pi-x|+1, & x\le 5\\ b|x-\pi|+3, & x>5\end{cases}$$

and it is also given that $f(x)$ is continuous at $x=5$.

Clearly, $\quad f(5)=a(5-\pi)+1 \quad \ldots(\text{i})$

$$\lim_{x\to 5^-} f(x)=\lim_{h\to 0}[a|\pi-(5-h)|+1]$$

$$=a(5-\pi)+1 \quad \ldots(\text{ii})$$

$$\text{and}\ \lim_{x\to 5^+} f(x)=\lim_{h\to 0}[b|(5+h)-\pi|+3]$$

$$=b(5-\pi)+3 \quad \ldots(\text{iii})$$

$\because$ Function $f(x)$ is continuous at $x=5$.

$$\therefore \quad f(5)=\lim_{x\to 5^+} f(x)=\lim_{x\to 5^-} f(x)$$

$$\Rightarrow a(5-\pi)+1=b(5-\pi)+3$$

$$\Rightarrow (a-b)(5-\pi)=2$$

$$\Rightarrow a-b=\frac{2}{5-\pi}$$

**13.** *(c)* Given function $f:[-1,3]\to R$ is defined as

$$f(x)=\begin{cases}|x|+[x], & -1\le x<1\\ x+|x|, & 1\le x<2\\ x+[x], & 2\le x\le 3\end{cases}$$

$$=\begin{cases}-x-1, & -1\le x<0\\ x, & 0\le x<1\\ 2x, & 1\le x<2\\ x+2, & 2\le x<3\\ 6, & x=3\end{cases}$$

$[\because$ if $n\le x<n+1,\ \forall\, n\in$ Integer, $[x]=n]$

$\because \lim_{x\to 0^-} f(x)=-1\ne f(0) \quad [\because f(0)=0]$

$\because \lim_{x\to 1^-} f(x)=1\ne f(1) \quad [\because f(1)=2]$

$\because \lim_{x\to 2^-} f(x)=4=f(2)=\lim_{x\to 2^+} f(x)=4$

$[\because f(2)=4]$

and $\lim_{x\to 3^-} f(x)=5\ne f(3) \quad [\because f(3)=6]$

$\therefore$ Function $f(x)$ is discontinuous at points 0, 1 and 3.

**14.** *(a)* Since, the function $f(x)$ is continuous at $x=2$.

$$\lim_{x\to 2} f(x)=f(2)$$

$$\Rightarrow \quad \lim_{x\to 2}(x-1)^{\frac{1}{2-x}}=k$$

$$\Rightarrow (e)^{\lim_{x\to 2}[(x-1)-1]\frac{1}{2-x}}=k$$

$[\because$ LHS is in the form of $1^\infty]$

$$\Rightarrow (e)^{\lim_{x\to 2}-\left(\frac{x-2}{x-2}\right)}=k$$

$$\Rightarrow \quad e^{-1}=k\Rightarrow k=\frac{1}{e}$$

**15.** *(b)* $\because f(x)$ is continuous at $x=\pi/2$.

$$\therefore \quad \lim_{x\to\frac{\pi}{2}}\left(\frac{4}{5}\right)^{\frac{\tan 4x}{\tan 5x}}=f\left(\frac{\pi}{2}\right)$$

$$\Rightarrow \quad \left(\frac{4}{5}\right)^0=k+\frac{2}{5}$$

$$\Rightarrow \quad k+\frac{2}{5}=1\Rightarrow k=1-\frac{2}{5}=\frac{3}{5}$$

**16.** *(c)* Since, $f(x)$ is continuous at $x=1$.

$\therefore \quad \frac{2}{a}=a \Rightarrow a=\pm\sqrt{2}$

Also $f(x)$ is continuous at $x=\sqrt{2}$

$\therefore a=\frac{2b^2-4b}{2\sqrt{2}} \quad [\because f(\sqrt{2})=\lim_{x\to\sqrt{2}} f(x)]$

When, $a=\sqrt{2}$, we get $2=b^2-2b$

$\Rightarrow \quad b^2-2b-2=0$

$\Rightarrow \quad b=\frac{2\pm\sqrt{4+4\times 2}}{2}=1\pm\sqrt{3}$

$\therefore (a,b)\equiv(\sqrt{2},1\pm\sqrt{3})$

When $a=-\sqrt{2}$, we get $-2=b^2-2b$

$\Rightarrow b^2-2b+2=0$

$\Rightarrow b=\frac{2\pm\sqrt{4-8}}{2}=1\pm i$ (Neglected)

(imaginary roots)

**17.** *(c)* Since $f(x)$ is a continuous function, therefore $\lim_{x\to 0} f(x)=f(0)$

$$\Rightarrow \lim_{x\to 0}\frac{\frac{(e^x-1)^2}{x^2}}{\frac{\sin\left(\frac{x}{k}\right)\cdot\log\left(1+\frac{x}{4}\right)}{k\cdot\frac{x}{k}\cdot 4\cdot\frac{x}{4}}}=12$$

$\Rightarrow \quad 4k=12 \Rightarrow k=3$

**18.** *(d)* Here, $f(x)=\begin{cases} x\cdot\sin(1/x), & x\neq 0\\ 0, & x=0\end{cases}$

To check continuity at $x=0$,

$\text{LHL}=\lim_{h\to 0}\left\{(-h)\sin\left(-\frac{1}{h}\right)\right\}=0$

$\text{RHL}=\lim_{h\to 0}\{h\sin(1/h)\}=0$

$\Rightarrow \quad f(0)=0$

So, $f(x)$ is continuous at $x=0$.

Hence, Statement I is correct.

$f_2(x)=\begin{cases}\sin\left(\frac{1}{x}\right), & x\neq 0\\ 0, & x=0\end{cases}$

Here, $\lim_{x\to 0} f_2(x)=\lim_{x\to 0}\sin\left(\frac{1}{x}\right)$

which does not exist. So, $f_2(x)$ is not continuous at $x=0$. Hence, Statement II is false.

**19.** *(d)* Now, $\lim_{x\to 0}\left(\frac{1}{x}-\frac{2}{e^{2x}-1}\right)$

$=\lim_{x\to 0}\frac{e^{2x}-1-2x}{x(e^{2x}-1)}$

$=\lim_{x\to 0}\frac{2e^{2x}-2}{(e^{2x}-1)+2xe^{2x}}$

[using L'Hospital rule]

$=\lim_{x\to 0}\frac{4e^{2x}}{4e^{2x}+4xe^{2x}}=1$

[using L'Hospital rule]

So, $f(x)$ is continuous at $x=0$, then

$\lim_{x\to 0} f(x)=f(0)$

$\Rightarrow \quad 1=f(0)$

**20.** *(c)* $\because \; f(x)=\frac{1-\tan x}{4x-\pi}$

$\therefore \lim_{x\to\pi/4} f(x)=\lim_{x\to\pi/4}\left(\frac{1-\tan x}{4x-\pi}\right)$

$=\lim_{x\to\pi/4}\left(\frac{-\sec^2 x}{4}\right)$

[using L'Hospital rule]

$=-\frac{2}{4}$

$\Rightarrow \lim_{x\to\pi/4} f(x)=-1/2$

Also, $f(x)$ is continuous in $[0,\pi/2]$, so $f(x)$ will be continuous at $\frac{\pi}{4}$.

$\therefore \quad f\left(\frac{\pi}{4}\right)=\lim_{x\to\pi/4} f(x)=-\frac{1}{2}$

**21.** *(d)* $\because f(2x)-f(x)=x$

Put $x\to\frac{x}{2} \quad f(x)-f\left(\frac{x}{2}\right)=\frac{x}{2}$

$x\to\frac{x}{2} \quad f\left(\frac{x}{2}\right)-f\left(\frac{x}{4}\right)=\frac{x}{2^2}$

$\vdots$

$f\left(\frac{x}{2^{n-1}}\right)-f\left(\frac{x}{2^n}\right)=\frac{x}{2^n}$

$$f(x)-f\left(\frac{x}{2^n}\right)=\frac{x}{2}\,\frac{\left(1-\left(\frac{1}{2}\right)^h\right)}{1-\frac{1}{2}}$$

$f(x)-f\left(\frac{x}{2^n}\right)=x\left(1-\left(\frac{1}{2}\right)^h\right)$

$\because \lim_{n\to\infty} f(x)-f\left(\frac{x}{2^n}\right)$

$\Rightarrow \lim_{n\to 0} x\left(1-\left(\frac{1}{2}\right)^n\right)=x$

$\therefore G(x)=x$

$\therefore \sum_{r=1}^{10} G(r^2)=\sum_{r=1}^{10} r^2=\frac{10}{6}(11)(21)$

$=385$

**22.** *(b)*

> **key idea** A function $f(x)$ is continuous at $x=a$,
> then $\lim_{x\to a^+} f(x)=\lim_{x\to a^-} f(x)=f(a)$

$f(x)=$

$$\begin{cases}\frac{2}{x}\{\sin(k_1+1)x+\sin(k_2-1)x\}, & x<0\\ 4, & x=0\\ \frac{2}{x}\log_e\left(\frac{2+k_1x}{2+k_2x}\right), & x>0\end{cases}$$

$\text{LHL}=\lim_{x\to 0^-} f(x)$

$=\lim_{x\to 0^-}\frac{2}{x}\{\sin(k_1+1)x+\sin(k_2-1)x\}$

$$\lim_{x\to 0^-} 2\left\{\frac{\sin(k_1+1)x}{(k_1+1)x}\cdot(k_1+1)+\frac{\sin(k_2-1)x}{(k_2-1)x}\times(k_2-1)\right\}$$

$=2[k_1+1+k_2-1]$

$=2[k_1+k_2]$

$\text{RHL}=\lim_{x\to 0^+} f(x)$

$=\lim_{x\to 0^+}\frac{2}{x}\log\left(\frac{2+k_1x}{2+k_2x}\right)$

$=\lim_{x\to 0^+}\frac{2}{x}\log\left[1+\frac{(k_1-k_2)x}{2+k_2x}\right]$

$k_1-k_2$ and since, we know that if $f(x)$ is continuous at $x=0$

$\Rightarrow \quad \text{LHL}=\text{RHL}=f(0)$

$\Rightarrow 2(k_1+k_2)=4$

$\Rightarrow \quad k_1+k_2=2 \quad \ldots$(i)

$\Rightarrow \quad k_1-k_2=4 \quad \ldots$(ii)

From Eqs (i) and (ii), we get

$k_1=3,\; k_2=-1$

Now $k_1^2+k_2^2=9+1=10$

**23.** *(c)* $f(x)=2x^2+x+[x^2]-[x]$

$f:[-1,2]\to R$

$\because [x]$ is discontinuous at integral point

$\therefore \quad -1\le x\le 2,\; 0\le x^2\le 4$

$\therefore$ Point where $f(x)$ can be discontinuous are

$\Rightarrow \quad x=-1, 0, 1, \sqrt{2}, \sqrt{3}, 2$

At end points we have to check

$\Rightarrow$ At $x=-1$

$\therefore \quad f(-1)=3$

$f(-1^+)=2-1+0-(-1)=2$

$\therefore$ Discontinuous at $x=-1$

at $\quad x=0,\ f(0)=0$

$f(0^-)=0+0+0-(-1)=1$

$\therefore$ Discontinuous at $x=0$

at $\quad x=2$

$f(2)=8+2+4-2=12$

$f(2^-)=8+2+3-1=12$

$\therefore$ Continuous at $x=2$

at $x=1$

$f(1)=3-1+1=3$

$f(1^+)=2+1-1+1=3$

$f(1^-)=2+1-0-0=3$

$\therefore$ Continuous at $x=1$

Similarly, $f(x)$ will be discontinuous at $x=\sqrt{2}$ and $\sqrt{3}$ also.

Number of points of discontinuity $=4$

> • Concept **Enhancer**
>
> Polynomial function is continuous everywhere and Greatest Integer Function is discontinuous at integer.

**24.** (a) Given,

$$f(x)=\begin{cases}\dfrac{72^x-9^x-8^x+1}{\sqrt{2}-\sqrt{1+\cos x}}, & x\neq 0\\ a\log_e 2\log_e 3, & x=0\end{cases}$$

$\because f(x)$ is continuous at $x=0$

$\therefore \lim\limits_{x\to 0} f(x)=f(0)$

$$\Rightarrow \lim_{x\to 0}\frac{72^x-9^x-8^x+1}{\sqrt{2}-\sqrt{1+\cos x}}=a\ln 2\cdot\ln 3$$

$$\Rightarrow \lim_{x\to 0}\frac{(8^x-1)(9^x-1)}{\sqrt{2}-\sqrt{1+\cos x}}\times\frac{\sqrt{2}+\sqrt{1+\cos x}}{\sqrt{2}+\sqrt{1+\cos x}}=a\ln 2\ln 3$$

$$\Rightarrow \lim_{x\to 0}\frac{(8^x-1)(9^x-1)}{1-\cos x}\times(\sqrt{2}+\sqrt{1+\cos x})=a\ln 2\cdot\ln 3$$

$$\Rightarrow \lim_{x\to 0}\frac{\left(\dfrac{8^x-1}{x}\right)\left(\dfrac{9^x-1}{x}\right)x^2(\sqrt{2}+\sqrt{1+\cos x})}{\dfrac{2\sin^2\frac{x}{2}}{\left(\frac{x}{2}\right)^2}\times\dfrac{x^2}{4}}=a\ln\cdot 2\ln 3$$

$\Rightarrow \ln 8\cdot\ln 9\cdot 2\cdot 2\sqrt{2}=a\ln 2\cdot\ln 3$

$$\Rightarrow \quad a=\frac{3\ln 2\cdot 2\ln 3\times 4\sqrt{2}}{\ln 2\cdot\ln 3}$$

$\Rightarrow \quad a=24\sqrt{2}$

$\Rightarrow \quad a^2=576\times 2=1152$

**25.** (b) $f(x)=\dfrac{\sqrt[7]{p(729+x)}-3}{\sqrt[3]{729+qx}-9}$

$\lim\limits_{x\to 0} f(x)$ is finite and equal to $f(0)$.

$\therefore \quad \sqrt[7]{p(729)}-3=0\Rightarrow p=3$

Now,

$$\lim_{x\to 0} f(x)=\lim_{x\to 0}\frac{\sqrt[7]{3(729+x)}-3}{\sqrt[3]{729+qx}-9}$$

$$=\lim_{x\to 0}\frac{1}{3}\times\frac{\left(1+\dfrac{x}{3^6}\right)^{1/7}-1}{\left(1+\dfrac{qx}{9^3}\right)^{\frac{1}{3}}-1}\quad\left[\because \frac{0}{0}\text{ form}\right]$$

$$=\frac{1}{3}\times\frac{\frac{1}{7}\times\frac{1}{3^6}}{\frac{1}{3}\times\frac{q}{9^3}}\Rightarrow\frac{1}{7q}=f(0)$$

$\Rightarrow \quad 7q\,f(0)=1$

$\Rightarrow \quad 63q\,f(0)=9=p^2 \quad [\because p=3]$

$\therefore \quad 63q\,f(0)-p^2=0$

**26.** (d) Given, $f(x)=\begin{cases}x+a, & x\le 0\\ |x-4|, & x>0\end{cases}$

and $g(x)=\begin{cases}x+1, & x<0\\ (x-4)^2+b, & x\ge 0\end{cases}$

As, $f(x)$ is continuous on $R$,

$\lim\limits_{x\to 0^-} f(x)=\lim\limits_{x\to 0^+} f(x)=f(0)$

$\Rightarrow \quad 0+a=|0-4|\Rightarrow a=4$

As $g(x)$ is continuous on $R$,

$\lim\limits_{x\to 0^-} g(x)=\lim\limits_{x\to 0^+} g(x)=g(0)$

$\Rightarrow \quad 0+1=(0-4)^2+b$

$\Rightarrow \quad 1=16+b\Rightarrow b=-15$

$\therefore \quad f(x)=\begin{cases}x+4, & x\le 0\\ |x-4|, & x>0\end{cases}$

and $\quad g(x)=\begin{cases}x+1, & x<0\\ (x-4)^2-15, & x\ge 0\end{cases}$

$gof(2)+fog(-2)=g[f(2)]+f[g(-2)]$

$=g(2)+f(-1)$

$=(2-4)^2-15+(-1+4)=8-16=-8$

**27.** (b) We have, $f(x)=\begin{cases}[x], & x<0\\ |1-x|, & x\ge 0\end{cases}$

and $g(x)=\begin{cases}e^x-x, & x<0\\ (x-1)^2-1, & x\ge 0\end{cases}$

Now, $fog\,(x)=\begin{cases}[g(x)], & g(x)<0\\ |1-g(x)|, & g(x)\ge 0\end{cases}$

$$=\begin{cases}1+x-e^x, & x<0\\ 1, & x=0\\ [(x-1)^2-1], & 0<x<2\\ |2-(x-1)^2|, & x\ge 2\end{cases}$$

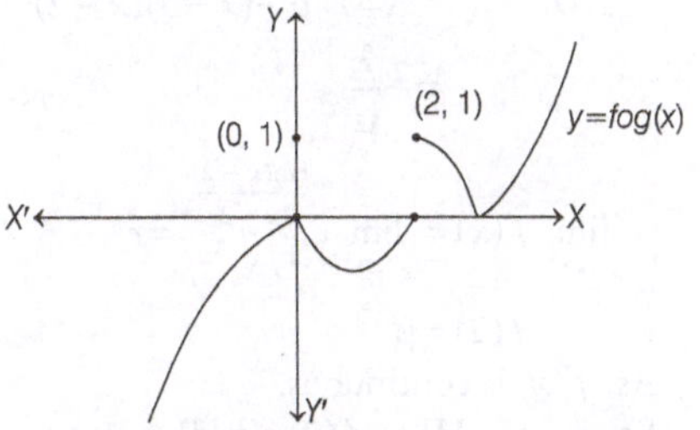


So, $x=0,\ 2$ are the two points where $fog$ is discontinuous.

**28.** (a) $f(x)$ is continuous at $x=0$

LHL at $x=0=f(0)=$ RHL at $x=0$

$$\Rightarrow \lim_{x\to 0^-}\frac{\ln\left(\dfrac{1+\frac{x}{a}}{1-\frac{x}{b}}\right)}{x}=\frac{\lim\limits_{x\to 0^-}\left(\frac{1}{a}\right)\ln\left(1+\frac{x}{a}\right)}{\left(\frac{1}{a}\right)x}$$

$$-\frac{\lim\limits_{x\to 0^-}\left(-\frac{1}{b}\right)\ln\left(1-\frac{x}{b}\right)}{\left(-\frac{1}{b}\right)x}=\left(\frac{1}{a}+\frac{1}{b}\right)$$

and $\quad f(0)=k$

and $\quad \lim\limits_{x\to 0^+}\dfrac{\cos^2 x-\sin^2 x-1}{\sqrt{x^2+1}-1}$

$$\Rightarrow \lim_{x\to 0^+}\frac{-2\sin^2 x}{\sqrt{x^2+1}-1}$$

$$=\lim_{x\to 0^+}-\left(\frac{2\sin^2 x}{x^2}\right)(\sqrt{x^2+1}+1)=-4$$

$$\Rightarrow \quad \frac{1}{a}+\frac{1}{b}=-4=k$$

$$\left(\frac{1}{a}+\frac{1}{b}\right)+\left(\frac{4}{k}\right)=-4-1=-5$$

**29.** (a) We have,

$$f(x)=\begin{cases}\dfrac{\lambda|x^2-5x+6|}{\mu(5x-x^2-6)}, & x<2\\ e^{\frac{\tan(x-2)}{x-[x]}}, & x>2\\ \mu, & x=2\end{cases}$$

$f(x)$ is continuous at $x=2$.

$$\therefore \lim_{x\to 2^-} f(x)=\lim_{x\to 2^+} f(x)=f(2)$$

$$\therefore \lim_{x\to 2^-} f(x)=\lim_{x\to 2^-}\frac{\lambda\,|(x-3)(x-2)|}{\mu\,-(x-3)(x-2)}$$

$$=-\frac{\lambda}{\mu}$$

$$\lim_{x\to 2^+} f(x)=\lim_{x\to 2^+} e^{\frac{\tan(x-2)}{x-2}}=e$$

$$f(2)=\mu$$

As, $f(x)$ is continuous.

So, $\text{LHL}=f(2)=\text{RHL}$

$$-\frac{\lambda}{\mu}=\mu=e$$

$$\lambda=-e^2 \Rightarrow \mu=e$$

$$\lambda+\mu=e(-e+1)$$

**30.** (a) $f(x)=$

$$\begin{cases}\dfrac{x^3}{(1-\cos 2x)^2}\log_e\left[\dfrac{1+2xe^{-2x}}{(1-xe^{-x})^2}\right], & x\neq 0\\ \alpha, & x=0\end{cases}$$

For continuity,

$$\lim_{x\to 0}\frac{x^3}{4\sin^4 x}[\log_e(1+2xe^{-2x})$$

$$-\log_e(1-xe^{-x})^2]=\alpha \text{ (by expansion)} \quad \ldots(i)$$

$$\because \log(1+2xe^{-2x})=2xe^{-2x}-\frac{(2xe^{-2x})^2}{2}+\ldots$$

$$\text{and } \log(1-xe^{-x})=-xe^{-x}-\frac{(xe^{-x})^2}{2}-\ldots$$

On putting the values in Eq. (i), we get

$$\lim_{x\to 0}\left(\frac{1}{4}\cdot\frac{x}{x}\right)\left(\frac{x^3}{\sin^4 x}\right)[2xe^{-2x}-2(-xe^{-x})]$$

$$[\because \log(1-xe^{-x})^2=2\log(1-xe^{-x})]$$

$$=\lim_{x\to 0}\left(\frac{1}{4x}\right)\left(\frac{x}{\sin x}\right)^4(2xe^{-2x}+2xe^{-x})$$

$$=\lim_{x\to 0}\left(\frac{1}{4x}\right)\cdot\left(\frac{x}{\sin x}\right)^4\cdot 2x\cdot(e^{-2x}+e^{-x})$$

$$=\left(\frac{1}{4}\right)(1)(2)(2)\Rightarrow \alpha=1$$

**31.** (d) Given,

$$f(x)=\begin{cases}\dfrac{\sin(a+1)x+\sin 2x}{2x}, & x<0\\ b, & x=0\\ \dfrac{\sqrt{x+bx^3}-\sqrt{x}}{bx^{5/2}}, & x>0\end{cases}$$

$\because$ $f(x)$ is continuous at $x=0$.

$$\therefore \lim_{x\to 0^-} f(x)=\lim_{x\to 0^+} f(x)=f(0) \quad \ldots(i)$$

$$\because f(0)=b \quad \ldots(ii)$$

Now, $\lim_{x\to 0^-} f(x)=\lim_{x\to 0^-}\left(\dfrac{\sin(a+1)x+\sin 2x}{2x}\right)$

$$=\lim_{x\to 0^-}\left(\frac{\sin(a+1)x}{(a+1)x}\times\left(\frac{a+1}{2}\right)+\frac{\sin 2x}{2x}\right)$$

$$=\frac{a+1}{2}+1 \quad \ldots(iii)$$

Again,

$$\lim_{x\to 0^+} f(x)=\lim_{x\to 0^+}\left(\frac{\sqrt{x+bx^3}-\sqrt{x}}{bx^{5/2}}\right)$$

$$=\lim_{x\to 0^+}\frac{(\sqrt{x+bx^3}-\sqrt{x})(\sqrt{x+bx^3}+\sqrt{x})}{bx^{5/2}(\sqrt{x+bx^3}+\sqrt{x})}$$

$$=\lim_{x\to 0^+}\frac{(x+bx^3-x)}{bx^{5/2}(\sqrt{x+bx^3}+\sqrt{x})}$$

$$=\lim_{x\to 0^+}\frac{\sqrt{x}}{\sqrt{x}(\sqrt{1+bx^2}+1)}$$

$$\Rightarrow \lim_{x\to 0^+} f(x)=\frac{1}{2} \quad \ldots(iv)$$

From Eqs. (i), (ii), (iii) and (iv),

$$\frac{1}{2}=b=\frac{a+1}{2}+1\Rightarrow b=\frac{1}{2},\ a=-2$$

$$\therefore \quad a+b=\frac{-3}{2}$$

**32.** (d) Given function

$$f(x)=\begin{cases}ae^x+be^{-x}, & -1\le x<1\\ cx^2, & 1\le x\le 3\\ ax^2+2cx, & 3<x\le 4\end{cases}$$

Since, given function is continuous for some $a, b, c\in R$, So

$$\lim_{x\to 1^-} f(x)=\lim_{x\to 1^+} f(x)$$

$$\Rightarrow ae+be^{-1}=c \quad \ldots(i)$$

and $\lim_{x\to 3^-} f(x)=\lim_{x\to 3^+} f(x)$

$$\Rightarrow c(3)^2=9a+6c$$

$$\Rightarrow 3c=9a\Rightarrow c=3a \quad \ldots(ii)$$

Now, $f'(0)+f'(2)=a-b+4c=e$ [given]... (iii)

From Eqs. (i), (ii) and (iii), we get

$$a-(3a-ae)e+12a=e$$

$$\Rightarrow (13-3e+e^2)a=e$$

$$\Rightarrow a=\frac{e}{e^2-3e+13}$$

**33.** (d) Given function

$$f(x)=\begin{cases}\dfrac{\sin(p+1)x+\sin x}{x}, & x<0\\ q, & x=0\\ \dfrac{\sqrt{x+x^2}-\sqrt{x}}{x^{3/2}}, & x>0\end{cases}$$

is continuous at $x=0$, then

$$f(0)=\lim_{x\to 0^-} f(x)=\lim_{x\to 0^+} f(x) \quad \ldots(i)$$

$$\lim_{x\to 0^-} f(x)=\lim_{x\to 0^-}\frac{\sin(p+1)x+\sin x}{x}$$

$$=p+1+1=p+2$$

and $\lim_{x\to 0^+} f(x)=\lim_{x\to 0^+}\dfrac{\sqrt{x+x^2}-\sqrt{x}}{x^{3/2}}$

$$=\lim_{x\to 0^+}\frac{\sqrt{x}[(1+x)^{1/2}-1]}{x\sqrt{x}}$$

$$=\lim_{x\to 0^+}\frac{\left(1+\frac{1}{2}x+\frac{\frac{1}{2}\left(\frac{1}{2}-1\right)}{2!}x^2+\ldots.-1\right)}{x}$$

$$\left[\because (1+x)^n=1+nx+\frac{n(n-1)}{1\cdot 2}x^2+\frac{n(n-1)(n-2)}{1\cdot 2\cdot 3}x^3+\ldots,\ |x|<1\right]$$

$$=\lim_{x\to 0^+}\left(\frac{1}{2}+\frac{\frac{1}{2}\left(\frac{1}{2}-1\right)}{2!}x+\ldots\right)=\frac{1}{2}$$

From Eq. (i), we get

$$f(0)=q=\frac{1}{2}$$

and $\lim_{x\to 0^-} f(x)=p+2=\dfrac{1}{2}\Rightarrow p=-3/2$

So, $(p, q)=\left(-\dfrac{3}{2},\dfrac{1}{2}\right)$

**34.** *(a)* Given, function $f: R \to R$ defined by $f(x) = [x] \cos \pi\left(x - \frac{1}{2}\right)$, where $[\cdot]$ denotes the greatest integer function.

Now, $\cos x$ is continuous, $\forall x \in R$.

$\Rightarrow \cos \pi\left(x - \frac{1}{2}\right)$ is also continuous, $\forall x \in R$.

Hence, the continuity of $f$ depends upon the continuity of $[x]$. Now, $[x]$ is discontinuous, $\forall x \in I$.

So, we should check the coitnuity of $f$ at $x = n$, $\forall n \in I$.

LHL at $x = n$ is given by

$$f(n^-) = \lim_{x \to n^-} f(x)$$

$$= (n-1) \cos \frac{(2n-1)\pi}{2} = 0$$

RHL at $x = n$ is given by

$$f(n^+) = \lim_{x \to n^+} f(x)$$

$$= (n) \cos \frac{(2n-1)\pi}{2} = 0$$

Also, value of the function at $x = n$ is

$$f(n) = [n] \cos \pi\left(n - \frac{1}{2}\right)$$

$$= (n) \cos \frac{(2n-1)\pi}{2} = 0$$

$$\therefore \quad f(n^+) = f(n^-) = f(n)$$

Hence, $f$ is continuous at $x = n$, $\forall n \in I$.

**35.** *(d)*

$$f(x) = \begin{cases} \dfrac{\tan((a+1)x) + b \tan x}{x}, & x < 0 \\ 3, & x = 0 \\ \dfrac{\sqrt{ax + b^2x^2} - \sqrt{ax}}{b\sqrt{a} \cdot x\sqrt{x}}, & x > 0 \end{cases}$$

$\because f(x)$ is continuous at $x = 0$

So, $\lim_{x \to 0} f(x) = f(0) = 3$

$$\Rightarrow \lim_{x \to 0^+} \frac{\sqrt{ax + b^2x^2} - \sqrt{ax}}{b\sqrt{a} \cdot x\sqrt{x}} = 3$$

$$\Rightarrow \lim_{x \to 0^+} \frac{(\sqrt{ax + b^2x^2} - \sqrt{ax})(\sqrt{ax + b^2x^2} + \sqrt{ax})}{(b\sqrt{a} \cdot x\sqrt{x})(\sqrt{ax + b^2x^2} + \sqrt{ax})} = 3$$

$$\Rightarrow \lim_{x \to 0^+} \frac{ax + b^2x^2 - ax}{b\sqrt{a}\, x^{3/2}(\sqrt{ax + b^2x^2} + \sqrt{ax})} = 3$$

$$\Rightarrow \lim_{x \to 0^+} \frac{b^2x^2}{b\sqrt{a}\, x^2 (\sqrt{a + b^2x} + \sqrt{a})} = 3$$

$$\Rightarrow \lim_{x \to 0^+} \frac{b}{\sqrt{a}(\sqrt{a + b^2x} + \sqrt{a})} = 3$$

At $x = 0$,

$$\frac{b}{\sqrt{a}(\sqrt{a} + \sqrt{a})} = 3$$

$$\Rightarrow \quad \frac{b}{2a} = 3 \Rightarrow \frac{b}{a} = 6$$

**36.** *(c)* Given,

$$f(x) = \begin{cases} \dfrac{a(7x - 12 - x^2)}{b|x^2 - 7x + 12|}, & x < 3 \\ 2^{\frac{\sin(x-3)}{x - [x]}}, & x > 3 \\ b, & x = 3 \end{cases}$$

Also, given $f(x)$ is continuous at $x = 3$.

So, LHL

$$= f(3^-) = \lim_{x \to 3^-} \left\{\frac{a(7x - 12 - x^2)}{b|x^2 - 7x + 12|}\right\}$$

$$= \frac{-a(x-3)(x-4)}{b(x-3)(x-4)}$$

[$\because$ when $x < 3$, then $|x^2 - 7x + 12| = x^2 - 7x + 12$ as $x^2 - 7x + 12 > 0$ $\forall x < 3$]

$$= -\frac{a}{b}$$

$$\text{RHL} = f(3^+) = \lim_{x \to 3^+} f(x)$$

$$= \lim_{x \to 3^+} \left(2^{\frac{\sin(x-3)}{x - [x]}}\right)$$

$$= 2^{\lim_{x \to 3^+} \left\{\frac{\sin(x-3)}{x - [x]}\right\}}$$

$$= 2^{\lim_{x \to 3^+} \frac{\sin(x-3)}{(x-3)}}$$

$$= 2 \qquad [\because \lim_{x \to 3^+} [x] = 3]$$

$f(3) = b$ = value of function

For $f(x)$ to be continuous at $x = 3$;

$$f(3^-) = f(3^+) = f(3)$$

$$\Rightarrow \quad -\frac{a}{b} = b = 2$$

$$\Rightarrow \quad b = 2 \text{ and } -\frac{a}{2} = 2 \Rightarrow a = -4$$

Hence, only one ordered pair $(-4, 2)$.

$\therefore n(S)$ = Number of elements in the set $S = 1$

**37.** *(a)* Let

$$f(x) = \begin{cases} (1 + |\cos x|)^{\frac{\lambda}{|\cos x|}}, & 0 < x < \frac{\pi}{2} \\ \mu, & x = \frac{\pi}{2} \\ e^{\left(\frac{\cot 6x}{\cot 4x}\right)}, & \frac{\pi}{2} < x < \pi \end{cases}$$

Now, LHL

$$= \lim_{x \to \frac{\pi}{2}^-} f(x) = \lim_{x \to \frac{\pi}{2}^-} (1 + |\cos x|)^{\frac{\lambda}{|\cos x|}}$$

[it's of $1^\infty$ form]

$$\text{So, LHL} = e^{\left(\lim_{x \to \frac{\pi}{2}^-} \lambda\right)} = e^\lambda$$

Now,

$$\text{RHL} = \lim_{x \to \frac{\pi}{2}^+} f(x) = \lim_{x \to \frac{\pi}{2}^+} e^{\left(\frac{\cot 6x}{\cot 4x}\right)}$$

$$\text{RHL} = e^{\lim_{x \to \frac{\pi}{2}^+} \frac{\cot 6x}{\cot 4x}}$$

$$\text{So,} \quad \text{RHL} = e^{\lim_{x \to \frac{\pi}{2}^+} \left(\frac{\tan 4x}{\tan 6x}\right)}$$

$$\Rightarrow \quad \text{RHL} = e^{\lim_{h \to 0} \left(\frac{\tan 4\left(\frac{\pi}{2} + h\right)}{\tan 6\left(\frac{\pi}{2} + h\right)}\right)}$$

$$\text{So,} \quad \text{RHL} = e^{\lim_{h \to 0} \left[\frac{\tan 4h}{\tan 6h}\right]}$$

Now, multiplying and divide $4h$ and $6h$ in power of $e$

$$\text{RHL} = e^{\lim_{h \to 0} \left[\left(\frac{\tan 4h}{4h}\right)\left(\frac{6h}{\tan 6h}\right)\left(\frac{4h}{6h}\right)\right]}$$

$$\left[\because \lim_{x \to 0} \frac{\tan x}{x} = 1\right]$$

So, RHL $= e^{2/3}$

For continuity of a function

LHL = RHL = Functional value

$$e^\lambda = e^{2/3} = \mu$$

By comparison, $\lambda = 2/3$ and $\mu = e^{2/3}$

Now, $9\lambda + 6\log_e \mu + \mu^6 - e^{6\lambda}$

$$= 6 + 6\left(\frac{2}{3}\right) + e^4 - e^4 = 10$$

**38.** *(b)*

$$f(x)=\lim_{n\to\infty}\frac{\cos(2\pi x)-x^{2n}\sin(x-1)}{1+x^{2n+1}-x^{2n}}$$

Case I $|x|<1$

If $|x|<1$, then $x\in(-1,1)$

$$\lim_{n\to\infty}|x|^{2n+1}=\lim_{n\to\infty}|x|^{2n}=0$$

$$f(x)=\lim_{n\to\infty}\cos 2\pi x$$

$f(x)$ is continuous everywhere.

Case II $|x|>1$

If $|x|>1$, then $x\in(-\infty,-1)\cup(1,\infty)$

$$\lim_{n\to\infty}|x|^{2n+1}=\infty$$

$$\lim_{n\to\infty}\frac{\cos 2\pi x-x^{2n}\sin(x-1)}{1+x^{2n+1}-x^{2n}}$$

$$\lim_{n\to\infty}\frac{|x|^{2n+1}\left[\frac{\cos 2\pi x}{|x|^{2n+1}}-\frac{\sin(x-1)}{|x|}\right]}{|x|^{2n+1}\left[\frac{1}{|x|^{2n+1}}+1-\frac{1}{|x|}\right]}$$

$$=\lim_{n\to\infty}\frac{-\sin(x-1)}{|x|-1}$$ which is continuous.

Case III When $|x|=1$

$$f(x)=\begin{cases}1 & , \text{ if } x=1\\ -(1+\sin 2) & , \text{ if } x=-1\end{cases}$$

Now, $\lim_{x\to1^+}f(x)=\lim_{x\to1^+}-\frac{\sin(x-1)}{|x|-1}$ $\left[\frac{0}{0}\text{ form}\right]$

$$\lim_{x\to1^+}-\frac{\cos(x-1)}{1}=-1$$

$\Rightarrow \lim_{x\to1^-}f(x)=\lim_{x\to1^-}\cos 2\pi x=1$

As $\lim_{x\to1^+}f(x)\neq\lim_{x\to1^-}f(x)$

$\therefore f(x)$ is discontinuous at $x=1$

Similarly, $f(x)$ is also discontinuous at $x=-1$. Hence, the function $f(x)$ is continuous at $R-\{-1,1\}$.

**39.** *(a)* Given,

$$f(x)=\begin{cases}\frac{\log_e(1-x+x^2)+\log_e(1+x+x^2)}{\sec x-\cos x}, & x\in\left(-\frac{\pi}{2},\frac{\pi}{2}\right)-\{0\}\\ k, & x=0\end{cases}$$

$f(x)$ is continuous at $x=0$

$$\lim_{x\to0}f(x)=f(0)$$

$$\Rightarrow \lim_{x\to0}\frac{\log_e(1-x+x^2)+\log_e(1+x+x^2)}{\sec x-\cos x}=k$$

$$\Rightarrow \lim_{x\to0}\frac{\log_e[(1+x^2)^2-x^2]}{\frac{1}{\cos x}-\cos x}=k$$

$$\Rightarrow \lim_{x\to0}\frac{\cos x\cdot\log_e[1+x^4+2x^2-x^2]}{\sin^2 x}=k$$

$$\Rightarrow \lim_{x\to0}\cos x\lim_{x\to0}\frac{x^2}{\sin^2 x}\times\lim_{x\to0}\frac{\log_e(x^4+x^2+1)}{x^2}=k$$

$$\Rightarrow k=1\times1\times\lim_{x\to0}\frac{\log_e(x^4+x^2+1)}{x^2}\quad\left[\frac{0}{0}\text{ form}\right]$$

$$=\lim_{x\to0}\frac{4x^3+2x}{(x^4+x^2+1)2x}$$

$$=\lim_{x\to0}\frac{2x^2+1}{x^4+x^2+1}$$

Now, substitute $x=0\Rightarrow k=1$

**40.** *(c)* $f:\left(-\frac{\pi}{4},\frac{\pi}{4}\right)\to R$

$$f(x)=\begin{cases}(1+|\sin x|)^{\frac{3a}{|\sin x|}}, & -\frac{\pi}{4}<x<0\\ b, & x=0\\ e^{\frac{\cot 4x}{\cot 2x}}, & 0<x<\frac{\pi}{4}\end{cases}$$

Given, $f(x)$ is continuous at $x=0$

LHL at $x=0$

Put $x=0-h$

We get $\lim_{h\to0}(1-\sin h)^{\frac{3a}{-\sin h}}$

$$e^{\lim_{h\to0}(1-\sin h-1)\frac{3a}{-\sin h}}=e^{3a}$$

RHL at $x=0\Rightarrow \lim_{x\to0^+}e^{\frac{\cot 4x}{\cot 2x}}$

Put $x=0+h$, we get

$$\lim_{h\to0}e^{\frac{\cot 4h}{\cot 2h}}=\lim_{h\to0}e^{\frac{\cos 4h}{\cos 2h}\times\frac{\sin 2h}{\sin 4h}}$$

$$=\lim_{h\to0}e^{\frac{\cos 4h}{\cos 2h}\times\frac{\frac{\sin 2h}{2h}\times 2h}{\frac{\sin 4h}{4h}\times 4h}}=e^{1/2}$$

As, $f(x)$ is continuous at $x=0$.

So, LHL $=f(0)=$ RHL

$$e^{3a}=b=e^{1/2}$$

$$\therefore\quad a=\frac{1}{6},\ b=\sqrt{e}$$

$$\therefore\quad 6a+b^2=6\left(\frac{1}{6}\right)+(\sqrt{e})^2$$

$$=1+e$$

**41.** *(c)* Given,

$$f(x)=\begin{cases}\frac{\cos^{-1}(1-\{x\}^2)\sin^{-1}(1-\{x\})}{\{x\}-\{x\}^3}, & x\neq0\\ \alpha, & x=0\end{cases}$$

$$\{x\}=x-[x]$$

So, when $x\to0^+$

$\Rightarrow \{x\}=x-0=x$

And, when $x\to0^-$

$\Rightarrow \{x\}=x+1$

LHL $=\lim_{x\to0^-}f(x)$

$$=\lim_{x\to0}\frac{\cos^{-1}[1-(1+x)^2]\sin^{-1}[1-(1+x)]}{(1+x)-(1+x)^3}$$

$$=\lim_{x\to0}\frac{\cos^{-1}(-x^2-2x)\sin^{-1}(-x)}{(1+x)(x+2)(-x)}$$

$$=\lim_{x\to0}\frac{\cos^{-1}(-x^2-2x)}{(1+x)(x+2)}=\frac{\cos^{-1}(0)}{1.2}=\frac{\pi}{4}$$

RHL $=\lim_{x\to0^+}f(x)$

$$=\lim_{x\to0}\frac{\cos^{-1}(1-x^2)\sin^{-1}(1-x)}{x(1-x)(1+x)}$$

$$=\frac{\pi}{2}\lim_{x\to0}\frac{\cos^{-1}(1-x^2)}{x}$$

Applying L-Hopital Rule,

$$\frac{\pi}{2}\lim_{x\to0}\frac{(-1)(-2x)}{\sqrt{1-(1-x^2)^2}}$$

$$=\frac{\pi}{2}\cdot2\lim_{x\to0}\frac{x}{\sqrt{2x^2-x^4}}$$

$$=\pi\lim_{x\to0}\frac{1}{\sqrt{2-x^2}}=\frac{\pi}{\sqrt2}$$

LHL $=\frac{\pi}{4}$ and RHL $=\frac{\pi}{\sqrt2}$

Hence, LHL $\neq$ RHL

So, the function will be discontinuous for every value of $\alpha\in R$.

$\therefore$No such $\alpha$ exist.

**42.** *(8)* Probable values of $x$ where $\left[\frac{x^2}{2}\right]$ may be discontinuous in $x\in[0,4]$ are $=1, 2, 3, 4, 5, 6, 7, 8$

$x=\sqrt{2}, 2, \sqrt{6}, 2\sqrt{2}, \sqrt{10}, 2\sqrt{3}, \sqrt{14}, 4$

And for $[\sqrt{x}]$ corresponding values are $x=1, 2$

On checking for continuity at these points we get the $f(x)$ is discontinuous at

$x=1, \sqrt{2}, 2, \sqrt{6}, 2\sqrt{2}, \sqrt{10}, 2\sqrt{3}, \sqrt{14}$

Hence, $f(x)$ is discontinuous for 8 values of $x \in [0, 4]$

**43.** *(2)* A function $f(x)$ is continuous at $x=0$, if $\lim_{x\to 0} f(x)=f(0)$

Use $\tan x = x+\frac{x^3}{3}+\frac{2}{15}x^5+\ldots$ and $\sin x = x-\frac{x^3}{3!}+\frac{x^5}{5!}-\ldots$

$\lim_{x\to 0} f(x)=f(0)$

$$=\lim_{x\to 0}\frac{\left(\tan x+\frac{(\tan x)^3}{3}+\frac{2}{15}(\tan x)^5+\ldots\right)-\left(\sin x-\frac{(\sin x)^3}{6}+\frac{(\sin x)^5}{120}-\ldots\right)}{x+\frac{x^3}{3}+\frac{2}{15}x^5+\ldots-\left(-\frac{x^3}{3!}+\frac{x^5}{5!}-\ldots\right)}$$

$$=\lim_{x\to 0}\frac{(\tan x-\sin x)+\left(\frac{(\tan x)^3}{3}+\frac{(\sin x)^3}{6}\right)+\left(\frac{2}{15}(\tan x)^5-\frac{(\sin x)^5}{120}\right)\ldots}{x^3\left(\frac{1}{3}+\frac{1}{6}\right)+x^5\left(\frac{2}{15}-\frac{1}{120}\right)+\ldots}$$

Divide numerator and denominator by $x^3$

$$=\lim_{x\to 0}\frac{\frac{(\tan x-\sin x)}{x^3}+\left(\frac{(\tan x)^3}{3x^3}+\frac{(\sin x)^3}{6x^3}\right)+\left(\frac{2}{15x^3}(\tan x)^5-\frac{(\sin x)^5}{120x^3}\right)\ldots}{\left(\frac{1}{3}+\frac{1}{6}\right)+x^2\left(\frac{2}{15}-\frac{1}{120}\right)+\ldots}$$

Now, use $\lim_{x\to 0}\frac{\tan x}{x}=1$ and $\lim_{x\to 0}\frac{\sin x}{x}=1$

$$=\lim_{x\to 0}\frac{\frac{(\tan x-\sin x)}{x^3}+\left(\frac{1}{3}+\frac{1}{6}\right)}{\left(\frac{1}{3}+\frac{1}{6}\right)}$$

Because $\lim_{x\to 0}\left(\frac{2}{15x^3}(\tan x)^5-\frac{(\sin x)^5}{120x^3}\right)$

$$=\lim_{x\to 0}\left(\frac{2}{15x^3}(\tan x)^3(\tan x)^2-\frac{(\sin x)^3}{120x^3}(\sin x)^2\right)=\left(\frac{2}{15}\times 1\times 0-\frac{1}{120}\times 1\times 0\right)=0$$

Likewise higher term will become zero.

$$\lim_{x\to 0} x^2\left(\frac{2}{15}-\frac{1}{120}\right)=0$$

$$\text{Now, }\lim_{x\to 0}\frac{\frac{(\tan x-\sin x)}{x^3}+\left(\frac{1}{3}+\frac{1}{6}\right)}{\left(\frac{1}{3}+\frac{1}{6}\right)}$$

$$=\lim_{x\to 0}\frac{\frac{\sin x}{x^3}\left(\frac{1-\cos x}{\cos x}\right)+\frac{1}{2}}{\frac{1}{2}}$$

$$=\lim_{x\to 0}\frac{\frac{\sin x}{x}\times 2\frac{\sin^2\frac{x}{2}}{x^2}\times\frac{1}{\cos x}+\frac{1}{2}}{1/2}$$

$$=\lim_{x\to 0}\frac{1\times 2\times\frac{\sin^2\frac{x}{2}}{4\left(\frac{x}{2}\right)^2}\times\frac{1}{\cos x}+\frac{1}{2}}{1/2}=\frac{1\times 2\times\frac{1}{4}+\frac{1}{2}}{1/2}=\frac{1}{1/2}=2$$

Thus, $f(0)=2$

**44.** *(81)* Given,

$$f(x)=\begin{cases}\left(\frac{8}{7}\right)^{\frac{\tan 8x}{\tan 7x}}, & 0<x<\frac{\pi}{2}\\ a-8, & x=\frac{\pi}{2}\\ (1+|\cot x|)^{\frac{b}{a}|\tan x|}, & \frac{\pi}{2}<x<\pi\end{cases}$$

LHL at $x=\frac{\pi}{2}$ i.e. $f\left(\frac{\pi^-}{2}\right)=\lim_{x\to\frac{\pi^-}{2}}\left(\frac{8}{7}\right)^{\frac{\tan 8x}{\tan 7x}}$

Putting $x=\frac{\pi}{2}-h$, as $x\to\frac{\pi^-}{2}$, then $h\to 0$

$$\text{So, LHL}=\lim_{h\to 0}\left(\frac{8}{7}\right)^{\frac{\tan\left[8\left(\frac{\pi}{2}-h\right)\right]}{\tan\left[7\left(\frac{\pi}{2}-h\right)\right]}}$$

$$=\lim_{h\to 0}\left(\frac{8}{7}\right)^{\frac{-\tan 8h}{\cot h}}=\left(\frac{8}{7}\right)^0=1$$

$$f\left(\frac{\pi}{2}\right)=a-8$$

$$\text{RHL}=\lim_{x\to\frac{\pi^+}{2}}(1+|\cot x|)^{\frac{b}{a}|\tan x|}$$

$$=e^{\lim_{x\to\frac{\pi^+}{2}}|\cot x|\cdot\frac{b}{a}|\tan x|}=e^{\frac{b}{a}}$$

Since, $f(x)$ is continuous at $x=\frac{\pi}{2}$

So, $a-8=1=e^{b/a}$

$\Rightarrow \quad a=9$ and $b=0$

Hence, $a^2+b^2=(9)^2+(0)^2=81$

**45.** (17) $\left[\frac{x}{2}+3\right]$ is discontinuous at

$x=2, 4, 6, 8$

$\sqrt{x}$ is discontinuous at $x=1, 4$

$\because$ Continuous $\pm$ continuous $=$ continuous

continuous $\pm$ discontinuous $=$ discontinuous

and discontinuous $\pm$ discontinuous $=$ Need to be checked

so, at $x=4$,

LHL = RHL = $f(4)=3$

that mean, $f(x)$ is continuous at $x=4$

$f(x)$ is discontinuous at $x=1, 2, 6, 8$

$S=\{1, 2, 6, 8\}$

$\sum_{a\in S} a=1+2+6+8=17$

**46.** (2) We have,

$f(x)=|[x]|+\sqrt{x-[x]},\ x\in(-2,1)$

Now, $[x]=\begin{cases}-2, & x\in(-2,-1)\\ -1, & x\in[-1,0)\\ 0, & x\in(0,1)\end{cases}$

$\therefore f(x)=\begin{cases}2+\sqrt{x+2}, & x\in(-2,-1)\\ 1+\sqrt{x+1}, & x\in[-1,0)\\ \sqrt{x}, & x\in(0,1)\end{cases}$

Since, square root function is continuous, we need to check on end points of intervals.

$f(-1^-)=\lim_{x\to 1^-} f(x)=2+\sqrt{-1+2}$

$=3$

$\lim_{x\to -1^+} f(x)=1+\sqrt{-1+1}=1$

Since LHL $\neq$ RHL at $x=-1$, it is a point of discontinuity.

$\lim_{x\to 0^-} f(x)=1+\sqrt{0+1}=2$

$\lim_{x\to 0^+} f(x)=\sqrt{0}=0$

Since LHL $\neq$ RHL at $x=0$, it is a point of discontinuity.

So, total 2 points of discontinuity of $f(x)$ in $(-2, 1)$.

**47.** (62) $f(x)=[2x^2+1]$,

$g(x)=\begin{cases}2x-3, & x<0\\ 2x+3, & x\geq 0\end{cases}$

$fog(x)=\begin{cases}[2(2x-3)^2]+1, & x<0\\ [2(2x+3)^2]+1, & x\geq 0\end{cases}$

The function is continuous at $x=0$

$fog(x)$ is discontinuous when

(i) $2(2x-3)^2\in I$ and $x\in(-1,0)$

(ii) $2(2x+3)^2\in I$ and $x\in(0,1)$

Now, when $-1<x<0$

$-5<2x-3<-3$

$9<(2x-3)^2<25$

$18<2(2x-3)^2<50$

$\therefore$ There are total 31 discontinuous points.

When $0<x<1$

$3<2x+3<5$

$9<(2x+3)^2<25$

$18<2(2x+3)^2<50$

$\therefore$ There are 31 point of discontinuous in this case

Hence, there are total 62 points of discontinuous.

**48.** (2) For $-2<x<-1$

$f(x)=-2\times(x^2-1)+\sin(\pi)-(-2+1)$

$=(3-2x^2)$

For $-1\leq x<0$,

$f(x)=-1\times(1-x^2)+\sin\left(\frac{\pi}{2}\right)-(-1+1)$

$=1-(1-x^2)=x^2$

For $0\leq x<1$,

$f(x)=0\cdot|x^2-1|+\sin\left(\frac{\pi}{3}\right)-(0+1)$

$=\sin\frac{\pi}{3}-1=\frac{\sqrt{3}}{2}-1$

For $1\leq x<2$,

$f(x)=(x^2-1)+\sin\left(\frac{\pi}{1+3}\right)-(1+1)$

$=x^2-1+\frac{1}{\sqrt{2}}-2$

$=x^2-3+\frac{1}{\sqrt{2}}$

$\therefore f(x)$ is defines as,

$f(x)=\begin{cases}3-2x^2, & x\in(-2,-1)\\ x^2, & x\in[-1,0)\\ \frac{\sqrt{3}}{2}-1, & x\in[0,1)\\ \frac{1}{\sqrt{2}}-3+x^2, & x\in[1,2)\end{cases}$

Now, $\lim_{x\to -1^-} f(x)=1=f(-1)$ and

$\lim_{x\to -1^+} f(x)=1$

$\therefore f(x)$ is continuous at $x=-1$.

$\lim_{x\to 0^-} f(x)=0$ and $\lim_{x\to 0^+} f(x)=\frac{\sqrt{3}}{2}-1$

$\therefore f(x)$ is discontinuous at $x=0$.

$\lim_{x\to 1^-} f(x)=\frac{\sqrt{3}}{2}-1$

and $\lim_{x\to 1^+} f(x)=\frac{1}{\sqrt{2}}-2$

$\therefore f(x)$ is discontinuous at $x=1$.

$\Rightarrow f(x)$ is discontinuous at 2 points.

**49.** (6) $f(x)=\frac{\cos(\sin x)-\cos x}{x^4}$

As, $f(x)$ is continuous everywhere, so

$f(0)=\lim_{x\to 0} f(x)$

$=\lim_{x\to 0}\frac{\cos(\sin x)-\cos x}{x^4}$

On expanding the numerator and only identifying the coefficient of $x^4$ will give us the required limit.

$\cos(\sin x)=\left(1-\frac{\sin^2 x}{2}+\frac{\sin^4 x}{24}\right)$

$=1-\frac{1}{2}\left(x-\frac{x^3}{6}\right)^2+\frac{1}{24}(x)^4$

$=1-\frac{1}{2}\left(x^2-\frac{x^4}{3}\right)+\frac{x^4}{24}$

$=1-\frac{x^2}{2}+\frac{x^4}{6}+\frac{x^4}{24}$

and $\cos x=1-\frac{x^2}{2}+\frac{x^4}{24}$

$\therefore \frac{\cos(\sin x)-\cos x}{x^4}$

$=\frac{\left(1-\frac{x^2}{2}+\frac{x^4}{6}+\frac{x^4}{24}\right)-\left(1-\frac{x^2}{2}+\frac{x^4}{24}\right)}{x^4}$

$=\frac{1}{6}$

$\therefore f(0)=\frac{1}{6}=\frac{1}{k}$

Hence, $k=6$.

**50.** (1) $f(x)=\begin{cases} x+a, & x<0 \\ |x-1|, & x\geq 0 \end{cases}$

$g(x)=\begin{cases} x+1, & x<0 \\ (x-1)^2+b, & x\geq 0 \end{cases}$

$g[f(x)]=\begin{cases} f(x)+1, & f(x)<0 \\ [f(x)-1]^2+b, & f(x)\geq 0 \end{cases}$

$f(x)<0$

Case I $x+a<0$ and $x<0 \Rightarrow x<-a$

Case II $|x-1|<0$ and $x\geq 0$

$\Rightarrow$ Not possible

$f(x)\geq 0$

Case I $x+a\geq 0$ and $x<0$

$\Rightarrow x\in[-a, 0)$

Case II $|x-1|\geq 0$ and $x\geq 0$

$\Rightarrow x\geq 0$

$g[f(x)]=$

$\begin{cases} x+a+1, & x<0 \text{ and } f(x)<0 \\ |x-1|+1, & x\geq 0 \text{ and } f(x)<0 \\ (x+a-1)^2+b, & x<0 \text{ and } f(x)\geq 0 \\ (|x-1|-1)^2+b, & x\geq 0 \text{ and } f(x)\geq 0 \end{cases}$

$g[f(x)]=\begin{cases} x+a+1, & x<-a \\ (x+a-1)^2+b, & -a\leq x<0 \\ (|x-1|-1)^2+b, & x\geq 0 \end{cases}$

Since, $g[f(x)]$ is continuous for all $x\in R$

So, $g(f(x))$ will be continuous at $x=-a$ and $x=0$. Now, at $x=-a$

LHL = RHL = value of function

$\Rightarrow \quad 1=1+b=1+b \Rightarrow b=0$

At $\quad x=0$

LHL = RHL = value of function

$\Rightarrow (a-1)^2+b=b$

$\Rightarrow (a-1)^2=0 \Rightarrow a=1$

Hence, $a+b=1$

**51.** (8) The given function, $f(x)=x\cdot\left[\frac{x}{2}\right]$, for $-10<x<10$ may be discontinuous at the points, where $x/2$ is an integer.

So, possible points of discontinuity are

$x=0, \pm 2, \pm 4, \pm 6, \pm 8$

But at $x=0$

$\lim_{x\to 0^+} f(x)=0 \quad \text{and} \quad \lim_{x\to 0^-} f(x)=0$

$\therefore x=0$ is the point of continuity.

So, the given function $f(x)$ is discontinuous at

$x=\pm 2, \pm 4, \pm 6, \pm 8$

Therefore, the number of points of discontinuity of $f$ is equal to eight.

## Topic 2

**01.** (a) $\lim_{x\to 0}[f(2+x)]^{\frac{3}{x}}=e^{\alpha}$

($1^{\infty}$ from)

$e^{\lim_{x\to 0}(f(2+x)-1)\frac{3}{x}}=e^{\alpha}$

$\Rightarrow 3\lim_{x\to 0}\frac{f(2+x)-1}{x}=\alpha$

$\Rightarrow 3\lim_{x\to 0}\frac{f'(2+x)}{1}=\alpha$

$\Rightarrow 3f'(2)=\alpha$

$\Rightarrow 3\times 4=\alpha$

$\Rightarrow \alpha=12$

$\therefore y=4x^3-4x^2-4(12-7)x-12$

$=4x^3-4x^2-20x-12$

$=4(x^3-x^2-5x-3)$

$=4(x+1)(x^2-2x-3)$

$=4(x+1)(x-3)(x+1)$

$=4(x+1)^2(x-3)$

$\therefore$ Meets $X$-axis at 2 points.

**02.** (b) Given, $f(x)=[x]+|x-2|$, $-2<x<3$

$\Rightarrow f(x)=\begin{cases} -x, & -2<x<-1 \\ -x+1, & -1\leq x<0 \\ -x+2, & 0\leq x<1 \\ -x+3, & 1\leq x<2 \\ x, & 2\leq x<3 \end{cases}$

Hints : $\begin{pmatrix} [x]=-2, |x-2|=-(x-2) \\ [x]=-1, |x-2|=-(x-2) \\ [x]=0, |x-2|=-(x-2) \\ [x]=1, |x-2|=-(x-2) \\ [x]=2, |x-2|=x-2 \end{pmatrix}$

We see that at $x=-1$, LHL $\neq$ RHL

at $x=0$, LHL $\neq$ RHL

at $x=1$, 2, LHL $\neq$ RHL

So, $f(x)$ is discontinuous at 4 points and hence obviously it is non-differentiable at these points.

So, $\quad m=n=4$

$\therefore \quad m+n=4+4=8$

**03.** (a) Given, $f:(-\infty,\infty)-\{0\}\to R$

$f'(1)=\lim_{a\to\infty}a^2 f\left(\frac{1}{a}\right)$

Now, we have to find

$\lim_{a\to\infty}\frac{a(a+1)}{2}\tan^{-1}\left(\frac{1}{a}\right)+a^2-2\ln(a)$

$=\lim_{a\to\infty}a^2\left(\frac{\left(1+\frac{1}{a}\right)}{2}\tan^{-1}\left(\frac{1}{a}\right)+1\frac{2}{a^2}\ln\frac{1}{a}\right)$

$\Rightarrow f(x)=\frac{1}{2}(1+x)\tan^{-1}x+1+2x^2\ln(x)$

$f'(x)=\frac{1}{2}\left(\frac{1+x}{1+x^2}+\tan^{-1}x\right)+4x\ln x+2x$

$f'(1)=\frac{1}{2}\left(1+\frac{\pi}{4}\right)+2$

$f'(1)=\frac{5}{2}+\frac{\pi}{8}$

**04.** (b) $f:R\to R$

$f(x)=|x-1|\cos|x-2|\sin|x-1|+(x-3)|x^2-5x+4|$

$f(x)=|x-1|\cos|x-2|\sin|x-1|+(x-3)|x-1||x-4|$

$=|x-1|[\cos|x-2|\sin|x-1|+(x-3)|x-4|]$

So, there are only two points $x=1$ and $x=4$, where $f(x)$ is not differentiable.

**05.** (b) Given, $f(x)=\min\{1, 1+x\sin x\}$, $0\leq x\leq 2\pi$

Number of non-differentiable points $=1(m)$

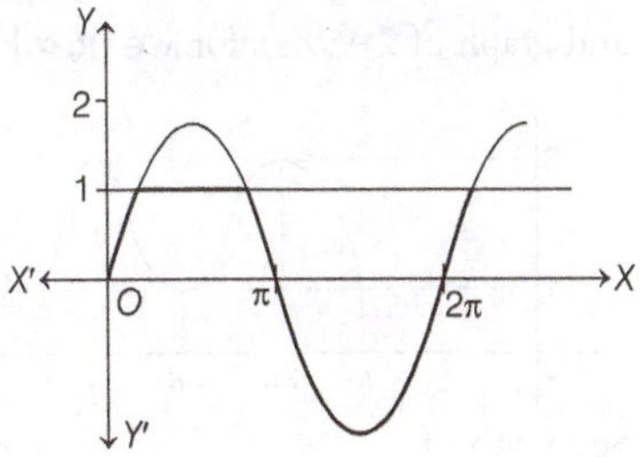


Number of non-continuous point $=0(n)$

$\therefore \quad (m, n)=(1, 0)$

**06.** (c) $f(x)=$

$\begin{cases} (x-3)(x+1)e^{(3x-2)^2}, & x>3 \\ -(x-3)(x+1)e^{(3x-2)^2}, & -1\leq x\leq 3 \\ (x-3)(x+1)e^{(3x-2)^2}, & x<-1 \end{cases}$

At $x=-1$, let LHD be $\alpha$, then its clear that RHD be $-\alpha$.

Similarly, at $x=3$, if LHD is $\beta$, then RHD at $x=3$ will be $-\beta$.

So, $f(x)$ is not differentiable at $x=-1$, $x=3$

At all other points $f(x)$ will be differentiable.

Hence, option (c) is correct.

**07.** *(a)* We have, $f(x)=x-[x]=\{x\}$ and $g(x)=1-x+[x]=1-\{x\}$

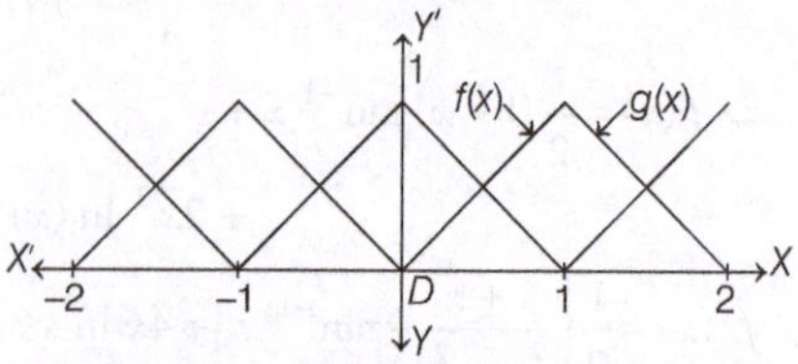


Again, $h(x)=\min[f(x), g(x)]$, so graph of $h(x)$ will be

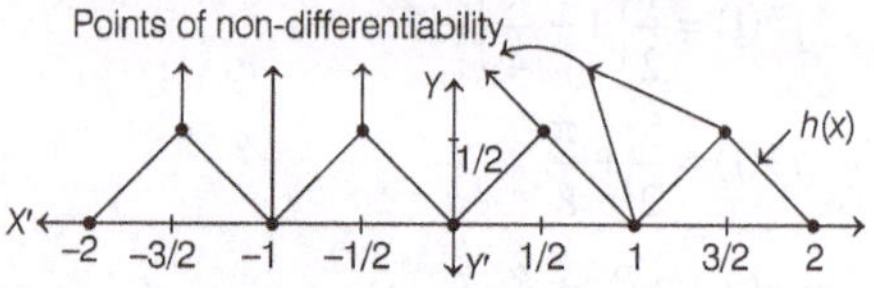


From graph, it is clear that $h(x)$ is continuous in $[-2, 2]$ but not differentiable at $x=\dfrac{-3}{2}$,

$$-1, \frac{-1}{2}, 0, \frac{1}{2}, 1, \frac{3}{2} \text{ in } (-2, 2).$$

**08.** *(b)* Graph of $\max(\sin t: 0 \le t \le x)$ in $x \in [0, \pi]$

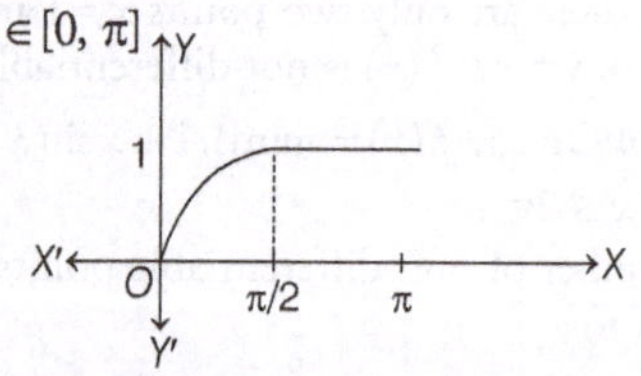


and graph of $2+\cos x$ for $x \in [\pi, \infty]$

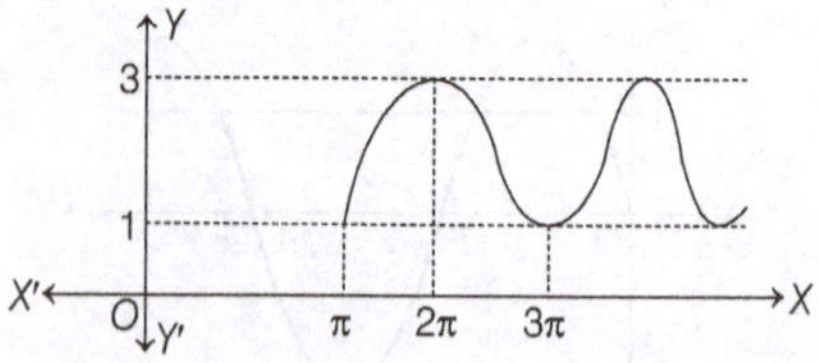


So, graph of

$$f(x)=\begin{cases}\max[\sin t: 0 \le t \le x], & 0 \le x \le \pi \\ 2+\cos x, & x>\pi\end{cases}$$

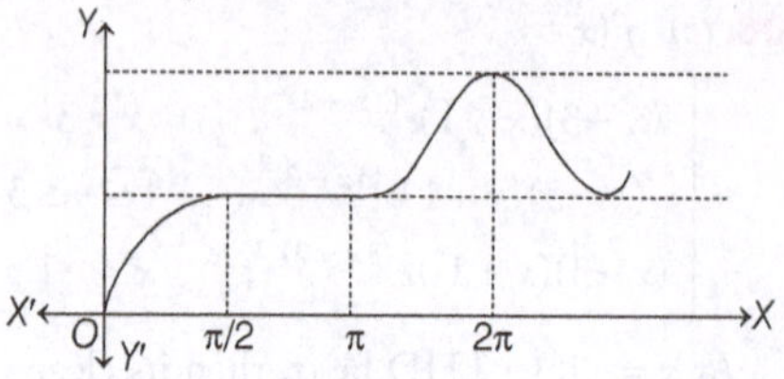


So, $f(x)$ is differentiable everywhere in $(0, \infty)$.

**09.** *(a)* The function $f: R \to R$ is defined by $f(x)=\max\{x, x^2\}$

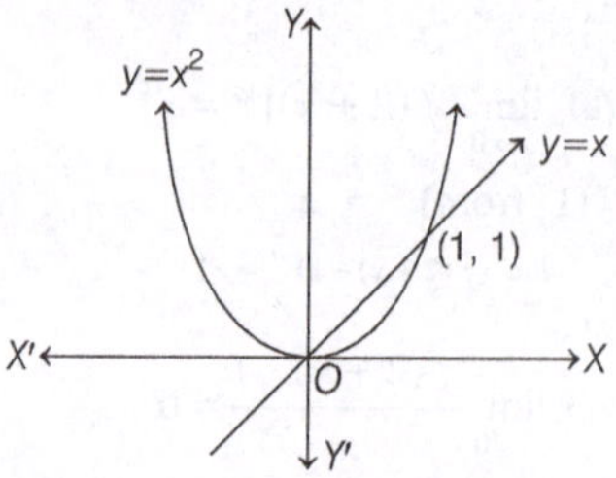


$$=\begin{cases}x^2, & x<0 \\ x, & 0 \le x<1 \\ x^2, & x \ge 1\end{cases}$$

$$\Rightarrow f'(x)=\begin{cases}2x, & x<0 \\ 1, & 0 \le x<1 \\ 2x, & x \ge 1\end{cases}$$

$\because f'(0^-)=0$, but $f'(0^+)=1$, so $f$ is not differentiable at $x=0$.

Similarly, as $f'(1^-)=1$, but $f'(1^+)=2$

$\Rightarrow f$ is not differentiable at $x=1$. So, the required set $S=\{0, 1\}$.

**10.** *(d)* Given function, $g(x)=|f(x)|$, where $f: R \to R$ be differentiable at $c \in R$ and $f(c)=0$, then for function $g$ at $x=c$

$$g'(c)=\lim_{h \to 0} \frac{g(c+h)-g(c)}{h} \quad \text{[Where, } h>0]$$

$$=\lim_{h \to 0} \frac{|f(c+h)|-|f(c)|}{h}$$

$$=\lim_{h \to 0} \frac{|f(c+h)|}{h} \quad [\text{as } f(c)=0 \text{ [given]}]$$

$$=\lim_{h \to 0} \left|\frac{f(c+h)-f(c)}{h}\right| \quad [\because h>0]$$

$$=\left|\lim_{h \to 0} \frac{f(c+h)-f(c)}{h}\right|$$

$$=|f'(c)|$$

$[\because f$ is differentiable at $x=c]$

Now, if $f'(c)=0$, then $g(x)$ is differentiable at $x=c$, otherwise LHD (at $x=c$) and RHD (at $x=c$) is different.

**11.** *(b)* Given function is $f(x)=15-|x-10|, x \in R$

and $g(x)=f(f(x))=f(15-|x-10|)$

$$=15-|15-|x-10|-10|$$

$$=15-|5-|x-10||$$

$$=\begin{cases}15-|5-(x-10)|, & x \ge 10 \\ 15-|5+(x-10)|, & x<10\end{cases}$$

$$=\begin{cases}15-|15-x|, & x \ge 10 \\ 15-|x-5|, & x<10\end{cases}$$

$$=\begin{cases}15+(x-5)=10+x, & x<5 \\ 15-(x-5)=20-x, & 5 \le x<10 \\ 15+(x-15)=x, & 10 \le x<15 \\ 15-(x-15)=30-x, & x \ge 15\end{cases}$$

From the above definition it is clear that $g(x)$ is not differentiable at $x=5, 10, 15$.

**12.** *(c)* Let us draw the graph of $y=f(x)$, as shown below

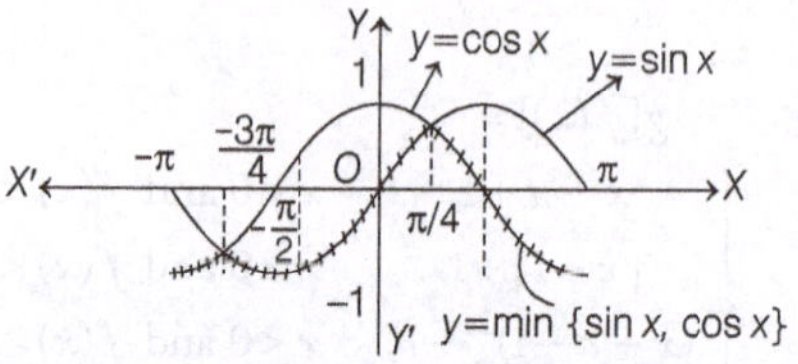


Clearly, the function $f(x)=\min\{\sin x, \cos x\}$ is not differentiable at $x=\dfrac{-3\pi}{4}$ and $\dfrac{\pi}{4}$

[these are point of intersection of graphs of $\sin x$ and $\cos x$ in $(-\pi, \pi)$, on which function has sharp edges]. So, $S=\left\{\dfrac{-3\pi}{4}, \dfrac{\pi}{4}\right\}$, which is a subset of $\left\{\dfrac{-3\pi}{4}, \dfrac{-\pi}{4}, \dfrac{3\pi}{4}, \dfrac{\pi}{4}\right\}$.

**13.** *(c)* Given,

$$f(t)=(|\lambda| e^{|t|}-\mu) \cdot \sin(2|t|)$$

$$\text{RHD (at } x=0)=\lim_{h \to 0} \frac{f(0+h)-f(0)}{h}$$

$$=\lim_{h \to 0} \frac{(|\lambda| e^h-\mu) \sin 2h}{h}$$

$$=\lim_{h \to 0} (|\lambda| e^h-\mu) \frac{2 \cosh \sinh}{h}$$

$$=2(|\lambda|-\mu)$$

$$\left[\because \lim_{h \to 0} \frac{\sin h}{h}=\lim_{h \to 0} \cos h=1\right]$$

$$\text{LHD (at } x=0)=\lim_{h \to 0} \frac{f(0-h)-f(0)}{-h}$$

$$= \lim_{h\to 0} \frac{(|\lambda|e^h - \mu)\sin 2h}{-h}$$

$$= 2(|\lambda| - \mu)$$

For $f$ to be differentiable;

RHD (at $x=0$) = LHD (at $x=0$)

$\Rightarrow \quad 4|\lambda| = 4\mu$

$\Rightarrow \quad |\lambda| = \mu \Rightarrow \mu \geq 0$ and $\lambda \in R$

**14.** *(b)* We have, $f(x) = |\log 2 - \sin x|$ and $g(x) = f(f(x)), x \in R$

Note that, for $x \to 0$, $\log 2 > \sin x$

$\therefore \quad f(x) = \log 2 - \sin x$

$\Rightarrow \quad g(x) = \log 2 - \sin(f(x))$

$\qquad = \log 2 - \sin(\log 2 - \sin x)$

Clearly, $g(x)$ is differentiable at $x=0$ as $\sin x$ is differentiable.

Now, $g'(x) = -\cos(\log 2 - \sin x)\,(-\cos x)$

$\qquad = \cos x \cdot \cos(\log 2 - \sin x)$

$\Rightarrow g'(0) = 1 \cdot \cos(\log 2)$

**15.** *(a)* Since, $g(x)$ is differentiable

$\Rightarrow g(x)$ must be continuous.

$$\therefore \quad g(x) = \begin{cases} k\sqrt{x+1}, & 0 \leq x \leq 3 \\ mx+2, & 3 < x \leq 5 \end{cases}$$

At $x=3$, RHL $= 3m+2$

and at $x=3$, LHL $= 2k$

$\therefore \quad 2k = 3m+2 \qquad \ldots(\text{i})$

$$\text{Also, } g'(x) = \begin{cases} \dfrac{k}{2\sqrt{x+1}}, & 0 \leq x < 3 \\ m, & 3 < x \leq 5 \end{cases}$$

$\therefore \quad L\{g'(3)\} = \dfrac{k}{4}$ and $R\{g'(3)\} = m$

$\Rightarrow \quad k/4 = m$ i.e. $k = 4m \qquad \ldots(\text{ii})$

On solving Eqs. (i) and (ii), we get

$$k + m = 2$$

**16.** *(d)* Here, $\displaystyle\lim_{x\to a} \frac{x^2 f(a) - a^2 f(x)}{x-a}$

$$= \lim_{x\to a} \frac{2xf(a) - a^2 f'(x)}{1-0}$$

[Using L' Hospital rule]

$$= 2af(a) - a^2 f'(a)$$

**17.** *(b)* We have, $f : (-1, 1) \longrightarrow R$

$f(0) = -1, \qquad f'(0) = 1$

$g(x) = [f(2f(x)+2)]^2$

$\Rightarrow \quad g'(x) = 2[f(2f(x)+2)] \times f'(2f(x)+2) \times 2f'(x)$

$\Rightarrow \quad g'(0) = 2[f\{2f(0)+2\}] \times f'\{2f(0)+2\} \times 2f'(0)$

$\qquad = 2\,[f(0)] \times f'(0) \times 2f'(0)$

$\qquad = 2 \times (-1) \times 1 \times 2 \times 1 = -4$

**18.** *(b)* Since, $f(x) = \dfrac{x}{1+|x|}$

Let $\quad f(x) = \dfrac{g(x)}{h(x)} = \dfrac{x}{1+|x|}$

It is clear that $g(x) = x$ and $h(x) = 1 + |x|$ are differentiable on $(-\infty, \infty)$ and $(-\infty, 0) \cup (0, \infty)$, respectively. Thus, $f(x)$ is differentiable on $(-\infty, 0) \cup (0, \infty)$.

Now, we have to check the differentiability at $x = 0$.

$$\therefore \quad \lim_{x\to 0} \frac{f(x) - f(0)}{x - 0} = \lim_{x\to 0} \frac{\dfrac{x}{1+|x|} - 0}{x}$$

$$= \lim_{x\to 0} \frac{1}{1+|x|} = 1$$

Hence, $f(x)$ is differentiable on $(-\infty, \infty)$.

**19.** *(b)* $f'(1) = \displaystyle\lim_{h\to 0} \frac{f(1+h) - f(1)}{h}$

$$= \lim_{h\to 0} \frac{f(1+h)}{h} - \lim_{h\to 0} \frac{f(1)}{h}$$

Since, $\displaystyle\lim_{h\to 0} \frac{f(1+h)}{h} = 5$, so $\displaystyle\lim_{h\to 0} \frac{f(1)}{h}$ must be finite as $f'(1)$ exists and $\displaystyle\lim_{h\to 0} \frac{f(1)}{h}$ can be finite only, if $f(1) = 0$ and $\displaystyle\lim_{h\to 0} \frac{f(1)}{h} = 0$.

$$\therefore \quad f'(1) = \lim_{h\to 0} \frac{f(1+h)}{h} = 5$$

**20.** *(a)* We have,

$$f(x) = |2x^2 + 5|x| - 3|$$

Firstly we draw the graph of $f(x)$

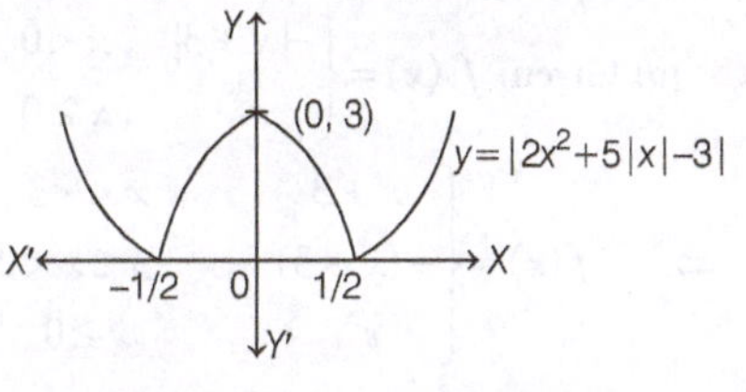


Clearly by graph of $y$, $f(x)$ is continuous for every $x \in R$ and not differentiable at $x = -\dfrac{1}{2}, \dfrac{1}{2}$ and 0

i.e. 3 points $m = 0$ and $n = 3$

$\qquad m + n = 3$

**21.** *(a)* $f : (0, \infty) \to R$, $f(x) = e^{-|\log_e x|}$

$$f(x) = \begin{cases} \dfrac{1}{e^{-\ln x}} = \dfrac{1}{e^{\ln \frac{1}{x}}} = \dfrac{1}{1/x} = x, & 0 < x < 1 \\ \dfrac{1}{e^{\ln x}} = \dfrac{1}{x}, & x \geq 1 \end{cases}$$

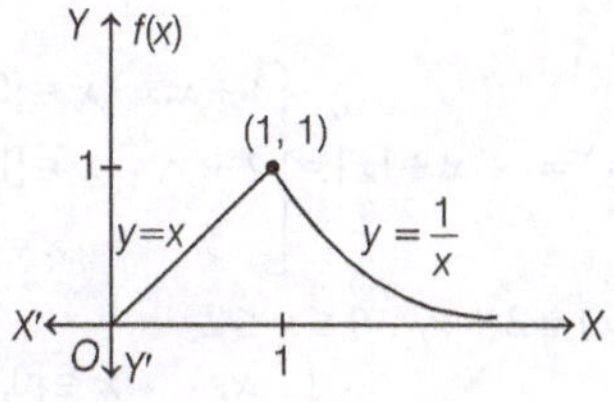


$m = 0$, (no point at which function is not continuous)

$n = 1$, (not differentiable)

$\therefore m + n = 1$

> **Concept Enhancer**
>
> A point on the curve at which, we can not draw a unique tangent is called the point of non-differentiability.

**22.** *(c)* Given, $f : (0, 2) \to R$ such that

$$f(x) = \frac{x}{2} + \frac{2}{x}$$

$$\Rightarrow \quad f'(x) = \frac{1}{2} - \frac{2}{x^2}$$

For $x \in (0, 2)$, $f(x)$ is decreasing.

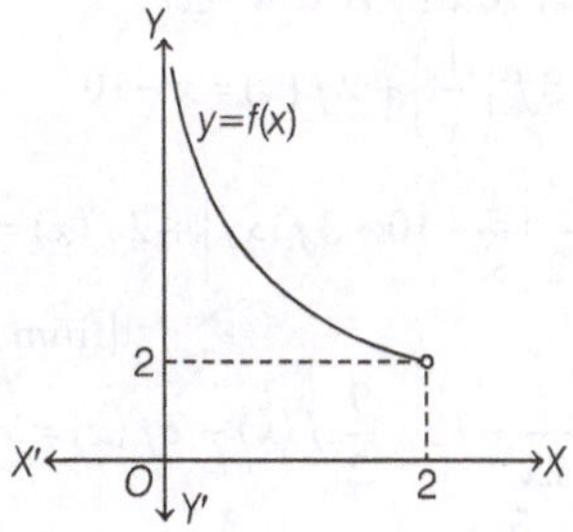


$$g(x) = \begin{cases} \dfrac{x}{2} + \dfrac{2}{x}, & 0 < x \leq 1 \\ \dfrac{3}{2} + x, & 1 < x < 2 \end{cases}$$

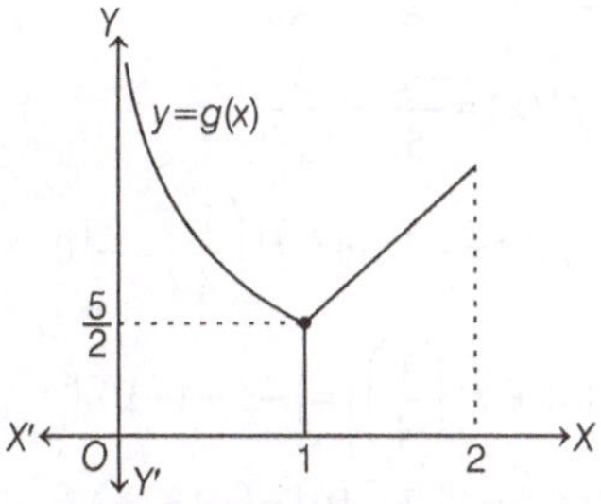

**23.** (b) Given,
$f(x) = \max\{1 + x + [x];\ 2 + x,\ x + 2[x]\}$
$0 \le x \le 2$
Given that, $m$ be denotes the number of points in [0, 2], where $f$ is not continuous and '$n$' be denotes the number of point in (0, 2) where $f$ is not differentiable.
Let

$$f_1(x) = 1 + x + [x] = \begin{cases} 1+x, & x \in [0,1) \\ 2+x, & x \in [1, 2) \\ 5 & x = 2 \end{cases}$$

$$f_2(x) = 2 + x, \quad 0 \le x \le 2$$

$$f_3(x) = x + 2[x] = \begin{cases} x, & x \in [0,1) \\ 2+x, & x \in [3, 2) \\ 6, & x = 2 \end{cases}$$

So, $f(x) = \max\{f_1(x), f_2(x), f_3(x)\}$ in $0 \le x \le 2$.

Then, $f(x) = \begin{cases} 2+x, & x \in [0, 2) \\ 6, & x = 2 \end{cases}$

Clearly, $f(x)$ is discontinuous only at $x = 2$
So, $m = 1$
and $f(x)$ is differentiable in (0, 2) so, $n = 0$
$\therefore (m+n)^2 + 2 = (1+0)^2 + 2 = 1 + 2 = 3$

**24.** (d) Given, $3f(x) + 2f\left(\frac{1}{x}\right) = \frac{1}{x} - 10$ ...(i)

Change $x$ by $1/x$, we get

$$3f\left(\frac{1}{x}\right) + 2f(x) = x - 10$$

$$\Rightarrow \frac{3}{2}\left[\frac{1}{x} - 10 - 3f(x)\right] + 2f(x) = x - 10$$ [from Eq. (i)]

$$\Rightarrow \frac{3}{2x} - 15 - \frac{9}{2}f(x) + 2f(x) = x - 10$$

$$\Rightarrow -\frac{5}{2}f(x) - 15 + \frac{3}{2x} = x - 10$$

$$\Rightarrow -\frac{5}{2}f(x) = x - \frac{3}{2x} + 5$$

$$\Rightarrow f(x) = -\frac{2}{5}x + \frac{3}{5x} - 2$$

$$\Rightarrow f'(x) = \frac{-2}{5} - \frac{3}{5x^2}$$

$$\therefore \quad f(3) = -3 \text{ and } f'\left(\frac{1}{4}\right) = -10$$

$$\left|f(3) + f'\left(\frac{1}{4}\right)\right| = |-3 + (-10)|$$
$$= |-3 - 10| = |-13| = 13$$

**25.** (a)
Given, $f(x) = \begin{cases} x+1, & x < 0 \\ |x-1|, & x \ge 0 \end{cases}$

and $g(x) = \begin{cases} x+1, & x < 0 \\ 1, & x \ge 0 \end{cases}$

Then, $gof(x) = g(f(x))$

$$= \begin{cases} f(x)+1, & f(x) < 0 \\ 1, & f(x) \ge 0 \end{cases}$$

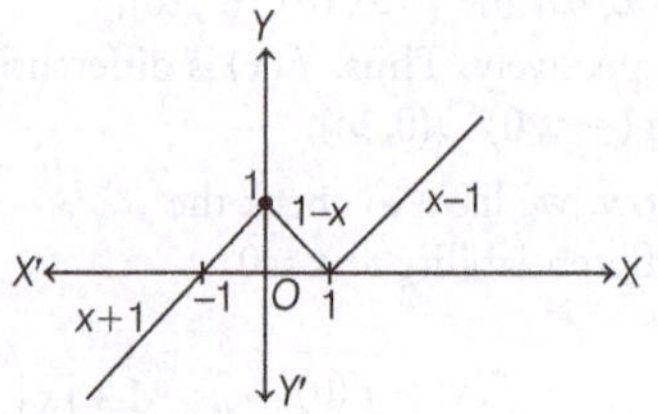


Here, $f(x) < 0$ for $x \in (-\infty, -1]$ and $f(x) \ge 0$ for $x \in [-1, \infty)$

Now, $g(f(x)) = \begin{cases} (x+1)+1, & x < -1 \\ 1, & x \ge -1 \end{cases}$

$$g(f(x)) = \begin{cases} x+2, & x < -1 \\ 1, & x \ge -1 \end{cases}$$

Check continuity at $x = -1$
LHL $= g(f(-1)) = -1 + 2 = 1$
RHL $= g(f(-1^+)) = 1$
continuous at $x = -1$

$$(g((f(x)))' = \begin{cases} 1, & x < -1 \\ 0, & x \ge -1 \end{cases}$$

For differentiability at $x = -1$
LHD $= g'(f(-1^-)) = 1$
RHD $= g'(f(-1^+)) = 0$
LHD $\ne$ RHD
$\Rightarrow$ not differentiable at $x = -1$
Hence, $gof(x)$ is continuous everywhere but not differentiable at $x = -1$ (exactly at one point)

**26.** (d) Given, $f(x) = \begin{cases} -|x+3| & , x < 0 \\ e^x & , x \ge 0 \end{cases}$

$$\Rightarrow \quad f(x) = \begin{cases} x+3 & , \quad x < -3 \\ -(x+3) & , \quad -3 \le x < 0 \\ e^x & , \quad x \ge 0 \end{cases}$$

and $g(x) = \begin{cases} x^2 + k_1 x & , x < 0 \\ 4x + k_2 & , x \ge 0 \end{cases}$

For $x < 0$, $(gof)(x) =$
$\{-(x+3)\}^2 + k_1(-(x+3))$
$= (x+3)^2 - k_1(x+3)$

For $x \ge 0$, $(gof)(x) = 4e^x + k_2$
It is given that, $(gof)(x)$ is differentiable at $x = 0$

$$(gof)(0^-) = (gof)(0^+)$$
$$\Rightarrow (0+3)^2 - k_1(0+3) = 4e^0 + k_2$$
$$\Rightarrow \quad 9 - 3k_1 = 4 + k_2 \quad \text{...(i)}$$

Also, LHD = RHD
$\Rightarrow 2(x+3) - k_1 = 4e^x$
$\Rightarrow 2(0+3) - k_1 = 4 \times e^0 = 4 \Rightarrow k_1 = 2$
Put $k_1 = 2$ in Eq. (i), we get
$9 - 6 = 4 + k_2$
$\Rightarrow \quad k_2 = -1$
Now, $(gof)(-4) = (x+3)^2 + k_1(x+3)$ [$\because -4 < -3$]
$= (-4+3)^2 + 2(-4+3) = 1 - 2 = -1$
and $(gof)(4) = 4e^4 - 1$
Then, $(gof)(4) + (gof)(-4)$
$= 4e^4 - 1 - 1 = 4e^4 - 2 = 2(2e^4 - 1)$

**27.** (b) We have,
$f(x) = \sin|x| - |x| + 2(x - \pi)\cos|x|$
$f(x) =$

$$\begin{cases} -\sin x + x + 2(x-\pi)\cos x, & \text{if } x < 0 \\ \sin x - x + 2(x-\pi)\cos x, & \text{if } x \ge 0 \end{cases}$$

[$\because \sin(-\theta) = -\sin\theta$ and $\cos(-\theta) = \cos\theta$]

$\therefore f'(x) =$

$$\begin{cases} -\cos x + 1 + 2\cos x - 2(x-\pi)\sin x; & \text{if } x < 0 \\ \cos x - 1 + 2\cos x - 2(x-\pi)\sin x, & \text{if } x \ge 0 \end{cases}$$

Clearly, $f(x)$ is differentiable everywhere except possibly at $x = 0$
[$\because f'(x)$ exist for $x < 0$ and $x > 0$]

Here, $Rf'(0) = \lim_{x \to 0^+} (3\cos x - 1 - 2(x-\pi)\sin x)$
$= 3 - 1 - 0 = 2$
and $Lf'(0) = \lim_{x \to 0^-} (\cos x + 1 - 2(x-\pi)\sin x)$
$= 1 + 1 - 0 = 2$
$\because \quad Rf'(0) = Lf'(0)$
So, $f(x)$ is differentiable at all values of $x$.
$\Rightarrow \quad K = \phi$

**28.** (a) We have,
$f(x) = |x - \pi| \cdot (e^{|x|} - 1)\sin|x|$

$$f(x) = \begin{cases} (x-\pi)(e^{-x} - 1)\sin x, & x < 0 \\ -(x-\pi)(e^x - 1)\sin x, & 0 \le x < \pi \\ (x-\pi)(e^x - 1)\sin x, & x \ge \pi \end{cases}$$

We check the differentiability at $x=0$ and $\pi$.

We have, $f'(x)=$

$$\begin{cases}(x-\pi)(e^{-x}-1)\cos x+(e^{-x}-1)\sin x \\ \quad +(x-\pi)\sin xe^{-x}(-1), x<0 \\ -[(x-\pi)(e^{x}-1)\cos x+(e^{x}-1)\sin x \\ \quad +(x-\pi)\sin xe^{x}], 0<x<\pi \\ (x-\pi)(e^{x}-1)\cos x+(e^{x}-1)\sin x \\ \quad +(x-\pi)\sin xe^{x}, x>\pi\end{cases}$$

Clearly, $\lim_{x\to0^-} f'(x)=0=\lim_{x\to0^+} f'(x)$

and $\lim_{x\to\pi^-} f'(x)=0=\lim_{x\to\pi^+} f'(x)$

$\therefore$ $f$ is differentiable at $x=0$ and $x=\pi$

Hence, $f$ is differentiable for all $x$.

**29.** *(b)* Check for the continuity at $x=0$,

$$\text{LHL}=\lim_{x\to0^-} f(x)$$

$$=\lim_{h\to0}(0-h)e^{-\left(\frac{1}{|-h|}+\frac{1}{(-h)}\right)}$$

$$=\lim_{h\to0}(-h)e^{-\left(\frac{1}{h}-\frac{1}{h}\right)}$$

$$=\lim_{h\to0}(-h)=0$$

$$\text{RHL}=\lim_{x\to0^+} f(x)$$

$$=\lim_{h\to0}(0+h)e^{-\left(\frac{1}{|h|}+\frac{1}{h}\right)}$$

$$=\lim_{h\to0} he^{-1/h+1/h}$$

$$=\lim_{h\to0}\frac{h}{e^{2/h}}=0$$

$$\Rightarrow \lim_{x\to0^-} f(x)=\lim_{x\to0^+} f(x)=f(0)$$

Therefore, $f(x)$ is continuous for all $x$.

Differentiability at $x=0$,

$$Lf'(0)=\lim_{h\to0}\frac{(-h)e^{-\left(\frac{1}{h}-\frac{1}{h}\right)}-0}{(-h)-0}$$

$$=\lim_{h\to0} e^{0}=1$$

$$Rf'(0)=\lim_{h\to0}\frac{he^{-(1/h+1/h)}-0}{h-0}$$

$$=\lim_{h\to0}\frac{1}{e^{2/h}}=0$$

$$\Rightarrow \quad Rf'(0)\neq Lf'(0)$$

Therefore, $f(x)$ is not differentiable at $x=0$.

**30.** *(b)* At $x=0$, RHL $=2$

$$\text{LHL}=\lim_{x\to0}\frac{a-b\cos2x}{x^2}$$

$$\Rightarrow \lim_{x\to0}\frac{a-b\left(1-\frac{(2x)^2}{2!}+\frac{(2x)^4}{4!}\dots\infty\right)}{x^2}=2$$

$\Rightarrow$ For LHL = RHL

$\Rightarrow$ $a-b=0$

$\Rightarrow$ $a=b$

$\Rightarrow$ $2b=2$

$\Rightarrow$ $b=1$

$\Rightarrow$ $a=b=1$

At $x=1$, $f(x)$ is continuous therefore

$$f(1^-)=f(1)=f(1^+)$$

$\Rightarrow$ $f(1)=3+c$

$\Rightarrow$ $f(1^+)=\lim_{h\to0}(2(1+h)+1)$

$\Rightarrow$ $f(1^+)=\lim_{h\to0}3+2h=3$

$\Rightarrow$ $3+c=3$

$\therefore$ $c=0$

$$f(x)=\begin{cases}\frac{1-\cos2x}{x^2}, & x<0\\ x^2+2, & 0\le x\le1\\ 2x+1, & x>1\end{cases}$$

At $x=1$, LHD $=2=$ RHD

$\Rightarrow f$ is differentiable at $x=1$

At $x=0$, RHD $=0$

$$\text{LHD}\Rightarrow \lim_{x\to0}\frac{1-\cos2x}{x^2}$$

$$=\lim_{h\to0}\frac{\frac{1-\cos2h}{h^2}-2}{-h}$$

$$=\lim_{h\to0}\frac{1-\left(1-\frac{4h^2}{2!}+\frac{16h^2}{4!}\dots\right)\dots2h^2}{-h^3}=0$$

$\Rightarrow$ LHD = RHD

$\Rightarrow f$ is differentiable at $x=0$

$\Rightarrow m=0$

$m+a+b+c=0+1+1+0=2$

**Alternate Solution**

$$f(x)=\begin{cases}\frac{a-b\cos2x}{x^2}, & x<0\\ x^2+c+2, & 0\le x\le1\\ 2x+1, & x>1\end{cases}$$

$f(x)$ is continuous every where

$$\lim_{x\to0^-}\frac{a-b\cos2x}{x^2}=2$$

$$\lim_{x\to0^-}\frac{2b\sin2x}{2x}=2$$

$\Rightarrow$ $b=1$

Also, $a-b=0$ $a=b=1$

$$\lim_{x\to1^-} x^2+c+2=\lim_{x\to1^+} 2x+1$$

$1+c+2=2+1$

$c=0$

$$f(x)=\begin{cases}\frac{1-\cos2x}{x^2}, & x<0\\ x^2+2, & 0\le x\le1\\ 2x+1, & x>1\end{cases}$$

$$f(x)=\begin{cases}\frac{2\sin^2x}{x^2}, & x<0\\ x^2+2, & 0\le x\le1\\ 2x+1, & x>1\end{cases}$$

$f(x)$ is differential for all value of $x$ is $m=0$

$m+a+b+c=0+1+1+0=2$

**31.** *(b)* Given function,

$$f(x)=\begin{cases}x^2\sin\left(\frac{1}{x}\right), & x\neq0\\ 0, & x=0\end{cases}$$

**To check continuity at $x=0$**

$$\text{LHL} \lim_{x\to0^-} f(x)=\lim_{h\to0} f(0-h)$$

$$=\lim_{h\to0}(-h)^2\cdot\sin\left(-\frac{1}{h}\right)=0$$

$$\text{RHL}=\lim_{x\to0^+} f(x)=\lim_{h\to0} f(0+h)$$

$$=\lim_{h\to0} h^2\sin\left(\frac{1}{h}\right)=0$$

Value of function $=f(0)=0$

Here, LHL = RHL $=f(0)$

$=f(x)$ is continuous at $x=0$.

**To check differentiability of $f(x)$ at $x=0$**

$$\text{RHD}=f'(0^+)=\lim_{h\to0}\frac{f(0+h)-f(0)}{h}$$

$$=\lim_{h\to0} h\sin\left(\frac{1}{h}\right)=0$$

$$\text{LHD}=f'(0^-)$$

$$=\lim_{h\to0}\frac{f(0-h)-f(0)}{-h}$$

$$=\lim_{h\to0} h\sin\left(\frac{1}{h}\right)=0$$

Here, LHD = RHD $=0$

$\therefore f(x)$ is differentiable at $x=0$.

Now, we differentiate $f(x)$,

$$\Rightarrow f'(x) = \begin{cases} 2x\sin\left(\frac{1}{x}\right) - \cos\left(\frac{1}{x}\right), & x \neq 0 \\ 0, & x = 0 \end{cases}$$

As, we know that $\cos(1/x)$ is highly oscillating at $x = 0$.

So, $f'(x)$ is not continuous at $x = 0$.

**32.** *(d)* Given, $f(x) = \begin{cases} \frac{1}{|x|}, & |x| \geq 1 \\ ax^2 + b, & |x| < 1 \end{cases}$

$$\Rightarrow f(x) = \begin{cases} \frac{1}{|x|}, & x \leq -1 \text{ or } x \geq 1 \\ ax^2 + b, & -1 < x < 1 \end{cases}$$

$$\Rightarrow f(x) = \begin{cases} \frac{-1}{x}, & x \leq -1 \\ ax^2 + b, & -1 < x < 1 \\ \frac{1}{x}, & x \geq 1 \end{cases}$$

Given, $f(x)$ is differentiable at every point of domain.

$$\therefore \quad f'(x) = \begin{cases} \frac{1}{x^2}, & x < -1 \\ 2ax, & -1 < x < 1 \\ \frac{-1}{x^2}, & x > 1 \end{cases}$$

$\because$ $f(x)$ is differentiatble at $x = 1$

$\therefore$ (LHD at $x = 1$) = (RHD at $x = 1$)

$\Rightarrow \quad f'(1^-) = f'(1^+)$

$\Rightarrow \quad 2a = -1 \Rightarrow a = -1/2$

As, we know that, a function is differentiable at $x = a$, if it is continuous at $x = a$.

Hence, $f(x)$ is also continuous at $x = 1$.

i.e., (LHL at $x = 1$) = (RHL at $x = 1$) $= f(1)$

$\Rightarrow \quad a + b = 1$

$\Rightarrow \quad \left(-\frac{1}{2}\right) + b = 1 \Rightarrow b = \frac{3}{2}$

Hence, $a = -\frac{1}{2}, b = \frac{3}{2}$

**Note** You can also (or apply) continuity and differentiability at $x = -1$.

**33.** *(b)*

> **key idea** (i) Use suitable method to find $f(g(x))$.
> (ii) If a function is discontinuous at any point, then it will be not differentiable at that point.
> (iii) A function will be differential at $x = a$ if LHD$|_{x = a^-}$ = RHD$|_{x = a^+}$ = Finite quantity

$$f(x) = \begin{cases} x + 2, & x < 0 \\ x^2, & x \geq 0 \end{cases}$$

$$g(x) = \begin{cases} x^3, & x < 1 \\ 3x - 2, & x \geq 1 \end{cases}$$

$$f[g(x)] = \begin{cases} g(x) + 2, & g(x) < 0 \\ g^2(x), & g(x) \geq 0 \end{cases}$$

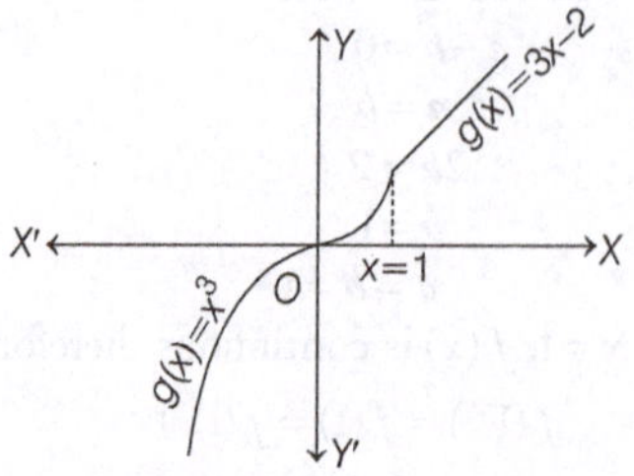


When $g(x) < 0 \Rightarrow g(x) = x^3, x < 0$

When $g(x) \geq 0$

$$\Rightarrow g(x) = \begin{cases} x^3, & 0 \leq x < 1 \\ 3x - 2, & x \geq 1 \end{cases}$$

$$f[g(x)] = \begin{cases} x^3 + 2, & x < 0 \\ x^6, & 0 \leq x < 1 \\ (3x-2)^2, & x \geq 1 \end{cases}$$

As, polynomial function is continuous everywhere in its domain. So, $f[g(x)]$ will be continuous everywhere at $x < 0$, $0 < x < 1$ and $x > 1$. We will check the behaviour of $fog(x)$ only at boundary points which is $x = 0$ and $x = 1$.

At $x = 0$, $\lim\limits_{x \to 0^-} (x^3 + 2) = 2$

$$\lim_{x \to 0^+} x^6 = 0$$

Clearly, LHL $\neq$ RHL at $x = 0$

So, $fog(x)$ is discontinuous at $x = 0$.

At $x = 1$, $\lim\limits_{x \to 1^-} x^6 = 1$

$$\lim_{x \to 1^+} (3x - 2)^2 = 1$$

Also $f(1) = 1$

$fog(x)$ is continuous at $x = 1$

Derivative test at $x = 1$,

$$\text{LHD} = \lim_{h \to 0} \frac{f(1) - f(1-h)}{h}$$

$$= \lim_{h \to 0} \frac{1 - (1-h)^6}{h}$$

$$= \lim_{h \to 0} 6(1-h)^5 = 6$$

$$\text{RHD} = \lim_{h \to 0} \frac{f(1+h) - f(1)}{h}$$

$$= \lim_{h \to 0} \frac{[3(1+h) - 2]^2 - 1}{h}$$

$$= \lim_{h \to 0} 2[3(1+h) - 2] \cdot 3$$

$$= 6$$

$\therefore$ $fog(x)$ is continuous and differentiable at $x = 1$.

$\therefore$ $fog(x)$ is discontinuous and non-differentiable at $x = 0$.

So, number of points of non-differentiability of $fog(x)$ is 1.

**34.** *(a)*

> **key idea** This type of problem can be solved graphically

We have,

$$f(x) = \begin{cases} -1, & -2 \leq x < 0 \\ x^2 - 1, & 0 \leq x \leq 2 \end{cases}$$

and $\quad g(x) = |f(x)| + f(|x|)$

$$\text{Clearly, } |f(x)| = \begin{cases} 1, & -2 \leq x < 0 \\ |x^2 - 1|, & 0 \leq x \leq 2 \end{cases}$$

$$= \begin{cases} 1, & -2 \leq x < 0 \\ -(x^2 - 1), & 0 \leq x < 1 \\ x^2 - 1, & 1 \leq x \leq 2 \end{cases}$$

and $f(|x|) = |x|^2 - 1, \ 0 \leq |x| \leq 2$

[$\because$ $f(|x|) = -1$ is not possible as $|x| \not< 0$]

$= x^2 - 1, \quad |x| \leq 2 \qquad [\because |x|^2 = x^2]$

$= x^2 - 1, \quad -2 \leq x \leq 2$

$\therefore$ $g(x) = |f(x)| + f(|x|)$

$$= \begin{cases} 1 + x^2 - 1, & -2 \leq x < 0 \\ -(x^2 - 1) + x^2 - 1, & 0 \leq x < 1 \\ x^2 - 1 + x^2 - 1, & 1 \leq x \leq 2 \end{cases}$$

$$= \begin{cases} x^2, & -2 \leq x < 0 \\ 0, & 0 \leq x < 1 \\ 2(x^2 - 1), & 1 \leq x \leq 2 \end{cases}$$

Now, let us draw the graph of $y = g(x)$, as shown in the figure.

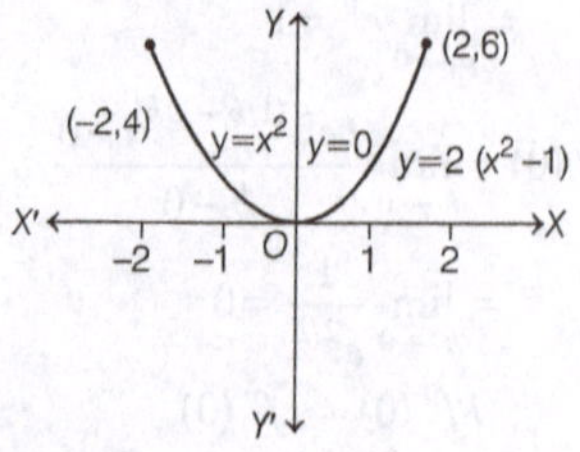


[Here, $y = 2(x^2 - 1)$

or $x^2 = \frac{1}{2}(y+2)$ represent a parabola with vertex (0, – 2) and it is open upward]

Note that there is a sharp edge at $x = 1$ only, so $g(x)$ is not differentiable at $x = 1$ only.

**35.** *(a)*

**key idea** This type of Questions can be solved graphically

Given, $f: (-1, 1) \to R$, such that

$$f(x) = \max\{-|x|, -\sqrt{1-x^2}\}$$

On drawing the graph, we get the follwong figure.

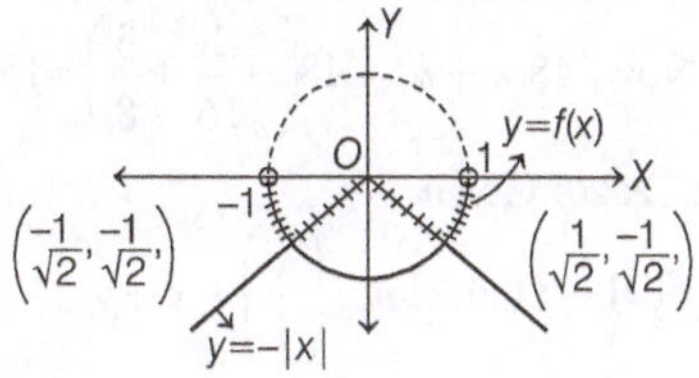


$\because$ Graph of $y = -|x|$ is

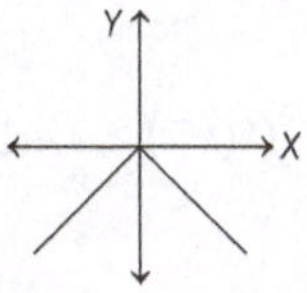


and graph of $y = -\sqrt{1-x^2}$

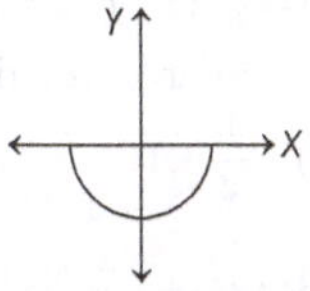


[$\because x^2 + y^2 = 1$ represent a complete circle]

$$\Rightarrow f(x) = \begin{cases} -\sqrt{1-x^2}, & -1 < x \le -\frac{1}{\sqrt{2}} \\ -|x|, & -\frac{1}{\sqrt{2}} < x \le \frac{1}{\sqrt{2}} \\ -\sqrt{1-x^2}, & \frac{1}{\sqrt{2}} < x < 1 \end{cases}$$

From the figure, it is clear that function have sharp edges, at $x = -\frac{1}{\sqrt{2}}, 0, \frac{1}{\sqrt{2}}$

$\therefore$ Function is not differentiable at 3 points.

**36.** *(a)* We have,

$$f(x) = \begin{cases} \max\{|x|, x^2\}, & |x| \le 2 \\ 8 - 2|x|, & 2 < |x| \le 4 \end{cases}$$

Let us draw the graph of $y = f(x)$

For $|x| \le 2$ $f(x) = \max\{|x|\ x^2\}$

Let us first draw the graph of $y = |x|$ and $y = x^2$ as shown in the following figure.

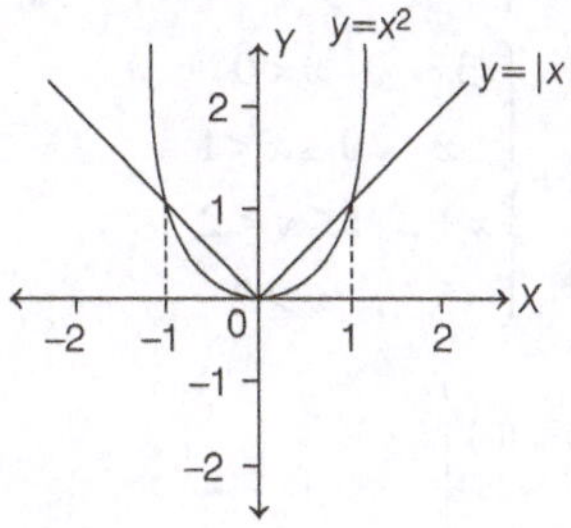


Clearly, $y = |x|$

and $y = x^2$ intersect at $x = -1, 0, 1$

Now, the graph of $y = \max\{|x|, x^2\}$ for $|x| \le 2$ is

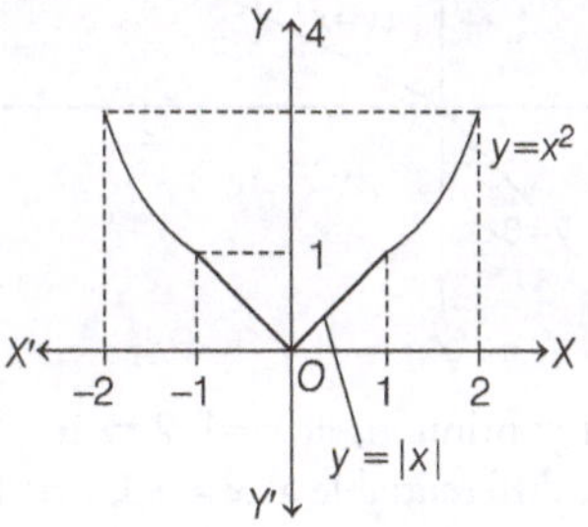


For $|x| \in (2, 4]$

$$f(x) = 8 - 2|x|$$
$$= \begin{cases} 8 - 2x, & x \in (2, 4] \\ 8 + 2x, & x \in [-4, -2) \end{cases}$$

[$\because 2 < |x| \le 4$
$\Rightarrow |x| > 2$ and $|x| \le 4$

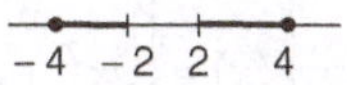

]

Hence, the graph of $y = f(x)$ is

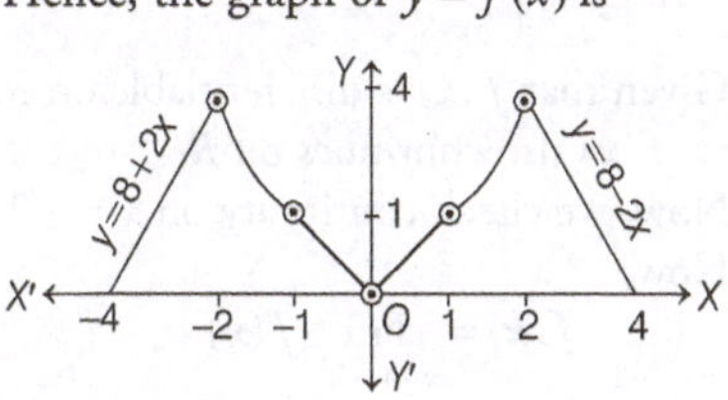


From the graph it is clear that at $x = -2, -1, 0, 1, 2$ the curve has sharp edges and hence at these points $f$ is not differentiable.

**37.** *(c)* **Given** A function $f$ such that $f(x) = |x-2| + |x-5|$.

**To discuss** Continuity and differentiability of $f$ in interval (2, 5).

We know that

$$|x-2| = \begin{cases} x-2 & x \ge 2 \\ 2-x, & x < 2 \end{cases}$$

and $|x-5| = \begin{cases} x-5, & x \ge 5 \\ 5-x, & x < 5 \end{cases}$

$$\Rightarrow \quad |x-2| = \begin{cases} 2-x, & x < 2 \\ x-2, & 2 \le x \le 5 \\ x-2, & x > 5 \end{cases}$$

and $|x-5| = \begin{cases} 5-x, & x < 2 \\ 5-x, & 2 \le x \le 5 \\ x-5, & x > 5 \end{cases}$

$$\Rightarrow f(x) = |x-2| + |x-5|$$
$$= \begin{cases} (2-x) + (5-x), & x < 2 \\ (x-2) + (5-x), & 2 \le x \le 5 \\ (x-2) + (x-5), & x > 5 \end{cases}$$
$$= \begin{cases} 7 - 2x, & x < 2 \\ 3, & 2 \le x \le 5 \\ 2x - 7, & x > 5 \end{cases}$$

Now, we can draw the graph of $f$ very easily.

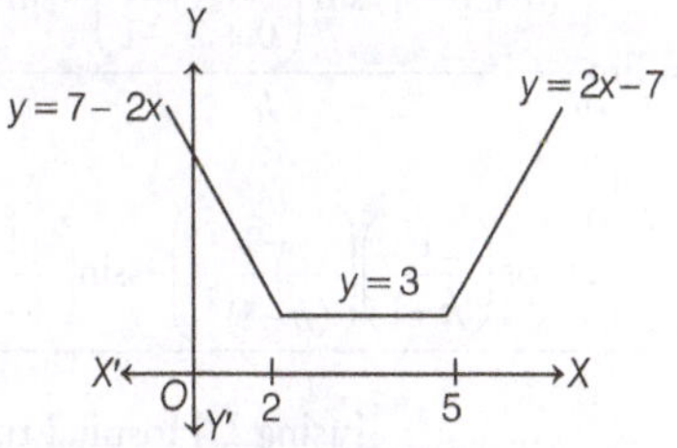


From the above graph, we can analyse all the required things.

**Statement I** $f'(4) = 0$

It is obviously clear that $f$ is constant around $x = 4$, hence $f'(4) = 0$. Hence, Statement I is correct.

**Statement II** It can be clearly seen that

(i) $f$ is continuous, $\forall x \in [2, 5]$.

(ii) $f$ is differentiable, $\forall x \in (2, 5)$.

(iii) $f(2) = f(5) = 3$

Hence, Statement II is also correct but obviously not a correct explanation of Statement I.

**38.** *(d)* Now,

$$f'(1^-) = \lim_{h \to 0} \frac{f(1-h) - f(1)}{-h}$$

$$= \lim_{h \to 0} \frac{(1-h-1) \cdot \sin\left(\frac{1}{1-h-1}\right) - 0}{-h}$$

$$= \lim_{h \to 0} \sin\left(-\frac{1}{h}\right) = -\lim_{h \to 0} \sin\frac{1}{h}$$

and $f'(1^+) = \lim_{h \to 0} \dfrac{f(1+h) - f(1)}{h}$

$$= \lim_{h \to 0} \frac{(1+h-1)\sin\left(\frac{1}{1+h-1}\right) - 0}{h}$$

$$= \lim_{h \to 0} \sin\frac{1}{h} \Rightarrow f'(1^-) \neq f'(1^+)$$

Hence, $f$ is not differentiable at $x = 1$.

Again, $f'(0)$

$$= \lim_{h \to 0} \frac{(0+h-1)\sin\left(\frac{1}{0+h-1}\right) - \sin 1}{-h}$$

$$= \lim_{h \to 0} \frac{\left[\left\{(h-1)\cos\left(\frac{1}{h-1}\right) \times \left(\frac{-1}{(h-1)^2}\right)\right\} + \sin\left(\frac{1}{h-1}\right)\right]}{-1}$$

[using L' Hospital rule]

$= \cos 1 - \sin 1$ and $f'(0^+)$

$$= \lim_{h \to 0} \frac{(0+h-1)\sin\left(\frac{1}{0+h-1}\right) - \sin 1}{h}$$

$$= \lim_{h \to 0} \frac{(h-1)\cos\left(\frac{1}{h-1}\right)\left(\frac{-1}{(h-1)^2}\right) + \sin\left(\frac{1}{h-1}\right)}{1}$$

[using L'Hospital rule]

$= \cos 1 - \sin 1$

$\Rightarrow \quad f'(0^-) = f'(0^+)$

Hence, $f$ is differentiable at $x = 0$.

**39.** *(3)* $f(x) = \max\{x, x^3, x^5, \ldots, x^{21}\}$

$$f(x) = \begin{cases} x & x \le -1 \\ x^{21} & -1 < x < 0 \\ x & 0 < x < 1 \\ x^{21} & x \ge 1 \end{cases}$$

$\because f(x)$ is continuous at everywhere now $n = 0$

$$f'(x) = \begin{cases} 1 & x < -1 \\ 21x^{20} & -1 < x < 0 \\ 1 & 0 < x < 1 \\ 21x^{20} & x > 1 \end{cases}$$

$\therefore f'(-1^-) \neq f'(-1^+)$ and $f'(1^+) \neq f'(1^-)$ and $f'(0^+) \neq f'(0^-)$

$\therefore$ Not differentiable at 3 points

$\therefore m = 3$

$\therefore m + n = 3$

**40.** *(5)*

$$f(x) = \begin{cases} 3x &, \quad x < 0 \\ \min\{1+x, [x]\} &, \quad 0 \le x < 1 \\ \min\{2+x, x+2\} &, \quad 1 \le x < 2 \\ 5 &, \quad x > 2 \end{cases}$$

$$f(x) = \begin{cases} 3x \;, & x < 0 \\ x \;, & 0 \le x < 1 \\ x+2 \;, & 1 \le x < 2 \\ 5 \;, & x > 2 \end{cases}$$

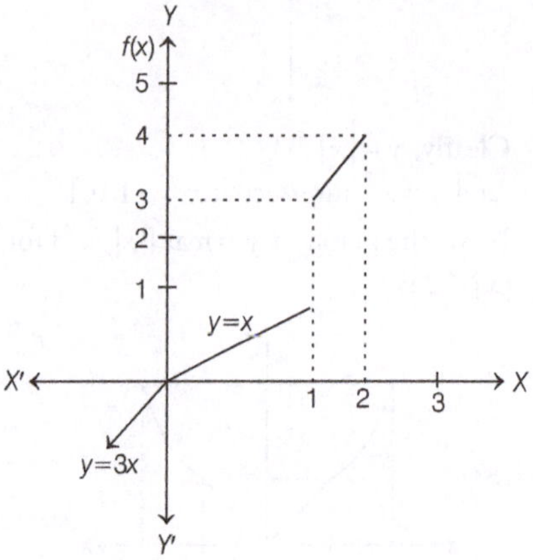


Not continuous at $x = 1, 2 \Rightarrow \alpha = 2$

Not differentiable at $x = 0, 1, 2 \Rightarrow \beta = 3$

$\therefore \quad \alpha + \beta = 5$

**41.** *(15)* Given,

$$f(x) = \begin{cases} \dfrac{1}{|x|}, & |x| \ge 2 \\ ax^2 + 2b, & |x| < 2 \end{cases}$$

$$\Rightarrow f(x) = \begin{cases} -\dfrac{1}{x}, & x < -2 \\ ax^2 + 2b, & -2 < x < 2 \\ 1/x, & x \ge 2 \end{cases}$$

Given that $f(x)$ is differentiable on $R$.

$\Rightarrow f(x)$ also continues on $R$.

Now, we check continuity on $x = -2$.

Now,

$$\lim_{x \to -2^-} f(x) = \lim_{x \to -2^+} f(x)$$

$$\Rightarrow \lim_{x \to -2} -\frac{1}{x} = \lim_{x \to -2^+} (ax^2 + 2b)$$

$$\Rightarrow 4a + 2b = \frac{1}{2} \qquad \ldots \text{(i)}$$

Now, LHD = RHD at $x = -2$

$$f'(x) = \begin{cases} 1/x^2, & x \le -2 \\ 2ax, & -2 < x < 2 \\ -1/x^2, & x \ge 2 \end{cases}$$

$$\lim_{x \to 2^-} f'(x) = \lim_{x \to 2^+} f'(x)$$

$$\Rightarrow \quad \lim_{x \to -2^-}\left(\frac{1}{x^2}\right) = \lim_{x \to -2^+} (2ax)$$

$$\Rightarrow \quad \frac{1}{4} = -4a$$

$$\Rightarrow \quad a = -1/16$$

Put $a = -1/16$ in Eq. (i),

$$4\left(-\frac{1}{16}\right) + 2b = \frac{1}{2}$$

$$\Rightarrow \quad b = 3/8$$

Now, $48(a+b) = 48\left(-\dfrac{1}{16} + \dfrac{3}{8}\right) = 15$

**42.** *(2890)* Given,

$$f(x) - f(y) \ge \log_e\left(\frac{x}{y}\right) + x - y$$

$$\Rightarrow \frac{f(x) - f(y)}{x - y} \ge \frac{\ln(x) - \ln(y)}{x - y} + 1$$

Let $x > y$

Then, $\lim_{y \to x} f'(x^-) \ge \dfrac{1}{x} + 1 \qquad \ldots$(i)

Let $x < y$

Then, $\lim_{y \to x} f'(x^+) \le \dfrac{1}{x} + 1 \qquad \ldots$(ii)

$f'(x^-) = f'(x^+)$

[$\because f$ is differentiable function in $(0, \infty)$]

So, $\quad f'(x) = \dfrac{1}{x} + 1$

$$\Rightarrow f'\left(\frac{1}{x^2}\right) = x^2 + 1$$

Now, $\displaystyle\sum_{x=1}^{20} (x^2 + 1) = \sum_{x=1}^{20} x^2 + \sum_{x=1}^{20} 1$

$$= \frac{20 \times 21 \times 41}{6} + 20 = 2890$$

**43.** *(4)* Given function is

$$f(x) = \begin{cases} x[x], & -2 < x < 0 \\ (x-1)[x], & 0 \le x < 2 \end{cases}$$

When $[x]$ is denotes greatest integer function

$$f(x) = \begin{cases} -2x; & -2 < x < -1 \\ -x; & -1 \le x < 0 \\ 0; & 0 \le x < 1 \\ x - 1; & 1 \le x < 2 \end{cases}$$

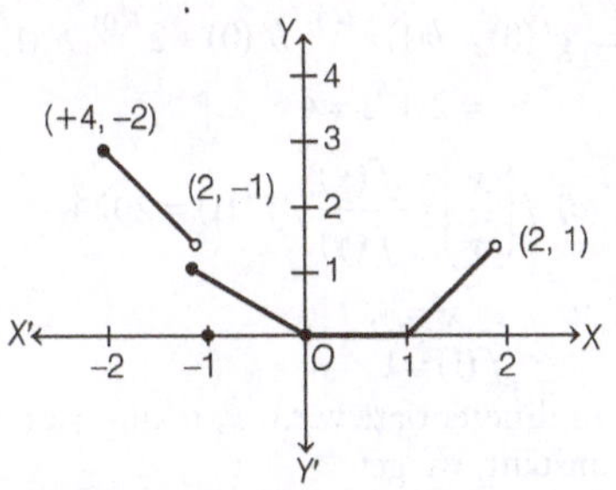


Clearly $|f(x)|$ remains same.
Given that, $m$ and $n$ respectively the number points in $(-2, 2)$ at which $y=|f(x)|$ is not continuous and not differentiable.
So, $m=1$ where $y=|f(x)|$ not continuous
and $n=3$ when $|f(x)|$ is not differentiable.
Thus $m+n=4$

**44.** (309) $\because f(x)$ is continuous at $x=1$

$\therefore \quad f(1^-)=f(1^+)$

$\Rightarrow \quad 3+k\sqrt{2}=m+k^2 \quad \ldots$ (i)

$$f'(1^-)=\left[6x+\frac{k}{2\sqrt{x+1}}\right]_{x=1^-}=6+\frac{k}{2\sqrt{2}}$$

$$f'(1^+)=[2mx]_{x=1^+}=2m$$

$f(x)$ is differentiability at $x=1$

$$6+\frac{k}{2\sqrt{2}}=2m \quad \ldots \text{(ii)}$$

From Eqs. (i) and (ii), we get

$$k=\frac{7}{4\sqrt{2}} \text{ and } m=\frac{103}{32}$$

$$f'(8)=[2mx]_{x=8}=16m$$

$$=16\times\frac{103}{32}=\frac{103}{2}$$

$$f'\left(\frac{1}{8}\right)=\left[6x+\frac{k}{2\sqrt{x+1}}\right]_{x=\frac{1}{8}}$$

$$=\left[6\times\frac{1}{8}+\frac{k}{2\sqrt{\frac{1}{8}+1}}\right]$$

$$=\frac{3}{4}+\frac{7}{4\sqrt{2}}\times\frac{\sqrt{2}}{3}=\frac{4}{3}$$

$$\therefore \quad \frac{8f'(8)}{f'\left(\frac{1}{8}\right)}=\frac{8\times\frac{103}{2}}{\frac{4}{3}}=4\times 103\times\frac{3}{4}$$

$$=309$$

**45.** (25) We have,
$f(x)=[a+13\sin x], x\in(0,\pi)$
$[13\sin x]$ is not differentiable, where $13\sin x$ is an integer.
$\therefore$ Total number of point where $f(x)$ is not differential $=2\times 13-1$
$=25$
$[\because f(x)=[n+p\sin x], 0<x<\pi, n\in I.$ and $p$ is a prime number, then the number of points where $f(x)$ is not differentiable $=2p-1]$

**46.** (79) $f(x)=4\,|2x+3|+9\left[x+\frac{1}{2}\right]$
$-12\,[x+20]\ ;\ x\in(-20, 20)$
$f(x)$ is not differentiating at
$x=I\in\{-19,-18,\ldots.0\ldots.19\}$ 39 points
At $x=I+\frac{1}{2}$, $f(x)$ is non-differentiating at 39 points.
Also $f\times N$ is non-differentiating at $x=-\frac{3}{2}$
So, total points of non-differentiability $=39+39+1=79$

**47.** (4) Given, $f$ and $g$ are twice differentiable even functions.

$$\therefore f\left(\frac{1}{4}\right)=f\left(-\frac{1}{4}\right)=f\left(\frac{1}{2}\right)$$
$$=f\left(-\frac{1}{2}\right)=0$$

$$g\left(\frac{3}{4}\right)=g\left(-\frac{3}{4}\right)=0$$

It means $f(x)$ has atleast 4 roots and $g(x)$ has atleast 2 roots as per the given information.
Let $h(x)=f(x)\,g''(x)\rightarrow 4$ roots
Let $p(x)=f'(x)g'(x)\rightarrow 4$ roots
$[\because f'(x)$ has atleast 3 roots, $g'(x)$ has atleast 1 root]
Hence, the given expression $f(x)g''(x)+f'(x)\,g'(x)$ has atleast 4 roots.

**48.** (5) Given, $f(x)=\min\{x-[x], 1+[x]-x\}$
$\therefore f(x)=\min(\{x\}, 1-\{x\})$

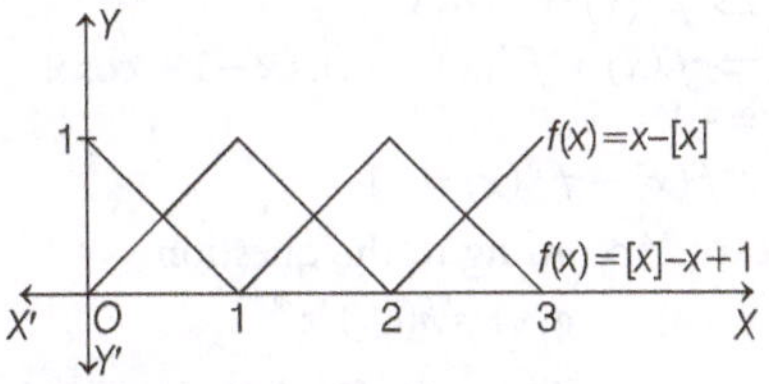


So, the graph of $f(x)$ will be

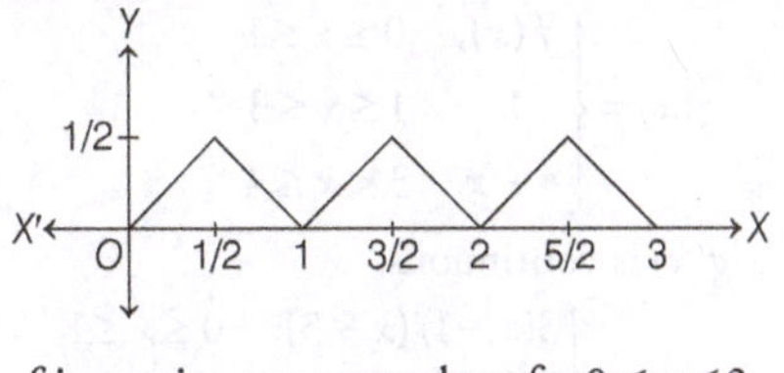


$f$ is continuous everywhere for $0\le x\le 3$.
But $f$ is non-differentiable at $x=\frac{1}{2},\frac{3}{2},\frac{5}{2}$ and $x=1, 2$. So, if set $P$ denotes the points of discontinuity, then $n(P)=0$.
And if set $Q$ denotes the points of non-differentiable, then
$$n(Q)=5$$
$$\therefore \quad n(P)+n(Q)=0+5=5$$

**49.** (4) $f(x+2)$

$$=\begin{cases}\frac{3x}{2}, & -4\le x\le -2\\ \frac{-3x}{2}, & -2<x\le 0\\ 0, & x\in(-\infty,-4)\cup(0,\infty)\end{cases}$$

amd $f(x-2)$

$$=\begin{cases}\frac{3x}{2}, & 0\le x\le 2\\ \frac{-3x}{2}+6, & 2<x\le 4\\ 0, & x\in(-\infty,0)\cup(4,\infty)\end{cases}$$

$\therefore g(x)=f(x+2)-f(x-2)$

$$=\begin{cases}\frac{3x}{2}+6, & -4\le x\le -2\\ \frac{-3x}{2}, & -2<x<2\\ \frac{3x}{2}-6, & 2\le x\le 4\\ 0, & x\in(-\infty,-4)\cup(0,\infty)\end{cases}$$

$n=0\Rightarrow m=4\Rightarrow(m+n=4)$

**50.** (1) Let $f(x)=x^3-6x^2+9x-3$
$$f'(x)=3x^2-12x+9$$
$f'(x)=0$ gives
$$3x^2-12x+9=0$$
$\Rightarrow \quad 3(x-1)(x-3)=0$
$\therefore \quad x=1$ or $x=3$
Now, $f(1)=1$ and $f(3)=-3$

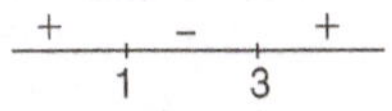

$$g(x)=\begin{cases} f(x), & 0\le x\le 1 \\ 1, & 1\le x\le 3 \\ 4-x, & 3<x\le 4 \end{cases}$$

$g(x)$ is continuous.

$$g'(x)=\begin{cases} 3(x-1)(x-3), & 0\le x\le 1 \\ 0, & 1\le x\le 3 \\ -1, & 3<x\le 4 \end{cases}$$

$g(x)$ is non-differentiable at $x=3$.

So, the number of points in (0, 4) where $g(x)$ is not differentiable at 1.

**51.** *(5)* For this particular problem, try to draw graph in the region $(-3, 3)$, it will be as follows,

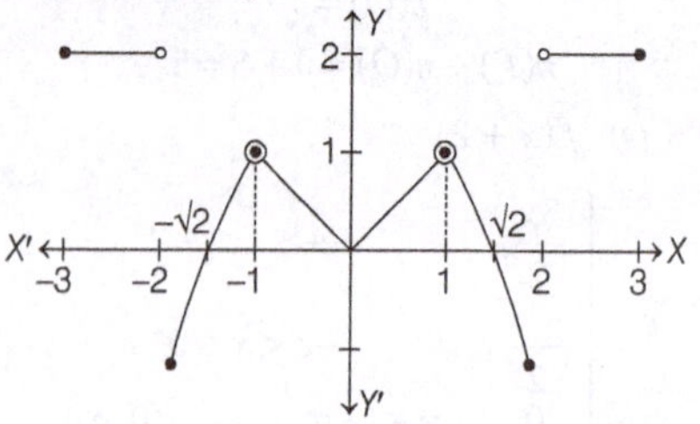


Thus, points of discontinuity will be at −2, 2 because the curve breaks at these points and at −1, 0, 1 $f(x)$ is not differentiable because curve has sharp points.

$\therefore$ Point where $f$ is not differentiable are −2, −1, 0, 1, 2 i.e. 5 points.

**52.** *(2)* Given,

$f(x)=|2x+1|-3|x+2|+|x^2+x-2|$
$=|2x+1|-3|x+2|+|x+2|\times|x-1|$

Here, critical points are $x=\dfrac{-1}{2}, -2, 1$

$\therefore f(x)$

$$=\begin{cases} x^2+2x+3, & x\le 2 \\ -x^2-6x-5, & -2<x<\dfrac{-1}{2} \\ -x^2-2x-3, & \dfrac{-1}{2}\le x<1 \\ x^2-7, & x\ge 1 \end{cases}$$

Now,

$$f'(x)=\begin{cases} 2x+2, & x\le -2 \\ -2x-6, & -2<x<\dfrac{-1}{2} \\ -2x-2, & \dfrac{-1}{2}\le x<1 \\ 2x, & x\ge 1 \end{cases}$$

Now, $f'(x)$ at 1, −2 and −1/2.

For $x=1$,

$f'(x)=2x=2\times 1=2$

and $-2x-2=-(2\times 1)-2=-4$

both are not equal.

$\therefore$ Non-differentiable at $x=1$

Similarly, for $x=-2$,

$f'(x)=2x+2=2\times(-2)+2=-2$

and $-2x-6=-2\times(-2)-6=-2$ both are equal.

$\therefore$ Differentiable at $x=-2$

and for $x=-1/2$, $f'(x)=-2x-6$

$=-2\times\left(\dfrac{-1}{2}\right)-6=-5$

and $-2x-2=-2\times\left(\dfrac{-1}{2}\right)-2=-1$

both are not equal.

$\therefore$ Non-differentiable at $x=-1/2$

$\therefore$ The number of points at which $f(x)$ is non-differentiable is 2.

**53.** *(3)* Given function $f(x)=|2-|x-3||$

$$=\begin{cases} |2+(x-3)|, & x<3 \\ |2-(x-3)|, & x\ge 3 \end{cases} = \begin{cases} |x-1|, & x<3 \\ |5-x|, & x\ge 3 \end{cases}$$

$$=\begin{cases} 1-x, & x<1 \\ x-1, & 1\le x<3 \\ 5-x, & 3\le x<5 \\ x-5, & x\ge 5 \end{cases}$$

$\because$ Function $f(x)$ is not differentiable at $x=1, 3, 5$

$\therefore \quad S=\{1, 3, 5\}$

Now $\sum_{x\in S} f(f(x))=f(f(1))+f(f(3))+f(f(5))$

$=f(1-1)+f(5-3)+f(5-5)$
$=f(0)+f(2)+f(0)$
$=(1-0)+(2-1)+(1-0)=1+1+1=3$

## Topic 3

**01.** *(b)*

$$f(x)=\begin{vmatrix} \sin x & \cos x & \sin x+\cos x+1 \\ 27 & 28 & 27 \\ 1 & 1 & 1 \end{vmatrix}$$

$f(x)=\sin x(1)-\cos x(0)$
$+(\sin x+\cos x+1)(-1)$
$\Rightarrow f(x)=\sin x-\sin x-\cos x-1$
$\Rightarrow f(x)=-\cos x-1 \Rightarrow f'(x)=\sin x$
$\Rightarrow f''(x)=-\cos x$
$\Rightarrow f(x)+f''(x)=-\cos x-1+\cos x$
$=-1$
$\therefore f(x)+f''(x)=-1$

**02.** *(c)* According to the question,

$g(x)=h(e^x)\cdot e^{h(x)}$

$\Rightarrow \quad g'(x)=h(e^x)\cdot e^{h(x)}\cdot h'(x)+e^{h(x)}\cdot h'(e^x)\cdot e^x$

$\Rightarrow g'(0)=h(1)\,e^{h(0)}\,h'(0)+e^{h(0)}\,h'(1)$
$=2+2=4$

**03.** *(d)* $f\left(\dfrac{x}{y}\right)=\dfrac{f(x)}{f(y)}$, $f'(1)=2024$

Put $\quad x=y,$
$f(1)=1$

On differentiate w.r.t. $x$, taking $y$ as constant, we get

$f'\left(\dfrac{x}{y}\right)\cdot\dfrac{1}{y}=\dfrac{f'(x)}{f(y)}$

Put $\quad x=y$

$f'(1)\dfrac{1}{x}=\dfrac{f'(x)}{f(x)}$

$\Rightarrow \quad \dfrac{2024}{x}=\dfrac{f'(x)}{f(x)}$

$\Rightarrow \quad 2024f(x)=xf'(x)$

$\Rightarrow x f'(x)-2024f(x)=0$

**04.** *(b)* It is given that

$(a+\sqrt{2}b\cos x)(a-\sqrt{2}b\cos y)$
$=a^2-b^2$, where $a>b>0$.

On differentiating w.r.t. $y$, we get

$(a-\sqrt{2}b\cos y)\left(0+\sqrt{2}b(-\sin x)\dfrac{dx}{dy}\right)$
$+(a+\sqrt{2}b\cos x)(0-\sqrt{2}b(-\sin y))=0$

$\Rightarrow \quad (a-\sqrt{2}b\cos y)\left(-\sqrt{2}b(\sin x)\dfrac{dx}{dy}\right)$
$+\sqrt{2}b(\sin y)(a+\sqrt{2}\,b\cos x)=0$

At $\left(\dfrac{\pi}{4}, \dfrac{\pi}{4}\right)$, we get

$(a-b)\left(-b\dfrac{dx}{dy}\right)+b(a+b)=0$

$\Rightarrow \quad \dfrac{dx}{dy}=\dfrac{a+b}{a-b}$

**05.** *(a)* It is given that $fog(x)=x$, then

$gof(x)=x$

$\Rightarrow \quad g'(f(x))f'(x)=1$

On putting $x=b$, we get

$g'(f(b))f'(b)=1$

$\Rightarrow \quad f'(b)=\dfrac{1}{g'(f(b))}$

$gof(x)=x$

$\Rightarrow \quad g\,f(b)=b \quad$ [On putting $x=b$]

$\Rightarrow \quad g(f(b))=g(a) \quad [\because g(a)=b]$

$\Rightarrow \quad f(b)=a$

$\therefore \quad f'(b)=\dfrac{1}{g'(a)}=\dfrac{1}{5} \quad$ [given]

**06.** (c) It is given that $x^k + y^k = a^k$, $(a, k > 0)$

On differentiating both sides w.r.t. $x$,

we get $kx^{k-1} + ky^{k-1}\dfrac{dy}{dx} = 0$

$$\Rightarrow \frac{dy}{dx} + \left(\frac{x}{y}\right)^{k-1} = 0, \text{ [as, } k \neq 0] \quad \ldots(i)$$

$$\text{Since, } \frac{dy}{dx} + \left(\frac{y}{x}\right)^{1/3} = 0 \text{ (given)} \quad \ldots(ii)$$

On comparing Eqs. (i) and (ii), we get

$$k - 1 = -\frac{1}{3} \Rightarrow k = \frac{2}{3}$$

**07.** (b) Given functions,

$f(x) = \log_e(\sin x)$, $(0 < x < \pi)$

and $g(x) = \sin^{-1}(e^{-x})$, $x \geq 0$.

Now, $fog(x) = f(g(x)) = f(\sin^{-1}(e^{-x}))$

$= \log_e(\sin(\sin^{-1}(e^{-x}))) = \log_e(e^{-x})$

$\{\because \sin(\sin^{-1} x) = x, \text{ if } x \in [-1, 1]\}$

$= -x \quad \ldots(i)$

and $(fog)'(x) = \dfrac{d}{dx}(-x) = -1 \quad \ldots(ii)$

According to the question,

$\because a = (fog)'(\alpha) = -1$ [from Eq. (ii)]

and $b = (fog)(\alpha) = -(\alpha)$ [from Eq. (i)]

for a positive real value '$\alpha$'.

Since, the value of $a = -1$ and $b = -\alpha$, satisfy the quadratic equation (from the given options)

$$a\alpha^2 - b\alpha - a = 1.$$

**08.** (d) Let $y = f(f(f(x))) + (f(x))^2$

On differentiating both sides w.r.t. $x$, we get

$$\frac{dy}{dx} = f'(f(f(x))) \cdot f'(f(x)) \cdot f'(x) + 2f(x)f'(x) \text{ [by chain rule]}$$

So,

$$\left.\frac{dy}{dx}\right|_{\text{at } x=1} = f'(f(f(1))) \cdot f'(f(1)) \cdot f'(1) + 2f(1)f'(1)$$

$$\therefore \left.\frac{dy}{dx}\right|_{x=1} = f'(f(1)) \cdot f'(1) \cdot (3) + 2(1)(3) \quad [\because f(1) = 1 \text{ and } f'(1) = 3]$$

$= f'(1) \cdot (3) \cdot (3) + 6 = (3 \times 9) + 6$

$= 27 + 6 = 33$

**09.** (b) Given, $f(x) = \begin{vmatrix} \cos x & x & 1 \\ 2\sin x & x^2 & 2x \\ \tan x & x & 1 \end{vmatrix}$

$f(x) = \cos x(x^2 - 2x^2)$

$\quad - x(2\sin x - 2x\tan x)$

$\quad + 1(2x\sin x - x^2\tan x)$

(expanding along $R_1$)

$= -x^2\cos x + x^2\tan x$

$= x^2(\tan x - \cos x)$

$f'(x) = 2x(\tan x - \cos x) + x^2(\sec^2 x + \sin x)$

(differentiating w.r.t. $x$)

$$\lim_{x \to 0} \frac{f'(x)}{x}$$

$$= \lim_{x \to 0}\left[\frac{2x(\tan x - \cos x)}{x} + \frac{x^2(\sec^2 x + \sin x)}{x}\right]$$

(taking limit as $x \to 0$)

$= 2(-1) + 0 = -2$

**10.** (a) $f(x) = \sin^{-1}\left(\dfrac{2 \times 3^x}{1 + 9^x}\right)$

Put $3^x = \tan t \Rightarrow t = \tan^{-1} 3^x$

$$\therefore f(x) = \sin^{-1}\left(\frac{2\tan t}{1 + \tan^2 t}\right)$$

$= \sin^{-1}(\sin 2t) = 2t = 2\tan^{-1}(3^x)$

$$\Rightarrow f'(x) = \frac{2}{1 + 9^x} \times 3^x \log_e 3$$

$$\Rightarrow f'\left(-\frac{1}{2}\right) = \sqrt{3} \times \frac{1}{2}\log_e 3$$

$= \sqrt{3}\log_e\sqrt{3}$

**11.** (d) Let $u = \tan^{-1}\left(\dfrac{6x\sqrt{x}}{1 - 9x^3}\right)$,

$$x \in \left(0, \frac{1}{4}\right)$$

$$= \tan^{-1}\left(\frac{2 \cdot (3x^{3/2})}{1 - (3x^{3/2})^2}\right)$$

$= 2\tan^{-1}(3x^{3/2})$

This holds as $3x^{3/2} \in \left(0, \dfrac{3}{8}\right)$

Differentiating w.r.t. $x$, we obtain

$$\frac{du}{dx} = 2 \cdot \frac{1}{1 + 9x^3} \cdot 3 \cdot \frac{3}{2} \cdot x^{\frac{1}{2}} = \frac{9\sqrt{x}}{1 + 9x^3}$$

$$\Rightarrow \sqrt{x} \cdot g(x) = \frac{9\sqrt{x}}{1 + 9x^3}$$

$$\Rightarrow g(x) = \frac{9}{1 + 9x^3}$$

**12.** (c) According to the question,

$f(g(x)) = x \Rightarrow f'(g(x))g'(x) = 1$

(differentiating w.r.t. $x$ on both sides)

$$\Rightarrow g'(x) = \frac{1}{f'(g(x))}$$

$$\Rightarrow g'(x) = 1 + \{g(x)\}^5$$

**13.** (a) Given, $y = \sec(\tan^{-1} x)$

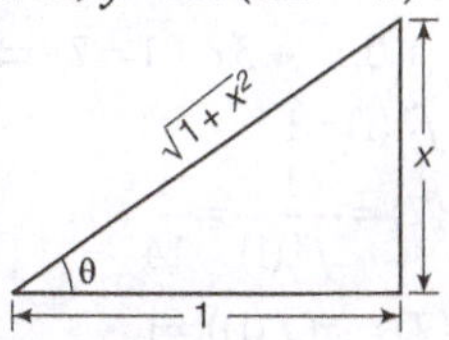


Let $\tan^{-1} x = \theta \Rightarrow x = \tan\theta$

$$\therefore \quad y = \sec\theta = \sqrt{1 + x^2}$$

On differentiating w.r.t. $x$, we get

$$\frac{dy}{dx} = \frac{1}{2\sqrt{1 + x^2}} \cdot 2x$$

$$\text{At } x = 1, \quad \frac{dy}{dx} = \frac{1}{\sqrt{2}}$$

**14.** (a) $x^m \times y^n = (x + y)^{m+n}$

Taking log both sides, we get

$m\log x + n\log y = (m + n)\log(x + y)$

Differentiating w.r.t $x$, we get

$$\frac{m}{x} + \frac{n}{y}\frac{dy}{dx} = \frac{m + n}{x + y}\left(1 + \frac{dy}{dx}\right)$$

$$\Rightarrow \frac{dy}{dx}\left(\frac{n}{y} - \frac{m + n}{x + y}\right) = \frac{m + n}{x + y} - \frac{m}{x}$$

$$\Rightarrow \frac{dy}{dx}\left(\frac{nx + ny - my - ny}{y(x + y)}\right) = \frac{mx + nx - mx - my}{x(x + y)}$$

$$\Rightarrow \frac{dy}{dx} = \left(\frac{nx - my}{nx - my}\right)\frac{y}{x}$$

$$\Rightarrow \frac{dy}{dx} = \frac{y}{x}$$

**15.** (d) Since, $x^y = e^{x-y}$

Taking log on both sides, we get

$y\log x = (x - y)\log_e e$

$$\Rightarrow y = \frac{x}{1 + \log x}$$

On differentiating w.r.t. $x$, we get

$$\frac{dy}{dx} = \frac{(1 + \log x) - x \cdot 1/x}{(1 + \log x)^2} = \frac{\log x}{(1 + \log x)^2}$$

**16.** (d) We have,

$f(x) = x^5 + 2x^3 + 3x + 1$

$\Rightarrow f'(x) = 5x^4 + 6x^2 + 3$

$\Rightarrow f'(1)=5(1)+6(1)+3=14$

Also, $g(f(x))=x$

$\Rightarrow g'(f(x))\,f'(x)=1$

Now, for $f(x)=7$

$\Rightarrow x^5+2x^3+3x+1=7 \Rightarrow x=1$

$\therefore g'(7)\,f'(1)=1$

$\Rightarrow g'(7)=\dfrac{1}{f'(1)}=\dfrac{1}{14}$

and $g(7)=g(f(1))=1$

$\therefore \dfrac{g(7)}{g'(7)}=\dfrac{1}{1/14}=14$

**17.** *(d)* Given, $f(x)=x^5+2e^{x/4}$ ... (i)

and $g(f(x))=x$ ... (ii)

Here, $f(0)=0+2\times 1=2$ ...(iii)

Differentiating w.r.t. $x$, we get

$g'(f(x))\cdot f'(x)=1$

$\Rightarrow g'(f(x))=1/f'(x)$

Putting $x=0$, we get

$g'(f(0))=\dfrac{1}{f'(0)} \Rightarrow g'(2)=\dfrac{1}{f'(0)}$ ...(iv)

Now, $f'(x)=5x^4+\dfrac{2}{4}.e^{x/4}$

$=5x^4+\dfrac{1}{2}e^{x/4}$

$\therefore f'(0)=0+\dfrac{1}{2}\times 1=\dfrac{1}{2} \Rightarrow g'(2)=2$

Hence, $8g'(2)=8\times 2=16$

**18.** *(a)* Given,

$$f(x)=\begin{vmatrix} x^3 & 2x^2+1 & 1+3x \\ 3x^2+2 & 2x & x^3+6 \\ x^3-x & 4 & x^2-2 \end{vmatrix}$$

$$f(0)=\begin{vmatrix} 0 & 1 & 1 \\ 2 & 0 & 6 \\ 0 & 4 & -2 \end{vmatrix}$$

$=-1(-4)+1(8)=12$

$$f'(x)=\begin{vmatrix} 3x^2 & 4x & 3 \\ 3x^2+2 & 2x & x^3+6 \\ x^3-x & 4 & x^2-2 \end{vmatrix}+\begin{vmatrix} x^3 & 2x^2+1 & 1+3x \\ 6x & 2 & 3x^2 \\ x^3-x & 4 & x^2-2 \end{vmatrix}+\begin{vmatrix} x^3 & 2x^2+1 & 1+3x \\ 3x^2+2 & 2x & x^3+6 \\ 3x^2-1 & 0 & 2x \end{vmatrix}$$

$$f'(0)=\begin{vmatrix} 0 & 0 & 3 \\ 2 & 0 & 6 \\ 0 & 4 & -2 \end{vmatrix}+\begin{vmatrix} 0 & 1 & 1 \\ 0 & 2 & 0 \\ 0 & 4 & -2 \end{vmatrix}+\begin{vmatrix} 0 & 1 & 1 \\ 2 & 0 & 6 \\ -1 & 0 & 0 \end{vmatrix}$$

$=3(8)+2(0)+(-1)(6)$

$=24-6=18$

$2f(0)+f'(0)=24+18=42$

**19.** *(d)* We have,

$$f(x)=\frac{\tan x\sqrt{\tan^{-1}(x^2-x+1)}\,(2^x+2^{-x})}{(7x^2+3x+1)^3}$$

$$f(x)=\frac{(\tan x)[(2^x+2^{-x})\sqrt{\tan^{-1}(x^2-x+1)}]}{(7x^2+3x+1)^3}$$

Let $g(x)=(2^x+2^{-x})\sqrt{\tan^{-1}(x^2-x+1)}$

$$\therefore f(x)=\frac{\tan x\cdot g(x)}{(7x^2+3x+1)^3}$$

$$\therefore f'(x)=\frac{-\tan x\cdot g(x)\dfrac{d}{dx}(7x^2+3x+1)^3}{(7x^2+3x+1)^6}$$

$$\therefore f'(0)=\frac{(1)^3\,g(0)\sec^2(0)-0}{(1)^6}$$

$f'(0)=(1+1)\sqrt{\tan^{-1}\cdot 1}=2\sqrt{\dfrac{\pi}{4}}$

$=\dfrac{2\sqrt{\pi}}{2}=\sqrt{\pi}$

**20.** *(b)*

> **key idea** If $f(a)\cdot f(b)<0$
> $\Rightarrow f(x)$ has at least one root in $x\in[a,b]$

$f(x)=4\sqrt{2}x^3-3\sqrt{2}x-1$

$\Rightarrow f'(x)=12\sqrt{2}x^2-3\sqrt{2}$

For $\left[\dfrac{1}{2},1\right]$, $f'(x)\geq 0$

Since, $f\left(\dfrac{1}{2}\right)<0$ and $f(1)>0$

$\Rightarrow y=f(x)$ intersect the $X$-axis exactly at one point

Now, $f(x)=\sqrt{2}(4x^3-3x)-1=0$

put $x=\cos\alpha$

$[\because 4\cos 3\alpha-3\cos\alpha=\cos 3\alpha]$

$\therefore \cos 3\alpha=\dfrac{1}{\sqrt{2}}=\cos\dfrac{\pi}{4}$

$\Rightarrow 3\alpha=\dfrac{\pi}{4} \Rightarrow \alpha=\dfrac{\pi}{12}$

$\Rightarrow x=\cos\dfrac{\pi}{12}$

Both Statement (I) and (II) are correct.

**21.** *(d)* Let $f(x)=$

$\log_e 2\dfrac{d}{dx}(\log_{\cos x}\operatorname{cosec} x)$

$$=\log_e 2\times\frac{d}{dx}\left[\frac{\log_e\operatorname{cosec} x}{\log_e\cos x}\right]$$

$$=\log_e 2\times\frac{d}{dx}\left(\frac{\log_e|\sin x|}{\log_e|\cos x|}\right)$$

$$=-\log_e 2\times\frac{d}{dx}\left(\frac{\log_e\sin x}{\log_e\cos x}\right)$$

$$=-\log_e 2\,\frac{\log_e\cos x\left[\dfrac{\cos x}{\sin x}\right]-\log_e\sin x\left[-\dfrac{\sin x}{\cos x}\right]}{(\log_e\cos x)^2}$$

$$=-\log_e 2\left[\frac{\cot x\log_e\cos x+\tan x\log_e\sin x}{(\log_e\cos x)^2}\right]$$

$$f\left(\frac{\pi}{4}\right)=-\log_e 2\left[\frac{\cot\dfrac{\pi}{4}\log_e\cos\dfrac{\pi}{4}+\tan\dfrac{\pi}{4}\log_e\sin\dfrac{\pi}{4}}{\left(\log_e\cos\dfrac{\pi}{4}\right)^2}\right]$$

$$=-\log_e 2\left[\frac{2}{\log_e\dfrac{1}{\sqrt{2}}}\right]$$

$$=-\log_e 2\times\frac{2}{-\dfrac{1}{2}\log_e 2}=4$$

**22.** *(c)* We have,

$$y(x)=\cot^{-1}\left(\frac{\sqrt{1+\sin x}+\sqrt{1-\sin x}}{\sqrt{1+\sin x}-\sqrt{1-\sin x}}\right)$$

$$=\cot^{-1}\frac{\left|\cos\dfrac{x}{2}+\sin\dfrac{x}{2}\right|+\left|\cos\dfrac{x}{2}-\sin\dfrac{x}{2}\right|}{\left|\cos\dfrac{x}{2}+\sin\dfrac{x}{2}\right|-\left|\cos\dfrac{x}{2}-\sin\dfrac{x}{2}\right|}$$

$\left[\text{as } \cos^2\frac{x}{2}+\sin^2\frac{x}{2}=1 \text{ and } \sin x = 2\sin\frac{x}{2}\cos\frac{x}{2}\right]$

$$=\cot^{-1}\left(\frac{\cos\frac{x}{2}+\sin\frac{x}{2}+\sin\frac{x}{2}-\cos\frac{x}{2}}{\cos\frac{x}{2}+\sin\frac{x}{2}-\sin\frac{x}{2}+\cos\frac{x}{2}}\right)$$

$$\forall x\in\left(\frac{\pi}{2},\pi\right)$$

$$=\cot^{-1}\left(\frac{\sin\frac{x}{2}}{\cos\frac{x}{2}}\right)=\cot^{-1}\left(\tan\frac{x}{2}\right)$$

$$=\frac{\pi}{2}-\tan^{-1}\left(\tan\frac{x}{2}\right)$$

$$\therefore \quad y(x)=\frac{\pi}{2}-\frac{x}{2}$$

$$\Rightarrow \frac{dy}{dx}=y'(x)=-\frac{1}{2}=y'\left(\frac{5\pi}{6}\right)$$

**23.** *(c)* Given,

$$f(x)=\cos\left(2\tan^{-1}\sin\left(\cot^{-1}\sqrt{\frac{1-x}{x}}\right)\right)$$

$$\cot^{-1}\sqrt{\frac{1-x}{x}}=\sin^{-1}\sqrt{x}$$

$$\therefore f(x)=\cos(2\tan^{-1}\sin\sin^{-1}\sqrt{x})$$

or $\quad f(x)=\cos(2\tan^{-1}\sqrt{x})$

$$=\cos\tan^{-1}\left(\frac{2\sqrt{x}}{1-x}\right)$$

$$f(x)=\cos\cos^{-1}\left(\frac{1-x}{1+x}\right)$$

$$f(x)=\frac{1-x}{1+x}$$

$$f'(x)=\frac{-(1+x)-(1-x)}{(1+x)^2}=\frac{-2}{(1+x)^2}$$

$$f'(x)(1-x)^2=-2\left(\frac{1-x}{1+x}\right)^2$$

$$(1-x)^2f'(x)+2[f(x)]^2=-2\left(\frac{1-x}{1+x}\right)^2+2\left(\frac{1-x}{1+x}\right)^2=0$$

**24.** *(d)* It is given that

$$y(\alpha)=\sqrt{2\left(\frac{\tan\alpha+\cot\alpha}{1+\tan^2\alpha}\right)+\frac{1}{\sin^2\alpha}},$$

$$\alpha\in\left(\frac{3\pi}{4},\pi\right)$$

$\Rightarrow y(\alpha)$

$$=\sqrt{2\cot\alpha\left(\frac{\tan^2\alpha+1}{1+\tan^2\alpha}\right)+\frac{1}{\sin^2\alpha}}$$

$$=\sqrt{2\cot\alpha+\operatorname{cosec}^2\alpha}$$

$$=\sqrt{2\cot\alpha+1+\cot^2\alpha}$$

$$=\sqrt{(1+\cot\alpha)^2}=|1+\cot\alpha|$$

$\because \cot\alpha\in(-\infty,-1)$, for $\alpha\in\left(\frac{3\pi}{4},\pi\right)$

$\therefore y(\alpha)=-(1+\cot\alpha)$ $[\because |x|=-x$, for $x<0]$

$$\therefore \frac{dy}{d\alpha}=-(0-\operatorname{cosec}^2\alpha)=\operatorname{cosec}^2\alpha$$

So, $\left.\frac{dy}{d\alpha}\right|_{\alpha=\frac{5\pi}{6}}=\operatorname{cosec}^2\left(\frac{5\pi}{6}\right)$

$$=\operatorname{cosec}^2\left(\pi-\frac{\pi}{6}\right)=\operatorname{cosec}^2\left(\frac{\pi}{6}\right)=(2)^2=4$$

**25.** *(b)* Given equation is

$$(2x)^{2y}=4\cdot e^{2x-2y} \quad \ldots(\text{i})$$

On applying $\log_e$ both sides, we get

$$\log_e(2x)^{2y}=\log_e 4+\log_e e^{2x-2y}$$

$$2y\log_e(2x)=\log_e(2)^2+(2x-2y)$$

$[\because \log_e n^m = m\log_e n$ and $\log_e e^{f(x)}=f(x)]$

$$\Rightarrow (2\log_e(2x)+2)y=2x+2\log_e(2)$$

$$\Rightarrow y=\frac{x+\log_e 2}{1+\log_e(2x)}$$

On differentiating '$y$' w.r.t. '$x$', we get

$$\frac{dy}{dx}=\frac{(1+\log_e(2x))1-(x+\log_e 2)\frac{2}{2x}}{(1+\log_e(2x))^2}$$

$$=\frac{1+\log_e(2x)-1-\frac{1}{x}\log_e 2}{(1+\log_e(2x))^2}$$

So, $(1+\log_e(2x))^2\frac{dy}{dx}$

$$=\left(\frac{x\log_e(2x)-\log_e 2}{x}\right)$$

**26.** *(b)* We have,

$x\log_e(\log_e x)-x^2+y^2=4$, which can be written as

$$y^2=4+x^2-x\log_e(\log_e x) \quad \ldots(\text{i})$$

Now, differentiating Eq. (i) w.r.t. $x$, we get

$$2y\frac{dy}{dx}=2x-x\frac{1}{\log_e x}\cdot\frac{1}{x}-1\cdot\log_e(\log_e x)$$

[by using product rule of derivative]

$$\Rightarrow \left(\frac{dy}{dx}\right)=\frac{2x-\frac{1}{\log_e x}-\log_e(\log_e x)}{2y} \quad \ldots(\text{ii})$$

Now, at $x=e$,

$$y^2=4+e^2-e\log_e(\log_e e)$$

[using Eq. (i)]

$$=4+e^2-e\log_e(1)=4+e^2-0$$

$$\Rightarrow \quad y=\sqrt{e^2+4} \quad [\because y>0]$$

$\therefore$ At $x=e$ and $y=\sqrt{e^2+4}$,

$$\frac{dy}{dx}=\frac{2e-1-0}{2\sqrt{e^2+4}}=\frac{2e-1}{2\sqrt{e^2+4}}$$

[using Eq. (ii)]

**27.** *(b)* $\because \sin y = x\sin(a+y)$

$$\Rightarrow \quad x=\frac{\sin y}{\sin(a+y)}$$

On differentiating w.r.t. $y$, we get

$$\frac{dx}{dy}=\frac{\sin(a+y)\cos y-\sin y\cos(a+y)}{\sin^2(a+y)}$$

$$\Rightarrow \frac{dx}{dy}=\frac{\sin a}{\sin^2(a+y)}\Rightarrow\frac{dy}{dx}=\frac{\sin^2(a+y)}{\sin a}$$

**28.** *(b)* $g(x)=$ linear function $=ax+b$ (let)

$$f(x)=\begin{cases} g(x), & x\le 0\\ \left(\frac{1+x}{2+x}\right)^{\frac{1}{x}}, & x>0\end{cases}$$

$$=\begin{cases} ax+b, & x\le 0\\ \left(\frac{1+x}{2+x}\right)^{\frac{1}{x}}, & x>0\end{cases}$$

$f(x)$ is continuous at $x=0$

$$f(0^+)=f(0)=f(0^-)\Rightarrow 0=b$$

$f'(x)$ for $x=1$ is $f(-1)$

Let $\quad y=f(x)=\left(\frac{1+x}{2+x}\right)^{1/x}$

$$\ln y=\frac{1}{x}\ln\left(\frac{1+x}{2+x}\right)$$

$$\Rightarrow \frac{1}{y}\frac{dy}{dx}=\frac{1}{x}\cdot\frac{1}{\frac{1+x}{2+x}}\cdot\frac{2+x-1-x}{(2+x)^2}+\ln\left(\frac{1+x}{2+x}\right)\left(\frac{-1}{x^2}\right)$$

$$\Rightarrow \frac{dy}{dx}=y\left[\frac{1}{x(1+x)(2+x)}-\frac{1}{x^2}\ln\left(\frac{1+x}{2+x}\right)\right]$$

$$f'(1)=\left.\frac{dy}{dx}\right|_{x=1}=y(1)\left[\frac{1}{6}-\ln\left(\frac{2}{3}\right)\right]$$

$$=\left(\frac{2}{3}\right)\left(\frac{1}{6}-\ln\frac{2}{3}\right)$$

Let $f(-1)=-a$

$\because f'(1)=f(-1)$

$$a=\frac{2}{3}\left(\ln\frac{2}{3}-\frac{1}{6}\right)$$

$$g(3)=3a=2\left[\ln\left(\frac{2}{3}\right)-\frac{1}{6}\right]$$

$$=\ln\left(\frac{4}{9}\right)-\frac{1}{3}$$

$$=\ln\left(\frac{4}{9}\right)-\ln e^{1/3}$$

$$=\ln\left(\frac{4}{9e^{1/3}}\right)$$

**29.** (a) $f(x)=\dfrac{\sin x+\cos x-\sqrt{2}}{\sin x-\cos x}$

On dividing numerator and denominator by $\sqrt{2}$,

$$\frac{\frac{1}{\sqrt{2}}\sin x+\frac{1}{\sqrt{2}}\cos x-1}{\frac{1}{\sqrt{2}}\sin x-\frac{1}{\sqrt{2}}\cos x}$$

$$f(x)=\frac{\sin\left(x+\frac{\pi}{4}\right)-1}{\sin\left(x-\frac{\pi}{4}\right)}$$

Differentiating $f(x)$ w.r.t. $x$, we get

$$f'(x)=\frac{\sin\left(x-\frac{\pi}{4}\right)\cos\left(x+\frac{\pi}{4}\right)-\left\{\cos\left(x-\frac{\pi}{4}\right)\right\}\left\{\sin\left(x+\frac{\pi}{4}\right)-1\right\}}{\sin^2\left(x-\frac{\pi}{4}\right)}$$

$$=\frac{\cos\left(x-\frac{\pi}{4}\right)-1}{\sin^2\left(x-\frac{\pi}{4}\right)}$$

$$=\frac{-1}{1+\cos\left(x-\frac{\pi}{4}\right)}$$

Again differentiating $f(x)$ w.r.t. $x$, we get

$$f''(x)=\frac{-\sin\left(x-\frac{\pi}{4}\right)}{\left[1+\cos\left(x-\frac{\pi}{4}\right)\right]^2}$$

Now, $f\left(\frac{7\pi}{12}\right)=-\frac{1}{\sqrt{3}}$

and $f''\left(\frac{7\pi}{12}\right)=-\frac{2\sqrt{3}}{9}$

$$\therefore f\left(\frac{7\pi}{12}\right)f''\left(\frac{7\pi}{12}\right)=\frac{2}{9}$$

**30.** (a) We have,

$$y=f(x)=\sin^3\left(\frac{\pi}{3}\left(\cos\left(\frac{\pi}{3\sqrt{2}}(-4x^3+5x^2+1)^{3/2}\right)\right)\right)$$

$$\therefore \sin^3\left(\frac{\pi}{3}\left(\cos\left(\frac{\pi}{3\sqrt{2}}(-4x^3+5x^2+1)^{3/2}\right)\right)\right).$$

$$\frac{dy}{dx}=3\sin^2\times\cos\left(\frac{\pi}{3}\left(\cos\left(\frac{\pi}{3\sqrt{2}}(-4x^3+5x^2+1)^{3/2}\right)\right)\right)\times\frac{\pi}{3}\times$$

$$\left(-\sin\left(\frac{\pi}{3\sqrt{2}}(-4x^3+5x^2+1)^{3/2}\right)\right)\times\frac{\pi}{3\sqrt{2}}\times\frac{3}{2}$$

$$\times(-4x^3+5x^2+1)^{1/2}\times(-12x^2+10x)$$

Now, at $x=1$ $\quad y=f(1)=\sin^3\left(\frac{\pi}{3}\cos\left(\frac{\pi}{3\sqrt{2}}(-4+5+1)^{3/2}\right)\right)$

$$=\sin^3\left(\frac{\pi}{3}\left(\cos\frac{2\pi}{3}\right)\right)=\sin^3\left(\frac{\pi}{3}\times\left(-\frac{1}{2}\right)\right)$$

$$=\left(-\frac{1}{2}\right)^3=-\frac{1}{8}$$

and $y'=3\left(-\frac{1}{2}\right)^2\cdot\left(\cos\frac{\sqrt{3}}{2}\right)$

$$\left(-\frac{\pi}{3}\right)\left(\frac{\sqrt{3}}{2}\right)\cdot\frac{\pi}{3\sqrt{2}}\left(\frac{3}{2}\sqrt{2}\right)(-2)=\frac{3\pi^2}{16}$$

Now, $2y'+3\pi^2y$

$$=2\left(\frac{3\pi^2}{16}\right)+3\pi^2\left(-\frac{1}{8}\right)$$

$$=\frac{3\pi^2}{8}-\frac{3\pi^2}{8}=0$$

**31.** (a) We have,

$$f(\theta)=3\left\{\sin^4\left(\frac{3\pi}{2}-\theta\right)+\sin^4(3\pi+\theta)\right\}-2(1-\sin^2(2\theta))$$

(Using $\sin\left(\frac{3\pi}{2}-\theta\right)=-\cos\theta$

and $\sin(3\pi+\theta)=\sin\theta$), we get

$$f(\theta)=3(\cos^4\theta+\sin^4\theta)-2(1-\sin^2(2\theta))$$

$$=3[(\cos^2\theta+\sin^2\theta)^2-2\sin^2\theta\cos^2\theta]-2[(1-\sin^2(2\theta)]$$

$$=3\left[1-\frac{(2\sin\theta\cos\theta)^2}{2}\right]-2+2\sin^2(2\theta)$$

$[\because \sin^2\theta+\cos^2\theta=1]$

$$=3-\frac{3\sin^2(2\theta)}{2}-2+2\sin^2(2\theta)$$

$(\because 2\sin\theta\cos\theta=\sin 2\theta)$

$$=1+\frac{\sin^2(2\theta)}{2} \quad \text{... (i)}$$

Now $S=\left\{\theta\in[0,\pi]: f'(\theta)=-\frac{\sqrt{3}}{2}\right\}$

Now, $f'(\theta)=\left[1+\frac{\sin^2(2\theta)}{2}\right]'$

$$=\frac{1}{2}\times[\sin^2(2\theta)]'$$

$$=\frac{1}{2}\times 2\sin(2\theta)\cdot[\sin(2\theta)]'$$

$$=\sin(2\theta)\cdot\cos(2\theta)\cdot 2=\sin(4\theta)$$

$\theta\in[0,\pi]\Rightarrow 4\theta\in[0,4\pi]$

According to question, $\sin(4\theta)=-\frac{\sqrt{3}}{2}$ in $[0,4\pi]$

$$\Rightarrow 4\theta=\frac{4\pi}{3},\frac{5\pi}{3},\frac{10\pi}{3},\frac{11\pi}{3}$$

$$\Rightarrow \theta=\frac{\pi}{3},\frac{5\pi}{12},\frac{5\pi}{6},\frac{11\pi}{12}$$

Given, $4\beta=\sum_{\theta\in S}\theta=\frac{\pi}{3}+\frac{5\pi}{12}+\frac{5\pi}{6}+\frac{11\pi}{12}$

$$=\frac{4\pi+5\pi+10\pi+11\pi}{12}=\frac{5\pi}{2}$$

$$\Rightarrow \beta=\frac{5\pi}{8}$$

$$\therefore f(\beta)=1+\frac{\sin^2(2\beta)}{2}=1+\frac{\sin^2\left(\frac{5\pi}{4}\right)}{2}$$

$$=1+\frac{\sin^2\left(\pi+\frac{\pi}{4}\right)}{2}$$

[Using $\sin(\pi+x)=-\sin x$]

$$= 1 + \frac{\sin^2\left(\frac{\pi}{4}\right)}{2} = 1 + \frac{\left(\frac{1}{\sqrt{2}}\right)^2}{2}$$

$$= 1 + \frac{1}{4} = \frac{5}{4}$$

**32.** *(b)* Given functional relation is

$$y\sqrt{1-x^2} = k - x\sqrt{1-y^2}$$

On differentiating both sides w.r.t. $x$, we get

$$\frac{-2yx}{2\sqrt{1-x^2}} + \sqrt{1-x^2}\frac{dy}{dx} = 0 + \frac{2xy\frac{dy}{dx}}{2\sqrt{1-y^2}} - \sqrt{1-y^2}$$

$$\Rightarrow \frac{dy}{dx}\left\{\sqrt{1-x^2} - \frac{xy}{\sqrt{1-y^2}}\right\} = \left\{\frac{yx}{\sqrt{1-x^2}} - \sqrt{1-y^2}\right\}$$

$$\Rightarrow \frac{dy}{dx}\left\{\frac{\sqrt{1-x^2}\sqrt{1-y^2} - xy}{\sqrt{1-y^2}}\right\} = \frac{xy - \sqrt{1-x^2}\sqrt{1-y^2}}{\sqrt{1-x^2}}$$

$$\Rightarrow \frac{dy}{dx} = -\frac{\sqrt{1-y^2}}{\sqrt{1-x^2}} \Rightarrow y\left(\frac{1}{2}\right) = -\frac{1}{4} \text{ (given)}$$

$$\therefore \left.\frac{dy}{dx}\right|_{x=\frac{1}{2}} = -\frac{\sqrt{1-\left(-\frac{1}{4}\right)^2}}{\sqrt{1-\left(\frac{1}{2}\right)^2}} = -\frac{\sqrt{\frac{16-1}{16}}}{\sqrt{\frac{4-1}{4}}}$$

$$= -\frac{\sqrt{15}}{2\sqrt{3}} = -\frac{\sqrt{5}}{2}$$

**33.** *(d)* Let $f(x) = \tan^{-1}\left(\frac{\sin x - \cos x}{\sin x + \cos x}\right)$

$$= \tan^{-1}\left(\frac{\tan x - 1}{\tan x + 1}\right)$$

[dividing numerator and denominator by $\cos x > 0$, $x \in \left(0, \frac{\pi}{2}\right)$]

$$= \tan^{-1}\left(\frac{\tan x - \tan\frac{\pi}{4}}{1 + \left(\tan\frac{\pi}{4}\right)(\tan x)}\right)$$

$$= \tan^{-1}\left[\tan\left(x - \frac{\pi}{4}\right)\right]$$

$$\left[\because \frac{\tan A - \tan B}{1 + \tan A \tan B} = \tan(A - B)\right]$$

Since, it is given that $x \in \left(0, \frac{\pi}{2}\right)$, so

$$x - \frac{\pi}{4} \in \left(-\frac{\pi}{4}, \frac{\pi}{4}\right)$$

Also, for $\left(x - \frac{\pi}{4}\right) \in \left(-\frac{\pi}{4}, \frac{\pi}{4}\right)$

Then,

$$f(x) = \tan^{-1}\left(\tan\left(x - \frac{\pi}{4}\right)\right) = x - \frac{\pi}{4}$$

$$\left[\because \tan^{-1} \tan\theta = \theta, \text{ for } \theta \in \left(-\frac{\pi}{2}, \frac{\pi}{2}\right)\right]$$

Now, derivative of $f(x)$ w.r.t. $x/2$ is

$$\frac{d(f(x))}{d(x/2)} = 2\frac{df(x)}{d(x)}$$

$$= 2 \times \frac{d}{dx}\left(x - \frac{\pi}{4}\right) = 2$$

**34.** *(b)* Given expression is

$$2y = \left(\cot^{-1}\left(\frac{\sqrt{3}\cos x + \sin x}{\cos x - \sqrt{3}\sin x}\right)\right)^2$$

$$= \left(\cot^{-1}\left(\frac{\sqrt{3}\cot x + 1}{\cot x - \sqrt{3}}\right)\right)^2$$

[dividing each term of numerator and denominator by $\sin x$]

$$= \left(\cot^{-1}\left(\frac{\cot\frac{\pi}{6}\cot x + 1}{\cot x - \cot\frac{\pi}{6}}\right)\right)^2$$

$$\left[\because \cot\frac{\pi}{6} = \sqrt{3}\right]$$

$$= \left(\cot^{-1}\left(\cot\left(\frac{\pi}{6} - x\right)\right)\right)^2$$

$$\left[\because \cot(A - B) = \frac{\cot A \cot B + 1}{\cot B - \cot A}\right]$$

$$= \begin{cases} \left(\frac{\pi}{6} - x\right)^2, & 0 < x < \frac{\pi}{6} \\ \left(\pi + \left(\frac{\pi}{6} - x\right)\right)^2, & \frac{\pi}{6} < x < \frac{\pi}{2} \end{cases}$$

$$\left[\because \cot^{-1}(\cot\theta) = \begin{cases} \pi + \theta, & -\pi < \theta < 0 \\ \theta, & 0 < \theta < \pi \\ \theta - \pi, & \pi < \theta < 2\pi \end{cases}\right]$$

$$\Rightarrow 2y = \begin{cases} \left(\frac{\pi}{6} - x\right)^2, & 0 < x < \frac{\pi}{6} \\ \left(\frac{7\pi}{6} - x\right)^2, & \frac{\pi}{6} < x < \frac{\pi}{2} \end{cases}$$

$$\Rightarrow 2\frac{dy}{dx} = \begin{cases} 2\left(\frac{\pi}{6} - x\right)(-1), & 0 < x < \frac{\pi}{6} \\ 2\left(\frac{7\pi}{6} - x\right)(-1), & \frac{\pi}{6} < x < \frac{\pi}{2} \end{cases}$$

$$\Rightarrow \frac{dy}{dx} = \begin{cases} x - \frac{\pi}{6}, & 0 < x < \frac{\pi}{6} \\ x - \frac{7\pi}{6}, & \frac{\pi}{6} < x < \frac{\pi}{2} \end{cases}$$

**35.** *(a)* $x = \sqrt{2^{\text{cosec}^{-1} t}}$ ...(i)

and $y = \sqrt{2^{\sec^{-1} t}}$ ...(ii)

Differentiating Eqs. (i) and (ii) w.r.t, '$t$', we get

$$\frac{dx}{dt} = \frac{1}{2\sqrt{2^{\text{cosec}^{-1} t}}} \times (2^{\text{cosec}^{-1} t} \log 2) \times \frac{-1}{t\sqrt{t^2 - 1}}$$

$$\Rightarrow \frac{dx}{dt} = \frac{-(2^{\text{cosec}^{-1} t})^{\frac{1}{2}} \log 2}{2t\sqrt{t^2 - 1}}$$

$$= \frac{-x \log 2}{2t\sqrt{t^2 - 1}}$$

and $\frac{dy}{dt} = \frac{1}{2\sqrt{2^{\sec^{-t} t}}} \times (2^{\sec^{-1} t} \log 2) \times \frac{1}{\sqrt{t^2 - 1}}$

$$= \frac{(2^{\sec^{-1} t})^{\frac{1}{2}} \log 2}{2t\sqrt{t^2 - 1}} = \frac{y \log 2}{2t\sqrt{t^2 - 1}}$$

$$\therefore \frac{dy}{dx} = \frac{\frac{dy}{dt}}{\frac{dx}{dt}} = \frac{y \log 2 / 2t\sqrt{t^2 - 1}}{-x \log 2 / 2t\sqrt{t^2 - 1}}$$

$$= -\frac{y}{x}$$

**36.** *(55)* $L = \lim_{x \to 0} 2$

$$\left(\frac{1 - \cos x (\cos 2x)^{1/2} (\cos 3x)^{1/3} \ldots (\cos 10x)^{1/10}}{x^2}\right)$$

At $x=0$, $L$ makes $=\frac{0}{0}$ form

So, we will use L-Hospital's rule

$\because \frac{d}{dx}(x^2)=2x$ and $\frac{d}{dx}(1)=0$

and for $\frac{d}{dx}(1+\cos x(\cos 2x)^{1/2}....$

$(\cos 10x)^{1/10}$

Let $y=\cos x\,(\cos 2x)^{\frac{1}{2}}\ ...(\cos 10x)^{\frac{1}{10}}$

Let $\log(y)=\log(\cos x)+\frac{1}{2}\log\cos(2x)$

$+...+\frac{1}{10}\log(\cos 10x)$

$\frac{1}{y}\left(\frac{dy}{dx}\right)=\tan x+\frac{1}{2}\times 2\tan 2x$

$+...+\frac{1}{10}\times 10\tan 10x$

$\frac{dy}{dx}=y(\tan x+\tan 2x+...+\tan 10x)$

So, $L=\lim\limits_{x\to 0} 2$

$\left(\frac{y(\tan x+\tan 2x+...+\tan 10x)}{2x}\right)$

$\because \lim\limits_{x\to 0}\cos x\,(\cos 2x)^{\frac{1}{2}}\ ....(\cos 10x)^{\frac{1}{10}}=1$

$=\lim\limits_{x\to 0} 1\times$

$\left(\frac{\tan x}{x}+\frac{2\tan 2x}{2x}+...+\frac{10\tan 10x}{10x}\right)$

$=1+2+3+...+10$

$=\frac{10\times 11}{2}=55$

**37.** *(105)* We have,

$y=\frac{(\sqrt{x}+1)(x^2-\sqrt{x})}{x\sqrt{x}+x+\sqrt{x}}$

$+\frac{1}{15}(3\cos^2 x-5)\cos^3 x$

$y=\frac{(\sqrt{x}+1)\cdot\sqrt{x}(x\sqrt{x}-1)}{x\sqrt{x}+x+\sqrt{x}}$

$+\frac{1}{15}(3\cos^2 x-5)\cos^3 x$

$y=\frac{(\sqrt{x}+1)\cdot(\sqrt{x}((\sqrt{x})^3-1^3)}{x\sqrt{x}+x+\sqrt{x}}$

$+\frac{1}{15}[3\cos^5 x-5\cos^3 x]$

$y=\frac{(\sqrt{x}+1)\cdot\sqrt{x}(\sqrt{x}-1)(x+\sqrt{x}+1)}{\sqrt{x}\,(x+\sqrt{x}+1)}$

$+\frac{1}{15}[3\cos^5 x-5\cos^3 x]$

$y=(x-1)+\frac{1}{15}[3\cos^5 x-5\cos^3 x]$

Now, differentiate w.r.t. $x$, we get

$y'=1-0+\frac{1}{15}[15\cos^4 x(-\sin x)$

$-15\cos^2 x(-\sin x)]$

$y'=1-\sin x[\cos^4 x-\cos^2 x]$

$y'\left(\frac{\pi}{6}\right)=1-\sin\frac{\pi}{6}\left[\cos^4\frac{\pi}{6}-\cos^2\frac{\pi}{6}\right]$

$=1-\frac{1}{2}\left[\frac{9}{16}-\frac{3}{4}\right]$

$=1-\frac{1}{2}\left[\frac{9-12}{16}\right]$

$\Rightarrow 1+\frac{3}{32}\Rightarrow\frac{35}{32}$

$\therefore 96y'\left(\frac{\pi}{6}\right)=96\times\frac{35}{32}=105$

**38.** *(5)* $f(x)=2^x-x^2$

$\Rightarrow f'(x)=2^x\log 2-2x$

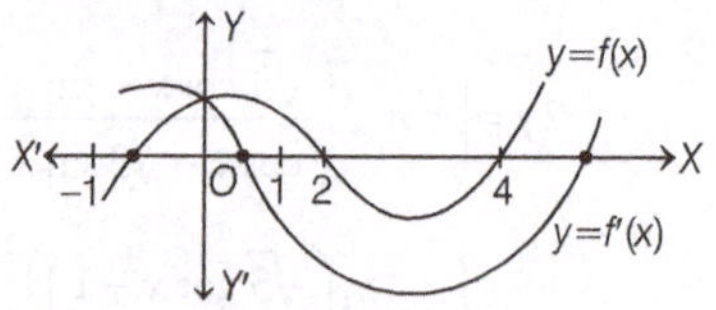


Clearly,

$m=3$ and $n=2$

$\therefore\ m+n=5$

**39.** *(248)* Given,

$f(x+y)=2^x f(y)+4^y f(x),\ \forall x, y\in R$

$f(2)=3$

To calculate : $14\frac{f'(4)}{f'(2)}$

Put $y=2$

$f(x+2)=2^x 3+4^2 f(x)$

$f(x+2)=2^x\cdot 3+16f(x)$

$f'(x+2)=16f'(x)+3\cdot 2^x\ln 2$

Put $x=2$,

$f'(4)=16f'(2)+12\ln 2$ ...(i)

$f(y+2)=4f(y)+4^y\cdot 3$

$f'(y+2)=4f'(y)+3\cdot 4^y\ln 4$

On putting $y=2$,

$f'(4)=4f'(2)+96\ln 2$ ...(ii)

Solving Eqs. (i) and (ii), we get

$16f'(2)+12\ln 2=4f'(2)+96\ln 2$

$\Rightarrow\quad 12f'(2)=84\ln 2$

$\Rightarrow\quad f'(2)=7\ln 2$ ...(iii)

On putting Eq. (iii) in Eq. (i), we get

$f'(4)=16(7\ln 2)+12\ln 2$

$\Rightarrow\quad f'(4)=112\ln 2+12\ln 2$

$\Rightarrow\quad f'(4)=124\ln 2$

$\therefore\quad 14\left(\frac{f'(4)}{f'(2)}\right)=\frac{14\times 124\ln 2}{7\ln 2}=248$

**40.** *(481)* $f(x)=\sin\left(\cos^{-1}\left(\frac{1-2^{2x}}{1+2^{2x}}\right)\right)$

Let $2^{2x}$ be $\tan^2\theta$.

$\therefore f(x)=\sin\left[\cos^{-1}\left(\frac{1-\tan^2\theta}{1+\tan^2\theta}\right)\right]$

$=\sin[\cos^{-1}(\cos 2\theta)]=\sin 2\theta$

$=\frac{2\tan\theta}{1+\tan^2\theta}=\frac{2\cdot 2^x}{1+2^{2x}}$

$f(x)=2\cdot\left(\frac{2^x}{1+2^{2x}}\right)$

$f'(x)=2\cdot$

$\left[\frac{(1+2^{2x})(2^x\log_e 2)-2^x(2^{2x}\log_e 2\cdot 2)}{(1+2^{2x})^2}\right]$

$f'(1)=2\left(\frac{5\cdot 2\log_e 2-2\cdot 8\log_e 2}{5^2}\right)$

$=\left(-\frac{12}{25}\right)\log_e 2=\frac{-b}{a}\log_e 2$

$\Rightarrow\quad a=25$ and $b=12$

$\therefore\quad |a^2-b^2|_{\min}=|25^2-12^2|=481$

## Topic 4

**01.** *(c)* $\ln y=3\sin^{-1}x$

Differentiating w.r.t. $x$, we get

$\frac{1}{y}\frac{dy}{dx}=\frac{3}{\sqrt{1-x^2}}$

$\sqrt{1-x^2}\,y'=3y$

Again, differentiating w.r.t. $x$, we get

$\sqrt{1-x^2}\,y''+y'\frac{1}{2\sqrt{1-x^2}}\times -2x=3y'$

$(1-x^2)y''-xy'=3y'(1-x^2)^{1/2}$

At $x=\frac{1}{2}$

$y=e^{\pi/2}\ \Rightarrow y'=2\sqrt{3}e^{\pi/2}$

$\therefore (1-x^2)y''-xy' = 3\times 2\sqrt{3}e^{\frac{\pi}{2}}\cdot\frac{\sqrt{3}}{2}$

$= 9e^{\pi/2}$

**02.** *(d)* According to the question,

$$f(x) = x^3 \sin\left(\frac{1}{x}\right)$$

Let's find first derivative

$$f'(x) = 3x^2 \sin\frac{1}{x} + x^3 \cos\frac{1}{x}\times\left(-\frac{1}{x^2}\right)$$

$$= 3x^2 \sin\frac{1}{x} - x\cos\frac{1}{x}$$

Now, $f''(x) = 6x\sin\frac{1}{x} + 3x^2\cos\frac{1}{x}\times\left(-\frac{1}{x^2}\right) - \cos\frac{1}{x} - x\times\left(-\sin\frac{1}{x}\right)\left(-\frac{1}{x^2}\right)$

$$= 6x\sin\frac{1}{x} - 4\cos\frac{1}{x} - \frac{1}{x}\sin\frac{1}{x};\ x\neq 0$$

Now, $f''(0) =$ Not defined,

$$f''\left(\frac{2}{\pi}\right) = 6\times\frac{2}{\pi}\sin\left(\frac{\pi}{2}\right) - 4\cos\frac{\pi}{2} - \frac{\pi}{2}\sin\frac{\pi}{2}$$

$$= \frac{12}{\pi} - \frac{\pi}{2} = \frac{24-\pi^2}{2\pi}$$

**03.** *(b)* Given,

$$y = \tan^{-1}(\sec x^3 - \tan x^3),$$

$$\frac{\pi}{2} < x^3 < \frac{3\pi}{2}$$

We know, $\sec\theta - \tan\theta = \tan\left(\frac{\pi}{4}-\frac{\theta}{2}\right)$

$\Rightarrow \quad y = \tan^{-1}\left\{\tan\left(\frac{\pi}{4}-\frac{x^3}{2}\right)\right\}$

$\Rightarrow \quad y = \frac{\pi}{4} - \frac{x^3}{2}$

Differentiating both sides w.r.t. $x$, we get

$$y' = -\frac{3x^2}{2} \Rightarrow y'' = -3x$$

Now check by option,

**Option (a)**

$$xy''+2y' = x(-3x^2)+2\left(-\frac{3x^2}{2}\right)$$

$$= -3x^3 - 6x^2 \neq 0$$

$\therefore$ Option (a) is false.

**Option (b)** $x^2y''-6y+\frac{3\pi}{2} = x^2(-3x)$

$$-6\left\{\frac{\pi}{4}-\frac{x^3}{2}\right\}+\frac{3\pi}{2}$$

$$= -3x^3 - \frac{3\pi}{2} + 3x^3 + \frac{3\pi}{2} = 0$$

**04.** *(b)* $y^{1/5} + y^{-1/5} = 2x$

$\Rightarrow \left(\frac{1}{5}y^{-\frac{4}{5}} - \frac{1}{5}y^{-\frac{6}{5}}\right)\frac{dy}{dx} = 2$

(differentiating w.r.t. $x$, we get)

$\Rightarrow y'(y^{1/5} - y^{-1/5}) = 10y$

$\Rightarrow \quad y(2\sqrt{x^2-1}) = 10y$

$[\because 2x = y^{1/5} + y^{-1/5}$

$\Rightarrow y^{1/5} - y^{-1/5} = 2\sqrt{x^2-1}]$

$\Rightarrow y''(2\sqrt{x^2-1}) + y'\left(2\frac{2x}{2\sqrt{x^2-1}}\right) = 10y'$

$\Rightarrow \quad y''(x^2-1) + xy' = 5\sqrt{x^2-1}(y')$

$\Rightarrow \quad y''(x^2-1) + xy' - 25y = 0$

On comparing, $\lambda = 1$, $k = -25$

So, $\quad \lambda + k = 1 - 25 = -24$

**05.** *(c)* Here, $\frac{dx}{dy} = \left(\frac{dy}{dx}\right)^{-1}$

On differentiating both sides w.r.t. $y$, we get

$$\frac{d^2x}{dy^2} = -\left(\frac{dy}{dx}\right)^{-2}\cdot\frac{d}{dy}\left(\frac{dy}{dx}\right)$$

$$= -\left(\frac{dy}{dx}\right)^{-2}\cdot\frac{d}{dy}\left(\frac{dy}{dx}\right)\cdot\frac{dx}{dx}$$

$$= -\left(\frac{dy}{dx}\right)^{-2}\cdot\frac{d}{dx}\cdot\left(\frac{dy}{dx}\right)\cdot\frac{dx}{dy}$$

$$= -\left(\frac{dy}{dx}\right)^{-2}\cdot\frac{d^2y}{dx^2}\cdot\left(\frac{dy}{dx}\right)^{-1}$$

$$= -\left(\frac{dy}{dx}\right)^{-3}\cdot\left(\frac{d^2y}{dx^2}\right)$$

**06.** *(a)* Let

$f(x) = a_nx^n + a_{n-1}x^{n-1} + \ldots + a_1x$

$\therefore \quad f(0) = 0$

Also, $f(\alpha) = 0$ (belongs to $x = \alpha$ is root of given equation)

and $f'(x) = na_nx^{n-1} + (n-1)a_{n-1}x^{n-2} + \ldots + a_1 = 0$ has atleast one root in $]0, \alpha[$

$\Rightarrow \quad na_n x^{n-1} + (n-1)a_{n-1}x^{n-2} + \ldots + a_1 = 0$

has a positive root smaller than $\alpha$.

**07.** *(c)* Let $f(x) = \frac{ax^3}{3} + \frac{bx^2}{2} + cx$

$f(0) = 0$ and $f(1) = \frac{a}{3} + \frac{b}{2} + c$

$$= \frac{2a+3b+6c}{6} = 0$$

Given, $2a + 3b + 6c = 0$

$\therefore x = 0$ and $x = 1$ are roots of

$$f(x) = \frac{ax^3}{3} + \frac{bx^2}{2} + cx = 0$$

$\therefore$ Atleast one root of the equation $ax^2 + bx + c = 0$ lies in (0, 1).

**08.** *(d)* Given, $f(x) = ax^3 + bx^2 + cx + 41$ ...(i)

$f(1) = 40$, $f'(1) = 2$, $f''(1) = 4$

Now, differentiating Eq. (i) w.r.t. $x$, we have

$f'(x) = 3ax^2 + 2bx + c$

$\therefore f'(1) = 3a + 2b + c = 2$ [given] ...(ii)

Again, differentiating,

$f''(x) = 6ax + 2b$

$\therefore \quad f''(1) = 6a + 2b = 4$ [given]

$\Rightarrow \quad 3a + b = 2$ ...(iii)

Subtracting in Eqs. (ii) and (iii), we get

$\Rightarrow \quad b + c = 0$ ...(iv)

Now, $f(1) = 40$

$\Rightarrow a + b + c + 41 = 40$

$\Rightarrow a + 0 = 40 - 41 \Rightarrow a = -1$ [using Eq. (iv)]

From Eq. (iii), $3\times(-1) + b = 2$

$\Rightarrow b = 5$

From Eq. (ii), $3\times(-1) + 2\times 5 + c = 2$

$\Rightarrow c = -5$

Hence, $a^2 + b^2 + c^2 = 1 + 25 + 25 = 51$

**09.** *(b)* Given,

$$g'\left(\frac{1}{2}\right) = g'\left(\frac{3}{2}\right)$$

Now, $f(x) = \frac{1}{2}[g(x) + g(2-x)]$

$$f'(x) = \frac{1}{2}(g'(x) - g'(2-x)]$$

Now, $f'\left(\frac{1}{2}\right) = \frac{1}{2}\left[g'\left(\frac{1}{2}\right) - g'\left(\frac{3}{2}\right)\right]$

$$=0 \quad \left[\because g'\left(\frac{1}{2}\right)=g'\left(\frac{3}{2}\right)\right]$$

Now, $f'\left(\frac{3}{2}\right)=\frac{1}{2}\left[g'\left(\frac{3}{2}\right)-g'\left(\frac{1}{2}\right)\right]=0$

$\therefore \quad f'\left(\frac{3}{2}\right)=f'\left(\frac{1}{2}\right)=0$

So roots in $\left(\frac{1}{2},1\right)$ and (1, 3/2).

$\Rightarrow f''(x)=0$ at least two $x$ in (1, 3/2)

So, $f''(x)=0$ for atleast two $x$ in (0, 2).

**10.** *(a)* We have, $y=x^x$

$\Rightarrow \quad \log y = x\log x$

On differentiating both the sides w.r.t. $x$, we get

$$\frac{1}{y}y'=\log x+\frac{x}{x}$$

$\Rightarrow \quad y'=y(\log x+1)$ ...(i)

Again, differentiating both the sides w.r.t. $x$, we get

$$y''=y'(\log x+1)+\frac{y}{x} \quad ...(ii)$$

Now, $y'(2)=y(2)(\log 2+1)$

$=2^2(\log 2+1)=4(\log 2+1)$

and $y''(2)=y'(2)(\log 2+1)+\frac{y(2)}{2}$

$=4(\log 2+1)^2+2$

Now, $y''(2)-2y'(2)=4(\log 2+1)^2+2-8(\log 2+1)$

$=4(\log 2+1)[\log 2+1-2]+2$

$=4(\log 2+1)(\log 2-1)+2$

$=4((\log 2)^2-1)+2$

$=4(\log 2)^2-4+2=4(\log 2)^2-2$

**11.** *(a)* $y=(1+x)(1+x^2)(1+x^4)(1+x^8)(1+x^{16})$

$$=\frac{(1-x)(1+x)(1+x^2)(1+x^4)(1+x^8)(1+x^{16})}{(1-x)}$$

$$=\frac{(1-x^2)(1+x^2)(1+x^4)(1+x^8)(1+x^{16})}{(1-x)}$$

$$=\frac{(1-x^4)(1+x^4)(1+x^8)(1+x^{16})}{(1-x)}$$

$$=\frac{(1-x^8)(1+x^8)(1+x^{16})}{(1-x)}$$

$$=\frac{(1-x^{16})(1+x^{16})}{1-x}=\frac{1-x^{32}}{1-x}$$

$\Rightarrow \quad y(1-x)=1-x^{32}$

On differentiating w.r.t. $x$, we get

$$y'(1-x)+y(-1)=-32x^{31}$$

Again differentiating w.r.t. $x$, we get

$$y''(1-x)+y'(-1)-y'=-32\times 31x^{30}$$

At $x=-1$,

$$2y''-2y'=-32\times 31$$

Hence, $\quad y'-y''=16\times 31=496$

**12.** *(a)* We have, $f:S\to S,\ S=(0,\infty)$

$$f(x+1)=x\cdot f(x)$$

$g:S\to R;\ g(x)=\log_e f(x)$

To find $|g''(5)-g''(1)|$

$\Rightarrow \quad g(x+1)=\log_e f(x+1)$

$\Rightarrow \quad g(x+1)=\log[x\cdot f(x)]$

$\Rightarrow \quad g(x+1)=\log x+\log f(x)$

$\Rightarrow \quad g(x+1)=\log x+g(x)$

$\Rightarrow \quad g(x+1)-g(x)=\log x$

$\Rightarrow \quad g'(x+1)-g'(x)=1/x$

$\Rightarrow \quad g''(x+1)-g''(x)=\frac{-1}{x^2}$

$x=1,\ g''(2)-g''(1)=-1$ ...(i)

$x=2,\ g''(3)-g''(2)=-1/4$ ...(ii)

$x=3,\ g''(4)-g''(3)=-1/9$ ...(iii)

$x=4,\ g''(5)-g''(4)=-1/16$ ...(iv)

On adding Eqs. (i), (ii), (iii) and (iv),

$$g''(5)-g''(1)=-1-\frac{1}{4}-\frac{1}{9}-\frac{1}{16}$$

$$=-\left(\frac{144+36+16+9}{144}\right)=\frac{-205}{144}$$

So, $|g''(5)-g''(1)|=\frac{205}{144}$

**13.** *(b)* The given function,

$$f(x)=\begin{cases}k_1(x-\pi)^2-1, & x\le\pi\\ k_2\cos x, & x>\pi\end{cases}$$

is twice differentiable, so $f(x)$ must be continuous at $x=\pi$, so

$$\lim_{x\to\pi^+}f(x)=f(\pi)$$

$\Rightarrow \quad -k_2=-1\Rightarrow k_2=1$ ...(i)

Now, $f'(x)=\begin{cases}2k_1(x-\pi), & x<\pi\\ -k_2\sin x, & x>\pi\end{cases}$

and $f''(x)=\begin{cases}2k_1, & x<\pi\\ -k_2\cos x, & x>\pi\end{cases}$

$\therefore \lim_{x\to\pi^+}f''(x)=\lim_{x\to\pi^-}f''(x)$

$\Rightarrow \quad k_2=2k_1$

$\Rightarrow \quad k_1=\frac{k_2}{2}=\frac{1}{2}$ [from Eq. (i)]

$\therefore$ The ordered pair $(k_1,k_2)=\left(\frac{1}{2},1\right)$

**14.** *(c)* Given equation,

$$y^2+\log_e(\cos^2 x)=y,\quad x\in\left(-\frac{\pi}{2},\frac{\pi}{2}\right) \quad ...(i)$$

On differentiating w.r.t. $x$, we get

$$2yy'-2\tan x=y' \quad ...(ii)$$

Again, on differentiating w.r.t. $x$, we get

$$2(y')^2+2yy''-2\sec^2 x=y''$$

$\Rightarrow 2(y')^2+2yy''=2\sec^2 x+y''$ ...(iii)

From Eq. (i), at $x=0\Rightarrow y=0$ or 1

Now, from Eq. (ii),

at $x=0,\ y=0\Rightarrow y'=0$,

or at $x=0,\ y=1\Rightarrow y'=0$

Now, from Eq. (iii),

at $x=0,\ y=0$ and $y'=0\ \ y''=-2$

So, at $x=0,\ y=1$ and $y'=0$

$\Rightarrow \quad y''=2$

$\Rightarrow \quad y''(0)=2;\ \therefore\ |y''(0)|=2$

**15.** *(c)* We have,

$$f(x)=x^3+x^2f'(1)+xf''(2)+f'''(3)$$

$\Rightarrow f'(x)=3x^2+2xf'(1)+f''(2)$ ...(i)

$\Rightarrow f''(x)=6x+2f'(1)$ ...(ii)

$\Rightarrow \quad f'''(x)=6$ ...(iii)

$\Rightarrow \quad f'''(3)=6$

Putting $x=1$ in Eq. (i), we get

$$f'(1)=3+2f'(1)+f''(2) \quad ...(iv)$$

and putting $x=2$ in Eq. (ii), we get

$$f''(2)=12+2f'(1) \quad ...(v)$$

From Eqs. (iv) and (v), we get

$$f'(1)=3+2f'(1)+(12+2f'(1))$$

$\Rightarrow \quad 3f'(1)=-15\Rightarrow f'(1)=-5$

$\Rightarrow \quad f''(2)=12+2(-5)=2$ [Using Eq. (v)]

$\therefore \quad f(x)=x^3+x^2f'(1)+xf''(2)+f'''(3)$

$\Rightarrow f(x)=x^3-5x^2+2x+6$

$\Rightarrow f(2)=2^3-5(2)^2+2(2)+6=-2$

**16.** *(b)* We have, $x=3\tan t$ and $y=3\sec t$

Clearly, $\frac{dy}{dx}=\frac{\frac{dy}{dt}}{\frac{dx}{dt}}=\frac{\frac{d}{dt}(3\sec t)}{\frac{d}{dt}(3\tan t)}$

$$=\frac{3\sec t\tan t}{3\sec^2 t}=\frac{\tan t}{\sec t}=\sin t$$

and $\frac{d^2y}{dx^2}=\frac{d}{dx}\left(\frac{dy}{dx}\right)=\frac{d}{dt}\left(\frac{dy}{dx}\right)\cdot\frac{dt}{dx}$

$$=\frac{\frac{d}{dt}\left(\frac{dy}{dx}\right)}{\frac{dx}{dt}}=\frac{\frac{d}{dt}(\sin t)}{\frac{d}{dt}(3\tan t)}$$

$$=\frac{\cos t}{3\sec^2 t}=\frac{\cos^3 t}{3}$$

Now, $\frac{d^2y}{dx^2}\left(\text{at } t=\frac{\pi}{4}\right)$

$$=\frac{\cos^3\frac{\pi}{4}}{3}=\frac{1}{3(2\sqrt{2})}=\frac{1}{6\sqrt{2}}$$

**17.** (a) Given,

$$x^2+y^2+\sin y=4 \quad \text{...(i)}$$

On differentiating Eq. (i) both sides w.r.t. $x$, we get

$$2x+2y\frac{dy}{dx}+\cos y\frac{dy}{dx}=0$$

$$\Rightarrow \frac{dy}{dx}=\frac{-2x}{2y+\cos y} \quad \text{...(ii)}$$

On differentiating Eq. (ii) both sides, we get

$$\frac{d^2y}{dx^2}=\frac{(2y+\cos y)(-2)-(-2x)(2-\sin y)\frac{dy}{dx}}{(2y+\cos y)^2}$$

$$\therefore \left.\frac{d^2y}{dx^2}\right|_{(-2,\,0)}=\frac{(1)(-2)-(4)(2)\left.\frac{dy}{dx}\right|_{(-2,\,0)}}{1}$$

$$=\frac{-2-8\times 4}{1}=-34$$

**18.** (c) Let $f(x)=ax^3+bx^2+cx+d$

$$\Rightarrow f(3x)=27ax^3+9bx^2+3cx+d$$

$$f'(x)=3ax^2+2bx+c,$$

$$f''(x)=6ax+2b$$

Now, $f(3x)=f'(x)\,f''(x)$ (By given information)

$$\Rightarrow 27ax^3+9bx^2+3cx+d$$
$$=18a^2x^3+(6ab+12ab)x^2+(4b^2+6ac)x+2bc$$

Comparing coefficient of like powers, we get

$27a=18a^2, 9b=18ab, 3c=4b^2+6ac,$
$d=2bc$

$$\Rightarrow a=\frac{3}{2}, b=0, c=0, d=0$$

$$\therefore f(x)=\frac{3}{2}x^3$$

Hence, $f'(x)=\frac{9}{2}x^2$, $f''(x)=9x$

$\therefore f(2)=12, f'(2)=18, f''(2)=18$

So, $f''(2)-f'(2)=0$

**19.** (d) $f(x)=x|x|$ and $g(x)=\sin x$

$$gof(x)=\sin(x|x|)=\begin{cases}-\sin x^2, & x<0\\ \sin x^2, & x\ge 0\end{cases}$$

$$\therefore (gof)'(x)=\begin{cases}-2x\cos x^2, & x<0\\ 2x\cos x^2, & x\ge 0\end{cases}$$

Clearly, $L(gof)'(0)=0=R(gof)'(0)$

So, $gof$ is differentiable at $x=0$ and also its derivative is continuous at $x=0$.

Now, $(gof)''(x)=$

$$\begin{cases}-2\cos x^2+4x^2\sin x^2, & x<0\\ 2\cos x^2-4x^2\sin x^2, & x\ge 0\end{cases}$$

$\therefore \quad L(gof)''(0)=-2$

and $\quad R(gof)''(0)=2$

$\therefore \quad L(gof)''(0)\ne R(gof)''(0)$

Hence, $gof(x)$ is not twice differentiable at $x=0$.

Therefore, Statement I is true, Statement II is false.

**20.** (c) $f(x)=x^n \Rightarrow f(1)=1$

$$f'(x)=nx^{n-1} \Rightarrow f'(1)=n$$

$$f''(x)=n(n-1)x^{n-2}$$

$$\Rightarrow f''(1)=n(n-1)$$

... ... ... ... ... ... ...

... ... ... ... ... ... ...

$$f^n(x)=n(n-1)(n-2)\ldots 2\cdot 1$$

$$\Rightarrow f^n(1)=n(n-1)(n-2)\ldots 2\cdot 1$$

Now,

$$f(1)-\frac{f'(1)}{1!}+\frac{f''(1)}{2!}-\frac{f'''(1)}{3!}+\ldots+\frac{(-1)^n f^n(1)}{n!}$$

$$=1-\frac{n}{1!}+\frac{n(n-1)}{2!}-\frac{n(n-1)(n-2)}{3!}+\ldots+\frac{(-1)^n\,n(n-1)(n-2)\ldots 2\cdot 1}{n!}$$

$[\because (1-x)^n=1-{}^nC_1x+{}^nC_2x^2-{}^nC_3x^3+\ldots+(-1)^n\,{}^nC_n]$

$$=(1-1)^n=0$$

**21.** (a) $\because\ y=(x+\sqrt{1+x^2})^n \quad \text{...(i)}$

On differentiating w.r.t. $x$, we get

$$\frac{dy}{dx}=n(x+\sqrt{1+x^2})^{n-1}\cdot\left(1+\frac{2x}{2\sqrt{1+x^2}}\right)$$

$$=\frac{n(x+\sqrt{1+x^2})^n}{\sqrt{1+x^2}}$$

$$\Rightarrow (1+x^2)\left(\frac{dy}{dx}\right)^2=n^2y^2 \text{ [from Eq. (i)]}$$

Again, differentiating w.r.t. $x$, we get

$$(1+x^2)\cdot 2\frac{dy}{dx}\cdot\frac{d^2y}{dx^2}+2x\left(\frac{dy}{dx}\right)^2=n^2 2y\frac{dy}{dx}$$

$$\Rightarrow (1+x^2)\frac{d^2y}{dx^2}+x\frac{dy}{dx}=n^2y$$

**22.** (d) $y=\dfrac{2\cos\theta+\cos 2\theta}{\cos 3\theta+4\cos 2\theta+5\cos\theta+2}$

$\because \cos 2\theta=2\cos^2\theta-1$

and $\cos 3\theta=4\cos^3\theta-3\cos\theta$

$$=\frac{(2\cos^2\theta+2\cos\theta-1)}{4\cos^3\theta-3\cos\theta+8\cos^2\theta-4+5\cos\theta+2}$$

$$=\frac{(2\cos^2\theta+2\cos\theta-1)}{(2\cos^2\theta+2\cos\theta-1)(2\cos\theta+2)}$$

$$y=\frac{1}{2\cos\theta+2}$$

$$y=\frac{1}{2}\left(\frac{1}{\cos\theta+1}\right)=\frac{1}{4}\cdot\sec^2\frac{\theta}{2}$$

$$\Rightarrow y_{\theta=\frac{\pi}{2}}=\frac{1}{4}\times 2=\frac{1}{2}$$

$$\Rightarrow y'=\frac{1}{4}\times 2\cdot\sec\frac{\theta}{2}\cdot\sec\frac{\theta}{2}\cdot\tan\frac{\theta}{2}\times\frac{1}{2}$$

$$=\frac{1}{2}\sec^2\frac{\theta}{2}\cdot\tan\frac{\theta}{2}\cdot\frac{1}{2}$$

$$y_{\theta=\frac{\pi}{2}}=\frac{1}{2}\times 2\times\frac{1}{2}=\frac{1}{2}$$

$$y''=\frac{1}{2}\left[\frac{1}{2}\left(2\sec\frac{\theta}{2}\cdot\sec\frac{\theta}{2}\cdot\tan\frac{\theta}{2}\right)+\frac{1}{2}\sec^2\frac{\theta}{2}\cdot\sec^2\frac{\theta}{2}\right]$$

$$\Rightarrow y''_{\theta=\pi/2}=1$$

$$\therefore y''+y'+y=2$$

**23.** (a) $y=\log_e\left(\frac{1-x^2}{1+x^2}\right), -1<x<1$

$$y'=\frac{1}{\left(\frac{1-x^2}{1+x^2}\right)}\cdot\frac{(1+x^2)(-2x)-(1-x^2)(2x)}{(1+x^2)^2}$$

$$=\frac{2x(-1-x^2-1+x^2)}{(1-x^2)(1+x^2)}=\frac{-4x}{1-x^4}$$

$$y''=\frac{(1-x^4)(-4)-(-4x)(-4x^3)}{(1-x^4)^2}$$

$$=4\frac{(-3x^4-1)}{(1-x^4)^2}=-4\left[\frac{3x^4+1}{(1-x^4)^2}\right]$$

At $x=\frac{1}{2}$

$$\Rightarrow y'=\frac{-4\times\frac{1}{2}}{1-\frac{1}{16}}=\frac{-2\times16}{15}=\frac{-32}{15}$$

and $y''=\frac{-4\left(\frac{3}{16}+1\right)}{\left(1-\frac{1}{16}\right)^2}$

$$=\frac{-4\times19}{16}\times\frac{16\times16}{15\times15}=\frac{-64\times19}{225}$$

Now,

$$225(y'-y'')=225\left(\frac{-32}{15}+\frac{64\times19}{225}\right)$$

$$=32(-15+38)=736$$

**24.** (b)

**key idea** Differentiating the given equation twice w.r.t. $x$.

Given equation is

$$e^y+xy=e \quad \ldots(i)$$

On differentiating both sides w.r.t. $x$, we get

$$e^y\frac{dy}{dx}+x\frac{dy}{dx}+y=0 \quad \ldots(ii)$$

$$\Rightarrow \frac{dy}{dx}=-\left(\frac{y}{e^y+x}\right) \quad \ldots(iii)$$

Again, differentiating Eq. (ii) w.r.t. '$x$', we get

$$e^y\frac{d^2y}{dx^2}+e^y\left(\frac{dy}{dx}\right)^2+x\frac{d^2y}{dx^2}+\frac{dy}{dx}+\frac{dy}{dx}=0 \quad \ldots(iv)$$

On putting $x=0$ in Eq. (i), we get

$$e^y=e^1 \Rightarrow y=1$$

On putting $x=0$, $y=1$ in Eq. (iii), we get

$$\frac{dy}{dx}=-\frac{1}{e+0}=-\frac{1}{e}$$

Now, on putting $x=0$, $y=1$ and $\frac{dy}{dx}=-\frac{1}{e}$ in Eq. (iv), we get

$$e^1\frac{d^2y}{dx^2}+e^1\left(-\frac{1}{e}\right)^2+0\left(\frac{d^2y}{dx^2}\right)+\left(-\frac{1}{e}\right)+\left(-\frac{1}{e}\right)=0$$

$$\Rightarrow \left.\frac{d^2y}{dx^2}\right|_{(0,1)}=\frac{1}{e^2}$$

So, $\left(\frac{dy}{dx},\frac{d^2y}{dx^2}\right)$ at (0, 1) is $\left(-\frac{1}{e},\frac{1}{e^2}\right)$.

**25.** (d) $\frac{dy}{dx}=15(x+\sqrt{x^2-1})^{14}\left(1+\frac{x}{\sqrt{x^2-1}}\right)$

$$+15(x-\sqrt{x^2-1})^{14}\left(1-\frac{x}{\sqrt{x^2-1}}\right)$$

$$=\frac{15}{\sqrt{x^2-1}}[(x+\sqrt{x^2-1})^{15}-(x-\sqrt{x^2-1})^{15}]$$

$$=\sqrt{x^2-1}\frac{d^2y}{dx^2}+\frac{dy}{dx}\left(\frac{x}{\sqrt{x^2-1}}\right)$$

$$=15\left[15(x+\sqrt{x^2-1})^{14}\left(1+\frac{x}{\sqrt{x^2-1}}\right)-15(x+\sqrt{x^2-1})^{14}\left(1-\frac{x}{\sqrt{x^2-1}}\right)\right]$$

$$=\frac{15}{\sqrt{x^2-1}}(15(x+\sqrt{x^2-1})^{15}+15(x+\sqrt{x^2-1})^{15})$$

$$=\frac{225y}{\sqrt{x^2-1}}$$

$$\Rightarrow (x^2-1)\frac{d^2y}{dx^2}+x\frac{dy}{dx}=225y$$

**26.** (5) $(3f'f''+ff''')(x)$

$$=[(ff''+(f')^2)(x)]'$$

$$=(ff'(x)')'=(f(x)-f'(x))''$$

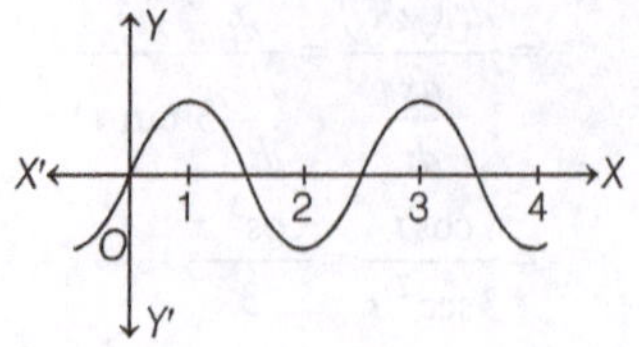


Minimum number of roots of $f(x)=4$
Minimum number of root of $f'(x)=3$
Minimum number of root of $f(x)\cdot f'(x)=7$
Minimum number of roots of $[f(x)\cdot f'(x)]''=5$

**27.** (202) Given equation is

$$f(x)=x^3+x^2f'(1)+xf''(2)+f'''(3)$$

$x\in R$ ...(i)

Differentiating

$$f'(x)=3x^2+2xf'(1)+f''(2) \quad \ldots(ii)$$

Again, differentiating

$$f''(x)=6x+2f'(1) \quad \ldots(iii)$$

Again, differentiating,

$$f'''(x)=6 \quad \ldots(iv)$$

Now, $f(1)=-5$, $f''(2)=2$, $f'''(3)=6$
[with the help of above Eqs. (ii), (iii) and (iv)]

$$f(x)=x^3+x^2(-5)+x(2)+6$$

$$f'(x)=3x^2-10x+2$$

Hence, $f'(10)=300-100+2$
$=202$

**28.** (10) Given, $f(x)=\sum_{k=1}^{10}k\cdot x^k$, $x\in R$

Now, $f(x)=x+2x^2+\ldots+10x^{10}$

$$\Rightarrow x\cdot f(x)=x^2+2x^3+\ldots+9x^{10}+10x^{11}$$

$$\Rightarrow (1-x)f(x)=(x+2x^2+\ldots+10x^{10})-(x^2+2x^3+\ldots+9x^{10}+10x^{11})$$

$$=x+x^2+x^3+\ldots+x^{10}-10x^{11}$$

$$=\frac{x(1-x^{10})}{(1-x)}-10x^{11}$$

$$\Rightarrow f(x)=\frac{x(1-x^{10})}{(1-x)^2}-\frac{10x^{11}}{(1-x)}$$

$$=\frac{x-x^{11}-10x^{11}+10x^{12}}{(1-x)^2}$$

$$=\frac{10x^{12}-11x^{11}+x}{(1-x)^2}$$

$$f'(x)=\frac{(10x^{12}-11x^{11}+x)\cdot 2\cdot(1-x)+(120x^{11}-121x^{10}+1)\cdot(1-x)^2}{(1-x)^4}$$

$$=\frac{20x^{12}-22x^{11}+2x+(120x^{11}-121x^{10}+1)(1-x)}{(1-x)^3}$$

Thus, $2f(2)+f'(2)=119\cdot 2^{10}+1$

So, $n=10$

**29.** *(16)* Given,

$C\ (x^2+y^2-3)+(x^2-y^2-1)^5=0$

Point $(\alpha, \alpha)$ lies on $C$ and $\alpha>0$

$\therefore\quad 2\alpha^2-3-1=0$

$\Rightarrow\quad \alpha=\sqrt{2}$

On differentiating w.r.t. $x$

$C:(x^2+y^2-3)+(x^2-y^2-1)^5=0$

$\Rightarrow\quad 2x+2yy'+5(x^2-y^2-1)^4(2x-2yy')=0$

On putting $x=\alpha,\ y=\alpha,\ \alpha=\sqrt{2}$,

$\sqrt{2}+\sqrt{2}y'+5(\sqrt{2}-\sqrt{2}y')=0$

$\Rightarrow\quad 6\sqrt{2}=4\sqrt{2}y'$

$\Rightarrow\quad y'=\frac{3}{2}$

Again differentiating $C$, we get

$2+2((y')^2+yy'')+20(x^2-y^2-1)^3(2x-2yy')^2+5(x^2-y^2-1)^4(2-2((y')^2+yy''))=0$

On putting $x=y=\sqrt{2},\ y'=\frac{3}{2}$,

$2+\frac{9}{2}+2\sqrt{2}y''-40+10-\frac{45}{2}-10\sqrt{2}y''=0$

$\Rightarrow\quad y''=\frac{-23}{4\sqrt{2}}$

$\therefore\quad 3y'-y^3y''$ at $(\sqrt{2},\sqrt{2})$

$3\times\frac{3}{2}-2\sqrt{2}\times\left(\frac{-23}{4\sqrt{2}}\right)=16$

**30.** *(17)* Given, $y^{1/4}+y^{-1/4}=2x$

$\Rightarrow\quad (y^{1/4}+y^{-1/4})^2=(2x)^2$

$\Rightarrow\quad (y^{1/4}+y^{-1/4})^2=4x^2$

On differentiating w.r.t. $x$, we get

$\frac{1}{4y}(y^{1/4}-y^{-1/4})\frac{dy}{dx}=2$

$\Rightarrow\quad \left(y^{\frac{1}{4}}-y^{-\frac{1}{4}}\right)\frac{dy}{dx}=8y\quad$ ...(i)

Now, $y^{1/4}-y^{-1/4}$

$=\sqrt{\left(y^{\frac{1}{4}}+y^{-\frac{1}{4}}\right)^2-4}$

$\Rightarrow\quad y^{1/4}-y^{-1/4}=2\sqrt{x^2-1}\quad$ ...(ii)

$\Rightarrow\quad (\sqrt{x^2-1})\frac{dy}{dx}=4y$

[using Eqs. (i) and (ii)]

On squaring on both sides,

$(x^2-1)\left(\frac{dy}{dx}\right)^2=16y^2$

Again, differentiating w.r.t. $x$,

$(x^2-1)\cdot 2\frac{dy}{dx}\cdot\frac{d^2y}{dx^2}+2x\left(\frac{dy}{dx}\right)^2=32y\frac{dy}{dx}$

On dividing by $\frac{2dy}{dx}$, we get

$(x^2-1)\frac{d^2y}{dx^2}+x\frac{dy}{dx}=16y$

or $(x^2-1)\frac{d^2y}{dx^2}+x\frac{dy}{dx}-16y=0$

On comparing with

$(x^2-1)\frac{d^2y}{dx^2}+\alpha x\frac{dy}{dx}+\beta y=0$

$\alpha=1, \beta=-16$

$\therefore\quad |\alpha-\beta|=|1+16|=17$

**31.** *(40)* We have, $\log(x+y)=4xy$

$\Rightarrow\quad x+y=e^{4xy}$

$\Rightarrow\quad 1+\frac{dy}{dx}=\left(4x\frac{dy}{dx}+4y\right)e^{4xy}$

If $x=0$, then $y=1$

At $(0, 1)$, $\frac{dy}{dx}=3$

$\frac{d^2y}{dx^2}=e^{4xy}\left(4x\frac{dy}{dx}+4y\right)^2+e^{4xy}\left(4x\frac{d^2y}{dx^2}+\frac{4dy}{dx}+4\frac{dy}{dx}\right)$

At $x=0$, $\frac{d^2y}{dx^2}=16+24=40$

# C-18

JEE Main Chapterwise Topicwise Mathematics

In JEE Main, there is usually 0-1 question per paper is visible from this chapter. In past 3 years, 39 conceptual and 7 numerical questions of moderate difficulty level were asked.

**For detailed analysis scan QR code**

# APPLICATIONS OF DERIVATIVES

## Topic 1 Rate of Change of Quantity

### Level 1

**01** A spherical chocolate ball has a layer of ice-cream of uniform thickness around it. When the thickness of the ice-cream layer is 1 cm, the ice-cream melts at the rate of 81 $cm^3$/min and the thickness of the ice-cream layer decreases at the rate of $\frac{1}{4\pi}$cm/min. The surface area (in $cm^2$) of the chocolate ball (without the ice-cream layer) is *[2025, 23 Jan Shift-II]*

(a) $128\pi$ (b) $196\pi$
(c) $256\pi$ (d) $225\pi$

**02** The position of a moving car at time $t$ is given by $f(t) = at^2 + bt + c, t > 0$, where a, $b$ and $c$ are real numbers greater than 1. Then, average speed of the car over the time interval $[t_1, t_2]$ is attained at the point *[2020, 6 Sep Shift-I]*

(a) $(t_2 - t_1)/2$ (b) $a(t_2 - t_1) + b$
(c) $(t_1 + t_2)/2$ (d) $2a(t_1 + t_2) + b$

**03** A spherical iron ball of 10 cm radius is coated with a layer of ice of uniform thickness that melts at a rate of 50 $cm^3$/min. When the thickness of ice is 5 cm, then the rate (in cm/ min.) at which of the thickness of ice decreases, is *[2020, 9 Jan Shift-I]*

(a) $\frac{5}{6\pi}$ (b) $\frac{1}{54\pi}$ (c) $\frac{1}{36\pi}$ (d) $\frac{1}{18\pi}$

**04** A 2m ladder leans against a vertical wall. If the top of the ladder begins to slide down the wall at the rate 25 cm/sec, then rate (in cm/sec) at which the bottom of the ladder slides away from the wall on the horizontal ground when the top of the ladder is 1m above the ground is *[2019, 12 April Shift-I]*

(a) $25/\sqrt{3}$ (b) $25\sqrt{3}$
(c) 25 (d) 25/3

**05** A point on the parabola $y^2 = 18x$ at which the ordinate increases at twice the rate of the abscissa, is *[AIEEE 2004]*

(a) (2, 4) (b) (2, – 4) (c) $\left(-\frac{9}{8}, \frac{9}{2}\right)$ (d) $\left(\frac{9}{8}, \frac{9}{2}\right)$

### Level 2

**06** If the surface area of a cube is increasing at a rate of 3.6 $cm^2$/sec, retaining its shape, then the rate of change of its volume (in $cm^3$/sec), when the length of a side of the cube is 10 cm, is *[2020, 3 Sep Shift-II]*

(a) 18 (b) 10 (c) 9 (d) 20

### Level 3

**07** The surface area of a balloon of spherical shape being inflated, increases at a constant rate. If initially, the radius of balloon is 3 units and after 5 s, it becomes 7 units, then its radius after 9 s is *[2022, 24 June Shift-I]*

(a) 9 (b) 10 (c) 11 (d) 12

**08** A spherical balloon is filled with $4500\pi$ cu m of helium gas. If a leak in the balloon causes the gas to escape at the rate of $72\pi$ cu m / min, then the rate (in m/min) at which the radius of the balloon decreases 49 min after the leakage began is *[AIEEE 2012]*

(a) 9/7 (b) 7/9 (c) 2/9 (d) 9/2

### Numerical Value Type Questions

**09** The number of points, where the curve $y = x^5 - 20x^3 + 50x + 2$ crosses the $X$-axis, is .......... *[2023, 6 April Shift-II]*

**10** A water tank has the shape of a right circular cone with axis vertical and vertex downwards. Its semi-vertical angle is $\tan^{-1} 3/4$. Water is poured in it at a constant rate of 6 cubic meter per hour. The rate (in square metre per hour), at which the wet curved surface area of the tank is increasing, when the depth of water in the tank is 4 m, is ............ . *[2022, 27 July Shift-II]*

## Topic 2 Monotonic Increasing and Decreasing Function

### Level 1

**01** Let the function $f(x)=\frac{x}{3}+\frac{3}{x}+3, x\neq 0$ be strictly increasing in $(-\infty,\alpha_1)\cup(\alpha_2,\infty)$ and strictly decreasing in $(\alpha_3,\alpha_4)\cup(\alpha_4,\alpha_5)$. Then $\sum_{i=1}^{5}\alpha_i^2$ is equal to [2025, 8 April Shift-II]
(a) 40 (b) 36 (c) 48 (d) 28

**02** Let (2, 3) be the largest open interval in which the function $f(x)=2\log_e(x-2)-x^2+ax+1$ is strictly increasing and $(b, c)$ be the largest open interval, in which the function $g(x)=(x-1)^3(x+2-a)^2$ is strictly decreasing. Then, $100(a+b-c)$ is equal to [2025, 24 Jan Shift-II]
(a) 420 (b) 160 (c) 360 (d) 280

**03** For the function $f(x)=(\cos x)-x+1, x\in R$, between the following two statements
(S1) $f(x)=0$ for only one value of $x$ in $[0,\pi]$.
(S2) $f(x)$ is decreasing in $\left[0,\frac{\pi}{2}\right]$ and increasing in $\left[\frac{\pi}{2},\pi\right]$. [2024, 8 April Shift-I]
(a) Only (S2) is correct.
(b) Both (S1) and (S2) are correct.
(c) Both (S1) and (S2) are incorrect.
(d) Only (S1) is correct.

**04** The interval in which the function $f(x)=x^x, x>0$, is strictly increasing is [2024, 6 April Shift-I]
(a) $\left[\frac{1}{e},\infty\right)$ (b) $\left(0,\frac{1}{e}\right]$ (c) $(0,\infty)$ (d) $\left[\frac{1}{e^2},1\right)$

**05** The function $f(x)=\frac{x}{x^2-6x-16}, x\in R-\{-2,8\}$ [2024, 29 Jan Shift-II]
(a) decreases in $(-\infty,-2)$ and increases in $(8,\infty)$
(b) increases in $(-\infty,-2)\cup(-2,8)\cup(8,\infty)$
(c) decreases in $(-\infty,-2)\cup(-2,8)\cup(8,\infty)$
(d) decreases in $(-2,8)$ and increases in $(-\infty,-2)\cup(8,\infty)$

**06** Let $f(x)=3\sin^4 x+10\sin^3 x+6\sin^2 x-3$ $x\in\left[-\frac{\pi}{6},\frac{\pi}{2}\right]$. Then, $f$ is [2021, 25 July Shift-I]
(a) increasing in $\left(-\frac{\pi}{6},\frac{\pi}{2}\right)$ (b) decreasing in $\left(0,\frac{\pi}{2}\right)$
(c) increasing in $\left(-\frac{\pi}{6},0\right)$ (d) decreasing in $\left(-\frac{\pi}{6},0\right)$

**07** The function
$$f(x)=\frac{4x^3-3x^2}{6}-2\sin x+(2x-1)\cos x$$ [2021, 24 Feb Shift-I]
(a) increases in $\left[\frac{1}{2},\infty\right)$ (b) decreases in $\left[\frac{1}{2},\infty\right)$
(c) increases in $\left(-\infty,\frac{1}{2}\right]$ (d) decreases in $\left(-\infty,\frac{1}{2}\right]$

**08** The function $f(x)=(3x-7)x^{2/3}, x\in R$, is increasing for all $x$ lying in [2020, 3 Sep Shift-I]
(a) $(-\infty,0)\cup\left[\frac{14}{15},\infty\right)$ (b) $(-\infty,0)\cup\left(\frac{3}{7},\infty\right)$
(c) $\left(-\infty,\frac{14}{15}\right)$ (d) $\left(-\infty,-\frac{14}{15}\right)\cup(0,\infty)$

**09** Let $f(x)=\frac{x}{\sqrt{a^2+x^2}}-\frac{d-x}{\sqrt{b^2+(d-x)^2}}, x\in R$, where $a, b$ and $d$ are non-zero real constants. Then, [2019, 11 Jan Shift-II]
(a) $f$ is an increasing function of $x$
(b) $f'$ is not a continuous function of $x$
(c) $f$ is a decreasing function of $x$
(d) $f$ is neither increasing nor decreasing function of $x$

**10** Let $f(x)=\sin^4 x+\cos^4 x$. Then, $f$ is an increasing function in the interval. [JEE Main 2016]
(a) $\left]\frac{5\pi}{8},\frac{3\pi}{4}\right[$ (b) $\left]\frac{\pi}{2},\frac{5\pi}{8}\right[$
(c) $\left]\frac{\pi}{4},\frac{\pi}{2}\right[$ (d) $\left]0,\frac{\pi}{4}\right[$

**11** The real number $k$ for which the equation $2x^3+3x+k=0$ has two distinct real roots in [0, 1] [JEE Main 2013]
(a) lies between 2 and 3 (b) lies between −1 and 0
(c) does not exist (d) lies between 1 and 2

**12** How many real solutions does the equation $x^7+14x^5+16x^3+30x-560=0$ have? [AIEEE 2008]
(a) 5 (b) 7 (c) 1 (d) 3

**13** The function $f(x)=\tan^{-1}(\sin x+\cos x)$ is an increasing function in [AIEEE 2007]
(a) $\left(\frac{\pi}{4},\frac{\pi}{2}\right)$ (b) $\left(\frac{-3\pi}{4},\frac{\pi}{4}\right)$
(c) $\left(0,\frac{\pi}{2}\right)$ (d) $\left(-\frac{\pi}{2},\frac{\pi}{2}\right)$

**14** The function $f(x)=\cot^{-1}x+x$ increases in the interval [AIEEE 2002]
(a) $(1,\infty)$ (b) $(-1,\infty)$
(c) $(-\infty,\infty)$ (d) $(0,\infty)$

## Level 2

**15** The number of critical points of the function $f(x)=(x-2)^{2/3}(2x+1)$ is **[2024, 8 April Shift-I]**

(a) 1 (b) 3 (c) 0 (d) 2

**16** For the function $f(x)=\sin x+3x-\frac{2}{\pi}(x^2+x)$, where $x\in\left[0,\frac{\pi}{2}\right]$, consider the following two statements: **[2024, 5 April Shift-I]**

I. $f$ is increasing in $\left(0,\frac{\pi}{2}\right)$.

II. $f'$ is decreasing in $\left(0,\frac{\pi}{2}\right)$.

Between the above two statements,

(a) Neither I nor II is true (b) Both I and II are true
(c) Only I is true (d) Only II is true

**Imp.** Sandwich Theorem

**17** Let $f:R\to(0,\infty)$ be strictly increasing function such that $\lim_{x\to\infty}\frac{f(7x)}{f(x)}=1$. Then, the value of $\lim_{x\to\infty}\left[\frac{f(5x)}{f(x)}-1\right]$ is equal to **[2024, 31 Jan Shift-II]**

(a) 7/5 (b) 1 (c) 0 (d) 4

**18** Let $g(x)=3f\left(\frac{x}{3}\right)+f(3-x)$ and $f''(x)>0$ for all $x\in(0,3)$. If $g$ is decreasing in $(0,\alpha)$ and increasing in $(\alpha,3)$, then $8\alpha$ is **[2024, 27 Jan Shift-II]**

(a) 0 (b) 20 (c) 18 (d) 24

**19** Let $g(x)=f(x)+f(1-x)$ and $f''(x)>0$, $x\in(0,1)$. If $g$ is decreasing in the interval $(0,\alpha)$ and increasing in the interval $(\alpha,1)$, then $\tan^{-1}(2\alpha)+\tan^{-1}\left(\frac{1}{\alpha}\right)+\tan^{-1}\left(\frac{\alpha+1}{\alpha}\right)$ is equal to **[2023, 12 April Shift-II]**

(a) $\frac{5\pi}{4}$ (b) $\frac{3\pi}{4}$
(c) $\frac{3\pi}{2}$ (d) $\pi$

**20** The function $f(x)=xe^{x(1-x)}$, $x\in R$ is **[2022, 28 July Shift-II]**

(a) increasing in $\left(-\frac{1}{2},1\right)$ (b) decreasing in $\left(\frac{1}{2},2\right)$
(c) increasing in $\left(-1,-\frac{1}{2}\right)$ (d) decreasing in $\left(-\frac{1}{2},\frac{1}{2}\right)$

**21** If the maximum value of $a$, for which the function $f_a(x)=\tan^{-1}2x-3ax+7$ is non-decreasing in $\left(-\frac{\pi}{6},\frac{\pi}{6}\right)$, is $\bar{a}$, then $f_{\bar{a}}(\pi/8)$ is equal to **[2022, 26 July Shift-II]**

(a) $8-\frac{9\pi}{4(9+\pi^2)}$ (b) $8-\frac{4\pi}{9(4+\pi^2)}$
(c) $8\left(\frac{1+x^2}{9+\pi^2}\right)$ (d) $8-\frac{\pi}{4}$

**22** Let $\lambda^*$ be the largest value of $\lambda$ for which the function $f_\lambda(x)=4\lambda x^3-36\lambda x^2+36x+48$ is increasing for all $x\in R$. Then, $f_{\lambda}^*(1)+f_{\lambda}^*(-1)$ is equal to **[2022, 24 June Shift-II]**

(a) 36 (b) 48 (c) 64 (d) 72

**23** The function $f(x)=x^3-6x^2+ax+b$ is such that $f(2)=f(4)=0$. Consider two statements. $(S_1)$ there exists $x_1,x_2\in(2,4)$, $x_1<x_2$, such that $f'(x_1)=-1$ and $f'(x_2)=0$. $(S_2)$ there exists $x_3,x_4\in(2,4)$, $x_3<x_4$, such that $f$ is decreasing in $(2,x_4)$, increasing in $(x_4,4)$ and $2f'(x_3)=\sqrt{3}f(x_4)$. Then, **[2021, 1 Sep Shift-II]**

(a) Both $(S_1)$ and $(S_2)$ are true
(b) $(S_1)$ is false and $(S_2)$ is true
(c) Both $(S_1)$ and $(S_2)$ are false
(d) $(S_1)$ is true and $(S_2)$ is false

**24** Let $f:R\to R$ be defined as

$$f(x)=\begin{cases}-55x, & \text{if } x<-5\\ 2x^3-3x^2-120x, & \text{if } -5\le x\le 4\\ 2x^3-3x^2-36x-336, & \text{if } x>4\end{cases}$$

Let $A=\{x\in R: f \text{ is increasing}\}$. Then, $A$ is equal to **[2021, 24 Feb Shift-II]**

(a) $(-\infty,-5)\cup(4,\infty)$ (b) $(-5,\infty)$
(c) $(-\infty,-5)\cup(-4,\infty)$ (d) $(-5,-4)\cup(4,\infty)$

## Level 3

**25** If $5f(x)+4f\left(\frac{1}{x}\right)=x^2-2$, $\forall x\ne0$ and $y=9x^2f(x)$, then $y$ is strictly increasing in **[2024, 1 Feb Shift-I]**

(a) $\left(0,\frac{1}{\sqrt{5}}\right)\cup\left(\frac{1}{\sqrt{5}},\infty\right)$

(b) $\left(-\frac{1}{\sqrt{5}},0\right)\cup\left(\frac{1}{\sqrt{5}},\infty\right)$

(c) $\left(-\frac{1}{\sqrt{5}},0\right)\cup\left(0,\frac{1}{\sqrt{5}}\right)$

(d) $\left(-\infty,\frac{1}{\sqrt{5}}\right)\cup\left(0,\frac{1}{\sqrt{5}}\right)$

**26** Let $f(x)=2x+\tan^{-1}x$ and $g(x)=\log_e(\sqrt{1+x^2}+x), x\in[0,3]$. Then, **[2023, 1 Feb Shift-I]**

(a) there exists $x\in[0,3]$ such that $f'(x)<g'(x)$
(b) there exist $0<x_1<x_2<3$ such that $f(x)<g(x)$, $\forall x\in(x_1,x_2)$
(c) $\max f(x)>\max g(x)$
(d) $\min f'(x)=1+\max g'(x)$

**27** Let $f:(0,1)\to R$ be a function defined by $f(x)=\dfrac{1}{1-e^{-x}}$ and $g(x)=(f(-x)-f(x))$. Consider two statements

(I) $g$ is an increasing function in (0, 1).
(II) $g$ is one-one in (0, 1).

Then, **[2023, 25 Jan Shift-I]**

(a) only (II) is true
(b) Both (I) and (II) are true
(c) only (I) is true
(d) Neither (I) nor (II) is true

**28** For the function $f(x)=4\log_e(x-1)-2x^2+4x+5, x>1$, which one of the following is not correct ? **[2022, 24 June Shift-I]**

(a) $f$ is increasing in (1, 2) and decreasing in $(2,\infty)$
(b) $f(x)=-1$ has exactly two solutions
(c) $f'(e)-f''(2)<0$
(d) $f(x)=0$ has a root in the interval $(e, e+1)$

**29** If $m$ is the minimum value of $k$ for which the function $f(x)=x\sqrt{kx-x^2}$ is increasing in the interval [0, 3] and $M$ is the maximum value of $f$ in the interval [0, 3], when $k=m$, then the ordered pair $(m, M)$ is equal to **[2019, 12 April Shift-I]**

(a) $(4,3\sqrt{2})$ (b) $(4,3\sqrt{3})$ (c) $(3,3\sqrt{3})$ (d) $(5,3\sqrt{6})$

### Numerical Value Type Questions

**30** Let the set of all values of $p$, for which $f(x)=(p^2-6p+8)(\sin^2 2x-\cos^2 2x)+2(2-p)x+7$ does not have any critical point, be the interval $(a, b)$. Then, $16ab$ is equal to ……… . **[2024, 9 April Shift-II]**

**31** If $R$ is the least value of $a$ such that the function $f(x)=x^2+ax+1$ is increasing on [1, 2] and $S$ is the greatest value of $a$ such that the function $f(x)=x^2+ax+1$ is decreasing on [1, 2], then the value of $|R-S|$ is …… . **[2021, 31 Aug Shift-I]**

**32** The number of elements in the set $\{n\in\{1,2,3,\ldots,100\}\mid(11)^n>(10)^n+(9)^n\}$ is …… . **[2021, 22 July Shift-II]**

## Topic 3 Maxima and Minima

### Level 1

**01** If the function $f(x)=2x^3-9ax^2+12a^2x+1$, where $a>0$, attains its local maximum and local minimum values at $p$ and $q$, respectively, such that $p^2=q$, then $f(3)$ is equal to **[2025, 2 April Shift-I]**

(a) 37 (b) 23 (c) 55 (d) 10

**02** Let $a>0$. If the function $f(x)=6x^3-45ax^2+108a^2x+1$ attains its local maximum and minimum values at the points $x_1$ and $x_2$ respectively such that $x_1x_2=54$, then $a+x_1+x_2$ is equal to **[2025, 4 April Shift-II]**

(a) 18 (b) 13 (c) 24 (d) 15

**03** Let $f:R\to R$ be a polynomial function of degree four having extreme values at $x=4$ and $x=5$. If $\lim\limits_{x\to 0}\dfrac{f(x)}{x^2}=5$, then $f(2)$ is equal to **[2025, 7 April Shift-II]**

(a) 10 (b) 12 (c) 8 (d) 14

**04** Let $f(x)=\int_0^x t(t^2-9t+20)\,dt, 1\le x\le 5$. If the range of $f$ is $[\alpha,\beta]$, then $4(\alpha+\beta)$ equals **[2025, 29 Jan Shift-II]**

(a) 253 (b) 154
(c) 125 (d) 157

**05** Let the sum of the maximum and the minimum values of the function $f(x)=\dfrac{2x^2-3x+8}{2x^2+3x+8}$ be $\dfrac{m}{n}$, where $gcd(m, n)=1$. Then, $m+n$ is equal to **[2024, 4 April Shift-I]**

(a) 201 (b) 217
(c) 195 (d) 182

**06** Let $f(x)=(x+3)^2(x-2)^3, x\in[-4,4]$. If $M$ and $m$ are the maximum and minimum values of $f$, respectively in $[-4,4]$, then the value of $M-m$ is **[2024, 30 Jan Shift-II]**

(a) 392 (b) 108
(c) 608 (d) 600

**07** The function $f(x)=2x+3(x)^{2/3}, x\in R$ has **[2024, 29 Jan Shift-II]**

(a) exactly one point of local minima and no point of local maxima
(b) exactly two points of local maxima and exactly one point of local minima
(c) exactly one point of local maxima and exactly one point of local minima
(d) exactly one point of local maxima and no point of local minima

**08** The sum of the absolute maximum and minimum values of the function $f(x)=|x^2-5x+6|-3x+2$ in the interval $[-1, 3]$ is equal to **[2023, 1 Feb Shift-II]**

(a) 12 (b) 24
(c) 13 (d) 10

**09** The abolute minimum value of the function $f(x)=|x^2-x+1|+[x^2-x+1]$, which $[t]$ denotes the greatest integer function, in the interval $[-1, 2]$, is **[2023, 31 Jan Shift-II]**

(a) 3/2 (b) 5/4
(c) 3/4 (d) 1/4

**10** If the functions $f(x)=\dfrac{x^3}{3}+2bx+\dfrac{ax^2}{2}$ and $g(x)=\dfrac{x^3}{3}+ax+bx^2$, $a\neq 2b$ have a common extreme point, then $a+2b+7$ is equal to **[2023, 30 Jan Shift-II]**

(a) 3 (b) $\dfrac{3}{2}$
(c) 6 (d) 4

**11** The minimum value of the sum of the squares of the roots of $x^2+(3-a)x+1=2a$ is **[2022, 26 July Shift-II]**

(a) 4 (b) 5
(c) 6 (d) 8

**Imp.** **AM-GM Inequality**

**12** Let $x, y>0$. If $x^3y^2=2^{15}$, then the least value of $3x+2y$ is **[2022, 24 June Shift-II]**

(a) 30 (b) 32 (c) 36 (d) 40

**13** Let $f(x)$ be a cubic polynomial with $f(1)=-10$, $f(-1)=6$, and has a local minima at $x=1$, and $f'(x)$ has a local minima at $x=-1$. Then $f(3)$ is equal to **[2021, 31 Aug Shift-II]**

(a) 22 (b) 24
(c) 26 (d) 20

**14** The local maximum value of the function $f(x)=\left(\dfrac{2}{x}\right)^{x^2}$, $x>0$ **[2021, 26 Aug Shift-II]**

(a) $(2\sqrt{e})^{\frac{1}{e}}$ (b) $(4/\sqrt{e})^{\frac{e}{4}}$
(c) $(e)^{\frac{2}{e}}$ (d) 1

**15** If $p(x)$ be a polynomial of degree three that has a local maximum value 8 at $x=1$ and a local minimum value 4 at $x=2$, then $p(0)$ is equal to **[2020, 2 Sep Shift-I]**

(a) −24 (b) 6
(c) 12 (d) −12

**16** If $f(x)$ is a non-zero polynomial of degree four, having local extreme points at $x=-1, 0, 1$, then the set $S=\{x\in R: f(x)=f(0)\}$ contains exactly **[2019, 9 April Shift-I]**

(a) four rational numbers
(b) two irrational and two rational numbers
(c) four irrational numbers
(d) two irrational and one rational number

**17** If the function $f$ given by $f(x)=x^3-3(a-2)x^2+3ax+7$, for some $a\in R$ is increasing in $(0, 1]$ and decreasing in $[1, 5)$, then a root of the equation, $\dfrac{f(x)-14}{(x-1)^2}=0\ (x\neq 1)$ is **[2019, 12 Jan Shift-I]**

(a) −7 (b) 6 (c) 7 (d) 5

**18** Let $f(x)=x^2+\dfrac{1}{x^2}$ and $g(x)=x-\dfrac{1}{x}$, $x\in R-\{-1, 0, 1\}$. If $h(x)=\dfrac{f(x)}{g(x)}$, then the local minimum value of $h(x)$ is **[JEE Main 2018]**

(a) 3 (b) −3 (c) $-2\sqrt{2}$ (d) $2\sqrt{2}$

**Imp.** **Absolute Maximum & Absolute Minimum**

**19** Let $M$ and $m$ be respectively the absolute maximum and the absolute minimum value of the function, $f(x)=2x^3-9x^2+12x+5$ in the interval $[0, 3]$. Then, $M-m$ is equal to **[JEE Main 2018]**

(a) 5 (b) 1 (c) 4 (d) 9

**20** If $x=-1$ and $x=2$ are extreme points of $f(x)=\alpha\log|x|+\beta x^2+x$, then **[JEE Main 2014]**

(a) $\alpha=-6, \beta=\dfrac{1}{2}$ (b) $\alpha=-6, \beta=-\dfrac{1}{2}$
(c) $\alpha=2, \beta=-\dfrac{1}{2}$ (d) $\alpha=2, \beta=\dfrac{1}{2}$

**21** Let $f: R\to R$ be a positive increasing function with $\lim\limits_{x\to\infty}\dfrac{f(3x)}{f(x)}=1$. Then, $\lim\limits_{x\to\infty}\dfrac{f(2x)}{f(x)}$ is equal to **[AIEEE 2010]**

(a) 1 (b) 2/3
(c) 3/2 (d) 3

**22** Let $f: R\to R$ be defined by
$$f(x)=\begin{cases}k-2x, & \text{if } x\leq -1\\ 2x+3, & \text{if } x>-1\end{cases}.$$
If $f$ has a local minimum at $x=-1$, then a possible value of $k$ is **[AIEEE 2010]**

(a) 1 (b) 0
(c) $-\dfrac{1}{2}$ (d) −1

**23** Let $f : R \to R$ be a continuous function defined by

$$f(x) = \frac{1}{e^x + 2e^{-x}}$$

**Statement I** $f(c) = 1/3$, for some $c \in R$.

**Statement II** $0 < f(x) \le \frac{1}{2\sqrt{2}}, \forall x \in R$ *[AIEEE 2010]*

(a) Statement I is true, Statement II is true; Statement II is a correct explanation of Statement I.

(b) Statement I is true, Statement II is true, Statement II is not a correct explanation for Statement I.

(c) Statement I is true, Statement II is false.

(d) Statement I is false, Statement II is true.

**24** The greatest value of

$f(x) = (x+1)^{1/3} - (x-1)^{1/3}$ on [0, 1] is *[AIEEE 2002]*

(a) 1 (b) 2 (c) 3 (d) 1/3

## Level 2

**25** Let $f : R \to R$ be a function defined by $f(x) = ||x+2| - 2|x||$. If $m$ is the number of points of local minima and $n$ is the number of points of local maxima of $f$, then $m + n$ is *[2025, 3 April Shift-II]*

(a) 2 (b) 5

(c) 3 (d) 4

**26** Let $x = -1$ and $x = 2$ be the critical points of the function $f(x) = x^3 + ax^2 + b\log_e|x| + 1, x \ne 0$. Let $m$ and $M$ respectively be the absolute minimum and the absolute maximum values of $f$ in the interval $\left[-2, -\frac{1}{2}\right]$. Then, $[M + m]$ is equal to (take $\log_e 2 = 0.7$) *[2025, 7 April Shift-I]*

(a) 19.8 (b) 20.9

(c) 21.1 (d) 22.1

**27** The sum of all local minimum values of the function

$$f(x) = \begin{cases} 1 - 2x, & x < -1 \\ \frac{1}{3}(7 + 2|x|), & -1 \le x \le 2 \\ \frac{11}{18}(x-4)(x-5), & x > 2 \end{cases}$$ is *[2025, 28 Jan Shift-I]*

(a) $\frac{131}{72}$ (b) $\frac{167}{72}$

(c) $\frac{157}{72}$ (d) $\frac{171}{72}$

**28** Let $f(x) = \int_0^{x^2} \frac{t^2 - 8t + 15}{e^t} dt, x \in R$. Then, the number of local maximum and local minimum points of $f$, respectively are *[2025, 22 Jan Shift-II]*

(a) 1 and 3 (b) 3 and 2

(c) 2 and 3 (d) 2 and 2

**Imp.** **Local Maxima & Local Minima**

**29** Let $f(x) = 4\cos^3 x + 3\sqrt{3}\cos^2 x - 10$. The number of points of local maxima of $f$ in interval $(0, 2\pi)$ is *[2024, 8 April Shift-I]*

(a) 3 (b) 1 (c) 2 (d) 4

**30** If the function $f(x) = \left(\frac{1}{x}\right)^{2x}; x > 0$ attains the maximum value at $x = \frac{1}{e}$, then *[2024, 6 April Shift-II]*

(a) $e^{2\pi} < (2\pi)^e$ (b) $e^\pi > \pi^e$

(c) $e^\pi < \pi^e$ (d) $(2e)^\pi > \pi^{(2e)}$

**31** Let a rectangle $ABCD$ of sides 2 and 4 be inscribed in another rectangle $PQRS$ such that the vertices of the rectangle $ABCD$ lie on the sides of the rectangle $PQRS$. Let $a$ and $b$ be the sides of the rectangle $PQRS$ when its area is maximum. Then, $(a+b)^2$ is equal to *[2024, 5 April Shift-I]*

(a) 72 (b) 64 (c) 60 (d) 80

**32** Let $f(x) = 3\sqrt{x-2} + \sqrt{4-x}$ be a real valued function. If $\alpha$ and $\beta$ are respectively the minimum and the maximum values of $f$, then $\alpha^2 + 2\beta^2$ is equal to *[2024, 4 April Shift-II]*

(a) 42 (b) 38 (c) 44 (d) 24

**33** Let the function $f(x) = 2x^3 + (2p-7)x^2 + 3(2p-9)x - 6$ have a maxima, for some value of $x < 0$ and a minima for some value of $x > 0$. Then, the set of all values of $p$ is *[2023, 25 Jan Shift-II]*

(a) $\left(0, \frac{9}{2}\right)$ (b) $\left(-\infty, \frac{9}{2}\right)$

(c) $\left(\frac{9}{2}, \infty\right)$ (d) $\left(-\frac{9}{2}, \frac{9}{2}\right)$

**34** Let $x = 2$ be a local minima of the function $f(x) = 2x^4 - 18x^2 + 8x + 12, x \in (-4, 4)$. If $M$ is local maximum value of the function $f$ in $(-4, 4)$, then $M =$ *[2023, 25 Jan Shift-I]*

(a) $12\sqrt{6} - \frac{33}{2}$ (b) $18\sqrt{6} - \frac{33}{2}$

(c) $18\sqrt{6} - \frac{31}{2}$ (d) $12\sqrt{6} - \frac{31}{2}$

**35** If the absolute maximum value of the function $f(x) = (x^2 - 2x + 7)e^{(4x^3 - 12x^2 - 180x + 31)}$ in the interval $[-3, 0]$ is $f(\alpha)$, then *[2022, 25 July Shift-I]*

(a) $\alpha = 0$ (b) $\alpha = -3$

(c) $\alpha \in (-1, 0)$ (d) $\alpha \in (-3, -1]$

**36** Let $f : R \to R$ be a function defined by $f(x) = (x-3)^{n_1}(x-5)^{n_2}, n_1, n_2 \in N$. Then, which of the following is not true? *[2022, 29 June Shift-II]*

(a) For $n_1 = 3, n_2 = 4$, there exists $\alpha \in (3, 5)$, where $f$ attains local maxima
(b) For $n_1 = 4, n_2 = 3$, there exists $\alpha \in (3, 5)$, where $f$ attains local minima
(c) For $n_1 = 3, n_2 = 5$, there exists $\alpha \in (3, 5)$ where, $f$ attains local maxima
(d) For $n_1 = 4, n_2 = 6$, there exists $\alpha \in (3, 5)$ where, $f$ attains local maxima

**37** The lengths of the sides of a triangle are $10 + x^2$, $10 + x^2$ and $20 - 2x^2$. If for $x = k$, the area of the triangle is maximum, then $3k^2$ is equal to *[2022, 27 June Shift-I]*

(a) 5 (b) 8 (c) 10 (d) 12

**38** The sum of the absolute minimum and the absolute maximum values of the function $f(x) = |3x - x^2 + 2| - x$ in the interval $[-1, 2]$ is *[2022, 26 June Shift-I]*

(a) $\frac{\sqrt{17}+3}{2}$ (b) $\frac{\sqrt{17}+5}{2}$
(c) 5 (d) $\frac{9-\sqrt{17}}{2}$

**39** Let $A$ be the set of all points $(\alpha, \beta)$ such that the area of triangle formed by the points $(5, 6)$, $(3, 2)$ and $(\alpha, \beta)$ is 12 sq units. Then, the least possible length of a line segment joining the origin to a point in $A$, is *[2021, 31 Aug Shift-II]*

(a) $4/\sqrt{5}$ (b) $16/\sqrt{5}$
(c) $8/\sqrt{5}$ (d) $12/\sqrt{5}$

**40** The sum of all the local minimum values of the twice diffrentiable function $f : R \to R$ defined by

$$f(x) = x^3 - 3x^2 - \frac{3f''(2)}{2}x + f''(1) \text{ is}$$ *[2021, 20 July Shift-II]*

(a) $-22$ (b) 5
(c) $-27$ (d) 0

**41** Let $a$ be a real number such that the function $f(x) = ax^2 + 6x - 15, x \in R$ is increasing in $\left(-\infty, \frac{3}{4}\right)$ and decreasing in $\left(\frac{3}{4}, \infty\right)$. Then, the function $g(x) = ax^2 - 6x + 15$, $x \in R$ has a *[2021, 20 July Shift-I]*

(a) local maximum at $x = -3/4$
(b) local minimum at $x = -3/4$
(c) local maximum at $x = 3/4$
(d) local minimum at $x = 3/4$

**42** If the curves, $\frac{x^2}{a} + \frac{y^2}{b} = 1$ and $\frac{x^2}{c} + \frac{y^2}{d} = 1$ intersest each other at an angle of 90°, then which of the following relations is true? *[2021, 25 Feb Shift-I]*

(a) $ab = \frac{c+d}{a+b}$ (b) $a - b = c - d$
(c) $a - c = b + d$ (d) $a + b = c + d$

**43** The set of all real values of $\lambda$ for which the function $f(x) = (1 - \cos^2 x).(\lambda + \sin x)$, $x \in \left(-\frac{\pi}{2}, \frac{\pi}{2}\right)$, has exactly one maxima and exactly one minima, is *[2020, 6 Sep Shift-II]*

(a) $\left(-\frac{1}{2}, \frac{1}{2}\right) - \{0\}$ (b) $\left(-\frac{3}{2}, \frac{3}{2}\right)$
(c) $\left(-\frac{1}{2}, \frac{1}{2}\right)$ (d) $\left(-\frac{3}{2}, \frac{3}{2}\right) - \{0\}$

**44** If $x = 1$ is a critical point of the function $f(x) = (3x^2 + ax - 2 - a)e^x$, then *[2020, 5 Sep Shift-II]*

(a) $x = 1$ and $x = -\frac{2}{3}$ are local minima of $f$
(b) $x = 1$ and $x = -\frac{2}{3}$ are local maxima of $f$
(c) $x = 1$ is a local maxima and $x = -\frac{2}{3}$ is a local minima of $f$
(d) $x = 1$ is a local minima and $x = -\frac{2}{3}$ is a local maxima of $f$

**45** The maximum area (in sq units) of a rectangle having its base on the $X$-axis and its other two vertices on the parabola, $y = 12 - x^2$ such that the rectangle lies inside the parabola, is *[2019, 12 Jan Shift-I]*

(a) 36 (b) $18\sqrt{3}$ (c) $20\sqrt{2}$ (d) 32

**46** The maximum volume (in cum) of the right circular cone having slant height 3m is *[2019, 9 Jan Shift-I]*

(a) $4/3\pi$ (b) $2\sqrt{3}\pi$ (c) $3\sqrt{3}\pi$ (d) $6\pi$

**47** If a right circular cone, having maximum volume, is inscribed in a sphere of radius 3 cm, then the curved surface area (in cm$^2$) of this cone is *[JEE Main 2018]*

(a) $6\sqrt{3}\pi$ (b) $6\sqrt{2}\pi$ (c) $8\sqrt{2}\pi$ (d) $8\sqrt{3}\pi$

**48** Let $f(x)$ be a polynomial of degree 4 having extreme values at $x = 1$ and $x = 2$. If $\lim_{x \to 0}\left(\frac{f(x)}{x^2} + 1\right) = 3$, then $f(-1)$ is equal to *[JEE Main 2018]*

(a) 5/2 (b) 9/2 (c) 1/2 (d) 3/2

**49** If 20 m of wire is available for fencing off a flower-bed in the form of a circular sector, then the maximum area (in sq m) of the flower-bed is *[JEE Main 2017]*

(a) 12.5 (b) 10 (c) 25 (d) 30

**50** Let $f$ be a function defined by

$$f(x) = \begin{cases} \frac{\tan x}{x}, & x \neq 0 \\ 1, & x = 0 \end{cases}$$

**Statement I** $x = 0$ is point of minima of $f$.
**Statement II** $f'(0) = 0$ *[AIEEE 2011]*

(a) Statement I is false, Statement II is true
(b) Statement I is true, Statement II is true; Statement II is correct explanation of Statement I
(c) Statement I is true, Statement II is true; Statement II is not a correct explanation of Statement I
(d) Statement I is true, Statement II is false

**51** The function $f(x) = \dfrac{x}{2} + \dfrac{2}{x}$ has a local minimum at **[AIEEE 2006]**

(a) $x = -2$ (b) $x = 0$
(c) $x = 1$ (d) $x = 2$

**52** If the function $f(x) = 2x^3 - 9ax^2 + 12a^2x + 1$, where $a > 0$, attains its maximum and minimum at $p$ and $q$ respectively, such that $p^2 = q$, then $a$ is equal to **[AIEEE 2003]**

(a) 3 (b) 1
(c) 2 (d) 1/2

## Level 3

**53** A variable line $L$ passes through the point (3, 5) and intersects the positive coordinate axes at the points $A$ and $B$. The minimum area of the $\Delta OAB$, where $O$ is the origin, is **[2024, 9 April Shift-I]**

(a) 35 (b) 30
(c) 25 (d) 40

**54** If the function $f(x) = 2x^3 - 9ax^2 + 12a^2x + 1, a > 0$ has a local maximum at $x = \alpha$ and a local minimum at $x = \alpha^2$, then $\alpha$ and $\alpha^2$ are the roots of the equation **[2024, 8 April Shift-II]**

(a) $x^2 - 6x + 8 = 0$ (b) $8x^2 - 6x + 1 = 0$
(c) $8x^2 + 6x - 1 = 0$ (d) $x^2 + 6x + 8 = 0$

**55** $\underset{0 \le x \le \pi}{\text{Max}} \left\{ x - 2\sin x \cos x + \dfrac{1}{3}\sin 3x \right\} =$ **[2023, 13 April Shift-I]**

(a) $\dfrac{5\pi + 2 + 3\sqrt{3}}{6}$ (b) $\dfrac{\pi + 2 - 3\sqrt{3}}{6}$
(c) $\pi$ (d) 0

**56** If the local maximum value of the function $f(x) = \left( \dfrac{\sqrt{3e}}{2\sin x} \right)^{\sin^2 x}, x \in \left(0, \dfrac{\pi}{2}\right)$ is $\dfrac{k}{e}$, then $\left(\dfrac{k}{e}\right)^8 + \dfrac{k^8}{e^5} + k^8$ is equal to **[2023, 12 April Shift-I]**

(a) $e^3 + e^6 + e^{10}$ (b) $e^3 + e^6 + e^{11}$
(c) $e^5 + e^6 + e^{11}$ (d) $e^3 + e^5 + e^{11}$

**57** A square piece of tin of side 30 cm is to be made into a box without top by cutting a square from each corner and folding up the flaps to form a box. If the volume of the box is maximum, then its surface area (in $cm^2$) is equal to **[2023, 10 April Shift-I]**

(a) 800 (b) 900
(c) 1025 (d) 675

**58** Let $f(x) = 3^{(x^2-2)^3+4}, x \in R$. Then, which of the following statements are true?
$P: x = 0$ is a point of local minima of $f$
$Q: x = \sqrt{2}$ is a point of inflection of $f$
$R: f$ is increasing for $x > \sqrt{2}$

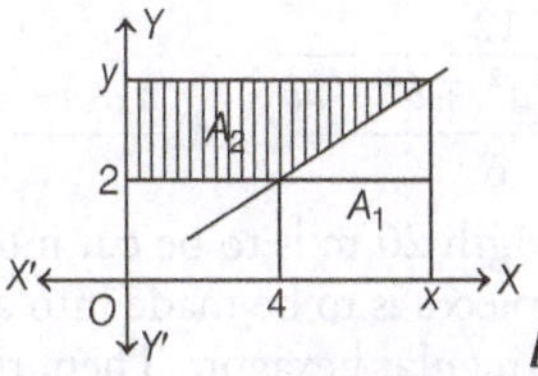


**[2022, 29 July Shift-I]**

(a) only $P$ and $Q$ (b) only $P$ and $R$
(c) only $Q$ and $R$ (d) All, $P$, $Q$ and $R$

**59** Let $P$ and $Q$ be any points on the curves $(x-1)^2 + (y+1)^2 = 1$ and $y = x^2$, respectively. The distance between $P$ and $Q$ is minimum for some value of the abscissa of $P$ in the interval **[2022, 26 July Shift-II]**

(a) $\left(0, \dfrac{1}{4}\right)$ (b) $\left(\dfrac{1}{2}, \dfrac{3}{4}\right)$
(c) $\left(\dfrac{1}{4}, \dfrac{1}{2}\right)$ (d) $\left(\dfrac{3}{4}, 1\right)$

**60** The curve $y(x) = ax^3 + bx^2 + cx + 5$ touches the $X$-axis at the point $P(-2, 0)$ and cuts the $Y$-axis at the point $Q$, where $y'$ is equal to 3. Then, the local maximum value of $y(x)$ is **[2022, 25 July Shift-I]**

(a) 27/4 (b) 29/4
(c) 37/4 (d) 9/2

**61** A wire of length 22 m is to be cut into two pieces. One of the pieces is to be made into a square and the other into an equilateral triangle. Then, the length of the side of the equilateral triangle, so that the combined area of the square and the equilateral triangle is minimum, is **[2022, 29 June Shift-I]**

(a) $\dfrac{22}{9 + 4\sqrt{3}}$ (b) $\dfrac{66}{9 + 4\sqrt{3}}$ (c) $\dfrac{22}{4 + 9\sqrt{3}}$ (d) $\dfrac{66}{4 + 9\sqrt{3}}$

**62** Consider a cuboid of sides $2x$, $4x$ and $5x$ and a closed hemisphere of radius $r$. If the sum of their surface areas is a constant $k$, then the ratio $x : r$, for which the sum of their volumes is maximum, is **[2022, 26 June Shift-II]**

(a) 2 : 5 (b) 19 : 45
(c) 3 : 8 (d) 19 : 15

**63** The sum of absolute maximum and absolute minimum values of the function $f(x) = |2x^2 + 3x - 2| + \sin x \cos x$ in the interval [0, 1] is **[2022, 24 June Shift-I]**

(a) $3 + \dfrac{\sin(1)\cos^2(1/2)}{2}$ (b) $3 + \dfrac{1}{2}(1 + 2\cos(1)|\sin(1))$
(c) $5 + \dfrac{1}{2}(\sin(1) + \sin(2))$ (d) $2 + \sin\left(\dfrac{1}{2}\right)\cos\left(\dfrac{1}{2}\right)$

**64** A box open from top is made from a rectangular sheet of dimension $a \times b$ by cutting squares each of side $x$ from each of the four corners and folding up the flaps. If the volume of the box is maximum, then $x$ is equal to **[2021, 27 Aug Shift-II]**

(a) $\dfrac{a+b-\sqrt{a^2+b^2-ab}}{12}$ (b) $\dfrac{a+b-\sqrt{a^2+b^2+ab}}{6}$

(c) $\dfrac{a+b-\sqrt{a^2+b^2-ab}}{6}$ (d) $\dfrac{a+b+\sqrt{a^2+b^2-ab}}{6}$

**65** A wire of length 20 m is to be cut into two pieces. One of the pieces is to be made into a square and the other into a regular hexagon. Then, the length of the side (in m) of the hexagon, so that the combined area of the square and the hexagon is minimum, is **[2021, 27 Aug Shift-I]**

(a) $\dfrac{5}{2+\sqrt{3}}$ (b) $\dfrac{10}{2+3\sqrt{3}}$

(c) $\dfrac{5}{3+\sqrt{3}}$ (d) $\dfrac{10}{3+2\sqrt{3}}$

**66** Let $A=[a_{ij}]$ be a $3\times 3$ matrix, where

$$a_{ij}=\begin{cases} 1, & \text{if } i=j \\ -x, & \text{if } |i-j|=1 \\ 2x+1, & \text{otherwise} \end{cases}$$

Let a function $f: R \to R$ be defined as $f(x)=\det(A)$. Then, the sum of maximum and minimum values of $f$ on $R$ is equal to **[2021, 20 July Shift-I]**

(a) $\dfrac{88}{27}$ (b) $\dfrac{20}{27}$

(c) $-\dfrac{88}{27}$ (d) $-\dfrac{20}{27}$

**67** The range of $a \in R$ for which the function $f(x)=(4a-3)(x+\log_e 5)+2(a-7)$ $\cot\left(\dfrac{x}{2}\right)\sin^2\left(\dfrac{x}{2}\right)$, $x \neq 2n\pi, n \in N$, has critical points, is **[2021, 16 March Shift-I]**

(a) $(-3, 1)$ (b) $\left[-\dfrac{4}{3}, 2\right]$

(c) $[1, \infty)$ (d) $(-\infty, -1]$

**68** A water tank has the shape of an inverted right circular cone, whose semi-vertical angle is $\tan^{-1}\left(\dfrac{1}{2}\right)$.

Water is poured into it at a constant rate of 5 cu m/min. Then, the rate (in m/min) at which the level of water is rising at the instant when the depth of water in the tank is 10 m is **[2019, 9 April Shift-II]**

(a) $\dfrac{2}{\pi}$ (b) $\dfrac{1}{5\pi}$

(c) $\dfrac{1}{15\pi}$ (d) $\dfrac{1}{10\pi}$

**69** If $\lambda \in R$ is such that the sum of the cubes of the roots of the equation, $x^2+(2-\lambda)x+(10-\lambda)=0$ is minimum, then the magnitude of the difference of the roots of this equation is **[JEE Main 2018]**

(a) $4\sqrt{2}$ (b) 20 (c) $2\sqrt{7}$ (d) $2\sqrt{5}$

**70** A wire of length 2 units is cut into two parts which are bent respectively to form a square of side $x$ units and a circle of radius $r$ units. If the sum of the areas of the square and the circle so formed is minimum, then **[JEE Main 2016]**

(a) $2x=(\pi+4)r$ (b) $(4-\pi)x=\pi r$

(c) $x=2r$ (d) $2x=r$

**71** Let $f(x)$ be a polynomial of degree four having extreme values at $x=1$ and $x=2$. If

$\lim_{x\to 0}\left[1+\dfrac{f(x)}{x^2}\right]=3$, then $f(2)$ is equal to **[JEE Main 2015]**

(a) $-8$ (b) $-4$ (c) 0 (d) 4

**72** Let $a, b \in R$ be such that the function $f$ given by $f(x)=\log|x|+bx^2+ax, x\neq 0$ has extreme values at $x=-1$ and $x=2$.

**Statement I** f has local maximum at $x=-1$ and at $x=2$.

**Statement II** $a=\dfrac{1}{2}$ and $b=\dfrac{-1}{4}$ **[AIEEE 2012]**

(a) Statement I is false, Statement II is true

(b) Statement I is true, Statement II is true; Statement II is a correct explanation of Statement I

(c) Statement I is true, Statement II is true; Statement II is not a correct explanation of Statement I

(d) Statement I is true, Statement II is false

**73** Given, $P(x)=x^4+ax^3+bx^2+cx+d$ such that $x=0$ is the only real root of $P'(x)=0$. If $P(-1)<P(1)$, then in the interval $[-1, 1]$, **[AIEEE 2009]**

(a) $P(-1)$ is the minimum and $P(1)$ is the maximum of $P$

(b) $P(-1)$ is not minimum but $P(1)$ is the maximum of $P$

(c) $P(-1)$ is the minimum and $P(1)$ is not the maximum of $P$

(d) Neither $P(-1)$ is the minimum nor $P(1)$ is the maximum of $P$

**74** Area of the greatest rectangle that can be inscribed in the ellipse $\dfrac{x^2}{a^2}+\dfrac{y^2}{b^2}=1$ is **[AIEEE 2005]**

(a) $ab$ (b) $2ab$

(c) $a/b$ (d) $\sqrt{ab}$

**75** If $u=\sqrt{a^2\cos^2\theta+b^2\sin^2\theta}+\sqrt{a^2\sin^2\theta+b^2\cos^2\theta}$, then the difference between the maximum and minimum values of $u^2$ is given by **[AIEEE 2004]**

(a) $2(a^2+b^2)$ (b) $2\sqrt{a^2+b^2}$

(c) $(a+b)^2$ (d) $(a-b)^2$

## Numerical Value Type Questions

**76** Let the set of all positive values of $\lambda$, for which the point of local minimum of the function $(1+x(\lambda^2-x^2))$ satisfies $\frac{x^2+x+2}{x^2+5x+6}<0$, be $(\alpha, \beta)$.
Then, $\alpha^2+\beta^2$ is equal to ............ **[2024, 9 April Shift-I]**

**77** Let $A$ be the region enclosed by the parabola $y^2=2x$ and the line $x=24$. Then, the maximum area of the rectangle inscribed in the region $A$ is .................... . **[2024, 8 April Shift-II]**

**78** Let the maximum and minimum values of $(\sqrt{8x-x^2-12}-4)^2+(x-7)^2, x \in R$ be $M$ and $m$, respectively. Then, $M^2-m^2$ is equal to ...... . **[2024, 5 April Shift-II]**

**79** Let $S=\{\sin^2 2\theta : (\sin^4\theta+\cos^4\theta)x^2+(\sin 2\theta)x +(\sin^6\theta+\cos^6\theta)=0$ has real roots$\}$.
If $\alpha$ and $\beta$ are the smallest and largest elements of the set $S$, respectively, then $3((\alpha-2)^2+(\beta-1)^2)$ equals ...... . **[2024, 4 April Shift-II]**

**80** If $a_\alpha$ is the greatest term in the sequence $a_n=\frac{n^3}{n^4+147}\cdot n=1,2,3...$, then $\alpha$ is equal to ...... . **[2023, 8 April Shift-I]**

**81** Let $f(x)=|(x-1)(x^2-2x-3)|+x-3, x \in R$. If $m$ and $M$ are respectively the number of points of local minimum and local maximum of $f$ in the interval $(0, 4)$, then $m+M$ is equal to ......... . **[2022, 25 June Shift-II]**

**82** The number of distinct real roots of the equation $3x^4+4x^3-12x^2+4=0$ is .... . **[2021, 27 Aug Shift-I]**

**83** The minimum value of $\alpha$ for which the equation $\frac{4}{\sin x}+\frac{1}{1-\sin x}=\alpha$ has at least one solution in $(0, \pi/2)$ is ......... . **[2021, 24 Feb Shift-I]**

**84** Let $AD$ and $BC$ be two vertical poles at $A$ and $B$, respectively on a horizontal ground. If $AD=8$ m, $BC=11$ m and $AB=10$ m; then the distance (in meters) of a point $M$ on $AB$ from the point $A$ such that $MD^2+MC^2$ is minimum is ......... . **[2020, 6 Sep Shift-I]**

# Key Index

### Topic 1

| | | | | | | | | | |
|---|---|---|---|---|---|---|---|---|---|
| 1 (c) | 2 (c) | 3 (d) | 4 (a) | 5 (d) | 6 (c) | 7 (a) | 8 (c) | 9 (5) | 10 (5) |

### Topic 2

| | | | | | | | | | |
|---|---|---|---|---|---|---|---|---|---|
| 1 (b) | 2 (c) | 3 (d) | 4 (*) | 5 (c) | 6 (d) | 7 (a) | 8 (a) | 9 (a) | 10 (c) |
| 11 (c) | 12 (c) | 13 (b) | 14 (c) | 15 (d) | 16 (b) | 17 (c) | 18 (c) | 19 (d) | 20 (a) |
| 21 (a) | 22 (d) | 23 (a) | 24 (d) | 25 (b) | 26 (c) | 27 (b) | 28 (c) | 29 (b) | 30 (252) |
| 31 (2) | 32 (96) | | | | | | | | |

### Topic 3

| | | | | | | | | | |
|---|---|---|---|---|---|---|---|---|---|
| 1 (a) | 2 (a) | 3 (a) | 4 (d) | 5 (a) | 6 (c) | 7 (c) | 8 (d) | 9 (c) | 10 (c) |
| 11 (c) | 12 (d) | 13 (a) | 14 (c) | 15 (d) | 16 (d) | 17 (c) | 18 (d) | 19 (d) | 20 (c) |
| 21 (a) | 22 (d) | 23 (a) | 24 (b) | 25 (c) | 26 (c) | 27 (c) | 28 (c) | 29 (c) | 30 (b) |
| 31 (a) | 32 (a) | 33 (b) | 34 (a) | 35 (b) | 36 (c) | 37 (c) | 38 (a) | 39 (c) | 40 (c) |
| 41 (a) | 42 (b) | 43 (d) | 44 (d) | 45 (d) | 46 (b) | 47 (d) | 48 (b) | 49 (c) | 50 (c) |
| 51 (d) | 52 (c) | 53 (b) | 54 (a) | 55 (a) | 56 (b) | 57 (a) | 58 (d) | 59 (c) | 60 (a) |
| 61 (b) | 62 (b) | 63 (b) | 64 (c) | 65 (d) | 66 (c) | 67 (b) | 68 (b) | 69 (d) | 70 (c) |
| 71 (c) | 72 (c) | 73 (b) | 74 (b) | 75 (d) | 76 (39) | 77 (128) | 78 (1600) | 79 (4) | 80 (5) |
| 81 (3) | 82 (4) | 83 (9) | 84 (5) | | | | | | |

# Detailed **Explanations**

## Topic 1

**01.** *(c)* Given, $\frac{dr}{dt}=\frac{1}{4\pi}$ cm/min

Let volume of sphere be $V$ at any time $t$.

$$V=\frac{4}{3}\pi r^3$$

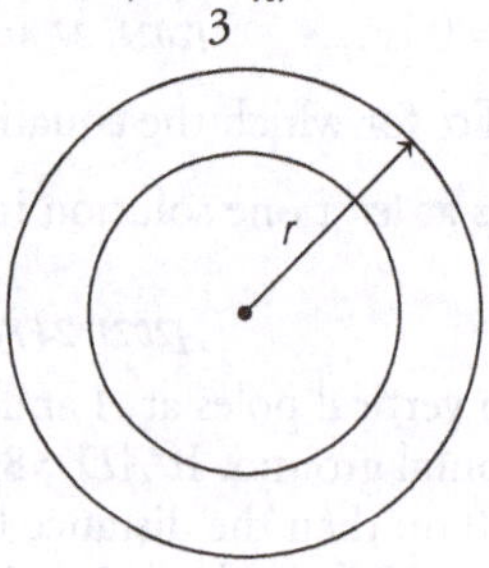


$$\frac{dV}{dt}=4\pi r^2\frac{dr}{dt}$$

$$\Rightarrow \quad 81=4\pi r^2\left(\frac{1}{4\pi}\right)$$

$$\Rightarrow \quad r^2=81$$

$$\Rightarrow \quad r=9$$

$\because$ Surface area of chocolate

$$S=4\pi(r-1)^2=4\pi(8)^2$$

$$=256\pi$$

**02.** *(c)* The average speed of the car, for time interval $[t_1, t_2]$ is

$$\frac{f(t_2)-f(t_1)}{t_2-t_1}$$

$$=\frac{a(t_2^2-t_1^2)+b(t_2-t_1)}{t_2-t_1}=\frac{d(f(t))}{dt}$$

$$\therefore \quad 2at+b=a(t_2+t_1)+b$$

$$\Rightarrow \quad t=(t_1+t_2)/2$$

$\therefore$ The average speed of the car over the time interval $[t_1, t_2]$ is attained at the point $\frac{t_1+t_2}{2}$.

**03.** *(d)* It is given that, a spherical iron ball of 10 cm radius is coated with a layer of ice of uniform thickness, let the thickness is $x$ cm, then volume of the ball is

$$V=\frac{4}{3}\pi(10+x)^3$$

On differentiating w.r.t. $t$, we get

$$\frac{dV}{dt}=4\pi(10+x)^2\frac{dx}{dt} \quad \ldots \text{(i)}$$

where, $t$ is time in min.

It is given, the $\frac{dV}{dt}=-50\text{ cm}^3/\text{min}$,

Now, when $x$ is 5 cm, then

$$-50=4\pi(10+5)^2\frac{dx}{dt} \quad \text{[from Eq. (i)]}$$

$$\Rightarrow \quad \frac{dx}{dt}=-\frac{50}{4\pi(225)}=-\frac{1}{18\pi}\text{ cm/min}$$

Negative sign indicates the thickness of ice layer decreases with time.

**04.** *(a)* Let $AB$ be the ladder.

Given, $\frac{dy}{dt}=-25$

We have, $x^2+y^2=4$

On differentiating w.r.t. $t$, we get

$$2x\frac{dx}{dt}+2y\frac{dy}{dt}=0$$

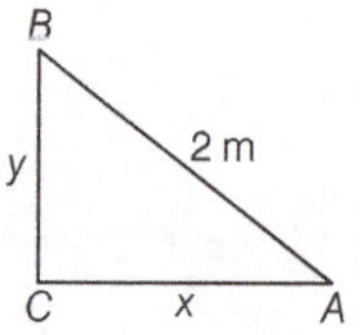


$$\Rightarrow \quad \frac{dx}{dt}=-\frac{y}{x}\frac{dy}{dt}$$

When top of ladder is 1 m above the ground, i.e., $y=1$, then $x=\sqrt{4-1}=\sqrt{3}$ (by Pythagoras theorem)

$$\therefore \quad \frac{dx}{dt}=\frac{-1}{\sqrt{3}}(-25)=\frac{25}{\sqrt{3}}\text{ cm/sec}$$

**05.** *(d)* Equation of parabola is $y^2=18x$.

On differentiating w.r.t. $t$, we get

$$2y\frac{dy}{dt}=18\frac{dx}{dt}$$

$$\Rightarrow \quad 2\cdot 2y=18 \quad \left[\because \frac{dy}{dt}=2\frac{dx}{dt}, \text{ given}\right]$$

$$\Rightarrow \quad y=\frac{9}{2}$$

From equation of parabola,

$$\left(\frac{9}{2}\right)^2=18x \Rightarrow \frac{81}{4}=18x \quad \Rightarrow x=\frac{9}{8}$$

Hence, required point is $\left(\frac{9}{8}, \frac{9}{2}\right)$.

**06.** *(c)* Since, surface area of cube,

$$A=6a^2\text{ cm}^2.$$

It is given, $\frac{dA}{dt}=3.6\text{ cm}^2/\text{sec}$

$$\Rightarrow \quad 12a\frac{da}{dt}=3.6\text{ cm}^2/\text{sec} \quad \ldots \text{(i)}$$

Now, as volume of cube, $V=a^3\text{ cm}^3$

$$\therefore \quad \frac{dV}{dt}=3a^2\frac{da}{dt}$$

$$=3a^2\frac{3.6}{12a} \quad \text{[from Eq. (i)]}$$

So, at $a=10$ cm, $\frac{dV}{dt}=0.9\times 10$

$$=9\text{ cm}^3/\text{sec}$$

**07.** *(a)* We know that surface area of a balloon of spherical shape is $S=4\pi r^2$, where $r$ is a radius.

Now, $S=4\pi r^2$

On differentiating both side w.r.t. $t$, we get

$$\frac{dS}{dt}=8\pi r\frac{dr}{dt}$$

Given that surface area of a balloon increases at a constant rate.

So, $dS/dt=$ constant ($k$ say),

Thus, $k=8\pi\left(r\frac{dr}{dt}\right)$

$$\Rightarrow r\frac{dr}{dt}=\lambda, \text{ where } \lambda=\frac{k}{8\pi}=\text{constant}$$

(say)

$$\Rightarrow \quad rdr=\lambda dt$$

Now, integrating on both sides, we get

$$\frac{r^2}{2}=\lambda t+C$$

At $r=3, t=0$, then $C=\frac{9}{2}$

Thus, $\frac{r^2}{2}=\lambda t+\frac{9}{2}$

Now, given when $t=5$, then $r=7$

Thus, $\frac{(7)^2}{2}=\lambda\cdot 5+\frac{9}{2}$

$$\Rightarrow \quad \frac{40}{2}=5\lambda \quad \Rightarrow \quad \lambda=4$$

Hence, $\frac{r^2}{2}=4t+\frac{9}{2}$

Now, at $t=9$; $\frac{r^2}{2}=4\times 9+\frac{9}{2}$

$\Rightarrow \quad r=9$

Hence, its radius after 9 s is 9 units.

**08.** *(c)* Given

Rate of decrease of volume of the balloon is

$$\frac{dV}{dt}=-72\,\pi\text{ m}^3/\text{min}$$

**To find** The rate of decrease of the radius of the balloon 49 min after the leakage started.

*i.e.,* $\frac{dr}{dt}$ at $t=49$ min

[Assuming that the leakage started at time $t=0$]

Now, the balloon is spherical in shape, hence the volume of the balloon is $V=\frac{4}{3}\pi r^3$.

On differentiating both sides w.r.t. $t$, we get

$$\frac{dV}{dt}=\frac{4}{3}\pi\left(3r^2\times\frac{dr}{dt}\right)$$

$$\Rightarrow \frac{dr}{dt}=\frac{dV/dt}{4\pi r^2} \quad \ldots\text{(i)}$$

Now, to find $\frac{dr}{dt}$ at $t=49$ min, we require $dV/dt$ and the radius ($r$) at that stage,

$$dV/dt=-72\,\pi\text{ m}^3/\text{min}$$

$\therefore$ Amount of volume lost in 49 min

$$=72\,\pi\times 49\text{ m}^3$$

$\therefore$ Final volume at the end of 49 min

$$=(4500\,\pi-3528\,\pi)\text{ m}^3=972\,\pi\text{ m}^3$$

If $r$ is the radius at the end of 49 min, then

$$\frac{4}{3}\pi r^3=972\pi$$

$$\Rightarrow r^3=729 \Rightarrow r=9$$

$\therefore$ Radius of the balloon at the end of 49 min $=9$ m

Hence, from Eq. (i), we get

$$\frac{dr}{dt}=\frac{dV/dt}{4\pi r^2}$$

$$\Rightarrow \left(\frac{dr}{dt}\right)_{t=49}=\frac{(dV/dt)_{t=49}}{4\pi(r^2)_{t=49}}$$

$$=\frac{72\pi}{4\pi(9^2)}=\frac{2}{9}\text{ m/min}$$

**09.** *(5)* Given equation of curve

$$y=x^5-20x^3+50x+2$$

$$\Rightarrow \frac{dy}{dx}=5x^4-60x^2+50$$

On putting $\frac{dy}{dx}=0$

$$\Rightarrow 5(x^4-12x^2+10)=0$$

$$\Rightarrow x^2=\frac{12\pm\sqrt{144-40}}{2}=6\pm\sqrt{26}$$

$$\Rightarrow x^2=6-\sqrt{26},\,6+\sqrt{26}$$

$$\Rightarrow x^2=6-5.10,\,6+5.10$$

$$\Rightarrow x^2=0.9,\,11.1$$

$$\Rightarrow x=\pm\sqrt{0.9},\,\pm\sqrt{11.1}$$

$$\Rightarrow x=-0.95,\,0.95,\,-3.33,\,3.33$$

Now, $y(0)=2(+\text{ve}) \Rightarrow y(1)=+\text{ve}$

$y(2)=-\text{ve} \Rightarrow y(3.3)=-\text{ve}$

$y(-1)=-\text{ve} \Rightarrow y(-2)=+\text{ve}$

$y(-3.3)=\text{ve}$

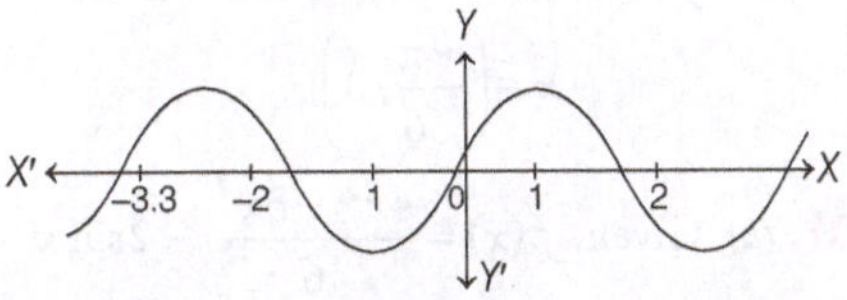


$\because$ Required number of points $=5$

**10.** *(5)*

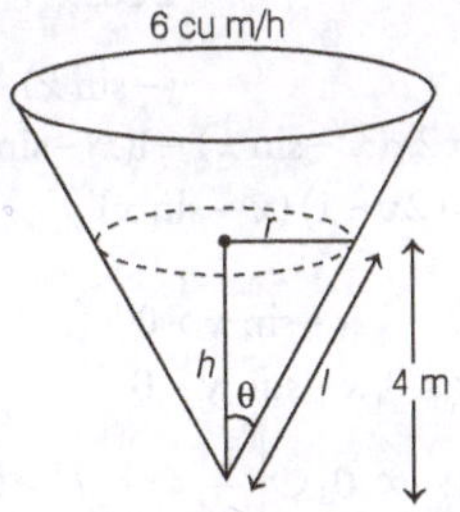


We have, $\tan\theta=\frac{3}{4} \Rightarrow \frac{r}{h}=\frac{3}{4}$

$$\therefore \quad h=\frac{4r}{3} \quad \ldots\text{(i)}$$

$$\Rightarrow \quad l=\sqrt{r^2+\frac{16r^2}{9}}=\frac{5r}{3}$$

Volume of water

$$=V=\frac{1}{3}\pi r^2 h=\frac{4\pi(r)^3}{9}$$

$$\frac{dV}{dt}=\frac{4\pi r^2}{3}\cdot\frac{dr}{dt}=6$$

$$\therefore \quad \frac{dr}{dt}=\frac{9}{2\pi r^2}$$

Now, curved surface area $=S=\pi rl$

$$\therefore \quad S=\frac{5\pi r^2}{3}$$

$$\therefore \frac{dS}{dt}=\frac{10\pi r}{3}\frac{dr}{dt}=\frac{10\pi r}{3}\times\frac{9}{2\pi r^2}=\frac{15}{r}$$

Now, $h=4$, then from Eq. (i), we get

$$r=3$$

$$dS/dt=5\text{ m}^2/\text{h}$$

## Topic 2

**01.** *(b)* Given, $f(x)=\frac{x}{3}+\frac{3}{x}+3$

$$\Rightarrow f'(x)=\frac{1}{3}-\frac{3}{x^2} \Rightarrow f'(x)=\frac{x^2-9}{3x^2}$$

| +ve | –ve | –ve | +ve |
|---|---|---|---|
| | –3 | 0 | 3 |

$f'(x)>0\;\forall\; x\in(-\infty,-3)\cup(3,\infty)$

$f'(x)<0\;\forall\; x\in(-3,0)\cup(0,3)$

$\Rightarrow \quad \alpha_1=-3, \quad \alpha_2=3$

$\alpha_3=-3, \quad \alpha_4=0$

$\alpha_5=3$

So, $\sum_{i=1}^{5}\alpha_i^2=\alpha_1^2+\alpha_2^2+\alpha_3^2+\alpha_4^2+\alpha_5^2$

$$=9+9+9+0+9=36$$

**02.** *(c)* Given,

$f(x)=2\log_e(x-2)-x^2+ax+1$ is increasing in (2, 3)

$$\therefore \quad f'(x)=\frac{2}{x-2}-2x+a\geq 0$$

and $\quad f''(x)=\frac{-2}{(x-2)^2}-2<0$

then we can say that $f'(x)$ is strictly decreasing.

and $\quad f'(3)\geq 0$

$$\Rightarrow \quad 2-6+a\geq 0 \Rightarrow a\geq 4$$

$$\therefore \quad a_{\min}=4$$

Also given, $g(x)=(x-1)^3\,(x+2-a)^2$ is strictly decreasing.

$$\Rightarrow \quad g(x)=(x-1)^3\,(x+2-4)^2$$

$$=(x-1)^3\,(x-2)^2$$

$$\Rightarrow g'(x)=(x-1)^3\,2\,(x-2)+(x-2)^2\cdot 3(x-1)^2$$

$$=(x-1)^2\,(x-2)\,(2x-2+3x-6)$$

$$=(x-1)^2\,(x-2)\,(5x-8)<0 \text{ in}$$

$x\in\left(\frac{8}{5},2\right)$

$$\therefore \quad b=8/5 \text{ and } c=2.$$

| $-\infty$ | + | + | – | + | $\infty$ |
|---|---|---|---|---|---|
| | 1 | | 8/5 | 2 | |

$\therefore 100\,(a+b-c)=100\,(4+8/5-2)$

$$=360$$

**03.** (d) $f(x)=\cos x-x+1$

So, $f'(x)=-\sin x-1$

Put $f'(x)=0$

$\Rightarrow -\sin x=1$

$\Rightarrow \quad \sin x=-1$

In $x\in[0,\pi]$, $\sin x=-1$ is not possible.

So, there is no critical point in $[0,\pi]$.

Now, $f(0)=2>0$

$f(\pi)=-1-\pi+1=-\pi<0$

That means, $f(x)$ will have only one zero in $[0,\pi]$.

and $f(x)$ is decreasing in $[0,\pi]$.

Hence, only (SI) is correct.

**04.** (*)

> **key idea** Firstly find $\frac{dy}{dx}$, then use $f'(x)>0$ for strictly increasing function.

According to the question,

$\Rightarrow \quad f(x)=x^x=e^{x\ln x}$

So, $f'(x)=e^{x\ln x}(\ln x+1)$

Now, according to the question,

$f'(x)>0$

$\Rightarrow \quad \ln x+1>0 \Rightarrow \ln x>-1$

$\Rightarrow \quad x>e^{-1}$

$\Rightarrow \quad x\in\left(\frac{1}{e},\infty\right)$

$\therefore$ Option not matched.

**05.** (c) $f(x)=\dfrac{x}{x^2-6x-16}$,

$x\in R-\{-2,8\}$

$$f'(x)=\frac{(x^2-6x-16)(1)-x(2x-6)}{(x^2-6x-16)^2}$$

$$=\frac{x^2-6x-16-2x^2+6x}{(x^2-6x-16)^2}$$

$$=\frac{-x^2-16}{(x^2-6x-16)^2}$$

$$=\frac{-(x^2+16)}{(x^2-6x-16)^2}<0$$

$\because f'(x)<0$ for $x\in$ domain.

$\therefore$ It is a decreasing function in its domain.

$x\in(-\infty,-2)\cup(-2,8)\cup(8,\infty)$

**06.** (d) $f(x)=3\sin^4 x+10\sin^3 x+6\sin^2 x-3$

$\Rightarrow f'(x)=\cos x(12\sin^3 x+30\sin^2 x+12\sin x)$

$\Rightarrow f'(x)=6\sin x\cos x(2\sin^2 x+5\sin x+2)$

$\Rightarrow f'(x)=3\sin 2x(2\sin x+1)(\sin x+2)$

$f'(x)=0$

$\sin 2x=0$ or $2\sin x+1=0$ $\quad[\because \sin x\neq-2]$

$\Rightarrow \quad x=0$ or $x=n\pi+(-1)^n\left(-\frac{\pi}{6}\right)$

As, $\quad x\in\left[\frac{-\pi}{6},\frac{\pi}{2}\right]\Rightarrow x=0,\frac{-\pi}{6}$

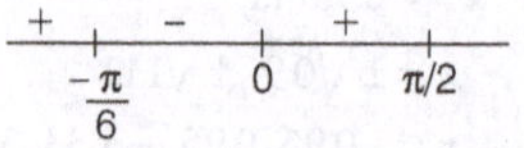


So, $f(x)$ is increasing in the interval

$x\in\left(0,\frac{\pi}{2}\right)$

And $f(x)$ is decreasing in the interval

$x\in\left(-\frac{\pi}{6},0\right)$

**07.** (a) Given, $f(x)=\dfrac{4x^3-3x^2}{6}-2\sin x+(2x-1)\cos x$

$f'(x)=\dfrac{12x^2-6x}{6}-2\cos x+(2x-1)(-\sin x)+\cos x(2)$

$=2x(x-\sin x)-1(x-\sin x)$

$f'(x)=(2x-1)(x-\sin x)$

For $x>0$

$x-\sin x>0$

$x<0, x-\sin x<0$

For $x\in(-\infty,0]\cup\left[\frac{1}{2},\infty\right)$, $f'\geq0$

For $x\in\left[0,\frac{1}{2}\right]$, $f'(x)\leq0$

Hence, $f(x)$ increases in $\left[\frac{1}{2},\infty\right)$.

**08.** (a) Since, the given function $f(x)=(3x-7)x^{2/3}$ is increasing for $x\in R$

$\therefore \quad f'(x)>0$

$\Rightarrow \quad \frac{2}{3}x^{-1/3}(3x-7)+x^{2/3}(3)\geq0,\ x\neq0$

$\Rightarrow \quad \dfrac{2(3x-7)+9x}{x^{1/3}}\geq0,\ x\neq0$

$\Rightarrow \quad \dfrac{15x-14}{x^{1/3}}>0$

$\Rightarrow x\in(-\infty,0)\cup\left[\frac{14}{15},\infty\right)$

**09.** (a) We have, $f(x)=\dfrac{x}{(a^2+x^2)^{1/2}}-\dfrac{(d-x)}{(b^2+(d-x)^2)^{1/2}}$

Differentiating above w.r.t. $x$, we get

$$f'(x)=\frac{(a^2+x^2)^{1/2}-x\dfrac{1}{2}\dfrac{2x}{(a^2+x^2)^{1/2}}}{(a^2+x^2)}-\frac{(b^2+(d-x)^2)^{1/2}(-1)-(d-x)\dfrac{2(d-x)(-1)}{2(b^2+(d-x)^2)^{1/2}}}{(b^2+(d-x)^2)}$$

[By using quotient rule of derivative]

$$=\frac{a^2+x^2-x^2}{(a^2+x^2)^{3/2}}+\frac{b^2+(d-x)^2-(d-x)^2}{(b^2+(d-x)^2)^{3/2}}$$

$$=\frac{a^2}{(a^2+x^2)^{3/2}}+\frac{b^2}{(b^2+(d-x)^2)^{3/2}}>0,$$

$\forall x\in R$

Hence, $f$ is an increasing function of $x$.

**10.** (c) We have, $f(x)=\sin^4 x+\cos^4 x$

On differentiating w.r.t $x$, we get

$f'(x)=4\sin^3 x\cos x+4\cos^3 x(-\sin x)$

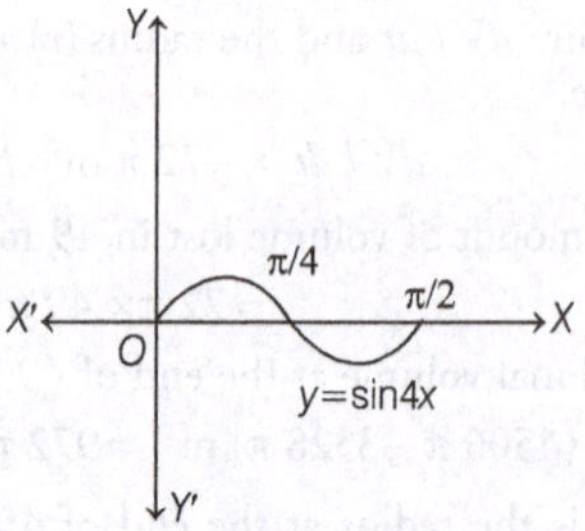


$=4\sin x\cos x(\sin^2 x-\cos^2 x)$

$=-2\sin 2x\cos 2x=-\sin 4x$

Since, $f(x)$ is increasing when $f'(x)>0$

$\Rightarrow \quad -\sin 4x>0\Rightarrow \sin 4x<0$

$\Rightarrow \quad x\in\left(\frac{\pi}{4},\frac{\pi}{2}\right)$

**11.** (c) Let $f(x)=2x^3+3x+k$

$\Rightarrow \quad f'(x)=6x^2+3>0$

Thus $f$ is strictly increasing. Hence, it has almost one real root. But a polynomial equation of odd degree has atleast one root. So, the equation has exactly one root.

Thus, the two distinct roots, in any interval what so ever is impossible. No such $k$ exists.

**12.** *(c)* Let $f(x)=x^7+14x^5+16x^3+30x-560$

$f'(x)=7x^6+70x^4+48x^2+30>0, \forall x \in R$

So, $f(x)$ is increasing.

Hence, $f(x)=0$ has only one solution.

**13.** *(b)* Since, $f(x)=\tan^{-1}(\sin x+\cos x)$

On differentiating w.r.t. $x$, we get

$$f'(x)=\frac{1}{1+(\sin x+\cos x)^2}(\cos x-\sin x)$$

$$=\frac{\sqrt{2}\left\{\cos x\cdot\cos\frac{\pi}{4}-\sin x\cdot\sin\frac{\pi}{4}\right\}}{1+(\sin x+\cos x)^2}$$

$$=\frac{\sqrt{2}\cos\left(x+\frac{\pi}{4}\right)}{1+(\sin x+\cos x)^2}$$

For $f(x)$ to be increasing,

$$-\frac{\pi}{2}<x+\frac{\pi}{4}<\frac{\pi}{2}$$

$$\Rightarrow \quad -\frac{3\pi}{4}<x<\frac{\pi}{4}$$

Hence, option (b) is correct, which lies in the above interval.

**14.** *(c)* Since, $f(x)=\cot^{-1}x+x$

On differentiating w.r.t. $x$, we get

$$f'(x)=-\frac{1}{1+x^2}+1=\frac{x^2}{1+x^2}\geq 0$$

Hence, $f(x)$ is increasing function for all $x\in(-\infty,\infty)$.

**15.** *(d)* $f(x)=(x-2)^{2/3}(2x+1)$

For critical points,

Put $f'(x)=0$

So, $f'(x)=\frac{d}{dx}[(x-2)^{2/3}(2x+1)]$

$$=2(x-2)^{2/3}+(2x+1)\left(\frac{2}{3}(x-2)^{-1/3}\right)$$

$$=\frac{6(x-2)^{2/3}\cdot(x-2)^{1/3}+2(2x+1)}{3(x-2)^{1/3}}\ \{x\neq 2\}$$

So, $x=2$ is critical point.

Put $f'(x)=0$

$$\Rightarrow \frac{2(3(x-2)+(2x+1))}{3(x-2)^{1/3}}=0$$

$\Rightarrow 3x-6+2x+1=0$

$\Rightarrow 5x=5 \Rightarrow x=1$

Hence, there is only two critical points.

**16.** *(b)* Given,

$$f(x)=\sin x+3x-\frac{2}{\pi}(x^2+x)$$

$$\Rightarrow f'(x)=\cos x+3-\frac{2}{\pi}(2x+1)$$

$$=\cos x-\frac{4x}{\pi}-\frac{2}{\pi}+3$$

When $x\in\left[0,\frac{\pi}{2}\right]$

$\Rightarrow \quad 4x\in[0,2\pi]$

$\Rightarrow \quad \frac{4x}{\pi}\in[0,2]$,

$\cos x>0$, when $x\in\left[0,\frac{\pi}{2}\right]$

$\therefore \quad f'(x)=\cos x-\frac{4x}{\pi}-\frac{2}{\pi}+3>0$

$\Rightarrow f(x)$ is increasing

Now, $f''(x)=-\sin x-\frac{4}{\pi}<0$

when $x\in\left(0,\frac{\pi}{2}\right)$

$\Rightarrow f'(x)$ is decreasing

Hence, both statements are true.

**17.** *(c)* $f:R\to(0,\infty)$, $\lim_{x\to\infty}\frac{f(7x)}{f(x)}=1$

Since, $f(x)$ is strictly increasing function.

$$f(x)<f(5x)<f(7x)$$

$$\frac{f(x)}{f(x)}<\frac{f(5x)}{f(x)}<\frac{f(7x)}{f(x)}$$

$$\lim_{x\to\infty}1<\lim_{x\to\infty}\frac{f(5x)}{f(x)}<\lim_{x\to\infty}\frac{f(7x)}{f(x)}$$

$$1<\lim_{x\to\infty}\frac{f(5x)}{f(x)}<1$$

So, by sandwich theorem,

$$\lim_{x\to\infty}\frac{f(5x)}{f(x)}=1$$

$$\text{Or }\lim_{x\to\infty}\left[\frac{f(5x)}{f(x)}-1\right]=1-1=0$$

**• Concept Enhancer**

**Sandwich theorem** Let $f(x), g(x)$ and $h(x)$ be real functions such that $f(x)\leq g(x)\leq h(x), \forall x$ in the common domain, If $\lim_{x\to a}f(x)=l=\lim_{x\to a}h(x)$

Then, $\lim g(x)=l$

**18.** *(c)* Given, $g(x)=3f\left(\frac{x}{3}\right)+f(3-x)$

On differentiating w.r.t. $x$ both sides, we get

$$g'(x)=3\times\frac{1}{3}f'\left(\frac{x}{3}\right)-f'(3-x)$$

$$=f'\left(\frac{x}{3}\right)-f'(3-x)\quad\text{...(i)}$$

Given $g(x)$ is decreasing in $(0,\alpha)$.

So, $g'(x)<0$

$$\Rightarrow \quad f'\left(\frac{x}{3}\right)<f'(3-x)$$

[from Eq. (i)] ...(ii)

$\because \quad f''(x)>0$ [given]

i.e. $f'(x)$ is an increasing function.

From Eq. (ii),

$$\frac{x}{3}<3-x \Rightarrow \frac{4x}{3}<3$$

$$\Rightarrow \quad x<\frac{9}{4}$$

So, $\alpha=\frac{9}{4}$

Hence, $8\alpha=8\times\frac{9}{4}=18$

**19.** *(d)* We have, $g(x)=f(x)+f(1-x)$

Differentiating both side, we get

$g'(x)=f'(x)-f'(1-x)$

As $f''(x)>0$, $f'(x)$ is an increasing function.

Also, $g(x)=f(x)+f(2a-x)$ is always symmetric about $x=a$

So, $g(x)=f(x)+f(1-x)$ is also symmetric about $x=1/2$

$\therefore g$ is decreasing in the interval $(0,1/2)$ and increasing in the interval $(1/2,1)$.

$\therefore \quad \alpha=1/2$

Now,

$$\tan^{-1}2\alpha+\tan^{-1}\left(\frac{1}{\alpha}\right)+\tan^{-1}\left(\frac{\alpha+1}{\alpha}\right)$$

$$=\tan^{-1}1+\tan^{-1}(2)+\tan^{-1}3=\pi$$

**20.** *(a)* $f(x)=xe^{x(1-x)}$

$$\log f(x)=\log x+\log e^{x(1-x)}$$

$$\log f(x)=\log x+x(1-x)$$

$$\log f(x)=\log x+x-x^2$$

On differentiating both sides w.r.t. $x$, we get

$$\frac{1}{f(x)}f'(x)=\frac{1}{x}+1-2x$$

$$\Rightarrow \quad f'(x)=xe^{x(1-x)}\left[\frac{1+x-2x^2}{x}\right]$$

$\Rightarrow f'(x) = -e^{x(1-x)}(2x^2 - x - 1)$

$= -e^{x(1-x)}(2x^2 - 2x + x - 1)$

$= -e^{x(1-x)}(2x + 1)(x - 1)$

$f'(x) = 0 \Rightarrow x = -\frac{1}{2}, 1$

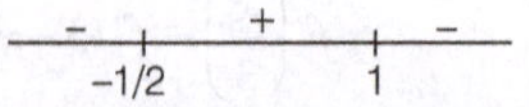


$\therefore f(x)$ is increasing in $\left(-\frac{1}{2}, 1\right)$ and decreasing $\left(-\infty, -\frac{1}{2}\right) \cup (1, \infty)$.

**21.** *(a)* Given, $f_a(x) = \tan^{-1} 2x - 3ax + 7$

$$f_a'(x) = \frac{2}{1+4x^2} - 3a$$

As the function $f_a'(x)$ is non-decreasing in $\left(-\frac{\pi}{6}, \frac{\pi}{6}\right)$,

$$f_a'(x) \geq 0$$

$$\Rightarrow \frac{2}{1+4x^2} - 3a \geq 0 \Rightarrow \frac{2}{1+4x^2} \geq 3a$$

$$\Rightarrow a \leq \frac{2}{3(1+4x^2)}, \text{ when } x \in \left(-\frac{\pi}{6}, \frac{\pi}{6}\right)$$

$\because a$ is maximum when $x^2 = \frac{\pi^2}{36}$,

$$a_{\max} = \frac{2}{3\left(1 + \frac{4\pi^2}{36}\right)} = \frac{2 \times 12}{36 + 4\pi^2}$$

$$= \frac{6}{9 + \pi^2}$$

$$\therefore \bar{a} = \frac{6}{9 + \pi^2}$$

$$\therefore f_{\bar{a}}(x) = \tan^{-1} 2x - \frac{18}{9+\pi^2}x + 7$$

$$f_{\bar{a}}\left(\frac{\pi}{8}\right) = \tan^{-1} 2\left(\frac{\pi}{8}\right) - \frac{18}{9+\pi^2} \times \frac{\pi}{8} + 7$$

$$= \tan^{-1}\frac{\pi}{4} - \frac{9\pi}{36 + 4\pi^2} + 7$$

$$= 8 - \frac{9\pi}{36 + 4\pi^2} = 8 - \frac{9\pi}{4(9+\pi^2)}$$

**22.** *(d)* Given that

$f_\lambda(x) = 4\lambda x^3 - 36\lambda x^2 + 36x + 48$

$\Rightarrow \quad f_\lambda'(x) = 12(\lambda x^2 - 6\lambda x + 3)$

For $f_\lambda(x)$ is increasing, thus

$(6\lambda)^2 - 12\lambda \leq 0$ and $\lambda > 0$

$\Rightarrow \quad \lambda \in [0, 1/3]$

Since, $\lambda^*$ be the largest value of $\lambda$, thus $\lambda^* = 1/3$

Hence, $f_\lambda{}^*(x) = \frac{4}{3}x^3 - 12x^2 + 36x + 48$

$\therefore f_\lambda{}^*(1) = \frac{4}{3}(1)^3 - 12(1)^2 + 36(1) + 48$

$= \frac{4}{3} - 12 + 36 + 48 = \frac{4}{3} + 72$

and $f_\lambda{}^*(-1) = \frac{4}{3}(-1)^3 - 12(-1)^2 + 36(-1) + 48$

$= -\frac{4}{3} + 0 = -\frac{4}{3}$

Thus, $f_\lambda{}^*(1) + f_\lambda{}^*(-1)$

$= \left(\frac{4}{3} + 72\right) + \left(-\frac{4}{3}\right) = 72$

**23.** *(a)* $f(x) = x^3 - 6x^2 + ax + b$

$\because \quad f(2) = 0 \Rightarrow 2a + b = 16$

and $\quad f(4) = 0 \Rightarrow 4a + b = 32$

On solving $a = 8, b = 0$

$\therefore f(x) = x^3 - 6x^2 + 8x$

$= x(x-2)(x-4)$

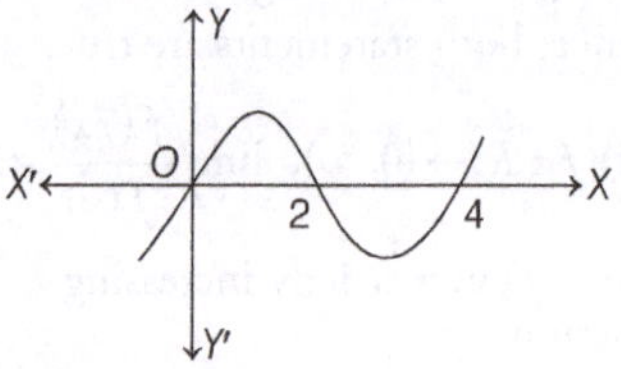


$f'(x) = 3x^2 - 12x + 8$

$f'(x) = 0 \Rightarrow x = 2 \pm \frac{2}{\sqrt{3}}$

$\because f'(x_2) = 0$ and $x_2 \in (2, 4)$

$\Rightarrow x_2 = 2 + \frac{2}{\sqrt{3}}$ and $f'(x_1) = -1$

$\Rightarrow \quad 3x_1^2 - 12x_1 + 8 = -1$

$\Rightarrow \quad x_1 = 1, 3$ ($S_1$ is true)

Again $f'(x) < 0$ for $x \in (2, x_4)$

and $f'(x) > 0$ for $x \in (x_4, 4)$

$\therefore \quad x_4 \in (3, 4)$

Now, $f(x) = x^3 - 6x^2 + 8x$

$f(3) = 27 - 54 + 24 = -3$

$f(4) = 64 - 96 + 32 = 0$

For $x_4 \in (3, 4)$

$f(x_4) < -3\sqrt{3}$

and $f'(x_3) > -4$

$2f'(x_3) > -8$

So, $2f'(x_3) = \sqrt{3} f(x_4)$

$S_2$ is true.

**24.** *(d)* Given,

$$f(x) = \begin{cases} -55x, & x < -5 \\ 2x^3 - 3x^2 - 120x, & -5 \leq x \leq 4 \\ 2x^3 - 3x^2 - 36x - 336, & x > 4 \end{cases}$$

$$\therefore f'(x) = \begin{cases} -55, & x < -5 \\ 6(x^2 - x - 20), & -5 \leq x \leq 4 \\ 6(x^2 - x - 6), & x > 4 \end{cases}$$

$$f'(x) = \begin{cases} -55, & x < -5 \\ 6(x-5)(x+4), & -5 \leq x \leq 4 \\ 6(x-3)(x+2), & x > 4 \end{cases}$$

For $f$ to be increasing, $f'(x) > 0$.

Now, $f'(x) = -55$ is always less than zero.

$f'(x) = 6(x-5)(x+4) < 0, -5 \leq x < 4$

Critical points $= 5, -4$

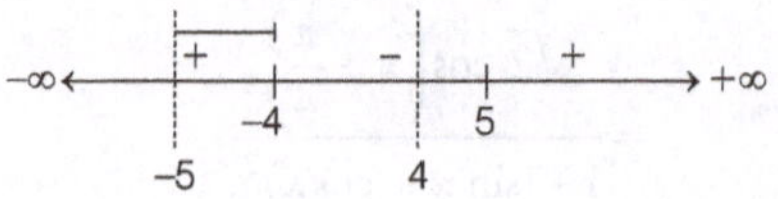


$$x \in (-5, -4) \quad \ldots\text{(i)}$$

and $f'(x) = 6(x-3)(x+2) < 0, x \geq 4$

Critical point, $= 3, -2$

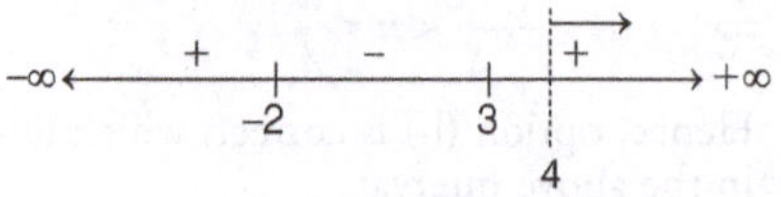


$$x \in (4, \infty) \quad \ldots\text{(ii)}$$

From Eqs. (i) and (ii), $f(x)$ is increasing in $x \in (-5, -4) \cup (4, \infty)$.

**25.** *(b)* Given,

$$5f(x) + 4f\left(\frac{1}{x}\right) = x^2 - 2 \quad \ldots\text{(i)}$$

Substitute $x \to \frac{1}{x}$, we get

$$5f\left(\frac{1}{x}\right) + 4f(x) = \frac{1}{x^2} - 2 \quad \ldots\text{(ii)}$$

On solving Eqs. (i) and (ii), we get

$$9f(x) = 5x^2 - \frac{4}{x^2} - 2$$

$$\because \quad y = 9x^2 f(x)$$

$$y = x^2\left(5x^2 - \frac{4}{x^2} - 2\right)$$

$$y = 5x^4 - 4 - 2x^2$$

$$\frac{dy}{dx} = 20x^3 - 4x = 4x(5x^2 - 1)$$

For increasing $4x(5x^2 - 1) > 0$

$$x = 0, \frac{1}{\sqrt{5}}, \frac{-1}{\sqrt{5}}$$

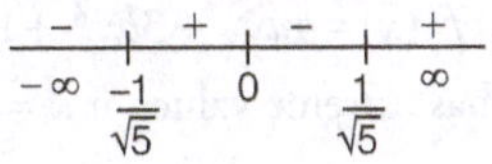


$$x \in \left(\frac{-1}{\sqrt{5}}, 0\right) \cup \left(\frac{1}{\sqrt{5}}, \infty\right)$$

**26.** *(c)* We have,

$$f(x) = 2x + \tan^{-1} x$$

and $g(x) = \log(x + \sqrt{1+x^2})$

Let $h(x) = f(x) - g(x)$

$$= 2x + \tan^{-1} x - \log(x + \sqrt{1+x^2})$$

$$\therefore h'(x) = 2 + \frac{1}{1+x^2} - \frac{1}{x+\sqrt{1+x^2}}\left[1 + \frac{1}{2\sqrt{1+x^2}}(2x)\right]$$

$$= 2 + \frac{1}{1+x^2} - \frac{1}{x+\sqrt{1+x^2}}\left[\frac{\sqrt{1+x^2}+x}{\sqrt{1+x^2}}\right]$$

$$= 2 + \frac{1}{1+x^2} - \frac{1}{\sqrt{1+x^2}}$$

Now, $1 + x^2 \geq 1$

$$\Rightarrow \sqrt{1+x^2} \geq 1 \Rightarrow \frac{1}{\sqrt{1+x^2}} \leq 1$$

$$\therefore h'(x) = 2 + \frac{1}{1+x^2} - \frac{1}{\sqrt{1+x^2}} \geq 0$$

So, $h'(x)$ is increasing on $R$.

$\therefore h(x) > h(0) \Rightarrow h(x) > f(0) - g(0)$

$\Rightarrow h(x) > 2(0) + \tan^{-1} 0 - \log(0 + \sqrt{1+0})$

$\Rightarrow h(x) > 0 \Rightarrow f(x) - g(x) > 0$

$\Rightarrow f(x) > g(x)$

So, $f_{max}(x) > g_{max}(x)$

**27.** *(b)* Given, $f:(0,1) \to R$,

$$f(x) = \frac{1}{1-e^{-x}}$$

$$g(x) = f(-x) - f(x) = \frac{1}{1-e^{x}} - \frac{1}{1-e^{-x}}$$

$$= \frac{1}{1-e^{x}} - \frac{1}{1-\frac{1}{e^{x}}}$$

$$g(x) = \frac{1}{1-e^{x}} - \frac{e^{x}}{e^{x}-1} = \frac{1+e^{x}}{1-e^{x}}$$

Now, differentiating w.r.t. $x$, we get

$$g'(x) = \frac{(1-e^{x})e^{x} - (1+e^{x})(-e^{x})}{(1-e^{x})^2}$$

$$= \frac{2e^{x}}{(1-e^{x})^2}$$

Now, $g'(x) = \dfrac{2e^{x}}{(1-e^{x})^2} > 0$

So, $g$ is increasing in (0, 1).

Now, for one-one or not $g(x_1) = g(x_2)$

$$\Rightarrow \frac{1+e^{x_1}}{1-e^{x_1}} = \frac{1+e^{x_2}}{1-e^{x_2}}$$

$$\Rightarrow 1 + e^{x_1} - e^{x_2} - e^{x_1} \cdot e^{x_2} = 1 + e^{x_2} - e^{x_1} - e^{x_1} \cdot e^{x_2}$$

$$\Rightarrow 2e^{x_1} = 2e^{x_2} \Rightarrow x_1 = x_2$$

So, $g(x)$ is one-one also.

**28.** *(c)* Given that

$$f(x) = 4\log_e(x-1) - 2x^2 + 4x + 5, \; x > 1$$

Differentiating both sides w.r.t. '$x$', we get

$$f'(x) = \frac{4}{x-1} - 4x + 4$$

$$= \frac{4(2x - x^2)}{x-1} = \frac{-4x(x-2)}{x-1} \quad \text{...(i)}$$

Since, if $1 < x < 2$ then $f'(x) > 0$, thus the function increasing but if $x \geq 2$, then $f'(x) < 0$, thus the function decreasing.

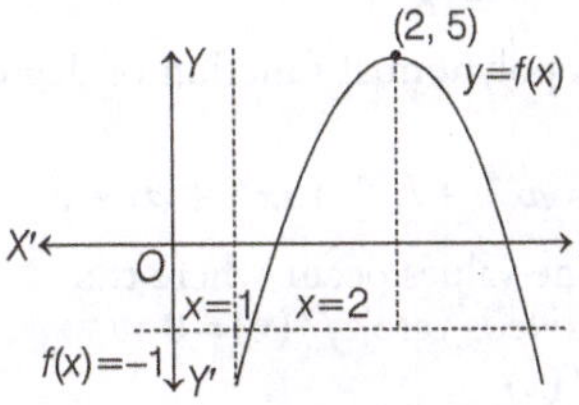


So that $f$ is increasing in (1, 2) and decreasing in (2, ∞).

From Eq. (i), $f(x)$ has maxima occurs at $x = 2$

$$f(2) = 4 \times 0 - 2 \times 2^2 + 4 \times 2 + 5 = 5$$

So clearly $f(x) = -1$ has exactly two solutions.

Now, differentiating Eq. (i) w.r.t. $x$, we get

$$f''(x) = \frac{4(x-1)(-2x+2) - 4(2x - x^2)}{(x-1)^2}$$

Now, $f'(e) = \dfrac{4(2e - e^2)}{(e-1)} < 0$ and

$$f''(2) = -8 < 0$$

From above,

$\therefore f'(e) - f''(2) > 0$

Thus, $f'(e) - f''(2) < 0$ is not correct.

**29.** *(b)* Given function,

$$f(x) = x\sqrt{kx - x^2} \quad \text{... (i)}$$

the function $f(x)$ is defined if

$$kx - x^2 \geq 0$$

$$\Rightarrow x^2 - kx \leq 0$$

$$\Rightarrow x \in [0, k] \quad \text{... (ii)}$$

because it is given that $f(x)$ is increasing in interval $x \in [0, 3]$, so $k$ should be positive.

Now, on differentiating the function $f(x)$ w.r.t. $x$, we get

$$f'(x) = \sqrt{kx - x^2} + \frac{x}{2\sqrt{kx - x^2}} \times (k - 2x)$$

$$= \frac{2(kx - x^2) + kx - 2x^2}{2\sqrt{kx - x^2}}$$

$$= \frac{3kx - 4x^2}{2\sqrt{kx - x^2}}$$

as $f(x)$ is increasing in interval $x \in [0, 3]$, so $f'(x) \geq 0 \; \forall \; x \in (0, 3)$

$$\Rightarrow 3kx - 4x^2 \geq 0$$

$$\Rightarrow 4x^2 - 3kx \leq 0$$

$$\Rightarrow 4x\left(x - \frac{3k}{4}\right) \leq 0$$

$$\Rightarrow x \in \left[0, \frac{3k}{4}\right] \quad \text{(as } k \text{ is positive)}$$

So, $3 \leq \dfrac{3k}{4} \Rightarrow k \geq 4$

$\Rightarrow$ Minimum value of $k = m = 4$ and the maximum value of $f$ in [0, 3] is $f(3)$.

$\because$ $f$ is increasing function in interval $x \in [0, 3]$

$$\therefore M = f(3) = 3\sqrt{4 \times 3 - 3^2} = 3\sqrt{3}$$

Therefore, ordered pair

$(m, M) = (4, 3\sqrt{3})$

**30.** *(252)*

$$f(x) = (p^2 - 6p + 8)(\sin^2 2x - \cos^2 2x) + 2(2-p)x + 7$$

$$f(x) = (p^2 - 6p + 8)(-\cos 4x) + 2(2-p)x + 7$$

$$f'(x) = (p^2 - 6p + 8)(\sin 4x)(4) + 2(2-p) \neq 0$$

$$\sin 4x \neq \frac{2p-4}{4(p-4)(p-2)}$$

$$\sin 4x \neq \frac{1}{2(p-4)}, \text{ where } p \neq 2$$

$$\therefore \quad \left|\frac{1}{2(p-4)}\right| > 1$$

$$|2(p-4)| < 1$$

$$-1 < 2(p-4) < 1$$

$$-\frac{1}{2} < p-4 < \frac{1}{2}$$

$$\frac{7}{2} < p < \frac{9}{2} \Rightarrow a = \frac{7}{2}, b = \frac{9}{2}$$

$$16ab = 16 \times \frac{7}{2} \times \frac{9}{2} = 252$$

**31.** *(2)* $f(x) = x^2 + ax + 1$

$$f'(x) = 2x + a$$

According to the question, $f'(x) \geq 0$ for $x \in [1, 2]$

For the least value $2x + a \geq 0$

$\Rightarrow \quad a \geq -2x \Rightarrow a \geq -2$

$\Rightarrow \quad R = -2$

For the greatest value $2x + a \leq 0$

$\Rightarrow \quad a \leq -2x \quad \{x \in [1, 2]\}$

$\Rightarrow \quad a \leq -4 \Rightarrow S = -4$

$$|R - S| = |-2 + 4| = 2$$

**32.** *(96)* Let $f(n) = \left(\frac{10}{11}\right)^n + \left(\frac{9}{11}\right)^n$, then

$$f'(n) = \left(\frac{10}{11}\right)^n \log\left(\frac{10}{11}\right) + \left(\frac{9}{11}\right)^n$$

$$\log\left(\frac{9}{11}\right) < 0 \qquad \left[\because \frac{d}{dx}(a^x) = a^x \log a\right]$$

$\Rightarrow f(n)$ is decreasing function.

Putting $n = 1, 2, 3, 4, \ldots\ldots$, we observe that

$$f(1) = \frac{10}{11} + \frac{9}{11} = \frac{19}{11} > 1$$

$$\text{and } f(2) = \left(\frac{10}{11}\right)^2 + \left(\frac{9}{11}\right)^2$$

$$= \frac{100}{121} + \frac{81}{121} = \frac{181}{121} > 1$$

Similarly, we find that $f(3)$ and $f(4) > 1$.

$$\text{But } f(5) = \left(\frac{10}{11}\right)^5 + \left(\frac{9}{11}\right)^5 < 1$$

Similarly, for $n = 6, 7, 8, \ldots 100$, $f(n) < 1$

$\therefore$Required number of elements $= 96$.

## Topic 3

**01.** *(a)* $f(x) = 2x^3 - 9ax^2 + 12a^2x + 1$ and $a > 0$

$$f'(x) = 6x^2 - 18ax + 12a^2$$
$$= 6(x^2 - 3ax + 2a^2)$$
$$= 6(x-a)(x-2a) = 0$$

$x = a, 2a$

$\therefore x = a$ is point of maxima and $x = 2a$ is point of minima.

$\therefore \quad p = a, q = 2a$

$p^2 = q$ [given]

$\Rightarrow \quad a^2 = 2a \Rightarrow a = 2$

$$f(x) = 2x^3 - 18x^2 + 48x + 1$$

$\Rightarrow \quad f(3) = 37$

**02.** *(a)* $f(x) = 6x^3 - 45ax^2 + 108a^2x + 1$

$$f'(x) = 18x^2 - 90ax + 108a^2$$
$$= 18(x^2 - 5ax + 6a^2)$$
$$= 18(x-3a)(x-2a)$$

$\therefore \quad f'(x) = 0$

$\Rightarrow \quad x = 2a, 3a$

$\because x_1 \cdot x_2 = 54$

$\therefore (2a)(3a) = 54$

$a^2 = 9$

$a = 3 \qquad (\because a > 0)$

$f'(x) = 0, x = 6, 9$

$\therefore \quad a + x_1 + x_2$
$= 3 + 6 + 9$
$= 18$

**03.** *(a)* A polynomial function of degree four

$$f(x) = ax^4 + bx^3 + cx^2 + dx + e$$

Extreme values occur where the derivative is zero, $f'(x) = 0$

$$\lim_{x \to 0} \frac{f(x)}{x^2} = 5$$

$$\lim_{x \to 0} \frac{ax^4 + bx^3 + cx^2 + dx + e}{x^2} = 5$$

$$\lim_{x \to 0} ax^2 + bx + c + \frac{d}{x} + \frac{e}{x^2} = 5$$

For the limit to be exist and equal to 5, then $d$ and $e$ must be zero.

Thus, $f(x) = ax^4 + bx^3 + cx^2$

$$\lim_{x \to 0} \frac{ax^4 + bx^3 + cx^2}{x^2} = 5$$

$$\lim_{x \to 0} ax^2 + bx + c = 5 \Rightarrow c = 5$$

So, $f(x) = ax^4 + bx^3 + 5x^2$

Now, $f'(x) = 4ax^3 + 3bx^2 + 10x$

$f(x)$ has extreme values at $x = 4$ and $x = 5$

$$f'(4) = 256a + 48b + 40 = 0$$

$\Rightarrow \quad 32a + 6b + 5 = 0$

$$f'(5) = 500a + 75b + 50 = 0$$

$\Rightarrow \quad 20a + 3b + 2 = 0$

Solving these equations, we get $a = \frac{1}{8}$, $b = -\frac{3}{2}$

The polynomial becomes

$$f(x) = \frac{1}{8}x^4 - \frac{3}{2}x^3 + 5x^2$$

$$\text{Now, } f(2) = \frac{1}{8} \times 16 - \frac{3}{2} \times 8 + 5 \times 4$$
$$= 2 - 12 + 20 = 10$$

**04.** *(d)* We have,

$$f(x) = \int_0^x t(t^2 - 9t + 20)\,dt, 1 \leq x \leq 5$$

$$\Rightarrow f'(x) = x^3 - 9x^2 + 20x$$
$$= x(x-4)(x-5)$$

– + – +
0 4 5

$f(x)$ is increasing in $(0, 4)$

$\therefore$ Minimum value of $f(x)$ at $x = 1$ and maximum value of $x = 4$

$$\therefore \quad f(x) = \frac{x^4}{4} - \frac{9x^3}{3} + \frac{20x^2}{2}$$

$$\Rightarrow f(1) = \frac{1}{4} - 3 + 10 = \frac{29}{4} = \alpha$$

$$\Rightarrow f(4) = \frac{256}{4} - 3(64) + 10 \times (16)$$
$$= 32 = \beta$$

$$\therefore 4(\alpha + \beta) = 4\left(\frac{29}{4} + 32\right) = 157$$

**05.** *(a)*

**key idea** First convert $y = \frac{2x^2 - 3x + 8}{2x^2 + 3x + 8}$ into quadratic equation in $x$ where, $y$ is constant. Then, use $D \geq 0$ to find minimum and maximum value of $y$.

$$\text{Let } y = \frac{2x^2 - 3x + 8}{2x^2 + 3x + 8} \quad \ldots \text{(i)}$$

Here, we observe that

$$2x^2 + 3x + 8 > 0, \forall x \in R$$

$(\because a > 0 \text{ and } D < 0)$

Now, Eq. (i) we get

$\Rightarrow 2x^2y + 3xy + 8y = 2x^2 - 3x + 8$

$\Rightarrow x^2(2y-2) + x(3y+3) + 8y - 8 = 0$

$\because x \in R$, so the equation has real roots.

i.e. discriminant $\geq 0$

$\Rightarrow (3y+3)^2 - 4(2y-2)(8y-8) \geq 0$

$\Rightarrow (11y-5)(5y-11) \leq 0$

$\Rightarrow y \in \left[\frac{5}{11}, \frac{11}{5}\right]$

Hence, sum of $y_{max}$ and

$y_{min} = \frac{11}{5} + \frac{5}{11} = \frac{121+25}{55} = \frac{146}{55} = \frac{m}{n}$ [given]

$\therefore \quad m + n = 146 + 55 = 201$

**06.** *(c)* $f(x) = (x+3)^2(x-2)^3,\ x \in [-4, 4]$

$\therefore \quad f'(x) = (x+3)^2\, 3(x-2)^2 + (x-2)^3 \cdot 2(x+3)$

$= (x+3)(x-2)^2\,[3(x+3) + 2(x-2)]$

$= (x-2)^2\,(x+3)(5x+5)$

$= 5(x-2)^2\,(x+3)(x+1)$

$\because \quad f'(x) = 0 \Rightarrow x = 2, -3, -1$

$\therefore \quad f(2) = 0,\ f(-3) = 0,$

$f(-1) = -4 \times 27 = -108,$

$f(-4) = -216$

and $f(4) = 392$

$\therefore \quad M = 392,\ m = -216$

$\Rightarrow M - m = 392 + 216 = 608$

**07.** *(c)* $f(x) = 2x + 3x^{2/3}$

$f'(x) = 2 + 2x^{\frac{-1}{3}} = 2 + \frac{2}{x^{1/3}}$

$= 2\left(\frac{x^{\frac{1}{3}} + 1}{x^{\frac{1}{3}}}\right)$

$f'(x) = 0 \Rightarrow x^{1/3} + 1 = 0 \Rightarrow x = -1$

Critical point $x = 0, -1$

Also at $x = 0$, $f(x) = 0$

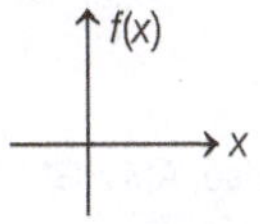


Using first order derivative test

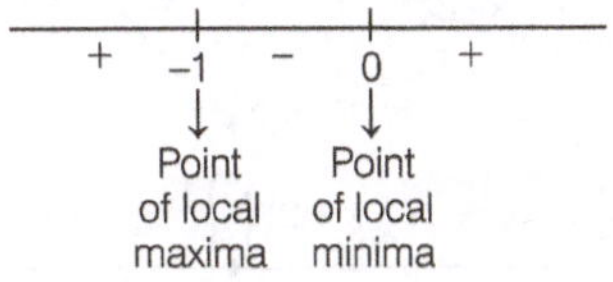


$\Rightarrow$ Exactly one point local maxima and exactly one point local minima.

**08.** *(d)* We have,

$f(x) = |x^2 - 5x + 6| - 3x + 2$

$= \begin{cases} (x^2 - 5x + 6) - 3x + 2, & x \in [-1, 2] \\ -(x^2 - 5x + 6) - 3x + 2, & x \in [2, 3] \end{cases}$

$= \begin{cases} x^2 - 8x + 8, & x \in [-1, 2] \\ -x^2 + 2x - 4, & x \in [2, 3] \end{cases}$

Now, we have to check in the interval $[-1, 3]$ only

$\therefore$ Absolute maximum

$= f(-1) = (-1)^2 - 8(-1) + 8 = 17$

and absolute minimum

$= f(3) = (3)^2 - 8(3) + 8 = -7$

$\therefore$ Required sum $= 17 - 7 = 10$

**09.** *(c)* We have, $f(x) = |x^2 - x + 1| + [x^2 - x + 1]$

Let $g(x) = x^2 - x + 1$

$= \left(x - \frac{1}{2}\right)^2 + \frac{3}{4}$

$\therefore g(x)$ attains minimum value at $x = \frac{1}{2}$.

So, $f(x)$ attains minimum value at $x = \frac{1}{2}$.

$\therefore f_{min} = f\left(\frac{1}{2}\right) = \left|\frac{3}{4}\right| + \left[\frac{3}{4}\right] = \frac{3}{4}$

**10.** *(c)* $\because f(x) = \frac{x^3}{3} + 2bx + \frac{ax^2}{2}$

and $g(x) = \frac{x^3}{3} + ax + bx^2$

For extreme points,

$f'(x) = x^2 + 2b + ax = 0$ ...(i)

and $g'(x) = x^2 + a + 2bx = 0$ ...(ii)

From Eqs. (i) and (ii), we get

$x = 1$ will be common root, then put $x = 1$

$\Rightarrow (1)^2 + 2b + a(1) = 0$

$\Rightarrow 2b + a + 1 = 0$

On adding 6 both sides, we get

$2b + a + 1 + 6 = 0 + 6 = 6$

$\Rightarrow 2b + a + 7 = 6$

**11.** *(c)* Given, $x^2 + (3-a)x + 1 - 2a = 0$

Let the roots of the given equation be $\alpha$ and $\beta$, respectively.

$\therefore \alpha + \beta = a - 3$ and $\alpha\beta = 1 - 2a$

Now, $\alpha^2 + \beta^2 = (\alpha + \beta)^2 - 2\alpha\beta$

$= (a-3)^2 - 2(1-2a)$

$= a^2 + 9 - 6a - 2 + 4a$

$= a^2 - 2a + 7$

$= a^2 - 2a + 1 + 6$

$= (a-1)^2 + 6$

$\therefore$ The minimum value of $\alpha^2 + \beta^2$ is 6.

[$\because$ Minimum value $(a-1)^2 = 0$]

**12.** *(d)* Given that $x, y > 0$ and $x^3y^2 = 2^{15}$

Now, $3x + 2y$ can be expressed as

$3\left(\frac{3x}{3}\right) + 2\left(\frac{2y}{2}\right)$.

As, $x, y > 0$. So, apply AM $\geq$ GM inequality here,

$\Rightarrow \frac{3\left(\frac{3x}{3}\right) + 2\left(\frac{2y}{2}\right)}{3+2} \geq \left\{\left(\frac{3x}{3}\right)^3\left(\frac{2y}{2}\right)^2\right\}^{\frac{1}{3+2}}$ [$\because$ AM$\geq$GM]

$\Rightarrow \frac{3x + 2y}{5} \geq (x^3y^2)^{1/5}$

$\Rightarrow \frac{3x + 2y}{5} \geq (2^{15})^{1/5}$

[Since, $x^3y^2 = 2^{15}$]

$\Rightarrow \frac{3x + 2y}{5} \geq 2^3 = 8 \Rightarrow 3x + 2y \geq 40$

Hence, the least value of $3x + 2y$ is 40.

**13.** *(a)* Let $f(x) = ax^3 + bx^2 + cx + d$

$f'(x) = 3ax^2 + 2bx + c$

$f''(x) = 6ax + 2b$

$f'(x)$ has local minima at $x = -1$.

$\therefore \quad f''(-1) = 0 \Rightarrow -6a + 2b = 0$

$\Rightarrow \quad b = 3a$ ...(i)

$f(x)$ has local minima at $x = 1$

$f'(1) = 0$

$\Rightarrow 3a + 6a + c = 0$

$\Rightarrow \quad c = -9a$ ...(ii)

$f(1) = -10$

$\Rightarrow \quad -5a + d = -10$ ...(iii)

$f(-1) = 6$

$\Rightarrow \quad 11a + d = 6$ ...(iv)

Solving Eqs. (iii) and (iv), we get

$a = 1,\ d = -5$

From Eqs. (i) and (ii), we get

$b = 3,\ c = -9$

$\therefore \quad f(x) = x^3 + 3x^2 - 9x - 5$

So, $\quad f(3) = 27 + 27 - 27 - 5 = 22$

**14.** *(c)* $f(x)=\left(\frac{2}{x}\right)^{x^2}$ ; $x>0$

$\therefore \log f(x)=x^2(\log 2-\log x)$

$f'(x)=f(x)[-x+(\log 2-\log x)2x]$

$f'(x)=f(x)\cdot x\,(2\log 2-2\log x-1)$

For maxima or minima put $f'(x)=0$, we get

$2\log 2-2\log x-1=\log\left(\frac{4}{x^2}\right)-1=0$

$\Rightarrow x=2/\sqrt{e}$

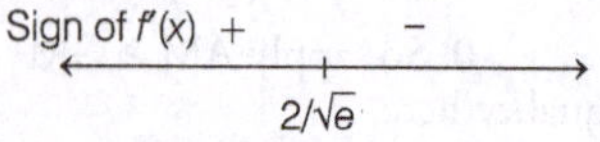


$\therefore$ At $x=\frac{2}{\sqrt{e}}$, $f(x)$ has maximum value.

Maximum $=\left(\frac{2\sqrt{e}}{2}\right)^{\frac{4}{e}}=e^{\frac{2}{e}}$

**15.** *(d)* Since, $p'(x)=0$ at $x=1$ and $x=2$ and $p(x)$ is cubic polynomial, so

$p'(x)=a(x-1)(x-2)$

$=a(x^2-3x+2)$

$\therefore p(x)=a\left(\frac{x^3}{3}-\frac{3}{2}x^2+2x\right)+b$

According to the question,

$p(1)=8 \Rightarrow a\left(\frac{1}{3}-\frac{3}{2}+2\right)+b=8$

$\Rightarrow a\left(\frac{1}{3}+\frac{1}{2}\right)+b=8$

$\Rightarrow 5a+6b=48$ ... (i)

and $p(2)=4 \Rightarrow a\left(\frac{8}{3}-6+4\right)+b=4$

$\Rightarrow 2a+3b=12$ ... (ii)

From Eqs. (i) and (ii), we get

$a=24,\ b=-12$

$\because p(0)=b=-12$

**16.** *(d)* The non-zero four degree polynomial $f(x)$ has extreme points at $x=-1, 0, 1$, so we can assume

$f'(x)=a(x+1)(x-0)(x-1)$

$=ax(x^2-1)$

where, $a$ is non-zero constant.

$f'(x)=ax^3-ax$

$\Rightarrow f(x)=\frac{a}{4}x^4-\frac{a}{2}x^2+C$

[integrating both sides]

where, $C$ is constant of integration.

Now, since $f(x)=f(0)$

$\Rightarrow \frac{a}{4}x^4-\frac{a}{2}x^2+C=C$

$\Rightarrow \frac{x^4}{4}=\frac{x^2}{2} \Rightarrow x^2(x^2-2)=0$

$\Rightarrow x=-\sqrt{2}, 0, \sqrt{2}$

Thus, $f(x)=f(0)$ has one rational and two irrational roots.

**17.** *(c)* Given that function, $f(x)=x^3-3(a-2)x^2+3ax+7$, for some $a\in R$ is increasing in (0, 1] and decreasing in [1, 5).

$f'(1)=0$ [$\because$ tangent at $x=1$ will be parallel to $X$-axis]

$\Rightarrow (3x^2-6(a-2)x+3a)_{x=1}=0$

$\Rightarrow 3-6(a-2)+3a=0$

$\Rightarrow 3-6a+12+3a=0$

$\Rightarrow 15-3a=0 \Rightarrow a=5$

So, $f(x)=x^3-9x^2+15x+7$

$\Rightarrow f(x)-14=x^3-9x^2+15x-7$

$\Rightarrow f(x)-14=(x-1)(x^2-8x+7)$

$=(x-1)(x-1)(x-7)$

$\Rightarrow \frac{f(x)-14}{(x-1)^2}=(x-7)$ ...(i)

Now, $\frac{f(x)-14}{(x-1)^2}=0, (x\neq 1)$

$\Rightarrow x-7=0$ [from Eq. (i)]

$\Rightarrow x=7$

**18.** *(d)* We have,

$f(x)=x^2+\frac{1}{x^2}$ and $g(x)=x-\frac{1}{x}$

$\Rightarrow h(x)=\frac{f(x)}{g(x)}$

$\therefore h(x)=\frac{x^2+\frac{1}{x^2}}{x-\frac{1}{x}}$

$\Rightarrow h(x)=\left(x-\frac{1}{x}\right)+\frac{2}{x-\frac{1}{x}}$

$x-\frac{1}{x}>0,$

$\left(x-\frac{1}{x}\right)+\frac{2}{x-\frac{1}{x}}\in[2\sqrt{2}, \infty)$

$x-\frac{1}{x}<0,$

$\left(x-\frac{1}{x}\right)+\frac{2}{x-\frac{1}{x}}\in(-\infty, 2\sqrt{2}]$

$\therefore$ Local minimum value is $2\sqrt{2}$.

**19.** *(d)* $f(x)=2x^3-9x^2+12x+5$, $x\in[0, 3]$

$\Rightarrow f'(x)=6x^2-18x+12$

$=6(x-1)(x-2)$

For maxima-minima, put $f'(x)=0$,

$(x-1)(x-2)=0 \Rightarrow x=1$ or $x=2$

Now, $f(1)=10$, $f(2)=9$, $f(0)=5$ and $f(3)=14$

$\therefore M=14$ and $m=5$

So, $M-m=14-5=9$

**20.** *(c)* Here, $x=-1$ and $x=2$ are extreme points of

$f(x)=\alpha\log|x|+\beta x^2+x$, then

$f'(x)=\frac{\alpha}{x}+2\beta x+1$

$\therefore f'(-1)=-\alpha-2\beta+1=0$ ...(i)

[at extreme point, $f'(x)=0$]

and $f'(2)=\frac{\alpha}{2}+4\beta+1=0$ ...(ii)

On solving Eqs. (i) and (ii), we get

$\alpha=2, \beta=-1/2$

**21.** *(a)*

> **key idea** **Sandwich Theorem** Let $f, g$ and $h$ be real functions such that $f(x)\le g(x)\le h(x)$, $\forall x$ in the common domain of definition for some real number $a$, if
> $\because \lim_{x\to a} f(x)=l=\lim_{x\to a} h(x)$, then $\lim_{x\to a} g(x)=l$

Since, $f(x)$ is a positive increasing function.

$\Rightarrow 0<f(x)<f(2x)<f(3x)$

$\Rightarrow 0<1<\frac{f(2x)}{f(x)}<\frac{f(3x)}{f(x)}$

$\Rightarrow \lim_{x\to\infty} 1\le \lim_{x\to\infty}\frac{f(2x)}{f(x)}\le \lim_{x\to\infty}\frac{f(3x)}{f(x)}$

By Sandwich theorem,

$\lim_{x\to\infty}\frac{f(2x)}{f(x)}=1$

**22.** *(d)*

> **key idea** Use, AM $\ge$ GM

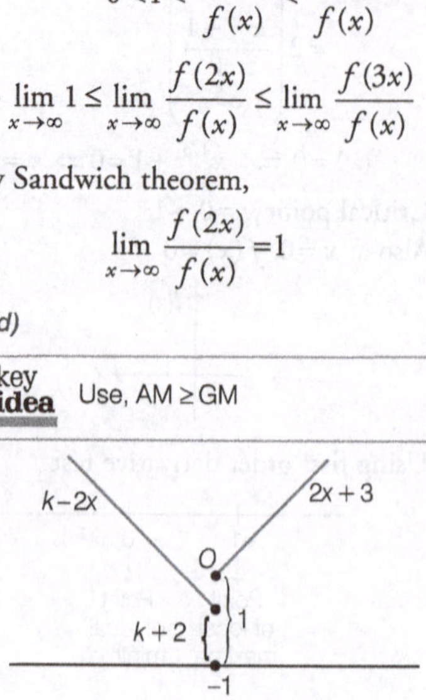

$\because \quad k+2 \le 1$

$\therefore \quad k \le -1$

**23.** *(a)* Using AM $\ge$ GM inequality

$$\frac{e^x + 2e^{-x}}{2} \ge \sqrt{e^x \cdot 2e^{-x}}$$

Thus, $e^x + 2e^{-x} \ge 2\sqrt{2}$.

Then, $\dfrac{1}{e^x + 2e^{-x}} \le \dfrac{1}{2\sqrt{2}}$

As, $\dfrac{1}{e^x + 2e^{-x}}$ is always positive,

we get

$$0 < \frac{1}{e^x + 2e^{-x}} \le \frac{1}{2\sqrt{2}}$$

Observe that $f(0) = 1/3$. Thus, $f(c) = 1/3$.

Using extreme-value theorem, we can say that as $f$ is continuous, $f$ will attain a value 1/3 at some point. Here we are able to identify the point as well.

**24.** *(b)* Given that,

$f(x) = (x+1)^{1/3} - (x-1)^{1/3}$

On differentiating w.r.t. $x$, we get

$$f'(x) = \frac{(x-1)^{2/3} - (x+1)^{2/3}}{3(x^2-1)^{2/3}}$$

Clearly, $f'(x)$ does not exist at $x = \pm 1$.

Also $(x+1)^{2/3} > (x-1)^{2/3} \le 0$

$\therefore f(x)$ is always decreasing.

If end point were included, the maximum occurs at $x = 0$

At $\qquad x = 0$

$f(0) = (0+1)^{1/3} - (0-1)^{1/3} = 2$

Hence, the greatest value of $f(x)$ is 2.

**25.** *(c)* $f(x) = \begin{cases} |-x-2+2x|, & x \le -2 \\ |x+2+2x|, & -2 \le x \le 0 \\ |x+2-2x|, & x \ge 0 \end{cases}$

$$f(x) = \begin{cases} |-x-2+2x|, & x \le -2 \\ |x+2+2x|, & -2 \le x \le 0 \\ |x+2-2x|, & x \ge 0 \end{cases}$$

$$f(x) = \begin{cases} |x-2|, & x \le -2 \\ |3x+2|, & -2 \le x \le 0 \\ |2-x|, & x > 0 \end{cases}$$

$$f(x) = \begin{cases} 2-x, & x \le -2 \\ -3x-2, & -2 \le x \le \frac{-2}{3} \\ 3x+2, & -\frac{2}{3} < x \le 0 \end{cases}$$

$$f(x) = \begin{cases} 2-x, & x \le -2 \\ -3x-2, & -2 < x \le \frac{-2}{3} \\ 3x+2, & \frac{-2}{3} < x \le 0 \\ 2-x, & \frac{-2}{3} < x < 2 \\ x-2, & x \ge 2 \end{cases}$$

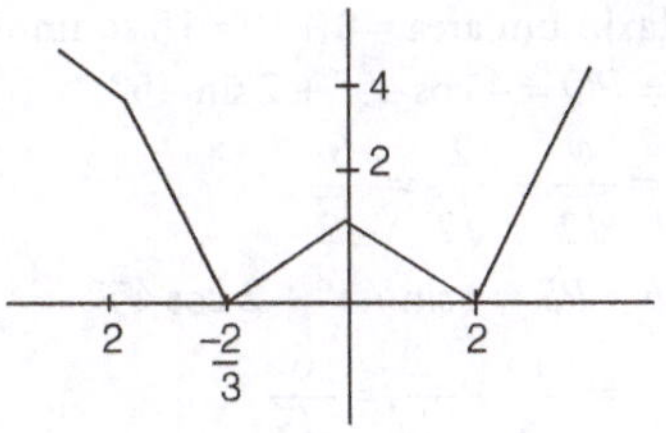


Number of maxima (n) = 1

Number of minima (m) = 2

$\therefore m + n = 3$

**26.** *(c)* $f(x) = x^3 + ax^2 + b\ln|x| + 1, x \ne 0$

$\Rightarrow f'(x) = 3x^2 + 2ax + \dfrac{b}{x}$

$\Rightarrow f'(-1) = 0 = 3 - 2a - b \qquad \ldots$(i)

$\Rightarrow f'(2) = 0 = 12 + 4a + \dfrac{b}{2} \qquad \ldots$(ii)

From Eqs. (i) and (ii), we get

$a = \dfrac{-9}{2}, b = 12$

$\Rightarrow f(x) = x^3 - \dfrac{9}{2}x^2 + 12\ln|x| + 1$

$\Rightarrow f''(x) = 6x - 9 - \dfrac{12}{x^2}$

$\Rightarrow f''(x = -1) \le 0$

$\therefore x = -1$ is the point of Maximum

$f''(x = 2) = 0$

Max of $f(x) = -1 - \dfrac{9}{2} + 1 = -\dfrac{9}{2} = -4.5$

$f(-2) = -8 - 18 + 12\ln 2 + 1$

$= -25 + 12\ln 2$

$= -25 + 8.4$

$= -16.6$

$f\left(-\dfrac{1}{2}\right) = -\dfrac{1}{8} - \dfrac{9}{8} - 12\ln 2 + 1$

$= \dfrac{-5}{4} + 1 - 8.4$

$= \dfrac{-5 + 4 - 33.6}{4}$

$= \dfrac{-34.6}{4}$

$= -8.65$

$|M + m| = |-4.5 - 16.6|$

$= 21.1$

**27.** *(c)* Given,

$$f(x) = \begin{cases} 1-2x, & x < -1 \\ \frac{1}{3}(7 + 2|x|), & -1 \le x \le 2 \\ \frac{11}{18}(x-4)(x-5), & x > 2 \end{cases}$$

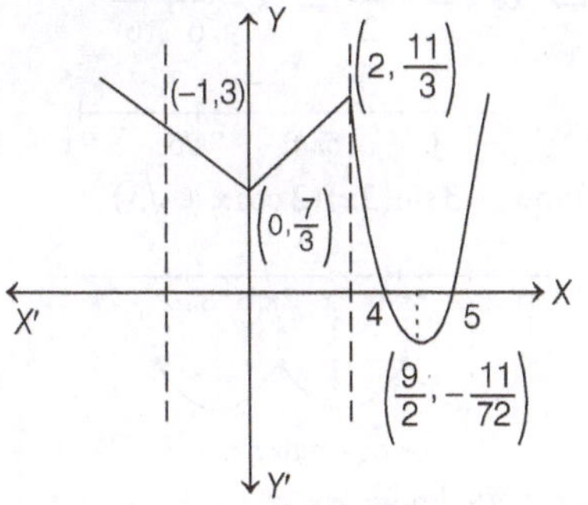


Local minimum at $\left(0, \dfrac{9}{2}\right)$

$f(0) = \dfrac{7}{3}$

$f\left(\dfrac{9}{2}\right) = \dfrac{11}{18}\left(\dfrac{1}{2}\right)\left(-\dfrac{1}{2}\right) = -\dfrac{11}{72}$

Sum $= \dfrac{7}{3} - \dfrac{11}{72} = \dfrac{168 - 11}{72} = \dfrac{157}{72}$

**28.** *(c)* $f(x) = \displaystyle\int_0^{x^2} \frac{t^2 - 8t + 15}{e^t} dt$

$f'(x) = \dfrac{2x(x^4 - 8x^2 + 15)}{e^{x^2}}$

For extreme points

$f'(x) = 0$

$2x(x^2 - 5)(x^2 - 3) = 0$

$x = 0, \pm\sqrt{5}, \pm\sqrt{3}$

Sign. of $f'(x)$

$$\underset{-\sqrt{5}}{-\;|}\;+\;\underset{-\sqrt{3}}{|}\;-\;\underset{0}{|}\;+\;\underset{\sqrt{3}}{|}\;-\;\underset{\sqrt{5}}{|}\;+$$

$\therefore$ Local maximum at $x = -\sqrt{3}, \sqrt{3}$

Local minimum at $x = -\sqrt{5}, 0, \sqrt{5}$

**Concept Enhancer**

$f(x) = \displaystyle\int_{g(x)}^{h(x)} F(t)dt$

$\Rightarrow f'(x) = h'(x)F(h(x)) - g'(x)F(g(x))$

**29.** *(c)* $f(x) = 4 \cdot \cos^3 x + 3\sqrt{3}\cos^2 x - 10$

and $x \in (0, 2\pi)$

So, $f'(x) = 12\cos^2 x(-\sin x)$

$+ 6\sqrt{3}\cos x(-\sin x)$

$= -6\sin x \cdot \cos x\,(2\cos x + \sqrt{3})$

$= -3\sin(2x)\,(2\cos x + \sqrt{3})$

Now, $-3\sin(2x) = 0,\ x = \frac{\pi}{2}, \frac{3\pi}{2}, \pi, 2\pi$

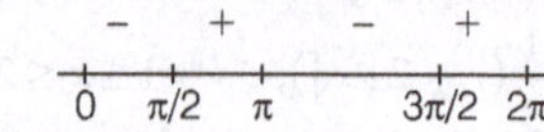


and $(2\cos x + \sqrt{3})$

$\Rightarrow \cos x = -\frac{\sqrt{3}}{2} \Rightarrow x = \frac{5\pi}{6}, \frac{7\pi}{6}$

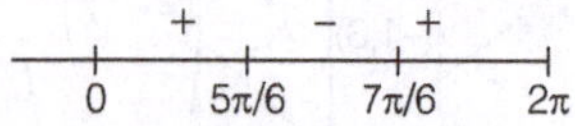


Now, $-3\sin(2x)(2\cos x + \sqrt{3})$

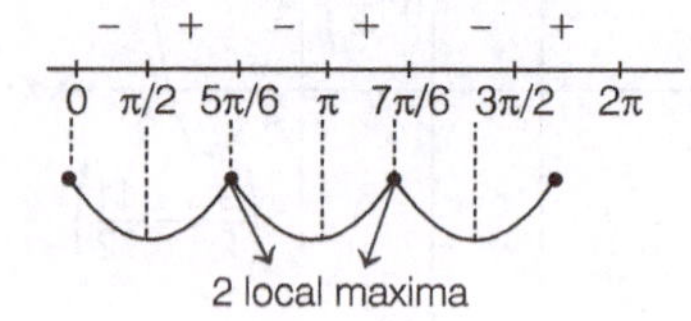


**30.** *(b)* We know, $e < \pi$

$\Rightarrow \quad \frac{1}{e} > \frac{1}{\pi}$

Then, according to the question,
If $f(x)$ attains maximum value at $x = 1/e$, then

$f(x)\big|_{x=\frac{1}{\pi}} < f(x)\big|_{x=\frac{1}{e}}$

$\Rightarrow \quad f\left(\frac{1}{\pi}\right) < f\left(\frac{1}{e}\right)$

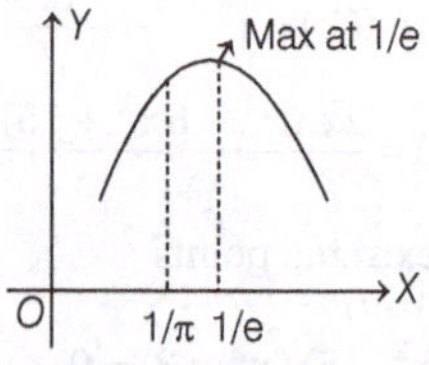


$\Rightarrow \quad (\pi)^{\frac{2}{\pi}} < e^{\frac{2}{e}} \Rightarrow (\pi)^{2e} < e^{2\pi}$

$\Rightarrow \quad \pi^e < e^\pi$

**31.** *(a)* Given, rectangle $ABCD$ of sides 2 and 4 be inscribed in another rectangle $PQRS$.

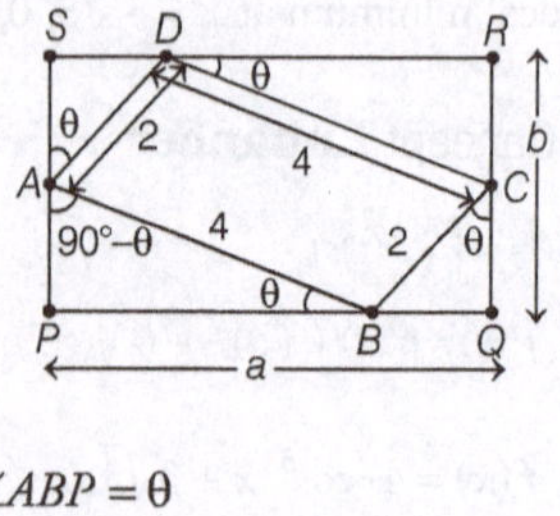


Let $\angle ABP = \theta$

$\therefore \quad \angle DAS = \angle CDR = \angle BCQ = \theta$

Now, $PB = 4\cos\theta$ and $BQ = 2\sin\theta$

$\therefore \quad PQ = 4\cos\theta + 2\sin\theta$

Also, $AP = 4\sin\theta$, $AS = 2\cos\theta$

$\therefore \quad PS = 4\sin\theta + 2\cos\theta$

$\therefore$ Area of rectangle $PQRS$

$= (4\cos\theta + 2\sin\theta)(4\sin\theta + 2\cos\theta)$

$= 4(2\cos^2\theta + 2\sin^2\theta + 5\sin\theta\cos\theta)$

$= 4(2 + 5\sin\theta\cos\theta) = 8 + 10\sin 2\theta$

Area will be maximum, when $\sin 2\theta = 1$

$\Rightarrow \theta = 45°$

$\therefore$ Maximum area $= 8 + 10 = 18$ sq units

$\therefore a = PQ = 4\cos 45° + 2\sin 45°$

$= \frac{4}{\sqrt{2}} + \frac{2}{\sqrt{2}} = \frac{6}{\sqrt{2}}$

and $b = PS = 4\sin 45° + 2\cos 45°$

$= \frac{4}{\sqrt{2}} + \frac{2}{\sqrt{2}} = \frac{6}{\sqrt{2}}$

$\therefore (a+b)^2 = \left(\frac{6}{\sqrt{2}} + \frac{6}{\sqrt{2}}\right)^2 = \left(\frac{12}{\sqrt{2}}\right)^2 = 72$

**32.** *(a)* $f(x) = 3\sqrt{x-2} + \sqrt{4-x}$

Domain of $f(x)$

$x - 2 \ge 0$ and $4 - x \ge 0$

$\Rightarrow \quad x \ge 2$ and $x \le 4$

$\Rightarrow \quad x \in [2, 4]$

$f'(x) = \frac{3}{2\sqrt{x-2}} + \frac{1}{2\sqrt{4-x}}(-1)$

$\Rightarrow \quad f'(x) = 0$

$\frac{3}{\sqrt{x-2}} = \frac{1}{\sqrt{4-x}}$

$\Rightarrow \quad \frac{9}{x-2} = \frac{1}{4-x}$

$\Rightarrow 36 - 9x = x - 2 \Rightarrow 10x = 38$

$\Rightarrow \quad x = \frac{38}{10} = \frac{19}{5}$

$\therefore \quad f(2) = \sqrt{2},\ f(4) = 3\sqrt{2}$

and $f\left(\frac{19}{5}\right) = 3\sqrt{\frac{19}{5} - 2} + \sqrt{4 - \frac{19}{5}}$

$= 3 \times \frac{3}{\sqrt{5}} + \frac{1}{\sqrt{5}} = \frac{10}{\sqrt{5}} = 2\sqrt{5}$

$f_{\min} = \alpha = \sqrt{2}$

$f_{\max} = \beta = 2\sqrt{5}$

$\therefore \quad \alpha^2 + 2\beta^2 = 2 + 40 = 42$

**33.** *(b)* $f(x) = 2x^3 + (2p-7)x^2 + 3(2p-9)x - 6$

On differentiate $f(x)$ w.r.t. $x$, we get

$f'(x) = 6x^2 + 2(2p-7)x + 3(2p-9)$

$f'(x)$ is a quadratic function.

So, Product of root's $= \frac{\text{Constant}}{\text{Lead coefficient}}$

if sign of maxima or minima are opposite, then

Product of roots $< 0$

$\frac{3(2p-9)}{6} < 0$

$\Rightarrow \quad 2p - 9 < 0 \Rightarrow p < \frac{9}{2}$

or $\quad p \in \left(-\infty, \frac{9}{2}\right)$

**34.** *(a)* $f(x) = 2x^4 - 18x^2 + 8x + 12;$
$x \in (-4, 4)$

On differentiating $f(x)$ w.r.t. $x$, we get

$f'(x) = 8x^3 - 36x + 8$

$f'(x) = 0 = 4(2x^3 - 9x + 2)$

$\Rightarrow \quad 2x^3 - 9x + 2 = 0$

$(x-2)(2x^2 + 4x - 1) = 0$

$x = 2,$

$x = \frac{-4 \pm \sqrt{16+8}}{2 \times 2} = \frac{\pm\sqrt{6} - 2}{2} \in (-4, 4)$

Now, $f(x) = 2x^4 - 18x^2 + 8x + 12$

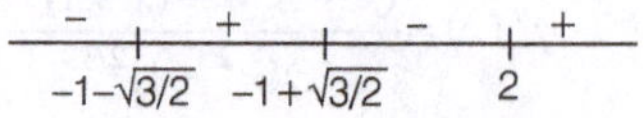


$\because f'(x)$ changes sign from *+ve* to *–ve* when passing through $-1 + \frac{\sqrt{3}}{2}$

$\therefore x = -1 + \sqrt{\frac{3}{2}}$ is point of local maxima.

$\therefore M = f\left(-1 + \sqrt{\frac{3}{2}}\right)$

$= 2\left(-1 + \sqrt{\frac{3}{2}}\right)^4 - 18\left(-1 + \sqrt{\frac{3}{2}}\right)^2 + 8\left(-1 + \sqrt{\frac{3}{2}}\right) + 12$

$= 2 - 4\sqrt{6} + 18 - 6\sqrt{6} + \frac{4}{2} - 18 + 18\sqrt{6} - 27 - 8 + 4\sqrt{6} + 12$

$= 12\sqrt{6} - \frac{33}{2}$

**35.** *(b)* Given, $f(x) = (x^2 - 2x + 7) \cdot e^{(4x^3 - 12x^2 - 180x + 31)}$

Let $f_1(x) = (x^2 - 2x + 7) > 0,\ \forall\, x \in R$

$f_1'(x) = 2x - 2 = 0$

$\Rightarrow \quad x = 1$

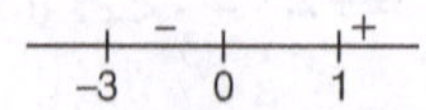

Thus, $f(x)$ is decreasing on $[-3, 0]$.

Also, let $f_2(x) = e^{4x^3 - 12x^2 - 180x + 31}$

$$\Rightarrow f_2'(x) = e^{4x^3 - 12x^2 - 180x + 31} \cdot (12x^2 - 24x - 180)$$

$$= 12(x^2 - 2x - 15) \cdot e^{4x^3 - 12x^2 - 180x + 31}$$

$$= 12(x+3)(x-5) e^{4x^3 - 12x^2 - 180x + 31}$$

Thus, sign of $f_2'(x)$: $+$ on $(-\infty, -3)$, $-$ on $(-3, 0)$, $-$ on $(0, 5)$, $+$ on $(5, \infty)$ (points $-3$, $0$, $5$).

$\therefore$ Clearly, $f_2(x)$ is also deceasing in $[-3, 0]$ and also positive, $\forall\, x \in R$.

$\therefore$ Absolute maximum value of $f(x)$ occurs at $x = -3$

Given that, absolute maximum value of the above function in $[-3, 0]$ is $f(\alpha)$.

Thus, $\alpha = -3$.

**36.** (c)

**key idea** (i) Find $f'(x)$ using product rule of differentiation and expression for $f'(x)$ for different value of $n_1$ and $n_2$ as per options and solve further using first derivative test.
(ii) $(uv)' = u'v + v'u$

$$f(x) = (x-3)^{n_1}(x-5)^{n_2}$$

Differentiating w.r.t. $x$, we get

$$f'(x) = n_1 (x-3)^{n_1 - 1}(x-5)^{n_2} + (x-3)^{n_1} \cdot n_2 (x-5)^{n_2 - 1}$$

$$= (x-3)^{n_1 - 1}(x-5)^{n_2 - 1}[n_1(x-5) + n_2(x-3)]$$

$$= (x-3)^{n_1 - 1}(x-5)^{n_1 - 1}[n_1 x + n_2 x - 5n_1 - 3n_2]$$

$$= (x-3)^{n_1 - 1}(x-5)^{n_2 - 1}[x(n_1 + n_2) - (5n_1 + 3n_2)]$$

$$= (x-3)^{n_1 - 1}(x-5)^{n_2 - 1}(n_1 + n_2)\left[x - \frac{5n_1 + 3n_2}{n_1 + n_2}\right]$$

Option (c) is incorrect.

For $n_1 = 3, n_2 = 5$

$$f'(x) = 8(x-3)^2 (x-5)^4 \left(x - \frac{30}{8}\right)$$

By $f'(x) = 0$

Critical point : $x = 30/8$

$$f''(x) = 8\left[2(x-3)(x-5)^4\left(x - \frac{30}{8}\right) + (x-3)^2 \cdot 4(x-5)^3\left(x - \frac{30}{8}\right) + (x-3)^2(x-5)^4(1-0)\right]$$

At $x = \frac{30}{8}$, $f''(x) = 8$ [0 + 0 + positive value]

= positive

$\therefore f$ attains local minima at $x = \alpha = 30/8$

**37.** (c) Let $a = 20 - 2x^2$

and $b = 10 + x^2, c = 10 + x^2$

$$s = \frac{a+b+c}{2} = 20$$

$$\Delta = \sqrt{20(2x^2)(10-x^2)(10-x^2)}$$

$$\Delta = 2\sqrt{10}\sqrt{x^2(10-x^2)^2}$$

$$\Delta = 2\sqrt{10}|10x - x^3|$$

Let $p = 10x - x^3$

$dp/dx = 10 - 3x^2$

For $\Delta$ to be maximum

$$\frac{dp}{dx} = 0 \Rightarrow 10 - 3x^2 = 0$$

$$\Rightarrow x^2 = \frac{10}{3} \Rightarrow x = \pm\sqrt{\frac{10}{3}}$$

$$\frac{d^2p}{dx^2} = -6x$$

At $x = \sqrt{\frac{10}{3}}$, $\frac{d^2p}{dx^2} = -6\sqrt{\frac{10}{3}} < 0$

So, $x = \sqrt{\frac{10}{3}}$ is point of maxima.

So, $k = \sqrt{\frac{10}{3}} \Rightarrow 3k^2 = 3 \times \frac{10}{3} = 10$

**38.** (a) $f(x) = |x^2 - 3x - 2| - x$, $x \in [-1, 2]$

For $x^2 - 3x - 2$, $-\frac{b}{2a} = \frac{3}{2}$ and $\frac{D}{4a} = \frac{17}{4}$

where, $D = b^2 - 4ac = 9 + 8 = 17$

and roots are $x = \frac{3 \pm \sqrt{17}}{2}$ out of which only

$$x = \frac{3 - \sqrt{17}}{2} \in [-1, 2]$$

So,

$$f(x) = \begin{cases} x^2 - 4x - 2, & -1 \le x \le \frac{3-\sqrt{17}}{2} \\ -x^2 + 2x + 2, & \frac{3-\sqrt{17}}{2} \le x \le 2 \end{cases}$$

$$f\left(\frac{3-\sqrt{17}}{2}\right) = 0 + \frac{\sqrt{17}-3}{2}$$

$$= \frac{\sqrt{17}-3}{2} = f(x)_{\min}$$

$$f'(x) = -2x + 2 = 0$$

$$\Rightarrow x = 1 \in \left[\frac{3-\sqrt{17}}{2} \le x \le 2\right]$$

$$\therefore f(x)_{\max} = f(1) = -1 + 2 + 2 = 3$$

So, $\frac{\sqrt{17}-3}{2} + 3 = \frac{\sqrt{17}+3}{2}$

**39.** (c) Area = 12 sq units

$$\Rightarrow \begin{vmatrix} \alpha & \beta & 1 \\ 5 & 6 & 1 \\ 3 & 2 & 1 \end{vmatrix} = \pm 24$$

$$\Rightarrow 4\alpha - 2\beta - 8 = \pm 24$$

$$\Rightarrow 4\alpha - 2\beta = 32, 4\alpha - 2\beta + 16 = 0$$

$$\Rightarrow 2\alpha - \beta - 16 = 0, 2\alpha - \beta + 8 = 0$$

Distance from origin when $\beta = 2\alpha + 8$ is

$$D = \sqrt{\alpha^2 + (2\alpha + 8)^2}$$

$$= \sqrt{5\alpha^2 + 32\alpha + 64}$$

$$\Rightarrow D^2 = 5\alpha^2 + 32\alpha + 64$$

Now, $\frac{d}{d\alpha}(D^2) = 0$

$$\Rightarrow 10\alpha + 32 = 0 \Rightarrow \alpha = -\frac{16}{5}$$

$$\Rightarrow \beta = -\frac{32}{5} + 8 = \frac{8}{5}$$

$$\therefore D = \sqrt{\left(\frac{-16}{5}\right)^2 + \left(\frac{8}{5}\right)^2} = \frac{8}{5}\sqrt{5} = \frac{8}{\sqrt{5}}$$

Similarly, if $\beta = 2\alpha - 16$, then $D = \frac{16}{\sqrt{5}}$

So, least possible length of line segment $= \frac{8}{\sqrt{5}}$

**40.** (c) Given,

$$f(x) = x^3 - 3x^2 - \frac{3}{2}f''(2) \cdot x + f''(1)$$

Then, $f'(x) = 3x^2 - 6x - \frac{3}{2}f''(2)$

$$f''(x) = 6x - 6$$

Put $x = 1$ and $x = 2$,

$$f''(2) = 12 - 6 = 6 \text{ and } f''(1) = 0$$

Therefore,

$$f(x) = x^3 - 3x^2 - 9x \quad \text{...(i)}$$

and $f'(x) = 3x^2 - 6x - 9$

and $f''(x) = 6x - 6$

Equate $f'(x) = 0$ gives,

$$\Rightarrow 3x^2 - 6x - 9 = 0$$

$$3(x+1)(x-3) = 0$$

$$\Rightarrow x = -1, x = 3$$

Now, $f''(-1) = 6(-1) - 6 = -12 < 0$

$f''(3) = 6(3) - 6 = 12 > 0$

$\therefore$ $(-1)$ is maxima and 3 is minima.

Local minimum value $= f(3)$

$f(3) = (3)^3 - 3(3)^2 - 9(3)$ [Using Eq. (i)]

$= 27 - 27 - 27$

$f(3) = -27$

**41.** *(a)* $f(x) = ax^2 + 6x - 15$

$f'(x) = 2ax + 6$

$f'(x) = 2(ax + 3)$

$f(x)$ is increasing for $\left(-\infty, \dfrac{3}{4}\right)$.

So, $x = -\dfrac{3}{a} = \dfrac{3}{4}$

$\Rightarrow a = -4$

$g(x) = ax^2 - 6x + 15$

$g'(x) = 2(ax - 3)$

If $g'(x) = 0$

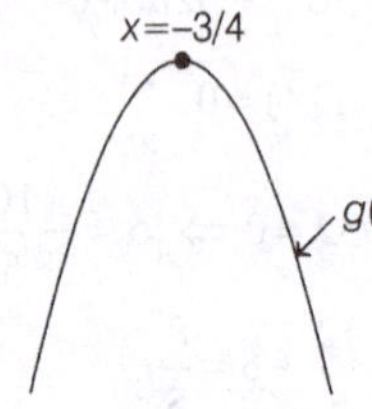


$x = \dfrac{3}{a} = -\dfrac{3}{4}$

$g(x)$ is maximum at $x = -\dfrac{3}{4}$.

**42.** *(b)* We have, $\dfrac{x^2}{a} + \dfrac{y^2}{b} = 1$ ...(i)

Differentiating Eq. (i) w.r.t. $x$, we get

$$\frac{dy}{dx} = -\frac{bx}{ay} = m_1 \text{ (say)}$$

Also, $\dfrac{x^2}{c} + \dfrac{y^2}{d} = 1$ ...(ii)

Differentiating Eq. (ii) w.r.t. $x$, we get

$$\frac{dy}{dx} = -\frac{dx}{cy} = m_2 \text{ (say)}$$

Now, $m_1 m_2 = -1$ [given]

$\Rightarrow \left(-\dfrac{bx}{ay}\right)\left(\dfrac{-dx}{cy}\right) = -1$

$\Rightarrow bdx^2 = -acy^2$ ...(iii)

Now, subtracting Eq.(ii) from Eq. (i), we get

$$\left(\frac{1}{a} - \frac{1}{c}\right)x^2 + \left(\frac{1}{b} - \frac{1}{d}\right)y^2 = 0$$

$\Rightarrow \left(\dfrac{c-a}{ac}\right)x^2 + \left(\dfrac{d-b}{bd}\right)y^2 = 0$

$\Rightarrow \dfrac{(c-a)}{ac}x^2 + \left(\dfrac{d-b}{bd}\right)\left(\dfrac{-bdx^2}{ac}\right) = 0$ [using Eq. (iii)]

$\Rightarrow (c-a) - (d-b) = 0$

$\Rightarrow \quad c - d = a - b$

**43.** *(d)* Given function,

$f(x) = (1 - \cos^2 x)(\lambda + \sin x)$

$= \sin^2 x(\lambda + \sin x)$

$\therefore \quad f'(x) = \sin 2x(\lambda + \sin x) + \sin^2 x(\cos x)$

$= \sin 2x\left[\lambda + \sin x + \dfrac{1}{2}\sin x\right]$

$= \sin 2x\left[\lambda + \dfrac{3}{2}\sin x\right]$

For maxima and minima, as

$f'(x) = 0$

$\Rightarrow \quad \sin 2x\left(\lambda + \dfrac{3}{2}\sin x\right) = 0$

So, either $\sin 2x = 0$

$\Rightarrow x = 0$ as $x \in \left(-\dfrac{\pi}{2}, \dfrac{\pi}{2}\right)$

or $\lambda + \dfrac{3}{2}\sin x = 0$ as there must exactly one maxima and exactly one minima, so

$\lambda \in \left(-\dfrac{3}{2}, \dfrac{3}{2}\right) - \{0\}$.

**44.** *(d)* It is given that $x = 1$ is a critical point of the function

$f(x) = (3x^2 + ax - 2 - a)e^x$

So, $f'(1) = e^x(6x + a) + e^x(3x^2 + ax - 2 - a)\big|_{x=1} = 0$

$\Rightarrow 6 + a + 3 + a - 2 - a = 0 \Rightarrow a = -7.$

$\therefore f'(x) = e^x[3x^2 - x - 2] = 0$

$\Rightarrow \quad x = 1$ or $\left(-\dfrac{2}{3}\right)$

and $f''(x) = e^x(6x - 1 + 3x^2 - x - 2)$

$= e^x(3x^2 + 5x - 3)$

$\because \quad f''(1) = 5e > 0$

$\Rightarrow x = 1$ is the point of local minima

and $f''\left(-\dfrac{2}{3}\right) = e^{-2/3}\left(\dfrac{4}{3} - \dfrac{10}{3} - 3\right)$

$= -5e^{-2/3} < 0$

$\Rightarrow x = -\dfrac{2}{3}$ is the point of local maxima.

**45.** *(d)* Let $(a, 0)$ be any point on the $X$-axis, which is the vertex of the rectangle.

So, the co-ordinates of the vertex of the rectangle lying on the parabola $y = 12 - x^2$ is $(a, 12 - a^2)$.

$\therefore$Area of rectangle, $f(a) = 2a(12 - a^2)$

$\therefore f'(a) = 2(12 - 3a^2)$

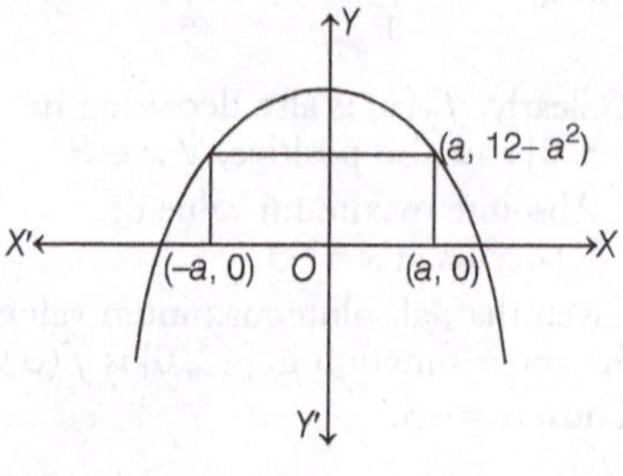


For maximum area

$f'(a) = 0$

$\Rightarrow 2(12 - 3a^2) = 0 \Rightarrow a = \pm 2$

$\therefore$Maximum area at $a = 2$ is $f(2)$

$= 32$ sq units

**46.** *(b)* Let $h$ = height of the cone,

$r$ = radius of circular base

$= \sqrt{(3)^2 - h^2}$

$= \sqrt{9 - h^2}$ ...(i)

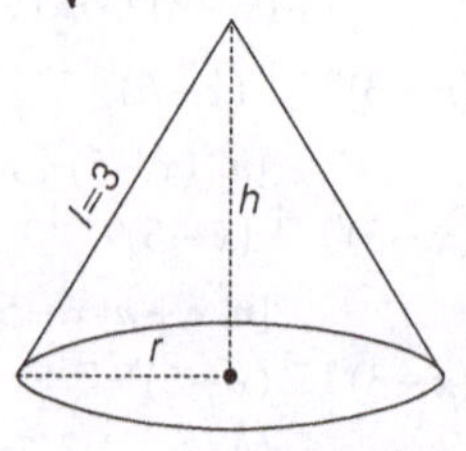


Now, volume $(V)$ of cone $= \dfrac{1}{3}\pi r^2 h$

$\Rightarrow V(h) = \dfrac{1}{3}\pi(9 - h^2)h$ [from Eq. (i)]

$= \dfrac{1}{3}\pi[9h - h^3]$ ...(ii)

For maximum volume $V'(h) = 0$ and $V''(h) < 0$.

Here, $\quad V'(h) = 0$

$\Rightarrow \quad (9 - 3h^2) = 0$

$\Rightarrow \quad h = \sqrt{3}$ [$\because h \not< 0$]

and $\quad V''(h) = \dfrac{1}{3}\pi(-6h) < 0$

for $\quad h = \sqrt{3}$

Thus, volume is maximum when $h = \sqrt{3}$

Now, maximum volume

$$V(\sqrt{3})=\frac{1}{3}\pi(9\sqrt{3}-3\sqrt{3})$$

[from Eq. (ii)]

$$=2\sqrt{3}\pi\ m^3$$

**47.** (d) Let cone of radius $r$ and height $h$ be inscribed in a sphere of radius, $R=3$

Now, $OD=AD-OA=h-3$

In $\Delta ODC$, $(OC)^2=(OD)^2+(CD)^2$

$\Rightarrow (3)^2=(h-3)^2+r^2$

$\Rightarrow\ 9=h^2+9-6h+r^2$

$\Rightarrow\ r^2=6h-h^2$ ...(i)

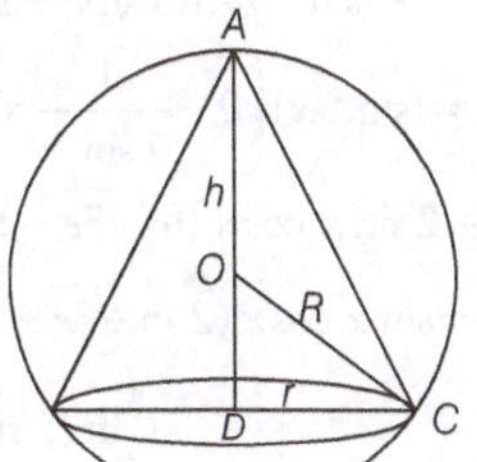


Volume of cone

$(V)=\frac{1}{3}\pi r^2h=\frac{1}{3}\pi h(6h-h^2)$

$=\frac{1}{3}\pi(6h^2-h^3)$

Now, differentiating it w.r.t. $h$, we get

$\frac{dV}{dh}=\frac{\pi}{3}\times(12h-3h^2)$

Put $\frac{dV}{dh}=0$

$\Rightarrow 3h^2-12h=0 \Rightarrow 3h(h-4)=0$

$\Rightarrow h=4$ ($\because h=0$ is not possible)

Also, $\frac{d^2V}{dh^2}<0$ at $h=4$

Put $h=4$ in Eq. (i), we get

$r^2=24-16=8$

$\therefore$ Slant height,

$l=\sqrt{h^2+r^2}=\sqrt{16+8}=\sqrt{24}$

Now, curved surface area of cone $=\pi rl$

$=\pi\times 2\sqrt{2}\times 2\sqrt{6}=8\sqrt{3}\ \pi$

**48.** (b) Let

$f(x)=ax^4+bx^3+cx^2+dx+e]$ ... (i)

Given, $\lim_{x\to 0}\left(\frac{f(x)}{x^2}+1\right)=3$

$\Rightarrow \lim_{x\to 0}\left(ax^2+bx+c+\frac{d}{x}+\frac{e}{x^2}+1\right)=3$

$\Rightarrow\ c+1=3$

$\Rightarrow\ c=2$

[$\because$ limit exists finitely, so $d=e=0$]

$\therefore\ f(x)=ax^4+bx^3+2x^2$

$\Rightarrow\ f'(x)=4ax^3+3bx^2+4x$

Given that $f(x)$ has extreme values at $x=1$ and $x=2$

$f'(1)=0$ and $f'(2)=0$

$\Rightarrow\ 4a+3b+4=0$ ...(ii)

and $32a+12b+8=0$ ...(iii)

From Eqs. (ii) and (iii), we get

$a=1/2, b=-2$

Thus, $f(x)=\frac{1}{2}x^4-2x^3+2x^2$

$\therefore\ f(-1)=\frac{1}{2}+2+2=\frac{9}{2}$

**49.** (c) Total length $=2r+r\theta=20$

$\Rightarrow\ \theta=\frac{20-2r}{r}$

Now, area of flower-bed,

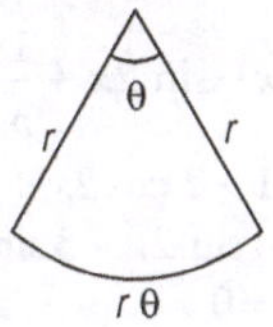


$A=\frac{1}{2}r^2\theta$

$\Rightarrow A=\frac{1}{2}r^2\left(\frac{20-2r}{r}\right)\Rightarrow A=10r-r^2$

$\therefore dA/dr=10-2r$

For maxima or minima, put $\frac{dA}{dr}=0$.

$\Rightarrow\ r=5$

$\therefore A_{max}=\frac{1}{2}(5)^2\left[\frac{20-2(5)}{5}\right]=25$ sq m

**50.** (c) $f(x)=\begin{cases}\frac{\tan x}{x}, & x\neq 0\\ 1, & x=0\end{cases}$

As, $\frac{\tan x}{x}>1, \forall\ x\neq 0$

$\therefore\ f(0+h)>f(0)$

and $f(0-h)>f(0)$

At $x=0$, $f(x)$ attains minima.

$\therefore\ f'(0)=\lim_{h\to 0}\frac{f(h)-f(0)}{h}$

$=\lim_{h\to 0}\frac{\tan h-h}{h^2}\left[\because \frac{0}{0}\text{ form}\right]$

$=\lim_{h\to 0}\frac{\sec^2 h-1}{2h}$

[using L' Hopital rule]

$=\lim_{h\to 0}\frac{\tan^2 h}{2h^2}\cdot h$

$=\frac{1}{2}\cdot 0=0\left[\because \lim_{\theta\to 0}\frac{\tan\theta}{\theta}=1\right]$

So, Statement II is true.

Hence, both statements are true but Statement II is not the correct explanation of Statement I.

**51.** (d) $\because\ f(x)=\frac{x}{2}+\frac{2}{x}$

$\therefore\ f'(x)=\frac{1}{2}-\frac{2}{x^2}$

For maxima or minima, put $f'(x)=0$

$\therefore \frac{1}{2}-\frac{2}{x^2}=0 \Rightarrow x^2=4\Rightarrow x=\pm 2$

Now, $f''(x)=\frac{4}{x^3}$

$\Rightarrow\ f''(2)=\frac{4}{8}=\frac{1}{2}>0$ [minima]

and $f''(-2)=-\frac{4}{8}=-\frac{1}{2}<0$ [maxima]

Hence, $f(x)$ is minimum at $x=2$.

**52.** (c) $\because f(x)=2x^3-9ax^2+12a^2x+1$

$\therefore\ f'(x)=6x^2-18ax+12a^2$

For maxima or minima, put $f'(x)=0$

$\Rightarrow\ x^2-3ax+2a^2=0$

$\Rightarrow\ (x-a)(x-2a)=0$

$\Rightarrow\ x=a,\ x=2a$

Now, $f''(x)=12x-18a$

At $x=a$

$f''(x)=12a-18a=-6a$

So, $f(x)$ will be maximum at $x=a$.

i.e., $p=a$

Again, at $x=2a$

$f''(x)=24a-18a=6a$

So, $f(x)$ will be minimum at $x=2a$.

i.e., $q=2a$

Given, $p^2=q \Rightarrow a^2=2a$

$\therefore\ a=2$

**53.** (b) Equation of line in intercept form is

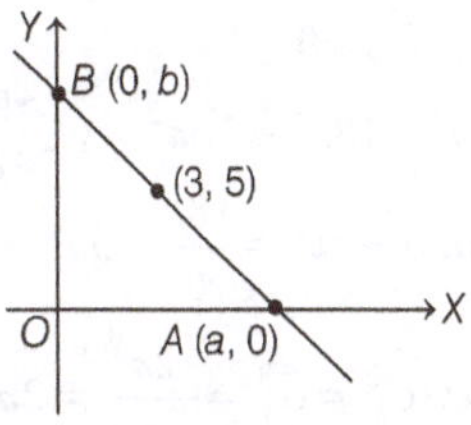

$$\frac{x}{a}+\frac{y}{b}=1 \quad \text{...(i)}$$

Since, Eq. (i) passes through (3, 5), then

$$\frac{3}{a}+\frac{5}{b}=1$$

$$\Rightarrow \quad b=\frac{5a}{a-3},\ a-3>0 \text{ is } a>3$$

Now, area of $\Delta OAB=\Delta=\frac{1}{2}ab$

$$\Rightarrow \quad \Delta=\frac{1}{2}\cdot a\cdot\frac{5a}{a-3}=\frac{5}{2}\cdot\frac{a^2}{a-3}$$

$$=\frac{5}{2}\left(\frac{a^2-9+9}{a-3}\right)$$

$$=\frac{5}{2}\left(a+3+\frac{9}{a-3}\right)$$

$$=\frac{5}{2}\left(a-3+\frac{9}{a-3}+6\right)\geq 30$$

**Alternate Solution**

(To find minimum area)

$$\Delta=\frac{1}{2}a\cdot\frac{5a}{a-3}=\frac{5}{2}\frac{a^2}{a-3}$$

$$=\frac{5}{2}\left(a+3+\frac{9}{a-3}\right)$$

Now, $\frac{d\Delta}{da}=\frac{5}{2}\left[1+0-\frac{9}{(a-3)^2}\right]=0$

$\Rightarrow (a-3)^2=9 \Rightarrow a-3=\pm 3 \Rightarrow a=0, 6$

Again, differentiating,

$$\frac{d^2\Delta}{da^2}=\frac{5}{2}\left[0+\frac{9\times 2}{(a-3)^3}\right]$$

Here, when $a=0$, then $\frac{d^2\Delta}{da^2}<0$ which gives maximum area

and when $a=6$, then $\frac{d^2\Delta}{da^2}>0$ which gives minimum area

$$\therefore \quad \Delta|_{a=6}=\frac{5}{2}\times\frac{(6)^2}{(6-3)}=\frac{5}{2}\times\frac{6\times 6}{3}=30$$

**54.** (a) $\because f(x)=2x^3-9ax^2+12a^2x+1$

$$\Rightarrow \quad f'(x)=6x^2-18ax+12a^2$$

Put $f'(x)=0$

$$\Rightarrow 6x^2-18ax+12a^2=0 \begin{matrix}\nearrow \alpha \\ \searrow \alpha^2\end{matrix}$$

Then, $\alpha+\alpha^2=\frac{18a}{6}=3a \quad$ ...(i)

and $\alpha\cdot\alpha^2=\alpha^3=\frac{12a^2}{6}=2a^2 \quad$ ...(ii)

From Eq. (i), we get

$$(\alpha+\alpha^2)^3=27a^3$$

$$\Rightarrow \alpha^3+(\alpha^2)^3+3\alpha^3(\alpha+\alpha^2)=27a^3$$

Put the values from Eqs. (i) and (ii), we get

$$2a^2+4a^4+3(2a^2)(3a)=27a^3$$

$$\Rightarrow a^2(4a^2+18a+2)=27a^3$$

$$\Rightarrow 4a^2-9a+2=0$$

$$\Rightarrow 4a^2-8a-a+2=0$$

$$\Rightarrow 4a(a-2)-1(a-2)=0$$

$$\Rightarrow (a-2)(4a-1)=0 \Rightarrow a=2$$

Then, $6x^2-36x+48=0$

$$\Rightarrow \quad x^2-6x+8=0$$

**55.** (a) Let

$$f(x)=x-2\sin x\cos x+\frac{1}{3}\sin 3x, \quad 0\leq x\leq\pi$$

$$=x-\sin 2x+\frac{1}{3}\sin 3x$$

$$\Rightarrow \quad f'(x)=1-2\cos 2x+\cos 3x$$

$$\Rightarrow f''(x)=4\sin 2x-3\sin 3x$$

Now, $f'(x)=0$

$$\Rightarrow 1-2\cos 2x+\cos 3x=0$$

$$\Rightarrow 1-2(2\cos^2 x-1)+4\cos^3 x-3\cos x=0$$

$$\Rightarrow 1-4\cos^2 x+2+4\cos^3 x-3\cos x=0$$

$$\Rightarrow 4\cos^3 x-4\cos^2 x-3\cos x+3=0$$

$$\Rightarrow 4\cos^2 x(\cos x-1)-3(\cos x-1)=0$$

$$\Rightarrow (\cos x-1)(4\cos^2 x-3)=0$$

$$\Rightarrow (\cos x-1)(2\cos x+\sqrt{3})(2\cos x-\sqrt{3})=0$$

$$\therefore \quad \cos x=-\frac{\sqrt{3}}{2},\ \cos x=\frac{\sqrt{3}}{2},\ \cos x=1$$

$$\therefore x=\frac{5\pi}{6},\ x=\frac{\pi}{6},\ x=0$$

$$f''\left(\frac{5\pi}{6}\right)=4\sin\frac{5\pi}{3}-3\sin\frac{5\pi}{2}$$

$$=-2\sqrt{3}-3<0$$

$$f''\left(\frac{\pi}{6}\right)=4\sin\frac{\pi}{3}-3\sin\frac{\pi}{2}$$

$$=2\sqrt{3}-3>0$$

$$f''(0)=0$$

So, $x=\frac{5\pi}{6}$ is point of maxima.

$$\therefore f\left(\frac{5\pi}{6}\right)=\frac{5\pi}{6}-\sin\frac{5\pi}{3}+\frac{1}{3}\sin\frac{5\pi}{2}$$

$$=\frac{5\pi}{6}+\frac{\sqrt{3}}{2}+\frac{1}{3}=\frac{5\pi+3\sqrt{3}+2}{6}$$

**56.** (b) Let $P(x)=y=\left(\frac{\sqrt{3e}}{2\sin x}\right)^{\sin^2 x}$, $x\in\left(0,\frac{\pi}{2}\right)$

$$\Rightarrow \ln(y)=(\sin^2 x)\ln\left(\frac{\sqrt{3e}}{2\sin x}\right)$$

$$=(\sin^2 x)(\ln\sqrt{3e}-\ln(2\sin x))$$

Differentiating w.r.t. $x$, we get

$$\frac{1}{y}\times y'=(\sin^2 x)(\ln\sqrt{3e}-\ln(2\sin x))'+(\sin^2 x)'(\ln\sqrt{3e}-\ln(2\sin x))$$

$$=(\sin^2 x)\left(0-\frac{1}{2\sin x}\times 2\cos x\right)+2\sin x\cos x(\ln\sqrt{3e}-\ln(2\sin x))$$

$$\Rightarrow \frac{y'}{y}=\sin x\cos x(2\ln\sqrt{3e}-2\ln(2\sin x)-1)$$

For local maxima, $y'=0$

$$\Rightarrow \sin x\cos x(2\ln\sqrt{3e}-2\ln(2\sin x)-1)=0$$

$$\Rightarrow \quad 2\ln\sqrt{3e}-2\ln(2\sin x)-1=0 \quad \left(\because x\in\left(0,\frac{\pi}{2}\right)\right)$$

$$\Rightarrow \quad \ln\left(\frac{3e}{e}\right)=2\ln(2\sin x)$$

$$\Rightarrow \quad \frac{1}{2}\ln(3)=\ln(2\sin x)\Rightarrow \sin x=\frac{\sqrt{3}}{2}$$

$$\therefore \text{Local maximum value}=\left(\frac{\sqrt{3e}}{2\times\left(\frac{\sqrt{3}}{2}\right)}\right)^{\frac{3}{4}}$$

$$=(\sqrt{e})^{\frac{3}{4}}=e^{\frac{3}{8}}=\frac{k}{e}\ \text{(given)}\Rightarrow k=e^{\frac{11}{8}}$$

$$\therefore \left(\frac{k}{e}\right)^8+\frac{k^8}{e^5}+k^8=e^3+e^6+e^{11}$$

**57.** (a) Given length of square sheet of side is 30 cm.

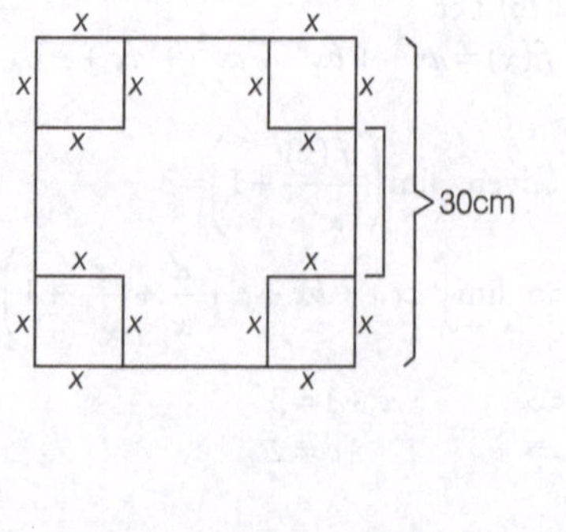

Let side of small square be $x$ cm. Since, a square piece of tin is to be made into a box without top by cutting square from each convex and folding up the flaps to from a box. Then, this shape framed a cuboidal shape.

Their, volume of the box ($V$)

$$= (30-2x)(30-2x)x$$

$$\Rightarrow \quad V = x(30-2x)^2$$

Differentiating both side w.r.t. $x$, we get

$$\frac{dV}{dx} = (30-2x)^2 + x\{2(30-2x)\cdot(-2)\}$$
$$= (30-2x)^2 - 4x(30-2x)$$
$$= (30-2x)\{30-6x\}$$

Therefore, maximum of $V$, $\frac{dV}{dx} = 0$

$$(30-2x)(30-6x) = 0$$

$$\Rightarrow \quad x = 15 \text{ or } x = 5$$

$\because x = 15$ is not possible $x = 5$

So, the surface area is

$$= 4 \times (30-2x) \times x + (30-2x)^2$$
$$= 4 \times (30-10) \times 5 + (30-10)^2 \quad (\text{Put } x = 5)$$
$$= 400 + 400 = 800$$

**58.** *(d)* $f(x) = 3^{(x^2-2)^3+4}, x \in R$

$$f'(x) = \ln 3 \cdot 3^{(x^2-2)^3+4} \cdot 3(x^2-2)^2(2x)$$
$$f'(x) = \ln 3 \cdot 3^4 \cdot 3^{(x^2-2)^3} \cdot 3(x^2-2)^2(2x)$$
$$= (3^4 \times 3 \times 2)\ln 3 \cdot 3^{(x^2-2)^3} x \cdot (x^2-2)^2$$

$3^{(x^2-2)^3}$ cannot be zero.

Put $f'(x) = 0 \Rightarrow x = 0, \pm\sqrt{2}$

$$\begin{array}{cccccccc} - & & - & & + & & + \\ \hline & -\sqrt{2} & & 0 & & \sqrt{2} & \end{array}$$

$f'(x) = 0$ at $x = 0, \pm\sqrt{2}$

$x = 0$ is point of local minima

$$f''(x) = (486 \ln 3)$$
$$[\ln 3 \cdot 3^{(x^2-2)^3} \cdot 3(x^2-2)^2 \cdot 2x \cdot x(x^2-2)^2] + 3^{(x^2-3)^3}$$
$$[(x^2-2)^2 + x \cdot 2(x^2-2) \cdot 2x]$$
$$f''(x) = (486 \ln 3) 3^{(x^2-2)^3}$$
$$[(\ln 3) 6x^2 \cdot (x^2-2)^4 + (x^2-2)^2 + 4x^2(x^2-2)]$$
$$= (486) \ln 3 \cdot 3^{(x^2-2)^3} (x^2-2)$$
$$[6x^2 \ln 3 (x^2-2)^3 + 5x^2 - 2]$$
$$f''(\sqrt{2}) = 0$$

$x = \sqrt{2}$ is point of inflection.

So, $x = 0$ is point of local minima

$x = \sqrt{2}$ is a point of inflection.

$f'$ is increasing for $x > \sqrt{2}$.

**59.** *(c)* Given, equation of circle is $(x-1)^2 + (y+1)^2 = 1$ and equation of parabola is $y = x^2$

Centre of circle is $(1, -1)$ and radius is 1 units.

$\because$ Circle touches the coordinate axes and the centre lies in 4th quadrant.

The parabola lies upwards in the first and second quadrant.

$\therefore$ The circle and parabola does not intersect each other.

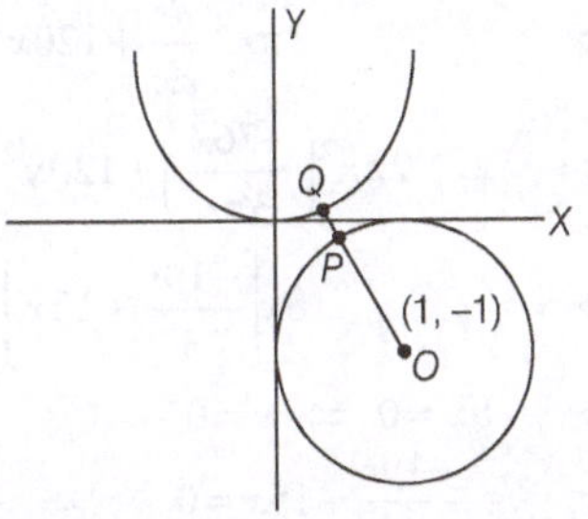


The distance $PQ$ is minimum when it passes through the centre of circle when extended.

Let coordinate of $Q$ be $(t, t^2)$.

$$PQ = OQ - OP$$
$$PQ = \sqrt{(t-1)^2 + (t^2+1)^2} - 1$$

$PQ$ is minimum when

$\sqrt{(t-1)^2 + (t^2+1)^2}$ is minimum or when $(t-1)^2 + (t^2+1)^2$ is minimum

Let $\quad f(t) = (t-1)^2 + (t^2+1)^2$

$$f(t) = t^2 - 2t + 1 + t^4 + 2t^2 + 1$$
$$= t^4 + 3t^2 - 2t + 2$$
$$f'(t) = 4t^3 + 6t - 2$$
$$f''(t) = 12t^2 + 6 > 0$$

$\therefore f(t)$ is minimum when $f'(t) = 0$

Now, take the values from options,

$$f'(0) = -2 < 0$$
$$f'\left(\frac{1}{4}\right) = 4\left(\frac{1}{4}\right)^3 + 6\left(\frac{1}{4}\right) - 2$$
$$= \frac{1}{16} + \frac{3}{2} - 2 < 0$$
$$f'\left(\frac{1}{2}\right) = 4\left(\frac{1}{2}\right)^3 + 6\left(\frac{1}{2}\right) - 2$$
$$= \frac{1}{2} + 3 - 2 > 0$$

$$\therefore \quad t \in \left(\frac{1}{4}, \frac{1}{2}\right)$$

$\therefore PQ$ is minimum in the interval $\left(\frac{1}{4}, \frac{1}{2}\right)$.

**60.** *(a)* Given,

$$f(x) = y = ax^3 + bx^2 + cx + 5 \quad \ldots \text{(i)}$$
$$\Rightarrow \frac{dy}{dx} = 3ax^2 + 2bx + c \quad \ldots \text{(ii)}$$

Touches $X$-axis at $P(-2, 0)$

$$\Rightarrow \left.\frac{dy}{dx}\right|_{x=-2} = 0 \Rightarrow 3a(-2)^2 + 2b(-2) + c = 0$$
$$\Rightarrow \quad 12a - 4b + c = 0 \quad \ldots \text{(iii)}$$

and also $y|_{x=-2} = 0$

$$\Rightarrow -8a + 4b - 2c + 5 = 0 \quad \ldots \text{(iv)}$$

and $y = f(x)$ cut $Y$-axis at point $Q$, where $y'$ is equal to 3.

Then, $\quad c = 3 \quad \ldots$ (v)

From Eqs. (iii), (iv) and (v), we get

$$a = -\frac{1}{2}, b = -\frac{3}{4}, c = 3$$

Thus, $f(x) = -\frac{x^3}{2} - \frac{3}{4}x^2 + 3x + 5$

$$\Rightarrow \quad f'(x) = -\frac{3}{2}x^2 - \frac{3}{2}x + 3$$
$$= \frac{-3}{2}(x+2)(x-1)$$

Now, $f'(x) = 0 \Rightarrow x = -2$ and 1

By first derivative test, $x = 1$ is the point of Local maximum.

Hence, the local maximum value of $f(x)$ at $x = 1$ is

$$f(1) = -\frac{1}{2} - \frac{3}{4} + 3 + 5 = \frac{27}{4}$$

**61.** *(b)*

**key idea**
(i) The area of equilateral triangle with side $a$ is $\frac{\sqrt{3}}{4}a^2$.
(ii) Final total area in terms of length of wire of triangle and find critical points.

Let the length of one piece be $x$, then the length of other piece will be $22 - x$.

Let first piece of length $x$ is formed into a square and let it's side length be $y$.

Then, $\quad x = 4y \Rightarrow y = \frac{x}{4} \quad \ldots$ (i)

From the second piece of length $(22-x)$ we make an equilateral triangle, then

Side of equilateral triangle $= \dfrac{22-x}{3}$

Combined area, $A = y^2 + \dfrac{\sqrt{3}}{4}(\text{side})^2$

$$A = \left(\frac{x}{4}\right)^2 + \frac{\sqrt{3}}{4}\left(\frac{22-x}{3}\right)^2$$

Differentiating w.r.t. $x$, we get

$$\frac{dA}{dx} = 2\left(\frac{x}{4}\right)\left(\frac{1}{4}\right) + \frac{\sqrt{3}}{4}.2\left(\frac{22-x}{3}\right)\left(\frac{-1}{3}\right)$$

$$= \frac{x}{8} - \frac{\sqrt{3}}{18}(22-x)$$

For maximum or minimum value,

putting $\dfrac{dA}{dx} = 0$, we get

$$\frac{x}{8} - \frac{\sqrt{3}}{18}(22-x) = 0$$

$$\Rightarrow \frac{x}{8} - \frac{11\sqrt{3}}{9} + \frac{\sqrt{3}}{18}x = 0$$

$$\Rightarrow x\left(\frac{3\sqrt{3}+4}{24\sqrt{3}}\right) = \frac{11\sqrt{3}}{9}$$

$$\Rightarrow x = \frac{11\times 24\times 3}{9(3\sqrt{3}+4)} = \frac{11\times 8}{3\sqrt{3}+4}$$

$$= \frac{88\sqrt{3}}{9+4\sqrt{3}}$$

$\therefore$ Side of equilateral triangle

$$= \frac{22 - \dfrac{88\sqrt{3}}{(9+4\sqrt{3})}}{3}$$

$$= \frac{198 + 88\sqrt{3} - 88\sqrt{3}}{3(9+4\sqrt{3})}$$

$$= \frac{66}{9+4\sqrt{3}}$$

**62.** *(b)* Dimensions of cuboid : $2x, 4x, 5x$

Radius of hemisphere $r$.

Let sum of surface area,

$S = S_1 + S_2$

$S_1$ = Surface area of hemisphere $= 3\pi r^2$

$S_2$ = Surface area of cuboid

$= 2[(2x)(4x) + (4x)(5x) + (2x)(5x)]$

$= (8x^2 + 20x^2 + 10x^2)\times 2$

$= 38x^2 \times 2 = 76x^2$

$\therefore\ S = 3\pi r^2 + 76x^2 = k$ (constant)

$$\frac{dS}{dx} = 0$$

$$\Rightarrow 6\pi r\frac{dr}{dx} + 152x = 0$$

$$\Rightarrow \frac{dr}{dx} = -\frac{152x}{6\pi r} = -\frac{76}{3\pi r}x \quad \ldots\text{(i)}$$

$V_1$ = Volume of cuboid

$= (2x)(4x)(5x) = 40x^3$

$V_2$ = Volume of hemisphere $= \dfrac{2}{3}\pi r^3$

Total volume

$= V_1 + V_2 = 40x^3 + \dfrac{2}{3}\pi r^3$

For maximum volume $dV/dx = 0$

$$\Rightarrow \frac{2}{3}\pi 3r^2\frac{dr}{dx} + 120x^2 = 0$$

$$\Rightarrow 2\pi r^2\frac{dr}{dx} + 120x^2 = 0$$

$$\Rightarrow 2\pi r^2\left(\frac{-76x}{3\pi r}\right) + 120x^2 = 0$$

$$\Rightarrow 8x\left[\frac{-19r}{3} + 15x\right] = 0$$

$$\Rightarrow 8x = 0 \Rightarrow x = 0$$

or $\quad \dfrac{-19r}{3} + 15x = 0$

$$15x = \frac{19}{3}r \Rightarrow \frac{x}{r} = \frac{19}{45}$$

$\therefore \quad x : r = 19 : 45$

**63.** *(b)* The given function is,

$f(x) = |2x^2 + 3x - 2| + \sin x \cos x,$

$x \in [0, 1]$

$= |(2x-1)(x+2)| + \dfrac{\sin 2x}{2},\ x \in [0, 1]$

$$\therefore f(x) = \begin{cases} -(2x^2+3x-2) + \dfrac{\sin 2x}{2}, & 0 \le x \le \dfrac{1}{2} \\ (2x^2+3x-2) + \dfrac{\sin 2x}{2}, & x > \dfrac{1}{2} \end{cases}$$

For $0 \le x \le \dfrac{1}{2}$, then

$$f(x) = -(2x^2+3x-2) + \frac{\sin 2x}{2}$$

$\Rightarrow f'(x) = (-4x-3) + \cos 2x$

Clearly, $f'(x) < 0$ if $0 \le x \le \dfrac{1}{2}$

Thus, $f(x)$ is decreasing on $\left[0, \dfrac{1}{2}\right]$.

For $x > \dfrac{1}{2}$, then

$$f(x) = (2x^2+3x-2) + \frac{\sin 2x}{2}$$

$\Rightarrow f'(x) = (4x+3) + \cos 2x$

Clearly, $f'(x) > 0$ if $x > \dfrac{1}{2}$

So, clearly from above, minima occurs at $x = 1/2$

$$f(x)|_{\min} = \left|2\left(\frac{1}{2}\right)^2 + \frac{3}{2} - 2\right| + \sin\left(\frac{1}{2}\right).\cos\left(\frac{1}{2}\right) = \frac{1}{2}\sin(1)$$

And maxima is possible at $x = 0$ or $x = 1$.

Now checking for $x = 0$ and $x = 1$, we can see it attains its value at $x = 1$.

$$f(x)|_{\max} = |2+3-2| + \frac{\sin 2}{2} = 3 + \frac{1}{2}\sin 2$$

$\therefore$ Sum of absolute maxima and absolute minimum value

$$= 3 + \frac{1}{2}(\sin(1) + \sin(2))$$

$$= 3 + \frac{1}{2}(\sin(1) + 2\sin(1)\cdot\cos(1))$$

$$= 3 + \frac{1}{2}(1 + 2\cos(1))\sin(1)$$

**64.** *(c)* Length of box $= a - 2x$

Breadth of box $= b - 2x$

Height of box $= x$

Volume of box, $V = (a-2x)(b-2x)x$

$\Rightarrow V = 4x^3 - 2ax^2 - 2bx^2 + abx$

Differentiating $V$ w.r.t. $x$,

$V'_{(x)} = 12x^2 - 4(a+b)x + ab$

Critical Point, $V'_{(x)} = 0$

$$\Rightarrow x = \frac{4(a+b) \pm \sqrt{16(a+b)^2 - 4\cdot 12\cdot ab}}{2(12)}$$

$$x = \frac{(a+b) \pm \sqrt{a^2+b^2-ab}}{6}$$

$V''(x) = 24x - 4(a+b)$

For $x = \dfrac{(a+b) - \sqrt{a^2+b^2-ab}}{6}$,

$V''(x) < 0$

Hence, for maximum volume

$$x = \frac{(a+b) - \sqrt{a^2+b^2-ab}}{6}$$

**65.** *(d)* Let two pieces of wire one of length $x$ and other of the length $20 - x$.

Wire of length $x$ is made into a square.

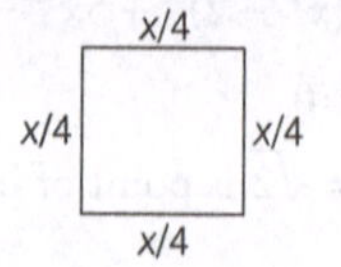

$\therefore$ Area of square $= \left(\frac{x}{4}\right)^2 = A_S$ (let)

Wire of length $(20 - x)$ is made into a regular hexagon.

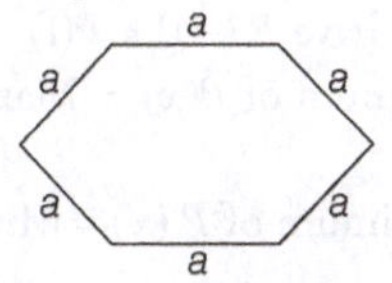


Area of hexagon $= 6 \times \frac{\sqrt{3}}{4} a^2$ (Let)

$$A_H = \frac{3\sqrt{3}}{2}\left(\frac{20 - x}{6}\right)^2$$

Sum of both area

$$A = A_S + A_H = \frac{x^2}{16} + \frac{\sqrt{3}}{24}(20 - x)^2$$

$$\frac{dA}{dx} = \frac{x}{8} - \frac{\sqrt{3}}{12}(20 - x)$$

$$= \frac{3x - 40\sqrt{3} + 2\sqrt{3}x}{24}$$

$$\frac{dA}{dx} = 0$$

$$\Rightarrow x = \frac{40\sqrt{3}}{3 + 2\sqrt{3}} = \frac{40}{\sqrt{3} + 2} = 40(2 - \sqrt{3})$$

$$\frac{d^2A}{dx^2} = \frac{3 + 2\sqrt{3}}{24} > 0$$

$\Rightarrow$ Area will be minimum, when

$x = 40(2 - \sqrt{3})$

$$\therefore \text{ Side of hexagon} = \frac{20 - 40(2 - \sqrt{3})}{6}$$

$$= \frac{20\sqrt{3} - 30}{3} = \frac{10(2\sqrt{3} - 3)}{3}$$

$$= \frac{10}{2\sqrt{3} + 3}$$

**66.** *(c)* We have,

$$A = \begin{bmatrix} 1 & -x & 2x+1 \\ -x & 1 & -x \\ 2x+1 & -x & 1 \end{bmatrix}$$

$\therefore f(x) = \det(A) = 4x^3 - 4x^2 - 4x$

$\Rightarrow f'(x) = 12x^2 - 8x - 4$

$= 4(x - 1)(3x + 1)$

Now, $f'(x) = 0 \Rightarrow x = 1, -\frac{1}{3}$

Also, $f''(x) = 24x - 8$

At $x = 1$, $f''(x) = 16 > 0$

$\therefore f(x)$ attains its minimum value at $x = 1$

At $x = -\frac{1}{3}$, $f''(x) = -16 < 0$

$\therefore f(x)$ attains its maximum value at $x = -\frac{1}{3}$.

Hence, sum of maximum and minimum values of $f$

$$= f\left(\frac{-1}{3}\right) + f(1) = \frac{20}{27} - 4 = \frac{-88}{27}$$

**67.** *(b)* Given,

$$f(x) = (4a - 3)(x + \log_e 5) + 2(a - 7)\cot\frac{x}{2}\cdot\sin^2\frac{x}{2}$$

$$\Rightarrow f(x) = (4a - 3)(x + \log_e 5) + 2(a - 7)\cos\frac{x}{2}\sin\frac{x}{2}$$

$$\Rightarrow f(x) = (4a - 3)(x + \log_e 5) + (a - 7)\sin x$$

$\Rightarrow f'(x) = (4a - 3)(1 + 0) + (a - 7)\cos x$

$\Rightarrow f'(x) = (4a - 3) + (a - 7)\cos x$

When $f'(x) = 0$,

$$\Rightarrow \cos x = \frac{4a - 3}{7 - a}$$

As, $-1 \le \cos x \le 1$

$$\text{So, } -1 \le \frac{4a - 3}{7 - a} \le 1 \Rightarrow \frac{4a - 3}{7 - a} + 1 \ge 0$$

$$\Rightarrow \frac{4a - 3 + 7 - a}{7 - a} \ge 0$$

$$\Rightarrow \frac{3a + 4}{7 - a} \ge 0$$

$-$ $+$ $-$ (sign scheme with points $-4/3$ and $7$)

$$a \in [-4/3, 7) \quad \ldots(i)$$

$$\text{Now, } \frac{4a - 3}{7 - a} - 1 \le 0$$

$$\Rightarrow \frac{4a - 3 - 7 + a}{7 - a} \le 0 \Rightarrow \frac{5a - 10}{7 - a} \le 0$$

$-$ $+$ $-$ (sign scheme with points $2$ and $7$)

$$a \in (-\infty, 2] \cup [7, \infty) \quad \ldots(ii)$$

From Eqs. (i) and (ii), we get

$$a \in \left[-\frac{4}{3}, 2\right]$$

**68.** *(b)* Given, semi-vertical angle of right circular cone

$$= \tan^{-1}\left(\frac{1}{2}\right)$$

$$\text{Let} \quad \alpha = \tan^{-1}\left(\frac{1}{2}\right)$$

$$\Rightarrow \tan\alpha = \frac{1}{2} \Rightarrow \frac{r}{h} = \frac{1}{2}$$

$$\Rightarrow r = h/2 \quad \ldots(i)$$

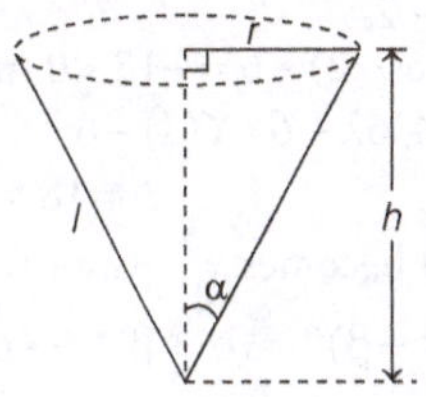


$\because$ Volume of cone is $(V) = \frac{1}{3}\pi r^2 h$

$$\therefore \quad V = \frac{1}{3}\pi\left(\frac{1}{2}h\right)^2(h) = \frac{1}{12}\pi h^3$$

[from Eq. (i)]

On differentiating both sides w.r.t. $t$, we get

$$\frac{dV}{dt} = \frac{1}{12}\pi(3h^2)\frac{dh}{dt}$$

$$\Rightarrow \quad \frac{dh}{dt} = \frac{4}{\pi h^2}\frac{dV}{dt}$$

$$\Rightarrow \quad \frac{dh}{dt} = \frac{4}{\pi h^2} \times 5$$

$$\left[\because \text{ given } \frac{dV}{dt} = 5 \text{ m}^3/\text{min}\right]$$

Now, at $h = 10$ m, the rate at which height of water level is rising $= \left.\frac{dh}{dt}\right|_{h=10}$

$$= \frac{4}{\pi(10)^2} \times 5 = \frac{1}{5\pi} \text{ m/min}$$

**69.** *(d)* Let $\alpha, \beta$ be the roots of

$$x^2 + (2 - \lambda)x + (10 - \lambda) = 0 \quad \ldots(i)$$

$\therefore \alpha + \beta = \lambda - 2$

and $\alpha\beta = (10 - \lambda)$ ...(ii)

Let, $A = \alpha^3 + \beta^3$

$$= (\alpha + \beta)^3 - 3\alpha\beta(\alpha + \beta)$$

$$= (\lambda - 2)^3 - 3(10 - \lambda)(\lambda - 2)$$

$$= \lambda^3 - 8 - 6\lambda^2 + 12\lambda + 3(\lambda^2 - 12\lambda + 20)$$

$$= \lambda^3 - 3\lambda^2 - 24\lambda + 52$$

Now, differentiating it w.r.t $\lambda$, we get

$$\text{Now, } \frac{dA}{d\lambda} = 3\lambda^2 - 6\lambda - 24$$

$$\text{On putting } \frac{dA}{d\lambda} = 0$$

$$\Rightarrow \quad \lambda^2 - 2\lambda - 8 = 0$$

$$\Rightarrow \quad (\lambda + 2)(\lambda - 4) = 0$$

$$\Rightarrow \quad \lambda = -2, 4$$

$$\text{Now, } \quad \frac{d^2A}{d\lambda^2} = 6\lambda - 6$$

For $\lambda = -2$,
$6\lambda - 6 = 6(-2) - 6 = -18 < 0$ (max.)
For $\lambda = 4$, $6\lambda - 6 = 6(4) - 6$
$= 18 > 0$ (min.)
$\therefore$ Eq. (i) becomes $x^2 - 2x + 6 = 0$
Thus, $(\alpha - \beta)^2 = (\alpha + \beta)^2 - 4\alpha\beta$
$= (2)^2 - 4(6) = 4 - 24 = -20$
$\Rightarrow \alpha - \beta = \pm 2\sqrt{5}i \Rightarrow |\alpha - \beta| = 2\sqrt{5}$

**70.** *(c)* According to given information, we have
Perimeter of square + Perimeter of circle = 2 units

$$\Rightarrow 4x + 2\pi r = 2 \Rightarrow r = \frac{1-2x}{\pi} \quad \ldots(i)$$

Now, let $A$ be the sum of the areas of the square and the circle. Then,

$$A = x^2 + \pi r^2$$

$$\Rightarrow A(x) = x^2 + \frac{(1-2x)^2}{\pi}$$

Now, for minimum value of $A(x)$, $dA/dx = 0$

$$\Rightarrow 2x + \frac{2(1-2x)}{\pi} \cdot (-2) = 0$$

$$\Rightarrow \pi x + 4x = 2$$

$$\Rightarrow x = \frac{2}{\pi + 4} \quad \ldots(ii)$$

Now, from Eq. (i), we get

$$r = \frac{1 - 2 \cdot \frac{2}{\pi+4}}{\pi} = \frac{1}{\pi + 4} \quad \ldots(iii)$$

From Eqs. (ii) and (iii), we get

$$x = 2r$$

**71.** *(c)* Since, the function have extreme values at $x = 1$ and $x = 2$.
$\therefore$ $f'(x) = 0$ at $x = 1$ and $x = 2$
Also it is given that

$$\lim_{x \to 0}\left[1 + \frac{f(x)}{x^2}\right] = 3$$

$$\Rightarrow \lim_{x \to 0} \frac{f(x)}{x^2} = 2$$

$\Rightarrow$ $f(x)$ will be of the form

$$ax^4 + bx^3 + 2x^2$$

[$\because f(x)$ is of four degree polynomial]

Let $f(x) = ax^4 + bx^3 + 2x^2$

$$\Rightarrow f'(x) = 4ax^3 + 3bx^2 + 4x$$

$$\Rightarrow f'(1) = 4a + 3b + 4 = 0 \quad \ldots(i)$$

and $f'(2) = 32a + 12b + 8 = 0$

$$\Rightarrow 8a + 3b + 2 = 0 \quad \ldots(ii)$$

On solving Eqs. (i) and (ii), we get

$$a = \frac{1}{2}, b = -2$$

$$\therefore f(x) = \frac{x^4}{2} - 2x^3 + 2x^2$$

$$\Rightarrow f(2) = 8 - 16 + 8 = 0$$

**72.** *(c)* Given,
(i) A function $f$, such that
$$f(x) = \log|x| + bx^2 + ax, x \neq 0$$
(ii) The function $f$ has extrema at $x = -1$ and $x = 2$, i.e., $f'(-1) = f'(2) = 0$ and $f''(-1) \neq 0 \neq f''(2)$.

Now, given function $f$ is

$$f(x) = \log|x| + bx^2 + ax$$

$$\Rightarrow f'(x) = \frac{1}{x} + 2bx + a$$

$$\Rightarrow f''(x) = \frac{-1}{x^2} + 2b$$

Since, $f$ has extrema at $x = -1$ and $x = 2$.
Hence,

$$f'(-1) = 0$$

$$\Rightarrow a - 2b = 1 \quad \ldots(i)$$

and $f'(2) = 0$

$$\Rightarrow a + 4b = \frac{-1}{2} \quad \ldots(ii)$$

Solving Eqs. (i) and (ii), we get

$$a = \frac{1}{2} \text{ and } b = \frac{-1}{4}$$

$$\Rightarrow f''(x) = \frac{-1}{x^2} + \frac{-1}{2} = -\left(\frac{x^2+2}{2x^2}\right)$$

$$\Rightarrow f''(-1) < 0 \text{ and } f''(2) < 0$$

Hence, $f$ has local maxima at both $x = -1$ and $x = 2$. Hence, Statement I is correct. Also, while solving for Statement I, we found the values of $a$ and $b$, which justify that Statement II is also correct.
However, Statement II does not explain Statement I in any way.

**73.** *(b)* Given,

$$P(x) = x^4 + ax^3 + bx^2 + cx + d$$

$$\Rightarrow P'(x) = 4x^3 + 3ax^2 + 2bx + c$$

Since, $x = 0$ is a solution for $P'(x) = 0$, then

$$c = 0$$

$$\therefore P(x) = x^4 + ax^3 + bx^2 + d \quad \ldots(i)$$

Also, we have $P(-1) < P(1)$

$$\Rightarrow 1 - a + b + d < 1 + a + b + d$$

$$\Rightarrow a > 0$$

Since, $P'(x) = 0$, only when $x = 0$ and $P(x)$ is differentiable in $(-1, 1)$, we should have the maximum and minimum at the points $x = -1$, 0 and 1.
Also, we have $P(-1) < P(1)$
$\therefore$ Maximum of $P(x) =$ Max $\{P(0), P(1)\}$
and minimum of $P(x) =$ Min $\{P(-1), P(0)\}$
In the interval $[0, 1]$,

$$P'(x) = 4x^3 + 3ax^2 + 2bx = x(4x^2 + 3ax + 2b)$$

Since, $P'(x)$ has only one root $x = 0$, then $4x^2 + 3ax + 2b = 0$ has no real roots.

$$\therefore (3a)^2 - 32b < 0$$

$$\Rightarrow \frac{9a^2}{32} < b$$

$$\therefore b > 0$$

Thus, we have $a > 0$ and $b > 0$.
$\therefore P'(x) = 4x^3 + 4ax^2 + 2bx > 0$, $\forall x \in (0, 1)$
Hence, $P(x)$ is increasing in $[0, 1]$.
$\therefore$ Maximum of $P(x) = P(1)$
Similarly, $P(x)$ is decreasing in $[-1, 0]$.
Therefore, minimum $P(x)$ does not occur at $x = -1$.

**74.** *(b)* Any point on the ellipse $\frac{x^2}{a^2} + \frac{y^2}{b^2} = 1$ is taken as $(a\cos\theta, b\sin\theta)$. So, the area of rectangle inscribed in the ellipse is given by

$$A = (2a\cos\theta)(2b\sin\theta)$$

$$\therefore A = 2ab\sin 2\theta$$

On differentiate it w.r.t., $\theta$ we get

$$\frac{dA}{d\theta} = 4ab\cos 2\theta$$

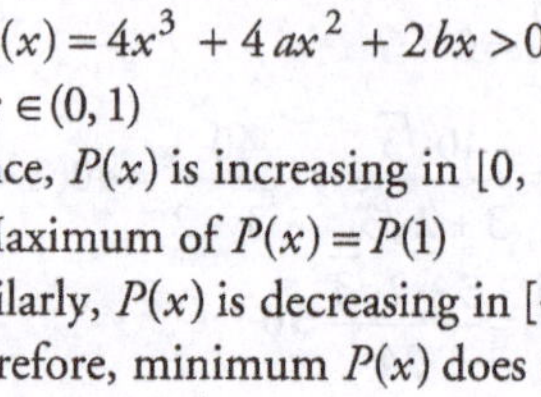


Now for maximum area,
putting $\frac{dA}{d\theta} = 0$

$$\Rightarrow \theta = \frac{\pi}{4} \text{ and } \left(\frac{d^2A}{d\theta^2}\right)_{\theta = \pi/4}$$

$$= -8ab\sin 2\theta$$

As, $\frac{d^2A}{d\theta^2}<0$, area is maximum for

$\theta=\frac{\pi}{4}$.

$\therefore$ Sides of rectangle are $\frac{2a}{\sqrt{2}}, \frac{2b}{\sqrt{2}}$

Required area $=2ab$.

**75.** *(d)* $\therefore u^2=a^2\cos^2\theta+b^2\sin^2\theta$

$$+a^2\sin^2\theta$$

$$+b^2\cos^2\theta+2\sqrt{(a^2\cos^2\theta+b^2\sin^2\theta)}$$

$$\times\sqrt{(a^2\sin^2\theta+b^2\cos^2\theta)}$$

$\Rightarrow u^2=a^2+b^2+2\sqrt{x(a^2+b^2-x)}$

[where, $x=a^2\cos^2\theta+b^2\sin^2\theta$]

$\Rightarrow u^2=(a^2+b^2)+2\sqrt{(a^2+b^2)x-x^2}$

On differentiating w.r.t. $\theta$, we get

$$\frac{du^2}{d\theta}=\frac{2}{2\sqrt{(a^2+b^2)x-x^2}}$$

$$(a^2+b^2-2x)\times\frac{dx}{d\theta}$$

and $\frac{dx}{d\theta}=(b^2-a^2)\sin 2\theta$

$$\Rightarrow \frac{du^2}{d\theta}=\frac{(a^2+b^2-2x)}{\sqrt{(a^2+b^2)x-x^2}}\times(b^2-a^2)\sin 2\theta$$

For maxima and minima, put $\frac{du^2}{d\theta}=0$

$\therefore a^2+b^2=2(a^2\cos^2\theta+b^2\sin^2\theta)$

and $\sin 2\theta=0$

$\Rightarrow \cos 2\theta\,(b^2-a^2)=0$

and $\sin 2\theta=0$

$\Rightarrow \cos 2\theta=0$

and $\theta=0$

$\Rightarrow 2\theta=\frac{\pi}{2}$ and $\theta=0$

$\Rightarrow \theta=\frac{\pi}{4}$ and $\theta=0$

So, $u^2$ will be minimum at $\theta=0$ and will be maximum at $\theta=\frac{\pi}{4}$.

$\therefore u^2_{\min}=(a+b)^2$

and $u^2_{\max}=2(a^2+b^2)$

Hence,

$$u^2_{\max}-u^2_{\min}=2(a^2+b^2)-(a+b)^2$$
$$=(a-b)^2$$

**76.** *(39)* Given, $\frac{x^2+x+2}{x^2+5x+6}<0$

$\Rightarrow \frac{1}{(x+2)(x+3)}<0$

$(\because x^2+x+2>0,\ \forall\, x\in R)$

$\Rightarrow x\in(-3,-2)$ ...(i)

Also, $f(x)=1+x(\lambda^2-x^2)$

$\Rightarrow f'(x)=(\lambda^2-x^2)+(-2x)\,x$

For finding local minima, we put $f'(x)=0$

$\lambda^2=3x^2\Rightarrow x=\pm\frac{\lambda}{\sqrt{3}}$

$$\underset{-\lambda/\sqrt{3}}{-}\quad + \quad\underset{\lambda/\sqrt{3}}{-}$$

$\therefore$Local minima at $x=\frac{-\lambda}{\sqrt{3}}$

So, from Eq. (i), $-3<x<-2$

$\Rightarrow -3<-\frac{\lambda}{\sqrt{3}}<-2$

$\Rightarrow 3\sqrt{3}>\lambda>2\sqrt{3}\Rightarrow 2\sqrt{3}<\lambda<3\sqrt{3}$

$\therefore \alpha=2\sqrt{3}$ and $\beta=3\sqrt{3}$

Hence, $\alpha^2+\beta^2=12+27=39$

**77.** *(128)*

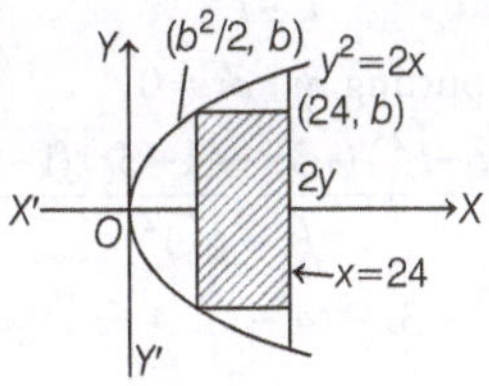


So, $A=\left(24-\frac{b^2}{2}\right)\times 2b$

Then, for maximum area

$$\frac{dA}{db}=0$$

$\Rightarrow 2b\left(0-\frac{2b}{2}\right)+\left(24-\frac{b^2}{2}\right)\times 2=0$

$\Rightarrow -b^2+48-2b^2=0$

$\Rightarrow 3b^2=48\Rightarrow b=4$

Hence, $A=\left(24-\frac{16}{2}\right)\times 2\times 4$

$=16\times 8=128$

**78.** *(1600)* $\because$ Let $y=\sqrt{8x-x^2-12}$

$\Rightarrow y^2=8x-x^2-12$

$(x-4)^2+y^2=4$ ...(i)

Centre $(4, 0)$, radius $=2$

$\therefore$ Let $P=(x-7)^2+(y-4)^2$ ...(ii)

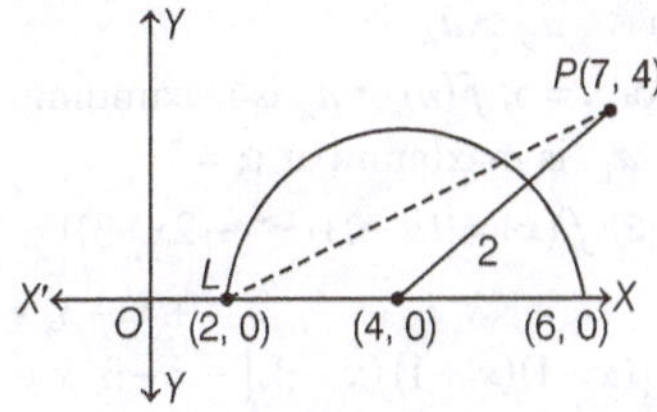


$M=PL^2=(7-2)^2+(4-0)^2$

$=25+16=41$

and $m=PR^2=\sqrt{(7-4)^2+(4-0)^2}$

$=(\sqrt{9+16}-2)^2=(5-2)^2=9$

$\therefore m=9$

and $M=41$

Then, $M^2-m^2=(41-9)(41+9)=1600$

**79.** *(4)* $D=\sin^2 2\theta-4\left(1-\frac{\sin^2 2\theta}{2}\right)$

$$\left(1-\frac{3}{4}\sin^2 2\theta\right)$$

$=(\sin 2\theta)^2-4\left(1-\frac{5}{4}\sin^2 2\theta+\frac{3}{8}\sin^4 2\theta\right)$

$\Rightarrow D=-\frac{3}{2}\sin^4 2\theta+6\sin^2 2\theta-4>0$

$\Rightarrow 3\sin^4 2\theta-12\sin^2 2\theta+8<0$

Critical point

$$\sin^2 2\theta=\frac{12\pm\sqrt{(12)^2-4(3)(8)}}{2\times 3}$$

$$=\frac{12\pm 4\sqrt{3}}{6}=\frac{6\pm 2\sqrt{3}}{3}$$

$\sin^2 2\theta=\frac{6\pm 2\sqrt{3}}{3}$ $[\because \sin^2 2\theta\in[0,1]]$

$\therefore \sin^2 2\theta\in\left[2-\frac{2}{\sqrt{3}},1\right]$

$\alpha=2-\frac{2}{\sqrt{3}},\ \beta=1$

$(\alpha-2)^2=\frac{4}{3},\ (\beta-1)^2=0$

$\therefore 3(\alpha-2)^2+(\beta-1)^2=4$

**80.** *(5)* Let $f(n)=\frac{n^3}{n^4+147}$

So, $f'(n)=\frac{3n^2(n^4+147)-n^3(4n^3)}{(n^4+147)^2}$

For maxima, $f'(n)=0$

$\Rightarrow 3n^2(n^4+147)-n^3(4n^3)=0$

$\Rightarrow n=\sqrt{21}$

$\because 4<\sqrt{21}<5\Rightarrow f(5)>f(4)$

or $\quad a_5 > a_4$

For $n = 5$, $f(n)$ or $a_n$ is maximum.

$\therefore a_\alpha$ is maximum at $\alpha = 5$

**81.** (3) $f(x) = |(x-1)(x^2-2x-3)| + x - 3, x \in R$

$= |(x-1)(x+1)(x-3)| + x - 3, x \in R$

$= |(x^2-1)(x-3)| + (x-3), x \in R$

$$f(x) = \begin{cases} (x^2-1)(x-3) + (x-3), & 3 \le x < 4 \\ -(x^2-1)(x-3) + (x-3), & 1 \le x < 3 \\ (x^2-1)(x-3) + (x-3), & 0 < x < 1 \end{cases}$$

$$f(x) = \begin{cases} (x-3)x^2, & 3 \le x < 4 \\ (2-x^2)(x-3), & 1 \le x < 3 \\ (x-3)x^2, & 0 < x < 1 \end{cases}$$

$$f'(x) = \begin{cases} 3x^2 - 6x, & 3 \le x < 4 \\ -3x^2 + 6x + 2, & 1 \le x < 3 \\ 3x^2 - 6x, & 0 < x < 1 \end{cases}$$

Clearly, $f'(3^+) > 0$ and $f(3^-) < 0$

$\therefore 3$ is local minima.

$$f'(1^+) > 0 \text{ and } f'(1^{-1}) < 0$$

$\therefore$ 1 is local minima.

When $x \in (1, 3)$

$$f'(x) = 0$$

$$-3x^2 + 6x + 2 = 0$$

$$x = \frac{-6 \pm \sqrt{36+24}}{2 \times (-3)}$$

$$= \frac{-6 \pm \sqrt{60}}{-6} = 1 \pm \sqrt{\frac{5}{3}}$$

$\therefore$ There are two minimum and one maximum point.

$\therefore m + M = 2 + 1 = 3$

**82.** (4) Let $f(x) = 3x^4 + 4x^3 - 12x^2 + 4$

Differentiating w.r.t. $x$,

$$f'(x) = 12x^3 + 12x^2 - 24x$$

$$= 12x(x^2 + x - 2) = 12x(x+2)(x-1)$$

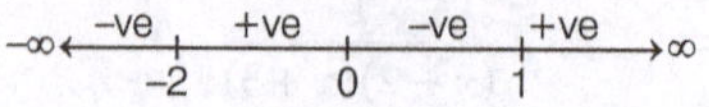


Sign of $f'(x)$

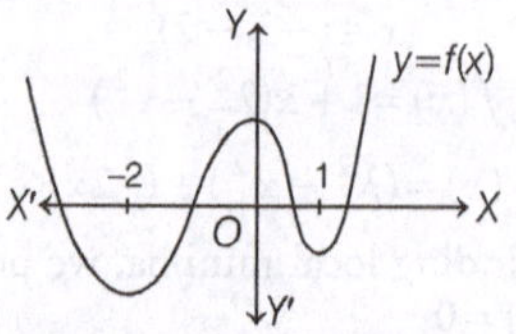


From graph we observe that, $f(x)$ has minima at $x = -2$ and $x = 1$ and maxima at $x = 0$.

$\Rightarrow f(x)$ cuts the $X$-axis at 4 distinct points.

So, there are 4 distinct real roots of the given equation.

**83.** (9) Let $y = \dfrac{4}{\sin x} + \dfrac{1}{1 - \sin x}$

$$= \frac{4 - 3\sin x}{\sin x(1 - \sin x)}$$

Let $\sin x = t$, where $t \in (0, 1)$

$$\therefore \quad y = \frac{4 - 3t}{t - t^2}$$

Now, putting $dy/dt = 0$

$$\frac{dy}{dt} = \frac{(t-t^2)(-3) - (4-3t)(1-2t)}{(t-t^2)^2} = 0$$

$$\Rightarrow 3t^2 - 3t - (4 - 11t + 6t^2) = 0$$

$$\Rightarrow \quad 3t^2 - 8t + 4 = 0$$

$$\Rightarrow \quad (3t-2)(t-2) = 0 \Rightarrow t = \frac{2}{3}$$

[$t = 2$ discarded because $t \in (0, 1)$ only]

$$\Rightarrow \quad \sin x = 2/3$$

$$\therefore y_{\min} = \frac{4}{2/3} + \frac{1}{1 - 2/3} = 9$$

$[\because y_{\max} \to \infty]$

$\Rightarrow \alpha_{\min} = 9$

**84.** (5) We have,

$AD = 8$ m, $BC = 11$ m, $AB = 10$ m.

Let $AM = x$ m, then

$BM = (10 - x)$ m

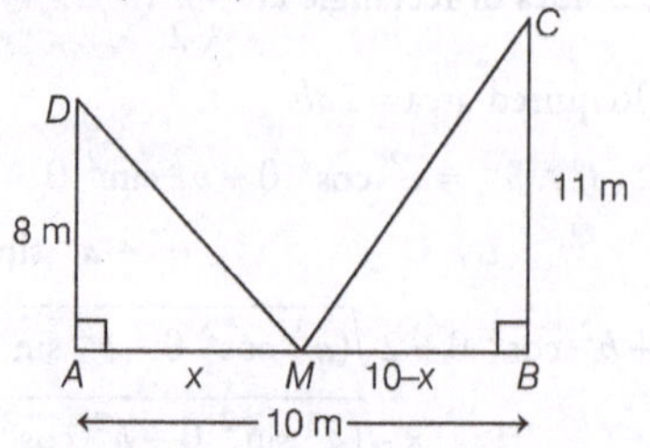


In $\Delta DAM$, by Pythagoras theorem, we get

$$MD^2 = AD^2 + AM^2 = 8^2 + x^2$$

$$\Rightarrow MD^2 = 64 + x^2 \quad \ldots\text{(i)}$$

In $\Delta CBM$, by Pythagoras theorem, we get

$$MC^2 = BC^2 + BM^2 = 121 + (10-x)^2$$

$$\Rightarrow MC^2 = 121 + 100 + x^2 - 20x$$

$$\Rightarrow MC^2 = 221 + x^2 - 20x \quad \ldots\text{(ii)}$$

From Eqs. (i) and (ii), we have

$$MD^2 + MC^2 = 64 + x^2 + 221 + x^2 - 20x$$

$$= 285 - 20x + 2x^2$$

Let $S = 285 - 20x + 2x^2$

$$\Rightarrow \quad \frac{dS}{dx} = -20 + 4x.$$

Now, $\dfrac{dS}{dx} = 0 \Rightarrow 4x - 20 = 0 \Rightarrow x = 5$

Also, $\dfrac{d^2S}{dx^2} = 4 > 0.$

So, $S$ is minimum for $x = 5$.

Thus, distance of $M$ from the point $A$ is 5m.

# C-19

JEE Main Chapterwise Topicwise Mathematics

In JEE Main, there are usually 1-2 questions per paper are visible from this chapter. In past 3 years, 75 conceptual and 37 numerical questions of moderate to hard difficulty level were asked from this chapter.

**For detailed analysis scan QR code**

# INTEGRALS

## Topic 1 Indefinite Integrals

### Level 1

**01** Let $f(x)=\int x^3\sqrt{3-x^2}\,dx$. If $5f(\sqrt{2})=-4$, then $f(1)$ is equal to *[2025, 3 April Shift-I]*

(a) $-\frac{8\sqrt{2}}{5}$ (b) $-\frac{4\sqrt{2}}{5}$ (c) $-\frac{2\sqrt{2}}{5}$ (d) $-\frac{6\sqrt{2}}{5}$

**Imp.** Integration by Parts

**02** Let $\int x^3 \sin x\,dx = g(x)+C$, where $C$ is the constant of integration.

If $8\left(g\left(\frac{\pi}{2}\right)+g'\left(\frac{\pi}{2}\right)\right)=\alpha\pi^3+\beta\pi^2+\gamma, \alpha, \beta, \gamma, \in Z$, then $\alpha+\beta-\gamma$ equals *[2025, 23 Jan Shift-II]*

(a) 55 (b) 48 (c) 62 (d) 47

**03** If $\int e^x\left(\frac{x\sin^{-1}x}{\sqrt{1-x^2}}+\frac{\sin^{-1}x}{(1-x^2)^{\frac{3}{2}}}+\frac{x}{1-x^2}\right)dx$ $=g(x)+C$, where $C$ is the constant of integration, then $g\left(\frac{1}{2}\right)$ equals *[2025, 22 Jan Shift-II]*

(a) $\frac{\pi}{6}\sqrt{\frac{e}{3}}$ (b) $\frac{\pi}{4}\sqrt{\frac{e}{2}}$

(c) $\frac{\pi}{4}\sqrt{\frac{e}{3}}$ (d) $\frac{\pi}{6}\sqrt{\frac{e}{2}}$

**04** Let $\int \frac{2-\tan x}{3+\tan x}dx=\frac{1}{2}(\alpha x+\log_e|\beta\sin x+\gamma\cos x|)$ $+C$, where $C$ is the constant of integration. Then, $\alpha+\frac{\gamma}{\beta}$ is equal to *[2024, 9 April Shift-I]*

(a) 1 (b) 3 (c) 7 (d) 4

**05** For $x\in\left(-\frac{\pi}{2},\frac{\pi}{2}\right)$, if

$$y(x)=\int\frac{\operatorname{cosec} x+\sin x}{\operatorname{cosec} x\sec x+\tan x\sin^2 x}dx$$ and

$\lim_{x\to\left(\frac{\pi}{2}\right)^-} y(x)=0$, then $y\left(\frac{\pi}{4}\right)$ is equal to *[2024, 29 Jan Shift-I]*

(a) $\frac{1}{2}\tan^{-1}\left(\frac{1}{\sqrt{2}}\right)$ (b) $\frac{1}{\sqrt{2}}\tan^{-1}\left(-\frac{1}{2}\right)$

(c) $\tan^{-1}\left(\frac{1}{\sqrt{2}}\right)$ (d) $-\frac{1}{\sqrt{2}}\tan^{-1}\left(\frac{1}{\sqrt{2}}\right)$

**06** If $I(x)=\int e^{\sin^2 x}(\cos x\sin 2x-\sin x)dx$ and $I(0)=1$, then $I\left(\frac{\pi}{3}\right)$ is equal to *[2023, 10 April Shift-I]*

(a) $e^{\frac{3}{4}}$ (b) $-e^{\frac{3}{4}}$

(c) $\frac{1}{2}e^{\frac{3}{4}}$ (d) $-\frac{1}{2}e^{\frac{3}{4}}$

**07** The integral $\int\left(\left(\frac{x}{2}\right)^x+\left(\frac{2}{x}\right)^x\right)\log_2 x\,dx$ is equal to *[2023, 8 April Shift-II]*

(a) $\left(\frac{x}{2}\right)^x\log_2\left(\frac{2}{x}\right)+C$ (b) $\left(\frac{x}{2}\right)^x\log_2\left(\frac{x}{2}\right)+C$

(c) $\left(\frac{x}{2}\right)^x+\left(\frac{2}{x}\right)^x+C$ (d) $\left(\frac{x}{2}\right)^x-\left(\frac{2}{x}\right)^x+C$

**08** Let $g:(0,\infty)\to R$ be a differentiable function such that $\int\left(\frac{x(\cos x-\sin x)}{e^x+1}+\frac{g(x)(e^x+1-xe^x)}{(e^x+1)^2}\right)dx$ $=\frac{xg(x)}{e^x+1}+c, \forall x>0$, where $c$ is an arbitrary constant. Then, *[2022, 25 June Shift-I]*

(a) $g$ is decreasing in $\left(0, \frac{\pi}{4}\right)$

(b) $g'$ is increasing in $\left(0, \frac{\pi}{4}\right)$

(c) $g + g'$ is increasing in $\left(0, \frac{\pi}{2}\right)$

(d) $g - g'$ is increasing in $\left(0, \frac{\pi}{2}\right)$

**09** The integral $\int \frac{1}{\sqrt[4]{(x-1)^3 (x+2)^5}} dx$ is equal to (where, $C$ is a constant of integration)

*[2021, 31 Aug Shift-I]*

(a) $\frac{3}{4}\left(\frac{x+2}{x-1}\right)^{\frac{1}{4}} + C$ (b) $\frac{3}{4}\left(\frac{x+2}{x-1}\right)^{\frac{5}{4}} + C$

(c) $\frac{4}{3}\left(\frac{x-1}{x+2}\right)^{\frac{1}{4}} + C$ (d) $\frac{4}{3}\left(\frac{x-1}{x+2}\right)^{\frac{5}{4}} + C$

**10** The integral $\int \frac{(2x-1)\cos\sqrt{(2x-1)^2+5}}{\sqrt{4x^2-4x+6}} dx$ is equal to (where, $C$ is a constant of integration)

*[2021, 18 March Shift-I]*

(a) $\frac{1}{2}\sin\sqrt{(2x-1)^2+5}+C$

(b) $\frac{1}{2}\cos\sqrt{(2x+1)^2+5}+C$

(c) $\frac{1}{2}\cos\sqrt{(2x-1)^2+5}+C$

(d) $\frac{1}{2}\sin\sqrt{(2x+1)^2+5}+C$

**11** The integral $\int \frac{2x^3-1}{x^4+x} dx$ is equal to (here $C$ is a constant of integration)

*[2019, 12 April Shift-I]*

(a) $\frac{1}{2}\log_e \frac{|x^3+1|}{x^2}+C$ (b) $\frac{1}{2}\log_e \frac{(x^3+1)^2}{|x^3|}+C$

(c) $\log_e \left|\frac{x^3+1}{x}\right|+C$ (d) $\log_e \frac{|x^3+1|}{x^2}+C$

**12** $\int \frac{\sin \frac{5x}{2}}{\sin \frac{x}{2}} dx$ is equal to (where, $C$ is a constant of integration)

*[2019, 8 April Shift-I]*

(a) $2x+\sin x+2\sin 2x+C$ (b) $x+2\sin x+2\sin 2x+C$

(c) $x+2\sin x+\sin 2x+C$ (d) $2x+\sin x+\sin 2x+C$

**13** The integral

$$\int \frac{\sin^2 x \cos^2 x}{(\sin^5 x+\cos^3 x \sin^2 x+\sin^3 x\cos^2 x+\cos^5 x)^2} dx$$

is equal to *[JEE Main 2018]*

(a) $\frac{1}{3(1+\tan^3 x)}+C$ (b) $\frac{-1}{3(1+\tan^3 x)}+C$

(c) $\frac{1}{1+\cot^3 x}+C$ (d) $\frac{-1}{1+\cot^3 x}+C$

(Where, $C$ is a constant of integration)

**14** If $f\left(\frac{x-4}{x+2}\right)=2x+1$, $(x \in R-\{1, 2\})$, then $\int (f(x))\, dx$ is equal to (where, $C$ is a constant of integration)

*[JEE Main 2018]*

(a) $12\log_e|1-x|-3x+C$

(b) $-12\log_e|1-x|-3x+C$

(c) $-12\log_e|1-x|+3x+C$

(d) $12\log_e|1-x|+3x+C$

**15** If $\int \frac{2x+5}{\sqrt{7-6x-x^2}} dx = A\sqrt{7-6x-x^2} + B\sin^{-1}\left(\frac{x+3}{4}\right)+C$

(where, $C$ is a constant of integration), then the ordered pair $(A, B)$ is equal to *[JEE Main 2018]*

(a) $(-2, -1)$ (b) $(2, -1)$ (c) $(-2, 1)$ (d) $(2, 1)$

**16** Let $I_n = \int \tan^n x\, dx$ $(n>1)$. If $I_4+I_6 = a\tan^5 x + bx^5 + C$, where $C$ is a constant of integration, then the ordered pair $(a, b)$ is equal to *[JEE Main 2017]*

(a) $\left(-\frac{1}{5}, 1\right)$ (b) $\left(\frac{1}{5}, 0\right)$ (c) $\left(\frac{1}{5}, -1\right)$ (d) $\left(-\frac{1}{5}, 0\right)$

**17** The integral $\int \sqrt{1+2\cot x(\operatorname{cosec} x+\cot x)}\, dx$ $\left(0<x<\frac{\pi}{2}\right)$ is equal to

(where $C$ is a constant of integration) *[JEE Main 2017]*

(a) $2\log\left(\sin\frac{x}{2}\right)+C$ (b) $2\log\left(\cos\frac{x}{2}\right)+C$

(c) $4\log\left(\cos\frac{x}{2}\right)+C$ (d) $4\log\left(\sin\frac{x}{2}\right)+C$

**18** The integral $\int \frac{dx}{x^2(x^4+1)^{\frac{3}{4}}}$ equals

*[JEE Main 2015]*

(a) $\left(\frac{x^4+1}{x^4}\right)^{1/4}+c$ (b) $(x^4+1)^{1/4}+c$

(c) $-(x^4+1)^{1/4}+c$ (d) $-\left(\frac{x^4+1}{x^4}\right)^{1/4}+c$

## Level 2

**19** If $f(x)=\int \frac{1}{x^{1/4}(1+x^{1/4})}dx$, $f(0)=-6$, then $f(1)$ is equal to *[2025, 28 Jan Shift-II]*

(a) $2-\log_e 2$ (b) $\log_e 2+2$
(c) $4(\log_e 2+2)$ (d) $4(\log_e 2-2)$

**20** Let $I(x)=\int \frac{dx}{(x-11)^{\frac{11}{13}}(x+15)^{\frac{15}{13}}}$.

If $I(37)-I(24)=\frac{1}{4}\left(\frac{1}{b^{1/13}}-\frac{1}{c^{1/13}}\right)$, $b, c, \in N$, then $3(b+c)$ is equal to *[2025, 23 Jan Shift-I]*

(a) 22 (b) 26 (c) 39 (d) 40

**21** Let $I(x)=\int \frac{6}{\sin^2 x(1-\cot x)^2}dx$. If $I(0)=3$, then $I\left(\frac{\pi}{12}\right)$ is equal to *[2024, 8 April Shift-I]*

(a) $2\sqrt{3}$ (b) $6\sqrt{3}$ (c) $\sqrt{3}$ (d) $3\sqrt{3}$

**Imp.** Maximum & Minimum Value

**22** If $\int \frac{1}{a^2\sin^2 x+b^2\cos^2 x}dx=\frac{1}{12}\tan^{-1}(3\tan x)+$ constant, then the maximum value of $a\sin x+b\cos x$, is *[2024, 6 April Shift-I]*

(a) $\sqrt{42}$ (b) $\sqrt{41}$
(c) $\sqrt{40}$ (d) $\sqrt{39}$

**23** If $\int \frac{\sin^{3/2} x+\cos^{3/2} x}{\sqrt{\sin^3 x\cos^3 x\sin(x-\theta)}}dx$
$=A\sqrt{\cos\theta\tan x-\sin\theta}+B\sqrt{\cos\theta-\sin\theta\cot x}+C$, where $C$ is the integration constant, then $AB$ is equal to *[2024, 29 Jan Shift-II]*

(a) $4\text{ cosec }(2\theta)$ (b) $4\sec\theta$
(c) $8\text{ cosec }(2\theta)$ (d) $2\sec\theta$

**24** The integral $\int \frac{(x^8-x^2)dx}{(x^{12}+3x^6+1)\tan^{-1}\left(x^3+\frac{1}{x^3}\right)}$ is equal to *[2024, 27 Jan Shift-II]*

(a) $\log_e\left(\left|\tan^{-1}\left(x^3+\frac{1}{x^3}\right)\right|\right)+C$

(b) $\log_e\left(\left|\tan^{-1}\left(x^3+\frac{1}{x^3}\right)\right|\right)^{1/2}+C$

(c) $\log_e\left(\left|\tan^{-1}\left(x^3+\frac{1}{x^3}\right)\right|\right)^{3}+C$

(d) $\log_e\left(\left|\tan^{-1}\left(x^3+\frac{1}{x^3}\right)\right|\right)^{1/3}+C$

**25** Let $I(x)=\int \frac{(x+1)}{x(1+xe^x)^2}dx$, $x>0$. If $\lim_{x\to\infty} I(x)=0$, then $I(1)$ is equal to *[2023, 8 April Shift-I]*

(a) $\frac{e+2}{e+1}-\log_e(e+1)$ (b) $\frac{e+2}{e+1}+\log_e(e+1)$
(d) $\frac{e+1}{e+2}+\log_e(e+1)$ (d) $\frac{e+1}{e+2}-\log_e(e+1)$

**26** Let $I(x)=\int \frac{x^2(x\sec^2 x+\tan x)}{(x\tan x+1)^2}dx$. If $I(0)=0$, then $I\left(\frac{\pi}{4}\right)$ is equal to *[2023, 6 April Shift-I]*

(a) $\log\frac{(\pi+4)^2}{16}-\frac{\pi^2}{4(\pi+4)}$ (b) $\log\frac{(\pi+4)^2}{16}+\frac{\pi^2}{4(\pi+4)}$

(c) $\log\frac{(\pi+4)^2}{32}+\frac{\pi^2}{4(\pi+4)}$ (d) $\log\frac{(\pi+4)^2}{32}-\frac{\pi^2}{4(\pi+4)}$

**27** Let $f(x)=\int \frac{2x}{(x^2+1)(x^2+3)}dx$. If $f(3)=\frac{1}{2}(\log_e 5-\log_e 6)$, then $f(4)$ is equal to *[2023, 25 Jan Shift-I]*

(a) $\frac{1}{2}(\log_e 19-\log_e 17)$ (b) $\frac{1}{2}(\log_e 17-\log_e 19)$
(c) $\log_e 19-\log_e 20$ (d) $\log_e 17-\log_e 18$

**28** For $I(x)=\int \frac{\sec^2 x-2022}{\sin^{2022} x}dx$, if $I\left(\frac{\pi}{4}\right)=2^{1011}$, then *[2022, 29 July Shift-II]*

(a) $3^{1010}I\left(\frac{\pi}{3}\right)-I\left(\frac{\pi}{6}\right)=0$ (b) $3^{1010}I\left(\frac{\pi}{6}\right)-I\left(\frac{\pi}{3}\right)=0$

(c) $3^{1011}I\left(\frac{\pi}{3}\right)-I\left(\frac{\pi}{6}\right)=0$ (d) $3^{1011}I\left(\frac{\pi}{6}\right)-I\left(\frac{\pi}{3}\right)=0$

**29** If $\int \frac{(x^2+1)e^x}{(x+1)^2}dx=f(x)e^x+C$, where $C$ is a constant, then $\frac{d^3f}{dx^3}$ at $x=1$ is equal to *[2022, 27 June Shift-I]*

(a) $-3/4$ (b) $3/4$
(c) $-3/2$ (d) $3/2$

**30** If $\int \frac{1}{x}\sqrt{\frac{1-x}{1+x}}dx=g(x)+c$, $g(1)=0$, then $g\left(\frac{1}{2}\right)$ is equal to *[2022, 26 June Shift-II]*

(a) $\log_e\left(\frac{\sqrt{3}-1}{\sqrt{3}+1}\right)+\frac{\pi}{3}$ (b) $\log_e\left(\frac{\sqrt{3}+1}{\sqrt{3}-1}\right)+\frac{\pi}{3}$

(c) $\log_e\left(\frac{\sqrt{3}+1}{\sqrt{3}-1}\right)-\frac{\pi}{3}$ (d) $\frac{1}{2}\log_e\left(\frac{\sqrt{3}-1}{\sqrt{3}+1}\right)-\frac{\pi}{6}$

**Imp.** Properties of Logarthim

**31** The integral

$$\int \frac{e^{3\log_e 2x}+5e^{2\log_e 2x}}{e^{4\log_e x}+5e^{3\log_e x}-7e^{2\log_e x}}\,dx,\ x>0,$$ is equal to

(Where, $C$ is a constant of integration) **[2021, 25 Feb Shift-II]**

(a) $\log_e|x^2+5x-7|+C$
(b) $4\log_e|x^2+5x-7|+C$
(c) $\frac{1}{4}\log_e|x^2+5x-7|+C$
(d) $\log_e\sqrt{x^2+5x-7}+C$

**32** If $\int \frac{\cos x-\sin x}{\sqrt{8-\sin 2x}}\,dx = a\sin^{-1}\left(\frac{\sin x+\cos x}{b}\right)+c$, where $c$ is a constant of integration, then the ordered pair $(a, b)$ is equal to **[2021, 24 Feb Shift-I]**

(a) (3, 1) (b) (1, 3)
(c) (−1, 3) (d) (1, −3)

**33** If $\int (e^{2x}+2e^x-e^{-x}-1)\,e^{(e^x+e^{-x})}\,dx = g(x)e^{(e^x+e^{-x})}+c$, where $c$ is a constant of integration, then $g(0)$ is equal to **[2020, 5 Sep Shift-I]**

(a) 2 (b) $e^2$
(c) $e$ (d) 1

**34** Let $f(x)=\int \frac{\sqrt{x}}{(1+x)^2}\,dx,\ (x\ge 0)$. Then, $f(3)-f(1)$ is equal to **[2020, 4 Sep Shift-I]**

(a) $-\frac{\pi}{6}+\frac{1}{2}+\frac{\sqrt{3}}{4}$ (b) $-\frac{\pi}{12}+\frac{1}{2}+\frac{\sqrt{3}}{4}$
(c) $\frac{\pi}{6}+\frac{1}{2}-\frac{\sqrt{3}}{4}$ (d) $\frac{\pi}{12}+\frac{1}{2}-\frac{\sqrt{3}}{4}$

**35** If $\int \sin^{-1}\left(\sqrt{\frac{x}{1+x}}\right)dx = A(x)\tan^{-1}(\sqrt{x})+B(x)+C$, where $C$ is a constant of integration, then the ordered pair $(A(x), B(x))$ can be **[2020, 3 Sep Shift-II]**

(a) $(x-1, \sqrt{x})$
(b) $(x+1, \sqrt{x})$
(c) $(x+1, -\sqrt{x})$
(d) $(x-1, -\sqrt{x})$

**36** If $\int \frac{d\theta}{\cos^2\theta(\tan 2\theta+\sec 2\theta)} = \lambda\tan\theta+2\log_e|f(\theta)|+C$, where $C$ is a constant of integration, then the ordered pair $(\lambda, f(\theta))$ is equal to **[2020, 9 Jan Shift-II]**

(a) $(1, 1+\tan\theta)$ (b) $(1, 1-\tan\theta)$
(c) $(-1, 1+\tan\theta)$ (d) $(-1, 1-\tan\theta)$

**37** If $\int x^5 e^{-x^2}\,dx = g(x)e^{-x^2}+C$, where $C$ is a constant of integration, then $g(-1)$ is equal to **[2019, 10 April Shift-II]**

(a) −1 (b) 1
(c) −1/2 (d) −5/2

**38** The integral $\int \frac{3x^{13}+2x^{11}}{(2x^4+3x^2+1)^4}\,dx$ is equal to

(Where, $C$ is a constant of integration) **[2019, 12 Jan Shift-II]**

(a) $\frac{x^4}{6(2x^4+3x^2+1)^3}+C$ (b) $\frac{x^{12}}{6(2x^4+3x^2+1)^3}+C$
(c) $\frac{x^4}{(2x^4+3x^2+1)^3}+C$ (d) $\frac{x^{12}}{(2x^4+3x^2+1)^3}+C$

**39** The integral $\int \cos(\log_e x)\,dx$ is equal to

(where, $C$ is a constant of integration) **[2019, 12 Jan Shift-I]**

(a) $\frac{x}{2}[\cos(\log_e x)+\sin(\log_e x)]+C$
(b) $x[\cos(\log_e x)+\sin(\log_e x)]+C$
(c) $x[\cos(\log_e x)-\sin(\log_e x)]+C$
(d) $x/2[\sin(\log_e x)-\cos(\log_e x)]+C$

**40** If $f\left(\frac{3x-4}{3x+4}\right)=x+2,\ x\ne\frac{-4}{3}$ and $\int f(x)\,dx = A\log|1-x|+Bx+C$, then the ordered pair $(A, B)$ is equal to

(Where, $C$ is a constant of integration) **[JEE Main 2017]**

(a) $\left(-\frac{8}{3},\frac{2}{3}\right)$ (b) $\left(\frac{8}{3},-\frac{2}{3}\right)$
(c) $\left(\frac{8}{3},\frac{2}{3}\right)$ (d) $\left(-\frac{8}{3},-\frac{2}{3}\right)$

**41** If $\int \frac{dx}{\cos^3 x\sqrt{2\sin 2x}} = (\tan x)^A + C(\tan x)^B + k$, where $k$ is a constant of integration, then $A+B+C$ equal **[JEE Main 2016]**

(a) 16/5 (b) 27/10 (c) 7/10 (d) 21/5

## Level 3

**42** Let $f$ and $g$ be twice differentiable functions on $R$ such that **[2023, 29 Jan Shift-II]**

$$f''(x)=g''(x)+6x,\ f'(1)=4g'(1)-3=9$$
$$f(2)=3g(2)=12.$$

Then, which of the following is not true?

(a) There exists $x_0\in\left(1,\frac{3}{2}\right)$ such that $f(x_0)=g(x_0)$
(b) $g(-2)-f(-2)=20$
(c) If $-1<x<2$, then $|f(x)-g(x)|<8$
(d) $|f'(x)-g'(x)|<6 \Rightarrow -1<x<1$

**43** The integral $\int \dfrac{\left(1-\dfrac{1}{\sqrt{3}}\right)(\cos x-\sin x)}{\left(1+\dfrac{2}{\sqrt{3}}\sin 2x\right)}dx$ is equal to **[2022, 26 July Shift-II]**

(a) $\dfrac{1}{2}\log_e\left|\dfrac{\tan\left(\dfrac{x}{2}+\dfrac{\pi}{12}\right)}{\tan\left(\dfrac{x}{2}+\dfrac{\pi}{6}\right)}\right|+C$ (b) $\dfrac{1}{2}\log_e\left|\dfrac{\tan\left(\dfrac{x}{2}+\dfrac{x}{6}\right)}{\tan\left(\dfrac{x}{2}+\dfrac{\pi}{3}\right)}\right|+C$

(c) $\log_e\left|\dfrac{\tan\left(\dfrac{x}{2}+\dfrac{\pi}{6}\right)}{\tan\left(\dfrac{x}{2}+\dfrac{\pi}{12}\right)}\right|+C$ (d) $\dfrac{1}{2}\log_e\left|\dfrac{\tan\left(\dfrac{x}{2}-\dfrac{\pi}{12}\right)}{\tan\left(\dfrac{x}{2}-\dfrac{\pi}{6}\right)}\right|+C$

**44** The value of the integral

$$\int\left[\frac{\sin\theta\cdot\sin 2\theta\,(\sin^6\theta+\sin^4\theta+\sin^2\theta)\sqrt{2\sin^4\theta+3\sin^2\theta+6}}{1-\cos 2\theta}\right]d\theta \text{ is}$$

(Where, $C$ is a constant of integration) **[2021, 25 Feb Shift-I]**

(a) $\dfrac{1}{18}[11-18\sin^2\theta+9\sin^4\theta-2\sin^6\theta]^{3/2}+C$

(b) $\dfrac{1}{18}[9-2\cos^6\theta\ 3\cos^4\theta-6\cos^2\theta]^{3/2}+C$

(c) $\dfrac{1}{18}[9-2\sin^6\theta-3\sin^4\theta-6\sin^2\theta]^{3/2}+C$

(d) $\dfrac{1}{18}[11-18\cos^2\theta+9\cos^4\theta-2\cos^6\theta]^{3/2}+C$

**45** If $\int e^{\sec x}(\sec x\tan x\, f(x)+(\sec x\tan x+\sec^2 x))\,dx=e^{\sec x}f(x)+C$, then a possible choice of $f(x)$ is **[2019, 9 April Shift-II]**

(a) $x\sec x+\tan x+\dfrac{1}{2}$ (b) $\sec x+\tan x+\dfrac{1}{2}$

(c) $\sec x+x\tan x-\dfrac{1}{2}$ (d) $\sec x-\tan x-\dfrac{1}{2}$

**46** Let $n\geq 2$ be a natural number and $0<\theta<\dfrac{\pi}{2}$. Then, $\int\dfrac{(\sin^n\theta-\sin\theta)^{1/n}\cos\theta}{\sin^{n+1}\theta}d\theta$ is equal to (Where, $C$ is a constant of integration) **[2019, 10 Jan Shift-I]**

(a) $\dfrac{n}{n^2-1}\left(1-\dfrac{1}{\sin^{n+1}\theta}\right)^{\frac{n+1}{n}}+C$

(b) $\dfrac{n}{n^2-1}\left(1+\dfrac{1}{\sin^{n-1}\theta}\right)^{\frac{n+1}{n}}+C$

(c) $\dfrac{n}{n^2-1}\left(1-\dfrac{1}{\sin^{n-1}\theta}\right)^{\frac{n+1}{n}}+C$

(d) $\dfrac{n}{n^2+1}\left(1-\dfrac{1}{\sin^{n-1}\theta}\right)^{\frac{n+1}{n}}+C$

## Numerical Value Type Questions

**47** If $\int\dfrac{(\sqrt{1+x^2}+x)^{10}}{(\sqrt{1+x^2}-x)^9}dx=\dfrac{1}{m}((\sqrt{1+x^2}+x)^n(n\sqrt{1+x^2}-x))+C$ where $C$ is the constant of integration and $m,n\in N$ then $m+n$ is equal to ............... . **[2025, 4 April Shift-II]**

**48** If $\int\left(\dfrac{1}{x}+\dfrac{1}{x^3}\right)(\sqrt[23]{3x^{-24}+x^{-26}})\,dx$

$=-\dfrac{\alpha}{3(\alpha+1)}(3x^{\beta}+x^{\gamma})^{\frac{\alpha+1}{\alpha}}+C,\ x>0,\ (\alpha,\beta,\gamma\in Z)$, where $C$ is the constant of integration, then $\alpha+\beta+\gamma$ is equal to ............... . **[2025, 7 April Shift-II]**

**49** If $\int\dfrac{2x^2+5x+9}{\sqrt{x^2+x+1}}dx=x\sqrt{x^2+x+1}+\alpha\sqrt{x^2+x+1}+\beta\log_e\left|x+\dfrac{1}{2}+\sqrt{x^2+x+1}\right|+C$, where $C$ is the constant of integration, then $\alpha+2\beta$ is equal to ............. . **[2025, 24 Jan Shift-II]**

**50** If $\int\dfrac{1}{\sqrt[5]{(x-1)^4(x+3)^6}}dx=A\left(\dfrac{\alpha x-1}{\beta x+3}\right)^B+C$, where $C$ is the constant of integration, then the value of $\alpha+\beta+20AB$ is ......... . **[2024, 8 April Shift-II]**

**51** If $\int\operatorname{cosec}^5 x\,dx=\alpha\cot x\operatorname{cosec} x\left(\operatorname{cosec}^2 x+\dfrac{3}{2}\right)+\beta\log_6\left|\tan\dfrac{x}{2}\right|+C$, where $\alpha,\beta\in R$ and $C$ is the constant of integration, then the value of $8(\alpha+\beta)$ equals ......... . **[2024, 4 April Shift-II]**

**52** Let $f(x)=\int\dfrac{dx}{(3+4x^2)\sqrt{4-3x^2}},\ |x|<\dfrac{2}{\sqrt{3}}$. If $f(0)=0$ and $f(1)=\dfrac{1}{\alpha\beta}\tan^{-1}\left(\dfrac{\alpha}{\beta}\right),\ \alpha,\beta>0$, then $\alpha^2+\beta^2$ is equal to ......... . **[2023, 15 April Shift-I]**

**53** Let $I(x)=\int\sqrt{\dfrac{x+7}{x}}dx$ and $I(9)=12+7\log_e 7$. If $I(1)=\alpha+7\log_e(1+2\sqrt{2})$, then $\alpha^4$ is equal to .......... **[2023, 12 April Shift-I]**

**54** If $\int \sqrt{\sec 2x - 1}\,dx = \alpha \log_e \left| \cos 2x + \beta + \sqrt{\cos 2x\left(1 + \cos \frac{1}{\beta} x\right)} \right|$ + constant, then $\beta - \alpha$ is equal to ........ . **[2023, 30 Jan Shift-II]**

**Imp.** Integration by Partial Fraction

**55** If $\int \frac{\sin x}{\sin^3 x + \cos^3 x} dx = \alpha \log_e |1 + \tan x| + \beta \log_e |1 - \tan x + \tan^2 x| + \gamma \tan^{-1}\left(\frac{2 \tan x - 1}{\sqrt{3}}\right) + C$, when $C$ is constant of integration, then the value of $18(\alpha + \beta + \gamma^2)$ is **[2021, 31 Aug Shift-II]**

**56** If $\int \frac{2e^x + 3e^{-x}}{4e^x + 7e^{-x}} dx = \frac{1}{14}(ux + v \log_e(4e^x + 7e^{-x})) + C$, where $C$ is a constant of integration, then $u + v$ is equal to ......... . **[2021, 27 Aug Shift-II]**

**57** If $\int \frac{dx}{(x^2 + x + 1)^2} = a \tan^{-1}\left(\frac{2x + 1}{\sqrt{3}}\right) + b\left(\frac{2x + 1}{x^2 + x + 1}\right) + C, x > 0$, where, $C$ is the constant of integration, then the value of $9(\sqrt{3}\,a + b)$ is equal to **[2021, 27 Aug Shift-I]**

**58** If $f(x) = \int \frac{5x^8 + 7x^6}{(x^2 + 1 + 2x^7)^2} dx, (x \geq 0), f(0) = 0$ and $f(1) = \frac{1}{K}$, then the value of $K$ is ........... . **[2021, 18 March Shift-I]**

## Topic 2 Definite Integrals

### Level 1

**01** The integral $\int_0^{\pi} \frac{8x\,dx}{4\cos^2 x + \sin^2 x}$ is equal to **[2025, 3 April Shift-II]**

(a) $\frac{3\pi^2}{2}$ (b) $\pi^2$
(c) $2\pi^2$ (d) $4\pi^2$

**02** The value of $\int_{e^2}^{e^4} \frac{1}{x}\left(\frac{e^{((\log_e x)^2 + 1)^{-1}}}{e^{((\log_e x)^2 + 1)^{-1}} + e^{((6 - \log_e x)^2 + 1)^{-1}}}\right) dx$ is **[2025, 23 Jan Shift-I]**

(a) $\log_e 2$ (b) $e^2$
(c) 2 (d) 1

**03** The integral $\int_{1/4}^{3/4} \cos\left(2\cot^{-1}\sqrt{\frac{1 - x}{1 + x}}\right) dx$ is equal to **[2024, 9 April Shift-II]**

(a) 1/4 (b) −1/4 (c) 1/2 (d) −1/2

**04** The value of the integral $\int_{-1}^{2} \log_e(x + \sqrt{x^2 + 1})\,dx$ is **[2024, 9 April Shift-II]**

(a) $\sqrt{2} - \sqrt{5} + \log_e\left(\frac{9 + 4\sqrt{5}}{1 + \sqrt{2}}\right)$

(b) $\sqrt{5} - \sqrt{2} + \log_e\left(\frac{9 + 4\sqrt{5}}{1 + \sqrt{2}}\right)$

(c) $\sqrt{5} - \sqrt{2} + \log_e\left(\frac{7 + 4\sqrt{5}}{1 + \sqrt{2}}\right)$

(d) $\sqrt{2} - \sqrt{5} + \log_e\left(\frac{7 + 4\sqrt{5}}{1 + \sqrt{2}}\right)$

**05** $\lim_{x \to \frac{\pi}{2}} \left(\frac{\int_{x^3}^{(\pi/2)^3} (\sin(2t^{1/3}) + \cos(t^{1/3}))\,dt}{\left(x - \frac{\pi}{2}\right)^2}\right)$ is equal to **[2024, 9 April Shift-II]**

(a) $\frac{3\pi^2}{2}$ (b) $\frac{9\pi^2}{8}$ (c) $\frac{11\pi^2}{10}$ (d) $\frac{5\pi^2}{9}$

**06** The value of $k \in N$ for which the integral $I_n = \int_0^1 (1 - x^k)^n\,dx, n \in N$, satisfies $147 I_{20} = 148 I_{21}$ is **[2024, 8 April Shift-I]**

(a) 14 (b) 7 (c) 10 (d) 8

**07** $\int_0^{\frac{\pi}{4}} \frac{\cos^2 x \sin^2 x}{(\cos^3 x + \sin^3 x)^2} dx$ is equal to **[2024, 8 April Shift-I]**

(a) 1/6 (b) 1/12 (c) 1/9 (d) 1/3

**08** Let $\beta(m, n) = \int_0^1 x^{m-1}(1 - x)^{n-1} dx, m, n > 0$. If $\int_0^1 (1 - x^{10})^{20} dx = a \times \beta(b, c)$, then $100(a + b + c)$ equals **[2024, 5 April Shift-II]**

(a) 2120 (b) 2012 (c) 1120 (d) 1021

**09** If the value of the integral $\int_{-1}^{1} \frac{\cos \alpha x}{1 + 3^x} dx$ is $\frac{2}{\pi}$. Then, a value of $\alpha$ is **[2024, 4 April Shift-II]**

(a) $\frac{\pi}{3}$ (b) $\frac{\pi}{4}$ (c) $\frac{\pi}{6}$ (d) $\frac{\pi}{2}$

**10** The value of $\int_0^1 (2x^3 - 3x^2 - x + 1)^{1/3} dx$ is equal to **[2024, 1 Feb Shift-II]**

(a) −1 (b) 0 (c) 2 (d) 1

**Imp.** Leibniz Rule

**11** $\lim_{x \to \frac{\pi}{2}} \left( \frac{1}{\left(x - \frac{\pi}{2}\right)^2} \int_{x^3}^{\left(\frac{\pi}{2}\right)^3} \cos(t^{1/3})\, dt \right)$ is equal to **[2024, 29 Jan Shift-I]**

(a) $\frac{3\pi}{4}$ (b) $\frac{3\pi^2}{4}$ (c) $\frac{3\pi^2}{8}$ (d) $\frac{3\pi}{8}$

**12** The value of $\frac{e^{-\frac{\pi}{4}} + \int_0^{\frac{\pi}{4}} e^{-x} \tan^{50} x\, dx}{\int_0^{\frac{\pi}{4}} e^{-x} (\tan^{49} x + \tan^{51} x)\, dx}$ is **[2023, 13 April Shift-II]**

(a) 50 (b) 25 (c) 51 (d) 49

**13** Let $f$ be a continuous function satisfying $\int_0^{t^2} (f(x) + x^2) dx = \frac{4}{3} t^3, \forall t > 0$. Then, $f\left(\frac{\pi^2}{4}\right)$ is equal to **[2023, 10 April Shift-II]**

(a) $-\pi\left(1 + \frac{\pi^3}{16}\right)$ (b) $\pi\left(1 - \frac{\pi^3}{16}\right)$

(c) $-\pi^2\left(1 + \frac{\pi^2}{16}\right)$ (d) $\pi^2\left(1 - \frac{\pi^2}{16}\right)$

**14** Let $f(x)$ be a function satisfying $f(x) + f(\pi - x) = \pi^2, \forall x \in R$. Then, $\int_0^{\pi} f(x) \sin x\, dx$ is equal to **[2023, 6 April Shift-II]**

(a) $\frac{\pi^2}{2}$ (b) $2\pi^2$ (c) $\pi^2$ (d) $\frac{\pi^2}{4}$

**15** Let $\alpha \in (0, 1)$ and $\beta = \log_e(1 - \alpha)$. Let $P_n(x) = x + \frac{x^2}{2} + \frac{x^3}{3} + \ldots + \frac{x^n}{n}, x \in (0, 1)$. Then, the integral $\int_0^{\alpha} \frac{t^{50}}{1-t} dt$ is equal to **[2023, 31 Jan Shift-I]**

(a) $P_{50}(\alpha) - \beta$ (b) $-(\beta + P_{50}(\alpha))$
(c) $\beta + P_{50}(\alpha)$ (d) $\beta - P_{50}(\alpha)$

**16** The value of the integral $\int_{1/2}^{2} \frac{\tan^{-1} x}{x} dx$ is equal to **[2023, 29 Jan Shift-II]**

(a) $\frac{1}{2} \log_e 2$ (b) $\frac{\pi}{2} \log_e 2$

(c) $\frac{\pi}{4} \log_e 2$ (d) $\pi \log_e 2$

**17** $\int_{\frac{3\sqrt{2}}{4}}^{\frac{3\sqrt{3}}{4}} \frac{48}{\sqrt{9 - 4x^2}} dx$ is equal to **[2023, 24 Jan Shift-II]**

(a) $\frac{\pi}{2}$ (b) $2\pi$

(c) $\frac{\pi}{3}$ (d) $\frac{\pi}{6}$

**18** The integral $\int_0^{\pi/2} \frac{1}{3 + 2\sin x + \cos x} dx$ is equal to **[2022, 29 July Shift-I]**

(a) $\tan^{-1}(2)$ (b) $\tan^{-1}(2) - \pi/4$

(c) $\frac{1}{2}\tan^{-1}(2) - \frac{\pi}{8}$ (d) $\frac{1}{2}$

**19** $\int_0^2 \left( |2x^2 - 3x| + \left[ x - \frac{1}{2} \right] \right) dx$, where $[t]$ is the greatest integer function is equal to **[2022, 27 July Shift-II]**

(a) 7/6 (b) 19/12
(c) 31/12 (d) 3/2

**20** Let $f(x) = 2 + |x| - |x - 1| + |x + 1| x \in R$. Consider

**Statement I**

$f'\left(-\frac{3}{2}\right) + f'\left(-\frac{1}{2}\right) + f'\left(\frac{1}{2}\right) + f'\left(\frac{3}{2}\right) = 2$

**Statement II** $\int_{-2}^{2} f(x)\, dx = 12$

Then, **[2022, 27 July Shift-II]**

(a) Both Statement I and Statement II are correct
(b) Both Statement I and Statement II are wrong
(c) only Statement I is correct
(d) only Statement II is correct

**21** Let $R \to R$ be function defined as $f(x) = a \sin\left(\frac{\pi[x]}{2}\right) + [2 - x], a \in R$, where $[t]$ is the greatest integer less than or equal to $t$. If $\lim_{x \to -1} f(x)$ exists, then the value of $\int_0^4 f(x)\, dx$ is equal to **[2022, 27 July Shift-I]**

(a) $-1$ (b) $-2$
(c) 1 (d) 2

**22** $\int_0^{20\pi} (|\sin x| + |\cos x|)^2\, dx$ is equal to **[2022, 26 July Shift-I]**

(a) $10(\pi + 4)$ (b) $10(\pi + 2)$
(c) $20(\pi - 2)$ (d) $20(\pi + 2)$

**23** $\int_0^5 \cos\left(\pi\left(x - \left[\frac{x}{2}\right]\right)\right) dx$, where $[t]$ denotes greatest integer less than or equal to $t$, is equal to **[2022, 29 June Shift-I]**

(a) $-3$ (b) $-2$
(c) 2 (d) 0

**24** If $m$ and $n$ respectively are the number of local maximum and local minimum points of the function

$f(x) = \int_0^{x^2} \frac{t^2 - 5t + 4}{2 + e^t} dt$, then the ordered pair $(m, n)$ is equal to **[2022, 27 June Shift-II]**

(a) (3, 2) (b) (2, 3) (c) (2, 2) (d) (3, 4)

**25** The value of the integral $\int_{-2}^{2} \frac{|x^3 + x|}{(e^{x|x|} + 1)} dx$ is equal to **[2022, 27 June Shift-I]**

(a) $5e^2$ (b) $3e^{-2}$ (c) 4 (d) 6

**26** The function $f(x)$, that satisfies the condition

$f(x) = x + \int_0^{\pi/2} \sin x \cdot \cos y\, f(y)\, dy$, is **[2021, 01 Sep Shift-II]**

(a) $x + \frac{2}{3}(\pi - 2)\sin x$ (b) $x + (\pi + 2)\sin x$

(c) $x + \frac{\pi}{2}\sin x$ (d) $x + (\pi - 2)\sin x$

**27** If $[x]$ is the greatest integer $\le x$, then $\pi^2 \int_0^2 \left(\sin \frac{\pi x}{2}\right)(x - [x])^{[x]} dx$ is equal to **[2021, 31 Aug Shift-II]**

(a) $2(\pi - 1)$ (b) $4(\pi - 1)$ (c) $4(\pi + 1)$ (d) $2(\pi + 1)$

**28** The value of $\int_{-1/\sqrt{2}}^{1/\sqrt{2}} \left(\left(\frac{x+1}{x-1}\right)^2 + \left(\frac{x-1}{x+1}\right)^2 - 2\right)^{1/2} dx$ is **[2021, 26 Aug Shift-I]**

(a) $\log_e 4$ (b) $\log_e 16$

(c) $2\log_e 16$ (d) $4\log_e(3 + 2\sqrt{2})$

**29** The value of the integral $\int_{-1}^{1} \log(x + \sqrt{x^2 + 1}) dx$ is **[2021, 25 July Shift-II]**

(a) 2 (b) 0 (c) $-1$ (d) 1

**30** If $[x]$ denotes the greatest integer less than or equal to $x$ then the value of the integral $\int_{-\pi/2}^{\pi/2} ([x] + [-\sin x])\, dx$ is equal to **[2021, 20 July Shift-II]**

(a) $-\pi$ (b) $\pi$ (c) 0 (d) 1

**31** Which of the following statements is correct for the function $g(\alpha)$ for $\alpha \in R$, such that

$g(\alpha) = \int_{\pi/6}^{\pi/3} \frac{\sin^\alpha x}{\cos^\alpha x + \sin^\alpha x} dx$ **[2021, 17 March Shift-I]**

(a) $g(\alpha)$ is a strictly increasing function.

(b) $g(\alpha)$ has an inflection point at $\alpha = -\frac{1}{2}$.

(c) $g(\alpha)$ is a strictly decreasing function.

(d) $g(\alpha)$ is an even function.

**32** Consider the integral $I = \int_0^{10} \frac{[x]e^{[x]}}{e^{x-1}} dx$, where $[x]$ denotes the greatest integer less than or equal to $x$. Then, the value of $I$ is equal to **[2021, 16 March Shift-II]**

(a) $9(e - 1)$ (b) $45(e + 1)$ (c) $45(e - 1)$ (d) $9(e + 1)$

**33** The value of $\int_{-1}^{1} x^2 e^{[x^3]} dx$, where $[t]$ denotes the greatest integer $\le t$, is **[2021, 25 Feb Shift-I]**

(a) $\frac{e-1}{3e}$ (b) $\frac{e+1}{3}$ (c) $\frac{e+1}{3e}$ (d) $\frac{1}{3e}$

**34** Let $f(x)$ be a differentiable function defined on $[0, 2]$, such that $f'(x) = f'(2 - x), \forall x \in (0, 2)$, $f(0) = 1$ and $f(2) = e^2$. Then, the value of $\int_0^2 f(x)\, dx$ is **[2021, 24 Feb Shift-II]**

(a) $1 - e^2$ (b) $1 + e^2$ (c) $2(1 - e^2)$ (d) $2(1 + e^2)$

**35** The value of the integral $\int_1^3 [x^2 - 2x - 2]\, dx$, where $[x]$ denotes the greatest integer less than or equal to $x$, is **[2021, 24 Feb Shift-II]**

(a) $-\sqrt{2} - \sqrt{3} + 1$ (b) $-\sqrt{2} - \sqrt{3} - 1$

(c) $-5$ (d) $-4$

**36** The integral $\int_1^2 e^x \cdot x^x (2 + \log_e x) dx$ equals **[2020, 6 Sep Shift-II]**

(a) $e(4e + 1)$ (b) $4e^2 - 1$

(c) $e(4e - 1)$ (d) $e(2e - 1)$

**37** The integral

$\int_{\pi/6}^{\pi/3} \tan^3 x \cdot \sin^2 3x (2\sec^2 x \cdot \sin^2 3x + 3\tan x \cdot \sin 6x)\, dx$ is equal to **[2020, 4 Sep Shift-II]**

(a) $-\frac{1}{9}$ (b) $\frac{7}{18}$

(c) $-\frac{1}{18}$ (d) $\frac{9}{2}$

**38** $\int_{-\pi}^{\pi} |\pi - |x||\, dx$ is equal to **[2020, 3 Sep Shift-I]**

(a) $\sqrt{2}\pi^2$ (b) $2\pi^2$

(c) $\pi^2$ (d) $\pi^2/2$

**39** Let a function $f: [0, 5] \to R$ be continuous, $f(1) = 3$ and $F$ be defined as:

$F(x) = \int_1^x t^2 g(t) dt$, where $g(t) = \int_1^t f(u) du$.

Then for the function $F$, the point $x = 1$ is **[2020, 9 Jan Shift-II]**

(a) not a critical point (b) a point of inflection

(c) a point of local maxima (d) a point of local minima

**40** $\lim_{x \to 0} \frac{\int_0^x t \sin(10t)\,dt}{x}$ is equal to **[2020, 8 Jan Shift-II]**

(a) 0 (b) $\frac{1}{10}$ (c) $-\frac{1}{10}$ (d) $-\frac{1}{5}$

**41** Let $f$ and $g$ be continuous functions on $[0, a]$ such that $f(x) = f(a-x)$ and $g(x) + g(a-x) = 4$, then $\int_0^a f(x)\,g(x)\,dx$ is equal to **[2019, 12 Jan Shift-I]**

(a) $4\int_0^a f(x)\,dx$ (b) $\int_0^a f(x)\,dx$

(c) $2\int_0^a f(x)\,dx$ (d) $-3\int_0^a f(x)\,dx$

**42** $\int_{\pi/4}^{3\pi/4} \frac{dx}{1+\cos x}$ is equal to **[JEE Main 2017]**

(a) $-2$ (b) 2 (c) 4 (d) $-1$

**43** If $2\int_0^1 \tan^{-1} x\,dx = \int_0^1 \cot^{-1}(1-x+x^2)\,dx$, then $\int_0^1 \tan^{-1}(1-x+x^2)\,dx$ is equal to **[JEE Main 2016]**

(a) $\frac{\pi}{2} + \log 2$ (b) $\log 2$ (c) $\frac{\pi}{2} - \log 4$ (d) $\log 4$

**44** For $x > 0$, let $f(x) = \int_1^x \frac{\log t}{1+t}\,dt$. Then, $f(x) + f\left(\frac{1}{x}\right)$ is equal to **[JEE Main 2015]**

(a) $\frac{1}{4}(\log x)^2$ (b) $\frac{1}{2}(\log x)^2$ (c) $\log x$ (d) $\frac{1}{4}\log x^2$

**45** Let $f : R \to R$ be a function such that $f(2-x) = f(2+x)$ and $f(4-x) = f(4+x), \forall x \in R$ and $\int_0^2 f(x)dx = 5$. Then, the value of $\int_{10}^{50} f(x)\,dx$ is **[JEE Main 2015]**

(a) 80 (b) 100 (c) 125 (d) 200

**46** Let $f : (-1, 1) \to R$ be a continuous function. If $\int_0^{\sin x} f(t)\,dt = \frac{\sqrt{3}}{2}x$, then $f\left(\frac{\sqrt{3}}{2}\right)$ is equal to **[JEE Main 2015]**

(a) $\sqrt{3}/2$ (b) $\sqrt{3}$ (c) $\sqrt{3/2}$ (d) 1/2

**47** The integral $\int_2^4 \frac{\log x^2}{\log x^2 + \log(36 - 12x + x^2)}dx$ is equal to **[JEE Main 2015]**

(a) 2 (b) 4 (c) 1 (d) 6

**48** The integral $\int_0^{\pi} \sqrt{1 + 4\sin^2\frac{x}{2} - 4\sin\frac{x}{2}}\,dx$ is equal to **[JEE Main 2014]**

(a) $\pi - 4$ (b) $\frac{2\pi}{3} - 4 - 4\sqrt{3}$

(c) $4\sqrt{3} - 4$ (d) $4\sqrt{3} - 4 - \frac{\pi}{3}$

**49** **Statement I** The value of the integral $\int_{\pi/6}^{\pi/3} \frac{dx}{1+\sqrt{\tan x}}$ is equal to $\pi/6$.

**Statement II** $\int_a^b f(x)\,dx = \int_a^b f(a+b-x)\,dx$ **[JEE Main 2013]**

(a) Statement I is true Statement II is true; Statement II is a correct explanation of Statement I
(b) Statement I is true, Statement II is true; Statement II is not a correct explanation of Statement I
(c) Statement I is true, Statement II is false
(d) Statement I is false, Statement II is true

**Imp.** Local Maximum & Local Minimum

**50** For $x \in \left(0, \frac{5\pi}{2}\right)$, define $f(x) = \int_0^x \sqrt{t}\sin t\,dt$. Then, $f$ has **[AIEEE 2011]**

(a) local minimum at $\pi$ and $2\pi$.
(b) local minimum at $\pi$ and local maximum at $2\pi$.
(c) local maximum at $\pi$ and local minimum at $2\pi$.
(d) local maximum at $\pi$ and $2\pi$.

**51** Let [ ] denotes the greatest integer function, then the value of $\int_0^{1.5} x[x^2]\,dx$ is **[AIEEE 2011]**

(a) 5/4 (b) 0
(c) 3/2 (d) 3/4

**52** $\int_0^{\pi} [\cot x]\,dx$, where [] denotes the greatest integer function, is equal to **[AIEEE 2009]**

(a) $\frac{\pi}{2}$ (b) 1

(c) $-1$ (d) $-\frac{\pi}{2}$

**53** If $I = \int_0^1 \frac{\sin x}{\sqrt{x}}\,dx$ and $J = \int_0^1 \frac{\cos x}{\sqrt{x}}\,dx$. Then, which one of the following is true? **[AIEEE 2008]**

(a) $I > \frac{2}{3}$ and $J < 2$ (b) $I > \frac{2}{3}$ and $J > 2$

(c) $I < \frac{2}{3}$ and $J < 2$ (d) $I < \frac{2}{3}$ and $J > 2$

**54** The solution for $x$ of the equation $\int_{\sqrt{2}}^{x} \frac{dt}{t\sqrt{t^2-1}} = \frac{\pi}{2}$ is **[AIEEE 2007]**

(a) $-\sqrt{2}$ (b) $\pi$

(c) $\frac{\sqrt{3}}{2}$ (d) $2\sqrt{2}$

**55** The value of $\int_3^6 \frac{\sqrt{x}}{\sqrt{9-x} + \sqrt{x}}\,dx$ is **[AIEEE 2006]**

(a) 3/2 (b) 2
(c) 1 (d) 1/2

**56** The value of $\int_1^a [x]f'(x)\,dx$, $a>1$, where $[x]$ denotes the greatest integer not exceeding $x$, is **[AIEEE 2006]**

(a) $[a]f(a)-\{f(1)+f(2)+\ldots+f([a])\}$
(b) $[a]f([a])-\{f(1)+f(2)+\ldots+f(a)\}$
(c) $af([a])-\{f(1)+f(2)+\ldots+f(a)\}$
(d) $af(a)-\{f(1)+f(2)+\ldots+f([a])\}$

**57** Let $f: R\to R$ be a differentiable function having $f(2)=6$, $f'(2)=\left(\dfrac{1}{48}\right)$. Then, $\lim\limits_{x\to 2}\int_6^{f(x)}\dfrac{4t^3}{x-2}\,dt$ is equal to **[AIEEE 2005]**

(a) 18 (b) 12
(c) 36 (d) 24

**58** If $I_1=\int_0^1 2^{x^2}dx$, $I_2=\int_0^1 2^{x^3}dx$, $I_3=\int_1^2 2^{x^2}dx$ and $I_4=\int_1^2 2^{x^3}dx$, then **[AIEEE 2005]**

(a) $I_3>I_4$ (b) $I_3=I_4$
(c) $I_1>I_2$ (d) $I_2>I_1$

**59** If $f(x)$ is a function satisfying $f'(x)=f(x)$ with $f(0)=1$ and $g(x)$ is a function that satisfies $f(x)+g(x)=x^2$. Then, the value of $\int_0^1 f(x)\,g(x)\,dx$, is **[AIEEE 2003]**

(a) $e-\dfrac{e^2}{2}-\dfrac{5}{2}$ (b) $e+\dfrac{e^2}{2}-\dfrac{3}{2}$

(c) $e-\dfrac{e^2}{2}-\dfrac{3}{2}$ (d) $e+\dfrac{e^2}{2}+\dfrac{5}{2}$

**60** $\int_0^{10\pi}|\sin x|\,dx$ is equal to **[AIEEE 2002]**

(a) 20 (b) 8
(c) 10 (d) 18

**61** $\int_0^{\pi/2}\dfrac{\sqrt{\sin x}}{\sqrt{\sin x}+\sqrt{\cos x}}\,dx$ is equal to **[AIEEE 2002]**

(a) $\pi/4$ (b) $\pi/2$
(c) 0 (d) 1

## Level 2

**Imp.** Properties of Greatest Integer Function

**62** Let the domain of the function $f(x)=\log_2\log_4\log_6(3+4x-x^2)$ be $(a,b)$. If $\int_0^{b-a}[x^2]dx=p-\sqrt{q}-\sqrt{r}$, $p,q,r\in N$, $\gcd(p,q,r)=1$, where $[\cdot]$ is the greatest integer function, then $p+q+r$ is equal to **[2025, 3 April Shift-I]**

(a) 8 (b) 10 (c) 11 (d) 9

**63** Let $f(x)+2f\left(\dfrac{1}{x}\right)=x^2+5$ and $2g(x)-3g\left(\dfrac{1}{2}\right)=x$, $x>0$. If $\alpha=\int_1^2 f(x)\,dx$ and $\beta=\int_1^2 g(x)\,dx$, then the value of $9\alpha+\beta$ is **[2025, 4 April Shift-II]**

(a) 1 (b) 10 (c) 11 (d) 0

**64** The value of $\int_{-1}^{1}\dfrac{(1+\sqrt{|x|-x})e^x+(\sqrt{|x|-x})e^{-x}}{e^x+e^{-x}}\,dx$ is equal to **[2025, 4 April Shift-I]**

(a) $1+\dfrac{2\sqrt{2}}{3}$ (b) $3-\dfrac{2\sqrt{2}}{3}$

(c) $1-\dfrac{2\sqrt{2}}{3}$ (d) $2+\dfrac{2\sqrt{2}}{3}$

**65** The integral $80\int_0^{\pi/4}\left(\dfrac{\sin\theta+\cos\theta}{9+16\sin 2\theta}\right)d\theta$ is equal to **[2025, 29 Jan Shift-I]**

(a) $2\log_e 3$ (b) $3\log_e 4$ (c) $6\log_e 4$ (d) $4\log_e 3$

**66** If $\int_{-\frac{\pi}{2}}^{\frac{\pi}{2}}\dfrac{96x^2\cos^2 x}{(1+e^x)}dx=\pi(\alpha\pi^2+\beta)$, $\alpha,\beta\in Z$, then $(\alpha+\beta)^2$ equals **[2025, 28 Jan Shift-I]**

(a) 144 (b) 196 (c) 100 (d) 64

**67** If $I=\int_0^{\frac{\pi}{2}}\dfrac{\sin^{\frac{3}{2}}x}{\sin^{\frac{3}{2}}x+\cos^{\frac{3}{2}}x}dx$, then $\int_0^{2I}\dfrac{x\sin x\cos x}{\sin^4 x+\cos^4 x}dx$ equals **[2025, 23 Jan Shift-II]**

(a) $\dfrac{\pi^2}{4}$ (b) $\dfrac{\pi^2}{16}$ (c) $\dfrac{\pi^2}{8}$ (d) $\dfrac{\pi^2}{12}$

**68** Let for $f(x)=7\tan^8 x+7\tan^6 x-3\tan^4 x-3\tan^2 x$, $I_1=\int_0^{\frac{\pi}{4}}f(x)\,dx$ and $I_2=\int_0^{\frac{\pi}{4}}xf(x)\,dx$. Then, $7I_1+12I_2$ is equal to **[2025, 22 Jan Shift-I]**

(a) 1 (b) $2\pi$ (c) 2 (d) $\pi$

**69** Let $\int_\alpha^{\log_e 4}\dfrac{dx}{\sqrt{e^x-1}}=\dfrac{\pi}{6}$. Then, $e^\alpha$ and $e^{-\alpha}$ are the roots of the equation **[2024, 8 April Shift-II]**

(a) $x^2-2x-8=0$ (b) $2x^2-5x-2=0$
(c) $2x^2-5x+2=0$ (d) $x^2+2x-8=0$

**70** The value of $\int_{-\pi}^{\pi} \frac{2y(1+\sin y)}{1+\cos^2 y} dy$ is **[2024, 5 April Shift-I]**

(a) $\pi^2$ (b) $2\pi^2$ (c) $\pi/2$ (d) $\pi^2/2$

**71** The integral $\int_0^{\pi/4} \frac{136\sin x}{3\sin x+5\cos x} dx$ is equal to **[2024, 5 April Shift-I]**

(a) $3\pi-50\log_e 2+20\log_e 5$
(b) $3\pi-25\log_e 2+10\log_e 5$
(c) $3\pi-10\log_e(2\sqrt{2})+10\log_e 5$
(d) $3\pi-30\log_e 2+20\log_e 5$

**72** Let $f(x)=\int_0^x (t+\sin(1-e^t))\,dt, x\in R$. Then, $\lim_{x\to 0}\frac{f(x)}{x^3}$ is equal to **[2024, 4 April Shift-II]**

(a) $\frac{1}{6}$ (b) $-\frac{2}{3}$ (c) $\frac{2}{3}$ (d) $-\frac{1}{6}$

**73** Let $f(x)=\begin{cases}-2, & -2\le x\le 0\\ x-2, & 0<x\le 2\end{cases}$ and $h(x)=f(|x|)+|f(x)|$. Then, $\int_{-2}^{2} h(x)\,dx$ is equal to **[2024, 4 April Shift-II]**

(a) 1 (b) 6 (c) 4 (d) 2

**74** If $\int_0^{\pi/3}\cos^4 x\,dx=a\pi+b\sqrt{3}$, where $a$ and $b$ are rational numbers, then $9a+8b$ is equal to **[2024, 1 Feb Shift-II]**

(a) 3 (b) 1 (c) 2 (d) 3/2

**75** Let $f, g:(0,\infty)\to R$ be two functions defined by $f(x)=\int_{-x}^{x}(|t|-t^2)e^{-t^2}dt$ and $g(x)=\int_0^{x^2} t^{1/2}e^{-t}dt$. Then, the value of $9(f(\sqrt{\log_e 9})+g(\sqrt{\log_e 9}))$ is equal to **[2024, 31 Jan Shift-II]**

(a) 8 (b) 10 (c) 9 (d) 6

**76** Let $a$ and $b$ be real constants such that the function $f$ defined by $f(x)=\begin{cases}x^2+3x+a, & x\le 1\\ bx+2, & x>1\end{cases}$ be differentiable on $R$. Then, the value of $\int_{-2}^{2} f(x)\,dx$ equals **[2024, 30 Jan Shift-II]**

(a) 15/6 (b) 17 (c) 21 (d) 19/6

**77** Let $f:R\to R$ be defined as $f(x)=ae^{2x}+be^x+cx$. If $f(0)=-1$, $f'(\log_e 2)=21$ and $\int_0^{\log_e 4}(f(x)-cx)\,dx=\frac{39}{2}$, then the value of $|a+b+c|$ equals **[2024, 30 Jan Shift-II]**

(a) 12 (b) 16 (c) 8 (d) 10

**78** Let $f:\left[-\frac{\pi}{2},\frac{\pi}{2}\right]\to R$ be a differentiable function such that $f(0)=\frac{1}{2}$. If the $\lim_{x\to 0}\frac{x\int_0^x f(t)\,dt}{e^{x^2}-1}=\alpha$, then $8\alpha^2$ is equal to **[2024, 30 Jan Shift-I]**

(a) 16 (b) 2 (c) 1 (d) 4

**79** If $\int_0^1 \frac{1}{\sqrt{3+x}+\sqrt{1+x}}dx=a+b\sqrt{2}+c\sqrt{3}$, where $a, b, c$ are rational numbers, then $2a+3b-4c$ is equal to **[2024, 27 Jan Shift-I]**

(a) 8 (b) 10 (c) 7 (d) 4

**80** $\int_0^{\infty}\frac{6}{e^{3x}+6e^{2x}+11e^x+6}dx$ is equal to **[2023, 13 April Shift-I]**

(a) $\log_e\left(\frac{256}{81}\right)$ (b) $\log_e\left(\frac{64}{27}\right)$
(c) $\log_e\left(\frac{32}{27}\right)$ (d) $\log_e\left(\frac{512}{81}\right)$

**81** Let the function $f:[0,2]\to R$ be defined as

$$f(x)=\begin{cases}e^{\min\{x^2,\, x-[x]\}}, & x\in[0,1)\\ e^{[x-\log_e x]}, & x\in[1,2]\end{cases}$$

Where, $[t]$ denotes the greatest integer less than or equal to $t$. Then, the value of the integral $\int_0^2 xf(x)\,dx$ is **[2023, 11 April Shift-I]**

(a) $2e-\frac{1}{2}$ (b) $2e-1$
(c) $1+\frac{3e}{2}$ (d) $(e-1)\left(e^2+\frac{1}{2}\right)$

**82** The value of the integral $\int_{-\log_e 2}^{\log_e 2} e^x(\log_e(e^x+\sqrt{1+e^{2x}}))\,dx$ is equal to **[2023, 11 April Shift-I]**

(a) $\log_e\left(\frac{(2+\sqrt{5})^2}{\sqrt{1+\sqrt{5}}}\right)+\frac{\sqrt{5}}{2}$

(b) $\log_e\left(\frac{2(2+\sqrt{5})}{\sqrt{1+\sqrt{5}}}\right)-\frac{\sqrt{5}}{2}$

(c) $\log_e\left(\frac{\sqrt{2}(3-\sqrt{5})^2}{\sqrt{1+\sqrt{5}}}\right)+\frac{\sqrt{5}}{2}$

(d) $\log_e\left(\frac{\sqrt{2}(2+\sqrt{5})^2}{\sqrt{1+\sqrt{5}}}\right)-\frac{\sqrt{5}}{2}$

**83** Let $5f(x)+4f\left(\frac{1}{x}\right)=\frac{1}{x}+3, x>0$. Then $18\int_1^2 f(x)\,dx$ is equal to [2023, 6 April Shift-I]

(a) $10\log_e 2-6$ (b) $10\log_e 2+6$
(c) $5\log_e 2-3$ (d) $5\log_e 2+3$

**84** The value of the integral $\int_{-\pi/4}^{\pi/4}\frac{x+\frac{\pi}{4}}{2-\cos 2x}\,dx$ is [2023, 1 Feb Shift-II]

(a) $\frac{\pi^2}{6}$ (b) $\frac{\pi^2}{12\sqrt{3}}$ (c) $\frac{\pi^2}{3\sqrt{3}}$ (d) $\frac{\pi^2}{6\sqrt{3}}$

**85** If $\phi(x)=\frac{1}{\sqrt{x}}\int_{\pi/4}^{x}(4\sqrt{2}\sin t-3\phi'(t))\,dt, x>0$, then $\phi'\left(\frac{\pi}{4}\right)$ is equal to [2023, 31 Jan Shift-II]

(a) $\frac{4}{6+\sqrt{\pi}}$ (b) $\frac{8}{6+\sqrt{\pi}}$
(c) $\frac{8}{\sqrt{\pi}}$ (d) $\frac{4}{6-\sqrt{\pi}}$

**86** If $[t]$ denotes the greatest integer $\le t$, then the value of $\frac{3(e-1)}{e}\int_1^2 x^2 e^{[x]+[x^3]}\,dx$ is [2023, 30 Jan Shift-I]

(a) $e^7-1$ (b) $e^8-1$ (c) $e^9-e$ (d) $e^8-e$

**87** The value of the integral $\int_1^2\left(\frac{t^4+1}{t^6+1}\right)dt$ is [2023, 29 Jan Shift-II]

(a) $\tan^{-1}2-\frac{1}{3}\tan^{-1}8+\frac{\pi}{3}$

(b) $\tan^{-1}2+\frac{1}{3}\tan^{-1}8-\frac{\pi}{3}$

(c) $\tan^{-1}\frac{1}{2}+\frac{1}{3}\tan^{-1}8-\frac{\pi}{3}$

(d) $\tan^{-1}\frac{1}{2}-\frac{1}{3}\tan^{-1}8+\frac{\pi}{3}$

**88** The integral $16\int_1^2\frac{dx}{x^3(x^2+2)^2}$ is equal to [2023, 25 Jan Shift-II]

(a) $\frac{11}{6}-\log_e 4$ (b) $\frac{11}{12}-\log_e 4$
(c) $\frac{11}{12}+\log_e 4$ (d) $\frac{11}{6}+\log_e 4$

**89** The minimum value of the function $f(x)=\int_0^2 e^{|x-t|}\,dt$ is [2023, 25 Jan Shift-I]

(a) $2(e-1)$ (b) $e(e-1)$ (c) 2 (d) $2e-1$

**90** If $f(\alpha)=\int_1^{\alpha}\frac{\log_{10}t}{1+t}\,dt, \alpha>0$, then $f(e^3)+f(e^{-3})$ is equal to [2022, 29 July Shift-I]

(a) 9 (b) $9/2$
(c) $9/\log_e(10)$ (d) $9/2\log_e(10)$

**91** Let $I_n(x)=\int_0^x\frac{1}{(t^2+5)^n}\,dt, n=1,2,3,\ldots.$ Then, [2022, 28 July Shift-II]

(a) $50I_6-9I_5=xI'(5)$ (b) $50I_6-11I_5=xI'(5)$
(c) $50I_6-9I_5=I'(5)$ (d) $50I_6-11I_5=I'(5)$

**92** Let $I=\int_{\pi/4}^{\pi/3}\left(\frac{8\sin x-\sin 2x}{x}\right)dx$. Then, [2022, 27 July Shift-I]

(a) $\frac{\pi}{2}<I<\frac{3\pi}{4}$ (b) $\frac{\pi}{5}<I<\frac{5\pi}{12}$
(c) $\frac{5\pi}{12}<I<\frac{\sqrt{2}\pi}{3}$ (d) $\frac{3\pi}{4}<I<\pi$

**93** Let $[t]$ denote the greatest integer less than or equal to $t$. Then, the value of the integral $\int_{-3}^{101}([\sin(\pi x)]+e^{[\cos(2\pi x)]})\,dx$ is equal to [2022, 25 July Shift-II]

(a) $\frac{52(1-e)}{e}$ (b) $\frac{52}{e}$
(c) $\frac{52(2+e)}{e}$ (d) $\frac{104}{e}$

**94** For any real number $x$, let $[x]$ denote the largest integer less than or equal to $x$. Let $f$ be a real valued function defined on the interval $[-10, 10]$ by

$$f(x)=\begin{cases}x-[x], & \text{if } [x] \text{ is odd}\\ 1+[x]-x, & \text{if } [x] \text{ is even}\end{cases}$$

Then, the value of $\frac{\pi^2}{10}\int_{-10}^{10}f(x)\cos\pi x\,dx$ [2022, 25 July Shift-I]

(a) 4 (b) 2 (c) 1 (d) 0

**95** Let $f: R\to R$ be a function defined by

$$f(x)=\begin{Bmatrix}\max\{t^3-3t\}; t\le x & x\le 2\\ x^2+2x-6; & 2<x<3\\ [x-3]+9; & 3\le x\le 5\\ 2x+1; & x>5\end{Bmatrix},$$

where $[t]$ is the greatest integer less than or equal to $t$. Let $m$ be the number of points where $f$ is not differentiable and $I=\int_{-2}^{2}f(x)\,dx$. Then, the ordered pair $(m, I)$ is equal to [2022, 29 June Shift-I]

(a) $\left(3, \frac{27}{4}\right)$ (b) $\left(3, \frac{23}{4}\right)$
(c) $\left(4, \frac{27}{4}\right)$ (d) $\left(4, \frac{23}{4}\right)$

**96** Let $f: R \to R$ be a differentiable function such that $f\left(\frac{\pi}{4}\right)=\sqrt{2}, f\left(\frac{\pi}{2}\right)=0$ and $f'\left(\frac{\pi}{2}\right)=1$ and let

$g(x)=\int_x^{\pi/4}(f'(t)\sec t+\tan t\sec t\, f(t))\,dt$ for

$x \in\left[\frac{\pi}{4}, \frac{\pi}{2}\right)$. Then, $\lim_{x \to \left(\frac{\pi}{2}\right)^-} g(x)$ is equal to **[2022, 28 June Shift-II]**

(a) 2 (b) 3 (c) 4 (d) $-3$

**97** Let $f: R \to R$ be continuous function satisfying $f(x)+f(x+k)=n$, for all $x \in R$ where $k>0$ and $n$ is a positive integer. If $I_1=\int_0^{4nk} f(x)\,dx$ and

$I_2=\int_{-k}^{3k} f(x)\,dx$, then **[2022, 28 June Shift-II]**

(a) $I_1+2I_2=4nk$ (b) $I_1+2I_2=2nk$
(c) $I_1+nI_2=4n^2k$ (d) $I_1+nI_2=6n^2k$

**98** Let $[t]$ denote the greatest integer less than or equal to $t$. Then, the value of the integral $\int_0^1[-8x^2+6x-1]dx$ is equal to **[2022, 28 June Shift-I]**

(a) $-1$ (b) $-\frac{5}{4}$ (c) $\frac{\sqrt{17}-13}{8}$ (d) $\frac{\sqrt{17}-16}{8}$

**99** Let $f$ be a differentiable function in $\left(0, \frac{\pi}{2}\right)$.

If $\int_{\cos x}^1 t^2 f(t)\,dt=\sin^3 x+\cos x$, then $\frac{1}{\sqrt{3}} f'\left(\frac{1}{\sqrt{3}}\right)$ is equal to **[2022, 27 June Shift-II]**

(a) $6-9\sqrt{2}$ (b) $6-\frac{9}{\sqrt{2}}$ (c) $\frac{9}{2}-6\sqrt{2}$ (d) $\frac{9}{\sqrt{2}}-6$

**100** The value of $\int_0^{\pi}\frac{e^{\cos x}\sin x}{(1+\cos^2 x)(e^{\cos x}+e^{-\cos x})}dx$ is equal to **[2022, 25 June Shift-I]**

(a) $\pi^2/4$ (b) $\pi^2/2$ (c) $\pi/4$ (d) $\pi/2$

**101** The value of the integral $\int_0^1\frac{\sqrt{x}\,dx}{(1+x)(1+3x)(3+x)}$ is **[2021, 27 Aug Shift-II]**

(a) $\frac{\pi}{8}\left(1-\frac{\sqrt{3}}{2}\right)$ (b) $\frac{\pi}{4}\left(1-\frac{\sqrt{3}}{6}\right)$
(c) $\frac{\pi}{8}\left(1-\frac{\sqrt{3}}{6}\right)$ (d) $\frac{\pi}{4}\left(1-\frac{\sqrt{3}}{2}\right)$

**102** $\int_6^{16}\frac{\log_e x^2}{\log_e x^2+\log_e(x^2-44x+484)}dx$ is equal to **[2021, 27 Aug Shift-I]**

(a) 6 (b) 8
(c) 5 (d) 10

**103** If the value of the integral $\int_0^5\frac{x+[x]}{e^{x-[x]}}dx=\alpha e^{-1}+\beta$, where $\alpha, \beta \in R$, $5\alpha+6\beta=0$ and $[x]$ denotes the greatest integer less than or equal to $x$, then the value of $(\alpha+\beta)^2$ is equal to **[2021, 26 Aug Shift-II]**

(a) 100 (b) 25
(c) 16 (d) 36

**104** The value of $\int_{-\pi/2}^{\pi/2}\left(\frac{1+\sin^2 x}{1+\pi^{\sin x}}\right)dx$ is **[2021, 26 Aug Shift-II]**

(a) $\pi/2$ (b) $5\pi/4$
(c) $3\pi/4$ (d) $3\pi/2$

**105** The value of the definite integral $\int_{\pi/24}^{5\pi/24}\frac{dx}{1+\sqrt[3]{\tan 2x}}$ is **[2021, 25 July Shift-I]**

(a) $\pi/3$ (b) $\pi/6$
(c) $\pi/12$ (d) $\pi/18$

**106** Let $g(t)=\int_{-\pi/2}^{\pi/2}\cos\left(\frac{\pi}{4}t+f(x)\right)dx$, where $f(x)=\log_e(x+\sqrt{x^2+1})$, $x \in R$. Then, which one of the following is correct? **[2021, 20 July Shift-II]**

(a) $g(1)=g(0)$ (b) $\sqrt{2}g(1)=g(0)$
(c) $g(1)=\sqrt{2}g(0)$ (d) $g(1)+g(0)=0$

**107** The value of the integral $\int_{-1}^1\log_e(\sqrt{1-x}+\sqrt{1+x})dx$ is equal to **[2021, 20 July Shift-I]**

(a) $\frac{1}{2}\log_e 2+\frac{\pi}{4}-\frac{3}{2}$

(b) $2\log_e 2+\frac{\pi}{4}-1$

(c) $\log_e 2+\frac{\pi}{2}-1$

(d) $2\log_e 2+\frac{\pi}{2}-\frac{1}{2}$

**108** Let $a$ be a positive real number, such that $\int_0^a e^{x-[x]}dx=10e-9$, where $[x]$ is the greatest integer less than or equal to $x$. Then, $a$ is equal to **[2021, 20 July Shift-I]**

(a) $10-\log_e(1+e)$ (b) $10+\log_e 2$
(c) $10+\log_e 3$ (d) $10+\log_e(1+e)$

**109** Let $f: R \to R$ be defined as $f(x) = e^{-x} \sin x$. If $F:[0, 1] \to R$ is a differentiable function, such that $F(x) = \int_0^x f(t)\, dt$, then the value of $\int_0^1 [F'(x) + f(x)] e^x \, dx$ lies in the interval **[2021, 17 March Shift-II]**

(a) $\left[\frac{327}{360}, \frac{329}{360}\right]$ (b) $\left[\frac{330}{360}, \frac{331}{360}\right]$

(c) $\left[\frac{331}{360}, \frac{334}{360}\right]$ (d) $\left[\frac{335}{360}, \frac{336}{360}\right]$

**110** If the integral $\int_0^{10} \frac{[\sin 2\pi x]}{e^{x-[x]}} dx = \alpha e^{-1} + \beta e^{-\frac{1}{2}} + \gamma$, where $\alpha, \beta, \gamma$ are integers and $[x]$ denotes the greatest integer less than or equal to $x$, then the value of $\alpha + \beta + \gamma$ is equal to **[2021, 17 March Shift-II]**

(a) 0 (b) 20 (c) 25 (d) 10

**111** For $x > 0$, if $f(x) = \int_1^x \frac{\log_e t}{(1+t)} dt$, then $f(e) + f\left(\frac{1}{e}\right)$ is equal to **[2021, 26 Feb Shift-II]**

(a) 1 (b) –1 (c) 1/2 (d) 0

**112** The value of $\int_{-\pi/2}^{\pi/2} \frac{\cos^2 x}{1+3^x} dx$ is **[2021, 26 Feb Shift-I]**

(a) $\pi/4$ (b) $4\pi$ (c) $\pi/2$ (d) $2\pi$

**113** If $I_n = \int_{\pi/4}^{\pi/2} \cot^n x \, dx$, then **[2021, 25 Feb Shift-II]**

(a) $\frac{1}{I_2 + I_4}, \frac{1}{I_3 + I_5}, \frac{1}{I_4 + I_6}$ are in AP

(b) $I_2 + I_4, I_3 + I_5, I_4 + I_6$ are in AP

(c) $\frac{1}{I_2 + I_4}, \frac{1}{I_3 + I_5}, \frac{1}{I_4 + I_6}$ are in GP

(d) $I_2 + I_4, (I_3 + I_5)^2, I_4 + I_6$ are in GP

**114** If $I_1 = \int_0^1 (1 - x^{50})^{100} dx$ and $I_2 = \int_0^1 (1 - x^{50})^{101} dx$ such that $I_2 = \alpha I_1$, then $\alpha$ equals to **[2020, 6 Sep Shift-I]**

(a) $\frac{5049}{5050}$ (b) $\frac{5050}{5049}$ (c) $\frac{5050}{5051}$ (d) $\frac{5051}{5050}$

**115** The value of $\int_{-\pi/2}^{\pi/2} \frac{1}{1+e^{\sin x}} dx$ is **[2020, 5 Sep Shift-I]**

(a) $\frac{\pi}{2}$ (b) $\frac{\pi}{4}$ (c) $\pi$ (d) $\frac{3\pi}{2}$

**116** The value of $\int_0^{2\pi} \frac{x \sin^8 x}{\sin^8 x + \cos^8 x} dx$ is equal to **[2020, 9 Jan Shift-I]**

(a) $2\pi$ (b) $4\pi$

(c) $2\pi^2$ (d) $\pi^2$

**117** If $\theta_1$ and $\theta_2$ be respectively the smallest and the largest values of $\theta$ in $(0, 2\pi) - \{\pi\}$, which satisfy the equation, $2\cot^2\theta - \frac{5}{\sin\theta} + 4 = 0$, then $\int_{\theta_1}^{\theta_2} \cos^2 3\theta \, d\theta$ is equal to **[2020, 7 Jan Shift-II]**

(a) $\frac{\pi}{3} + \frac{1}{6}$ (b) $\frac{\pi}{3}$ (c) $\frac{2\pi}{3}$ (d) $\frac{\pi}{9}$

**118** If $f(a + b + 1 - x) = f(x)$, $\forall x$, where $a$ and $b$ are fixed positive real numbers, then $\frac{1}{a+b}\int_a^b x(f(x) + f(x+1))dx$ is equal to **[2020, 7 Jan Shift-I]**

(a) $\int_{a+1}^{b+1} f(x+1)dx$ (b) $\int_{a+1}^{b+1} f(x)dx$

(c) $\int_{a-1}^{b-1} f(x+1)dx$ (d) $\int_{a-1}^{b-1} f(x)dx$

**119** A value of $\alpha$ such that $\int_\alpha^{\alpha+1} \frac{dx}{(x+\alpha)(x+\alpha+1)} = \log_e\left(\frac{9}{8}\right)$ is **[2019, 12 April Shift-II]**

(a) $-2$ (b) $\frac{1}{2}$ (c) $-\frac{1}{2}$ (d) 2

**120** Let $f: R \to R$ be a continuously differentiable function such that $f(2) = 6$ and $f'(2) = \frac{1}{48}$. If $\int_6^{f(x)} 4t^3 \, dt = (x-2)g(x)$, then $\lim_{x \to 2} g(x)$ is equal to **[2019, 12 April Shift-I]**

(a) 18 (b) 24

(c) 12 (d) 36

**121** The value of $\int_0^{2\pi} [\sin 2x(1 + \cos 3x)] \, dx$, where $[t]$ denotes the greatest integer function, is **[2019, 10 April Shift-I]**

(a) $-\pi$ (b) $2\pi$

(c) $-2\pi$ (d) $\pi$

**122** The value of the integral $\int_0^1 x \cot^{-1}(1 - x^2 + x^4)dx$ is **[2019, 9 April Shift-II]**

(a) $\frac{\pi}{4} - \frac{1}{2}\log_e 2$ (b) $\frac{\pi}{2} - \frac{1}{2}\log_e 2$

(c) $\frac{\pi}{4} - \log_e 2$ (d) $\frac{\pi}{2} - \log_e 2$

**123** The value of $\int_0^{\pi/2} \frac{\sin^3 x}{\sin x + \cos x} dx$ is **[2019, 9 April Shift-I]**

(a) $\frac{\pi - 1}{2}$ (b) $\frac{\pi - 2}{8}$

(c) $\frac{\pi - 1}{4}$ (d) $\frac{\pi - 2}{4}$

**124** Let $f(x)=\int_0^x g(t)\,dt$, where $g$ is a non-zero even function. If $f(x+5)=g(x)$, then $\int_0^x f(t)\,dt$ equals *[2019, 8 April Shift-II]*

(a) $5\int_{x+5}^{5} g(t)\,dt$ (b) $\int_{5}^{x+5} g(t)\,dt$

(c) $2\int_{5}^{x+5} g(t)\,dt$ (d) $\int_{x+5}^{5} g(t)\,dt$

**125** If $f(x)=\frac{2-x\cos x}{2+x\cos x}$ and $g(x)=\log_e x$, $(x>0)$, then the value of the integral $\int_{-\pi/4}^{\pi/4} g(f(x))\,dx$ is *[2019, 8 April Shift-I]*

(a) $\log_e 3$ (b) $\log_e e$ (c) $\log_e 2$ (d) $\log_e 1$

**126** The integral $\int_1^e \left\{\left(\frac{x}{e}\right)^{2x}-\left(\frac{e}{x}\right)^{x}\right\}\log_e x\,dx$ is equal to *[2019, 12 Jan Shift-II]*

(a) $\frac{3}{2}-e-\frac{1}{2e^2}$ (b) $-\frac{1}{2}+\frac{1}{e}-\frac{1}{2e^2}$

(c) $\frac{1}{2}-e-\frac{1}{e^2}$ (d) $\frac{3}{2}-\frac{1}{e}-\frac{1}{2e^2}$

**127** The integral $\int_{\pi/6}^{\pi/4}\frac{dx}{\sin 2x(\tan^5 x+\cot^5 x)}$ equals *[2019, 11 Jan Shift-II]*

(a) $\frac{1}{5}\left(\frac{\pi}{4}-\tan^{-1}\left(\frac{1}{3\sqrt{3}}\right)\right)$ (b) $\frac{1}{20}\tan^{-1}\left(\frac{1}{9\sqrt{3}}\right)$

(c) $\frac{1}{10}\left(\frac{\pi}{4}-\tan^{-1}\left(\frac{1}{9\sqrt{3}}\right)\right)$ (d) $\frac{\pi}{40}$

**128** The value of $\int_{-\pi/2}^{\pi/2}\frac{dx}{[x]+[\sin x]+4}$, where $[t]$ denotes the greatest integer less than or equal to $t$, is *[2019, 10 Jan Shift-II]*

(a) $\frac{1}{12}(7\pi-5)$ (b) $\frac{1}{12}(7\pi+5)$

(c) $\frac{3}{10}(4\pi-3)$ (d) $\frac{3}{20}(4\pi-3)$

**129** The value of $\int_0^\pi |\cos x|^3\,dx$ is *[2019, 9 Jan Shift-I]*

(a) $\frac{2}{3}$ (b) $-\frac{4}{3}$ (c) 0 (d) $\frac{4}{3}$

**130** The value of $\int_{-\pi/2}^{\pi/2}\frac{\sin^2 x}{1+2^x}\,dx$ is *[JEE Main 2018]*

(a) $\frac{\pi}{8}$ (b) $\frac{\pi}{2}$ (c) $4\pi$ (d) $\frac{\pi}{4}$

**131** The value of integral $\int_{\pi/4}^{3\pi/4}\frac{x}{1+\sin x}\,dx$ is *[JEE Main 2018]*

(a) $\pi\sqrt{2}$ (b) $2\pi(\sqrt{2}-1)$ (c) $\frac{\pi}{2}(\sqrt{2}+1)$ (d) $\pi(\sqrt{2}-1)$

**132** The integral $\int_{\pi/12}^{\pi/4}\frac{8\cos 2x}{(\tan x+\cot x)^3}\,dx$ equals *[JEE Main 2017]*

(a) $\frac{15}{128}$ (b) $\frac{13}{32}$

(c) $\frac{13}{256}$ (d) $\frac{15}{64}$

**133** The value of $\int_0^1\frac{8\log(1+x)}{1+x^2}\,dx$ is *[AIEEE 2011]*

(a) $\frac{\pi}{8}\log 2$ (b) $\frac{\pi}{2}\log 2$

(c) $\log 2$ (d) $\pi\log 2$

**134** Let $p(x)$ be a function defined on $R$ such that $\lim_{x\to\infty}\frac{f(3x)}{f(x)}=1$, $p'(x)=p'(1-x), \forall\, x\in[0,1]$, $p(0)=1$ and $p(1)=41$. Then, $\int_0^1 p(x)\,dx$ is equal to *[AIEEE 2010]*

(a) $\sqrt{41}$ (b) 21

(c) 41 (d) 42

**135** If $F(x)=f(x)+f\left(\frac{1}{x}\right)$, where $f(x)=\int_1^x\frac{\log t}{1+t}\,dt$. Then, $F(e)$ is equal to *[AIEEE 2007]*

(a) 1/2 (b) 0

(c) 1 (d) 2

**136** $\int_0^\pi x f(\sin x)\,dx$ is equal to *[AIEEE 2006]*

(a) $\pi\int_0^\pi f(\sin x)\,dx$ (b) $\frac{\pi}{2}\int_0^{\pi/2} f(\sin x)\,dx$

(c) $\pi\int_0^{\pi/2} f(\cos x)\,dx$ (d) $\pi\int_0^\pi f(\cos x)\,dx$

**137** $\int_{-\pi}^{\pi}\frac{2x(1+\sin x)}{1+\cos^2 x}\,dx$ is equal to *[AIEEE 2002]*

(a) $\pi^2/4$ (b) $\pi^2$

(c) 0 (d) $\pi/2$

## Level 3

**Imp.** Properties of Definite Integrals

**138** The integral $\int_0^\pi\frac{(x+3)\sin x}{1+3\cos^2 x}\,dx$ is equal to *[2025, 7 April Shift-I]*

(a) $\frac{\pi}{2\sqrt{3}}(\pi+4)$ (b) $\frac{\pi}{\sqrt{3}}(\pi+1)$

(c) $\frac{\pi}{\sqrt{3}}(\pi+2)$ (d) $\frac{\pi}{3\sqrt{3}}(\pi+6)$

**139** Let $f: R \to R$ be a twice differentiable function such that $f(2) = 1$. If $f(x) = xf'(x)\ \forall x \in R$, $\int_0^2 xF'(x)dx = 6$ and $\int_0^2 x^2 F''(x)dx = 40$, then $F'(2) + \int_0^2 F(x)dx$ is equal to **[2025, 28 Jan Shift-II]**

(a) 13 (b) 11
(c) 15 (d) 9

**140** If $I(m, n) = \int_0^1 x^{m-1}(1-x)^{n-1}dx$, $m, n > 0$, then $I(9, 14) + I(10, 13)$ is **[2025, 24 Jan Shift-I]**

(a) $I(9, 1)$ (b) $I(1, 13)$
(c) $I(19, 27)$ (d) $I(9, 13)$

**141** The value of the integral $\int_0^{\pi/4} \frac{xdx}{\sin^4(2x)+\cos^4(2x)}$ equals **[2024, 1 Feb Shift-II]**

(a) $\frac{\sqrt{2}\pi^2}{32}$ (b) $\frac{\sqrt{2}\pi^2}{8}$
(c) $\frac{\sqrt{2}\pi^2}{16}$ (d) $\frac{\sqrt{2}\pi^2}{64}$

**142** Let $a$ be the sum of all coefficients in the expansion of $(1-2x+2x^2)^{2023}(3-4x^2+2x^3)^{2024}$ and $b = \lim_{x \to 0}\left(\frac{\int_0^x \frac{\log(1+t)}{t^{2024}+1}dt}{x^2}\right)$. If the equations $cx^2 + dx + e = 0$ and $2bx^2 + ax + 4 = 0$ have a common root, where $c, d, e \in R$, then $d:c:e$ equals **[2024, 31 Jan Shift-I]**

(a) $4:1:4$ (b) $2:1:4$
(c) $1:2:4$ (d) $1:1:4$

**143** Let $f: R \to R$ be a function defined by $f(x) = \frac{x}{(1+x^4)^{1/4}}$ and $g(x) = f(f(f(f(x))))$. Then, $18\int_0^{\sqrt{2\sqrt{5}}} x^2 g(x)\,dx$ is equal to **[2024, 30 Jan Shift-II]**

(a) 42 (b) 36 (c) 39 (d) 33

**144** If the value of the integral $\int_{-\pi/2}^{\pi/2}\left(\frac{x^2\cos x}{1+\pi^x} + \frac{1+\sin^2 x}{1+e^{\sin x^{2023}}}\right)dx = \frac{\pi}{4}(\pi + a) - 2$, then the value of $a$ is **[2024, 29 Jan Shift-I]**

(a) $-\frac{3}{2}$ (b) $\frac{3}{2}$ (c) 3 (d) 2

**145** For $0 < a < 1$, the value of the integral $\int_0^{\pi} \frac{dx}{1-2a\cos x + a^2}$ is **[2024, 27 Jan Shift-II]**

(a) $\frac{\pi^2}{\pi + a^2}$ (b) $\frac{\pi^2}{\pi - a^2}$
(c) $\frac{\pi}{1-a^2}$ (d) $\frac{\pi}{1+a^2}$

**146** If $\int_0^1 \frac{1}{(5+2x-2x^2)(1+e^{(2-4x)})}dx = \frac{1}{\alpha}\log_e\left(\frac{\alpha+1}{\beta}\right)$, $\alpha, \beta > 0$, then $\alpha^4 - \beta^4$ is equal to **[2023, 15 April Shift-I]**

(a) $-21$ (b) 21
(c) 0 (d) 19

**147** If $f: R \to R$ be a continuous function satisfying $\int_0^{\pi/2} f(\sin 2x)\sin x\,dx + \alpha\int_0^{\pi/4} f(\cos 2x)\cos x\,dx = 0$, then the value of $\alpha$ is **[2023, 11 April Shift-II]**

(a) $\sqrt{2}$ (b) $\sqrt{3}$
(c) $-\sqrt{3}$ (d) $-\sqrt{2}$

**148** Let $\alpha > 0$. If $\int_0^{\alpha} \frac{x}{\sqrt{x+\alpha} - \sqrt{x}}dx = \frac{16+20\sqrt{2}}{15}$, then $\alpha$ is equal to **[2023, 31 Jan Shift-II]**

(a) 4 (b) 2
(c) $\sqrt{2}$ (d) $2\sqrt{2}$

**149** The value of $\int_{\pi/3}^{\pi/2} \frac{(2+3\sin x)}{\sin x(1+\cos x)}dx$ is equal to **[2023, 31 Jan Shift-I]**

(a) $\frac{7}{2} - \sqrt{3} - \log_e\sqrt{3}$ (b) $\frac{10}{3} - \sqrt{3} + \log_e\sqrt{3}$
(c) $\frac{10}{3} - \sqrt{3} - \log_e\sqrt{3}$ (d) $-2 + 3\sqrt{3} + \log_e\sqrt{3}$

**150** Let $f(x) = x + \frac{a}{\pi^2-4}\sin x + \frac{b}{\pi^2-4}\cos x$, $x \in R$ be a function which satisfies $f(x) = x + \int_0^{\pi/2}\sin(x+y)f(y)\,dy$. Then, $(a+b)$ is equal to **[2023, 29 Jan Shift-I]**

(a) $-\pi(\pi+2)$ (b) $-2\pi(\pi-2)$
(c) $-2\pi(\pi+2)$ (d) $-\pi(\pi-2)$

**151** The minimum value of the twice differentiable function $f(x) = \int_0^x e^{x-t}f'(t)dt - (x^2-x+1)e^x$, $x \in R$ is **[2022, 28 July Shift-I]**

(a) $-\frac{2}{\sqrt{e}}$ (b) $-2\sqrt{e}$
(c) $-\sqrt{e}$ (d) $\frac{2}{\sqrt{e}}$

**152** If $\int_0^2 (\sqrt{2x} - \sqrt{2x - x^2})\, dx =$

$$\int_0^1 \left(1 - \sqrt{1 - y^2} - \frac{y^2}{2}\right) dy + \int_1^2 \left(2 - \frac{y^2}{2}\right) dy + I,$$ then $I$

is equal to **[2022, 29 June Shift-II]**

(a) $\int_0^1 (1 + \sqrt{1 - y^2})\, dy$ (b) $\int_0^1 \left(\frac{y^2}{2} - \sqrt{1 - y^2} + 1\right) dy$

(c) $\int_0^1 (1 - \sqrt{1 - y^2})\, dy$ (d) $\int_0^1 \left(\frac{y^2}{2} + \sqrt{1 - y^2} + 1\right) dy$

**153** The integral $\int_0^1 \frac{1}{7^{\left[\frac{1}{x}\right]}}\, dx$, where [.] denotes the greatest integer function is equal to **[2022, 27 June Shift-II]**

(a) $1 + 6\log_e\left(\frac{6}{7}\right)$ (b) $1 - 6\log_e\left(\frac{6}{7}\right)$

(c) $\log_e\left(\frac{7}{6}\right)$ (d) $1 - 7\log_e\left(\frac{6}{7}\right)$

**154** The value of the integral

$\int_{-\pi/2}^{\pi/2} \frac{dx}{(1 + e^x)(\sin^6 x + \cos^6 x)}$ is equal to **[2022, 24 June Shift-II]**

(a) $2\pi$ (b) 0 (c) $\pi$ (d) $\pi/2$

**155** If $\int_0^{100\pi} \frac{\sin^2 x}{e^{\left(\frac{x}{\pi} - \left[\frac{x}{\pi}\right]\right)}}\, dx = \frac{\alpha\pi^3}{1 + 4\pi^2}, \alpha \in R,$

where $[x]$ is the greatest integer less than or equal to $x$, then the value of $\alpha$ is **[2021, 22 July Shift-II]**

(a) $200(1 - e^{-1})$ (b) $100(1 - e)$

(c) $50(e - 1)$ (d) $150(e^{-1} - 1)$

**156** The value of the definite integral

$\int_{-\pi/4}^{\pi/4} \frac{dx}{(1 + e^{x\cos x})(\sin^4 x + \cos^4 x)}$ is equal to **[2021, 27 July Shift-I]**

(a) $-\frac{\pi}{2}$ (b) $\frac{\pi}{2\sqrt{2}}$ (c) $-\frac{\pi}{4}$ (d) $\frac{\pi}{\sqrt{2}}$

**157** The value of the integral

$\int_{-\pi/2}^{\pi/2} \sin^4 x \left(1 + \log\left(\frac{2 + \sin x}{2 - \sin x}\right)\right) dx$ is **[JEE Main 2018]**

(a) $\frac{3}{8}\pi$ (b) 0 (c) $\frac{3}{16}\pi$ (d) $\frac{3}{4}$

**158** If $f(x) = \int_0^x t(\sin x - \sin t)\, dt$, then **[JEE Main 2018]**

(a) $f'''(x) + f'(x) = \cos x - 2x\sin x$
(b) $f'''(x) + f''(x) - f'(x) = \cos x$
(c) $f'''(x) - f''(x) = \cos x - 2x\sin x$
(d) $f'''(x) + f''(x) = \sin x$

**159** For $x \in R, x \neq 0$, if $y(x)$ is differentiable function such that $x\int_1^x y(t)\, dt = (x + 1)\int_1^x t y(t)\, dt$, then $y(x)$ equals (where, $C$ is a constant) **[JEE Main 2016]**

(a) $Cx^3 e^{1/x}$ (b) $\frac{C}{x^2} e^{-1/x}$ (c) $\frac{C}{x} e^{-1/x}$ (d) $\frac{C}{x^3} e^{-1/x}$

## Numerical Value Type Questions

**160** Let $[\cdot]$ denote the greatest integer function. If $\int_0^{e^3} \left[\frac{1}{e^{x-1}}\right] dx = \alpha - \log_e 2$, then $\alpha^3$ is equal to ......... **[2025, 2 April Shift-I]**

**161** Let $f(x)$ be a positive function and

$I_1 = \int_{-\frac{1}{2}}^{1} 2x\, f(2x(1 - 2x))\, dx$ and

$I_2 = \int_{-1}^{2} f(x(1 - x))\, dx$. Then, the value of $\frac{I_2}{I_1}$ is equal to ............ **[2025, 8 April Shift-II]**

**Imp.** Properties of Modulus and Greatest Integer Function

**162** If $24\int_0^{\frac{\pi}{4}} \left(\sin\left|4x - \frac{\pi}{12}\right| + [2\sin x]\right) dx = 2\pi + \alpha$, where [.] denotes the greatest integer function, then $\alpha$ is equal to.......... **[2025, 29 Jan Shift-II]**

**163** Consider the matrices : $A = \begin{bmatrix} 2 & -5 \\ 3 & m \end{bmatrix}, B = \begin{bmatrix} 20 \\ m \end{bmatrix}$, $X = \begin{bmatrix} x \\ y \end{bmatrix}$. Let the set of all $m$, for which the system of equations $AX = B$ has a negative solution (i.e., $x < 0$ and $y < 0$), be the interval $(a, b)$. Then, $8\int_a^b |A|\, dm$ is equal to ............ . **[2024, 9 April Shift-II]**

**164** Let $\lim_{n \to \infty} \left(\frac{n}{\sqrt{n^4 + 1}} - \frac{2n}{(n^2 + 1)\sqrt{n^4 + 1}} + \frac{n}{\sqrt{n^4 + 16}} - \frac{8n}{(n^2 + 4)\sqrt{n^4 + 16}} + \ldots + \frac{n}{\sqrt{n^4 + n^4}} - \frac{2n \cdot n^2}{(n^2 + n^2)\sqrt{n^4 + n^4}}\right)$ be $\frac{\pi}{k}$, using only the principal values of the inverse trigonometric functions. Then, $k^2$ is equal to .......... . **[2024, 9 April Shift-I]**

**165** Let $[t]$ denote the largest integer less than or equal to $t$. If $\int_0^3 \left([x^2]+\left[\frac{x^2}{2}\right]\right)dx = a+b\sqrt{2}-\sqrt{3}-\sqrt{5}$ $+c\sqrt{6}-\sqrt{7}$, where $a,b,c \in Z$ then $a+b+c$ is equal to ......... . **[2024, 6 April Shift-II]**

**166** Let $r_k = \dfrac{\int_0^1 (1-x^7)^k\, dx}{\int_0^1 (1-x^7)^{k+1}\, dx}$, $k \in N$. Then, the value of $\sum_{k=1}^{10} \frac{1}{7(r_k-1)}$ is equal to .......... . **[2024, 6 April Shift-II]**

**167** If $f(t) = \int_0^{\pi} \frac{2x\, dx}{1-\cos^2 t \sin^2 x}$, $0<t<\pi$, then the value of $\int_0^{\pi/2} \frac{\pi^2 dt}{f(t)}$ equals ........ **[2024, 5 April Shift-II]**

**168** If $\int_0^{\pi/4} \frac{\sin^2 x}{1+\sin x \cos x} dx = \frac{1}{a}\log_e\left(\frac{a}{3}\right)+\frac{\pi}{b\sqrt{3}}$, where $a, b \in N$, then $a+b$ is equal to ..... . **[2024, 4 April Shift-I]**

**169** If $\int_{-\pi/2}^{\pi/2} \frac{8\sqrt{2}\cos x\, dx}{(1+e^{\sin x})(1+\sin^4 x)} = \alpha\pi+\beta\log_e(3+2\sqrt{2})$, where $\alpha, \beta$ are integers, then $\alpha^2+\beta^2$ equals ........ . **[2024, 1 Feb Shift-I]**

**170** $\left|\frac{120}{\pi^3}\int_0^{\pi} \frac{x^2 \sin x \cos x}{\sin^4 x+\cos^4 x} dx\right|$ is equal to ......... . **[2024, 31 Jan Shift-II]**

**171** Let $f: R \to R$ be a function defined by $f(x) = \frac{4^x}{4^x+2}$ and $M = \int_{f(a)}^{f(1-a)} x\sin^4(x(1-x))dx$, $N = \int_{f(a)}^{f(1-a)} \sin^4(x(1-x))dx$; $a \neq \frac{1}{2}$. If $\alpha M = \beta N$, $\alpha, \beta \in N$, then the least value of $\alpha^2+\beta^2$ is equal to .......... . **[2024, 31 Jan Shift-I]**

**172** The value of $9\int_0^9 \left[\sqrt{\frac{10x}{x+1}}\right] dx$, where $[t]$ denotes the greatest integer less than or equal to $t$, is ......... . **[2024, 30 Jan Shift-I]**

**173** If $\int_{\pi/6}^{\pi/3} \sqrt{1-\sin 2x}\, dx = \alpha+\beta\sqrt{2}+\gamma\sqrt{3}$, where $\alpha, \beta$ and $\gamma$ are rational numbers, then $3\alpha+4\beta-\gamma$ is equal to .......... . **[2024, 29 Jan Shift-II]**

**174** Let the slope of the line $45x+5y+3=0$ be $27r_1+\frac{9r_2}{2}$ for some $r_1, r_2 \in R$. Then, $\lim_{x\to 3}\left(\int_3^x \frac{8t^2}{\frac{3r_2 x}{2}-r_2x^2-r_1x^3-3x} dt\right)$ is equal to .... . **[2024, 29 Jan Shift-II]**

**175** Let $f(x) = \int_0^x g(t)\log_e\left(\frac{1-t}{1+t}\right)dt$, where $g$ is a continuous odd function. If $\int_{-\pi/2}^{\pi/2}\left(f(x)+\frac{x^2\cos x}{1+e^x}\right)dx = \left(\frac{\pi}{\alpha}\right)^2-\alpha$, then $\alpha$ is equal to ......... . **[2024, 27 Jan Shift-II]**

**176** Let for $x \in R$, $S_0(x) = x$, $S_k(x) = C_k x + k\int_0^x S_{k-1}(t)dt$, where $C_0 = 1$, $C_k = 1-\int_0^1 S_{k-1}(x)dx$, $k = 1, 2, 3, \ldots$ Then $S_2(3)+6C_3$ is equal to ...... . **[2023, 13 April Shift-I]**

**177** If $\int_{-0.15}^{0.15} |100x^2-1|\, dx = \frac{k}{3000}$, then $k$ is equal to ............ . **[2023, 12 April Shift-I]**

**178** For $m, n > 0$, let $\alpha(m,n) = \int_0^2 t^m(1+3t)^n\, dt$. If $11\alpha(10,6)+18\alpha(11,5) = p(14)^6$, then $p$ is equal to ............ . **[2023, 11 April Shift-I]**

**179** Let $[t]$ denote the greatest integer function. If $\int_0^{2.4}[x^2]dx = \alpha+\beta\sqrt{2}+\gamma\sqrt{3}+\delta\sqrt{5}$, then $\alpha+\beta+\gamma+\delta$ is equal to .......... . **[2023, 8 April Shift-II]**

**180** Let $[t]$ denote the greatest integer $\leq t$. Then, $\frac{2}{\pi}\int_{\pi/6}^{5\pi/6}(8[\operatorname{cosec} x]-5[\cot x])dx$ is equal to .......... . **[2023, 8 April Shift-I]**

**181** Let $f(x) = \frac{x}{(1+x^n)^{\frac{1}{n}}}$, $x \in R-\{-1\}$, $n \in N$, $n > 2$. If $f^n(x) = (fofof\ldots \text{upto } n \text{ times})(x)$, then $\lim_{n\to\infty}\int_0^1 x^{n-2}(f^n(x))dx$ is equal to .......... . **[2023, 6 April Shift-II]**

**182** If $\int_0^{\pi} \frac{5^{\cos x}(1+\cos x\cos 3x+\cos^2 x+\cos^3 x\cos 3x)dx}{1+5^{\cos x}} = \frac{k\pi}{16}$, then $k$ is equal to ......... . **[2023, 1 Feb Shift-II]**

**183** If $\int_0^1 (x^{21}+x^{14}+x^7)(2x^{14}+3x^7+6)^{1/7}\,dx$

$=\frac{1}{l}(11)^{m/n}$, where $l, m, n \in N$, $m$ and $n$ are coprime, then $l+m+n$ is equal to ........... . **[2023, 1 Feb Shift-I]**

**184** $\lim_{x\to 0} \frac{48}{x^4}\int_0^x \frac{t^3}{t^6+1}\,dt$ is equal to ......... **[2023, 30 Jan Shift-I]**

**185** If $\int_{1/3}^3 |\log_e x|\,dx = \frac{m}{n}\log_e(n^2/e)$, where $m$ and $n$ are coprime natural numbers, then $m^2+n^2-5$ is equal to ........ **[2023, 25 Jan Shift-II]**

**186** Let $f$ be *a* differentiable function defined on $\left[0, \frac{\pi}{2}\right]$ such that $f(x)>0$ and

$f(x)+\int_0^x f(t)\sqrt{1-(\log_e f(t))^2}\,dt = e, \forall x \in \left[0, \frac{\pi}{2}\right]$.

Then, $\left(6\log_e f\left(\frac{\pi}{6}\right)\right)^2$ is equal to **[2023, 24 Jan Shift-II]**

**187** The value of $12\int_0^3 |x^2-3x+2|\,dx$ is ....... **[2023 24 Jan Shift-I]**

**188** The value of $\frac{8}{\pi}\int_0^{\pi/2}\frac{(\cos x)^{2023}}{(\sin x)^{2023}+(\cos x)^{2023}}\,dx$ is ........ **[2023, 24 Jan Shift-I]**

**189** The value of the integral $\int_0^{\pi/2} 60\,\frac{\sin(6x)}{\sin x}\,dx$ is equal to ............ . **[2022, 28 July Shift-II]**

**190** If $\int_0^{\pi/3}\frac{15x^3}{\sqrt{1+x^2+\sqrt{(1+x^2)^3}}}\,dx = \alpha\sqrt{2}+\beta\sqrt{3}$, where $\alpha, \beta$ are integers, then $\alpha+\beta$ is equal to ............ . **[2022, 28 July Shift-I]**

**191** Let $f(x)=\min\{[x-1],[x-2],.....,[x-10]\}$ where $[t]$ denotes the greatest integer $\le t$.

Then, $\int_0^{10} f(x)\,dx + \int_0^{10}(f(x))^2\,dx + \int_0^{10}|f(x)|\,dx$ is equal to ............ . **[2022, 27 July Shift-II]**

**192** Let $f$ be a differential function satisfying

$f(x)=\frac{2}{\sqrt{3}}\int_0^{\sqrt{3}} f\left(\frac{\lambda^2 x}{3}\right)d\lambda, x>0$ and $f(1)=\sqrt{3}$. If $y=f(x)$ passes through the point $(\alpha, 6)$, then $\alpha$ is equal to ............ . **[2022, 27 July Shift-II]**

**193** If $n(2n+1)\int_0^1 (1-x^n)^{2n}\,dx = 1177\int_0^1 (1-x^n)^{2n+1}\,dx$, then $n \in N$ is equal to ............ . **[2022, 26 July Shift-I]**

**194** Let $f$ be a twice differentiable function on $R$. If $f'(0)=4$ and $f(x)+\int_0^x (x-t)f'(t)\,dt$

$=(e^{2x}+e^{-2x})\cos 2x+\frac{2}{a}x$, then $(2a+1)^5\,a^2$ is equal to ............ . **[2022, 25 July Shift-II]**

**195** Let $a_n=\int_{-1}^{n}\left(1+\frac{x}{2}+\frac{x^2}{3}+\ldots+\frac{x^{n-1}}{n}\right)dx$ for every $n\in N$. Then, the sum of all the elements of the set $\{n\in N : a_n \in (2, 30)\}$ is ......... . **[2022, 25 July Shift-II]**

**196** The integral $\frac{24}{\pi}\int_0^{\sqrt{2}}\frac{(2-x^2)\,dx}{(2+x^2)\sqrt{4+x^4}}$ is equal to .... . **[2022, 26 June Shift-II]**

**197** The value of the integral $\frac{48}{\pi^4}\int_0^{\pi}\left(\frac{3\pi x^2}{2}-x^3\right)\frac{\sin x}{1+\cos^2 x}\,dx$ is equal to ...... . **[2022, 26 June Shift-I]**

**198** Let $f(x)=\max\{|x+1|,|x+2|,...,|x+5|\}$. Then, $\int_{-6}^{0} f(x)\,dx$ is equal to ...... . **[2022, 26 June Shift-I]**

**199** The value of $b>3$ for which

$12\int_3^b \frac{1}{(x^2-1)(x^2-4)}\,dx = \log_e\left(\frac{49}{40}\right)$ is equal to ...... . **[2022, 25 June Shift-II]**

**200** Let $f(\theta)=\sin\theta+\int_{-\pi/2}^{\pi/2}(\sin\theta+t\cos\theta)f(t)\,dt$. Then, the value of $\left|\int_0^{\pi/2} f(\theta)\,d\theta\right|$ is ...... . **[2022, 24 June Shift-I]**

**201** Let $\max_{0\le x\le 2}\left\{\frac{9-x^2}{5-x}\right\}=\alpha$ and $\min_{0\le x\le 2}\left\{\frac{9-x^2}{5-x}\right\}=\beta$.

If $\int_{\beta-8/3}^{2\alpha-1}\max\left\{\frac{9-x^2}{5-x}, x\right\}dx = \alpha_1+\alpha_2\log_e\left(\frac{8}{15}\right)$, then $\alpha_1+\alpha_2$ is equal to ...... . **[2022, 24 June Shift-I]**

**202** Let $[t]$ denote the greatest integer $\le t$. Then, the value of $8\cdot\int_{-1/2}^{1}([2x]+|x|)\,dx$ is ......... . **[2021, 31 Aug Shift-I]**

**203** If $x\phi(x)=\int_5^x (3t^2-2\phi'(t))\,dt,\ x>-2$, and $\phi(0)=4$, then $\phi(2)$ is ......... **[2021, 31 Aug Shift-I]**

**204** If $\int_0^{\pi}(\sin^3 x)e^{-\sin^2 x}\,dx=\alpha-\frac{\beta}{e}\int_0^1\sqrt{t}\,e^t\,dt$, then $(\alpha+\beta)$ is equal to ........... **[2021, 27 July Shift-II]**

**205** Let $f:[3,5]\to R$ be a twice differentiable function on (3, 5) such that

$$f(x)=e^{-x}\int_3^x[3t^2+2t+4f'(t)]\,dt.$$

If $f'(4)=\frac{\alpha e^{\beta}-224}{(e^{\beta}-4)^2}$, then $\alpha+\beta$ is equal to ......... **[2021, 27 July Shift-I]**

**206** Let the domain of the function $f(x)=\log_4[\log_5(\log_3(18x-x^2-77))]$ be $(a,b)$. Then, the value of the integral $\int_a^b\frac{\sin^3 x}{[\sin^3 x+\sin^3(a+b-x)]}\,dx$ is equal to ........... **[2021, 27 July Shift-I]**

**207** Let $P(x)$ be a real polynomial of degree 3 which vanishes at $x=-3$. Let $P(x)$ have local minima at $x=1$, local maxima at $x=-1$ and $\int_{-1}^{1}P(x)\,dx=18$, then the sum of all the coefficients of the polynomial $P(x)$ is equal to ............... **[2021, 18 March Shift-II]**

**208** Let $f(x)$ and $g(x)$ be two functions satisfying $f(x^2)+g(4-x)=4x^3$ and $g(4-x)+g(x)=0$, then the value of $\int_{-4}^{4}f(x^2)\,dx$ is ...... **[2021, 18 March Shift-I]**

**209** Let $I_n=\int_1^e x^{19}(\log|x|)^n\,dx$, where $n\in N$. If $(20)\,I_{10}=\alpha I_9+\beta I_8$, for natural numbers $\alpha$ and $\beta$, then $\alpha-\beta$ is equal to ............... **[2021, 17 March Shift-II]**

**210** If the normal to the curve $y(x)=\int_0^x(2t^2-15t+10)\,dt$ at a point $(a,b)$ is parallel to the line $x+3y=-5$, $a>1$, then the value of $|a+6b|$ is equal to......... **[2021, 16 March Shift-I]**

**211** Let $f:R\to R$ be a continuous function such that $f(x)+f(x+1)=2$, for all $x\in R$. If $I_1=\int_0^8 f(x)\,dx$ and $I_2=\int_{-1}^3 f(x)\,dx$, then the value of $I_1+2I_2$ is equal to......... **[2021, 16 March Shift-I]**

**212** If $I_{m\cdot n}=\int_0^1 x^{m-1}(1-x)^{n-1}\,dx$, for $m,n\ge 1$ and $\int_0^1\frac{x^{m-1}+x^{n-1}}{(1+x)^{m+n}}\,dx=\alpha I_{m\cdot n}$, $\alpha\in R$, then $\alpha$ equals ......... **[2021, 26 Feb Shift-II]**

**213** The value of the integral $\int_0^{\pi}|\sin 2x|\,dx$ is ......... **[2021, 26 Feb Shift-I]**

**214** The value of $\int_{-2}^{2}|3x^2-3x-6|\,dx$ is .......... **[2021, 25 Feb Shift-II]**

**215** If $\int_{-a}^{a}(|x|+|x-2|)\,dx=22$, $(a>2)$ and $[x]$ denotes the greatest integer $\le x$, then $\int_a^{-a}(x+[x])\,dx$ is equal to ......... **[2021, 24 Feb Shift-I]**

**216** Let $\{x\}$ and $[x]$ denote the fractional part of $x$ and the greatest integer $\le x$ respectively of a real number $x$. If $\int_0^n(\{x\})\,dx$, $\int_0^n[x]\,dx$ for $n\in[0,n]$ and $10(n^2-n)$, $(n\in N, n>1)$ are three consecutive terms of a GP, then $n$ is equal to ......... **[2020, 4 Sep Shift-II]**

## Key Index

### Topic 1

| | | | | | | | | | |
|---|---|---|---|---|---|---|---|---|---|
| **1** (d) | **2** (a) | **3** (a) | **4** (d) | **5** (b) | **6** (c) | **7** (*) | **8** (d) | **9** (c) | **10** (a) |
| **11** (c) | **12** (c) | **13** (b) | **14** (b) | **15** (a) | **16** (b) | **17** (a) | **18** (d) | **19** (d) | **20** (c) |
| **21** (d) | **22** (c) | **23** (c) | **24** (d) | **25** (a) | **26** (d) | **27** (b) | **28** (a) | **29** (b) | **30** (a) |
| **31** (b) | **32** (b) | **33** (a) | **34** (d) | **35** (c) | **36** (c) | **37** (d) | **38** (b) | **39** (a) | **40** (a) |
| **41** (a) | **42** (c) | **43** (a) | **44** (d) | **45** (b) | **46** (c) | **47** (379) | **48** (19) | **49** (16) | **50** (7) |
| **51** (1) | **52** (28) | **53** (64) | **54** (1) | **55** (3) | **56** (7) | **57** (15) | **58** (4) | | |

### Topic 2

| | | | | | | | | | |
|---|---|---|---|---|---|---|---|---|---|
| **1** (c) | **2** (d) | **3** (b) | **4** (a) | **5** (b) | **6** (b) | **7** (a) | **8** (a) | **9** (d) | **10** (b) |
| **11** (c) | **12** (a) | **13** (b) | **14** (c) | **15** (b) | **16** (b) | **17** (b) | **18** (b) | **19** (b) | **20** (d) |
| **21** (b) | **22** (d) | **23** (d) | **24** (b) | **25** (d) | **26** (d) | **27** (b) | **28** (b) | **29** (b) | **30** (a) |
| **31** (d) | **32** (c) | **33** (c) | **34** (b) | **35** (b) | **36** (c) | **37** (c) | **38** (c) | **39** (d) | **40** (a) |
| **41** (c) | **42** (b) | **43** (b) | **44** (b) | **45** (b) | **46** (b) | **47** (c) | **48** (d) | **49** (d) | **50** (c) |
| **51** (d) | **52** (d) | **53** (c) | **54** (a) | **55** (a) | **56** (a) | **57** (a) | **58** (c) | **59** (c) | **60** (a) |
| **61** (a) | **62** (b) | **63** (c) | **64** (a) | **65** (d) | **66** (c) | **67** (b) | **68** (a) | **69** (c) | **70** (a) |
| **71** (a) | **72** (d) | **73** (d) | **74** (c) | **75** (a) | **76** (b) | **77** (c) | **78** (b) | **79** (a) | **80** (c) |
| **81** (a) | **82** (d) | **83** (a) | **84** (d) | **85** (b) | **86** (d) | **87** (b) | **88** (a) | **89** (a) | **90** (d) |
| **91** (a) | **92** (c) | **93** (b) | **94** (a) | **95** (c) | **96** (b) | **97** (c) | **98** (c) | **99** (b) | **100** (c) |
| **101** (a) | **102** (c) | **103** (b) | **104** (c) | **105** (c) | **106** (b) | **107** (c) | **108** (b) | **109** (b) | **110** (a) |
| **111** (c) | **112** (a) | **113** (a) | **114** (c) | **115** (a) | **116** (d) | **117** (b) | **118** (c) | **119** (a) | **120** (a) |
| **121** (a) | **122** (a) | **123** (c) | **124** (d) | **125** (d) | **126** (a) | **127** (c) | **128** (d) | **129** (d) | **130** (d) |
| **131** (d) | **132** (a) | **133** (d) | **134** (b) | **135** (a) | **136** (c) | **137** (b) | **138** (d) | **139** (d) | **140** (d) |
| **141** (a) | **142** (d) | **143** (c) | **144** (c) | **145** (c) | **146** (b) | **147** (d) | **148** (b) | **149** (b) | **150** (c) |
| **151** (a) | **152** (c) | **153** (a) | **154** (c) | **155** (a) | **156** (b) | **157** (a) | **158** (a) | **159** (d) | **160** (8) |
| **161** (4) | **162** (12) | **163** (450) | **164** (32) | **165** (23) | **166** (65) | **167** (1) | **168** (8) | **169** (8) | **170** (15) |
| **171** (5) | **172** (155) | **173** (6) | **174** (12) | **175** (2) | **176** (18) | **177** (575) | **178** (32) | **179** (6) | **180** (14) |
| **181** (0) | **182** (13) | **183** (63) | **184** (12) | **185** (20) | **186** (27) | **187** (22) | **188** (2) | **189** (104) | **190** (10) |
| **191** (385) | **192** (12) | **193** (24) | **194** (8) | **195** (5) | **196** (3) | **197** (6) | **198** (21) | **199** (6) | **200** (1) |
| **201** (34) | **202** (5) | **203** (4) | **204** (5) | **205** (16) | **206** (1) | **207** (8) | **208** (512) | **209** (1) | **210** (406) |
| **211** (16) | **212** (1) | **213** (2) | **214** (19) | **215** (3) | **216** (21) | | | | |

# Detailed **Explanations**

## Topic 1

**01.** (d) $f(x)=\int x^3\sqrt{3-x^2}\,dx$

Let $3-x^2=u$

$du=-2x\,dx$

$\frac{-1}{2}du=x\,dx$

Also, $x^2=3-u$

$\therefore f(x)=\int x^2\sqrt{3-x^2}\,x\,dx$

$=\int(3-u)\sqrt{u}\left(-\frac{1}{2}du\right)$

$=\frac{-1}{2}\int(3u^{1/2}-u^{3/2})du$

$=-\frac{1}{2}\left(3\frac{u^{3/2}}{\frac{3}{2}}-\frac{u^{5/2}}{\frac{5}{2}}\right)+C$

$=-\frac{1}{2}\left(2u^{3/2}-\frac{2}{5}u^{5/2}\right)+C$

$=-u^{\frac{3}{2}}+\frac{1}{5}u^{\frac{5}{2}}+C$

$\Rightarrow f(x)=-(3-x^2)^{\frac{3}{2}}+\frac{1}{5}(3-x^2)^{\frac{5}{2}}+C$

$\Rightarrow 5f(\sqrt{2})=-4\Rightarrow f(\sqrt{2})=\frac{-4}{5}.$

$\Rightarrow f(\sqrt{2})=-(3-(\sqrt{2})^2)^{\frac{3}{2}}$

$+\frac{1}{5}(3-(\sqrt{2})^2)^{\frac{5}{2}}+C=-1+\frac{1}{5}+C$

$\Rightarrow -\frac{4}{5}=-1+\frac{1}{5}+C$

$\Rightarrow C=-\frac{4}{5}+1-\frac{1}{5}=0$

$\Rightarrow f(1)=(-3-1)^{\frac{3}{2}}+\frac{1}{5}(3-1)^{\frac{5}{2}}$

$=-2^{3/2}+\frac{1}{5}2^{5/2}$

$\Rightarrow f(1)=-2\sqrt{2}+\frac{1}{5}4\sqrt{2}$

$=-2\sqrt{2}+\frac{4\sqrt{2}}{5}$

$\Rightarrow f(1)=\frac{-10\sqrt{2}+4\sqrt{2}}{5}$

$=\frac{-6\sqrt{2}}{5}$

**02.** (a)

> **key idea** Successive integration by part
>
> $\int \underset{\text{I}}{U}\cdot \underset{\text{II}}{V}=UV_1-U'V_2+U''V_3$
>
> $-U'''V_4$ ......
>
> Where $U'$ = derivative of $U$
>
> $V_n=\int V_{n-1}$

$I=\int x^3\sin x\,dx$

Using successive integration

$=x^3(-\cos x)-3x^2(-\sin x)$

$+6x(\cos x)-6(\sin x)+C$

$=-x^3\cos x+3x^2\sin x+6x\cos x$

$-6\sin x+C$

$\therefore g(x)=-x^3\cos x+3x^2\sin x$

$+6x\cos x-6\sin x$

$g\left(\frac{\pi}{2}\right)=0+3\left(\frac{\pi}{2}\right)^2(1)+0-6(1)$

$=\frac{3\pi^2}{4}-6$

$\because\ g'(x)=x^3\sin x$

$\Rightarrow\ g'\left(\frac{\pi}{2}\right)=\frac{\pi^3}{8}$

$\therefore 8\left(g\left(\frac{\pi}{2}\right)+g'\left(\frac{\pi}{2}\right)\right)=6\pi^2-48+\pi^3$

$\therefore\ \alpha=1,\beta=6$ and $\gamma=-48$

$\therefore\ \alpha+\beta-\gamma=55$

**03.** (a) $I=\int e^x\left(\frac{x\sin^{-1}x}{\sqrt{1-x^2}}+\frac{\sin^{-1}x}{(1-x^2)^{3/2}}+\frac{x}{1-x^2}\right)dx$

Let $f(x)=\frac{x\sin^{-1}x}{\sqrt{1-x^2}}$

Differentiating w.r.t. $x$, we get

$$f'(x)=\frac{\sqrt{1-x^2}\left(x\times\frac{1}{\sqrt{1-x^2}}+\sin^{-1}x\right)-x\sin^{-1}x\,\frac{1}{2\sqrt{1-x^2}}(-2x)}{(1-x^2)}$$

$=\frac{x}{1-x^2}+\frac{\sin^{-1}x}{(1-x^2)^{3/2}}$

$\therefore\ I=\int e^x(f(x)+f'(x))\,dx$

$=e^x\,f(x)+c$

$\therefore\ I=e^x\left(\frac{x\sin^{-1}x}{\sqrt{1-x^2}}\right)+c$

$g(x)=e^x\left(\frac{x\sin^{-1}x}{\sqrt{1-x^2}}\right)$

$g\left(\frac{1}{2}\right)=\frac{e^{1/2}\,\frac{1}{2}\left(\frac{\pi}{6}\right)}{\frac{\sqrt{3}}{2}}=\frac{\pi}{6}\sqrt{\frac{e}{3}}$

> • Concept **Enhancer**
>
> $\int e^x(f(x)+f'(x))dx=e^xf(x)+C$

**04.** (d) $\int\frac{2-\tan x}{3+\tan x}dx=\int\frac{2\cos x-\sin x}{3\cos x+\sin x}dx$

$2\cos x-\sin x=A(3\cos x+\sin x)$

$+B(\cos x-3\sin x)$

On comparing coefficient of $\sin x$ and $\cos x$ from both sides, we get

$3A+B=2$ ...(i)

$A-3B=-1$ ...(ii)

On solving Eqs. (i) and (ii), we get

$A=\frac{1}{2},B=\frac{1}{2}$

So, $\int\frac{2\cos x-\sin x}{3\cos x+\sin x}dx$

$=\frac{x}{2}+\frac{1}{2}\ln|3\cos x+\sin x|+C$

$=\frac{1}{2}(x+\ln|3\cos x+\sin x|)+C$

$=\frac{1}{2}(\alpha x+\ln|\beta\sin x+\gamma\cos x|+C$

$\alpha=1,\beta=1,\gamma=3$

$\therefore\ \alpha+\frac{\gamma}{\beta}=1+\frac{3}{1}=4$

**05.** (b) Given,

$y(x)=\int\frac{\operatorname{cosec}x+\sin x}{\operatorname{cosec}x\sec x+\tan x\sin^2 x}dx$

$$= \int \frac{(1+\sin^2 x)\cos x}{1+\sin^4 x}\,dx$$

On putting, $\sin x = t$

$$\Rightarrow \cos x\,dx = dt = \int \frac{1+t^2}{t^4+1}\,dt$$

$$= \frac{1}{\sqrt{2}} \tan^{-1}\left(\frac{t-\frac{1}{t}}{\sqrt{2}}\right) + C$$

When, $x = \frac{\pi}{2}$, then $t = 1$

$\therefore \quad c = 0$

$\therefore \quad y\left(\frac{\pi}{4}\right) = \frac{1}{\sqrt{2}} \tan^{-1}\left(-\frac{1}{2}\right)$

**06.** *(c)* Given,

$$I(x) = \int e^{\sin^2 x}(\cos x \sin 2x - \sin x)\,dx$$

$$= \int e^{\sin^2 x} \cdot \cos x \cdot \sin 2x dx - \int \sin x e^{\sin^2 x} dx$$

$$= \int \underset{\text{I}}{[(\cos x)} \cdot \underset{\text{II}}{e^{\sin^2 x} \cdot \sin 2x]} dx - \int \sin x \cdot e^{\sin^2 x} dx$$

$$= \cos x \cdot e^{\sin^2 x} - \int (-\sin x) e^{\sin^2 x} dx - \int \sin x e^{\sin^2 x} dx$$

$$= \cos x \cdot e^{\sin^2 x} + \int \sin x\, e^{\sin^2 x} \cdot dx - \int \sin x \cdot e^{\sin^2 x} dx + C$$

$I(x) = e^{\sin^2 x} \cdot \cos x + C$

Given, $I(0) = 1$

$\Rightarrow \quad C = 0$

So, $I(x) = e^{\sin^2 x} \cdot \cos x$

$$\Rightarrow I(\pi/3) = e^{3/4} \cdot \frac{1}{2} = \frac{e^{3/4}}{2}$$

**07.** *(*)* $I = \int \left(\left(\frac{x}{2}\right)^x + \left(\frac{2}{x}\right)^x\right) \log_2(x) \cdot dx$

$$= \int \left(e^{\ln (x/2)^x} + e^{\ln (2/x)^x}\right) \log_2(x) dx$$

$$= \int (e^{x(\ln x - \ln 2)} + e^{x(\ln 2 - \ln x)}) \log_2(x) \cdot dx$$

Further solution in not possible.
Here, it is BONUS.

**08.** *(d)* $\int \left[\frac{x(\cos x - \sin x)}{e^x + 1} + \frac{g(x)[e^x + 1 - xe^x]}{(e^x+1)^2}\right] dx$

$$= \frac{xg(x)}{e^x + 1} + c$$

On differentiating both side w.r.t. $x$,

$$\frac{x(\cos x - \sin x)}{e^x + 1} + \frac{g(x)(e^x + 1 - xe^x)}{(e^x+1)^2}$$

$$= \frac{xg'(x)}{(e^x+1)} + \frac{g(x)[e^x + 1 - xe^x]}{(e^x+1)^2}$$

Now, comparing both parts, we get

$g'(x) = \cos x - \sin x,$

$g'(x) = 0$

$\cos x - \sin x = 0 \Rightarrow x = \pi/4$

$g(x) = \sin x + \cos x$

$g(x) - g'(x) = (\sin x + \cos x) - (\cos x - \sin x) = 2\sin x$

$\sin x$ is an increasing function in $\left(0, \frac{\pi}{2}\right)$, so $g(x) - g'(x)$ is also an increasing function in $\left(0, \frac{\pi}{2}\right)$.

**09.** *(c)* $\int \frac{1}{(x-1)^{3/4}(x+2)^{5/4}}\,dx$

$$= \int \frac{dx}{\left(\frac{x+2}{x-1}\right)^{5/4}(x-1)^2}$$

Let $\frac{x+2}{x-1} = t$

$$\Rightarrow \left(\frac{(x-1)-(x+2)}{(x-1)^2}\right) dx = dt$$

$$\Rightarrow \quad -\frac{3}{(x-1)^2}\,dx = dt$$

$$\Rightarrow -\frac{1}{3}\int \frac{dt}{t^{5/4}} = \frac{4}{3} \cdot \frac{1}{t^{1/4}} + C$$

$$= \frac{4}{3}\left(\frac{x-1}{x+2}\right)^{1/4} + C$$

**10.** *(a)* Let

$$I = \int \frac{(2x-1)\cos\sqrt{(2x-1)^2+5}}{\sqrt{4x^2-4x+6}}\,dx$$

$$= \int \frac{(2x-1)\cos\sqrt{(2x-1)^2+5}}{\sqrt{(2x-1)^2+5}}\,dx$$

Putting $\quad (2x-1)^2 + 5 = z^2$

$\Rightarrow \quad 2(2x-1) \times 2 \cdot dx = 2z\,dz$

$\Rightarrow \quad (2x-1)\,dx = \frac{1}{2} z\,dz$

$$\therefore \quad I = \int \frac{\cos z}{z} \cdot \frac{1}{2} z \cdot dz$$

$$= \frac{1}{2}\int \cos z\,dz$$

$$= \frac{1}{2}\sin z + C$$

$$= \frac{1}{2}\sin\sqrt{(2x-1)^2+5} + C$$

**11.** *(c)*

> **key idea** (i) Divide each term of numerator and denominator by $x^2$.
> (ii) Let $x^2 + \frac{1}{x} = t$

$$\text{Let } I = \int \frac{2x^3 - 1}{x^4 + x}\,dx = \int \frac{2x - 1/x^2}{x^2 + \frac{1}{x}}\,dx$$

On dividing each term of numerator and denominator by $x^2$

Put $x^2 + \frac{1}{x} = t$

$$\Rightarrow \left(2x + \left(-\frac{1}{x^2}\right)\right) dx = dt$$

$$\therefore \quad I = \int \frac{dt}{t} = \log_e |(t)| + C$$

$$= \log_e \left|\left(x^2 + \frac{1}{x}\right)\right| + C$$

$$= \log_e \left|\frac{x^3+1}{x}\right| + C$$

**12.** *(c)* Let $I = \int \frac{\sin \frac{5x}{2}}{\sin \frac{x}{2}}\,dx$

$$= \int \frac{2\sin\frac{5x}{2}\cos\frac{x}{2}}{2\sin\frac{x}{2}\cos\frac{x}{2}} dx$$

$$= \int \frac{\sin 3x + \sin 2x}{\sin x} dx$$

$$= \int \frac{(3\sin x - 4\sin^3 x) + 2\sin x\cos x}{\sin x}\,dx$$

$[\because \sin 3x = 3\sin x - 4\sin^3 x]$

$$= \int (3 - 4\sin^2 x + 2\cos x)\,dx$$

$$= \int [3 - 2(1 - \cos 2x) + 2\cos x]\,dx$$

$[\because 2\sin^2 x = 1 - \cos 2x]$

$$= \int [1 + 2\cos 2x + 2\cos x]\,dx$$

$$= x + 2\sin x + \sin 2x + C$$

**13.** *(b)* We have,

$$I = \int \frac{\sin^2 x \cdot \cos^2 x}{(\sin^5 x + \cos^3 x \cdot \sin^2 x + \sin^3 x \cdot \cos^2 x + \cos^5 x)^2}\,dx$$

$$= \int \frac{\sin^2 x \cos^2 x}{\{\sin^3 x(\sin^2 x + \cos^2 x) + \cos^3 x(\sin^2 x + \cos^2 x)\}^2} dx$$

$$= \int \frac{\sin^2 x \cos^2 x}{(\sin^3 x + \cos^3 x)^2} dx$$

$$= \int \frac{\sin^2 x \cos^2 x}{\cos^6 x(1 + \tan^3 x)^2} dx$$

$$= \int \frac{\tan^2 x \sec^2 x}{(1 + \tan^3 x)^2} dx$$

Put $\tan^3 x = t$

$\Rightarrow 3\tan^2 x \sec^2 x dx = dt$

$\therefore \quad I = \frac{1}{3}\int \frac{dt}{(1+t)^2}$

$\Rightarrow \quad I = \frac{-1}{3(1+t)} + C$

$\Rightarrow \quad I = \frac{-1}{3(1+\tan^3 x)} + C$

**14.** *(b)* Given, $f\left(\frac{x-4}{x+2}\right) = 2x + 1 \quad \ldots$(i)

Put $\frac{x-4}{x+2} = t \Rightarrow x - 4 = 2t + xt$

$\Rightarrow x - xt = 2t + 4 \Rightarrow x = \frac{2(t+2)}{1-t}$

$\therefore$ Eq. (i) becomes

$f(t) = 2\left(\frac{2(t+2)}{1-t}\right) + 1 = \frac{-4(t+2)}{t-1} + 1$

or $f(x) = \frac{-4(x+2)}{x-1} + 1 \quad \ldots$(ii)

On integrating both sides of Eq. (ii), we get

$\int f(x)\,dx = -4\int \frac{(x+2)}{x-1} dx + \int 1 dx$

$= -4\int \frac{x+3-1}{x-1} dx + x$

$= -4\int 1\,dx - 12\int \frac{1}{x-1} dx + x$

$= -3x - 12\log_e |1 - x| + C$

**15.** *(a)* Let $I = \int \frac{2x+5}{\sqrt{7-6x-x^2}} dx$

$\Rightarrow \quad I = \int \frac{2x+6-1}{\sqrt{7-6x-x^2}} dx$

$\Rightarrow \quad I = \int \frac{2x+6}{\sqrt{7-6x-x^2}} dx - \int \frac{1}{\sqrt{(4)^2 - (x+3)^2}} dx$

{Put $t = 7 - 6x - x^2$

$\Rightarrow -dt = (2x+6)\,dx$}

$\Rightarrow \quad I = \int \frac{-dt}{\sqrt{t}} - \int \frac{1}{\sqrt{4^2 - (x+3)^2}} dx$

$\Rightarrow \quad I = -2(7 - 6x - x^2)^{1/2} - \sin^{-1}\left(\frac{x+3}{4}\right) + C$

$\Rightarrow \quad I = -2(7 - 6x - x^2)^{1/2} - \sin^{-1}\left(\frac{x+3}{4}\right) + C$

Now, on comparing $I$ with the given solution, we get

$A = -2, \; B = -1$

**16.** *(b)* We have, $I_n = \int \tan^n x\,dx$

$\therefore \; I_n + I_{n+2} = \int \tan^n x\,dx + \int \tan^{n+2} x\,dx$

$= \int \tan^n x(1 + \tan^2 x)\,dx$

$= \int \tan^n x \sec^2 x\,dx$

$= \frac{\tan^{n+1} x}{n+1} + C$

On putting $n = 4$, we get

$I_4 + I_6 = \frac{\tan^5 x}{5} + C$

$\therefore \quad a = \frac{1}{5}$ and $b = 0$

**17.** *(a)* Let

$I = \int \sqrt{1 + 2\cot x(\operatorname{cosec} x + \cot x)}\,dx$

$= \int \sqrt{1 + 2\cot x \operatorname{cosec} x + 2\cot^2 x}\,dx$

Convert all the quantity in $\cos x$ and $\sin x$ form.

$= \int \sqrt{1 + 2\frac{\cos x}{\sin^2 x} + 2\frac{\cos^2 x}{\sin^2 x}}\,dx$

$= \int \sqrt{\frac{\sin^2 x + 2\cos x + 2\cos^2 x}{\sin^2 x}}\,dx$

$= \int \sqrt{\frac{\sin^2 x + \cos^2 x + 2\cos x + \cos^2 x}{\sin^2 x}}\,dx$

$= \int \sqrt{\frac{(1+\cos x)^2}{\sin^2 x}}\,dx = \int \frac{(1+\cos x)}{\sin x}\,dx$

$= \int \frac{2\cos^2(x/2)}{2\sin(x/2)\cos(x/2)}\,dx$

$= \int \cot(x/2)\,dx = 2\log\left|\sin\frac{x}{2}\right| + C$

**18.** *(d)* $\int \frac{dx}{x^2(x^4+1)^{\frac{3}{4}}} = \int \frac{dx}{x^5\left(1 + \frac{1}{x^4}\right)^{\frac{3}{4}}}$

Put $1 + \frac{1}{x^4} = t^4$

$\Rightarrow \quad -\frac{4}{x^5}dx = 4t^3\,dt \Rightarrow \frac{dx}{x^5} = -t^3\,dt$

$= \int \frac{-t^3\,dt}{t^3} = -\int dt = -t + c$

$= -\left(1 + \frac{1}{x^4}\right)^{1/4} + c$

**19.** *(d)* Let $x^{1/4} = t$

$\Rightarrow \quad x = t^4 \Rightarrow dx = 4t^3\,dt$

$\Rightarrow \quad I = 4\int \frac{t^3\,dt}{t(t+1)}$

$= 4\int\left(\frac{t^2-1}{t+1} + \frac{1}{t+1}\right)dt$

$I = 4\left[\frac{t^2}{2} - t + \log_e(t+1)\right] + C$

Given, $f(x) = 2x^{1/2} - 4x^{1/4} + 4\log_e(x^{1/4} + 1) + C$

$f(0) = -6 \Rightarrow C = -6$

$f(1) = 2 - 4 + 4\log_e 2 - 6$

$= 4(\log_e 2 - 2)$

**20.** *(c)* $I(x) = \int \frac{dx}{(x-11)^{\frac{11}{13}}(x+15)^{\frac{15}{13}}}$

Put $\frac{x-11}{x+15} = t; \quad \frac{x-11+15-15}{x+15} = t$

$\Rightarrow dt = \frac{26}{(x+15)^2}dx$

$I(x) = \frac{1}{26}\int \frac{(x+15)^2}{\left(\frac{x-11}{x+15}\right)^{\frac{11}{13}}(x+15)^2}dt$

$= \frac{1}{26}\int \frac{dt}{(t)^{\frac{11}{13}}} = \frac{1}{26}\frac{t^{\frac{2}{13}}}{\frac{2}{13}} + C$

$I(x) = \frac{1}{4}\left(\frac{x-11}{x+15}\right)^{\frac{2}{13}} + C$

$$\therefore I(37)-I(24)=\frac{1}{4}\left[\left(\frac{26}{52}\right)^{\frac{2}{13}}-\left(\frac{13}{39}\right)^{\frac{2}{13}}\right]$$

$$=\frac{1}{4}\left(\frac{1}{2^{\frac{2}{13}}}-\frac{1}{3^{\frac{2}{13}}}\right)$$

$$=\frac{1}{4}\left[\frac{1}{b^{\frac{1}{13}}}-\frac{1}{c^{\frac{1}{13}}}\right]=\frac{1}{4}\left[\frac{1}{4^{\frac{1}{13}}}-\frac{1}{9^{\frac{1}{13}}}\right]$$

$$\Rightarrow b=4, c=9$$

$$\therefore 3(b+c)=39$$

**21.** (d)

> **key idea** First convert given integration in the terms of cot $x$ and cosec $x$, then put $1-\cot x=t$.

$$I(x)=\int\frac{6}{\sin^2 x(1-\cot x)^2}dx$$

$$=6\int\frac{\operatorname{cosec}^2 x}{(1-\cot x)^2}\cdot dx$$

Let $(1-\cot x)=t$

$$\Rightarrow \operatorname{cosec}^2 x\cdot dx=dt\Rightarrow I(x)=6\int\frac{1}{t^2}\cdot dt$$

$$=6\times\left(-\frac{1}{t}\right)=-\frac{6}{t}+C$$

$$\Rightarrow I(x)=\frac{-6}{(1-\cot x)}+C$$

$$\because I(0)=3$$

$$\Rightarrow 3=\frac{-6}{(1-\infty)}+C\Rightarrow C=3$$

So, $I(x)=\frac{-6}{(1-\cot x)}+3$ then,

Hence, $I\left(\frac{\pi}{12}\right)=\frac{-6}{\left(1-\cot\frac{\pi}{12}\right)}+3$

$$=\frac{-6}{(1-2-\sqrt{3})}+3=\frac{-6-3-3\sqrt{3}}{(\sqrt{3}-1)}$$

$$=\frac{3\sqrt{3}(1+\sqrt{3})}{(\sqrt{3}+1)}$$

$$=3\sqrt{3}$$

**22.** (c)

> **key idea** First convert given integration in the terms of $\tan x$ and $\sec x$, then let $\tan x=t$ and solve it. After this apply identity, $=\int\frac{1}{x^2+a^2}dx=\frac{1}{a}\tan^{-1}\frac{x}{a}+c$

Let $I=\int\frac{dx}{a^2\sin^2 x+b^2\cos^2 x}$

Dividing numerator and denominator by $\cos^2 x$

$$I=\int\frac{\sec^2 x\,dx}{a^2\tan^2 x+b^2}$$

Now, let $\tan x=t$

$$\Rightarrow \sec^2 x\,dx=dt$$

$$I=\int\frac{dt}{a^2t^2+b^2}$$

$$I=\frac{1}{a^2}\int\frac{dt}{t^2+\left(\frac{b}{a}\right)^2}$$

Using identity

$$\int\frac{1}{x^2+P^2}dx=\frac{1}{P}\tan^{-1}\frac{x}{P}+C$$

So, $I=\frac{1}{a^2\left(\frac{b}{a}\right)}\tan^{-1}\left(\frac{t}{b/a}\right)+C$

$$I=\frac{1}{ab}\tan^{-1}\left(\frac{a}{b}\tan x\right)+C$$

Comparing with RHS,

$ab=12$ and $\frac{a}{b}=3$

Solving both, $b^2=4$ and $a^2=36$

Now, maximum value of

$a\sin x+b\cos x$ is $\sqrt{a^2+b^2}$

$$=\sqrt{36+4}=\sqrt{40}$$

**23.** (c) $\int\frac{\sin^{3/2}x+\cos^{3/2}x}{\sqrt{\sin^3 x\cdot\cos^3 x\cdot\sin(x-\theta)}}dx$

$$=A\sqrt{\cos\theta\cdot\tan x-\sin\theta}+B\sqrt{\cos\theta-\sin\theta\cot x}+C$$

Let

$$I=\int\frac{\sin^{3/2}x+\cos^{3/2}x}{\sqrt{\sin^3 x\cdot\cos^3 x\cdot\sin(x-\theta)}}dx$$

$$=\int\frac{dx}{\sqrt{\cos^3 x(\sin x\cos\theta-\cos x\sin\theta)}}+\int\frac{dx}{\sqrt{\sin^3 x(\sin x\cos\theta-\cos x\sin\theta)}}$$

$$=\int\frac{\sec^2 x\,dx}{\sqrt{\tan x\cos\theta-\sin\theta}}+\int\frac{\operatorname{cosec}^2 x\,dx}{\sqrt{\cos\theta-\cot x\sin\theta}}$$

$\tan x=y$

$$\Rightarrow -\cot x=z$$

$$\sec^2 x\,dx=dy\quad \operatorname{cosec}^2 x\,dx=dz$$

$$=\int\frac{dy}{\sqrt{y\cos\theta-\sin\theta}}+\int\frac{dz}{\sqrt{\cos\theta+z\sin\theta}}$$

$$=\frac{2\sqrt{y\cos\theta-\sin\theta}}{\cos\theta}+\frac{2\sqrt{\cos\theta+z\sin\theta}}{\sin\theta}$$

$$=2\sec\theta\sqrt{\cos\theta\tan x-\sin\theta}+2\operatorname{cosec}\theta\sqrt{\cos\theta-\sin\theta\cot x}+C$$

$$\therefore A=2\sec\theta, B=2\operatorname{cosec}\theta$$

$$AB=2\sec\theta\cdot 2\operatorname{cosec}\theta$$

$$=\frac{4}{\sin\theta\cos\theta}=\frac{8}{\sin 2\theta}$$

$$=8\operatorname{cosec}2\theta$$

**24.** (d) Let

$$I=\int\frac{(x^8-x^2)dx}{(x^{12}+3x^6+1)\tan^{-1}\left(x^3+\frac{1}{x^3}\right)}$$

Put $\tan^{-1}\left(x^3+\frac{1}{x^3}\right)=z$

$$\Rightarrow \frac{1}{1+\left(x^3+\frac{1}{x^3}\right)^2}\times\left(3x^2-\frac{3}{x^4}\right)dx=dz$$

$$\Rightarrow \frac{1}{1+\left(\frac{x^6+1}{x^3}\right)^2}\times\left(\frac{3x^6-3}{x^4}\right)dx=dz$$

$$\Rightarrow \frac{(x^3)^2}{(x^3)^2+(x^{12}+1+2\times 1\times x^6)}\times\frac{3(x^6-1)}{x^4}dx=dz$$

$$\Rightarrow \frac{x^6\times 3(x^6-1)}{x^4[x^6+x^{12}+1+2x^6]}dx=dz$$

$$\Rightarrow \frac{(x^8-x^2)}{(x^{12}+3x^6+1)}dx=\frac{1}{3}dz$$

So, $I=\frac{1}{3}\int\frac{1}{z}dz=\frac{1}{3}\ln|z|+C$

$$=\frac{1}{3}\log_e\left(\left|\tan^{-1}\left(x^3+\frac{1}{x^3}\right)\right|\right)+C$$

$$\Rightarrow I=\log_e\left(\left|\tan^{-1}\left(x^3+\frac{1}{x^3}\right)\right|\right)^{1/3}+C$$

**25.** (a) $I(x)=\int\frac{x+1}{x(1+xe^x)^2}dx$

$$=\int\frac{(x+1)e^x\,dx}{xe^x(1+xe^x)^2}$$

Let $1+xe^x=t \Rightarrow (x+1)e^x\,dx=dt$

$$I(x)=\int\frac{dt}{(t-1)t^2} \quad \ldots\text{(i)}$$

$$\therefore \quad I(x)=\int\left(\frac{-1}{t}+\frac{-1}{t^2}+\frac{1}{t-1}\right)dt$$

$$=-\ln(t)+\frac{1}{t}+\ln(t-1)+K$$

$$I(x)=\frac{1}{1+xe^x}+\ln\left(\frac{xe^x}{1+xe^x}\right)+K$$

Now, for $x\to\infty$, $I(x)\to 0$

$$\therefore \lim_{x\to\infty} I(x)=\lim_{x\to\infty}\frac{1}{1+xe^x}$$

$$+\lim_{x\to\infty}\ln\left(\frac{xe^x}{1+xe^x}\right)+K=0$$

$\Rightarrow 0+0+K=0 \Rightarrow K=0$

Now, $I(1)=\dfrac{1}{1+e}+\ln\left(\dfrac{e}{1+e}\right)$

$$=\frac{1}{1+e}+1-\ln(1+e)$$

$$=\frac{2+e}{1+e}-\ln(e+1)$$

**26.** *(d)* We have,

$$I(x)=\int\frac{x^2(x\sec^2x+\tan x)}{(x\tan x+1)^2}dx$$

Integration by parts,

$$x^2\int\frac{x\sec^2x+\tan x}{(x\tan x+1)^2}dx$$

$$-\int\left\{\frac{d}{dx}(x^2)\int\frac{x\sec^2x+\tan x}{(x\tan x+1)^2}\right\}dx$$

$$=x^2\left(\frac{-1}{x\tan x+1}\right)+\int\frac{2x}{x\tan x+1}dx$$

Now, let $I_1=2\displaystyle\int\frac{x}{x\tan x+1}dx$

$$I_1=2\int\frac{x\cos x}{x\sin x+\cos x}dx$$

On putting $x\sin x+\cos x=t$

$\Rightarrow (x\cos x+\sin x-\sin x)\,dx=dt$

$\Rightarrow x\cos x\,dx=dt$

$$\therefore \quad I_1=2\int\frac{dt}{t}=2\log t+c$$

$$=2\log(x\sin x+\cos x)+c$$

$$\therefore I(x)=\frac{-x^2}{x\tan x+1}$$

$$+2\log(x\sin x+\cos x)+c \quad \ldots\text{(i)}$$

When $x=0$,

then $I(0)=0+2\log(1)+c=0$

$\Rightarrow \quad c=0$

$$\therefore I(x)=\frac{-x^2}{x\tan x+1}+2\log(x\sin x+\cos x)$$

$$\Rightarrow I\left(\frac{\pi}{4}\right)=\frac{-\frac{\pi^2}{16}}{\frac{\pi}{4}+1}+2\log\left(\frac{\pi}{4}\times\frac{1}{\sqrt2}+\frac{1}{\sqrt2}\right)$$

$$=\log\left(\frac{(\pi+4)^2}{32}\right)-\frac{\pi^2}{4(\pi+4)}$$

**27.** *(b)* Given,

$$f(x)=\int\frac{2x}{(x^2+1)(x^2+3)}dx$$

Put $x^2=t \Rightarrow 2x\cdot dx=dt$

$$f(t)=\int\frac{dt}{(t+1)(t+3)}$$

$$=\frac{1}{2}\int\frac{1}{t+1}-\frac{1}{t+3}\,dt$$

$$f(t)=\frac{1}{2}(\ln|t+1|-\ln|t+3|)+c$$

$$f(t)=\frac{1}{2}\ln\left|\frac{t+1}{t+3}\right|+c=\frac{1}{2}\ln\left|\frac{x^2+1}{x^2+3}\right|+c$$

$$f(3)=\frac{1}{2}\ln\left|\frac{9+1}{9+3}\right|+c=\frac{1}{2}\ln\frac{5}{6}+c$$

$$\frac{1}{2}(\ln5-\ln6)=\frac{1}{2}(\ln5-\ln6)+c$$

$$\left[\because f(3)=\frac{1}{2}(\log5-\log6)\right]$$

$\Rightarrow \quad c=0$

So, $f(x)$ becomes

$$f(x)=\frac{1}{2}\ln\left|\frac{x^2+1}{x^2+3}\right|$$

$$f(4)=\frac{1}{2}\ln\left(\frac{17}{19}\right)$$

$$=\frac{1}{2}(\log_e17-\log_e19)$$

**28.** *(a)* $I(x)=\displaystyle\int\frac{\sec^2x-2022}{\sin^{2022}x}dx$

$$\Rightarrow I(x)=\int\underset{\text{II}}{\sec^2x}\,\underset{\text{I}}{\sin^{-2022}x}\,dx$$

$$-\int 2022\sin^{-2022}x\,dx$$

Using integration by parts,

$$I(x)=\sin^{-2022}x\int\sec^2x\,dx$$

$$-\int\left(\left(\frac{d}{dx}\sin^{-2022}x\right)\int\sec^2x\,dx\right)dx$$

$$-\int 2022\sin^{-2022}x\,dx$$

$$\Rightarrow I(x)=\tan x\sin^{-2022}x$$

$$+\int 2022\sin^{-2023}x\cdot\cos x\cdot\tan x\,dx$$

$$-\int 2022\sin^{-2022}x\,dx+C$$

$$\Rightarrow I(x)=\tan x\cdot\sin^{-2022}x$$

$$+\int 2022\sin^{-2022}x\,dx$$

$$-\int 2022\sin^{-2022}x\,dx+C$$

Given, $I\left(\dfrac{\pi}{4}\right)=2^{1011}$

$$2^{1011}=\tan\frac{\pi}{4}\cdot\left(\sin\frac{\pi}{4}\right)^{-2022}+C$$

$\Rightarrow \quad 2^{1011}=2^{1011}+C \Rightarrow C=0$

$$I(x)=\tan x\cdot\sin^{-2022}x$$

Option (a) : $3^{1010}I\left(\dfrac{\pi}{3}\right)-I\left(\dfrac{\pi}{6}\right)$

$$=3^{1010}\tan\frac{\pi}{3}\left(\sin\frac{\pi}{3}\right)^{-2022}$$

$$-\tan\frac{\pi}{6}\left(\sin\frac{\pi}{6}\right)^{-2022}$$

$$=3^{1010}\sqrt3\left(\frac{\sqrt3}{2}\right)^{-2022}-\frac{1}{\sqrt3}\left(\frac{1}{2}\right)^{-2022}$$

$$=3^{1011}\cdot\frac{1}{\sqrt3}\left(\frac{3^{-1011}}{2^{-2022}}\right)-\frac{1}{\sqrt3}\cdot\left(\frac{1}{2}\right)^{-2022}$$

$=0$

**29.** *(b)* $\displaystyle\int\frac{(x^2+1)}{(x+1)^2}e^x\,dx=f(x)e^x+C$

$$=\int\frac{(x^2-1+2)}{(x+1)^2}e^x\,dx$$

$$=\int\left(\frac{x^2-1}{(x+1)^2}+\frac{2}{(x+1)^2}\right)e^x\,dx$$

$$=\int\left(\frac{x-1}{x+1}+\frac{2}{(x+1)^2}\right)e^x\,dx$$

Let $f(x)=\dfrac{x-1}{x+1}$,

then $f'(x)=\dfrac{2}{(x+1)^2}$

The above integral is of the form

$\int(f(x)+f'(x))e^x\,dx=e^x f(x)+C$

$$=\int\left(\frac{x-1}{x+1}+\frac{2}{(x+1)^2}\right)e^x\,dx$$

$$=e^x\left(\frac{x-1}{x+1}\right)+C$$

On comparing with answer by taking

$$f(x)=\frac{x-1}{x+1},\ f'(x)=\frac{2}{(x+1)^2},$$

$$f''(x)=\frac{-4}{(x+1)^3},$$

$$f'''(x)=\frac{12}{(x+1)^4}$$

and $f'''(1)=\frac{12}{16}=\frac{3}{4}$

**30.** *(a)* Let $I=\int\frac{1}{x}\sqrt{\frac{1-x}{1+x}}\,dx$

Now, put $x=\cos 2\theta$

$\Rightarrow\quad dx=-2\sin 2\theta\, d\theta$

$$\therefore I=\int\frac{1}{\cos 2\theta}\sqrt{\frac{1-\cos 2\theta}{1+\cos 2\theta}}(-2\sin 2\theta)d\theta$$

$$=-2\int\frac{\tan\theta}{\cos 2\theta}(2\sin\theta\cdot\cos\theta)d\theta$$

$$=-2\int\frac{(1-\cos 2\theta)}{\cos 2\theta}d\theta$$

$$=-2\int\sec 2\theta\, d\theta+2\int d\theta$$

$$=\frac{-2}{2}\ln|\sec 2\theta+\tan 2\theta|+2\theta+c$$

$$=\ln|\sec 2\theta-\tan 2\theta|+2\theta+c$$

$$=\ln\left|\frac{1-\sin 2\theta}{\cos 2\theta}\right|+\cos^{-1}x+c$$

$[2\theta=\cos^{-1}x]$

$$=\ln\left|\frac{1-\sqrt{1-x^2}}{x}\right|+\cos^{-1}x+c$$

$\because\quad g(1)=0$ (given)

$$=\ln\left|\frac{1-\sqrt{1-1}}{1}\right|+\cos^{-1}1+c=0$$

$\Rightarrow 0+0+c=0$

$\Rightarrow\quad c=0$

$$\Rightarrow\ g(x)=\ln\left|\frac{1-\sqrt{1-x^2}}{x}\right|+\cos^{-1}x$$

Now,

$$g\left(\frac{1}{2}\right)=\ln\left|\frac{1-\sqrt{1-\frac{1}{4}}}{\frac{1}{2}}\right|+\cos^{-1}\left(\frac{1}{2}\right)$$

$$g\left(\frac{1}{2}\right)=\ln|2-\sqrt{3}|+\frac{\pi}{3}$$

$$g\left(\frac{1}{2}\right)=\ln\left|\frac{\sqrt{3}-1}{\sqrt{3}+1}\right|+\frac{\pi}{3}$$

**31.** *(b)*

$$I=\int\frac{e^{3\log_e(2x)}+5e^{2\log_e(2x)}}{\left(e^{4\log_e(x)}+5e^{3\log_e(x)}-7e^{2\log_e(x)}\right)}dx$$

$$=\int\frac{e^{\log_e(2x)^3}+5e^{\log_e(2x)^2}}{e^{\log x^4}+5e^{\log_e(x)^3}-7e^{\log_e(x)^2}}dx$$

[Using property $a\log x=\log x^a$]

$$=\int\frac{8x^3+5(2x)^2}{x^4+5(x)^3-7x^2}dx$$

[Using property $a^{\log_a x}=x$]

$$=\int\frac{8x^3+20x^2}{x^4+5x^3-7x^2}dx$$

$$=\int\frac{4x^2(2x+5)}{x^2(x^2+5x-7)}dx$$

$$=\int\frac{4(2x+5)}{x^2+5x-7}dx$$

Let $x^2+5x-7=t$, then

$(2x+5)dx=dt$

$$I=\int\frac{4dt}{t}=4\log_e t+C$$

Put $t=x^2+5x-7$

$$I=4\log_e|x^2+5x-7|+C$$

**32.** *(b)* Given, $\int\frac{\cos x-\sin x}{\sqrt{8-\sin 2x}}dx$

$$=a\sin^{-1}\left(\frac{\sin x+\cos x}{b}\right)+c$$

Put $\sin x+\cos x=t$

Also, $\sin^2 x+\cos^2 x+2\sin x\cos x=t^2$

$\Rightarrow\quad 2\sin x\cos x=t^2-1$

$\Rightarrow\quad \sin 2x=(t^2-1)$

and $(\cos x-\sin x)\,dx=dt$

Now,

$$\int\frac{\cos x-\sin x}{\sqrt{8-\sin 2x}}dx=\int\frac{dt}{\sqrt{8-(t^2-1)}}$$

$$=\int\frac{dt}{\sqrt{9-t^2}}=\sin^{-1}\left(\frac{t}{3}\right)+c$$

$$\Rightarrow\ \sin^{-1}\left(\frac{t}{3}\right)+c$$

$$=a\sin^{-1}\frac{(\sin x+\cos x)}{b}+c\text{(given)}$$

$$\Rightarrow\ 1\sin^{-1}\left(\frac{\sin x+\cos x}{3}\right)+c$$

$$=a\sin^{-1}\left(\frac{\sin x+\cos x}{b}\right)+c$$

$[\because t=\sin x+\cos x]$

On comparing, we get $a=1$ and $b=3$

Hence, $(a,b)=(1,3)$

> **Shortcut Method**
>
> Put $\sin x+\cos x=t\Rightarrow\sin 2x=t^2-1$
>
> $(\cos x-\sin x)dx=dt$
>
> $$\therefore\ I=\int\frac{dt}{\sqrt{9-t^2}}=\sin^{-1}\left(\frac{t}{3}\right)+c$$
>
> $$=\sin^{-1}\left(\frac{\sin x+\cos x}{3}\right)+c$$
>
> $\Rightarrow\ a=1$ and $b=3$

**33.** *(a)* Given integral,

$$I=\int(e^{2x}+2e^x-e^{-x}-1)e^{(e^x+e^{-x})}dx$$

Let $e^x=t\Rightarrow e^x\,dx=dt$

$$=\int\left(t+2-\frac{1}{t^2}-\frac{1}{t}\right)e^{\left(t+\frac{1}{t}\right)}dt$$

$$=\int\left[\left(t-\frac{1}{t}\right)+\left(1-\frac{1}{t^2}\right)+1\right]e^{(t+1/t)}dt$$

$$=\int\left[(1+t)\left(1-\frac{1}{t^2}\right)+1\right]e^{\left(t+\frac{1}{t}\right)}dt$$

$$=(1+t)e^{\left(t+\frac{1}{t}\right)}-\int 1\cdot e^{\left(t+\frac{1}{t}\right)}dt+\int e^{\left(t+\frac{1}{t}\right)}dt$$

$$=(1+t)e^{\left(t+\frac{1}{t}\right)}+c$$

$$=(1+e^x)e^{(e^x+e^{-x})}+c$$

$$=g(x)e^{(e^x+e^{-x})}+c\qquad\text{[given]}$$

$\therefore\ g(x)=1+e^x\Rightarrow g(0)=1+1=2$

**34.** *(d)*

> **key idea** (i) Put $x=\tan^2 t$ in integral form
>
> (ii) $\int(1-\cos 2t)\,dt=t-\frac{\sin 2t}{2}+C$

It is given that,

$$f(x)=\int\frac{\sqrt{x}}{(1+x)^2}dx,\ (x\geq 0)$$

Put $x=\tan^2 t$

$\Rightarrow\quad dx=2\tan t\sec^2 t\,dt$

$$\therefore\ f(t)=\int\frac{2\tan^2 t\sec^2 t}{\sec^4 t}dt$$

$$=\int 2\sin^2 t\,dt=\int(1-\cos 2t)dt$$

$$=t-\frac{\sin 2t}{2}+C$$

So, $f(t)=t-\frac{\tan t}{1+\tan^2 t}+C$

$f(x)=\tan^{-1}\sqrt{x}-\dfrac{\sqrt{x}}{1+x}+C$

$\therefore f(3)-f(1)$

$=\left(\tan^{-1}\sqrt{3}-\dfrac{\sqrt{3}}{1+3}+C\right)-\left(\tan^{-1}(1)-\dfrac{1}{1+1}+C\right)$

$=\dfrac{\pi}{3}-\dfrac{\sqrt{3}}{4}-\dfrac{\pi}{4}+\dfrac{1}{2}=\dfrac{\pi}{12}+\dfrac{1}{2}-\dfrac{\sqrt{3}}{4}$

**35.** (c) The integral $I=\int \sin^{-1}\sqrt{\dfrac{x}{1+x}}dx$

Let $x=\tan^2\theta$

$\Rightarrow dx=2\tan\theta\sec^2\theta d\theta$

So, $I=\int\left(\sin^{-1}\sqrt{\dfrac{\tan^2\theta}{1+\tan^2\theta}}\right)2\tan\theta\sec^2\theta d\theta$

$=2\int\left(\sin^{-1}\left(\dfrac{\tan\theta}{\sec\theta}\right)\right)\tan\theta\sec^2\theta d\theta$

$=2\int(\sin^{-1}(\sin\theta))\tan\theta\sec^2\theta d\theta$

$=2\int\theta\tan\theta\sec^2\theta d\theta$

By integration by parts

$I=2\left[\theta\int\tan\theta\sec^2\theta d\theta-\int\left(\dfrac{d}{d\theta}\theta\int\tan\theta\sec^2\theta d\theta\right)d\theta\right]$

$=2\left[\theta\dfrac{\tan^2\theta}{2}-\int\dfrac{\tan^2\theta}{2}d\theta\right]$

$=\theta\tan^2\theta-\int(\sec^2\theta-1)d\theta$

$=\theta\tan^2\theta-\tan\theta+\theta+C$

$=(x+1)\tan^{-1}\sqrt{x}-\sqrt{x}+C$

$=A(x)\tan^{-1}\sqrt{x}+B(x)+C$ [given]

On comparing, we get

$A(x)=x+1$ and $B(x)=-\sqrt{x}$

So, $(A(x),B(x))=(x+1,-\sqrt{x})$

**36.** (c) Given integral,

$I=\int\dfrac{d\theta}{\cos^2\theta(\tan 2\theta+\sec 2\theta)}$

$=\int\dfrac{\sec^2\theta\cos 2\theta}{(\sin 2\theta+1)}d\theta$

$=\int\dfrac{\sec^2\theta(1-\tan\theta)}{1+\tan\theta}d\theta$

Let $\tan\theta=t$

$\Rightarrow \sec^2\theta d\theta=dt$

Then,

$I=\int\dfrac{1-t}{1+t}dt=\int\left(-1+\dfrac{2}{1+t}\right)dt$

$=-t+2\log_e|1+t|+C$

$=-\tan\theta+2\log_e|1+\tan\theta|+C$

$=\lambda\tan\theta+2\log_e|f(\theta)|+C$ [Given]

On comparing,

$\therefore (\lambda,f(\theta))=(-1,1+\tan\theta)$

**37.** (d) Let $I=\int x^5e^{-x^2}dx$

Put $x^2=t$

$\Rightarrow 2xdx=dt$

So, $I=\dfrac{1}{2}\int t^2e^{-t}dt$

$=\dfrac{1}{2}[(-t^2e^{-t})+\int e^{-t}(2t)dt]$

[Integration by parts]

$=\dfrac{1}{2}[-t^2e^{-t}+2t(-e^{-t})+\int 2e^{-t}dt]$

$=\dfrac{1}{2}[-t^2e^{-t}-2te^{-t}-2e^{-t}]+C$

$=-\dfrac{e^{-x^2}}{2}(x^4+2x^2+2)+C$

$[\because t=x^2]$ ...(i)

$\because$ It is given that

$I=\int x^5e^{-x^2}dx=g(x)\cdot e^{-x^2}+C$

By Eq. (i), comparing both sides, we get

$g(x)=-\dfrac{1}{2}(x^4+2x^2+2)$

So, $g(-1)=-\dfrac{1}{2}(1+2+2)=-\dfrac{5}{2}$

**38.** (b) Let $I=\int\dfrac{3x^{13}+2x^{11}}{(2x^4+3x^2+1)^4}dx$

On dividing numerator and denominator by $x^{16}$.

$=\int\dfrac{\dfrac{3}{x^3}+\dfrac{2}{x^5}}{\left(2+\dfrac{3}{x^2}+\dfrac{1}{x^4}\right)^4}dx$

Now, put $2+\dfrac{3}{x^2}+\dfrac{1}{x^4}=t$

$\Rightarrow \left(\dfrac{-6}{x^3}-\dfrac{4}{x^5}\right)dx=dt$

$\Rightarrow \left(\dfrac{3}{x^3}+\dfrac{2}{x^5}\right)dx=-\dfrac{dt}{2}$

So, $I=\int\dfrac{-dt}{2t^4}=-\dfrac{1}{2}\times\dfrac{t^{-4+1}}{-4+1}+C$

$=\dfrac{1}{6t^3}+C=\dfrac{1}{6\left(2+\dfrac{3}{x^2}+\dfrac{1}{x^4}\right)^3}+C$

$\left[\because t=2+\dfrac{3}{x^2}+\dfrac{1}{x^4}\right]$

$=\dfrac{x^{12}}{6(2x^4+3x^2+1)^3}+C$

**39.** (a) Let $I=\int\cos(\log_e x)dx$

$=x\cos(\log_e x)-\int x(-\sin(\log_e x))\dfrac{1}{x}\cdot dx$

[Using integration by parts]

$=x\cos(\log_e x)+\int\sin(\log_e x)dx$

$=x\cos(\log_e x)+x\sin(\log_e x)-\int x(\cos(\log_e x))\dfrac{1}{x}dx$

[Again, using integration by parts]

$\Rightarrow I=x\cos(\log_e x)+x\sin((\log_e x)-I$

$\Rightarrow I=\dfrac{x}{2}[\cos(\log_e x)+\sin(\log_e x)]+C$

**40.** (a) Given, $f\left(\dfrac{3x-4}{3x+4}\right)=x+2, x\neq-\dfrac{4}{3}$

Let $\dfrac{3x-4}{3x+4}=y\Rightarrow 3x-4=3yx+4y$

$\Rightarrow x=\dfrac{4y+4}{3-3y}$

Now, $f(y)=\dfrac{4y+4}{3-3y}+2=\dfrac{10-2y}{3-3y}$

or $f(x)=\dfrac{2x-10}{3x-3}$

Now, $\int f(x)dx=\int\dfrac{2x-10}{3x-3}dx$

$=\int\dfrac{2x}{3x-3}dx-10\int\dfrac{dx}{3x-3}$

$=\dfrac{2}{3}\int\dfrac{x-1}{x-1}dx+\dfrac{2}{3}\int\dfrac{dx}{x-1}-\dfrac{10}{3}\int\dfrac{dx}{x-1}$

$=\dfrac{2}{3}x+\dfrac{2}{3}\log(x-1)-\dfrac{10}{3}\log(x-1)+C$

$=\dfrac{2x}{3}-\dfrac{8}{3}\log(x-1)+C.$

On comparing to the given solution, we get $A=\dfrac{-8}{3}, B=\dfrac{2}{3}$

**41.** (a) Consider,

$I=\int\dfrac{dx}{\cos^3 x\sqrt{2\sin 2x}}$

$$= \int \frac{dx}{\cos^3 x \sqrt{4 \sin x \cos x}}$$

$$= \int \frac{dx}{2 \cos^4 x \sqrt{\tan x}} = \int \frac{\sec^4 x}{2\sqrt{\tan x}} dx$$

On putting $\tan x = t^2$

$\Rightarrow \sec^2 x \, dx = 2t \, dt$

Also $\sec^2 x = 1 + t^4$

Now, $I = \int \frac{(1+t^4) 2t \, dt}{2t}$

$$= \int (1+t^4) \, dt = t + \frac{t^5}{5} + k$$

$$= \sqrt{\tan x} + \frac{1}{5} \tan^{5/2} x + k$$

On comparing with

$(\tan x)^A + C (\tan x)^B + k$ [Given]

$$A = \frac{1}{2}, B = \frac{5}{2}, C = \frac{1}{5}$$

Now, $A + B + C = 16/5$

**42.** *(c)* Given that $f''(x) = g''(x) + 6x$

$$f'(1) = 4g'(1) - 3 = 9$$

$$f(2) = 3g(2) = 12$$

Here, $f''(x) = g''(x) + 6x$

On integrating both side, we get

$$f'(x) = g'(x) + 3x^2 + c_1 \quad \ldots\text{(i)}$$

Again integrating both sides, we get

$$f(x) = g(x) + x^3 + c_1 x + c_2 \quad \ldots\text{(ii)}$$

Now, use $f'(1) = 4g'(1) - 3 = 9$ in Eq. (i), we get

$\therefore \quad 9 = 3 + 3 + c_1 \Rightarrow c_1 = 3$

Again use $f(2) = 3g(2) = 12$ in Eq. (ii), we get

$$12 = 4 + 8 + 6 + c_2 \Rightarrow c_2 = -6$$

So, $f(x) = g(x) + x^3 + 3x - 6$

Now, $f(x) - g(x) = x^3 + 3x - 6$

Let at $x = x_0$, $f(x_0) = g(x_0)$

So, $x_0^3 + 3x_0 - 6 = 0$

Now, let $h(x) = x^3 + 3x - 6$

In the interval $(1, 3/2)$,

$$h(1) = -2 \text{ (– ve)}$$

$$h\left(\frac{3}{2}\right) = \frac{15}{8} \text{ (+ ve)}$$

So, $h(x)$ has a root between $\left(1, \frac{3}{2}\right)$.

$\Rightarrow f(x_0) \in \left(1, \frac{3}{2}\right)$ such that

$f(x_0) = g(x_0)$

Option (a) is true.

$$h(-2) = f(-2) - g(-2) = -20$$

$$\Rightarrow \quad g(-2) - f(-2) = 20$$

Option (b) is true.

Now, $|f'(x) - g'(x)| < 6$

$\Rightarrow \quad |3x^2 + 3| < 6 \Rightarrow 3x^2 + 3 < 6$

$\Rightarrow \quad x^2 - 1 < 0 \Rightarrow x \in (-1, 1)$

Option (d) is true.

**43.** *(a)* Let

$$I = \int \frac{\left(1 - \frac{1}{\sqrt{3}}\right)(\cos x - \sin x)}{\left(1 + \frac{2}{\sqrt{3}} \sin 2x\right)} dx$$

On multiply and divide by $\sqrt{3}/2$, we get

$$\int \frac{\left(\frac{\sqrt{3}}{2} - \frac{1}{2}\right)(\cos x - \sin x)}{\frac{\sqrt{3}}{2} + \sin 2x} dx$$

$$= \int \frac{\left(\cos \frac{\pi}{6} - \sin \frac{\pi}{6}\right)(\cos x - \sin x)}{\sin \frac{\pi}{3} + \sin 2x} dx$$

$$= \int \frac{\left[\cos \frac{\pi}{6} \cos x + \sin \frac{\pi}{6} \sin x - \cos \frac{\pi}{6} \sin x - \sin \frac{\pi}{6} \cos x\right]}{2 \sin\left(\frac{\pi}{6} + x\right) \cos\left(\frac{\pi}{6} - x\right)} dx$$

$$= \frac{1}{2} \int \frac{\cos\left(\frac{\pi}{6} - x\right) - \sin\left(\frac{\pi}{6} + x\right)}{\sin\left(\frac{\pi}{6} + x\right) \cos\left(\frac{\pi}{6} - x\right)} dx$$

$$= \frac{1}{2} \int \left[\text{cosec}\left(\frac{\pi}{6} + x\right) - \sec\left(\frac{\pi}{6} - x\right)\right] dx$$

$$= \frac{1}{2} \int \left[\text{cosec}\left(\frac{\pi}{6} + x\right) - \text{cosec}\left(\frac{\pi}{3} + x\right)\right] dx$$

$$\left[\because \sec\left(\frac{\pi}{2} - \theta\right) = \text{cosec}\, \theta\right]$$

$$= \frac{1}{2} \left[\log \left| \tan\left(\frac{\pi}{12} + \frac{x}{2}\right) \right| - \log \left| \tan\left(\frac{\pi}{6} + \frac{x}{2}\right) \right| \right] + C$$

$$= \frac{1}{2} \log_e \left| \frac{\tan\left(\frac{\pi}{12} + \frac{x}{2}\right)}{\tan\left(\frac{\pi}{6} + \frac{x}{2}\right)} \right| + C$$

**44.** *(d)* Let

$$I = \int \left[\frac{\sin\theta \cdot \sin 2\theta (\sin^6\theta + \sin^4\theta + \sin^2\theta) \sqrt{2\sin^4\theta + 3\sin^2\theta + 6}}{1 - \cos 2\theta}\right] d\theta$$

$$I = \int \frac{\left(\sin\theta \cdot 2\sin\theta \cos\theta (\sin^6\theta + \sin^4\theta + \sin^2\theta) \sqrt{2\sin^4\theta + 3\sin^2\theta + 6}\right)}{2\sin^2\theta} d\theta$$

$$I = \int [\cos\theta (\sin^6\theta + \sin^4\theta + \sin^2\theta) \sqrt{2\sin^4\theta + 3\sin^2\theta + 6}] \, d\theta$$

Let $\sin\theta = t \Rightarrow \cos\theta \, d\theta = dt$

$$= \int (t^6 + t^4 + t^2) \sqrt{2t^4 + 3t^2 + 6} \, dt$$

$$= \int (t^5 + t^3 + t) \sqrt{2t^6 + 3t^4 + 6t^2} \, dt$$

Let $2t^6 + 3t^4 + 6t^2 = z$

$\therefore \; dz = (12t^5 + 12t^3 + 12t) \, dt$

$\therefore \; dz = 12(t^5 + t^3 + t) \, dt$

Now, $I = \frac{1}{12} \int \sqrt{z} \, dz$

$$= \frac{1}{12} \times \frac{z^{3/2}}{3/2} + C = \frac{1}{18} z^{3/2} + C$$

$$= \frac{1}{18} [2t^6 + 3t^4 + 6t^2]^{3/2} + C$$

$$= \frac{1}{18} [2\sin^6\theta + 3\sin^4\theta + 6\sin^2\theta]^{3/2} + c$$

$$= \frac{1}{18} [(1 - \cos^2\theta)\{2(1 - \cos^2\theta)^2 + 3 - 3\cos^2\theta + 6\}]^{3/2} + C$$

$$= \frac{1}{18} [(1 - \cos^2\theta)(2\cos^4\theta - 7\cos^2\theta + 11)]^{3/2} + C$$

$$= \frac{1}{18} [-2\cos^6\theta + 9\cos^4\theta - 18\cos^2\theta + 11]^{3/2} + C$$

$$= \frac{1}{18} [11 - 18\cos^2\theta + 9\cos^4\theta - 2\cos^6\theta]^{3/2} + C$$

**45.** *(b)* Given,

$$\int e^{\sec x} [(\sec x \tan x) f(x) + (\sec x \tan x + \sec^2 x)] dx$$

$$= e^{\sec x} \cdot f(x) + C$$

On differentiating both sides w.r.t. $x$, we get

$$e^{\sec x} [(\sec x \tan x) f(x) + (\sec x \tan x + \sec^2 x)]$$

$$= e^{\sec x} f'(x) + e^{\sec x}(\sec x \tan x) f(x)$$

$$\Rightarrow e^{\sec x}(\sec x \tan x + \sec^2 x)$$

$$= e^{\sec x} f'(x)$$

On comparing

$$\Rightarrow \quad f'(x) = \sec x \tan x + \sec^2 x$$

So, $f(x) = \int f'(x)dx$

$$= \int (\sec x \tan x + \sec^2 x)dx$$

$$= \sec x + \tan x + C$$

Possible value of $f(x)$ from options, is

$$f(x) = \sec x + \tan x + \frac{1}{2}$$

**46.** *(c)* Let

$$I = \int \frac{(\sin^n\theta - \sin\theta)^{1/n} \cos\theta}{\sin^{n+1}\theta} d\theta$$

Put $\sin\theta = t \Rightarrow \cos\theta\, d\theta = dt$

$$\therefore \quad I = \int \frac{(t^n - t)^{1/n}}{t^{n+1}} dt$$

$$= \int \frac{\left[t^n\left(1 - \frac{t}{t^n}\right)\right]^{1/n}}{t^{n+1}} dt$$

$$= \int \frac{t(1 - 1/t^{n-1})^{1/n}}{t^{n+1}} dt$$

$$= \int \frac{(1 - 1/t^{n-1})^{1/n}}{t^n} dt$$

Put $1 - \frac{1}{t^{n-1}} = u$ or $1 - t^{-(n-1)} = u$

$$\Rightarrow \frac{(n-1)}{t^n} dt = du \Rightarrow \frac{dt}{t^n} = \frac{du}{n-1}$$

$$\Rightarrow I = \int \frac{u^{1/n} du}{n-1} = \frac{u^{1/n+1}}{(n-1)\left(\frac{1}{n} + 1\right)} + C$$

$$= \frac{n\left(1 - \frac{1}{t^{n-1}}\right)^{(n+1)/n}}{(n-1)(n+1)} + C$$

$$= \frac{n\left(1 - \frac{1}{\sin^{n-1}\theta}\right)^{(n+1)/n}}{n^2 - 1} + C$$

$$\left[\because u = 1 - \frac{1}{t^{n-1}} \text{ and } t = \sin\theta\right]$$

**47.** *(379)* $I = \int \frac{(\sqrt{1+x^2} + x)^{10}}{(\sqrt{1+x^2} - x)^9} dx$

$$\sqrt{1+x^2} - x = \frac{\sqrt{1+x^2} - x}{\sqrt{1+x^2} + x} \times \sqrt{1+x^2} + x$$

$$= \frac{1}{\sqrt{1+x^2} + x}$$

$$\int (\sqrt{1+x^2} + x)^{19} dx$$

$$\Rightarrow \sqrt{1+x^2} + x = t \Rightarrow \sqrt{1+x^2} - x = \frac{1}{t}$$

$$\Rightarrow \left(\frac{1}{2\sqrt{1+x^2}} 2x + 1\right) dx = dt$$

$$\therefore 2\sqrt{1+x^2} = t + \frac{1}{t}$$

$$\frac{(x + \sqrt{1+x^2})}{\sqrt{1+x^2}} dx = dt$$

$$dx = \sqrt{1+x^2}\frac{dt}{t}$$

$$\therefore I = \frac{1}{2}\int t^{19} \cdot \left(t + \frac{1}{t}\right)\frac{dt}{t}$$

$$= \frac{1}{2}\int (t^{19} + t^{17})\, dt$$

$$= \frac{1}{2}\left(\frac{t^{20}}{20} + \frac{t^{18}}{18}\right) + C$$

$$= \frac{1}{2}\left(\frac{9t^{20} + 10t^{18}}{180}\right) + C$$

$$= \frac{t^{19}}{360}\left(9t + \frac{10}{t}\right) + C$$

$$= \frac{t^{19}}{360}\left(9t + \frac{9}{t} + \frac{1}{t}\right) + C$$

$$= \frac{t^{19}}{360}\left(9\left(t + \frac{1}{t}\right) + \frac{1}{t}\right) + C$$

$$= \frac{t^{19}}{360}(9(2\sqrt{1+x^2}) + \sqrt{1+x^2} - x) + C$$

$$= \frac{(\sqrt{1+x^2} + x)^{19}}{360}(19\sqrt{1+x^2} - x) + C$$

$m = 360$, $n = 19$

$\therefore m + n = 379$

**48.** *(19)* $\int\left(\frac{1}{x} + \frac{1}{x^3}\right)(\sqrt[23]{3x^{-24} + x^{-26}})dx$

$$= \int\left(\frac{1}{x} + \frac{1}{x^3}\right)\left(\frac{3}{x^{24}} + \frac{1}{x^{26}}\right)^{\frac{1}{23}} dx$$

$$= \int \frac{1}{x}\left(\frac{1}{x} + \frac{1}{x^3}\right)\left(\frac{3}{x} + \frac{1}{x^3}\right)^{\frac{1}{23}} dx$$

$$= \int\left(\frac{1}{x^2} + \frac{1}{x^4}\right)\left(\frac{3}{x} + \frac{1}{x^3}\right)^{\frac{1}{23}} dx$$

Let $\frac{3}{x} + \frac{1}{x^3} = t \Rightarrow \left(-\frac{3}{x^2} - \frac{3}{x^4}\right)dx = dt$

$$\Rightarrow -3\left(\frac{1}{x^2} + \frac{1}{x^4}\right)dx = dt$$

$$= -\frac{1}{3}\int t^{1/23} dt = -\frac{t^{\frac{24}{23}}}{\frac{24}{23}} + C$$

$$= -\frac{1}{3}\frac{23}{24}\left(\frac{3}{x} + \frac{1}{x^3}\right)^{\frac{24}{23}} + C$$

$$= -\frac{1}{3}\frac{23}{24}\left(\frac{3x^2 + 1}{x^3}\right)^{\frac{24}{23}} + C$$

$$= -\frac{1}{3}\frac{23}{24}(3x^{-1} + x^{-3})^{\frac{24}{23}} + C$$

Compare with

$$-\frac{\alpha}{3(\alpha+1)}(3x^\beta + x^\gamma)^{\frac{\alpha+1}{\alpha}} + C$$

We get $\alpha = 23$, $\beta = -1$, $\gamma = -3$

Thus, $\alpha + \beta + \gamma = 23 - 1 - 3 = 19$

**49.** *(16)* Let $I = \int \frac{2x^2 + 5x + 9}{\sqrt{x^2 + x + 1}} dx$ which is of the form $\int \frac{ax^2 + bx + c}{\sqrt{px^2 + qx + r}} dx$

So, $2x^2 + 5x + 9 = A(x^2 + x + 1) + B\,(x^2 + x + 1)' + C$

$$\Rightarrow 2x^2 + 5x + 9 = A(x^2 + x + 1) + B(2x + 1) + C$$

On comparing the coefficient from both sides and then solving, we get

$A = 2$, $B = 3/2$

and $C = 11/2$

So,

$$I = \int \frac{2(x^2 + x + 1) + \frac{3}{2}(2x + 1) + \frac{11}{2}}{\sqrt{x^2 + x + 1}} dx$$

$$= 2\int \frac{x^2 + x + 1}{\sqrt{x^2 + x + 1}} dx + \frac{3}{2}\int \frac{2x + 1}{\sqrt{x^2 + x + 1}} dx + \frac{11}{2}\int \frac{1}{\sqrt{x^2 + x + 1}} dx$$

$$= 2\int \sqrt{x^2 + x + 1}\, dx + \frac{3}{2} \times 2\sqrt{x^2 + x + 1} + \frac{11}{2}\int \frac{1}{\sqrt{x^2 + x + 1}} dx$$

$$= 2\int \sqrt{\left(x+\frac{1}{2}\right)^2 + \left(\frac{\sqrt{3}}{2}\right)^2}\, dx$$

$$+ 3\sqrt{x^2+x+1}$$

$$+ \frac{11}{2}\int \frac{dx}{\sqrt{\left(x+\frac{1}{2}\right)^2 + \left(\frac{\sqrt{3}}{2}\right)^2}}$$

$$= 2\left[\frac{x+1/2}{2}\sqrt{x^2+x+1} + \frac{3}{8} \ln\left(x+\frac{1}{2}+\sqrt{x^2+x+1}\right)\right]$$

$$+ 3\sqrt{x^2+x+1} + \frac{11}{2} \ln\left(x+\frac{1}{2}+\sqrt{x^2+x+1}\right) + C,$$

where $C$ is constant of integration.

$$= x\sqrt{x^2+x+1} + \frac{7}{2}\sqrt{x^2+x+1}$$

$$+ \frac{25}{4}\ln\left(x+\frac{1}{2}+\sqrt{x^2+x+1}\right) + C$$

So, $\alpha = \frac{7}{2}, \beta = \frac{25}{4}$

Hence, $\alpha + 2\beta = 16$

> **Concept Enhancer**
>
> To evaluate $\int \frac{px+q}{\sqrt{ax^2+bx+c}}\,dx$,
>
> Put $px + q = \lambda \frac{d}{dx}(ax^2+bx+c) + \mu$, then find values of $\lambda$, $\mu$ and then integrate it.

**50.** (7) Let $I = \int \frac{1}{(x-1)^{4/5}(x+3)^{6/5}}\,dx$

$$= \int \frac{1}{\left(\frac{x-1}{x+3}\right)^{4/5}\cdot(x+3)^2}\,dx$$

Let $\frac{x-1}{x+3} = t \Rightarrow \frac{4}{(x+3)^2}\,dx = dt$

$$\Rightarrow \quad I = \frac{1}{4}\int \frac{1}{t^{4/5}}\,dt$$

$$= \frac{1}{4}\left[\frac{t^{1/5}}{1/5}\right] + C$$

$$= \frac{5}{4}t^{1/5} + C$$

$$\Rightarrow \quad I = \frac{5}{4}\left(\frac{x-1}{x+3}\right)^{1/5} + C$$

Comparing with $A\left(\frac{\alpha x - 1}{\beta x + 3}\right)^B + C$,

we get

$$A = \frac{5}{4}, \alpha = \beta = 1, B = \frac{1}{5}$$

Then, $\alpha + \beta + 20AB$

$$= 2 + 20 \times \frac{5}{4}\times\frac{1}{5} = 7$$

**51.** (1) $I = \int \operatorname{cosec}^5 x\,dx$

$$= \int \operatorname{cosec}^3 x \cdot \operatorname{cosec}^2 x\,dx$$

Using by parts

$$I = \operatorname{cosec}^3 x \int \operatorname{cosec}^2 x$$

$$- \int (3\operatorname{cosec}^2 x\cdot(-\operatorname{cosec} x \cot x)$$

$$\int \operatorname{cosec}^2 x\,dx)\,dx$$

$$\Rightarrow I = -\operatorname{cosec}^3 x \cdot \cot x$$

$$- \int 3\operatorname{cosec}^3 x \cot^2 x\,dx$$

$$\Rightarrow I = -\operatorname{cosec}^3 x\cdot\cot x$$

$$-3\int \operatorname{cosec}^3 x(\operatorname{cosec}^2 x - 1)$$

$$\Rightarrow I = -\operatorname{cosec}^3 x\cdot\cot x - 3\int \operatorname{cosec}^5 x$$

$$+ 3\int \operatorname{cosec}^3 x \ldots \text{(i)}$$

$$\Rightarrow I = -\operatorname{cosec}^3 x\cdot\cot x - 3I$$

$$+ 3\int \operatorname{cosec} x\cdot\operatorname{cosec}^2 x\,dx$$

$$\Rightarrow 4I = -\operatorname{cosec}^3 x\cdot\cot x$$

$$+ 3\operatorname{cosec} x\cdot(-\cot x)$$

$$-3\int(-\operatorname{cosec} x\cot x)\cdot(-\cot x)\,dx$$

$$\Rightarrow$$

$$4I = -\operatorname{cosec}^3 x\cdot\cot x - 3\operatorname{cosec} x\cot x$$

$$-3\int \operatorname{cosec} x(\operatorname{cosec}^2 x - 1)$$

$$\Rightarrow$$

$$4I = -\operatorname{cosec}^2 x\cot x - 3\operatorname{cosec} x\cdot\cot x$$

$$-3\int \operatorname{cosec}^3 x\,dx + 3\int \operatorname{cosec} x\,dx$$

$$\Rightarrow$$

$$4I = -\operatorname{cosec}^3 x\cot x - 3\operatorname{cosec} x\cdot\cot x$$

$$-[I + \operatorname{cosec}^3 x\cot x + 3I)]$$

$$+ 3\ln\left|\operatorname{cosec}\frac{x}{2} + \cot\frac{x}{2}\right| + C$$

[by Eq. (i)]

$$\Rightarrow \quad 8I = -2\operatorname{cosec}^3 x\cot x$$

$$-3\operatorname{cosec} x\cdot\cot x + 3\ln\left|\tan\frac{x}{2}\right| + C$$

$$\Rightarrow I = -\frac{1}{4}\operatorname{cosec} x\cdot\cot x\left(\operatorname{cosec}^2 x + \frac{3}{2}\right)$$

$$+ \frac{3}{8}\ln\left|\tan\frac{x}{2}\right| + C$$

$$\alpha = -\frac{1}{4}, \beta = \frac{3}{8}$$

$$\therefore 8(\alpha+\beta) \Rightarrow 8\alpha + 8\beta = -2 + 3 = 1$$

**52.** (28) Given,

$$f(x) = \int \frac{dx}{(3+4x^2)\sqrt{4-3x^2}}, |x| < \frac{2}{\sqrt{3}}$$

Put $x = \frac{1}{t} \Rightarrow dx = -\frac{1}{t^2}\,dt$

$$\therefore f(x) = \int \frac{\left(-\frac{1}{t^2}\right)dt}{\left(3+\frac{4}{t^2}\right)\sqrt{4-\frac{3}{t^2}}}$$

$$= \int \frac{-t\,dt}{(3t^2+4)\sqrt{4t^2-3}}$$

Put $4t^2 - 3 = z^2$

$$\Rightarrow \quad 8t\,dt = 2z\,dz$$

$$\Rightarrow \quad t\,dt = \frac{1}{4}z\,dz$$

$$= \int \frac{\left(-\frac{1}{4}z\,dz\right)}{\left[3\left(\frac{z^2+3}{4}\right)+4\right]z}$$

$$= -\frac{1}{4}\int\frac{4\,dz}{3z^2+25} = -\int\frac{dz}{3\left\{z^2+\frac{25}{3}\right\}}$$

$$= -\frac{1}{3}\int\frac{dz}{(z)^2 + \left(\frac{5}{\sqrt{3}}\right)^2}$$

$$= -\frac{1}{3}\cdot\frac{\sqrt{3}}{5}\tan^{-1}\left(\frac{z}{5/\sqrt{3}}\right) + C$$

$$= \frac{-1}{5\sqrt{3}}\tan^{-1}\left(\frac{\sqrt{3}}{5}\sqrt{4t^2-3}\right) + C$$

$$\therefore f(x) = -\frac{1}{5\sqrt{3}}\tan^{-1}\left(\frac{\sqrt{3}}{5}\sqrt{\frac{4-3x^2}{x^2}}\right) + C$$

Since, $f(0) = 0$

$$\Rightarrow \quad C = \frac{\pi}{10\sqrt{3}}$$

$\therefore f(x)=-\dfrac{1}{5\sqrt{3}}\tan^{-1}\left(\dfrac{\sqrt{3}}{5}\dfrac{\sqrt{4-3x^2}}{x}\right)+\dfrac{\pi}{10\sqrt{3}}$

$\therefore f(1)=-\dfrac{1}{5\sqrt{3}}\tan^{-1}\left(\dfrac{\sqrt{3}}{5}\right)+\dfrac{\pi}{10\sqrt{3}}$

$=\dfrac{1}{5\sqrt{3}}\left(\dfrac{\pi}{2}-\tan^{-1}\left(\dfrac{\sqrt{3}}{5}\right)\right)$

$=\dfrac{1}{5\sqrt{3}}\cot^{-1}\left(\dfrac{\sqrt{3}}{5}\right)$

$f(1)=\dfrac{1}{5\sqrt{3}}\tan^{-1}\left(\dfrac{5}{\sqrt{3}}\right)$

$=\dfrac{1}{\alpha\beta}\tan^{-1}\left(\dfrac{\alpha}{\beta}\right)$ [given]

So, $\alpha=5, \beta=\sqrt{3}$ $(\because \alpha,\beta>0)$

$\therefore \alpha^2+\beta^2=(5)^2+(\sqrt{3})^2$

$=25+3=28$

**53.** *(64)* $I(x)=\int\sqrt{\dfrac{x+7}{x}}\,dx=\int\sqrt{1+\dfrac{7}{x}}\,dx$

Let $\sqrt{1+\dfrac{7}{x}}=\sec\theta$

$\Rightarrow 1+\dfrac{7}{x}=\sec^2\theta \Rightarrow x=7\cot^2\theta$

$\Rightarrow dx=-14\cot\theta\cdot(-\operatorname{cosec}^2\theta)d\theta$

$=14\cot\theta\operatorname{cosec}^2\theta\, d\theta$

$\therefore I(\theta)=14\int\sec\theta\cdot\cot\theta\operatorname{cosec}^2\theta\,d\theta$

$=14\int\operatorname{cosec}^3\theta\,d\theta$

$=14\left(-\dfrac{1}{2}\ln\left|\dfrac{1-\cos\theta}{\sin\theta}\right|+\dfrac{1}{2}\operatorname{cosec}\theta\cot\theta\right)+c$

$=-7\ln\left|\dfrac{1-\cos\theta}{\sin\theta}\right|+\dfrac{7\cos\theta}{\sin^2\theta}+c$

As, $\sec\theta=\sqrt{\dfrac{x+7}{x}}$

$\Rightarrow \cos\theta=\sqrt{\dfrac{x}{x+7}}$

$\Rightarrow \sin^2\theta=1-\cos^2\theta=1-\dfrac{x}{x+7}$

$=\dfrac{7}{x+7}$

$\Rightarrow \sin\theta=\dfrac{\sqrt{7}}{\sqrt{x+7}}$

$\therefore I(x)=-7\ln\left|\dfrac{1-\dfrac{\sqrt{x}}{\sqrt{x+7}}}{\left(\dfrac{\sqrt{7}}{\sqrt{x+7}}\right)}\right|+\dfrac{\left(\dfrac{7\sqrt{x}}{\sqrt{x+7}}\right)}{\left(\dfrac{7}{x+7}\right)}+c$

$\Rightarrow I(x)=-7\ln\left|\dfrac{\sqrt{x+7}-\sqrt{x}}{\sqrt{7}}\right|+\sqrt{x(x+7)}+c$

$I(9)=12+7\ln 7$ (Given)

$\Rightarrow -7\ln\left(\dfrac{\sqrt{9+7}-\sqrt{9}}{\sqrt{7}}\right)+\sqrt{9(9+7)}+c=12+7\ln 7$

$\Rightarrow 7\ln(\sqrt{7})+12+c=12+7\ln(7)$

$\Rightarrow \dfrac{7}{2}\ln(7)+c=7\ln(7)$

$\Rightarrow c=\dfrac{7}{2}\ln(7)$

Now, $I(1)=\alpha+7\ln(2\sqrt{2}+1)$

$\Rightarrow -7\ln\left(\dfrac{\sqrt{1+7}-\sqrt{1}}{\sqrt{7}}\right)+\sqrt{1(1+7)}+\dfrac{7}{2}\ln(7)$

$=\alpha+7\ln(2\sqrt{2}+1)$

$\Rightarrow -7\ln\left(\dfrac{2\sqrt{2}-1}{\sqrt{7}}\right)+2\sqrt{2}+7\ln(\sqrt{7})-7\ln(2\sqrt{2}+1)=\alpha$

$\Rightarrow \alpha=2\sqrt{2}+7\ln\left(\dfrac{\sqrt{7}\times\sqrt{7}}{(2\sqrt{2}-1)(2\sqrt{2}+1)}\right)$

$\Rightarrow \alpha=2\sqrt{2}+7\ln\left(\dfrac{7}{8-1}\right)=2\sqrt{2}+7\ln(1)$

$\Rightarrow \alpha=2\sqrt{2} \Rightarrow \alpha^4=64$

**54.** *(1)* $I=\int\sqrt{\sec 2x-1}\,dx$

$=\int\sqrt{\dfrac{1-\cos 2x}{\cos 2x}}\,dx$

$=\int\dfrac{\sqrt{2}\sin x}{\sqrt{2\cos^2 x-1}}\,dx$

Let $\sqrt{2}\cos x=t$

$\Rightarrow -\sqrt{2}\sin x\cdot dx=dt$

Then, $I=\int\dfrac{-1}{\sqrt{t^2-1}}\,dt$

$=-\ln|t+\sqrt{t^2-1}|+C$

Put $t=\sqrt{2}\cos x$

$=-\ln|\sqrt{2}\cos x+\sqrt{2\cos^2 x-1}|+C$

$=-\dfrac{1}{2}\ln|(\sqrt{2}\cos x+\sqrt{\cos 2x})^2|+C$

$=-\dfrac{1}{2}\ln\left|2\cos^2 x+\cos 2x+2\sqrt{2}\sqrt{\cos 2x}\sqrt{\dfrac{1+\cos 2x}{2}}\right|+C$

$=-\dfrac{1}{2}\ln|1+\cos 2x+\cos 2x+2\sqrt{\cos 2x(1+\cos 2x)}|+C$

$=-\dfrac{1}{2}\log_e\left|\cos 2x+\dfrac{1}{2}+\sqrt{\cos 2x(1+\cos 2x)}\right|+C$

Comparing to the given solution

So, $\alpha=-\dfrac{1}{2}, \beta=\dfrac{1}{2}$

$\beta-\alpha=\dfrac{1}{2}-\left(-\dfrac{1}{2}\right)=1$

**55.** *(3)* Let $I=\int\dfrac{\sin x}{\sin^3 x+\cos^3 x}\,dx$

$=\dfrac{\tan x\sec^2 x}{\tan^3 x+1}\,dx$

Put $\tan x=t$

$\Rightarrow \sec^2 x\,dx=dt$

$I=\int\dfrac{t\,dt}{t^3+1}=\int\dfrac{t}{(t+1)(t^2-t+1)}\,dt$

$=-\dfrac{1}{3}\int\dfrac{1}{t+1}\,dt+\dfrac{1}{3}\int\dfrac{t+1}{t^2-t+1}\,dt$

$=-\dfrac{1}{3}\ln(t+1)+\dfrac{1}{3}\int\dfrac{\frac{1}{2}(2t-1)+\frac{3}{2}}{t^2-t+1}\,dt$

$=-\dfrac{1}{3}\ln(t+1)+\dfrac{1}{6}\ln(t^2-t+1)+\dfrac{1}{2}\int\dfrac{dt}{\left(t-\frac{1}{2}\right)^2+\left(\frac{\sqrt{3}}{2}\right)^2}$

$=-\dfrac{1}{3}\ln(t+1)+\dfrac{1}{6}\ln(t^2-t+1)+\dfrac{1}{2}\cdot\dfrac{2}{\sqrt{3}}\tan^{-1}\left(\dfrac{2t-1}{\sqrt{3}}\right)+C$

$$= -\frac{1}{3}\ln(\tan x + 1) + \frac{1}{6}\ln(\tan^2 x - \tan x + 1) + \frac{1}{\sqrt{3}}\tan^{-1}\left(\frac{2\tan x - 1}{\sqrt{3}}\right) + C$$

Comparing to the given solution

$$\Rightarrow \alpha = \frac{-1}{3}, \beta = \frac{1}{6}, \gamma = \frac{1}{\sqrt{3}}$$

So, $18(\alpha + \beta + \gamma^2)$

$$= 18\left(\frac{-1}{3} + \frac{1}{6} + \frac{1}{3}\right) = 3$$

**56.** (7) $I = \int \frac{2e^x + 3e^{-x}}{4e^x + 7e^{-x}} dx$

$$= \int \frac{2e^{2x} + 3}{4e^{2x} + 7} dx$$

Let $2e^{2x} + 3 = A\frac{d}{dx}(4e^{2x} + 7) + B(4e^{2x} + 7)$

$$\Rightarrow 2e^{2x} + 3 = (8A + 4B)e^{2x} + 7B$$

On comparing both sides,

$$B = \frac{3}{7} \text{ and } A = \frac{1}{28}$$

$$\therefore I = \int \frac{\frac{1}{28}(8e^{2x}) + \frac{3}{7}(4e^{2x} + 7)}{4e^{2x} + 7} dx$$

$$= \frac{1}{28}\ln|4e^{2x} + 7| + \frac{3}{7}x + C$$

$$= \frac{1}{28}\ln|e^x(4e^x + 7e^{-x})| + \frac{3}{7}x + C$$

$$= \frac{1}{28}x + \frac{1}{28}\ln|4e^x + 7e^{-x}| + \frac{3}{7}x + C$$

$$= \frac{1}{14}\left(\frac{13}{2}x + \frac{1}{2}\ln|4e^x + 7e^{-x}|\right) + C$$

Comparing to the given solution

$u = 13/2$ and $v = 1/2$

$$\therefore \quad u + v = \frac{13}{2} + \frac{1}{2} = 7$$

**57.** (15) $\int \frac{dx}{(x^2 + x + 1)^2}$

$$= \int \frac{dx}{\left[\left(x + \frac{1}{2}\right)^2 + \left(\frac{\sqrt{3}}{2}\right)^2\right]^2}$$

Let $x + \frac{1}{2} = \frac{\sqrt{3}}{2}\tan\theta$

$$\Rightarrow dx = \frac{\sqrt{3}}{2}\sec^2\theta\, d\theta$$

$$\therefore \int \frac{\frac{\sqrt{3}}{2}\sec^2\theta\, d\theta}{\frac{9}{16}(\tan^2\theta + 1)^2} = \frac{8}{3\sqrt{3}}\int \frac{\sec^2\theta\, d\theta}{\sec^4\theta}$$

$$= \frac{8}{3\sqrt{3}}\int \cos^2\theta\, d\theta$$

$$= \frac{8}{3\sqrt{3}}\int \frac{1 + \cos 2\theta}{2} d\theta$$

$$= \frac{4}{3\sqrt{3}}\left(\theta + \frac{\sin 2\theta}{2}\right) + C$$

$$= \frac{4}{3\sqrt{3}}\tan^{-1}\left(\frac{2x+1}{\sqrt{3}}\right) + \frac{4}{3\sqrt{3}}\frac{\frac{2x+1}{\sqrt{3}}}{1 + \left(\frac{2x+1}{\sqrt{3}}\right)^2} + C$$

$$= \frac{4}{3\sqrt{3}}\tan^{-1}\left(\frac{2x+1}{3}\right) + \frac{1}{3}\frac{2x+1}{(x^2 + x + 1)} + C$$

Comparing to the given solution

$$\therefore \quad a = \frac{4}{3\sqrt{3}}, b = \frac{1}{3}$$

Hence, $9(\sqrt{3}a + b) = 9\left(\frac{4}{3} + \frac{1}{3}\right) = 15$

**58.** (4) Let $I = \int \frac{5x^8 + 7x^6}{(x^2 + 1 + 2x^7)^2} dx$

$$= \int \frac{5x^8 + 7x^6}{x^{14}(x^{-5} + x^{-7} + 2)^2} dx$$

$$= \int \frac{\frac{5x^8 + 7x^6}{x^{14}}}{\frac{x^{14}(x^{-5} + x^{-7} + 2)^2}{x^{14}}} dx$$

$$\Rightarrow \quad I = \int \frac{5x^{-6} + 7x^{-8}}{(x^{-5} + x^{-7} + 2)^2} dx$$

Putting $x^{-5} + x^{-7} + 2 = z$

$$\Rightarrow \quad -(5x^{-6} + 7x^{-8})dx = dz$$

$$\therefore \quad I = -\int \frac{dz}{z^2} = -\left(\frac{1}{-z}\right) + c$$

$$\Rightarrow \quad I = \frac{1}{x^{-5} + x^{-7} + 2} + c$$

$$\Rightarrow \quad f(x) = \frac{x^7}{x^2 + 1 + 2x^7} + c$$

Given, $f(0) = 0 \Rightarrow c = 0$

$$\therefore \quad f(x) = \frac{x^7}{x^2 + 1 + 2x^7}$$

$$\therefore \quad f(1) = \frac{1}{1 + 1 + 2} = \frac{1}{4} = \frac{1}{K}$$

[given]

Hence, $K = 4$

## Topic 2

**01.** (c) $I = \int_0^{\pi} \frac{8x}{4\cos^2 x + \sin^2 x} dx \quad \ldots(i)$

$$I = \int_0^{\pi} \frac{8(\pi - x)}{4\cos^2(\pi - x) + \sin^2(\pi - x)} dx$$

$$I = \int_0^{\pi} \frac{8(\pi - x)}{4\cos^2 x + \sin^2 x} dx \quad \ldots(ii)$$

On adding Eqs. (i) and (ii), we get

$$2I = 8\pi\int_0^{\pi} \frac{1}{4\cos^2 x + \sin^2 x} dx$$

$$I = 4\pi \times 2\int_0^{\pi} \frac{1}{4\cos^2 x + \sin^2 x} dx$$

$$I = 4\pi \times 2\int_0^{\pi/2} \frac{\sec^2 x}{4\tan^2 x} dx$$

Let $\tan^2 x = t$

$\sec^2 x\, dx = dt$

$$I = 8\pi\int_0^{\infty} \frac{dt}{4 + t^2}$$

$$I = 8\pi\frac{1}{2}\left(\tan^{-1}\frac{t}{2}\right)_0^{\infty} \Rightarrow I = 4\pi\left(\frac{\pi}{2}\right)$$

$$I = 2\pi^2$$

**02.** (d)

> **key idea** King property
>
> $\int_a^b f(x)dx = \int_a^b f(a + b - x)dx$

Let $I =$

$$\int_{e^2}^{e^4} \frac{1}{x}\left(\frac{e^{((\log_e x)^2 + 1)^{-1}}}{e^{((\log_e x)^2 + 1)^{-1}} + e^{((6 - \log_e x)^2 + 1)^{-1}}}\right) dx$$

Put $\log_e x = t \Rightarrow \frac{1}{x}dx = dt$

When $x = e^2, t = 2$

and $x = e^4, t = 4$

$$\therefore I = \int_2^4 \frac{e^{\frac{1}{1+t^2}}}{e^{\frac{1}{1+t^2}} + e^{\frac{1}{1+(6-t)^2}}} dt \quad \ldots(i)$$

$$I = \int_2^4 \frac{e^{\frac{1}{1+(6-t)^2}}}{e^{\frac{1}{1+(6-t)^2}} + e^{\frac{1}{1+t^2}}} dt \quad \ldots(ii)$$

Use property

$$\int_a^b f(x)dx = \int_a^b f(a+b-x)dx$$

From Eqs. (i) and (ii), we get

$$2I = \int_2^4 dt \Rightarrow 2I = 2 \Rightarrow I = 1$$

**03.** *(b)* $\int_{1/4}^{3/4} \cos\left(2\cot^{-1}\sqrt{\frac{1-x}{1+x}}\right)dx$

$$= \int_{1/4}^{3/4} \cos\left(2\tan^{-1}\sqrt{\frac{1+x}{1-x}}\right)dx$$

$$= \int_{1/4}^{3/4} \left(\frac{1-\tan^2\left(\tan^{-1}\sqrt{\frac{1+x}{1-x}}\right)}{1+\tan^2\left(\tan^{-1}\sqrt{\frac{1+x}{1-x}}\right)}\right)dx$$

$$= \int_{1/4}^{3/4} \left(\frac{1-\frac{1+x}{1-x}}{1+\frac{1+x}{1-x}}\right)dx = \int_{1/4}^{3/4}(-x)dx$$

$$= -\left(\frac{x^2}{2}\right)_{1/4}^{3/4} = -\frac{1}{2}\left(\frac{9}{16}-\frac{1}{16}\right) = \frac{-1}{4}$$

**04.** *(a)* $I = \int_{-1}^{2} \ln(x+\sqrt{x^2+1})dx$

Using by parts

$$\left| x\ln(x+\sqrt{x^2+1}) - \int \frac{1}{x+\sqrt{x^2+1}}\left(1+\frac{2x}{2\sqrt{1+x^2}}\right)x\,dx \right|_{-1}^{2}$$

$$= \left| x\ln(x+\sqrt{x^2+1}) - \int \frac{x}{\sqrt{1+x^2}}dx \right|_{-1}^{2}$$

$$= [x\ln(x+\sqrt{1+x^2}) - \sqrt{1+x^2}\,]_{-1}^{2}$$

$$= [2\ln(2+\sqrt{5}) - \sqrt{5}] - [-\ln(\sqrt{2}-1) - \sqrt{2}]$$

$$= \ln(2+\sqrt{5})^2 - \sqrt{5} + \ln(\sqrt{2}-1) + \sqrt{2}$$

$$= \sqrt{2} - \sqrt{5} + \ln(9+4\sqrt{5}) - \ln(\sqrt{2}+1)$$

$$= \sqrt{2} - \sqrt{5} + \ln\left(\frac{9+4\sqrt{5}}{\sqrt{2}+1}\right)$$

**05.** *(b)*

$$\lim_{x\to\frac{\pi}{2}} \frac{\int_{x^3}^{(\pi/2)^3} (\sin 2t^{1/3} + \cos t^{1/3})dt}{\left(x-\frac{\pi}{2}\right)^2}\left(\frac{0}{0}\text{ form}\right)$$

Using L' hospital rule,

$$\lim_{x\to\frac{\pi}{2}} \frac{0-3x^2(\sin 2x+\cos x)}{2\left(x-\frac{\pi}{2}\right)}\left(\frac{0}{0}\text{ form}\right)$$

$$-6x(\sin 2x + \cos x)$$

$$= \lim_{x\to\frac{\pi}{2}} + \frac{-3x^2(2\cos 2x - \sin x)}{2}$$

$$= -3\times\frac{\pi^2}{4}\times\frac{(-2-1)}{2} = \frac{9\pi^2}{8}$$

**06.** *(b)* $I_n = \int_0^1 (1-x^k)^n dx$

and $147 I_{20} = 148 I_{21}$

Let $I_{21} = \int_0^1 (1-x^k)^{21} \times 1\cdot dx$

Using integration by parts, we get

$$I_{21} = [(1-x^k)^{21}\cdot x]_0^1 + 21k\int_0^1 (1-x^k)^{20}\cdot x^{k-1}\cdot x\cdot dx$$

$$= 0 + 21k\int_0^1 (1-x^k)^{20}(x^k - 1 + 1)dx$$

$$= 21k\int_0^1 -(1-x^k)^{21}dx + \int (1-x^k)^{20}dx$$

$$I_{21} = 21k[-I_{21} + I_{20}]$$

$$\Rightarrow I_{21} = -21k\,I_{21} + 21k\,I_{20}$$

Comparing with $148 I_{21} = 147 I_{20}$, we get $k = 7$

**07.** *(a)*

> **key idea** Firstly convert the integral in terms of $\tan x$ and $\sec x$, then use substitution $1+\tan^3 x = t$

Let $I = \int_0^{\pi/4} \frac{\cos^2 x \sin^2 x}{(\cos^3 x + \sin^3 x)^2}dx$

Dividing numerator and denominator by $\cos^6 x$,

$$\Rightarrow I = \int_0^{\pi/4} \frac{(\cos^2 x\sin^2 x/\cos^6 x)}{\left(\frac{\cos^3 x+\sin^3 x}{\cos^3 x}\right)^2}dx$$

$$I = \int_0^{\pi/4} \frac{\tan^2 x\sec^2 x}{(1+\tan^3 x)^2}dx$$

Let $1+\tan^3 x = t$

$\Rightarrow 3\tan^2 x\sec^2 x\,dx = dt$

and at $x = 0$, $t = 1$ and $x = \frac{\pi}{4}$, $t = 2$

$$\Rightarrow I = \frac{1}{3}\int_1^2 \frac{dt}{t^2} = \frac{1}{3}\left[-\frac{1}{t}\right]_1^2$$

$$= \frac{1}{3}\left[1-\frac{1}{2}\right] = \frac{1}{6}$$

**08.** *(a)* $I = \int_0^1 1\cdot(1-x^{10})^{20}dx$

$$= a\times\beta(b,c)$$

$$= a\int_0^1 x^{b-1}\cdot(1-x)^{c-1}dx$$

Let $x^{10} = t$

$\Rightarrow 10x^9 dx = dt$

$$I = \int_0^1 (1-t)^{20}\cdot\frac{1}{10}\cdot t^{-9/10}dt$$

$$= \frac{1}{10}\cdot\int_0^1 (1-x)^{20}\cdot x^{-9/10}dx$$

$\therefore a = \frac{1}{10}, b = \frac{1}{10}, c = 21$

$\therefore 100(a+b+c) = 10+10+2100 = 2120$

**09.** *(d)* Using property,

$$\int_{-a}^{a} f(x)dx = \int_0^a f(x)dx + \int_0^a f(-x)dx$$

$$I = \int_{-1}^{1} \frac{\cos\alpha x}{1+3^x}dx$$

$$I = \int_0^1 \frac{\cos\alpha x}{1+3^x} + \frac{\cos\alpha(-x)}{1+3^{-x}}$$

$$= \int_0^1 \frac{\cos\alpha x(1+3^x)}{1+3^x}$$

$$= \int_0^1 (\cos\alpha x)dx$$

$$= \left[\frac{\sin\alpha x}{\alpha}\right]_0^1 = \frac{\sin\alpha}{\alpha}$$

$\because \quad I = \frac{2}{\pi}$

$$\frac{\sin\alpha}{\alpha} = \frac{2}{\pi} \Rightarrow \alpha = \frac{\pi}{2}$$

**10.** *(b)* $I = \int_0^1 (2x^3 - 3x^2 - x + 1)^{\frac{1}{3}}dx$ ...(i)

$$= \int_0^1 [2(1-x)^3 - 3(1-x)^2 - (1-x) + 1]^{\frac{1}{3}}dx$$

$$= \int_0^1 [2(1-x^3+3x^2-3x) - 3(1+x^2-2x) - 1 + x + 1]^{\frac{1}{3}}dx$$

$$= \int_0^1 [2-2x^3+6x^2-6x-3-3x^2+6x+x]^{\frac{1}{3}}dx$$

$$= \int_0^1 [-2x^3+3x^2+x-1]^{\frac{1}{3}}dx$$

$$= -\int_0^1 (2x^3-3x^2-x+1)^{\frac{1}{3}}dx$$

$= -I$ [from Eq. (i)]

$\Rightarrow 2I = 0; \quad \therefore I = 0$

**11.** *(c)*

**key idea** Use Leibnitz rule

$$\frac{d}{dx}\left[\int_{g(x)}^{h(x)} f(t)dt\right] = f(h(x))h'(x) - f(g(x))g'(x)$$

We have,

$$\lim_{x\to\frac{\pi}{2}} \left( \frac{1}{\left(x-\frac{\pi}{2}\right)^2} \int_{x^3}^{\left(\frac{\pi}{2}\right)^3} \cos(t^{1/3})\, dt \right)$$

By using L' Hospital rule, we get

$$\lim_{x\to\frac{\pi}{2}} \frac{\cos(\pi/2)(0)-(\cos x)(3x^2)}{2\left(x-\frac{\pi}{2}\right)}$$

$$= \lim_{x\to\frac{\pi}{2}} \frac{\sin\left(x-\frac{\pi}{2}\right)\left(\frac{3\pi^2}{4}\right)}{2\left(x-\frac{\pi}{2}\right)} = \frac{3\pi^2}{8}$$

**12.** *(a)* Let

$$I = \frac{e^{\frac{-\pi}{4}} + \int_0^{\pi/4} e^{-x} \tan^{50} x\, dx}{\int_0^{\pi/4} e^{-x}(\tan^{49} x + \tan^{51} x)\, dx}$$

Now, $\int_0^{\pi/4} e^{-x} \tan^{50} x\, dx$

$$= [-e^{-x} \tan^{50} x]_0^{\pi/4} + \int_0^{\pi/4} e^{-x}(50) \tan^{49} x \sec^2 x\, dx$$

$$= -e^{\frac{-\pi}{4}}(1)^{50} + 0 + 50\int_0^{\pi/4} e^{-x} \tan^{49} x(\tan^2 x+1)\, dx$$

$$= -e^{\frac{-\pi}{4}} + 50\int_0^{\pi/4} e^{-x} [\tan^{49} x + \tan^{51} x]\, dx \quad \ldots\text{(i)}$$

Now, $I = \dfrac{e^{\frac{-\pi}{4}} + \int_0^{\pi/4} e^{-x} \tan^{50} x dx}{\int_0^{\pi/4} e^{-x}(\tan^{49} x + \tan^{51} x)\, dx}$

Using Eq. (i), we get

$$I = \frac{50\int_0^{\pi/4} e^{-x}(\tan^{49} x + \tan^{51} x)\, dx}{\int_0^{\pi/4} e^{-x}(\tan^{49} x + \tan^{51} x)\, dx}$$

$\therefore\ I = 50$

**13.** *(b)* We have, $\int_0^{t^2} (f(x) + x^2)dx = \frac{4}{3}t^3$, $\forall\, t > 0$

Differentiate using Leibnitz rule, we get

$$(f(t^2)+t^4)\times 2t = 4t^2$$

$$f(t^2)+t^4 = 2t$$

$$f(t^2) = 2t - t^4$$

Substitute $\frac{\pi}{2}$ for $t$, we get

$$f\left(\frac{\pi}{2}\right)^2 = 2\left(\frac{\pi}{2}\right) - \frac{\pi^4}{16} = \pi\left(1-\frac{\pi^3}{16}\right)$$

**14.** *(c)* Let $I = \int_0^{\pi} f(x)\sin x\, dx \quad \ldots\text{(i)}$

$$I = \int_0^{\pi} f(\pi - x)\sin(\pi - x)\, dx$$

$$I = \int_0^{\pi} f(\pi - x)\sin x\, dx \quad \ldots\text{(ii)}$$

On adding Eqs. (i) and (ii), we get

$$2I = \int_0^{\pi} [f(x) + f(\pi - x)]\sin x\, dx$$

$$\Rightarrow\ 2I = \int_0^{\pi} \pi^2 \sin x\, dx$$

$$= \pi^2 \int_0^{\pi} \sin x\, dx$$

$$= \pi^2(-\cos x)_0^{\pi} = -\pi^2(-1-1) = 2\pi^2$$

$$\Rightarrow\ I = \pi^2$$

**15.** *(b)* Let

$$I = \int_0^{\alpha} \frac{t^{50}}{1-t}\, dt = -\int_0^{\alpha} \frac{-t^{50}}{1-t}\, dt$$

$$= -\int_0^{\alpha} \frac{1-t^{50}}{1-t}\, dt + \int_0^{\alpha} \frac{1}{1-t}\, dt$$

$$= -\int_0^{\alpha} (1+t+t^2+\ldots+t^{49})\, dt - [\log(1-t)]_0^{\alpha}$$

$$= -\left[t + \frac{t^2}{2} + \frac{t^3}{3} + \ldots + \frac{t^{50}}{50}\right]_0^{\alpha} - [\log(1-\alpha) - \log 1]$$

$$= -\left[\alpha + \frac{\alpha^2}{2} + \frac{\alpha^3}{3} + \ldots + \frac{\alpha^{50}}{50}\right] - \log(1-\alpha)$$

$$= -P_{50}(\alpha) - \beta = -(\beta + P_{50}(\alpha))$$

**16.** *(b)* Let $I = \int_{1/2}^{2} \frac{\tan^{-1} x}{x}\, dx \quad \ldots\text{(i)}$

Put $x = \frac{1}{t}$, then $dx = \frac{-1}{t^2}\, dt$

$$\therefore\ I = \int_2^{1/2} \frac{\tan^{-1}\left(\frac{1}{t}\right)}{\left(\frac{1}{t}\right)}\left(-\frac{1}{t^2}\right)dt$$

$$= \int_{1/2}^{2} \frac{\tan^{-1}\left(\frac{1}{t}\right)}{t}\, dt = \int_{1/2}^{2} \frac{\cot^{-1} t}{t}\, dt \quad \ldots\text{(ii)}$$

$\therefore$ Adding Eqs. (i) and (ii), we get

$$2I = \int_{1/2}^{2} \frac{\tan^{-1} x + \cot^{-1} x}{x}\, dx = \int_{1/2}^{2} \frac{\left(\frac{\pi}{2}\right)}{x}\, dx$$

$$\Rightarrow\ 2I = \frac{\pi}{2}[\log x]_{\frac{1}{2}}^{2} = \frac{\pi}{2}(\log 2 + \log 2)$$

$$\Rightarrow\ I = \frac{\pi}{2}\log_e 2$$

**17.** *(b)* Given, $= \int_{\frac{3\sqrt{2}}{4}}^{\frac{3\sqrt{3}}{4}} \frac{48}{\sqrt{9-4x^2}}\, dx$

$$= \int_{\frac{3\sqrt{2}}{4}}^{\frac{3\sqrt{3}}{4}} \frac{48}{2\sqrt{\left(\frac{3}{2}\right)^2 - x^2}}\, dx$$

$$= \frac{48}{2}\int_{\frac{3\sqrt{2}}{4}}^{\frac{3\sqrt{3}}{4}} \frac{1}{\sqrt{\left(\frac{3}{2}\right)^2 - x^2}}\, dx$$

$$= 24\left\{\sin^{-1}\left(\frac{x}{\frac{3}{2}}\right)\right\}_{\frac{3\sqrt{2}}{4}}^{\frac{3\sqrt{3}}{4}}$$

$$= 24\left\{\sin^{-1}\left(\frac{2}{3}\cdot\frac{3\sqrt{3}}{4}\right) - \sin^{-1}\left(\frac{2}{3}\cdot\frac{3\sqrt{2}}{4}\right)\right\}$$

$$= 24\left\{\sin^{-1}\left(\frac{\sqrt{3}}{2}\right) - \sin^{-1}\left(\frac{1}{\sqrt{2}}\right)\right\}$$

$$= 24\left(\frac{\pi}{3} - \frac{\pi}{4}\right)$$

$$= 24\left(\frac{4\pi - 3\pi}{12}\right) = 2\pi$$

**18.** *(b)* $\int_0^{\pi/2} \frac{1}{3 + 2\sin x + \cos x}\, dx$

$$= \int_0^{\pi/2} \frac{1}{3 + 2\frac{2\tan x/2}{1+\tan^2 x/2} + \frac{1-\tan^2 x/2}{1+\tan^2 x/2}}\, dx$$

$$= \int_0^{\pi/2} \frac{1/2\sec^2 x/2\, dx}{\tan^2 x/2 + 2\tan x/2 + 2}$$

On putting $\tan\frac{x}{2} = t$, we get

$1/2\sec^2 x/2\,dx = dt$

$$\Rightarrow \int_0^1 \frac{dt}{t^2+2t+2} = \int_0^1 \frac{dt}{(t+1)^2+1}$$

$$= \frac{1}{1}\left[\tan^{-1}\frac{(t+1)}{1}\right]_0^1$$

$$= \tan^{-1}2 - \tan^{-1}1 = \tan^{-1}2 - \frac{\pi}{4}$$

**19** (b) $I = \int_0^2 |2x^2-3x|\,dx + \int_0^2 \left[x-\frac{1}{2}\right]dx$

$$= \int_0^{3/2}(-2x^2+3x)dx + \int_{3/2}^2 (2x^2-3x)\,dx$$

$$+ \int_0^{1/2}(-1)\,dx + \int_{1/2}^{3/2}(0)dx + \int_{3/2}^2 1\,dx$$

$$= \left(\frac{-2x^3}{3}+\frac{3x^2}{2}\right)_0^{3/2} + \left(\frac{2x^3}{3}-\frac{3x^2}{2}\right)_{3/2}^2$$

$$+(-x)_0^{1/2} + (x)_{3/2}^2$$

$$= \frac{-9}{4}+\frac{27}{8}+\frac{16}{3}-6-\frac{9}{4}+\frac{27}{8}=\frac{19}{12}$$

**20.** (d) $f(x) = 2+|x|-|x-1|+|x+1|$, $x \in R$

$$\therefore \quad f(x) = \begin{cases} -x, & x \le -1 \\ x+2, & -1 < x \le 0 \\ 3x+2, & 0 < x \le 1 \\ x+4, & 1 < x \end{cases}$$

$$f'\left(\frac{-3}{2}\right)+f'\left(\frac{-1}{2}\right)+f'\left(\frac{1}{2}\right)+f'\left(\frac{3}{2}\right)$$

$$= -1+1+3+1 = 4$$

$$\int_{-2}^2 f(x)dx = \int_{-2}^{-1}(-x)dx + \int_{-1}^0 (x+2)dx$$

$$+ \int_0^1 (3x+2)dx + \int_1^2 (x+4)dx$$

$$\therefore \quad = \frac{3}{2}-\frac{1}{2}+2+\frac{3}{2}+2+\frac{3}{2}+4 = 12$$

Statement I is wrong and Statement II is correct.

**21.** (b) We have, $f: R \to R$

$$f(x) = a\sin\left(\frac{\pi[x]}{2}\right)+[2-x],\ a \in R$$

At $x=-1$

$$\text{LHL} = a\sin\left(\frac{-2\pi}{2}\right)+3$$

$$\text{RHL} = a\sin\left(-\frac{\pi}{2}\right)+2$$

Now, LHL = RHL

$$\Rightarrow \quad 3 = -a+2 \Rightarrow a = -1$$

$$\text{Now, } \int_0^4 f(x)dx = \int_0^1 (-\sin(0)+1)\,dx$$

$$+\int_1^2 (-1+0)dx + \int_2^3 (0-1)dx$$

$$+\int_3^4 (1-2)dx$$

$$= 1-1-1-1 = -2$$

**22.** (d) Let $I = \int_0^{20\pi} (|\sin x| + |\cos x|)^2\,dx$

Let $f(x) = (|\sin x| + |\cos x|)^2$

$$\because f(\pi+x) = (|\sin(\pi+x)| + |\cos(\pi+x)|)^2$$

$$= (|\sin x| + |\cos x|)^2 = f(x)$$

$$I = 20\int_0^{\pi}(|\sin x| + |\cos x|)^2\,dx$$

$$= 20\int_0^{\pi}(1+|\sin 2x|)\,dx$$

$$I = 40\int_0^{\pi/2}(1+\sin 2x)dx$$

$$= 40\left[x-\frac{\cos 2x}{2}\right]_0^{\pi/2}$$

$$= 40\left[\left(\frac{\pi}{2}-\frac{1}{2}\cos 2\left(\frac{\pi}{2}\right)\right)-\left(0-\frac{\cos 0}{2}\right)\right]$$

$$= 40\left[\frac{\pi+2}{2}\right] = 20(\pi+2)$$

**23.** (d) Let $I = \int_0^2 \cos\left(\pi\left(x-\left[\frac{x}{2}\right]\right)\right)dx$

$$+\int_2^4 \cos\left(\pi\left(x-\left[\frac{x}{2}\right]\right)\right)dx$$

$$+\int_4^5 \cos\left(\pi\left(x-\left[\frac{x}{2}\right]\right)\right)dx$$

$$= \int_0^2 \cos(\pi x - 0)\,dx + \int_2^4 \cos(\pi x - \pi)\,dx$$

$$+\int_4^5 \cos(\pi x - 2\pi)\,dx$$

$$= \int_0^2 \cos(\pi x)\,dx - \int_2^4 \cos(\pi x)\,dx$$

$$+\int_4^5 \cos(\pi x)\,dx$$

$$= \left[\frac{\sin \pi x}{\pi}\right]_0^2 - \left[\frac{\sin \pi x}{\pi}\right]_2^4 + \left[\frac{\sin \pi x}{\pi}\right]_4^5$$

$$= [0-0]-[0-0]+[0-0]$$

$$= 0$$

**Shortcut Method**

Let $I = \int_0^5 \cos\left(\pi\left(x-\left[\frac{\pi}{2}\right]\right)\right)dx$

$$\int_0^2 \cos(\pi x)\,dx + \int_2^4 \cos(\pi x - \pi)$$

$$+\int_4^5 \cos(\pi x - 2\pi)\,dx$$

$\Rightarrow I = 0$

**24.** (b) $f(x) = \int_0^{x^2} \frac{t^2-5t+4}{2+e^t}\,dt$

$$f'(x) = \frac{x^4-5x^2+4}{2+e^{x^2}}\cdot 2x$$

For maxima or minima

$$f'(x) = 0$$

$$\Rightarrow \quad x(x^4-5x^2+4) = 0$$

$$\Rightarrow \quad x(x^2-4)(x^2-1) = 0$$

$$\Rightarrow \quad x = 0, \pm 2, \pm 1$$

$-\ +\ -\ +\ -\ +$ (sign scheme on number line: $-2, -1, 0, 1, 2$)

Point of local maxima $= -1$ and $1$

Point of local minima $= -2, 0, 2$

$\therefore (m, n) = (2, 3)$

**25.** (d) $f(x) = \frac{|x^3+x|}{(e^{x|x|}+1)}$

$$\int_{-2}^2 f(x)dx = \int_0^2 (f(x)+f(-x))dx$$

$$= \int_0^2 \left(\frac{|x^3+x|}{(e^{x|x|}+1)}+\frac{|-x^3-x|}{(e^{-x|-x|}+1)}\right)dx$$

$$= \int_0^2 \frac{|x^3+x|}{(e^{x^2}+1)}+\frac{|x^3+x|}{(e^{-x|x|}+1)}dx$$

$$= \int_0^2 \left(\frac{x^3+x}{e^{x^2}+1}+\frac{x^3+x}{e^{-x^2}+1}\right)dx$$

$$= \int_0^2 \left(\frac{x^3+x}{e^{x^2}+1}+\frac{e^{x^2}(x^3+x)}{1+e^{x^2}}\right)dx$$

$$= \int_0^2 (x^3+x)\,dx = \left[\frac{x^4}{4}+\frac{x^2}{2}\right]_0^2$$

$$= 4+2 = 6$$

**26.** (d) $f(x) = x + \int_0^{\pi/2} \sin x \cdot \cos y\, f(y)\,dy$ ...(i)

Let $k = \int_0^{\pi/2} \cos y\, f(y)\,dy$

Then, $f(x) = x + k\sin x$ ...(ii)

From Eqs. (i) and (ii), we get

$$f(x)=x+\int_0^{\pi/2}\sin x\cos y(y+k\sin y)\,dy$$

$$=x+\sin x\int_0^{\pi/2} y\cos y\,dy$$

$$+\frac{k}{2}\sin x\int_0^{\pi/2}\sin 2y\,dy$$

$$f(x)=x+\sin x\cdot\frac{\pi-2}{2}+\frac{k\sin x}{2}\quad\ldots\text{(iii)}$$

From Eqs. (ii) and (iii),

$$k=\frac{\pi-2}{2}+\frac{k}{2}\Rightarrow k=\pi-2$$

$$\therefore\quad f(x)=x+(\pi-2)\sin x$$

**27.** (b) $I=\pi^2\int_0^2\sin\left(\frac{\pi x}{2}\right)(x-|x|)^{[x]}\,dx$

$$=\pi^2\int_0^1\sin\left(\frac{\pi x}{2}\right)x^0\,dx$$

$$+\pi^2\int_1^2\sin\left(\frac{\pi x}{2}\right)(x-1)\,dx$$

$$=\pi^2\left[\frac{-2}{\pi}\cos\frac{\pi x}{2}\right]_0^1$$

$$+\pi^2\left[(x-1)\frac{2}{\pi}\left(-\cos\frac{\pi x}{2}\right)\right]_1^2$$

$$+\pi^2\int_1^2\frac{2}{\pi}\cos\frac{\pi x}{2}\,dx$$

$$=2\pi+2\pi+4(0-1)$$

$$=4\pi-4=4(\pi-1)$$

**28.** (b) Let

$$I=\int_{\frac{-1}{\sqrt2}}^{\frac{1}{\sqrt2}}\left(\left(\frac{x+1}{x-1}\right)^2+\left(\frac{x-1}{x+1}\right)^2-2\right)^{1/2}dx$$

$$I=\int_{\frac{-1}{\sqrt2}}^{\frac{1}{\sqrt2}}\left[\left(\frac{x+1}{x-1}-\frac{x-1}{x+1}\right)^2\right]^{\frac12}dx$$

$$=\int_{-1/\sqrt2}^{1/\sqrt2}\left|\frac{x+1}{x-1}-\frac{x-1}{x+1}\right|dx$$

$$=\int_{\frac{-1}{\sqrt2}}^{\frac{1}{\sqrt2}}\left|\frac{(x+1)^2-(x-1)^2}{(x-1)(x+1)}\right|dx$$

$$=\int_{-1/\sqrt2}^{1/\sqrt2}\left|\frac{4x}{(x-1)(x+1)}\right|dx$$

$$=2\cdot4\int_0^{1/\sqrt2}\left|\frac{x}{(x-1)(x+1)}\right|dx$$

$$=4\int_0^{1/\sqrt2}\frac{-2x}{x^2-1}\,dx$$

$$=-4\left[\log(x^2-1)\right]_0^{1/\sqrt2}$$

$$=-4\log\left(\frac12\right)$$

$$=4\ln 2=\ln 16$$

**29.** (b) Let $I=\int_{-1}^{1}\log(x+\sqrt{x^2+1})dx$

Let $f(x)=\log(x+\sqrt{x^2+1})$

Now, $f(-x)=\log(-x+\sqrt{x^2+1})$

$$=\log\left((\sqrt{x^2+1}-x)\times\left(\frac{\sqrt{x^2+1}+x}{\sqrt{x^2+1}+x}\right)\right)$$

$$=\log\left(\frac{x^2+1-x^2}{\sqrt{x^2+1}+x}\right)=\log\left(\frac{1}{x+\sqrt{x^2+1}}\right)$$

$$=\log 1-\log(x+\sqrt{x^2+1})$$

$$=-\log(x+\sqrt{x^2+1})$$

$$\Rightarrow\quad f(-x)=-f(x)$$

$\because\log(x+\sqrt{x^2+1})$ is an odd function.

$$\therefore\int_{-1}^{1}\log(x+\sqrt{x^2+1})dx=0$$

**30.** (a) $I=\int_{-\pi/2}^{\pi/2}([x]+[-\sin x])dx\quad\ldots\text{(i)}$

Using property,

$$\int_a^b f(x)dx=\int_a^b f(a+b-x)dx$$

$$I=\int_{-\pi/2}^{\pi/2}([-x]+[\sin x])dx\quad\ldots\text{(ii)}$$

On adding Eqs. (i) and (ii), we get

$$2I=\int_{-\pi/2}^{\pi/2}([x]+[-x])$$

$$+([\sin x]+[-\sin x])dx$$

$$=\int_{-\pi/2}^{\pi/2}(-2)dx=-2x]_{-\pi/2}^{\pi/2}$$

[$\because[-x]+[x]=-1$, if $x$ is not an integer]

$$2I=(-2)\left(\frac{\pi}{2}\right)+(2)\left(-\frac{\pi}{2}\right)=-2\pi$$

$$\Rightarrow\quad I=-\pi$$

**31.** (d) $g(\alpha)=\int_{\pi/6}^{\pi/3}\frac{\sin^\alpha x}{\cos^\alpha x+\sin^\alpha x}dx\quad\ldots\text{(i)}$

On applying $\int_a^b f(x)dx=\int_a^b f(a+b-x)dx$

$$g(\alpha)=\int_{\pi/6}^{\pi/3}\frac{\sin^\alpha(\pi/2-x)}{\cos^\alpha\left(\frac{\pi}{2}-x\right)+\sin^\alpha\left(\frac{\pi}{2}-x\right)}dx$$

$$g(\alpha)=\int_{\pi/6}^{\pi/3}\frac{\cos^\alpha x}{\cos^\alpha x+\sin^\alpha x}dx\quad\ldots\text{(ii)}$$

On adding Eqs. (i) and (ii), we get

$$2g(\alpha)=\int_{\pi/6}^{\pi/3}\frac{\sin^\alpha x+\cos^\alpha x}{\sin^\alpha x+\cos^\alpha x}dx$$

$$2g(\alpha)=\int_{\pi/6}^{\pi/3}1\cdot dx=[x]_{\pi/6}^{\pi/3}$$

$$\therefore\quad g(\alpha)=\frac{\pi}{12}$$

$g(\alpha)$ is constant function.

$\therefore$ It is even function.

**32.** (c) We have, $\int_0^{10}\frac{[x]e^{[x]}}{e^{x-1}}dx$

$$=e\int_0^{10}\frac{[x]e^{[x]}}{e^x}dx=e\int_0^1\frac{0}{e^x}dx$$

$$+e\int_1^2\frac{e}{e^x}dx+e\int_2^3\frac{2e^2}{e^x}dx$$

$$+\ldots+e\int_9^{10}\frac{9x^9}{e^x}dx$$

$$\left[\int_a^b e^{-x}dx=\frac{e^{-x}}{-1}\Big|_a^b=(e^{-a}-e^{-b})\right]$$

$$=e^2\left(\frac1e-\frac1{e^2}\right)+2e^3\left(\frac1{e^2}-\frac1{e^3}\right)$$

$$+3e^4\left(\frac1{e^3}-\frac1{e^4}\right)\quad+\ldots+9e^{10}\left(\frac1{e^9}-\frac1{e^{10}}\right)$$

$$=(e-1)+2(e-1)+3(e-1)+\ldots+9(e-1)$$

$$=(1+2+3+\ldots+9)(e-1)$$

$$=\left(\frac{9\times10}{2}\right)(e-1)=45(e-1)$$

**33.** (c) Given, $\int_{-1}^{1}x^2e^{[x^3]}dx$, where $[t]$ is greatest integer function.

$\because\quad[x^3]=0,\ \forall\ x\in(0,1)$

and $[x^3]=-1,\ \forall\ x\in(-1,0)$

So, $\int_{-1}^{1} x^2 e^{[x^3]} dx = \int_{-1}^{0} x^2 e^{-1} dx + \int_{0}^{1} x^2 e^{0} dx$

$= \frac{1}{e}\int_{-1}^{0} x^2 dx + \int_{0}^{1} x^2 dx$

$= \frac{1}{e}\left[\frac{x^3}{3}\right]_{-1}^{0} + \left[\frac{x^3}{3}\right]_{0}^{1} = \frac{1}{e}\left[0+\frac{1}{3}\right] + \left[\frac{1}{3}\right]$

$= \frac{1}{3e} + \frac{1}{3} = \frac{1+e}{3e}$

**34.** *(b)* Given, $f(0) = 1$ ... (i)

$f(2) = e^2$ ... (ii)

$f'(x) = f'(2-x)$

On integrating w.r.t. $x$, we get

$f(x) = -f(2-x) + C$

Putting $x = 0$

$f(0) = -f(2) + C$

$\Rightarrow 1 = -e^2 + C$ [From Eqs. (i) and (ii)]

$\Rightarrow C = 1 + e^2$

$\therefore f(x) = -f(2-x) + 1 + e^2$

or $f(x) + f(2-x) = 1 + e^2$ ...(iii)

Let $I = \int_0^2 f(x)\, dx$ ...(iv)

Also, $I = \int_0^2 f(2-x)\, dx$ ...(v)

Now, adding Eqs. (iv) and (v), we get

$2I = \int_0^2 [f(x) + f(2-x)]\, dx$

$2I = \int_0^2 (1+e^2)\, dx$ [From Eq. (iii)]

$2I = 2(1+e^2)$

$\therefore I = 1 + e^2$

**35.** *(b)* Let $I = \int_1^3 [x^2 - 2x - 2]\, dx$

$= \int_1^3 [x^2 - 2x + 1 - 3]\, dx$

$= \int_1^3 [(x-1)^2 - 3] dx$

$= \int_1^3 [(x-1)^2]\, dx + \int_1^3 (-3)\, dx$

Putting $x - 1 = t$, $dx = dt$,

when $x = 1$, $t = 0$ and when $x = 3$, $t = 2$

$\therefore I = -3[x]_1^3 + \int_0^2 [t^2]\, dt$

$= -6 + \int_0^1 0\, dt + \int_1^{\sqrt{2}} 1\, dt + \int_{\sqrt{2}}^{\sqrt{3}} 2\, dt + \int_{\sqrt{3}}^{2} 3\, dt$

$= -6 + (0) + (\sqrt{2} - 1) + 2(\sqrt{3} - \sqrt{2}) + 3(2 - \sqrt{3})$

$= -6 + \sqrt{2} - 1 + 2\sqrt{3} - 2\sqrt{2} + 6 - 3\sqrt{3}$

$I = -1 - \sqrt{2} - \sqrt{3}$

**36.** *(c)*

> **key idea** (i) Take $e^x \cdot x^x = t$
> $\Rightarrow e^x x^x (2 + \log_e x) dx = dt$
> (ii) $\int_a^b dt = [t]_a^b$

Given integral,

$$I = \int_1^2 e^x x^x (2 + \log_e x)\, dx$$

Let $e^x \cdot x^x = t$

$\Rightarrow [e^x x^x + e^x x^x (1 + \log_e x)]\, dx = dt$

$\Rightarrow e^x x^x (2 + \log_e x)\, dx = dt$

and at $x = 1$, $t = e$

at $x = 2$, $t = 4e^2$

$\therefore I = \int_e^{4e^2} dt = [t]_e^{4e^2} = 4e^2 - e$

$= e(4e - 1)$

**37.** *(c)* Given integral,

$I = \int_{\pi/6}^{\pi/3} \tan^3 x \cdot \sin^2 3x (2 \sec^2 x \sin^2 3x + 3 \tan x \cdot \sin 6x) dx$

$= \int_{\pi/6}^{\pi/3} [2 \tan^3 x \sec^2 x \sin^4 3x + 3 \tan^4 x (2 \sin^3 3x \cos 3x)] dx$

$= \int_{\pi/6}^{\pi/3} \left[\frac{d}{dx}\left(\frac{\tan^4 x \sin^4 3x}{2}\right)\right] dx$

$= \left[\frac{1}{2} \tan^4 x \sin^4 3x\right]_{\pi/6}^{\pi/3} = -\frac{1}{18}$

**38.** *(c)* Given integral, $\int_{-\pi}^{\pi} |\pi - |x||\, dx$

$= 2\int_0^{\pi} |\pi - |x||\, dx = 2\int_0^{\pi} |\pi - x|\, dx$

$= 2\int_0^{\pi} (\pi - x)\, dx$ [$\because x \in (0, \pi)$]

$= 2\left[\pi x - \frac{x^2}{2}\right]_0^{\pi} = 2\left(\frac{\pi^2}{2}\right) = \pi^2$

**39.** *(d)* It is given, $F(x) = \int_1^x t^2 g(t) dt$, then

$F'(x) = x^2 g(x)$

[By Newton-Leibnitz rule]

$\because g(t) = \int_1^t f(u) du$

$\Rightarrow g(1) = 0$

$\therefore F'(x=1) = 1^2 \cdot g(1) = 0$

Now, $F''(x) = 2x\, g(x) + x^2 g'(x)$

$= 2x \int_1^x f(u) du + x^2 f(x)$

Now, $F''(1) = 2\int_1^1 f(u) du + f(1)$

$= 2 \times 0 + 3$ ($\because f(1) = 3$ given)

$= 3$

$\because F''(1) = 3 > 0$ and $F'(1) = 0$

$\therefore$ The point $x = 1$ is point of local minima for the function $F$.

**40.** *(a)* Given limit $\lim_{x \to 0} \frac{\int_0^x t \sin(10t) dt}{x}$

On applying L'Hospital rule, we get

$\lim_{x \to 0} \frac{x \sin(10x)}{1}$

(on applying Newton Leibnitz rule)

$= 0 \sin(10 \times 0) = 0$

Hence, option (a) is correct.

**41.** *(c)* Let $I = \int_0^a f(x)\, g(x)\, dx$ ... (i)

$= \int_0^a f(a-x)\, g(a-x)\, dx$

$\left[\because \int_0^a f(x)\, dx = \int_0^a f(a-x)\, dx\right]$

$I = \int_0^a f(x)\, [4 - g(x)]\, dx$

[$\because f(x) = f(a-x)$ and $g(x) + g(a-x) = 4$]

$= \int_0^a 4 f(x)\, dx - \int_0^a f(x)\, g(x)\, dx$

$\Rightarrow I = 4\int_0^a f(x)\, dx - I$ [from Eq. (i)]

$\Rightarrow 2I = 4\int_0^a f(x) dx$

$\Rightarrow I = 2\int_0^a f(x)\, dx$

**42.** *(b)* Let $I = \int_{\pi/4}^{3\pi/4} \frac{dx}{1 + \cos x}$

$= \int_{\pi/4}^{3\pi/4} \frac{1 - \cos x}{1 - \cos^2 x} dx$

$= \int_{\pi/4}^{3\pi/4} \frac{1 - \cos x}{\sin^2 x} dx$

$= \int_{\pi/4}^{3\pi/4} (\text{cosec}^2 x - \text{cosec}\, x \cot x) dx$

$= [-\cot x + \text{cosec}\, x]_{\pi/4}^{3\pi/4}$

$= [(1 + \sqrt{2}) - (-1 + \sqrt{2})] = 2$

**43.** *(b)* Given, $2\int_0^1 \tan^{-1} x\, dx$

$= \int_0^1 \cot^{-1} (1 - x + x^2)\, dx$

$= \int_0^1 \left(\frac{\pi}{2} - \tan^{-1} (1 - x + x^2)\right) dx$

$$\Rightarrow 2\int_0^1 \tan^{-1} x\,dx = \int_0^1 \frac{\pi}{2}\,dx - \int_0^1 \tan^{-1}(1-x+x^2)\,dx$$

$$\Rightarrow \int_0^1 \tan^{-1}(1-x+x^2)\,dx = \frac{\pi}{2} - 2\int_0^1 \tan^{-1} x\,dx$$

$$= \frac{\pi}{2} - 2\left[ [(\tan^{-1} x)\,x]_0^1 - \int_0^1 \frac{1}{1+x^2}\,x\,dx \right]$$

$$= \frac{\pi}{2} - 2\left(\frac{\pi}{4}\right) + \int_0^1 \frac{2x}{1+x^2}\,dx$$

$$= \frac{\pi}{2} - \frac{\pi}{2} + [\log(1+x^2)]_0^1$$

$$= \log(2) - \log(1) = \log 2$$

**44.** *(b)* Given,

$$f\left(\frac{1}{x}\right) = \int_1^{1/x} \frac{\log t}{1+t}\,dt \quad \ldots(\text{i})$$

Put $t = \dfrac{1}{z}$

$$\Rightarrow dt = -\frac{1}{z^2}\,dz$$

$$f(x) = \int_1^x \frac{\log z}{z^2\left(1+\dfrac{1}{z}\right)}\,dz$$

$$= \int_1^x \frac{\log z}{z(1+z)}\,dz \quad \ldots(\text{ii})$$

On adding Eqs. (i) and (ii),

$$f(x) + f\left(\frac{1}{x}\right) = \int_1^x \log z\left[\frac{1}{1+z} + \frac{1}{z(1+z)}\right]dz$$

$$= \int_1^x \frac{\log z}{z}\,dz$$

$$= \left[\frac{(\log z)^2}{2}\right]_1^x = \frac{(\log x)^2}{2}$$

**45.** *(b)* $f(2-x) = f(2+x)$

$\Rightarrow$ Function is symmetrical about $x = 2$

and $f(4-x) = f(4+x)$

$\Rightarrow$ Function is symmetrical about $x = 4$

Hence, $f(x)$ is periodic function and the period of $f(x)$ is 2.

$$\int_{10}^{50} f(x)\,dx = \int_{2(5)}^{2(25)} f(x)\,dx$$

$$= (25-5)\int_0^2 f(x)\,dx$$

$$= 20 \times 5 = 100$$

**46.** *(b)* $\displaystyle\int_0^{\sin x} f(t)\,dt = \frac{\sqrt{3}}{2}x \quad \ldots(\text{i})$

On differentiating Eq. (i) w.r.t. $x$, we get

$$f(\sin x)\cos x = \frac{\sqrt{3}}{2} \quad \ldots(\text{ii})$$

On putting $x = \dfrac{\pi}{3}$ in Eq. (ii), we get

$$f\left(\frac{\sqrt{3}}{2}\right) \times \frac{1}{2} = \frac{\sqrt{3}}{2}$$

$$\Rightarrow f\left(\frac{\sqrt{3}}{2}\right) = \sqrt{3}$$

**47.** *(c)*

**key idea** Apply the property, $\displaystyle\int_a^b f(x)\,dx = \int_a^b f(a+b-x)\,dx.$

Let

$$I = \int_2^4 \frac{\log x^2}{\log x^2 + \log(36 - 12x + x^2)}\,dx$$

$$= \int_2^4 \frac{2\log x}{2\log x + \log(6-x)^2}\,dx$$

$$= \int_2^4 \frac{2\log x\,dx}{2[\log x + \log(6-x)]}$$

$$\Rightarrow I = \int_2^4 \frac{\log x\,dx}{[\log x + \log(6-x)]} \quad \ldots(\text{i})$$

$$\Rightarrow I = \int_2^4 \frac{\log(6-x)}{\log(6-x) + \log x}\,dx \quad \ldots(\text{ii})$$

$$\left[\because \int_a^b f(x)dx = \int_a^b f(a+b-x)dx\right]$$

On adding Eqs. (i) and (ii), we get

$$2I = \int_2^4 \frac{\log x + \log(6-x)}{\log x + \log(6-x)}\,dx$$

$$\Rightarrow \quad 2I = \int_2^4 dx = [x]_2^4 \Rightarrow 2I = 2$$

$$\Rightarrow \quad I = 1$$

**48.** *(d)* $\displaystyle\int_0^{\pi} \sqrt{\left(1 - 2\sin\frac{x}{2}\right)^2}\,dx$

$$= \int_0^{\pi} \left|1 - 2\sin\frac{x}{2}\right| dx$$

$$= \int_0^{\frac{\pi}{3}} \left(1 - 2\sin\frac{x}{2}\right)dx - \int_{\frac{\pi}{3}}^{\pi} \left(1 - 2\sin\frac{x}{2}\right)dx$$

$$= \left[x + 4\cos\frac{x}{2}\right]_0^{\frac{\pi}{3}} - \left[x + 4\cos\frac{x}{2}\right]_{\frac{\pi}{3}}^{\pi}$$

$$= 4\sqrt{3} - 4 - \frac{\pi}{3}$$

**49.** *(d)* Let $\displaystyle I = \int_{\pi/6}^{\pi/3} \frac{dx}{1+\sqrt{\tan x}} \quad \ldots(\text{i})$

$$\therefore \quad I = \int_{\pi/6}^{\pi/3} \frac{dx}{1 + \sqrt{\tan\left(\dfrac{\pi}{2} - x\right)}}$$

$$= \int_{\pi/6}^{\pi/3} \frac{dx}{1+\sqrt{\cot x}}$$

$$\Rightarrow \quad I = \int_{\pi/6}^{\pi/3} \frac{\sqrt{\tan x}\,dx}{1+\sqrt{\tan x}} \quad \ldots(\text{ii})$$

On adding Eqs. (i) and (ii), we get

$$2I = \int_{\pi/6}^{\pi/3} dx \Rightarrow 2I = [x]_{\pi/6}^{\pi/3}$$

$$\therefore \quad I = \frac{1}{2}\left[\frac{\pi}{3} - \frac{\pi}{6}\right] = \frac{\pi}{12}$$

Hence, Statement I is false.

But $\displaystyle\int_a^b f(x)\,dx = \int_a^b f(a+b-x)\,dx$ is a true statement by property of definite integrals.

**50.** *(c)* Here, $f(x) = \displaystyle\int_0^x \sqrt{t}\sin t\,dt$, where $x \in \left(0, \dfrac{5\pi}{2}\right)$

$$f'(x) = \{\sqrt{x}\sin x - 0\} \quad \ldots(\text{i})$$

[Using Newton-Leibnitz formula]

$$\therefore \quad f'(x) = \sqrt{x}\sin x = 0$$

$$\Rightarrow \quad \sin x = 0$$

$$\therefore \quad x = \pi, 2\pi$$

$$f''(x) = \sqrt{x}\cos x + \frac{1}{2\sqrt{x}}\sin x$$

$$f''(\pi) = -\sqrt{\pi} < 0$$

So, $f(x)$ has local maximum at $x = \pi$.

$$f''(2\pi) = \sqrt{\pi} > 0$$

So, $f(x)$ has local minimum at $x = 2\pi$.

**51.** *(d)* Here, $\displaystyle\int_0^{1.5} x[x^2]\,dx$

$$I = \int_0^1 x\cdot 0\,dx + \int_1^{\sqrt{2}} x\cdot 1\,dx + \int_{\sqrt{2}}^{1.5} x\cdot 2\,dx$$

$$= 0 + \left[\frac{x^2}{2}\right]_1^{\sqrt{2}} + [x^2]_{\sqrt{2}}^{1.5}$$

$$= \frac{1}{2}\{2-1\} + \{(1.5)^2 - 2\} = \frac{1}{2} + \frac{1}{4} = \frac{3}{4}$$

**52.** *(d)* Let $\displaystyle I = \int_0^{\pi} [\cot x]\,dx \quad \ldots(\text{i})$

$$\Rightarrow \quad I = \int_0^{\pi} [\cot(\pi - x)]\,dx$$

$= \int_0^{\pi} [-\cot x]\, dx$ ...(ii)

On adding Eqs. (i) and (ii), we get

$2I = \int_0^{\pi} [\cot x]\, dx + \int_0^{\pi} [-\cot x]\, dx$

$= \int_0^{\pi} (-1)\, dx$

$\left[\because [x] + [-x] = \begin{cases} -1 \text{ if } x \notin Z \\ 0, \text{ if } x \in Z \end{cases}\right]$

$= [-x]_0^{\pi} = -\pi$

$\therefore \quad I = -\dfrac{\pi}{2}$

**53.** *(c)* Since,

$I = \int_0^1 \dfrac{\sin x}{\sqrt{x}} dx < \int_0^1 \dfrac{x}{\sqrt{x}} dx,$

because in $x \in (0, 1)$, $x > \sin x$

$\Rightarrow I < \int_0^1 \sqrt{x}\, dx = \dfrac{2}{3}[x^{3/2}]_0^1 \Rightarrow I < \dfrac{2}{3}$

and $J = \int_0^1 \dfrac{\cos x}{\sqrt{x}} dx < \int_0^1 x^{-\frac{1}{2}} dx$

$= 2[x^{1/2}]_0^1 = 2$

$\therefore \quad J < 2$

**54.** *(a)* Given, $\int_{\sqrt{2}}^{x} \dfrac{dt}{t\sqrt{t^2 - 1}} = \dfrac{\pi}{2}$

$\Rightarrow \quad [\sec^{-1} t]_{\sqrt{2}}^{x} = \dfrac{\pi}{2}$

$\Rightarrow \quad \sec^{-1} x - \sec^{-1} \sqrt{2} = \dfrac{\pi}{2}$

$\Rightarrow \quad \sec^{-1} x = \dfrac{\pi}{2} + \dfrac{\pi}{4} = \dfrac{3\pi}{4}$

$\therefore \quad x = \sec \dfrac{3\pi}{4} = -\sqrt{2}$

**55.** *(a)* Let $I = \int_3^6 \dfrac{\sqrt{x}}{\sqrt{9 - x} + \sqrt{x}} dx$ ...(i)

$= \int_3^6 \dfrac{\sqrt{9 - x}}{\sqrt{9 - 9 + x} + \sqrt{9 - x}} dx$

$\Rightarrow I = \int_3^6 \dfrac{\sqrt{9 - x}}{\sqrt{x} + \sqrt{9 - x}} dx$ ...(ii)

On adding Eqs. (i) and (ii), we get

$2I = \int_3^6 \dfrac{\sqrt{x} + \sqrt{9 - x}}{\sqrt{x} + \sqrt{9 - x}} dx \Rightarrow I = \dfrac{3}{2}$

**56.** *(a)* Since, $\int_1^a [x] f'(x)\, dx$

$= \int_1^2 f'(x)\, dx + \int_2^3 2f'(x)\, dx$

$+ \ldots + \int_{[a]}^{a} [a] f'(x)\, dx$

$= [f(x)]_1^2 + 2[f(x)]_2^3 + \ldots + [a][f(x)]_{[a]}^{a}$

$= f(2) - f(1) + 2f(3) - 2f(2) + \ldots$

$+ [a] f(a) - [a] f([a])$

$= [a] f(a) - \{f(1) + f(2) + \ldots + f([a])\}$

**57.** *(a)*

> **key idea** (i) Use L' Hospital rule as the given equation is of the form $\left(\dfrac{0}{0}\right)$.
>
> (ii) And then use the value of $f(2)$ and $f'(2)$.

$\lim_{x \to 2} \int_6^{f(x)} \dfrac{4t^3}{x - 2} dt$

$= \lim_{x \to 2} \dfrac{\int_6^{f(x)} 4t^3\, dt}{(x - 2)} \quad \left[\text{form } \dfrac{0}{0}\right]$

[By Leibnitz's rule]

$= \lim_{x \to 2} \dfrac{4\{f(x)\}^3}{1} f'(x)$

$= 4\{f(2)\}^3 f'(2) = 4 \times (6)^3 \times \dfrac{1}{48}$

$\left[\because f(2) = 6 \text{ and } f'(2) = \dfrac{1}{48}, \text{ given}\right]$

$= 18$

**58.** *(c)* Given that

$I_1 = \int_0^1 2^{x^2} dx$

$I_2 = \int_0^1 2^{x^3} dx,$

$I_3 = \int_1^2 2^{x^2} dx$ and $I_4 = \int_1^2 2^{x^3} dx$

Since, $\quad 2^{x^3} < 2^{x^2}$ for $0 < x < 1$

and $\quad 2^{x^3} > 2^{x^2}$ for $x > 1$

$\therefore \quad \int_0^1 2^{x^3} dx < \int_0^1 2^{x^2} dx$

and $\int_1^2 2^{x^3} dx > \int_1^2 2^{x^2} dx$

$\Rightarrow \quad I_2 < I_1$ and $I_4 > I_3$

**59.** *(c)* Given, $f'(x) = f(x)$ and $f(0) = 1$

Let $\quad f(x) = e^x$ ...(i)

Also, $\quad f(x) + g(x) = x^2$

$\Rightarrow \quad g(x) = x^2 - e^x$ ...(ii)

Now,

$\int_0^1 f(x) g(x)\, dx = \int_0^1 e^x (x^2 - e^x)\, dx$

From Eqs. (i) and (ii), we get

$\int_0^1 x^2 e^x\, dx - \int_0^1 e^{2x}\, dx$

$= [x^2 e^x - \int 2xe^x\, dx]_0^1 - \dfrac{1}{2}[e^{2x}]_0^1$

$= [x^2 e^x - 2xe^x + 2e^x]_0^1 - \dfrac{1}{2}(e^2 - 1)$

$= [(x^2 - 2x + 2)e^x]_0^1 - \dfrac{1}{2}e^2 + \dfrac{1}{2}$

$= e - 2 - \dfrac{e^2}{2} + \dfrac{1}{2} = e - \dfrac{e^2}{2} - \dfrac{3}{2}$

**60.** *(a)* Since, $|\sin x|$ is a periodic function with period $\pi$.

$\therefore \quad \int_0^{10\pi} |\sin x|\, dx = 10 \int_0^{\pi} |\sin x|\, dx$

$= 10 \int_0^{\pi} \sin x\, dx = 10[-\cos x]_0^{\pi}$

$= 10[-\cos \pi + \cos 0] = 10[1 + 1] = 20$

**61.** *(a)*

> **key idea** (i) Directly use the property of integration
>
> $\int_a^b f(x)\, dx = \int_a^b f(a + b - x)\, dx$
>
> (ii) And then solve it.

Let

$I = \int_0^{\pi/2} \dfrac{\sqrt{\sin x}}{\sqrt{\cos x} + \sqrt{\sin x}} dx$ ...(i)

$I = \int_0^{\pi/2} \dfrac{\sqrt{\sin\left(\frac{\pi}{2} - x\right)}}{\sqrt{\cos\left(\frac{\pi}{2} - x\right)} + \sqrt{\sin\left(\frac{\pi}{2} - x\right)}} dx$

$I = \int_0^{\pi/2} \dfrac{\sqrt{\cos x}}{\sqrt{\sin x} + \sqrt{\cos x}} dx$ ...(ii)

On adding Eqs. (i) and (ii), we get

$2I = \int_0^{\pi/2} 1\, dx = [x]_0^{\pi/2} \Rightarrow I = \dfrac{\pi}{4}$

**62.** *(b)* $\log_4 \log_6 (3 + 4x - x^2) > 0$

$\log_6(3 + 4x - x^2) > 4^0$

$3 + 4x - x^2 > 6$

$-x^2 + 4x - 3 > 0$

$x \in (1, 3)$ ...(i)

Also, $\log_6(3 + 4x - x^2) > 0$

$-x^2 + 4x + 2 > 0$

$\Rightarrow x \in (2 - \sqrt{6}, 2 + \sqrt{6})$ ...(ii)

and $3 + 4x - x^2 > 0$

$\Rightarrow x \in (2 - \sqrt{7}, 2 + \sqrt{7})$ ...(iii)

From Eqs. (i), (ii), (iii), we get

$a = 1, b = 3 \Rightarrow b - a = 2$

$\therefore \int_0^2 [x^2] dx = \int_0^1 0 + \int_1^{\sqrt{2}} 1\, dx$

$+ \int_{\sqrt{2}}^{\sqrt{3}} 1\, dx + \int_{\sqrt{3}}^{2} 3\, dx$

$= 5 - \sqrt{2} - \sqrt{3} = p - \sqrt{q} - \sqrt{r}$ (given)

$\Rightarrow p = 5, q = 2, r = 3 \Rightarrow p + q + r = 10$

**63.** (c) $f(x) + 2f\left(\frac{1}{x}\right) = x^2 + 5$ ...(i)

Replace $x \to \frac{1}{x}$

$f\left(\frac{1}{x}\right) + 2f(x) = \frac{1}{x^2} + 5$ ...(ii)

Solving Eqs. (i) and (ii), we get

$3f(x) = \frac{2}{x^2} + 10 - x^2 - 5$

$f(x) = \frac{2}{3x^2} + \frac{5}{3} - \frac{x^2}{3}$

$\alpha = \int_1^2 f(x)\, dx$

$= \int_1^2 \left(\frac{2}{3x^2} + \frac{5}{3} - \frac{x^2}{3}\right) dx$

$= \left[-\frac{2}{3x} + \frac{5x}{3} - \frac{x^3}{9}\right]_1^2$

$= \frac{-2}{3}\left(\frac{1}{2} - 1\right) + \frac{5}{3}(2-1) - \frac{1}{9}(2^3 - 1^3)$

$= \frac{1}{3} + \frac{5}{3} - \frac{1}{9}(7) = 2 - \frac{7}{9}$

$9\alpha = 18 - 7 = 11$

and $2g(x) - 3g\left(\frac{1}{2}\right) = x$

$x = \frac{1}{2} \Rightarrow 2g\left(\frac{1}{2}\right) - 3g\left(\frac{1}{2}\right) = \frac{1}{2}$

$\Rightarrow -g\left(\frac{1}{2}\right) = \frac{1}{2} \Rightarrow g\left(\frac{1}{2}\right) = \frac{-1}{2}$

$\therefore 2g(x) - 3\left(\frac{-1}{2}\right) = x \Rightarrow 2g(x) = x - \frac{3}{2}$

$\Rightarrow g(x) = \frac{x}{2} - \frac{3}{4} \Rightarrow \beta = \int_1^2 \left(\frac{x}{2} - \frac{3}{4}\right) dx$

$= \left[\frac{x^2}{4} - \frac{3x}{4}\right]_1^2 = \frac{1}{4}(2^2 - 1^2) - \frac{3}{4}(2-1)$

$= \frac{3}{4} - \frac{3}{4} = 0$

**64.** (a)

$I = \int_{-1}^{1} \frac{(1 + \sqrt{|x| - x})\, e^x (1 + \sqrt{|x| + x})\, e^x}{e^x + e^{-x}}$ ...(i)

Using kings property

$I = \int_{-1}^{1} \frac{1 + (\sqrt{|x| + x})\, e^{-x} \sqrt{|x| + x\, e^x}}{e^{-x} + e^x} dx$ ...(ii)

Adding Eqs. (i) and (ii), we get

$2I = \int_{-1}^{1} \frac{(e^{-x} + e^x)}{e^x + e^{-x}}$

$(1 + \sqrt{|x| + x} + \sqrt{|x| - x})dx$

$I = \frac{1}{2}\int_{-1}^{1} (1 + \sqrt{|x| + x} + \sqrt{|x| - x})dx$

$= \frac{1}{2} \times 2 \int_0^1 (1 + \sqrt{|x| + x} + \sqrt{|x| - x})dx$

$= \int_0^1 (1 + \sqrt{2x} + 0)dx$

$= \left[x + \frac{2\sqrt{2}}{3} x^{\frac{3}{2}}\right]_0^1$

$= 1 + \frac{2\sqrt{2}}{3}$

**65.** (d) Let $I = 80 \int_0^{\frac{\pi}{4}} \frac{\sin\theta + \cos\theta}{9 + 16\sin 2\theta} d\theta$

$= 80 \int_0^{\frac{\pi}{4}} \frac{\sin\theta + \cos\theta}{9 + 16(2\sin\theta \cdot \cos\theta)} d\theta$

$= 80 \int_0^{\frac{\pi}{4}} \frac{\sin\theta + \cos\theta}{9 + 16(1 + 2\sin\theta\cos\theta - 1)} d\theta$

$= 80 \int_0^{\frac{\pi}{4}} \frac{\sin\theta + \cos\theta}{9 + 16 - 16(\sin\theta - \cos\theta)^2} d\theta$

On putting $\sin\theta - \cos\theta = z$ so that $(\cos\theta + \sin\theta)\, d\theta = dz$

| $\theta$ | 0 | $\pi/4$ |
|---|---|---|
| $z$ | $-1$ | 0 |

$I = 80 \int_{-1}^{0} \frac{dz}{25 - 16z^2}$

$= 80 \int_{-1}^{0} \frac{dz}{16\left(\frac{25}{16} - z^2\right)}$

$= \frac{80}{16} \int_{-1}^{0} \frac{dz}{\left(\frac{5}{4}\right)^2 - (z)^2}$

$= 5 \times \frac{1}{2 \times \frac{5}{4}} \log_e \left(\frac{\frac{5}{4} + z}{\frac{5}{4} - z}\right)_{-1}^{0}$

$= 2\log_e(1) + 4\log_e(3)$

$= 2 \times 0 + 4\log_e(3) = 4\log_e(3)$

**66.** (c)

**key idea** $\int_{-a}^{a} f(x)\,dx = \int_0^a [f(x) + f(-x)]\,dx$

Let $I = \int_{-\pi/2}^{\pi/2} \frac{96x^2 \cos^2 x}{1 + e^x} dx$

$I = \int_0^{\pi/2} \left(\frac{96x^2 \cos^2 x}{1 + e^x} + \frac{96x^2 \cos^2 x}{1 + e^{-x}}\right) dx$

$= \int_0^{\pi/2} 96x^2 \cos^2 x\, dx$

$= 96 \int_0^{\pi/2} x^2 \left(\frac{1 + \cos 2x}{2}\right) dx$

$= 48 \left[\frac{x^3}{3} + \frac{x^2}{2}\sin 2x - \int 2x \cdot \frac{\sin 2x}{2} dx\right]_0^{\pi/2}$

$= 48 \left[\frac{x^3}{3} + \frac{x^2}{2}\sin 2x + \frac{x}{2}\cos 2x - \frac{\sin 2x}{4}\right]_0^{\pi/2}$

$= 48\left[\frac{\pi^3}{24} + 0 + \frac{\pi}{4}(-1) - 0\right]$

$= 2\pi^3 - 12\pi = \pi(2\pi^2 - 12)$

$\Rightarrow \alpha = 2, \beta = -12$

$\therefore (\alpha + \beta)^2 = (2 - 12)^2 = 100$

**67.** (b)

**key idea** Use property $\int_0^a f(x)dx = \int_0^a f(a - x)dx$

$I = \int_0^{\pi/2} \frac{\sin^{3/2} x}{\sin^{3/2} x + \cos^{3/2} x} dx$ ...(i)

Using property,

$I = \int_0^{\pi/2} \frac{\sin^{3/2}\left(\frac{\pi}{2} - x\right)}{\sin^{3/2}\left(\frac{\pi}{2} - x\right) + \cos^{3/2}\left(\frac{\pi}{2} - x\right)} dx$

$I = \int_0^{\pi/2} \frac{\cos^{3/2}\left(\frac{\pi}{2} - x\right)}{\cos^{3/2}\left(\frac{\pi}{2} - x\right) + \sin^{3/2}\left(\frac{\pi}{2} - x\right)}$ ...(ii)

On adding Eqs. (i) and (ii), we get

$$2I = \int_0^{\pi/2} 1\,dx$$

$$I = \frac{\pi}{4}$$

Let $I_2 = \int_0^{2(\pi/4)} \frac{x \sin x \cos x}{\sin^4 x + \cos^4 x} dx$ ...(iii)

Using property,

$$I_2 = \int_0^{\pi/2} \frac{\left(\frac{\pi}{2} - x\right) \cos x \sin x}{\sin^4 x + \cos^4 x} \quad \text{...(iv)}$$

On adding Eqs. (iii) and (iv), we get

$$2I_2 = \int_0^{\pi/2} \frac{\frac{\pi}{2} \sin x \cos x}{\sin^4 x + \cos^4 x} dx$$

$$\Rightarrow \quad 2I_2 = \frac{\pi}{2} \int_0^{\pi/2} \frac{\sin x \cos x\, dx}{\sin^4 x + \cos^4 x}$$

$$\Rightarrow \quad 2I_2 = \frac{\pi}{2} \int_0^{\pi/2} \frac{\tan x \sec^2 x\, dx}{\tan^4 x + 1}$$

let $\tan^2 x = t$

$$\Rightarrow \quad 2 \tan x \sec^2 x\, dx = dt$$

$$\Rightarrow \quad 2I_2 = \frac{\pi}{4} \int_0^{\infty} \frac{dt}{1 + t^2}$$

$$\Rightarrow \quad I_2 = \frac{\pi}{8} [\tan^{-1} t]_0^{\infty} = \frac{\pi^2}{16}$$

**68.** (a) $f(x) = 7 \tan^8 x + 7 \tan^6 x - 3 \tan^4 x - 3 \tan^2 x$

$$= 7 \tan^6 x (\tan^2 x + 1) - 3 \tan^2 x (\tan^2 x + 1)$$

$$= \sec^2 x (7 \tan^6 x - 3 \tan^2 x)$$

$$= \sec^2 x \tan^2 x (7 \tan^4 x - 3)$$

$$I_1 = \int_0^{\frac{\pi}{4}} f(x) dx$$

$$= \int_0^{\frac{\pi}{4}} \sec^2 x (7 \tan^6 x - 3 \tan^2 x) dx$$

Let $\tan x = t$

$\Rightarrow \sec^2 x\, dx = dt$

$$= \int_0^1 (7t^6 - 3t^2) dt$$

$$= [t^7 - t^3]_0^1 = 0$$

$$I_2 = \int_0^{\frac{\pi}{4}} x\, f(x) dx$$

$$= x \int_0^{\frac{\pi}{4}} f(x) - \int_0^{\frac{\pi}{4}} \left(1 \cdot \int f(x) dx\right) dx$$

$$= 0 - \int_0^{\frac{\pi}{4}} \tan^7 x - \tan^3 x$$

$$I_2 = -\int_0^{\frac{\pi}{4}} \tan^3 x (\tan^4 x - 1) dx$$

$$= -\int_0^{\frac{\pi}{4}} \tan^3 x (\tan^2 x - 1)(\tan^2 x + 1) dx$$

$$= -\int_0^{\frac{\pi}{4}} \tan^3 x (\tan^2 x - 1) \sec^2 x\, dx$$

Let $\tan x = t$

$\sec^2 x dx = dt$

when $x = 0,\ t = 0$

and $x = \frac{\pi}{4},\ t = 1$

$$= \int_0^1 t^3 (1 - t^2) dt$$

$$= \int_0^1 (t^3 - t^5) dt$$

$$= \left[\frac{t^4}{4} - \frac{t^6}{6}\right]_0^1 = \frac{1}{4} - \frac{1}{6} = \frac{1}{12}$$

$$\therefore \quad 7I_1 + 12 I_2 = 1$$

**69.** (c) $\int_{\alpha}^{\log_e(4)} \frac{dx}{\sqrt{e^x - 1}} = \frac{\pi}{6}$

Let $e^x - 1 = t^2$

$\Rightarrow e^x \cdot dx = 2t \cdot dt$

$$\Rightarrow \int \frac{2}{t^2 + 1} \cdot dt = \frac{\pi}{6}$$

$$\Rightarrow \ 2 [\tan^{-1}(t)] = \frac{\pi}{6}$$

Put $t = \sqrt{e^x - 1}$,

$$2 [\tan^{-1} (\sqrt{e^x - 1})]_{\alpha}^{\log_e(4)} = \frac{\pi}{6}$$

$$\Rightarrow 2 [\tan^{-1}(\sqrt{3}) - \tan^{-1} (\sqrt{e^{\alpha} - 1})] = \frac{\pi}{6}$$

$$\Rightarrow \frac{\pi}{3} - \tan^{-1} \sqrt{e^{\alpha} - 1} = \frac{\pi}{12}$$

$$\Rightarrow \tan^{-1}(\sqrt{e^{\alpha} - 1}) = \frac{\pi}{4}$$

$$\Rightarrow e^{\alpha} - 1 = 1$$

$$\Rightarrow \quad e^{\alpha} = 2 \text{ and } e^{-\alpha} = \frac{1}{2}$$

$$e^{\alpha} + e^{-\alpha} = 2 + \frac{1}{2} = \frac{5}{2}$$

and $\quad e^{\alpha} \cdot e^{-\alpha} = 1$

Then, the equation is

$$x^2 - \left(\frac{5}{2}\right) x + 1 = 0$$

$$\Rightarrow \quad 2x^2 - 5x + 2 = 0$$

**70.** (a)

> **key idea** Use first
>
> $$\int_{-a}^{a} f(x) dx = \begin{cases} 2\int_0^a f(x) dx, & \text{if } f(-x) = f(x) \\ 0, & \text{if } f(-x) = -f(x) \end{cases}$$
>
> and then $\int_0^a f(x) dx = \int_0^a f(a - x) dx$

We have,

$$\int_{-\pi}^{\pi} \frac{2y(1 + \sin y)}{1 + \cos^2 y} dy$$

$$= \int_{-\pi}^{\pi} \frac{2y}{1 + \cos^2 y} dy + \int_{-\pi}^{\pi} \frac{2y \sin y}{1 + \cos^2 y} dy$$

Here, $\int_{-\pi}^{\pi} \frac{2y}{1 + \cos^2 y} dy = 0$

$$\left\{\text{Since, } \frac{2y}{1 + \cos^2 y} \text{ is odd function}\right\}$$

Let $I = 2 \times 2 \int_0^{\pi} \frac{y \sin y}{1 + \cos^2 y} dy$ ...(i)

$$\left\{\text{Since, } \frac{y \sin y}{1 + \cos^2 y} \text{ is even}\right\}$$

$$= 4 \int_0^{\pi} \frac{(\pi - y) \sin (\pi - y)}{1 + \cos^2 (\pi - y)} dy$$

$$= 4 \int_0^{\pi} \frac{(\pi - y) \sin y}{1 + \cos^2 y} \quad \text{...(ii)}$$

On adding Eqs. (i) and (ii), we get

$$2I = 4 \int_0^{\pi} \frac{\pi \sin y}{1 + \cos^2 y} dy$$

$$\Rightarrow I = 2\pi \int_0^{\pi} \frac{\sin y}{1 + \cos^2 y}$$

$$= 2\pi [-\tan^{-1} (\cos y)]_0^{\pi}$$

$$= -2\pi [\tan^{-1} (\cos \pi) - \tan^{-1} (\cos 0)]$$

$$= -2\pi \left[-\frac{\pi}{4} - \frac{\pi}{4}\right]$$

$$= (-2\pi)\left(-\frac{\pi}{2}\right) = \pi^2$$

**71.** *(a)* Let $I = \int_0^{\pi/4} \frac{136 \sin x}{3 \sin x + 5 \cos x} dx$

Now, $136 \sin x = A(3 \sin x + 5 \cos x) + B(3 \cos x - 5 \sin x)$

$\Rightarrow 136 \sin x = (3A - 5B) \sin x + (5A + 3B) \cos x$

On comparing both sides, we get

$3A - 5B = 136$ ...(i)

and $5A + 3B = 0$ ...(ii)

On solving Eqs. (i) and (ii), we get

$A = 12$ and $B = -20$

$\therefore \quad I = \int_0^{\pi/4} \frac{12(3 \sin x + 5 \cos x) - 20(3 \cos x - 5 \sin x)}{(3 \sin x + 5 \cos x)} dx$

$= \int_0^{\pi/4} 12 dx - 20 \int_0^{\pi/4} \frac{3 \cos x - 5 \sin x}{3 \sin x + 5 \cos x} dx$

$\Rightarrow I = 12[x]_0^{\pi/4} - 20[\log_e (3 \sin x + 5 \cos x)]_0^{\pi/4}$

$= 12\left(\frac{\pi}{4}\right) - 20\left[\log_e\left(3 \sin \frac{\pi}{4} + 5 \cos \frac{\pi}{4}\right) - \log_e (3 \times 0 + 5 \times 1)\right]$

$= 3\pi - 20\left[\log_e\left(\frac{8}{\sqrt{2}}\right) - \log_e (5)\right]$

$= 3\pi - 20 [\log_e (4\sqrt{2}) - \log_e (5)]$

$= 3\pi - 20 \log_e (4\sqrt{2}) + 20 \log_e (5)$

$= 3\pi - 20 \log_e (\sqrt{32}) + 20 \log_e 5$

$= 3\pi - 20 \times \frac{5}{2} \log_e 2 + 20 \log_e 5$

$= 3\pi - 50 \log_e 2 + 20 \log_e 5$

**72.** *(d)*

**key idea** Use Leibnitz's rule

$F(x) = \int_{u(x)}^{v(x)} f(t) dt$

$\Rightarrow F'(x) = f(v(x)) \cdot v'(x) - f(u(x)) \cdot u'x$

$f(x) = \int_0^x [t + \sin (1 - e^t)] dt$

$\lim_{x \to 0} \frac{f(x)}{x^3} = \left(\frac{0}{0}\right)$

Using L'Hospital rule,

$\lim_{x \to 0} \frac{f(x)}{3x^2} = \lim_{x \to 0} \frac{1(x + \sin (1 - e^x))}{3x^2}$ $\left[\frac{0}{0} \text{ form}\right]$

(Using Leibnitz rule)

$= \lim_{x \to 0} \frac{1 + \cos (1 - e^x) \cdot (-e^x)}{6x}$

[again L'Hospital rule]

$\Rightarrow \lim_{x \to 0} \frac{0 + \sin (1 - e^x) e^x (-e^x) - e^x (0) (1 - e^x)}{6}$

[again L'Hospital rule]

$= -\frac{1}{6}$

**73.** *(d)* Given,

$f(x) = \begin{cases} -2, & -2 \le x \le 0 \\ x - 2, & 0 < x \le 2 \end{cases}$

$f(|x|) = \begin{cases} -x - 2, & -2 \le x \le 0 \\ x - 2, & 0 < x \le 2 \end{cases}$

($\because f(x) = -2, -2 \le x \le 0$ is not possible for $f(|x|)$)

and $|f(x)| = \begin{cases} 2, & -2 \le x \le 0 \\ -(x - 2), & 0 < x \le 2 \end{cases}$

So, $h(x) = f(|x|) + |f(x)|$

$= \begin{cases} (-x - 2) + 2 = -x, & -2 \le x \le 0 \\ x - 2 - x + 2 = 0, & 0 < x \le 2 \end{cases}$

Hence, $\int_{-2}^2 h(x) dx = \int_{-2}^0 h(x) dx + \int_0^2 h(x) dx = \int_{-2}^0 (-x) dx + 0 = 2 + 0 = 2$

**Alternate method**

(Graphical Approach)

First of all we draw the graph of $f(x)$.

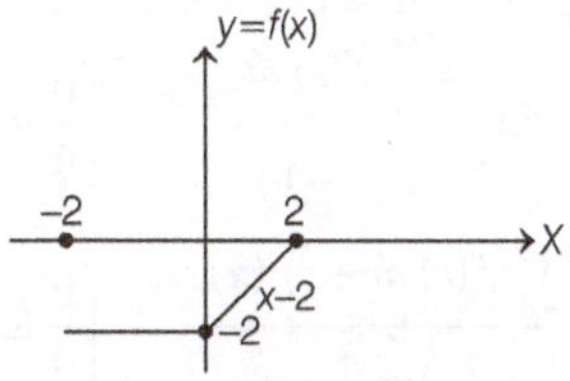


Now, we draw the graph of $f(|x|)$ and $|f(x)|$ as below→

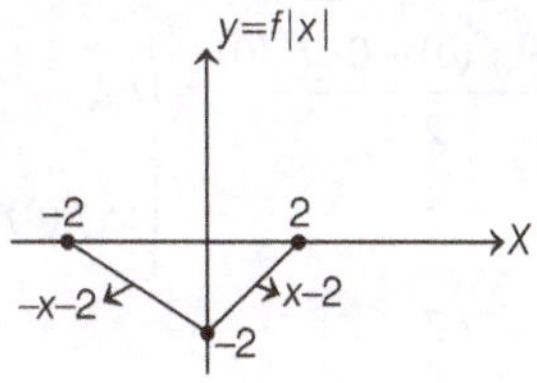


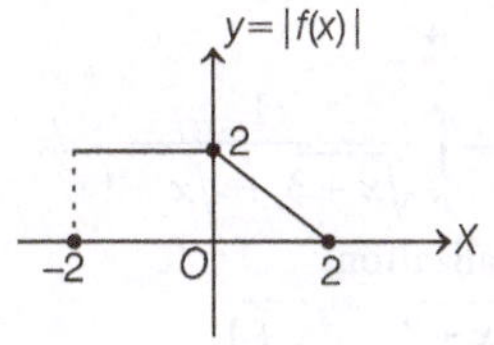


So, $h(x) = \begin{cases} -x - 2 + 2 = -x, & -2 \le x \le 0 \\ x - 2 + 2 - x = 0, & 0 < x \le 2 \end{cases}$

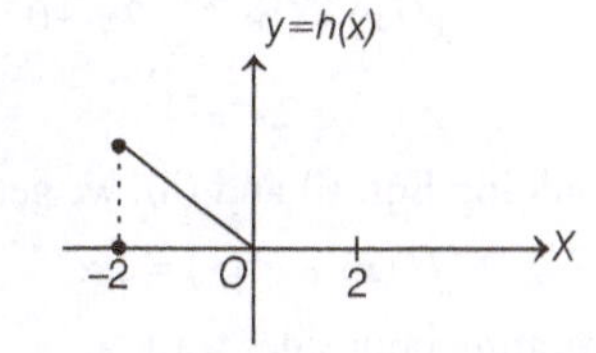


Hence, $\int_{-2}^2 h(x) dx = \int_{-2}^0 h(x) dx + \int_0^2 h(x) dx$

$= 2 + 0 = 2$

**74.** *(c)* We have,

$I = \int_0^{\frac{\pi}{3}} \cos^4 x \, dx = a\pi + b\sqrt{3}$

$= \int_0^{\frac{\pi}{3}} [\cos^2 x]^2 dx$

$= \int_0^{\frac{\pi}{3}} \left[\frac{\cos 2x + 1}{2}\right]^2 dx$

$= \frac{1}{4} \int_0^{\frac{\pi}{3}} (1 + \cos^2 2x + 2 \cos 2x) dx$

$= \frac{1}{4}\left[\frac{\pi}{3} + \int_0^{\frac{\pi}{3}} \cos^2 2x \, dx + [\sin 2x]_0^{\frac{\pi}{3}}\right]$

$= \frac{1}{4}\left[\frac{\pi}{3} + \int_0^{\frac{\pi}{3}} \left[\frac{1 + \cos 4x}{2}\right] dx + \frac{\sqrt{3}}{2}\right]$

$= \frac{1}{4}\left[\frac{\pi}{3} + \frac{1}{2}\left(\frac{\pi}{3} + \frac{1}{4} (\sin 4x)_0^{\frac{\pi}{3}}\right) + \frac{\sqrt{3}}{2}\right]$

$= \frac{1}{4}\left[\frac{\pi}{3} + \frac{\pi}{6} + \frac{1}{8} \sin \frac{4\pi}{3} + \frac{\sqrt{3}}{2}\right]$

$= \frac{1}{4}\left[\frac{\pi}{2} + \frac{1}{8}\left(\frac{-\sqrt{3}}{2}\right) + \frac{\sqrt{3}}{2}\right]$

$= \frac{\pi}{8} + \frac{7\sqrt{3}}{64} \therefore a = \frac{1}{8}, b = \frac{7}{64}$

Now, $9a + 8b = \frac{9}{8} + \frac{7}{8} = 2$

**75.** *(a)* Let $f, g : (0, \infty) \to R$

$f(x) = \int_{-x}^x (|t| - t^2) e^{-t^2} dt$

Differentiating w.r.t. $x$ both side, we get

$f'(x) = (|x| - x^2) e^{-x^2} - (|x| - x^2) e^{-x^2} (-1)$

$f'(x) = 2(|x| - x^2) e^{-x^2}$

$f'(x) = 2|x| e^{-x^2} - 2x^2 e^{-x^2}$ ...(i)

$g : (0, \infty) \to R$

$\Rightarrow \quad g(x) = \int_0^{x^2} t^{1/2} e^{-t} dt$

$\Rightarrow \quad g'(x) = (xe^{-x^2})\, 2x - 0$

$\Rightarrow \quad g'(x) = 2x^2 e^{-x^2}$ ...(ii)

On adding Eqs. (i) and (ii), we get

$$f'(x) + g'(x) = 2xe^{-x^2}$$

Integrating both sides w.r.t. $x$,

$$f(x) + g(x) = \int_0^{\sqrt{\log_e 9}} 2xe^{-x^2} dx$$

Put $x^2 = t \Rightarrow 2x\,dx = dt$

$$f(x) + g(x) = \int_0^{\log_e 9} e^{-t} dt = [-e^{-t}]_0^{\log_e 9}$$

$$f(x) + g(x) = -[e^{-\log_e 9} - e^0]$$

$$= 1 - \frac{1}{9} = \frac{8}{9}$$

$\Rightarrow \quad 9(f(x) + g(x)) = 8$

$\Rightarrow 9(f(\sqrt{\log_e 9} + g(\sqrt{\log_e 9})) = 8$

**76.** *(b)* Given, $f(x) = \begin{cases} x^2 + 3x + a, & x \le 1 \\ bx + 2, & x > 1 \end{cases}$

$f(x)$ is continuous and differentiable on $R$.

$\therefore \quad f(1) = f(1^+) = f(1^-)$

$\Rightarrow \quad 1 + 3 + a = b + 2$

$\Rightarrow \quad a - b + 2 = 0$ ...(i)

Now, on differentiating w.r.t. $x$, we get

$$f'(x) = \begin{cases} 2x + 3, & x < 1 \\ b, & x > 1 \end{cases}$$

$$f'(1^-) = f'(1^+)$$

$$5 = b \Rightarrow b = 5$$

$\therefore$ By Eq. (i), $\quad a = 3$

Now, $\quad \int_{-2}^{2} f(x) = \int_{-2}^{1} (x^2 + 3x + 3) dx + \int_1^2 (5x + 2) dx$

$$= \left[\frac{x^3}{3} + \frac{3x^2}{2} + 3x\right]_{-2}^{1} + \left[\frac{5x^2}{2} + 2x\right]_1^2$$

$$= \frac{1}{3}(1 + 8) + \frac{3}{2}[1 - 4] + 3(1 + 2) + \frac{5}{2}(4 - 1) + 2(2 - 1)$$

$$= 3 - \frac{9}{2} + 9 + \frac{15}{2} + 2 = 14 + 3 = 17$$

**77.** *(c)* Given, $f(x) = ae^{2x} + be^x + cx$

and $\quad f'(x) = 2ae^{2x} + be^x + c$

Now, $\quad f(0) = -1$

$\Rightarrow \quad a + b = -1$ ...(i)

$\because \quad f'(\ln 2) = 21$

$\Rightarrow 2ae^{2\ln 2} + be^{\ln 2} + c = 21$

$\Rightarrow \quad 8a + 2b + c = 21$ ...(ii)

and $\quad \int_0^{\log_e 4} [f(x) - cx] dx = \frac{39}{2}$

$\Rightarrow \quad \int_0^{\ln 4} (ae^{2x} + be^x) dx = \frac{39}{2}$

$\Rightarrow \quad \left[\frac{ae^{2x}}{2} + be^x\right]_0^{\ln 4} = \frac{39}{2}$

$\Rightarrow \left[\frac{ae^{2\ln 4}}{2} + be^{\ln 4}\right] - \left(\frac{a}{2} + b\right) = \frac{39}{2}$

$\Rightarrow \quad 8a + 4b - \frac{a}{2} - b = \frac{39}{2}$

$\Rightarrow \quad 16a + 8b - a - 2b = 39$

$\Rightarrow \quad 15a + 6b = 39$

$\Rightarrow \quad 5a + 2b = 13$ ...(iii)

By Eqs. (i) and (iii), $a = 5$, $b = -6$

by Eq. (ii), $c = -7$

$\therefore \quad |a + b + c| = |5 - 6 - 7| = 8$

**78.** *(b)*

> **key idea** Consider Newton Leibnitz formula for limits
>
> $\frac{d}{dx}\int_{h(x)}^{g(x)} f(t) dt = f(g(x)) \times g'(x) - f(h(x)) \times h'(x)$

We have, $\lim_{x \to 0} \frac{x\int_0^x f(t)\,dt}{e^{x^2} - 1} = \alpha$

or $\alpha = \lim_{x \to 0} \frac{x\int_0^x f(t)\,dt}{e^{x^2} - 1}$ $\quad \left[\frac{0}{0} \text{ form}\right]$

$= \lim_{x \to 0} \frac{\int_0^x f(t)\,dt + xf(x)}{2xe^{x^2}}$ $\quad \left[\frac{0}{0} \text{ form}\right]$

$= \lim_{x \to 0} \frac{f(x) + f(x) + xf'(x)}{2e^{x^2} + 4x^2 e^{x^2}}$

$= \frac{f(0) + f(0) + 0 \cdot f'(0)}{2} = 1/2$ $\quad \left[\because f(0) = \frac{1}{2}\right]$

So, $\quad \alpha = 1/2$

$\therefore \quad 8\alpha^2 = 8 \cdot \frac{1}{4} = 2$

**79.** *(a)* Let $I = \int_0^1 \frac{1}{\sqrt{x+3} + \sqrt{x+1}} dx$

On rationalization,

$$I = \int_0^1 \frac{\sqrt{x+3} - \sqrt{x+1}}{(x+3) - (x+1)} dx$$

$$= \int_0^1 \frac{(x+3)^{1/2} - (x+1)^{1/2}}{2} dx$$

$$= \frac{1}{2}\left[\int_0^1 (x+3)^{\frac{1}{2}} dx - \int_0^1 (x+1)^{\frac{1}{2}} dx\right]$$

$$= \frac{1}{2}\left[\frac{2}{3}(x+3)^{3/2}\Big|_0^1 - \frac{2}{3}(x+1)^{3/2}\Big|_0^1\right]$$

$$= \frac{2}{2 \times 3}[\{(4)^{3/2} - (3)^{3/2}\} - \{2^{3/2} - (1)^{3/2}\}]$$

$$= \frac{1}{3}[(4^3)^{1/2} - (3^3)^{1/2} - (2^3)^{1/2} + 1]$$

$$= \frac{1}{3}[8 - 3\sqrt{3} - 2\sqrt{2} + 1]$$

$$= 3 - \sqrt{3} - \frac{2}{3}\sqrt{2} = a + b\sqrt{2} + c\sqrt{3}$$ [given]

So, $a = 3$, $b = -\frac{2}{3}$, $c = -1$

$\Rightarrow \quad 2a = 6,\ 3b = -2$

and $\quad 4c = -4$

Hence, $2a + 3b - 4c = 6 + (-2) - (-4)$

$= 4 + 4 = 8$

**80.** *(c)* Let

$$I = \int_0^\infty \frac{6}{e^{3x} + 6e^{2x} + 11e^x + 6} dx$$

$$= \int_0^\infty \frac{6}{(e^x + 1)(e^x + 2)(e^x + 3)} dx$$

[Using partial decomposition]

$$= 6\int_0^\infty \left[\frac{1}{2} \cdot \frac{1}{e^x + 1} + \frac{(-1)}{e^x + 2} + \frac{1/2}{e^x + 3}\right] dx$$

$$= \int_0^\infty 3 \cdot \frac{1}{e^x + 1} dx - 6\int_0^\infty \frac{dx}{e^x + 2} + 3\int_0^\infty \frac{dx}{e^x + 3}$$

$$= 3\int_0^\infty \frac{e^{-x} dx}{e^{-x} + 1} - 6\int_0^\infty \frac{e^{-x} dx}{1 + 2e^{-x}} + 3\int_0^\infty \frac{e^{-x} dx}{1 + 3e^{-x}}$$

$$= -3[\ln(1 + e^{-x})]_0^\infty + 6 \cdot \frac{1}{2}[\ln(1 + 2e^{-x})]_0^\infty + 3\left(\frac{-1}{3}\right)[\ln(1 + 3e^{-x})]_0^\infty$$

$$= 3\ln 2 - 3\ln 3 + \ln 4 = 3\ln\frac{2}{3} + \ln 4$$

$$= \ln\frac{8}{27} + \ln 4 = \ln\frac{32}{27}$$

**81.** (a) Given: $f:[0,2]\to R$

$$f(x)=\begin{cases} e^{\min\{x^2,\, x-[x]\}}, & x\in[0,1) \\ e^{[x-\log_e x]}, & x\in[1,2]\end{cases}$$

where $[t]=$ G.I.F.

Now, $\int_0^2 x f(x)\,dx$

$$=\int_0^1 xf(x)dx+\int_1^2 x\,f(x)dx$$

$$=\int_0^1 xe^{\min\{x^2,\,x-[x]\}}\,dx$$

$$\Rightarrow \int_0^2 x\,f(x)dx=\int_0^1 x\,e^{x^2}\,dx+\int_1^2 xe^1\,dx$$

$\left[\because x^2<\{x\}\text{ when, } x\in(0,0)\right.$
$\left.\text{and } x-\log_e x\in[1,2),\text{ when } x\in 1\right]$

$$\Rightarrow \int_0^2 xf(x)\,dx=\frac{1}{2}\int_0^1 (2x)\cdot e^{x^2}\,dx+e\int_1^2 x\,dx$$

$$=\frac{1}{2}e^{x^2}\Big|_0^1+e\frac{x^2}{2}\Big|_1^2$$

$$=\frac{1}{2}(e^1-1)+\frac{e}{2}(4-1)$$

$$=\frac{e-1}{2}+\frac{3e}{2}$$

$$=\frac{4e-1}{2}=2e-\frac{1}{2}$$

Hence, $\int_0^1 xf(x)\,dx=2e-\frac{1}{2}$

**82.** (d) Let $I=\int_{-\log_e 2}^{\log_e 2} e^x(\log_e(e^x+\sqrt{1+e^{2x}}))dx$

Let $e^x=t\Rightarrow e^x\,dx=dt$

When $x\to-\log_e 2$, then $t\to\frac{1}{2}$

When $x\to\log_e 2$, then $t\to 2$

$$I=\int_{\frac{1}{2}}^{2}[\log_e(t+\sqrt{1+t^2})]dt \quad \ldots(i)$$

Applying integration by part method,

$$I=[t\log_e(t+\sqrt{1+t^2})]_{1/2}^{2}-\int_{1/2}^{2}\frac{t}{t+\sqrt{1+t^2}}\left(1+\frac{2t}{2\sqrt{1+t^2}}\right)dt$$

$$=2\log_e(2+\sqrt{5})-\frac{1}{2}\log_e\left(\frac{1+\sqrt{5}}{2}\right)-\int_{1/2}^{2}\frac{t}{\sqrt{1+t^2}}\,dt$$

$$=\log_e\left(\frac{(2+\sqrt{5})^2}{\left(\frac{1+\sqrt{5}}{2}\right)^{1/2}}\right)-\frac{1}{2}\int_{1/2}^{2}\frac{2t}{\sqrt{1+t^2}}\,dt \quad \ldots(ii)$$

Let $I_1=\int_{1/2}^{2}\frac{2t}{\sqrt{1+t^2}}\,dt$

Let $1+t^2=\omega$

$\Rightarrow \quad 2tdt=d\omega$

When $t\to\frac{1}{2}$, then $\omega=\frac{5}{4}$

When $t\to 2$, then $\omega=5$

$$I_1=\int_{5/4}^{5}\frac{1}{\sqrt{\omega}}\,d\omega=[2\sqrt{\omega}]_{5/4}^{5}$$

$$=2\left[\sqrt{5}-\frac{\sqrt{5}}{2}\right]=\sqrt{5}$$

Substitute value of $I_1$ in Eq. (ii), we get

$$I=\log_e\left(\frac{\sqrt{2}(2+\sqrt{5})^2}{\sqrt{1+\sqrt{5}}}\right)-\frac{\sqrt{5}}{2}$$

**83.** (a) We have,

$$5f(x)+4f\left(\frac{1}{x}\right)=\frac{1}{x}+3,\ x>0 \quad \ldots(i)$$

On replacing $x$ by $\frac{1}{x}$ in Eq. (i), we get

$$5f\left(\frac{1}{x}\right)+4f(x)=x+3 \quad \ldots(ii)$$

Now using Eq. (i) $\times 5-$ Eq. (ii) $\times 4$, we get

$$25f(x)-16f(x)=\left(\frac{5}{x}+15\right)-(4x+12)$$

$$\Rightarrow \quad 9f(x)=\frac{5}{x}-4x+3$$

$$\Rightarrow \quad f(x)=\frac{1}{9}\left(\frac{5}{x}-4x+3\right) \quad \ldots(iii)$$

$$\therefore 18\int_1^2 f(x)\,dx=18\int_1^2\frac{1}{9}\left(\frac{5}{x}-4x+3\right)dx$$

[using Eq. (iii)]

$$=2\int_1^2\left(\frac{5}{x}-4x+3\right)dx$$

$$=2\left[5\log_e x-4\frac{x^2}{2}+3x\right]_1^2$$

$$=2[5\log_e 2-2(2)^2+3(2)-(5\log 1-2(1)^2+3(1)]$$

$$=2[5\log_e 2-8+6+2-3]$$

$(\because \log 1=0)$

$$=10\log_e 2-6$$

**84.** (d) Let $I=\int_{-\frac{\pi}{4}}^{\frac{\pi}{4}}\frac{x+\frac{\pi}{4}}{2-\cos 2x}\,dx \quad \ldots(i)$

$$I=\int_{-\frac{\pi}{4}}^{\frac{\pi}{4}}\frac{\left(\frac{\pi}{4}-\frac{\pi}{4}-x\right)+\frac{\pi}{4}}{2-\cos 2\left(\frac{\pi}{4}-\frac{\pi}{4}-x\right)}\,dx$$

$$\left[\because \int_a^b f(x)\,dx=\int_a^b f(a+b-x)\,dx\right]$$

$$\Rightarrow \quad I=\int_{-\frac{\pi}{4}}^{\frac{\pi}{4}}\frac{-x+\frac{\pi}{4}}{2-\cos 2x}\,dx \quad \ldots(ii)$$

On adding Eqs. (i) and (ii), we get

$$2I=\int_{-\frac{\pi}{4}}^{\frac{\pi}{4}}\frac{\frac{\pi}{2}}{2-\cos 2x}\,dx$$

$$\Rightarrow \quad I=\frac{\pi}{4}\int_{-\frac{\pi}{4}}^{\frac{\pi}{4}}\frac{1}{2-\cos 2x}\,dx$$

$$=\frac{\pi}{4}\int_{-\frac{\pi}{4}}^{\frac{\pi}{4}}\frac{1}{2-\left(\frac{1-\tan^2 x}{1+\tan^2 x}\right)}\,dx$$

$$=\frac{\pi}{4}\int_{-\frac{\pi}{4}}^{\frac{\pi}{4}}\frac{\sec^2 x}{2+2\tan^2 x-1+\tan^2 x}\,dx$$

$$=\frac{\pi}{4}\int_{-\frac{\pi}{4}}^{\frac{\pi}{4}}\frac{\sec^2 x}{1+3\tan^2 x}\,dx$$

Put $\tan x=t\Rightarrow \sec^2 x\,dx=dt$

When, $x=-\frac{\pi}{4}\Rightarrow \quad t=\tan\left(-\frac{\pi}{4}\right)=-1$

and when $x=\frac{\pi}{4}\Rightarrow t=\tan\frac{\pi}{4}=1$

$$\therefore I=\frac{\pi}{4}\int_{-1}^{1}\frac{dt}{1+3t^2}$$

$$=\frac{\pi}{4}\int_{-1}^{1}\frac{dt}{1+(\sqrt{3}t)^2}$$

$$=\frac{\pi}{4}\cdot\frac{1}{\sqrt{3}}[\tan^{-1}\sqrt{3}t]_{-1}^{1}$$

$$=\frac{\pi}{4\sqrt{3}}[\tan^{-1}\sqrt{3}-\tan^{-1}(-\sqrt{3})]$$

$$=\frac{\pi}{4\sqrt{3}}\left[\frac{\pi}{3}-\left(-\frac{\pi}{3}\right)\right]$$

$$=\frac{\pi}{4\sqrt{3}}\times\frac{2\pi}{3}=\frac{\pi^2}{6\sqrt{3}}$$

**85.** *(b)* We have,

$$\phi(x)=\frac{1}{\sqrt{x}}\int_{\frac{\pi}{4}}^{x}(4\sqrt{2}\sin t-3\phi'(t))\,dt,$$

$x>0$

$$\Rightarrow \sqrt{x}\,\phi(x)=\int_{\frac{\pi}{4}}^{x}(4\sqrt{2}\sin t-3\phi'(t))\,dt \quad \ldots(i)$$

On differentiating both the sides w.r.t. $x$, we get

$$\frac{1}{2\sqrt{x}}\phi(x)+\sqrt{x}\,\phi'(x)=(4\sqrt{2}\sin x-3\phi'(x)).1$$

$$\Rightarrow \frac{1}{2\sqrt{x}}\phi(x)+\sqrt{x}\,\phi'(x)=4\sqrt{2}\sin x-3\phi'(x) \quad \ldots(ii)$$

On putting $x=\frac{\pi}{4}$ in Eq. (i), we get

$$\sqrt{\frac{\pi}{4}}\,\phi\left(\frac{\pi}{4}\right)=\int_{\frac{\pi}{4}}^{\frac{\pi}{4}}(4\sqrt{2}\sin t-3\phi'(t))\,dt$$

$$\Rightarrow \sqrt{\frac{\pi}{4}}\,\phi\left(\frac{\pi}{4}\right)=0\Rightarrow \phi\left(\frac{\pi}{4}\right)=0 \quad \ldots(iii)$$

Again, putting $x=\frac{\pi}{4}$ in Eq. (ii), we get

$$\frac{1}{2\sqrt{\frac{\pi}{4}}}\phi\left(\frac{\pi}{4}\right)+\sqrt{\frac{\pi}{4}}\,\phi'\left(\frac{\pi}{4}\right)=4\sqrt{2}\sin\frac{\pi}{4}-3\phi'\left(\frac{\pi}{4}\right)$$

$$\Rightarrow \sqrt{\frac{\pi}{4}}\,\phi'\left(\frac{\pi}{4}\right)=4-3\phi'\left(\frac{\pi}{4}\right) \quad \left[\because \phi\left(\frac{\pi}{4}\right)=0\right]$$

$$\Rightarrow \phi'\left(\frac{\pi}{4}\right)\left[3+\frac{\sqrt{\pi}}{2}\right]=4$$

$$\Rightarrow \phi'\left(\frac{\pi}{4}\right)=\frac{8}{6+\sqrt{\pi}}$$

> **Concept Enhancer**
>
> If function $\phi(x)$ and $\psi(x)$ are defined on $[\alpha,\beta]$ and differentiable on $[\alpha,\beta]$ and $f(t)$ is continuous on $[\psi(\alpha),\phi(\beta)]$, then
>
> $$\frac{d}{dx}\left[\int_{\phi(x)}^{\psi(x)}f(t)dt\right]=\left\{\frac{d}{dx}(\psi(x))\right\}f(\psi(x))-\left\{\frac{d}{dx}(\phi(x))\right\}f(\phi(x))$$
>
> This is called Leibnitz theorem.

**86.** *(d)* Let $I=\frac{3(e-1)}{e}\int_1^2 x^2e^{[x]+[x^3]}\,dx,$

Since, integration limit is from 1 to 2.

So, $[x]=1$, when $1<x<2$

$$\text{So, } \int_1^2 x^2e^{1+[x^3]}\,dx=\int_1^2 x^2e\,e^{[x^3]}dx$$

$$=e\int_1^2 x^2e^{[x^3]}\,dx$$

$$=e\int_1^8\frac{1}{3}e^{[t]}\,dt, \text{ where } t=x^3$$

| $x$ | 1 | 2 |
|---|---|---|
| $t$ | 1 | 8 |

$$=\frac{e}{3}\left[\int_1^2 e^{[t]}\,dt+\int_2^3 e^{[t]}dt+\int_3^4 e^{[t]}dt+\ldots+\int_7^8 e^{[t]}dt\right]$$

$$=\frac{e}{3}\left[\int_1^2 e\,dt+\int_2^3 e^2dt+\int_3^4 e^3\,dt+\ldots+\int_7^8 e^7\,dt\right]$$

$$=\frac{e^2}{3}(1+e+e^2+\ldots+e^6)=\frac{e^2}{3}\left[\frac{e^7-1}{e-1}\right]$$

$$\left(\because S_n=\frac{a(r^n-1)}{r-1}, r>1 \quad \text{Here, } a=1, r=e, n=7\right)$$

$$\therefore \ I=\frac{3(e-1)}{e}\times\frac{e^2}{3}\times\left(\frac{e^7-1}{e-1}\right)$$

$$=e(e^7-1)=e^8-e$$

> ***Alternate Solution***
>
> $$\text{Let } I=\frac{3(e-1)}{e}\int_1^2 x^2e^{[x]+[x^3]}dx$$
>
> $$=\frac{3(e-1)}{e}\int_1^2 x^2e^{1+[x^3]}dx$$
>
> Let $x^3=u \Rightarrow \quad 3x^2dx=du$
>
> $$\therefore \quad I=\frac{3(e-1)}{e}\int_1^8\frac{e\cdot e^{[u]}}{3}dt=(e-1)\int_1^8 e^{[u]}$$
>
> $$=(e-1)[e+e^2+e^3+\ldots+e^7]$$
>
> $$=(e-1)e[1+e+e^2+\ldots+e^6]$$
>
> $$=e(e-1)\left(\frac{e^7-1}{e-1}\right)=e^8-e$$

**87.** *(b)* Let

$$I=\int_1^2\frac{t^4+1}{t^6+1}\,dt=\int_1^2\frac{t^4+1}{(t^2)^3+1^3}\,dt$$

$$=\int_1^2\frac{t^4+1}{(t^2+1)(t^4-t^2+1)}\,dt$$

$$=\int_1^2\frac{dt}{t^2+1}+\int_1^2\frac{t^2\,dt}{(t^2+1)(t^4-t^2+1)}$$

$$=\int_1^2\frac{dt}{t^2+1}+\frac{1}{3}\int_1^2\frac{3t^2\,dt}{(t^3)^2+1}$$

Let $t^3=u\Rightarrow 3t^2\,dt=du$

When, $t=1\Rightarrow u=1$

$t=2\Rightarrow u=8$

$$\therefore \quad I=\int_1^2\frac{dt}{t^2+1}+\frac{1}{3}\int_1^8\frac{du}{u^2+1}$$

$$=[\tan^{-1}t]_1^2+\frac{1}{3}[\tan^{-1}u]_1^8$$

$$=\tan^{-1}2-\frac{\pi}{4}+\frac{1}{3}\tan^{-1}8-\frac{\pi}{12}$$

$$=\tan^{-1}2+\frac{1}{3}\tan^{-1}8-\frac{\pi}{3}$$

**88.** *(a)* Let $I=16\int_1^2\frac{dx}{x^3(x^2+2)^2}$

$$I=16\int_1^2\frac{dx}{x^3\cdot x^4\left(1+\frac{2}{x^2}\right)^2}$$

$\Rightarrow$ Now, put $1+\frac{2}{x^2}=t$

$$\Rightarrow (-2)(2)x^{-3}\,dx=dt$$

$$\Rightarrow \quad \frac{dx}{x^3}=-\frac{dt}{4}\Rightarrow 1+\frac{2}{x^2}=t$$

$$\Rightarrow \quad \frac{1}{x^2}=\left(\frac{t-1}{2}\right)$$

Now,

| $x$ | 1 | 2 |
|---|---|---|
| $t$ | 3 | 3/2 |

$$\text{So, } I=16\int_3^{3/2}\left(\frac{t-1}{2}\right)^2\left(\frac{1}{t^2}\right)\left(\frac{-dt}{4}\right)$$

$$I=-\int_3^{3/2}\left(\frac{t-1}{t}\right)^2dt$$

$$I=\int_{3/2}^{3}\left(1+\frac{1}{t^2}-\frac{2}{t}\right)dt$$

$$I=\left[t+\frac{t^{-2+1}}{(-2+1)}-2\ln t\right]_{3/2}^{3}$$

$$I=\frac{8}{3}-2\ln 3-\left(\frac{3}{2}\right)+\frac{2}{3}+2\,[\ln 3-\ln 2]$$

$$I=\frac{11}{6}-2\ln 2\Rightarrow I=\frac{11}{6}-\ln 4$$

**89.** *(a)* $f(x)=\int_0^2 e^{|x-t|}\,dt$

$$|x-t|=\begin{cases}x-t & : \ x>t>0\\ t-x & : \ 2>t>x\end{cases}$$

$$f(x)=\int_0^x e^{x-t}\,dt+\int_x^2 e^{t-x}\,dt$$

$= [-e^{x-t}]_0^x + [e^{t-x}]_x^2$

$f(x) = -1 + e^x + \frac{e^2}{e^x} - 1$

$= e^x + \frac{e^2}{e^x} - 2$

Now, we know that AM $\ge$ GM

$$\frac{e^x + \frac{e^2}{e^x}}{2} \ge \sqrt{e^x \cdot \frac{e^2}{e^x}} \Rightarrow e^x + \frac{e^2}{e^x} \ge 2e$$

$$\left(e^x + \frac{e^2}{e^x}\right)_{\min} = 2e$$

$[(f(x)]_{\min} = 2e - 2 = 2(e-1)$

**90.** (d) $f(\alpha) = \int_1^\alpha \frac{\log_{10} t}{1+t}\, dt,\ \alpha > 0$

$$f(e^3) = \int_0^{e^3} \frac{\log_e t}{\log_e 10\,(1+t)}\, dt \quad \ldots\text{(i)}$$

Put $t = \frac{1}{x} \Rightarrow x = \frac{1}{t} \Rightarrow dt = \frac{-1}{x^2}dx$

$$f(\alpha) = \int_1^{\frac{1}{\alpha}} \frac{-\ln x}{\ln 10\left(1+\frac{1}{x}\right)}\left(\frac{-1}{x^2}\right)dx$$

$$= \frac{1}{\ln 10}\int_1^{1/\alpha}\frac{\ln x}{x(x+1)}\,dx$$

$$f(e^{-3}) = \frac{1}{\ln 10}\int_1^{e^3}\frac{\ln t}{t(t+1)}\,dt \quad \ldots\text{(ii)}$$

$\therefore\ f(e^3) + f(e^{-3})$

$$= \frac{1}{\ln 10}\int_1^{e^3}\frac{\ln t}{1+t}\left(1+\frac{1}{t}\right)dt$$

$$= \frac{1}{\ln 10}\int_1^{e^3}\frac{\ln t}{t}\,dt = \frac{1}{\ln 10}\left[\frac{(\ln t)^2}{2}\right]_1^{e^3}$$

$$= \frac{1}{2\ln 10}(9-0) = \frac{9}{2\ln 10}$$

**91.** (a) $I_n(x) = \int_0^x \frac{1}{(t^2+5)^n}\,dt$

Using integration by parts,

$$\int_0^x \underbrace{\frac{1}{(t^2+5)^n}}_{u} \times \underset{v}{1}\, dt$$

$$I_n(x) = \left[\frac{t}{(t^2+5)^n}\right]_0^x - \int_0^x \frac{-n(2t)}{(t^2+5)^{n+1}}\times t\,dt$$

$$= \frac{x}{(x^2+5)^n} + 2n\int_0^x \frac{t^2+5\,dt}{(t^2+5)^{n+1}} - 10n\int_0^x \frac{1}{(t^2+5)^{n+1}}\,dt$$

$$I_n = \frac{x}{(x^2+5)^n} + 2n\, I_n - 10n\, I_{n+1}$$

$$I_n(1-2n) + 10n(I_{n+1}) = \frac{x}{(x^2+5)^n}$$

$$I_n(1-2n) + 10n(I_{n+1}) = xI'(n)$$

Now, on substituting $n = 5$

$I_5(-9) + 50I_6 = xI'(5)$

$50I_6 - 9I_5 = xI'(5)$

**92.** (c) $I = \int_{\pi/4}^{\pi/3}\left(\frac{8\sin x - \sin 2x}{x}\right)dx$

Consider $f(x) = 8\sin x - \sin 2x$

$\Rightarrow f(0) = 0$

Now, $f'(x) = 8\cos x - 2\cos 2x$

$f''(x) = -8\sin x + 4\sin 2x$

$= -8\sin x(1-\cos x) < 0,\ \forall\, x \in (0, \pi)$

$\therefore\quad f'\left(\frac{\pi}{3}\right) < f'(x) < f'\left(\frac{\pi}{4}\right)$

$\Rightarrow\quad 5 < f'(x) < 4\sqrt{2}$

$\therefore\ \int_0^x 5\,dx < \int_0^x f'(x)dx < \int_0^x 4\sqrt{2}\,dx$

$\Rightarrow\quad 5x < f(x) - f(0) < 4\sqrt{2}x$

$\Rightarrow\quad 5 < \frac{f(x)}{x} < 4\sqrt{2}\quad [\because f(0) = 0]$

$\Rightarrow\quad \int_{\pi/4}^{\pi/3} 5\,dx < \int_{\pi/4}^{\pi/3}\frac{f(x)}{x}\,dx < \int_{\pi/4}^{\pi/3}4\sqrt{2}\,dx$

$\therefore\quad \frac{5\pi}{12} < I < \frac{\sqrt{2}\pi}{3}$

**93.** (b) Given that $[t]$ denote the greatest integer less than or equal to $t$.

So, $[\sin \pi x]$ is periodic function with period 2 and $e^{[\cos(2\pi x)]}$ is periodic also with period 1.

Thus, $\int_{-3}^{101}([\sin \pi x] + e^{[\cos(2\pi x)]}]\,dx$

For property of periodic function,

$\int_0^{104}([\sin \pi x] + e^{[\cos(2\pi x)}])\,dx$

$= \int_0^{2\times 52}([\sin \pi x] + e^{[\cos 2\pi x]})\,dx$

Since, $\int_0^{na} f(x)dx = n\int_0^a f(x)dx$

Using this property,

$52\int_0^2([\sin(\pi x)] + e^{[\cos 2\pi x]})]$

$$= 52\int_0^2 [\sin(\pi x)]\,dx + 52\int_0^2 e^{[\cos 2\pi x]}\,dx$$

$$= 52\left[\int_0^1 0\cdot dx + \int_1^2(-1)dx\right]$$

$$+52\left[\int_0^{\frac{1}{4}} e^0\,dx + \int_{\frac{1}{4}}^{\frac{1}{2}} e^{-1}\,dx + \int_{\frac{1}{2}}^{\frac{3}{4}} e^{-1}\,dx + \int_{\frac{3}{4}}^{1} e^0\,dx + \int_1^{\frac{5}{4}} e^0\,dx + \int_{5/4}^{3/2} e^{-1}\,dx + \int_{3/2}^{7/4} e^{-1}\,dx + \int_{7/4}^{2} e^0\,dx\right]$$

$$= -52 + 52\left(1+\frac{1}{e}\right) = \frac{52}{e}$$

**94.** (a) Given, $f$ is a real valued function defined on the interval $[-10, 10]$ by

$$f(x) = \begin{cases} x - [x], & \text{if } [x] \text{ is odd} \\ 1 + [x] - x, & \text{if } [x] \text{ is even} \end{cases}$$

The graph of $f(x)$ is

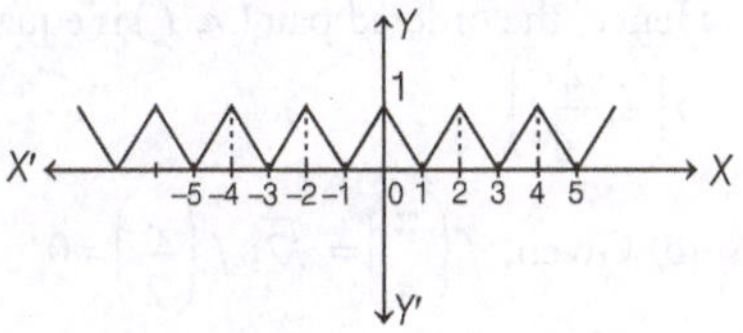


Clearly, $f(x)$ is an even function and also period is 2.

Thus, $\frac{\pi^2}{10}\int_{-10}^{10} f(x)\cos \pi x\,dx$

$$= \frac{\pi^2}{10}\cdot 20\int_0^1 f(x)\cos \pi x\,dx$$

Since, $f(x) = \begin{cases} -1-x, & x \in [-2, 1) \\ x+1, & x \in [-1, 0) \\ 1-x, & x \in [0, 1) \\ x-1, & x \in [1, 2) \\ 3-x, & x \in [2, 3) \\ x-3, & x \in [3, 4) \end{cases}$

$$= \frac{\pi^2}{10}\cdot 20\int_0^1 (1-x)\cos \pi x\,dx$$

$$= 2\pi^2\left\{\left((1-x)\frac{\sin \pi x}{\pi}\right)_0^1 - \left(\frac{\cos \pi x}{\pi^2}\right)_0^1\right\}$$

$$= 2\pi^2\left\{\frac{2}{\pi^2}\right\} = 4$$

**95.** (c)

$$f(x) = \begin{cases} \max\{t^2 - 3t\} & x \le 2\ \ t \le x \\ x^2 + 2x - 6 & 2 < x < 3 \\ [x-3] + 9 & 3 \le x \le 5 \\ 2x + 1 & x > 5 \end{cases}$$

$$f(x)=\begin{cases} x^3-3x & x\le -1\\ 2 & -1<x<2\\ x^2+2x-6 & 2<x<3\\ 9 & 3\le x<4\\ 10 & 4\le x<5\\ 11; & x=5\\ 2x+1; & x>5\end{cases}$$

Using the simplified form of the function, we find that $f(x)$ is not differentiable at $x = 2, 3, 4, 5$ when we check the continuity and differentiability.

$\therefore\ m=4$

$$I=\int_{-2}^{2} f(x)\,dx=\int_{-2}^{-1}(x^3-3x)\,dx+\int_{-1}^{2}2\,dx$$

$$=\frac{3}{4}+6=\frac{27}{4}$$

Hence, the ordered pair $(m, I)$ is equal to $\left(4,\frac{27}{4}\right)$.

**96.** *(b)* Given, $f\left(\frac{\pi}{4}\right)=\sqrt{2},\ f\left(\frac{\pi}{2}\right)=0$

and $f'\left(\frac{\pi}{2}\right)=1$

$$g(x)=\int_x^{\pi/4}[f'(t)\sec t+\tan t\sec t f(t)]\,dt$$

$$=\int_x^{\pi/4}\left[\frac{d}{dt}(t)\sec t\right]dt$$

$$=[\sec t\cdot f(t)]_x^{\pi/4}=2-\sec x\, f(x)$$

Now, $\lim\limits_{x\to\frac{\pi}{2}^-} g(x)=\lim\limits_{h\to 0} g\left(\frac{\pi}{2}-h\right)$

$$=\lim_{h\to 0}\left[2-\operatorname{cosec} h\, f\left(\frac{\pi}{2}-h\right)\right]$$

$$=\lim_{h\to 0}\left[2-\frac{f\left(\frac{\pi}{2}-h\right)}{\sin h}\right]$$

$$=\lim_{h\to 0}\left[2+\frac{f'\left(\frac{\pi}{2}-h\right)}{\cos h}\right]=2+\frac{1}{1}=3$$

**97.** *(c)* Given, $f: R\to R$ and

$f(x)+f(x+k)=n,\ \forall\ x\in R$

$x\to x+k$

$f(x+k)+f(x+2k)=n$

$\therefore\quad f(x+2k)=f(x)$

So, period of $f(x)$ is $2k$.

Now, $I_1=\int_0^{4nk} f(x)\,dx=2n\int_0^{2k} f(x)\,dx$

On putting

$$2n\left[\int_0^k f(x)\,dx+\int_k^{2k} f(x)\,dx\right]$$

Putting $x=t+k \Rightarrow dx=dt$ in second integral

$$I_1=2n\left[\int_0^k f(x)\,dx+\int_0^k f(t+k)\,dt\right]$$

$$=2n^2k$$

Now, $I_2=\int_{-k}^{3k} f(x)\,dx=2\int_0^{2k} f(x)\,dx$

$I_2=2(nk);$

$\therefore\ I_1+nI_2=4n^2k$

**98.** *(c)* Let $I=\int_0^1[-8x^2+6x-1]\,dx$

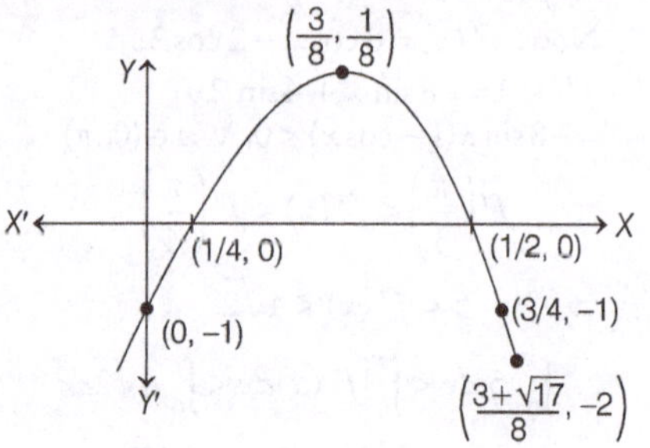


$$I=\int_0^{1/4}(-1)dx+\int_{1/4}^{1/2}0\,dx+\int_{1/2}^{3/4}(-1)\,dx$$

$$+\int_{3/4}^{(3+\sqrt{17})/8}(-2)dx+\int_{3+\sqrt{17}/8}^{1}(-3)dx$$

$$=-\left(\frac{1}{4}-0\right)-\left(\frac{3}{4}-\frac{1}{2}\right)-2\left(\frac{3+\sqrt{17}}{8}-\frac{3}{4}\right)$$

$$-3\left(1-\frac{3+\sqrt{17}}{8}\right)$$

$$=\frac{\sqrt{17}-13}{8}$$

**99.** *(b)* Given,

$$\int_{\cos x}^{1} t^2 f(t)\,dt=\sin^3 x+\cos x$$

On differentiating both sides w.r.t. $x$,

$$0-\cos^2 x\, f(\cos x)\cdot(-\sin x)$$

$$=3\sin^2 x\cdot\cos x-\sin x$$

$$\Rightarrow\ \cos^2 x\cdot\sin x\cdot f(\cos x)$$

$$=\sin x(3\sin x\cdot\cos x-1)$$

$$\Rightarrow\ \cos^2 x\cdot f(\cos x)=\frac{3}{2}\sin 2x-1 \quad \ldots(\text{i})$$

Again differentiating both sides w.r.t. $x$, we get

$$2\cos x(-\sin x)\cdot f(\cos x)+\cos^2 x\cdot f'(\cos x)(-\sin x)$$

$$=\frac{3}{2}\cos 2x\cdot 2 \quad \ldots(\text{ii})$$

On putting $\cos x=\frac{1}{\sqrt{3}}$ in Eq. (i), we get

$$\sin x=\frac{\sqrt{2}}{\sqrt{3}}$$

$$\Rightarrow\ \left(\frac{1}{\sqrt{3}}\right)^2\cdot f\left(\frac{1}{\sqrt{3}}\right)=\frac{3}{2}\left(2\cdot\frac{\sqrt{2}}{\sqrt{3}}\cdot\frac{1}{\sqrt{3}}\right)-1$$

$$\Rightarrow\ f\left(\frac{1}{\sqrt{3}}\right)=3(\sqrt{2}-1)$$

From Eq. (ii),

$$2\left(\frac{1}{\sqrt{3}}\right)\left(-\frac{\sqrt{2}}{\sqrt{3}}\right)\cdot f\left(\frac{1}{\sqrt{3}}\right)+\left(\frac{1}{\sqrt{3}}\right)^2\cdot f'\left(\frac{1}{\sqrt{3}}\right)\left(-\frac{\sqrt{2}}{\sqrt{3}}\right)$$

$$=3\left[2\left(\frac{1}{\sqrt{3}}\right)^2-1\right]$$

$$\Rightarrow\ \frac{\sqrt{2}}{3\sqrt{3}}\cdot f'\left(\frac{1}{\sqrt{3}}\right)=2\sqrt{2}-3$$

$$\frac{1}{\sqrt{3}}f'\left(\frac{1}{\sqrt{3}}\right)=6-\frac{9}{\sqrt{2}}$$

**100.** *(c)* Let

$$I=\int_0^{\pi}\frac{e^{\cos x}\sin x}{(1+\cos^2 x)(e^{\cos x}+e^{-\cos x})}dx \quad \ldots(\text{i})$$

Using the property,

$$\int_0^a f(x)\,dx=\int_0^a f(a-x)\,dx$$

$$I=\int_0^{\pi}\frac{e^{\cos(\pi-x)}\sin(\pi-x)}{(1+\cos^2(\pi-x))[e^{\cos(\pi-x)}+e^{-\cos(\pi-x)}]}dx$$

$$I=\int_0^{\pi}\frac{e^{-\cos x}\sin x}{(1+\cos^2 x)(e^{-\cos x}+e^{\cos x})}dx \quad \ldots(\text{ii})$$

On adding Eqs. (i) and (ii), we get

$$2I=\int_0^{\pi}\frac{\sin x(e^{\cos x}+e^{-\cos x})}{(1+\cos^2 x)(e^{\cos x}+e^{-\cos x})}dx$$

$$=\int_0^{\pi}\frac{\sin x}{(1+\cos^2 x)}dx$$

Let $\cos x = t$

$\Rightarrow \quad -\sin x\, dx = dt$ or $\sin x\, dx = -dt$

When $x \to 0$ then $t \to 1$ and $x \to \pi$ then $t \to -1$

$$\Rightarrow \quad 2I = -\int_1^{-1} \frac{1}{1+t^2} dt$$

$$\Rightarrow \quad 2I = [\tan^{-1} t]_{-1}^{1}$$

$$\Rightarrow \quad 2I = \tan^{-1}(1) - [\tan^{-1}(-1)]$$

$$= \frac{\pi}{4} - \left(\frac{-\pi}{4}\right)$$

$$\Rightarrow \quad I = \pi/4$$

**101.** *(a)* Let $I = \int_0^1 \frac{\sqrt{x}\, dx}{(1+x)(1+3x)(3+x)}$

Put $\sqrt{x} = t \Rightarrow x = t^2$ and $dx = 2t\, dt$

$$\therefore \quad I = \int_0^1 \frac{2t^2\, dt}{(t^2+1)(3t^2+1)(t^2+3)}$$

$$= \int_0^1 \frac{(3t^2+1)-(t^2+1)}{(t^2+1)(3t^2+1)(t^2+3)} dt$$

$$= \int_0^1 \left[\frac{1}{(t^2+3)(t^2+1)} - \frac{1}{(t^2+3)(3t^2+1)}\right] dt$$

$$= \int_0^1 \left[\frac{1}{2(t^2+1)} - \frac{1}{2(t^2+3)} - \frac{3}{8(3t^2+1)} + \frac{1}{8(t^2+3)}\right] dt$$

$$= \int_0^1 \frac{dt}{2(t^2+1)} - \int_0^1 \frac{3}{8}\frac{dt}{(3t^2+1)} - \int_0^1 \frac{3}{8}\frac{dt}{(t^2+3)}$$

$$= \left(\frac{1}{2}\tan^{-1} t\right)_0^1 - \left(\frac{3\sqrt{3}}{8\ 3}\tan^{-1}\sqrt{3}t\right)_0^1 - \left(\frac{3}{8\sqrt{3}}\tan^{-1}\frac{t}{\sqrt{3}}\right)_0^1$$

$$= \frac{\pi}{8}\left(1 - \frac{\sqrt{3}}{2}\right)$$

**102.** *(c)* Let

$$I = \int_6^{16} \frac{\ln_e(x^2)}{\ln_e(x^2) + \ln_e(484 - 44x + x^2)} dx$$

$$= \int_6^{16} \frac{\ln_e(x^2)}{\ln_e(x^2) + \ln_e(22-x)^2} dx$$

$$= \int_6^{16} \frac{2\ln_e x\, dx}{2\ln_e x + 2\ln_e(22-x)}$$

$$I = \int_6^{16} \frac{\ln_e x\, dx}{\ln_e x + \ln_e(22-x)} \quad \ldots(i)$$

$$\because \int_a^b f(x)dx = \int_a^b f(a+b-x)dx$$

$$\therefore \quad I = \int_6^{16} \frac{\ln_e(22-x)}{\ln_e(22-x) + \ln_e x} dx \quad \ldots(ii)$$

On adding Eqs. (i) and (ii), we get

$$2I = \int_6^{16} \frac{\ln_e x + \ln_e(22-x)}{\ln_e x + \ln_e(22-x)} dx$$

$$2I = \int_6^{16} dx = x\Big|_6^{16} = 10 \text{ or } I = 5$$

**103.** *(b)* $I = \int_0^5 \frac{x+[x]}{e^{x-[x]}} dx = \alpha e^{-1} + \beta$

$$I = \int_0^1 \frac{x}{e^x} dx + \int_1^2 \frac{x+1}{e^{x-1}} dx + \int_2^3 \frac{x+2}{e^{x-2}} dx + \int_3^4 \frac{x+3}{e^{x-3}} dx + \int_4^5 \frac{x+4}{e^{x-4}} dx$$

Let $\quad I = I_1 + I_2 + I_3 + I_4 + I_5$

Here, $I_2 = \int_1^2 \frac{x+1}{e^{x-1}} dx$ Put $x = t+1$

$\Rightarrow dx = dt$

$$= \int_0^1 \frac{t+2}{e^t} dt = \int_0^1 \frac{t}{e^t} dt + \int_0^1 \frac{2}{e^t} dt$$

$$I_2 = I_1 + 2\int_0^1 e^{-t}\, dt = I_1 + 2(1 - e^{-1})$$

Similarly, $I_3 = I_1 + 4(1 - e^{-1})$

$I_4 = I_1 + 6(1 - e^{-1})$

$I_5 = I_1 + 8(1 - e^{-1})$

$I = I_1 + I_2 + I_3 + I_4 + I_5$

$= 5I_1 + 20(1 - e^{-1})$

$I_1 = \int_0^1 xe^{-x} dx = -[e^{-x}(x+1)]_0^1$

$= 1 - 2e^{-1}$

$\therefore 5I_1 + 20(1 - e^{-1}) = 5(1 - 2e^{-1}) + 20(1 - e^{-1})$

$= 25 - 30e^{-1}$

$\therefore \quad \alpha = -30, \beta = 25$

Also, it satisfy $5\alpha + 6\beta = 0$

Now, $(\alpha+\beta)^2 = (-30+25)^2$

$= (-5)^2 = 25$

**104.** *(c)* $I = \int_{-\frac{\pi}{2}}^{\frac{\pi}{2}} \frac{1+\sin^2 x}{1+\pi^{\sin x}} dx \quad \ldots(i)$

$$I = \int_{-\pi/2}^{\pi/2} \frac{1+\sin^2(-x)}{1+\pi^{\sin(-x)}} dx$$

$$\left[\because \int_a^b f(x)\, dx = \int_a^b f(a+b-x)\, dx\right]$$

$$= \int_{-\pi/2}^{\pi/2} \frac{1+\sin^2 x}{1+\pi^{(-\sin x)}} dx$$

$$I = \int_{-\frac{\pi}{2}}^{\frac{\pi}{2}} \frac{\pi^{\sin x}(1+\sin^2 x)}{1+\pi^{\sin x}} dx \quad \ldots(ii)$$

On adding Eqs. (i) and (ii), we get

$$2I = \int_{\frac{-\pi}{2}}^{\frac{\pi}{2}} \frac{(1+\pi^{\sin x})(1+\sin^2 x)}{(1+\pi^{\sin x})} dx$$

$$\Rightarrow \quad 2I = \int_{-\pi/2}^{\pi/2} (1+\sin^2 x)\, dx$$

$$\Rightarrow \quad 2I = [x]_{\frac{-\pi}{2}}^{\frac{\pi}{2}} + 2\int_0^{\frac{\pi}{2}} \sin^2 x\, dx$$

[$\because \sin^2 x$ is an even function, so $\int_{-\pi/2}^{\pi/2} \sin^2 dx = 2\int_0^{\pi/2} \sin^2 x\, dx$]

$$\Rightarrow \quad I = \frac{1}{2}\left(\frac{\pi}{2} - \left(-\frac{\pi}{2}\right)\right) + \frac{1}{2}\int_0^{\frac{\pi}{2}} (1 - \cos 2x)\, dx$$

$$\Rightarrow \quad I = \frac{\pi}{2} + \frac{\pi}{4} = \frac{3\pi}{4}$$

**105.** *(c)* Let $I = \int_{\frac{\pi}{24}}^{\frac{5\pi}{24}} \frac{dx}{1+\sqrt[3]{\tan 2x}} \quad \ldots(i)$

$$\because \int_a^b f(x)dx = \int_a^b f(a+b-x)dx$$

So, $I = \int_{\frac{\pi}{24}}^{\frac{5\pi}{24}} \frac{dx}{1+\left[\tan 2\left(\frac{\pi}{4} - x\right)\right]^{\frac{3}{2}}}$

$$I = \int_{\frac{\pi}{24}}^{\frac{5\pi}{24}} \frac{dx}{1+\sqrt[3]{\cot 2x}} = \int_{\frac{\pi}{24}}^{\frac{5\pi}{24}} \frac{\sqrt[3]{\tan 2x}}{1+\sqrt[3]{\tan 2x}} \quad \ldots(ii)$$

Adding Eqs. (i) and (ii), we get

$$2I = \int_{\pi/24}^{5\pi/24} \frac{1+\sqrt[3]{\tan 2x}}{1+\sqrt[3]{\tan 2x}}\, dx$$

$$\Rightarrow\ 2I = \int_{\pi/24}^{5\pi/24} 1\, dx = [x]_{\pi/24}^{5\pi/24} \Rightarrow 2I = \frac{4\pi}{24}$$

$$\therefore\quad I = \frac{\pi}{12}$$

**106.** *(b)* We have,

$$g(t) = \int_{-\pi/2}^{\pi/2} \cos\left(\frac{\pi}{4}t + f(x)\right) dx$$

$$g(t) = \int_{-\frac{\pi}{2}}^{\frac{\pi}{2}} \left[\cos\left(\frac{\pi}{4}\right)t \cos(f(x)) - \sin\frac{\pi}{4}t \sin f(x)\right] dx$$

Given, $f(x) = \log(x + \sqrt{1+x^2})$

$f(-x) = \log(-x + \sqrt{1+x^2})$

$= -\log(x + \sqrt{1+x^2})$

$f(-x) = -f(x)$

$f(x)$ is an odd function.

$$\therefore g(t) = \int_{-\pi/2}^{\pi/2} \cos\frac{\pi}{4}t \cos f(x) dx$$

[$\because \sin f(x)$ is an odd function]

$$\Rightarrow\quad g(1) = \int_{-\pi/2}^{\pi/2} \frac{1}{\sqrt{2}} \cos f(x) dx$$

$$\Rightarrow\quad \sqrt{2}\, g(1) = \int_{-\pi/2}^{\pi/2} \cos f(x) dx$$

and $$g(0) = \int_{-\pi/2}^{\pi/2} \cos f(x)\, dx$$

$$\therefore\quad \sqrt{2}\, g(1) = g(0)$$

**107.** *(c)* $\displaystyle\int_{-1}^{1} \log_e(\sqrt{1-x} + \sqrt{1+x}) dx$

$$\left[\because \int_{-a}^{a} f(x)\, dx = 2\int_0^a f(x) dx, \text{ if } f(-x) = f(x)\right]$$

So, $$2\int_0^1 \log_e(\sqrt{1-x} + \sqrt{1+x}) dx$$

$$I = 2\int_0^1 \log_e(\sqrt{1-x} + \sqrt{1+x}) \cdot 1\, dx$$

$$\Rightarrow \frac{I}{2} = [\log_e(\sqrt{1-x} + \sqrt{1+x}) \cdot x]_0^1 - \int_0^1 \frac{\frac{1}{2\sqrt{1+x}} - \frac{1}{2\sqrt{1-x}}}{\sqrt{1-x} + \sqrt{1+x}} \cdot x\, dx$$

$$\Rightarrow \frac{I}{2} = \log_e \sqrt{2} - \frac{1}{2}\int_0^1 \left(\frac{\sqrt{1-x} - \sqrt{1+x}}{\sqrt{1-x} + \sqrt{1+x}}\right) \frac{x}{\sqrt{1-x^2}}\, dx$$

$$\Rightarrow\quad \frac{I}{2} = \log_e \sqrt{2} - \frac{1}{2} \int_0^1 \left(\frac{(1-x) + (1+x) - 2\sqrt{1-x^2}}{(1-x) - (1+x)}\right) \frac{x}{\sqrt{1-x^2}}\, dx$$

$$\Rightarrow \frac{I}{2} = \log_e \sqrt{2} + \frac{1}{2}\int_0^1 \left(\frac{1}{\sqrt{1-x^2}} - 1\right) dx$$

$$\Rightarrow \frac{I}{2} = \log_e \sqrt{2} + \frac{1}{2}[\sin^{-1} x - x]_0^1$$

$$\Rightarrow \frac{I}{2} = \log_e \sqrt{2} + \frac{1}{2}\left(\frac{\pi}{2} - 1\right)$$

$$\therefore\quad I = \log_e 2 + \frac{\pi}{2} - 1$$

**108.** *(b)* $\displaystyle\int_0^a e^{x-[x]} dx = 10e - 9$

Let $x = I + f$ and $n \le a < n+1$

$[x] = I,\ [a] = n$

$$\int_0^1 e^{x-0} dx + \int_1^2 e^{x-1} dx + \int_2^3 e^{x-2} dx + \ldots + \int_{n-1}^{n} e^{x-n+1} dx + \int_n^a e^{x-n} dx$$

$$(e^1 - e^0) + \left(\frac{e^2 - e^1}{e}\right) + \left(\frac{e^3 - e^2}{e^2}\right) + \ldots + \left(\frac{e^n - e^{n-1}}{e^{n-1}}\right) + \left(\frac{e^a - e^n}{e^n}\right)$$

$\Rightarrow n(e-1) + [e^{a-n} - 1] = 10e - 10 + 1$

$\Rightarrow\quad n = 10 \Rightarrow e^{a-10} - 1 = 1$

$\Rightarrow\quad e^{a-10} = 2 \Rightarrow a - 10 = \log_e 2$

$\therefore\quad a = 10 + \log_e 2$

**109.** *(b)* Given, $f(x) = e^{-x} \cdot \sin x$

and $\quad F(x) = \int_0^x f(t)\, dt$

$\because F(x)$ is differentiable function.

$\therefore F'(x) = f(x) \times 1 - f(0) \times 0$

(Using Newton-Leibnitz rule)

$\Rightarrow\quad F'(x) = f(x)$ ...(i)

Let $I = \int_0^1 [F'(x) + f(x)] e^x dx$

$= \int_0^1 [f(x) + f(x)]\, e^x\, dx$

[From Eq. (i)]

$\Rightarrow\quad I = 2 \cdot \int_0^1 f(x) \cdot e^x dx$

$= 2 \cdot \int_0^1 e^{-x} \sin x \cdot e^x\, dx$

$= 2\int_0^1 \sin x dx = 2[-\cos x]_0^1$

$= 2[-(\cos 1 - \cos 0)] = 2(1 - \cos 1)$

$$\Rightarrow I = 2 \cdot \left[1 - \left(1 - \frac{(1)^2}{2!} + \frac{(1)^4}{4!} - \frac{(1)^6}{6!} + \frac{(1)^8}{8!} - \ldots\right)\right]$$

[Using expansion of $\cos x$ i.e.,

$$\cos x = 1 - \frac{x^2}{2!} + \frac{x^4}{4!} - \frac{x^6}{6!} + \frac{x^8}{8!} - \ldots]$$

$$\Rightarrow\quad I = 2\left(\frac{1}{2} - \frac{1}{24} + \frac{1}{720} - \ldots\right)$$

Now, $I < 2\left(\dfrac{1}{2} - \dfrac{1}{24} + \dfrac{1}{720}\right)$

$$\Rightarrow\quad I < \left(1 - \frac{1}{12} + \frac{1}{360}\right)$$

$$\Rightarrow\quad I < \frac{360 - 30 + 1}{360} \Rightarrow I < \frac{331}{360} \quad \ldots\text{(ii)}$$

Also, $I > 2\left(\dfrac{1}{2} - \dfrac{1}{24}\right)$

$$\Rightarrow\quad I > \left(1 - \frac{1}{12}\right) \Rightarrow\quad I > \frac{11}{12}$$

$$\Rightarrow\quad I > \frac{11 \times 30}{12 \times 30} \Rightarrow I > \frac{330}{360} \quad \ldots\text{(iii)}$$

From Eqs. (ii) and (iii), we get

$$\frac{330}{360} < I < \frac{331}{360}$$

**110.** *(a)* Let $I = \displaystyle\int_0^{10} \frac{[\sin 2\pi x]}{e^{x-[x]}}\, dx$

$$\Rightarrow\quad I = \int_0^{10} \frac{[\sin 2\pi x]}{e^{\{x\}}}\, dx$$

$[\because x - [x] = \{x\}]$

From above integrand, we observe that $\dfrac{[\sin 2\pi x]}{e^{\{x\}}}$ is a periodic function with period 1.

$$\therefore\quad I = 10\int_0^1 \frac{[\sin 2\pi x]}{e^{\{x\}}}$$

[By the property of definite integral,

$$\int_0^{nT} f(x)\,dx = n\int_0^{T} f(x)\,dx,$$

where $f(x)$ is a periodic function with period $=T$]

$$\Rightarrow \quad I = 10\cdot\int_0^1 \frac{[\sin 2\pi x]}{e^x}\,dx$$

$[\because \{x\} = x, 0 \le x < 1]$

$$\Rightarrow \quad I = 10\left(\int_0^{1/2}\frac{[\sin 2\pi x]}{e^x}\,dx + \int_{1/2}^{1}\frac{[\sin 2\pi x]}{e^x}\,dx\right)$$

$$\Rightarrow \quad I = 10\left(\int_0^{1/2}\frac{0}{e^x}\,dx + \int_{1/2}^{1}\frac{(-1)}{e^x}\,dx\right)$$

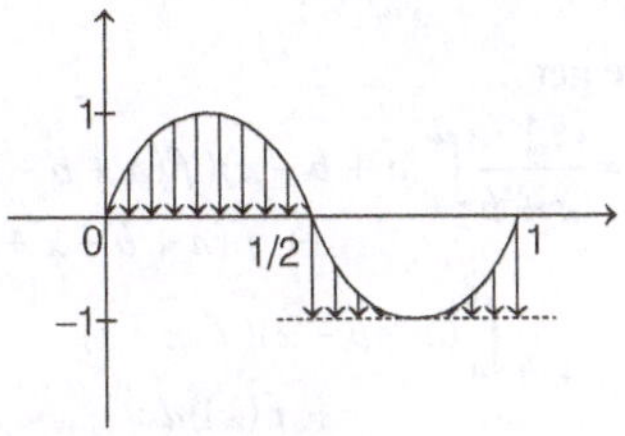


$$\Rightarrow \quad I = 10\left(0 - \int_{1/2}^{1} e^{-x}\,dx\right)$$

$$\Rightarrow \quad I = -10\left[\frac{e^{-x}}{-1}\right]_{1/2}^{1}$$

$$\Rightarrow \quad I = 10\,[e^{-1} - e^{-1/2}]$$

$$\Rightarrow \quad I = 10e^{-1} - 10e^{\frac{-1}{2}}$$

$$\Rightarrow I = 10e^{-1} + (-10)\cdot e^{\frac{-1}{2}} + 0 \quad \ldots(\text{i})$$

On comparing Eq. (i) by

$\alpha e^{-1} + \beta e^{\frac{-1}{2}} + \gamma$, we get

$\alpha = 10, \beta = -10$ and $\gamma = 0$

Hence, $\alpha + \beta + \gamma = 10 - 10 + 0$

$\Rightarrow \quad \alpha + \beta + \gamma = 0$

**111.** (c) Given, $f(x) = \int_1^x \frac{\log_e t}{(1+t)}\,dt$

So, $f(e) = \int_1^e \frac{\log_e t}{1+t}\,dt \quad \ldots(\text{i})$

and $f\left(\frac{1}{e}\right) = \int_1^{\frac{1}{e}} \frac{\log_e t}{1+t}\,dt \quad \ldots(\text{ii})$

Let $t = \frac{1}{u}$ and putting in Eq. (ii), we get

$$f\left(\frac{1}{e}\right) = \int_1^e \frac{\log\left(\frac{1}{u}\right)}{1+\frac{1}{u}}\cdot\frac{-1}{u^2}\,du$$

$$= \int_1^e \frac{\log u}{u(u+1)}\,du$$

Using change of variable

$$f\left(\frac{1}{e}\right) = \int_1^e \frac{\log t}{t(t+1)}\,dt \quad \ldots(\text{iii})$$

From Eqs. (i) and (iii), we get

$$f(e) + f\left(\frac{1}{e}\right) = \int_1^e \frac{\log t}{1+t}\,dt + \int_1^e \frac{\log t}{t(1+t)}\,dt$$

$$= \int_1^e \frac{\log t}{t}\,dt$$

Let $\log t = v$, then $\frac{1}{t}\,dt = dv$

$$f(e) + f\left(\frac{1}{e}\right) = \int_0^1 v\,dv = \left[\frac{v^2}{2}\right]_0^1 = \frac{1}{2}$$

$$\therefore \quad f(e) + f\left(\frac{1}{e}\right) = \frac{1}{2}$$

**112.** (a) Let $I = \int_{-\pi/2}^{\pi/2} \frac{\cos^2 x}{1+3^x}\,dx \quad \ldots(\text{i})$

Using the property,

$$\int_a^b f(x)\,dx = \int_a^b f(a+b-x)\,dx$$

$$I = \int_{-\pi/2}^{\pi/2} \frac{\cos^2(\pi/2 - \pi/2 - x)}{1+3^{\pi/2-\pi/2-x}}\,dx$$

$$= \int_{-\pi/2}^{\pi/2} \frac{\cos^2 x}{1+3^{-x}}\,dx$$

$[\because \cos(-x) = \cos x]$

$$I = \int_{-\pi/2}^{\pi/2} \frac{3^x \cos^2 x}{(1+3^x)}\,dx \quad \ldots(\text{ii})$$

On adding Eqs. (i) and (ii), we get

$$2I = \int_{-\pi/2}^{\pi/2} \frac{\cos^2 x}{1+3^x}\,dx + \int_{-\pi/2}^{\pi/2} \frac{3^x \cos^2 x}{1+3^x}\,dx$$

$$= \int_{-\pi/2}^{\pi/2} \frac{(1+3^x)\cos^2 x}{1+3^x}\,dx$$

$$= \int_{-\pi/2}^{\pi/2} \frac{1+\cos 2x}{2}\,dx$$

$[\because \cos 2x = 2\cos^2 x - 1]$

$$= \frac{1}{2}\left[x + \frac{\sin 2x}{2}\right]_{-\pi/2}^{\pi/2} = \frac{1}{2}[\pi]$$

$$\Rightarrow \quad 2I = \pi/2 \Rightarrow I = \frac{\pi}{4}$$

**113.** (a) $I_n = \int_{\pi/4}^{\pi/2} \cot^n x\,dx$

$$= \int_{\pi/4}^{\pi/2} \cot^{n-2} x(\cot^2 x)\,dx$$

$$I_n = \int_{\pi/4}^{\pi/2} \cot^{n-2} x \operatorname{cosec}^2 x\,dx - \int_{\pi/4}^{\pi/2} \cot^{n-2} x\,dx$$

$$I_n + I_{n-2} = \int_{\pi/4}^{\pi/2} \cot^{n-2} x \cdot \operatorname{cosec}^2 x\,dx$$

Now, let $\cot x = t$, then

$\operatorname{cosec}^2 x\,dx = -dt$, limit will be

$$I_n + I_{n-2} = \int_1^0 -t^{n-2}\,dt = \left[\frac{-(t)^{n-1}}{n-1}\right]_1^0$$

$$I_n + I_{n-2} = \frac{1}{n-1}$$

Now, putting $n = 4$

$\Rightarrow I_2 + I_4 = \frac{1}{3}$, then

$$\frac{1}{I_2 + I_4} = 3 \quad \ldots(\text{i})$$

Putting $n = 5$

$\Rightarrow \quad I_5 + I_3 = \frac{1}{4}$, then $\frac{1}{I_3 + I_5} = 4 \quad \ldots(\text{ii})$

Putting $n = 6$

$\Rightarrow I_6 + I_4 = \frac{1}{5}$, then $\frac{1}{I_4 + I_6} = 5 \quad \ldots(\text{iii})$

Here, from Eqs. (i), (ii) and (iii), we conclude

$\frac{1}{I_2 + I_4}, \frac{1}{I_3 + I_5}$ and $\frac{1}{I_4 + I_6}$ are in AP with common difference 1.

**114.** (c) It is given that

$$I_1 = \int_0^1 (1 - x^{50})^{100}\,dx$$

and $I_2 = \int_0^1 (1 - x^{50})^{101}\,dx$

$$= \int_0^1 (1 - x^{50})(1 - x^{50})^{100}\,dx$$

$$= \int_0^1 (1 - x^{50})^{100}\,dx - \int_0^1 x \cdot x^{49}(1 - x^{50})^{100}\,dx$$

Now, let $I = \int_0^1 x \cdot x^{49}(1 - x^{50})^{100}\,dx$

$$=[x\int x^{49}(1-x^{50})^{100}\,dx - \int [1\int x^{49}(1-x^{50})^{100}\,dx]\,dx]_0^1$$

$$=\left[t^{1/50}\int \frac{(1-t)^{100}}{50}\,dt - \int\left(\int \frac{(1-t)^{100}}{50}\,dt\right)dt\right]_0^1$$

[Where, $x^{50}=t$]

$$I = 0 - \frac{I_2}{50\times 101}$$

$$\therefore \quad I_2 = I_1 - \frac{I_2}{5050}$$

$$\Rightarrow \frac{5051}{5050} I_2 = I_1$$

$$\because \quad I_2 = \alpha I_1$$

$$\therefore \quad \alpha = \frac{5050}{5051}$$

**115.** *(a)* Given integral,

$$I = \int_{-\pi/2}^{\pi/2} \frac{1}{1+e^{\sin x}}\,dx \qquad \ldots(i)$$

As we know that $\int_a^b f(x)dx$

$$= \int_a^b f(a+b-x)dx, \text{ so}$$

$$I = \int_{-\pi/2}^{\pi/2} \frac{1}{1+e^{-\sin x}}\,dx$$

$$= \int_{-\pi/2}^{\pi/2} \frac{e^{\sin x}}{e^{\sin x}+1}\,dx \qquad \ldots(ii)$$

On adding Eqs. (i) and (ii), we get

$$2I = \int_{-\pi/2}^{\pi/2} \frac{1+e^{\sin x}}{1+e^{\sin x}}\,dx = \int_{-\pi/2}^{\pi/2} dx = \pi$$

$$\Rightarrow \quad I = \frac{\pi}{2}$$

**116.** *(d)* Given integral is

$$I = \int_0^{2\pi} \frac{x\sin^8 x}{\sin^8 x + \cos^8 x}\,dx \qquad \ldots(i)$$

On applying property,

$$\int_0^a f(x)\,dx = \int_0^a f(a-x)\,dx,$$

we get

$$I = \int_0^{2\pi} \frac{(2\pi - x)\sin^8 x}{\sin^8 x + \cos^8 x}\,dx \qquad \ldots(ii)$$

On adding integrals Eqs. (i) and (ii), we get

$$2I = \int_0^{2\pi} \frac{2\pi \sin^8 x}{\sin^8 x + \cos^8 x}\,dx$$

$$\Rightarrow \quad I = \pi\int_0^{2\pi} \frac{\sin^8 x}{\sin^8 x + \cos^8 x}\,dx$$

On applying property,

$$\int_0^{2a} f(x)\,dx = 2\int_0^a f(x)dx,$$

if $f(2a-x) = f(x)$, we get

$$I = 2\pi\int_0^{\pi} \frac{\sin^8 x}{\sin^8 x + \cos^8 x}\,dx$$

Again, if we apply the same property, we get

$$I = 4\pi\int_0^{\pi/2} \frac{\sin^8 x}{\sin^8 x + \cos^8 x}\,dx \qquad \ldots(iii)$$

Now, on applying property,

$$\int_0^a f(x)\,dx = \int_0^a f(a-x)\,dx, \text{ we get}$$

$$I = 4\pi\int_0^{\pi/2} \frac{\cos^8 x}{\cos^8 x + \sin^8 x}\,dx \qquad \ldots(iv)$$

On adding integrals (iii) and (iv), we get

$$2I = 4\pi\int_0^{\pi/2} \frac{\sin^8 x + \cos^8 x}{\cos^8 x + \sin^8 x}\,dx$$

$$\Rightarrow \quad I = 2\pi\int_0^{\pi/2} dx = 2\pi\left(\frac{\pi}{2}\right)$$

$$\Rightarrow \quad I = \pi^2$$

**117.** *(b)* Given trigonometric equation for $\theta \in (0, 2\pi) - \{\pi\}$ is

$$2\cot^2\theta - \frac{5}{\sin\theta} + 4 = 0$$

$$\Rightarrow \quad 2\cos^2\theta - 5\sin\theta + 4\sin^2\theta = 0$$

$$\Rightarrow \quad 2\sin^2\theta - 4\sin\theta - \sin\theta + 2 = 0$$

$$\Rightarrow 2\sin\theta(\sin\theta - 2) - 1(\sin\theta - 2) = 0$$

$$\Rightarrow \quad (\sin\theta - 2)(2\sin\theta - 1) = 0$$

$\because$ $\sin\theta \in [-1, 1] - \{0\}$, for $\theta \in (0, 2\pi) - \{\pi\}$

$\therefore$ $\sin\theta - 2 \neq 0$

So, $\sin\theta = \dfrac{1}{2} \Rightarrow \theta = \dfrac{\pi}{6}, \dfrac{5\pi}{6}$

$$\therefore \quad \theta_1 = \frac{\pi}{6} \text{ and } \theta_2 = \frac{5\pi}{6}$$

Now, the integral $\int_{\theta_1}^{\theta_2} \cos^2 3\theta\, d\theta$

$$= \int_{\pi/6}^{5\pi/6} \left(\frac{1+\cos 6\theta}{2}\right) d\theta$$

$$= \frac{1}{2}\left[\theta + \frac{\sin 6\theta}{6}\right]_{\pi/6}^{5\pi/6}$$

$$= \frac{1}{2}\left(\frac{4\pi}{6}\right) = \frac{\pi}{3}$$

**118.** *(c)* Let the integral

$$I = \frac{1}{a+b}\int_a^b x(f(x) + f(x+1))dx \qquad \ldots(i)$$

On applying property

$$\int_a^b f(x)dx = \int_a^b f(a+b-x)dx,$$

we get

$$I = \frac{1}{a+b}\int_a^b (a+b-x)(f(a+b-x) + f(a+b-x+1))dx$$

$$= \frac{1}{a+b}\int_a^b (a+b-x)(f(x+1) + f(x))dx \qquad \ldots(ii)$$

$$\left[\begin{array}{l}\because f(a+b+1-x) = f(x)\\ \therefore f(a+b-x) = f(x+1)\end{array}\right]$$

On adding Eqs. (i) and (ii), we get

$$2I = \frac{a+b}{a+b}\int_a^b (f(x) + f(x+1))dx$$

$$\Rightarrow \quad I = \frac{1}{2}\left[\int_a^b f(x)dx + \int_a^b f(x+1)dx\right]$$

$$= \frac{1}{2}\left[\int_a^b f(x)dx + \int_a^b (a+b-x+1)dx\right]$$

$$= \frac{1}{2}\left[\int_a^b f(x)dx + \int_a^b f(x)dx\right]$$

$$\Rightarrow \quad I = \int_a^b f(x)dx$$

Putting $x = t+1$, then upper limit $t = b-1$ and lower limit $t = a-1$ and $dx = dt$, so

$$I = \int_{a-1}^{b-1} f(t+1)dt = \int_{a-1}^{b-1} f(x+1)dx$$

**119.** *(a)* Let $I = \displaystyle\int_\alpha^{\alpha+1} \frac{dx}{(x+\alpha)(x+\alpha+1)}$

$$= \int_\alpha^{\alpha+1} \frac{(x+\alpha+1)-(x+\alpha)}{(x+\alpha)(x+\alpha+1)}\,dx$$

$$= \int_\alpha^{\alpha+1} \left(\frac{1}{x+\alpha} - \frac{1}{x+\alpha+1}\right)dx$$

$$= [\log_e(x+\alpha) - \log_e(x+\alpha+1)]_\alpha^{\alpha+1}$$

$$= \left[\log_e \left(\frac{x+\alpha}{x+\alpha+1}\right)\right]_{\alpha}^{\alpha+1}$$

$$= \log_e \frac{2\alpha+1}{2\alpha+2} - \log_e \frac{2\alpha}{2\alpha+1}$$

$$= \log_e \left(\frac{2\alpha+1}{2\alpha+2} \times \frac{2\alpha+1}{2\alpha}\right)$$

$$= \log_e \left(\frac{9}{8}\right) \quad \text{[Given]}$$

$$\Rightarrow \frac{(2\alpha+1)^2}{4\alpha(\alpha+1)} = \frac{9}{8}$$

$$\Rightarrow 8[4\alpha^2 + 4\alpha + 1] = 36(\alpha^2 + \alpha)$$

$$\Rightarrow \quad \alpha^2 + \alpha - 2 = 0$$

$$\Rightarrow \quad (\alpha+2)(\alpha-1) = 0 \Rightarrow \alpha = 1, -2$$

From the options, we get

$$\alpha = -2$$

**120.** *(a)* Given, $\int_6^{f(x)} 4t^3\, dt = (x-2)\, g(x)$

$$\Rightarrow \quad g(x) = \frac{\int_6^{f(x)} 4t^3\, dt}{(x-2)}$$

[Provided $x \neq 2$]

$$\text{So,} \quad \lim_{x\to 2} g(x) = \lim_{x\to 2} \frac{\int_6^{f(x)} 4t^3\, dt}{x-2}$$

$$\left[\because \frac{0}{0} \text{ form as } x \to 2 \Rightarrow f(2) = 6\right]$$

$$\lim_{x\to 2} g(x) = \lim_{x\to 2} \frac{4(f(x))^3 f'(x)}{1}$$

$$\left[\because \frac{d}{dx}\int_{\phi_1(x)}^{\phi_2(x)} f(t)\, dt = f(\phi_2(x))\cdot \phi_2'(x) - f(\phi_1(x))\cdot \phi_1'(x)\right]$$

On applying limit, we get

$$\lim_{x\to 2} g(x) = 4(f(2))^3 f'(2)$$

$$= 4 \times (6)^3 \frac{1}{48} = 18$$

**121.** *(a)* Given integral,

$$I = \int_0^{2\pi} [\sin 2x \cdot (1 + \cos 3x)]dx$$

$$= \int_0^{\pi} [\sin 2x \cdot (1 + \cos 3x)]dx + \int_{\pi}^{2\pi} [\sin 2x \cdot (1 + \cos 3x)]dx$$

$$= I_1 + I_2 \text{ (let)} \quad \dots \text{(i)}$$

Now, $I_2 = \int_{\pi}^{2\pi} [\sin 2x \cdot (1 + \cos 3x)]dx$

Let $2\pi - x = t$, upper limit $t = 0$ and lower limit $t = \pi$

and $\quad dx = -dt$

$$\text{So,} \quad I_2 = -\int_{\pi}^{0} [-\sin 2t \cdot (1 + \cos 3t)]dt$$

$$= \int_0^{\pi} [-\sin 2t \cdot (1 + \cos 3t)]dt \quad \dots \text{(ii)}$$

$$\therefore \quad I = \int_0^{\pi} [\sin 2t \cdot (1 + \cos 3t)]dt + \int_0^{\pi} [-\sin 2t \cdot (1 + \cos 3t)]dt$$

[from Eqs. (i) and (ii) and change variable of Eq. (i) from $t$ to $x$.]

$$= \int_0^{\pi} (-1)\, dx$$

$$[\because [x] + [-x] = -1,\ x \notin \text{Integer}]$$

$$= -\pi$$

**122.** *(a)* Let $I = \int_0^1 x \cot^{-1}(1 - x^2 + x^4)\, dx$

Now, putting $x^2 = t \Rightarrow 2x dx = dt$

Lower limit at $x = 0,\ t = 0$

Upper limit at $x = 1,\ t = 1$

$$\therefore \quad I = \frac{1}{2}\int_0^1 \cot^{-1}(1 - t + t^2)\, dt$$

$$= \frac{1}{2}\int_0^1 \tan^{-1}\left(\frac{1}{1 - t + t^2}\right) dt$$

$$[\because \cot^{-1} x = \tan^{-1} 1/x]$$

$$= \frac{1}{2}\int_0^1 \tan^{-1}\left(\frac{t - (t-1)}{1 + t(t-1)}\right) dt$$

$$= \frac{1}{2}\left[\int_0^1 (\tan^{-1} t - \tan^{-1}(t-1)\, dt\right]$$

$$\left[\because \tan^{-1} \frac{x - y}{1 + xy} = \tan^{-1} x - \tan^{-1} y\right]$$

$$\because \quad \int_0^1 \tan^{-1}(t-1)dt$$

$$= \int_0^1 \tan^{-1}(1 - t - 1)\, dt$$

$$= -\int_0^1 \tan^{-1}(t)\, dt$$

$$\text{because} \int_0^a f(x)dx = \int_0^a f(a - x)\, dx$$

$$\text{So,} \ I = \frac{1}{2}\int_0^1 (\tan^{-1} t + \tan^{-1} t)\, dt$$

$$= \int_0^1 \tan^{-1} t dt$$

$$= [t \tan^{-1} t]_0^1 - \int_0^1 \frac{t}{1+t^2}\, dt$$

[using integration by parts method]

$$= \frac{\pi}{4} - \frac{1}{2}[\log_e(1 + t^2)]_0^1$$

$$= \frac{\pi}{4} - \frac{1}{2}\log_e 2$$

**123.** *(c)* Let $I = \int_0^{\frac{\pi}{2}} \frac{\sin^3 x}{\sin x + \cos x}\, dx \quad \dots \text{(i)}$

On applying the property

$$\int_a^b f(x)dx = \int_a^b f(a + b - x)\, dx,$$

we get

$$I = \int_0^{\pi/2} \frac{\cos^3 x}{\cos x + \sin x}\, dx \quad \dots \text{(ii)}$$

On adding integrals (i) and (ii), we get

$$2I = \int_0^{\pi/2} \frac{\sin^3 x + \cos^3 x}{\sin x + \cos x}\, dx$$

$$= \int_0^{\pi/2} \frac{(\sin x + \cos x)(\sin^2 x + \cos^2 x - \sin x \cos x)}{\sin x + \cos x}\, dx$$

$$= \int_0^{\pi/2} \left[1 - \frac{1}{2}(2\sin x \cos x)\right] dx$$

$$= \left[x + \frac{1}{4}\cos 2x\right]_0^{\pi/2}$$

$$\Rightarrow \quad I = \frac{\pi}{4} - \frac{1}{4} = \frac{\pi - 1}{4}$$

**124.** *(d)* Given, $f(x) = \int_0^x g(t)\, dt$

On replacing $x$ by $(-x)$, we get

$$f(-x) = \int_0^{-x} g(t)dt$$

Now, putting $t = -u$, so

$$f(-x) = -\int_0^x g(-u)du$$

$$= -\int_0^x g(u)du = -f(x)$$

[$\because g$ is an even function]

$\Rightarrow f(-x) = -f(x)$

$\Rightarrow f$ is an odd function.

Now, it is given that $f(x + 5) = g(x)$

$$\therefore \quad f(5 - x) = g(-x)$$

$$= g(x) = f(x + 5)$$

[$\because g$ is an even function]

$$\Rightarrow \quad f(5 - x) = f(x + 5) \quad \dots \text{(i)}$$

Let $I = \int_0^x f(t)dt$

Putting $t = u + 5$

$\Rightarrow \quad t - 5 = u \Rightarrow dt = du$

$\therefore \quad I = \int_{-5}^{x-5} f(u+5)du = \int_{-5}^{x-5} g(u)du$

On putting $u = -t$

$\Rightarrow \quad du = -dt$, we get

$$I = -\int_{5}^{5-x} g(-t)dt = \int_{5-x}^{5} g(t)dt$$

$\left[\because -\int_a^b f(x)dx = \int_b^a f(x)dx \text{ and } g \text{ is an even function}\right]$

$$I = \int_{5-x}^{5} f'(t)dt$$

[By Leibnitz rule $f'(x) = g(x)$]

$= f(5) - f(5-x)$

$= f(5) - f(5+x)$ [From Eq. (i)]

$$= \int_{5+x}^{5} f'(t)dt = \int_{5+x}^{5} g(t)dt$$

**125.** *(d)*

**key idea** (i) First calculate the value of $g(f(x))$
(ii) Then use the property of definite integration
$I = \int_a^b f(x)\,dx = \int_a^b f(a+b-x)\,dx$

The given functions are

$g(x) = \log_e x, x > 0$

and $f(x) = \dfrac{2 - x\cos x}{2 + x\cos x}$

Let $\quad I = \int_{-\pi/4}^{\pi/4} g(f(x))dx$

Then,

$$I = \int_{-\pi/4}^{\pi/4} \log_e\left(\frac{2 - x\cos x}{2 + x\cos x}\right)dx \quad \ldots\text{(i)}$$

Now, by using the property

$\int_a^b f(x)\,dx = \int_a^b f(a+b-x)\,dx$, we get

$$I = \int_{-\pi/4}^{\pi/4} \log_e\left(\frac{2 + x\cos x}{2 - x\cos x}\right)dx \quad \ldots\text{(ii)}$$

On adding Eqs. (i) and (ii), we get

$$2I = \int_{-\pi/4}^{\pi/4}\left[\log_e\left(\frac{2 - x\cos x}{2 + x\cos x}\right) + \log_e\left(\frac{2+\cos x}{2 - x\cos x}\right)\right]dx$$

$$= \int_{-\pi/4}^{\pi/4}\log_e\left(\frac{2 - x\cos x}{2 + x\cos x} \times \frac{2 + x\cos x}{2 - x\cos x}\right)dx$$

$\Rightarrow \quad 2I = \int_{-\pi/4}^{\pi/4} \log_e(1)dx = 0$

$\Rightarrow \quad I = 0 = \log_e(1)$

**126.** *(a)* Let

$$I = \int_1^e \left\{\left(\frac{x}{e}\right)^{2x} - \left(\frac{e}{x}\right)^x\right\}\log_e x\,dx$$

Now, putting $\left(\dfrac{x}{e}\right)^x = t$

$\Rightarrow x\log_e\left(\dfrac{x}{e}\right) = \log t$

$\Rightarrow x(\log_e x - \log_e e) = \log t$

$\Rightarrow \left[x\left(\dfrac{1}{x}\right) + (\log_e x - \log_e e)\right]dx = \dfrac{1}{t}dt$

$\Rightarrow (1 + \log_e x - 1)\,dx = \dfrac{1}{t}dt$

$\Rightarrow (\log_e x)\,dx = \dfrac{1}{t}dt$

Also, upper limit $x = e \Rightarrow t = 1$

and lower limit

$x = 1 \Rightarrow t = 1/e$

$$\therefore I = \int_{1/e}^{1}\left(t^2 - \frac{1}{t}\right)\cdot\frac{1}{t}dt$$

$$\Rightarrow I = \int_{1/e}^{1}(t - t^{-2})\,dt = \left[\left(\frac{t^2}{2} + \frac{1}{t}\right)\right]_{1/e}^{1}$$

$$= \frac{3}{2} - e - \frac{1}{2e^2}$$

**127.** *(c)* Let

$$I = \int_{\pi/6}^{\pi/4}\frac{dx}{\sin 2x(\tan^5 x + \cot^5 x)}$$

$$= \int_{\pi/6}^{\pi/4}\frac{(1+\tan^2 x)\tan^5 x}{2\tan x(\tan^{10} x + 1)}dx$$

$$\left[\because \sin 2x = \frac{2\tan x}{1 + \tan^2 x}\right]$$

$$= \frac{1}{2}\int_{\pi/6}^{\pi/4}\frac{\tan^4 x\sec^2 x}{(\tan^{10} x + 1)}dx$$

Putting $\tan^5 x = t$

$\Rightarrow \quad 5\tan^4 x\sec^2 x\,dx = dt$

| $x$ | $\pi/6$ | $\pi/4$ |
|---|---|---|
| $t$ | $(1/\sqrt{3})^5$ | 1 |

$$\therefore I = \frac{1}{2}\cdot\frac{1}{5}\int_{(1/\sqrt{3})^5}^{1}\frac{dt}{t^2 + 1}$$

$$= \frac{1}{10}(\tan^{-1}(t))_{(1/\sqrt{3})^5}^{1}$$

$$= \frac{1}{10}\left(\tan^{-1}(1) - \tan^{-1}\left(\frac{1}{9\sqrt{3}}\right)\right)$$

$$= \frac{1}{10}\left(\frac{\pi}{4} - \tan^{-1}\left(\frac{1}{9\sqrt{3}}\right)\right)$$

**128.** *(d)*

**key idea** (i) Define the function $[x]$ and $[\sin x]$ in the internal $\left(-\dfrac{\pi}{2}, \dfrac{\pi}{2}\right)$
(ii) Break the integration into four part accordingly and then integrate.

Let $I = \int_{-\frac{\pi}{2}}^{\frac{\pi}{2}}\dfrac{dx}{[x] + [\sin x] + 4}$

$$= \int_{-\frac{\pi}{2}}^{-1}\frac{dx}{[x]+[\sin x]+4} + \int_{-1}^{0}\frac{dx}{[x]+[\sin x]+4}$$

$$+ \int_0^1\frac{dx}{[x]+[\sin x]+4} + \int_1^{\pi/2}\frac{dx}{[x]+[\sin x]+4}$$

$$\because \quad [x] = \begin{cases} -2, & -\pi/2 < x < -1 \\ -1, & -1 \le x < 0 \\ 0, & 0 \le x < 1 \\ 1, & 1 \le x < \pi/2 \end{cases}$$

and

$$[\sin x] = \begin{cases} -1, & \dfrac{-\pi}{2} < x < -1; \quad \dfrac{-\pi}{2} \le x < 0 \\ -1, & -1 < x < 0; \quad -1 \le \sin x < 0 \\ 0, & 0 < x < 1; \quad 0 < x \le \dfrac{\pi}{2} \\ 0, & 1 < x < \dfrac{\pi}{2}; \quad 0 < \sin x \le 1 \end{cases}$$

So, $I = \int_{-\frac{\pi}{2}}^{-1}\dfrac{dx}{-2-1+4} + \int_{-1}^{0}\dfrac{dx}{-1-1+4}$

$$+\int_0^1\frac{dx}{0+0+4} + \int_1^{\pi/2}\frac{dx}{1+0+4}$$

$$= \int_{-\frac{\pi}{2}}^{-1}\frac{dx}{1} + \int_{-1}^{0}\frac{dx}{2} + \int_0^1\frac{dx}{4} + \int_1^{\frac{\pi}{2}}\frac{dx}{5}$$

$$= \left(-1 + \frac{\pi}{2}\right) + \frac{1}{2}(0+1) + \frac{1}{4}(1-0) + \frac{1}{5}\left(\frac{\pi}{2} - 1\right)$$

$$= -\frac{9}{20} + \frac{3\pi}{5}$$

$$= \frac{3}{20}(4\pi - 3)$$

**129.** *(d)* We know, graph of $y = \cos x$ is

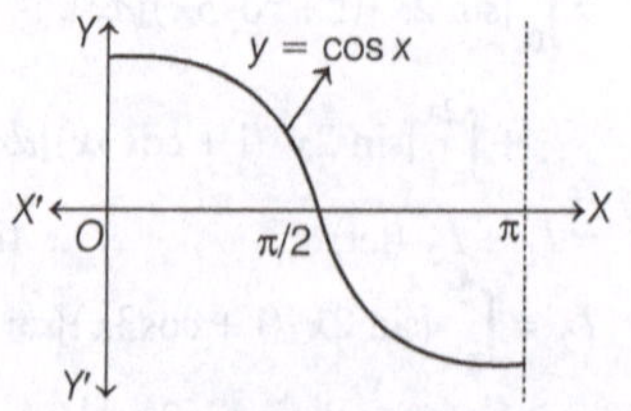

∴ The graph of $y = |\cos x|$ is

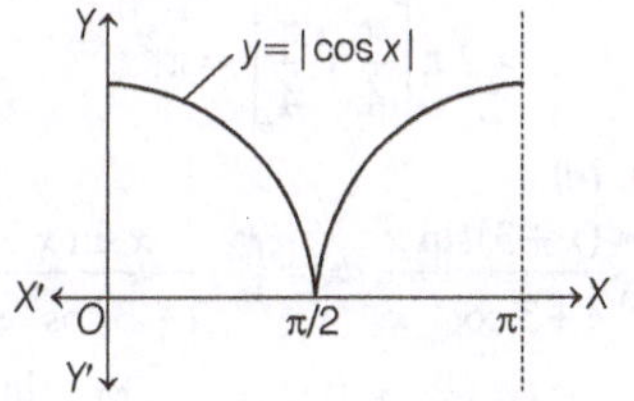


$$\therefore\ I = \int_0^{\pi} |\cos x|^3 = 2\int_0^{\frac{\pi}{2}} |\cos x|^3\,dx$$

[∵ $y = |\cos x|$ is symmetric about $x = \pi/2$]

$$= 2\int_0^{\frac{\pi}{2}} \cos^3 x\,dx$$

$$\left[\because \cos x \geq 0 \text{ for } x \in \left[0, \frac{\pi}{2}\right]\right]$$

Now, as $\cos 3x = 4\cos^3 x - 3\cos x$

$$\therefore\ \cos^3 x = \frac{1}{4}(\cos 3x + 3\cos x)$$

$$\therefore\quad I = \frac{2}{4}\int_0^{\frac{\pi}{2}} (\cos 3x + 3\cos x)\,dx$$

$$= \frac{1}{2}\left[\frac{\sin 3x}{3} + 3\sin x\right]_0^{\frac{\pi}{2}}$$

$$= \frac{1}{2}\left\{\left[\frac{1}{3}\sin\frac{3\pi}{2} + 3\sin\frac{\pi}{2}\right] - \left[\frac{1}{3}\sin 0 + 3\sin 0\right]\right\}$$

$$= \frac{1}{2}\left[-\frac{1}{3} + 3\right] = \frac{4}{3}$$

**130.** *(d)*

> **key idea** Use property : $\int_a^b f(x)dx = \int_a^b f(a+b-x)\,dx$

Let $I = \int_{-\pi/2}^{\pi/2} \frac{\sin^2 x}{1+2^x}\,dx$ ...(i)

$$\Rightarrow\quad I = \int_{-\pi/2}^{\pi/2} \frac{\sin^2\left(-\frac{\pi}{2} + \frac{\pi}{2} - x\right)}{1 + 2^{-\frac{\pi}{2}+\frac{\pi}{2}-x}}\,dx$$

$$\left[\because \int_a^b f(x)dx = \int_a^b f(a+b-x)dx\right]$$

$$\Rightarrow\quad I = \int_{-\pi/2}^{\pi/2} \frac{\sin^2 x}{1+2^{-x}}\,dx$$

$$\Rightarrow\quad I = \int_{-\pi/2}^{\pi/2} \frac{2^x \sin^2 x}{2^x + 1}\,dx \quad \text{...(ii)}$$

On adding Eqs. (i) and (ii), we get

$$2I = \int_{-\pi/2}^{\pi/2} (\sin^2 x)\left(\frac{2^x+1}{2^x+1}\right)dx$$

$$\Rightarrow\quad 2I = \int_{-\pi/2}^{\pi/2} \sin^2 x\,dx$$

$$\Rightarrow\quad 2I = 2\int_0^{\pi/2} \sin^2 x\,dx$$

[∵ $\sin^2 x$ is an even function]

$$\Rightarrow\ I = \int_0^{\pi/2} \sin^2 x dx \Rightarrow I = \int_0^{\pi/2} \cos^2 x dx$$

$$\left[\because \int_0^a f(x)dx = \int_0^a f(a-x)dx\right]$$

$$\Rightarrow\ 2I = \int_0^{\pi/2} dx \Rightarrow 2I = [x]_0^{\pi/2} \Rightarrow I = \frac{\pi}{4}$$

**131.** *(d)* Let $I = \int_{\pi/4}^{3\pi/4} \frac{x}{1+\sin x}\,dx$

$$\Rightarrow I = \int_{\pi/4}^{3\pi/4}\left(\frac{x}{1+\sin x} \times \frac{1-\sin x}{1-\sin x}\right)dx$$

$$\Rightarrow I = \int_{\pi/4}^{3\pi/4} \frac{x\,(1-\sin x)}{\cos^2 x}\,dx$$

$$\{\because 1 - \sin^2 x = \cos^2 x\}$$

$$\Rightarrow I = \int_{\pi/4}^{3\pi/4} x \cdot \sec^2 x\,dx - \int_{\pi/4}^{3\pi/4} x\,(\sec x \tan x)\,dx$$

$$\Rightarrow I = |x\tan x|_{\pi/4}^{3\pi/4} - |\log|\sec x||_{\pi/4}^{3\pi/4} - \left|\frac{x}{\cos x}\right|_{\pi/4}^{3\pi/4} - \int_{\pi/4}^{3\pi/4} \frac{1}{\cos x}\,dx$$

$$\Rightarrow I = |x\tan x|_{\pi/4}^{3\pi/4} - \left|\frac{x}{\cos x}\right|_{\pi/4}^{3\pi/4} - \log|\sec x + \tan x|_{\pi/4}^{3\pi/4}$$

After putting the limits, we get

$$I = -\frac{3\pi}{4} - \frac{\pi}{4} + \frac{3\pi}{4}(\sqrt{2}) + \frac{\pi}{4}\sqrt{2} - \log(\sqrt{2}+1) + \log(\sqrt{2}+1)$$

$$\Rightarrow I = -\pi + \sqrt{2}\pi$$

$$\Rightarrow I = \pi(\sqrt{2} - 1)$$

**132.** *(a)* We have,

$$\int_{\pi/12}^{\pi/4} \frac{8\cos 2x}{(\tan x + \cot x)^3}\,dx$$

$$= \int_{\pi/12}^{\pi/4} \frac{8\cos 2x}{8\left(\frac{\sin^2 x + \cos^2 x}{2\sin x \cos x}\right)^3}\,dx$$

$$= \int_{\pi/12}^{\pi/4} \frac{\cos 2x}{\left(\frac{1}{\sin 2x}\right)^3}\,dx$$

$$= \int_{\pi/12}^{\pi/4} \cos 2x \times \sin 2x \times \sin^2(2x)\,dx$$

$$= \frac{1}{4}\int_{\pi/12}^{\pi/4} \sin 4x \cdot (1 - \cos 4x)\,dx$$

$$= \frac{1}{4}\int_{\pi/12}^{\pi/4} \sin 4x\,dx - \frac{1}{8}\int_{\pi/12}^{\pi/4} \sin 8x\,dx$$

$$= \frac{1}{4}\left[\frac{-\cos 4x}{4}\right]_{\pi/12}^{\pi/4} - \frac{1}{8}\left(-\frac{\cos 8x}{8}\right)_{\pi/12}^{\pi/4}$$

$$= \frac{-1}{16}\left(\cos\pi - \cos\frac{\pi}{3}\right) + \frac{1}{64}\left(\cos 2\pi - \cos\frac{2\pi}{3}\right)$$

$$= \frac{3}{32} + \frac{3}{128} = \frac{15}{128}$$

**133.** *(d)* Let $I = \int_0^1 \frac{8\log(1+x)}{(1+x^2)}\,dx$

Put $x = \tan\theta \Rightarrow dx = \sec^2\theta d\theta$

When, $x = 0 \Rightarrow \tan\theta = 0$

$\therefore\quad \theta = 0$

When, $x = 1 = \tan\theta \Rightarrow \theta = \frac{\pi}{4}$

$$\therefore I = \int_0^{\pi/4} \frac{8\log(1+\tan\theta)}{1+\tan^2\theta} \cdot \sec^2\theta d\theta$$

$$= 8\int_0^{\pi/4} \log(1+\tan\theta)\,d\theta \quad \text{...(i)}$$

Using $\int_0^a f(x)\,dx = \int_0^a f(a-x)\,dx$

$$\therefore I = 8\int_0^{\pi/4} \log\left\{1 + \tan\left(\frac{\pi}{4} - \theta\right)\right\}d\theta$$

$$= 8\int_0^{\pi/4} \log\left\{1 + \frac{1-\tan\theta}{1+\tan\theta}\right\}d\theta$$

$$= 8\int_0^{\pi/4} \log\left\{\frac{2}{1+\tan\theta}\right\}d\theta \quad \text{...(ii)}$$

On adding Eqs. (i) and (ii), we get

$$2I = 8\int_0^{\pi/4}\left[\log\{1+\tan\theta\} + \log\left\{\frac{2}{1+\tan\theta}\right\}\right]d\theta$$

$$\Rightarrow I = 4\int_0^{\pi/4} \log 2\,d\theta$$

$$= 4 \cdot \log 2(\theta)_0^{\pi/4}$$

$$= 4\log 2 \cdot \left[\frac{\pi}{4} - 0\right]$$

$$= \pi \log 2$$

**134.** (b) We have,

$p'(x)=p'(1-x),\ \forall\ x\in[0,1],$

$p(0)=1,\ p(1)=41$ [Given]

On integrating, we get

$$\int p'(x)\,dx=\int p'(1-x)\,dx$$

$$\Rightarrow \quad p(x)=-p(1-x)+C$$

Put $x=1$; $p(1)=-p(0)+C$

$$\Rightarrow \quad 41=-1+C\Rightarrow C=42$$

$$\therefore p(x)+p(1-x)=42$$

Now, $I=\int_0^1 p(x)\,dx$ ...(i)

$$\Rightarrow \quad I=\int_0^1 p(1-x)\,dx \quad ...(ii)$$

$$\left[\because \int_0^a f(x)\,dx=\int_0^a f(a-x)\,dx\right]$$

On adding, Eqs. (i) and (ii), we get

$$2I=\int_0^1 [(p(x)+p(1-x)]\,dx$$

$$=\int_0^1 42\,dx=42$$

$$\therefore \quad I=21$$

**135.** (a) Since, $f(x)=\int_1^x \frac{\log t}{1+t}\,dt$

and $F(e)=f(e)+f\left(\frac{1}{e}\right)$

$$\Rightarrow F(e)=\int_1^e \frac{\log t}{1+t}\,dt+\int_1^{1/e}\frac{\log t}{1+t}\,dt$$

Put $t=1/t$ in second integration.

$$\therefore F(e)=\int_1^e \frac{\log t}{1+t}\,dt+\int_1^e \frac{-\log t}{1+\frac{1}{t}}\,d\left(\frac{1}{t}\right)$$

$$=\int_1^e \frac{\log t}{1+t}\,dt-\int_1^e \frac{t\log t}{(1+t)}\times\left(\frac{-dt}{t^2}\right)$$

$$=\int_1^e \frac{\log t}{1+t}\,dt+\int_1^e \frac{\log t}{t(1+t)}\,dt$$

$$=\int_1^e \frac{\log t}{1+t}\,dt+\int_1^e \frac{\log t}{t}\,dt$$

$$-\int_1^e \frac{\log t}{(1+t)}\,dt\left[\because \frac{1}{t(1+t)}=\frac{1}{t}-\frac{1}{t+1}\right]$$

$$=\int_1^e \frac{\log t}{t}\,dt=\left[\frac{(\log t)^2}{2}\right]_1^e$$

$$=\frac{1}{2}[(\log e)^2-(\log 1)^2]=\frac{1}{2}$$

**136.** (c) Let $I=\int_0^\pi x\,f(\sin x)\,dx$ ...(i)

$$\Rightarrow I=\int_0^\pi (\pi-x)f[\sin(\pi-x)]\,dx$$

$$\Rightarrow I=\int_0^\pi (\pi-x)\,f(\sin x)\,dx \quad ...(ii)$$

On adding Eqs. (i) and (ii), we get

$$2I=\int_0^\pi \pi f(\sin x)\,dx$$

$$\Rightarrow \quad I=\frac{\pi}{2}\int_0^\pi f(\sin x)\,dx \quad ...(iii)$$

$$\left[\because \int_0^{2a} f(x)\,dx=\begin{cases}2\int_0^a f(x)\,dx, & \text{if } f(2a-x)=f(x)\\ 0, & \text{if } f(2a-x)=-f(x)\end{cases}\right]$$

$$\Rightarrow \quad I=\pi\int_0^{\pi/2} f(\sin x)\,dx$$

Put $\frac{\pi}{2}-x=t\Rightarrow x=\frac{\pi}{2}-t$

$$\Rightarrow \quad dx=-dt$$

we get $I=\frac{\pi}{2}\int_{-\pi/2}^{\pi/2} f(\cos t)\,dt$

$$=\frac{\pi}{2}\int_{-\pi/2}^{\pi/2} f(\cos x)\,dx$$

$$=\pi\int_0^{\pi/2} f(\cos x)\,dx$$

[$\because f(\cos x)$ is an even function]

**137.** (b) Let $I=\int_{-\pi}^{\pi}\frac{2x(1+\sin x)}{1+\cos^2 x}\,dx$

$$=\int_{-\pi}^{\pi}\frac{2x}{1+\cos^2 x}\,dx+\int_{-\pi}^{\pi}\frac{2x\sin x}{1+\cos^2 x}\,dx$$

$$\Rightarrow \ I=0+4\int_0^\pi \frac{x\sin x}{1+\cos^2 x}\,dx \quad ...(i)$$

$\left[\because \frac{2x}{1+\cos^2 x}\right.$ is an odd function

and $\frac{2x\sin x}{1+\cos^2 x}$ is an even function$\left.\right]$

$$\Rightarrow I=4\int_0^\pi \frac{(\pi-x)\sin(\pi-x)}{1+\cos^2(\pi-x)}\,dx$$

$$\Rightarrow I=4\int_0^\pi \frac{\pi\sin x}{1+\cos^2 x}\,dx-4\int_0^\pi \frac{x\sin x}{1+\cos^2 x}\,dx$$

$$\Rightarrow I=4\pi\int_0^\pi \frac{\sin x}{1+\cos^2 x}\,dx-I$$

[from Eq. (i)]

$$\Rightarrow \quad I=2\pi\int_0^\pi \frac{\sin x}{1+\cos^2 x}\,dx$$

Put $\cos x=t\Rightarrow -\sin x\,dx=dt$

$$\therefore \ I=-2\pi\int_1^{-1}\frac{1}{1+t^2}\,dt$$

$$=2\pi[\tan^{-1} t]_{-1}^{1}$$

$$=2\pi\left[\frac{\pi}{4}+\frac{\pi}{4}\right]=\pi^2$$

**138.** (d)

$$\int_0^\pi \frac{(x+3)\sin x}{1+3\cos^2 x}\,dx=\int_0^\pi \frac{x\sin x}{1+3\cos^2 x}\,dx+3\int_0^\pi \frac{\sin x\,dx}{1+3\cos^2 x}$$

$$=I_1+I_2$$

$$\therefore I_1=\int_0^\pi \frac{x\sin x\,dx}{1+3\cos^2 x}$$

$$=\int_0^\pi \frac{(\pi-x)\sin(\pi-x)dx}{1+3(\cos(\pi-x))^2}$$

$$\left(\because \int_0^a f(x)dx=\int_0^a f(a-x)dx\right)$$

$$=\pi\int_0^\pi \frac{\sin x dx}{1+3\cos^2 x}-I_1$$

$$\Rightarrow 2I_1=2\pi\int_0^{\frac{\pi}{2}}\frac{\sin x dx}{1+3\cos^2 x}$$

$$\left(\because \int_0^{2a} f(x)dx=2\int_0^a f(x)dx \text{ iff } (2a-x)=f(x)\right)$$

$$\Rightarrow I_1=\pi\int_0^{\frac{\pi}{2}}\frac{\sin x\,dx}{1+3\cos^2 x}$$

$$\Rightarrow I_1=\pi\int_0^1 \frac{dt}{1+3t^2}$$

$$\Rightarrow I_1=\frac{\pi}{3}\int_0^1 \frac{dt}{t^2+\left(\frac{1}{3}\right)^2}$$

$$=\frac{\pi}{3}\frac{1}{\frac{1}{\sqrt{3}}}\tan^{-1}(t\sqrt{3})|_0^1$$

$$=\frac{\pi}{\sqrt{3}}\cdot\frac{\pi}{3}=\frac{\pi^2}{3\sqrt{3}}$$

$$\therefore I_1+I_2=\frac{\pi^2}{3\sqrt{3}}+\frac{2\pi}{\sqrt{3}}=\frac{\pi}{3\sqrt{3}}(\pi+6)$$

**139.** (d) Given, $\int_0^2 x^2F''(x)dx=40$

$$\Rightarrow [x^2F'(x)]_0^2-2\int_0^2 xF'(x)dx=40$$

$$\Rightarrow \quad 4F'(2)-0-2(6)=40$$

$$\left(\text{given} \int_0^2 xF'(x)dx=6\right)$$

$$\Rightarrow \quad F'(2)=\frac{52}{4}=13$$

Also, $\int_0^2 xF'(x)dx=6$

$\Rightarrow [xF(x)]_0^2 - \int_0^2 F(x)dx = 6$

$\Rightarrow 2F(2) - 0 - \int_0^2 F(x)dx = 6$

$\Rightarrow \quad 2(1) - \int_0^2 F(x)dx = 6$

$\Rightarrow \quad \int_0^2 F(x)dx = -4$

$\therefore F'(2) + \int_0^2 F(x)dx = 13 - 4 = 9$

**140.** *(d)* As, we know that beta function

$\beta(p, q) = \int_0^1 x^{p-1}(1-x)^{q-1}dx$

$= \dfrac{(p-1)!(q-1)!}{(p+q-1)!}$, where $p, q \in N$

So, $I(9, 14) = \dfrac{(8!)(13!)}{(22!)}$

and $I(10, 13) = \dfrac{(9!)(12!)}{(22!)}$

Now, $I(9, 14) + I(10, 13)$

$= \dfrac{1}{22!}(8!13! + 9!12!)$

$= \dfrac{1}{22!}(8!)(12!)(13+9)$

$= \dfrac{(8!)(12!)}{(21!)}$

$= \dfrac{(9-1)!(13-1)!}{(9+13-1)!} = I(9, 13)$

**141.** *(a)* Let $I = \int_0^{\pi/4} \dfrac{x\,dx}{\sin^4(2x) + \cos^4(2x)}$

Put $2x = t \Rightarrow 2dx = dt$

$\Rightarrow x = 0, t = 0, x = \dfrac{\pi}{4}, t = \dfrac{\pi}{2}$

$I = \dfrac{1}{4}\int_0^{\pi/2} \dfrac{t\,dt}{\sin^4 t + \cos^4 t}$

$I = \dfrac{1}{4}\int_0^{\pi/2} \dfrac{\left(\dfrac{\pi}{2} - t\ dt\right)}{\cos^4 t + \sin^4 t}dt$

$2I = \dfrac{\pi}{8}\int_0^{\pi/2} \dfrac{dt}{\cos^4 t + \sin^4 t}$

$I = \dfrac{\pi}{16}\int_0^{\pi/2} \dfrac{dt}{\sin^4 t + \cos^4 t}$

$= 2 \times \dfrac{\pi}{16}\int_0^{\pi/4} \dfrac{dt}{\sin^4 t + \cos^4 t}$

$= \dfrac{\pi}{8}\int_0^{\pi/4} \dfrac{dt}{(\sin^2 t + \cos^2 t)^2 - 2\sin^2 t\cos^2 t}$

$= \dfrac{\pi}{8}\int_0^{\pi/4} \dfrac{dt}{1 - 2(\sin t \cdot \cos t)^2}$

$I = \dfrac{\sqrt{2}\pi^2}{32} = \dfrac{\pi}{4}\int_0^{\pi/4} \dfrac{dt}{2 - \sin^2 2t}$

Put $2t = u \Rightarrow 2t\ dt = du$

$= \dfrac{\pi}{8}\int_0^{\pi/2} \dfrac{du}{2 - \sin^2 u}$

$= \dfrac{\pi}{8}\int_0^{\pi/2} \dfrac{\sec^2 u}{2 + \tan^2 u}du$

Let $\tan u = r$

$I = \dfrac{\pi}{8\sqrt{2}}\left(\tan^{-1}\dfrac{r}{\sqrt{2}}\right)_0^\infty$

$\Rightarrow \quad I = \dfrac{\pi^2}{16\sqrt{2}} = \dfrac{\sqrt{2}\pi^2}{32}$

**142.** *(d)*

> **key idea** Use Leibnitz rule,
> $\dfrac{d}{dx}\int_{h(x)}^{g(x)} f(t)dt = f(g(x))g'(x) - f(h(x))h'(x)$

$a$ = Sum of all coefficient in the expansion of

$(1 - 2x + 2x^2)^{2023}\ (3 - 4x^2 + 2x^3)^{2024}$

For a, put $x = 1$ in expression

$a = (1 - 2 + 2)^{2023}\ (3 - 4 + 2)^{2024}$

$\Rightarrow a = 1$

$b = \lim\limits_{x \to 0} \dfrac{\displaystyle\int_0^x \dfrac{\ln(1+t)}{t^{2024}+1}dt}{x^2} \quad \left(\dfrac{0}{0}\text{ form}\right)$

Using $L$ Hospital rule,

$= \lim\limits_{x \to 0} \dfrac{\dfrac{\ln(1+x)}{x^{2024}+1}}{2x}$ [using Leibnitz rule]

$= \lim\limits_{x \to 0} \dfrac{\ln(1+x)}{2x(x^{2024}+1)}$

$= \dfrac{1}{2 \times 1} = \dfrac{1}{2} \quad \left[\because \lim\limits_{x \to 0} \dfrac{\ln(1+x)}{x} = 1\right]$

$2b = 1$

Equations $2bx^2 + ax + 4 = 0$

$\Rightarrow x^2 + x + 4 = 0$

and $cx^2 + dx + e$ have one common root

But discriminant of $x^2 + x + 4 = 0$ is

$D = -15 < 0$

$\therefore$ Roots are imaginary.

$\Rightarrow$ both roots are common

$\dfrac{1}{c} = \dfrac{1}{d} = \dfrac{4}{e} = k$ (let)

$\Rightarrow \quad d = \dfrac{1}{k} \Rightarrow c = \dfrac{1}{k} \Rightarrow e = \dfrac{4}{k}$

$\therefore \quad d : c : e$

$= \dfrac{1}{k} : \dfrac{1}{k} : \dfrac{4}{k} = 1 : 1 : 4$

**143.** *(c)* Given, $f : R \to R$,

$f(x) = \dfrac{x}{(1+x^4)^{1/4}}$ and

$g(x) = f(f(f(f(x))))$

Now, $fof(x) = \dfrac{f}{(1+f^4)^{1/4}}$

$= \dfrac{\dfrac{x}{(1+x^4)^{1/4}}}{\left[1 + \left(\dfrac{x}{(1+x^4)^{1/4}}\right)^4\right]^{1/4}}$

$= \dfrac{\dfrac{x}{(1+x^4)^{1/4}}}{\left[1 + \dfrac{x^4}{(1+x^4)}\right]^{1/4}} = \dfrac{x}{(1+2x^4)^{1/4}}$

Similarly, $fofof = \dfrac{x}{(1+3x^4)^{1/4}}$

and $\quad fofofof = \dfrac{x}{(1+4x^4)^{1/4}} = g(x)$

Now, according to the question,

$18\int_0^{\sqrt{2\sqrt{5}}} x^2 \cdot g(x)dx$

$= 18\int_0^{\sqrt{2\sqrt{5}}} x^2 \cdot \dfrac{x}{(1+4x^4)^{1/4}}dx$

$= 18\int_0^{\sqrt{2\sqrt{5}}} \dfrac{x^3}{(1+4x^4)^{1/4}}dx$

Let $1 + 4x^4 = t^4$

$\Rightarrow \quad 16x^3\,dx = 4t^3\,dt$

$\Rightarrow \quad x^3\,dx = \dfrac{1}{4}t^3\,dt$

When $x = 0, t = 1$ and $x = \sqrt{2\sqrt{5}}, t = 3$

$= 18\int_1^3 \dfrac{t^3}{4} \cdot \dfrac{dt}{t} = \dfrac{18}{4}\int_1^3 t^2\,dt$

$= \dfrac{9}{2} \times \left[\dfrac{t^3}{3}\right]_1^3 = \dfrac{3}{2}(27 - 1)$

$= \dfrac{3}{2} \times 26 = 3 \times 13 = 39$

**144.** *(c)* Let

$I = \int_{-\pi/2}^{\pi/2} \left(\dfrac{x^2\cos x}{1+\pi^x} + \dfrac{1+\sin^2 x}{1+e^{\sin x^{2023}}}\right)dx$

$$= \int_0^{\pi/2} \left[\left(\frac{x^2 \cos x}{1+\pi^x} + \frac{1+\sin^2 x}{1+e^{(\sin x^{2023})}}\right) + \left(\frac{x^2 \cos x}{1+\pi^{-x}} + \frac{1+\sin^2 x}{1+e^{-(\sin x^{2023})}}\right)\right] dx$$

$$= \int_0^{\pi/2} (x^2 \cos x + 1 + \sin^2 x) dx$$

$$= \int_0^{\pi/2} x^2 \cos x \, dx + \int_0^{\pi/2} (1+\sin^2 x)\, dx$$

Now,

let $I_1 = \int_0^{\pi/2} (1+\sin^2 x)\, dx$

$$= \int_0^{\pi/2} 1\, dx + \int_0^{\pi/2} \left(\frac{1-\cos 2x}{2}\right) dx$$

$$= \frac{\pi}{2} + \frac{1}{2}\left(\frac{\pi}{2} + 0\right)$$

$$\Rightarrow I_1 = \frac{\pi}{2} + \frac{\pi}{4} = \frac{3\pi}{4}$$

and $I_2 = \int_0^{\pi/2} x^2 \cos x dx$

$$= [(x^2 \sin x) - \int 2x \int \cos x\, dx]_0^{\pi/2}$$

$$= [x^2 (\sin x) - 2\int x \sin x dx]_0^{\pi/2}$$

$$= [x^2 \sin x - 2(-x\cos x + \sin x dx)]_0^{\pi/2}$$

$$= \frac{\pi^2}{4} - 2$$

$$\therefore \quad I = \frac{3\pi}{4} + \frac{\pi^2}{4} - 2$$

$$= \frac{\pi}{4}(\pi + a) - 2 \qquad \text{[given]}$$

$\therefore \quad a = 3$

**145.** *(c)*

> **key idea** By using the properties of definite integrals, we replace $x$ into $(a+b-x)$ e.g. $\int_a^b f(x)\,dx = \int_a^b f(a+b-x)\,dx$

Let $I = \int_0^{\pi} \frac{1}{1-2a\cos x + a^2} dx$ ...(i)

$0 < a < 1$

$$\Rightarrow \quad I = \int_0^{\pi} \frac{1}{1-2a\cos(\pi - x) + a^2} dx$$

$$\Rightarrow \quad I = \int_0^{\pi} \frac{1}{1+2a\cos x + a^2} dx \quad \text{...(ii)}$$

On adding Eqs. (i) and (ii), we get

$$2I = \int_0^{\pi} \left[\frac{1}{(1+a^2) - 2a\cos x} + \frac{1}{(1+a^2)+2a\cos x}\right] dx$$

$$\Rightarrow 2I = 2\int_0^{\pi/2} \frac{2(1+a^2)}{(1+a^2)^2 - 4a^2\cos^2 x} dx$$

$$\Rightarrow I = \int_0^{\pi/2} \frac{2(1+a^2)\cdot \sec^2 x}{(1+a^2)^2 \sec^2 x - 4a^2} dx$$

$$= \int_0^{\pi/2} \frac{2\cdot(1+a^2)\cdot \sec^2 x}{(1+a^2)^2 \tan^2 x + (1-a^2)^2} dx$$

$[\because \sec^2 x = 1 + \tan^2 x]$

$$I = \int_0^{\pi/2} \frac{\frac{2\cdot \sec^2 x}{1+a^2}}{\tan^2 x + \left(\frac{1-a^2}{1+a^2}\right)^2} dx$$

$$= \frac{2}{1-a^2}\left[\frac{\pi}{2} - 0\right] = \frac{\pi}{1-a^2}$$

**146.** *(b)* We have,

$$I = \int_0^1 \left(\frac{dx}{(5+2x-2x^2)(1+e^{(2-4x)})}\right) \quad \text{...(i)}$$

$$= \int_0^1 \frac{dx}{\{5+2(1-x)-2(1-x)^2\}(1+e^{2-4(1-x)})}$$

$$I = \int_0^1 \frac{dx}{(5+2x-2x^2)(1+e^{4x-2})} \quad \text{...(ii)}$$

On adding Eqs. (i) and (ii), we get

$$2I = \int_0^1 \frac{dx}{(5+2x-2x^2)} \left\{\frac{1}{(1+e^{2-4x})} + \frac{1}{(1+e^{4x-2})}\right\}$$

$$= \int_0^1 \frac{dx}{(5+2x-2x^2)}$$

$$\left[\because \left\{\frac{1}{e^{2-4x}} + \frac{1}{e^{4x-2}}\right\} = 1\right]$$

$$\Rightarrow I = \frac{1}{2}\int_0^1 \frac{dx}{(5+2x-2x^2)}$$

$$I = \frac{1}{2}\int_0^1 \frac{dx}{-2\left(x^2 - x - \frac{5}{2}\right)}$$

$$= -\frac{1}{4}\int_0^1 \frac{dx}{\left[\left(x-\frac{1}{2}\right)^2 - \left(\frac{\sqrt{11}}{2}\right)^2\right]}$$

$$= \left(\frac{-1}{4} \times \frac{1}{2\times\frac{\sqrt{11}}{2}} \log\left|\frac{\left(x-\frac{1}{2}\right)-\left(\frac{\sqrt{11}}{2}\right)}{\left(x-\frac{1}{2}\right)+\left(\frac{\sqrt{11}}{2}\right)}\right|\right)_0^1$$

$$= \frac{2}{4} \times \frac{1}{\sqrt{11}} \log\left|\frac{\sqrt{11}+1}{\sqrt{11}-1}\right|$$

$$= \frac{1}{\sqrt{11}} \log\left(\frac{1+\sqrt{11}}{\sqrt{10}}\right) = \frac{1}{\alpha}\log_e\left(\frac{\alpha+1}{\beta}\right)$$

$$\Rightarrow \alpha = \sqrt{11} \text{ and } \beta = \sqrt{10}$$

$$\Rightarrow \alpha^4 - \beta^4 = (\sqrt{11})^4 - (\sqrt{10})^4$$

$$= (11)^{4/2} - (10)^{4/2}$$

$$= 121 - 100 = 21$$

**147.** *(d)* Given: $f : R \to R$

and $\int_0^{\frac{\pi}{2}} f(\sin 2x) \sin x\, dx$

$$+ \alpha \int_0^{\frac{\pi}{4}} f(\cos 2x)\cdot \cos x\, dx = 0$$

Let $I_1 = \int_0^{\frac{\pi}{2}} f(\sin 2x)\cdot \sin x\, dx$ ...(i)

and $I_2 = \int_0^{\frac{\pi}{4}} f(\cos 2x)\cdot \cos x\, dx$

Now, use $x \to \frac{\pi}{2} - x$ in $I_1$

$$\Rightarrow I_1 = \int_0^{\frac{\pi}{2}} f\left(\sin\left(2\left(\frac{\pi}{2}-x\right)\right)\right)\cdot \sin\left(\frac{\pi}{2}-x\right) dx$$

$$I_1 = \int_0^{\frac{\pi}{2}} f(\sin 2x)\cdot \cos x\, dx \quad \text{...(ii)}$$

On adding Eqs. (i) and (ii), we get

$$2I_1 = \int_0^{\frac{\pi}{2}} f(\sin 2x)(\sin x + \cos x) dx$$

$$\Rightarrow 2I_1 = 2\int_0^{\frac{\pi}{4}} f(\sin 2x)\cdot(\sin x + \cos x) dx$$

$$\left[\because \int_0^{2a} f(2a-x)dx = 2\int_0^{a} f(2a-x)\, dx\right]$$

If $f(2a-x)=f(x)$

$$\Rightarrow I_1=\int_0^{\frac{\pi}{4}} f(\sin 2x)\cdot(\sin x+\cos x)dx \quad \text{...(iii)}$$

Now, $I_2=\int_0^{\frac{\pi}{4}} f(\cos 2x)\cdot \cos x\, dx$

Use $x \to \frac{\pi}{4}-x$

$$I_2=\int_0^{\frac{\pi}{4}} f(\sin 2x)\cdot\left(\frac{\sin x+\cos x}{\sqrt{2}}\right)dx$$

$$\Rightarrow I_2=\frac{1}{\sqrt{2}}\int_0^{\frac{\pi}{4}} f(\sin 2x)\cdot(\sin x+\cos x)dx$$

$$\Rightarrow I_2=\frac{1}{\sqrt{2}}I_1 \quad \text{[Use eq. (iii)]}$$

$$\Rightarrow \sqrt{2}\,I_2=I_1$$

$$\Rightarrow \quad I_1=\sqrt{2}\,I_2 \quad [\because I_1=-\alpha I_2]$$

$$\Rightarrow \quad \alpha=-\sqrt{2}$$

Hence, value of $\alpha$ is $-\sqrt{2}$.

**148.** *(b)* Let $I=\int_0^{\alpha}\frac{x}{\sqrt{x+\alpha}-\sqrt{x}}dx$

$$=\int_0^{\alpha}\frac{x[\sqrt{x+\alpha}+\sqrt{x}]}{\alpha}dx$$

$$=\frac{1}{\alpha}\int_0^{\alpha}x\sqrt{x+\alpha}\,dx+\frac{1}{\alpha}\int_0^{\alpha}x^{3/2}dx$$

$$=\frac{1}{\alpha}\int_0^{\alpha}(x+\alpha-\alpha)\sqrt{x+\alpha}\,dx+\frac{1}{\alpha}\int_0^{\alpha}x^{3/2}dx$$

$$=\frac{1}{\alpha}\left[\int_0^{\alpha}(x+\alpha)^{3/2}dx\right]-\int_0^{\alpha}\sqrt{x+\alpha}\,dx+\frac{1}{\alpha}\int_0^{\alpha}x^{3/2}dx$$

$$=\left[\frac{2(x+\alpha)^{5/2}}{5\alpha}\right]_0^{\alpha}-\left[\frac{2}{3}(x+\alpha)^{3/2}\right]_0^{\alpha}+\left[\frac{2x^{5/2}}{5\alpha}\right]_0^{\alpha}$$

$$=\left[\frac{8\sqrt{2}\alpha^{5/2}}{5\alpha}-\frac{2\alpha^{5/2}}{5\alpha}\right]-\left[\frac{2\times 2\sqrt{2}\alpha^{3/2}}{3}-\frac{2\alpha^{3/2}}{3}\right]+\frac{2\alpha^{3/2}}{5}$$

$$=\frac{(24\sqrt{2}-6-20\sqrt{2}+10+6)\,\alpha^{3/2}}{15}$$

$$=\left(\frac{4\sqrt{2}+10}{15}\right)\alpha^{3/2}$$

But, it is given, $I=\frac{16+20\sqrt{2}}{15}$

$$\therefore \frac{4\sqrt{2}+10}{15}\alpha^{3/2}=\frac{16+20\sqrt{2}}{15}$$

$$\Rightarrow \quad \alpha^{3/2}=\frac{2\sqrt{2}\,[10+4\sqrt{2}]}{10+4\sqrt{2}}$$

$$\Rightarrow \quad \alpha^{3/2}=2\sqrt{2}$$

$$\Rightarrow \quad \alpha=2$$

**149.** *(b)* Let $I=\int_{\pi/3}^{\pi/2}\frac{2+3\sin x}{\sin x\,(1+\cos x)}dx$

$$=\int_{\pi/3}^{\pi/2}\frac{2}{\sin x(1+\cos x)}dx+\int_{\pi/3}^{\pi/2}\frac{3}{1+\cos x}dx$$

$$=\int_{\pi/3}^{\pi/2}\frac{2(1-\cos x)}{\sin x(1-\cos^2 x)}dx+\int_{\pi/3}^{\pi/2}\frac{3}{1+\cos x}dx$$

$$=\int_{\pi/3}^{\pi/2}\frac{2(1-\cos x)}{\sin^3 x}dx+\int_{\pi/3}^{\pi/2}\frac{3}{2\cos^2\frac{x}{2}}dx$$

$$=\int_{\pi/3}^{\pi/2}\frac{2}{\sin^3 x}dx-\int_{\pi/3}^{\pi/2}2\cot x\,\text{cosec}^2 x\,dx+\int_{\pi/3}^{\pi/2}\frac{3}{2}\sec^2\frac{x}{2}dx$$

$$=2I_1-2I_2+\frac{3}{2}I_3$$

Now, $I_1=\int_{\pi/3}^{\pi/2}\text{cosec}^3 x\,dx$

$$=\int_{\pi/3}^{\pi/2}\sqrt{1+\cot^2 x}\,\text{cosec}^2 x\,dx$$

Put $\cot x=t \Rightarrow -\text{cosec}^2 x\,dx=dt$

When $x=\frac{\pi}{3}\Rightarrow t=\frac{1}{\sqrt{3}}$

and when $x=\frac{\pi}{2}\Rightarrow t=0$

$$\therefore I_1=-\int_{\frac{1}{\sqrt{3}}}^{0}\sqrt{1+t^2}\,dt$$

$$=\int_0^{\frac{1}{\sqrt{3}}}\sqrt{1+t^2}\,dt$$

$$=\left[\frac{t}{2}\sqrt{1+t^2}+\frac{1}{2}\log[t+\sqrt{1+t^2}]\right]_0^{\frac{1}{\sqrt{3}}}$$

$$=\frac{1}{3}+\frac{1}{2}\log\sqrt{3}$$

$$I_2=\int_{\pi/3}^{\pi/2}\cot x\,\text{cosec}^2 x\,dx$$

Put $\cot x=t\Rightarrow \text{cosec}^2 x\,dx=-dt$

When, $x=\frac{\pi}{3}\Rightarrow t=\frac{1}{\sqrt{3}}$

and $x=\frac{\pi}{2}\Rightarrow t=0$

$$\therefore \quad I_2=-\int_{\frac{1}{\sqrt{3}}}^{0}t\,dt$$

$$=\int_0^{\frac{1}{\sqrt{3}}}t\,dt=\left[\frac{t^2}{2}\right]_0^{\frac{1}{\sqrt{3}}}=\frac{1}{6}$$

$$I_3=\int_{\frac{\pi}{3}}^{\frac{\pi}{2}}\sec^2\frac{x}{2}dx$$

$$=2\left(\tan\frac{\pi}{4}-\tan\frac{\pi}{6}\right)=2\left(1-\frac{1}{\sqrt{3}}\right)$$

$$\therefore \quad I=2I_1-2I_2+\frac{3}{2}I_3$$

$$=2\left(\frac{1}{3}+\frac{1}{2}\log\sqrt{3}\right)-2\left(\frac{1}{6}\right)+\frac{3}{2}\left(2\left(1-\frac{1}{\sqrt{3}}\right)\right)$$

$$=\frac{2}{3}+\log\sqrt{3}-\frac{1}{3}+3-\sqrt{3}$$

$$=\frac{10}{3}-\sqrt{3}+\log_e\sqrt{3}$$

**150.** *(c)* Given,

$$f(x)=x+\int_0^{\pi/2}(\sin x\cos y+\cos x\sin y)\,f(y)dy$$

$$\Rightarrow f(x)=x+\int_0^{\pi/2}[(\cos y f(y)dy)\sin x+(\sin y f(y)dy)\cos x] \quad \text{...(i)}$$

On comparing with

$$f(x)=x+\frac{a}{\pi^2-4}\sin x+\frac{b}{\pi^2-4}\cos x \quad \text{...(A)}$$

$x\in R$, we get

$$\frac{a}{\pi^2-4}=\int_0^{\pi/2}\cos y\,f(y)\,dy \quad \text{... (ii)}$$

and $\frac{b}{\pi^2-4}=\int_0^{\pi/2}\sin y\, f(y)\,dy$ ...(iii)

On adding Eqs. (ii) and (iii), we get

$$\frac{a+b}{\pi^2-4}=\int_0^{\pi/2}(\sin y+\cos y)\,f(y)\,dy \quad \text{...(iv)}$$

$$\Rightarrow \frac{a+b}{\pi^2-4}=\int_0^{\pi/2}\left[\sin\left(\frac{\pi}{2}-y\right)+\cos\left(\frac{\pi}{2}-y\right)f\left(\frac{\pi}{2}-y\right)dy\right]$$

$$\Rightarrow \frac{a+b}{\pi^2-4}=\int_0^{\pi/2}(\sin y+\cos y)f\left(\frac{\pi}{2}-y\right)dy \quad \text{... (v)}$$

On adding Eqs. (iv) and (v), we get

$$2\left(\frac{a+b}{\pi^2-4}\right)=\int_0^{\pi/2}(\sin y+\cos y)\left[f(y)+f\left(\frac{\pi}{2}-y\right)\right]dy \quad \text{... (vi)}$$

Now,

$$f(y)=y+\frac{a}{\pi^2-4}\sin y+\frac{b}{\pi^2-4}\cos y$$

[From Eq. (A)] ...(vii)

and $f\left(\frac{\pi}{2}-y\right)=\left(\frac{\pi}{2}-y\right)+\frac{a}{\pi^2-4}\sin\left(\frac{\pi}{2}-y\right)+\frac{b}{\pi^2-4}\cos\left(\frac{\pi}{2}-y\right)$

$$\Rightarrow f\left(\frac{\pi}{2}-y\right)=\frac{\pi}{2}-y+\frac{a}{\pi^2-4}\cos y+\frac{b}{\pi^2-4}\sin y \quad \text{...(viii)}$$

On adding Eqs. (vii) and (viii), we get

$$f(y)+f\left(\frac{\pi}{2}-y\right)=\frac{\pi}{2}+\left(\frac{a+b}{\pi^2-4}\right)(\sin y+\cos y) \quad \text{...(ix)}$$

On putting the value from Eq. (ix) into Eq. (vi), we get

$$2\left(\frac{a+b}{\pi^2-4}\right)=\int_0^{\pi/2}(\sin y+\cos y)\left[\frac{\pi}{2}+\left(\frac{a+b}{\pi^2-4}\right)(\sin y+\cos y)\right]dy$$

$$=\frac{\pi}{2}\int_0^{\pi/2}(\sin y+\cos y)\,dy+\left(\frac{a+b}{\pi^2-4}\right)\int_0^{\pi/2}(\sin y+\cos y)^2\,dy$$

$$=\frac{\pi}{2}[(-\cos y)_0^{\pi/2}+(\sin y)_0^{\pi/2}]+\left(\frac{a+b}{\pi^2-4}\right)\int_0^{\pi/2}(1+\sin 2y)\,dy$$

$$=\frac{\pi}{2}\left[-\left(\cos\frac{\pi}{2}-\cos 0\right)+\sin\frac{\pi}{2}-\sin 0\right]+\frac{a+b}{\pi^2-4}\left[y+\frac{-\cos 2y}{2}\right]_0^{\pi/2}$$

$$=\frac{\pi}{2}\times 2+\left(\frac{a+b}{\pi^2-4}\right)\left(\frac{\pi}{2}+1\right)$$

$$\Rightarrow 2\left(\frac{a+b}{\pi^2-4}\right)=\pi+\left(\frac{a+b}{\pi^2-4}\right)\left(\frac{\pi}{2}+1\right)$$

$\therefore\ a+b=-2\pi(\pi+2)$

**151.** (a) As,

$$f(x)=\int_0^x e^{x-t}f'(t)\,dt-(x^2-x+1)e^x$$

$$f(x)=e^x\int_0^x e^{-t}f'(t)\,dt-(x^2-x+1)e^x$$

$$e^{-x}f(x)=\int_0^x e^{-t}f'(t)\,dt-(x^2-x+1)$$

On differentiate both sides w.r.t. $x$, we get

$$e^{-x}f'(x)+\{-f(x)e^{-x}\}=e^{-x}f'(x)-2x+1$$

$$f(x)=e^x(2x-1)$$

For maxima or minima we should first differentiate function with w.r.t. $x$, we get

$$f'(x)=e^x(2)+e^x(2x-1)=e^x(2x+1)$$

Now, on putting $f'(x)=0$

$\Rightarrow\quad x=-1/2$

Now, $f''(x)=e^x(2)+(2x+1)e^x$

$$f''\left(-\frac{1}{2}\right)=e^{-1/2}(2)+\left[2\left(\frac{-1}{2}\right)+1\right]e^x$$

$$=\frac{2}{\sqrt{e}}>0$$

$\Rightarrow\quad x=-\frac{1}{2}$ is point of minima.

$$\therefore\ f\left(-\frac{1}{2}\right)=e^{-1/2}\left[2\times\left(-\frac{1}{2}\right)-1\right]$$

$$=-\frac{2}{\sqrt{e}}$$

**152.** (c)

> **key idea** (i) Simplify using method of substitution and solve further.
> (ii) Use $\int_{-a}^{a}f(x)\,dx=2\int_0^a f(x)\,dx$; $f(-x)=f(x)$
> (iii) $\int_a^b f(x)\,dx=F(b)-F(a)$, where $F'(x)=f(x)$

$$\int_0^2(\sqrt{2x}-\sqrt{2x-x^2})\,dx=\int_0^1\left(1-\sqrt{1-y^2}-\frac{y^2}{2}\right)dy+\int_1^2\left(2-\frac{y^2}{2}\right)dy+I$$

$$\text{LHS}=\sqrt{2}\int_0^2(\sqrt{x}-\int_0^2\sqrt{1-(x-1)^2})\,dx$$

$$=\sqrt{2}\left[\frac{x^{3/2}}{3/2}\right]_0^2-\left[\frac{x-1}{2}\sqrt{1-(x-1)^2}+\frac{1}{2}\sin^{-1}(x-1)\right]_0^2$$

$$=\frac{8}{3}-\frac{\pi}{2} \quad \text{... (i)}$$

$$\text{RHS}=\int_0^1 1\,dy-\int_0^1\sqrt{1-y^2}\,dy-\int_0^2\frac{y^2}{2}\,dy+\int_1^2 2\,dy-\int_1^2\frac{y^2}{2}\,dy+I$$

$$=[y]_0^1-\left[\frac{y}{2}\sqrt{1-y^2}+\frac{1}{2}\sin^{-1}y\right]_0^1-\frac{1}{2}\left[\frac{y^3}{3}\right]_0^1+2[y]_1^2-\frac{1}{2}\cdot\left[\frac{y^3}{3}\right]_1^2+I$$

$$=1-\frac{\pi}{4}-\frac{1}{6}+2-\frac{7}{6}+I$$

$$=\frac{5}{3}-\frac{\pi}{4}+I \quad \text{... (ii)}$$

From Eqs. (i) and (ii), we get

$$\frac{8}{3}-\frac{\pi}{2}=\frac{5}{3}-\frac{\pi}{4}+I\Rightarrow I=1-\pi/4$$

Option (c) : Let $I=\int_0^1 1-\sqrt{1-y^2}\,dy$

$$=[y]_0^1-\left[\frac{y}{2}\sqrt{1-y^2}+\frac{1}{2}\sin^{-1}y\right]_0^1$$

$$=1-\frac{\pi}{4}$$

Hence, option (c) is the correct expression for $I$.

**153.** (a) $I=\int_0^1 \frac{dx}{7^{\left[\frac{1}{x}\right]}}$

Let $\frac{1}{x}=t \Rightarrow x=\frac{1}{t}$

$\Rightarrow dx=-\frac{1}{t^2}dt$

$x\to 0, t\to\infty$

$x\to 1, t\to 1$

$I=\int_\infty^1 \frac{1}{7^{[t]}}\left(-\frac{1}{t^2}dt\right)$

$=\int_1^\infty \frac{1}{7^{[t]}t^2}dt$ ...(i)

Let $I_n=\int_n^{n+1}\frac{dt}{7^{[t]}t^2}, n=1, 2, 3, \ldots$

$=\int_n^{n+1}\frac{dt}{7^n t^2}, n=1, 2, 3, \ldots$

$\left[\because n\le t<n+1 \\ \therefore [t]=n\right]$

$I_n=\frac{1}{7^n}\left[-\frac{1}{t}\right]_n^{n+1}$

$I_n=\frac{1}{7^n}\left[\frac{1}{n}-\frac{1}{n+1}\right]$

$=\frac{1}{n\cdot 7^n}-\frac{7}{(n+1)7^{n+1}}$

$I=I_1+I_2+I_3+\ldots$ to $\infty$

$=\left[\frac{1}{1\cdot 7}+\frac{1}{2\cdot 7^2}+\frac{1}{3\cdot 7^3}+\ldots\infty\right]$

$-7\left[\frac{1}{2\cdot 7^2}+\frac{1}{3\cdot 7^3}+\ldots\infty\right]$

$=-\log\left(1-\frac{1}{7}\right)+7\left[\log\left(1-\frac{1}{7}\right)+\frac{1}{7}\right]$

$\left[\because \log(1-x)=-x-\frac{x^2}{2}-\frac{x^3}{3}-\ldots\infty\right]$

$I=-\log\frac{6}{7}+7\log\frac{6}{7}+1=1+6\log\frac{6}{7}$

**154.** (c) Let

$I=\int_{-\pi/2}^{\pi/2}\frac{dx}{(1+e^x)(\sin^6 x+\cos^6 x)}$ ...(i)

$\Rightarrow I=\int_{-\pi/2}^{\pi/2}\frac{dx}{(1+e^{-x})(\sin^6 x+\cos^6 x)}$

$\left[\text{using} \int_a^b f(x)dx=\int_a^b f(a+b-x)dx\right]$

$\Rightarrow I=\int_{-\pi/2}^{\pi/2}\frac{e^x dx}{(1+e^x)(\sin^6 x+\cos^6 x)}$ ...(ii)

On adding Eqs. (i) and (ii), we get

$2I=\int_{-\pi/2}^{\pi/2}\frac{(e^x+1)dx}{(e^x+1)(\sin^6 x+\cos^6 x)}$

$\Rightarrow 2I=\int_{-\pi/2}^{\pi/2}\frac{dx}{(\sin^6 x+\cos^6 x)}$

$\Rightarrow 2I=2\int_0^{\pi/2}\frac{dx}{(\sin^6 x+\cos^6 x)}$

$\Rightarrow I=\int_0^{\pi/2}\frac{dx}{(\sin^2 x)^3+(\cos^2 x)^3}$

$\Rightarrow I=\int_0^{\pi/2}\frac{dx}{(\sin^2 x+\cos^2 x)^3 -3\sin^2 x\cos^2 x(\sin^2 x+\cos^2 x)}$

Since, $\sin^2 x+\cos^2 x=1$

$\Rightarrow I=\int_0^{\pi/2}\frac{dx}{1-3\sin^2 x\cos^2 x}$

$=\int_0^{\pi/2}\frac{4dx}{4-3\cdot\{4\sin^2 x\cos^2 x\}}$

$=\int_0^{\pi/2}\frac{4dx}{4-3\sin^2 2x}$

$=\int_0^{\pi/2}\frac{4\sec^2 2x\, dx}{4\sec^2 2x-3\tan^2 2x}$

$=2\int_0^{\pi/4}\frac{4\sec^2 2x\, dx}{4\sec^2 2x-3\tan^2 2x}$

$\left[\text{We know, }\int_0^{2a}f(x)dx=2\int_0^a f(x)dx\right]$

Let $\tan 2x=z \Rightarrow 2\sec^2 2xdx=dz$

$=\int_0^\infty\frac{4dz}{4+z^2}$

| $x$ | 0 | $\pi/4$ |
|---|---|---|
| $z$ | 0 | $\infty$ |

$=\frac{4}{2}\left[\tan^{-1}\frac{z}{2}\right]_0^\infty=2\tan^{-1}(\infty)=2\cdot\frac{\pi}{2}=\pi$

Thus, $I=\pi$

**155.** (a) Given,

$\int_0^{100\pi}\frac{\sin^2 x}{e^{\left(\frac{x}{\pi}-\left[\frac{x}{\pi}\right]\right)}}dx=\frac{\alpha\pi^3}{1+4\pi^2}, \alpha\in R$

$\Rightarrow \int_0^{100\pi}\frac{\sin^2 x}{e^{\left\{\frac{x}{\pi}\right\}}}dx=\frac{\alpha\pi^2}{1+4\pi^2}$

where, $\{x\}$ is fractional part function.

If $f(x)=\frac{\sin^2 x}{e^{\left\{\frac{x}{\pi}\right\}}}$

$f(x+\pi)=\frac{\sin^2(x+\pi)}{e^{\left\{\frac{x+\pi}{\pi}\right\}}}=\frac{\sin^2 x}{e^{\left\{\frac{x}{\pi}\right\}}}$

Now, $\int_0^{100\pi}\frac{\sin^2 x}{e^{\left\{\frac{x}{\pi}\right\}}}dx=100\int_0^\pi\frac{\sin^2 x}{e^{\frac{x}{\pi}}}dx$

$=100\int_0^\pi e^{\frac{-x}{\pi}}\sin^2 xdx$

$=50\int_0^\pi e^{\frac{-x}{\pi}}(1-\cos 2x)dx$

$=50\int_0^\pi\left(\underbrace{e^{\frac{-x}{\pi}}}_{A}-\underbrace{e^{\frac{-x}{\pi}}\cos 2xdx}_{B}\right)$

$A=\int_0^\pi e^{\frac{-x}{\pi}}dx=\left[\frac{e^{\frac{-x}{\pi}}}{\frac{-1}{\pi}}\right]_0^\pi$

$=\left(\frac{e^{-1}-1}{\frac{-1}{\pi}}\right)=-\pi(e^{-1}-1)$

Using $\int e^{ax}\cos bx\,dx$

$=\frac{e^{ax}}{a^2+b^2}(a\cos bx+b\sin bx)$

$B=\int_0^\pi e^{\left(\frac{-1}{\pi}\right)^x}\cos 2xdx$

$=\left(\frac{e^{\frac{-x}{\pi}}}{4+\frac{1}{\pi^2}}\right)\left[\left(\frac{-1}{\pi}\cos 2x+2\sin 2x\right)\right]_0^\pi$

$=\left(\frac{e^{-1}}{4+\frac{1}{\pi^2}}\right)\left(\frac{-1}{\pi}\right)-\left(\frac{1}{4+\frac{1}{\pi^2}}\right)\left(\frac{-1}{\pi}\right)$

$B=\frac{e^{-1}-1}{-\pi\left(4+\frac{1}{\pi^2}\right)}$

$\therefore A-B$

$=[-\pi(e^{-1}-1)]-\left[\frac{(e^{-1}-1)}{-\pi\left(4+\frac{1}{\pi^2}\right)}\right]$

$$= -\pi(e^{-1}-1) + \frac{\pi(e^{-1}-1)}{(4\pi^2+1)}$$

$$= -(e^{-1}-1)\left(\frac{4\pi^3}{4\pi^2+1}\right) = \frac{4(1-e^{-1})\pi^3}{4\pi^2+1}$$

$$50(A-B) = 200(1-e^{-1})\left(\frac{\pi^3}{4\pi^2+1}\right)$$

$$\therefore \quad \alpha = 200(1-e^{-1})$$

**156.** *(b)*

$$I = \int_{-\pi/4}^{\pi/4} \frac{dx}{(1+e^{x\cos x})(\sin^4 x + \cos^4 x)} \quad \text{...(i)}$$

Using $\int_a^b f(x)dx = \int_a^b f(a+b-x)dx$

$$I = \int_{-\frac{\pi}{4}}^{\frac{\pi}{4}} \frac{dx}{(1+e^{-x\cos x})(\sin^4 x+\cos^4 x)}$$

$$I = \int_{-\frac{\pi}{4}}^{\frac{\pi}{4}} \frac{e^{x\cos x}\,dx}{(1+e^{x\cos x})(\sin^4 x+\cos^4 x)} \quad \text{...(ii)}$$

On adding Eqs. (i) and (ii),

$$2I = \int_{-\frac{\pi}{4}}^{\frac{\pi}{4}} \frac{(1+e^{x\cos x})dx}{(1+e^{x\cos x})(\sin^4 x+\cos^4 x)}$$

$$\Rightarrow \quad 2I = \int_{-\frac{\pi}{4}}^{\frac{\pi}{4}} \frac{dx}{\sin^4 x+\cos^4 x}$$

$$\Rightarrow \quad 2I = 2\int_0^{\frac{\pi}{4}} \frac{dx}{\sin^4 x+\cos^4 x}$$

$$\left[\because \frac{1}{\sin^4 x+\cos^4 x} \text{ is an even function}\right]$$

$$\Rightarrow \quad I = \int_0^{\frac{\pi}{4}} \frac{\sec^4 x}{\tan^4 x+1}dx$$

$$\Rightarrow \quad I = \int_0^{\frac{\pi}{4}} \frac{(1+\tan^2 x)\sec^2 x}{1+\tan^4 x}dx$$

Let $\tan x = t$

$\Rightarrow \sec^2 x\, dx = dt$

When, $x = 0, t = 0$ and when $x = \pi/4$, $t = 1$

$$\Rightarrow I = \int_0^1 \frac{1+t^2}{1+t^4}dt = \int_0^1 \frac{1+\frac{1}{t^2}}{\left(t-\frac{1}{t}\right)^2+2}dt$$

Let $t - \frac{1}{t} = u \Rightarrow \left(1+\frac{1}{t^2}\right)dt = du$

When, $t = 0, u \to -\infty$

When, $t = 1, u = 0$

$$I = \int_{-\infty}^0 \frac{du}{u^2+2}$$

$$= \frac{1}{\sqrt{2}}\left[\tan^{-1}\left(\frac{u}{\sqrt{2}}\right)\right]_{-\infty}^0$$

$$= \frac{1}{\sqrt{2}}\left[0-\left(-\frac{\pi}{2}\right)\right] = \frac{\pi}{2\sqrt{2}}$$

**157.** *(a)* Let

$$I = \int_{-\pi/2}^{\pi/2} \sin^4 x\left(1+\log\left(\frac{2+\sin x}{2-\sin x}\right)\right)dx$$

$$= \int_{-\pi/2}^{\pi/2} \sin^4 x\, dx + \int_{-\pi/2}^{\pi/2} \sin^4 x \log\left(\frac{2+\sin x}{2-\sin x}\right)dx$$

$$= I_1 + I_2$$

Now, $I_1 = \int_{-\pi/2}^{\pi/2} \sin^4 x\, dx$

$$= 2\int_0^{\pi/2} \sin^4 x\, dx \quad \text{...(i)}$$

We know that

$$\left[\because \int_{-a}^{a} f(x)\,dx = 2\int_0^a f(x)\,dx \text{ if } f(x) = f(-x)\right]$$

$$I_2 = \int_{-\pi/2}^{\pi/2} \sin^4 x \cdot \log\left(\frac{2+\sin x}{2-\sin x}\right)dx$$

$$= 0 \quad \text{...(ii)}$$

$$\left[\because \int_{-a}^{a} f(x)\,dx = 0 \text{ if } f(x) = -f(-x)\right]$$

On adding Eqs. (i) and (ii), we get

$$I = 2\int_0^{\pi/2} \sin^4 x$$

$$= 2\int_0^{\pi/2} (\sin^2 x)^2\, dx$$

$$= 2\int_0^{\pi/2} \left(\frac{1-\cos 2x}{2}\right)^2 dx$$

$$= \frac{1}{2}\int_0^{\pi/2} [1+\cos^2 2x - 2\cos 2x]\, dx$$

$$= \frac{1}{2}\int_0^{\pi/2}\left[1+\left(\frac{1+\cos 4x}{2}\right)-2\cos 2x\right]dx$$

$$= \frac{1}{2}\int_0^{\pi/2}\left(\frac{3}{2}+\frac{\cos 4x}{2}-2\cos 2x\right)dx$$

$$= \frac{1}{2}\left[\frac{3}{2}|x|_0^{\pi/2} + \frac{1}{2}\left|\frac{\sin 4x}{4}\right|_0^{\pi/2} - 2\left|\frac{\sin 2x}{2}\right|_0^{\pi/2}\right]$$

After putting the limits, we get

$$\frac{1}{2}\left[\left(\frac{3}{2}\times\frac{\pi}{2}\right)+\frac{1}{8}(\sin 2\pi - \sin 0) - \sin\pi + \sin 0\right] = \frac{3\pi}{8}$$

**158.** *(a)* $f(x) = \int_0^x t(\sin x - \sin t)\,dt$

$$\Rightarrow f(x) = \int_0^x t\sin x \cdot dt - \int_0^x t\sin t\, dt$$

$$\Rightarrow f(x) = \sin x\left[\frac{t^2}{2}\right]_0^x - \left[|t(-\cos t)|_0^x - \int_0^x \frac{d(t)}{dt}\left(\int \sin t\cdot dt\right)dt\right]$$

After putting the value of limits, we get

$$\Rightarrow f(x) = \sin x\cdot\frac{x^2}{2} - \left[-x\cos x - \int_0^x 1\cdot(-\cos t)\cdot dt\right]$$

$$\Rightarrow f(x) = \sin x\cdot\frac{x^2}{2} - [-x\cos x + |\sin t|_0^x]$$

$$\Rightarrow f(x) = \sin x\cdot\frac{x^2}{2} + x\cos x - \sin x \quad \text{...(i)}$$

On differentiating Eq. (i) w.r.t. $x$, we get

$$f'(x) = \sin x\cdot\left(\frac{2x}{2}\right) + \frac{x^2}{2}\cdot(\cos x) - x\sin x + \cos x - \cos x$$

$$\Rightarrow f'(x) = \frac{x^2}{2}\cdot\cos x \quad \text{...(ii)}$$

On differentiating Eq. (ii) w.r.t. $x$, we get

$$f''(x) = \frac{x^2}{2}\cdot(-\sin x) + \left(\frac{2x}{2}\right)\cdot\cos x$$

$$\Rightarrow f''(x) = x\cos x - \frac{x^2}{2}\cdot\sin x \quad \text{...(iii)}$$

On differentiating Eq. (iii) w.r.t. $x$, we get

$$f'''(x) = x(-\sin x) + \cos x - \frac{x^2}{2}(\cos x) - \left(\frac{2x}{2}\right)\sin x$$

$$\Rightarrow f'''(x) = \cos x - 2x \sin x - \frac{x^2}{2}\cos x \quad \ldots(\text{iv})$$

On adding Eqs. (ii) and (iv), we get

$$f'''(x) + f'(x) = \cos x - 2x \sin x$$

**159.** *(d)* $x\int_1^x y(t)\,dt$

$$= x\int_1^x ty(t)\,dt + \int_1^x ty(t)\,dt$$

On differentiating w.r.t $x$, we get,

$$\int_1^x y(t)dt + x[y(x) - 0\,y(1)]$$

$$= \int_1^x ty(t)\,dy + x[xy(x) - 0\,y(1)] + xy(x) - 0\,y(1)$$

$$\Rightarrow \int_1^x (t)\,dt = \int_1^x ty(t)\,dt + x^2 y(x)$$

Again differentiating the above equation w.r.t. $x$, we get

$$y(x) - 0\,y(1) = xy(x) - 0\,y(1) + 2xy(x) + x^2 y'(x)$$

$$\Rightarrow \quad (1-3x)y(x) = x^2 y'(x)$$

$$\Rightarrow \frac{y'(x)}{y(x)} = \frac{1-3x}{x^2} \Rightarrow \frac{1}{y}\frac{dy}{dx} = \frac{1-3x}{x^2}$$

$$\Rightarrow \int \frac{1}{y}\,dy = \int \frac{1}{x^2}\,dx - \int \frac{3}{x}\,dx$$

$$\Rightarrow \ln y = -\frac{1}{x} - 3\ln x + \ln C$$

$$\Rightarrow \ln\left|\frac{yx^3}{C}\right| = -\frac{1}{x}$$

$$\Rightarrow \quad \frac{yx^3}{C} = e^{-1/x}$$

$$\Rightarrow \quad y = \frac{Ce^{-1/x}}{x^3}$$

**160.** *(8)* $I = \int_0^{e^3}\left[\frac{1}{e^{x-1}}\right]dx$

$$= \int_0^{1-\ln 2} 2dx + \int_{1-\ln 2}^{1} dx + \int_1^{e^3} 0\cdot dx$$

$$= 2(1-\ln 2) + (1-(1-\ln 2)) + 0$$

$$= 2 - \ln 2 \Rightarrow \alpha = 2 \Rightarrow \alpha^3 = 8$$

**161.** *(4)* Given

$$I_1 = \int_{-1/2}^{1} 2xf(2x(1-2x))\,dx$$

Put $t = 2x$

$$\Rightarrow \quad dt = 2dx$$

$$\Rightarrow \quad I_1 = \int_{-1}^{2} tf(t(1-t))\frac{dt}{2}$$

$$\Rightarrow \quad 2I_1 = \int_{-1}^{2} tf(t(1-t))\,dt \quad \ldots(\text{i})$$

$$\Rightarrow \quad 2I_1 = \int_{-1}^{2} (1-t)f(1-t)(t))\,dt \quad \ldots(\text{ii})$$

$$\left[\because \int_a^b f(x)dx = \int_a^b f(a+b-x)\,dx\right]$$

Adding Eqs. (i) and (ii), we get

$$4I_1 = \int_{-1}^{2} f(t(1-t))\,dt$$

$$\Rightarrow \quad 4I_1 = \int_{-1}^{2} f(x(1-x))\,dx$$

$$\left[\because \int_a^b f(t)\,dt = \int_a^b f(x)\,dx\right]$$

$$I_2 = \int_{-1}^{2} f(x(1-x))\,dx$$

So, $4I_1 = I_2 \Rightarrow \dfrac{I_2}{I_1} = \dfrac{4}{1}$

**162.** *(12)* We have,

$$\Rightarrow 24\int_0^{\pi/4} (\sin|4x - \pi/12|) + [2\sin x])dx = 2\pi + \alpha$$

$$\Rightarrow 24\int_0^{\pi/48} -\sin\left(4x - \frac{\pi}{12}\right)dx + \int_{\pi/48}^{\pi/4} \sin\left(4x - \frac{\pi}{12}\right)dx + \int_0^{\frac{\pi}{6}} [0]\,dx + \int_{\frac{\pi}{6}}^{\frac{\pi}{4}} [2\sin x]dx$$

$$= 24\left[\frac{\left(1-\cos\frac{\pi}{12}\right)}{4} - \frac{\left(-\cos\frac{\pi}{12} - 1\right)}{4}\right] + \frac{\pi}{4} - \frac{\pi}{6}$$

$$= 24\left(\frac{1}{2} + \frac{\pi}{12}\right) = 2\pi + 12$$

Compare with $2\pi + \alpha$, we get

$$\Rightarrow \quad \alpha = 12$$

**163.** *(450)* $A = \begin{bmatrix} 2 & -5 \\ 3 & m \end{bmatrix}$

$$B = \begin{bmatrix} 20 \\ m \end{bmatrix},\ X = \begin{bmatrix} x \\ y \end{bmatrix}$$

$$AX = B$$

$$2x - 5y = 20$$

$$3x + my = m$$

$$\Rightarrow \quad y = \frac{2m - 60}{2m + 15}$$

$$y < 0,\ m \in \left(\frac{-15}{2}, 30\right) \quad \ldots(\text{i})$$

Now, $\quad x = \dfrac{25m}{2m+15}$

$$\because \quad x < 0$$

$$m \in \left(-\frac{15}{2}, 0\right) \quad \ldots(\text{ii})$$

By Eqs. (i) and (ii), we get

$$m \in \left(-\frac{15}{2}, 0\right)$$

$$|A| = 2m + 15$$

$$\therefore 8\int_{-\frac{15}{2}}^{0} (2m+15) = 450$$

**164.** *(32)*

> **key idea** Use $\lim\limits_{x\to\infty} \sum\limits_{r=0}^{n-1} f\left(\frac{r}{n}\right) = \int_0^1 f(x)dx$

$$\sum_{r=1}^{\infty} \frac{n}{\sqrt{n^4 + r^4}} - \frac{2nr^2}{(n^2 + r^2)\sqrt{n^4 + r^4}}$$

$$= \sum_{r=1}^{\infty} \frac{\frac{1}{n}}{\sqrt{1 + \left(\frac{r}{n}\right)^4}} - \frac{2\left(\frac{1}{n}\right)\left(\frac{r}{n}\right)^2}{\left[1 + \left(\frac{r}{n}\right)^2\right]\sqrt{1 + \left(\frac{r}{n}\right)^4}}$$

$$= \int_0^1 \left[\frac{1}{\sqrt{1+x^4}} - \frac{2x^2\,dx}{(1+x^2)\sqrt{1+x^4}}\right]$$

$$= \int_0^1 \frac{(1-x^2)dx}{(1+x^2)\sqrt{1+x^4}}$$

$$= \int_0^1 \frac{\frac{1}{x^2} - 1}{\left(x + \frac{1}{x}\right)\sqrt{x^2 + \frac{1}{x^2}}}\,dx$$

$$= -\int_0^1 \frac{1 - \frac{1}{x^2}}{\left(x + \frac{1}{x}\right)\sqrt{\left(x + \frac{1}{x}\right)^2 - 2}}\,dx$$

Put $x + \dfrac{1}{x} = z$

$$\Rightarrow 1 - \frac{1}{x^2}\,dx = dz$$

So, $-\int_{\infty}^{2}\frac{dz}{z\sqrt{z^2-2}}$

Using property

$$=\frac{-1}{\sqrt{2}}\left[\sec^{-1}\frac{z}{\sqrt{2}}\right]_{\infty}^{2}$$

$$=\frac{-1}{\sqrt{2}}\left(\frac{\pi}{4}-\frac{\pi}{2}\right)=\frac{\pi}{4\sqrt{2}}=\frac{\pi}{k}$$

$$k=4\sqrt{2} \Rightarrow k^2=32$$

$\therefore\ k^2=32$

**165.** *(23)* By applying property of greatest integer function given

$\int_0^3\left([x^2]+\left[\frac{x^2}{2}\right]\right)dx$ can be written as

$$\text{LHS}\Rightarrow\int_0^1(0+0)\,dx+\int_1^{\sqrt{2}}(1+0)\,dx$$

$$+\int_{\sqrt{2}}^{\sqrt{3}}(2+1)\,dx+\int_{\sqrt{3}}^{2}(3+1)\,dx$$

$$+\int_2^{\sqrt{5}}(4+2)\,dx+\int_{\sqrt{5}}^{\sqrt{6}}(5+2)\,dx$$

$$+\int_{\sqrt{6}}^{\sqrt{7}}(6+3)\,dx+\int_{\sqrt{7}}^{\sqrt{8}}(7+3)\,dx$$

$$+\int_{\sqrt{8}}^{3}(8+4)\,dx$$

$$\Rightarrow\int_0^3\left([x^2]+\left[\frac{x^2}{2}\right]\right)dx$$

$$=\sqrt{2}-1+3\sqrt{3}$$

$$-3\sqrt{2}+8-4\sqrt{3}+6\sqrt{5}-12$$

$$+7\sqrt{6}-7\sqrt{5}+9\sqrt{7}-9\sqrt{6}+10\sqrt{8}$$

$$-10\sqrt{7}+36-12\sqrt{8}$$

$$=31-6\sqrt{2}-\sqrt{3}-\sqrt{5}-2\sqrt{6}-\sqrt{7}\ \ldots\text{(i)}$$

Comparing Eq. (i) with $a+b\sqrt{2}-\sqrt{3}-\sqrt{5}+c\sqrt{6}-\sqrt{7}$, we get

$$a=31,\ b=-6,\ c=-2$$

$\therefore a+b+c=31-6-2=23$

> **• Concept Enhancer**
>
> Properties of greatest integer function are given below
>
> | $x$ | $[x]$ |
> |---|---|
> | $0\le x<1$ | 0 |
> | $1\le x<2$ | 1 |
> | $-1\le x<0$ | $-1$ |
> | $-2\le x<-1$ | $-2$ |

**166.** *(65)* According to the question,

Here, we will use reduction method.

Let $I_k=\int_0^1(1-x^7)^k\,dx$, then

$$I_{k+1}=\int_0^1(1-x^7)^{k+1}dx\Rightarrow r_k=\frac{I_k}{I_{k+1}}$$

Now, applying by part in $I_{k+1}$

$$I_{k+1}=\int_0^1(1-x^7)^{k+1}\cdot 1\,dx$$

$$I_{k+1}=[(1-x^7)^{k+1}\cdot x]_0^1+\int_0^1(k+1)(1-x^7)^k\cdot 7x^6x\,dx$$

$$I_{k+1}=0+7(k+1)\int_0^1(1-x^7)^k(1-(1-x^7))\,dx$$

$$I_{k+1}=7(k+1)[I_k-I_{k+1}]$$

$$\Rightarrow(7k+8)I_{k+1}=(7k+7)I_k$$

$$\Rightarrow\frac{I_k}{I_{k+1}}=\frac{7k+8}{7k+7}=r_k$$

Now, $7(r_{k-1})=7\left(\frac{7k+8}{7k+7}-1\right)=\frac{1}{(k+1)}$

$$\Rightarrow\frac{1}{7(r_{k-1})}=k+1$$

Now, $\sum_{k=1}^{10}\frac{1}{7(r_{k-1})}=\sum_{k=1}^{10}k+1$

$$=\frac{10\times 11}{2}+10=65$$

**167.** *(1)* $f(t)=\int_0^{\pi}\frac{2x}{1-\cos^2t\sin^2x}dx\ \ldots\text{(i)}$

$$f(t)=2\int_0^{\pi}\frac{\pi-x}{1-\cos^2t\sin^2x}dx\ \ldots\text{(ii)}$$

On adding Eqs. (i) and (ii), we get

$$2f(t)=2\pi\int_0^{\pi}\frac{1}{1-\cos^2t\sin^2x}dx$$

$$2f(t)=2\pi\int_0^{\pi}\frac{\sec^2x}{\sec^2x-\cos^2t\tan^2x}dx$$

$$f(t)=2\pi\int_0^{\pi/2}\frac{\sec^2x\,dx}{\sec^2x-\cos^2t\tan^2x}$$

Let $\tan x=z\Rightarrow\sec^2x\,dx=dz$

$$f(t)=2\pi\int_0^{\pi/2}\frac{dz}{1+\sin^2t\cdot z^2}=\frac{\pi^2}{\sin t}$$

$$\therefore\int_0^{\pi/2}\frac{\pi^2}{f(t)}dt=\int_0^{\pi/2}\sin t\,dt=1$$

**168.** *(8)* $\int_0^{\pi/4}\frac{\sin^2x}{1+\frac{1}{2}\sin 2x}dx$

$$=\int_0^{\pi/4}\frac{1-\cos 2x}{2+\sin 2x}dx$$

Let $\underbrace{\int_0^{\pi/4}\frac{1}{2+\sin 2x}dx}_{I_1\text{ (say)}}-\underbrace{\int_0^{\pi/4}\frac{\cos 2x}{2+\sin 2x}dx}_{I_2\text{ (say)}}$

$$I_1=\int_0^{\pi/4}\frac{dx}{2+\frac{2\tan x}{1+\tan^2x}}$$

$$=\int_0^{\pi/4}\frac{\sec^2x}{2\tan^2x+2\tan x+2}dx$$

Put $\tan x=z$

$$I=\frac{1}{2}\int_0^1\frac{dz}{\left(z+\frac{1}{2}\right)^2+\frac{3}{4}}=\frac{\pi}{6\sqrt{3}}$$

Also, $I_2=\int_0^{\frac{\pi}{4}}\frac{\cos 2x}{2+\sin 2x}dx=\frac{1}{2}\left(\ln\frac{3}{2}\right)$

$$\therefore I_1-I_2=\frac{1}{\sqrt{3}}\times\frac{\pi}{6}+\frac{1}{2}\ln\frac{2}{3}$$

Hence, $a=2$, $b=6$ and its sum $=a+b=8$

**169.** *(8)* $I=\int_{-\frac{\pi}{2}}^{\frac{\pi}{2}}\frac{8\sqrt{2}\cos x\cdot dx}{(1+e^{\sin x})(1+\sin^4x)}$

$$=\alpha\pi+\beta\log_e(3+2\sqrt{2})$$

$$\Rightarrow\quad I=\int_{-\frac{\pi}{2}}^{\frac{\pi}{2}}\frac{8\sqrt{2}\cos x\cdot dx}{(1+e^{-\sin x})(1+\sin^4x)}$$

$$\Rightarrow\quad 2I=\int_{-\frac{\pi}{2}}^{\frac{\pi}{2}}\frac{8\sqrt{2}\cos x(1+e^{\sin x})}{(1+e^{\sin x})(1+\sin^4x)}dx$$

$$\Rightarrow 2I=2\int_0^{\frac{\pi}{2}}\frac{8\sqrt{2}\cos x}{(1+\sin^4x)}dx$$

Put $\sin x=t$

$\therefore\ \cos x\,dx=dt$

$$\Rightarrow\quad I=8\sqrt{2}\int_0^1\frac{dt}{1+t^4}$$

$$\Rightarrow\quad I=4\sqrt{2}\int_0^1\frac{2}{1+t^4}dt$$

$$\Rightarrow\quad I=4\sqrt{2}\int_0^1\frac{(1+t^2)-(t^2-1)}{t^2\left(\frac{1}{t^2}+t^2\right)}\cdot dt$$

$$\Rightarrow I=4\sqrt{2}$$

$$\left[\int_0^1 \frac{\left(\frac{1}{t^2}+1\right)}{\left(t-\frac{1}{t}\right)^2+2}dt-\int_0^1 \frac{\left(1-\frac{1}{t^2}\right)}{\left(1+\frac{1}{t}\right)^2-2}dt\right]$$

Put $t-\frac{1}{t}=z$, put $t+\frac{1}{t}=s$

$\left(1+\frac{1}{t^2}\right)dt=dz;\left(1-\frac{1}{t^2}\right)dt=ds$

$$I=4\sqrt{2}\left[\int_{-\infty}^{0}\frac{dz}{z^2+(\sqrt{2})^2}-\int_{\infty}^{2}\frac{ds}{s^2-(\sqrt{2})^2}\right]$$

$$I=4\sqrt{2}\left[\left(\frac{1}{\sqrt{2}}\tan^{-1}\frac{z}{\sqrt{2}}\right)_{-\infty}^{0}-\left(\frac{1}{2\sqrt{2}}\log_e\left|\frac{s-\sqrt{2}}{s+\sqrt{2}}\right|\right)_{\infty}^{2}\right]$$

$I=2\pi+2\ln(3+2\sqrt{2})\Rightarrow\alpha=2,\beta=2$

Hence, $\alpha^2+\beta^2=4+4=8$

**170.** *(15)*

Let $I=\int_0^{\pi}\frac{x^2\sin x\cos x}{\sin^4x+\cos^4x}dx$ ...(i)

Using $\int_0^a f(x)\,dx=\int_0^a f(a-x)\,dx$

$$I=\int_0^{\pi}\frac{(\pi-x)^2\sin(\pi-x)\cos(\pi-x)}{\sin^4(\pi-x)+\cos^4(\pi-x)}dx$$

$$I=-\int_0^{\pi}\frac{(\pi^2+x^2-2\pi x)\sin x\cos x}{\sin^4x+\cos^4x}dx \quad ...(ii)$$

On adding Eqs. (i) and (ii), we get

$$2I=\int_0^{\pi}\frac{(2\pi x-\pi^2)\sin x\cos x}{\sin^4x+\cos^4x}dx$$

$$2I=2\int_0^{\pi/2}\frac{\pi(2x-\pi)\sin x\cos x}{\sin^4x+\cos^4x}dx$$

$$I=\pi\int_0^{\pi/2}\frac{(2x-\pi)\sin x\cos x}{\sin^4x+\cos^4x}dx \quad ...(iii)$$

$$I=\pi\int_0^{\pi/2}\frac{\sin\left(\frac{\pi}{2}-x\right)\cos\left(\frac{\pi}{2}-x\right)}{\sin^4\left(\frac{\pi}{2}-x\right)+\cos^4\left(\frac{\pi}{2}-x\right)}dx$$

$$I=\pi\int_0^{\pi/2}\frac{-2x\sin x\cos x}{\sin^4x+\cos^4x}dx \quad ...(iv)$$

On adding Eqs. (iii) and (iv), we get

$$2I=\pi\int_0^{\pi/2}\frac{-\pi\sin x\cos x}{\sin^4x+\cos^4x}dx$$

$$I=-\frac{\pi^2}{4}\int_0^{\pi/2}\frac{2\sin x\cos x}{\sin^4x+\cos^4x}dx$$

$$I=-\frac{\pi^2}{4}\int_0^{\pi/2}\frac{2\tan x\sec^2x}{\tan^4x+1}dx$$

Put $\tan^2x=t$

$2\tan x\sec^2x\,dx=dt$

$$I=-\frac{\pi^2}{4}\int_0^{\pi/2}\frac{dt}{1+t^2}$$

$$=-\frac{\pi^2}{4}[\tan^{-1}t]_0^{\pi/2}$$

$$I=-\frac{\pi^2}{4}[\tan^{-1}(\tan^2x)]_0^{\pi/2}$$

$$I=-\frac{\pi^2}{4}[\tan^{-1}(\infty)-\tan^{-1}(0)]$$

$$I=-\frac{\pi^2}{4}\times\left(\frac{\pi}{2}-0\right)=-\frac{\pi^3}{8}$$

So $\left|\frac{120}{\pi^3}\int_0^{\pi}\frac{x^2\sin x\cos x}{\sin^4x+\cos^4x}dx\right|$

$$=\left|\frac{120}{\pi^3}\times\frac{-\pi^3}{8}\right|=15$$

**171.** *(5)* $f(x)=\frac{4^x}{4^x+2}$

$$f(1-x)=\frac{4^{1-x}}{4^{1-x}+2}=\frac{\frac{4}{4^x}}{\frac{4}{4^x}+2}$$

$$=\frac{4}{4+2\cdot4^x}=\frac{2}{4^x+2}$$

$\Rightarrow f(x)+f(1-x)=1$

$$M=\int_{f(a)}^{f(1-a)}x\sin^4(x(1-x)dx$$

$$=\int_{f(a)}^{f(1-a)}(1-x)\sin^4(1-x)\cdot x\,dx$$

$$=\int_{f(a)}^{f(1-a)}\sin^4[(1-x)x]\,dx$$

$$-\int_{f(a)}^{f(1-a)}x\sin^4[(1-x)x]\,dx$$

$\Rightarrow \quad M=N-M$

$\Rightarrow \quad 2M=N$

$\Rightarrow \quad \alpha=2,\ \beta=1$

$\therefore \alpha^2+\beta^2=5$

**172.** *(155)* We have,

$$I=9\int_0^9\left[\sqrt{\frac{10x}{x+1}}\right]dx$$

Let $f(x)=\sqrt{\frac{10x}{x+1}}$

If $\frac{10x}{x+1}=1\Rightarrow x=1/9$

If $\frac{10x}{x+1}=4\Rightarrow x=2/3$

If $\frac{10x}{x+1}=9\Rightarrow x=9$

$$\therefore f(x)=\begin{cases}0, & 0\le x<1/9\\ 1, & 1/a\le x<2/3\\ 2, & 2/3\le x<9\end{cases}$$

$$\therefore \quad I=9\left[\int_0^{1/9}0\,dx+\int_{1/9}^{2/3}dx+\int_{2/3}^{9}2\,dx\right]$$

$$=9\left[\frac{2}{3}-\frac{1}{9}+18-\frac{4}{3}\right]$$

$$=9[18-7/9]=155$$

**173.** *(6)* $\int_{\pi/6}^{\pi/3}\sqrt{1-\sin 2x}\,dx=\alpha+\beta\sqrt{2}+\gamma\sqrt{3}$

Let $I=\int_{\pi/6}^{\pi/3}\sqrt{1-\sin 2x}\,dx$

$$=\int_{\pi/6}^{\pi/3}\sqrt{(\sin x-\cos x)^2}\,dx$$

$$=\int_{\pi/6}^{\pi/3}|\sin x-\cos x|\,dx$$

$$=\int_{\pi/6}^{\pi/4}(\cos x-\sin x)\,dx+\int_{\pi/4}^{\pi/3}(\sin x-\cos x)\,dx$$

$$=[\sin x+\cos x]_{\frac{\pi}{6}}^{\frac{\pi}{4}}-[\cos x+\sin x]_{\frac{\pi}{4}}^{\frac{\pi}{3}}$$

$$=\left(\sqrt{2}-\frac{1}{2}-\frac{\sqrt{3}}{2}\right)-\left(\frac{1}{2}+\frac{\sqrt{3}}{2}-\sqrt{2}\right)$$

$$=2\sqrt{2}-1-\sqrt{3}$$

$\alpha=-1,\beta=2$ and $\gamma=-1$

$3\alpha+4\beta-\gamma=-3+8+1=6$

**174.** *(12)* $45x+5y+3=0$

Slope $=\frac{-45}{5}=27r_1+\frac{9r_2}{2}$

$-9=27r_1+\frac{9r_2}{2}$

$-2=6r_1+r_2=6r_1+r_2+2=0$

$$= \lim_{x\to 3}\left(\int_3^x \frac{8t^2}{\frac{3r_2x}{2} - r_2x^2 - r_1x^3 - 3x}\,dt\right)$$

$$= \lim_{x\to 3}\frac{\frac{8}{3}(x^3-27)}{\frac{3r_2x}{2} - r_2x^2 - r_1x^3 - 3x}$$

L'Hospital rule,

$$\lim_{x\to 3}\frac{\frac{8}{3}\cdot 3x^2}{\frac{3r_2}{2} - 2r_2x - 3r_1x^2 - 3}$$

$$= \frac{8(9)}{\frac{3r_2}{2} - 6r_2 - 27r_1 - 3}$$

$$= \frac{8\times 9\times 2}{3r_2 - 12r_2 - 54r_1 - 6}$$

$$= \frac{144}{-9(r_2+6r_1)-6} = \frac{144}{12} = 12$$

**175.** *(2)* Given,

$$f(x) = \int_0^x g(t)\log_e\left(\frac{1-t}{1+t}\right)dt$$

$$\Rightarrow f'(x) = g(x)\log_e\left(\frac{1-x}{1+x}\right)$$

Since, $g(x)$ and $\log_e\left(\frac{1-x}{1+x}\right)$ both are odd.

So, $f'(x)$ is an even function.

Hence, $f(x)$ is an odd function.

So, $f(x) + f(-x) = 0$ ...(i)

Now, $\int_{-\frac{\pi}{2}}^{\frac{\pi}{2}}\left(f(x) + \frac{x^2\cos x}{1+e^x}\right)dx$

$$= \int_0^{\pi/2}\left\{f(x) + f(-x) + x^2\cos x\left(\frac{1}{1+e^x} + \frac{1}{1+e^{-x}}\right)dx\right\}$$

$\int_0^{\pi/2} x^2\cos x\,dx$ $\left[\because f(x)+f(-x)=0 \text{ and } \frac{1}{1+e^x}+\frac{1}{1+e^{-x}}=1\right]$

$$= x^2\sin x\,\Big|_0^{\pi/2} - \int_0^{\pi/2} 2x\sin x\,dx$$

$$= \frac{\pi}{4} - 2\left\{[-x\cos x]_0^{\pi/2} - \int_0^{\pi/2}(-\cos x)\,dx\right\}$$

$$= \frac{\pi^2}{4} - 2$$

Given that, $\frac{\pi^2}{\alpha^2} - \alpha = \frac{\pi^2}{4} - 2$

$\Rightarrow \alpha = 2$

**176.** *(18)* Given, $S_0(x) = x$

$$S_k(x) = C_k x + k\int_0^x S_{k-1}(t)\,dt$$

On putting $k = 1$,

$$S_1(x) = C_1x + \int_0^x t\,dt = C_1x + \frac{x^2}{2}$$

On putting $k = 2$, we have

$$S_2(x) = C_2x + 2\int_0^x S_1(t)\,dt$$

$$= C_2x + 2\int_0^x\left\{C_1t + \frac{t^2}{2}\right\}dt$$

$$= C_2x + 2\left[C_1\frac{t^2}{2} + \frac{t^3}{6}\right]_0^x$$

$$= C_2x + 2C_1\frac{x^2}{2} + \frac{2x^3}{6}$$

$$S_2(x) = C_2x + C_1x^2 + \frac{x^3}{3}$$

$$S_2(3) = 3C_2 + 9C_1 + 9$$

Now, $C_0 = 1$ and

$C_k = 1 - \int_0^1 S_{k-1}(x)\,dx$ [Given]

On putting $k = 3$, we get

$$C_3 = 1 - \int_0^1 S_2(x)\,dx$$

$$= 1 - \int_0^1\left\{C_2x + C_1x^2 + \frac{x^3}{3}\right\}dx$$

$$= 1 - \left[\frac{C_2x^2}{2} + \frac{C_1x^3}{3} + \frac{x^4}{12}\right]_0^1$$

$$C_3 = 1 - \frac{C_2}{2} - \frac{C_1}{3} - \frac{1}{12}$$

$$6C_3 = 6 - 3C_2 - 2C_1 - \frac{1}{2}$$

$$S_2(3) + 6C_3 = (3C_2 + 9C_1 + 9) + \left(6 - 3C_2 - 2C_1 - \frac{1}{2}\right)$$

$$= \left(15 - \frac{1}{2}\right) + 7C_1$$

$$= \frac{29}{2} + 7C_1$$

Again, $C_1 = 1 - \int_0^1 S_0(x)\,dx = 1 - \frac{1}{2} = \frac{1}{2}$

$$\therefore S_2(3) + 6C_3 = \frac{29}{2} + \frac{7}{2} = \frac{36}{2} = 18$$

**177.** *(575)* Let $f(x) = |100x^2 - 1|$

$$\Rightarrow f(-x) = |100(-x^2) - 1| = |100x^2 - 1| = f(x)$$

i.e. $f(x)$ is even function.

$$\therefore I = \int_{-0.15}^{0.15} f(x)\,dx = 2\int_0^{0.15} f(x)dx \quad \text{...(i)}$$

Now,

$$|100x^2 - 1| = \begin{cases}100x^2 - 1, & 100x^2 - 1 \ge 0\\ 1 - 100x^2, & 100x^2 - 1 < 0\end{cases}$$

$$= \begin{cases}100x^2 - 1, & x \in R - (-0.1, 0.1)\\ 1 - 100x^2, & x \in (-0.1, 0.1)\end{cases}$$

$$\therefore I = 2\left[\int_0^{0.1}(1 - 100x^2)dx + \int_{0.1}^{0.15}(100x^2 - 1)dx\right]$$

$$= 2\left[\left(x - \frac{100x^3}{3}\right)_0^{0.1} + \left(\frac{100x^3}{3} - x\right)_{0.1}^{0.15}\right]$$

$$= 2\left(0.1 - \frac{100\times 0.001}{3} + \frac{100\times 0.003375}{3} - 0.15 - \frac{100\times 0.001}{3} + 0.1\right)$$

$$= \frac{23}{120} = \frac{k}{3000} \text{ [Given]} \Rightarrow k = 575$$

**178.** *(32)* We have,

$$\alpha(m, n) = \int_0^2 t^m(1+3t)^n\,dt$$

Also, $11\alpha(10, 6) + 18\,\alpha(11, 5) = p(14)^6$

$$\Rightarrow 11\int_0^2 t^{10}(1+3t)^6\,dt + 18\int_0^2 t^{11}(1+3t)^5\,dt = p(14)^6$$

Using integration by part for expression $t^{10}(1+3t)^6$

$$\Rightarrow 11\left[\left[(1+3t)^6\times\frac{t^{11}}{11}\right]_0^2 - \int_0^2 6(1+3t)^5\times 3\times\frac{t^{11}}{11}\,dt\right]$$

$$+ 18\int_0^2 t^{11}(1+3t)^5\,dt = P(14)^6$$

$$\Rightarrow 7^6\times 2^{11} - 18\int_0^2 t^{11}(1+3t)^5\,dt + 18\int_0^2 t^{11}(1+3t)^5\,dt = p(14)^6$$

$$\Rightarrow \quad 7^6 \times 2^{11} = p(14)^6$$

$$\Rightarrow (7\times 2)^6 \times 2^5 = p(14)^6$$

$$\Rightarrow \quad p = 2^5 = 32$$

**179.** *(6)* $\int_0^{2.4} [x^2]\, dx$

$$= \int_0^1 [x^2] dx + \int_1^{\sqrt{2}} [x^2] dx + \int_{\sqrt{2}}^{\sqrt{3}} [x^2] dx + \int_{\sqrt{3}}^{2} [x^2]\, dx + \int_2^{\sqrt{5}} [x^2] dx + \int_{\sqrt{5}}^{2.4} [x^2] dx$$

$$= \int_0^1 0 dx + \int_1^{\sqrt{2}} 1\, dx + \int_{\sqrt{2}}^{\sqrt{3}} 2 dx + \int_{\sqrt{3}}^{2} 3 dx + \int_2^{\sqrt{5}} 4 dx + \int_{\sqrt{5}}^{2.4} 5 dx$$

$$= (\sqrt{2}-1) + 2(\sqrt{3}-\sqrt{2}) + 3(2-\sqrt{3}) + 4(\sqrt{5}-2) + 5(2.4-\sqrt{5})$$

$$= 9 - \sqrt{2} - \sqrt{3} - \sqrt{5}$$

$$\Rightarrow \alpha = 9, \beta = -1, \gamma = -1, \delta = -1$$

$$\therefore \alpha + \beta + \gamma + \delta = 6$$

**180.** *(14)* Let

$$I = \frac{2}{\pi} \int_{\pi/6}^{5\pi/6} (8[\text{cosec}\, x] - 5[\cot x])\, dx$$

For $x \in \left(\frac{\pi}{6}, \frac{5\pi}{6}\right)$, $\text{cosec}\, x \in [1, 2)$

$\Rightarrow [\text{cosec}\, x] = 1$

For $x \in \left(\frac{\pi}{6}, \frac{\pi}{4}\right)$, $[\cot x] = 1$

For $x \in \left(\frac{\pi}{4}, \frac{\pi}{2}\right)$, $[\cot x] = 0$

For $x \in \left(\frac{\pi}{2}, \frac{3\pi}{4}\right)$, $[\cot x] = -1$

For $x \in \left(\frac{3\pi}{4}, \frac{5\pi}{6}\right)$, $[\cot x] = -2$

$$\therefore I = \frac{2}{\pi}\int_{\pi/6}^{5\pi/6} 8(1)\, dx - \int_{\pi/6}^{\pi/4} 5(1)\, dx - \int_{\pi/4}^{\pi/2} 5(0)\, dx - \int_{\pi/2}^{3\pi/4} 5(-1)\, dx - \int_{3\pi/4}^{5\pi/6} 5(-2)\, dx$$

$$= \frac{2}{\pi} \times 7\pi = 14$$

**181.** *(0)* We have, $f(x) = \dfrac{x}{(1+x^n)^{1/n}}$

$$\therefore f(f(x)) = \frac{f(x)}{(1+(f(x))^n)^{1/n}}$$

$$= \frac{\dfrac{x}{(1+x^n)^{1/n}}}{\left(1+\dfrac{x^n}{1+x^n}\right)^{1/n}} = \frac{x}{(1+2x^n)^{1/n}}$$

$$\Rightarrow f(f(f(x))) = \frac{x}{(1+3x^n)^{1/n}}$$

Similarly, $f^n(x) = \dfrac{x}{(1+nx^n)^{1/n}}$

Now, $\lim\limits_{n\to\infty} \int_0^1 x^{n-2}(f^n(x))dx$

$$\lim_{n\to\infty} \int_0^1 x^{n-2} \frac{x}{(1+nx^n)^{1/n}}\, dx$$

$$= \lim_{n\to\infty} \int_0^1 \frac{x^{n-1}}{(1+nx^n)^{1/n}}\, dx$$

Let $1+nx^n = t \Rightarrow n^2 x^{n-1}\, dx = dt$

When, $x = 0$, then $t = 1$

When, $x = 1$, then $t = 1+n$

$$\therefore \lim_{n\to\infty} \int_1^{1+n} \frac{dt}{n^2 (t)^{1/n}} = \lim_{n\to\infty} \frac{1}{n^2} \int_1^{1+n} \frac{dt}{(t)^{1/n}}$$

$$= \lim_{n\to\infty} \frac{1}{n(n-1)} [(1+n)^{1-\frac{1}{n}} - 1]$$

On putting $n = 1/h$, when $n \to \infty$, then $h \to 0$

$$\therefore \lim_{h\to 0} \frac{1}{\frac{1}{h}\left(\frac{1}{h}-1\right)} \left[\left(1+\frac{1}{h}\right)^{1-h} - 1\right]$$

$$\lim_{h\to 0} \frac{1}{\frac{1}{h}\left(\frac{1}{h}-1\right)} \left[1-(1-h)\left(1+\frac{1}{h}\right) + \ldots - 1\right]$$

$$\lim_{h\to 0} \frac{h^2}{1-h}\left[-(1-h)\left(1+\frac{1}{h}\right) + \ldots\right]$$

$$= \lim_{h\to 0} -h[h+1+\ldots] = 0$$

**182.** *(13)* Let

$$I = \int_0^{\pi} \frac{5^{\cos x}(1+\cos x \cos 3x + \cos^2 x + \cos^3 x \cos 3x)}{1+5^{\cos x}}\, dx \quad \ldots\text{(i)}$$

$$= \int_0^{\pi} \frac{5^{\cos(\pi - x)}(1+\cos(\pi - x) \cos 3(\pi - x) + \cos^2(\pi - x) + \cos^3(\pi - x)\cos 3(\pi - x)}{1+5^{\cos(\pi - x)}}\, dx$$

$$= \int_0^{\pi} \frac{5^{-\cos x}(1+\cos x \cos 3x + \cos^2 x + \cos^3 x \cos 3x}{1+5^{-\cos x}}\, dx$$

$$\therefore I = \int_0^{\pi} \frac{(1+\cos x \cos 3x + \cos^2 x + \cos^3 x \cos 3x)}{5^{\cos x}+1}\, dx \quad \ldots\text{(ii)}$$

On adding Eqs. (i) and (ii), we get

$$2I = \int_0^{\pi} (1+\cos x \cos 3x + \cos^2 x + \cos^3 x \cos 3x) dx$$

$$\Rightarrow 2I = 2\int_0^{\pi/2} (1+\cos x \cos 3x + \cos^2 x + \cos^3 x \cos 3x)\, dx$$

$$\Rightarrow I = \int_0^{\pi/2} (1+\cos x \cos 3x + \cos^2 x + \cos^3 x \cos 3x)\, dx$$

$$\Rightarrow I = \frac{1}{2}\int_0^{\pi/2} 2 + 2\cos x \cos 3x + 2\cos^2 x + 2\cos^3 x \cos 3x\, dx$$

$$= \frac{1}{2}\int_0^{\pi/2} 2 + \cos 4x + \cos 2x + 1 + \cos 2x + 2\cos 3x \left[\frac{\cos 3x + 3\cos x}{4}\right] dx$$

$$= \frac{1}{2}\int_0^{\pi/2} \left(3 + \cos 4x + 2\cos 2x + \frac{1}{2}\cos^2 3x + \frac{3}{2}\cos 3x \cos x\right) dx$$

$$= \frac{1}{2}\int_0^{\pi/2} \left[3 + \cos 4x + 2\cos 2x + \frac{1}{4}(1+\cos 6x) + \frac{3}{4}(\cos 4x + \cos 2x)\right] dx$$

$$= \frac{1}{2}\int_0^{\pi/2} \left(\frac{13}{4} + \frac{11}{4}\cos 2x + \frac{7}{4}\cos 4x + \frac{1}{4}\cos 6x\right) dx$$

$$= \frac{1}{2}\left[\frac{13}{4}x + \frac{11}{8}\sin 2x + \frac{7}{16}\sin 4x + \frac{1}{24}\sin 6x\right]_0^{\pi/2}$$

$$= \frac{1}{2}\left[\left(\frac{13}{4} \times \frac{\pi}{2} + \frac{11}{8}\sin \pi + \frac{7}{16}\sin 2\pi + \frac{1}{24}\sin 3\pi\right) - 0\right]$$

$$= \frac{1}{2} \times \frac{13}{4} \times \frac{\pi}{2} = \frac{13\pi}{16}$$

$$\therefore k = 13$$

**183.** *(63)* Let $I = \int_0^1 [(x^{21} + x^{14} + x^7)(2x^{14} + 3x^7 + 6)^{1/7}]\, dx$

$$= \int_0^1 [(x^{20} + x^{13} + x^6)x(2x^{14} + 3x^7 + 6)^{1/7}]dx$$

$$= \int_0^1 [(x^{20} + x^{13} + x^6)(2x^{21} + 3x^{14} + 6x^7)^{1/7}]dx$$

Put $2x^{21} + 3x^{14} + 6x^7 = t$

$\Rightarrow (42x^{20} + 42x^{13} + 42x^6)dx = dt$

$\Rightarrow (x^{20} + x^{13} + x^6)\, dx = \dfrac{1}{42} dt$

Also, when $x = 0 \Rightarrow t = 0$

and when $x = 1$

$\Rightarrow t = 2 + 3 + 6 = 11$

$$\therefore \quad I = \frac{1}{42}\int_0^{11} t^{\frac{1}{7}} dt = \frac{1}{42}\left[\frac{t^{\frac{1}{7}+1}}{\frac{1}{7}+1}\right]_0^{11}$$

$$= \frac{7}{42 \times 8}[11^{8/7}]$$

$$= \frac{1}{48}(11)^{8/7}$$

$\therefore \quad l = 48, m = 8, n = 7$

So, $l + m + n = 48 + 8 + 7 = 63$

**184.** *(12)* Let $L = \lim\limits_{x \to 0} \dfrac{48}{x^4} \displaystyle\int_0^x \dfrac{t^3}{t^6 + 1} dt$

$$= \lim_{x \to 0} 48 \frac{\int_0^x \frac{t^3}{t^6+1} dt}{x^4} \quad \left[\frac{0}{0} \text{ form}\right]$$

On applying L'Hopital rule, we get

$$L = 48 \lim_{x \to 0} \frac{\frac{x^3}{x^6+1}}{4x^3}$$

$$= 48 \times \lim_{x \to 0} \frac{x^3}{x^6+1} \times \frac{1}{4x^3}$$

$$= 48 \times \frac{1}{4} \times \frac{1}{1} = 12$$

**185.** *(20)* Let $I = \displaystyle\int_{1/3}^{3} |\log_e x|\, dx$

$$\therefore \ |\log_e x| = \begin{cases} -\log_e x, & x \in (0, 1) \\ \log_e x, & x > 1 \end{cases}$$

$$= \int_{1/3}^{1} |\log_e x|\, dx + \int_1^3 |\log_e x|\, dx$$

$$= -\int_{1/3}^{1} \log_e x\, dx + \int_1^3 \log_e x\, dx$$

$$= \frac{2}{3} - \frac{1}{3}\ln 3 + 3\ln 3 - 2$$

$$= \frac{8}{3}\ln 3 - \frac{4}{3}$$

$$= \frac{4}{3}(2\ln 3 - \ln e)$$

$$= \frac{4}{3}\ln\left(\frac{3^2}{e}\right) = \frac{m}{n}\log_e\left(\frac{n^2}{e}\right)$$

$\therefore \quad m = 4, n = 3$

$\therefore m^2 + n^2 - 5 = 20$

**186.** *(27)* Given, $x \in \left[0, \dfrac{\pi}{2}\right]$

and $f(x) + \displaystyle\int_0^x f(t)\sqrt{1 - (\ln f(t))^2}\, dt = e$ ... (i)

On differentiating both sides, we get

$$f'(x) + f(x)\sqrt{1 - (\ln f(x))^2} = 0$$

$$\Rightarrow \ f'(x) = -f(x)\sqrt{1 - (\ln f(x))^2}$$

$$\frac{-f'(x)}{f(x)\sqrt{1 - (\ln f(x))^2}} = 1$$

On integrating both sides, we get

$$-\int \frac{f'(x)}{f(x)\sqrt{1 - (\ln f(x))^2}} dx = \int 1\, dx$$

Let $\ln f(x) = t$

$\therefore \quad \dfrac{1}{f(x)} \cdot f'(x)\, dx = dt$

$$\therefore \quad -\int \frac{dt}{\sqrt{1 - t^2}} = \int 1 \cdot dx$$

$\Rightarrow \quad \cos^{-1} t = x + c$

$\Rightarrow \quad \cos^{-1}(\ln f(x)) = x + c$

$\Rightarrow \quad \cos^{-1}(\ln f(0)) = 0 + c$

$\because$ From Eq. (i),

$$f(0) + \int_0^x f(t)\sqrt{1 - (\ln f(t))^2}\, dt = e$$

$f(0) = e$

$\Rightarrow \quad \cos^{-1}(\ln e) = c \Rightarrow \cos^{-1}(1) = c$

$\Rightarrow \quad c = 0$

$\therefore \quad \cos^{-1}(\ln f(x)) = x$

$$\therefore \quad \left(6\ln f\left(\frac{\pi}{6}\right)\right)^2 = \left(6\cos\left(\frac{\pi}{6}\right)\right)^2$$

$$= \left(6 \times \frac{\sqrt{3}}{2}\right)^2 = (3\sqrt{3})^2 = 27$$

**187.** *(22)* Let $I = \displaystyle\int_0^3 |x^2 - 3x + 2|\, dx$

Put $x^2 - 3x + 2 = 0$

$\Rightarrow \quad (x - 1)(x - 2) = 0 \Rightarrow x = 1, 2$

So, limits of integration breaks at $x = 1$, 2.

i.e. $I = \displaystyle\int_0^1 |x^2 - 3x + 2|\, dx$

$$+ \int_1^2 |x^2 - 3x + 2|\, dx + \int_2^3 |x^2 - 3x + 2|\, dx$$

$$\Rightarrow \quad I = \left(\frac{x^3}{3} - \frac{3x^2}{2} + 2x\right)_0^1 - \left(\frac{x^3}{3} - \frac{3x^2}{2} + 2x\right)_1^2 + \left(\frac{x^3}{3} - \frac{3x^2}{2} + 2x\right)_2^3$$

$$\Rightarrow \quad I = \frac{5}{6} - \left(\frac{2}{3} - \frac{5}{6}\right) + \left(\frac{3}{2} - \frac{2}{3}\right) = \frac{11}{6}$$

$\therefore \quad 12I = 12 \times \dfrac{11}{6} = 22$

**188.** *(2)*

> **key idea** Use property of definite integral
>
> $$\int_0^{2a} f(x)\, dx = \begin{cases} 2\int_0^a f(x)\, dx, & \text{If } f(2a - x) = f(x) \\ 0, & \text{If } f(2a - x) = -f(x) \end{cases}$$

Let

$$I = \frac{8}{\pi}\int_0^{\pi/2} \frac{(\cos x)^{2023}}{(\sin x)^{2023} + (\cos x)^{2023}} dx \quad \text{...(i)}$$

As we know that

$$\int_0^a f(x)\, dx = \int_0^a f(a - x) dx$$

$$\text{So, } I = \frac{8}{\pi}\int_0^{\frac{\pi}{2}} \frac{\left[\cos\left(\frac{\pi}{2} - x\right)\right]^{2023}}{\left[\sin\left(\frac{\pi}{2} - x\right)\right]^{2023} + \left[\cos\left(\frac{\pi}{2} - x\right)\right]^{2023}} dx$$

$$\Rightarrow I = \frac{8}{\pi}\int_0^{\frac{\pi}{2}} \frac{(\sin x)^{2023}}{(\cos x)^{2023} + (\sin x)^{2023}} dx \quad ...(ii)$$

On adding Eqs. (i) and (ii), we get

$$2I = \frac{8}{\pi}\left[\int_0^{\frac{\pi}{2}} \frac{(\cos x)^{2023} + (\sin x)^{2023}}{(\sin x)^{2023} + (\cos x)^{2023}} dx\right]$$

$$\Rightarrow 2I = \frac{8}{\pi}\int_0^{\pi/2} 1 \cdot dx$$

$$\Rightarrow I = \frac{4}{\pi}[x]_0^{\frac{\pi}{2}}$$

$$\Rightarrow I = \frac{4}{\pi}\left[\frac{\pi}{2} - 0\right] \Rightarrow I = 2$$

**189.** *(104)* Let $I = 60\int_0^{\frac{\pi}{2}} \frac{\sin 6x}{\sin x} dx$

$$= 60 \times 2\int_0^{\frac{\pi}{2}} \frac{\sin 3x \cos 3x}{\sin x} dx$$

$$= 120\int_0^{\frac{\pi}{2}} \frac{(3\sin x - 4\sin^3 x)(4\cos^3 x - 3\cos x)\, dx}{\sin x}$$

$$= 120\int_0^{\frac{\pi}{2}} (3 - 4\sin^2 x)(4\cos^2 x - 3) \cos x\, dx$$

$$= 120\int_0^{\frac{\pi}{2}} (3 - 4\sin^2 x)(4(1 - \sin^2 x) - 3) \cos x\, dx$$

$$= 120\int_0^{\frac{\pi}{2}} (3 - 4\sin^2 x)(1 - 4\sin^2 x) \cos x\, dx$$

Let $\sin x = t$, then $\cos x\, dx = dt$

When, $x \to 0$, then $t \to 0$

When, $x \to \frac{\pi}{2}$, then $t \to 1$

$$= 120\int_0^1 (3 - 4t^2)(1 - 4t^2) dt$$

$$= 120\int_0^1 (3 - 16t^2 + 16t^4)\, dt$$

$$= 120\left[3t - \frac{16t^3}{3} + \frac{16t^5}{5}\right]_0^1$$

$$= 120\left[3 - \frac{16}{3} + \frac{16}{5}\right]$$

$$= 120\left[\frac{45 - 80 + 48}{15}\right]$$

$$= 8 \times 13 = 104$$

**190.** *(10)* Given,

$$\int_0^{\pi/3} \frac{15x^3}{\sqrt{1 + x^2 + \sqrt{(1 + x^2)^3}}} dx = \alpha\sqrt{2} + \beta\sqrt{3}$$

$\therefore$ On putting $x = \tan\theta$

$$\Rightarrow dx = \sec^2\theta\, d\theta$$

$$\Rightarrow I = \int_0^{\pi/3} \frac{15\tan^3\theta \sec^2\theta\, d\theta}{\sqrt{1 + \tan^2\theta + \sqrt{\sec^6\theta}}}$$

$$\Rightarrow I = \int_0^{\pi/3} \frac{15\tan^3\theta \sec^2\theta}{\sec\theta\sqrt{1 + \sec\theta}} d\theta$$

$$I = \int_0^{\pi/3} \frac{15(\sec^2\theta - 1)\sec\theta\tan\theta\, d\theta}{\sqrt{1 + \sec\theta}}$$

Now, on putting $1 + \sec\theta = t^2$

$\sec\theta\tan\theta\, d\theta = 2t\, dt$

$$I = \int_{\sqrt{2}}^{\sqrt{3}} \frac{15[(t^2 - 1)^2 - 1]\, 2t\, dt}{t}$$

$$I = 30\int_{\sqrt{2}}^{\sqrt{3}} \frac{(t^4 - 2t^2 + 1 - 1)t}{t} dt$$

$$I = 30\int_{\sqrt{2}}^{\sqrt{3}} (t^4 - 2t^2)\, dt$$

$$\Rightarrow I = 30\left(\frac{t^5}{5} - \frac{2t^3}{3}\right)_{\sqrt{2}}^{\sqrt{3}}$$

$$= 30\left[\left(\frac{9}{5}\sqrt{3} - 2\sqrt{3}\right) - \left(\frac{4\sqrt{2}}{5} - \frac{4\sqrt{2}}{3}\right)\right]$$

$$= 16\sqrt{2} - 6\sqrt{3}$$

$\alpha = 16$ and $\beta = -6$

$\therefore \alpha + \beta = 10$

**191.** *(385)* Given,

$$f(x) = \min\{[x - 1], [x - 2], \ldots, [x - 10]\}$$

$$\Rightarrow \quad f(x) = [x - 10] = [x] - 10$$

$$I = \int_0^{10} f(x) dx + \int_0^{10} |f(x)|\, dx + \int_0^{10} (f(x))^2\, dx$$

$\therefore \quad I = \int_0^{10} (f(x))^2\, dx$ as for $x \in (0, 10)$

$|f(x)| = -f(x)$

$$\therefore \quad I = \int_0^1 (-10)^2\, dx + \int_1^2 (-9)^2\, dx + \int_2^3 (-8)^2\, dx + \ldots\ldots + \int_9^{10} (-1)^2\, dx$$

[as $f(x) < 0$]

$$\therefore I = (10)^2 + 9^2 + 8^2 + \ldots.. + 1^2 = \frac{10 \times 11 \times 21}{6} = 385$$

**192.** *(12)* $f(x) = \frac{2}{\sqrt{3}}\int_0^{\sqrt{3}} f\left(\frac{\lambda^2 x}{3}\right) d\lambda,\ x > 0$

On putting $\frac{\lambda^2 x}{3} = t,\ \lambda = \frac{\sqrt{3}\sqrt{t}}{\sqrt{x}}$

$$\Rightarrow \quad d\lambda = \frac{\sqrt{3}}{2\sqrt{x}\sqrt{t}} dt$$

$$f(x) = \frac{2}{\sqrt{3}} \times \frac{\sqrt{3}}{2\sqrt{x}}\int_0^x \frac{f(t)}{\sqrt{t}} dt$$

$$\sqrt{x} f(x) = \int_0^x \frac{f(t)}{\sqrt{t}} dt$$

$$\Rightarrow \sqrt{x} \cdot f'(x) + \frac{f(x)}{2\sqrt{x}} = \frac{f(x)}{\sqrt{x}}$$

$$\Rightarrow \quad \sqrt{x} f'(x) = \frac{f(x)}{2\sqrt{x}}$$

$$\Rightarrow \quad \frac{dy}{y} = \frac{dx}{2x} \quad (\because y = f(x))$$

$$\therefore \quad \ln|y| = \frac{1}{2}\ln x + C$$

Now, $f(1) = \sqrt{3}\ \therefore\ \ln\sqrt{3} = C$

$$\therefore \quad \ln|y| = \ln(\sqrt{3x})$$

$\Rightarrow y = \sqrt{3x}$, this curve passes through $(\alpha, 6)$.

$$\therefore \quad 6 = \sqrt{3\alpha}$$

$$\Rightarrow \quad \alpha = 12$$

**193.** *(24)* Given, $n(2n + 1)\int_0^1 (1 - x^n)^{2n} dx = 1177\int_0^1 (1 - x^n)^{2n+1} dx$

Let $I = \int_0^1 (1 - x^n)^{2n+1} dx$

$$I = \int_0^1 (1 - x^n)^{2n+1} \times 1\, dx$$

$$= [x(1 - x^n)^{2n+1}]_0^1 + \int_0^1 n(2n + 1)(1 - x^n)^{2n} x^{n-1} \times x\, dx$$

$$= n(2n + 1)\int_0^1 (1 - x^n)^{2n} \times x^n\, dx$$

$$= n(2n + 1)\int_0^1 (1 - x^n)^{2n} (1 - (1 - x^n))dx$$

$$= n(2n + 1)\left[\int_0^1 (1 - x^n)^{2n} dx - \int_0^1 (1 - x^n)^{2n+1} dx\right]$$

$$= n(2n+1)\int_0^1 (1-x^n)^{2n}\,dx - n(2n+1)\,I$$

$$I(1+n(2n+1)) = n(2n+1)\int_0^1 (1-x^n)^{2n}\,dx$$

$$\Rightarrow (2n^2+n+1)\int_0^1 (1-x)^{2n+1}\,dx = n(2n+1)\int_0^1 (1-x^n)^{2n}\,dx$$

On comparing it with given equation, we get

$$2n^2+n+1 = 1177$$

$$\Rightarrow (2n+49)(n-24) = 0$$

$$\Rightarrow n = 24, -49/2$$

$$\therefore n = 24, \; n \neq -49/2$$

**194.** *(8)* Since, $f(x)+\int_0^x (x-t)\,f'(t)\,dt$

$$= (e^{2x}+e^{-2x})\cos 2x + \frac{2}{a}x \quad \ldots\text{(i)}$$

On putting $x=0$, we get

Here, $f(0) = 2$ ...(ii)

$$\therefore f(x) + x\int_0^x f'(t)\,dt - \int_0^x t\,f'(t)\,dt = (e^{2x}+e^{-2x})\cos 2x + \frac{2}{a}x$$

Now, on differentiating Eq. (i) w.r.t. $x$, we get

$$f'(x) + \int_0^x f'(t)\,dt + xf'(x) - xf'(x) = 2(e^{2x}-e^{-2x})\cos 2x - 2(e^{2x}+e^{-2x})\sin 2x + \frac{2}{a}$$

$$\Rightarrow f'(x) + (f(x) - f(0)) = (e^{2x}-e^{-2x})(2\cos 2x) - (e^{2x}+e^{-2x})(2\sin 2x) + \frac{2}{a}$$

On putting $x=0$, we get

$$f'(0) = 0 + 0 + \frac{2}{a}$$

$$\Rightarrow 4 = 2/a \quad [\because f'(0) = 4, \text{given}]$$

$$\Rightarrow a = 1/2$$

Thus, $(2a+1)^5\,a^2 = 2^5 \cdot \frac{1}{2^2} = 2^3 = 8$

**195.** *(5)* Since,

$$a_n = \int_{-1}^{n}\left(1+\frac{x}{2}+\frac{x^2}{3}+\ldots+\frac{x^{n-1}}{n}\right)dx$$

$$= \left[x + \frac{x^2}{2^2} + \frac{x^3}{3^2} + \ldots + \frac{x^n}{n^2}\right]_{-1}^{n}$$

$$= \left(n+\frac{n^2}{2^2}+\frac{n^3}{3^2}+\ldots+\frac{n^n}{n^2}\right) - \left(-1+\frac{1}{2^2}-\frac{1}{3^2}+\ldots+\frac{(-1)^{n+1}}{n^2}\right)$$

$$a_n = \frac{n+1}{1^2} + \frac{n^2-1}{2^2} + \frac{n^3+1}{3^2} + \ldots + \frac{n^n+(-1)^{n+1}}{n^2}$$

Hence, $a_1 = 2$,

$$a_2 = \frac{2+1}{1} + \frac{2^2-1}{2^2} = \frac{15}{4}$$

$$a_3 = 4 + 2 + \frac{28}{9} = \frac{82}{9} > 2$$

$$a_4 = 5 + \frac{15}{4} + \frac{65}{9} + \frac{255}{16} > 31$$

Since, $a_n \in (2, 30)$, $n \in N$

Thus, the required set is {2, 3}.

Sum of the elements is $2+3=5$

**196.** *(3)* $\dfrac{24}{\pi}\displaystyle\int_0^{\sqrt{2}} \frac{(2-x^2)}{(2+x^2)(\sqrt{4+x^4})}\,dx$

$$= \frac{24}{\pi}\int_0^{\sqrt{2}} \frac{\left(\frac{2}{x^2}-1\right)}{\left(\frac{2}{x}+x\right)\sqrt{\left(\frac{2}{x}+x\right)^2-4}}\,dx$$

Let $\frac{2}{x}+x = t \Rightarrow \left(\frac{-2}{x^2}+1\right)dx = dt$

At $x=0$, $t=\infty$

and at $x=\sqrt{2}$, $t=2\sqrt{2}$

$$I = \frac{-24}{\pi}\int_{\infty}^{2\sqrt{2}} \frac{dt}{t\sqrt{t^2-4}}$$

$$= \frac{-12}{\pi}\int_{\infty}^{2\sqrt{2}} \frac{2t\,dt}{t^2\sqrt{t^2-4}}$$

Let $t^2-4 = z^2 \Rightarrow 2t\,dt = 2z\,dz$

or $t\,dt = z\,dz$

At $t\to\infty$, $z\to\infty$

and at $t = 2\sqrt{2} \Rightarrow z = 2$

$$I = \frac{-12}{\pi}\int_{\infty}^{2} \frac{2z\,dz}{z(z^2+4)}$$

$$I = \frac{-24}{\pi}\int_{\infty}^{2}\frac{dz}{z^2+4} = \frac{-24}{\pi}\frac{1}{2}\left(\tan^{-1}\frac{z}{2}\right)_{\infty}^{2}$$

$$= \frac{-24}{\pi}\left(\frac{\pi}{8}-\frac{\pi}{4}\right) = 3$$

**197.** *(6)* We will use King's rule,

$$\int_0^a f(x)\,dx = \int_0^a f(a-x)\,dx$$

$$I = \frac{48}{\pi^4}\int_0^{\pi}\left(\frac{3\pi x^2}{2}-x^3\right)\frac{\sin x}{1+\cos^2 x}\,dx \quad \ldots\text{(i)}$$

$$I = \frac{48}{\pi^4}\int_0^{\pi}\left\{\frac{3\pi}{2}(\pi-x)^2-(\pi-x)^3\right\}\frac{\sin x}{1+\cos^2 x}\,dx \quad \ldots\text{(ii)}$$

On adding Eqs. (i) and (ii), we get

$$2I = \frac{48}{\pi^4}\int_0^{\pi}\frac{\sin x}{1+\cos^2 x}\left[\frac{3\pi x^2}{2}-x^3+\frac{3\pi^3}{2}+\frac{3\pi x^2}{2} - 3\pi^2 x - (\pi^3 - x^3 - 3\pi^2 x + 3\pi x^2)\right]dx$$

$$2I = \frac{48}{\pi^4}\int_0^{\pi}\frac{\sin x}{1+\cos^2 x}\times\frac{\pi^3}{2}\,dx$$

$$\Rightarrow 2I = \frac{24}{\pi}\int_0^{\pi}\frac{\sin x}{1+\cos^2 x}\,dx$$

$$\Rightarrow I = \frac{12}{\pi}\int_0^{\pi}\frac{\sin x}{1+\cos^2 x}\,dx$$

$$\Rightarrow I = -\frac{12}{\pi}\left[\tan^{-1}(\cos x)\right]_0^{\pi}$$

$$\Rightarrow \; = -\frac{12}{\pi}\left[-\frac{\pi}{4}-\frac{\pi}{4}\right]$$

$$\therefore I = \frac{12}{2} = 6$$

**198.** *(21)*

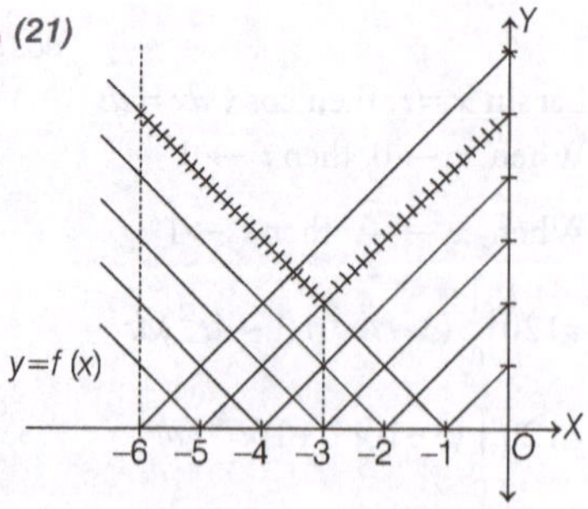


Clearly, $\int_{-6}^{0} f(x)\,dx$

$$= \int_{-6}^{-3} -(x+1)\,dx + \int_{-3}^{0}(x+5)\,dx$$

$$= \left[-\frac{(x+1)^2}{2}\right]_{-6}^{-3} + \left[\frac{(x+5)^2}{2}\right]_{-3}^{0}$$

$$= -\frac{1}{2}[4-25] + \frac{1}{2}[25-4]$$

$$= \frac{1}{2} \times 2[25-4] = 21$$

**199.** *(6)* $12\int_3^b \frac{1}{(x^2-1)(x^2-4)}\, dx$

$$= \log_e\left(\frac{49}{40}\right)$$

$$\Rightarrow \frac{12}{3}\int_3^b \frac{(x^2-1)-(x^2-4)}{(x^2-1)(x^2-4)}\, dx$$

$$= \log_e\left(\frac{49}{40}\right)$$

$$\Rightarrow 4\int_3^b \left(\frac{1}{x^2-4} - \frac{1}{x^2-1}\right) dx$$

$$= \log_e\left(\frac{49}{40}\right)$$

$$\Rightarrow 4\left[\frac{1}{4}\log\left|\frac{x-2}{x+2}\right| - \frac{1}{2}\left|\frac{x-1}{x+1}\right|\right]_3^b$$

$$= \log_e\left(\frac{49}{40}\right)$$

$$\Rightarrow \left[\log\left|\frac{x-2}{x+2}\right| - \log\left|\frac{x-1}{x+1}\right|^2\right]_3^b$$

$$= \log_e\left(\frac{49}{40}\right)$$

$$\Rightarrow \left[\log_e\left|\left(\frac{x-2}{x+2}\right) \times \left(\frac{x+1}{x-1}\right)^2\right|\right]_3^b$$

$$= \log_e\left(\frac{49}{40}\right)$$

$$\Rightarrow \log_e\left[\frac{(b-2)(b+1)^2}{(b+2)(b-1)^2}\right] - \log_e\left[\frac{4}{5}\right] = \log_e\left(\frac{49}{40}\right)$$

$$\Rightarrow \frac{5}{4}\left[\frac{(b-2)(b+1)^2}{(b+2)(b-1)^2}\right] = \frac{49}{40}$$

$$\Rightarrow \frac{(b-2)(b^2+2b+1)}{(b+2)(b^2-2b+1)} = \frac{49}{50}$$

$$\Rightarrow \frac{b^3+2b^2+b-2b^2-4b-2}{b^3-2b^2+b+2b^2-4b+2} = \frac{49}{50}$$

$$\Rightarrow 50b^3 - 150b - 100 = 49b^3 - 147b + 98$$

$$\Rightarrow b^3 - 3b - 198 = 0$$

$$\therefore \quad b = 6$$

**200.** *(1)* Given,

$$f(\theta) = \sin\theta + \int_{-\pi/2}^{\pi/2} (\sin\theta + t\cos\theta) f(t)\, dt$$

$$= \sin\theta\left(1 + \int_{-\pi/2}^{\pi/2} f(t)\, dt\right) + \cos\theta\left(\int_{-\pi/2}^{\pi/2} t f(t)\, dt\right)$$

Clearly, $f(\theta) = a\sin\theta + b\cos\theta$, where

$$a = 1 + \int_{-\pi/2}^{\pi/2} f(t)\, dt$$

$$= 1 + \int_{-\pi/2}^{\pi/2} (a\sin t + b\cos t)\, dt$$

$$= 1 - a\cos t\Big|_{-\pi/2}^{\pi/2} + b\sin t\Big|_{-\pi/2}^{\pi/2}$$

$$\therefore \quad a = 1 + 2b \qquad \ldots \text{(i)}$$

$$\text{and } b = \int_{-\pi/2}^{\pi/2} t f(t)\, dt$$

$$= \int_{-\pi/2}^{\pi/2} (at\sin t + bt\cos t)\, dt$$

$$b = 2a \qquad \ldots \text{(ii)}$$

Now, solving Eqs. (i) and (ii), we get

$$a = -\frac{1}{3} \text{ and } b = -\frac{2}{3}$$

$$\therefore \quad f(\theta) = -\frac{1}{3}(\sin\theta + 2\cos\theta)$$

$$\Rightarrow \left|\int_0^{\pi/2} f(\theta)\, d\theta\right|$$

$$= \frac{1}{3}\left|\int_0^{\pi/2} (\sin\theta + 2\cos\theta)\, d\theta\right|$$

$$= \frac{1}{3}(1 + 2\times 1) = 1$$

**201.** *(34)* Given, $\max\limits_{0\le x\le 2}\left\{\frac{9-x^2}{5-x}\right\} = \alpha$

and $\min\limits_{0\le x\le 2}\left\{\frac{9-x^2}{5-x}\right\} = \beta$

Let $f(x) = \frac{9-x^2}{5-x}$

$$\Rightarrow f'(x) = \frac{(x-1)(x-9)}{(x-5)^2}$$

Clearly, if $0 \le x < 1$, then $f'(x) > 0$ but $1 < x \le 2$, then $f'(x) < 0$. Thus, maximum value occurs at $x = 1$ in [0, 2].

Thus, $\alpha = f(1) = 2$

Similarly, $f(x)$ gets minimum value at $x = 0$ or $x = 2$

Thus, $\beta = \min\{f(0), f(2)\} = \left\{\frac{9}{5}, \frac{5}{3}\right\}$

$$= 5/3$$

Thus, $\alpha = 2$ and $\beta = 5/3$

$$\text{Now, } \int_{\beta-8/3}^{2\alpha-1} \max\left\{\frac{9-x^2}{5-x}, x\right\} dx$$

$$= \int_{-1}^{3} \max\left\{\frac{x^2-9}{x-5}, x\right\} dx$$

$$= \int_{-1}^{9/5} \frac{x^2-9}{x-5}\, dx + \int_{9/5}^{3} x\, dx$$

$$= \int_{-1}^{9/5}\left(x + 5 + \frac{16}{x-5}\right) dx + \left(\frac{x^2}{2}\right)_{9/5}^{3}$$

$$= \frac{28}{25} + 14 + 16\, ln_e\left(\frac{8}{15}\right) + \frac{72}{25}$$

$$= 18 + 16\log_e\left(\frac{8}{15}\right)$$

$$\therefore \quad \alpha_1 = 18 \text{ and } \alpha_2 = 16$$

$$\therefore \quad \alpha_1 + \alpha_2 = 18 + 16 = 34$$

**202.** *(5)* $8\int_{-1/2}^{1} ([2x] + |x|\, dx$

$$\because \ -\frac{1}{2} \le x \le 0 \Rightarrow [2x] = -1$$

$$0 \le x < \frac{1}{2} \Rightarrow [2x] = 0$$

$$\frac{1}{2} \le x < 1 \Rightarrow [2x] = 1$$

$$I = \int_{-1/2}^{0} (-1-x)\, dx + \int_{0}^{1/2} (0+x) dx + \int_{1/2}^{1} (1+x)\, dx$$

$$= \left[-x - \frac{x^2}{2}\right]_{-1/2}^{0} + \left[\frac{x^2}{2}\right]_0^{1/2} + \left[x + \frac{x^2}{2}\right]_{1/2}^{1}$$

$$= -\left(\frac{1}{2} - \frac{1}{8}\right) + \left(\frac{1}{8}\right) + \left(1 + \frac{1}{2}\right) - \left(\frac{1}{2} + \frac{1}{8}\right)$$

$$= \frac{5}{8} \Rightarrow \therefore 8I = 8\cdot\frac{5}{8} = 5$$

**203.** *(4)*

**key idea** Use property of definite integral $\int_0^a f(x)dx = \int_0^a f(a-x)dx$

$$x\,\phi(x) = \int_5^x 3t^2 - 2\phi'(t)\, dt$$

$\Rightarrow x\,\phi(x) = [t^3 - 2\phi(t)]_5^x$

$\Rightarrow x\,\phi(x) = (x^3 - 125) - 2[\phi(x) - \phi(5)]$

Now, $\phi(0) = 4$

$\Rightarrow \quad 0 = -125 - 2[4 - \phi(5)]$

$\Rightarrow \quad \phi(5) = 133/2$

For $\phi(2)$,

$\Rightarrow 2\phi(2) = (8 - 125) - 2\left[\phi(2) - \dfrac{133}{2}\right]$

$\Rightarrow 4\phi(2) = 16 \Rightarrow \phi(2) = 4$

**204.** *(5)* Let $I = \int_0^{\pi} (\sin^3 x)\, e^{-\sin^2 x}\, dx$

$= \int_0^{\pi} (1 - \cos^2 x) \sin x\, e^{-(1-\cos^2 x)} dx$

If $\quad f(x) = f(2a - x)$, then

$\int_0^{2a} f(x)\, dx = 2\int_0^a f(x)\, dx$

$I = 2\int_0^{\pi/2} (1 - \cos^2 x)\sin x\, e^{-(1-\cos^2 x)}\, dx$

Let $\cos^2 x = t \Rightarrow -2\cos x \sin x\, dx = dt$

$\Rightarrow \sin x\, dx = -\dfrac{dt}{2\sqrt{t}}$

$= -2\int_0^1 (1-t)e^{t-1} \dfrac{dt}{-2\sqrt{t}}$

$= \dfrac{1}{e}\int_0^1 \dfrac{(1-t)e^t}{\sqrt{t}}\, dt$

$= \dfrac{1}{e}\left\{\int_0^1 \dfrac{1}{\sqrt{t}} e^t\, dt - \int_0^1 \sqrt{t}\, e^t\, dt\right\}$

$= \dfrac{1}{e}\left\{\begin{matrix}(2\sqrt{t}\, e^t)_0^1 - 2\int_0^1 \sqrt{t}\, e^t\, dt \\ -\int_0^1 \sqrt{t}\, e^t\, dt\end{matrix}\right\}$

$= \dfrac{1}{e}\left\{2e - 3\int_0^1 \sqrt{t}\, e^t\, dt\right\}$

$= 2 - \dfrac{3}{e}\int_0^1 \sqrt{t}\, e^t\, dt$

On comparing with $\alpha - \dfrac{\beta}{e}\int_0^1 \sqrt{t}\, e^t\, dt$

$\Rightarrow \alpha = 2$ and $\beta = 3$

$\therefore \alpha + \beta = 2 + 3 = 5$

**205.** *(16)* Given, $f(x) = e^{-x}\int_3^x [3t^2 + 2t + 4f'(t)]\, dt$

$f(x) = e^{-x}[t^3 + t^2 + 4f(x)]_3^x$

$f(x) = e^{-x}[(x^3 - 27) + (x^2 - 9) + 4[f(x) - f(3)]]$

Now, $f(3) = e^{-3}\int_3^3 3t^2 + 2t + 4f'(t)\, dt = 0$

$\Rightarrow f(x) = e^{-x}(x^3 + x^2 + 4f(x) - 36)$

$\Rightarrow e^x f(x) = x^3 + x^2 + 4f(x) - 36 \quad \ldots(\text{i})$

On putting $x = 4$, we get

$e^4 f(4) = 44 + 4f(4)$

$\Rightarrow \quad f(4) = \left(\dfrac{44}{e^4 - 4}\right)$

On differentiating Eq. (i) w.r.t. $x$, we get

$e^x[f(x) + f'(x)] = 3x^2 + 2x + 4f'(x)$

$f'(x)(e^x - 4) = 3x^2 + 2x - e^x f(x)$

$\Rightarrow f'(4)(e^4 - 4) = 48 + 8 - e^4\left(\dfrac{44}{e^4 - 4}\right)$

$\Rightarrow f'(4)(e^4 - 4) = \dfrac{56(e^4 - 4) - 44e^4}{e^4 - 4}$

$f'(4) = \dfrac{12e^4 - 224}{(e^4 - 4)^2}$

According to the question, by comparing

$\alpha = 12, \beta = 4$

So, $\quad \alpha + \beta = 16$

**206.** *(1)* $\log_4[\log_5\{\log_3(18x - x^2 - 77)\}]$

$\Rightarrow \log_5[\log_3(18x - x^2 - 77)] > 0$

$\Rightarrow \log_3(18x - x^2 - 77) > 1$

$\Rightarrow \quad (18x - x^2 - 77) > 3$

$\Rightarrow \quad x^2 - 18x + 80 < 0$

$\Rightarrow \quad (x - 8)(x - 10) < 0$

$\Rightarrow \quad 8 < x < 10 \Leftrightarrow a < x < b$

$I = \int_a^b \dfrac{\sin^3 x}{\sin^3 x + \sin^3(a + b - x)}\, dx \ldots(\text{i})$

Using king's rule,

$\int_a^b f(x)dx = \int_a^b f(a + b - x)dx$

$I = \int_a^b \dfrac{\sin^3(a + b - x)}{\sin^3(a + b - x)\sin^3 x}\, dx \quad \ldots(\text{ii})$

On adding Eqs. (i) and (ii), we get

$2I = \int_a^b \dfrac{\sin^3 x}{\sin^3 x + \sin^3(a + b - x)}\, dx + \int_a^b \dfrac{\sin^3(a + b - x)}{\sin^3 x + \sin^3(a + b - x)}\, dx$

$\Rightarrow 2I = \int_a^b 1\, dx \Rightarrow 2I = [x]_a^b$

$\Rightarrow 2I = [b - a] \Rightarrow I = \left(\dfrac{b - a}{2}\right)$

As $\; b = 10 \;$ and $\; a = 8$

So, $\; I = \dfrac{10 - 8}{2} = 1$

**207.** *(8)* Let $\quad P'(x) = a(x - 1)(x + 1)$

$\Rightarrow \quad P'(x) = a(x^2 - 1)$

$\therefore \quad P(x) = a\int (x^2 - 1)\, dx$

$\Rightarrow \quad P(x) = a\left(\dfrac{x^3}{3} - x\right) + C$

According to the question, $P(-3) = 0$

$a\left(-\dfrac{27}{3} + 3\right) + C = 0$

$\Rightarrow \quad -6a + C = 0 \quad \ldots(\text{i})$

Now, $\int_{-1}^{1}\left(a\left(\dfrac{x^3}{3} - x\right) + C\right) dx = 18$ [Given]

$\Rightarrow \quad 2C = 18 \Rightarrow C = 9 \quad \ldots(\text{ii})$

From Eqs. (i) and (ii), we get

$-6a + 9 = 0 \Rightarrow a = 3/2$

$\therefore \; P(x) = \dfrac{3}{2}\left(\dfrac{x^3}{3} - x\right) + 9$

$\therefore$Sum of the all coefficients

$= \dfrac{1}{2} - \dfrac{3}{2} + 9 = 8$

**208.** *(512)* Given, $f(x^2) + g(4 - x) = 4x^3$

and $\quad g(4 - x) + g(x) = 0$

Let $\; I = \int_{-4}^{4} f(x^2)\, dx$

$= 2\int_0^4 f(x^2)dx \quad$ [Even function]

$\Rightarrow \; I = 2\cdot\int_0^4 [4x^3 - g(4 - x)]\, dx$

$= 8\int_0^4 x^3\, dx - 2\int_0^4 g(4 - x)\, dx$

$= 8\left(\dfrac{x^4}{4}\right)_0^4 - 2I_1$

$= 2(4^4 - 0^4) - 2I_1 = 2^9 - 2I_1$

Where, $\; I_1 = \int_0^4 g(4 - x)\, dx$

Now, $\quad I_1 = \int_0^4 g(4 - x)\, dx \quad \ldots(\text{i})$

$\Rightarrow \quad I_1 = \int_0^4 g[4 - (0 + 4 - x]\, dx$

$\left[\because \int_a^b f(x)\, dx = \int_a^b f(a + b - x)\, dx\right]$

$\Rightarrow \quad I_1 = \int_0^4 g(x)\,dx$ ...(ii)

On adding Eqs. (i) and (ii), we get

$2I_1 = \int_0^4 [g(x) + g(4-x)]\,dx \Rightarrow 2I_1 = 0$

$\Rightarrow I_1 = 0$ ($\because g(x) + g(4-x) = 0$, given)

$\therefore \quad I = 2^9 - 2I_1 \Rightarrow I = 2^9 = 512$

**209.** *(1)* Given, $I_n = \int_1^e x^{19} (\log|x|)^n\,dx$

$\Rightarrow \quad I_n = \left[\frac{x^{20}}{20}(\ln|x|^n)\right]_1^e - \int_1^e n \cdot \frac{(\ln|x|)^{n-1}}{x} \cdot \frac{x^{20}}{20}\,dx$

(using integration by parts)

$\Rightarrow I_n = \frac{e^{20}}{20} - \frac{n}{20}\int_1^e (\ln|x|)^{n-1} \cdot x^{19}\,dx$

$\Rightarrow \quad I_n = \frac{e^{20}}{20} - \frac{n}{20} \cdot I_{n-1}$

$\Rightarrow \quad 20I_n + nI_{n-1} = e^{20}$

Put $n = 10$ and $n = 9$, we get

$20\,I_{10} + 10\,I_9 = e^{20}$ ... (i)

and $\quad 20\,I_9 + 9\,I_8 = e^{20}$ ... (ii)

From Eqs. (i) and (ii),

$20\,I_{10} - 10\,I_9 - 9\,I_8 = 0$

$\Rightarrow 20I_{10} = 10I_9 + 9I_8$

On comparing with

$20(I_{10}) = \alpha I_9 + \beta I_8$, we get

$\alpha = 10, \beta = 9 \quad \therefore \quad \alpha - \beta = 1$

**210.** *(406)* Given,

$y(x) = \int_0^x (2t^2 - 15t + 10)\,dt$

$\Rightarrow y'(x) = 2x^2 - 15x + 10$

Equation of normal $= x + 3y = -5$

$\frac{-1}{[y'(x)]_{a,b}} = \frac{-1}{3}$ or $[y'(x)]_{a,b} = 3$

$2a^2 - 15a + 10 = 3$

$\Rightarrow \quad 2a^2 - 15a + 7 = 0$

$\Rightarrow (2a-1)(a-7) = 0 \;\therefore a = 1/2$ or $7$

As, $a > 1$, so $a = 7$

Now, $(7, b)$ lies on $y(x)$.

$\therefore \quad b = \int_0^a (2t^2 - 15t + 10)\,dt$

$\Rightarrow \quad b = \frac{2}{3}a^3 - \frac{15}{2}a^2 + 10a$

$\Rightarrow \quad b = \frac{2}{3}(7)^3 - \frac{15}{2}(7)^2 + 10(7)$

$\Rightarrow \quad b = \frac{-413}{6}$

So, $\quad a + 6b = 7 - 6\left(\frac{413}{6}\right) = -406$

$\therefore \quad |a + 6b| = 406$

**211.** *(16)* Given, $f(x) + f(x+1) = 2$

$I_1 = \int_0^8 f(x)$ and $I_2 = \int_{-1}^3 f(x)\,dx$

Let $\quad f(0) = a$

$f(0) + f(1) = 2 \Rightarrow f(1) = 2 - a$

$f(1) + f(2) = 2$

$f(2) = a$ and so on

So, $\quad f(0) = f(2) = f(4) \ldots = a$

$f(1) = f(3) = f(5) \ldots = 2 - a$

Clearly, $f(x)$ is periodic with its period 2 units.

So, $I_1 = 2\int_0^4 f(x)\,dx \Rightarrow I_1 = 4\int_0^2 f(x)\,dx$

Now, $\quad I_2 = \int_{-1}^3 f(x)\,dx$

Put $x = x + 1$

$I_2 = \int_0^4 f(x+1)\,dx = \int_0^4 [2 - f(x)]\,dx$

$\Rightarrow I_2 = 8 - 2\int_0^2 f(x)\,dx$

$\Rightarrow 2I_2 = 16 - 4\int_0^2 f(x)\,dx$

$\Rightarrow 2I_2 = 16 - I_1$

$\therefore \quad I_1 + 2I_2 = 16$

**212.** *(1)* Given, $I_{mn} = \int_0^1 x^{m-1}(1-x)^{n-1}\,dx$

Using substitution method, $x = \frac{1}{t+1}$

Then, $dx = \frac{-1}{(t+1)^2}\,dt$

$I_{mn} = \int_\infty^0 (-1)\frac{1}{(t+1)^{m-1}} \cdot \frac{t^{n-1}}{(t+1)^{n-1}} \cdot \frac{1}{(t+1)^2}\,dt$

$= -\int_\infty^0 \frac{t^{n-1}}{(t+1)^{m+n}}\,dt$

$I_{mn} = \int_0^\infty \frac{t^{n-1}}{(t+1)^{m-n}}\,dt$ ...(i)

Similarly, $I_{mn} = \int_0^1 x^{n-1}(1-x)^{m-1}\,dx$

$\Rightarrow I_{mn} = \int_0^\infty \frac{t^{m-1}}{(t+1)^{m+n}}$ ...(ii)

From Eqs. (i) and (ii), we get

$2I_{mn} = \int_0^\infty \frac{t^{n-1} + t^{m-1}}{(t+1)^{m+n}}\,dt$

$2I_{mn} = \int_0^1 \frac{t^{n-1} + t^{m-1}}{(t+1)^{m+n}}\,dt + \int_1^\infty \frac{t^{n-1} + t^{m-1}}{(t+1)^{m+n}}\,dt$

Let $I_1 = \int_1^\infty \frac{t^{n-1} + t^{m-1}}{(t+1)^{m+n}}\,dt$

Let $t = \frac{1}{z}$, then $dt = \frac{-1}{z^2}\,dz$

$I_1 = \int_1^0 (-1)\frac{\left(\frac{1}{z}\right)^{n-1} + \left(\frac{1}{z}\right)^{m-1}}{\left(\frac{1}{z}+1\right)^{m+n}} \cdot \frac{1}{z^2}\,dz$

$= -\int_1^0 \frac{z^{n-1} + z^{m-1}}{(z+1)^{m+n}}\,dz$

Put $I_1$ in $2I_{mn}$

$2I_{mn} = \int_0^1 \frac{t^{n-1} + t^{m-1}}{(t+1)^{m+n}}\,dt - \int_1^0 \frac{z^{n-1} + z^{m-1}}{(z+1)^{m+n}}\,dz$

$= 2\int_0^1 \frac{t^{m-1} + t^{n-1}}{(t+1)^{m+n}}\,dt$

$\Rightarrow \alpha = 1$

**213.** *(2)* Let $I = \int_0^\pi |\sin 2x|\,dx$

$= 2\int_0^{\pi/2} |\sin 2x|\,dx$

[$\because \sin 2x$ is periodic function]

$= 2\int_0^{\pi/2} \sin 2x\, dx$ [sin 2x is positive in range $(0, \pi/2)$]

$= 2\left[\dfrac{-\cos 2x}{2}\right]_0^{\pi/2}$

$= -[\cos\pi - \cos 0]$

$= -(-1-1) = 2$

$I = 2$

**214.** *(19)* $\int_{-2}^{2} |3x^2 - 3x - 6|\, dx = I$ (say)

$I = 3\int_{-2}^{2} |x^2 - x - 2|\, dx$

$= 3\left[\int_{-2}^{-1}(x^2 - x - 2)dx + \int_{-1}^{2}(-x^2 + x + 2)dx\right]$

$= 3\left[\left(\dfrac{x^3}{3} - \dfrac{x^2}{2} - 2x\right)_{-2}^{-1} - \left(\dfrac{x^3}{3} - \dfrac{x^2}{2} - 2x\right)_{-1}^{2}\right] = 19$

**215.** *(3)* Given, $\int_{-a}^{a} (|x| + |x-2|)\, dx = 22$

$\Rightarrow \int_{-a}^{0}(-2x+2)\, dx + \int_0^2 (x + 2 - x)\, dx + \int_2^a (2x-2)\, dx = 22$

$\Rightarrow (x^2 - 2x)|_0^{-a} + (2x)|_0^2 + (x^2 - 2x)\,|_2^a = 22$

$\Rightarrow a^2 + 2a + 4 + a^2 - 2a - (4-4) = 22$

$\Rightarrow 2a^2 = 18$

$\Rightarrow a^2 = 9$

$\Rightarrow a = 3$

$\therefore \int_a^{-a} x + [x]\, dx = \int_3^{-3} (x + [x])\, dx$

$= -\int_{-3}^{3} (x + [x])\, dx$

$= -[-3 - 2 - 1 + 1 + 2]$

$= -(-3) = 3$

**216.** *(21)* Since, $\int_0^n \{x\}dx = n\int_0^1 x dx$ (as $n \in N, n > 1$)

$= n[x^2/2]_0^1 = n/2$

and $\int_0^n [x]dx = \int_0^1 0\, dx + \int_1^2 1 dx + \int_2^3 2dx + \ldots + \int_{n-1}^{n} (n-1)dx$

$= 0 + 1(2-1) + 2(3-2) + \ldots + (n-1)(n-(n-1))$

$= 1 + 2 + 3 + \ldots + (n-1) = \dfrac{n(n-1)}{2}$

$\because \dfrac{n}{2}, \dfrac{n(n-1)}{2}, 10n(n-1)$ are in GP

$\Rightarrow \dfrac{1}{2}, \dfrac{n-1}{2}, 10(n-1)$ are also in GP [as $n > 1$]

$\Rightarrow \left(\dfrac{n-1}{2}\right)^2 = 5(n-1)$

$\Rightarrow n - 1 = 20$ [As, $n > 1$]

$\Rightarrow n = 21$

# C-20

JEE Main Chapterwise Topicwise Mathematics

In JEE Main, there is usually 0-1 question per paper from this chapter. In past 3 years, 37 conceptual and 25 numericals questions of moderate to difficulty level were asked from this chapter.

**For detailed analysis scan QR code**

# APPLICATIONS OF INTEGRALS

## Topic 1 Area Under Simple Curves in Standard Forms

### Level 1

**01** The area (in sq. units) of the region described by $\{(x, y): y^2 \leq 2x \text{ and } y \geq 4x - 1\}$ is *[2024, 4 April Shift-II]*

(a) 11/12 (b) 9/32 (c) 8/9 (d) 11/32

**02** The area bounded by the curves $y = |x - 1| + |x - 2|$ and $y = 3$ is equal to *[2023, 6 April Shift-I]*

(a) 4 (b) 5 (c) 6 (d) 3

**03** The odd natural number $a$, such that the area of the region bounded by $y = 1$, $y = 3$, $x = 0$, $x = y^a$ is $364/3$, is equal to *[2022, 26 July Shift-I]*

(a) 3 (b) 5 (c) 7 (d) 9

**04** Let the locus of the centre $(\alpha, \beta)$, $\beta > 0$, of the circle which touches the circle $x^2 + (y-1)^2 = 1$ externally and also touches the $X$-axis be $L$. Then, the area bounded by $L$ and the line $y = 4$ is *[2022, 25 July Shift-I]*

(a) $\frac{32\sqrt{2}}{3}$ (b) $\frac{40\sqrt{2}}{3}$ (c) $\frac{64}{3}$ (d) $\frac{32}{3}$

**05** The area of the bounded region enclosed by the curve $y = 3 - \left|x - \frac{1}{2}\right| - |x + 1|$ and the $X$-axis is *[2022, 28 June Shift-II]*

(a) $\frac{9}{4}$ (b) $\frac{45}{16}$ (c) $\frac{27}{8}$ (d) $\frac{63}{16}$

**06** Let $f:[-2, 3] \to [0, \infty)$ be a continuous function such that $f(1-x) = f(x)$ for all $x \in [-2, 3]$.
If $R_1$ is the numerical value of the area of the region bounded by $y = f(x)$, $x = -2$, $x = 3$ and the axis of $X$ and $R_2 = \int_{-2}^{3} x\, f(x)\, dx$, then *[JEE Main 2013]*

(a) $3R_1 = 2R_2$ (b) $2R_1 = 3R_2$
(c) $R_1 = R_2$ (d) $R_1 = 2R_2$

**07** The area under the curve $y = |\cos x - \sin x|$, $0 \leq x \leq \frac{\pi}{2}$ and above $X$-axis is *[JEE Main 2013]*

(a) $2\sqrt{2}$ (b) $2\sqrt{2} - 2$
(c) $2\sqrt{2} + 2$ (d) 0

**08** The area of the region bounded by the curve $y = x^3$, and the lines, $y = 8$ and $x = 0$, is *[AIEEE 2012]*

(a) 8 (b) 12
(c) 10 (d) 16

**09** If $f(x)$ is a non-negative continuous function such that the area bounded by the curve $y = f(x)$, $X$-axis and the ordinates $x = \frac{\pi}{4}$ and $x = \beta > \frac{\pi}{4}$ is $\left(\beta \sin \beta + \frac{\pi}{4} \cos \beta + \sqrt{2}\beta\right)$. Then, $f\left(\frac{\pi}{2}\right)$ is equal to *[AIEEE 2005]*

(a) $\left(1 - \frac{\pi}{4} + \sqrt{2}\right)$ (b) $\left(1 - \frac{\pi}{4} - \sqrt{2}\right)$
(c) $\left(\frac{\pi}{4} - \sqrt{2} + 1\right)$ (d) $\left(\frac{\pi}{4} + \sqrt{2} - 1\right)$

### Level 2

**10** Let $f: R \to R$ be a twice differentiable function such that $f(x + y) = f(x) f(y)$, $\forall x, y \in R$. If $f'(0) = 4a$ and $f$ satisfies $f''(x) - 3af'(x) - f(x) = 0$, $a > 0$, then the area of the region $R = \{|(x, y)|, 0 \leq y \leq f(ax), 0 \leq x \leq 2\}$ is *[2025, 22 Jan Shift-I]*

(a) $e^2 - 1$ (b) $e^4 + 1$ (c) $e^4 - 1$ (d) $e^2 + 1$

**11** The area (in square units) of the region enclosed by the ellipse $x^2 + 3y^2 = 18$ in the first quadrant below the line $y = x$ is *[2024, 9 April Shift-II]*

(a) $\sqrt{3}\pi$ (b) $\sqrt{3}\pi - \frac{3}{4}$
(c) $\sqrt{3}\pi + \frac{3}{4}$ (d) $\sqrt{3}\pi + 1$

**12** The area of the region enclosed by the curve $y = x^3$ and its tangent at the point $(-1, -1)$ is **[2023, 12 April Shift-I]**

(a) $\frac{23}{4}$ (b) $\frac{19}{4}$ (c) $\frac{31}{4}$ (d) $\frac{27}{4}$

**Imp.** Area between Line and Parabola

**13** The area bounded by the curves $y = |x^2 - 1|$ and $y = 1$ is **[2022, 27 July Shift-II]**

(a) $\frac{2}{3}(\sqrt{2}+1)$ (b) $\frac{4}{3}(\sqrt{2}-1)$

(c) $2(\sqrt{2}-1)$ (d) $\frac{8}{3}(\sqrt{2}-1)$

**14** The area of the region $A = \{(x, y); 0 \le y \le x|x| + 1$ and $-1 \le x \le 1\}$ in sq. units, is **[2019, 9 Jan Shift-II]**

(a) 2 (b) $\frac{4}{3}$ (c) $\frac{1}{3}$ (d) $\frac{2}{3}$

**15** The area of the region bounded by the curves $y = |x - 2|$, $x = 1$, $x = 3$ and the $X$-axis is **[AIEEE 2004]**

(a) 1 sq unit (b) 2 sq units

(c) 3 sq units (d) 4 sq units

## Level 3

**16** The area bounded by the curve $y = |x^2 - 9|$ and the line $y = 3$ is **[2022, 26 June Shift-I]**

(a) $4(2\sqrt{3} + \sqrt{6} - 4)$ (b) $4(4\sqrt{3} + \sqrt{6} - 4)$

(c) $8(4\sqrt{3} + 3\sqrt{6} - 9)$ (d) $8(4\sqrt{3} + \sqrt{6} - 9)$

**17** The area (in sq units) of the region $A = \{(x, y) \in R \times R | 0 \le x \le 3, 0 \le y \le 4, y \le x^2 + 3x\}$ is **[2019, 8 April Shift-I]**

(a) $\frac{53}{6}$ (b) 8 (c) $\frac{59}{6}$ (d) $\frac{26}{3}$

**18** Let $g(x) = \cos x^2$, $f(x) = \sqrt{x}$ and $\alpha, \beta$ $(\alpha < \beta)$ be the roots of the quadratic equation $18x^2 - 9\pi x + \pi^2 = 0$. Then, the area (in sq units) bounded by the curve $y = (gof)(x)$ and the lines $x = \alpha$, $x = \beta$ and $y = 0$, is **[JEE Main 2018]**

(a) $\frac{1}{2}(\sqrt{3}-1)$ (b) $\frac{1}{2}(\sqrt{3}+1)$ (c) $\frac{1}{2}(\sqrt{3}-\sqrt{2})$ (d) $\frac{1}{2}(\sqrt{2}-1)$

### Numerical Value Type Questions

**19** Let the area of the region enclosed by the curve $y = \min\{\sin x, \cos x\}$ and the $X$-axis between $x = -\pi$ to $x = \pi$ be $A$. Then, $A^2$ is equal to .......... . **[2024, 8 April Shift-I]**

**20** Let $A$ be the area bounded by the curve $y = x|x - 3|$, the $X$-axis and the ordinates $x = -1$ and $x = 2$. Then, $12A$ is equal to ......... **[2023, 1 Feb Shift-I]**

**21** Let $\alpha$ be the area of the larger region bounded by the curve $y^2 = 8x$ and the lines $y = x$ and $x = 2$, which lies in the first quadrant. Then, the value of $3\alpha$ is equal to ........ **[2023, 30 Jan Shift-I]**

**22** Let $a$ and $b$ respectively be the points of local maximum and local minimum of the function $f(x) = 2x^3 - 3x^2 - 12x$. If $A$ is the total area of the region bounded by $y = f(x)$, the $X$-axis and the lines $x = a$ and $x = b$, then $4A$ is equal to ...... . **[2021, 26 Aug Shift-II]**

**23** If the area of the triangle formed by the positive $X$-axis, the normal and the tangent to the circle $(x-2)^2 + (y-3)^2 = 25$ at the point (5, 7) is $A$, then $24A$ is equal to .... . **[2021, 24 Feb Shift-II]**

# Topic 2 Area Bounded between The Curves

## Level 1

**01** The area of the region $\{(x, y): |x - y| \le y \le 4\sqrt{x}\}$ is **[2025, 3 April Shift-II]**

(a) $\frac{1024}{3}$ (b) $\frac{2048}{3}$ (c) 512 (d) $\frac{512}{3}$

**02** If the area of the region bounded by the curves $y = 4 - \frac{x^2}{4}$ and $y = \frac{x-4}{2}$ is equal to $\alpha$, then $6\alpha$ equals **[2025, 7 April Shift-I]**

(a) 250 (b) 220 (c) 240 (d) 210

**03** If the area of the region $\{(x, y): 1 + x^2 \le y \le \min(x + 7, 11 - 3x)\}$ is $A$, then $3A$ is equal to **[2025, 7 April Shift-II]**

(a) 49 (b) 47 (c) 46 (d) 50

**04** Let the area enclosed between the curves $|y| = 1 - x^2$ and $x^2 + y^2 = 1$ be $\alpha$. If $9\alpha = \beta\pi + \gamma$; $\beta, \gamma$ are integers, then the value of $|\beta - \gamma|$ equals **[2025, 29 Jan Shift-II]**

(a) 27 (b) 33 (c) 15 (d) 18

**05** Let the area of the region $\{(x, y): 2y \le x^2 + 3, y + |x| \le 3, y \ge |x - 1|\}$ be A. Then, 6A is equal to **[2025, 29 Jan Shift-I]**

(a) 18 (b) 12 (c) 16 (d) 14

**06** The area (in sq. units) of the region $\{(x, y): 0 \le y \le 2|x| + 1, 0 \le y \le x^2 + 1, |x| \le 3\}$ is **[2025, 28 Jan Shift-I]**

(a) $\frac{80}{3}$ (b) $\frac{64}{3}$ (c) $\frac{17}{3}$ (d) $\frac{32}{3}$

**07** The area of the region enclosed by the curves $y = e^x$, $y = |e^x - 1|$ and $Y$-axis is **[2025, 24 Jan Shift-II]**

(a) $1 + \log_e 2$ (b) $\log_e 2$

(c) $1 - \log_e 2$ (d) $2\log_e 2 - 1$

**08** The area of the region $\{(x, y): x^2 + 4x + 2 \le y \le |x + 2|\}$ is equal to *[2025, 24 Jan Shift-I]*

(a) 7 (b) 5 (c) 20/3 (d) 24/3

**09** The area of the region in the first quadrant inside the circle $x^2 + y^2 = 8$ and outside the parabola $y^2 = 2x$ is equal to *[2024, 8 April Shift-II]*

(a) $\pi - \frac{2}{3}$ (b) $\frac{\pi}{2} - \frac{2}{3}$ (c) $\frac{\pi}{2} - \frac{1}{3}$ (d) $\pi - \frac{1}{3}$

**10** If the area of the region $\left\{(x, y): \frac{a}{x^2} \le y \le \frac{1}{x}, 1 \le x \le 2, 0 < a < 1\right\}$ is $(\log_e 2) - \frac{1}{7}$, then the value of $7a - 3$ is equal to *[2024, 6 April Shift-II]*

(a) 1 (b) 0
(c) 2 (d) −1

**11** The area enclosed by the curves $xy + 4y = 16$ and $x + y = 6$ is equal to *[2024 Ist Feb Shift I]*

(a) $30 - 32 \log_e 2$ (b) $32 - 30 \log_e 2$
(c) $30 - 28 \log_e 2$ (d) $28 - 30 \log_e 2$

**12** The area (in sq units) of the region bounded by the parabola $y^2 = 4(x - 2)$ and the line $y = 2x - 8$, is *[2024, 30 Jan , Shift - II]*

(a) 6 (b) 9
(c) 7 (d) 8

**13** Area of the region $\{(x, y): x^2 + (y - 2)^2 \le 4, x^2 \ge 2y\}$ is *[2023, 11 April Shift-I]*

(a) $\pi - \frac{8}{3}$ (b) $2\pi + \frac{16}{3}$
(c) $2\pi - \frac{16}{3}$ (d) $\pi + \frac{8}{3}$

**14** The area of the region $\{(x, y): x^2 \le y \le 8 - x^2, y \le 7\}$ is *[2023, 8 April Shift-I]*

(a) 21 (b) 20
(c) 18 (d) 24

**15** The area of the region given by $\{(x, y): xy \le 8, 1 \le y \le x^2\}$ is *[2023, 1 Feb Shift-II]*

(a) $16 \log_e 2 + \frac{7}{3}$ (b) $16 \log_e 2 - \frac{14}{3}$
(c) $8 \log_e 2 + \frac{7}{6}$ (d) $8 \log_e 2 - \frac{13}{3}$

**16** The area of the region $A = \left\{(x, y): |\cos x - \sin x| \le y \le \sin x, 0 \le x \le \frac{\pi}{2}\right\}$ is *[2023, 29 Jan Shift-II]*

(a) $\sqrt{5} + 2\sqrt{2} - 4.5$ (b) $\frac{3}{\sqrt{5}} - \frac{3}{\sqrt{2}} + 1$
(c) $1 - \frac{3}{\sqrt{2}} + \frac{4}{\sqrt{5}}$ (d) $\sqrt{5} - 2\sqrt{2} + 1$

**17** The area of the region enclosed by $y \le 4x^2$, $x^2 \le 9y$ and $y \le 4$, is equal to *[2022, 27 July Shift-II]*

(a) 40/3 (b) 56/3 (c) 112/3 (d) 80/3

**18** The area of the smaller region enclosed by the curves $y^2 = 8x + 4$ and $x^2 + y^2 + 4\sqrt{3}x - 4 = 0$ is equal to *[2022, 26 July Shift-I]*

(a) $\frac{1}{3}(2 - 12\sqrt{3} + 8\pi)$ (b) $\frac{1}{3}(2 - 12\sqrt{3} + 6\pi)$
(c) $\frac{1}{3}(4 - 12\sqrt{3} + 8\pi)$ (d) $\frac{1}{3}(4 - 12\sqrt{3} + 6\pi)$

**19** The area enclosed by $y^2 = 8x$ and $y = \sqrt{2}x$ that lies outside the triangle formed by $y = \sqrt{2}x$, $x = 1$, $y = 2\sqrt{2}$, is equal to *[2022, 29 June Shift-I]*

(a) $\frac{16\sqrt{2}}{6}$ (b) $\frac{11\sqrt{2}}{6}$ (c) $\frac{13\sqrt{2}}{6}$ (d) $\frac{5\sqrt{2}}{6}$

**20** The area, enclosed by the curves $y = \sin x + \cos x$ and $y = |\cos x - \sin x|$ and the lines $x = 0$, $x = \frac{\pi}{2}$, is *[2021, 01 Sep Shift-II]*

(a) $2\sqrt{2}(\sqrt{2} - 1)$ (b) $2(\sqrt{2} + 1)$
(c) $4(\sqrt{2} - 1)$ (d) $2\sqrt{2}(\sqrt{2} + 1)$

**21** The area (in sq. units) of the region, given by the set $\{(x, y) \in R \times R | x \ge 0, 2x^2 \le y \le 4 - 2x\}$ is *[2021, 25 July Shift-I]*

(a) 8/3 (b) 17/3 (c) 13/3 (d) 7/3

**22** The area of the region $R = \{(x, y): 5x^2 \le y \le 2x^2 + 9\}$ is *[2021, 24 Feb Shift-II]*

(a) $11\sqrt{3}$ sq units (b) $12\sqrt{3}$ sq units
(c) $9\sqrt{3}$ sq units (d) $6\sqrt{3}$ sq units

**23** The area (in sq. units) of the part of the circle $x^2 + y^2 = 36$, which is outside the parabola $y^2 = 9x$, is *[2021, 24 Feb Shift-I]*

(a) $24\pi + 3\sqrt{3}$ (b) $24\pi - 3\sqrt{3}$
(c) $12\pi + 3\sqrt{3}$ (d) $12\pi - 3\sqrt{3}$

**24** The area (in sq. units) of the region $\{(x, y): 0 \le y \le x^2 + 1, 0 \le y \le x + 1, \frac{1}{2} \le x \le 2\}$ is *[2020, 3 Sep Shift-I]*

(a) $\frac{23}{16}$ (b) $\frac{79}{24}$
(c) $\frac{79}{16}$ (d) $\frac{23}{6}$

**25** Consider region $R = \{(x, y) \in R^2 : x^2 \le y \le 2x\}$. If a line $y = \alpha$ divides the area of region $R$ into two equal parts, then which of the following is true? *[2020, 2 Sep Shift-II]*

(a) $\alpha^3 - 6\alpha^2 + 16 = 0$ (b) $3\alpha^2 - 8\alpha^{3/2} + 8 = 0$
(c) $3\alpha^2 - 8\alpha + 8 = 0$ (d) $\alpha^3 - 6\alpha^{3/2} - 16 = 0$

**26** The area (in sq. units) of the region $\{(x, y) \in R^2 : x^2 \le y \le 3 - 2x\}$, is **[2020, 8 Jan Shift-II]**

(a) 31/3 (b) 32/3 (c) 29/3 (d) 34/3

**27** If the area of the region bounded by the curves, $y = x^2$, $y = \frac{1}{x}$ and the lines $y = 0$ and $x = t\ (t > 1)$ is 1 sq. unit, then $t$ is equal to **[JEE Main 2018]**

(a) 4/3 (b) $e^{2/3}$ (c) 3/2 (d) $e^{3/2}$

**28** The area (in sq units) of the region bounded by the curves $y + 2x^2 = 0$ and $y + 3x^2 = 1$, is equal to **[JEE Main 2015]**

(a) 3/5 (b) 1/3 (c) 4/3 (d) 3/4

**29** The area (in sq units) bounded by the curves $y = \sqrt{x}$, $2y - x + 3 = 0$, $X$-axis and lying in the first quadrant is **[JEE Main 2013]**

(a) 9 (b) 36 (c) 18 (d) 27/4

**30** The area bounded by the curves $y^2 = 4x$ and $x^2 = 4y$ is **[AIEEE 2011]**

(a) 0 (b) 32/3 (c) 16/3 (d) 8/3

**31** The area bounded by the curves $y = \cos x$ and $y = \sin x$ between the ordinates $x = 0$ and $x = \frac{3\pi}{2}$, is **[AIEEE 2009]**

(a) $(4\sqrt{2} - 2)$ sq units (b) $(4\sqrt{2} + 2)$ sq units
(c) $(4\sqrt{2} - 1)$ sq units (d) $(4\sqrt{2} + 1)$ sq units

**32** The area enclosed between the curves $y^2 = x$ and $y = |x|$ is **[AIEEE 2007]**

(a) 2/3 sq unit (b) 1 sq unit
(c) 1/6 sq unit (d) 1/3 sq unit

**33** The parabolas $y^2 = 4x$ and $x^2 = 4y$ divide the square region bounded by the lines $x = 4$, $y = 4$ and the coordinate axes. If $S_1, S_2$ and $S_3$ are respectively the areas of these parts numbered from top to bottom, then $S_1 : S_2 : S_3$ is equal to **[AIEEE 2005]**

(a) 1 : 1 : 1 (b) 2 : 1 : 2 (c) 1 : 2 : 3 (d) 1 : 2 : 1

## Level 2

**34** The area of the region bounded by the curves $x(1 + y^2) = 1$ and $y^2 = 2x$ is **[2025, 28 Jan Shift-II]**

(a) $\frac{\pi}{4} - \frac{1}{3}$ (b) $2\left(\frac{\pi}{2} - \frac{1}{3}\right)$ (c) $\frac{\pi}{2} - \frac{1}{3}$ (d) $\frac{1}{2}\left(\frac{\pi}{2} - \frac{1}{3}\right)$

**35** Consider the region

$R = \{(x, y) : x \le y \le 9 - \frac{11}{3}x^2, x \ge 0\}$.

The area of the largest rectangle of sides parallel to coordinate axes and inscribed in $R$ is **[2025, 24 Jan Shift-I]**

(a) $\frac{625}{111}$ (b) $\frac{567}{121}$ (c) $\frac{821}{123}$ (d) $\frac{730}{119}$

**36** If the area of the region $\{(x, y) : -1 \le x \le 1, 0 \le y \le a + e^{|x|} - e^{-x}, a > 0\}$ is $\frac{e^2 + 8e + 1}{e}$, then the value of $a$ is **[2025, 23 Jan Shift-II]**

(a) 6 (b) 8 (c) 5 (d) 7

**37** The area of the region enclosed by the curves $y = x^2 - 4x + 4$ and $y^2 = 16 - 8x$ is **[2025, 22 Jan Shift-II]**

(a) 5 (b) 8 (c) 4/3 (d) 8/3

**38** The area of the region, inside the circle $(x - 2\sqrt{3})^2 + y^2 = 12$ and outside the parabola $y^2 = 2\sqrt{3}x$ is **[2025, 22 Jan Shift-I]**

(a) $6\pi - 8$ (b) $3\pi + 8$ (c) $6\pi - 16$ (d) $3\pi - 8$

**39** The parabola $y^2 = 4x$ divides the area of the circle $x^2 + y^2 = 5$ in two parts. The area of the smaller part of equal to **[2024, 9 April Shift-I]**

(a) $\frac{1}{3} + 5\sin^{-1}\left(\frac{2}{\sqrt{5}}\right)$ (b) $\frac{2}{3} + 5\sin^{-1}\left(\frac{2}{\sqrt{5}}\right)$

(c) $\frac{2}{3} + \sqrt{5}\sin^{-1}\left(\frac{2}{\sqrt{5}}\right)$ (d) $\frac{1}{3} + \sqrt{5}\sin^{-1}\left(\frac{2}{\sqrt{5}}\right)$

**40** Let the area of the region enclosed by the curves $y = 3x$, $2y = 27 - 3x$ and $y = 3x - x\sqrt{x}$ be $A$. Then, $10A$ is equal to **[2024, 6 April Shift-I]**

(a) 154 (b) 162 (c) 172 (d) 184

**41** The area of the region

$\left\{(x, y) : y^2 \le 4x, x < 4, \frac{xy(x-1)(x-2)}{(x-3)(x-4)} > 0, x \ne 3\right\}$ is **[2024, 31 Jan Shift I]**

(a) 32/2 (b) 8/3 (c) 64/3 (d) 16/3

**42** The area of the region $\{(x, y) : x^2 \le y \le |x^2 - 4|\}$ and $y \ge 1$ is **[2023, 13 April Shift-II]**

(a) $\frac{3}{4}(4\sqrt{2} - 1)$ (b) $\frac{3}{4}(4\sqrt{2} + 1)$

(c) $\frac{4}{3}(4\sqrt{2} - 1)$ (d) $\frac{4}{3}(4\sqrt{2} + 1)$

**43** The area of the region enclosed by the curve $f(x) = \max\{\sin x, \cos x\}$, $-\pi \le x \le \pi$ and the $X$-axis is **[2023, 13 April Shift-I]**

(a) $2\sqrt{2}(\sqrt{2} + 1)$ (b) $4(\sqrt{2})$
(c) 4 (d) $2(\sqrt{2} + 1)$

**44** The area enclosed by the curves $y^2 + 4x = 4$ and $y - 2x = 2$ is **[2023, 24 Jan Shift-I]**

(a) $\frac{25}{3}$ (b) $\frac{23}{3}$

(c) 9 (d) $\frac{22}{3}$

**45** The area enclosed by the curves $y=\log_e(x+e^2)$, $x=\log_e\left(\frac{2}{y}\right)$ and $x=\log_e 2$, above the line $y=1$ is **[2022, 28 July Shift-II]**

(a) $2+e-\log_e 2$ (b) $1+e-\log_e 2$
(c) $e-\log_e 2$ (d) $1+\log_e 2$

**46** The area of the region given by $A=\{(x, y); x^2 \le y \le \min\{x+2, 4-3x\}\}$ is **[2022, 25 July Shift-I]**

(a) $\frac{31}{8}$ (b) $\frac{17}{6}$ (c) $\frac{19}{6}$ (d) $\frac{27}{8}$

**47** The area of the region bounded by $y^2=8x$ and $y^2=16(3-x)$ is equal to **[2022, 26 June Shift-II]**

(a) $\frac{32}{3}$ (b) $\frac{40}{3}$ (c) 19 (d) 16

**48** The area of the region enclosed between the parabolas $y^2=2x-1$ and $y^2=4x-3$ is **[2022, 25 June Shift-II]**

(a) 1/3 (b) 1/6 (c) 2/3 (d) 3/4

**49** Let $A_1$ be the area of the region bounded by the curves $y=\sin x$, $y=\cos x$ and $Y$-axis in the first quadrant. Also, let $A_2$ be the area of the region bounded by the curves $y=\sin x$, $y=\cos x$, $X$-axis and $x=\frac{\pi}{2}$ in the first quadrant. Then, **[2021, 26 Feb Shift-II]**

(a) $A_1 : A_2 = 1:\sqrt{2}$ and $A_1 + A_2 = 1$
(b) $A_1 = A_2$ and $A_1 + A_2 = \sqrt{2}$
(c) $2A_1 = A_2$ and $A_1 + A_2 = 1+\sqrt{2}$
(d) $A_1 : A_2 = 1:2$ and $A_1 + A_2 = 1$

**50** The area (in sq. units) of the region $A=\{(x, y): |x|+|y| \le 1, 2y^2 \ge |x|\}$ is **[2020, 6 Sep Shift-I]**

(a) $\frac{1}{3}$ (b) $\frac{7}{6}$
(c) $\frac{1}{6}$ (d) $\frac{5}{6}$

**51** The area of the region, enclosed by the circle $x^2+y^2=2$ which is not common to the region bounded by the parabola $y^2=x$ and the straight line $y=x$, is **[2020, 7 Jan Shift-I]**

(a) $\frac{1}{3}(12\pi-1)$ (b) $\frac{1}{6}(12\pi-1)$
(c) $\frac{1}{6}(24\pi-1)$ (d) $\frac{1}{3}(6\pi-1)$

**52** If the area (in sq units) bounded by the parabola $y^2=4\lambda x$ and the line $y=\lambda x$, $\lambda>0$, is $1/9$, then $\lambda$ is equal to **[2019, 12 April Shift-II]**

(a) $2\sqrt{6}$ (b) 48
(c) 24 (d) $4\sqrt{3}$

**53** The area (in sq units) of the region $A=\left\{(x, y): \frac{y^2}{2} \le x \le y+4\right\}$ is **[2019, 9 April Shift-II]**

(a) 30 (b) 53/2 (c) 16 (d) 18

**54** The area (in sq units) of the region bounded by the curve $x^2=4y$ and the straight line $x=4y-2$ is **[2019, 11 Jan Shift-I]**

(a) 7/8 (b) 9/8 (c) 5/4 (d) 3/4

**55** The area (in sq. units) of the smaller portion enclosed between the curves, $x^2+y^2=4$ and $y^2=3x$, is **[JEE Main 2017]**

(a) $\frac{1}{2\sqrt{3}}+\frac{\pi}{3}$ (b) $\frac{1}{\sqrt{3}}+\frac{2\pi}{3}$
(c) $\frac{1}{2\sqrt{3}}+\frac{2\pi}{3}$ (d) $\frac{1}{\sqrt{3}}+\frac{4\pi}{3}$

**56** The area (in sq units) of the region $\{(x, y): y^2 \ge 2x$ and $x^2+y^2 \le 4x, x \ge 0, y \ge 0\}$ is **[JEE Main 2016]**

(a) $\pi-\frac{4}{3}$ (b) $\pi-\frac{8}{3}$
(c) $\pi-\frac{4\sqrt{2}}{3}$ (d) $\frac{\pi}{2}-\frac{2\sqrt{2}}{3}$

**57** The area enclosed by the curves $y=x^2$, $y=x^3$, $x=0$ and $x=p$, where $p>1$, is 1/6. The $p$ equals **[AIEEE 2012]**

(a) 8/3 (b) 16/3 (c) 2 (d) 4/3

**58** The area of the region bounded by the curves $y=|x-1|$ and $y=3-|x|$ is **[AIEEE 2003]**

(a) 2 sq units (b) 3 sq units
(c) 4 sq units (d) 6 sq units

## Level 3

**Imp.** Area bounded by Two Curves

**59** One of the points of intersection of the curves $y=1+3x-2x^2$ and $y=\frac{1}{x}$ is $\left(\frac{1}{2}, 2\right)$. Let the area of the region enclosed by these curves be $\frac{1}{24}(l\sqrt{5}+m)-n\log_e(1+\sqrt{5})$, where $l, m, n \in N$. Then, $l+m+n$ is equal to **[2024, 4 April Shift-I]**

(a) 32 (b) 29 (c) 30 (d) 31

**60** The area of the region enclosed by the parabolas $y=4x-x^2$ and $3y=(x-4)^2$ is equal to **[2024, 31 Jan Shift-II]**

(a) $\frac{14}{3}$ (b) 6
(c) 4 (d) $\frac{32}{9}$

**61** Let $q$ be the maximum integral value of $p$ in [0, 10] for which the roots of the equation $x^2 - px + \frac{5}{4}p = 0$ are rational. Then, the area of the region $\{(x, y): 0 \le y \le (x-q)^2, 0 \le x \le q\}$ is **[2023, 30 Jan Shift-II]**

(a) $\frac{125}{3}$ (b) 243 (c) 25 (d) 164

**62** Let $A = \{(x, y) \in R^2 : y \ge 0, 2x \le y \le \sqrt{4-(x-1)^2}\}$ and $B = \{(x, y) \in R \times R : 0 \le y \le \min\{2x, \sqrt{4-(x-1)^2}\}\}$.

Then, the ratio of the area of $A$ to the area of $B$ is **[2023, 29 Jan Shift-I]**

(a) $\frac{\pi}{\pi-1}$ (b) $\frac{\pi+1}{\pi-1}$ (c) $\frac{\pi-1}{\pi+1}$ (d) $\frac{\pi}{\pi+1}$

**63** Let $\Delta$ be the area of the region $\{(x, y) \in R^2 : x^2 + y^2 \le 21, y^2 \le 4x, x \ge 1\}$. Then, $\frac{1}{2}\left(\Delta - 21\sin^{-1}\frac{2}{\sqrt{7}}\right)$ is equal to **[2023, 29 Jan Shift-I]**

(a) $2\sqrt{3} - \frac{2}{3}$ (b) $\sqrt{3} - \frac{2}{3}$

(c) $\sqrt{3} - \frac{4}{3}$ (d) $2\sqrt{3} - \frac{1}{3}$

**64** If the area of the bounded region $R = \left\{(x, y): \max\{0, \log_e x\} \le y \le 2^x, \frac{1}{2} \le x \le 2\right\}$ is $\alpha(\log_e 2)^{-1} + \beta(\log_e 2) + \gamma$, then the value of $(\alpha + \beta - 2\gamma)^2$ is equal to **[2021, 27 July Shift-I]**

(a) 8 (b) 2 (c) 4 (d) 1

**65** The area bounded by the curve $4y^2 = x^2(4-x)(x-2)$ is equal to **[2021, 18 March Shift-II]**

(a) $\frac{\pi}{8}$ (b) $\frac{3\pi}{8}$ (c) $\frac{3\pi}{2}$ (d) $\frac{\pi}{16}$

**66** The area (in sq units) of the region described by $\{(x, y): y^2 \le 2x \text{ and } y \ge 4x - 1\}$ is **[JEE Main 2015]**

(a) $\frac{7}{32}$ units (b) $\frac{5}{64}$ units

(c) $\frac{15}{64}$ units (d) $\frac{9}{32}$ units

## Numerical Value Type Questions

**67** If the area of the region $\{(x, y): |4 - x^2| \le y \le x^2, y \le 4, x \ge 0\}$ is $\left(\frac{80\sqrt{2}}{\alpha} - \beta\right)$, $\alpha, \beta \in N$, then $\alpha + \beta$ is equal to .......... **[2025, 2 April Shift-I]**

**68** The area of the region bounded by the curve $y = \max\{|x|, x|x-2|\}$, then $X$-axis and the lines $x = -2$ and $x = 4$ is equal to .......... **[2025, 3 April Shift-I]**

**69** If the area of the region $\{(x, y): |x-5| \le y \le 4\sqrt{x}\}$ is $A$, then $3A$ is equal to ........ **[2025, 4 April Shift-I]**

**70** Let the area of the bounded region $\{(x, y): 0 \le 9x \le y^2, y \ge 3x - 6\}$ be $A$. Then, $6A$ is equal to ........... **[2025, 8 April Shift-II]**

**Imp.** Area bounded by Two Curves

**71** If the area of the larger portion bounded between the curves $x^2 + y^2 = 25$ and $y = |x-1|$ is $\frac{1}{4}(b\pi + c)$, $b, c \in N$, then $b + c$ is equal to .......... . **[2025, 23 Jan Shift-I]**

**72** Let the function $f(x) = \begin{cases} -3ax^2 - 2, & x < 1 \\ a^2 + bx, & x \ge 1 \end{cases}$ be differentiable, $\forall x \in R$, where $a > 1$, $b \in R$. If the area of the region enclosed by $y = f(x)$ and the line $y = -20$ is $\alpha + \beta\sqrt{3}$, $\alpha, \beta \in Z$, then the value of $\alpha + \beta$ is ...... **[2025, 22 Jan Shift-I]**

**73** The area of the region enclosed by the parabolas $y = x^2 - 5x$ and $y = 7x - x^2$ is ......... **[2024, 4 April Shift-I]**

**74** The area of the region enclosed by the parabola $(y-2)^2 = x - 1$, the line $x - 2y + 4 = 0$ and the positive coordinate axes is ......... . **[2024, 30 Jan Shift-II]**

**75** Let the area of the region $\{(x, y): 0 \le x \le 3, 0 \le y \le \min\{x^2 + 2, 2x + 2\}\}$ be $A$. Then, $12A$ is equal to ......... . **[2024 29 Jan Shift - II]**

**76** The area (in sq. units) of the part of the circle $x^2 + y^2 = 169$ which is below the line $5x - y = 13$ is $\frac{\pi\alpha}{2\beta} - \frac{65}{2} + \frac{\alpha}{\beta}\sin^{-1}\left(\frac{12}{13}\right)$, where $\alpha, \beta$ are coprime numbers. Then, $\alpha + \beta$ is equal to .... **[2024, 29 Jan Shift I]**

**77** If the area of the region $\{(x, y): 0 \le y \le \min\{2x, 6x - x^2\}\}$ is $A$, then $12A$ is equal to ......... **[2024, 27 Jan shift II]**

**78** Let the area of the region $\{(x, y): x - 2y + 4 \ge 0, x + 2y^2 \ge 0, x + 4y^2 \le 8, y \ge 0\}$ be $\frac{m}{n}$, where $m$ and $n$ are coprime numbers. Then, $m + n$ is equal to .... . **[2024, 27 Jan shift - I]**

**79** If the area bounded by the curve $2y^2 = 3x$, lines $x + y = 3$, $y = 0$ and outside the circle $(x-3)^2 + y^2 = 2$ is $A$, then $4(\pi + 4A)$ is equal to ......... . **[2023, 15 April Shift-I]**

**80** If $A$ is the area in the first quadrant enclosed by the curve $C: 2x^2 - y + 1 = 0$, the tangent to $C$ at the point (1, 3) and the line $x + y = 1$, then the value of $60A$ is ........... .

*[2023, 11 April Shift-II]*

**81** If the area of the region $\{(x, y): |x^2 - 2| \le y \le x\}$ is $A$, then $6A + 16\sqrt{2}$ is equal to .........

*[2023, 10 April Shift-II]*

**82** Let $y = p(x)$ be the parabola passing through the points $(-1, 0)$, $(0, 1)$ and $(1, 0)$. If the area of the region $\{(x, y): (x+1)^2 + (y-1)^2 \le 1, y \le p(x)\}$ is $A$, then $12(\pi - 4A)$ is equal to .........

*[2023, 10 April Shift-I]*

**83** Let the area enclosed by the lines $x + y = 2$, $y = 0$, $x = 0$ and the curve $f(x) = \min\left\{x^2 + \frac{3}{4}, 1 + [x]\right\}$, where $[x]$ denotes the greatest integer $\le x$, be $A$. Then, the value of $12A$ is ..............

*[2023, 8 April Shift-II]*

**84** If the area of the region $S = \{(x, y): 2y - y^2 \le x^2 \le 2y, x \ge y\}$ is equal to $\frac{n+2}{n+1} - \frac{\pi}{n-1}$, then the natural number $n$ is equal to ........

*[2023, 6 April Shift-I]*

**85** Let the area of the region $\{(x, y): |2x - 1| \le y \le |x^2 - x|, 0 \le x \le 1\}$ be $A$. Then, $(6A + 11)^2$ is equal to ..........

*[2023, 31 Jan Shift-II]*

**86** Let for $x \in R$, $f(x) = \frac{x + |x|}{2}$ and $g(x) = \begin{cases} x, & x < 0 \\ x^2, & x \ge 0 \end{cases}$, then area bounded by the curve $y = (fog)(x)$ and the lines $y = 0$, $2y - x = 15$ is equal to ......... .

*[2023, 31 Jan Shift-I]*

**87** Let $A$ be the area of the region $\{(x, y): y \ge x^2, y \ge (1-x)^2, y \le 2x(1-x)\}$. Then, $540A$ is equal to .......

*[2023, 30 Jan Shift-II]*

**88** If the area enclosed by the parabolas $P_1: 2y = 5x^2$ and $P_2: x^2 - y + 6 = 0$ is equal to the area enclosed by $P_1$ and $y = \alpha x$, $\alpha > 0$, then $\alpha^3$ is equal to ........

*[2023, 25 Jan Shift-I]*

**89** If the area of the region $\left\{(x, y): x^{\frac{2}{3}} + y^{\frac{2}{3}} \le 1, x + y \ge 0, y \ge 0\right\}$ is $A$, then $\frac{256A}{\pi}$ is equal to ......... .

*[2022, 27 June Shift-II]*

**90** Let $A_1 = \{(x, y): |x| \le y^2, |x| + 2y \le 8\}$ and $A_2 = \{(x, y): |x| + |y| \le k\}$. If 27 (Area $A_1$) = 5 (Area $A_2$), then $k$ is equal to ........... .

*[2022, 27 June Shift-I]*

**91** The area (in sq units) of the region enclosed between the parabola $y^2 = 2x$ and the line $x + y = 4$ is ...... .

*[2022, 24 June Shift-II]*

**92** Let $S$ be the region bounded by the curves $y = x^3$ and $y^2 = x$. The curve $y = 2|x|$ divides $S$ into two regions of areas $R_1$ and $R_2$. If $\max\{R_1, R_2\} = R_2$, then $\frac{R_2}{R_1}$ is equal to ...... .

*[2022, 24 June Shift-I]*

**93** If the line $y = mx$ bisects the area enclosed by the lines $x = 0$ and $y = 0$, $x = \frac{3}{2}$ and the curve $y = 1 + 4x - x^2$, then $12m$ is equal to ...... .

*[2021, 31 Aug Shift-II]*

**94** The area of the region $S = \{(x, y): 3x^2 \le 4y \le 6x + 24\}$ is ...... .

*[2021, 26 Aug Shift-I]*

**95** The area (in sq units) of the region bounded by the curves $x^2 + 2y - 1 = 0$, $y^2 + 4x - 4 = 0$ and $y^2 - 4x - 4 = 0$, in the upper half plane is ......... .

*[2021, 22 July Shift-II]*

**96** Let $f: [-3, 1] \to R$ be given as

$$f(x) = \begin{cases} \min\{(x+6), x^2\}, & -3 \le x \le 0 \\ \max\{\sqrt{x}, x^2\}, & 0 \le x \le 1 \end{cases}.$$

If the area bounded by $y = f(x)$ and $X$-axis is $A$, then the value of $6A$ is equal to ........... .

*[2021, 17 March Shift-II]*

**97** The graphs of sine and cosine functions, intersect each other at a number of points and between two consecutive points of intersection, the two graphs enclose the same area $A$. Then $A^4$ is equal to ............ .

*[2021, 25 Feb Shift-I]*

## • Key Index

### Topic 1

| | | | | | | | | | |
|---|---|---|---|---|---|---|---|---|---|
| 1 (b) | 2 (a) | 3 (b) | 4 (c) | 5 (c) | 6 (d) | 7 (b) | 8 (b) | 9 (a) | 10 (a) |
| 11 (a) | 12 (d) | 13 (d) | 14 (a) | 15 (a) | 16 (*) | 17 (c) | 18 (a) | 19 (9) | 20 (62) |
| 21 (22) | 22 (114) | 23 (1225) | | | | | | | |

### Topic 2

| | | | | | | | | | |
|---|---|---|---|---|---|---|---|---|---|
| 1 (a) | 2 (a) | 3 (d) | 4 (b) | 5 (d) | 6 (b) | 7 (c) | 8 (c) | 9 (a) | 10 (d) |
| 11 (a) | 12 (b) | 13 (b) | 14 (b) | 15 (b) | 16 (d) | 17 (d) | 18 (c) | 19 (c) | 20 (a) |
| 21 (d) | 22 (b) | 23 (b) | 24 (b) | 25 (b) | 26 (b) | 27 (b) | 28 (c) | 29 (a) | 30 (c) |
| 31 (a) | 32 (c) | 33 (a) | 34 (c) | 35 (b) | 36 (c) | 37 (d) | 38 (c) | 39 (b) | 40 (b) |
| 41 (a) | 42 (c) | 43 (c) | 44 (c) | 45 (b) | 46 (b) | 47 (d) | 48 (a) | 49 (a) | 50 (d) |
| 51 (b) | 52 (c) | 53 (d) | 54 (b) | 55 (d) | 56 (b) | 57 (d) | 58 (c) | 59 (c) | 60 (b) |
| 61 (b) | 62 (c) | 63 (c) | 64 (b) | 65 (c) | 66 (d) | 67 (22) | 68 (12) | 69 (368) | 70 (15) |
| 71 (77) | 72 (34) | 73 (72) | 74 (5) | 75 (164) | 76 (171) | 77 (304) | 78 (119) | 79 (42) | 80 (16) |
| 81 (27) | 82 (16) | 83 (17) | 84 (5) | 85 (125) | 86 (72) | 87 (25) | 88 (600) | 89 (36) | 90 (6) |
| 91 (18) | 92 (19) | 93 (26) | 94 (27) | 95 (2) | 96 (41) | 97 (64) | | | |

# Detailed **Explanations**

### Topic 1

**01.** *(b)* $y^2 \le 2x,\ y \ge 4x-1$

Point of intersection

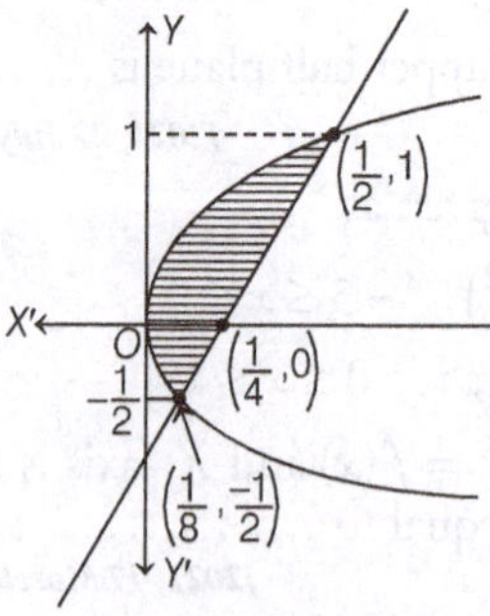


$y^2 = 2x$ and $y = 4x-1$

$y = 2y^2 - 1$

$\Rightarrow 2y^2 - y - 1 = 0 \Rightarrow 2y^2 - 2y + y - 1 = 0$

$\Rightarrow 2y(y-1) + 1(y-1) = 0$

$\Rightarrow \quad (y-1)(2y+1) = 0$

$\Rightarrow \quad y = 1, -\frac{1}{2} \Rightarrow x = \frac{1}{2}, \frac{1}{8}$

$\therefore$ Point of intersection

$\left(\frac{1}{2},1\right),\left(\frac{1}{8},-\frac{1}{2}\right)$

$\therefore$ Required area $= \int_{-1/2}^{1} (x_1 - x_2)\, dy$

$$= \int_{-1/2}^{1} \left(\left(\frac{y+1}{4}\right) - \left(\frac{y^2}{2}\right)\right) dy$$

$$= \left[\frac{(y+1)^2}{8} - \frac{y^3}{6}\right]_{-1/2}^{1}$$

$$= \left(\frac{4}{8} - \frac{1}{6}\right) - \left(\frac{1}{32} + \frac{1}{48}\right) = \left(\frac{2}{6}\right) - \left(\frac{5}{96}\right)$$

$$= \frac{32-5}{96} = \frac{27}{96} = \frac{9}{32}$$

**02.** *(a)* Given equation of curves are

$y = |x-1| + |x-2|$ and $y = 3$

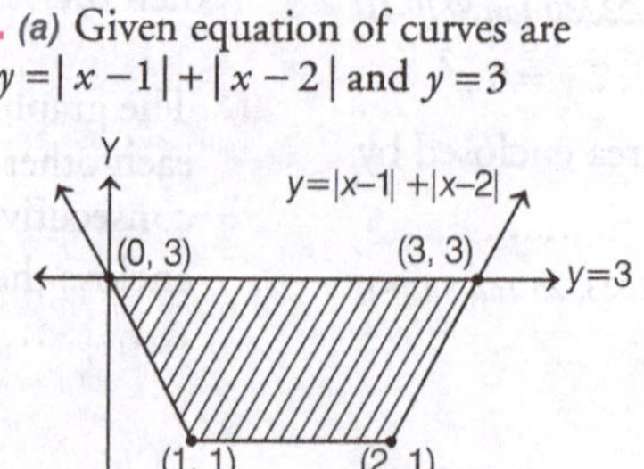


Now, the required area

$= $ Area of trapezium

$= \frac{1}{2}(1+3) \times 2 = 4$

**03.** *(b)* Area bounded by the curve $x = y^a$, $x = 0,\ y = 1,\ y = 3$ is $\left|\int_1^3 y^a\, dy\right|$.

But the area of region is given $364/3$

$$\Rightarrow \left|\int_1^3 y^a\, dy\right| = \frac{364}{3} \Rightarrow \left[\frac{y^{a+1}}{a+1}\right]_1^3 = \frac{364}{3}$$

$a$ can be odd natural number when $a = 1, 3, 5, 7, \ldots$

By hit and trial method,

$$\frac{3^{5+1}-1}{5+1} = \frac{3^6-1}{6} = \frac{729-1}{6} = \frac{728}{6} = \frac{364}{3}$$

Hence, $a = 5$

**04.** *(c)* Since, the locus of the centre $(\alpha, \beta)$, $\beta > 0$ of the circle which touches the circle $x^2 + (y-1)^2 = 1$ externally and also touches the the $X$-axis be $L$.

Thus, $\alpha^2 + (\beta - 1)^2 = (|\beta| + 1)^2$

$\Rightarrow \alpha^2 + \beta^2 - 2\beta + 1 = \beta^2 + 1 + 2|\beta|$

$\Rightarrow \quad \alpha^2 = 4\beta$ as $\beta > 0$

Thus, the Locus $L$ is $x^2 = 4y,\ y > 0$

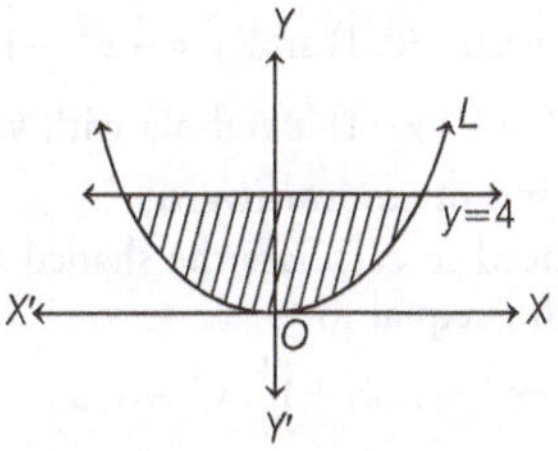


Hence, the area bounded by $L$ and the line $y = 4$ is

$$= 2\int_0^4 2\sqrt{y}\,dy = 4\left[\frac{2}{3}y^{\frac{3}{2}}\right]_0^4 = \frac{64}{3} \text{ sq units}$$

**05.** *(c)* Given, $y = 3 - \left|x - \frac{1}{2}\right| - |x+1|$

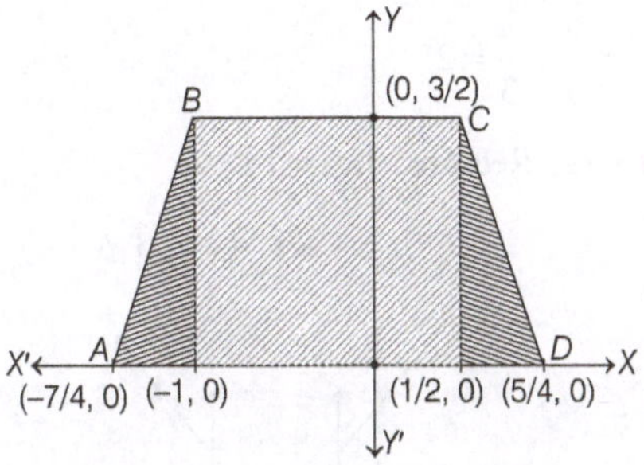


$$y = \begin{cases} 3 + (x+1) + \left(x - \frac{1}{2}\right), & x < -1 \\ 3 - (x+1) + \left(x - \frac{1}{2}\right), & -1 \le x < \frac{1}{2} \\ 3 - (x+1) - \left(x - \frac{1}{2}\right), & \frac{1}{2} \le x \end{cases}$$

$$y = \begin{cases} 2x + \frac{7}{2}, & x < -1 \\ \frac{3}{2}, & -1 \le x < \frac{1}{2} \\ \frac{5}{2} - 2x, & x \ge \frac{1}{2} \end{cases}$$

Area of shaded region

$$ABCD = \frac{1}{2}\left(3 + \frac{3}{2}\right)\frac{3}{2} = \frac{27}{8} \text{ sq units}$$

**06.** *(d)* We have,

$$R_2 = \int_{-2}^{3} x\, f(x)dx = \int_{-2}^{3} (1-x)\, f(1-x)dx$$

$$\left[\text{Using} \int_a^b f(x)\,dx = \int_a^b f(a+b-x)dx\right]$$

$$\Rightarrow R_2 = \int_{-2}^{3} (1-x)\, f(x)dx$$

$$[\because f(x) = f(1-x) \text{ on } [-2, 3]]$$

$$\therefore R_2 + R_2 = \int_{-2}^{3} x\, f(x)\,dx + \int_{-2}^{3} (1-x)\, f(x)\,dx$$

$$= \int_{-2}^{3} f(x)dx = R_1 \Rightarrow 2R_2 = R_1$$

**07.** *(b)* Given, $y = |\cos x - \sin x|$

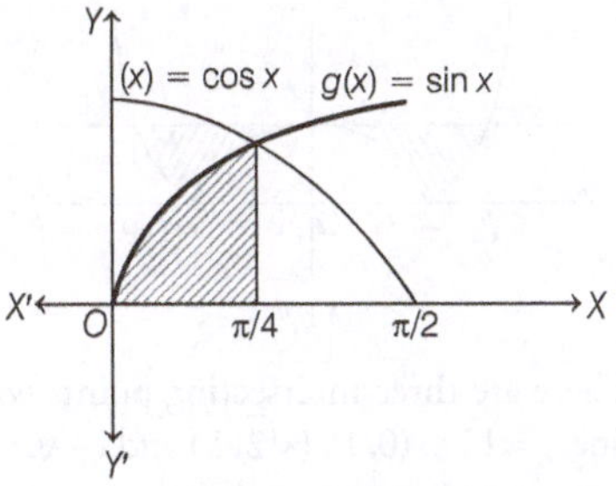


$$\text{The required area} = 2\int_0^{\pi/4} (\cos x - \sin x)dx$$

$$= 2[\sin x + \cos x]_0^{\pi/4}$$

$$= 2\left[\frac{1}{\sqrt{2}} + \frac{1}{\sqrt{2}} - 0 - 1\right] = 2\left[\frac{2}{\sqrt{2}} - 1\right]$$

$$= (2\sqrt{2} - 2) \text{ sq units}$$

**08.** *(b)* Given, $f(x) = x^3$,

$g(x) = 8,\ f(x) > g(x)$

$$\text{The required area} = \int_{y=0}^{8} y^{\frac{1}{3}}\,dy$$

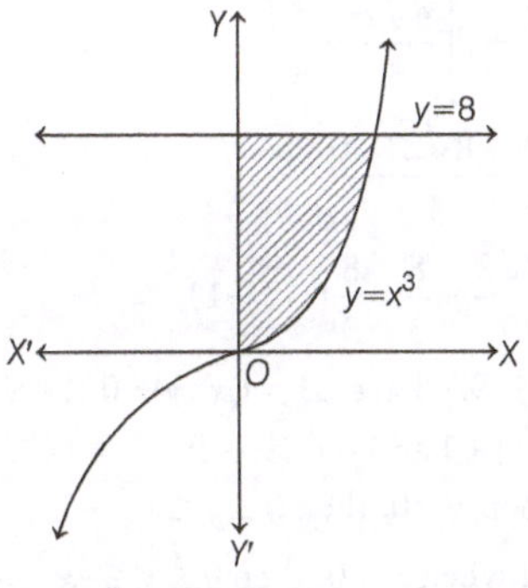


$$= \left.\frac{y^{\frac{1}{3}+1}}{\frac{1}{3}+1}\right|_0^8 = \left.\frac{3}{4}\left(y^{\frac{4}{3}}\right)\right|_0^8$$

$$= \frac{3}{4}\left[(8)^{\frac{4}{3}} - 0\right] = \frac{3}{4}[2^4] = \frac{3}{4} \times 16$$

$= 12$ sq units

**09.** *(a)* According to the given condition,

$$\int_{\pi/4}^{\beta} f(x)\,dx = \beta \sin\beta + \frac{\pi}{4}\cos\beta + \sqrt{2}\,\beta$$

On differentiating both sides w.r.t. β, we get

$$f(\beta) = \sin\beta + \beta\cos\beta - \frac{\pi}{4}\sin\beta + \sqrt{2}$$

$$\therefore f\left(\frac{\pi}{2}\right) = 1 + 0 - \frac{\pi}{4} + \sqrt{2} = 1 - \frac{\pi}{4} + \sqrt{2}$$

**10.** *(a)* $f(x+y) = f(x) \cdot f(y),\ \forall\, x, y \in R$

$$f(x) = k^x$$

$$f'(x) = k^x \ln k$$

$$\Rightarrow f'(0) = \ln k = 4a \Rightarrow k = e^{4a}$$

$$f(x) = (e^{4a})^x = e^{4ax}$$

$$f'(x) = 4a\, e^{4ax}$$

$$f''(x) = 16a^2\, e^{4ax}$$

Now, $f''(x) - 3af'(x) - f(x) = 0$

$$16a^2 e^{4ax} - 12a^2 e^{4ax} - e^{4ax} = 0$$

$$(4a^2 - 1)e^{4ax} = 0 \Rightarrow 4a^2 = 1$$

$$a = 1/2 \qquad [\because a > 0]$$

$$\therefore \quad y = e^{2x} \Rightarrow f(ax) = e^x$$

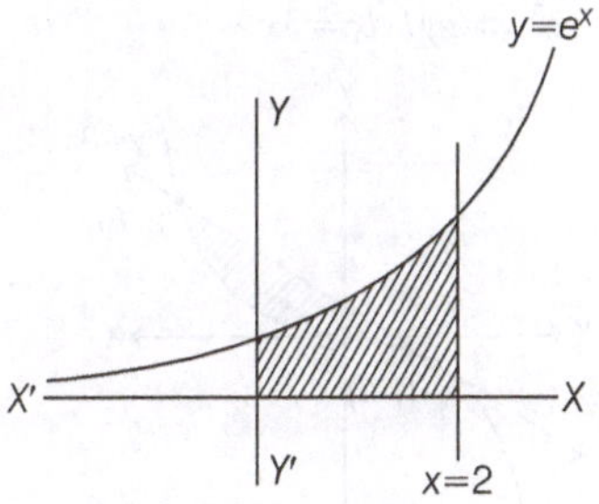


Area of region

$$R = \{(x, y);\ 0 \le y \le f(ax),\ 0 \le x \le 2\}$$

$$A = \int_0^2 e^x\,dx = [e^x]_0^2 = e^2 - 1$$

**11.** *(a)* Equation of ellipse : $\frac{x^2}{18} + \frac{y^2}{6} = 1$

Line : $y = x$

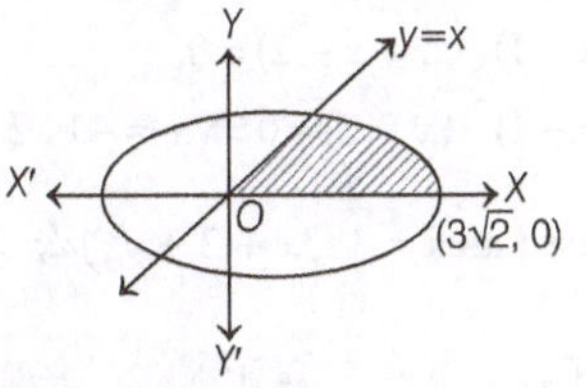


Since, $y = x$, we have

$$\frac{x^2}{18} + \frac{x^2}{6} = 1 \Rightarrow \frac{4x^2}{18} = 1$$

$$\Rightarrow \quad x = \frac{3}{\sqrt{2}} = y$$

∴Required area

$$= \frac{1}{2} \times \frac{3}{\sqrt{2}} \times \frac{3}{\sqrt{2}} + \int_{3/\sqrt{2}}^{3\sqrt{2}} \frac{\sqrt{18 - x^2}}{\sqrt{3}}\,dx$$

$$= \frac{9}{4} + \frac{1}{\sqrt{3}} \left( \frac{x\sqrt{18-x^2}}{2} + \frac{18}{2}\sin^{-1}\frac{x}{3\sqrt{2}} \right)_{3/\sqrt{2}}^{3\sqrt{2}}$$

$$= \frac{9}{4} + \frac{1}{\sqrt3}\left[\left(0+\frac{18}{2}\sin^{-1}\frac{3\sqrt2}{3\sqrt2}\right) - \frac{\frac{3}{\sqrt2}\cdot\frac{3\sqrt3}{\sqrt2}}{2} - \frac{18}{2}\sin^{-1}\frac{\frac{3}{\sqrt2}}{3\sqrt2}\right]$$

$$= \frac{9}{4} + \frac{1}{\sqrt3}\left(9\times\frac{\pi}{2} - \frac{9\sqrt3}{4} - 9\times\frac{\pi}{6}\right)$$

$$= \frac{9}{4} + \frac{1}{\sqrt3}\left(\frac{9\pi}{2} - \frac{3\pi}{2} - \frac{9\sqrt3}{4}\right)$$

$$= \frac{9}{4} + \frac{1}{\sqrt3}\times 3\pi - \frac{9}{4} = \sqrt3\pi$$

**12.** *(d)* Given region is,

$y = x^3 \Rightarrow dy/dx = 3x^2$

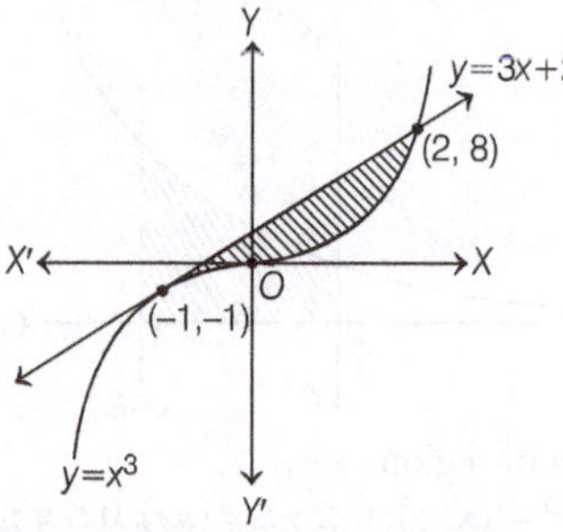


Slope of tangent at $(-1,-1) = 3(-1)^2 = 3$

So, equation of tangent : $y + 1 = 3(x+1)$

$\Rightarrow 3x + 2 = y$

Solving with $y = x^3$,

$3x + 2 = x^3 \Rightarrow x^3 - 3x - 2 = 0$

$\Rightarrow \quad (x+1)(x^2 - x - 2) = 0$

$\Rightarrow \quad (x+1)^2(x-2) = 0 \Rightarrow x = -1, 2$

$\therefore$ Required area $= \int_{-1}^{2}(3x + 2 - x^3)\,dx$

$$= \left[\frac{3x^2}{2} + 2x - \frac{x^4}{4}\right]_{-1}^{2}$$

$$= \left[\frac{3\cdot 2^2}{2} + 2\cdot 2 - \frac{2^4}{4}\right] - \left(\frac{3\cdot(-1)^2}{2} + 2(-1) - \frac{(-1)^4}{4}\right)$$

$= 27/4$ sq. units

**13.** *(d)* Given, $y = |x^2 - 1|$ and line $y = 1$

$$y = \begin{cases} 1 - x^2, & \text{if } x^2 - 1 < 0 \\ x^2 - 1, & \text{if } x^2 - 1 \geq 0 \end{cases}$$

When $x^2 - 1 < 0$ and $y = 1$

Then, $x^2 = 0 \Rightarrow x = 0$

When $x^2 - 1 \geq 0$ and $y = 1$

$\Rightarrow \quad x^2 = 2 \Rightarrow x = \pm\sqrt2$

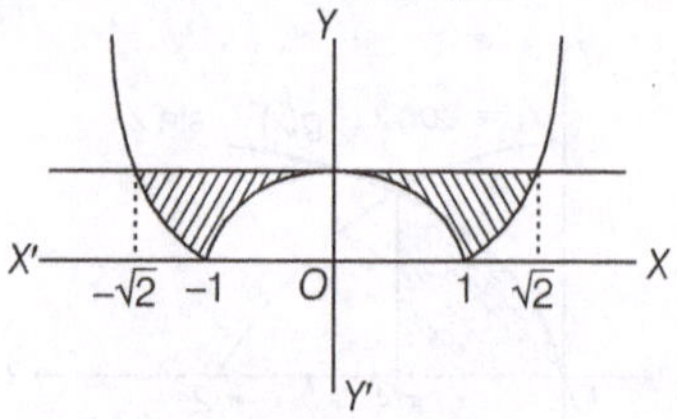


There are three intersecting points with line $y = 1$ i.e, $(0, 1)$, $(\sqrt2, 1)$ and $(-\sqrt2, 1)$

The area bounded by the lines and curve is shown in the figure.

Area bounded by the lines and curve

$$= 2\int_0^{\sqrt2}(1 - |x^2 - 1|)\,dx$$

$$= 2\left[\int_0^1 1 - (1 - x^2)\,dx + \int_1^{\sqrt2}(1 - x^2 + 1)\,dx\right]$$

$$= 2\left[\frac{x^3}{3}\right]_0^1 + 2\left[2x - \frac{x^3}{3}\right]_1^{\sqrt2}$$

$$= \frac{2}{3} + 2\left[2\sqrt2 - \frac{2\sqrt2}{3} - 2 + \frac{1}{3}\right]$$

$$= \frac{2}{3} + 2\left[\frac{4\sqrt2}{3} - \frac{5}{3}\right]$$

$$= \frac{2}{3} + \frac{8\sqrt2}{3} - \frac{10}{3}$$

$$= \frac{8\sqrt2}{3} - \frac{8}{3} = \frac{8}{3}(\sqrt2 - 1)$$

**14.** *(a)* We have, $A = \{(x, y) : 0 \leq y \leq x|x| + 1$ and $-1 \leq x \leq 1\}$

When $x \geq 0$, then $0 \leq y \leq x^2 + 1$

and when $x < 0$, then $0 \leq y \leq -x^2 + 1$

Now, the required region is the shaded region.

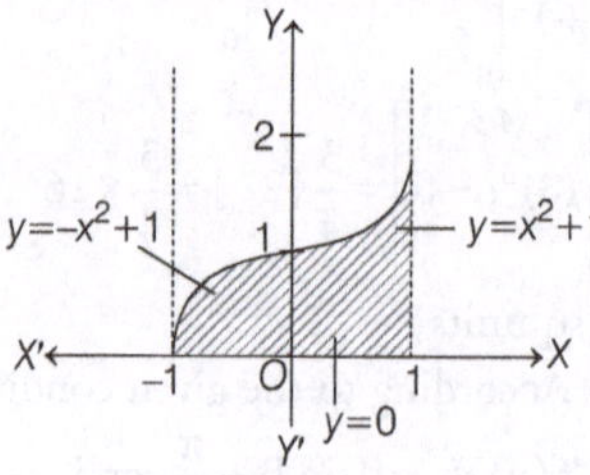


[$\because y = x^2 + 1 \Rightarrow x^2 = (y-1)$, parabola with vertex (0, 1) and $y = -x^2 + 1$

$\Rightarrow x^2 = -(y-1)$ parabola with vertex (0,1) but open downward]

We need to calculate the shaded area, which is equal to

$$\int_{-1}^{0}(-x^2+1)\,dx + \int_0^1 (x^2+1)\,dx$$

$$= \left[-\frac{x^3}{3} + x\right]_{-1}^{0} + \left[\frac{x^3}{3} + x\right]_0^1$$

$$= \left(0 - \left[-\frac{(-1)^3}{3} + (-1)\right]\right) + \left(\left[\frac{1}{3} + 1\right] - 0\right)$$

$$= -\left(\frac{1}{3} - 1\right) + \frac{4}{3}$$

$$= \frac{2}{3} + \frac{4}{3} = 2$$

**15.** *(a)* Required area $= \int_1^3 y\,dx$

$$= \int_1^3 |x - 2|\,dx$$

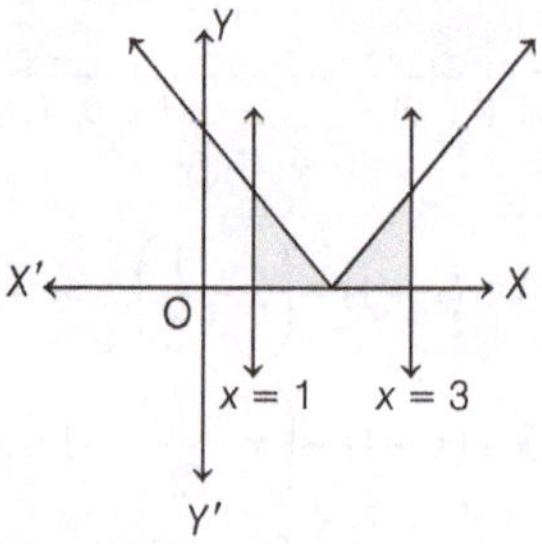


$$= \int_1^2 -(x-2)\,dx + \int_2^3 (x-2)\,dx$$

$$= \int_1^2 (2-x)\,dx + \int_2^3 (x-2)\,dx$$

$$= \left[2x - \frac{x^2}{2}\right]_1^2 + \left[\frac{x^2}{2} - 2x\right]_2^3$$

$$= \left[4 - 2 - \left(2 - \frac{1}{2}\right)\right] + \left[\frac{9}{2} - 6 - (2-4)\right]$$

$$= 2 - \frac{3}{2} - \frac{3}{2} + 2 = 1 \text{ sq unit}$$

**16.** (*) $y = |x^2 - 9| = |(x+3)(x-3)|$

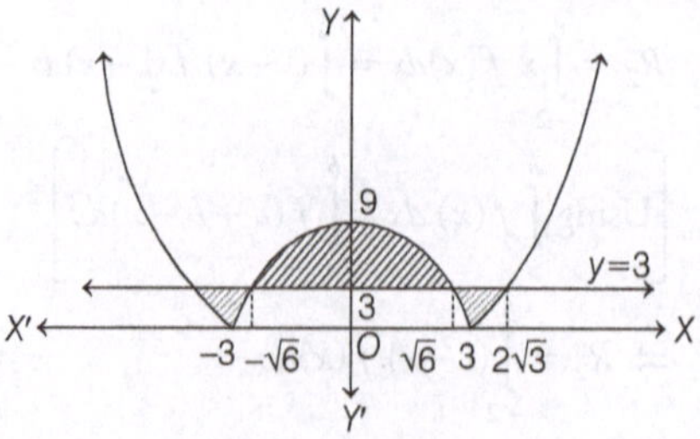

and $y=3$

On solving above equations, we get

$$x=\pm\sqrt{6} \text{ and } \pm 2\sqrt{3}$$

Required area

$$=2\left[\int_0^{\sqrt{6}}(9-x^2-3)\,dx+\int_{\sqrt{6}}^{3}(3-9+x^2)\,dx+\int_3^{2\sqrt{3}}(12-x^2)dx\right]$$

$$=2\left[\left[6x-\frac{x^3}{3}\right]_0^{\sqrt{6}}+\left[\frac{x^3}{3}-6x\right]_{\sqrt{6}}^{3}+\left[12x-\frac{x^3}{3}\right]_3^{2\sqrt{3}}\right]$$

$$=2\,[6\sqrt{6}-2\sqrt{6}+(9-18)-(2\sqrt{6}-6\sqrt{6})+(24\sqrt{3}-8\sqrt{3})-(36-9)]$$

$$=8\,[2\sqrt{6}-9+4\sqrt{3}]$$

**17.** *(c)* Given, $y\le x^2+3x$

$$\Rightarrow\quad y\le\left(x+\frac{3}{2}\right)^2-\frac{9}{4}$$

$$\Rightarrow\quad \left(x+\frac{3}{2}\right)^2\ge\left(y+\frac{9}{4}\right)$$

Since, $0\le y\le 4$ and $0\le x\le 3$

$\therefore$ The diagram for the given inequalities is

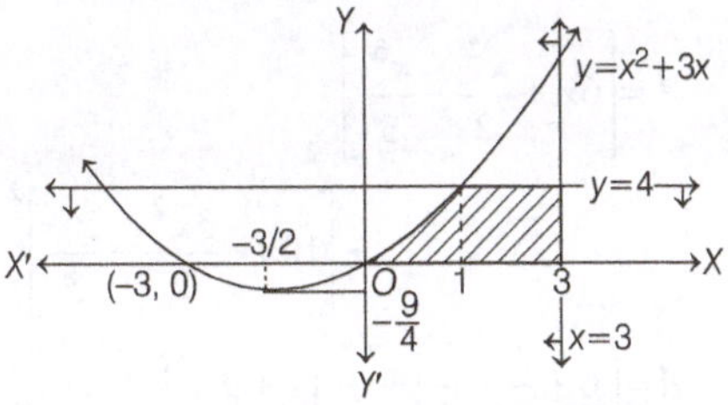


and points of intersection of curves $y=x^2+3x$ and $y=4$ are (1, 4) and (−4, 4)

Now, required area

$$=\int_0^1(x^2+3x)dx+\int_1^3 4\,dx$$

$$=\left[\frac{x^3}{3}+\frac{3x^2}{2}\right]_0^1+[4x]_1^3$$

$$=\frac{1}{3}+\frac{3}{2}+4(3-1)=\frac{2+9}{6}+8=\frac{11}{6}+8$$

$=59/6$ sq units

**18.** *(a)* We have,

$$\Rightarrow\quad 18x^2-9\pi x+\pi^2=0$$

$$(6x-\pi)(3x-\pi)=0$$

$$\Rightarrow\quad x=\frac{\pi}{6},\frac{\pi}{3}$$

Now, $\alpha<\beta$

$$\therefore\quad \alpha=\frac{\pi}{6},\ \beta=\frac{\pi}{3}$$

Given, $g(x)=\cos x^2$ and $f(x)=\sqrt{x}$

$$y=gof(x)$$

$$\therefore\quad y=g(f(x))=\cos x$$

Area of region bounded by $x=\alpha$, $x=\beta$, $y=0$ and curve $y=g(f(x))$ is

$$A=\int_{\pi/6}^{\pi/3}\cos x\,dx\ \Rightarrow\ A=[\sin x]_{\pi/6}^{\pi/3}$$

$$A=\sin\frac{\pi}{3}-\sin\frac{\pi}{6}=\frac{\sqrt{3}}{2}-\frac{1}{2}$$

$$A=\left(\frac{\sqrt{3}-1}{2}\right)$$

**Shortcut Method**

We have, $18x^2-9\pi x+\pi^2=0$

$x=\frac{\pi}{6},\frac{\pi}{3}$ $\quad\alpha=\frac{\pi}{6},\beta=\frac{\pi}{3}$ $\quad[\because\alpha<\beta]$

$y=g(f(x))=\cos x$

$\therefore$ Required area $=\int_{\pi/6}^{\pi/3}\cos x\,dx=\frac{\sqrt{3}}{2}-\frac{1}{2}=\frac{\sqrt{3}-1}{2}$

**19.** *(9)* $y=\min\{\sin x,\cos x\}$ and $x\in[-\pi,\pi]$

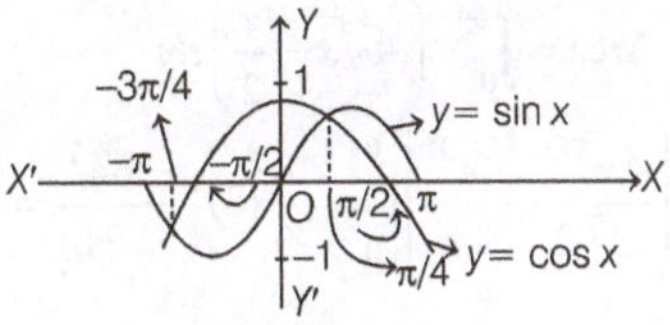


So, $\min\{\sin x,\cos x\}$

$$=\begin{cases}\cos x, & -\pi\le x<-\frac{3\pi}{4}\\ \sin x, & -\frac{3\pi}{4}<x<\frac{\pi}{4}\\ \cos x, & \frac{\pi}{4}<x<\pi\end{cases}$$

Therefore, the region enclosed

$$A=\int_{-\pi}^{-3\pi/4}-\cos x\cdot dx+\int_{-3\pi/4}^{0}-\sin x\cdot dx+\int_0^{\pi/4}\sin x\cdot dx+\int_{\pi/4}^{\pi/2}\cos x\cdot dx+\int_{\pi/2}^{\pi}-\cos x\cdot dx$$

$$=[-\sin x]_{-\pi}^{\frac{-3\pi}{4}}+[\cos x]_{-\frac{3\pi}{4}}^{0}+[-\cos x]_0^{\pi/4}+[\sin x]_{\pi/4}^{\pi/2}+[-\sin x]_{\pi/2}^{\pi}$$

$$=\left(\frac{1}{\sqrt{2}}-1\right)+\left(1+\frac{1}{\sqrt{2}}\right)+\left(-\frac{1}{\sqrt{2}}+1\right)+\left(1-\frac{1}{\sqrt{2}}\right)+(-0+1)$$

$$=-1+1+1+1+1=3$$

So, $A^2=9$

**20.** *(62)* We have, $y=x|x-3|$

$$=\begin{cases}-x(x-3), & x<3\\ x(x-3), & x\ge 3\end{cases}$$

$$=\begin{cases}3x-x^2, & x<3\\ x^2-3x, & x\ge 3\end{cases}$$

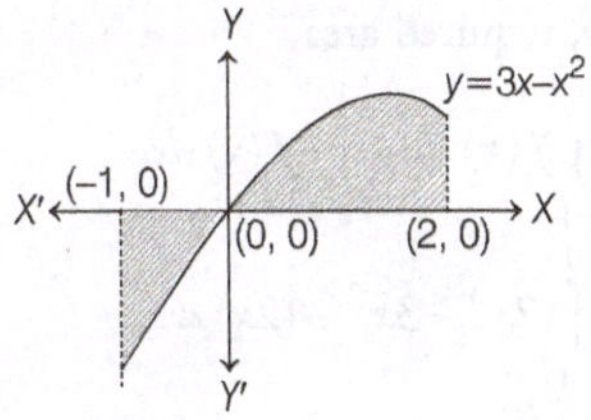


$\therefore$ Required area,

$$A=\int_{-1}^{0}-(3x-x^2)dx+\int_0^2(3x-x^2)\,dx$$

$$=-\left[\frac{3x^2}{2}-\frac{x^3}{3}\right]_{-1}^{0}+\left[\frac{3x^2}{2}-\frac{x^3}{3}\right]_0^2$$

$$=-\left[0-\left(\frac{3}{2}+\frac{1}{3}\right)\right]+\left[\left(6-\frac{8}{3}\right)-0\right]$$

$$=\frac{3}{2}+\frac{1}{3}+6-\frac{8}{3}=\frac{9+2+36-16}{6}=\frac{31}{6}$$

$$\therefore\ 12A=12\times\frac{31}{6}=62$$

**21.** *(22)* Given equation of curves are

$$y^2=8x \quad \ldots\text{(i)}$$

$$y=x \quad \ldots\text{(ii)}$$

and $$x=2 \quad \ldots\text{(iii)}$$

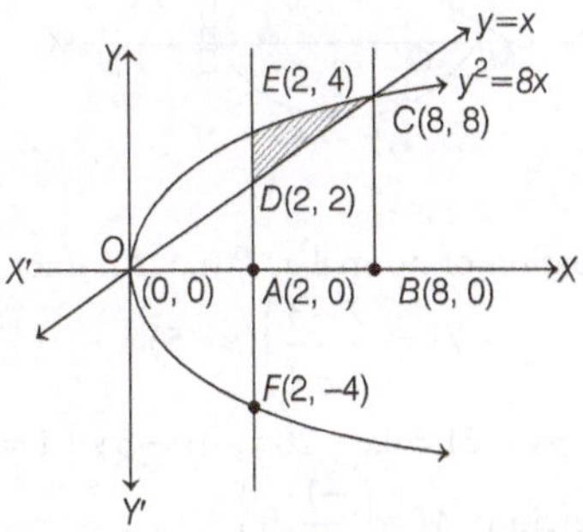


Hence, area of shaded portion is

$$\int_2^8(\sqrt{8x}-x)\,dx=\int_2^8 2\sqrt{2}\sqrt{x}\ dx-\int_2^8 x\,dx$$

$$=2\sqrt{2}\left[\frac{x^{3/2}}{3/2}\right]_2^8-\left[\frac{x^2}{2}\right]_2^8$$

$$= \frac{4\sqrt{2}}{3}[(8)^{3/2} - (2)^{3/2}] - \frac{1}{2}[(8)^2 - (2)^2]$$

$$= \frac{22}{3} = \alpha \quad \text{[given]}$$

Hence, $3\alpha = 22$

**22.** *(114)* We have,

$$f(x) = 2x^3 - 3x^2 - 12x$$

$$\therefore f'(x) = 6x^2 - 6x - 12$$

$$= 6(x^2 - x - 2) = 6(x+1)(x-2)$$

$$f'(x) = 0 \Rightarrow x = -1 \text{ and } 2$$

$\therefore x = -1$ and 2 are critical points

$\therefore a = -1$ and $b = 2$

Now, required area,

$$A = \int_{-1}^{0} f(x)\,dx + \int_{0}^{2} -f(x)\,dx$$

$$= \int_{-1}^{0} (2x^3 - 3x^2 - 12x)\,dx + \int_{0}^{2} (12x + 3x^2 - 2x^3)\,dx$$

$$= \left[\frac{x^4}{2} - x^3 - 6x^2\right]_{-1}^{0} + \left[6x^2 + x^3 - \frac{x^4}{2}\right]_{0}^{2} = \frac{114}{4}$$

$\therefore 4A = 114$

**23.** *(1225)* Given, circle

$$(x-2)^2 + (y-3)^2 = 5^2$$

$$C = (2, 3) \quad r = 5$$

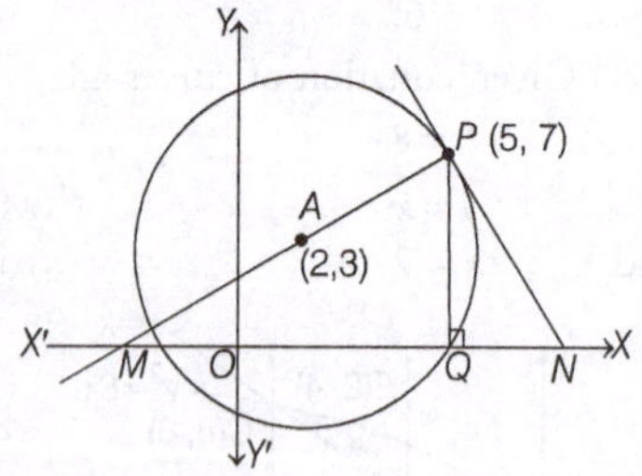


Equation of normal at $P$ (i.e. $PA$ line)

$$\Rightarrow \quad (y-7) = \left(\frac{7-3}{5-2}\right)(x-5)$$

$$\Rightarrow 3y - 21 = 4x - 20 \Rightarrow 4x - 3y + 1 = 0$$

Therefore, $M = \left(\frac{-1}{4}, 0\right)$

[Put $y = 0$ in above equation]

Now, equation of tangent at $P$.

$$y - 7 = \frac{-3}{4}(x-5)$$

$$\Rightarrow \quad 4y - 28 = -3x + 15$$

$$\Rightarrow \quad 3x + 4y = 43$$

Therefore, $N = \left(\frac{43}{3}, 0\right)$

$$\therefore \text{ Area } (A) = \frac{1}{2} \times MN \times PQ$$

$$= \frac{1}{2} \times \left(\frac{43}{3} + \frac{1}{4}\right) \times 7 = \frac{1}{2} \times \frac{175}{12} \times 7$$

$$\therefore 24A = 24 \times \frac{1}{2} \times \frac{175}{12} \times 7 = 1225$$

But this question is wrong as in question it is mentioned that the triangle is formed with the positive $X$-axis which contradicts the solution.

## Topic 2

**01.** *(a)* $|x - y| \le y \le 4\sqrt{x}$

$x = 2y, \qquad x - y = -y \ (x \le y)$

$x - y = y \ (x > y) \qquad x = 0$

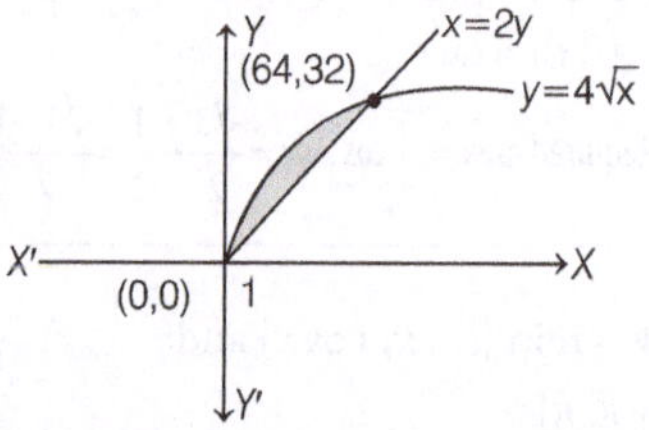


$$\therefore \text{Area} = \int_0^{64} \left(4\sqrt{x} - \frac{x}{2}\right) dx$$

$$= \left[\frac{4x^{3/2}}{3/2} - \frac{x^2}{4}\right]_0^{64} = \frac{8}{3}(8)^3 - \frac{64^2}{4} = \frac{1024}{3}$$

**02.** *(a)* $y = 4 - \frac{x^2}{4}$ and $y = \frac{x-4}{2}$

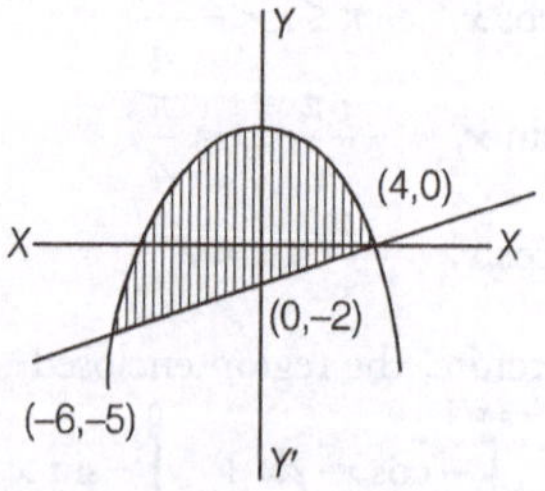


$$\text{Area} = \int_{-6}^{4} \left(4 - \frac{x^2}{4} - \frac{x}{2} + 2\right) dx$$

$$= \left[6x - \frac{x^3}{12} - \frac{x^2}{4}\right]_{-6}^{4}$$

$$= \left(6 \cdot 4 - \frac{4^3}{12} - \frac{4^2}{4}\right) - \left(6 \cdot (-6) - \frac{(-6)^3}{12} - \frac{(-6)^2}{4}\right)$$

$$\alpha = \frac{125}{3}$$

$$\therefore 6\alpha = 250$$

**03.** *(d)* Find the intersection of

$x + 7 = 11 - 3x$

$\Rightarrow x = 1, y = 1 + 7 = 8$

The point of intersection is (1, 8)

The point of intersection of

$1 + x^2 = x + 7$

$\Rightarrow x = 3, -2$

Points are (3, 10), (– 2, 5)

The point of intersection

$11 - 3x = 1 + x^2 \Rightarrow x = -5, 2$

Points are (2, 5), (–5, 26)

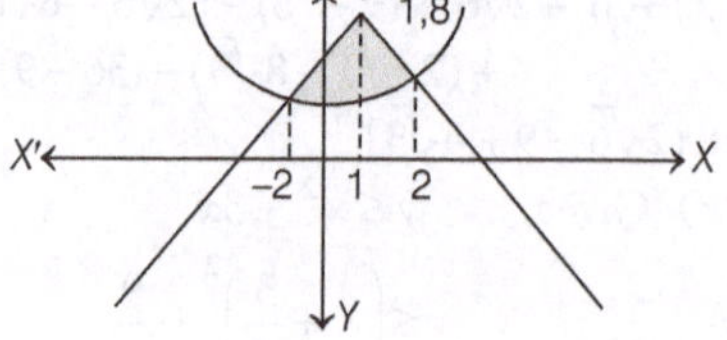


$$\text{Area} = \int_{-2}^{1} (x + 7 - (1 + x^2))dx + \int_{1}^{2} (11 - 3x - (1 + x^2))dx$$

$$A = \int_{-2}^{1} (6 + x - x^2)dx + \int_{1}^{2} (10 - 3x - x^2)dx$$

$$A = \left[6x + \frac{x^2}{2} - \frac{x^3}{3}\right]_{-2}^{1} + \left[10x - \frac{3x^2}{2} - \frac{x^3}{3}\right]_{1}^{2}$$

$$A = \left(6 + \frac{1}{2} - \frac{1}{3}\right) - \left(-12 + 2 + \frac{8}{3}\right) + \left(20 - 6 - \frac{8}{3}\right) - \left(10 - \frac{3}{2} - \frac{1}{3}\right)$$

$$A = 50/3$$

Thus, $3A = 50$

**04.** *(b)* We have,

$$c_1 : |y| = 1 - x^2$$

$$c_2 : x^2 + y^2 = 1$$

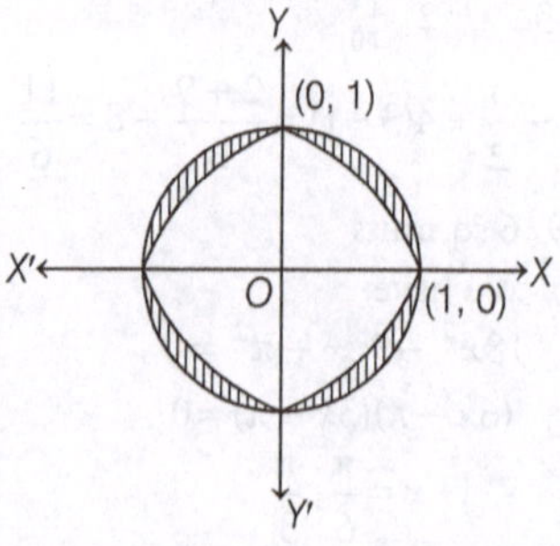

$\because$ Required area

$= \alpha = 4$ [area of circle is 1st quadrant $-\int_0^1 (1 - x^2)dx$]

$$= 4\left[\frac{\pi}{4} - \left[x - \frac{x^3}{3}\right]_0^1\right]$$

$$\alpha = \pi - \frac{8}{3}$$

$\therefore \quad 3\alpha = 3\pi - 8 \Rightarrow 9\alpha = 9\pi - 24$

$\therefore \quad \beta = 9, \gamma = -24$

$\therefore \quad |\beta - \gamma| = 33$

**05.** *(d)* $\therefore$ Required area $= A$

ar (rectangle $ABDE$) − ar (region $EDC$)

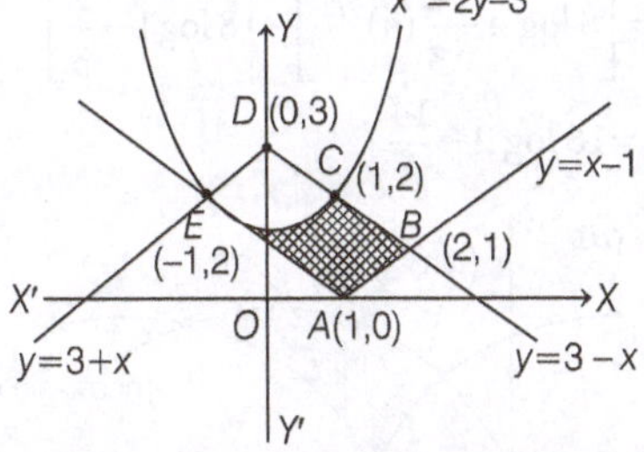


$$\Rightarrow A = 4 - 2\int_0^1 (3 - x) - \left(\frac{x^2 + 3}{2}\right) dx$$

$$= 4 - 2\left[3x - \frac{x^2}{2} - \frac{x^3}{6} - \frac{3}{2}x\right]_0^1$$

$$= 4 - 2\left[3 - \frac{1}{2} - \frac{1}{6} - \frac{3}{2}\right] = \frac{7}{3}$$

$\therefore 6A = 6 \times \frac{7}{3} = 14$ sq units

**06.** *(b)* $R = \{(x, y): 0 \le y \le 2|x| + 1, 0 \le y \le x^2 + 1, |x| \le 3\}$

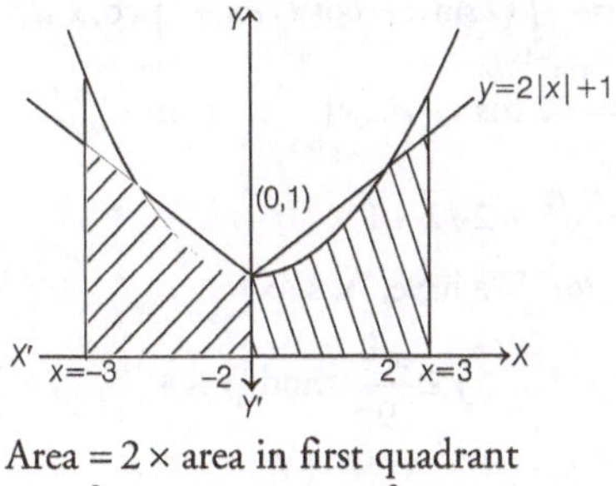


Area $= 2 \times$ area in first quadrant

$$= 2\int_0^2 (x^2 + 1)\,dx + 2\int_2^3 (2x + 1)\,dx$$

$$= 2\left[\frac{x^3}{3} + x\right]_0^2 + 2[x^2 + x]_2^3$$

$$= \frac{64}{3}$$

**07.** *(c)* Given, $y = e^x$ ...(i)

and $\quad y = |e^x - 1|$ ...(ii)

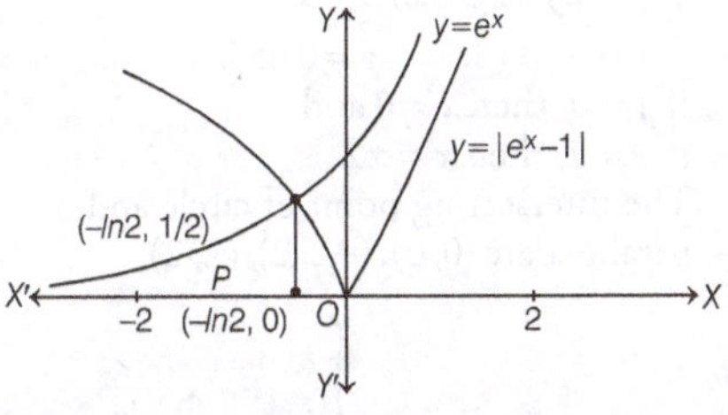


To get coordinates of point $P$, we have to solve Eqs. (i) and (ii)

So, $\quad e^x = |e^x - 1|$

$\Rightarrow \quad e^x - 1 = \pm e^x \quad [\because e^x > 0]$

$\therefore \quad e^x - 1 = e^x$ or $e^x - 1 = -e^x$

$\Rightarrow \quad e^x - e^x = 1$ (which is false).

So, $2e^x = 1 \Rightarrow x = \ln 1/2 \Rightarrow x = -\ln 2$.

$\therefore \quad y = e^{-\log_e 2} = 1/2$

So, $\quad P \equiv (-\log_e 2, 0)$

Now, area $= \int_{-\log_e 2}^0 (e^x - |e^x - 1|)\,dx$

$$= \int_{-\log_e 2}^0 (e^x - (1 - e^x))\,dx$$

$[\because$ when $-\log_e 2 < x < 0, |e^x - 1| = 1 - e^x]$

$$= \int_{-\log_e 2}^0 (2e^x - 1)\,dx = (2e^x - x)_{-\ln 2}^0$$

$$= 2 - (1 + \log_e 2) = 1 - \log_e 2$$

**08.** *(c)* Given, $x^2 + 4x + 2 \le y \le |x + 2|$

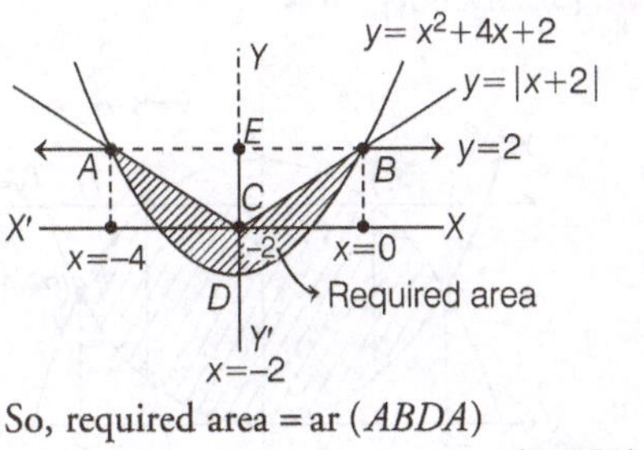


So, required area = ar $(ABDA)$ − ar $(\Delta ACB)$

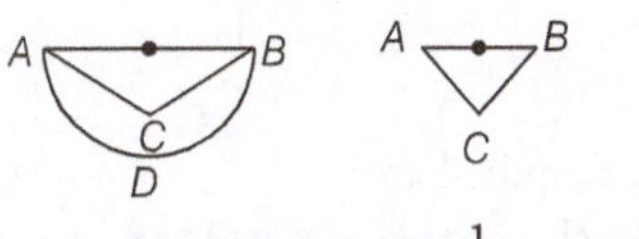


$$= \int_{-4}^0 [2 - (x^2 + 4x + 2)]dx - \frac{1}{2} \times AB \times CE$$

$$= \left(\frac{-x^3}{3} - 2x^2\right)\Bigg|_{-4}^0 - \frac{1}{2} \times 4 \times 2$$

$$= 0 - \left(\frac{64}{3} - 32\right) - 4 = 32 - \frac{64}{3} - 4$$

$$= \frac{20}{3} \text{ sq units}$$

**09.** *(a)* $x^2 + y^2 = 8$

$\Rightarrow \quad x^2 + y^2 = (2\sqrt{2})^2$

and $\quad y^2 = 2x$

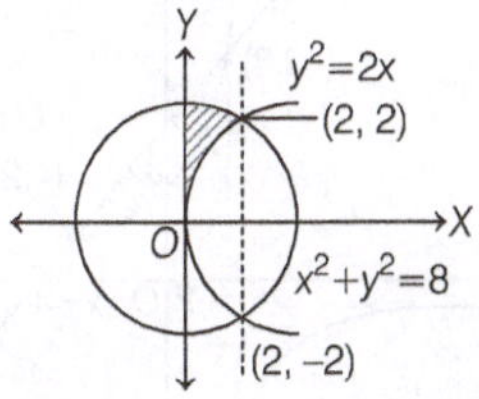


$\therefore$ Required area

$$= \int_0^2 \sqrt{8 - x^2} \cdot dx - \int_0^2 \sqrt{2x} \cdot dx$$

$$= \left[\frac{x}{2}\sqrt{8 - x^2} + \frac{8}{2}\sin^{-1}\left(\frac{x}{2\sqrt{2}}\right)\right]_0^2 - \left[\frac{2\sqrt{2}}{3}(x\sqrt{x})\right]_0^2$$

$$= \left[2 + 4\sin^{-1}\left(\frac{1}{\sqrt{2}}\right) - 0\right] - \frac{8}{3}$$

$$= 2 + 4 \cdot \frac{\pi}{4} - \frac{8}{3} = \pi - \frac{2}{3}$$

**10.** *(d)* Given, $\frac{a}{x^2} \le y \le \frac{1}{x}$,

$1 \le x \le 2$ and $0 < a < 1$

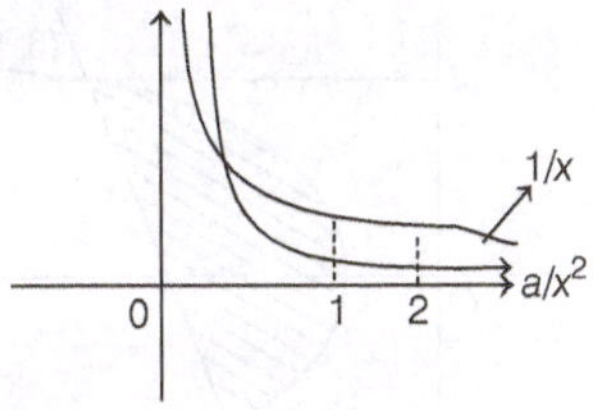


According to the question,

$$\int_1^2 \left(\frac{1}{x} - \frac{a}{x^2}\right) dx = \log_e 2 - \frac{1}{7}$$

$$\Rightarrow \left[\ln|x| + \frac{a}{x}\right]_1^2 = \ln 2 - \frac{1}{7}$$

$$\Rightarrow \ln 2 - \ln 1 + \frac{a}{2} - a = \ln 2 - \frac{1}{7}$$

$$\Rightarrow -\frac{a}{2} = -\frac{1}{7}$$

$$\Rightarrow a = \frac{2}{7}$$

$\therefore$The value of $7a - 3 = 7 \times \frac{2}{7} - 3$

$$= 2 - 3 = -1$$

**11.** *(a)* Given, curves $xy + 4y = 16$ and $x + y = 6$

On solving both equation, we get

$$(6 - x)(x + 4) = 16$$

$\Rightarrow \quad x^2 - 2x - 8 = 0$

$\Rightarrow \quad x = -2, 4$

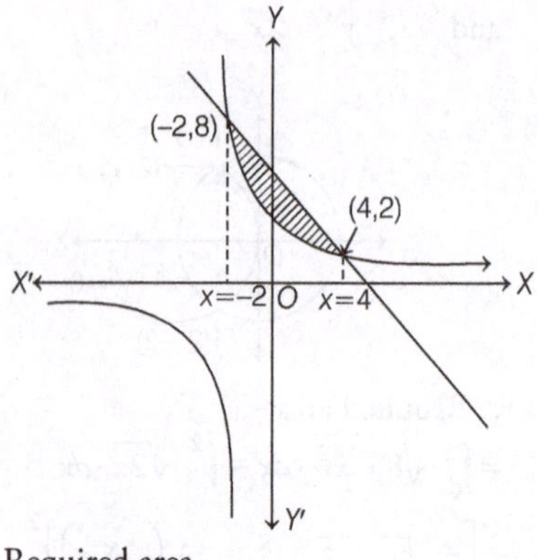


$\therefore$ Required area

$$= \int_{-2}^{4} \left[ (6 - x) - \left( \frac{16}{x+4} \right) \right] dx$$

$$= (24 - 8 - 16 \log 8) - (-12 - 2 - 16 \log 2)$$

$$= 30 - 32 \log 2$$

**12.** *(b)* Given, curve

$y^2 = 4(x - 2)$ and line $y = 2x - 8$

Firstly, we draw the graph.

$\therefore$ Required area

$$= \int_{-2}^{4} x_l \, dy - \int_{-2}^{4} x_P \, dy$$

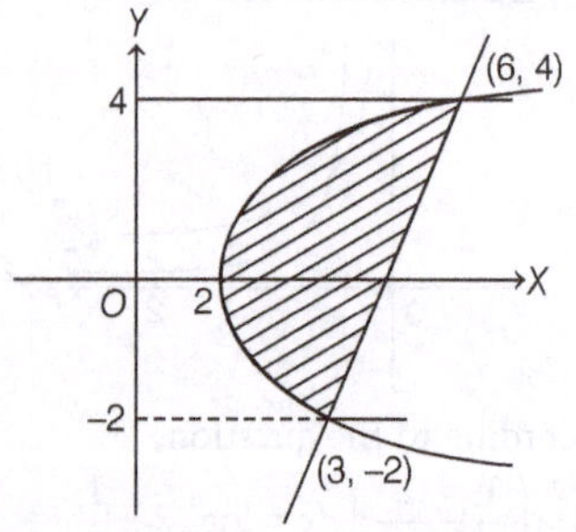


$$= \int_{-2}^{4} \left[ \left( \frac{y+8}{2} \right) - \left( \frac{y^2}{4} + 2 \right) \right] dy$$

$$= \left[ \frac{1}{2} \left( \frac{y^2}{2} + 8y \right) - \left( \frac{y^3}{12} + 2y \right) \right]_{-2}^{4}$$

$$= \left[ \frac{y^2}{4} - \frac{y^3}{12} + 2y \right]_{-2}^{4}$$

$$= \left[ \left( 4 - \frac{16}{3} + 8 \right) - \left( 1 + \frac{2}{3} - 4 \right) \right]$$

$$= [9] = 9 \text{ sq units}$$

**13.** *(b)* We have,

$$x^2 + (y-2)^2 \le 2^2$$

and $\quad x^2 \ge 2y$

Solving the given equation of parabola and circle, we get

$$2y + (y-2)^2 = 4$$

$$y = 0 \text{ or } 2$$

If $y = 0$, then $x = 0$ and

If $y = 2$, then $x = \pm 2$

The intersecting point of circle and parabola are $(0, 0), (-2, 2), (2, 2)$.

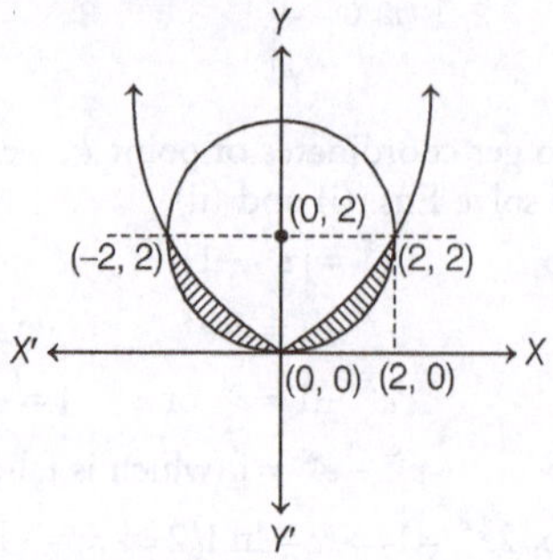


Area of the shaded region

$$= 2\int_0^2 \left[ \frac{x^2}{2} - (\sqrt{4 - x^2} + 2) \right] dx$$

$$= 2 \left| \left[ \frac{x^3}{6} - \frac{x}{2}\sqrt{4 - x^2} - 2\sin^{-1}\left( \frac{x}{2} \right) - 2x \right]_0^2 \right|$$

$$= 2 \left| \left[ \frac{8}{6} - 2 \times \frac{\pi}{2} - 4 \right] \right| = 2 \left| \left[ \frac{4}{3} - \pi - 4 \right] \right|$$

$$= 2\pi + \frac{16}{3} \text{ sq units}$$

**14.** *(b)* Given region is,

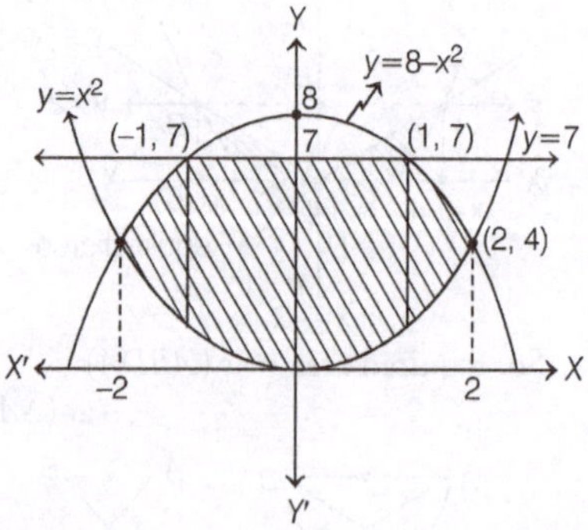


$\therefore$ Required area

$$= 2\int_0^1 (7 - x^2) \, dx + 2\int_1^2 (8 - x^2 - x^2) \, dx$$

$$= 2 \left[ 7x - \frac{x^3}{3} \right]_0^1 + 2 \left[ 8x - \frac{2x^3}{3} \right]_1^2$$

$$= 2\left( \frac{20}{3} \right) + 2\left( \frac{10}{3} \right)$$

$$\Rightarrow \frac{40}{3} + \frac{20}{3} = \frac{60}{3} = 20$$

**15.** *(b)* We have, $xy \le 8, 1 \le y \le x^2$

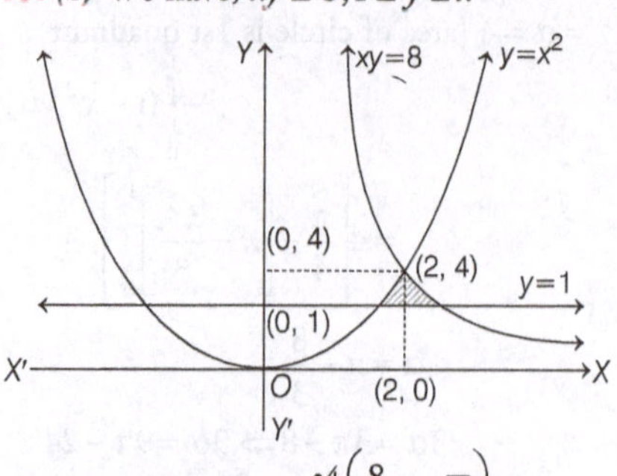


$$\therefore \text{ Required area} = \int_1^4 \left( \frac{8}{y} - \sqrt{y} \right) dy$$

$$= \left[ 8 \log y - \frac{2}{3} y^{3/2} \right]_1^4$$

$$= \left[ 8 \log 4 - \frac{2}{3}(4)^{3/2} \right] - \left[ 8 \log 1 - \frac{2}{3} \right]$$

$$= 16 \log 2 - \frac{14}{3}$$

**16.** *(d)*

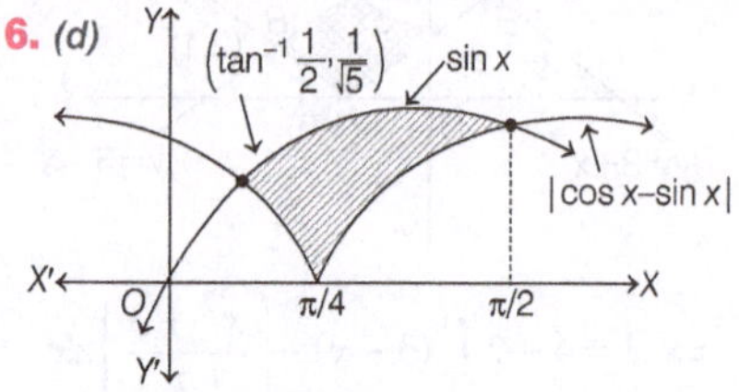


$\therefore \cos x - \sin x = \sin x,$

$\cos x - \sin x = -\sin x$

$\Rightarrow \quad \cos x = 2 \sin x, \cos x = 0$

$\Rightarrow \quad \tan x = \frac{1}{2}, x = \frac{\pi}{2}$

$$\therefore \int_{\tan^{-1} 1/2}^{\pi/4} (\sin x - \cos x + \sin x) \, dx + \int_{\pi/4}^{\pi/2} (\sin x + \cos x - \sin x) \, dx$$

$$= \int_{\tan^{-1} 1/2}^{\pi/4} (2 \sin x - \cos x) \, dx + \int_{\pi/4}^{\pi/2} \cos x \, dx$$

$$= -2 \cos x - \sin x \Big|_{\tan^{-1}\frac{1}{2}}^{\pi/4} + \sin x \Big|_{\pi/4}^{\pi/2}$$

$$= \sqrt{5} - 2\sqrt{2} + 1$$

**17.** *(d)* We have, $y \le 4x^2$

$$y \ge \frac{x^2}{9} \quad \text{and} \quad y \le 4$$

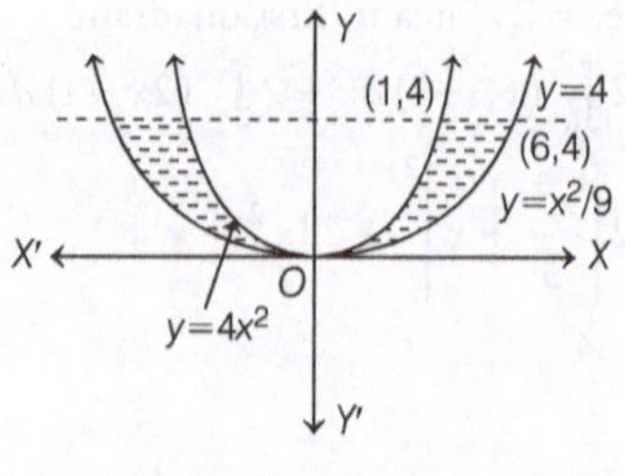

Area of shaded region

$$= 2\int_0^4 \left(3\sqrt{y} - \frac{\sqrt{y}}{2}\right) dy$$

$$= 2\left[2y^{3/2} - \frac{1}{3}y^{3/2}\right]_0^4$$

$$= \frac{10}{3}[y^{3/2}]_0^4 = \frac{80}{3}$$

**18.** *(c)* Given, $y^2 = 8x + 4$ and

$x^2 + y^2 + 4\sqrt{3}x - 4 = 0$

∴ Curves are

$x = \frac{y^2 - 4}{8}$ and $(x + 2\sqrt{3})^2 + y^2 = 16$

For point of intersection, put $y^2 = 8x + 4$ in the second equation,

$x^2 + 8x + 4 + 4\sqrt{3}x - 4 = 0$

$\Rightarrow x = 0$ or $x = -8 - 4\sqrt{3}$ (rejected)

$\Rightarrow y = 2$ or $y = -2$

Take the area about $Y$-axis,

$$A = 2\int_0^2 (\sqrt{16 - y^2} - 2\sqrt{3})dy - \int_0^2 \left(\frac{y^2 - 4}{8}\right) dy$$

$$= 2\left[\frac{y}{2}\sqrt{16 - y^2} + \frac{16}{2}\sin^{-1}\left(\frac{y}{4}\right) - 2\sqrt{3}y - \frac{y^3}{24} + \frac{y}{2}\right]_0^2$$

$$= \frac{1}{3}[8\pi + 4 - 12\sqrt{3}]$$

**19.** *(c)* Given, $y^2 = 8x$ and $y = \sqrt{2x}$

Lines for triangle,

$y = \sqrt{2x}, x = 1, y = 2\sqrt{2}$

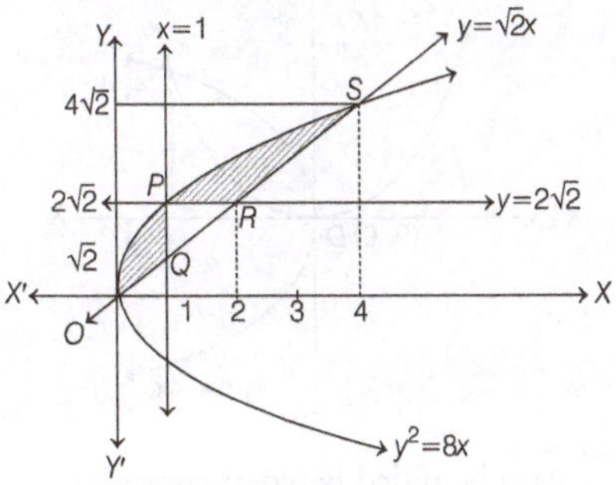


Required area

$$= \int_0^4 (y_{\text{curve}} - y_{\text{line}})\, dx - \text{Area of } \Delta PQR$$

$$= \int_0^4 (\sqrt{8x} - \sqrt{2x})\, dx - \frac{1}{2} \times \sqrt{2} \times 1$$

$$= 2\sqrt{2}\left(\frac{16}{3} - \frac{0}{3}\right) - \sqrt{2}\left(\frac{16}{2} - \frac{0}{2}\right) - \frac{\sqrt{2}}{2}$$

$$= \frac{32\sqrt{2}}{3} - 8\sqrt{2} - \frac{\sqrt{2}}{2}$$

$$= \frac{64\sqrt{2} - 48\sqrt{2} - 3\sqrt{2}}{6} = \frac{13\sqrt{2}}{6}$$

**20.** *(a)*

**key idea** (i) First define the function according as modulus.

(ii) Then, break the integration into two part as $I = \int_0^{\pi/4} + \int_{\pi/4}^{\pi/2}$ and then solve it further.

Area

$$= \int_0^{\pi/2} ((\cos x + \sin x) - |\cos x - \sin x|)\, dx$$

$$= \int_0^{\pi/4} ((\cos x + \sin x) - (\cos x - \sin x))\, dx$$

$$+ \int_{\pi/4}^{\pi/2} ((\cos x + \sin x) - (\sin x - \cos x))\, dx$$

$$= 2\int_0^{\pi/4} \sin x\, dx + 2\int_{\pi/4}^{\pi/2} \cos x\, dx$$

$$= 2\left(\frac{-1}{\sqrt{2}} + 1\right) + 2\left(1 - \frac{1}{\sqrt{2}}\right)$$

$$= 2\sqrt{2}(\sqrt{2} - 1)$$

**21.** *(d)* The given region is $\begin{cases} x \ge 0 \\ y \ge 2x^2 \\ y \le 4 - 2x \end{cases}$

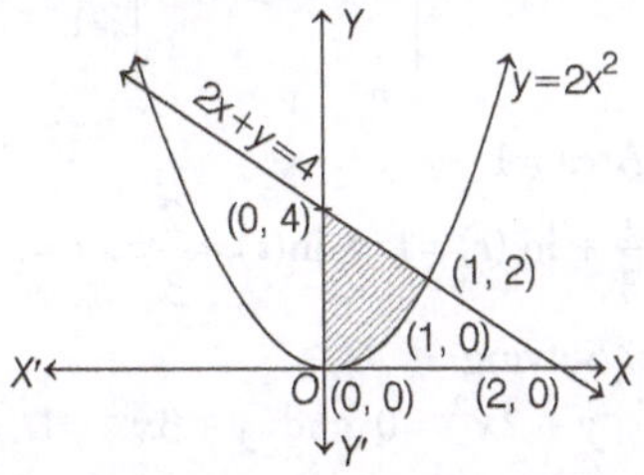


$4 - 2x = 2x^2 \Rightarrow 2x^2 + 2x = 4$

$\Rightarrow \quad x = 1, -2 \quad$ [Reject $-2$]

Area of shaded region

$$= \int_0^1 (4 - 2x)\, dx - \int_0^1 2x^2\, dx$$

$$= [4x - x^2]_0^1 - \left[\frac{2}{3}x^3\right]_0^1$$

$$= (4 - 1) - \left(\frac{2}{3}\right) = \frac{7}{3}$$

**22.** *(b)* Given,

$R = \{(x, y) : 5x^2 \le y \le 2x^2 + 9\}$

Here, we have two curves $y = 5x^2$ and $y = 2x^2 + 9$, point of intersection of both curves is find by solving both equations i.e.

$$5x^2 = 2x^2 + 9$$

$$\Rightarrow \quad x^2 = 3 \Rightarrow x = \pm\sqrt{3}$$

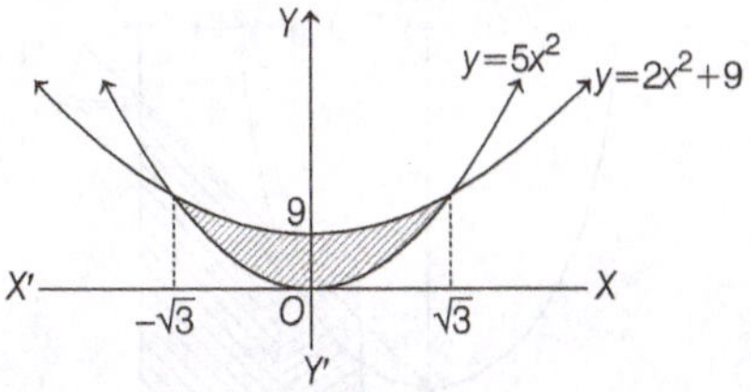


$$\therefore \text{Area} = \int_{-\sqrt{3}}^{\sqrt{3}} (2x^2 + 9 - 5x^2)\, dx$$

$$= 2\int_0^{\sqrt{3}} (9 - 3x^2)\, dx = 2[9x - x^3]_0^{\sqrt{3}}$$

$$= 2[9\sqrt{3} - 3\sqrt{3}]$$

$= 12\sqrt{3}$ sq units.

**23.** *(b)*

**key idea** (i) $\int \sqrt{a^2 - x^2}\,dx = \frac{x}{2}\sqrt{a^2 - x^2} + \frac{a^2}{2}\sin^{-1}\left(\frac{x}{a}\right) + C$

(ii) $\int \sqrt{x}\, dx = \frac{2}{3}x^{3/2} + C$

Given, equation of circle

$\Rightarrow \quad x^2 + y^2 = 36$

Equation of parabola $\Rightarrow y^2 = 9x$

We have to find area of shaded region.

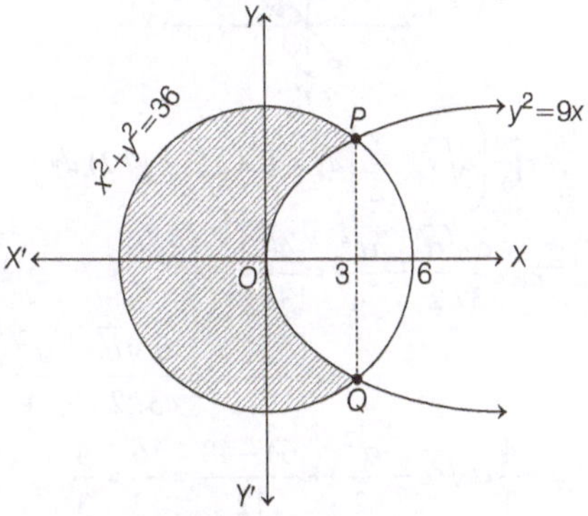


Required area

$$= \pi r^2 - 2\left[\int_0^3 \sqrt{9x}\, dx + \int_3^6 \sqrt{36 - x^2}\, dx\right]$$

$$= 36\pi - 12\sqrt{3} - 2\left|\frac{x}{2}\sqrt{36 - x^2} + 18\sin^{-1}\left(\frac{x}{6}\right)\right|_3^6$$

$$= 36\pi - 12\sqrt{3} - 2\left[9\pi - \left(\frac{9\sqrt{3}}{2} + 3\pi\right)\right]$$

$$= (24\pi - 3\sqrt{3})$$

**24.** *(b)* Given region $\{(x, y): 0 \le y \le x^2 + 1, 0 \le y \le x + 1,$ $\frac{1}{2} \le x \le 2\}$ = Shaded region in the figure

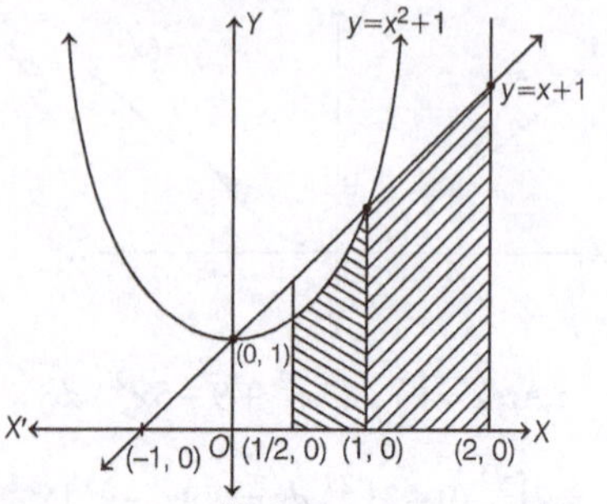


$\therefore$ Area of required region

$$= \int_{1/2}^{1} (x^2 + 1)\, dx + \int_{1}^{2} (x + 1) dx$$

$$= \left[\frac{x^3}{3} + x\right]_{1/2}^{1} + \left[\frac{x^2}{2} + x\right]_{1}^{2}$$

$$= \frac{1}{3} + 1 - \frac{1}{24} - \frac{1}{2} + \frac{4}{2} + 2 - \frac{1}{2} - 1$$

$$= \frac{79}{24} \text{ sq units}$$

**25.** *(b)* According to the question,

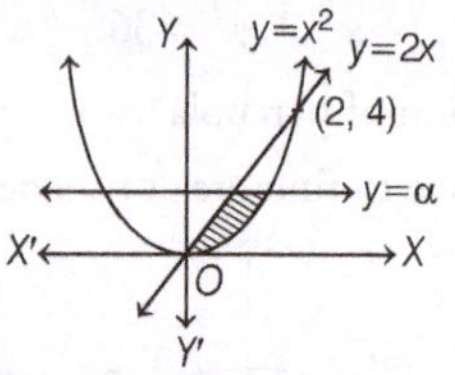


$$\int_0^{\alpha} \left(\sqrt{y} - \frac{y}{2}\right) dy = \int_{\alpha}^{4} (\sqrt{y} - y/2)\, dy$$

$$\Rightarrow \frac{\alpha\sqrt{\alpha}}{3/2} - \frac{\alpha^2}{4} = \frac{4(2)}{3/2} - \frac{16}{4} - \frac{\alpha\sqrt{\alpha}}{3/2} + \frac{\alpha^2}{4}$$

$$\Rightarrow \frac{4}{3}\alpha\sqrt{\alpha} - \frac{\alpha^2}{2} + \frac{64 - 48}{12} = \frac{16}{12} = \frac{4}{3}$$

$$\Rightarrow 3\alpha^2 - 8\alpha^{3/2} + 8 = 0$$

**26.** *(b)* The area of the region $\{(x, y) \in R^2 : x^2 \le y \le 3 - 2x\}$ is the area of region bounded by curves $y = x^2$ and $y = 3 - 2x$.

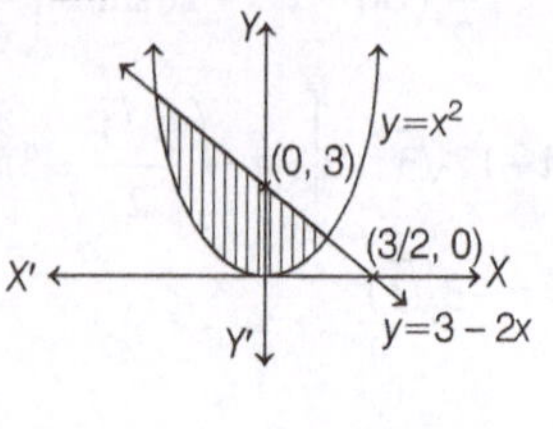


On solving $y = 3 - 2x$ and $y = x^2$,

we have $x^2 = 3 - 2x \Rightarrow x^2 + 2x - 3 = 0$

$\Rightarrow x = -3, 1$, then $(x, y) = (-3, 9)$ and $(1, 1)$

So, required area $= \int_{-3}^{1} [(3 - 2x) - x^2] dx$

$$= \left[3x - \frac{2x^2}{2} - \frac{x^3}{3}\right]_{-3}^{1}$$

$$= \left(3 - 1 - \frac{1}{3}\right) - (-9 - 9 + 9)$$

$$= 9 + \frac{5}{3} = \frac{32}{3} \text{ sq units}$$

**27.** *(b)* Given $y = x^2$ and $y = 1/x$ is (1, 1).

Area bounded by the curves is the region *MNOP* is given as

$$\text{Area} = \int_0^1 x^2 dx + \int_1^t \frac{1}{x} dx$$

$$= \left[\frac{x^3}{3}\right]_0^1 + [\ln(x)]_1^t = \frac{1}{3} + \ln(t)$$

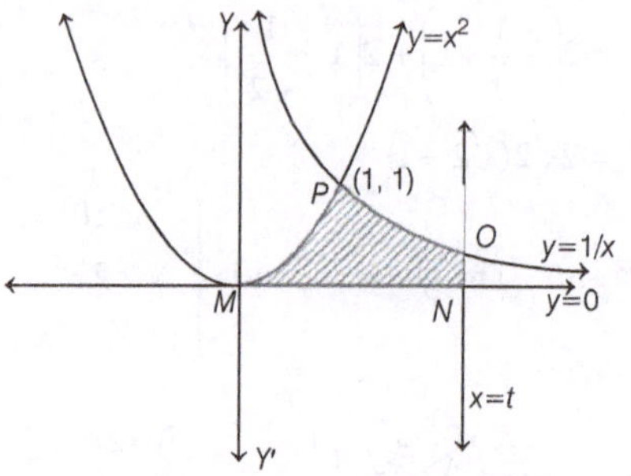


$\because$ Area = 1

$$\Rightarrow \frac{1}{3} + \ln(t) = 1 \Rightarrow \ln(t) = \frac{2}{3} \Rightarrow t = e^{2/3}$$

**28.** *(c)* Solving

$$y + 2x^2 = 0 \text{ and } y + 3x^2 = 1$$

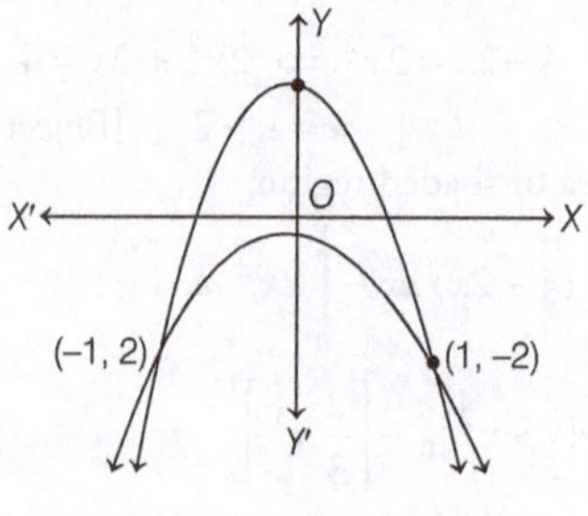


Point of intersection (1, – 2) and (–1, – 2).

The required area

$$= 2\int_0^1 ((1 - 3x^2) - (-2x^2))\, dx$$

$$= 2\int_0^1 (1 - x^2) dx$$

$$= 2\left(x - \frac{x^3}{3}\right)_0^1 = \frac{4}{3}$$

**29.** *(a)* Given curves are

$$y = \sqrt{x} \quad \text{...(i)}$$

and $\quad 2y - x + 3 = 0 \quad$ ...(ii)

On solving Eqs. (i) and (ii), we get

$$2\sqrt{x} - (\sqrt{x})^2 + 3 = 0$$

$$\Rightarrow \quad (\sqrt{x})^2 - 2\sqrt{x} - 3 = 0$$

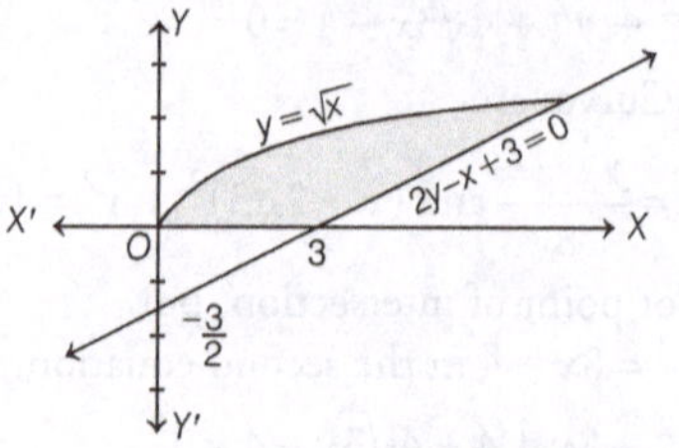


$\Rightarrow (\sqrt{x} - 3)(\sqrt{x} + 1) = 0$

$\Rightarrow \sqrt{x} = 3 \quad [\because \sqrt{x} = -1$ is not possible$]$

$\therefore \quad y = 3$

$\therefore$ Required area $= \int_0^3 (x_2 - x_1)\, dy$

$$= \int_0^3 \{(2y + 3) - y^2\}\, dy$$

$$= \left[y^2 + 3y - \frac{y^3}{3}\right]_0^3$$

$$= 9 + 9 - 9 = 9$$

**30.** *(c)* For the point of intersection, solve $y^2 = 4x$ and $x^2 = 4y$.

$$\Rightarrow \quad \left(\frac{x^2}{4}\right)^2 = 4x \Rightarrow x^4 = 4^3 x$$

$$\Rightarrow \quad x = 0, 4$$

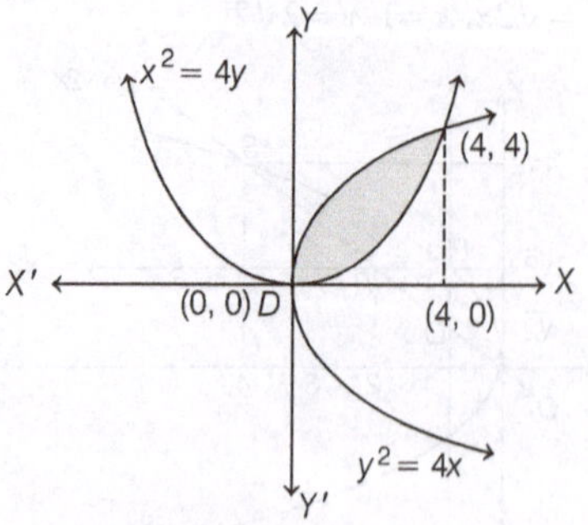


$\therefore$ Area bounded between curves

$$= \int_0^4 \left(\sqrt{4x} - \frac{x^2}{4}\right) dx = \left[2 \cdot \frac{x^{3/2}}{3/2} - \frac{x^3}{12}\right]_0^4$$

$$= \frac{4}{3} \cdot (4)^{3/2} - \frac{(4)^3}{12} = \frac{16}{3}$$

**31.** (a)

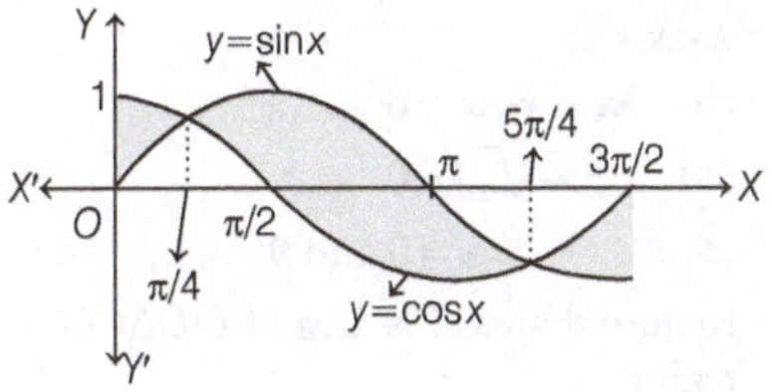


$\therefore$ Required area

$= \int_0^{\pi/4} (\cos x - \sin x)\, dx + \int_{\pi/4}^{5\pi/4} (\sin x - \cos x)\, dx + \int_{5\pi/4}^{3\pi/2} (\cos x - \sin x)\, dx$

$= [\sin x + \cos x]_0^{\pi/4} + [-\cos x - \sin x]_{\pi/4}^{5\pi/4} + [\sin x + \cos x]_{5\pi/4}^{3\pi/2}$

$= \left(\frac{1}{\sqrt{2}} + \frac{1}{\sqrt{2}} - 1\right) + \left[\left(\frac{1}{\sqrt{2}} + \frac{1}{\sqrt{2}}\right) - \left(\frac{-1}{\sqrt{2}} - \frac{1}{\sqrt{2}}\right)\right] + \left[(-1+0) - \left(\frac{-1}{\sqrt{2}} - \frac{1}{\sqrt{2}}\right)\right]$

$= \frac{8}{\sqrt{2}} - 2 = (4\sqrt{2} - 2)$ sq units

**32.** (c) Required area, $A = \int_0^1 (\sqrt{x} - x)\, dx$

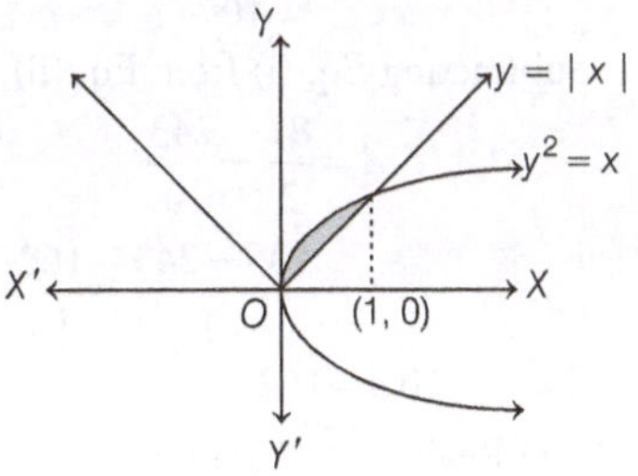


$= \left[\frac{2}{3}x^{3/2} - \frac{x^2}{2}\right]_0^1 = \frac{2}{3} - \frac{1}{2}$

$= \frac{1}{6}$ sq unit

**33.** (a)

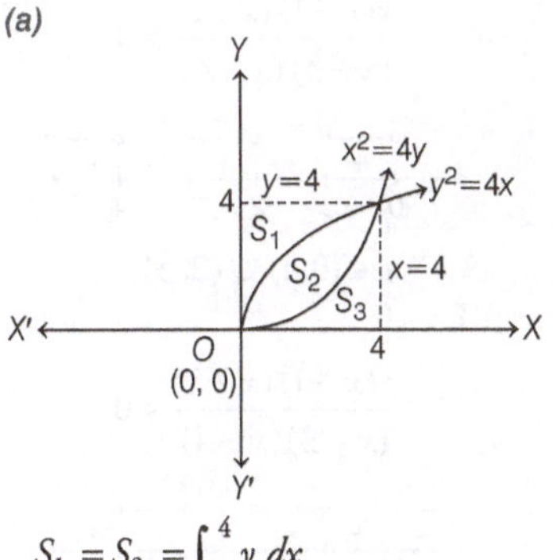


$S_1 = S_3 = \int_0^4 y\, dx$

$= \int_0^4 \frac{x^2}{4}\, dx = \frac{1}{4}\left[\frac{x^3}{3}\right]_0^4$

$\Rightarrow S_1 = S_3 = \frac{1}{12} \times 64 = \frac{16}{3}$ sq units ...(i)

and $S_2 + S_3 = \int_0^4 \sqrt{4x}\, dx$

$= 2\left[\frac{x^{3/2}}{3/2}\right]_0^4 = \frac{4}{3} \times 8$

$\Rightarrow S_2 = \frac{32}{3} - \frac{16}{3}$ [from Eq. (i)]

$\Rightarrow S_2 = \frac{16}{3}$ sq units

$\therefore S_1 : S_2 : S_3 = \frac{16}{3} : \frac{16}{3} : \frac{16}{3} = 1:1:1$

**Shortcut Method**

$S_1 = S_3 = \int_0^4 \frac{x^2}{4}\, dx = \frac{1}{4}\left[\frac{x^3}{3}\right]_0^4 = \frac{16}{3}$ sq units

and $S_2 + S_3 = \int_0^4 \sqrt{4x}\, dx = \frac{32}{3}$ sq units

$S_2 = \frac{16}{3}$ sq units

$\therefore S_1 : S_2 : S_3 = \frac{16}{3} : \frac{16}{3} : \frac{16}{3} = 1:1:1$

**34.** (c) Given, $x(1 + y^2) = 1$ and $y^2 = 2x$

$\Rightarrow x(1 + 2x) = 1$

$\Rightarrow x + 2x^2 - 1 = 0$

$\Rightarrow 2x^2 + x - 1 = 0$

$\Rightarrow 2x^2 + 2x - x - 1 = 0$

$\Rightarrow 2x(x + 1) - (x + 1) = 0$

$\Rightarrow (2x - 1)(x + 1) = 0$

$\Rightarrow x = -1, \frac{1}{2}$

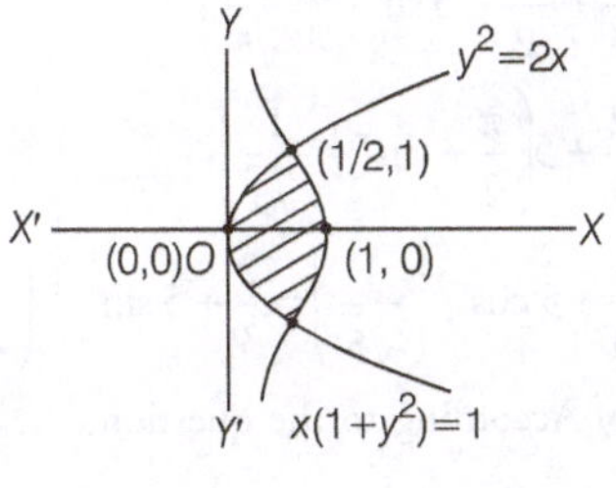


$\therefore$ Area $= 2\int_0^1 (x_2 - x_1)\, dy$

$= 2\int_0^1 \left(\frac{1}{y^2 + 1} - \frac{y^2}{2}\right) dy$

$= 2\left[\tan^{-1} y - \frac{y^3}{6}\right]_0^1$

$= 2\left[\left(\frac{\pi}{4} - \frac{1}{6}\right) - 0\right] = \frac{\pi}{2} - \frac{1}{3}$

**35.** (b) Let $Q(t, t)$ be any point on $y = x$.

and $P$ be any point on the curve

$y = 9 - \frac{11}{3}x^2$

So, coordinates of $P$ is $\left(t, 9 - \frac{11}{3}t^2\right)$

Let us consider $PQRS$ be the required rectangle

So, ar($\Delta PQRS$) $= A = QR \times PQ$

$= QR \times (SO - RO)$

$= t\left(9 - \frac{11t^2}{3} - t\right)$

$= 9t - t^2 - \frac{11t^3}{3}$

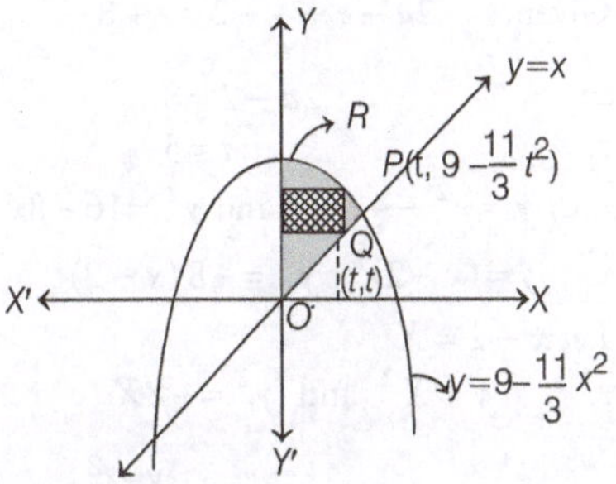


Now, $\frac{dA}{dt} = 9 - 2t - 11t^2$

For maxima on minima put $\frac{dA}{dt} = 0$

$\Rightarrow 11t^2 + 2t - 9 = 0$

$\Rightarrow 11t^2 + 11t - 9t - 9 = 0$

$\Rightarrow 11t(t + 1) - 9(t + 1) = 0$

$\Rightarrow (t + 1)(11t - 9) = 0$

$\Rightarrow t = -1, 9/11$

Hence, at $t = 9/11$ gives us maximum value.

So, put $t = 9/11$ in $A$.

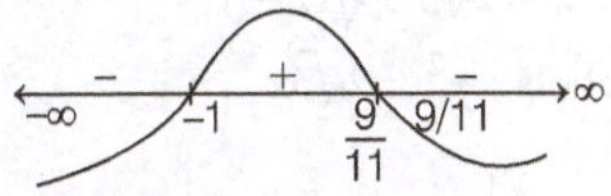


$\therefore$ Largest area

$= \frac{9}{11}\left(9 - \frac{11}{3} \times \frac{81}{121} - \frac{9}{11}\right)$

$= \frac{9}{11} \times \frac{63}{11} = \frac{567}{121}$ sq units

**36.** (c) $R = \{(x, y) : -1 \le x \le 1,$

$0 \le y \le a + e^{|x|} - e^{-x}, a > 0\}$

$y = a + e^{|x|} - e^{-x}$

$y = \begin{bmatrix} a + e^{-x} - e^{-x} = a, & -1 < x < 0 \\ a + e^x - e^{-x}, & 0 \le x < 1 \end{bmatrix}$

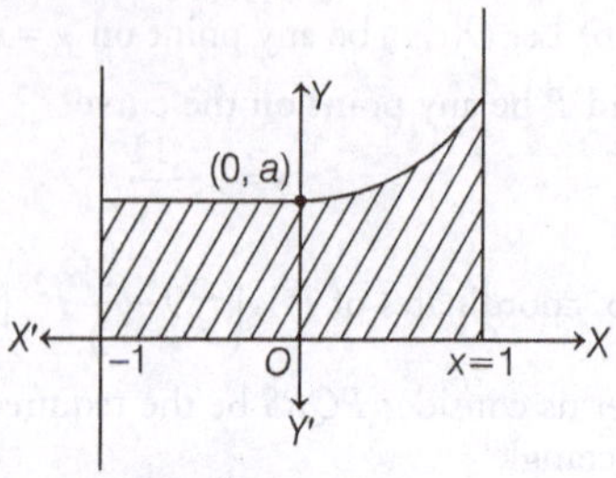


Required area

$$= a + \int_0^1 (a + e^x - e^{-x})\, dx$$

$$= a + a + (e^1 - 1) + (e^{-1} - 1)$$

$$= 2a + e + \frac{1}{e} - 2$$

Given, $\quad 2a + e + \frac{1}{e} - 2 = e + 8 + \frac{1}{e}$

$\Rightarrow \quad 2a - 2 = 8$

$\therefore \quad a = 5$

**37.** *(d)* $y = x^2 - 4x + 4$ and $y^2 = 16 - 8x$

$y = (x-2)^2, \; y^2 = -8(x-2)$

Let $x - 2 = X$

$\therefore \quad y = X^2$ and $y^2 = -8X$

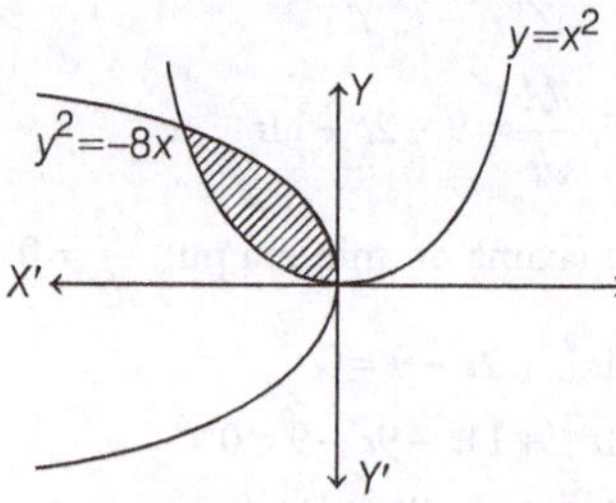


$\therefore$ Area enclosed $= \dfrac{16ab}{3} = \dfrac{16 \times \frac{1}{4} \times \frac{8}{4}}{3} = \dfrac{8}{3}$

**38.** *(c)* Circle : $(x - 2\sqrt{3})^2 + y^2 = 12$

$\Rightarrow \quad x = 2\sqrt{3} \pm \sqrt{12 - y^2}$

Parabola : $y^2 = 2\sqrt{3}x$

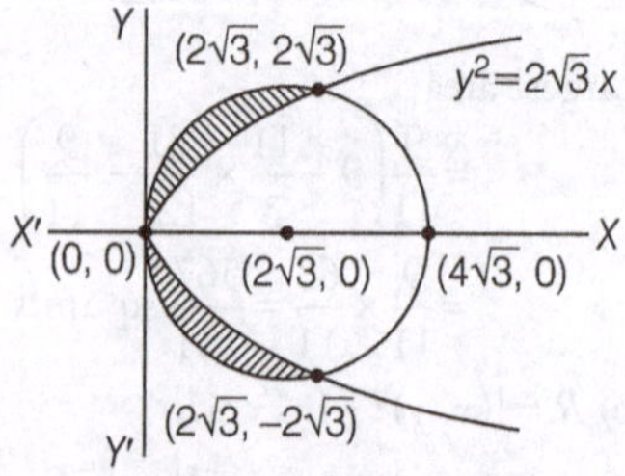


Point of intersection $\Rightarrow y^2 = 12$

$\Rightarrow \quad y = \pm 2\sqrt{3}$

$\therefore \quad x = 2\sqrt{3}$

Area of required region

$$= 2\int_0^{2\sqrt{3}} (x_1 - x_2)\, dy$$

$$= 2\int_0^{2\sqrt{3}} \left( \frac{y^2}{2\sqrt{3}} - 2\sqrt{3} + \sqrt{12 - y^2} \right) dy$$

$$= 2\left[ \frac{y^3}{3\sqrt{3}} - 2\sqrt{3}y + \frac{y}{2}\sqrt{12 - y^2} + \frac{12}{2}\sin^{-1}\frac{y}{2\sqrt{3}} \right]_0^{2\sqrt{3}}$$

$$= 2\left[ 4 - 12 + 0 + 6 \times \frac{\pi}{2} \right] = 6\pi - 16$$

**39.** *(b)* Given equation of parabola $y^2 = 4x$ and circle is $x^2 + y^2 = 5$

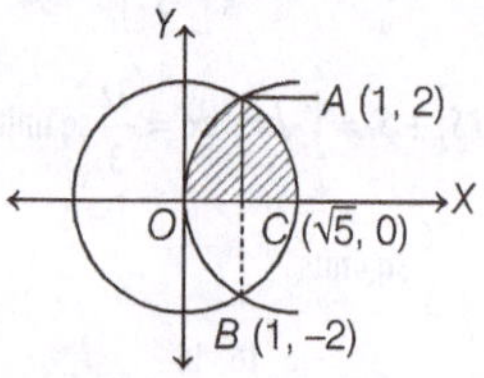


Now, area of shaded region

$$= A_1 = \int_0^1 \sqrt{4x}\, dx + \int_1^{\sqrt{5}} \sqrt{5 - x^2}\, dx$$

$$\Rightarrow A_1 = \frac{4}{3}[x^{3/2}]_0^1 + \left[ \frac{x}{2}\sqrt{5 - x^2} + \frac{5}{2}\sin^{-1}\frac{x}{\sqrt{5}} \right]_1^{\sqrt{5}}$$

$$= \frac{1}{3} + \frac{5\pi}{4} - \frac{5}{2}\sin^{-1}\left(\frac{1}{\sqrt{5}}\right)$$

$\therefore$ Required area $= 2A_1$

$$= \frac{2}{3} + \frac{5\pi}{2} - 5\sin^{-1}\left(\frac{1}{\sqrt{5}}\right)$$

$$= \frac{2}{3} + 5\left(\frac{\pi}{2} - \sin^{-1}\frac{1}{\sqrt{5}}\right)$$

$$= \frac{2}{3} + 5\cos^{-1}\left(\frac{1}{\sqrt{5}}\right) = \frac{2}{3} + 5\sin^{-1}\left(\frac{2}{\sqrt{5}}\right)$$

**40.** *(b)* According to the question,

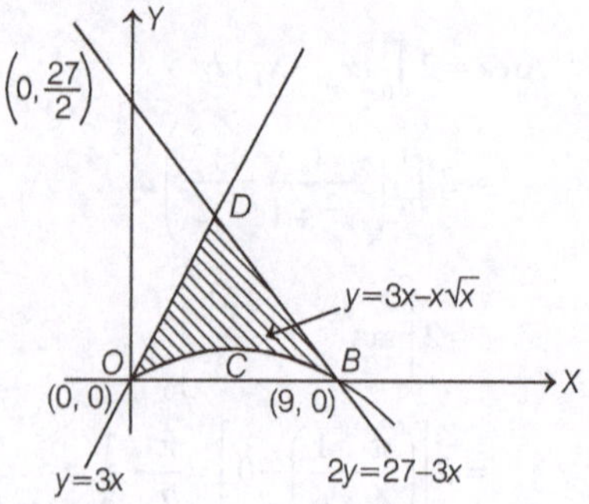


Given curve, $y = 3x - x\sqrt{x}$ intersect $X$-axis at

$\Rightarrow \quad 3x - x\sqrt{x} = 0$

$\Rightarrow \quad x(3 - \sqrt{x}) = 0$

$\Rightarrow \quad x = 0$ and $9$

Required area $A$ is area of $ODBCO$ region.

Coordinates of $D$ can be find by solving $y = 3x$ and $2y = 27 - 3x$

$\Rightarrow \quad 2y = 27 - y$

$\Rightarrow \quad 3y = 27$

$\Rightarrow \quad y = 9$ and $x = 3$

$\therefore \quad D \equiv (3, 9)$

Area of $\Delta ODB = \frac{1}{2} \times 9 \times 9 = \frac{81}{2}$ ...(i)

Area of $OCBO = \int_0^9 3x - x\sqrt{x}\, dx$

$$= \left[ \frac{3x^2}{2} - \frac{2x^{5/2}}{5} \right]_0^9$$

$$= \left[ \frac{243}{2} - \frac{2}{5} \times 243 \right]$$

$$= \frac{243}{10} \quad \text{...(ii)}$$

Subtracting Eq. (i) from Eq. (ii), we get

$$A = \frac{81}{2} - \frac{243}{10}$$

$$= \frac{405 - 243}{10} = \frac{162}{10}$$

So, $\quad 10A = 162$

**41.** *(a)* Region $y^2 \le 4x, \; x < 4$

$$\frac{xy(x-1)(x-2)}{(x-3)(x-4)} > 0$$

**Case I** $y > 0$

The inequality becomes,

$$\frac{x(x-1)(x-2)}{(x-3)(x-4)} > 0$$

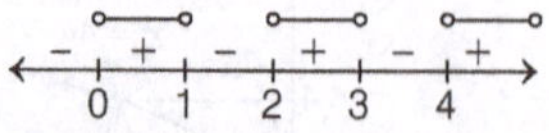


$x \in (0, 1) \cup (2, 3)$

**Case II** $y < 0$

$$\Rightarrow \quad \frac{x(x-1)(x-2)}{(x-3)(x-4)} < 0$$

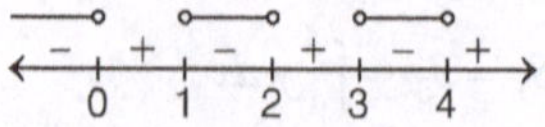


$x \in (1, 2) \cup (3, 4)$

Graph

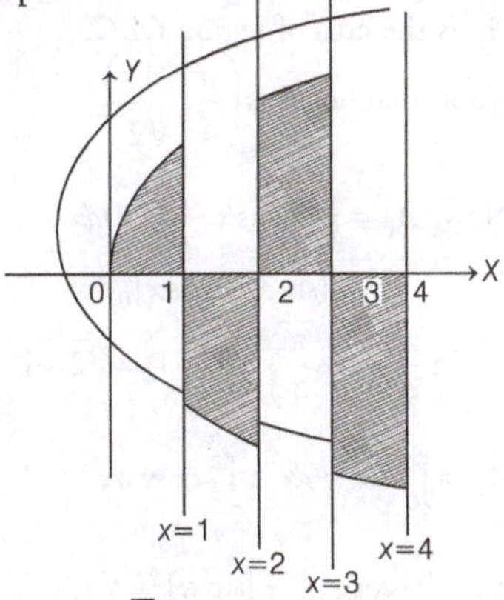


$$y = 2\sqrt{x}\, dx$$

$\therefore$ Area $= \int_0^4 2\sqrt{x}$ [due to symmetry]

$$= 2\frac{x^{3/2}}{3/2} = \frac{4}{3}(4)^{3/2} = \frac{32}{3}$$

**42.** *(c)* The given region is

$\{(x, y), x^2 \le y \le |x^2 - 4|, y \ge 1\}$

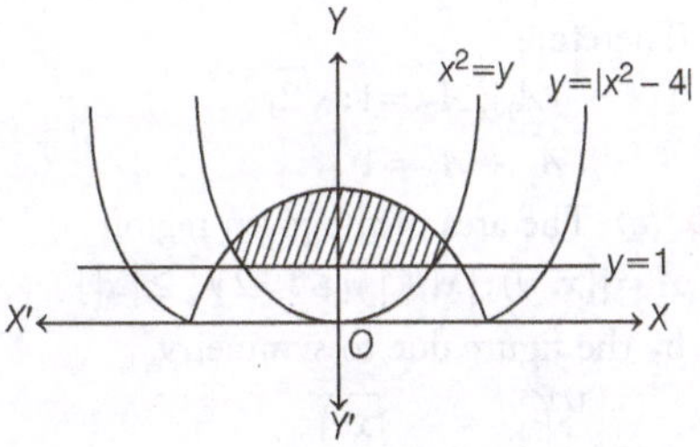


Then, area of required shaded region is

$$2\left[\int_1^2 \sqrt{y}dy + \int_2^4 \sqrt{4-y}dy\right]$$

$$= 2\left[\frac{2}{3}(y^{3/2})_1^2 - \left(\frac{2}{3}(4-y)\right)^{3/2}\right]_2^4$$

$$= \frac{4}{3}[\{(2)^{3/2} - (1)^{3/2}\}$$
$$-\{(4-4)^{3/2} - (4-2)^{3/2}\}]$$

$$= \frac{4}{3}[2\sqrt{2} - 1 - 0 + 2\sqrt{2}] = \frac{4}{3}[4\sqrt{2} - 1]$$

**43.** *(c)* Given, $f(y) = \max\{\sin x, \cos x\}$, $-\pi \le x \le \pi$

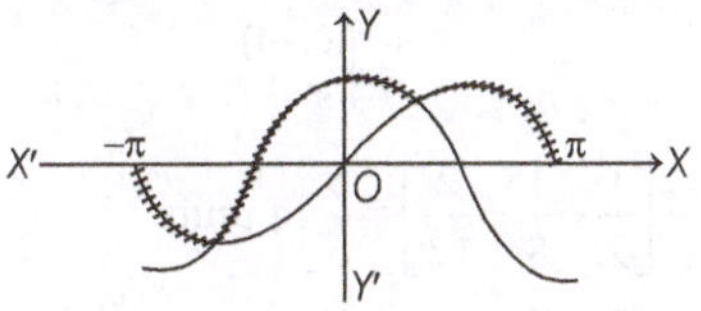


Thus, required area of the region enclosed by the curve $f(x)$, $-\pi \le x \le \pi$, and the $X$-axis is

$$\int_{-\pi}^{-3\pi/4}(-\sin x)\, dx + \int_{-3\pi/4}^{-\pi/2}(-\cos x)\, dv$$
$$+ \int_{-\pi 2}^{\pi 4}(\cos x)\, dx + \int_{\pi/4}^{\pi}\sin x\, dx$$
$$= (\cos x)_{-\pi}^{-3\pi/4} - (\sin x)_{-3\pi/4}^{-\pi/2}$$
$$+ (\sin x)_{-\pi/2}^{\pi/4} - (\cos x)_{\pi/4}^{\pi}$$
$$= \left(-\frac{1}{\sqrt{2}} + 1\right) - \left(-1 + \frac{1}{\sqrt{2}}\right) + \left(\frac{1}{\sqrt{2}} + 1\right)$$
$$- \left(-1 - \frac{1}{\sqrt{2}}\right)$$
$$= -\frac{1}{\sqrt{2}} + 1 + 1 - \frac{1}{\sqrt{2}} + \frac{1}{\sqrt{2}} + 1 + 1 + \frac{1}{\sqrt{2}}$$
$$= 4$$

**44.** *(c)* Given equation of curves are

$$y^2 + 4x = 4 \quad \text{...(i)}$$

and

$$y - 2x = 2 \quad \text{...(ii)}$$

$\Rightarrow \quad y = 2x - 2$

On putting in Eq. (i), we get

$$(2x + 2)^2 + 4x = 4$$

$\Rightarrow 4x^2 + 4 + 8x + 4x = 4$

$\Rightarrow 4x^2 + 12x = 0 \Rightarrow 4x(x + 3) = 0$

$\Rightarrow \quad x = 0, -3$

From Eq. (ii), we get

When $x = 0$, $y = 2$

When $x = -3$, $y = -4$

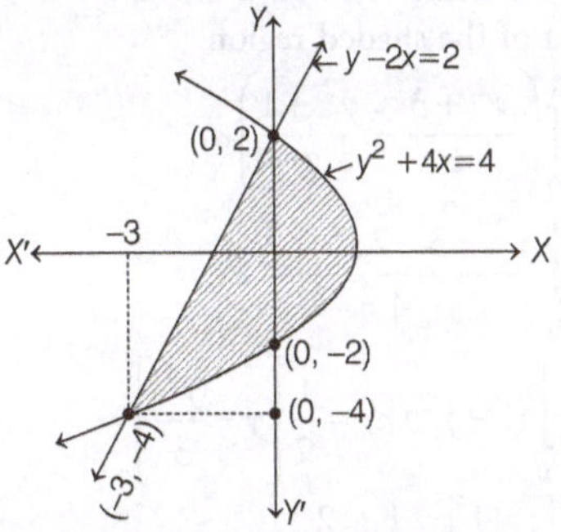


So, points of intersection of Eqs. (i) and (ii) are (0, 2) and (−3, −4).

Now, required area

$$= \int_{y=-4}^{y=2}(x_{\text{right}} - x_{\text{left}})\, dy$$

$$= \int_{-4}^{2}\left(\frac{4-y^2}{4} - \frac{y-2}{2}\right)dy$$

$$= \int_{-4}^{2}\left(2 - \frac{y^2}{4} - \frac{y}{2}\right)dy = \left[2y - \frac{y^3}{12} - \frac{y^2}{4}\right]_{-4}^{2}$$

$$= \left(4 - \frac{8}{12} - 1\right) - \left(-8 + \frac{16}{3} - 4\right)$$

$$= \left(3 - \frac{2}{3}\right) + 12 - \frac{16}{3} = 9 \text{ sq units}$$

**45.** *(b)*

> **key idea** (i) Find the point of intersection by solving both the curve.
> (ii) And then take the area about Y-axis.

Given, three curves $y = \log(x + e^2)$, $x = \log_e(2/y)$ and $x = \log_e 2$.

When, $x = 0$ is substituted in both the curves, $x = \log_e(2/y)$ and $y = \log(x + e^2)$, we get $y = 2$

$\therefore$The intersecting point of two curves is (0, 2).

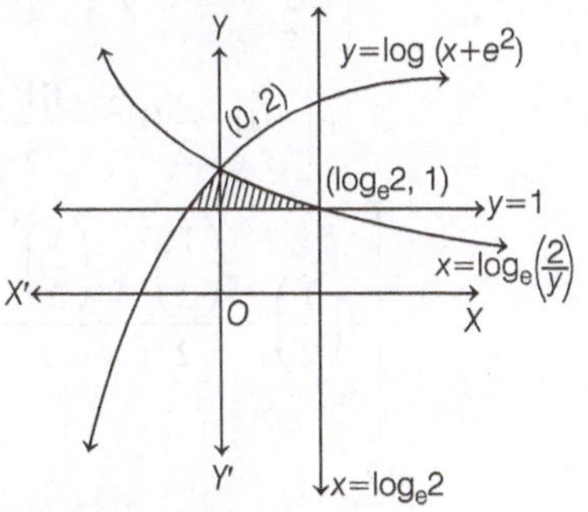


Area of the shaded portion

$$= \int_1^2 \left[\log\left(\frac{2}{y}\right) - (e^y - e^2)\right] dy$$

$$= \left[y\log\frac{2}{y}\right]_1^2 - \int_1^2 \frac{y}{2} \times \left(-\frac{2}{y^2}\right) \times y dy$$
$$- [e^y - ye^2]_1^2$$

$$= 2\log 1 - \log 2 + [y]_1^2$$
$$- [e^2 - 2e^2 - e + e^2]$$

$$= -\log 2 + [2 - 1] + e$$
$$= 1 + e - \log 2$$

**46.** *(b)* Given that

$$A = \{(x, y); x^2 \le y \le \min\{x + 2, 4 - 3x\}\}$$

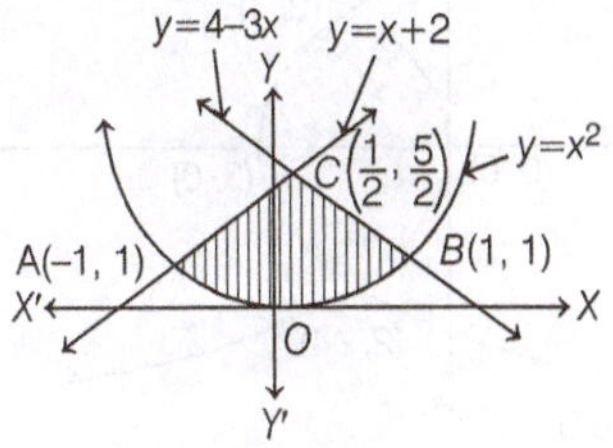


Thus, the point of intersection of $y = x^2$ and $y = 4 - 3x$ is $B(1, 1)$ and the point of intersection of $y = x + 2$ and $y = x^2$ is $A(-1, 1)$ and also the point of intersection of $y = x + 2$ and $y = 4 - 3x$ is $C\left(\frac{1}{2}, \frac{5}{2}\right)$.

So, the required region $ACBA$ is

$$\int_{-1}^{1/2}(x+2-x^2)dx+\int_{1/2}^{1}(4-3x-x^2)dx$$

$$=\left[\frac{x^2}{2}+2x-\frac{x^3}{3}\right]_{-1}^{1/2}+\left[4x-\frac{3x^2}{2}-\frac{x^3}{3}\right]_{1/2}^{1}$$

$$=\left[\frac{1}{2}\left(\frac{1}{4}\right)+2\left(\frac{1}{2}\right)-\frac{1}{3}\left(\frac{1}{2}\right)^3\right]-\left[\frac{(-1)^2}{2}+2(-1)-\frac{(-1)^3}{3}\right]$$

$$+\left[4(1)-\frac{3}{2}(1)^2-\frac{(1)^3}{3}\right]-\left[4\left(\frac{1}{2}\right)-\frac{3\left(\frac{1}{2}\right)^2}{2}-\frac{\left(\frac{1}{2}\right)^3}{3}\right]$$

$$=\frac{17}{6}$$

**47.** *(d)* Given, curves $y^2=8x$ ... (i)

$y^2=16(3-x)$ ...(ii)

Now, from Eqs. (i) and (ii), we get

$8x=16(3-x)$

$\Rightarrow\ 24x=48$

$\Rightarrow\ x=2$

and $y^2=16$

$\Rightarrow\ y=\pm 4$

Common points of both the graphs $(2, 4)$ : $(2, -4)$

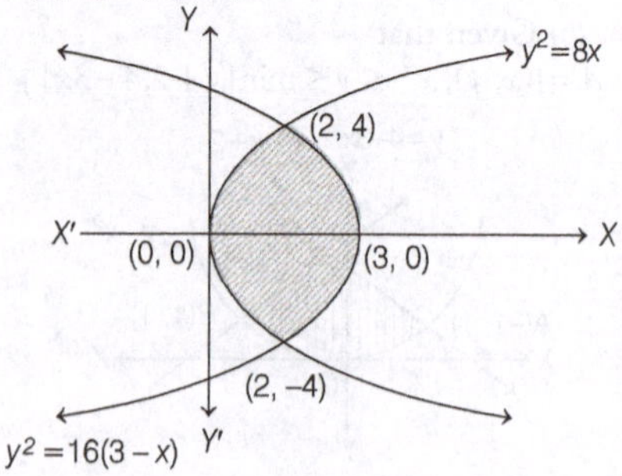


$$\text{Required area}=\left[\int_{-4}^{4}\left[\left(3-\frac{y^2}{16}\right)-\frac{y^2}{8}\right]\right]dy$$

$$=2\int_0^4\left[\left(3-\frac{y^2}{16}\right)-\frac{y^2}{8}\right]dy$$

$$=2\left[3y-\frac{y^3}{48}-\frac{y^3}{24}\right]_0^4$$

$$=2\left[12-\frac{64}{48}-\frac{64}{24}\right]=16$$

**48.** *(a)* Given $y^2=2x-1$ ... (i)

and $y^2=4x-3$ ... (ii)

$$y^2=2\left(x-\frac{1}{2}\right)\text{ and }y^2=4\left(x-\frac{3}{4}\right)$$

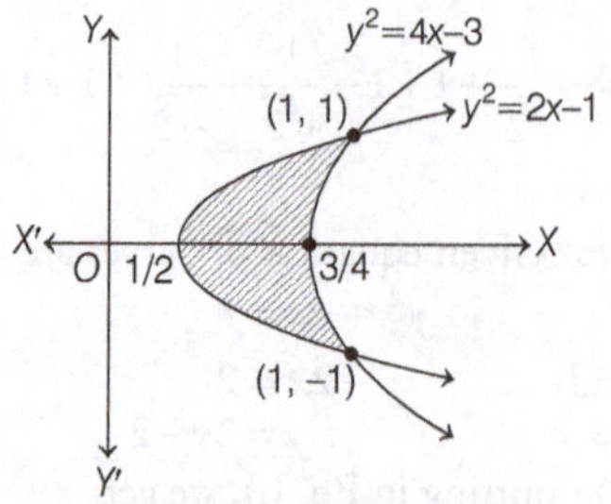


On comparing both the equations, we get

$2x-1=4x-3$

$\Rightarrow\ 2x=2\Rightarrow x=1$

$\Rightarrow\ y^2=2\times 1-1$

$\Rightarrow\ y^2=1\Rightarrow y=\pm 1$

The two intersection point of the parabolas are $(1, 1)$ and $(1, -1)$.

Area of the shaded region

$$=2\int_0^1\left(\frac{y^2+3}{4}-\frac{y^2+1}{2}\right)dy$$

$$=2\int_0^1\frac{y^2+3-2y^2-2}{4}dy$$

$$=\frac{1}{2}\int_0^1(1-y^2)\,dy=\frac{1}{2}\left[y-\frac{y^3}{3}\right]_0^1$$

$$=\frac{1}{2}\left[1-\frac{1}{3}\right]=\frac{1}{2}\times\frac{2}{3}$$

$$=\frac{1}{3}\text{ sq units}$$

**49.** *(a)*

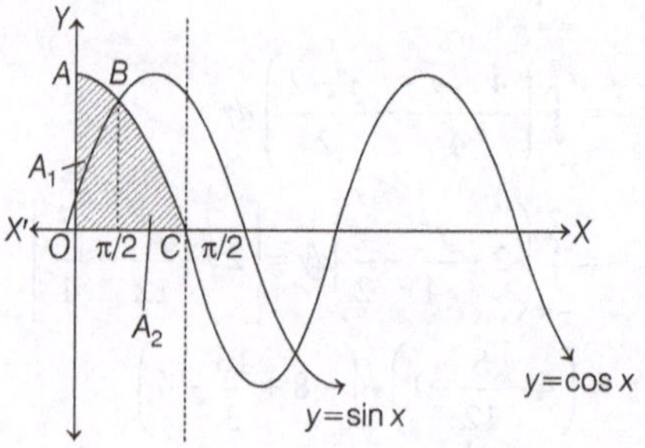


$A_1$ is the area of region $OAB$.

$A_2$ is the area of region $OBC$.

Coordinate of $B$ is $\left(\frac{\pi}{4},\frac{1}{\sqrt{2}}\right)$.

$$\text{Now, }A_1=\int_0^{\frac{\pi}{4}}(\cos x-\sin x)dx$$

$$=[\sin x+\cos x]_0^{\pi/4}$$

$$=\left(\sin\frac{\pi}{4}+\cos\frac{\pi}{4}\right)-(0+1)=\sqrt{2}-1$$

$$A_2=\int_0^{\frac{\pi}{4}}\sin x\,dx+\int_{\frac{\pi}{4}}^{\frac{\pi}{2}}\cos x\,dx$$

$$A_2=[-\cos x]_0^{\frac{\pi}{4}}+[\sin x]_{\frac{\pi}{4}}^{\frac{\pi}{2}}$$

$$=\sqrt{2}(\sqrt{2}-1)$$

Now, $A_1:A_2=(\sqrt{2}-1):\sqrt{2}(\sqrt{2}-1)$

$A_1:A_2=1:\sqrt{2}$

and $A_1+A_2=(\sqrt{2}-1)+\sqrt{2}(\sqrt{2}-1)$

$=(\sqrt{2}-1)(\sqrt{2}+1)$

$A_1+A_2=2-1=1$

Therefore,

$A_1:A_2=1:\sqrt{2}$,

$A_1+A_2=1$

**50.** *(d)* The area of the given region $A=\{(x, y):|x|+|y|\le 1, 2y^2\ge |x|\}$ by the figure due to symmetry

$$A=4\int_0^{1/2}\left(1-x-\sqrt{\frac{x}{2}}\right)dx$$

$$[\because 2y^2+y-1=0\Rightarrow(2y-1)(y+1)=0\Rightarrow y=1/2]$$

$$=4\left[x-\frac{x^2}{2}-\frac{2x^{3/2}}{3\sqrt{2}}\right]_0^{1/2}$$

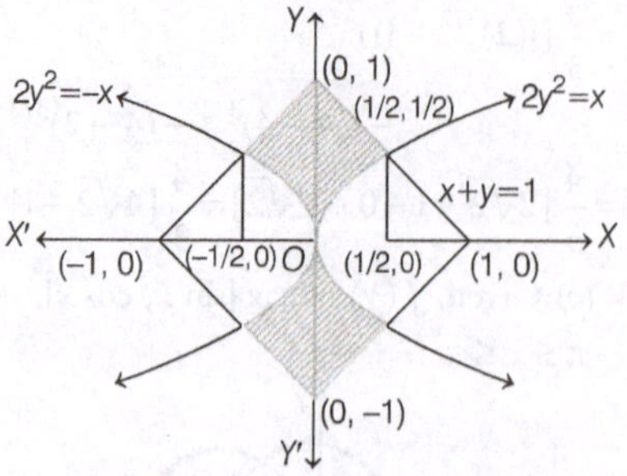


$$=4\left[\frac{1}{2}-\frac{1}{8}-\frac{2}{12}\right]=\frac{5}{6}\text{ sq units}$$

**51.** *(b)* Let the area of the region, enclosed by the parabola $y^2=x$ and straight line $y=x$ is

$$A=\int_0^1(\sqrt{x}-x)dx$$

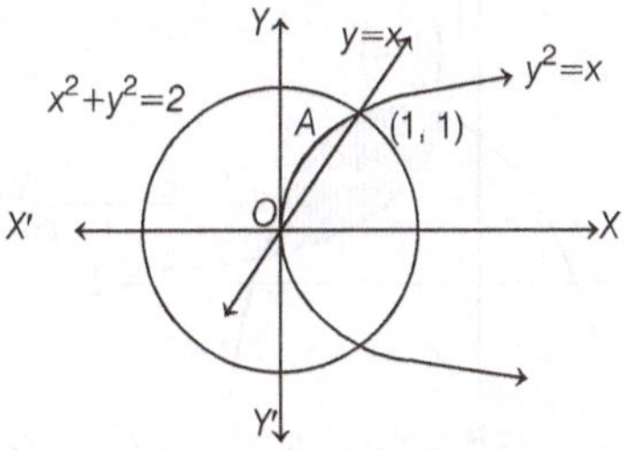


$$=\left[\frac{x^{3/2}}{3/2}-\frac{x^2}{2}\right]_0^1=\left(\frac{1}{3/2}-\frac{1}{2}\right)$$

$$=\left(\frac{2}{3}-\frac{1}{2}\right)=\frac{4-3}{6}=\frac{1}{6}$$

$\because$ Area of circle having radius $r=\sqrt{2}$ unit is $\pi r^2=2\pi$

$\therefore$ The area of the region, enclosed by the circle $x^2+y^2=2$, which is not common to the region bounded by the parabola $y^2=x$ and the straight line $y=x$, is $2\pi-\frac{1}{6}=\frac{1}{6}(12\pi-1)$.

**52.** *(c)* Given, equation of curves are

$$y^2=4\lambda x \quad \ldots\text{(i)}$$

and $$y=\lambda x \quad \ldots\text{(ii)}$$

$$\lambda>0$$

Area bounded by above two curve is, as per figure

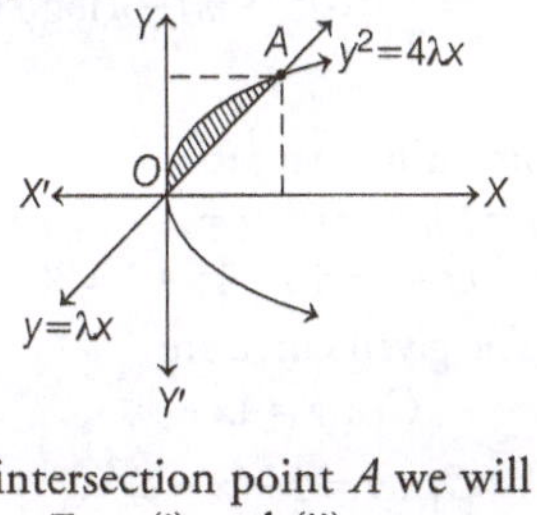


The intersection point $A$ we will get on solving Eqs. (i) and (ii), we get

$$\lambda^2x^2=4\lambda x \Rightarrow x=\frac{4}{\lambda}, \text{ so } y=4$$

So, $$A\left(\frac{4}{\lambda}, 4\right)$$

Now, required area $=\int_0^{4/\lambda}(2\sqrt{\lambda x}-\lambda x)\,dx$

$$=2\sqrt{\lambda}\left[\frac{x^{3/2}}{3/2}\right]_0^{4/\lambda}-\lambda\left[\frac{x^2}{2}\right]_0^{4/\lambda}$$

$$=\frac{4}{3}\sqrt{\lambda}\,\frac{4\sqrt{4}}{\lambda\sqrt{\lambda}}-\frac{\lambda}{2}\left(\frac{4}{\lambda}\right)^2=\frac{32}{3\lambda}-\frac{8}{\lambda}$$

$$=\frac{32-24}{3\lambda}=\frac{8}{3\lambda}$$

It is given that area $=\frac{1}{9} \Rightarrow \frac{8}{3\lambda}=\frac{1}{9}$

$$\Rightarrow \quad \lambda=24$$

**53.** *(d)* Given region

$$A=\left\{(x, y): \frac{y^2}{2}\le x\le y+4\right\}$$

$$\therefore \quad y^2/2=x$$

$$\Rightarrow \quad y^2=2x \quad \ldots\text{(i)}$$

and $x=y+4 \Rightarrow y=x-4 \quad \ldots$(ii)

Graphical representation of $A$ is

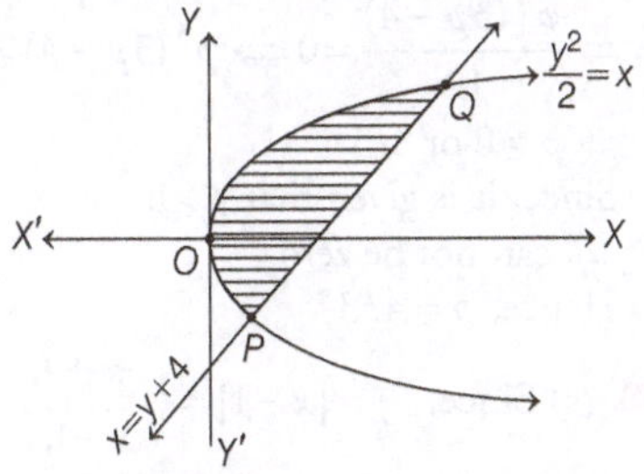


On substituting $y=x-4$ from Eq. (ii) to Eq. (i), we get

$$(x-4)^2=2x \Rightarrow x^2-8x+16=2x$$

$$\Rightarrow x^2-10x+16=0 \Rightarrow (x-2)(x-8)=0$$

$$\Rightarrow \quad x=2, 8$$

$\therefore \quad y=-2, 4$ [from Eq. (ii)]

So, the point of intersection of Eqs. (i) and (ii) are $P(2,-2)$ and $Q(8, 4)$.

Now, the area enclosed by the region $A$

$$=\int_{-2}^{4}\left[(y+4)-\frac{y^2}{2}\right]dy$$

$$=\left[\frac{y^2}{2}+4y-\frac{y^3}{6}\right]_{-2}^{4}$$

$$=\left(\frac{16}{2}+16-\frac{64}{6}\right)-\left(\frac{4}{2}-8+\frac{8}{6}\right)$$

$$=8+16-\frac{32}{3}-2+8-\frac{4}{3}$$

$$=30-12=18 \text{ sq units}$$

**54.** *(b)* Given equation of curve is $x^2=4y$, which represent a parabola with vertex (0, 0) and it open upward

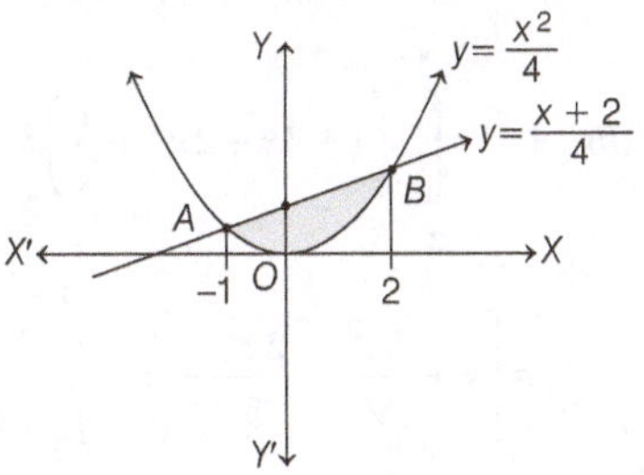


Now, let us find the points of intersection of $x^2=4y$ and $4y=x+2$

For this consider, $x^2=x+2$

$$\Rightarrow x^2-x-2=0$$

$$\Rightarrow (x-2)(x+1)=0 \Rightarrow x=-1, x=2$$

When $x=-1$, then $y=\frac{1}{4}$

and when $x=2$, then $y=1$

Thus, the points of intersection are $A\left(-1, \frac{1}{4}\right)$ and $B(2, 1)$.

Now, required area = area of shaded region

$$=\int_{-1}^{2}\{y\,(\text{line})-y\,(\text{parabola})\}\,dx$$

$$=\int_{-1}^{2}\left(\frac{x+2}{4}-\frac{x^2}{4}\right)dx$$

$$=\frac{1}{4}\left[\frac{x^2}{2}+2x-\frac{x^3}{3}\right]_{-1}^{2}$$

$$=\frac{1}{4}\left[\left(2+4-\frac{8}{3}\right)-\left(\frac{1}{2}-2+\frac{1}{3}\right)\right]$$

$$=\frac{1}{4}\left[5-\frac{1}{2}\right]=\frac{9}{8} \text{ sq units}$$

**55.** *(d)*

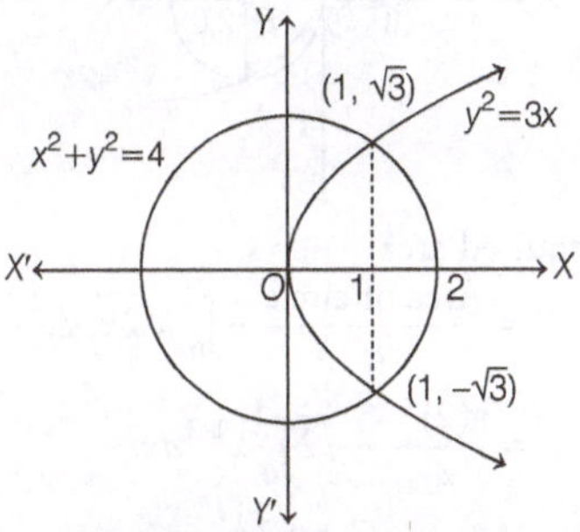


On putting $y^2=3x$, we get

$$x^3+3x-4=0$$

$$\Rightarrow (x+4)(x-1)=0 \Rightarrow x=-4, x=1$$

$$x=1, y=\sqrt{3}$$

Area

$$=\left(\int_0^1\sqrt{3}\cdot\sqrt{x}\,dx+\int_1^2\sqrt{4-x^2}\cdot dx\right)\times 2$$

$$=\left[\sqrt{3}\left(\frac{x^{3/2}}{3/2}\right)_0^1+\left(\frac{x}{2}\sqrt{4-x^2}+2\sin^{-1}\frac{x}{2}\right)_1^2\right]\times 2$$

$$=\left(\sqrt{3}\left(\frac{2}{3}\right)+\left\{2\cdot\frac{\pi}{2}-\left(\frac{\sqrt{3}}{2}+\frac{\pi}{3}\right)\right\}\right)\times 2$$

$$=\left(\frac{1}{2\sqrt{3}}+\frac{2\pi}{3}\right)\times 2$$

$$=\frac{1}{\sqrt{3}}+\frac{4\pi}{3}$$

**56.** *(b)* Given equations of curves are

$$y^2 = 2x \quad ...(i)$$

which is a parabola with vertex (0, 0) and axis parallel to $X$-axis.

And $\quad x^2+y^2=4x \quad ...(ii)$

which is a circle with centre (2, 0) and radius = 2

On substituting $y^2=2x$ in Eq. (ii), we get

$$x^2+2x=4x$$

$$\Rightarrow \quad x^2 = 2x$$

$$\Rightarrow \quad x=0 \text{ or } x=2$$

$$\Rightarrow \quad y=0 \text{ or } y=\pm 2 \quad \text{[using Eq. (i)]}$$

Now, the required area is the area of shaded region, i.e.

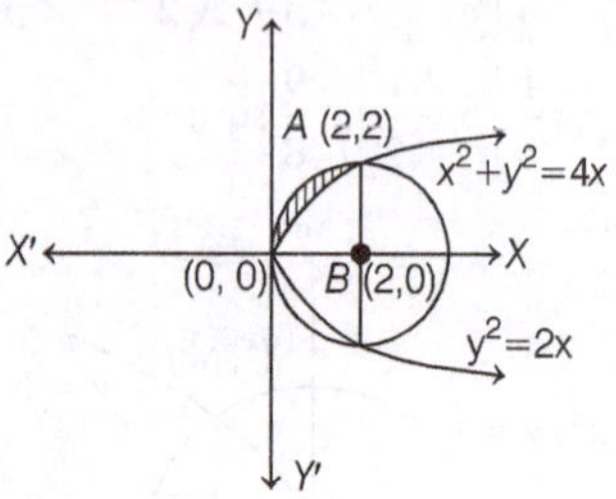


Required area

$$=\frac{\text{Area of circle}}{4}-\int_0^2 \sqrt{2x}\,dx$$

$$=\frac{\pi(2)^2}{4}-\sqrt{2}\int_0^2 x^{1/2}dx$$

$$=\pi-\sqrt{2}\left[\frac{x^{3/2}}{3/2}\right]_0^2$$

$$=\pi-\frac{2\sqrt{2}}{3}[2\sqrt{2}-0]$$

$$=\pi-\frac{8}{3} \text{ sq units}$$

**57.** *(d)* Given, $y=x^2$ and $y=x^3$

$x=0$ and $x=p,\ p>1$

Now, intersecting point is (1, 1).

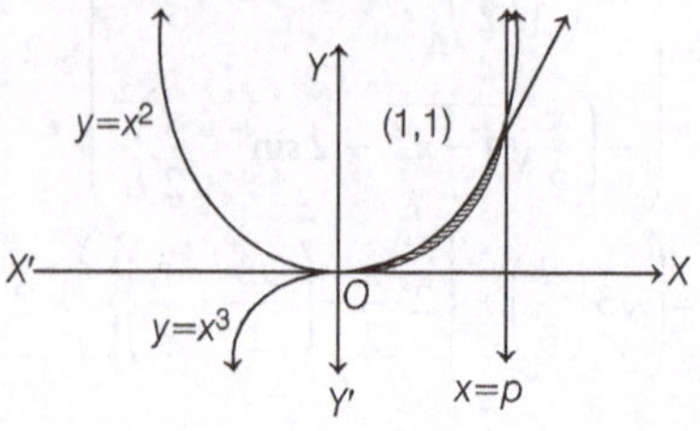


The required area

$$=\int_0^1 (x^2-x^3)dx+\int_1^p (x^3-x^2)dx$$

$$\Rightarrow \frac{1}{6}=\left.\frac{x^3}{3}-\frac{x^4}{4}\right|_0^1+\left.\frac{x^4}{4}-\frac{x^3}{3}\right|_1^p$$

$$\Rightarrow \frac{1}{6}=\left(\frac{1}{3}-\frac{1}{4}\right)+\left(\frac{p^4}{4}-\frac{p^3}{3}-\frac{1}{4}+\frac{1}{3}\right)$$

$$\Rightarrow \frac{1}{6}-\frac{1}{3}+\frac{1}{4}+\frac{1}{4}-\frac{1}{3}=\frac{3p^4-4p^3}{12}$$

$$\Rightarrow \frac{p^3(3p-4)}{12}=0 \Rightarrow p^3(3p-4)=0$$

$$\Rightarrow p=0 \text{ or } 4/3$$

Since, it is given that $p>1$.

$\therefore p$ can not be zero.

Hence, $p=4/3$

**58.** *(c)* Since, $\quad y=|x-1|=\begin{cases}-x+1, & x\le 1\\ x-1, & x>1\end{cases}$

and $\quad y=3-|x|=\begin{cases}3+x, & x\le 0\\ 3-x, & x>0\end{cases}$

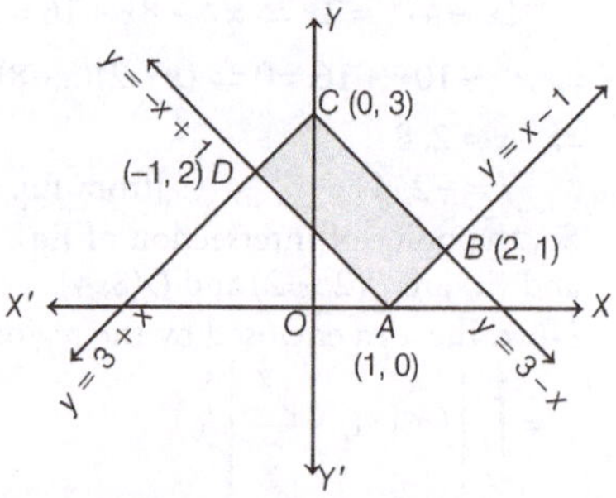


On solving, we get

$$y=x-1 \text{ and } y=3-x$$

$$\Rightarrow \quad x-1=3-x \Rightarrow x=2$$

and $\quad y=3-2 \Rightarrow y=1$

Now, $\quad AB^2=(2-1)^2+(1-0)^2$

$$=1+1=2$$

$$\Rightarrow \quad AB=\sqrt{2}$$

and $BC^2=(0-2)^2+(3-1)^2=4+4=8$

$$\Rightarrow \quad BC=2\sqrt{2}$$

$\therefore$ Area of rectangle $ABCD = AB\times BC$

$$=\sqrt{2}\times 2\sqrt{2}=4 \text{ sq units}$$

**59.** *(c)* $A=\displaystyle\int_{\frac{1}{2}}^{\frac{1+\sqrt{5}}{2}}\left(1+3x-2x^2-\frac{1}{x}\right)dx$

$$=\left[x+\frac{3x^2}{2}-\frac{2x^3}{3}-\ln x\right]_{\frac{1}{2}}^{\frac{1+\sqrt{5}}{2}}$$

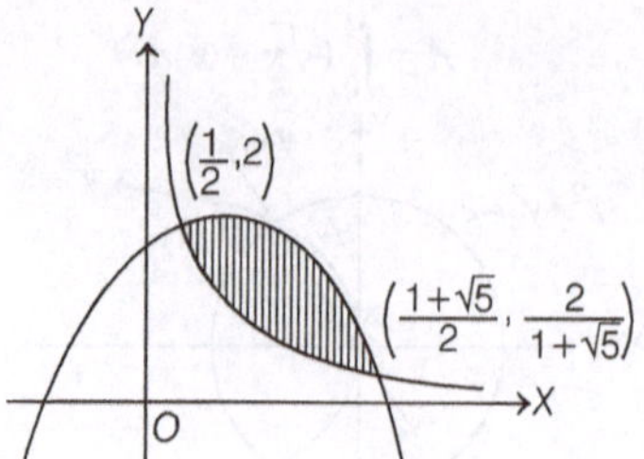


$$=\frac{1+\sqrt{5}}{2}+\frac{3}{2}\left(\frac{1+\sqrt{5}}{2}\right)^2-\frac{2}{3}\left(\frac{1+\sqrt{5}}{2}\right)^3-\ln\left(\frac{1+\sqrt{5}}{2}\right)-\frac{1}{2}-\frac{3}{2}\left(\frac{1}{4}\right)+\frac{2}{3}\left(\frac{1}{8}\right)+\ln\left(\frac{1}{2}\right)$$

$$\Rightarrow A=\frac{1}{2}+\frac{\sqrt{5}}{2}+\frac{3}{8}+\frac{3}{4}\sqrt{5}+\frac{15}{8}-\frac{4}{3}-\frac{2}{3}\sqrt{5}-\frac{1}{2}-\frac{3}{8}+\frac{1}{12}-\ln(1+\sqrt{5})$$

$$\Rightarrow A=\sqrt{5}\left(\frac{1}{2}+\frac{3}{4}-\frac{2}{3}\right)+\frac{15}{8}-\frac{4}{3}+\frac{1}{12}-\ln(1+\sqrt{5})$$

$$\Rightarrow \quad \frac{14}{24}\sqrt{5}+\frac{15}{24}-\ln(1+\sqrt{5})$$

$$=\frac{1}{24}(l\sqrt{5}+m)-n\log_e(1+\sqrt{5}) \quad \text{[given]}$$

By comparing, we get

$$l=14,\ m=15,\ n=1$$

$$\therefore \quad l+m+n=14+15+1=30$$

**60.** *(b)* The given curve are

$$C_1: y=4x-x^2$$

So, $\quad C_1: y=4-(x-2)^2$

$$C_2: (x-4)^2=3y$$

$$C_1: y=-(x^2-4x)$$

$$C_1: y=-(x^2-4x+4-4)$$

$$C_1: y=4-(x-2)^2$$

$$C_1: (x-2)^2=-(y-4)$$

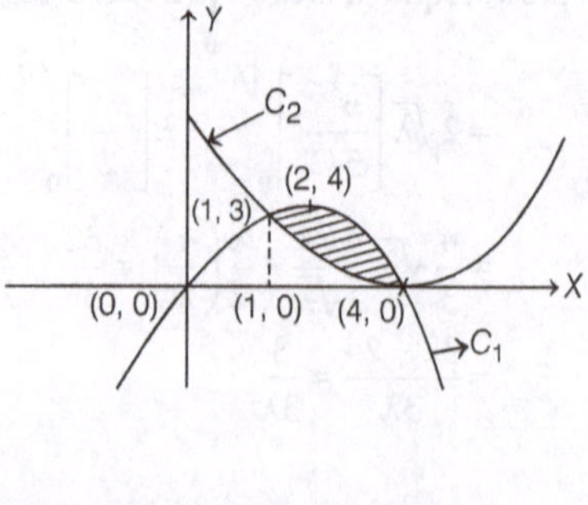

Solving two curves

$$3(4x - x^2) = (x-4)^2$$
$$4x^2 - 20x + 16 = 0$$
$$x^2 - 5x + 4 = 0$$
$$x = 1, 4$$

Both curves intersect at (1, 3) and (4, 0).

$\therefore$ Shaded area $= \int_1^4 (y_{C_1} - y_{C_2})\, dx$

$$= \int_1^4 \left((4x - x^2) - \frac{(x-4)^2}{3}\right) dx$$

$$= \left[2x^2 - \frac{x^3}{3} - \frac{(x-4)^3}{9}\right]_1^4$$

$$= \left(32 - \frac{64}{3} - 0\right) - \left(2 - \frac{1}{3} + 3\right)$$

$= 6$ units

**61.** *(b)* Given that $q$ is the maximum integral value of $p$ in $[0, 10]$ for which the roots of the equation $x^2 - px + \frac{5}{4}p = 0$ are rational.

Now, $D = (-p)^2 - 4 \times 1 \times \frac{5}{4}p$

$$= p^2 - 5p = p(p-5)$$

In $\quad p \in [0, 10]$

Let $\quad p = 10$

$D = 10 \times (10 - 5) = 50$ (reject)

($\because D$ is not a perfect square)

Let $\quad p = 9$

$D = 9 \times (9 - 5) = 36$ (select)

($D$ is a perfect square)

So, $\quad q = 9$

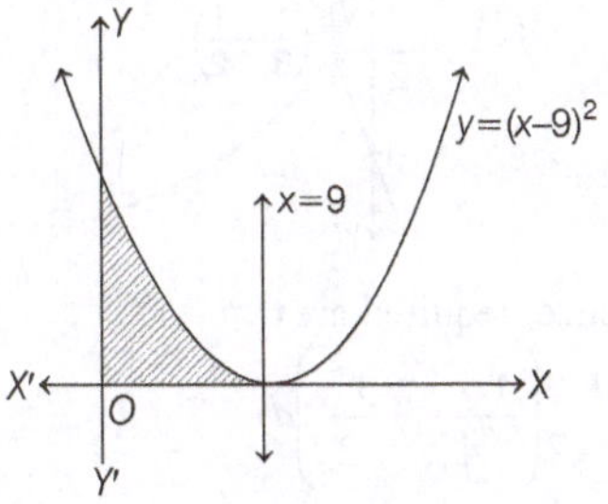


Now, Area $A = \int_0^9 (x-9)^2\, dx$

$$= \left|\frac{(x-9)^3}{3}\right|_0^9$$

$$= 0 - \frac{(-9)^3}{3}$$

$= 243$ unit$^2$

**62.** *(c)* We have, $A = \{(x, y) \in R^2 : y \geq 0,\ 2x \leq y \leq \sqrt{4 - (x-1)^2}\}$

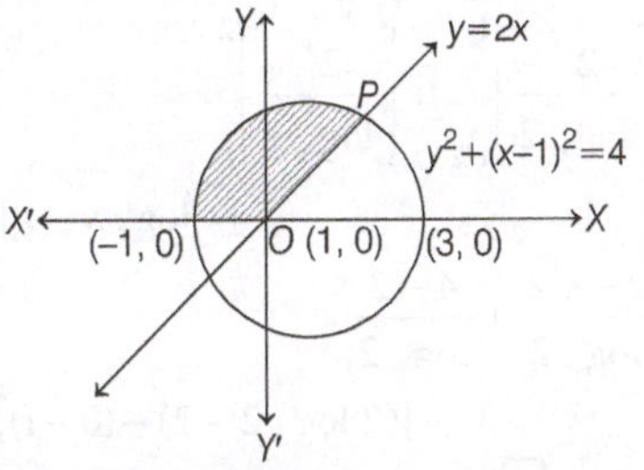


On solving $y^2 + (x-1)^2 = 4$ and $y = 2x$, we get

$$(x-1)^2 + (2x)^2 = 4$$
$$\Rightarrow \quad x^2 - 2x + 1 + 4x^2 = 4$$
$$\Rightarrow \quad 5x^2 - 2x - 3 = 0$$
$$\Rightarrow \quad x = \frac{2 \pm \sqrt{(-2)^2 - 4 \cdot 5(-3)}}{2 \cdot 5}$$
$$= \frac{2 \pm \sqrt{4 + 60}}{2 \cdot 5} = \frac{2 \pm 8}{10} = 1, -\frac{3}{5}$$

For $P$, $x = 1$; $\therefore y = 2$

i.e $P(1, 2)$

$A$ is the shaded region

$$= \left|\int_{-1}^{1} \sqrt{4 - (x-1)^2}\, dx - \int_0^1 (2x)\, dx\right|$$

$$= \left|\left(\frac{(x-1)\sqrt{4-(x-1)^2} + 4\sin^{-1}\left(\frac{x-1}{2}\right)}{2}\right)_{-1}^{1} - (x^2)_0^1\right|$$

$$\left[\because \int \sqrt{a^2 - x^2}\, dx = \frac{x\sqrt{a^2 - x^2} + a^2 \sin^{-1}\left(\frac{x}{a}\right)}{2} + c\right]$$

$$= \left|0 - \left(\frac{(-2)\sqrt{4-(-2)^2} + 4\sin^{-1}\left(\frac{-2}{2}\right)}{2}\right) - (1^2 - 0^2)\right|$$

$= (\pi - 1)$ sq units

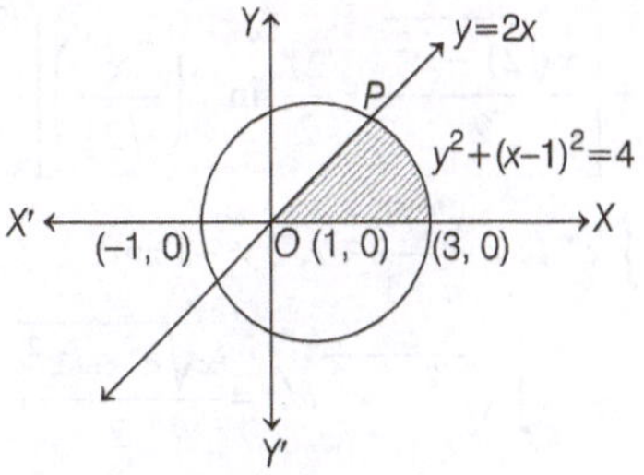


$B$ is the shaded region

$=$ Area of semi-circle $- A$

$$= \left[\frac{\pi(4)}{2} - (\pi - 1)\right] \text{ sq units}$$

$= (\pi + 1)$ sq units

So, required ratio $= \frac{\pi - 1}{\pi + 1}$

**63.** *(c)* For $x^2 + y^2 \leq 21$, $y^2 \leq 4x$, $x \geq 1$

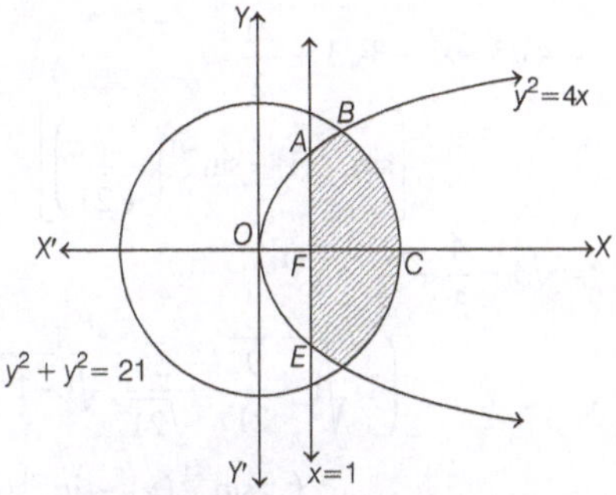


Required area, triangle is the shaded region $ABCDEFA$.

Since, all are symmetric about $X$-axis.

$\therefore \Delta = 2\Delta_1$, where $\Delta_1$ is the shaded region $ABCF$.

On solving $x^2 + y^2 = 21$ and $y^2 = 4x$, we get

i.e. $x^2 + 4x = 21 \Rightarrow x^2 + 4x - 21 = 0$

$$\Rightarrow \quad x = \frac{-4 \pm \sqrt{4^2 - 4 \cdot 1(-21)}}{2 \cdot 1}$$
$$= \frac{-4 \pm 10}{2} = 3, -7$$

For $B$, we get $B(3, \sqrt{4 \cdot 3})$

i.e. $B(3, 2\sqrt{3})$

Also, putting $y = 0$ in $x^2 + y^2 = 21$, we get

$$x = \pm\sqrt{21}$$

$\therefore C(\sqrt{21}, 0)$

Now, $\Delta_1 = \int_1^3 2\sqrt{x}\, dx + \int_3^{\sqrt{21}} \sqrt{21 - x^2}\, dx = \frac{2 \cdot 2}{3}(x^{\frac{3}{2}})_1^3$

$$+\left[\frac{x\sqrt{21-x^2}}{2}+\frac{21}{2}\sin^{-1}\left(\frac{x}{\sqrt{21}}\right)\right]_3^{\sqrt{21}}$$

$\left[\because \int x^n dx = \frac{x^{n+1}}{n+1}+c, n \neq -1 \text{ and}\right.$

$$\left.\int\sqrt{a^2-x^2}\,dx = \frac{x\sqrt{a^2-x^2}}{2}+\frac{a^2}{2}\sin^{-1}\left(\frac{x}{a}\right)+c\right]$$

$$\Rightarrow \Delta_1 = \frac{4}{3}(3\sqrt{3}-1)+\left[\frac{21}{2}\sin^{-1}(1)-\frac{3\sqrt{21-3^2}}{2}-\frac{21}{2}\sin^{-1}\left(\frac{3}{\sqrt{21}}\right)\right]$$

$$=4\sqrt{3}-\frac{4}{3}-3\sqrt{3}+\frac{21}{2}\left[\sin^{-1}(1)-\sin^{-1}\left(\frac{3}{\sqrt{21}}\right)\right]$$

$$=\sqrt{3}-\frac{4}{3}+\frac{21}{2}\sin^{-1}\left(1\cdot\sqrt{1-\frac{9}{21}}-\frac{3}{\sqrt{21}}\cdot\sqrt{1-1^2}\right)$$

$$(\because \sin^{-1}(a)-\sin^{-1}(b) = \sin^{-1}(a\sqrt{1-b^2}-b\sqrt{1-a^2}))$$

$$=\sqrt{3}-\frac{4}{3}+\frac{21}{2}\sin^{-1}\left(\frac{2}{\sqrt{7}}\right)$$

$$\therefore \Delta = 2\Delta_1 = 2\sqrt{3}-\frac{8}{3}+21\sin^{-1}\left(\frac{2}{\sqrt{7}}\right)$$

Thus, $\frac{1}{2}\left[\Delta-21\sin^{-1}\left(\frac{2}{\sqrt{7}}\right)\right]=\sqrt{3}-\frac{4}{3}$

**64.** *(b)* $R=\{(x,y): \max\{0, \log_e x\} \le y \le 2^x, \frac{1}{2}\le x \le 2\}$

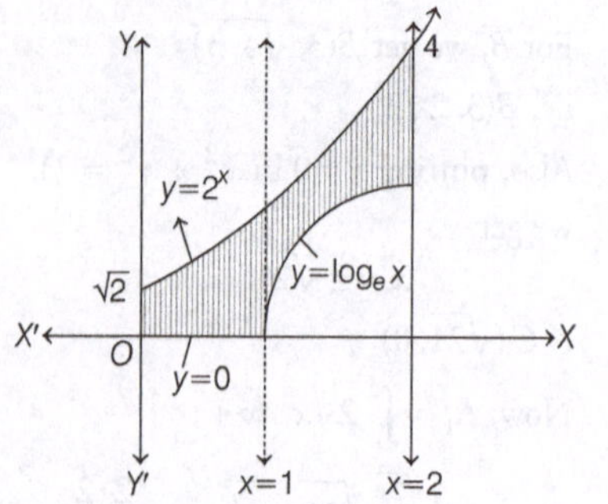


$$\max\{0, \log_e x\} = \begin{cases} 0 & , \frac{1}{2}\le x<1 \\ \log_e x & , 1\le x\le 2\end{cases}$$

$$\text{Area} = \int_{1/2}^{1} 2^x dx + \int_1^2 (2^x - \log_e x)dx$$

$$=\left[\frac{2^x}{\log_e 2}\right]_{1/2}^1+\left[\frac{2^x}{\log_e 2}\right]_1^2-[x\log_e x - x]_1^2$$

$$=\frac{2-\sqrt{2}}{\log_e 2}+\frac{4-2}{\log_e 2}-[(2\log_e 2-2)-(0-1)]$$

$$=\frac{4-\sqrt{2}}{\log_e 2}-2\log_e 2+1$$

On comparing with expression,

$$\alpha(\log_e 2)^{-1}+\beta\log_e 2+\gamma$$

$$\alpha = 4-\sqrt{2}, \beta=-2, \gamma=1$$

$$\therefore (\alpha+\beta-2\gamma)^2 = (4-\sqrt{2}-2-2)^2 = 2$$

**65.** *(c)* Given, equation of curve is

$$4y^2 = x^2(4-x)(x-2)$$

$$\Rightarrow \sqrt{4y^2}=\sqrt{x^2(4-x)(x-2)}$$

$$\Rightarrow |y| = \frac{|x|}{2}\sqrt{(4-x)(x-2)}$$

$$y=\begin{cases}\frac{x}{2}\sqrt{(4-x)(x-2)}, & x>0\\ -\frac{x}{2}\sqrt{(4-x)(x-2)}, & x<0\end{cases}$$

Let $y_1 = \frac{x}{2}\sqrt{(4-x)(x-2)}$

$$y_2 = -\frac{x}{2}\sqrt{(4-x)(x-2)}$$

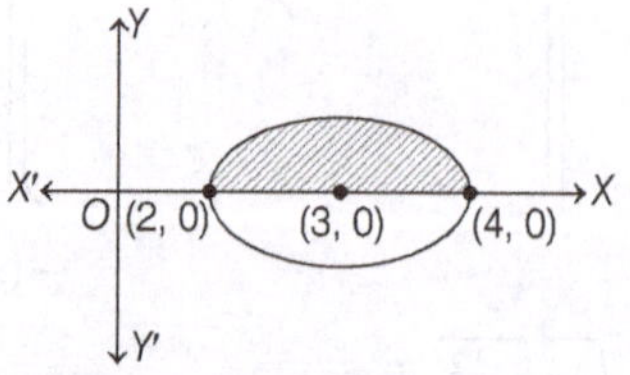


and domain $x\in[2,4]$

$$\therefore \text{Required area} = \int_2^4 (y_1-y_2)\,dx$$

$$=\int_2^4 x\sqrt{(4-x)(x-2)}\,dx \quad \ldots\text{(i)}$$

Using property,

$$\int_a^b f(x)\,dx = \int_a^b f(a+b-x)\,dx$$

From Eq. (i),

Area

$$=\int_2^4 (6-x)\sqrt{(4-x)(x-2)}\,dx \quad \ldots\text{(ii)}$$

From Eqs. (i) and (ii), we get

$$2A = 6\int_2^4\sqrt{(4-x)(x-2)}\,dx$$

$$\Rightarrow A = 3\int_2^4\sqrt{1-(x-3)^2}\,dx$$

$$\Rightarrow A = 3\cdot\left(\frac{\pi}{2}\right)\cdot(1)^2 = \frac{3\pi}{2}$$

**66.** *(d)* Given region is $\{(x,y): y^2\le 2x$ and $y\ge 4x-1\}$

$y^2 \le 2x$ represents a region inside the parabola

$$y^2 = 2x \quad \ldots\text{(i)}$$

and $y\ge 4x-1$ represents a region to the left of the line

$$y=4x-1 \quad \ldots\text{(ii)}$$

The point of intersection of the curves (i) and (ii) is

$$(4x-1)^2 = 2x$$

$$\Rightarrow 16x^2+1-8x = 2x$$

$$\Rightarrow 16x^2-10x+1 = 0$$

$$\Rightarrow x = \frac{1}{2}, \frac{1}{8}$$

$\therefore$ The points where these curves intersect, are $\left(\frac{1}{2},1\right)$ and $\left(\frac{1}{8},-\frac{1}{2}\right)$.

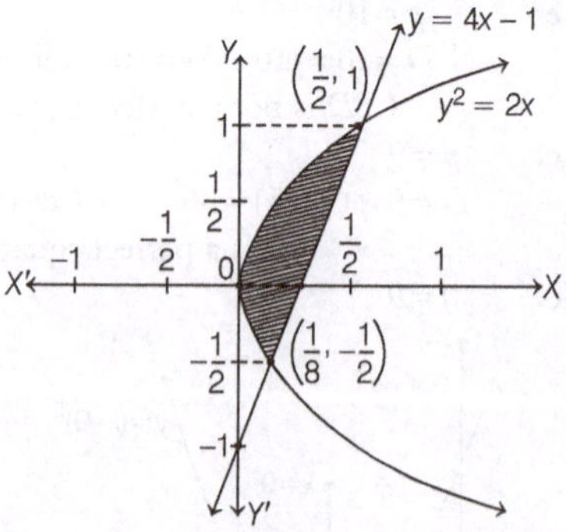


Hence, required area

$$=\int_{-1/2}^{1}\left(\frac{y+1}{4}-\frac{y^2}{2}\right)dy$$

$$=\frac{1}{4}\left(\frac{y^2}{2}+y\right)_{-1/2}^{1}-\frac{1}{6}(y^3)_{-1/2}^{1}$$

$$=\frac{1}{4}\left\{\left(\frac{1}{2}+1\right)-\left(\frac{1}{8}-\frac{1}{2}\right)\right\}-\frac{1}{6}\left\{1+\frac{1}{8}\right\}$$

$$=\frac{1}{4}\left\{\frac{3}{2}+\frac{3}{8}\right\}-\frac{1}{6}\left\{\frac{9}{8}\right\}$$

$$=\frac{1}{4}\times\frac{15}{8}-\frac{3}{16}=\frac{9}{32}$$

**67.** *(22)*

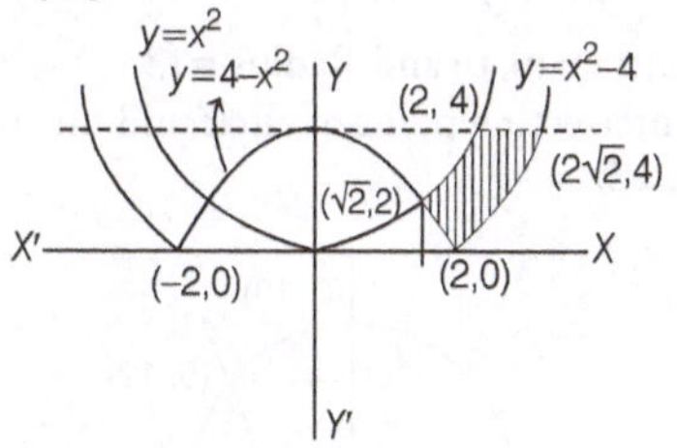


$$\text{Area} = \int_{\sqrt{2}}^{2} (x^2 - (4 - x^2))\, dx + (2\sqrt{2} - 2) \times 4 - \int_{2}^{2\sqrt{2}} (x^2 - 4)\, dx$$

$$= \left[\frac{2x^3}{3} - 4x\right]_{\sqrt{2}}^{2} + 8\sqrt{2} - 8 - \left[\frac{x^3}{3} - 4x\right]_{2}^{2\sqrt{2}}$$

$$= \frac{40\sqrt{2}}{3} - 16$$

$\Rightarrow \alpha = 6, \beta = 16 \Rightarrow \alpha + \beta = 22$

**68.** *(12)* Max $\{|x|, x, |x-2|\}$

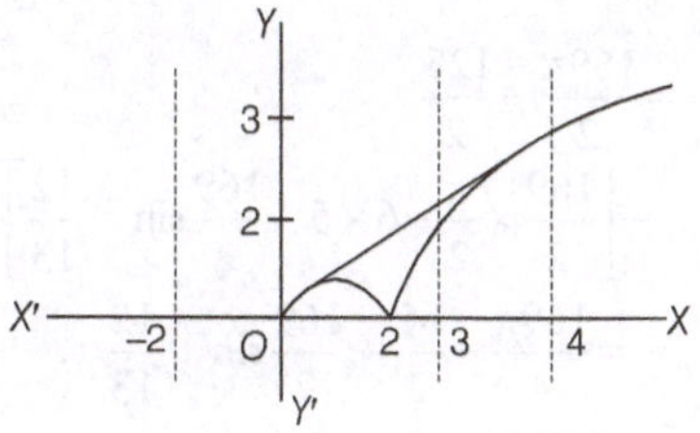


$$\therefore \text{Required area} = \frac{1}{2} \times 2 \times 2 + \frac{1}{2} \times 3 \times 3 + \frac{1}{2} \times 1 \times 11$$

**69.** *(368)* Region $= \{(x, y) : |x - 5| \le y \le 4\sqrt{x}\}$

Graph

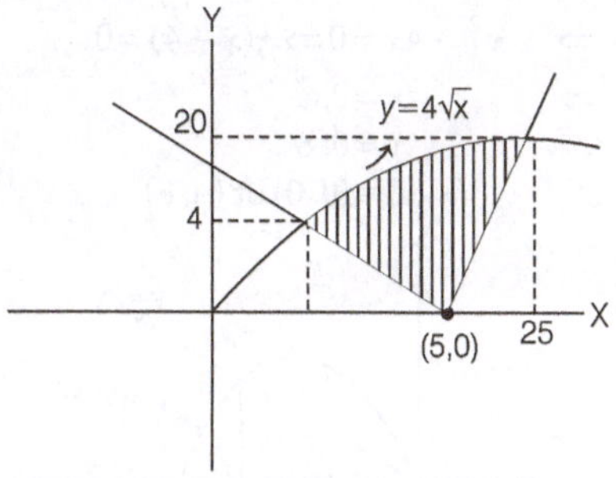


Solving $|x - 5| = 4\sqrt{x}$

We get $x = 1$ and $x = 25$

Area required

$$= \int_{1}^{25} 4\sqrt{x}\, dx - \frac{1}{2} \times 4 \times 4 - \frac{1}{2} \times 20 \times 20$$

$$= \frac{8}{3} [x^{3/2}]_1^{25} - 208$$

$$\Rightarrow A = \frac{8}{3}(124) - 208 \Rightarrow A = \frac{368}{3}$$

$\therefore 3A = 368$

**70.** *(15)* $0 \le 9x \le y^2 \Rightarrow y \ge 3x - 6$

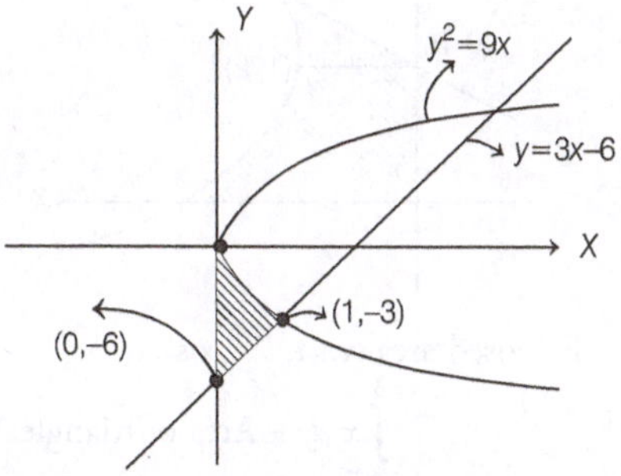


Solving $y^2 = 9x$ and $y = 3x - 6$

$y^2 = 3y + 18 \;\; x^{3/2} \;\; y^2 - 3y - 18 = 0$

$(y - 6)(y + 3) = 0$

$y = 6$ and $y = -3$

$x = 4$ and $x = 1$

$$A = \int_{-6}^{-3} \frac{y + 6}{3}\, dy + \int_{-3}^{0} \frac{y^2}{9}\, dy$$

$$A = \frac{1}{6}[y^2]_{-6}^{-3} + 2(6 - 3) + \frac{1}{9}\left[\frac{y^3}{3}\right]_{-3}^{0}$$

$$= \frac{1}{6}[9 - 36] + 6 + \frac{1}{9}\left[\frac{27}{3}\right]$$

$$= \frac{-27}{6} + 6 + 1 = \frac{-27 + 42}{6} \Rightarrow 6A = 15$$

**71.** *(77)*

> **key idea** $\int \sqrt{a^2 - x^2}\, dx$
> $$= \frac{1}{2}\left[x\sqrt{a^2 - x^2} + a^2 \sin^{-1}\left(\frac{x}{a}\right)\right] + C$$

We have, $x^2 + y^2 = 25$

$$y = \begin{cases} x - 1, & x \ge 1 \\ -x + 1 & x < 1 \end{cases}$$

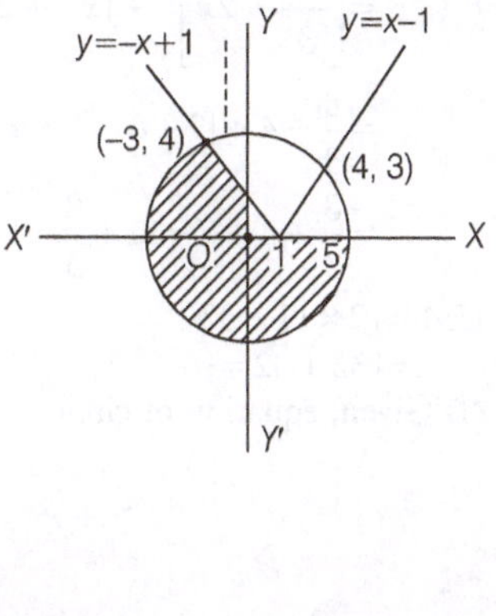


$\Rightarrow \quad x^2 + (x - 1)^2 = 25$

$\Rightarrow \quad x = 4$

$x^2 + (-x + 1)^2 = 25$

$\Rightarrow \quad x = -3$

Area of shaded region

$$= 25\pi - \int_{-3}^{4} \sqrt{25 - x^2}\, dx + \frac{1}{2} \times 4 \times 4 + \frac{1}{2} \times 3 \times 3$$

$$= 25\pi - \left[\frac{x}{2}\sqrt{25 - x^2} + \frac{25}{2}\sin^{-1}\frac{x}{5}\right]_{-3}^{4} + \frac{25}{2}$$

$$A = 25\pi + \frac{25}{2} - \left[(6 + 3) + \frac{25}{2}\frac{\pi}{2}\right]$$

$$= \frac{75\pi}{4} + \frac{1}{2}$$

$$\Rightarrow \frac{75\pi}{4} + \frac{1}{2} = \frac{1}{4}(b\pi + c)$$

$\Rightarrow \quad b = 75, c = 2$

$\therefore \quad b + c = 75 + 2 = 77$

**72.** *(34)* $f(x) = \begin{cases} -3ax^2 - 2, & x < 1 \\ a^2 + bx, & x \ge 1 \end{cases}$

$\because f(x)$ is differentiable.

$$f'(x) = \begin{cases} -6ax, & x < 1 \\ b, & x > 1 \end{cases}$$

$f(1^+) = f(1^-) = f(1)$

$\therefore -3a - 2 = a^2 + b$

$a^2 + 3a + b = -2 \quad$ ...(i)

and $\quad f'(1^+) = f'(1^-)$

$-6a = b \quad$ ...(ii)

Solving Eqs. (i) and (ii), we get

$a^2 - 3a + 2 = 0$

$(a - 1)(a - 2) = 0$

$\Rightarrow \quad a = 1, 2 \quad [\because a > 1]$

$\Rightarrow \quad a = 2$

$b = -12$

Now, $y = f(x) = \begin{cases} -6x^2 - 2, & x < 1 \\ -12x + 4, & x \ge 1 \end{cases}$

$-6x^2 - 2 = -20$

$6x^2 = 18$

$x^2 = 3$

$\Rightarrow \quad x = -\sqrt{3}$

or $\quad -12x + 4 = -20$

$12x = 24$

$x = 2$

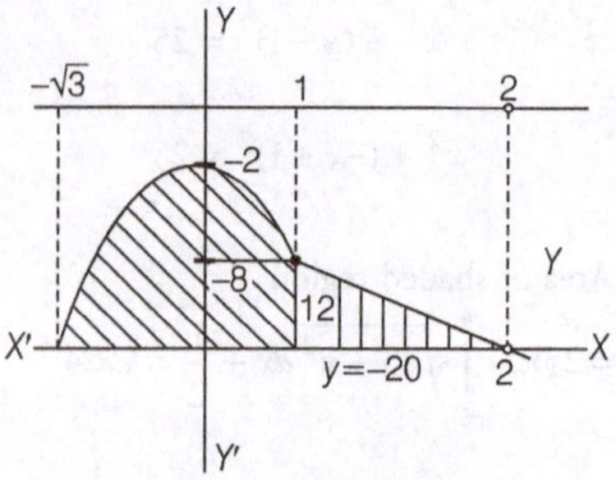


$\therefore$ Required area $= 20(\sqrt{3}+1)$

$$-\left|\int_{-\sqrt{3}}^{1} y\,dx\right| + \frac{1}{2}\times 1\times 12$$

$$= 20(\sqrt{3}+1) - \left|\int_{-\sqrt{3}}^{1} (-6x^2-2)\,dx\right| + 6$$

$$= 20\sqrt{3} + 26 - |[2x^3+2x]_{-\sqrt{3}}^{1}|$$

$$= 20\sqrt{3} + 26 - |4 + 8\sqrt{3}|$$

$$= 20\sqrt{3} + 26 - 8\sqrt{3} - 4 = 22 + 12\sqrt{3}$$

$\alpha = 22$ and $\beta = 12$

$\therefore\ \alpha + \beta = 34$

**73.** *(72)* Given, equation of curves

$y = x^2 - 5x$ and $y = 7x - x^2$

Intersecting point of given curves

$x^2 - 5x = 7x - x^2 \Rightarrow\ 2x^2 = 12x$

$\Rightarrow\ 2x(x-6) = 0 \Rightarrow\ x = 0, 6$

When $x = 0$, then $y = 0$

and $x = 6$, then $y = 6$

$\therefore$ Required area

$$= \int_0^6 (7x - x^2) - (x^2 - 5x)\,dx$$

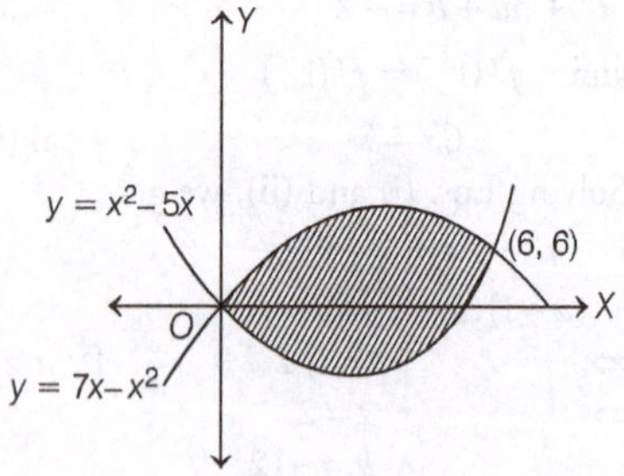


$$= \int_0^6 (12x - 2x^2)\,dx$$

$$= \left[12\left(\frac{x^2}{2}\right) - 2\left(\frac{x^3}{3}\right)\right]_0^6$$

$$= 12\left(\frac{36}{2}\right) - 2\left(\frac{216}{3}\right)$$

$= 216 - 144 = 72$ sq units

**74.** *(5)* Parabola $(y-2)^2 = x - 1$

Line $\quad x - 2y + 4 = 0 \quad$ ...(i)

$\Rightarrow \quad x = 2(y-2)$

Let $\quad y - 2 = Y$

$\Rightarrow \quad Y^2 = x - 1$ and $x = 2Y$

Solving $Y^2 = 2Y - 1 \Rightarrow (Y-1)^2 = 0$

$\Rightarrow \quad Y = 1 \Rightarrow y - 2 = 1 \Rightarrow y = 3$

and $x = 2$ $\quad$ [from Eq (i)]

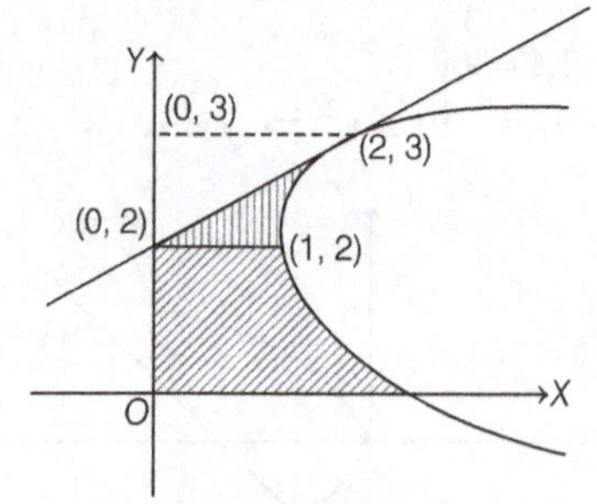


Enclosed area w.r.t. $X$-axis

$$= \int_0^3 x\,dy - \text{Area of triangle}$$

$$= \int_0^3 [(y-2)^2 + 1] - \frac{1}{2}\times 1\times 2$$

$$= \left[\frac{(y-2)^3}{3} + y\right]_0^3 - 1$$

$$= \frac{1}{3} + 3 - \left(-\frac{8}{3}\right) - 1$$

$= 6 - 1 = 5$

**75.** *(164)* Region $= \{(x, y): 0 \le x \le 3,$
$0 \le y \le \min(x^2 + 2, 2x + 2)\}$

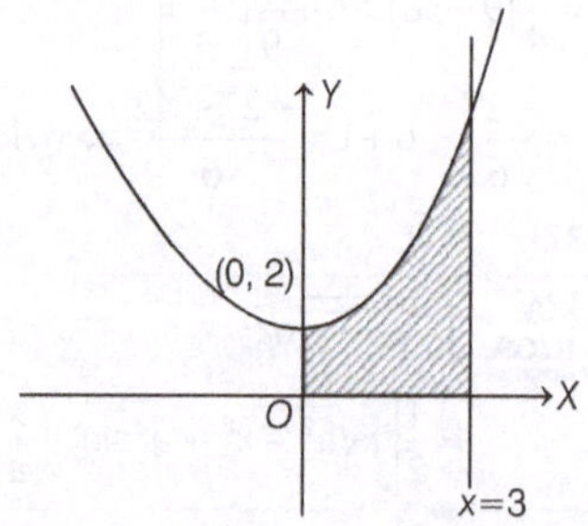


$$A = \int_0^2 (x^2+2) + \int_2^3 (2x+2)$$

$$= \left[\frac{x^3}{3} + 2x\right]_0^2 + [x^2 + 2x]_2^3$$

$$= \frac{8}{3} + 4 + [9 + 6 - 4 - 4]$$

$$= \frac{8}{3} + 4 + 7 = 11 + \frac{8}{3}$$

$\therefore\ 12A = 12\times 11 + 32$

$= 132 + 32 = 164$

**76.** *(171)* Given, equation of circle,

$$x^2 + y^2 = 169$$

Centre = (0, 0) and Radius = 13

$\therefore$ Intersecting point of circle and line is (5, 12)

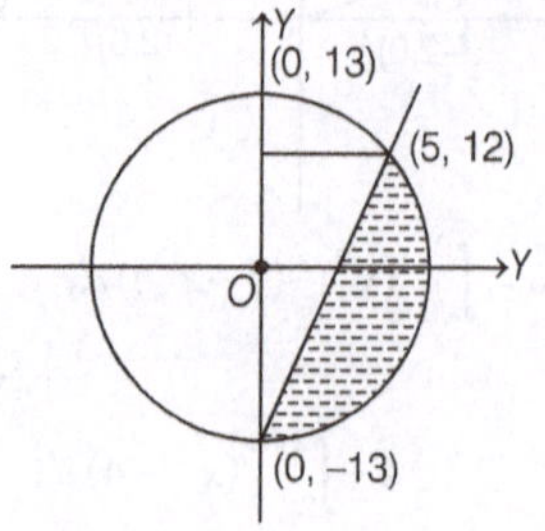


$\therefore$ Required area

$$= \frac{\pi(13)^2}{2} - \left[\frac{1}{2}\times 25\times 5 + \int_{12}^{13}\sqrt{(169-y^2)}\,dy\right]$$

$$= \frac{169\pi}{2} - \left[\frac{125}{2} + \left\{\frac{y}{2}\sqrt{169-y^2} + \frac{169}{2}\sin^{-1}\left(\frac{y}{13}\right)\right\}_{12}^{13}\right]$$

$$= \frac{169\pi}{2} - \frac{125}{2} - \left[\frac{169}{2}\times\frac{\pi}{2} - 6\times 5 - \frac{169}{2}\sin^{-1}\frac{12}{13}\right]$$

$$= \frac{169\pi}{4} - \frac{65}{2} + \frac{169}{2}\sin^{-1}\frac{12}{13}$$

$\therefore\ \alpha = 169, \beta = 2$

$\therefore\ \alpha + \beta = 171$

**77.** *(304)*

Given, curves

$y = 2x$ $\quad$ ...(i)

and $\quad y = 6x - x^2$ $\quad$ ...(ii)

On solving Eqs. (i) and (ii), we get

$2x = 6x - x^2$

$\Rightarrow\ x^2 - 4x = 0 \Rightarrow x(x-4) = 0$

$\Rightarrow\ x = 0, 4$

$\therefore\ y = 0, 8$

$\therefore\ (x, y) \equiv (0, 0)$ or $(4, 8)$

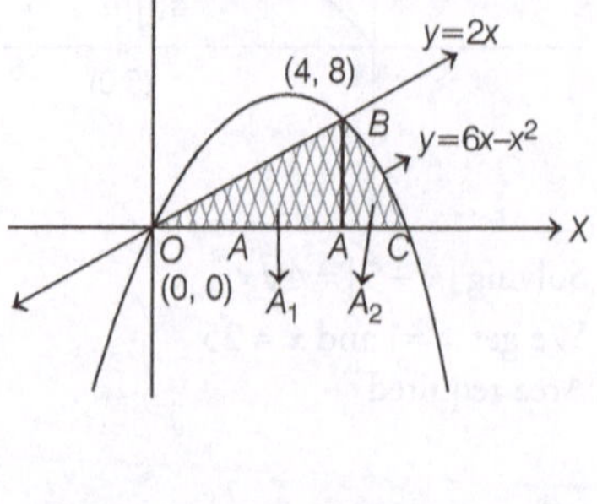

Now, required area $= A_1 + A_2$

$\therefore$ Required area

$$= \frac{1}{2} \times 4 \times 8 + \int_4^6 (6x - x^2)dx$$

$$\Rightarrow A = 16 + \left[3x^2 - \frac{x^3}{3}\right]_4^6 \Rightarrow$$

$$A = 16 + \left(\frac{28}{3}\right)$$

$$\therefore \quad 12A = (12 \times 16 + 28 \times 4) = 304 \text{ sq units}$$

**78.** *(119)* To drawn the region, we consider equations as

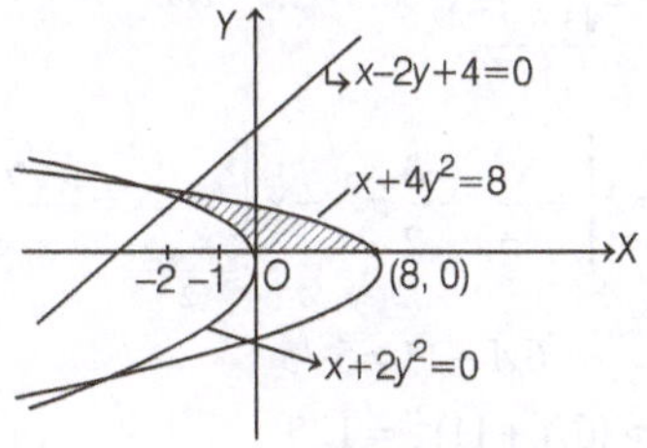


$$x - 2y + 4 = 0 \quad \text{...(i)}$$

and $\quad x + 2y^2 = 0$

$$\Rightarrow \quad y^2 = -\frac{1}{2}x \quad \text{...(ii)}$$

and $\quad x + 4y^2 = 8$

$$\Rightarrow \quad y^2 = -\frac{1}{4}(x-8) \quad \text{...(iii)}$$

$$\text{Area, } A = \int_0^1 [(8-4y^2) - (-2y^2)]dy + \int_1^{3/2} [(8-4y^2) - (2y-4)]dy$$

$$= \left[8y - \frac{2y^3}{3}\right]_0^1 + \left[12y - y^2 - \frac{4y^3}{3}\right]_1^{3/2}$$

$$= \frac{107}{12} = \frac{m}{n}$$

Hence, $m + n = 107 + 12 = 119$

**79.** *(42)*

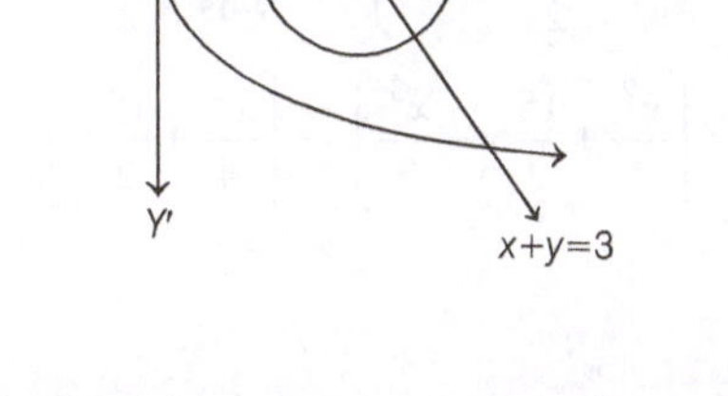


$A =$ Required area

$$= \int_0^{3/2} \left[(3-y) - \left(\frac{2y^2}{3}\right)\right] dy - \pi[\sqrt{2}]^2 \cdot \frac{1}{8}$$

$$= \left[\left(3y - \frac{y^2}{2} - \frac{2}{9}y^3\right)\right]_0^{3/2} - \frac{\pi}{4}$$

$$= 3 \times \frac{3}{2} - \frac{9}{8} - \frac{2}{9} \times \frac{27}{8} - \frac{\pi}{4}$$

$$= \frac{36-9-6}{8} - \frac{\pi}{4} = \frac{21}{8} - \frac{\pi}{4}$$

So, $4(\pi + 4A)$

$$= 4\left(\pi + 4\left(\frac{21}{8} - \frac{\pi}{4}\right)\right) = 42$$

**80.** *(16)* Given, $C = 2x^2 - y + 1 = 0$

$\Rightarrow y = 3x + 1$

Tangent to $c$ at point (1, 3)

and Line $x + y = 1$

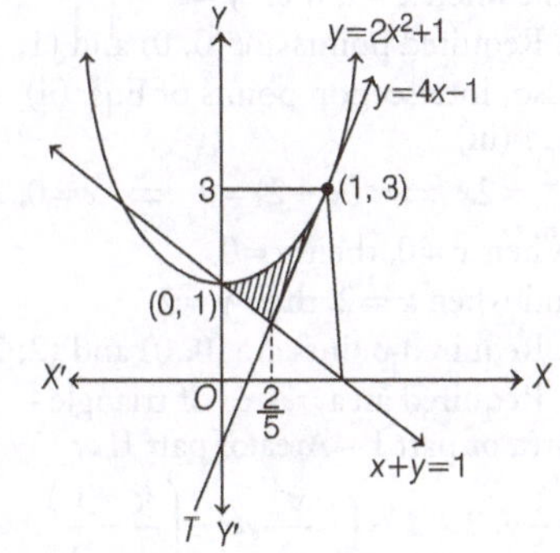


$$\text{Tangent} = \frac{dy}{dx} = 4x$$

$$y - 3 = 4(x-1)$$

$$\Rightarrow \quad y = 4x - 1$$

Hence, shaded area $= A$

$$A = \int_0^{2/5} [(2x^2+1) - (1-x)]dx + \int_{2/5}^1 [(2x^2+1) - (4x-1)]dx$$

$$= \int_0^{2/5} (2x^2 + x)dx + \int_{2/5}^1 (2x^2 - 4x + 2)dx$$

$$= \left(\frac{2}{3}x^3 + \frac{x^2}{2}\right)\Bigg|_0^{\frac{2}{5}} + \left(\frac{2}{3}x^3 - 2x^2 + 2x\right)\Bigg|_{2/5}^1$$

$$= \frac{16}{3 \times 5^3} + \frac{2}{5^2} + \frac{2}{3} - 2 + 2 - \frac{16}{3 \times 5^3} + \frac{8}{5^2} - \frac{4}{5}$$

$$A = \frac{10}{25} + \frac{2}{3} - \frac{4}{5} = \frac{2}{5} + \frac{2}{3} - \frac{4}{5} = \frac{2}{3} - \frac{2}{5}$$

$$\Rightarrow A = \frac{4}{15}; \quad 60A = 60 \times \frac{4}{15} = 16$$

**81.** *(27)* We have, $\{(x, y): |x^2 - 2| \le y \le x\}$

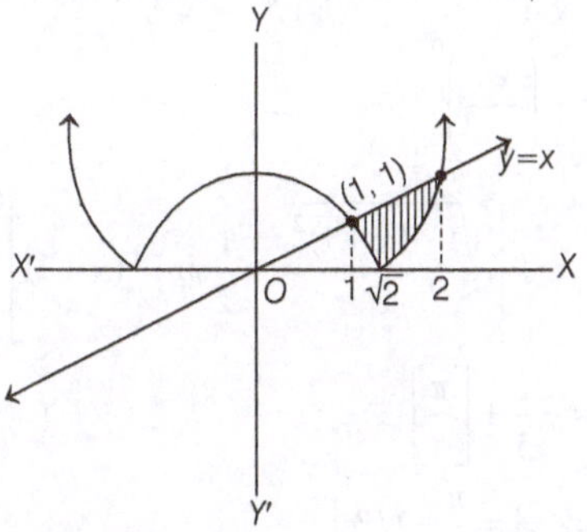


Solving $|x^2 - 2| = y$ and $y = x$, we get (1, 1) and (2, 2)

$$A = \int_1^{\sqrt{2}} (x - (2 - x^2))dx + \int_{\sqrt{2}}^2 (x - (x^2 - 2))dx$$

$$= \left[\frac{x^2}{2} - 2x + \frac{x^3}{3}\right]_1^{\sqrt{2}} + \left[\frac{x^2}{2} - \frac{x^3}{3} + 2x\right]_{\sqrt{2}}^2$$

$$= \left(1 - 2\sqrt{2} + \frac{2\sqrt{2}}{3}\right) - \left(\frac{1}{2} - 2 + \frac{1}{3}\right) + \left(2 - \frac{8}{3} + 4\right) - \left(1 - \frac{2\sqrt{2}}{3} + 2\sqrt{2}\right)$$

$$= \frac{-8\sqrt{2}}{3} + \frac{9}{2} = \frac{-16\sqrt{2} + 27}{6}$$

$$\Rightarrow 6A = -16\sqrt{2} + 27 \Rightarrow 6A + 16\sqrt{2} = 27$$

**82.** *(16)* Let $y = p(x)$ be the parabola passing through the points $(-1, 0)\ (0, 1)\ (1, 0)$.

Now, to find the area of the region $\{(x, y), (x+1)^2 + (y-1)^2 \le 1, y \le p(x)\}$

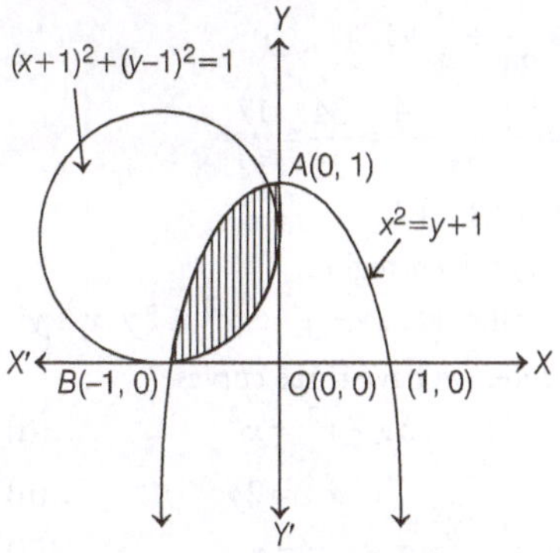

The area of shaded region is

$$A = \int_{-1}^{0} \{(1-x^2) - (1-\sqrt{1-(x+1)^2}\}dx$$

$$= \int_{-1}^{0} \{-x^2 + \sqrt{1-(x+1)^2}\}dx$$

$$= \int_{-1}^{0} (-x^2)\,dx + \int_{-1}^{0} \sqrt{1-(x+1)^2}\,dx$$

$$= -\left(\frac{x^3}{3}\right)_{-1}^{0}$$

$$+ \left[\frac{\sqrt{1-(x+1)^2}}{2} + \frac{1}{2}\sin^{-1}(x+1)\right]_{-1}^{0}$$

$$= -\frac{1}{3} + \left[\frac{\pi}{4}\right]$$

$$\Rightarrow A = \frac{\pi}{4} - 1/3$$

$$\text{Now, } 12(\pi - 4A) = 12\left(\pi - 4\left(\frac{\pi}{4} - \frac{1}{3}\right)\right)$$

$$= 16$$

**83.** *(17)*

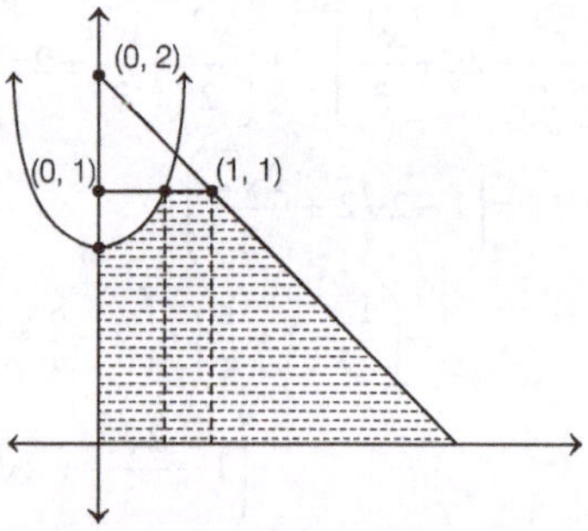


When $0 \le x < 1$, then $1 + [x] = 1 + 0 = 1$

$$A = \int_0^{1/2}\left(x^2 + \frac{3}{4}\right)dx + \left(\frac{1}{2}\times 1\right) + \left(\frac{1}{2}\times 1\times 1\right)$$

$$= \left[\frac{x^3}{3}\right]_0^{1/2} + \frac{3}{4}[x]_0^{1/2} + 1$$

$$= \frac{1}{3}\cdot\frac{1}{8} + \frac{3}{4}\times\frac{1}{2} + 1$$

$$= \frac{1}{24} + \frac{3}{8} + 1$$

$$= \frac{1+9+24}{24} = \frac{34}{24} = \frac{17}{12}$$

$\therefore\ 12A = 17$

**84.** *(5)* Given region,

$$S = \{(x, y) : 2y - y^2 \le x^2 \le 2y, x \ge y\}$$

Here, we have three curves

$$2y - y^2 = x^2 \qquad \ldots(i)$$

$$x^2 = 2y \qquad \ldots(ii)$$

and $$x = y \qquad \ldots(iii)$$

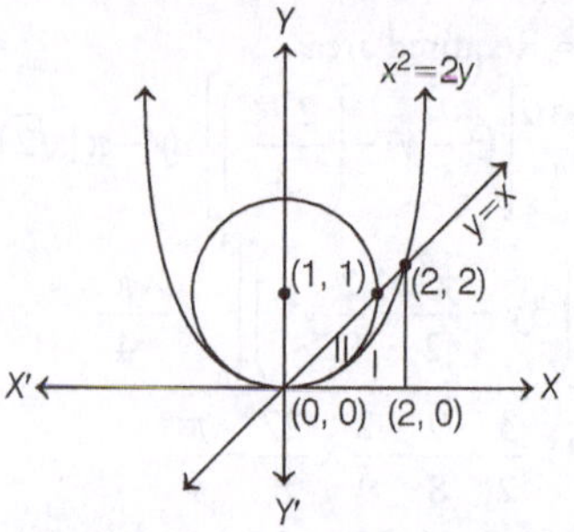


From Eq. (i),

$$x^2 + y^2 - 2y = 0$$

$$\Rightarrow \quad x^2 + (y-1)^2 = 1$$

is a circle with centre (0, 1) and radius is 1.

Intersection points of Eqs. (i) and (iii) are

$$2x - x^2 = x^2$$

$\Rightarrow 2x^2 = 2x \Rightarrow x(x-1) = 0 \Rightarrow x = 0, 1$

When $x = 0$, then $y = 0$

and when $x = 1$, then $y = 1$

$\therefore$ Required points are (0, 0) and (1, 1).

Also, intersection points of Eqs. (ii) and (iii)

$x^2 = 2x \Rightarrow x(x-2) = 0 \quad \Rightarrow x = 0, 2$

When $x = 0$, then $y = 0$

and when $x = 2$, then $y = 2$

$\therefore$ Required points are (0, 0) and (2, 2)

$\therefore$ Required area = Area of triangle – Area of part I – Area of part II

$$= \frac{1}{2}\times 2\times 2 - \int_0^2 \frac{x^2}{2}\,dx - \left(\frac{\pi}{4} - \frac{1}{2}\right)$$

$$= 2 - \frac{1}{2}\left(\frac{x^3}{3}\right)_0^2 - \frac{\pi}{4} + \frac{1}{2}$$

$$= 2 - \frac{1}{2}\times\frac{8}{3} - \frac{\pi}{4} + \frac{1}{2} = \frac{7}{6} - \frac{\pi}{4}$$

$$= \frac{5+2}{5+1} - \frac{\pi}{5-1}$$

$\therefore \quad n = 5$

**85.** *(125)* We have, $|2x-1| \le y \le |x^2 - x|$, $0 \le x \le 1$

$\therefore y \ge |2x-1|$ and $y \le |x^2 - x|$, $0 \le x \le 1$

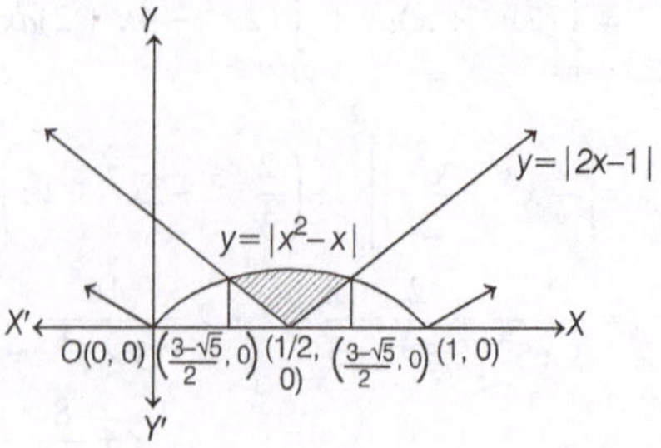


Now, $y = x^2 - x$

$\Rightarrow 2x - 1 = x^2 - x \Rightarrow x^2 - 3x + 1 = 0$

$$\Rightarrow \qquad x = \frac{3 \pm \sqrt{9-4}}{2} = \frac{3 \pm \sqrt{5}}{2}$$

So, both curve intersect at $x = \dfrac{3+\sqrt{5}}{2}$

and $x = \dfrac{3-\sqrt{5}}{2}$.

$\therefore$ Required area,

$$A = 2\int_{\frac{3-\sqrt{5}}{2}}^{1/2} [(x - x^2) - (1 - 2x)]\,dx$$

$$= 2\int_{\frac{3-\sqrt{5}}{2}}^{1/2} (-x^2 + 3x - 1)dx$$

$$= 2\left[-\frac{x^3}{3} + \frac{3}{2}x^2 - x\right]_{\frac{3-\sqrt{5}}{2}}^{\frac{1}{2}} = \frac{5\sqrt{5} - 11}{6}$$

$$\therefore \qquad 6A + 11 = 5\sqrt{5}$$

$$\Rightarrow (6A + 11)^2 = 125$$

**86.** *(72)* We have,

$$f(x) = \frac{x + |x|}{2} = \begin{cases} x, & x \ge 0 \\ 0, & x < 0 \end{cases}$$

and $$g(x) = \begin{cases} x, & x < 0 \\ x^2, & x \ge 0 \end{cases}$$

$$\therefore \qquad fog(x) = f(g(x))$$

$$= \begin{cases} g(x), & g(x) \ge 0 \\ 0, & g(x) < 0 \end{cases}$$

$$= \begin{cases} x^2, & x \ge 0 \\ 0, & x < 0 \end{cases}$$

Also, equation of line is $2y - x = 15$

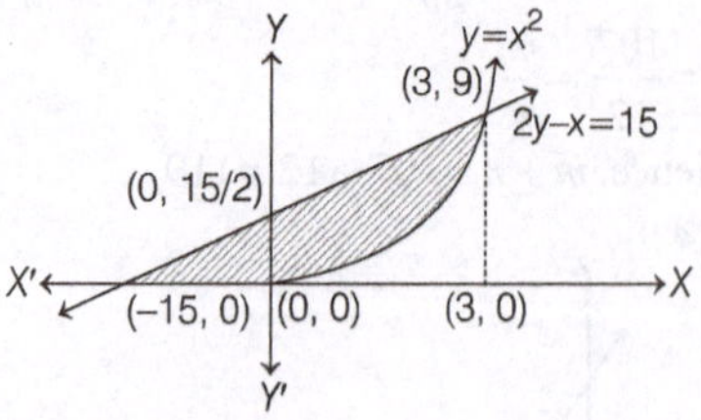


Intersecting points of $y = x^2$ and $2y - x = 15$ is (3, 9).

$\therefore$ Required area

$$= \int_0^3 \left(\frac{x+15}{2} - x^2\right)dx + \int_{-15}^{0} \frac{x+15}{2}\,dx$$

$$= \left[\frac{x^2}{4} + \frac{15}{2}x - \frac{x^3}{3}\right]_0^3 + \left[\frac{x^2}{4} + \frac{15}{2}x\right]_{-15}^{0}$$

$$=\left(\frac{9}{4}+\frac{45}{2}-9\right)+\left(-\frac{225}{4}+\frac{225}{2}\right)=72$$

**87.** *(25)* We have, $\{(x, y): y \geq x^2,$ $y \geq (1-x)^2,\ y \leq 2x(1-x)\}$

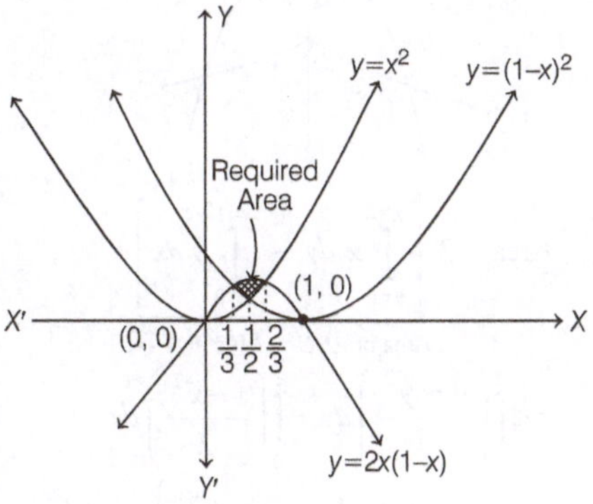


So, $A=\int_{1/3}^{1/2}[(2x-2x^2)-(1-x)^2]\,dx$

$$+\int_{1/2}^{2/3}(2x-2x^2)-x^2]\,dx$$

$$=\left[x^2-\frac{2}{3}x^3+\frac{(1-x)^3}{3}\right]_{1/3}^{1/2}$$

$$+[x^2-x^3]_{1/2}^{2/3}$$

$$=\left(\frac{1}{4}-\frac{1}{12}+\frac{1}{24}\right)-\left(\frac{1}{9}-\frac{2}{81}+\frac{8}{81}\right)$$

$$+\left(\frac{4}{9}-\frac{8}{27}\right)-\left(\frac{1}{4}-\frac{1}{8}\right)$$

$$=-\frac{1}{24}-\frac{1}{9}-\frac{2}{27}-\frac{8}{27}+\frac{4}{9}+\frac{1}{8}=\frac{-10}{27}$$

$$+\frac{1}{3}+\frac{1}{12}=\frac{5}{108}$$

Therefore, $540A=540\times\frac{5}{108}=25$

**88.** *(600)* Parabola $P_1: 2y=5x^2$,

Vertex : (0, 0)

Parabola $P_2: x^2-y+6=0$

$\Rightarrow x^2=(y-6)$, Vertex (0, 6)

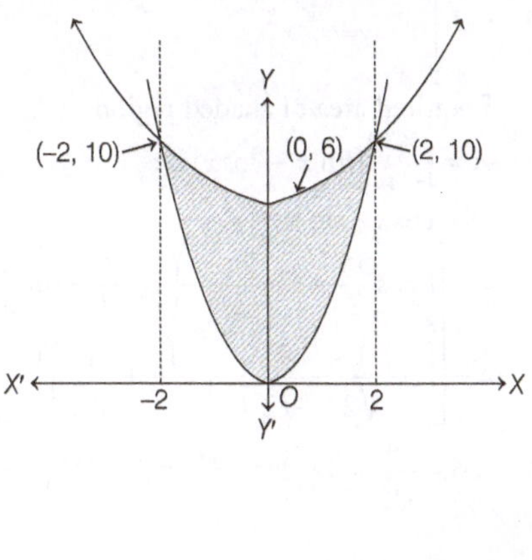


For their intersection, $x^2-\frac{5}{2}x^2+6=0$

$$\frac{-3x^2}{2}+6=0 \Rightarrow x=\pm 2$$

$$\text{Area}=2\int_0^2 (x^2+6)-\frac{5x^2}{2}\,dx$$

$$=2\int_0^2\left(\frac{-3x^2}{2}+6\right)dx=2\left[-\frac{x^3}{2}+6x\right]_0^2$$

$$=2[-4+12]=16 \qquad \ldots(\text{i})$$

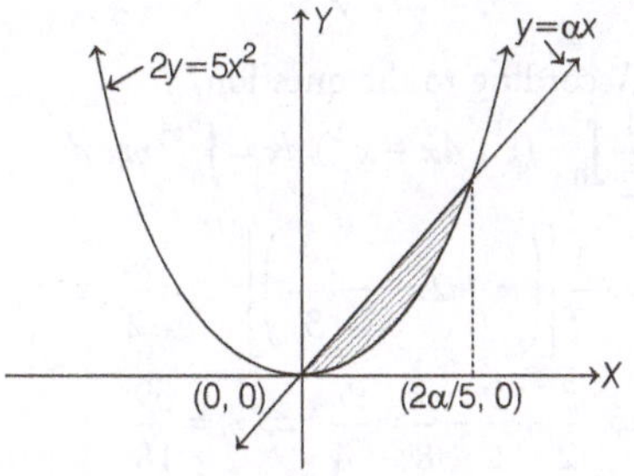


Solving $2y=5x^2$ and $y=\alpha x$,

$$2\alpha x=5x^2 \Rightarrow x=0,\ x=\frac{2\alpha}{5}$$

$$A=\int_0^{2\alpha/5}\left(\alpha x-\frac{5}{2}x^2\right)dx$$

$$A=\left[\frac{\alpha x^2}{2}-\frac{5x^3}{6}\right]_0^{2\alpha/5}$$

$$=\left[\frac{\alpha}{2}\cdot\frac{4\alpha^2}{25}-\frac{5}{6}\cdot\frac{8\alpha^3}{125}\right] \qquad \ldots(\text{ii})$$

Given that area by Eqs. (i) and (ii) are same.

So, $\alpha^3\left[\frac{4}{50}-\frac{40}{750}\right]=16$

$$\alpha^3\times\frac{20}{750}=16 \Rightarrow \alpha^3=600$$

**89.** *(36)*

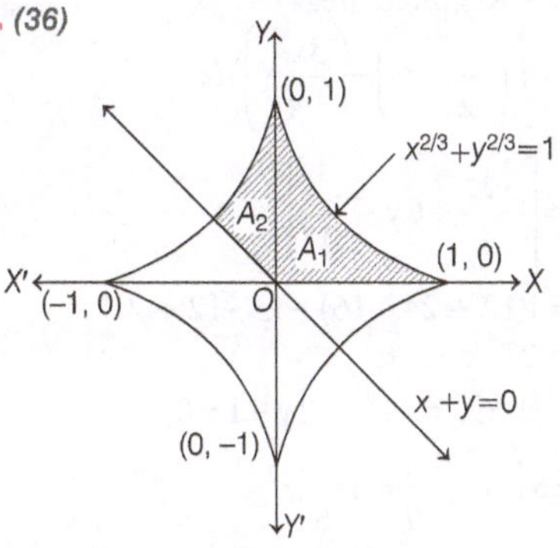


$$A_1=\int_0^1 y\,dx=\int_0^1 (1-x^{2/3})^{3/2}\,dx$$

Putting $x=\sin^3\theta$

$\Rightarrow$ $dx=3\sin^2\theta\cos\theta\,d\theta$

$x\to 0,\ \theta\to 0$

$x\to 1,\ \theta=\pi/2$

$$A_1=\int_0^{\frac{\pi}{2}}(1-\sin^2\theta)^{\frac{3}{2}}\cdot 3\sin^2\theta\cdot\cos\theta\,d\theta$$

$$=3\int_0^{\frac{\pi}{2}}\sin^2\theta\cos^4\theta\,d\theta$$

$$=3\,.\,\frac{1\times 3\times 1}{6\times 4\times 2}\times\frac{\pi}{2}=\frac{3\pi}{32}$$

$$\therefore\ A_2=\frac{1}{2}\times\frac{3\pi}{32}$$

$$A=A_1+A_2=\frac{3\pi}{32}+\frac{1}{2}\cdot\frac{3\pi}{32}$$

$$=\frac{3\pi}{32}\times\frac{3}{2}=\frac{9\pi}{64}$$

$$\therefore\ \frac{256A}{\pi}=\frac{256}{\pi}\times\frac{9\pi}{64}=36$$

**90.** *(6)* $A_1=\{(x, y);\ |x|\leq y^2,\ |x|+2y\leq 8\}$

$A_2=\{(x, y);\ |x|+|y|\leq k\}$

For $A_1$ the line and curves are

$x\geq 0;\ x+2y\leq 8,\ x\leq y^2$

$x<0;\ -x+2y\leq 8,\ -x\leq y^2$

Solving $y^2=x$ and $x+2y=8$

Put $x=8-2y$ in $y^2=x$

$$y^2=8-2y \Rightarrow y^2+2y-8=0$$

$$(y+4)(y-2)=0\quad y=-4,\ y=2$$

$$\text{Area }(A_1)=2\left[\int_0^2 y^2\,dy+\int_0^4 (8-2y)\,dy\right]2$$

$$=2\left[\left(\frac{y^3}{3}\right)_0^2+(8y-y^2)_2^4\right]$$

$$=2\left[\left(\frac{8}{3}-0\right)+((32-16)-(16-4))\right]=\frac{40}{3}$$

For area, $A_2$

$x+y\leq k,\ x-y\leq k$

$-x+y\leq k,\ -x-y\leq k$

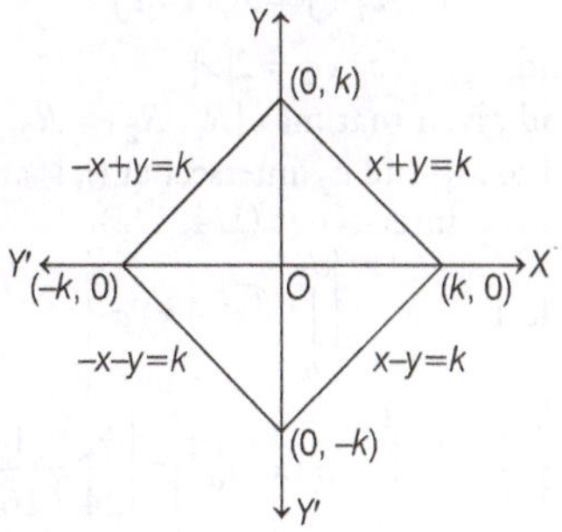


$$\text{Area }(A_2)=4\times\frac{1}{2}\times k\times k=2k^2$$

Given, $27(\text{Area }A_1)=5\,(\text{Area }A_2)$

$$\Rightarrow 27\times\frac{40}{3}=5\times 2k^2 \Rightarrow k=6$$

**91.** *(18)* Given that,

Equation of the parabola $y^2 = 2x$ and equation of the line $x + y = 4$

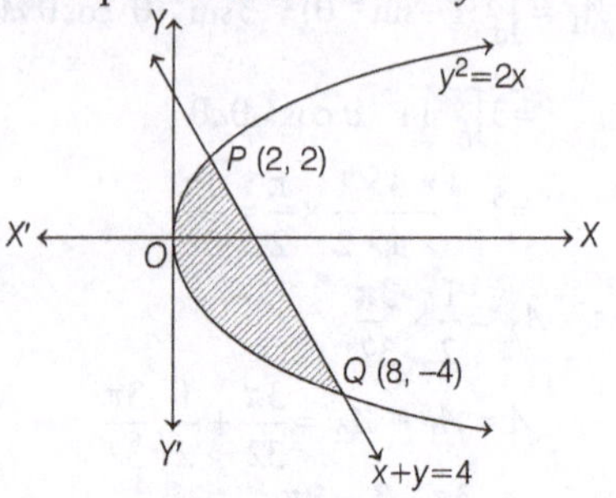


Now solving above two equations $y^2 = 2x$ and $x + y = 4$ to get intersecting points $P(2, 2)$ and $Q(8, -4)$.

Thus, the required area is

$$= \int_{-4}^{2} \left( (4 - y) - \frac{y^2}{2} \right) dy$$

$$= \left[ 4y - \frac{y^2}{2} - \frac{y^3}{6} \right]_{-4}^{2}$$

$$= \left( 8 - 2 - \frac{4}{3} \right) - \left( -16 - 8 + \frac{32}{3} \right) = 18$$

**92.** *(19)* The given curve are

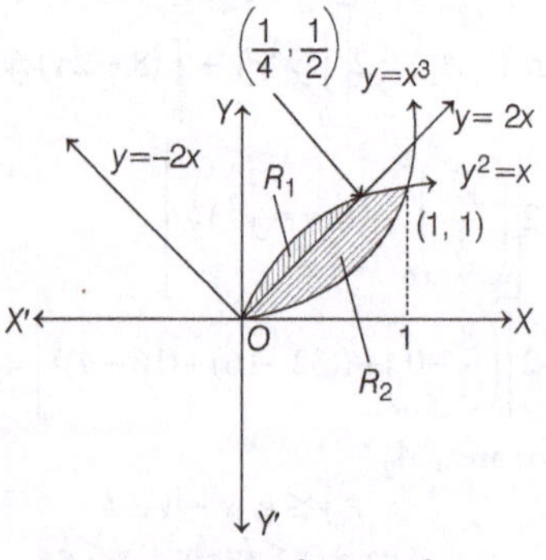


$$c_1 : y = x^3, c_2 : y^2 = x$$

and $\quad c_3 : y = 2|x|$

and given that $\max\{R_1, R_2\} = R_2$

Thus, $c_1$ and $c_2$ intersect at (1,1) and $c_2$ and $c_3$ intersect at (1/4, 1/2).

Clearly, $R_1 = \left| \int_0^{1/4} (\sqrt{x} - 2x)\, dx \right|$

$$= \left| \left( \frac{2}{3} x^{3/2} \right)_0^{1/4} - (x^2)_0^{1/4} \right| = \left| \frac{2}{24} - \frac{1}{16} \right| = \frac{1}{48}$$

and $R_1 + R_2 = \int_0^1 (\sqrt{x} - x^3)\, dx$

$$= \frac{2}{3} - \frac{1}{4} = \frac{5}{12}$$

So, $\dfrac{R_2 + R_1}{R_1} = \dfrac{5/12}{1/48} = 20$

$$\Rightarrow \frac{R_2}{R_1} + 1 = 20 \Rightarrow \frac{R_2}{R_1} = 19$$

**93.** *(26)*

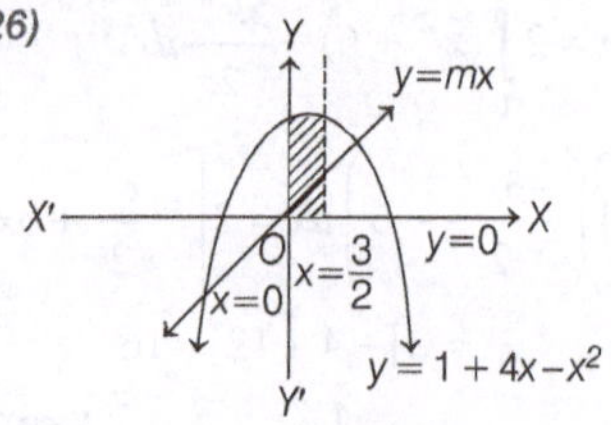


According to the question,

$$\frac{1}{2} \int_0^{3/2} (1 + 4x - x^2)\, dx = \int_0^{3/2} mx\, dx$$

$$\Rightarrow \frac{1}{2} \left[ \left( x + 2x^2 - \frac{x^3}{3} \right) \right]_0^{3/2} = \frac{m}{2} [x^2]_0^{3/2}$$

$$\Rightarrow \frac{3}{2} + \frac{9}{2} - \frac{9}{8} = \frac{9m}{4} \Rightarrow m = \frac{39}{18}$$

$$\Rightarrow \quad 12m = 26$$

**94.** *(27)*

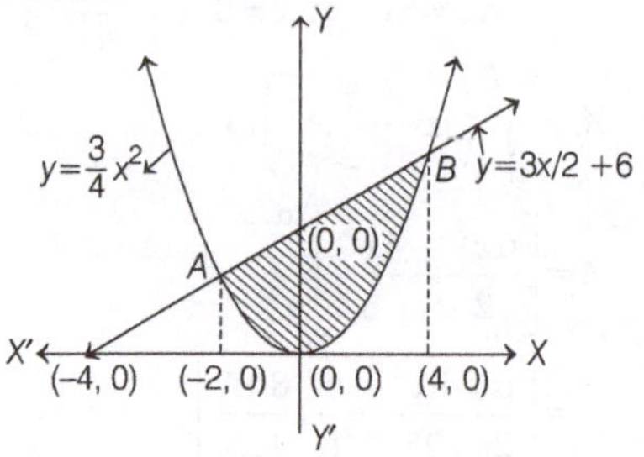


We have, $y = \left( \dfrac{3}{4} \right) x^2$ and $y = \left( \dfrac{3x}{2} \right) + 6$

$$\Rightarrow \frac{3x^2}{4} = \frac{3x}{2} + 6 \Rightarrow (x - 4)(x + 2) = 0$$

$$\Rightarrow \quad x = -2, 4 \Rightarrow y = 3, 12$$

$A(-2, 3)$ and $B(4, 12)$

∴ Required area

$$= \int_{-2}^{4} \left( \frac{3x}{2} + 6 \right) - \left( \frac{3x^2}{4} \right) dx$$

$$= \left[ \frac{3x^2}{4} + 6x - \frac{x^3}{4} \right]_{-2}^{4}$$

$$= [(12 + 24 - 16) - (3 - 12 + 2)]$$

= 27 sq units

**95.** *(2)* $C_1 \Rightarrow x^2 + 2y - 1 = 0$

$$\Rightarrow \left( y - \frac{1}{2} \right) = 4 \cdot \frac{1}{8} x^2$$

$$C_2 \Rightarrow y^2 + 4x - 4 = 0$$

$$\Rightarrow y^2 = -4(x - 1)$$

$$C_3 \Rightarrow y^2 - 4x - 4 = 0$$

$$\Rightarrow y^2 = 4(x + 1)$$

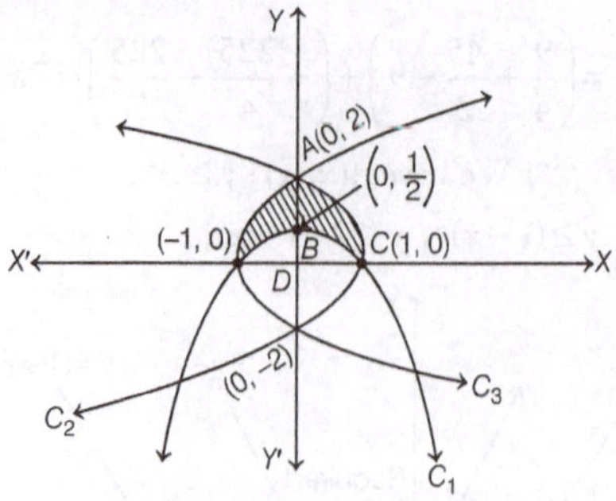


$$\text{Area} = 2 \left[ \underbrace{\int_{y=0}^{y=2} x\, dy}_{\text{Area of } ADC} - \underbrace{\int_{x=0}^{x=1} y\, dx}_{\text{Area of } BCD} \right]$$

$$= 2 \int_0^2 \left( \frac{4 - y^2}{4} \right) dy - \int_0^1 \left( \frac{1 - x^2}{2} \right) dx$$

$$= 2 \left\{ \left[ y - \frac{y^3}{12} \right]_0^2 - \left[ \frac{x}{2} - \frac{x^3}{6} \right]_0^1 \right\}$$

$$= 2 \left( \frac{4}{3} - \frac{1}{3} \right) = 2$$

**96.** *(41)* $A$ = Area bounded by $y = f(x)$ and $X$-axis.

$$= \int_{-3}^{-2} (x + 6)\, dx + \int_{-2}^{0} x^2\, dx + \int_0^1 \sqrt{x}\, dx$$

$$= \left[ \frac{x^2}{2} \right]_{-3}^{-2} + 6[x]_{-3}^{-2} + \left[ \frac{x^3}{3} \right]_{-2}^{0} + \left[ \frac{2}{3} x^{3/2} \right]_0^1 = \frac{41}{6}$$

$$\therefore \quad 6A = 6 \times \frac{41}{6} \Rightarrow 6A = 41$$

**97.** *(64)*

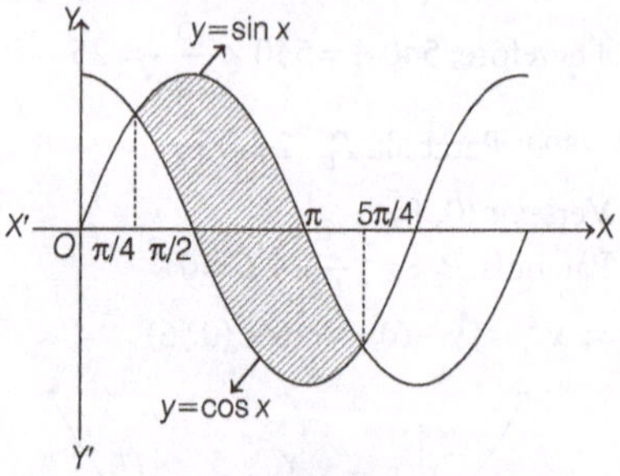


Required area of shaded region

$$A = \int_{\pi/4}^{5\pi/4} (\sin x - \cos x)\, dx$$

$$= [-\cos x - \sin x]_{\pi/4}^{5\pi/4}$$

$$= -\left[ \left( \cos \frac{5\pi}{4} + \sin \frac{5\pi}{4} \right) - \left( \cos \frac{\pi}{4} + \sin \frac{\pi}{4} \right) \right]$$

$$= -\left[ \left( -\frac{1}{\sqrt{2}} - \frac{1}{\sqrt{2}} \right) - \left( \frac{1}{\sqrt{2}} + \frac{1}{\sqrt{2}} \right) \right]$$

$$\therefore A = \frac{4}{\sqrt{2}} = 2\sqrt{2} \Rightarrow A^4 = (2\sqrt{2})^4 = 64$$

# C-21

JEE Main Chapterwise Topicwise Mathematics

In JEE Main, there is usually 0-1 question per paper is visible from this chapter. In past 3 years, 59 conceptual and 28 numerical questions were asked from this chapter.

**For detailed analysis scan QR code**

# DIFFERENTIAL EQUATIONS

## Topic 1 Ordinary Differential Equation, Their Order and Degree

### Level 1

**01** The differential equation of the family of circles passing through the origin and having centre at the line $y = x$ is *[2024, 5 April Shift-II]*

(a) $(x^2 - y^2 + 2xy)dx = (x^2 - y^2 + 2xy)dy$

(b) $(x^2 + y^2 - 2xy)dx = (x^2 + y^2 + 2xy)dy$

(c) $(x^2 + y^2 + 2xy)dx = (x^2 + y^2 - 2xy)dy$

(d) $(x^2 - y^2 + 2xy)dx = (x^2 - y^2 - 2xy)dy$

**Imp.** Distance of a Point from a Line

**02** A differential equation representing the family of parabolas with axis parallel to $Y$-axis and whose length of latus rectum is the distance of the point $(2, -3)$ from the line $3x + 4y = 5$, is given by *[2021, 27 Aug Shift-II]*

(a) $10\dfrac{d^2y}{dx^2} = 11$ (b) $11\dfrac{d^2x}{dy^2} + 10$

(c) $10\dfrac{d^2x}{dy^2} = 11$ (d) $11\dfrac{d^2y}{dx^2} = 10$

**03** If $y = \left(\dfrac{2}{\pi}x - 1\right)\operatorname{cosec} x$ is the solution of the differential equation, $\dfrac{dy}{dx} + p(x)y = \dfrac{2}{\pi}\operatorname{cosec} x$, $0 < x < \dfrac{\pi}{2}$, then the function $p(x)$ is equal to *[2020, 6 Sep Shift-II]*

(a) $\cot x$ (b) $\operatorname{cosec} x$

(c) $\sec x$ (d) $\tan x$

**04** The differential equation whose solution is $Ax^2 + By^2 = 1$, where $A$ and $B$ are arbitrary constant, is of *[AIEEE 2006]*

(a) first order and second degree

(b) first order and first degree

(c) second order and first degree

(d) second order and second degree

**05** The differential equation representing the family of curves $y^2 = 2c(x + \sqrt{c})$, where $c > 0$, is a parameter, is of order and degree as follows *[AIEEE 2005]*

(a) order 2, degree 2

(b) order 1, degree 3

(c) order 1, degree 1

(d) order 1, degree 2

**06** The degree and order of the differential equation of the family of all parabolas whose axis is $X$-axis, are respectively *[AIEEE 2003]*

(a) 2, 1 (b) 1, 2

(c) 3, 2 (d) 2, 3

**07** The order and degree of the differential equation $\left(1 + 3\dfrac{dy}{dx}\right)^{\frac{2}{3}} = 4\dfrac{d^3y}{dx^3}$ are *[AIEEE 2002]*

(a) $\left(1, \dfrac{2}{3}\right)$ (b) (3, 1)

(c) (3, 3) (d) (1, 2)

### Level 2

**08** Let $f(x)$ be a positive function such that the area bounded by $y = f(x)$, $y = 0$ from $x = 0$ to $x = a > 0$ is $e^{-a} + 4a^2 + a - 1$. Then, the differential equation, whose general solution is $y = c_1 f(x) + c_2$, where $c_1$ and $c_2$ are arbitrary constants, is *[2024, 8 April Shift-I]*

(a) $(8e^x-1)\frac{d^2y}{dx^2}+\frac{dy}{dx}=0$ (b) $(8e^x+1)\frac{d^2y}{dx^2}-\frac{dy}{dx}=0$

(c) $(8e^x+1)\frac{d^2y}{dx^2}+\frac{dy}{dx}=0$ (d) $(8e^x-1)\frac{d^2y}{dx^2}-\frac{dy}{dx}=0$

**09** The differential equation of the family of circles passing through the points (0, 2) and (0, – 2) is **[2022, 28 July Shift-II]**

(a) $2xy\frac{dy}{dx}+(x^2-y^2+4)=0$

(b) $2xy\frac{dy}{dx}+(x^2+y^2-4)=0$

(c) $2xy\frac{dy}{dx}+(y^2-x^2+4)=0$

(d) $2xy\frac{dy}{dx}-(x^2-y^2+4)=0$

**10** The differential equation satisfied by the system of parabolas $y^2=4a(x+a)$ is **[2021, 18 March Shift-I]**

(a) $y\left(\frac{dy}{dx}\right)^2-2x\left(\frac{dy}{dx}\right)-y=0$

(b) $y\left(\frac{dy}{dx}\right)^2-2x\left(\frac{dy}{dx}\right)+y=0$

(c) $y\left(\frac{dy}{dx}\right)^2+2x\left(\frac{dy}{dx}\right)-y=0$

(d) $y\left(\frac{dy}{dx}\right)+2x\left(\frac{dy}{dx}\right)-y=0$

**11** If the differential equation representing the family of all circles touching X-axis at the origin is $(x^2-y^2)\frac{dy}{dx}=g(x)\,y$, then $g(x)$ equals **[JEE Main 2014]**

(a) $\frac{1}{2}x$ (b) $2x^2$

(c) $2x$ (d) $\frac{1}{2}x^2$

## Numerical Value Type Questions

**12** Let $f:R\to R$ be a thrice differentiable odd function satisfying $f'(x)\ge 0$, $f''(x)=f(x)$, $f(0)=0$, $f'(0)=3$. Then, $9f(\log_e 3)$ is equal to .......... **[2025, 02 April Shift-I]**

**13** For the curve $C:(x^2+y^2-3)+(x^2-y^2-1)^5=0$, the value of $3y'-y^3y''$, at the point $(\alpha,\alpha)$, $\alpha>0$, on $C$, is equal to .......... . **[2022, 27 July Shift-II]**

**14** The difference between degree and order of a differential equation that represents the family of curves given by $y^2=a\left(x+\frac{\sqrt{a}}{2}\right)$, $a>0$ is .......... . **[2021, 26 Feb Shift-I]**

## Topic 2 Solution of Differential Equations by Method of Variables Form and Homogeneous

### Level 1

**01** The solution curve, of the differential equation $2y\frac{dy}{dx}+3=5\frac{dy}{dx}$, passing through the point (0, 1) is a conic, whose vertex lies on the line. **[2024, 9 April Shift-I]**

(a) $2x+3y=9$ (b) $2x+3y=-9$

(c) $2x+3y=6$ (d) $2x+3y=-6$

**02** Let $y=y(x)$ be the solution of the differential equation $(1+y^2)e^{\tan x}dx+\cos^2 x(1+e^{2\tan x})dy=0$, $y(0)=1$. Then, $y\left(\frac{\pi}{4}\right)$ is equal to **[2024, 8 April Shift-I]**

(a) $2/e$ (b) $2/e^2$ (c) $1/e^2$ (d) $1/e$

**03** The temperature $T(t)$ of a body at time $t=0$ is 160°F and it decreases continuously as per the differential equation $\frac{dT}{dt}=-K(T-80)$, where $K$ is a positive constant. If $T(15)=120$°F, then $T(45)$ is equal to **[2024, 31 Jan Shift-II]**

(a) 85°F (b) 95°F (c) 80°F (d) 90°F

**Imp.** Homogeneous Differential Equation

**04** The solution curve of the differential equation $y\frac{dx}{dy}=x(\log_e x-\log_e y+1)$, $x>0$, $y>0$ passing through the point $(e, 1)$ is **[2024, 31 Jan Shift-I]**

(a) $\left|\log_e\frac{y}{x}\right|=x$ (b) $\left|\log_e\frac{x}{y}\right|=y$

(c) $\left|\log_e\frac{y}{x}\right|=y^2$ (d) $2\left|\log_e\frac{x}{y}\right|=y+1$

**05** If $\sin\left(\frac{y}{x}\right)=\log_e|x|+\frac{\alpha}{2}$ is the solution of the differential equation $x\cos\left(\frac{y}{x}\right)\frac{dy}{dx}=y\cos\left(\frac{y}{x}\right)+x$ and $y(1)=\pi/3$, then $\alpha^2$ is equal to **[2024, 22 Jan Shift-I]**

(a) 3 (b) 12 (c) 9 (d) 4

**06** Let $y=f(x)$ be the solution of the differential equation $y(x+1)dx-x^2dy=0$, $y(1)=e$. Then, $\lim_{x\to 0^+} f(x)$ is equal to **[2023, 29 Jan Shift-I]**

(a) 0 (b) $e^2$ (c) $\frac{1}{e^2}$ (d) $\frac{1}{e}$

**07** Let $x = x(y)$ be the solution of the differential equation $2ye^{x/y^2}\,dx + (y^2 - 4xe^{x/y^2})\,dy = 0$ such that $x(1) = 0$. Then, $x(e)$ is equal to **[2022, 28 June Shift-II]**

(a) $e\log_e(2)$ (b) $-e\log_e(2)$
(c) $e^2\log_e(2)$ (d) $-e^2\log_e(2)$

**08** Let $g:(0,\infty)\to R$ be a differentiable function such that $\int\left(\frac{x(\cos x - \sin x)}{e^x+1} + \frac{g(x)(e^x+1-xe^x)}{(e^x+1)^2}\right)dx = \frac{xg(x)}{e^x+1} + c, \forall x > 0$, where $c$ is an arbitrary constant.

Then, **[2022, 25 June Shift-I]**

(a) $g$ is decreasing in $\left(0, \frac{\pi}{4}\right)$

(b) $g'$ is increasing in $\left(0, \frac{\pi}{4}\right)$

(c) $g + g'$ is increasing in $\left(0, \frac{\pi}{2}\right)$

(d) $g - g'$ is increasing in $\left(0, \frac{\pi}{2}\right)$

**09** Let $f$ be a non-negative function in [0, 1] and twice differentiable in (0, 1). If $\int_0^x \sqrt{1-(f'(t))^2}\,dt = \int_0^x f(t)\,dt, 0 \le x \le 1$ and $f(0) = 0$, then $\lim_{x\to 0}\frac{1}{x^2}\int_0^x f(t)\,dt$ **[2021, 31 Aug Shift-I]**

(a) equals 0 (b) equals 1
(c) does not exist (d) equals 1/2

**10** If $\frac{dy}{dx} = \frac{2^{x+y} - 2^x}{2^y}$, $y(0) = 1$, then $y(1)$ is equal to **[2021, 31 Aug Shift-I]**

(a) $\log_2(2+e)$ (b) $\log_2(1+e)$
(c) $\log_2(2e)$ (d) $\log_2(1+e^2)$

**11** Let $y = y(x)$ be solution of the differential equation $\log_e\left(\frac{dy}{dx}\right) = 3x + 4y$, with $y(0) = 0$.

If $y\left(-\frac{2}{3}\log_e 2\right) = \alpha\log_e 2$, then the value of $\alpha$ is equal to **[2021, 27 July Shift-I]**

(a) $-1/4$ (b) $1/4$ (c) 2 (d) $-1/2$

**12** Let $y = y(x)$ be the solution of the differential equation $x\,dy = (y + x^3\cos x)\,dx$ with $y(\pi) = 0$, then $y(\pi/2)$ is equal to **[2021, 25 July Shift-II]**

(a) $\frac{\pi^2}{4} + \frac{\pi}{2}$ (b) $\frac{\pi^2}{2} + \frac{\pi}{4}$ (c) $\frac{\pi^2}{2} - \frac{\pi}{4}$ (d) $\frac{\pi^2}{4} - \frac{\pi}{2}$

**13** Let $y = y(x)$ be the solution of the differential equation $e^x\sqrt{1-y^2}\,dx + \left(\frac{y}{x}\right)dy = 0$, $y(1) = -1$. Then, the value of $[y(3)]^2$ is equal to **[2021, 20 July Shift-I]**

(a) $1 - 4e^3$
(b) $1 - 4e^6$
(c) $1 + 4e^3$
(d) $1 + 4e^6$

**14** Which of the following is true for $y(x)$, that satisfies the differential equation $\frac{dy}{dx} = xy - 1 + x - y$; $y(0) = 0$ **[2021, 17 March Shift-I]**

(a) $y(1) = e^{-\frac{1}{2}} - 1$ (b) $y(1) = e^{\frac{1}{2}} - e^{-\frac{1}{2}}$
(c) $y(1) = 1$ (d) $y(1) = e^{\frac{1}{2}} - 1$

**15** If $y = y(x)$ is the solution of the differential equation $\frac{5+e^x}{2+y}\cdot\frac{dy}{dx} + e^x = 0$ satisfying $y(0) = 1$, then a value of $y(\log_e 13)$ is **[2020, 5 Sep Shift-I]**

(a) 0 (b) $-1$
(c) 1 (d) 2

**16** The solution of the differential equation $\frac{dy}{dx} - \frac{y+3x}{\log_e(y+3x)} + 3 = 0$ is
(Where, $C$ is a constant of integration) **[2020, 4 Sep Shift-II]**

(a) $x - \log_e(y+3x) = C$ (b) $y + 3x - \frac{1}{2}(\log_e x)^2 = C$
(c) $x - 2\log_e(y+3x) = C$ (d) $x - \frac{1}{2}(\log_e(y+3x))^2 = C$

**17** If $p(x)$ be a polynomial of degree three that has a local maximum value 8 at $x = 1$ and a local minimum value 4 at $x = 2$, then $p(0)$ is equal to **[2020, 2 Sep Shift-I]**

(a) 6 (b) $-12$ (c) $-24$ (d) 12

**18** Let $f:[0,1]\to R$ be such that $f(xy) = f(x).f(y)$, for all $x, y \in [0,1]$ and $f(0) \ne 0$. If $y = y(x)$ satisfies the differential equation, $\frac{dy}{dx} = f(x)$ with $y(0) = 1$, then $y\left(\frac{1}{4}\right) + y\left(\frac{3}{4}\right)$ is equal to **[2019, 9 Jan Shift-II]**

(a) 5 (b) 3
(c) 2 (d) 4

**19** The solution of the differential equation $y\,dx - (x + 2y^2)\,dy = 0$ is $x = f(y)$. If $f(-1) = 1$, then $f(1)$ is equal to **[JEE Main 2015]**

(a) 4 (b) 3 (c) 2 (d) 1

**20** At present, a firm is manufacturing 2000 items. It is estimated that the rate of change of production $P$ w.r.t. additional number of workers $x$ is given by $\frac{dP}{dx} = 100 - 12\sqrt{x}$. If the firm employees 25 more workers, then the new level of production of items is **[JEE Main 2013]**

(a) 2500 (b) 3000 (c) 3500 (d) 4500

**21** If $y(x)$ is the solution of the differential equation $(x+2)\frac{dy}{dx} = x^2 + 4x - 9, x \neq -2$ and $y(0) = 0$, then $y(-4)$ is equal to **[JEE Main 2010]**

(a) 0 (b) 1 (c) $-1$ (d) 2

**22** The solution of the differential equation $y\,dx + (x + x^2 y)\,dy = 0$ is **[AIEEE 2004]**

(a) $-\frac{1}{xy} = C$ (b) $-\frac{1}{xy} + \log y = C$

(c) $\frac{1}{xy} + \log y = C$ (d) $\log y = Cx$

**23** The solution of the equation $\frac{d^2y}{dx^2} = e^{-2x}$ is **[AIEEE 2002]**

(a) $\frac{e^{-2x}}{4}$ (b) $\frac{e^{-2x}}{4} + Cx + D$

(c) $\frac{1}{4}e^{-2x} + Cx^2 + D$ (d) $\frac{1}{4}e^{-2x} + C + D$

## Level 2

**24** Let for some function $y = f(x)$, $\int_0^x t f(t)\,dt = x^2 f(x)$, $x > 0$ and $f(2) = 3$. Then, $f(6)$ is equal to **[2025, 28 Jan Shift-I]**

(a) 6 (b) 3 (c) 2 (d) 1

**25** Let $f:(0, \infty) \to R$ be a function which is differentiable at all points of its domain and satisfies the condition $x^2 f'(x) = 2xf(x) + 3$ with $f(1) = 4$. Then, $2f(2)$ is equal to **[2025, 24 Jan Shift-II]**

(a) 29 (b) 23

(c) 19 (d) 39

**26** Let $f(x)$ be a real differentiable function such that $f(0) = 1$ and $f(x+y) = f(x)f'(y) + f'(x)f(y)$ $\forall$ $x, y \in R$. Then, $\sum_{n=1}^{100} \log_e f(n)$ is equal to **[2025, 22 Jan Shift-I]**

(a) 2525 (b) 2384

(c) 5220 (d) 2406

**27** The solution of the differential equation $(x^2 + y^2)dx - 5xy\,dy = 0$, $y(1) = 0$, is **[2024, 9 April Shift-I]**

(a) $|x^2 - 2y^2|^6 = x$ (b) $|x^2 - 4y^2|^6 = x$

(c) $|x^2 - 4y^2|^5 = x^2$ (d) $|x^2 - 2y^2|^5 = x^2$

**28** If the solution $y = y(x)$ of the differential equation $(x^4 + 2x^3 + 3x^2 + 2x + 2)dy - (2x^2 + 2x + 3)dx = 0$ satisfies $y(-1) = -\frac{\pi}{4}$, then $y(0)$ is equal to **[2024, 4 April Shift-I]**

(a) $-\frac{\pi}{12}$ (b) $\frac{\pi}{4}$

(c) $\frac{\pi}{2}$ (d) 0

**29** Let $\alpha$ be a non-zero real number. Suppose $f: R \to R$ is a differentiable function such that $f(0) = 2$ and $\lim_{x \to -\infty} f(x) = 1$. If $f'(x) = \alpha f(x) + 3$, for all $x \in R$, then $f(-\log_e 2)$ is equal to ...... **[2024, 1 Feb Shift-I]**

(a) 5 (b) 9 (c) 7 (d) 3

**Imp.** **Separation of Variables**

**30** Let $y = y(x)$ be the solution of the differential equation $\sec x\,dy + [2(1-x)\tan x + x(2-x)]\,dx = 0$ such that $y(0) = 2$. Then, $y(2)$ is equal to **[2024, 30 Jan Shift-I]**

(a) 2 (b) $2\{1 - \sin(2)\}$

(c) 1 (d) $2\{\sin(2) + 1\}$

**31** Let $x = x(t)$ and $y = y(t)$ be solutions of the differential equations $\frac{dx}{dt} + ax = 0$ and $\frac{dy}{dt} + by = 0$ respectively, $a, b \in R$. Given that $x(0) = 2$; $y(0) = 1$ and $3y(1) = 2x(1)$, the value of $t$, for which $x(t) = y(t)$, is **[2024, 27 Jan Shift-I]**

(a) $\log_{4/3} 2$ (b) $\log_4 3$

(c) $\log_{2/3} 2$ (d) $\log_3 4$

**32** Let $y = y_1(x)$ and $y = y_2(x)$ be the solution curves of the differential equation $\frac{dy}{dx} = y + 7$ with initial conditions $y_1(0) = 0$ and $y_2(0) = 1$ respectively. Then, the curves $y = y_1(x)$ and $y = y_2(x)$ intersect at **[2023, 13 April Shift-I]**

(a) infinite number of points

(b) two points

(c) one point

(d) no point

**33** Let $y = y(x)$ be a solution curve of the differential equation $(1 - x^2y^2)dx = y\,dx + x\,dy$.

If the line $x = 1$ intersects the curve $y = y(x)$ at $y = 2$ and the line $x = 2$ intersects the curve $y = y(x)$ at $y = \alpha$, then a value of $\alpha$ is **[2023, 11 April Shift-I]**

(a) $\frac{3e^2}{2(3e^2+1)}$ (b) $\frac{3e^2}{2(3e^2-1)}$

(c) $\frac{1-3e^2}{2(3e^2+1)}$ (d) $\frac{1+3e^2}{2(3e^2-1)}$

**34** The area enclosed by the closed curve $C$ given by the differential equation $\frac{dy}{dx}+\frac{x+a}{y-2}=0$, $y(1)=0$ is $4\pi$.

Let $P$ and $Q$ be the points of intersection of the curve $C$ and the $Y$-axis. If normals at $P$ and $Q$ on the curve $C$ intersect $X$-axis at points $R$ and $S$ respectively, then the length of the line segment $RS$ is **[2023, 1 Feb Shift-I]**

(a) $\frac{4\sqrt{3}}{3}$ (b) $2\sqrt{3}$

(c) 2 (d) $\frac{2\sqrt{3}}{3}$

**35** Let $y=y(x)$ be the solution of the differential equation $(3y^2-5x^2)y\,dx+2x(x^2-y^2)\,dy=0$ such that $y(1)=1$. Then, $|(y(2))^3-12y(2)|$ is equal to **[2023, 31 Jan Shift-II]**

(a) $16\sqrt{2}$ (b) 32

(c) $32\sqrt{2}$ (d) 64

**36** The solution of the differential equation $\frac{dy}{dx}=-\left(\frac{x^2+3y^2}{3x^2+y^2}\right)$, $y(1)=0$ is **[2023, 30 Jan Shift-II]**

(a) $\log_e|x+y|-\frac{2xy}{(x+y)^2}=0$

(b) $\log_e|x+y|-\frac{xy}{(x+y)^2}=0$

(c) $\log_e|x+y|+\frac{2xy}{(x+y)^2}=0$

(d) $\log_e|x+y|+\frac{xy}{(x+y)^2}=0$

**37** Let $y=y(x)$ be the solution of the differential equation $(x^2-3y^2)dx+3xy\,dy=0$, $y(1)=1$. Then, $6y^2(e)$ is equal to **[2023, 24 Jan Shift-II]**

(a) $e^2$ (b) $2e^2$

(c) $3e^2$ (d) $\frac{3}{2}e^2$

**38** Let the solution curve of the differential equation $xdy=(\sqrt{x^2+y^2}+y)\,dx$, $x>0$, intersect the line $x=1$ at $y=0$ and the line $x=2$ at $y=\alpha$. Then, the value of $\alpha$ is **[2022, 28 July Shift-I]**

(a) $\frac{1}{2}$ (b) $\frac{3}{2}$ (c) $-\frac{3}{2}$ (d) $\frac{5}{2}$

**39** Consider a curve $y=y(x)$ in the first quadrant as show in the figure. Let the area $A_1$ is twice the area $A_2$. Then, the normal to the curve perpendicular to the line $2x-12y=15$ does not pass through the point. **[2022, 27 July Shift-II]**

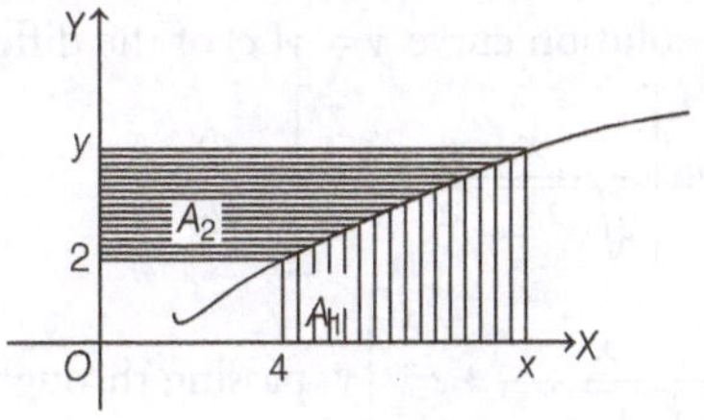


(a) (6, 21) (b) (8, 9)

(c) (10, –4) (d) (12, –15)

**40** Let a smooth curve $y=f(x)$ be such that the slope of the tangent at any point $(x, y)$ on it is directly proportional to $\left(-\frac{y}{x}\right)$. If the curve passes through the points (1, 2) and (8, 1), then $\left|y\left(\frac{1}{8}\right)\right|$ is equal to **[2022, 25 July Shift-II]**

(a) $2\log_e 2$ (b) 4

(c) 1 (d) $4\log_e 2$

**41** The slope of the tangent to a curve $C: y=y(x)$ at any point $(x, y)$ on it is $\frac{2e^{2x}-6e^{-x}+9}{2+9e^{-2x}}$. If $C$ passes through the points $\left(0, \frac{1}{2}+\frac{\pi}{2\sqrt{2}}\right)$ and $\left(\alpha, \frac{1}{2}e^{2\alpha}\right)$, then $e^\alpha$ is equal to **[2022, 25 July Shift-I]**

(a) $\frac{3+\sqrt{2}}{3-\sqrt{2}}$ (b) $\frac{3}{\sqrt{2}}\left(\frac{3+\sqrt{2}}{3-\sqrt{2}}\right)$

(c) $\frac{1}{\sqrt{2}}\left(\frac{\sqrt{2}+1}{\sqrt{2}-1}\right)$ (d) $\frac{\sqrt{2}+1}{\sqrt{2}-1}$

**42** The general solution of the differential equation $(x-y^2)dx+y(5x+y^2)\,dy=0$ is **[2022, 25 July Shift-I]**

(a) $(y^2+x)^4=C\,|(y^2+2x)^3|$

(b) $(y^2+2x)^4=C\,|(y^2+x)^3|$

(c) $|(y^2+x)^3|=C\,(2y^2+x)^4$

(d) $|(y^2+2x)^3|=C\,(2y^2+x)^4$

**43** If $y=y\,(x)$ is the solution of the differential equation $(1+e^{2x})\frac{dy}{dx}+2(1+y^2)e^x=0$ and $y(0)=0$, then $6\{y'(0)+(y(\log_e\sqrt{3}))^2\}$ is equal to **[2022, 29 June Shift-II]**

(a) 2 (b) $-2$

(c) $-4$ (d) $-1$

**44** Let the solution curve of the differential equation $x\frac{dy}{dx}-y=\sqrt{y^2+16x^2}$, $y(1)=3$ be $y=y(x)$. Then, $y(2)$ is equal to **[2022, 29 June Shift-I]**

(a) 15 (b) 11 (c) 13 (d) 17

**45** Let the solution curve $y = y(x)$ of the differential equation $\left[\frac{x}{\sqrt{x^2 - y^2}} + e^{\frac{y}{x}}\right] x\frac{dy}{dx} = x + \left[\frac{x}{\sqrt{x^2 - y^2}} + e^{\frac{y}{x}}\right] y$ passing through the points (1, 0) and $(2\alpha, \alpha)$, $\alpha > 0$. Then, $\alpha$ is equal to *[2022, 28 June Shift-I]*

(a) $\frac{1}{2}\exp\left(\frac{\pi}{6} + \sqrt{e} - 1\right)$ (b) $\frac{1}{2}\exp\left(\frac{\pi}{3} + e - 1\right)$

(c) $\exp\left(\frac{\pi}{6} + \sqrt{e} + 1\right)$ (d) $2\exp\left(\frac{\pi}{3} + \sqrt{e} - 1\right)$

**46** If $\frac{dy}{dx} + \frac{2^{x-y}(2^y - 1)}{2^x - 1} = 0$, $x, y > 0$, $y(1) = 1$, then $y(2)$ is equal to *[2022, 27 June Shift-I]*

(a) $2 + \log_2 3$ (b) $2 + \log_3 2$

(c) $2 - \log_3 2$ (d) $2 - \log_2 3$

**47** Let $\frac{dy}{dx} = \frac{ax - by + a}{bx + cy + a}$, where $a, b, c$ are constants, represent a circle passing through the point (2, 5). Then, the shortest distance of the point (11, 6) from this circle is *[2022, 27 June Shift-I]*

(a) 10 (b) 8 (c) 7 (d) 5

**48** If $y = y(x)$ is the solution of the differential equation $2x^2\frac{dy}{dx} - 2xy + 3y^2 = 0$ such that $y(e) = e/3$, then $y(1)$ is equal to *[2022, 25 June Shift-II]*

(a) 1/3 (b) 2/3 (c) 3/2 (d) 3

**49** If the solution curve $y = y(x)$ of the differential equation $y^2 dx + (x^2 - xy + y^2) dy = 0$, which passes through the point (1, 1) and intersects the line $y = \sqrt{3}\,x$ at the point $(\alpha, \sqrt{3}\,\alpha)$, then value of $\log_e(\sqrt{3}\,\alpha)$ is equal to *[2022, 25 June Shift-I]*

(a) $\frac{\pi}{3}$ (b) $\frac{\pi}{2}$

(c) $\frac{\pi}{12}$ (d) $\frac{\pi}{6}$

**50** The slope of normal at any point $(x, y)$, $x > 0$, $y > 0$ on the curve $y = y(x)$ is given by $\frac{x^2}{xy - x^2y^2 - 1}$. If the curve passes through the point (1, 1), then $e \cdot y(e)$ is equal to *[2022, 24 June Shift-II]*

(a) $\frac{1 - \tan(1)}{1 + \tan(1)}$ (b) $\tan(1)$

(c) 1 (d) $\frac{1 + \tan(1)}{1 - \tan(1)}$

**51** A particle is moving in the $XY$-plane along a curve $C$ passing through the point (3, 3). The tangent to the curve $C$ at the point $P$ meets the $X$-axis at $Q$. If the $Y$-axis bisects the segment $PQ$, then $C$ is a parabola with *[2022, 24 June Shift-II]*

(a) length of latusrectum 3 (b) length of latusrectum 6

(c) focus $\left(\frac{4}{3}, 0\right)$ (d) focus $\left(0, \frac{3}{4}\right)$

**52** If $y\frac{dy}{dx} = x\left[\frac{y^2}{x^2} + \frac{\phi\left(\frac{y^2}{x^2}\right)}{\phi'\left(\frac{y^2}{x^2}\right)}\right]$, $x > 0$, $\phi > 0$, and $y(1) = -1$, then $\phi\left(\frac{y^2}{4}\right)$ is equal to *[2021, 31 Aug Shift-II]*

(a) $4\phi(2)$ (b) $4\phi(1)$ (c) $2\phi(1)$ (d) $\phi(1)$

**53** Let $y = y(x)$ be the solution of the differential equation $\frac{dy}{dx} = 2(y + 2\sin x - 5)x - 2\cos x$ such that, $y(0) = 7$. Then, $y(\pi)$ is equal to *[2021, 27 Aug Shift-I]*

(a) $2e^{\pi^2} + 5$ (b) $e^{\pi^2} + 5$ (c) $3e^{\pi^2} + 5$ (d) $7e^{\pi^2} + 5$

**54** Let $y(x)$ be the solution of the differential equation $2x^2 dy + (e^y - 2x)\,dx = 0$, $x > 0$. If $y(e) = 1$, then $y(1)$ is equal to *[2021, 26 Aug Shift-II]*

(a) 0 (b) 2 (c) $\log_e 2$ (d) $\log_e(2e)$

**55** Let $y = y(x)$ be the solution of the differential equation $(x - x^3)\,dy = (y + yx^2 - 3x^4)\,dx$, $x > 2$. If $y(3) = 3$, then $y(4)$ is equal to *[2021, 27 July Shift-II]*

(a) 4 (b) 12 (c) 8 (d) 16

**56** Let $y = y(x)$ be the solution of the differential equation $dy/dx = 1 + xe^{y-x}$, $-\sqrt{2} < x < \sqrt{2}$, $y(0) = 0$, then the minimum value of $y(x)$, $x \in (-\sqrt{2}, \sqrt{2})$ is equal to *[2021, 25 July Shift-I]*

(a) $(2 - \sqrt{3}) - \log_e 2$ (b) $(2 + \sqrt{3}) - \log_e 2$

(c) $(1 + \sqrt{3}) - \log_e(\sqrt{3} - 1)$ (d) $(1 - \sqrt{3}) - \log_e(\sqrt{3} - 1)$

**57** Let $C_1$ be the curve obtained by the solution of differential equation $2xy\frac{dy}{dx} = y^2 - x^2$, $x > 0$. Let the curve $C_2$ be the solution of $\frac{2xy}{x^2 - y^2} = \frac{dy}{dx}$. If both the curves pass through (1, 1), then the area enclosed by the curves $C_1$ and $C_2$ is equal to *[2021, 16 March Shift-II]*

(a) $\pi - 1$ (b) $\frac{\pi}{2} - 1$

(c) $\pi + 1$ (d) $\frac{\pi}{4} + 1$

**58** Let $f(x)=\int_0^x e^t f(t)\,dt+e^x$ be a differentiable function for all $x\in R$. Then, $f(x)$ equals **[2021, 26 Feb Shift-II]**

(a) $2e^{(e^x-1)}-1$ (b) $e^{e^x}-1$ (c) $2e^{e^x}-1$ (d) $e^{(e^x-1)}$

**59** Let slope of the tangent line to a curve at any point $P(x, y)$ be given by $\frac{xy^2+y}{x}$. If the curve intersects the line $x+2y=4$ at $x=-2$, then the value of $y$, for which the point $(3, y)$ lies on the curve, is **[2021, 26 Feb Shift-II]**

(a) $\frac{18}{35}$ (b) $-\frac{4}{3}$

(c) $-\frac{18}{19}$ (d) $-\frac{18}{11}$

**60** The rate of growth of bacteria in a culture is proportional to the number of bacteria present and the bacteria count is 1000 at initial time $t=0$. The number of bacteria is increased by 20% in 2 h. If the population of bacteria is 2000 after $\frac{k}{\log_e(6/5)}$ h, then $\left(\frac{k}{\log_e 2}\right)^2$ is equal to **[2021, 26 Feb Shift-I]**

(a) 4 (b) 8 (c) 2 (d) 16

**61** Let $f:(0,\infty)\to(0,\infty)$ be a differentiable function such that $f(1)=e$ and $\lim_{t\to x}\frac{t^2f^2(x)-x^2f^2(t)}{t-x}=0$. If $f(x)=1$, then $x$ is equal to **[2020, 4 Sep Shift-II]**

(a) $1/2e$ (b) $1/e$

(c) $e$ (d) $2e$

**62** If $x^3\,dy+xy\,dx=x^2\,dy+2y\,dx$; $y(2)=e$ and $x>1$, then $y(4)$ is equal to **[2020, 3 Sep Shift-II]**

(a) $\frac{\sqrt{e}}{2}$ (b) $\frac{3}{2}+\sqrt{e}$ (c) $\frac{3}{2}\sqrt{e}$ (d) $\frac{1}{2}+\sqrt{e}$

**63** If a curve $y=f(x)$, passing through the point $(1, 2)$, is the solution of the differential equation, $2x^2dy=(2xy+y^2)dx$, then $f\left(\frac{1}{2}\right)$ is equal to **[2020, 2 Sep Shift-II]**

(a) $\frac{1}{1+\log_e 2}$ (b) $\frac{1}{1-\log_e 2}$ (c) $1+\log_e 2$ (d) $\frac{-1}{1+\log_e 2}$

**64** Let $y=y(x)$ be a solution of the differential equation, $\sqrt{1-x^2}\frac{dy}{dx}+\sqrt{1-y^2}=0, |x|<1$.

If $y\left(\frac{1}{2}\right)=\frac{\sqrt{3}}{2}$, then $y\left(\frac{-1}{\sqrt{2}}\right)$ is equal to **[2020, 8 Jan Shift-I]**

(a) $\frac{\sqrt{3}}{2}$ (b) $-\frac{\sqrt{3}}{2}$ (c) $-\frac{1}{\sqrt{2}}$ (d) $\frac{1}{\sqrt{2}}$

**65** Let $f$ be a differentiable function such that $f(1)=2$ and $f'(x)=f(x)$ for all $x\in R$. If $h(x)=f(f(x))$, then $h'(1)$ is equal to **[2019, 12 Jan Shift-II]**

(a) $4e^2$ (b) $4e$ (c) $2e$ (d) $2e^2$

**66** If a curve $y=f(x)$ passes through the point $(1,-1)$ and satisfies the differential equation $y(1+xy)dx=x\,dy$, then $f\left(-\frac{1}{2}\right)$ is equal to **[JEE Main 2016]**

(a) $-\frac{2}{5}$ (b) $-\frac{4}{5}$ (c) $\frac{2}{5}$ (d) $\frac{4}{5}$

**67** The population $p(t)$ at time $t$ of a certain mouse species satisfies the differential equation. $\frac{dp(t)}{dt}=0.5(t)-450$. If $p(0)=850$, then the time at which the population becomes zero is **[AIEEE 2012]**

(a) $2\log 18$ (b) $\log 9$ (c) $\frac{1}{2}\log 18$ (d) $\log 18$

## Level 3

**68** Let $x=x(y)$ be the solution of the differential equation $y=\left(x-y\frac{dx}{dy}\right)\sin\left(\frac{x}{y}\right)$, $y>0$ and $x(1)=\frac{\pi}{2}$

Then, $\cos(x(2))$ is equal to **[2025, 23 Jan Shift-II]**

(a) $2(\log_e 2)^2-1$ (b) $2(\log_e 2)-1$

(c) $1-2(\log_e 2)$ (d) $1-2(\log_e 2)^2$

**69** Suppose the solution of the differential equation $\frac{dy}{dx}=\frac{(2+\alpha)x-\beta y+2}{\beta x-2\alpha y-(\beta\gamma-4\alpha)}$ represents a circle passing through origin. Then, the radius of this circle is **[2024, 6 April Shift-II]**

(a) 2 (b) $\frac{\sqrt{17}}{2}$ (c) $\frac{1}{2}$ (d) $\sqrt{17}$

**70** Let $y=y(x)$ be the solution of the differential equation $\frac{dy}{dx}=2x(x+y)^3-x(x+y)-1$, $y(0)=1$.

Then, $\left(\frac{1}{\sqrt{2}}+y\left(\frac{1}{\sqrt{2}}\right)\right)^2$ equals **[2024, 1 Feb Shift-I]**

(a) $\frac{3}{3-\sqrt{e}}$ (b) $\frac{4}{4+\sqrt{e}}$ (c) $\frac{1}{2-\sqrt{e}}$ (d) $\frac{2}{1+\sqrt{e}}$

**71** If $y=y(x)$ is the solution curve of the differential equation $(x^2-4)dy-(y^2-3y)dx=0$, $x>2$, $y(4)=\frac{3}{2}$ and the slope of the curve is never zero, then the value of $y(10)$ equals. **[2024, 27 Jan Shift II]**

(a) $\frac{3}{1-(8)^{1/4}}$ (b) $\frac{3}{1+2\sqrt{2}}$

(c) $\frac{3}{1-2\sqrt{2}}$ (d) $\frac{3}{1+(8)^{1/4}}$

**72** If the solution curve of the differential equation $\frac{dy}{dx}=\frac{x+y-2}{x-y}$ passes through the point (2, 1) and $(k+1, 2), k>0$, then *[2022, 29 July Shift-II]*

(a) $2\tan^{-1}\left(\frac{1}{k}\right)=\log_e(k^2+1)$

(b) $\tan^{-1}\left(\frac{1}{k}\right)=\log_e(k^2+1)$

(c) $2\tan^{-1}\left(\frac{1}{k+1}\right)=\log_e(k^2+2k+2)$

(d) $2\tan^{-1}\left(\frac{1}{k}\right)=\log_e\left(\frac{k^2+1}{k^2}\right)$

**73** Let $y=y(x)$ be the solution of the differential equation

$$x\tan\left(\frac{y}{x}\right)dy=\left(y\tan\left(\frac{y}{x}\right)-x\right)dx, -1\le x\le 1,$$

$y\left(\frac{1}{2}\right)=\frac{\pi}{6}$. Then, the area of the region bounded by the curves $x=0$, $x=\frac{1}{\sqrt{2}}$ and $y=y(x)$ in the upper half plane is *[2021, 20 July Shift-I]*

(a) $\frac{1}{8}(\pi-1)$ (b) $\frac{1}{12}(\pi-3)$ (c) $\frac{1}{4}(\pi-2)$ (d) $\frac{1}{6}(\pi-1)$

**74** Let $f$ be a twice differentiable function defined on $R$, such that $f(0)=1$, $f'(0)=2$ and $f''(x)\neq 0$ for all $x\in R$. If $\begin{vmatrix} f(x) & f'(x) \\ f'(x) & f''(x)\end{vmatrix}=0$, for all $x\in R$, then the value of $f(1)$ lies in the interval *[2021, 24 Feb Shift-II]*

(a) (9, 12) (b) (6, 9)

(c) (0, 3) (d) (3, 6)

**75** The general solution of the differential equation $\sqrt{1+x^2+y^2+x^2y^2}+xy\frac{dy}{dx}=0$ is (Where, $C$ is a constant of integration) *[2020, 6 Sep Shift-I]*

(a) $\sqrt{1+y^2}+\sqrt{1+x^2}=\frac{1}{2}\log_e\left(\frac{\sqrt{1+x^2}+1}{\sqrt{1+x^2}-1}\right)+C$

(b) $\sqrt{1+y^2}-\sqrt{1+x^2}=\frac{1}{2}\log_e\left(\frac{\sqrt{1+x^2}+1}{\sqrt{1+x^2}-1}\right)+C$

(c) $\sqrt{1+y^2}+\sqrt{1+x^2}=\frac{1}{2}\log_e\left(\frac{\sqrt{1+x^2}-1}{\sqrt{1+x^2}+1}\right)+C$

(d) $\sqrt{1+y^2}-\sqrt{1+x^2}=\frac{1}{2}\log_e\left(\frac{\sqrt{1+x^2}-1}{\sqrt{1+x^2}+1}\right)+C$

**76** Given that the slope of the tangent to a curve $y=y(x)$ at any point $(x, y)$ is $2y/x^2$. If the curve passes through the centre of the circle $x^2+y^2-2x-2y=0$, then its equation is *[2019, 8 April Shift-II]*

(a) $x^2\log_e|y|=-2(x-1)$

(b) $x\log_e|y|=x-1$

(c) $x\log_e|y|=2(x-1)$

(d) $x\log_e|y|=-2(x-1)$

## Numerical Value Type Questions

**77** Let $f:(0,\infty)\to R$ be a twice differentiable function. If for some $a\neq 0$, $\int_0^1 f(\lambda x)\,d\lambda=af(x)$, $f(1)=1$ and $f(16)=\frac{1}{8}$, then $16-f'\left(\frac{1}{16}\right)$ is equal to ............ . *[2025, 29 Jan Shift-I]*

**78** Let $f$ be a differentiable function such that

$$2(x+2)^2 f(x)-3(x+2)^2=10\int_0^x (t+2)f(t)\,dt, x\ge 0,$$ then $f(2)$ is equal to *[2025, 24 Jan Shift-I]*

**79** Let $\alpha|x|=|y|e^{xy-\beta}$, $\alpha,\beta\in N$ be the solution of the differential equation $xdy-ydx+xy(xdy+ydx)=0$, $y(1)=2$. Then, $\alpha+\beta$ is equal to ......... . *[2024, 8 April Shift-II]*

**80** If the solution $y(x)$ of the given differential equation $(e^y+1)\cos x\,dx+e^y\sin x\,dy=0$ passes through the point $\left(\frac{\pi}{2},0\right)$, then the value of $e^{y\left(\frac{\pi}{6}\right)}$ is equal to ...... . *[2024, 6 April Shift-II]*

**Imp.** **Maximum and Minimum Value of Function**

**81** Let $y=y(x)$ be the solution of the differential equation $(x+y+2)^2dx=dy$, $y(0)=-2$. Let the maximum and minimum values of the function $y=y(x)$ in $\left[0,\frac{\pi}{3}\right]$ be $\alpha$ and $\beta$, respectively. If $(3\alpha+\pi)^2+\beta^2=\gamma+\delta\sqrt{3}$, $\gamma,\delta\in Z$, then $\gamma+\delta$ equals........ *[2024, 4 April Shift-II]*

**82** Let $f(x)=\sqrt{\lim_{r\to x}\left\{\frac{2r^2[(f(r))^2-f(x)f(r)]}{r^2-x^2}-r^3e^{\frac{f(r)}{r}}\right\}}$ be differentiable in $(-\infty,0)\cup(0,\infty)$ and $f(1)=1$. Then, the value of $ea$, such that $f(a)=0$ is equal to ......... . *[2024, 29 Jan Shift-I]*

**83** If the solution curve $y = y(x)$ of the differential equation $(1+y^2)(1+\log_e x)\,dx + x\,dy = 0,\ x > 0$ passes through the point (1, 1) and

$$y(e) = \frac{\alpha - \tan\left(\frac{3}{2}\right)}{\beta + \tan\left(\frac{3}{2}\right)},$$ then $\alpha + 2\beta$ is ......... .

*[2024, 29 Jan Shift-I]*

**84** If the solution curve of the differential equation $\frac{dy}{dx} = \frac{x+y-2}{x-y}$ passing through the point (2, 1) is

$$\tan^{-1}\left(\frac{y-1}{x-1}\right) - \frac{1}{\beta}\log_e\left(\alpha + \left(\frac{y-1}{x-1}\right)^2\right)$$

$= \log_e |x-1|$, then $5\beta + \alpha$ is equal to ......... .

*[2024, 27 Jan Shift-II]*

**85** If the solution of the differential equation $(2x+3y-2)\,dx + (4x+6y-7)\,dy = 0,\ y(0) = 3$, is $\alpha x + \beta y + 3\log_e |2x+3y-\gamma| = 6$, then $\alpha + 2\beta + 3\gamma$ is equal to ......... .

*[2024, 27 Jan Shift-I]*

**86** Let the tangent at any point $P$ on a curve passing through the points (1, 1) and $\left(\frac{1}{10}, 100\right)$, intersect positive $X$-axis and $Y$-axis at the points $A$ and $B$ respectively. If $PA : PB = 1 : k$ and $y = y(x)$ is the solution of the differential equation $e^{dy/dx} = kx + \frac{k}{2}$, $y(0) = k$, then $4y(1) - 5\log_e 3$ is equal to......... .

*[2023, 10 April Shift-II]*

**87** Let $y = y(x)$ be the solution of the differential equation $\frac{dy}{dx} = \frac{4y^3 + 2yx^2}{3xy^2 + x^3}$, $y(1) = 1$.

If for some $n \in N$, $y(2) \in [n-1, n]$, then $n$ is equal to ...... .

*[2022, 25 July Shift-II]*

**88** Let $y = y(x)$ be the solution of the differential equation $dy = e^{\alpha x + y}\,dx;\ \alpha \in N$. If $y(\log_e 2) = \log_e 2$ and $y(0) = \log_e\left(\frac{1}{2}\right)$, then the value of $\alpha$ is equal to ........... .

*[2021, 27 July Shift-II]*

**89** If $y = y(x),\ y \in \left[0, \frac{\pi}{2}\right)$ is the solution of the differential equation

$$\sec y \frac{dy}{dx} - \sin(x+y) - \sin(x-y) = 0$$ with $y(0) = 0$,

then $5y'\left(\frac{\pi}{2}\right)$ is equal to ......... .

*[2021, 27 July Shift-I]*

**90** Let a curve $y = f(x)$ pass through the point $[2, (\log_e 2)^2]$ and have slope $\frac{2y}{x\log_e x}$ for all positive real value of $x$. Then, the value of $f(e)$ is equal to ...... .

*[2021, 25 July Shift-II]*

**91** Let $y = y(x)$ be the solution of the differential equation $\left((x+2)e^{\left(\frac{y+1}{x+2}\right)} + (y+1)\right)dx = (x+2)\,dy$, $y(1) = 1$. If the domain of $y = y(x)$ is an open interval $\alpha, \beta$, then $|\alpha + \beta|$ is equal to ...... .

*[2021, 22 July Shift-II]*

**92** Let a curve $y = y(x)$ be given by the solution of the differential equation

$$\cos\left(\frac{1}{2}\cos^{-1}(e^{-x})\right)dx = \sqrt{e^{2x} - 1}\,dy.$$

If it intersects $Y$-axis at $y = -1$ and the intersection point of the curve with $X$-axis is $(\alpha, 0)$, then $e^{\alpha}$ is equal to ...... .

*[2021, 20 July Shift-II]*

**93** Let $y = y(x)$ be the solution of the differential equation $x\,dy - y\,dx = \sqrt{(x^2 - y^2)}\,dx,\ x \geq 1$, with $y(1) = 0$. If the area bounded by the line $x = 1$, $x = e^{\pi}$, $y = 0$ and $y = y(x)$ is $\alpha e^{2\pi} + \beta$, then the value of $10(\alpha + \beta)$ is equal to ........... .

*[2021, 18 March Shift-II]*

**94** Let the curve $y = y(x)$ be the solution of the differential equation $\frac{dy}{dx} = 2(x+1)$. If the numerical value of area bounded by the curve $y = y(x)$ and $X$-axis is $\frac{4\sqrt{8}}{3}$, then the value of $y(1)$ is equal to......... .

*[2021, 16 March Shift-I]*

## Topic 3 Linear Differential Equations

### Level 1

**01** Let $y = y(x)$ be the solution of the differential equation $(x^2+1)y' - 2xy = (x^4 + 2x^2 + 1)\cos x$, $y(0) = 1$. Then, $\int_{-3}^{3} y(x)\,dx$ is

*[2025, 07 April Shift-II]*

(a) 30
(b) 18
(c) 36
(d) 24

**02** Let $y = y(x)$ be the solution of the differential equation
$(xy - 5x^2\sqrt{1+x^2})dx + (1+x^2)dy = 0,\ y(0) = 0.$
Then, $y(\sqrt{3})$ is equal to **[2025, 24 Jan Shift-I]**
(a) $\frac{5\sqrt{3}}{2}$ (b) $2\sqrt{2}$
(c) $\sqrt{\frac{15}{2}}$ (d) $\sqrt{\frac{14}{3}}$

**03** A function $y = f(x)$ satisfies
$f(x)\sin 2x + \sin x - (1+\cos^2 x) f'(x) = 0$ with condition $f(0) = 0$. Then, $f\left(\frac{\pi}{2}\right)$ is equal to **[2024, 29 Jan Shift-I]**
(a) 0 (b) $-1$ (c) 2 (d) 1

**04** Let $y = y_1(x)$ and $y = y_2(x)$ be two distinct solution of the differential equation $\frac{dy}{dx} = x + y$, with $y_1(0) = 0$ and $y_2(0) = 1$, respectively. Then, the number of points of intersection of $y = y_1(x)$ and $y = y_2(x)$ is **[2022, 27 July Shift-I]**
(a) 0 (b) 1 (c) 2 (d) 3

**05** If $x = x(y)$ is the solution of the differential equation $y\frac{dx}{dy} = 2x + y^3(y+1)e^y$, $x(1) = 0$; then $x(e)$ is equal to **[2022, 24 June Shift-I]**
(a) $e^3(e^e - 1)$ (b) $e^e(e^3 - 1)$
(c) $e^2(e^e + 1)$ (d) $e^e(e^2 - 1)$

**06** If the solution curve of the differential equation $(2x - 10y^3)dy + ydx = 0$, passes through the points $(0, 1)$ and $(2, \beta)$, then $\beta$ is a root of the equation **[2021, 27 Aug Shift-II]**
(a) $y^5 - 2y - 2 = 0$ (b) $2y^5 - 2y - 1 = 0$
(c) $2y^5 - y^2 - 2 = 0$ (d) $y^5 - y^2 - 1 = 0$

**07** Let us consider a curve, $y = f(x)$ passing through the point $(-2, 2)$ and the slope of the tangent to the curve at any point $(x, f(x))$ is given by $f(x) + xf'(x) = x^2$. Then, **[2021, 27 Aug Shift-I]**
(a) $x^2 + 2x\,f(x) - 12 = 0$ (b) $x^3 + 2x\,f(x) + 12 = 0$
(c) $x^3 - 3x\,f(x) - 4 = 0$ (d) $x^2 + 2x\,f(x) + 4 = 0$

**08** If $y = y(x)$ is the solution of the differential equation $\frac{dy}{dx} + (\tan x)y = \sin x$, $0 \le x \le \frac{\pi}{3}$, with $y(0) = 0$, then $y\left(\frac{\pi}{4}\right)$ is equal to **[2021, 16 March Shift-II]**
(a) $\frac{1}{4}\log_e 2$ (b) $\left(\frac{1}{2\sqrt{2}}\right)\log_e 2$ (c) $\log_e 2$ (d) $\frac{1}{2}\log_e 2$

**09** Let $y = y(x)$ be the solution of the differential equation $\cos x\frac{dy}{dx} + 2y\sin x = \sin 2x$, $x \in \left(0, \frac{\pi}{2}\right)$. If $y\left(\frac{\pi}{3}\right) = 0$, then $y\left(\frac{\pi}{4}\right)$ is equal to **[2020, 5 Sep Shift-II]**
(a) $2 - \sqrt{2}$ (b) $2 + \sqrt{2}$
(c) $\sqrt{2} - 2$ (d) $\frac{1}{\sqrt{2}} - 1$

**10** The curve satisfying the differential equation, $ydx - (x + 3y^2)dy = 0$ and passing through the point $(1, 1)$, also passes through the point **[JEE Main 2017]**
(a) $\left(\frac{1}{4}, \frac{1}{2}\right)$ (b) $\left(\frac{1}{4}, -\frac{1}{2}\right)$
(c) $\left(\frac{1}{3}, -\frac{1}{3}\right)$ (d) $\left(-\frac{1}{3}, \frac{1}{3}\right)$

**11** If $(2 + \sin x)\frac{dy}{dx} + (y + 1)\cos x = 0$ and $y(0) = 1$, then $y\left(\frac{\pi}{2}\right)$ is equal to **[JEE Main 2017]**
(a) $\frac{1}{3}$ (b) $-\frac{2}{3}$
(c) $-\frac{1}{3}$ (d) $\frac{4}{3}$

**12** Let $y(x)$ be the solution of the differential equation $(x\log x)\frac{dy}{dx} + y = 2x\log x$, $(x \ge 1)$. Then, $y(e)$ is equal to **[JEE Main 2015]**
(a) $e$ (b) 0
(c) 2 (d) $2e$

**13** The general solution of the differential equation, $\sin 2x\left(\frac{dy}{dx} - \sqrt{\tan x}\right) - y = 0$, is **[JEE Main 2014]**
(a) $y\sqrt{\tan x} = x + c$ (b) $y\sqrt{\cot x} = \tan x + c$
(c) $y\sqrt{\tan x} = \cot x + c$ (d) $y\sqrt{\cot x} = x + c$

**14** The integrating factor of the differential equation $(x^2 - 1)\frac{dy}{dx} + 2xy = x$ is **[AIEEE 2012]**
(a) $\frac{1}{x^2 - 1}$ (b) $x^2 - 1$
(c) $\frac{x^2 - 1}{x}$ (d) $\frac{x}{x^2 - 1}$

**15** The solution of the differential equation $\frac{dy}{dx} = \frac{x + y}{x}$ satisfying the condition $y(1) = 1$ is **[AIEEE 2008]**
(a) $y = x\log x + x$ (b) $y = \log x + x$
(c) $y = x\log x + x^2$ (d) $y = xe^{(x-1)}$

## Level 2

**Imp.** Leibniz Integral Rule

**16** Let $f:[1,\infty)\to[2,\infty)$ be a differentiable function. If $10\int_1^x f(t)\,dt = 5xf(x) - x^5 - 9$ for all $x \ge 1$, then the value of $f(3)$ is **[2025, 02 April Shift-II]**

(a) 22 (b) 26 (c) 18 (d) 32

**17** Let $g$ be a differentiable function such that $\int_0^x g(t)dt = x - \int_0^x tg(t)dt$, $x \ge 0$ and let $y = y(x)$ satisfy the differential equation $\frac{dy}{dx} - y\tan x = 2(x+1)\sec x\, g(x)$, $x \in \left[0, \frac{\pi}{2}\right)$. If $y(0) = 0$, then $y\left(\frac{\pi}{3}\right)$ is equal to **[2025, 03 April Shift-I]**

(a) $\frac{4\pi}{3}$ (b) $\frac{2\pi}{3}$ (c) $\frac{4\pi}{3\sqrt{3}}$ (d) $\frac{2\pi}{3\sqrt{3}}$

**18** Let $y = y(x)$ be the solution of the differential equation $\frac{dy}{dx} + 3(\tan^2 x)y + 3y = \sec^2 x$, $y(0) = \frac{1}{3} + e^3$. Then, $y\left(\frac{\pi}{4}\right)$ is equal to **[2025, 03 April Shift-II]**

(a) $\frac{2}{3} + e^3$ (b) $\frac{4}{3} + e^3$ (c) $\frac{2}{3}$ (d) $\frac{4}{3}$

**19** Let $f:[0,\infty)\to R$ be a differentiable function such that $f(x) = 1 - 2x + \int_0^x e^{x-t} f(t)\,dt$ for all $x \in [0,\infty)$. Then, the area of the region bounded by $y = f(x)$ and the coordinate axes is **[2025, 04 April Shift-I]**

(a) $\sqrt{2}$ (b) 1/2 (c) $\sqrt{5}$ (d) 2

**20** If a curve $y = y(x)$ passes through the point $\left(1, \frac{\pi}{2}\right)$ and satisfies the differential equation $(7x^4\cot y - e^x \text{cosec} y)\frac{dx}{dy} = x^5$, $x \ge 1$, then at $x = 2$, the value of $\cos y$ is **[2025, 04 April Shift-II]**

(a) $\frac{2e^2+e}{128}$ (b) $\frac{2e^2-e}{128}$ (c) $\frac{2e^2+e}{64}$ (d) $\frac{2e^2-e}{64}$

**21** Let $y = y(x)$ be the solution curve of the differential equation $x(x^2+e^x)dy + (e^x(x-2)y - x^3)dx = 0$, $x > 0$, passing through the point (1, 0). Then, $y(2)$ is equal **[2025, 07 April Shift-I]**

(a) $\frac{4}{4-e^2}$ (b) $\frac{2}{2+e^2}$ (c) $\frac{4}{4+e^2}$ (d) $\frac{2}{2-e^2}$

**22** Let $f(x) = x - 1$ and $g(x) = e^x$ for $x \in R$. If $\frac{dy}{dx} = \left(e^{-2\sqrt{x}} g(f(f(x))) - \frac{y}{\sqrt{x}}\right)$, $y(0) = 0$, then $y(1)$ is **[2025, 08 April Shift-II]**

(a) $\frac{e-1}{e^4}$ (b) $\frac{2e-1}{e^3}$ (c) $\frac{1-e^3}{e^4}$ (d) $\frac{1-e^2}{e^4}$

**23** Let a curve $y = f(x)$ passes through the point (0, 5) and $(\log_e 2, k)$. If the curve satisfies the differential equation $2(3+y)e^{2x}dx - (7+e^{2x})dy = 0$, then $k$ is equal to **[2025, 23 Jan Shift-I]**

(a) 8 (b) 16 (c) 4 (d) 32

**24** Let $x = x(y)$ be the solution of the differential equation $y^2 dx + \left(x - \frac{1}{y}\right)dy = 0$. If $x(1) = 1$, then $x(1/2)$ is **[2025, 22 Jan Shift-I]**

(a) $\frac{1}{2} + e$ (b) $3 + e$ (c) $\frac{3}{2} + e$ (d) $3 - e$

**25** If $x = f(y)$ is the solution of the differential equation $(1+y^2) + (x - 2e^{\tan^{-1} y})\frac{dy}{dx} = 0$, $y \in \left(-\frac{\pi}{2}, \frac{\pi}{2}\right)$ with $f(0) = 1$, then $f\left(\frac{1}{\sqrt{3}}\right)$ is equal to **[2025, 22 Jan Shift-II]**

(a) $e^{\pi/4}$ (b) $e^{\pi/6}$ (c) $e^{\pi/12}$ (d) $e^{\pi/3}$

**Imp.** Linear Differential Equation

**26** Let $y = y(x)$ be the solution curve of the differential equation $\sec y\frac{dy}{dx} + 2x\sin y = x^3\cos y$, $y(1) = 0$. Then, $y(\sqrt{3})$ is equal to **[2024, 8 April Shift-I]**

(a) $\frac{\pi}{3}$ (b) $\frac{\pi}{12}$ (c) $\frac{\pi}{6}$ (d) $\frac{\pi}{4}$

**27** Let $y = y(x)$ be the solution of the differential equation $(1+x^2)\frac{dy}{dx} + y = e^{\tan^{-1} x}$, $y(1) = 0$. Then, $y(0)$ is **[2024, 6 April Shift-I]**

(a) $\frac{1}{2}\left(1 - e^{\frac{\pi}{2}}\right)$ (b) $\frac{1}{2}\left(e^{\frac{\pi}{2}} - 1\right)$

(c) $\frac{1}{4}\left(1 - e^{\frac{\pi}{2}}\right)$ (d) $\frac{1}{4}\left(e^{\frac{\pi}{2}} - 1\right)$

**28** If $y = y(x)$ is the solution of the differential equation $\frac{dy}{dx} + 2y = \sin(2x)$, $y(0) = \frac{3}{4}$, then $y\left(\frac{\pi}{8}\right)$ is equal to **[2024, 5 April Shift-I]**

(a) $e^{\pi/4}$ (b) $e^{-\pi/8}$ (c) $e^{\pi/8}$ (d) $e^{-\pi/4}$

**29** Let $y = y(x)$ be the solution of the differential equation $(x^2+4)^2 dy + (2x^3 y + 8xy - 2) dx = 0$. If $y(0) = 0$, then $y(2)$ is equal to **[2024, 4 April Shift-II]**

(a) $\frac{\pi}{16}$ (b) $\frac{\pi}{32}$ (c) $2\pi$ (d) $\frac{\pi}{8}$

**30** Let $x = x(y)$ be the solution of the differential equation $2(y+2)\log_e(y+2)\, dx + (x + 4 - 2\log_e(y+2))dy = 0$, $y > -1$ with $x(e^4 - 2) = 1$. Then, $x(e^9 - 2)$ is equal to **[2023, 15 April Shift-II]**

(a) 4/9 (b) 10/3 (c) 32/9 (d) 3

**31** Let $y = y(x)$ be the solution of the differential equation $\frac{dy}{dx} + \frac{5}{x(x^5+1)} y = \frac{(x^5+1)^2}{x^7}$, $x > 0$. If $y(1) = 2$, then $y(2)$ is equal to **[2023, 11 April Shift-II]**

(a) $\frac{693}{128}$ (b) $\frac{697}{128}$ (c) $\frac{637}{128}$ (d) $\frac{679}{128}$

**32** Let $f$ be a differentiable function such that $x^2 f(x) - x = 4\int_0^x tf(t)dt$, $f(1) = \frac{2}{3}$. Then, $18f(3)$ is equal to **[2023, 10 April Shift-I]**

(a) 210 (b) 160
(c) 180 (d) 130

**33** If the solution curve $f(x, y) = 0$ of the differential equation $(1 + \log_e x)\frac{dx}{dy} - x\log_e x = e^y$, $x > 0$, passes through the points (1, 0) and $(\alpha, 2)$, then $\alpha^\alpha$ is equal to **[2023, 6 April Shift-II]**

(a) $e^{2e^{\sqrt{2}}}$ (b) $e^{\sqrt{2}e^2}$ (c) $e^{2e^2}$ (d) $e^{e^2}$

**Imp.** Reducible to the Linear Differential Equation

**34** Let $\alpha x = \exp(x^\beta y^\gamma)$ be the solution of the differential equation $2x^2 y dy - (1 - xy^2)\, dx = 0$, $x > 0$, $y(2) = \sqrt{\log_e 2}$. Then, $\alpha + \beta - \gamma$ equals **[2023, 1 Feb Shift-II]**

(a) $-1$ (b) 3 (c) 0 (d) 1

**35** If $y = y(x)$ is the solution curve of the differential equation $\frac{dy}{dx} + y\tan x = x\sec x$, $0 \le x \le \frac{\pi}{3}$, $y(0) = 1$, then $y\left(\frac{\pi}{6}\right)$ is equal to **[2023, 1 Feb Shift-I]**

(a) $\frac{\pi}{12} + \frac{\sqrt{3}}{2}\log_e\left(\frac{2}{e\sqrt{3}}\right)$ (b) $\frac{\pi}{12} - \frac{\sqrt{3}}{2}\log_e\left(\frac{2\sqrt{3}}{e}\right)$

(c) $\frac{\pi}{12} - \frac{\sqrt{3}}{2}\log_e\left(\frac{2}{e\sqrt{3}}\right)$ (d) $\frac{\pi}{12} + \frac{\sqrt{3}}{2}\log_e\left(\frac{2\sqrt{3}}{e}\right)$

**36** Let $y = y(x)$ be the solution of the differential equation $x\log_e x\frac{dy}{dx} + y = x^2\log_e x$, $(x > 1)$. If $y(2) = 2$, then $y(e)$ is equal to **[2023, 29 Jan Shift-II]**

(a) $\frac{1+e^2}{2}$ (b) $\frac{2+e^2}{2}$ (c) $\frac{4+e^2}{4}$ (d) $\frac{1+e^2}{4}$

**37** Let $y = y(t)$ be a solution of the differential equation $\frac{dy}{dt} + \alpha y = \gamma e^{-\beta t}$, where $\alpha > 0$, $\beta > 0$ and $\gamma > 0$. Then, $\lim_{t\to\infty} y(t)$ **[2023, 25 Jan Shift-II]**

(a) does not exist (b) is 0
(c) is 1 (d) is $-1$

**38** Let $y = y(x)$ be the solution of the differential equation $x^3 dy + (xy - 1)dx = 0$, $x > 0$, $y\left(\frac{1}{2}\right) = 3 - e$. Then, $y(1)$ is equal to **[2023, 24 Jan Shift-I]**

(a) $e$ (b) $2 - e$ (c) 1 (d) 3

**39** Let the solution curve $y = y(x)$ of the differential equation $(1 + e^{2x})\left(\frac{dy}{dx} + y\right) = 1$ passes through the point $\left(0, \frac{\pi}{2}\right)$. Then, $\lim_{x\to\infty} e^x y(x)$ is equal to **[2022, 29 July Shift-I]**

(a) $\frac{\pi}{4}$ (b) $\frac{3\pi}{4}$
(c) $\frac{\pi}{2}$ (d) $\frac{3\pi}{2}$

**40** Let the slope of the tangent to a curve $y = f(x)$ at $(x, y)$ be given by $2\tan x(\cos x - y)$. If the curve passes through the point $\left(\frac{\pi}{4}, 0\right)$, then the value of $\int_0^{\pi/2} y\, dx$ is equal to **[2022, 28 June Shift-II]**

(a) $(2 - \sqrt{2}) + \frac{\pi}{\sqrt{2}}$ (b) $2 - \frac{\pi}{\sqrt{2}}$
(c) $(2 + \sqrt{2}) + \frac{\pi}{\sqrt{2}}$ (d) $2 + \frac{\pi}{\sqrt{2}}$

**41** Let $y = y(x)$ be the solution of the differential equation $x(1 - x^2)\frac{dy}{dx} + (3x^2 y - y - 4x^3) = 0$, $x > 1$, with $y(2) = -2$. Then, $y(3)$ is equal to **[2022, 28 June Shift-I]**

(a) $-18$ (b) $-12$ (c) $-6$ (d) $-3$

**42** If the solution curve of the differential equation $((\tan^{-1} y) - x)dy = (1 + y^2)dx$ passes through the point (1, 0), then the abscissa of the point on the curve whose ordinate is tan (1) is **[2022, 27 June Shift-II]**

(a) $2e$ (b) $2/e$
(c) 2 (d) $1/e$

**43** If $y = y(x)$ is the solution of the differential equation $x\dfrac{dy}{dx} + 2y = xe^x$, $y(1) = 0$, then the local maximum value of the function $z(x) = x^2 y(x) - e^x$, $x \in R$ is **[2022, 26 June Shift-II]**

(a) $1-e$ (b) 0 (c) $\dfrac{1}{2}$ (d) $\dfrac{4}{e} - e$

**44** If the solution of the differential equation $\dfrac{dy}{dx} + e^x (x^2 - 2)\, y = (x^2 - 2x)(x^2 - 2)\, e^{2x}$ satisfies $y(0) = 0$, then the value of $y(2)$ is ……… . **[2022, 26 June Shift-II]**

(a) $-1$ (b) 1 (c) 0 (d) $e$

**45** Let $y = y(x)$ be the solution of the differential equation $(x+1)\, y' - y = e^{3x} (x+1)^2$, with $y(0) = \dfrac{1}{3}$. Then, the point $x = -\dfrac{4}{3}$ for the curve $y = y(x)$ is **[2022, 25 June Shift-I]**

(a) not a critical point
(b) a point of local minima
(c) a point of local maxima
(d) a point of inflection

**46** Let $y = y(x)$ be a solution curve of the differential equation $(y+1)\tan^2 x\, dx + \tan x\, dy + y\, dx = 0$, $x \in \left(0, \dfrac{\pi}{2}\right)$. If $\lim\limits_{x \to 0^+} xy(x) = 1$, , then the value of $y(\pi/4)$ is **[2021, 26 Aug Shift-I]**

(a) $-\dfrac{\pi}{4}$ (b) $\dfrac{\pi}{4} - 1$ (c) $\dfrac{\pi}{4} + 1$ (d) $\dfrac{\pi}{4}$

**47** Let $y = y(x)$ be the solution of the differential equation $\operatorname{cosec}^2 x\, dy + 2dx = (1 + y\cos 2x)\operatorname{cosec}^2 x dx$, with $y\left(\dfrac{\pi}{4}\right) = 0$. Then, the value of $(y(0)+1)^2$ is equal to **[2021, 22 July Shift-II]**

(a) $e^{1/2}$ (b) $e^{-1/2}$ (c) $e^{-1}$ (d) $e$

**48** Let $y = y(x)$ satisfies the equation $\dfrac{dy}{dx} - |A| = 0$, for all $x > 0$, where $A = \begin{bmatrix} y & \sin x & 1 \\ 0 & -1 & 1 \\ 2 & 0 & \dfrac{1}{x} \end{bmatrix}$. If $y(\pi) = \pi + 2$, then the value of $y\left(\dfrac{\pi}{2}\right)$ is **[2021, 20 July Shift-II]**

(a) $\dfrac{\pi}{2} + \dfrac{4}{\pi}$ (b) $\dfrac{\pi}{2} - \dfrac{1}{\pi}$
(c) $\dfrac{3\pi}{2} - \dfrac{1}{\pi}$ (d) $\dfrac{\pi}{2} - \dfrac{4}{\pi}$

**49** If $y = y(x)$ is the solution of the differential equation, $\dfrac{dy}{dx} + 2y\tan x = \sin x$, $y\left(\dfrac{\pi}{3}\right) = 0$, then the maximum value of the function $y(x)$ over $R$ is equal to **[2021, 16 March Shift-I]**

(a) 8 (b) 1/2
(c) $-15/4$ (d) 1/8

**50** Let $y = y(x)$ be the solution of the differential equation, $xy' - y = x^2(x\cos x + \sin x)$, $x > 0$. If $y(\pi) = \pi$, then $y''\left(\dfrac{\pi}{2}\right) + y\left(\dfrac{\pi}{2}\right)$ is equal to **[2020, 4 Sep Shift-I]**

(a) $2 + \dfrac{\pi}{2} + \dfrac{\pi^2}{4}$ (b) $1 + \dfrac{\pi}{2}$ (c) $1 + \dfrac{\pi}{2} + \dfrac{\pi^2}{4}$ (d) $2 + \dfrac{\pi}{2}$

**51** Consider the differential equation, $y^2 dx + \left(x - \dfrac{1}{y}\right) dy = 0$. If value of $y$ is 1 when $x = 1$, then the value of $x$ for which $y = 2$, is **[2019, 12 April Shift-I]**

(a) $\dfrac{5}{2} + \dfrac{1}{\sqrt{e}}$ (b) $\dfrac{3}{2} - \dfrac{1}{\sqrt{e}}$ (c) $\dfrac{1}{2} + \dfrac{1}{\sqrt{e}}$ (d) $\dfrac{3}{2} - \sqrt{e}$

**52** Let $y = y(x)$ be the solution of the differential equation, $\dfrac{dy}{dx} + y\tan x = 2x + x^2 \tan x$, $x \in \left(-\dfrac{\pi}{2}, \dfrac{\pi}{2}\right)$, such that $y(0) = 1$. Then, **[2019, 10 April Shift-II]**

(a) $y'\left(\dfrac{\pi}{4}\right) - y'\left(-\dfrac{\pi}{4}\right) = \pi - \sqrt{2}$

(b) $y'\left(\dfrac{\pi}{4}\right) + y'\left(-\dfrac{\pi}{4}\right) = -\sqrt{2}$

(c) $y\left(\dfrac{\pi}{4}\right) + y\left(-\dfrac{\pi}{4}\right) = \dfrac{\pi^2}{2} + 2$

(d) $y\left(\dfrac{\pi}{4}\right) - y\left(-\dfrac{\pi}{4}\right) = \sqrt{2}$

**53** If $y = y(x)$ is the solution of the differential equation $\dfrac{dy}{dx} = (\tan x - y)\sec^2 x$, $x \in \left(-\dfrac{\pi}{2}, \dfrac{\pi}{2}\right)$, such that $y(0) = 0$, then $y\left(-\dfrac{\pi}{4}\right)$ is equal to **[2019, 10 April Shift-I]**

(a) $\dfrac{1}{e} - 2$ (b) $\dfrac{1}{2} - e$ (c) $2 + \dfrac{1}{e}$ (d) $e - 2$

**54** Let $y = y(x)$ be the solution of the differential equation, $x\dfrac{dy}{dx} + y = x\log_e x$, $(x > 1)$. If $2y(2) = \log_e 4 - 1$, then $y(e)$ is equal to **[2019, 12 Jan Shift-I]**

(a) $-\dfrac{e}{2}$ (b) $-\dfrac{e^2}{2}$ (c) $\dfrac{e}{4}$ (d) $\dfrac{e^2}{4}$

**55** Let $f$ be a differentiable function such that $f'(x)=7-\frac{3}{4}\frac{f(x)}{x},(x>0)$ and $f(1)\neq 4$. Then, $\lim_{x\to 0^+} x f\left(\frac{1}{x}\right)$ **[2019, 10 Jan Shift-II]**

(a) does not exist (b) exists and equals 4/7
(c) exists and equals 0 (d) exists and equals 4

**56** If $y=y(x)$ is the solution of the differential equation, $x\frac{dy}{dx}+2y=x^2$ satisfying $y(1)=1$, then $y\left(\frac{1}{2}\right)$ is equal to **[2019, 9 Jan Shift-I]**

(a) 13/16 (b) 1/4 (c) 49/16 (d) 7/64

**57** Let $y=y(x)$ be the solution of the differential equation $\sin x\frac{dy}{dx}+y\cos x=4x,\ x\in(0,\pi)$.

If $y\left(\frac{\pi}{2}\right)=0$, then $y\left(\frac{\pi}{6}\right)$ is equal to **[JEE Main 2018]**

(a) $\frac{4}{9\sqrt{3}}\pi^2$ (b) $\frac{-8}{9\sqrt{3}}\pi^2$ (c) $-\frac{8}{9}\pi^2$ (d) $-\frac{4}{9}\pi^2$

**58** If $f(x)$ is a differentiable function in the interval $(0,\infty)$ such that $f(1)=1$ and $\lim_{t\to x}\frac{t^2f(x)-x^2f(t)}{t-x}=1$, for each $x>0$, then $f(3/2)$ is equal to **[JEE Main 2016]**

(a) 23/18 (b) 13/6 (c) 25/9 (d) 31/18

**59** The solution of the differential equation $\frac{dy}{dx}+\frac{y}{2}\sec x=\frac{\tan x}{2y}$, where $0\le x<\frac{\pi}{2}$, and $y(0)=1$, is given by **[JEE Main 2016]**

(a) $y^2=1+\frac{x}{\sec x+\tan x}$ (b) $y=1+\frac{x}{\sec x+\tan x}$

(c) $y=1-\frac{x}{\sec x+\tan x}$ (d) $y^2=1-\frac{x}{\sec x+\tan x}$

**60** Consider the differential equation $y^2dx+\left(x-\frac{1}{y}\right)dy=0$. If $y(1)=1$, then $x$ is given by **[AIEEE 2011]**

(a) $1-\frac{1}{y}+\frac{e^{1/y}}{e}$ (b) $4-\frac{2}{y}-\frac{e^{1/y}}{e}$

(c) $3-\frac{1}{y}+\frac{e^{1/y}}{e}$ (d) $1+\frac{1}{y}-\frac{e^{1/y}}{e}$

**61** The solution of the differential equation $(1+y^2)+(x-e^{\tan^{-1}y})\frac{dy}{dx}=0$, is **[AIEEE 2003]**

(a) $(x-2)=Ce^{-2\tan^{-1}y}$ (b) $2xe^{\tan^{-1}y}=e^{2\tan^{-1}y}+C$

(c) $xe^{\tan^{-1}y}=\tan^{-1}y+C$ (d) $xe^{2\tan^{-1}y}=e^{\tan^{-1}y}+C$

## Level 3

**62** If for the solution curve $y=f(x)$ of the differential equation $\frac{dy}{dx}+(\tan x)y=\frac{2+\sec x}{(1+2\sec x)^2}$, $x\in\left(\frac{-\pi}{2},\frac{\pi}{2}\right)$, $f\left(\frac{\pi}{3}\right)=\frac{\sqrt{3}}{10}$, then $f\left(\frac{\pi}{4}\right)$ is equal to **[2025, 29 Jan Shift-II]**

(a) $\frac{\sqrt{3}+1}{10(4+\sqrt{3})}$ (b) $\frac{5-\sqrt{3}}{2\sqrt{2}}$

(c) $\frac{9\sqrt{3}+3}{10(4+\sqrt{3})}$ (d) $\frac{4-\sqrt{2}}{14}$

**63** Let $y=y(x)$ be the solutions of the differential equation $\cos x\,(\log_e(\cos x))^2dy+(\sin x-3y\sin x\log_e(\cos x))\,dx=0,\ x\in\left(0,\frac{\pi}{2}\right)$. If $y\left(\frac{\pi}{4}\right)=\frac{-1}{\log_e 2}$, then $y\left(\frac{\pi}{6}\right)$ is equal to **[2025, 29 Jan Shift-I]**

(a) $\frac{2}{\log_e(3)-\log_e(4)}$ (b) $\frac{1}{\log_e(3)-\log_e(4)}$

(c) $\frac{1}{\log_e(4)-\log_e(3)}$ (d) $-\frac{1}{\log_e(4)}$

**64** Let $y=y(x)$ be the solution of the differential equation $(2x\log_e x)\frac{dy}{dx}+2y=\frac{3}{x}\log_e x,\ x>0$ and $y(e^{-1})=0$. Then, $y(e)$ is equal to **[2024, 6 April Shift-I]**

(a) $-\frac{2}{3e}$ (b) $-\frac{3}{e}$

(c) $-\frac{3}{2e}$ (d) $-\frac{2}{e}$

**65** Let $y=y(x)$ be the solution of the differential equation $\frac{dy}{dx}=\frac{(\tan x)+y}{\sin x(\sec x-\sin x\tan x)},\ x\in\left(0,\frac{\pi}{2}\right)$ satisfying the condition $y\left(\frac{\pi}{4}\right)=2$. Then, $y\left(\frac{\pi}{3}\right)$ is **[2024, 31 Jan Shift-I]**

(a) $\sqrt{3}(2+\log_e\sqrt{3})$ (b) $\sqrt{3}(1+2\log_e 3)$

(c) $\frac{\sqrt{3}}{2}(2+\log_e 3)$ (d) $\sqrt{3}(2+\log_e 3)$

**66** Let $y=y(x)$, $y>0$, be a solution curve of the differential equation $(1+x^2)dy=y(x-y)\,dx$. If $y(0)=1$ and $y(2\sqrt{2})=\beta$, then **[2023, 12 April Shift-I]**

(a) $e^{\beta^{-1}}=e^{-2}(5+\sqrt{2})$ (b) $e^{\beta^{-1}}=e^{-2}(3+2\sqrt{2})$

(c) $e^{3\beta^{-1}}=e(3+2\sqrt{2})$ (d) $e^{3\beta^{-1}}=e(5+\sqrt{2})$

**67** Let the solution curve $y = y(x)$ of the differential equation

$$\frac{dy}{dx} - \frac{3x^5 \tan^{-1}(x^3)}{(1+x^6)^{\frac{3}{2}}} y = 2x \exp\left\{\frac{x^3 - \tan^{-1} x^3}{\sqrt{(1+x^6)}}\right\}$$

pass through the origin. Then, $y(1)$ is equal to *[2023, 30 Jan Shift-I]*

(a) $\exp\left(\frac{4+\pi}{4\sqrt{2}}\right)$ (b) $\exp\left(\frac{1-\pi}{4\sqrt{2}}\right)$

(c) $\exp\left(\frac{4-\pi}{4\sqrt{2}}\right)$ (d) $\exp\left(\frac{\pi-4}{4\sqrt{2}}\right)$

**68** Let $y = y(x)$ be the solution curve of the differential equation $\frac{dy}{dx} = \frac{y}{x}(1 + xy^2(1 - \log_e x)), x > 0, y(1) = 3.$

Then, $\frac{y^2(x)}{9}$ is equal to *[2023, 25 Jan Shift-I]*

(a) $\frac{x^2}{3x^3(1+\log_e x^2) - 2}$ (b) $\frac{x^2}{2x^3(2+\log_e x^3) - 3}$

(c) $\frac{x^2}{5 - 2x^3(2+\log_e x^3)}$ (d) $\frac{x^2}{7 - 3x^3(2+\log_e x^2)}$

**69** Let $y = y(x)$ be the solution curve of the differential equation $\frac{dy}{dx} + \frac{1}{x^2 - 1} y = \left(\frac{x-1}{x+1}\right)^{1/2}, x > 1$ passing through the point $\left(2, \sqrt{\frac{1}{3}}\right)$. Then, $\sqrt{7}\, y(8)$ is *[2022, 28 July Shift-II]*

(a) $11 + 6\log_e 3$ (b) 19
(c) $12 - 2\log_e 3$ (d) $19 - 6\log_e 3$

**70** Let the solution curve $y = f(x)$ of the differential equation $\frac{dy}{dx} + \frac{xy}{x^2 - 1} = \frac{x^4 + 2x}{\sqrt{1 - x^2}}, x \in (-1, 1)$ pass through the origin. Then, $\int_{-\sqrt{3}/2}^{\sqrt{3}/2} f(x)\, dx$ is *[2022, 26 July Shift-II]*

(a) $\frac{\pi}{3} - \frac{1}{4}$ (b) $\frac{\pi}{3} - \frac{\sqrt{3}}{4}$ (c) $\frac{\pi}{6} - \frac{\sqrt{3}}{4}$ (d) $\frac{\pi}{6} - \frac{\sqrt{3}}{2}$

**71** Let $y = y(x)$ be the solution of the differential equation $\frac{dy}{dx} = (y+1)[(y+1)e^{x^2/2} - x], 0 < x < 2.1,$ with $y(2) = 0$. Then, the value of $dy/dx$ at $x = 1$ is equal to *[2021, 18 March Shift-II]*

(a) $\frac{-e^{3/2}}{(e^2+1)^2}$ (b) $-\frac{2e^2}{(1+e^2)^2}$ (c) $\frac{e^{5/2}}{(1+e^2)^2}$ (d) $\frac{5e^{1/2}}{(e^2+1)^2}$

**72** Let $y = y(x)$ be the solution of the differential equation $\cos x\,(3\sin x + \cos x + 3)dy = [1 + y\sin x\,(3\sin x + \cos x + 3)]dx, 0 \le x \le \frac{\pi}{2}, y(0) = 0.$

Then, $y\left(\frac{\pi}{3}\right)$ is equal to *[2021, 17 March Shift-II]*

(a) $2\log_e\left(\frac{2\sqrt{3}+9}{6}\right)$ (b) $2\log_e\left(\frac{2\sqrt{3}+10}{11}\right)$

(c) $2\log_e\left(\frac{\sqrt{3}+7}{2}\right)$ (d) $2\log_e\left(\frac{3\sqrt{3}-8}{4}\right)$

**73** If the curve $y = y(x)$ is the solution of the differential equation $2(x^2 + x^{5/4})\, dy - y(x + x^{1/4})\, dx = 2x^{9/4}\, dx, x > 0$ which passes through the point $\left(1, 1 - \frac{4}{3}\log_e 2\right)$, then the value of $y(16)$ is equal to *[2021, 17 March Shift-II]*

(a) $4\left(\frac{31}{3} + \frac{8}{3}\log_e 3\right)$ (b) $\left(\frac{31}{3} + \frac{8}{3}\log_e 3\right)$

(c) $4\left(\frac{31}{3} - \frac{8}{3}\log_e 3\right)$ (d) $\left(\frac{31}{3} - \frac{8}{3}\log_e 3\right)$

## Numerical Value Type Questions

**74** Let $y = y(x)$ be the solution of the differential equation $\frac{dy}{dx} + 2y\sec^2 x = 2\sec^2 x + 3\tan x \cdot \sec^2 x$ such that $y(0) = \frac{5}{4}$. Then, $12\left(y\left(\frac{\pi}{4}\right) - e^{-2}\right)$ is equal to ......... *[2025, 02 April Shift-II]*

**75** Let $y = f(x)$ be the solution of the differential equation $\frac{dy}{dx} + \frac{xy}{x^2 - 1} = \frac{x^6 + 4x}{\sqrt{1 - x^2}}, -1 < x < 1$ such that $f(0) = 0$.

If $6\int_{-\frac{1}{2}}^{1/2} f(x)\, dx = 2\pi - \alpha$, then $\alpha^2$ is equal to ...... . *[2025, 22 Jan Shift-II]*

**76** Let $y = y(x)$ be the solution of the differential equation $2\cos x \frac{dy}{dx} = \sin 2x - 4y\sin x, x \in \left(0, \frac{\pi}{2}\right)$.

If $y\left(\frac{\pi}{3}\right) = 0$, then $y'\left(\frac{\pi}{4}\right) + y\left(\frac{\pi}{4}\right)$ is equal to ...... . *[2025, 24 Jan Shift-II]*

**77** If $y = y(x)$ is the solution of the differential equation,

$$\sqrt{4 - x^2}\frac{dy}{dx} = \left(\left(\sin^{-1}\left(\frac{x}{2}\right)\right)^2 - y\right)\sin^{-1}\left(\frac{x}{2}\right),$$

$-2 \le x \le 2, y(2) = \frac{\pi^2 - 8}{4}$, then $y^2(0)$ is equal to ......... *[2025, 28 Jan Shift-II]*

**78** For a differentiable function $f: R \to R$, suppose $f'(x) = 3f(x) + \alpha$, where $\alpha \in R$, $f(0) = 1$ and $\lim_{x \to -\infty} f(x) = 7$. Then, $9f(-\log_e 3)$ is equal to ......... **[2024, 9 April Shift-II]**

**79** Let $y = y(x)$ the solution of the differential equation

$$\frac{dy}{dx} + \frac{2x}{(1+x^2)^2} y = xe^{\left(\frac{1}{1+x^2}\right)}; \; y(0) = 0$$ Then, the area enclosed by the curve $f(x) = y(x)e^{-\frac{1}{(1+x^2)}}$ and the line $y - x = 4$ is ......... **[2024, 5 April Shift-II]**

**80** Let the solution $y = y(x)$ of differential equation $\frac{dy}{dx} - y = 1 + 4\sin x$ satisfy $y(\pi) = 1$. Then, $y\left(\frac{\pi}{2}\right) + 10$ is equal to ......... . **[2024, 4 April Shift-I]**

**81** If $\frac{dx}{dy} = \frac{1 + x - y^2}{y}$, $x(1) = 1$, then $5x(2)$ is equal to ......... . **[2024, 1 Feb Shift-II]**

**82** If $x = x(t)$ is the solution of the differential equation $(t+1)dx = (2x + (t+1)^4)dt$, $x(0) = 2$, then, $x(1)$ equals .......... . **[2024, 1 Feb Shift-I]**

**83** Let $y = y(x)$ be the solution of the differential equation $\sec^2 x \, dx + (e^{2y}\tan^2 x + \tan x)\, dy = 0$, $0 < x < \frac{\pi}{2}$, $y(\pi/4) = 0$. If $y(\pi/6) = \alpha$, then $e^{8\alpha}$ is equal to ......... . **[2024, 31 Jan Shift-II]**

**84** Let $Y = Y(X)$ be a curve lying in the first quadrant such that the area enclosed by the line $Y - y = Y'(x)(X - x)$ and the coordinate axes, where $(x, y)$ is any point on the curve is always $\frac{-y^2}{2Y'(x)} + 1$, $Y'(x) \neq 0$. If $Y(1) = 1$, then $12Y(2)$ equals ......... . **[2024, 30 Jan Shift-II]**

**85** Let $y = y(x)$ be the solution of the differential equation $(1 - x^2)\, dy = [xy + (x^3 + 2)\sqrt{3(1 - x^2)}]\, dx$, $-1 < x < 1$, $y(0) = 0$. If $y\left(\frac{1}{2}\right) = \frac{m}{n}$, $m$ and n are coprime numbers, then $m + n$ is equal to ......... . **[2024, 30 Jan Shift-I]**

**86** If $y = y(x)$ is the solution of the differential equation $\frac{dy}{dx} + \frac{4x}{(x^2 - 1)} y = \frac{x + 2}{(x^2 - 1)^{5/2}}$, $x > 1$ such that $y(2) = \frac{2}{9}\log_e(2 + \sqrt{3})$ and $y(\sqrt{2}) = \alpha \log_e(\sqrt{\alpha} + \beta) + \beta - \sqrt{\gamma}$, $\alpha, \beta, \gamma \in N$, then $\alpha\beta\gamma$ is equal to ......... . **[2023, 13 April Shift-II]**

**87** Let the solution curve $x = x(y)$, $0 < y < \frac{\pi}{2}$ of the differential equation $(\log_e(\cos y))^2 \cos y \, dx - (1 + 3x\log_e(\cos y))\sin y \, dy = 0$ satisfy $x\left(\frac{\pi}{3}\right) = \frac{1}{2\log_e 2}$.
If $x\left(\frac{\pi}{6}\right) = \frac{1}{\log_e m - \log_e n}$, where $m$ and $n$ are coprime, then $mn$ is equal to...... . **[2023, 8 April Shift-II]**

**88** If the solution curve of the differential equation $(y - 2\log_e x)dx + (x\log_e x^2)dy = 0$, $x > 1$ passes through the points $\left(e, \frac{4}{3}\right)$ and $(e^4, \alpha)$, then $\alpha$ is equal to ....... **[2023, 8 April Shift-I]**

**89** Let $y = y(x)$ be a solution of the differential equation $(x\cos x)dy + (xy\sin x + y\cos x - 1)dx = 0$, $0 < x < \frac{\pi}{2}$. If $\frac{\pi}{3} y\left(\frac{\pi}{3}\right) = \sqrt{3}$, then $\left|\frac{\pi}{6} y''\left(\frac{\pi}{6}\right) + 2y'\left(\frac{\pi}{6}\right)\right|$ is equal to ......... **[2023, 6 April Shift-I]**

**90** Let $f: R \to R$ be a differentiable function such that $f'(x) + f(x) = \int_0^2 f(t)dt$. If $f(0) = e^{-2}$, then $2f(0) - f(2)$ is equal to ............ . **[2023, 1 Feb Shift-I]**

**91** Let $y = y(x)$ be the solution curve of the differential equation
$$\sin(2x^2)\log_e(\tan x^2)dy + \left(4xy - 4\sqrt{2}x\sin\left(x^2 - \frac{\pi}{4}\right)\right)dx = 0,$$
$0 < x < \sqrt{\frac{\pi}{2}}$, which passes through the point $\left(\sqrt{\frac{\pi}{6}}, 1\right)$. Then, $\left|y\left(\sqrt{\frac{\pi}{3}}\right)\right|$ is equal to ...... . **[2022, 27 July Shift-I]**

**92** Suppose, $y = y(x)$ be the solution curve to the differential equation $\frac{dy}{dx} - y = 2 - e^{-x}$ such that $\lim_{x \to \infty} y(x)$ is finite. If $a$ and $b$ are respectively the $x$ and $y$-intercepts of the tangent to the curve at $x = 0$, then the value of $a - 4b$ is equal to ........... . **[2022, 26 July Shift-II]**

**93** Let $y = y(x)$, $x > 1$, be the solution of the differential equation

$$(x-1)\frac{dy}{dx} + 2xy = \frac{1}{x-1} \text{ with } y(2) = \frac{1+e^4}{2e^4}. \text{ If}$$

$$y(3) = \frac{e^{\alpha}+1}{\beta e^{\alpha}}, \text{ then the value of } \alpha + \beta \text{ is equal to}$$

......... *[2022, 29 June Shift-II]*

**94** Let $y = y(x)$ be the solution of the differential equation $\frac{dy}{dx} + \frac{\sqrt{2}y}{2\cos^4 x - \cos 2x} = xe^{\tan^{-1}(\sqrt{2}\cot 2x)}$,

$0 < x < \frac{\pi}{2}$ with $y\left(\frac{\pi}{4}\right) = \frac{\pi^2}{32}$.

If $y\left(\frac{\pi}{3}\right) = \frac{\pi^2}{18}e^{-\tan^{-1}(\alpha)}$, then the value of $3\alpha^2$ is equal to ...... . *[2022, 29 June Shift-I]*

**95** Let $y = y(x)$ be the solution of the differential equation $(1-x^2)dy = (xy + (x^3+2)\sqrt{1-x^2})\,dx$, $-1 < x < 1$ and $y(0) = 0$ if $\int_{-1/2}^{1/2} \sqrt{1-x^2}\; y(x)\, dx = k$,

then $k^{-1}$ is equal to ......... . *[2022, 27 June Shift-II]*

**96** Let $y = y(x)$ be solution of the following differential equation $e^y \frac{dy}{dx} - 2e^y \sin x + \sin x \cos^2 x = 0$, $y(\pi/2) = 0$. If $y(0) = \log_e(\alpha + \beta e^{-2})$, then $4(\alpha + \beta)$ is equal to ......... . *[2021, 25 July Shift-I]*

**97** If $y = y(x)$ is the solution of the equation

$$e^{\sin y}\cos y\frac{dy}{dx} + e^{\sin y}\cos x = \cos x, \quad y(0) = 0, \text{ then}$$

$$1 + y\left(\frac{\pi}{6}\right) + \frac{\sqrt{3}}{2}\,y\left(\frac{\pi}{3}\right) + \frac{1}{\sqrt{2}}\,y\left(\frac{\pi}{4}\right) \text{ is } \ldots\ldots\ldots .$$

*[2021, 26 Feb Shift-I]*

**98** If the curve $y = y(x)$ represented by the solution of the differential equation $(2xy^2 - y)\,dx + x\,dy = 0$, passes through the intersection of the lines $2x - 3y = 1$ and $3x + 2y = 8$, then $|y(1)|$ is equal to ............ . *[2021, 25 Feb Shift-II]*

## • Key Index

### Topic 1

| | | | | | | | | | |
|---|---|---|---|---|---|---|---|---|---|
| 1 (d) | 2 (d) | 3 (a) | 4 (c) | 5 (b) | 6 (b) | 7 (c) | 8 (c) | 9 (a) | 10 (c) |
| 11 (c) | 12 (36) | 13 (16) | 14 (2) | | | | | | |

### Topic 2

| | | | | | | | | | |
|---|---|---|---|---|---|---|---|---|---|
| 1 (a) | 2 (d) | 3 (d) | 4 (b) | 5 (a) | 6 (a) | 7 (d) | 8 (d) | 9 (d) | 10 (b) |
| 11 (a) | 12 (a) | 13 (b) | 14 (a) | 15 (b) | 16 (d) | 17 (b) | 18 (b) | 19 (b) | 20 (c) |
| 21 (a) | 22 (b) | 23 (b) | 24 (d) | 25 (d) | 26 (a) | 27 (c) | 28 (b) | 29 (b) | 30 (a) |
| 31 (a) | 32 (d) | 33 (d) | 34 (a) | 35 (c) | 36 (c) | 37 (b) | 38 (b) | 39 (c) | 40 (b) |
| 41 (b) | 42 (a) | 43 (c) | 44 (a) | 45 (a) | 46 (d) | 47 (b) | 48 (b) | 49 (c) | 50 (d) |
| 51 (a) | 52 (b) | 53 (a) | 54 (c) | 55 (b) | 56 (d) | 57 (b) | 58 (a) | 59 (c) | 60 (a) |
| 61 (b) | 62 (c) | 63 (a) | 64 (d) | 65 (b) | 66 (d) | 67 (a) | 68 (a) | 69 (b) | 70 (c) |
| 71 (d) | 72 (a) | 73 (a) | 74 (b) | 75 (a) | 76 (c) | 77 (112) | 78 (19) | 79 (4) | 80 (3) |
| 81 (31) | 82 (2) | 83 (3) | 84 (11) | 85 (29) | 86 (5) | 87 (3) | 88 (2) | 89 (2) | 90 (1) |
| 91 (4) | 92 (2) | 93 (4) | 94 (2) | | | | | | |

### Topic 3

| | | | | | | | | | |
|---|---|---|---|---|---|---|---|---|---|
| 1 (d) | 2 (a) | 3 (d) | 4 (a) | 5 (a) | 6 (d) | 7 (c) | 8 (b) | 9 (c) | 10 (d) |
| 11 (a) | 12 (c) | 13 (d) | 14 (b) | 15 (a) | 16 (d) | 17 (a) | 18 (d) | 19 (b) | 20 (b) |
| 21 (c) | 22 (a) | 23 (a) | 24 (d) | 25 (b) | 26 (d) | 27 (a) | 28 (d) | 29 (b) | 30 (c) |
| 31 (a) | 32 (b) | 33 (c) | 34 (d) | 35 (c) | 36 (c) | 37 (b) | 38 (c) | 39 (b) | 40 (b) |
| 41 (a) | 42 (b) | 43 (d) | 44 (c) | 45 (b) | 46 (d) | 47 (c) | 48 (a) | 49 (d) | 50 (d) |
| 51 (b) | 52 (a) | 53 (d) | 54 (c) | 55 (d) | 56 (c) | 57 (c) | 58 (d) | 59 (d) | 60 (d) |
| 61 (b) | 62 (d) | 63 (b) | 64 (b) | 65 (a) | 66 (c) | 67 (c) | 68 (c) | 69 (d) | 70 (b) |
| 71 (a) | 72 (b) | 73 (c) | 74 (21) | 75 (27) | 76 (1) | 77 (4) | 78 (61) | 79 (18) | 80 (7) |
| 81 (5) | 82 (14) | 83 (9) | 84 (20) | 85 (97) | 86 (6) | 87 (12) | 88 (3) | 89 (2) | 90 (1) |
| 91 (1) | 92 (3) | 93 (14) | 94 (2) | 95 (320) | 96 (4) | 97 (1) | 98 (1) | | |

# Detailed **Explanations**

## Topic 1

**01.** (d) $C : x^2 + y^2 + gx + gy = 0$ ...(i)

Differentiating w.r.t. $x$,

$$2x + 2yy' + g + gy' = 0$$

$$g = -\frac{2x + 2yy'}{1 + y'}$$

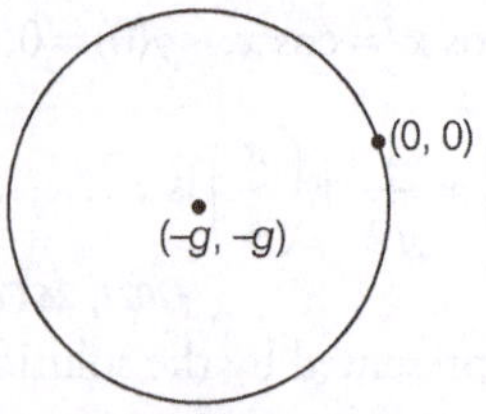


Put in Eq. (i), we get

$$x^2 + y^2 - \left(\frac{2x + 2yy'}{1 + y'}\right)(x + y) = 0$$

$$\Rightarrow (x^2 - y^2 - 2xy)\, y' = x^2 - y^2 + 2xy$$

$$\Rightarrow (x^2 - y^2 - 2xy)\, dy = (x^2 - y^2 + 2xy)\, dx$$

**02.** (d) Let $(h, k)$ be the vertex of parabola. Then, equation of parabola parallel to $Y$-axis is

$$(x - h)^2 = 4a(y - k) \quad \text{... (i)}$$

Also, Length of latusrectum = Distance of point $(2, -3)$ from the line $3x + 4y = 5$

$$\Rightarrow 4a = \frac{|6 - 12 - 5|}{\sqrt{3^2 + 4^2}} \Rightarrow 4a = \frac{11}{5}$$

$\therefore$ From Eq. (i), $(x - h)^2 = \frac{11}{5}(y - k)$

On differentiating w.r.t. $x$, we get

$$2(x - h) = \frac{11}{5}\frac{dy}{dx}$$

Again, differentiating w.r.t. $x$, we get

$$2 = \frac{11}{5}\frac{d^2y}{dx^2} \Rightarrow 11\frac{d^2y}{dx^2} = 10$$

**03.** (a) Since, $y = \left(\frac{2}{\pi}x - 1\right)\operatorname{cosec} x$ ...(i)

On differentiating w.r.t. $x$, we get

$$\frac{dy}{dx} = \frac{2}{\pi}\operatorname{cosec} x - \left(\frac{2}{\pi}x - 1\right)\operatorname{cosec} x \cot x$$

$$\Rightarrow \frac{dy}{dx} + y \cot x = \frac{2}{\pi}\operatorname{cosec} x$$

[from Eq. (i)] ...(ii)

It is given that

$$\frac{dy}{dx} + p(x)y = \frac{2}{\pi}\operatorname{cosec} x,\ 0 < x < \frac{\pi}{2}$$

On comparing with Eq. (ii), we get

$$p(x) = \cot x$$

**04.** (c) The given equation is

$$Ax^2 + By^2 = 1.$$

On differentiating w.r.t. $x$, we get

$$2\,Ax + 2\,By\frac{dy}{dx} = 0 \quad \text{...(i)}$$

Again differentiating, we get

$$2A + 2B\left\{\left(\frac{dy}{dx}\right)^2 + y\frac{d^2y}{dx^2}\right\} = 0 \quad \text{...(ii)}$$

On eliminating $A$ and $B$ from Eqs. (i) and (ii), we get

$$y\frac{d^2y}{dx^2} + \left(\frac{dy}{dx}\right)^2 - \frac{y}{x}\cdot\frac{dy}{dx} = 0$$

This is the required differential equation whose order is two and degree is one.

**05.** (b) Given, equation of family of curves is

$$y^2 = 2c(x + \sqrt{c}) \quad \text{...(i)}$$

On differentiating Eq. (i) w.r.t. $x$, we get

$$2yy_1 = 2c \Rightarrow c = yy_1$$

On putting the value of $c$ in Eq. (i), we get

$$y^2 = 2yy_1(x + \sqrt{yy_1})$$

$$\therefore \quad (y^2 - 2yy_1x)^2 = 4(yy_1)^3$$

Hence, the degree and order of above equation are three and one, respectively.

**06.** (b) General equation of parabola whose axis is $X$-axis, is

$$y^2 = 4a(x + h)$$

On differentiating w.r.t. $x$, we get

$$2y\frac{dy}{dx} = 4a$$

Again differentiating w.r.t. $x$, we get

$$\left(\frac{dy}{dx}\right)^2 + y\frac{d^2y}{dx^2} = 0$$

This is a differential equation whose degree and order are one and two, respectively.

**07.** (c) The given differential equation can be rewritten as

$$\left(1 + 3\frac{dy}{dx}\right)^2 = 4\left(\frac{d^3y}{dx^3}\right)^3$$

This shows that the order and degree of given equation are 3 and 3, respectively.

**08.** (c) According to the question,

$$A = \int_0^a f(x)\cdot dx = e^{-a} + 4a^2 + a - 1$$

$$\Rightarrow f(a) = -e^{-a} + 8a + 1$$

Then, $f(x) = -e^{-x} + 8x + 1$

Let $\quad y = c_1 f(x) + c_2$

$$\frac{dy}{dx} = c_1 f'(x) = c_1(e^{-x} + 8) \quad \text{...(i)}$$

$$\frac{d^2y}{dx^2} = -c_1e^{-x} \Rightarrow -e^x\left(\frac{d^2y}{dx^2}\right) = c_1$$

So, $\quad \frac{dy}{dx} = -e^x\left(\frac{d^2y}{dx^2}\right)(e^{-x} + 8)$

$$\Rightarrow \frac{dy}{dx} = -e^x\left(\frac{d^2y}{dx^2}\right)\left(\frac{1 + 8e^x}{e^x}\right)$$

$$\Rightarrow (8e^x + 1)\left(\frac{d^2y}{dx^2}\right) + \frac{dy}{dx} = 0$$

**09.** (a) Let the equation of circle be

$$(x - h)^2 + (y - k)^2 = r^2$$

$\because$ The circle passes through (0, 2),

$$(0 - h)^2 + (2 - k)^2 = r^2$$

$$\Rightarrow \quad h^2 + 4 + k^2 - 4k = r^2 \quad \text{...(i)}$$

And the circle passes through (0, – 2),

$$(0 - h)^2 + (-2 - k)^2 = r^2$$

$$\Rightarrow \quad h^2 + 4 + k^2 + 4k = r^2 \quad \text{...(ii)}$$

On subtracting Eq. (i) from Eq. (ii), we get

$$8k = 0 \Rightarrow k = 0$$

$$\therefore \quad h^2 + 4 = r^2$$

Now, the equation of circle is

$$(x - h)^2 + y^2 = h^2 + 4$$

$$x^2 - 2hx + y^2 = 4 \quad \text{...(iii)}$$

On differentiating both sides w.r.t. $x$, we get

$$2h = 2x + 2y\frac{dy}{dx}$$

On substituting the value of $2h$ in Eq. (iii), we get

$$x^2 - 2x^2 - 2xy\frac{dy}{dx} + y^2 = 4$$

$$\Rightarrow 2xy\frac{dy}{dx} + x^2 - y^2 + 4 = 0$$

**10.** *(c)* Given, equation of curve is

$$y^2 = 4a(x + a)$$

$$\Rightarrow \quad y^2 = 4ax + 4a^2 \quad \ldots(i)$$

On differentiating Eq. (i) w.r.t. $x$, we get

$$2y\frac{dy}{dx} = 4a \Rightarrow a = \left(\frac{y}{2}\right)\cdot\frac{dy}{dx} \quad \ldots(ii)$$

$\therefore$ Required differential equation is

$$y^2 = 4 \times \frac{y}{2} \times \frac{dy}{dx} x + 4\left(\frac{y}{2}\cdot\frac{dy}{dx}\right)^2$$

[From Eqs. (i) and (ii)]

$$\Rightarrow \quad y^2\left(\frac{dy}{dx}\right)^2 + 2xy\left(\frac{dy}{dx}\right) - y^2 = 0$$

$$\Rightarrow \quad y\left[y\left(\frac{dy}{dx}\right)^2 + 2x\left(\frac{dy}{dx}\right) - y\right] = 0$$

As, $y \neq 0$

$$\therefore y\left(\frac{dy}{dx}\right)^2 + 2x\left(\frac{dy}{dx}\right) - y = 0$$

**11.** *(c)*

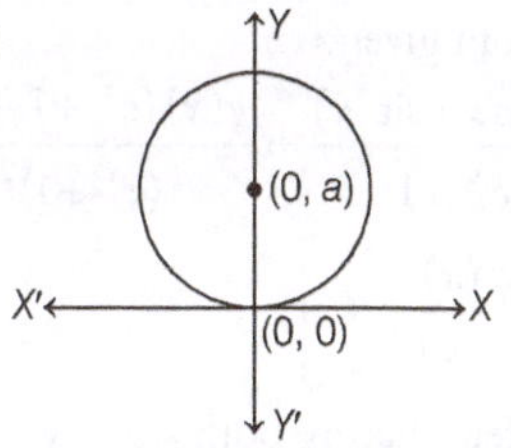


Equation of family is

$$x^2 + (y - a)^2 = a^2$$

where, $(0, a)$ is the centre of circle.

$$\Rightarrow \quad x^2 + y^2 - 2ay = 0 \quad \ldots(i)$$

On differentiating both side w.r.t. $x$, we get

$$x + y\frac{dy}{dx} - a\frac{dy}{dx} = 0$$

$$\Rightarrow \quad \frac{x + y\frac{dy}{dx}}{dy/dx} = a$$

On putting $a$ in original equation, we get

$$x^2 + y^2 - 2y\left[\frac{y\frac{dy}{dx} + x}{\frac{dy}{dx}}\right] = 0$$

$$\Rightarrow \quad (x^2 + y^2 - 2y^2)\frac{dy}{dx} = 2xy$$

Hence, $\quad g(x) = 2x$.

By solving this equation,

$$(x^2 - y^2)\frac{dy}{dx} = 2xy = g(x)y$$

$$\Rightarrow \quad g(x) = 2x$$

**12.** *(36)* $f'(x) \geq 0$

$$f''(x) = f(x)$$

Second order differential equation

$$f(x) = Ae^x + Be^{-x}$$

$$f(0) = 0 \Rightarrow A = -B$$

$$f(x) = A(e^x - e^{-x})$$

$$\Rightarrow \quad f'(x) = A(e^x + e^{-x})$$

$$f'(0) = 3 = A(e^0 + e^{-0}) \Rightarrow A = \frac{3}{2}$$

$$f(x) = \frac{3}{2}(e^x - e^{-x})$$

$$\Rightarrow f(\ln(3)) = \frac{3}{2}(e^{\ln 3} - e^{-\ln 3})$$

$$\Rightarrow 9f(\ln 3) = \frac{27}{2}\left(3 - \frac{1}{3}\right) = 36$$

**13.** *(16)* Given a circle, $C : (x^2 + y^2 - 3) + (x^2 - y^2 - 1)^5 = 0$

On putting the point $(\alpha, \alpha)$

$$2\alpha^2 - 3 + (-1)^5 = 0 \Rightarrow \alpha = \sqrt{2}$$

Now, differentiating w.r.t. $x$, we get

$$2x + 2y \cdot y' + 5(x^2 - y^2 - 1)^4 (2x - 2yy') = 0 \quad \ldots(i)$$

At $(\sqrt{2}, \sqrt{2})$,

$$\sqrt{2} + \sqrt{2}y' + 5(-1)^4 (\sqrt{2} - \sqrt{2}y') = 0$$

$$\Rightarrow \quad y' = \frac{3}{2} \quad \ldots(ii)$$

Again, differentiating Eq. (i) w.r.t. $x$, we get

$$1 + (y')^2 + yy'' + 20(x^2 - y^2 - 1)^3 (x - yy')^2 \cdot 2 + 5(x^2 - y^2 - 1)^4 (1 - (y')^2 - yy'') = 0$$

At $(\sqrt{2}, \sqrt{2})$ and $y' = \frac{3}{2}$

We have,

$$\left(1 + \frac{9}{4}\right) + \sqrt{2}y'' - 40\left(\sqrt{2} - \sqrt{2}\cdot\frac{3}{2}\right)^2 + 5(1)\left(1 - \frac{9}{4} - \sqrt{2}y''\right) = 0$$

$$\Rightarrow \quad 4\sqrt{2}y'' = -23$$

$$\because 3y' - y^3 y'' = \frac{9}{2} + \frac{23}{2} = 16$$

**14.** *(2)* Given, $y^2 = a\left[x + \frac{\sqrt{a}}{2}\right], a > 0 \ldots(i)$

On differentiating both sides w.r.t. $x$, we get

$$2y\frac{dy}{dx} = a[1 + 0] = a \quad \ldots(ii)$$

Using Eqs. (i) and (ii) to eliminate the constant $a$.

$$y^2 = 2y\frac{dy}{dx}\left(x + \sqrt{2y}\sqrt{\frac{dy}{dx}}\right)$$

$$\Rightarrow y^2 - 2xy\frac{dy}{dx} = \frac{2\sqrt{2}}{2}\cdot y\sqrt{y}\cdot\frac{dy}{dx}\sqrt{\frac{dy}{dx}}$$

On squaring both sides, we get

$$y^4 + 4x^2y^2\left(\frac{dy}{dx}\right)^2 - 4xy^3\frac{dy}{dx} = 2y^3\left(\frac{dy}{dx}\right)^3$$

Thus, degree of above differential equation is 3 and its order is 1.

Difference between degree and order $= 3 - 1 = 2$

## Topic 2

**01.** *(a)* Given, differential equation is

$$2y\frac{dy}{dx} + 3 = 5\frac{dy}{dx}$$

$$\Rightarrow \quad (2y - 5)\frac{dy}{dx} = -3$$

$$\Rightarrow \quad (2y - 5)dy = -3dx$$

On integrating, we get

$$2\cdot\frac{y^2}{2} - 5y = -3x + C \quad \ldots(i)$$

$\because$ Curve passes through (0, 1) i.e. putting $x = 0$ and $y = 1$ in Eq. (i)

$\therefore \quad C = -4$

From Eq. (i) $\Rightarrow y^2 - 5y = -3x - 4$

$$\Rightarrow y^2 - 2\left(\frac{5}{2}\right)y + \frac{25}{4} = -3x - 4 + \frac{25}{4}$$

$\Rightarrow \left(y - \frac{5}{2}\right)^2 = -3\left(x - \frac{3}{4}\right)$ which is the equation of shifted parabola whose vertex is $\left(\frac{3}{4}, \frac{5}{2}\right)$ and this vertex satisfies the equation $2x + 3y = 9$

**02.** *(d)* $(1 + y^2)e^{\tan x}\, dx + \cos^2 x\,(1 + e^{2\tan x})dy = 0$

$$\Rightarrow -\frac{1}{1 + y^2}\cdot dy = \frac{e^{\tan x}\cdot\sec^2 x}{1 + (e^{\tan x})^2}\cdot dx$$

On integrating both sides,

$$\int \frac{1}{1+y^2}dy = -\int \frac{e^{\tan x}}{1+(e^{\tan x})^2}\sec^2 x \cdot dx$$

Let $\tan x = t$

$\Rightarrow \sec^2 x \cdot dx = dt$

$\Rightarrow \tan^{-1}(y) = -\int \frac{e^t}{1+(e^t)^2}\cdot dt$

Let $e^t = u, e^t \cdot dt = du$

$\Rightarrow \tan^{-1}(y) = -\int \frac{1}{1+u^2}dx$

$\Rightarrow \tan^{-1}(y) = -\tan^{-1}(u) + c$

$\Rightarrow \tan^{-1}(y) = -\tan^{-1}(e^{\tan x}) + c$

$\because y(0) = 1$

$\Rightarrow \frac{\pi}{4} = -\tan^{-1}(1) + c$

$\Rightarrow c = \frac{\pi}{4} + \frac{\pi}{4} = \frac{\pi}{2}$

So, $y = \tan\left(\frac{\pi}{2} - \tan^{-1}(e^{\tan x})\right)$

$= \cot(\tan^{-1}(e^{\tan x}))$

At $x = \frac{\pi}{4}, y\left(\frac{\pi}{4}\right) = \cot(\tan^{-1}(e))$

$= \cot\left(\cot^{-1}\left(\frac{1}{e}\right)\right) = \frac{1}{e}$

**03.** *(d)* Given, $\frac{dT}{dt} = -K(T-80)$

$$\int_{160}^{T} \frac{dT}{T-80} = \int_0^t -K\,dt$$

$\Rightarrow [\ln|T-80|]_{160}^{T} = -K[t]_0^t$

$\Rightarrow \ln|T-80| - \ln(80) = -K(t-0)$

$\Rightarrow \ln\left|\frac{T-80}{80}\right| = -Kt$

$\Rightarrow \frac{T-80}{80} = e^{-Kt}$

$T = 80 + 80e^{-Kt}$

Given, $T(15) = 120°\text{F}$

$120 = 80 + 80e^{-15K}$

$\Rightarrow 40 = 80e^{-15K}$

$\Rightarrow e^{-15K} = \frac{1}{2}$

At $t = 45, T = 80 + 80e^{-45K}$

$T = 80 + 80(e^{-15K})^3$

$= 80 + 80\left(\frac{1}{2}\right)^3 = 90°\text{F}$

**04.** *(b)*

**key idea** The given differential equation is a homogenous differential equation. To find the solution of such type of differential equation substitute $x = vy$

Then, solve it by integration.

Given, $y\frac{dx}{dy} = x(\ln x - \ln y + 1)$

$x > 0, y > 0$

$\Rightarrow \frac{dx}{dy} = \frac{x}{y}\left(\ln\frac{x}{y} + 1\right)$

$\Rightarrow \frac{x}{y} = v$

$\Rightarrow x = vy$

$\Rightarrow \frac{dx}{dy} = v + y\frac{dv}{dy}$

$\therefore$ Equation becomes,

$v + y\frac{dv}{dy} = v(\ln v + 1)$

$\Rightarrow y\frac{dv}{dy} = v\ln v$

Integrating both sides, we get

$$\int \frac{dv}{v\ln v} = \int \frac{dy}{y}$$

$\ln(\ln v) = \ln y + C$

$\ln v = ky$

$\because \ln\left(\frac{x}{y}\right) = ky$

$x = e$ when $y = 1$ [given]

$\therefore \ln e = k \Rightarrow k = 1$

$\ln\frac{x}{y} = y$ or $\left|\ln\frac{x}{y}\right| = y$

**05.** *(a)* $\sin\frac{y}{x} = \ln x + \frac{\alpha}{2}$

$\because y(1) = \frac{\pi}{3}$

$\therefore \sin\left(\frac{\pi}{3}\right) = \ln(1) + \frac{\alpha}{2}$

$\left(\because x = 1 \text{ and } y = \frac{\pi}{3}\right)$

$\Rightarrow \frac{\sqrt{3}}{2} = 0 + \frac{\alpha}{2} \Rightarrow \alpha = \sqrt{3}$

$\alpha^2 = 3$

**06.** *(a)* We have, $y(x+1)dx = x^2 dy$

$\Rightarrow \left(\frac{x+1}{x^2}\right)dx = \frac{dy}{y}$

On integrating both the sides, we get

$\log x - \frac{1}{x} = \log y + c$

When, $x = 1, y = e$, then

$c = 0 - 1 - 1 = -2$

$\therefore \log y = \log x - \frac{1}{x} + 2 \Rightarrow y = xe^{2-\frac{1}{x}}$

$\therefore \lim_{x\to 0^+} y = 0 \times e^{-\infty} = 0$

**07.** *(d)* Given, differential equation

$2ye^{x/y^2}dx + (y^2 - 4xe^{x/y^2})dy = 0,$

$x(1) = 0$

$\Rightarrow e^{x/y^2}[2y\,dx - 4x\,dy] = -y^2\,dy$

$\Rightarrow e^{x/y^2}\left[\frac{2y^2\,dx - 4xy\,dy}{y^4}\right] = -\frac{1}{y}dy$

$\Rightarrow 2e^{x/y^2}\,d\left(\frac{x}{y^2}\right) = -\frac{1}{y}dy$

$\Rightarrow 2e^{x/y^2} = -\ln y + c$ ... (i)

Now, using $x(1) = 0$, then $c = 2$

So, for $x(e)$, put $y = e$ in Eq. (i),

$2e^{x/e^2} = -1 + 2$

$\Rightarrow \frac{x}{e^2} = \ln\left(\frac{1}{2}\right) \Rightarrow x(e) = -e^2\ln 2$

**08.** *(d)* From given,

$$\int\left(\frac{x(\cos x - \sin x)}{e^x + 1} + \frac{g(x)(e^x + 1 - xe^x)}{(e^x+1)^2}\right)dx = \left[\frac{xg(x)}{e^x+1} + C\right]$$

On differentiating both sides, we get

$$\frac{x(\cos x - \sin x)}{e^x+1} + \frac{g(x)(e^x + 1 - xe^x)}{(e^x+1)^2} = \frac{d}{dx}\left[\frac{xg(x)}{e^x+1} + C\right]$$

We get, $g'(x) = \cos x - \sin x$

$g(x) = \sin x + \cos x \therefore g - g' = 2\sin x$

We know, $\sin x$ is in increasing in $\left(0, \frac{\pi}{2}\right)$.

So, $g - g'$ will be increasing in $\left(0, \frac{\pi}{2}\right)$.

**09.** *(d)*

**key idea** Use L' Hospital Rule to remove $\frac{0}{0}, \frac{\infty}{\infty}$ indeterminant form and find the limit by differentiating numerator and denominator upto finite value.

$$\int_0^x [1-(f'(t))^2]^{1/2} dt = \int_0^x f(t)\, dt$$

On differentiating both sides, we get

$$\sqrt{1-[f'(x)]^2} = [f(x)]$$

$$\Rightarrow \quad 1-[f'(x)]^2 = [f(x)]^2$$

$$\Rightarrow \quad 1-[f(x)]^2 = [f'(x)]^2$$

$$\Rightarrow \quad f'(x) = \sqrt{1-[f(x)]^2}$$

$$\Rightarrow \quad \int \frac{f'(x)\, dx}{\sqrt{1-[f(x)]^2}} = \int 1\, dx$$

$$\Rightarrow \quad \sin^{-1} f(x) = x + C$$

$$\because \quad f(0) = 0 \Rightarrow C = 0$$

$$f(x) = \sin x$$

$$\Rightarrow \lim_{x\to 0} \frac{\int_0^x f(t)\, dt}{x^2} = \lim_{x\to 0} \frac{\int_0^x \sin t\, dt}{x^2}$$

On applying L'Hospital rule, we get

$$\lim_{x\to 0}\left(\frac{\sin x}{2x}\right) = \frac{1}{2}$$

**10.** (b) $\frac{dy}{dx} = \frac{2^{x+y}-2^x}{2^y} = \frac{2^x(2^y-1)}{2^y}$

$$\int \frac{2^y}{2^y-1}\, dy = \int 2^x\, dx$$

$$\frac{\ln(2^y-1)}{\ln 2} = \frac{2^x}{\ln 2} + C$$

As, $y(0) = 1 \Rightarrow 0 = \frac{1}{\log 2} + C$

For $y(1)$, $\ln(2^y-1) = 2^1 - 1$

$$\Rightarrow \quad 2^y - 1 = e$$

$$\Rightarrow \quad y = \log_2(e+1)$$

**11.** (a) $\log_e\left(\frac{dy}{dx}\right) = 3x + 4y$

$$\Rightarrow \frac{dy}{dx} = e^{3x}\cdot e^{4y} \Rightarrow \int \frac{dy}{e^{4y}} = \int e^{3x}\, dx$$

$$\Rightarrow \frac{e^{-4y}}{-4} = \frac{e^{3x}}{3} + C$$

Now, $y(0) = 0$

So, $\frac{e^0}{-4} = \frac{e^0}{3} + C \Rightarrow C = -\frac{1}{3} - \frac{1}{4} = -\frac{7}{12}$

$$\therefore \quad \frac{e^{-4y}}{-4} = \frac{e^{3x}}{3} - \frac{7}{12}$$

At $x = \left(-\frac{2}{3}\log 2\right)$, $y = \alpha \log 2$

$$\Rightarrow \frac{e^{-4y}}{-4} = \frac{2^{-2}}{3} - \frac{7}{12}$$

$$\Rightarrow \frac{e^{-4y}}{-4} = \frac{1}{12} - \frac{7}{12} \Rightarrow \frac{e^{-4y}}{-4} = \frac{-1}{2}$$

$$\Rightarrow e^{-4y} = 2 \qquad \therefore e^{-4\alpha\log 2} = 2$$

$$\Rightarrow -4\alpha \log 2 = \log 2 \Rightarrow -4\alpha = 1$$

$$\therefore \quad \alpha = \frac{-1}{4}$$

**12.** (a) Given, differential equation

$$xdy = (y + x^3 \cos x)\, dx$$

$$\Rightarrow xdy = ydx + x^3 \cos x\, dx$$

$$\Rightarrow xdy - ydx = x^3 \cos x\, dx$$

$$\Rightarrow \frac{xdy - ydx}{x^2} = x\cos x\, dx$$

$$\Rightarrow \frac{d}{dx}\left(\frac{y}{x}\right) = \int x \cos x\, dx$$

$$\Rightarrow \frac{y}{x} = x\sin x - \int 1\cdot \sin x\, dx$$

$$\Rightarrow \frac{y}{x} = x\sin x + \cos x + c \qquad \text{...(i)}$$

At $x = \pi$ and $y = 0$

$$0 = -1 + c \Rightarrow c = 1$$

On substituting the value of $c$ in Eq. (i), we get

$$\frac{y}{x} = x\sin x + \cos x + 1$$

$$\Rightarrow y = x^2 \sin x + x\cos x + x$$

At, $x = \pi/2$

So, $y\left(\frac{\pi}{2}\right) = \frac{\pi^2}{4}(1) + 0 + \frac{\pi}{2} = \frac{\pi^2}{4} + \frac{\pi}{2}$

**13.** (b) $e^x\sqrt{(1-y^2)}dx + \frac{y}{x}dy = 0$

$$\frac{dy}{dx} = \frac{-e^x\sqrt{1-y^2}}{\left(\frac{y}{x}\right)}$$

$$\Rightarrow \int \frac{ydy}{\sqrt{1-y^2}} = -\int xe^x\, dx$$

$$\Rightarrow \sqrt{1-y^2} = e^x(x-1) + c \Rightarrow c = 0$$

$x = 1, y = -1$ $\qquad 0 = 0 + c$

$$\therefore \quad 1 - y^2 = e^{2x}(x-1)^2$$

At $\quad x = 3, 1-y^2 = e^6\cdot 4$

$$y^2 = 1 - 4e^6$$

**14.** (a) $\frac{dy}{dx} = xy - 1 + x - y$

$$\frac{dy}{dx} = x(y+1) - 1(y+1)$$

$$\frac{dy}{dx} = (x-1)(y+1)$$

$$\int \frac{dy}{y+1} = \int (x-1)\, dx$$

$$\log_e(y+1) = \frac{x^2}{2} - x + c$$

$$y(0) = 0$$

$$\log_e 1 = 0 + c \Rightarrow c = 0$$

$$y(1) \Rightarrow \log_e(y+1) = \frac{1}{2} - 1$$

$$y + 1 = e^{-1/2} \Rightarrow y(1) = -1 + e^{-1/2}$$

**15.** (b) Given differential equation

$$\frac{5+e^x}{2+y}\frac{dy}{dx} + e^x = 0$$

$$\Rightarrow \int \frac{dy}{2+y} + \int \frac{e^x}{5+e^x}\, dx = 0$$

$$\Rightarrow \log_e|2+y| + \log_e|5+e^x| = \log_e C$$

$$\Rightarrow \quad |2+y|\cdot|5+e^x| = C$$

$$\because \quad y(0) = 1 \Rightarrow C = 18$$

$\therefore$ At, $x = \log_e 13$ $\quad |2+y||5+13| = 18$

$$\Rightarrow \quad |2+y| = 1 \Rightarrow y + 2 = \pm 1$$

$$\Rightarrow \quad y = -1 \text{ or } -3$$

**16.** (d) Given differential equation is

$$\frac{dy}{dx} - \frac{y+3x}{\log_e(y+3x)} + 3 = 0$$

On puttting, $y + 3x = t \Rightarrow \frac{dy}{dx} + 3 = \frac{dt}{dx}$

$$\therefore \frac{dt}{dx} - \frac{t}{\log_e(t)} = 0 \Rightarrow \int \frac{\log(t)}{t}\, dt = \int dx$$

$$\Rightarrow \frac{1}{2}(\log_e t)^2 = x + c'$$

$$\Rightarrow x - \frac{1}{2}(\log_e(y+3x))^2 = C,$$

where $c' = -C$

**17.** (b) Let $p'(x) = \lambda(x-1)(x-2)$, where $\lambda > 0$

$$\Rightarrow p'(x) = \lambda(x^2 - 3x + 2)$$

$$\Rightarrow p(x) = \lambda\left(\frac{x^3}{3} - \frac{3x^2}{2} + 2x\right) + C$$

$$\because \quad p(1) = 8 \Rightarrow 8 = \lambda\left(\frac{1}{3} - \frac{3}{2} + 2\right) + C$$

$$\Rightarrow \quad 8 = \frac{5}{6}\lambda + C \qquad \text{...(i)}$$

Also, $p(2) = 4 \Rightarrow \lambda\left(\frac{8}{3} - 6 + 4\right) + C = 4$

$$\Rightarrow \quad 4 = \frac{2}{3}\lambda + C \qquad \text{...(ii)}$$

On solving Eqs. (i) and (ii), we get

$C = -12$

$\therefore \quad p(0) = -12$

**18.** *(b)* $f(xy) = f(x)\cdot f(y)$

Put $x = 0,\ y = 0$

$f(0) = (f(0))^2$

$\Rightarrow f(0) = 1$ as $f(0) \neq 0$

Now, put $y = 0$

$f(0) = f(x)\cdot f(0)$

$\Rightarrow f(x) = 1$

As given, $\dfrac{dy}{dx} = f(x)$

$\Rightarrow y = \int f(x)\,dx = \int 1\,dx$

$\Rightarrow y = x + c$

At $x = 0,\ y(0) = 1$

$\Rightarrow c = 1$

$\therefore y(x) = x + 1$

$y\left(\dfrac{1}{4}\right) + y\left(\dfrac{3}{4}\right) = \dfrac{1}{4} + 1 + \dfrac{3}{4} + 1 = 3$

**19.** *(b)* Given, $ydx - (x + 2y^2)dy = 0$

$\Rightarrow ydx - xdy - 2y^2dy = 0$

$\Rightarrow \quad \dfrac{ydx - xdy}{y^2} = 2dy$

$\Rightarrow \quad d(x/y) = 2dy$

On integrating both sides, we get

$\dfrac{x}{y} = 2y + c$

Using $f(-1) = 1$, we get

$c = 1 \Rightarrow x/y = 2y + 1$

On putting $y = 1$, we get $f(1) = 3$

**20.** *(c)* Given, $\dfrac{dP}{dx} = (100 - 12\sqrt{x})$

$\Rightarrow \quad dP = (100 - 12\sqrt{x})\,dx$

On integrating both sides, we get

$\int dP = \int (100 - 12\sqrt{x})\,dx$

$\Rightarrow \quad P = 100x - 8x^{3/2} + C$

When, $x = 0$, then $P = 2000$

$\Rightarrow \quad C = 2000$

Now, when $x = 25$, then

$P = 100 \times 25 - 8 \times (25)^{3/2} + 2000$

$= 2500 - 8 \times 125 + 2000$

$= 4500 - 1000 = 3500$

**21.** *(a)* Given,

$(x+2)\dfrac{dy}{dx} = x^2 + 4x - 9,\ x \neq -2$

$\Rightarrow \quad \dfrac{dy}{dx} = \dfrac{x^2 + 4x - 9}{x + 2}$

$\Rightarrow \quad dy = \dfrac{x^2 + 4x - 9}{x + 2}\,dx$

$\Rightarrow \int dy = \int \dfrac{x^2 + 4x - 9}{x + 2}\,dx$

$\Rightarrow \quad y = \int \left(x + 2 - \dfrac{13}{x + 2}\right) dx$

$\Rightarrow \quad y = \int (x+2)dx - 13\int \dfrac{1}{x+2}\,dx$

$\Rightarrow y = \dfrac{x^2}{2} + 2x - 13\log|x + 2| + c$

Since, $y(0) = (0),\quad 0 = -13\log 2 + c$

$\Rightarrow c = 13\log 2$

$y = \dfrac{x^2}{2} + 2x - 13\log|x + 2| + 13\log 2$

$\therefore y(-4) = 8 - 8 - 13\log 2 + 13\log 2 = 0$

**22.** *(b)* Given that

$ydx + (x + x^2y)\,dy = 0$

$\therefore \quad \dfrac{y\,dx + x\,dy}{x^2y^2} = -\dfrac{1}{y}\,dy$

$\Rightarrow \quad \int d\left(-\dfrac{1}{xy}\right) = -\int \dfrac{1}{y}\,dy$

On integrating both sides, we get

$-\dfrac{1}{xy} = -\log y + C \Rightarrow -\dfrac{1}{xy} + \log y = C$

**23.** *(b)* $\because \quad \dfrac{d^2y}{dx^2} = e^{-2x}$

On integrating both sides, we get

$\dfrac{dy}{dx} = \dfrac{e^{-2x}}{-2} + C$

Again integrating, we get

$y = \dfrac{e^{-2x}}{4} + Cx + D$

**24.** *(d)*

> **key idea** If $f(x) = \int_{g(x)}^{h(x)} F(x)\,dx$
>
> $\Rightarrow f'(x) = h'(x)\,F(h(x)) - g'(x)\,F(g(x))$

$\int_0^x t\,f(t)\,dt = x^2 f(x)$

On differentiating w.r.t $x$, wet get

$xf(x) = 2xf(x) + x^2 f'(x)$

$\Rightarrow \quad x^2 f'(x) = -xf(x)$

$\Rightarrow \quad \dfrac{f'(x)}{f(x)} = -\dfrac{1}{x}$

$\Rightarrow \quad \int \dfrac{f'(x)}{f(x)}\,dx = -\int \dfrac{1}{x}$

$\Rightarrow \quad \log(f(x)) = -\log(x) + \log(c)$

$f(x) = \dfrac{c}{x}$

$\because \quad f(2) = 3 \Rightarrow 3 = \dfrac{c}{2} \Rightarrow c = 6$

$\Rightarrow \quad f(x) = \dfrac{6}{x}$

$\therefore \quad f(6) = \dfrac{6}{6} = 1$

**25.** *(d)*

> **key idea** To solve this problem, use formula
>
> $\dfrac{d}{dx}\left(\dfrac{f(x)}{g(x)}\right) = \dfrac{f'(x)g(x) - g'(x)f(x)}{(f(x))^2}$

Given, $\quad x^2 f'(x) - 2xf(x) = 3$

$\Rightarrow \quad \dfrac{x^2 f'(x) - 2xf(x)}{(x^2)^2} = \dfrac{3}{(x^2)^2}$

$\Rightarrow \quad \dfrac{d}{dx}\left(\dfrac{f(x)}{x^2}\right) = \dfrac{3}{x^4}$

Integrating both sides, we get

$\dfrac{f(x)}{x^2} = -\dfrac{1}{x^3} + C$, where $C$ is constant of integration.

$\Rightarrow \quad f(x) = -\dfrac{1}{x} + Cx^2$

Given, $f(1) = 4$ i.e. when $x = 1$, then $f(x) = 4$.

So, $4 = -\dfrac{1}{1} + C \times (1)^2 \Rightarrow C = 5$

$\therefore f(x) = -\dfrac{1}{x} + 5x^2$

Hence, $2f(2) = 2 \times \left[-\dfrac{1}{2} + 5(2)^2\right] = 39$

**26.** *(a)* Given, $f(x + y) = f(x)\cdot f'(y) + f'(x)\cdot f(y)$

Put $y = 0$

$f(x) = f(x)\cdot f'(0) + f'(x)\,f(0)$

$f(x) = f(x)\cdot f'(0) + f'(x) \quad$ ...(i) $\quad [\because f(0) = 1]$

Put $x = y = 0$

$f(0) = f'(0) + f'(0) \Rightarrow f'(0) = \dfrac{1}{2}$

Hence, equation (i) becomes

$f(x) = \dfrac{f(x)}{2} + f'(x)$

$f'(x) = \dfrac{f(x)}{2} \Rightarrow \dfrac{f'(x)}{f(x)} = \dfrac{1}{2}$

On integrating both sides, we get

$$\ln f(x) = \frac{x}{2} + C$$

Put $x = 0 \Rightarrow C = 0$

$$\Rightarrow \quad \ln f(x) = \frac{x}{2} \Rightarrow f(x) = e^{\frac{x}{2}}$$

$$\therefore \quad f(n) = e^{\frac{n}{2}}$$

$$\log_e f(n) = \frac{n}{2}$$

$$\therefore \quad \sum_{n=1}^{100} \log_e f(n) = \frac{1}{2} \sum_{n=1}^{100} n$$

$$= \frac{1}{2} \sum_{n=1}^{100} n = \frac{1}{2}\left(\frac{100}{2}(100+1)\right)$$

$$= 25 \times 101 = 2525$$

**27.** *(c)* Given, the differential equation is

$$(x^2 + y^2)dx - 5xydy = 0$$

$$\Rightarrow \quad \frac{dy}{dx} = \frac{x^2 + y^2}{5xy} \quad \text{...(i)}$$

which is homogeneous differential equation.

Put, $y = vx \Rightarrow \frac{dy}{dx} = v + x\frac{dv}{dx}$

From Eq. (i), $v + x\frac{dv}{dx} = \frac{1+v^2}{5v}$

$$\Rightarrow \quad x\frac{dv}{dx} = \frac{1-4v^2}{5v}$$

So, $\int \frac{v}{1-4v^2} dv = \int \frac{dx}{5x}$ ...(ii)

Let $1 - 4v^2 = z \Rightarrow -8vdv = dz$

From Eq. (ii) $\Rightarrow \int \frac{dz}{-8z} = \int \frac{dx}{5x}$

$$\Rightarrow -\frac{1}{8} \ln|z| = \frac{1}{5} \ln|x| + \ln C_1$$

$\Rightarrow -5 \ln|z| = 8 \ln|x| + \ln C_1$, where $\ln C_1 = 8 \ln C$

$$\Rightarrow \ln(x^8) + \ln|z^5| + \ln C_1 = 0$$

$\Rightarrow x^8 |z^5| = C_2$, $\quad$ [where $C_2 = \frac{1}{C_1}$]

$$\Rightarrow x^8 |1 - 4v^2|^5 = C_2$$

$$\Rightarrow x^8 \cdot \left|\frac{x^2 - 4y^2}{x^2}\right|^5 = C_2$$

$$\Rightarrow |x^2 - 4y^2|^5 = C_2 x^2$$

Given, $y(1) = 0$

$\therefore |1|^5 = C_2 \Rightarrow C_2 = 1$

Hence, $|x^2 - 4y^2|^5 = x^2$

**28.** *(b)* Given differential equation is

$$(x^4 + 2x^3 + 3x^2 + 2x + 2)dy - (2x^2 + 2x + 3)dx = 0$$

$$\Rightarrow$$

$$\int dy = \int \frac{2x^2 + 2x + 3}{x^4 + 2x^3 + 3x^2 + 2x + 2} dx$$

$$\Rightarrow \quad y = \int \frac{2x^2 + 2x + 3}{(x^2+1)(x^2 + 2x + 2)} dx$$

$$= \int \frac{dx}{x^2 + 2x + 2} + \int \frac{dx}{x^2 + 1}$$

$$= \int \frac{dx}{(x+1)^2 + 1} + \int \frac{dx}{x^2 + 1}$$

$$\Rightarrow y = \tan^{-1}(x+1) + \tan^{-1} x + C \text{ ...(i)}$$

$$\because \; y(-1) = -\frac{\pi}{4}$$

From Eq. (i), $-\frac{\pi}{4} = 0 - \frac{\pi}{4} + c$

$\Rightarrow C = 0$

$\therefore \; y = \tan^{-1}(x+1) + \tan^{-1} x$

Hence, $y(0) = \tan^{-1}(1) = \pi/4$

**29.** *(b)* We have, $f'(x) = \alpha f(x) + 3$

Let $f(x) = y \Rightarrow \frac{dy}{dx} = f'(x)$

$$\therefore \frac{dy}{dx} = \alpha y + 3$$

On variable separable and integrating

$$\int \frac{1}{\alpha y + 3} dy = \int dx$$

$$\Rightarrow \frac{1}{\alpha} \ln|\alpha y + 3| = x + c$$

$$\Rightarrow \ln|\alpha y + 3| = \alpha x + \alpha c$$

$\ln|\alpha y + 3| = \alpha x + \ln k \quad [\alpha c = \ln k]$

$\alpha y + 3 = ke^{\alpha x}$ ...(i)

Use $f(0) = 2$ in Eq. (i),

$2\alpha + 3 = ke^0 \Rightarrow k = 2\alpha + 3$

$$\therefore f(x) = \frac{(2\alpha + 3)e^{\alpha x} - 3}{\alpha}$$

Given, $\lim_{x \to -\infty} f(x) = 1$

$$\therefore \lim_{x \to -\infty} \frac{(2\alpha + 3)e^{\alpha x} - 3}{\alpha} = 1$$

$$= -\frac{3}{\alpha} = 1$$

$$\Rightarrow \quad \alpha = -3$$

$$\therefore f(x) = \frac{-3e^{-3x} - 3}{-3} = e^{-3x} + 1$$

Now, $f(-\log_e 2) = e^{-(-3 \log_e 2)} + 1$

$$= e^{\log_e 2^3} + 1$$

$$= 2^3 + 1 = 9$$

**30.** *(a)* We have,

$$\sec x \, dy + \{2(1-x) \tan x + x(2-x)\} dx = 0$$

$$\Rightarrow \frac{dy}{dx} = -[2(1-x) \sin x + x(2-x) \cos x]$$

On variable separable and integration,

$$\int dy = -\int (2(1-x) \sin x + x(2-x) \cos x) \, dx$$

$$y = 2\int (x-1) \sin x \, dx + \int (x-2)x \cos x \, dx$$

$$= 2[-(x-1) \cos x + \int \cos x \, dx] + \int (x-2)x \cos x \, dx$$

$$= 2[\sin x - (x-1) \cos x] + (x-2)x \sin x - \int (2x-2) \sin x \, dx$$

$$= 2(\sin x - (x-1) \cos x) + (x-2)x \sin x - 2[-(x-1) \cos x + \sin x] + C$$

$$y = (x-2)x \sin x + C$$

Given, $y(0) = 2$

$\Rightarrow \quad 2 = c$

$\therefore \quad y(x) = (x-2)x \sin x + 2$

So, $\quad y(2) = 2$

**31.** *(a)* Given, differential equations are

$$\frac{dx}{dt} + ax = 0 \quad \text{...(i)}$$

and $\quad \frac{dy}{dt} + by = 0 \quad$ ...(ii)

From Eq. (i), $\frac{dx}{x} = -adt$

$$\Rightarrow \quad \int \frac{dx}{x} = -a\int dt$$

$$\Rightarrow \quad \ln|x| = -at + C,$$

where $C$ is constant of integration

At $t = 0$, $x = 2$ [given]

$$\ln 2 = 0 + C \Rightarrow C = \ln(2)$$

So, $\quad \ln|x| = -at + \ln(2)$

$$\Rightarrow \quad \frac{x}{2} = e^{-at}$$

$$\Rightarrow \quad x = 2 \cdot e^{-at} \quad \text{...(iii)}$$

From Eq. (ii), we get

$$\frac{dy}{y} = -bdt \Rightarrow \int \frac{dy}{y} = -b\int dt$$

$\Rightarrow \ln|y| = -bt + k,$

where $k$ = constant of integration.

Given, $\quad t = 0, y = 1$

$\therefore \quad \ln|1| = -b \times 0 + k$

$\Rightarrow \quad k = 0$

So, $\quad y = e^{-bt}$ ...(iv)

According to the question,

$3y(1) = 2 \times x(1)$

$\Rightarrow \quad 3e^{-b} = 2(2e^{-a})$

$\Rightarrow \quad e^{a-b} = \frac{4}{3}$ ...(v)

[using multiplication in Eqs. (iii) and (iv)]

For $x(t) = y(t) \Rightarrow 2e^{-at} = e^{-bt}$

$\Rightarrow \quad 2 = e^{(a-b)t} = \left(\frac{4}{3}\right)^t$ [from Eq. (v)]

$\Rightarrow \quad \log_{4/3} 2 = t$

**32.** *(d)* Given, $\frac{dy}{dx} = y + 7$ with initial conditions $y_1(0) = 0$, $y_2(0) = 1$

$\Rightarrow \int \frac{dy}{y+7} = \int dx$

[Taking integration on both sides]

$\Rightarrow \quad \log(y+7) = x + \log c$

$\Rightarrow \quad (y+7) = ce^x$ ($c$ is constant)

Since, $y_1(x)$ and $y_2(x)$ satisfied the given equation, we get

$y_1 + 7 = c_1 e^x$

and $\quad y_2 + 7 = c_2 e^x$

Given, $y_1(0) = 0$, $y_2(0) = 1$

$\therefore \quad 0 + 7 = c_1$ and $1 + 7 = c_2$

$c_1 = 7$ and $c_2 = 8$

$\therefore \quad y_1(x) = 7(e^x - 1)$

and $\quad y_2(x) = 8e^x - 7$

Hence, its point of intersection at

$7e^x - 7 = 8e^x - 7$

$\Rightarrow e^x = 0$, which is impossible.

Thus, $y_1(x)$ and $y_2(x)$ curves not intersect at any point.

So, no point.

**33.** *(d)* We have,

$(1 - x^2y^2)dx = ydx + xdy$, $y(1) = 2$

$\Rightarrow \quad dx = \frac{ydx + xdy}{1-(xy)^2}$

On integrating both sides, we get

$\int dx = \int \frac{d(xy)}{1-(xy)^2}$

$x = \frac{1}{2}\log\left|\frac{1+xy}{1-xy}\right| + C$

As $y(1) = 2$

$1 = \frac{1}{2}\log\left|\frac{1+2}{1-2}\right| + C \Rightarrow C = 1 - \frac{1}{2}\log 3$

Now, substituting $x = 2$ as $y(2) = \alpha$

$2 = \frac{1}{2}\log\left|\frac{1+2\alpha}{1-2\alpha}\right| + 1 - \frac{1}{2}\log 3$

$\Rightarrow 1 + \frac{1}{2}\log 3 = \frac{1}{2}\log\left|\frac{1+2\alpha}{1-2\alpha}\right|$

$\Rightarrow \quad 2 + \log 3 = \log\left|\frac{1+2\alpha}{1-2\alpha}\right|$

$\Rightarrow \left|\frac{1+2\alpha}{1-2\alpha}\right| = 3e^2 \Rightarrow \frac{1+2\alpha}{1-2\alpha} = \pm\frac{3e^2}{1}$

$\Rightarrow \quad \frac{1}{2\alpha} = \frac{\pm 3e^2 + 1}{\pm 3e^2 - 1} \Rightarrow 2\alpha = \frac{\pm 3e^2 - 1}{\pm 3e^2 + 1}$

$\Rightarrow \quad \alpha = \frac{1}{2}\left(\frac{\pm 3e^2 - 1}{1 \pm 3e^2}\right)$

$\therefore \quad \alpha = \frac{3e^2 - 1}{2(1+3e^2)}$ or $\frac{3e^2+1}{2(3e^2-1)}$

**34.** *(a)* Given, $\frac{dy}{dx} + \frac{x+a}{y-2} = 0$

$\Rightarrow \quad (y-2)\,dy + (x+a)\,dx = 0$

$\Rightarrow \quad \frac{y^2}{2} - 2y + \frac{x^2}{2} + ax = c$

$\Rightarrow \quad x^2 + y^2 + 2ax - 4y = c$

Again, $\quad y(1) = 0$

$\therefore \quad 1 + 0 + 2a = c$

$\Rightarrow \quad c = 2a + 1$

So, equation of curve become

$x^2 + y^2 + 2ax - 4y - 2a - 1 = 0$

which is equation of a circle.

Now, radius

$= \sqrt{(-a)^2 + (2)^2 - (-2a-1)}$

$= \sqrt{a^2 + 2a + 5}$

It is given that area of circle is $4\pi$

$\therefore \quad \pi(a^2 + 2a + 5) = 4\pi$

$\Rightarrow (a+1)^2 = 0$

$\Rightarrow \quad a = -1$

Now, curve become

$x^2 + y^2 - 2x - 4y + 1 = 0$ ...(i)

On putting $x = 0$ in Eq. (i), we get

$y^2 - 4y + 1 = 0$

$\Rightarrow y = \frac{4 \pm \sqrt{16-4}}{2} = \frac{4 \pm 2\sqrt{3}}{2}$

$= 2 \pm \sqrt{3}$

So, intersecting point of curve with $Y$-axis are $(0, 2-\sqrt{3})$ and $(0, 2+\sqrt{3})$

$\therefore P(0, 2-\sqrt{3})$ and $Q(0, 2+\sqrt{3})$

Normal at $P$ and $Q$ are

$y - 2 = \sqrt{3}(x-1)$ and

$y - 2 = -\sqrt{3}(x-1)$

Now, these normals intersect $X$-axis at $\left(1 - \frac{2}{\sqrt{3}}, 0\right)$ and $\left(1 + \frac{2}{\sqrt{3}}, 0\right)$, respectively.

$\therefore R\left(1 - \frac{2}{\sqrt{3}}, 0\right)$ and $S\left(1 + \frac{2}{\sqrt{3}}, 0\right)$

So, distance $RS = \frac{4}{\sqrt{3}} = \frac{4\sqrt{3}}{3}$

**35.** *(c)* We have,

$(3y^2 - 5x^2)y\,dx + 2x(x^2 - y^2)dy = 0$

$\Rightarrow \quad \frac{dy}{dx} = \frac{y(5x^2 - 3y^2)}{2x(x^2 - y^2)}$ ...(i)

which is a homogeneous differential equation.

On putting $y = vx$ and $\frac{dy}{dx} = v + x\frac{dv}{dx}$ in Eq. (i), we get

$v + x\frac{dv}{dx} = \frac{vx(5x^2 - 3v^2x^2)}{2x(x^2 - v^2x^2)}$

$\Rightarrow \quad v + x\frac{dv}{dx} = \frac{v(5-3v^2)}{2(1-v^2)}$

$\Rightarrow \quad x\frac{dv}{dx} = \frac{3v - v^3}{2(1-v^2)}$

$\Rightarrow \quad \frac{2(1-v^2)}{3v - v^3}dv = \frac{dx}{x}$

On integrating both the sides, we get

$2\int \frac{(1-v^2)}{3v - v^3}dv = \int \frac{dx}{x}$

$\Rightarrow \quad \frac{2}{3}\int \frac{3 - 3v^2}{3v - v^3}dv = \int \frac{dx}{x}$

$\Rightarrow \quad \log|3v - v^3| = \log(cx)^{3/2}$

$\Rightarrow \quad 3v - v^3 = c^{3/2}x^{3/2}$

$\Rightarrow \quad \frac{3y}{x} - \frac{y^3}{x^3} = c^{3/2}x^{3/2}$

[$\because y = vx$]

When, $x = 1$, $y = 1$

$\therefore \quad 3 - 1 = c^{3/2}$

$\Rightarrow \quad c^{3/2} = 2$

$\therefore \quad \frac{3y}{x} - \frac{y^3}{x^3} = 2x^{3/2}$

$\Rightarrow \quad 3x^2y - y^3 = 2x^{9/2}$

Now, putting $x = 2$, we get

$$12y(2) - y^3(2) = 2 \times 2^{9/2}$$

$$\Rightarrow \quad -(y^3(2) - 12y(2)) = 2^{11/2}$$

$$\Rightarrow \quad |y^3_{(2)} - 12y_{(2)}| = 32\sqrt{2}$$

**36.** *(c)*

> **key idea** The given differential equation is a homogeneous differential equation. To find the solution of such type of differential equation substitute $y = vx$. Then, solve it using integration.

The given differential equation is

$$\frac{dy}{dx} = -\left(\frac{x^2 + 3y^2}{3x^2 + y^2}\right) \quad \ldots(i)$$

On substituting $y = vx$

$$\Rightarrow \quad \frac{dy}{dx} = v + x\frac{dv}{dx}$$

Then, $v + x\dfrac{dv}{dx} = -\left(\dfrac{x^2 + 3v^2x^2}{3x^2 + v^2x^2}\right)$

$$\Rightarrow \quad v + x\frac{dv}{dx} = -\left[\frac{1 + 3v^2}{3 + v^2}\right]$$

$$\Rightarrow \quad x\frac{dv}{dx} = -\frac{(1 + 3v^2)}{3 + v^2} - v$$

$$\Rightarrow \quad x\frac{dy}{dx} = \frac{-(1 + 3v^2 + 3v + v^3)}{3 + v^2}$$

On variables separable and integrating,

$$\int \frac{3 + v^2}{(1 + v)^3} dv = -\int \frac{1}{x} dx$$

We can write on using partial fraction

$$\frac{3 + v^2}{(1 + v)^3} = \frac{1}{v + 1} - \frac{2}{(v + 1)^2} + \frac{4}{(v + 1)^3}$$

$$\therefore \int \left[\frac{1}{v + 1} - \frac{2}{(v + 1)^2} + \frac{4}{(v + 1)^3}\right] dv$$

$$= -\int \frac{1}{x} dx$$

$$\Rightarrow \ln|v + 1| + \frac{2}{(v + 1)} - \frac{2}{(v + 1)^2}$$

$$= -\ln x + C$$

$$\Rightarrow \ln\left|\frac{y}{x} + 1\right| + \frac{2[y/x + 1 - 1]}{(y/x + 1)^2}$$

$$= -\ln x + C$$

$$\Rightarrow \ln\left(\frac{y + x}{x}\right) + \frac{2xy}{(x + y)^2} = -\ln x + C$$

or $\ln(x + y) + \dfrac{2xy}{(x + y)^2} = C \quad \ldots(ii)$

Given, $y(1) = 0$, put in Eq. (ii), we get

$$c = 0$$

Hence, $\ln|x + y| + \dfrac{2xy}{(x + y)^2} = 0$

**37.** *(b)*

> **key idea** This is homogeneous differential equation form $\dfrac{dy}{dx} = f(x, y)$. This equation can be solved by putting $y = vx$, where $v = V(x)$ a function of $x$.

Given differential equation is

$$(x^2 - 3y^2)dx + 3xy\, dy = 0, \; y(1) = 1$$

$$\frac{3y^2 - x^2}{3xy} = \frac{dy}{dx} \quad \ldots(i)$$

which is a homogeneous type.

Let $\quad y = vx$

$$\therefore \quad \frac{dy}{dx} = v + x\frac{dv}{dx}$$

From Eq. (i),

$$\frac{3(vx)^2 - x^2}{3x(vx)} = v + x\frac{dv}{dx}$$

$$\Rightarrow \frac{3v^2 - 1}{3v} - v = \frac{xdv}{dx} \Rightarrow \frac{-1}{3v} = \frac{xdv}{dx}$$

$$\int \frac{dx}{x} = -\int 3v\,dv$$

On integrating both sides, we get

$$\ln x = -3.\frac{v^2}{2} + c$$

which satisfies $y(1) = 1$

$$\therefore \; \ln(1) = -\frac{3}{2}\left(\frac{1}{1}\right)^2 + c \Rightarrow c = \frac{3}{2}$$

$\therefore$ Equation is

$$\ln x = \frac{-3}{2}\left(\frac{y}{x}\right)^2 + \frac{3}{2}$$

$$\Rightarrow \quad 3y^2 = 2x^2\left(\frac{3}{2} - \ln x\right)$$

$$6y^2 = 6x^2 - 4x^2 \ln x$$

$\therefore$ For $x = e$

$$6y^2(e) = 6e^2 - 4e^2 \cdot \ln e$$

$$\Rightarrow \quad 6e^2 - 4e^2 \times 1 = 2e^2$$

**38.** *(b)* Given, $x\,dy = \sqrt{x^2 + y^2}\,dx + ydx$

$$\Rightarrow \quad xdy - ydx = \sqrt{x^2 + y^2}\,dx$$

$$\Rightarrow \quad \frac{x\,dy - ydx}{x^2} = \frac{\sqrt{x^2 + y^2}\,dx}{x^2}$$

On dividing by $x^2$ both sides, we get

$$d\left(\frac{y}{x}\right) = \frac{x\sqrt{1 + \dfrac{y^2}{x^2}}\,dx}{x^2}$$

$$= \sqrt{1 + \left(\frac{y}{x}\right)^2} \cdot \frac{dx}{x}$$

$$\Rightarrow \quad \frac{d\left(\dfrac{y}{x}\right)}{\sqrt{1 + \left(\dfrac{y}{x}\right)^2}} = \frac{dx}{x}$$

On integrating both sides, we get

$$\ln\left(\frac{y}{x} + \sqrt{1 + \left(\frac{y}{x}\right)^2}\right) = \ln(x) + \ln C$$

$\because$ Using $\displaystyle\int \frac{dx}{\sqrt{1 + x^2}} = \ln(x + \sqrt{1 + x^2})$

$$\Rightarrow \ln\left(\frac{y}{x} + \frac{\sqrt{x^2 + y^2}}{x}\right) = \ln(Cx)$$

$$\Rightarrow \quad y + \sqrt{x^2 + y^2} = Cx^2 \quad \ldots(i)$$

According to given condition; when $x = 1$, then $y = 0$

From Eq. (i), $0 + \sqrt{1 + 0}$

$= C \times (1)^2 \Rightarrow C = 1$

So, equation of curve is

$$y + \sqrt{x^2 + y^2} = x^2 \quad \text{[from Eq. (i)]}$$

Now, when $x = 2$, then $y = \alpha$ [given]

So, $\alpha + \sqrt{4 + \alpha^2} = 4$

$$\Rightarrow \quad 4 + \alpha^2 = 16 + \alpha^2 - 8\alpha$$

$$\Rightarrow \quad 8\alpha = 12 \Rightarrow \alpha = 3/2$$

**39.** *(c)* From given $A_1 = 2A_2$

From the graph, $A_1 + A_2 = xy - 8$

$$\Rightarrow \frac{3}{2}A_1 = xy - 8 \Rightarrow A_1 = \frac{2}{3}xy - \frac{16}{3}$$

$$\Rightarrow \int_4^x f(x)dx = \frac{2}{3}xy - \frac{16}{3}$$

On differentiating both sides w.r.t. $x$, we get

$$f(x) = \frac{2}{3}\left[x \cdot \frac{dy}{dx} + y\right] - 0$$

$$\Rightarrow \frac{2}{3}x\frac{dy}{dx} = \frac{y}{3} \Rightarrow 2\int \frac{dy}{y} = \int \frac{dx}{x}$$

$\Rightarrow 2\ln y = \ln x + \ln c$

$\Rightarrow \quad y^2 = cx,$

As, $f(4) = 2 \Rightarrow c = 1$

$\therefore$ The required curve is, $y^2 = x$

Slope of normal will be $-6$.

So, equation of normal is

$$y = -6x - \frac{1}{2}(-6) - \frac{1}{4}(-6)^3$$

$\Rightarrow y + 6x = 57$

$\therefore$ The point (10, – 4) will not lies on the above equation.

**40.** *(b)* Let the slope of the tangent at any point of $P(x, y)$ is $dy/dx$.

According to the question, $\frac{dy}{dx} \propto \left(-\frac{y}{x}\right)$

$$\Rightarrow \frac{dy}{dx} = \frac{-ky}{x}$$

[$k$ be a constant]

$$\Rightarrow \frac{dy}{y} = -k\frac{dx}{x}$$

On integrating on both sides, we get

$$\int \frac{dy}{y} = -k\int \frac{dx}{x}$$

$\Rightarrow \ln|y| = -k\ln|x| + C$

Since, the above equation, satisfy the points (1, 2) and (8, 1),

$\ln 2 = -k \times 0 + C \Rightarrow C = \ln 2,$

then $\ln 1 = -k\ln 8 + \ln 2$

$$\Rightarrow \quad k = \frac{1}{3}$$

Thus, at $x = 1/8$

$$\ln|y| = -\frac{1}{3}\ln\left(\frac{1}{8}\right) + \ln 2$$

$$= \ln 2 + \ln 2 = 2\ln 2$$

$\Rightarrow \quad |y| = 4$

**41.** *(b)* We have,

$$\frac{dy}{dx} = \frac{2e^{2x} - 6e^{-x} + 9}{2 + 9e^{-2x}}$$

$$\frac{dy}{dx} = e^{2x} - \frac{6e^x}{2e^{2x} + 9}$$

$$dy = \left(e^{2x} - \frac{6e^x}{2e^{2x} + 9}\right)dx$$

On integrating both sides, we get

$$y = \frac{e^{2x}}{2} - \int \frac{6e^x}{2e^{2x} + 9}dx$$

$$y = \frac{e^{2x}}{2} - \sqrt{2}\tan^{-1}\left(\frac{\sqrt{2}e^x}{3}\right) + c$$

$\because$ Given curve passes through $\left(0, \frac{1}{2} + \frac{\pi}{2\sqrt{2}}\right)$.

$$\therefore \frac{1}{2} + \frac{\pi}{2\sqrt{2}} = \frac{1}{2} - \sqrt{2}\tan^{-1}\left(\frac{\sqrt{2}}{3}\right) + c$$

$$c = \frac{\pi}{2\sqrt{2}} + \sqrt{2}\tan^{-1}\left(\frac{\sqrt{2}}{3}\right)$$

$$\Rightarrow c = \sqrt{2}\left[\frac{\pi}{4} + \tan^{-1}\left(\frac{\sqrt{2}}{3}\right)\right]$$

So, the curve passes through $\left(\alpha, \frac{1}{2}e^{2\alpha}\right)$.

$$\therefore \frac{1}{2}e^{2\alpha} = \frac{e^{2\alpha}}{2} - \sqrt{2}\tan^{-1}\left(\frac{\sqrt{2}e^\alpha}{3}\right) + \sqrt{2}\left[\frac{\pi}{4} + \tan^{-1}\left(\frac{\sqrt{2}}{3}\right)\right]$$

$$\Rightarrow e^\alpha = \frac{3}{\sqrt{2}}\left(\frac{3+\sqrt{2}}{3-\sqrt{2}}\right)$$

**42.** *(a)* Given,

$(x - y^2)\,dx + y\,(5x + y^2)\,dy = 0$

$$\Rightarrow \quad y\frac{dy}{dx} = \frac{y^2 - x}{5x + y^2}$$

Let $\quad y^2 = z$

$$\Rightarrow \quad 2y\frac{dy}{dx} = \frac{dz}{dx} \Rightarrow y\frac{dy}{dx} = \frac{1}{2}\frac{dz}{dx}$$

$$\Rightarrow \quad \frac{1}{2}\frac{dz}{dx} = \frac{z - x}{5x + z}$$

Now, substitute $z = vx$

$$\Rightarrow \quad \frac{dz}{dx} = v + x\frac{dv}{dx}$$

$$\Rightarrow \quad \frac{1}{2}\left\{v + x\frac{dv}{dx}\right\} = \frac{v - 1}{5 + v}$$

$$\Rightarrow \quad v + x\frac{dv}{dx} = \frac{2v - 2}{5 + v}$$

$$\Rightarrow x\left(\frac{dv}{dx}\right) = \frac{2v - 2}{5 + v} - v$$

$$= \frac{2v - 2 - 5v - v^2}{5 + v}$$

$$\Rightarrow \quad x\frac{dv}{dx} = \frac{-3v - 2 - v^2}{5 + v}$$

$$\Rightarrow \int \frac{(5 + v)\,dv}{v^2 + 3v + 2} = -\int \frac{dx}{x}$$

$$\Rightarrow \int \frac{4dv}{v + 1} - 3\int \frac{dv}{v + 2} = -\int \frac{dx}{x}$$

On integrating both sides, we get

$4\log|v + 1| - 3\log|v + 2|$

$= -\log|x| + \log C$

$$\Rightarrow \log|(v + 1)^4| - \log|(v + 2)^3| = \log\frac{C}{x}$$

$$\Rightarrow \log\left|\frac{(v + 1)^4}{(v + 2)^3}\right| = \log\frac{C}{x}$$

$$\Rightarrow \frac{\left|\left(\frac{y^2}{x} + 1\right)^4\right|}{\left|\left(\frac{y^2}{x} + 2\right)^3\right|} = \frac{C}{x} \quad \left[\because v = \frac{z}{x} = \frac{y^2}{x}\right]$$

$\Rightarrow |(y^2 + x)^4| = C\,|(y^2 + 2x)^3|$

**43.** *(c)*

> **key idea** (i) Apply variable separable method for solving given differential education.
> (ii) Use $\int \frac{1}{x^2 + a^2}dx = \frac{1}{a}\tan^{-1}\frac{x}{a}$

$$(1 + e^{2x})\frac{dy}{dx} + 2(1 + y^2)e^x = 0 \quad \ldots\text{(i)}$$

$$\Rightarrow \quad \frac{dy}{(1 + y^2)} = -\frac{2e^x}{1 + e^{2x}}dx$$

$$\Rightarrow \quad \int \frac{dy}{1 + y^2} = -2\int \frac{e^x}{1 + (e^x)^2}dx$$

$\Rightarrow \tan^{-1}y = -2\tan^{-1}e^x + c \quad \ldots\text{(ii)}$

On putting $x = 0$ and $y = 0$ as $y(0) = 0$, we get

$$0 = -2\tan^{-1}(1) + c \Rightarrow c = \frac{\pi}{2}$$

$$\therefore \tan^{-1}y = -2\tan^{-1}e^x + \frac{\pi}{2}$$

[from Eq. (ii)]

$$\Rightarrow y = \tan\left(\frac{\pi}{2} - 2\tan^{-1}e^x\right)$$

$\Rightarrow \quad y = \cot(2\tan^{-1}e^x)$

When, $x = \log_e\sqrt{3}$

$$y = \cot(2\tan^{-1}\sqrt{3}) = \cot\left(2 \times \frac{\pi}{3}\right)$$

$$= -\cot\frac{\pi}{3} = -\frac{1}{\sqrt{3}} \quad \ldots\text{(iii)}$$

From Eq. (i), when $x = 0$

$$(1 + 1)\frac{dy}{dx} + 2(1 + 0) \times 1 = 0$$

$$\Rightarrow \quad \frac{dy}{dx} = -1 \quad \ldots\text{(iv)}$$

Therefore, $6[y'(0)+(y(\ln\sqrt{3})^2]$

$$=6\left[(-1)+\left(-\frac{1}{\sqrt{3}}\right)^2\right]$$

[from Eqs. (iii) and (iv)]

$$=6\left[-1+\frac{1}{3}\right]=6\left[-\frac{2}{3}\right]=-4$$

**44.** (a) $x\frac{dy}{dx}-y=\sqrt{y^2+16x^2}$

$$\frac{dy}{dx}=\frac{y+\sqrt{y^2+16x^2}}{x}$$

Let $y=vx \Rightarrow \frac{dy}{dx}=v+x\frac{dv}{dx}$

$$\therefore \quad v+x\frac{dv}{dx}=\frac{vx+\sqrt{v^2x^2+16x^2}}{x}$$

$$\Rightarrow \quad v+x\frac{dv}{dx}=v+\sqrt{v^2+16}$$

$$\Rightarrow \int\frac{dv}{\sqrt{v^2+16}}=\int\frac{dx}{x}$$

$$\Rightarrow \quad \ln|v+\sqrt{v^2+16}|=\ln x+\ln c$$

$$\Rightarrow \quad \frac{y}{x}+\frac{\sqrt{y^2+16x^2}}{x}=xc$$

$$\Rightarrow \quad y+\sqrt{y^2+16x^2}=cx^2 \quad \dots \text{(i)}$$

$\because \quad y(1)=3$

$\therefore \quad 3+\sqrt{9+16}=c$

$\Rightarrow \quad 3+5=c$

$\Rightarrow \quad c=8$

$y+\sqrt{y^2+16x^2}=8x^2$ [from Eq. (i)]

On putting $x=2$

$$\sqrt{y^2+64}=32-y$$

$\Rightarrow \quad y^2+64=(32)^2+y^2-64y$

$\Rightarrow \quad 64(1+y)=32\times 32$

$\Rightarrow \quad y=16-1=15$

$y(2)=15$

**45.** (a) Given, $\left[\frac{x}{\sqrt{x^2-y^2}}+e^{y/x}\right]x\frac{dy}{dx}$

$$=x+\left[\frac{x}{\sqrt{x^2-y^2}}+e^{y/x}\right]y$$

$$\Rightarrow \left[\frac{1}{\sqrt{1-\frac{y^2}{x^2}}}+e^{y/x}\right]\frac{dy}{dx}$$

$$=1+\left[\frac{1}{\sqrt{1-\frac{y^2}{x^2}}}+e^{y/x}\right]\frac{y}{x}$$

On putting $y=vx$

$$\left(\frac{1}{\sqrt{1-v^2}}+e^v\right)\left(v+x\frac{dv}{dx}\right)$$

$$=1+\left(\frac{1}{\sqrt{1-v^2}}+e^v\right)v$$

$$\Rightarrow \quad x\left(\frac{1}{\sqrt{1-v^2}}+e^v\right)\frac{dv}{dx}=1$$

$$\Rightarrow \sin^{-1}v+e^v=\ln x+C$$

$$\Rightarrow \quad \sin^{-1}\left(\frac{y}{x}\right)+e^{y/x}=\ln x+C$$

Since, curve passes through (1, 0).

$\therefore$ At $x=1$, $y=0$

$$0+e^0=0+C$$

$\Rightarrow \quad C=1$

At $(2\alpha,\alpha)$ $\sin^{-1}\left(\frac{y}{x}\right)+e^{y/x}=\ln x+1$

$$\Rightarrow \quad \frac{\pi}{6}+e^{1/2}-1=\ln(2\alpha)$$

$$\Rightarrow \quad \alpha=\frac{1}{2}e^{\left(\frac{\pi}{6}+\sqrt{e}-1\right)}$$

**46.** (d) $\frac{dy}{dx}+\frac{2^{x-y}(2^y-1)}{2^x-1}=0$

$$\Rightarrow \quad \frac{dy}{dx}+\frac{2^x}{2^x-1}\times\frac{2^y-1}{2^y}=0$$

$$\Rightarrow \quad \frac{dy}{dx}=\frac{-2^x(2^y-1)}{2^y(2^x-1)}$$

$$\Rightarrow \quad \int\frac{2^y}{2^y-1}dy=-\int\frac{2^x}{2^x-1}dx$$

$$\Rightarrow \frac{1}{\ln 2}\int\frac{2^y\ln 2}{2^y-1}dy=\frac{-1}{\ln 2}\int\frac{2^x\ln 2}{2^x-1}dx$$

$$\Rightarrow \ln|2^y-1|=-\ln|2^x-1|+c$$

At $x=1$, $y=1$,

[Given]

$$\ln 1=-\ln 1+c$$

$$0=-0+c \Rightarrow c=0$$

$$\Rightarrow \ln|2^y-1|=-\ln|2^x-1|$$

$$\Rightarrow \ln((2^x-1)(2^y-1))=0$$

$$\Rightarrow (2^x-1)(2^y-1)=1$$

At $\quad x=2$

$$(2^y-1)(4-1)=1 \Rightarrow 2^y=1+\frac{1}{3}=\frac{4}{3}$$

$$y=\log_2 4/3=\log_2 4-\log_2 3$$

$$=2-\log_2 3$$

**47.** (b) $\frac{dy}{dx}=\frac{ax-by+a}{bx+cy+a}$

$$\Rightarrow bx\,dy+cy\,dy+a\,dy=ax\,dx-by\,dx+a\,dx$$

$$\Rightarrow cy\,dy+a\,dy-ax\,dx-a\,dx+b(xdy+y\,dx)=0$$

By integrating both sides, we get

$$c\int ydy+a\int dy-a\int xdx-a\int dx+b\int d(xy)=k$$

$$\Rightarrow \frac{cy^2}{2}+ay-\frac{ax^2}{2}-ax+bxy=k$$

$$\Rightarrow ax^2-cy^2+2ax-2ay-2bxy=K$$

(Where, $K=-2k$)

Above equation represents circle, if $a=-c$ and $b=0$

$$\Rightarrow ax^2+ay^2+2ax-2ay=K$$

$$\Rightarrow x^2+y^2+2x-2y=\lambda$$

$\left(\text{Where, } \lambda=\frac{K}{a}\right)$

It passes through (2, 5). So, $\lambda=23$

Equation of circle is

$$x^2+y^2+2x-2y-23=0$$

Centre of circle is $O(-1, 1)$ and radius is

$$r=\sqrt{(-1)^2+(1)^2+23}=5$$

Minimum distance of point $P(11, 6)$ from circle is,

$$PQ=OP-r=\sqrt{(12)^2+5^2}-5$$

$$=13-5=8 \text{ (as, } Q \text{ is point on circle)}$$

**48.** (b) Given, $2x^2\frac{dy}{dx}-2xy+3y^2=0$

$$\frac{dy}{dx}=\frac{2xy-3y^2}{2x^2} \quad \dots \text{(i)}$$

Let $y=vx$

$$\Rightarrow \quad \frac{dy}{dx}=v+x\frac{dv}{dx}$$

On substituting the value of $y$ and $\frac{dy}{dx}$ in Eq. (i), we get

$$v+x\frac{dv}{dx}=\frac{2xvx-3v^2x^2}{2x^2}$$

$$\Rightarrow \quad v+x\frac{dv}{dx}=v-\frac{3}{2}v^2$$

$\Rightarrow x\dfrac{dv}{dx}=-\dfrac{3}{2}v^2 \Rightarrow \int \dfrac{1}{v^2}dv=-\dfrac{3}{2}\int\dfrac{1}{x}dx$

$\Rightarrow \quad -\dfrac{1}{v}=-\dfrac{3}{2}\log x+c$

$\dfrac{x}{y}-\dfrac{3}{2}\log x+c=0$

$\therefore\ y(e)=\dfrac{e}{3}\Rightarrow \dfrac{e}{e/3}-\dfrac{3}{2}\log e+c=0$

$\Rightarrow\ 3-\dfrac{3}{2}+c=0$

$\Rightarrow \quad c=-\dfrac{3}{2}$

Now, $\dfrac{x}{y}-\dfrac{3}{2}\log x-\dfrac{3}{2}=0$

When, $x=1$, then

$\dfrac{1}{y}-\dfrac{3}{2}\log 1-\dfrac{3}{2}=0$

$\Rightarrow \quad \dfrac{1}{y}=\dfrac{3}{2}\ \Rightarrow\ y=\dfrac{2}{3}$

**49.** *(c)* Given,

$y^2\,dx+(x^2-xy+y^2)\,dy=0$

$\dfrac{dy}{dx}=\dfrac{-y^2}{x^2-xy+y^2}$ ...(i)

Let $\quad y=vx$

$\Rightarrow \dfrac{dy}{dx}=v+x\dfrac{dv}{dx}$

On substituting the value of $y$ and $\dfrac{dy}{dx}$ in Eq. (i), we get

$v+x\dfrac{dv}{dx}=-\dfrac{v^2x^2}{x^2-xvx+v^2x^2}$

$\Rightarrow v+x\dfrac{dv}{dx}=\dfrac{-v^2}{1-v+v^2}$

$\Rightarrow \quad x\dfrac{dv}{dx}=\dfrac{-v^2}{1-v+v^2}-v$

$\Rightarrow \quad x\dfrac{dv}{dx}=\dfrac{-v^2-v+v^2-v^3}{1-v+v^2}$

$\Rightarrow \int\dfrac{(1-v+v^2)}{v+v^3}dv=-\int\dfrac{1}{x}dx$

$\Rightarrow \int\left(\dfrac{1+v^2}{v(1+v^2)}-\dfrac{v}{v(1+v^2)}\right)dv=-\int\dfrac{1}{x}dx$

$\Rightarrow \quad -\log v+\tan^{-1}v=\log_e x+c$

$\Rightarrow -\log_e\dfrac{y}{x}+\tan^{-1}\dfrac{y}{x}=\log_e x+c$

When, $x=1$, $y=1$

$0+\dfrac{\pi}{4}=C\Rightarrow C=\dfrac{\pi}{4}$

$\Rightarrow \tan^{-1}\left(\dfrac{y}{x}\right)=\log_e\dfrac{y}{x}+\log_e x+\dfrac{\pi}{4}$

$\Rightarrow \tan^{-1}\left(\dfrac{y}{x}\right)=\log_e y+\dfrac{\pi}{4}$

The lines passes through the point $(\alpha, \sqrt{3}\alpha)$

$\tan^{-1}\left(\dfrac{\sqrt{3}\alpha}{\alpha}\right)=\log_e\sqrt{3}\alpha+\dfrac{\pi}{4}$

$\Rightarrow \quad \dfrac{\pi}{3}-\dfrac{\pi}{4}=\log_e\sqrt{3}\alpha$

$\Rightarrow \quad \log_e(\sqrt{3}\alpha)=\pi/12$

**50.** *(d)* Given, slope of normal

$=\dfrac{x^2}{xy-x^2y^2-1}$

$\Rightarrow \dfrac{-dx}{dy}=\dfrac{x^2}{xy-x^2y^2-1}$

$\Rightarrow \quad \dfrac{dy}{dx}=\dfrac{x^2y^2-xy+1}{x^2}$ ...(i)

On putting $xy=t\Rightarrow \dfrac{xdy}{dx}+y=\dfrac{dt}{dx}$

From Eq. (i), we get

$\dfrac{1}{x}\left(\dfrac{dt}{dx}-\dfrac{t}{x}\right)=\dfrac{t^2-t+1}{x^2}$

$\Rightarrow \quad \dfrac{dt}{dx}-\dfrac{t}{x}=\dfrac{t^2}{x}-\dfrac{t}{x}+\dfrac{1}{x}$

$\Rightarrow \quad \dfrac{dt}{dx}=\dfrac{t^2+1}{x}\Rightarrow \int\dfrac{dt}{t^2+1}=\int\dfrac{dx}{x}+C$

$\Rightarrow \quad \tan^{-1}t=\log x+C$

$\Rightarrow \tan^{-1}(xy)=\log x+C$ ...(ii)

Now, Eq. (ii) passes through (1, 1)

$\therefore \quad C=\dfrac{\pi}{4}$

So, Eq. (ii) becomes,

$\tan^{-1}(xy)=\log x+\dfrac{\pi}{4}$

On putting $x=e$

$\therefore\ e\cdot y(e)=\tan\left(1+\dfrac{\pi}{4}\right)=\dfrac{1+\tan(1)}{1-\tan(1)}$

**51.** *(a)* Equation of tangent to the curve $C$ at point

$P(x, y)$ is $Y-y=m(X-x)$ ...(i)

Now, Eq. (i) passes through $X$-axis at $Q$.

$\therefore$Co-ordinates of $Q=\left(x-\dfrac{y}{m}, 0\right)$

Similarly, Eq. (i) passes through $Y$-axis at $M$ (mid-point of $P$ and $Q$)

$\therefore$Co-ordinates of $M=(0, y-mx)$

($\because M$ bisects $PQ$)

$\therefore\ x-\dfrac{y}{m}+x=0\Rightarrow\ 2x=y\dfrac{dy}{dx}$

$\left(\because m=\dfrac{dy}{dx}\right)$

$\Rightarrow \dfrac{dx}{x}=\dfrac{2dy}{y}\Rightarrow 2\log y=\log x+\log C$

$\Rightarrow \quad y^2=xC$ ...(ii)

Now, Eq. (ii) passes through (3, 3).

$\therefore \quad C=3$

So, the curve $C$ is $y^2=3x$

$\Rightarrow$ Length of latus rectum $=3$ and focus $=(3/4, 0)$

**52.** *(b)* Given,

$y\dfrac{dy}{dx}=x\left[\dfrac{y^2}{x^2}+\dfrac{\phi(y^2/x^2)}{\phi'(y^2/x^2)}\right]$ ...(i)

Let $t=\dfrac{y}{x}\Rightarrow y=xt\Rightarrow \dfrac{dy}{dx}=t+x\dfrac{dt}{dx}$

$\therefore$ Eq. (i) becomes

$t\left(t+x\dfrac{dt}{dx}\right)=\left(t^2+\dfrac{\phi(t^2)}{\phi'(t^2)}\right)$

$\Rightarrow xt\dfrac{dt}{dx}=\dfrac{\phi(t^2)}{\phi'(t^2)}\Rightarrow \dfrac{t\,\phi'(t^2)}{\phi(t^2)}dt=\dfrac{dx}{x}$

On integrating both sides, we get

$\int\dfrac{t\,\phi'(t^2)}{\phi(t^2)}dt=\int\dfrac{dx}{x}$

Let $\phi(t^2)=u\Rightarrow t\phi'(t^2)\,dt=\dfrac{du}{2}$

$\therefore \quad \dfrac{1}{2}\int\dfrac{du}{u}=\int\dfrac{dx}{x}$

$\Rightarrow \quad \dfrac{1}{2}\ln u=\ln x+C$

$\Rightarrow \quad \dfrac{1}{2}\ln\phi(t^2)=\ln x+C$

$\Rightarrow \quad \dfrac{1}{2}\ln\left(\phi\left(\dfrac{y^2}{x^2}\right)\right)=\ln x+C$

If $x=1$, $y=-1$, then $C=\dfrac{1}{2}\ln(\phi(1))$

$\therefore \quad \dfrac{1}{2}\ln\left(\phi\left(\dfrac{y^2}{x^2}\right)\right)=\ln x+\dfrac{1}{2}\ln(\phi(1))$

If $x=2$, then

$$\ln\left(\phi\left(\frac{y^2}{4}\right)\right)=\ln 4+\ln[\phi(1)]$$

or $\phi\left(\frac{y^2}{4}\right)=4\phi(1)$

**53.** (a) Given,

$dy/dx=2(y+2\sin x-5)x-2\cos x$, $y(0)=7$

$$\Rightarrow \frac{dy}{dx}+2\cos x=2(y+2\sin x-5)x \quad \ldots\text{(i)}$$

Let $y+2\sin x-5=t$

$$\Rightarrow \frac{dy}{dx}+2\cos x=\frac{dt}{dx}$$

Then, Eq. (i) becomes

$$\frac{dt}{dx}=2tx \Rightarrow \frac{dt}{t}=2x\,dx$$

On integrating both sides, we get

$\ln t=x^2+C$

$$\Rightarrow \ln(y+2\sin x-5)=x^2+C \quad \ldots\text{(ii)}$$

Given, $y(0)=7$

$\Rightarrow \ln(7+0-5)=0+C$

$\Rightarrow C=\ln 2$

$\therefore$ From Eq. (ii), we get

$\ln(y+2\sin x-5)=x^2+\ln 2$

Now, at $x=\pi$

$\ln(y(\pi)+2\sin\pi-5)=\pi^2+\ln 2$

$\Rightarrow y(\pi)-5=e^{\pi^2+\ln 2}$

$\Rightarrow y(\pi)=2e^{\pi^2}+5$

**54.** (c) We have, $2x^2\,dy+(e^y-2x)\,dx=0$

$$\Rightarrow \frac{dy}{dx}+\frac{e^y-2x}{2x^2}=0$$

$$\Rightarrow \frac{dy}{dx}+\frac{e^y}{2x^2}-\frac{1}{x}=0$$

$$e^{-y}\frac{dy}{dx}-\frac{e^{-y}}{x}=-\frac{1}{2x^2} \quad \ldots\text{(i)}$$

$$e^{-y}=t \quad \ldots\text{(ii)}$$

$-e^{-y}\,dy=dt$

$$dy=-\left(\frac{dt}{t}\right) \quad \ldots\text{(iii)}$$

$$\Rightarrow \frac{-dt}{dx}-\frac{t}{x}=-\frac{1}{2x^2}$$

[from Eq. (i)]

$$\Rightarrow x\,dt+t\,dx=\frac{dx}{2x}$$

$$\Rightarrow \int d(xt)=\int\frac{dx}{2x}$$

$$\Rightarrow xt=\frac{1}{2}\log(x)+\frac{c}{2}$$

$\Rightarrow 2xe^{-y}=\log x+c$

When, $x=e$, $y=1$

$2e\cdot e^{-1}=\log e+c \Rightarrow c=1$

$\therefore 2xe^{-y}=\log x+1$

When, $x=1$,

$2e^{-y}=0+1 \Rightarrow e^y=2$

$\Rightarrow y=\log_e 2$

**55.** (b) Given differential equation,

$(x-x^3)\,dy=(y+yx^2-3x^4)\,dx$

$\Rightarrow x\,dy-x^3\,dy=y\,dx+yx^2\,dx-3x^4\,dx$

$\Rightarrow x\,dy-y\,dx=x^2(y\,dx+x\,dy-3x^2\,dx)$

$$\Rightarrow \frac{x\,dy-y\,dx}{x^2}=(y\,dx+x\,dy)-3x^2\,dx$$

$$\Rightarrow d\left(\frac{y}{x}\right)=d(xy)-d(x^3)$$

On integrating both sides, we get

$$\frac{y}{x}=xy-x^3+c \quad \ldots\text{(i)}$$

Given, $f(3)=3$

$$\frac{3}{3}=3\times3-3^3+c$$

$\Rightarrow 1=-18+c \Rightarrow c=19$

From Eq. (i),

$$\frac{y}{x}=xy-x^3+19$$

At, $x=4$

$$\frac{y}{4}=4y-64+19$$

$\Rightarrow y=16y-180 \Rightarrow 15y=180$

$\Rightarrow y=12$

Hence, $y(4)=12$.

**56.** (d) $\frac{dy}{dx}=1+xe^{y-x}$, $-\sqrt{2}<x<\sqrt{2}$

Let $y-x=t \Rightarrow \frac{dy}{dx}-1=\frac{dt}{dx}$

$$\Rightarrow \frac{dt}{dx}=xe^t \Rightarrow \int e^{-t}\,dt=\int x\,dx$$

$$-e^{-t}=\frac{x^2}{2}+C \Rightarrow y(0)=0$$

$\Rightarrow t=0, x=0 \Rightarrow -1=C$

$-2e^{-t}=x^2-2$

$$\Rightarrow 2e^{x-y}=2-x^2 \Rightarrow e^{x-y}=\frac{2-x^2}{2}$$

$$\Rightarrow x-y=\log_e\left(\frac{2-x^2}{2}\right)$$

$$\Rightarrow y=x-\log_e\left(\frac{2-x^2}{2}\right)$$

$$\frac{dy}{dx}=1+\left(\frac{2}{2-x^2}\right)x=\frac{-x^2+2x+2}{2-x^2}$$

$$=\frac{-(x^2-2x+1)+3}{2-x^2}=\frac{3-(x-1)^2}{2-x^2}$$

$-\sqrt{2}$   $-$   $1-\sqrt{3}$   $+$   $\sqrt{2}$

Minimum value

Minimum value $=x=(1-\sqrt{3})$

$$\Rightarrow y=(1-\sqrt{3})-\log_e\left(\frac{2-(1-\sqrt{3})^2}{2}\right)$$

$y=(1-\sqrt{3})-\log_e(\sqrt{3}-1)$

**57.** (b) Given, $2xy\frac{dy}{dx}=y^2-x^2$, $x>0$

$$\frac{dy}{dx}=\left(\frac{y^2-x^2}{2xy}\right), x>0$$

Let $y=vx$

$$\frac{dy}{dx}=v+x\frac{dv}{dx}$$

$$v+x\frac{dv}{dx}=\frac{v^2x^2-x^2}{2x\cdot vx}=\frac{v^2-1}{2v}$$

$$\Rightarrow x\frac{dv}{dx}=\frac{v^2-1}{2v}-v$$

$$\Rightarrow x\frac{dv}{dx}=\frac{-(v^2+1)}{2v}$$

$$\Rightarrow \int-\left(\frac{2v}{v^2+1}\right)dv=\int\frac{dx}{x}$$

$\Rightarrow -\log_e|v^2+1|=\log_e x+c$

$\Rightarrow \log_e|(v^2+1)x|=c$

$$\Rightarrow \left(\frac{y^2}{x^2}+1\right)x=c$$

$$\Rightarrow \frac{(y^2+x^2)}{x}=c$$

$\Rightarrow y^2+x^2=cx$

Similarly, for second curve,

$x^2+y^2=cy$

Both passes through (1, 1),

$C_1 \Rightarrow 1+1=c_1 \Rightarrow c_1=2$

$C_1 \Rightarrow x^2+y^2=2x$

$C_2 \Rightarrow x^2+y^2=2y$

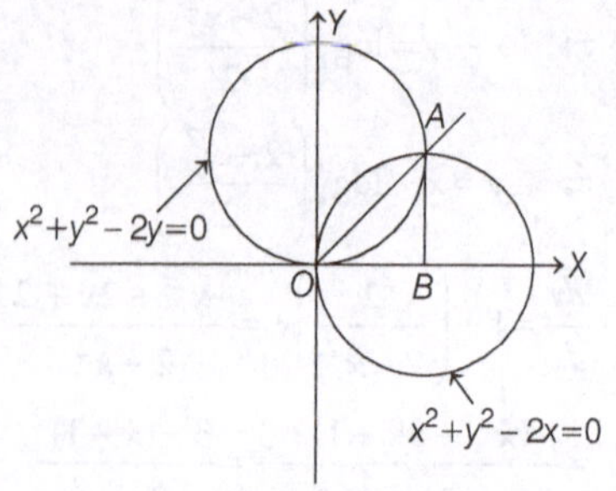


$\therefore$ Required area $= 2\left[\dfrac{\pi \cdot 1^2}{4} - \dfrac{1}{2}\cdot 1 \cdot 1\right]$

$= 2\left[\left(\dfrac{\pi - 2}{4}\right)\right] = \dfrac{\pi}{2} - 1$

**58.** *(a)* Given, $f(x) = \int_0^x e^t f(t)dt + e^x$ ...(i)

Since, $f(x)$ is differentiable function.
Differentiate Eq. (i),

$f'(x) = e^x f(x) + e^x$

[Using Newton's Leibnitz theorem]

$f'(x) = e^x(f(x) + 1)$

$\Rightarrow \dfrac{f'(x)}{f(x)+1} = e^x$

On integrating both sides, we get

$\int \dfrac{f'(x)}{f(x)+1}dx = \int e^x dx + C$

$\log(f(x)+1) = e^x + C \quad [\because u = f(x)+1]$

Now, $f(x) = \int_0^x e^t f(t)dt + e^x$

$\Rightarrow f(0) = e^0 = 1$

On putting $x = 0$ in Eq. (i), we get

$\log(2) = e^0 + C \Rightarrow C = \log(2) - 1$

From Eq. (i), we get

$\log(f(x)+1) = e^x + \log 2 - 1$

$f(x) + 1 = e^{e^x + \log 2 - 1}$

$= e^{e^x} \cdot e^{\log 2} \cdot e^{-1}$

$f(x) + 1 = 2e^{e^x} \cdot e^{-1} = 2e^{e^x - 1}$

$\therefore \quad f(x) = 2e^{e^x - 1} - 1$

**59.** *(c)* Given, slope of tangent line to curve at $(x, y)$ is $\dfrac{xy^2 + y}{x}$

i.e $\dfrac{dy}{dx} = \dfrac{xy^2 + y}{x} \Rightarrow \dfrac{dy}{dx} = y^2 + \dfrac{y}{x}$

$\Rightarrow \quad xdy = xy^2 dx + ydx$

$\Rightarrow \quad xdy - ydx = xy^2 dx$

$\Rightarrow \quad \dfrac{xdy - ydx}{y^2} = xdx$

On integrating, both sides, we get

$\dfrac{-x}{y} = \dfrac{x^2}{2} + C$ ...(i)

The curve intersect line at $x = -2$
Then, $x = -2$, is satisfied by $x + 2y = 4$
Hence, $(-2) + 2y = 4$
Gives, $y = 3$

$\therefore$ Curve passes through $(-2, 3)$.
Use $(-2, 3)$ is Eq. (i), we get

$\dfrac{2}{3} = \dfrac{(2)^2}{2} + C \Rightarrow C = \dfrac{-4}{3}$

$\therefore$ The curve is $\dfrac{-x}{y} = \dfrac{x^2}{2} + \dfrac{-4}{3}$ ...(ii)

It also passes through $(3, y)$.
Put $(3, y)$ in Eq. (ii), we get

$\dfrac{-3}{y} = \dfrac{(3)^2}{2} - \dfrac{4}{3} \Rightarrow y = -\dfrac{18}{19}$

**60.** *(a)* Let $x$ be the number of bacteria at any time $t$.

Given that $\dfrac{dx}{dt} \propto x$

$\left[\because \text{Rate of growth} = \dfrac{dx}{dt}\right]$

$\Rightarrow \quad \dfrac{dx}{dt} = \lambda x$

$\Rightarrow \quad \dfrac{dx}{x} = \lambda dt$

After integrating it, we get

$\log x = \lambda t + c$ ... (i)

Given, when $t = 0$, $x = 1000$ which gives

$\log 1000 = 0 + c$

$\Rightarrow \quad c = \log 1000$

From Eq. (i), we get

$\log x - \log 1000 = \lambda t$

or $\quad \log\left(\dfrac{x}{1000}\right) = \lambda t$ ... (ii)

Given that in 2h, number of bacteria increased by 20% i.e. when $t = 2$h, $x = 1200$
On putting, $t = 2$ and $x = 1200$ in Eq. (ii),

$\log\left(\dfrac{1200}{1000}\right) = 2\lambda$ gives, $\lambda = \dfrac{1}{2}\log\left(\dfrac{6}{5}\right)$

Again, from Eq. (ii),

$\log\left(\dfrac{x}{1000}\right) = \dfrac{1}{2}\log\left(\dfrac{6}{5}\right)\cdot t$

or $\quad \dfrac{x}{1000} = e^{\frac{t}{2}\log\left(\frac{6}{5}\right)}$ ... (iii)

Given, $x = 2000$ at $t = k/\log_e(6/5)$, put in Eq. (iii),

$\dfrac{2000}{1000} = e^{\frac{k}{2}\log\left(\frac{6}{5}\right)/\log\left(\frac{6}{5}\right)}$

$2 = e^{k/2}$ or $\log 2 = k/2$

or $\quad k/\log 2 = 2$

$\Rightarrow \quad (k/\log 2)^2 = (2)^2 = 4$

**61.** *(b)* Given, $\lim\limits_{t \to x} \dfrac{t^2 f^2(x) - x^2 f^2(t)}{t - x}$

$= 0 \quad \left[\dfrac{0}{0} \text{ form}\right]$

On applying L' Hospital rule, we get

$\lim\limits_{t \to x} \dfrac{2tf^2(x) - x^2 \cdot 2f(t)\, f'(t)}{1} = 0$

$\Rightarrow 2xf^2(x) - x^2 2f(x)\, f'(x) = 0$

$\Rightarrow \quad 2xf(x)\,[f(x) - xf'(x)] = 0$

$\because xf(x) \neq 0$, so $f(x) - xf'(x) = 0$

$\Rightarrow \quad xf'(x) = f(x) \Rightarrow \dfrac{f'(x)}{f(x)} = \dfrac{1}{x}$

On integrating both sides, we get

$\log f(x) = \log x + \log c \Rightarrow f(x) = cx$

When $x = 1$, $f(1) = e \Rightarrow c = e$

So, $f(x) = xe \Rightarrow 1 = xe \Rightarrow x = 1/e$

$[\because f(x) = 1 \text{ (given)}]$

**62.** *(c)* Given differential equation

$x^3 dy + xydx = x^2 dy + 2ydx$

$\Rightarrow \quad (x^3 - x^2)dy = y(2 - x)dx$

$\Rightarrow \quad \int \dfrac{dy}{y} = \int \dfrac{2 - x}{x^2(x-1)}dx$

On making partial fractions for RHS,

we get $\dfrac{2-x}{x^2(x-1)} = \dfrac{A}{x} + \dfrac{B}{x^2} + \dfrac{C}{x-1}$

$\Rightarrow 2 - x = Ax(x-1) + B(x-1) + Cx^2$

On comparing $A + C = 0$, $B - A = -1$ and $B = -2$

$\therefore \quad A = -1$ and $C = 1$

$\therefore \int \dfrac{dy}{y} = -\int \dfrac{dx}{x} - 2\int \dfrac{dx}{x^2} + \int \dfrac{dx}{x-1}$

$\Rightarrow \quad \log_e y = -\log_e x + \dfrac{2}{x} + \log_e(x-1) + C$

$\because \quad y(2) = e \Rightarrow 1 = -\log_e 2 + 1 + 0 + C$

$\Rightarrow \quad C = \log_e 2$

$\therefore$ At $x = 4$,

$$\log_e y = -\log_e (4) + \frac{1}{2} + \log_e (3) + \log_e 2$$

$$\Rightarrow \log_e\left(\frac{2y}{3}\right) = \frac{1}{2} \Rightarrow y = \frac{3}{2}\sqrt{e}$$

$$\Rightarrow y(4) = \frac{3}{2}\sqrt{e}$$

**63.** *(a)* Given differential equation is

$$2x^2 dy = (2xy + y^2)\, dx$$

$$\Rightarrow \frac{dy}{dx} = \frac{2xy + y^2}{2x^2}$$

[Homogeneous differential equation]

Let $y + vx = \frac{dy}{dx} = v + x\frac{dv}{dx}$

$\therefore$ The differential equation becomes

$$v + x\frac{dv}{dx} = \frac{2vx^2 + v^2x^2}{2x^2}$$

$$\Rightarrow x\frac{dv}{dx} = \frac{2v + v^2 - 2v}{2}$$

$$\Rightarrow x\frac{dv}{dx} = \frac{v^2}{2} \Rightarrow 2\frac{dv}{v^2} = \frac{dx}{x}$$

$$\Rightarrow 2\left(\frac{-1}{v}\right) = \log_e x + C$$

$$\Rightarrow \log_e x + \frac{2x}{y} + C = 0 \quad \ldots(i)$$

$\because$ The curve (i) passes through the point (1, 2).

So, $C = -1 \quad \therefore \log_e x + \frac{2x}{y} = 1$

Now, at $x = \frac{1}{2}$

$$\log_e\left(\frac{1}{2}\right) + \frac{1}{y} = 1 \Rightarrow \frac{1}{y} = 1 + \log_e 2$$

$$\Rightarrow y = f\left(\frac{1}{2}\right) = \frac{1}{1 + \log_e 2}$$

**64.** *(d)* Given differential equation

$$\sqrt{1-x^2}\frac{dy}{dx} + \sqrt{1-y^2} = 0, |x| < 1$$

$$\Rightarrow \int \frac{dy}{\sqrt{1-y^2}} = -\int \frac{dx}{\sqrt{1-x^2}}$$

$$\Rightarrow \sin^{-1} y = -\sin^{-1} x + C$$

$$\because y\left(\frac{1}{2}\right) = \frac{\sqrt{3}}{2} \quad \text{[given]}$$

$$\therefore \sin^{-1}\frac{\sqrt{3}}{2} = -\sin^{-1}\frac{1}{2} + C$$

$$\Rightarrow \frac{\pi}{3} = -\frac{\pi}{6} + C \Rightarrow C = \frac{\pi}{2}$$

$\therefore$ at $x = -\frac{1}{\sqrt{2}}$

$$\sin^{-1} y = -\sin^{-1}\left(-\frac{1}{\sqrt{2}}\right) + \frac{\pi}{2}$$

$$\sin^{-1} y = \sin^{-1}\left(\frac{1}{\sqrt{2}}\right) + \frac{\pi}{2} = \frac{\pi}{4} + \frac{\pi}{2} = \frac{3\pi}{4}$$

$$\Rightarrow \sin^{-1} y = \pi - \frac{\pi}{4}$$

$$\Rightarrow y = \sin\left(\pi - \frac{\pi}{4}\right) = \sin\frac{\pi}{4}$$

$$\Rightarrow y = \frac{1}{\sqrt{2}}$$

**65.** *(b)* $\int \frac{f'(x)}{f(x)} = \int 1\, dx$

$$\Rightarrow \log|f(x)| = x + C$$

$$\Rightarrow \log 2 = 1 + C$$

$$\Rightarrow C = \log 2 - 1$$

$$\Rightarrow \log|f(x)| = x + \log 2 - 1$$

If $f(x) > 0$

$$\log f(x) = (x-1) + \log 2$$

$$\frac{\log f(x)}{2} = x - 1$$

$$\Rightarrow f(x) = 2\cdot e^{x-1}$$

$$\Rightarrow f(x) = \frac{2}{e}(e^x)$$

Now, $f(f(x)) = \frac{2}{e}(e^{f(x)})$

$$\Rightarrow h(x) = \frac{2}{e}\left(e^{\frac{2}{e}e^x}\right)$$

$$\Rightarrow h'(x) = \frac{2}{e}\left(e^{\frac{2}{e}e^x} \times \frac{2}{e}\cdot e^x\right)$$

$$\Rightarrow h'(x) = \frac{2}{e}\left[e^2 \times \frac{2}{e} e\right] = 4e$$

If $f(x) < 0$

$$|f(x)| = 2e^{x-1}$$

$$f(x) = -2e^{x-1} < 0$$

Then, $h'(1)$ will be negative.

**66.** *(d)* Given differential equation is

$$y(1 + xy)\, dx = x\, dy$$

$$\Rightarrow y\, dx + xy^2\, dx = x\, dy$$

$$\Rightarrow \frac{x\, dy - y\, dx}{y^2} = x\, dx$$

$$\Rightarrow -\frac{(y\, dx - x\, dy)}{y^2} = x\, dx$$

$$\Rightarrow -d(x/y) = x\, dx$$

On integrating both sides, we get

$$-\frac{x}{y} = \frac{x^2}{2} + C \quad \ldots(i)$$

$\because$ It passes through (1, –1).

$$\therefore 1 = \frac{1}{2} + C \Rightarrow C = \frac{1}{2}$$

Now, from Eq. (i), $-\frac{x}{y} = \frac{x^2}{2} + \frac{1}{2}$

$$\Rightarrow x^2 + 1 = -\frac{2x}{y} \Rightarrow y = -\frac{2x}{x^2 + 1}$$

$$\therefore f(-1/2) = 4/5$$

**67.** *(a)* Given,

(i) The population of mouse at time $t$ satisfies the differential equation

$$p'(t) = \frac{dp(t)}{dt} = 0.5p(t) - 450$$

(ii) Population of mouse at time $t = 0$ is $p(0) = 850$

**To find** : The time at which the population of the mouse will become zero, *i.e.*, to find the value of '$t$' at which $p(t) = 0$.

Let's solve the differential equation first,

$$p'(t) = \frac{dp(t)}{dt} = 0.5p(t) - 450$$

$$\Rightarrow \frac{2dp(t)}{p(t) - 900} = dt \Rightarrow \int \frac{2dp(t)}{p(t) - 900} = \int dt$$

$$\Rightarrow 2\log|p(t) - 900| = t + C$$

Where, $C$ is the constant of integration. To find the value of $C$, let's substitute $t = 0$.

$$\Rightarrow 2\log|p(0) - 900| = 0 + C$$

$$\Rightarrow C = 2\log|850 - 900| \; [\because p(0) = 850]$$

$$\therefore C = 2\log 50$$

Now, substituting the value of $C$ back in the solution, we get

$$2\log|p(t) - 900| = t + 2\log 50$$

On substituting $p(t) = 0$, we get

$$2\log|0 - 900| = t + 2\log 50$$

$$\Rightarrow t = 2\log\left|\frac{900}{50}\right|$$

$$\therefore t = 2\log 18$$

**68.** *(a)* $y = \left(x - y\frac{dx}{dy}\right)\sin\left(\frac{x}{y}\right)$

$$\Rightarrow 1 = \left(\frac{x}{y} - \frac{dx}{dy}\right)\sin\left(\frac{x}{y}\right)$$

$$\Rightarrow \frac{dx}{dy} = \frac{x}{y} - \text{cosec}\left(\frac{x}{y}\right)$$

Put $x = vy$

$\Rightarrow \quad \frac{dx}{dy} = v + y\frac{dv}{dy}$

$\therefore$ Then, equation

$$v + y\frac{dv}{dy} = v - \operatorname{cosec} v$$

$\Rightarrow \quad \sin v\, dv = -\frac{dy}{y}$

Integrating both sides, we get

$$-\cos v = -\log_e y - C$$

$\Rightarrow \quad \cos v = \log_e y + C$

$\Rightarrow \quad \cos\left(\frac{x}{y}\right) = \log_e y + C$

$\therefore \quad y = 1$ and $x = \frac{\pi}{2}$

$$\cos\frac{\pi}{2} = \log_e 1 + C \Rightarrow C = 0$$

$\therefore \quad \log_e y = \cos\left(\frac{x}{y}\right)$

$$\frac{x}{y} = \cos^{-1}(\log_e y)$$

$$x = y(\cos^{-1}\log_e y)$$

$$x(2) = 2\cos^{-1}\log_e 2$$

$$\cos(x(2)) = \cos(2\cos^{-1}\log_e 2)$$

$$= 2(\log_e 2)^2 - 1$$

$$= 2(\log_e 2)^2 - 1$$

**69.** *(b)* $\frac{dy}{dx} = \frac{(2+\alpha)x - \beta y + 2}{\beta x - 2\alpha y - (\beta y - 4\alpha)}$

$\Rightarrow (\beta x - 2\alpha y - (\beta y - 4\alpha))\, dy$

$= [(2+\alpha)x - \beta y + 2]\, dx$

$\Rightarrow \beta x\, dy - 2\alpha y\, dy - [\beta y - 4\alpha]\, dy$

$= (2+\alpha)\, x\, dy - \beta y\, dx + 2\, dx$

$\Rightarrow \beta[x\, dy + y\, dx] - 2\alpha y\, dy - [\beta y - 4\alpha]\, dy$

$= [2+\alpha]\, x dx + 2dx$

$\Rightarrow \beta[d(xy)] - 2\alpha y\, dy - [\beta y - 4\alpha]\, dy$

$= [2+\alpha]\, x dx + 2\, dx$

On integrating

$\beta xy - \alpha y^2 - [\beta y - 4\alpha]y$

$= [2+\alpha]\frac{x^2}{2} + 2x + C$

This passes through (0, 0),

$\therefore \quad C = 0$

$\beta xy - \alpha y^2 - [\beta y - 4\alpha]y$

$= [2+\alpha]\frac{x^2}{2} + 2x$

$\Rightarrow \quad \frac{[2+\alpha]x^2}{2} + \alpha y^2 + 2x + [\beta y - 4\alpha]y$

$- \beta xy = 0$

Put $\beta = 0$ for equation of circle and $\frac{2+\alpha}{2} = \alpha$

$\Rightarrow \quad \alpha = 2$ and $\beta = 0$

$\Rightarrow \quad 2x^2 + 2y^2 + 2x - 8y = 0$

$\Rightarrow \quad x^2 + y^2 + x - 4y = 0$

$\Rightarrow \quad \left(x + \frac{1}{2}\right)^2 + (y-2)^2 = \frac{1}{4} + 4$

$\Rightarrow \quad \left(x + \frac{1}{2}\right)^2 + (y-2)^2 = \frac{17}{4}$

$\Rightarrow \quad \left(x + \frac{1}{2}\right)^2 + (y-2)^2 = \left(\frac{\sqrt{17}}{2}\right)^2$

Hence, radius of circle is $\frac{\sqrt{17}}{2}$.

**70.** *(c)* Given differential equation

$$\frac{dy}{dx} = 2x(x+y)^3 - x(x+y) - 1$$

$$\frac{dy}{dx} + 1 = x(2(x+y)^3 - x + y)$$

Put $x + y = v$

$\Rightarrow \quad 1 + \frac{dy}{dx} = \frac{dv}{dx}$

$$\frac{dv}{dx} = x(2v^3 - v)$$

$$\frac{dv}{2v^3 - v} = x\, dx$$

$\Rightarrow \int \frac{dv}{v^3\left(2 - \frac{1}{v^2}\right)} = \int x\, dx$

Put $2 - \frac{1}{v^2} = t \Rightarrow \frac{2}{v^3}dv = dt$

$\therefore \quad \frac{1}{2}\int\frac{dt}{t} = \int x\, dx$

$$\frac{1}{2}\log t = \frac{x^2}{2} + \frac{c}{2}$$

$\Rightarrow \quad \log t = x^2 + c$

$\Rightarrow \quad t = e^{x^2 + c} = Ae^{x^2}$

$$2 - \frac{1}{v} = Ae^{x^2}$$

$$2 - \frac{1}{(x+y)^2} = Ae^{x^2}$$

Put $\quad x = 0, y = 1,$

we get $A = 1$

$\Rightarrow \quad 2 - \frac{1}{(x+y)^2} = e^{x^2}$

$\Rightarrow 2 - e^{x^2} = \frac{1}{(x+y)^2}$

$\Rightarrow (x+y)^2 = \frac{1}{2 - e^{x^2}}$

Put $x = \frac{1}{\sqrt{2}}$

$$\left(\frac{1}{\sqrt{2}} + y\right)^2 = \frac{1}{2 - \sqrt{e}}$$

$\therefore \quad \left(\frac{1}{\sqrt{2}} + y\left(\frac{1}{\sqrt{2}}\right)\right)^2 = \frac{1}{2 - \sqrt{e}}$

**71.** *(d)* Given, differential equation is $(x^2 - 4)dy - (y^2 - 3y)dx = 0$, where $x > 2$ and $y(4) = \frac{3}{2}$

$\Rightarrow \quad (x^2 - 4)dy = (y^2 - 3y)\, dx$

$= y(y-3)dx$

$\Rightarrow \quad \frac{1}{y(y-3)}dy = \frac{1}{x^2 - 4}dx$

$= \frac{1}{(x-2)(x+2)}dx$

$\Rightarrow \frac{1}{3}\left[\frac{y - (y-3)}{y(y-3)}\right]dy$

$= \frac{1}{4}\left[\frac{(x+2) - (x-2)}{(x+2)(x-2)}\right]dx$

$\Rightarrow \frac{1}{3}\left[\frac{1}{y-3} - \frac{1}{y}\right]dy$

$= \frac{1}{4}\left[\frac{1}{x-2} - \frac{1}{x+2}\right]dx$

On integrating both sides, we get

$\frac{1}{3}[\ln|y-3| - \ln|y|]$

$= \frac{1}{4}[\ln|x-2| - \ln|x+2|] + C$

$\Rightarrow \quad \frac{1}{3}\ln\left|\frac{y-3}{y}\right| = \frac{1}{4}\ln\left|\frac{x-2}{x+2}\right| + C \quad$ ...(i)

As given, $y(4) = \frac{3}{2}$, i.e. we put $x = 4$ and $y = \frac{3}{2}$ in Eq. (i), we get

$$\frac{1}{3}\ln\left|\frac{\frac{3}{2} - 3}{\frac{3}{2}}\right| = \frac{1}{4}\ln\left|\frac{4-2}{4+2}\right| + C$$

$\Rightarrow \quad \frac{1}{3}\ln|-1| = \frac{1}{4}\ln\left|\frac{1}{3}\right| + C$

$\Rightarrow \quad \frac{1}{3}\ln(1) = -\frac{1}{4}\ln\left(\frac{1}{3}\right) + C$

$\Rightarrow \quad C = \frac{1}{4}\ln(3) \quad [\because \ln(1)=0]$

From Eq. (i),

$$\frac{1}{3}\ln\left|\frac{y-3}{y}\right| = \frac{1}{4}\ln\left|\frac{x-2}{x+2}\right| + \frac{1}{4}\ln(3) \quad \text{...(ii)}$$

We have to find $y(10)$, i.e when $x=10$, then

From Eq. (ii),

$$\frac{1}{3}\ln\left|\frac{y-3}{y}\right| = \frac{1}{4}\ln\left|\frac{2}{3}\right| + \frac{1}{4}\ln(3)$$

$$\Rightarrow \frac{1}{3}\ln\left|\frac{y-3}{y}\right| = \frac{1}{4}\ln(2) - \frac{1}{4}\ln(3) + \frac{1}{4}\ln(3) = \frac{1}{4}\ln(2)$$

$$\Rightarrow \ln\left|\frac{y-3}{y}\right| = \ln(2)^{\frac{3}{4}}$$

$$\Rightarrow \left|\frac{y-3}{y}\right| = 2^{\frac{3}{4}}, \forall x > 2, \frac{dy}{dx} < 0$$

As, $\quad y(4) = \frac{3}{2} \Rightarrow y \in (0, 3)$

So, $\quad -y + 3 = 8^{\frac{1}{4}} \cdot y$

$$\Rightarrow \quad y = \frac{3}{1+(8)^{\frac{1}{4}}}$$

**72.** (a) $\frac{dy}{dx} = \frac{x+y-2}{x-y} = \frac{(x-1)+(y-1)}{(x-1)-(y-1)}$

On putting $x-1 = X$, $y-1 = Y$

$$\frac{dY}{dX} = \frac{X+Y}{X-Y} \quad \text{...(i)}$$

Now, this is a homogeneous differential function

Put $\quad Y = VX$

On differentiating w.r.t. $X$, we get

$$\frac{dY}{dX} = V + X\frac{dV}{dX}$$

On putting in Eq (i), we get

$$V + X\frac{dV}{dX} = \frac{X+VX}{X-VX} = \frac{1+V}{1-V}$$

$$\Rightarrow X\frac{dV}{dX} = \frac{1+V-V+V^2}{1-V}$$

$$\Rightarrow \frac{1-V}{1+V^2}dV = \frac{dX}{X}$$

$$\Rightarrow \int\frac{1}{1+V^2}dV - \int\frac{1}{2}\frac{2V}{1+V^2}dV = \int\frac{dX}{X}$$

$$\tan^{-1}V - \frac{1}{2}\ln(1+V^2) = \ln X + c$$

On putting $V = Y/X$,

$$\tan^{-1}\frac{Y}{X} - \frac{1}{2}\ln\left(1+\frac{Y^2}{X^2}\right) = \ln X + c$$

On putting $Y = y-1$, $X = x-1$

$$\tan^{-1}\left(\frac{y-1}{x-1}\right) - \frac{1}{2}\ln\left(1+\frac{(y-1)^2}{(x-1)^2}\right) = \ln(x-1) + c$$

Given that curve passes through (2, 1)

$$\tan^{-1}(0) - \frac{1}{2}\ln(1+0) = \ln 1 + c$$

$$0 - 0 = 0 + c \Rightarrow c = 0$$

So, curve is

$$\tan^{-1}\left(\frac{y-1}{x-1}\right) - \frac{1}{2}\ln\left(1+\frac{(y-1)^2}{(x-1)^2}\right) = \ln(x-1)$$

Curve also passes through $(k+1, 2)$, $k>0$

$$\tan^{-1}\left(\frac{1}{k}\right) - \frac{1}{2}\ln\left(1+\frac{1}{k^2}\right) = \ln k$$

$$\Rightarrow 2\tan^{-1}\left(\frac{1}{k}\right) = \ln\left(\frac{k^2+1}{k^2}\right) + 2\ln k$$

$$\Rightarrow 2\tan^{-1}\left(\frac{1}{k}\right) = \ln(1+k^2)$$

**73.** (a) $y = y(x)$

$$x\tan\left(\frac{y}{x}\right)dy = \left[y\tan\left(\frac{y}{x}\right) - x\right]dx$$

$$\frac{dy}{dx} = \frac{y\tan\frac{y}{x} - x}{x\tan\frac{y}{x}}$$

On putting $y = vx$

$$\Rightarrow \frac{dy}{dx} = v + x\frac{dv}{dx}$$

$$\therefore \quad v + x\frac{dv}{dx} = \frac{vx\tan v - x}{x\tan v}$$

$$\Rightarrow v + x\frac{dv}{dx} = \frac{v\tan v - 1}{\tan v}$$

$$\Rightarrow v + x\frac{dv}{dx} = v - \cot v$$

$$\Rightarrow \int\frac{dv}{-\cot v} = \int\frac{dx}{x}$$

$$\Rightarrow \log_e \cos v = \log_e x + \log_e c$$

$$\Rightarrow \cos\left(\frac{y}{x}\right) = cx$$

When, $x = \frac{1}{2}$, $y = \frac{\pi}{6}$

$$\therefore \cos\left(\frac{\pi}{3}\right) = \frac{c}{2} \Rightarrow c = 1 \quad \cos\left(\frac{y}{x}\right) = x$$

$$\Rightarrow y = x\cos^{-1}x$$

$$\therefore \text{ Area} = \int_0^{\frac{1}{\sqrt{2}}} y\,dy = \int_0^{\frac{1}{\sqrt{2}}} x\cos^{-1}x\,dx$$

$$= \left[\cos^{-1}x \cdot \frac{x^2}{2}\right]_0^{\frac{1}{\sqrt{2}}} - \int_0^{\frac{1}{\sqrt{2}}}\frac{-1}{\sqrt{1-x^2}} \cdot \frac{x^2}{2}dx$$

$$= \frac{\pi}{16} + \frac{1}{2}\int_0^{1/\sqrt{2}}\frac{x^2-1+1}{\sqrt{1-x^2}}dx$$

$$= \frac{\pi}{16} + \frac{1}{2}\int\left(\frac{1}{\sqrt{1-x^2}} - \sqrt{1-x^2}\right)dx$$

$$= \frac{\pi}{16} + \frac{1}{2}\left[\sin^{-1}x - \left(\frac{x}{2}\sqrt{1-x^2} + \frac{1}{2}\sin^{-1}x\right)\right]_0^{\frac{1}{\sqrt{2}}}$$

$$= \frac{\pi}{16} + \frac{1}{2}\left(\frac{\pi}{4} - \frac{1}{4} - \frac{\pi}{8}\right)$$

$$= \frac{\pi}{16} + \frac{\pi}{16} - \frac{1}{8} = \frac{\pi}{8} - \frac{1}{8}$$

**74.** (b) Given, $f(0) = 1$,

$$f'(0) = 2,$$

$$f''(x) \neq 0$$

$$\begin{vmatrix} f(x) & f'(x) \\ f'(x) & f''(x) \end{vmatrix} = 0$$

$$\Rightarrow f(x)f''(x) - f'(x)f'(x) = 0$$

$$\Rightarrow \frac{f''(x)}{f'(x)} = \frac{f'(x)}{f(x)}$$

$$\Rightarrow \int\frac{f''(x)}{f'(x)}dx = \int\frac{f'(x)}{f(x)}dx$$

$$\Rightarrow \log f'(x) = \log f(x) + \log c \quad \text{...(i)}$$

or $\quad f'(x) = cf(x)$

Now, putting $x = 0$, we get

$$f'(0) = cf(0)$$

$$\Rightarrow \quad 2 = c \times 1 \Rightarrow c = 2$$

On putting the value of $c = 2$ in Eq. (i), we get

$$f'(x) = 2f(x)$$

$$\Rightarrow \int \frac{f'(x)}{f(x)}dx = \int 2dx$$
$$\Rightarrow \log f(x) = 2x + D$$
$$\Rightarrow f(x) = e^{2x+D}$$
$$\Rightarrow f(x) = e^D \cdot e^{2x}$$
$$\Rightarrow f(x) = k\cdot e^{2x} \quad [\text{Let } k = e^D]$$

On putting $x = 0$, we get

$$f(0) = k\cdot e^0$$
$$\Rightarrow 1 = k$$
$$\therefore f(x) = e^{2x}$$

On putting $x = 1$, we get, $f(1) = e^2$

Clearly, $e^2$ lies in (6, 9).

**75.** *(a)* The given differential equation

$$\sqrt{1+x^2+y^2+x^2y^2} + xy\frac{dy}{dx} = 0$$
$$\Rightarrow \sqrt{(1+x^2)(1+y^2)} + xy\frac{dy}{dx} = 0$$
$$\Rightarrow \int \frac{\sqrt{1+x^2}}{x}dx + \int \frac{y}{\sqrt{1+y^2}}dy = 0 \quad \text{...(i)}$$

Now, let $I_1 = \int \frac{\sqrt{1+x^2}}{x}dx$

On putting, $x = \tan\theta$

$$\Rightarrow dx = \sec^2\theta\, d\theta$$
$$\therefore I_1 = \int \frac{\sec\theta}{\tan\theta}\sec^2\theta\, d\theta$$
$$= \int \frac{d\theta}{\sin\theta\cos^2\theta}$$
$$= \int \frac{\sin\theta}{\cos^2\theta(1-\cos^2\theta)}d\theta$$

On puttting, $\cos\theta = t$

$$\Rightarrow -\sin\theta d\theta = dt$$
$$\therefore I_1 = \int \frac{-dt}{t^2(1-t^2)}$$
$$= \int\left(\frac{-1}{1-t^2} - \frac{1}{t^2}\right)dt$$
$$= -\frac{1}{2}\log_e\left|\left(\frac{1+t}{1-t}\right)\right| + \frac{1}{t} + C_1$$
$$= \frac{-1}{2}\log_e\left|\frac{1+\cos\theta}{1-\cos\theta}\right| + \frac{1}{\cos\theta} + C_1$$
$$= \frac{-1}{2}\log_e\left|\frac{1+\frac{1}{\sqrt{1+x^2}}}{1-\frac{1}{\sqrt{1+x^2}}}\right| + \sqrt{1+x^2} + C_1$$
$$\Rightarrow I_1 = \frac{-1}{2}\log_e\left|\frac{\sqrt{1+x^2}+1}{\sqrt{1+x^2}-1}\right| + \sqrt{1+x^2} + C_1 \quad \text{...(ii)}$$

And $I_2 = \frac{1}{2}\int \frac{2y}{\sqrt{1+y^2}}dy$

$$= \frac{1}{2}\frac{\sqrt{1+y^2}}{1/2} + C_2$$
$$= \sqrt{1+y^2} + C_2 \quad \text{...(iii)}$$

From Eqs. (i), (ii) and (iii), we get

$$-\frac{1}{2}\log_e\left(\frac{\sqrt{1+x^2}+1}{\sqrt{1+x^2}-1}\right) + \sqrt{1+x^2} + \sqrt{1+y^2} - C = 0$$
$$\Rightarrow \sqrt{1+x^2} + \sqrt{1+y^2} = \frac{1}{2}\log_e\left(\frac{\sqrt{1+x^2}+1}{\sqrt{1+x^2}-1}\right) + C$$

[Where, $-C = C_1 + C_2$]

**76.** *(c)* Given, $\frac{dy}{dx} = \frac{2y}{x^2}$

On variable separable and integrating,

$$\int \frac{1}{2y}dy = \int \frac{1}{x^2}dx$$
$$= \frac{1}{2}\ln|y| = -\frac{1}{x} + C$$

The centre of given circle is (1, 1).

$$\therefore \frac{1}{2}\ln 1 = -\frac{1}{1} + C$$
$$\Rightarrow C = 1$$
$$\therefore \frac{1}{2}\ln|y| = -\frac{1}{x} + 1$$
$$\Rightarrow \ln|y| = -\frac{2}{x} + 2$$
$$\Rightarrow x\ln|y| = -2 + 2x$$
$$x\ln|y| = 2(x-1)$$

**77.** *(112)* Given $\int_0^1 f(\lambda x)\, d\lambda = af(x)$ ...(i)

In LHS of Eq. (i), we get

Put $\lambda x = t$

$$\Rightarrow d\lambda = \frac{1}{x}dt$$

| $\lambda$ | 0 | 1 |
|---|---|---|
| $t$ | 0 | $x$ |

So, Eq. (i) becomes,

$$\frac{1}{x}\int_0^x f(t)\,dt = af(x)$$
$$\Rightarrow \int_0^x f(t)\,dt = ax\, f(x)$$

Differentiating both sides, we get

$$(f(x)\times 1 - 0) = a\,(xf'(x) + f(x)\times 1)$$
$$\Rightarrow f(x)(1-a) = a\cdot x\cdot f'(x)$$
$$\Rightarrow \frac{f'(x)}{f(x)} = \frac{1-a}{a}\times\frac{1}{x}$$
$$\Rightarrow \int \frac{f'(x)}{f(x)}dx = \int\left(\frac{1-a}{a}\right)\times\frac{1}{x}dx$$
$$\Rightarrow \ln f(x) = \frac{1-a}{a}\ln x + C \quad \text{...(ii)}$$

where $C$ is constant of integration.

Given initial conditions as

$$f(1) = 1$$

So, put $x = 1$ and $f(1) = 1$ in Eq. (ii) we get,

$$\ln f(1) = \frac{1-a}{a}\ln 1 + C$$
$$\Rightarrow 0 = 0 + C$$
$$\Rightarrow C = 0 \quad [\because \ln f(1) = \ln 1 = 0]$$

For Eq. (ii), we get

$$\Rightarrow \ln f(x) = \frac{1-a}{a}\ln x + 0$$
$$= \frac{1-a}{a}\ln x \quad \text{...(iii)}$$

Again, putting $x = 16$ and $f(16) = \frac{1}{8}$ [as data given]

$$\ln f(16) = \frac{1-a}{a}\ln 16$$
$$\Rightarrow \ln\frac{1}{8} = \ln(16)^{\frac{1-a}{a}}$$
$$\Rightarrow \frac{1}{8} = (16)^{\frac{1-a}{a}} \Rightarrow -3 = \frac{4-4a}{a}$$
$$\Rightarrow a = 4$$

Hence, $f(x) = x^{-\frac{3}{4}}$

$$\Rightarrow f'(x) = \frac{-3}{4}\cdot x^{-7/4}$$

Thus,

$$16 - f'\left(\frac{1}{16}\right) = 16 - \left(\frac{-3}{4}\times(2^{-4})^{-\frac{7}{4}}\right)$$
$$= 16 - \left(\frac{-3}{4}\times 2^7\right)$$
$$= 16 + \frac{3}{2^2}\times 2^7$$
$$= 16 + 96 = 112$$

**78.** *(19)* Given,

$$2(x+2)^2\cdot f(x) - 3(x+2)^2 = 10\int_0^x (t+2)f(t)dt \quad \text{... (i)}$$

On differentiating Eq. (i) both side, we get

$4(x+2)f(x)+2(x+2)^2\cdot f'(x)-6(x+2)$
$=10(x+2)f(x)$

$\Rightarrow 2(x+2)^2\cdot f'(x)-6(x+2)f(x)$
$=6(x+2)$

$\Rightarrow (x+2)\dfrac{dy}{dx}-3y=3$, where $y=f(x)$

and $\dfrac{dy}{dx}=f'(x)$

$\Rightarrow (x+2)\dfrac{dy}{dx}=3(y+1)$

$\Rightarrow \int\dfrac{1}{y+1}\,dy=\int\dfrac{3}{x+2}\,dx$

$\Rightarrow \ln(y+1)=3\ln(x+2)+\ln c$

$=\ln(c(x+2)^3)$

$\Rightarrow (y+1)=c(x+2)^3$

For initial condition put $x=0$ in Eq. (i) we get

$2\times(0+2)^2\cdot f(0)-3(0+2)^2$
$=10\int_0^0(t+2)f(t)\,dt=0$

$\Rightarrow f(0)=\dfrac{3}{2}$

$\therefore \quad c=\dfrac{5}{16}$

So, $y+1=\dfrac{5}{16}(x+2)^3$ ...(ii)

$\therefore f(2)=\dfrac{5}{16}\times 64-1=20-1=19$

**79.** (4) $\alpha|x|=|y|e^{yx-\beta}$, $\alpha,\beta\in N$

$x\,dy-y\,dx+xy(x\,dy+y\,dx)=0$

$\Rightarrow \dfrac{dy}{y}-\dfrac{dx}{x}+(x\,dy+y\,dx)=0$

$\Rightarrow \ln|y|-\ln|x|+xy=C$

$\because \quad y(1)=2$

So, $\ln(2)-0+2=C$

$\Rightarrow C=2+\ln(2)$

Then, $\ln|y|-\ln|x|+xy=2+\ln 2$

$\Rightarrow \ln|x|=\ln\left|\dfrac{y}{2}\right|-2+xy$

$\Rightarrow |x|=\left|\dfrac{y}{2}\right|e^{xy-2}\Rightarrow 2|x|=|y|e^{xy-2}$

Comparing with $\alpha|x|=|y|e^{xy-\beta}$,

we get $\quad \alpha=2,\beta=2$

Therefore, $\alpha+\beta=4$

**80.** (3) $(e^y+1)\cos x\,dx+e^y\sin x\,dy=0$

$\Rightarrow d((e^y+1)\sin x)=0$

$\Rightarrow \quad (e^y+1)\sin x=C$

It passes through $\left(\dfrac{\pi}{2},0\right)\Rightarrow C=2$

Now, at $\quad x=\dfrac{\pi}{6}$

$e^y=3$

**81.** (31) $\dfrac{dy}{dx}=(x+y+2)^2$ ...(i)

$y(0)=-2$

Let $x+y+2=t$

$1+\dfrac{dy}{dx}=\dfrac{dt}{dx}$

Eq. (i) becomes

$\dfrac{dt}{dx}-1=t^2\Rightarrow\dfrac{dt}{dx}=t^2+1$

$\int\dfrac{dt}{t^2+1}=\int dx$

$\Rightarrow \quad \tan^{-1}t=x+C$

$\Rightarrow \quad t=\tan(x+C)$

$\Rightarrow \quad x+y+2=\tan(x+C)$

$\Rightarrow \quad x=0,\ y=-2$

$\Rightarrow \quad 0+0=\tan C$

$\Rightarrow \quad C=0$

$\therefore \quad x+y+2=\tan x$

$y=\tan x-x-2$

$f(x)=\tan x-x-2$

$f'(x)=\sec^2x-1>0$

$f(x)$ is increasing function

$f_{\min}=f(0)=-2=\beta$

$f_{\max}=f\left(\dfrac{\pi}{3}\right)=\sqrt{3}-\dfrac{\pi}{3}-2=\alpha$

Now, $(3\alpha+\pi)^2+\beta^2$

$=(3\sqrt{3}-6)^2+4$

$=67-36\sqrt{3}$

$\therefore \quad \gamma=67,\ \delta=-36$

$\gamma+\delta=31$

**82.** (2)

**key idea** Use, $\dfrac{xdy-ydx}{x^2}=\dfrac{d}{dx}\left(\dfrac{y}{x}\right)$

$$f(x)=\sqrt{\lim_{r\to x}\frac{2r^2[(f(r))^2-f(x)\cdot f(r)]}{r^2-x^2}-r^3e^{\frac{f(r)}{r}}}$$

$$=\sqrt{\lim_{r\to x}\frac{2r^2f(r)[f(r)-f(x)]}{(r-x)\cdot(r+x)}-\lim_{r\to x}r^3e^{\frac{f(r)}{r}}}$$

$$=\sqrt{\lim_{r\to x}\frac{2r^2\cdot f(r)}{r+x}\cdot\left(\frac{f(r)-f(x)}{r-x}\right)-x^3\cdot e^{\frac{f(x)}{x}}}$$

$$=\sqrt{\frac{2x^2\cdot f(x)}{2x}\cdot f'(x)-x^3e^{\frac{f(x)}{x}}}$$

On squaring both sides, we get

$f^2(x)=x\cdot f(x)\cdot f'(x)-x^3e^{\frac{f(x)}{x}}$

$\Rightarrow y^2=xy\dfrac{dy}{dx}-x^3e^{y/x}$

$\Rightarrow x^3e^{y/x}\,dx=xy\,dy-y^2dx$

$\Rightarrow x^3dx=y\dfrac{(xdy-ydx)}{e^{y/x}}$

$\Rightarrow x^3dx=y(xdy-ydx)\cdot e^{-y/x}$

$\Rightarrow \dfrac{x}{y}dx=\dfrac{(xdy-ydx)}{x^2}\cdot e^{\frac{-y}{x}}$ ...(i)

$e^{\frac{-y}{x}}=t$

$\Rightarrow \dfrac{-y}{x}=\ln t$

$\Rightarrow e^{\frac{-y}{x}}\cdot(-1)\left(\dfrac{xdy-ydx}{x^2}\right)=dt$

$\Rightarrow e^{\frac{-y}{x}}\left(\dfrac{xdy-ydx}{x^2}\right)=-dt$

Now, Eq. (i) become,

$-\dfrac{dx}{\ln t}=-dt$

$dx=\ln t\cdot dt$

Integrating both sides, we get

$\int dx=\int\ln t\,dt$

$x=t[\ln t-1]+C$

$\Rightarrow x=e^{\frac{-y}{x}}\left(\dfrac{-y}{x}-1\right)+C$

$\Rightarrow x+\left(1+\dfrac{y}{x}\right)e^{\frac{-y}{x}}=C$ ...(ii)

$f(1)=1$

$\Rightarrow x=1$ and $y=1$ satisfies Eq. (ii),

$1+(1+1)e^{-1}=C$

$\Rightarrow \quad 1+2e^{-1}=C$

Now, $f(a)=0$

$x=a$ and $y=0$

$a+(1+0)e^{-0}=C$

$\Rightarrow a+1=C$

$\Rightarrow \quad a=C-1$

$\Rightarrow \quad a = 1 + 2e^{-1} - 1$

$\Rightarrow \quad a = 2e^{-1} \quad \Rightarrow \quad ae = 2$

**83.** *(3)* We have,

$$(1+y^2)(1+\log_e x)\,dx + x\,dy = 0$$

$$\Rightarrow (1+y^2)\left(\frac{1}{x}+\frac{\log x}{x}\right)dx + dy = 0$$

$$\Rightarrow \int\left(\frac{1}{x}+\frac{\log x}{x}\right)dx + \int\frac{dy}{1+y^2} = 0$$

$$\Rightarrow \log x + \frac{(\log x)^2}{2} + \tan^{-1} y = c$$

On putting $x = y = 1$, we get $c = \frac{\pi}{4}$

$$\therefore \log x + \frac{(\log x)^2}{2} + \tan^{-1} y = \frac{\pi}{4}$$

At $x = e$

$$y = \tan\left(\frac{\pi}{4}-\frac{3}{2}\right) = \frac{1-\tan\frac{3}{2}}{1+\tan\frac{3}{2}}$$

$\therefore \quad \alpha = 1, \beta = 1$

$\therefore \quad \alpha + 2\beta = 1 + 2 = 3$

**84.** *(11)* Given differential equation is

$$\frac{dy}{dx} = \frac{x+y-2}{x-y}$$

Let $\quad X = x-1$

and $\quad Y = y-1$

$$\Rightarrow \quad \frac{dY}{dX} = \frac{X+Y}{X-Y}$$

Now, let $Y = VX$

$$\Rightarrow \quad \frac{dY}{dX} = V + X\cdot\frac{dV}{dX}$$

$$\Rightarrow V + X\frac{dV}{dX} = \frac{1+V}{1-V}$$

$$\Rightarrow \quad \frac{(1-V)dV}{(1+V^2)} = \frac{dX}{X}$$

On integrating, we have

$$\tan^{-1} V - \ln(\sqrt{1+V^2}) = \ln|X| + C$$

$$\Rightarrow \tan^{-1}\left(\frac{y-1}{x-1}\right) - \ln\sqrt{1+\left(\frac{y-1}{x-1}\right)^2} = \ln|(x-1)| + C$$

$$\Rightarrow \tan^{-1}\left(\frac{y-1}{x-1}\right) - \frac{1}{2}\ln\left[1+\left(\frac{y-1}{x-1}\right)^2\right] = \ln|x-1| + C \quad ...(i)$$

Since, Eq. (i) passes through (2, 1).

i.e. when $x = 2$, then $y = 1$

So, put in Eq. (i) we get $C = 0$

Thus, $$\tan^{-1}\left(\frac{y-1}{x-1}\right) - \frac{1}{2}\ln\left[1+\left(\frac{y-1}{x-1}\right)^2\right] = \ln|x-1|$$

On comparing with

$$\tan^{-1}\left(\frac{y-1}{x-1}\right) - \frac{1}{\beta}\log_e\left[\alpha+\left(\frac{y-1}{x-1}\right)^2\right] = \log_e|x-1|,$$

we get $\quad \beta = 2$ and $\alpha = 1$

Hence, $5\beta + \alpha = 5\times 2 + 1 = 11$

**85.** *(29)* $2x + 3y - 2 = t$

$$2 + \frac{3dy}{dx} = \frac{dt}{dx}$$

$$4x + 6y - 4 = 2t$$

$$4x + 6y - 7 = 2t$$

$$\frac{dy}{dx} = \frac{-(2x+3y-2)}{4x+6y-7}$$

$$\frac{dt}{dx} = \frac{-3t+4t-6}{2t-3} = \frac{t-6}{2t-3}$$

$$\int\frac{2t-3}{t-6}\,dt = \int dx$$

$$\Rightarrow \int\left(\frac{(2t-12)}{t-6}+\frac{9}{t-6}\right)dt = x$$

$$\Rightarrow 2t + 9\log(t-6) = x + C$$

$$\Rightarrow 2(2x+3y-2) + 9\log(2x+3y-8) = x + C$$

$x = 0, \; y = 3, \; C = 14$

$$\Rightarrow \quad 4x + 6y - 4 + 9\log(2x+3y-8) = x + 14$$

$$\Rightarrow x + 2y + 3\log(2x+3y-8) = 6$$

$\Rightarrow \alpha = 1, \; \beta = 2, \; \gamma = 8$

$\therefore \alpha + 2\beta + 3\gamma = 1 + 4 + 24 = 29$

**86.** *(5)* Let the coordinate of a point $P$ be $(x, y)$.

Let the tangent intersect $X$-axis and $Y$-axis at a point $A(\alpha, 0)$ and $(0, \beta)$ respectively.

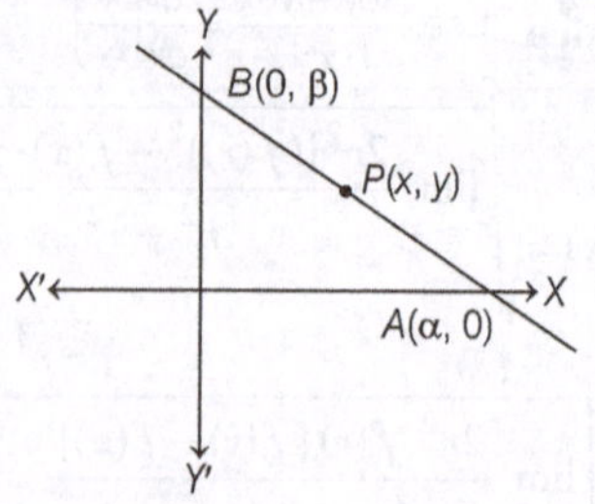


Point $P$ divides $AB$ in the ratio $1 : K$.

$$x = \frac{K\alpha}{K+1} \text{ and } y = \frac{\beta}{K+1}$$

$$\Rightarrow \alpha = \frac{x(K+1)}{K} \text{ and } \beta = y(K+1)$$

Slope of tangent $= \frac{0-\beta}{\alpha-0}$

$$\frac{dy}{dx} = -\frac{y(K+1)}{\frac{x(K+1)}{K}} \Rightarrow \frac{dy}{dx} = -\frac{Ky}{x}$$

$$\frac{1}{y}dy = -\frac{K}{x}dx$$

$$\log y = -K\log x + \log C$$

$$y\cdot x^K = C$$

The curve passes through $(1, 1) \Rightarrow C = 1$

The curve also passes through $\left(\frac{1}{10}, 100\right)$

$$\Rightarrow \quad 100\times\left(\frac{1}{10}\right)^K = 1$$

$\Rightarrow \quad K = 2$

Also, $e^{\frac{dy}{dx}} = 2x + 1$

$$\frac{dy}{dx} = \log_e(2x+1)$$

$$dy = \log_e(2x+1)dx$$

Let $t = 2x + 1 \Rightarrow \frac{1}{2}dt = dx$

$$\Rightarrow \int dy = \frac{1}{2}\int \log_e t\,dt$$

$$y = \frac{1}{2}[t\log_e t - t]$$

$$y = \frac{2x+1}{2}[\log_e(2x+1) - 1] + C$$

$y(0) = 2$ is its solution

$$2 = \frac{1}{2}[\log 1 - 1] + C$$

$$\Rightarrow C = \frac{5}{2}$$

$$\therefore \quad y = \frac{2x+1}{2}[\log_e(2x+1)-1] + \frac{5}{2}$$

$$y(1) = \frac{3}{2}[\log_e 3 - 1] + \frac{5}{2}$$

$$= \frac{3}{2}\log_e 3 + 1$$

$$4y(1) - 5\log_e 3 = 6\log_e 3 + 4 - 5\log_e 3$$

$$= \log_e 3 + 4$$

$$= 1.09 + 4 = 5.09 \approx 5$$

**87.** *(3)* Given, $\frac{dy}{dx} = \frac{y}{x}\frac{(4y^2+2x^2)}{(3y^2+x^2)}$

On putting $y = vx$

$$\Rightarrow \quad \frac{dy}{dx} = v + x\cdot\frac{dv}{dx}$$

$$\Rightarrow v + x\cdot\frac{dv}{dx} = v\cdot\left(\frac{4v^2+2}{3v^2+1}\right)$$

$$\Rightarrow x\frac{dv}{dx} = v\left(\frac{4v^2+2}{3v^2+1} - 1\right) = v\left(\frac{v^2+1}{3v^2+1}\right)$$

$$\Rightarrow \int\frac{(3v^2+1)dv}{v^3+v} = \int\frac{dx}{x}$$

[∵ Integration on both sides]

$$\Rightarrow \log|v^3+v| = \log|x| + C$$

$$\Rightarrow \log\left|\left(\frac{y}{x}\right)^3 + \left(\frac{y}{x}\right)\right| = \log|x| + C$$

Since, $y(1)=1 \Rightarrow C = \ln 2$

$$\Rightarrow \ln\left(\frac{y^3}{x^3} + \frac{y}{x}\right) = \ln|x| + \ln 2$$

At $x = 2$

$$\ln\left(\frac{y^3}{8} + \frac{y}{2}\right) = \ln 2 + \ln 2 = 2\ln 2$$

$$\Rightarrow \frac{y^3}{8} + \frac{y}{2} = 4 \Rightarrow y^3 + 4y - 32 = 0$$

Thus, clearly $y(2) \in [2, 3]$.

$\therefore n = 3$

**88.** (2) Given differential equation,

$$dy = e^{\alpha x + y}\ dx$$

$$e^{-y}dy = e^{\alpha x}\ dx$$

On integrating both sides, we get

$$\int e^{-y}dy = \int e^{\alpha x}\ dx$$

$$\Rightarrow \quad -e^{-y} = \frac{e^{\alpha x}}{\alpha} + c \quad \ldots\text{(i)}$$

On putting $(x, y) = (\log_e 2, \log_e 2)$

$$\Rightarrow \quad -\frac{1}{2} = \frac{2^\alpha}{\alpha} + c \quad \ldots\text{(ii)}$$

On putting $(x, y) = (0, -\log_e 2)$ in Eq. (i), we get $-2 = \frac{1}{\alpha} + c$ ...(iii)

On subtracting Eq. (iii) from Eq. (ii), we get

$$-\frac{1}{2} + 2 = \frac{2^\alpha - 1}{\alpha} \Rightarrow \frac{2^\alpha - 1}{\alpha} = \frac{3}{2}$$

$$\Rightarrow \quad \alpha = 2$$

(as $\alpha \in N$)

**89.** (2) $\sec y\frac{dy}{dx} - \sin(x+y) - \sin(x-y) = 0$

$$\sec y\frac{dy}{dx} = \sin(x+y) + \sin(x-y)$$

$$\Rightarrow \quad \sec y\frac{dy}{dx} = 2\sin x\cos y$$

$$\Rightarrow \int\sec^2 y\ dy = \int 2\sin x\ dx$$

$$\Rightarrow \quad \tan y = -2\cos x + c$$

$$y(0) = 0$$

$$\Rightarrow \quad \tan 0 = -2 + c$$

$$\therefore \quad c = 2$$

$$y\left(\frac{\pi}{2}\right) \Rightarrow \tan y = -2\cos\frac{\pi}{2} + 2$$

$$\tan y = 2$$

$$\Rightarrow \sec^2 y = 1 + \tan^2 y = 1 + 4 = 5$$

$$y'\left(\frac{\pi}{2}\right) \Rightarrow \sec^2 y\frac{dy}{dx} = 2\sin x$$

$$\Rightarrow 5\frac{dy}{dx} = 2\sin\frac{\pi}{2} \quad \Rightarrow 5\left(\frac{dy}{dx}\right)_{x=\frac{\pi}{2}} = 2$$

**90.** (1) We have,

$$\frac{dy}{dx} = \frac{2y}{x\log_e x} \Rightarrow \frac{dy}{y} = \frac{2dx}{x\log_e x}$$

On integrating both sides, we get

$$\int\frac{dy}{y} = \int\frac{2dx}{x\log_e x}$$

$$\Rightarrow \log_e|y| = 2\log_e|\log_e x| + c \quad \ldots\text{(i)}$$

Putting $x = 2,\ y = (\log_e 2)^2$

$$\Rightarrow \log_e|(\log_e 2)^2| = 2\log_e|\log_e 2| + c$$

$$\Rightarrow 2\log_e|\log_e 2| = 2\log_e|\log_e 2| + c$$

$$\Rightarrow \quad c = 0$$

On substituting in Eq. (i), we get

$$\log_e|y| = 2\log_e|\log_e x|$$

$$\Rightarrow \quad y = (\log_e x)^2$$

Now, value of $y$ at $x = e$

$$f(e) = (\log_e e)^2 = 1$$

**91.** (4) Given,

$$\left[(x+2)e^{\left(\frac{y+1}{x+2}\right)} + (y+1)\right]dx = (x+2)dy$$

$$y(1) = 1 \quad \therefore \frac{dy}{dx} = e^{\frac{y+1}{x+2}} + \left(\frac{y+1}{x+2}\right)$$

Let $y + 1 = Y \Rightarrow x + 2 = X$

$$\Rightarrow \frac{dY}{dX} = e^{\frac{Y}{X}} + \frac{Y}{X} \Rightarrow \frac{dY}{dX} - e^{\frac{Y}{X}} = \frac{Y}{X}$$

Let $Y = VX \quad \Rightarrow \frac{dY}{dX} = V + X\frac{dV}{dX}$

$$\Rightarrow V + X\frac{dV}{dX} - e^V = V$$

$$\Rightarrow \int\frac{dV}{e^V} = \int\frac{dX}{X}$$

$$\Rightarrow \frac{e^{-V}}{-1} = \log_e X + c$$

$$\Rightarrow -e^{\frac{-Y}{X}} = \log_e|X| + c$$

$$\Rightarrow \quad x + 2 = X \Rightarrow 1 + 2 = 3$$

$$\Rightarrow \quad y + 1 = Y \Rightarrow 1 + 1 = 2$$

$$\Rightarrow -e^{\frac{-2}{3}} = \log_e 3 + c$$

$$\Rightarrow c = -e^{\frac{-2}{3}} - \log_e 3$$

$$\Rightarrow e^{\frac{-Y}{X}} = e^{\frac{-2}{3}} + \log_e 3 - \log_e|X| > 0$$

$$\log_e|X| < e^{\frac{-2}{3}} + \log_e 3$$

$$|x+2| < e^{(e^{-2/3} + \log_e 3)}$$

$$\Rightarrow |x+2| < k \Rightarrow -k < x + 2 < k$$

$$\Rightarrow -k - 2 < x < k - 2$$

$$\therefore \alpha + \beta = -4$$

$$\Rightarrow |\alpha + \beta| = 4$$

**92.** (2) Given equation,

$$\cos\left(\frac{1}{2}\cos^{-1}(e^{-x})\right)dx = \sqrt{e^{2x} - 1}\ dy \quad \ldots\text{(i)}$$

Let $\cos^{-1}(e^{-x}) = \theta$, then $e^{-x} = \cos\theta$

$$\Rightarrow 2\cos^2\frac{\theta}{2} - 1 = e^{-x}$$

$$\Rightarrow \cos\frac{\theta}{2} = \sqrt{\frac{e^{-x}+1}{2}} \quad \ldots\text{(ii)}$$

From Eq. (i),

$$\cos\left(\frac{1}{2}\cdot\theta\right)dx = \sqrt{e^{2x} - 1}\ dy$$

From Eqs. (i) and (ii), we get

$$\sqrt{\frac{e^{-x}+1}{2}}\cdot dx = \sqrt{e^{2x} - 1}\ dy$$

$$\text{or} \quad \frac{\sqrt{e^{-x}+1}}{\sqrt{e^{2x}-1}}dx = \sqrt{2}\ dy$$

$$\text{or} \quad \frac{\sqrt{1+e^x}}{\sqrt{e^x}\cdot\sqrt{e^{2x}-1}}dx = \sqrt{2}\ dy$$

On putting $e^x = t$, then $e^x\ dx = dt$

$$\text{or } dx = \frac{dt}{t}, \ \frac{\sqrt{1+t}}{\sqrt{t}\sqrt{t^2-1}}\frac{dt}{t} = \sqrt{2}\ dy$$

$$\Rightarrow \frac{\sqrt{1+t}}{\sqrt{t}\sqrt{t+1}\cdot\sqrt{t-1}}\frac{dt}{t} = \sqrt{2}\ dy$$

$$\Rightarrow \frac{dt}{t\sqrt{t}\sqrt{t-1}} = dy\cdot\sqrt{2}$$

On putting $t = \frac{1}{z}$, then $dt = -\frac{1}{z^2}dz$

$$\frac{-\frac{dz}{z^2}}{\frac{1}{z}\sqrt{\frac{1}{z^2}-\frac{1}{z}}}=\sqrt{2}\,dy-\frac{dz}{\sqrt{1-z}}=\sqrt{2}dy$$

On integrating both sides, we get

$$-\int\frac{dz}{\sqrt{1-z}}=\sqrt{2}\int dy$$

$$\Rightarrow \quad -2\frac{(1-z)^{1/2}}{-1}=\sqrt{2}y+c$$

$$\Rightarrow \quad 2\left(1-\frac{1}{t}\right)^{1/2}=\sqrt{2}y+c$$

$$\Rightarrow \quad 2(1-e^{-x})^{\frac{1}{2}}=\sqrt{2}y+c \quad \text{...(iii)}$$

Given condition, $y(0)=-1$

From Eq. (iii),

$$2(1-1)^{\frac{1}{2}}=\sqrt{2}(-1)+c \Rightarrow c=\sqrt{2}$$

From Eq. (iii),

$$2\sqrt{1-e^{-x}}=\sqrt{2}(y+1) \quad \text{...(iv)}$$

It passes through $(\alpha, 0)$, putting in Eq. (iv),

$$2\sqrt{1-e^{-\alpha}}=\sqrt{2}(0+1)$$

$$\Rightarrow \sqrt{1-e^{-\alpha}}=\frac{1}{\sqrt{2}} \Rightarrow 1-e^{-\alpha}=\frac{1}{2}$$

$$\Rightarrow \quad e^{-\alpha}=\frac{1}{2} \quad \Rightarrow \quad e^{\alpha}=2$$

**93.** *(4)* $x\,dy-y\,dx=\sqrt{x^2-y^2}\,dx$

$$\Rightarrow \frac{xdy-ydx}{x^2}=\frac{1}{x}\sqrt{1-\frac{y^2}{x^2}}\,dx$$

$$\Rightarrow \frac{d(y/x)}{\sqrt{1-(y/x)^2}}=\frac{1}{x}\,dx$$

On integrating both sides, we get

$$\int\frac{1}{\sqrt{1-\left(\frac{y}{x}\right)^2}}\cdot d(y/x)=\int\frac{1}{x}\,dx$$

$$\Rightarrow \quad \sin^{-1}(y/x)=\log|x|+C$$

Now, at $x=1$, $y=0$

$\Rightarrow \quad \therefore C=0$

Hence, $y=x\sin(\log x)$

$$\therefore \quad A=\int_1^{e^{\pi}}x\sin(\log x)\,dx$$

On putting $x=e^t$

$$\Rightarrow \quad dx=e^t\,dt$$

$$\therefore \quad A=\int_0^{\pi}e^{2t}\sin(t)\,dt$$

$$\left[\text{using} \int e^{ax}\cdot\sin bx=\frac{e^{ax}}{a^2+b^2}(a\sin bx-b\cos bx)\right]$$

$$\Rightarrow \alpha e^{2\pi}+\beta=\left(\frac{e^{2t}}{5}(2\sin t-\cos t)\right)_0^{\pi}$$

$$=\frac{1+e^{2\pi}}{5}; \quad \therefore \alpha=\frac{1}{5}, \beta=\frac{1}{5}$$

$$\therefore 10(\alpha+\beta)=4$$

**94.** *(2)* We have, $dy/dx=2(x+1)$

$$y=\int 2(x+1)\,dx=2\left(\frac{x^2}{2}+x\right)+c$$

$$y=x^2+2x+c$$

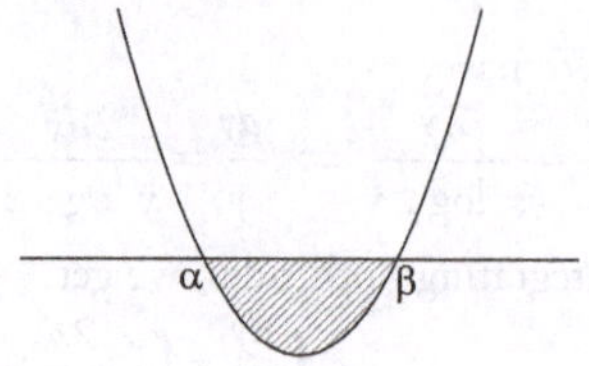

$$\text{Now,} \int_{\alpha}^{\beta}y\,dx=\frac{4\sqrt{8}}{3}$$

$$\int_{\alpha}^{\beta}(x^2+2x+c)\,dx=\frac{4\sqrt{8}}{3}$$

$$\Rightarrow \left[\frac{x^3}{3}+x^2+cx\right]_{\alpha}^{\beta}=\frac{4\sqrt{8}}{3}$$

$$\Rightarrow \left(\frac{\beta^3-\alpha^3}{3}\right)+(\beta^2-\alpha^2)+c(\beta-\alpha)=\frac{4\sqrt{8}}{3}$$

$$\alpha+\beta=-2 \text{ and } \alpha\beta=c$$

$$\therefore \beta-\alpha=\sqrt{(\alpha+\beta)^2-4\alpha\beta}=\sqrt{4-4c}=2\sqrt{1-c}$$

$$\Rightarrow \frac{1}{3}(\beta-\alpha)(\alpha^2+\beta^2+\alpha\beta)+(\beta-\alpha)(\beta+\alpha)+c(\beta-\alpha)=\frac{4\sqrt{8}}{3}$$

$$\Rightarrow \frac{1}{3}(2\sqrt{1-c})(4-c)+(2\sqrt{1-c})(-2+c)=\frac{4\sqrt{8}}{3}$$

$$\Rightarrow \quad (2\sqrt{1-c})[4-c+(-6+3c)]=4\sqrt{8}$$

$$\Rightarrow \quad (2\sqrt{1-c})[2(c-1)]=4\sqrt{8}$$

$$\Rightarrow \quad (1-c)^{3/2}=-\sqrt{8}$$

$$\Rightarrow \quad (1-c)^3=8 \Rightarrow 1-c=2$$

$$\therefore \quad c=-1$$

Now, $y=x^2+2x-1$

$\therefore \quad y(1)=1^2+2\cdot 1-1=2$

## Topic 3

**01.** *(d)* We have,

$$(x^2+1)y'-2xy=(x^4+2x^2+1)\cos x$$

$$y'-\frac{2x}{1+x^2}y=\frac{(1+x^2)^2}{1+x^2}\cos x$$

$$y'-\frac{2x}{1+x^2}y=(1+x^2)\cos x$$

This is linear differential equation of the form $\frac{dy}{dx}+Py=Q$

Where $P=-\frac{2x}{1+x^2}$

$$\Rightarrow \int Pdx=\int-\frac{2x}{1+x^2}dx=-\ln(1+x^2)$$

Find integrating factor

$$\text{IF}=e^{\int Pdx}=e^{-\ln(1+x^2)}=\frac{1}{1+x^2}$$

The solution is given by

$$y\times\text{IF}=\int Q\times\text{IF}dx+C$$

$$y\frac{1}{1+x^2}=\int(1+x^2)\cos x\times\frac{1}{1+x^2}\,dx$$

$$\frac{y}{1+x^2}=\int\cos x\,dx \Rightarrow \text{At } y(0)=1$$

$$\frac{1}{1+0}=\sin 0+C \Rightarrow C=1$$

$$\Rightarrow \frac{y}{1+x^2}=\sin x+1$$

$$\Rightarrow \quad y=(1+x^2)(\sin x+1)$$

Now, $\int_{-3}^{3}y(x)dx$

$$=\int_{-3}^{3}(1+x^2)(\sin x+1)dx$$

$$=\int_{-3}^{3}(\sin x+1+x^2\sin x+x^2)dx$$

Here, $\sin x$ and $x^2\sin x$ are odd function, so their integral is zero.

$$\int_{-3}^{3}1+x^2dx=\left(x+\frac{x^3}{3}\right)_{-3}^{3}$$

$$=(3+9)-(-3-9)=12+12=24$$

**02.** *(a)* Given, differential equation is

$$(xy-5x^2\sqrt{1+x^2})dx+(1+x^2)dy=0$$

$$\Rightarrow \left(\frac{xy}{1+x^2}-\frac{5x^2\sqrt{1+x^2}}{1+x^2}\right)dx=-dy$$

$\Rightarrow \frac{dy}{dx} = \frac{5x^2}{\sqrt{1+x^2}} - \frac{x}{1+x^2}y$

$\Rightarrow \frac{dy}{dx} + \frac{x}{1+x^2}y = \frac{5x^2}{\sqrt{1+x^2}}$ which is linear differential equation

$\text{IF} = e^{\int \frac{x}{1+x^2}dx} = e^{\frac{1}{2}\int \frac{2x}{1+x^2}dx}$

$= e^{1/2\ln(1+x^2)} = e^{\ln(1+x^2)^{1/2}} = \sqrt{1+x^2}$

So, its solution is

$y\times(\text{IF}) = \int Q\times(\text{IF})dx$

$\Rightarrow y\times\sqrt{1+x^2} = \int \frac{5x^2}{\sqrt{1+x^2}}\times\sqrt{1+x^2}\,dx$

$= 5\int x^2 dx$

$= 5\times\frac{x^3}{3} + C$ ...(i)

But given $y(0)=0$

From (i), we get

$0\times\sqrt{1+0^2} = 5\times\frac{(0)^3}{3} + C$

$\Rightarrow C = 0$

$\therefore y = \frac{5x^3}{3\sqrt{1+x^2}}$

$\therefore y(\sqrt{3}) = \frac{5(\sqrt{3})^3}{3\sqrt{1+3}} = \frac{5\sqrt{3}}{2}$

**03.** *(d)* We have,

$f(x)\sin 2x + \sin x - (1+\cos^2 x)f'(x) = 0$

$\Rightarrow y\sin 2x + \sin x - (1+\cos^2 x)\frac{dy}{dx} = 0$

$\Rightarrow \frac{dy}{dx} - \left(\frac{\sin 2x}{1+\cos^2 x}\right)y = \frac{\sin x}{1+\cos^2 x}$

$\therefore \text{IF} = e^{\int P dx} = e^{\int \frac{-\sin 2x}{1+\cos^2 x}dx}$

$= e^{\log(1+\cos^2 x)} = 1+\cos^2 x$

$\therefore y(\text{IF}) = \int (Q\times\text{IF})\,dx + C$

$\Rightarrow y(1+\cos^2 x)$

$= \int \frac{\sin x}{(1+\cos^2 x)}(1+\cos^2 x)\,dx + c$

$= \int \sin x\,dx + c$

$= -\cos x + c$

Given when, $x=0$, then $y=0$

$\Rightarrow 0 = -1 + c$

$\Rightarrow c = 1$

$\therefore y(1+\cos^2 x) = -\cos x + 1$

$\Rightarrow y(\pi/2) = 1$

**04.** *(a)* $\frac{dy}{dx} - y = x$, $\text{IF} = e^{-x}$

$\therefore y\cdot e^{-x} = \int (x\cdot e^{-x})dx + C$

$\Rightarrow ye^{-x} = -xe^{-x} - e^{-x} + C$

$\therefore y = -x - 1 + C\cdot e^x$

$y_1(0) = 0 \Rightarrow 0 = -1 + C \Rightarrow C = 1$

$\therefore y_1(x) = -x - 1 + e^x$

$y_2(0) = 1 \Rightarrow 1 = -1 + C \Rightarrow C = 2$

$\therefore y_2(x) = -x - 1 + 2e^x$

For point of intersection,

$-x - 1 + e^x = -x - 1 + 2e^x$

$\Rightarrow e^x = 0$, which is not possible.

$\therefore$Number of points of intersection is 0.

**05.** *(a)* The given differential equation is

$\frac{dx}{dy} = \frac{2x}{y} + y^2(y+1)e^y$, $x(1)=0$

$\Rightarrow \frac{dx}{dy} - \frac{2x}{y} = y^2(y+1)e^y$

This is the form of linear differential equation thus,

$\text{IF} = e^{\int \frac{-2}{y}dy} = e^{-2\ln y} = \frac{1}{y^2}$

Thus, the solution is

$x\cdot\frac{1}{y^2} = \int y^2(y+1)\cdot e^y\cdot\frac{1}{y^2}dy$

$= \int (y+1)e^y dy \Rightarrow \frac{x}{y^2} = ye^y + C$

Thus, $x(1) = 0$

So, $0 = 1(e^1) + C \Rightarrow C = -e$

Now, $\frac{x}{y^2} = (ye^y - e)$

At $y = e$

$x = e^2(ee^e - e) = e^3(e^e - 1)$

**06.** *(d)* Given, differential equation

$(2x - 10y^3)\,dy + ydx = 0$

$\Rightarrow \frac{dx}{dy} + \frac{2x}{y} = 10y^2$ ... (i)

This is Linear differential equation

Integrating factor $\text{IF} = e^{2\int\frac{1}{y}dy} = y^2$

Solution of differential Eq. (i),

$x\cdot y^2 = \int 10y^2\cdot y^2 dy + C$

$\Rightarrow xy^2 = 2y^5 + C$ ... (ii)

Solution Eq. (ii) passes through (0, 1)

$\Rightarrow 0\cdot 1^2 = 2\cdot 1^5 + C$

$\Rightarrow C = -2$

$\therefore$ Solution of Eq. (i) is

$xy^2 = 2y^5 - 2$

Now, this equation passes through $(2, \beta)$.

$\therefore 2\cdot\beta^2 = 2\beta^5 - 2$

$\Rightarrow \beta^5 - \beta^2 - 1 = 0$

$\Rightarrow \beta$ is root of the equation

$y^5 - y^2 - 1 = 0$

**07.** *(c)* Given, $f(x) + xf'(x) = x^2$

$\Rightarrow f'(x) + \frac{f(x)}{x} = x$

$\Rightarrow \frac{dy}{dx} + \frac{1}{x}y = x$

$\left[\because y = f(x) \Rightarrow \frac{dy}{dx} = f'(x)\right]$

This is linear differential equation.

$\therefore$Integrating factor $\text{IF} = e^{\int\frac{1}{x}dx} = x$

Solution, $y\cdot x = \int x\cdot x\,dx + C$

or $xy = \frac{x^3}{3} + C$

$\because$ It passes through $(-2, 2)$.

$\therefore -2\cdot 2 = \frac{(-2)^3}{3} + C \Rightarrow C = -\frac{4}{3}$

Hence, $xf(x) = \frac{x^3}{3} - \frac{4}{3}$

or $x^3 - 3f(x)\cdot x - 4 = 0$

**08.** *(b)* Given, $\frac{dy}{dx} + (\tan x)y = \sin x$, $x\in[0, \pi/3]$

which is a linear differential equation of the form of $\frac{dy}{dx} + Py = Q$

$\therefore \text{IF} = e^{\int P dx}$

$\Rightarrow e^{\int \tan x\,dx} = e^{\log(\sec x)} = \sec x$

$\therefore$ Its solution is

$y\sec x = \int \tan x\,dx$

$\Rightarrow y\sec x = \log(\sec x) + c$

$y = \cos x\log(\sec x) + c\cdot\cos x$

$y(0) = 0 \Rightarrow 0 = 1\cdot 0 + c\cdot 1$

$\Rightarrow c = 0$

$\therefore y = \cos x\cdot\log(\sec x)$

$\Rightarrow y\left(\frac{\pi}{4}\right) = \cos\left(\frac{\pi}{4}\right)\cdot\log\left(\sec\frac{\pi}{4}\right)$

$= \frac{1}{\sqrt{2}}\log(\sqrt{2}) = \frac{1}{2\sqrt{2}}\log 2$

**09.** *(c)* Given differential equation

$(\cos x)\frac{dy}{dx}+2y\sin x=\sin 2x,\ x\in\left(0,\frac{\pi}{2}\right)$

$\Rightarrow \frac{dy}{dx}+2(\tan x)y=2\sin x,$

is in the form of linear differential equation, so

$IF=e^{2\int \tan x dx}=\sec^2 x$

$\therefore$ Solution is

$y(\sec^2 x)=\int 2\tan x\sec x dx+c$

$=2\sec x+c$

$\because y\left(\frac{\pi}{3}\right)=0\Rightarrow c=-4$

So, the required equation is

$y(\sec^2 x)=2\sec x-4$

At $x=\frac{\pi}{4}$, we get $2y=2\sqrt{2}-4$

$\therefore y\left(\frac{\pi}{4}\right)=\sqrt{2}-2$

**10.** *(d)* From given, $ydx-(x+3y^2)dy=0$

$\Rightarrow ydx=(x+3y^2)dy$

$\Rightarrow \frac{dx}{dy}=\frac{x}{y}+3y$

$\Rightarrow \frac{dx}{dy}-\frac{x}{y}=3y$

Above equation is in the form of linear differential equation.

$\therefore IF\ e^{-\int\frac{1}{y}dy}=e^{-\ln y}=\frac{1}{y}$

$\therefore \frac{x}{y}=\int 3y\cdot\frac{1}{y}dy$

$\Rightarrow \frac{x}{y}=3y+c$ ...(i)

Given, Eq. (i) passes through (1, 1).

$\therefore 1=3+c$

$\Rightarrow c=-2$

$\therefore$ Required curve is $x=3y^2-2y$

This curve also passes through the point $\left(-\frac{1}{3},\frac{1}{3}\right)$

**11.** *(a)* We have,

$(2+\sin x)\frac{dy}{dx}+(y+1)\cos x=0$

$\Rightarrow \frac{dy}{dx}+\frac{\cos x}{2+\sin x}y=\frac{-\cos x}{2+\sin x}$

which is a linear differential equation.

$\therefore IF=e^{\int\frac{\cos x}{2+\sin x}dx}$

$=e^{\log(2+\sin x)}$

$=2+\sin x$

$\therefore$ Required solution is given by

$y\cdot(2+\sin x)=\int\frac{-\cos x}{2+\sin x}\cdot(2+\sin x)dx+C$

$\Rightarrow y(2+\sin x)=-\sin x+C$

Also, $y(0)=1$

$\therefore 1(2+\sin 0)=-\sin 0+C$

$\Rightarrow C=2$

$\therefore y=\frac{2-\sin x}{2+\sin x}$

$\Rightarrow y\left(\frac{\pi}{2}\right)=\frac{2-\sin\frac{\pi}{2}}{2+\sin\frac{\pi}{2}}=\frac{1}{3}$

**12.** *(c)* Given differential equation is

$(x\log x)\frac{dy}{dx}+y=2x\log x,$ $(x\geq 1)$

$\Rightarrow \frac{dy}{dx}+\frac{y}{x\log x}=2$

This is a linear differential equation.

$\therefore IF=e^{\int\frac{1}{x\log x}dx}=e^{\log(\log x)}=\log x$

Now, the solution of given differential equation is given by

$y\cdot\log x=\int\log x\cdot 2dx$

$\Rightarrow y\cdot\log x=2[x\log x-x]+c$

At $x=1$, $c=2$

$\Rightarrow y\cdot\log x=2[x\log x-x]+2$

At $x=e$,

$y=2(e-e)+2$

$\Rightarrow y=2$

**13.** *(d)* $\sin 2x\left(\frac{dy}{dx}-\sqrt{\tan x}\right)-y=0$

[from given]

or, $\frac{dy}{dx}=\frac{y}{\sin 2x}+\sqrt{\tan x}$

or, $\frac{dy}{dx}-y\ \text{cosec}\ 2x=\sqrt{\tan x}$ ...(i)

Now, $IF=e^{\int-\text{cosec}\ 2x}$

or, $IF=e^{-\frac{1}{2}\log|\tan x|}=e^{\log(\sqrt{\tan x})^{-1}}$

$=\frac{1}{\sqrt{\tan x}}=\sqrt{\cot x}$

Now, general solution of Eq. (i)

$y(IF)=\int Q(IF)dx+c$

$\therefore y\sqrt{\cot x}=\int\sqrt{\tan x}\sqrt{\cot x}\,dx+c$

$\therefore y\sqrt{\cot x}=\int 1\cdot dx+c$

$\therefore y\sqrt{\cot x}=x+c$

**14.** *(b)* Given, $(x^2-1)\frac{dy}{dx}+2xy=x$

$\Rightarrow \frac{dy}{dx}+\frac{2x}{x^2-1}\cdot y=\frac{x}{x^2-1}$

This is linear differential equation

$\therefore IF=e^{\int\frac{2x}{x^2-1}dx}=e^{\int\frac{dt}{t}}$

where, $t=x^2-1$

$=e^{\log t}=x^2-1$

Hence, required $IF=x^2-1$.

**15.** *(a)* Given equation can be rewritten as

$\frac{dy}{dx}-\frac{1}{x}\cdot y=1$

Now, $IF=e^{-\int\frac{1}{x}dx}=e^{-\log x}=\frac{1}{x}$

$\therefore$ Required solution,

$y\left(\frac{1}{x}\right)=\int\frac{1}{x}dx=\log x+C$

$\because y(1)=1$

$\Rightarrow 1=\log 1+C$

$\Rightarrow C=1$

$\therefore y=x\log x+x$

**16.** *(d)* $f:[1,\infty)\rightarrow[2,\infty)$

Given, $10\int_1^x f(t)dt=5x\,f(x)-x^5-9$

Put $x=1$

$\Rightarrow 0=5f(1)-10$

$\Rightarrow f(1)=2$ ...(i)

Now, on differentiating w.r.t $x$ of given equation,

$10[1\cdot f(x)-0]=5f(x)+5xf'(x)-5x^4$

$\Rightarrow f(x)=xf'(x)-x^4$

Put $f(x)=y$ and $f'(x)=\frac{dy}{dx}$

$\Rightarrow y=x\cdot\frac{dy}{dx}-x^4$

$\Rightarrow \frac{dy}{dx}-\frac{y}{x}=x^3$

Integrating factor,

$IF=e^{-\int\frac{1}{x}dx}=e^{-\ln x}=\frac{1}{x}$

Solution of differential equation,

$y\cdot\left(\frac{1}{x}\right)=\int\frac{1}{x}(x^3)\cdot dx+C$

$$\Rightarrow \quad \frac{y}{x} = \frac{x^3}{3} + C$$

$$\Rightarrow \quad f(x) = \frac{x^4}{3} + Cx$$

Put $x = 1$

$$f(1) = \frac{1}{3} + C$$

From Eq. (i), we get

$$2 = \frac{1}{3} + C \Rightarrow C = \frac{5}{3}$$

$$\therefore \quad f(x) = \frac{x^4}{3} + \frac{5}{3}\cdot x$$

$$\Rightarrow \quad f(3) = 27 + 5 = 32$$

**17.** (a) $\int_0^x g(t)\,dt = x - \int^x t\,g(t)\,dt$

Differentiating both sides, we get

$g(x) = 1 - x\,g(x)$

$$\Rightarrow g(x)(1+x) = 1 \Rightarrow g(x) = \frac{1}{1+x}$$

$$\Rightarrow \frac{dy}{dx} - y\tan x = 2(x+1)\sec x \cdot g(x)$$

$$\Rightarrow \text{IF} = e^{-\int \tan x\,dx} = e^{-(-\log\cos x)}$$

$\text{IF} = \cos x$

$$\Rightarrow y\cos x = \int 2(x+1)\sec x \frac{\cos x}{(1+x)}\,dx$$

$$\Rightarrow y\cos x = \int 2dx \quad \text{...(i)}$$

$$\Rightarrow y\cos x = 2x + C \quad \Rightarrow \quad y(0) = 0$$

$C = 0$

From Eq., (i), we get

$$y\cos x = dx$$

$$x = \frac{\pi}{3} \Rightarrow y\cdot\frac{1}{2} = \frac{2\pi}{3}$$

$$y = \frac{4\pi}{3}$$

**18.** (d) $\frac{dy}{dx} + 3y(1+\tan^2 x) = \sec^2 x$

$$\Rightarrow \quad \frac{dy}{dx} + y(3\sec^2 x) = \sec^2 x$$

$$\text{IF} = e^{\int 3\sec^2 x\,dx} = e^{3\tan x}$$

$$\Rightarrow \quad y(e^{3\tan x}) = \int e^{3\tan x}\cdot\sec^2 x\,dx + C$$

$$= \frac{e^{\tan x}}{3} + C = \frac{1}{3} + e^{-3\tan x}\cdot C$$

$y(0) = \frac{1}{3} + e^x$ [given]

Put $x = 0$

$$\frac{1}{3} + e^3 = \frac{1}{3} + C\cdot e^{-3\tan(0)}$$

$$\frac{1}{3} + e^3 = \frac{1}{3} + C$$

$$C = e^3$$

$$\therefore \quad y(x) = \frac{1}{3} + e^3\cdot e^{-3\tan x}$$

$$y(x) = \frac{1}{3} + e^{3-3\tan x}$$

put $x = \frac{\pi}{4}$ $\quad y\left(\frac{\pi}{4}\right) = \frac{1}{3} + e^{3-3\tan\left(\frac{\pi}{4}\right)}$

$$= \frac{1}{3} + e^{3-3} = \frac{1}{3} + 1 = \frac{4}{3}$$

**19.** (b) $f(x) = 1 - 2x + \int_0^x e^{x-t} f(t)\,dt$

$$f(x) = 1 - 2x + e^x \int_0^x e^{-t} f(t)\,dt$$

On differentiating w.r.t. $x$, we get

$$f'(x) = -2 + e^x\{e^{-x} f(x)\} + e^x \int e^{-t} f(t)\,dt$$

$$\Rightarrow \quad \frac{dy}{dx} = -2 + y + (y + 2x - 1)$$

$$\Rightarrow \quad \frac{dy}{dx} = -2 + 2y + 2x - 1$$

$$\Rightarrow \frac{dy}{dx} - 2y = 2x - 3$$

Linear differential equation

$$\text{IF} = e^{\int -2dx} = e^{-2x}$$

Solution of differential equation

$$ye^{-2x} = \int (2x-3)e^{-2x}\,dx + C$$

$$\Rightarrow \quad ye^{-2x} = -\frac{(2x-3)}{2}e^{-2x} - \frac{1}{2}e^{-2x} + C$$

$x = 0,\ y = 1$

$$1 = \frac{3}{2} - \frac{1}{2} + C$$

$C = 0$

$$\therefore \quad y = -\frac{2x-3}{2} - \frac{1}{2}$$

$y = -x + 1$

$x + y = 1$

$$\therefore \text{Required area} = \frac{1}{2}\times|x| = \frac{1}{2}$$

**20.** (b) $(7x^4\cot y - e^x\,\text{cosec}\,y)\frac{dx}{dy} = x^5$

$$(7x^4\cos y - e^x) = x^5\sin y\frac{dy}{dx}$$

$$x^5\sin y\frac{dy}{dx} - 7x^4\cos y = -e^x$$

Put $\quad -\cos y = t$

$$\sin y\frac{dy}{dx} = \frac{dt}{dx}$$

$$x^5\frac{dt}{dx} + (7x^4)t = -e^x$$

$$\frac{dt}{dx} + \frac{7}{x}t = \frac{-e^x}{x^5}$$

It is linear differential equation.

$$\text{IF} = e^{\int\frac{7}{x}dx} = e^{7\ln x} = x^7$$

$\therefore$ Solution of differential equation

$$\Rightarrow \quad t(x^7) = -\int\frac{e^x}{x^5}(x^7)\,dx + c$$

$$\Rightarrow \quad tx^7 = -\int x^2e^x\,dx + c$$

$$\Rightarrow \quad tx^7 = -e^x(x^2 - 2x + 2) + c$$

$$\Rightarrow (-\cos y)x^7 = -e^x(x^2 - 2x + 2) + c$$

$\therefore$ Curve passing through $\left(1, \frac{\pi}{2}\right)$

$$0 = -e(1) + c \Rightarrow c = e$$

$$\therefore (-\cos y)x^7 = -e^x(x^2 - 2x + 2) + e$$

$$\therefore \quad x = 2 \Rightarrow -\cos y(2^7) = -2e^2 + e$$

$$\cos y = \frac{2e^2 - e}{128}$$

**21.** (c) $x(x^2 + e^x)dy + (e^x(x-2)y - x^3)dx = 0,\ x > 0$

$$\Rightarrow \quad \frac{dy}{dx} + \frac{e^x(x-2)}{x(x^2+e^x)}\cdot y = \frac{x^3}{x(x^2+e^x)}$$

$$\text{IF} = e^{\int\frac{e^x(x-2)}{x(x^2+e^x)}dx} = e^{\int\frac{e^x+2x}{x^2+e^x}dx - \int\frac{2}{x}dx}$$

$$= e^{\left(\frac{e^x+x^2}{x^2}\right)} = \frac{e^x + x^2}{x^2}$$

$$\therefore \quad y\left(\frac{e^x+x^2}{x^2}\right) = \int dx + c$$

$$\Rightarrow \quad y\left(\frac{e^x+x^2}{x^2}\right) = x + c$$

Also, $y(1) = 0$ given

$$\Rightarrow \quad c = -1$$

$$\therefore \quad y\left(\frac{e^x+x^2}{x^2}\right) = x - 1$$

Hence, $\quad y(2) = \frac{4(1)}{e^2+4} = \frac{4}{e^2+4}$

**22.** (a) Given, and $\left.\begin{array}{l} f(x) = x - 1 \\ g(x) = e^x \end{array}\right\} x \in R$

Also, $\frac{dy}{dx} = e^{-2\sqrt{x}} \times g[f(f(x))] - \frac{y}{\sqrt{x}}$

Also, $y(0) = 0$

$f(f(x))=f(x)-1=x-1-1=x-2$

$g(x-2)=e^{x-2}$

$\Rightarrow \quad \dfrac{dy}{dx}=e^{-2\sqrt{x}}\times e^{x-2}-\dfrac{y}{\sqrt{x}}$

$\Rightarrow \quad \dfrac{dy}{dx}+y\cdot x^{-1/2}=e^{x-2-2\sqrt{x}}$

$\Rightarrow y\times \text{IF}=\int(Q\times\text{IF})\,dx$

$\text{IF}=e^{\int Pdx=e^{\int x^{-1/2}dx}}$

$\text{IF}=e^{2\sqrt{x}}$

$\Rightarrow \quad y\times e^{2\sqrt{x}}=\int\dfrac{e^{x-2}}{e^{2\sqrt{x}}}\times e^{2\sqrt{x}}\,dx$

$\Rightarrow \quad y\times e^{2\sqrt{x}}=\dfrac{1}{e^2}\int e^x\,dx$

$\Rightarrow \quad y\times e^{2\sqrt{x}}=\dfrac{e^x}{e^2}+C$

At $x=0,\ y=0$

$0=\dfrac{1}{e^2}+C \Rightarrow C=-\dfrac{1}{e^2}$

$\Rightarrow y\times e^{2\sqrt{x}}=e^{x-2}-\dfrac{1}{e^2}$

$\Rightarrow \quad y(1)=\dfrac{e^{-1}-\dfrac{1}{e^2}}{e^2}=\dfrac{\dfrac{1}{e}-\dfrac{1}{e^2}}{e^2}=\dfrac{e-1}{e^4}$

**23.** *(a)* Given that

$2(3+y)e^{2x}\,dx-(7+e^{2x})dy=0$

$\Rightarrow \quad \dfrac{dy}{dx}=\dfrac{2(3+y)e^{2x}}{7+e^{2x}}$

$\dfrac{dy}{dx}-\dfrac{2ye^{2x}}{7+e^{2x}}=\dfrac{6\cdot e^{2x}}{7+e^{2x}}$ ...(i)

This is linear differential equation and it's form is

$\dfrac{dy}{dx}+py=Q \qquad [\because \text{IF}=e^{\int pdx}]$

$\text{IF}=e^{\int\frac{-2e^{2x}}{7+e^{2x}}dx}$

Put $7+e^{2x}=t$

$\Rightarrow \quad 2e^{2x}\,dx=dt=e^{-\log t}=\dfrac{1}{t}=\dfrac{1}{7+e^{2x}}$

Solution of Eq. (i), we get

$\therefore \quad y\cdot\dfrac{1}{7+e^{2x}}=\int\dfrac{6e^{2x}}{(7+e^{2x})^2}dx$

$\Rightarrow \quad y\cdot\dfrac{1}{7+e^{2x}}=\dfrac{-3}{7+e^{2x}}+c$

This curve passes through (0, 5), then

$\dfrac{5}{8}=\dfrac{-3}{8}+c \Rightarrow c=1$

$\therefore \quad y=-3+7+e^{2x}$

$\Rightarrow \quad y=e^{2x}+4$

Now, put $x=\log_e 2,\ y=k$, we get

$k=e^{2\log_e 2}+4\Rightarrow k=8$

> • Concept **Enhancer**
>
> Solution linear differential equation $\dfrac{dy}{dx}+py=Q$ is given by
>
> $y(\text{IF})=\int Q\cdot(\text{IF})dx+C,$
>
> where $\text{IF}=e^{\int Pdx}$

**24.** *(d)* $y^2dx+\left(x-\dfrac{1}{y}\right)dy=0$

$\Rightarrow \quad y^2\dfrac{dx}{dy}+x=\dfrac{1}{y}$

$\Rightarrow \quad \dfrac{dx}{dy}+\left(\dfrac{1}{y^2}\right)x=\dfrac{1}{y^3}$

$P=\dfrac{1}{y^2}$ and $Q=\dfrac{1}{y^3}$

$\text{IF}=e^{\int Pdy}=e^{\int\frac{1}{y^2}dy}=e^{-\frac{1}{y}}$

Solution of differential equation

$x(\text{IF})=\int Q\cdot(\text{IF})dy+C$

$xe^{-\frac{1}{y}}=\int\dfrac{1}{y^3}e^{-\frac{1}{y}}\,dy+C \qquad$ ...(i)

Put $\quad e^{-\frac{1}{y}}=t\Rightarrow -\dfrac{1}{y}=\log t$

$e^{-\frac{1}{y}}\cdot\dfrac{1}{y^2}\,dy=dt$

From Eq. (i), we get

$\therefore \quad xe^{-\frac{1}{y}}=\int-\ln t\,dt+C$

$xe^{-\frac{1}{y}}=-t(\ln t-1)+C$

$xe^{-\frac{1}{y}}=-e^{-\frac{1}{y}}\left(-\dfrac{1}{y}-1\right)+C$

$x=\dfrac{1}{y}+1+Ce^{\frac{1}{y}}$

Put $x=y=1$

$1=1+1+Ce$

$\Rightarrow C=-\dfrac{1}{e}$

$\therefore \quad x\left(\dfrac{1}{2}\right)=2+1-\left(\dfrac{1}{e}\right)\cdot e^2=3-e$

> • Concept **Enhancer**
>
> Solution of linear differential equation $\dfrac{dy}{dx}+py=Q$ is $y(\text{IF})=\int Q(\text{IF})dx+C,$
>
> where $\text{IF}=e^{\int Pdx}$

**25.** *(b)* $(1+y^2)+(x-2e^{\tan^{-1}y})\dfrac{dy}{dx}=0$

$\Rightarrow \quad (1+y^2)\dfrac{dx}{dy}+x=2e^{\tan^{-1}y}$

$\Rightarrow \quad \dfrac{dx}{dy}+\dfrac{x}{1+y^2}=\dfrac{2e^{\tan^{-1}y}}{1+y^2}$

$P=\dfrac{1}{1+y^2},\ Q=\dfrac{2e^{\tan^{-1}y}}{1+y^2}$

$\text{IF}=e^{\int P\,dy}=e^{\int\frac{1}{1+y^2}dy}=e^{\tan^{-1}y}$

Solution of differential equation

$x\,(\text{IF})=\int Q\cdot(\text{IF})\,dy+C$

$\Rightarrow \quad xe^{\tan^{-1}y}=2\int\dfrac{(e^{\tan^{-1}y})^2}{1+y^2}dy+C$

Put $e^{\tan^{-1}y}=t\Rightarrow e^{\tan^{-1}y}\dfrac{1}{1+y^2}dy=dt$

$\Rightarrow \quad xe^{\tan^{-1}y}=2\int t\,dt+C$

$\Rightarrow \quad xe^{\tan^{-1}y}=t^2+C$

$\Rightarrow \quad xe^{\tan^{-1}y}=e^{2\tan^{-1}y}+C$

$\Rightarrow \quad x=e^{\tan^{-1}y}+Ce^{-\tan^{-1}y}$

$x=1,\ y=0\Rightarrow 1=e^0+Ce^0$

$C=0$

$\therefore \quad x=e^{\tan^{-1}y}$

at $y=\dfrac{1}{\sqrt{3}},\ x=e^{\tan^{-1}\frac{1}{\sqrt{3}}}=e^{\pi/6}$

**26.** *(d)* $\sec y\dfrac{dy}{dx}+2x\sin y=x^3\cos y$

$\Rightarrow \sec^2 y\cdot\dfrac{dy}{dx}+2x\sin y\sec y=x^3$

$\Rightarrow \sec^2 y\cdot\dfrac{dy}{dx}+2x\cdot\tan y=x^3$

Let $\tan y=t$

$\Rightarrow \sec^2 y\cdot dy=dt \Rightarrow \dfrac{dt}{dx}+2xt=x^3$

$\text{IF}=e^{\int 2x\cdot dx}=e^{x^2}$

So, $t \cdot e^{x^2} = \int x^3 \cdot e^{x^2}\, dx + C$

Let $x^2 = u \Rightarrow x \cdot dx = \frac{1}{2} du$

$\Rightarrow t \cdot e^u = \int \frac{1}{2} u \cdot e^u \cdot du + C$

$\Rightarrow t \cdot e^u = \frac{1}{2}\left[e^u \cdot u - e^u \cdot \frac{d}{du}(u)\right] + C$

$= \frac{1}{2}[u \cdot e^u - e^u] + C$

$\Rightarrow \tan y\, e^{x^2} = \frac{1}{2} e^{x^2}(x^2 - 1) + C$

$\because y(1) = 0$

$0 = \frac{1}{2}(1-1) + C \Rightarrow C = 0$

So, $y(\sqrt{3}) = \tan^{-1}\left(\frac{e^3}{e^3}\right) = \frac{\pi}{4}$

**27.** *(a)*

> **key idea** First convert the given differential equation in the form of
> $\frac{dy}{dx} + Py = Q$
> Then, use formula for
> $y \cdot IF = \int IF \times Q dx + c$
> where $IF = e^{\int P dx}$

According to the question,

$(1 + x^2)\frac{dy}{dx} + y = e^{\tan^{-1} x}$

$\Rightarrow \frac{dy}{dx} + \frac{1}{1+x^2} y = \frac{e^{\tan^{-1} x}}{1+x^2}$

Above equation is linear differential equation of form

$\frac{dy}{dx} + Py = Q$

Now, $IF = e^{\int P dx} = e^{\int \frac{1}{1+x^2} dx}$

$= e^{\tan^{-1} x}$

Now, solution of LDE is

$\Rightarrow y \cdot e^{\tan^{-1} x} = \int e^{\tan^{-1} x} \times \frac{e^{\tan^{-1} x}}{1+x^2} dx$

Let $\tan^{-1} x = t \Rightarrow \frac{dx}{1+x^2} = dt$

$\Rightarrow y \cdot e^{\tan^{-1} x} = \int e^{2t} dt = \frac{e^{2t}}{2} + C$

$\Rightarrow y \cdot e^{\tan^{-1} x} = \frac{e^{2\tan^{-1} x}}{2} + C$

Now, $y(1) = 0$ [given]

$\therefore \quad 0 = \frac{e^{2\tan^{-1}(1)}}{2} + C \Rightarrow C = -\frac{e^{\pi/2}}{2}$

$\Rightarrow y \cdot e^{\tan^{-1} x} = \frac{e^{2\tan^{-1} x}}{2} - \frac{e^{\pi/2}}{2}$

Now, value of $y$ at $x = 0$

$\Rightarrow y \cdot e^0 = \frac{e^0}{2} - \frac{e^{\pi/2}}{2} \Rightarrow y = \frac{1}{2}(1 - e^{\pi/2})$

**28.** *(d)* Given, $\frac{dy}{dx} + 2y = \sin(2x)$

which is differential equation of the form $\frac{dy}{dx} + Py = Q$

$\therefore IF = e^{\int P dx} = e^{\int 2dx} = e^{2x}$

Solution is $y(IF) = \int Q(IF)\, dx + C$

$\Rightarrow ye^{2x} = \int (\sin 2x)(e^{2x})\, dx + C$ ...(i)

Let $I = \int \sin 2x \cdot e^{2x} dx + C$

$= \sin 2x \int e^{2x} dx - \int \left(\frac{d}{dx}(\sin 2x) \int e^{2x} dx\right) dx$

$= \sin 2x \left(\frac{e^{2x}}{2}\right) - \int (2\cos 2x)\left(\frac{e^{2x}}{2}\right) dx$

$= \frac{\sin 2x\, e^{2x}}{2} - \left(\frac{\cos 2x\, e^{2x}}{2} + I\right)$

$\Rightarrow 2I = \frac{e^{2x}}{2}(\sin 2x - \cos 2x) + C$

$\Rightarrow \quad I = \frac{e^{2x}}{4}(\sin 2x - \cos 2x) + C$

From Eq. (i), we have

$ye^{2x} = \frac{e^{2x}}{4}(\sin 2x - \cos 2x) + C$

$\Rightarrow y = \frac{1}{4}(\sin 2x - \cos 2x) + C(e^{-2x})$ ...(ii)

When $x = 0$, then $y = 3/4$

$\therefore \frac{3}{4} = \frac{1}{4}(0 - 1) + C \Rightarrow C = \frac{3}{4} + \frac{1}{4} = 1$

$\therefore y\left(\frac{\pi}{8}\right) = \frac{1}{4}\left(\sin\frac{\pi}{4} - \cos\frac{\pi}{4}\right) + 1(e^{-\pi/4})$ (using Eq. (ii))

$= \frac{1}{4}\left(\frac{1}{\sqrt{2}} - \frac{1}{\sqrt{2}}\right) + e^{-\pi/4} = e^{-\pi/4}$

**29.** *(b)*

> **key idea** First convert the given differential equation in the form $\frac{dy}{dx} + Py = Q$
> Then, $IF = e^{\int P dx}$
> So, the solution of differential equation is
> $y\,(IF) = \int Q(IF) dx + C$

$(x^2 + 4)^2\, dy + (2x^3 y + 8xy - 2)\, dx = 0$

$\Rightarrow (x^2 + 4)^2 \frac{dy}{dx} + (2x^3 + 8x)\, y = 2$

$\Rightarrow \frac{dy}{dx} + \frac{2x(x^2 + 4)}{(x^2 + 4)^2} y = \frac{2}{(x^2 + 4)^2}$

$\Rightarrow \frac{dy}{dx} + \left(\frac{2x}{x^2 + 4}\right) y = \frac{2}{(x^2 + 4)^2}$

$P = \frac{2x}{x^2 + 4}, Q = \frac{2}{(x^2 + 4)^2}$

$IF = e^{\int P\, dx} = e^{\int \frac{2x}{x^2 + 4} dx}$

$= e^{\ln x^2 + 4} = x^2 + 4$

Solution of differential equation is

$y(IF) = \int Q \cdot (IF)\, dx + C$

$\Rightarrow y(x^2 + 4) = \int \frac{2}{(x^2 + 4)^2}(x^2 + 4) + C$

$\Rightarrow y(x^2 + 4) = 2 \times \frac{1}{2}\tan^{-1}\left(\frac{x}{2}\right) + C$

$\Rightarrow y(x^2 + 4) = \tan^{-1}\left(\frac{x}{2}\right) + C$

$\therefore \quad y(0) = 0 \Rightarrow x = 0, y = 0$

$0(0 + 4) = \tan^{-1}(0) + C$

$\Rightarrow \quad 0 = 0 + C \Rightarrow C = 0$

$\Rightarrow \quad 8y = \frac{\pi}{4} \Rightarrow C = 0$

$\therefore y(x^2 + 4) = \tan^{-1}\left(\frac{x}{2}\right)$

$\therefore$ At $x = 2$

$\Rightarrow y(4 + 4) = \tan^{-1}(1) \Rightarrow 8y = \frac{\pi}{4}$

$\Rightarrow \quad y = \frac{\pi}{32}$

$\therefore y(2) = \frac{\pi}{32}$

**30.** *(c)* The given differential equation is

$2(y + 2)\log_e(y + 2)\, dx$
$+ (x + 4 - 2\log_e(y + 2))\, dy = 0$

$\Rightarrow 2\ln(y + 2) + (x + 4 - 2\ln(y + 2))$
$\frac{1}{y + 2}\frac{dy}{dx} = 0$

Let $\ln(y + 2) = z$

$\Rightarrow \quad \frac{1}{y + 2} \cdot \frac{dy}{dx} = \frac{dz}{dx}$

$\Rightarrow \quad 2z + (x + 4 - 2z)\frac{dz}{dx} = 0$

$\Rightarrow \quad \frac{dz}{dx} = -\frac{2z}{x + 4 + 2z}$

$$\Rightarrow \frac{dx}{dz} = -\frac{x+4-2z}{2z}$$

$$\Rightarrow \frac{dx}{dz} + \frac{x}{2z} = \frac{2z-4}{2z}$$

On comparing $\frac{dx}{dz} + Px = Q$, whose

$$P = \frac{1}{2z}, Q = \frac{2z-4}{2z}$$

$$\text{IF} = e^{\int \frac{1}{2z} dz} = z^{1/2}$$

Thus, solution is

$$x \cdot z^{1/2} = \int \left\{ \frac{2z-4}{2z} \cdot z^{1/2} \right\} dt$$

$$= \int \left\{ \frac{2z-4}{2z^{1/2}} \right\} dt$$

$$= \int z^{1/2} dz - 2\int z^{-1/2} dz$$

$$x = \frac{2}{3} z^{3/2} - 4z^{1/2} + c$$

$$\Rightarrow xz^{1/2} = \frac{2z}{3} - 4 + cz^{-1/2}$$

On putting $z = \ln(y+2)$ we get

$$x(z) = \frac{2}{3}\ln(y+2) - 4 + c(\ln(y+2))^{-1/2}$$

$$\Rightarrow x(e^4 - 2) = 1$$

$$\Rightarrow 1 = \frac{2}{3}\ln(e^4) - 4 + c \cdot (4)^{-1/2}$$

$$\Rightarrow 1 = \frac{8}{3} - 4 + \frac{c}{2} \Rightarrow c = \frac{14}{3}$$

$$\therefore x(y) = \frac{2}{3}\ln(y+2) - 4 + \frac{14}{3}\{\ln(y+2)\}^{-1/2}$$

$$\Rightarrow x(e^9 - 2) = \frac{2}{3} \times 9 - 4 + \frac{14}{3} \times \frac{1}{3}$$

$$= 6 - 4 + \frac{14}{9} = 2 + \frac{14}{9} = \frac{32}{9}$$

$$\therefore x(e^9 - 2) = \frac{32}{9}$$

**31.** *(a)* Given,

$$\frac{dy}{dx} + \frac{5}{x(x^2+1)} y = \frac{(x^5+1)^2}{x^7}, x > 0$$

and $y(1) = 2$

$\therefore$ Given differential equation is linear

$$\text{IF} = e^{\int \frac{5}{x(x^5+1)} dx} = e^{\int \frac{5x^4 dx}{x^5(x^5+1)}}$$

Let $x^5 = t$, then $5x^4 dx = dt$

$$\text{IF} = e^{\int \frac{dt}{t(t+1)}} = e^{\int \left(\frac{1}{t} - \frac{1}{t+1}\right) dt}$$

$$= e^{\ln\left(\frac{t}{t+1}\right)} = e^{\ln\left(\frac{x^5}{x^5+1}\right)}$$

$$\text{IF} = \frac{x^5}{x^5+1}$$

Now, $\left(\frac{x^5}{x^5+1}\right) y(x)$

$$= \int \frac{x^5}{(x^5+1)} \cdot \frac{(x^5+1)^2}{x^7} dx$$

$$= \int \frac{x^5+1}{x^2} dx = \int \left(x^3 + \frac{1}{x^2}\right) dx$$

$$\Rightarrow \left(\frac{x^5}{x^5+1}\right) y(x) = \frac{x^4}{4} - \frac{1}{x} + C$$

When, $x = 1$ and $y = 2$, then $C = \frac{7}{4}$

$$\therefore \left(\frac{x^5}{x^5+1}\right) y(x) = \frac{x^4}{4} - \frac{1}{x} + \frac{7}{4} \quad \text{...(i)}$$

On putting $x = 2$, in Eq. (i), we get

$$\left(\frac{32}{33}\right) y(2) = \frac{21}{4} \Rightarrow y(2) = \frac{693}{128}$$

**32.** *(b)* Given that,

$$x^2 f(x) - x = 4\int_0^x tf(t) dt$$

Differentiating both side w.r.t. $x$, we get

$$2xf(x) + x^2 f'(x) - 1 = 4xf(x)$$

$$\Rightarrow x^2 f'(x) - 2xf(x) - 1 = 0$$

$$\Rightarrow x^2 \frac{dy}{dx} - 2xy = 1 \quad [\text{Let } y = f(x)]$$

Comparing above equation with

$\frac{dy}{dx} + Py = Q$ where $P = -\frac{2}{x}, Q = \frac{1}{x^2}$

$$\text{Now, IF} = e^{\int (-2/x) dx} = e^{-2\log x} = 1/x^2$$

$$\text{Solution is } y \cdot \left(\frac{1}{x^2}\right) = \int \left(\frac{1}{x^2}\right) \frac{1}{x^2} dx$$

$$\Rightarrow \frac{y}{x^2} = \int \frac{1}{x^4} dx = -\frac{1}{3x^3} + C$$

$$\Rightarrow y = -\frac{1}{3x} + Cx^2$$

Given, $f(1) = \frac{2}{3}$

So, $\frac{2}{3} = -\frac{1}{3} + C \Rightarrow C = 1$

Thus, $y = f(x) = -\frac{1}{3x} + x^2$

$$\therefore 18f(3) = 18\left\{9 - \frac{1}{9}\right\}$$

$$= 18 \times \frac{80}{9} = 160$$

**33.** *(c)* We have,

$$(1 + \log_e x)\frac{dx}{dy} - x\log_e x = e^y, x > 0$$

Let $x \log_e x = t$

$$\left(x \cdot \frac{1}{x} + \log_e x\right)\frac{dx}{dy} = \frac{dt}{dy}$$

$$\Rightarrow (1 + \log_e x)\frac{dx}{dy} = \frac{dt}{dy}$$

$$\therefore \frac{dt}{dy} - t = e^y$$

$$\text{IF} = e^{\int(-1)dy} = e^{-y}$$

$\therefore$ General solution,

$$t(e^{-y}) = \int (e^y \cdot e^{-y}) dy + c$$

$$\Rightarrow te^{-y} = \int dy + c \Rightarrow te^{-y} = y + c$$

$$\Rightarrow (x\log_e x)e^{-y} = y + c \quad \text{...(i)}$$

Eq. (i) passes through the point (1, 0).

$$\therefore 0 = 0 + c \Rightarrow c = 0$$

$$\therefore (x\log_e x)e^{-y} = y$$

$$\Rightarrow x\log_e x = ye^y$$

which passes through $(\alpha, 2)$.

$$\therefore \alpha \log_e \alpha = 2e^2$$

$$\Rightarrow \log_e \alpha^\alpha = 2e^2$$

$$\Rightarrow \alpha^\alpha = e^{2e^2}$$

**34.** *(d)* We have, $2x^2 y\, dy - (1 - xy^2) dx = 0$

$$\Rightarrow 2x^2 y\, dy = (1 - xy^2)\, dx$$

$$\Rightarrow 2x^2 y \frac{dy}{dx} = 1 - xy^2$$

$$\Rightarrow 2y\frac{dy}{dx} + \frac{1}{x}y^2 = \frac{1}{x^2}$$

$$\text{Let } y^2 = v \Rightarrow 2y\frac{dy}{dx} = \frac{dv}{dx}$$

$$\therefore \frac{dv}{dx} + \frac{1}{x}v = \frac{1}{x^2}$$

Which is a linear differential equation.

$$\therefore \text{IF} = e^{\int \frac{1}{x} dx} = e^{\log_e x} = x$$

Now, solution of differential equation is

$$vx = \int \frac{1}{x^2} \cdot x\, dx + C$$

$$\Rightarrow vx = \int \frac{1}{x} dx + C \Rightarrow y^2 x = \log x + C$$

Now, we have $y(2) = \sqrt{\log 2}$

$$\therefore 2\log 2 = \log 2 + c$$

$$\Rightarrow c = \log 2$$

$\therefore$ Solution is $y^2 x = \log x + \log 2$

$$\Rightarrow y^2 x = \log 2x$$

$\Rightarrow \quad 2x = e^{xy^2}$

But, solution given is $\alpha x = e^{x^\beta y^\gamma}$

$\therefore \quad \alpha = 2, \beta = 1, \gamma = 2$

So, $\quad \alpha + \beta - \gamma = 2 + 1 - 2 = 1$

**35.** *(c)* We have, $\frac{dy}{dx} + y \tan x = x \sec x$

Which is a linear differential equation.

Now, IF $= e^{\int \tan x \, dx} = e^{\log \sec x} = \sec x$

$\therefore$ Solution of differential equation is given by

$y \cdot \sec x = \int x \sec x \cdot \sec x \, dx + c$

$y \sec x = \int x \sec^2 x \, dx + c$

$y \sec x = x \tan x - \int 1 \cdot \tan x \, dx + c$

$y \sec x = x \tan x - \log \sec x + c$

Now, $y(0) = 1$

$\therefore \quad 1 \times \sec 0 = 0 \times \tan 0 - \log \sec 0 + c$

$\Rightarrow \quad 1 = c$

$\therefore y \sec x = x \tan x - \log \sec x + 1$

$\Rightarrow y = x \sin x - \cos x \log \sec x + \cos x$

$\therefore y\left(\frac{\pi}{6}\right) = \frac{\pi}{6} \sin \frac{\pi}{6} - \cos \frac{\pi}{6} \log \sec \frac{\pi}{6} + \cos \frac{\pi}{6}$

$= \frac{\pi}{12} - \frac{\sqrt{3}}{2} \log_e \left(\frac{2}{\sqrt{3}}\right) + \frac{\sqrt{3}}{2}$

$= \frac{\pi}{12} - \frac{\sqrt{3}}{2} \left(\log_e \left(\frac{2}{\sqrt{3}}\right) - 1\right)$

$= \frac{\pi}{12} - \frac{\sqrt{3}}{2} \log_e \left(\frac{2}{\sqrt{3}e}\right)$

**36.** *(c)* $x \log x \frac{dy}{dx} + y = x^2 \log x$

$\Rightarrow \frac{dy}{dx} + \frac{y}{x \log x} = x$

IF $= e^{\int \frac{dx}{x \log x}} = e^{\log |\log x|} = |\log x|$

Solution of equation is

$y(\text{IF}) = \int x \cdot \log x | \, dx$

$\Rightarrow \quad y |\log x| = |\log x| \frac{x^2}{2} - \frac{x^2}{4} + c$

Put $x = 2$

$2 \log 2 = \log 2 \cdot 2 - 1 + c$

$\Rightarrow \quad c = 1$

Put $x = e, \; y = \frac{e^2}{2} - \frac{e^2}{4} + 1$

$\Rightarrow \quad y(e) = 1 + \frac{e^2}{4} = \frac{4 + e^2}{4}$

**37.** *(b)* Let $y = y(t) \frac{dy}{dt} + \alpha y = \gamma e^{-Bt}$,

where $\alpha > 0, B > 0, \gamma > 0$,

where, $\quad$ IF $= e^{\int P \, dx}$

So, $\quad$ IF $= e^{\int \alpha \, dt} = e^{\alpha t}$

Thus, solution is

$y \cdot (e^{\alpha t}) = \int (e^{\alpha t}) \cdot \gamma e^{-\beta t} \, dt$

$y \cdot e^{\alpha t} = \gamma \int e^{(\alpha - \beta)t} \cdot dt$

Now, let $\quad (\alpha - \beta)t = x$

$\Rightarrow (\alpha - \beta) dt = dx \Rightarrow dt = dx / \alpha - \beta$

$\therefore \quad y e^{\alpha t} = \gamma \int e^x \frac{dx}{(\alpha - \beta)}$

$\Rightarrow \quad y e^{\alpha t} = \frac{\gamma}{\alpha - \beta} (e^x) + C$

$\Rightarrow \quad y e^{\alpha t} = \frac{\gamma}{(\alpha - \beta)} e^{(\alpha - \beta)t} + C$

On dividing both the sides by $(e^{\alpha t})$, we get

$y = \left(\frac{\gamma}{\alpha - \beta}\right) e^{-\beta t} + c \cdot e^{-\alpha t} = y(t)$

Now, $\lim_{t \to \infty} y(t) =$

$\lim_{t \to \infty} \left[\left(\frac{\gamma}{\alpha - \beta}\right) e^{-\beta t} + c \cdot e^{-\alpha t}\right]$

$\left[\because \lim_{x \to \infty} e^{-x} = 0\right]$

So, $\quad \lim_{t \to \infty} y(t) = 0 + 0 = 0$

**38.** *(c)* Given differential equation is

$x^3 dy + (xy - 1) dx = 0, \; x > 0$

and $\quad y\left(\frac{1}{2}\right) = 3 - e \quad$ ...(i)

We have to find $y(1)$.

For this, we can write

$x^3 dy = (xy - 1) dx = 0$ as

$\frac{dy}{dx} = \frac{1 - xy}{x^3} = \frac{1}{x^3} - \frac{y}{x^2}$

$\Rightarrow \frac{dy}{dx} + \left(\frac{1}{x^2}\right) y = \frac{1}{x^3}$, which is linear differential equation of the form

$\frac{dy}{dx} + Py = Q$, where $P$ and $Q$ are either constants or functions of $x$.

So, integrating factor (IF)

$= e^{\int P dx} = e^{\int \frac{1}{x^2} dx} = e^{-\frac{1}{x}}$

Its solution is

$y \times (\text{IF}) = \int Q \times (\text{IF}) \, dx$

$\Rightarrow \quad y \times e^{-\frac{1}{x}} = \int \frac{1}{x^3} \times e^{-\frac{1}{x}} dx$

$\Rightarrow \quad y e^{-1/x} = -\int e^t \cdot t dt$

(Where, $-\frac{1}{x} = t$, so that $\frac{1}{x^2} dx = dt$)

$\Rightarrow \quad y = \frac{1}{x} + 1 + Ce^{\frac{1}{x}}$,

where, $C$ is constant. ...(ii)

From Eq. (i), we put $x = \frac{1}{2}$ and $y = 3 - e$ in Eq. (ii), we get

$3 - e = 2 + 1 + Ce^2$

$\Rightarrow \quad C = -\frac{1}{e}$

$\therefore \quad y(1) = 1$

**39.** *(b)* Given, $(1 + e^{2x})\left(\frac{dy}{dx} + y\right) = 1$

$\frac{dy}{dx} + y = \frac{1}{1 + e^{2x}}$

On comparing with standard equation of differential equation $\frac{dy}{dx} + Py = Q$,

we get $P = 1$ and $Q = \frac{1}{1 + e^{2x}}$

Integrating factor (IF)

$= e^{\int P dx} = e^{\int 1 dx} = e^x$

General solution $y$ (IF)

$= \int Q \, (\text{IF}) \, dx + C$

$y \cdot e^x = \int e^x \left(\frac{1}{1 + (e^x)^2}\right) dx$

$= \tan^{-1}(e^x) + C$

Above curve passes through $\left(0, \frac{\pi}{2}\right)$

$\frac{\pi}{2}(e^0) = \tan^{-1}(e^0) + C$

$\Rightarrow \frac{\pi}{2} = \frac{\pi}{4} + C \quad \Rightarrow C = \frac{\pi}{4}$

Curve $y \cdot e^x = \tan^{-1}(e^x) + \frac{\pi}{4}$

$\lim_{x \to \infty} (y \cdot e^x) = \lim_{x \to \infty} \tan^{-1} e^x + \frac{\pi}{4}$

$= \tan^{-1} e^\infty + \frac{\pi}{4} = \frac{\pi}{2} + \frac{\pi}{4} = \frac{3\pi}{4}$

**40.** *(b)* Given, $dy / dx = 2 \tan x (\cos x - y)$

$\Rightarrow \frac{dy}{dx} + 2y \tan x = 2 \sin x$

$\therefore$ IF $= e^{\int 2 \tan x\, dx} = \sec^2 x$

$\therefore$Solution of differential equation will be $y \sec^2 x = \int 2 \sin x \sec^2 x\, dx$

$\Rightarrow y \sec^2 x = 2 \sec x + c$

$\because$ Curve passes through $\left(\frac{\pi}{4}, 0\right)$.

$0 = 2\sqrt{2} + c \quad \therefore \quad c = -2\sqrt{2}$

$\Rightarrow y = 2 \cos x - 2\sqrt{2} \cos^2 x$

$\therefore \int_0^{\pi/2} y\, dx = \int_0^{\pi/2} (2 \cos x - 2\sqrt{2} \cos^2 x)\, dx$

$= 2\int_0^{\pi/2} \cos x\, dx - 2\sqrt{2} \int_0^{\pi/2} \cos^2 x\, dx$

$= 2[\sin x]_0^{\pi/2} - 2\sqrt{2}\left[\frac{x}{2} + \frac{\sin 2x}{4}\right]_0^{\pi/2}$

$= 2 - 2\sqrt{2} \cdot \left(\frac{\pi}{4}\right) = 2 - \frac{\pi}{\sqrt{2}}$

**41.** *(a)* Given,

$x(1 - x^2)\frac{dy}{dx} + (3x^2 y - y - 4x^3) = 0$

$\Rightarrow \frac{dy}{dx} + \frac{y(3x^2 - 1)}{x(1 - x^2)} = \frac{4x^3}{x(1 - x^2)}$

IF $= e^{\int \frac{3x^2 - 1}{x - x^3} dx} = e^{-\ln|x^3 - x|}$

$= e^{-\ln(x^3 - x)} = \frac{1}{x^3 - x}$

Solution of differential equation

$y \times \text{IF} = \int (Q \times \text{IF})\, dx + C$

$y.\left(\frac{1}{x^3 - x}\right) = \int \frac{4x^3}{x(1 - x^2)} \cdot \frac{1}{x(x^2 - 1)}\, dx + C$

$\Rightarrow \frac{y}{x^3 - x} = \int \frac{-4x}{(x^2 - 1)^2} dx + C$

$\Rightarrow \frac{y}{x^3 - x} = \frac{2}{x^2 - 1} + C$

At $x = 2, y = -2$

$-\frac{2}{6} = \frac{2}{3} + C$

$\Rightarrow \quad C = -1$

At $x = 3, \Rightarrow \frac{y}{24} = \frac{2}{8} - 1$

$\Rightarrow y = -18$

**42.** *(b)* Given,

$(\tan^{-1} y - x)\, dy = (1 + y^2)\, dx$

$\Rightarrow \frac{\tan^{-1} y}{1 + y^2} - \frac{x}{1 + y^2} = \frac{dx}{dy}$

$\Rightarrow \frac{dx}{dy} + \left(\frac{1}{1 + y^2}\right) x = \frac{\tan^{-1} y}{1 + y^2}$

IF $= e^{\int \frac{1}{1+y^2} dy} = e^{\tan^{-1} y}$

Solution of the above differential equation is given by

$xe^{\tan^{-1} y} = \int \left[e^{\tan^{-1} y} \cdot \frac{\tan^{-1} y}{1 + y^2}\right] dy + C_1$

On putting $\tan^{-1} y = t$

$\Rightarrow \frac{1}{1 + y^2} dy = dt$

$\Rightarrow \int \left[e^{\tan^{-1} y} \frac{\tan^{-1} y}{1 + y^2}\right] dy$

$= \int te^t\, dt = e^t (t - 1) + c_2$

$= e^{\tan^{-1} y} (\tan^{-1} y - 1) + c_2$

$\therefore x \cdot e^{\tan^{-1} y} = e^{\tan^{-1} y} (\tan^{-1} y - 1) + c$

When, $x = 1, y = 0$

$\Rightarrow 1 \times e^0 = e^0 (0 - 1) + c$

$\Rightarrow \quad c = 2$

When, $y = \tan 1$

$xe^{\tan^{-1}(\tan 1)} = e^{\tan^{-1}(\tan 1)}$

$(\tan^{-1} (\tan 1) - 1) + 2$

$\Rightarrow x \cdot e = e(1 - 1) + 2$

$\Rightarrow \quad x = \frac{2}{e}$

**43.** *(d)* $x\frac{dy}{dx} + 2y = xe^x$

Given, $y(1) = 0$ and $\frac{dy}{dx} + \frac{2y}{x} = e^x$

IF $= e^{\int \frac{2}{x} dx} = e^{2 \ln x} = e^{\ln x^2} = x^2$

Solution is $yx^2 = \int e^x (x^2) dx + c \quad \ldots$(i)

$\int e^x f(x) dx = e^x [f(x) - f'(x) + f''(x)$

$- f'''(x) + \ldots + (-1)^n f^n(x)] + c$

$\therefore \int e^x (x^2) dx = e^x [x^2 - 2x + 2] \quad \ldots$(ii)

On putting Eq. (ii) in Eq. (i), we get

$yx^2 = e^x [x^2 - 2x + 2] + c$

$0 = e[1 - 2 + 2] + c$

$\Rightarrow \quad c = -e$

Thus, $y = \frac{e^x}{x^2}[x^2 - 2x + 2] - \frac{e}{x^2}$

So, $z(x) = x^2 \frac{e^x}{x^2}[x^2 - 2x + 2] - \frac{ex^2}{x^2} - e^x$

$= e^x [x^2 - 2x + 1] - e$

$z'(x) = e^x (x^2 - 2x + 1) + e^x (2x - 2)$

$= e^x x^2 - 2xe^x + e^x + 2xe^x - e^x 2$

$= e^x x^2 - e^x$

$z'(x) = e^x (x^2 - 1)$

$\Rightarrow z'(x) = 0 \Rightarrow e^x (x^2 - 1) = 0$

$\therefore e^x > 0, \forall x \in R$

$\therefore \quad x^2 - 1 = 0$

$(x - 1)(x + 1) = 0$

$\Rightarrow \quad x = 1, -1$

$z''(x) = e^x (2x) + (x^2 - 1)e^x$

$= e^x (x^2 + 2x - 1)$

$z''(-1) = (e^{-1})(1 - 2 - 1)$

$= \frac{1}{e}(1 - 3) = \frac{-2}{e} < 0$

$\therefore$Maxima at $x = -1$

Thus, $z(-1) = e^{-1}(1 + 2 + 1) - e$

$= \frac{4}{e} - e$

**44.** *(c)* Given,

$\frac{dy}{dx} + e^x (x^2 - 2)y = (x^2 - 2x)$

$(x^2 - 2)e^{2x}; \; y(0) = 0$

IF $= e^{\int (x^2 - 2) e^x dx}$

$e^{[\int x^2 e^x dx - \int 2e^x dx]}$

$= e^{x^2 e^x - 2\int x e^x dx - 2e^x}$

$= e^{x^2 e^x - 2[xe^x - e^x] - 2e^x} = e^{(x^2 - 2x)e^x}$

$\therefore$ Solution is

$ye^{(x^2 - 2x)e^x} = \int e^{(x^2 - 2x)e^x} \times (x^2 - 2x)(x^2 - 2)e^{2x}\, dx$

Let $(x^2 - 2x)e^x = t$

$e^x (x^2 - 2x)dx + e^x (2x - 2)dx = dt$

$\Rightarrow \; e^x (x^2 - 2)dx = dt$

$\Rightarrow \quad ye^{(x^2 - 2x)e^x} = \int e^t \cdot t\, dt$

$\Rightarrow \quad ye^{(x^2 - 2x)e^x} = e^t t - e^t + c$

$\Rightarrow \quad ye^{(x^2 - 2x)e^x} = e^t (t - 1) + c$

$\Rightarrow \quad ye^{(x^2 - 2x)e^x} = e^{(x^2 - 2x)e^x} [(x^2 - 2x)e^x - 1] + c$

$y(0) = 0$

$\therefore \; 0 = e^0 [0 - 1] + c$

$0 = (-1) + c \Rightarrow c = 1$

Thus, $ye^{(x^2 - 2x)e^x} = e^{(x^2 - 2x)e^x} [(x^2 - 2x)e^x - 1] + 1$

To find $y(2)$

$ye^{(4-4)e^2} = e^{(4-4)e^2}[(4-4)e^2 - 1] + 1$

$\Rightarrow \quad ye^0 = e^0[-1] + 1 \Rightarrow y = -1 + 1$

$\Rightarrow \quad y = 0$

**45.** *(b)* Given, $(x+1)y' - y = e^{3x}(x+1)^2$

$$y' - \frac{1}{x+1}y = e^{3x}(x+1)$$

which is linear differential equation.

$\therefore$ IF $= e^{\int P\,dx} = e^{\int -\frac{1}{x+1}dx} = e^{-\log(x+1)}$

$$= \frac{1}{x+1}$$

Now, $\dfrac{y}{x+1} = \displaystyle\int e^{3x}(x+1) \times \frac{1}{x+1}\,dx$

$\Rightarrow \quad \dfrac{y}{x+1} = \dfrac{e^{3x}}{3} + C$

$\Rightarrow \quad y = \left(\dfrac{e^{3x}}{3} + C\right)(x+1)$

When, $x = 0$, $y = 1/3$

$\dfrac{1}{3} = \dfrac{1}{3} + C \Rightarrow C = 0$

$\therefore \quad y = \dfrac{e^{3x}}{3}(x+1)$

$y'(x) = \dfrac{e^{3x}}{3} + (x+1)e^{3x}$

$y'(x) = 0 \Rightarrow e^{3x}\left[\dfrac{1}{3} + x + 1\right] = 0$

$\Rightarrow \quad x = \dfrac{-4}{3}$

$\Rightarrow y''(x) = e^{3x} + e^{3x} + 3(x+1)e^{3x}$

$\Rightarrow y''(x) = e^{3x}[5 + 3x]$

$\Rightarrow y''\left[-\dfrac{4}{3}\right] > 0$

$\therefore -\dfrac{4}{3}$ is a point of local minima.

**46.** *(d)* We have,

$(y+1)\tan^2 x\,dx + \tan x\,dy + y\,dx = 0$

$\Rightarrow [(y+1)\tan^2 x + y]\,dx + \tan x\,dy = 0$

$\Rightarrow \dfrac{dy}{dx} + (y+1)\tan x + \dfrac{y}{\tan x} = 0$

$\Rightarrow \dfrac{dy}{dx} + \dfrac{y\tan^2 x + \tan^2 x + y}{\tan x} = 0$

$\Rightarrow \dfrac{dy}{dx} + \dfrac{y\sec^2 x}{\tan x} + \tan x = 0$

$\Rightarrow \dfrac{dy}{dx} + \left(\dfrac{\sec^2 x}{\tan x}\right)y = -\tan x$

This is a linear differential equation

$\therefore$ IF $= e^{\int \frac{\sec^2 x}{\tan x}dx} = e^{\ln(\tan x)} = \tan x$

So, solution is given by

$(y\tan x) = \displaystyle\int -\tan^2 x\,dx$

$= \displaystyle\int (1 - \sec^2 x)\,dx = x - \tan x + C$

$$y = x\cot x - 1 + C\cot x$$

Now, $\displaystyle\lim_{x\to 0^+} x \cdot y = 1$

$\Rightarrow \displaystyle\lim_{x\to 0^+}(x^2\cot x - x + Cx\cot x) = 1$

$\Rightarrow \displaystyle\lim_{x\to 0^+}\left(x\cdot\frac{x}{\tan x} - x + \frac{Cx}{\tan x}\right) = 1$

$\Rightarrow \quad 0 - 0 + C = 1$

$\Rightarrow \quad C = 1$

$\therefore \quad y = x\cot x - 1 + \cot x$

Now, $x = \dfrac{\pi}{4}$

$$y = \frac{\pi}{4} - 1 + 1 = \frac{\pi}{4}$$

**47.** *(c)* $\operatorname{cosec}^2 x\,dy + 2\,dx$

$= (1 + y\cos 2x)\operatorname{cosec}^2 x\,dx$

On dividing LHS and RHS by $\operatorname{cosec}^2 x\,dx$,

$\dfrac{dy}{dx} + \left(\dfrac{2}{\operatorname{cosec}^2 x}\right) = (1 + y\cos 2x)$

$\Rightarrow \dfrac{dy}{dx} + 2\sin^2 x - y\cos 2x = 1$

$\Rightarrow \dfrac{dy}{dx} - y\cos 2x = 1 - 2\sin^2 x$

$\therefore$IF $= e^{\int -\cos 2x dx} = e^{\left(\frac{-1}{2}\right)\sin 2x}$

So, IF$\cdot y = \displaystyle\int Q\cdot$IF $dx$

$\Rightarrow$

$ye^{\frac{-1}{2}\sin 2x} = \displaystyle\int e^{\frac{-1}{2}\sin 2x}(1 - 2\sin^2 x)dx$

$= \displaystyle\int e^{\frac{-1}{2}\sin 2x}\cos 2x dx$

Let $\dfrac{-1}{2}\sin 2x = t \Rightarrow \cos 2x dx = -dt$

$\Rightarrow \quad ye^{\frac{-1}{2}\sin 2x} = -e^{\frac{-\sin 2x}{2}} + C$

Now, $y\left(\dfrac{\pi}{4}\right) = 0$

$\Rightarrow \quad 0 = -e^{\frac{-1}{2}} + C \Rightarrow C = \dfrac{1}{\sqrt{e}}$

$\therefore \quad y(0) \Rightarrow \quad y\cdot 1 = -1 + \dfrac{1}{\sqrt{e}}$

$\Rightarrow \quad y + 1 = \dfrac{1}{\sqrt{e}} \Rightarrow (y+1)^2 = \dfrac{1}{e}$

**48.** *(a)* $|A| = \begin{vmatrix} y & \sin x & 1 \\ 0 & -1 & 1 \\ 2 & 0 & \frac{1}{x} \end{vmatrix}$

$= y\left(-\dfrac{1}{x}\right) + \sin x(2) + 1(0+2)$

$= -\dfrac{y}{x} + 2\sin x + 2$

Now, $\dfrac{dy}{dx} - |A| = 0$ gives

$\dfrac{dy}{dx} + \dfrac{y}{x} - 2\sin x - 2 = 0$

or $\quad \dfrac{dy}{dx} + \dfrac{y}{x} = 2\sin x + 2$

$$\text{IF} = e^{\int \frac{1}{x}dx} = e^{\log x} = x$$

$\therefore y\cdot$IF $= \displaystyle\int (2\sin x + 2)\cdot$IF $dx$

$y\cdot x = \displaystyle\int (2x\sin x + 2x)\,dx$

$yx = 2[-x\cos x + \sin x] + x^2 + c$ ...(i)

Given, $y(\pi) = \pi + 2$, putting in Eq. (i)

$(\pi + 2)\pi = 2(-\pi\cos\pi + \sin\pi) + \pi^2 + c$

$\Rightarrow \quad \pi^2 + 2\pi = \pi^2 + 2\pi + c$

$\Rightarrow \quad c = 0$

From Eq. (i),

$$yx = x^2 - 2x\cos x + 2\sin x$$

Putting $x = \dfrac{\pi}{2}$,

$y\cdot\left(\dfrac{\pi}{2}\right) = \left(\dfrac{\pi}{2}\right)^2 - 2\left(\dfrac{\pi}{2}\right)\cos\dfrac{\pi}{2} + 2\sin\left(\dfrac{\pi}{2}\right)$

$\Rightarrow y\cdot\left(\dfrac{\pi}{2}\right) = \dfrac{\pi}{2} - 2(0) + \dfrac{2}{\pi/2}$

$\Rightarrow y\cdot\left(\dfrac{\pi}{2}\right) = \dfrac{\pi}{2} + \dfrac{4}{\pi}$

**49.** *(d)*

> **key idea** The range of $\cos x$ is $-1 \le \cos x \le 1$.
> So, $\cos x_{\min} = -1$. But, $\left(\cos x - \dfrac{1}{4}\right)^2_{\min} = 0$

Given, $\dfrac{dy}{dx} + 2y\tan x = \sin x$

$\therefore$ IF $= e^{\int 2\tan x\,dx} = e^{\int \frac{2\sin x}{\cos x}dx}$

$= e^{-2\log(\cos x)} = \sec^2 x$

Its solution is

$$y\sec^2 x = \int \sin x \sec^2 x\, dx$$

$$\Rightarrow y\sec^2 x = \int \sec x \tan x\, dx$$

$$\Rightarrow y\sec^2 x = \sec x + c$$

Now, $x = \pi/3,\ y = 0$

$$0 = 2 + c$$

$$\Rightarrow c = -2$$

So, $y\sec^2 x = \sec x - 2$

$$y = \cos^2 x\left(\frac{1}{\cos x} - 2\right)$$

$$= -2\cos^2 x + \cos x$$

$$\Rightarrow y = -2\left(\cos^2 x - \frac{\cos x}{2}\right)$$

$$\Rightarrow y = -2\left[\left(\cos^2 x - \frac{\cos x}{2} + \frac{1}{16}\right) - \frac{1}{16}\right]$$

$$\Rightarrow y = -2\left[(\cos x - 1/4)^2 - \frac{1}{16}\right]$$

$$\Rightarrow y = \frac{1}{8} - 2(\cos x - 1/4)^2$$

So, $y_{max} = 1/8$

**50.** *(d)* Given differential equation

$$xy' - y = x^2(x\cos x + \sin x),\ x > 0$$

$$\Rightarrow \frac{dy}{dx} - \frac{1}{x}y = x(x\cos x + \sin x),$$

it is in the form of linear differential equation, of the form $\frac{dy}{dx} + py = Q$

So, $\quad \text{IF} = e^{-\int \frac{dx}{x}} = \frac{1}{x}$

∴Solution of given differential equation is

$$\frac{y}{x} = \int (x\cos x + \sin x)dx + C$$

$$= x\sin x + C$$

$$[\because \int (xf'(x) + f(x))dx = xf(x) + C]$$

$\because \quad y(\pi) = \pi \Rightarrow C = 1$

$\therefore \quad y = x^2 \sin x + x$

$\therefore \quad y' = 2x\sin x + x^2\cos x + 1$

$\Rightarrow \quad y'' = 2x\cos x + 2\sin x + 2x\cos x - x^2 \sin x$

$$\text{So, } y''\left(\frac{\pi}{2}\right) + y\left(\frac{\pi}{2}\right) = 2 - \frac{\pi^2}{4} + \frac{\pi^2}{4} + \frac{\pi}{2}$$

$$= 2 + \frac{\pi}{2}$$

**51.** *(b)*

> **key idea** Using integration by parts,
> $\int f_1 \cdot f_2\, dx = f_1 \int f_2 dx - \int f'_1 f_2 dx$

Given differential equation is

$$y^2 dx + \left(x - \frac{1}{y}\right) dy = 0$$

$$\Rightarrow \frac{dx}{dy} + \frac{1}{y^2}x = \frac{1}{y^3}$$

which is the linear differential equation of the form $\frac{dx}{dy} + Px = Q$.

$$\text{Now, IF} = e^{\int \frac{1}{y^2}dy} = e^{-\frac{1}{y}}$$

∴ The solution of linear differential equation is

$$x\cdot(\text{IF}) = \int Q\,(\text{IF})\, dy + C$$

$$\Rightarrow x e^{-1/y} = \int \frac{1}{y^3} e^{-1/y} dy + C$$

$$\therefore x e^{-1/y} = \int (-t)\, e^t\, dt + C$$

$$\left[\because \text{let } -\frac{1}{y} = t \Rightarrow +\frac{1}{y^2}dy = dt\right]$$

$$= -te^t + \int e^t dt + C$$

[Integration by parts]

$$= -te^t + e^t + C$$

$$\Rightarrow x e^{-1/y} = \frac{1}{y} e^{-1/y} + e^{-1/y} + C \quad \ldots\text{(i)}$$

Now, at $y = 1$, the value of $x = 1$, so

$$1\cdot e^{-1} = e^{-1} + e^{-1} + C$$

$$\Rightarrow C = -\frac{1}{e}$$

On putting the value of $C$, in Eq. (i),

$$\text{we get } x = \frac{1}{y} + 1 - \frac{e^{1/y}}{e}$$

So, at $y = 2$, the value of

$$x = \frac{1}{2} + 1 - \frac{e^{\frac{1}{2}}}{e} = \frac{3}{2} - \frac{1}{\sqrt{e}}$$

**52.** *(a)* Given differential equation is $\frac{dy}{dx} + y\tan x = 2x + x^2 \tan x$, which is linear differential equation in the form of $\frac{dy}{dx} + Py = Q$.

$$\therefore \text{IF} = e^{\int \tan x\, dx} = e^{\log_e(\sec x)} = \sec x$$

Now, solution of linear differential equation is given as

$$y \times \text{IF} = \int (Q \times \text{IF})dx + C$$

$$\therefore y(\sec x) = \int (2x + x^2\tan x) \sec x\, dx + C$$

$$= \int (2x\sec x)\, dx + \int x^2 \sec x \tan x\, dx + C$$

$$\because \int x^2 \sec x \tan x\, dx = x^2 \sec x - \int (2x\sec x)\, dx$$

Therefore, solution is

$$y\sec x = 2\int x\sec x\, dx + x^2 \sec x - 2\int x\sec x\, dx + C$$

$$\Rightarrow y\sec x = x^2 \sec x + C \quad \ldots\text{(i)}$$

$$\because y(0) = 1 \Rightarrow 1(1) = 0(1) + C \Rightarrow C = 1$$

Now, $\quad y = x^2 + \cos x$ [from Eq. (i)]

and $\quad y' = 2x - \sin x$

Now,

$$y'\left(\frac{\pi}{4}\right) - y'\left(\frac{-\pi}{4}\right) = \left(2\left(\frac{\pi}{4}\right) - \frac{1}{\sqrt{2}}\right) - \left(2\left(-\frac{\pi}{4}\right) + \frac{1}{\sqrt{2}}\right)$$

$$= \pi - \sqrt{2}$$

**53.** *(d)* Given differential equation

$$\frac{dy}{dx} = (\tan x - y)\sec^2 x$$

$$\Rightarrow \frac{dy}{dx} + (\sec^2 x)y = \sec^2 x \tan x,$$

which is linear differential equation of the form $\frac{dy}{dx} + Py = Q$,

$$\text{IF} = e^{\int \sec^2 x\, dx} = e^{\tan x}$$

So, solution of given differential equation is

$$y(e^{\tan x}) = \int e^{\tan x} \cdot \sec^2 x \tan x\, dx + C$$

Let $\tan x = t \Rightarrow \sec^2 x\, dx = dt$

$$ye^{\tan x} = \int e^t \cdot t\, dt + C$$

$$= te^t - \int e^t dt + C$$

[Using integration by parts method]

$$= e^t(t - 1) + C$$

$$\Rightarrow y\cdot e^{\tan x} = e^{\tan x}(\tan x - 1) + C$$

$[\because t = \tan x]$

$\because \quad y(0) = 0$

$\Rightarrow \quad 0 = 1(0 - 1) + C$

$\Rightarrow \quad C = 1$

$\therefore \quad y\cdot e^{\tan x} = e^{\tan x}(\tan x - 1) + 1$

Now, at $x = -\frac{\pi}{4}$

$$\Rightarrow ye^{-1} = e^{-1}(-1 - 1) + 1$$

$$\Rightarrow ye^{-1} = -2e^{-1} + 1$$

$$\Rightarrow y = e - 2$$

**54.** (c) Given differential equation is

$$x\frac{dy}{dx}+y=x\log_e x,\ (x>1)$$

$$\Rightarrow \quad \frac{dy}{dx}+\frac{1}{x}y=\log_e x \quad \ldots(i)$$

Which is a linear differential equation.

So, $\text{IF}=e^{\int\frac{1}{x}dx}=e^{\log_e x}=x$

Now, solution of differential Eq. (i), is

$$y\times x=\int(\log_e x)\,x\,dx+C$$

$$\Rightarrow yx=\frac{x^2}{2}\log_e x-\int\frac{x^2}{2}\times\frac{1}{x}\,dx+C$$

[Using integration by parts]

$$\Rightarrow \quad yx=\frac{x^2}{2}\log_e x-\frac{x^2}{4}+C \quad \ldots(ii)$$

Given that $2y(2)=\log_e 4-1 \quad \ldots(iii)$

On substituting, $x=2$, in Eq. (ii), we get

$$2y(2)=\frac{4}{2}\log_e 2-\frac{4}{4}+C,$$

[Where, $y(2)$ represents value of $y$ at $x=2$]

$$\Rightarrow \quad 2y(2)=\log_e 4-1+C \ldots(iv)$$

$[\because m\log a=\log a^m]$

From Eqs. (iii) and (iv), we get

$$C=0$$

So, required solution is

$$yx=\frac{x^2}{2}\log_e x-\frac{x^2}{4}$$

Now, at $\quad x=e,\ ey(e)=\frac{e^2}{2}\log_e e-\frac{e^2}{4}$

[Where, $y(e)$ represents value of $y$ at $x=e$]

$$\Rightarrow \quad y(e)=e/4 \quad [\because \log_e e=1]$$

**55.** (d) Given,

$$f'(x)=7-\frac{3}{4}\frac{f(x)}{x},\ (x>0)$$

On putting $f(x)=y$ and $f'(x)=\frac{dy}{dx}$,

we get

$$\frac{dy}{dx}=7-\frac{3}{4}\frac{y}{x}\Rightarrow\frac{dy}{dx}+\frac{3}{4x}y=7 \quad \ldots(i)$$

Which is a linear differential equation of the form $\frac{dy}{dx}+Py=Q$,

Now, integrating factor (IF)

$$=e^{\int\frac{3}{4x}dx}=e^{\frac{3}{4}\log x}=e^{\log x^{3/4}}=x^{3/4}$$

And solution of differential Eq. (i) is given by

$$y(\text{IF})=\int(Q\cdot(\text{IF}))\,dx+C$$

$$yx^{3/4}=\int 7x^{3/4}\,dx+C$$

$$\Rightarrow \quad y=4x+C\,x^{-3/4}$$

So, $\quad y=f(x)=4x+C\cdot x^{-3/4}$

Now, $\quad f\left(\frac{1}{x}\right)=\frac{4}{x}+C\cdot x^{3/4}$

$$\therefore \lim_{x\to0^+}x\,f\left(\frac{1}{x}\right)=\lim_{x\to0^+}x\left(\frac{4}{x}+Cx^{3/4}\right)$$

$$=\lim_{x\to0^+}(4+Cx^{7/4})=4$$

**56.** (c) Given differential equation can be rewritten as $\frac{dy}{dx}+\left(\frac{2}{x}\right)\cdot y=x$, which is a linear differential equation of the form

$$\frac{dy}{dx}+Py=Q$$

Now, integrating factor

$$(\text{IF})=e^{\int\frac{2}{x}dx}=e^{2\log x}=e^{\log x^2}=x^2$$

$[\because e^{\log f(x)}=f(x)]$

And the solution is given by

$$y(\text{IF})=\int(Q\times\text{IF})\,dx+C$$

$$\Rightarrow \quad yx^2=\int x^3\,dx+C$$

$$\Rightarrow \quad yx^2=\frac{x^4}{4}+C \quad \ldots(i)$$

Since, it is given that $y=1$ when $x=1$

$\therefore$ From Eq. (i), we get

$$1=\frac{1}{4}+C$$

$$\Rightarrow \quad C=\frac{3}{4} \quad \ldots(ii)$$

$$\therefore \quad 4x^2y=x^4+3$$

[Using Eqs. (i) and (ii)]

$$\Rightarrow \quad y=\frac{x^4+3}{4x^2}$$

Now, $\quad y\left(\frac{1}{2}\right)=\frac{\frac{1}{16}+3}{4\times\frac{1}{4}}=\frac{49}{16}$

**57.** (c) We have,

$$\sin x\frac{dy}{dx}+y\cos x=4x$$

$$\Rightarrow \quad \frac{dy}{dx}+y\cot x=4x\operatorname{cosec} x$$

This is a linear differential equation of form

$$\frac{dy}{dx}+Py=Q$$

Now, $\text{IF}=e^{\int Pdx}=e^{\int\cot x\,dx}=e^{\log\sin x}$

$$=\sin x$$

Solution of the differential equation is

$$y\cdot\sin x=\int 4x\operatorname{cosec} x\sin x\,dx+C$$

$$\Rightarrow y\sin x=\int 4x\,dx+C=2x^2+C$$

On putting $x=\frac{\pi}{2}$, $y=0$, we get

$$C=-\frac{\pi^2}{2}\Rightarrow y\sin x=2x^2-\frac{\pi^2}{2}$$

On putting $x=\frac{\pi}{6}$, we get

$$y\left(\frac{1}{2}\right)=2\left(\frac{\pi^2}{36}\right)-\frac{\pi^2}{2}$$

$$\Rightarrow \quad y=\frac{\pi^2}{9}-\pi^2\Rightarrow y=-\frac{8\pi^2}{9}$$

**58.** (d) Let $L=\lim_{t\to x}\frac{t^2f(x)-x^2f(t)}{t-x}=1$

On applying L' Hospital rule, we get

$$L=\lim_{t\to x}\frac{2t\,f(x)-x^2\,f'(t)}{1}=1$$

or $\quad 2xf(x)-x^2f'(x)=1$

$$\Rightarrow \quad f'(x)-\frac{2}{x}f(x)=\frac{-1}{x^2}$$

$$\text{IF}=e^{\int\frac{-2}{x}dx}=e^{-2\log x}=\frac{1}{x^2}$$

Solution of differential equation

$$[f(x)]\frac{1}{x^2}=\int\frac{1}{x^2}\left(-\frac{1}{x^2}\right)dx$$

$$\Rightarrow \frac{f(x)}{x^2}=\frac{1}{3x^3}+C$$

$$\because f(1)=1\Rightarrow1=\frac{1}{3}+C\ \Rightarrow\ C=\frac{2}{3}$$

$$\therefore\ f(x)=\frac{2}{3}x^2+\frac{1}{3x}$$

$$\therefore\ f\left(\frac{3}{2}\right)=\frac{2}{3}\times\left(\frac{3}{2}\right)^2+\frac{1}{3}\times\frac{2}{3}=\frac{31}{18}$$

**59.** (d) $\frac{dy}{dx}+\frac{y}{2}\sec x=\frac{\tan x}{2y}$

$$2y\frac{dy}{dx}+y^2\sec x=\tan x$$

Put $y^2=t\ \Rightarrow\ 2y\frac{dy}{dx}=\frac{dt}{dx}$

$$\frac{dt}{dx}+t\sec x=\tan x$$

$$\text{IF}=e^{\int\sec x\,dx}=e^{\ln(\sec x+\tan x)}$$

$$=\sec x+\tan x$$

$\therefore$ Solution is $t\cdot(\sec x+\tan x)$
$=\int \tan x\cdot(\sec x+\tan x)dx+c$
$\Rightarrow t\,(\sec x+\tan x)=\int \sec x\cdot\tan x\,dx + \int \tan^2 x dx + c$
$\Rightarrow t\cdot(\sec x+\tan x) = \sec x+\tan x - x + c$
On putting $x=0$
$\Rightarrow$ $y=1$, we get
$c=0$
$\Rightarrow t(\sec x+\tan x)=\sec x+\tan x - x$
$\Rightarrow y^2 = 1-\dfrac{x}{\sec x+\tan x}$

**60.** *(d)* Here, $\dfrac{dx}{dy}+\dfrac{1}{y^2}\cdot x=\dfrac{1}{y^3}$
[Linear differential equation in $x$]
$\therefore$ IF $=e^{\int \frac{1}{y^2}dy}=e^{-\frac{1}{y}}$
Complete solution is
$x\cdot e^{-\frac{1}{y}}=\int \dfrac{1}{y^3}\cdot e^{-\frac{1}{y}}dy$
$\Rightarrow x\cdot e^{-\frac{1}{y}}=\int \dfrac{1}{y}\cdot\dfrac{1}{y^2}\cdot e^{-\frac{1}{y}}dy$
On putting $-\dfrac{1}{y}=t \Rightarrow \dfrac{1}{y^2}dy=dt$
$\Rightarrow xe^{-\frac{1}{y}}=\int -t\cdot e^t dt$
$\Rightarrow xe^{-\frac{1}{y}}=-\{t\cdot e^t-\int 1\cdot e^t dt\}+C$
$\Rightarrow xe^{-\frac{1}{y}}=-te^t+e^t+C$
$\Rightarrow xe^{-\frac{1}{y}}=\dfrac{1}{y}\cdot e^{-\frac{1}{y}}+e^{-\frac{1}{y}}+C$
$\because y(1)=1 \;\therefore\; e^{-1}=e^{-1}+e^{-1}+C$
$\Rightarrow C=-\dfrac{1}{e}$
$\therefore xe^{-\frac{1}{y}}=\dfrac{1}{y}\cdot e^{-\frac{1}{y}}+e^{-\frac{1}{y}}-\dfrac{1}{e}$
$\Rightarrow x=\dfrac{1}{y}+1-\dfrac{1}{e}\cdot e^{\frac{1}{y}}$

**61.** *(b)* Given differential equation can be rewritten as
$(1+y^2)\dfrac{dx}{dy}+x=e^{\tan^{-1}y}$
or $\dfrac{dx}{dy}+\dfrac{1}{1+y^2}x=\dfrac{e^{\tan^{-1}y}}{1+y^2}$
$\therefore$ IF $=e^{\int P\,dy}=e^{\int\frac{1}{1+y^2}dy}=e^{\tan^{-1}y}$
$\therefore$ Required solution,
$xe^{\tan^{-1}y}=\int \dfrac{e^{2\tan^{-1}y}}{1+y^2}dy+C_1$
Put $t=\tan^{-1}y$
$\Rightarrow dt=\dfrac{dy}{1+y^2}$
$\therefore$ $xe^{\tan^{-1}y}=\int e^{2t}dt+C_1$
$\Rightarrow xe^{\tan^{-1}y}=\dfrac{1}{2}e^{2t}+C_1$
$\Rightarrow xe^{\tan^{-1}y}=\dfrac{1}{2}e^{2\tan^{-1}y}+C_1$
$\Rightarrow 2xe^{\tan^{-1}y}=e^{2\tan^{-1}y}+C$
[$\because$ put $2C_1=C$]

**62.** *(d)* IF $=e^{\int \tan x\,dx}=e^{\ln(\sec x)}$
$=\sec x$
$\therefore$ $y.\sec x=\int\left\{\dfrac{2+\sec x}{(1+2\sec x)^2}\right\}\sec x\,dx$
$=\int \dfrac{2\cos x+1}{(\cos x+2)^2}dx$
Let $t=\tan x$
$\therefore$ $\cos x=\dfrac{1-t^2}{1+t^2}$
Now, differentiating both sides w.r.t $x$ we get
$-\sin x dx=\dfrac{(1+t^2)(-2t)-(1-t^2)(2t)}{(1+t^2)^2}dt$
$\Rightarrow -\sin x dx=\dfrac{-2t(1+t^2+1-t^2)}{(1+t^2)^2}dt$
$\Rightarrow -\sin x dx=-\dfrac{4tdt}{(1+t^2)^2}$
$\Rightarrow dx=\dfrac{4tdt}{(1+t^2)^2\sin x}$
If $\cos x=\dfrac{1-t^2}{1+t^2}$,
then $\sin x=\dfrac{2t}{1+t^2}$
$\therefore$ $dx=\dfrac{2}{1+t^2}dt$
$=\int \dfrac{2\left(\dfrac{1-t^2}{1+t^2}\right)+1}{\left(\dfrac{1-t^2}{1+t^2}+2\right)^2}\cdot\dfrac{2}{(1+t^2)}dt$
$=\int \dfrac{\dfrac{(2-2t^2+1+t^2)}{(1-t^2+2+2t^2)^2}\cdot\dfrac{2}{(1+t^2)^2}dt}{(1+t^2)^2}$
$=\int \dfrac{(3-t^2)}{(t^2+3)}\cdot 2dt \Rightarrow 2\int\left(\dfrac{3-t^2}{t^2+3}\right)dt$
Let $t+\dfrac{3}{t}=u$
$\Rightarrow \left(1-\dfrac{3}{t^2}\right)dt=du=-2\int\dfrac{du}{u^2}$
$y.(\sec x)=\dfrac{2}{u}+C$
$y\cdot\sec x=\dfrac{2}{t+\dfrac{3}{t}}+C$ ... (i)
at $x=\dfrac{\pi}{3}$, $t=\tan\dfrac{x}{2}=\dfrac{1}{\sqrt{3}}$
$2\cdot\dfrac{\sqrt{3}}{10}=\dfrac{2}{\dfrac{1}{\sqrt{3}}+3\sqrt{3}}+C$
$\dfrac{2\sqrt{3}}{10}=\dfrac{2\sqrt{3}}{10}+C\Rightarrow C=0$
at $x=\dfrac{\pi}{4}$, $t=\tan\dfrac{x}{2}=\sqrt{2}-1$
$\therefore$ $y.\sqrt{2}=\dfrac{2}{\sqrt{2}-1+\dfrac{3}{\sqrt{2}-1}}$
$\Rightarrow y\cdot\sqrt{2}=\dfrac{2(\sqrt{2}-1)}{6-2\sqrt{2}}\Rightarrow y=\dfrac{\sqrt{2}(\sqrt{2}-1)}{2(3-\sqrt{2})}$
$=\dfrac{1}{\sqrt{2}}\times\dfrac{(2\sqrt{2}-1)}{7}=\dfrac{4-\sqrt{2}}{14}$

**63.** *(b)* Given differential equation is
$\cos x\,(\log_e(\cos x))^2dy+(\sin x-3y(\sin x)\log_e(\cos x))\,dx=0$
$\Rightarrow \cos x\,(\log_e(\cos x))^2\cdot dy/dx - 3\sin x\cdot\log_e(\cos x)y = -\sin x$
$\Rightarrow \dfrac{dy}{dx}+\dfrac{3\tan x}{\log_e(\sec x)}y=\dfrac{-\tan x}{(\log_e(\sec x))^2}$
($\because \log_e(\sec x)=-\log_e(\cos x)$)
The above differential equation is linear differential equation
So, integrating factor (IF)
$=e^{\int\frac{3\tan x}{\log_e(\sec x)}dx}=(\log_e(\sec x))^3$
Its solution is
$y\times(\log_e(\sec x))^3=\int\dfrac{-\tan x}{(\log_e(\sec x))^2}(\log_e(\sec x))^3dx$

$$\Rightarrow y \times (\log_e(\sec x))^3 = -\frac{1}{2}(\log_e(\sec x))^2 + C \text{ ...(i)}$$

where $C$ is constant of integration.
Given that when $x = \pi/4$, then

$$y = -\frac{1}{\log_e 2}$$

$$\text{So, } \frac{-1}{\log_e 2} \times (\log_e \sqrt{2})^3 = \frac{-1}{2}(\log_e \sqrt{2})^2 + C$$

$$\Rightarrow \frac{-1}{8\log_e 2} \times (\log_e 2)^3 = \frac{-1}{2} \times \frac{1}{4}(\log_e 2)^2 + C$$

$$\Rightarrow \frac{-1}{8}(\log_e 2)^2 = \frac{-1}{8}(\log_e 2)^2 + C$$

$$\Rightarrow C = 0$$

So, Eq. (i), we get

$$\Rightarrow y \times (\log_e(\sec x))^3 = \frac{-1}{2}(\log_e(\sec x))^3$$

$$\Rightarrow y = \frac{-1}{2\log_e(\sec x)} = \frac{1}{2\log_e(\cos x)}$$

Hence,

$$y\left(\frac{\pi}{6}\right) = \frac{1}{2\log_e\left(\cos\frac{\pi}{6}\right)} = \frac{1}{2\log_e\left(\frac{\sqrt{3}}{2}\right)}$$

$$= \frac{1}{2\left(\frac{1}{2}\log_e 3 - \log_e 2\right)}$$

$$= \frac{1}{\log_e 3 - \log_e 4}$$

**64.** *(b)* According to the question,

$$2x\log x\frac{dy}{dx} + 2y = \frac{3}{x}\log x$$

$$\Rightarrow \frac{dy}{dx} + \frac{1}{x\log x}\cdot y = \frac{3}{2x^2}$$

Above equation is linear differential equation of form

$$\frac{dy}{dx} + Py = Q$$

$$\text{Now, IF} = e^{\int P\,dx} = e^{\int \frac{1}{x\log x}dx}$$

Putting $\log x = t$

$$\Rightarrow \frac{1}{x}dx = dt$$

$$\text{IF} = e^{\int\frac{dt}{t}} = e^{\ln t} = t = \log x$$

Now, solution of LDE is

$$y \times (\log x) = \int (\log x) \times \frac{3}{2x^2}dx$$

$$y(\log x) = \frac{3}{2}\left[(\log x)\int \frac{1}{x^2}dx - \int\left(\frac{1}{x} \times \int\frac{1}{x^2}dx\right)dx\right]$$

$$y(\log x) = \frac{3}{2}\left[-\frac{1}{x}(\log x) + \int\frac{1}{x^2}dx\right]$$

$$y(\log x) = \frac{3}{2}\left[-\frac{1}{x}(\log x) - \frac{1}{x}\right] + C$$

Now, $y(e^{-1}) = 0$

$$\Rightarrow 0 = \frac{3}{2}[-e(\log e^{-1}) - e] + C$$

$$\Rightarrow C = 0$$

$$y(\log x) = \frac{3}{2}\left[-\frac{1}{x}(\log x) - \frac{1}{x}\right]$$

Now, $y(e)$

$$\Rightarrow y \times 1 = \frac{3}{2}\left[-\frac{1}{e} - \frac{1}{e}\right] = \frac{3}{2}\left(-\frac{2}{e}\right) = -\frac{3}{e}$$

**65.** *(a)*

$$\frac{dy}{dx} = \frac{\tan x + y}{\sin x(\sec x - \sin x \cdot \tan x)}$$

$$= \frac{\sin x + y\cos x}{\sin x(1 - \sin^2 x)}$$

$$= \frac{\sin x + y\cos x}{\sin x \cdot \cos^2 x}$$

$$\frac{dy}{dx} = \sec^2 x + \frac{y}{\sin x\cos x}$$

$$\frac{dy}{dx} = \sec^2 x + \frac{2y}{\sin 2x}$$

$$\frac{dy}{dx} - (2\operatorname{cosec} 2x)y = \sec^2 x$$

$$\text{IF} = e^{-\int 2\operatorname{cosec} 2x\,dx}$$

$$= e^{-\frac{2\ln|\operatorname{cosec} 2x - \cot 2x|}{2}}$$

$$= \frac{1}{\operatorname{cosec} 2x - \cot 2x}$$

$$= \operatorname{cosec} 2x + \cot 2x$$

$$= \frac{1}{\sin 2x} + \frac{\cos 2x}{\sin 2x}$$

$$= \frac{1 + \cos 2x}{\sin 2x} = \frac{2\cos^2 x}{2\sin x\cos x}$$

$$= \cot x$$

$$\text{Then, } y(\cot x) = \int(\cot x)\sec^2 x\,dx + C$$

$$\Rightarrow \frac{y}{\tan x} = \int\frac{\sec^2 x}{\tan x}dx + C$$

$$\Rightarrow \frac{y}{\tan x} = \ln|\tan x| + C$$

$$\Rightarrow y = \tan x\ln|\tan x| + C\tan x$$

Given, $y\left(\frac{\pi}{4}\right) = 2$

When $x = \pi/4$, $y = 2$

$$\therefore 2 = 1 \times 0 + C,\ C = 2$$

$$\because y\left(\frac{\pi}{3}\right) = \tan\frac{\pi}{3}\ln\tan\frac{\pi}{3} + 2\tan\frac{\pi}{3}$$

$$= \sqrt{3}\ln\sqrt{3} + 2\sqrt{3}$$

$$= \sqrt{3}(\ln\sqrt{3} + 2)$$

**66.** *(c)* We have $(1 + x^2)dy = y(x - y)dx$ ...(i)

$$\text{Let } y = \frac{1}{t} \Rightarrow dy = \frac{-1}{t^2}dt$$

So, Eq. (i) becomes

$$\frac{-(1 + x^2)}{t^2}dt = \frac{1}{t}\left(x - \frac{1}{t}\right)dx$$

$$\Rightarrow \frac{-(1 + x^2)}{t^2}dt = \frac{1}{t^2}(tx - 1)dx$$

$$\Rightarrow \frac{dt}{dx} + \frac{tx}{1 + x^2} = \frac{1}{1 + x^2} \quad \text{...(ii)}$$

$$\text{Let } P(x) = \frac{x}{1 + x^2}$$

$$\Rightarrow \int P(x)dx = \frac{1}{2}\int\frac{2x\,dx}{1 + x^2} = \frac{1}{2}\ln(1 + x^2)$$

$$= \ln\sqrt{1 + x^2}$$

Now integrating factor in Eq. (ii) is

$$e^{\int P(x)dx} = e^{\ln\sqrt{1 + x^2}} = \sqrt{1 + x^2}$$

$\therefore$ Solution of Eq. (ii) is

$$t \cdot \sqrt{1 + x^2} = \int\frac{1}{1 + x^2}\cdot\sqrt{1 - x^2}\,dx$$

$$\Rightarrow t \cdot \sqrt{1 + x^2} = \int\frac{1}{\sqrt{1 + x^2}}dx \quad \text{...(iii)}$$

$$\Rightarrow \frac{\sqrt{1 + x^2}}{y} = \ln|x + \sqrt{1 + x^2}| + c$$

Now $y(0) = 1$

$$\frac{\sqrt{1 + 0^2}}{1} = \ln|0 + \sqrt{1 + 0^2}| + c$$

$$\Rightarrow c = 1$$

$$\therefore y = \frac{\sqrt{1 + x^2}}{\ln|x + \sqrt{1 + x^2} + 1|}$$

Now, $y(2\sqrt{2})=\beta$

$$\Rightarrow \quad \beta=\frac{\sqrt{1+(2\sqrt{2})^2}}{\ln|2\sqrt{2}+\sqrt{1+(2\sqrt{2})^2}|+1}$$

$$=\frac{3}{\ln|3+2\sqrt{2}|+1}$$

$$\Rightarrow \quad \ln|3+2\sqrt{2}|+1=3\beta^{-1}$$

$$\Rightarrow \quad \ln(3+2\sqrt{2})+\ln e=3\beta^{-1}$$

$$\Rightarrow \quad \ln[(e)(3+2\sqrt{2})]=3\beta^{-1}$$

$$\Rightarrow \quad e^{3\beta^{-1}}=e(3+2\sqrt{2})$$

**67.** *(c)* Given differential equation is

$$\frac{dy}{dx}-\frac{3x^5\tan^{-1}(x^3)}{(1+x^6)^{3/2}}y$$

$$=2x\exp\left(\frac{x^3-\tan^{-1}(x^3)}{\sqrt{(1+x)^6}}\right)$$

$$\Rightarrow \quad \frac{dy}{dx}+\left(\frac{-3x^5\tan^{-1}(x^3)}{(1+x^6)^{3/2}}\right)y$$

$$=2xe^{\left[\frac{x^3-\tan^{-1}(x^3)}{\sqrt{(1+x)^6}}\right]}$$

which is a linear differential equation of the form $\frac{dy}{dx}+Py=Q$, whose integrating factor (IF) is $e^{\int Pdx}$ and its solution is

$$y\times(\text{IF})=\int Q\times(\text{IF})\,dx \quad \text{...(i)}$$

So, $\text{IF}=e^{\int\frac{-3x^5\tan^{-1}(x^3)}{(1+x^6)^{3/2}}dx}$

$$=e^{\frac{\tan^{-1}(x^3)-x^3}{\sqrt{1+x^6}}}$$

(you can solve it by setting $x^3=z$ and then again putting $\tan^{-1}z=u$ and then proceed)

From Eq. (i),

$$y\times e^{\frac{\tan^{-1}(x^3)-x^3}{\sqrt{1+x^6}}}$$

$$=\int 2x\,e^{\left(\frac{x^3-\tan^{-1}x^3}{\sqrt{1+x^6}}\right)}\cdot e^{\left(\frac{\tan^{-1}(x^3)-x^3}{\sqrt{1+x^6}}\right)}dx$$

$$\Rightarrow \quad y\times e^{\left(\frac{\tan^{-1}x^3-x^3}{\sqrt{1+x^6}}\right)}=\int 2x\,dx$$

$$\Rightarrow y\times e^{\left(\frac{\tan^{-1}(x^3)-x^3}{\sqrt{1+x^6}}\right)}=2\cdot\frac{x^2}{2}+C$$

$$=x^2+C \quad \text{... (ii)}$$

Also (ii) passes through origin as question said.

So, put $x=y=0$ in Eq. (ii), we get

$$C=0$$

Hence, $y\cdot e^{(\tan^{-1}x^3-x^3/\sqrt{1+x^6})}=x^2$

$$\therefore \quad y(1)\cdot e^{\tan^{-1}(1)-1/\sqrt{1+1}}=1$$

$$\Rightarrow y(1)\cdot e^{\frac{\pi}{4}-1/\sqrt{2}}=1$$

$$\Rightarrow y(1)=\frac{1}{e^{\pi-4/4\sqrt{2}}}=e^{4-\pi/4\sqrt{2}}$$

$$=\exp\left(\frac{4-\pi}{4\sqrt{2}}\right)$$

**68.** *(c)*

> **key idea** Convert the given differential equation in the standard form of linear differential equation, $\frac{dy}{dx}+Py=Q$, where $P$ and $Q$ are function of $x$, then solve the linear differential equation using integration.

$$\frac{dy}{dx}=\frac{y}{x}(1+xy^2(1+\log x_e)),\ x>0,$$

$$y(1)=3$$

$$\frac{dy}{dx}-\frac{y}{x}=y^3(1+\log_e x)$$

$$\frac{1}{y^3}\frac{dy}{dx}-\frac{1}{y^2}\frac{1}{x}=1+\log_e x$$

Let $\quad -\frac{1}{y^2}=t$

On differentiating w.r.t. $x$, we get

$$\frac{2}{y^3}\frac{dy}{dx}=\frac{dt}{dx}$$

$$\Rightarrow \quad \frac{1}{y^3}\frac{dy}{dx}=\frac{1}{2}\frac{dt}{dx}$$

Differential equation becomes,

$$\frac{1}{2}\frac{dt}{dx}+\frac{t}{x}=1+\log_e x$$

$$\Rightarrow \quad \frac{dt}{dx}+\frac{2}{x}t=2(1+\log_e x)$$

On comparing with standard equation of linear differential equation,

$$\frac{dy}{dx}+Py=Q$$

$$P=\frac{2}{x},\ Q=2(1+\log_e x)$$

$\Rightarrow$ Integrating factor,

$$\text{IF}=e^{\int P\,dx}=e^{2\int\frac{1}{x}dx}$$

$$\text{IF}=e^{2\ln x}=e^{\ln x^2}=x^2$$

General solution $t\cdot\text{IF}=\int Q\cdot\text{IF}\,dx$

$$t\cdot x^2=\int 2(1+\log_e x)\,x^2dx$$

$$\Rightarrow \frac{-x^2}{y^2}=2\left[(1+\log_e x)\frac{x^3}{3}-\int\frac{1}{x}\cdot\frac{x^3}{3}dx\right]$$

$$\Rightarrow \quad \frac{-x^2}{y^2}=\frac{2}{3}[x^3(1+\log_e x)-\int x^2dx]$$

$$\Rightarrow \quad \frac{-x^2}{y^2}=\frac{2}{3}\left[x^3(1+\log_e x)-\frac{x^3}{3}\right]+C$$

Given, $y(1)=3$

$$\frac{-1}{9}=\frac{2}{3}\left[1(1+\log_e 1)-\frac{1}{3}\right]+C$$

$$\Rightarrow \quad \frac{-1}{9}=\frac{4}{9}+C\Rightarrow C=\frac{-5}{9}$$

Solution becomes,

$$-\frac{x^2}{y^2}=\frac{2}{3}\left[x^3(1+\log_e x)-\frac{x^3}{3}\right]-\frac{5}{9}$$

$$\Rightarrow -\frac{x^2}{y^2}=\frac{2x^3}{9}[2+\log_e x^3]-\frac{5}{9}$$

$$\Rightarrow \quad \frac{9}{y^2}=-2x[2+\log_e x^3]+\frac{5}{x^2}$$

$$\Rightarrow \quad \frac{9}{y^2}=\frac{5-2x^3\ [2+\log_e x^3]}{x^2}$$

$$\Rightarrow \quad \frac{y^2}{9}=\frac{x^2}{5-2x^3(2+\log_e x^3)}$$

**69.** *(d)* Given, differential equation

$$\frac{dy}{dx}+\frac{1}{x^2-1}y=\left(\frac{x-1}{x+1}\right)^{\frac{1}{2}},$$

$$x>1$$

which is of first order linear differential equation.

Here, $P=\frac{1}{x^2-1},\ Q=\sqrt{\frac{x-1}{x+1}}$

$$\therefore \quad \text{IF}=e^{\int Pdx}=e^{\int\frac{1}{x^2-1}dx}$$

$$=e^{\frac{1}{2}\log\left|\frac{x-1}{x+1}\right|}=\sqrt{\frac{x-1}{x+1}}$$

Now, solution is given by,

$$y\cdot\text{IF}=\int Q\cdot\text{IF}\,dx$$

$$\Rightarrow \quad y\sqrt{\frac{x-1}{x+1}} = \int \sqrt{\frac{x-1}{x+1}}\sqrt{\frac{x-1}{x+1}}\,dx$$

$$\Rightarrow \quad y\sqrt{\frac{x-1}{x+1}} = \int \frac{x-1}{x+1}\,dx$$

$$\Rightarrow \quad y\sqrt{\frac{x-1}{x+1}} = \int \left(1-\frac{2}{x+1}\right)dx$$

$$\Rightarrow \quad y\sqrt{\frac{x-1}{x+1}} = x - 2\log(x+1) + C$$

$\because$ The curve passes through $\left(2, \sqrt{\frac{1}{3}}\right)$.

$$\therefore \quad \sqrt{\frac{1}{3}}\sqrt{\frac{1}{3}} = 2 - 2\log 3 + C$$

$$\Rightarrow \quad \frac{1}{3} - 2 + 2\log 3 = C$$

$$\Rightarrow \quad C = 2\log 3 - \frac{5}{3}$$

$$\therefore y\sqrt{\frac{x-1}{x+1}} = x - 2\log(x+1) + 2\log 3 - \frac{5}{3}$$

$$y = \sqrt{\frac{x+1}{x-1}}\left[x - 2\log(x+1) + 2\log 3 - \frac{5}{3}\right]$$

$$\Rightarrow y(8) = \frac{3}{\sqrt{7}}\left[8 - 2\log 9 + 2\log 3 - \frac{5}{3}\right]$$

$$\Rightarrow \sqrt{7}y(8) = 3\left[\frac{19}{3} - 2\log 3^2 + 2\log 3\right]$$

$$= 3\left[\frac{19 - 12\log 3 + 6\log 3}{3}\right]$$

$$= 19 - 6\log 3$$

**70.** *(b)*

> **key idea** If $f(-x) = -f(x)$, then it is odd function and if $f(-x) = f(x)$, then even function.

Given, $\dfrac{dy}{dx} + \dfrac{xy}{x^2-1} = \dfrac{x^4+2x}{\sqrt{1-x^2}}$,

$x \in (-1, 1)$ which is of first order differential equation.

Here, $P = \dfrac{x}{x^2-1}$, $Q = \dfrac{x^4+2x}{\sqrt{1-x^2}}$

$$\text{IF} = e^{\int P dx} = e^{\frac{1}{2}\int \frac{2x}{x^2-1}dx}$$

$$= e^{\frac{1}{2}\log|x^2-1|}$$

$$= \sqrt{1-x^2},\ x \in (-1, 1)$$

Now, solution is given by

$$y \cdot \text{IF} = \int Q\,\text{IF}\,dx + C$$

$$\Rightarrow y\cdot\sqrt{1-x^2} = \int \frac{x^4+2x}{\sqrt{1-x^2}} \times \sqrt{1-x^2}\,dx + C$$

$$\Rightarrow y\sqrt{1-x^2} = \frac{x^5}{5} + x^2 + C$$

The given curve passes through origin.

$$\therefore \quad C = 0$$

$$y\sqrt{1-x^2} = \frac{x^5}{5} + x^2 \Rightarrow y = \frac{x^5+5x^2}{5\sqrt{1-x^2}}$$

$$\text{Now, let } I = \int_{-\frac{\sqrt{3}}{2}}^{\frac{\sqrt{3}}{2}} \frac{x^5+5x^2}{5\sqrt{1-x^2}}dx$$

$$= \int_{-\frac{\sqrt{3}}{2}}^{\frac{\sqrt{3}}{2}} \frac{x^5}{\sqrt{1-x^2}}dx + \int_{-\frac{\sqrt{3}}{2}}^{\frac{\sqrt{3}}{2}} \frac{x^2}{\sqrt{1-x^2}}dx$$

As, $\dfrac{x^5}{5\sqrt{1-x^2}}$ is an odd function and

$\dfrac{x^2}{\sqrt{1-x^2}}$ is an even function.

$$\therefore \quad I = 0 + 2\int_0^{\sqrt{3}/2} \frac{x^2}{\sqrt{1-x^2}}$$

On putting $x = \sin\theta \Rightarrow dx = \cos\theta\,d\theta$, when $x = 0$, then $\theta = 0$

and when $x = \dfrac{\sqrt{3}}{2}$, then $\theta = \dfrac{\pi}{3}$

$$\therefore \quad I = 2\int_0^{\pi/3} \frac{\sin^2\theta\cos\theta\,d\theta}{\cos\theta}$$

$$= \int_0^{\pi/3}(1-\cos 2\theta)\,d\theta$$

$$= \left(\theta - \frac{\sin 2\theta}{2}\right)_0^{\pi/3} = \frac{\pi}{3} - \frac{\sqrt{3}}{4}$$

**71.** *(a)* Given,

$$\frac{dy}{dx} = (y+1)\left[(y+1)e^{x^2/2} - x\right] \quad \ldots\text{(i)}$$

$$\Rightarrow \frac{dy}{dx} = (y+1)^2 e^{x^2/2} - x(y+1)$$

$$\Rightarrow \frac{dy}{dx} + x(y+1) = (y+1)^2 e^{x^2/2}$$

$$\Rightarrow \frac{1}{(y+1)^2}\cdot\frac{dy}{dx} + \frac{1}{y+1}\cdot x = e^{x^2/2} \quad \ldots\text{(ii)}$$

$$\text{Let } \frac{1}{y+1} = t \Rightarrow \frac{-1}{(y+1)^2}\cdot\frac{dy}{dx} = \frac{dt}{dx}$$

$$\Rightarrow -\frac{dt}{dx} + tx = e^{x^2/2}$$

$$\Rightarrow \frac{dt}{dx} + (-x)t = -e^{x^2/2}$$

which is linear differential equation

$$\text{IF} = e^{\int -x\,dx} = e^{-x^2/2}$$

Now, solution of differential equation is,

$$t\cdot e^{-x^2/2} = -\int e^{-x^2/2}\cdot e^{x^2/2}dx$$

$$\Rightarrow \left(\frac{1}{y+1}\right)e^{-x^2/2} = -x + c, \text{ where } c$$

is constant of integration. ...(iii)

Given, $y(2) = 0$ i.e. when $x = 2$, then $y = 0$.

From Eq. (iii),

$$e^{-2} = -2 + c \Rightarrow c = 2 + e^{-2}$$

Now, at $x = 1$

$$\frac{1}{y+1}e^{-1/2} = -1 + e^{-2} + 2$$

$$\Rightarrow \quad (y+1) = \frac{e^{-1/2}}{1+e^{-2}}$$

Now, putting the value of $(y+1)$ in Eq. (i), we get

$$y'(1) = \frac{e^{-1/2}}{1+e^{-2}}\left(\frac{e^{-1/2}}{1+e^{-2}}\cdot e^{1/2} - 1\right)$$

$$= \frac{e^{-1/2}}{1+e^{-2}}\left(\frac{1-1-e^{-2}}{1+e^{-2}}\right)$$

$$= \frac{-e^{-5/2}}{(1+e^{-2})^2} = \frac{-e^{-5/2}}{\frac{(e^2+1)^2}{e^4}} = \frac{-e^{3/2}}{(e^2+1)^2}$$

**72.** *(b)* Given, $\cos x(3\sin x + \cos x + 3)\,dy$
$= [1 + y\sin x(3\sin x + \cos x + 3)]\,dx$

$$\Rightarrow \cos x\,dy = \left(\frac{1}{3\sin x + \cos x + 3} + y\sin x\right)dx$$

$$\Rightarrow \frac{dy}{dx} = y\frac{\sin x}{\cos x} + \frac{1}{\cos x(3\sin x + \cos x + 3)}$$

$$\Rightarrow \frac{dy}{dx} - (\tan x)\,y = \frac{1}{\cos x(3\sin x + \cos x + 3)} \quad \ldots\text{(i)}$$

Which is linear differential equation.

$\therefore$ Integrating factor (IF) $= e^{\int(-\tan x)\,dx}$

$$= e^{\log|\cos x|} = |\cos x|$$

$\because |\cos x| > 0,\ \forall\, x \in [0, \pi/2)$

$\therefore |\cos x| = \cos x$

Hence, solution of Eq. (i) is

$$y(\cos x) = \int (\cos x)\cdot\frac{1}{\cos x(3\sin x + \cos x + 3)}\,dx$$

$$\Rightarrow y\cos x = \int \frac{1}{3\sin x + \cos x + 3}\,dx$$

$$\Rightarrow y\cos x = \int \frac{\sec^2\frac{x}{2}\,dx}{2\tan^2\frac{x}{2} + 6\tan\frac{x}{2} + 4}\,dx$$

$$= \int \frac{\sec^2\frac{x}{2}\,dx}{2\left(\tan^2\frac{x}{2} + 3\tan\frac{x}{2} + 2\right)}\,dx$$

On putting $\tan x/2 = z$

$$\Rightarrow \frac{1}{2}\sec^2\frac{x}{2}\,dx = dz$$

$$\therefore\ y\cos x = \int \frac{dz}{z^2 + 3z + 2}$$

$$= \int \frac{dz}{(z+1)(z+2)}$$

$$= \int \frac{1}{z+1}\,dz - \int \frac{1}{z+2}\,dz$$

$$= \log(z+1) - \log(z+2) + c$$

$$\Rightarrow y\cos x = \log\left|\frac{z+1}{z+2}\right| + c$$

$$= \log\left|\frac{\tan\frac{x}{2} + 1}{\tan\frac{x}{2} + 2}\right| + c \quad \ldots\text{(i)}$$

Since, $y(0) = 0$

$$\therefore\quad c = -\log\left(\frac{1}{2}\right) = \log 2$$

From Eq. (i),

$$y\cos x = \log\left|\frac{\tan\frac{x}{2} + 1}{\tan\frac{x}{2} + 2}\right| + \log 2$$

$$\therefore\ y\left(\frac{\pi}{3}\right) = 2\left[\log\left|\frac{\frac{1}{\sqrt{3}} + 1}{\frac{1}{\sqrt{3}} + 2}\right| + \log 2\right]$$

$$= 2\log\left|2\left(\frac{\sqrt{3}+1}{2\sqrt{3}+1}\right)\right|$$

$$= 2\log\left|2\left(\frac{\sqrt{3}+1}{2\sqrt{3}+1} \times \frac{2\sqrt{3}-1}{2\sqrt{3}-1}\right)\right|$$

$$= 2\log\left|\frac{2\sqrt{3}+10}{11}\right|$$

**73.** *(c)* Given,

$2(x^2 + x^{5/4})dy - y\cdot(x + x^{1/4})\,dx = 2x^{9/4}\,dx$, where $x > 0$

After rearranging, we get

$$\frac{dy}{dx} - \frac{y}{2x} = \frac{x^{9/4}}{x^{5/4}(x^{3/4}+1)}$$

$$= e^{\int -\frac{1}{2x}dx} = e^{\frac{-1}{2}\int\frac{1}{x}dx}$$

$$= e^{-\frac{1}{2}\log x} = e^{\log(x)^{-1/2}} = \frac{1}{x^{1/2}}$$

Its solution is

$$y \times (\text{IF}) = \int Q \times (\text{IF})\,dx$$

$$\Rightarrow\quad y \times \frac{1}{x^{1/2}} = \int \frac{x^{9/4}}{x^{5/4}(x^{3/4}+1)} \times x^{-1/2}\,dx$$

$$= \int \frac{(x)^{\frac{9}{4}-\frac{5}{4}-\frac{1}{2}}}{(x^{3/4}+1)}\,dx$$

$$\Rightarrow\quad y \times \frac{1}{x^{1/2}} = \int \frac{x^{1/2}\,dx}{(x^{3/4}+1)} \quad \ldots\text{(i)}$$

On putting $x = z^4 \Rightarrow dx = 4z^3\cdot dz$

RHS of Eq. (i) becomes,

$$\int \frac{z^2\cdot 4z^3}{(z^3+1)}\cdot dz = 4\int \frac{z^2(z^3+1-1)}{(z^3+1)}\,dz$$

$$= 4\left[\int \frac{z^2(z^3+1)}{(z^3+1)}\,dz - \int \frac{z^2}{(z^3+1)}\,dz\right]$$

$$= 4\left[\frac{z^3}{3} - \frac{1}{3}\cdot\int \frac{3z^2}{z^3+1}\,dz\right]$$

$$= 4\left[\frac{z^3}{3} - \frac{1}{3}\cdot\log|z^3+1|\right]$$

$$\left(\because \int \frac{f'(x)}{f(x)}\,dx = \log|f(x)| + C\right)$$

$$= \frac{4z^3}{3} - \frac{4}{3}\log|z^3+1| + C$$

$$= \frac{4x^{3/4}}{3} - \frac{4}{3}\log|x^{3/4}+1| + C$$

$$\left(\begin{array}{l}\because x = z^4\\ \therefore x^{3/4} = z^3\end{array}\right)$$

$\therefore$ Eq. (i) becomes, $y \times x^{-1/2}$

$$= \frac{4x^{3/4}}{3} - \frac{4}{3}\log|x^{3/4}+1| + C$$

Since, this passes through $\left(1, 1 - \frac{4}{3}\log_e 2\right)$.

Then, $\left(1 - \frac{4}{3}\log_e 2\right) \times 1$

$$= \frac{4\times 1}{3} - \frac{4}{3}\log|1+1| + C$$

$$\Rightarrow\quad C = 1 - \frac{4}{3} \Rightarrow C = \frac{-1}{3}$$

Hence,

$$y = \frac{4}{3}x^{5/4} - \frac{4}{3}\sqrt{x}\log|x^{3/4}+1| - \frac{\sqrt{x}}{3}$$

$\because\ x > 0$

$\therefore\ x^{3/4} > 0 \Rightarrow x^{3/4} + 1 > 0$

i.e., $|x^{3/4}+1| = x^{3/4}+1$

$$\therefore y = \frac{4}{3}x^{5/4} - \frac{4}{3}\sqrt{x}\log(x^{3/4}+1) - \frac{\sqrt{x}}{3}$$

Now, putting $x = 16$, we get

$$y(16) = \frac{4}{3}\times 32 - \frac{4}{3}\times 4\log 9 - \frac{4}{3}$$

$$= 4\left(\frac{31}{3} - \frac{8}{3}\log 3\right)$$

**74.** *(21)* $\dfrac{dy}{dx} + 2y\sec^2 x = 2\sec^2 x + 3\tan x\cdot\sec^2 x$

$$\text{IF} = e^{\int 2\sec^2 x\,dx} = e^{2\tan x}$$

$$\Rightarrow y\cdot e^{2\tan x} = \int e^{2\tan x}(2 + 3\tan x)\sec^2 x\,dx$$

Put $\tan x = u$

$\sec^2 x\,dx = du$

$$y\cdot e^{2u} = \int e^{2u}(2+3u)dx$$

$$y\cdot e^{2u} = 2\cdot\frac{e^{2u}}{2} + 3\int e^{2u}\cdot u\,dx$$

$$y\cdot e^{2u} = e^{2u} + 3\left[\frac{ue^{2u}}{2} - \int \frac{e^{2u}}{2}\cdot du\right]$$

$$y\cdot e^{2u} = e^{2u} + \frac{3}{2}u\cdot e^{2u} - \frac{3}{4}e^{2u} + C$$

$$F(0) = \frac{5}{4}$$

$$\frac{5}{4} - \frac{1}{4} = C \Rightarrow C = 1$$

$$\Rightarrow\quad y = 1 + 3\left(\frac{\tan x}{2} - \frac{1}{4}\right) + 1\cdot e^{-2\tan x}$$

$$\Rightarrow y\left(\frac{\pi}{4}\right) = 1 + 3\left(\frac{1}{2} - \frac{1}{4}\right) + \frac{1}{e^2}$$

$$\Rightarrow 12\left(y\left(\frac{\pi}{4}\right) - \frac{1}{e^2}\right)$$

$$= 12\left(\frac{7}{4} + \frac{1}{e^2} - \frac{1}{e^2}\right) = 21$$

**75.** (27) $\frac{dy}{dx}+\left(\frac{x}{x^2-1}\right)y$

$$=\frac{x^6+4x}{\sqrt{1-x^2}}\quad -1<x<1$$

$$P=\frac{-x}{1-x^2},\quad Q=\frac{x^6+4x}{\sqrt{1-x^2}}$$

$$\text{IF}=e^{\int P\,dx}=e^{\int\frac{-x}{1-x^2}}$$

$$=e^{-\frac{1}{2}\ln(1-x^2)}=\sqrt{1-x^2}$$

Solution of differential equation

$$y\sqrt{1-x^2}=\int\frac{x^6+4x}{\sqrt{1-x^2}}\cdot\sqrt{1-x^2}\,dx+C$$

$$y\sqrt{1-x^2}=\frac{x^7}{7}+2x^2+C$$

$$x=0,\quad y=0\Rightarrow C=0$$

$$y=\frac{\frac{x^7}{7}+2x^2}{\sqrt{1-x^2}}=f(x)$$

$$\therefore\int_{-1/2}^{1/2}f(x)\,dx=\int_{-1/2}^{1/2}\frac{\frac{x^7}{7}+2x^2}{\sqrt{1-x^2}}\,dx$$

$$=\int_{-1/2}^{1/2}\frac{x^7}{7\sqrt{1-x^2}}\,dx+2\int_{-1/2}^{1/2}\frac{x^2}{\sqrt{1-x^2}}\,dx$$

↓ odd function ↓ even function

$$\therefore\quad =4\int_0^{1/2}\frac{x^2}{\sqrt{1-x^2}}\,dx$$

$$=-4\int_0^{1/2}\frac{1-x^2-1}{\sqrt{1-x^2}}\,dx$$

$$=-4\int_0^{1/2}\sqrt{1-x^2}\,dx+4\int_0^{1/2}\frac{1}{\sqrt{1-x^2}}\,dx$$

$$=-4\left[\frac{x}{2}\sqrt{1-x^2}+\frac{1}{2}\sin^{-1}x\right]_0^{1/2}+4(\sin^{-1}x)_0^{1/2}$$

$$=-4\left(\frac{\sqrt{3}}{8}+\frac{1}{2}\times\frac{\pi}{6}\right)+4\left(\frac{\pi}{6}\right)$$

$$=-\frac{\sqrt{3}}{2}+\frac{4\pi}{6}\left(-\frac{1}{2}+1\right)=-\frac{\sqrt{3}}{2}+\frac{\pi}{3}$$

$\therefore$ Required value $=2\pi-3\sqrt{3}$

$\therefore\quad \alpha=3\sqrt{3}$

$\Rightarrow\quad \alpha^2=27$

> **Concept Enhancer**
>
> The solution of the differential equation $\frac{dy}{dx}+Py=Q$ is
>
> $y(\text{IF})=\int Q\cdot(\text{IF})dx+C$, where $\text{IF}=e^{\int P dx}$

**76.** (1) Given differential equation is

$$2\cos x\frac{dy}{dx}=\sin 2x-4y\sin x,$$

$$x\in\left(0,\frac{\pi}{2}\right)$$

$$\Rightarrow\quad \frac{dy}{dx}=\frac{2\sin x\cos x}{2\cos x}-\frac{4y\sin x}{2\cos x}$$

$$\Rightarrow\quad \frac{dy}{dx}+(2\tan x)\,y=\sin x,$$

which is linear differential equation.

So, $\text{IF}=e^{\int 2\tan x\,dx}=e^{2\log_e\sec x}$

$$=e^{\log_e\sec^2 x}=\sec^2 x$$

Its solution is

$$y\times\sec^2 x=\int\sin x\times\sec^2 x\,dx$$

$$=\int\frac{\sin x}{\cos^2 x}\,dx=\int\tan x\cdot\sec x\,dx$$

$$=\sec x+C$$

Where, $C$ is constant of integration.

$\because\quad y(\pi/3)=0$ [given]

$\therefore\quad 0\times\sec^2\pi/3=\sec\pi/3+C$

$\Rightarrow\quad C=-2$

Hence, $\quad y=\cos x-2\cos^2 x$

$\Rightarrow\quad y'=-\sin x+4\cos x\sin x$

Now, $y(\pi/4)=\frac{1}{\sqrt{2}}-1$ and

$$y'(\pi/4)=-\frac{1}{\sqrt{2}}+2$$

So, $y\left(\frac{\pi}{4}\right)+y'\left(\frac{\pi}{4}\right)$

$$=\frac{1}{\sqrt{2}}-1-\frac{1}{\sqrt{2}}+2=1$$

**77.** (4) Since, $\sqrt{4-x^2}\frac{dy}{dx}$

$$=\left\{\left(\sin^{-1}\frac{x}{2}\right)^2-y\right\}\sin^{-1}\frac{x}{2}$$

$$=\left(\sin^{-1}\frac{x}{2}\right)^3-\sin^{-1}\frac{x}{2}\cdot y$$

$$\Rightarrow\frac{dy}{dx}+\frac{\sin^{-1}\frac{x}{2}\cdot y}{\sqrt{4-x^2}}=\frac{\left(\sin^{-1}\frac{x}{2}\right)^3}{\sqrt{4-x^2}}$$

which is liner differential equation

$$\therefore\ \text{IF}=e^{\int\frac{\sin^{-1}\frac{x}{2}}{\sqrt{4-x^2}}dx}$$

Put $t=\sin^{-1}\frac{x}{2}\Rightarrow dt=\frac{1}{\sqrt{4-x^2}}dx$

$$\Rightarrow e^{\int t\,dt}=e^{\frac{t^2}{2}}=e^{\frac{\left(\sin^{-1}\frac{x}{2}\right)^2}{2}}$$

$$\Rightarrow y\cdot e^{\frac{\left(\sin^{-1}\frac{x}{2}\right)^2}{2}}$$

$$=\int\frac{\left(\sin^{-1}\frac{x}{2}\right)^3}{\sqrt{4-x^2}}\cdot e^{\frac{\left(\sin^{-1}\frac{x}{2}\right)^2}{2}}dx+C$$

Put $\frac{\left(\sin^{-1}\frac{x}{2}\right)^2}{2}=z$

$$\Rightarrow\quad ye^z=2\int z\,e^z\,dz+C$$

$$=2[e^z(z-1)]+C$$

$$\Rightarrow 2\sin^{-1}\frac{x}{2}\times\frac{1}{\sqrt{4-x^2}}dx=2dz$$

$$ye^{\frac{\left(\sin^{-1}\frac{x}{2}\right)^2}{2}}=2\left[e^{\frac{\left(\sin^{-1}\frac{x}{2}\right)^2}{2}}\frac{\left(\sin^{-1}\frac{x}{2}\right)^2}{2}-1\right]+c$$

Given, $y(2)=\frac{\pi^2-8}{4}$

$$\Rightarrow\left(\frac{\pi^2-8}{4}\right)e^{\frac{\pi^2}{8}}=2e^{\frac{\pi^2}{8}}\left[\frac{\pi^2-8}{8}\right]+C$$

$\Rightarrow C=0$

$$\therefore\ ye^{\frac{\left(\sin^{-1}\frac{x}{2}\right)^2}{2}}=2e^{\frac{\left[\sin^{-1}\left(\frac{x}{2}\right)\right]^2}{2}}\left[\frac{\left(\sin^{-1}\frac{x}{2}\right)^2}{2}-1\right]+C$$

Put $x=0\Rightarrow ye^0=2e^0[0-1]$

$\Rightarrow y(0)=-2\Rightarrow y^2(0)=4$

**78.** (61) Let $y=f(x)$

Thus, $\frac{dy}{dx}=3y+\alpha\Rightarrow\frac{dy}{dx}-3y=\alpha$

$$e^{\int -3\,dx} = e^{-3x} = \text{IF}$$

Solution is $y(e^{-3x}) = \int \alpha \cdot e^{-3x}\,dx + C$

$$ye^{-3x} = \frac{\alpha e^{-3x}}{-3} + C$$

$$\Rightarrow \quad y = \frac{\alpha}{-3} + Ce^{3x} \Rightarrow x = 0,\ y = 1$$

$$\Rightarrow \quad 1 = \frac{\alpha}{-3} + C \quad \text{...(i)}$$

$$x \to -\infty,\ y = 7$$

$$7 = \frac{\alpha}{-3} + 0 \Rightarrow \alpha = -21$$

By Eq. (i), $C = -6 \Rightarrow y = 7 - 6e^{3x}$

$$9y(-\ln 3) = 9(7 - 6e^{-3\ln 3})$$

$$= 9\left(7 - \frac{6}{27}\right) = 9\left(7 - \frac{2}{9}\right) = 63 - 2 = 61$$

**79.** *(18)* $\frac{dy}{dx} + \frac{2x}{(1+x^2)^2} y = x\,e^{\frac{1}{1+x^2}}$

$$P = \frac{2x}{(1+x^2)^2},\ Q = x\,e^{\frac{1}{1+x^2}}$$

$$\text{IF} = e^{\int \frac{2x}{(1+x^2)^2} dx} = e^{\frac{-1}{1+x^2}}$$

$\therefore$ Solution is $y(\text{IF}) = \int Q\cdot(\text{IF})\,dx + C$

$$y\left(e^{\frac{-1}{1+x^2}}\right) = \int x\,e^{\frac{1}{1+x^2}} \cdot e^{\frac{-1}{1+x^2}} \cdot dx + C$$

$$\frac{y}{e^{\frac{1}{1+x^2}}} = \frac{x^2}{2} + C$$

$\therefore y(0) = 0 \Rightarrow 0 = 0 + C \Rightarrow C = 0$

$$\therefore \quad y = \frac{x^2}{2} e^{\frac{1}{1+x^2}}$$

$$f(x) = \frac{x^2}{2} e^{\left(\frac{1}{1+x^2}\right)} e^{\left(\frac{-1}{1+x^2}\right)} = \frac{x^2}{2}$$

$f(x) = \frac{x^2}{2}$ and line $y - x = 4$

$$A = \int_{-2}^{4} (x+4) - \frac{x^2}{2}$$

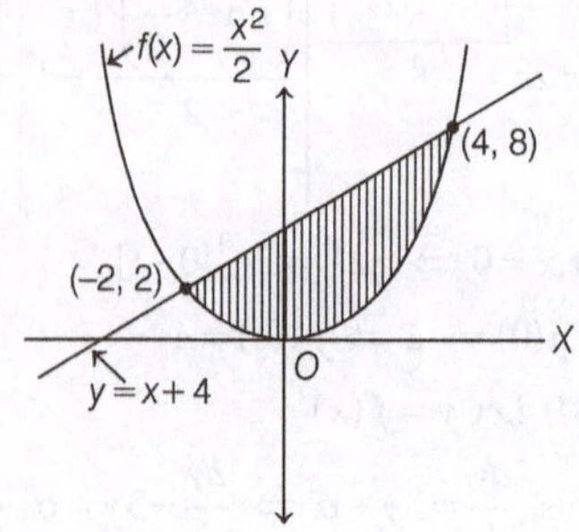


= 18 sq units.

**80.** *(7)* Given differential equation is

$$\frac{dy}{dx} - y = 1 + 4\sin x$$

$$\text{IF} = e^{\int (-1)\,dx} = e^{-x}$$

So, its solution is

$$y \times \text{IF} = \int Q \times \text{IF}\,dx$$

$$\Rightarrow y \times e^{-x} = \int (e^{-x} + 4e^{-x}\sin x)\,dx$$

$$= -e^{-x} - 2(e^{-x}\sin x + e^{-x}\cos x) + c$$

$$\Rightarrow y = -1 - 2(\sin x + \cos x) + ce^{x}$$

$\because y(\pi) = 1$ so, $c = 0$

Hence, $y\left(\frac{\pi}{2}\right) = -1 - 2 = -3$ and thus

$$y\left(\frac{\pi}{2}\right) + 10 = -3 + 10 = 7$$

**81.** *(5)* We have,

$$\frac{dx}{dy} = \frac{1 + x - y^2}{y}$$

$$\Rightarrow \quad \frac{dx}{dy} - \frac{x}{y} = \frac{1}{y} - y \quad \text{...(i)}$$

This form of

$$\frac{dx}{dy} + Px = Q(y) \quad \text{...(ii)}$$

So, IF $e^{\int P dy} = e^{-\int \frac{1}{y} dy} = e^{-\ln y} = \frac{1}{y}$

Now, solution of Eq. (i) is given by

$x \cdot \text{IF} = \int Q \cdot \text{IF}\,dy + c$

$$= \int \left(\frac{1}{y^2} - 1\right) dy + c$$

$$\frac{x}{y} = -\frac{1}{y} - y + c$$

$$x = -1 - y^2 + cy$$

[use $x(1) = 1$]

$\Rightarrow \quad 1 = -1 - 1 + c$

$\Rightarrow \quad c = 3$

$\therefore \quad x = 3y - y^2 - 1$

$\Rightarrow \quad 5x(2) = 5(3 \times 2 - 4 - 1) = 5$

**82.** *(14)*

> **key idea** First, form linear differential equation in *x*. Then, conclude the solution of differential equation by $x \times \text{I.F} = \int Q \times \text{I.F}\,dx$

$$\frac{dx}{dt} = \frac{2x}{(t+1)} + (t+1)^3$$

$$\Rightarrow \quad \frac{dx}{dt} - \frac{2x}{t+1} = (t+1)^3 \quad \text{[LDE]}$$

$$\text{IF} \Rightarrow e^{-2\int \frac{dt}{t+1}} = e^{-2\ln(t+1)} = \frac{1}{(t+1)^2}$$

$$\Rightarrow \quad x \cdot \frac{1}{(1+t)^2} = \int (t+1)^3 \cdot \frac{1}{(t+1)^2}\,dt$$

$$\Rightarrow \quad \frac{x}{(t+1)^2} = \frac{(t+1)^2}{2} + c$$

$$\Rightarrow \quad = \frac{(t+1)^4}{2} + c(t+1)^2$$

Put $t = 0$, $x = 2$

$$2 = \frac{1}{2} + c \Rightarrow c = \frac{3}{2}$$

$$\Rightarrow x = \frac{(t+1)^4}{2} + \frac{3}{2}(t+1)^2$$

Put $t = 1$, $x = \frac{16}{2} + \frac{3}{2} \cdot 4 = 14$

**83.** *(9)* $\sec^2 x \frac{dx}{dy} + e^{2y}\tan^2 x + \tan x = 0$

Put $\tan x = t$

$$\sec^2 x \frac{dx}{dy} = \frac{dt}{dy}$$

$$\frac{dt}{dy} + t^2 e^{2y} + t = 0$$

$$\frac{dt}{dy} + t = -t^2 e^{2y}$$

$$\frac{1}{t^2}\frac{dt}{dy} + \frac{1}{t} = -e^{2y}$$

$$\frac{1}{t} = z$$

$$\Rightarrow \quad \frac{1}{t^2}\frac{dt}{dy} = -\frac{dz}{dy}$$

$$-\frac{dz}{dy} + z = -e^{2y}$$

$$\frac{dz}{dy} - z = e^{2y}$$

$$P = -1,\ Q = e^{2y}$$

$$\text{IF} = e^{\int P\,dy} = e^{-\int dy} = e^{-y}$$

General solution

$$z \cdot \text{IF} = \int Q \cdot \text{IF}\,dy$$

$$z \cdot e^{-y} = \int e^{2y} e^{-y}\,dy = e^{y} + C$$

$$ze^{-y} = e^{y} + C$$

$$\frac{1}{\tan x} \cdot e^{-y} = e^{y} + C$$

Given, $y\left(\frac{\pi}{4}\right) = 0$

$$\frac{1}{1}e^{0} = e^{0} + C \Rightarrow C = 0$$

Curve, $\tan x \cdot e^{2y} = 1$

at $x = \frac{\pi}{6}$, $y = \alpha$

$$\tan \frac{\pi}{6} e^{2\alpha} = 1 \Rightarrow e^{2\alpha} = \sqrt{3}$$

$$\Rightarrow \quad e^{8\alpha} = 9$$

**84.** *(20)* Equation of line

$(Y - y) = Y'(x)[X - x]$

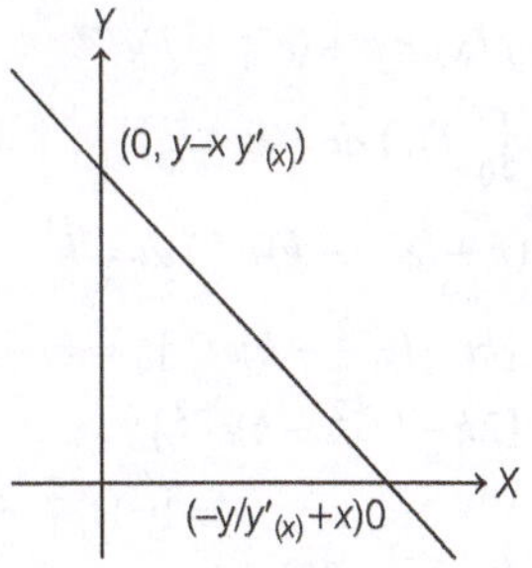


$$\text{Area} = \frac{1}{2} \times \left(x - \frac{y}{Y'(x)}\right) \cdot [y - xY'(x)]$$

$$\frac{-y^2}{2Y'(x)} + 1 = \frac{1}{2}\left(\frac{xY'(x) - y}{Y'(x)}\right)(y - xY'(x))$$

$$\Rightarrow -y^2 + 2Y'(x) = (xY'(x) - y)(y - xY'(x))$$

$$\Rightarrow -y^2 + 2Y'(x) = xyY'(x) - x^2$$

$$\Rightarrow (Y'(x))^2 - y^2 + xyY'(x)$$

$$\Rightarrow \quad 2Y'(x) = 2xyY'(x) - x^2[Y'(x)]^2$$

$$\Rightarrow \quad 2 = 2xy - x^2 Y'(x)$$

$$\Rightarrow \quad x^2 \frac{dy}{dx} = 2xy - 2$$

$$\Rightarrow \frac{dy}{dx} - 2\frac{y}{x} = -\frac{2}{x^2} \Rightarrow \text{IF} = e^{\int -\frac{2}{x} dx} = \frac{1}{x^2}$$

$\therefore$ Solution

$$y\left(\frac{1}{x^2}\right) = \int -\frac{2}{x^2} \times \frac{1}{x^2} dx + C$$

$$\Rightarrow \quad \frac{y}{x^2} = \frac{2}{3x^3} + C$$

Given, $\quad f(1) = 1 \Rightarrow x = 1$ and $y = 1$

$$\therefore \quad 1 = \frac{2}{3} + C \Rightarrow C = \frac{1}{3}$$

Equation is

$$Y = \frac{2}{3X} + \frac{X^2}{3} \Rightarrow Y(2) = \frac{2}{3 \times 2} + \frac{4}{3} = \frac{5}{3}$$

$$\Rightarrow \quad 12Y(2) = 12 \times \frac{5}{3} = 20$$

**85.** *(97)* We have,

$$(1 - x^2)dy = [xy + (x^3 + 2)\sqrt{3(1 - x^2)}]\, dx$$

$$\Rightarrow \frac{dy}{dx} - \frac{x}{1 - x^2} y = \frac{(x^3 + 2)\sqrt{3}}{\sqrt{1 - x^2}}$$

This form of

$$\frac{dy}{dx} + Py = Q$$

$$\text{So, IF} = e^{-\int \frac{x}{1 - x^2} dx} = \sqrt{1 - x^2}$$

Solution is given by

$$y \cdot \text{IF} = \int Q \cdot \text{IF} + C$$

$$y \cdot \sqrt{1 - x^2} = \sqrt{3}\int (x^3 + 2)dx + C$$

$$y(\sqrt{1 - x^2}) = \sqrt{3}\left[\frac{x^4}{4} + 2x\right] + C$$

Now, $y(0) = 0 \Rightarrow 0 = C$

$$\therefore \quad y(\sqrt{1 - x^2}) = \sqrt{3}\left(\frac{x^4}{4} + 2x\right)$$

$$\text{Now, } y\left(\frac{1}{2}\right) \cdot \frac{\sqrt{3}}{2} = \sqrt{3}\left[\frac{1}{64} + 1\right]$$

$$y\left(\frac{1}{2}\right) = \frac{1}{32} + 2 = \frac{65}{32}$$

$$\therefore \quad m = 65, n = 32$$

$$\therefore \quad m + n = 97$$

**86.** *(6)* The given differential equation is

$$\frac{dy}{dx} + \frac{4x}{x^2 - 1} y = \frac{x + 2}{(x^2 - 1)^{5/2}}, x > 1$$

$$\text{IF} = e^{\int \frac{4x}{x^2 - 1} dx} = e^{2\int \frac{2x}{x^2 - 1} dx}$$

$$= e^{2\log (x^2 - 1)} = (x^2 - 1)^2$$

Solution is

$$y \cdot (x^2 - 1)^2 = \int (x^2 - 1)^2 \frac{x + 2}{(x^2 + 1)^{5/2}} dx$$

$$= \int \frac{x + 2}{\sqrt{x^2 - 1}} dx$$

$$= \frac{1}{2}\int \frac{2x\, dx}{\sqrt{x^2 - 1}} + 2\int \frac{dx}{\sqrt{x^2 - 1}}$$

$$y(x^2 - 1)^2 = \sqrt{x^2 - 1} + 2\ln(x + \sqrt{x^2 - 1}) + C$$

Given, $y(2) = \frac{2}{9} \log_e (2 + \sqrt{3})$

$$\frac{2}{9} \log_e (2 + \sqrt{3})(3)^2 = \sqrt{3} + 2\ln|2 + \sqrt{3}| + C$$

$$\Rightarrow 2\ln(2 + \sqrt{3}) = \sqrt{3} + 2\ln|2 + \sqrt{3}| + C$$

$$\Rightarrow \quad C = -\sqrt{3}$$

Hence, $y(x^2 - 1)^2 = \sqrt{x^2 - 1} + 2\ln|x + \sqrt{x^2 - 1}| - \sqrt{3}$

Put $x = \sqrt{2}$

$$y = 1 + 2\ln|\sqrt{2} + 1| - \sqrt{3}$$

$$\Rightarrow y = 2\ln(\sqrt{2} + 1) + 1 - \sqrt{3}$$

$$y = \alpha \log_e (\sqrt{\alpha} + \beta) + \beta - \gamma$$

$$\therefore \quad \alpha = 2, \beta = 1, \gamma = 3$$

Thus, $\alpha\beta\gamma = 6$

**87.** *(12)* Given, $(\ln(\cos y))^2 \cos y\, dx = (1 + 3x \ln \cos y) \sin y\, dy$

$$\frac{dx}{dy} = \tan y\left(\frac{1}{\ln^2 \cos y} + \frac{3\ln \cos y}{\ln^2 \cos y} x\right)$$

$$\Rightarrow \frac{dx}{dy} - \left(\frac{3\tan y}{\ln \cos y}\right)x = \frac{\tan y}{\ln^2 \cos y}$$

$$\text{IF} = e^{\int \frac{-3\tan y}{\ln \cos y} dy} = e^{\int -\frac{3}{\ln \cos y} \frac{\sin y}{\cos y} dy}$$

On putting $\ln \cos y = t$

$$\Rightarrow \frac{1}{\cos y}(-\sin y)dy = dt$$

$$\text{IF} = e^{\int \frac{3}{t} dt} = e^{\ln t^3} = t^3 = \ln^3(\cos y)$$

Solution : $x \cdot \ln^3 \cos y$

$$= \int \left[\frac{\sin y}{\cos y} \cdot \frac{1}{\ln^2 \cos y} \times \ln^3 \cos y\right] dy + C$$

$$\text{or } x \ln^3 \cos y = \int (\ln \cos y)\frac{\sin y}{\cos y} \cdot dy + C$$

$$\text{or } x \ln^3 \cos y = -\frac{\ln^2 \cos y}{2} + C$$

$$\because x\left(\frac{\pi}{3}\right) = \frac{1}{2\ln 2}$$

$$\therefore \frac{1}{2\ln 2} \times \ln^3 \cos\frac{\pi}{3} = -\frac{1}{2}\ln^2 \cos\frac{\pi}{3} + C$$

$$\Rightarrow -\frac{1}{2\ln 2} \times \left(\ln\left(\frac{1}{2}\right)\right)^3 = -\frac{1}{2}\left(\ln\frac{1}{2}\right)^2 + C$$

$$\Rightarrow -\frac{1}{2} \cdot \frac{1}{\ln\left(\frac{1}{2}\right)} \cdot \left(\ln\left(\frac{1}{2}\right)\right)^3 = -\frac{1}{2}\left(\ln\frac{1}{2}\right)^2 + C$$

$$C = 0$$

When, $y = \frac{\pi}{6}$

$$x \ln^3\left(\frac{\sqrt{3}}{2}\right) = -\frac{1}{2}\ln^2\left(\frac{\sqrt{3}}{2}\right) + 0$$

$$\Rightarrow \quad x = -\frac{1}{2\ln\left(\frac{\sqrt{3}}{2}\right)}$$

$$= -\frac{1}{\ln\frac{3}{4}} = \frac{1}{\ln\left(\frac{4}{3}\right)}$$

$$\Rightarrow \quad x = \frac{1}{\ln 4 - \ln 3}$$

$\Rightarrow m = 4$ and $n = 3 \Rightarrow mn = 12$

**88.** *(3)* Given, differential equation

$$(y - 2\log_e x)\,dx + (x \log_e x^2)\,dy = 0$$

$$\Rightarrow 2x \log x \frac{dy}{dx} + y = 2\log_e x$$

$$\Rightarrow \frac{dy}{dx} + \frac{y}{2x\log_e x} = \frac{1}{x}$$

$$\text{IF} = e^{\int \frac{1}{2x\log_e x}dx} = e^{\frac{1}{2}\log(\log x)} = \sqrt{\log x}$$

$\therefore$ Solution of differential equation

$$y \cdot \sqrt{\log_e x} = \int \frac{1}{x}\sqrt{\log x}\,dx + C$$

$$y\sqrt{\log_e x} = \frac{(\log x)^{3/2}}{3/2} + C$$

It passes through $\left(e, \frac{4}{3}\right)$

$$\Rightarrow \quad \frac{4}{3} = \frac{2}{3} + C$$

$$\Rightarrow \quad C = \frac{2}{3}$$

It also passes through $(e^4, \alpha)$,

$$\therefore \quad \alpha \cdot (2) = \frac{16}{3} + \frac{2}{3} = 6$$

$$\Rightarrow \quad \alpha = 3$$

**89.** *(2)* Given, differential equation

$$(x\cos x)\,dy + (xy\sin x + y\cos x - 1)\,dx = 0$$

$$0 < x < \frac{\pi}{2}$$

$$\Rightarrow \quad \frac{dy}{dx} + \frac{xy\sin x + y\cos x - 1}{x\cos x} = 0$$

$$\Rightarrow \quad \frac{dy}{dx} + \left(\frac{xy\sin x + y\cos x}{x\cos x}\right) = \frac{1}{x\cos x}$$

$$\Rightarrow \quad \frac{dy}{dx} + \left(\tan x + \frac{1}{x}\right)y = \frac{1}{x\cos x}$$

Which is linear differential equation in the form of $\frac{dy}{dx} + Py = Q$

$$\therefore \text{IF} = e^{\int\left(\tan x + \frac{1}{x}\right)dx}$$

$$= e^{(\log\sec x + \log x)} = e^{\log(x\sec x)}$$

$$= x\sec x$$

$\therefore$ The general solution of the given differential equation

$$y \cdot \text{IF} = \int (Q \times \text{IF})\,dx + c$$

$$\Rightarrow y\,(x\sec x) = \int\left(\frac{1}{x\cos x} x\sec x\right)dx + c$$

$$\Rightarrow \quad xy\sec x = \int \sec^2 x\,dx + c$$

$$\Rightarrow \quad xy\sec x = \tan x + c \quad \ldots\text{(i)}$$

Since, $\frac{\pi}{3} y\left(\frac{\pi}{3}\right) = \sqrt{3}$

$$\Rightarrow \quad y\left(\frac{\pi}{3}\right) = \frac{3\sqrt{3}}{\pi}$$

$$\therefore \frac{\pi}{3}\left(\frac{3\sqrt{3}}{\pi}\right)\sec\left(\frac{\pi}{3}\right) = \tan\left(\frac{\pi}{3}\right) + c$$

$$\Rightarrow \quad \sqrt{3}\,(2) = \sqrt{3} + c \Rightarrow c = \sqrt{3}$$

On putting value of $c$ in Eq. (i), we get

$$xy\sec x = \tan x + \sqrt{3}$$

$$\Rightarrow \quad y = \frac{1}{x}(\sin x + \sqrt{3}\cos x)$$

$$y' = \frac{1}{x}(\cos x - \sqrt{3}\sin x) - \frac{1}{x^2}(\sin x + \sqrt{3}\cos x)$$

and $y'' = \frac{1}{x}(-\sin x - \sqrt{3}\cos x)$

$$- \frac{1}{x^2}(\cos x - \sqrt{3}\sin x) - \frac{1}{x^2}(\cos x - \sqrt{3}\sin x) + \frac{2}{x^3}(\sin x - \sqrt{3}\cos x)$$

$$\therefore \left|\frac{\pi}{6}y''\left(\frac{\pi}{6}\right) + 2y'\left(\frac{\pi}{6}\right)\right| = |-2| = 2$$

**90.** *(1)* We have,

$$f'(x) + f(x) = \int_0^2 f(t)\,dt$$

Let $\int_0^2 f(t)\,dt = k$

Then, $f'(x) + f(x) = k$

$$\Rightarrow \quad \frac{d}{dx}f(x) + f(x) = k$$

Which is a linear differential equation

Now, $\text{IF} = e^{\int 1\,dx} = e^x$

$\therefore$ Solution is

$$e^x f(x) = \int k \cdot e^x\,dx + c$$

$$e^x f(x) = ke^x + c$$

Now, $f(0) = e^{-2}$

$$\therefore \quad e^0 f(0) = ke^0 + c$$

$$\Rightarrow \quad e^{-2} = k + c \Rightarrow c = e^{-2} - k$$

$$\therefore\ e^x f(x) = ke^x + e^{-2} - k$$

$$\Rightarrow \quad f(x) = k + (e^{-2} - k)e^{-x}$$

Now, $\int_0^2 f(t)\,dt = k$

$$\Rightarrow \int_0^2 (k + (e^{-2} - k)e^{-t})\,dt = k$$

$$\Rightarrow \quad [kt - (e^{-2} - k)e^{-t}]_0^2 = k$$

$$\Rightarrow \quad [2k - (e^{-2} - k)e^{-2}] - [-(e^{-2} - k)] = k$$

$$\Rightarrow \quad k = e^{-2} - 1$$

Now, $2f(0) - f(2) = 2e^{-2} - [k + (e^{-2} - k)e^{-2}]$

$$= e^{-2} + 1 - e^{-2} = 1$$

**91.** *(1)*

$$\frac{dy}{dx} + \frac{4xy}{\sin(2x^2)\log(\tan x^2)} = \frac{4\sqrt{2}\,x\sin\left(x^2 - \frac{\pi}{4}\right)}{\sin(2x^2)\log(\tan x^2)}$$

Here, $\text{IF} = e^{\int \frac{4x\,dx}{\sin(2x^2)\log(\tan x^2)}}$

$$= e^{\log\{\log(\tan x^2)\}} = \log(\tan x^2)$$

Now, $y \cdot \log(\tan x^2)$

$$= \int \frac{4\sqrt{2}x\sin\left(x^2 - \frac{\pi}{4}\right)}{\sin(2x^2)}dx$$

$$\Rightarrow \quad y\log(\tan x^2) = \log\left|\frac{\sec x^2 + \tan x^2}{\operatorname{cosec} x^2 - \cot x^2}\right| + C \quad \ldots\text{(i)}$$

Since, Eq. (i) passes through $\left(\sqrt{\frac{\pi}{6}}, 1\right)$.

$$\therefore \log\left(\frac{1}{\sqrt{3}}\right) = \log\left(\frac{3/\sqrt{3}}{2 - \sqrt{3}}\right) + C$$

$$\Rightarrow \quad C = \log\left(\frac{1}{\sqrt{3}}\right) - \log\left(\frac{\sqrt{3}}{2 - \sqrt{3}}\right)$$

For $y\left(\sqrt{\frac{\pi}{3}}\right)$, we get from Eq. (i),

$$y\log(\sqrt{3})=\log\left|\frac{2+\sqrt{3}}{\frac{2}{\sqrt{3}}-\frac{1}{\sqrt{3}}}\right|+\log\left(\frac{1}{\sqrt{3}}\right)-\log\left(\frac{\sqrt{3}}{2-\sqrt{3}}\right)$$

$$\Rightarrow y\log\sqrt{3}=\log\left(\frac{1}{\sqrt{3}}\right)$$

$$\Rightarrow \frac{y}{2}\log 3=-\frac{1}{2}\log 3$$

$$\Rightarrow \frac{y}{2}=-\frac{1}{2}$$

$$\Rightarrow y=-1$$

Hence, $\left|y\left(\sqrt{\frac{\pi}{3}}\right)\right|=1$

**92.** *(3)* Given, differential equation $\frac{dy}{dx}-y=2-e^{-x}$, which is of the first order linear differential equation.

Here, $P=-1,\ Q=2-e^{-x}$

$$\therefore \quad \text{IF}=e^{\int P dx}=e^{\int -1\,dx}=e^{-x}$$

Now, solution is given by

$$y\cdot\text{IF}=\int Q\times\text{IF}\,dx+C$$

$$\Rightarrow y\times e^{-x}=\int(2-e^{-x})e^{-x}+C$$

$$\Rightarrow ye^{-x}=\int(2e^{-x}-e^{-2x})dx+C$$

$$\Rightarrow ye^{-x}=-2e^{-x}+\frac{e^{-2x}}{2}+C$$

$$\Rightarrow y=-2+\frac{e^{-x}}{2}+ce^{x}$$

$\lim_{x\to\infty} y(x)$ is finite, then

$\lim_{x\to\infty}\left(-2+\frac{e^{-x}}{2}+ce^{-x}\right)$ is finite

This is possible when $c=0$

$$y=-2+\frac{e^{-x}}{2}\Rightarrow\frac{dy}{dx}=-\frac{e^{-x}}{2}$$

$$\left.\frac{dy}{dx}\right|_{x=0}=-\frac{1}{2}$$

$\therefore$ Slope $=-\frac{1}{2}$

When, $x=0$, then $y=-2+\frac{1}{2}=-\frac{3}{2}$

Equation of tangent

$$y+\frac{3}{2}=-\frac{1}{2}(x-0)$$

$$\Rightarrow 2y+3=-x$$

$$\Rightarrow x+2y=-3$$

$$\frac{x}{-3}+\frac{y}{-\frac{3}{2}}=1$$

$$\therefore a=-3,\ b=-\frac{3}{2}$$

Now,

$$a-4b=-3-4\left(-\frac{3}{2}\right)$$

$$=-3+6=3$$

**93.** *(14)* Given, $(x-1)\frac{dy}{dx}+2xy=\frac{1}{x-1}$ ... (i),

$$y(2)=\frac{1+e^4}{2e^4}\text{ and }y(3)=\frac{e^{\alpha}+1}{\beta e^{\alpha}}$$

From Eq. (i),

$$\frac{dy}{dx}+\frac{2x.y}{(x-1)}=\frac{1}{(x-1)^2}$$

$$\text{IF}=e^{\int\frac{2x}{x-1}.dx}=e^{2\int\frac{x-1+1}{x-1}.dx}$$

$$=e^{2\int 1+\frac{1}{x-1}dx}=e^{2x+2\ln(x-1)}$$

$$=e^{2x}\cdot e^{\ln(x-1)^2}=(x-1)^2\cdot e^{2x}$$

$\therefore$ Solution is

$$y\cdot(x-1)^2e^{2x}=\int\frac{1}{(x-1)^2}(x-1)^2e^{2x}\,dx+c$$

$$\Rightarrow y(x-1)^2e^{2x}=\frac{e^{2x}}{2}+c \quad \text{... (ii)}$$

On putting $x=2$, $y=\frac{1+e^4}{2e^4}$ in Eq. (ii),

we get

$$\left(\frac{1+e^4}{2e^4}\right)\cdot(1)\cdot(e^4)=\frac{e^4}{2}+c\Rightarrow c=\frac{1}{2}$$

$$\therefore\ y(x-1)^2e^{2x}=\frac{e^{2x}}{2}+\frac{1}{2}$$

On putting $x=3$ in above equation

$$y(2)^2e^6=\frac{e^6}{2}+\frac{1}{2}$$

$$\Rightarrow y=\frac{e^6+1}{8e^6}=\frac{e^{\alpha}+1}{\beta e^{\alpha}}$$

$$\Rightarrow \alpha+\beta=14$$

**94.** *(2)* $\frac{dy}{dx}+\frac{\sqrt{2}y}{2\cos^4x-\cos 2x}=xe^{\tan^{-1}(\sqrt{2}\cot 2x)}$

$$\text{IF}=e^{\sqrt{2}\int\frac{dx}{2\cos^4x-\cos 2x}}$$

$$=e^{\sqrt{2}\int\frac{dx}{2(\cos^2x)^2-(2\cos^2x-1)}}$$

$$=e^{\sqrt{2}\int\frac{dx}{2\cos^2x\ (\cos^2x-1)+1}}$$

$$=e^{\sqrt{2}\int\frac{dx}{-2\cos^2x\sin^2x+1}}$$

$$=e^{\sqrt{2}\int\frac{dx}{\cos^4x+\sin^4x}}$$

$$\because \cos^4x+\sin^4x=(\cos^2x+\sin^2x)^2-2\cos^2x\cdot\sin^2x$$

$$=1-2\cos^2x\sin^2x$$

$$\therefore\ \text{IF}=e^{\sqrt{2}\int\frac{\text{cosec}^4x\,dx}{1+\cot^4x}}=e^{\sqrt{2}\int-\frac{(t^2+1)}{1+t^4}dt}$$

[Assuming $\cot x=t$]

$$=e^{\sqrt{2}\int\frac{\left(1+\frac{1}{t^2}\right)}{t^2+\frac{1}{t^2}}.dt}=e^{-\sqrt{2}\int\frac{\left(1+\frac{1}{t^2}\right)}{\left(t-\frac{1}{t}\right)^2+2}dt}$$

$$=e^{\frac{-\sqrt{2}}{\sqrt{2}}\tan^{-1}\left(\frac{t-\frac{1}{t}}{\sqrt{2}}\right)}$$

$$=e^{-\tan^{-1}\left(\frac{\cot^2x-1}{\sqrt{2}\cot x}\right)}=e^{-\tan^{-1}\left(\frac{\sqrt{2}(\cot^2x-1)}{2\cot x}\right)}$$

$$=e^{-\tan^{-1}(\sqrt{2}\cot 2x)}$$

$$y\cdot e^{-\tan^{-1}(\sqrt{2}\cot 2x)}=\int x\,dx$$

$$\Rightarrow y\cdot e^{-\tan^{-1}(\sqrt{2}\cot 2x)}=\frac{x^2}{2}+c$$

At $x=\frac{\pi}{4},\ y=\frac{\pi^2}{32}$

$$\therefore\ \frac{\pi^2}{32}.e^{-0}=\frac{\pi^2}{32}+c\Rightarrow c=0$$

$$\therefore\ y=\frac{x^2}{2}.e^{\tan^{-1}}(\sqrt{2}\cot 2x)$$

At $x=\frac{\pi}{3},\ y=\frac{\pi^2}{18}e^{\tan^{-1}\left(\sqrt{2}\times\left(\frac{-1}{\sqrt{3}}\right)\right)}$

$$\therefore\ y\left(\frac{\pi}{3}\right)=\frac{\pi^2}{18}e^{-\tan^{-1}\left(\sqrt{\frac{2}{3}}\right)}$$

Therefore, $\alpha=\sqrt{\frac{2}{3}}\Rightarrow 3\alpha^2=2$

**95.** *(320)* Given equation is

$$(1-x^2)dy=(xy+(x^3+2)\sqrt{1-x^2})\,dx$$

$$\frac{dy}{dx}=\frac{xy}{1-x^2}+\frac{x^3+2}{\sqrt{1-x^2}}$$

$$\Rightarrow \frac{dy}{dx}+\left(\frac{-x}{1-x^2}\right)y=\frac{(x^3+2)}{\sqrt{1-x^2}}$$

$$\text{IF}=e^{\int\frac{-x}{1-x^2}dx}$$

$$=e^{\frac{1}{2}\int\frac{-2x}{1-x^2}dx}$$

$$=e^{\frac{1}{2}\log(1-x^2)}$$

$$=\sqrt{1-x^2}$$

$\therefore$ Solution is given by $y\times\sqrt{1-x^2}$

$$=\int\left(\sqrt{1-x^2}\cdot\frac{x^3+2}{\sqrt{1-x^2}}\right)dx+c$$

or $\quad y\sqrt{1-x^2}=\frac{x^4}{4}+2x+c$

When, $x=0$, $y=0$

$\Rightarrow \quad c=0$

$\therefore \quad y\sqrt{1-x^2}=\frac{x^4}{4}+2x$

Now, $k=\int_{-1/2}^{1/2}\sqrt{1-x^2}\cdot y\,dx$

$$=\int_{-1/2}^{1/2}\left(\frac{x^4}{4}+2x\right)dx$$

$$=\frac{1}{4}\int_{-1/2}^{1/2}x^4\,dx+2\int_{-1/2}^{1/2}x\,dx$$

$$=\frac{1}{4}\times2\int_0^{1/2}x^4\,dx+0$$

$$=\frac{1}{2}\left[\frac{x^5}{5}\right]_0^{1/2}=\frac{1}{320}$$

$\therefore \quad k^{-1}=320$

**96.** *(4)* $y=y(x)$

$$e^y\frac{dy}{dx}-2e^y\sin x+\sin x\cos^2 x=0$$

Let $e^y=t$

$$\Rightarrow e^y\frac{dy}{dx}=\frac{dt}{dx}$$

$$\therefore \frac{dt}{dx}-2t\sin x+\sin x\cos^2 x=0$$

$$\frac{dt}{dx}+(-2\sin x)t=-\sin x\cos^2 x$$

$$\text{IF}=e^{\int-2\sin x dx}=e^{2\cos x}$$

$$(e^{2\cos x}.t)=\int e^{2\cos x}.(-\sin x\cos^2 x)dx$$

$$\Rightarrow te^{2\cos x}=\int e^{2z}.z^2dz$$

$$\Rightarrow te^{2\cos x}=\frac{z^2e^{2z}}{2}-\int\frac{2ze^{2z}}{2}dz$$

$$\Rightarrow te^{2\cos x}=\frac{z^2e^{2z}}{2}-\frac{ze^{2z}}{2}+\frac{e^{2z}}{4}+C$$

$$\therefore e^ye^{2\cos x}=\frac{\cos^2 xe^{2\cos x}}{2}-\cos x\frac{e^{2\cos x}}{2}+\frac{e^{2\cos x}}{4}+C$$

$$x=\frac{\pi}{2},\ y=0,\ 1=\frac{1}{4}+C$$

$$\Rightarrow \quad C=\frac{3}{4}$$

$$\Rightarrow \quad x=0,$$

$$e^y.e^2=\frac{e^2}{2}-\frac{e^2}{2}+\frac{e^2}{4}+\frac{3}{4}$$

$$\Rightarrow 4e^{y+2}=e^2+3$$

$$\Rightarrow \quad e^y=\frac{1}{4}+\frac{3}{4}e^{-2}$$

$$y=\log_e\left(\frac{1}{4}+\frac{3}{4}e^{-2}\right)$$

$\Rightarrow \quad \alpha+\beta=1$

$\therefore \quad 4(\alpha+\beta)=4$

**97.** *(1)* Given $e^{\sin y}\cos y\frac{dy}{dx}+e^{\sin y}\cos x$

$$=\cos x \quad \ldots\text{(i)}$$

Let $e^{\sin y}=t$, then $e^{\sin y}\cdot\cos y\cdot\frac{dy}{dx}=\frac{dt}{dx}$,

putting in Eq. (i),

$\frac{dt}{dx}+t\cos x=\cos x$ (Linear form) … (ii)

Then, IF $=e^{\int\cos x\,dx}=e^{\sin x}$

Solution of differential Eq. (ii) is

$$t\cdot\text{IF}=\int\cos x\cdot\text{IF}\,dx+C$$

$$t\cdot e^{\sin x}=\int\cos x\cdot e^{\sin x}dx+C$$

Let $e^{\sin x}=e^u$ i.e. let $\sin x=u$ then $\cos x\,dx=du$

$$\Rightarrow \quad t\cdot e^{\sin x}=\int e^u\,du+C=e^u+C$$

On putting $u=\sin x$ and $t=e^{\sin y}$

$$\Rightarrow \quad e^{\sin y}\cdot e^{\sin x}=e^{\sin x}+C$$

Given, $y(0)=0$, this gives $C=0$

$$\Rightarrow e^{\sin y}\cdot e^{\sin x}=e^{\sin x}$$

$$\Rightarrow e^{\sin y+\sin x}=e^{\sin x}$$

$$\Rightarrow \quad \sin y+\sin x=\sin x$$

$$\Rightarrow \quad \sin y=0$$

$$\Rightarrow \quad y=0$$

$$\therefore y\left(\frac{\pi}{6}\right)=y\left(\frac{\pi}{3}\right)=y\left(\frac{\pi}{4}\right)=0$$

Hence, $1+y(\pi/6)+\frac{\sqrt{3}}{2}y(\pi/3)+\frac{1}{\sqrt{2}}y(\pi/4)$

$$=1+0+0+0=1$$

**98.** *(1)* Given, $(2xy^2-y)dx+xdy=0$

$$\Rightarrow \quad \frac{dy}{dx}-\frac{y}{x}=-2y^2$$

$$\Rightarrow \quad \frac{-1}{y^2}\frac{dy}{dx}+\frac{1}{xy}=2 \quad \ldots\text{(i)}$$

[divide by $y^2$]

Let $\frac{1}{y}=v$, then $-\frac{1}{y^2}\cdot\frac{dy}{dx}=\frac{dv}{dx}$,

On putting in Eq. (i), we get

$\frac{dv}{dx}+v\left(\frac{1}{x}\right)=2$ [This is a linear form]

Now, integrating factor (IF)

$$=e^{\int\frac{1}{x}dx}=e^{\log x}=x$$

$$\therefore \quad (\text{IF})v=\int 2(\text{IF})dx=\int 2xdx$$

$$=2\frac{x^2}{2}+C$$

$$\therefore \quad (\text{IF})v=x^2+C$$

On putting $v=\frac{1}{y}$, this gives

$$x^2+c=x/y$$

Now, first find point of intersection of lines $2x-3y=1$ and $3x=-2y+8$ by elimination method, we get $x=2$, $y=1$

$\therefore$ The curve $x^2+c=\frac{x}{y}$ passes through (2, 1).

putting $x=2$, $y=1$, we get $c=-2$

$$\frac{x}{y}=x^2-2$$

or $\quad y=\frac{x}{x^2-2}$

Put $x=1$, we get

$$y(1)=\frac{1}{1-2}=-1$$

$$\therefore |y(1)|=1$$

# C-22

JEE Main Chapterwise Topicwise Mathematics

In JEE Main, there are usually 1-2 questions per paper are visible from this chapter. In past 3 years, 68 conceptual and 19 numerical questions of moderate difficulty level were asked from this chapter.

**For detailed analysis scan QR code**

# VECTOR ALGEBRA

## Topic 1 Algebra & Modulus of Vectors

### Level 1

**01** If the components of $\mathbf{a} = \alpha\hat{i} + \beta\hat{j} + \gamma\hat{k}$ along and perpendicular to respectively, are $\frac{16}{11}(3\hat{i} + \hat{j} - \hat{k})$ and $\frac{1}{11}(-4\hat{i} - 5\hat{j} - 17\hat{k})$, then $\alpha^2 + \beta^2 + \gamma^2$ is equal to *[2025, 28 Jan Shift-I]*

(a) 23 (b) 16 (c) 26 (d) 18

**02** Let **a**, **b** and **c** be three non-zero vectors such that **b** and **c** are non-collinear. If $\mathbf{a} + 5\mathbf{b}$ is collinear with **c**, $\mathbf{b} + 6\mathbf{c}$ is collinear with **a** and $\mathbf{a} + \alpha\mathbf{b} + \beta\mathbf{c} = \mathbf{0}$, then $\alpha + \beta$ is equal to *[2024, 29 Jan Shift-I]*

(a) –25 (b) 35 (c) 30 (d) –30

**03** Let $O$ be the origin and the position vectors of $A$ and $B$ be $2\hat{i} + 2\hat{j} + \hat{k}$ and $2\hat{i} + 4\hat{j} + 4\hat{k}$, respectively. If the internal bisector of $\angle AOB$ meets the line $AB$ at $C$, then the length of $OC$ is *[2024, 29 Jan Shift-I]*

(a) $\frac{3}{2}\sqrt{34}$ (b) $\frac{2}{3}\sqrt{34}$ (c) $\frac{3}{2}\sqrt{31}$ (d) $\frac{2}{3}\sqrt{31}$

**04** Let $ABCD$ be a quadrilateral. If $E$ and $F$ are the mid-points of the diagonals $AC$ and $BD$ respectively and $(\mathbf{AB} - \mathbf{BC}) + (\mathbf{AD} - \mathbf{DC}) = k\mathbf{FE}$, then $k$ is equal to *[2023, 15 April Shift-I]*

(a) 4 (b) – 4
(c) – 2 (d) 2

**05** For any vector $\mathbf{a} = a_1\hat{i} + a_2\hat{j} + a_3\hat{k}$, with $10|a_i| < 1$, $i = 1, 2, 3$, consider the following statements

(A) $\max\{|a_1|, |a_2|, |a_3|\} \le |\mathbf{a}|$

(B) $|\mathbf{a}| \le 3\max\{|a_1|, |a_2|, |a_3|\}$ *[2023, 11 April Shift-I]*

(a) Neither (A) nor (B) is true
(b) Both (A) and (B) are true
(c) Only (B) is true
(d) Only (A) is true

**06** If the points $P$ and $Q$ are respectively the circumcenter and the orthocentre of a $\Delta ABC$, then $\mathbf{PA} + \mathbf{PB} + \mathbf{PC}$ is equal to *[2023, 10 April Shift-II]*

(a) **PQ** (b) 2**QP** (b) **QP** (d) 2**PQ**

**Imp.** Collinear Points

**07** If the points with position vectors $\alpha\hat{i} + 10\hat{j} + 13\hat{k}$, $6\hat{i} + 11\hat{j} + 11\hat{k}$, $\frac{9}{2}\hat{i} + \beta\hat{j} - 8\hat{k}$ are collinear, then $(19\alpha - 6\beta)^2$ is equal to *[2023, 8 April Shift-I]*

(a) 36 (b) 25 (c) 49 (d) 16

**08** Let the vectors $\mathbf{a} = (1+t)\hat{i} + (1-t)\hat{j} + \hat{k}$, $\mathbf{b} = (1-t)\hat{i} + (1+t)\hat{j} + 2\hat{k}$ and $\mathbf{c} = t\hat{i} - t\hat{j} + \hat{k}$, $t \in R$ be such that for $\alpha, \beta, \gamma \in R$, $\alpha\mathbf{a} + \beta\mathbf{b} + \gamma\mathbf{c} = \mathbf{0}$ $\Rightarrow \alpha = \beta = \gamma = 0$. Then, the set of all values of $t$ is *[2022, 28 July Shift-I]*

(a) a non-empty finite set (b) equal to $N$
(c) equal to $R - \{0\}$ (d) equal to $R$

**09** A vector **a** has components $3p$ and 1 with respect to rectangular cartesian system. This system is rotated through a certain angle about the origin in the counter clockwise sense. If, with respect to new system, **a** has components $p + 1$ and $\sqrt{10}$, then a value of $p$ is equal to *[2021, 18 March Shift-I]*

(a) 1 (b) $-\frac{5}{4}$ (c) $\frac{4}{5}$ (d) –1

**10** Let **a**, **b** and **c** be three non-zero vectors which are pairwise non-collinear. If $\mathbf{a} + 3\mathbf{b}$ is collinear with **c** and $\mathbf{b} + 2\mathbf{c}$ is collinear with **a**, then $\mathbf{a} + 3\mathbf{b} + 6\mathbf{c}$ is *[AIEEE 2011]*

(a) **a**+ **c** (b) **a** (c) **c** (d) **0**

**11** The non-zero vectors **a**, **b** and **c** are related by $\mathbf{a} = 8\mathbf{b}$ and $\mathbf{c} = -7\mathbf{b}$. Then, the angle between **a** and **c** is *[AIEEE 2008]*

(a) $\pi$ (b) 0 (c) $\pi/4$ (d) $\pi/2$

**12** If $C$ is the mid-point of $AB$ and $P$ is any point outside $AB$, then **[AIEEE 2005]**
(a) $\mathbf{PA}+\mathbf{PB}+\mathbf{PC}=0$ (b) $\mathbf{PA}+\mathbf{PB}+2\mathbf{PC}=0$
(c) $\mathbf{PA}+\mathbf{PB}=\mathbf{PC}$ (d) $\mathbf{PA}+\mathbf{PB}=2\mathbf{PC}$

**13** If $\mathbf{a}$, $\mathbf{b}$ and $\mathbf{c}$ are three non-zero vectors such that no two of these are collinear. If the vector $\mathbf{a}+2\mathbf{b}$ is collinear with $\mathbf{c}$ and $\mathbf{b}+3\mathbf{c}$ is collinear with $\mathbf{a}$ ($\lambda$ being some non-zero scalar), then $\mathbf{a}+2\mathbf{b}+6\mathbf{c}$ equal to **[AIEEE 2004]**
(a) $\lambda\mathbf{a}$ (b) $\lambda\mathbf{b}$ (c) $\lambda\mathbf{c}$ (d) 0

**14** If $\begin{vmatrix} a & a^2 & 1+a^3 \\ b & b^2 & 1+b^3 \\ c & c^2 & 1+c^3 \end{vmatrix}=0$ and vectors $(1, a, a^2)$, $(1, b, b^2)$ and $(1, c, c^2)$ are non- coplanar, then the product $abc$ equal to **[AIEEE 2003]**
(a) 2 (b) −1 (c) 1 (d) 0

**15** The vector $\hat{i}+x\hat{j}+3\hat{k}$ is rotated through an angle $\theta$ and doubled in magnitude, then it becomes $4\hat{i}+(4x-2)\hat{j}+2\hat{k}$. The value of $x$ are **[AIEEE 2002]**
(a) $\left\{-\frac{2}{3}, 2\right\}$ (b) $\left(\frac{1}{3}, 2\right)$
(c) $\left\{\frac{2}{3}, 0\right\}$ (d) $\{2, 7\}$

## Level 2

**16** Let the position vectors of three vertices of a triangle be $4\mathbf{p}+\mathbf{q}-3\mathbf{r}$, $-5\mathbf{p}+\mathbf{q}+2\mathbf{r}$ and $2\mathbf{p}-\mathbf{q}+2\mathbf{r}$. If the position vectors of the orthocenter and the circumcenter of the triangle are $\frac{\mathbf{p}+\mathbf{q}+\mathbf{r}}{4}$ and $\alpha\mathbf{p}+\beta\mathbf{q}+\gamma\mathbf{r}$ respectively, then $\alpha+2\beta+5\gamma$ is equal to **[2025, 24 Jan Shift-II]**
(a) 1 (b) 3 (c) 6 (d) 4

**17** The position vectors of the vertices $A$, $B$ and $C$ of a triangle are $2\hat{i}-3\hat{j}+3\hat{k}$, $2\hat{i}+2\hat{j}+3\hat{k}$ and $-\hat{i}+\hat{j}+3\hat{k}$, respectively. Let $l$ denotes the length of the angle bisector $AD$ of $\angle BAC$, where $D$ is on the line segment $BC$, then $2l^2$ equals **[2024, 27 Jan Shift-II]**
(a) 49 (b) 42 (c) 45 (d) 50

**18** Let the position vectors of the vertices $A$, $B$ and $C$ of a triangle be $2\hat{i}+2\hat{j}+\hat{k}$, $\hat{i}+2\hat{j}+2\hat{k}$ and $2\hat{i}+\hat{j}+2\hat{k}$, respectively. Let $l_1$, $l_2$ and $l_3$ be the lengths of perpendiculars drawn from the orthocenter of the triangle on the sides $AB$, $BC$ and $CA$ respectively, then $l_1^2+l_2^2+l_3^2$ equals **[2024, 27 Jan Shift-II]**
(a) 1/5 (b) 1/4 (c) 1/3 (d) 1/2

**19** An arc $PQ$ of a circle subtends a right angle at its centre $O$. The mid-point of the arc $PQ$ is $R$. If $\mathbf{OP}=\mathbf{u}$, $\mathbf{OR}=\mathbf{v}$ and $\mathbf{OQ}=\alpha\mathbf{u}+\beta\mathbf{v}$, then $\alpha$, $\beta^2$ are the roots of the equation **[2023, 10 April Shift-I]**
(a) $3x^2-2x-1=0$ (b) $3x^2+2x-1=0$
(c) $x^2-x-2=0$ (d) $x^2+x-2=0$

**20** Let $A$, $B$, $C$ be three points whose position vectors respectively are $\mathbf{a}=\hat{i}+4\hat{j}+3\hat{k}$; $\mathbf{b}=2\hat{i}+\alpha\hat{j}+4\hat{k}$, $\alpha\in R$ and $\mathbf{c}=3\hat{i}-2\hat{j}+5\hat{k}$
If $\alpha$ is the smallest positive integer for which $\mathbf{a}$, $\mathbf{b}$, $\mathbf{c}$ are non-collinear, then the length of the median in $\Delta ABC$, through $A$ is **[2022, 29 June Shift-II]**
(a) $\frac{\sqrt{82}}{2}$ (b) $\frac{\sqrt{62}}{2}$ (c) $\frac{\sqrt{69}}{2}$ (d) $\frac{\sqrt{66}}{2}$

**21** Let $\alpha=(\lambda-2)\mathbf{a}+\mathbf{b}$ and $\beta=(4\lambda-2)\mathbf{a}+3\mathbf{b}$ be two given vectors where vectors $\mathbf{a}$ and $\mathbf{b}$ are non- collinear. The value of $\lambda$ for which vectors $\alpha$ and $\beta$ are collinear, is **[2019, 10 Jan Shift-II]**
(a) 4 (b) −3 (c) 3 (d) −4

**22** If the vectors $\mathbf{AB}=3\hat{i}+4\hat{k}$ and $\mathbf{AC}=5\hat{i}-2\hat{j}+4\hat{k}$ are the sides of a $\Delta ABC$, then the length of the median through $A$ is **[JEE Main 2013, 2003]**
(a) $\sqrt{18}$ (b) $\sqrt{72}$ (c) $\sqrt{33}$ (d) $\sqrt{45}$

### Numerical Value Type Questions

**23** Let the three sides of a $\Delta ABC$ be given by the vectors $2\hat{i}-\hat{j}+\hat{k}$, $\hat{i}-3\hat{j}-5\hat{k}$ and $3\hat{i}-4\hat{j}-4\hat{k}$. Let $G$ be the centroid of the $\Delta ABC$. Then $6(|\mathbf{AG}|^2+|\mathbf{BG}|^2+|\mathbf{CG}|^2)$ is equal to .............. . **[2025, 4 April Shift-II]**

# Topic 2 Scalar or Dot Product of Two Vectors and Its Applications

## Level 1

**01** Let $\mathbf{a}$ and $\mathbf{b}$ be the vectors of the same magnitude such that $\frac{|\mathbf{a}+\mathbf{b}|+|\mathbf{a}-\mathbf{b}|}{|\mathbf{a}+\mathbf{b}|-|\mathbf{a}-\mathbf{b}|}=\sqrt{2}+1$. Then, $\frac{|\mathbf{a}+\mathbf{b}|^2}{|\mathbf{a}|^2}$ is **[2025, 7 April Shift-II]**
(a) $2+\sqrt{2}$ (b) $2+4\sqrt{2}$ (c) $4+2\sqrt{2}$ (d) $1+\sqrt{2}$

**Imp.** Projection on One Vector on Another Vector

**02** If $\mathbf{a}$ is a non-zero vector such that its projections on the vectors $2\hat{i}-\hat{j}+2\hat{k}$, $\hat{i}+2\hat{j}-2\hat{k}$ and $\hat{k}$ are equal, then a unit vector along $\mathbf{a}$ is **[2025, 2 April Shift-I]**
(a) $\frac{1}{\sqrt{155}}(7\hat{i}+9\hat{j}-5\hat{k})$ (b) $\frac{1}{\sqrt{155}}(-7\hat{i}+9\hat{j}-5\hat{k})$
(c) $\frac{1}{\sqrt{155}}(7\hat{i}+9\hat{j}+5\hat{k})$ (d) $\frac{1}{\sqrt{155}}(-7\hat{i}+9\hat{j}+5\hat{k})$

**03** Let **a** and **b** be two unit vectors such that the angle between them is $\pi/3$. If $\lambda\mathbf{a}+2\mathbf{b}$ and $3\mathbf{a}-\lambda\mathbf{b}$ are perpendicular to each other, then the number of values of $\lambda$ in $[-1, 3]$ is *[2025, 22 Jan Shift-II]*
(a) 1 (b) 2 (c) 0 (d) 3

**Imp.** Scalar Product of Two Vector

**04** Let $\mathbf{a}=\hat{i}+2\hat{j}+3\hat{k}$, $\mathbf{b}=2\hat{i}+3\hat{j}-5\hat{k}$ and $\mathbf{c}=3\hat{i}-\hat{j}+\lambda\hat{k}$ be three vectors. Let **r** be a unit vector along $\mathbf{b}+\mathbf{c}$. If $\mathbf{r}\cdot\mathbf{a}=3$, then $3\lambda$ is equal to *[2024, 8 April Shift-II]*
(a) 27 (b) 30 (c) 25 (d) 21

**05** For $\lambda>0$, let $\theta$ be the angle between the vectors $\mathbf{a}=\hat{i}+\lambda\hat{j}-3\hat{k}$ and $\mathbf{b}=3\hat{i}-\hat{j}+2\hat{k}$.
If the vectors $\mathbf{a}+\mathbf{b}$ and $\mathbf{a}-\mathbf{b}$ are mutually perpendicular, then the value of $(14\cos\theta)^2$ is equal to *[2024, 4 April Shift-II]*
(a) 20 (b) 50 (c) 40 (d) 25

**06** Let $\mathbf{a}=\hat{i}+\alpha\hat{j}+\beta\hat{k}$, $\alpha,\beta\in R$. Let a vector **b** be such that the angle between **a** and **b** is $\frac{\pi}{4}$ and $|\mathbf{b}|^2=6$. If $\mathbf{a}\cdot\mathbf{b}=3\sqrt{2}$, then the value of $(\alpha^2+\beta^2)|\mathbf{a}\times\mathbf{b}|^2$ is equal to *[2024, 30 Jan Shift-II]*
(a) 90 (b) 85 (c) 95 (d) 75

**07** Let $P(3, 2, 3)$, $Q(4, 6, 2)$ and $R(7, 3, 2)$ be the vertices of $\Delta PQR$. Then, the $\angle QPR$ is *[2024, 30 Jan Shift-II]*
(a) $\frac{\pi}{6}$ (b) $\cos^{-1}\left(\frac{1}{18}\right)$
(c) $\cos^{-1}\left(\frac{7}{18}\right)$ (d) $\frac{\pi}{3}$

**08** Let $\mathbf{a}=5\hat{i}-\hat{j}-3\hat{k}$ and $\mathbf{b}=\hat{i}+3\hat{j}+5\hat{k}$ be two vectors. Then, which one of the following is true? *[2023, 1 Feb Shift-II]*
(a) Projection of **a** on **b** is $\frac{-17}{\sqrt{35}}$ and the direction of the projection vector is opposite to the direction of **b**.
(b) Projection of **a** on **b** is $\frac{-17}{\sqrt{35}}$ and the direction of the projection vector is same as of **b**.
(c) Projection of **a** on **b** is $\frac{17}{\sqrt{35}}$ and the direction of the projection vector is same as of **b**.
(d) Projection of **a** on **b** is $\frac{17}{\sqrt{35}}$ and the direction of the projection vector is opposite to the direction of **b**.

**09** Let $S$ be the set of all $a\in R$ for which the angle between the vectors $\mathbf{u}=a(\log_e b)\hat{i}-6\hat{j}+3\hat{k}$ and $\mathbf{v}=(\log_e b)\hat{i}+2\hat{j}+2a(\log_e b)\hat{k}$, $(b>1)$ is acute. Then, a is equal to *[2022, 28 July Shift-II]*
(a) $\left(-\infty,-\frac{4}{3}\right)$ (b) $\phi$ (c) $\left(-\frac{4}{3},0\right)$ (d) $\left(\frac{12}{7},\infty\right)$

**10** A hall has a square floor of dimension $10\text{m}\times10\text{m}$ (see the figure) and vertical walls. If the $\angle GPH$ between the diagonals $AG$ and $BH$ is $\cos^{-1}\frac{1}{5}$, then the height of the hall (in m) is *[2021, 26 Aug Shift-II]*

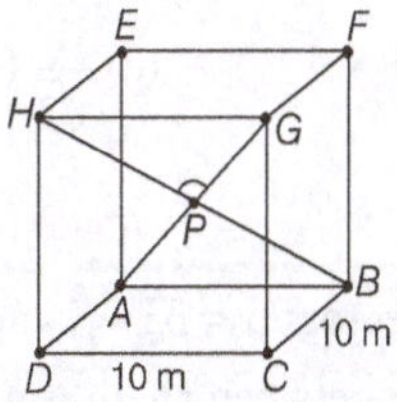


(a) 5 (b) $2\sqrt{10}$ (c) $5\sqrt{3}$ (d) $5\sqrt{2}$

**Imp.** Projection of a Vector

**11** In $\Delta ABC$, if $|\mathbf{BC}|=3$, $|\mathbf{CA}|=5$ and $|\mathbf{BA}|=7$, then the projection of the vector **BA** on **BC** is equal to *[2021, 20 July Shift-II]*
(a) $\frac{19}{2}$ (b) $\frac{13}{2}$ (c) $\frac{11}{2}$ (d) $\frac{15}{2}$

**12** Let **a** and **b** be two unit vectors. If the vectors $\mathbf{c}=\mathbf{a}+2\mathbf{b}$ and $\mathbf{d}=5\mathbf{a}-4\mathbf{b}$ are perpendicular to each other, then the angle between **a** and **b** is *[AIEEE 2012]*
(a) $\pi/6$ (b) $\pi/2$
(c) $\pi/3$ (d) $\pi/4$

**13** If the vectors $\mathbf{a}=\hat{i}-\hat{j}+2\hat{k}$, $\mathbf{b}=2\hat{i}+4\hat{j}+\hat{k}$ and $\mathbf{c}=\lambda\hat{i}+\hat{j}+\mu\hat{k}$ are mutually orthogonal, then $(\lambda,\mu)$ is equal to *[AIEEE 2010]*
(a) $(-3, 2)$ (b) $(2, -3)$ (c) $(-2, 3)$ (d) $(3, -2)$

**14** The value of $a$, for which the points, $A, B, C$ with position vectors $2\hat{i}-\hat{j}+\hat{k}$, $\hat{i}-3\hat{j}-5\hat{k}$ and $a\hat{i}-3\hat{j}+\hat{k}$ respectively are the vertices of a right angled triangle with $C=\frac{\pi}{2}$ are *[AIEEE 2006]*
(a) $-2$ and $-1$ (b) $-2$ and 1 (c) 2 and $-1$ (d) 2 and 1

**15** A particle is acted upon by constant forces $4\hat{i}+\hat{j}-3\hat{k}$ and $3\hat{i}+\hat{j}-\hat{k}$ which displace it from a point $\hat{i}+2\hat{j}+3\hat{k}$ to the point $5\hat{i}+4\hat{j}+\hat{k}$. The work done in standard units by the forces is given by *[AIEEE 2004]*
(a) 40 units (b) 30 units (c) 25 units (d) 15 units

**16** $\mathbf{a}, \mathbf{b}, \mathbf{c}$ are three vectors, such that $\mathbf{a}+\mathbf{b}+\mathbf{c}=\mathbf{0}, |\mathbf{a}|=1, |\mathbf{b}|=2, |\mathbf{c}|=3$, then $\mathbf{a}\cdot\mathbf{b}+\mathbf{b}\cdot\mathbf{c}+\mathbf{c}\cdot\mathbf{a}$ is equal to *[AIEEE 2003]*

(a) 0 (b) −7
(c) 7 (d) 1

## Level 2

**17** Let $\mathbf{a}=\hat{\mathbf{i}}+2\hat{\mathbf{j}}+\hat{\mathbf{k}}$ and $\mathbf{b}=2\hat{\mathbf{i}}+\hat{\mathbf{j}}-\hat{\mathbf{k}}$. Let $\mathbf{c}$ be a unit vector in the plane of the vectors $\mathbf{a}$ and $\mathbf{b}$ and be perpendicular to $\mathbf{a}$. Then, such *a* vector $\mathbf{c}$ is *[2025, 8 April Shift-II]*

(a) $\frac{1}{\sqrt{2}}(-\hat{\mathbf{i}}+\hat{\mathbf{k}})$ (b) $\frac{1}{\sqrt{5}}(\hat{\mathbf{j}}-2\hat{\mathbf{k}})$
(c) $\frac{1}{\sqrt{3}}(\hat{\mathbf{i}}-\hat{\mathbf{j}}+\hat{\mathbf{k}})$ (d) $\frac{1}{\sqrt{3}}(-\hat{\mathbf{i}}+\hat{\mathbf{j}}-\hat{\mathbf{k}})$

**Imp.** **Angle Between Two Vectors**

**18** Consider two vectors $\mathbf{u}=3\hat{\mathbf{i}}-\hat{\mathbf{j}}$ and $\mathbf{v}=2\hat{\mathbf{i}}+\hat{\mathbf{j}}-\lambda\hat{\mathbf{k}}$, $\lambda>0$. The angle between them is given by $\cos^{-1}\left(\frac{\sqrt{5}}{2\sqrt{7}}\right)$ Let $\mathbf{v}=\mathbf{v}_1+\mathbf{v}_2$, where $\mathbf{v}_1$ is parallel to $\mathbf{u}$ and $\mathbf{v}_2$ is perpendicular to $\mathbf{u}$. Then, the value $|\mathbf{v}_1|^2+|\mathbf{v}_2|^2$ is equal to *[2025, 4 April Shift-I]*

(a) 10 (b) $\frac{23}{2}$ (c) 14 (d) $\frac{25}{2}$

**19** The set of all $\alpha$, for which the vectors $\mathbf{a}=\alpha t\hat{\mathbf{i}}+6\hat{\mathbf{j}}-3\hat{\mathbf{k}}$ and $\mathbf{b}=t\hat{\mathbf{i}}-2\hat{\mathbf{j}}-2\alpha t\hat{\mathbf{k}}$ are inclined at an obtuse angle for all $t\in R$ is *[2024, 8 April Shift-I]*

(a) $(-2, 0]$ (b) $\left(-\frac{4}{3}, 0\right]$
(c) $[0, 1)$ (d) $\left(-\frac{4}{3}, 1\right)$

**20** Let a unit vector which makes an angles of 60° with $2\hat{\mathbf{i}}+2\hat{\mathbf{j}}-\hat{\mathbf{k}}$ and an angle of 45° with $\hat{\mathbf{i}}-\hat{\mathbf{k}}$ be $\mathbf{c}$. Then, $\mathbf{c}+\left(-\frac{1}{2}\hat{\mathbf{i}}+\frac{1}{3\sqrt{2}}\hat{\mathbf{j}}-\frac{\sqrt{2}}{3}\hat{\mathbf{k}}\right)$ is *[2024, 4 April Shift-I]*

(a) $\left(\frac{1}{\sqrt{3}}+\frac{1}{2}\right)\hat{\mathbf{i}}+\left(\frac{1}{\sqrt{3}}-\frac{1}{3\sqrt{2}}\right)\hat{\mathbf{j}}+\left(\frac{1}{\sqrt{3}}+\frac{\sqrt{2}}{3}\right)\hat{\mathbf{k}}$
(b) $-\frac{\sqrt{2}}{3}\hat{\mathbf{i}}+\frac{\sqrt{2}}{3}\hat{\mathbf{j}}+\left(\frac{1}{2}+\frac{2\sqrt{2}}{3}\right)\hat{\mathbf{k}}$
(c) $\frac{\sqrt{2}}{3}\hat{\mathbf{i}}-\frac{1}{2}\hat{\mathbf{k}}$
(d) $\frac{\sqrt{2}}{3}\hat{\mathbf{i}}+\frac{1}{3\sqrt{2}}\hat{\mathbf{j}}-\frac{1}{2}\hat{\mathbf{k}}$

**21** Consider a $\Delta ABC$ where $A(1, 3, 2)$, $B(-2, 8, 0)$ and $C(3, 6, 7)$. If the angle bisector of $\angle BAC$ meets the line $BC$ at $D$, then the length of the projection of the vector $\mathbf{AD}$ on the vector $\mathbf{AC}$ is *[2024, 1 Feb Shift-I]*

(a) $\frac{39}{2\sqrt{38}}$ (b) $\frac{\sqrt{38}}{2}$
(c) $\sqrt{19}$ (d) $\frac{37}{2\sqrt{38}}$

**22** Let a unit vector $\mathbf{u}=x\hat{\mathbf{i}}+y\hat{\mathbf{j}}+z\hat{\mathbf{k}}$ make angles $\frac{\pi}{2}, \frac{\pi}{3}$ and $\frac{2\pi}{3}$ with the vectors $\frac{1}{\sqrt{2}}\hat{\mathbf{i}}+\frac{1}{\sqrt{2}}\hat{\mathbf{k}}, \frac{1}{\sqrt{2}}\hat{\mathbf{j}}+\frac{1}{\sqrt{2}}\hat{\mathbf{k}}$ and $\frac{1}{\sqrt{2}}\hat{\mathbf{i}}+\frac{1}{\sqrt{2}}\hat{\mathbf{j}}$, respectively. If $\mathbf{v}=\frac{1}{\sqrt{2}}\hat{\mathbf{i}}+\frac{1}{\sqrt{2}}\hat{\mathbf{j}}+\frac{1}{\sqrt{2}}\hat{\mathbf{k}}$, then $|\mathbf{u}-\mathbf{v}|^2$ is equal to *[2024, 29 Jan Shift-I]*

(a) 9 (b) $\frac{5}{2}$ (c) 7 (d) $\frac{11}{2}$

**23** Let $\mathbf{a}=2\hat{\mathbf{i}}+\hat{\mathbf{j}}+\hat{\mathbf{k}}$ and $\mathbf{b}$ and $\mathbf{c}$ be two non-zero vectors such that $|\mathbf{a}+\mathbf{b}+\mathbf{c}|=|\mathbf{a}+\mathbf{b}-\mathbf{c}|$ and $\mathbf{b}\cdot\mathbf{c}=0$. Consider the following two statements

(A) $|\mathbf{a}+\lambda\mathbf{c}|\geq|\mathbf{a}|, \forall\lambda\in R$ *[2023, 31 Jan Shift-I]*
(B) $\mathbf{a}$ and $\mathbf{c}$ are always parallel.

Then,
(a) Neither (A) nor (B) is correct
(b) Both (A) and (B) are correct
(c) only (B) is correct
(d) only (A) is correct

**24** Let $\alpha=4\hat{\mathbf{i}}+3\hat{\mathbf{j}}+5\hat{\mathbf{k}}$ and $\beta=\hat{\mathbf{i}}+2\hat{\mathbf{j}}-4\hat{\mathbf{k}}$. Let $\beta_1$ be parallel to $\alpha$ and $\beta_2$ be perpendicular to $\alpha$. If $\beta=\beta_1+\beta_2$, then the value of $5\beta_2\cdot(\hat{\mathbf{i}}+\hat{\mathbf{j}}+\hat{\mathbf{k}})$ is *[2023, 24 Jan Shift-II]*

(a) 6 (b) 9 (c) 7 (d) 11

**25** Let $\mathbf{a}=\hat{\mathbf{i}}+\hat{\mathbf{j}}+2\hat{\mathbf{k}}$, $\mathbf{b}=2\hat{\mathbf{i}}-3\hat{\mathbf{j}}+\hat{\mathbf{k}}$ and $\mathbf{c}=\hat{\mathbf{i}}-\hat{\mathbf{j}}+\hat{\mathbf{k}}$ be three given vectors. Let $\mathbf{v}$ be a vector in the plane of $\mathbf{a}$ and $\mathbf{b}$ whose projection on $\mathbf{c}$ is $\frac{2}{\sqrt{3}}$, if $\mathbf{v}\cdot\hat{\mathbf{j}}=7$, then $\mathbf{v}\cdot(\hat{\mathbf{i}}+\hat{\mathbf{k}})$ is equal to *[2022, 26 June Shift-II]*

(a) 6 (b) 7 (c) 8 (d) 9

**26** Let $\mathbf{a}$ and $\mathbf{b}$ be two vectors such that $|2\mathbf{a}+3\mathbf{b}|=|3\mathbf{a}+\mathbf{b}|$ and the angle between $\mathbf{a}$ and $\mathbf{b}$ is 60°. If $\frac{1}{8}\mathbf{a}$ is a unit vector, then $|\mathbf{b}|$ is equal to *[2021, 31 Aug Shift-I]*

(a) 4 (b) 6 (c) 5 (d) 8

**27** Let $a, b, c \in R$ be such that $a^2 + b^2 + c^2 = 1$. If $a\cos\theta = b\cos\left(\theta + \frac{2\pi}{3}\right) = c\cos\left(\theta + \frac{4\pi}{3}\right)$, where $\theta = \frac{\pi}{9}$, then the angle between the vectors $a\hat{i} + b\hat{j} + c\hat{k}$ and $b\hat{i} + c\hat{j} + a\hat{k}$ is *[2020, 3 Sep Shift-II]*

(a) $\frac{\pi}{2}$ (b) 0 (c) $\frac{2\pi}{3}$ (d) $\frac{\pi}{9}$

**28** Let $\mathbf{a} = \hat{i} + \hat{j} + \sqrt{2}\hat{k}$, $\mathbf{b} = b_1\hat{i} + b_2\hat{j} + \sqrt{2}\hat{k}$ and $\mathbf{c} = 5\hat{i} + \hat{j} + \sqrt{2}\hat{k}$ be three vectors such that the projection vector of **b** on **a** is **a**. If **a** + **b** is perpendicular to **c**, then $|\mathbf{b}|$ is equal to *[2019, 9 Jan Shift-II]*

(a) 6 (b) 4 (c) $\sqrt{22}$ (d) $\sqrt{32}$

**29** Let **u** be a vector coplanar with the vectors $\mathbf{a} = 2\hat{i} + 3\hat{j} - \hat{k}$ and $\mathbf{b} = \hat{j} + \hat{k}$. If **u** is perpendicular to **a** and $\mathbf{u} \cdot \mathbf{b} = 24$, then $|\mathbf{u}|^2$ is equal to *[JEE Main 2018]*

(a) 336 (b) 315 (c) 256 (d) 84

**30** Let *ABCD* be a parallelogram such that **AB** = **q**, **AD** = **p** and $\angle BAD$ be an acute angle. If **r** is the vector that coincides with the altitude directed from the vertex *B* to the side *AD*, then **r** is given by *[AIEEE 2012]*

(a) $\mathbf{r} = 3\mathbf{q} - \frac{3(\mathbf{p}\cdot\mathbf{q})}{(\mathbf{p}\cdot\mathbf{p})}\mathbf{p}$ (b) $\mathbf{r} = -\mathbf{q} + \left(\frac{\mathbf{p}\cdot\mathbf{q}}{\mathbf{p}\cdot\mathbf{p}}\right)\mathbf{p}$

(c) $\mathbf{r} = \mathbf{q} - \left(\frac{\mathbf{p}\cdot\mathbf{q}}{\mathbf{p}\cdot\mathbf{p}}\right)\mathbf{p}$ (d) $\mathbf{r} = -3\mathbf{q} + \frac{3(\mathbf{p}\cdot\mathbf{q})}{(\mathbf{p}\cdot\mathbf{p})}\mathbf{p}$

**31** Given, two vectors are $\hat{i} - \hat{j}$ and $\hat{i} + 2\hat{j}$, the unit vector coplanar with the two vectors and perpendicular to first is *[AIEEE 2002]*

(a) $\frac{1}{\sqrt{2}}(\hat{i} + \hat{j})$ (b) $\frac{1}{\sqrt{5}}(2\hat{i} + \hat{j})$

(c) $\pm\frac{1}{\sqrt{2}}(\hat{i} + \hat{j})$ (d) None of these

## Numerical Value Type Questions

**32** The least positive integral value of $\alpha$, for which the angle between the vectors $\alpha\hat{i} - 2\hat{j} + 2\hat{k}$ and $\alpha\hat{i} + 2\alpha\hat{j} - 2\hat{k}$ is acute, is ......... *[2024, 27 Jan Shift-I]*

**33** If $(\mathbf{a} + 3\mathbf{b})$ is perpendicular to $(7\mathbf{a} - 5\mathbf{b})$ and $(\mathbf{a} - 4\mathbf{b})$ is perpendicular to $(7\mathbf{a} - 2\mathbf{b})$, then the angle between **a** and **b** (in degrees) is ......... *[2021, 25 July Shift-II]*

**34** For $p > 0$, a vector $\mathbf{V}_2 = 2\hat{i} + (p+1)\hat{j}$ is obtained by rotating the vector $\mathbf{V}_1 = \sqrt{3}\cdot p\,\hat{i} + \hat{j}$ by an angle $\theta$ about origin in counter clockwise direction. If $\tan\theta = \frac{(\alpha\sqrt{3} - 2)}{(4\sqrt{3} + 3)}$, then the value of $\alpha$ is equal to ......... *[2021, 20 July Shift-II]*

**35** If **x** and **y** be two non-zero vectors such that $|\mathbf{x} + \mathbf{y}| = |\mathbf{x}|$ and $2\mathbf{x} + \lambda\mathbf{y}$ is perpendicular to **y**, then the value of $\lambda$ is ......... *[2020, 6 Sep Shift-II]*

**36** **a**, **b** and **c** be three unit vectors such that $|\mathbf{a} - \mathbf{b}|^2 + |\mathbf{a} - \mathbf{c}|^2 = 8$. Then, $|\mathbf{a} + 2\mathbf{b}|^2 + |\mathbf{a} - 2\mathbf{c}|^2$ is equal to ...... *[2020, 2 Sep Shift-I]*

## Topic 3 Vector or Cross Product of Two Vectors & Its Applications

### Level 1

**01** Let the angle $\theta, 0 < \theta < \frac{\pi}{2}$ between two unit vectors $\hat{a}$ and $\hat{b}$ be $\sin^{-1}\left(\frac{\sqrt{65}}{9}\right)$. If the vector $\hat{c} = 3\hat{a} + 6\hat{b} + 9(\hat{a} \times \hat{b})$, then the value of $9(\mathbf{c}\cdot\hat{a}) - 3(\mathbf{c}\cdot\hat{b})$ is *[2025, 7 Sep Shift-I]*

(a) 29 (b) 31
(c) 24 (d) 27

**02** Let $\mathbf{a} = 2\hat{i} - 3\hat{j} + \hat{k}$, $\mathbf{b} = 3\hat{i} + 2\hat{j} + 5\hat{k}$ and a vector **c** be such that $(\mathbf{a} - \mathbf{c}) \times \mathbf{b} = -18\hat{i} - 3\hat{j} + 12\hat{k}$ and $\mathbf{a}\cdot\mathbf{c} = 3$. If $\mathbf{b} \times \mathbf{c} = \mathbf{d}$, then $|\mathbf{a}\cdot\mathbf{d}|$ is equal to *[2025, 2 Sep Shift-II]*

(a) 12 (b) 15
(c) 18 (d) 9

**03** Let $\mathbf{a} = 3\hat{i} - \hat{j} + 2\hat{k}$, $\mathbf{b} = \mathbf{a} \times (\hat{i} - 2\hat{k})$ and $\mathbf{c} = \mathbf{b} \times \hat{k}$. Then, the projection of $\mathbf{c} - 2\hat{j}$ on **a** is *[2025, 24 Jan Shift-II]*

(a) $3\sqrt{7}$ (b) $\sqrt{14}$ (c) $2\sqrt{7}$ (d) $2\sqrt{14}$

**04** Let $\mathbf{OA} = 2\mathbf{a}$, $\mathbf{OB} = 6\mathbf{a} + 5\mathbf{b}$ and $\mathbf{OC} = 3\mathbf{b}$, where *O* is the origin. If the area of the parallelogram with adjacent sides **OA** and **OC** is 15 sq units, then the area (in sq units) of the quadrilateral *OABC* is equal to *[2024, 9 April Shift-I]*

(a) 38 (b) 32 (c) 40 (d) 35

**Imp.** Cross Product of Two Vectors

**05** Let $\mathbf{a} = 4\hat{i} - \hat{j} + \hat{k}$, $\mathbf{b} = 11\hat{i} - \hat{j} + \hat{k}$ and **c** be a vector such that $(\mathbf{a} + \mathbf{b}) \times \mathbf{c} = \mathbf{c} \times (-2\mathbf{a} + 3\mathbf{b})$. If $(2\mathbf{a} + 3\mathbf{b})\cdot\mathbf{c} = 1670$, then $|\mathbf{c}|^2$ is equal to *[2024, 8 April Shift-II]*

(a) 1627 (b) 1618 (c) 1600 (d) 1609

**06** Let $\mathbf{a}=6\hat{i}+\hat{j}-\hat{k}$ and $\mathbf{b}=\hat{i}+\hat{j}$. If $\mathbf{c}$ is a vector such that $|\mathbf{c}|\geq 6, \mathbf{a}\cdot\mathbf{c}=6|\mathbf{c}|, |\mathbf{c}-\mathbf{a}|=2\sqrt{2}$ and the angle between $\mathbf{a}\times\mathbf{b}$ and $\mathbf{c}$ is $60°$, then $|(\mathbf{a}\times\mathbf{b})\times\mathbf{c}|$ is equal to *[2024, 6 April Shift-II]*

(a) $\frac{9}{2}(6+\sqrt{6})$ (b) $\frac{3}{2}\sqrt{3}$ (c) $\frac{3}{2}\sqrt{6}$ (d) $\frac{9}{2}(6-\sqrt{6})$

**07** Consider three vectors $\mathbf{a}, \mathbf{b}, \mathbf{c}$. Let $|\mathbf{a}|=2, |\mathbf{b}|=3$ and $\mathbf{a}=\mathbf{b}\times\mathbf{c}$. If $\alpha\in\left[0,\frac{\pi}{3}\right]$ is the angle between the vectors $\mathbf{b}$ and $\mathbf{c}$, then the minimum value of $27|\mathbf{c}-\mathbf{a}|^2$ is equal to *[2024, 5 April Shift-II]*

(a) 124 (b) 121 (c) 110 (d) 105

**08** Let $\mathbf{a}=2\hat{i}+5\hat{j}-\hat{k}, \mathbf{b}=2\hat{i}-2\hat{j}+2\hat{k}$ and $\mathbf{c}$ be three vectors such that $(\mathbf{c}+\hat{i})\times(\mathbf{a}+\mathbf{b}+\hat{i})=\mathbf{a}\times(\mathbf{c}+\hat{i})$. If $\mathbf{a}\cdot\mathbf{c}=-29$, then $\mathbf{c}\cdot(-2\hat{i}+\hat{j}+\hat{k})$ is equal to *[2024, 5 April Shift-II]*

(a) 12 (b) 15 (c) 10 (d) 5

**09** If $A(1,-1,2), B(5,7,-6), C(3,4,-10)$ and $D(-1,-4,-2)$ are the vertices of a quadrilateral $ABCD$, then its area is *[2024, 5 April Shift-I]*

(a) $24\sqrt{7}$ (b) $48\sqrt{7}$
(c) $24\sqrt{29}$ (d) $12\sqrt{29}$

**10** Let $\mathbf{a}=3\hat{i}+\hat{j}-2\hat{k}, \mathbf{b}=4\hat{i}+\hat{j}+7\hat{k}$ and $\mathbf{c}=\hat{i}-3\hat{j}+4\hat{k}$ be three vectors. If a vectors $\mathbf{p}$ satisfies $\mathbf{p}\times\mathbf{b}=\mathbf{c}\times\mathbf{b}$ and $\mathbf{p}\cdot\mathbf{a}=0$, then $\mathbf{p}\cdot(\hat{i}-\hat{j}-\hat{k})$ is equal to *[2024, 31 Jan Shift-I]*

(a) 36 (b) 32 (c) 24 (d) 28

**11** Let $\mathbf{a}$ and $\mathbf{b}$ be two vectors such that $|\mathbf{b}|=1$ and $|\mathbf{b}\times\mathbf{a}|=2$. Then, $|(\mathbf{b}\times\mathbf{a})-\mathbf{b}|^2$ is equal to *[2024, 30 Jan Shift-II]*

(a) 1 (b) 3 (c) 4 (d) 5

**12** Let $\mathbf{a}=a_1\hat{i}+a_2\hat{j}+a_3\hat{k}$ and $\mathbf{b}=b_1\hat{i}+b_2\hat{j}+b_3\hat{k}$ be two vectors such that $|\mathbf{a}|=1, \mathbf{a}\cdot\mathbf{b}=2$ and $|\mathbf{b}|=4$. If $\mathbf{c}=2(\mathbf{a}\times\mathbf{b})-3\mathbf{b}$, then the angle between $\mathbf{b}$ and $\mathbf{c}$ is equal to *[2024, 30 Jan Shift-I]*

(a) $\cos^{-1}\left(\frac{2}{\sqrt{3}}\right)$ (b) $\cos^{-1}\left(-\frac{1}{\sqrt{3}}\right)$
(c) $\cos^{-1}\left(\frac{2}{3}\right)$ (d) $\cos^{-1}\left(-\frac{\sqrt{3}}{2}\right)$

**13** Let $A(2,3,5)$ and $C(-3,4,-2)$ be opposite vertices of a parallelogram $ABCD$. If the diagonal $\mathbf{BD}=\hat{i}+2\hat{j}+3\hat{k}$, then the area of the parallelogram is equal to *[2024, 30 Jan Shift-I]*

(a) $\frac{1}{2}\sqrt{410}$ (b) $\frac{1}{2}\sqrt{586}$ (c) $\frac{1}{2}\sqrt{306}$ (d) $\frac{1}{2}\sqrt{474}$

**14** Let $\mathbf{a}=\hat{i}+2\hat{j}+\hat{k}, \mathbf{b}=3(\hat{i}-\hat{j}+\hat{k})$. Let $\mathbf{c}$ be the vector such that $\mathbf{a}\times\mathbf{c}=\mathbf{b}$ and $\mathbf{a}\cdot\mathbf{c}=3$. Then, $\mathbf{a}\cdot((\mathbf{c}\times\mathbf{b})-\mathbf{b}-\mathbf{c})$ is equal to *[2024, 27 Jan Shift-I]*

(a) 32 (b) 24
(c) 20 (d) 36

**15** Let $\mathbf{a}=\hat{i}+4\hat{j}+2\hat{k}, \mathbf{b}=3\hat{i}-2\hat{j}+7\hat{k}$ and $\mathbf{c}=2\hat{i}-\hat{j}+4\hat{k}$. If a vector $\mathbf{d}$ satisfies $\mathbf{d}\times\mathbf{b}=\mathbf{c}\times\mathbf{b}$ and $\mathbf{d}\cdot\mathbf{a}=24$, then $|\mathbf{d}|^2$ is equal to *[2023, 13 April Shift-I]*

(a) 323 (b) 313 (c) 423 (d) 413

**16** Let $\mathbf{a}$ be a non-zero vector parallel to the line of intersection of the two planes described by $\hat{i}+\hat{j}, \hat{i}+\hat{k}$ and $\hat{i}-\hat{j}, \hat{j}-\hat{k}$. If $\theta$ is the angle between the vector $\mathbf{a}$ and the vector $\mathbf{b}=2\hat{i}-2\hat{j}+\hat{k}$ and $\mathbf{a}\cdot\mathbf{b}=6$, then the ordered pair $(\theta, |\mathbf{a}\times\mathbf{b}|)$ is equal to *[2023, 11 April Shift-I]*

(a) $\left(\frac{\pi}{3}, 6\right)$ (b) $\left(\frac{\pi}{4}, 3\sqrt{6}\right)$ (c) $\left(\frac{\pi}{3}, 3\sqrt{6}\right)$ (d) $\left(\frac{\pi}{4}, 6\right)$

**17** Let $\mathbf{a}=2\hat{i}+3\hat{j}+4\hat{k}, \mathbf{b}=\hat{i}-2\hat{j}-2\hat{k}$ and $\mathbf{c}=-\hat{i}+4\hat{j}+3\hat{k}$.
If $\mathbf{d}$ is a vector perpendicular to both $\mathbf{b}$ and $\mathbf{c}$, and $\mathbf{a}\cdot\mathbf{d}=18$, then $|\mathbf{a}\times\mathbf{d}|^2$ is equal to *[2023, 6 April Shift-I]*

(a) 720 (b) 680 (c) 640 (d) 760

**18** Let $\mathbf{a}$ and $\mathbf{b}$ be two vectors. Let $|\mathbf{a}|=1, |\mathbf{b}|=4$ and $\mathbf{a}\cdot\mathbf{b}=2$. If $\mathbf{c}=(2\mathbf{a}\times\mathbf{b})-3\mathbf{b}$, then the value of $\mathbf{b}\cdot\mathbf{c}$ is *[2023, 30 Jan Shift-II]*

(a) $-84$ (b) $-60$ (c) $-24$ (d) $-48$

**19** If $\mathbf{a}=\hat{i}+2\hat{k}, \mathbf{b}=\hat{i}+\hat{j}+\hat{k}, \mathbf{c}=7\hat{i}-3\hat{j}+4\hat{k}$, $\mathbf{r}\times\mathbf{b}+\mathbf{b}\times\mathbf{c}=\mathbf{0}$ and $\mathbf{r}\cdot\mathbf{a}=0$. Then, $\mathbf{r}\cdot\mathbf{c}$ is equal to *[2023, 29 Jan Shift-II]*

(a) 34 (b) 36 (c) 30 (d) 32

**20** Let $\mathbf{a}=\alpha\hat{i}+3\hat{j}-\hat{k}, \mathbf{b}=3\hat{i}-\beta\hat{j}+4\hat{k}$ and $\mathbf{c}=\hat{i}+2\hat{j}-2\hat{k}$ where $\alpha, \beta\in R$, be three vectors. If the projection of $\mathbf{a}$ on $\mathbf{c}$ is $\frac{10}{3}$ and $\mathbf{b}\times\mathbf{c}=-6\hat{i}+10\hat{j}+7\hat{k}$, then the value of $\alpha+\beta$ is equal to *[2022, 29 June Shift-I]*

(a) 3 (b) 4 (c) 5 (d) 6

**21** If $|\mathbf{a}|=2, |\mathbf{b}|=5$ and $|\mathbf{a}\times\mathbf{b}|=8$, then $|\mathbf{a}\cdot\mathbf{b}|$ is equal to *[2021, 25 July Shift-II]*

(a) 6 (b) 4 (c) 3 (d) 5

**22** Let $\mathbf{a}=3\hat{i}+2\hat{j}+2\hat{k}$ and $\mathbf{b}=\hat{i}+2\hat{j}-2\hat{k}$ be two vectors. If a vector perpendicular to both the vectors $\mathbf{a}+\mathbf{b}$ and $\mathbf{a}-\mathbf{b}$ has the magnitude 12, then one such vector is *[2019, 12 April Shift-I]*

(a) $4(2\hat{i}+2\hat{j}+\hat{k})$ (b) $4(2\hat{i}-2\hat{j}-\hat{k})$
(c) $4(2\hat{i}+2\hat{j}-\hat{k})$ (d) $4(-2\hat{i}-2\hat{j}+\hat{k})$

**23** The vectors **a** and **b** are not perpendicular and **c** and **d** are two vectors satisfying $\mathbf{b}\times\mathbf{c}=\mathbf{b}\times\mathbf{d}$ and $\mathbf{a}\cdot\mathbf{b}=0$. Then, the vector **d** is equal to **[AIEEE 2011]**

(a) $\mathbf{c}+\left(\frac{\mathbf{a}\cdot\mathbf{c}}{\mathbf{a}\cdot\mathbf{b}}\right)\mathbf{b}$ (b) $\mathbf{b}+\left(\frac{\mathbf{b}\cdot\mathbf{c}}{\mathbf{a}\cdot\mathbf{b}}\right)\mathbf{c}$

(c) $\mathbf{c}-\left(\frac{\mathbf{a}\cdot\mathbf{c}}{\mathbf{a}\cdot\mathbf{b}}\right)\mathbf{b}$ (d) $\mathbf{b}-\left(\frac{\mathbf{b}\cdot\mathbf{c}}{\mathbf{a}\cdot\mathbf{b}}\right)\mathbf{c}$

**24** If **u** and **v** are unit vectors and $\theta$ is the acute angle between them, then $2\,\mathbf{u}\times 3\,\mathbf{v}$ is a unit vector for **[AIEEE 2007]**

(a) exactly two values of $\theta$ (b) more than two values of $\theta$
(c) No value of $\theta$ (d) exactly one value of $\theta$

**25** For any vector **a**, the value of $(\mathbf{a}\times\hat{\mathbf{i}})^2+(\mathbf{a}\times\hat{\mathbf{j}})^2+(\mathbf{a}\times\hat{\mathbf{k}})^2$ is equal to **[AIEEE 2005]**

(a) $4a^2$ (b) $2a^2$
(c) $a^2$ (d) $3a^2$

**26** If $\mathbf{a}\times\mathbf{b}=\mathbf{b}\times\mathbf{c}=\mathbf{c}\times\mathbf{a}$, then $\mathbf{a}+\mathbf{b}+\mathbf{c}$ equal to **[AIEEE 2003]**

(a) abc (b) $-1$
(c) 0 (d) 2

**27** If the vectors **a**, **b** and **c** form the sides $BC, CA$ and $AB$ respectively of a $\Delta ABC$, then **[AIEEE 2002]**

(a) $\mathbf{a}\cdot\mathbf{b}=\mathbf{b}\cdot\mathbf{c}=\mathbf{c}\cdot\mathbf{b}=0$
(b) $\mathbf{a}\times\mathbf{b}=\mathbf{b}\times\mathbf{c}=\mathbf{c}\times\mathbf{a}$
(c) $\mathbf{a}\cdot\mathbf{b}=\mathbf{b}\cdot\mathbf{c}=\mathbf{c}\cdot\mathbf{a}=0$
(d) $\mathbf{a}\times\mathbf{a}+\mathbf{a}\times\mathbf{c}+\mathbf{a}\times\mathbf{a}=0$

**28** If the vectors **c**, $\mathbf{a}=x\,\hat{\mathbf{i}}+y\,\hat{\mathbf{j}}+z\,\hat{\mathbf{k}}$ and $\mathbf{b}=\hat{\mathbf{j}}$ are such that **a**, **c** and **b** form a right handed system, then **c** is **[AIEEE 2002]**

(a) $z\,\hat{\mathbf{i}}-x\,\hat{\mathbf{k}}$ (b) **0**
(c) $y\,\hat{\mathbf{j}}$ (d) $-z\,\hat{\mathbf{i}}+x\,\hat{\mathbf{k}}$

**29** If $|\mathbf{a}|=4$, $|\mathbf{b}|=2$ and the angle between **a** and **b** is $\frac{\pi}{6}$, then $(\mathbf{a}\times\mathbf{b})^2$ is equal to **[AIEEE 2002]**

(a) 48 (b) 16
(c) **a** (d) None of these

## Level 2

**Imp.** Area of Triangle

**30** Let $ABCD$ be a tetrahedron such that the edges $AB$, $AC$ and $AD$ are mutually perpendicular. Let the areas of the triangles $ABC$, $ACD$ and $ADB$ be 5, 6 and 7 square units respectively. Then, the area (in square units) of the $\Delta BCD$ is equal to **[2025, 2 April Shift-I]**

(a) $\sqrt{340}$ (b) 12
(c) $7\sqrt{3}$ (d) $\sqrt{110}$

**31** Let **a** be a unit vector perpendicular to the vectors $\mathbf{b}=\hat{\mathbf{i}}-2\hat{\mathbf{j}}+3\hat{\mathbf{k}}$ and $\hat{\mathbf{c}}=2\hat{\mathbf{i}}+3\hat{\mathbf{j}}-\hat{\mathbf{k}}$ and makes an angle of $\cos^{-1}\left(-\frac{1}{3}\right)$ with the vector $\hat{\mathbf{i}}+\hat{\mathbf{j}}+\hat{\mathbf{k}}$. If $\hat{\mathbf{a}}$ makes an angle of $\frac{\pi}{3}$ with the vector $\hat{\mathbf{i}}+\alpha\hat{\mathbf{j}}+\hat{\mathbf{k}}$, then the vlue of $\alpha$ is **[2025, 29 Jan Shift-II]**

(a) $\sqrt{6}$ (b) $-\sqrt{6}$ (c) $-\sqrt{3}$ (d) $\sqrt{3}$

**32** Let $\mathbf{a}=2\hat{\mathbf{i}}-\hat{\mathbf{j}}+3\hat{\mathbf{k}}$, $\mathbf{b}=3\hat{\mathbf{i}}-5\hat{\mathbf{j}}+\hat{\mathbf{k}}$ and **c** be a vector such that $\mathbf{a}\times\mathbf{c}=\mathbf{c}\times\mathbf{b}$ and $(\mathbf{a}+\mathbf{c})\cdot(\mathbf{b}+\mathbf{c})=168$. Then, the maximum value of $|\mathbf{c}|^2$ is **[2025, 29 Jan Shift-I]**

(a) 154 (b) 308 (c) 462 (d) 77

**33** Let $\mathbf{a}=\hat{\mathbf{i}}+2\hat{\mathbf{j}}+3\hat{\mathbf{k}}$, $\mathbf{b}=3\hat{\mathbf{i}}+\hat{\mathbf{j}}-\hat{\mathbf{k}}$ and **c** be three vectors such that **c** is coplanar with **a** and **b**. If the vector **c** is perpendicular to **b** and $\mathbf{a}\cdot\mathbf{c}=5$, then $|\mathbf{c}|$ is equal to **[2025, 24 Jan Shift-I]**

(a) 18 (b) 16 (c) $\sqrt{\frac{11}{6}}$ (d) $\frac{1}{3\sqrt{2}}$

**34** Let $\mathbf{a}=2\hat{\mathbf{i}}+\alpha\hat{\mathbf{j}}+\hat{\mathbf{k}}$, $\mathbf{b}=-\hat{\mathbf{i}}+\hat{\mathbf{k}}$, $\mathbf{c}=\beta\hat{\mathbf{j}}-\hat{\mathbf{k}}$, where $\alpha$ and $\beta$ are integers and $\alpha\beta=-6$. Let the values of the ordered pair $(\alpha,\beta)$, for which the area of the parallelogram of diagonals $\mathbf{a}+\mathbf{b}$ and $\mathbf{b}+\mathbf{c}$ is $\frac{\sqrt{21}}{2}$, be $(\alpha_1,\beta_1)$ and $(\alpha_2,\beta_2)$. Then, $\alpha_1^2+\beta_1^2-\alpha_2\beta_2$ is equal to **[2024, 9 April Shift-II]**

(a) 24 (b) 19 (c) 21 (d) 17

**35** Let three vectors $\mathbf{a}=\alpha\hat{\mathbf{i}}+4\hat{\mathbf{j}}+2\hat{\mathbf{k}}$, $\mathbf{b}=5\hat{\mathbf{i}}+3\hat{\mathbf{j}}+4\hat{\mathbf{k}}$, $\mathbf{c}=x\hat{\mathbf{i}}+y\hat{\mathbf{j}}+z\hat{\mathbf{k}}$ form a triangle such that $\mathbf{c}=\mathbf{a}-\mathbf{b}$ and the area of the triangle is $5\sqrt{6}$. If $\alpha$ is a positive real number, then $|\mathbf{c}|^2$ is equal to **[2024, 9 April Shift-I]**

(a) 12 (b) 10 (c) 16 (d) 14

**36** Let $\mathbf{a}=2\hat{\mathbf{i}}+\hat{\mathbf{j}}-\hat{\mathbf{k}}$, $\mathbf{b}=((\mathbf{a}\times(\hat{\mathbf{i}}+\hat{\mathbf{j}}))\times\hat{\mathbf{i}})\times\hat{\mathbf{i}}$. Then, the square of the projection of **a** on **b** is **[2024, 6 April Shift-II]**

(a) $\frac{1}{5}$ (b) $\frac{2}{3}$
(c) 2 (d) $\frac{1}{3}$

**37** If $A(3,1,-1)$, $B\left(\frac{5}{3},\frac{7}{3},\frac{1}{3}\right)$, $C(2,2,1)$ and $D\left(\frac{10}{3},\frac{2}{3},\frac{-1}{3}\right)$ are the vertices of a quadrilateral $ABCD$, then its area is **[2024, 6 April Shift-I]**

(a) $\frac{2\sqrt{2}}{3}$ (b) $2\sqrt{2}$
(c) $\frac{5\sqrt{2}}{3}$ (d) $\frac{4\sqrt{2}}{3}$

**38** Let $\mathbf{a}=\hat{i}+\hat{j}+\hat{k}$, $\mathbf{b}=2\hat{i}+4\hat{j}-5\hat{k}$ and $\mathbf{c}=x\hat{i}+2\hat{j}+3\hat{k}$, $x \in R$.

If $\mathbf{d}$ is the unit vector in the direction of $\mathbf{b}+\mathbf{c}$ such that $\mathbf{a}\cdot\mathbf{d}=1$, then $(\mathbf{a}\times\mathbf{b})\cdot\mathbf{c}$ is equal to *[2024, 4 April Shift-II]*

(a) 9 (b) 11 (c) 6 (d) 3

**39** Let $\mathbf{a}=-5\hat{i}+\hat{j}-3\hat{k}$, $\mathbf{b}=\hat{i}+2\hat{j}-4\hat{k}$ and $\mathbf{c}=(((\mathbf{a}\times\mathbf{b})\times\hat{i})\times\hat{i})\times\hat{i}$. Then, $\mathbf{c}\cdot(-\hat{i}+\hat{j}+\hat{k})$ is equal to *[2024, 1 Feb Shift-I]*

(a) −13 (b) −12 (c) −10 (d) −15

**Imp.** **Area of Parallelogram**

**40** Let $OA=\mathbf{a}$, $OB=12\mathbf{a}+4\mathbf{b}$ and $OC=\mathbf{b}$, where $O$ is the origin. If $S$ is the parallelogram with adjacent sides OA and OC, then $\dfrac{\text{area of the quadrilateral } OABC}{\text{area of } S}$ is equal to *[2024, 29 Jan Shift-II]*

(a) 8 (b) 10 (c) 7 (d) 6

**41** Let $|\mathbf{a}|=2$, $|\mathbf{b}|=3$ and the angle between the vectors $\mathbf{a}$ and $\mathbf{b}$ be $\frac{\pi}{4}$. Then, $|(\mathbf{a}+2\mathbf{b})\times(2\mathbf{a}-3\mathbf{b})|^2$ is equal to *[2023, 13 April Shift-II]*

(a) 841 (b) 882 (c) 482 (d) 441

**42** Let $\lambda=Z$, $\mathbf{a}=\lambda\hat{i}+\hat{j}-\hat{k}$ and $\mathbf{b}=3\hat{i}-\hat{j}+2\hat{k}$. Let $\mathbf{c}$ be a vector such that $(\mathbf{a}+\mathbf{b}+\mathbf{c})\times\mathbf{c}=\mathbf{0}$, $\mathbf{a}\cdot\mathbf{c}=-17$ and $\mathbf{b}\cdot\mathbf{c}=-20$. Then, $|\mathbf{c}\times(\lambda\hat{i}+\hat{j}+\hat{k})|^2$ is equal to *[2023, 12 April Shift-I]*

(a) 53 (b) 46 (c) 62 (d) 49

**43** Let $O$ be the origin and the position vector of the point $P$ be $-\hat{i}-2\hat{j}+3\hat{k}$. If the position vectors of the points $A$, $B$ and $C$ are $-2\hat{i}+\hat{j}-3\hat{k}$, $2\hat{i}+4\hat{j}-2\hat{k}$ and $-4\hat{i}+2\hat{j}-\hat{k}$ respectively, then the projection of the vector OP on a vector perpendicular to the vectors AB and AC is *[2023, 10 April Shift-I]*

(a) $\frac{7}{3}$ (b) 3 (c) $\frac{8}{3}$ (d) $\frac{10}{3}$

**44** The area of the quadrilateral $ABCD$ with vertices $A(2,1,1)$, $B(1,2,5)$, $C(-2,-3,5)$ and $D(1,-6,-7)$ is equal to *[2023, 8 April Shift-II]*

(a) 54 (b) $8\sqrt{38}$ (c) 48 (d) $9\sqrt{38}$

**45** Let $\mathbf{a}=2\hat{i}-7\hat{j}+5\hat{k}$, $\mathbf{b}=\hat{i}+\hat{k}$ and $\mathbf{c}=\hat{i}+2\hat{j}-3\hat{k}$ be three given vectors. If $\mathbf{r}$ is a vector such that $\mathbf{r}\times\mathbf{a}=\mathbf{c}\times\mathbf{a}$ and $\mathbf{r}\cdot\mathbf{b}=0$, then $|\mathbf{r}|$ is equal to *[2023, 1 Feb Shift-II]*

(a) $\frac{11}{7}$ (b) $\frac{11}{5}\sqrt{2}$ (c) $\frac{\sqrt{914}}{7}$ (d) $\frac{11}{7}\sqrt{2}$

**46** Let $\mathbf{a}=\hat{i}+2\hat{j}+3\hat{k}$, $\mathbf{b}=\hat{i}-\hat{j}+2\hat{k}$ and $\mathbf{c}=5\hat{i}-3\hat{j}+3\hat{k}$ be three vectors. If $\mathbf{r}$ is a vector such that, $\mathbf{r}\times\mathbf{b}=\mathbf{c}\times\mathbf{b}$ and $\mathbf{r}\cdot\mathbf{a}=0$, then $25|\mathbf{r}|^2$ is equal to *[2023, 31 Jan Shift-II]*

(a) 560 (b) 449 (c) 339 (d) 336

**47** Let $\mathbf{a}=4\hat{i}+3\hat{j}$ and $\mathbf{b}=3\hat{i}-4\hat{j}+5\hat{k}$. If $\mathbf{c}$ is a vector such that $\mathbf{c}\cdot(\mathbf{a}\times\mathbf{b})+25=0$, $\mathbf{c}\cdot(\hat{i}+\hat{j}+\hat{k})=4$ and projection of $\mathbf{c}$ on $\mathbf{a}$ is 1, then the projection of $\mathbf{c}$ on $\mathbf{b}$ equals *[2023, 29 Jan Shift-II]*

(a) $\frac{1}{5}$ (b) $\frac{3}{\sqrt{2}}$ (c) $\frac{5}{\sqrt{2}}$ (d) $\frac{1}{\sqrt{2}}$

**48** Let $\mathbf{a}$, $\mathbf{b}$ and $\mathbf{c}$ be three coplanar concurrent vectors such that angles between any two of them is same. If the product of their magnitudes is 14 and $(\mathbf{a}\times\mathbf{b})\cdot(\mathbf{b}\times\mathbf{c})+(\mathbf{b}\times\mathbf{c})\cdot(\mathbf{c}\times\mathbf{a})+(\mathbf{c}\times\mathbf{a})\cdot(\mathbf{a}\times\mathbf{b})=168$, then $|\mathbf{a}|+|\mathbf{b}|+|\mathbf{c}|$ is equal to *[2022, 29 July Shift-II]*

(a) 10 (b) 14
(c) 16 (d) 18

**49** Let $\mathbf{a}=\alpha\hat{i}+\hat{j}-\hat{k}$ and $\mathbf{b}=2\hat{i}+\hat{j}-\alpha\hat{k}$, $\alpha>0$. If the projection of $\mathbf{a}\times\mathbf{b}$ on the vector $-\hat{i}+2\hat{j}-2\hat{k}$ is 30, then $\alpha$ is equal to *[2022, 26 July Shift-I]*

(a) 15/2 (b) 8
(c) 13/2 (d) 7

**50** Let $\mathbf{a}=\alpha\hat{i}+2\hat{j}-\hat{k}$ and $\mathbf{b}=-2\hat{i}+\alpha\hat{j}+\hat{k}$, where $\alpha\in R$. If the area of the parallelogram whose adjacent sides are represented by the vectors $\mathbf{a}$ and $\mathbf{b}$ is $\sqrt{15(\alpha^2+4)}$, then the value of $2|\mathbf{a}|^2+(\mathbf{a}\cdot\mathbf{b})|\mathbf{b}|^2$ is equal to *[2022, 28 June Shift-II]*

(a) 10 (b) 7
(c) 9 (d) 14

**51** Let $\mathbf{a}$ and $\mathbf{b}$ be the vectors along the diagonals of a parallelogram having area $2\sqrt{2}$. Let the angle between $\mathbf{a}$ and $\mathbf{b}$ be acute, $|\mathbf{a}|=1$ and $|\mathbf{a}\cdot\mathbf{b}|=|\mathbf{a}\times\mathbf{b}|$. If $\mathbf{c}=2\sqrt{2}(\mathbf{a}\times\mathbf{b})-2\mathbf{b}$, then an angle between $\mathbf{b}$ and $\mathbf{c}$ is *[2022, 27 June Shift-II]*

(a) $\frac{\pi}{4}$ (b) $-\frac{\pi}{4}$
(c) $\frac{5\pi}{6}$ (d) $\frac{3\pi}{4}$

**52** Let $\mathbf{a}$, $\mathbf{b}$ and $\mathbf{c}$ be three unit vectors such that $\mathbf{a}+\mathbf{b}+\mathbf{c}=\mathbf{0}$. If $\lambda=\mathbf{a}\cdot\mathbf{b}+\mathbf{b}\cdot\mathbf{c}+\mathbf{c}\cdot\mathbf{a}$ and $\mathbf{d}=\mathbf{a}\times\mathbf{b}+\mathbf{b}\times\mathbf{c}+\mathbf{c}\times\mathbf{a}$, then the ordered pair, $(\lambda,\mathbf{d})$ is equal to *[2020, 7 Jan Shift-II]*

(a) $\left(\frac{3}{2}, 3\mathbf{b}\times\mathbf{c}\right)$ (b) $\left(-\frac{3}{2}, 3\mathbf{c}\times\mathbf{b}\right)$
(c) $\left(\frac{3}{2}, 3\mathbf{a}\times\mathbf{c}\right)$ (d) $\left(-\frac{3}{2}, 3\mathbf{a}\times\mathbf{b}\right)$

**53** Let $\mathbf{a}=3\hat{i}+2\hat{j}+x\hat{k}$ and $\mathbf{b}=\hat{i}-\hat{j}+\hat{k}$, for some real $x$. Then, $|\mathbf{a}\times\mathbf{b}|=r$ is possible if *[2019, 8 April Shift-II]*

(a) $0<r\le\sqrt{\frac{3}{2}}$ (b) $\sqrt{\frac{3}{2}}<r\le 3\sqrt{\frac{3}{2}}$

(c) $3\sqrt{\frac{3}{2}}<r<5\sqrt{\frac{3}{2}}$ (d) $r\ge 5\sqrt{\frac{3}{2}}$

**54** Let $\mathbf{a}=2\hat{i}+\hat{j}-2\hat{k}$, $\mathbf{b}=\hat{i}+\hat{j}$ and $\mathbf{c}$ be a vector such that $|\mathbf{c}-\mathbf{a}|=3$, $|(\mathbf{a}\times\mathbf{b})\times\mathbf{c}|=3$ and the angle between $\mathbf{c}$ and $\mathbf{a}\times\mathbf{b}$ is 30°. Then, $\mathbf{a}\cdot\mathbf{c}$ is equal to *[JEE Main 2017]*

(a) 25/8 (b) 2 (c) 5 (d) 1/8

**55** Let $\mathbf{a}$ and $\mathbf{b}$ be two unit vectors such that $|\mathbf{a}+\mathbf{b}|=\sqrt{3}$. If $\mathbf{c}=\mathbf{a}+2\mathbf{b}+3(\mathbf{a}\times\mathbf{b})$, then $2|\mathbf{c}|$ is equal to *[JEE Main 2015]*

(a) $\sqrt{55}$ (b) $\sqrt{51}$ (c) $\sqrt{43}$ (d) $\sqrt{37}$

**56** Let $\mathbf{u}=\hat{i}+\hat{j}$, $\mathbf{v}=\hat{i}-\hat{j}$ and $\mathbf{w}=\hat{i}+2\hat{j}+3\hat{k}$. If $\mathbf{n}$ is a unit vector such that $\mathbf{u}\cdot\mathbf{n}=0$ and $\mathbf{v}\cdot\mathbf{n}=0$, then $|\mathbf{w}\cdot\mathbf{n}|$ is equal to *[AIEEE 2003]*

(a) 0 (b) 1 (c) 2 (d) 3

## Level 3

**57** Let the point A divide the line segment joining the points $P(-1,-1,2)$ and $Q(5,5,10)$ internally in the ratio $r:1(r>0)$, If $O$ is the origin and $(\mathbf{OQ}\cdot\mathbf{OA})-\frac{1}{5}|\mathbf{OP}\times\mathbf{OA}|^2=10$, then the value of $r$ is *[2025, 23 Jan Shift-II]*

(a) 7 (b) 3 (c) 14 (d) $\sqrt{7}$

**58** Between the following two statements :

**Statement I** Let $\mathbf{a}=\hat{i}+2\hat{j}-3\hat{k}$ and $\mathbf{b}=2\hat{i}+\hat{j}-\hat{k}$. Then, the vector $\mathbf{r}$ satisfying $\mathbf{a}\times\mathbf{r}=\mathbf{a}\times\mathbf{b}$ and $\mathbf{a}\cdot\mathbf{r}=0$ is of magnitude $\sqrt{10}$.

**Statement II** In a $\Delta ABC$, $\cos 2A+\cos 2B+\cos 2C\ge-\frac{3}{2}$. *[2024, 9 April Shift-II]*

(a) Statement I is incorrect but Statement II is correct.
(b) Statement I is correct but Statement II is incorrect.
(c) Both Statement I and Statement II are correct.
(d) Both Statement I and Statement II are incorrect.

**59** Let for a $\Delta ABC$,

$$\mathbf{AB}=-2\hat{i}+\hat{j}+3\hat{k}$$
$$\mathbf{CB}=\alpha\hat{i}+\beta\hat{j}+\gamma\hat{k}$$
$$\mathbf{CA}=4\hat{i}+3\hat{j}+\delta\hat{k}$$

If $\delta>0$ and the area of the $\Delta ABC$ is $5\sqrt{6}$, then $\mathbf{CB}\cdot\mathbf{CA}$ is equal to *[2023, 13 April Shift-II]*

(a) 60 (b) 54
(c) 108 (d) 120

**60** The vector $\mathbf{a}=-\hat{i}+2\hat{j}+\hat{k}$ is rotated through a right angle, passing through the $Y$-axis in its way and the resulting vector is $\mathbf{b}$. Then, the projection of $3\mathbf{a}+\sqrt{2}\mathbf{b}$ on $\mathbf{c}=5\hat{i}+4\hat{j}+3\hat{k}$ is *[2023, 25 Jan Shift-I]*

(a) 1 (b) $\sqrt{6}$ (c) $3\sqrt{2}$ (d) $2\sqrt{3}$

**61** Let $ABC$ be a triangle such that $\mathbf{BC}=\mathbf{a}$, $\mathbf{CA}=\mathbf{b}$, $\mathbf{AB}=\mathbf{c}$, $|\mathbf{a}|=6\sqrt{2}$, $|\mathbf{b}|=2\sqrt{3}$ and $\mathbf{b}\cdot\mathbf{c}=12$. Consider the statements

**Statement I** $|(\mathbf{a}\times\mathbf{b})+(\mathbf{c}\times\mathbf{b})|-|\mathbf{c}|=6(2\sqrt{2}-1)$.

**Statement II** $\angle ACB=\cos^{-1}\left(\sqrt{\frac{2}{3}}\right)$.

Then, *[2022, 25 July Shift-I]*

(a) Both statement I and Statement II are true
(b) only Statement I is true
(c) only Statement II is true
(d) Both Statement I and Statement II are false

## Numerical Value Type Questions

**62** Let $\mathbf{a}=\hat{i}+2\hat{j}+\hat{k}$, $\mathbf{b}=3\hat{i}-3\hat{j}+3\hat{k}$, $\mathbf{c}=2\hat{i}-\hat{j}+2\hat{k}$ and $\mathbf{d}$ be a vector such that $\mathbf{b}\times\mathbf{d}=\mathbf{c}\times\mathbf{d}$ and $\mathbf{a}\cdot\mathbf{d}=4$. Then, $|(\mathbf{a}\times\mathbf{d})|^2$ is equal to ……… . *[2025, 3 April Shift-II]*

**63** Let $\mathbf{a}=\hat{i}+\hat{j}+\hat{k}$, $\mathbf{b}=3\hat{i}+2\hat{j}-\hat{k}$, $\mathbf{c}=\lambda\hat{j}+\mu\hat{k}$ and $\hat{\mathbf{d}}$ be a unit vector such that $\mathbf{a}\times\hat{\mathbf{d}}=\mathbf{b}\times\hat{\mathbf{d}}$ and $\mathbf{c}\cdot\hat{\mathbf{d}}=1$. If $\mathbf{c}$ is perpendicular to $\mathbf{a}$, then $|3\lambda\hat{\mathbf{d}}+\mu\hat{\mathbf{c}}|^2$ is equal to ………. *[2025, 3 April Shift-I]*

**64** Let $\mathbf{a}=\hat{i}+\hat{j}+\hat{k}$, $\mathbf{b}=2\hat{i}+2\hat{j}+\hat{k}$ and $\mathbf{d}=\mathbf{a}\times\mathbf{b}$. If $\mathbf{c}$ is a vector such that $\mathbf{a}\cdot\mathbf{c}=|\mathbf{c}|$, $|\mathbf{c}-2\mathbf{a}|^2=8$ and the angle between $\mathbf{d}$ and $\mathbf{c}$ is $\frac{\pi}{4}$, then $|10-3\mathbf{b}\cdot\mathbf{c}|+|\mathbf{d}\times\mathbf{c}|^2$ is equal to ……… *[2025, 28 Jan Shift-I]*

**65** Let $\mathbf{c}$ be the projection vector of $\mathbf{b}=\lambda\hat{i}+4\hat{k}$, $\lambda>0$, on the vector $\mathbf{a}=\hat{i}+2\hat{j}+2\hat{k}$. If $|\mathbf{a}+\mathbf{c}|=7$, then the area of the parallelogram formed by the vectors $\mathbf{b}$ and $\mathbf{c}$ is ……. *[2025, 22 Jan Shift-I]*

**66** Let $\mathbf{a}=9\hat{i}-13\hat{j}+25\hat{k}$, $\mathbf{b}=3\hat{i}+7\hat{j}-13\hat{k}$ and $\mathbf{c}=17\hat{i}-2\hat{j}+\hat{k}$ be three given vectors. If $\mathbf{r}$ is a vector such that $\mathbf{r}\times\mathbf{a}=(\mathbf{b}\times\mathbf{c})\times\mathbf{a}$ and $\mathbf{r}\cdot(\mathbf{b}-\mathbf{c})=0$, then $\frac{|593\mathbf{r}+67\mathbf{a}|^2}{(593)^2}$ is equal to ………. . *[2024, 8 April Shift-I]*

**67** Let $\mathbf{a}=2\hat{i}-3\hat{j}+4\hat{k}$, $\mathbf{b}=3\hat{i}+4\hat{j}-5\hat{k}$ and a vector $\mathbf{c}$ be such that $\mathbf{a}\times(\mathbf{b}+\mathbf{c})+\mathbf{b}\times\mathbf{c}=\hat{i}+8\hat{j}+13\hat{k}$. If $\mathbf{a}\cdot\mathbf{c}=13$, then $(24-\mathbf{b}\cdot\mathbf{c})$ is equal to ………. *[2024, 6 April Shift-I]*

**68** Let $\mathbf{a} = \hat{i} - 3\hat{j} + 7\hat{k}$, $\mathbf{b} = 2\hat{i} - \hat{j} + \hat{k}$ and $\mathbf{c}$ be a vector such that $(\mathbf{a} + 2\mathbf{b}) \times \mathbf{c} = 3(\mathbf{c} \times \mathbf{a})$. If $\mathbf{a} \cdot \mathbf{c} = 130$, then $\mathbf{b} \cdot \mathbf{c}$ is equal to ......... . *[2024, 5 April Shift-I]*

**69** Let $ABC$ be a triangle of area $15\sqrt{2}$ and the vectors $\mathbf{AB} = \hat{i} + 2\hat{j} - 7\hat{k}$, $\mathbf{BC} = a\hat{i} + b\hat{j} + c\hat{k}$ and $\mathbf{AC} = 6\hat{i} + d\hat{j} - 2\hat{k}$, $d > 0$. Then, the square of the length of the largest side of the $\Delta ABC$ is. *[2024, 9 April Shift-II]*

**70** Let $\mathbf{a} = \hat{i} + \hat{j} + \hat{k}$, $\mathbf{b} = -\hat{i} - 8\hat{j} + 2\hat{k}$ and $\mathbf{c} = 4\hat{i} + c_2\hat{j} + c_3\hat{k}$ be three vectors such that $\mathbf{b} \times \mathbf{a} = \mathbf{c} \times \mathbf{a}$. If the angle between the vector $\mathbf{c}$ and the vector $3\hat{i} + 4\hat{j} + \hat{k}$ is $\theta$, then the greatest integer less than or equal to $\tan^2\theta$ is ......... . *[2024, 1 Feb Shift-II]*

**71** Let $\mathbf{a} = 3\hat{i} + 2\hat{j} + \hat{k}$, $\mathbf{b} = 2\hat{i} - \hat{j} + 3\hat{k}$ and $\mathbf{c}$ be a vector such that $(\mathbf{a} + \mathbf{b}) \times \mathbf{c} = 2(\mathbf{a} \times \mathbf{b}) + 24\hat{j} - 6\hat{k}$ and $(\mathbf{a} - \mathbf{b} + \hat{i}) \cdot \mathbf{c} = -3$, Then, $|\mathbf{c}|^2$ is equal to ......... . *[2024, 31 Jan Shift-II]*

**72** Let $\mathbf{a}$ and $\mathbf{b}$ be two vectors such that $|\mathbf{a}| = 1$, $|\mathbf{b}| = 4$ and $\mathbf{a} \cdot \mathbf{b} = 2$. If $\mathbf{c} = (2\mathbf{a} \times \mathbf{b}) - 3\mathbf{b}$ and the angle between $\mathbf{b}$ and $\mathbf{c}$ is $\alpha$, then $192 \sin^2 \alpha$ is equal to ....... . *[2024, 31 Jan Shift-I]*

**73** Let $\mathbf{a} = 3\hat{i} + \hat{j} - \hat{k}$ and $\mathbf{c} = 2\hat{i} - 3\hat{j} + 3\hat{k}$. If $\mathbf{b}$ is a vector such that $\mathbf{a} = \mathbf{b} \times \mathbf{c}$ and $|\mathbf{b}|^2 = 50$, then $|72 - |\mathbf{b} + \mathbf{c}|^2|$ is equal to ........... . *[2023, 13 April Shift-I]*

**74** Let $\mathbf{a} = \hat{i} + 2\hat{j} + 3\hat{k}$ and $\mathbf{b} = \hat{i} + \hat{j} - \hat{k}$. If $\mathbf{c}$ is a vector such that $\mathbf{a} \cdot \mathbf{c} = 11$, $\mathbf{b} \cdot (\mathbf{a} \times \mathbf{c}) = 27$ and $\mathbf{b} \cdot \mathbf{c} = -\sqrt{3}|\mathbf{b}|$, then $|\mathbf{a} \times \mathbf{c}|^2$ is equal to ........... . *[2023, 11 April Shift-II]*

**75** Let $\mathbf{a} = 6\hat{i} + 9\hat{j} + 12\hat{k}$, $\mathbf{b} = \alpha\hat{i} + 11\hat{j} - 2\hat{k}$ and $\mathbf{c}$ vectors such that $\mathbf{a} \times \mathbf{c} = \mathbf{a} \times \mathbf{b}$. If $\mathbf{a} \cdot \mathbf{c} = -12$, $\mathbf{c} \cdot (\hat{i} - 2\hat{j} + \hat{k}) = 5$, then $\mathbf{c} \cdot (\hat{i} + \hat{j} + \hat{k})$ is equal to ........ . *[2023, 8 April Shift-I]*

**76** Let $\mathbf{a}$, $\mathbf{b}$, $\mathbf{c}$ be three vectors such that $|\mathbf{a}| = \sqrt{31}$, $4|\mathbf{b}| = |\mathbf{c}| = 2$ and $2(\mathbf{a} \times \mathbf{b}) = 3(\mathbf{c} \times \mathbf{a})$. If the angle between $\mathbf{b}$ and $\mathbf{c}$ is $\frac{2\pi}{3}$, then $\left(\frac{\mathbf{a} \times \mathbf{c}}{\mathbf{a} \cdot \mathbf{b}}\right)^2$ is equal to ........ *[2023, 31 Jan Shift-II]*

**77** Let $\mathbf{a}$ and $\mathbf{b}$ be two vectors such that $|\mathbf{a}| = \sqrt{14}$, $|\mathbf{b}| = \sqrt{6}$ and $|\mathbf{a} \times \mathbf{b}| = \sqrt{48}$. Then, $(\mathbf{a} \cdot \mathbf{b})^2$ is equal to .......... . *[2023, 31 Jan Shift-I]*

**78** Let $\mathbf{a} = \hat{i} + 2\hat{j} + \lambda\hat{k}$, $\mathbf{b} = 3\hat{i} - 5\hat{j} - \lambda\hat{k}$, $\mathbf{a} \cdot \mathbf{c} = 7$, $2\mathbf{b} \cdot \mathbf{c} + 43 = 0$, $\mathbf{a} \times \mathbf{c} = \mathbf{b} \times \mathbf{c}$. Then, $|\mathbf{a} \cdot \mathbf{b}|$ is equal to *[2023, 24 Jan Shift-II]*

**79** Let $\mathbf{a}$ and $\mathbf{b}$ be two vectors such that $|\mathbf{a} + \mathbf{b}|^2 = |\mathbf{a}|^2 + 2|\mathbf{b}|^2$, $\mathbf{a} \cdot \mathbf{b} = 3$ and $|\mathbf{a} \times \mathbf{b}|^2 = 75$. Then, $|\mathbf{a}|^2$ is equal to ............ . *[2022, 29 July Shift-II]*

**80** Let $\mathbf{a} = \hat{i} - 2\hat{j} + 3\hat{k}$, $\hat{\mathbf{b}} = \hat{i} + \hat{j} + \hat{k}$ and $\mathbf{c}$ be a vector such that $\mathbf{a} + (\mathbf{b} \times \mathbf{c}) = \mathbf{0}$ and $\mathbf{b} \cdot \mathbf{c} = 5$. Then, the value of $3(\mathbf{c} \cdot \mathbf{a})$ is equal to ......... . *[2022, 29 June Shift-II]*

**81** Let $\mathbf{b} = \hat{i} + \hat{j} + \lambda\hat{k}$, $\lambda \in R$. If $\mathbf{a}$ is a vector such that $\mathbf{a} \times \mathbf{b} = 13\hat{i} - \hat{j} - 4\hat{k}$ and $\mathbf{a} \cdot \mathbf{b} + 21 = 0$, then $(\mathbf{b} - \mathbf{a}) \cdot (\hat{k} - \hat{j}) + (\mathbf{b} + \mathbf{a}) \cdot (\hat{i} - \hat{k})$ is equal to ......... . *[2022, 25 June Shift-II]*

**82** Let $\theta$ be the angle between the vectors $\mathbf{a}$ and $\mathbf{b}$ where $|\mathbf{a}| = 4$, $|\mathbf{b}| = 3$, $\theta \in \left(\frac{\pi}{4}, \frac{\pi}{3}\right)$. Then, $|(\mathbf{a} - \mathbf{b}) \times (\mathbf{a} + \mathbf{b})|^2 + 4(\mathbf{a} \cdot \mathbf{b})^2$ is equal to ...... . *[2022, 25 June Shift-I]*

**83** Let $\mathbf{a} = \hat{i} + 5\hat{j} + \alpha\hat{k}$, $\mathbf{b} = \hat{i} + 3\hat{j} + \beta\hat{k}$ and $\mathbf{c} = -\hat{i} + 2\hat{j} - 3\hat{k}$ be three vectors such that, $|\mathbf{b} \times \mathbf{c}| = 5\sqrt{3}$ and $\mathbf{a}$ is perpendicular to $\mathbf{b}$. Then, the greatest amongst the values of $|\mathbf{a}|^2$ is ........... . *[2021, 27 Aug Shift-I]*

**84** Let $\mathbf{p} = 2\hat{i} + 3\hat{j} + \hat{k}$ and $\mathbf{q} = \hat{i} + 2\hat{j} + \hat{k}$ be two vectors. If a vector $\mathbf{r} = (\alpha\hat{i} + \beta\hat{j} + \gamma\hat{j})$ is perpendicular to each of the vectors $(\mathbf{p} + \mathbf{q})$ and $(\mathbf{p} - \mathbf{q})$, and $|\mathbf{r}| = \sqrt{3}$, then $|\alpha| + |\beta| + |\gamma|$ is equal to ...... . *[2021, 25 July Shift-I]*

**85** Let $\mathbf{a} = \hat{i} + \alpha\hat{j} + 3\hat{k}$ and $\mathbf{b} = 3\hat{i} - \alpha\hat{j} + \hat{k}$. If the area of the parallelogram whose adjacent sides are represented by the vectors $\mathbf{a}$ and $\mathbf{b}$ is $8\sqrt{3}$ square units, then $\mathbf{a} \cdot \mathbf{b}$ is equal to ............ . *[2021, 25 Feb Shift-II]*

# Key Index

## Topic 1

| | | | | | | | | | |
|---|---|---|---|---|---|---|---|---|---|
| 1 (c) | 2 (b) | 3 (b) | 4 (b) | 5 (b) | 6 (a) | 7 (a) | 8 (c) | 9 (d) | 10 (d) |
| 11 (a) | 12 (d) | 13 (d) | 14 (b) | 15 (a) | 16 (b) | 17 (c) | 18 (d) | 19 (c) | 20 (a) |
| 21 (d) | 22 (c) | 23 (164) | | | | | | | |

## Topic 2

| | | | | | | | | | |
|---|---|---|---|---|---|---|---|---|---|
| 1 (a) | 2 (c) | 3 (c) | 4 (c) | 5 (d) | 6 (a) | 7 (d) | 8 (*) | 9 (b) | 10 (d) |
| 11 (c) | 12 (c) | 13 (a) | 14 (d) | 15 (a) | 16 (b) | 17 (a) | 18 (c) | 19 (b) | 20 (c) |
| 21 (d) | 22 (b) | 23 (d) | 24 (c) | 25 (d) | 26 (c) | 27 (a) | 28 (a) | 29 (a) | 30 (b) |
| 31 (a) | 32 (5) | 33 (60) | 34 (6) | 35 (1) | 36 (2) | | | | |

## Topic 3

| | | | | | | | | | |
|---|---|---|---|---|---|---|---|---|---|
| 1 (a) | 2 (b) | 3 (d) | 4 (d) | 5 (b) | 6 (a) | 7 (a) | 8 (d) | 9 (d) | 10 (b) |
| 11 (d) | 12 (d) | 13 (d) | 14 (b) | 15 (d) | 16 (d) | 17 (a) | 18 (d) | 19 (a) | 20 (a) |
| 21 (a) | 22 (b) | 23 (c) | 24 (d) | 25 (b) | 26 (c) | 27 (b) | 28 (a) | 29 (b) | 30 (d) |
| 31 (b) | 32 (b) | 33 (c) | 34 (b) | 35 (d) | 36 (c) | 37 (d) | 38 (b) | 39 (b) | 40 (a) |
| 41 (b) | 42 (b) | 43 (b) | 44 (b) | 45 (d) | 46 (c) | 47 (c) | 48 (c) | 49 (d) | 50 (d) |
| 51 (d) | 52 (d) | 53 (d) | 54 (b) | 55 (a) | 56 (d) | 57 (a) | 58 (a) | 59 (a) | 60 (c) |
| 61 (c) | 62 (128) | 63 (5) | 64 (6) | 65 (16) | 66 (569) | 67 (46) | 68 (30) | 69 (54) | 70 (38) |
| 71 (38) | 72 (48) | 73 (66) | 74 (285) | 75 (11) | 76 (3) | 77 (36) | 78 (8) | 79 (14) | 80 (10) |
| 81 (14) | 82 (576) | 83 (90) | 84 (3) | 85 (2) | | | | | |

# Detailed **Explanations**

## Topic 1

**01.** *(c)* Given $\mathbf{a} = \mathbf{a}$ along $\mathbf{b} + \mathbf{a} \perp \mathbf{b}$

$$\Rightarrow \alpha\hat{i} + \beta\hat{j} + \gamma\hat{k} = \frac{16}{11}(3\hat{i} + \hat{j} - \hat{k}) + \frac{1}{11}(-4\hat{i} - 5\hat{j} - 17\hat{k}) = 4\hat{i} + \hat{j} - 3\hat{k}$$

$\Rightarrow \alpha = 4, \beta = 1$ and $\gamma = -3$

$\Rightarrow \alpha^2 + \beta^2 + \gamma^2 = 16 + 1 + 9 = 26$

**02.** *(b)* We have,

$$\mathbf{a} + 5\mathbf{b} = \lambda\mathbf{c} \quad \ldots\text{(i)}$$

and
$$\mathbf{b} + 6\mathbf{c} = \mu\mathbf{a} \quad \ldots\text{(ii)}$$

From Eq. (i), we have,

$$\mathbf{a} = \lambda\mathbf{c} - 5\mathbf{b}$$

$\therefore \quad \mathbf{b} + 6\mathbf{c} = \mu(\lambda\mathbf{c} - 5\mathbf{b})$ [from Eq. (ii)]

$$\Rightarrow \frac{\mathbf{b}}{\mu} + \frac{6}{\mu}\mathbf{c} = \lambda\mathbf{c} - 5\mathbf{b}$$

$$\therefore \frac{6}{\mu} = \lambda \text{ and } \frac{1}{\mu} = -5$$

$$\Rightarrow \mu = -\frac{1}{5} \Rightarrow \lambda = -30$$

$$\therefore \mathbf{a} + 5\mathbf{b} + 30\mathbf{c} = 0$$

$$\Rightarrow \alpha = 5 \text{ and } \beta = 30$$

$$\therefore \alpha + \beta = 5 + 30 = 35$$

**03.** *(b)* We have, $\mathbf{OA} = 2\hat{i} + 2\hat{j} + \hat{k}$

and $\mathbf{OB} = 2\hat{i} + 4\hat{j} + 4\hat{k}$

$$|\mathbf{OA}| = \sqrt{4 + 4 + 1} = \sqrt{9} = 3$$

$$|\mathbf{OB}| = \sqrt{4 + 16 + 16} = 6$$

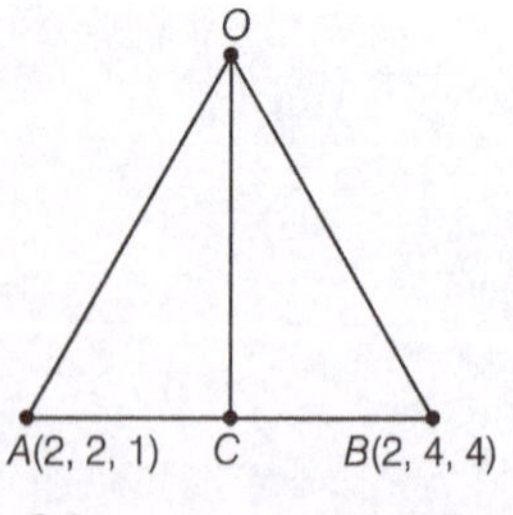


$$\Rightarrow \frac{\mathbf{OA}}{\mathbf{OB}} = \frac{3}{6} = \frac{1}{2} \Rightarrow \frac{\mathbf{AC}}{\mathbf{BC}} = \frac{1}{2}$$

$$\therefore \mathbf{OC} = \frac{1 \times (2\hat{i} + 4\hat{j} + 4\hat{k}) + 2(2\hat{i} + 2\hat{j} + \hat{k})}{1 + 2} = \frac{6\hat{i} + 8\hat{j} + 6\hat{k}}{3}$$

$$\therefore |\mathbf{OC}| = \frac{\sqrt{36 + 64 + 36}}{3} = \frac{\sqrt{136}}{3} = \frac{2\sqrt{34}}{3}$$

**04.** *(b)* Let the position vectors of vertices *A, B,* C and *D* of quadrilateral *ABCD* be **a, b, c** and **d** respectively.

*E* is the mid point of *AC*

$\therefore$ Position vector of $E = \frac{1}{2}(\mathbf{a} + \mathbf{c})$

*F* is the mid-point of *BD*

$\therefore$ Position vector of $F = \frac{1}{2}(\mathbf{b} + \mathbf{d})$

$\therefore$ **EF** = (position vector of *F*) − (position vector of *E*)

$$= \frac{1}{2}(\mathbf{b} + \mathbf{d}) - \frac{1}{2}(\mathbf{a} + \mathbf{c}) = \frac{1}{2}(\mathbf{b} + \mathbf{d} - \mathbf{a} - \mathbf{c})$$

Now,

$$\mathbf{AB} + \mathbf{AD} + \mathbf{CB} + \mathbf{CD} = (\mathbf{b} - \mathbf{a}) + (\mathbf{d} - a) + (\mathbf{b} - \mathbf{c}) + (\mathbf{d} - \mathbf{c})$$
$$= 2(\mathbf{b} + \mathbf{d} - \mathbf{a} - \mathbf{c}) = 2 \times 2\mathbf{EF}$$

$$\mathbf{AB} + \mathbf{AD} + \mathbf{CB} + \mathbf{CD} = 4\mathbf{EF}$$

$$\mathbf{AB} + (-\mathbf{BC}) + \mathbf{AD} + (-\mathbf{DC}) = 4\mathbf{EF} = -4\mathbf{FE}$$

$$(\mathbf{AB} - \mathbf{BC}) + (\mathbf{AD} - \mathbf{DC}) = -4\mathbf{FE} = k\mathbf{FE}$$

$$\Rightarrow k = -4$$

**05.** *(b)* We have,

$$10|a_i| < 1, i = 1, 2, 3$$

Let $\quad |a_1| \geq |a_2| \geq |a_3|$

$$|\mathbf{a}| = \sqrt{a_1^2 + a_2^2 + a_3^2} \geq \sqrt{a_1^2}$$

$$\therefore \quad |\mathbf{a}| \geq |a_1|$$

or $\quad \max\{|a_1|, |a_2|, |a_3|\}$

Hence, (A) is true.

$$|\mathbf{a}| = \sqrt{a_1^2 + a_2^2 + a_3^2} \leq \sqrt{a_1^2 + a_1^2 + a_1^2} = \sqrt{3}|a_1|$$

$$\therefore |\mathbf{a}| = \sqrt{3}|a_1| < 3|a_1|$$

$$\therefore |\mathbf{a}| < 3\max\{|a_1|, |a_2|, |a_3|\}$$

Hence, (B) is also true.

**06.** *(a)*

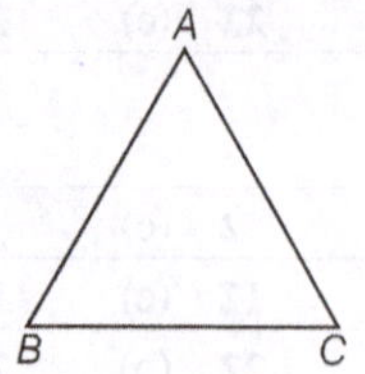


Let position vector of *A*, *B* and *C* are **a**, **b** and **c**, respectively.

*P* and *Q* are respectively circumcentre and orthocentre of $\Delta ABC$, whose position vector are **p** and **q**.

Let *G* be the centroid of triangle whose position vector is

$$\mathbf{G} = \frac{\mathbf{a} + \mathbf{b} + \mathbf{c}}{3} \quad \ldots\text{(i)}$$

Centroid divides the line joining the circumcenter and orthocenter in the ratio 1 : 2.

$$\mathbf{G} = \frac{\mathbf{q} + 2\mathbf{p}}{3} \quad \ldots\text{(ii)}$$

On comparing Eqs. (i) and (ii), we get

$$\mathbf{q} + 2\mathbf{p} = \mathbf{a} + \mathbf{b} + \mathbf{c} \quad \ldots\text{(iii)}$$

Now, $\mathbf{PA} + \mathbf{PB} + \mathbf{PC} = \mathbf{a} - \mathbf{p} + \mathbf{b} - \mathbf{p} + \mathbf{c} - \mathbf{p}$

$= \mathbf{q} + 2\mathbf{p} - 3\mathbf{p}$ [from Eq. (iii)]

$= \mathbf{q} - \mathbf{p} = \mathbf{PQ}$

**07.** *(a)* Let $\mathbf{a} = \alpha\hat{i} + 10\hat{j} + 13\hat{k}$

$$\mathbf{b} = 6\hat{i} + 11\hat{j} + 11\hat{k}$$

$$\mathbf{c} = \frac{9}{2}\hat{i} + \beta\hat{j} - 8\hat{k}$$

If $P(\mathbf{a})$, $Q(\mathbf{b})$, $R(\mathbf{c})$ are collinear, then **PQ || QR || PR**

Now, $\mathbf{PQ} = (6 - \alpha)\hat{i} + \hat{j} - 2\hat{k}$

$$\mathbf{QR} = -\frac{3}{2}\hat{i} + (\beta - 11)\hat{j} - 19\hat{k}$$

$$\mathbf{PQ} \,||\, \mathbf{QR} \Rightarrow \frac{6 - \alpha}{-3/2} = \frac{1}{\beta - 11} = \frac{2}{19}$$

$$\Rightarrow 19\alpha = 117 \Rightarrow 6\beta = 123$$

$$\therefore (19\alpha - 6\beta)^2 = (117 - 123)^2 = 36$$

**08.** *(c)* According to given condition in question; **a, b** and c are linearly independent vectors.

So, scalar triple product of **a, b** *and* **c** are non-zero, i.e. $[\mathbf{a}\ \mathbf{b}\ \mathbf{c}] \neq 0$.

$$\Rightarrow \begin{vmatrix} 1+t & 1-t & 1 \\ 1-t & 1+t & 2 \\ t & -t & 1 \end{vmatrix} \neq 0$$

$$\Rightarrow \begin{vmatrix} 1+t & 2 & 1 \\ 1-t & 2 & 2 \\ t & 0 & 1 \end{vmatrix} \neq 0$$

[Applying $C_2 \to C_1 + C_2$]

$$\Rightarrow 2\begin{vmatrix} 1+t & 1 & 1 \\ 1-t & 1 & 2 \\ t & 0 & 1 \end{vmatrix} \neq 0$$

[Taking 2 as common from $C_2$]

$\Rightarrow \quad 2[(1+t)-(1-t-2t)-t] \neq 0$

$\Rightarrow \quad 2(3t) \neq 0 \quad \Rightarrow t \neq 0$

Hence, $t \in R-\{0\}$

**09.** *(d)* After counter clockwise (or anti-clockwise) rotation, the length of the vector **a** remains constant.

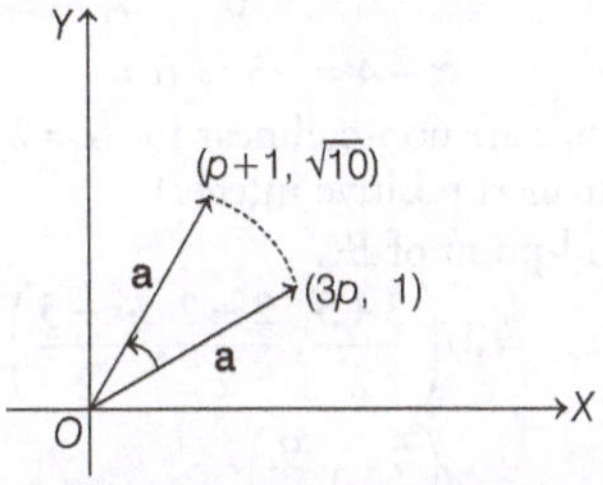


i.e. $|\mathbf{a}|$ at old position $= |\mathbf{a}|$ at new position

$\Rightarrow (3p)^2 + (1)^2 = (p+1)^2 + (\sqrt{10})^2$

$\Rightarrow \quad 9p^2 + 1 = p^2 + 1 + 2p + 10$

$\Rightarrow \quad 8p^2 - 2p - 10 = 0$

$\Rightarrow \quad 4p^2 - p - 5 = 0$

$\Rightarrow \quad (p+1)(4p-5) = 0 \Rightarrow p = 5/4, -1$

**10.** *(d)* As, $\mathbf{a}+3\mathbf{b}$ is collinear with **c**.

$\therefore \quad \mathbf{a}+3\,\mathbf{b} = \lambda\mathbf{c}$ ...(i)

Also $\mathbf{b}+2\mathbf{c}$ is collinear with **a**.

$\Rightarrow \quad \mathbf{b}+2\mathbf{c} = \mu\,\mathbf{a}$ ...(ii)

From Eq. (i), we get

$\mathbf{a}+3\mathbf{b}+6\,\mathbf{c} = (\lambda+6)\mathbf{c}$ ...(iii)

From Eq. (ii), we get

$\mathbf{a}+3\,\mathbf{b}+6\,\mathbf{c} = (1+3\mu)\,\mathbf{a}$ ...(iv)

From Eqs. (iii) and (iv), we get

$\therefore \quad (\lambda+6)\,\mathbf{c} = (1+3\,\mu)\,\mathbf{a}$

Since, **a** is not collinear with **c**.

$\Rightarrow \quad \lambda+6 = 1+3\mu = 0$

From Eq. (iv), we get

$\mathbf{a}+3\,\mathbf{b}+6\,\mathbf{c} = 0$

**11.** *(a)* Since, $\mathbf{a} = 8\mathbf{b}$ and $\mathbf{c} = -7\,\mathbf{b}$

So, **a** is parallel to **b** and **c** is anti-parallel to **b**.

$\Rightarrow$ **a** and **c** are anti-parallel.

So, the angle between **a** and **c** is $\pi$.

**12.** *(d)* Let $P$ be the origin outside of $AB$ and $C$ is mid-point of $AB$, then

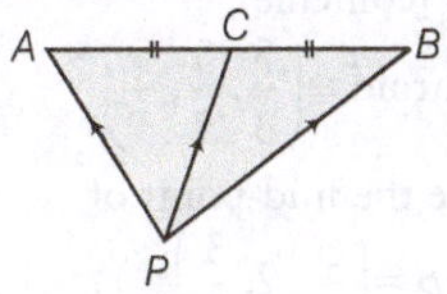


$$\mathbf{PC} = \frac{\mathbf{PA}+\mathbf{PB}}{2} \Rightarrow 2\,\mathbf{PC} = \mathbf{PA}+\mathbf{PB}$$

**13.** *(d)* If $\mathbf{a}+2\mathbf{b}$ is collinear with **c**, then

$\mathbf{a}+2\mathbf{b} = t\mathbf{c}$ ...(i)

Also, $\mathbf{b}+3\,\mathbf{c}$ is collinear with **a**, then

$\mathbf{b}+3\,\mathbf{c} = \lambda\mathbf{a}$

$\Rightarrow \quad \mathbf{b} = \lambda\mathbf{a}-3\mathbf{c}$ ...(ii)

From Eqs. (i) and (ii), we get

$\mathbf{a}+2(\lambda\mathbf{a}-3\,\mathbf{c}) = t\mathbf{c}$

$\Rightarrow \quad (\mathbf{a}-6\,\mathbf{c}) = t\mathbf{c}-2\lambda\mathbf{a}$

On comparing the coefficients of **a** and **c**, we get

$1 = -2\lambda \Rightarrow \lambda = -\dfrac{1}{2}$

and $\quad -6 = t \Rightarrow t = -6$

From Eq. (i), we get

$\mathbf{a}+2\,\mathbf{b} = -6\mathbf{c} \Rightarrow \mathbf{a}+2\mathbf{b}+6\mathbf{c} = 0$

**14.** *(b)* Since,

$$\begin{vmatrix} a & a^2 & 1+a^3 \\ b & b^2 & 1+b^3 \\ c & c^2 & 1+c^3 \end{vmatrix} = \begin{vmatrix} a & a^2 & 1 \\ b & b^2 & 1 \\ c & c^2 & 1 \end{vmatrix} + \begin{vmatrix} a & a^2 & a^3 \\ b & b^2 & b^3 \\ c & c^2 & c^3 \end{vmatrix} = 0$$

$$\Rightarrow \begin{vmatrix} a & a^2 & 1 \\ b & b^2 & 1 \\ c & c^2 & 1 \end{vmatrix} + abc\begin{vmatrix} a & a^2 & 1 \\ b & b^2 & 1 \\ c & c^2 & 1 \end{vmatrix} = 0$$

$$\Rightarrow (1+abc)\begin{vmatrix} a & a^2 & 1 \\ b & b^2 & 1 \\ c & c^2 & 1 \end{vmatrix} = 0$$

$$\left[\because \begin{vmatrix} a & a^2 & 1 \\ b & b^2 & 1 \\ c & c^2 & 1 \end{vmatrix} \neq 0\right]$$

$\Rightarrow \quad 1+abc = 0 \Rightarrow abc = -1$

**15.** *(a)* Since, the vector $\hat{\mathbf{i}}+x\,\hat{\mathbf{j}}+3\,\hat{\mathbf{k}}$ is doubled in magnitude, then it becomes

$4\,\hat{\mathbf{i}}+(4x-2)\,\hat{\mathbf{j}}+2\,\hat{\mathbf{k}}$

$\therefore \quad 2|\hat{\mathbf{i}}+x\,\hat{\mathbf{j}}+3\,\hat{\mathbf{k}}| = |4\,\hat{\mathbf{i}}+(4x-2)\,\hat{\mathbf{j}}+2\hat{\mathbf{k}}|$

$\Rightarrow 2\sqrt{1+x^2+9} = \sqrt{16+(4x-2)^2+4}$

$\Rightarrow \quad 40+4x^2 = 20+(4x-2)^2$

$\Rightarrow \quad 3x^2-4x-4 = 0$

$\Rightarrow \quad (x-2)(3x+2) = 0 \Rightarrow x = 2, -\dfrac{2}{3}$

**16.** *(b)* Given, position vectors of vertices of triangle are

$4\mathbf{p}+\mathbf{q}-3\mathbf{r}, -5\,\mathbf{p}+\mathbf{q}+2\mathbf{r}$

and $2\mathbf{p}-\mathbf{q}+2\mathbf{r}$.

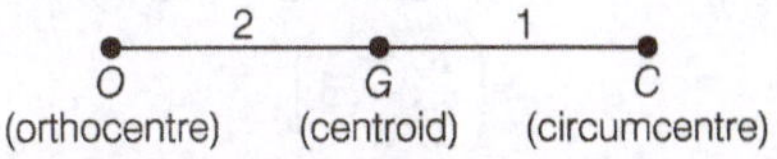


As we know that in a triangle, centroid divides join of orthocentre and circumcentre in 2 : 1

$\therefore \quad O = \dfrac{\mathbf{p}+\mathbf{q}+\mathbf{r}}{4}$ [given]

$C = \alpha\mathbf{p}+\beta\mathbf{q}+\gamma\mathbf{r}$ [given]

$$\therefore G = \frac{(4\mathbf{p}+\mathbf{q}-3\mathbf{r})+(-5\mathbf{p}+\mathbf{q}+2\mathbf{r})+(2\mathbf{p}-\mathbf{q}+2\mathbf{r})}{3}$$

$$= \frac{\mathbf{p}+\mathbf{q}+\mathbf{r}}{3}$$

Since, $G$ divides $OC$ in 2 : 1

$$\therefore \frac{2(\alpha\mathbf{p}+\beta\mathbf{q}+\gamma\mathbf{r})+\dfrac{\mathbf{p}+\mathbf{q}+\mathbf{r}}{4}}{(2+1)} = \frac{\mathbf{p}+\mathbf{q}+\mathbf{r}}{3}$$

$\Rightarrow \quad 8\,(\alpha\mathbf{p}+\beta\mathbf{q}+\gamma\mathbf{r}) = 3\,(\mathbf{p}+\mathbf{q}+\mathbf{r})$

$$\Rightarrow \left.\begin{aligned} 8\alpha &= 3 \\ 8\beta &= 3 \\ 8\gamma &= 3 \end{aligned}\right\} \Rightarrow \alpha = \beta = \gamma = 3/8$$

$$\therefore \alpha+2\beta+5\gamma = \frac{3}{8}+\frac{6}{8}+\frac{15}{8} = \frac{24}{8} = 3$$

**17.** *(c)* Given position vectors of the vertices $A, B, C$ of $\Delta ABC$ are

$2\hat{\mathbf{i}}-3\hat{\mathbf{j}}+3\hat{\mathbf{k}}, 2\hat{\mathbf{i}}+2\hat{\mathbf{j}}+3\hat{\mathbf{k}}$

and $-\hat{\mathbf{i}}+\hat{\mathbf{j}}+3\hat{\mathbf{k}}$ respectively.

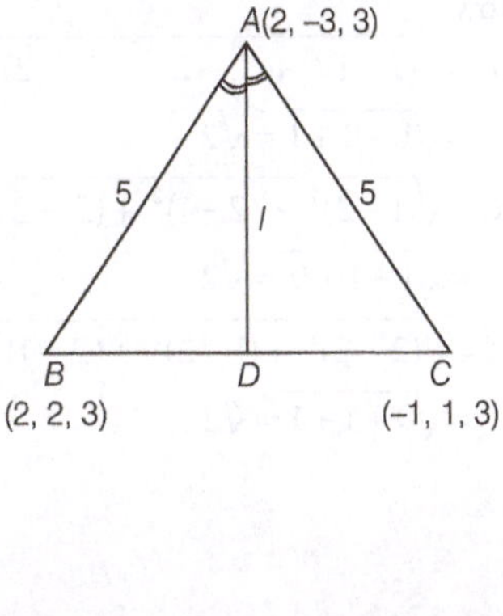

So, coordinates of $A \equiv (2, -3, 3)$,

$B \equiv (2, 2, 3)$

and $C \equiv (-1, 1, 3)$

Now, $AB = 5$ and $AC = 5$

As we know that

$$\frac{BD}{DC} = \frac{AB}{AC} \Rightarrow \frac{BD}{DC} = \frac{5}{5} = \frac{1}{1}$$

$\Rightarrow \quad BD = DC$

i.e. $D$ is mid-point of $BC$.

Now, $D \equiv \left(\frac{2-1}{2}, \frac{2+1}{2}, \frac{3+3}{2}\right)$

$$\equiv \left(\frac{1}{2}, \frac{3}{2}, 3\right)$$

Hence, $l = AD$

$$= \sqrt{\left(2-\frac{1}{2}\right)^2 + \left(-3-\frac{3}{2}\right)^2 + (3-3)^2}$$

$$= \sqrt{\frac{9}{4} + \frac{81}{4}} = \sqrt{\frac{90}{4}} = \sqrt{\frac{45}{2}}$$

$\therefore \quad 2l^2 = 2 \times \frac{45}{2} = 45$

**18.** *(d)* Given, position vectors of the vertices $A, B, C$ are $2\hat{i} + 2\hat{j} + \hat{k}$, $\hat{i} + 2\hat{j} + 2\hat{k}$ and $2\hat{i} + \hat{j} + 2\hat{k}$, respectively.

So, coordinates of $A \equiv (2, 2, 1)$,

$B \equiv (1, 2, 2)$ and $C \equiv (2, 1, 2)$

$l_1$ : Length of perpendicular drawn from orthocentre of $\Delta ABC$ on $AB$.

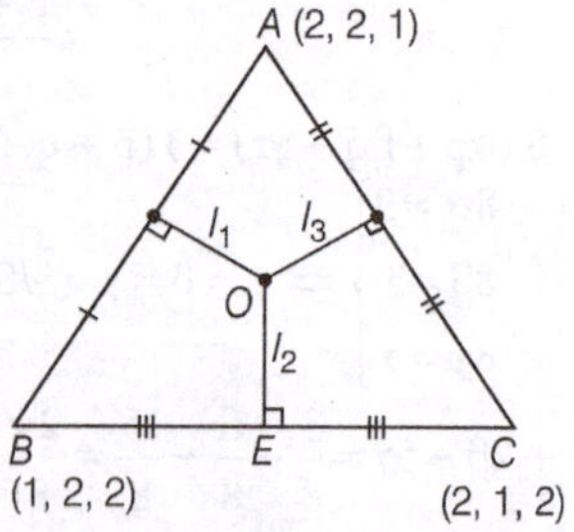


$l_2$ : Length of perpendicular drawn from orthocentre of $\Delta ABC$ on $BC$.

$l_3$ : Length of perpendicular drawn from orthocentre of $\Delta ABC$ on $CA$.

Now,

$$AB = \sqrt{(2-1)^2 + (2-2)^2 + (1-2)^2}$$

$$= \sqrt{1+0+1} = \sqrt{2}$$

$$BC = \sqrt{(1-2)^2 + (2-1)^2 + (2-2)^2}$$

$$= \sqrt{1+1+0} = \sqrt{2}$$

$$CA = \sqrt{(2-2)^2 + (1-2)^2 + (2-1)^2}$$

$$= \sqrt{0+1+1} = \sqrt{2}$$

This represents $\Delta ABC$ is an equilateral triangle.

As we known that, in an equilateral triangle, orthocentre and centroid will be same [coincide].

So, centroid $\equiv \left(\frac{5}{3}, \frac{5}{3}, \frac{5}{3}\right)$

Let $D$ be the mid-point of

$$AB = \left(\frac{3}{2}, 2, \frac{3}{2}\right)$$

$$\therefore \quad l_1 = \sqrt{\frac{1}{36} + \frac{1}{9} + \frac{1}{36}}$$

$$= \frac{1}{\sqrt{6}}$$

$$= l_2 = l_3$$

Hence, $l_1^2 + l_2^2 + l_3^2 = \frac{1}{6} + \frac{1}{6} + \frac{1}{6}$

$$= 3\left(\frac{1}{6}\right) = \frac{1}{2}$$

**19.** *(c)* An arc $PQ$ of a circle subtends a right angle at its centre $O$. The mid point of an arc $PQ$ is $R$. So, $\mathbf{PR} = \mathbf{RQ}$ also given that, $\mathbf{OP} = \mathbf{u}$, $\mathbf{OR} = \mathbf{v}$

$\mathbf{OQ} = \alpha\mathbf{u} + \beta\mathbf{v}$

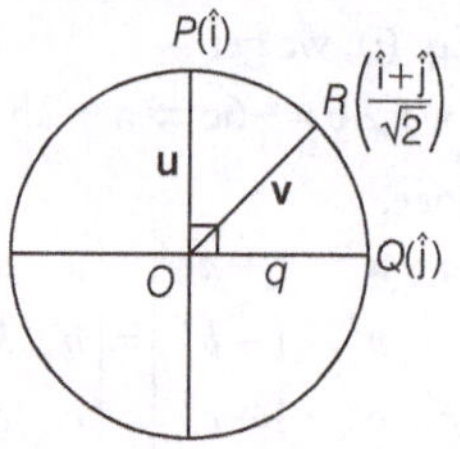


Let, $\mathbf{OP} = \mathbf{u}$

$\mathbf{OQ} = \mathbf{q} = \alpha\mathbf{u} + \beta\mathbf{v}$

$\vec{OR} = \mathbf{v}$

Clearly $\mathbf{q} = \hat{\mathbf{j}}$, $\mathbf{u} = \hat{\mathbf{i}}$, $\mathbf{v} = \frac{\hat{\mathbf{i}} + \hat{\mathbf{j}}}{\sqrt{2}}$

Now, given, $\mathbf{q} = \alpha\mathbf{u} + \beta\mathbf{v}$

$$\Rightarrow \hat{j} = \alpha(\hat{i}) + \beta\left(\frac{\hat{i} + \hat{j}}{\sqrt{2}}\right)$$

$$\Rightarrow \hat{j} = \hat{i}\left(\alpha + \frac{\beta}{\sqrt{2}}\right) + \frac{\beta}{\sqrt{2}}\hat{j}$$

On comparing both sides, we get

$$\alpha + \beta/\sqrt{2} = 0 \text{ and } \frac{\beta}{\sqrt{2}} = 1$$

$\Rightarrow \quad \alpha = -\beta/\sqrt{2}, \ \beta = \sqrt{2}$

$\Rightarrow \quad \alpha = -1, \ \beta^2 = 2$

Now, $\alpha$ and $\beta^2$ are roots of the quadratic equation is $x^2 - x - 2 = 0$

**20.** *(a)*

> **key idea** Find the value of $\alpha$ for which points are collinear using $\mathbf{AB} \parallel \mathbf{AC}$ and choose smallest positive integer value of $\alpha$ for which given points are non-collinear.

$\mathbf{OA} = \mathbf{a} = \hat{\mathbf{i}} + 4\hat{\mathbf{j}} + 3\hat{\mathbf{k}}$

$\mathbf{OB} = \mathbf{b} = 2\hat{\mathbf{i}} + \alpha\hat{\mathbf{j}} + 4\hat{\mathbf{k}}, \mathbf{a} \in \mathbf{R}$

$\mathbf{OC} = \mathbf{c} = 3\hat{\mathbf{i}} - 2\hat{\mathbf{j}} + 5\hat{\mathbf{k}}$

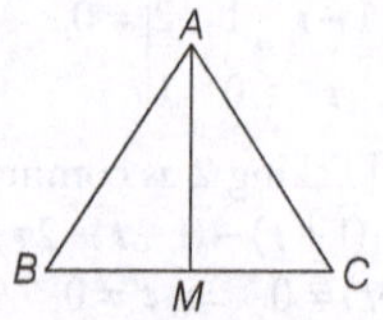


$\mathbf{AB} = \mathbf{b} - \mathbf{a} = \hat{\mathbf{i}} + (\alpha - 4)\hat{\mathbf{j}} + \hat{\mathbf{k}}$

$\mathbf{AC} = \mathbf{c} - \mathbf{a} = 2\hat{\mathbf{i}} - 6\hat{\mathbf{j}} + 2\hat{\mathbf{k}}$

$\mathbf{AB} \parallel \mathbf{AC}$, if $\frac{1}{2} = \frac{\alpha - 4}{-6} = \frac{1}{2}$

$\Rightarrow \quad \alpha - 4 = -3 \Rightarrow \alpha = 1$

**a, b, c** are non-collinear for $\alpha = 2$ (smallest positive integer)

Mid-point of $BC$

$$= M\left(\frac{2+3}{2}, \frac{2-2}{2}, \frac{4+5}{2}\right)$$

$$= M\left(\frac{5}{2}, 0, \frac{9}{2}\right)$$

$$AM = \sqrt{\left(\frac{5}{2} - 1\right)^2 + (0-4)^2 + \left(\frac{9}{2} - 3\right)^2}$$

$$= \sqrt{\frac{9}{4} + 16 + \frac{9}{4}} = \sqrt{\frac{82}{4}} = \frac{\sqrt{82}}{2}$$

**21.** *(d)* Two vectors **c** and **b** are said to be collinear if we can write $\mathbf{c} = \lambda\mathbf{b}$ for some non-zero scalar $\lambda$.

Let the vectors $\alpha = (\lambda - 2)\mathbf{a} + \mathbf{b}$ and $\beta = (4\lambda - 2)\mathbf{a} + 3\mathbf{b}$ are collinear, where **a** and **b** are non-collinear.

$\therefore$ We can write $\alpha = k\beta$, for some $k \in R - \{0\}$

$\Rightarrow \ (\lambda - 2)\mathbf{a} + \mathbf{b} = k[(4\lambda - 2)\mathbf{a} + 3\mathbf{b}]$

$\Rightarrow [(\lambda - 2) - k(4\lambda - 2)]\mathbf{a} + (1 - 3k)\mathbf{b} = \mathbf{0}$

Now, as **a** and **b** are non-collinear, therefore they are linearly independent and hence

$(\lambda - 2) - k(4\lambda - 2) = 0$ and $1 - 3k = 0$

$\Rightarrow \ \lambda - 2 = k(4\lambda - 2)$ and $3k = 1$

$\Rightarrow \lambda - 2 = \frac{1}{3}(4\lambda - 2) \left[\because 3k = 1 \Rightarrow k = \frac{1}{3}\right]$

$\Rightarrow 3\lambda - 6 = 4\lambda - 2 \Rightarrow \lambda = -4$

**22.** *(c)* We know that the sum of three vectors of a triangle is zero.

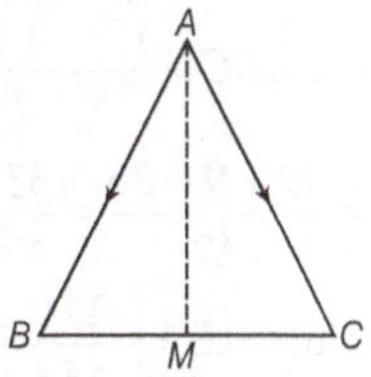


$\therefore \quad \mathbf{AB}+\mathbf{BC}+\mathbf{CA}=\mathbf{0}$

$\Rightarrow \quad \mathbf{BC}=\mathbf{AC}-\mathbf{AB} \quad [\because \mathbf{AC}=-\mathbf{CA}]$

$\Rightarrow \quad \mathbf{BM}=\dfrac{\mathbf{AC}-\mathbf{AB}}{2}$

[$\because M$ is a mid-point of $BC$]

Also, $\mathbf{AB}+\mathbf{BM}+\mathbf{MA}=\mathbf{0}$

[by properties of a triangle]

$\Rightarrow \mathbf{AB}+\dfrac{\mathbf{AC}-\mathbf{AB}}{2}=\mathbf{AM} \; [\because \mathbf{AM}=-\mathbf{MA}]$

$\Rightarrow \mathbf{AM}=\dfrac{\mathbf{AB}+\mathbf{AC}}{2}$

$=\dfrac{3\hat{\mathbf{i}}+4\hat{\mathbf{k}}+5\hat{\mathbf{i}}-2\hat{\mathbf{j}}+4\hat{\mathbf{k}}}{2}$

$=4\hat{\mathbf{i}}-\hat{\mathbf{j}}+4\hat{\mathbf{k}}$

$\Rightarrow \quad |\mathbf{AM}|=\sqrt{4^2+1^2+4^2}$

$=\sqrt{33}$

**23.** *(164)* let $A$ be origin

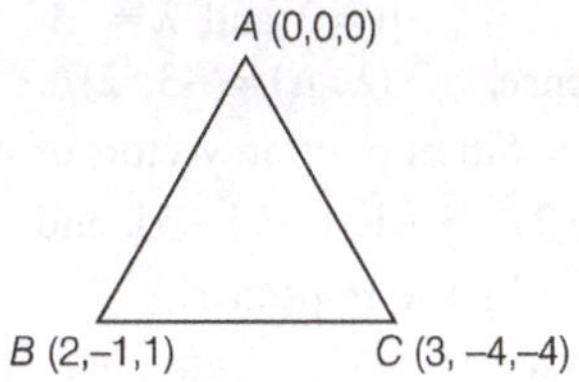


$\therefore \; G=\left(\dfrac{5}{3}, \dfrac{-5}{3}, -1\right)$

$|AG|=\sqrt{\dfrac{25}{9}+\dfrac{25}{9}+1}=\dfrac{\sqrt{59}}{3}$

$|BG|=\sqrt{\dfrac{1}{9}+\dfrac{4}{9}+4}=\dfrac{\sqrt{41}}{3}$

$|CG|=\sqrt{\dfrac{16}{9}+\dfrac{49}{9}+9}=\dfrac{\sqrt{146}}{3}$

$\therefore 6\,(|AG|^2+|BG|^2+|CG|^2)$

$=6\left(\dfrac{59}{9}+\dfrac{41}{9}+\dfrac{146}{9}\right)$

$=6\left(\dfrac{246}{9}\right)$

$=2\times 82=164$

## Topic 2

**01.** *(a)* $\dfrac{|\mathbf{a}+\mathbf{b}|+|\mathbf{a}-\mathbf{b}|}{|\mathbf{a}+\mathbf{b}|-|\mathbf{a}-\mathbf{b}|}=\dfrac{\sqrt{2}}{1}$

Using componendo and dividendo,

$\dfrac{2|\mathbf{a}+\mathbf{b}|}{2|\mathbf{a}-\mathbf{b}|}=\dfrac{\sqrt{2}+1}{\sqrt{2}-1}$

$\Rightarrow \dfrac{|\mathbf{a}+\mathbf{b}|^2}{|\mathbf{a}-\mathbf{b}|^2}=(1+\sqrt{2})^2$

$=3+2\sqrt{2}$

$\dfrac{a^2+b^2+2ab\cos\theta}{a^2+b^2-2ab\cos\theta}=3+2\sqrt{2}$

$\dfrac{a^2+a^2+2aa\cos\theta}{a^2+a^2-2aa\cos\theta}=3+2\sqrt{2}$

$[\because |\mathbf{a}|=|\mathbf{b}|]$

$\dfrac{2|\mathbf{a}|^2(1+\cos\theta)}{2|\mathbf{a}|^2(1-\cos\theta)}=3+2\sqrt{2}$

$(1+\cos\theta)=(3+2\sqrt{2})(1-\cos\theta)$

$\Rightarrow \quad \cos\theta=\dfrac{2+2\sqrt{2}}{4+2\sqrt{2}}$

$\cos\theta=\dfrac{1+\sqrt{2}}{2+\sqrt{2}}=\dfrac{\sqrt{2}}{2}$

$\Rightarrow \cos\theta=\dfrac{1}{\sqrt{2}}$

Thus, $\dfrac{|\mathbf{a}+\mathbf{b}|^2}{|\mathbf{a}|^2}=\dfrac{a^2+a^2+2aa\cos\theta}{a^2}$

$=\dfrac{2a^2(1+\cos\theta)}{a^2}=2+\sqrt{2}$

**02.** *(c)* Let $\mathbf{a}=a_1\hat{\mathbf{i}}+a_2\hat{\mathbf{j}}+a_3\hat{\mathbf{k}}$

$a_1^2+a_2^2+a_3^2=1$

Let $\quad \mathbf{b}=2\hat{\mathbf{i}}-\hat{\mathbf{j}}+2\hat{\mathbf{k}},$

$\mathbf{c}=\hat{\mathbf{i}}+2\hat{\mathbf{j}}-2\hat{\mathbf{k}},$

$\mathbf{d}=\hat{\mathbf{k}}$

$\dfrac{\mathbf{a}\cdot\mathbf{b}}{|b|}=\dfrac{\mathbf{a}\cdot\mathbf{c}}{|c|}=\dfrac{\mathbf{a}\cdot\mathbf{d}}{|d|}$

$\dfrac{2a_1-a_2+2a_3}{3}=\dfrac{a_1+2a_2-2a_3}{3}=a_3$

By solving

$a_1=\dfrac{7}{\sqrt{155}}, a_2=\dfrac{9}{\sqrt{155}}, a_3=\dfrac{5}{\sqrt{155}}$

Unit vector along a is

$\dfrac{1}{\sqrt{155}}(7\hat{\mathbf{i}}+9\hat{\mathbf{j}}+5\hat{\mathbf{k}})$

**03.** *(c)*

> **key idea** If two vectors **a** and **b** are perpendicular to each other than $\mathbf{a}\cdot\mathbf{b}=0$

Given,

$|\mathbf{a}|=|\mathbf{b}|=1, \; \theta=\dfrac{\pi}{3}$

$(\lambda\mathbf{a}+2\mathbf{b})\perp^r 3\mathbf{a}-\lambda\mathbf{b}$

$\therefore \quad (\lambda\mathbf{a}+2\mathbf{b})\cdot(3\mathbf{a}-\lambda\mathbf{b})=0$

$3\lambda|\mathbf{a}|^2-\lambda^2\mathbf{a}\cdot\mathbf{b}+6\mathbf{b}\cdot\mathbf{a}-2\lambda|\mathbf{b}|^2=0$

$\Rightarrow \quad 3\lambda-\lambda^2(\cos\theta)+6\cos\theta-2\lambda=0$

$\Rightarrow \quad \lambda-\lambda^2\left(\cos\dfrac{\pi}{3}\right)+6\cos\dfrac{\pi}{3}=0$

$\Rightarrow \lambda-\dfrac{\lambda^2}{2}+3=0 \Rightarrow \lambda^2-2\lambda-6=0$

$\lambda=1\pm\sqrt{7}$

$\therefore$ Number of values of $\lambda$ in $[-1, 3]$ is 0.

**04.** *(c)* $\mathbf{r}=k(\mathbf{b}+\mathbf{c})$ and $\mathbf{r}\cdot\mathbf{a}=3$

$\Rightarrow \quad \mathbf{r}\cdot\mathbf{a}=k(\mathbf{b}\cdot\mathbf{a}+\mathbf{c}\cdot\mathbf{a})$

$\Rightarrow \quad 3=k(2+6-15+3-2+3\lambda)$

$\Rightarrow \quad 3=k(3\lambda-6)$

$\Rightarrow \quad k(\lambda-2)=1 \quad$ ...(i)

Now,

$|\mathbf{r}|=k\sqrt{25+4+25+\lambda^2-10\lambda}=1$

Put $k=\dfrac{1}{(\lambda-2)}$

$\Rightarrow \dfrac{1}{(\lambda-2)}\sqrt{\lambda^2-10\lambda+54}=1$

$\Rightarrow \lambda^2-10\lambda+54=\lambda^2-4\lambda+4$

[on squaring both sides]

$\Rightarrow \quad 6\lambda=50$

$\Rightarrow \quad 3\lambda=25$

**05.** *(d)* $\quad \mathbf{a}=\hat{\mathbf{i}}+\lambda\hat{\mathbf{j}}-3\hat{\mathbf{k}}$

and $\quad \mathbf{b}=3\hat{\mathbf{i}}-\hat{\mathbf{j}}+2\hat{\mathbf{k}}$

$\therefore \; \mathbf{a}+\mathbf{b}=4\hat{\mathbf{i}}+(\lambda-1)\hat{\mathbf{j}}-\hat{\mathbf{k}}$

and $\mathbf{a}-\mathbf{b}=-2\hat{\mathbf{i}}+(\lambda+1)\hat{\mathbf{j}}-5\hat{\mathbf{k}}$

$\because \quad (\mathbf{a}+\mathbf{b})\perp(\mathbf{a}-\mathbf{b})$

$\therefore \quad (\mathbf{a}+\mathbf{b})\cdot(\mathbf{a}-\mathbf{b})=0$

$\Rightarrow (4\hat{\mathbf{i}}+(\lambda-1)\hat{\mathbf{j}}-\hat{\mathbf{k}})$

$\cdot(-2\hat{\mathbf{i}}+(\lambda+1)\hat{\mathbf{j}}-5\hat{\mathbf{k}})=0$

$\Rightarrow \quad -8+(\lambda-1)(\lambda+1)+5=0$

$\Rightarrow \quad \lambda^2-1=3$

$\Rightarrow \quad \lambda^2=4$

$\lambda=2 \quad [\because \lambda>0]$

$\Rightarrow \quad \cos\theta=\dfrac{\mathbf{a}\cdot\mathbf{b}}{|\mathbf{a}|\,|\mathbf{b}|}$

$$= \frac{(\hat{i} + 2\hat{j} - 3\hat{k})\cdot(3\hat{i} - \hat{j} + 2\hat{k})}{\sqrt{1+4+9}\sqrt{9+1+4}}$$

$$= \frac{3-2-6}{\sqrt{14}\sqrt{14}} = \frac{-5}{14}$$

$\Rightarrow \quad 14\cos\theta = -5$

$\Rightarrow \quad (14\cos\theta)^2 = 25$

**06.** *(a)* Given, $\mathbf{a} = \hat{i} + \alpha\hat{j} + \beta\hat{k}$, $\alpha, \beta \in R$.

Angle between $\mathbf{a}$ and $\mathbf{b} = \theta = \frac{\pi}{4}$

and $\quad |\mathbf{b}|^2 = 6$

$\because \quad \mathbf{a}\cdot\mathbf{b} = 3\sqrt{2}$

$\therefore \quad |\mathbf{a}||\mathbf{b}|\cos\theta = 3\sqrt{2}$

$\Rightarrow |\mathbf{a}|^2|\mathbf{b}|^2\cos^2\frac{\pi}{4} = 18$

$\Rightarrow \quad |\mathbf{a}|^2 \times 6 \times \frac{1}{2} = 18$

$\Rightarrow \quad |\mathbf{a}|^2 = 6$

$\therefore \quad 1 + \alpha^2 + \beta^2 = 6$

$\Rightarrow \quad \alpha^2 + \beta^2 = 5$

Now, $(\alpha^2 + \beta^2)|\mathbf{a}\times\mathbf{b}|^2$

$= 5|\mathbf{a}|^2|\mathbf{b}|^2\sin^2\theta$

$= 5\times 6\times 6\times\frac{1}{2} = 90$

**07.** *(d)* $P(3, 2, 3)$, $Q(4, 6, 2)$ and $R(7, 3, 2)$ be the vertices of $\Delta PQR$

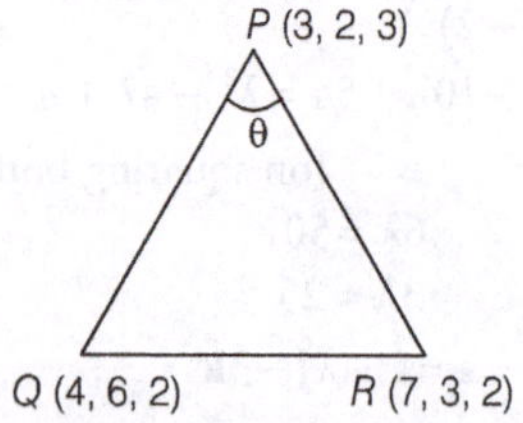


Let $\angle QPR = \theta$

$\mathbf{QP} = -\hat{i} - 4\hat{j} + \hat{k}$

$\mathbf{RP} = -4\hat{i} - \hat{j} + \hat{k}$

$$\therefore \quad \cos\theta = \left|\frac{\mathbf{QP}\cdot\mathbf{RP}}{|\mathbf{QP}||\mathbf{RP}|}\right|$$

$$= \left|\frac{4+4+1}{\sqrt{18}\sqrt{18}}\right| = \frac{9}{18} = \frac{1}{2}$$

So, $\quad \theta = \frac{\pi}{3}$

**08.** *(*)* We have, $\mathbf{a} = 5\hat{i} - \hat{j} - 3\hat{k}$ and $\mathbf{b} = \hat{i} + 3\hat{j} + 5\hat{k}$

$\therefore$Projection of $\mathbf{a}$ on $\mathbf{b} = \frac{\mathbf{a}\cdot\mathbf{b}}{|\mathbf{b}|}$

$$= \frac{(5\hat{i} - \hat{j} - 3\hat{k})\cdot(\hat{i} + 3\hat{j} + 5\hat{k})}{|\hat{i} + 3\hat{j} + 5\hat{k}|}$$

$$= \frac{5-3-15}{\sqrt{(1)^2 + (3)^2 + (5)^2}} = \frac{-13}{\sqrt{35}}$$

Again, projection vector $= \left(\frac{\mathbf{a}\cdot\mathbf{b}}{|\mathbf{b}|^2}\right)\cdot\mathbf{b}$

$$= \frac{-13}{35}\mathbf{b}$$

So, direction of projection vector is opposite of $\mathbf{b}$.

**09.** *(b)* Given, $\mathbf{u} = a(\log_e b)\hat{i} - 6\hat{j} + 3\hat{k}$

and $\mathbf{v} = (\log_e b)\hat{i} + 2\hat{j} + 2a(\log_e b)\hat{k}$

The angle between $\mathbf{u}$ and $\mathbf{v}$ is acute if $\mathbf{u}\cdot\mathbf{v} > 0$

$a(\log_e b)^2 - 12 + 6a\log_e b > 0$

$\Rightarrow \quad a(\log_e b)^2 + 6a(\log_e b) - 12 > 0$

Let $\log_e b = \alpha$

As $\quad b > 1$ and $\alpha > 0$

$a\alpha^2 + 6a\alpha - 12 > 0$

$\Rightarrow \quad a(\alpha^2 + 6\alpha) > 12$

$\Rightarrow \quad a \in \phi$

**10.** *(d)* Let $\quad A = (0, 0, 0)$

$B = (10, 0, 0)$

$G = (10, 10, h)$

$H = (0, 10, h)$

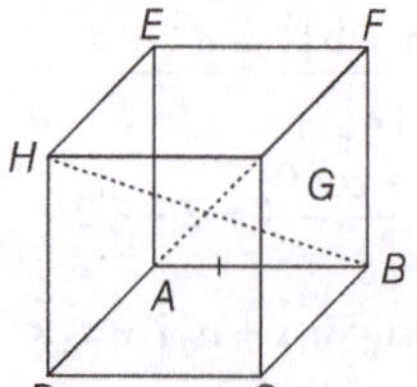


$\mathbf{AG} = 10\hat{i} + 10\hat{j} + h\hat{k}$

$\mathbf{BH} = -10\hat{i} + 10\hat{j} + h\hat{k}$

Since, $\mathbf{AG}\cdot\mathbf{BH} = |\mathbf{AG}||\mathbf{BH}|\cos\theta$

$\Rightarrow (-100 + 100 + h^2)$

$= \sqrt{h^2 + 200}\cdot\sqrt{h^2 + 200}.\left(\frac{1}{5}\right)$

$\Rightarrow \quad 5h^2 = h^2 + 200 \Rightarrow h^2 = 50$

$\Rightarrow \quad h = 5\sqrt{2}$

**11.** *(c)* Given, $\Delta ABC$, $|\mathbf{BC}| = 3$, $|\mathbf{CA}| = 5$, $|\mathbf{BA}| = 7$

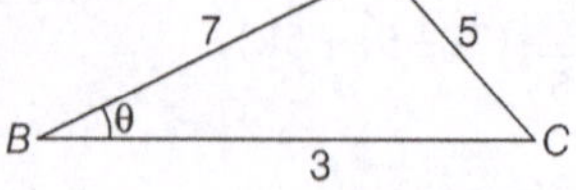


Now, projection of $\mathbf{BA}$ on

$\mathbf{BC} = |\mathbf{BA}|\cos\angle ABC$

Now, $\cos\angle ABC = \frac{(7)^2 + (3)^2 - (5)^2}{2\cdot 7\cdot 3}$

$$= \frac{49 + 9 - 25}{42} = \frac{33}{42} = \frac{11}{14}$$

$\therefore$ Projection $= |\mathbf{BA}|\cdot\frac{11}{14} = 7\times\frac{11}{14} = \frac{11}{2}$

**12.** *(c)* Given that

(i) $\mathbf{a}$ and $\mathbf{b}$ are unit vectors, *i.e.*, $|\mathbf{a}| = |\mathbf{b}| = 1$

(ii) $\mathbf{c} = \mathbf{a} + 2\mathbf{b}$ and $\mathbf{d} = 5\mathbf{a} - 4\mathbf{b}$

(iii) $\mathbf{c}$ and $\mathbf{d}$ are perpendicular to each other i.e., $\mathbf{c}\cdot\mathbf{d} = 0$

To find Angle between $\mathbf{a}$ and $\mathbf{b}$.

Now,

$\mathbf{c}\cdot\mathbf{d} = 0 \Rightarrow (\mathbf{a} + 2\,\mathbf{b})\cdot(5\,\mathbf{a} - 4\,\mathbf{b}) = 0$

$\Rightarrow 5\,\mathbf{a}\cdot\mathbf{a} - 4\,\mathbf{a}\cdot\mathbf{b} + 10\,\mathbf{b}\cdot\mathbf{a} - 8\,\mathbf{b}\cdot\mathbf{b} = 0$

$\Rightarrow \quad 6\,\mathbf{a}\cdot\mathbf{b} = 3 \Rightarrow \mathbf{a}\cdot\mathbf{b} = 1/2$

So, the angle between $\mathbf{a}$ and $\mathbf{b}$ is $\pi/3$.

**13.** *(a)* Since, the given vectors are mutually orthogonal, therefore

$\mathbf{a}\cdot\mathbf{b} = 2 - 4 + 2 = 0$

$\mathbf{a}\cdot\mathbf{c} = \lambda - 1 + 2\mu = 0 \quad$ ...(i)

and $\quad \mathbf{b}\cdot\mathbf{c} = 2\lambda + 4 + \mu = 0 \quad$ ...(ii)

On solving Eqs. (i) and (ii), we get

$\mu = 2$ and $\lambda = -3$

Hence, $\quad (\lambda, \mu) = (-3, 2)$

**14.** *(d)* Since, position vectors of $A$, $B$, $C$ are $2\hat{i} - \hat{j} + \hat{k}$, $\hat{i} - 3\hat{j} - 5\hat{k}$ and $a\hat{i} - 3\hat{j} + \hat{k}$, respectively.

Now,

$\mathbf{AC} = (a\hat{j} - 3\hat{j} + \hat{k}) - (2\hat{i} - \hat{j} + \hat{k})$

$= (a-2)\hat{i} - 2\hat{j}$

and $\mathbf{BC} = (a\hat{i} - 3\hat{j} + \hat{k}) - (\hat{i} - 3\hat{j} - 5\hat{k})$

$= (a-1)\hat{i} + 6\hat{k}$

Since, the $\Delta ABC$ is right angled at $C$, then

$\mathbf{AC}\cdot\mathbf{BC} = 0$

$\Rightarrow \{(a-2)\hat{i} - 2\hat{j}\}\cdot\{(a-1)\hat{i} + 6\hat{k}\} = 0$

$\Rightarrow \quad (a-2)(a-1) = 0$

$\therefore \quad a = 1$ and $a = 2$

**15.** *(a)* Total force,

$\mathbf{F} = (4\hat{i} + \hat{j} - 3\hat{k}) + (3\hat{i} + \hat{j} - \hat{k})$

$\therefore \quad \mathbf{F} = 7\hat{i} + 2\hat{j} - 4\hat{k}$

The particle is displaced from $A(\hat{i} + 2\hat{j} + 3\hat{k})$ to $B(5\hat{i} + 4\hat{j} + \hat{k})$.

Now, displacement,

$AB = (5\hat{i}+4\hat{j}+\hat{k})-(\hat{i}+2\hat{j}+3\hat{k})$

$= 4\hat{i}+2\hat{j}-2\hat{k}$

$\therefore$ Work done $= \mathbf{F}\cdot \mathbf{AB}$

$= (7\hat{i}+2\hat{j}-4\hat{k})\cdot(4\hat{i}+2\hat{j}-2\hat{k})$

$= 28+4+8 = 40$ units

**16.** *(b)* Given that, $|\mathbf{a}|=1, |\mathbf{b}|=2, |\mathbf{c}|=3$

and $\quad \mathbf{a}+\mathbf{b}+\mathbf{c}=0$

Now, $(\mathbf{a}+\mathbf{b}+\mathbf{c})^2 = |\mathbf{a}|^2+|\mathbf{b}|^2+|\mathbf{c}|^2 + 2(\mathbf{a}\cdot\mathbf{b}+\mathbf{b}\cdot\mathbf{c}+\mathbf{c}\cdot\mathbf{a})$

$\Rightarrow \quad 0 = 1^2+2^2+3^2 + 2(\mathbf{a}\cdot\mathbf{b}+\mathbf{b}\cdot\mathbf{c}+\mathbf{c}\cdot\mathbf{a})$

$\Rightarrow \quad 2(\mathbf{a}\cdot\mathbf{b}+\mathbf{b}\cdot\mathbf{c}+\mathbf{c}\cdot\mathbf{a}) = -14$

$\Rightarrow \quad \mathbf{c}\cdot\mathbf{b}+\mathbf{b}\cdot\mathbf{c}+\mathbf{c}\cdot\mathbf{a} = -7$

**17.** *(a)* Let $\mathbf{c} = x\,\mathbf{a} + y\,\mathbf{b}$

$\mathbf{c}\cdot\mathbf{a} = 0$

$\lambda a^2 + y\mathbf{b}\cdot\mathbf{a} = 0$

$a^2 = 2^2+1^2+1^2 = 6$

$\mathbf{a}\cdot\mathbf{b} = 2+2-1 = 3$

$\Rightarrow \quad 6x+3y = 0$

$\Rightarrow \quad y = -2x \qquad \ldots(\text{i})$

Also $|\mathbf{c}| = 1$

$\mathbf{c} = x(\hat{i}+2\hat{j}+\hat{k}) + y(2\hat{i}+\hat{j}-\hat{k})$

$= (x+2y)\,\hat{i} + (2x+y)\,\hat{j} + (x-y)\,\hat{k}$

$\mathbf{c} = (x-4x)\,\hat{i} + (2x-2x)\,\hat{j} + (x+2x)\,\hat{k}$

$[\because y = -2x]$

$\mathbf{c} = -3x\,\hat{i} + 3x\,\hat{k}$

$|\mathbf{c}|^2 = 9x^2 + 9x^2 = 1$

$18x^2 = 1$

$x = \pm\dfrac{1}{3\sqrt{2}}$

So, $\mathbf{c} = 3x\,\hat{k} - 3x\,\hat{i}$

$\mathbf{c} = \pm\dfrac{1}{\sqrt{2}}(\hat{k}-\hat{i})$

So, $\mathbf{c} = \dfrac{1}{\sqrt{2}}(\hat{k}-\hat{i})$

**18.** *(c)* $\mathbf{u} = 3\hat{i}-\hat{j}$

$\mathbf{v} = 2\hat{i}+\hat{j}-\lambda\hat{k}$

$\because$ Let angle between $\mathbf{u}$ and $\mathbf{v}$ be $\theta$.

$\cos\theta = \dfrac{\mathbf{u}\cdot\mathbf{v}}{|\mathbf{u}||\mathbf{v}|}$

$\Rightarrow \dfrac{\sqrt{5}}{2\sqrt{7}} = \dfrac{6-1}{\sqrt{10}\,\sqrt{4+1+\lambda^2}}$

$\Rightarrow \dfrac{\sqrt{5}}{2\sqrt{7}} = \dfrac{5}{\sqrt{10}\,\sqrt{5+\lambda^2}}$

$\Rightarrow 5+\lambda^2 = 14 \Rightarrow \lambda^2 = 9$

$\Rightarrow \qquad \lambda = 3 \qquad (\because \lambda > 0)$

$\because \mathbf{v}_1$ is parallel to $\mathbf{u}$ and $\mathbf{v}_2$ is perpendicular $\mathbf{u}$

$\therefore \qquad \mathbf{v}_1 = k\mathbf{u} \qquad v_2\cdot u = 0$

$\because \qquad \mathbf{v} = \mathbf{v}_1 + \mathbf{v}_2$

$\therefore \quad (\mathbf{v}-\mathbf{v}_1)\cdot\mathbf{u} = 0$

$(\mathbf{v}-k\mathbf{u})\cdot\mathbf{u} = 0$

$\mathbf{v}\cdot\mathbf{u} = k\mathbf{u}^2$

$k = \dfrac{\mathbf{v}\cdot\mathbf{u}}{\mathbf{u}^2} = \dfrac{5}{10} = \dfrac{1}{2}$

$\mathbf{v}_1 = \dfrac{1}{2}(3\hat{i}-\hat{j})$

$\mathbf{v}_2 = \mathbf{v}-\mathbf{v}_1$

$= (2\hat{i}+\hat{j}-3\hat{k}) - \left(\dfrac{3}{2}\hat{i}-\dfrac{1}{2}\hat{j}\right)$

$= \dfrac{1}{2}(\hat{i}+3\hat{j}-6\hat{k})$

$|\mathbf{v}_1|^2\,|\mathbf{v}_2|^2 = \left|\dfrac{1}{2}(3\hat{i}-\hat{j})\right|^2 + \dfrac{1}{2}\left|\hat{i}+3\hat{j}-6\hat{k}\right|^2$

$= \dfrac{1}{4}\times 10 + \dfrac{1}{4}\times 46 = \dfrac{56}{4} = 14$

**19.** *(b)*

> **key idea** $\mathbf{a}$ and $\mathbf{b}$ inclined at an obtuse angle iff $\mathbf{a}\cdot\mathbf{b} < 0$

$\mathbf{a} = \alpha t\hat{i} + 6\hat{j} - 3\hat{k}$ and $\mathbf{b} = t\hat{i} - 2\hat{j} - 2\alpha t\hat{k}$

$\because$ $\mathbf{a}$ and $\mathbf{b}$ are inclined at an obtuse angle.

So, $\qquad \mathbf{a}\cdot\mathbf{b} < 0$

$\Rightarrow \alpha t^2 - 12 + 6\alpha t < 0$

$\Rightarrow \alpha t^2 + 6\alpha t - 12 < 0$

To satisfy this condition

$D < 0$

$\Rightarrow (6\alpha)^2 - 4(\alpha)(-12) < 0$

$\Rightarrow \quad 36\alpha^2 + 48\alpha < 0 \Rightarrow 3\alpha^2 + 4\alpha < 0$

$\Rightarrow \quad \alpha(3\alpha+4) < 0$

If $\alpha < 0$, $3\alpha + 4 > 0$

$\Rightarrow \alpha < 0$ and $\alpha > -\dfrac{4}{3}$

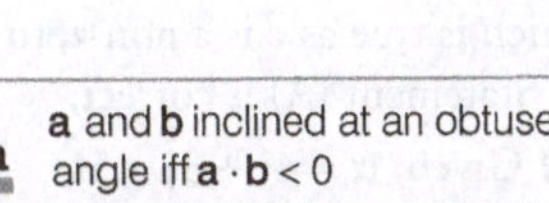


$\alpha \in \left(-\dfrac{4}{3}, 0\right)$

If $\alpha > 0$, $3\alpha + 4 < 0$

$\Rightarrow \alpha > 0, \alpha < -\dfrac{4}{3}$

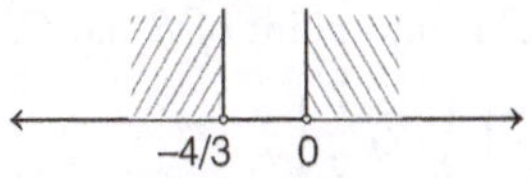


No solution

and when $\alpha = 0$

$0t^2 + 6t\times 0 - 12 = -12 < 0$

Therefore, $\alpha \in \left(-\dfrac{4}{3}, 0\right]$

**20.** *(c)* Let $\mathbf{c} = \alpha\hat{i} + \beta\hat{j} + \gamma\hat{k}$

$\Rightarrow \quad |\mathbf{c}| = \sqrt{\alpha^2+\beta^2+\gamma^2} = 1$

$\Rightarrow \quad \alpha^2+\beta^2+\gamma^2 = 1 \qquad \ldots(\text{i})$

Now, $\mathbf{c}\cdot(2\hat{i}+2\hat{j}-\hat{k}) = |\mathbf{c}|\sqrt{9}\cdot\cos 60°$

$\Rightarrow \; 2\alpha + 2\beta - \gamma = 1\times\dfrac{3}{2} \qquad \ldots(\text{ii})$

Also, $\mathbf{c}\cdot(\hat{i}-\hat{k}) = |\mathbf{c}|\sqrt{2}\cos 45°$

$\Rightarrow \alpha - \gamma = 1\times\sqrt{2}\times\dfrac{1}{\sqrt{2}} = 1 \qquad \ldots(\text{iii})$

Subtracting Eq. (iii) from Eq. (ii), we get

$\alpha + 2\beta = 1/2 \qquad \ldots(\text{iv})$

and hence we get $\alpha = \dfrac{\sqrt{2}}{3} + \dfrac{1}{2}$

$\beta = \dfrac{-1}{3\sqrt{2}}$ and $\gamma = \dfrac{\sqrt{2}}{3} - \dfrac{1}{2}$

(by solving equation)

Thus, $\mathbf{c} + \left(-\dfrac{1}{2}\hat{i} + \dfrac{1}{3\sqrt{2}}\hat{j} - \dfrac{\sqrt{2}}{3}\hat{k}\right)$

$= \dfrac{\sqrt{2}}{3}\hat{i} - \dfrac{1}{2}\hat{k}$

**21.** *(d)* We have, $D$ divides $BC$ in the ratio of

$|\mathbf{AB}| : |\mathbf{AC}|$

$\therefore \quad \mathbf{AB} = -3\hat{i} + 5\hat{j} - 2\hat{k}$

and $\mathbf{AC} = 2\hat{i} + 3\hat{j} + 5\hat{k}$

$\therefore |\mathbf{AB}| = \sqrt{9+25+4} = \sqrt{38}$

and $|\mathbf{AC}| = \sqrt{4+9+25} = \sqrt{38}$

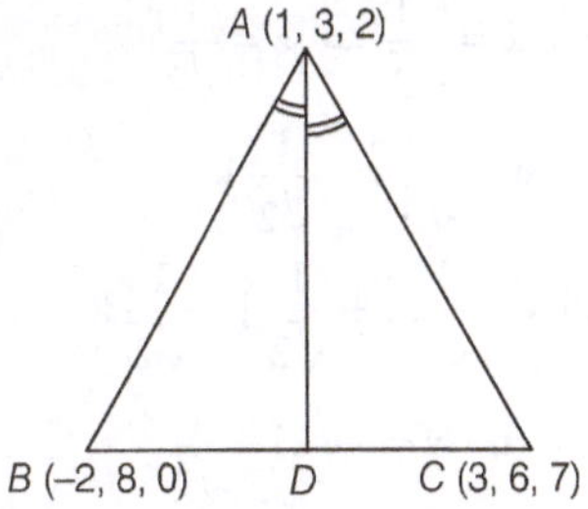

$\therefore \quad \dfrac{BD}{DC}=\dfrac{\sqrt{38}}{\sqrt{38}}=\dfrac{1}{1}$

So, $D$ is mid-point of $B$ and $C$.

So, $D\left(\dfrac{1}{2}, 7, \dfrac{7}{2}\right)$

Now, $\mathbf{AD}=-\dfrac{1}{2}\hat{\mathbf{i}}+4\hat{\mathbf{j}}+\dfrac{3}{2}\hat{\mathbf{k}}$

$\therefore$ Projection of **AD** on **AC**

$$=\frac{\mathbf{AD}\cdot\mathbf{AC}}{|\mathbf{AC}|}=\frac{-1+12+\frac{15}{2}}{\sqrt{38}}=\frac{37}{2\sqrt{38}}$$

**22.** *(b)* $\because$ Unit vector $\mathbf{u}=x\hat{\mathbf{i}}+y\hat{\mathbf{j}}+z\hat{\mathbf{k}}$ makes $\dfrac{\pi}{2}, \dfrac{\pi}{3}$ and $\dfrac{2\pi}{3}$ with the vectors $\dfrac{1}{\sqrt{2}}\hat{\mathbf{i}}+\dfrac{1}{\sqrt{2}}\hat{\mathbf{k}}, \dfrac{1}{\sqrt{2}}\hat{\mathbf{j}}+\dfrac{1}{\sqrt{2}}\hat{\mathbf{k}}$ and $\dfrac{1}{\sqrt{2}}\hat{\mathbf{i}}+\dfrac{1}{\sqrt{2}}\hat{\mathbf{j}}$, respectively.

Taking dot product of unit vector **u** with given vectors respectively, we have

$$\therefore \frac{x}{\sqrt{2}}+\frac{z}{\sqrt{2}}=|\mathbf{u}|\sqrt{\frac{1}{2}+\frac{1}{2}}\cdot\cos\frac{\pi}{2}=0$$

$$\Rightarrow \quad x+z=0 \quad \ldots\text{(i)}$$

$$\text{and} \quad \frac{y}{\sqrt{2}}+\frac{z}{\sqrt{2}}=|\mathbf{u}|\sqrt{\frac{1}{2}+\frac{1}{2}}\cos\frac{\pi}{3}$$

$$\frac{y+z}{\sqrt{2}}=\frac{1}{2}$$

$$\Rightarrow \quad y+z=\frac{1}{\sqrt{2}} \quad \ldots\text{(ii)}$$

$$\text{and} \quad \frac{x}{\sqrt{2}}+\frac{y}{\sqrt{2}}=|\mathbf{u}|\sqrt{\frac{1}{2}+\frac{1}{2}}\cos\frac{2\pi}{3}$$

$$\frac{x+y}{\sqrt{2}}=1\times1\times\frac{-1}{2}$$

$$\Rightarrow \quad x+y=\frac{-1}{\sqrt{2}} \quad \ldots\text{(iii)}$$

On solving Eqs. (i), (ii) and (iii), we get

$$y=0$$

$$\Rightarrow \quad x=\frac{-1}{\sqrt{2}} \Rightarrow z=\frac{1}{\sqrt{2}}$$

$$\Rightarrow \quad \mathbf{u}=\frac{-1}{\sqrt{2}}\hat{\mathbf{i}}+\frac{1}{\sqrt{2}}\hat{\mathbf{k}}$$

$$\text{and } \mathbf{v}=\frac{1}{\sqrt{2}}\hat{\mathbf{i}}+\frac{1}{\sqrt{2}}\hat{\mathbf{j}}+\frac{1}{\sqrt{2}}\hat{\mathbf{k}}$$

$$\therefore \quad \mathbf{u}-\mathbf{v}=-\sqrt{2}\hat{\mathbf{i}}-\frac{1}{\sqrt{2}}\hat{\mathbf{j}}$$

$$\Rightarrow \quad |\mathbf{u}-\mathbf{v}|^2=(\sqrt{2})^2+\left(\frac{1}{\sqrt{2}}\right)^2$$

$$=2+\frac{1}{2}=\frac{5}{2}$$

**23.** *(d)* We have,

$$|\mathbf{a}+\mathbf{b}+\mathbf{c}|=|\mathbf{a}+\mathbf{b}-\mathbf{c}|$$

$$\Rightarrow \quad |\mathbf{a}+\mathbf{b}+\mathbf{c}|^2=|\mathbf{a}+\mathbf{b}-\mathbf{c}|^2$$

$$\Rightarrow \quad (\mathbf{a}+\mathbf{b}+\mathbf{c})\cdot(\mathbf{a}+\mathbf{b}+\mathbf{c})=(\mathbf{a}+\mathbf{b}-\mathbf{c})\cdot(\mathbf{a}+\mathbf{b}-\mathbf{c})$$

$$\Rightarrow \quad |\mathbf{a}|^2+|\mathbf{b}|^2+|\mathbf{c}|^2+2\,\mathbf{a}\cdot\mathbf{b}+2\mathbf{b}\cdot\mathbf{c}+2\mathbf{c}\cdot\mathbf{a}=|\mathbf{a}|^2+|\mathbf{b}|^2+|\mathbf{c}|^2+2\mathbf{a}\cdot\mathbf{b}-2\mathbf{b}\cdot\mathbf{c}-2\mathbf{c}\cdot\mathbf{a}$$

$$\Rightarrow \quad 4\,\mathbf{b}\cdot\mathbf{c}+4\,\mathbf{c}\cdot\mathbf{a}=0$$

$$\Rightarrow \quad 4\times0+4\,\mathbf{c}\cdot\mathbf{a}=0 \quad [\because \mathbf{b}.\mathbf{c}=0]$$

$$\Rightarrow \quad 4\mathbf{c}\cdot\mathbf{a}=0 \Rightarrow \mathbf{c}\cdot\mathbf{a}=0$$

So, **a** and **c** are perpendicular to each other. So, Statement (B) is incorrect.

Again, we have,

$$|\mathbf{a}+\lambda\mathbf{c}|\geq|\mathbf{a}|$$

$$\Rightarrow \quad |\mathbf{a}+\lambda\mathbf{c}|^2\geq|\mathbf{a}|^2$$

$$\Rightarrow \quad (\mathbf{a}+\lambda\mathbf{c})\cdot(\mathbf{a}+\lambda\mathbf{c})\geq|\mathbf{a}|^2$$

$$\Rightarrow \quad |\mathbf{a}|^2+\lambda^2|\mathbf{c}|^2+2\lambda\,\mathbf{a}.\mathbf{c}\geq|\mathbf{a}|^2$$

$$\Rightarrow \quad \lambda^2|\mathbf{c}|^2+0\geq0 \quad [\because \mathbf{a}.\mathbf{c}=0]$$

$$\Rightarrow \quad |\mathbf{c}|\geq0$$

Which is true as **c** is a non-zero vector.

So, Statement (A) is correct.

**24.** *(c)* Given, $\alpha=4\hat{\mathbf{i}}+3\hat{\mathbf{j}}+5\hat{\mathbf{k}}$

$$\beta=\hat{\mathbf{i}}+2\hat{\mathbf{j}}-4\hat{\mathbf{k}}$$

$\because \beta_1$ parallel to $\alpha$

$$\therefore \quad \beta_1=\mu(4\hat{\mathbf{i}}+3\hat{\mathbf{j}}+5\hat{\mathbf{k}})$$

$$\because \quad \beta=\beta_1+\beta_2$$

$$\Rightarrow \beta_2=\beta-\beta_1$$

$$\therefore \beta_2=(\hat{\mathbf{i}}+2\hat{\mathbf{j}}-4\hat{\mathbf{k}})-\mu(4\hat{\mathbf{i}}+3\hat{\mathbf{j}}+5\hat{\mathbf{k}})$$

$$\beta_2=(1-4\mu)\hat{\mathbf{i}}+(2-3\mu)\hat{\mathbf{j}}-(4+5\mu)\hat{\mathbf{k}}$$

$\because \beta_2$ perpendicular to $\alpha$, so its dot product $=0$

$$\therefore (1-4\mu)4+(2-3\mu)3-(4+5\mu)5=0$$

$$\Rightarrow \quad 4-16\mu+6-9\mu-20-25\mu=0$$

$$\Rightarrow \quad -10-50\mu=0$$

$$\therefore \mu=\frac{-1}{5}$$

$\therefore$ The value of $5\beta_2.(\hat{\mathbf{i}}+\hat{\mathbf{j}}+\hat{\mathbf{k}})$

$$=5\left\{\left(1+\frac{4}{5}\right)\hat{\mathbf{i}}+\left(2+\frac{3}{5}\right)\hat{\mathbf{j}}-\left(4-\frac{5}{5}\right)\hat{\mathbf{k}}\right\}\cdot(\hat{\mathbf{i}}+\hat{\mathbf{j}}+\hat{\mathbf{k}})$$

$$=(9\hat{\mathbf{i}}+13\hat{\mathbf{j}}-15\hat{\mathbf{k}})\cdot(\hat{\mathbf{i}}+\hat{\mathbf{j}}+\hat{\mathbf{k}})$$

$$=9+13-15=7$$

**25.** *(d)* Let $\mathbf{v}=\lambda\mathbf{a}+\mu\mathbf{b}$

$$\mathbf{v}=\lambda(1,1,2)+\mu(2,-3,1)$$

$$\mathbf{v}=(\lambda+2\mu, \lambda-3\mu, 2\lambda+\mu)$$

$$\mathbf{v}\cdot\hat{\mathbf{j}}=7 \text{ and } \mathbf{v}\cdot\frac{\mathbf{c}}{|\mathbf{c}|}=\frac{2}{\sqrt{3}}$$

$$\Rightarrow \quad \lambda-3\mu=7 \quad \ldots\text{(i)}$$

$$\text{and} \quad \mathbf{v}\cdot\frac{\mathbf{c}}{\sqrt{3}}=\frac{2}{\sqrt{3}}$$

$$\mathbf{v}\cdot\mathbf{c}=2$$

$$\Rightarrow \quad (\lambda+2\mu, \lambda-3\mu, 2\lambda+\mu)\cdot(1,-1,1)=2$$

$$\Rightarrow \quad \lambda+2\mu-\lambda+3\mu+2\lambda+\mu=2$$

$$\Rightarrow 2\lambda+6\mu=2 \Rightarrow \lambda+3\mu=1 \quad \ldots\text{(ii)}$$

$$\text{and} \quad \lambda-3\mu=7 \quad \text{[from Eq. (i)]}$$

$$\therefore \quad 2\lambda=8 \Rightarrow \lambda=4 \text{ and } \mu=-1$$

Thus, $\mathbf{v}=(2,7,7)$

Hence, $\mathbf{v}\cdot(\hat{\mathbf{i}}+\hat{\mathbf{k}})$

$$=(2,7,7)\cdot(1,0,1)=2+7=9$$

**26.** *(c)* $|2\mathbf{a}+3\mathbf{b}|=|3\mathbf{a}+\mathbf{b}|$

$$\Rightarrow \quad |2\mathbf{a}+3\mathbf{b}|^2=|3\mathbf{a}+\mathbf{b}|^2$$

$$\Rightarrow \quad 4|\mathbf{a}|^2+9|\mathbf{b}|^2+12\,\mathbf{a}\cdot\mathbf{b}=9|\mathbf{a}|^2+|\mathbf{b}|^2+6\mathbf{a}\cdot\mathbf{b}$$

$$\Rightarrow \quad 5|\mathbf{a}|^2-6\mathbf{a}\cdot\mathbf{b}-8|\mathbf{b}|^2=0$$

$\mathbf{a}/8$ is a unit vector.

So, $|\mathbf{a}|=8$

And, $5\cdot64-6\cdot8|\mathbf{b}|\left(\dfrac{1}{2}\right)-8|\mathbf{b}|^2=0$

$$\Rightarrow \quad |\mathbf{b}|^2+3|\mathbf{b}|-40=0$$

$$\Rightarrow \quad (|\mathbf{b}|+8)(|\mathbf{b}|-5)=0$$

$$|\mathbf{b}|=5$$

As, $|\mathbf{b}|=-8$ Not possible.

**27.** *(a)* It is given that

Since, $a\cos\theta=b\cos\left(\theta+\dfrac{2\pi}{3}\right)$

$$=c\cos\left(\theta+\frac{4\pi}{3}\right)=k \text{ (let)}$$

$$\Rightarrow \quad a=\frac{k}{\cos\theta}$$

$$b=\frac{k}{\cos\left(\theta+\frac{2\pi}{3}\right)} \quad \ldots\text{(i)}$$

$$\text{and } c=\frac{k}{\cos\left(\theta+\frac{4\pi}{3}\right)}$$

Now angle between vectors $a\hat{\mathbf{i}}+b\hat{\mathbf{j}}+c\hat{\mathbf{k}}$ and $b\hat{\mathbf{i}}+c\hat{\mathbf{j}}+a\hat{\mathbf{k}}$, where $a^2+b^2+c^2=1$ for $a, b, c\in R$ is

$$\alpha=\cos^{-1}|(ab+bc+ca)|$$

$$\because \quad ab + bc + ca = k^2$$

$$\left[\frac{1}{\cos\theta\cos\left(\theta+\frac{2\pi}{3}\right)} + \frac{1}{\cos\left(\theta+\frac{2\pi}{3}\right)\cos\left(\theta+\frac{4\pi}{3}\right)} + \frac{1}{\cos\left(\frac{4\pi}{3}+\theta\right)\cos\theta}\right]$$

$$= k^2 \left[\frac{\cos\left(\frac{4\pi}{3}+\theta\right) + \cos\left(\frac{2\pi}{3}+\theta\right) + \cos\theta}{\cos\theta\cos\left(\frac{2\pi}{3}+\theta\right)\cos\left(\frac{4\pi}{3}+\theta\right)}\right]$$

$$= k^2 \frac{2\cos(\pi+\theta)\cos\left(\frac{\pi}{3}\right) + \cos\theta}{\cos\theta\cos\left(\frac{2\pi}{3}+\theta\right)\cos\left(\frac{4\pi}{3}+\theta\right)}$$

$$= k^2 \frac{-\cos\theta + \cos\theta}{\cos\theta\cos\left(\frac{2\pi}{3}+\theta\right)\cos\left(\frac{4\pi}{3}+\theta\right)} = 0$$

$\therefore \alpha = \cos^{-1}(0) = \pi/2$

**28.** *(a)* According to given information, we have the following figure.

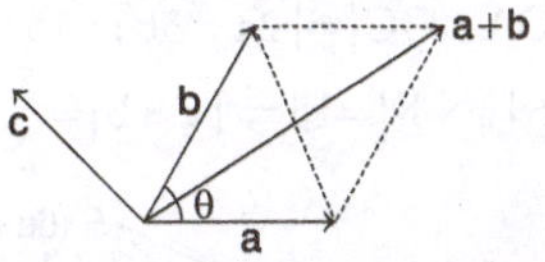


Clearly, projection of **b** on **a** $= \dfrac{\mathbf{b}\cdot\mathbf{a}}{|\mathbf{a}|}$

$$= \frac{(b_1\hat{\mathbf{i}} + b_2\hat{\mathbf{j}} + \sqrt{2}\hat{\mathbf{k}})\cdot(\hat{\mathbf{i}} + \hat{\mathbf{j}} + \sqrt{2}\,\hat{\mathbf{k}})}{\sqrt{1^2 + 1^2 + (\sqrt{2})^2}}$$

$$= \frac{b_1 + b_2 + 2}{\sqrt{4}} = \frac{b_1 + b_2 + 2}{2}$$

But, projection of **b** on **a** $= |\mathbf{a}|$

$$\therefore \quad \frac{b_1 + b_2 + 2}{2} = \sqrt{1^2 + 1^2 + (\sqrt{2})^2}$$

$$\Rightarrow \quad \frac{b_1 + b_2 + 2}{2} = 2$$

$$\Rightarrow \quad b_1 + b_2 = 2 \quad \text{...(i)}$$

Now, $\mathbf{a} + \mathbf{b} = (\hat{\mathbf{i}} + \hat{\mathbf{j}} + \sqrt{2}\hat{\mathbf{k}}) + (b_1\hat{\mathbf{i}} + b_2\hat{\mathbf{j}} + \sqrt{2}\hat{\mathbf{k}})$

$$= (b_1 + 1)\hat{\mathbf{i}} + (b_2 + 1)\hat{\mathbf{j}} + 2\sqrt{2}\hat{\mathbf{k}}$$

$\because$ $(\mathbf{a} + \mathbf{b}) \perp \mathbf{c}$, therefore $(\mathbf{a} + \mathbf{b})\cdot\mathbf{c} = 0$

$$\Rightarrow \{(b_1 + 1)\hat{\mathbf{i}} + (b_2 + 1)\hat{\mathbf{j}} + 2\sqrt{2}\hat{\mathbf{k}}\}\cdot(5\hat{\mathbf{i}} + \hat{\mathbf{j}} + \sqrt{2}\,\hat{\mathbf{k}}) = 0$$

$$\Rightarrow \quad 5(b_1 + 1) + 1(b_2 + 1) + 2\sqrt{2}(\sqrt{2}) = 0$$

$$\Rightarrow \quad 5b_1 + b_2 = -10 \quad \text{...(ii)}$$

From Eqs. (i) and (ii), we get

$$b_1 = -3 \text{ and } b_2 = 5$$

$$\Rightarrow \quad \mathbf{b} = -3\hat{\mathbf{i}} + 5\hat{\mathbf{j}} + \sqrt{2}\hat{\mathbf{k}}$$

$$\Rightarrow \quad |\mathbf{b}| = \sqrt{(-3)^2 + (5)^2 + (\sqrt{2})^2} = \sqrt{36} = 6$$

**29.** *(a)*

> **key idea** If any vector **x** is coplanar with the vector **y** and **z**, then
> **x** = λ**y** + μ**z**
> Here, **u** is coplanar with **a** and **b**.

$\therefore \mathbf{u} = \lambda\,\mathbf{a} + \mu\,\mathbf{b}$

Dot product with **a**, we get

$$\mathbf{u}\cdot\mathbf{a} = \lambda(\mathbf{a}\cdot\mathbf{a}) + \mu(\mathbf{b}\cdot\mathbf{a})$$

$$\Rightarrow \quad 0 = 14\lambda + 2\mu \quad \text{...(i)}$$

$$[\because \mathbf{a} = 2\hat{\mathbf{i}} + 3\hat{\mathbf{j}} - \hat{\mathbf{k}},\ \mathbf{b} = \hat{\mathbf{j}} + \hat{\mathbf{k}},\ \mathbf{u}\cdot\mathbf{a} = 0]$$

Dot product with **b**, we get

$$\mathbf{u}\cdot\mathbf{b} = \lambda(\mathbf{a}\cdot\mathbf{b}) + \mu(\mathbf{b}\cdot\mathbf{b})$$

$$24 = 2\lambda + 2\mu \quad \text{...(ii)}$$

$$[\because \mathbf{u}\cdot\mathbf{b} = 24]$$

On solving Eqs. (i) and (ii), we get

$$\lambda = -2,\ \mu = 14$$

Dot product with **u**, we get

$$|\mathbf{u}|^2 = \lambda(\mathbf{u}\cdot\mathbf{a}) + \mu(\mathbf{u}\cdot\mathbf{b})$$

$$|\mathbf{u}|^2 = -2(0) + 14(24) \Rightarrow |\mathbf{u}|^2 = 336$$

**30.** *(b)* Given

(i) A parallelogram $ABCD$ such that $\mathbf{AB} = \mathbf{q}$ and $\mathbf{AD} = \mathbf{p}$.

(ii) The altitude from vertex $B$ to side $AD$ coincides with a vector **r**.

To find the vector **r** in terms of **p** and **q**.

Let $E$ be the foot of perpendicular from $B$ to side AD.

AE = Projection of vector **q** on **p** $= \dfrac{\mathbf{q}\cdot\mathbf{p}}{|\mathbf{p}|}$

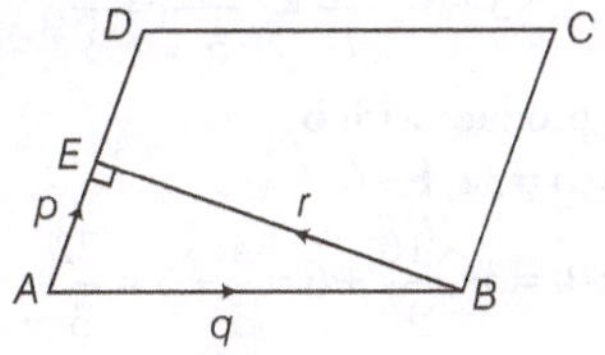


**AE** = Vector along $AE$ of length $AE$

$$= |\mathbf{AE}|\,\mathbf{AE} = \left(\frac{\mathbf{q}\cdot\mathbf{p}}{|\mathbf{p}|}\right)\mathbf{p} = \frac{(\mathbf{q}\cdot\mathbf{p})\,\mathbf{p}}{|\mathbf{p}|^2}$$

Now, applying triangles law in $\Delta ABE$, we get

$$\mathbf{AB} + \mathbf{BE} = \mathbf{AE}$$

$$\Rightarrow \quad \mathbf{q} + \mathbf{r} = \frac{(\mathbf{q}\cdot\mathbf{p})\,\mathbf{p}}{|\mathbf{p}|^2}$$

$$\Rightarrow \quad \mathbf{r} = \frac{(\mathbf{q}\cdot\mathbf{p})\,\mathbf{p}}{|\mathbf{p}|^2} - \mathbf{q}$$

$$\Rightarrow \quad \mathbf{r} = -\mathbf{q} + \left(\frac{\mathbf{q}\cdot\mathbf{p}}{\mathbf{p}\cdot\mathbf{p}}\right)\mathbf{p}$$

**31.** *(a)* Given two vectors lie in *xy*-plane. So, a vector coplanar with them is

$$\mathbf{a} = x\,\hat{\mathbf{i}} + y\,\hat{\mathbf{j}}$$

Since, $\quad \mathbf{a} \perp (\hat{\mathbf{i}} - \hat{\mathbf{j}})$

$$\Rightarrow (x\,\hat{\mathbf{i}} + y\,\hat{\mathbf{j}})\cdot(\hat{\mathbf{i}} - \hat{\mathbf{j}}) = 0$$

$$\Rightarrow \quad x - y = 0$$

$$\Rightarrow \quad x = y$$

$$\therefore \quad \mathbf{a} = x\,\hat{\mathbf{i}} + x\,\hat{\mathbf{j}}$$

and $|\mathbf{a}| = \sqrt{x^2 + x^2} = x\sqrt{2}$

$\therefore$ Required unit vector

$$= \frac{\mathbf{a}}{|\mathbf{a}|} = \frac{x(\hat{\mathbf{i}} + \hat{\mathbf{j}})}{x\sqrt{2}} = \frac{1}{\sqrt{2}}(\hat{\mathbf{i}} + \hat{\mathbf{j}})$$

**32.** *(5)* Let θ be the acute angle between the vectors $\alpha\hat{\mathbf{i}} - 2\hat{\mathbf{j}} + 2\hat{\mathbf{k}}$ and $\alpha\hat{\mathbf{i}} + 2\alpha\hat{\mathbf{j}} - 2\hat{\mathbf{k}}$.

Now,

$$\cos\theta = \frac{(\alpha\hat{\mathbf{i}} - 2\hat{\mathbf{j}} + 2\hat{\mathbf{k}})\cdot(\alpha\hat{\mathbf{i}} + 2\alpha\hat{\mathbf{j}} - 2\hat{\mathbf{k}})}{\sqrt{\alpha^2 + 4 + 4}\,\sqrt{\alpha^2 + 4\alpha^2 + 4}}$$

$\Rightarrow \cos\theta$

$$= \frac{\alpha^2 - 4\alpha - 4}{\sqrt{\alpha^2 + 4 + 4}\,\sqrt{\alpha^2 + 4\alpha^2 + 4}}$$

$$\Rightarrow \cos\theta = \frac{\alpha^2 - 4\alpha - 4}{\sqrt{\alpha^2 + 8}\,\sqrt{5\alpha^2 + 4}}$$

$$\Rightarrow \alpha^2 - 4\alpha - 4 > 0$$

$$[\because \cos\theta > 0 \text{ for acute angle}]$$

$$\Rightarrow (\alpha - 2)^2 > 8$$

$$\Rightarrow \quad \alpha - 2 > 2\sqrt{2}$$

or $\quad \alpha - 2 < -2\sqrt{2}$

$$\Rightarrow \quad \alpha > 2 + 2\sqrt{2}$$

or $\quad \alpha < 2 - 2\sqrt{2}$

$$\Rightarrow x \in (-\infty, -0.82) \cup (4.82, \infty)$$

Hence, least positive integral value of $\alpha = 5$.

**33.** *(60)* If $(\mathbf{a}+3\mathbf{b})$ is perpendicular to $(7\mathbf{a}-5\mathbf{b})$, then $(\mathbf{a}+3\mathbf{b})\cdot(7\mathbf{a}-5\mathbf{b})=0$

$\Rightarrow 7|\mathbf{a}|^2-15|\mathbf{b}|^2+16\mathbf{a}\cdot\mathbf{b}=0$ ...(i)

Also, $(\mathbf{a}-4\mathbf{b})$ is perpendicular to $(7\mathbf{a}-2\mathbf{b})$.

$$(\mathbf{a}-4\mathbf{b})\cdot(7\mathbf{a}-2\mathbf{b})=0$$

$\Rightarrow 7|\mathbf{a}|^2+8|\mathbf{b}|^2-30\mathbf{a}\cdot\mathbf{b}=0$ ...(ii)

On adding Eq. (i) × 15 and Eq. (ii) × 8, we get

$105|\mathbf{a}|^2-225|\mathbf{b}|^2+240\,\mathbf{a}\cdot\mathbf{b}+$
$56|\mathbf{a}|^2+64|\mathbf{b}|^2-240\mathbf{a}\cdot\mathbf{b}=0$

$$161|\mathbf{a}|^2-161|\mathbf{b}|^2=0$$

$\Rightarrow |\mathbf{a}|=|\mathbf{b}|$ ...(ii)

From (i), we get

$$7|\mathbf{a}|^2-15|\mathbf{a}|^2+16\mathbf{a}\cdot\mathbf{b}=0$$

$\Rightarrow 16\mathbf{a}\cdot\mathbf{b}=8|\mathbf{a}|^2$

$\Rightarrow 16|\mathbf{a}||\mathbf{b}|\cos\theta=8|\mathbf{a}|^2$

$\Rightarrow \cos\theta=\dfrac{8|\mathbf{a}|^2}{16|\mathbf{a}||\mathbf{b}|}$

$\Rightarrow \cos\theta=\dfrac{1}{2}\Rightarrow\theta=60^\circ$ $[\because|\mathbf{a}|=|\mathbf{b}|]$

**34.** *(6)* Since, $V_2$ is obtained by rotating $V_1$.

$\therefore |V_1|^2=|V_2|^2$

$\Rightarrow 3p^2+1=4+(p+1)^2$

$\Rightarrow 3p^2+1=4+p^2+1+2p$

$\Rightarrow 2p^2-2p-4=0\Rightarrow p^2-p-2=0$

$\Rightarrow (p+1)(p-2)=0\Rightarrow p=-1$

or $p=2$

Since, $p>0$ given, then $p=-1$ is discarded.

Now, $\cos\theta=\dfrac{V_1\cdot V_2}{|V_1||V_2|}$

$$=\frac{4\sqrt{3}+3}{\sqrt{13}\cdot\sqrt{13}}=\frac{4\sqrt{3}+3}{13}$$

$\sin\theta=\sqrt{1-\cos^2\theta}=\sqrt{1-\dfrac{(4\sqrt{3}+3)^2}{(13)^2}}$

$$=\frac{\sqrt{(13)^2-(4\sqrt{3}+3)^2}}{13}$$

$\Rightarrow \sin\theta=\dfrac{\sqrt{112-24\sqrt{3}}}{13}$

$\therefore \tan\theta=\dfrac{\sin\theta}{\cos\theta}=\dfrac{\sqrt{112-24\sqrt{3}}}{4\sqrt{3}+3}$

Given, $\tan\theta=\dfrac{\alpha\sqrt{3}-2}{4\sqrt{3}+3}$

$\therefore \sqrt{112-24\sqrt{3}}=\alpha\sqrt{3}-2$

$\Rightarrow 6\sqrt{3}-2=\alpha\sqrt{3}-2$

$\therefore \alpha=6$

**35.** *(1)* For two non-zero vectors **x** and **y**, it is given

$$|\mathbf{x}+\mathbf{y}|=|\mathbf{x}|$$

On squaring both sides, we get

$$|\mathbf{x}|^2+2\cdot\mathbf{x}\cdot\mathbf{y}+|\mathbf{y}|^2=|\mathbf{x}|^2$$

$\Rightarrow |\mathbf{y}|^2+2\mathbf{x}\cdot\mathbf{y}=0$ ...(i)

Now, as $(2\mathbf{x}+\lambda\mathbf{y})$ is perpendicular to **y**, so

$$(2\mathbf{x}+\lambda\mathbf{y})\cdot\mathbf{y}=0$$

$\Rightarrow 2\mathbf{x}\cdot\mathbf{y}+\lambda|\mathbf{y}|^2=0$ ...(ii)

From Eqs.(i) and (ii), we get

$$\lambda=1$$

**36.** *(2)* Given, for unit vectors **a**, **b** and **c**

$$|\mathbf{a}-\mathbf{b}|^2+|\mathbf{a}-\mathbf{c}|^2=8$$

$\Rightarrow |\mathbf{a}|^2+|\mathbf{b}|^2-2\mathbf{a}\cdot\mathbf{b}+|\mathbf{a}|^2+|\mathbf{c}|^2-2\mathbf{a}\cdot\mathbf{c}=8$

$\Rightarrow -2[\mathbf{a}\cdot\mathbf{b}+\mathbf{a}\cdot\mathbf{c}]=4$

$\Rightarrow \mathbf{a}\cdot\mathbf{b}+\mathbf{a}\cdot\mathbf{c}=-2$ ...(i)

Now, $|\mathbf{a}+2\mathbf{b}|^2+|\mathbf{a}+2\mathbf{c}|^2$

$=|\mathbf{a}|^2+4|\mathbf{b}|^2+4\mathbf{a}\cdot\mathbf{b}+|\mathbf{a}|^2+4|\mathbf{c}|^2+4\mathbf{a}\cdot\mathbf{c}$

$=10+4[\mathbf{a}\cdot\mathbf{b}+\mathbf{a}\cdot\mathbf{c}]$

$=10+4(-2)$ [from Eq. (i)]

$=10-8=2$

## Topic 3

**01.** *(a)* $\because 0<\theta<\dfrac{\pi}{2}$

Angle between **a** and **b** be $\sin^{-1}\left(\dfrac{\sqrt{65}}{9}\right)$

$$\mathbf{c}=3\mathbf{a}+6\mathbf{b}+9(\mathbf{a}\times\mathbf{b})$$

dot product with **a**

$$\mathbf{a}\cdot\mathbf{c}=3+6\,\mathbf{a}\cdot\mathbf{b}$$

$\cos\theta=\dfrac{\sqrt{16}}{9}$ $\left(\because\sin\theta=\dfrac{\sqrt{65}}{9}\right)$

$$=3+6\cdot\frac{\sqrt{16}}{9}=\frac{9+8}{3}=\frac{17}{3}$$

dot product with **b**

$$\mathbf{c}\cdot\mathbf{b}=3\mathbf{a}\cdot\mathbf{b}+6$$

$$\mathbf{c}\cdot\mathbf{b}=3\cdot\frac{\sqrt{16}}{9}+6=\frac{4}{3}+6=\frac{22}{3}$$

$9(\mathbf{c}\cdot\mathbf{a})-3(\mathbf{c}\cdot\mathbf{b})$

$$=9\cdot\left(\frac{17}{3}\right)-3\left(\frac{22}{3}\right)$$

$=51-22$

$=29$

**02.** *(b)* $(\mathbf{a}-\mathbf{c})\times\mathbf{b}=-18\hat{i}-3\hat{j}+12\hat{k}$

$\mathbf{a}\times\mathbf{b}-\mathbf{c}\times\mathbf{b}=-18\hat{i}-3\hat{j}+12\hat{k}$

$\Rightarrow \mathbf{a}\times\mathbf{b}+\hat{\mathbf{d}}=-18\hat{i}-3\hat{j}+12\hat{k}$

$\because \mathbf{b}\times\mathbf{c}=\mathbf{d}$ [given]

Take dot with **a**

$$\mathbf{a}\cdot\mathbf{d}=-36+9+12=-15$$

$\therefore |\mathbf{a}\cdot\mathbf{d}|=15$

**03.** *(d)*

> **key idea** Projection of **a** on **b** is $\mathbf{a}\cdot\hat{\mathbf{b}}=\dfrac{\mathbf{a}\cdot\mathbf{b}}{|\mathbf{b}|}$

Given, $\mathbf{a}=3\hat{i}-\hat{j}+2\hat{k}$

$$\mathbf{b}=\mathbf{a}\times(\hat{i}-2\hat{k}),\ \mathbf{c}=\mathbf{b}\times\hat{k}$$

$$\because \mathbf{b}=\mathbf{a}\times(\hat{i}-2\hat{k})=\begin{vmatrix}\hat{i}&\hat{j}&\hat{k}\\3&-1&2\\1&0&-2\end{vmatrix}$$

$$=2\hat{i}+8\hat{j}+\hat{k}$$

and $\mathbf{c}=\mathbf{b}\times\hat{k}=8\hat{i}-2\hat{j}$

$\therefore \mathbf{c}-2\hat{j}=8\hat{i}-4\hat{j}$

Hence, projection of $(\mathbf{c}-2\hat{j})$ on $\mathbf{a}=(\mathbf{c}-2\hat{j})\cdot\mathbf{a}$

$$=\frac{(8\hat{i}-4\hat{j})\cdot(3\hat{i}-\hat{j}+2\hat{k})}{\sqrt{3^2+(-1)^2+(2)^2}}=\frac{28}{\sqrt{14}}=2\sqrt{14}$$

**04.** *(d)* Area of parallelogram having sides *OA* and *OC*

$=|\mathbf{OA}\times\mathbf{OC}|=|2\mathbf{a}\times3\mathbf{b}|=15$ [given]

$\Rightarrow 6|\mathbf{a}\times\mathbf{b}|=15\Rightarrow|\mathbf{a}\times\mathbf{b}|=\dfrac{5}{2}$ ...(i)

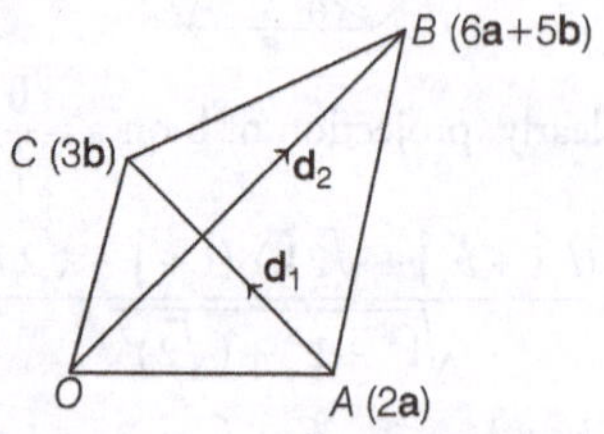


Also, area of quadrilateral *OABC*

$=\dfrac{1}{2}|\mathbf{d}_1\times\mathbf{d}_2|=\dfrac{1}{2}|(\mathbf{AC}\times\mathbf{OB})|$

$=\dfrac{1}{2}|(3\mathbf{b}-2\mathbf{a})\times(6\mathbf{a}+5\mathbf{b})|$

$=\dfrac{1}{2}|18\mathbf{b}\times\mathbf{a}-10\mathbf{a}\times\mathbf{b}|=14|\mathbf{a}\times\mathbf{b}|$

$=14\times\dfrac{5}{2}$ [using Eq. (i)]

$=35$

**05.** (b) $(\mathbf{a}+\mathbf{b})\times\mathbf{c}=\mathbf{c}\times(-2\mathbf{a}+3\mathbf{b})$

$\Rightarrow (\mathbf{a}+\mathbf{b})\times\mathbf{c}-\mathbf{c}\times(-2\mathbf{a}+3\mathbf{b})=0$

$\Rightarrow (\mathbf{a}+\mathbf{b})\times\mathbf{c}+(-2\mathbf{a}+3\mathbf{b})\times\mathbf{c}=0$

$\Rightarrow (\mathbf{a}+\mathbf{b}-2\mathbf{a}+3\mathbf{b})\times\mathbf{c}=0$

$\Rightarrow (4\mathbf{b}-\mathbf{a})\times\mathbf{c}=0$

Therefore, $\mathbf{c}=\lambda(4\mathbf{b}-\mathbf{a})$

$\Rightarrow \mathbf{c}=\lambda(44\hat{i}-4\hat{j}+4\hat{k}-4\hat{i}+\hat{j}-\hat{k})$

$\Rightarrow \mathbf{c}=\lambda(40\hat{i}-3\hat{j}+3\hat{k})$

Now, $(2\mathbf{a}+3\mathbf{b})\cdot\mathbf{c}=1670$

$\Rightarrow (8\hat{i}-2\hat{j}+2\hat{k}+33\hat{i}-3\hat{j}+3\hat{k})\cdot\lambda(40\hat{i}-3\hat{j}+3\hat{k})=1670$

$\Rightarrow \lambda(40\times 41+15+15)=1670$

$\Rightarrow \lambda=\dfrac{1670}{1670}=1$

So, $\mathbf{c}=40\hat{i}-3\hat{j}+3\hat{k}$

$|\mathbf{c}|^2=(40)^2+(-3)^2+(3)^2=1618$

**06.** (a) $|(\mathbf{a}\times\mathbf{b})\times\mathbf{c}|=|\mathbf{a}\times\mathbf{b}||\mathbf{c}|\times\dfrac{\sqrt{3}}{2}$

Now, $|\mathbf{c}-\mathbf{a}|=2\sqrt{2}$

$\Rightarrow |\mathbf{c}|^2+|\mathbf{a}|^2-2\mathbf{c}\cdot\mathbf{a}=8$

$|\mathbf{c}|^2+38-12|\mathbf{c}|=8$

$\Rightarrow |\mathbf{c}|^2-12|\mathbf{c}|+30=0$

$\Rightarrow |\mathbf{c}|=\dfrac{12\pm\sqrt{144-120}}{2}$

$|\mathbf{c}|=\dfrac{12\pm 2\sqrt{6}}{2}$

$|\mathbf{c}|=6+\sqrt{6}$ $\quad[\because d\geq 6]$

$$\mathbf{a}\times\mathbf{b}=\begin{vmatrix}\hat{i}&\hat{j}&\hat{k}\\6&1&-1\\1&1&0\end{vmatrix}=\hat{i}-\hat{j}+5\hat{k}$$

$|\mathbf{a}\times\mathbf{b}|=\sqrt{27}$

$|(\mathbf{a}\times\mathbf{b})\times\mathbf{c}|=\sqrt{27}\,(6+\sqrt{6})\dfrac{\sqrt{3}}{2}$

$=\dfrac{9}{2}(6+\sqrt{6})$

**07.** (a) $|\mathbf{c}-\mathbf{a}|^2=|\mathbf{c}|^2+|\mathbf{a}|^2-2\mathbf{c}\cdot\mathbf{a}$

$=|\mathbf{c}|^2+4-0=|\mathbf{c}|^2+4$

$\because \quad \mathbf{a}=\mathbf{b}\times\mathbf{c}$

$|\mathbf{a}|=|\mathbf{b}\times\mathbf{c}|$

$2=|\mathbf{b}||\mathbf{c}|\sin\alpha$

$2=3|\mathbf{c}|\sin\alpha$

$|\mathbf{c}|=\dfrac{2}{3\sin\alpha}=\dfrac{2}{3}\operatorname{cosec}\alpha$

$\because \quad \alpha\in\left[0,\dfrac{\pi}{3}\right]$

$\operatorname{cosec}\alpha\in\left[\dfrac{2}{\sqrt{3}},\infty\right)$

$|\mathbf{c}|_{\min}=\dfrac{2}{3}\times\dfrac{2}{\sqrt{3}}=\dfrac{4}{3\sqrt{3}}$

$\therefore\ 27|\mathbf{c}-\mathbf{a}|^2_{\min}=27\left(\dfrac{16}{27}+4\right)$

$=16+108=124$

**08.** (d) $\mathbf{a}=2\hat{i}+5\hat{j}-\hat{k}$

$\mathbf{b}=2\hat{i}-2\hat{j}+2\hat{k}$

$\mathbf{a}+\mathbf{b}+\hat{i}=5\hat{i}+3\hat{j}+\hat{k}$

$\therefore$ According to the question,

$(\mathbf{c}+\hat{i})\times(\mathbf{a}+\mathbf{b}+\hat{i})=\mathbf{a}\times(\mathbf{c}+\hat{i})$

$(\mathbf{c}+\hat{i})\times(5\hat{i}+3\hat{j}+\hat{k})$

$=(2\hat{i}+5\hat{j}-\hat{k})\times(\mathbf{c}+\hat{i})$

$(\mathbf{c}+\hat{i})\times(5\hat{i}+3\hat{j}+\hat{k}+2\hat{i}+5\hat{j}-\hat{k})=0$

$(\mathbf{c}+\hat{i})\times(7\hat{i}+8\hat{j})=0$

$(\mathbf{c}+\hat{i})=\lambda(7\hat{i}+8\hat{j})$

$\mathbf{c}=\lambda(7\hat{i}+8\hat{j})-\hat{i}$

$\mathbf{c}=(7\lambda-1)\,\hat{i}+8\lambda\hat{j}$

$\mathbf{a}\cdot\mathbf{c}=-29$

$2(7\lambda-1)+5\times 8\lambda=-29\Rightarrow \lambda=-\dfrac{1}{2}$

$\mathbf{c}\cdot(-2\hat{i}+\hat{j}+\hat{k})=-2(7\lambda-1)+8\lambda(1)$

$=-14\lambda+2+8\lambda$

$=-6\lambda+2=3+2=5$

**09.** (d) We have, vertices of a quadrilateral *ABCD*

$A(1,-1,2)$, $B(5,7,-6)$, $C(3,4,-10)$ and $D(-1,-4,-2)$

$\therefore\ \mathbf{AC}=2\hat{i}+5\hat{j}-12\hat{k}$

and $\mathbf{BD}=-6\hat{i}-11\hat{j}+4\hat{k}$

$\therefore$Area of quadrilateral *ABCD*

$=\dfrac{1}{2}|\mathbf{AC}\times\mathbf{BD}|$

$$\text{Now, }\mathbf{AC}\times\mathbf{BD}=\begin{vmatrix}\hat{i}&\hat{j}&\hat{k}\\2&5&-12\\-6&-11&4\end{vmatrix}$$

$=(20-132)\,\hat{i}-(8-72)\hat{j}+(-22+30)\,\hat{k}$

$=-112\hat{i}+64\hat{j}+8\hat{k}$

$\therefore$Area of quadrilateral *ABCD*

$=\dfrac{1}{2}|112\hat{i}-64\hat{j}-8\hat{k}|=4|14\hat{i}-8\hat{j}-\hat{k}|$

$=4\sqrt{196+64+1}\ =4\sqrt{261}$ sq units

$=12\sqrt{29}$ sq units

**10.** (b) Given, $\mathbf{a}=3\hat{i}+\hat{j}-2\hat{k}$,

$\mathbf{b}=4\hat{i}+\hat{j}+7\hat{k}$

and $\quad \mathbf{c}=\hat{i}-3\hat{j}+4\hat{k}$

$\therefore\quad \mathbf{p}\times\mathbf{b}=\mathbf{c}\times\mathbf{b}$

$\Rightarrow\quad \mathbf{p}\times\mathbf{b}-\mathbf{c}\times\mathbf{b}=0$

$\Rightarrow\quad (\mathbf{p}-\mathbf{c})\times\mathbf{b}=0$

$\therefore\quad \mathbf{p}-\mathbf{c}\,||\,\mathbf{b}$

$\Rightarrow\quad \mathbf{p}-\mathbf{c}=\lambda\mathbf{b}$

$\Rightarrow\quad \mathbf{p}=\mathbf{c}+\lambda\mathbf{b}$

$\mathbf{p}\cdot\mathbf{a}=0$ $\quad$ [given]

$0=\mathbf{c}\cdot\mathbf{a}+\lambda\mathbf{a}\cdot\mathbf{b}$

$\lambda=-\dfrac{\mathbf{a}\cdot\mathbf{c}}{\mathbf{a}\cdot\mathbf{b}}=-\dfrac{(3-3-8)}{12+1-14}$

$=\dfrac{8}{-1}=-8$

$\therefore\ \mathbf{p}=\mathbf{c}-8\mathbf{b}=-31\hat{i}-11\hat{j}-52\hat{k}$

$\mathbf{p}\cdot(\hat{i}-\hat{j}-\hat{k})=-31+11+52=32$

**11.** (d) $|\mathbf{b}|=1$ and $|\mathbf{b}\times\mathbf{a}|=2$

Let $\quad \mathbf{d}=\mathbf{b}\times\mathbf{a}$

Now, for $|(\mathbf{b}\times\mathbf{a})-\mathbf{b}|^2=|\mathbf{d}-\mathbf{b}|^2$

$=|\mathbf{d}|^2+|\mathbf{b}|^2-2\mathbf{d}\cdot\mathbf{b}$

$=|\mathbf{b}\times\mathbf{a}|^2+|\mathbf{b}|^2-2(\mathbf{b}\times\mathbf{a})\cdot\mathbf{b}$

$=4+1-0=5$

**12.** (d) Given, $|\mathbf{a}|=1$, $|\mathbf{b}|=4$ and $\mathbf{a}\cdot\mathbf{b}=2$

and $\mathbf{c}=2(\mathbf{a}\times\mathbf{b})-3\mathbf{b}$

Dot product with $\mathbf{a}$ on both sides

$\mathbf{c}\cdot\mathbf{a}=-6$ $\quad$ ...(i)

Dot product with $\mathbf{b}$ on both sides

$\mathbf{b}\cdot\mathbf{c}=-48$ $\quad$ ...(ii)

$\mathbf{c}\cdot\mathbf{c}=4|\mathbf{a}\times\mathbf{b}|^2+9|\mathbf{b}|^2$

$|\mathbf{c}|^2=4[|\mathbf{a}|^2|\mathbf{b}|^2-(\mathbf{a}\cdot\mathbf{b})^2]+9|\mathbf{b}|^2$

$=4[(1)(4)^2-(4)]+9(16)$

$=4[12]+144=48+144$

$|\mathbf{c}|^2=192$

$\therefore\ \cos\theta=\dfrac{\mathbf{b}\cdot\mathbf{c}}{|\mathbf{b}||\mathbf{c}|}\Rightarrow\cos\theta=\dfrac{-48}{\sqrt{192}\cdot 4}$

$\therefore\quad \theta=\cos^{-1}\left(\dfrac{-\sqrt{3}}{2}\right)$

**13.** (d) Now, find diagonal

$\mathbf{AC}=(-3-2)\hat{i}+(4-3)\hat{j}+(-2-5)\hat{k}$

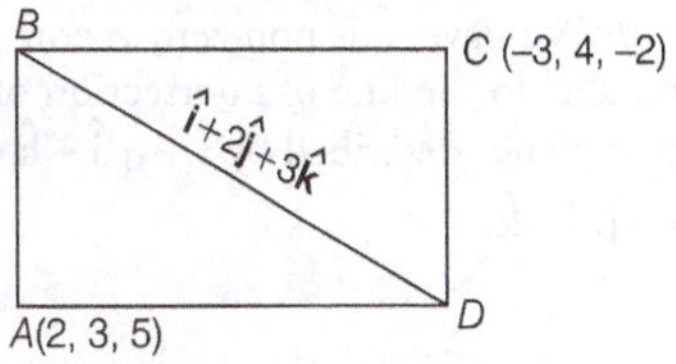

$= -5\hat{i} + \hat{j} - 7\hat{k}$

and $BD = \hat{i} + 2\hat{j} + 3\hat{k}$

Hence, area of parallelogram *ABCD*

$= \frac{1}{2} |AC \times BD|$

Now, $AC \times BD = \begin{vmatrix} \hat{i} & \hat{j} & \hat{k} \\ -5 & 1 & -7 \\ 1 & 2 & 3 \end{vmatrix}$

$= (3+14)\hat{i} - (-15+7)\hat{j} + (-10-1)\hat{k}$

$= 17\hat{i} + 8\hat{j} - 11\hat{k}$

So, $|AC \times BD|$

$= \sqrt{17^2 + 8^2 + 11^2} = \sqrt{474}$

Hence, area of parallelogram is $\frac{1}{2}\sqrt{474}$.

**14.** *(b)* Given, $a = \hat{i} + 2\hat{j} + \hat{k}$

$b = 3(\hat{i} - \hat{j} + \hat{k})$

and $a \times c = b$

$\Rightarrow (a \times c) \cdot b = b \cdot b = |b|^2 = 27$

Now $a \cdot (c \times b) = [a \quad c \quad b]$

$= (a \times c) \cdot b = 27$ ...(i)

Now, $a \cdot b = 3 - 6 + 3 = 0$ ...(ii)

and $a \cdot c = 3$ [given] ...(iii)

Now, $a \cdot ((c \times b) - b - c)$

$= a \cdot (c \times b) - a \cdot b - a \cdot c$

$= 27 - 0 - 3 = 24$

**15.** *(d)* $a = \hat{i} + 4\hat{j} + 2\hat{k}$,

$b = 3\hat{i} - 2\hat{j} + 7\hat{k}$

and $c = 2\hat{i} - \hat{j} + 4\hat{k}$

Also, given $d \cdot a = 24$ and $d \times b = c \times b$, then $(d - c) \times b = 0$

$\Rightarrow d - c = \lambda b$

$\Rightarrow d = c + \lambda b$

Now, $a \cdot d = a \cdot (c + \lambda b) = a \cdot c + \lambda(a \cdot b)$

$= 6 + \lambda(9)$

$\therefore 24 = 6 + 9\lambda$

$\Rightarrow \lambda = 2$

$\therefore d = c + 2b$

$= (2\hat{i} - \hat{j} + 4\hat{k}) + 2(3\hat{i} - 2\hat{j} + 7\hat{k})$

$= (8\hat{i} - 5\hat{j} + 18\hat{k})$

$\therefore |d|^2 = 64 + 25 + 324 = 413$

**16.** *(d)* We have, a is non-zero vector parallel to the line of intersection of the two planes described by $\hat{i} + \hat{j}$; $\hat{i} + \hat{k}$ and $\hat{i} - \hat{j}$; $\hat{j} - \hat{k}$.

Let $x_1$ and $x_2$ are the normal vector to the plane $\hat{i} + \hat{j}$; $\hat{i} + \hat{k}$ and $\hat{i} - \hat{j}$; $\hat{j} - \hat{k}$ respectively

$x_1 = \begin{vmatrix} \hat{i} & \hat{j} & \hat{k} \\ 1 & 1 & 0 \\ 1 & 0 & 1 \end{vmatrix} = \hat{i} - \hat{j} - \hat{k}$

and $x_2 = \begin{vmatrix} \hat{i} & \hat{j} & \hat{k} \\ 1 & -1 & 0 \\ 0 & 1 & -1 \end{vmatrix} = \hat{i} + \hat{j} + \hat{k}$

$a_1 = \lambda |x_1 \times x_2|$

[$\because$ a is parallel to intersection of both planes]

$\lambda = \begin{vmatrix} \hat{i} & \hat{j} & \hat{k} \\ 1 & -1 & -1 \\ 1 & 1 & 1 \end{vmatrix}$

$= \lambda(-2\hat{j} + 2\hat{k})$

$a \cdot b = 6$ [given]

$\lambda(0 + 4 + 2) = 6$

$\lambda = 1$

$\therefore a = -2\hat{j} + 2\hat{k}$

$\cos\theta = \frac{a \cdot b}{|a||b|} = \frac{6}{\sqrt{4+4}\sqrt{4+4+1}}$

$= \frac{1}{\sqrt{2}}$

$\because \theta = \frac{\pi}{4} \Rightarrow a \times b = \begin{vmatrix} \hat{i} & \hat{j} & \hat{k} \\ 0 & -2 & 2 \\ 2 & -2 & 1 \end{vmatrix}$

$= 2\hat{i} + 4\hat{j} + 4\hat{k}$

$|a \times b| = \sqrt{4+16+16} = 6$

Hence, $(\theta, |a \times b|) = \left(\frac{\pi}{4}, 6\right)$

**17.** *(a)* Given, $a = 2\hat{i} + 3\hat{j} + 4\hat{k}$

$b = \hat{i} - 2\hat{j} - 2\hat{k}$

and $c = -\hat{i} + 4\hat{j} + 3\hat{k}$

$b \times c = \begin{vmatrix} \hat{i} & \hat{j} & \hat{k} \\ 1 & -2 & -2 \\ -1 & 4 & 3 \end{vmatrix}$

$= \hat{i}(-6+8) - \hat{j}(3-2) + \hat{k}(4-2)$

$= 2\hat{i} - \hat{j} + 2\hat{k}$

$\therefore d = \lambda(b \times c)$

$\Rightarrow d = \lambda(2\hat{i} - \hat{j} + 2\hat{k})$

Since, $a \cdot d = 18$

$\Rightarrow \lambda(4 - 3 + 8) = 18$

$\Rightarrow \lambda = 2$

$\therefore d = 2(2\hat{i} - \hat{j} + 2\hat{k})$

Now, $|a \times d|^2 = |a|^2 |d|^2 - (a \cdot d)^2$

$= (29)(36) - (18)^2$

$= 1044 - 324 = 720$

**18.** *(d)* Given that $|a| = 1$, $|b| = 4$ and $a \cdot b = 2$

and $c = (2a \times b) - 3b$

Now, $b \cdot c = b \cdot [(2a \times b) - 3b]$

$= b \cdot (2a \times b) - 3b \cdot b = 0 - 3|b|^2$

[$\because b \cdot (2a \times b) = 0$]

$= -3(4)^2$

$= -48$

**19.** *(a)* Given that $a = \hat{i} + 2\hat{k}$

$b = \hat{i} + \hat{j} + \hat{k}$

$c = 7\hat{i} - 3\hat{j} + 4\hat{k}$

$r \times b + b \times c = 0$

and $r \cdot a = 0$

$\Rightarrow r \times b - c \times b = 0$

$\Rightarrow (r - c) \times b = 0$

$\Rightarrow r = c + \lambda b$

$\because r \cdot a = 0$

So, $(c + \lambda b) \cdot a = 0$

$\Rightarrow c \cdot a + \lambda b \cdot a = 0$

$\Rightarrow (8 + 7) + \lambda \cdot 3 = 0$

$\Rightarrow \lambda = -5$

$\Rightarrow r = c - 5b$

Now, $r \cdot c = (c - 5b) \cdot c = c \cdot c - 5b \cdot c$

$= 34$

**20.** *(a)*

**key idea**
(i) The projection of **x** on **y** is $\frac{x \cdot y}{|y|}$
(ii) If $x = x_1\hat{i} + y_1\hat{j} + z_1\hat{k}$
and $y = x_2\hat{i} + y_2\hat{j} + z_2\hat{k}$,
then $x \times y = \begin{vmatrix} \hat{i} & \hat{j} & \hat{k} \\ x_1 & y_1 & z_1 \\ x_2 & y_2 & z_2 \end{vmatrix}$

$a = \alpha\hat{i} + 3\hat{j} - \hat{k}$, $b = 3\hat{i} - \beta\hat{j} + 4\hat{k}$ and $c = \hat{i} + 2\hat{j} - 2\hat{k}$

$$\frac{\mathbf{a}\cdot\mathbf{c}}{|\mathbf{c}|}=\frac{10}{3}$$

$$\Rightarrow \frac{\alpha+6+2}{\sqrt{1+4+4}}=\frac{10}{3}\Rightarrow \alpha=2$$

$$\mathbf{b}\times\mathbf{c}=-6\hat{i}+10\hat{j}+7\hat{k}$$

$$\Rightarrow \begin{vmatrix}\hat{i} & \hat{j} & \hat{k}\\ 3 & -\beta & 4\\ 1 & 2 & -2\end{vmatrix}=-6\hat{i}+10\hat{j}+7\hat{k}$$

$$\Rightarrow (2\beta-8)\hat{i}-(-6-4)\hat{j}+(6+\beta)\hat{k}=-6\hat{i}+10\hat{j}+7\hat{k}$$

$$\therefore\quad 2\beta-8=-6\Rightarrow \beta=1$$

Therefore, $\alpha+\beta=3$.

**21.** *(a)* Given that $|\mathbf{a}|=2, |\mathbf{b}|=5$ and $|\mathbf{a}\times\mathbf{b}|=8$

$$\Rightarrow |\mathbf{a}||\mathbf{b}|\sin\theta=\pm 8$$

$$\Rightarrow 10\sin\theta=\pm 8$$

$$\Rightarrow \sin\theta=\pm\frac{4}{5}$$

So, $\cos\theta=\pm\frac{3}{5}$ $(\because \sin^2\theta+\cos^2\theta=1)$

Now, $\mathbf{a}\cdot\mathbf{b}=|\mathbf{a}||\mathbf{b}|\cos\theta$

$$=2\times5\times\left(\pm\frac{3}{5}\right)$$

$$\Rightarrow \mathbf{a}\cdot\mathbf{b}=\pm6;\ \therefore |\mathbf{a}\cdot\mathbf{b}|=6$$

**22.** *(b)* Given vectors are

$$\mathbf{a}=3\hat{i}+2\hat{j}+2\hat{k}$$

and $$\mathbf{b}=\hat{i}+2\hat{j}-2\hat{k}$$

Now, vectors $\mathbf{a}+\mathbf{b}=4\hat{i}+4\hat{j}$

and $\mathbf{a}-\mathbf{b}=2\hat{i}+4\hat{k}$

$\therefore$ A vector which is perpendicular to both the vectors $\mathbf{a}+\mathbf{b}$ and $\mathbf{a}-\mathbf{b}$ is

$$(\mathbf{a}+\mathbf{b})\times(\mathbf{a}-\mathbf{b})=\begin{vmatrix}\hat{i} & \hat{j} & \hat{k}\\ 4 & 4 & 0\\ 2 & 0 & 4\end{vmatrix}$$

$$=\hat{i}(16)-\hat{j}(16)+\hat{k}(-8)$$

$$=8(2\hat{i}-2\hat{j}-\hat{k})$$

Then, the required vector along $(\mathbf{a}+\mathbf{b})\times(\mathbf{a}-\mathbf{b})$ having magnitude 12 is

$$\pm12\times\frac{8(2\hat{i}-2\hat{j}-\hat{k})}{8\times\sqrt{4+4+1}}$$

$$=\pm4(2\hat{i}-2\hat{j}-\hat{k})$$

**23.** *(c)* Given, $\mathbf{a}\cdot\mathbf{b}\neq0,\ \mathbf{a}\cdot\mathbf{b}=0$ ...(i)

and $$\mathbf{b}\times\mathbf{c}=\mathbf{b}\times\mathbf{d}$$

$$\Rightarrow \mathbf{b}\times(\mathbf{c}-\mathbf{d})=0$$

$$\therefore \mathbf{b}\,||\,(\mathbf{c}-\mathbf{d})$$

$$\Rightarrow \mathbf{c}-\mathbf{d}=\lambda\mathbf{b}$$

$$\Rightarrow \mathbf{d}=\mathbf{c}-\lambda\mathbf{b} \quad \text{...(ii)}$$

On taking dot product with $\mathbf{a}$, we get

$$\mathbf{a}\cdot\mathbf{d}=\mathbf{a}\cdot\mathbf{c}-\lambda\mathbf{a}\cdot\mathbf{b}$$

$$\Rightarrow 0=\mathbf{a}\cdot\mathbf{c}-\lambda(\mathbf{a}\cdot\mathbf{b})$$

$$\therefore \lambda=\frac{\mathbf{a}\cdot\mathbf{c}}{\mathbf{a}\cdot\mathbf{b}} \quad \text{...(iii)}$$

$$\therefore \mathbf{d}=\mathbf{c}-\left(\frac{\mathbf{a}\cdot\mathbf{c}}{\mathbf{a}\cdot\mathbf{b}}\right)\mathbf{b}$$

**24.** *(d)* Since, $(2\mathbf{u}\times3\mathbf{v})$ is a unit vector.

$$\Rightarrow |2\mathbf{u}\times3\mathbf{v}|=1$$

$$\Rightarrow 6|\mathbf{u}||\mathbf{v}||\sin\theta|=1$$

$$\Rightarrow \sin\theta=\frac{1}{6} \quad [\because |\mathbf{u}|=|\mathbf{v}|=1]$$

Since, $\theta$ is an acute angle, then there is exactly one value of $\theta$ for which $(2\mathbf{u}\times3\mathbf{v})$ is a unit vector.

**25.** *(b)* Let $\mathbf{a}=a_1\hat{i}+a_2\hat{j}+a_3\hat{k}$

Then, $$\mathbf{a}\times\hat{i}=-a_2\hat{k}+a_3\hat{j}$$

$$\mathbf{a}\times\hat{j}=a_1\hat{k}-a_3\hat{i}$$

$$\mathbf{a}\times\mathbf{k}=-a_1\mathbf{j}+a_2\mathbf{i}$$

$$\therefore (\mathbf{a}\times\mathbf{i})^2+(\mathbf{a}\times\mathbf{j})^2+(\mathbf{a}\times\mathbf{k})^2$$

$$=a_2^2+a_3^2+a_1^2+a_3^2+a_1^2+a_2^2$$

$$=2(a_1^2+a_2^2+a_3^2)=2a^2$$

**26.** *(c)* If possible say $\mathbf{a}+\mathbf{b}+\mathbf{c}=0$

$$\mathbf{a}\times(\mathbf{a}+\mathbf{b}+\mathbf{c})$$

$$\Rightarrow \mathbf{a}\times\mathbf{b}+\mathbf{a}\times\mathbf{c}\Rightarrow \mathbf{a}\times\mathbf{b}=\mathbf{c}\times\mathbf{a}$$

Now, $\mathbf{b}\times(\mathbf{a}+\mathbf{b}+\mathbf{c})=\mathbf{b}\times\mathbf{a}$

$$\Rightarrow \mathbf{b}\times\mathbf{a}+\mathbf{b}\times\mathbf{c}=0$$

$$\Rightarrow \mathbf{a}\times\mathbf{b}=\mathbf{b}\times\mathbf{c}$$

**27.** *(b)* Since, $\mathbf{a}+\mathbf{b}+\mathbf{c}=0$

$$\Rightarrow \mathbf{a}+\mathbf{b}=-\mathbf{c}$$

$$\Rightarrow (\mathbf{a}+\mathbf{b})\times\mathbf{c}=-\mathbf{c}\times\mathbf{c}$$

$$\Rightarrow \mathbf{b}\times\mathbf{c}=\mathbf{c}\times\mathbf{a}$$

Similarly, $\mathbf{a}\times\mathbf{b}=\mathbf{b}\times\mathbf{c}$

Hence, $\mathbf{a}\times\mathbf{b}=\mathbf{b}\times\mathbf{c}=\mathbf{c}\times\mathbf{a}$

**28.** *(a)* Since, the vectors $\mathbf{a}=x\hat{i}+y\hat{j}+z\hat{k}$ and $\mathbf{b}=\hat{j}$ are such that $\mathbf{a}$, $\mathbf{c}$ and $\mathbf{b}$ form a right handed system.

$$\therefore \mathbf{c}=\mathbf{b}\times\mathbf{a}=\begin{vmatrix}\hat{i} & \hat{j} & \hat{k}\\ 0 & 1 & 0\\ x & y & z\end{vmatrix}=z\hat{i}-x\hat{k}$$

**29.** *(b)* Using fact :

$$(\mathbf{a}\times\mathbf{b})^2=\mathbf{a}^2\mathbf{b}^2-(\mathbf{a}\cdot\mathbf{b})^2$$

$$=\mathbf{a}^2\mathbf{b}^2-\mathbf{a}^2\mathbf{b}^2\cos^2\theta=\mathbf{a}^2\mathbf{b}^2\sin^2\theta$$

$$=64\times\sin^2\frac{\pi}{6}=64\times\frac{1}{4}=16$$

**30.** *(d)* $ar(\Delta ABC)=5$

$$\frac{1}{2}\times bc=5$$

$$\Rightarrow bc=10$$

$$ar(\Delta ACD)=6$$

$$=\frac{1}{2}\times cd=6\Rightarrow cd=12$$

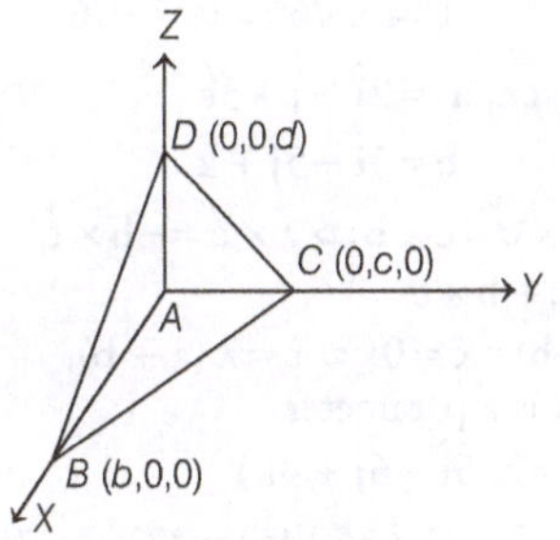


$$ar(\Delta ABD)=7$$

$$\Rightarrow bd=14$$

$$\text{Area}(\Delta BCD)=\frac{1}{2}(\mathbf{BC}\times\mathbf{BD})$$

$$=<-b,c,0>$$

$$=<-b,0,d>$$

$$|\mathbf{BC}\times\mathbf{BD}|=\sqrt{c^2d^2+b^2d^2+b^2c^2}$$

$$=\sqrt{440}$$

$$ar(\Delta BCD)=\sqrt{110}$$

**31.** *(b)*

> **key idea** The angle $\theta$ between $\mathbf{a}$ and $\mathbf{b}$ is given by
> $$\cos\theta=\frac{\mathbf{a}\cdot\mathbf{b}}{|\mathbf{a}||\mathbf{b}|}$$

Let $\mathbf{P}=\hat{i}+\hat{j}+\hat{k}$

and $$\mathbf{b}\times\mathbf{c}=\begin{vmatrix}\hat{i} & \hat{j} & \hat{k}\\ 1 & -2 & 3\\ 2 & 3 & -1\end{vmatrix}$$

$$=-7\hat{i}+7\hat{j}+7\hat{k}=-7(\hat{i}-\hat{j}-\hat{k})$$

Now, $$\mathbf{a}=\frac{\hat{i}-\hat{j}-\hat{k}}{\sqrt{3}} \text{ or } \frac{-\hat{i}+\hat{j}+\hat{k}}{\sqrt{3}}$$

$$\cos\theta=\frac{\mathbf{a}\cdot\mathbf{p}}{|\mathbf{p}|}=\frac{1-1-1}{\sqrt{3}\sqrt{3}}=-\frac{1}{3}$$

or $$\cos\theta=\frac{-1+1+1}{3}=\frac{-1}{3}=\frac{1}{3}$$

(rejected)

$$\Rightarrow \mathbf{a}=\frac{\hat{i}-\hat{j}-\hat{k}}{\sqrt{3}}$$

Now, $$\cos\frac{\pi}{3}=\frac{\mathbf{a}\cdot(\hat{i}+\alpha\hat{j}+\hat{k})}{\sqrt{1+\alpha^2+1}}$$

$\Rightarrow \quad \frac{1}{2}=\frac{1-\alpha-1}{\sqrt{3}\sqrt{\alpha^2+2}}$

$\Rightarrow \quad \frac{\sqrt{3}}{2}\sqrt{\alpha^2+2}=-\alpha \qquad [\because \alpha<0]$

On squaring both sides, we get

$3\alpha^2+6=4\alpha^2$

$\Rightarrow \quad \alpha=\pm\sqrt{6}\Rightarrow \alpha=-\sqrt{6}$

**32.** *(b)* Given, $\mathbf{a}=2\hat{\mathbf{i}}-\hat{\mathbf{j}}+3\hat{\mathbf{k}}$

and $\quad \mathbf{b}=3\hat{\mathbf{i}}-5\hat{\mathbf{j}}+\hat{\mathbf{k}}$

Also, $\mathbf{a}\times\mathbf{c}=\mathbf{c}\times\mathbf{b}\Rightarrow \mathbf{a}\times\mathbf{c}=-\mathbf{b}\times\mathbf{c}$

$\Rightarrow \mathbf{a}\times\mathbf{c}+\mathbf{b}\times\mathbf{c}=0$

$\Rightarrow (\mathbf{a}+\mathbf{b})\times\mathbf{c}=\mathbf{0}\Rightarrow \mathbf{c}=\lambda(\mathbf{a}+\mathbf{b})$,

where $\lambda$ is a parameter

$\Rightarrow \quad \mathbf{c}=\lambda(5\hat{\mathbf{i}}-6\hat{\mathbf{j}}+4\hat{\mathbf{k}}) \qquad \ldots(i)$

$\Rightarrow |\mathbf{c}|^2=\lambda^2(25+36+16)=77\lambda^2 \qquad \ldots(ii)$

Also, given, $(\mathbf{a}+\mathbf{c})\cdot(\mathbf{b}+\mathbf{c})=168$

$\Rightarrow \mathbf{a}\cdot\mathbf{b}+\mathbf{a}\cdot\mathbf{c}+\mathbf{c}\cdot\mathbf{b}+|\mathbf{c}|^2=168$

$\Rightarrow 14+\mathbf{c}\cdot(\mathbf{a}+\mathbf{b})+77\lambda^2=168$

[using Eq. (ii), $\mathbf{a}\cdot\mathbf{b}=14$]

$\Rightarrow \lambda|5\hat{\mathbf{i}}-6\hat{\mathbf{j}}+4\hat{\mathbf{k}}|^2+77\lambda^2=154$

[(using Eq. (i))]

$\Rightarrow 77\lambda+77\lambda^2-154=0$

$\Rightarrow \quad \lambda^2+\lambda-2=0$

$\Rightarrow \quad \lambda=-2, 1$

$\therefore$ Maximum value of $|\mathbf{c}|^2$ occurs when $\lambda=-2$

$\therefore \quad (|\mathbf{c}|^2)_{\max}=77\times(-2)^2$

$=77\times4=308$

**33.** *(c)*

> **key idea** Three vectors **a**, **b** and **c** are coplanar iff $\mathbf{a}\cdot(\mathbf{b}\times\mathbf{c})=0$

Given, $\mathbf{a}=\hat{\mathbf{i}}+2\hat{\mathbf{j}}+3\hat{\mathbf{k}}$

$\mathbf{b}=3\hat{\mathbf{i}}+\hat{\mathbf{j}}-\hat{\mathbf{k}}$

and **c** is coplanar with **a** and **b**

$\therefore [\mathbf{c}\ \mathbf{a}\ \mathbf{b}]=0$

$\Rightarrow \begin{vmatrix} c_1 & c_2 & c_3\\ 1 & 2 & 3\\ 3 & 1 & -1\end{vmatrix}=0,$

where $\mathbf{c}=c_1\hat{\mathbf{i}}+c_2\hat{\mathbf{j}}+c_3\hat{\mathbf{k}}$ (say)

$\Rightarrow \quad c_1-2c_2+c_3=0 \qquad \ldots(i)$

Also, $\mathbf{a}\cdot\mathbf{c}=5$ (given)

$\Rightarrow \quad c_1+2c_2+3c_3=5 \qquad \ldots(ii)$

and $\quad \mathbf{b}\cdot\mathbf{c}=0$ (given)

$\Rightarrow \quad 3c_1+c_2-c_3=0 \qquad \ldots(iii)$

On solving Eqs. (i), (ii) and (iii), we get

$c_1=\frac{1}{6}, c_2=\frac{4}{6}, c_3=\frac{7}{6}$

$\therefore \quad |\mathbf{c}|=\sqrt{\frac{1}{36}+\frac{16}{36}+\frac{49}{36}}$

$(\because |\mathbf{c}|=\sqrt{c_1^2+c_2^2+c_3^2})$

$=\sqrt{\frac{\pi}{6}}$

**34.** *(b)*

> **key idea** Use area of parallelogram in terms of diagonalsis $=\frac{1}{2}|\mathbf{d}_1\times\mathbf{d}_2|$
>
> where $\mathbf{d}_1$ and $\mathbf{d}_2$ are diagonals of the parallelogram.

$A=\frac{1}{2}|(\mathbf{a}+\mathbf{b})\times(\mathbf{b}+\mathbf{c})|=\frac{\sqrt{21}}{2}$

$\Rightarrow \quad |(\mathbf{a}+\mathbf{b})\times(\mathbf{b}+\mathbf{c})|=\sqrt{21}$

Now, $\quad \mathbf{a}+\mathbf{b}=\hat{\mathbf{i}}+\alpha\hat{\mathbf{j}}+2\hat{\mathbf{k}}$

$\mathbf{b}+\mathbf{c}=-\hat{\mathbf{i}}+\beta\hat{\mathbf{j}}$

$(\mathbf{a}+\mathbf{b})\times(\mathbf{b}+\mathbf{c})=\begin{vmatrix}\hat{\mathbf{i}} & \hat{\mathbf{j}} & \hat{\mathbf{k}}\\ 1 & \alpha & 2\\ -1 & \beta & 0\end{vmatrix}$

$=\hat{\mathbf{i}}(-2\beta)-\hat{\mathbf{j}}(2)+\hat{\mathbf{k}}(\beta+\alpha)$

$\therefore |(\mathbf{a}+\mathbf{b})\times(\mathbf{b}+\mathbf{c})|$

$=\sqrt{4\beta^2+4+(\alpha+\beta)^2}=\sqrt{21}$

$\Rightarrow 4\beta^2+4+\alpha^2+\beta^2+2\alpha\beta=21$

$\Rightarrow \quad \alpha^2+5\beta^2=29 \quad [\because \alpha\beta=-6]$

$\therefore$ $\alpha$ and $\beta$ are integers

$\alpha=-3, \beta=2$ or $\alpha=3, \beta=-2$

$(\alpha_1,\beta_1)=(-3,2), (\alpha_2,\beta_2)=(3,-2)$

$\therefore \alpha_1^2+\beta_1^2-\alpha_2\beta_2=9+4+6=19$

**35.** *(d)* Given, $\mathbf{a}=\alpha\hat{\mathbf{i}}+4\hat{\mathbf{j}}+2\hat{\mathbf{k}}$

$\mathbf{b}=5\hat{\mathbf{i}}+3\hat{\mathbf{j}}+4\hat{\mathbf{k}}$, $\mathbf{c}=x\hat{\mathbf{i}}+y\hat{\mathbf{j}}+z\hat{\mathbf{k}}$

and $\mathbf{c}=\mathbf{a}-\mathbf{b}=(\alpha-5)\hat{\mathbf{i}}+\hat{\mathbf{j}}-2\hat{\mathbf{k}}$

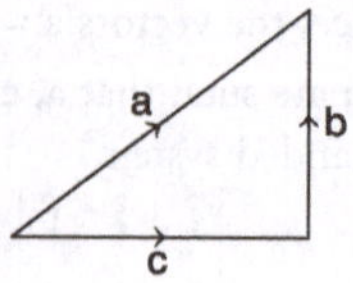


So, $x=\alpha-5, y=1, z=-2 \qquad \ldots(i)$

Also, given area of $\Delta=5\sqrt{6}$

$\Rightarrow \quad \frac{1}{2}|\mathbf{a}\times\mathbf{c}|=5\sqrt{6}$

$\Rightarrow \left\|\begin{matrix}\hat{\mathbf{i}} & \hat{\mathbf{j}} & \hat{\mathbf{k}}\\ \alpha & 4 & 2\\ x & y & z\end{matrix}\right\|=10\sqrt{6}$

$\Rightarrow \left\|\begin{matrix}\hat{\mathbf{i}} & \hat{\mathbf{j}} & \hat{\mathbf{k}}\\ \alpha & 4 & 2\\ \alpha-5 & 1 & -2\end{matrix}\right\|=10\sqrt{6}$

$\Rightarrow |-10\hat{\mathbf{i}}-\hat{\mathbf{j}}(-2\alpha-2\alpha+10)+\hat{\mathbf{k}}(\alpha-4\alpha+20)|$

$=10\sqrt{6}$

$\Rightarrow \sqrt{(-10)^2+(2\alpha+2\alpha-10)^2+(\alpha-4\alpha+20)^2}$

$=10\sqrt{6}$

$\Rightarrow (4\alpha-10)^2+(20-3\alpha)^2$

$=600-100=500$

$\Rightarrow 25\alpha^2-80\alpha-120\alpha=0$

$\Rightarrow \alpha(25\alpha-200)=0\Rightarrow \alpha=0, 8$

But $\alpha>0$ [given]

$\therefore \quad \alpha=0$ [rejected]

$\therefore \quad \alpha=8$

$\Rightarrow \quad x=\alpha-5=3$

Hence, $|\mathbf{c}|^2=x^2+y^2+z^2$

$=9+1+4=14$

**36.** *(c)*

> **key idea** Firstly find the vector *b*, then use formula for projection of **a** on **b** $=\frac{\mathbf{a}\cdot\mathbf{b}}{|\mathbf{b}|}$

According to the question,

$\mathbf{a}=2\hat{\mathbf{i}}+\hat{\mathbf{j}}-\hat{\mathbf{k}}$

and $\mathbf{b}=((\mathbf{a}\times(\hat{\mathbf{i}}+\hat{\mathbf{j}}))\times\hat{\mathbf{i}})\times\hat{\mathbf{i}}$

$\Rightarrow \mathbf{a}\times(\hat{\mathbf{i}}+\hat{\mathbf{j}})=\begin{vmatrix}\hat{\mathbf{i}} & \hat{\mathbf{j}} & \hat{\mathbf{k}}\\ 2 & 1 & -1\\ 1 & 1 & 0\end{vmatrix}=\hat{\mathbf{i}}-\hat{\mathbf{j}}+\hat{\mathbf{k}}$

$\Rightarrow ((\mathbf{a}\times(\hat{\mathbf{i}}+\hat{\mathbf{j}}))\times\hat{\mathbf{i}})$

$=\begin{vmatrix}\hat{\mathbf{i}} & \hat{\mathbf{j}} & \hat{\mathbf{k}}\\ 1 & -1 & 1\\ 1 & 0 & 0\end{vmatrix}=\hat{\mathbf{j}}+\hat{\mathbf{k}}$

$\Rightarrow ((\mathbf{a}\times(\hat{\mathbf{i}}+\hat{\mathbf{j}}))\times\hat{\mathbf{i}})\times\hat{\mathbf{i}}$

$=\begin{vmatrix}\hat{\mathbf{i}} & \hat{\mathbf{j}} & \hat{\mathbf{k}}\\ 0 & 1 & 1\\ 1 & 0 & 0\end{vmatrix}=\hat{\mathbf{j}}-\hat{\mathbf{k}}$

$\therefore \quad \mathbf{b}=\hat{\mathbf{j}}-\hat{\mathbf{k}}$

Now, projection of **a** on $\mathbf{b}=\frac{\mathbf{a}\cdot\mathbf{b}}{|\mathbf{b}|}$

$=\frac{(2\hat{\mathbf{i}}+\hat{\mathbf{j}}-\hat{\mathbf{k}})\cdot(\hat{\mathbf{j}}-\hat{\mathbf{k}})}{\sqrt{1+1}}=\frac{1+1}{\sqrt{2}}=\sqrt{2}$

So, square of projection **a** on $\mathbf{b}=2$

**37.** *(d)* According to the question,

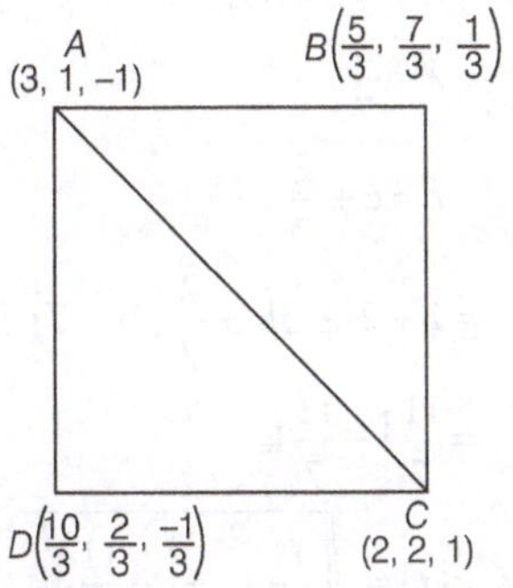


Area of quadrilateral $ABCD = \frac{1}{2} \times$ area of $\Delta ADC + \frac{1}{2} \times$ area of $\Delta ABC$

Area of $\Delta ADC = \frac{1}{2}|\mathbf{AD} \times \mathbf{AC}|$

$$= \frac{1}{2}\begin{vmatrix} \hat{i} & \hat{j} & \hat{k} \\ \frac{1}{3} & -\frac{1}{3} & \frac{2}{3} \\ -1 & 1 & 2 \end{vmatrix}$$

$$= \frac{1}{2}\left| -\frac{4}{3}\hat{i} - \frac{4}{3}\hat{j} + 0\hat{k} \right|$$

$$= \frac{1}{2} \times \sqrt{2 \times \frac{16}{9}}$$

$$= \frac{1}{2} \times \frac{4}{3}\sqrt{2} = \frac{2}{3}\sqrt{2}$$

Area of $\Delta ABC = \frac{1}{2}|\mathbf{AB} \times \mathbf{AC}|$

$$= \frac{1}{2}\begin{vmatrix} \hat{i} & \hat{j} & \hat{k} \\ -\frac{4}{3} & \frac{4}{3} & \frac{4}{3} \\ -1 & 1 & 2 \end{vmatrix}$$

$$= \frac{1}{2}\left| \frac{4}{3}\hat{i} - \frac{4}{3}\hat{j} + 0\hat{k} \right|$$

$$= \frac{1}{2}\sqrt{\frac{16}{9} + \frac{16}{9}}$$

$$= \frac{1}{2}\sqrt{\frac{2 \times 16}{9}}$$

$$= \frac{1}{2} \times \frac{4}{3}\sqrt{2}$$

$$= \frac{2}{3}\sqrt{2}$$

Area of $ABCD = \frac{2}{3}\sqrt{2} + \frac{2}{3}\sqrt{2}$

$$= \frac{4}{3}\sqrt{2}$$

**Alternate Solution**

Clearly, *ABCD* is a parallelogram.

∴ Area of quadrilateral

$$\mathbf{AB} \times \mathbf{AD} = \begin{vmatrix} \hat{i} & \hat{j} & \hat{k} \\ -4/3 & 4/3 & 4/3 \\ 1/3 & -1/3 & 2/3 \end{vmatrix}$$

Now, $|\mathbf{AB} \times \mathbf{AD}| = \left|\frac{1}{9}(12\hat{i} - 12\hat{j})\right|$

$$= \frac{12\sqrt{2}}{9} = \frac{4\sqrt{2}}{3}$$

**38.** *(b)* $\mathbf{a} = \hat{i} + \hat{j} + \hat{k}$,

$\mathbf{b} = 2\hat{i} + 4\hat{j} - 5\hat{k}$

and $\mathbf{c} = x\hat{i} + 2\hat{j} + 3\hat{k}$

$\mathbf{d}$ = unit vector in the direction of $\mathbf{b} + \mathbf{c}$

$\mathbf{d} = \lambda(\mathbf{b} + \mathbf{c})$

$\Rightarrow \quad |\mathbf{d}| = 1$

Now, $\quad \mathbf{a} \cdot \mathbf{d} = 1$

$\mathbf{a} \cdot \lambda(\mathbf{b} + \mathbf{c}) = 1$

$\Rightarrow \lambda(\mathbf{a} \cdot \mathbf{b} + \mathbf{a} \cdot \mathbf{c}) = 1$

$\Rightarrow \lambda(2 + 4 - 5 + x + 2 + 3) = 1$

$\Rightarrow \quad \lambda(x + 6) = 1$

$\because \quad |\mathbf{d}| = 1$

$|\lambda(\mathbf{b} + \mathbf{c})| = 1$

$\Rightarrow \quad \lambda^2(\mathbf{b} + \mathbf{c})^2 = 1$

$\lambda^2((x + 2)^2 + 6^2 + 2^2) = 1$

$\Rightarrow x^2 + 4x + 4 + 36 + 4 = \frac{1}{\lambda^2}$

$\Rightarrow \quad x^2 + 4x + 44 = (x + 6)^2$

$\Rightarrow \quad 4x + 44 = 12x + 36$

$\Rightarrow \quad 8x = 8$

$x = 1 \Rightarrow \lambda = 1/7$

$\therefore \quad \mathbf{c} = \hat{i} + 2\hat{j} + \hat{k}$

$$\therefore (\mathbf{a} \times \mathbf{b}) \cdot \mathbf{c} = \begin{vmatrix} 1 & 1 & 1 \\ 2 & 4 & -5 \\ 1 & 2 & 3 \end{vmatrix}$$

$= 1(12 + 10) - 1(6 + 5) + 1(4 - 4)$

$= 22 - 6 - 5 = 11$

**39.** *(b)* Given, $\mathbf{a} = -5\hat{i} + \hat{j} - 3\hat{k}$,

$\mathbf{b} = \hat{i} + 2\hat{j} - 4\hat{k}$

and $\quad \mathbf{c} = (((\mathbf{a} \times \mathbf{b}) \times \hat{i}) \times \hat{i}) \times \hat{i}$

$$\mathbf{a} \times \mathbf{b} = \begin{vmatrix} \hat{i} & \hat{j} & \hat{k} \\ -5 & 1 & -3 \\ 1 & 2 & -4 \end{vmatrix}$$

$= (-4 + 6)\,\hat{i} - (20 + 3)\,\hat{j} + (-10 - 1)\,\hat{k}$

$\mathbf{a} \times \mathbf{b} = 2\hat{i} - 23\hat{j} - 11\hat{k}$

$$(\mathbf{a} \times \mathbf{b}) \times \hat{i} = \begin{vmatrix} \hat{i} & \hat{j} & \hat{k} \\ 2 & -23 & -11 \\ 1 & 0 & 0 \end{vmatrix}$$

$= -11\hat{j} + 23\hat{k}$

$$((\mathbf{a} \times \mathbf{b}) \times \hat{i}) \times \hat{i} = \begin{vmatrix} \hat{i} & \hat{j} & \hat{k} \\ 0 & -11 & 23 \\ 1 & 0 & 0 \end{vmatrix}$$

$= 23\hat{j} + 11\hat{k}$

$$(((\mathbf{a} \times \mathbf{b}) \times \hat{i}) \times \hat{i}) \times \hat{i} = \begin{vmatrix} \hat{i} & \hat{j} & \hat{k} \\ 0 & 23 & 11 \\ 1 & 0 & 0 \end{vmatrix}$$

$= 11\hat{j} - 23\hat{k}$

$\therefore \quad \mathbf{c} = 11\hat{j} - 23\hat{k}$

$\mathbf{c} \cdot (-\hat{i} + \hat{j} + \hat{k}) = (11\hat{j} - 23\hat{k}) \cdot (-\hat{i} + \hat{j} + \hat{k})$

$= 11 - 23 = -12$

**40.** *(a)* $\quad \mathbf{OA} = \mathbf{a}$, $\mathbf{OB} = 12\mathbf{a} + 4\mathbf{b}$

and $\quad \mathbf{OC} = \mathbf{b}$

$\because$ *S* is parallelogram with adjacent sides *OA* and *OC*.

Area of quadrilateral

$OABC = \frac{1}{2}|\mathbf{d}_1 \times \mathbf{d}_2|$

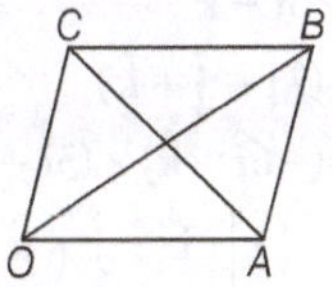


$= \frac{1}{2}|\mathbf{OB} \times \mathbf{AC}|$

$= \frac{1}{2}|(12\mathbf{a} + 4\mathbf{b}) \times (\mathbf{b} - \mathbf{a})|$

$= \frac{1}{2}|12(\mathbf{a} \times \mathbf{b}) - 4(\mathbf{b} \times \mathbf{a})|$

$= \frac{1}{2}|16(\mathbf{a} \times \mathbf{b})|$

$= 8|\mathbf{a} \times \mathbf{b}|$

Area of $S = |\mathbf{a} \times \mathbf{b}|$

$\therefore \frac{\text{Area of quadrilateral } OABC}{\text{Area of } S} = 8$

**41.** *(b)* Given that $|\mathbf{a}| = 2$ and $|\mathbf{b}| = 3$ and angle between the vector $\mathbf{a}$ and $\mathbf{b}$ be $\frac{\pi}{4}$

$\therefore |(\mathbf{a} + 2\mathbf{b}) \times (2\mathbf{a} - 3\mathbf{b})|^2$

$= |-3(\mathbf{a} \times \mathbf{b}) + 4(\mathbf{b} \times \mathbf{a})|^2$

$= |-3(\mathbf{a} \times \mathbf{b}) - 4(\mathbf{a} \times \mathbf{b})|^2$

$$=\left|-7|\mathbf{a}|\times|\mathbf{b}|\times\sin\left(\frac{\pi}{4}\right)\right|^2$$

$$=\left|-7\times2\times3\times\frac{1}{\sqrt{2}}\right|^2$$

$$=49\times4\times9\times\frac{1}{2}=882$$

**42.** *(b)* We have, $(\mathbf{a}+\mathbf{b}+\mathbf{c})\times\mathbf{c}=0$

$\Rightarrow (\mathbf{a}+\mathbf{b})\times\mathbf{c}+\mathbf{c}\times\mathbf{c}=0$

$\Rightarrow \quad (\mathbf{a}+\mathbf{b})\times\mathbf{c}=0 \; (\because \mathbf{c}\times\mathbf{c}=0)$

$\Rightarrow \quad \mathbf{c}=k(\mathbf{a}+\mathbf{b})$

$=k((\lambda\hat{i}+\hat{j}-\hat{k})+(3\hat{i}-\hat{j}+2\hat{k}))$

$=k((\lambda+3)\hat{i}+\hat{k}$ ...(i)

Now, $\mathbf{a}\cdot\mathbf{c}=-17$

$\Rightarrow \lambda k(\lambda+3)-k=-17$ ...(ii)

$\mathbf{b}\cdot\mathbf{c}=-20$

$\Rightarrow 3k(\lambda+3)+2k=-20$

$\Rightarrow \lambda=\dfrac{-20-11k}{3k}$ ...(iii)

On putting in Eq. (ii),

$13k^2+413k+400=0$

$\Rightarrow \quad k=-1,-\dfrac{400}{13}\Rightarrow\lambda=3,\dfrac{-207}{60}$

$\because \quad \lambda\in Z\Rightarrow\lambda=3$ and $k=-1$

$\therefore \quad \mathbf{c}=-6\hat{i}-\hat{k}$

Now, $\mathbf{c}\times(\lambda\hat{i}+\hat{j}+\hat{k})$

$=(-6\hat{i}-\hat{k})\times(3\hat{i}+\hat{j}+\hat{k})$

$$=\begin{vmatrix}\hat{i} & \hat{j} & \hat{k}\\ -6 & 0 & -1\\ 3 & 1 & 1\end{vmatrix}$$

$=\hat{i}+3\hat{j}-6\hat{k}$

$\therefore \left|\mathbf{c}\times(\lambda\hat{i}+\hat{j}+\hat{k})\right|^2=\left|\hat{i}+3\hat{j}-6\hat{k}\right|^2$

$=1^2+3^2+(-6)^2=46$

**43.** *(b)* Given $\mathbf{OP}=-\hat{i}-2\hat{j}+3\hat{k}$

$\mathbf{OA}=-2\hat{i}+\hat{j}-3\hat{k}$

$\mathbf{OB}=2\hat{i}+4\hat{j}-2\hat{k}$

$\mathbf{OC}=-4\hat{i}+2\hat{j}-\hat{k}$

Now, $\mathbf{AB}=\mathbf{OB}-\mathbf{OA}$

$=(2\hat{i}+4\hat{j}-2\hat{k})-(-2\hat{i}+\hat{j}-3\hat{k})$

$=4\hat{i}+3\hat{j}+\hat{k}$

$\mathbf{AC}=\mathbf{OC}-\mathbf{OA}$

$=(-4\hat{i}+2\hat{j}-\hat{k})-(-2\hat{i}+\hat{j}-3\hat{k})$

The projection of vector OP on a vector perpendicular to the vector AB and AC is

$$\frac{\mathbf{OP}\cdot(\mathbf{AB}\times\mathbf{AC})}{|\mathbf{AB}\times\mathbf{AC}|}$$

$$=\frac{(-\hat{i}-2\hat{j}+3\hat{k})\cdot(5\hat{i}-10\hat{j}+10\hat{k})}{|5\hat{i}-10\hat{j}+10\hat{k}|}$$

$$=\frac{-5+20+30}{\sqrt{25+100+100}}=\frac{45}{15}=3$$

**44.** *(b)* $D(1,-6,-7)$

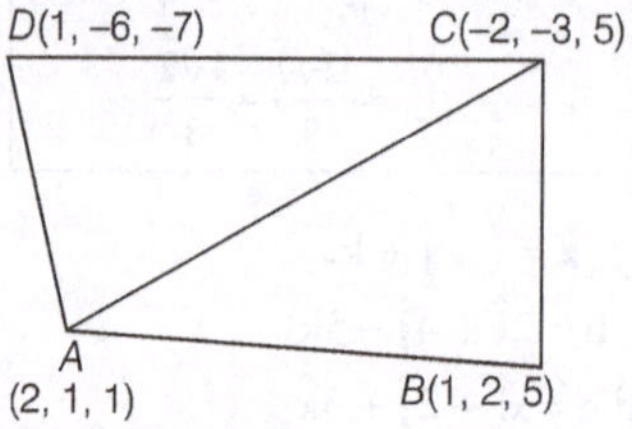


$\therefore$ Required Area

$$=\left|\frac{1}{2}(\mathbf{AB}\times\mathbf{AC})+\frac{1}{2}(\mathbf{AC}\times\mathbf{AD})\right|$$

$$=\frac{1}{2}|(\mathbf{AB}\times\mathbf{AC})-(\mathbf{AD}\times\mathbf{AC})|$$

$$=\frac{1}{2}|(\mathbf{AB}-\mathbf{AD})\times\mathbf{AC}|$$

$$=\frac{1}{2}|\{(-\hat{i}+\hat{j}+4\hat{k})-(-\hat{i}-7\hat{j}-8\hat{k})\}\times(-4\hat{i}-4\hat{j}+4\hat{k})|$$

$$=\frac{1}{2}|(8\hat{j}+12\hat{k})\times(-4\hat{i}-4\hat{j}+4\hat{k})|$$

$$=\frac{1}{2}\begin{vmatrix}\hat{i} & \hat{j} & \hat{k}\\ 0 & 8 & 12\\ -4 & -4 & 4\end{vmatrix}$$

$$=\frac{1}{2}|(32+48)\hat{i}-(0+48)\hat{j}+(0+32)\hat{k}|$$

$$=\frac{1}{2}|(80\hat{i}-48\hat{j}+32\hat{k}|$$

$$=\frac{1}{2}\times16|5\hat{i}-3\hat{j}+2\hat{k})|$$

$$=8\times\sqrt{25+9+4}=8\sqrt{38}$$

**45.** *(d)* We have,

$\mathbf{a}=2\hat{i}-7\hat{j}+5\hat{k},\ \mathbf{b}=\hat{i}+\hat{k},$

$\mathbf{c}=\hat{i}+2\hat{j}-3\hat{k}$

Now, $\quad \mathbf{r}\times\mathbf{a}=\mathbf{c}\times\mathbf{a}$

$\mathbf{r}\times\mathbf{a}-\mathbf{c}\times\mathbf{a}=0$

$\Rightarrow \quad (\mathbf{r}-\mathbf{c})\times\mathbf{a}=0$

$\therefore \quad \mathbf{r}-\mathbf{c}\,||\,\mathbf{a}$

$\Rightarrow \quad \mathbf{r}-\mathbf{c}=\lambda\mathbf{a}$

$\Rightarrow \quad \mathbf{r}=\mathbf{c}+\lambda\mathbf{a}$

Now, $\quad \mathbf{r}=\mathbf{c}+\lambda\mathbf{a}$

$\Rightarrow \quad \mathbf{r}\cdot\mathbf{b}=\mathbf{c}\cdot\mathbf{b}+\lambda\mathbf{a}\cdot\mathbf{b}$

$\Rightarrow \quad 0=(1+0-3)+\lambda(2+0+5)$

$\Rightarrow \quad 0=-2+7\lambda$

$\Rightarrow \quad \lambda=\dfrac{2}{7}$

$\therefore \quad \mathbf{r}=\mathbf{c}+\dfrac{2}{7}\mathbf{a}$

$=\hat{i}+2\hat{j}-3\hat{k}+\dfrac{2}{7}(2\hat{i}-7\hat{j}+5\hat{k})$

$=\dfrac{11}{7}\hat{i}-\dfrac{11}{7}\hat{k}$

Now, $\quad |\mathbf{r}|=\sqrt{\left(\dfrac{11}{7}\right)^2+\left(-\dfrac{11}{7}\right)^2}$

$=\sqrt{\dfrac{121}{49}+\dfrac{121}{49}}$

$=\sqrt{\dfrac{2\times121}{49}}=\dfrac{11}{7}\sqrt{2}$

> **Shortcut Method**
>
> $\mathbf{r}=\mathbf{c}+\lambda\mathbf{a}$
>
> $\mathbf{r}\cdot\mathbf{b}=\mathbf{c}\cdot\mathbf{b}+\alpha\cdot\mathbf{b}=0$
>
> $-2+\lambda(7)=0\Rightarrow\lambda=\dfrac{2}{7}$
>
> Now, $r=(\hat{i}+2\hat{j}-3\hat{k})+\dfrac{2}{7}(2\hat{i}-7\hat{j}+5\hat{k})$
>
> $=\dfrac{11}{7}\hat{i}+0\hat{j}-\dfrac{11}{7}\hat{k}$
>
> $\therefore \quad |r|=\sqrt{\left(\dfrac{11}{7}\right)^2+\left(\dfrac{-11}{7}\right)^2}=\dfrac{11}{7}\sqrt{2}$

**46.** *(c)* We have,

$\mathbf{a}=\hat{i}+2\hat{j}+3\hat{k},\ \mathbf{b}=\hat{i}-\hat{j}+2\hat{k}$ and $\mathbf{c}=5\hat{i}-3\hat{j}+3\hat{k}.$

Again, $\quad \mathbf{r}\times\mathbf{b}=\mathbf{c}\times\mathbf{b}$

$\Rightarrow \quad \mathbf{r}\times\mathbf{b}-\mathbf{c}\times\mathbf{b}=0$

$\Rightarrow \quad (\mathbf{r}-\mathbf{c})\times\mathbf{b}=0$

$\therefore \quad (\mathbf{r}-\mathbf{c})\,||\,\mathbf{b}$

$\therefore \quad \mathbf{r}-\mathbf{c}=\lambda\mathbf{b}$

$\Rightarrow \quad \mathbf{r}=\lambda\mathbf{b}+\mathbf{c}$

$=\lambda(\hat{i}-\hat{j}+2\hat{k})+5\hat{i}-3\hat{j}+3\hat{k}$

$=(\lambda+5)\hat{i}+(-\lambda-3)\hat{j}+(2\lambda+3)\hat{k}$

Also, $\quad \mathbf{r}\cdot\mathbf{a}=0$

$\Rightarrow ((\lambda+5)\hat{i}+(-\lambda-3)\hat{j}+(2\lambda+3)k)\cdot(\hat{i}+2\hat{j}+3\hat{k})=0$

$\Rightarrow \quad \lambda+5-2\lambda-6+6\lambda+9=0$

$\Rightarrow \quad 5\lambda+8=0$

$\Rightarrow \quad \lambda=-\dfrac{8}{5}$

$\therefore \quad r = -\frac{8}{5}b + c$

$\Rightarrow \quad 5r = -8b + 5c$

$= -8(i - j + 2k) + 5(5i - 3j + 3k)$

$= 17\hat{i} - 7\hat{j} - \hat{k}$

$\therefore \; 25|r|^2 = (17)^2 + (-7)^2 + (-1)^2$

$= 289 + 49 + 1 = 339$

**47.** *(c)* Let $a = 4\hat{i} + 3\hat{j}$

and $b = 3\hat{i} - 4\hat{j} + 5\hat{k}$

Given, $c \cdot (a \times b) + 25 = 0$,

$c \cdot (\hat{i} + \hat{j} + \hat{k}) = 4$

And projection of c on a is 1.

Let $c = c_1\hat{i} + c_2\hat{j} + c_3\hat{k}$

$$a \times b = \begin{vmatrix} \hat{i} & \hat{j} & \hat{k} \\ 4 & 3 & 0 \\ 3 & -4 & 5 \end{vmatrix} = 15\hat{i} - 20\hat{j} - 25\hat{k}$$

$\therefore \; c \cdot (a \times b) = 15c_1 - 20c_2 - 25c_3$

Given, $c \cdot (a \times b) + 25 = 0$

$\Rightarrow 15c_1 - 20c_2 - 25c_3 + 25 = 0 \quad \ldots$ (i)

Also, $c \cdot (\hat{i} + \hat{j} + \hat{k}) = 4$

$\Rightarrow \quad c_1 + c_2 + c_3 - 4 = 0 \quad \ldots$(ii)

And projection of c on a is 1

$\Rightarrow \frac{c \cdot a}{|a|} = 1 \Rightarrow 4c_1 + 3c_2 - 5 = 0 \quad \ldots$(iii)

On solving Eqs. (i), (ii) and (iii), we get

$c_1 = 2, c_2 = -1, c_3 = 3$

So, $c = 2\hat{i} - \hat{j} + 3\hat{k}$

Now, projection of c on b equals,

$$\frac{c \cdot b}{|b|} = \frac{6 + 4 + 15}{5\sqrt{2}} = \frac{25}{5\sqrt{2}} = \frac{5}{\sqrt{2}}$$

**48.** *(c)* a, b and c be three coplanar concurrent vectors.

Given, $|a||b||c| = 14$

Angle between any two of them is same

$\theta_1 = \theta_2 = \theta_3 = \theta$

$\theta_1 + \theta_2 + \theta_3 = 360°$

$3\theta = 360° \Rightarrow \theta = 120°$

So, $a \cdot b = |a|\,|b| \cos\theta = (14)^2 \cos 120°$

$= -\frac{(14)^2}{2}$

Similarly, $b \cdot c = c \cdot a = -\frac{(14)^2}{2}$

$(a \times b) \cdot (b \times c) = (a \cdot b)(b \cdot c) - (a \cdot c)(b \cdot b)$

$= (|a||b|\cos 120°)(|b||c|\cos 120°) - (|a||c|\cos 120° \times |b|^2)$

$= \frac{|a||b|^2|c|}{4} + \frac{1}{2}|a||b|^2|c|$

$= \frac{3}{4}|a||b|^2|c|$

Similarly, $(b \times c) \cdot (c \times a) = \frac{3}{4}|a||b||c|^2$

$\Rightarrow \quad (c \times a) \cdot (a \times b) = \frac{3}{4}|a|^2|b||c|$

Given, $(a \times b) \cdot (b \times c) + (b \times c) \cdot (c \times a) + (c \times a) \cdot (a \times b) = 168$

$\Rightarrow \frac{3}{4}|a||b|^2|c| + \frac{3}{4}|a||b||c|^2 + \frac{3}{4}|a|^2|b||c| = 168$

$\Rightarrow \frac{3}{4}|a||b||c|(|a| + |b| + |c|) = 168$

$\Rightarrow 14(|a| + |b| + |c|) = 168 \times \frac{4}{3}$

$\Rightarrow \quad |a| + |b| + |c| = \frac{168}{14} \times \frac{4}{3} = 16$

**49.** *(d)* Given, $a = \alpha\hat{i} + \hat{j} - \hat{k}$ and

$b = 2\hat{i} + \hat{j} - \alpha\hat{k}$

$$a \times b = \begin{vmatrix} \hat{i} & \hat{j} & \hat{k} \\ \alpha & 1 & -1 \\ 2 & 1 & -\alpha \end{vmatrix}$$

$= \hat{i}(-\alpha + 1) - \hat{j}(-\alpha^2 + 2) + \hat{k}(\alpha - 2)$

$= \hat{i}(1 - \alpha) + \hat{j}(\alpha^2 - 2) + \hat{k}(\alpha - 2)$

Projection of $a \times b$ on vector $-\hat{i} + 2\hat{j} - 2\hat{k}$

$$= \frac{|[\hat{i}(1-\alpha) + \hat{j}(\alpha^2 - 2) + \hat{k}(\alpha - 2)] \cdot [-\hat{i} + 2\hat{j} - 2\hat{k}]|}{\sqrt{(-1)^2 + (2)^2 + (-2)^2}}$$

$\Rightarrow \frac{|\alpha - 1 + 2\alpha^2 - 4 - 2\alpha + 4|}{3} = 30$

$\Rightarrow \quad |2\alpha^2 - \alpha - 1| = 90$

$\Rightarrow \quad 2\alpha^2 - \alpha - 1 = 90$

or $\quad 2\alpha^2 - \alpha - 1 = -90$

$\Rightarrow \quad 2\alpha^2 - \alpha - 91 = 0$

or $\quad 2\alpha^2 - \alpha + 91 = 0$

$\Rightarrow \quad 2\alpha^2 - 14\alpha + 13\alpha - 91 = 0$

[$\because 2\alpha^2 - \alpha + 91$ has no real roots]

$\Rightarrow \quad \alpha = 7, \alpha \neq \frac{-13}{2}$ [as $\alpha > 0$]

**50.** *(d)* Given, $a = \alpha\hat{i} + 2\hat{j} - \hat{k}$ and

$b = -2i + \alpha\hat{j} + \hat{k}$

$$\therefore a \times b = \begin{vmatrix} \hat{i} & \hat{j} & \hat{k} \\ \alpha & 2 & -1 \\ -2 & \alpha & 1 \end{vmatrix}$$

$= (2 + \alpha)\hat{i} - (\alpha - 2)\hat{j} + (\alpha^2 + 4)\hat{k}$

Now, $|a \times b| = \sqrt{15(\alpha^2 + 4)}$

$\Rightarrow \quad (2 + \alpha)^2 + (\alpha - 2)^2 + (\alpha^2 + 4)^2 = 15(\alpha^2 + 4)$

$\Rightarrow \quad \alpha^4 - 5\alpha^2 - 36 = 0 \Rightarrow \alpha^2 = 9$

$\therefore \quad \alpha = \pm 3$

$|a|^2 = 9 + 4 + 1 = 14$

$|b|^2 = 4 + 9 + 1 = 14$

$a \cdot b = -2(9) + 2(9) - 1 = -1$

Now,

$2|a|^2 + (a \cdot b)|b|^2 = 2(14) - 14 = 14$

**51.** *(d)* Given, $|a \cdot b| = |a \times b|$

$\Rightarrow \quad |a||b| \cos\theta = |a||b| \sin\theta$

$\Rightarrow \quad \tan\theta = 1$

$\Rightarrow \quad \theta = \frac{\pi}{4}$ [$\because \theta$ is an acute angle]

According to question, Area of parallelogram $= 2\sqrt{2}$

$\Rightarrow \quad |a \times b| = 2\sqrt{2}$

$\Rightarrow \quad |a||b| \sin\theta = 2\sqrt{2}$

$\Rightarrow \quad 1 \times |b| \times \sin\frac{\pi}{4} = 2\sqrt{2} \Rightarrow |b| = 4$

Let $\alpha$ be the angle between b and c.

$\therefore \quad \cos\alpha = \frac{b \cdot c}{|b||c|} \quad \ldots$(i)

Now, $b \cdot c = b \cdot (2\sqrt{2}(a \times b) - 2b)$

$= 2\sqrt{2}\, b \cdot (a \times b) - 2b \cdot b$

$= 0 - 2|b|^2$

[$\because b \cdot (a \times b) = 0$]

$= -2 \times 4^2 = -32$

$|c|^2 = |2\sqrt{2}(a \times b)|^2 + 4|b|^2 - 2 \cdot 2\sqrt{2}(a \times b) 2 \cdot b$

$= 8|a \times b|^2 + 4 \times 4^2 - 0$

[$\because (a \times b) \cdot b = 0$]

$= 8 \times 8 + 64 = 128 \Rightarrow |c| = 8\sqrt{2}$

From Eq. (i), we get

$\cos\alpha = \frac{-32}{4 \times 8\sqrt{2}} = -\frac{1}{\sqrt{2}}$

$\therefore \quad \alpha = 3\pi/4$

**52.** *(d)* Given three unit vectors a, b and c such that $a + b + c = 0 \quad \ldots$(i)

If we do dot product in Eq. (i) with a, b and c, then we get relations

$$a\cdot a+a\cdot b+a\cdot c=0$$

$\Rightarrow \quad a\cdot b+a\cdot c=-1,\ b\cdot a+b\cdot c=-1$

and $\quad c\cdot a+c\cdot b=-1$

On adding above three obtained relations, we get

$$2(a\cdot b+b\cdot c+c\cdot a)=-3 \quad (\because a\cdot b=b\cdot a)$$

$$\Rightarrow \lambda=a\cdot b+b\cdot c+c\cdot a=-\frac{3}{2}$$

If we do cross product in Eq. (i) with a, b and c, then we get relation

$$0+a\times b+a\times c=0$$

$\Rightarrow \quad a\times b+a\times c=0$

and $\quad b\times a+b\times c=0$

and $\quad c\times a+c\times b=0$

$\therefore \quad a\times b+b\times c+c\times a=3(a\times b)=3(b\times c)=3(c\times a)$

$\therefore \quad d=3a\times b=3b\times c=3c\times a$

From the given options the ordered pair,

$$(\lambda, d)=\left(-\frac{3}{2}, 3a\times b\right)$$

**53.** *(d)* Given vectors are $a=3\hat{i}+2\hat{j}+x\hat{k}$ and $b=\hat{i}-\hat{j}+\hat{k}$

$$\therefore a\times b=\begin{vmatrix}\hat{i} & \hat{j} & \hat{k}\\ 3 & 2 & x\\ 1 & -1 & 1\end{vmatrix}$$

$$=\hat{i}(2+x)-\hat{j}(3-x)+\hat{k}(-3-2)$$

$$=(x+2)\hat{i}+(x-3)\hat{j}-5\hat{k}$$

$$\Rightarrow |a\times b|=\sqrt{(x+2)^2+(x-3)^2+25}$$

$$=\sqrt{2x^2-2x+4+9+25}$$

$$=\sqrt{2\left(x^2-x+\frac{1}{4}\right)-\frac{1}{2}+38}$$

$$=\sqrt{2\left(x-\frac{1}{2}\right)^2+\frac{75}{2}}$$

So, $|a\times b|\geq\sqrt{75/2}$

[at $x=1/2$, $|a\times b|$ is minimum]

$\Rightarrow \quad r\geq 5\sqrt{3/2}$

**54.** *(b)* We have, $a=2\hat{i}+\hat{j}-2\hat{k}$

$\Rightarrow \quad |a|=\sqrt{4+1+4}=3$

and $\quad b=\hat{i}+\hat{j}$

$\Rightarrow \quad |b|=\sqrt{1+1}=\sqrt{2}$

Now, $|c-a|=3\Rightarrow|c-a|^2=9$

$\Rightarrow \quad (c-a)\cdot(c-a)=9$

$\Rightarrow \quad |c|^2+|a|^2-2c\cdot a=9 \quad \ldots(i)$

Again, $\quad |(a\times b)\times c|=3$

$\Rightarrow \quad |a\times b||c|\sin 30^\circ=3$

$\Rightarrow \quad |c|=\dfrac{6}{|a\times b|}$

But, $a\times b=\begin{vmatrix}\hat{i} & \hat{j} & \hat{k}\\ 2 & 1 & -2\\ 1 & 1 & 0\end{vmatrix}=2\hat{i}-2\hat{j}+\hat{k}$

$\therefore \quad |c|=\dfrac{6}{\sqrt{4+4+1}}=2 \quad \ldots(ii)$

From Eqs. (i) and (ii), we get

$$(2)^2+(3)^2-2c\cdot a=9$$

$\Rightarrow \quad 4+9-2c\cdot a=9\Rightarrow c\cdot a=2$

**55.** *(a)* Given that $|a+b|=\sqrt{3}$ $\quad \ldots(i)$

On squaring Eq. (i) both sides, we get

$$|a+b|^2=3$$

$\Rightarrow |a|^2+|b|^2+2a\cdot b=3$

$\Rightarrow \quad 1+1+2\cos\theta=3$

$\Rightarrow \quad \cos\theta=1/2$

$\Rightarrow \quad \theta=60^\circ$

$a\times b$ is perpendicular to plane containing a and b

$$c=a+2b+3(a\times b) \quad \ldots(ii)$$

On squaring Eq. (ii) both sides, we get

$|c|^2=||a|^2+4|b|^2+9(a\times b)^2$
$+4(a\cdot b)+6a\cdot(a\times b)+12b\cdot(a\times b)|$

$=|1+4+9\sin^2\theta+4\cos\theta+0+0|$

$=\left|5+9\times\dfrac{3}{4}+4\times\dfrac{1}{2}\right|$

$\Rightarrow |c|^2=7+\dfrac{27}{4}=\dfrac{55}{4}$

$\Rightarrow 2|c|=\sqrt{55}$

**56.** *(d)* Given that $u=\hat{i}+\hat{j},\quad v=\hat{i}-\hat{j},$

$$w=\hat{i}+2\hat{j}+3\hat{k},$$

$$u\cdot n=0$$

and $\quad v\cdot n=0$

i.e., $\quad n=\dfrac{u\times v}{|u\times v|}$

Now, $\quad u\times v=\begin{vmatrix}\hat{i} & \hat{j} & \hat{k}\\ 1 & 1 & 0\\ 1 & -1 & 0\end{vmatrix}$

$$=0\hat{i}-0\hat{j}-2\hat{k}$$

$$=-2\hat{k}$$

$\therefore \quad |w\cdot n|=\dfrac{|w\cdot u\times v|}{|u\times v|}=\dfrac{|-6\hat{k}|}{|-2\hat{k}|}=3$

$[\because w\cdot(u\times v)=(\hat{i}+2\hat{j}+3\hat{k})\cdot(-2\hat{k})=-6\hat{k}]$

Hence, $\quad |w\cdot n|=3$

**57.** *(a)*

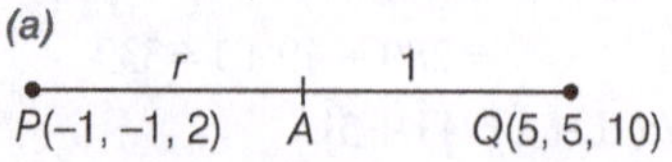


$\therefore$ Coordinates of $A$

$$=\left(\frac{5r-1}{r+1},\frac{5r-1}{r+1},\frac{10r+2}{r+1}\right)$$

$$\because (OQ\cdot OA)-\frac{1}{5}|OP\times OA|^2=10 \ldots(i)$$

Now, $OP\times OA$

$$=\begin{vmatrix}\hat{i} & \hat{j} & \hat{k}\\ -1 & -1 & 2\\ \dfrac{5r-1}{r+1} & \dfrac{5r-1}{r+1} & \dfrac{10r+2}{r+1}\end{vmatrix}$$

$$=\left(\frac{-10r-2-10r+2}{r+1}\right)\hat{i}-\left(\frac{-10r-2-10r+2}{r+1}\right)\hat{j}+(0)\hat{k}$$

$$=\left(\frac{-20r}{r+1}\right)\hat{i}+\left(\frac{20r}{r+1}\right)\hat{j}$$

$$=\left(\frac{20r}{r+1}\right)(-\hat{i}+\hat{j})$$

$\Rightarrow |OP\times OA|^2$

$$=\frac{400r^2}{(r+1)^2}\cdot 2=\frac{800r^2}{(r+1)^2} \quad \ldots(ii)$$

and $\quad OQ\cdot OA=(5\hat{i}+5\hat{j}+10\hat{k})\cdot\left(\dfrac{5r-1}{r+1}\hat{i}+\dfrac{5r-1}{r+1}\hat{j}+\dfrac{10r+2}{r+1}\hat{k}\right)$

$$=\frac{5}{r+1}(5r-1+5r-1+20r+4)$$

$$=\frac{5}{r+1}(30r+2) \quad \ldots(iii)$$

From Eqs. (i) and (ii), we get

$$\frac{10}{r+1}(15r+1)-\frac{1}{5}\frac{r^2(800)}{(r+1)^2}=10$$

$\Rightarrow (15r+1)(r+1)-16r^2=(r+1)^2$

$\Rightarrow \quad 15r^2+15r+r+1-16r^2=r^2+2r+1$

$\Rightarrow \quad 2r^2-14r=0$

$\Rightarrow \qquad r = 0, 7$

$\because \qquad r > 0 \Rightarrow r = 7$

**58.** *(a)* $\mathbf{a} = \hat{i} + 2\hat{j} - 3\hat{k}$

$\mathbf{b} = 2\hat{i} + \hat{j} - \hat{k}$

$\mathbf{a} \times \mathbf{r} - \mathbf{a} \times \mathbf{b} = 0$

$\mathbf{a} \times (\mathbf{r} - \mathbf{b}) = 0$

$\mathbf{r} - \mathbf{b} = \lambda \mathbf{a}$

$\mathbf{r} = \mathbf{b} + \lambda \mathbf{a}$

$\mathbf{a} \cdot \mathbf{r} = \mathbf{a} \cdot \mathbf{b} + \lambda(\mathbf{a} \cdot \mathbf{a}) = 0$ [given $\mathbf{a} \cdot \mathbf{r} = 0$]

$\Rightarrow \lambda = -\frac{(\mathbf{a} \cdot \mathbf{b})}{\mathbf{a} \cdot \mathbf{a}} = -\frac{(2+2+3)}{1+4+9} = -\frac{7}{14}$

$\lambda = -\frac{1}{2}$

$\therefore \mathbf{r} = \mathbf{b} - \frac{\mathbf{a}}{2} = \frac{2\mathbf{b} - \mathbf{a}}{2} = \frac{3\hat{i} + \hat{k}}{2}$

Now, $|\mathbf{r}| = \frac{\sqrt{10}}{2}$

Statement I is incorrect.

Let $|\mathbf{OA}| = |\mathbf{a}|$, $\mathbf{OB} = |\mathbf{b}|$ and $|\mathbf{OC}| = |\mathbf{c}|$

Here, $O$ is the circumcentre:

$\therefore \mathbf{OA} = \mathbf{OB} = \mathbf{OC}$

$\Rightarrow |\mathbf{a}| = |\mathbf{b}| = |\mathbf{c}| \qquad \ldots(i)$

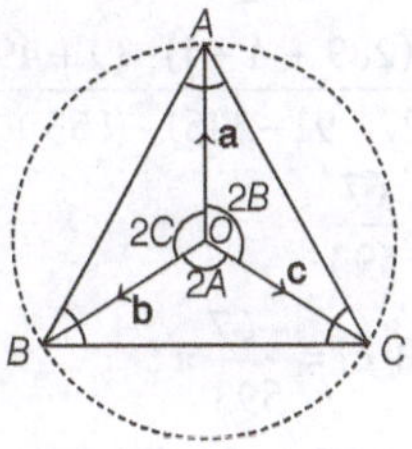


Now,

$(\mathbf{OA} + \mathbf{OB} + \mathbf{OC})^2 = (\mathbf{a} + \mathbf{b} + \mathbf{c})^2 \ge 0$

$(\mathbf{a} + \mathbf{b} + \mathbf{c}) \cdot (\mathbf{a} + \mathbf{b} + \mathbf{c}) \ge 0$

$|\mathbf{a}|^2 + |\mathbf{b}|^2 + |\mathbf{c}|^2 + 2\mathbf{a} \cdot \mathbf{b} + 2\mathbf{b} \cdot \mathbf{c} + 2\mathbf{c} \cdot \mathbf{a} \ge 0$

Or $3|\mathbf{a}|^2 + 2|\mathbf{a}|^2 \cos 2c + 2|\mathbf{b}|^2 \cos 2A + 2|\mathbf{c}|^2 \cos 2B \ge 0$ [from Eq. (i)]

or

$|\mathbf{a}|^2 (3 + 2(\cos 2A + \cos 2B + \cos 2C)) \ge 0$

$\therefore 3 + 2(\cos 2A + \cos 2B + \cos 2C) \ge 0$

$\cos 2A + \cos 2B + \cos 2C \ge -\frac{3}{2}$

$\therefore$ Statement II is correct.

**59.** *(a)* Given that for a triangle

$\mathbf{AB} = -2\hat{i} + \hat{j} + 3\hat{k}$

$\mathbf{CB} = \alpha\hat{i} + \beta\hat{j} + \gamma\hat{k}$

$\mathbf{CA} = 4\hat{i} + 3\hat{j} + \delta\hat{k}$

We know, $\mathbf{AB} + \mathbf{BC} + \mathbf{CA} = 0$

$\Rightarrow (-2\hat{i} + \hat{j} + 3\hat{k}) + (-\alpha\hat{i} - \beta\hat{j} - \gamma\hat{k}) + (4\hat{i} + 3\hat{j} + \delta\hat{k}) = \mathbf{0}$

$\Rightarrow (-2 - \alpha + 4)\hat{i} + (1 - \beta + 3)\hat{j} + (3 - \gamma + \delta)\hat{k} = \mathbf{0}$

$\Rightarrow (-\alpha + 2)\hat{i} + (-\beta + 4)\hat{j} + (-\gamma + \delta + 3)\hat{k} = \mathbf{0}$

On comparing both sides, we get

$-\alpha + 2 = 0 \quad \Big| \quad -\beta + 4 = 0$

$\alpha = 2 \quad \Big| \quad \beta = 4$

and $-\gamma + \delta + 3 = 0 \Rightarrow -\gamma + \delta = -3$

Also given the area of $\Delta ABC$ is $5\sqrt{6}$.

Thus, $\frac{1}{2}|\mathbf{AB} \times \mathbf{AC}| = 5\sqrt{6}$

$\Rightarrow |\mathbf{AB} \times \mathbf{AC}| = 10\sqrt{6} \qquad \ldots(i)$

Now, $\mathbf{AB} \times \mathbf{AC}$

$= \begin{vmatrix} \hat{i} & \hat{j} & \hat{k} \\ -2 & 1 & 3 \\ 4 & 3 & \delta \end{vmatrix} = \hat{i}(\delta - 9) + \hat{j}(12 + 2\delta) + \hat{k}(-6 - 4)$

$= \hat{i}(\delta - 9) + \hat{j}(2\delta + 12) - 10\hat{k}$

From Eq. (i), we get

$|\mathbf{AB} \times \mathbf{AC}| = 10\sqrt{6}$

$\Rightarrow \sqrt{(\delta - 9)^2 + (2\delta + 12)^2 + 100} = 10\sqrt{6}$

On squaring both sides, we get

$(\delta - 9)^2 + (2\delta + 12)^2 + 100 = 600$

$\Rightarrow \delta^2 - 18\delta + 81 + 4\delta^2 + 48\delta + 144 - 500 = 0$

$\Rightarrow 5\delta^2 + 30\delta - 275 = 0$

$\therefore \delta = 5 \; (\delta > 0)$

So, $\delta - \gamma = -3$

$-\gamma = -8 \Rightarrow \gamma = 8$

Now, $\mathbf{CB} = 2\hat{i} + 4\hat{j} + 8\hat{k}$

$\mathbf{CA} = 4\hat{i} + 3\hat{j} + 5\hat{k}$

$\mathbf{CB} \cdot \mathbf{CA} = 8 + 12 + 40 = 60$

**60.** *(c)* $\mathbf{a} = -\hat{i} + 2\hat{j} + \hat{k}$ is rotated through a right angle, passing through $Y$-axis, and resulted vector is $\mathbf{b}$.

$\mathbf{b} = \lambda \mathbf{a} \times (\hat{a} \times \hat{j})$

$\mathbf{a} \times \hat{j} = \begin{vmatrix} \hat{i} & \hat{j} & \hat{k} \\ -1 & 2 & 1 \\ 0 & 1 & 0 \end{vmatrix} = -\hat{i} + 0\hat{j} - \hat{k}$

$\mathbf{a} \times (\mathbf{a} \times \hat{j}) = \begin{vmatrix} \hat{i} & \hat{j} & \hat{k} \\ -1 & 2 & 1 \\ -1 & 0 & -1 \end{vmatrix}$

$= -2\hat{i} - 2\hat{j} + 2\hat{k}$

$\mathbf{b} = \lambda(-2\hat{i} - 2\hat{j} + 2\hat{k})$

Also, $|\mathbf{a}| = |\mathbf{b}| \Rightarrow \sqrt{1 + 4 + 1}$

$= \sqrt{(-2\lambda)^2 + (-2\lambda)^2 + (2\lambda)^2}$

$6 = 12\lambda^2 \Rightarrow \lambda^2 = \frac{1}{2} \Rightarrow \lambda = \pm \frac{1}{\sqrt{2}}$

[($\lambda = \frac{1}{\sqrt{2}}$ rejected, $\because$ $\mathbf{b}$ makes acute angle with $Y$-axis)]

$\mathbf{b} = \frac{-1}{\sqrt{2}}(-2\hat{i} - 2\hat{j} + 2\hat{k})$

$= -\sqrt{2}(-\hat{i} - \hat{j} + \hat{k})$

Projection of $(3\mathbf{a} + \sqrt{2}\mathbf{b})$ on $\mathbf{c}$

$= (3\mathbf{a} + \sqrt{2}\mathbf{b}) \cdot \frac{\mathbf{c}}{|\mathbf{c}|}$

$= (-3\hat{i} + 6\hat{j} + 3\hat{k} + 2\hat{i} + 2\hat{j} - 2\hat{k}) \cdot \frac{(5\hat{i} + 4\hat{j} + 3\hat{k})}{\sqrt{25 + 16 + 9}}$

$= \frac{(-\hat{i} + 8\hat{j} + \hat{k}) \cdot (5\hat{i} + 4\hat{j} + 3\hat{k})}{5\sqrt{2}}$

$= \frac{-5 + 32 + 3}{5\sqrt{2}} = \frac{30}{5\sqrt{2}} = 3\sqrt{2}$

**61.** *(c)* Given, $ABC$ is a triangle such that $\mathbf{BC} = \mathbf{a}$, $\mathbf{CA} = \mathbf{b}$, $\mathbf{AB} = \mathbf{c}$, $|\mathbf{a}| = 6\sqrt{2}$, $|\mathbf{b}| = 2\sqrt{3}$ and $\mathbf{b} \cdot \mathbf{c} = 12$

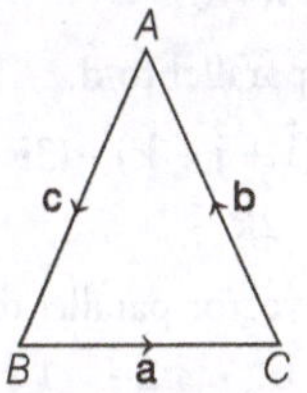


Since, $\mathbf{a} + \mathbf{b} + \mathbf{c} = 0 \qquad \ldots(i)$

$\Rightarrow \mathbf{a} + \mathbf{c} = -\mathbf{b}$

$\Rightarrow (\mathbf{a} + \mathbf{c}) \times \mathbf{b} = -\mathbf{b} \times \mathbf{b}$

$\Rightarrow (\mathbf{a} \times \mathbf{b}) + \mathbf{c} \times \mathbf{b} = 0 \qquad \ldots(ii)$

Now, from Statement I

$|(\mathbf{a} \times \mathbf{b}) + (\mathbf{c} \times \mathbf{b})| - |\mathbf{c}| = 6(2\sqrt{2} - 1)$

$\Rightarrow |0| - |\mathbf{c}| = (12\sqrt{2} - 6)$

$\Rightarrow |\mathbf{c}| = 6 - 12\sqrt{2} < 0$ (which is impossible)

Hence, Statement I is not correct.

For Statement II, from Eq. (i), we get

$\mathbf{b} + \mathbf{c} = -\mathbf{a}$

$\Rightarrow \mathbf{b} \cdot (\mathbf{b} + \mathbf{c}) = -\mathbf{b} \cdot \mathbf{a}$

$\Rightarrow \mathbf{b} \cdot \mathbf{b} + \mathbf{b} \cdot \mathbf{c} = -\mathbf{a} \cdot \mathbf{b}$

$\Rightarrow |\mathbf{b}|^2 + \mathbf{b} \cdot \mathbf{c} = -\mathbf{a} \cdot \mathbf{b}$

$\Rightarrow 12 + 12 = -|\mathbf{a}||\mathbf{b}| \cos(\pi - \angle ACB)$

$\Rightarrow 24 = -6\sqrt{2} \cdot 2\sqrt{3}\,[-\cos(\angle ACB)]$

$\Rightarrow \cos(\angle ACB) = \frac{24}{12\sqrt{6}} = \frac{2}{\sqrt{6}} = \sqrt{\frac{2}{3}}$

$\Rightarrow \angle ACB = \cos^{-1}\left(\sqrt{\frac{2}{3}}\right)$

Hence, Statement II is correct.

**62.** *(128)* $\mathbf{b} \times \mathbf{d} = \mathbf{c} \times \mathbf{d} \Rightarrow (\mathbf{b} - \mathbf{c}) \times \mathbf{d}$

$\Rightarrow \mathbf{b} - \mathbf{c} \,||\, \mathbf{d}$

$\Rightarrow \mathbf{b} - \mathbf{c} = (3\hat{i} - 3\hat{j} + 3\hat{k}) - (2\hat{i} - \hat{j} + 2\hat{k})$

$= \hat{i} - 2\hat{j} + \hat{k}$

Since $\mathbf{d}$ is parallel to $\mathbf{b} - \mathbf{c}$

$\Rightarrow \mathbf{d} = \lambda(\mathbf{b} - \mathbf{c}) = \lambda(\hat{i} - 2\hat{j} + \hat{k})$

Also, $\mathbf{a} \cdot \mathbf{d} = 4$

$\Rightarrow (\hat{i} + 2\hat{j} + \hat{k}) \cdot \lambda(\hat{i} - 2\hat{j} + \hat{k}) = 4$

$\Rightarrow \lambda(1 - 4 + 1) = 4$

$\Rightarrow 2\lambda = -4 \Rightarrow \lambda = -2$

$\therefore \mathbf{d} = -2\hat{i} + 4\hat{j} - 2\hat{k}$

$\therefore |\mathbf{a} \times \mathbf{d}|^2 = \begin{vmatrix} \hat{i} & \hat{j} & \hat{k} \\ 1 & 2 & 1 \\ -2 & 4 & -2 \end{vmatrix}^2$

$= \left|\hat{i}(-8) - \hat{j}(0) + \hat{k}(8)\right|^2$

$= (\sqrt{64 + 64})^2 = 128$

**63.** *(5)* $\mathbf{a} \times \hat{\mathbf{d}} = \mathbf{b} \times \hat{\mathbf{d}}$

$\Rightarrow (\mathbf{a} - \mathbf{b}) \times \hat{\mathbf{d}} = 0$

$\Rightarrow \mathbf{a} - \mathbf{b}$ is parallel to $\hat{\mathbf{d}}$.

$\Rightarrow \mathbf{a} - \mathbf{b} = (\hat{i} + \hat{j} + \hat{k}) - (3\hat{i} + 2\hat{j} - \hat{k})$

$= -2\hat{i} - \hat{j} + 2\hat{k}$

$\because \hat{\mathbf{d}}$ is unit vector parallel to $\mathbf{a} - \mathbf{b}$

$\Rightarrow \hat{\mathbf{d}} = \frac{\mathbf{a} - \mathbf{b}}{|\mathbf{a} - \mathbf{b}|} = -\frac{2}{3}\hat{i} - \frac{1}{3}\hat{j} + \frac{2}{3}\hat{k}$

Since, $\mathbf{c}$ is perpendicular to $\mathbf{a} \Rightarrow \mathbf{a} \cdot \mathbf{c} = 0$

$\Rightarrow \lambda + \mu = 0 \Rightarrow \lambda = -\mu$ ...(i)

Also, $\mathbf{c} \cdot \mathbf{d} = 1$

$\Rightarrow (\lambda\hat{i} + \mu\hat{k}) \cdot \left(-\frac{2}{3}\hat{i} - \frac{1}{3}\hat{j} + \frac{2}{3}\hat{k}\right) = 1$

$\Rightarrow -\lambda + 2\mu = 3$ ...(ii)

From Eqs. (i) and (ii) we get,

$\lambda = -1, \mu = 1$

$\Rightarrow \mathbf{c} = -\hat{j} + \hat{k}$

$\therefore |3\lambda\mathbf{d} + \mu\mathbf{c}|^2$

$= \left| 3(-1)\left(-\frac{2}{3}\hat{i} - \frac{1}{3}\hat{j} + \frac{2}{3}\hat{k}\right) + (1)(-\hat{j} + \hat{k})^2 \right|$

$= |2\hat{i} + \hat{j} - 2\hat{k} - \hat{j} + \hat{k}|^2$

$= |2\hat{i} - \hat{k}|^2 = (\sqrt{4 + 1})^2 = 5$

**64.** *(6)* $\mathbf{d} = \mathbf{a} \times \mathbf{b} = -\hat{i} + \hat{j}$

$\because |\mathbf{c} - 2\mathbf{a}|^2 = 8$

$\Rightarrow |\mathbf{c}|^2 + 4|\mathbf{a}|^2 - 4(\mathbf{a} \cdot \mathbf{c}) = 8$

$\Rightarrow |\mathbf{c}| = 2 \quad [\because |\mathbf{a}| = \sqrt{3}, \mathbf{a} \cdot \mathbf{c} = |\mathbf{c}|]$

We have, $\mathbf{d} = \mathbf{a} \times \mathbf{b}$

Or $\mathbf{d} \times \mathbf{c} = (\mathbf{a} \times \mathbf{b}) \times \mathbf{c}$

$\left[|\mathbf{d}||\mathbf{c}|\sin\frac{\pi}{4}\right]^2 = [(\mathbf{a} \cdot \mathbf{c})\mathbf{b} - (\mathbf{b} \cdot \mathbf{c})\mathbf{a}]^2$

$4 = 4b^2 + (\mathbf{b} \cdot \mathbf{c})^2 (\mathbf{a})^2 - 2(\mathbf{a} \cdot \mathbf{c})(\mathbf{a} \cdot \mathbf{b})(\mathbf{b} \cdot \mathbf{c})$

Let $\mathbf{b} \cdot \mathbf{c} = x$

$4 = 36 + 3x^2 - 20x$

$\Rightarrow 3x^2 - 20x + 32 = 0$

$\Rightarrow x = \frac{8}{3}, 4$

$\Rightarrow \mathbf{b} \cdot \mathbf{c} = \frac{8}{3}$

Now, $|10 - 3(\mathbf{b} \cdot \mathbf{c})| + |\mathbf{b} \times \mathbf{c}|^2$

$= |10 - 8| + (2)^2 = 6$

**65.** *(16)* $\mathbf{b} = \lambda\hat{i} + 4\hat{k}, \lambda > 0$

$\mathbf{c}$ = Projection of $\mathbf{b}$ on vector $\mathbf{a} = \hat{i} + 2\hat{j} + 2\hat{k}$

$= (|\mathbf{b}|\cos\theta)\,\hat{\mathbf{a}}$

$= \left(\frac{\mathbf{b} \cdot \mathbf{a}}{|a|}\right)\hat{\mathbf{a}}$

$\Rightarrow \mathbf{c} = \frac{(\lambda + 8)}{9} \cdot (\hat{i} + 2\hat{j} + 2\hat{k})$

$\because |\mathbf{a} + \mathbf{c}| = 7$

$|\mathbf{a}|^2 + |\mathbf{c}|^2 + 2\mathbf{a} \cdot \mathbf{c} = 49$

$\Rightarrow 9 + \left(\frac{\lambda + 8}{9}\right)^2 (9) + 2\frac{(\lambda + 8)}{9}(9) = 49$

$\Rightarrow 9 + \frac{(\lambda + 8)^2}{9} + 2(\lambda + 8) = 49$

Let $\lambda + 8 = t$

$t^2 + 81 + 18t = 49 \times 9$

$\Rightarrow t^2 + 18t - 360 = 0$

$\Rightarrow t^2 + 30t - 12t - 360 = 0$

$\Rightarrow (t + 30)(t - 12) = 0$

$t = -30, 12$

$\lambda = -38, 4$

$\because \lambda > 0, \lambda = 4$

$\mathbf{c} = \frac{12}{9}(\hat{i} + 2\hat{j} + 2\hat{k}) = \frac{4}{3}(\hat{i} + 2\hat{j} + 2\hat{k})$

$\therefore$ Area of parallelogram $= |\mathbf{b} \times \mathbf{c}|$

$= \left\| \begin{matrix} \hat{i} & \hat{j} & \hat{k} \\ 4 & 0 & 4 \\ \frac{4}{3} & \frac{8}{3} & \frac{8}{3} \end{matrix} \right\| = 4 \times \frac{4}{3} \left\| \begin{matrix} \hat{i} & \hat{j} & \hat{k} \\ 1 & 0 & 1 \\ 1 & 2 & 2 \end{matrix} \right\|$

$= \frac{16}{3}\sqrt{(2)^2 + (1)^2 + (2)^2} = 16$

**66.** *(569)* $\mathbf{a} = 9\hat{i} - 13\hat{j} + 25\hat{k}$,

$\mathbf{b} = 3\hat{i} + 7\hat{j} - 13\hat{k}$

and $\mathbf{c} = 17\hat{i} - 2\hat{j} + \hat{k}$

$\therefore \mathbf{b} + \mathbf{c} = 20\hat{i} + 5\hat{j} - 12\hat{k}$

and $\mathbf{b} - \mathbf{c} = -14\hat{i} + 9\hat{j} - 14\hat{k}$

$\because \mathbf{r} \times \mathbf{a} = (\mathbf{b} + \mathbf{c}) \times \mathbf{a}$

$\Rightarrow [\mathbf{r} - (\mathbf{b} + \mathbf{c})] \times \mathbf{a} = 0$

$\Rightarrow \mathbf{r} - (\mathbf{b} + \mathbf{c}) = \lambda\mathbf{a}$

$\Rightarrow \mathbf{r} = \lambda\mathbf{a} + \mathbf{b} + \mathbf{c}$ ...(i)

Also, $\mathbf{r} \cdot (\mathbf{b} - \mathbf{c}) = 0$

$\Rightarrow (\lambda\mathbf{a} + \mathbf{b} + \mathbf{c}) \cdot (\mathbf{b} - \mathbf{c}) = 0$

$\Rightarrow \lambda\mathbf{a} \cdot \mathbf{b} + \mathbf{b} \cdot \mathbf{b} + \mathbf{c} \cdot \mathbf{b} - \lambda\mathbf{a} \cdot \mathbf{c} - \mathbf{b} \cdot \mathbf{c} - \mathbf{c} \cdot \mathbf{c} = 0$

$\Rightarrow \lambda = \frac{\mathbf{c} \cdot \mathbf{c} - \mathbf{b} \cdot \mathbf{b}}{\mathbf{a} \cdot \mathbf{b} - \mathbf{a} \cdot \mathbf{c}}$

$= \frac{(289 + 4 + 1) - (9 + 49 + 169)}{(27 - 91 - 325) - (153 + 26 + 25)}$

$= -\frac{67}{593}$

$\therefore \mathbf{r} - (\mathbf{b} + \mathbf{c}) = \frac{-67}{593}\mathbf{a}$

$\Rightarrow 593\mathbf{r} + 67\mathbf{a} = 593(\mathbf{b} + \mathbf{c})$

So, $\frac{|593\mathbf{r} + 67\mathbf{a}|^2}{(593)^2} = \frac{(593)^2|\mathbf{b} + \mathbf{c}|^2}{(593)^2}$

$= |\mathbf{b} + \mathbf{c}|^2 = 20^2 + 5^2 + 12^2$

$= 569$

**67.** *(46)* We have, $\mathbf{a} \times \mathbf{b} + \mathbf{a} \times \mathbf{c} + \mathbf{b} \times \mathbf{c} = \hat{i} + 8\hat{j} + 13\hat{k}$

Taking cross product with $\mathbf{a}$, we get

$\mathbf{a} \times (\mathbf{a} \times \mathbf{b}) + \mathbf{a} \times (\mathbf{a} \times \mathbf{c}) + \mathbf{a} \times (\mathbf{b} \times \mathbf{c}) = \mathbf{a} \times (\hat{i} + 8\hat{j} + 13\hat{k})$

$\Rightarrow (\mathbf{a} \cdot \mathbf{b})\,\mathbf{a} - \mathbf{a}^2\mathbf{b} + (\mathbf{a} \cdot \mathbf{c})\,\mathbf{a} - \mathbf{a}^2\mathbf{c} + (\mathbf{a} \cdot \mathbf{c})\,\mathbf{b} - (\mathbf{a} \cdot \mathbf{b})\,\mathbf{c} = \mathbf{a} \times (\hat{i} + 8\hat{j} + 13\hat{k})$

$\Rightarrow -26\mathbf{a} - 29\mathbf{b} + 13\mathbf{a} - 29\mathbf{c} + 13\mathbf{b} + 26\mathbf{c} = \mathbf{a} \times (\hat{i} + 8\hat{j} + 13\hat{k})$

$[\because \mathbf{a} \cdot \mathbf{b} = -26]$

$|\mathbf{a}|^2 = 29$

$\Rightarrow -13\mathbf{a} - 16\mathbf{b} - 3\mathbf{c} = \mathbf{a} \times (\hat{i} + 8\hat{j} + 13\hat{k})$

Taking dot product with **b**, we get

$-13\mathbf{a}\cdot\mathbf{b}-16\mathbf{b}^2-3\mathbf{b}\cdot\mathbf{c}$

$=\{\mathbf{a}\times(\hat{\mathbf{i}}+8\hat{\mathbf{j}}+13\hat{\mathbf{k}})\}\cdot\mathbf{b}$

$\Rightarrow(-13)(-26)-16\times 50-3\mathbf{b}\cdot\mathbf{c}$

$=\begin{vmatrix}2&-3&4\\1&8&13\\3&4&-5\end{vmatrix}$ $[|\mathbf{b}^2|=50]$

$\Rightarrow -462-3\mathbf{b}\cdot\mathbf{c}=-396\Rightarrow \mathbf{b}\cdot\mathbf{c}=-22$

$\Rightarrow \quad 24-\mathbf{b}\cdot\mathbf{c}=46$

**68.** *(30)* We have, $\mathbf{a}=\hat{\mathbf{i}}-3\hat{\mathbf{j}}+7\hat{\mathbf{k}}$,

$\mathbf{b}=2\hat{\mathbf{i}}-\hat{\mathbf{j}}+\hat{\mathbf{k}}$

and $(\mathbf{a}+2\mathbf{b})\times\mathbf{c}=3(\mathbf{c}\times\mathbf{a})$

$\Rightarrow \quad \mathbf{a}\times\mathbf{c}+2(\mathbf{b}\times\mathbf{c})=3(\mathbf{c}\times\mathbf{a})$

$\Rightarrow \quad 2(\mathbf{b}\times\mathbf{c})+4(\mathbf{a}\times\mathbf{c})=\mathbf{0}$

$\Rightarrow \quad (\mathbf{b}\times\mathbf{c})+2(\mathbf{a}\times\mathbf{c})=\mathbf{0}$

$\Rightarrow \quad (\mathbf{b}+2\mathbf{a})\times\mathbf{c}=\mathbf{0}$

$\Rightarrow \quad \mathbf{c}=\lambda(\mathbf{b}+2\mathbf{a})$

$\Rightarrow \quad \mathbf{c}\cdot\mathbf{a}=\lambda(\mathbf{b}\cdot\mathbf{a}+2\mathbf{a}\cdot\mathbf{a})$

$\Rightarrow \quad \mathbf{c}\cdot\mathbf{a}=\lambda(12+2\times 59)$

$\Rightarrow \quad \mathbf{c}\cdot\mathbf{a}=\lambda(130)=130$ [given]

$\Rightarrow \quad \lambda=1$

$\therefore \quad \mathbf{c}=4\hat{\mathbf{i}}-7\hat{\mathbf{j}}+15\hat{\mathbf{k}}$

$\therefore \quad \mathbf{b}\cdot\mathbf{c}=(2\hat{\mathbf{i}}-\hat{\mathbf{j}}+\hat{\mathbf{k}})\cdot(4\hat{\mathbf{i}}-7\hat{\mathbf{j}}+15\hat{\mathbf{k}})$

$=8+7+15=30$

**69.** *(54)* Given, area of $\Delta ABC=15\sqrt{2}$

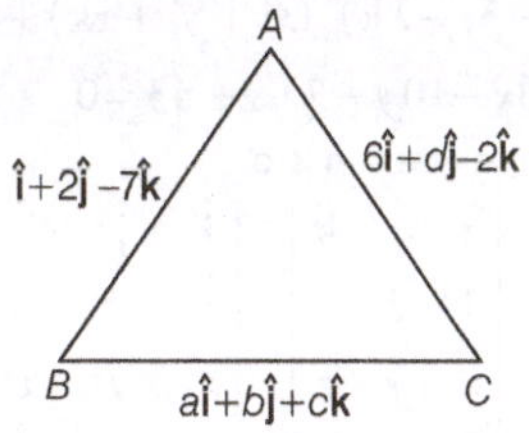


$\Rightarrow \frac{1}{2}\left\|\begin{matrix}\hat{\mathbf{i}}&\hat{\mathbf{j}}&\hat{\mathbf{k}}\\1&2&-7\\6&d&-2\end{matrix}\right\|=15\sqrt{2}$

$\Rightarrow |(-4+7d)\hat{\mathbf{i}}-\hat{\mathbf{j}}(-2+42)+\hat{\mathbf{k}}(d-12)|$

$=(15\sqrt{2})(2)$

$\Rightarrow (7d-4)^2+(40)^2+(d-12)^2=1800$

$\Rightarrow \quad 50d^2-80d-40=0$

$\Rightarrow \quad 5d^2-8d-4=0$

$\Rightarrow 5d^2-10d+2d-4=0$

$\Rightarrow \quad 5d^2-8a-4=0$

$\Rightarrow \quad d=2$ or $d=-2/5$

But $d>0$

$\therefore \quad d=2$ only

So, $(a+1)\hat{\mathbf{i}}+(b+2)\hat{\mathbf{j}}+(c-7)\hat{\mathbf{k}}$

$=6\hat{\mathbf{i}}+2\hat{\mathbf{j}}-2\hat{\mathbf{k}}$

$\Rightarrow \quad a+1=6$ and $b+2=2$

and $\quad c-7=-2$

$\Rightarrow \quad a=5, b=0, c=5$

Now, $|AB|=\sqrt{1+4+49}=\sqrt{54}$,

$|BC|=\sqrt{25+25}=\sqrt{50}$

and $\quad |AC|=\sqrt{36+4+4}=\sqrt{44}$

Thus, square of the length of the largest side $=(\sqrt{54})^2=54$

**70.** *(38)* We have,

$\mathbf{a}=\hat{\mathbf{i}}+\hat{\mathbf{j}}+\hat{\mathbf{k}}, \mathbf{b}=-\hat{\mathbf{i}}-8\hat{\mathbf{j}}+2\hat{\mathbf{k}}$

and $\mathbf{c}=4\hat{\mathbf{i}}+c_2\hat{\mathbf{j}}+c_3\hat{\mathbf{k}}$

Given, $\mathbf{b}\times\mathbf{a}=\mathbf{c}\times\mathbf{a}$

$-10\hat{\mathbf{i}}+3\hat{\mathbf{j}}+7\hat{\mathbf{k}}=(c_2-c_3)\hat{\mathbf{i}}+(c_3-4)\hat{\mathbf{j}}$
$+(4-c_2)\hat{\mathbf{k}}$

$c_2-c_3=-10, c_3-4=3, 4-c_2=7$

$c_3=7, c_2=-3$

So, $\mathbf{c}=4\hat{\mathbf{i}}-3\hat{\mathbf{j}}+7\hat{\mathbf{k}}$

Angle between $c$ and $(3\hat{\mathbf{i}}+4\hat{\mathbf{j}}+\hat{\mathbf{k}})$ is $\theta$.

$\therefore \cos\theta=\dfrac{12-12+7}{\sqrt{74}\cdot\sqrt{26}}\Rightarrow\dfrac{7}{\sqrt{74}\cdot\sqrt{26}}$

Now, $\tan^2\theta=\sec^2\theta-1\Rightarrow\dfrac{74\times 26}{49}-1$

$=\dfrac{1924-49}{49}=\dfrac{1875}{49}=38.26$

$\therefore \quad \tan^2\theta=38$

**71.** *(38)* Given, $\mathbf{a}=3\hat{\mathbf{i}}+2\hat{\mathbf{j}}+\hat{\mathbf{k}}$

$\mathbf{b}=2\hat{\mathbf{i}}-\hat{\mathbf{j}}+3\hat{\mathbf{k}}$

$\mathbf{a}\times\mathbf{b}=\begin{vmatrix}\hat{\mathbf{i}}&\hat{\mathbf{j}}&\hat{\mathbf{k}}\\3&2&1\\2&-1&3\end{vmatrix}$

$=7\hat{\mathbf{i}}-7\hat{\mathbf{j}}-7\hat{\mathbf{k}}$

Let $\mathbf{c}=x\hat{\mathbf{i}}+y\hat{\mathbf{j}}+z\hat{\mathbf{k}}$

$(\mathbf{a}+\mathbf{b})\times\mathbf{c}=2(\mathbf{a}\times\mathbf{b})+24\hat{\mathbf{j}}-6\hat{\mathbf{k}}$

$(5\hat{\mathbf{i}}+\hat{\mathbf{j}}+4\hat{\mathbf{k}})\times\mathbf{c}$

$=2(7\hat{\mathbf{i}}-7\hat{\mathbf{j}}-7\hat{\mathbf{k}})+24\hat{\mathbf{j}}-6\hat{\mathbf{k}}$

$(5\hat{\mathbf{i}}+\hat{\mathbf{j}}+4\hat{\mathbf{k}})\times(x\hat{\mathbf{i}}+y\hat{\mathbf{j}}+z\hat{\mathbf{k}})$

$=14\hat{\mathbf{i}}+10\hat{\mathbf{j}}-20\hat{\mathbf{k}}$

$\begin{vmatrix}\hat{\mathbf{i}}&\hat{\mathbf{j}}&\hat{\mathbf{k}}\\5&1&4\\x&y&z\end{vmatrix}=14\hat{\mathbf{i}}+10\hat{\mathbf{j}}-20\hat{\mathbf{k}}$

$(z-4y)\hat{\mathbf{i}}+(4x-5z)\hat{\mathbf{j}}+(5y-x)\hat{\mathbf{k}}$

$=14\hat{\mathbf{i}}+10\hat{\mathbf{j}}-20\hat{\mathbf{k}}$

$z-4y=14$

$4x-5z=10$

$5y-x=-20$

$x=5, y=-3$ and $z=2$

$|\mathbf{c}|^2=x^2+y^2+z^2$

$=25+9+4=38$

**72.** *(48)* Given, $|\mathbf{a}|=1, |\mathbf{b}|=4, \mathbf{a}\cdot\mathbf{b}=2$

$\mathbf{c}=(2\mathbf{a}\times\mathbf{b})-3\mathbf{b}$, angle between **b** and **c** is $\alpha$

Now, $\mathbf{a}\cdot\mathbf{b}=2$

$|a||b|\cos\theta=2$

$1\times 4\cos\theta=2$

$\cos\theta=\dfrac{1}{2}$

$\therefore$ Angle between **a** and **b**

$\theta=\dfrac{\pi}{3}$

$\mathbf{c}=(2\mathbf{a}\times\mathbf{b})-3\mathbf{b}$

$\mathbf{b}\cdot\mathbf{c}=\mathbf{b}\cdot(2\mathbf{a}\times\mathbf{b})-3\cdot\mathbf{b}\cdot\mathbf{b}$

$|\mathbf{b}||\mathbf{c}|\cos\alpha=0-3|\mathbf{b}|^2$

$|\mathbf{c}|\cos\alpha=-3|\mathbf{b}|=-12$ ...(i)

Again,

$\mathbf{c}=2\mathbf{a}\times\mathbf{b}-3\mathbf{b}$

$|\mathbf{c}|^2=(2\mathbf{a}\times\mathbf{b}-3\mathbf{b})^2$

$=|2\mathbf{a}\times\mathbf{b}|^2+9|b|^2-6(2\mathbf{a}\times\mathbf{b})\cdot\mathbf{b}$

$=4|\mathbf{a}|^2|\mathbf{b}|^2\sin^2\dfrac{\pi}{3}+9|\mathbf{b}|^2-0$

$=4\times 1\times 16\times\dfrac{3}{4}+9\times 16$

$=48+144=192$

By Eq. (i), we get

$|\mathbf{c}|^2\cos^2\alpha=144$

$\cos^2\alpha=\dfrac{144}{192}$

$\sin^2\alpha=1-\dfrac{144}{192}=\dfrac{192-144}{192}$

$\therefore 192\sin^2\alpha=48$

**73.** *(66)* Given, $\mathbf{a}=3\hat{\mathbf{i}}+\hat{\mathbf{j}}-\hat{\mathbf{k}}$

$\mathbf{c}=2\hat{\mathbf{i}}-3\hat{\mathbf{j}}+3\hat{\mathbf{k}}$

and given $\mathbf{a}=\mathbf{b}\times\mathbf{c}$

$|\mathbf{a}|=\sqrt{11}, |\mathbf{c}|=\sqrt{22}$

$|\mathbf{a}|=|\mathbf{b}\times\mathbf{c}|=|\mathbf{b}|\,|\mathbf{c}|\sin\theta$

$\sqrt{11}=\sqrt{50}\sqrt{22}\sin\theta$

$\Rightarrow \quad \sin\theta=\dfrac{1}{10}$

$|\mathbf{b}+\mathbf{c}|^2=|\mathbf{b}|^2+|\mathbf{c}|^2+2\mathbf{b}\cdot\mathbf{c}$

$|\mathbf{b}|^2+|\mathbf{c}|^2+2|\mathbf{b}||\mathbf{c}|\cos\theta$

$=50+22+2\times\sqrt{50}\times\sqrt{22}\times\frac{\sqrt{99}}{10}$

$=72+66$

Now, $|72-|\mathbf{b}+\mathbf{c}|^2|=66$

**74.** *(285)* Given: $\mathbf{a}=\hat{i}+2\hat{j}+3\hat{k}$,

$\mathbf{b}=\hat{i}+\hat{j}-\hat{k}$

$\mathbf{a}\cdot\mathbf{c}=11,\ \mathbf{b}\cdot(\mathbf{a}\times\mathbf{c})=27,\ \mathbf{b}\cdot\mathbf{c}=-\sqrt{3}\,|b|$

$\Rightarrow \mathbf{b}\cdot\mathbf{c}=-\sqrt{3}\times\sqrt{3}$

$[\because |b|=\sqrt{1^2+1^2+(-1)^2}=\sqrt{3}]$

$\Rightarrow \mathbf{b}\cdot\mathbf{c}=-3$

Let $\mathbf{c}=x\hat{i}+y\hat{j}+z\hat{k}$

Then, using $\mathbf{a}\cdot\mathbf{c}=11$

$\Rightarrow x+2y+3z=11$ ...(i)

and Using $\mathbf{b}\cdot\mathbf{c}=-3$

$\Rightarrow x+y-z=-3$ ...(ii)

and Using $\mathbf{b}\cdot(\mathbf{a}\times\mathbf{c})=27$

$\Rightarrow (\mathbf{a}\times\mathbf{b})\cdot\mathbf{c}=27$

$$\therefore \mathbf{a}\times\mathbf{b}=\begin{vmatrix}\hat{i}&\hat{j}&\hat{k}\\1&2&3\\1&1&-1\end{vmatrix}=-5\hat{i}+4\hat{j}-\hat{k}$$

$\Rightarrow -5x+4y-z=-27$

$\Rightarrow 5x-4y+z=27$ ...(iii)

On adding Eqs. (ii) and (iii), we get

$6x-3y=24$

$\Rightarrow \quad 2x-y=8$

$\Rightarrow \quad y=2x-8$ ...(iv)

On multiplying Eq. (i) with 2 times of Eq. (ii), we get

$4x+5y=2$

Using Eq. (iv), we get

$4x+5(2x-8)=2$

$\Rightarrow x=3$

From Eq. (iv), $y=2\times3-8=-2$

Now, from Eq. (i),

$3+2(-2)+3z=11$

$\Rightarrow \quad z=4$

So, $\mathbf{c}=3\hat{i}-2\hat{j}+4\hat{k}$

Now, $|\mathbf{a}\times\mathbf{c}|^2=-(\mathbf{a}\cdot\mathbf{c})^2+a^2c^2$

$=-(3-4+12)^2+14\times29$

$=-121+406=285$

**75.** *(11)* $\mathbf{a}=6\hat{i}+9\hat{j}+12\hat{k}$

$\mathbf{b}=\alpha\hat{i}+11\hat{j}-2\hat{k}$

$\mathbf{a}\cdot\mathbf{b}=6\alpha+99-24=6\alpha+75$

$|\mathbf{a}|=\sqrt{36+81+144}=\sqrt{261}$

Since, $\mathbf{a}\times\mathbf{c}=\mathbf{a}\times\mathbf{b}$, $\mathbf{a}\cdot\mathbf{c}=-12$

and, $\mathbf{c}\cdot(\hat{i}\hat{i}-2\hat{j}+\hat{k})=5$

$\mathbf{a}\times\mathbf{c}=\mathbf{a}\times\mathbf{b}$

$\Rightarrow \mathbf{a}\times(\mathbf{c}-\mathbf{b})=0$

$\Rightarrow \mathbf{c}-\mathbf{b}\,||\,\mathbf{a}$

$\Rightarrow \mathbf{c}=\mathbf{b}+\mathbf{a}\lambda$

Now, $\mathbf{c}\cdot\mathbf{a}=\mathbf{b}\cdot\mathbf{a}+|\mathbf{a}|^2\lambda$

$\Rightarrow 6\alpha+75+261\lambda=-12$

$\Rightarrow 6\alpha+261\lambda=-87$ ...(i)

Also, $(\mathbf{b}+\lambda\mathbf{a})\cdot(\hat{i}-2\hat{j}+\hat{k})=5$

$\Rightarrow (\alpha-22-2)+\lambda(6-18+12)=5$

$\Rightarrow \alpha=29$

From Eq. (i),

$\lambda=-1$

Hence,

$\mathbf{c}(\hat{i}+\hat{j}+\hat{k})=(\mathbf{b}-\mathbf{a})\cdot(\hat{i}+\hat{j}+\hat{k})=11$

**76.** *(3)* Given, $|\mathbf{a}|=\sqrt{31},\ 4|\mathbf{b}|=|\mathbf{c}|=2$, $2(\mathbf{a}\times\mathbf{b})=3(\mathbf{c}\times\mathbf{a})$ and angle between $\mathbf{b}$ and $\mathbf{c}$ is $\frac{2\pi}{3}$.

Now, $\quad 2(\mathbf{a}\times\mathbf{b})=3(\mathbf{c}\times\mathbf{a})$

$\Rightarrow \quad 3(\mathbf{c}\times\mathbf{a})+2(\mathbf{b}\times\mathbf{c})=0$

$\Rightarrow \quad (3\mathbf{c}+2\mathbf{b})\times\mathbf{a}=0$

$\therefore \quad (3\mathbf{c}+2\mathbf{b})\,||\,\mathbf{a} \Rightarrow 3\mathbf{c}+2\mathbf{b}=\lambda\mathbf{a}$

$\Rightarrow \quad |3\mathbf{c}+2\mathbf{b}|=|\lambda\mathbf{a}|$

$\Rightarrow \quad |3\mathbf{c}+2\mathbf{b}|^2=\lambda^2|\mathbf{a}|^2$

$\Rightarrow \quad (3\mathbf{c}+2\mathbf{b})\cdot(3\mathbf{c}+2\mathbf{b})=\lambda^2|\mathbf{a}|^2$

$\Rightarrow 9|\mathbf{c}|^2+4|\mathbf{b}|^2+12\mathbf{c}\cdot\mathbf{b}=\lambda^2|\mathbf{a}|^2$

$\Rightarrow \quad 9|\mathbf{c}|^2+4|\mathbf{b}|^2+12|\mathbf{c}||\mathbf{b}|\cos\frac{2\pi}{3}=\lambda^2|\mathbf{a}|^2$

$\Rightarrow 9(2)^2+4\left(\frac{1}{2}\right)^2+12\times2\times\frac{1}{2}\times\left(-\frac{1}{2}\right)=\lambda^2(\sqrt{31})^2$

$\Rightarrow \quad 36+1-6=31\lambda^2 \Rightarrow 31=31\lambda^2$

$\Rightarrow \quad \lambda^2=1 \Rightarrow \lambda=\pm1$

$\therefore \quad 3\mathbf{c}+2\mathbf{b}=\pm\mathbf{a}$

Now, $\quad 3\mathbf{c}+2\mathbf{b}=\pm\mathbf{a}$

$\Rightarrow \quad (3\mathbf{c}+2\mathbf{b})\cdot\mathbf{b}=\pm\mathbf{a}\cdot\mathbf{b}$

$\Rightarrow \quad 3\mathbf{c}\cdot\mathbf{b}+2|\mathbf{b}|^2=\pm\mathbf{a}\cdot\mathbf{b}$

$\Rightarrow \quad 3|\mathbf{c}||\mathbf{b}|\cos\frac{2\pi}{3}+2|\mathbf{b}|^2=\mathbf{a}\cdot\mathbf{b}$

$\Rightarrow 3\times2\times\frac{1}{2}\times\left(-\frac{1}{2}\right)+2\left(\frac{1}{2}\right)^2=\pm\mathbf{a}\cdot\mathbf{b}$

$\Rightarrow \quad -\frac{3}{2}+\frac{1}{2}=\pm\mathbf{a}\cdot\mathbf{b}$

$\Rightarrow \quad \mathbf{a}\cdot\mathbf{b}=\pm1$

$\Rightarrow \quad (\mathbf{a}\cdot\mathbf{b})^2=1$

Again, $|\mathbf{a}\times\mathbf{b}|^2=|\mathbf{a}|^2|\mathbf{b}|^2-|\mathbf{a}\cdot\mathbf{b}|^2$

$=(\sqrt{31})^2\left(\frac{1}{2}\right)^2-(1)^2=\frac{31}{4}-1=\frac{27}{4}$

Now, $3(\mathbf{c}\times\mathbf{a})=2(\mathbf{a}\times\mathbf{b})$

$\Rightarrow \mathbf{c}\times\mathbf{a}=\frac{2}{3}\times(\mathbf{a}\times\mathbf{b})$

$\Rightarrow |\mathbf{c}\times\mathbf{a}|=\frac{2}{3}|\mathbf{a}\times\mathbf{b}|$

$\Rightarrow |\mathbf{c}\times\mathbf{a}|^2=\frac{4}{9}|\mathbf{a}\times\mathbf{b}|^2=\frac{4}{9}\times\frac{27}{4}=3$

$\therefore \quad \left(\frac{\mathbf{a}\times\mathbf{c}}{\mathbf{a}\cdot\mathbf{b}}\right)^2=\frac{3}{1}=3$

**77.** *(36)* We know that

$|\mathbf{a}\cdot\mathbf{b}|^2+|\mathbf{a}\times\mathbf{b}|^2=|\mathbf{a}|^2|\mathbf{b}|^2$

$\Rightarrow \quad |\mathbf{a}\cdot\mathbf{b}|^2+(\sqrt{48})^2=(\sqrt{14})^2(\sqrt{6})^2$

$\Rightarrow \quad |\mathbf{a}\cdot\mathbf{b}|^2+48=14\times6$

$\Rightarrow |\mathbf{a}\cdot\mathbf{b}|^2+48=84 \Rightarrow |\mathbf{a}\cdot\mathbf{b}|^2=36$

**78.** *(8)* Given, $\mathbf{a}=\hat{i}+2\hat{j}+\lambda\hat{k}$,

$\mathbf{b}=3\hat{i}-5\hat{j}-\lambda\hat{k}$

Let $\quad \mathbf{c}=x\hat{i}+y\hat{j}+z\hat{k}$

$\because$ Given, $\mathbf{a}\cdot\mathbf{c}=7$

$\therefore \quad (\hat{i}+2\hat{j}+\lambda\hat{k})\cdot(x\hat{i}+y\hat{j}+z\hat{k})=7$

$x+2y+\lambda z=7$ ...(i)

Also, $2\mathbf{b}\cdot\mathbf{c}+43=0$

$2(3\hat{i}-5\hat{j}-\lambda\hat{k}).(x\hat{i}+y\hat{j}+z\hat{k})+43=0$

$6x-10y-2\lambda z+43=0$ ...(ii)

Again, $\mathbf{a}\times\mathbf{c}=\mathbf{b}\times\mathbf{c}$

$$\therefore \begin{vmatrix}\hat{i}&\hat{j}&\hat{k}\\1&2&\lambda\\x&y&z\end{vmatrix}=\begin{vmatrix}\hat{i}&\hat{j}&\hat{k}\\3&-5&-\lambda\\x&y&z\end{vmatrix}$$

$\Rightarrow \hat{i}(2z-\lambda y)-\hat{j}(z-\lambda x)+\hat{k}(y-2x)$

$=\hat{i}(-5z+\lambda y)-\hat{j}(3z+\lambda x)+\hat{k}(3y+5x)$

$\therefore$ On comparing

$2z-\lambda y=-5z+\lambda y \Rightarrow 7z=2\lambda y$ ...(iii)

$z-\lambda x=3z+\lambda x \Rightarrow -2z=2\lambda x$ ...(iv)

and $y-2x=3y+5x \Rightarrow -2y=7x$

On putting, $x=\frac{-2y}{7}$ in Eqs. (i) and (ii), we get $\frac{-2y}{7}+2y+\lambda z=7$

$\Rightarrow \quad 12y+7\lambda z=49$ ...(A)

Also, $6\left(\frac{-2y}{7}\right)-10y-2\lambda z+43=0$

$$\frac{-12y-70y-14\lambda z}{7}=-43$$

$\Rightarrow$ $82y+14\lambda z=301$ ...(B)

$\because$ $7z=2\lambda y \Rightarrow z=\frac{2\lambda y}{7}$

From Eqs. (A) and (B), we get

$$12y+7\lambda\left(\frac{2\lambda y}{7}\right)=49$$

$$12y+2\lambda^2 y=49 \quad ...(C)$$

Also, $82y+14\lambda\left(\frac{2\lambda y}{7}\right)=301$

$$82y+4\lambda^2 y=301 \quad ...(D)$$

Multiply by 2 in Eq. (C), then subtract Eq. (D) from it

$$24y+4\lambda^2 y=98$$
$$-82y-4\lambda^2 y=-301$$
$$58y=203$$

$\Rightarrow$ $y=\frac{203}{58}=\frac{7}{2}$

$\therefore$ $x=\frac{-2y}{7}=\frac{-2}{7}\times\frac{7}{2}=-1$

and $z=\frac{2\lambda}{7}\times\frac{7}{2}=\lambda \Rightarrow z=\lambda$

From Eq. (i), we get

$$-1+7+\lambda^2=7$$

$\Rightarrow$ $\lambda^2=1 \Rightarrow \lambda=\pm 1$

$\therefore |\mathbf{a}\cdot\mathbf{b}|=|(\hat{i}+2\hat{j}+\lambda\hat{k}).(3\hat{i}-5\hat{j}-\lambda\hat{k})|$

$=|3-10-\lambda^2|=|3-10-1|=|-8|=8$

**79.** *(14)* $|\mathbf{a}+\mathbf{b}|^2=|\mathbf{a}|^2+2|\mathbf{b}|^2$

$\Rightarrow \mathbf{a}\cdot\mathbf{b}=3$ and $|\mathbf{a}\times\mathbf{b}|^2=75$

$\therefore$ $|\mathbf{a}|^2+|\mathbf{b}|^2+2\mathbf{a}\cdot\mathbf{b}=|\mathbf{a}|^2+2|\mathbf{b}|^2$

$\Rightarrow$ $2\mathbf{a}\cdot\mathbf{b}=|\mathbf{b}|^2$

$\Rightarrow$ $2\times 3=|\mathbf{b}|^2 \Rightarrow |\mathbf{b}|^2=6$

Also, $(\mathbf{a}\cdot\mathbf{b})^2+|\mathbf{a}\times\mathbf{b}|^2=|\mathbf{a}|^2|\mathbf{b}|^2$

$\Rightarrow 9+75=6|\mathbf{a}|^2 \Rightarrow |\mathbf{a}|^2=\frac{84}{6}=14$

**80.** *(10)*

> **key idea** Find $a\cdot b$ using dot product formula and also find $a\cdot b$ using given education $a+(b\times c)$ and analyse both result.

$\mathbf{a}=\hat{i}-2\hat{j}+3\hat{k}, \mathbf{b}=\hat{i}+\hat{j}+\hat{k}$

$\mathbf{a}+(\mathbf{b}\times\mathbf{c})=0$ and $\mathbf{b}\cdot\mathbf{c}=5$

$\mathbf{a}+(\mathbf{b}\times\mathbf{c})=0$

$\Rightarrow$ $\mathbf{a}\times\mathbf{b}+(\mathbf{b}\times\mathbf{c})\times\mathbf{b}=0\times\mathbf{b}$

$\Rightarrow$ $\mathbf{a}\times\mathbf{b}+(\mathbf{b}\cdot\mathbf{b})\mathbf{c}-(\mathbf{c}\cdot\mathbf{b})\mathbf{b}=0$

$[\because (\mathbf{p}\times\mathbf{q})\times\mathbf{r}=(\mathbf{p}\cdot\mathbf{r})\mathbf{q}-(\mathbf{q}\cdot\mathbf{r})\mathbf{p}]$

$\Rightarrow$ $\mathbf{a}\times\mathbf{b}+|\mathbf{b}|^2\mathbf{c}-5\mathbf{b}=0$

$\Rightarrow \mathbf{c}=\frac{5\mathbf{b}-\mathbf{a}\times\mathbf{b}}{1+1+1}$

$=\frac{((5\hat{i}+5\hat{j}+5\hat{k})-\mathbf{a}\times\mathbf{b}}{3}$

$$\therefore \mathbf{a}\times\mathbf{b}=\begin{vmatrix}\hat{i} & \hat{j} & \hat{k}\\ 1 & -2 & 3\\ 1 & 1 & 1\end{vmatrix}$$

$=\hat{i}(-2-3)-\hat{j}(1-3)+\hat{k}(1+2)$

$=-5\hat{i}+2\hat{j}+3\hat{k}$

$\therefore$ $\mathbf{c}=\frac{10\hat{i}+3\hat{j}+2\hat{k}}{3}$

$\therefore$ $3\mathbf{a}\cdot\mathbf{c}=(1)(10)+(-2)(3)+(3)(2)$

$=10-6+6=10$

**81.** *(14)* Let $\mathbf{a}=x\hat{i}+y\hat{j}+z\hat{k}$

$$\therefore \mathbf{a}\times\mathbf{b}=\begin{vmatrix}\hat{i} & \hat{j} & \hat{k}\\ x & y & z\\ 1 & 1 & \lambda\end{vmatrix}=13\hat{i}-\hat{j}-4\hat{k}$$

$\Rightarrow$ $\hat{i}(\lambda y-z)-\hat{j}(x\lambda-z)+\hat{k}(x-y)$

$=13\hat{i}-\hat{j}-4\hat{k}$

$\Rightarrow$ $\lambda y-z=13$ ...(i)

and $x\lambda-z=1$ ...(ii)

and $x-y=-4$ ...(iii)

On subtracting Eq. (i) from Eq. (ii), we get

$\lambda(x-y)=-12$

$\Rightarrow$ $\lambda\times(-4)=-12 \Rightarrow \lambda=3$

$3y-z=13, 3x-z=1$ and $x-y=-4$

Also, $\mathbf{ab}+21=0$

$x+y+\lambda z+21=0$

$\Rightarrow y-4+y+3(3y-13)-21=0$

$\Rightarrow$ $11y-22=0 \Rightarrow y=2$

$\Rightarrow$ $x=-2$ and $z=-7$

$\mathbf{a}=-2\hat{i}+2\hat{j}-7\hat{k}$

$\mathbf{b}-\mathbf{a}=(\hat{i}+\hat{j}+3\hat{k})-(-2\hat{i}+2\hat{j}-7\hat{k})$

$=3\hat{i}-\hat{j}+13\hat{k}$

$\therefore (\mathbf{b}-\mathbf{a})\cdot(\hat{k}-\hat{j})+(\mathbf{b}+\mathbf{a})(\hat{i}-\hat{k})$

$=(3\hat{i}-\hat{j}+10\hat{k})\cdot(\hat{k}-\hat{j})$

$+(-\hat{i}+3\hat{j}-4\hat{k})(\hat{i}-\hat{k})$

$=1+10+(-1)+4=14$

**82.** *(576)* $|(\mathbf{a}-\mathbf{b})\times(\mathbf{a}+\mathbf{b})|^2+4(\mathbf{a}\cdot\mathbf{b})^2$

$=|\mathbf{a}\times\mathbf{a}-\mathbf{b}\times\mathbf{a}+\mathbf{a}\times\mathbf{b}-\mathbf{b}\times\mathbf{b}|^2$

$+4(\mathbf{a}\cdot\mathbf{b})^2$

$=|2\mathbf{a}\times\mathbf{b}|^2+4(\mathbf{a}\cdot\mathbf{b})^2$

$[\because \mathbf{a}\times\mathbf{a}=0$ and $\mathbf{b}\times\mathbf{a}=-(\mathbf{a}\times\mathbf{b})]$

$=4|\mathbf{a}|^2|\mathbf{b}|^2\sin^2\theta+4|\mathbf{a}|^2|\mathbf{b}|^2\cos^2\theta$

$=4|\mathbf{a}|^2|\mathbf{b}|^2[\sin^2\theta+\cos^2\theta]$

$=4\times 4^2\times 3^2=4\times 16\times 9=576$

**83.** *(90)* Given, $\mathbf{a}=\hat{i}+5\hat{j}+\alpha\hat{k}$

$\mathbf{b}=\hat{i}+3\hat{j}+\beta\hat{k}$

and $\mathbf{c}=-\hat{i}+2\hat{j}-3\hat{k}$

$\because$ $\mathbf{a}\perp\mathbf{b}\Rightarrow\mathbf{a}\cdot\mathbf{b}=0$

$\Rightarrow$ $(\hat{i}+5\hat{j}+\alpha\hat{k})\cdot(\hat{i}+3\hat{j}+\beta\hat{k})=0$

$\Rightarrow$ $1+15+\alpha\beta=0$

or $\alpha\beta=-16$ ...(i)

$$\text{Now, } \mathbf{b}\times\mathbf{c}=\begin{vmatrix}\hat{i} & \hat{j} & \hat{k}\\ 1 & 3 & \beta\\ -1 & 2 & -3\end{vmatrix}$$

$=\hat{i}(-9-2\beta)-\hat{j}(-3+\beta)+\hat{k}(2+3)$

$\mathbf{b}\times\mathbf{c}=\hat{i}(-9-2\beta)+\hat{j}(3-\beta)+5\hat{k}$

Given, $|\mathbf{b}\times\mathbf{c}|=5\sqrt{3}\Rightarrow|\mathbf{b}\times\mathbf{c}|^2=75$

$\Rightarrow (-9-2\beta)^2+(3-\beta)^2+25=75$

$\Rightarrow \beta^2+6\beta+8=0\Rightarrow\beta=-2,-4$

From Eq. (i), we get

For $\beta=-2, \alpha=8$

For $\beta=-4, \alpha=4$

For maximum value of $|\mathbf{a}|^2, \alpha=8$

$\therefore |\mathbf{a}|^2=1+25+64=90$

**84.** *(3)* $\mathbf{p}=(2,3,1), \mathbf{q}=(1,2,1)$

$\mathbf{r}$ is perpendicular to $\mathbf{p}+\mathbf{q}$ and $\mathbf{p}-\mathbf{q}$.

$\mathbf{p}+\mathbf{q}=(3,5,2)$

$\mathbf{p}-\mathbf{q}=(1,1,0)$

$\mathbf{r}$ is parallel to $(\mathbf{p}+\mathbf{q})\times(\mathbf{p}-\mathbf{q})$

$$\Rightarrow \begin{vmatrix}i & j & k\\ 3 & 5 & 2\\ 1 & 1 & 0\end{vmatrix}=(-2,2,-2)$$

$\Rightarrow$ $\mathbf{r}=\lambda(-1,1,-1)\Rightarrow|\mathbf{r}|=\sqrt{3}$

$\Rightarrow$ $\lambda\sqrt{1+1+1}=\sqrt{3}\Rightarrow\lambda=1$

$\therefore$ $\mathbf{r}=(-1,1,-1)\Rightarrow\alpha=-1,\beta=1$

and $\gamma=-1$

$\therefore$ $|\alpha|+|\beta|+|\gamma|=3$

**85.** *(2)* Area of parallelogram $=|\mathbf{a}\times\mathbf{b}|$

$=|(\hat{i}+\alpha\hat{j}+3\hat{k})\times(3\hat{i}-\alpha\hat{j}+\hat{k})|$

$\Rightarrow 8\sqrt{3}=|4\alpha\hat{i}+8\hat{j}-4\alpha\hat{k}|$

[given, area $=8\sqrt{3}$]

On squaring on both sides, we get

$(64)(3)=16\alpha^2+64+16\alpha^2\Rightarrow\alpha^2=4$

Now, $\mathbf{a}\cdot\mathbf{b}=3-\alpha^2+3$

$=6-\alpha^2=6-4=2$

# C-23

JEE Main Chapterwise Topicwise Mathematics

In JEE Main, there are usually 0-3 questions per paper are visible from this chapter. In past 3 years, 64 conceptual and 23 numerical questions were asked from this chapter.

**For detailed analysis scan QR code**

# THREE-DIMENSIONAL GEOMETRY

## Topic 1 System of Co-ordinates, Direction Cosines and Direction Ratios of a Line

### Level 1

**01** If two straight line whose direction cosines are given by relations $l+m-n=0$, $3l^2+m^2+cnl=0$ are parallel, then the positive value of $c$ is *[2022, 27 June Shift-I]*

(a) 6 (b) 4
(c) 3 (d) 2

**02** If a point $R(4, y, z)$ lies on the line segment joining the points $P(2,-3,4)$ and $Q(8,0,10)$, then distance of $R$ from the origin is *[2019, 08 April Shift-II]*

(a) $2\sqrt{14}$ (b) $2\sqrt{21}$
(c) 6 (d) $\sqrt{53}$

**03** A line in the 3-dimensional space makes an angle $\theta$ $\left(0<\theta\leq\frac{\pi}{2}\right)$ with both the $X$ and $Y$-axes. Then, the set of all values of $\theta$ is the interval. *[JEE Main 2014]*

(a) $\left[0,\frac{\pi}{4}\right]$ (b) $\left[\frac{\pi}{6},\frac{\pi}{3}\right]$ (c) $\left[\frac{\pi}{4},\frac{\pi}{2}\right]$ (d) $\left[\frac{\pi}{3},\frac{\pi}{2}\right]$

**04** The angle between the lines whose direction cosines satisfy the equation $l+m+n=0$ and $l^2=m^2+n^2$ is *[JEE Main 2014]*

(a) $\frac{\pi}{3}$ (b) $\frac{\pi}{4}$ (c) $\frac{\pi}{6}$ (d) $\frac{\pi}{2}$

**05** A line $AB$ in three-dimensional space makes angles 45° and 120° with the positive $X$-axis and the positive $Y$-axis, respectively. If $AB$ makes an acute angle $\theta$ with the positive $Z$-axis, then $\theta$ is equal to *[AIEEE 2010]*

(a) 30° (b) 45° (c) 60° (d) 75°

**06** The projections of a vector on the three coordinate axes are 6, $-3$ and 2, respectively. The direction cosines of the vector are *[AIEEE 2009]*

(a) $6,-3,2$ (b) $\frac{6}{5},-\frac{3}{5},\frac{2}{5}$
(c) $\frac{6}{7},-\frac{3}{7},\frac{2}{7}$ (d) $-\frac{6}{7},-\frac{3}{7},\frac{2}{7}$

**07** If a line makes an angle of $\pi/4$ with the positive directions of each of $X$-axis and $Y$-axis, then the angle that the line makes with the positive direction of the $Z$-axis is *[AIEEE 2007]*

(a) $\pi/6$ (b) $\pi/3$ (c) $\pi/4$ (d) $\pi/2$

**08** A line makes the same angle $\theta$ with each of the $X$ and $Z$-axes. If the angle $\beta$, which it makes with $Y$-axis, is such that $\sin^2\beta=3\sin^2\theta$, then $\cos^2\theta$ is equal to *[AIEEE 2004]*

(a) 2/3 (b) 1/5 (c) 3/5 (d) 2/5

### Level 2

**09** Let $A(x, y, z)$ be a point in $XY$ plane, which is equidistant from three points $(0,3,2)$, $(2,0,3)$ and $(0,0,1)$. Let $B=(1,4,-1)$ and $C=(2,0,-2)$. Then, among the statements $(S_1):\Delta ABC$ is an isosceles right angled triangle and $(S_2)$ : the area of $\Delta ABC$ is $\frac{9\sqrt{2}}{2}$, *[2025, 28 Jan Shift-I]*

(a) only $(S2)$ is true (b) both are false
(c) only $(S1)$ is true (d) both are true

**10** Let $A$ be the point of intersection of the lines $3x+2y=14$, $5x-y=6$ and $B$ be the point of intersection of the line $4x+3y=8$, $6x+y=5$. The distance of the point $P(5,-2)$ from the line $AB$ is *[2024, 29 Jan Shift-I]*

(a) 13/2 (b) 5/2 (c) 8 (d) 6

**11** $ABC$ is triangle in a plane with vertices $A(2, 3, 5)$, $B(-1, 3, 2)$ and $C(\lambda, 5, \mu)$. If the median through $A$ is equally inclined to the coordinate axes, then the value of $(\lambda^3 + \mu^3 + 5)$ is *[JEE Main 2016]*

(a) 1130 (b) 1348 (c) 1077 (d) 676

## Level 3

**12** The angle between the straight lines, whose direction cosines are given by the equations $2l + 2m - n = 0$ and $mn + nl + lm = 0$, is *[2021, 27 Aug Shift-II]*

(a) $\pi/2$ (b) $\pi - \cos^{-1}(4/9)$

(c) $\cos^{-1}(8/9)$ (d) $\pi/3$

**13** Let $\alpha$ be the angle between the lines whose direction cosines satisfy the equations $l + m - n = 0$ and $l^2 + m^2 - n^2 = 0$. Then, the value of $\sin^4 \alpha + \cos^4 \alpha$ is *[2021, 25 Feb Shift-I]*

(a) $\frac{3}{4}$ (b) $\frac{3}{8}$ (c) $\frac{5}{8}$ (d) $\frac{1}{2}$

## Numerical Value Type Questions

**14** A line with direction ratios 2, 1, 2 meets the lines $x = y + 2 = z$ and $x + 2 = 2y = 2z$ respectively at the points $P$ and $Q$. If the length of the perpendicular from the point (1, 2, 12) to the line $PQ$ is $l$, then $l^2$ is. *[2024, 29 Jan Shift-I]*

**15** Let $P(-2, -1, 1)$ and $Q\left(\frac{56}{17}, \frac{43}{17}, \frac{111}{17}\right)$ be the vertices of the rhombus $PRQS$. If the direction ratios of the diagonal $RS$ are $\alpha, -1, \beta$ where both $\alpha$ and $\beta$ are integers of minimum absolute values, then $\alpha^2 + \beta^2$ is equal to ………… . *[2022, 28 June Shift-I]*

**16** The projection of the line segment joining the points $(1, -1, 3)$ and $(2, -4, 11)$ on the line joining the points $(-1, 2, 3)$ and $(3, -2, 10)$ is……… . *[2020, 09 Jan Shift-I]*

# Topic 2 Lines

## Level 1

**01** If the equation of the line passing through the point $\left(0, -\frac{1}{2}, 0\right)$ and perpendicular to the lines $\mathbf{r} = \lambda(\hat{\mathbf{i}} + a\hat{\mathbf{j}} + b\hat{\mathbf{k}})$ and $\mathbf{r} = (\hat{\mathbf{i}} - \hat{\mathbf{j}} - 6\hat{\mathbf{k}}) + \mu(-b\hat{\mathbf{i}} + a\hat{\mathbf{j}} + 5\hat{\mathbf{k}})$ is $\frac{x-1}{-2} = \frac{y+4}{d} = \frac{z-c}{-4}$, then $a + b + c + d$ is equal to *[2025, 07 April Shift-II]*

(a) 13 (b) 12 (c) 10 (d) 14

**02** Consider the lines $L_1: x - 1 = y - 2 = z$ and $L_2: x - 2 = y = z - 1$. Let the feet of the perpendiculars from the point $P(5, 1, -3)$ on the lines $L_1$ and $L_2$ be $Q$ and $R$ respectively. If the area of the $\Delta PQR$ is $A$, then $4A^2$ is equal to *[2025, 07 April Shift-II]*

(a) 147 (b) 151 (c) 143 (d) 139

**Imp.** Shortest Distance between Two Lines

**03** Line $L_1$ passes through the point (1, 2, 3) and is parallel to $Z$-axis. Line $L_2$ passes through the point $(\lambda, 5, 6)$ and is parallel to $Y$-axis. Let for $\lambda = \lambda_1, \lambda_2$, $\lambda_2 < \lambda_1$, the shortest distance between the two lines be 3. Then, the square of the distance of the point $(\lambda_1, \lambda_2, 7)$ from the line $L_1$ is *[2025, 03 April Shift-I]*

(a) 37 (b) 25

(c) 32 (d) 40

**04** The line $L_1$ is parallel to the vector $\mathbf{a} = -3\hat{\mathbf{i}} + 2\hat{\mathbf{j}} + 4\hat{\mathbf{k}}$ and passes through the point (7, 6, 2) and the line $L_2$ is parallel to the vector $\mathbf{b} = 2\hat{\mathbf{i}} + \hat{\mathbf{j}} + 3\hat{\mathbf{k}}$ and passes through the point (5, 3, 4). The shortest distance between the lines $L_1$ and $L_2$ is *[2025, 02 April Shift-II]*

(a) $\frac{23}{\sqrt{57}}$ (b) $\frac{21}{\sqrt{57}}$ (c) $\frac{21}{\sqrt{38}}$ (d) $\frac{23}{\sqrt{38}}$

**05** Let the vertices $Q$ and $R$ of the $\Delta PQR$ lie on the line $\frac{x+3}{5} = \frac{y-1}{2} = \frac{z+4}{3}$, $QR = 5$ and the coordinates of the point $P$ be $(0, 2, 3)$. If the area of the $\Delta PQR$ is $\frac{m}{n}$, then *[2025, 02 April Shift-I]*

(a) $5m - 2\sqrt{21}n = 0$ (b) $2m - 5\sqrt{21}n = 0$

(c) $5m - 21\sqrt{2}n = 0$ (d) $m - 5\sqrt{21}n = 0$

**06** Let a straight line $L$ pass through the point $P(2, -1, 3)$ and be perpendicular to the lines $\frac{x-1}{2} = \frac{y+1}{1} = \frac{z-3}{-2}$ and $\frac{x-3}{1} = \frac{y-2}{3} = \frac{z+2}{4}$. If the line $L$ intersects the $YZ$-plane at the point $Q$. Then, the distance between the points $P$ and $Q$ is *[2025, 29 Jan Shift-II]*

(a) $\sqrt{10}$ (b) $2\sqrt{3}$ (c) 2 (d) 3

**07** Let $\mathbf{a} = \hat{\mathbf{i}} + 2\hat{\mathbf{j}} + \hat{\mathbf{k}}$ and $\mathbf{b} = 2\hat{\mathbf{i}} + 7\hat{\mathbf{j}} + 3\hat{\mathbf{k}}$.

Let $L_1: \mathbf{r} = (-\hat{\mathbf{i}} + 2\hat{\mathbf{j}} + \hat{\mathbf{k}}) + \lambda\mathbf{a}, \lambda \in R$ and $L_2: \mathbf{r} = (\hat{\mathbf{j}} + \hat{\mathbf{k}}) + \mu\mathbf{b}, \mu \in R$ be two lines. If the line $L_3$ passes through the point of intersection of $L_1$ and $L_2$ and is parallel to $\mathbf{a} + \mathbf{b}$, then $L_3$ passes through the point *[2025, 29 Jan Shift-I]*

(a) (5, 17, 4) (b) (8, 26, 12)

(c) (2, 8, 5) (d) (−1, −1, 1)

**08** If the image of the point (4, 4, 3) in the line $\frac{x-1}{2}=\frac{y-2}{1}=\frac{z-1}{3}$ is $(\alpha, \beta, \gamma)$, then $\alpha+\beta+\gamma$ is equal to **[2025, 28 Jan Shift-I]**

(a) 12 (b) 8 (c) 9 (d) 7

**09** Let the line passing through the points (– 1, 2, 1) and parallel to the line $\frac{x-1}{2}=\frac{y+1}{3}=\frac{z}{4}$ intersect the line $\frac{x+2}{3}=\frac{y-3}{2}=\frac{z-4}{1}$ at the point $P$. Then, the distance of $P$ from the point $Q$ (4, – 5, 1) is **[2025, 24 Jan Shift-I]**

(a) 10 (b) $5\sqrt{5}$ (c) 5 (d) $5\sqrt{6}$

**10** If the square of the shortest distance between the lines $\frac{x-2}{1}=\frac{y-1}{2}=\frac{z+3}{-3}$ and $\frac{x+1}{2}=\frac{y+3}{4}=\frac{z+5}{-5}$ is $\frac{m}{n}$, where $m, n$ are coprime number, then $m+n$ is equal to **[2025, 23 Jan Shift-II]**

(a) 9 (b) 14 (c) 6 (d) 21

**11** The distance of the line $\frac{x-2}{2}=\frac{y-6}{3}=\frac{z-3}{4}$ from the point (1, 4, 0) along the line $\frac{x}{1}=\frac{y-2}{2}=\frac{z+3}{3}$ is **[2025, 23 Jan Shift-II]**

(a) $\sqrt{13}$ (b) $\sqrt{15}$ (c) $\sqrt{17}$ (d) $\sqrt{14}$

**Imp.**

## Distance of a Point from a Line

**12** Let $P$ be the foot of the perpendicular from the point $Q(10, -3, -1)$ on the line $\frac{x-3}{7}=\frac{y-2}{-1}=\frac{z+1}{-2}$.

Then, the area of the right angled $\Delta PQR$, where $R$ is the point (3, – 2, 1) is **[2025, 23 Jan Shift-I]**

(a) $\sqrt{30}$ (b) $9\sqrt{15}$ (c) $8\sqrt{15}$ (d) $3\sqrt{30}$

**13** The perpendicular distance of the line $\frac{x-1}{2}=\frac{y+2}{-1}=\frac{z+3}{2}$ from the point $P(2, -10, 1)$ is **[2025, 22 Jan Shift-II]**

(a) $5\sqrt{2}$ (b) $3\sqrt{5}$ (c) $4\sqrt{3}$ (d) 6

**14** Let a line passing through two distinct points $P(-2, -1, 3)$ and $Q$ and be parallel to the vector $3\hat{i}+2\hat{j}+2\hat{k}$. If the distance of the point $Q$ from the point $R(1, 3, 3)$ is 5, then the square of the area of $\Delta PQR$ is equal to **[2025, 22 Jan Shift-II]**

(a) 144 (b) 136 (c) 148 (d) 140

**15** The shortest distance between the lines $\frac{x-3}{4}=\frac{y+7}{-11}=\frac{z-1}{5}$ and $\frac{x-5}{3}=\frac{y-9}{-6}=\frac{z+2}{1}$ is **[2024, 9 April Shift I]**

(a) $\frac{178}{\sqrt{563}}$ (b) $\frac{185}{\sqrt{563}}$ (c) $\frac{179}{\sqrt{563}}$ (d) $\frac{187}{\sqrt{563}}$

**16** If the shortest distance between the lines
$L_1: \mathbf{r}=(2+\lambda)\hat{i}+(1-3\lambda)\hat{j}+(3+4\lambda)\hat{k}., \lambda \in R$
$L_2: \mathbf{r}=2(1+\mu)\hat{i}+3(1+\mu)\hat{j}+(5+\mu)\hat{k}, \mu \in R$
is $m/\sqrt{n}$, where $\gcd(m, n)=1$, then the value of $m+n$ equals **[2024, 9 April Shift I]**

(a) 387 (b) 377 (c) 384 (d) 390

**17** The shortest distance between the lines $\frac{x-3}{2}=\frac{y+15}{-7}=\frac{z-9}{5}$ and $\frac{x+1}{2}=\frac{y-1}{1}=\frac{z-9}{-3}$ is **[2024, 6 April Shift I]**

(a) $5\sqrt{3}$ (b) $4\sqrt{3}$ (c) $8\sqrt{3}$ (d) $6\sqrt{3}$

**18** Let $(\alpha, \beta, \gamma)$ be the image of the point (8, 5, 7) in the line $\frac{x-1}{2}=\frac{y+1}{3}=\frac{z-2}{5}$. Then, $\alpha+\beta+\gamma$ is equal to **[2024, 5 April Shift II]**

(a) 16 (b) 14 (c) 18 (d) 20

**19** If the line $\frac{2-x}{3}=\frac{3y-2}{4\lambda+1}=4-z$ makes a right angle with the line $\frac{x+3}{3\mu}=\frac{1-2y}{6}=\frac{5-z}{7}$, then $4\lambda+9\mu$ is equal to **[2024, 5 April Shift I]**

(a) 13 (b) 6 (c) 4 (d) 5

**20** Let the point, on the line passing through the points $P(1, -2, 3)$ and $Q(5, -4, 7)$, farther from the origin and at a distance of 9 units from the point $P$ be $(\alpha, \beta, \gamma)$. Then, $\alpha^2+\beta^2+\gamma^2$ is equal to **[2024, 4 April Shift I]**

(a) 155 (b) 165 (c) 150 (d) 160

**21** If the shortest distance between the lines $\frac{x-\lambda}{-2}=\frac{y-2}{1}=\frac{z-1}{1}$ and $\frac{x-\sqrt{3}}{1}=\frac{y-1}{-2}=\frac{z-2}{1}$ is 1, then the sum of all possible values of $\lambda$ is **[2024, 1 Feb Shift-I]**

(a) 0 (b) $2\sqrt{3}$ (c) $3\sqrt{3}$ (d) $2\sqrt{3}$

**22** Let $(\alpha, \beta, \gamma)$ be the mirror image of the point (2, 3, 5) in the line $\frac{x-1}{2}=\frac{y-2}{3}=\frac{z-3}{4}$. Then, $2\alpha+3\beta+4\gamma$ is equal to : **[2024, 31 Jan Shift-II]**

(a) 34 (b) 31 (c) 33 (d) 32

**23** Let $(\alpha, \beta, \gamma)$ be the foot of perpendicular from the point (1, 2, 3) on the line $\frac{x+3}{5}=\frac{y-1}{2}=\frac{z+4}{3}$. Then, $19(\alpha+\beta+\gamma)$ is equal to **[2024, 30 Jan Shift-I]**

(a) 101 (b) 99 (c) 100 (d) 102

**24** Let $\mathbf{a}=\hat{\mathbf{i}}+\hat{\mathbf{j}}+2\hat{\mathbf{k}}$, $\mathbf{b}=2\hat{\mathbf{i}}-3\hat{\mathbf{j}}+\hat{\mathbf{k}}$ and $\mathbf{c}=\hat{\mathbf{i}}-\hat{\mathbf{j}}+\hat{\mathbf{k}}$ be the three given vectors. Let $\mathbf{v}$ be a vector in the plane of $\mathbf{a}$ and $\mathbf{b}$ whose projection on $\mathbf{c}$ is $\frac{2}{\sqrt{3}}$. If $\mathbf{v}\cdot\hat{\mathbf{j}}=7$, then $\mathbf{v}\cdot(\hat{\mathbf{i}}+\hat{\mathbf{k}})$ is equal to *[2022, 26 June Shift-II]*

(a) 6 (b) 7 (d) 8 (d) 9

**25** The equation of the line through the point (0,1,2) and perpendicular to the line $\frac{x-1}{2}=\frac{y+1}{3}=\frac{z-1}{-2}$ is *[2021, 25 Feb Shift-I]*

(a) $\frac{x}{3}=\frac{y-1}{4}=\frac{z-2}{3}$ (b) $\frac{x}{3}=\frac{y-1}{-4}=\frac{z-2}{3}$

(c) $\frac{x}{3}=\frac{y-1}{4}=\frac{z-2}{-3}$ (d) $\frac{x}{-3}=\frac{y-1}{4}=\frac{z-2}{3}$

**26** If $(a, b, c)$ is the image of the point $(1, 2, -3)$ in the line $\frac{x+1}{2}=\frac{y-3}{-2}=\frac{z}{-1}$, then $a+b+c$ is equal to *[2020, 5 Sep Shift-I]*

(a) 3 (b) 2 (c) $-1$ (d) 1

**27** The two lines $x=ay+b, z=cy+d$ and $x=a'y+b', z=c'y+d'$ are perpendicular to each other, if *[AIEEE 2006, 2003]*

(a) $aa'+cc'=1$ (b) $\frac{a}{a'}+\frac{c}{c'}=-1$

(c) $\frac{a}{a'}+\frac{c}{c'}=1$ (d) $aa'+cc'=-1$

**28** The angle between the lines $2x=3y=-z$ and $6x=-y=-4z$ is *[AIEEE 2005]*

(a) 30° (b) 45°

(c) 90° (d) 0°

## Level 2

**29** Let the line $L$ pass through (1, 1, 1) and intersect the lines $\frac{x-1}{2}=\frac{y+1}{3}=\frac{z-1}{4}$ and $\frac{x-3}{1}=\frac{y-4}{2}=\frac{z}{1}$. Then, which of the following points lies on the line $L$? *[2025, 07 April Shift-I]*

(a) (7, 15, 13) (b) (5,4,3)

(c) (4,22, 7) (d) (10, −29, −50)

**30** If the shortest distance between the lines $\frac{x-1}{2}=\frac{y-2}{3}=\frac{z-3}{4}$ and $\frac{x}{1}=\frac{y}{\alpha}=\frac{z-5}{1}$ is $\frac{5}{\sqrt{6}}$, then the sum of all possible values of $\alpha$ is *[2025, 07 April Shift-I]*

(a) 3 (b) −3

(c) $\frac{3}{2}$ (d) $-\frac{3}{2}$

**Imp.** Shortest Distance and Length of Letus Reaction

**31** Let the values of $p$, for which the shortest distance between the lines $\frac{x+1}{3}=\frac{y}{4}=\frac{z}{5}$ and $\mathbf{r}=(p\hat{\mathbf{i}}+2\hat{\mathbf{j}}+\hat{\mathbf{k}})+\lambda(2\hat{\mathbf{i}}+3\hat{\mathbf{j}}+4\hat{\mathbf{k}})$ is $\frac{1}{\sqrt{6}}$, be $a, b$, $(a<b)$. Then, the length of the latus rectum of the ellipse $\frac{x^2}{a^2}+\frac{y^2}{b^2}=1$ is *[2025, 4 April Shift-II]*

(a) $\frac{3}{2}$ (b) $\frac{2}{3}$ (c) 9 (d) 18

**32** Let the shortest distance between the lines $\frac{x-3}{3}=\frac{y-\alpha}{-1}=\frac{z-3}{1}$ and $\frac{x+3}{-3}=\frac{y+7}{2}=\frac{z-\beta}{4}$ be $3\sqrt{30}$. Then, the positive value of $5\alpha+\beta$ is *[2025, 4 April Shift-I]*

(a) 40 (b) 48 (c) 46 (d) 42

**33** Let $A$ and $B$ be two distinct points on the line $L:\frac{x-6}{3}=\frac{y-7}{2}=\frac{z-7}{-2}$. Both $A$ and $B$ are at a distance $2\sqrt{17}$ from the foot of perpendicular drawn from the point (1, 2, 3) on the line $L$. If $O$ is the origin, then $\mathbf{OA}\cdot\mathbf{OB}$ is equal to *[2025, 4 April Shift-I]*

(a) 62 (b) 21 (c) 47 (d) 49

**Imp.** Distance between Lines

**34** The distance of the point (7, 10, 11) from the line $\frac{x-4}{1}=\frac{y-4}{0}=\frac{z-2}{3}$ along the line $\frac{x-9}{2}=\frac{y-13}{3}=\frac{z-17}{6}$ is *[2025, 3 April Shift-II]*

(a) 12 (b) 18 (c) 16 (d) 14

**35** Let a line passing through the point (4, 1, 0) intersect the line $L_1:\frac{x-1}{2}=\frac{y-2}{3}=\frac{z-3}{4}$ at the point $A(\alpha,\beta,\gamma)$ and the $L_2: x-6=y=-z+4$ at the point $B(a,b,c)$. Then, $\begin{vmatrix}1 & 0 & 1\\ \alpha & \beta & \gamma\\ a & b & c\end{vmatrix}$ is equal to *[2025, 3 April Shift-I]*

(a) 6 (b) 12 (c) 16 (d) 8

**36** If the image of the point $P$ (1, 0, 3) in the line joining the points $A$(4, 7, 1) and $B$(3, 5, 3) is $Q(\alpha,\beta,\gamma)$, then $\alpha+\beta+\gamma$ is equal to *[2025, 2 April Shift-II]*

(a) 13 (b) $\frac{47}{3}$ (c) $\frac{46}{3}$ (d) 18

**37** Let $P$ be the foot of the perpendicular from the point $(1,2,2)$ on the line $L: \frac{x-1}{1}=\frac{y+1}{-1}=\frac{z-2}{2}$. Let the line $\hat{r}=(-\hat{i}+\hat{j}-2\hat{k})+\lambda(\hat{i}-\hat{j}+\hat{k}), \lambda \in R$, intersect the line $L$ at $Q$. Then, $2(PQ)^2$ is equal to. **[2025, 29 Jan Shift-II]**

(a) 25 (b) 19 (c) 29 (d) 27

**38** Let $L_1: \frac{x-1}{1}=\frac{y-2}{-1}=\frac{z-1}{2}$ and $L_2: \frac{x+1}{-1}=\frac{y-2}{2}=\frac{z}{1}$ be two lines.

Let $L_3$ be a line passing through the point $(\alpha, \beta, \gamma)$ and be perpendicular to both $L_1$ and $L_2$. If $L_3$ intersects $L_1$, then $|5\alpha-11\beta-8\gamma|$ equals **[2025, 29 Jan Shift-I]**

(a) 16 (b) 20 (c) 18 (d) 25

**39** The square of the distance of the point $\left(\frac{15}{7}, \frac{32}{7}, 7\right)$ from the line $\frac{x+1}{3}=\frac{y+3}{5}=\frac{z+5}{7}$ in the direction of the vector $\hat{i}+4\hat{j}+7\hat{k}$ is **[2025, 28 Jan Shift-II]**

(a) 44 (b) 41 (c) 54 (d) 66

**40** Let in a $\Delta ABC$, the length of the side $AC$ be 6, the vertex $B$ be $(1, 2, 3)$ and the vertices $A, C$ lie on the line $\frac{x-6}{3}=\frac{y-7}{2}=\frac{z-7}{-2}$. Then, the area (in sq. units) of $\Delta ABC$ is **[2025, 24 Jan Shift-I]**

(a) 21 (b) 56 (c) 42 (d) 17

**41** Let the position vectors of the vertices $A$, $B$ and $C$ of a tetrahedron $ABCD$ be $\hat{i}+2\hat{j}+\hat{k}$, $\hat{i}+3\hat{j}-2\hat{k}$ and $2\hat{i}+\hat{j}-\hat{k}$, respectively. The altitude from the vertex $D$ to the opposite face $ABC$ meets the median line segment through $A$ of the $\Delta ABC$ at the point $E$. If the length of $AD$ is $\frac{\sqrt{110}}{3}$ and the volume of the tetrahedron is $\frac{\sqrt{805}}{6\sqrt{2}}$, then the position vector of $E$ is **[2025, 23 Jan Shift-I]**

(a) $\frac{1}{2}(\hat{i}+4\hat{j}+7\hat{k})$ (b) $\frac{1}{12}(7\hat{i}+4\hat{j}+3\hat{k})$

(c) $\frac{1}{6}(7\hat{i}+12\hat{j}+\hat{k})$ (d) $\frac{1}{6}(12\hat{i}+12\hat{j}+\hat{k})$

**42** Let $L_1: \frac{x-1}{2}=\frac{y-2}{3}=\frac{z-3}{4}$ and $L_2: \frac{x-2}{3}=\frac{y-4}{4}=\frac{z-5}{5}$ be two lines. Then, which of the following points lies on the line of the shortest distance between $L_1$ and $L_2$? **[2025, 22 Jan Shift-I]**

(a) $\left(\frac{8}{3}, -1, \frac{1}{3}\right)$ (b) $\left(-\frac{5}{3}, -7, 1\right)$

(c) $\left(\frac{14}{3}, -3, \frac{22}{3}\right)$ (d) $\left(2, 3, \frac{1}{3}\right)$

**43** Consider the line $L$ passing through the points $(1, 2, 3)$ and $(2, 3, 5)$. The distance of the point $\left(\frac{11}{3}, \frac{11}{3}, \frac{19}{3}\right)$ from the line $L$ along the line $\frac{3x-11}{2}=\frac{3y-11}{1}=\frac{3z-19}{2}$ is equal to **[2024, 9 April Shift-II]**

(a) 3 (b) 5 (c) 6 (d) 4

**44** If the shortest distance between the lines $\frac{x-\lambda}{2}=\frac{y-4}{3}=\frac{z-3}{4}$ and $\frac{x-2}{4}=\frac{y-4}{6}=\frac{z-7}{8}$ is $\frac{13}{\sqrt{29}}$, then a value of $\lambda$ is **[2024, 8 April Shift-II]**

(a) $\frac{13}{25}$ (b) $-\frac{13}{25}$ (c) 1 (d) $-1$

**45** Let $d$ be the distance of the point of intersection of the lines $\frac{x+6}{3}=\frac{y}{2}=\frac{z+1}{1}$ and $\frac{x-7}{4}=\frac{y-9}{3}=\frac{z-4}{2}$ from the point $(7, 8, 9)$. Then, $d^2+6$ is equal to **[2024, 8 April Shift-II]**

(a) 72 (b) 78 (c) 69 (d) 75

**46** The shortest distance between lines $L_1$ and $L_2$, where $L_1: \frac{x-1}{2}=\frac{y+1}{-3}=\frac{z+4}{2}$ and $L_2$ is the line passing through the points $A(-4, 4, 3)$, $B(-1, 6, 3)$ and perpendicular to the line $\frac{x-3}{-2}=\frac{y}{3}=\frac{z-1}{1}$ is **[2024, 31 Jan Shift-II]**

(a) $\frac{121}{\sqrt{221}}$ (b) $\frac{24}{\sqrt{117}}$ (c) $\frac{42}{\sqrt{117}}$ (d) $\frac{141}{\sqrt{221}}$

**47** If the shortest distance between the lines $\frac{x-4}{1}=\frac{y+1}{2}=\frac{z}{-3}$ and $\frac{x-\lambda}{2}=\frac{y+1}{4}=\frac{z-2}{-5}$ is $\frac{6}{\sqrt{5}}$, then the sum of all possible values of $\lambda$ is **[2024, 27 Jan Shift-I]**

(a) 7 (b) 8 (c) 5 (d) 10

**48** The distance of the point $(7, -2, 11)$ from the line $\frac{x-6}{1}=\frac{y-4}{0}=\frac{z-8}{3}$ along the line $\frac{x-5}{2}=\frac{y-1}{-3}=\frac{z-5}{6}$, is **[2024, 27 Jan Shift-I]**

(a) 14 (b) 12

(c) 18 (d) 21

**49** Let $S$ be the set of all values of $\lambda$, for which the shortest distance between the lines $\frac{x-\lambda}{0}=\frac{y-3}{4}=\frac{z+6}{1}$ and $\frac{x+\lambda}{3}=\frac{y}{-4}=\frac{z-6}{0}$ is 13.

Then, $8\left|\sum_{\lambda \in S} \lambda\right|$ is equal to *[2023, 15 April shift-I]*

(a) 304 (b) 302 (c) 306 (d) 308

**50** The shortest distance between the lines $\frac{x-1}{2}=\frac{y+8}{-7}=\frac{z-4}{5}$ and $\frac{x-1}{2}=\frac{y-2}{1}=\frac{z-6}{-3}$ is *[2023, 29 Jan Shift-II]*

(a) $5\sqrt{3}$ (b) $3\sqrt{3}$ (c) $4\sqrt{3}$ (d) $2\sqrt{3}$

**51** The distance of the point $P(4, 6, -2)$ from the line passing through the point $(-3, 2, 3)$ and parallel to a line with direction ratios 3, 3, −1 is equal to *[2023, 25 Jan Shift-I]*

(a) $\sqrt{14}$ (b) 3
(c) $2\sqrt{3}$ (d) $\sqrt{6}$

**52** If the two lines $l_1 : \frac{x-2}{3}=\frac{y+1}{-2}, z-2$ and $l_2 : \frac{x-1}{1}=\frac{2y+3}{\alpha}=\frac{z+5}{2}$ are perpendicular, then an angle between the lines $l_2$ and $l_3 : \frac{1-x}{3}=\frac{2y-1}{-4}=\frac{z}{4}$ is : *[2022, 26 June Shift-I]*

(a) $\cos^{-1}\left(\frac{29}{4}\right)$ (b) $\sec^{-1}\left(\frac{29}{4}\right)$
(c) $\cos^{-1}\left(\frac{2}{29}\right)$ (d) $\cos^{-1}\left(\frac{2}{\sqrt{29}}\right)$

**53** The vertices $B$ and $C$ of a $\Delta ABC$ lie on the line, $\frac{x+2}{3}=\frac{y-1}{0}=\frac{z}{4}$ such that $BC = 5$ units. Then, the area (in sq units) of this triangle, given that point $A(1, -1, 2)$ is *[2019, 09 April Shift-II]*

(a) $\sqrt{34}$ (b) $2\sqrt{34}$ (c) $5\sqrt{17}$ (d) 6

**54** If the line $\frac{x-1}{2}=\frac{y+1}{3}=\frac{z-1}{4}$ and $\frac{x-3}{1}=\frac{y-k}{2}=\frac{z}{1}$ intersect, then $k$ is equal to *[AIEEE 2012]*

(a) −1 (b) $\frac{2}{9}$ (c) $\frac{9}{2}$ (d) 0

**55** A line with direction cosines proportional to 2, 1, 2 meets each of the lines $x = y + a = z$ and $x + a = 2y = 2z$. The coordinates of each of the points of intersection are given by *[AIEEE 2004]*

(a) $(3a, 3a, 3a), (a, a, a)$ (b) $(3a, 2a, 3a), (a, a, a)$
(c) $(3a, 2a, 3a), (a, a, 2a)$ (d) $(2a, 3a, 3a), (2a, a, a)$

## Level 3

**56** Let the values of $\lambda$, for which the shortest distance between the lines $\frac{x-1}{2}=\frac{y-2}{3}=\frac{z-3}{4}$ and $\frac{x-\lambda}{3}=\frac{y-4}{4}=\frac{z-5}{5}$ is $\frac{1}{\sqrt{6}}$ be $\lambda_1$ and $\lambda_2$. Then, the radius of the circle passing through the points $(0, 0), (\lambda_1, \lambda_2)$ and $(\lambda_2, \lambda_1)$ is *[2025, 08 April Shift-II]*

(a) 4 (b) $\frac{5\sqrt{2}}{3}$ (c) 3 (d) $\frac{\sqrt{2}}{3}$

**Imp.** Area of Triangle

**57** Let $A$ be the point of intersection of the lines $L_1 : \frac{x-7}{1}=\frac{y-5}{0}=\frac{z-3}{-1}$ and $L_2 : \frac{x-1}{3}=\frac{y+3}{4}=\frac{z+7}{5}$. Let $B$ and $C$ be the points on the lines $L_1$ and $L_2$ respectively such that $AB = AC = \sqrt{15}$. Then, the square of the area of the $\Delta ABC$ is *[2025, 04 April Shift-II]*

(a) 57 (b) 54 (c) 63 (d) 60

**58** Let the line $L$ intersect the lines $x-2=-y=z-1$, $2(x+1)=2(y-1)=z+1$ and be parallel to the lines $\frac{x-2}{3}=\frac{y-1}{1}=\frac{z-2}{2}$. Then, which of the following points lies on $L$? *[2024, 09 April Shift-I]*

(a) $\left(-\frac{1}{3}, -1, -1\right)$ (b) $\left(-\frac{1}{3}, -1, 1\right)$
(c) $\left(-\frac{1}{3}, 1, 1\right)$ (d) $\left(-\frac{1}{3}, 1, -1\right)$

**59** Let $P(\alpha, \beta, \gamma)$ be the image of the point $Q(3, -3, 1)$ in the line $\frac{x-0}{1}=\frac{y-3}{1}=\frac{z-1}{-1}$ and $R$ be the point $(2, 5, -1)$. If the area of the $\Delta PQR$ is $\lambda$ and $\lambda^2 = 14K$, then $K$ equal to *[2024, 06 April Shift-II]*

(a) 18 (b) 72 (c) 36 (d) 81

**60** Let $P$ be the point of intersection of the lines $\frac{x-2}{1}=\frac{y-4}{5}=\frac{z-2}{1}$ and $\frac{x-3}{2}=\frac{y-2}{3}=\frac{z-3}{2}$. Then, the shortest distance of $P$ from the line $4x = 2y = z$ is *[2024, 04 April Shift-II]*

(a) $\frac{3\sqrt{14}}{7}$ (b) $\frac{\sqrt{14}}{7}$ (c) $\frac{5\sqrt{14}}{7}$ (d) $\frac{6\sqrt{14}}{7}$

**61** If the mirror image of the point $P(3, 4, 9)$ in the line $\frac{x-1}{3}=\frac{y+1}{2}=\frac{z-2}{1}$ is $(\alpha, \beta, \gamma)$, then $14(\alpha+\beta+\gamma)$ is *[2024, 01 Feb Shift-II]*

(a) 132 (b) 138 (c) 108 (d) 102

**62** The distance of the point $Q(0, 2, -2)$ form the line passing through the point $P(5, -4, 3)$ and perpendicular to the lines $\mathbf{r} = (-3\hat{i} + 2\hat{k}) + \lambda(2\hat{i} + 3\hat{j} + 5\hat{k}), \lambda \in R$ and $\mathbf{r} = (\hat{i} - 2\hat{j} + \hat{k}) + \mu(-\hat{i} + 3\hat{j} + 2\hat{k}), \mu \in R$ is *[2024, 31 Jan Shift-I]*

(a) $\sqrt{20}$ (b) $\sqrt{86}$ (c) $\sqrt{54}$ (d) $\sqrt{74}$

**63** Let the image of the point (1, 0, 7) in the line $\frac{x}{1} = \frac{y-1}{2} = \frac{z-2}{3}$ be the point $(\alpha, \beta, \gamma)$. Then, which one of the following points lies on the line passing through $(\alpha, \beta, \gamma)$ and making angles $\frac{2\pi}{3}$ and $\frac{3\pi}{4}$ with $Y$-axis and $Z$-axis respectively and an acute angle with $X$-axis? *[2024, 27 Jan Shift-II]*

(a) $(1, -2, 1+\sqrt{2})$ (b) $(1, 2, 1-\sqrt{2})$

(c) $(3, -4, 3+2\sqrt{2})$ (d) $(3, 4, 3-2\sqrt{2})$

**64** Let the line $\frac{x}{1} = \frac{6-y}{2} = \frac{z+8}{5}$ intersect the lines $\frac{x-5}{4} = \frac{y-7}{3} = \frac{z+2}{1}$ and $\frac{x+3}{6} = \frac{3-y}{3} = \frac{z-6}{1}$ at the points $A$ and $B$ respectively. Then, the distance of the mid-point of the line segment $AB$ from the plane $2x - 2y + z = 14$ is *[2023, 10 April shift-II]*

(a) 4 (b) 3

(c) 11/3 (d) 10/3

**65** The shortest distance between the lines $\frac{x+2}{1} = \frac{y}{-2} = \frac{z-5}{2}$ and $\frac{x-4}{1} = \frac{y-1}{2} = \frac{z+3}{0}$ is *[2023, 10 April shift-I]*

(a) 6 (b) 8

(c) 9 (d) 7

**66** The shortest distance between the lines $\frac{x-4}{4} = \frac{y+2}{5} = \frac{z+3}{3}$ and $\frac{x-1}{3} = \frac{y-3}{4} = \frac{z-4}{2}$ is *[2023, 8 April shift-I]*

(a) $3\sqrt{6}$ (b) $6\sqrt{3}$

(c) $2\sqrt{6}$ (d) $6\sqrt{2}$

**67** The shortest distance between the lines $\frac{x-5}{1} = \frac{y-2}{2} = \frac{z-4}{-3}$ and $\frac{x+3}{1} = \frac{y+5}{4} = \frac{z-1}{-5}$ is *[2023, 1 Feb Shift-I]*

(a) $6\sqrt{3}$ (b) $5\sqrt{3}$

(c) $4\sqrt{3}$ (d) $7\sqrt{3}$

**68** Let the shortest distance between the lines $L: \frac{x-5}{-2} = \frac{y-\lambda}{0} = \frac{z+\lambda}{1}, \lambda \geq 0$ and $L_1: x+1 = y-1 = 4-z$ be $2\sqrt{6}$. If $(\alpha, \beta, \gamma)$ lies on $L$, then which of the following is not possible? *[2023, 31 Jan Shift-I]*

(a) $\alpha - 2\gamma = 19$ (b) $\alpha + 2\gamma = 24$

(c) $2\alpha - \gamma = 9$ (d) $2\alpha + \gamma = 7$

**69** The foot of perpendicular of the point (2, 0, 5) on the line $\frac{x+1}{2} = \frac{y-1}{5} = \frac{z+1}{-1}$ is $(\alpha, \beta, \gamma)$. Then, which of the following is not correct? *[2023, 25 Jan Shift-II]*

(a) $\frac{\beta}{\gamma} = -5$ (b) $\frac{\gamma}{\alpha} = \frac{5}{8}$ (c) $\frac{\alpha}{\beta} = -8$ (d) $\frac{\alpha\beta}{\gamma} = \frac{4}{15}$

**70** The shortest distance between the lines $x+1 = 2y = -12z$ and $x = y+2 = 6z-6$ is *[2023, 25 Jan Shift-II]*

(a) 3 (b) $\frac{3}{2}$ (c) 2 (d) $\frac{5}{2}$

**71** Consider the lines $L_1$ and $L_2$ given by $L_1: \frac{x-1}{2} = \frac{y-3}{1} = \frac{z-2}{2}$; $L_2: \frac{x-2}{1} = \frac{y-2}{2} = \frac{z-3}{3}$

A line $L_3$ having direction ratios $1, -1, -2$, intersects $L_1$ and $L_2$ at the points $P$ and $Q$, respectively. Then, the length of line segment $PQ$ is *[2023, 25 Jan Shift-I]*

(a) $4\sqrt{3}$ (b) 4 (c) $2\sqrt{6}$ (d) $3\sqrt{2}$

**72** The shortest distance between the lines $\frac{x+7}{-6} = \frac{y-6}{7} = z$ and $\frac{7-x}{2} = y-2 = z-6$ is *[2022, 25 July Shift-II]*

(a) $2\sqrt{29}$ (b) 1 (c) $\sqrt{\frac{37}{29}}$ (d) $\frac{\sqrt{29}}{2}$

**73** If the shortest distance between the straight lines $3(x-1) = 6(y-2) = 2(z-1)$ and $4(x-2) = 2(y-\lambda) = (z-3), \lambda \in R$ is $\frac{1}{\sqrt{38}}$, then the integral value of $\lambda$ is equal to *[2021, 22 July Shift-II]*

(a) 3 (b) 2 (c) 5 (d) $-1$

**74** The length of the perpendicular from the point $(2, -1, 4)$ on the straight line, $\frac{x+3}{10} = \frac{y-2}{-7} = \frac{z}{1}$ is

(a) greater than 3 but less than 4. *[2019, 8 April Shift-I]*

(b) less than 2.

(c) greater than 2 but less than 3.

(d) greater than 4.

## Numerical Value Type Questions

**75** Let the area of the triangle formed by the lines $x+2=y-1=z, \frac{x-3}{5}=\frac{y}{-1}=\frac{z-1}{1}$ and $\frac{x}{-3}=\frac{y-3}{3}=\frac{z-2}{1}$ be $A$. Then, $A^2$ is equal to ………… [2025, 08 April Shift-II]

**76** Let $P$ be the image of the point $Q(7,-2,5)$ in the line $L: \frac{x-1}{2}=\frac{y+1}{3}=\frac{z}{4}$ and $R(5, p, q)$ be a point on $L$. Then, the square of the area of $\Delta PQR$ is ……… . [2025, 25 Jan Shift-II]

**77** Let $L_1: \frac{x-1}{3}=\frac{y-1}{-1}=\frac{z+1}{0}$ and $L_2: \frac{x-2}{2}=\frac{y}{0}=\frac{z+4}{\alpha}, \alpha \in R$ be two lines, which intersect at the point $B$.
If $P$ is the foot of perpendicular from the point $A(1,1,-1)$ on $L_2$, then the value of $26\,\alpha(PB)^2$ is ……… . [2025, 22 Jan Shift-I]

**78** The square of the distance of the image of the point $(6, 1, 5)$ in the line $\frac{x-1}{3}=\frac{y}{2}=\frac{z-2}{4}$, from the origin is ……… . [2024, 9 April Shift-II]

**79** Let $P(\alpha, \beta, \gamma)$ be the image of the point $Q(1, 6, 4)$ in the line $\frac{x}{1}=\frac{y-1}{2}=\frac{z-2}{3}$. Then, $2\alpha+\beta+\gamma$ is equal to ……… . [2024, 8 April Shift-II]

**80** If the shortest distance between the lines $\frac{x-\lambda}{3}=\frac{y-2}{-1}=\frac{z-1}{1}$ and $\frac{x+2}{-3}=\frac{y+5}{2}=\frac{z-4}{4}$ is $\frac{44}{\sqrt{30}}$, then the largest possible value of $|\lambda|$ is equal to ………. [2024, 6 April Shift-II]

**81** Let $P$ be the point $(10,-2,-1)$ and $Q$ be the foot of the perpendicular drawn from the point $R(1,7,6)$ on the line passing through the points $(2,-5,11)$ and $(-6,7,-5)$. Then, the length of the line segment $PQ$ is equal to ……… . [2024, 6 April Shift-I]

**82** Let the point $(-1, \alpha, \beta)$ lie on the line of the shortest distance between the lines $\frac{x+2}{-3}=\frac{y-2}{4}=\frac{z-5}{2}$ and $\frac{x+2}{-1}=\frac{y+6}{2}=\frac{z-1}{0}$. Then, $(\alpha-\beta)^2$ is equal to ………. [2024, 5 April Shift-II]

**83** Consider a line $L$ passing through the points $P(1,2,1)$ and $Q(2,1,-1)$. If the mirror image of the point $A(2,2,2)$ in the line $L$ is $(\alpha, \beta, \gamma)$, then $\alpha+\beta+6\gamma$ is equal to…… . [2024, 4 April Shift-II]

**84** If the shortest distance between the lines $\frac{x+2}{2}=\frac{y+3}{3}=\frac{z-5}{4}$ and $\frac{x-3}{1}=\frac{y-2}{-3}=\frac{z+4}{2}$ is $\frac{38}{3\sqrt{5}}k$, and $\int_0^k [x^2]\,dx=\alpha-\sqrt{\alpha}$, where $[x]$ denotes the greatest integer function, then $6\alpha^3$ is equal to … . [2024, 4 April Shift-II]

**85** Let the line of the shortest distance between the lines
$L_1: \mathbf{r}=(\hat{i}+2\hat{j}+3\hat{k})+\lambda(\hat{i}-\hat{j}+\hat{k})$ and
$L_2: \mathbf{r}=(4\hat{i}+5\hat{j}+6\hat{k})+\mu(\hat{i}+\hat{j}-\hat{k})$
intersect $L_1$ and $L_2$ at $P$ and $Q$, respectively. If $(\alpha, \beta, \gamma)$ is the mid-point of the line segment $PQ$, then $2(\alpha+\beta+\gamma)$ is equal to ……. . [2024, 1st Feb Shift-I]

**86** Let $Q$ and $R$ be the feet of perpendiculars from the point $P(a, a, a)$ on the lines $x=y, z=1, x=-y$ and $z=-1$, respectively. If $\angle QPR$ is a right angle, then $12a^2$ is equal to …… . [2024, 31 Jan Shift-I]

**87** Let a line passing through the point $(-1, 2, 3)$ intersect the lines $L_1: \frac{x-1}{3}=\frac{y-2}{2}=\frac{z+1}{-2}$ at $M(\alpha, \beta, \gamma)$ and $L_2: \frac{x+2}{-3}=\frac{y-2}{-2}=\frac{z-1}{4}$ at $N(a, b, c)$. Then, the value of $\frac{(\alpha+\beta+\gamma)^2}{(a+b+c)^2}$ equals ……… . [2024, 30 Jan Shift-II]

**88** If $d_1$ is the shortest distance between the lines $x+1=2y=-12z, x=y+2=6z-6$ and $d_2$ is the shortest distance between the lines $\frac{x-1}{2}=\frac{y+8}{-7}=\frac{z-4}{5}, \frac{x-1}{2}=\frac{y-2}{1}=\frac{z-6}{-3}$, then the value of $\frac{32\sqrt{3}d_1}{d_2}$ is ……… . [2024, 30 Jan Shift-I]

**89** Let $O$ be the origin, and $M$ and $N$ be the points on the lines $\frac{x-5}{4}=\frac{y-4}{1}=\frac{z-5}{3}$ and $\frac{x+8}{12}=\frac{y+2}{5}=\frac{z+11}{9}$ respectively, such that $MN$ is the shortest distance between the given lines. Then, $\mathbf{OM}\cdot\mathbf{ON}$ is equal to ……… . [2024, 29 Jan Shift-II]

**90** The lines $\frac{x-2}{2}=\frac{y}{-2}=\frac{z-7}{16}$ and $\frac{x+3}{4}=\frac{y+2}{3}=\frac{z+2}{1}$ intersect at the point $P$. If the distance of $P$ from the line $\frac{x+1}{2}=\frac{y-1}{3}=\frac{z-1}{1}$ is $l$, then $14l^2$ is equal to ......... . [2024, 27 Jan Shift-II]

**91** Let a line $l$ pass through the origin and be perpendicular to the line $\mathbf{r}=(\hat{\mathbf{i}}-11\hat{\mathbf{j}}-7\hat{\mathbf{k}})+\lambda(\hat{\mathbf{i}}+2\hat{\mathbf{j}}+3\hat{\mathbf{k}}), \lambda \in R$ and $l_2$ : $\mathbf{r}=(-\hat{\mathbf{i}}+\hat{\mathbf{k}})+\mu(2\hat{\mathbf{i}}+2\hat{\mathbf{j}}+\hat{\mathbf{k}}), \mu \in R$.

If $P$ is the point of intersection of $l$ and $l_1$, and $Q(\alpha, \beta, \gamma)$ is the foot of perpendicular from $P$ on $l_2$, then $9(\alpha+\beta+\gamma)$ is equal to ...... . [2023, 11 April Shift-I]

**92** If the lines $\frac{x-1}{2}=\frac{2-y}{-3}=\frac{z-3}{\alpha}$ and $\frac{x-4}{5}=\frac{y-1}{2}=\frac{z}{\beta}$ intersect, then the magnitude of the minimum value of $8\alpha\beta$ is ......... [2023, 6 April Shift-II]

**93** Let a line $L$ pass through the point $P(2, 3, 1)$ and be parallel to the line $x+3y-2z-2=0=x-y+2z$. If the distance of $L$ from the point (5,3,8) is $\alpha$, then $3\alpha^2$ is equal to ......... [2023, 30 Jan Shift-II]

**94** Let the co-ordinates of one vertex of $\Delta ABC$ be $A(0, 2, \alpha)$ and the other two vertice lie on the line $\frac{x+\alpha}{5}=\frac{y-1}{2}=\frac{z+4}{3}$. For $\alpha \in Z$, if the area of $\Delta ABC$ is 21 sq units and the line segment $BC$ has length $2\sqrt{21}$ units, then $\alpha^2$ is equal to ....... [2023, 29 Jan Shift-I]

**95** The shortest distance between the lines $\frac{x-2}{3}=\frac{y+1}{2}=\frac{z-6}{2}$ and $\frac{x-6}{3}=\frac{1-y}{2}=\frac{z+8}{0}$ is equal to. [2023, 24 Jan Shift-I]

**96** Consider a $\Delta ABC$ whose vertices are $A(0, \alpha, \alpha)$, $B(\alpha, 0, \alpha)$ and $C(\alpha, \alpha, 0)$, $\alpha>0$. Let $D$ be a point moving on the line $x+z-3=0=y$ and $G$ be the centroid of $\Delta ABC$. If the minimum length of $GD$ is $\sqrt{\frac{57}{2}}$, then $\alpha$ is equal to ......... [2022, 30 June Shift-I]

**Imp.** **Image of Point on a Line**

**97** Let the image of the point $P(1, 2, 3)$ in the line $L: \frac{x-6}{3}=\frac{y-1}{2}=\frac{z-2}{3}$ be $Q$.

Let $R(\alpha, \beta, \gamma)$ be a point that divides internally the line segment $PQ$ in the ratio 1:3. Then, the value of $22(\alpha+\beta+\gamma)$ is equal to ........ [2022, 28 June Shift-II]

**98** If the lines $\frac{x-k}{1}=\frac{y-2}{2}=\frac{z-3}{3}$ and $\frac{x+1}{3}=\frac{y+2}{2}=\frac{z+3}{1}$ are co-planar, then the value of $k$ is ......... . [2021, 25 July Shift-II]

**99** If the foot of the perpendicular drawn from the point (1, 0, 3) on a line passing through $(\alpha, 7, 1)$ is $\left(\frac{5}{3}, \frac{7}{3}, \frac{17}{3}\right)$, then $\alpha$ is equal to .... . [2020, 7 Jan Shift-II]

## Key Index

### Topic 1

| | | | | | | | | | |
|---|---|---|---|---|---|---|---|---|---|
| 1 (a) | 2 (a) | 3 (c) | 4 (a) | 5 (c) | 6 (c) | 7 (d) | 8 (c) | 9 (c) | 10 (d) |
| 11 (b) | 12 (a) | 13 (c) | 14 (65) | 15 (450) | 16 (8) | | | | |

### Topic 2

| | | | | | | | | | |
|---|---|---|---|---|---|---|---|---|---|
| 1 (a) | 2 (a) | 3 (b) | 4 (d) | 5 (b) | 6 (d) | 7 (b) | 8 (c) | 9 (b) | 10 (a) |
| 11 (d) | 12 (d) | 13 (b) | 14 (b) | 15 (d) | 16 (a) | 17 (b) | 18 (b) | 19 (b) | 20 (a) |
| 21 (b) | 22 (c) | 23 (a) | 24 (d) | 25 (d) | 26 (b) | 27 (d) | 28 (c) | 29 (a) | 30 (b) |
| 31 (b) | 32 (c) | 33 (c) | 34 (d) | 35 (a) | 36 (c) | 37 (d) | 38 (d) | 39 (d) | 40 (a) |
| 41 (c) | 42 (c) | 43 (a) | 44 (c) | 45 (d) | 46 (d) | 47 (b) | 48 (a) | 49 (c) | 50 (c) |
| 51 (a) | 52 (b) | 53 (a) | 54 (c) | 55 (b) | 56 (b) | 57 (b) | 58 (d) | 59 (d) | 60 (a) |
| 61 (c) | 62 (d) | 63 (d) | 64 (a) | 65 (c) | 66 (a) | 67 (a) | 68 (b) | 69 (a) | 70 (c) |
| 71 (c) | 72 (a) | 73 (a) | 74 (a) | 75 (56) | 76 (957) | 77 (216) | 78 (62) | 79 (11) | 80 (43) |
| 81 (13) | 82 (25) | 83 (6) | 84 (48) | 85 (21) | 86 (12) | 87 (196) | 88 (16) | 89 (9) | 90 (108) |
| 91 (5) | 92 (18) | 93 (158) | 94 (9) | 95 (14) | 96 (6) | 97 (125) | 98 (1) | 99 (4) | |

# Detailed Explanations

## Topic 1

**01.** (a) Given, $l+m-n=0$,

$$3l^2+m^2+cl(l+m)=0 \quad [\because n=l+m]$$

$$\Rightarrow \quad 3l^2+m^2+cl^2+clm=0$$

$$\Rightarrow \quad (3+c)l^2+clm+m^2=0$$

$$(3+c)\left(\frac{l}{m}\right)^2+c\left(\frac{l}{m}\right)+1=0 \quad \ldots\text{(i)}$$

$\because$ Lines are parallel.

Roots of quadratic equation in Eq. (i) must be equal

$$\Rightarrow \quad D=0 \Rightarrow c^2-4(3+c)=0$$

$$\Rightarrow c^2-4c-12=0$$

$$\Rightarrow \quad c=6 \text{ or } c=-2$$

Positive value of $c=6$

**02.** (a) Given, $P$, $Q$, $R$ are collinear

$\therefore PR=\lambda PQ$, where $\lambda$ be any constant

$$2\hat{i}+(y+3)\hat{j}+(z-4)\hat{k}=\lambda[6\hat{i}+3\hat{j}+6\hat{k}]$$

$$\Rightarrow 6\lambda=2,\ y+3=3\lambda,\ z-4=6\lambda$$

$$\Rightarrow \quad \lambda=\frac{1}{3},\ y=-2,\ z=6$$

$\therefore$Point $R(4,-2,6)$

Now, $OR=\sqrt{(4)^2+(-2)^2+(6)^2}$

$$=\sqrt{56}=2\sqrt{14}$$

**03.** (c) It makes $\theta$ with $X$ and $Y$-axis.

Let given line makes $\alpha$ angle with $Z$-axis. $\alpha \in \left[0, \frac{\pi}{2}\right]$

$$l=\cos\theta,\ m=\cos\theta,\ n=\cos\alpha$$

$$\because \quad l^2+m^2+n^2=1$$

$$\therefore \cos^2\theta+\cos^2\theta+\cos^2\alpha=1$$

$$\Rightarrow 2\cos^2\theta=1-\cos^2\alpha$$

$$\Rightarrow \quad \cos^2\theta=\frac{\sin^2\alpha}{2} \Rightarrow \cos\theta=\frac{\sin\alpha}{\sqrt{2}}$$

As $\alpha \in \left[0,\frac{\pi}{2}\right]$, then $\theta \in \left[\frac{\pi}{4},\frac{\pi}{2}\right]$

**04.** (a) Given, $l+m+n=0$

$$\Rightarrow -(m+n)=l$$

$$\Rightarrow \quad (m+n)^2=l^2$$

$$\Rightarrow \quad m^2+n^2+2mn=m^2+n^2 \quad [\because l^2=m^2+n^2, \text{ given}]$$

$$\Rightarrow \quad 2mn=0$$

Case I When, $m=0$, then $l=-n$

Hence, $(l, m, n)$ is $(1, 0, -1)$.

Case II When, $n=0$, then $l=-m$

Hence, $(l, m, n)$ is $(1, -1, 0)$.

So, the direction ratios of the lines are $(1, 0, -1)$ and $(1, -1, 0)$.

$$\therefore \quad \cos\theta=\frac{1+0+0}{\sqrt{2}\times\sqrt{2}}=\frac{1}{2} \Rightarrow \theta=\frac{\pi}{3}$$

**05.** (c) We know that

$$\cos^2 45°+\cos^2 120°+\cos^2\theta=1$$

$$\Rightarrow \quad \frac{1}{2}+\frac{1}{4}+\cos^2\theta=1$$

$$\Rightarrow \quad \cos^2\theta=\frac{1}{4} \Rightarrow \cos\theta=\pm\frac{1}{2}$$

$$\therefore \quad \theta=60° \text{ or } 120°$$

**06.** (c) Projection of a vector on co-ordinate axes are

$$x_2-x_1,\ y_2-y_1,\ z_2-z_1$$

$$\Rightarrow x_2-x_1=6,\ y_2-y_1=-3,\ z_2-z_1=2$$

Now,

$$\sqrt{(x_2-x_1)^2+(y_2-y_1)^2+(z_2-z_1)^2}$$

$$=\sqrt{36+9+4}=7$$

So, the DC's of the vector are $6/7, -3/7$ and $2/7$.

**07.** (d) Since, a line makes an angle of $\frac{\pi}{4}$ with positive directions of each of $X$-axis and $Y$-axis, therefore

$$\alpha=\frac{\pi}{4}, \quad \beta=\frac{\pi}{4}$$

We know that

$$\cos^2\alpha+\cos^2\beta+\cos^2\gamma=1$$

$$\therefore \quad \cos^2\frac{\pi}{4}+\cos^2\frac{\pi}{4}+\cos^2\gamma=1$$

$$\Rightarrow \quad \frac{1}{2}+\frac{1}{2}+\cos^2\gamma=1 \Rightarrow \cos^2\gamma=0$$

$$\therefore \quad \gamma=\pi/2$$

**08.** (c) A line makes angle $\theta$ with $X$-axis and $Z$-axis and $\beta$ with $Y$-axis.

$$\therefore \quad l=\cos\theta, \quad m=\cos\beta, \quad n=\cos\theta$$

$$\because \quad l^2+m^2+n^2=1$$

$$\therefore \cos^2\theta+\cos^2\beta+\cos^2\theta=1$$

$$\Rightarrow \quad 2\cos^2\theta=1-\cos^2\beta$$

$$\Rightarrow \quad 2\cos^2\theta=\sin^2\beta \quad \ldots\text{(i)}$$

But, it is given that

$$\sin^2\beta=3\sin^2\theta \quad \ldots\text{(ii)}$$

From Eqs. (i) and (ii), we get

$$3\sin^2\theta=2\cos^2\theta$$

$$\Rightarrow \quad 3(1-\cos^2\theta)=2\cos^2\theta$$

$$\Rightarrow \quad 3=5\cos^2\theta$$

$$\Rightarrow \quad \cos^2\theta=3/5$$

**09.** (c) Let point $A(x, y, 0)$

Let $P(0, 3, 2)$, $Q(2, 0, 3)$ and $R(0, 0, 1)$

$$AP=AQ=AR$$

$$x^2+(y-3)^2+4=(x-2)^2+y^2+9$$

$$=x^2+y^2+1$$

$$AQ=AR \Rightarrow -4x+13=1$$

$$x=3$$

$$AP=AR \Rightarrow y=2$$

$$A\equiv(3, 2, 0)$$

$$AB=3$$

$$BC=3\sqrt{2}$$

$$AC=3$$

$\because AB^2+AC^2=BC^2$ and $AB=AC$

Area of $\Delta ABC$ is

$$=\frac{1}{2}\times3\times3=\frac{9}{2} \text{ sq unit}$$

**10.** (d) Point of intersection of the lines

$$3x+2y=14 \quad \ldots\text{(i)}$$

$$5x-y=6 \quad \ldots\text{(ii)}$$

On multiplying by 2 in Eq. (ii) and then adding Eqs. (i) and (ii),

$$3x+2y=14$$

$$10x-2y=12$$

$$\overline{13x \quad =26}$$

$$\Rightarrow \quad x=2 \Rightarrow y=4$$

$$\therefore \quad A(2, 4)$$

Point of intersection of lines

$$4x+3y=8 \quad \ldots\text{(iii)}$$

$$6x+y=5 \quad \ldots\text{(iv)}$$

On multiply by 3 in Eq. (iv) and then subtracting Eq. (iv) from Eq. (iii),

$$4x+3y=8$$

$$18x+3y=15$$

$$- \quad - \quad -$$

$$\overline{-14x=-7}$$

$$x=\frac{1}{2} \Rightarrow y=2$$

$$\therefore \quad B=\left(\frac{1}{2}, 2\right)$$

Equation of line $AB$

$\Rightarrow \quad y-4=\dfrac{2-4}{\frac{1}{2}-2}(x-2)$

$\Rightarrow \quad y-4=\dfrac{4}{3}(x-2)$

$\Rightarrow \quad 3y-12=4x-8$

$\Rightarrow \quad 4x-3y+4=0$

Distance of line $4x-3y+4=0$ from $P(5,-2)$ is

$=\dfrac{4(5)-3(-2)+4}{5}=\dfrac{30}{5}=6$

**11.** *(b)* $M$ is mid-piont of $BC$.

Co-ordinates of $M$ is $\left(\dfrac{\lambda-1}{2}, 4, \dfrac{\mu+2}{2}\right)$

DR's of $AM$ are $\dfrac{\lambda-1}{2}-2, 4-3,$

$\dfrac{\mu+2}{2}-5$

i.e. $\quad \dfrac{\lambda-5}{2}, 1, \dfrac{\mu-8}{2}$

$\because$ The median makes equal angles with co-ordinate axes, therefore,

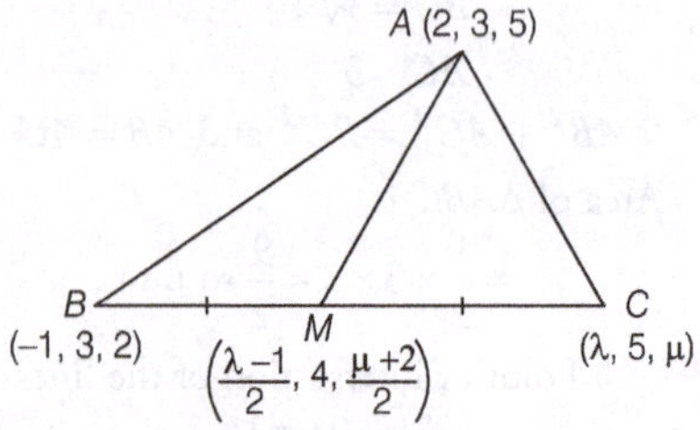


$\dfrac{\lambda-5}{2}=1=\dfrac{\mu-8}{2}$

$\Rightarrow \quad \lambda=7$ and $\mu=10$

$\therefore \lambda^3+\mu^3+5=1348$

**12.** *(a)* Given,

$2l+2m-n=0 \quad \ldots$ (i)

$mn+nl+lm=0 \quad \ldots$(ii)

From Eq. (i), we get

$n=2l+2m \quad \ldots$(iii)

On substituting, $n=2l+2m$ in Eq. (ii), we get

$m(2l+2m)+l(2l+2m)+lm=0$

$\Rightarrow 2lm+2m^2+2l^2+2lm+lm=0$

$\Rightarrow \quad 2l^2+4lm+lm+2m^2=0$

$\Rightarrow \quad 2l(l+2m)+m(l+2m)=0$

$\Rightarrow \quad (2l+m)(l+2m)=0$

When, $2l=-m$

From Eq (iii),

$n=m$

$\Rightarrow \quad \dfrac{2l}{-1}=\dfrac{m}{1}=\dfrac{n}{1} \Rightarrow \dfrac{l}{-1/2}=\dfrac{m}{1}=\dfrac{n}{1}$

or $\quad \dfrac{l}{1}=\dfrac{m}{-2}=\dfrac{n}{-2}$

$\Rightarrow (l, m, n)=(1,-2,-2)$

When, $l=-2m$

From Eq. (iii),

$n=-2m \Rightarrow l=-2m=n$

$\Rightarrow \quad \dfrac{l}{-2}=\dfrac{m}{1}=\dfrac{n}{-2}$

$\Rightarrow \quad (l, m, n)=(-2,1,-2)$

$\therefore$ Angles between straight lines

$\cos\theta=\dfrac{(\hat{i}-2\hat{j}-2\hat{k})\cdot(-2\hat{i}+\hat{j}-2\hat{k})}{|\hat{i}-2\hat{j}-2\hat{k}||-2\hat{i}+\hat{j}-2\hat{k}|}$

$\cos\theta=\dfrac{-2-2+4}{9}=0 \Rightarrow \theta=\dfrac{\pi}{2}$

**13.** *(c)* Given, $l+m-n=0 \quad \ldots$(i)

and $\quad l^2+m^2-n^2=0 \quad \ldots$(ii)

On squaring Eq. (i), we get

$(l+m)^2=n^2$

$\Rightarrow \quad l^2+m^2+2lm=n^2 \quad \ldots$(iii)

From Eqs. (ii) and (iii), we get

$l^2+m^2-n^2=0$

$l^2+m^2+2lm=n^2$

$\underline{- \quad - \quad - \quad -}$

$-n^2-2lm=-n^2$

$\Rightarrow \quad 2lm=0$

$\Rightarrow \quad lm=0$

$\Rightarrow \quad l=0$

or $\quad m=0$

Case I When, $l=0$

$\Rightarrow \quad 0+m-n=0$

$\Rightarrow \quad m=n$

and $l^2+m^2+n^2=1$

$\Rightarrow \quad m^2+m^2=1\ [\because n=m$ and $l=0]$

$\Rightarrow \quad m^2=\dfrac{1}{2}$

$\therefore \quad m=\pm\dfrac{1}{\sqrt{2}}=n$

$\therefore \quad (l, m, n)=\left(0, \dfrac{1}{\sqrt{2}}, \dfrac{1}{\sqrt{2}}\right)$

or $\left(0, \dfrac{-1}{\sqrt{2}}, \dfrac{-1}{\sqrt{2}}\right)$

Case II When, $m=0$,

then $\quad l+m-n=0$

$\Rightarrow \quad l=n$

and $l^2+m^2+n^2=1$

$\Rightarrow \quad l^2+0+l^2=1\ [\because n=l$ and $m=0]$

$\Rightarrow \quad l=\pm\dfrac{1}{\sqrt{2}}$

$\therefore\ (l, m, n)=\left(\dfrac{1}{\sqrt{2}}, 0, \dfrac{1}{\sqrt{2}}\right)$

or $\left(\dfrac{-1}{\sqrt{2}}, 0, \dfrac{-1}{\sqrt{2}}\right) \Rightarrow \mathbf{a}=\left(0, \dfrac{1}{\sqrt{2}}, \dfrac{1}{\sqrt{2}}\right)$

and $\quad \mathbf{b}=\left(\dfrac{1}{\sqrt{2}}, 0, \dfrac{1}{\sqrt{2}}\right)$

Then, $\cos\alpha=\dfrac{\mathbf{a}\cdot\mathbf{b}}{|\mathbf{a}||\mathbf{b}|}=\dfrac{1}{2}$

$\therefore \quad \sin\alpha=\pm\dfrac{\sqrt{3}}{2}$

Now, $\cos^4\alpha+\sin^4\alpha=\dfrac{1}{16}+\dfrac{9}{16}$

$=\dfrac{10}{16}=\dfrac{5}{8}$

**14.** *(65)* Given, line are

$x=y+2=z$

$\Rightarrow \quad \dfrac{x}{1}=\dfrac{y-(-2)}{1}=\dfrac{z}{1}=t$ (let) $\ldots$ (i)

Let any point on the line (i) be

$P(t, t-2, t)$

and any point on the $x+2=2y=2z$ is $Q(2s-2, s, s)$

$\therefore$Direction ratio of $PQ$ is

$2s-2-t, s-t+2, s-t$

$\therefore \quad \dfrac{2s-2-t}{2}=\dfrac{s-t+2}{1}=\dfrac{s-t}{2}$

$\Rightarrow \ t=6$ and $s=2$

$\therefore \quad P=(6, 4, 6)$ and $Q=(2, 2, 2)$

Equation of line $PQ$ is

$\dfrac{x-2}{2}=\dfrac{y-2}{1}=\dfrac{z-2}{2}=\lambda$ (say) $\ldots$(ii)

Let any point on the line (ii) be $F(2\lambda+2, \lambda+2, 2\lambda+2)$ and let $A(1, 2, 12)$.

$\therefore$Direction ratio of $AF$ is

$2\lambda+1, \lambda, 2\lambda-10$

Here, $\mathbf{AF}\cdot\mathbf{PQ}=0$

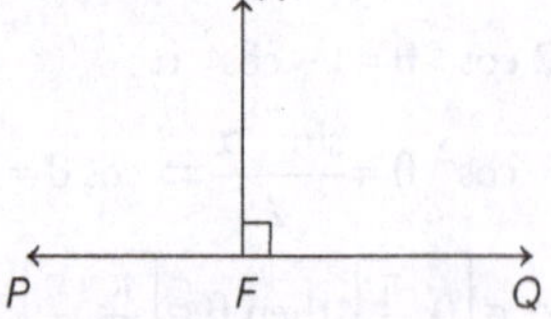


$\Rightarrow \quad 4\lambda+2+\lambda+4\lambda-20=0$

$\Rightarrow \quad 9\lambda=18$

$\Rightarrow \quad \lambda=2$

$F\equiv(6, 4, 6)$

$\therefore\ AF=\sqrt{(6-1)^2+(4-2)^2+(6-12)^2}$

$=\sqrt{25+4+36}=\sqrt{65}$

$\therefore \quad l^2=65$

**15.** *(450)* According to the question,

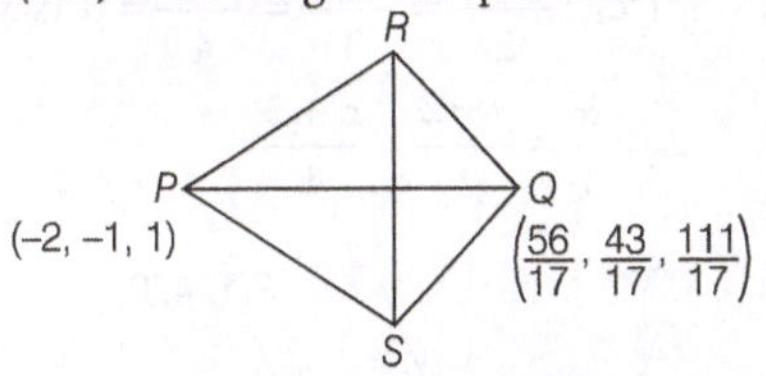


$$\mathbf{PQ} = \frac{90}{17}\hat{i} + \frac{60}{17}\hat{j} + \frac{94}{17}\hat{k}$$

$\Rightarrow \quad \mathbf{RS} = \alpha\hat{i} - 1\hat{j} + \beta\hat{k}$

As, we know that diagonals of rhombus are perpendicular to each other.

So, we can use $\mathbf{a} \cdot \mathbf{b} = 0 \Rightarrow \mathbf{PQ} \cdot \mathbf{RS} = 0$

After solving, we get

$$\alpha = \frac{30 - 47\beta}{45}$$

For minimum integral value

$$\alpha = -15 \text{ and } \beta = 15$$

$\therefore \quad \alpha^2 + \beta^2 = 450$

**16.** *(8)* Given points $A(1, -1, 3)$, $B(2, -4, 11)$, $C(-1, 2, 3)$ and $D(3, -2, 10)$, then

$$\mathbf{AB} = \hat{i} - 3\hat{j} + 8\hat{k}$$

and $\quad \mathbf{CD} = 4\hat{i} - 4\hat{j} + 7\hat{k}$

So, the length of projection of line segment vector **AB** on the line vector joining points $C$ and $D$ is

$$= \frac{|\mathbf{AB} \cdot \mathbf{CD}|}{|\mathbf{CD}|}$$

$$= \frac{4 + 12 + 56}{\sqrt{16 + 16 + 49}} = \frac{72}{\sqrt{81}} = \frac{72}{9} = 8$$

## Topic 2

**01.** *(a)* Line $\mathbf{r} = \lambda(\hat{i} + a\hat{j} + b\hat{k})$ has direction ratios $d_1 = <1, a, b>$

Line $\mathbf{r} = (\hat{i} - \hat{j} - 6\hat{k}) + \mu(-b\hat{i} + a\hat{j} + 5\hat{k})$ has direction ratios $d_2 = <-b, a, 5>$

$$\mathbf{l} = (1, a, b) \times (-b, a, 5)$$

$$= (5a - ab, -b^2 - 5, a + ab)$$

$$\Rightarrow \frac{5a - ab}{-2} = \frac{-b^2 - 5}{d} = \frac{a + ab}{-4}$$

$$\Rightarrow \frac{5a - ab}{-2} = \frac{a + ab}{-4}$$

$$\Rightarrow \frac{a(5 - b)}{-2} = \frac{a(1 + b)}{-4}$$

$\Rightarrow 2(5 - b) = 1 + b$

$\Rightarrow 10 - 2b = 1 + b \Rightarrow b = 3$

Then, $\dfrac{5a - 3a}{-2} = \dfrac{-9 - 5}{d}$

$$\Rightarrow -a = -\frac{14}{d} \Rightarrow a = \frac{14}{d}$$

Also, line passing through the point $\left(0, -\frac{1}{2}, 0\right)$ then

$$\frac{0 - 1}{-2} = \frac{-\frac{1}{2} + 4}{d} = \frac{0 - c}{-4}$$

$$\frac{1}{2} = \frac{7}{2d} = \frac{c}{4} \Rightarrow d = 7, \text{ then } a = \frac{14}{7} = 2$$

$$\frac{7}{2d} = \frac{c}{4} \Rightarrow \frac{7}{2 \times 7} = \frac{c}{2} \Rightarrow c = 1$$

Thus, $a + b + c + d = 2 + 3 + 1 + 7 = 13$

**02.** *(a)* The parametric equation of $L_1$ is

$$x - 1 = y - 2 = z = \lambda_1$$

$\Rightarrow Q(\lambda_1 + 1, \lambda_1 + 2, \lambda_1)$

Direction ratio of $PQ$ is

$(\lambda_1 + 1 - 5, \lambda_1 - 1 + 2, \lambda_1 + 3)$

$= (\lambda_1 - 4, \lambda_1 + 1, \lambda_1 + 3)$

Direction ratio of $L_1$ is $(1, 1, 1)$

Since $PQ \perp L_1$

$\Rightarrow (\lambda_1 - 4)1 + (\lambda_1 + 1)1 + (\lambda_1 + 3)1 = 0$

$\Rightarrow \quad 3\lambda_1 = 0 \Rightarrow \lambda_1 = 0$

Thus, $Q(1, 2, 0)$

The parametric equation of $L_2$ is

$$x - 2 = y = z - 1 = \lambda_2$$

$\Rightarrow R(\lambda_2 + 2, \lambda_2, \lambda_2 + 1)$

Direction ratio of $PR$ is

$(\lambda_2 + 2 - 5, \lambda_2 + 1 + 3)$

$= (\lambda_2 - 3, \lambda_2 - 1, \lambda_2 + 4)$

Direction ratio of $L_2$ is $(1, 1, 1)$

Since $PR \perp L_2$

$\Rightarrow (\lambda_2 - 3)1 + (\lambda_2 - 1)1 + (\lambda_2 + 4)1 = 0$

$\Rightarrow \quad 3\lambda_2 = 0 \Rightarrow \lambda_2 = 0$

Thus, $R(2, 0, 1)$

Now, $\mathbf{PQ} = <-4, 1, 3>$

and $\quad \mathbf{PR} = <-4, -1, 4>$

Area of $\Delta PQR = \frac{1}{2}|\mathbf{PQ} \times \mathbf{PR}|$

$$= \frac{1}{2}\begin{vmatrix} \hat{i} & \hat{j} & \hat{k} \\ -4 & 1 & 3 \\ -4 & -1 & 4 \end{vmatrix}$$

$$= \frac{1}{2}\sqrt{7^2 + 7^2 + 7^2} = \frac{7\sqrt{3}}{2}$$

Thus, $4A^2 = \dfrac{49 \times 3}{4} = 147$

**03.** *(b)* $\mathbf{a}_1 \equiv (1, 2, 3)$, $\mathbf{b}_1 = (0, 0, 1)$,

$\mathbf{a}_2 \equiv (\lambda, 5, 6)$, $\mathbf{b}_2 \equiv (0, 1, 0)$

$$\mathbf{b}_1 \times \mathbf{b}_2 = \begin{vmatrix} \hat{i} & \hat{j} & \hat{k} \\ 0 & 0 & 1 \\ 0 & 1 & 0 \end{vmatrix} = -\hat{i}$$

$$d = \frac{|(\lambda - 1, 3, 3) - (-1, 0, 0)|}{|(-1, 0, 0)|} = |\lambda - 1|$$

$\Rightarrow |\lambda - 1| = 3$

$\therefore \quad \lambda = 4, -2$

Since, $\lambda_2 < \lambda_1 \Rightarrow \lambda_1 = 4$ and $\lambda_2 = -2$ and point $P = (4, -2, 7)$

Line $L_1$ passes through $(1, 2, 3)$ and parallel to the $Z$-axis

$$L: \frac{x - 1}{0} = \frac{y - 2}{0} = \frac{z - 3}{1} = k$$

$(3, -4), (4 - k)1k$

$4 - k = 0 \Rightarrow k = 4$

Now, distance between $(4, -2, 7)$ and $(1, 2, 7)$

$$\Rightarrow \quad d^2 = (4 - 1)^2 + (-2 - 2)^2 + (7 - 7)^2 = 25$$

**04.** *(d)* Equation of $L_1$

$$\mathbf{r} = 7\hat{i} + 6\hat{j} + 2\hat{k} + \lambda(-3\hat{i} + 2\hat{j} + 4\hat{k})$$

Equation of $L_2$

$$\mathbf{r} = 5\hat{i} + 3\hat{j} + 4\hat{k} + u(2\hat{i} + \hat{j} + 3\hat{k})$$

Shortest distance between $L_1$ and $L_2$

$$= \left|\frac{(\mathbf{a}_2 - \mathbf{a}_1) \cdot (\mathbf{b}_1 \times \mathbf{b}_2)}{|\mathbf{b}_1 \times \mathbf{b}_2|}\right|$$

$$= \left|\frac{(-2\hat{i} - 3\hat{j} + 2\hat{k}) \cdot (2\hat{i} + 17\hat{j} - 7\hat{k})}{\sqrt{2^2 + (+17)^2 + (-7)^2}}\right|$$

$$= \left|\frac{-4 - 51 - 14}{\sqrt{342}}\right| = \frac{69}{\sqrt{342}} = \frac{23}{\sqrt{38}}$$

**05.** *(b)* $H: (5\lambda - 3, 2\lambda + 1, 3\lambda - 4)$

<Direction ratios of $PH$>

$<5\lambda - 3, 2\lambda - 1, 3\lambda - 7>$

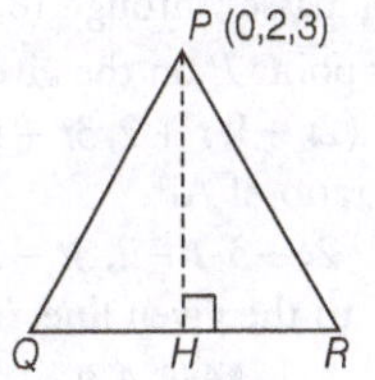


$\mathbf{PH} \cdot \mathbf{QR} = 0$

$\Rightarrow 5(5\lambda - 3) + 2(2\lambda - 1) + 3(3\lambda - 7) = 0$

$\Rightarrow \quad \lambda = 1$

$\therefore \quad H \equiv (2, 3, -1)$

$PH = \sqrt{4 + 1 + 16} = \sqrt{21}$

$$\text{Area} = \frac{1}{2}(PH)(QR)$$

$$= \frac{1}{2}\sqrt{21} \times 5$$

$$= \frac{5}{2}\sqrt{21} = \frac{m}{n}$$

$\Rightarrow 2m - 5\sqrt{21}\,n = 0$

**06.** *(d)* Vector parallel to '$L$'

$$=\begin{vmatrix}\hat{i} & \hat{j} & \hat{k}\\ 2 & 1 & -2\\ 1 & 3 & 4\end{vmatrix}=10\hat{i}-10\hat{j}+5\hat{k}$$

$$=5(2\hat{i}-2\hat{j}+\hat{k})$$

Equation of '$L$'

$$\frac{x-2}{2}=\frac{y+1}{-2}=\frac{z-3}{1}=\lambda \text{ (say)}$$

Let $Q(2\lambda+2, -2\lambda-1, \lambda+3)$

$\Rightarrow 2\lambda+2=0$

$\Rightarrow \quad \lambda=-1$

$\Rightarrow \quad Q(0, 1, 2)$

$\therefore$ Perpendicular distance between $P$ and $Q=3$

**07.** *(b)*

$L_1: \mathbf{r}=(-\hat{i}+2\hat{j}+\hat{k})+\lambda(\hat{i}+2\hat{j}+\hat{k})$

$\Rightarrow \mathbf{r}=(\lambda-1)\hat{i}+2(\lambda+1)\hat{j}+(\lambda+1)\hat{k}$

$L_2: \mathbf{r}=(\hat{j}+\hat{k})+\mu(2\hat{i}+7\hat{j}+3\hat{k})$

$\Rightarrow \mathbf{r}=(2\mu)\hat{i}+(1+7\mu)\hat{j}+(1+3\mu)\hat{k}$

For point of intersection, we equating respective components

i.e. $\lambda-1=2\mu \quad \ldots$(i)

$2(\lambda+1)=1+7\mu \quad \ldots$(ii)

$\lambda+1=1+3\mu \quad \ldots$(iii)

On solving these equation simultaneously, we get

$\lambda=3$ and $\mu=1$

$\Rightarrow$ Point of intersection $=(2, 8, 4)$

Given, $\mathbf{a}+\mathbf{b}=3\hat{i}+9\hat{j}+4\hat{k}$

$L_3: \mathbf{r}=2\hat{i}+8\hat{j}+4\hat{k}+\alpha(3\hat{i}+9\hat{j}+4\hat{k})$

For $\alpha=2$, $\mathbf{r}=8\hat{i}+26\hat{j}+12\hat{k}$

Hence, $L_3$ passes through (8, 26, 12)

**08.** *(c)* Any point $P$ on the given line is

$(2t+1, t+2, 3t+1)$

direction ratio of AP

$2t-3, t-2, 3t-2$

$\because$ AP is $\perp$ to the given line,

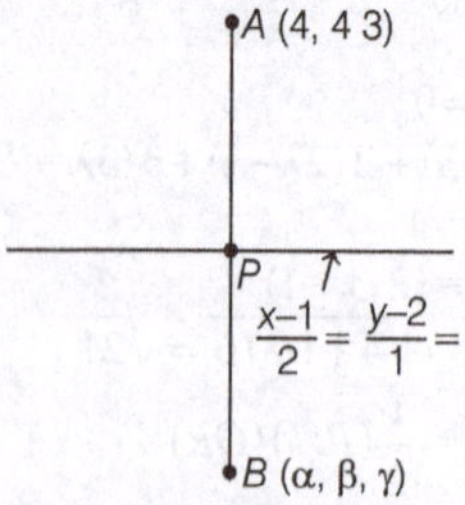


$2(2t-3)+1(t-2)+3(3t-2)=0$

$4t-6+t-2+9t-6=0$

$14t=14 \Rightarrow t=1$

$P:(3, 3, 4)$

$\because P$ is the mid-point of $AB$, then

$$(3, 3, 4)=\left(\frac{4+\alpha}{2}, \frac{4+\beta}{2}, \frac{3+\gamma}{2}\right)$$

$\alpha=2, \quad \beta=2$ and $\gamma=5$

$\therefore \quad \alpha+\beta+\gamma=2+2+5=9$

**09.** *(b)* Equation of any line passing through (–1, 2, 1) and parallel to the line $\frac{x-1}{2}=\frac{y+1}{3}=\frac{z}{4}$ ...(i) is given by

$$\frac{x+1}{2}=\frac{y-2}{3}=\frac{z-1}{4}=\lambda \text{ (say)} \quad \ldots\text{(ii)}$$

Any point on line (ii), we get

$x=2\lambda-1, y=3\lambda+2, z=4\lambda+1$

$$\frac{x+2}{3}=\frac{y-3}{2}=\frac{z-4}{1}=\mu \text{ (say)} \quad \ldots\text{(iii)}$$

Any point on line (iii)

$x=3\mu-2, y=2\mu+3, z=\mu+4$

Since, lines (ii) and (iii) intersect at point $P$.

So, $\quad x=2\lambda-1=3\mu-2 \quad \ldots$ (iv)

$y=3\lambda+2=2\mu+3 \quad \ldots$(v)

$z=4\lambda+1=\mu+4 \quad \ldots$(vi)

On solving Eqs. (iv), (v) and (vi), we get

$\lambda=1, \mu=1$

$\therefore$ Coordinates of point $P\equiv(1, 5, 5)$

Hence, distance of point $P$ from $Q(4, -5, 1)$ is

$$PQ=\sqrt{(4-1)^2+(-5-5)^2+(1-5)^2}$$

$$=\sqrt{9+100+16}$$

$$=\sqrt{125}=5\sqrt{5}$$

**10.** *(a)* $d=$ Shortest distance between line

$$=\left|\frac{(\mathbf{a}_2-\mathbf{a}_1)\cdot(\mathbf{b}_1\times\mathbf{b}_2)}{\mathbf{b}_1\times\mathbf{b}_2}\right|$$

$$\mathbf{b}_1\times\mathbf{b}_2=\begin{vmatrix}\hat{i} & \hat{j} & \hat{k}\\ 1 & 2 & -3\\ 2 & 4 & -5\end{vmatrix}$$

$$=\hat{i}(2)-\hat{j}(1)+\hat{k}(0)=2\hat{i}-\hat{j}$$

$\Rightarrow |\mathbf{b}_1\times\mathbf{b}_2|=\sqrt{5}$

$$\therefore (\mathbf{a}_2-\mathbf{a}_1)\cdot(\mathbf{b}_1\times\mathbf{b}_2)=\begin{vmatrix}-3 & -4 & -2\\ 1 & 2 & -3\\ 2 & 4 & -5\end{vmatrix}$$

$$=(-3(2)+4(1)-20)=-6+4=-2$$

$$\therefore \quad |d|=\left|\frac{-2}{\sqrt{5}}\right|=\frac{2}{\sqrt{5}}$$

$\Rightarrow \quad d^2=\frac{4}{5}=\frac{m}{n}$ [given]

$\therefore \; m+n=4+5=9$

**11.** *(d)* $L_1: \frac{x-2}{2}=\frac{y-6}{3}=\frac{z-3}{4}=\mu$ (say)

$L_2: \frac{x}{1}=\frac{y-2}{2}=\frac{z+3}{3}$

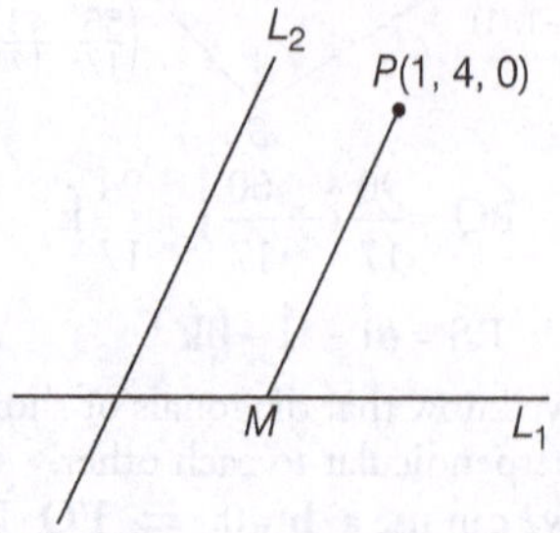


Equation of line passing through (1, 4, 0) and parallel to line $L_2$ is

$$\frac{x-1}{1}=\frac{y-4}{2}=\frac{z-0}{3}=\lambda \text{ (say)}$$

Let $M(\lambda+1, 2\lambda+4, 3\lambda)$

or $M(2\mu+2, 3\mu+6, 4\mu+3)$

$\lambda+1=2\mu+2$

$\Rightarrow \quad \lambda=2\mu+1,$

and $\quad 2\lambda+4=3\mu+6$

$\Rightarrow \quad 2\lambda=3\mu+2$

and $\quad 3\lambda=4\mu+3$

$\therefore \quad \mu=0$

$\therefore \quad M(2, 6, 3)$

$\Rightarrow PM=\sqrt{(2-1)^2+(6-4)^2+(3-0)^2}$

$=\sqrt{1+4+9}=\sqrt{14}$

**12.** *(d)* Given, line

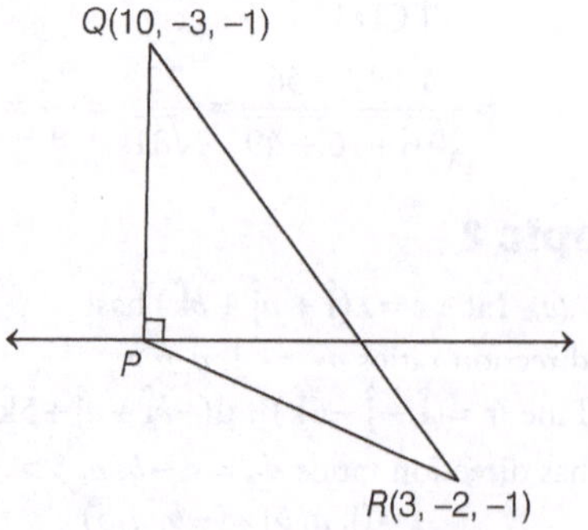


$$L: \frac{x-3}{7}=\frac{y-2}{-1}=\frac{z+1}{-2}=\lambda \text{ (say)}$$

Any point on line be ($P$)

$(7\lambda+3, -\lambda+2, -2\lambda-1)$

Direction ratio of

$QP\equiv 7\lambda-7, -\lambda+5, -2\lambda$

Now, $PQ \perp$ line (i)

$7(7\lambda-7)-1(-\lambda+5)-2(-2\lambda)=0$

$\Rightarrow \quad \lambda=1$

$P\equiv(10, 1, -3)$

$PQ=-4\hat{j}+2\hat{k}$

$PR=-7\hat{i}-3\hat{j}+4\hat{k}$

$$\therefore \text{Required area} = \frac{1}{2}\left|\begin{vmatrix} \hat{i} & \hat{j} & \hat{k} \\ 0 & -4 & 2 \\ -7 & -3 & 4 \end{vmatrix}\right|$$

$$= 3\sqrt{30}$$

**13.** *(b)* $L : \dfrac{x-1}{2} = \dfrac{y+2}{-1} = \dfrac{z+3}{2} = \lambda$ (say)

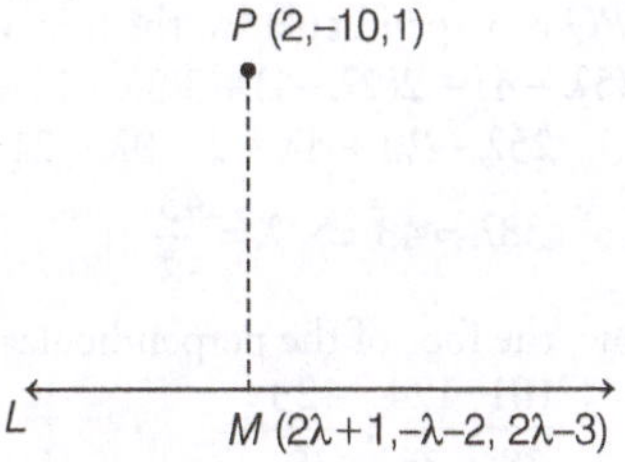


Let $M = (2\lambda+1, -\lambda-2, 2\lambda-3)$

$$\mathbf{PM} = (2\lambda-1)\,\hat{i} + (8-\lambda)\,\hat{j} + (2\lambda-4)\,\hat{k}$$

$$\mathbf{b} = 2\hat{i} - \hat{j} + 2\hat{k}$$

$\because \quad \mathbf{PM} \perp \mathbf{b}$

$\Rightarrow \quad \mathbf{PM} \cdot \mathbf{b} = 0$

$\Rightarrow \quad 4\lambda - 2 - 8 + \lambda + 4\lambda - 8 = 0$

$\Rightarrow \quad \lambda = 2$

$$PM = \sqrt{(4-1)^2 + (8-2)^2 + (4-4)^2}$$

$$= \sqrt{3^2 + 6^2} = 3\sqrt{5}$$

**Alternate Method**

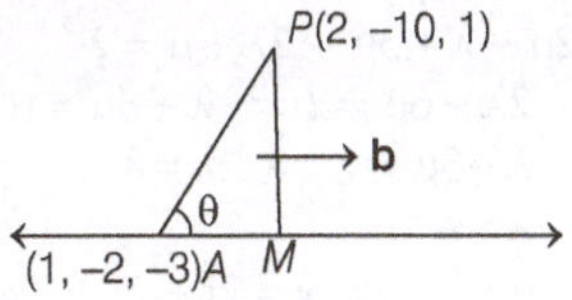


$$PM = PA\sin\theta = \frac{|\mathbf{PA} \times \mathbf{b}|}{|\mathbf{b}|}$$

$$= \frac{|(-\hat{i} + 8\hat{j} - 4\hat{k}) \times (2\hat{i} - \hat{j} + 2\hat{k})|}{\sqrt{4+1+4}}$$

$$= \frac{\sqrt{(12)^2 + (6)^2 + (15)^2}}{3} = 3\sqrt{5}$$

**14.** *(b)* $\mathbf{PQ} \,||\, 3\hat{i} + 2\hat{j} + 2\hat{k}$

Equation of line *PQ*

$$\frac{x+2}{3} = \frac{y+1}{2} = \frac{z-3}{2} = \lambda \quad \text{(say)}$$

Let $Q\,(3\lambda-2, 2\lambda-1, 2\lambda+3)$

$R\,(1, 3, 3)$

$\because \quad QR = 5$

$\Rightarrow \quad QR^2 = 25$

$\Rightarrow (3\lambda-3)^2 + (2\lambda-4)^2 + (2\lambda)^2 = 25$

$\Rightarrow \quad 17\lambda^2 - 34\lambda = 0$

$17\lambda(\lambda - 2) = 0$

$\lambda = 0, 2$

Area of $PQR = \dfrac{1}{2}|\mathbf{PQ} \times \mathbf{QR}|$

$$A = \frac{1}{2}\begin{vmatrix} \hat{i} & \hat{j} & \hat{k} \\ 6 & 4 & 4 \\ 3 & 4 & 0 \end{vmatrix}$$

$$A = \sqrt{64+36+36}$$

$$= \sqrt{136}$$

$$A^2 = 136$$

**15.** *(d)* Given equation of lines are

$$\frac{x-3}{4} = \frac{y+7}{-11} = \frac{z-1}{5}$$

and $\quad \dfrac{x-5}{3} = \dfrac{y-9}{-6} = \dfrac{z+2}{1}$

From figure,

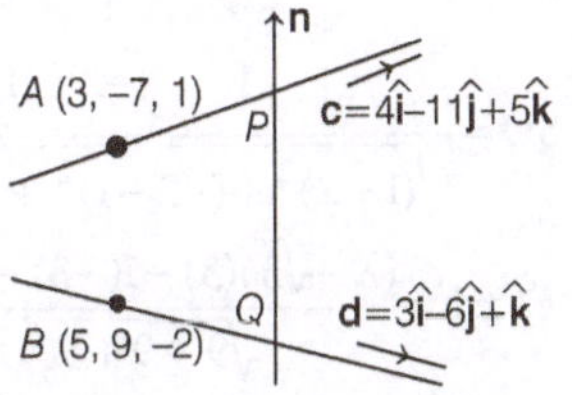


$$\mathbf{n} = \mathbf{c} \times \mathbf{d}$$

$$= \begin{vmatrix} \hat{i} & \hat{j} & \hat{k} \\ 4 & -11 & 5 \\ 3 & -6 & 1 \end{vmatrix} = 19\hat{i} + 11\hat{j} + 9\hat{k}$$

$\therefore$ Shortest distance (SD) = Projection of **AB** on **n**

$$= \left|\frac{\mathbf{AB} \cdot \mathbf{n}}{|\mathbf{n}|}\right|$$

$$= \left|\frac{(2\hat{i} + 16\hat{j} - 3\hat{k}) \cdot (19\hat{i} + 11\hat{j} + 9\hat{k})}{\sqrt{361 + 121 + 81}}\right|$$

$$= \frac{38 + 176 - 27}{\sqrt{563}} = \frac{187}{\sqrt{563}}$$

**16.** *(a)*

**key idea** If $\mathbf{r} = \mathbf{a}_1 + \lambda\mathbf{b}_1$ and $\mathbf{r} = \mathbf{a}_2 + \mu\mathbf{b}_2$ be two skew lines, then the shortest distance between two lines is

$$d = \left|\frac{(\mathbf{a}_2 - \mathbf{a}_1) \cdot (\mathbf{b}_1 \times \mathbf{b}_2)}{|\mathbf{b}_1 \times \mathbf{b}_2|}\right|$$

$$L_1 : \mathbf{r} = (2+\lambda)\hat{i} + (1-3\lambda)\hat{j} + (3+4\lambda)\hat{k}$$

$$L_2 : \mathbf{r} = 2(1+\mu)\hat{i} + 3(1+\mu)\hat{j} + (5+\mu)\hat{k}$$

$$= (2+2\mu)\hat{i} + (3+3\mu)\hat{j} + (5+\mu)\hat{k}$$

Shortest distance $(d)$

$$= \left|\frac{\begin{vmatrix} x_2 - x_1 & y_2 - y_1 & z_2 - z_1 \\ a_1 & b_1 & c_1 \\ a_2 & b_2 & c_2 \end{vmatrix}}{\left|\begin{vmatrix} \hat{i} & \hat{j} & \hat{k} \\ a_1 & a_2 & a_3 \\ b_1 & b_2 & b_3 \end{vmatrix}\right|}\right|$$

$$= \frac{\begin{vmatrix} 0 & -2 & -2 \\ 1 & -3 & 4 \\ 2 & 3 & 1 \end{vmatrix}}{\left|\begin{vmatrix} \hat{i} & \hat{j} & \hat{k} \\ 1 & -3 & 4 \\ 2 & 3 & 1 \end{vmatrix}\right|}$$

$$= \left|\frac{2(1-8) - 2(3+6)}{\hat{i}(-15) - \hat{j}(1-8) + \hat{k}(3+6)}\right|$$

$$= \left|\frac{-14-18}{-15\hat{i} + 7\hat{j} + 9\hat{k}}\right|$$

$$= \frac{32}{\sqrt{(-15)^2 + 7^2 + 9^2}}$$

$$= \frac{32}{\sqrt{225 + 49 + 81}}$$

$$= \frac{32}{\sqrt{355}} = \frac{m}{\sqrt{n}}$$

So, $m = 32$ and $n = 355$

Therefore, $m + n = 32 + 355 = 387$

**17.** *(b)*

**key idea** First write the given equation of lines in the form of

$\mathbf{r}_1 = \mathbf{a}_1 + \lambda\mathbf{b}_1$ and $\mathbf{r}_2 = \mathbf{a}_2 + \mu\mathbf{b}_2$

then use formula to find shortest distance,

$$d = \left|\frac{(\mathbf{a}_2 - \mathbf{a}_1) \cdot (\mathbf{b}_1 \times \mathbf{b}_2)}{|\mathbf{b}_1 \times \mathbf{b}_2|}\right|$$

Given lines are,

$$\frac{x-3}{2} = \frac{y+15}{-7} = \frac{z-9}{5}$$

and $\quad \dfrac{x+1}{2} = \dfrac{y-1}{1} = \dfrac{z-9}{-3}$

So, $\quad \mathbf{a}_1 = 3\hat{i} - 15\hat{j} + 9\hat{k}$,

$\mathbf{a}_2 = -\hat{i} + \hat{j} + 9\hat{k}$

and $\quad \mathbf{b}_1 = 2\hat{i} - 7\hat{j} + 5\hat{k}$,

$\mathbf{b}_2 = 2\hat{i} + \hat{j} - 3\hat{k}$

Now, $\mathbf{a}_2 - \mathbf{a}_1 = -4\hat{i} + 16\hat{j}$

$$(\mathbf{b}_1 \times \mathbf{b}_2) = \begin{vmatrix} \hat{i} & \hat{j} & \hat{k} \\ 2 & -7 & 5 \\ 2 & 1 & -3 \end{vmatrix}$$

$$= 16\hat{i} + 16\hat{j} + 16\hat{k}$$

$$|\mathbf{b}_1 \times \mathbf{b}_2| = 16\sqrt{3}$$

$$d = \left| \frac{(-4\hat{i} + 16\hat{j}) \cdot 16(\hat{i} + \hat{j} + \hat{k})}{16\sqrt{3}} \right|$$

$$= \left| \frac{-4 + 16}{\sqrt{3}} \right| = \frac{12}{\sqrt{3}} = 4\sqrt{3}$$

**18.** *(b)* Let $R = 2\lambda + 1, 3\lambda - 1, 5\lambda + 2$

$\mathbf{PR} \perp \mathbf{b}$

$$(2\lambda + 1 - 8)\,2 + (3\lambda - 1 - 5)\,3 + (5\lambda + 2 - 7)\,5 = 0$$

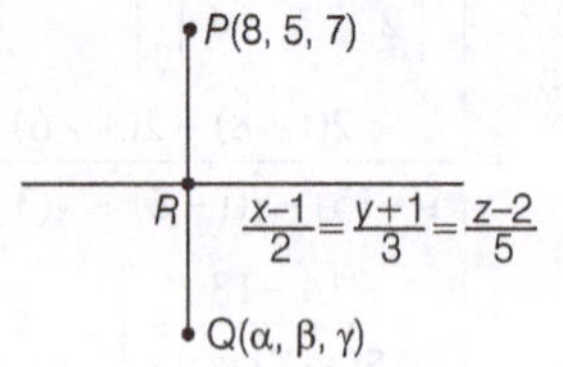


$$\lambda = \frac{57}{38} = \frac{3}{2}$$

$$\because \frac{\alpha + 8}{2} = 2\lambda + 1 \Rightarrow \alpha = 4\lambda + 2 - 8$$

$$\frac{\beta + 5}{2} = 3\lambda - 1 \Rightarrow \beta = 6\lambda - 2 - 5$$

$$\frac{\gamma + 7}{2} = 5\lambda + 2 \Rightarrow \gamma = 10\lambda + 4 - 7$$

$$\alpha + \beta + \gamma = 20\lambda - 16$$

$$= 20 \times \frac{3}{2} - 16 = 14$$

**19.** *(b)* Given lines are

$$\frac{2 - x}{3} = \frac{3y - 2}{4\lambda + 1} = 4 - z$$

$$\Rightarrow \frac{x - 2}{-3} = \frac{y - \frac{2}{3}}{(4\lambda + 1)/3} = \frac{z - 4}{-1} \quad \text{...(i)}$$

and $\frac{x + 3}{3\mu} = \frac{1 - 2y}{6} = \frac{5 - z}{7}$

$$\Rightarrow \frac{x + 3}{3\mu} = \frac{y - \frac{1}{2}}{-3} = \frac{z - 5}{-7} \quad \text{...(ii)}$$

Since, lines (i) and (ii) are perpendicular

$$\therefore \quad (-3)(3\mu) + \frac{(4\lambda + 1)}{3}(-3) + (-1)(-7) = 0$$

$$\Rightarrow \quad -9\mu - 4\lambda - 1 + 7 = 0$$

$$\Rightarrow \quad 4\lambda + 9\mu = 6$$

**20.** *(a)* Equation of line

$PQ$ is $\frac{x - 1}{4} = \frac{y + 2}{-2} = \frac{z - 3}{4} = \lambda$ (say)

So, coordinates of point is

$(4\lambda + 1, -2\lambda - 2, 4\lambda + 3)$

Now, $(\text{Distance})^2 = 16\lambda^2 + 4\lambda^2 + 16\lambda^2$

$$\Rightarrow \quad 36\lambda^2 = (9)^2 = 81 \Rightarrow \lambda = \pm\frac{3}{2}$$

So, coordinates of point is

$(7, -5, 9) \equiv (\alpha, \beta, \gamma)$

Hence, $\alpha^2 + \beta^2 + \gamma^2$

$$= (7)^2 + (-5)^2 + (9)^2 = 155$$

**21.** *(b)* Given equation

$$\frac{x - \lambda}{-2} = \frac{y - 2}{1} = \frac{z - 1}{1}$$

and $\frac{x - \sqrt{3}}{1} = \frac{y - 1}{-2} = \frac{z - 2}{1}$

$$SD = \frac{\begin{vmatrix} \lambda - \sqrt{3} & 1 & -1 \\ -2 & 1 & 1 \\ 1 & -2 & 1 \end{vmatrix}}{\sqrt{(1+2)^2 + (-2-1)^2 + (4-1)^2}}$$

$$\Rightarrow \quad \pm 1 = \frac{(\lambda - \sqrt{3})(3) - 1(-3) - 1(3)}{\sqrt{9 + 9 + 9}}$$

$$\Rightarrow \pm 3\sqrt{3} = 3\lambda - 3\sqrt{3} + 3 - 3$$

$$\Rightarrow \quad 3\lambda = 3\sqrt{3} \pm 3\sqrt{3}$$

$$\Rightarrow \quad \lambda = \sqrt{3} \pm \sqrt{3}$$

$$\Rightarrow \quad \lambda = 2\sqrt{3}, 0$$

Sum of all $\lambda = 2\sqrt{3} + 0 = 2\sqrt{3}$

**22.** *(c)*

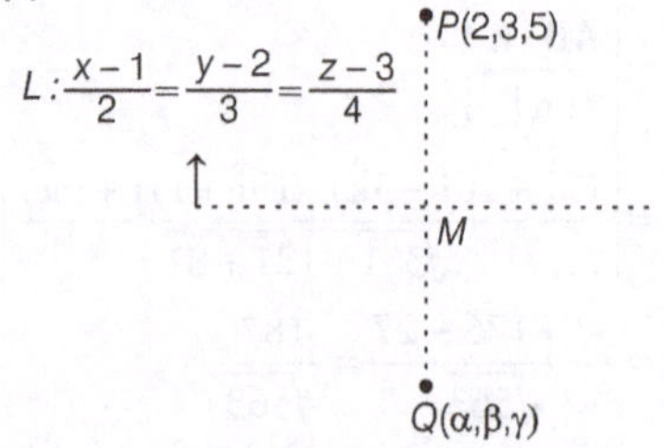


Let $\frac{x - 1}{2} = \frac{y - 2}{3} = \frac{z - 3}{4} = \lambda$ (let)

and $P(2, 3, 5)$ and $Q(\alpha, \beta, \gamma)$.

$\therefore$ DR's of $PQ: (\alpha - 2, \beta - 3, \gamma - 5)$

DR's of the line $L$: 2, 3, 4

$\because$ $L$ is perpendicular to $PQ$.

$$\therefore \quad 2(\alpha - 2)\ 3(\beta - 3) + 4(\gamma - 5) = 0$$

$$\Rightarrow \quad 2\alpha + 3\beta + 4\gamma = 33$$

**23.** *(a)* We have,

$$\frac{x + 3}{5} = \frac{y - 1}{2} = \frac{z + 4}{3} = \lambda \text{ (say)}$$

Now, the coordinates of any point $Q$ on the line are

$$x = 5\lambda - 3,\ y = 2\lambda + 1,\ z = 3\lambda - 4$$

and given point is $P(1, 2, 3)$.

Then, direction ratios of $PQ$ are $5\lambda - 4$, $2\lambda - 1$, $3\lambda - 7$

and the direction ratios of the given line are 5, 2, 3.

If $PQ$ is perpendicular to the line, then

$$5(5\lambda - 4) + 2(2\lambda - 1) + 3(3\lambda - 7) = 0$$

$$\Rightarrow \quad 25\lambda - 20 + 4\lambda - 2 + 9\lambda - 21 = 0$$

$$\Rightarrow \quad 38\lambda = 43 \Rightarrow \lambda = \frac{43}{38}$$

Now, the foot of the perpendicular is

$$\frac{101}{38}, \frac{124}{38}, \frac{-23}{38}$$

So, $\alpha = \frac{101}{38}, \beta = \frac{62}{19},$

$$\gamma = \frac{-23}{38}$$

So, $19(\alpha + \beta + \gamma) = 101$

**24.** *(d)* Here, $\mathbf{v}$ can be written as,

$$\mathbf{v} = \lambda\mathbf{a} + \mu\mathbf{b}$$

$$\mathbf{v} = \lambda(\hat{i} + \hat{j} + 2\hat{k}) + \mu(2\hat{i} - 3\hat{j} + \hat{k})$$

$$\mathbf{v} = (\lambda + 2\mu)\hat{i} + (\lambda - 3\mu)\hat{j} + (2\lambda + \mu)\hat{k}$$

Given, $\mathbf{v} \cdot \hat{j} = 7$ and $\mathbf{v} \cdot \frac{\mathbf{c}}{|\mathbf{c}|} = \frac{2}{\sqrt{3}}$

$$= \mathbf{v} \cdot \mathbf{c} = 2$$

$$\Rightarrow \quad \lambda - 3\mu = 7 \Rightarrow \mathbf{v} \cdot \mathbf{c} = 2$$

$$\lambda + 2\mu - \lambda + 3\mu + 2\lambda + \mu = 2$$

$$2\lambda + 6\mu = 2 \Rightarrow \lambda + 3\mu = 1$$

and $\lambda - 3\mu = 7 \Rightarrow 2\lambda = 8$

or $\lambda = 4$

$$\Rightarrow \quad \mu = -1 \Rightarrow \mathbf{v} = 2\hat{i} + 7\hat{j} + 7\hat{k}$$

$$\mathbf{v} \cdot (\hat{i} + \hat{k}) = 2 + 7 = 9$$

**25.** *(d)* Given,

line $\Rightarrow \frac{x - 1}{2} = \frac{y + 1}{3} = \frac{z - 1}{-2} = \lambda$ (let)

Any point on this line is $B(2\lambda + 1, 3\lambda - 1, -2\lambda + 1)$ and direction ratios of this line $= (2, 3, -2) = \mathbf{d}_1$

Let given point be $A(0, 1, 2)$.

Then direction ratio of

$$AB = (2\lambda + 1, 3\lambda - 2, -2\lambda - 1) = \mathbf{d}_2$$

$\because$ Both lines are perpendicular to each other.

$$\therefore \quad \mathbf{d}_1 \cdot \mathbf{d}_2 = 0$$

$$2(2\lambda + 1) + 3(3\lambda - 2) - 2(-2\lambda - 1) = 0$$

$$\Rightarrow \quad 4\lambda + 2 + 9\lambda - 6 + 4\lambda + 2 = 0$$

$$\Rightarrow \quad 17\lambda = 2 \Rightarrow \lambda = 2/17$$

$\therefore$ Direction ratio of required line

$$\mathbf{d}_2 = (21, -28, -21)$$

$$= (3, -4, -3) = (-3, 4, 3)$$

This line passes through $A(0, 1, 2)$.

$\therefore$ Required equation of line

$$\Rightarrow \frac{x-0}{-3}=\frac{y-1}{4}=\frac{z-2}{3}$$

**26.** *(b)* As, $P(a, b, c)$ is the image of the point $Q(1, 2, -3)$ in the line,

$$\frac{x+1}{2}=\frac{y-3}{-2}=\frac{z}{-1} \quad \ldots(i)$$

So, the mid-point of $P$ and $Q$ lies in the line,

$$\frac{\frac{a+1}{2}+1}{2}=\frac{\frac{b+2}{2}-3}{-2}=\frac{\frac{c-3}{2}}{-1}$$

$$\Rightarrow \frac{a+3}{4}=\frac{b-4}{-4}=\frac{c-3}{-2}$$

$$\Rightarrow \frac{a+3}{-2}=\frac{b-4}{2}=\frac{c-3}{1}=k \quad \text{(let)}$$

$$\Rightarrow a=-2k-3,$$
$$b=2k+4 \text{ and } c=k+3 \quad \ldots(ii)$$

As the line $PQ$ is perpendicular to line (i).

$$\therefore\ 2(a-1)-2(b-2)-1(c+3)=0$$

$$\Rightarrow 2a-2b-c=1 \quad \ldots(iii)$$

From Eqs. (ii) and (iii), we get

$$-4k-6-4k-8-k-3=1$$

$$\Rightarrow 9k=-18 \Rightarrow k=-2$$

$$\therefore\ a+b+c=k+4=-2+4=2$$

**27.** *(d)* Given equations can be rewritten as

$$\frac{x-b}{a}=\frac{y-0}{1}=\frac{z-d}{c}$$

and
$$\frac{x-b'}{a'}=\frac{y-0}{1}=\frac{z-d'}{c'}$$

These lines will be perpendicular, if

$$aa'+cc'+1=0 \Rightarrow aa'+cc'=-1$$
$$[\because l_1l_2+m_1m_2+n_1n_2=0]$$

**28.** *(c)* The given equations of lines can be rewritten as

$$\frac{x}{3}=\frac{y}{2}=\frac{z}{-6} \text{ and } \frac{x}{2}=\frac{y}{-12}=\frac{z}{-3}$$

$\therefore$ Angle between the lines is

$$\theta=\cos^{-1}\left(\frac{3\times 2+2(-12)-6(-3)}{\sqrt{3^2+2^2+(-6)^2}\sqrt{(2)^2+(-12)^2+(-3)^2}}\right)$$

$$\left[\because \cos\theta=\frac{a_1\cdot a_2+b_2\cdot b_2+c_1\cdot c_2}{\sqrt{a_1^2+b_1^2+c_1^2}\sqrt{a_2^2+b_2^2+c_2^2}}\right]$$

$$=\cos^{-1}(0)=90^\circ$$

**29.** *(a)*

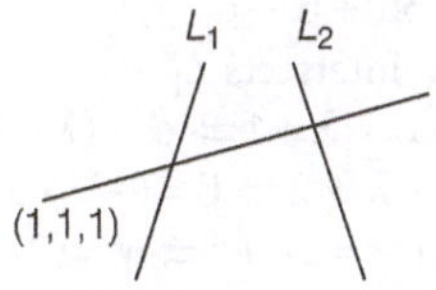


$$L: \frac{x-1}{a}=\frac{y-1}{b}=\frac{z-1}{c}$$

$$L_1: \frac{x-1}{2}=\frac{y+1}{3}=\frac{z-1}{4}=\lambda \text{ (say)}$$

Any point on $L_1$ be $A(2\lambda+1, 3\lambda-1, 4\lambda+1)$

$$L_2: \frac{x-3}{1}=\frac{y-4}{2}=\frac{z}{1}=\mu \text{ (say)}$$

Any point on $L_2$ be $B(\mu+3, 2\mu+4, \mu)$

Direction ratio of $L$ be

$$<2\lambda, 3\lambda-2, 4\lambda>$$

or
$$<\mu+2, 2\mu+3, \mu-1>$$

Now,
$$\frac{2\lambda}{\mu+2}=\frac{3\lambda-2}{2\mu+3}=\frac{4\lambda}{\mu-1}$$

$$\Rightarrow \lambda=-\frac{6}{5} \text{ and } \mu=-5$$

$$\therefore\ <a, b, c>\equiv<-3, -7, -6>$$

or
$$<3, 7, 6>$$

$$\therefore\ L: \frac{x-1}{3}=\frac{y-1}{7}=\frac{z-1}{6}$$

$\therefore$ (7, 15, 13) lies on the line

**30.** *(b)*
$$\frac{x-1}{2}=\frac{y-2}{3}=\frac{z-3}{4}$$

and
$$\frac{x}{1}=\frac{y}{\alpha}=\frac{z-5}{1}$$

Shortest distance between two lines

$$=\left|\frac{(\mathbf{a}_1-\mathbf{a}_2)\cdot(\mathbf{b}_1\times\mathbf{b}_2)}{|\mathbf{b}_1\times\mathbf{b}_2|}\right|=\frac{5}{\sqrt{6}} \quad \text{[given]}$$

$$\left|\frac{(\hat{i}+2\hat{j}-2\hat{k})\cdot[\hat{i}(3-4\alpha)+2\hat{j}+\hat{k}(2\alpha-3)]}{\sqrt{(3-4\alpha)^2+2^2+(2\alpha-3)^2}}\right|=\frac{5}{\sqrt{6}}$$

$$\Rightarrow \left|\frac{3-4\alpha+4-2(2\alpha-3)}{\sqrt{22-36\alpha+20\alpha^2}}\right|=\frac{5}{\sqrt{6}}$$

$$\Rightarrow \frac{(13-8\alpha)^2}{22-36\alpha+20\alpha^2}=\frac{25}{6}$$

$$\Rightarrow 3(169+64\alpha^2-208\alpha)=25(11-18\alpha+10\alpha^2)$$

$$\Rightarrow (250-192)\alpha^2+(624-450)\alpha+275-507=0$$

$$\Rightarrow 58\alpha^2+174\alpha-232=0$$

$$\Rightarrow 29\alpha^2+87\alpha-116=0$$

$$\Rightarrow 29\alpha^2+116\alpha-29\alpha-116=0$$

$$\Rightarrow \alpha(29\alpha+116)-1(29\alpha+116)=0$$

$$\Rightarrow \alpha=1, -\frac{116}{29}$$

Hence, sum of all possible values of

$$\alpha=\frac{1-116}{29}=\frac{-87}{29}=-3$$

**31.** *(b)*
$$L_1: \frac{x+1}{3}=\frac{y}{4}=\frac{z}{5}$$

$$L_2: (P\hat{i}+2\hat{j}+\hat{k})+\lambda(2\hat{i}+3\hat{j}+4\hat{k})$$

$$\Rightarrow \frac{x-P}{2}=\frac{y-2}{3}=\frac{z-1}{4}$$

Shortest distance between lines

$$=\left|\frac{(\mathbf{a}_2-\mathbf{a}_1)\cdot(\mathbf{b}_1\times\mathbf{b}_2)}{|\mathbf{b}_1\times\mathbf{b}_2|}\right|$$

$$\Rightarrow \frac{1}{\sqrt{6}}=\left|\frac{\begin{vmatrix} P+1 & 2-0 & 1-0 \\ 3 & 4 & 5 \\ 2 & 3 & 4 \end{vmatrix}}{|\mathbf{b}_1\times\mathbf{b}_2|}\right|$$

$$\Rightarrow \frac{1}{\sqrt{6}}=\left|\frac{(P+1)(1)-2(2)+1(1)}{\sqrt{1^2+2^2+1}}\right|$$

$$1=|P+1-4+1|$$
$$1=|P-2| \Rightarrow P-2=\pm 1$$
$$P=2\pm 1=3, 1 \quad \therefore a=1, b=3$$

$\therefore$ Length of LR of $\frac{x^2}{a^2}+\frac{y^2}{b^2}=1$ is

$$\Rightarrow \frac{2a^2}{b}=\frac{2\times 1}{3}=\frac{2}{3}$$

**32.** *(c)*
$$L_1: \frac{x-3}{3}=\frac{y-\alpha}{-1}=\frac{z-3}{1}$$

$$L_2: \frac{x+3}{-3}=\frac{y+7}{2}=\frac{z-\beta}{4}$$

Shortest distance between $L_1$ and $L_2$ is $3\sqrt{30}$.

$$\left|\frac{(\mathbf{a}_2-\mathbf{a}_1)\cdot(\mathbf{b}_1\times\mathbf{b}_2)}{|\mathbf{b}_1\times\mathbf{b}_2|}\right|=3\sqrt{30}$$

$$\left|\frac{\begin{vmatrix} 3+3 & \alpha+7 & 3-\beta \\ 3 & -1 & 1 \\ -3 & 2 & 4 \end{vmatrix}}{|\mathbf{b}_1\times\mathbf{b}_2|}\right|=3\sqrt{30}$$

$$\Rightarrow \frac{|6(-4-2)-(\alpha+7)(12+3)+(3-\beta)(6-3)|}{\sqrt{36+225+9}}=3\sqrt{30}$$

$$\Rightarrow |-36-15\alpha-105+9-3\beta|=3\sqrt{30}\sqrt{270}$$

$$\Rightarrow |15\alpha+3\beta+132|=270$$

$$\Rightarrow 15\alpha+3\beta=270-132$$

$$\Rightarrow 5\alpha+\beta=\frac{138}{3}=46$$

**33.** *(c)* $L: \frac{x-6}{3} = \frac{y-7}{2} = \frac{z-7}{-2} = k$

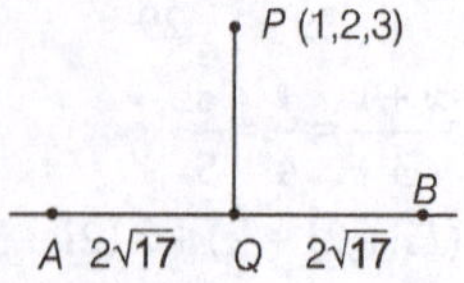


Let point $Q(3k+6, 2k+7, -2k+7)$

$\because \quad \mathbf{PQ} \perp 3\hat{i} + 2\hat{j} - 2\hat{k}$

$\Rightarrow 3(3k+6-1) + 2(2k+7-2) - 2(-2k+7-3) = 0$

$\Rightarrow 17k + 17 = 0 \Rightarrow k = -1$

$\therefore \quad Q = (3, 5, 9)$

Let $A = (3\lambda+6, 2\lambda+7, -2\lambda+7)$

$AQ = 2\sqrt{17}$

$\Rightarrow (3\lambda+3)^2 + (2\lambda+2)^2 + (-2\lambda-2)^2 = 4\times 17$

$(9+4+4)(\lambda+1)^2 = 4 \times 17$

$(\lambda+1)^2 = 4$

$\lambda + 1 = \pm 2, \lambda = -3, 1$

$\therefore A(-3, 1, 13), B(9, 9, 5)$

$\mathbf{OA} - \mathbf{OB} = -27 + 9 + 65 = 47$

**34.** *(d)* $\frac{x-4}{1} = \frac{y-4}{0} = \frac{z-2}{3}$

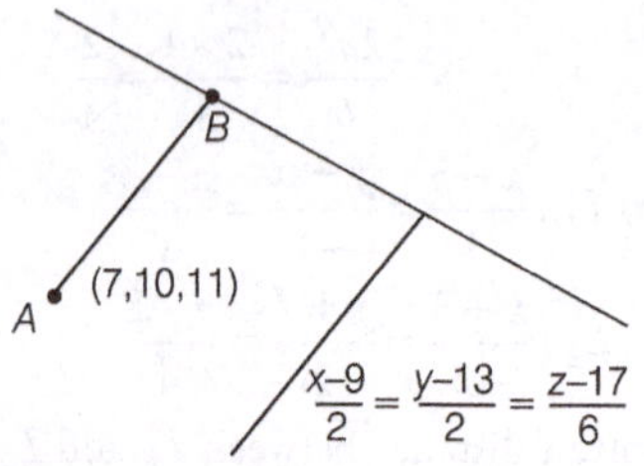


equation of $AB$

$\frac{x-7}{2} = \frac{y-10}{3} = \frac{z-11}{6} = \lambda$ (say)

for point $B \Rightarrow (2\lambda+7, 3\lambda+10, 6\lambda+11)$

According to the question point $B$ lies *m* the line $\frac{x-4}{1} = \frac{y-4}{0} = \frac{z-2}{3}$

$\frac{2\lambda+7-4}{1} = \frac{3\lambda+10-4}{0} = \frac{6\lambda+11-2}{3}$

$3\lambda + 6 = 0 \Rightarrow \lambda = -2$

Point $B = (3, 4, -1)$

Point $A = (7, 10, 11)$

$AB = \sqrt{4^2 + 6^2 + 12^2} = 14$

**35.** *(a)* $L_1 = \frac{x-1}{2} = \frac{y-2}{3} = \frac{z-3}{4} = t_1$ (Let)

$\Rightarrow x = 2t_1 + 1, y = 3t_1 + 2, z = 4t_1 + 3$

$L_2 = x - 6 = y = -z + 4 = t_2$ (let)

$x = t_2 + 6, y = t_2, z = -t_2 + 4$

Let the line passing through $P(4, 1, 0)$ and $A(\alpha, \beta, \gamma)$ be $L$

$\Rightarrow$ Point $P$, $A$ and $B$ are collinear

Directions ratios of line $PA$ are $(\alpha - 4, \beta - 1, \gamma)$ since $A$ lies on $L_1$, we have

$\alpha = 2t_1 + 1, \beta = 3t_1 + 2, \gamma = 4t_1 + 3$

Since, $B(a, b, c)$ lies on $L_2$ we have,

$a = t_2 + 6, b = t_2, c = -t_2 + 4$

Directio ratios of line $PB$ are

$(a-4, b-1, c) = (t_2 + 2, t_2 - 1, -t_2 + 4)$

$P$, $A$ and $B$ are collinear

$\Rightarrow t_1 = -1, t_2 = 2$

$A \equiv [-1, -1, -1]$ and $B \equiv [8, 2, 2]$

$\therefore \quad \alpha = -1, \beta = -1, \gamma = -1$

$a = 8, b = 2, c = 2$

$$\therefore \begin{vmatrix} 1 & 0 & 1 \\ \alpha & \beta & \gamma \\ a & b & c \end{vmatrix} = \begin{vmatrix} 1 & 0 & 1 \\ -1 & -1 & -1 \\ 8 & 2 & 2 \end{vmatrix}$$

$= 1(-2+2) + 1(-2+8) = 6$

**36.** *(c)*

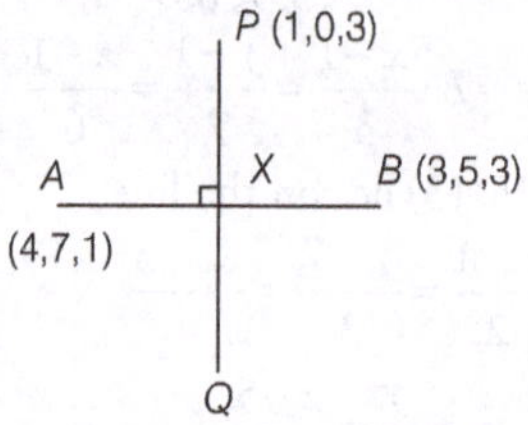


Let $x$ be mid-point of $P$ and $Q$, which would be also feet of perpendicular.

Let $x$ divides $A$ and $B$ in $\lambda : 1, \lambda \neq -1$

$$X \equiv \left(\frac{3\lambda+4}{\lambda+1}, \frac{5\lambda+7}{\lambda+1}, \frac{3\lambda+1}{\lambda+1}\right)$$

Now, $PX \perp AB \Rightarrow \mathbf{PX} \cdot \mathbf{AB} = 0$

$$\left(\frac{3\lambda+4}{\lambda+1} - 1\right)(4-3) + \left(\frac{5\lambda+7}{\lambda+1} - 0\right)(7-5) + \left(\frac{3\lambda+1}{\lambda+1} - 3\right)(1-3) = 0$$

$\Rightarrow \quad 12\lambda + 21 = 0 \Rightarrow \quad \lambda = -\frac{7}{4}$

$X \equiv \left(\frac{5}{3}, \frac{7}{3}, \frac{17}{3}\right)$, Since $X$ is the mid-point of $PQ$

$$Q \equiv \left(2 \cdot \frac{5}{3} - 1, 2 \cdot \frac{7}{3} - 0, 2 \cdot \frac{17}{3} - 3\right) \equiv (\alpha, \beta, \gamma)$$

$\Rightarrow \alpha + \beta + \gamma = \frac{46}{3}$

**37.** *(d)*

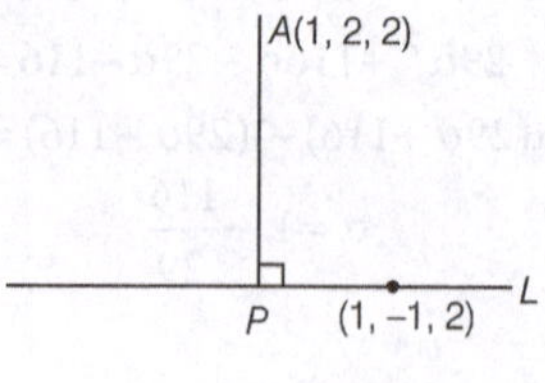


$(1, -1, 2) =$ for of given line

$L: \frac{x-1}{1} = \frac{y+1}{-1} = \frac{z-2}{2} = \mu$(say)

Put $(\mu + 1, -\mu - 1, 2\mu + 2)$

$\mathbf{AP} \cdot \mathbf{d} = 0$

$\Rightarrow (\mu, -\mu - 3, 2\mu) \cdot (1, -1, 2) = 0$

$\Rightarrow \quad \mu + \mu + 3 + 4\mu = 0$

$\Rightarrow \quad \mu = \frac{-1}{2}$

$$\therefore \quad P\left(\frac{-1}{2} + 1, +\frac{1}{2} - 1, 2\left(\frac{-1}{2}\right) + 2\right)$$

$$P\left(\frac{1}{2}, -\frac{1}{2}, 1\right)$$

Now, general point on $L'$ is

$a(-1+\lambda, 1-\lambda, -2+\lambda)$

For point of intersection of line $L$ and $L'$

$\mu + 1 = -1 + \lambda \quad | \quad -\mu - 1 = 1 - \lambda$

$\mu = \lambda - 2 \quad | \quad \mu = \lambda - 2$

$2\mu + 2 = -2 + \lambda$

$2(\lambda - 2) + 2 = -2 + \lambda$

$\therefore \quad \mu = -2, \lambda = 0$

$\Rightarrow \quad Q \equiv (-1, 1, -2)$

So,

$$PQ = \sqrt{\left(\frac{1}{2} + 1\right)^2 + \left(\frac{-1}{2} - 1\right)^2 + (1+2)^2}$$

$$= \sqrt{\frac{9}{4} + \frac{9}{4} + 9} = \sqrt{\frac{54}{4}}$$

$$\therefore 2(PQ)^2 = 2\left(\frac{54}{4}\right) = 27$$

**38.** *(d)* Given

$L_1: \frac{x-1}{1} = \frac{y-2}{-1} = \frac{z-1}{2} = \lambda \quad$ (say)

$\therefore$ Any point taken on $L_1$ is

$(\lambda + 1, -\lambda + 2, 2\lambda + 1)$

Let equation of $L_3$ be

$$\frac{x-\alpha}{a} = \frac{y-\beta}{b} = \frac{z-\gamma}{c} \quad \ldots(i)$$

where, $(a\hat{i} + b\hat{j} + c\hat{k})$ is parallel to

$[(\hat{i} - \hat{j} + 2\hat{k}) \times (-\hat{i} + 2\hat{j} + \hat{k})]$

So, $(a, b, c) \equiv (5, 3, -2)$

From Eq. (i), we get

$$\Rightarrow \frac{x-\alpha}{5} = \frac{y-\beta}{3} = \frac{z-\gamma}{-1} = \mu \text{ (say)} \ldots\text{(ii)}$$

Any point taken on Eq. (ii) i.e. $L_3$ is

$(5\mu + \alpha, 3\mu + \beta, -\mu + \gamma)$

Since, $L_3$ intersects $L_1$

So, $5\mu + \alpha = \lambda + 1 \Rightarrow \alpha = (\lambda + 1 - 5\mu)$

$3\mu + \beta = -\lambda + 2 \Rightarrow \beta = (-\lambda + 2 - 3\mu)$

and $-\mu + \gamma = 2\lambda + 1 \Rightarrow \gamma = (2\lambda + 1 + \mu)$

Hence, $|5\alpha - 11\beta - 8\gamma| = |-25| = 25$

**39.** *(d)*

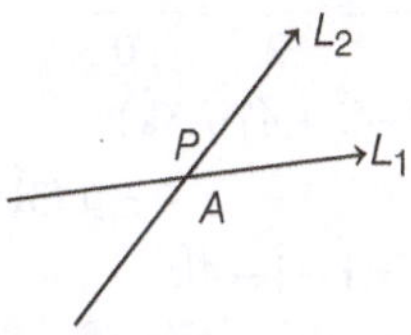


Let $L_1 : \frac{x+1}{3}=\frac{y+3}{5}=\frac{z+5}{7}=t$ (say)

$L_2 : \frac{x-\frac{15}{7}}{1}=\frac{y-\frac{32}{7}}{4}=\frac{z-7}{7}=r$ (say)

$\Rightarrow P(3t-1, 5t-3, 7t-5)$

$= P\left(\frac{15}{7}+r, \frac{32}{7}+4r, 7+7r\right)$

$\Rightarrow \quad 3t-1=\frac{15}{7}+r \quad \ldots(\text{i})$

$5t-3=\frac{32}{7}+4r \quad \ldots(\text{ii})$

Multiplying by 4 in Eq. (i) and subtract Eq. (ii), we get

$12t-4=\frac{60}{7}+4r$

$5t-3=\frac{32}{7}+4r$

$\;-\quad+\quad-\quad-$

$7t-1=4$

$\Rightarrow \quad 7t=5 \Rightarrow t=\frac{5}{7}$

$\therefore$ Coordinate of point

$P \equiv \left(3\times\frac{5}{7}-1, 5\times\frac{5}{7}-3, 7\times\frac{5}{7}-5\right)$

$=\left(\frac{8}{7}, \frac{4}{7}, 0\right)$

and coordinate of point $A \equiv \left(\frac{15}{7}, \frac{32}{7}, 7\right)$

$\therefore \quad AP=\sqrt{\left(\frac{15}{7}-\frac{8}{7}\right)^2+\left(\frac{32}{7}-\frac{4}{7}\right)^2+(7-0)^2}$

$=\sqrt{1^2+4^2+7^2}$

$=\sqrt{1+16+49}=\sqrt{66}$

$\Rightarrow AP^2=66$

**40.** *(a)* Given, $B \equiv (1, 2, 3)$

$AC=6$

Equation of line

$\frac{x-6}{3}=\frac{y-7}{2}=\frac{z-7}{-2} \quad \ldots(\text{i})$

We have to find ar ($\Delta ABC$).

So, ar$(\Delta ABC)=\frac{1}{2}\times AC\times BD$

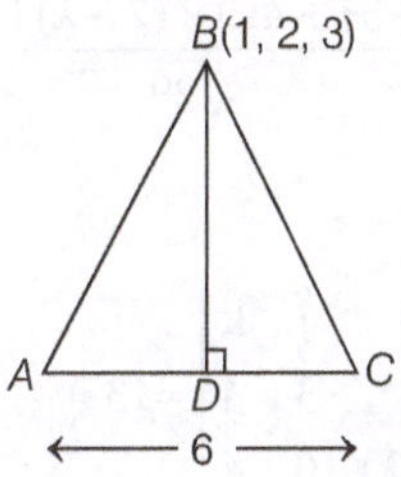


$=\frac{1}{2}\times 6\times BD=3(BD) \quad \ldots(\text{ii})$

Now, any point taken on line (i) is $D(3\lambda+6, 2\lambda+7, -2\lambda+7)$

So, $\mathbf{BD}=(3\lambda+5)\hat{\mathbf{i}}+(2\lambda+5)\hat{\mathbf{j}}+(-2\lambda+4)\hat{\mathbf{k}}$

$\because \mathbf{AC}\perp\mathbf{BD}$

$\therefore \mathbf{AC}\cdot\mathbf{BD}=0$

($\because$ AC is parallel to $3\hat{\mathbf{i}}+2\hat{\mathbf{j}}-2\hat{\mathbf{k}}$)

$\Rightarrow 3(3\lambda+5)+2(2\lambda+5)-2(-2\lambda+4)=0$

$\Rightarrow 9\lambda+4\lambda+4\lambda+15+10-8=0$

$\Rightarrow 17\lambda=-17 \Rightarrow \lambda=-1$

$\therefore \mathbf{BD}=2\hat{\mathbf{i}}+3\hat{\mathbf{j}}+6\hat{\mathbf{k}}$

$\Rightarrow |\mathbf{BD}|=\sqrt{(2)^2+(3)^2+(6)^2}$

$=\sqrt{4+9+36}=\sqrt{49}=7$

Hence, ar$(\Delta ABC)=3(BD)$ [from Eq. (ii)]

$=3\times 7=21$

**41.** *(c)*

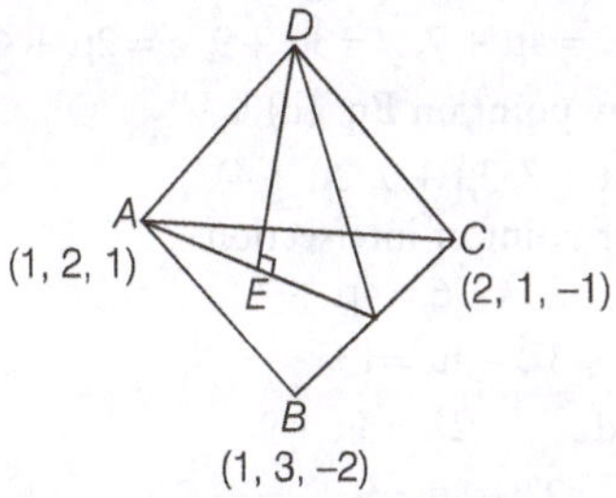


Area of $\Delta ABC=\frac{1}{2}|\mathbf{AB}\times\mathbf{AC}|$

$=\frac{1}{2}|5\hat{\mathbf{i}}+3\hat{\mathbf{j}}+\hat{\mathbf{k}}|=\frac{1}{2}\sqrt{35}$

$\therefore$ Volume of tetrahedron

$=\frac{1}{3}$ Base area $\times$ Height

$\Rightarrow \frac{\sqrt{805}}{6\sqrt{2}}=\frac{1}{3}\times\frac{1}{2}\sqrt{35}\times h$

$\therefore \quad h=\sqrt{\frac{23}{2}}$

Again, $AE^2=AD^2-DE^2$

$\Rightarrow AE^2=\frac{110}{9}-\frac{23}{2}=\frac{13}{18} \Rightarrow AE=\sqrt{\frac{13}{18}}$

$\mathbf{AE}=|AE|\left|\frac{\hat{\mathbf{i}}-5\hat{\mathbf{k}}}{\sqrt{26}}\right|$

$=\sqrt{\frac{13}{18}}\left(\frac{\hat{\mathbf{i}}-5\hat{\mathbf{k}}}{\sqrt{26}}\right)=\frac{\hat{\mathbf{i}}-5\hat{\mathbf{k}}}{6}$

$\therefore$ Position vector of

$E=\frac{\hat{\mathbf{i}}-5\hat{\mathbf{k}}}{6}+\hat{\mathbf{i}}+2\hat{\mathbf{j}}+\hat{\mathbf{k}}$

$=\frac{1}{6}(7\hat{\mathbf{i}}+12\hat{\mathbf{j}}+\hat{\mathbf{k}})$

**42.** *(c)* $L_1 : \frac{x-1}{2}=\frac{y-2}{3}=\frac{z-3}{4}=\lambda$ (say)

$L_2 : \frac{x-2}{3}=\frac{y-4}{4}=\frac{z-5}{5}=\mu$ (say)

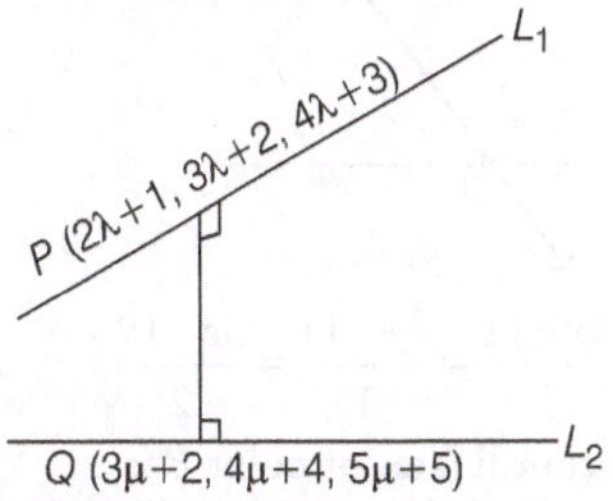


$\mathbf{PQ}=(3\mu-2\lambda+1)\hat{\mathbf{i}}+(4\mu-3\lambda+2)\hat{\mathbf{j}}+(5\mu-4\lambda+2)\hat{\mathbf{k}}$

$\because \mathbf{PQ}\perp\mathbf{b}_1$ and $\mathbf{PQ}\perp\mathbf{b}_2$

$\therefore \mathbf{PQ}$ is parallel to $\mathbf{b}_1\times\mathbf{b}_2$

$\mathbf{b}_1\times\mathbf{b}_2=\begin{vmatrix}\hat{\mathbf{i}} & \hat{\mathbf{j}} & \hat{\mathbf{k}}\\ 2 & 3 & 4\\ 3 & 4 & 5\end{vmatrix}=-\hat{\mathbf{i}}+2\hat{\mathbf{j}}-\hat{\mathbf{k}}$

$\mathbf{PQ}\,||\,\mathbf{b}_1\times\mathbf{b}_2$

$\frac{3\mu-2\lambda+1}{-1}=\frac{4\mu-3\lambda+2}{2}$

$=\frac{5\mu-4\lambda+2}{-1}$

$6\mu-4\lambda+2=-4\mu+3\lambda-2$

$\Rightarrow \quad 10\mu-7\lambda+4=0\times 7$

$-4\mu+3\lambda-2=10\mu-8\lambda+4$

$\Rightarrow \quad -14\mu+11\lambda-6=0\times 5$

$70\mu-49\lambda+28=0$

$-70\mu+55\lambda-30=0$

$6\lambda-2=0$

$\lambda=\frac{1}{3}, \mu=\left(\frac{7}{3}-4\right)\frac{1}{10}=-\frac{5}{3\times 10}=-\frac{1}{6}$

$P=\left(2\left(\frac{1}{3}\right)+1, 3\left(\frac{1}{3}\right)+2, 4\left(\frac{1}{3}\right)+3\right)$

$=\left(\frac{5}{3}, 3, \frac{13}{3}\right)$

Equation of $PQ$

$$\frac{x-\frac{5}{3}}{-1}=\frac{y-3}{2}=\frac{z-\frac{13}{3}}{-1}$$

For $\left(\frac{14}{3},-3,\frac{22}{3}\right)$ satisfies the equation.

**43.** *(a)* $L: \frac{x-1}{1}=\frac{y-2}{1}=\frac{z-3}{2}=\lambda$ (say)

Point $A\left(\frac{11}{3},\frac{11}{3},\frac{19}{3}\right)$ lies on

the given line $\frac{x-\frac{11}{3}}{2}=\frac{y-\frac{11}{3}}{1}=\frac{z-\frac{19}{3}}{2}$

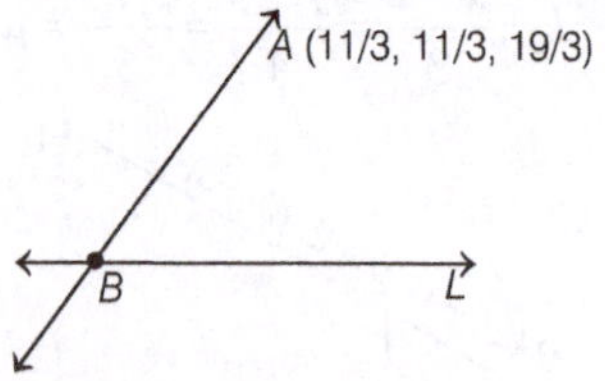


$$\Rightarrow \frac{3x-11}{2}=\frac{3y-11}{1}=\frac{3z-19}{2} \quad \text{...(i)}$$

Let coordinates of point $B$ is $(\lambda+1, \lambda+2, 2\lambda+3)$ which lies on line (i)

$$\frac{3\lambda-8}{2}=\frac{3\lambda-5}{1}=\frac{6\lambda-10}{2}$$

$$\Rightarrow \quad 3\lambda-8=6\lambda-10 \Rightarrow 3\lambda=2$$

$$\lambda=\frac{2}{3}$$

$\therefore$ Coordinates of point $B \equiv \left(\frac{5}{3},\frac{8}{3},\frac{13}{3}\right)$

Now, $AB=\sqrt{(2)^2+(1)^2+(2)^2}=3$

**44.** *(c)*

> **key idea** To find the shortest distance between parallel lines $r_1 = a_1 + \lambda b$ and $r_2 = a_2 + \mu b$ use the formula
>
> $$d=\left|\frac{(a_2-a_1)\times b}{|b|}\right|$$

From the given equations

$$r_1=(\lambda\hat{i}+4\hat{j}+3\hat{k})+\alpha(2\hat{i}+3\hat{j}+4\hat{k})$$

$$r_2=(2\hat{i}+4\hat{j}+7\hat{k})+\beta(2\hat{i}+3\hat{j}+4\hat{k})$$

So, $a_1=\lambda\hat{i}+4\hat{j}+3\hat{k}$

$a_2=2\hat{i}+4\hat{j}+7\hat{k}$

$b=2\hat{i}+3\hat{j}+4\hat{k}$

Shortest distance

$$=\frac{|b\times(a_2-a_1)|}{|b|}=\frac{13}{\sqrt{29}}$$

$$\Rightarrow \frac{|(2\hat{i}+3\hat{j}+4\hat{k})\times(2-\lambda)\hat{i}+4\hat{k})|}{\sqrt{29}}=\frac{13}{\sqrt{29}}$$

$$\Rightarrow \begin{vmatrix}\hat{i} & \hat{j} & \hat{k}\\ 2 & 3 & 4\\ (2-\lambda) & 0 & 4\end{vmatrix}=13$$

$$\Rightarrow |\hat{i}(12)-\hat{j}(8-4(2-\lambda))+\hat{k}(-3(2-\lambda))|=13$$

$$\Rightarrow \quad |12\hat{i}-4\lambda\hat{j}+(3\lambda-6)\hat{k}|=13$$

$$\Rightarrow \quad (12)^2+(4\lambda)^2+(3\lambda-6)^2=(13)^2$$

$$\Rightarrow 144+16\lambda^2+9\lambda^2+36-36\lambda=169$$

$$\Rightarrow \quad 25\lambda^2-36\lambda+11=0$$

Hence, $\lambda=1$

**45.** *(d)* Given lines are

$$\frac{x+6}{3}=\frac{y}{2}=\frac{z+1}{1}=\lambda \quad \text{...(i)}$$

and $\frac{x-7}{4}=\frac{y-9}{3}=\frac{z-4}{2}=\mu$ ...(ii)

From Eq. (i), we get

$x=3\lambda-6,\ y=2\lambda,\ z=\lambda-1$

$\therefore$ Any point on Eq. (i) is

$(3\lambda-6, 2\lambda, \lambda-1)$

From Eq. (ii), we get

$x=4\mu+7,\ y=3\mu+9,\ z=2\mu+4$

Any point on Eq. (ii) is

$(4\mu+7, 3\mu+9, 2\mu+4)$

For point of intersection,

$3\lambda-6=4\mu+7$

$\Rightarrow \quad 3\lambda-4\mu=13$ ...(iii)

and $\quad 2\lambda=3\mu+9$

$\Rightarrow \quad 2\lambda-3\mu=9$ ...(iv)

On solving Eqs. (iii) and (iv), we get

$\lambda=3$ and $\mu=-1$

$\therefore$ Point of intersection is (3, 6, 2).

$\therefore$ Distance between (3, 6, 2) and (7, 8, 9)

$d=\sqrt{16+4+49}=\sqrt{69}$

$\therefore \quad d^2+6=69+6=75$

**46.** *(d)* $L_1: \frac{x-1}{2}=\frac{y+1}{-3}=\frac{z+4}{2}$

$L_1: r=(\hat{i}-\hat{j}-4\hat{k})+\lambda(2\hat{i}-3\hat{j}+2\hat{k})$

$L_2$ passes through $A(-4, 4, 3)$ and $B(-1, 6, 3)$.

DR's of

$L_2: -1-(-4), 6-4, 3-3 \equiv 3, 2, 0$

$L_2: \frac{x+4}{3}=\frac{y-4}{2}=\frac{z-3}{0}$

$L_2: r=(-4\hat{i}+4\hat{j}+3\hat{k})+\mu(3\hat{i}+2\hat{j}+0\hat{k})$

$\therefore \quad a_1=\hat{i}-\hat{j}-4\hat{k}$

and $\quad a_2=-4\hat{i}+4\hat{j}+3\hat{k}$

$\therefore \quad a_2-a_1=-5\hat{i}+5\hat{j}+7\hat{k}$

$b_1=2\hat{i}-3\hat{j}+2\hat{k},$

$b_2=3\hat{i}+2\hat{j}+0\hat{k}$

$$\text{SD}=\left|\frac{(a_2-a_1)\cdot(b_1\times b_2)}{|b_1\times b_2|}\right|$$

and $b_1\times b_2=\begin{vmatrix}\hat{i} & \hat{j} & \hat{k}\\ 2 & -3 & 2\\ 3 & 2 & 0\end{vmatrix}$

$=-4\hat{i}+6\hat{j}+13\hat{k}$

$|b_1\times b_2|=\sqrt{16+36+169}=\sqrt{221}$

$$\text{SD}=\left|\frac{(-5\hat{i}+5\hat{j}+7\hat{k})\cdot(-4\hat{i}+6\hat{j}+13\hat{k})}{\sqrt{221}}\right|$$

$$\text{SD}=\frac{20+30+91}{\sqrt{221}}=\frac{141}{\sqrt{221}}$$

**47.** *(b)* Given, lines are

$$\frac{x-4}{1}=\frac{y+1}{2}=\frac{z}{-3} \quad \text{...(i)}$$

and $\frac{x-\lambda}{2}=\frac{y+1}{4}=\frac{z-2}{-5}$ ...(ii)

Since, line (i) passes through (4, −1, 0) having DR's < 1, 2, −3 >.

So, $\quad a_1=4\hat{i}-\hat{j}-0\hat{k}$

and $\quad b_1=\hat{i}+2\hat{j}-3\hat{k}$

and line (ii) passes through $(\lambda, -1, 2)$ having DR's < 2, 4, −5 >

So, $\quad a_2=\lambda\hat{i}-\hat{j}+2\hat{k}$

and $\quad b_2=2\hat{i}+4\hat{j}-5\hat{k}$

Now, shortest distance (SD) $=\frac{6}{\sqrt{5}}$ [given]

$$\Rightarrow \left|\frac{(a_2-a_1)\cdot(b_1\times b_2)}{|b_1\times b_2|}\right|=\frac{6}{\sqrt{5}}$$

$$\Rightarrow \left|\frac{\begin{vmatrix}\lambda-4 & 0 & 2\\ 1 & 2 & -3\\ 2 & 4 & -5\end{vmatrix}}{\left|\begin{vmatrix}\hat{i} & \hat{j} & \hat{k}\\ 1 & 2 & -3\\ 2 & 4 & -5\end{vmatrix}\right|}\right|=\frac{6}{\sqrt{5}}$$

$$\Rightarrow \left|\frac{(\lambda-4)(-10+12)-0+2(4-4)}{|2\hat{i}-1\cdot\hat{j}+0\hat{k}|}\right| = \frac{6}{\sqrt{5}}$$

$$\Rightarrow \left|\frac{2(\lambda-4)}{\sqrt{5}}\right| = \frac{6}{\sqrt{5}}$$

$$\Rightarrow 3=|\lambda-4| \Rightarrow \lambda-4=\pm 3 \Rightarrow \lambda=7,1$$

Hence, sum of all possible values of $\lambda=8$

**48.** (a) Given equation of lines are

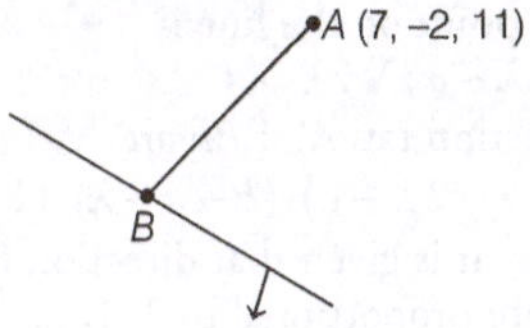


$$\frac{x-6}{1}=\frac{y-4}{0}=\frac{z-8}{3}$$

$$\frac{x-6}{1}=\frac{y-4}{0}=\frac{z-8}{3} \quad \ldots\text{(i)}$$

and $\frac{x-5}{2}=\frac{y-1}{-3}=\frac{z-5}{6}=\lambda$ (say)...(ii)

Let $A \equiv (7,-2,11)$

Any point $B$ taken on Eq. (ii) as

$$(2\lambda+5, -3\lambda+1, 6\lambda+5)$$

$\because$ Point $B$ lies on Eq. (i).

$$\therefore \frac{2\lambda+5-6}{1}=\frac{-3\lambda+1-4}{0}=\frac{6\lambda+5-8}{3}$$

(iii) (iv) (v)

From Eqs. (iii) and (iv), we get

$$-3\lambda-3=0 \Rightarrow \lambda=-1$$

So, $B \equiv (3,4,-1)$

Hence,

$$AB=\sqrt{(7-3)^2+(-2-4)^2+(11+1)^2}$$

$$=\sqrt{16+36+144}=\sqrt{196}=14$$

**49.** (c) The given linear are,

$$\frac{x-\lambda}{0}=\frac{y-3}{4}=\frac{z+6}{1}$$

$$\frac{x+\lambda}{3}=\frac{y}{-4}=\frac{z-6}{0}$$

and shortest distance is 13

We know, formula for shortest distance,

$$SD=\frac{\begin{vmatrix} x_2-x_1 & y_2-y_1 & z_2-z_1 \\ a_1 & b_1 & c_1 \\ a_2 & b_2 & c_2 \end{vmatrix}}{\begin{vmatrix} \hat{i} & \hat{j} & \hat{k} \\ a_1 & b_1 & c_1 \\ a_2 & b_2 & c_2 \end{vmatrix}}$$

$$=\left|\frac{\begin{vmatrix} 2\lambda & 3 & -12 \\ 0 & 4 & 1 \\ 3 & -4 & 0 \end{vmatrix}}{\begin{vmatrix} \hat{i} & \hat{j} & \hat{k} \\ 0 & 4 & 1 \\ 3 & -4 & 0 \end{vmatrix}}\right|$$

$$=\left|\frac{2\lambda\times(4)+3(3)-12(-12)}{|4\hat{i}+3\hat{j}-12\hat{k}|}\right|$$

$$=\frac{|8\lambda+153|}{\sqrt{16+9+144}}=\frac{|8\lambda+153|}{13}$$

According to the question,

$$\frac{|8\lambda+153|}{13}=13$$

$$\Rightarrow |8\lambda+153|=169$$

$$\therefore \text{ either } 8\lambda+153=169 \Rightarrow \lambda=\frac{16}{8}=2$$

or $8\lambda+153=-169$

$$\Rightarrow \quad 8\lambda=-322$$

$$\Rightarrow \quad \lambda=\frac{-322}{8}=\frac{-161}{4}$$

Let $S$ be the set of one value of $\lambda$.

$$\text{So, } 8\left|\sum_{\lambda\in S}\lambda\right|=8\left|2-\frac{161}{4}\right|$$

$$=8\left|\frac{-153}{4}\right|=306$$

**50.** (c) $L_1: \frac{x-1}{2}=\frac{y+8}{-7}=\frac{(z-4)}{5}$

$$\Rightarrow \quad a_1=\hat{i}-8\hat{j}+4\hat{k};\ b_1=2\hat{i}-7\hat{j}+5\hat{k}$$

$$L_2: \frac{x-1}{2}=\frac{y-2}{1}=\frac{z-6}{-3}$$

$$\Rightarrow a_2=\hat{i}+2\hat{j}+6\hat{k};\ \mathbf{b}_2=2\hat{i}+\hat{j}-3\hat{k}$$

$$\mathbf{b}_1\times\mathbf{b}_2=\begin{vmatrix} \hat{i} & \hat{j} & \hat{k} \\ 2 & -7 & 5 \\ 2 & 1 & -3 \end{vmatrix}$$

$$=\hat{i}(21-5)-\hat{j}(-6-10)+\hat{k}(2+14)$$

$$=16\hat{i}+16\hat{j}+16\hat{k}$$

$$\therefore (\mathbf{a}_2-\mathbf{a}_1)\cdot(\mathbf{b}_1\times\mathbf{b}_2)=(10\hat{j}+2\hat{k})\cdot 16(\hat{i}+\hat{j}+\hat{k})$$

$$=32(5\hat{j}+\hat{k})\cdot(\hat{i}+\hat{j}+\hat{k})$$

$$=32(5+1)=192$$

$\therefore$ Shortest distance

$$=\left|\frac{(a_2-a_1)\cdot(b_1\times b_2)}{|b_1\times b_2|}\right|$$

$$=\frac{192}{16\sqrt{3}}=\frac{12}{\sqrt{3}}=4\sqrt{3}$$

**51.** (a) Equation of line passing through $(-3, 2, 3)$ and parallel to a line with direction ratios $3, 3, -1$

$$L: \frac{x+3}{3}=\frac{y-2}{3}=\frac{z-3}{-1}=\lambda \text{ (say)}$$

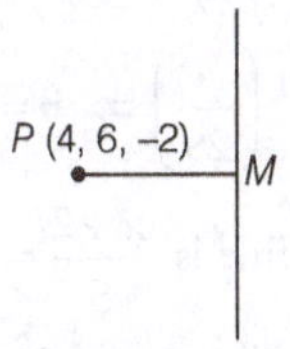


$$L: x=3\lambda-3,\ y=3\lambda+2,\ z=-\lambda+3$$

Let $M(3\lambda-3, 3\lambda+2, -\lambda+3)$

Direction ratio of line

$$PM: (3\lambda-3-4, 3\lambda+2-6, -\lambda+3+2)$$

$$\equiv(3\lambda-7, 3\lambda-4, -\lambda+5)$$

Direction ratio of line $L: 3, 3, -1$

We know that if two lines are perpendicular, then

$$a_1a_2+b_1b_2+c_1c_2=0$$

Here, $PM \perp L$

$$\Rightarrow 3(3\lambda-7)+3(3\lambda-4)-1(-\lambda+5)=0$$

$$19\lambda-38=0 \Rightarrow \lambda=2$$

So, point $M(3, 8, 1)$

Distance

$$PM=\sqrt{(3-4)^2+(8-6)^2+(1+2)^2}$$

$$=\sqrt{1+4+9}=\sqrt{14}$$

**52.** (b) $l_1: \frac{x-2}{3}=\frac{y+1}{-2}=\frac{z-2}{0}$

$$a_1=3, b_1=-2, c_1=0$$

$$l_2: \frac{x-1}{1}=\frac{y+\left(\frac{3}{2}\right)}{\left(\frac{\alpha}{2}\right)}=\frac{z+5}{2}$$

$$a_2=1, b_2=\frac{\alpha}{2},$$

$$c_2=2$$

Given that $l_1$ perpendicular to $l_2$.

$$\therefore \quad a_1a_2+b_1b_2+c_1c_2=0$$

$$\Rightarrow \quad 3-\alpha+0=0$$

$$\Rightarrow \quad \alpha=3$$

$$\Rightarrow \quad a_2=1, b_2=\frac{3}{2}, c_2=2$$

$$\text{Now, } l_3: \frac{x-1}{-3}=\frac{y-\left(\frac{1}{2}\right)}{-2}=\frac{z-0}{4}$$

$$a_3=-3, b_3=-2, c_3=4$$

Angle between $l_2$ and $l_3$ is given by,

$$\cos\theta=\frac{|a_2a_3+b_2b_3+c_2c_3|}{\sqrt{a_2^2+b_2^2+c_2^2}\sqrt{a_3^2+b_3^2+c_3^2}}$$

$$=\frac{|-3-3+8|}{\sqrt{1+\frac{9}{4}+4\sqrt{9+4+16}}}$$

$$\cos\theta=\frac{4}{\sqrt{29}\sqrt{29}}$$

$$\Rightarrow \theta=\cos^{-1}\left(\frac{4}{29}\right) \Rightarrow \theta=\sec^{-1}\left(\frac{29}{4}\right)$$

**53.** *(a)* Given line is $\frac{x+2}{3}=\frac{y-1}{0}=\frac{z}{4}$

Vector along line is, $\mathbf{a}=3\hat{\mathbf{i}}+4\hat{\mathbf{k}}$ and vector joining the points (1, −1, 2) to (−2, 1, 0) is

$$\mathbf{b}=(1+2)\hat{\mathbf{i}}+(-1-1)\hat{\mathbf{j}}+(2-0)\hat{\mathbf{k}}$$
$$=3\hat{\mathbf{j}}-2\hat{\mathbf{j}}+2\hat{\mathbf{k}}$$

and $|\mathbf{BC}|=5$ units

Now area of required $\Delta ABC$

$$=\frac{1}{2}|\mathbf{BC}||\mathbf{b}||\sin\theta| \quad \ldots\text{(i)}$$

[where, $\theta$ is angle between vectors **a** and **b**]

$$\because \quad |\mathbf{b}|\sin\theta=\frac{|\mathbf{a}\times\mathbf{b}|}{|\mathbf{a}|},$$

$$\because \quad |\mathbf{a}\times\mathbf{b}|=\begin{vmatrix}\hat{\mathbf{i}} & \hat{\mathbf{j}} & \hat{\mathbf{k}}\\ 3 & 0 & 4\\ 3 & -2 & 2\end{vmatrix}$$
$$=8\hat{\mathbf{i}}+6\hat{\mathbf{j}}-6\hat{\mathbf{k}}$$

$$\Rightarrow \quad |\mathbf{a}\times\mathbf{b}|=\sqrt{64+36+36}$$
$$=\sqrt{136}=2\sqrt{34}$$

and $\quad |\mathbf{a}|=\sqrt{9+16}=5$

$$\therefore \quad |\mathbf{b}|\sin\theta=\frac{2\sqrt{34}}{5}$$

On substituting these values in Eq. (i), we get

$$\text{Required area}=\frac{1}{2}\times5\times\frac{2\sqrt{34}}{5}$$
$$=\sqrt{34}\text{ sq. units}$$

**Alternate Method**

Given line is $\frac{x+2}{3}=\frac{y-1}{0}=\frac{z}{4}=\lambda$ (let) ...(i)

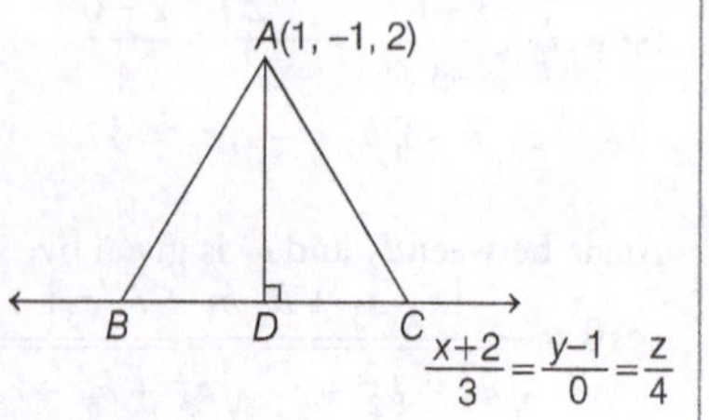


Since, point $D$ lies on the line $BC$.

$\therefore$ Coordinates of $D=(3\lambda-2,1,4\lambda)$

Now, DR's of $BC \Rightarrow a_1=3, b_1=0, c_1=4$

and DR of $AD \Rightarrow a_2=3\lambda-3, b_2=2, c_2=4\lambda-2$

Since, $AD\perp BC, a_1a_2+b_1b_2+c_1c_2=0$

$$3\times(3\lambda-3)+0(2)+4(4\lambda-2)=0$$
$$\Rightarrow \quad 9\lambda-9+0+16\lambda-8=0$$
$$\Rightarrow \quad 25\lambda-17=0 \Rightarrow \lambda=\frac{17}{25}$$

$$\therefore \text{Coordinates of } D=\left(\frac{1}{25},1,\frac{68}{25}\right).$$

$$\text{Now, } AD=\sqrt{\left(1-\frac{1}{25}\right)^2+(-1-1)^2+\left(2-\frac{68}{25}\right)^2}$$
$$=\sqrt{\left(\frac{24}{25}\right)^2+(-2)^2+\left(\frac{-18}{25}\right)^2}$$
$$=\sqrt{\frac{576}{625}+4+\frac{324}{625}}=\frac{2}{5}\sqrt{34}$$

$$\therefore \text{Area of } \Delta ABC=\frac{1}{2}BC\times AD$$
$$=\frac{1}{2}\times5\times\frac{2}{5}\sqrt{34}=\sqrt{34}\text{ sq units}$$

**54.** *(c)* Given two lines,

$$L_1:\frac{x-1}{2}=\frac{y+1}{3}=\frac{z-1}{4}$$

and $\quad L_2:\frac{x-3}{1}=\frac{y-k}{2}=\frac{z-0}{1}$

**To find** The value of $k$ of the given lines $L_1$ and $L_2$ are intersecting each other.

Let $L_1:\frac{x-1}{2}=\frac{y+1}{3}=\frac{z-1}{4}=p$

and $L_2:\frac{x-3}{1}=\frac{y-k}{2}=\frac{z-0}{1}=q$

$\Rightarrow$ Any point $P$ on line $L_1$ is of type $P(2p+1, 3p-1, 4p+1)$ and any point $Q$ on line $L_2$ is of type $Q(q+3, 2q+k, q)$.

Since, $L_1$ and $L_2$ are intersecting each other, hence both points $P$ and $Q$ should coincide at the point of intersection, i.e., corresponding coordinates of $P$ and $Q$ should be same.

$2p+1=q+3, 3p-1=2q+k$

and $\quad 4p+1=q$

On solving $2p+1=q+3$ and $4p+1=q$, we get the values of $p$ and $q$ as

$$p=-3/2 \text{ and } q=-5$$

On substituting the values of $p$ and $q$ in the third equation $3p-1=2q+k$, we get

$$3\left(\frac{-3}{2}\right)-1=2(-5)+k$$

$$\Rightarrow \quad k=9/2$$

**55.** *(b)* Let the equation of line $AB$ be

$$\frac{x-0}{1}=\frac{y+a}{1}=\frac{z-0}{1}=k \quad \text{[say]}$$

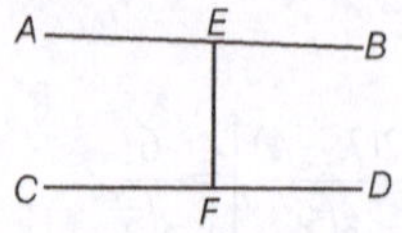


Any point on the line is $E(k, k-a, k)$.

Also, the equation of other line $CD$ is

$$\frac{x+a}{2}=\frac{y-0}{1}=\frac{z-0}{1}=\lambda \quad \text{[say]}$$

Any point on the line is $F(2\lambda-a, \lambda, \lambda)$.

Direction ratios of $EF$ are

$$[(k-2\lambda+a),(k-a-\lambda),(k-\lambda)].$$

Since, it is given that direction ratios of $EF$ are proportional to 2, 1, 2.

$$\therefore \quad \frac{k-2\lambda+a}{2}=\frac{k-\lambda-a}{1}=\frac{k-\lambda}{2}$$

On solving first and second fractions, we get

$$k-2\lambda+a=2k-2\lambda-2a$$
$$\Rightarrow \quad k=3a \quad \ldots\text{(i)}$$

On solving second and third fractions, we get

$$2k-2\lambda-2a=k-\lambda$$
$$\Rightarrow \quad k-\lambda=2a \Rightarrow \lambda=3a-2a$$

From Eq. (i),

$$\lambda=a$$

Hence, co-ordinates of $E$ are $(3a, 2a, 3a)$ and coordinates of $F$ are $(a, a, a)$.

**56.** *(b)* $\mathbf{r}_1=\overbrace{\hat{\mathbf{i}}+2\hat{\mathbf{j}}+3\hat{\mathbf{k}}}^{\mathbf{a}}+k_1\overbrace{(2\hat{\mathbf{i}}+3\hat{\mathbf{j}}+4\hat{\mathbf{k}})}^{\mathbf{b}}$

$\mathbf{r}_2=\underbrace{\lambda\hat{\mathbf{i}}+4\hat{\mathbf{j}}+5\hat{\mathbf{k}}}_{\mathbf{c}}+k_2\underbrace{(3\hat{\mathbf{i}}+4\hat{\mathbf{j}}+5\hat{\mathbf{k}})}_{\mathbf{d}}$

$$\frac{1}{\sqrt{6}}=\left|\frac{(\mathbf{a}-\mathbf{c})\cdot(\mathbf{b}\times\mathbf{d})}{\mathbf{b}\times\mathbf{d}}\right|$$

$$\mathbf{a}-\mathbf{c}=(1-\lambda)\hat{\mathbf{i}}-2\hat{\mathbf{j}}-2\hat{\mathbf{k}}$$

$$\mathbf{b}\times\mathbf{d}=\begin{vmatrix}\hat{\mathbf{i}} & \hat{\mathbf{j}} & \hat{\mathbf{k}}\\ 2 & 3 & 4\\ 3 & 4 & 5\end{vmatrix}$$
$$=-\hat{\mathbf{i}}+2\hat{\mathbf{j}}-\hat{\mathbf{k}}$$

$$\Rightarrow \quad |\mathbf{b}\times\mathbf{d}|=\sqrt{6}$$

$$(\mathbf{a}-\mathbf{c})\cdot(\mathbf{b}\times\mathbf{d})=\lambda-1-4+2=\lambda-3$$

$$\text{So,} \quad \frac{1}{\sqrt{6}}=\left|\frac{\lambda-3}{\sqrt{6}}\right|$$

$$|\lambda-3|=1$$

$\Rightarrow \quad \lambda - 3 = \pm 1$

$\Rightarrow \quad \lambda_1 = 4$ and $\lambda_2 = 2$

Equation of circle passing through (0, 0) (4, 2) and (2, 4)

$$x^2 + y^2 + 2gx + 2fy + c = 0$$

$\Rightarrow \quad c = 0$

$$20 + 8g + 4f = 0$$

$\Rightarrow g = f$

$$20 + 4g + 8f = 0$$

$\Rightarrow \quad 12g = -20$

$\Rightarrow \quad -g = \dfrac{5}{3}$

$\Rightarrow \quad -f = \dfrac{5}{3}$

$$r = \sqrt{\frac{25}{9} \times 2} = \frac{5\sqrt{2}}{3}$$

So, option (b) is correct.

**57.** *(b)* $L_1: \dfrac{x-7}{1} = \dfrac{y-5}{0} = \dfrac{z-3}{-1} = \lambda$ (say)

$L_2: \dfrac{x-1}{3} = \dfrac{y+3}{4} = \dfrac{z+7}{5} = \mu$ (say)

Let any point on $L_1$ and $L_2$ are $(\lambda + 7, 5, -\lambda + 3)$ and $(3\mu + 1, 4\mu - 3, 5\mu - 7)$

For point of intsersection of $L_1$ and $L_2$

$$\lambda + 7 = 3\mu + 1$$
$$\lambda = 0$$
$$5 = 4\mu - 3$$
$$\mu = 2$$
$$-\lambda + 3 = 5\mu - 7$$
$$\lambda = 0$$

$\therefore A\,(7, 5, 3)$

$\because$ $B$ and $C$ lies on $L_1$ and $L_2$ respectively.

$\because \quad AB = \sqrt{15}$ and $AC = \sqrt{15}$

$$\sqrt{\lambda^2 + 0 + \lambda^2} = \sqrt{15}$$
$$2\lambda^2 = 15$$
$$\lambda^2 = \frac{15}{2}$$

and $\sqrt{(3\mu - 6)^2 + (4\mu - 8)^2 + (5\mu - 10)^2} = \sqrt{15}$

$\Rightarrow \quad (\mu - 2)^2 (9 + 16 + 25) = 15$

$\Rightarrow \quad (\mu - 2)^2 = \dfrac{15}{50}$

$\Rightarrow \quad (\mu - 2)^2 = \dfrac{3}{10}$

Area of $\Delta\, ABC = A$

$$A = \frac{1}{2} |\mathbf{AB} \times \mathbf{BC}|$$

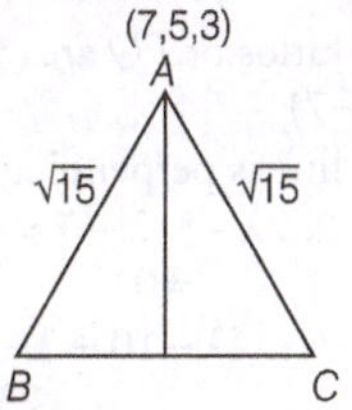


$$= \frac{1}{2} \left\| \begin{matrix} \hat{i} & \hat{j} & \hat{k} \\ \lambda & 0 & -\lambda \\ 3(\mu - 2) & 4(\mu - 2) & 5(\mu - 2) \end{matrix} \right\|$$

$$= \frac{1}{2} |\mu - 2| |\lambda| \left\| \begin{matrix} \hat{i} & \hat{j} & \hat{k} \\ 1 & 0 & -1 \\ 3 & 4 & 5 \end{matrix} \right\|$$

$$= \frac{1}{2} \left(\sqrt{\frac{3}{10}}\right)\left(\sqrt{\frac{15}{2}}\right)\sqrt{4^2 + 8^2 + 4^2}$$

$$= \frac{1}{2} \times \frac{3}{2} \times 4\sqrt{1 + 4 + 1} = 3\sqrt{6}$$

$$A^2 = 9 \times 6 = 54$$

**58.** *(d)* Given equation of lines can be rewritten (in standard form) as—

$L_1 : \dfrac{x-2}{1} = \dfrac{y}{-1} = \dfrac{z-1}{1} = r$ (say)

$L_2 : \dfrac{x+1}{1/2} = \dfrac{y-1}{1/2} = \dfrac{z+1}{1} = s$ (say)

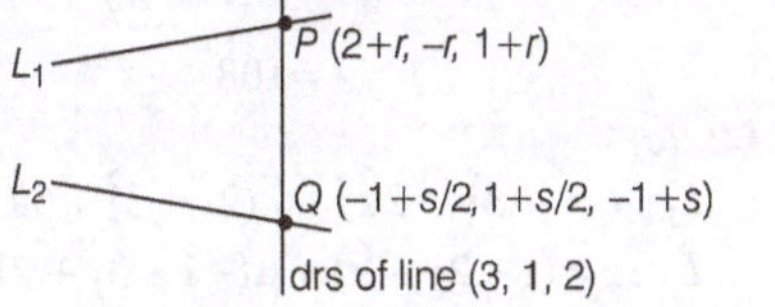


Let $P$ and $Q$ be two points on the lines $L_1$ and $L_2$, respectively.

So, $P \equiv (r + 2, -r, r + 1)$

$$Q \equiv \left(-1 + \frac{1}{2}s, 1 + \frac{1}{2}s, -1 + s\right)$$

Now, d.r.'s of the line $PQ$ is

$$< 3 + r - \frac{s}{2}, -1 - r - \frac{s}{2}, 2 + r - s >$$

and this will be proportional to <3, 1, 2>

So, $\underset{\text{(I)}}{\dfrac{3 + r - \frac{s}{2}}{3}} = \underset{\text{(II)}}{\dfrac{-1 - r - \frac{s}{2}}{1}} = \underset{\text{(III)}}{\dfrac{2 + r - s}{2}}$

From (I) and (II),

$$4r + s = -6 \qquad \text{...(i)}$$

From (II) and (III), we get

$$4 + 3r = 0 \Rightarrow r = -\frac{4}{3} \text{ and hence, } s = \frac{-2}{3}$$

(from Eq. (i))

$$\therefore P \equiv \left(\frac{2}{3}, \frac{4}{3}, -\frac{1}{3}\right)$$

Now, equation of required line will be

$$\frac{x - \frac{2}{3}}{3} = \frac{y - \frac{4}{3}}{1} = \frac{z + \frac{1}{3}}{2} = \lambda \quad \text{(say)}$$

Coordinates of any point taken on it will be $\left(\dfrac{2}{3} + 3\lambda, \dfrac{4}{3} + \lambda, -\dfrac{1}{3} + 2\lambda\right)$

$\because \quad \dfrac{2}{3} + 3\lambda = -\dfrac{1}{3} \Rightarrow \lambda = -\dfrac{1}{3}$

$\therefore$Point lie on the line for $\lambda = -\dfrac{1}{3}$ is

$$\left(-\frac{1}{3}, 1, -1\right)$$

**59.** *(d)* According to the question,

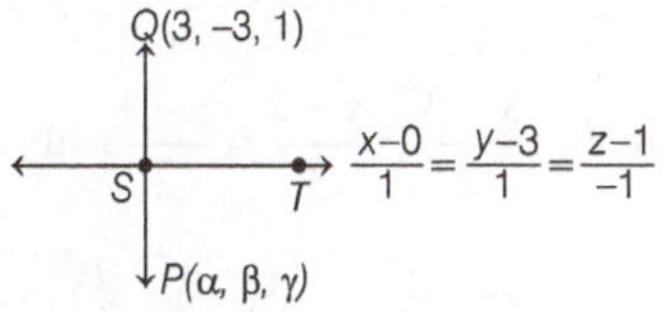


Let $S$ be the foot of perpendicular, it can be written as

$S \equiv (\mu, \mu + 3, -\mu + 1)$,

where $\mu$ be any real number for which $S$ be the foot of perpendicular

According to the conditions,

$SQ \cdot ST = 0$

$\Rightarrow [(\mu - 3)\,\hat{i} + (\mu + 3 + 3)\,\hat{j} + (-\mu)\,\hat{k}] \cdot [\hat{i} + \hat{j} - \hat{k}] = 0$

$\Rightarrow \mu - 3 + \mu + 6 + \mu = 0$

$\Rightarrow \quad 3\mu = -3 \Rightarrow \mu = -1$

$\therefore \quad S \equiv (-1, 2, 2)$

So, $\quad P \equiv (-5, 7, 3)$

Now, point $R(2, 5, -1)$ lies on the given line.

Now, area of $\Delta PQR$

$$\lambda = \frac{1}{2} |PQ \times PR|$$

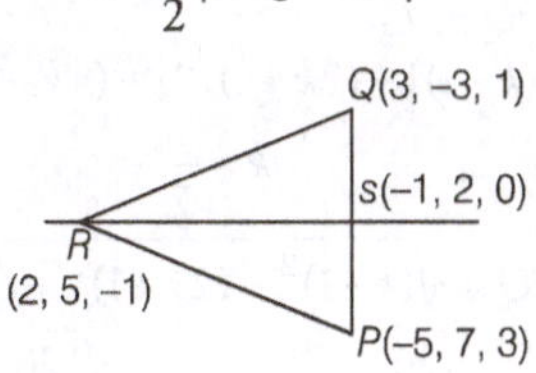


$$\lambda = \frac{1}{2} \left| \begin{matrix} \hat{i} & \hat{j} & \hat{k} \\ 8 & -10 & -2 \\ 7 & -2 & -4 \end{matrix} \right|$$

$= \frac{1}{2}|36\hat{i}+18\hat{j}+54\hat{k}|$

$\lambda = \frac{\sqrt{36^2+18^2+54^2}}{2} = \frac{\sqrt{4536}}{2}$

According to the question,

$\lambda^2 = 14k \Rightarrow \frac{4536}{4} = 14k \Rightarrow k = 81$

**Concept Enhancer**

Area of triangle $\Delta ABC, \Delta = \frac{1}{2}|\mathbf{AB} \times \mathbf{AC}|$

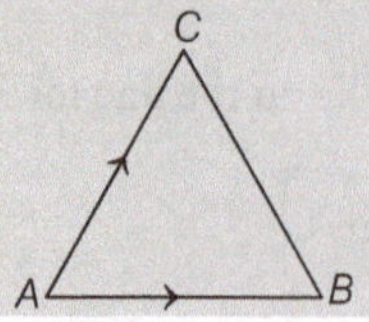


**60.** *(a)* $L_1 : \frac{x-2}{1} = \frac{y-4}{5} = \frac{z-2}{1} = \lambda$

and $L_2 : \frac{x-3}{2} = \frac{y-2}{3} = \frac{z-3}{2} = \mu$

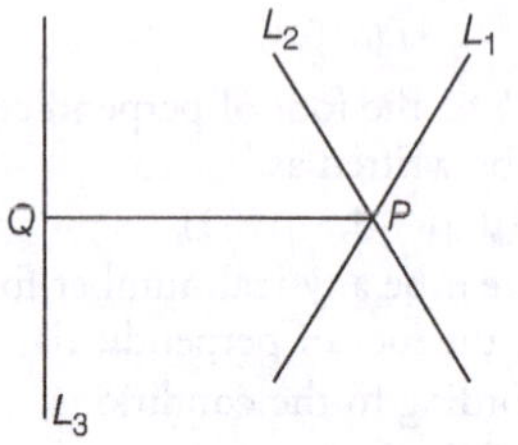


Let $P$ be the point of intersection, then

$\lambda + 2 = 2\mu + 3,$
$5\lambda + 4 = 3\mu + 2,$
$\lambda + 2 = 2\mu + 3$

$\Rightarrow \quad \lambda - 2\mu = 1$ and $5\lambda - 3\mu = -2$

By solving, we get $\lambda = -1, \mu = -1$

$\therefore \; P = (1, -1, 1)$

Now, $L_3 \Rightarrow \frac{x}{1} = \frac{y}{2} = \frac{z}{4} = k$

DR of $L_3$ is proportional to 1, 2, 4.

Let $Q$ be $(k, 2k, 4k)$ on line $L_3$.

$\therefore \; PQ \perp L_3$

$\Rightarrow (k-1)1 + (2k+1)(2) + (4k-1)(4) = 0$

$\Rightarrow \quad k = \frac{1}{7}$

$\therefore \; PQ = \sqrt{(k-1)^2 + (2k+1)^2 + (4k-1)^2}$

$= \sqrt{\frac{36}{49} + \frac{81}{49} + \frac{9}{49}} = \frac{\sqrt{126}}{7} = \frac{3}{7}\sqrt{14}$

**61.** *(c)* Given, equation of line

$\frac{x-1}{3} = \frac{y+1}{2} = \frac{z-2}{1} = \lambda$

Then, any point on given line $Q(3\lambda+1, 2\lambda-1, \lambda+2)$ and $P(3, 4, 9)$.

Direction ratios of $PQ$ are $(3\lambda-2, 2\lambda-5, \lambda-7)$

and given line is perpendicular to $PQ$.

So, $< 3\lambda-2, 2\lambda-5, \lambda-7 > \cdot < 3, 2, 1 > = 0$

$\Rightarrow \quad 9\lambda - 6 + 4\lambda - 10 + \lambda - 7 = 0$

$\Rightarrow \quad 14\lambda - 23 = 0 \Rightarrow \lambda = \frac{23}{14}$

So, $Q\left(\frac{83}{14}, \frac{32}{14}, \frac{51}{14}\right)$

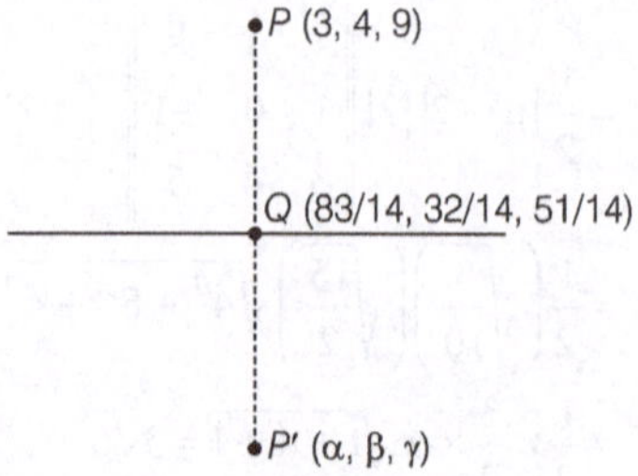


$Q$ is mid-point of $P$ and $P'$.

$\therefore \; \frac{\alpha+3}{2} = \frac{83}{14} \Rightarrow \alpha = \frac{62}{7}$

$\frac{\beta+4}{2} = \frac{32}{14} \Rightarrow \beta = \frac{4}{7}$

$\frac{\gamma+9}{2} = \frac{51}{14} \Rightarrow \gamma = -\frac{12}{7}$

Now, $14(\alpha+\beta+\gamma) = 14\left(\frac{62+4-12}{7}\right)$

$= 108$

**62.** *(d)*

$L_1 : \mathbf{r} = (-3\hat{i} + 2\hat{k}) + \lambda(2\hat{i} + 3\hat{j} + 5\hat{k})$

$L_2 : \mathbf{r} = (\hat{i} - 2\hat{j} + \hat{k}) + \mu(-\hat{i} + 3\hat{j} + 2\hat{k})$

Let $L$ be a line which passes through $(5, -4, 3)$ and perpendicular to $L_1, L_2$ both

$\mathbf{b} = \begin{vmatrix} \hat{i} & \hat{j} & \hat{k} \\ 2 & 3 & 5 \\ -1 & 3 & 2 \end{vmatrix} = \hat{i}(-9) - \hat{j}(9) + \hat{k}(9)$

$\mathbf{b} = -9\hat{i} - 9\hat{j} + 9\hat{k}$

$\therefore$ Equation of line $L$

$\mathbf{r} = 5\hat{i} - 4\hat{j} + 3\hat{k} + \lambda(-9\hat{i} - 9\hat{j} + 9\hat{k})$

$= 5\hat{i} - 4\hat{j} + 3\hat{k} - 9\lambda(\hat{i} + \hat{j} - \hat{k})$

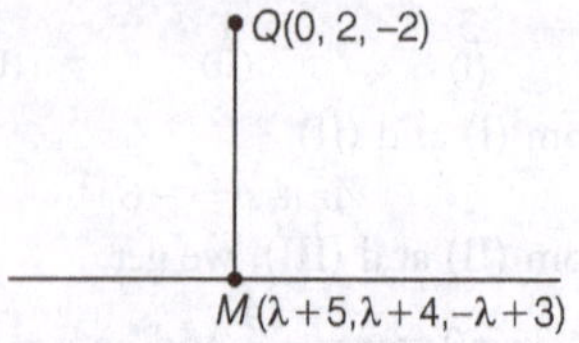


$\frac{x-5}{1} = \frac{y+4}{1} = \frac{z-3}{-1}$

$\mathbf{PM} = (\lambda+5)\hat{i} + (\lambda-6)\hat{j} + (-\lambda+5)\hat{k}$ (say)

$\mathbf{PM} \cdot (\hat{i} + \hat{j} - \hat{k}) = 0$

$\lambda + 5 + \lambda - 6 + \lambda - 5 = 0$

$\Rightarrow \quad 3\lambda = 6 \Rightarrow \lambda = 2$

$\mathbf{PM} = 7\hat{i} - 4\hat{j} + 3\hat{k}$

$\therefore$ Required distance $= |\mathbf{PM}|$

$= \sqrt{49+16+9} = \sqrt{74}$

**63.** *(d)*

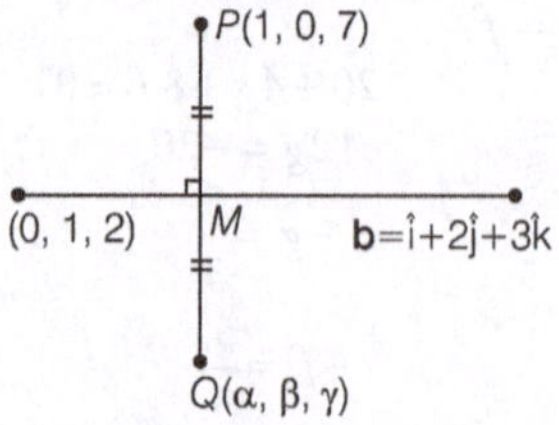


Let $L_1 : \frac{x-0}{1} = \frac{y-1}{2} = \frac{z-2}{3} = k$

$\therefore$ Coordinates of $M$ can be taken as

$x = k, \; y = 2k+1, \; z = 3k+2$

i.e. $\quad M \equiv (k, 2k+1, 3k+2)$

Now,

$\mathbf{PM} = (k-1)\hat{i} + (2k+1)\hat{j} + (3k-5)\hat{k}$

$\because \quad \mathbf{PM} \perp L_1$

$\therefore \quad \mathbf{PM} \cdot \mathbf{b} = 0$

$\Rightarrow \; k - 1 + 4k + 2 + 9k - 15 = 0$

(as $\mathbf{b} = \hat{i} + 2\hat{j} + 3\hat{k}$) $\Rightarrow k = 1;$

$\therefore \quad M \equiv (1, 3, 5)$

Now, here $M$ is mid-point of P and Q.

$\therefore \quad \mathbf{Q} = 2\mathbf{M} - \mathbf{P}$

$\Rightarrow \quad \mathbf{Q} = (2\hat{i} + 6\hat{j} + 10\hat{k}) - (\hat{i} + 7\hat{k})$

$\Rightarrow \quad \mathbf{Q} = \hat{i} + 6\hat{j} + 3\hat{k}$

$\therefore \quad (\alpha, \beta, \gamma) = (1, 6, 3)$

Let required line having direction cosines $< l, m, n >$.

Given, $m = -\frac{1}{2}$ and $n = -\frac{1}{\sqrt{2}}$

According to the question,

$\because l^2 + m^2 + n^2 = 1 \Rightarrow l^2 = \frac{1}{4} \Rightarrow l = \frac{1}{2}$

[$\because$ line makes an acute angle with $X$-axis.]

Therefore, equation of line passes through (1, 6, 3) will be

$\mathbf{r} = (\hat{i} + 6\hat{j} + 3\hat{k}) + \mu\left(\frac{1}{2}\hat{i} - \frac{1}{2}\hat{j} - \frac{1}{\sqrt{2}}\hat{k}\right)$

Option (d) satisfying for $\mu = 4$.

**64.** *(a)* We have, $\frac{x}{1} = \frac{6-y}{2} = \frac{z+8}{5}$

intersect the line

$\frac{x-5}{4} = \frac{y-7}{3} = \frac{z+2}{1}$

and $\quad \dfrac{x+3}{6}=\dfrac{3-y}{3}=\dfrac{z-6}{1}$

Now, $\dfrac{x}{1}=\dfrac{6-y}{2}=\dfrac{z+8}{5}=\lambda \quad \text{...(i)}$

$\Rightarrow x=\lambda,\ y=6-2\lambda,\ z=5\lambda-8$

Also, $\dfrac{x-5}{4}=\dfrac{y-7}{3}=\dfrac{z+2}{1}=k \quad \text{...(ii)}$

$\Rightarrow \quad x=4k+5,\ y=3k+7,\ z=k-2$

$\dfrac{x+3}{6}=\dfrac{3-y}{3}=\dfrac{z-6}{1}=\mu \quad \text{...(iii)}$

$\Rightarrow x=6\mu-3,\ y=3-3\mu,\ z=\mu+6$

On solving Eqs. (i) and (ii), we get

$\lambda=1,\ k=-1$

$\therefore$Co-ordinate of $A$ is $(1, 4, -3)$

On solving Eqs. (i) and (ii), we get

$\lambda=3,\ k=1$

$\therefore$Co-ordinate of $\beta$ is $(3, 0, 7)$

Co-ordinate of mid-point of $AB$ is

$$\left(\frac{1+3}{2}, \frac{4+0}{2}, \frac{-3+7}{2}\right)=(2, 2, 2)$$

Perpendicular-distance of mid-point of $AB$ form the plane $2x-2y+z=14$ is

$$\frac{|2(2)-2(2)+2-14|}{\sqrt{2^2+(-2)^2+1^2}}=4$$

**65.** *(c)* Given lines are

$$\frac{x+2}{1}=\frac{y}{-2}=\frac{z-5}{2} \quad \text{...(i)}$$

and $\quad \dfrac{x-4}{1}=\dfrac{y-1}{2}=\dfrac{z+3}{0} \quad \text{...(ii)}$

$\because$ Shortest distance between two skew-line, $SD=\left|\dfrac{(\mathbf{a}_2-\mathbf{a}_1)\cdot(\mathbf{b}_2\times\mathbf{b}_2)}{|\mathbf{b}_1\times\mathbf{b}_2|}\right|$

Vector form of Eqs. (i) and (ii), we get

$\mathbf{n}_1=(-2\hat{i}+5\hat{k})+\lambda(\hat{i}-2\hat{j}+2\hat{k})$

and $\mathbf{n}_2=(4\hat{i}+\hat{j}-3\hat{k})+\mu(\hat{i}+2\hat{j})$,

where $\mathbf{a}_1=-2\hat{i}+5\hat{k}$

$\mathbf{b}_1=\hat{i}-2\hat{j}+2\hat{k}$

$\mathbf{a}_2=4\hat{i}+\hat{j}-3\hat{k},\ \mathbf{b}_2=\hat{i}+2\hat{j}$

Now,

$\mathbf{a}_2-\mathbf{a}_1=(4\hat{i}+\hat{j}-3\hat{k})-(-2\hat{i}+5\hat{k})$

$=(6\hat{i}+\hat{j}-8\hat{k})$

$$\mathbf{b}_1\times\mathbf{b}_2=\begin{vmatrix}\hat{i} & \hat{j} & \hat{k}\\ 1 & -2 & 2\\ 1 & 2 & 0\end{vmatrix}$$

$=\hat{i}(-4)+\hat{j}(2)+\hat{k}(4)=-4\hat{i}+2\hat{j}+4\hat{k}$

$|\mathbf{b}_1\times\mathbf{b}_2|=\sqrt{(-4)^2+(2)^2+(4)^2}$

$=\sqrt{16+4+16}=\sqrt{36}=6$

and $(\mathbf{a}_2-\mathbf{a}_1)\cdot(\mathbf{b}_1\times\mathbf{b}_2)=-24+2-32$

$=-54$

Now, required shortest distance,

$$SD=\left|\frac{-54}{6}\right|=|-9|=9$$

**Alternate Solution**

We know, formula for shortest distance,

$$SD=\left|\frac{\begin{vmatrix}x_2-x_1 & y_2-y_1 & z_2-z_1\\ a_1 & b_1 & c_1\\ a_2 & b_2 & c_2\end{vmatrix}}{\left|\begin{vmatrix}\hat{i} & \hat{j} & \hat{k}\\ a_1 & b_1 & c_1\\ a_2 & b_2 & c_2\end{vmatrix}\right|}\right|$$

$$=\left|\frac{\begin{vmatrix}6 & 1 & -8\\ 1 & -2 & 2\\ 1 & 2 & 0\end{vmatrix}}{\left|\begin{vmatrix}\hat{i} & \hat{j} & \hat{k}\\ 1 & -2 & 2\\ 1 & 2 & 0\end{vmatrix}\right|}\right|=\left|\frac{6(-4)+1(-2)+8(4)}{|-4\hat{i}+2\hat{j}+4\hat{k}|}\right|$$

$$=\left|\frac{-24+2-32}{\sqrt{36}}\right|=\left|\frac{-54}{6}\right|=|-9|=9$$

**66.** *(a)* Given lines

$$\frac{x-4}{4}=\frac{y+2}{5}=\frac{z+3}{3}=\lambda \quad \text{(let)}$$

$$\frac{x-1}{3}=\frac{y-3}{4}=\frac{z-4}{2}=\mu \quad \text{(let)}$$

Vector form of lines

$L_1: \mathbf{r}=(4\hat{i}-2\hat{j}-3\hat{k})+\lambda(4\hat{i}+5\hat{j}+3\hat{k})$

and $L_2=\mathbf{r}=(\hat{i}+3\hat{j}+4\hat{k})$

$+\mu(3\hat{i}+4\hat{j}+2\hat{k})$

Here, $\mathbf{a}_1=4\hat{i}-2\hat{j}-3\hat{k}$,

$\mathbf{b}_1=4\hat{i}+5\hat{j}+3\hat{k}$

$\mathbf{a}_2=\hat{i}+3\hat{j}+4\hat{k},\ \mathbf{b}_2=3\hat{i}+4\hat{j}+2\hat{k}$

Now, $\mathbf{a}_2-\mathbf{a}_1=-3\hat{i}+5\hat{j}+7\hat{k}$

$$(\mathbf{a}_2-\mathbf{a}_1)\cdot(\mathbf{b}_1\times\mathbf{b}_2)=\begin{vmatrix}-3 & 5 & 7\\ 4 & 5 & 3\\ 3 & 4 & 2\end{vmatrix}$$

$=18$

$$\mathbf{b}_1\times\mathbf{b}_2=\begin{vmatrix}\hat{i} & \hat{j} & \hat{k}\\ 4 & 5 & 3\\ 3 & 4 & 2\end{vmatrix}=-2\hat{i}+\hat{j}+\hat{k}$$

$|\mathbf{b}_1\times\mathbf{b}_2|=\sqrt{6}$

Shortest distance

$$=\left|\frac{(\mathbf{a}_2-\mathbf{a}_1)\cdot(\mathbf{b}_1\times\mathbf{b}_2)}{|\mathbf{b}_1\times\mathbf{b}_2|}\right|=\frac{18}{\sqrt{6}}=3\sqrt{6}$$

**67.** *(a)* Given equation of lines are

$$L_1: \frac{x-5}{1}=\frac{y-2}{2}=\frac{z-4}{-3}$$

$$L_2: \frac{x+3}{1}=\frac{y+5}{4}=\frac{z-1}{-5}$$

So, equation of lines in vector form are

$L_1: \mathbf{r}=5\hat{i}+2\hat{j}+4\hat{k}+\lambda(\hat{i}+2\hat{j}-3\hat{k})$

$L_2: \mathbf{r}=-3\hat{i}-5\hat{j}+\hat{k}+\mu(\hat{i}+4\hat{j}-5\hat{k})$

$\therefore\ \mathbf{a}_1=5\hat{i}+2\hat{j}+4\hat{k},\ \mathbf{b}_1=\hat{i}+2\hat{j}-3\hat{k}$

$\mathbf{a}_2=-3\hat{i}-5\hat{j}+\hat{k},\ \mathbf{b}_2=\hat{i}+4\hat{j}-5\hat{k}$

Now,

$$\mathbf{b}_1\times\mathbf{b}_2=\begin{vmatrix}\hat{i} & \hat{j} & \hat{k}\\ 1 & 2 & -3\\ 1 & 4 & -5\end{vmatrix}=2\hat{i}+2\hat{j}+2\hat{k}$$

$\mathbf{a}_2-\mathbf{a}_1=(-3\hat{i}-5\hat{j}+\hat{k})-(5\hat{i}+2\hat{j}+4\hat{k})$

$=-8\hat{i}-7\hat{j}-3\hat{k}$

Again,

$|\mathbf{b}_1\times\mathbf{b}_2|=\sqrt{(2)^2+(2)^2+(2)^2}=2\sqrt{3}$

and

$(\mathbf{a}_2-\mathbf{a}_1).(\mathbf{b}_1\times\mathbf{b}_2)=(-8\hat{i}-7\hat{j}-3\hat{k})$

$.(2\hat{i}+2\hat{j}+2\hat{k})$

$=-16-14-6=-36$

Now, required distance

$$=\left|\frac{(\mathbf{a}_2-\mathbf{a}_1).(\mathbf{b}_1\times\mathbf{b}_2)}{|\mathbf{b}_1\times\mathbf{b}_2|}\right|=\left|\frac{-36}{2\sqrt{3}}\right|=6\sqrt{3}$$

**68.** *(b)* Given equations of lines are

$$L: \frac{x-5}{-2}=\frac{y-\lambda}{0}=\frac{z+\lambda}{1}$$

and $L_1: \dfrac{x+1}{1}=\dfrac{y-1}{1}=\dfrac{4-z}{1}$

or $\quad \dfrac{x+1}{1}=\dfrac{y-1}{1}=\dfrac{z-4}{-1}$

We have, $\mathbf{a}_1=5\hat{i}+\lambda\hat{j}-\lambda\hat{k}$,

$\mathbf{b}_1=-2\hat{i}+0\hat{j}+\hat{k}$

$\mathbf{a}_2=-\hat{i}+\hat{j}+4\hat{k},\ \mathbf{b}_2=\hat{i}+\hat{j}-\hat{k}$

Now,

$\mathbf{a}_2-\mathbf{a}_1=-6\hat{i}+(1-\lambda)\hat{j}+(4+\lambda)\hat{k}$

$$\mathbf{b}_1\times\mathbf{b}_2=\begin{vmatrix}\hat{i} & \hat{j} & \hat{k}\\ -2 & 0 & 1\\ 1 & 1 & -1\end{vmatrix}=-\hat{i}-\hat{j}-2\hat{k}$$

$\therefore (\mathbf{a}_2-\mathbf{a}_1)\cdot(\mathbf{b}_1\times\mathbf{b}_2)$

$=[-6\hat{i}+(1-\lambda)\hat{j}+(4+\lambda)\hat{k}]$

$\cdot[-\hat{i}-\hat{j}-2\hat{k}]$

$=6-1+\lambda-8-2\lambda=-3-\lambda$

and $|\mathbf{b}_1\times\mathbf{b}_2|=\sqrt{(-1)^2+(-1)^2+(-2)^2}$

$=\sqrt{6}$

Again, according to the question,

$$\left|\frac{(\mathbf{a}_2-\mathbf{a}_1)\cdot(\mathbf{b}_1\times\mathbf{b}_2)}{|\mathbf{b}_1\times\mathbf{b}_2|}\right| = 2\sqrt{6}$$

$$\Rightarrow \left|\frac{-3-\lambda}{\sqrt{6}}\right| = 2\sqrt{6} \Rightarrow |-3-\lambda| = 12$$

$\Rightarrow -(3+\lambda) = \pm 12 \Rightarrow \lambda = 9, -15$

But, $\lambda \geq 0$; $\lambda = 9$

So, equation of line $L$ is

$$\frac{x-5}{-2} = \frac{y-9}{0} = \frac{z+9}{1}$$

Since, $(\alpha, \beta, \gamma)$ lies on $L$, so $(\alpha, \beta, \gamma) = (-2k+5, 9, k-9)$, where $k \in R$

$\therefore \quad \alpha = -2k+5$ and $\gamma = k-9$

Again, $\alpha + 2\gamma = -2k+5+2k-18$
$= -13$

So, $\alpha + 2\gamma = 24$ is not possible.

**69.** *(a)* Given, equation of line

$$\frac{x+1}{2} = \frac{y-1}{5} = \frac{z+1}{-1}$$

Foot of perpendicular from $P(2, 0, 5)$ to given lines is $(\alpha, \beta, \gamma)$.

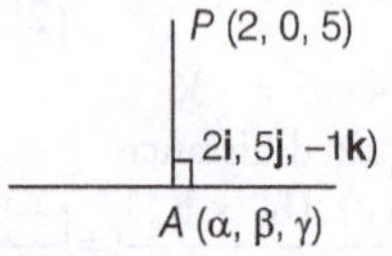


Direction ratio of given line is $(2, 5, -1)$

Direction vector $\mathbf{a} = 2\hat{i} + 5\hat{j} - \hat{k}$

and line passes through $= (-1, 1, -1)$

Now, let's suppose

$$\frac{x+1}{2} = \frac{y-1}{5} = \frac{z+1}{-1} = \lambda$$

or $x = 2\lambda - 1$, $y = 5\lambda + 1$, $z = -\lambda - 1$

Let the foot of perpendicular from point $P(2, 0, 5)$ on the line is $(\alpha, \beta, \gamma) \equiv A(2\lambda-1, 5\lambda+1, -\lambda-1)$

Now, $\mathbf{PA}$ = Position vector of $A$ − Position vector of $P$

$= [(2\lambda-1)\hat{i} + (5\lambda+1)\hat{j} + (-\lambda-1)\hat{k}] - (2\hat{i} + 5\hat{k})$

$\mathbf{PA} = (2\lambda-3)\hat{i} + (5\lambda+1)\hat{j} + (-\lambda-6)\hat{k}$

Now, $\mathbf{PA} \perp$ (direction of line)

Now, $\mathbf{PA} \perp \mathbf{a}$. So, $\mathbf{PA}\cdot\mathbf{a} = 0$

$[(2\lambda-3)\hat{i} + (5\lambda+1)\hat{j} + (-\lambda-6)\hat{k}] \cdot [2\hat{i} + 5\hat{j} - \hat{k}] = 0$

So, $2(2\lambda-3) + 5(5\lambda+1) + (\lambda+6) = 0$

$$30\lambda = -5 \Rightarrow \lambda = -\frac{1}{6}$$

So, $\quad A \equiv (2\lambda-1, 5\lambda+1, -\lambda-1)$

$$A \equiv \left(2\left(\frac{-1}{6}\right)-1, 5\left(-\frac{1}{6}\right)+1, -\left(\frac{-1}{6}\right)-1\right)$$

$$A \equiv \left(-\frac{4}{3}, \frac{1}{6}, -\frac{5}{6}\right) \equiv (\alpha, \beta, \gamma)$$

or $\alpha = -\frac{4}{3}, \beta = \frac{1}{6}, \gamma = \frac{-5}{6}$

$$\therefore \quad \frac{\beta}{\gamma} = \frac{\left(\frac{1}{6}\right)}{\left(-\frac{5}{6}\right)} = \frac{-1}{5}$$

**70.** *(c)* Given, line (i), $x + 1 = 2y = -12z$

$$\text{or} \quad \frac{x+1}{1} = \frac{y}{\left(\frac{1}{2}\right)} = \frac{z}{\left(-\frac{1}{12}\right)} \quad \text{...(i)}$$

Line (ii), $x = y + 2 = 6z - 6$

$$\text{or} \quad \frac{x}{1} = \frac{y+2}{1} = \frac{(z-1)}{(1/6)} \quad \text{...(ii)}$$

Shortest distance between line (i) and (ii)

Now, point lie on $L_1 = (-1, 0, 0)$

Position vector of point $= \mathbf{a}_1 = -\hat{i}$

$\Rightarrow$ Direction ratio on line

$$L_1 \equiv \left(1, \frac{1}{2}, -\frac{1}{12}\right)$$

or direction vector of

$$L_1 = \mathbf{b}_1 = \hat{i} + \frac{\hat{j}}{2} - \frac{\hat{k}}{12}$$

Now, point lie on $L_2 = (0, -2, 1)$

Position vector of the point

$$= \mathbf{a}_2 = -2\hat{j} + \hat{k}$$

$\Rightarrow$ Direction ratio of line

$$L_2 = \left(1, 1, \frac{1}{6}\right)$$

or direction vector of $L_2 = b_2 = \hat{i} + \hat{j} + \frac{\hat{k}}{6}$

Now, shortest distance between two skew lines

$$d = \left|\frac{(\mathbf{a}_1-\mathbf{a}_2)\cdot(\mathbf{b}_1\times\mathbf{b}_2)}{|\mathbf{b}_1\times\mathbf{b}_2|}\right|$$

$\mathbf{a}_1 - \mathbf{a}_2 = -\hat{i} + 2\hat{j} - \hat{k}$

$$\mathbf{b}_1 \times \mathbf{b}_2 = \begin{vmatrix} \hat{i} & \hat{j} & \hat{k} \\ 1 & \frac{1}{2} & -\frac{1}{12} \\ 1 & 1 & \frac{1}{6} \end{vmatrix}$$

$$= \hat{i}\left[\frac{1}{12} - \left(-\frac{1}{12}\right)\right] - \hat{j}\left[\frac{1}{6} - \left(-\frac{1}{12}\right)\right] + \hat{k}\left(1 - \frac{1}{2}\right)$$

$$\mathbf{b}_1 \times \mathbf{b}_2 = \frac{1}{6}\hat{i} - \frac{1}{4}\hat{j} + \frac{1}{2}\hat{k}$$

$$|\mathbf{b}_1 \times \mathbf{b}_2| = \sqrt{\left(\frac{1}{6}\right)^2 + \left(-\frac{1}{4}\right)^2 + \left(\frac{1}{2}\right)^2}$$

$$|\mathbf{b}_1 \times \mathbf{b}_2| = \sqrt{\frac{1}{36} + \frac{1}{16} + \frac{1}{4}}$$

$$|\mathbf{b}_1 \times \mathbf{b}_2| = \frac{7}{12}$$

Now, $(\mathbf{a}_1 - \mathbf{a}_2)\cdot(\mathbf{b}_1 \times \mathbf{b}_2)$

$$= (-\hat{i} + 2\hat{j} - \hat{k}) \cdot \left(\frac{\hat{i}}{6} - \frac{\hat{j}}{4} + \frac{\hat{k}}{2}\right)$$

$$= -\frac{1}{6} - \frac{1}{2} - \frac{1}{2} = -\frac{7}{6}$$

$$\text{So } d = \frac{\left|\left(-\frac{7}{6}\right)\right|}{\frac{7}{12}} = 2 \text{ units}$$

**71.** *(c)*

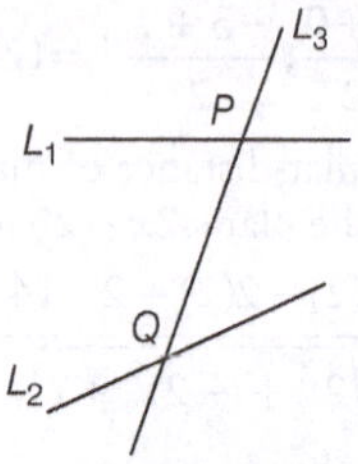


We have, $L_1 : \frac{x-1}{2} = \frac{y-3}{1} = \frac{z-2}{2} = \lambda$

$x = 2\lambda + 1$, $y = \lambda + 3$, $z = 2\lambda + 2$

Let $P(2\lambda+1, \lambda+3, 2\lambda+2)$

$L_2 : \frac{x-2}{1} = \frac{y-2}{2} = \frac{z-3}{3} = \mu$

$x = \mu + 2$, $y = 2\mu + 2$, $z = 3\mu + 3$

Let $Q(\mu+2, 2\mu+2, 3\mu+3)$

Direction ratio of line $PQ$

$= (2\lambda + 1 - \mu - 2, \lambda + 3 - 2\mu - 2, 2\lambda + 2 - 3\mu - 3)$

$= (2\lambda - \mu - 1, \lambda - 2\mu + 1, 2\lambda - 3\mu - 1)$

Given, direction ratio of $L_3 : 1, -1, -2$

and $L_3$ and $PQ$ are same line

$$\text{So,} \frac{2\lambda-\mu-1}{1} = \frac{\lambda-2\mu+1}{-1} = \frac{2\lambda-3\mu-1}{-2}$$

$$\text{On taking,} \frac{2\lambda-\mu-1}{1} = \frac{\lambda-2\mu+1}{-1}$$

$2\lambda - \mu - 1 = -\lambda + 2\mu - 1$

$3\lambda = 3\mu \Rightarrow \lambda = \mu \quad$ ...(i)

$$\text{On taking,} \frac{\lambda-2\mu+1}{-1} = \frac{2\lambda-3\mu-1}{-2}$$

$2\lambda - 4\mu + 2 = 2\lambda - 3\mu - 1$

$\mu = 3$

From Eq. (i), $\lambda = \mu = 3$

Now, $P(7,6,8)$ and $Q(5,8,12)$

Length of segment

$$PQ=\sqrt{(7-5)^2+(6-8)^2+(8-12)^2}$$
$$=\sqrt{4+4+16}=\sqrt{24}=2\sqrt{6}$$

**72.** *(a)* Given that first line

$$L_1:\frac{x+7}{-6}=\frac{y-6}{7}=\frac{z-0}{1}$$

Any point on the line $\mathbf{a}_1\equiv(-7,6,0)$ and $L_1$ parallel to $\mathbf{b}_1\equiv(-6,7,1)$ and second line,

$$L_2:\frac{x-7}{-2}=\frac{y-2}{1}=\frac{z-6}{1}$$

Then, any point on it $\mathbf{a}_2$ is $(7,2,6)$ and the line $L_2$ is parallel to $\mathbf{b}_2$ is $(-2,1,1)$.

Thus, shortest distance between line $L_1$ and $L_2$ is

$$\left|\frac{(\mathbf{a}_2-\mathbf{a}_1)\cdot(\mathbf{b}_1\times\mathbf{b}_2)}{|\mathbf{b}_1\times\mathbf{b}_2|}\right|$$

$a_2-a_1=14,-4,6$

and $b_1\times b_2=\begin{vmatrix}\hat{\mathbf{i}}&\hat{\mathbf{j}}&\hat{\mathbf{k}}\\-6&7&1\\-2&1&1\end{vmatrix}$

$$=6\hat{\mathbf{i}}+4\hat{\mathbf{j}}+8\hat{\mathbf{k}}=6,4,8$$

$$=\left|\frac{(14,-4,6)\cdot(6,4,8)}{\sqrt{6^2+4^2+8^2}}\right|$$

$$=\frac{116}{\sqrt{116}}=2\sqrt{29}$$

**73.** *(a)* $L_1\Rightarrow 3(x-1)=6(y-2)=2(z-1)$`

$L_2\Rightarrow 4(x-2)=2(y-\lambda)=(z-3)$

$$L_1\Rightarrow\frac{x-1}{2}=\frac{y-2}{1}=\frac{z-1}{3}$$

$L_1\Rightarrow\mathbf{r}=(\hat{\mathbf{i}}+2\hat{\mathbf{j}}+\hat{\mathbf{k}})+p(2\hat{\mathbf{i}}+\hat{\mathbf{j}}+3\hat{\mathbf{k}})$

$\Rightarrow\mathbf{r}=\mathbf{a}_1+p\mathbf{b}_1$

Similarly, $L_2\Rightarrow\dfrac{x-2}{1}=\dfrac{y-\lambda}{2}=\dfrac{z-3}{4}$

$L_2\Rightarrow\mathbf{r}=(2\hat{\mathbf{i}}+\lambda\hat{\mathbf{j}}+3\hat{\mathbf{k}})+q(\hat{\mathbf{i}}+2\hat{\mathbf{j}}+4\hat{\mathbf{k}})$

$\Rightarrow\quad\mathbf{r}=\mathbf{a}_2+q\mathbf{b}_2$

Shortest distance will be the projection of $(\mathbf{a}_1-\mathbf{a}_2)$ along $(\mathbf{b}_1\times\mathbf{b}_2)$.

or $\quad(\mathbf{a}_1-\mathbf{a}_2).\dfrac{(\mathbf{b}_1\times\mathbf{b}_2)}{|\mathbf{b}_1\times\mathbf{b}_2|}$

$$\mathbf{b}_1\times\mathbf{b}_2=\begin{vmatrix}\hat{\mathbf{i}}&\hat{\mathbf{j}}&\hat{\mathbf{k}}\\2&1&3\\1&2&4\end{vmatrix}$$

$$=-2\hat{\mathbf{i}}-5\hat{\mathbf{j}}+3\hat{\mathbf{k}}$$

$\therefore\quad|\mathbf{b}_1\times\mathbf{b}_2|=|-2\hat{\mathbf{i}}-5\hat{\mathbf{j}}+3\hat{\mathbf{k}}|$

$$=\sqrt{4+25+9}=\sqrt{38}$$

Now,

$(\mathbf{a}_1-\mathbf{a}_2)=(\hat{\mathbf{i}}+2\hat{\mathbf{j}}+\hat{\mathbf{k}})-(2\hat{\mathbf{i}}+\lambda\hat{\mathbf{j}}+3\hat{\mathbf{k}})$

$$=-\hat{\mathbf{i}}+(2-\lambda)\hat{\mathbf{j}}-2\hat{\mathbf{k}}$$

$\therefore\ (\mathbf{a}_1-\mathbf{a}_2).\dfrac{(\mathbf{b}_1\times\mathbf{b}_2)}{|\mathbf{b}_1\times\mathbf{b}_2|}$

$$=\frac{(-\hat{\mathbf{i}}+(2-\lambda)\hat{\mathbf{j}}-2\hat{\mathbf{k}})\cdot(-2\hat{\mathbf{i}}-5\hat{\mathbf{j}}+3\hat{\mathbf{k}})}{\sqrt{38}}$$

$$=\frac{2+5\lambda-10-6}{\sqrt{38}}=\frac{5\lambda-14}{\sqrt{38}}$$

$$\Rightarrow\frac{5\lambda-14}{\sqrt{38}}=\frac{1}{\sqrt{38}}\Rightarrow\lambda=3$$

**74.** *(a)* Equation of given line is

$$\frac{x+3}{10}=\frac{y-2}{-7}=\frac{z}{1}=r\text{ (let)}\quad\ldots\text{(i)}$$

Co-ordinates of a point on line (i) is

$$A(10r-3,-7r+2,r)$$

Now, let the line joining the points $P(2,-1,4)$ and $A(10r-3,-7r+2,r)$ is perpendicular to line (i).

Then, $\mathbf{PA}\cdot(10\hat{\mathbf{i}}-7\hat{\mathbf{j}}+\hat{\mathbf{k}})=0$

[$\because$ Vector along line (i) is $(10\hat{\mathbf{i}}-7\hat{\mathbf{j}}+\hat{\mathbf{k}})$]

$\Rightarrow$

$[(10r-5)\hat{\mathbf{i}}+(-7r+3)\hat{\mathbf{j}}+(r-4)\hat{\mathbf{k}}]\cdot[10\hat{\mathbf{i}}-7\hat{\mathbf{j}}+\hat{\mathbf{k}}]=0$

$\Rightarrow 10(10r-5)-7(3-7r)+(r-4)=0$

$\Rightarrow\quad 100r-50-21+49r+r-4=0$

$\Rightarrow\quad 150r=75\Rightarrow r=1/2$

So, the foot of perpendicular is

$$A\left(2,-\frac{3}{2},\frac{1}{2}\right)$$

[Put $r=\dfrac{1}{2}$ in the coordinates of point $A$]

Now, perpendicular distance of point $P(2,-1,4)$ from the line (i) is

$$PA=\sqrt{(2-2)^2+\left(-\frac{3}{2}+1\right)^2+\left(\frac{1}{2}-4\right)^2}$$

$$=\sqrt{\frac{1}{4}+\frac{49}{4}}=\sqrt{\frac{50}{4}}=\frac{5}{\sqrt{2}}$$

which lies in (3, 4).

**75.** *(56)* Given lines are

Equation of $BC$

$$k_1=\frac{x+2}{1}=\frac{y-1}{1}=\frac{z}{1}\quad\ldots\text{(i)}$$

Equation of $CA$

$$k_2=\frac{x-3}{5}=\frac{y}{-1}=\frac{z-1}{1}\quad\ldots\text{(ii)}$$

Equation of $AB$

$$k_3=\frac{x}{-3}=\frac{y-3}{3}=\frac{z-2}{1}\quad\ldots\text{(iii)}$$

Solving Eqs. (i) and (ii), we get point of interection

$$k_1-2=5k_2+3$$
$$k_1+1=-k_2$$
$$-3=6k_2+3\Rightarrow k_2=-1$$
$$k_1=0\Rightarrow C=(-2,1,0)$$

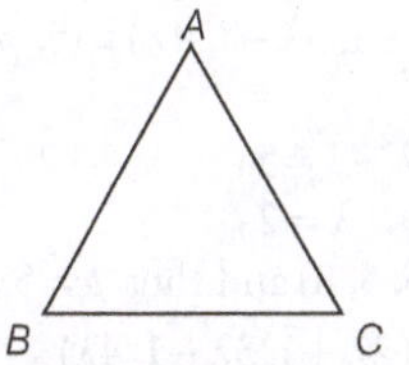


Solving Eqs. (ii) and (iii), we get point of interection $A$

$$5k_2+3=-3k_3$$
$$-k_2=3k_3+3$$
$$4k_2+3=3\Rightarrow k_2=0$$
$$A=(3,0,1)$$

Solving Eq. (i) and (iii), we get point of interection $B$

$$k_1-2=-3k_3$$
$$k_1+1=3k_3+3\Rightarrow k_3=0'$$
$$B\equiv(0,3,2)$$
$$A=(3,0,1)$$
$$B=(0,3,2)$$
$$C=(-2,1,0)$$

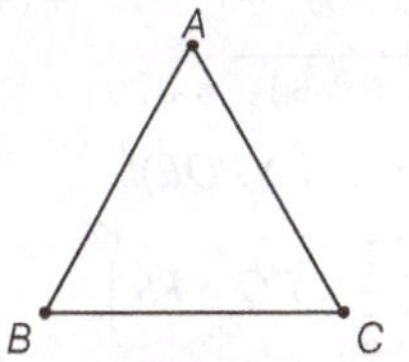


$$\mathbf{BC}=-2\hat{\mathbf{i}}-2\hat{\mathbf{j}}-2\hat{\mathbf{k}}$$
$$\mathbf{CA}=5\hat{\mathbf{i}}-\hat{\mathbf{j}}+\hat{\mathbf{k}}$$

Area of $\Delta ABC=\dfrac{1}{2}|\mathbf{BC}\times\mathbf{CA}|=A$

$$\mathbf{BC}\times\mathbf{CA}=\begin{vmatrix}\hat{\mathbf{i}}&\hat{\mathbf{j}}&\hat{\mathbf{k}}\\-2&-2&-2\\5&-1&1\end{vmatrix}$$

$$=\hat{\mathbf{i}}(-2-2)-\hat{\mathbf{j}}(-2+10)+\hat{\mathbf{k}}(2+10)$$

$$=-4\hat{\mathbf{i}}-8\hat{\mathbf{j}}+12\hat{\mathbf{k}}$$

$$A=|-2\hat{\mathbf{i}}-4\hat{\mathbf{j}}+6\hat{\mathbf{k}}|$$

$$A^2=2^2+4^2+6^2=56$$

**76.** *(957)* Given equation of line,

$$\frac{x-1}{2}=\frac{y+1}{3}=\frac{z}{4}=\lambda\text{ (say)}\quad\ldots\text{(i)}$$

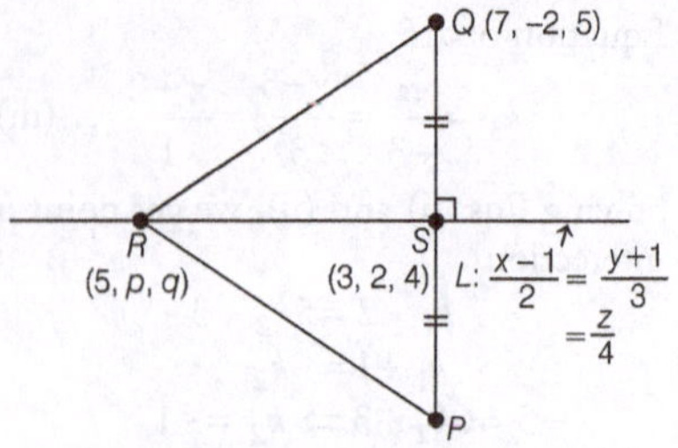


Any point taken on Eq. (i) is

$R \equiv (2\lambda+1, 3\lambda-1, 4\lambda) \equiv (5, p, q)$ [given]

$\therefore \quad 2\lambda+1=5$

$\Rightarrow \quad \lambda=2$

$\therefore R \equiv (5, 5, 8)$ and thus $p=5$ and $q=8$

Let $S \equiv (2\lambda+1, 3\lambda-1, 4\lambda)$

$\therefore$ **QS** $=(2\lambda-6)\hat{i}+(3\lambda+1)\hat{j}+(4\lambda-5)\hat{k}$

and let $\mathbf{b}=2\hat{i}+3\hat{j}+4\hat{k}$

$\therefore \quad \mathbf{QS}\cdot\mathbf{b}=0 \quad [\because \mathbf{QS} \perp \mathbf{RS}]$

$\Rightarrow 4\lambda-12+9\lambda+3+16\lambda-20=0$

$\Rightarrow \quad 29\lambda-29=0 \Rightarrow \lambda=1$

$\therefore \quad S \equiv (3, 2, 4)$

Hence,

$QS=\sqrt{(7-3)^2+(-2-2)^2+(5-4)^2}$

$=\sqrt{16+16+1}=\sqrt{33}$

and

$RS=\sqrt{(5-3)^2+(5-2)^2+(8-4)^2}$

$=\sqrt{4+9+16}=\sqrt{29}$

Now, (area of $\Delta PQR)^2$

$=\left(\frac{1}{2}\times PQ \times RS\right)^2$

$=\left(\frac{1}{2}\times\sqrt{29}\times 2\sqrt{33}\right)^2=957$

**77.** *(216)* $L_1:\frac{x-1}{3}=\frac{y-1}{-1}=\frac{z+1}{0}=\lambda$ (say)

$\Rightarrow P(3\lambda+1, -\lambda+1, -1)$

$L_2:\frac{x-2}{2}=\frac{y}{0}=\frac{z+4}{\alpha}=\mu$ (say)

$\Rightarrow Q(2\mu+2, 0, \alpha\mu-4)$

$\because$ $L_1$ and $L_2$ intersect.

$\therefore \quad 3\lambda+1=2\mu+2$

and $\quad -\lambda+1=0$

$\lambda=1$

and $\quad -1=\alpha\mu-4$

$\Rightarrow \quad -1=\alpha-4 \Rightarrow \alpha=3$

$4=2\mu+2$

$\Rightarrow \quad \mu=1 \Rightarrow B=(4, 0, -1)$

$\therefore \ L_2:\frac{x-2}{2}=\frac{y}{0}=\frac{z+4}{3}$

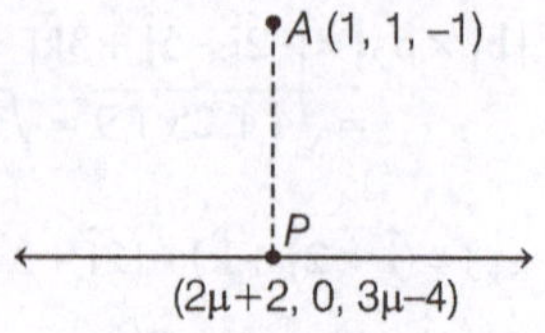


**AM** $\perp (2\hat{i}+0\hat{j}+3\hat{k})$

$\Rightarrow \quad \mathbf{AM}\cdot(2\hat{i}+3\hat{k})=0$

$\Rightarrow \quad 2(2\mu+1)+3(3\mu-3)=0$

$\Rightarrow \quad 4\mu+2+9\mu-9=0$

$\Rightarrow \quad 13\mu=7$

$\Rightarrow \quad \mu=\frac{7}{13}$

$PB=\sqrt{(2\mu-2)^2+0^2+(3\mu-3)^2}$

$=\sqrt{(\mu-1)^2(4+9)}=\sqrt{\left(\frac{6}{13}\right)^2(13)}$

$=\frac{6}{\sqrt{13}}$

$\therefore \quad 26\alpha(PB)^2=26(3)\frac{36}{13}$

$6\times 36=216$

**78.** *(62)* Let $M(3\lambda+1, 2\lambda, 4\lambda+2)$

$\mathbf{AM}=(3\lambda-5)\hat{i}+(2\lambda-1)\hat{j}+(4\lambda-3)\hat{k}$

$\mathbf{AM}\cdot\mathbf{b}=0$

$(3\lambda-5)3+(2\lambda-1)2+(4\lambda-3)4=0$

$29\lambda=29$

$\lambda=1$

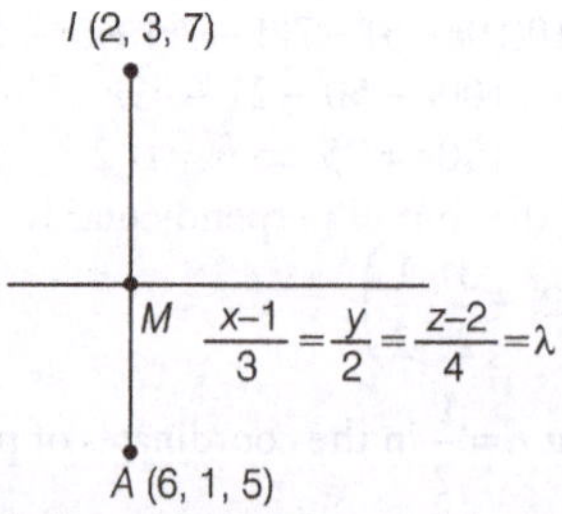


$M(4, 2, 6)$, Image of $M$ is $I=(2, 3, 7)$

Required distance $=\sqrt{4+9+49}=\sqrt{62}$

$\therefore$Square $=62$

**79.** *(11)*

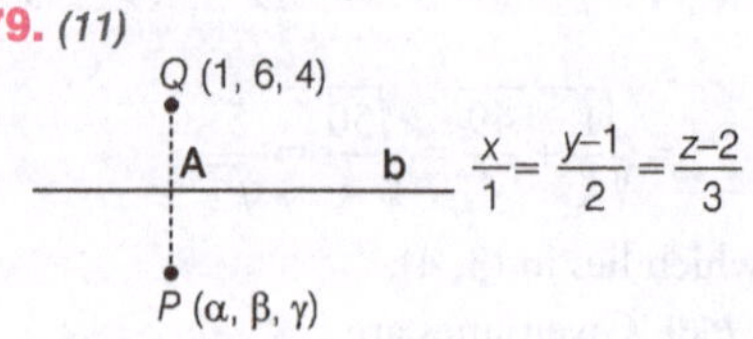


$A(t, 2t+1, 3t+2)$

**QA** $=(t-1)\hat{i}+(2t-5)\hat{j}+(3t-2)\hat{k}$

**QA** $\cdot$ **b** $=0$

$\Rightarrow \quad (t-1)\times 1+(2t-5)\times 2+(3t-2)\times 3=0$

$\Rightarrow \quad t-1+4t-10+9t-6=0$

$\Rightarrow \quad 14t-17=0 \Rightarrow t=17/14$

So, $A \equiv \left(\frac{17}{14}, \frac{48}{14}, \frac{79}{14}\right)$

$\because A$ is mid-point of $Q$ and $P$

$\frac{\alpha+1}{2}=\frac{17}{14}\Rightarrow \alpha=\frac{20}{14}$

Similarly, $\beta=\frac{12}{14}$ and $\gamma=\frac{102}{14}$

Now, $2\alpha+\beta+\gamma=\frac{20}{7}+\frac{12}{14}+\frac{102}{14}$

$=\frac{40+12+102}{14}=\frac{154}{14}=11$

**80.** *(43)*

> **key idea** Firstly convert the given equation of lines in the form of $\mathbf{r}_1=\mathbf{a}_1+\lambda\mathbf{b}_1$ and $\mathbf{r}_2=\mathbf{a}_2+\mu\mathbf{b}_2$, then use formula for shortest distance
> $d=\left|\frac{(\mathbf{a}_2-\mathbf{a}_1)\cdot(\mathbf{b}_1\times\mathbf{b}_2)}{|\mathbf{b}_1\times\mathbf{b}_2|}\right|$

$\mathbf{a}_1=\lambda\hat{i}+2\hat{j}+\hat{k}$

$\mathbf{a}_2=-2\hat{i}-5\hat{j}+4\hat{k}$

$\mathbf{b}_1=3\hat{i}-\hat{j}+\hat{k}$

$\mathbf{b}_2=-3\hat{i}+2\hat{j}+4\hat{k}$

$\Rightarrow \quad \mathbf{a}_1-\mathbf{a}_2=(\lambda+2)\hat{i}+7\hat{j}-3\hat{k}$

$\Rightarrow \quad \mathbf{b}_1\times\mathbf{b}_2=\begin{vmatrix}\hat{i} & \hat{j} & \hat{k}\\ 3 & -1 & 1\\ -3 & 2 & 4\end{vmatrix}$

$=-6\hat{i}-15\hat{j}+3\hat{k}$

Now, $\quad d=\left|\frac{(a_2-a_1)\cdot(b_1\times b_2)}{|b_1\times b_2|}\right|$

$\frac{44}{\sqrt{30}}=\frac{|6\lambda+126|}{3\sqrt{30}}$

$\Rightarrow \quad 132=|6\lambda+126|$

$\Rightarrow \quad \lambda=1, \lambda=-43$

$\Rightarrow \quad |\lambda|=43$

**81.** *(13)* According to the question,

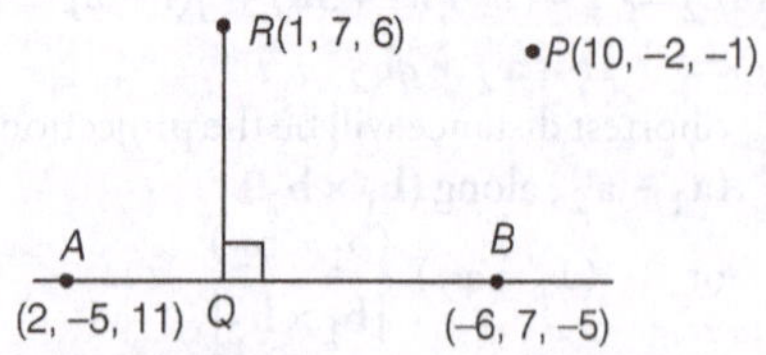


Equation of **AB**

$\Rightarrow \quad \frac{x-2}{-8}=\frac{y+5}{12}=\frac{z-11}{-16}$

Now, coordinates of

$Q \equiv (-8\lambda+2, 12\lambda-5, -16\lambda+11)$

According to the question,

$$RQ \cdot AB = 0$$

$$\Rightarrow [(-8\lambda + 2 - 1)\hat{i} + (12\lambda - 5 - 7)\hat{j} + (-16\lambda + 11 - 6)\hat{k}] \cdot [-8\hat{i} + 12\hat{j} - 16\hat{k}] = 0$$

$$\Rightarrow [(-8\lambda + 1)\hat{i} + (12\lambda - 12)\hat{j} + (-16\lambda + 5)\hat{k}] \cdot [-2\hat{i} + 3\hat{j} - 4\hat{k}] = 0$$

$$\Rightarrow 16\lambda - 2 + 36\lambda - 36 + 64\lambda - 20 = 0$$

$$116\lambda = 58$$

$$\Rightarrow \quad \lambda = \frac{1}{2}$$

$$Q \equiv \left(-\frac{8}{2} + 2, 12 \times \frac{1}{2} - 5, -16 \times \frac{1}{2} + 11\right)$$

$$\therefore Q \equiv (-2, 1, 3)$$

Now, $PQ = \sqrt{(10+2)^2 + (-2-1)^2 + (-1-3)^2}$

$$= \sqrt{144 + 9 + 16} = \sqrt{169} = 13$$

**82.** *(25)* $L_1 : \dfrac{x+2}{-3} = \dfrac{y-2}{4} = \dfrac{z-5}{2} = \lambda$ (say)

$L_2 : \dfrac{x+2}{-1} = \dfrac{y+6}{2} = \dfrac{z-1}{0} = \mu$ (say)

Let $P = -3\lambda - 2, 4\lambda + 2, 2\lambda + 5$

and $Q = -\mu - 2, 2\mu - 6, 1$

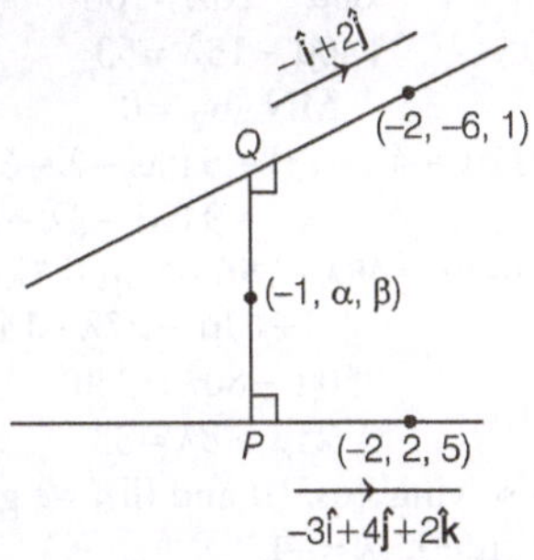


$\because$ **PQ** $\perp$ $\mathbf{b}_1$ and $\mathbf{b}_2$

$\therefore$ **PQ** perpendicular $\mathbf{b}_1 \times \mathbf{b}_2$

DR's of $PQ \propto$ DR's of $\mathbf{b}_1 \times \mathbf{b}_2$

$$\mathbf{b}_1 \times \mathbf{b}_2 = \begin{vmatrix} \hat{i} & \hat{j} & \hat{k} \\ -1 & 2 & 0 \\ -3 & 4 & 2 \end{vmatrix}$$

$$\Rightarrow 4\hat{i} + 2\hat{j} + 2\hat{k}$$

$\therefore$ DR's of $PQ = 2, 1, 1$

$$\therefore \quad \frac{3\lambda - \mu}{2} = \frac{2\mu - 4\lambda - 8}{1} = \frac{-2\lambda - 4}{1}$$

$$\Rightarrow \quad \lambda = -1, \mu = 1$$

Line $PQ : \dfrac{x+3}{2} = \dfrac{y+4}{1} = \dfrac{z-1}{1}$

$\because (-1, \alpha, \beta)$ satisfying line $PQ$

$$1 = \frac{\alpha + 4}{1} = \frac{\beta - 1}{1} \Rightarrow \alpha = -3, \beta = 2$$

$$\therefore (\alpha - \beta)^2 = 25$$

**83.** *(6)* Equation of $PQ$

$$\frac{x-1}{1} = \frac{y-2}{-1} = \frac{z-1}{-2} = \lambda$$

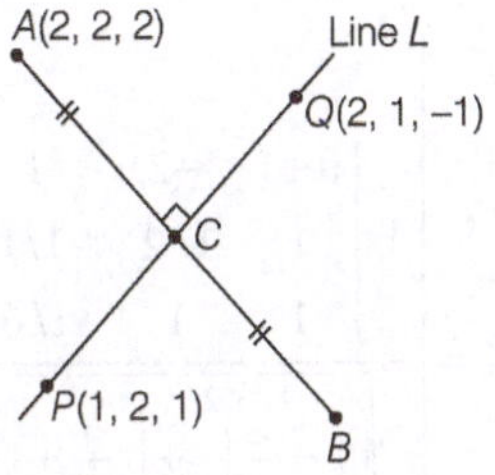


Let $C = (\lambda + 1, -\lambda + 2, -2\lambda + 1)$

$AC \perp$ line $PQ$

$$\therefore (\lambda + 1 - 2) \cdot 1 + (-\lambda + 2 - 2)(-1) + (-2\lambda + 1 - 2)(-2) = 0$$

$$\Rightarrow \quad \lambda - 1 + \lambda + 4\lambda + 2 = 0$$

$$\Rightarrow \quad 6\lambda = -1$$

$$\because \quad \lambda = -\frac{1}{6}$$

$\beta$ = image of $A = (\alpha, \beta, \gamma)$

$$\frac{\alpha + 2}{2} = \lambda + 1,$$

$$\frac{\beta + 2}{2} = -\lambda + 2,$$

$$\frac{\gamma + 2}{2} = -2\lambda + 1$$

$$\alpha = 2\lambda + 2 - 2,$$

$$\beta = -2\lambda + 4 - 2$$

and $\quad \gamma = -4\lambda + 2 - 2$

$$\alpha = 2\lambda, \beta = -2\lambda + 2, \gamma = -4\lambda$$

$$\therefore \quad \alpha + \beta + 6\gamma = 2\lambda - 2\lambda + 2 - 24\lambda$$

$$= -24\lambda + 2$$

$$= -24 \times \frac{-1}{6} + 2$$

$$= 4 + 2 = 6$$

**84.** *(48)*

> **key idea** Use, Shortest distance $= \dfrac{(\mathbf{a}_2 - \mathbf{a}_1) \cdot (\mathbf{b}_1 \times \mathbf{b}_2)}{|\mathbf{b}_1 \times \mathbf{b}_2|}$
>
> between lines $\mathbf{r} = \mathbf{a}_1 + \lambda \mathbf{b}_1$
>
> and $\mathbf{r} = \mathbf{a}_2 + \mu \mathbf{b}_2$

According to the question,

$$\frac{38}{3\sqrt{5}} k = \frac{5\hat{i} + 5\hat{j} - 9\hat{k}}{\sqrt{5}} \begin{vmatrix} \hat{i} & \hat{j} & \hat{k} \\ 2 & 3 & 4 \\ 1 & -3 & 2 \end{vmatrix}$$

$$= \frac{19}{\sqrt{5}}$$

$$\Rightarrow \quad k = \frac{19}{\sqrt{5}} \times \frac{3\sqrt{5}}{38} = \frac{3}{2}$$

So, $\displaystyle\int_0^{\frac{3}{2}} [x^2] dx = \int_0^1 0\, dx + \int_1^{\sqrt{2}} 1 \cdot dx + \int_{\sqrt{2}}^{\frac{3}{2}} 2 \cdot dx$

$$= \sqrt{2} - 1 + 2\left(\frac{3}{2} - \sqrt{2}\right)$$

$$= 2 - \sqrt{2}$$

Hence, $\alpha = 2$ and thus $6\alpha^3 = 48$

**85.** *(21)*

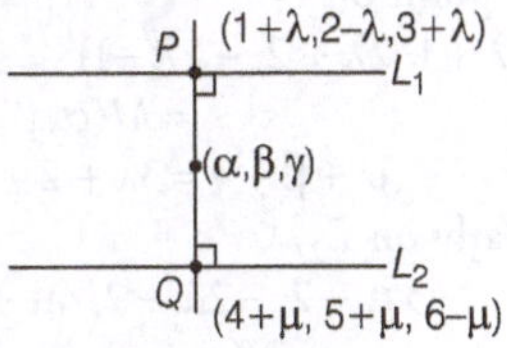


$PQ = 3 + \mu - \lambda, 3 + \mu + \lambda, 3 - \mu - \lambda$

**PQ** is perpendicular to $L_1$ and $L_2$

$$(3 + \mu - \lambda)(1) + (3 + \mu + \lambda)(-1) + (3 - \mu - \lambda)(1) = 0$$

$$\Rightarrow \quad -3\lambda - \mu + 3 = 0 \quad \text{...(i)}$$

$$(3 + \mu - \lambda)(1) + (3 + \mu + \lambda)(1) + (3 - \mu - \lambda)(-1) = 0$$

$$\Rightarrow \quad \lambda + 3\mu + 3 = 0 \quad \text{...(ii)}$$

On solving Eqs. (i) and (ii), we get

$$8\mu = -12, \; \mu = \frac{-3}{2}, \lambda = \frac{3}{2}$$

$$P\left(\frac{5}{2}, \frac{1}{2}, \frac{9}{2}\right) \text{ and } Q\left(\frac{5}{2}, \frac{7}{2}, \frac{15}{2}\right)$$

$$\alpha = \frac{5}{2}, \beta = 2, \gamma = 6$$

Hence, $2(\alpha + \beta + \gamma) = 21$

**86.** *(12)* $x = y, z = 1 \rightarrow$ Line $[L_1]$

$$\frac{x}{1} = \frac{y}{1} = \frac{z-1}{0}$$

Let point $Q$ on line $(\lambda, \lambda, 1)$

$x = -y, z = -1 \rightarrow$ Line $[L_2]$

$$\frac{x}{1} = \frac{y}{-1} = \frac{z+1}{0}$$

Let point $R$ on line $L_2$ $(\mu, -\mu, -1)$.

Given, point $P(a, a, a)$

$$\mathbf{PQ} \cdot \mathbf{b}_1 = 0, \; \mathbf{PR} \cdot \mathbf{b}_2 = 0$$

$$(a - \lambda) + (a - \lambda) + 0 = 0$$

and $(a - \mu) + (a + \mu)(-1) + 0 = 0$

$$a = \lambda, \mu = 0$$

$$\because \quad \angle QPR = 90^\circ$$

$$\mathbf{PQ} \cdot \mathbf{PR} = 0$$

$$(\lambda - a)(\mu - a) + (\lambda - a)$$
$$\Rightarrow \quad (-\mu + a) + (a-1)(a+1) = 0$$
$$\Rightarrow \quad 0 + 0 + a^2 - 1 = 0 \Rightarrow a^2 = 1$$
$$\therefore \quad 12a^2 = 12$$

**87.** *(196)* $L_1 : \dfrac{x-1}{3} = \dfrac{y-2}{2} = \dfrac{z+1}{-2}$

$L_2 : \dfrac{x+2}{-3} = \dfrac{y-2}{-2} = \dfrac{z-1}{4}$

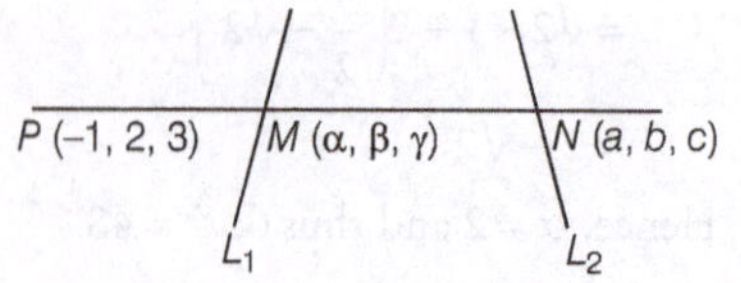


Let a point on $L_1$

$$\Rightarrow (3\lambda + 1, 2\lambda + 2, -2\lambda - 1) = M(\alpha, \beta, \gamma)$$
$$\alpha + \beta + \gamma = 3\lambda + 2 \quad \text{...(i)}$$

Similarly on $L_2$,

$$(-3\mu - 2, -2\mu + 2, 4\mu + 1) = N(a, b, c)$$
$$a + b + c = -\mu + 1 \quad \text{...(ii)}$$

∴ PM || PN

$$\frac{3\lambda + 1 + 1}{-3\mu - 2 + 1} = \frac{2\lambda + 2 - 2}{-2\mu + 2 - 2} = \frac{-2\lambda - 1 - 3}{4\mu + 1 - 3}$$
$$\Rightarrow \frac{3\lambda + 2}{-3\mu - 1} = \frac{2\lambda}{-2\mu} = \frac{-2\lambda - 4}{4\mu - 2}$$
$$\Rightarrow \quad -3\lambda\mu - 2\mu = -3\lambda\mu - \lambda;$$
$$\lambda = 2\mu$$

and $4\lambda\mu - 2\lambda = 4\lambda\mu - 8$

$$2\lambda = 8, \lambda = 4$$
$$\therefore \quad \lambda = 4 \text{ and } \mu = 2$$

Required value $\dfrac{(\alpha + \beta + \gamma)^2}{(a + b + c)^2}$

$$= \frac{(3\lambda + 2)^2}{(-\mu + 1)^2} \text{ [from Eqs. (i) and (ii)]}$$
$$= \frac{14^2}{1^2} = 196$$

**88.** *(16)* Given, lines are

$$\frac{x+1}{1} = \frac{y-0}{1/2} = \frac{z-0}{-1/12} : L_1$$

and $\dfrac{x-0}{1} = \dfrac{y+2}{1} = \dfrac{z-1}{1/6} : L_2$

Here, $x_1 = -1, y_1 = 0, z_1 = 0$

$$a_1 = 1, b_1 = \frac{1}{2}, c_1 = -\frac{1}{12}$$

and $x_2 = 0, y_2 = -2, z_2 = 1$

$$a_2 = 1, b_2 = 1, c_2 = \frac{1}{6}$$

Now,

$$d = \left| \frac{\begin{vmatrix} x_2 - x_1 & y_2 - y_1 & z_2 - z_1 \\ a_1 & b_1 & c_1 \\ a_2 & b_2 & c_2 \end{vmatrix}}{\sqrt{(a_1b_2 - a_2b_1)^2 + (b_1c_2 - b_2c_1)^2 + (c_1a_2 - c_2a_1)^2}} \right|$$

So, $d_1 = \left| \dfrac{\begin{vmatrix} 0+1 & -2 & 1 \\ 1 & 1/2 & -1/12 \\ 1 & 1 & 1/6 \end{vmatrix}}{\sqrt{\left(1 - \frac{1}{2}\right)^2 + \left(\frac{1}{12} + \frac{1}{12}\right)^2 + \left(-\frac{1}{12} - \frac{1}{6}\right)^2}} \right|$

$$= \left| \frac{\begin{vmatrix} 1 & -2 & 1 \\ 1 & 1/2 & -1/12 \\ 1 & 1 & 1/6 \end{vmatrix}}{\sqrt{\frac{1}{4} + \frac{1}{36} + \frac{1}{16}}} \right|$$

$$\left| \frac{\left(\frac{1}{12} + \frac{1}{12}\right) + 2\left(\frac{1}{6} + \frac{1}{12}\right) + (1 - 1/2)}{\sqrt{\frac{36 + 4 + 9}{144}}} \right|$$

$$= \frac{12}{7} \times \left[\frac{1}{6} + \frac{1}{2} + \frac{1}{2}\right] = 2$$

Similarly,

$$d_2 = \left| \frac{\begin{vmatrix} 0 & 10 & 2 \\ 2 & -7 & 5 \\ 2 & 1 & -3 \end{vmatrix}}{\sqrt{256 + 256 + 256}} \right|$$

$$= \frac{1}{16\sqrt{3}} [0 - 10(-6 - 10) + 2(2 + 14)]$$
$$= \frac{1}{16\sqrt{3}} (160 + 32) = \frac{12}{\sqrt{3}}$$

Now, $\dfrac{32\sqrt{3}d_1}{d_2} = \dfrac{32\sqrt{3} \times 2}{12/\sqrt{3}} = 16$

**Alternate Solution**

We have,

$$L_1 : \frac{x+1}{1} = \frac{y-0}{1/2} = \frac{z-0}{-1/12}$$
$$L_2 : \frac{x-0}{1} = \frac{y+2}{1} = \frac{z-1}{1/6}$$
$$d_1 = \left| \frac{(\mathbf{a}_2 - \mathbf{a}_1) \cdot (\mathbf{b}_1 \times \mathbf{b}_2)}{|(\mathbf{b}_1 \times \mathbf{b}_2)|} \right| = 2$$

Similarly, we can find $d_2 = \dfrac{12}{\sqrt{3}}$

So, $\dfrac{d_1 \times 32\sqrt{3}}{d_2} = \dfrac{2 \times 32 \times \sqrt{3} \times \sqrt{3}}{12} = 16$

**89.** *(9)*

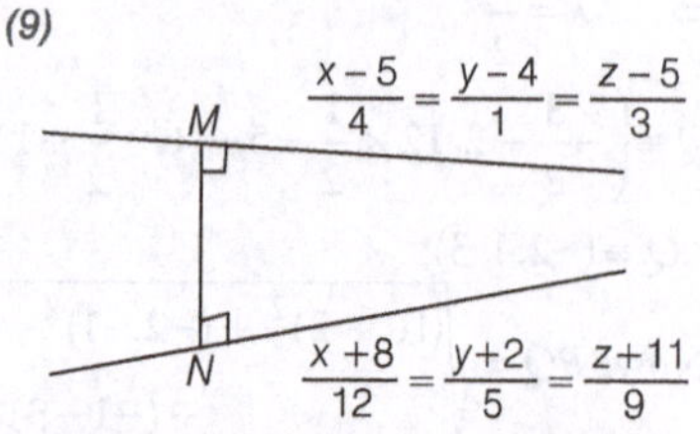


Let $M(4\lambda + 5, \lambda + 4, 3\lambda + 5)$ and $N(12\mu - 8, 5\mu - 2, 9\mu - 11)$ be two points on given lines.

∴ $\mathbf{MN} \perp \mathbf{b}_1$ and $\mathbf{MN} \perp \mathbf{b}_2$

$$\mathbf{MN} \cdot \mathbf{b}_1 = 0$$
$$4(12\mu - 4\lambda - 13) + 1(5\mu - \lambda - 6) + 3(9\mu - 3\lambda - 16) = 0$$
$$\Rightarrow 48\mu - 16\lambda - 52 + 5\mu - \lambda - 6 + 27\mu - 9\lambda - 48 = 0$$
$$\Rightarrow \quad 80\mu - 26\lambda = 106$$
$$40\mu - 13\lambda = 53 \quad \text{...(i)}$$
$$\mathbf{MN} \cdot \mathbf{b}_2 = 0$$
$$12(12\mu - 4\lambda - 13) + 5(5\mu - \lambda - 6) + 9(9\mu - 3\lambda - 16) = 0$$
$$\Rightarrow 144\mu - 48\lambda - 156 + 25\mu - 5\lambda - 30 + 81\mu - 27\lambda - 144 = 0$$
$$\Rightarrow \quad 250\mu - 80\lambda = 330$$
$$25\mu - 8\lambda = 33 \quad \text{...(ii)}$$

On solving Eqs. (i) and (ii), we get

$$\mu = 1, \lambda = -1$$
$$\therefore \quad M = (1, 3, 2) \text{ and } N = (4, 3, -2)$$
$$\mathbf{OM} = \hat{i} + 3\hat{j} + 2\hat{k}$$
$$\mathbf{ON} = 4\hat{i} + 3\hat{j} - 2\hat{k}$$
$$\mathbf{OM} \cdot \mathbf{ON} = 4 + 9 - 4 = 9$$

**90.** *(108)* Let $L_1 : \dfrac{x-2}{2} = \dfrac{y}{-2} = \dfrac{z-7}{16} = \lambda$ (say)

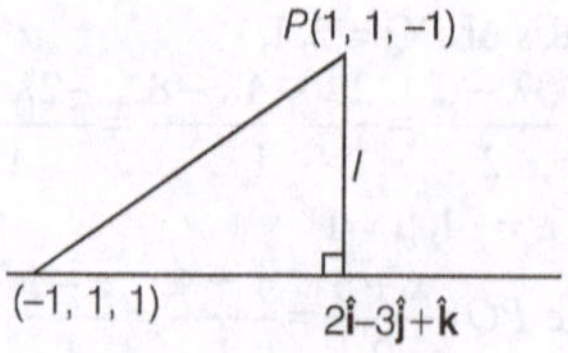

and $L_2 : \dfrac{x+3}{4} = \dfrac{y+2}{3} = \dfrac{z+2}{1} = \mu$ (say)

any point taken on $L_1$ as $(2\lambda+2, -2\lambda, 16\lambda+7)$

and any point taken on $L_2$ as $(4\mu-3, 3\mu-2, \mu-2)$

According to the question,

$$2\lambda+2 = 4\mu-3 \quad \text{...(i)}$$
$$-2\lambda = 3\mu-2 \quad \text{...(ii)}$$
$$16\lambda+7 = \mu-2 \quad \text{...(iii)}$$

On solving Eqs. (i) and (ii), we get

$\mu = 1$ and $\lambda = -\dfrac{1}{2}$ and these values satisfies Eq. (iii).

So, $P \equiv (1, 1, -1)$

Now, line $\dfrac{x+1}{2} = \dfrac{y-1}{3} = \dfrac{z-1}{1}$ passes through $(-1, 1, 1)$ and having DR's $\langle 2, 3, 1 \rangle$ [as shown in figure]

Now, projection of $2\hat{i}+0\hat{j}+(-2)\hat{k}$ on $2\hat{i}+3\hat{j}+\hat{k}$ is

$$= \frac{4-2}{\sqrt{4+9+1}} = \frac{2}{\sqrt{14}}$$

$$\therefore \quad l^2 = 8 - \frac{4}{14} = \frac{108}{14} \Rightarrow 14l^2 = 108$$

**91.** *(5)* We have, $l_1 : \mathbf{r} = (\hat{i} - 11\hat{j} - 7\hat{k}) + \lambda(\hat{i}+2\hat{j}+3\hat{k}), \lambda \in R$

and $l_2 : \mathbf{r} = (-\hat{i}+\hat{k}) + \mu(2\hat{i}+2\hat{j}+\hat{k}), \quad \mu \in R$

Let direction ratio of line $l$ be $a, b$ and $c$

Equation of line $l$

$$\mathbf{r} = (0\hat{i}+0\hat{j}+0\hat{k}) + \delta(a\hat{i}+b\hat{j}+c\hat{k})$$
$$= \delta(a\hat{i}+b\hat{j}+c\hat{k})$$

As line $l$ is perpendicular to $l_1$ and $l_2$,

$$a\hat{i}+b\hat{j}+c\hat{k} = \begin{vmatrix} \hat{i} & \hat{j} & \hat{k} \\ 1 & 2 & 3 \\ 2 & 2 & 1 \end{vmatrix} = -4\hat{i}+5\hat{j}-2\hat{k}$$

Equation line $l$ : $\mathbf{r} = \delta(-4\hat{i}+5\hat{j}-2\hat{k})$

As $P$ is the intersecting point of $l$ and $l_1$

$-4\delta = 1+\lambda, 5\delta = -11+2\lambda, -2\delta = -7+3\lambda$

After solving the above three equation, we get

$$\delta = -1 \text{ and } \lambda = 3$$

Co-ordinate of point $P$ is $(4, -5, 2)$.

$Q$ is a point on line $l_2$.

Let co-ordinate of $Q$ be $(-1+2\mu, 2\mu, 1+\mu)$

$\mathbf{PQ} = (-5+2\mu)\hat{\mathbf{i}} + (2\mu+5)\hat{\mathbf{j}} + (\mu-1)\hat{\mathbf{k}}$

$\mathbf{PQ}\cdot(2\hat{i}+2\hat{j}+\hat{k}) = 0 \quad [\because PQ \perp l_2]$

$2(-5+2\mu) + 2(2\mu+5) + \mu - 1 = 0$

$$9\mu - 1 = 0 \Rightarrow \mu = 1/9$$

$$\therefore \alpha = -1 + \frac{2}{9} = \frac{-7}{9}, \beta = 2 \times \frac{1}{9} = \frac{2}{9},$$

$$r = 1 + \frac{1}{9} = \frac{10}{9}$$

Hence,

$$9(\alpha+\beta+\gamma) = 9\left(-\frac{7}{9}+\frac{2}{9}+\frac{10}{9}\right) = 5$$

**92.** *(18)* Given lines $\dfrac{x-1}{2} = \dfrac{2-y}{-3} = \dfrac{z-3}{\alpha}$

$$\frac{x-1}{2} = \frac{y-2}{3} = \frac{z-3}{\alpha} = \lambda \quad \text{...(i)}$$

Any point on the line (i)

$x = 2\lambda+1, y = 3\lambda+2, z = \alpha\lambda+3$

and line $\dfrac{x-4}{5} = \dfrac{y-1}{2} = \dfrac{z}{\beta} = \mu \quad \text{...(ii)}$

any point on line (ii)

$\Rightarrow \quad x = 5\mu+4, y = 2\mu+1, z = \beta\mu$

Since, given lines intersects

$$\therefore \quad 2\lambda+1 = 5\mu+4 \quad \text{...(iii)}$$
$$3\lambda+2 = 2\mu+1 \quad \text{...(iv)}$$
$$\text{and} \quad \alpha\lambda+3 = \beta\mu \quad \text{...(v)}$$

On solving (iii) and (iv), we get

$$\lambda = -1, \mu = -1$$

On putting value of $\lambda$ and $\mu$ in Eq. (v), we get

$$\alpha(-1)+3 = -\beta \Rightarrow \alpha = \beta+3$$

Now, $8\alpha\beta = 8(\beta+3)(\beta) = 8(\beta^2+3\beta)$

$$= 8\left(\beta^2+3\beta+\frac{9}{4}-\frac{9}{4}\right)$$

$$= 8\left(\left(\beta+\frac{3}{2}\right)^2 - \frac{9}{4}\right) = 8\left(\beta+\frac{3}{2}\right)^2 - 18$$

Here, minimum value $= -18$

$\therefore$ Magnitude of the minimum value of $8\alpha\beta$ is 18.

**93.** *(158)* We have to find direction ratio of line

$$\begin{vmatrix} \hat{i} & \hat{j} & \hat{k} \\ 1 & 3 & -2 \\ 1 & -1 & 2 \end{vmatrix} = \hat{i}(6-2) - \hat{j}(2+2) + \hat{k}(-1-3)$$

$$= 4\hat{i} - 4\hat{j} - 4\hat{k}$$

Then, equation of line

$$\frac{x-2}{1} = \frac{y-3}{-1} = \frac{z-1}{-1} = \lambda$$

Let $M(\lambda+2, -\lambda+3, -\lambda+1)$

Direction ratio of

$MQ = \langle \lambda+2-5, -\lambda+3-3, -\lambda+1-8 \rangle$

$= \langle \lambda-3, -\lambda, -\lambda-7 \rangle$

$\because L \perp MQ$

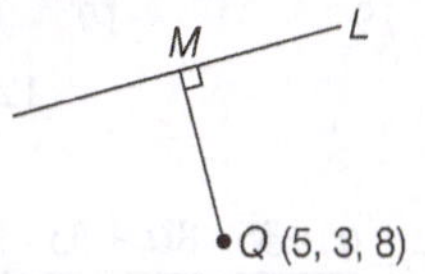


Then, $(\lambda-3)(1) + (-\lambda)(-1) + (-\lambda-7)(-1) = 0$

$\Rightarrow \lambda - 3 + \lambda + \lambda + 7 = 0$

$\Rightarrow \; 3\lambda = -4 \Rightarrow \lambda = -4/3$

$$\therefore \quad M = \left(-\frac{4}{3}+2, \frac{4}{3}+3, \frac{4}{3}+1\right)$$
$$= \left(\frac{2}{3}, \frac{13}{3}, \frac{7}{3}\right)$$

and $MQ = a$, then

$$3a^2 = 3\left[\left(5-\frac{2}{3}\right)^2 + \left(3-\frac{13}{3}\right)^2 + \left(8-\frac{7}{3}\right)^2\right]$$

$$= 3\left[\frac{169}{9} + \frac{16}{9} + \frac{289}{9}\right]$$

$$= 3\left(\frac{474}{9}\right) = \frac{474}{3} = 158$$

**94.** *(9)* We have, $B$ and $C$ lying on $\dfrac{x+\alpha}{5} = \dfrac{y-1}{2} = \dfrac{z+4}{3}$

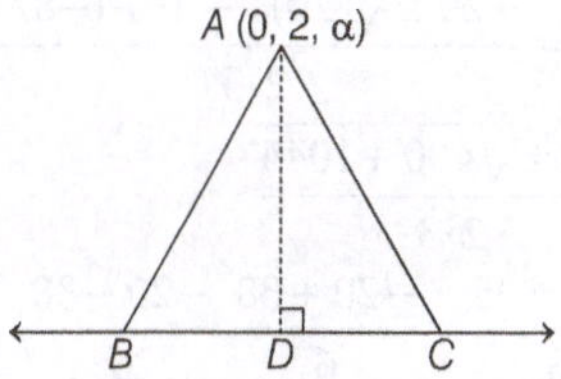


Also, $A(0, 2, \alpha)$, ar$(ABC) = 21$ sq units, $BC = 2\sqrt{21}$ units

$$\Rightarrow \quad \frac{1}{2} \times BC \times AD = 21$$

$$\Rightarrow \quad \frac{1}{2} \times 2\sqrt{21} \times AD = 21$$

$\Rightarrow AD = \sqrt{21}$ units

Let $D(5h-\alpha, 2h+1, 3h-4)$

$\Rightarrow \quad \langle AD \rangle = \langle 5h-\alpha-0, 2h+1-2, 3h-4-\alpha \rangle$

$= \langle 5h-\alpha, 2h-1, 3h-\alpha-4 \rangle$

Direction ratios of line $BC$ are $\langle 5, 2, 3 \rangle$

Since, $AD \perp BC$.

$\therefore$ Dot product of direction ratios $= 0$

$\Rightarrow \quad 5(5h-\alpha) + 2(2h-1) + 3(3h-\alpha-4) = 0$

$\Rightarrow (5^2+2^2+3^2)h = 5\alpha + 2 + 3\alpha + 12$

$\Rightarrow \quad 38h = 8\alpha + 14$

$$\Rightarrow \quad h = \frac{4\alpha+7}{19} \quad \text{... (i)}$$

$\therefore$Co-ordinates of $D$

$$= \left(\frac{20\alpha+35}{19}-\alpha, \frac{8\alpha+14}{19}+1, \frac{12\alpha+21}{19}-4\right)$$

i.e., $D\left(\frac{\alpha+35}{19}, \frac{8\alpha+33}{19}, \frac{12\alpha-55}{19}\right)$

Using distance formula for points i.e., if $P(x_1, y_1, z_1)$ and $Q(x_2, y_2, z_2)$, then

$$PQ = \sqrt{(x_2-x_1)^2+(y_2-y_1)^2+(z_2-z_1)^2}$$

We get,

$$AD^2 = \left(\frac{\alpha+35}{19}-0\right)^2+\left(\frac{8\alpha+33}{19}-2\right)^2+\left(\frac{12\alpha-55}{19}-\alpha\right)^2$$

$$\Rightarrow (\sqrt{21})^2 = \frac{(\alpha+35)^2}{19^2}+\frac{(8\alpha+33-38)^2}{19^2}+\frac{(12\alpha-55-19\alpha)^2}{19^2}$$

$$\Rightarrow 114\alpha^2+760\alpha-3306=0$$

$$\Rightarrow 3\alpha^2+20\alpha-87=0$$

$$\Rightarrow \alpha = \frac{-20\pm\sqrt{(20)^2-4\cdot 3\cdot(-87)}}{2\cdot 3}$$

$$= \frac{-20\pm\sqrt{400+1044}}{2\cdot 3}$$

$$= \frac{-20\pm 38}{6} = \frac{-20+38}{6}, \frac{-20-38}{6}$$

$$= 3, \frac{-58}{6}$$

Rejecting $\alpha = \frac{-58}{6} \Rightarrow \alpha^2 = 9$

**95.** *(14)*

> **key idea** To find the shortest distance between the lines firstly find the vector form of lines. Then, use of formula $\text{SD} = \left|\frac{(\mathbf{a}_2-\mathbf{a}_1)\cdot(\mathbf{b}_1\times\mathbf{b}_2)}{|\mathbf{b}_1\times\mathbf{b}_2|}\right|$.

Given equations of line are

$$\frac{x-2}{3} = \frac{y+1}{2} = \frac{z-6}{2} \quad \text{...(i)}$$

and $$\frac{x-6}{3} = \frac{1-y}{2} = \frac{z+8}{0} \quad \text{...(ii)}$$

Eqs. (i) and (ii) can be re-written in original standard form as,

$$\frac{x-2}{3} = \frac{y-(-1)}{2} = \frac{z-6}{2}$$

and $$\frac{x-6}{3} = \frac{y-1}{-2} = \frac{z-(-8)}{0}$$

So, shortest distance (SD)

$$= \frac{\left|\begin{vmatrix} 4 & 2 & -14 \\ 3 & 2 & 2 \\ 3 & -2 & 0 \end{vmatrix}\right|}{\left\|\begin{vmatrix} \hat{i} & \hat{j} & \hat{k} \\ 3 & 2 & 2 \\ 3 & -2 & 0 \end{vmatrix}\right\|}$$

$$= \frac{16+12+168}{|4\hat{i}+6\hat{j}-12\hat{k}|} = \frac{196}{14} = 14$$

$$[\because |x\hat{i}+y\hat{j}+z\hat{k}| = \sqrt{x^2+y^2+z^2}]$$

**96.** *(6)* Centroid of

$$\Delta ABC = G\left(\frac{2\alpha}{3}, \frac{2\alpha}{3}, \frac{2\alpha}{3}\right)$$

Given equation of line is

$$\frac{x}{1} = \frac{z-3}{-1} = \frac{y}{0} = \lambda \text{ (say)}$$

$$x = \lambda, y = 0, z = -\lambda+3$$

$\therefore D(\lambda, 0, -\lambda+3)$ be any point on given line

$$\therefore GD = \sqrt{\left(\lambda-\frac{2\alpha}{3}\right)^2+\left(\frac{2\alpha}{3}\right)^2+\left(-\lambda+3-\frac{2\alpha}{3}\right)^2}$$

Let $$P = GD^2 = \left(\lambda-\frac{2\alpha}{3}\right)^2+\left(\frac{-2\alpha}{3}\right)^2+\left(-\lambda+3-\frac{2\alpha}{3}\right)^2$$

$$\frac{dP}{d\lambda} = 2\left(\lambda-\frac{2\alpha}{3}\right)-2\left(-\lambda+3-\frac{2\alpha}{3}\right)$$

$$= 4\lambda-6 = 0 \Rightarrow \lambda = 3/2$$

$\therefore$ Minimum $GD$

$$= \sqrt{\left(\frac{3}{2}-\frac{2\alpha}{3}\right)^2+\left(\frac{2\alpha}{3}\right)^2+\left(-\frac{3}{2}+3-\frac{2\alpha}{3}\right)^2}$$

$$\sqrt{\frac{57}{2}} = \sqrt{\left(\frac{9-4\alpha}{6}\right)^2+\frac{4\alpha^2}{9}+\left(\frac{9-4\alpha}{6}\right)^2}$$

$$\Rightarrow \frac{57}{2} = \frac{24\alpha^2-72\alpha+81}{18}$$

$$\Rightarrow \alpha^2-3\alpha-18=0 \Rightarrow \alpha = -3, 6$$

$\therefore \quad \alpha = 6 \quad (\because \alpha > 0)$

**97.** *(125)*

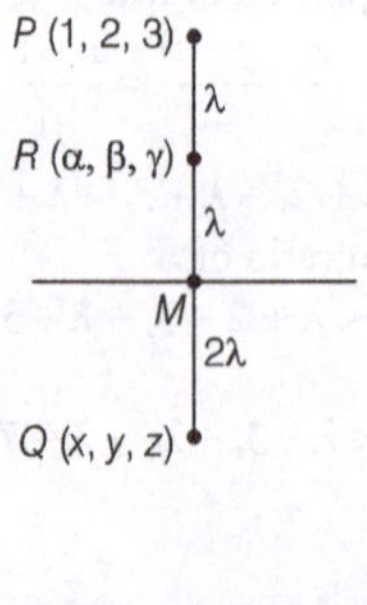


Let $$\frac{x-6}{3} = \frac{y-1}{2} = \frac{z-2}{3} = \lambda \text{ (say)}$$

$M$ be the mid-point of $PQ$.

$\therefore$ Co-ordinate of $M$ is $(3\lambda+6, 2\lambda+1, 3\lambda+2)$

$\therefore PM = (3\lambda+5)\hat{i}+(2\lambda-1)\hat{j}+(3\lambda-1)\hat{k}$

Here, DR's of normal to the line $L$ is $(3, 2, 3)$.

$\therefore$ Normal vector of the line $L$ is : $(3\hat{i}+2\hat{j}+3\hat{k})$

$\because \mathbf{PM} \perp (3\hat{i}+2\hat{j}+3\hat{k})$

$\therefore 3(3\lambda+5)+2(2\lambda-1)+3(3\lambda-1) = 0$

$\therefore \quad \lambda = \frac{-5}{11}$,

$\therefore \quad M\left(\frac{51}{11}, \frac{1}{11}, \frac{7}{11}\right)$

Since, $R$ is mid-point of $PM$,

co-ordinate of $R$ is $\left(\frac{31}{11}, \frac{23}{22}, \frac{20}{11}\right)$.

$\therefore \alpha = \frac{31}{11}, \beta = \frac{23}{22}, \gamma = \frac{20}{11}$

$\therefore \quad 22(\alpha+\beta+\gamma) = 125$

**98.** *(1)* Given lines,

$$\frac{x-k}{1} = \frac{y-2}{2} = \frac{z-3}{3}$$

and $$\frac{x+1}{3} = \frac{y+2}{2} = \frac{z+3}{1}$$

are coplanar.

So, $$\begin{vmatrix} k+1 & 4 & 6 \\ 1 & 2 & 3 \\ 3 & 2 & 1 \end{vmatrix} = 0$$

$\Rightarrow (k+1)(2-6)-4(1-9)+6(2-6) = 0$

$\Rightarrow \quad -4k-4+32-24 = 0$

$\Rightarrow \quad -4k+4 = 0$

$\Rightarrow \quad k = 1$

**99.** *(4)* It is given that point $Q\left(\frac{5}{3}, \frac{7}{3}, \frac{17}{3}\right)$ is foot of perpendicular of point $P(1, 0, 3)$ to a line passes through point $A(\alpha, 7, 1)$, so $PQ \perp AQ$.

$\because$ DR's of line segment $PQ$ is $\left(\frac{2}{3}, \frac{7}{3}, \frac{8}{3}\right)$ and DR's of line $AQ$ is $\left(\alpha-\frac{5}{3}, \frac{14}{3}, -\frac{14}{3}\right)$.

$$\therefore \quad \frac{2}{3}\left(\alpha-\frac{5}{3}\right)+\frac{7}{3}\left(\frac{14}{3}\right)+\frac{8}{3}\left(-\frac{14}{3}\right) = 0$$

$\Rightarrow \quad (3\alpha-5)+49-56 = 0$

$\Rightarrow \quad 3\alpha-5-7 = 0$

$\Rightarrow \quad \alpha = 4$

# C-24

JEE Main Chapterwise Topicwise Mathematics

In JEE Main, there is usually 0-1 question per paper is visible from this chapter. In past 3 years, 32 conceptual and 9 numerical value type questions of moderate to difficulty level were asked from this chapter.

**For detailed analysis scan QR code**

# PROBABILITY

## Topic 1 Conditional Probability, Multiplication Theorem on Probability, Baye's Theorem

### Level 1

**01** Given three identical bags each containing 10 balls, whose colours are as follows

| | Red | Blue | Green |
|---|---|---|---|
| Bag I | 3 | 2 | 5 |
| Bag II | 4 | 3 | 3 |
| Bag III | 5 | 1 | 4 |

A person chooses a bag at random and takes out a ball. If the ball is Red, the probability that it is from bag I is $p$ and if the ball is Green, then probability that it is from bag III is $q$, then the value of $\left(\frac{1}{p}+\frac{1}{q}\right)$ is *[2025, 2 April Shift-II]*

(a) 6 (b) 9 (c) 7 (d) 8

**02** A bag contains 19 unbiased coins and one coin with head on both sides. One coin drawn at random is tossed and head turns up. If the probability that the drawn coin was unbiased is $\frac{m}{n}$, gcd $(m, n) = 1$, then $n^2 - m^2$ is equal to *[2025, 7 April Shift-II]*

(a) 64 (b) 72
(c) 80 (d) 60

**03** Bag 1 contain 4 white balls and 5 black balls and Bag 2 contains $n$ white balls and 3 black balls. One ball is drawn randomly from Bag 1 and transferred to Bag 2. A ball is then drawn randomly from Bag 2. If the probability that the ball drawn is white is 29/45, then $n$ is equal to *[2025, 29 Jan Shift-II]*

(a) 6 (b) 3
(c) 5 (d) 4

**04** Bag $B_1$ contains 6 white and 4 blue balls, Bag $B_2$ contains 4 white and 6 blue balls and Bag $B_3$ contains 5 white and 5 blue balls. One of the bags is selected at random and a ball is drawn from it, If the ball is white, then the probability that the ball is drawn from Bag $B_2$ is *[2025, 28 Jan Shift-II]*

(a) 2/5 (b) 2/3 (c) 1/3 (d) 4/15

**Imp.** Baye's Theorem

**05** Two balls are selected at random one by one without replacement from a bag containing 4 white and 6 black balls. If the probability that the first selected ball is black given that the second selected ball is also black is $m / n$, where $gcd\,(m, n) = 1$, then $m + n$ is equal to *[2025, 22 Jan Shift-I]*

(a) 14 (b) 4 (c) 13 (d) 11

**06** Three urns $A$, $B$ and $C$ contain 7 red, 5 black; 5 red, 7 black and 6 red, 6 black balls, respectively. One of the urn is selected at random and a ball is drawn from it. If the ball drawn is black, then the probability that it is drawn from urn $A$ is *[2024, 4 April Shift-I]*

(a) 4/17 (b) 5/16 (c) 5/18 (d) 7/18

**07** A bag contains 6 white and 4 black balls. A die is rolled once and the number of balls equal to the number obtained on the die are drawn from the bag at random. The probability that all the balls drawn are white is *[2023, 15 April Shift-I]*

(a) $\frac{9}{50}$ (b) $\frac{1}{4}$ (c) $\frac{11}{50}$ (d) $\frac{1}{5}$

**08** Two dice are thrown independently. Let $A$ be the event that the number appeared on the 1st die is less than the number appeared on the 2nd die. $B$ be the event that the number appeared on the 1st die is even and that on the second die is odd and $C$ be the event that the number appeared on the 1st die is odd and that on the 2nd is even. Then, *[2023, 1 Feb Shift-II]*

(a) $B$ and $C$ are independent
(b) the number of favourable cases of the event $(A \cup B) \cap C$ is 6
(c) the number of favourable cases of the events $A$, $B$ and $C$ are 15, 6 and 6, respectively
(d) $A$ and $B$ are mutually exclusive

**09** If an unbiased die, marked with $-2, -1, 0, 1, 2, 3$ on its faces, is thrown five times, then the probability that the product of the outcomes is positive, is *[2023, 30 Jan Shift-I]*

(a) $\frac{27}{288}$ (b) $\frac{521}{2592}$
(c) $\frac{440}{2592}$ (d) $\frac{881}{2592}$

**10** Bag I contains 3 red, 4 black and 3 white balls and bag II contains 2 red, 5 black and 2 white balls. One ball is transferred from bag I to bag II and then a ball is draw from bag II. The ball so drawn is found to be black in colour. Then, the probability, that the transferred ball is red, is *[2022, 29 July Shift-II]*

(a) 4/9 (b) 5/18
(c) 1/6 (d) 3/10

**11** Out of 60% female and 40% male candidates appearing in an exam, 60% candidates qualify it. The number of females qualifying the exam is twice the number of males qualifying it. A candidate is randomly chosen from the qualified candidates. The probability, that the chosen candidate is a female, is *[2022, 28 July Shift-I]*

(a) 3/4 (b) 11/16
(c) 23/32 (d) 13/16

**12** The probability that a relation $R$ from $\{x,y\}$ to $\{x,y\}$ is both symmetric and transitive is equal to *[2022, 29 June Shift-II]*

(a) $\frac{5}{16}$ (b) $\frac{9}{16}$
(c) $\frac{11}{16}$ (d) $\frac{13}{16}$

**13** The probability that a randomly chosen $2 \times 2$ matrix with all the entries from the set of first 10 primes, is singular, is equal to *[2022, 29 June Shift-I]*

(a) $\frac{133}{10^4}$ (b) $\frac{18}{10^3}$ (c) $\frac{19}{10^3}$ (d) $\frac{271}{10^4}$

**Imp.** Type of Functions

**14** The probability that a randomly chosen one-one function from the set $\{a, b, c, d\}$ to the set $\{1, 2, 3, 4, 5\}$ satisfies $f(a) + 2f(b) - f(c) = f(d)$ is *[2022, 28 June Shift-II]*

(a) $\frac{1}{24}$ (b) $\frac{1}{40}$ (c) $\frac{1}{30}$ (d) $\frac{1}{20}$

**15** Let $A$ and $B$ be independent events such that $P(A) = p$ and $P(B) = 2p$. The largest value of $p$, for which $P$ (exactly one of $A$, $B$ occurs) $= \frac{5}{9}$, is *[2021, 26 Aug Shift-I]*

(a) $\frac{1}{3}$ (b) $\frac{2}{9}$ (c) $\frac{4}{9}$ (d) $\frac{5}{12}$

**16** When a missile is fired from a ship, the probability that it is intercepted is $1/3$ and the probability that the missile hits the target, given that it is not intercepted, is $3/4$. If three missiles are fired independently from the ship, then the probability that all three hit the target, is *[2021, 25 Feb Shift-I]*

(a) $\frac{1}{27}$ (b) $\frac{3}{4}$ (c) $\frac{1}{8}$ (d) $\frac{3}{8}$

**17** In a game two players A and B take turns in throwing a pair of fair dice starting with player A and total of scores on the two dice, in each throw is noted. A wins the game, if he throws a total of 6 before B throws a total of 7 and $B$ wins the game, if he throws a total of 7 before A throws a total of six. The game stops as soon as either of the players wins. The probability of A winning the game is *[2020, 4 Sep Shift-II]*

(a) $\frac{31}{61}$ (b) $\frac{30}{61}$ (c) $\frac{5}{31}$ (d) $\frac{5}{6}$

**18** A die thrown two times and the sum of the scores appearing on the die is observed to be a multiple of 4. Then the conditional probability that the score 4 has appeared atleast once is *[2020, 3 Sep Shift-I]*

(a) $\frac{1}{4}$ (b) $\frac{1}{3}$ (c) $\frac{1}{8}$ (d) $\frac{1}{9}$

**19** Box I contains 30 cards numbered 1 to 30 and Box II contains 20 cards numbered 31 to 50. A box is selected at random and a card is drawn from it. The number on the card is found to be a non-prime number. The probability that the card was drawn from Box I is *[2020, 2 Sep Shift-I]*

(a) 8/17 (b) 2/3 (c) 2/5 (d) 4/17

**20** Let $A$ and $B$ be two events such that the probability that exactly one of them occurs is 2/5 and the probability that $A$ or $B$ occurs is 1/2, then the probability of both of them occur together is *[2020, 8 Jan Shift-II]*

(a) 0.10 (b) 0.20 (c) 0.01 (d) 0.02

**21** Assume that each born child is equally likely to be a boy or a girl. If two families have two children each, then the conditional probability that all children are girls given that at least two are girls, is *[2019, 10 April Shift-I]*

(a) $\frac{1}{17}$ (b) $\frac{1}{12}$ (c) $\frac{1}{10}$ (d) $\frac{1}{11}$

**22** Let $A$ and $B$ be two non-null events such that $A \subset B$. Then, which of the following statements is always correct. *[2019, 8 April Shift-I]*

(a) $P(A/B) = P(B) - P(A)$ (b) $P(A/B) \geq P(A)$
(c) $P(A/B) \leq P(A)$ (d) $P(A/B) = 1$

**23** In a class of 60 students, 40 opted for NCC, 30 opted for NSS and 20 opted for both NCC and NSS. If one of these students is selected at random, then the probability that the student selected has opted neither for NCC nor for NSS is *[2019, 12 Jan Shift-II]*

(a) 1/6 (b) 1/3 (c) 2/3 (d) 5/6

**24** A pot contain 5 red and 2 green balls. At random a ball is drawn from this pot. If a drawn ball is green then put a red ball in the pot and if a drawn ball is red, then put a green ball in the pot, while drawn ball is not replaced in the pot. Now we draw another ball randomly, the probability of second ball to be red is *[2019, 9 Jan Shift-II]*

(a) $\frac{27}{49}$ (b) $\frac{26}{49}$ (c) $\frac{21}{49}$ (d) $\frac{32}{49}$

**25** A box $A$ contains 2 white, 3 red and 2 black balls. Another box $B$ contains 4 white, 2 red and 3 black balls. If two balls are drawn at random, without replacement, from a randomly selected box and one ball turns out to be white while the other ball turns out to be red, then the probability that both balls are drawn from box $B$ is *[JEE Main 2018]*

(a) 7/16 (b) 7/8 (c) 9/16 (d) 9/32

**26** Let $A$, $B$ and $C$ be three events which are pair-wise independent and $\overline{E}$ denotes the complement of an event $E$. If $P(A \cap B \cap C) = 0$ and $P(C) > 0$, then $P[(\overline{A} \cap \overline{B})|C]$ is equal to *[JEE Main 2018]*

(a) $P(A) + P(\overline{B})$ (b) $P(\overline{A}) - P(\overline{B})$
(c) $P(\overline{A}) - P(B)$ (d) $P(\overline{A}) + P(\overline{B})$

**Imp.** Independent Events

**27** Let $A$ and $B$ be two events such that $P(\overline{A \cup B}) = \frac{1}{6}$, $P(A \cap B) = \frac{1}{4}$ and $P(\overline{A}) = \frac{1}{4}$, where $\overline{A}$ stands for the complement of the event $A$. Then, the events $A$ and $B$ are *[JEE Main 2014, AIEEE 2005]*

(a) independent but not equally likely
(b) independent and equally likely
(c) mutually exclusive and independent
(d) equally likely but not independent

**28** Let $A$, $B$ and $C$ be pairwise independent events with $P(C) > 0$ and $P(A \cap B \cap C) = 0$. Then, $P(A^C \cap B^C / C)$ is equal to *[AIEEE 2011]*

(a) $P(A^C) - P(B)$ (b) $P(A) - P(B^C)$
(c) $P(A^C) + P(B^C)$ (d) $P(A^C) - P(B^C)$

**29** One ticket is selected at random from 50 tickets numbered 00, 01, 02, ..., 49. Then, the probability that the sum of the digits on the selected ticket is 8, given that the product of these digits is zero equal to *[AIEEE 2009]*

(a) $\frac{1}{14}$ (b) $\frac{1}{7}$ (c) $\frac{5}{14}$ (d) $\frac{1}{50}$

**30** It is given that the events $A$ and $B$ are such that $P(A) = \frac{1}{4}$, $P(A / B) = \frac{1}{2}$ and $P(B / A) = \frac{2}{3}$. Then, $P(B)$ is equal to *[AIEEE 2008]*

(a) $\frac{1}{2}$ (b) $\frac{1}{6}$ (c) $\frac{1}{3}$ (d) $\frac{2}{3}$

**31** Two aeroplanes I and II bomb a target in succession. The probabilities of I and II scoring a hit correctly are 0.3 and 0.2, respectively. The second plane will bomb only if the first misses the target. The probability that the target is hit by the second plane, is *[AIEEE 2007]*

(a) 0.06 (b) 0.14
(c) 0.32 (d) 0.7

**32** Five horses are in a race. Mr $A$ selects two of the horses at random and bets on them. The probability that Mr $A$ selected the winning horse, is *[AIEEE 2003]*

(a) $\frac{4}{5}$ (b) $\frac{3}{5}$
(c) $\frac{1}{5}$ (d) $\frac{2}{5}$

**33** The probability of India winning a test match against the West Indies is 1/2 assuming independence from match-to-match. The probability that in a match series India's second win occurs at the third test, is *[AIEEE 2002]*

(a) $\frac{1}{8}$ (b) $\frac{1}{4}$ (c) $\frac{1}{2}$ (d) $\frac{2}{3}$

## Level 2

**34** $A$ and $B$ alternately throw a pair of dice. A wins if he throws a sum of 5 before $B$ throws a sum of 8 and $B$ wins if he throws a sum of 8 before $A$ throws a sum of 5. The probability that $A$ wins if $A$ makes the first throw, is *[2025, 24 Jan Shift-I]*

(a) $\frac{8}{17}$ (b) $\frac{8}{19}$ (c) $\frac{9}{19}$ (d) $\frac{9}{17}$

**35** If $A$ and $B$ are two events such that $P(A \cap B) = 0.1$ and $P(A|B)$ and $P(B|A)$ are the roots of the equation $12x^2 - 7x + 1 = 0$, then the value of $\dfrac{P(\overline{A} \cup \overline{B})}{P(\overline{A} \cap \overline{B})}$ is **[2025, 22 Jan Shift-II]**

(a) 5/3 (b) 7/4 (c) 4/3 (d) 9/4

**36** A company has two plants $A$ and $B$ to manufacture motorcycles. 60% motorcycles are manufactured at plant $A$ and the remaining are manufactured at plant $B$. 80% of the motorcycles manufactured at plant $A$ are rated of the standard quality, while 90% of the motorcycles manufactured at plant $B$ are rated of the standard quality. A motorcycle picked up randomly from the total production is found to be of the standard quality. If $p$ is the probability that it was manufactured at plant $B$, then $126p$ is **[2024, 6 April Shift-I]**

(a) 66 (b) 56 (c) 64 (d) 54

**37** A bag contains 8 balls, whose colours are either white or black. 4 balls are drawn at random without replacement and it was found that 2 balls are white and other 2 balls are black. The probability that the bag contains equal number of white and black balls is **[2024, 1 Feb Shift-I]**

(a) $\frac{2}{7}$ (b) $\frac{1}{5}$ (c) $\frac{2}{5}$ (d) $\frac{1}{7}$

**38** Bag $A$ contains 3 white, 7 red balls and Bag $B$ contains 3 white, 2 red balls. One bag is selected at random and a ball is drawn from it. The probability of drawing the ball from the bag $A$, if the ball drawn is white, is **[2024, 30 Jan Shift-II]**

(a) $\frac{1}{4}$ (b) $\frac{1}{3}$ (c) $\frac{3}{10}$ (d) $\frac{1}{9}$

**39** A coin is biased so that the head is 3 times as likely to occur as tail. This coin is tossed until a head or three tails occur. If $X$ denotes the number of tosses of the coin. then the mean of $X$ is **[2023, 13 April Shift-I]**

(a) $\frac{21}{16}$ (b) $\frac{81}{64}$ (c) $\frac{37}{16}$ (d) $\frac{15}{16}$

**40** In a bolt factory, machines $A$, $B$ and $C$ manufacture respectively 20%, 30% and 50% of the total bolts. Of their output 3, 4 and 2 percent are respectively defective bolts. A bolt is drawn at random from the product. If the bolt drawn is found the defective, then the probability that it is manufactured by the machine $C$ is **[2023, 8 April Shift-I]**

(a) $\frac{2}{7}$ (b) $\frac{9}{28}$

(c) $\frac{3}{7}$ (d) $\frac{5}{14}$

**41** A bag contains 6 balls. Two balls are drawn from it at random and both are found to be black. The probability that the bag contains at least 5 black balls is **[2023, 31 Jan Shift-I]**

(a) $\frac{5}{7}$ (b) $\frac{2}{7}$ (c) $\frac{5}{6}$ (d) $\frac{3}{7}$

**42** Let $A$ and $B$ be two events such that $P(B/A) = \frac{2}{5}$, $P(A/B) = \frac{1}{7}$ and $P(A \cap B) = 1/9$.

Consider

**Statement I** $P(A' \cup B) = 5/6$,

**Statement II** $P(A' \cap B') = 1/18$. Then, **[2022, 28 July Shift-II]**

(a) Both statement I and statement II are true
(b) Both statement I and statement II are false
(c) only statement I is true
(d) only statement II is true

**43** If a point $A(x, y)$ lies in the region bounded by the $Y$-axis, straight lines $2y + x = 6$ and $5x - 6y = 30$, then the probability that $y < 1$ is **[2022, 27 June Shift-II]**

(a) 1/6 (b) 5/6
(c) 2/3 (d) 6/7

**44** Let $E_1$ and $E_2$ be two events such that the conditional probabilities $P(E_1 | E_2) = 1/2$, $P(E_2|E_1) = 3/4$ and $P(E_1 \cap E_2) = 1/8$. Then, **[2022, 25 June Shift-I]**

(a) $P(E_1 \cap E_2) = P(E_1) \cdot P(E_2)$
(b) $P(E'_1 \cap E'_2) = P(E'_1) \cdot P(E_2)$
(c) $P(E_1 \cap E'_2) = P(E_1) \cdot P(E_2)$
(d) $P(E'_1 \cap E_2) = P(E_1) \cdot P(E_2)$

**45** A fair die is tossed until six is obtained on it. Let $X$ be the number of required tosses, then the conditional probability $P(X \geq 5 | X > 2)$ is **[2021, 26 Aug Shift-II]**

(a) $\frac{125}{216}$ (b) $\frac{11}{36}$ (c) $\frac{5}{6}$ (d) $\frac{25}{36}$

**Imp.** Singular and Non-singular Matrices

**46** Four dice are thrown simultaneously and the numbers shown on these dice are recorded in $2 \times 2$ matrices. The probability that such formed matrices have all different entries and are non singular, is **[2021, 22 July Shift-II]**

(a) $\frac{45}{162}$ (b) $\frac{23}{81}$ (c) $\frac{22}{81}$ (d) $\frac{43}{162}$

**47** Let $A$, $B$ and $C$ be three events such that the probability that exactly one of $A$ and $B$ occurs is $(1-k)$, the probability that exactly one of $B$ and $C$ occurs is $(1-2k)$, the probability that exactly one of $C$ and $A$ occurs is $(1-k)$ and the probability of all $A$, $B$ and $C$ occur simultaneously is $k^2$, where $0 < k < 1$.

Then the probability that at least one of $A$, $B$ and $C$ occur is [2021, 20 July Shift-II]

(a) greater than 1/8 but less than 1/4
(b) greater than 1/2
(c) greater than 1/4 but less than 2/2
(d) exactly equal to 1/2

**48** In a group of 400 peoples, 160 are smokers and non-vegetarian; 100 are smokers and vegetarian and the remaining 140 are non-smokers and vegetarian. Their chances of getting a particular chest disorder are 35%, 20% and 10%, respectively. A person is chosen from the group at random and is found to be suffering from the chest disorder. The probability that the selected person is a smoker and non-vegetarian is [2021, 25 Feb Shift-II]

(a) $\frac{7}{45}$ (b) $\frac{8}{45}$ (c) $\frac{28}{45}$ (d) $\frac{14}{45}$

**49** Let $A$ and $B$ be two independent events such that $P(A)=\frac{1}{3}$ and $P(B)=\frac{1}{6}$. Then, which of the following is true? [2020, 8 Jan Shift-I]

(a) $P(A/(A\cup B))=\frac{1}{4}$ (b) $P(A'/B')=\frac{1}{3}$
(c) $P(A/B')=\frac{1}{3}$ (d) $P(A/B)=\frac{2}{3}$

**50** An unbiased coin is tossed. If the outcome is a head, then a pair of unbiased dice is rolled and the sum of the numbers obtained on them is noted. If the toss of the coin results in tail, then a card from a well-shuffled pack of nine cards numbered 1, 2, 3, ..., 9 is randomly picked and the number on the card is noted. The probability that the noted number is either 7 or 8 is [2019, 10 Jan Shift-II]

(a) $\frac{15}{72}$ (b) $\frac{13}{36}$ (c) $\frac{19}{72}$ (d) $\frac{19}{36}$

**51** A player $X$ has a biased coin whose probability of showing heads is $p$ and a player $Y$ has a fair coin. They start playing a game with their own coins and play alternately. The player who throws a head first is a winner. If $X$ starts the game, and the probability of winning the game by both the players is equal, then the value of $p$ is [JEE Main 2018]

(a) 1/3 (b) 2/5 (c) 1/4 (d) 1/5

**52** Let $E$ and $F$ be two independent events. The probability that both $E$ and $F$ happen is $\frac{1}{12}$ and the probability that neither $E$ nor $F$ happens is $\frac{1}{2}$, then a value of $\frac{P(E)}{P(F)}$ is [JEE Main 2017]

(a) $\frac{1}{3}$ (b) $\frac{5}{12}$ (c) $\frac{3}{2}$ (d) $\frac{4}{3}$

**53** Let two fair six-faced dice $A$ and $B$ be thrown simultaneously. If $E_1$ is the event that die $A$ shows up four, $E_2$ is the event that die $B$ shows up two and $E_3$ is the event that the sum of numbers on both dice is odd, then which of the following statements is not true? [JEE Main 2016]

(a) $E_1$ and $E_2$ are independent
(b) $E_2$ and $E_3$ are independent
(c) $E_1$ and $E_3$ are independent
(d) $E_1$, $E_2$ and $E_3$ are not independent

## Level 3

**54** Let the sum of two positive integers be 24. If the probability, that their product is not less than 3/4 times their greatest possible product is $\frac{m}{n}$, where gcd $(m, n) = 1$, then $n-m$ equals [2024, 8 April Shift-I]

(a) 8 (b) 10 (c) 11 (d) 9

**55** Two integers $x$ and $y$ are chosen with replacement from the set [0, 1, 2, 3, ..., 10]. Then, the probability that $|x-y|\geq 5$, is [2024, 30 April Shift-I]

(a) $\frac{62}{121}$ (b) $\frac{30}{121}$
(c) $\frac{60}{121}$ (d) $\frac{31}{121}$

**56** Bag $A$ contains 2 white, 1 black and 3 red balls and bag $B$ contains 3 black, 2 red and $n$ white balls. One bag is chosen at random and 2 balls drawn from it at random, are found to be 1 red and 1 black. If the probability that both balls come from bag A is 6/11, then $n$ is equal to .... [2022, 24 June Shift-I]

(a) 13 (b) 6
(c) 4 (d) 3

**57** A pack of cards has one card missing. Two cards are drawn randomly and are found to be spades. The probability that the missing card is not a spade, is [2021, 16 March Shift-I]

(a) $\frac{3}{4}$ (b) $\frac{52}{867}$
(c) $\frac{39}{50}$ (d) $\frac{22}{425}$

### Numerical Value Type Questions

**58** A card from a pack of 52 cards is lost. From the remaining 51 cards, $n$ cards are drawn and are found to be spades. If the probability of the lost card to be a spade is $\frac{11}{50}$, then $n$ is equal to ........... . [2025, 4 April Shift-II]

**59** Let the probability of getting head for a biased coin be 1/4. It is tossed repeatedly until a head appears. Let $N$ be the number of tosses required. If the probability that the equation $64x^2 + 5Nx + 1 = 0$ has no real root is $p/q$, where $p$ and $q$ are co-prime, then $q - p$ is equal to ......... . *[2023, 11 April Shift-II]*

**60** 25% of the population are smokers. A smoker has 27 times more chances to develop lung cancer than a non smoker. A person is diagnosed with lung cancer and the probability that this person is a smoker is $k/10$. Then, the value of $k$ is .......... *[2023, 25 Jan Shift-II]*

**61** Three urns $A$, $B$ and C contain 4 red, 6 black, 5 red, 5 black and $\lambda$ red, 4 black balls, respectively. One of the urns is selected at random and a ball is drawn. If the ball drawn is red and the probability that it is drawn from urn C is 0.4, then the square of the length of the side of the largest equilateral triangle, inscribed in the parabola $y^2 = \lambda x$ with one vertex at the vertex of the parabola, is ......... . *[2023, 24 Jan Shift-II]*

**62** In an examination, there are 10 true-false type questions. Out of 10, a student can guess the answer of 4 questions correctly with probability 3/4 and the remaining 6 questions correctly with probability 1/4. If the probability that the student guesses the answers of exactly 8 questions correctly out of 10 is $\frac{27k}{4^{10}}$, then $k$ is equal to ...... . *[2022, 24 June Shift-II]*

**63** An electric instrument consists of two units. Each unit must function independently for the instrument to operate. The probability that the first unit functions is 0.9 and that of the second unit is 0.8. The instrument is switched on and it fails to operate. If the probability that only the first unit failed and second unit is functioning is $P$, then 98 $P$ is equal to ......... . *[2021, 31 Aug Shift-I]*

**64** Let there be three independent events $E_1$, $E_2$ and $E_3$. The probability that only $E_1$ occurs is $\alpha$, only $E_2$ occurs is $\beta$ and only $E_3$ occurs is $\gamma$. Let $p$ denote the probability of none of events occur that satisfies the equations $(\alpha - 2\beta)\, p = \alpha\beta$ and $(\beta - 3\gamma)\, p = 2\beta\gamma$. All the given probabilities are assumed to lie in the interval (0, 1). Then, $\frac{\text{probability of occurrence of } E_1}{\text{probability of occurrence of } E_3}$ is equal to ......... . *[2021, 17 March Shift-I]*

## Topic 2 Probability Distribution

### Level 1

**Imp.** Sum of Infinite G.P.

**01** If the probability that the random variable $X$ takes the value $x$ is given by $P(X = x) = k(x+1)3^{-x}$, $x = 0, 1, 2, 3, \ldots$, where $k$ is a constant, then $P(X \geq 3)$ is equal to *[2025, 03 April Shift-II]*

(a) $\frac{4}{9}$ (b) $\frac{1}{9}$ (c) $\frac{7}{27}$ (d) $\frac{8}{27}$

**02** Three defective oranges are accidently mixed with seven good ones and on looking at them, it is not possible to differentiate between them. Two oranges are drawn at random from the lot. If $x$ denote the number of defective oranges, then the variance of $x$ is *[2025, 28 Jan Shift-I]*

(a) $\frac{14}{25}$ (b) $\frac{18}{25}$ (c) $\frac{26}{75}$ (d) $\frac{28}{75}$

**03** A coin is tossed three times. Let $X$ denote the number of times a tail follows a head. If $\mu$ and $\sigma^2$ denote the mean and variance of $X$, then the value of $64(\mu + \sigma^2)$ is *[2025, 22 Jan Shift-I]*

(a) 48 (b) 32 (c) 64 (d) 51

**04** A fair die is thrown until 2 appears. Then, the probability that 2 appears in even number of throws, is *[2024, 29 Jan Shift-I]*

(a) $\frac{5}{11}$ (b) $\frac{1}{6}$ (c) $\frac{5}{6}$ (d) $\frac{6}{11}$

***05** Let a die be rolled $n$ times. Let the probability of getting odd numbers seven times be equal to the probability of getting odd numbers nine times. If the probability of getting even numbers twice is $\frac{k}{2^{15}}$ then, $k$ is equal to *[2023, 10 April Shift-II]*

(a) 90 (b) 60 (c) 30 (d) 15

**06** Let $N$ denote the sum of the numbers obtained when two dice are rolled. If the probability that $2^N < N!$ is $m/n$, where $m$ and $n$ are coprime, then $4m - 3n$ is equal to *[2023, 10 April Shift-I]*

(a) 10 (b) 12 (c) 6 (d) 8

***07** A pair of dice is thrown 5 times. For each throw, a total of 5 is considered a success. If the probability of at least 4 successes is $\frac{k}{3^{11}}$, then $k$ is equal to *[2023, 6 April Shift-I]*

(a) 75 (b) 123 (c) 164 (d) 82

**(Marked Qns. are from topics, not included in latest JEE Main syllabus.)*

***08** In a binomial distribution $B(n, p)$ the sum and the product of the mean and the variance are 5 and 6 respectively, then $6(n+p-q)$ is equal to **[2023, 1 Feb Shift-I]**

(a) 53 (b) 50 (c) 51 (d) 52

***09** Let $X$ have a binomial distribution $B(n, p)$ such that the sum and the product of the mean and variance of $X$ are 24 and 128, respectively. If $P(X > n-3) = \frac{k}{2^n}$, then $k$ is equal to **[2022, 27 July Shift-II]**

(a) 528 (b) 529 (c) 629 (d) 630

***10** Let $X$ be a random variable having binomial distribution $B(7, p)$. If $P(X=3) = 5P(X=4)$, then the sum of the mean and the variance of $X$ is **[2022, 27 June Shift-II]**

(a) $\frac{105}{16}$ (b) $\frac{7}{16}$ (c) $\frac{77}{36}$ (d) $\frac{49}{16}$

**11** A random variable $x$ has the following probability distribution

| $x$ | 0 | 1 | 2 | 3 | 4 |
|---|---|---|---|---|---|
| $P(x)$ | $k$ | $2k$ | $4k$ | $6k$ | $8k$ |

The value of $P(1 < x < 4 \mid x \le 2)$ is equal to **[2022, 24 June Shift-II]**

(a) $\frac{4}{7}$ (b) $\frac{2}{3}$ (c) $\frac{3}{7}$ (d) $\frac{4}{5}$

**12** Each of the persons $A$ and $B$ independently tosses three fair coins. The probability that both of them get the same number of heads is **[2021, 27 Aug Shift-II]**

(a) $\frac{1}{8}$ (b) $\frac{5}{8}$ (c) $\frac{5}{16}$ (d) 1

***13** An ordinary dice is rolled for a certain number of times. If the probability of getting an odd number 2 times is equal to the probability of getting an even number 3 times, then the probability of getting an odd number for odd number of times is **[2021, 24 Feb Shift-I]**

(a) $\frac{1}{32}$ (b) $\frac{3}{16}$ (c) $\frac{5}{16}$ (d) $\frac{1}{2}$

**14** A random variable $X$ has the following probability distribution

| $X$ | 1 | 2 | 3 | 4 | 5 |
|---|---|---|---|---|---|
| $P(X)$ | $K^2$ | $2K$ | $K$ | $2K$ | $5K^2$ |

Then $P(X > 2)$ is equal to **[2020, 9 Jan Shift-II]**

(a) $\frac{1}{6}$ (b) $\frac{23}{36}$ (c) $\frac{1}{36}$ (d) $\frac{7}{12}$

**15** An unbiased coin is tossed 5 times. Suppose that a variable $X$ is assigned the value $k$ when $k$ consecutive heads are obtained for $k = 3, 4, 5$, otherwise $X$ takes the value $-1$. Then, the expected value of $X$, is **[2020, 7 Jan Shift-I]**

(a) $-\frac{3}{16}$ (b) $\frac{3}{16}$ (c) $\frac{1}{8}$ (d) $-\frac{1}{8}$

***16** Let a random variable $X$ have a binomial distribution with mean 8 and variance 4. If $P(X \le 2) = \frac{k}{2^{16}}$, then $k$ is equal to **[2019, 12 April Shift-I]**

(a) 17 (b) 121 (c) 1 (d) 137

**17** The minimum number of times one has to toss a fair coin so that the probability of observing atleast one head is atleast 90% is **[2019, 8 April Shift-II]**

(a) 2 (b) 3 (c) 5 (d) 4

***18** A box contains 15 green and 10 yellow balls. If 10 balls are randomly drawn one-by-one with replacement, then the variance of the number of green balls drawn is **[JEE Main 2017]**

(a) 12/5 (b) 6 (c) 4 (d) 6/25

**19** If $A$ and $B$ are any two events such that $P(A) = \frac{2}{5}$ and $P(A \cap B) = \frac{3}{20}$, then the conditional probability, $P(A \mid A' \cup B'))$, where $A'$ denotes the complement of $A$, is equal to **[JEE Main 2016]**

(a) 11/20 (b) 5/17 (c) 8/17 (d) 1/4

***20** A multiple choice examination has 5 questions. Each question has three alternative answers of which exactly one is correct. The probability that a student will get 4 or more correct answers just by guessing is **[JEE Main 2013]**

(a) $\frac{17}{3^5}$ (b) $\frac{13}{3^5}$ (c) $\frac{11}{3^5}$ (d) $\frac{10}{3^5}$

***21** Consider 5 independent Bernoulli's trials each with probability of success $p$. If the probability of atleast one failure is greater than or equal to 31/32, then $p$ lies in the interval **[AIEEE 2011]**

(a) $\left(\frac{3}{4}, \frac{11}{12}\right]$ (b) $\left[0, \frac{1}{2}\right]$ (c) $\left(\frac{11}{12}, 1\right]$ (d) $\left(\frac{1}{2}, \frac{3}{4}\right]$

***22** In a binomial distribution $B\left(n, p = \frac{1}{4}\right)$, if the probability of atleast one success is greater than or equal to 9/10, then $n$ is greater than **[AIEEE 2009]**

(a) $\frac{1}{\log_{10} 4 - \log_{10} 3}$ (b) $\frac{1}{\log_{10} 4 + \log_{10} 3}$

(c) $\frac{9}{\log_{10} 4 - \log_{10} 3}$ (d) $\frac{4}{\log_{10} 4 - \log_{10} 3}$

***23** A pair of fair dice is thrown independently three times. The probability of getting a score of exactly 9 twice is **[AIEEE 2007]**

(a) 1/729 (b) 8/9 (c) 8/729 (d) 8/243

**24** At a telephone enquiry system, the number of phone calls regarding relevant enquiry follow. Poisson distribution with an average of 5 phone calls during 10 min time intervals. The probability that there is atmost one phone call during a 10 min time period, is **[AIEEE 2006]**

(a) $\frac{5}{6}$ (b) $\frac{6}{55}$ (c) $\frac{6}{e^5}$ (d) $\frac{6}{5^e}$

**25** A random variable $X$ has Poisson distribution with mean 2. Then, $P(X > 1.5)$ equal to **[AIEEE 2005]**

(a) $\frac{3}{e^2}$ (b) $1-\frac{3}{e^2}$ (c) 0 (d) $\frac{2}{e^2}$

***26** The mean and variance of a random variable $X$ having a binomial distribution are 4 and 2 respectively, then $P(X = 1)$ is **[AIEEE 2003]**

(a) $\frac{1}{32}$ (b) $\frac{1}{16}$ (c) $\frac{1}{8}$ (d) $\frac{1}{4}$

**27** A biased coin with probability $p, 0 < p < 1$, of heads is tossed until a head appears for the first time. If the probability that the number of tosses required is even, is 2/5, then $p$ equal to **[AIEEE 2002]**

(a) $\frac{1}{3}$ (b) $\frac{2}{3}$ (c) $\frac{2}{5}$ (d) $\frac{3}{5}$

## Level 2

**Imp.**

### Variance of Random Variable

**28** A box contains 10 pens of which 3 are defective. A sample of 2 pens is drawn at random and let $X$ denote the number of defective pens. Then, the variance of $X$ is **[2025, 4 April Shift-I]**

(a) $\frac{3}{5}$ (b) $\frac{11}{15}$ (c) $\frac{2}{15}$ (d) $\frac{28}{75}$

**29** Let $A = [a_{ij}]$ be a $2 \times 2$ matrix such that $a_{ij} \in \{0, 1\}$ for all $i$ and $j$. Let the random variable $X$ denote the possible values of the determinant of the matrix $A$. Then, the variance of $X$ is **[2025, 29 Jan Shift-II]**

(a) 3/4 (b) 5/8 (c) 3/8 (d) 1/4

***30** The random variable $X$ follows binomial distribution $B(n, p)$, for which the difference of the mean and the variance is 1. If $2P(X = 2) = 3P(X = 1)$, then $n^2 P(X > 1)$ is equal to **[2023, 13 April Shift-II]**

(a) 15 (b) 12 (c) 16 (d) 11

**31** If the probability that the random variable $X$ takes values $x$ is given by $P(X = x) = k(x+1)3^{-x}$, $x = 0, 1, 2, 3, \ldots$, where $k$ is a constant, then $P(X \geq 2)$ is equal to **[2023, 8 April Shift-II]**

(a) 11/18 (b) 7/27 (c) 20/27 (d) 7/18

**32** Three rotten apples are mixed accidently with seven good apples and four apples are drawn one by one without replacement. Let the random variable $X$ denote the number of rotten apples. If $\mu$ and $\sigma^2$ represent mean and variance of $X$ respectively, then $10(\mu^2 + \sigma^2)$ is equal to **[2023, 29 Jan Shift-I]**

(a) 20 (b) 250 (c) 25 (d) 30

***33** A six faced die is biased such that $3 \times P$ (a prime number) $= 6 \times P$ (a composite number) $= 2 \times P(1)$. Let $X$ be a random variable that counts the number of times one gets a perfect square on some throws of this die. If the die is thrown twice, then the mean of $X$ is **[2022, 27 July Shift-II]**

(a) $\frac{3}{11}$ (b) $\frac{5}{11}$ (c) $\frac{7}{11}$ (d) $\frac{8}{11}$

***34** Let $X$ be a binomially distributed random variable with mean 4 and variance $\frac{4}{3}$. Then, $54\, P(X \leq 2)$ is equal to **[2022, 26 July Shift-II]**

(a) $\frac{73}{27}$ (b) $\frac{146}{27}$ (c) $\frac{146}{81}$ (d) $\frac{126}{81}$

***35** If the sum and the product of mean and variance of a binomial distribution are 24 and 128, respectively, then the probability of one or two successes is **[2022, 25 July Shift-I]**

(a) $\frac{33}{2^{32}}$ (b) $\frac{33}{2^{29}}$ (c) $\frac{33}{2^{28}}$ (d) $\frac{33}{2^{27}}$

***36** Let a biased coin be tossed 5 times. If the probability of getting 4 heads is equal to the probability of getting 5 heads, then the probability of getting atmost two heads is **[2022, 26 June Shift-I]**

(a) $\frac{275}{6^5}$ (b) $\frac{36}{5^4}$ (c) $\frac{181}{5^5}$ (d) $\frac{46}{6^4}$

***37** If a random variable $X$ follows the Binomial distribution $B(33, p)$ such that $3P(X = 0) = P(X = 1)$, then the value of $\frac{P(X=15)}{P(X=18)} - \frac{P(X=16)}{P(X=17)}$ is equal to **[2022, 24 June Shift-I]**

(a) 1320 (b) 1088 (c) $\frac{120}{1331}$ (d) $\frac{1088}{1089}$

***38** A student appeared in an examination consisting of 8 true-false type questions. The student guesses the answers with equal probability. The smallest value of $n$, so that the probability of guessing at least '$n$' correct answers is less than 1/2, is **[2021, 27 July Shift-II]**

(a) 5 (b) 6 (c) 3 (d) 4

**39** Let $x$ be a random variable such that the probability function of a distribution is given by $P(X = 0) = \frac{1}{2}$,

$P(X=j)=\frac{1}{3^j}(j=1,2,3,....,\infty)$. Then, the mean of the distribution and $P$ ($X$ is positive and even) respectively are *[2021, 25 July Shift-II]*

(a) $\frac{3}{8}$ and $\frac{1}{8}$ (b) $\frac{3}{4}$ and $\frac{1}{8}$ (c) $\frac{3}{4}$ and $\frac{1}{9}$ (d) $\frac{3}{4}$ and $\frac{1}{16}$

***40** A fair coin is tossed a fixed number of times. If the probability of getting 7 heads is equal to probability of getting 9 heads, then the probability of getting 2 heads is *[2021, 26 Feb Shift-I]*

(a) $\frac{15}{2^{13}}$ (b) $\frac{15}{2^{12}}$ (c) $\frac{15}{2^{8}}$ (d) $\frac{15}{2^{14}}$

***41** For an initial screening of an admission test, a candidate is given fifty problems to solve. If the probability that the candidate can solve any problem is 4/5, then the probability that he is unable to solve less than two problem is *[2019, 12 April Shift-II]*

(a) $\frac{201}{5}\left(\frac{1}{5}\right)^{49}$ (b) $\frac{316}{25}\left(\frac{4}{5}\right)^{48}$ (c) $\frac{54}{5}\left(\frac{4}{5}\right)^{49}$ (d) $\frac{164}{25}\left(\frac{1}{5}\right)^{48}$

***42** In a game, a man wins ₹100 if he gets 5 or 6 on a throw of a fair die and loses ₹50 for getting any other number on the die. If he decides to throw the die either till he gets a five or a six or to a maximum of three throws, then his expected gain/loss (in rupees) is *[2019, 12 Jan Shift-II]*

(a) $\frac{400}{3}$ loss (b) $\frac{400}{9}$ loss (c) 0 (d) $\frac{400}{3}$ gain

***43** A bag contains 30 white balls and 10 red balls. 16 balls are drawn one by one randomly from the bag with replacement. If $X$ be the number of white balls drawn, then $\left(\frac{\text{mean of } X}{\text{standard deviation of } X}\right)$ is equal to *[2019, 11 Jan Shift-II]*

(a) $\frac{4\sqrt{3}}{3}$ (b) 4 (c) $3\sqrt{2}$ (d) $4\sqrt{3}$

***44** Two cards are drawn successively with replacement from a well shuffled deck of 52 cards. Let $X$ denote the random variable of number of aces obtained in the two drawn cards. Then, $P(X=1)+P(X=2)$ equals *[2019, 9 Jan Shift-I]*

(a) $\frac{25}{169}$ (b) $\frac{52}{169}$ (c) $\frac{49}{169}$ (d) $\frac{24}{169}$

**45** A random variable $X$ has the probability distribution

| X | 1 | 2 | 3 | 4 | 5 | 6 | 7 | 8 |
|---|---|---|---|---|---|---|---|---|
| P(X) | 0.15 | 0.23 | 0.12 | 0.10 | 0.20 | 0.08 | 0.07 | 0.05 |

For the events, $E=\{X \text{ is a prime number}\}$ and $F=\{X<4\}$, the probability $P(E\cup F)$ is *[AIEEE 2004]*

(a) 0.87 (b) 0.77 (c) 0.35 (d) 0.50

## Numerical Value Type Questions

**46** Three balls are drawn at random from a bag containing 5 blue and 4 yellow balls. Let the random variables $X$ and $Y$, respectively denote the number of blue and yellow balls. If $\bar{X}$ and $\bar{Y}$ are the means of $X$ and $Y$ respectively. Then, $7\bar{X}+4\bar{Y}$ is equal to ......... . *[2024, 8 April Shift-I]*

**47** Let the mean and the standard deviation of the probability distribution

| $X$ | $\alpha$ | 1 | 0 | $-3$ |
|---|---|---|---|---|
| $P(X)$ | $\frac{1}{3}$ | $K$ | $\frac{1}{6}$ | $\frac{1}{4}$ |

be $\mu$ and $\alpha$, respectively. If $\sigma-\mu=2$, then $\sigma+\mu$ is equal to ...... . *[2024, 5 April Shift-II]*

**48** From a lot of 10 items which include 3 defective items, a sample of 5 items is drawn at random. Let the random variable $X$ denote the number of defective items in the sample. If the variance of $X$ is $\sigma^2$, then $96\sigma^2$ is equal to ............ . *[2024, 5 April Shift-I]*

**49** In a tournament, a team plays 10 matches with probabilities of winning and losing each match as $\frac{1}{3}$ and $\frac{2}{3}$ respectively. Let $x$ be the number of matches that the team wins and $y$ be the number of matches that team loses. If the probability $P(|x-y|\le 2)$ is $p$, then $3^9\ p$ equals...... . *[2024, 4 April Shift-II]*

**50** A fair die is tossed repeatedly until a six is obtained. Let $X$ denote the number of tosses required and let $a=P(X=3)$, $b=P(X\ge 3)$ and $c=P(X\ge 6\,|\,X>3)$. Then, $\frac{b+c}{a}$ is equal to ......... . *[2024, 27 Jan Shift-I]*

**51** The probability distribution of random variable $X$ is given by

| $X$ | 1 | 2 | 3 | 4 | 5 |
|---|---|---|---|---|---|
| $P(X)$ | $K$ | $2K$ | $2K$ | $3K$ | $K$ |

Let $p=P(1<X<4\,|\,X<3)$. If $5p=\lambda K$ then, $\lambda$ equals to. *[2021, 27 Aug Shift-II]*

**52** A fair coin is tossed $n$ times such that the probability of getting at least one head is at least 0.9. Then the minimum value of $n$ is ...... . *[2021, 25 July Shift-II]*

**53** The probability of a man hitting a target is 1/10. The least number of shots required, so that the probability of his hitting the target at least once is greater than 1/4, is .......... . *[2020, 4 Sep Shift-I]*

## Key Index

### Topic 1

| | | | | | | | | | | | | | | | | | | | |
|---|---|---|---|---|---|---|---|---|---|---|---|---|---|---|---|---|---|---|---|
| 1 | (c) | 2 | (c) | 3 | (a) | 4 | (d) | 5 | (a) | 6 | (c) | 7 | (d) | 8 | (b) | 9 | (b) | 10 | (b) |
| 11 | (*) | 12 | (a) | 13 | (c) | 14 | (d) | 15 | (d) | 16 | (c) | 17 | (b) | 18 | (d) | 19 | (a) | 20 | (a) |
| 21 | (d) | 22 | (b) | 23 | (a) | 24 | (d) | 25 | (a) | 26 | (c) | 27 | (a) | 28 | (a) | 29 | (a) | 30 | (c) |
| 31 | (c) | 32 | (d) | 33 | (b) | 34 | (c) | 35 | (d) | 36 | (d) | 37 | (a) | 38 | (b) | 39 | (a) | 40 | (d) |
| 41 | (a) | 42 | (a) | 43 | (b) | 44 | (c) | 45 | (d) | 46 | (d) | 47 | (b) | 48 | (c) | 49 | (c) | 50 | (c) |
| 51 | (a) | 52 | (d) | 53 | (d) | 54 | (b) | 55 | (b) | 56 | (c) | 57 | (c) | 58 | (2) | 59 | (27) | 60 | (9) |
| 61 | (432) | 62 | (479) | 63 | (28) | 64 | (6) | | | | | | | | | | | | |

### Topic 2

| | | | | | | | | | | | | | | | | | | | |
|---|---|---|---|---|---|---|---|---|---|---|---|---|---|---|---|---|---|---|---|
| 1 | (c) | 2 | (d) | 3 | (a) | 4 | (a) | 5 | (b) | 6 | (d) | 7 | (b) | 8 | (d) | 9 | (b) | 10 | (c) |
| 11 | (a) | 12 | (c) | 13 | (d) | 14 | (b) | 15 | (c) | 16 | (d) | 17 | (d) | 18 | (a) | 19 | (b) | 20 | (c) |
| 21 | (b) | 22 | (a) | 23 | (d) | 24 | (c) | 25 | (b) | 26 | (a) | 27 | (a) | 28 | (d) | 29 | (c) | 30 | (d) |
| 31 | (b) | 32 | (a) | 33 | (d) | 34 | (b) | 35 | (c) | 36 | (d) | 37 | (a) | 38 | (a) | 39 | (b) | 40 | (a) |
| 41 | (c) | 42 | (c) | 43 | (d) | 44 | (a) | 45 | (b) | 46 | (17) | 47 | (5) | 48 | (56) | 49 | (8288) | 50 | (12) |
| 51 | (30) | 52 | (4) | 53 | (3) | | | | | | | | | | | | | | |

# Detailed **Explanations**

### Topic 1

**01.** *(c)*

$$P\left(\frac{B_1}{R}\right)=\frac{P(B_1)\cdot P\left(\frac{R}{B_1}\right)}{P(B_1)\cdot P\left(\frac{R}{B_1}\right)+P(B_2)\cdot P\left(\frac{R}{B_2}\right)+P(B_3)\cdot P\left(\frac{R}{B_3}\right)}$$

$$=\frac{\frac{1}{3}\times\frac{3}{10}}{\frac{1}{3}\times\frac{3}{10}+\frac{1}{3}\times\frac{4}{10}+\frac{1}{3}\times\frac{5}{10}}$$

$$=\frac{1}{4}=p$$

$$P\left(\frac{B_3}{G}\right)=\frac{P(B_3)\,P\left(\frac{G}{B_3}\right)}{P(G)}$$

$$=\frac{\frac{1}{3}\times\frac{4}{10}}{\frac{1}{3}\times\frac{5}{10}+\frac{1}{3}\times\frac{3}{10}+\frac{1}{3}\times\frac{4}{10}}=\frac{1}{3}=q$$

$$\therefore \frac{1}{p}+\frac{1}{q}=7$$

**02.** *(c)* Total number of coins = 20 (19 unbiased and one with two head)

$U$ = Event of unbiased coin $H$ = Event that head turns up

$$P(U)=\frac{19}{20} \text{ and } P(H)=\frac{1}{2}$$

$P(H/U)$ = Probability of head from unbiased coin $=\frac{1}{2}$

$P(U/X)$ = Probability of head from two head coin $=1$

We need to find to get a head from drawn coin was unbiased

Using Baye's Theorem $P(U/H)$

$$=\frac{\frac{19}{20}\times\frac{1}{2}}{\frac{19}{20}\times\frac{1}{2}+\frac{1}{20}\times 1}=\frac{\frac{19}{40}}{\frac{21}{40}}=\frac{19}{21}=\frac{m}{n}$$

gcd$(m,n)=1$ this means $m$ and $n$ are co prime, so here 19 and 21 are co prime.

Thus, $n^2-m^2=21^2-19^2$

$$=441-361=80$$

**03.** *(a)* Bag 1 $=\{4W, 5B\}$

Bag 2 $=\{nW, 3B\}$

$$P\left(\frac{W}{\text{Bag2}}\right)=\frac{29}{45}$$

$$\Rightarrow P\left(\frac{W}{B_1}\right)\times P\left(\frac{W}{B_2}\right)+P\left(\frac{W}{B_1}\right)\times P\left(\frac{W}{B_2}\right)$$

$$=\frac{29}{45}$$

$$\Rightarrow \frac{4}{9}\times\frac{n+1}{n+4}+\frac{5}{9}\times\frac{n}{n+4}=\frac{29}{45}\Rightarrow n=6$$

**04.** *(d)*

| Bag $B_1$ | Bag $B_2$ | Bag $B_3$ |
|---|---|---|
| 6W 4B | 4W 6B | 5W 5B |

$$P(B_1)=P(B_2)=P(B_3)=\frac{1}{3}$$

$$P\left(\frac{W}{B_1}\right)=\frac{6}{10},\; P\left(\frac{W}{B_2}\right)=\frac{4}{10},$$

$$P\left(\frac{W}{B_3}\right)=\frac{5}{10}$$

$$\therefore P\left(\frac{B_2}{W}\right)=\frac{P(B_2)\,P\left(\frac{W}{B_2}\right)}{P(W)}$$

$$=\frac{\frac{1}{3}\times\frac{4}{10}}{\frac{1}{3}\times\frac{6}{10}+\frac{1}{3}\times\frac{4}{10}+\frac{1}{3}\times\frac{5}{10}}$$

$$= \frac{\frac{4}{30}}{\frac{15}{30}} = \frac{4}{30} \times \frac{30}{15} = \frac{4}{15}$$

**05.** *(a)*

| 4 White balls |
|---|
| 6 Black balls |

$E_1$ = First selected ball is white.
$E_2$ = First selected ball is black.
$A$ = Second selected ball is black.

$$P\left(\frac{E_2}{A}\right) = \frac{P(E_2) \cdot P\left(\frac{A}{E_2}\right)}{P(E_1) \cdot P\left(\frac{A}{E_1}\right) + P(E_2)\, P\left(\frac{A}{E_2}\right)}$$

$$= \frac{\frac{6}{10} \cdot \frac{5}{9}}{\frac{4}{10} \cdot \frac{6}{9} + \frac{6}{10} \cdot \frac{5}{9}} = \frac{30}{24+30} = \frac{30}{54} = \frac{5}{9} = \frac{m}{n}$$

$\therefore \quad m + n = 5 + 9 = 14$

**06.** *(c)*

| A | B | C |
|---|---|---|
| 7 Red 5 Black | 5 Red 7 Black | 6 Red 6 Black |

Now, if ball drawn is black, then probability that it is chosen from urn

$$A = \frac{\frac{1}{3} \times \frac{5}{12}}{\frac{1}{3} \times \frac{5}{12} + \frac{1}{3} \times \frac{7}{12} + \frac{1}{3} \times \frac{6}{12}}$$

$$= \frac{\frac{5}{36}}{\frac{5}{36} + \frac{7}{36} + \frac{6}{36}} = \frac{5}{18}$$

**07.** *(d)* $\left.\begin{matrix} 6-\text{White} \\ 4-\text{Black} \end{matrix}\right] \rightarrow$ Total number of balls $= 10$

Probability getting all the balls drawn are white is

$$= \frac{1}{6} \times \frac{{}^6C_1}{{}^{10}C_1} + \frac{1}{6} \times \frac{{}^6C_2}{{}^{10}C_2} + \frac{1}{6} \frac{{}^6C_3}{{}^{10}C_3} + \frac{1}{6} \frac{{}^6C_4}{{}^{10}C_4} + \frac{1}{6} \frac{{}^6C_5}{{}^{10}C_5} + \frac{1}{6} \times \frac{{}^6C_6}{{}^{10}C_6}$$

$$= \frac{1}{6} \times \frac{6}{10} + \frac{1}{6} \times \frac{15}{45} + \frac{1}{6} \times \frac{20}{120} + \frac{1}{6} \times \frac{15}{210} + \frac{1}{6} \times \frac{6}{252} + \frac{1}{6} \times \frac{1}{210}$$

$$= \frac{1}{6}\left[\frac{3}{5} + \frac{1}{3} + \frac{1}{6} + \frac{1}{14} + \frac{1}{42} + \frac{1}{210}\right]$$

$$= \frac{1}{6} \times \frac{6}{5} = \frac{1}{5}$$

**08.** *(b)* We have,

$A = \{(1, 2), (1, 3), (1,4), (1, 5), (1, 6), (2, 3), (2, 4), (2, 5), (2, 6), (3, 4), (3, 5), (3, 6), (4, 5), (4, 6), (5, 6)\}$

$B = \{(2, 1), (2, 3), (2, 5), (4, 1), (4, 3), (4, 5), (6, 1), (6, 3), (6, 5)\}$

$C = \{(1, 2), (1, 4), (1, 6), (3, 2), (3, 4), (3, 6), (5, 2), (5, 4), (5, 6)$

$A \cup B = \{(1, 2), (1, 3), (1, 4), (1, 5), (1, 6), (2, 1), (2, 3), (2, 4), (2, 5), (2, 6), (3, 4), (3, 5), (3, 6), (4, 1), (4, 3), (4, 5), (4, 6), (5, 6), (6, 1), (6, 3), (6, 5)\}$

$(A \cup B) \cap C = \{(1, 2), (1, 4), (1, 6), (3, 4), (3, 6), (5, 6)\}$

Clearly, $n((A \cup B) \cap C) = 6$

**09.** *(b)* Numbers marked on the faces of a die are $-2, -1, 0, 1, 2, 3$.

$\therefore \quad n(s) = 6.$

Either all outcomes are positive or any two are negative or any four are negative.

Now, $P$ (positive number) $= \frac{3}{6} = \frac{1}{2}$

$P$ (negative number) $= \frac{2}{6} = \frac{1}{3}$

and $E$ = Product is positive

$\therefore E$ = [(5 positive) or (3 positive and 2 negative) or (1 positive 4 negative)

$$\therefore \ P(E) = {}^5C_5 \left(\frac{1}{2}\right)^5 + {}^5C_3 \left(\frac{1}{2}\right)^3 \left(\frac{1}{3}\right)^2 + {}^5C_1 \left(\frac{1}{2}\right)^1 \left(\frac{1}{3}\right)^4$$

$$= \frac{1}{32} + \frac{5}{36} + \frac{5}{81 \times 2} = \frac{521}{2592}$$

**10.** *(b)*

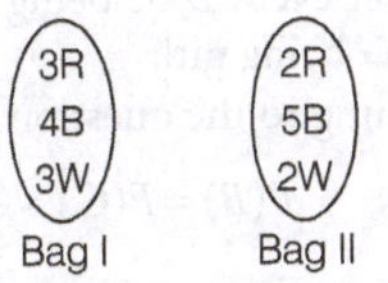


Let $A$ = Transferred ball is red.
$B$ = Black ball is drawn from Bag II.

$$\therefore \ P\left(\frac{A}{B}\right) = \frac{P(A \cap B)}{P(B)}$$

$$= \frac{\frac{3}{10} \times \frac{5}{10}}{\frac{3}{10} \times \frac{5}{10} + \frac{4}{10} \times \frac{6}{10} + \frac{3}{10} \times \frac{5}{10}} = \frac{5}{18}$$

**11.** *(*)*

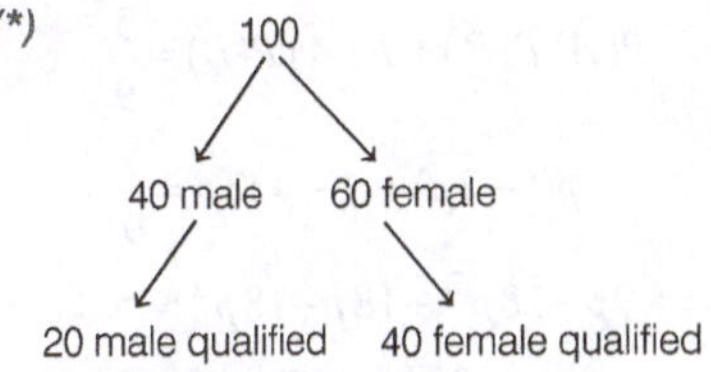


$\therefore$ Probability that chosen candidate is female $= \frac{40}{60} = \frac{2}{3}$

**12.** *(a)* $R$ is relation from $\{x, y\}$ to $\{x, y\}$.

Total number of relation $= 2^{2 \times 2} = 16$

The relation which are symmetric as well as transitive are

$\phi, \{(x, x)\}, \{(y, y)\}, \{(x, x), (y, y)\}$ and $\{(x, x), (y, y), (x, y), (y, x)\}$.

$\therefore$ Required probability $= \frac{5}{16}$

**13.** *(c)* All entries of a $2 \times 2$ matrix belongs to first 10 primes.

$\therefore$ Total number of outcomes $= 10 \times 10 \times 10 \times 10 = 10^4$

Now, let $A$ is singular, then $|A| = 0$

Determinant of matrix will be zero when

(i) All entries are same.

or (ii) diagonal elements are same i.e. only two prime numbers are used in the matrix.

Number of favourable singular matrix $= 10 + 10 \times 9 \times (2!) = 190$

$\therefore$ Required probability $= \frac{190}{10^4} = \frac{19}{10^3}$

**14.** *(d)* Number of one-one function from $\{a, b, c, d\}$ to set $\{1, 2, 3, 4, 5\}$ is

${}^5P_4 = 120$

$\therefore \quad n(S) = 120$

The required possible set of value $(f(a), f(b), f(c), f(d))$ such that $f(a) + 2f(b) - f(c) = f(d)$ are $(5, 1, 3, 4), (5, 1, 4, 3), (4, 2, 3, 5)$ $(4, 2, 5, 3), (1, 3, 2, 5)$ and $(1, 3, 5, 2)$

$\therefore \quad n(E) = 6$

$\therefore$ Required probability

$$= \frac{n(E)}{n(S)} = \frac{6}{120} = \frac{1}{20}$$

**15.** *(d)* $P$(Exactly one out of $A$ or $B$ occurs) $= \frac{5}{9}$

$$P(A \cap \bar{B}) + P(\bar{A} \cap B) = \frac{5}{9}$$

$$P(A)P(\bar{B})+P(\bar{A})P(B)=\frac{5}{9}$$

$$\therefore \quad p(1-2p)+(1-p)2p=\frac{5}{9}$$

$$\Rightarrow 9p-18p^2+18p-18p^2=5$$

$$\Rightarrow \quad 36^2p-27p+5=0$$

$$\Rightarrow \quad (3p-1)(12p-5)=0$$

$$p=\frac{1}{3} \text{ or } p=\frac{5}{12}$$

$$\Rightarrow \quad p_{\max}=5/12$$

**16.** *(c)* Probability of missile to get intercepted $=1/3$

$\therefore$ Probability of missile to not get intercepted $=1-\frac{1}{3}=\frac{2}{3}$

Probability of missile to hit the target given that it is not intercepted $=\frac{3}{4}$

$\therefore$ Probability of three missiles to hit the target

$$=\left(\frac{2}{3}\times\frac{3}{4}\right)\times\left(\frac{2}{3}\times\frac{3}{4}\right)\times\left(\frac{2}{3}\times\frac{3}{4}\right)=\frac{1}{8}$$

**17.** *(b)* The possibility to throw the dice such that sum is 6 are (1, 5), (2, 4), (3, 3), (4, 2), (5, 1)

$\therefore$Probability of getting sum 6 is $P(6)=5/36$

And, similarly the possibility to throw the dice such that sum is 7 are (1, 6), (2, 5), (3, 4), (4, 3), (5, 2), (6, 1).

$\therefore$Probability of getting 7 is $P(7)=\frac{6}{36}$

$\therefore$The required probability

$$=P(6)+P(6)P(7)P(6)+\ldots\ldots$$

$$=\frac{5}{36}+\left(\frac{31}{36}\times\frac{30}{36}\times\frac{5}{36}\right)+\ldots\ldots$$

$$=\frac{5/36}{1-\frac{31}{36}\times\frac{30}{36}}=\frac{30}{61}$$

**18.** *(d)* Let define the events

$F$ is the event that score 4 has appeared once and $E$ is the event the sum of scores appearing on the die is a multiple of 4.

So, $F=\{(1, 4), (2, 4), (3, 4), (4, 4), (5, 4), (6, 4), (4, 6), (4, 5), (4, 3), (4, 2), (4, 1)\}$

and $E=\{(1, 3), (2, 2), (2, 6), (3, 5), (4, 4), (5, 3), (6, 2), (3, 1), (6, 6)\}$

So, required conditional probability is

$$P(F/E)=\frac{P(F\cap E)}{P(E)}=\frac{1}{9}$$

**19.** *(a)* According to the question, let us define the events

$E_1=$ the event of selection of Box-I

$E_2=$ the event of selection of Box-II

$$\therefore \quad P(E_1)=P(E_2)=\frac{1}{2}$$

and $A=$ the event of drawn a non-prime numbered card

$$\therefore \quad P\left(\frac{A}{E_1}\right)=\frac{20}{30}=\frac{2}{3}$$

$$\text{and} \quad P\left(\frac{A}{E_2}\right)=\frac{15}{20}=\frac{3}{4}$$

Now, according to Baye's Theorem

$$P(E_1/A)=\frac{P(E_1)P(A/E_1)}{P(E_1)P(A/E_1)+P(E_2)P(A/E_2)}$$

$$=\frac{1/2\times 2/3}{\left(\frac{1}{2}\times\frac{2}{3}\right)+\left(\frac{1}{2}\times\frac{3}{4}\right)}=\frac{8}{17}$$

**20.** *(a)* For two events $A$ and $B$ it is given that probability of occurrence of exactly one of them is 2/5.

$$\text{So, } P(A)+P(B)-2P(A\cap B)=\frac{2}{5} \quad \ldots(i)$$

and probability that $A$ or $B$ occurs is $\frac{1}{2}$,

$$\text{so} \quad P(A\cup B)=1/2$$

$$\Rightarrow P(A)+P(B)-P(A\cap B)=\frac{1}{2} \quad \ldots(ii)$$

From Eqs. (i) and (ii), we get

$$P(A\cap B)=\frac{1}{2}-\frac{2}{5}=\frac{5-4}{10}=\frac{1}{10}$$

Probability of both of them occur together $=\frac{1}{10}=0.10$

**21.** *(d)* Let event $B$ be being boy while event $G$ being girl.

According to the question,

$$P(B)=P(G)=\frac{1}{2}$$

Now, required conditional probability that all children are girls given that at least two are girls, is

$$=\frac{\text{All 4 girls}}{(\text{All 4 girls})+(\text{exactly 3 girls}+1\text{ boy})+(\text{exactly 2 girls }+2\text{ boys})}$$

$$=\frac{\left(\frac{1}{2}\right)^4}{\left(\frac{1}{2}\right)^4+{}^4C_3\left(\frac{1}{2}\right)^3\left(\frac{1}{2}\right)+{}^4C_2\left(\frac{1}{2}\right)^2\left(\frac{1}{2}\right)^2}$$

$$=\frac{1}{1+4+6}=\frac{1}{11}$$

**22.** *(b)* We know that $P(A/B)=\frac{P(A\cap B)}{P(B)}$

[By the definition of conditional probability]

$$\because \quad A\subset B$$

$$\Rightarrow \quad A\cap B=A$$

$$\therefore \quad P(A/B)=\frac{P(A)}{P(B)} \quad \ldots(i)$$

As we know that, $0\le P(B)\le 1$

$$\therefore \quad 1\le\frac{1}{P(B)}<\infty\Rightarrow P(A)\le\frac{P(A)}{P(B)}<\infty$$

$$\Rightarrow \quad \frac{P(A)}{P(B)}\ge P(A) \quad \ldots(ii)$$

Now, from Eqs (i) and (ii), we get

$$P(A/B)\ge P(A)$$

**23.** *(a)* Let $C$ and $S$ represent the set of students who opted for NCC and NSS, respectively.

Then,

$n(C)=40$, $n(S)=30$, $n(C\cap S)=20$

and $\quad n(U)=60$

Now, $n(\bar{C}\cap\bar{S})=n(\overline{C\cup S})$

$$=n(U)-n(C\cup S)$$

$$=60-[n(C)+n(S)-n(C\cap S)]$$

$$=60-[40+30-20]=10$$

So, required probability $=\frac{10}{60}=\frac{1}{6}$

**24.** *(d)* $P(G1)=$ Probability that Ist ball drawn is green $=2/7$

$P(R1)=$ Probability that Ist ball drawn in red $=5/7$

$P(R)=$ Probability that 2nd ball drawn is red.

$$=P(G1)\cdot\frac{6}{7}+P(R1)\cdot\frac{4}{7}$$

$$=\frac{2}{7}\times\frac{6}{7}+\frac{5}{7}\times\frac{4}{7}=\frac{32}{49}$$

**25.** *(a)*

> **key idea** Poisson Distribution with mean $\lambda$ is
>
> $$P(X=k)=e^{-\lambda}\cdot\frac{\lambda^k}{k!}$$

Let $E_1$ denotes the event of drawing a white and a red ball and $E$ denotes the event that ball is drawn from box $B$.

$\therefore \quad P(E) = 1/2$

Now, $P$ (ball in drawn from box $A$)

$= P(\bar{E}) = \frac{1}{2}$

$P\left(\frac{E_1}{E}\right) = \frac{{}^4C_1 \times {}^2C_1}{{}^9C_2} = \frac{2}{9}$ and

$P(E_1 / \bar{E}) = \frac{{}^2C_1 \times {}^3C_1}{{}^7C_2} = \frac{2}{7}$

Using Bayes' Theorem,

$$P\left(\frac{E}{E_1}\right) = \frac{P(E)\, P\left(\frac{E_1}{E}\right)}{P(E)\, P\left(\frac{E_1}{E}\right) + P(\bar{E})\, P\left(\frac{E_1}{\bar{E}}\right)}$$

$$= \frac{7}{16}$$

**26.** (c) $P[(\bar{A} \cap \bar{B}) \mid C] = \frac{P[(\bar{A} \cap \bar{B}) \cap C]}{P(C)}$

$$= \frac{P[(\overline{A \cup B}) \cap C]}{P(C)} = \frac{P[C - (A \cup B)]}{P(C)}$$

$$= \frac{[P(C) - P(A \cap C) - P(B \cap C) + P(A \cap B \cap C)]}{P(C)}$$

$$= \frac{P(C) - P(A \cap C) - P(B \cap C)}{P(C)}$$

$[\because P(A \cap B \cap C) = 0]$

$$= \frac{P(C) - P(A) \cdot P(C) - P(B) \cdot P(C)}{P(C)}$$

[$\because A$, $B$ and $C$ are independent events]

$= 1 - P(A) - P(B) = P(\bar{A}) - P(B)$ or $P(\bar{B}) - P(A)$

**27.** (a) $P(\overline{A \cup B}) = \frac{1}{6}, P(A \cap B) = \frac{1}{4},$

$P(\bar{A}) = \frac{1}{4}$

$\therefore P(A \cup B) = 1 - P(\overline{A \cup B}) = 1 - \frac{1}{6} = \frac{5}{6}$

and $\quad P(A) = 1 - P(\bar{A}) = 1 - \frac{1}{4} = \frac{3}{4}$

$P(A \cup B) = P(A) + P(B) - P(A \cap B)$

$\Rightarrow \quad \frac{5}{6} = \frac{3}{4} + P(B) - \frac{1}{4}$

$\Rightarrow \quad P(B) = \frac{1}{3}$

$\Rightarrow A$ and $B$ are not equally likely.

Also, $P(A \cap B) = P(A) \cdot P(B) = 1/4$

So, events are independent.

**28.** (a) $P\left(\frac{A^C \cap B^C}{C}\right)$

$$= \frac{P(A^C \cap B^C \cap C)}{P(C)}$$

$$= \frac{P(C) - P(A \cap C) - P(B \cap C) + P(A \cap B \cap C)}{P(C)} \quad \ldots(i)$$

Given, $P(A \cap B \cap C) = 0$ and $A$, $B$, $C$ are pairwise independent.

$\therefore \quad P(A \cap C) = P(A) \cdot P(C)$

and $\quad P(B \cap C) = P(B) \cdot P(C)$

$\therefore \quad P\left(\frac{A^C \cap B^C}{C}\right)$

$$= \frac{P(C) - P(A) \cdot P(C) - P(B) \cdot P(C) + 0}{P(C)}$$

$= 1 - P(A) - P(B)$

$= P(A^C) - P(B)$

**29.** (a) $S = \{00, 01, 02, \ldots, 49\}$

Let $A$ be the event that sum of the digits on the selected ticket is 8, then

$A = \{08, 17, 26, 35, 44\}$

Let $B$ be the event that the product of the digits is zero.

$B = \{00, 01, 02, 03, \ldots, 09, 10, 20, 30, 40\}$

$\therefore \quad A \cap B = \{08\}$

$\therefore \quad$ Required probability $= P\left(\frac{A}{B}\right)$

$$= \frac{P(A \cap B)}{P(B)} = \frac{1/50}{14/50} = \frac{1}{14}$$

**30.** (c) Given that $P(A) = \frac{1}{4},\ P\left(\frac{A}{B}\right) = \frac{1}{2}$

and $P\left(\frac{B}{A}\right) = \frac{2}{3}$

We know that $P\left(\frac{A}{B}\right) = \frac{P(A \cap B)}{P(B)} \quad \ldots(i)$

and $\quad P\left(\frac{B}{A}\right) = \frac{P(B \cap A)}{P(A)} \quad \ldots(ii)$

$$\therefore P(B) = \frac{P\left(\frac{B}{A}\right) \cdot P(A)}{P\left(\frac{A}{B}\right)} = \frac{\left(\frac{2}{3}\right)\left(\frac{1}{4}\right)}{\left(\frac{1}{2}\right)} = \frac{1}{3}$$

**31.** (c) Let the events, $A =$ Ist aeroplane hit the target

$B =$ IInd aeroplane hit the target

And their corresponding probabilities are

$P(A) = 0.3$ and $P(B) = 0.2$

$\Rightarrow \quad P(\bar{A}) = 0.7$ and $P(\bar{B}) = 0.8$

$\therefore$ Required probability

$= P(\bar{A})\, P(B) + P(\bar{A})\, P(\bar{B})\, P(\bar{A})\, P(B) + \ldots$

$= (0.7)(0.2) + (0.7)(0.8)(0.7)(0.2) + (0.7)(0.8)(0.7)(0.8)(0.7)(0.2) + \ldots$

$= 0.14\,[1 + (0.56) + (0.56)^2 + \ldots]$

$= 0.14\left(\frac{1}{1 - 0.56}\right) = \frac{0.14}{0.44} = \frac{7}{22} \approx 0.32$

**32.** (d) The probability that Mr A selected the loosing horse $= \frac{4}{5} \times \frac{3}{4} = \frac{3}{5}$

The probability that Mr A selected the winning horse $= 1 - \frac{3}{5} = \frac{2}{5}$

**33.** (b) Let $A_1$, $A_2$ and $A_3$ be the events of match winning in first, second and third matches respectively and whose probabilities are

$P(A_1) = P(A_2) = P(A_3) = 1/2$

$\therefore$ Required probability

$= P(A_1\, A_2'\, A_3) + P(A_1'\, A_2\, A_3)$

$= P(A_1)\, P(A_2')\, P(A_3) + P(A_1')\, P(A_2)\, P(A_3)$

$= \left(\frac{1}{2}\right)^3 + \left(\frac{1}{2}\right)^3 = \frac{1}{8} + \frac{1}{8} = \frac{1}{4}$

**34.** (c) For sum $= 5$, $\{(1, 4), (2, 3), (3, 2), (4, 1)\}$

$\therefore \quad P(A) = \frac{4}{36} = \frac{1}{9}$

For sum $= 8$

$\Rightarrow \{(2, 6), (3, 5), (4, 4), (5, 3), (6, 2)\}$

$\therefore P(B) = 5/36$

Now, $P(\bar{A}) = 1 - P(A) = 1 - \frac{4}{36} = \frac{32}{36}$

and $P(\bar{B}) = 1 - P(B) = 1 - \frac{5}{36} = \frac{31}{36}$

$P(A \text{ win } S) = P(A) + P(\bar{A}) \cdot P(\bar{B}) \cdot P(A) + \overbrace{P(\bar{A})\, P(\bar{B}) P(\bar{A}) P(\bar{B})}^{(P(\bar{A}) \cdot P(\bar{B}))^2} \cdot P(A) \ldots$ (takes the form sum of infinite GP)

$\left(S_\infty = \frac{a}{1 - r}\right)$

$$= \frac{P(A)}{1 - P(\bar{A}) P(\bar{B})} = \frac{\frac{4}{36}}{1 - \frac{32}{36} \times \frac{31}{36}} = \frac{9}{19}$$

**35.** *(d)* Given, $P(A \cap B) = 0.1$

$$12x^2 - 7x + 1 = 0$$

$$\Rightarrow (3x-1)(4x-1) = 0$$

$$x = \frac{1}{3}, x = \frac{1}{4}$$

$$P\left(\frac{A}{B}\right) = \frac{1}{3}, \quad P\left(\frac{B}{A}\right) = \frac{1}{4}$$

$$\frac{P(A \cap B)}{P(B)} = \frac{1}{3} \text{ or } \frac{P(B \cap A)}{P(A)} = \frac{1}{4}$$

$P(B) = 0.3$ or $P(A) = 0.4$

Now, $P(\overline{A} \cup \overline{B}) = P(\overline{A \cap B})$

$= 1 - P(A \cap B)$

$= 1 - 0.1 = 0.9$

and $P(\overline{A} \cap \overline{B}) = P(\overline{A \cup B})$

$= 1 - P(A \cup B)$

$= 1 - [P(A) + P(B) - P(A \cap B)]$

$= 1 - (0.3 + 0.4 - 0.1) = 0.4$

$$\therefore \frac{P(\overline{A} \cup \overline{B})}{P(\overline{A} \cap \overline{B})} = \frac{0.9}{0.4} = \frac{9}{4}$$

**36.** *(d)* Let $E_1 \to$ Motorcycle produced by company $A$

$E_2 \to$ Motorcycle produced by company $B$

$A \to$ Motorcycle of standard quality

So, according to the question,

$P(E_1) = 0.6; P(E_2) = 0.4$

$P(A/E_1) = 0.8; P(A/E_2) = 0.9$

So, by Bayes theorem, probability of motorcycle produced at $B$ and of standard quality

$P(E_2/A)$

$$= \frac{P(E_2)\,P(A/E_2)}{P(E_1)\,P(A/E_1) + P(E_2)\,P(A/E_2)}$$

$$= \frac{0.4 \times 0.9}{0.6 \times 0.8 + 0.4 \times 0.9} = \frac{36}{84} = \frac{3}{7}$$

$$\therefore p = \frac{3}{7}$$

and $126p = 126 \times \frac{3}{7} = 54$

**37.** *(a)* $E_1 = 2$ black, 6 white

$E_2 = 3$ black, 5 white

$E_3 = 4$ black, 4 white

$E_4 = 5$ black, 3 white

$E_5 = 6$ black, 2 white,

$A = 2$ white, 2 black ball drawn

$P(E_1) = P(E_2) = P(E_3) = P(E_4)$

$= P(E_5) = \frac{1}{5}$

$$P\left(\frac{A}{E_1}\right) = \frac{{}^2C_2\,{}^6C_2}{8C_4},$$

$$P\left(\frac{A}{E_2}\right) = \frac{{}^3C_2 \times {}^5C_2}{{}^8C_4},$$

$$P\left(\frac{A}{E_3}\right) = \frac{{}^4C_2 \times {}^4C_2}{{}^8C_4}$$

$$P\left(\frac{A}{E_4}\right) = \frac{{}^5C_2\,{}^3C_2}{{}^8C_3}$$

$$P\left(\frac{A}{E_5}\right) = \frac{{}^6C_2\,{}^2C_2}{{}^8C_4}$$

Required proabality $= P\left(\frac{E_3}{A}\right)$

$$= \frac{\frac{1}{5} \times \frac{{}^4C_2 \times {}^4C_2}{{}^8C_4}}{\frac{1}{5}\left(\frac{{}^2C_2 \times {}^6C_2}{{}^8C_4} \times \frac{{}^3C_2 \times {}^5C_2}{{}^8C_4} + \frac{{}^4C_2 \times {}^4C_2}{{}^8C_4} + \frac{{}^5C_2\,{}^3C_2}{{}^8C_4} + \frac{{}^6C_2\,{}^3C_2}{{}^8C_4}\right)}$$

$$= \frac{36}{15 + 30 + 36 + 30 + 15} = \frac{36}{126} = \frac{2}{7}$$

**38.** *(b)* Use having formula

$$P(E_i/A) = \frac{P(E_i)P(A/E_i)}{\sum_{i=1}^{n} P(E_i)P(A/E_i)},$$

$$i = 1, 2, \ldots, n$$

$E_1 =$ Bag $A$ is selected

$E_2 =$ Bag $B$ is selected

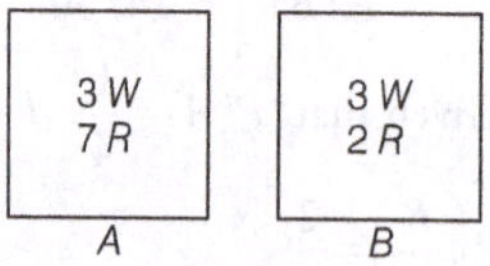


$E =$ Drawing ball from bag is white.

According to the question, we have to find $P\left[\frac{E_1}{E}\right]$.

Using Baye's theorem,

$$P\left(\frac{E_1}{E}\right) = \frac{P(E_1) \cdot P\left(\frac{E}{E_1}\right)}{P(E_1) \cdot P\left(\frac{E}{E_1}\right) + P(E_2)P\left(\frac{E}{E_2}\right)}$$

$$= \frac{\frac{1}{2} \times \frac{3}{10}}{\frac{1}{2} \times \frac{3}{10} + \frac{1}{2} \times \frac{3}{5}} = \frac{3}{3+6} = \frac{1}{3}$$

**39.** *(a)* Clearly, probability of head $= 3/4$ and probability of tails $= 1/4$

Since, coin is tossed untill a head or 3 tails occurs this process will be end in the maximum 3 thrown.

So, the probability distribution is

If $x$ denote, the number of toss of the coin, then

| $x_i$ | 1 | 2 | 3 |
|---|---|---|---|
| $p_i$ | $\frac{3}{4}$ | $\frac{1}{4} \cdot \frac{3}{4}$ | $\frac{1}{4} \cdot \frac{1}{4} \cdot \frac{3}{4} + \frac{1}{4} \cdot \frac{1}{4} \cdot \frac{1}{4}$ |

Hence, mean $= \sum p_i x_i$

$$p_1x_1 + p_2x_2 + p_3x_3$$

$$= \left(\frac{3}{4}\right) \cdot 1 + \left(\frac{1}{4} \cdot \frac{3}{4}\right) \cdot 2 + \left(\frac{1}{4} \cdot \frac{1}{4} \cdot \frac{3}{4} + \frac{1}{4} \cdot \frac{1}{4} \cdot \frac{1}{4}\right) \cdot 3$$

$$= \frac{3}{4} + \frac{3}{16} \cdot 2 + \frac{1}{16} \cdot 3$$

$$= \frac{12 + 6 + 3}{16} = \frac{21}{16}$$

**40.** *(d)* Let $D$ : Bolt is defective

$A$ : Bolt produced by machine $A$

B : Bolt produced by machine $B$

$C$ : Bolt produced by machine $C$

Then, $P(A) = \frac{20}{100}, P(B) = \frac{30}{100},$

$P(C) = \frac{50}{100}$

$P(D/A) = \frac{3}{100}, P(D/B) = \frac{4}{100},$

$P(D/C) = \frac{2}{100}$

$$P(C/D) = \frac{P(C) \cdot P(D/C)}{P(A) \cdot P(D/A) + P(B) \cdot P(D/B) + P(C) \cdot P(D/C)}$$

$$= \frac{\frac{50}{100} \times \frac{2}{100}}{\frac{20}{100} \times \frac{3}{100} + \frac{30}{100} \times \frac{4}{100} + \frac{50}{100} \times \frac{2}{100}}$$

$$= \frac{5}{14}$$

**41.** *(a)* We have, there are 6 balls in a bag. Total possibilities are

(2 black + 4 others), (3 black + 3 others), (4 black + 2 others), (5 black + 1 others) and (6 black).

Let $A$ : The event that 2 drawn balls are black.

$B$ : The event that there is atleast 5 black balls.

So, required probability $= P(B/A)$

$$= \frac{P(A\cap B)}{P(A)}$$

$$= \frac{\dfrac{^5C_2 + {^6C_2}}{^6C_2}}{\dfrac{^2C_2 + {^3C_2} + {^4C_2} + {^5C_2} + {^6C_2}}{^6C_2}}$$

$$= \frac{5\times4+6\times5}{2\times1+3\times2+4\times3+5\times4+6\times5}$$

$$= \frac{50}{70} = \frac{5}{7}$$

**42.** (a) Given, $P\left(\frac{B}{A}\right) = \frac{2}{5},\ P\left(\frac{A}{B}\right) = \frac{1}{7}$

and $P(A\cap B) = \frac{1}{9}$

$$\therefore\ P\left(\frac{B}{A}\right) = \frac{P(B\cap A)}{P(A)}$$

$$\Rightarrow\quad P(A) = \frac{P(B\cap A)}{P\left(\frac{B}{A}\right)} = \frac{1/9}{2/5} = \frac{5}{18}$$

Similarly, $P(B) = \frac{P(B\cap A)}{P\left(\frac{A}{B}\right)} = \frac{1/9}{1/7} = \frac{7}{9}$

**Statement I** $P(A'\cup B) = 1 - P(A-B)$

$$= 1 - P(A) + P(A\cap B)$$

$$= 1 - \frac{5}{18} + \frac{1}{9} = \frac{15}{18} = \frac{5}{6}$$

**Statement II** $P(A'\cap B') = P(A\cup B)'$

$$= 1 - P(A\cup B)$$

$$= 1 - P(A) - P(B) + P(A\cap B)$$

$$= 1 - \frac{5}{18} - \frac{7}{9} + \frac{1}{9} = \frac{1}{18}$$

Hence, both Statements I and II are true.

**43.** (b)

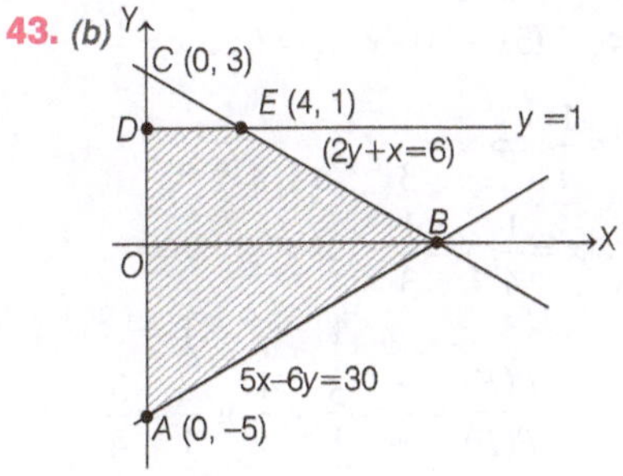


Probability

$$(y<1) = \frac{\text{ar}\,(DEBAD)}{\text{ar}\,(ABC)} \quad \text{...(i)}$$

$$\text{ar}\,(ABC) = \frac{1}{2}\times8\times6 = 24$$

$$\text{ar}\,(DEBAD) = \text{ar}\,(AOB) + \text{ar}\,(DEBO)$$

$$= \frac{1}{2}\times6\times5 + \int_0^1 (6-2y)\,dy$$

$$= 15 + [6y - y^2]_0^1$$

$$= 15 + 5 = 20$$

$\therefore$ Required probability $= \frac{20}{24} = \frac{5}{6}$ [from Eq. (i)]

**44.** (c) Given, $P(E_1 \mid E_2) = \frac{1}{2}$

$$\Rightarrow\quad \frac{P(E_1\cap E_2)}{P(E_2)} = \frac{1}{2}$$

$$P(E_2) = \frac{1}{8}\times2 = \frac{1}{4}$$

Again, $P(E_2 \mid E_1) = \frac{3}{4}$

$$\Rightarrow\quad \frac{P(E_2\cap E_1)}{P(E_1)} = \frac{3}{4}$$

$$P(E_1) = \frac{4}{3}\times\frac{1}{8} = \frac{1}{6}$$

$$P(E_1)\cdot P(E_2) = \frac{1}{4}\cdot\frac{1}{6} = \frac{1}{24}$$

$$P(E_1\cap E_2') = P(E_1) - P(E_1\cap E_2)$$

$$= \frac{1}{6} - \frac{1}{8} = \frac{4-3}{24} = \frac{1}{24}$$

$$\therefore\ P(E_1\cap E_2') = P(E_1)\cdot P(E_2)$$

**45.** (d) $P(x\ge5 \mid x>2) = \frac{P(x\ge5 \cap x>2)}{P(x>2)}$

$$= \frac{P(x\ge5)}{P(x>2)}$$

$$= \frac{\left(\frac{5}{6}\right)^4\left(\frac{1}{6}\right) + \left(\frac{5}{6}\right)^5\left(\frac{1}{6}\right) + \ldots + \infty}{\left(\frac{5}{6}\right)^2\left(\frac{1}{6}\right) + \left(\frac{5}{6}\right)^3\left(\frac{1}{6}\right) + \ldots + \infty}$$

$$= \frac{\dfrac{\left(\frac{5}{6}\right)^4\cdot\left(\frac{1}{6}\right)}{1-\frac{5}{6}}}{\dfrac{\left(\frac{5}{6}\right)^2\cdot\left(\frac{1}{6}\right)}{1-\frac{5}{6}}} = \left(\frac{5}{6}\right)^2 = \frac{25}{36}$$

**46.** (d) Let $A = \begin{vmatrix} a & b \\ c & d \end{vmatrix},\ |A| = ad - bc$

$$n(s) = 6\times6\times6\times6 = 6^4$$

For non-singular matrix $|A| \ne 0$

$$\Rightarrow\quad ad \ne bc$$

Now, $a, b, c$ and $d$ should be different.

If $ad = bc$

$$\Rightarrow\quad \frac{a}{b} = \frac{c}{d}$$

If $a = 1, \frac{1}{2} = \frac{3}{6}$ and $\frac{1}{3} = \frac{2}{6}$

$\{a, b, c, d\} = \{1, 2, 3, 6\}$ or $\{1, 3, 2, 6\}$

If $a = 2$

$$\frac{2}{1} = \frac{6}{3}, \frac{2}{3} = \frac{4}{6}, \frac{2}{4} = \frac{3}{6} = \frac{1}{2}, \frac{2}{6} = \frac{1}{3}$$

So, $\{a, b, c, d\} = \{2, 1, 6, 3\}, \{2, 3, 4, 6\}, \{2, 4, 3, 6\}, \{2, 4, 1, 2\}\ \{2, 6, 1, 3\}$

If $a = 3, \frac{3}{1} = \frac{6}{2}, \frac{3}{2} = \frac{6}{4}, \frac{3}{6} = \frac{1}{2}$

$\{a, b, c, d\} = \{3, 1, 6, 2\}, \{3, 2, 6, 4\}$ $\{3, 6, 1, 2\}$

If $a = 4, \frac{4}{2} = \frac{6}{3}, \frac{4}{6} = \frac{2}{3}$, $\{a, b, c, d\}$

$= \{4, 2, 6, 3\}, \{4, 6, 2, 3\}$

If $a = 5 \Rightarrow$ Not possible

If $a = 6, \frac{6}{2} = \frac{3}{1}, \frac{6}{3} = \frac{2}{1}, \frac{6}{4} = \frac{3}{2}, \frac{6}{3} = \frac{4}{2}$

$\{a, b, c, d\} = \{6, 2, 3, 1\}, \{6, 3, 2, 1\}, \{6, 4, 3, 2\}, \{6, 3, 4, 2\}$

So, $n(\overline{E}) = 2 + 5 + 3 + 2 + 0 + 4 = 16$

So, $n(E) = {^6C_4}\,4! - 16$

$$\text{Probability} = \frac{^6C_4\,4! - 16}{6^4} = \frac{43}{162}$$

**47.** (b) There are three events $A, B, C$.

Probability that exactly one of $A$ and $B$ occur is

$$P(\overline{A}\cap B) + P(A\cap\overline{B}) = 1 - k \quad \text{...(i)}$$

Probability that exactly one of $B$ and $C$ occur is

$$P(\overline{B}\cap C) + P(B\cap\overline{C}) = 1 - 2k \quad \text{...(ii)}$$

Probability that exactly one of $C$ and $A$ occur is

$$P(\overline{A}\cap C) + P(A\cap\overline{C}) = 1 - k \quad \text{...(iii)}$$

Probability of all $A, B, C$ occur simultaneously, is

$$P(A\cap B\cap C) = k^2 \quad \text{...(iv)}$$

From Eqs. (i), (ii) and (iii),

$$P(A) + P(B) - 2P(A\cap B) = 1 - k \quad \text{...(v)}$$

$$P(B) + P(C) - 2P(B\cap C) = 1 - 2k \quad \text{...(vi)}$$

$$P(A) + P(C) - 2P(A\cap C) = 1 - k \quad \text{...(vii)}$$

On adding Eqs. (v), (vi) and (vii), we get

$$2[P(A) + P(B) + P(C) - P(A\cap B) - P(B\cap C) - P(C\cap A)]$$

$$= (1-k) + (1-2k) + (1-k) = 3 - 4k$$

$\therefore P(A)+P(B)+P(C)-P(A\cap B)$
$-P(B\cap C)-P(A\cap C)=\dfrac{3-4k}{2}$

$\Rightarrow P(A\cup B\cup C)=\dfrac{3-4k}{2}+P(A\cap B\cap C)$

$=\dfrac{3-4k}{2}+k^2=\dfrac{2k^2-4k+3}{2}$

$\Rightarrow P(A\cup B\cup C)=\dfrac{2(k-1)^2+1}{2}$

$=(k-1)^2+\dfrac{1}{2}>\dfrac{1}{2}$

$\therefore\quad P(A\cup B\cup C)>\dfrac{1}{2}$

**48.** *(c)* Let $A$ be the event of smokers and non-vegetarian.
Let $B$ be the event of smokers and vegetarian.
Let $C$ be the event of non-smokers and vegetarian.
Let $E$ be the event of chest disorders.
According to question,

$P(A)=\dfrac{160}{400}=\dfrac{2}{5},\ P(B)=\dfrac{100}{400}=\dfrac{1}{4},$

$P(C)=\dfrac{140}{400}=\dfrac{7}{20};\ P\left(\dfrac{E}{A}\right)=35\%=\dfrac{35}{100}$

$P\left(\dfrac{E}{B}\right)=\dfrac{20}{100},\ P\left(\dfrac{E}{C}\right)=\dfrac{10}{100}$

Thus,

$$P\left(\dfrac{A}{E}\right)=\dfrac{P(A)\cdot P\left(\dfrac{E}{A}\right)}{P(A)\cdot P\left(\dfrac{E}{A}\right)+P(B)\cdot P\left(\dfrac{E}{B}\right)+P(C)\cdot P\left(\dfrac{E}{C}\right)}$$

$$=\dfrac{\dfrac{2}{5}\cdot\dfrac{35}{100}}{\dfrac{2}{5}\cdot\dfrac{35}{100}+\dfrac{1}{4}\cdot\dfrac{20}{100}+\dfrac{7}{20}\cdot\dfrac{10}{100}}$$

$$=\dfrac{14}{14+5+\dfrac{7}{2}}=\dfrac{28}{45}$$

**49.** *(c)* It is given that $A$ and $B$ are two independent events such that $P(A)=\dfrac{1}{3}$ and $P(B)=1/6$, then

$P(A\cap B)=P(A)\cdot P(B)=\dfrac{1}{18}$

Now, $P(A|(A\cup B))=\dfrac{P(A\cap(A\cup B))}{P(A\cup B)}$

$=\dfrac{P(A)}{P(A)+P(B)-P(A\cap B)}$

$=\dfrac{\dfrac{1}{3}}{\dfrac{1}{3}+\dfrac{1}{6}-\dfrac{1}{18}}=\dfrac{\dfrac{1}{3}}{\dfrac{6+3-1}{18}}=\dfrac{6}{8}=\dfrac{3}{4}$

and $P(A'|B')=\dfrac{P(A'\cap B')}{P(B')}$

$=\dfrac{P((A\cup B)')}{P(B')}=\dfrac{1-P(A\cup B)}{1-P(B)}$

$=\dfrac{1-\dfrac{6+3-1}{18}}{1-\dfrac{1}{6}}=\dfrac{\dfrac{10}{18}}{\dfrac{5}{6}}=\dfrac{2}{3}$

and $P(A|B')=\dfrac{P(A\cap B')}{P(B')}$

$=\dfrac{P(A)\cdot P(B')}{P(B')}=P(A)=\dfrac{1}{3}$

and $P(A|B)=\dfrac{P(A\cap B)}{P(B)}$

$=\dfrac{P(A)P(B)}{P(B)}=\dfrac{1}{3}$

**50.** *(c)* Clearly, $P(H)=$ Probability of getting head $=1/2$

and $P(T)=$ Probability of getting tail $=\dfrac{1}{2}$

Now, let $E_1$ be the event of getting a sum 7 or 8, when a pair of dice is rolled.
Then, $E_1=\{(6,1),(5,2),(4,3),(3,4),(2,5),$
$(1,6),(6,2),(5,3),(4,4),(3,5),$
$(2,6)\}$

$\Rightarrow P(E_1)=$ Probability of getting 7 or 8, when a pair of dice is thrown $=\dfrac{11}{36}$

Also, let $P(E_2)=$ Probability of getting 7 or 8 when a card is picked from cards numbered 1, 2, ...., 9 $=\dfrac{2}{9}$

$\therefore$Probability that the noted number is 7 or 8

$=P((H\cap E_1)\text{ or }(T\cap E_2))$

$=P(H\cap E_1)+P(T\cap E_2)$

$[\because (H\cap E_1)$ and $(T\cap E_2)$ are mutually exclusive]

$=P(H)\cdot P(E_1)+P(T)\cdot P(E_2)$

$[\because \{H, E_1\}$ and $\{T, E_2\}$ both are sets of independent events]

$=\dfrac{1}{2}\times\dfrac{11}{36}+\dfrac{1}{2}\times\dfrac{2}{9}$

$=\dfrac{19}{72}$

**51.** *(a)* Given, $P(X$ getting head$)=p$

$\therefore P(X$ not getting head$)=1-p$

$P(Y$ getting head$)=P(Y$ not getting head$)=1/2$

$P(X$ wins$)$

$=p+(1-p)\dfrac{1}{2}\cdot p+(1-p)\cdot\dfrac{1}{2}\cdot(1-p)\cdot\dfrac{1}{2}p+....$

$=\dfrac{p}{1-\left(\dfrac{1-p}{2}\right)}=\dfrac{2p}{1+p}$

$P(Y$ wins$)$

$=(1-p)\dfrac{1}{2}+(1-p)\dfrac{1}{2}\cdot(1-p)\dfrac{1}{2}+...$

$=\left(\dfrac{1-p}{2}\right)\cdot\dfrac{1}{\left(1-\dfrac{1-p}{2}\right)}=\dfrac{1-p}{1+p}$

It is given that $P(X$ wins$)=P(Y$ wins$)$

$\therefore \dfrac{2p}{1+p}=\dfrac{1-p}{1+p}\Rightarrow 3p=1\Rightarrow p=\dfrac{1}{3}$

**52.** *(d)* Let $P(E)=x$ and $P(F)=y$

$P(E\cap F)=P(E)\cdot P(F)=\dfrac{1}{12}\Rightarrow xy=\dfrac{1}{12}$

$P(\bar{E}\cap\bar{F})=P(\bar{E})\cdot P(\bar{F})=\dfrac{1}{2}$

$\Rightarrow\quad (1-P(E))(1-P(F))=\dfrac{1}{2}$

$\Rightarrow\quad (1-x)(1-y)=\dfrac{1}{2}$

$\Rightarrow\quad 1-x-y+\dfrac{1}{12}=\dfrac{1}{2}\Rightarrow x+y=\dfrac{7}{12}$

$\Rightarrow\quad x+\dfrac{1}{12x}=\dfrac{7}{12}\Rightarrow\dfrac{12x^2+1}{12x}=\dfrac{7}{12}$

$\Rightarrow\quad 12x^2-4x-3x+1=0$

$\Rightarrow\quad (3x-1)(4x-1)=0\Rightarrow x=\dfrac{1}{3},$

$x=\dfrac{1}{4}\quad\Rightarrow y=\dfrac{1}{3}$

$\therefore\ x=\dfrac{1}{4}, y=\dfrac{1}{3}$

$\therefore\quad \dfrac{P(E)}{P(F)}=\dfrac{x}{y}=\dfrac{\dfrac{1}{3}}{\dfrac{1}{4}}=\dfrac{4}{3}$ or $\dfrac{\dfrac{1}{4}}{\dfrac{1}{3}}=\dfrac{3}{4}$

**53.** *(d)* Clearly,

$E_1=\{(4,1),(4,2),(4,3),(4,4),(4,5),(4,6)\}$

$E_2=\{(1,2),(2,2),(3,2),(4,2),(5,2),(6,2)\}$

and $E_3 = \{(1, 2), (1, 4), (1, 6), (2, 1), (2, 3), (2, 5), (3, 2), (3, 4), (3, 6), (4, 1), (4, 3), (4, 5), (5, 2), (5, 4), (5, 6), (6, 1), (6, 3), (6, 5)\}$

$\Rightarrow \quad P(E_1) = \frac{6}{36} = \frac{1}{6}, \quad P(E_2) = \frac{6}{36} = \frac{1}{6}$

and $P(E_3) = \frac{18}{36} = \frac{1}{2}$

Now, $P(E_1 \cap E_2) = P$

(Probability of getting 4 on die $A$ and 2 on die $B$) $= \frac{1}{36} = P(E_1) \cdot P(E_2)$

$P(E_2 \cap E_3) = P$

(Probability of getting 2 on die $B$ and sum of numbers on both dice is odd)

$= \frac{3}{36} = P(E_2) \cdot P(E_3)$

$P(E_1 \cap E_3) = P$

(Getting 4 on die $A$ and sum of numbers on both dice is odd)

$= \frac{3}{36} = P(E_1) \cdot P(E_3)$

and $P(E_1 \cap E_2 \cap E_3) = P$

(Getting 4 on die $A$, 2 on die $B$ and sum of numbers is odd)

$= P$ (impossible event) $= 0$

Hence, $E_1$, $E_2$ and $E_3$ are not independent.

**54.** *(b)* Let the numbers be $x$ and $y$.

So, $x + y = 24$; $x, y \in N$

and $x \cdot y \geq \frac{3}{4}$ (maximum product)

Maximum product will be found.

When $x = y = \frac{24}{2} = 12$

So, $xy \geq \frac{3}{4}(12 \times 12) \Rightarrow xy \geq \frac{3}{4} \times 144$

$\Rightarrow xy \geq 108$

Possible cases

| $x$ | $y$ | $xy$ |
|---|---|---|
| 1 | 23 | 23 |
| 2 | 22 | 44 |
| ⋮ | ⋮ | ⋮ |
| 6 | 18 | 108 |
| ⋮ | ⋮ | ⋮ |
| 11 | 13 | 143 |
| 12 | 12 | 144 |
| 13 | 11 | 143 |
| ⋮ | ⋮ | ⋮ |
| 18 | 6 | 108 |

Total combination $= 13$

So, $P(xy \geq 108) = \frac{13}{23} = \frac{m}{n}$

So, $n - m = 23 - 13 = 10$

**55.** *(b)* Since, $x$ and $y$ each can take values from 0 to 10, so the total number of ways of selecting $x$ and $y$ is

$11 \times 11 = 121$. Now, $|x - y| > 5$

$\Rightarrow x - y < -5$ or $x - y > 5$

When $x - y > 5$, we have following cases

When $x = 0$, $y = 0 \rightarrow 1$ case

$x = 7$, $y = 0, 1 \rightarrow 2$ cases

$x = 8$, $y = 0, 1, 2, \rightarrow 3$ cases

$x = 9$, $y = 0, 1, 2, 3 \rightarrow 4$ cases

$x = 10$, $0, 1, 2, 3, 4 \rightarrow 5$ cases

Hence, the total number of cases 15.

Similarly, we have 15 cases for $x - y < 5$.

There are 30 pairs values of $x$ and $y$ satisfying these two in equalities.

So, favourable number of ways is 30.

Hence, required probability is 30/121.

**56.** *(c)* Given that, Bag $A$ contains 2 white, 1 black and 3 red balls and Bag $B$ contains 3 black, 2 red and $n$ white balls.

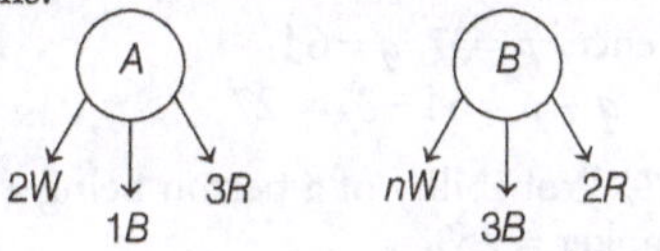


Thus, $P(1B \text{ and } 1R) = P(A) \cdot P\left(\frac{1R1B}{A}\right) + P(B) \cdot P\left(\frac{1R1B}{B}\right) \quad \ldots \text{(i)}$

$\therefore \quad P(A) = \frac{1}{2}$ and $P(B) = \frac{1}{2}$

$P\left(\frac{1R1B}{A}\right) = \frac{{}^3C_1 \cdot {}^1C_1}{{}^6C_2}$

and $P\left(\frac{1R1B}{B}\right) = \frac{{}^3C_1 \cdot {}^2C_1}{{}^{n+5}C_2}$

From Eq. (i), we get

$$p\,(1R \text{ and } 1B) = \frac{1}{2} \cdot \frac{{}^3C_1 \cdot {}^1C_1}{{}^6C_2} + \frac{1}{2} \cdot \frac{{}^3C_1 \cdot {}^2C_1}{{}^{n+5}C_2}$$

$$\therefore P\left(\frac{1R \text{ and } 1B}{A}\right) = \frac{\frac{1}{2} \cdot \frac{{}^3C_1 \cdot {}^1C_1}{{}^6C_2}}{\frac{1}{2} \cdot \frac{{}^3C_1 \cdot {}^1C_1}{{}^6C_2} + \frac{1}{2} \cdot \frac{{}^3C_1 \cdot {}^2C_1}{{}^{n+5}C_2}}$$

Given that $P\left(\frac{1R \text{ and } 1B}{A}\right) = \frac{6}{11}$

$$\Rightarrow \frac{\frac{1}{2} \cdot \frac{{}^3C_1 \cdot {}^1C_1}{{}^6C_2}}{\frac{1}{2} \cdot \frac{{}^3C_1 \cdot {}^1C_1}{{}^6C_2} + \frac{1}{2} \cdot \frac{{}^3C_1 \cdot {}^2C_1}{{}^{n+5}C_2}} = \frac{6}{11}$$

$$\Rightarrow \frac{\frac{1}{2} \cdot \frac{3}{15}}{\frac{1}{2} \cdot \frac{3}{15} + \frac{1}{2} \cdot \frac{6 \cdot 2}{(n+5)(n+4)}} = \frac{6}{11}$$

$\Rightarrow \quad \frac{6}{10} + \frac{36}{(n+5)(n+4)} = \frac{11}{10}$

$\Rightarrow \quad (n+5)(n+4) - 72 = 0$

$\Rightarrow \quad n^2 + 9n - 52 = 0$

$\Rightarrow \quad (n+13)(n-4) = 0$

$\Rightarrow \quad n = -13$ or $n = 4$

$n = -13$ not possible case.

So, the only possibility case is $n = 4$.

**57.** *(c)* Let $S$ and $NS$ be the respective events of choosing a spade card and a card which is not a spade.

$P(S) = \frac{{}^{13}C_1}{{}^{52}C_1} = \frac{1}{4}$; $P(NS) = \frac{{}^{39}C_1}{{}^{52}C_1} = \frac{3}{4}$

Let $A$ be the event when both the chosen cards are spade.

Let $\bar{A}$ be the event when both the chosen cards are not spade.

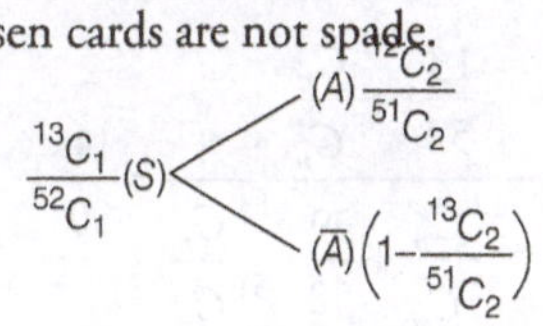


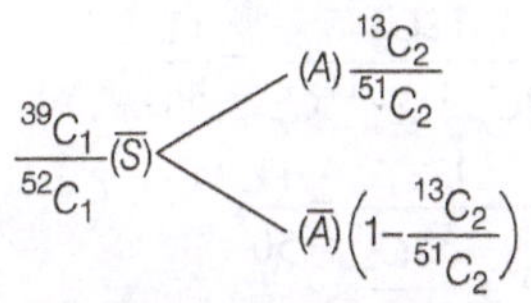


We have to find the probability of missing card not being spade when both the chosen cards are spade $= P(\bar{S} \mid A)$

By using Bayes theorem,

$$P(\bar{S} \mid A) = \frac{P(\bar{S} \cap A)}{P(A)}$$

$$= \frac{P(A \mid \bar{S})\, P(\bar{S})}{P(A \mid S)\, P(S) + P(A \mid \bar{S})\, P(\bar{S})}$$

$$= \frac{\left(\frac{{}^{13}C_2}{{}^{51}C_2}\right)\left(\frac{{}^{39}C_1}{{}^{52}C_1}\right)}{\left(\frac{{}^{13}C_1}{{}^{52}C_1}\right)\left(\frac{{}^{12}C_2}{{}^{51}C_2}\right) + \left(\frac{{}^{13}C_2}{{}^{51}C_2}\right)\left(\frac{{}^{39}C_1}{{}^{52}C_1}\right)}$$

$$=\frac{\left(\frac{13\cdot 12}{51\cdot 50}\right)\left(\frac{3}{4}\right)}{\left(\frac{1}{4}\right)\left(\frac{12\cdot 11}{51\cdot 50}\right)+\left(\frac{3}{4}\right)\left(\frac{13\cdot 12}{51\cdot 50}\right)}$$

$$=\frac{13\times 12\times 3}{(12\times 11)+(3\times 12\times 13)}=\frac{39}{50}$$

**Shortcut Method**

$P(\bar{S}$ missing/Both found spade)

$$=\frac{P(\bar{S}_m \cap \text{Both found spade})}{P(\text{Both found spade})}$$

$$=\frac{\left(1-\frac{13}{52}\right)\left(\frac{13}{51}\right)\left(\frac{12}{50}\right)}{\left(1-\frac{13}{52}\right)\left(\frac{13}{51}\times\frac{12}{50}\right)+\left(\frac{13}{52}\times\frac{12}{51}\right)\times\frac{11}{50}}=\frac{39}{50}$$

**58.** *(2)* $E_1$ = lost card is spade

$E_2$ = lost card is not spade

$A = n$ drawing cards are spade

$$P\left(\frac{E_1}{A}\right)=\frac{11}{50}$$

$$\Rightarrow \frac{P(E_1)\,P\left(\frac{A}{E_1}\right)}{P(E_1)\cdot P\left(\frac{A}{E_1}\right)+P(E_2)\cdot P\left(\frac{A}{E_2}\right)}=\frac{11}{50}$$

$$\Rightarrow \frac{\frac{13}{52}\cdot\frac{^{12}C_n}{^{51}C_n}}{\frac{13}{52}\cdot\frac{^{12}C_n}{^{51}C_n}+\frac{39}{52}\cdot\frac{^{13}C_n}{^{51}C_n}}=\frac{11}{50}$$

$$\Rightarrow \frac{12C_n}{12C_n+3\cdot{}^{13}C_n}=\frac{11}{50}$$

$$\Rightarrow \frac{1}{1+3\frac{^{13}C_n}{^{12}C_n}}=\frac{11}{50}$$

$$\Rightarrow 1+3\frac{^{13}C_n}{^{12}C_n}=\frac{50}{11}$$

$$\Rightarrow \frac{^{13}C_n}{^{12}C_n}=\frac{39}{11}\times\frac{1}{3}\Rightarrow\frac{^{13}C_n}{^{12}C_n}=\frac{13}{11}$$

$\therefore n=2$

**59.** *(27)* Given, $P(H)=1/4$ then,

$$P(T)=1-\frac{1}{4}=\frac{3}{4}$$

If $N$ = number of tosses required when the process is repeated until a head appears.

Then, probability

$$=\frac{1}{4}+\left(\frac{3}{4}\right)\left(\frac{1}{4}\right)+\left(\frac{3}{4}\right)^2\left(\frac{1}{4}\right)+\ldots+\left(\frac{3}{4}\right)^{N-1}\left(\frac{1}{4}\right)$$

$$P=\frac{1}{4}\left[1+\frac{3}{4}+\left(\frac{3}{4}\right)^2+\ldots+\left(\frac{3}{4}\right)^{N-1}\right]$$

$$P=\frac{1}{4}\left\{\frac{1\cdot\left\{1-\left(\frac{3}{4}\right)^N\right\}}{1-\frac{3}{4}}\right\}\Rightarrow P=1-\left(\frac{3}{4}\right)^N$$

Now, quadratic equation

$64x^2-5Nx+1=0$ has no real root

$$\Rightarrow D<0$$

$$25N^2-4\times 64<0$$

$$\Rightarrow (5N)^2<(16)^2\Rightarrow 5N<16$$

$$\Rightarrow N<\frac{16}{5}\Rightarrow N<3.2$$

Then, $p=1-\left(\frac{3}{4}\right)^3$ ; $p=1-\frac{27}{64}=\frac{37}{64}$

Hence, $p=37$, $q=64$

$q-p=64-37=27$

**60.** *(9)* Probability of a person being a smoker = 25%

$$P(S)=\frac{25}{100}\Rightarrow P(S)=\frac{1}{4}$$

So, probability of a person being non smoker $=P(N)=1-\frac{1}{4}=\frac{3}{4}$

And probability of a man to be a smoker to get a lung cancer $=P\left(\frac{L}{S}\right)$

$\because$ Probability of a non-smoker to get a lung cancer $=P\left(\frac{L}{N}\right)$

and $P\left(\frac{L}{S}\right)=27P\left(\frac{L}{N}\right)$ and $P\left(\frac{S}{L}\right)=\frac{k}{10}$

$$P\left(\frac{S}{L}\right)=\frac{P(S)\cdot P\left(\frac{L}{S}\right)}{P(L)}=\frac{k}{10}$$

Where,

$$P(L)=P(N)\cdot P\left(\frac{L}{N}\right)+P(S)\cdot P\left(\frac{L}{S}\right)$$

$$=\frac{3}{4}P\left(\frac{L}{N}\right)+\frac{1}{4}\left\{27\,P\left(\frac{L}{N}\right)\right\}$$

So, $P(L)=\frac{30}{4}P\left(\frac{L}{N}\right)$

So, $P\left(\frac{S}{L}\right)=\frac{\frac{1}{4}\left\{27\,P\left(\frac{L}{N}\right)\right\}}{\frac{30}{4}P\left(\frac{L}{N}\right)}=\frac{k}{10}$

$$\Rightarrow P\left(\frac{S}{L}\right)=\frac{9}{10}=\frac{k}{10}\Rightarrow 9$$

**61.** *(432)*

**Key idea** Use the Baye's Theorem

$$P\left(\frac{C}{E}\right)=\frac{P(C)\cdot\left(\frac{E}{C}\right)}{P(A)\cdot P\left(\frac{E}{A}\right)+P(B)\cdot\left(\frac{E}{B}\right)+P(C)\cdot P\left(\frac{E}{C}\right)}$$

The given three urns are $A, B, C$ which contains balls like below

$A\to 4$ red and 6 black

$B\to 5$ red and 5 black

and $C\to\lambda$ red and 4 black

$E\to$ Ball drawn is red

$\because$ Probability of selecting $A$, $B$ and $C$ will be $P(A)=P(B)=P(C)=1/3$

Also, probability of selecting red ball are

$$P\left(\frac{E}{A}\right)=\frac{4}{10},\ P\left(\frac{E}{B}\right)=\frac{5}{10}$$

and $P\left(\frac{E}{C}\right)=\frac{\lambda}{\lambda+4}$

$\because$ From Baye's theorem,

$$P\left(\frac{C}{E}\right)=\frac{P(C)\cdot P\left(\frac{E}{C}\right)}{P(A)\cdot P\left(\frac{E}{A}\right)+P(B)\cdot P\left(\frac{E}{B}\right)+P(C)\cdot P\left(\frac{E}{C}\right)}$$

$$0.4=\frac{\frac{1}{3}\times\frac{\lambda}{\lambda+4}}{\frac{1}{3}\times\frac{4}{10}+\frac{1}{3}\times\frac{5}{10}+\frac{1}{3}\times\frac{\lambda}{\lambda+4}}$$

$$\Rightarrow 0.4=\frac{\frac{\lambda}{\lambda+4}}{\frac{9}{10}+\frac{\lambda}{\lambda+4}}$$

$$\Rightarrow \frac{2}{5}\left(\frac{9\lambda+36+10\lambda}{10(\lambda+4)}\right)=\frac{\lambda}{\lambda+4}$$

$$\Rightarrow 38\lambda+72=50\lambda\Rightarrow\lambda=6$$

$\therefore$ Equation of parabola is $y^2 = 6x$

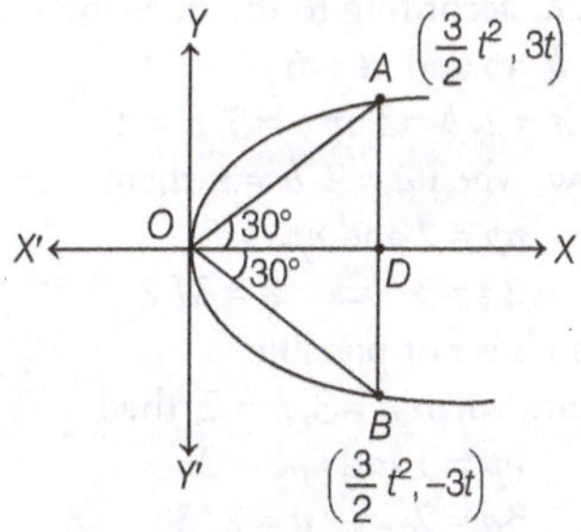


$\because$ $OAB$ is an equilateral triangle, and $OD$ bisects $AB$ at point $D$.

$\therefore \quad \angle AOD = 30°$

In $\Delta OAD$,

$\because \quad \tan 30° = \frac{AD}{OD} \Rightarrow \frac{1}{\sqrt{3}} = \frac{3t}{\frac{3}{2}t^2}$

$\Rightarrow \quad \frac{1}{\sqrt{3}} = \frac{2}{t} \quad \therefore t = 2\sqrt{3}$

$\therefore \quad A\left(\frac{3}{2} \times (2\sqrt{3})^2, 3 \times 2\sqrt{3}\right)$

$\Rightarrow \quad A(18, 6\sqrt{3})$

$\therefore$Square of side of an equilateral triangle

$= (OA)^2 = 18^2 + (6\sqrt{3})^2 = 432$

**62.** *(479)* Required probability

$$= {}^4C_2\left(\frac{3}{4}\right)^2\left(\frac{1}{4}\right)^2\left(\frac{1}{4}\right)^6 + {}^6C_2\left(\frac{3}{4}\right)^2\left(\frac{1}{4}\right)^4\left(\frac{3}{4}\right)^4 + {}^4C_1\,{}^6C_1\left(\frac{3}{4}\right)\left(\frac{1}{4}\right)\left(\frac{3}{4}\right)^3\left(\frac{1}{4}\right)^5$$

$$= \frac{1}{4^{10}}[6 \times 9 + 15 \times 9^3 + 24 \times 9^2]$$

$$= \frac{27}{4^{10}} \times 479 = \frac{27}{4^{10}}k$$

Thus, the value of $k$ is 479.

**63.** *(28)* $P(A) =$ Probability of first unit to function and $P(B) =$ Probability of second unit to function

$P(A) = 0.9, P(\bar{A}) = 0.1$

$P(B) = 0.8, P(\bar{B}) = 0.2$

$AB \rightarrow$ Operate

$A\bar{B} \rightarrow$ Not operate

$\bar{A}B \rightarrow$ Not operate

$\bar{A}\bar{B} \rightarrow$ Not operate

According to question, first failed and second functioning, given that instrument has failed, So according to Bayes theorem

$$\therefore \quad P = \frac{P(\bar{A}) \cdot P(B)}{P(A)P(\bar{B}) + P(\bar{A})P(B) + P(\bar{A})P(\bar{B})}$$

$$= \frac{(0.1 \times 0.8)}{(0.9 \times 0.2) + (0.1 \times 0.8) + (0.1 \times 0.2)}$$

$$= \frac{8}{28}$$

$$\therefore 98P = \frac{8}{28} \times 98 = 28$$

**64.** *(6)*

> **key idea** (i) $P(X) = p$
> $\Rightarrow P(\bar{X}) = 1 - p$
>
> (ii) If $A, B, C$ are independent events, then $P$ (only $A$ occurs) $= P(A\bar{B}\bar{C}) = P(A) \cdot P(\bar{B}) \cdot P(\bar{C})$
> Similarly, for $P(B)$ and $P(C)$ and
> $P$(none occurs) $= P(\bar{A}) \cdot P(\bar{B}) \cdot P(\bar{C})$.

Let $P(E_1) = x$, $P(E_2) = y$, $P(E_3) = z$

Then, $(1 - x)(1 - y)(1 - z) = p \quad \ldots$ (A)

Probability of only $E_1 = P(E_1 \cap \bar{E}_2 \cap \bar{E}_3)$

$\alpha = x(1 - y)(1 - z) \quad \ldots$ (B)

Similarly, $\beta = (1 - x)y(1 - z) \quad \ldots$ (C)

$\gamma = (1 - x)(1 - y)z \quad \ldots$ (D)

On dividing Eq. (A) by Eq. (B), we get

$$\frac{p}{\alpha} = \frac{1 - x}{x}$$

$$\Rightarrow \quad xp = \alpha - \alpha x \Rightarrow x = \frac{\alpha}{\alpha + p}$$

Similarly, $y = \frac{\beta}{\beta + p} \Rightarrow z = \frac{\gamma}{\gamma + p}$

$$\frac{P(E_1)}{P(E_3)} = \frac{x}{z} = \left(\frac{\alpha}{\alpha + p}\right) \times \left(\frac{\gamma + p}{\gamma}\right)$$

$$\frac{x}{z} = \left(\frac{\alpha}{\gamma}\right)\left(\frac{\gamma + p}{\alpha + p}\right) = \left(\frac{1 + p/\gamma}{1 + p/\alpha}\right)$$

Also given, $p = \frac{\alpha\beta}{\alpha - 2\beta} = \frac{2\beta\gamma}{\beta - 3\gamma}$

$$\alpha p - 2\beta p = \alpha\beta$$

$$\Rightarrow \quad \frac{\alpha p}{\alpha + 2p} = \beta \quad \ldots\text{(i)}$$

$$\beta p - 3p\gamma = 2\beta\gamma$$

$$\Rightarrow \quad \beta(p - 2\gamma) = 3p\gamma$$

$$\Rightarrow \quad \beta = \frac{3p\gamma}{p - 2\gamma} \quad \ldots\text{(ii)}$$

From Eqs. (i) and (ii), we get

$$\frac{\alpha p}{\alpha + 2p} = \frac{3p\gamma}{p - 2\gamma}$$

$$\Rightarrow \quad \frac{1}{1 + 2\left(\frac{p}{\alpha}\right)} = \frac{3}{\left(\frac{p}{\gamma}\right) - 2}$$

$$\Rightarrow \quad (p/\gamma) - 2 = 3 + 6(p/\alpha)$$

$$\Rightarrow \quad \frac{p}{\gamma} = 5 + 6\left(\frac{p}{\alpha}\right)$$

$$\frac{x}{z} = \frac{1 + p/\gamma}{1 + p/\alpha} = \frac{1 + 5 + 6(p/\alpha)}{1 + p/\alpha}$$

$$= 6\left(\frac{1 + p/\alpha}{1 + p/\alpha}\right) = 6$$

## Topic 2

**01.** *(c)* $\sum_{x=0}^{\infty} P(X = x) = 1$

$k(1 + 2 \cdot 3^{-1} + 3 \cdot 3^{-2} + 4 \cdot 3^{-3} + \ldots + \infty)$

Let $s = 1 + \frac{2}{3} + \frac{3}{3^2} + \frac{4}{3^3} + \ldots + \infty$

$\frac{s}{3} = \frac{1}{3} + \frac{2}{3^2} + \frac{3}{3^3} + \frac{4}{3^4} + \ldots + \infty$

$\frac{2s}{3} = 1 + \frac{1}{3} + \frac{1}{3^2} + \frac{1}{3^3} + \ldots + \infty$

$\frac{2s}{3} = \frac{1}{1 - \frac{1}{3}} = \frac{3}{2}$

$s = \frac{9}{4}$

$k = \frac{4}{9}$

$P(X \geq 2) = 1 - p(x = 0) - p(x = 1)$

$= 1 - \frac{4}{9}\left(1 + \frac{2}{3}\right) = \frac{7}{27}$

**02.** *(d)* Since, two oranges are drawn.

Then, $x = 0$; both oranges are good.

$x = 1$; one orange defective and one good.

$x = 2$; both oranges are defective.

Total number of ways to choose 2 oranges from 10 is $= {}^{10}C_2 = \frac{10 \times 9}{2} = 45$

$$X = 0 \Rightarrow {}^7C_2 = \frac{7 \times 6}{2} = 21$$

$$\therefore \quad P(X = 0) = \frac{21}{45} = \frac{7}{15}$$

$$X = 1 \Rightarrow {}^7C_1 \times {}^3C_1 = 21$$

$$\therefore \quad P(X = 1) = \frac{21}{45} = \frac{7}{15}$$

$$X = 2 \Rightarrow {}^3C_2 = \frac{3 \times 2}{2} = 3$$

$\therefore\ P(X=2)=\frac{3}{45}=\frac{1}{15}$

$\therefore E(X)=\Sigma x\,P(X=x)$

$=0\times\frac{7}{15}+1\times\frac{7}{15}+2\times\frac{1}{15}$

$=\frac{7+2}{15}=\frac{9}{15}=\frac{3}{5}$

$E(X^2)=\Sigma x^2 P(X=x)$

$=0+1^2\times\frac{7}{15}+2^2\times\frac{1}{15}=\frac{11}{15}$

$\therefore Var(X)=E(X^2)-[(E(X)]^2$

$=\frac{11}{15}-\left(\frac{3}{5}\right)^2=\frac{55-27}{75}=\frac{28}{75}$

**03.** (a)

**key idea** Mean $(\mu)=\Sigma P_i x_i$, Variance $(\sigma^2)=\Sigma P_i x_i^2-(\Sigma P_i x_i)^2$

$X=$ Number of times a tail follows a head.

| Outcomes | $X_i$ | $P_i$ |
|---|---|---|
| HHH | 0 | 1/8 |
| HHT | 1 | 1/8 |
| HTH | 1 | 1/8 |
| THH | 0 | 1/8 |
| HTT | 1 | 1/8 |
| THT | 1 | 1/8 |
| TTH | 0 | 1/8 |
| TTT | 0 | 1/8 |

$P(X=0)=\frac{4}{8}=\frac{1}{2}$

$P(X=1)=\frac{4}{8}=\frac{1}{2}$

$\mu=\sum P_i x_i=0+\frac{1}{2}=\frac{1}{2}$

$\sigma^2=\sum P_i x_i^2-\mu^2=0+\frac{1}{2}-\frac{1}{4}=\frac{1}{4}$

$\therefore\ 64(\mu+\sigma^2)=64\left(\frac{3}{4}\right)=48$

**04.** (a) We have,

$P(2)=\frac{1}{6}$ and $P(\overline{2})=\frac{5}{6}$

$\therefore P$[2 obtained on even numbered toss]

$=\frac{5}{6}\times\frac{1}{6}+\left(\frac{5}{6}\right)^3\times\frac{1}{6}+\left(\frac{5}{6}\right)^5\times\frac{1}{6}+\ldots$

$=\frac{\frac{5}{6}\times\frac{1}{6}}{1-\left(\frac{5}{6}\right)^2}=\frac{5}{36}\times\frac{36}{11}=\frac{5}{11}$

***05.** (b) Here, $p=\frac{1}{2}$ and $q=\frac{1}{2}$

We have, $P$ (getting an odd number 7 times)

$=P$ (getting an odd numbers nine time)

$\Rightarrow {}^nC_7\left(\frac{1}{2}\right)^7\left(\frac{1}{2}\right)^{n-7}={}^nC_9\left(\frac{1}{2}\right)^9\left(\frac{1}{2}\right)^{n-9}$

$\Rightarrow {}^nC_7={}^nC_9 \Rightarrow n=7+9=16$ $[\because 7\neq 9]$

$P$ (getting an even number twice) $=\frac{k}{2^{15}}$

${}^{16}C_2\times\left(\frac{1}{2}\right)^2\times\left(\frac{1}{2}\right)^{16-2}=\frac{k}{2^{15}}$

$\frac{16\times15}{1\times2}\times\frac{1}{2^{16}}=\frac{k}{2^{15}}\Rightarrow k=60$

**06.** (d) N denote the sum of the number obtained when two dice are rolled such that $2^N<N$.

i.e., $4\le N\le 12$

i.e., $N\in\{4,5,6,\ldots 12\}$

Now, $P(N=2)+P(N=3)$

$=\frac{1}{36}+\frac{2}{36}=\frac{3}{36}=\frac{1}{12}$

So, required probability

$=1-\frac{1}{12}=\frac{11}{12}=\frac{m}{n}$

$4m-3n=4\times11-3\times12$

$=44-36=8$

***07.** (b) Given, a pair of dice is thrown once then number of outcomes $=6\times6=36$. A total of 5 are $\{(1,4),(2,3),(4,1),(3,2)\}$

$\therefore p=P(\text{sucess})=\frac{4}{36}=\frac{1}{9}$

$q=1-\frac{1}{9}=\frac{8}{9}$

Let $X$ represent the number of success

$\therefore P$ (at least 4 success)

$=P(X=4)+P(X=5)$

$={}^5C_4\,p^4q+{}^5C_5p^5q^{5-5}$

$=5\left(\frac{1}{9}\right)^4\left(\frac{8}{9}\right)+\left(\frac{1}{9}\right)^5=\frac{40+1}{(3^2)^5}=\frac{41}{3^{10}}$

$=\frac{41\times3}{3^{11}}=\frac{123}{3^{11}}\Rightarrow k=123$

***08.** (d) Let mean be $a$ and variance be $b$. Then, according to the question

$a+b=5$ and $ab=6$

$\therefore\ a=2, b=3$ or $a=3, b=2$

Now, when $a=2, b=3$, then

$np=2$ and $npq=3$

$\therefore\ 2q=3\Rightarrow q=3/2$

Which is not possible.

Again, when $a=3, b=2$, then

$np=3$ and $npq=2$

$\therefore\ 3q=2\Rightarrow q=2/3$

Also, $p=1-q=1-\frac{2}{3}=\frac{1}{3}$

Again, $np=3$

$\Rightarrow\ n\times\frac{1}{3}=3\Rightarrow n=9$

So, $6(n+p-q)=6\left(9+\frac{1}{3}-\frac{2}{3}\right)$

$=54+2-4=52$

***09.** (b) Mean $=np$

Variance $=np(1-p)$

Given, Mean + Variance

$=np+np(1-p)=24$ and

Mean $\times$ Variance $=(np)(np(1-p))=128$

$\Rightarrow\ np=16,\ (np)(1-p)=8$

$\Rightarrow 1-p=\frac{1}{2}\Rightarrow p=\frac{1}{2}$

$\Rightarrow\ n=32$

$P(X>n-3)=P(X=n-2)+P(X=n-1)+P(X=n)$

$={}^{32}C_{30}\left(\frac{1}{2}\right)^{32}+{}^{32}C_{31}\left(\frac{1}{2}\right)^{32}+{}^{32}C_{32}\left(\frac{1}{2}\right)^{32}$

$=\frac{16\times31+32+1}{2^{32}}=\frac{529}{2^{32}}$

$k=529, n=32$

***10.** (c) $B(7,p)\Rightarrow n=7$

Given, $P(x=3)=5P(x=4)$

${}^7C_3p^3(1-p)^4=5\,{}^7C_4p^4(1-p)^3$

$\frac{{}^7C_3}{5\,{}^7C_4}=\frac{p}{1-p}\Rightarrow 1-p=5p$

$\Rightarrow\ p=\frac{1}{6}, q=\frac{5}{6}$

Mean $(np)=7\times\frac{1}{6}=\frac{7}{6}$

Var $(npq)=7\times\frac{1}{6}\times\frac{5}{6}=\frac{35}{36}$

Sum $=\frac{7}{6}+\frac{35}{36}=\frac{77}{36}$

**11.** *(a)* Given that

| $x$ | 0 | 1 | 2 | 3 | 4 |
|---|---|---|---|---|---|
| $P(x)$ | $k$ | $2k$ | $4k$ | $6k$ | $8k$ |

Now, $P(1<x<4)\,|\,x\le 2)=\dfrac{P(1<x\le 2)}{p(x\le 2)}$

$$=\frac{P(x=2)}{P(x=0)+P(x=1)+P(x=2)}$$

$$=\frac{4k}{k+2k+4k}=\frac{4k}{7k}=\frac{4}{7}$$

**12.** *(c)* Let $X=$ Number of heads

| $X=x$ | 0 | 1 | 2 | 3 |
|---|---|---|---|---|
| $P(x)$ | 1/8 | 3/8 | 3/8 | 1/8 |

Required probability

$=P_A(0)\times P_B(0)+P_A(1)\times P_B(1)+P_A(2)\times P_B(2)+P_A(3)\times P_B(3)$

$$=\frac{1}{8}\times\frac{1}{8}+\frac{3}{8}\times\frac{3}{8}+\frac{3}{8}\times\frac{3}{8}+\frac{1}{8}\times\frac{1}{8}=\frac{5}{16}$$

***13.** *(d)*

**key idea** Use this concept $^nC_r={}^nC_q \Rightarrow n=r+q$.

Given, $P$(odd number 2 times)

$=P$(even number 3 times)

$$\Rightarrow\quad {}^nC_2\left(\frac{1}{2}\right)^n={}^nC_3\left(\frac{1}{2}\right)^n$$

$$n=3+2=5$$

$[\because {}^nC_r={}^nC_q$ only when $n=r+q]$

$\therefore$ Probability of getting an odd number, odd number of times

$=P(1)+P(3)+P(5)$

$$={}^5C_1\left(\frac{1}{2}\right)^5+{}^5C_3\left(\frac{1}{2}\right)^5+{}^5C_5\left(\frac{1}{2}\right)^5$$

$$=\frac{16}{2^5}=\frac{16}{32}=\frac{1}{2}$$

**14.** *(b)* Given probability distribution for a random variable $X$ is

| $X$ | 1 | 2 | 3 | 4 | 5 |
|---|---|---|---|---|---|
| $P(X)$ | $K^2$ | $2K$ | $K$ | $2K$ | $5K^2$ |

Then,

$P(X>2)=1-\{P(X=1)+P(X=2)\}$

$=1-K^2-2K$

$$\therefore\quad \sum_{X=1}^{5} P(X)=1$$

$$\Rightarrow\quad 6K^2+5K=1$$

$$\Rightarrow\quad 6K(K+1)-1(K+1)=0$$

$$\Rightarrow\quad K=\frac{1}{6},-1\ \{\because P(X)\ge 0,\ \text{so } K\nless 0\}$$

$$\Rightarrow\quad K=1/6$$

$$\therefore\quad P(X>2)=1-\frac{1}{36}-\frac{2}{6}$$

$$=\frac{36-1-12}{36}=\frac{23}{36}$$

**Shortcut Method**

$$\sum_{X=1}^{5}P(x)=1$$

$6K^2+5K=1\Rightarrow K=1/6,-1$

We take, $K=1/6$

$P(X>2)=1-[P(X=1)+P(X=2)]$

$$=1-\left(\frac{1}{36}+\frac{2}{6}\right)=\frac{23}{36}$$

**15.** *(c)*

**key idea** Use the expected value of random variableX is $\Sigma x, P(x_i)$

If $k$ represents the number of consecutive heads and $P(k)$ is the corresponding probability while an unbiased coin is tossed 5 times, then

| $k$ | 0 | 1 | 2 | 3 | 4 | 5 |
|---|---|---|---|---|---|---|
| $P(k)$ | $\frac{1}{2^5}$ | $\frac{12}{2^5}$ | $\frac{11}{2^5}$ | $\frac{5}{2^5}$ | $\frac{2}{2^5}$ | $\frac{1}{2^5}$ |

Now, the expected value of variable $X$ is

$$(-1)\frac{1}{2^5}+(-1)\frac{12}{2^5}+(-1)\frac{11}{2^5}+(3)\frac{5}{2^5}+(4)\frac{2}{2^5}+(5)\frac{1}{2^5}$$

$$=\frac{1}{2^5}[-1-12-11+15+8+5]=\frac{1}{8}$$

***16.** *(d)* According to the question, Mean $=np=8$ and variance $=npq=4$

$$\therefore\quad q=\frac{1}{2}\Rightarrow p=1-q=\frac{1}{2}$$

Now, $\quad n\times\dfrac{1}{2}=8$

$$\Rightarrow\quad n=16$$

$$P(X=r)={}^{16}C_r\left(\frac{1}{2}\right)^{16}$$

$\therefore\ P(X\le 2)=P(X=0)+P(X=1)+P(X=2)$

$$={}^{16}C_0\left(\frac{1}{2}\right)^{16}+{}^{16}C_1\left(\frac{1}{2}\right)^{16}+{}^{16}C_2\left(\frac{1}{2}\right)^{16}$$

$$=\frac{1+16+120}{2^{16}}=\frac{137}{2^{16}}=\frac{k}{2^{16}}\quad\text{[given]}$$

$$\Rightarrow\quad k=137$$

**17.** *(d)* The required probability of observing atleast one head $=1-P$ (no head)

$$=1-\frac{1}{2^n}\qquad\text{[Let number of toss are } n]$$

$$\left[\because P(\text{Head})=P(\text{Tail})=\frac{1}{2}\right]$$

According to the question,

$$1-\frac{1}{2^n}\ge\frac{90}{100}$$

$$\Rightarrow\quad \frac{1}{2^n}\le\frac{1}{10}\Rightarrow 2^n\ge 10\Rightarrow n\ge 4$$

So, minimum number of times one has to toss a fair coin so that the probability of observing atleast one head is atleast 90% is 4.

***18.** *(a)* Given box contains 15 green and 10 yellow balls.

$\therefore$ Total number of balls $=15+10=25$

$$P(\text{green balls})=\frac{15}{25}=\frac{3}{5}=p$$

$=$ Probability of success

$$P(\text{yellow balls})=\frac{10}{25}=\frac{2}{5}=q$$

Probability of unsuccess and $n=10$

$=$ Number of trials.

$$\therefore \text{Variance}=npq=10\times\frac{3}{5}\times\frac{2}{5}=\frac{12}{5}$$

**19.** *(b)* We have, $P(A)=2/5$

and $\quad P(A\cap B)=3/20$

$P(A'\cup B')=P((A\cap B)')=1$

$$-P(A\cap B)=1-\frac{3}{20}=\frac{17}{20}$$

Now, $A\cap(A'\cup B')=A\cap(A\cap B)'$

$=A-[A\cap(A\cap B)]=A-(A\cap B)$

$$\therefore\ P(A-(A\cap B))=\frac{2}{5}-\frac{3}{20}=\frac{1}{4}$$

$$\therefore p(A\,|\,(A'\cup B'))=\frac{P(A-(A\cap B))}{P(A'\cup B')}$$

$$=\frac{1/4}{17/20}=\frac{5}{17}$$

***20.** *(c)* Probability of guessing a correct answer, $p=1/3$ and probability of guessing a wrong answer, $q=\dfrac{2}{3}$.

So, the probability of guessing 4 or more correct answers

$= {}^5C_4\left(\frac{1}{3}\right)^4\cdot\frac{2}{3}+{}^5C_5\left(\frac{1}{3}\right)^5$

$=5\cdot\frac{2}{3^5}+\frac{1}{3^5}=\frac{11}{3^5}$

***21.** *(b)* Here, $n=5$ and $r\geq 1$

$\therefore \quad P(X=r)={}^nC_r\,p^{n-r}\,q^r$

$P(X\geq 1)=1-P(X=0)$

$=1-{}^5C_0\cdot p^5\cdot q^0\geq\frac{31}{32}$ [given]

$\Rightarrow \quad p^5\leq 1-\frac{31}{32}=\frac{1}{32}$

$\therefore \quad p\leq\frac{1}{2}$ and $p\geq 0$

$\Rightarrow \quad p\in\left[0,\frac{1}{2}\right]$

***22.** *(a)* According to the given condition,

$1-\left(\frac{3}{4}\right)^n\geq\frac{9}{10}$

$\Rightarrow \quad \left(\frac{3}{4}\right)^n\leq 1-\frac{9}{10}=\frac{1}{10}$

$\Rightarrow \quad \left(\frac{4}{3}\right)^n\geq 10$

On taking log on both sides, we get

$n\,[\log 4-\log 3]\geq\log_{10}10=1$

$\Rightarrow \quad n\geq\frac{1}{\log_{10}4-\log_{10}3}$

***23.** *(d)* Probability of getting score 9 in a single throw $=\frac{4}{36}=\frac{1}{9}$

Probability of getting score 9 exactly in double throw $={}^3C_2\times\left(\frac{1}{9}\right)^2\times\frac{8}{9}=\frac{8}{243}$

**24.** *(c)* Required probability

$=P(X=0)+P(X=1)$

$=\frac{e^{-5}}{0!}\cdot 5^0+\frac{e^{-5}}{1!}\cdot 5^1=e^{-5}+5e^{-5}=\frac{6}{e^5}$

**25.** *(b)* $\therefore P(X>1.5)=P(2)+P(3)+\ldots+\infty$

$=1-[P(0)+P(1)]$

$=1-\left(e^{-2}+\frac{e^{-2}\times 2}{1}\right)=1-\frac{3}{e^2}$

***26.** *(a)* Given that for binomial distribution mean, $np=4$ and variance, $npq=2$.

$\therefore q=1/2$, but $p+q=1 \Rightarrow p=1/2$

and $\quad n\times\frac{1}{2}=4 \Rightarrow n=8$

We know that

$P(X=r)={}^nC_r\,p^r\,q^{n-r}$

$\therefore \quad P(X=1)={}^8C_1\left(\frac{1}{2}\right)^7\left(\frac{1}{2}\right)^1$

$=8\times\frac{1}{2^8}=\frac{1}{2^5}=\frac{1}{32}$

**27.** *(a)* Let the probability of getting a head be $p$ and not getting a head be $q$. Since, head appears first time in an even throw 2 or 4 or 6. . .

$\therefore \quad \frac{2}{5}=qp+q^3p+q^5p+\ldots$

$\Rightarrow \quad \frac{2}{5}=\frac{qp}{1-q^2}$

$\Rightarrow \quad \frac{2}{5}=\frac{(1-p)p}{1-(1-p)^2} \quad [\because q=1-p]$

$\Rightarrow \quad \frac{2}{5}=\frac{1-p}{2-p}\Rightarrow 4-2p=5-5p$

$\Rightarrow \quad p=\frac{1}{3}$

**28.** *(d)* $\because$ A box contain

10 Pens → 3 defective, 7 Non-defective

$X=$ Number of defective

| $X$ | $X^2$ | $P(x)$ | $P_ix_i$ | $P_ix_i^2$ |
|---|---|---|---|---|
| 0 | 0 | $\frac{^7C_2}{^{10}C_2}$ | 0 | 0 |
| 1 | 1 | $^7C_1\cdot\frac{^3C_1}{^{10}C_2}$ | $\frac{^7C_1\cdot{}^3C_1}{^{10}C_2}$ | $\frac{^7C_1\,{}^3C_1}{^{10}C_2}$ |
| 2 | 4 | $\frac{^3C_2}{^{10}C_2}$ | $2\cdot\frac{^3C_2}{^{10}C_2}$ | $4\cdot\frac{^3C_2}{^{10}C_2}$ |

Mean $=\sum P_ix_i=\frac{21}{45}+\frac{6}{45}=\frac{27}{45}=\frac{3}{5}$

Variance $=\sum P_ix_i^2-(\Sigma P_ix_i)^2$

$=\left(\frac{21}{45}+\frac{12}{45}\right)-\frac{9}{25}$

$=\frac{11}{15}-\frac{9}{25}$

$=\frac{55-27}{75}=\frac{28}{75}$

**29.** *(c)* We have, $A=\begin{bmatrix}a_{11} & a_{12}\\ a_{21} & a_{22}\end{bmatrix}$

$|A|=a_{11}a_{22}-a_{12}a_{21}=\{-1,0,1\}$

| $X$ | $P_i$ | $P_iX_i$ | $P_iX_i^2$ |
|---|---|---|---|
| −1 | 3/16 | −3 /16 | 3/16 |
| 0 | 10/16 | 0 | 0 |
| 1 | 3/16 | 3/16 | 3/16 |
| | | $\Sigma P_iX_i=0$ | $\Sigma P_iX_i^2=3/8$ |

$\therefore \mathrm{Var}(x)=\Sigma P_iX_i^2-(\Sigma P_iX_i)^2$

$=\frac{3}{8}-0=\frac{3}{8}$

***30.** *(d)* Given that the binomial distribution $B(n, p)$ for which the difference of the mean and variance is 1.

So, mean $=n\cdot p$ and variance $=npq$, where $p+q=1$

According to the question,

$np-npq=1 \Rightarrow np\,(1-q)=1$

$\Rightarrow \quad np^2=1 \quad \ldots(i)$

Given that, $2P(X=2)=3P(X=1)$

$\Rightarrow \quad 2^n\cdot C_2p^2q^{n-2}=3\cdot{}^nc_1\,pq^{n-1}$

$\Rightarrow \quad (n-1)\,p=3q$

$\Rightarrow \quad np-p=3q=3(1-p)$

$\Rightarrow np+2p=3 \Rightarrow p=\frac{3}{n+2} \quad \ldots(ii)$

On solving Eqs. (i) and (ii), we get

$n=4,\ p=\frac{1}{2}$

Now, $P(X>1)=1-\{P(X=0)+P(X=1)\}$

$=1-\left\{{}^4C_0\left(\frac{1}{2}\right)^4+{}^4C_1\left(\frac{1}{2}\right)\left(\frac{1}{2}\right)^3\right\}$

$=1-\left\{\frac{1}{16}+\frac{1}{4}\right\}=1-\frac{5}{16}=\frac{11}{16}$

Thus, $n^2\cdot P(X>1)=4^2\,P(X>1)$ $(\because n=4)$

$=16\times\frac{11}{16}=11$

**31.** *(b)* $P(X=0)+P(X=1)+P(X=2)+\ldots\text{to }\infty=1\ (\because\Sigma p_i=1)$

or, $k\left[1+\frac{2}{3}+\frac{3}{3^2}+\frac{4}{3^3}+\ldots\text{to }\infty\right]=1 \quad \ldots(i)$

Let $S=1+\frac{2}{3}+\frac{3}{3^2}+\frac{4}{3^3}+\ldots\text{to }\infty \quad \ldots(ii)$

$$\Rightarrow \frac{1}{3}S = \frac{1}{3} + \frac{2}{3^2} + \frac{3}{3^3} + \ldots \text{to } \infty$$

On subtracting Eq. (iii) from Eq. (ii), we get

$$\frac{2S}{3} = 1 + \frac{1}{3} + \frac{1}{3^2} + \frac{1}{3^3} + \ldots \text{to } \infty \ldots \text{(iii)}$$

$$\text{or, } \frac{2S}{3} = \frac{1}{1-\frac{1}{3}} = \frac{3}{2} \Rightarrow S = \frac{9}{4}$$

From Eq. (i),

$$k\left(\frac{9}{4}\right) = 1 \Rightarrow k = \frac{4}{9}$$

$$P(X \geq 2) = 1 - P(X=0) - P(X=1)$$

$$= 1 - \frac{4}{9} - \frac{4}{9} \times \frac{2}{3} = \frac{7}{27}$$

**32.** (a) We have 3 rotten apples and 7 good apples, i.e. total 10 apples, mixed. Let $X_i$ be the number of rotten apples, when 4 apples are drawn one-by-one without replacement.

Then, $X_i = 0, 1, 2, 3$

$$\text{Now, } P(X_i = 0) = \frac{^7C_4}{^{10}C_4} = \frac{\left(\frac{7!}{4!3!}\right)}{\left(\frac{10!}{4!6!}\right)} = \frac{1}{6}$$

$$P(X_i = 1) = \frac{^3C_1 \times {^7C_3}}{^{10}C_4} = \frac{\frac{3!}{2!1!} \times \frac{7!}{4!3!}}{\frac{10!}{4!6!}} = \frac{1}{2}$$

$$P(X_i = 2) = \frac{^3C_2 \times {^7C_2}}{^{10}C_4} = \frac{\frac{3!}{2!1!} \times \frac{7!}{2!5!}}{\frac{10!}{4!6!}} = \frac{3}{10}$$

$$P(X_i = 3) = \frac{^3C_3 \times {^7C_1}}{^{10}C_4} = \frac{\frac{3!}{3!0!} \times \frac{7!}{1!6!}}{\frac{10!}{4!6!}} = \frac{1}{30}$$

For mean ($\mu$) and variance ($\sigma^2$),

$$\text{we have, } \sigma^2 + \mu^2 = \sum_{i=0}^{3} P(X_i) \cdot X_i^2$$

$$\Rightarrow \sigma^2 + \mu^2 = 0^2 \cdot \frac{1}{6} + 1^2 \cdot \frac{1}{2} + 2^2 \cdot \frac{3}{10} + 3^2 \cdot \frac{1}{30}$$

$$= \frac{1}{2} + \frac{12}{10} + \frac{3}{10} = 2$$

$$\therefore \quad 10(\sigma^2 + \mu^2) = 20$$

***33.** (d) Let $P(1) = k$

$$\Rightarrow P(a \text{ prime number}) = \frac{2k}{3}$$

$$\text{and } P(\text{a composite number}) = \frac{k}{3}$$

$$\therefore P(1) = k, P(2) = P(3) = P(5) = \frac{2k}{3},$$

$$P(4) = P(6) = \frac{k}{3}$$

$$k + 3\left(\frac{2k}{3}\right) + 2\left(\frac{k}{3}\right) = 1$$

$$\Rightarrow k + 2k + \frac{2k}{3} = 1 \Rightarrow k = \frac{3}{11}$$

$$\therefore P(1) = \frac{3}{11}, P(2) = P(3) = P(5) = \frac{2}{11},$$

$$P(4) = P(6) = \frac{1}{11}$$

$p$ (number obtained is perfect square)

$$= P(1) + P(4) = \frac{4}{11}$$

A die is tossed twice.

$$\therefore \quad n = 2, P = \frac{4}{11};$$

$$\text{Mean} = np = \frac{8}{11}$$

***34.** (b) Given,

$$\text{mean} \quad (E(X)) = np = 4 \quad \ldots\text{(i)}$$

$$\text{Variance } (\sigma^2) = npq = \frac{4}{3} \quad \ldots\text{(ii)}$$

Here, $n$ = number of trials, $p$ and $q$ are the probability of success and failure in a single throw, respectively.

$$q = \frac{npq}{np} = \frac{4}{3 \times 4} = \frac{1}{3}$$

$$\text{Now, } p = 1 - q = 1 - \frac{1}{3} = \frac{2}{3}$$

$$\therefore \quad n \times \frac{2}{3} = 4 \quad [\because \text{from Eq. (i)}]$$

$$\Rightarrow \quad n = 6$$

$$\text{Now, } 54\, P(X \leq 2) = 54[P(X=0) + P(X=1) + P(X=2)]$$

$$= 54\left[{^6C_0}\left(\frac{2}{3}\right)^0\left(\frac{1}{3}\right)^6 + {^6C_1}\left(\frac{2}{3}\right)\left(\frac{1}{3}\right)^5 + {^6C_2}\left(\frac{2}{3}\right)^2\left(\frac{1}{3}\right)^4\right]$$

$$= 54 \times \frac{73}{729} = \frac{146}{27}$$

***35.** (c)

> **key idea** For Binomial Distribution $B(n, p)$
>
> $P(X = x) = {^nC_x} q^{n-x} \cdot p^x$ where, $x = 0, 1, \ldots n$ and $q = 1 - p$.

Let $n$ be the number of trials.
$p$ be the probability of success
and $q$ be the probability of unsuccess.
Thus, mean $= np$ and variance $= npq$

$$[\because q = 1 - p]$$

$$\text{Here, } np + npq = 24 \quad \ldots\text{(i)}$$

$$\text{and} \quad np \cdot npq = 128 \quad \ldots\text{(ii)}$$

Let $\alpha = np$ and $\beta = npq$.

Thus, $\alpha + \beta = 24$ and $\alpha\beta = 128$

$$\therefore \quad x^2 - (\alpha + \beta)x + \alpha\beta = 0$$

$$\Rightarrow \quad x^2 - (24)x + 128 = 0$$

$$\Rightarrow \quad (x-8)(x-16) = 0 \Rightarrow x = 8, 16$$

$\because \alpha = np$ and $\beta = npq$, $np > npq$

So, $np = 16$, $npq = 8$

Thus, $q = 1/2$ so,

$$p = 1 - q = 1 - \frac{1}{2} = \frac{1}{2} \text{ and } n = \frac{16}{1/2} = 32$$

Thus, required probability is

$$p(X = 1 \text{ or } 2) = p(X=1) + p(X=2)$$

$$= {^{32}C_1}\left(\frac{1}{2}\right)^1\left(\frac{1}{2}\right)^{31} + {^{32}C_2}\left(\frac{1}{2}\right)^2 \cdot \left(\frac{1}{2}\right)^{30}$$

$$= {^{32}C_1}\left(\frac{1}{2}\right)^{32} + {^{32}C_2}\left(\frac{1}{2}\right)^{32}$$

$$= ({^{32}C_1} + {^{32}C_2})\left(\frac{1}{2}\right)^{32}$$

$$= \left(32 + \frac{32 \times 31}{2}\right) \times \frac{1}{2^{32}} = \frac{32 \cdot 33}{2} \times \frac{1}{2^{32}}$$

$$= 33\left(\frac{1}{2^{32} \cdot 2^{-4}}\right) = \frac{33}{2^{28}}$$

***36.** (d) Given, $n = 5$

Let $p$ be the probability of getting head.
$q$ be the probability of getting tail.
According to the question,

$$^5C_4\, p^4 q^1 = {^5C_5}\, p^5 q^0$$

$$\Rightarrow \quad 5p^4 q = p^5 \Rightarrow q = \frac{p}{5}$$

$$\text{Now, } p + q = 1 \Rightarrow p + \frac{p}{5} = 1$$

$$\Rightarrow \quad p = \frac{5}{6} \text{ and } q = \frac{1}{6}$$

Then,

$$P(X \leq 2) = {^5C_0}\, p^0 q^5 + {^5C_1}\, p^1 q^4 + {^5C_2}\, p^2 q^3$$

$$=\left(\frac{1}{6}\right)^5+5\left(\frac{5}{6}\right)\left(\frac{1}{6}\right)^4+10\left(\frac{5}{6}\right)^2\left(\frac{1}{6}\right)^3$$

$$=\left(\frac{1}{6}\right)^5+\frac{25}{6^5}+\frac{250}{6^5}=\frac{276}{6^5}$$

$$p(\le 2)=46/6^4$$

***37.** (a) We know that $P(X=r)$ for a binomial distribution is ${}^nC_r p^r(1-p)^{n-r}$ or ${}^nC_r\,p^r q^{n-r}$, where $p+q=1$

Given, $n=33$ such that $3P(X=0)=P(X=1)$

$$\Rightarrow 3\,{}^nC_0p^0(1-p)^n={}^nC_1p^1(1-p)^{n-1}$$

$$\Rightarrow 3\,{}^{33}C_0p^0(1-p)^{33}={}^{33}C_1p^1(1-p)^{32}$$

$$\Rightarrow \frac{3}{33}=\frac{p}{(1-p)} \Rightarrow p=\frac{1}{12}$$

and $q=\dfrac{11}{12}$ $[\because q=1-p]$

Given that

$$\frac{P(X=15)}{P(X=18)}-\frac{P(X=16)}{P(X=17)}$$

$$\Rightarrow \frac{{}^{33}C_{15}\,p^{15}q^{18}}{{}^{33}C_{18}\,p^{18}q^{15}}-\frac{{}^{33}C_{16}\,p^{16}q^{17}}{{}^{33}C_{17}\,p^{17}q^{16}}$$

$$=\left(\frac{q}{p}\right)^3-\left(\frac{q}{p}\right)$$

$$=11^3-11=1320$$

***38.** (a) The examination consist 8 true-false type questions.

$$P(\text{true})=P(\text{false})=\frac{1}{2}$$

The probability of guessing atleast $n$ correct answer is $P(E)$.

$$P(E)<\frac{1}{2}$$

$$\Rightarrow \sum_{r=n}^{8}{}^8C_r\left(\frac{1}{2}\right)^{8-r}\left(\frac{1}{2}\right)^r<\frac{1}{2}$$

$$\Rightarrow \sum_{r=n}^{8}{}^8C_r\left(\frac{1}{2}\right)^8<\frac{1}{2}$$

$$\Rightarrow {}^8C_n+{}^8C_{n+1}+{}^8C_{n+2}+\ldots\ldots+{}^8C_8<128$$

$$\Rightarrow 256-({}^8C_0+{}^8C_1+\ldots\ldots+{}^8C_{n-1})<128$$

$$\Rightarrow {}^8C_0+{}^8C_1+\ldots\ldots+{}^8C_{n-1}>128$$

$$\Rightarrow n-1\ge 4\Rightarrow n\ge 5$$

Hence, least value of $n$ is 5.

**39.** (b) We have,

| $x_i$ | 0 | 1 | 2 | 3 | 4 | ... |
|---|---|---|---|---|---|---|
| $P_i$ | $\frac{1}{2}$ | $\frac{1}{3}$ | $\frac{1}{3^2}$ | $\frac{1}{3^3}$ | $\frac{1}{3^4}$ | ... |

Mean of distribution $=\sum P_ix_i$

$$M=\frac{1}{3}+\frac{2}{3^2}+\frac{3}{3^3}+\frac{4}{3^4}+\ldots \quad \ldots(\text{i})$$

$$\frac{M}{3}=\frac{1}{3^2}+\frac{2}{3^3}+\frac{3}{3^4}+\ldots. \quad \ldots(\text{ii})$$

On subtracting Eq. (ii) from Eq. (i), we get

$$\frac{2M}{3}=\frac{1}{3}+\frac{1}{3^2}+\frac{1}{3^3}+\ldots=\frac{\frac{1}{3}}{1-\frac{1}{3}}$$

$$\left[\because \text{Sum of infinite GP}=\frac{a}{1-r}\right]$$

$$M=\frac{3}{4}$$

$P$ ($x$ is positive and even)

$$=\frac{1}{3^2}+\frac{1}{3^4}+\frac{1}{3^6}+\ldots$$

$$=\frac{\frac{1}{3^2}}{1-\frac{1}{3^2}}=\frac{\frac{1}{3^2}}{\frac{8}{3^2}}=\frac{1}{8}$$

***40.** (a) Let the coin be tossed '$n$' times. Also, let $x$ denote number of times head occurs.

According to the question,

$$P(X=7)=P(X=9)$$

Using formula $P(X=r)={}^nC_rp^rq^{n-r}$

Where, $p=$ probability of getting head in tossing a coin $=1/2$

and $q=1-p=1-1/2=1/2$

$\therefore P(X=7)=P(X=9)$

$$\Rightarrow {}^nC_7\left(\frac{1}{2}\right)^7\left(\frac{1}{2}\right)^{n-7}={}^nC_9\left(\frac{1}{2}\right)^9\left(\frac{1}{2}\right)^{n-9}$$

$$\Rightarrow {}^nC_7\left(\frac{1}{2}\right)^{7+n-7}={}^nC_9\left(\frac{1}{2}\right)^{9+n-9}$$

$$\Rightarrow {}^nC_7={}^nC_9$$

$$\Rightarrow n=7+9=16$$

$[\because {}^nC_r={}^nC_q$ only when $n=r+q]$

$$\therefore \quad P(X=2)={}^{16}C_2\left(\frac{1}{2}\right)^2\left(\frac{1}{2}\right)^{14}$$

$$P(X=2)=\frac{16!}{2!14!}\left(\frac{1}{2}\right)^{16}$$

$$=\frac{2^3\times 15}{2^{16}}=\frac{15}{2^{13}}$$

***41.** (c) Given that there are 50 problems to solve in an admission test and probability that the candidate can solve any problem is $4/5=q$ (say). So, probability that the candidate cannot solve a problem is $p=1-q=1-\frac{4}{5}=\frac{1}{5}$.

Now, let $X$ be a random variable which denotes the number of problems that the candidate is unable to solve. Then, $X$ follows binomial distribution with parameters $n=50$ and $p=1/5$.

Now, according to binomial probability distribution concept

$$P(X=r)={}^{50}C_r\left(\frac{1}{5}\right)^r\left(\frac{4}{5}\right)^{50-r},$$

$r=0,1,\ldots,50$

$\therefore$ Required probability

$$=P(X<2)=P(X=0)+P(X=1)$$

$$={}^{50}C_0\left(\frac{4}{5}\right)^{50}+{}^{50}C_1\frac{4^{49}}{(5)^{50}}$$

$$=\left(\frac{4}{5}\right)^{49}\left(\frac{4}{5}+\frac{50}{5}\right)=\frac{54}{5}\left(\frac{4}{5}\right)^{49}$$

***42.** (c) Let $p$ and $q$ represents the probability of success and failure in a trial respectively. Then,

$p=P(5 \text{ or } 6)=2/6=1/3$

and $q=1-p=\frac{4}{6}=\frac{2}{3}$.

So, the expected gain/loss (in ₹)

$$=(p\times 100)+qp(-50+100)+q^2p(-50-50+100)+q^3(-50-50-50)$$

$$=\left(\frac{1}{3}\times 100\right)+\left(\frac{2}{3}\times\frac{1}{3}\right)(50)+\left(\frac{2}{3}\right)^2\left(\frac{1}{3}\right)(0)+\left(\frac{2}{3}\right)^3(-150)$$

$$=\frac{100}{3}+\frac{100}{9}+0-\frac{1200}{27}$$

$$=\frac{900+300-1200}{27}=\frac{1200-1200}{27}=0$$

***43.** (d) Number of white balls $=30$ and number of red balls $=10$

Let $p=$ probability of success in a trial $=$ probability of getting a white ball in a trial $=\frac{30}{40}=\frac{3}{4}$.

$$q = 1 - p = 1 - \frac{3}{4} = \frac{1}{4}$$

Here, $n$ = number of trials = 16.

Clearly, $X$ follows binomial distribution with parameter $n = 16$ and $p = \frac{3}{4}$.

$\therefore$ Mean of $X = np, = 16 \cdot \frac{3}{4} = 12$

and variance of $X = npq = 16 \cdot \frac{3}{4} \cdot \frac{1}{4} = 3$

Now, $\frac{\text{mean of } X}{\text{standard deviation of } X}$

$= \frac{12}{\sqrt{3}} = 4\sqrt{3}$ [$\because$ SD $= \sqrt{\text{variance}}$]

***44.** (a) Let $p$ = probability of getting an ace in a draw = probability of success and $q$ = probability of not getting an ace in a draw = probability of failure

Then, $p = \frac{4}{52} = \frac{1}{13}$ and $q = 1 - p$

$= 1 - \frac{1}{13} = \frac{12}{13}$

Here, number of trials, $n = 2$

Clearly, $X$ follows binomial distribution with parameter $n = 2$ and $p = \frac{1}{13}$.

Now, $P(X = x) = {}^2C_x \left(\frac{1}{13}\right)^x \left(\frac{12}{13}\right)^{2-x}$

$\therefore P(X = 1) + P(X = 2)$

$= {}^2C_1 \left(\frac{1}{13}\right)^1 \left(\frac{12}{13}\right) + {}^2C_2 \left(\frac{1}{13}\right)^2 \left(\frac{12}{13}\right)^0$

$= \frac{25}{169}$

**45.** (b) Given, $E = \{X \text{ is a prime number}\}$

$= \{2, 3, 5, 7\}$

$P(E) = P(X = 2) + P(X = 3) + P(X = 5) + P(X = 7)$

$= 0.23 + 0.12 + 0.20 + 0.07 = 0.62$

and $F = \{X < 4\} = \{1, 2, 3\}$

$\Rightarrow P(F) = P(X = 1) + P(X = 2) + P(X = 3)$

$= 0.15 + 0.23 + 0.12 = 0.5$

and $E \cap F = \{X$ is prime number as well as $< 4\} = \{2, 3\}$

$P(E \cap F) = P(X = 2) + P(X = 3)$

$= 0.23 + 0.12 = 0.35$

$\therefore$ Required probability,

$P(E \cup F) = P(E) + P(F) - P(E \cap F)$

$= 0.62 + 0.5 - 0.35$

$= 0.77$

**46.** (17)

| Blue balls | Probability |
|---|---|
| $\frac{{}^5C_0 \cdot {}^4C_1}{{}^9C_3}$ | 0 |
| $\frac{{}^5C_1 \cdot {}^4C_2}{{}^9C_3}$ | 1 |
| $\frac{{}^5C_2 \cdot {}^4C_1}{{}^9C_3}$ | 2 |
| $\frac{{}^5C_3 \cdot {}^4C_0}{{}^9C_3}$ | 3 |
| 0 | 4 |
| 0 | 5 |

$$7\bar{X} = \frac{({}^5C_0 \cdot {}^4C_1 \times 0) + {}^5C_1 \cdot {}^4C_2 + {}^5C_2 \cdot {}^4C_1 \times 2 + {}^5C_3 \cdot {}^4C_0 \times 3}{{}^9C_3} \times 7$$

$$= \frac{30 + 80 + 30}{84} \times 7 = \frac{35}{3}$$

| Yellow | Probability |
|---|---|
| ${}^5C_3 \cdot {}^4C_0$ | 0 |
| ${}^5C_2 \cdot {}^4C_1$ | 1 |
| ${}^5C_1 \cdot {}^4C_2$ | 2 |
| ${}^5C_0 \cdot {}^4C_3$ | 3 |
| 0 | 4 |

$$4\bar{Y} = \frac{0 + {}^5C_2 \cdot {}^4C_1 + {}^5C_1 \cdot {}^4C_2 \times 2 + {}^5C_0 \cdot {}^4C_3 \times 3}{{}^9C_3} \times 4$$

$$= \frac{40 + 60 + 12}{84} \times 4 = \frac{112}{21} = \frac{16}{3}$$

Then, $7\bar{X} + 4\bar{Y} = \frac{35}{3} + \frac{16}{3} = 17$

**47.** (5)

| $X$ | $\alpha$ | 1 | 0 | $-3$ |
|---|---|---|---|---|
| $P(X)$ | 1/3 | $k$ | 1/6 | 1/4 |

$\Sigma P_i = 1$

$\frac{1}{3} + k + \frac{1}{6} + \frac{1}{4} = 1 \Rightarrow k = \frac{1}{4}$

Mean $= \Sigma P_i x_i = \frac{\alpha}{3} + k + 0 - \frac{3}{4}$

$= \frac{\alpha}{3} + \frac{1}{4} - \frac{3}{4} = \frac{\alpha}{3} - \frac{1}{2}$

Variance $= \Sigma P_i x_i^2 - (\Sigma P_i x_i)^2$

$= \left(\frac{\alpha^2}{3} + k + 0 + \frac{9}{4}\right) - \left(\frac{\alpha}{3} - \frac{1}{2}\right)^2$

$\sigma^2 = \frac{\alpha^2}{3} + \frac{1}{4} + \frac{9}{4} - \frac{\alpha^2}{9} + \frac{\alpha}{3} - \frac{1}{4}$

$\sigma^2 = \frac{2\alpha^2}{9} + \frac{\alpha}{3} + \frac{9}{4}$

Now, $\sigma - \mu = 2$

$\Rightarrow \sigma = \mu + 2$

$\sigma^2 = (\mu + 2)^2$

$\frac{2\alpha^2}{9} + \frac{\alpha}{3} + \frac{9}{4} = \left(\frac{\alpha}{3} + \frac{3}{2}\right)^2$

$\frac{\alpha^2}{9} - \frac{2\alpha}{3} = 0$

$\alpha = 0$ (reject) or 6

$\therefore \sigma + \mu = \mu + 2 + \mu = 2\mu + 2$

$= \frac{2\alpha}{3} - 1 + 2 = \frac{2 \times 0}{3} + 1 = 5$

**48.** (56) Total number of items = 10

Number of defective items = 3

and number of non-defective items = 7

$P(X = 0) = \frac{{}^7C_5}{{}^{10}C_5} = \frac{1}{12}$

$P(X = 1) = \frac{{}^7C_4 \times {}^3C_1}{{}^{10}C_5} = \frac{5}{12}$

$P(X = 2) = \frac{{}^7C_3 \times {}^3C_2}{{}^{10}C_5} = \frac{5}{12}$

$P(X = 3) = \frac{{}^7C_2 \times {}^3C_3}{{}^{10}C_5} = \frac{1}{12}$

$\therefore$ Probability distribution

| $X$ | 0 | 1 | 2 | 3 |
|---|---|---|---|---|
| $P(X)$ | 1/12 | 5/12 | 5/12 | 1/12 |

$\therefore \mu = \Sigma X \cdot P(X)$

$= 0 \times \frac{1}{12} + 1 \times \frac{5}{12} + 2 \times \frac{5}{12} + 3 \times \frac{1}{12}$

$= \frac{18}{12} = \frac{3}{2}$

$\therefore \sigma^2 = (0)^2 \times \left(\frac{1}{12}\right) + (1)^2 \times \frac{5}{12} + (2)^2 \times \frac{5}{12} + (3)^2 \times \frac{1}{12} - \left(\frac{3}{2}\right)^2$

$= \frac{5}{12} + \frac{20}{12} + \frac{9}{12} - \frac{9}{4} = \frac{34 - 9 \times 3}{12} = \frac{7}{12}$

$\therefore 96\sigma^2 = 96 \times \frac{7}{12} = 56$

**49.** *(8288)* $P(W)=\frac{1}{3}, P(L)=\frac{2}{3}$

$x$ = Number of matches that team wins

$y$ = Number of matches that team loses

$|x-y|\le 2$ and $x+y=10$

$|x-y|=0, 1, 2\ x, y\in N$

Case 1 $|x-y|=0 \Rightarrow x=y$

$\because x+y=10 \Rightarrow x=y=5$

$\therefore P(|x-y|=0)$

$$= {}^{10}C_5\left(\frac{1}{3}\right)^5\left(\frac{2}{3}\right)^5 \quad ...(i)$$

Case 2 $|x-y|=1 \Rightarrow x-y=\pm 1$

| +1 | −1 |
|---|---|
| $x-y=1$ | $x-y=-1$ |
| $x+y=10$ | $x+y=10$ |
| $x=\frac{11}{2}$ | $2x=\frac{9}{2}$ |
| NP | NP |

Case 3 $|x-y|=2$

$x-y=\pm 2$

| +2 | −2 |
|---|---|
| $x+y=10$ | $x+y=10$ |
| $x-y=2$ | $x-y=-2$ |
| $x=6$ | $x=4$ |
| $y=4$ | $y=6$ |

$$\therefore P(|x-4|=2) = {}^{10}C_6\left(\frac{1}{3}\right)^6\left(\frac{2}{3}\right)^4 + {}^{10}C_4\left(\frac{1}{3}\right)^4\left(\frac{2}{3}\right)^6 \quad ...(ii)$$

Required probability

$$= {}^{10}C_5\left(\frac{1}{3}\right)^5\left(\frac{2}{3}\right)^5 + {}^{10}C_6\left(\frac{1}{3}\right)^6\left(\frac{2}{3}\right)^4 + {}^{10}C_4\left(\frac{1}{3}\right)^4\left(\frac{2}{3}\right)^6$$

$$P=\frac{{}^{10}C_5\,2^5}{3^{10}}+\frac{{}^{10}C_6\,2^4}{3^{10}}+\frac{{}^{10}C_4\,2^6}{3^{10}}$$

$$\Rightarrow 3^9 P=\frac{1}{3}({}^{10}C_5 2^5+{}^{10}C_6 2^4 + {}^{10}C_4 2^6)=8288$$

**50.** *(12)* $a=P(X=3)=\frac{5}{6}\times\frac{5}{6}\times\frac{1}{6}=\frac{25}{216}$

and $b=P(X\ge 3)$

$$=\frac{5}{6}\times\frac{5}{6}\times\frac{1}{6}+\left(\frac{5}{6}\right)^3\times\frac{1}{6}+\left(\frac{5}{6}\right)^4\times\frac{1}{6}+...$$

$$=\frac{25/216}{1-\frac{5}{6}}=\frac{25}{216}\times\frac{6}{1}=\frac{25}{36}$$

$$P(X\ge 6)=\left(\frac{5}{6}\right)^5\times\frac{1}{6}+\left(\frac{5}{6}\right)^6\times\frac{1}{6}+...$$

$$=\frac{\left(\frac{5}{6}\right)^5\times\frac{1}{6}}{1-\frac{5}{6}}=\left(\frac{5}{6}\right)^5$$

$$c=\frac{\left(\frac{5}{6}\right)^5}{\left(\frac{5}{6}\right)^3}=\frac{25}{36}$$

Hence, $\frac{b+c}{a}=\frac{\left(\frac{5}{6}\right)^2+\left(\frac{5}{6}\right)^2}{\left(\frac{5}{6}\right)^2\times\frac{1}{6}}=12$

**51.** *(30)* We know that $\Sigma P(X)=1$

$\Rightarrow\ K+2K+2K+3K+K=1$

$\Rightarrow\ K=1/9$

$$p=P(1<X<4|X<3)$$

$$=\frac{P((1<X<4)\cap(X<3))}{P(X<3)}$$

$$=\frac{P(X=2)}{P(X=1)+P(X=2)}$$

$$=\frac{2K}{K+2K}=\frac{2}{3}$$

$\because\ 5p=\lambda K$

$5\times\frac{2}{3}=\lambda\times\frac{1}{9}\Rightarrow\lambda=30$

**52.** *(4)* A fair coin is tossed, $P(\text{Head})=1/2$

Probability of getting atleast one head is atleast 0.9.

$P(\text{atleast one head})\ge 0.9$

$\Rightarrow\ 1-P(\text{all tail})\ge 0.9$

$\Rightarrow\ 1-\left(\frac{1}{2}\right)^n\ge 0.9$

$\Rightarrow\ \left(\frac{1}{2}\right)^n\le 0.1$

$\Rightarrow\ \left(\frac{1}{2}\right)^n\le\frac{1}{10}$

$\Rightarrow\ n_{min}=4$

**53.** *(3)* Let probability of hitting a target

$p=1/10$

So, probability of not hitting a target

$q=9/10$

So, probability of not hitting target in $n$ trials $=(9/10)^n$

$\therefore$ Probability of hitting target at least one $=1-\left(\frac{9}{10}\right)^n>\frac{1}{4}\Rightarrow(0.9)^n<0.75$s

$\therefore$ Least number of shots required $=n_{min}=3$